2 BREEDING EVIDENCE HISTOGRAM

The various categories of breeding evidence are shown in code form along the bottom of the histogram; the farther right the code, the more conclusive the breeding evidence. The number at the top of each column represents the number of squares in which the category of breeding evidence in question was reported (only the highest level of evidence reported in each Atlas square was used). The height of the column corresponds to the percentage of squares in which the category of breeding evidence was found relative to the total number of squares in which the species was reported. The aggregate totals for the three levels of breeding evidence (possible, probable, and confirmed) are given below the histogram. Confirmed breeding evidence is shown in black and possible and probable breeding evidence is shown in ochre in both the histogram and the squares on the map.

Possible breeding

- H Species present in suitable nesting habitat during its breeding season

Probable breeding

- P Pair observed in suitable habitat during its breeding season
- T Territorial behaviour (song, territorial disputes, etc.) observed on at least two days, a week or more apart, at the same place
- C Courtship behaviour: display, copulation, or food exchange between adults
- V Visiting a probable nest site, as distinct from a roosting site
- A Agitated behaviour or anxiety calls of an adult indicating that a nest site or young are nearby
- N Excavation of holes by woodpeckers

Confirmed breeding

- CN Nest building or carrying of nesting materials (except woodpeckers)
- NU Used nest or egg shells found (nest occupied or eggs laid during the current season)
- AT Adult carrying food for young during the breeding season
- PH Physiological evidence (i.e., highly vascularized incubation patch or egg in oviduct) observed in bird in hand
- DD Distraction display or injury feigning by such birds as ducks, grouse, shorebirds, etc.
- NO Adult leaving, occupying, or entering nest site in circumstances indicating that the nest is occupied (high nests, nest holes, or nest boxes where the contents of the nest cannot be seen)
- FE Adult carrying fecal sac
- JE Downy young or recently fledged young incapable of sustained flight
- NJ Nest containing eggs or young (seen or heard)

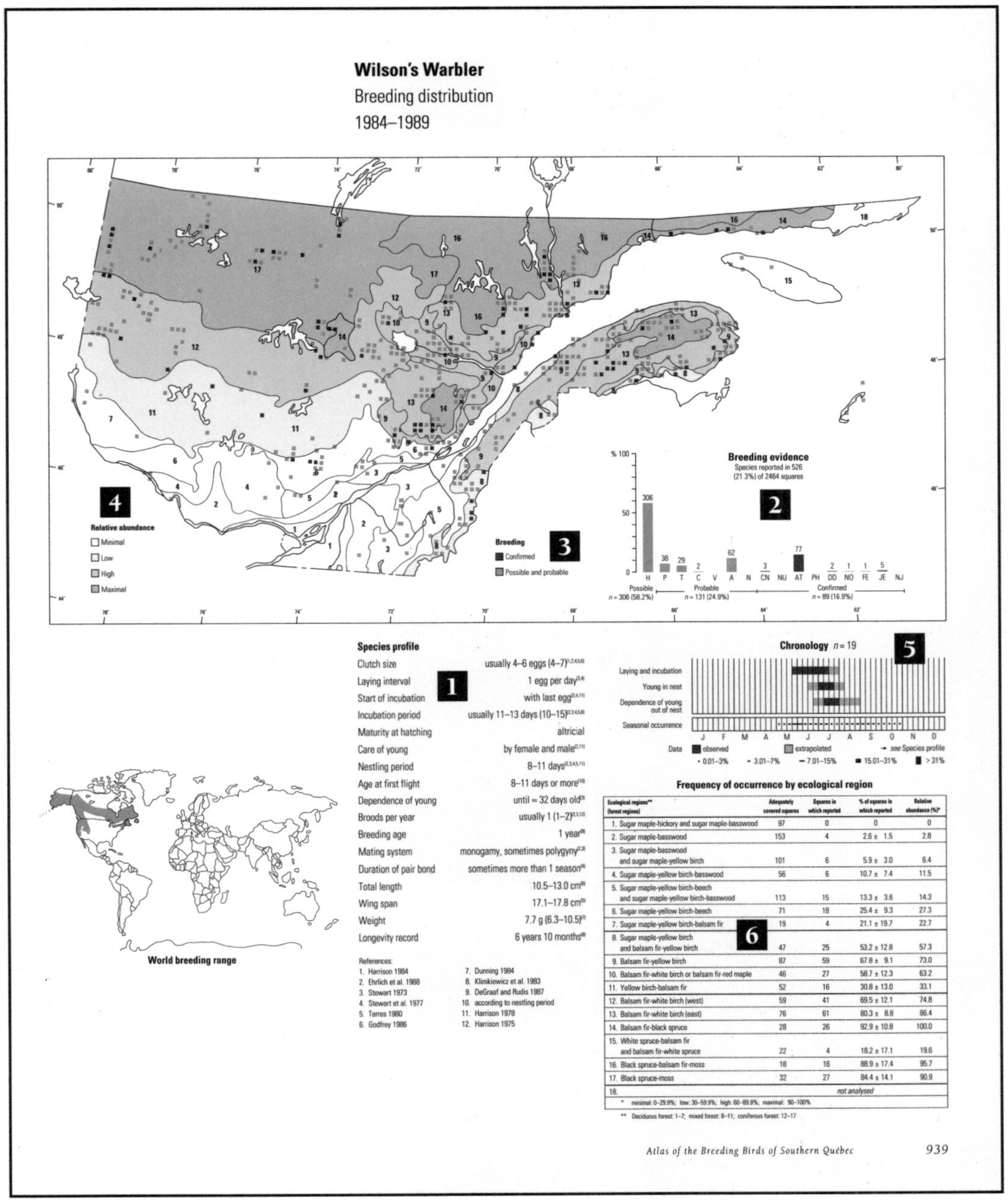

Species profile 1

Clutch size	usually 4–6 eggs (4–7)[1,2,4,5,6]
Laying interval	1 egg per day[3,4]
Start of incubation	with last egg[3,4,11]
Incubation period	usually 11–13 days (10–15)[2,3,4,5,6]
Maturity at hatching	altricial
Care of young	by female and male[2,11]
Nestling period	8–11 days[2,3,4,5,11]
Age at first flight	8–11 days or more[10]
Dependence of young	until ≈ 32 days old[3]
Broods per year	usually 1 (1–2)[2,3,12]
Breeding age	1 year[9]
Mating system	monogamy, sometimes polygyny[2,3]
Duration of pair bond	sometimes more than 1 season[4]
Total length	10.5–13.0 cm[6]
Wing span	17.1–17.8 cm[5]
Weight	7.7 g (6.3–10.5)[7]
Longevity record	6 years 10 months[8]

References:
1. Harrison 1984
2. Ehrlich et al. 1988
3. Stewart 1973
4. Stewart et al. 1977
5. Terres 1980
6. Godfrey 1986
7. Dunning 1984
8. Klimkiewicz et al. 1983
9. DeGraaf and Rudis 1987
10. according to nestling period
11. Harrison 1978
12. Harrison 1975

Frequency of occurrence by ecological region 6

Ecological regions** (forest regions)	Adequately covered squares	Squares in which reported	% of squares in which reported	Relative abundance (%)*
1. Sugar maple-hickory and sugar maple-basswood	97	0	0	0
2. Sugar maple-basswood	153	4	2.6 ± 1.5	2.8
3. Sugar maple-basswood and sugar maple-yellow birch	101	6	5.9 ± 3.0	6.4
4. Sugar maple-yellow birch-basswood	56	6	10.7 ± 7.4	11.5
5. Sugar maple-yellow birch-beech and sugar maple-yellow birch-basswood	113	15	13.3 ± 3.6	14.3
6. Sugar maple-yellow birch-beech	71	18	25.4 ± 9.3	27.3
7. Sugar maple-yellow birch-balsam fir	19	4	21.1 ± 19.7	22.7
8. Sugar maple-yellow birch and balsam fir-yellow birch	47	25	53.2 ± 12.8	57.3
9. Balsam fir-yellow birch	87	59	67.8 ± 9.1	73.0
10. Balsam fir-white birch or balsam fir-red maple	46	27	58.7 ± 12.3	63.2
11. Yellow birch-balsam fir	52	16	30.8 ± 13.0	33.1
12. Balsam fir-white birch (west)	59	41	69.5 ± 12.1	74.8
13. Balsam fir-white birch (east)	76	61	80.3 ± 8.8	86.4
14. Balsam fir-black spruce	28	26	92.9 ± 10.8	100.0
15. White spruce-balsam fir and balsam fir-white spruce	22	4	18.2 ± 17.1	19.6
16. Black spruce-balsam fir-moss	18	16	88.9 ± 17.4	95.7
17. Black spruce-moss	32	27	84.4 ± 14.1	90.9
18.			*not analysed*	

* minimal: 0–29.9%; low: 30–59.9%; high: 60–89.9%; maximal: 90–100%

** Deciduous forest: 1–7; mixed forest: 8–11; coniferous forest: 12–17

Atlas of the Breeding Birds of Southern Québec 939

3 DISTRIBUTION MAP – *BREEDING EVIDENCE*

The small squares on the map represent the 10 km × 10 km Atlas squares in which breeding evidence for the species was obtained during the Atlas survey period (1984–89). Black squares represent confirmed breeding and ochre squares, possible or probable breeding.

4 DISTRIBUTION MAP – *RELATIVE ABUNDANCE*

The map shows the species' relative abundance in each ecological region (forest region) of southern Québec during the Atlas period. The numbers on the map refer to the ecological regions listed in the table entitled "Frequency of occurrence by ecological region." The background colours indicate the species' relative abundance: white for 0–29.9% (minimal abundance), light ochre for 30–59.9% (low abundance), medium ochre for 60–89.9% (high abundance), and dark ochre for 90–100% (maximal abundance).

Readers should note that the confidence interval shown in the table must be taken into account in interpreting the map.

5 CHRONOLOGY

This bar graph shows the timing of the various stages of the species' breeding cycle in Québec. For the purposes of the graph, each month is divided into four parts: the 1st to the 7th day, the 8th to the 14th day, the 15th to the 21st day, and the 22nd to the last day. The symbol *n* represents the total number of nests or broods used to produce the graph or, in the case of colonial species, the number of sources used (QNRCP records, ÉPOQ checklists, various studies, personal communications, etc.). The black lines represent actual observations from all over Québec, while the ochre-coloured lines represent data extrapolated from these observations and from the species profiles. When a range is given in the species profile, it was calculated conservatively by taking the least extreme figure.

In species where "Dependence of young out of nest" lasts for a number of months or cannot be determined precisely (e.g., until autumn or winter), the ochre-coloured line is replaced by an arrow, indicating that the reader should refer to the species profile for more information.

The line entitled "Seasonal occurrence" shows the bird's frequency of occurrence throughout the year in southern Québec based on ÉPOQ and other data; the thicker the line, the greater the frequency, indicating when the species is easiest to find.

6 TABLE – *FREQUENCY OF OCCURRENCE BY ECOLOGICAL REGION*

This table shows the data used to calculate the percentage of squares in which the species was reported and, by extension, its relative abundance by ecological region. The percentage of squares in which the species was reported was arrived at by dividing the number of squares with observations by the number of squares with adequate coverage. Each percentage is accompanied by a confidence interval (with a risk coefficient of 5%) which gives a general idea of its precision: the greater the confidence interval, the less precise the percentage.

The data on relative abundance are indicative of the ecological regions most frequented by the species. A species' relative abundance in a given region is calculated by dividing the percentage of squares in which it was reported in the region in question by the highest percentage in any region (in this case 92.9) and multiplying the result by 100. Consequently, relative abundance is always 100% in the region with the highest frequency of occurrence (92.9/92.9 × 100 = 100%). This allows for comparison of the same species' relative abundance in different regions. The relative abundance of different species cannot be compared, however, and data on the percentage of squares with observations must be used instead.

The term "ecological region," also called "forest region" in the *Atlas*, should not be construed as referring to a single habitat, but rather to a region with distinctive bioclimactic characteristics and a variety of different habitats.

ABREVIATIONS USED

Adj.	adjective
Ant.	antonym
AOU	American Ornithologists' Union
AQGO	Association québécoise des groupes d'ornithologues (Québec Association of Birdwatching Clubs)
BAPE	Bureau des audiences publiques sur l'environnement (Québec Office for Public Hearings on the Environment)
BBS	Breeding Bird Survey
BDJV	Black Duck Joint Venture
BIOMQ	Banque informatisée des oiseaux marins du Québec (Québec Seabird Database)
C	Celsius (centigrade)
c.	*circa* (approximately)
CBC	Christmas Bird Count
CGIS	Canada Geographic Information System
cm	centimetre(s)
CMN	Canadian Museum of Nature
COQ	Club des ornithologues de Québec (Québec City Birdwatchers Club)
COSEWIC	Committee on the Status of Endangered Wildlife in Canada
CWS	Canadian Wildlife Service
dbh	diameter at breast height
DDT	dichloro-diphenyl-trichloroethane
DSS	Department of Supply and Services
DU	Ducks Unlimited
DUC	Ducks Unlimited Canada
E	east
ed.	edition, editor
e.g.	for example
ÉPOQ	Étude des populations d'oiseaux du Québec (Studies of Bird Populations in Québec)
et al.	and others
F	female
g	gram(s)
GREBE	Groupe de recherche et d'étude en biostatistique de l'environnement (Environmental Biostatistics Research and Study Group)
i.e.	that is
in prep.	in preparation
JBEC	James Bay Energy Corporation
kg	kilogram(s)
kHz	kilohertz
kJ	kilojoule(s)
km	kilometre(s)
km^2	square kilometre(s)
M	male
m	metre(s)
m^2	square metre(s)
MAPAQ	Ministère de l'Agriculture, des Pêcheries et de l'Alimentation du Québec (Québec Department of Agriculture, Fisheries and Food)
Max.	maximum
MEF	Ministère de l'Environnement et de la Faune (Québec Ministry of Environment and Wildlife)
MENVIQ	Ministère de l'Environnement du Québec (Québec Ministry of the Environment); merged with MLCP to become MEF
min.	minimum
MLCP	Ministère du Loisir, de la Chasse et de la Pêche (Québec Ministry of Recreation, Fish and Game); merged with MENVIQ to become MEF
mm	millimetre(s)
MRC	Municipalités régionales de comté (regional county municipalities)
MRN	Ministère des Ressources naturelles du Québec (Québec Department of Natural Resources)
MTFQ	Ministère des Terres et Forêts du Québec (Québec Department of Lands and Forests); merged with MRN
N	north
n	number; sample size
NAWMP	North American Waterfowl Management Plan
NGS	National Geographic Society
NRCan	Natural Resources Canada
NWRC	National Wildlife Research Centre
p.	page(s)
pers. comm.	personal communication
pers. corr.	personal correspondence
pers. obs.	personal observation(s)
PQSPB	Province of Quebec Society for the Protection of Birds
QLF	Quebec–Labrador Foundation
QNRCP	Quebec Nest Record Card Program
rev. ed.	revised edition
S	south
s	second(s)
SARCEL	Société d'aménagement récréatif et de conservation du lac Saint-Pierre (Lake Saint-Pierre Recreation and Conservation Association)
SDBJ	Société de développement de la Baie James (James Bay Development Corporation)
SÉPAQ	Société des établissements de plein air du Québec (Québec Association of Outdoor Pursuits Centres)
Syn.	synonym
t	metric ton(s); syn.: tonne
U	University
unpubl. data	unpublished data
UQCN	Union québécoise de conservation de la nature (Québec Nature Conservancy)
UQTR	Université du Québec à Trois-Rivières
USDA	United States Department of Agriculture
USFWS	United States Fish and Wildlife Service
UTM	Universal Transverse Mercator
Vb.	verb
VIREO	Visual Resources for Ornithology
vol.	volume
W	west
ZEC	zone d'exploitation contrôlée (controlled zone)
≈	approximately
≤	less than or equal to
≥	greater than or equal to
–	not applicable
?	data unavailable

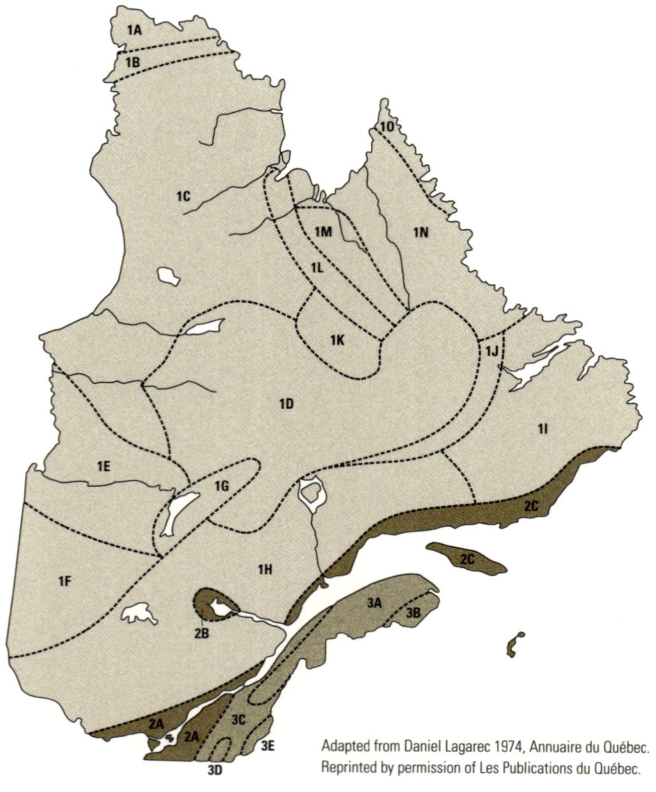

Adapted from Daniel Lagarec 1974, Annuaire du Québec.
Reprinted by permission of Les Publications du Québec.

Major geological and physiographic regions

The Shield
1A Saglouc Plateau
1B Povungnituk Hills
1C Larch Plateau
1D Lake Plateau
1E Eastmain Lowland
1F Abitibi Upland
1G Mistassini Hills
1H Laurentian Highlands
1I Mécatina Plateau
1J Hamilton Plateau
1K Caniapiscau Plateau
1L Labrador Hills
1M Whale Lowland
1N George Plateau
1O Labrador Highlands

St. Lawrence Lowlands (Southern Québec Lowlands)
2A Central St. Lawrence Lowland (St. Lawrence Plain)
2B Lake Saint-Jean Lowland
2C East St. Lawrence Lowland

Appalachian Region
3A Notre-Dame Mountains
3B Chaleur Uplands
3C Eastern Québec Uplands
3D Sutton Mountains
3E Mégantic Hills

The Breeding Birds of Québec

ATLAS OF THE BREEDING BIRDS OF SOUTHERN QUÉBEC

The Breeding Birds of Québec

Edited by
Jean Gauthier and Yves Aubry

Canadian Wildlife Service
Environment Canada
Québec Region

ALL PROCEEDS FROM THE SALE OF THIS BOOK WILL GO TO THE ASSOCIATION QUÉBÉCOISE DES GROUPES D'ORNITHOLOGUES FOR BIRD CONSERVATION AND ADVOCACY IN QUÉBEC.

ASSOCIATION QUÉBÉCOISE DES GROUPES D'ORNITHOLOGUES

THE PROVINCE OF QUEBEC SOCIETY FOR THE PROTECTION OF BIRDS

CANADIAN WILDLIFE SERVICE, ENVIRONMENT CANADA, QUÉBEC REGION

1996
Montréal (Québec)
Canada

Available in bookstores or by order from:

L'Association québécoise des groupes d'ornithologues
4545 Pierre de Coubertin Street
P. O. Box 1000, Station M
Montréal, Québec
Canada H1V 3R2

Aussi disponible en français sous le titre:
Les Oiseaux nicheurs du Québec: Atlas des oiseaux nicheurs du Québec méridional
ISBN: 2-9801553-1-4

Canadian Cataloguing in Publication Data

Main entry under title:

The Breeding Birds of Québec: Atlas of the Breeding Birds of Southern Québec

Co-published with the Province of Quebec Society for the Protection of Birds and the Canadian Wildlife Service, Québec Region.

Issued also in French under the title: *Les Oiseaux nicheurs du Québec: Atlas des oiseaux nicheurs du Québec méridional*

Includes bibliographical references.

Translation, with minor corrections, of *Les Oiseaux nicheurs du Québec: Atlas des oiseaux nicheurs du Québec méridional,* published in 1995.

ISBN: 2-9801553-2-2

1. Birds — Québec (Province) — Geographical distribution. 2. Birds — Nests — Québec (Province). I. Gauthier, Jean. (1947–). II. Aubry, Yves. (1958–). III. Association québécoise des groupes d'ornithologues. IV. Province of Quebec Society for the Protection of Birds. V. Canadian Wildlife Service, Québec Region.
VI. Title: The Breeding Birds of Québec.

QL685.5Q82 1996 598.29714 C95-980149-9

Art editor:
Norman Dupuis

Computer graphics
and colour separation:
Compélec Inc.

Published by L'Association québécoise
des groupes d'ornithologues,
the Province of Quebec Society for
the Protection of Birds,
and the Canadian Wildlife Service,
Environment Canada, Québec Region

Editors

Legal deposit – 1996
Bibliothèque nationale du Québec
National Library of Canada
ISBN 2-9801553-2-2 1996

Correct citation for this work:

Gauthier, J. and Y. Aubry (eds.). 1996. The Breeding Birds of Québec: Atlas of the Breeding Birds of Southern Québec. Association québécoise des groupes d'ornithologues, Province of Quebec Society for the Protection of Birds, Canadian Wildlife Service, Environment Canada, Québec Region, Montréal, 1302 pp.

Printing and binding:
Métropole Litho Inc.

Cartography (computerized maps
of southern Québec):
Environmental Information Systems Division,
State of the Environment Reporting,
Environment Canada, Ottawa

Correct citation for a section of this work:

Example:

Alvo, R. 1996. Common Loon, pp. 206-209 *in* J. Gauthier and Y. Aubry, eds. The Breeding Birds of Québec: Atlas of the Breeding Birds of Southern Québec. Association québécoise des groupes d'ornithologues, Province of Quebec Society for the Protection of Birds, Canadian Wildlife Service, Environment Canada, Québec Region, Montréal, 1302 pp.

To all those who ventured
forth to look for birds
and all those whose commitment
and generosity have brought
this project to fruition.

Main Partners

Association québécoise des groupes d'ornithologues

The Province of Quebec Society for the Protection of Birds was founded in 1917 and the Club des ornithologues de Québec in 1955, but amateur birdwatching really became popular in the 1970s. It was a hobby that took root among all segments of society, and other clubs were founded in various areas. In October 1981, five clubs banded together to found the Association québécoise des groupes d'ornithologues (AQGO). Today, the AQGO has 28 affiliates with nearly 5000 members.

The common goal of the AQGO and its affiliates is to foster the development of amateur birdwatching, promote the study, observation, and protection of birds in Québec, and ensure conservation of bird habitat.

The role of the affiliates is to promote amateur birdwatching by organizing activities locally, while the AQGO works on publicizing information on Québec birds and lobbying on issues affecting their protection. As part of this thrust, the AQGO produced *The Threatened Birds of Quebec*, and between 1987 and 1992 conducted research on the breeding of the Piping Plover in the Magdalen Islands.

In furtherance of its main goals, the AQGO collaborated on the production of the Atlas by having its members participate in data gathering, by acting in an advisory capacity and through representation on the advisory committee and the editorial board.

Canadian Wildlife Service

The Canadian Wildlife Service (CWS), a component of Environment Canada, is responsible for wildlife matters within the jurisdiction of the federal government and for administration of three pieces of legislation: the 1994 Migratory Birds Convention Act, the Canada Wildlife Act, and the Wild Animal and Plant Protection and Regulation of International and Interprovincial Trade Act. Its responsibilities include protection and management of migratory species, habitat of national importance, and threatened species, as well as work on wildlife issues of national and international scope. The CWS acts directly by establishing and developing national wildlife areas and migratory bird sanctuaries. Wildlife biologists also monitor the effects of toxic chemicals that get into ecosystems.

The CWS is particularly concerned with understanding migratory bird populations and conducts research on waterfowl, seabirds, and passerines. It was thus natural for the CWS to act as project manager and co-ordinator for the Atlas, drawing on its broad experience and expertise in the study of avian fauna.

The Province of Quebec Society for the Protection of Birds

The Province of Quebec Society for the Protection of Birds (PQSPB), founded in 1917, is the oldest ornithological society in Québec. Today its membership is mostly concentrated in the Montréal area.

After 75 years, PQSPB is still flourishing and has over 700 members. Its objective is twofold: to initiate its members and the larger community into the world of birds, and to promote the protection of birds and their environment.

Thus, in addition to offering a wide range of services and activities—field trips, lectures, and Bird Fairs—the PQSPB is also often called on to lobby lawmakers in defence of birds and their habitat. Its most important role, however, is in conservation, where it works closely with other national and international conservation groups. As it has financial resources of its own, it is able to acquire and manage bird sanctuaries.

The PQSPB is also active in the furtherance of scientific knowledge of birdlife, both through the research of its own members and the bursaries and grants it awards to other ornithologists.

Ministère de l'Environnement et de la Faune du Québec

The mission of the Ministère de l'Environnement et de la Faune (Québec Ministry of Environment and Wildlife) is to oversee the conservation and development of Québec wildlife and its habitat.

Its chief responsibility is to ensure the survival of species. It therefore monitors various forms of wildlife exploitation in the interests of sustainable development and maintenance of biodiversity. To achieve this, it conducts analyses of the status of wildlife populations and adopts various conservation measures as and when appropriate. Habitat preservation is also a key element in its species conservation strategy. It protects exceptional wildlife sites through its network of parks and ecological and wildlife reserves. It operates a number of surveillance, education, and awareness programs relating to the protection of wildlife and its habitat.

Under the Act respecting the conservation and development of wildlife, the Ministère protects bird species not covered by federal legislation on migratory birds. The Act mandates the Ministère to regulate hunting, trapping, habitat, animals in captivity, and wildlife territory. The Act respecting threatened or vulnerable species charges it with designating species at risk and extending additional protection to them. Wildlife conservation officers also co-operate in applying the 1994 Migratory Birds Convention Act.

The Ministère de l'Environnement et de la Faune encourages all initiatives intended to further knowledge of bird populations and their habitat, and was accordingly an enthusiastic contributor to the Atlas.

Table of Contents

Preface

Publication of *The Breeding Birds of Québec: Atlas of the Breeding Birds of Southern Québec* is a true milestone in the history of ornithology in Québec. This work, the first of its kind in Québec, embodies the efforts of hundreds of people over more than 10 years.

Large-scale volunteer participation has been the key to the success of this undertaking. It has also demonstrated that a number of ornithological organizations, amateur as well as professional, working together, can make a noteworthy contribution to science.

As is now widely recognized, management of the environment and of wildlife is everybody's business, not just government's. However, Environment Canada's Canadian Wildlife Service can facilitate the work of people and organizations seeking to collaborate on managing migratory bird populations, which know no borders and reach every part of Québec.

In taking on the task of co-ordinating a wide range of expertise and input, the Canadian Wildlife Service has grown to be more attentive to the public and to its partners, while continuing to fulfill the duties mandated to it. Among the environmental sciences, in Québec as elsewhere, ornithology attracts thousands of enthusiasts. As it is the pastime and passion of some and the profession and vocation of others, only a spark, a word was needed to launch this joint project and rally all interested parties, but the crucial factor was the dedication of committed people who would carry the torch and keep it alight. Through their tenacity and perseverance, this shared initiative has produced not only distribution maps for each species breeding in Québec, but also a monumental synthesis of their breeding ecology. The Atlas is thus more than an inventory of all the species breeding in Québec, whose capacity for regeneration depends, of course, on the state of ecosystems. It is a fund of essential data for any land use project, since Québec's breeding bird species are associated with habitats which are highly diversified, essential to their survival, and necessary to the maintenance of the quality of our environment.

In short, this work is a gold mine of information for readers seeking to compare the distribution of one species with that of some other or to study the use of various habitats. Professionals too will be able to use it for project planning and environmental impact assessments, proposals for natural parks or wildlife reserves, or any other program with the potential to affect the environment. As breeding birds are an excellent ecological barometer, their presence or absence can shed much light on the state of the environment. Breeding bird atlases are also useful for comparing the breeding populations of different epochs. One day, naturalists will produce another atlas showing changes over time in bird populations and their habitats. Current pressures on the environment caused by demographic growth and industrialization, as well as habitat loss, mean that this future atlas will cover a different set of species, as some species, more tolerant of environmental change, will become more numerous, while others will decline in numbers.

Throughout this work, the wealth of maps and photographs will reveal to the reader not just the variety of niches occupied by birds, but also the richness of the land that is Québec. Proper management of our environment represents a major challenge that requires co-operation and commitment on everybody's part, but involvement and determination will help overcome great obstacles. This Atlas is incontrovertible evidence of that. The many volunteers who have contributed and participated as well as project managers Jean Gauthier and Yves Aubry of the Canadian Wildlife Service can be rightfully proud of this accomplishment. I offer them my sincere congratulations.

Jean Cinq-Mars
Director, Canadian Wildlife Service, Québec Region
1984–94

Foreword

Birds, visible, varied, and diverse in their range of habitat, arouse people's interest in a way that few other creatures can match. This popular interest in birdwatching and environmental issues, together with wildlife management imperatives, is what prompted Québec ornithologists to produce a breeding bird atlas comparable to those of other provinces, states, and countries. Since knowledge of how species are distributed—especially data on ecological factors influencing distribution—is fundamental to sound management, the appearance of this work is very timely.

A breeding bird atlas also provides baseline data by which to assess changes to our environment. Unlike birds of passage, breeders spend several months in intimate relationship with an environment which they share with us. Québec's mosaic of habitats is home to almost 300 species, each of which occupies, year after year, a particular niche suited to raising a healthy brood. If the species dependent on each niche fail to find their preferred habitat or to adapt to a new one, their distribution and abundance will suffer accordingly. Just how sensitive birds are to environmental change is shown, for instance, by the way certain warblers react to spruce budworm outbreaks by breeding more prolifically. The state of health of bird populations thus depends on the environment and reflects the state of the habitat they occupy, and this is why birds make good environmental indicators. One need only recall the baleful effects of DDT use on bird populations, especially raptors like the Peregrine Falcon, though others were also affected; Chapdelaine et al. (1987), for example, demonstrated a connection between pesticide use and the decline in breeding success among Bonaventure Island gannets noted in the 1960s and 1970s.

During the 1970s, in the years following the start of work on the British breeding bird atlas, several Québec ornithologists conceived the idea of launching a similar project. However, few ornithologists were then available for the immense area to be covered, and hence the project had to be deferred.

Less than 10 years later, some biologists of the Canadian Wildlife Service began work outside of their official duties on a book on the ecology and distribution of Québec waterfowl. Meeting weekly, they found it hard to complete this task on a purely voluntary basis and without logistical resources of any kind, even though they were limiting themselves to waterfowl.

The project died, but the idea did not. In the early 1980s, the Canadian Wildlife Service, Québec Region, tasked its biologists with drawing up a report on what was known about Québec's birdlife. Of the several hundred species for which the Canadian Wildlife Service was responsible in Québec under the terms of the Migratory Birds Convention Act, very little was known, except in the case of a few colonial or game species and the flocks of duck and goose species that gather on the St. Lawrence during migration.

At that time, Ontario biologists and ornithologists, like their counterparts in a number of U.S. states, were preparing to launch their own breeding bird atlas project. In 1981, at a symposium at which the eastern states and provinces of North America were represented, Québec was invited to join this international community.

Two years later, in the late fall of 1983, the Canadian Wildlife Service decided to join in work on a breeding bird atlas, on condition that birding associations also be involved. During that winter, an advisory committee composed of specialists from the biology departments of Québec universities and representatives of the Québec Ministry of Environment and Wildlife and the Canadian Museum of Nature was established, and birding associations, clubs, and societies geared up under the leadership of the Association québécoise des groupes d'ornithologues; the Province of Quebec Society for the Protection of Birds made a financial commitment and pledged the efforts of its members. The first field season began six months later, in May 1984.

Six field seasons were required, one more than for most other atlas projects. As the work proceeded, an editorial board started work, and it quickly became apparent that there was a dearth of information in French on Québec breeding species. There were many complaints, as people needed information to make their field work more effective. It was at that point that the needs and expectations of various resource conservation groups began to crystallize. At the same time, the number of requests for information from the public—teachers, biologists, consultants, students, hunters—addressed to the Canadian Wildlife Service began to escalate, making it necessary to publish reasonably complete guides. Thus, while it was clear that this went well beyond the scope of traditional atlases, the editorial board decided to add to the project a synthesis of bird biology, listing major reference works so that members of the public could go back to the sources themselves to study in depth those aspects of interest to them. Special attention has been given to the various stages of breeding—laying, incubation, and raising the young, both in and off the nest; such information will be useful for land-use planners and consultants in preparing wildlife impact studies. Similarly, under the heading "Notes on territory," we discuss each species' home range requirements. Density data by habitat are also given, so that the reader will have an idea of how abundant a species is in different habitats.

In addition, to provide a better understanding of the distribution of breeding birds, we thought it useful to give an outline sketch of the biogeography of Québec. This should not be seen as anything more than a synopsis of data with a bearing on present bird distribution, and it is in no sense intended as a rigorous exposition. We suspect that the bird communities occupying the province today reflect Québec's changing landscape in recent centuries; it would thus be utopian to think of turning the clock back, as it would be hard to establish a reference period, human impact on wildlife having set in long before the arrival of European settlers, though it has escalated since. As demographic pressures on the environment are much stronger than in the past and are still growing, it is important to understand the biology of the species around us so that we can minimize the effects on fauna and flora.

Access to information is the key to grassroots involvement in environmental management. Indeed, this is not viable without community support. Public interest grows to the extent that information is widely available. Broadly based participation is what makes real conservation possible, since we know that the most effective resource management mechanisms are those that stimulate preventive practices. On the basis of the number of Québecers (85% in 1991) who indulge in wildlife-related pursuits (CWS and Statistics Canada 1993), there is every justification for believing that the extra effort made to produce a non-traditional atlas will have been worth the cost.

Completion of this work on the breeding birds of Québec has taken much longer and has proved very much more demanding than could have been foreseen. The sheer scale of the undertaking was compounded by the unexpected complexity of the compilation and analysis. Innumerable problems arose for which there were no evident or immediate solutions, but none of those involved was deterred, and this steadfastness meant that problems could be addressed one by one.

This atlas would never have appeared had we not had the support of Environment Canada management and the confidence and generosity of the members of the editorial board. We benefited from the dedication and efficiency of the entire team: publishing, editing, translation, data capture, layout, not to mention the essential contribution of all the birders. This book would never have seen the light of day without the participation of numerous photographers and the contributions of the various agencies, government departments, and public and private corporations to which we turned for help.

This book embodies both the efforts and the aspirations of a community of people anxious to find a way to achieve better management, understanding, and awareness of our environment. It is the fruit of the pooled experience and knowledge of all those involved. For our part, we hope to reach as wide a readership as possible with all that we have learned about birds, especially the new insights gained in the course of this project.

Jean Gauthier and Yves Aubry

Acknowledgments

The Breeding Birds of Québec: Atlas of the Breeding Birds of Southern Québec is the product of the largest ornithological undertaking ever in Québec. By virtue of the co-operative and original nature of the project, it brought together with a single purpose ornithologists, amateur and professional photographers, and various institutions, both public and private. Thousands of bird lovers came together and shared the many tasks involved in the project. The *Atlas* is testimony to the co-operation that prevailed throughout the project and to which it owes its success.

The editors and publishers enjoyed the unqualified support of birdwatching clubs, the Québec ornithological community, and Environment Canada, Québec Region, in particular Jean-Pierre Gauthier, Regional Director General, as well as Michel Lamontagne, Jean Cinq-Mars, Charles-A. Drolet, Georges Arsenault, and Isabelle Ringuet.

Through the 11 years that it ran, the project was able to rely on invaluable assistance from many people who devoted to it their time, interest, talents, gifts, and enthusiasm. Without them, the *Atlas* would not have seen the light of day.

Partners

The following partners provided invaluable assistance in furtherance of environmental awareness objectives.

Alcan
Association québécoise des groupes d'ornithologues
Canadian Forest Service, Natural Resources Canada, Québec Region
Canadian Wildlife Service, Environment Canada, Québec Region
Fondation de la faune du Québec
Hydro-Québec
Ministère de l'Environnement et de la Faune du Québec
Mouvement des caisses Desjardins
Province of Quebec Society for the Protection of Birds

Birdwatching clubs

Centre d'interprétation de la nature du lac Boivin
Club des ornithologues amateurs du Saguenay–Lac-Saint-Jean
Club des ornithologues de Brome-Missisquoi
Club des ornithologues de la Gaspésie
Club des ornithologues de l'Outaouais
Club des ornithologues de Québec
Club des ornithologues des Bois-Francs
Club des ornithologues du Bas-Saint-Laurent
Club d'observateurs d'oiseaux de la Haute-Yamaska
Club d'observateurs d'oiseaux de Laval
Club d'ornithologie de la Manicouagan
Club d'ornithologie de Longueuil
Club d'ornithologie de Trois-Rivières
Club d'ornithologie des îles de la Madeleine
Club d'ornithologie du Cégep de La Pocatière
Club d'ornithologie Sorel-Tracy
Club du loisir ornithologique maskoutain
Club Marie-Victorin
Club Natural de Granby
Club ornithologique de la Mauricie
Club ornithologique des Hautes-Laurentides
Georgeville Nature Society
Groupe d'animation en sciences naturelles du Québec
Province of Quebec Society for the Protection of Birds
Société de biologie de Montréal
Société d'ornithologie de Lanaudière
Société du loisir ornithologique de l'Abitibi
Société du loisir ornithologique de l'Estrie
Société ornithologique du centre du Québec
Société zoologique de la Mauricie
Société zoologique de Québec
St. Francis Valley Nature Society
St. Lawrence Valley Naturalist Club

Editorial Board

Yves Aubry, Canadian Wildlife Service, Environment Canada, Québec Region
Pierre Bannon, Province of Quebec Society for the Protection of Birds
Normand David, Association québécoise des groupes d'ornithologues
Jean Gauthier, Canadian Wildlife Service, Environment Canada, Québec Region
Daniel Jauvin, Association québécoise des groupes d'ornithologues
Pierre Lamothe, Hydro-Québec
Raymond Lemieux, Canadian Wildlife Service, Environment Canada, Québec Region
Pierre Normand, Environment Canada, Québec Region
Réginald Ouellet, Ministère de l'Environnement et de la Faune du Québec
Desneige Perreault, Association québécoise des groupes d'ornithologues
Michael Spencer, Province of Quebec Society for the Protection of Birds
Jacqueline Vincent, Canadian Wildlife Service, Environment Canada, Québec Region
Richard Yank, Province of Quebec Society for the Protection of Birds

Advisory Committee

Yves Aubry, Canadian Wildlife Service, Environment Canada, Québec Region
Pierre Bannon, Province of Quebec Society for the Protection of Birds
Roger Bider, McGill University
David M. Bird, McGill University
Normand David, Association québécoise des groupes d'ornithologues
Pierre Drapeau
Jean Gauthier, Canadian Wildlife Service, Environment Canada, Québec Region
Jean-François Giroux, University of Québec at Montréal
Christian Houle, Collège de Sherbrooke (Cégep)
Daniel Jauvin, Association québécoise des groupes d'ornithologues
Alain Leduc
Michel Lepage, Ministère de l'Environnement et de la Faune du Québec
Raymond McNeil, University of Montréal
François Morneau, Groupe de recherche et d'étude en biostatistique de l'environnement
Henri Ouellet, Canadian Museum of Nature
Bruno Scherrer, University of Québec at Montréal
Rodger D. Titman, McGill University

Regional co-ordinators and atlassers

By their dedication and dynamism, the 42 regional co-ordinators recruited, motivated and trained atlassers and acted as liaison with the Canadian Wildlife Service. The tasks they accomplished over a period of seven years—distributing supplies, passing on information, checking, compiling, and validating data—expedited communications and decentralized the work.

The invaluable contribution of birders to the science of ornithology cannot pass without mention. What prompts all these volunteer sleuths to spend hours in the field looking for clues to breeding is their devotion to science and their commitment to their avocation.

All these volunteers, with close to 70 000 field hours to their credit, are the foundation of the Atlas. Without them, we would never have obtained so much new information on bird distribution in southern Québec. We acknowledge their efforts and their contribution to Québec ornithology:

Regional co-ordinators

Pierre Bannon
Claire Bédard
Marc-André Bélanger
Gérard Bellavance
André-Guy Bernier
Diane Berthiaume
Rachel Blais
Mario Boisvert
Adalbert Bouchard
Marc Boudreau
Geneviève Caron
Jean-Pierre Caron
Louis-Sarto Carrier
Jean-Marc Côté
Michel Côté
Gérard Cyr
Normand David
Gilles Desrochers
Alain Desrosiers
Bruno Duchesne
Pierre Fradette
Jean Gaudreault
Guy Gendron
Rémy Gilbert
Christian Houle
Daniel Jauvin
Pierre Laporte
Vincent Létourneau
Yvan Lévesque
Kathleen Lord
Alain Pelletier
Réal Pelletier
Desneige Perreault
Pierre Poulin
Reynald Rivard
Laval Roy
Danielle Ruest
Pascal Samson
Marc Simard
Daniel St-Hilaire
Serge Tanguay
Yvon Villemure

ATLASSERS

Sylvie Abbott
René Adam-Chaumet
David Ainley
Marianne G. Ainley
Jocelyne Aird-Bélanger
Sylvie Alain
Sylvie Allard
Yvon Allen
Lucie Amyot
André Anctil
Fred Anderson
Raynald Anderson
Rhéal Angers
Robert Angers
Teresa Aniskowicz
Tristan Ap Rheinallt
Dany April
Claude Arbour
Sylvain Arbour
François Arcand
Suzanne Armstrong
Christian Arnaudin
Élise Arnaudin
Margaret Arnaudin
Adèle Arsenault
Hélène Arsenault
Jocelyne Arsenault
Stéphane Arteau
Sheila Arthur
Anne Artigau
Étienne Artigau
Jean-Pierre Artigau
Francine Asselin
Gary Atkinson
Bernard Aubé
Denis Aubé
Martin Aubé
Michel Aubé
Clément Aubert
Yves Aubry
Claude Auchu
Francine Aucoin
Guy Aucoin
Guylain Aucoin
Gilbert Audet
Robert Audet
Faith Avis
Guy Avoine
Alain Ayotte
Andrée Ayotte
Camille Ayotte
Yves Bachand
René Baillargeon
Pierre Bannon
René Bannon
Anne-Marie Banville
Daniel Banville
Suzanne Banville
A. Baril
Claude Baril
Robert Barnhurst
John Barny
Robert Baronet
Armand Barré
Lucie Barré
François Barrette
Gaétan Barrette
Simon Barrette
Jean-Pierre Barry
Raymond Barry
Michelle Bastien
Ruth Bastien
Charles Baumann
Catherine Bayne
Gladys Beattie
Nancy Beattie
S. Beaudet
Annie Beaudoin
Luc Beaudoin
Michelle Beaudoin
Ghislaine R. Beaudry
Rosanne Beaulé
Gaétan Beaulieu
Michel Beaulieu
Denis Beaumier
Marc Beaupré
Michel Beaupré
Normand Beauregard
Ghislain Beauséjour
Michel Beauséjour
Sylvain Beauséjour
Claude Beausoleil
Michel Beausoleil
Denise Beauvais
Emil Beck
Gregor Beck
Tony Beck
Christian Bédard
Claire Bédard
Daniel Bédard
Jean Bédard
Josée Bédard
Yves Bédard
Colombe Béland
Marie-Paule Béland
Claude Bélanger
France Bélanger
Marc-André Bélanger
Martin Bélanger
Renaud Bélanger
Yvon Bélanger
Luc Bélisle
Marc Bélisle
Jean Béliveau
Jean-Marc Béliveau
Gérard Bellavance
Sébastien Bellefeuille
Jean Bellerose
Gérald Benoît
Gilles Benoît
Laurie Benoît
Réjean Benoit
Yves Benoît
Marg Benson
Alain Bergeron
Clément Bergeron
Daniel Bergeron
Maxime Bergeron S.G.
Michel Bergeron
Paul Bergeron
Pierre Bergeron
Marc-André Bernard
Normand Bernatchez
Joyce Berndt
André-Guy Bernier
Ghislaine Bernier
Hélène Bernier
Martine Bernier
Carole Berruel
Sylvie Berson
Walter Bertacchi
Pascal Berthelot
Diane Berthiaume
Jean-Louis Berthiaume
Sylvie Berthiaume
Lucille Bertrand
Michel Bertrand
André Bérubé
Denise D. Bérubé
Huguette Bérubé
Jacques Bérubé
Simon Bérubé
Laurier Bessette
Marjolaine Bezeau
Roger Bider
Claude Bilodeau
Georges Bilodeau
B. Bird
David M. Bird
Dominique Biron
Réal Bisson
Danielle Blackburn
Yves Blackburn
Philippe Blain
Denyse Blais
Johanne Blais
Rachel Blais
Stéphane Blais
Claudette Blanchard
Diane Blanchard
Kathleen A. Blanchard
Marie Blanchet
Luc Blanchette
Pierre Blanchette
Pauline Blaquière
Daniel Blouin
François Blouin
Jeremy Bodycomb
Camille Boily
Félix-Antoine Boily
Monique Boily
Paul Boily
Alain Boisclair
Claude Boisclair
Jeannette Boisclerc
Jean Boisvert
Marc Boisvert
Mario Boisvert
André Boivin
Thérèse Boivin
Thierry Boivin
Daniel Bolduc
François Bolduc
Louis Bolduc
Denis Bombardier
Robert Bonneau
Daniel Bordage
Donald Bordeleau
Elisabeth Bossert
Adalbert Bouchard
Albert Bouchard
André Bouchard
Andrée Bouchard
Cécile Bouchard
François Bouchard
Gilbert Bouchard
Jean-Luc Bouchard
Lisianne Bouchard
Louise Bouchard
M. Bouchard
Rémy Bouchard
Roger Bouchard
Steven Bouchard
Sylvie Bouchard
Claudette Boucher
Marc Boudreau
Steve Boudreault
Michel Boulard
Michelle Boulard
Michel Boulé
Ghislaine Boulet
Jacinthe Boulet
Marie Boulet
Réal Boulet
André Bouliane
Paul Bouliane
Danny Boulianne
Michel Boulianne
France Bourassa
Michel Bourassa
André-François Bourbeau
Jean-Claude Bourgeois
André Bourget
Michel Bourgon
François Bourguignon
Gérard Bourque
François Bourret
Alain Bousquet
Lucie Bousquet
Marcel Bousquet
Jacques Bouvier
Thelma Bowers
Reed Bowman
Rama Boyd
John Boynton
Louise Brabant
Daniele Bracke
Robert Bracken
L. Brais
Claude Brassard
Léo Brassard
Ernest Brault
Robert Brault
Noël Breton
Christian Briand
Céline Brien
Charles Brière
Lucie Brière
John Brisbois
Jacques Brisson
Steve Brisson
Denise Brochu
Gaston Brochu
Louise Brochu
Claire Brodeur
Serge Brodeur
Pierrette Brossard
Gaétan Brouillette
Serge Brouillette
Laurent Brousseau
Pierre Brousseau
Albert Bruggeman
Jean D. Bruneau
Lauraine Bruneau
Michel Brunette
Denise Bruno
Hugues Brunoni
Jackie Budd
Henry Budger
Charles Bury
Joëlle Bussière
Bruno Bussières
Desneiges Bussières
Paul Bussières
Hegin Buttom
François Cadieux
Jean Cadieux
Louise Cadieux
Réal Cadoret
Roger Caissy
Harry Calvert
Erwin Camber
A. Camerone
Margaret Campbell
Agnès Caouette
Florienne Caouette
Martin Caplan
Robert Capozi
Juliette Cardinal
Monique Cardinal
Céline Caron
Danielle Caron
Geneviève Caron
Ghislain Caron
Gilbert Caron
Gisèle Caron
Jean Caron
Jean-Pierre Caron
Jocelyne-Maude Caron
Julien Caron
Louise Caron
Madeleine Caron
Martine Caron
Raymond Caron
France Carreau
Alain Carrier
André Carrier
Annie Carrier
Daniel Carrier
Louis-Sarto Carrier
Gary Carrigan
Gary Carruthers
Jean Carruthers
John Cartwright
Rosemary Cartwright
Martin Carver
Marcel Castilloux
Julien Cauvier
Stéphanie Cazelais
Jacques Chabot
Patricia Chabot
Pierre Chagnon
Cynthia Chalk
Julien Chamberland
Pauline Chamberland
Andrée Champoux
Gilles Chapdelaine
J. Chapman
B. Chaput
Édouard Charbonneau
Jean-Luc Charbonneau
Michel Charbonneau
Noël Charbonneau
Alain Charest
Jocelyne Charest
Denis Charlebois
David Charlton
Luc Chartré
Jacinthe Châteauvert
Sylvain Châteauvert
Ted Cheskey
Odilon Chiasson
Roland Chiasson
Alain Chouinard
Hélène Chouinard
Rosemarie Chrétien
Bill Clark
Colin Clarke
Geoffrey Clarke
Janet Clarke
Benoît Clément
Madeleine Clément
Jean-Paul Clermont
Brooke Clibbon
Murielle Cliff
André Cloutier
Danielle Cloutier
Elizabeth Cloutier
François Cloutier
Marcel Cloutier
Michèle Cloutier
Nancy Cloutier
Club des ornithologues de Québec
Pierre Codère
Philippe Coggin
Isabelle Collin
Sophie Collin
Éloi Colpron
Eric Comtois
Sylvie Comtois
Sandra Conciatori
Claudette Corbeil
M. Corbin
Aurore Cormier
Benoît Cormier
Charles Cormier
Claudette Cormier
Denis Cormier
Ronald Cormier
Serge Cormier
Ernest Cornes
Charles Corriveau
Lucie Corriveau
Louis Cossette
Adèle Côté
Alain Côté
Christian Côté
Élaine Côté
Francine Côté
Jean-Marc Côté
Jeannine Côté
Loraine Côté
Martin Côté
Michel Côté
Michel Côté
Réjean Côté
Robert Côté
Jean Couchouron
Daniel Coulombe
Josette Coulombe
Jean-Guy Courchesne
Trefflé Courchesne
Élyse Cournoyer
Sylvie Cournoyer
Yvan Cournoyer
Louise Cousineau
Céline Coutu
Albini Couture
Gaétan Couture
Guy Couture
Léo Couture
John Cowans
Mary Cowans
Jean Crépeau
Linda Crête
Benoît Crevier
William J. Crins
Bruno Croft
Réal Croteau
Ron Curtis
André Cyr
Gérard Cyr
Huguette Cyr
Linda Cyr
Marielle Cyr
André Dagenais
Pauline Daigle
Daniel Daigneault
Pierre Daigneault
Donald Dallaire
Lise Dallaire
Lucie Dalpé
Lucie Dame
Lucette D'Amour
Lucie D'Amour
Marcel Darveau
Claude Dassylva
Pierre Dassylva
Suzanne Dassylva
Francine D'Astous
Diane Dauphin
Susan Davey
Normand David
Sharon David
Nathan Davis
R. Decelles
Manon Declos
Jean De Joncas
Paul De La Chevrotière
Philippe Delacrétaz
Gladys Delafield
Danièle Delisle
Alain Demers
Alain Demers
Benoît Demers
M.G. Demers
Bernard Denault
Isabelle Denault
Jean-Bernard Denault
Léo-Guy de Repentigny
Jean Desaulniers
Serge Desbiens
Claude Deschênes
Marc Deschênes
Steeve Desfossés
Suzanne Desgagné
Jean-Luc DesGranges
Serge Desgreniers
Stéphane Deshaies
Jean Désilet
Daniel Desjardins
Gérard Desjardins
Gilles Desjardins
Luc Desjardins
Patrick Deslauriers
Paul Desmarais
Bernard Desmeules
Normand Desormiers
Alice Desrochers
Alphonse Desrochers
André Desrochers
Gilles Desrochers
Gilles Desrochers
Robert Desrochers
Sylvie Desrochers
Alain Desrosiers
Anatole Desrosiers
Denis Dessin
Pierre De Tonnancour
Bruce DiLabio
Joe Dilena
Jean Dillon
Claude Dion
François Dion
Jacques Dion
Paul-Émile Dion
Richard Dion
Robert Dion
Roger Dion
Lyne Dionne
Micheline Dionne
Michel Dixon
Daniel Dolan
Jean Domingue
June Donovan
Daniel Dorais
Pierre Dorais
Hélène Dorval
Robert Doyon
Louis Drainville
Pierre Drapeau
Michel Drew
Bernard Drolet
Bruno Drolet
Charles-A. Drolet
Éric Drolet
Gemma Drolet
René Drolet
Silvie Drouin
Armand Dubé
Bruno Dubé
Gilbert Dubé
Jean Dubois
John Dubois
Louise Dubois
Réjean Dubois
Rodolphe-Louis Dubois
Sonia Ducasse
Claude Duchaine
Marc Duchemin
Marguerite Duchesnay
Bruno Duchesne
René Duchesne
Yvan Duchesne
Camille Dufresne
Carlo Dufresne
Pierre Dufresne
Henri-Paul Dugré
François Duguay
Yvon Duguay
Christiane Duhamel
Serge Dulac
Carole Dulude
Claudette Dumais
Alain Dumont
Bruno Dumont
Robert Dumont
Serge Dumont
Guy Dumoulin
Raymond Dumoulin
André Dupéré
Cécile Dupuis
Pierre Dupuis
Serge Dupuis
Thérèse Dupuis
Yvette Dupuis
Pierre Dupuy
Richard Dupuys
Gaétan Duquette
Michel Dussureault
Brigitte Dutil
Anne Duval
Cécile T. Duval
Chantale Duval
Jim Eastman
Cameron Eckert
Beth Edmonds
Margaret Elliot
Kyra Emo
Hélène Essiambre
Pierre Ethier
Ronald Ewing
Jean-Pierre Fabien
Gilles Falardeau
Marie-Josée Falardeau
Patrick Fallu
Harry Farfan

Renée Faubert
Ilay C. Ferrier
Lionel Filiatrault
Luc Filiatrault
Muriel Filiatrault
Patrick Filiatrault
Gaétan Fillion
Guy Fitzgerald
Ethel Fogarty
Christiane Foley
Pierre Fontaine
Pierre Fontayne
Forillon National
Park naturalists
Alain Fort
Liette Fortier
Fernand Fortin
Francine Fortin
Guy Fortin
Jean-Luc Fortin
Jean-Pierre Fortin
Mélanie Fortin
Roger Fortin
Claude Fournier
Denis Fournier
Pierre Fournier
Yvan Fournier
Yvon Fournier
Daniel Fowler
Heather Fox
Pierre Fradette
Fernande Fragnier
Jérôme Fragnier
Philippe Fragnier
Claudel Francoeur
Betty Fransham
Doreen Fraser
Raymond Fredette
Christiane Frenette
Rolland Fuoco
Teddy Furter
Mireille Fyfe
Hercule Gaboury
Patrick Gadbois
Andrée Gagné
Clarence Gagné
Denise Gagné
Diane Gagné
Hélène Gagné
Jacques Gagné
Jacques Gagné
Johanne Gagné
Paul Gagné
Paul-Aimé Gagné
Richard Gagné
Serge Gagné
Sophie Gagné
Alain Gagnon
Benoît Gagnon
Christian Gagnon
François Gagnon
Gilles Gagnon
Jean Gagnon
Jocelyne Gagnon
Léon Gagnon
Louis-Gilles Gagnon
Marjolaine Gagnon
Monique Gagnon
Rachelle Gagnon
Suzanne Gagnon
Sylvie Gagnon
Dee Gaiger
Sylvia Galbac
Robert Galbraith
Normand Galipeau
Alexis Gallant
Denis Gallant
Yves Garant
N. Gard
Carole Gariépy
Céline Gariépy
Louis Gariépy
Claude Garneau
Lucie Garneau
Brian Garvin
Collen Gaskill
Martin Gaudet
René Gaudet
André Gaudreau
Claude Gaudreau
Ghyslain Gaudreau
Jean-Claude Gaudreau
Alain Gaudreault
Jean Gaudreault
Giselle Gaumont
Alain Gauthier
Alain Gauthier
Christiane Gauthier
Claude Gauthier
Claude P. Gauthier
Diane Gauthier
Doug Gauthier
Gérald Gauthier
Jean Gauthier
Jean-Claude Gauthier
Jean L. Gauthier
Marc Gauthier
Michelle Gauthier
Scott Gauthier
Serge Gauthier
Steeve Gauthier
Yves Gauthier
Clémence Gauvin
Michèle Gauvin
Lyse Gauvin-Surprenant
Bernard Gauvreau
Cécile Gauvreau
Marie-Christine Gauvreau
Brian Gawn
Mark Gawn
Simon Gawn
Stephen Gawn
Bernard Gendron
Guy Gendron
Paul Gendron
Marc Genest
Sandra Genest
Serge Genest
David Gervais
Denis Gervais
Lise Gervais
Luc Gervais
Alain Giard
Louis Giard
Line Giasson
Ginette Giguère
Jean-Roch Giguère
Gaston Gilbert
Martin Gilbert
Pierre Gilbert
Rémy Gilbert
Sylvain Gilbert
André Gingras
Daniel Gingras
Noëlla Gingras
Normand Gingras
Réjean Gingras
Jim Ginns
Gilles Giraldeau
Jacques Giraldeau
Sylvain Giraldeau
Alain Girard
Christiane Girard
Lynda Girard
Pierre Girard
Sylvie Girard
Jean-Marc Giroux
Michel Giroux
Robert Giroux
Roger Giroux
M. Gladu
Fred Goodrow
Bob Gorman
Angèle Gosselin
France Gosselin
Luc Gosselin
Michel Gosselin
Julie Gosselin-Gendron
Rachel Gosselin-Gendron
Alain Gouge
Bernard Goulet
Robert Goulet
Céline Goyette
Douglas Graham
Dominique Gravel
Roger Gravel
Yvon Gravel
Marion Greenlay
Mario Grégoire
Alain Grenier
Clermont Grenier
Julie Grenier
Lise Grenier
Mario Grenier
François Grenon
Catherine Grohe
Conrad Groleau
Michel Groleau
Ghislaine Groulx
Denis Guay
Gilles Guay
Sylvie Guénette
Jacqueline Guillemette
Jean Guillemette
Jocelyn Guillemette
Michel Guillemette
Renée Guillemette
Andrée Guimond
Suzanne Guimond
Guillaume Guitard
Dorothy Gurden
Jean-Paul Habel
Richard Habel
Alison Hackney
Raymond Haddad
John Haemmerli
Pierre Hainault
Jean-Claude Hallé
Bertrand Hamel
Martin Hamel
Nathalie Hamel
Pascale Hamel
Stéphanie Hamel
Steve Hamel
Jean-Pierre Hameury
Christine Hanrahan
William Hansen
Jean Hardy
Jean-Marc Hardy
Yves Harnois
Jim Harris
Jeff Harrison
Carol Harvey
Conrad Harvey
Léo Haubrich
Eve Hauser
P. Hays
Carmen Hébert
Gina Hébert
Martial Hébert
Nathalie Hébert
Normand Hébert
Roger Heirman
Steve Heiter
Pierre Hénault
Denis Henri
Kathleen Héon
Marlène Heppel
Madeleine Héroux
Claude Hétu
Tom Hickcox
Jean Higginson
Sarah Jane Hills
T. Hills
Félix Hilton
P. Hilton
Gaston Hogue
Charles Hogues
Phillis Holtz
Bill Horsnall
Benoît Houde
Marie-Josée Houde
Michel Houde
Jim S. Houghton
Christian Houle
Gabriel Houle
J. G. Houle
Patrick Houle
Yvette Houle
Marguerite Hoyes
Jean-Marie Hudon
Andy Hugessen
Jane K. Hugessen
Claude Huot
Guy Huot
Laurent Huot
Suzette Huot
Robert Huxley
Jacques Ibarzabal
Claude Imbeau
Louis Imbeau
Louis Imbeault
René Isabel
Guy Isabelle
Robert Isabelle
Henri Jacob
Sylvie Jacob
Jocelyne Jacques
Marielle Jacques
Raymond Jalbert
S. Jalbert
Hélène Jargaille
Daniel Jauvin
Jocelyne Jauvin
Renald Jauvin
Serge Jauvin
Colette Jean
Louis Jean
Benoît Jobin
Elizabeth John
Wayne Johnson
Ann Johnston
Audrey Johnston
Austen Johnston
Diane Jolin
Hélène Jolin
Jean Jolin
Elizabeth Joly
Georges Joly
P. Jones
Mario Julien
Michel Julien
Yvan Julien
Marc Juteau
Denis Jutras
Jacques Jutras
Paul Kauler
Charles Kavanagh
Carole Kearny
Paul Kennedy
Marie-Françoise Kennes
Marg Kenney
Martin Kuhnigk
Anne Labbé
Francis Labbé
Denis Labonté
Serge Labonté
Gilles Labranche
Damien Labrecque
Georges Lachaîne
Claire Lachance
Claude Lachance
Jacques Lachance
Roger Lachance
Micheline Lachman
Roger Lachman
Diane Lacombe
André Lacroix
Bruno Lacroix
Michel Lacroix
Micheline Lacroix
Bernard Ladouceur
Lucien Laferté
André Laflamme
Gino Laflamme
Yoland Laflamme
Anne-Marie Lafleur
Bibiane Lafleur
Maryse Lafleur
Michelle Lafleur
Yvan Lafleur
Richard Lafond
B. Lafontaine
Mario Laforest
Charles Lafortune
Alfred Lagrenade
Jacques Lajoie
Liliane Lajoie
Yvon Lajoie
Michel Lalancette
Lucien Laliberté
Pierre Laliberté
Ghislain Lallier
Paul-Émile Lallier
Andrée Lalonde
Bruno Lalongé
Jean-Paul Lamarre
Mrs. Jean-Paul Lamarre
La Mauricie National Park
wardens
Daniel Lambert
François Lambert
Gilbert Lambert
Steeve Lambert
Claude Lamirande
André Lamontagne
Josette Lamontagne
Denis Lamothe
Marie Landreville
Gilles Landry
Jean Landry
Marie B. Landry
Michèle Landry
Nicole Landry
Peter Landry
Réjean Landry
Peter Lane
Jacques Langelier
B. Langevin
Natacha Langevin
J. Langlois
Jean-Yves Langlois
Yves Lanthier
Alain Lapierre
Jean Lapierre
Antoinette Laplante
Jean-Guy Laplante
Donald Lapointe
Jean Lapointe
Paul Lapointe
Sylvain Lapointe
Pierre Laporte
Gaétan Laprise
Jean-Marc Laprise
Pierre Laprise
Hugo Laquerre
Madeleine Laramée
Elaine Laraway
Gilles Larivée
Jacques Larivée
Yvon Larochelle
Armand Larocque
Gérard Larocque
Jean-Claude Larocque
Robert Larocque
Roger Larose
Ginette Larouche
Pierre Larue
Daniel Latendresse
Normand Latour
Francis Laurencelle
Bakiss Laurent
Arlette Laurin
Jacques-Rémi Lauzon
Geneviève Lavallée
Jocelyne Lavallée
Michel Lavallée
Jacques Laverdière
Mario Laverdière
Richard Laverdière
Victor Lavie
Jean-Guy Lavigne
Sylvie Lavigne
Lance Laviolette
Andrée Lavoie
Bertrand Lavoie
Denise Lavoie
Hélène Lavoie
Jocelyne Lavoie
Clark Lawton
Elizabeth Lawton
Hélène Lebeau
Normand Lebeau
Eric Le Bel
Jean-Pierre le Bel
Pauline Lebel
Bernise Leblanc
Edouard Leblanc
Jocelyn Leblanc
Luc Le Blanc
Manon Leblanc
Marie-Anne Leblanc
Yves Leblanc
Mario Lebrun
Robert Lebrun
Daniel Leclerc
Jocelyn Leclerc
Marc Leclerc
Mario Leclerc
Richard Leclerc
Viateur Leclerc
Johanne Lecomte
Marc Lecomte
André L'Écuyer
Lise L'Écuyer
François Leduc
Gilbert Leduc
Louise Lefebvre
Luc Légaré
Michel Légaré
Antoine Legault
Jean-Guy Legault
Pierre Legault
Robert Léger
Serge Lehouiller
Patrick Lehouillier
Denis Lehoux
Jeanne Lehoux
Gilles Lejeune
Jean-Pierre Lemay
Guy Lemelin
Thérèse Lemerise O.P.
Jacques Lemieux
Michel Lemieux
Serge Lemieux
Jean Lemire
Luc Le Moyne
Peter Lenton
Denis Lepage
Jacques Lepage
Jean-Raymond Lepage
Michel Lepage
Ronald Lepage
Claire Lépine
Nicole Lépine
Gill Leroux
Suzanne Leroux
Les Frères du Sacré-Coeur
Les Jeunes Explos
Kay Leslie
Carole Lessard
Daniel Lessard
François Lessard
Isabelle Lessard
Sylvain Lessard
Sylvie Lessard
Anita Létourneau
Jacinthe Létourneau
Jacques Létourneau
Simon Létourneau
Vincent Létourneau
Andrée A. Lévesque
Fernand Lévesque
Georgette Lévesque
Pierre Lévesque
Yvan Lévesque
Denise Lieutenant
Daniel Lizotte
Kathleen Lord
Robert Lord
Gordon Lorimer
Angela Losito
John Luce
Patrice Luce
Gilles Lupien
Isabelle Lupien
Mario Lussier
Léo MacBriem
Mary Lou MacDonald
Barbara MacDuff
Don MacDuff
Katherine MacKenzie
E. Mackey
Marlène MacKinnon
Suzzanne MacLean
Elaine MacMillian
Marc MacQuart
Lise Magnan
Claude Maheu
Yvon Maheu
Nicole Mailhot
Suzanne Maillé
France Maloney
Jacques Malouin
Jean Maltais
Maurice Maltais
Dennis Mann
Stéphane Marchand
Louise Marcheterre
Ianik Marcil
Christian Marcotte
René Marcouiller
Sophie Marcouiller
Yves Marcouiller
France Marcoux
Pierre Marcoux
Pierre Marinier
André Marleau
Jeanne Marler
Robert Marsan
Daniel Martel
Denise Martel
François Martel
Jean-Louis Martel
Lisette Martel
Barb Martin
Daniel Martin
Jean-Jacques Martin
Michel Martin
Paul Martin
Serge Martin
Bertrand Martineau
Guy Martineau
Marc Martineau
Marielle Martineau
Christina Martinez
Robert Masella
Barbara Mason
André Massé
Josée Massé
Serge Massicotte
Michèle Masson
Yvon Masson
Sylvain Mathieu
Claude Matte
Céline Mayer
Paul Mayer
Anne McCormick
Donald McCutcheon
Donald McEwen
Mabel W. McIntosh
Raymond McNicoll
Harvey Mead
Barbara Meadacroft
Jacques Meloche
Josée Ménard
Judith Ménard
Suzanne Ménard

Atlassers (continued)

Daniel Mercier
Danielle Mercier
Denis Mercier
Jacques Mercier
Yvon Mercier
Ken Merring
Peggy Merring
Louis Messely
F. Messier
Paul Messier
Marie Meunier
Joan Meyer
Mary Meyer
Chris Michener
Maryse Michetti
Pierre Mignot
Mark Mills
Emmanuel Milot
Renaud Mimeault
Diane Mitchell
Elsie Mitchell
Peter Mitchell
Alain Mochon
Léo Moisan
Michel Moisan
Chantal Monette
Nicole Monette
Michel Mongeon
George H. Montgomery
Mary Montgomery
Phyllis Montgomery
Hélène Montminy
Annie Montour
Peter Mordy
Jean-René Moreau
Pierre Morency
Robert Morency
André Morin
Christiane Morin
Claude Morin
Danielle Morin
Denis Morin
Jean-Paul Morin
Josée Morin
Nathalie Morin
Pierre Morin
Gaétan Morissette
Marie-France Morissette
Pierre Morissette
Jo Morisson
François Morneau
David J. N. Mulholland
André Munger
Donal Murnaghan
Christine Murphy
Gaby Murphy
Ross Murphy
Claude Myette
Claude Nadeau
Roch Nadon
Bernard Napert
Hélène Napert
Patricia Nash
Malcom Neil
Jacques Nelson
Daniel Néron
Dan Nickerson
Dominique Nicol
Michelle Niquet
Éric Niskanen
Jean Niskanen
William Niven
Marguerite A. Normandeau
Yves Normandeau
Ed Norsworthy
Mrs. Norsworthy
Phillip Norton
Gisèle Olivier
Hélène Olivier
Georges O'Neil
Shirley Orichefski
Earl Orr
Pierre Otis
Clément Ouellet
Gilles Ouellet
Jean Ouellet
Jean-Pierre Ouellet
Louise Ouellet
Michael Ouellet
Paul-Henri Ouellet
Réginald Ouellet
H. Pagé
Claudia Pageau
Michel Pageau
Louis Panneton
Madeleine Papineau
Bernard Paquet
Gilles Paquet
Danielle Paquette
Jaqueline Paquette
Michel Paquette
Éric Paquin
Jean Paquin
Patrice Paquin
Sophie Paquin
Carol Paradis
Jean Paradis
Sylvain Paradis
Daniel Paré
Diane Paré
François Paré
Louis Paré
Josée Parent
Yves Parent
Jean Pari
Joanne Pari
Daniel Pariseau
Robert Patenaude
G. Patriquin
Michel Patry
Robert Patry
Flora Patterson
Gilles Pauzé
Mario Pearson
Patrick Pednault
Robert Pelland
Céline Pellerin
Alain Pelletier
Armand Pelletier
Martin Pelletier
Pauline Pelletier
Rachel Pelletier
Raymond Pelletier
Réal Pelletier
Richard Pelletier
Roger Pelletier
Brian Penney
Louise Pépin
Danielle Perras
François Perras
Christine Perreault
Desneige Perreault
Paul Perreault
Roger Perreault
Alain Perron
Brigitte Perron
Daniel Perron
François Perron
Henry Perry
Charlie Petch
Denis Petit
Shannon Pew
Cynthia Phinney
Jean-Guy Picard
Stéphane Picard
Sylvain Picard
Raymond Piché
Richard Piché
Jean Pillon
Claude Pilon
Diane Pilon
Chantal Pineau
Laurette Pintal
B. Piper
Lucien Pitre
Jean Piuze
Guy Plamondon
Marie Plamondon
Gaétane Plante
Michel Plante
Steeve Plante
Michel Plourde
Roger Plourde
Serge Plourde
Antoine Poirier
Lorraine Poirier
Louis Poirier
Pierre Poirier
Réjean Poirier
Sylvain Poirier
André Poisson
Jacques Poitras
Louis Poitras
Vivian Pollock
Richard Poole
Micheline Pothier
Alain Potvin
Anne Potvin
Francine Potvin
Gaétan Potvin
Roberto Potvin
Eric Poulin
François Poulin
Guy Poulin
Pierre Poulin
Fred Poulter
Phyllis Poulter
Serge Poussier
John Pratt
Jean-Pierre Pratte
Huguette Préfontaine
Robert Préfontaine
Sylvain Primeau
Gordon Pringle
Roger Proteau
Christian Proulx
Gérard Proulx
Johanne Provancher
Léo Provancher
Michel Provencher
Sébastien Provencher
Brigitte Prud'homme
Raymond Prud'homme
Raymond Quenneville
Jacques Quintin
Michel Quintin
Joshua Radu
Jean-Paul Raîche
Jean-François Rail
Arne Rasmussen
Austin Reed
Daniel Rémillard
Brigitte Renaud
François Renaud
Michel Renaud
Solange Renaud
Daniel René
T. Rhicard
Dan Rhodes
Alain Richard
Christian Richard
Lise Richard
Steve Richard
Marcel Riopel
Mrs. Marcel Riopel
François Riou
Steve Rioux
Anita Rivard
Christian Rivard
Micheline Rivard
René Rivard
Reynald Rivard
Réal Rivest
Yvon Rivest
Donald Roarke
Carmel Roberge
Alain Robert
Francine Robert
Michel Robert
Yolande Robert
Marcel Robillard
Ruby Robinson
Stewart Robinson
Lise Robitaille
Line Rochefort
Louise Rochefort
Charles Rochette
Laurette Rochette
Mrs. Rodier
Jean Rodrigue
Nicolas Rodrigue
John Rodriguez
Keith Rogers
Suzanne Rompré
Guy Rondeau
Richard Rondeau
Claude Rony
Catherine Rooney
Jacques Rosa
Heather Rose
Daphne Roseby
Ian Roseby
Eva Rosenthal
Yves Rouillard
Jocelyn Rouleau
Richard Rouleau
Daniel Rousseau
Jean-François Rousseau
Thérèse Rousseau
Yvan Rousseau
Yvon E. Roussel
Roseline Roux
Alice Roy
Clémence Roy
Clément Roy
Clodin Roy
Danielle Roy
Danny Roy
Hélène Roy
Huguette Roy
Jean-Simon Roy
Jean-Yves Roy
Laval Roy
Madeleine Roy
Marc-André Roy
Marcel Roy
Michel Roy
Rénald Roy
René Roy
Roger Roy
Alain Royer
Louis Ruelland
Carl Ruest
Danielle Ruest
Julien Ruest
Anne Russell
Gordon Russell
Helen Russell
Denise Sabourin
Monique Salathé
Jane Saltman
André Samson
Céline Samson
Denis Samson
Pascal Samson
George Samuel
Mat Samuel
Guy Sanche
Jean Sanscartier
Daniel Sanson
Reta Sarrasin
Isabelle Saucier
Daniel Sauvageau
Thérèse Sauvé
Germain Savard
Jean-Pierre L. Savard
Jocelyne Savard
Michel Savard
Carl Savignac
Bruno Scherrer
Roland Schryer
Nancy Schumann
John Schwinghamer
Mario Scrosati
Brian Scullion
Danny Scullion
Agathe B. Séguin
René Séguin
Raymonde Sélesse
David Sergeant
Gilles Seutin
Marco Sévigny
François Shaffer
Julie Shaffer
Marc Sherrer
Rodrigue Shooner
Luciano Silicani
Claude Simard
Gilbert Simard
Lise Simard
Marc Simard
Odette Simard
Sylvain Simard
D. Simley
Claude Simoneau
Claudie Simoneau
Denis Simoneau
Françoise Simoneau
Michel Simoneau
Dan Simpson
Langis Sirois
Tim Skelton
Matthew Smar
Charles Smith
Donald Smith
J. Smith
Margaret Smith
P. Smith
Peter K. Smith
Richard Smith
Robert Smith
Michel Sokolyk
Peter Somerville
Bruno Soucie
Denise Soucie
Josée Soucie
Louis-Marc Soyez
Michael Spencer
Patrice St-Amand
Robert St-Amand
Michel St-Amant
Betty Stanger
Jean-Claude St-Arnault
Richard St-Arneault
Guy St-Germain
Daniel St-Hilaire
Lyne St-Jacques
René St-Laurent
Claude St-Louis
Margo St-Louis
Serge St-Louis
Nathalie St-Onge
Sylvain St-Onge
Alain St-Pierre
André St-Pierre
Iann St-Pierre
Louiselle St-Pierre
Tony Straessle
Daniel Strasbourg
E. Strathdee
Marc Surprenant
Michel Surprenant
Soren Suse
N. Swan
Claire Sylvestre
Louise Sylvestre
André Taillon
Denis Talbot
Marc Talbot
Pierre Tanguay
Pierre Tanguay
Serge Tanguay
Pete Tarassoff
Claude Tardif
Fernand Tardif
Céline Taylor
Christine Taylor
Mark Taylor
Roger Taylor
Marc Tellier
Gilles Tessier
Michel Tessier
Normand Tessier
Wellie Théorêt
Michel Thériault
Réal Thériault
Dominique Therrien
Raymond Therrien
Yves Therrien
Marc Thibault
Marius Thibault
Germain Thibodeau
Julie Thibodeau
Maurice Thibodeau
Heiki Thoen
Gérald Thomassin
Linda Thomassin
Art Thompson
Kenneth Thorpe
Francine Thouin
Germain Thuot
Rodger D. Titman
Mark Tomalty
Pat Tottenham
Richard Tourangeau
Élise Tousignant
Daniel Toussaint
Jean-Marc Trahan
Normand Traversy
Aline Tremblay
Aurélien Tremblay
Cécile Tremblay
Chantal Tremblay
Daniel Tremblay
Dany Tremblay
Gaston Tremblay
Hervé Tremblay, O.B.S.
J. Tremblay
Jean-François Tremblay
Monique Tremblay
Nelson Tremblay
Nicolas Tremblay
Pierre Tremblay
Gaston Trépanier
René Trépanier
Benoît Truax
Denis Trudeau
Michel Trudeau
Monique Trudeau
Aimé Trudel
France Trudel
François Trudel
Gilles Trudel
Hugo Trudel
Pierre-David Trudel
Robert Trudel
Suzanne Trudel
Eric Tull
Jean-Luc Turbide
Yves Turbide
Cécile Turcotte
Donald Turcotte
Simon Turcotte
Sylvain Turcotte
Yves Turcotte
Andrée Turenne
Louise Turgeon
Michel Turgeon
Réjean Turgeon
Stéphane Turgeon
Ian Turner
Willie Ule
Charles Vachon
Line Vadeboncoeur
Jean-Pierre Vallée
Axel van de Walle
Édith van de Walle
Albert Van Dijk
Benoît Vanier
Carmen Vanier
Noël Vanier
Pierre Vanier
Monique Veillette
Pauline Veillette
Raynald Veilleux
Laurence Vergnole
Julie Verner
Guy Verreault
Marie-Claude Vezeau
Lucie Vézina
Alain Vié
Guy Vigneault
Louis Vigneault
Marc-André Villard
Yvon Villemure
A. Villeneuve
Jean Villeneuve
Jacqueline Vincent
Johanne Vincent
Pierre Vincent
Carol Vlassoff
Kathy Waghorn
Terry Wagner
Dorothy Walker
Peter Walker
Don Wallace
Jan Wallace
John Walsh
Jeff Webster
Robert Weeds
Craig Weil
Ken Weir
Lewis Wels
Pierre Wéry
Bernice Whitney
Gilbert Whitney
Dorothy Williams
Jim Williams
Sandy Willis
David Wilson
Roger Woodworth
Robert M. Worona
Jean Yanakis
Richard Yank
Linda Yearwood
Ryan Young
Jane Zdansky
Ed Zellen
Guy Zenaitis
Wilma Zomer

Data on breeding birds in Québec were also provided by the following organizations:

Association québécoise des groupes d'ornithologues
Banque informatisée des oiseaux marins du Québec (BIOMQ), Canadian Wildlife Service, Environment Canada, Québec Region
Canadian Museum of Nature
Ducks Unlimited Canada
Étude des populations d'oiseaux du Québec (ÉPOQ)
Hydro-Québec
Ministère de l'Environnement et de la Faune du Québec
Ontario Breeding Bird Atlas
Parks Canada, Department of Canadian Heritage
Société Duvetnor
Zones d'exploitation contrôlée (ZEC) du Québec

Authors

Réjean Benoit
Rémy Bouchard
Pierre Chagnon
Alain Côté
André Dion
Gaétan Duquette
Philippe Fragnier
Christian Gagnon
Alain Gauthier
Yves Gauthier
Claire Gauvreau
Jean-Denis Gauvreau
Christiane Girard
Douglas Graham
Louise Gratton
Pierre Hainault
Jacques Ibarzabal
Benoît Jobin
Lucie Landry
Maryse Lepage
Sylvain Lessard
Paul Messier
Gaston Moisan
Gilles Pauzé
Éric Razurel
Reynald Rivard
Laval Roy
Germain Savard
Michel Savard
René Séguin
Serge Tanguay
Nicolas Tremblay
Marianne G. Ainley, University of Northern British Columbia
Johanne Baril, State of the Environment Reporting, Environment Canada, Ottawa
Kathleen A. Blanchard, Quebec-Labrador Foundation
Robert Bonneau, Club d'observateurs d'oiseaux Marie-Victorin
Jacques Bouvier, La cité collégiale, Hawkesbury
Raymond Cayouette, Société zoologique de Québec
André Cyr, University of Sherbrooke
André Desrochers, Laval University
Pierre Drapeau, University of Montréal
Jean-François Giroux, University of Québec at Montréal
Christian Houle, Collège de Sherbrooke (Cégep)
Pierre Lamothe, Hydro-Québec
Jacques Larivée, Cégep de Rimouski
Guy Lemieux, Ministère des Ressources naturelles du Québec
Serge Lemieux, Environment Canada, Québec Region
Hélène Lévesque, Canadian Wildlife Service, Environment Canada, Hull
François Morneau, Groupe de recherche et d'étude en biostatistique de l'environnement
Pierre Mousseau, St. Lawrence Centre, Environment Canada, Québec Region
Jean Munro, Fisheries and Oceans Canada
Sylvain Paradis, Parks Canada, Department of Canadian Heritage, Québec Region
Gilles Seutin, Smithsonian Tropical Research Institute, Panama
Dan Strickland, Ontario Ministry of Natural Resources
Association québécoise des groupes d'ornithologues
Normand David
Pierre Fradette
Daniel Jauvin
Jean Paquin
Desneige Perreault
Canadian Forest Service, Natural Resources Canada, Québec Region
Jean-Louis Bélair
Réal Bisson
Canadian Museum of Nature
Michel Gosselin
Henri Ouellet
Canadian Wildlife Service, Environment Canada, Québec Region
Robert Alvo
Yves Aubry
Simon Barrette
Luc Bélanger
Manon Bombardier
Daniel Bordage
André Bourget
Louis Breton
Pierre Brousseau
Louise Champoux
Gilles Chapdelaine
Line Choinière
Richard C. Cotter
Marcel Darveau
Diane Dauphin
Alain Demers
Léo-Guy de Repentigny
Jean-Luc DesGranges
Mireille Desponts
Marie-Renée Doyon
Pierre Dupuis
Gilles Falardeau
Nicolas Gagnon
Jean Gauthier
Serge Gauthier
Marcelle Grenier
Derin Henderson
Serge Labonté
Pierre Lafontaine
Christine Langevin
Alain Lanoue
Pierre Laporte
Denis Lehoux
Vincent Létourneau
Benoît Limoges
Yvon Mercier
Louis Messely
Alain Morrier
Claude Nadeau
Richard Pelletier
Jean-François Rail
Austin Reed
Michel Robert
Jean Rodrigue
Jean-Pierre L. Savard
François Shaffer
Bernard Tardif
Josée Tardif
Jacqueline Vincent
McGill University
David M. Bird
Rodger D. Titman
Ministère de l'Environnement et de la Faune du Québec
Daniel Banville
Jean-Claude Bourgeois
Alain Desrosiers
Jean Lapointe
Michel Lepage
Raymond McNicoll
Réginald Ouellet
Daniel St-Hilaire
Province of Quebec Society for the Protection of Birds
Pierre Bannon
Bob Barnhurst
Sharon David
Mabel W. McIntosh
Michael Spencer
Richard Yank
University of Moncton
David Cairns
Marc-André Villard

Photographers

No other Québec work of ornithology contains as many photographs as The Breeding Birds of Québec. *The talent, patience, and generosity of all contributing photographers have made it possible to include pictures, often previously unpublished, of every species. Anyone who has taken up wildlife photography knows how long it takes to get the right results. This means that each photograph in this Atlas is the fruit of unflagging effort, in many cases involving interminable hours of waiting and much experimentation.*

Arthur A. Allen
Robert Alvo
R. M. Anderson
Yves Aubry
Jean Bachand
Pierre Bannon
Robert Baronet
Jean Beaubien
Napoléon Beaudet
Michel Beaudoin
Jean Bédard
Luc Bélanger
Réjean Benoit
André-Guy Bernier
André Bilodeau
David M. Bird
Réal Bisson
Bruno Blais
Kathleen A. Blanchard
Daniel Bordage
Rémy Bouchard
Michel Boulianne
André Bourget
Lina Breton
Jacques Brisson
Serge Brodeur
Pierre Brousseau
David Cairns
André Cayouette
Raymond Cayouette
Jacques Chabot
Gilles Chapdelaine
Alain Côté
Rob Curtis/VIREO
André Cyr
Marcel Darveau
Normand David
Sharon David
Léo-Guy de Repentigny
Jean-Luc DesGranges
Patricia Desjardins
Mireille Desponts
André Desrochers
Alain Desrosiers
Pierre Dupuis
Gaétan Duquette
M. Durand
Gilles Falardeau
Denis Faucher
Bernard Filion
Jim N. Flynn
Pierre Fradette
Alain Gagnon
Tony Gaston
Jean Gauthier
Yves Gauthier
Denis Gingras
Jean Giroux
Jean-Marc Giroux
Marcel Giroux
Douglas Graham
Robertson Graham
Gaétan Guertin
Ross Harris
Christian Harvey
Steven Holt/VIREO
Benoît Houde
Jack Hughes
Daniel Jauvin
Serge Jauvin
Paul Laliberté
Pierre Lamothe
Renée Langevin
Jean Lapointe
Pierre Laporte
Gaétan Laprise
Roger Larose
Mario Laverdière
Eric Le Bel
Jean-Pierre le Bel
Luc Leclerc
Jeanne Lehoux
Guy Lemieux
Renée Lemieux
Christine LePage
Gaston Le Page
Ronald Lepage
Chantal Lepire
M. Letendre
Vincent Létourneau
Christian Marcotte
Robert Michaud
Pierre Mineau
Robert Morin
François Morneau
A. and E. Morris/VIREO
Henry Mousley
Pierre Mousseau
Jean Munro
Claude Nadeau
Pierre Normand
W. S. Odell
Henri Ouellet
Réginald Ouellet
Michel Papageorges
Jean Paquin
Sylvain Paradis
Robert Parent
Richard Pelletier
Desneige Perreault
Paul Perreault
Jacques Poitras
Austin Reed
Alain Richard
Michel Robert
Jean Rodrigue
Guy Rondeau
Jacques Rosa
Jean-Pierre L. Savard
Michel Savard
B. Schorre/VIREO
François Shaffer
Guy Simard
Richard Sloss
Michel Sokolyk
Daniel St-Hilaire
Lyne St-Jacques
Sylvain St-Onge
Dan Strickland
Jean Sylvain
P. A. Taverner
Stéphane Turgeon
Jacqueline Vincent
Doug Wechsler/VIREO
D. and M. Zimmerman/VIREO

Other picture credits (photographs, aerial and satellite images)

American Museum of Natural History
Association touristique régionale de la Gaspésie
Canadian Forest Service, Natural Resources Canada
Canadian Museum of Nature
Canadian Wildlife Service, Environment Canada, Québec Region
Fisheries and Oceans Canada
Laval University Collection
Ministère de l'Environnement et de la Faune du Québec
Ministère des Ressources naturelles du Québec
Musée François-Pilote
National Air Photo Library, Natural Resources Canada
Natural Resources Canada
Parks Canada, Department of Canadian Heritage, Québec Region
Québec City Archives
St. Lawrence Centre, Environment Canada, Québec Region

Special collaboration

Dominique Aubry
Raymond Cayouette
Christine Ellefsen
Wilfrid Gaboriault
Paul-André Leclerc
Ronald Lepage
Claire Lépine
Yvan Lévesque
André Perron
Yvon Villemure
Gérard Barbe, Ministère de l'Environnement et de la Faune du Québec
Jean Beaubien, Canadian Forest Service, Natural Resources Canada, Québec Region
Jean Bédard, Laval University
Michel Bédard, Parks Canada, Department of Canadian Heritage, Québec Region
Jean-Louis Bélair, Canadian Forest Service, Natural Resources Canada, Québec Region
Michel Bourassa, Ministère des Ressources naturelles du Québec
Denis Brière, Ministère de l'Agriculture, des Pêcheries et de l'Alimentation du Québec
Denis Burelle, Environment Canada, Québec Region
Michael D. Cadman, Canadian Wildlife Service, Ontario Region
Manon Carignan, Ministère de l'Agriculture, des Pêcheries et de l'Alimentation du Québec
Nicole Chartrand, State of the Environment Reporting, Environment Canada, Ottawa
Pierre-Denis Cloutier, Canadian Wildlife Service, Environment Canada, Québec Region
Robert Décarie, Groupe de recherche et d'étude en biostatistique de l'environnement
Pierre Demers, Parks Canada, Department of Canadian Heritage, Québec Region
Léo-Guy de Repentigny, Canadian Wildlife Service, Environment Canada, Québec Region
Paul F. J. Eagles, University of Waterloo
Gilles Gaboury, Ministère des Ressources naturelles du Québec
Léopold Gaudreault, Ministère de l'Environnement et de la Faune du Québec
Marcelle Grenier, Canadian Wildlife Service, Environment Canada, Québec Region
Pierre Grondin, Ministère des Ressources naturelles du Québec
Luc Jobin, Canadian Forest Service, Natural Resources Canada, Québec Region
Peter Landry, Georgeville Nature Society
Jean Lapointe, Ministère de l'Environnement et de la Faune du Québec
Jacques Larivée, Cégep de Rimouski
Sarah B. Laughlin, Vermont Institute of Natural Science
Jean-Pierre le Bel, Ministère de l'Environnement et de la Faune du Québec
Mary LeCroy, American Museum of Natural History
Martin Léveillé, Ministère de l'Environnement et de la Faune du Québec
Alain Lévesque, Ministère des Ressources naturelles du Québec
Patricia Logan, Canadian Wildlife Service, Environment Canada, Ottawa
Raymond McNeil, University of Montréal
Michel Melançon, Canadian Wildlife Service, Environment Canada, Québec Region
Yvon Mercier, Canadian Wildlife Service, Environment Canada, Québec Region
Jennifer Milton, Atmospheric Environment Branch, Environment Canada, Ottawa
William A. Montevecchi, Memorial University

Special collaboration (continued)

Diane Nolet, Terminology and Documentation Branch, Translation Bureau, Public Works and Government Services Canada
Serge Pelletier, Environment Canada, Québec Region
Jean-Marie Perron, Laval University
Jean Piuze, Fisheries and Oceans Canada
André Robitaille, Parks Canada, Department of Canadian Heritage, Québec Region
Daniel Rochefort, Ministère des Ressources naturelles du Québec
Claude Saint-Charles, Environment Canada, Québec Region
Raymond Sarrazin, Canadian Wildlife Service, Environment Canada, Québec Region
Jean-Louis Scantland, State of the Environment Reporting, Environment Canada, Ottawa
Richard Sloss, American Museum of Natural History
Jocelyn Tessier, Imprimerie Le Laurentien
Michel Thibault, Ministère des Ressources naturelles du Québec
Jean Thuot, Environment Canada, Québec Region
Doug Wechsler, VIREO
Daniel A. Welsh, Canadian Wildlife Service, Ontario Region
Agriculture and Agri-food Canada
American Museum of Natural History
Atmospheric Environment Branch, Environment Canada, Ottawa
Canadian Forest Service, Natural Resources Canada
Canadian Museum of Nature
Canadian Wildlife Service, Environment Canada, Ottawa
Daishowa
Ducks Unlimited Canada
Environment Canada Library Service, Québec Region
Environmental Protection Branch, Environment Canada, Québec Region
Fapel-Faune
Federation of Ontario Naturalists
Fisheries and Oceans Canada
Groupe de recherche et d'étude en biostatistique de l'environnement (GREBE)
International Paper Company
Landsat, Canada Centre for Remote Sensing, Geomatics Canada (satellites images)
La Presse
Laval University Collection
Les forages Groleau
Les zones d'exploitation contrôlée (ZEC) du Québec
Ministère de l'Agriculture, des Pêcheries et de l'Alimentation du Québec
Ministère de l'Environnement et de la Faune du Québec
Ministère des Affaires municipales du Québec
Ministère des Ressources naturelles du Québec
Ministère des Transports du Québec
National Air Photo Library, Natural Resources Canada
National Wildlife Research Centre, Environment Canada, Ottawa
Québec and Ontario Paper Company
Québec City Archives
Société des établissements de plein air du Québec (SÉPAQ)
Société Duvetnor
Société zoologique de Québec
St. Lawrence Centre, Environment Canada, Québec Region
Visual Resources for Ornithology (VIREO)

Editing, verification, and standardization of scientific content

Jean Giroux
Association québécoise des groupes d'ornithologues
Normand David
Canadian Museum of Nature
Michel Gosselin
Henri Ouellet
Canadian Wildlife Service, Environment Canada, Québec Region
Robert Alvo
Yves Aubry
Simon Barrette
Luc Bélanger
Claire Boismenu
Manon Bombardier
André Bourget
Louis Breton
Pierre Brousseau
Gilles Chapdelaine
Line Choinière
Richard C. Cotter
Marcel Darveau
Diane Dauphin
Marie-Renée Doyon
Pierre Dupuis
Gilles Falardeau
Nicolas Gagnon
Jean Gauthier
Serge Gauthier
Derin Henderson
Pierre Lafontaine
Christine Langevin
Alain Lanoue
Pierre Laporte
Christine LePage
Vincent Létourneau
Benoît Limoges
Louis Messely
Alain Morrier
Richard Pelletier
Jean-François Rail
Austin Reed
Michel Robert
Jean-Pierre L. Savard
Josée Tardif
Luce Venne-Forcione
Laval University
Gilles Gauthier
McGill University
David M. Bird
Ministère de l'Environnement et de la Faune du Québec
Gildo Lavoie
Ministère des Ressources naturelles du Québec
Guy Lemieux
University of Montréal
Raymond McNeil
University of Québec at Trois-Rivières
Richard Couture

Wildlife illustrators

Carole Bérubé
Ghislain Caron
Pierre Dupuis
Christiane Girard
Jean-Marc Giroux
Michel Gosselin
Luc Leclerc
Marie-France Lelièvre
Guy Michaud
Clodin Roy
René Roy
Louis Ruelland
Jean-Pierre Vallée

Computerization of data

Canadian Wildlife Service, Environment Canada, Québec Region
Vital Boulé
Pierre Chagnon
Marcel Darveau
Jasmine Dion
Gilles Falardeau
Alain Gouge
Lyne Grégoire
Yvon Laplante
Josée Lessard
Yvan Lévesque
Jean Maltais
Louise Rioux
Line Rochefort
Suzanne Smith
Charles Vachon

Compilation and validation of data

Canadian Wildlife Service, Environment Canada, Québec Region
Yves Aubry
Dominique Biron
Louis Breton
Gilles Falardeau
Jean-Marc Giroux
François Lambert
Christine LePage
Christian Marcotte
Louis Messely
Shirley Orichefsky
Mario Paquin
Chantale Pineau
Francine Rousseau
Monique Salathé
Stéphane Turgeon
Frédéric Willème
Fondation Les oiseleurs du Québec Inc.

Statistical analysis (frequency of occurrence by ecological region)

Canadian Wildlife Service, Environment Canada, Québec Region
Yves Aubry
Gilles Falardeau
Jean Gauthier
Nathalie Plante

Data processing and analysis – waterfowl abundance maps

Canadian Wildlife Service, Environment Canada, Québec Region
Daniel Bordage
Marcelle Grenier
Le Groupe-conseil Argus

Bibliographical research

Jean-François Bélanger, Environment Canada, Hull
Mireille Boudreault, Canadian Museum of Nature
Marcel Darveau, Canadian Wildlife Service, Environment Canada, Québec Region
Normand David, Association québécoise des groupes d'ornithologues
Gilles Falardeau, Canadian Wildlife Service, Environment Canada, Québec Region
Michel Gosselin, Canadian Museum of Nature
Julia Innes, Environment Canada, Québec Region
Carmen Joseph, Environment Canada, Québec Region

Secretarial services

Françoise Lapointe, Canadian Wildlife Service, Environment Canada, Québec Region

Word processing

Canadian Wildlife Service, Environment Canada, Québec Region
Françoise Lapointe, co-ordinator
Marie-Thérèse Guertin
Sophie Guillemette
Karolyn Vachon

Typographical standardization

Françoise Lapointe, Canadian Wildlife Service, Environment Canada, Québec Region

Art editor

Norman Dupuis

Publishing advisers

Normand David, Association québécoise des groupes d'ornithologues
Norman Dupuis
Jean-Marc Gagnon, Communications Science-Impact
Pierre Normand, Environment Canada, Québec Region
Austin Reed, Canadian Wildlife Service, Environment Canada, Québec Region

Graphic Design

Norman Dupuis
Boissonneault Gagné
Parallèle Design

Page layout co-ordinator

Christine LePage, Canadian Wildlife Service, Environment Canada, Québec Region

Selection of photographs

Canadian Wildlife Service, Environment Canada, Québec Region
Yves Aubry
Jean Gauthier
Christine LePage
Jacqueline Vincent

Preliminary graphics

Dendrek

Graphics and map production

Compélec Inc.
Project manager: Benoit Yergeau
Project co-ordinator: André Robitaille

Camera-ready and colour separation

Compélec Inc.

Typographical revision and correction of proofs

Canadian Wildlife Service, Environment Canada, Québec Region
Richard C. Cotter
Françoise Lapointe
Christine LePage
Communications Science-Impact

Promotion and marketing

Canadian Wildlife Service, Environment Canada, Québec Region
Jean Gauthier
Environment Canada, Québec Region
Pierre Normand
Danielle-Josée Pelletier

English version

Translation

Montreal English Translation Section, Translation Bureau, Public Works and Government Services Canada

Translation co-ordinator

Joan Dyer

Translators

Margaret Arnaudin
Barbara Chunn
Jim Connelly
Debby Dubrofsky
Hilles Hamersley
Betsy McFarlane
Karin Montin
Lou Nelson
Susan Reed

Revisers (species accounts)

Jim Connelly
Betsy McFarlane
Karin Montin

Technical verification committee

Canadian Wildlife Service, Environment Canada, Québec Region
Richard C. Cotter, co-ordinator and liaison
Robert Alvo
Manon Bombardier
Derin Henderson
R. J. Hughes
Angela Losito
Austin Reed
Jean-Pierre L. Savard

Editorial committee

Yves Aubry
Richard C. Cotter
Diane Dauphin
Joan Dyer
Jean Gauthier
Françoise Lapointe
Christine LePage

We would also like to thank the staff of the Corporate Services and Corporate Affairs branches of Environment Canada, Québec Region for their professionalism and dedication throughout work on The Atlas of the Breeding Birds of Southern Québec.

INTRODUCTION

THE QUÉBEC ATLAS

The Breeding Birds of Québec: Atlas of the Breeding Birds of Southern Québec is both a synthesis of the breeding biology of almost 300 Québec bird species and an atlas of the breeding birds of southern Québec. It is a collective work and contains a wealth of information, much of it recondite or hitherto unpublished, on the province's birdlife. It was written for resource managers, biologists, both professional and amateur ornithologists, students, teachers, and researchers, indeed for anyone wishing to know more about birds or seeking answers to hitherto unresolved questions.

The book addresses three main themes: (1) the changes that have transformed the Québec landscape, the home of our breeding birds; (2) breeding biology and distribution; and (3) the biodiversity of the birdlife of southern Québec in relation to the major ecological components of the region.

The first part identifies and describes Québec bird habitats, these being crucial to distribution. An outline is given of the province's physical geography and its main characteristics, especially the importance of water in Québec (the St. Lawrence River, lakes, and wetlands such as peatlands, marshes, and swamps). The influence of the last ice age is covered, elucidating how it has affected present-day soils and vegetation. The habitats thus formed then determine the distribution of many bird species.

Habitats are also subject to human and natural pressures. The impact of human activity on the natural environment since European settlement has profoundly and irreversibly altered the landscape of southern Québec. In the Central St. Lawrence Lowland upstream from Québec City, for example, changes to habitat have resulted in changes in bird populations, in that species typical of open country and forest edges have certainly thrived. Throughout southern Québec, logging, the development or abandonment of farming, regulation and straightening of watercourses, and urbanization are among the human factors that have shaped and are still shaping the landscape. There are natural agents that are undeniably sources of habitat disturbance: forest fires, insect epidemics, and dieback. Their scale and duration affect the lives of many birds. The dynamics of these disturbances is of great importance for maintenance of high regional species diversity. In short, the combination of human and natural factors directly and indirectly impinges upon current species composition and the balance of bird populations.

The nucleus of the work consists of individual accounts of all species considered breeders in Québec. It thus includes all birds known and confirmed to have bred in Québec within the Atlas area between 1984 and 1989, in addition to some species selected as prior breeders, putative breeders that could not be confirmed during the Atlas period (1984–89), or species where there were indications that breeding might be imminent. Each species account generally includes a brief presentation with characteristics and status, details of its habitat and general behaviour during breeding, Québec distribution, and, where there is sufficient information, population history and trends. This is complemented by a species profile, a map of its world breeding range, a chronology of its annual occurrence and breeding cycle in Québec, and a histogram of the breeding evidence codes used to determine its breeding status in the Atlas squares and its frequency of occurrence by ecological region. The frequency data were used to map preferred breeding areas.

The last part of the book defines the concept of biodiversity and applies it to Québec's birdlife. This part takes a global rather than individual approach to the avian fauna of the various regions and ecological segments of southern Québec.

Lastly, there are appendices containing information to round out the species accounts. Here are found detailed statistical tables (analyses described in the Introduction) on all species reported in the course of the project, a compilation of hunters' kills over the previous two decades, and a listing giving the status of each species found in Québec. For computer mavens, details are provided of the computer-graphics processing of all Atlas data.

Bird atlases

The purpose of a bird atlas is to present in detail the breeding ranges of the species found in the area covered. The first work of this kind, *The Atlas of Breeding Birds in Britain and Ireland* (Sharrock 1976), sparked other comparable projects. The approach was simple, involving division of the whole area into a grid of squares and determination of distribution by surveying each square for occurrence and breeding evidence. The same approach was followed in subsequent works: for France (Yeatman 1976), Denmark (Dybbro 1976), West Germany (Rheinwald 1977), The Netherlands (Teixeira 1979), Switzerland (Schifferli et al. 1980), and several others. In North America, a number of provinces and states have also produced atlases in recent years, including Vermont (Laughlin and Kibbe 1985), Maine (Adamus 1987), Michigan (Brewer et al. 1991), Alberta (Semenchuk 1992), Ontario (Cadman et al. 1987), and the Maritime provinces (Erskine 1992).

The Atlas of Australian Birds, published in 1984, is based on a different concept: it presents the year-round distribution of birds, their breeding status, and historical maps of their distribution (Blakers et al. 1984). In *The Birds of British Columbia* (Campbell et al. 1990a,b), bird distribution throughout the province is shown for each season; the book also contains a synthesis of accumulated records of several decades about the birds of the province.

The *Atlas of the Breeding Birds of Southern Québec* continues this atlas series and is part of an international effort to map the distribution of all avian species worldwide. However, it goes further than traditional atlases in its treatment of the breeding biology of Québec's birds and their relative abundance in each of the ecological regions of southern Québec.

HISTORY OF ORNITHOLOGY

The development of ornithology in Québec: from its origins until 1960

The study of birds in Québec goes back at least to the voyages of the first explorers, who recorded in their writings, some of which remain unpublished, their impressions and observations concerning the flora and fauna of the lands they discovered. The first observers of Québec's avifauna were, of course, the Indians and Inuit, the original inhabitants. Their ornithological knowledge derives mainly from their oral tradition. Some of their artifacts stand as a testimony to their acute sense of observation, but since their knowledge was unstructured, it did not lend itself to being handed down in writing (Jenness 1958). Although there are a number of travel and natural history accounts dating from the 17th and 18th centuries, only those from the 19th century attempt to approach the study of birds in Québec in a structured manner. This occurred a few decades later than in other parts of the world, Europe in particular (Stresemann 1975). Despite this delay, the body of early information on birds in Québec, dispersed as it is in a host of works of a general nature, gives an indication of how ornithology would develop and the interest it would arouse, particularly after the middle of the 19th century.

Discovery to European settlement, 1534–1608

This period, which covers almost a century and begins with the first voyage of Jacques Cartier to Canada in 1534, is somewhat lacking in information on the birds found in the lands newly claimed by the French king, François I. Scattered throughout the writings of Jacques Cartier[1] are a few descriptions of sea and aquatic birds observed on the islands off the east coast of Canada and in the Gulf of St. Lawrence. The species described are for the most part fairly large birds that were used to supplement the diet of the sailors, who were avid for fresh food after their lengthy crossing. In 1534 and 1535, Jacques Cartier sailed in the Blanc-Sablon region and in the Gulf of St. Lawrence (Lewis 1934a), where he stopped off at Bird Rocks off the Magdalen Islands. Aside from the Great Auk, which he captured and used for food while on the coast of Newfoundland, more specifically on Funk Island, he mentions species he sighted elsewhere in the gulf, particularly near Blanc-Sablon "... where there are a great number of tinkers [Razorbill] and puffins [Atlantic Puffin] which have red beaks and feet and make their nests in holes under the earth like rabbits...".[2] During his second trip, in 1535, he sailed up the St. Lawrence to Montréal, where he saw "...great numbers of birds, to wit: cranes, bustards, swans, white and gray wild geese, ducks, drakes, blackbirds, thrushes, turtle-doves, wood-pigeons, goldfinches, tarins, canaries, linnets, nightingales, sparrows and other birds the same as in France...".[3] These enumerations are only of relative value (Laporte 1974a), as the species cited are identified in an approximate fashion—except in the case of more characteristic species such as the Great Auk, Razorbill, Atlantic Puffin, and the Northern Gannet—for the names he used are those of similar European species that he was familiar with.

Aside from Jacques Cartier, few explorers at that time (Sieur de Roberval, Jean Alfonse and Charles Leight) recorded sightings that had actually been made in Québec (Laporte 1974a). The ornithological elements contained in the writings of Jacques Cartier and the other chroniclers who came to what is now Québec, before the days of European settlement, consist not so much of detailed lists of plants and animals but rather an overall view of the ecological conditions.

Ornithology in New France, 1608–1759

The information on Québec birds that appeared after the founding of Québec City by Samuel de Champlain in 1608 was more complete. Early information was scattered through the writings of the explorers, visitors, and settlers, but later on data were recorded in a more systematic fashion. The first work from this period containing carefully recorded observations was the account of the voyages of Champlain in New France, published in Paris in 1613 (MacNamara 1926; Giguère 1973). These observations are still valid today. Champlain was both a seaman and a geographer and gave meticulous accounts of his explorations, with precise descriptions of the habitats and birds that he observed. For example, on 7 June 1611, when sailing up the St. Lawrence, he reported a large colony of herons on one of the islands in the Lachine Rapids. This island no doubt was Île aux Hérons, and the birds must have been the Great Blue Heron and the Black-crowned Night-Heron. To this day, both species still breed on the island during the breeding season. This type of information is useful to the biogeographer for the analysis and study of evolution of regional faunas, but is rare in the writings of the early settlers.

The *Jesuit Relations* (Thwaites 1895–1901) also provide valuable information on the flora and fauna of Québec. The authors of this vast work were not as a general rule as rigorous in their descriptions as were Champlain and some of the other explorers of that period (Fournier 1932). The accounts of Fr. Paul LeJeune, who lived from 1591 to 1664, and Marc Lescarbot (1928) are particularly interesting, although the latter deals mainly with Acadia at the beginning of the 17th century. The Récollet brother Gabriel Sagard left an impressive account in 1632 of his trips through the interior of North America (Sagard 1865), but it contains little ornithological information that originates unquestionably from Québec, the bulk of it being from the "Land of the Hurons," which is now Ontario.

The first book devoted to the natural history of New France, *L'Histoire véritable et naturelle des mœurs et production du pays de la Nouvelle-France, vulgairement dite le Canada*, was written by Pierre Boucher and was published in Paris in 1664 (Boucher 1664). Born in France, Pierre Boucher arrived in Canada at the age of 13. His many trips through the continent enabled him to acquire a thorough knowledge of the country. He describes mainly the flora and fauna, while giving what might be termed an ecological view of the areas with which he was familiar, which extended well beyond the borders of present-day Québec. He devoted an entire chapter to the birds he saw while travelling in Québec and elsewhere in Canada; this he called *Noms des Oyseaux qui se voyent en la Nouvelle-France* (Boucher 1664). This part of the book contains first-hand information on the species he encountered and is very useful from an ecological point of view, as it gives an understanding of what habitats in the St. Lawrence Valley were like before intensive colonization began.

Other sources of ornithological information from this period include an unpublished and little-known manuscript entitled *Traité des animaux à quatre pieds, terrestres et amphibies, qui se trouvent dans les Indes occidentales ou Amérique septentrionale*. Housed at the Bibliothèque nationale de France, it is followed by a text on birds entitled *Traité des oyseaux*, apparently written around 1677 by a Jesuit missionary, probably Fr. Louis Nicolas. This manuscript contains descriptions and notes on about fifty species observed by the author during his stay in New France between 1664 and 1675. Although several species are not rigorously described and precise information as to the origin of the sightings is lacking, the manuscript is worth mentioning as it is the first document devoted to the birds of Québec. An annotated edition is in preparation.

Around the same period, in 1672, Nicolas Denys[4] published in Paris an important document on the natural resources and geography of Acadia. He gave a description of various birds that he saw, revealing a keen sense of observation. However, as he did not specify whether the birds mentioned were seen in Québec or elsewhere in New France, the importance of the document as concerns the development of ornithological knowledge in Québec is difficult to establish, although it no doubt achieved a certain acclaim at the time (Allen

1 Biggar, H. P. 1924. The Voyages of Jacques Cartier. F. A. Acland, Ottawa, 314 pp.
2 Idem p. 16.
3 Idem p. 198.
4 Denys, N. The description and natural history of the coast of North America (Acadia). Translated from the original of 1672, published in 1908 by the Champlain Society, Toronto, 624 pp.

1939, 1940). A less important document by Fr. Louis Hennepin (1903) is nonetheless interesting for the pertinent ornithological details it contains. Its geographical descriptions of Québec give an idea of what some habitats were like at the mid and late 17th century.

The travel accounts of Baron Louis de Lahontan (1703), who sojourned in New France from 1683 to 1703, contain useful and interesting information on the distribution and behaviour of many birds. Since he travelled widely throughout Canada, mainly in Acadia, it is sometimes difficult to determine where he made his observations, but the information nonetheless applies to species in Québec. Although severely criticized by some historians, these accounts do contain information of historical and ornithological value.

Fr. François-Xavier Charlevoix (1744, 1761), in his *Histoire et description générale de la Nouvelle France* [History and Description of New France], gives a number of details on the flora and fauna of Canada, the Mississippi Valley, and Acadia. It is nonetheless difficult to determine which sightings were made within the boundaries of Québec as they stand now. This work is especially useful for its descriptions of places and habitats as they were at the time, rather than for its ornithological content.

Nearly half a century later, Roland-Michel Barrin de La Galissonière, governor of New France from 1747 to 1749, made a considerable effort to promote science during his stay in Québec (Groulx 1970). Not only did he send his observations to his correspondents and to learned societies in France, but he also sent specimens, mostly plants but also birds. He also encouraged a number of naturalists in their work and urged them to publish their results. Moreover, he asked his administrators to make natural history inventories in all parts of the territory under his jurisdiction (Lamontagne 1960a,b). His contribution to ornithology is difficult to assess, as he left no documents devoted exclusively to the birds of New France. Nonetheless, his unpublished observations and the specimens sent to France were used by J. M. Brisson (1760) in his book entitled *Ornithologie*. This monumental work was the first ornithological treatise to be published in France and was widely quoted, including the information he gave on the birds of New France, by many other authors such as Buffon in France, Latham (1781–85, 1809) and Pennant (1785) in England, and Gmelin and Müller in Germany.

The first naturalist to visit Québec under the French Regime was Pehr Kalm. Kalm, a Swede, was a follower and collaborator of Carolus Linnaeus. He travelled throughout New France and elsewhere in northeastern North America from June to October 1749. In Québec, he visited the areas around Lake Champlain, the Richelieu Valley, and the Montréal and Québec City regions. His writings were published between 1753 and 1761 and contain interesting observations on many species of birds; the English version appeared a few years later, in 1770 (Kalm 1770–71). His travel account, however, which contained a large number of ornithological observations, was only translated and annotated in 1977 (Rousseau et al. 1977). Aside from his observations on birds, Kalm gave a number of descriptions of areas that can be readily identified today. Because of his work, it is often possible to have a precise idea of what certain habitats were like at that time and even to establish comparisons which are useful in studying population dynamics and the historical distribution of certain birds, animals, and plants. The specimens he collected and his observations were no doubt also used by Linnaeus, who incorporated this information and that of his other collaborators into the 10th edition of his *Systema Naturae*, published in 1758 and which remains the basis for zoological nomenclature.

During the 17th and 18th centuries, until the conquest, British explorers and others sent by Britain and the Hudson's Bay Company published a considerable amount of information on northern birds in various British publications (Allen 1951). The writings and reports of these explorers offer a wealth of historical and scientific data.

Ornithology at the end of the 18th century, 1760–1800

Important as was the contribution of the British explorers and writers of this period, at least as concerns the regions of Hudson and James bays, the context of their work is difficult to situate, as most of the Hudson's Bay Company trading posts lay beyond the borders of Québec as we know them today. These writings were by people like George Edwards (1776), Andrew Graham (Williams 1978), Thomas Pennant (Pennant 1785; Allen 1908), and Johan R. Forster (1771, 1772; Willoughby Society 1882). Moreover, some works of lesser importance published by the Royal Society of London also contain interesting observations on the birds of northern Québec, although here again it is often difficult to determine where the specimens were collected and the observations made.

The beginnings of structured ornithology, 1771–1900

The first part of the 19th century was marked by the appearance of several major works that constitute the foundation of modern ornithology. Although most contain only a small amount of data from Québec, they nonetheless deserve mention because of their influence on Québec ornithology. In the 18th century, Mark Catesby (1771) had published a work whose influence would be felt in Québec and Canada for many years thereafter. The work of Alexander Wilson (1808–14) had an even broader influence, because information from his life histories was used by many authors up until the end of the 19th century, often without verification or even acknowledgment. The major works of C. J. Temminck (1807, 1813–15, 1817), L. J. P. Vieillot (1807, 1816, 1817), and G. Cuvier (1826–28) would serve as a basis for the ornithological classifications used in the years thereafter, including species found in Québec.

In the 19th century, the first work that contained original observations on the birds of Québec was *The Birds of America* (Double Elephant Folio) by John James Audubon (1827–38, 1840). Audubon's monumental work contains a host of data on a variety of species that he observed in 1833 on the Lower North Shore of the Gulf of St. Lawrence, particularly in the Blanc-Sablon region (Audubon 1827–38; Audubon and Coues 1898; Deane 1910; Chancellor 1978). Audubon also described a new species, Lincoln's Sparrow, which he saw for the first time near the mouth of the Natashquan River. The works of Wilson and Audubon can be considered the cornerstones of North American ornithology, and they had a significant influence on Québec ornithology. Audubon's work comprises not only colour plates of all known species, executed in a highly innovative style, but a host of information on the ecology, behaviour, abundance, and distribution of these species in North America, including Québec. Audubon's descriptions of the places and nest sites he visited are so accurate that often they can be located today, as is the case with the Peregrine Falcon nest found in Brador Bay, west of Blanc-Sablon (Audubon 1827–38).

It was not until 1857 that the first authentically Québec work on ornithology appeared. This was a short article on the winter distribution of land birds in the Montréal region in the winter of 1856–57, written by William D'Urban. A few years later, in 1862, Dr. Archibald Hall published a unique document on the distribution and abundance of birds he had seen in about 1830 in the Montréal region and elsewhere in the St. Lawrence Valley (Hall 1862).

Between 1859 and 1901, James MacPherson Le Moine (1860, 1861, 1864, 1866, 1874, 1889, 1892, 1901), a historian and member of the Québec City bar, published several works on the birds of Québec; these contain little original material, but the most important are: *Ornithologie du Canada; quelques groupes d'après la nomenclature de Baird* (Le Moine 1860) and *Ornithologie du Canada d'après la nomenclature du Smithsonian Institution de Washington* (2nd edition, part 2) (Le Moine 1861). Other publications on the birds of Québec from this period were written by Montague Chamberlain (1887, 1888, 1891, 1894, 1895, 1903), whose original work is limited to New Brunswick (Taverner 1925). Chamberlain's principal merit is to have published the first overall work on the birds of Canada; most of his material, however, is derived from the writings of Thomas Nuttall, who wrote on the birds of the United States.

Around this time, Charles-Eusèbe Dionne, who might be considered Québec's first ornithologist (Déry 1925; Gaboriault 1974), published two works in French, one on the birds of Canada, *Les oiseaux du Canada* (1883) and the other more narrowly focused, *Catalogue des oiseaux de la province de Québec, avec des notes sur leur distribution géographique* (1889). These works, which contained both original and hitherto unpublished information on the birds of Québec, along with data from contemporary literature (Déry 1925; Gaboriault 1974), mark a new stage in the history of ornithology in Québec. It was, however, his *Les oiseaux de la province de Québec* (Dionne 1906) that did the most to further the development of ornithology in Québec, as it provided a broad readership with a reliable source of information and thus stimulated interest in birds.

Charles-Eusèbe Dionne is considered Québec's first ornithologist. Associated mainly with Laval University, he worked as a taxidermist, librarian, and curator of the zoological collections. He held the degrees of Master and Doctor of Science and was one of the founding members of the Provancher Society of Natural History of Canada in 1919. *Laval University Collection.*

Toward the end of the 19th century, other books were published that contributed to the understanding of the distribution of Québec birds. Among these was the publication by John Rae (1888), containing interesting information gathered from the Hudson Bay region. Still more important was the book by Lucien M. Turner (1885), who described for the first time the distribution of birds in northern Québec, mainly in the areas around Kuujjuaq (Fort Chimo), Hudson and James bays, and on Anticosti Island. The observations he made and the specimens he collected constitute an impressive source of material for the study and analysis of the birdlife of these regions, and their importance is more than historical. Another important work, also published around that time, was *The Birds of Montreal* (Wintle 1896), which for a number of years was the main reference for naturalists in the Montréal region. The book was a synthesis of Wintle's work published in periodicals since 1883, and it replaced an incomplete, preliminary edition that had appeared in 1893 (Wintle 1893).

The 19th century also saw the introduction of journals and periodicals on natural history and the creation of societies for bird lovers, their main activity being birdwatching. The most important periodicals were: the *Canadian Naturalist and Geologist* (Montréal 1856–1925), the *Canadian Sportsman and Naturalist* (Montréal 1881–83), *Le Naturaliste canadien* (Québec City, 1868–), and *The Canadian Field-Naturalist* (Ottawa, 1879–). These periodicals made a major contribution to the development of ornithology in Québec and Canada, as they published articles on the birds of Québec and promoted correspondence between readers and authors. The natural history societies of the time, such as The Montreal Natural History Society and La Société d'histoire naturelle de Québec, founded by Fr. Léon Provancher in 1870, also fulfilled these functions, often against remarkable odds. Naturalists at that time also published the results of their observations in periodicals elsewhere in Canada and the United States, often at the invitation of editors and colleagues outside Québec.

The 19th century can be considered as a period of consolidation for Québec ornithology. There was not a great amount of ornithological activity, given no doubt the more immediate socio-economic preoccupations of those times. On the other hand, intensive ornithological research was being carried out in the United States, especially during the second half of the century. This research had a profound influence on several Québec ornithologists, notably Charles-Eusèbe Dionne and Napoléon-A. Comeau. The Americans who had the greatest influence on Québec ornithology at that time were Spencer F. Baird, Elliott Coues of the Smithsonian Institution, and C. Hart Merriam of the United States Biological Survey. Napoléon-A. Comeau, the "naturalist of the North Shore," sent his sight records and specimens to C. H. Merriam for over 30 years (Merriam 1882, 1883a,b, 1884, 1885a,b,c). These ornithologists not only sent numerous publications to Québec naturalists, but they also took an interest in their observations and gave them personal encouragement over the years.

Salmon fishing guide, postmaster, telegraph operator, fisheries officer, amateur physician, and the "Grenfell of the Canadian Labrador," Napoléon-A. Comeau was a man of many talents, besides being an accomplished naturalist. His listings of the birdlife on the North Shore of the St. Lawrence River testify to his dedication and the importance of his contribution to Québec ornithology. *Photograph by R. M. Anderson, c. 1925. Photo No. 68112, by permission of the Canadian Museum of Nature, Ottawa, Canada.*

A period of expansion, 1900–60

The early years of the 20th century brought the publication of two major works in the history of Québec ornithology. The first was *Les oiseaux de la province de Québec* (Dionne 1906), the first work devoted entirely to the birds of Québec. Although it contained little information that was original or specific to Québec, it was nonetheless the major work of a French-speaking writer. A few years later, the naturalists John Macoun and James M. Macoun, published their *Catalogue of Canadian Birds* (Macoun 1900–04; Macoun and Macoun 1909; see also Macoun 1920 and Macoun 1979), the French version of which appeared only in 1915 (Macoun and Macoun 1915). This major work on the birds of Canada included data from Dionne (1906), along with unpublished information from other naturalists. It treated the distribution of Québec birds from a wider perspective, describing each bird in the context of its distribution in Canada. Other interesting works were also published during this period, by authors such as C. W. G. Eifrig, whose article on the birds of Hudson Bay (Eifrig 1905) has a certain historical importance, as does his series of articles on the birds of Ottawa (Eifrig 1910–11), which, despite its title, contains several sight records from the Québec side of the Ottawa Valley. Few species were described from specimens taken in Québec; for information of this type, authors such as Déry (1933), Stevens (1936), and Houston (1983) should be consulted.

A few years later, C. W. Townsend made several visits to the Lower North Shore of the Gulf of St. Lawrence, then known as the Canadian Labrador, and published several articles on his observations (Townsend and Allen 1907; Townsend and Bent 1910; Townsend 1913a,b, 1914, 1916, 1917a,b, 1918). He later became interested in the birds of the Gaspé (Townsend 1920, 1923a,b).

In 1882, John Macoun became the first botanist to work with the Geological Survey of Canada. Between 1870 and 1920, his numerous expeditions took him throughout Canada. With his son, James M. Macoun, he produced the *Catalogue of Canadian Birds*, published for the first time in 1900–04.
Photo No. 69358, by permission of the Canadian Museum of Nature, Ottawa, Canada.

In 1911, Percy A. Taverner became the first professional ornithologist in Canada; he also worked for the Geological Survey of Canada. Taverner published hundreds of articles on birds, many about birds in Québec. His work in the field of conservation contributed among other things to the survival of the colony of Northern Gannets on Bonaventure Island. Petrie Islands, Ottawa River, Ontario.
Photograph by W. S. Odell. Photo No. 44459, by permission of the Canadian Museum of Nature, Ottawa, Canada.

In 1911, following upon recommendations by John Macoun, of the Geological Survey of Canada, and pressure from naturalists all across Canada, Percy Algernon Taverner was appointed naturalist and assistant curator of natural history at the Museum of the Geological Survey of Canada (later the National Museum of Canada). Taverner thus became the first professional ornithologist in the country, which marked an important step in the history of ornithology in Canada and Québec. He remained in this position until 1943 (Ouellet and Godfrey 1980; Ouellet 1987, 1988, 1995). During his career, he published nearly 300 articles, several of which dealt with the birds of Québec, conservation of specimens, taxonomy, migration, and behaviour (McAtee 1949). It was partly through his efforts in the area of conservation that the Bonaventure Island Migratory Bird Sanctuary was established, thus protecting the colony of Northern Gannets that nests there. Aside from establishing, as curator of the ornithology section of the National Museum, the foundation of the national ornithological collection, his greatest achievement remains his books on the birds of Canada (Taverner 1919, 1922, 1934). Although all his books contain important data on the birds of Québec, only the one on the birds of eastern Canada was translated into French, several years later as *Les Oiseaux de l'Est du Canada, deuxième édition* (Taverner 1922). Around the beginning of the century, Taverner had begun marking birds using small metallic leg bands, an innovation that others have taken credit for (Bub and Oelke 1989). Taverner also kept in close contact with Québec naturalists and encouraged them in their work, urging the few French-speaking ornithologists to create a French nomenclature for the birds of Canada. He corresponded regularly with Charles-Eusèbe Dionne, Gustave Langelier, and D.-A. Déry in Québec City, Willie Labrie in Kamouraska, Henry Mousley in the Eastern Townships, and Lewis McIver Terrill in Montréal (Ouellet 1995).

It was also during the first quarter of the 20th century that steps were first taken to regulate bird conservation in Québec. One of the pioneers in this area was Hoyes Lloyd (Munro 1979), whose vision and wide-ranging influence have benefited ornithology to this day. As the official in charge of the protection of migratory birds in Canada, he drafted the first wildlife management plans (Solman 1974). His influence was first felt in Québec when he appointed Harrison F. Lewis to be in charge of enforcing the regulations under the Migratory Birds Convention Act, further to the Migratory Birds Convention between Canada and the United States signed in 1916. Lewis himself contributed greatly to bird protection, especially on the Middle and Lower North Shore. He wrote extensively on distribution and ecology, publishing more than 60 notes and articles between 1920 and 1943. He also implemented the five-year seabird surveys, held since 1925, on the North Shore of the Gulf of St. Lawrence, and established several migratory bird sanctuaries on the North Shore and elsewhere in Québec. The Provancher Society of Natural History of Canada, for example, following his initiative and that of its directors, notably Dr. D.-A. Déry, proposed the designation of the Razade Islands and Île aux Basques, near Trois-Pistoles, as the Île aux Basques Migratory Bird Sanctuary. Furthermore, in 1937 at the urging of Harrison F. Lewis, the Province of Quebec Society for the Protection of Birds (PQSPB) purchased Île aux Perroquets, which was incorporated into the Baie de Brador Migratory Bird Sanctuary.

Mention should also be made of the immense contribution of amateur naturalists who devoted a considerable amount of their leisure time to ornithology over the years and who produced most of the major books and articles on Québec birds published up until the end of the 1950s. Lewis McIver Terrill was unquestionably the most important of these. Throughout his long career he not only assembled a huge collection of eggs (now at the Canadian Museum of Nature), but he also published a number of articles on the birds of southern Québec (Terrill 1910a,b, 1911a,b, 1913, 1915, 1916, 1917, 1921a,b, 1922, 1923, 1924, 1929, 1931a,b, 1938, 1942, 1943a,b, 1944, 1945, 1946, 1947, 1951, 1961; Terrill and Smith 1930). He was also one of the founders and first President of the Province of Quebec Society for the Protection of Birds (PQSPB), in which he was particularly active. More information on this remarkable naturalist can be found in a short biography published in the 1982 PQSPB annual report, *Tchebec* (Ainley 1982). Another eminent naturalist of this period was Henry Mousley, who studied the life histories of several species of birds in the Eastern Townships. Using an innovative approach, he produced detailed studies that gave a wealth of information on species that are difficult to study in the field. Mousley published widely in scientific periodicals (Mousley 1916a,b, 1917a,b,c,d,e,f,g,h,i, 1918a,b,c,d, 1919a,b, 1921a,b,c,d,e,f,g, 1924a,b,c, 1925a,b, 1926, 1927, 1928, 1930a,b, 1931a,b,c, 1932a,b, 1934a,b, 1935, 1936, 1937a,b, 1938, 1939a,b, 1940, 1945a, 1946; Ainley 1981). His article on territoriality, which appeared in 1921 (Mousley 1921d), made him a forerunner in the study of this type of bird behaviour (Lack 1944).

During the 1950s, Stanley C. Ball, an American ornithologist, published several articles on the birds of the eastern part of the Gaspé Peninsula. His major work, however, was a study on migration in the Forillon Peninsula (Ball 1952), which can be considered a classic of its type. Another ornithologist, Victor Gaboriault, began a major study on the birds of Québec that he unfortunately was unable to complete (David 1978b). Fr. René Tanguay, of the Collège de Sainte-Anne-de-la-Pocatière, built up a representative collection of the birds of the region, besides publishing a number of articles; he also encouraged students to take up careers in biology and ornithology. The few French-speaking ornithologists in Québec at that time were mostly part of a small group of birdwatchers in the Québec City area, to whom the Club des ornithologues de Québec, founded in 1955, owes its existence. A few dedicated members, among them Raymond Cayouette, began producing the *Bulletin ornithologique*, which became the organ par excellence for communicating ornithological news within and even beyond Québec. The Club des ornithologues de Québec and the *Bulletin ornithologique* have made a sizable contribution to the store of knowledge on birds in Québec, and have built up an information base that will continue to be useful to ornithologists and biologists alike.

Beginning in 1901, W. E. Clyde Todd, an American ornithologist with the Carnegie Museum in Pittsburgh, Pennsylvania, began a series of trips to Labrador and the interior of northern Québec, expeditions that could almost be termed heroic. He published the results of his observations and taxonomic studies in most of the major ornithological periodicals, but it was only in 1963, after a long period of preparation, that his monumental work, *Birds of the Labrador Peninsula and Adjacent Areas,* appeared (Todd 1963). A few years earlier, F. Harper (1958) had published a lesser work on northern Québec, *Birds of the Ungava Peninsula*, which mostly contained sightings from the Schefferville region and along the railway line between Sept-Îles and Schefferville. The birdlife of the western part of Québec and Labrador was better known because of the work of naturalist and explorer Thomas H. Manning (Manning 1946, 1949; Manning and Coates 1952; Manning and MacPherson 1952).

Fr. René Tanguay of the Collège de Sainte-Anne-de-la-Pocatière did much to promote ornithology, through his collection of birds, his publications, and his work with birdwatchers. *Courtesy of Musée François-Pilote, La Pocatière.*

Appointed Curator of Ornithology at the National Museum of Canada in 1947, W. Earl Godfrey conducted research in several localities in Québec, publishing his findings on birds of Lakes Mistassini and Albanel (Godfrey 1949), Lake Saint-Jean (Godfrey and Wilk 1948), and Abitibi (Godfrey 1968). Besides giving information on species distribution and ecology, these writings provide data on geographical variation and the taxonomy of most of the species found in these regions. Godfrey conducted studies in almost all of Canada (Godfrey 1956), and he used his regional studies and a vast quantity of unpublished data to produce his major work, *The Birds of Canada* (Godfrey 1966, 1986). All of the species found in Québec when the book was published are described therein, with relevant comments on their status here.

Québec has hosted numerous conferences at which ornithologists, amateur and professional, have gathered to exchange ideas and present research findings. Perhaps the most prestigious ornithological conference in North America is the annual meeting of the American Ornithologists' Union, held twice in Québec before 1960. The first time was in Québec City in 1932, when it was hosted by the Provancher Society of Natural History of Canada (attendance 250). The next was in Montréal in 1951 (attendance 262), hosted by the PQSPB.

During the period between 1900 and 1960, considerable effort was made in the area of conservation, especially with regard to the protection of aquatic game birds. This led to the publication of several important works on population management. Among these is the study by Louis Lemieux on the Snow Goose (Lemieux 1959a,b).

The contribution of ornithologists to knowledge of Québec birds was particularly important since the end of the Second World War, when results of their work appeared in various natural history periodicals in Canada, the United States, and elsewhere in the world. This type of publication, sporadic for the most part, indicates that studies were being done in an isolated manner, rather than according to an overall plan for the study of Québec birds.

Despite the progress made after the Second World War, ornithology in Québec did not advance at the same pace as in the United States and the rest of Canada. Aside from the social and economic context, there are other factors which might explain this situation, at least in part. Until the early 1960s, none of the French-speaking universities offered courses in ornithology, even at the undergraduate level, which meant that French-speaking biologists who wanted to specialize in ornithology had to attend other Canadian or American universities.

Also, few Québec libraries at that time had books on ornithology. The library of the former provincial museum, which around 1945 had acquired most of P. A. Taverner's collection—books that after the mid-1960s were difficult to obtain—and the library of the Société zoologique de Québec filled the gap in the Québec City region to a certain extent. The core collection of the Blacker-Wood Biology Library at McGill University, which began as the Blacker-Wood Library of Zoology and Ornithology, was one of the best ornithology libraries in North America and certainly the best in Canada. The vast collection of books on ornithology was amassed in the 1920s and 1930s primarily through the efforts of Dr. Casey A. Wood, an ophthalmologist of world renown and McGill graduate. He acquired all of the works on birds available at that time, along with many unique works and periodicals that were difficult to come by. McGill University has attempted to maintain this extraordinary collection, inasmuch as its resources have allowed. There were also a few private collections that reflected the tastes of their owners; among the most noteworthy collections of French-speaking ornithologists were those of Fr. René Tanguay of the Collège de Sainte-Anne-de-la-Pocatière and Gustave A. and Bernadette Langelier of Québec City.

All over the world, private and public museums that possess ornithological collections have stimulated the study of birds by becoming involved in research and by maintaining and building up their collections. The provincial museums have never held to any sustained mission in the area of ornithology, except for brief periods of activity. Their nascent efforts proved fruitless and caused their collections to become dispersed among other institutions. This is what happened to the ornithological collections of the former Musée de la Province in Québec City and the Redpath Museum at McGill University. This lack of a resource, so crucial and fundamental to ornithological research may to some extent explain a certain lack of interest on the part of most Québec ornithologists in systematics, taxonomy, and bird evolution.

Louis Lemieux was a pioneer in the area of large scale studies on the ecology of waterfowl populations in Québec. His work included studies on the Snow Goose, Canada Goose, and American Black Duck. He held a number of different positions: researcher with the Canadian Wildlife Service, head of the Ministère du Tourisme, de la Chasse et de la Pêche du Québec, and director of the National Museum of Natural Sciences. Florida. *October 1952, Raymond Cayouette.*

Since the early 1960s, most ornithological work in Québec has been conducted by professional ornithologists or by biologists with an interest in the discipline, either in provincial or federal government departments, universities, or the private sector. One needs only to consult periodicals that publish articles on ornithology to appreciate the contribution of Québec ornithologists to the study of birds in a global and general perspective. Articles by Ouellet and Godfrey (1980), Cooke (1981), and Ouellet (1988, 1995) are helpful in giving an overall view of the progress of ornithology in Canada. To situate Québec ornithology in a broader context, Bartholomew (1986), Farber (1977, 1982), Mayr (1975), and Stresemann (1975) can be consulted for a comparison of the progress of ornithology in Québec and other parts of the world, keeping in mind the context in which ornithology has developed here.

Henri Ouellet

Memoirs of a pioneer birdwatcher in Québec

It is no exaggeration to say that the history of birdwatching in Québec began with the arrival of the first explorers.

On 21 May 1534, Jacques Cartier recorded having seen the now-extinct Great Auk and the Northern Gannet, which he named *Margaulx*, on Bird Rocks, off the Magdalen Islands. On 9 June, off Blanc-Sablon, he found the *Richar*, which, from his description, was an Atlantic Puffin. He also saw *Godez*, which may have been the Razorbill—a bird that until recently was called *Gode* in French—or it may have been one of the two murres of the genus *Uria*.

Cartier mentions several other birds in his travel accounts; the names he gave them were for the most part inspired by European ones. Even today, the Canada Goose is often wrongly referred to in French as *oultarde* or *outarde* (bustard) in popular parlance (Laporte 1974a).

Samuel de Champlain, the founder of Québec City, left fascinating accounts in which he reported several birds, including the *huard* (1613), or loon, which he thus named because of its call. The bird still bears its mimetic name to this day (Anonymous 1968).

The forerunners

Several other travellers and explorers took an interest in Québec birds: the Récollet brother Gabriel Sagard, who spent the years 1624–28 in Canada; Pierre Boucher, onetime governor of Trois-Rivières, published in 1664 a well-documented work entitled *Histoire véritable et naturelle des mœurs et production du pays de la Nouvelle-France, vulgairement dite le Canada*. In this book, which has been analysed by Jacques Rousseau, Boucher, in his chapter on birds, named some of the species he had observed (Rousseau and Legendre 1966). Pehr Kalm, while travelling through the American colonies and Canada between 1749 and 1751, also described several animal and plant species (Rousseau et al. 1977).

Fr. Léon Provancher was a naturalist who specialized in entomology. In 1868, he founded Le Naturaliste canadien and in 1870 the Société d'histoire naturelle de Québec. *Laval University Collection.*

In the 19th century, William Stewart D'Urban (1857) reported on the birds of the Montréal area. In 1868, the well-known entomologist Fr. Léon Provancher founded the journal *Le Naturaliste canadien*. Although primarily an entomologist, Provancher was interested in all branches of natural science; his contribution to ornithology, however, was rather limited. After Provancher's death in 1892, Canon V.-A. Huard edited *Le Naturaliste canadien* for 36 years. It was then taken over by Laval University until 1991. In 1994, the Provancher Society of Natural History of Canada took over the publication in place of its own journal, *L'Euskarien*, which had been a forum for many ornithologists and birdwatchers, including Charles-E. Dionne, Fr. F.-X. Burque, James M. Le Moine, Raoul Lavoie, J. Gérald Coote, Edmond Yourans, Fr. René Tanguay, D.-A. Déry, and Gustave and Bernadette Langelier, all of whom were to some extent the forerunners of ornithology in Québec today.

The famous American naturalist and painter, John James Audubon (1785–1851), explored the Gulf of St. Lawrence in 1833 and made a number of interesting discoveries, being the first to describe Lincoln's Sparrow at Natashquan. He spent several weeks in Québec City in 1844 (Cayouette 1966).

Also in the 19th century, Sir James MacPherson Le Moine published in Québec City several brochures on birds, including the *Tableau synoptique de l'ornithologie du Canada*, the *Liste des oiseaux du Canada*, and *L'Ornithologie du Canada* (Potvin 1952). Napoléon-A. Comeau, an amateur naturalist who made a study of the game and bird life on the North Shore of the Gulf of St. Lawrence, published in English in 1909 and subsequently in French *Life and Sport on the North Shore of the Lower St. Lawrence and Gulf*.

In 1827, in Montréal, the first natural history society was founded, the Montreal Natural History Society. The Montréal journal *Canadian Naturalist and Geologist* appeared between 1856 and 1868. From 1881 to 1883, William Couper, who at the time was the editor of *The Canadian Sportsman and Naturalist*, published extensive notes on the distribution of birds in Québec. The Ottawa Field-Naturalists' Club published *The Transactions of the Ottawa Field-Naturalists' Club* from 1879 to 1886, *The Ottawa Naturalist* from 1887 to 1919, and, since 1919, *The Canadian Field-Naturalist*. These journals contain a number of significant studies on Québec birds.

During the same period, Charles-Eusèbe Dionne (1849–1925) had three volumes on birds published: *Les oiseaux du Canada* in 1883, *Catalogue des oiseaux de la province de Québec, avec des notes sur leur distribution géographique* in 1889, and his last and most important work, *Les oiseaux de la province de Québec*, in 1906. Former curator of the zoological museum of Laval University, he worked in various capacities at the Séminaire de Québec (Gaboriault 1974). Dionne observed and collected birds for the most part around Québec City, but also in Saint-Denis, Kamouraska County, his native village.

Many other ornithologists in Québec City and Montréal and in other parts of Québec made significant contributions to the study of birds: Archibald Hall, Ernest D. Wintle, who published *The Birds of Montreal* in 1896, Percy A. Taverner, the first professional ornithologist in Canada, who published in English in 1919 and in French in 1922 *Birds of Eastern Canada* and in 1934 *Birds of Canada*, Hoyes Lloyd, Henry Mousley, Stanley C. Ball, Lewis McIver Terrill, Harrison F. Lewis, the first director of what is today the Canadian Wildlife Service, W. Earl Godfrey, who published *The Birds of Canada* in 1966, and many others (Ouellet 1988).

All these major figures in ornithology helped set the stage for what was to develop into the tremendous modern popular interest in birdwatching, fostered and encouraged in the French-speaking community by the special contribution of some noteworthy leaders.

Marie-Victorin's École de la Route

Interest in natural science reached new heights in the early 1930s, stimulated by the activities of Bro. Marie-Victorin and his followers. Marie-Victorin founded the Montréal Botanical Gardens, in 1931, but the enthusiasm and encouragement of this great botanist and his followers also contributed to the founding of the Québec City zoo in 1931, the Cercles des jeunes naturalistes (CJN) by Bro. Adrien Rivard in 1931, and the 4-H Clubs in 1942. In a lecture in Québec City on 22 November 1931, Marie-Victorin put forward the idea of a new type of nature education, the *École de la Route*. The following year, he organized summer courses in natural science for primary and secondary school teachers. At that time, Fr. Ovila Fournier was the Director General of the Cercles des jeunes naturalistes (Audet 1949). I myself gave courses in Québec City in 1948, 1949, and 1950 as part of this program, which was designed to create a structure for the teaching of natural science in the schools. The religious orders were behind this initiative, which subsequently led to the creation of the CJNs.

Interest in natural science was slower to develop among French-speaking Quebecers than among English Canadians and Americans, but, as in other fields, a lot of ground was gained after the 1940s.

In Québec City and Montréal there were a number of associations dedicated to conservation and the study of natural science.

THE PQSPB, THE OLDEST BIRDWATCHING CLUB IN QUÉBEC

Founded in Montréal in 1917, the Province of Quebec Society for the Protection of Birds was the first birdwatching club formed in Québec. This society, whose name is clearly derived from that of the Royal Society for the Protection of Birds in the United Kingdom, had as its first president and co-founder Lewis McIver Terrill. Mr. Terrill was well known in the province as an ornithologist, and he spared no effort to gain acceptance for the organization and the importance of its role. By 1919, the PQSPB had 100 adult members and 2500 youth members; within 10 years of its founding, membership had grown to 300 and 24000, respectively. Well-known contemporary naturalists joined the PQSPB, among them Bro. Marie-Victorin, the renowned Montréal botanist; Henry Mousley, an ornithologist active in the Eastern Townships and author of many scientific articles; and Harrison F. Lewis, first director of what is now the Canadian Wildlife Service. Terrill corresponded actively with prominent ornithologists in the United States and elsewhere in Canada, such as Arthur Cleveland Bent and Edward Howe Forbush, and gave the Society standing and respect in North American ornithological circles.

The Society has been guided by distinguished presidents, including the Honourable Mr. Justice G. H. Montgomery, who was instrumental in the founding of the Society's Philipsburg Migratory Bird Sanctuary, Dr. Max Dunbar, a professor at McGill University for many years, and Dr. Ian McLaren, of Dalhousie University's Biology Department.

The PQSPB set itself various objectives, among them creation of a forum for the exchange of information on birds, teaching ornithology in schools, and arousing the interest of the general public in protecting birds by means of lectures, exhibits, and field trips. From the mid-1930s on, publication of reports on birds observed in the Montréal area further contributed to the spread of information about birds.

One of the PQSPB's objectives is to protect bird habitat, and it has made noteworthy and sustained efforts over the years to convert various sites into bird sanctuaries. In particular, it fought during the 1960s to save the Common Tern colony on Moffat Island (now Île Notre-Dame), but the site was filled in during work on Expo 67. The effort was not in vain, however, as the pressures brought to bear led to the creation of other bird sanctuaries in the Montréal area, including that on Îles de la Paix.

The PQSPB, since 1979 also known as the *Société québécoise de protection des oiseaux (SQPO)*, is still very much involved today in the protection of birds.

Field trips are still the main activity of a birdwatching society. Here a group from the Société de biologie de Montréal is seen with Wilfrid Gaboriault, president (upper right), and Roger Perreault (centre). Île aux Noix, Richelieu River. *Spring 1967, courtesy of Wilfrid Gaboriault.*

Among them were the Provancher Society of Natural History of Canada, the Société lévisienne d'histoire naturelle, the Province of Quebec Society for the Protection of Birds (PQSPB), the Société linnéenne de Québec, the Société canadienne d'histoire naturelle, the Société de biologie de Montréal, and the Société zoologique de Québec.

One of the many activities organized by the Provancher Society, founded in Québec City in April 1919, was a provincial bird festival, the Fête provinciale des oiseaux, held in the late 1920s and early 1930s in various localities around the province. One aspect of this annual festival was a birdhouse-building contest. On 28 May 1931, I received a modest prize for a Purple Martin house I had built, one of 400 entries. Three pairs of Purple Martins took up residence in my birdhouse on the very day that I set it up in my parents' backyard! This event marked the beginning of my interest in birds.

One of the dreams of the founders of the Provancher Society, the purchase of three islands in the St. Lawrence, was realized less than 10 years after the society's inception. These islands, the two Razade Islands, acquired in 1927, and Île aux Basques, acquired in 1929, are well known for their birdlife. In Lévis, on 12 November 1931, the Société lévisienne d'histoire naturelle came into being. I became a member in 1932 and took part that year in a natural science exhibit at the Académie de Québec. This society ceased its activities in the mid-1940s.

Since its founding in 1917, the Province of Quebec Society for the Protection of Birds (PQSPB) has been very active in the field of conservation. Through its petitions and influence, it acquired lands containing valuable habitat which were subsequently turned into bird sanctuaries—Philipsburg, Montréal area, Eastern Townships, estuary, and Gulf of St. Lawrence on Île aux Perroquets (Lourdes-de-Blanc-Sablon)—and helped in the protection of many other sites. Since 1935, the knowledge and enthusiasm of its first president Lewis McIver Terrill and its founding members have been passed on to thousands of young people. Its annual reports, published in *Tchebec*, testify to the dedication of PQSPB members and contain valuable information on birds.

The PQSPB has inspired many birders to become ornithologists. Other associations, such as the Société canadienne d'histoire naturelle (1929), the Société de biologie de Montréal (1922), the Société linnéenne de Québec (1929), and the Société zoologique de Québec (1932) have at various times during their existence and to varying degrees promoted the study and observation of birds.

One of the PQSPB's active members, Bro. Victor Gaboriault, C.S.V. (1909–52) also inspired many people in the Montréal area to take up birdwatching. He began recording his sightings in 1933 (David 1978b) and encouraged many bird lovers in Québec to begin collecting data on birds.

Victor Gaboriault played an important role in promoting birdwatching in Québec. He created the first French checklist for daily bird sightings, thus promoting the spread of knowledge about birds. *Courtesy of Wilfrid Gaboriault.*

Bro. Gaboriault, a solitary man by nature, was a highly active and methodical worker. He devised a system of data compilation, checklists, exchanges of notes, and bird distribution maps in preparation for the eventual publication of his projected *Le catalogue annoté des oiseaux du Québec*.

In 1947, he completed a thesis on Charles-Eusèbe Dionne (Gaboriault 1974) and published several articles in periodicals, notably in the journal of the Clerics of Saint Viateur, *Le Viateur naturaliste*, and *Le Jeune naturaliste*, two publications edited by Bro. Léo Brassard, C.S.V. (David 1978b). In 1948, he began exchanging notes on sightings with several colleagues in various parts of Québec. He had also acquired a considerable body of data by methodically gleaning material from books, journals, and collections in Québec museums. At the time of his premature death, on 22 March 1952, his work remained unfinished.

In 1948, I had a very friendly discussion with Bro. Gaboriault about collecting and compiling information on birds in Québec. I had previously devised a similar system of my own, and I proposed a kind of affiliation and even financial assistance from the Société zoologique de Québec. For entirely valid reasons, he preferred to compile his notes on his own, although he did not by any means eschew collaboration with me and other birdwatchers. In fact, he found such co-operation essential, and it was freely given (pers. corr.).

A group of adult students at Camp Marie-Victorin in Sainte-Félicité. *August 1956, Raymond Cayouette.*

Louis-Amédée Lord (left) and Fr. Robert Plante birdwatching at the edge of a beaver pond. Sainte-Félicité, near Montmagny. *August 1956, Raymond Cayouette.*

Nature camps

In June 1955, Bro. Léo Brassard founded the Camp des jeunes explorateurs, or Jeunes Explos. Every year, schoolchildren from all over Québec attended this nature camp. It was first held in Saint-Fulgence, on the Saguenay, and then in Grandes-Bergeronnes and Sainte-Luce-sur-Mer, on the north and south shores respectively of the St. Lawrence estuary. This camp aroused considerable interest among young people and inspired many to become naturalists and even take up careers in science (Brassard 1962).

Other similar camps were inaugurated at this time: the Camp de Saint-Fidèle de Charlevoix, founded by Fr. Louis Genest, C.S.V. and known for its location and the quality of the instruction, the camp at Lake Chicobi in Abitibi, and the Camp Rolland-Dumais at Rivière-à-Pierre in Portneuf. In June 1947, Fr. Raoul Cloutier inaugurated the Camp-École Trois Saumons at L'Islet. This co-educational camp offered a variety of activities, including natural science, dear to the heart of its founder. One site, known as "Plantation" and run by Fr. Robert Plante, was very popular among young people (Blanchet 1963).

Camp Marie-Victorin

In 1952, Fr. Robert Plante, a teacher at the Collège de Lévis, was sponsored by the Provancher Society to attend the Maine Audubon Nature Camp. The Société zoologique de Québec also gave a bursary to Rolland Dumais, the guide for school tours of the zoo, to attend a 15-day intensive course there. The idea of founding a similar camp for adults in Québec began to take root. The next year, on 2 August 1953, Rolland Dumais and Robert Plante founded the Camp Marie-Victorin.

The new camp held its activities at Fr. Raoul Cloutier's camp grounds at Lake Trois Saumons. Every June since it was founded, groups of 20–40 counsellors or monitors from playgrounds, camps, 4-H Clubs, and other organizations all over Québec have attended courses in natural science there. Teachers had their own camp in August.

The room and board provided by the camp afforded adult participants the freedom they required to carry out their activities. Using their individual teaching methods, naturalists gave courses in the field in their own spheres of activity: entomology, ornithology, botany, meteorology, and astronomy (Dumais 1962). By the time of his death in October 1972, the camp director, Rolland Dumais, had welcomed more than 1800 adult trainees. I was privileged to give courses there in birdwatching for 15 years, along with Robert Plante, Louis-Amédée Lord, Jacques Normandin, Paul-Étienne Sirois, Jacques Roberge, and Ronald Lepage.

Jacques Normandin was one of the founding members of the Club des ornithologues de Québec (COQ). This club was founded by ornithologists who taught at the summer camp at Lake Trois Saumons near Montmagny. *July 1959, Napoléon Beaudet.*

The Club des ornithologues de Québec

François Hamel, the first president of the Club des ornithologues de Québec, wrote in 1961: "There were several natural history societies in Québec City, but we wanted a group devoted exclusively to ornithology. During the August session of the Camp Marie-Victorin at Lake Trois Saumons in 1955, the idea of forming such an association in Québec City took shape. Four teachers at this nature camp, Raymond Cayouette, Rolland Dumais, Louis-Amédée Lord, and Jacques Normandin, set the agenda for the first meeting, to be held the following September 22..." (Hamel 1961).

The first meeting brought together 40 or so birdwatchers who at that time were only slightly acquainted, if at all. In those days, a group of people in the field watching birds through binoculars was still considered a strange sight.

The Club des ornithologues de Québec, whose purpose was simply to observe birds in their natural setting, organized field trips in Québec City and the surrounding area: the Plains of Abraham, the beach at Saint-Augustin, Cap-Rouge, the Maizerets tidal flats in Beauport, and the woodland of the Québec City zoo. The leaders then began organizing field trips to Cap Tourmente, with its wealth of wildlife, particularly geese, the Sorel Islands, and in 1950, in conjunction with the PQSPB, to Covey Hill. Since 1959, Île aux Basques, off Trois-Pistoles, has been the site of the annual field trip at the end of May.

A smaller-scale program offered visits to the zoo and the Musée de la Province, which then had large zoological collections; Rolland Dumais was at that time in charge of the collections, and Louis-Amédée Lord was the resident taxidermist. During the winter months, there were lectures and slide shows.

Birdwatching is a year-round activity, and from the beginning the club took part in the Christmas Bird Count each year. It still participates in the Count, which began in Québec City in 1943, under the auspices of the Société zoologique de Québec. On 21 December 1943, Bernadette Langelier, Louis-Amédée Lord, and Raymond Cayouette identified 534 birds of 13 species, the humble beginnings of an activity which in its 50th year, on 20 December 1992, brought together 68 people who sighted 15 180 birds of 71 species.

A quarterly newsletter was sent to members as of 1956. In the beginning a modest publication, the *Bulletin ornithologique* is now in its 40th year. Produced by volunteers, it carries members' sightings, announcements of activities, and background articles.

Birdwatchers use checklists to keep track of the data they collect. Around 1948, they began using the checklist drawn up by Bro. Victor Gaboriault. In 1959, Bro. Léo Brassard's list was published and distributed by *Le Jeune naturaliste* and *Le Jeune scientifique*, and later, the list of the Club des ornithologues de Québec was used. Since 1980, people have been using checklists of the Association québécoise des groupes d'ornithologues. Unless data are processed locally, checklists are generally sent to Rimouski for entry in what has grown into an impressive and much-valued data bank on bird populations, known today as ÉPOQ (*Étude des populations d'oiseaux du Québec / Studies of Bird Populations in Québec*).

Along with modern field glasses and spotting scopes, birdwatchers used Roger Tory Peterson's popular *Field Guide to the Birds of North America* (1934) and the Robbins, Brunn, and Zim *Guide to the Birds of North America*. These illustrated guides, with their fine colour illustrations, were translated into French during the 1970s and 1980s, as were many other American books.

Among the French books for amateur birdwatchers are *Charmants voisins* by Claude Mélançon (1940), *Les oiseaux du Québec* by Raymond Cayouette and Jean-Luc Grondin (1972), and *Nichoirs d'oiseaux* by Raymond Cayouette. There are also the recordings of birdsongs by Jean Bédard and Pierre Morency and their more modern adaptations. Books and brochures by Normand David and Gaétan Duquette (1982), Peter Lane (1987), André Dion, Julien Boisclair, and several other authors have all been widely and enthusiastically received. The radio talks and books by Pierre Morency also achieved popular acclaim.

Many birdwatchers have taken an interest in bird banding, a fascinating ornithological activity that has contributed to our understanding of the mysteries of bird migration. The Breeding Bird Survey (BBS), organized by the U.S. and Canadian governments, has been enthusiastically taken up by experienced birdwatchers since 1966. Finally, the phenomenal increase in the number of birdwatchers has spurred publication of birding site guides. The Club des ornithologues de Québec, which today has 600 members, has over the years stimulated the founding of several other birdwatching clubs and associations in Québec. The thousands of members of these clubs do not all indulge in the same activities, but they all share the same interest in birds. Some enjoy identifying birds and listing their sightings. Others, increasing in number, attract birds with nest boxes and feeders. Many practise photography, painting, and sculpture. There are now a number of excellent photographers, filmmakers, and wildlife artists of international renown in Québec—J.-L. Grondin, P. Leduc, and J.-L. Fründ, the leading international nature filmmaker. There are others, mainly professionals, who carry out studies on bird behaviour and habits or conduct population surveys.

Ornithology, like astronomy, is one of those rare sciences that depend heavily on data gleaned by amateurs.

We can only hope that Quebecers' interest in birds will continue to grow and thus contribute to preserving bird populations for future generations.

Raymond Cayouette

A founding member of the Club des ornithologues de Québec, Raymond Cayouette compiled the sightings of Québec birdwatchers for many years. These compilations are published each season in the Bulletin ornithologique, which first appeared in 1956. *April 1988, André Cayouette.*

Birdwatching in Québec since the 1970s

In the late 1970s, birdwatching was booming. In Québec, the Club des ornithologues du Québec (Province of Québec) (COQ), known as the Club des ornithologues de Québec (Québec City) between 1955 and 1975 and again since 1991, had taken on an environmental battle in an effort to save the Beauport flats, which were threatened by the construction of the Dufferin-Montmorency expressway, the prelude to a vast project for industrial development and harbour facilities.

In May 1978, in conjunction with the Association des biologistes du Québec, the Société linnéenne de Québec, and several other groups, COQ set up a vigorous protest and brought the debate into the public forum. During the public hearings, the club pointed to the historical importance of these broad tidal flats, the "marais de la Canardière," located at the threshold of Québec City and long known for its wealth of game birds.

In its submissions, COQ demonstrated that the site was one of the best areas in the Québec City region for studying birds. It also described its importance as a stopover for large numbers of migrating waterfowl and shorebirds and the enjoyment they brought to birdwatchers.

COQ pointed out that the project would not only have a negative effect on birdlife, but would cut off the inhabitants of the greater Québec City region from their last significant area of direct access to the river. Thus the debate took on broader environmental implications than simply the protection of a birdwatching site.

This was the first time that a Québec bird club had undertaken such a large-scale public battle, and it threw all its energy into the endeavour. It was then that the associations involved realized what an enormous amount of work was required to carry on a public debate on major environmental issues—the Beauport flats, Québec conservation and recreational parks policy (1978–82), the Kamouraska dikes (1980). Thus, in the early 1980s, they decided to join forces and unite to form the Union québécoise de conservation de la nature (UQCN).

The late 1970s in Québec also saw the emergence of several regional clubs. In many cases, as in Pointe-au-Père, near Rimouski, environmental issues served as a backdrop. It was during the struggle to protect the marshes in that area that the local birdwatchers met in 1979 and founded the Club des ornithologues du Bas-Saint-Laurent.

Growth in the regions

In the late 1970s, the state of birdwatching in Québec was much as it had been in the mid-1950s. The established natural science associations pursued their educational activities, and two main bird societies served people interested in watching birds. These were the Province of Quebec Society for the Protection of Birds (PQSPB) in Montréal, founded in 1917, and the Club des ornithologues du Québec (COQ), founded in 1955.

Despite this apparent stability, things were happening. Since about 1975, COQ had widened its recruitment area and was enrolling more and more members from outside the Greater Québec City region. The result was that in 1978, the club had 650 members, for the most part from practically all over southern Québec.

At this time, nuclei of birdwatchers would appear in given areas where there was the potential for forming separate bird clubs. Such was the case in the Hull region, in 1978, when the Club des ornithologues de l'Outaouais was formed.

This trend continued, and by the end of the decade four new clubs had been founded: the Club des ornithologues de l'Outaouais, the Club des ornithologues amateurs du Saguenay–Lac-Saint-Jean, the Club des ornithologues du Bas-Saint-Laurent, and the Club des ornithologues de la Gaspésie.

A federation of regional clubs

As these new clubs emerged, a debate arose over the relationship between the regional clubs and COQ, which at that time was the mouthpiece for Québec birdwatchers.

In the spring of 1980, at a convention at Lake Trois Saumons, it was suggested that a COQ committee be set up to study potential ways of co-ordinating the activities of COQ and the regional clubs. The COQ board of directors delegated one of its members to look into the matter, and in early autumn an interregional committee was formed. Made up of two representatives from each of the existing clubs, the committee, which reported to the COQ board of directors, met for the first time a few weeks after its inception.

At this first meeting, held on 18 October 1980, all of the committee members agreed that a provincial ornithological association should be set up, but since the time was not ripe for creating such an organization, COQ was assigned the role of co-ordinator.

The situation developed during the following months to the point where, at the second meeting in April 1981, the committee members expressed the wish to have a federation of Québec bird clubs set up as soon as possible. Finally, at the beginning of September, the COQ board of directors presented its members with the idea of forming a provincial federation of birdwatching clubs. This idea was accepted, not without some lively debate, at a special general meeting held in early October 1981.

The birth of the AQGO

The founding of the Association québécoise des groupes d'ornithologues (AQGO) in October 1981 marked the end of the debate on the need for a federation to represent birdwatchers. From that point on, the new association would represent the birdwatching clubs, promote birdwatching, encourage the study and protection of birds in Québec, and advocate conservation of bird habitat.

This is how the founding members of the AQGO defined the objectives of the Association, which were set out in the letters patent. The first meeting of the AQGO took place on 10 October 1981. Five clubs, representing the regions of Québec City, the Ottawa Valley, Saguenay–Lac-Saint-Jean, the Lower St. Lawrence, and the Gaspé, took part in the meeting. By the end of the year the Société de loisir ornithologique de l'Estrie and the Province of Quebec Society for the Protection of Birds had joined the ranks of the AQGO.

By the end of 1981, Québec birdwatching societies possessed the structures that would enable them to deal with the dramatic development of birdwatching in Québec in the years to come.

New growth

In 1980, the publication of the French version of the field guide *Birds of North America* (Robbins et al. 1966, 1980) was an important event for French-speaking ornithologists. This translation gave Québec birdwatchers their first field guide in French. The names used were those of the Société zoologique de Québec, first published in 1983.

Also in 1980, Roger Tory Peterson published the fourth edition of his famous field guide. This was translated as well, and the French version came out in 1984 (Peterson 1984). The new French names used in the guide were those recommended by Ouellet and Gosselin (1983). The use of two French nomenclatures caused some confusion in the ornithological community, but in any case, anyone wanting to birdwatch in French now had access to French field guides.

These books gave new life to birdwatching, and its development in subsequent years had much to do with the availability of these new guides.

BIRD NAMES: DIVERGENT VIEWS ON NOMENCLATURE

Québec is certainly not the only place where conflict over bird nomenclature exists, whether over the common names or scientific ones. The problem is by no means a new one, and it is still not settled, given the dynamics of the process involved and the relatively recent interest in ornithology in the French-speaking community.

French bird names in Québec have been changed several times this century, notably by Dionne (1906), Taverner (1922), the Société zoologique de Québec (1983; CWS 1972), Ouellet and Gosselin (1983), and, more recently, by Devillers et al. (1993). Despite numerous attempts to establish an accord on English names (AOU 1983; Sibley and Monroe 1990; Inskipp and Sharrock 1992), there is still disagreement between the two main authorities, the American Ornithologists' Union and the British Ornithologists' Union, as, for example, in the case of "jaeger" as opposed to "skua" and "loon" as opposed to "diver" (Sibley and Monroe 1990).

Scientific nomenclature, which is governed by the International Code of Zoological Nomenclature, is revised in the light of developments in systematics, based on new techniques such as DNA analysis. The classification of species is being continually reviewed. Subspecies become species, species change families or become subspecies. The taxonomic order is being continually updated. Such changes are made to maintain clear relationships and thereby avoid confusion; they inevitably have a bearing on French and English nomenclature.

The increasing number of clubs

Although no new clubs joined the AQGO in the year after its inception, 1983 gave rise to a trend that led to the founding of new clubs in many parts of Québec.

This trend began in the Drummondville region, with the founding of the Société ornithologique du centre du Québec. During the next year, four other clubs were formed, in the Mauricie, Manicouagan (North Shore), Sorel–Tracy, and Lanaudière regions. By the end of 1984, the AQGO had 12 member clubs representing birdwatchers from as many regions in Québec. It was during this year that the AQGO, the Canadian Wildlife Service, the Ministère du Loisir, de la Chasse et de la Pêche, and the PQSPB agreed to initiate a breeding bird atlas for southern Québec, thus mobilizing the entire ornithological community.

Only three clubs were founded during the following few years: the Bois-Francs club in 1986 and the upper Laurentians and Trois-Rivières clubs in 1988. However, after 1989, there was a virtual explosion in the number of new clubs. Seven clubs joined the AQGO in 1989: the Brome–Missisquoi, Haute-Yamaska (Granby), Saint-Hyacinthe, Longueuil, Laval, and Magdalen Islands clubs, and the Société de biologie de Montréal, although the last was an existing organization.

Since 1986, new birdwatching clubs have tended to limit their sphere of endeavour to activities of a local nature. Moreover, some of the new clubs have been founded by people not affiliated with COQ, which had not been the case previously.

Between 1990 and 1993, the AQGO acquired six more member clubs—the Abitibi, Ahuntsic (Montréal), Moulins region (Terrebonne), Valleyfield, Châteauguay, and Témiscamingue clubs—bringing AQGO membership to 28.

Since the early 1980s, there has been not only an increase in the number of birdwatching clubs, but a major change in the way people birdwatch in Québec. Although primarily a hobby of a scientific nature, birdwatching has become a multi-faceted activity, practised by people of all ages, for a greater variety of reasons than in the past. People are motivated to join a birdwatching club not just for the scientific activity it provides, but also because they are seeking a means of relaxation, the enjoyment of seeing birds, and the social rewards of belonging to a club.

The scientific dimension

The scientific dimension nonetheless remains an important aspect of birdwatching, and even if it is not their primary motivation, birdwatchers are still collecting data on Québec birds, be it no more than recording their sightings on checklists. These sightings are compiled regionally before being entered into Québec's computerized database, known as ÉPOQ.

Conceived as a practical means of collecting data, checklists have been used by Québec birdwatchers since the 1950s. This simply requires indicating the number of individuals for each species sighted, using one list per site per day.

The system for gathering data has evolved considerably since the mid-1970s. At that time, COQ commissioned Jacques Larivée, a computer scientist interested in ornithology, to computerize data collected up until that point. Thus began the lengthy process that led to the creation of a provincial database and the development of the software necessary for the regional input of checklists, thus allowing AQGO member clubs to use their own data on the birds in their particular area.

The provincial database, managed by the AQGO since 1988, is consulted regularly, and the information on Québec bird populations is used for both scientific studies and popular publications on birds.

Another data gathering activity is the Christmas Bird Count, which is attracting a growing number of Québec birdwatchers. This activity, held during a two-week period around Christmas, consists in counting all the birds seen within a prescribed area 25 km in diameter. It also gives birdwatchers a chance to get together and exchange sightings. The information is published by the National Audubon Society. In 1970, there were only seven CBC territories in Québec; 23 years later, in 1993, there were nearly 30.

This increasing interest in birds on the part of Quebecers has contributed enormously to the collecting of ornithological data. The Québec Nest Record Card Program (QNRCP), the Breeding Bird Survey (BBS), the Woodcock Singing Ground Survey, Feederwatch, and the Eastern Bluebird, swallow, and Wood Duck nest box projects are among the various activities that birdwatchers pursue which at the same time enable scientists to benefit from the data accumulated.

The Québec Atlas, moreover, with which the AQGO has been associated since the beginning, has benefited enormously from the contribution of birdwatchers, who have provided data and even written some of the texts. Most of the photographs in the Atlas were taken by birdwatchers and professional ornithologists, who generously agreed to share them. They were obtained from all over Québec through solicitation of club members and by awareness campaigns involving announcements on the radio, in local papers, and in some of the major daily newspapers. During the early years of the project, regional co-ordinators hired and trained young birdwatchers recruited through the bird clubs and as part of student employment programs. Several regional associations grew out of participation in the Atlas field work, especially in areas where none had previously existed.

The publishing of bird books in Québec

The 1980s saw a surge in publication of bird books, as publishers strove to meet the needs of the ever-growing numbers of birdwatchers.

Aside from a few books on bird identification, little documentation existed in French in the 1970s for Québec birdwatchers. The main books in French were those by Taverner (1922), Godfrey (1967), and Ouellet (1974a), but a large part of the information on birds was distributed through notes and newsletters put out by the various bird clubs.

For example, in the late 1970s, the Club des ornithologues du Québec published the *Cahiers d'ornithologie—Victor-Gaboriault*, which provided the latest data on Québec birds (David 1978b, 1980a,b; Fortin et al. 1978; Boily 1983). In fact, the most recent annotated list of the birds of Québec, which dates from 1980, appeared as part of this collection (David 1980a,b).

In the early 1980s, a few French books on birdwatching written in Québec became available (Cayouette and Grondin 1977; Huot 1980; David and Gosselin 1984), including a small book on nest boxes (Cayouette 1978b) and a few books on feeders (Lane 1980, 1987; David and Duquette 1982). Add to this the two French field guides (Robbins et al. 1980; Peterson 1984) and the French version of *Birds of Canada* (Godfrey 1967), and you have a fair idea of what the library of a French-speaking birdwatcher was like at that time.

In the following years the number of books available increased considerably (Boisclair 1980; Dion 1988; Delaunois 1990; Surprenant 1993). More American books were translated, and in 1987 Éditions Marcel Broquet published the French version of the National Geographic Society field guide (1983, 1987).

Conferences organized by the Association québécoise des groupes d'ornithologues offer excellent opportunities to make new acquaintances and exchange information. They also enable birdwatchers to meet outstanding ornithologists, as was the case at the 1990 convention, when the honorary chairman was none other than Roger Tory Peterson (eighth from the left).

The world-famous field guides by Roger Tory Peterson have promoted the development of birdwatching, especially in North America. The French version of his *Guide to the Birds of Eastern North America* did more than anything else to stimulate interest in birdwatching in Québec. The death of Mr. Peterson in July 1996 leaves a great void in North American birding circles. Saint-Augustin-de-Desmaures, west of Québec City. *May 1990, Patricia Desjardins.*

Not to be outdone, Québec birdwatching clubs also began publishing, specializing in regional site guides, which describe the best locations for watching birds in a given area.

The first such site guide was the *Guide d'observation de l'Outaouais*, put out by the Club des ornithologues de l'Outaouais, the Ottawa Valley club, in 1985. In 1988, the Gaspé club published its *Birdwatching Itinerary of the Gaspé Peninsula* (*Itinéraire ornithologique de la Gaspésie*) (Girard 1988) and the Société ornithologique du centre du Québec brought out its *L'observation des oiseaux au lac Saint-Pierre* (Société ornithologique du centre du Québec 1988).

Other site guides include *Birdfinding in the Montreal Area* (*Où et quand observer les oiseaux dans la région de Montréal*) (Bannon 1991a), *Les oiseaux des Îles-de-la-Madeleine : populations et sites d'observation* (Fradette 1992), and the *Guide des sites de la Côte-Nord* (Club d'ornithologie de la Manicouagan 1992). Finally, in 1993 *L'observation des oiseaux en Estrie: les meilleurs sites, les périodes favorables* (Lepage 1993) appeared, along with the *Guide des sites ornithologiques de la grande région de Québec* (Otis et al. 1993). There is also a book on the best birdwatching sites in the province (David 1990) and another on the natural sites of Québec (Tanguay 1988), with the result that people now have access to all the information they need to birdwatch in southern Québec.

These books show how useful the computerized ÉPOQ data have been, as most contain informative annotated lists of the birds in the regions covered, usually shown in graphs or tables, as is the case with the annotated list for the Lower St. Lawrence (Larivée 1993) or for the Saguenay–Lac-Saint-Jean (Savard and Cormier 1995).

All of these books, published for the most part by the regional clubs, show how much birdwatchers can contribute to the spread of information about birds.

The AQGO joins in

The Association québécoise des groupes d'ornithologues (AQGO) also contributed to this spate of publishing when it began putting out its own periodical, *QuébecOiseaux*.

First published in 1989, *QuébecOiseaux* is a popular magazine that provides information on birds and birdwatching. The only periodical of its kind in Québec, it is not only informative but has forged bonds among birdwatchers.

QuébecOiseaux also popularizes research carried out by Québec ornithologists. Many photographers use this periodical to display their work to the birdwatching community.

The AQGO has also published, in conjunction with others, various technical works, including a study on the threatened birds of Québec (Robert 1989b). This study illustrates how successful the AQGO has been in using the data collected on Québec birds. The AQGO has participated in research projects in conjunction with other organizations, notably the Québec Departments of Higher Education and Science and of the Environment and Wildlife, the Québec Wildlife Foundation, and Environment Canada's Canadian Wildlife Service, such as research on the Piping Plover in the Magdalen Islands.

The AQGO is very active in the area of threatened species. Besides participating in studies, it has also developed a computerized database for compiling relevant data. Sight records contributed by volunteers from member clubs allow for the monitoring of certain species.

The AQGO is the mouthpiece for birdwatching clubs in their relations with a variety of organizations and government departments. It plays an important role in the various consultation processes involved in planning wildlife management programs. Thanks to the AQGO, birdwatchers now have a say in these matters.

Years of growth

This overview of the recent history of birdwatching in Québec shows the phenomenal growth of this activity over the last 15 years, a period when many clubs were founded all over Québec, in response to the needs of the ever-increasing number of people who are interested in birds.

Birds are now the subject of increased media coverage, which today is not limited to major environmental battles, as was the case in the late 1970s. Since the mid-1980s, two major Québec newspapers have been publishing regular columns on birds, *À tire-d'aile...* (P. Gingras since 1986) in *La Presse* and *Bird's eye view* (D. M. Bird since 1984) in *The Gazette*.

Yet beyond this institutional framework, birdwatching itself has undergone profound changes and is now carried out in a multiplicity of ways, which explains its ever-growing popularity. Whether observed at home, in the forest, or on vacation, birds remain a never-ending source of delight and fascination for the birdwatcher.

Jean Paquin

THE BIOGEOGRAPHY OF QUÉBEC

Introduction

The composition of the fauna and flora of a region is conditioned by the region's physical characteristics, its degree of isolation, and the evolution of the landscape over time, all factors that influence bird distribution and abundance. After dicussion of Québec as a whole, regional variations are considered through description of the major vegetation regions, which incorporate most other components.

Geographically, Québec has two major physiognomic characteristics. First, it is a vast peninsula surrounded by salt water on three sides: the Gulf of St. Lawrence to the east, the Labrador Sea and Hudson Strait to the north, and James Bay and Hudson Bay to the west. Second, southern Québec is crosscut by the St. Lawrence River and its estuary, one of the largest river systems in the world (Gratton and Dubreuil 1990).

Québec covers 1 667 926 km² (1 496 400 km² excluding coastal waters). It runs 1900 km south to north between latitudes 44°59' and 62°34' N, and 1500 km east to west between longitudes 56°55' and 79°45' W (Dugas et al. 1989). This is an immense landmass, understandably diverse in its geology and climate as well as in the ecological communities that it supports. The following pages briefly outline the main biogeographic features of Québec.

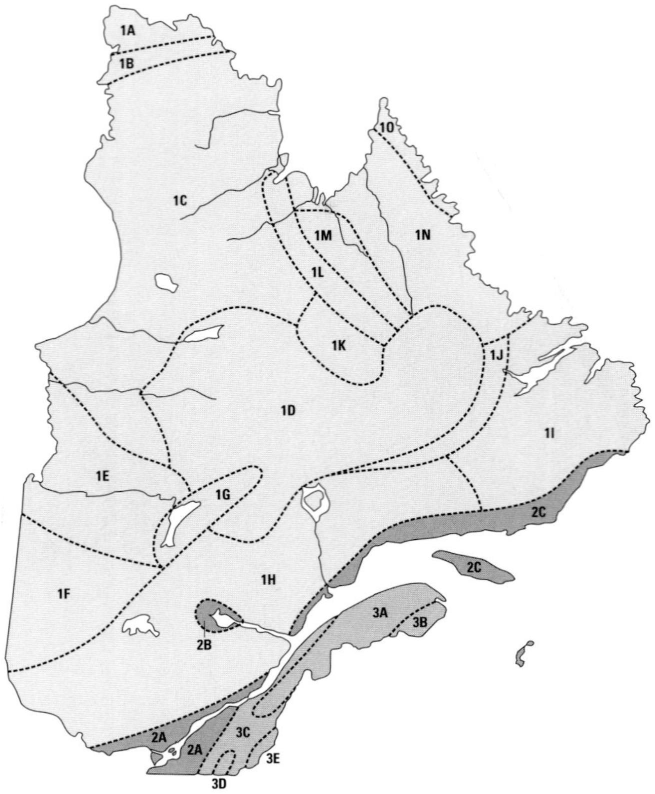

The Shield
1A Saglouc Plateau
1B Povungnituk Hills
1C Larch Plateau
1D Lake Plateau
1E Eastmain Lowland
1F Abitibi Upland
1G Mistassini Hills
1H Laurentian Highlands
1I Mécatina Plateau
1J Hamilton Plateau
1K Caniapiscau Plateau
1L Labrador Hills
1M Whale Lowland
1N George Plateau
1O Labrador Highlands

St. Lawrence Lowlands (Southern Québec Lowlands)
2A Central St. Lawrence Lowland (St. Lawrence Plain)
2B Lake Saint-Jean Lowland
2C East St. Lawrence Lowland

Appalachian Region
3A Notre-Dame Mountains
3B Chaleur Uplands
3C Eastern Québec Uplands
3D Sutton Mountains
3E Mégantic Hills

Figure 1
Major geological and physiographic regions
Adapted from Daniel Lagarec 1974, Annuaire du Québec. Reprinted by permission of Les Publications du Québec.

Major physiographic regions

There are three main geological regions in Québec: the Canadian Shield in the north, the Appalachians in the southeast, and the St. Lawrence Lowlands between them (Figure 1).

The Canadian Shield, regionally called the Laurentian Shield, covers 92% of Québec. It is composed of igneous rock originating in the Precambrian era, generally dated at more than one billion years before the present (Landry and Mercier 1983). The Canadian Shield is a vast, relatively monotonous peneplain comprising a number of groups of plateaus, hills, uplands, and lowlands (Dugas et al. 1989). Elevation generally ranges from 250 to 750 m, though there are massifs more than 1000 m high: the Torngat Mountains, at the northeastern end of Ungava Bay (including Mount D'Iberville, 1622 m, the highest peak in Québec), the Otish Mountains (including Mount Yapeitso, 1128 m) and the Groulx Mountains (including Mount Veyrier, 1104 m, and Mount Jauffret, 1065 m) in central Québec, and the Laurentians on either side of the Saguenay River. There are several high peaks, belonging to the Laurentians, just north of Québec City in Charlevoix (including Mount Raoul Blanchard, 1166 m, and Mount Bleu, 1052 m) and in the Grands-Jardins Conservation Park (Mount L.-A. Richard, 975 m, Mont du Lac des Cygnes, 960 m, and Mont du Lac à l'Empêche, 930 m).

The rock of the Canadian Shield is often exposed or partially covered with morainic material of glacial origin. The topography of the shield has been altered in places by large deposits of boulders, gravel, or sand. Postglacial seawater intrusion or large proglacial lakes have left thick layers of clay in the lowlands, flattening out the relief (the Eastmain Lowland and the Abitibi–Témiscamingue region, for example).

Compared to the Laurentian Shield, the Appalachians seem no more than a narrow strip of land (50–150 km wide) in southeastern Québec. Actually, the Appalachians are a long mountain range running from Alabama to Newfoundland, extending across more than 800 km of Québec from Montérégie to the Gaspé Peninsula, with peaks just breaking through the waters of the gulf in the Magdalen Islands.

The bedrock here is of sedimentary or volcanic origin and dates from the Paleozoic (about 250–500 million years ago). Mean elevation is 500 m, but relief nonetheless divides the Québec Appalachians into three distinct areas: the Eastern Townships, the Lower St. Lawrence, and the Shickshock Mountains. The Eastern Townships, in the south, are characterized by hills and valleys from which rise a few high peaks (Mount Gosford, 1186 m, and Mount Mégantic, 1100 m). In the Lower St. Lawrence the relief varies little, and elevation rarely exceeds 600 m, though there are a number of deep valleys, including those of the Matapédia River and Lake Témiscouata. Farther north, in the Notre-Dame Mountains of the Gaspé (the Shickshock Mountains), many of the highest peaks in Québec are found (Mount Jacques Cartier, 1268 m, Mont de la Table, 1173 m, and Mount Albert, 1151 m, for example). This, of course, affects the birdlife: the area supports several Golden Eagle pairs, and the American Pipit, an arctic species, is found on Mount Albert.

Inconsequential in size (about 17 280 km²), the St. Lawrence Lowlands are nonetheless home to the vast majority of the population of Québec. Of sedimentary origin and dating from the Lower Paleozoic (400–500 million years ago), the rock here is a geologic formation composed of three discontinuous zones: the Central St. Lawrence Lowland, bordering the St. Lawrence River from Ontario to Cap Tourmente, just north of Québec City; the East St. Lawrence Lowland (encompassing Anticosti Island); and the Lake Saint-Jean Lowland, surrounded by the Canadian Shield but connected to the St. Lawrence by the Saguenay Fjord.

The Central St. Lawrence Lowland, or St. Lawrence Plain, is a vast, very flat, triangular plain where elevation rarely exceeds 150 m. The relief is broken only by the St. Lawrence River and its tributaries (the Richelieu, Ottawa, and Saint-Maurice rivers and others) and by the Monteregian Hills, composed of intrusive rock of an older geologic formation. The highest of these hills is Mount Saint-Hilaire (411 m).

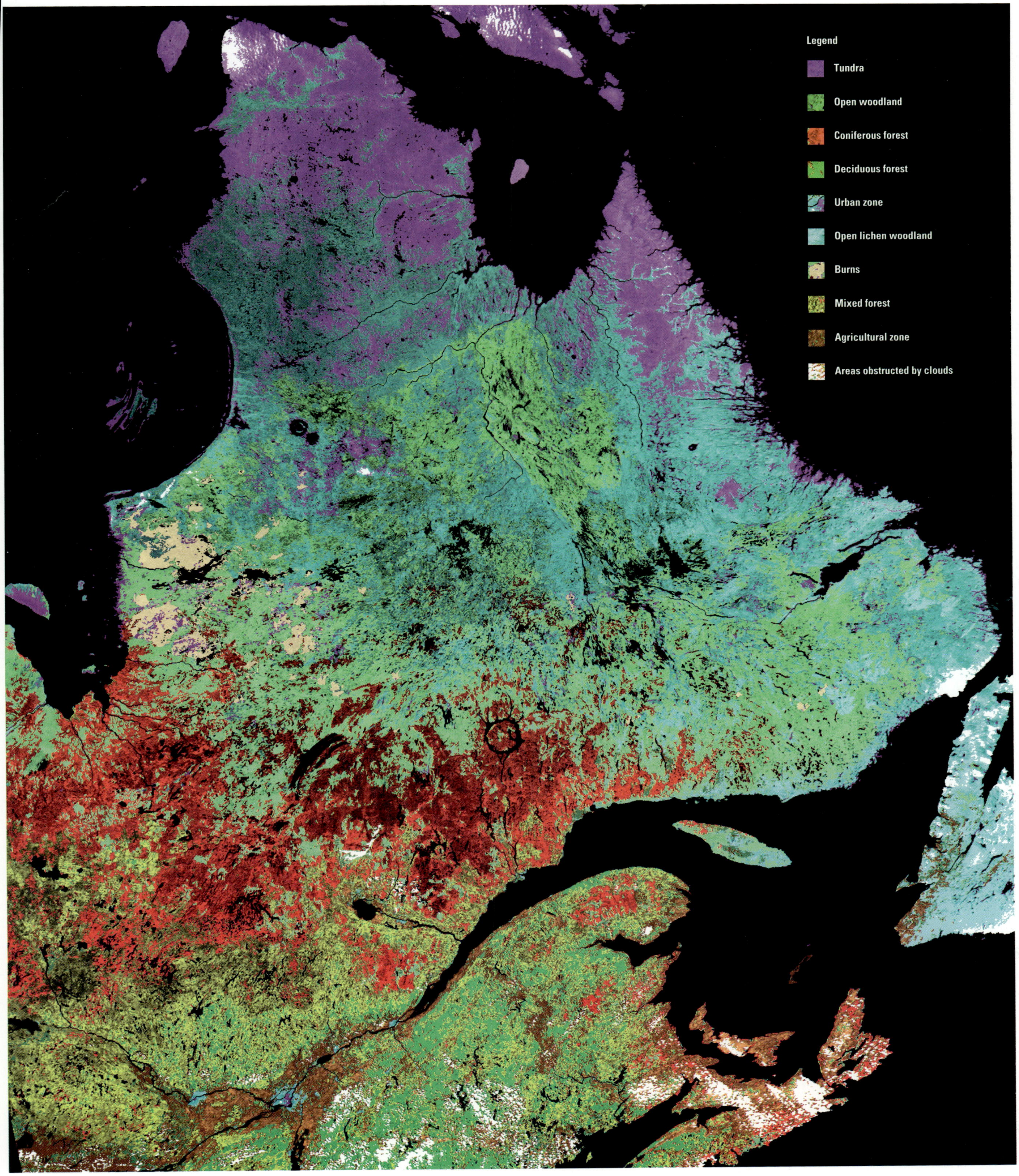

Satellite image of Québec

Composite false-colour digital image showing the latitudinal stratification of vegetation in Québec.
Source: Ministère de l'Énergie et des Ressources du Québec, photomap library. Produced in 1992 from images taken in August and September 1988 and 1989.

View of the Otish Mountains, which include some of the highest peaks in Québec. *July 1971, Henri Ouellet.*

The Central St. Lawrence Lowland is widest (about 100 km) around Montréal. The St. Lawrence Plain is singularly flat, largely because of the marine transgression at the end of the last glacial epoch. The Champlain Sea, which covered the St. Lawrence Valley, deposited a thick layer of clay and in places sand. This occurred in the Lake Saint-Jean Lowland as well (Laflamme Sea, Figure 7).

Mireille Desponts

Hydrographic system

Fresh water covers 183 890 km², that is 12.3% of the landmass of Québec (Dugas et al. 1989). It is therefore not surprising that close to a third of Québec's breeding birds are waterbirds. Québec's freshwater reserves are enormous: they constitute one third of Canada's freshwater reserves and are equivalent to half those of the entire United States. Not all of this fresh water is directly accessible to the population in the south, however, since many of Québec's 130 000 rivers flow north.

The waters flow through three large drainage basins: that of the St. Lawrence system (742 648 km²), that of James Bay and Hudson Bay (541 181 km²), and that of Ungava Bay (384 097 km²), including coastal waters (Figure 2). Lake and river abundance and morphology differ from one drainage basin to the next.

The basins of James Bay, Hudson Bay, Ungava Bay, and the north side of the St. Lawrence include tens of thousands of small lakes, a few very large lakes (Lake Mistassini, 2019 km², Lac à l'Eau Claire, 1243 km², and Lake Saint-Jean, 1041 km²), and a number of immense artificial reservoirs (the Caniapiscau Reservoir, 4285 km², the La Grande-2 Reservoir, 2836 km², and the La Grande-3 Reservoir, 2460 km², among others). Many of the associated rivers are long, with very large discharges: the Ottawa, Saguenay, La Grande, and Caniapiscau rivers, to name just a few. The south side of the St. Lawrence River basin is smaller and has fewer lakes, though some are fairly large: Lake Memphrémagog (95 km²), Lake Saint-François, and Lake Témiscouata (66 km²), among others. The rivers here are also shorter and their discharges smaller; mean annual discharge of the five largest rivers on the south side of the St. Lawrence is 150 m³/s, compared to 850 m³/s for the main rivers of the other drainage basins (Dugas et al. 1989).

Mireille Desponts

La Grande-2 Reservoir. View of the main rockfill dike and the spillway, which was carved right out of the rock. James Bay area. *June 1989, Jean-Luc DesGranges.*

Climate

Climate type changes several times north to south in Québec. In general, temperatures diminish northward, because of the decrease in sunshine and the increasing proximity of the polar front. The trend is the same for precipitation, because of the influence of arctic air masses, which generally have a low moisture content. The following paragraphs briefly summarize the main climatic trends in Québec.

Mean annual temperature ranges from –7.2°C (Quaqtaq) in northern Québec to +7.3°C (Montréal) in the south (Dugas et al. 1989). Isotherms pretty much coincide with latitudes (Figure 3). North of Trois-Rivières, mean temperature drops to less than +5°C, descending to 0°C at latitude 50° N. On the peaks of the Laurentide Wildlife Reserve and the Notre-Dame Mountains there are pockets where mean annual temperature is considerably lower than in the surrounding region. In Quaqtaq, mean temperature is –22°C in January and +6.2°C in July; in Montréal, mean temperature is –8.7°C in January and +21.8°C in July.

Figure 2
Drainage basins and primary watersheds
Adapted from the figure "Bassins versants et régions hydrographiques," Ministère de l'Environnement du Québec, Direction du milieu hydrique, 1989.

Figure 3
Mean annual temperature, 1951–80
Adapted from Environment Canada 1984, Climatic Atlas–Canada and from Ministère de l'Environnement du Québec 1987, Climatologie du Québec méridional.

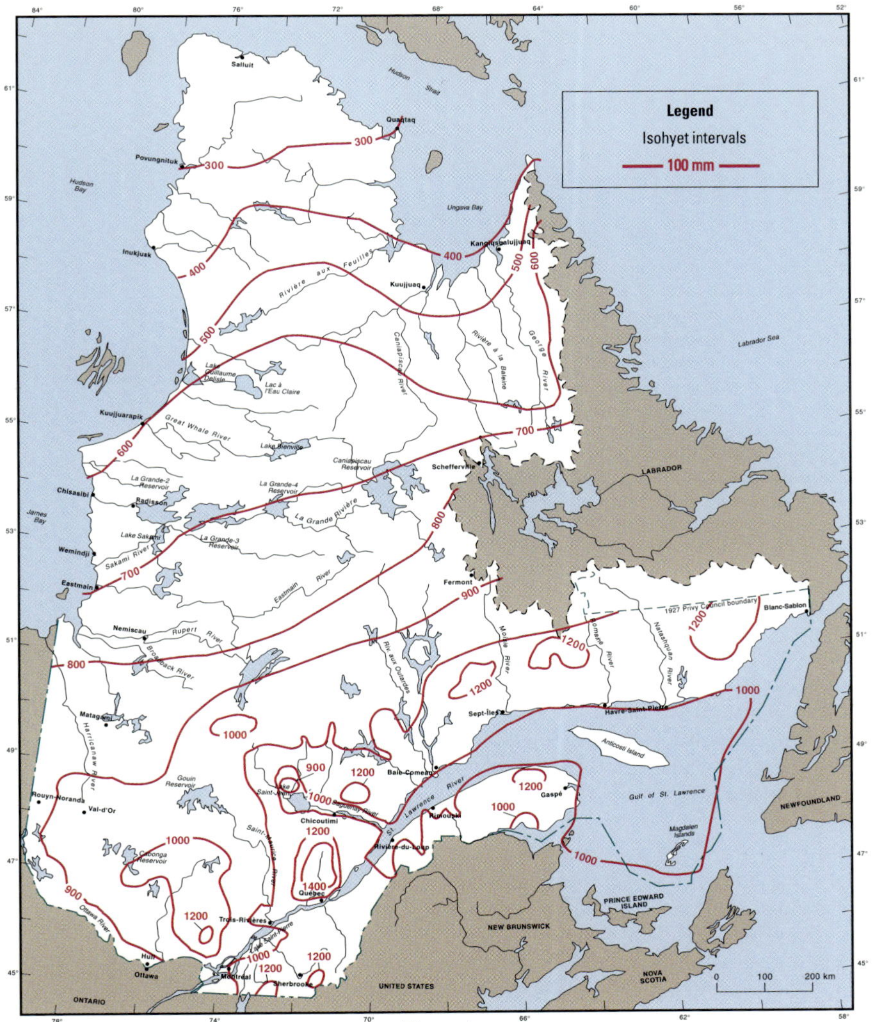

Figure 4
Mean annual precipitation, 1951–80
Adapted from Environment Canada 1986, Climatic Atlas–Canada and from Ministère de l'Environnement du Québec 1987, Climatologie du Québec méridional.

Mean annual precipitation generally increases north to south in Québec; extremes range from a mere 296 mm in Quaqtaq to more than 1500 mm in the Grands-Jardins Conservation Park in Charlevoix (Figure 4). The precipitation gradient is quite irregular in southern Québec, where the relief has a substantial influence. The air masses that rise to high altitudes, for example, to cross the Laurentians leave much more precipitation there than falls on the lowlands of the Saguenay–Lac-Saint-Jean region. In Montréal, annual precipitation is about 1000 mm; in Québec City it is close to 1200 mm. Precipitation values are of this order throughout the whole of southern Québec, except on the high peaks of the Laurentians, the Eastern Townships, and the Notre-Dame Mountains, where they are higher. The snow fraction generally ranges from 20 to 55%, Montréal receiving no more than one fifth of its precipitation in the form of snow, compared to more than half in the northernmost parts of Québec (Houde 1978).

The frost-free period diminishes south to north, dropping from 177 days in Montréal to 35 days in Quaqtaq (Dugas et al. 1989). The same gradient applies to ice break-up in the rivers, which begins around 10 April in southern Québec and only around 20 June in the Ungava Peninsula (Proulx et al. 1987). Break-up generally starts 10 days later on lakes than on rivers, and this is probably crucial for Ospreys; in northern regions, the only potential fishing areas for Ospreys in early spring are open-water areas in the rivers (Poole 1989). In fact, Osprey breeding success in northern regions of North America is largely dependent on the timing of the break-up (Wetmore and Gillespie 1976; Hughes 1986). In the fall, lakes start to freeze up in early November in the north but only in late December in the south.

The snow-free period is twice as long in southern Québec as it is in northern Québec, lasting an average 240 days in the south (15 April–15 November) compared to 120 days in the north (15 June–30 September) (Potter 1965). In some parts of southern Québec—the high plateaus of the Laurentians, for example—the snow-free period averages only 150 days. Many bird species that feed entirely on the ground (Killdeer and Horned Lark, for example) are forced to move south when snow covers the ground.

In southern Québec, the climate varies mainly as a function of relief and proximity to the estuary or the Gulf of St. Lawrence. A detailed map published by Ministère de l'Environnement du Québec (1987) summarizes these variations and identifies eight climate types based on mean annual temperature (°C), annual precipitation (mm), and continentality index (based on latitude and annual temperature range and expressed as a percentage) (Figure 5). On the whole, the climate is subpolar (mean annual temperatures range from –6.0°C to +4.15°C), subhumid (precipitation ranges from 800 to 1360 mm), and continental (continentality index ranges from 50 to 62%). In southern Québec, mean temperatures are higher (moderate climate, +4.15°C to +8.50°C) on the Magdalen Islands and in a small pocket around Île d'Orléans.

The harsh climate has shaped the coastal landscape along the Lower North Shore. Exposed areas and hilltops are barren, but vegetation colonizes wind-sheltered hollows. North of Tête-à-la-Baleine. *August 1979, Jean Beaubien and Guy Simard, Canadian Forest Service.*

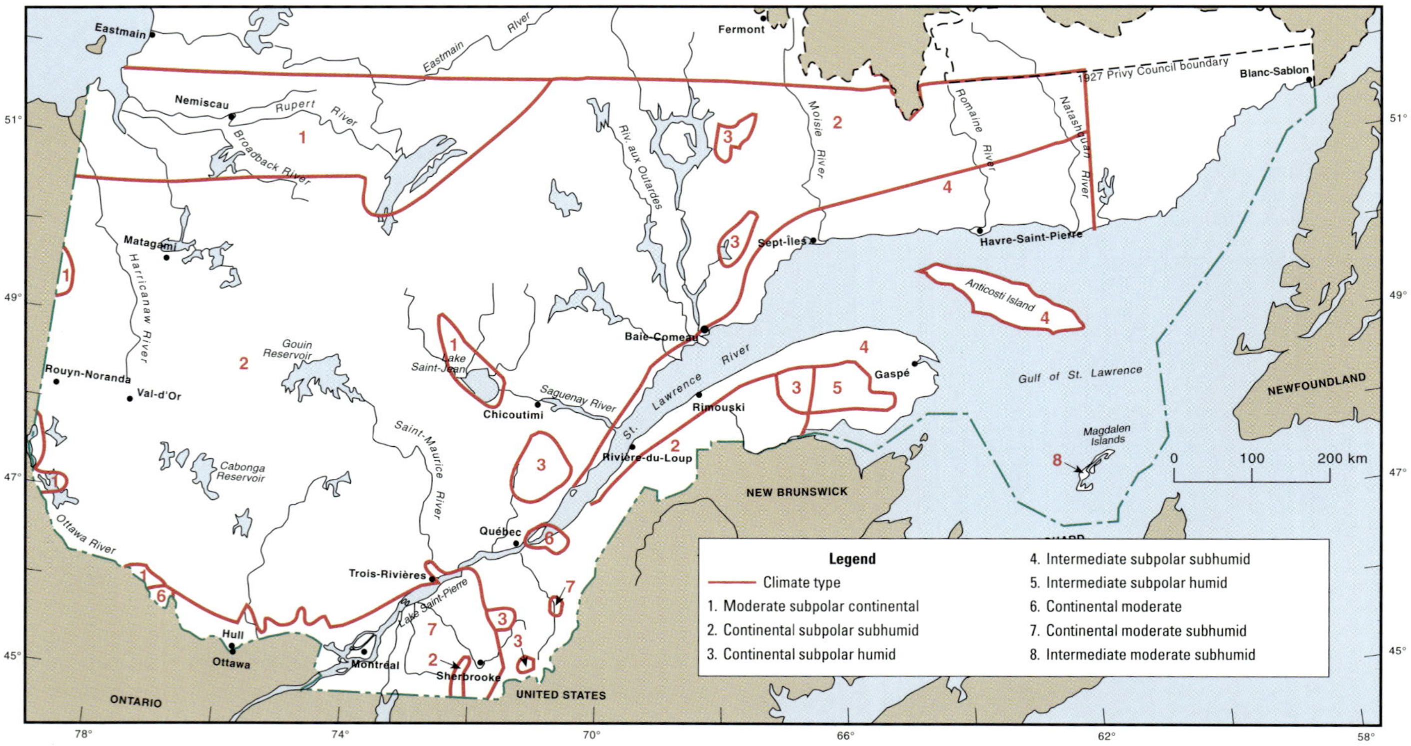

Figure 5
Climate types—southern Québec
Adapted from Ministère de l'Environnement du Québec 1987, Climatologie du Québec méridional.

As for precipitation, values are lower in the James Bay area (moderately humid climate, 680–800 mm) and in certain pockets, the northern parts of Lac-Saint-Jean and the Ottawa Valley (Île aux Allumettes), for example. Precipitation is greatest on the peaks of the Shickshock Mountains, the Groulx Mountains, and the Laurentide Wildlife Reserve (humid climate, 1360–2350 mm). The influence of the maritime air masses extends north of Québec City, along the St. Lawrence Valley, and into the eastern part of the Gaspé, including the Shickshock Mountains. The climate here is intermediate between continental and maritime (continentality index ranges from 33 to 55%).

Number of growing degree-days is an index commonly used to define vegetation growing seasons. Since plant growth essentially stops below +5°C, this base temperature has been selected as a reference value for calculating annual growing degree-days. A growing degree-day is a positive deviation expressed in degrees between the mean temperature of a day and the reference temperature (+5°C) (Environment Canada 1984). Though this index does not take into account all climate elements, it is widely used by phytogeographers. Temperature data are generally available, and the index, which is easy to calculate, can be used to predict vegetation growing seasons with considerable accuracy.

As Figure 6 shows, the number of growing degree-days decreases south to north in Québec from 2000 to 250. The length of the growing season has a major impact on bird breeding range limits, since it affects the plant life and the availability of food, mainly insects (Rodenhouse 1992). In fact, because of their homeothermy and behaviour mechanisms, birds are directly affected only by extremely low temperatures. The number of growing degree-days required for

Figure 6
Annual total growing degree-days above 5°C
Adapted from Environment Canada 1984, Climatic Atlas–Canada.

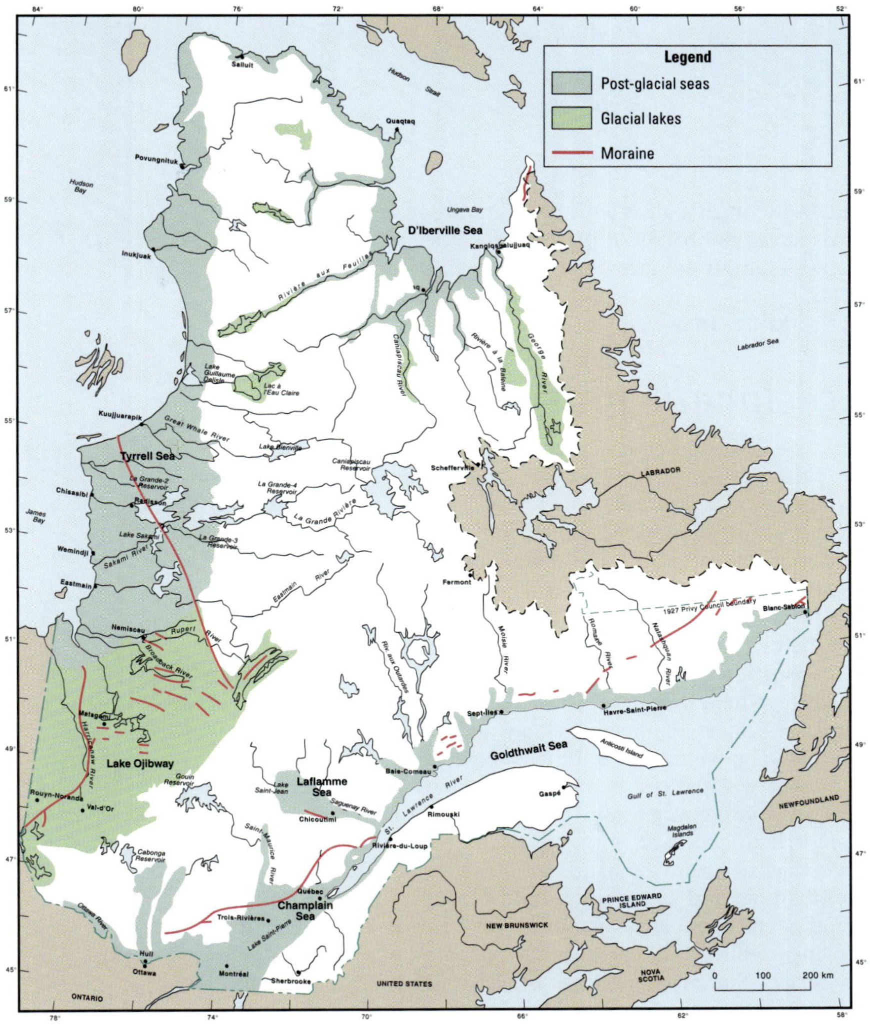

Figure 7
Major moraines and areas inundated by glacial lakes and marine waters during the last deglaciation in Québec
The glacial Lake Ojibway correspond to the Clay Belt that extends as far as Lake Nipigon in Ontario.
Adapted from Vincent (1989). By permission of the Geological Survey of Canada and the Minister of Public Works and Government Services.

insect development depends on the species. This explains the time lag in the spring arrival of insectivorous birds at different latitudes. Most warblers, for example, reach southern Québec around mid-May, the Laurentians around the end of May, the southern part of James Bay in early June, and Hudson Bay around mid-June, the males generally a few days ahead of the females (Francis and Cooke 1986).

Mireille Desponts

Glacial epochs

The climate and the physiognomy of Québec have undergone major changes in recent millennia. During the last million years, the whole of Québec except the Magdalen Islands was covered several times by continental ice sheets as much as 2000 m thick. The last glacial epoch, which began around 100 000 years ago, ended only 5500 years ago, when the last ice sheets disappeared from northern Québec (Richard, P. J. H. 1987).

The alternation of glacial and interglacial epochs during the Quaternary caused geographic displacement of animal and plant species in North America (Davis 1981, 1986). All bird species now living in Québec arrived in recent millennia, and the forest communities we know today only stabilized about 5000 years ago. The last glaciation also profoundly marked the geomorphology of the land. The glaciers formed large moraines, for example, the Sakami, Harricanaw, and Saint-Narcisse moraines (from north to south; Figure 7), and swept along countless erratic blocks, or glacial boulders, depositing them when the ice melted. Even more important are the bottom deposits of the lakes and proglacial seas that formed on the edges of the ice sheets during their retreat north. As the immense ice masses melted, large lakes formed in depressions such as Abitibi–Témiscamingue. Blocked by ice and with no outlet to the north, the meltwaters formed Lake Ojibway, which covered the drainage basins of the upper Ottawa River and James Bay (Figure 7) for about 2000 years (Vincent 1989). The deposits left by these lakes are the source of the Clay Belt in this region.

When the ice retreated, depressions and low-lying land—such as the St. Lawrence Valley, the girdle around Lake Saint-Jean, and, farther north, the lowlands of the James Bay coastal plain—were invaded by sea water because of isostatic depression of the land under the weight of the glaciers, coupled with a rise in sea level during the melt. Five large regions were submerged for several thousand years (Dionne 1972). Thus, 12 000 years ago the Champlain Sea covered the St. Lawrence Valley to a depth of about 200 m, the Laflamme Sea occupied the Lake Saint-Jean Lowland, and 7900 years ago the Tyrrell Sea covered the James Bay and Hudson Bay coastal plains to a depth of about 300 m. In fact, the site of the city of Montréal emerged only 7200 years ago. These seawater intrusions gradually retreated as a result of isostatic rebound of the land, leaving the drainage basins we know today. These maritime and lacustrian episodes are also the origin of the clay and sand deposits on the best farmland in Québec (the St. Lawrence Valley, the area around Lakes Timiskaming and Abitibi in the Clay Belt, and the Lake Saint-Jean Lowland)

and on the staging areas used by migrating geese, ducks, and shorebirds along the banks of the riverine lakes and their islands and along the intertidal marshes of the St. Lawrence estuary (Lehoux et al. 1985).

MIREILLE DESPONTS

Ecoclimatic provinces

Geology and climate determine the regional physical context in which soil-plant complexes form and develop. Regional variations in Québec's ecosystems can be understood by looking at the province's ecoclimatic regions (Canada Committee on Ecological Land Classification 1989; Figure 8). The different ecoclimatic provinces can be identified on a large-scale satellite image of Québec (page 19). Colour contrasts are obtained by image processing in the visible, infrared, and thermal spectral ranges.

North to south, the first province is the Arctic Ecoclimatic Province, where rock dominates and stunted vegetation grows on a carpet of lichens. Summers are cool and moist here and about four months long; winters are extremely cold and dry (less than 300 mm of precipitation a year). Next comes the Subarctic Ecoclimatic Province. Here the forest cover gradually becomes denser southward, moving from lichen woodland with a few stunted black spruce along the waterways to open woodland. Summers are cool, about four to five months long and especially humid between July and September. Winters are very cold with little snow. In the Boreal Ecoclimatic Province, which comes next, coniferous forest of spruce or fir clearly dominates. Summers are hot and relatively humid, and winters are cold and drier. In the most southerly province, the Cool Temperate Ecoclimatic Province, mixed or deciduous forest thrives. Summers are hot, with precipitation mainly at the end of the season. Winters are cold and snowy. This is also where human activity has concentrated. The satellite photo shows the urban centres and farmlands of the St. Lawrence Valley, the Ottawa Valley, and the Lake Saint-Jean Lowland, as well as the idle lands east of Lake Abitibi. The terrestrial and wetland ecosystems are analysed in detail in the sections that follow.

MIREILLE DESPONTS

Figure 8
Ecoclimatic provinces of Québec
Adapted from Canada Committee on Ecological Land Classification (1989), Environment Canada.

Soil

The origin or development of mineral soils results from the combined effect of diverse factors on material deposited by natural occurrences (mainly glaciers and the seas) that subsequently invaded the areas most deeply depressed by the ice sheets. Organic soils, on the other hand, of which peatlands are composed, are formed continuously wherever the rate of production of organic matter exceeds the rate of decomposition (Moore and Bellamy 1974). The Canadian system classifies soils by order, great group, subgroup, family, series, and phase (Soil Survey of Canada 1987). For purposes of this introduction, only order is considered.

Organic soils

The boreal forest is one region where the climate fosters development of organic soils. The cold and rainy weather, coupled with slightly permeable and flat or gently sloping terrain, helps keep the soil saturated with water. This inhibits decomposition of plant residues, which then accumulate and form peatlands. There is only one order for this soil type, the Organic order. Because of their acidity, peatlands support invertebrates on which certain bird species depend, such as sandpipers and the Common Snipe.

Mineral soils

Factors influencing mineral soil development are surficial deposits, relief, climate, biotic components (plants and animals, including human beings), and time. Soils are classified according to certain diagnostic A and B horizons (a horizon is a soil layer showing specific signs of transformation by soil forming factors). Among the light soils (sandy and loamy), Regosols are the most weakly developed; they have no B horizon. Podzols are the most extensively developed, with a leached Ae horizon and a Bf horizon enriched in iron and aluminum. Brunisols are classified as intermediate between Regosols and Podzols, with a slightly altered horizon (Bm) and other weakly expressed B horizons (Bfj and Btj). When drainage is poor, the term "gleyed" is added to the subgroup level. Well drained clayey soils, called Luvisols, have a horizon enriched in fine clay (Bt), whereas Gleysols formed on poorly drained sites have a grey horizon (Bg). There are also hardened horizons. They are most frequently found as ortsteins in the Laurentians and fragipans in the Appalachians.

Soil locations and classifications

Soil classification and general location information in the following paragraphs is based on the work of Gilbert et al. (1985) for the northern part of Québec and on the soil landscape map of Canada (Land Resource Research Centre 1990) for land south of 52° N and west of 68° W.

Soil on clay, sand, or silt deposits in the Central St. Lawrence Lowland makes excellent farmland. In the vicinity of Berthierville, near Lake Saint-Pierre. *October 1993, Pierre Normand.*

In the most northerly regions of Québec, there is no soil development properly speaking, organic matter production and chemical processes are reduced. The loose sediments of glacial and fluvioglacial origin are found mainly in the valleys and along the coasts and become more frequent and thicker in the southern part of this region. In arctic environments, above the tree line, the permafrost is continuous, and the 1–2 m surface layer subject to thawing in summer shows signs of disturbance attributable to freeze-thaw action with disrupted, mixed, or broken layers, and sorted polygons on the surface. Such "soils" are called Cryosols.

Farther south, the permafrost is more sporadic and restricted to fine-textured mineral soils or organic soils. The soils found here are mainly Regosols and Fibrisols. The mineral soil is thin and weakly developed in Regosols, a feature attributable to the youthfulness of the material and the cold, dry climate. Fibrisols are found mainly at poorly drained sites and are composed principally of relatively undecomposed organic material.

At the latitude of James Bay, Brunisols and Podzols appear. These soils are more developed than Regosols, but Brunisols still do not have well-defined horizons. Podzols are the dominant soil type under coniferous forests in a cool, moist climate. They can be easily recognized because of their well-defined horizons: an organic surface horizon composed of both litter that is more or less decomposed and of humus of variable thickness overlying a pale, eluvial, mineral horizon (leached of iron and aluminum oxides) atop one or more reddish-brown to black accumulation horizons which in turn lie on a yellowish or grey soil barely affected by soil development processes. On the clayey deposits of Abitibi and Témiscamingue, Gray Luvisols are found at well drained sites, and Gleysols interspersed with peatlands at poorly drained sites.

Podzols are the most common soils found in the Laurentian and Appalachian tills south of the 51st parallel and right to the U.S. border. Where slopes are very steep, there are soilless areas slightly back from the coast, such as along the North Shore. East of these soilless areas are large expanses of thin and very thin soils classified as Regosols. These soils comprise a layer of litter above an Ae horizon lying directly on rock.

In the Central St. Lawrence Lowland, in the vicinity of Montréal and farther south, we find richer soils, such as Melanic Brunisols and Gray Brown Luvisols typical of a deciduous forest. A characteristic feature of these soils is the presence of forest mull, a homogeneous forest humus where organic litter does not accumulate on the surface. Mull is formed by the intense biological activity of the microfauna (earthworms in particular), which are food for a variety of bird species, including the Wood Thrush, Veery, Hermit Thrush, and Ovenbird (Sabo and Holmes 1983). Soils on deposits of clay, sand, or silt are very good for agriculture and are generally very much disturbed by it.

Jean-Louis Bélair

Vegetation formations

There are six major vegetation zones in Québec: the arctic tundra, forest tundra, taiga, coniferous forest, mixed forest, and deciduous forest. These regions are distinguished less by their composition than by the structure of the forest landscape. The descriptions below are based on Rousseau (1974), Gérardin (1980), Payette (1983), and P. J. H. Richard (1987).

The arctic or shrub tundra is the most northerly vegetation zone in Québec (Figure 9). It corresponds to the Arctic Ecoclimatic Province, and the number of growing degree-days is less than 600. The landscape is rocky, with a discontinuous plant cover composed of herbaceous plants and lichens (*Cladina* spp.), with a few low shrubs (willow) and some green alder and scrub birch in wind-sheltered hollows heavily covered by snow in winter. At favourable sites, prostrate black spruce sometimes form krummholz, which often supports the White-crowned Sparrow (Godfrey 1986). These spruce are above the tree line, the northern limit for woody vegetation at least 5 m high (Payette 1983).

Often considered a simple transition between the tundra and the taiga, the forest tundra is located approximately between the northern and southern limits (summer and winter) of the predominance of the polar air mass (Bryson 1966). Unlike any other North American transition zone between tundra and forest properly speaking, the forest tundra of Québec covers 17% of the province, sometimes spanning 300–500 km between latitudes 55° and 58° N. The forest tundra corresponds to the northern part of the Subarctic Ecoclimatic Province (600–800 growing degree-days). The forest cover here is discontinuous. Black spruce grows in the valleys, as shrubs in the north and as trees farther south. Prostrate spruce are found on the more exposed slopes, and hilltops are generally barren. Because of these features, more breeding bird species probably live in the forest tundra than in the two zones flanking it. The American Pipit, for example, can be found on the mountain tops of the forest tundra, and the Blackpoll Warbler in its wooded valleys (Cons. G. G. – GREBE 1991a). White spruce grows along the coast of Hudson Bay, the most northerly part of its range, as well as in the eastern part of the forest tundra. Tamarack and balsam poplar are also found in the forest tundra, at riparian sites in particular.

In the taiga, the forest cover is continuous but relatively open, composed mainly of black spruce stands with lichen (reindeer, alpine reindeer, and yellow reindeer lichens). The taiga corresponds approximately to the southern part of the Subarctic Ecoclimatic Province, where there are 800–1000 growing degree-days. Though black spruce clearly dominates, the forest communities become more diversified here. White birch, trembling aspen, and balsam fir become more and more frequent southward, and jack pine grows on dry, sandy terraces. There are no bird species whose breeding ranges are contained by the taiga alone. The Black Scoter and Surf Scoter are nonetheless characteristic species of the biome, to which they are largely confined (Savard and Lamothe 1991; Cons. G. G. – GREBE 1992u).

South of the 52nd parallel, the forest cover gradually becomes denser and more diversified, and it has been altered by logging. Presence of a species often depends on site topography, microclimate, or past disturbances. The description and map of the forest regions of southern Québec come from Thibault (1985, 1988; Figure 10). There are three main vegetation zones—the coniferous forest, mixed forest, and deciduous forest—which subdivide into 17 forest regions. These 17 regions are distinguished on the basis of the dominant forest cover at mesic sites that have not been recently disturbed.

The coniferous forest essentially includes communities dominated by black spruce in the north and by balsam fir farther south and to the east. It occupies 575 000 km², or close to three quarters of the commercial forest. This zone corresponds to the Boreal Ecoclimatic Province, and total annual growing degree-days range from 750 in the north to 1250 in the south.

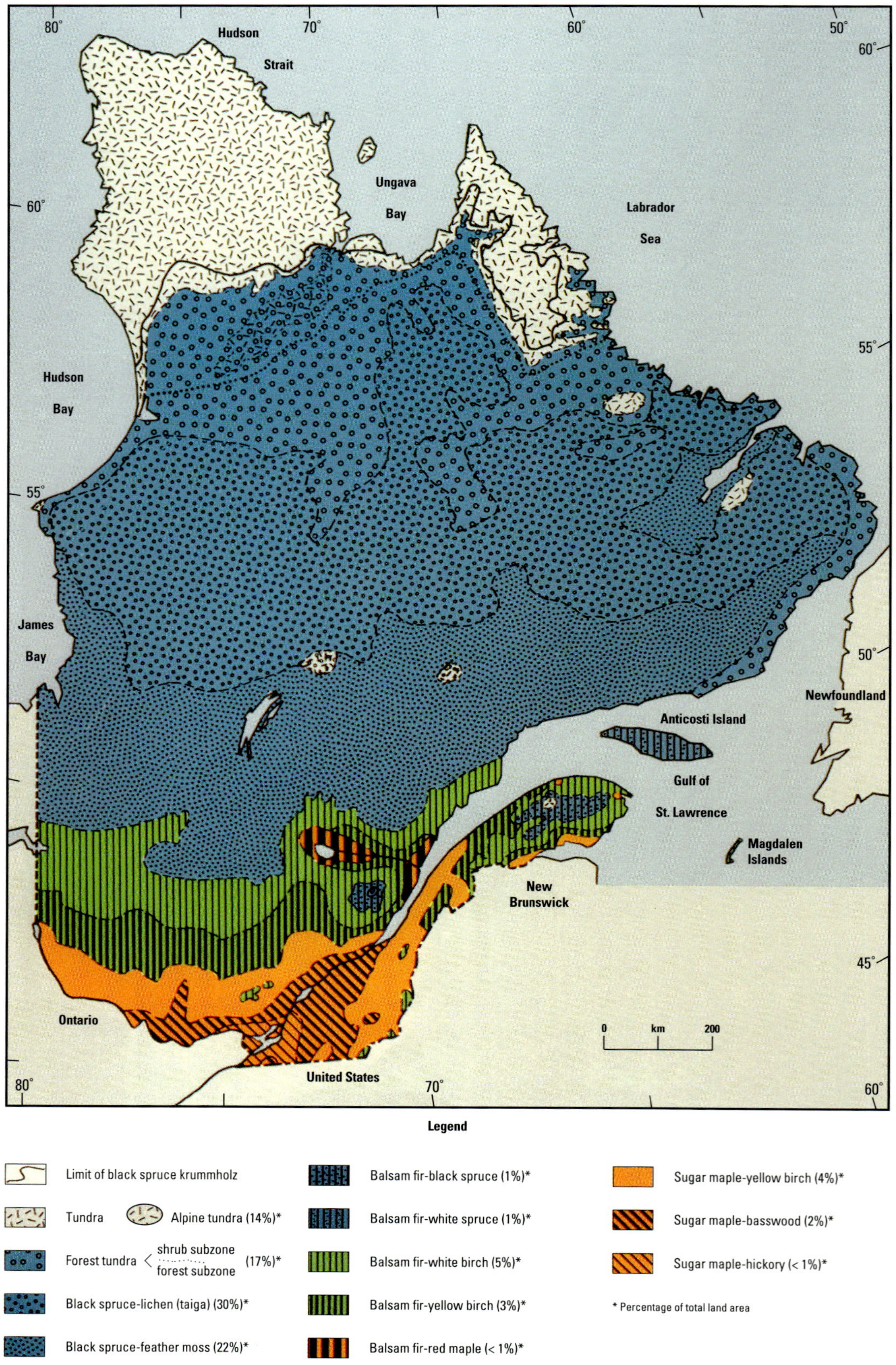

Figure 9
Major vegetation of Québec–Labrador

Produced in 1987 by Pierre J. H. Richard, Laboratoire Jacques-Rousseau, and available from the Geography Department, University of Montréal.
Sources: Grandtner 1966; Wilson 1971; Rowe 1972; Energy, Mines, and Resources Canada 1973; Jurdant et al. 1977; Payette 1983; Thibault 1985; Dyke and Prest 1987

In the Arctic, along the coast and at higher elevations, spruce are stunted when exposed to winds and the harsh climate. Near the Boniface River, Hudson Bay area. *September 1987, Mireille Desponts.*

Ornithologically, the coniferous forest is relatively monotonous. There are, however, many birds whose breeding ranges are largely or totally confined to the western part of the coniferous forest, that is the Eastmain Lowland and the Abitibi Upland (Figure 1). The assemblage—Sharp-tailed Grouse, Marbled Godwit, Sandhill Crane, Bonaparte's Gull, Connecticut Warbler, and Le Conte's Sparrow—shows similarities to the birdlife of western Canada.

Northwest of Lake Saint-Jean, the Black spruce-moss forest region (17) covers an immense, very uniform area. At mesic sites, the black spruce is mixed with balsam fir, jack pine, and, more rarely, tamarack. White birch and trembling aspen are found mainly at recently disturbed sites. Balsam fir becomes more dominant in the northeastern part of the coniferous forest zone, since the climate is wetter close to the Gulf of St. Lawrence. The main forest regions are Balsam fir-black spruce (14) and Black spruce-balsam fir-moss (16). Fir stands are characteristically found on uplands and generally dominate above 700 m (the Laurentians, the Notre-Dame Mountains, and the hilltops of the Manicouagan region). The Black spruce-balsam fir-moss forest region (16) constitutes a transition between the Balsam fir-black spruce (14) and the Black spruce-moss (17) forest regions. Pockets of alpine tundra are found on the peaks of the Groulx Mountains and other nearby mountains. Communities of mosses, lichens, and shrubs girdled by black spruce, balsam fir, and white spruce occupy these summits.

Forest tundra in the lowlands around Lac à l'Eau Claire. Barrens alternate with woodland, which occupies depressions and corridors. Hudson Bay area. *August 1979, Pierre Lamothe.*

Alpine tundra is generally associated with the harsher climatic conditions at high elevations (Wardle 1974). Alpine tundra communities are found in the more southerly parts of Québec—as far south as Les Grands Jardins, where their presence is, however, attributable to gradual opening of the forest cover after fires during climatic periods unfavourable to tree regeneration (Bussières 1992).

The temperate, maritime climate along the coast of the gulf and the estuary, on Anticosti Island, on the Magdalen Islands, and along the coast of Rupert Bay has promoted development of white spruce-balsam fir and balsam fir-white spruce forests (*cf.* region 15). Balsam fir-white spruce-feather moss forest covers 80% of Anticosti Island; fir mixed with deciduous species such as white birch, red maple, American mountain-ash, and trembling aspen is found along a strip of wind-sheltered coast less than 8 km wide (Desloges and Émond 1974). On the Magdalen Islands, the forest cover is composed mainly of balsam fir-white spruce stands (sometimes mixed with white birch, mountain-ash, and serviceberry) on hillsides, and of black spruce-laurel stands on the north shore of the islands. The communities most exposed to prevailing winds are often stunted (Grandtner 1967; Fradette 1992). This forest region covers a relatively small area, just a very thin strip along the Gaspé coast, and does not show on the map of the ecological regions. White spruce is also frequent on the North Shore, in the Balsam fir-black spruce forest region (14).

A relatively open subarctic forest formation composed mainly of conifers or broken by wetlands, the taiga is the transition between the commercial boreal forest and the tundra. North of Great Whale River, Hudson Bay area. *August 1989, Mireille Desponts.*

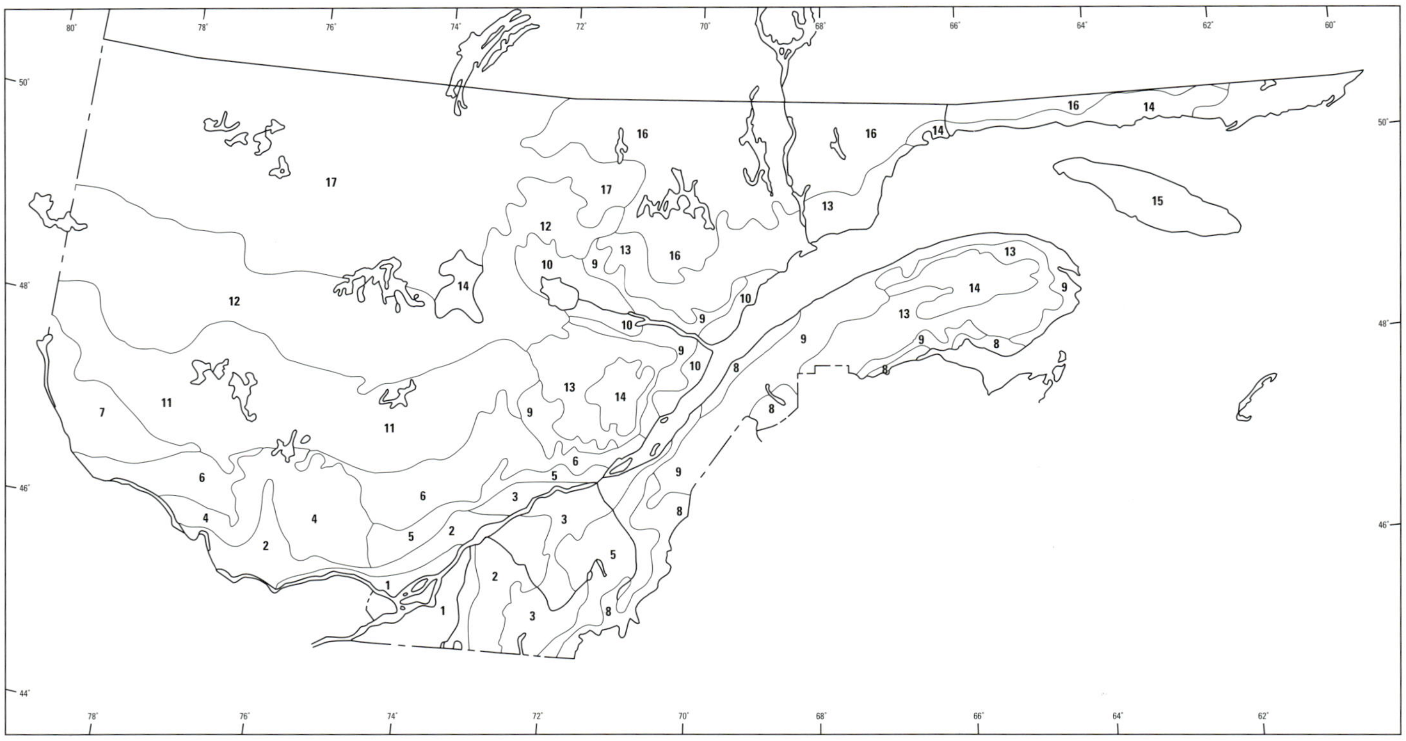

Ecological regions (forest regions)
Deciduous forest zone
1. Sugar maple-hickory and sugar maple-basswood
2. Sugar maple-basswood
3. Sugar maple-basswood and sugar maple-yellow birch
4. Sugar maple-yellow birch-basswood
5. Sugar maple-yellow birch-beech and sugar maple-yellow birch-basswood
6. Sugar maple-yellow birch-beech
7. Sugar maple-yellow birch-balsam fir
Mixed forest zone
8. Sugar maple-yellow birch and balsam fir-yellow birch
9. Balsam fir-yellow birch
10. Balsam fir-white birch or balsam fir-red maple
11. Yellow birch-balsam fir
Coniferous forest zone
12. Balsam fir-white birch (west)
13. Balsam fir-white birch (east)
14. Balsam fir-black spruce
15. White spruce-balsam fir and balsam fir-white spruce
16. Black spruce-balsam fir-moss
17. Black spruce-moss

Figure 10

Ecological regions (forest regions)—southern Québec

Adapted from Thibault 1985, Map "Les régions écologiques du Québec méridional, deuxième approximation."

Balsam fir-white birch forest region forms a wide band between latitudes 48° and 49° N, from Ontario to the Gaspé. It subdivides into two parts (considered subregions by Thibault 1985), one to the east (13) and the other to the west (12). In the western part of its range, however, balsam fir stands are confined to well-drained till deposits and actually occupies only a small area. The clay deposits of Abitibi are more favourable to black spruce. In the lowlands, other secondary communities occur, trembling aspen and jack pine stands, for example. Balsam fir forest clearly dominates to the east, however, on the Laurentian massifs and the Gaspé Peninsula. Here, Eastern white-cedar and Eastern white pine can also be found on well-drained sites.

The mixed forest (101 000 km²) marks the transition between the coniferous forest and the deciduous forest. It runs between latitudes 47° and 48° N, encompassing the Lake Saint-Jean Lowland and the foothills of the Laurentians along the Saguenay and St. Lawrence rivers from Baie-Saint-Paul to Forestville and girdling the Gaspé Peninsula. The mixed forest zone corresponds to the Cool Temperate Ecoclimatic Province, where total annual growing degree-days range from 1100 to 1400. The transitional nature of this forest is exemplified in the complexity of its communities. The change from dominance of deciduous species to dominance of conifers in any given area, for example, frequently depends on exposure or drainage conditions. Balsam fir-yellow birch forest (*cf.* region 9) is the largest community in the eastern part of the mixed forest, though it is confined to rich, relatively moist sites; it is found on the foothills of the Laurentians, in the lowlands of the northwestern part of the Gaspé, and along the Saguenay River. At sheltered sites, the balsam fir-yellow birch forest gives way to red maple-yellow birch or yellow birch-balsam fir stands. The balance depends on climatic or disturbance factors that are difficult to identify. The drier conditions west of La Tuque in the Laurentians massif explain the dominance of yellow birch over balsam fir in the Yellow birch-balsam fir forest region (11).

In the Upper Saguenay Valley and the Lake Saint-Jean Lowland, balsam fir-white birch forest dominates, sometimes with red maple communities (*cf.* region 10). Jack pine stands, as well as dry heath with ericaceous shrubs and lichens (*Cladina* spp.), often occupy well-drained sites, but Eastern white pine and red pine are also found here. Red maple and trembling aspen occupy disturbed, cool sites, and black spruce often occupies wet sites.

The Sugar maple-yellow birch and balsam fir-yellow birch forest region (8) constitutes the transition between the deciduous and the mixed forest. This region is typically a mosaic of the two communities at Appalachian mesic sites. It is a forest that is sometimes deciduous, sometimes mixed, and dominated by balsam fir, yellow birch, and red spruce. Red maple, trembling aspen, and white spruce are common at disturbed sites.

On the Gaspé Peninsula, valleys and mountains follow hard upon one another, breaking up the landscape. Bird communities therefore change rapidly between the farmland in a valley and the nearby conifer-covered mountain. Mont-Saint-Pierre. *Gaspé Tourist Association.*

The deciduous forest occupies the southernmost part of Québec, the area southwest of Cap Tourmente and west of La Pocatière (with pockets in Baie-Saint-Paul, on Île aux Coudres, and at La Malbaie). This forest zone is home to a number of bird species at the northern limit of their breeding range, mainly inhabitants of mature deciduous forest and often insectivores, such as the Scarlet Tanager, Northern Oriole, and Wood Thrush. The deciduous forest zone occupies 104 000 km² (7% of the province of Québec); it covers the St. Lawrence Plain and the Laurentian and Appalachian piedmonts. It is composed of sugar maple stands mixed with increasingly thermophile species toward the Central St. Lawrence Lowland. Annual growing degree-days can range from 1300 to 2000. This region is largely dominated by sugar maple on mesic soils, but it also supports a very large number of northern hardwoods, especially in the Sugar maple-hickory and sugar maple-basswood forest region (1). In addition to the northern species mentioned, the deciduous forest includes many species of oak, elm, and hickory, as well as Eastern

In the alpine tundra on certain peaks of the Shickshock Mountains in the Gaspé, the American Pipit finds barren heath like the arctic habitat where this species generally breeds. Top of Mount Jacques Cartier. *Summer 1985, R. Baronet, Ministère du Loisir, de la Chasse et de la Pêche.*

hemlock, ironwood, white ash, and other species. More than 40 tree species occupy this forest zone, and in any single wooded area 10 different species can easily be identified.

Forest regions dominated by sugar maple and yellow birch are located in the Laurentian and Appalachian piedmonts, from Lake Timiskaming to La Malbaie, and from the Eastern Townships to Saint-Pascal. There are four forest regions in this area (considered subregions by Thibault 1985), within which diverse forest communities associated with microclimates or soil drainage conditions can be identified. Mesic communities are dominated by sugar maple, American beech, and yellow birch mixed with red spruce or balsam fir.

Sugar maple-yellow birch forest, one of the forest regions of the Laurentians and Appalachians in southern Québec. Ottawa Valley. *September 1986, Daniel St-Hilaire, Ministère du Loisir, de la Chasse et de la Pêche.*

Sugar maple-yellow birch-balsam fir forest region (7) is found in Témiscamingue. In this region (considered a subregion by Thibault 1985), the maple stands are frequently accompanied by yellow birch-balsam fir stands at mesic sites. Balsam fir stands are also very abundant. Black spruce, balsam fir, yellow birch, and black ash vegetate wetter sites, and elsewhere white birch, red maple, trembling aspen, and black spruce stands form secondary communities. Sugar maple-yellow birch-beech forest (*cf.* region 6) is found at elevations in the Laurentian piedmont. At the highest or the most northerly sites, balsam fir is sometimes more abundant than sugar maple. Dry sites mainly support white birch, red maple, balsam fir, and beech stands, whereas wet sites are colonized by balsam fir, black spruce, and black ash stands. Trembling aspen stands are the main transition groups.

The Sugar maple-yellow birch-beech and sugar maple-yellow birch-basswood forest region (5) is typical of the Eastern Townships and the Beauce. Such forest is also found on a thin strip of the Laurentian and Appalachian piedmonts. Sugar maple-yellow birch-beech forest dominates mesic sites, whereas sugar maple-yellow birch-basswood forest is confined to sites where conditions are particularly favourable. Balsam fir and red maple stands are very frequent in this forest region, with red maple especially apt to colonize thin soils. White birch and trembling aspen form secondary communities. Sugar maple-yellow birch-basswood forest region (4) occupies both sides of the Gatineau Valley below elevation 300 m. Sugar maple forest largely dominates this area, mixed with beech, ironwood, red oak, basswood, and white elm. The wettest sites are colonized by balsam fir, Eastern white-cedar, black spruce, and black ash, whereas dry sites are covered by red oak and Eastern white pine stands in the southwest and beech stands in the southeast. Transition communities are dominated by red maple, white birch, and trembling or largetooth aspen.

The Sugar maple-basswood forest region (2) and Sugar maple-basswood and sugar maple-yellow birch forest region (3) occupy the Central St. Lawrence Lowland upstream of Québec City (Bois-Francs, the Eastern Townships, and central Québec right to the U.S. border), as well as the Ottawa and Gatineau river valleys. These regions are clearly dominated by sugar maple-basswood forest, but in part of the Appalachian piedmont, such forest is found only on the most fertile land, while poorer sites are colonized by sugar maple-yellow birch stands. Wet sites support communities of silver maple, Eastern white-cedar, black ash, balsam fir, or black spruce.

The Sugar maple-hickory and sugar maple-basswood forest region (1) is restricted to the St. Lawrence Lowlands from the Montréal islands to the Great Lakes and up the Ottawa Valley as far as Hull. Sugar maple dominates mesic sites, mixed with bitternut hickory and basswood. On well-drained soil, American beech, red oak, Eastern hemlock, and Eastern white pine are also found; cooler sites are colonized by butternut, white elm, red ash, and bur oak. Red maple often dominates sites composed of organic sandy or clayey soil.

In this Atlas, especially in the species accounts, these forest regions may be referred to as ecological regions. This is explained in the chapter entitled *The Atlas data*.

MIREILLE DESPONTS

Wetlands

Wetlands are sites saturated or flooded with water for periods long enough to affect soil and vegetation components. Québec wetlands fall into three categories: tidal wetlands, or wetlands affected by tides; riparian wetlands, or wetlands bordering an expanse of fresh water with mainly seasonal variations in water level, including poorly drained areas nourished by precipitation; and peatlands (bogs and fens) (Couillard and Grondin 1986; Buteau et al. 1994).

Tidal wetlands

Tidal wetlands are flooded daily by tides—freshwater, brackish, or saltwater—whose range varies with the season or month. Often water-logged, the soil sometimes dries out considerably between periods of submersion, especially if it is flooded only by very high tides. Tidal wetlands are found along all coastlines in Québec, notably along the shores of the St. Lawrence from the gulf to Grondines. The physiognomy of these wetlands can vary, however, depending on substrate, water salinity, tidal range, latitude, and shoreline slope and exposure. Four types of tidal wetlands have been identified: two in the gulf, one typically on coarse substrate and the other on deltas and at bay heads; and two in the estuary, one in the saltwater stretch and the other in the freshwater and brackish stretches.

The Gulf of St. Lawrence, an arm of the North Atlantic Ocean, is a vast expanse of water where salinity exceeds 2.5% and tide level ranges from 0.65 m (at the Magdalen Islands) to just over 1 m (along the Gaspé coast and the Lower and Middle North Shore). The coastline is generally exposed, with a coarse substrate of sand, pebbles, and rock. As a result, the salt marshes of the gulf are often small (called wet or salt meadows) and occupy only the highest coastal zone. Large marshes are found only at a few favourable sites, generally at river mouths, where finer sediment accumulates, and more particularly in deep bays or at sites protected by long bars. In the Gaspé, areas where salt marshes develop are called barachois and are found mainly at the eastern end of the peninsula and in Chaleur Bay. The largest salt marshes (200–400 ha) are found at Gaspé, Malbaie, Paspébiac, and Restigouche. Salt marshes are more numerous on the North Shore, east of Havre-Saint-Pierre, but they are still relatively small (100–200 ha); the largest are those of Baie-Saint-Laurent, Kegaska, Washicoutai Bay, Coacoachou Bay, Baie-des-Rochers, and Saint-Augustin. There is a comparable marsh on the Magdalen Islands, but it is associated with the lagoons and sand bars that connect the islands and is very large (occupying about 50% of the 4700 ha covered by the marshes of the gulf).

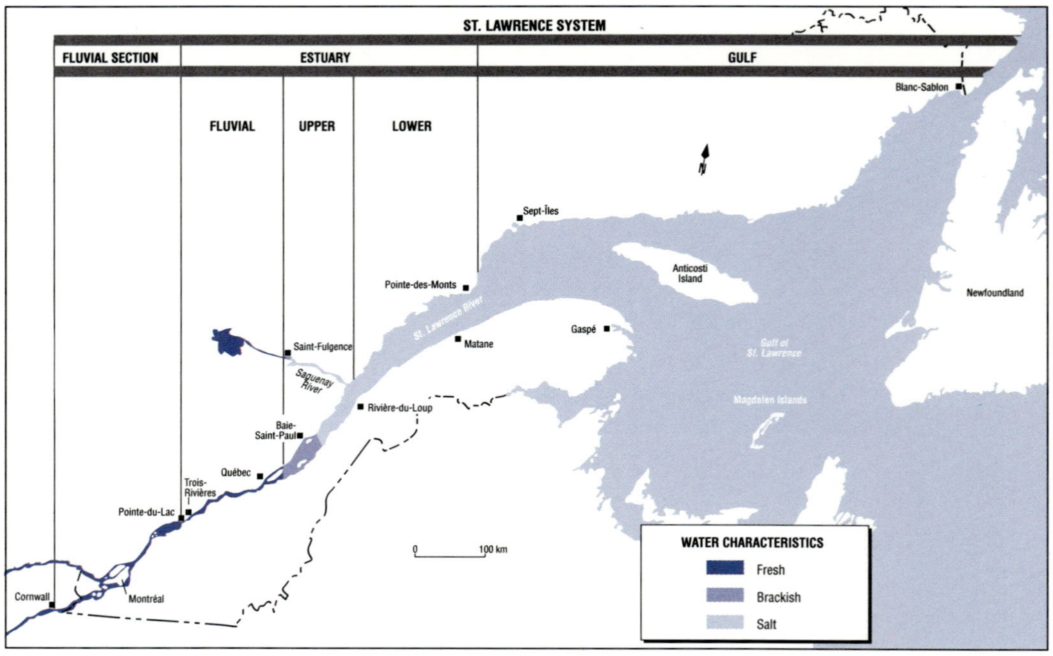

Hydrographic Divisions of the St. Lawrence

Source: Bouchard and Millet 1993, St. Lawrence Centre, Environment Canada

View of the barachois at Grand Pabos in the Gaspé. The salt marsh behind the sand bar attracts shorebirds and ducks from spring to winter. *Summer 1992, Christian Harvey for Fisheries and Oceans Canada.*

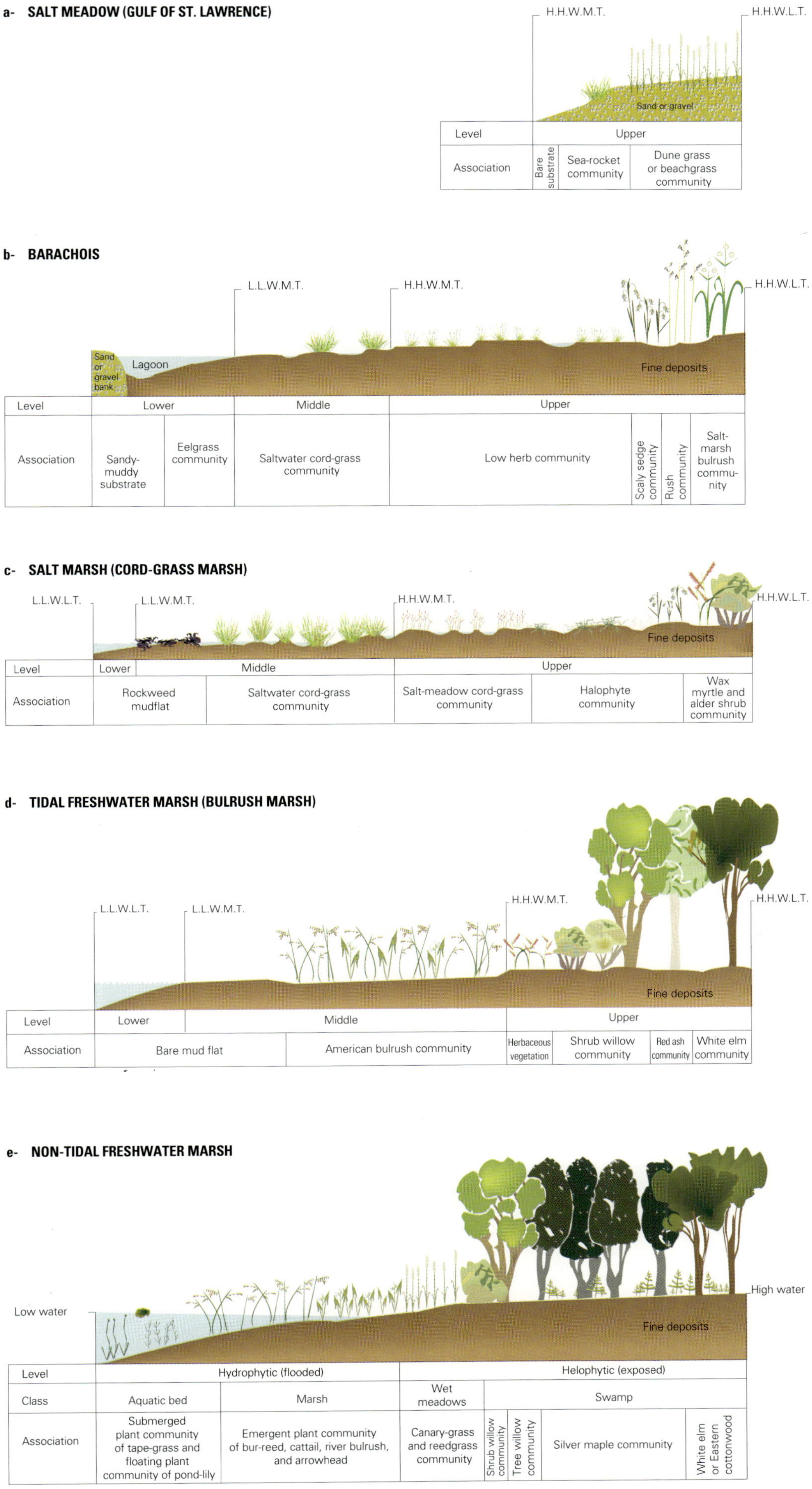

Figure 11

Cross-section of the principal wetland types found along the St. Lawrence River

Adapted from St. Lawrence Centre and Laval University (1991) modified from Couillard and Grondin (1986).

L.L.W.L.T. : Lower low water, large tide
L.L.W.M.T. : Lower low water, mean tide
H.H.W.M.T. : Higher high water, mean tide
H.H.W.L.T. : Higher high water, large tide

Wet or salt meadows have a very simple structure; there are just two parallel levels of vegetation (Figure 11a). The lower level, which is affected by equinoctial tides and storm surges, is discontinuous and composed mainly of sea-rocket, seabeach sandwort, and sea-lungwort. A little higher up, there is a denser band of American dune grass (mainly on the North Shore and Anticosti Island) or of beachgrass (on the sandy deposits of the Magdalen Islands and the eastern tip of the Gaspé). In the Mingan Islands, a community of cow-parsnip or wild strawberry sometimes develops behind the dune grass.

The large salt marshes of the gulf typically have an especially well-developed upper level (Figure 11b). The composition of the many halophytic herbaceous communities that colonize the upper level of the shoreline varies not only with submersion rate but also with microrelief, soil texture, and soil salinity. Communities of glasswort and salt-meadow cord-grass are found close to mean high-water level (the latter in the southern part of the gulf in particular). A little higher, the low herbaceous communities are dominated by seaside plantain and sea-milkwort, then by red fescue and poor goose-grass. The highest level of the marsh is dominated by Baltic rush, salt-marsh bulrush, or scaly sedge, the last mainly on sites with freshwater runoff.

Where the climate is milder, on the rare fine deposits of Chaleur Bay, the Magdalen Islands, and the east coast of the Gaspé, saltwater cord-grass colonizes the middle level regularly flooded by tides. On the lower level, in depressions and channels that remain flooded even at low tide, eelgrass grows.

Salt marshes typical of the St. Lawrence estuary are found between Saint-Jean-Port-Joli and Matane on the south shore and between Baie-Saint-Paul and Pointe-des-Monts on the north shore. These marshes are subjected to tides of more than 3 m, and the water in the marshes ranges from brackish (≈1.5%) to salty (≈2.5–3.5%). There are many more sites favourable to salt marshes in the estuary than in the gulf, especially along the south shore, where very large areas are covered with the fine deposits that foster salt marsh development. Among the largest marshes are those at L'Isle-Verte (1369 ha), Rivière-du-Loup (325 ha), Trois-Pistoles (252 ha), Saint-André (200 ha), and Kamouraska (150 ha). Generally smaller and less abundant, the marshes on the north shore mainly support high marsh communities, primarily because the coast is steeper. The main north shore salt marshes are at Baie aux Outardes (593 ha), in the Baie-Saint-Paul area (304 ha), at Baie de Mille-Vaches (249 ha), and on the Jérémie Islets (121 ha), southwest of Baie-Comeau.

The dominant estuarine marsh type is the cord-grass salt marsh (Figure 11c). The higher temperature of the water, the larger quantity of fine sediment, and the increasing range of the tide promote saltwater cord-grass colonization of vast expanses of the middle level regularly flooded by tides. The cord-grass first forms small patches sheltered behind communities of wrack or eelgrass, which colonize the low marsh and the lower part of the middle marsh. Eelgrass is found in residual waters at low tide wherever slopes are very gentle, the substrate is fine and, above all, currents are very slow (less than 20 cm/s) (Lalumière and Lemieux 1993). The saltwater cord-grass cover becomes continuous higher up. The marshes are intercut by pools, generally of glacial origin, that run parallel to the estuary. In the highest part of the middle marsh, saltwater cord-grass is mixed with a dozen other species, most frequently glasswort and Canada sand-spurry (Cantin 1974). Above mean high-water level, in the high marsh, salt-meadow cord-grass replaces saltwater cord-grass and becomes dominant. Several other plant communities (about 20 species in all) colonize this level. Here, ponds and pans maintained by physicobiochemical processes occupy about 25% of the marsh area (Gauthier and Goudreau 1983). These are generally shallow (10–25 cm) and rarely cover more than 0.5 ha (Argus 1989). Some of them, especially those that do not dry up completely in the summer, are colonized by sea-grass. The highest part of the marsh is the most diversified, colonized by 35 plant species. Freshwater cord-grass, glasswort, red fescue, hastate atriplex, and foxtail barley are the dominant species. Ponds are rarer and not as deep here (less than 15 cm) and they dry up in summer, sometimes developing into barrens with a high saline content that only glasswort can tolerate.

American bulrush marshes are typical along the freshwater to brackish stretches of the St. Lawrence estuary, from Grondines to Saint-Jean-Port-Joli on the south shore and to Saint-Tite-des-Caps on the north shore. The water is fresh to Montmagny, becoming more and more salty thereafter (to ≈1.5%). The tide is still felt in these stretches of the estuary, and its amplitude can reach 4–5 m. Upstream of Québec City, the coast is steep and often rocky and can support only small, very narrow bulrush marshes on the thin deposits between rock outcrops. The largest bulrush marshes have developed close to Grondines, Portneuf, and Donnacona, where fine deposits have accumulated. These marshes occupy some 1500 ha.

Downstream of Québec City, vast bulrush marshes (Figure 11d) have developed on the large muddy intertidal flats of the North Channel of Île d'Orléans, Cap Tourmente, and the Montmagny Islands, occupying as much as 2500 ha. This stretch of the St. Lawrence estuary alone includes close to 60% of the bulrush marshes in Québec (Dignard 1990). The dominant species in these marshes is American bulrush, mixed with other species to form communities that grow in parallel layers on the middle marsh subjected to tides. The largest of these communities is generally the one that includes wild rice. The zone straddling the middle and high marsh is colonized by a great bulrush–giant bur-reed community. In the Montmagny Islands, American bulrush is sometimes replaced by saltwater cord-grass, at the southern limit of its range in these slightly saline waters. The vegetation of the high marsh is highly variable, depending on many factors—including the height of the water table, the substrate, and the microrelief. The most typical communities where there has been no disturbance are as follows, in ascending order: freshwater cord-grass meadows, a shrub stratum of heartleaf willow or pussy willow, a treed stratum of crack willow, and a red ash or white elm association.

Ponds in cord-grass marshes are especially attractive to migrating shorebirds and American Black Ducks and their broods. The ponds offer abundant food rich in invertebrates, and the herbaceous communities provide cover. Kamouraska, southwest of Rivière-du-Loup. *Summer 1970, Austin Reed.*

Exposed at low tide, the Alouettes flats (on the left), west of the mouth of the Saguenay River, are a staging area for thousands of shorebirds and ducks, especially in late summer and fall. *October 1993, Pierre Normand.*

View of a bulrush marsh subject to tides. The different vegetation zones are well defined: an exposed mudflat is succeeded by an American bulrush marsh followed by the different herbaceous and willow communities that precede the agricultural zone. The fields and marshes of the Cap Tourmente National Wildlife Area attract hundreds of thousands of ducks and migrating Snow Geese. Today a popular destination for birders and hunters, this wildlife area was the site of the first farm in North America, cleared by Samuel de Champlain in 1626. Northeast of Québec City. *September 1987, Léo-Guy de Repentigny.*

Riparian wetlands

Riparian wetlands are wetlands bordering fresh water (rivers, lakes, streams or ponds) subjected to spring and sometimes fall flooding and to a summer low-water period provoked by a decline in precipitation and major evapotranspiration. Buteau et al. (1994) includes in this category poorly drained, non riverine areas nourished by precipitation, like cedar bottomlands. Four physiognomic classes can be identified based on vegetation and degree of immersion (Figure 11e). In permanently submerged shallow zones, aquatic beds of submergent and floating plants are generally found. The dominant species are tape-grass, Eurasian water-milfoil, and Canada waterweed. When broad-leaved arrowhead, giant bur-reed, cattail, or bulrush emerge, the aquatic bed becomes a marsh. Wet meadows (freshwater ones in this case) appear at sites which dry up during the summer low-water period. The vegetation here is still herbaceous, consisting of reed canary-grass, purple loosestrife, blue-joint, and freshwater cord-grass. When the shrub or tree cover exceeds 25%, the wet meadow becomes a swamp. Speckled alder, sweet gale, meadow willow, and red-osier dogwood compose the shrub swamp, and silver maple, red ash, crack willow, and Eastern cottonwood dominate the treed swamp. Couillard and Grondin (1986) subdivide the riparian system according to the physiognomy of the shoreline, the physicochemical quality of the water, and the composition of the plant communities. They divide the St. Lawrence River and two of its main tributaries—the Ottawa and Richelieu rivers—into four regions.

The Lake Saint-Pierre region, where tidal influence is no longer felt, runs from Grondines to the Montréal islands. In this vast widening of the St. Lawrence River, the waters are relatively calm, and there is substantial sedimentation of clay and silt from the river and its tributaries (the Richelieu and Saint-François rivers, for example). These conditions promote development of very large wetlands around the lake and on the Berthier and Sorel islands. A characteristic feature of these wetlands is their large aquatic beds (Figure 11e), where submerged-plant communities sometimes develop to a depth of 2.5 m. These aquatic beds are dominated by wild celery downstream of Sorel and by Eurasian water-milfoil upstream. Around Gentilly and the Contrecoeur and Verchères islands, American water-plantain is also very frequent. Shoreward, the different emergents comprising

Silver maple forest along the north shore of Lake Saint-Pierre. A true swamp, this maple stand harbours a variety of nesting birds, such as Great Blue Heron, Green Heron, and Wood Duck, as well as many passerines. Highway 40, near Yamachiche. *October 1993, Pierre Normand.*

This picture shows the extent and diversity of the marshes of the Berthier and Sorel islands. This large complex at the mouth of Lake Saint-Pierre extends out from the shoreline for several kilometres. The aquatic beds are large, especially in the eastern part of the complex. *Aerial image, 26 July and 21 August 1990, St. Lawrence Centre, Environment Canada, Québec Region. Approximate scale 1:125 000.*

Aerial view of Île du Moine. A large part of this island is occupied by marsh, which is frequented by many bird species. This hot spot attracts large numbers of birders in the summer and fall. Sorel Islands, west of Lake Saint-Pierre. *June 1988, Michel Beaudoin, Ministère du Loisir, de la Chasse et de la Pêche.*

the marsh appear in succession as the water becomes shallower. Great bulrush, American bulrush, and narrow-leaved cattail occupy the outer edges of the marsh, which are more exposed to wind and waves. At the low-water line, broad-leaved arrowhead and erect arrowhead colonize sheltered sites, such as protected bays and channels between islands. Communities of giant bur-reed, river bulrush, or common cattail generally mark the end of the marsh. The wet meadows, which are especially large on the Berthier, Sorel, Verchères, and Contrecoeur islands, are dominated by reed canarygrass and blue-joint. Low herbaceous communities dominated by redtop sometimes occupy the tops of inundable islands, vestiges of an agricultural past. The largest communities of purple loosestrife, an introduced species considered a threat to the quality of wildlife habitat, are found along the south shore of Lake Saint-Pierre. The shrub swamp is composed of several species of willow (crack, black, shining, meadow, and sandbar willow). The treed swamp is composed of communities that appear in succession as the elevation of the banks increases; black or crack willow stands appear first, followed by silver maple stands. From the wettest sites to the driest, there appear in succession amphibious yellow cress, sensitive fern, wood-nettle, and ostrich-fern.

The large aquatic beds near the eastern tip of Île Perrot (at the top) as well as between Îles de la Paix and Saint-Bernard Island (Île des Soeurs Grises) at the mouth of the Châteauguay River (bottom right of image) are excellent feeding grounds, not only for breeding waterbirds but also for the thousands of pochards that gather here in the fall. *Aerial image, 21 August 1990, St. Lawrence Centre, Environment Canada, Québec Region. Approximate scale 1:65 000.*

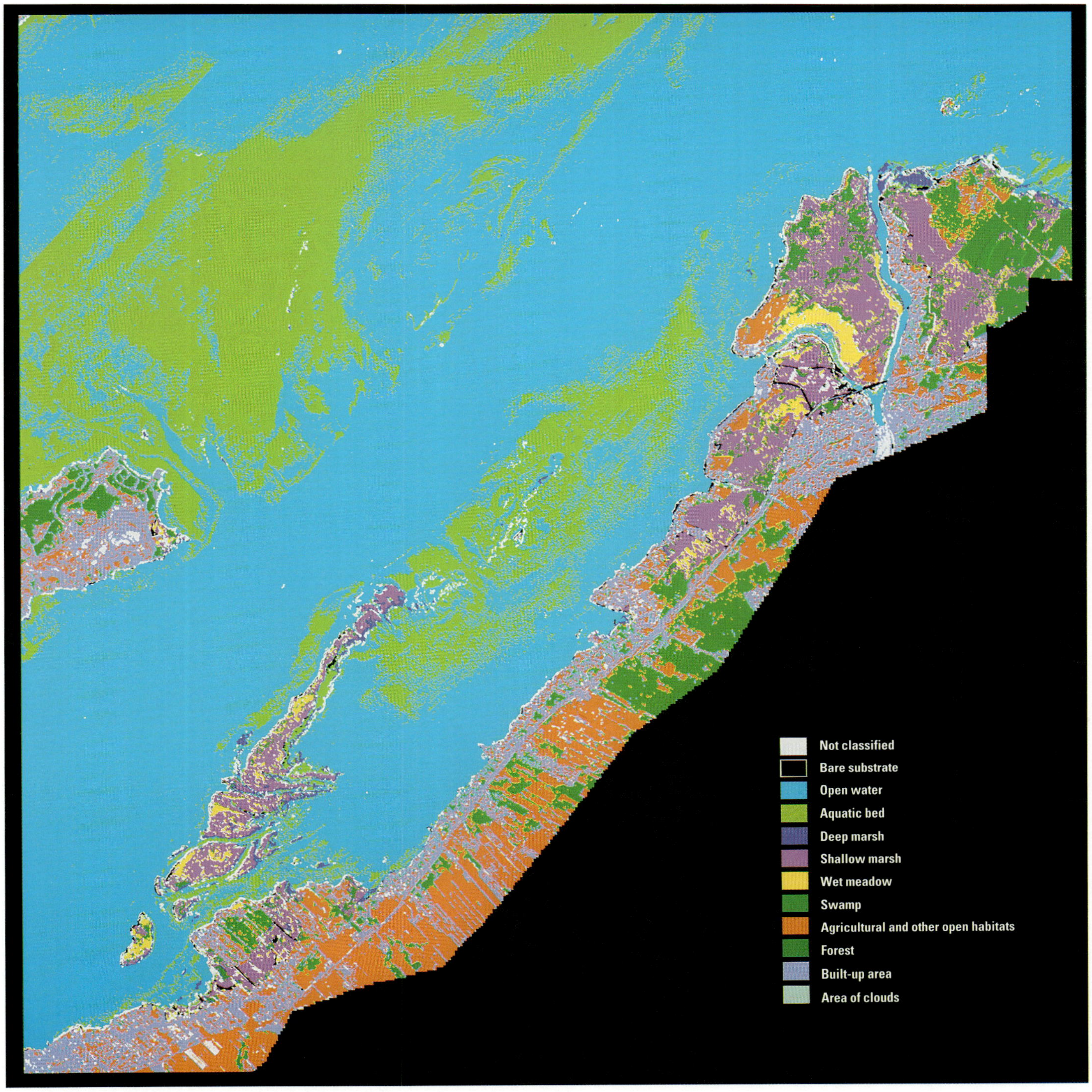

The Lake Saint-Louis region occupies the river corridor between Cornwall and Sainte-Thérèse Island, northeast of Montréal. This region in fact comprises three widenings of the St. Lawrence River separated by rapids—Lake Saint-François, Lake Saint-Louis, and the La Prairie Basin. In Lake Saint-Louis, the clear, green waters of the St. Lawrence mix with the more turbid brown waters of the Ottawa River, which run along the northeast shore of the river. The weak current and large fine deposits together with the shallow, clear, green waters in these large basins promote development of large aquatic beds, which, together with marshes, occupy 15 000 ha. These wetlands have, however, been subjected to major disturbances—of which artificial water-level regulation was not the least, considering that it upset the annual flood-low water cycle and altered the size of wet meadows and swamps bordering the river.

The aquatic beds are dominated by tape-grass, water-stargrass, sago pondweed, and Eurasian water-milfoil in this region. In the brown water corridor, only tape-grass is present. In the marshes, communities of giant bur-reed, narrow-leaved cattail, and river bulrush dominate. Great bulrush mainly occupies the shallows. Except on a few islands (Îles de la Paix and Saint-Bernard Island), wet meadows and swamps have generally disappeared from the shoreline as a result of urbanization. Where these vegetation zones are still present, reed canary-grass, purple loosestrife, and rice cut grass occupy the lowest ground, once colonized by stands of silver maple and sensitive fern—which almost completely disappeared in the 1970s because water level was kept too high during the growing season. Shrub swamp colonized by buttonbush and shrub willow occupies a few permanently flooded depressions. All that remains of the treed swamps is a discontinuous band of crack willow and sometimes

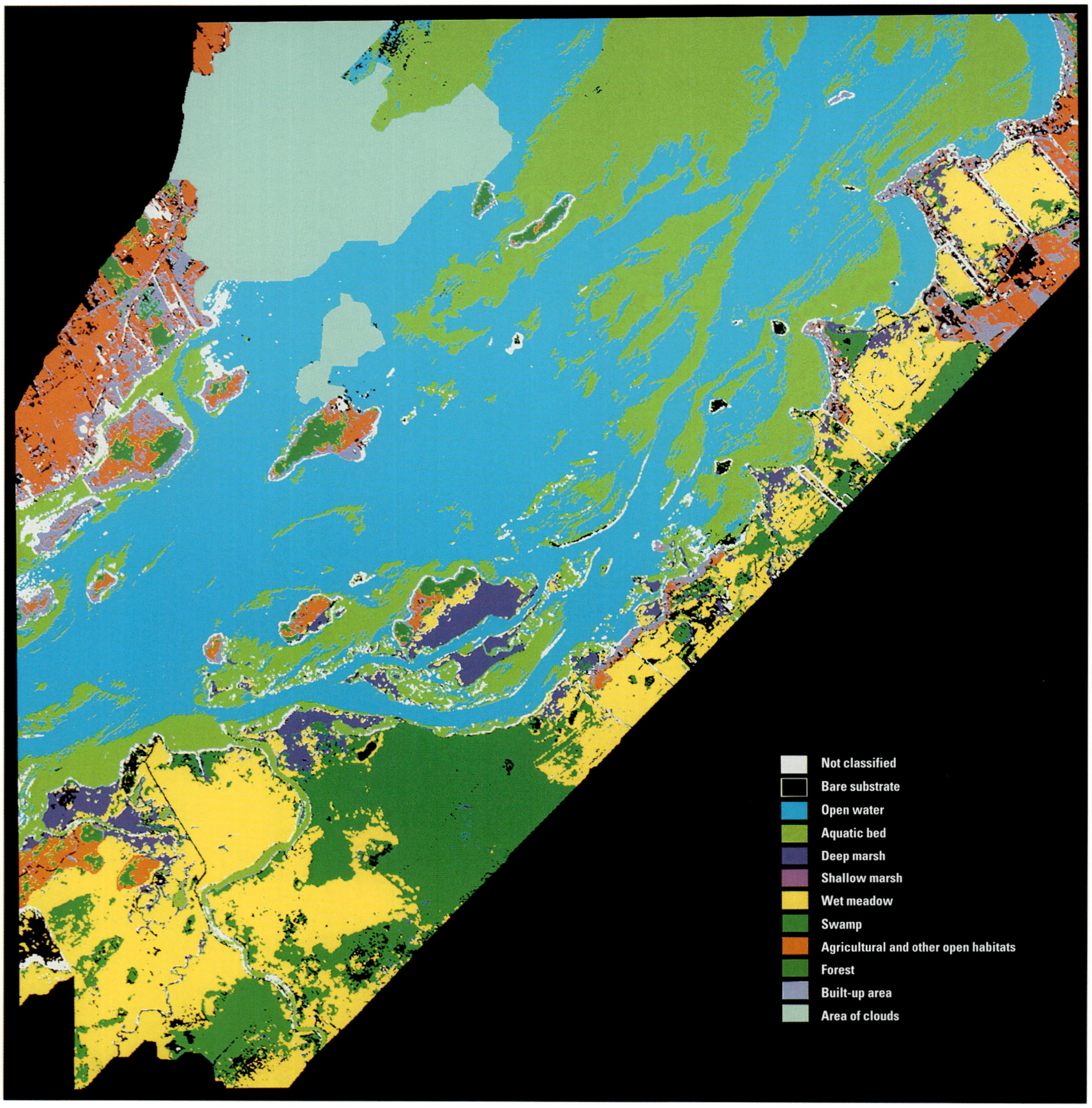

Digital processing of this aerial image makes it possible to show the extent of the large aquatic beds, which are often hidden from view. In late summer and fall, these aquatic beds attract large flocks of diving ducks that feed on invertebrates. Western part of Lake Saint-François.
Aerial image, 26 July 1990, St. Lawrence Centre, Environment Canada, Québec Region. Approximate scale 1:65 000.

silver maple with wood-nettle. This is not true around Lake Saint-François, however, where water levels have been controlled since the 1930s. In the Dundee region, large marshes and wet meadows grow on a thick layer of well-decomposed peat. The marshes are colonized by blue-joint, cattail, lake sedge, and aquatic sedge. The shrub swamp is composed of a variety of willow species together with speckled alder, and the treed swamp is dominated by tamarack, red maple, white elm, and red ash.

The Lac des Deux Montagnes region includes Lac des Deux Montagnes as well as the Ottawa River upstream of the lake and the Rivière des Mille Îles and the Rivière des Prairies downstream of it. Through this region flow the brown waters of the Ottawa River, which are more turbid and acidic than those of the St. Lawrence. The most obvious effect of this is the disappearance of certain less tolerant species, such as American water-plantain and sago pondweed, as well as the smaller size of the aquatic beds. Marshes are more abundant here, however, than in the Lake Saint-Louis region, because of the large bays favourable for their development at Oka, Rigaud, and Carillon, among other places. There are also large wetlands along Rivière des Mille Îles and bordering Lac des Deux Montagnes in Vaudreuil Bay.

The vegetation of the wetlands in this region is very similar to that of the wetlands in the Lake Saint-Louis region, though there are a number of distinctive local plant communities—such as the threadfoot beds in the rapids of Rivière des Mille Îles and Rivière des Prairies, and the marshes dominated by sweetflag and branching bur-reed. Buttonbush shrub communities sometimes occupy vast expanses in sheltered bays, and treed swamps, less affected by urbanization here, are more numerous. Dominated by silver maple, treed swamps cover large areas around Carillon and Oka as well as along Rivière des Mille Îles.

Marshes and swamps are characteristic of the mouth of the Saint-François River. The aquatic beds do not show up on this photo, but they can be seen on the aerial image of the Berthier and Sorel islands (on the far right about half way down). **La Grande Commune.** *October 1993, Pierre Normand.*

Various species of dabbling ducks breed in the marshes around Plaisance on the Ottawa River and various species of waterbirds also stage through here on migration, making it a favourite spot for birders. **Dubé Bay.** *July 1988, Jacques Chabot, Ministère du Loisir, de la Chasse et de la Pêche.*

The Richelieu Valley region includes the Richelieu River as well as Rivière du Sud, its main tributary, and Missisquoi Bay. Marsh and swamp cover very large areas in the upper Richelieu because of the relatively slow river current and the substantial spring flooding of the bottomlands. In this region, the marsh vegetation is rich, especially at Île aux Noix, at Ash Island, in Missisquoi Bay, and along Rivière du Sud. The aquatic beds are dominated by tape-grass, Canada waterweed, and magnolia water-lily, the last particularly abundant in coves and bays. Along Rivière du Sud, communities of common coontail, water-stargrass, and lesser duckweed add to the diversity. The marshes are dominated by communities of great bulrush and giant bur-reed, except along Rivière du Sud, where wild rice and narrow-leaved cattail are more prevalent. Wet meadows are generally composed of reed canary-grass and purple loosestrife, but buttonbush is also a frequent shrub. The willow stands are generally only small islands of black willow along the edge of the shoreline, and the silver maple stands have partly deteriorated from many years of high water. At the back of the wetlands, right at the end of the floodplain, swamp white oak stands have developed, another distinctive feature of this region.

Many species use this large freshwater marsh for a variety of purposes. Wood Duck and Hooded Merganser, for example, nest in the trees and bring their broods to feed in the aquatic beds. **Lac Saint-François National Wildlife Area, southwest of Montréal.** *June 1974. Léo-Guy de Repentigny.*

Peatlands

Peatlands are found throughout Québec. The abundance and size of the peatlands vary regionally depending on relief and climate and locally depending on the nature of the sediment. By definition, peatland is a place where peat accumulates because plant growth is too rapid for the environment to manage to decompose it. Factors that promote peatland development include level ground, a wet, cool climate, a relatively long growing season, and fine, impermeable sediment that inhibits drainage. The main peatland areas in Québec are the St. Lawrence, Saguenay–Lac-Saint-Jean, Abitibi, and Eastmain lowlands.There are several physiognomic types of peatland (Grondin and Ouzilleau 1980; Couillard and Grondin 1986), but the primary distinction is based on the system's trophic regime, that is, the method of water supply. Ombrotrophic peatlands, or bogs, receive their water mainly from precipitation. They are impoverished, strongly acidic environments with very poor water circulation. The peat decomposes very slowly in a bog. Bogs are colonized mainly by sphagnum, lichens, ericaceous shrubs, and black spruce. Minerotrophic peatlands, or fens, have a lateral water supply from runoff. Fens have better water circulation than bogs, and they are generally richer and less acidic. Mosses, sedges, and herbaceous plants dominate fens, together with a few shrubs and some tamarack.

Distribution of the different physiognomic types of peatland also depends on relief and climate (Couillard and Grondin 1986; Winter and Woo 1990; Larson 1991). In arctic environments, the climate is rather dry, the growing season is short, and permafrost largely inhibits the development of peatland. Peatlands in these environments are limited to small, thin fens associated with lakes, waterways, or snow beds which can provide enough water in summer for their development. Peatlands are not very extensive in subarctic environments either. They appear mainly along Hudson Bay and Ungava Bay, where the climate is humid and deposits are fine. Permafrost is sometimes found in peatlands, giving rise to a particular formation, the palsa bog. In these ombrotrophic peatlands, ice lenses grow, forming mounds of perennially frozen peat and mineral soil sometimes several metres high. In the Rivière aux Feuilles valley, another physiognomic type is characteristic, the peat plateau bog, composed of perennially frozen lichenous or wooded raised areas generally located along low terraces bordering the river. Fens have developed almost everywhere in this region, but they are small.

Peatlands are more numerous south of Hudson Bay and at the latitude of James Bay, where they sometimes occupy as much as 50% of the terrain. This area is largely dominated by bogs, frequently shrub-covered or with their peripheries invaded by black spruce. These bogs are typically string bogs, a regular succession of shrub-covered hummocks and of wet hollows carpeted by sphagnum and invaded by herbaceous plants. In the lowlands of Rupert Bay and Chisasibi, however, fens dominate, and peatland covers close to 75% of the ground. The fine sediment of the Tyrrell Sea that was deposited on this flat land has greatly contributed to the development of peatland. These lowlands also harbour feather fens, in which a feather-shaped shrub bog sits in the centre of a fen. Farther east of James Bay, in the Laurentian Shield, the till-covered soil and the rolling hills are less favourable for peatland, which covers no more than 6–10% of the terrain. Here peatland is most often found where drainage has been slowed down by glacial deposits such as eskers, drumlins, blocks, and so forth.

Peatlands also cover much of the ground in Abitibi (26–50%), appearing mainly in the form of treed, shrub, or sedge bogs. Farther east, however, in the Laurentian Shield, the uneven ground inhibits peatland development. Here peatlands occupy hollows and are generally small and very uniform, covering no more than 5% of the terrain.

The condition of the peatlands in the St. Lawrence Lowlands varies appreciably from one region to the next. Along the North Shore, peatlands are concentrated on the sandy deltaic deposits in the Sainte-Marguerite, Romaine, Natashquan, and Little Mecatina rivers. The bogs here are vast and an average 2 m thick, their development fostered by the level ground, the relatively impermeable deposits, and the very humid climate. At Lac-Saint-Jean, the glacial deposits and alluvions accumulated for millennia by the Ashuapmushuan, Péribonka, Mistassini, and other rivers have left an undulating relief with flat areas favourable to development of peatlands, generally bogs, which occupy 11–25% of the terrain. Most of the bogs are small, but east of Sainte-Marguerite-Marie there is a bog that covers an estimated 50 km^2. In the St. Lawrence Lowlands, between Québec City and Baie-Comeau and right to Matane on the south shore, peatlands have developed in depressions, such as at Plée de Saint-Charles near Lévis. But the largest peatlands are found close to Rivière-du-Loup and at Manicouagan Point; some of these are used for horticulture. Small peatlands used to be found everywhere in the Greater Montréal region, but most have been drained and are now used for agriculture. South of Montréal, this organic black muck soil is particularly prized for market gardening. Peatlands still remaining include those at Lanoraie, Clarenceville, Napierville, Lac-à-la-Tortue, and Saint-Gilles. The Appalachians, on the other hand, are not very suitable for peatland development. Peatlands here are small and dispersed, confined to valleys and depressions. Small concentrations are, however, found close to Saint-Fabien-de-Panet in Montmagny County, and west of Sherbrooke.

Mireille Desponts, Denis Lehoux, and Louise Gratton

Bog near the Eastmain River in the James Bay region. *June 1982, Yves Aubry.*

Fen near the Eastmain River in the James Bay region. *June 1982, Yves Aubry.*

HUMAN IMPRINT ON THE ENVIRONMENT

The immense size and apparently inexhaustible natural resources of Québec have profoundly marked its history and culture. In 1535, Jacques Cartier expressed his wonderment at "this country that was the finest and most excellent one could find anywhere" (Biggar 1924), as he sailed up "*the Great River of Canada*." That was the name he gave the river that would later be called the "St. Lawrence" by Champlain. The name "Baye Sainct Laurens," which was what Cartier called the strait between Anticosti Island and the North Shore, was later attributed (due to errors in translation of maps into Spanish and then into Italian) to the gulf and then to the river itself by Champlain in his book *Les Voyages de la Nouvelle France occidentale, dicte Canada* of 1632 (Frenette 1990). The potential for establishing a colony and exploiting the land's resources seemed highly promising in this still largely untouched valley. Quite naturally, the first attempts to set up a colony (apart from Acadia) and establish towns took place on the fertile and accessible lands along the St. Lawrence River, gateway to the continent.

The clearing of land, followed by industrial and urban development, have inevitably had an impact on the environment. Its main components—the soil, waterways, and forests—have been profoundly altered, triggering changes in the abundance and composition of wildlife, particularly birdlife.

Nature, as a dynamic system, has evolved continuously over millennia in response to climatic change and natural disturbance. However, human impact on the environment has greatly accelerated this process. Anthropogenic disturbances often threaten the balance of nature and the capacity of ecosystems to maintain or regenerate themselves, causing catastrophes that sometimes affect humans as well as other species.

To survive, every species requires a minimum of resources from its environment so that it can feed and reproduce; it occupies a habitat and exploits a particular ecological niche. If the habitat disappears, so does the species. Any major change in food supply will lead to an adjustment in population or even to local extirpations. On the other hand, altered habitats may create suitable conditions for species that were previously less abundant, or attract species that were once absent. For instance, few warblers nest in sites that are totally barren of trees and shrubs, but such habitats may be suitable for the Killdeer.

Humans, as predators, have also had a direct impact on the abundance of certain wildlife species. Some, such as the Passenger Pigeon and Labrador Duck, have been hunted to extinction, and others, like the wapiti, have been extirpated from Québec. In the 20th century, hunting and profound habitat changes threaten the disappearance of the woodland caribou from southern Québec. On the other hand, human activity has helped increase populations of Ring-billed Gulls, Snow Geese, and white-tailed deer, for example.

Throughout its short settled history, the environment in southern Québec has gradually been transformed. Over the years, an observer attentive to the landscape and natural environment would have noticed changes in the composition of certain bird communities and would have been able to correlate this sometimes with major environmental changes of the recent past. This brief overview of the human activities and socioeconomic contexts that have led to this transformation opens with a short discussion of the environment today in the late 20th century, considered under four main aspects: the urban, agricultural, forest, and aquatic environments.

The environment in the late 20th century

Human occupation in Québec

The St. Lawrence Valley has been central to Québec's development since the start of European settlement, for reasons that have changed little in 300 years. Until the 1800s, the river and its main tributaries—the Richelieu, Ottawa, Saint-Maurice, and Saguenay rivers—formed the only real network of communication and trade among regions, and between the regions and other countries. Also, the lowlands along these waterways contain, with few exceptions, the only arable land. The thin, acidic soils of the Canadian Shield, which covers 92% of Québec, have little or no potential for agriculture. The climate of southwestern Québec allows for higher yields and a wider range of crops, necessarily determining the pattern of settlement, as the following statistics show.

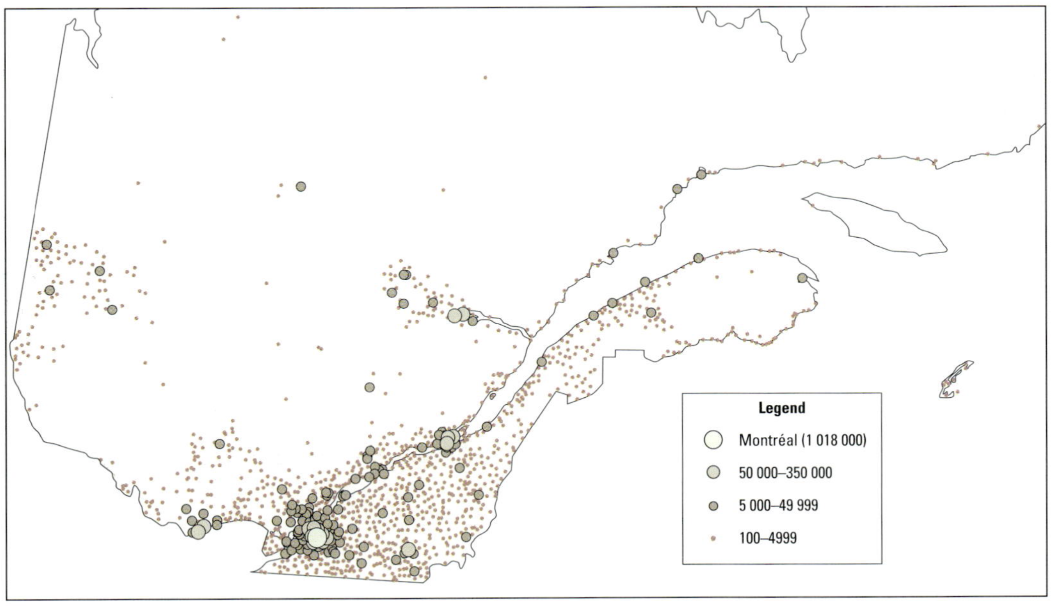

Figure 1
Population distribution in southern Québec, 1991
Adapted from the TOPOS database, Commission de toponymie du Québec.
Produced by M. Melançon, Environment Canada.

In 1986, 50% of the population (6 540 276 inhabitants) was concentrated in 0.7% of the province in the urban communities of Montréal and Québec City, the Outaouais Regional Community, and the regional county municipalities (MRC) of Laval and Champlain. Moreover, 80% of the population lives along the narrow corridor of the St. Lawrence Valley (Joe and Duchesne 1989). Figure 1 clearly shows this uneven population distribution. The heavily built-up environment (agricultural and urban zones) covers 2.5% of Québec, about the same size as The Netherlands. These inhabited areas roughly correspond to the distribution of arable land. Arable land totals 2 360 000 ha (2% of Québec); the largest zones are located in the St. Lawrence Plain (1 240 000 ha), the Appalachians (660 000 ha), and the enclaves of Abitibi–Témiscamingue and Saguenay–Lac-Saint-Jean (380 000 ha). To this could be added about 121 400 ha of organic soils with good potential located mainly in the St. Lawrence Lowlands (MENVIQ 1988; Figure 2). Creation of open habitat has encouraged the expansion of species such as the Killdeer, various sparrows, and the icterines (blackbird family), to name a few.

The human imprint on the environment is also apparent in the transportation networks that form a web across Québec. In 1988, there were 63 263 km of roads under provincial jurisdiction (slightly more than 60% of the total), not including 38 500 km of logging roads managed by the Ministère de l'Énergie et des Ressources, the Ministère des Forêts, and the Ministère du Loisir, de la Chasse et de la Pêche, as well as by Hydro-Québec and the ZECs (M. Béland,

Ministère des Transports du Québec, pers. comm.). Expansion of species such as the Barn Swallow into the hinterland parallelled expansion of the transportation network. The Barn Swallow likes to nest on man-made structures such as bridges, cottages, and other buildings; today its range extends as far north as James Bay. The construction of roads and logging roads involved opening many borrow (sand and gravel) pits, visibly marking the landscape. In a number of locations open sand deposits provide suitable nest sites for colonies of Bank Swallows.

Borrow pits are good habitats for Bank Swallow colonies. Martin-Valin ZEC, north of the Saguenay River. *August 1991, Jean Gauthier.*

In 1986, the railroad network had a total trackage of 7971 km, and pipelines totalled 1105 km. In 1991, Hydro-Québec had 33 067 km of high-tension power lines, for a total land area of about 1178 km^2 (MENVIQ 1988, 1993). The new habitat created by such developments is suited to species associated with scrubland, such as the Song Sparrow, Yellow Warbler, and Mourning Warbler.

Forest

Concentration of human activity in the St. Lawrence Valley has inevitably had an impact on the forest. The landscape has changed so much since 1535 that Jacques Cartier would have difficulty recognizing the countryside he so admired. Only a few forest fragments remain along the St. Lawrence, and stand composition corresponds very little to that of the virgin forest. Most stands have been selectively cut more than once, eliminating many sugar maple companion species. In effect, the fragments are almost all second-growth forests. The mature maple-dominated northern hardwoods forest community in the Sugar maple-hickory and sugar maple-basswood forest region would have included, on mesic sites, basswood, beech, bitternut hickory, ironwood, ash, black cherry, butternut, white elm, slippery elm, and hemlock. The extent of beech dominance may have been much greater than it is today, since this species is more often found in a climax formation (Nicholson et al. 1979; Leak 1987; Brisson et al. 1988, 1992, 1994).

Actually, the dominance of sugar maple shows that it is opportunistic in occupying disturbed sites. A study of forests farther north along the river, on the south shore (Montmagny–L'Islet), concluded that in the clayey plain, the climax association ought to be white spruce and red maple with ash, oak, and elm (Mathieu and Courville 1987). Sugar maple-yellow birch would have dominated on better-drained terraces, with Eastern white pine occupying gravel deposits. In fact, almost the entire commercial forest has been affected to varying degrees, although the impact of harvesting is not immediately obvious since, with regrowth over the long term, the forest landscape does not retain the bare look of urban and agricultural areas.

Commercial forest covers 46% of Québec, almost the entire area south of the 52nd parallel (Dugas et al. 1989). Its timber harvest potential is immense and was long considered inexhaustible. All of the deciduous and mixed forest has already been logged at least once, except for occasional small pockets. The coniferous forest has been heavily logged as far north as Chibougamau. The following summary of the situation in the major regions is based on data in Québec's 1988 state of the environment report (MENVIQ 1988).

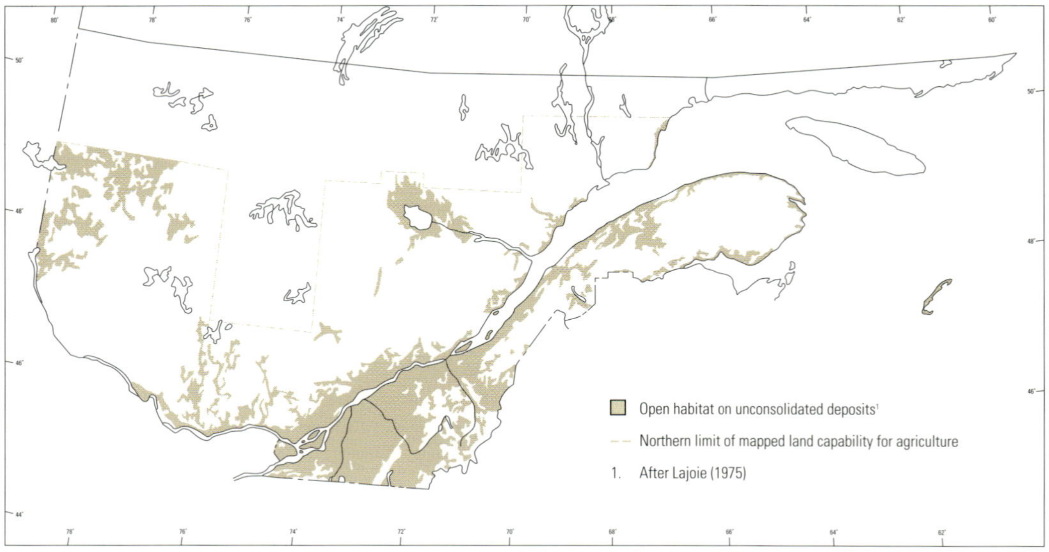

Figure 2
Farmland in southern Québec

Source: Compilation and cartography by the Soil Research Institute (now the Centre for Land and Biological Resources Research –CLBRR), Research Branch, Agriculture Canada, based on maps prepared by the Québec Soil Service (now the Québec Land Resource Unit of the CLBRR, and the Service des sols, Ministère de l'Agriculture, des Pêcheries et de l'Alimentation du Québec), with assistance from the Lands Directorate, Environment Canada.

In southwestern Québec, including the Eastern Townships and the southern Ottawa Valley, the forest has largely given way to farming and urbanization. In the Sugar maple-hickory and sugar maple-basswood forest region (*sensu* Thibault 1988), species once harvested for lumber—such as hickory, butternut, elm, yellow birch, ash, and pine—have sharply declined. Sugar maple stands have typically been developed for maple syrup production, to the detriment of young maples and other species in general. Diseased and overmature trees and snags have been removed, and this has had an effect on woodpeckers, nuthatches, and other cavity nesters. Eighty-five percent of the remaining forest, mainly in sugar maple-yellow birch associations, is composed of young mixed stands. In the Appalachians, idle land and sugar maple stands that have been repeatedly cut over tend to regenerate in conifers. Today, softwood stands are the most heavily logged, regenerating in their turn in shade-intolerant hardwoods such as poplar, birch, pin cherry, and red maple.

In the northern Ottawa Valley, Mauricie, and lower Laurentians, basically in the Sugar maple-yellow birch-balsam fir forest region, poplars and white birch tend to invade after intensive cutting or burns. Only local remnants remain of the vast tracts of white and red pine after overcutting in the last century. This explains the present patchy distribution of the Pine Warbler,

Strip cut/clear cut mix. The large uncut expanses allow forest species that need large territories to thrive. Maintaining major tracts of forest at the regional level is now considered essential. Awareness of the negative effects of forest fragmentation is increasing. It has even been suggested that large clearcut areas be alternated with large untouched forest tracts. The conservation of buffer strips along shorelines protects aquatic habitats and the animals that live in them. West of Lake Saint-Jean. *May 1985, Daniel Bordage.*

Strip cutting makes it possible for some of the original wildlife to remain in a habitat, but it also attracts species that like forest edges. The Brown-headed Cowbird moves in, taking advantage of the openings, to parasitize populations once protected deep in the forest. Nests also become more vulnerable to predation by some mammal and bird species. Near Leclercville, Lotbinière. *October 1993, Pierre Normand.*

a species that only colonizes areas where white and red pine are found in varying combinations (Collins et al. 1982). In some sites, selective cutting of yellow birch has left degraded stands with a high number of poor-quality stems, and the Pine Warbler is becoming increasingly rare in these areas. Lastly, the La Vérendrye Wildlife Reserve still has mature softwood stands, although these were logged to some extent in the 1980s.

In the Lower St. Lawrence and Gaspé, where balsam fir with yellow birch or white birch communities dominate, 80–90% of forests are young mixed or coniferous stands. Mature communities of balsam fir and spruce are now found only in the heart of the Notre-Dame Mountains, in the Gaspé, and are still in the throes of a severe spruce budworm outbreak in the 1990s. Deciduous species such as white birch and trembling aspen now dominate clearcut areas, while balsam fir has flourished in partial cuts. Throughout the Notre-Dame Mountains, former farmlands and repeatedly logged woodlands are regenerating in conifers, replacing the mixed deciduous and coniferous stands of the virgin forest. On the Magdalen Islands, forest cover has declined from 60 to 30% of total land area, largely due to farming and urban development (Fradette 1992). Many formerly wooded areas are now pastureland.

In Abitibi, the Laurentians, and on the North Shore, basically the Balsam fir-white birch forest region, mature stands are rare and mainly found north of the Gouin Reservoir and on the North Shore. Young mixed or softwood stands occupy 60–70% of central and eastern Québec; in Abitibi, trembling aspen has become widespread, particularly in burns and clearcuts. The Abitibi region has clayey soils, and clearcutting has raised the water table, causing waterlogging of flat or gently sloping land; this has slowed regeneration.

Forest fragmentation started with the arrival of the first settlers, who gradually turned the St. Lawrence Plain into farms and towns. In the 19th century, logging reached into and fragmented the hinterland forest. Kamouraska, southwest of Rivière-du-Loup. *October 1993, Pierre Normand.*

Landsat images

The following 1:1 000 000-scale images were taken by Landsat-5 (acronym for *land satellite*). This satellite was launched in March 1985 and orbits the earth at a height of approximately 900 km. It is equipped with a Thematic Mapper sensor, which obtains non-photographic images with a pixel size (from a contraction of the term "picture element"), or spatial resolution, of 30 m × 30 m. The sensor's spectral range comprises seven bands ranging from blue (0.450 µm) to thermal infrared (2.35 µm). The physical signal received by the sensor (electromagnetic radiation measured as a voltage) is processed using special mathematical techniques (Bonn and Rochon 1992).

To represent digital values as images that approximate natural colours, a combination of three bands is used, each of which is assigned to a different colour. In these images, band TM3 (0.63 µm–0.69 µm) is assigned to blue, band TM4 (0.76 µm–0.90 µm) to green, and band TM5 (1.55 µm–1.75 µm) to red. The hues associated with a specific habitat may vary from image to image, however, depending on the degree of development of the vegetation when the image was taken and, in some cases, the method used to accentuate contrasts. Each image is identified by the orbit, or trajectory, which is basically north-south, and the swath, which is east-west. For example, a typical image would be identified as follows: "Image from orbit 14, swath 28."

MARCELLE GRENIER

Coniferous forest
Clearcut
Mixed forest
Agriculture
Forest regenerating after logging
Built-up area or bare substrate
Peatland

Landsat image of the Abitibi–Témiscamingue region

In the lower left-hand corner of the image are Lake Timiskaming and its outlet, the Ottawa River. In the centre of the image is Lake Simard, also part of the river's headwaters, upstream on the Laurentian Shield.

The image clearly shows the region's economic basis: mining and agriculture. In the Témiscamingue area, the farms and villages around Lake Timiskaming can be seen, while in the Abitibi area, mining towns such as Val-d'Or, Malartic, and Rouyn-Noranda appear at the top of the image (from east to west). The nearby mines and their tailings are the same colour as the towns themselves since the nature of the radiation produced is the same, whether asphalt, the roofs of buildings, or crushed rock—they are all bare ground. Openings in the forest created by logging and agriculture attract birds of forest edges and open habitat.

The extensive peatlands southeast of Val-d'Or appear in light blue, and the scars from logging operations can be seen south of the town and north and east of Lake Simard.

The veiled bluish areas in the lower right are thin clouds. If these clouds were thicker, they would be white on the image.

Image from orbit 18, swath 27, taken on 29-05-1990, courtesy of the Canada Centre for Remote Sensing, Geomatics Canada. Approximate scale 1:1 060 000.

On Anticosti Island, logging on the west side in the 20th century has produced a thriving population of the introduced white-tailed deer, whose feeding habits promote regeneration mainly of white spruce, but also trembling aspen and white birch, instead of balsam fir (Desloges and Émond 1974; J. Huot, Laval U., pers. comm.). The last major mature softwood stands in Québec are northwest of the Gouin Reservoir and north of the Saguenay–Lac-Saint-Jean region. Already, north of the Saguenay, the black spruce forest is being heavily logged for lumber. Regeneration here is a major problem, especially on the North Shore, where barren heath often replaces logged forest.

In short, the forest landscape today looks quite different from what it was in the time of the first European settlers, with only small fragments remaining in the once heavily wooded St. Lawrence Valley. Birds like the Red-shouldered Hawk, Cerulean Warbler, and Cooper's Hawk have become so rare that they are now on the list of species likely to be designated vulnerable or threatened in Québec (MLCP 1992). In forests throughout southern Québec, tree species diversity has been reduced by repeated selective cutting.

The BAPE (Bureau des audiences publiques sur l'environnement—environmental hearings board) report on forest protection strategy consultations underlines a number of concerns for the future of the Québec lumber industry (Commission sur la protection des forêts 1991). The following examples are taken from the report and illustrate the state of the forest in the 1990s: additional supplies of commercial hardwoods such as oak, butternut, ash, black cherry, and yellow birch now have to be imported from the northeastern United States for the furniture industry in places like the Bois-Francs; the Eastern Townships and Ottawa Valley still have potential in quality hardwoods, but stocks are forecast to run out in the Laurentians by the year 2000. Added to this is the fact that more softwoods are growing on soils that used to support hardwoods. The Beauce is perhaps the most striking example of this.

In Québec's mixed and coniferous forests, depending on the extent of logging, original stands have been replaced by white birch and trembling aspen or very dense stands of balsam fir. Coniferous forests are slow to regenerate. As shown in Figure 3, cutting between 1974 and 1988 was mainly concentrated in Abitibi, northwest of Lake Saint-Jean, and in the Lower St. Lawrence (Fragnier 1992). Logging in the late 1980s mainly occurred in these last great reserves of mature softwoods, north of the 49th parallel and bordering the open woodland, where trees are of small diameter and widely spaced. Productivity here is relatively low, which leads to logging of larger areas.

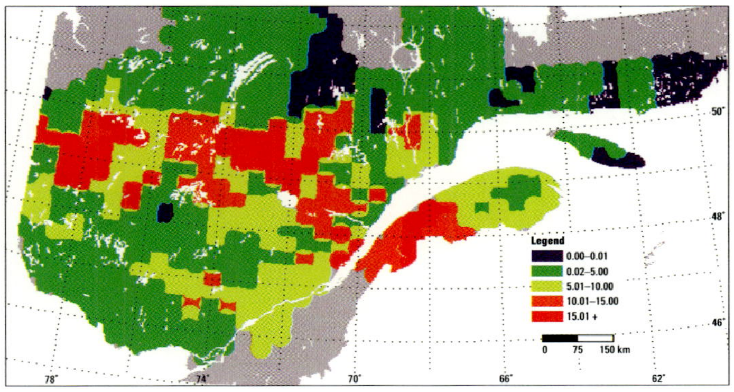

Figure 3
Areas logged between 1974 and 1988, expressed as a percentage (%) by survey unit
Source: Fragnier (1992), Canadian Wildlife Service, Environment Canada

Waterways and wetlands

The shorelines and wetlands of Québec have been subject to extensive disturbance. Very little of the shoreline in urban and suburban areas is still in its natural undisturbed state. Most marshes and swamps, long considered valueless, if not insalubrious or even noxious, have been filled or drained for farming, road construction, or residential housing. The demands of farming and industry have already resulted in the draining of a large proportion of peatland.

It is estimated that in 1966 only 20% of the original wetlands remained in the Montréal region (Kessel-Taylor 1984 *in* Champagne and Melançon 1985). The areas most affected by urban development have been the rich wet meadows. From 1945 to 1976, in the stretch between Cornwall and Matane, at least 3600 ha of wetlands were converted into farmland, roads, and harbour facilities (Lands Directorate 1986). In 1972, Reed and Smith estimated that 60% of the estuarine marshes had disappeared (Wildlife Habitat Canada 1986). Despite the much-reduced size of wetlands in the Montréal area, another 392 ha, or 6.7% of the remaining wetlands, were lost between 1966 and 1981—including 118 ha of peatland—which represented 11.3% of the 1966 wetland area. The most famous cases were the construction of the Dufferin-Montmorency expressway on the Beauport flats (1979), construction of a dock at Bécancour (1980), diking of part of the Kamouraska marshes (1980), diking of part of the Lake Saint-Pierre floodplain (1982), and construction of the deep water harbour at Cacouna (1965–82). The loss of these southern Québec wetlands is cause for serious concern, given the rarity and biological importance of such habitat and the fact that several species of ducks and shorebirds depend on these wetlands for food during the breeding season or in migration.

Aerial view of the marshes along the Verdun shoreline. The Aqueduct Canal outlet was in this area, near Marguerite Bourgeois Park. This was once a prime site for aquatic birds, but today it has disappeared. In the background is Mount Royal, which still serves as a migratory stopover for many passerines. *September 1923, National Air Photo Library, Natural Resources Canada.*

Landsat image of the eastern Ottawa Valley, southern Laurentians, and Montréal area

The complex topography of the Canadian Shield, wrinkled and indented by intense folding and faulting, contrasts with the uniform pattern of farm fields in the Central St. Lawrence Lowland. Just south of Montréal is the Châteauguay Valley, which has some of the best farmland in Québec. The mature deciduous forests in the Ottawa Valley north of Hull, the result of extensive logging at the turn of the century, appear in light green. The Rivière Rouge, and its sandy river valley, meander through the centre of the image. Northwest of Montréal, the runways of Mirabel Airport show up clearly in pink.

The image gives a good idea of the extent of Montréal's urban sprawl. In the centre of the island, tree-covered Mount Royal (Summit Park, Mount Royal Park, and the various cemeteries) appears in dark green, as do some of the well-treed surrounding municipalities and neighbourhoods. The Lachine and Vaudreuil-Soulanges canals cut grooves through the landscape.

In places where the forests have been razed for urban or agricultural development, the forest birds have been replaced by birds of open habitat. The increased diversity of species has not been without consequences, however; the remaining forest species found in the islands of woodland become more vulnerable to predation and parasitism, which further reduce their numbers.

The scattered clouds cast a few shadows on what must be a beautiful summer day in the Laurentians north of Montréal.

Image from orbit 15, swath 28, taken on 27-07-1990, courtesy of the Canada Centre for Remote Sensing, Geomatics Canada. Approximate scale 1:1 060 000.

Aerial view of part of Kamouraska Bay behind one of the Saint-Denis–Saint-Louis dikes showing the small dike and fields in the upper level of the salt marsh. *October 1979, Ministère du Loisir, de la Chasse et de la Pêche.*

In centuries past, rivers were open sewers and everything was thrown into them, much of it, such as household garbage, corpses, and carcasses, ending up in the marshes. Water, and especially stagnant water, was the source of many diseases and epidemics. Despite decades of public hygiene, in the public imagination marshes have remained ingalubrious or, at best, useless. Through ignorance, man sought to improve upon nature by converting marshland to other uses. Québec was no exception. Several major Québec marshes became expressways, like the La Prairie marshes at Boucherville, near Montréal, or the Beauport flats seen in this photo, taken at low tide. Near Québec City. *September 1978, Jacques Rosa.*

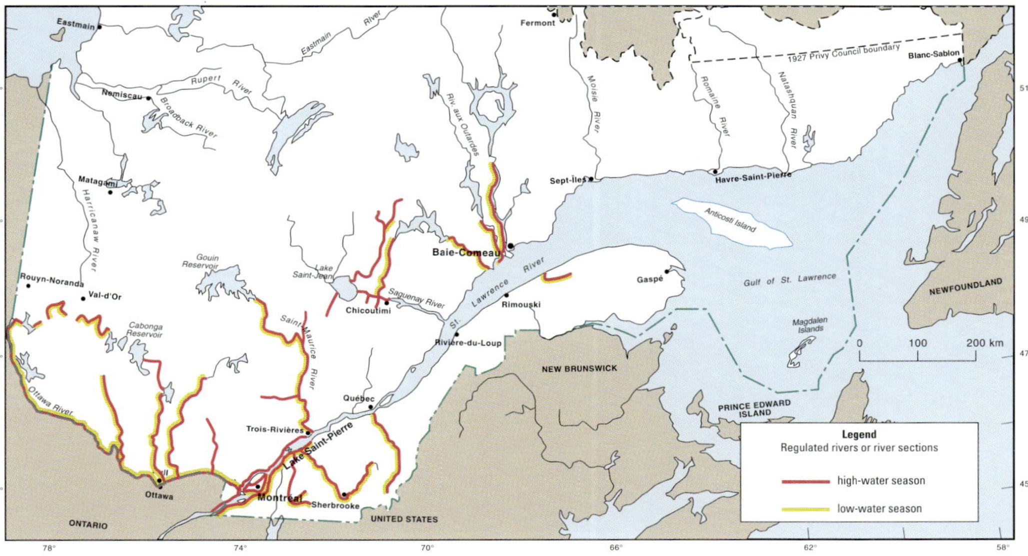

Figure 4

Regulated rivers in southern Québec

Source: Ministère de l'Environnement du Québec, Direction du milieu hydrique 1987, *in* Ministère de l'Environnement du Québec 1988, L'environnement au Québec : Un premier bilan : Document technique

In Québec, a large number of reservoirs have been created to generate electricity. Such reservoirs have altered the landscape of many regions. Dozois Reservoir. *May 1991, Yves Aubry.*

The flows of major tributaries in the St. Lawrence River drainage basin, as of other rivers elsewhere in Québec, have also been altered (Figure 4). Over time, the St. Lawrence basin has become highly regulated, with control structures to help prevent extreme fluctuations in water level (floods or low water), to assist in transporting timber, to generate hydro-electric power, and to control the St. Lawrence Seaway. The provincial government owns 220 dams, in addition to the 420 dams operated by Hydro-Québec (MENVIQ 1988). The reservoirs created by the dams cover 1.7% of the province, and a number of them are located in southern Québec, with 70% of them in the Ottawa–Gatineau, Saint-Maurice, and Saguenay River and Lake Saint-Jean watersheds (MENVIQ 1993). These include the Gouin Reservoir (1303 km^2), Pipmuacan Reservoir (780 km^2), and Lake Saint-Jean (1041 km^2), which is also a reservoir for generating hydro-electricity. The largest reservoirs in northern Québec are the Caniapiscau, with 4285 km^2 (Dugas et al. 1989), the Opinaca (1040 km^2), La Grande-2 (2836 km^2), and La Grande-3 (2420 km^2).

The provincial government also maintains about 100 small dams, most of which were built in the 1800s to help float logs downriver, in parks and wildlife reserves, for the purpose of preserving the wildlife habitats and recreational sites that have developed in these man-made habitats (MENVIQ 1988). In all, the Ministère de l'Environnement du Québec has counted over 9450 structures, mainly in southern Québec (Sarrazin et al. 1983), which gives some idea of the highly regulated nature of Québec's lakes and rivers. In 1980, no fewer than 3595 dams were counted in the Saint-Maurice River watershed alone. Most were small, privately owned, and are now used mainly for recreation and fish farming.

All these structures have altered the landscape by enlarging bodies of water or, in some river reaches, by reducing floodplains. Here and there, small swamps have dried up or been flooded. This effect is particularly marked in reservoirs used to generate electricity, especially when the drawdown zone is large and the cycle of variation in water level is reversed. On La Grande Rivière, shoreline vegetation has difficulty establishing itself at the maximum drawdown elevation, since that water level is reached only irregularly during the growing season (P. Lamothe and R. Denis, Hydro-Québec, pers. comm.).

This impoverishment of shoreline plant communities naturally leads to a parallel reduction in resident wildlife (Sarrazin et al. 1983). However, it is estimated that nest loss by early-nesting waterfowl in plains that would have once flooded has decreased. In southern Québec, flow regulation on the main rivers in the St. Lawrence basin has undoubtedly contributed to the increasing rarity of several marsh birds, four of which—the Least Bittern, Yellow Rail, Sedge Wren, and Sharp-tailed Sparrow—are on the list of species likely to be designated threatened or vulnerable in Québec (MLCP 1992).

MIREILLE DESPONTS

Landsat image of the Central St. Lawrence Lowland between Valleyfield and La Pérade
Here, in the heart of the St. Lawrence Lowlands, we get a good picture of the typical layout of agricultural lots in the region perpendicular to the nearest river or stream. The image also shows the extent of forest fragmentation in southwestern Québec, which affects species that require extensive tracts of woodland. East of the Chambly Basin, in the centre of the image, are the Monteregian Hills, which appear in relief despite the small scale. As we can see, these islands of woodland in the agricultural landscape provide crucial habitat for various species of breeding birds and are used as staging areas during migration. The reputation of birding hotspots such as Mount Royal and Mount Saint-Hilaire is firmly established.
The Berthier–Sorel Islands form a delta in mid-course along the St. Lawrence River. The myriad islands and islets are the result of sediments deposited by the St. Lawrence before it broadens into Lake Saint-Pierre. On the islands, flourishing stands of willows and silver maples grow. In spring and summer, these floodplain swamps are sought out by several species of ducks and other birds. The Richelieu, Yamaska, and Saint-François rivers also converge here. In the lower right-hand corner of the image, the Sutton Mountains and the peaks of the Eastern Townships frame Lake Memphrémagog and Brome Lake. The peatlands in Lotbinière County (south of the St. Lawrence; top right of photo) and in the Mauricie (north of Trois-Rivières) show up in dark green on the map, since they are forested almost exclusively with conifers. These isolated stands of conifers attract species of the northern forests.

Image from orbit 14, swath 28, taken on 21-08-1990, courtesy of the Canada Centre for Remote Sensing, Geomatics Canada. Approximate scale 1:1 060 000.

Changing activity, changing landscape

Over the past several centuries, logging and farming have greatly altered the landscape of southern Québec. We must look to the past to see how and in what circumstances this environment was shaped by its inhabitants. This brief history highlights the phases of development and natural resource exploitation that have made Québec what it is today. Past actions are hard to quantify and are often reported in anecdotal form only, yet such material can reveal much about the methods and intensity of environmental exploitation, as well as the major impact it has had on plants, wildlife, and human settlement.

The first inhabitants

The first traces of human occupation in northeastern North America date back about 10 000 years, in the areas south of Lake Erie and Georgian Bay (McGhee 1989), the end of a glacial epoch that covered most of Canada with ice. It is thought that about 15 000 to 20 000 years ago, lower sea levels allowed bands of hunters from Asia to cross the Bering Strait in the extreme northwest corner of the continent. Despite the ice caps, these Paleoindians managed to migrate south and spread along the southern margin of the ice sheet. In a land of steppe and tundra, they hunted the large herbivores such as caribou, giant bison, horse, camel, and mammoth. About 10 000 years ago they moved north as the glaciers retreated. This was a period of major warming, marked by the disappearance of all the large herbivores except the caribou. Since these species had already survived several glacial epochs, their first contact with humans was probably decisive. Their disappearance likely marks the first major human impact on the environment in this region (McGhee 1989).

Be that as it may, over subsequent millennia the landscape gradually evolved into the one Europeans saw when they first arrived. Aboriginal hunting communities adapted to their environments and developed unique cultures that led to the formation of today's nations. Along the St. Lawrence Valley, traces of occupation date back about 5000 years, particularly in the Montréal and Cap-Tourmente areas. When Jacques Cartier arrived, two very distinct societies inhabited the shores of the St. Lawrence River (Douville and Casanova 1982). At Tadoussac, Cartier met the Algonquins, the name given to all the nomadic nations that lived by hunting and fishing: the Montagnais, Micmac, Abenaki, Cree, Naskapi, and several other groups. To the south lived the Huron-Iroquois, sedentary and largely agricultural communities.

Cartier recounts his welcome in 1535 at Hochelaga by over 1000 Iroquois and describes his approach to the village through a forest of oak and fields of corn stretching as far as the eye could see (McGhee 1989). A century later, missionaries noted the presence of many villages of 1000 to 2000 inhabitants, mainly concentrated in Huronia, between Georgian Bay and Lake Simcoe, and mentioned that numerous clearings had been opened in the forest. These aboriginal peoples practised slash and burn agriculture, moving every 10–20 years when the soil became exhausted. They had cultivated corn for 500–1000 years and squash and beans since about 1400 (McGhee 1989). It is thus possible that birds associated with farmland habitat were already established in Québec by then.

The French Regime

The first attempt at permanent settlement by Europeans in the St. Lawrence Valley was in 1608, with the establishment of a fur trading post at Québec City, followed a few decades later by others at Trois-Rivières (1634) and Montréal (1642). Despite some efforts on the part of authorities to promote colonization, the first arrivals were mainly interested in the lucrative fur trade (Douville and Casanova 1982). Within a few decades, the French had dispersed across the continent, mainly in search of new sources of pelts, since beaver populations near trading posts had been decimated. By 1723, one had to travel 600–800 leagues to hunt fur-bearing animals (Martin 1990). This was to become typical of the situation with beaver all across the continent. The beaver population, estimated to have been somewhere between 60 and 400 million before the arrival of Europeans, was practically extinct by about 1900. The beaver population is now estimated at 6–12 million (Naiman et al. 1986). The increasing scarcity of beaver no doubt affected several bird species, such as the Wood Duck and Great Blue Heron, that make use of beaver ponds (Drapeau 1981; Pelletier and Scherrer 1983). Flooded trees in beaver ponds die, and the holes pecked in them by woodpeckers are used subsequently by cavity nesting birds.

During the French Regime, little land was cleared except near settled areas. By the British Conquest, in 1759, New France had only about 70 000 inhabitants, and a good many of them were not very interested in farming, according to Governor Murray in 1762. It was in the 1700s, however, that the area between Montréal and Québec City first became settled, and new seigneuries were established in the Beauce, Charlevoix, and Côte-du-Sud areas (Mathieu and Courville 1987; Figure 5). It is estimated that in 1765, less than 100 000 ha of land was under cultivation (Courville 1980).

The forest in the St. Lawrence Valley was partially cleared during the French Regime, but the greatest impact of colonization was on wildlife. Settlers soon became accustomed to obtaining fresh meat by hunting local game, a right recognized from the early days of the colony (Martin 1990). Large game such as moose, caribou, wapiti, and white-tailed deer soon disappeared from the vicinity of settlements. Pierre Boucher commented, in 1663, that one had to go 10–12 leagues from Québec City to find game (Guay 1983). In the latter half of the 17th century, the settler in the St. Lawrence Valley would mainly become a hunter of small game, taking advantage of the annual migrations of geese, ducks, Passenger Pigeons, woodcock, and so forth. In the spring, settlers harvested the eggs of dabbling ducks, eiders, guillemots, Razorbills, and gulls (Martin 1990). In the 1600s, hunters hunted on the St. Lawrence islands and at the mouth of the Saint-Charles River. By 1627, Robert Giffard, seigneur of Beauport, already had a blind on the marsh called "La Canardière" (duck ponds). The early explorers habitually stopped in the gulf to fill their holds with birds and eggs. Cartier recounts how in 1534 at Bird Rocks on the Magdalen Islands birds were so plentiful they might have loaded 30 longboats in just one hour (Biggar 1924).

In addition to local hunting, the *coureurs de bois* and Indians continued to trade beaver pelts, of course, as well as any larger game they caught. Exporting beaver pelts remained the colony's main economic activity until the early 19th century. Figures from Guay (1983) for 1786 give some idea of the scale of the fur trade. That year, the pelts of 116 623 beaver, 48 436 marten, 202 719 muskrat, and 108 521 raccoons, among others, were shipped out of the Port of Québec. Revealingly, in 1674 Frontenac issued an edict proclaiming moose pelts legal tender, to compensate for the lack of coin currency. Hunting pressure was so strong that species began to sharply decline or disappear, to the point where aboriginal peoples still living a traditional way of life in the 19th century were affected by the scarcity of game and had to turn to fishing, small game hunting, and long seasonal treks to survive (Cole Harris 1987).

Early logging operations

The British Conquest marks the real beginning of land clearing and settlement in the St. Lawrence Valley. Having become subjects of the British Empire, and cut off from their traditional economic networks, a large number of "Canadiens" (as French Canadians called themselves) took up farming. Until the early 1800s, timber was cut mainly to meet local requirements for building materials and firewood. However, some choice Eastern white pine and oak was cut for the British navy after trade broke off with New England during the American Revolution (Johnstone 1991).

Pine began to be logged on a massive scale after 1806, when Napoleon blockaded the Baltic ports to prevent trade between them and the British Isles, and the Royal Navy, deprived of timber from the Baltic countries, turned to the British colonies for the giant pines it needed for its masts and spars. That same year, the first timber rafting operation, led by the American immigrant Philemon Wright, set out from the mouth of the Gatineau River and descended the Ottawa and St. Lawrence rivers to Québec City (Dunn 1975). This event marked the beginning of wholesale logging of the forest. According to Dunn, the timber rafts could contain anywhere from 1300 to 5400 squared pine logs, and sometimes as many as 300 such rafts arrived at Québec City each year. The logging industry grew rapidly, largely due to a series of measures adopted by Britain between 1805 and 1814 to open the British market to Canadian timber, which had previously been too remote to be competitive. The measures remained in force until 1842, and when they were completely abolished in 1846 timber exports to England declined sharply. However, about that time the British market was replaced by the American market, where demand was flourishing for lumber to build its towns. Over time, sawtimber gradually came to replace square timber. For instance, it is estimated that between 1826 and 1881, the upper Ottawa River in Québec (west of the Rivière du Lièvre) supplied over 4 million logs of square pine and 19 500 000 of sawtimber pine (Hardy et al. 1980; Gaudreau 1986). In the first half of the 19th century, the Port of Québec was the shipping centre for the Atlantic, and shipyards were eventually built there, greatly contributing to the city's growth (Ouellet 1971).

Coniferous forest
Forest regenerating after logging
Recent clearcut
Peatland
Banks of reservoirs above water

Landsat image of the Laurentian Shield between the Rivière du Lièvre and the Saint-Maurice River
In the lower part of the image, north of Taureau Reservoir, the mixed forest begins to give way to coniferous forest as we move farther north on the Laurentian Shield. This image was obtained in early autumn and water levels are low, baring the banks of several reservoirs dammed for hydroelectric power. The exposed shoreline appears in pink. Huge areas of the forest along the upper Saint-Maurice River, at the top of the image, have been recently clearcut. On the right, the industrial portions of the Saint-Maurice Valley near La Tuque appear in purple, while nearby agricultural areas along the Saint-Maurice and Croche rivers appear in pink. The rights-of-way for the transmission lines from the La Grande Complex can also be seen clearly in the lower left of the image.
This region, with its extensive open habitat and large reservoirs, is transitional in nature, and attracts species that take advantage of a wide range of habitats.

Image from orbit 15, swath 27, taken on 05-09-1987, courtesy of the Canada Centre for Remote Sensing, Geomatics Canada. Approximate scale 1:1 060 000.

Logging and regional development

With the expansion of the sawtimber industry, logging quickly moved beyond the St. Lawrence Valley. First, the sites that had the most sought-after species and were most accessible by water were logged, along the St. Lawrence River, the Richelieu River to Lake Champlain, and in the Ottawa Valley. The cutting was highly selective: loggers took the finest specimens closest to the riverbanks. Logging operations continually relocated to new camps along rivers and major tributaries, wherever the flow was sufficient to float logs. Many river basins were logged in the early 1800s. Settlement followed the opening up of the outlying regions by logging, spreading into the Mauricie, Saguenay–Lac-Saint-Jean, Gaspé, and finally Abitibi–Témiscamingue. From the 1830s, a resource-based economy centred on farming and forestry developed, typical of this period of Québec settlement (Séguin 1977, 1980; Hardy and Séguin 1983; Gélinas 1984; Jean 1985).

Several factors played a role in the wave of settlement that started in the 1830s. Several decades of relative prosperity after the British Conquest were followed by several years of famine: agriculture was in crisis (Ouellet 1971). The main causes were the rapid population increase and the growing scarcity of land, especially since some seigneurs were unwilling to grant new lots because of higher land values. Numbering 60 000 in 1760, the population had grown to 1 000 000 a hundred years later (Mathieu and Courville 1987). Until 1820–30, farmers had been restricted to the old seigneurial parishes of the French Regime, but these were now becoming overpopulated. Some inhabitants, mightily encouraged in their endeavours by the clergy and seeking new opportunities for the younger generation, set

Logging in the mid-1900s had major repercussions on the environment. Timber covered the entire surface of some waterways, ready for transport to the mill. Many waterways were diked to ensure there was enough flow to raft timber. Saint-Maurice River. *September 1927, National Air Photo Library, Natural Resources Canada.*

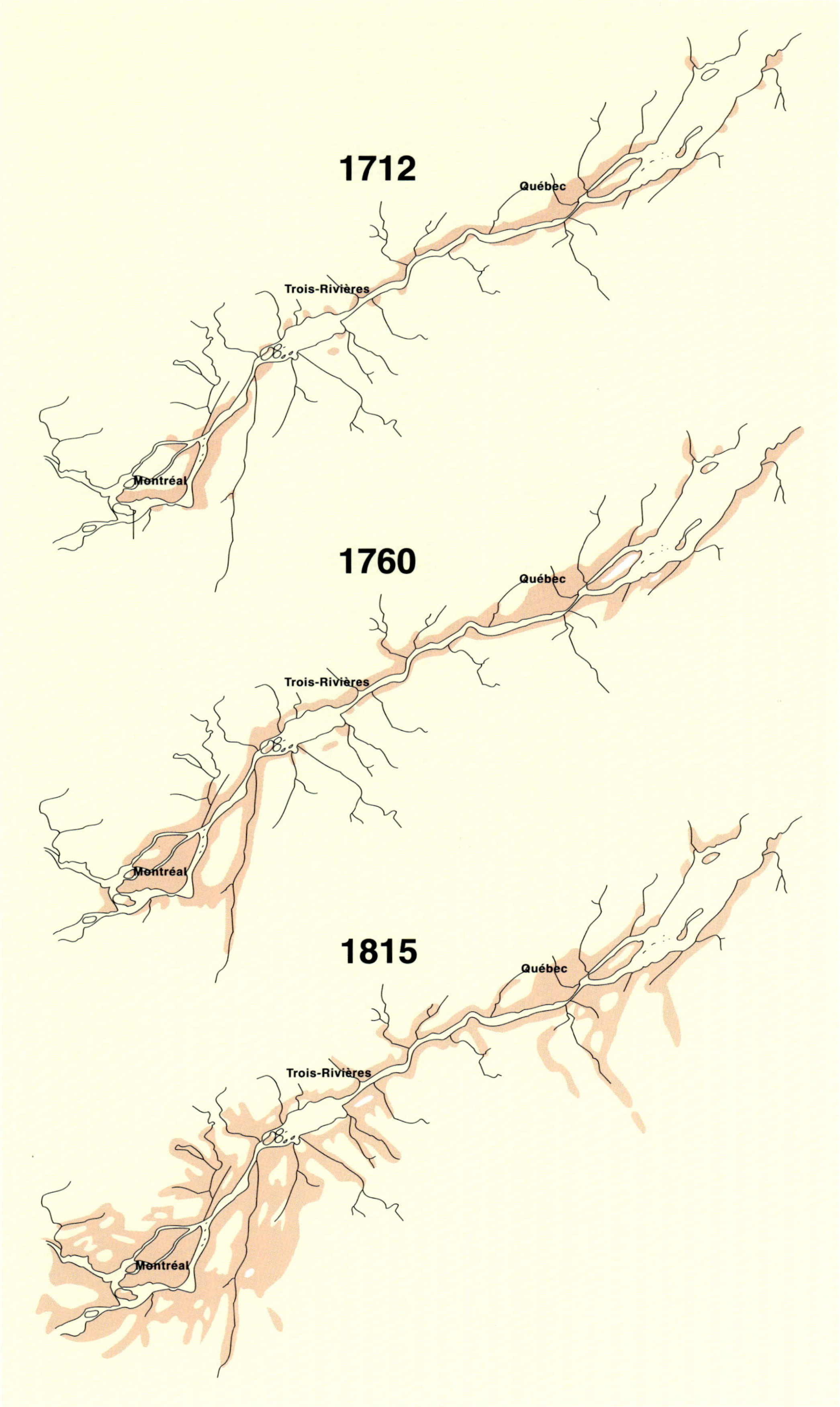

Figure 5
Settlement in the St. Lawrence Valley, 1712, 1760, and 1815

Until the early 19th century, settlement was limited to the shores of the St. Lawrence and the lower reaches of its tributaries. Settlers only ventured into the hinterland to hunt, and the landscape was yet unmarked by logging.

Source: Historical Atlas of Canada, Volume I. From the Beginning to 1800. R. Cole Harris (ed.). University of Toronto Press, 1987

out to found new parishes rather than abandon their country for the United States. Indicative of the prevailing attitude was the watchword of the St. Jean Baptiste Society, founded in 1834: "*Emparons-nous du sol*," which roughly translates as "We shall claim the land."

The 1820s and 1830s saw settlement reach into the upper Beauce, Bois-Francs, and the Laurentians north of Montréal, expand into the Mauricie and Saguenay–Lac-Saint-Jean regions, driven by logging in the 1840s and 1850s, then on into the Témiscouata and Gaspé regions in the 1860s. The Eastern Townships had been settled by the United Empire Loyalists starting in 1790. By the end of the 19th century, there were a number of settlements in the foothills of the Laurentians. In 1844, more than 900 000 ha of land was under cultivation; by 1861 this had risen to 1 600 000 ha (Courville 1980). The expansion occurred along the valleys of Québec's main waterways, partly because settlement followed on logging, but also because the best land was along the rivers, and the rivers were, until the late 19th century, almost the only means of communication and trade with the outside world. Roads, when there were any, were generally impassable, even with the sturdiest wagon.

The move to settle new areas was spontaneous and largely unplanned. The government did not take over from the clergy and local elites until 1888, when it set up the Department of Colonization. Symbolizing this transition, the Deputy Minister appointed was the "King of the North," Curé Labelle, one of the main proponents of new settlement. Initially, a complementary relationship was established between settlers and timber merchants, with the former providing labour and surplus produce to local logging operations, while the latter gave settlers the opportunity to earn additional seasonal income that was essential to their survival and establishment. Villages grew up around sawmills, but tension soon arose between the two partners. On the one hand, the seasonal work and continual relocation of logging camps led settlers to practise rather archaic subsistence farming; this limited the development and organization of parishes, which the clergy denounced on more than one occasion (Séguin 1977). On the other, the timber merchants feared the encroachment of agriculture and demanded the creation of forest reserves (Gélinas 1984).

From the 1850s onward, the forest was under attack. Loggers were accused of wasting timber, since they took only part of the trunk and left 30% of the tree behind (Gélinas 1984). Also, no steps were taken to preserve the seedlings that would ensure regeneration. The settlers, for their part, sold the finest specimens and burnt the rest. It was reported that settlers set fire to entire hardwood forests to obtain potash to sell as a bleaching agent to textile mills. It was also common, in the 19th century, for hemlock bark to be used in tanning leather, and it was even exported to the United States. Logging of hemlock was so extensive that it almost disappeared from Québec and Ontario (Perry 1986).

More serious still was the phenomenon of "forest pillagers," men who bought non-arable land at low prices, logged it out and then disappeared, only to return again later under another name to repeat the cycle. The government was unable to exercise adequate control over clearing and logging, so that anarchy reigned. Some settlers, strengthened by the common sentiment that clearing land was above all a patriotic act, illegally settled on land ceded to logging companies, a practice promoted by Curé Labelle. In fact, the settlers who moved to the upper Laurentians in the 1870s and 1880s were occupying land situated in forest reserves (Laurin 1989). Logging company monopolies over this region were openly denounced by Curé Labelle, who wasted no time, once appointed Deputy Minister, in trying to have their monopolies repealed. Caught between the settlers and the timber merchants, who paid considerable sums—for that time—into government coffers, the government chose the path of least resistance. Nothing was done for decades. Knowing that no action would be taken against the squatters and fearing the gradual invasion of their domain, the timber merchants responded by stripping lots of all valuable timber as quickly as possible; they then applied for new concessions.

Coniferous forest | Recent clearcut
Mixed forest | Agriculture
Deciduous forest | Built-up area
Ploughed field | Regenerating forest

Landsat image of the upper Saint-Maurice Valley and the Laurentide Wildlife Reserve
While mixed forests—and south of the Wayagamac lakes, deciduous forests—can be found at lower elevations on both sides of the Saint-Maurice Valley, the summits of the mountains in the Laurentide Wildlife Reserve are covered by dense coniferous forest, which appears very dark green in the image. Between the Reserve and the Saint-Maurice Valley is mixed forest dominated by conifers (mainly balsam fir-white birch), shown in dark olive green. The Jacques Cartier River Valley appears as a light green strip, indicating a mature deciduous forest. The large clearcuts around the lake of the same name (pink surrounded by dark green) are also very apparent, as is the extensive network of logging roads branching off from Route 175.
In this heterogeneous landscape, the composition of avian communities varies depending on whether they occur on the conifer-covered summits or in the deep valleys with their deciduous trees. Each enclave is associated with a specific habitat, resulting in a checkerboard of habitats and equally diverse avifauna.

Image from orbit 14, swath 27, taken on 10-06-1987, courtesy of the Canada Centre for Remote Sensing, Geomatics Canada. Approximate scale 1:1 060 000.

In the 1880s, Ontario and Québec shorelines produced an average of 64 414 square pine and 3 807 800 sawtimber pine logs a year. On an annual basis, Québec as a whole exported about 3 500 000 pine and spruce logs (Gaudreau 1986). In 1901, the grand total was 3 333 633 pine logs and 7 186 041 white spruce logs (Séguin 1977). By 1890, the Saguenay and Mauricie regions produced more spruce than pine, as the latter was becoming scarce. Oak had already been logged out for several decades (Ouellet 1971).

By the late 1800s, the extent of the destruction was beginning to arouse concern. Sir John A. Macdonald, Prime Minister of Canada, had watched the immense rafts float down the Ottawa River; in 1871 he wrote: "We are recklessly destroying the timber of Canada, and there is scarcely a possibility of replacing it" (Johnstone 1991). In 1889, the pamphleteer Arthur Buies, secretary to Curé Labelle, also eloquently denounced the situation: "Until now, many lumber merchants have practised not intelligent industry, but devastation of the forests, veritable pillaging in blind, brutal, furious extermination of some of the most beautiful species of wood in the world" (National Museum of Man 1983). In 1889, the former premier of Québec, Henri-Gustave Joly de Lotbinière, at the request of then premier Honoré Mercier, attended the American Forestry Congress. In his report he emphasized the disastrous results of wanton cutting practices. A man ahead of his time, Joly de Lotbinière insisted on the importance of learning from past mistakes and applying a more rational approach to logging of the remaining forests in the Crown domain. He also discussed natural resource concepts that were new to the time, such as preserving the common heritage and responsibility toward future generations (Martin 1990). His ideas were taken up again when forest reserves were created, but saw limited application in the 20th century.

It was not only the forest that was wantonly exploited in this period. The arrival of thousands of settlers and loggers brought increased hunting pressure on wildlife in general, and on deer, moose, and elk in particular. Government guides published for new settlers suggested that they could economize by hunting game and keep their cost of living down (Martin 1990). Logging camps supplied themselves more or less openly with local game. In the 1870s, the railroad began to reach settled areas and logging camps, increasing pressure on resources by making more land accessible to more people (Martin 1990). The behaviour of the hunters, settlers, and loggers of that era was based on the myth—still strong today—that these resources were inexhaustible. Hunting occurred all year round, with no bag limits.

The "exploits" of hunters were recounted in newspapers, and this helped attract hunters from afar, on the promise of miraculous bags. One report detailed a drive organized in 1856 in the Eastern Townships to "protect agriculture." In one week, 150 hunters killed 82 844 animals and birds, including 100 foxes, 75 hawks, 73 770 squirrels and chipmunks, 300 owls, and so forth. Similarly, in 1861, a certain De Niverville took 120 ducks and 360 snipe, plovers, and other birds at Lake Saint-Pierre. That same year, 50 moose trapped in deep snow were killed near Joly in Lotbinière. Three years later, hunters shot 480 ducks on a trip along the Ottawa River (Guay 1983). Reports like these were not rare.

However, by the second half of the 19th century, concerns were being raised about the declining numbers of migratory birds and the dearth of large game in various areas. This was when the Labrador Duck became extinct; the last specimen was reported in 1878. Warnings about the possible extinction of moose, decimated by unrelenting hunting, were issued in 1857–59 by the Montreal

View of the Lachine Rapids in the foreground, with part of Île aux Hérons on the right and the run-of-river power plant. This part of the Island of Montréal was quite heavily wooded. Mount Royal is in the background. *September 1923, National Air Photo Library, Natural Resources Canada.*

Natural History Society, which led to the hiring of the province's first game wardens in 1867 (Martin 1990). The moose was finally saved thanks to protection introduced under the Game Act of 1884. One of the last wapiti was killed in 1899, near the Causapscal River. Passenger Pigeons, whose primary range was in the United States and which had been under relentless attack since the early days of the colony because the birds invaded croplands by the thousands, were finally exterminated in the early 1900s. In fact, as early as 1686, Monseigneur de Laval, the Bishop of Québec, had tried to exorcise the Passenger Pigeon with bell, book, and candle and liberal asperging with holy water in an attempt to protect crops from these "legions of the devil" (Guay 1983). The clouds of Passenger Pigeons that filled the skies nevertheless represented manna from heaven for the settlers, who consumed large quantities of them. The last individual died in captivity in 1914. The list lengthens when one adds the Great Auk, bobcat, caribou, Snow Bunting, and a number of other species. In 1900, it was estimated that the total population of Greater Snow Geese was only 2000–3000.

A series of regulations establishing a closed season for hunting and setting bag limits helped at least partially reestablish populations in this period. In 1845, an act was passed prohibiting the killing of wildfowl from May to August. This law also made it illegal to destroy the eggs of partridge, duck, or snipe anywhere in the territory. Egg harvesting, which had always been practised by aboriginals and Europeans, had taken on alarming proportions, especially in the gulf, where many ships stopped to shamelessly pillage eggs from seabird colonies. The prohibition against gathering or even touching eggs was reiterated in 1876 in the Act to Amend and Consolidate the Game Laws in the Province of Québec. Despite these regulations, it is estimated that in 1891, 2 500 000 eggs a year were gathered by smugglers on the North Shore of the gulf and sold in the Boston and New York markets (Guay 1983). In 1884, legislation was passed to protect insectivorous and other birds useful to agriculture, the only way known at the time to reduce losses caused by insects and vermin. Raptors, however, were excluded from protection because they were considered dangerous to poultry and were killed by the thousands.

The pulp and paper industry

The late 19th century marked a turning point in the Québec forest industry. Eastern white pine had become rare by around 1880 and had to be replaced with white spruce (Hardy et al. 1980). In the Mauricie, sites were logged as far upriver as the headwaters of the Saint-Maurice River tributaries. Moreover, starting in 1875–76, Canadian forest products had to compete with American timber producers in a depressed market. Several logging companies and sawmills declared bankruptcy (Perry 1986). However, in the same period, it was realized that spruce and fir were ideal for making paper. With the economic recovery of the early 1900s, logging operations in Québec took a new direction.

In the decade from 1860 to 1870, the technique of making paper from wood pulp became more common, and American demand for newsprint shot to new heights. The first mill to produce wood pulp by mechanical means opened in Valleyfield in 1866, and by 1881 there were four more (Perry 1986). The pulp and paper industry's period of greatest expansion was from 1908 to 1920, when Québec production increased fivefold. This rapid growth was largely due to adoption of a process for making kraft paper at East Angus, in 1907, and the fact that the United States abolished customs duty on newsprint in 1911 (Perry 1986; Charland 1990). In addition, Québec was able to provide an abundance of fresh water, electricity, and above all fir and spruce logs. These species, considered of no commercial value just a few years before, now became highly prized. The pulp and paper industry infused new life into the forest industry, giving rise to extensive logging of spruce and fir forests in regions ever-farther removed from the St. Lawrence Valley. In the late 1800s, the *Progrès du Saguenay* had proclaimed that the forests could meet the demand for timber for the next 2000 years (Charland 1990). The lumber industry is still alive today, but was supplanted, in the early 1900s, by the pulp and paper industry, which had a voracious appetite. In 1929, pulp and paper companies held 91% of the province's total timber licences.

Hydro-electricity

The hydro-electric era was inaugurated in Québec in 1883, when a small power plant was built in Valleyfield to supply power to the Montreal Cotton Mill. Apart from that, however, hydro-electric development in the late 1800s was limited to construction of a few power plants on the St. Lawrence (Lachine Rapids in 1895 and 1897), Ottawa (Chaudières Falls in 1889), and Saint-Maurice rivers (Shawinigan Falls in 1899). From 1900 to 1930 use of this new resource expanded rapidly, and generation rose from 61 816 kW to 2 027 725 kW (Bush 1986). Awareness grew of the enormous potential of the St. Lawrence River, whose flow was naturally regulated by the Great Lakes and whose tributaries often descended abruptly from the Laurentian Shield. Until 1944, power stations were almost all privately owned. In 1944, the Hydro-electric Commission of Québec (commonly called Hydro-Québec) was created, with a mandate to provide municipalities, businesses, and citizens with electricity at the lowest possible cost.

Landsat image of the Laurentian Mountains north of Québec City

At Québec City, the St. Lawrence River broadens to form the estuary. A number of islands used for agriculture can be found here: Île d'Orléans, the Montmagny Islands with its two main islands (Île aux Grues and Île aux Oies), Île aux Coudres, and others. A thick layer of fine sediment has accumulated around these islands, appearing blue in the image. Most of the farming areas in the L'Islet–Kamouraska region and in the valleys of the Charlevoix region show up in pink, since most of the fields had already been harvested when this image was obtained at the end of August. On the south shore of the St. Lawrence River, the contrasting colours allow the St. Lawrence Lowlands to be distinguished from the lower slopes of the Notre-Dame Mountains, covered with mixed forest. North of Québec City and just west of the river, a series of river valleys cut through the outer slopes of the Laurentians (Jacques Cartier, Montmorency, and Sainte-Anne rivers) and are very conspicuous in pale green. Landmarks closer to Québec City include, on the south shore of the river, the Grande Plée Bleue peatland south of Lévis (dark green) and, on the north shore, a thin line of green extending from the Plains of Abraham and the Bois-de-Coulonge Park west into the wooded residential areas of Sillery and Sainte-Foy.

In this region, too, the diverse landscape is matched by an equally diverse birdlife.

Image from orbit 13, swath 27, taken on 27-08-1989, courtesy of the Canada Centre for Remote Sensing, Geomatics Canada. Approximate scale 1:1 060 000.

Reservoir impoundment not only affects ecological balance, it also causes changes in regional infrastructure. Near Saint-Coeur-de-Marie, Lac-Saint-Jean. *October 1927, National Air Photo Library, Natural Resources Canada.*

Reservoir impoundment causes erosion. Woods along the shore disappear, as do the species that depend on them. Saint-Coeur-de-Marie, Lac-Saint-Jean. *October 1927, National Air Photo Library, Natural Resources Canada.*

Infrared aerial photograph colours show checkerboard clearcutting near Petite-Rivière-Saint-François. As compared to black and white photography, infrared photography produces images showing vegetation type and structure in greater detail. This zone is between the ski centre and Highway 138. Unlike regular clearcutting, this cut promotes habitat diversity and edge species (such as the Mourning Warbler, Chestnut-sided Warbler, Sharp-shinned Hawk, and snowshoe hare). Vegetation recovers more quickly, even in the centre of cut lots, because of the presence of seed trees. The openings allow hunting by a number of birds of prey and also attract the Brown-headed Cowbird. Checkerboard clearcutting is also visible in the satellite images of the Laurentians north of Québec City, in the small green and pink squares mid-way between Saint-Tite-des-Caps and Baie-Saint-Paul. *Québec photomap library, Ministère de l'Énergie et des Ressources du Québec, Photo Q90419–47, 01-08-1990.*

Construction of impoundment dams and reservoirs had been supervised by the Québec Streams Commission since 1910. Such structures became necessary on rivers with large seasonal flow variations. Thus, in 1921 the Gouin Reservoir was built, followed by the Kénogami Reservoir in 1925, and the Baskatong Reservoir in 1929. Projects on the Saguenay River started in 1923, and shortly thereafter the level of Lake Saint-Jean was raised 32 feet above low-water level (Charland 1990). In 1940, Québec had 17 reservoirs. When the government took control of hydro-electric developments in 1963, the energy potential of the south sector of the St. Lawrence basin was almost fully harnessed. In the 1960s, development pushed into the North Shore (Rivière-aux-Outardes and Manicouagan), and then on to mammoth projects at James Bay in the 1970s (La Grande Rivière and Caniapiscau River). In the 1970s and 1980s, the creation of five immense reservoirs in the La Grande Rivière region added a total of 11 355 km² to areas under water, doubling the extent to 21 280 km², or 1.4% of the area of Québec (MENVIQ 1988).

In the early 20th century, industrialization in the pulp and paper industry fuelled development of regional centres. This in turn sparked a restructuring in agriculture, as farms, especially ones near towns, tended to become more specialized so that they could supply urban markets. The regional dairy industry, for example, underwent remarkable expansion. In this period of prosperity there was less interest in clearing marginal lands than in intensifying cultivation of good arable land for maximum economic return (Séguin 1977; Jean 1985). Unproductive land was abandoned, though in this period that trend was limited. In the 1920s, the pace of settlement declined, particularly in the Saguenay–Lac-Saint-Jean, Beauce, and Laurentians (Séguin 1977; Laurin 1989; Bélanger et al. 1990). Subsistence farming continued in more remote areas until the early 1950s.

Coniferous forest | Deciduous forest | Forest defoliated by the spruce budworm | Recent clearcut | Agriculture | Worked peatland | Built-up area

Landsat image of the Notre-Dame Mountains

Unlike the Laurentians, where the logged areas are quite large but fewer in number, here the forests are logged in numerous small parcels, creating a checkerboard effect. Closer to the St. Lawrence, the strips of different shades of green running parallel to the Notre-Dame Mountains illustrate the succession of ridges and valleys typical of the Appalachian Mountains. The nearby islands in the St. Lawrence are continuations of the same ridge pattern (Les Pèlerins, Île aux Lièvres, Île aux Coudres, Île Verte, and Île aux Basques). The peatlands being worked near Rivière-du-Loup and on Île aux Coudres show up in pink. These peatlands are home to several species of birds that usually inhabit more northerly forests. In the upper left-hand corner, the Saguenay River flows into the St. Lawrence estuary at Tadoussac.

Image from orbit 12, swath 27, taken on 27-07-1988, courtesy of the Canada Centre for Remote Sensing, Geomatics Canada. Approximate scale 1:1 060 000.

In the early part of the century, many deciduous trees were left standing in cutovers, and little attention was paid to leaving buffer strips along the shoreline. Mauricie region.
October 1927, National Air Photo Library, Natural Resources Canada.

Port of Québec in April 1929. Previously intense harbour activity declined after the advent of rail transport and development of rail networks. The Canardière marsh, west of the Beauport flats, stretched to the mouth of the Saint-Charles River, where a paper mill was already located. Note how close the farms are to the city. *National Air Photo Library, Natural Resources Canada.*

By 1993, as the city had spread out, the meandering course of the Saint-Charles River had been straightened, and the Canardière marsh had become a port-industrial complex. Québec City. *October 1993, Pierre Normand.*

Coniferous forest
Mixed forest
Regenerating forest
Forest defoliated by the spruce budworm
Recent clearcut
Built-up area

Landsat image of the Upper North Shore
On the Upper North Shore, a number of points of land jut into the estuary; they are (upstream to downstream) Grandes-Bergeronnes Point (bottom of the image), Boisvert Point, Portneuf sandbank, Betsiamites Point (centre), the Manicouagan Peninsula, and Pointe des Monts (extreme right). Logging companies are active inland from Baie-Comeau and Forestville, and the forest in this area shows different stages of succession after logging: the most recent clearcuts are pink (forbs and shrubs), older clearcuts are in a greenish tan, and the oldest clearcuts are in light green (trembling aspen and birch). The transmission line rights-of-way from the Manic-1, Manic-2, and, to a lesser extent, Manic-5 hydro-electric complexes are clearly visible. These corridors allow some species that frequent open habitat to colonize the back country. Logging roads and borrow pits provide suitable habitat for Bank Swallows and other species, while logging camps, outfitting operations, and cottaging areas attract Barn Swallows. Logging brings about even more complex changes in avian communities.
Inland from Les Escoumins, large areas of forest appear as brownish ochre, indicating that they have been defoliated by the spruce budworm.
On the south shore, the lowlands narrow progressively and disappear almost altogether east of Rimouski.

Image from orbit 12, swath 26, taken on 27-07-1986, courtesy of the Canada Centre for Remote Sensing, Geomatics Canada. Approximate scale 1:1 060 000.

View of Chicoutimi (right) in October 1927. The landscape is dominated by farms. *National Air Photo Library, Natural Resources Canada.*

View of Québec City in 1929. In the foreground, the Plains of Abraham. Wooded areas were rather small at that time. *National Air Photo Library, Natural Resources Canada.*

Protected sites like the Plains of Abraham, which now has mature stands, support local flora and fauna and provide a stopover for migrating passerines. Québec City. *October 1993, Pierre Normand.*

Coniferous forest | Recent clearcut | Deciduous forest | Agriculture | Regenerating forest | Peatland | Blueberry field | Clouds

Landsat image of the Saguenay–Lac-Saint-Jean region
Although the colours in this image are different from those in the preceding ones, the image shows quite clearly the distinction between agricultural areas (in yellow and pink) and forest (different shades of green). South of Lake Saint-Jean, a fault separates the lake basin from the Laurentian Highlands, indicated by a clear demarcation between the yellow and green. These hills support the most northerly sugar maple stands in Québec. The milder microclimate in the Lake Saint-Jean basin and the resulting habitats, including openings created for agriculture, have allowed a number of species with a generally more southerly distribution to colonize the region. Along the same fault line is Lake Kénogami, a vestige of the end of the last Ice Age, when Lake Saint-Jean emptied through Métabetchouan into Baie des Ha! Ha!, now the head of the fjord. Blueberries are one of the region's main crops, and the extensive blueberry fields show up as orange areas north of Alma and south of Dolbeau–Mistassini. A succession of marsh and peatland, shown in light blue, occurs at the mouths of the Mistassini and Péribonka rivers and at Taillon Point. Along the Saguenay River, the cities of Chicoutimi and Jonquière are clearly visible in purplish tones, as is the city of La Baie.

Image from orbit 14, swath 26, taken on 21-08-1990, courtesy of the Canada Centre for Remote Sensing, Geomatics Canada. Approximate scale 1:1 060 000.

However, for Québec as a whole, settlement peaked only after the crash of 1929. Development of Témiscamingue, in the Ville-Marie region, started in 1881; settlement of Abitibi took place from 1911 to 1921, as the land was opened up by the Québec–Winnipeg rail line and construction of a depot at Amos (Boileau and Dumont 1979). Acreage under cultivation in Québec peaked at 2 431 676 ha in 1941, compared to 1 881 758 ha in 1901 (Jean 1985). As in the recession of the late 1800s, the lack of jobs in urban areas forced young people to stay in rural communities. It also led those who could not find jobs in the city to leave for settlement zones. In the 1930s, this expansion was channelled toward Abitibi–Témiscamingue and the Gaspé Peninsula, regions with extremely limited agricultural potential given their climates and soil quality, though it must have been known that the best land had been occupied since the late 19th century. The federal government's Gordon plan in 1932 and the provincial government's Vautrin plan in 1934 both encouraged the urban unemployed to relocate in new settlement areas. Between 1931 and 1941, the population of Témiscamingue doubled and that of Abitibi tripled (Boileau and Dumont 1979).

The late 19th century and first half of the 20th century were both a good and bad period for wildlife. Expansion of the pulp and paper industry was largely responsible for the transformation of the virgin forest into a second-growth forest. Extensive logging, creation of reservoirs and, of course, settlement that moved ever deeper into the interior resulted in massive habitat loss. Indeed, settlement of more remote areas intensified with expansion of the railroad, particularly after 1870, and then construction of the road network in the early decades of the 20th century. Every improvement in transportation brought in more hunters and encouraged a flourishing trade in game meat (Martin 1990). Caribou from the Gaspé were shipped to towns by the cartload, and the herd in Laurentide Provincial Park, estimated at 10 000 individuals in the early 1900s, was decimated in less than 15 years and finally vanished completely (Martin 1990). Waterfowl was also becoming more scarce as a result of the wanton hunting of previous decades. As early as 1845, a Québec Act to protect wildfowl mentioned that bird populations had declined so much that "their entire disappearance from the country [was] threatened, unless protected by some Legislative enactment" (Guay 1983).

In 1929, the countryside west of Québec City was mainly fields, with few wooded areas. *National Air Photo Library, Natural Resources Canada.*

Several decades later, the riverbanks have disappeared under an expressway, and urban woodland can be seen in the residential neighbourhoods of the towns of Sainte-Foy and Sillery. As in many places, urban trees are the same age as the neighbourhoods they occupy. *October 1993, Pierre Normand.*

Coniferous forest — Recent clearcut — Regenerating forest — Banks of reservoirs above water

Landsat image of the Manicouagan region
Three rivers in this region have been dammed to produce hydro-electric power: from west to east, the Outardes, Manicouagan, and Sainte-Marguerite rivers. The Manicouagan originates in the huge reservoir of the same name; its southern part can be seen in the top left-hand corner of the image. The banks of the reservoirs are above water (pink) due to low water levels caused by a drought in early summer. This region is also logged extensively, as can be seen by the large clearcuts in different stages of regeneration, appearing in shades of pink, orange, and light green. The orange area near the centre of the image is probably a burn, given its configuration and the fact that it is far from a logging road. Logging and burns have a negative effect on certain local bird communities, but a positive effect on others.

Image from orbit 12, swath 25, taken on 03-07-1989, courtesy of the Canada Centre for Remote Sensing, Geomatics Canada. Approximate scale 1:1 060 000.

This situation provoked a number of reactions from groups that supported hunting or conservation. Even though the government issued regulations and appointed game wardens in 1867, it could not control the entire province. One solution was provided by groups of wealthier hunters, who around 1880 started to set up private hunting clubs on leased private shooting reserves, in order to preserve game and hunt in a sportsmanlike way (Martin 1990). This was doubly advantageous for the government because it could simultaneously collect ground rents and hand surveillance of the reserves over to the game clubs.

In 1866, the Montreal Fish and Game Protection Club started demanding protection for waterfowl to forestall their mass destruction. Although first introduced in 1878, various attempts to impose hunting licences came to nought because all regulation was stubbornly opposed by many citizens, for whom hunting was a necessity and not a recreational activity (Martin 1990). On the other hand, powerful American pressure groups such as the Audubon Society, American Ornithologists' Union, and the Sierra Club, benefiting from the support of several presidents, including T. D. Roosevelt, fought for the creation of wildlife refuges in the United States. They also lobbied hard for the signing in 1916 of the Migratory Birds Convention between Canada and the United States (Elman 1977), which was followed by passage of the Migratory Birds Convention Act in Canada the following year. This bilateral agreement marked a turning point for waterfowl and birds. Threatened species received protection, and spring hunting was prohibited. The Convention also put an end to the centuries-old trade in duck meat and eggs. Breeding sites and staging areas were now protected.

View of the area west of Trois-Rivières in April 1929. Trees and hedgerows were rare because as much land as possible was being farmed. *National Air Photo Library, Natural Resources Canada.*

Trois-Rivières-Ouest, 64 years later. Urban expansion and port-related industrial activities have taken over the farmland. Abandoned land has reverted to forest. These changes in the landscape have also altered bird communities. *October 1993, Pierre Normand.*

Landsat image of the Gaspé Peninsula
The Shickshock Mountains, with an average elevation of 800 m, are covered with dense coniferous forest, which appear as a very dark green. The tallest summits, such as Mount Jacques Cartier and Mount Albert, have no trees at all and therefore are the same colour as recent clearcuts (pink), although their edges are more blurred and uneven. The tundra-like conditions found on the summits of some of these mountains support some species that normally frequent arctic tundra, such as the American Pipit. West of the town of Gaspé, in the York River Valley, the forest has been logged to salvage a burned area. The entire peninsula, in fact, is fragmented by clearcuts. The narrow strip of agricultural land around Chaleur Bay between Port-Daniel and Maria can also be seen.

Image from orbit 10, swath 26, taken on 14-08-1986, courtesy of the Canada Centre for Remote Sensing, Geomatics Canada. Approximate scale 1:1 060 000.

Concern about the disappearance of wildlife had already led Mr. Parent, the Minister of Lands, Forests and Fisheries, to organize a meeting in Montréal in 1900 of representatives from Ontario, New Brunswick, and several American states to study the possibility of co-ordinating laws on protection of game and consider how such laws could be improved (Guay 1983). The measures subsequently adopted contributed substantially to the survival of several species. However, poaching remained a major problem for a long time, with poachers often taking advantage of the network of logging roads to "clean out" areas that still had game. In 1935, 787 offenders were arrested for poaching, an indication of how prevalent it was. In a 1942 review of natural resource harvesting, it was estimated that wildlife resources had declined 70% between 1900 and 1939 (Martin 1990).

Urban development and farm abandonment

After World War II, agriculture all across Québec went through a major restructuring. The pace of industrialization and urban development accelerated, causing an exodus from rural communities. In addition, the demands of the market economy meant that farmers had to abandon subsistence agriculture. To be profitable and provide families with access to consumer goods, farms had to become mechanized. These socioeconomic changes resulted in massive abandonment of lands in outlying regions (outside the Central St. Lawrence Lowland) and the loss of good farmland to suburban expansion. Leaving the land was therefore not simply an issue of abandoning marginal land. On the other hand, good arable land was cultivated more intensively, and the average area of a farming operation more than doubled, rising from 16 ha to about 36 ha between 1941 and 1981 (Jean 1985). The subsistence farmer had become an agricultural producer.

View of Saint-Hubert Airport in 1930, in the zeppelin era. The landscape was largely dominated by farms. *August 1930, National Air Photo Library, Natural Resources Canada.*

Saint-Hubert Airport in October 1993. The city has expanded over much of the former farmland.
Pierre Normand.

Coniferous forest

Mixed forest

Deciduous forest

Forest defoliated by the spruce budworm

Forest defoliated by the hemlock looper (Anticosti Island)

Recent clearcuts and open lichen woodland (Anticosti Island)

Landsat image of the tip of the Gaspé Peninsula and part of Anticosti Island
In this image, the dark green of the white spruce forests on Anticosti Island contrasts with the lighter green of the mixed and deciduous forests at the tip of the Gaspé Peninsula. On Anticosti Island, the broad rivers flowing over the island's characteristic pale limestone substrate show up clearly in pink. These rivers have their headwaters in an area ravaged by an insect epidemic in 1971, leaving a forest of snags that stands out in pale ochre on the image.
The pinkish area at the very top of the image consists of open lichen woodland.
The eastern coast of the Gaspé Peninsula is characterized by a deeply incised shoreline with numerous points and barachois, the latter showing up in pink. A barachois is a lagoon partly obstructed by a sand spit. On the image, you can see Forillon Peninsula, the bays at the mouths of the Dartmouth and York rivers, the sand spits at the entrance to Gaspé harbour, the barachois at the mouths of the Saint-Jean and Malbaie rivers, Saint-Pierre Point, Percé, Bonaventure Island, Cap d'Espoir, and the Chandler barachois. Large areas of forest behind Gaspé defoliated by the spruce budworm appear in brown on the image.
The rocky coastline provides a haven for seabirds, while the sheltered barachois provide refuge for several species of ducks and migrating shorebirds. Several species of passerines flourish in the conifer-covered hills ravaged by the spruce budworm, where this plentiful insect food improves breeding success. The narrow belt of agricultural land along the coast attracts numerous species that frequent open habitat.
The horizontal line near the bottom of the image is an artefact.

Image from orbit 09, swath 26, taken on 22-07-1986, courtesy of the Canada Centre for Remote Sensing, Geomatics Canada. Approximate scale 1:1 060 000.

It was during this period that the Turkey Vulture became established in Québec (Chabot and Barrette 1989). The consolidation and larger size of livestock farms undoubtedly played a role in this development, since dead livestock are a major part of the Turkey Vulture's diet (Coleman and Fraser 1989).

In outlying areas, the farms abandoned were often on poor soils. In some regions the decrease in land under cultivation was dramatic, dropping in the Gaspé, for example, from 63 720 ha to 18 196 ha between 1941 and 1981. Also, many villages shut down altogether in this period, a phenomenon that came to be called the destructuring of rural communities (Jean 1985). In Abitibi–Témiscamingue, it is estimated that from 1956 to 1976, over 80% of farms were abandoned; in 1975 only about 10 000 ha was under cultivation out of 220 000 ha of cleared land (Boileau and Dumont 1979). Between 1941 and 1981, for Québec as a whole, farmland area dropped from 2 431 676 ha to 1 732 317 ha.

The decline of farming continued to have a major impact in the 1980s and 1990s. Data from the 1991 census show that improved land, or cropland, including land under forage crops or lying fallow, declined by at least 300 000 ha in 10 years (MENVIQ 1993). A full evaluation based on satellite images conducted by the Québec Department of Agriculture, Fisheries and Food (MAPAQ) showed the extent of abandonment province-wide: in 8 out of 12 farming regions, about 8% or 138 290 ha of farmland was idle in the period 1987–90. Furthermore, hedgerows between fields have been widely sacrificed to intensive farming practices, with consequent impact on certain bird species such as the Vesper Sparrow (Jobin et al. 1994).

Landsat image of the Mingan Islands and part of Anticosti Island

The colour composition on this satellite image differs from that used on the others. The TM3, TM5, and TM4 bands have been assigned to blue, green, and red respectively. This breakdown makes for a clearer representation of the forest.

In the middle of Anticosti Island, the various stages of regeneration following the 1952 fire can be seen. The still largely barren areas appear bluish, while those beginning to regenerate are pink. Cutovers growing back in white spruce are orange.

Image from orbit 09, swath 25, taken on 26-10-1986, courtesy of the Canada Centre for Remote Sensing, Geomatics Canada. Approximate scale 1:1 060 000.

View of the south shore of Montréal at Charron Island. Farmland dominates, with hedgerows between fields. In the lower left is a star-shaped airfield. *May 1930, National Air Photo Library, Natural Resources Canada.*

Aerial view taken in 1982 of Charron Island and the surrounding area, between Boucherville and Longueuil, several years after construction of the St. Lawrence Seaway and the Louis-Hippolyte LaFontaine Bridge-Tunnel. Farmland has largely given way to urban development. *October 1982, Service de la cartographie, Ministère des Ressources naturelles du Québec.*

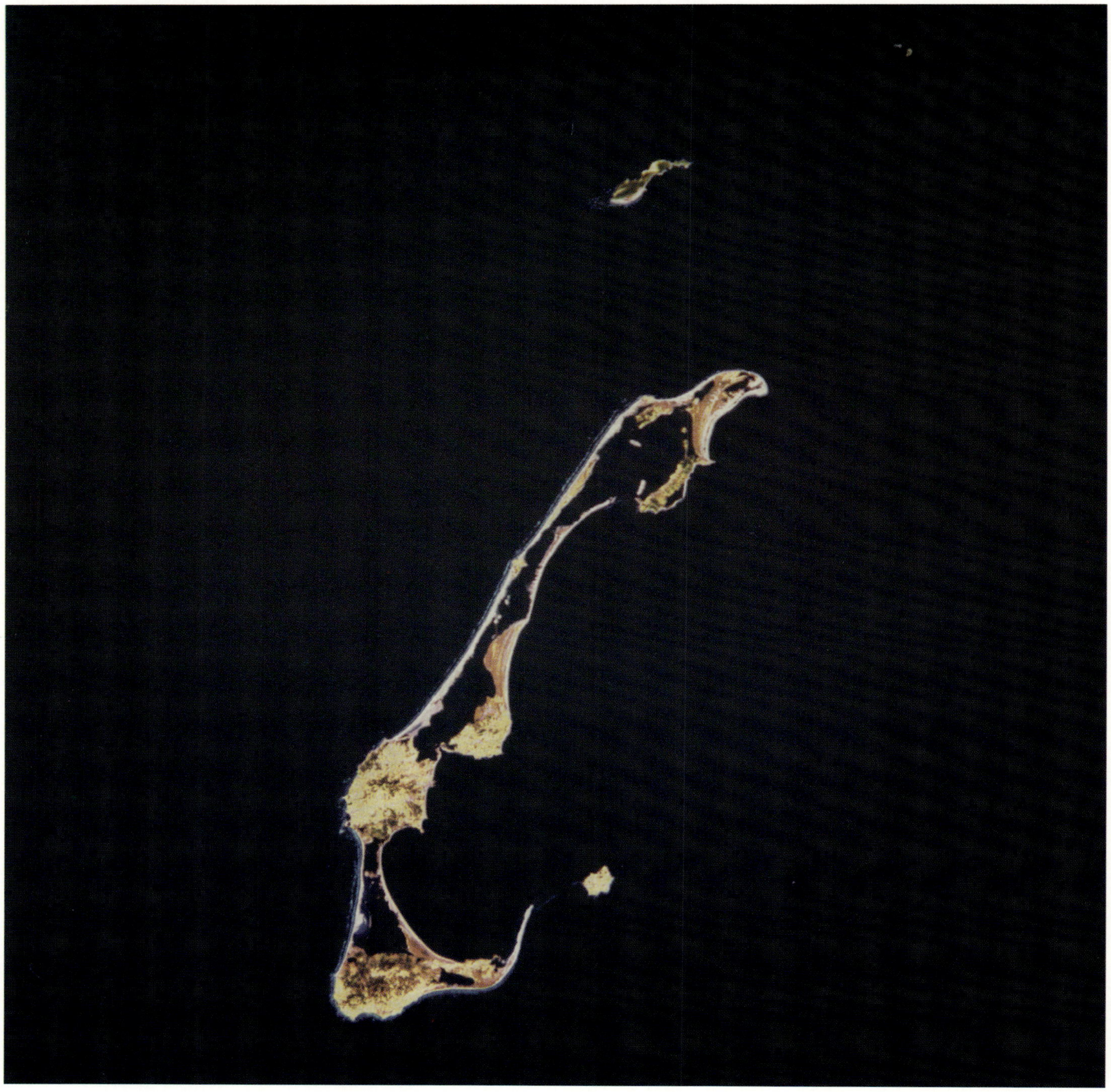

Aerial infrared colour photograph of the Magdalen Islands
Unlike the Landsat images, this photograph records the spectral signature of infrared radiation on various surfaces. The rocky islands show up as yellowish areas: Île d'Entrée, Île du Havre Aubert, Île du Cap aux Meules, Île du Havre aux Maisons, Île aux Loups, La Grosse Île, Grande Entrée Island, and Brion Island. The Bird Rocks can be barely seen in the distance (upper right) in the Gulf of St. Lawrence, northeast of Brion Island.
Sand dunes and spits covered with vegetation appear in orange and beaches in white. This photograph shows how the sand spits are formed between the islands by the currents and tides, including Sandy Hook Dune, which extends from Île du Havre Aubert toward Île d'Entrée.
The mud flats and extensive marshes inside the lagoons provide foraging areas for thousands of shorebirds. The Magdalen Islands are at a strategic point on these birds' fall migration routes; the birds come to feed and replenish their energy reserves so that they can continue their trip to the West Indies and South America.

Fisheries and Oceans Canada, 1980. Approximate scale 1:500000.

Abandoned marginal farmland in remote areas generally reverted to forest; urban areas expanded. Between 1966 and 1986, urban development claimed over 174 835 ha of arable land (Government of Canada 1991). Curiously, in recent years, in the most heavily built-up areas, wooded areas (or at least areas with some trees) have increased in both agricultural and urban environments. In the oldest suburbs built on cleared farmland, tree cover is greater today than before urbanization. In a related development, species such as the Northern Cardinal, House Finch, and Mourning Dove have recently become established in Québec towns and villages, undoubtedly aided by the presence of bird feeders.

Mireille Desponts

NATURAL DISTURBANCES

Forest fires

Fires caused by lightning in the boreal forest are a natural and recurrent phenomenon. In the boreal forest, fire helps stands regenerate, mainly by destroying old formations, which are less vigorous and more susceptible to epidemics, and by creating conditions favourable to new growth. At the biome level, fire ensures community diversity by creating a mosaic of populations of differing ages and species. Jack pine and black spruce forests are well adapted to regeneration by fire, and in the absence of human interference jack pine stands burn on average every 70 years and black spruce, every 130 years. However, in southern deciduous or mixed communities, a single cycle may last for several hundred years (Heinselman 1973; Wein and MacLean 1983; Cogbill 1985; Frelich and Lorimer 1991), with climax being reached only after a seral stage characterized by the presence of shade-intolerant species.

Fire ensures the regeneration of habitats attractive to many species of birds. Some birds, such as the Least Flycatcher, Red-eyed Vireo, and Ovenbird only live in habitat dominated by shade-intolerant species, mainly white birch and trembling aspen. Few birds depend solely on coniferous forests (Erskine 1977; Cons. G. G. – GREBE 1991a). In young burns, the snags serve as singing posts for a number of songbirds and as hunting perches for birds of prey such as the Red-tailed Hawk, Great Horned Owl, or American Kestrel. The kestrel also nests in natural cavities or woodpecker holes in standing dead trees. The Common Nighthawk lays its eggs on bare ground or in low ground cover. Poplars, and to a lesser extent white birch, tend to have natural cavities or woodpecker holes which are used by tree-nesting ducks, such as the Wood Duck and goldeneyes, and owls such as the Barred Owl, Boreal Owl, or Northern Saw-whet Owl (Meehan and Ritchie 1982).

In fact, fire is helpful to much birdlife in the boreal forest. Through its impact on vegetation dynamics, it increases avian diversity.

Fire burns the forest in an irregular pattern, leaving islands of green in wet depressions or along shorelines. Local bird populations are affected. Each succession stage after a fire attracts different species of birds. Near the Eastmain River, James Bay. *July 1983, Daniel Bordage.*

Early accounts of explorers mention a large forest fire had occurred in the interior as early as 1661, and smaller fires were quite common in the summer months, leaving whole forests of charred stumps (Cogbill 1985). Major fires occur mainly in exceptionally prolonged dry spells. Since recorded history they have occurred regularly in Québec: 1661, 1785, 1870, and 1923 (Lortie 1977; Cogbill 1985). In our era, there was a major fire on the North Shore in 1991. Such fires can burn up to hundreds of square kilometres of forest, and eventually result in major formations of white birch, jack pine, or black spruce, and other shade-intolerant species, followed by more stable climax formations.

According to contemporary accounts, forest fires became disturbingly frequent as of the mid-1800s. As settlement spread, the number of fires increased radically, with sometimes devastating results. Fire annually destroyed more timber than loggers did. Settlers and loggers mutually decried each other's practices. Settlers lit fires to clear land without taking proper precautions, and loggers left large amounts of slash on logging sites, which fed fires; it also made them more destructive, since the humus layer was sometimes destroyed to full depth (Gélinas 1984), which impeded regeneration (pers. obs.).

The great fires, like the one in the Saguenay region in 1870, caught the public imagination and moved authorities to impose regulations. For example, the period for burning slash was restricted, and railroads were obliged to install screens on locomotive smokestacks to prevent sparks from flying out. However, these measures proved ineffective, largely because they were not enforced. In 1923, fire ravaged close to 775 km^2 in the Mauricie alone and perhaps 2300 km^2 throughout Québec, possibly extending as far as James Bay or Hudson Bay (Lortie 1977). In 1923, there were 90 fires in Témiscamingue, 75 in the Lower St. Lawrence, and 100 in the Gaspé (Martin 1990). This catastrophic year triggered a response from the provincial government, which decided to set up a protection service in 1924. This agency appointed fire rangers and required anyone who wanted access to Crown lands to obtain a permit, allowing it to restrict and even forbid public access when weather conditions created a fire hazard. In addition, various means were used to sensitize citizens to the hazards of fire, the main one being the Sunday sermon delivered by the parish priest; thus, no one could be ignorant of the virtues of preventing forest fires.

In the end, fire was only successfully combatted by better weather forecasting and fire-watch networks. The advent of the telegraph, mechanized transportation, and above all aerial surveillance after World War II made the fight against forest fires increasingly effective. Between 1981 and 1991, an average of 34 000 ha of forest burned each year (MENVIQ 1993). However, in six of those years less than 8000 ha were lost, while in three of them more than 150 000 ha were lost, including a disastrous year when fire destroyed 380 000 ha (on the North Shore in 1991). This indicates that except in years when climatic conditions are particularly conducive to fire, losses are fewer. Forest fire control has been so successful that the cycle of natural regeneration by fire, which was once generally less than 150 years in the boreal forest, is now much longer. In fact, logging has in some ways contributed to revitalizing the forest, but conditions for regeneration after cutting are often much poorer than those created by fire. Also, current forestry practices in the boreal zone do not usually result in as complex a mosaic of habitats as that created by fire. This may lead to local declines in the number of bird species (DesGranges and Rondeau 1993).

Insect epidemics

Like fire, insect epidemics have always occurred in natural forests. Severe infestations generally occur in older monospecific forests and are therefore a factor in the revitalization and diversification of forest cover. Most forest species are attacked by one or more insects. For example, the Bruce spanworm attacks sugar maple, the large aspen tortrix and forest tent caterpillar attack mainly trembling aspen, the gypsy moth infests red or white oak, and the Swaine jack pine sawfly attacks jack pine. The current star of this hit parade is without question the spruce budworm, whose favourite target is the young leaves of balsam fir, though it is not averse to white spruce also (as it happens, the insect was first identified in a spruce forest).

There are few mentions of defoliator insects before the 20th century. It is known, however, that in Ontario epidemics of spruce budworm occurred in 1704, 1748, 1760, 1802, and 1863 (Lortie 1977). In Québec, infestations occurred in the Laurentians north of Québec City around 1704, again in 1748, and throughout southern Québec in the 1760s (Blais 1983). The Eastern Townships had their turn in 1877 and central and eastern Québec in 1911. In those days, however, other insect species did as much damage. In the period 1920–30, for instance, tamarack almost completely disappeared after a sawfly infestation. In the late 1930s and early 1940s, spruce forests in the Gaspé were decimated by another sawfly outbreak. In the 1950s, jack pine was attacked by Swaine jack pine sawfly (Lortie 1977).

In 1908, Québec produced its first two forest engineers, two new Yale graduates. One of them, Avila Bédard, is remembered for his commitment to Québec forestry and for founding the Association des ingénieurs forestiers du Québec (Gélinas 1984; Martin 1990). They were immediately alive to the dangers posed by insect epidemics and right from the start, they recommended that regulations to protect insectivorous birds be strengthened. In 1938, Québec established a forest insect survey service. During the 1940s, these surveys helped relate spruce budworm infestations to stands of old, overmature coniferous forests of balsam fir and spruce (L. Jobin, NRCan, Canadian Forest Service, pers. comm.). They also showed that mixed species forests were most resistant to infestations. Recommendations were issued proposing a forest management program to create more diverse regeneration after cutting, but the recommendations were never implemented. Research into biological controls, mainly focussed on identifying parasites of the insects, continued until the outbreak of World War II.

In the late 1940s, another spruce budworm outbreak occurred, predominantly in the Gaspé. A solution was then provided by the post-war chemical industry, which found a whole new outlet in the manufacture of pesticides. From 1950 to 1958, 36 000 km² of forest in Québec, New Brunswick, and Maine were sprayed with DDT (Gélinas 1984). The "treatment" was terribly effective, to the point where it was believed the problem of insect infestations had finally been solved, but the solution only addressed the effect, not the underlying cause, the proliferation of fir trees.

Balsam fir generally grows in the understorey of spruce forests and has difficulty regenerating after fire. Thus, before human interference, few communities were totally dominated by balsam fir, except in the humid climate of the North Shore and Gaspé where the natural fire cycle is particularly long. Two factors helped markedly increase the preponderance of fir. First, fire control sometimes led to the replacement of spruce by fir in old forests. Second, harvesting of dominant spruce allowed fir saplings to grow and dominate. In the 1960s and 1970s, favourable conditions for the development of spruce budworm larvae soon produced a catastrophe. The first outbreaks were discovered in 1967, and by 1975 it was estimated that over 30 million ha across the province were infected. This time it was no longer a local phenomenon, limited in time and space, but a pandemic: even the trees in cities were attacked. The 17-year infestation claimed a volume equivalent to 10 years of logging (MENVIQ 1988). Despite massive pesticide spraying and use of the biological insecticide *B.t.* (*Bacillus thuringiensis* var. Kurstaki) for a number of years, the infestation still rages in the heart of the Gaspé, on the North Shore between Forestville and Baie-Comeau, and on the Magdalen Islands. In 1990, spruce budworm still infested 1.26 million ha (Bordeleau and Auger 1991). Fire did, however, destroy a large part of the forests infested with spruce budworm on the North Shore in 1991 (MENVIQ 1993).

View of damage caused by the spruce budworm. Gaspé Peninsula. *August 1985, Lina Breton.*

The abundance of the Cape May Warbler, Bay-breasted Warbler, and Tennessee Warbler correlates strongly with spruce budworm outbreaks, since the budworm is their main food source in the breeding season. Abundance of these species may be low over large areas for several years until a spruce budworm outbreak occurs; they may then become highly abundant (MacArthur 1958; Morse 1978). It is possible that spruce budworm control programs in northeastern North America have stabilized abundance of these warblers for the past 45 years (Morse 1989a).

Forest dieback

Since the 1920s, it has been observed that certain forest species are declining in Québec. Causes are still not well understood, although there is increasing acknowledgment that sugar maple dieback is due to a combination of anthropogenic and natural factors.

The first reported cases of dieback concerned ash in 1927, and white and yellow birch between 1937 and 1960 (Lortie 1977; MENVIQ 1988). Since 1978, sugar maple has shown a disquieting decline not only in Québec, but also in Ontario, New Brunswick, and New England. In Québec, sugar maple stands in the Appalachians, particularly in the Eastern Townships, Beauce, and Bois-Francs, have the highest levels of dieback. The stands most affected are sugar maple-beech associations on poorly drained sites and sugar maple-yellow birch-ash stands colonizing thin well-drained soils. However, other sugar maple associates such as beech, ash, red maple, and yellow birch are also affected (Robitaille 1985). A high percentage of root systems in such habitat are shallow. This characteristic, in conjunction with the fact that dieback increased sharply in 1981 and 1982 after a winter marked by a thaw, loss of snow cover, and then a precipitous drop in temperature, suggests that abnormal climatic fluctuations may be a determining factor in dieback; freezing and deterioration of root systems, which leads to inadequate water uptake in summer, could also be a cause (Pomerleau 1991). Acid rain remains, however, the most-studied factor, and it is generally thought that sulphur dioxide deposits cause soil acidification and a mineral deficiency in trees (Parent and Lebrun 1988). In the late 1980s, the total area of sugar maple stands affected by dieback was estimated at 1 060 763 ha, of which close to 6% were moderately or seriously affected (MENVIQ 1993). In the hardest hit sectors, annual mortality of sugar maples was estimated at 4.5 stems/ha for the period 1979–85.

Mireille Desponts

CONSERVATION OF WILDLIFE HABITAT

Mixed use forest reserves

The idea that government should protect forests grew out of increasing concern on the part of elites close to government about the wholesale destruction of the forests. The Forest Reserves Act passed in 1875 initially had two major goals. First, it sought to prevent colonization on lands unsuited to agriculture to preserve forest resources (at the pressing request of timber merchants) and to maintain some control over forestry operations. Second, it regulated timber cutting at heads of rivers in an attempt to restore regular flows on rivers that, for several years, had shown alarming variations in flood and low-water patterns. In addition, there was a growing awareness of the need for conservation in the United States, which was spreading into Québec. The conservation movement sought to preserve natural areas of great interest for the public good, resulting in the creation of Yellowstone National Park in the United States in 1872 and Rocky Mountain Park (now Banff National Park) in western Canada in 1887.

When Québec decided to act in the late 1800s, it created conservation areas intended for several somewhat contradictory purposes. The Québec provincial parks, Trembling Mountain Park, established in 1894, and Laurentide Provincial Park, established in 1895, were forest reserves doubling as public park and recreation areas (Comité des parcs en forêt 1974; Martin 1990). These conservation areas were set aside for rational use of the forest, yet the presence of sportsmen was encouraged by the leasing of exclusive hunting and fishing rights to private clubs or individuals. In 1903, 93% of Laurentide Park was under timber licence to logging companies. Paradoxically, the forest roads helped increase the number of hunting and fishing camps. By the 1920s, the development of private clubs had become a significant source of revenue, which the government sought to promote by building camps (Camp Mercier in 1920, Grand Lac à l'Épaule and Lac du Bec Croche in 1922, Camp Devlin in 1923, and Grand Lac Jacques-Cartier in 1930) (Martin 1990). A number were built in Les Grands Jardins and along the road between Québec City and Lac-Saint-Jean. In 1930, there were no less than 555 clubs and private leaseholders in Québec, with access to 321 rivers and 3938 lakes; they contributed $2 500 000 to the regional economy and $200 000 directly to the government through ground rents (Comité des parcs en forêt 1974). The year 1977 culminated many years of public pressure for Québec's wildlife to be managed as a public resource (Martin 1990). The 1164 private clubs in existence that year were replaced by 60 ZECs (zones d'exploitation contrôlée/controlled zones).

In addition to the laws and programs aimed at protecting and managing Québec's forests, the government established experimental forests, research forests, and one superb forestry station located just north of Québec City—the Duchesnay Forestry Station. There are also forest educative centres, whose mandate has recently been expanded when they were given over to private, local control.

Recreational parks

It was not until the 1930s that Québec established its first real recreation and conservation parks. Gaspé Provincial Park was created in 1937 and Mont-Orford Provincial Park in 1938, with no mining or logging allowed in either. The following year, a further step was taken with the creation of a public wildlife reserve intended solely for hunting, fishing, and conservation, along the highway being built between Mont-Laurier and Senneterre (this reserve would become La Vérendrye Park in 1950) (Martin 1990). Between 1943 and 1945, the Petite Rivière Cascapédia and Matane River, among others, were opened to salmon fishing, in response to increasing public pressure from anglers.

After World War II, Québec set up mainly recreational and tourist-oriented parks, as well as public game and fish preserves in response to steadily increasing demand for such facilities from residents who now had cars and roads to travel on, which made remote regions more accessible. In 1946, there were already complaints about excessive pressure on the camps in Laurentide Provincial Park after the opening of Talbot Boulevard between Québec City and Chicoutimi. In the 1950s and 1960s, the government concentrated mainly on developing campgrounds, picnic sites, golf courses, and ski centres. In 1958, it finally decided to develop Mont-Tremblant Provincial Park, 64 years after the park (Trembling Mountain Park) had been established.

Conservation parks

In the 1970s and 1980s a new approach to conservation developed, one oriented toward protecting natural habitats for their intrinsic ecological value, and not just for their usefulness as recreation, hunting, or fishing areas. Thus, the government passed the Cultural Property Act in 1970, the Ecological Reserves Act in 1974, and the Parks Act in 1977, thereby creating sites with different levels of protection (MENVIQ 1983). Sites with exceptionally beautiful landscapes were protected for their aesthetic value, while others were off limits to all human activity, and access to still others was severely restricted.

Along with these provincial measures, the federal government created a network of national wildlife areas and migratory bird sanctuaries along the St. Lawrence to protect critical waterfowl habitat and wetland ecosystems increasingly threatened by residential development and farming (Ringuet and de Repentigny 1982). The federal government also created four large national parks: La Mauricie National Park, Forillon National Park, Mingan Archipelago National Park Reserve, and the Saguenay–St. Lawrence Marine Park, as part of a program to protect major natural environments representative of Canada's natural regions (Government of Canada 1991). As a signatory to the International Convention on Biological Diversity in 1992, Canada is planning to add four new national parks in southern Québec and four others in northern Québec. The percentage of protected territory should therefore soon rise from 0.4 to 4.1% (Gagné 1993).

In the past 20 years, Québec has established a complex network of parks and reserves of varying sizes and uses (Boucher 1992). In all, these sites represent about 10% of the area of the province, if wildlife and forest reserves are included. The conservation movement of the 1980s took a new direction, with private environmental organizations and individuals becoming concerned about the preservation of small sites of great ecological importance. By 1992, 15 environmental organizations had acquired, for the purposes of preservation and public education, over 4000 ha of habitat. Another recent concept is that of stewardship of private lands (Mercier 1992), and it is interesting that this initiative is coming from landowners aware of the ecological or heritage value of the sites they own; government merely provides support and advice in the management and protection of such habitat. It is clear that Québec's attitude toward conservation has been steadily evolving over the past 100 years. Now a new awareness is emerging, an awareness that should endure, largely because it fosters a sense of individual responsibility for the environment.

Forest protection

Government policy on commercial forests also shifted in the 1980s in response to the desperate situation of the forest industry, one of the province's main industries. Until recently, provincial governments have always hesitated to intervene in the logging industry. They simply granted concessions and left forest management up to the companies, which conducted forest inventories, built roads, and logged trees. Indeed, by the end of the 19th century, the tendency was for the government to grant larger and larger concessions to avoid the expense of building infrastructure or fighting fires. This left the forest wide open to abuse such as using dry stream beds as logging roads and clearcutting enormous tracts of forest with no reforestation program.

Initial attempts to change logging practices were made by the first forest engineers, who tried to convince logging companies that different methods would produce higher yields, and to persuade them that the emphasis should be on long-term rather than short-term returns. These efforts bore little fruit, and it was only when restrictive legislation was introduced that certain practices altered. To limit waste, regulations were passed controlling the height of the cut and the width of skidding roads. Measures to protect seedlings were also imposed, to ensure forest regeneration (Gélinas 1984).

Finally, in the early 1980s, the inherent problems of leaving forest management to concession holders became all too apparent. Public hearings on aerial spraying of insecticides and herbicides revealed grave deficiencies in forest management practices, which led the provincial government to act. The Forest Act passed in 1986 set out a new forest regime designed to obtain sustained timber yields through better management (Commission sur la protection des forêts 1991; MENVIQ 1993). Since 1986, logging operations on public lands have been bound by timber supply and forest management agreements with the government. These agreements allow companies to log a certain volume of timber as long as they agree to replant areas they harvest.

After 1986, there was a substantial increase in the number of trees planted in both public and private forests, largely due to subsidies from the Ministère des Forêts du Québec. In the 1980s, the total number of trees planted annually rose from 35 million to 200 million, and plantation areas from 13828 ha in 1980–81 to 74545 ha in 1987–88 (Parent 1989). Areas logged also increased in that period, from 240000 ha to 314000 ha. However, 97% of the trees replanted in the deciduous forest are conifers, mostly black spruce (38%), white spruce (22%), and jack pine (20%) (G. Couture, NRCan, Canadian Forest Service, pers. comm.). About 80% of private woodlots are in deciduous or mixed forests, while 80% of government funds for reforestation are allocated to softwoods (Commission sur la protection des forêts 1991). This "substitution" of communities could lead to decreased soil fertility and a decline in biodiversity, as well as greater susceptibility to disease and insect epidemics. In addition, these small private woodlot operators use large amounts of phytocides.

Mireille Desponts

Protecting birds and their habitat

In 1916, Canada and the United States signed the Migratory Birds Convention, by which they undertook to protect migratory species against indiscriminate hunting and preserve populations. The Migratory Birds Convention Act of 1917 provided for regulated hunting, prevention of illegal trading, controlled harvesting, and establishment of migratory bird sanctuaries. In Québec, the first such sanctuaries were created in 1919; these were islands heavily used by colonially nesting seabirds, for example the Bonaventure Island and Percé Rock sanctuary and the Bird Rocks sanctuary, the two principal breeding sites of the Northern Gannet. Several sanctuaries were also established at that time (1925) on the North Shore, for instance the St. Mary Islands Bird Sanctuary (now known as the Îles Sainte-Marie Migratory Bird Sanctuary), protecting an important breeding site for such species as the Common Murre and Razorbill.

Even then, private conservation organizations were active in purchasing islands to be turned into sanctuaries; thus, the Province of Quebec Society for the Protection of Birds acquired Île aux Perroquets near Blanc-Sablon, and the Provancher Society of Natural History of Canada bought Île aux Basques and the Razade Islands in the St. Lawrence estuary.

Although hunting and other forms of disturbance were subject to a comprehensive statutory ban, the legislation applied only to activities taking place within sanctuary boundaries while the birds were there; it did not protect bird habitat.

Though very important, the protection afforded to birds by the sanctuaries was not sufficient, and in the 1960s it quickly became apparent that loss and degradation of habitat were the greatest threat facing birds and other wildlife. Critical habitat for wild species, both plant and animal, especially wetlands, was indeed disappearing at an alarming rate.

The year 1973 saw enactment of the Canada Wildlife Act, under which the federal government is empowered to establish protected zones known as "national wildlife areas." The purpose of these areas is to conserve critical habitat for migratory birds. They enjoy fuller protection, and there is provision for surveillance year-round, even when no migratory species are present. For each national wildlife area, the Canadian Wildlife Service, as manager, draws up a management plan specifying which activities are to be controlled. Thus, grazing and farming may be authorized by a permit incorporating strict conditions. Plans may also specify how and where habitat is to be maintained. The Act also covers public awareness and research. In most wildlife areas, hiking, photography, and birdwatching are permitted. Some management plans also allow traditional activities like hunting, but these are carefully regulated to maintain bird and animal populations and prevent degradation of protected habitat. In Québec, national wildlife areas are located along the shores or on the islands of the St. Lawrence corridor between Lake Saint-François and the Magdalen Islands. In 1995, there were eight national wildlife areas protecting some 5500 ha of critical migratory bird habitat. Each of these areas protects specific habitat for various bird species and is thus very important to the habitat protection system, even though they may vary considerably in size.

The best known of these is without doubt Cap Tourmente National Wildlife Area, created to protect the bulrush marshes used as a food resource by the Greater Snow Goose. It is one of the most famous sites in ornithology, with a species count of over 250. It was also the first site in North America to be designated, in 1981, as a wetland of international importance under the Ramsar Convention, a status shared since 1987 with the Lac Saint-François and Baie de L'Isle-Verte national wildlife areas.

Since the mid-1980s, the Canadian Wildlife Service has no longer had the resources to purchase huge tracts for management as sanctuaries or wildlife areas, but other players have since become active, using new and innovative methods. Financing and project implementation are shared among partners according to their respective means. The North American Waterfowl Management Plan, which brings together both governmental and private partners, is a good example of this new *modus operandi* for protecting and managing migratory bird habitat. The Baie-du-Febvre site, established and protected by a group of private and public interests and now managed by the SARCEL private conservation group, is one of the best examples of what can be achieved when budgets and expertise are pooled.

A number of non-profit organizations have also taken up the cause of protecting sites that are home to migratory birds or other significant wildlife. Prominent among these are the Nature Conservancy of Canada, which protects numerous islands between Montréal and Lake Saint-Pierre, and Duvetnor, which has purchased island habitat in the St. Lawrence estuary.

Other joint environmental protection projects such as the St. Lawrence Action Plan also protect bird habitat.

Also to be taken into account is the protection afforded by Canada's national parks and Québec's conservation parks, ecological reserves, wildlife sanctuaries, and wildlife reserves. Through their highly important role in safeguarding the environment deriving from their strategic location or their extent, these areas also help maintain local bird populations.

The urgency of protecting bird habitat is about to receive even greater recognition, as the signatories to the Migratory Birds Convention are preparing to amend the treaty to enshrine management principles encompassing protection of critical migratory bird habitat. The signatories are also expected to agree on joint habitat conservation programs. This will be a milestone in the development of bird and habitat protection.

Yet regardless of all this official, legislated protection, joint habitat protection and management planning and the unflagging work of volunteers in private conservation associations, each and every one of us is responsible for protecting birds, and we should ensure that the ways we use the land remain compatible in the long term with their basic survival needs.

Jacqueline Vincent and Yvon Mercier

THE ATLAS DATA

Survey methods

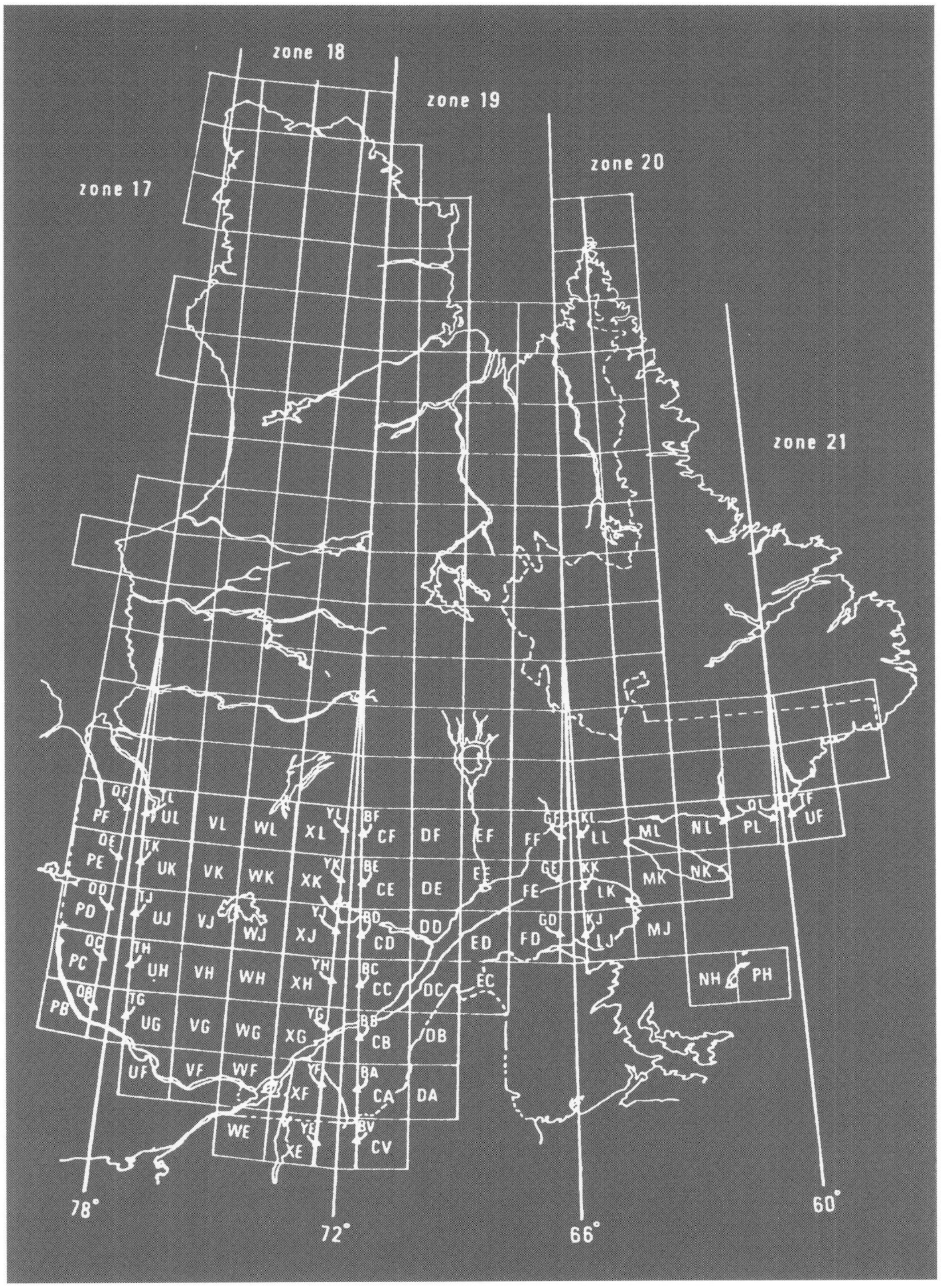

Figure 1
UTM grid for Québec and identification of blocks in southern Québec

Atlas area and survey units

To map the distribution of breeding birds in southern Québec, the Atlas area was divided into 10 km × 10 km survey units, or squares, according to the Universal Transverse Mercator (UTM) grid system. This grid is easy to identify on Canadian topographic maps, appearing in the form of blue lines on the 1:250 000- and 1:50 000-scale maps. In the UTM system, squares are grouped into 100 km × 100 km blocks, which are identified by two letters. In turn, blocks are organized into zones corresponding to 6° increments in longitude (Figure 1). This type of grid system has numerous advantages over latitude-longitude systems since the boundaries of the units are already shown on topographic maps and all the units are the same size except for those on the edge of the zones. Although a number of different survey units have been used in breeding bird atlas projects, the 10 km × 10 km square is the most common one. Squares are identified by the UTM co-ordinate of their southwest corner.

The UTM grid system has only one drawback: it is impossible to superimpose a rectangular plane on a spherical surface like the earth. This results in incomplete sliver- or pie-shaped squares along the edges of the zones, or every 6° of longitude. Figure 2 shows the method used to avoid having squares very different in size from the normal 100 km^2 squares.

Given Québec's huge size and the difficulty of accessing almost all of the northern part of the province, it was decided that surveying the entire province would not be feasible. Instead, the survey was limited to southern Québec, i.e., the area below latitude 50°30' N or, more precisely, UTM co-ordinate 5 600 000 N. The field work for the Atlas, which was originally scheduled to take place between 1984 and 1988, was extended until 1989 to ensure better coverage of the Atlas area.

Several sampling options were examined before total coverage was selected. One option was to apply to all of southern Québec the stratified sampling method used by Scherrer (1982) in Gatineau Park, whereby the Atlas area would be divided into strata, based on habitat characteristics, and then a certain number of sampling units in each stratum would be randomly selected. However, this method was rejected as unfeasible because of the difficulty of accessing some regions in the northern part of the Atlas area. Instead of a strict random sample, another option was to use a cluster sampling technique based on the random selection of logging roads. This idea was also rejected because many of these roads had become impassable, making it difficult to construct the sample. Therefore, a pragmatic approach was finally adopted: to aim for total coverage in the accessible parts of southern Québec, but in more remote and less accessible regions, to cover squares near access points, which in fact resembles cluster sampling.

Division of Atlas area and contact with participants

Breeding bird atlases are generally a volunteer effort involving the participation of hundreds, if not thousands, of local volunteers. The Québec Atlas area was divided into 35 regions, many corresponding roughly to the area covered by the local birdwatching club. The field work in each region was supervised by a regional co-ordinator, who was generally chosen because he or she was an active, experienced birder well known in the region. The regional co-ordinators' role was to act as a liaison between atlassers and Atlas officials. Their responsibilities included recruiting atlassers and encouraging them to search for evidence of breeding by the species found in the region. Therefore, co-ordinators had to make sure that atlassers thoroughly understood the goals of the project and the data collection methods used. Co-ordinators also had to ensure that the

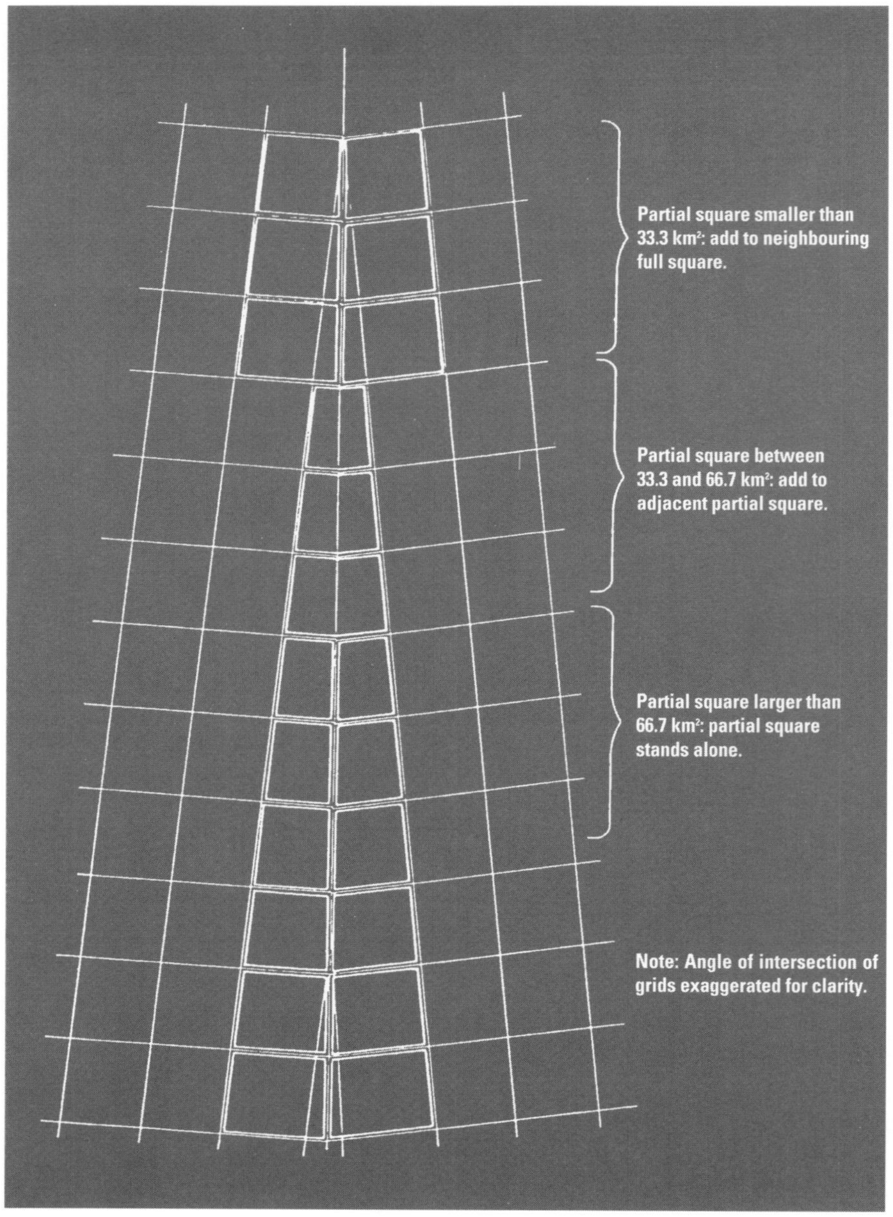

Figure 2
UTM grid and zone system, showing incomplete squares along edges of zones

squares in their regions were assigned to produce optimum coverage of the region as a whole and that the coverage in each square was adequate. At the end of each field season, the co-ordinators collected all the data cards from the atlassers in their region, checked them, and sent them to the Atlas office. They also subsequently checked and validated this information again when they received it in compiled form. Some co-ordinators held meetings to inform participants of the objectives of the project and its progress, as well as any interesting observations that had been made in the region. The regional co-ordinators played a key role in ensuring the successful completion of the field work.

Atlas officials promoted the project every spring with the help of local birdwatching clubs. Group presentations and articles in regional weekly newspapers and Québec birding publications celebrated the pleasures of observing birds in summer and the usefulness of this activity in conservation. A bilingual newsletter, *Info Atlas*, was also published to inform participants of the project's progress and some of the preliminary results. It also contained articles on the biology of some groups of birds and tips on spotting some of the more elusive species and on upgrading breeding evidence.

Data collection methods

Atlasser's kit

Data collection for the Atlas was done according to a well-established procedure, based mainly on the methods employed in the Ontario breeding bird atlas project (Cadman et al. 1987), which in turn were adapted from the methods used in the Great Britain and Ireland atlas (Sharrock 1976). Instructions for atlassers were provided in the *Guide for participants* (Gauthier and Aubry 1985), which was modelled on the pamphlet used for the Ontario atlas (Eagles and Cadman 1983). Participants also received *Québec Breeding Birds: Nesting and Rearing Chronology* (Aubry and Lambert 1984) and, for each square to be covered, a data card to record their observations, a 1:50 000-scale topographic map illustrating the square, and a 1:250000-scale map of the surrounding area.

An initial version of the data card, which could only contain data on four visits, was used during the first field season in 1984. It nevertheless proved useful in assessing the minimum atlassing effort required to cover each square.

From 1985 to 1989, a revised version of the card was used, containing space for data on up to 10 visits to the same square and a single box for recording the most conclusive breeding evidence obtained for each species (Figure 3). Atlassers generally filled out only one card per square annually, unless they had made more than 10 visits to the square that year. Occasionally, several participants each sent in a data card for the same square in the same year. On the first page, participants filled in their names and addresses; identified the square surveyed by providing the complete set of UTM co-ordinates and a place name (city, lake, mountain, etc.); and indicated the Atlas region where the square was located, the regional co-ordinator, the survey year, and the dates and times when the field work was done. When the card reached the Atlas office, Atlas staff completed the bottom part by inserting the square co-ordinates, assigning the atlasser a participant number, and inserting the code for the region, the survey year, the dates of the first and last visits, the total number of potential breeding species in the square, and the total number of hours of field work. The rest of the card was in the form of a checklist of the bird species known or likely to breed in Québec, against which participants recorded the most conclusive breeding evidence obtained for the species.

Breeding evidence

Breeding evidence levels were adapted from those used in the Ontario breeding bird atlas project (Cadman et al. 1987). The 17 categories of breeding evidence were divided into four different levels: species observed (1 category), possible breeding (1 category), probable breeding (6 categories), and confirmed breeding (9 categories). On the data card, the categories were shown in ascending order from the least certain to the most certain. The "Species observed" code (*X*) was reserved for reports of non-breeding individuals seen during the breeding season, in other words, species that could not possibly have been nesting in the area where they were seen for various reasons, including age, habitat, and distribution. For example, this code would apply to a Parasitic Jaeger seen along the St. Lawrence River, an area far-removed both in terms of habitat type and distance from

YEAR | ZONE | BLOCK | SQUARE (UTM) 000 000

QUEBEC BREEDING BIRD ATLAS

NAME

ADDRESS

POSTAL CODE | AREA CODE | PHONE –

SITE

REGION | COORDINATOR

DATA CARD no: ____ of ________

DAY	MONTH	TIME OF DAY		TIME (HOURS) ×	NO. OF OBSERVERS =	TOTAL HOURS
		START	END			

Environnement Canada — Environment Canada

Service canadien de la faune — Canadian Wildlife Service

DO NOT WRITE IN THIS SECTION

Z | UTM 000 000 | NP

R | A | PV | DV | P | NH

SPECIES	CODE	AOU	FINAL CODE	SPECIES	CODE	AOU	FINAL CODE
Red-throated Loon		0110		Red-breast. Merganser		1300	
Common Loon		0070		Ruddy Duck		1670	
Pied-billed Grebe		0060		Turkey Vulture		3250	
Horned Grebe		0030		Osprey		3640	
Red-necked Grebe		0020		Bald Eagle		3520	
				Northern Harrier		3310	
Leach's Storm-Petrel		1060		Sharp-shinned Hawk		3320	
Northern Gannet		1170		Cooper's Hawk		3330	
Great Cormorant		1190		Northern Goshawk		3340	
Double-cr. Cormorant		1200		Red-shouldered Hawk		3390	
				Broad-winged Hawk		3430	
American Bittern		1900		Red-tailed Hawk		3370	
Least Bittern		1910		Rough-legged Hawk		3470	
Great Blue Heron		1940		Golden Eagle		3490	
Great Egret		1960		American Kestrel		3600	
Green-backed Heron		2010		Merlin		3570	
Black-cr. Night-Heron		2020		Peregrine Falcon		3560	
Canada Goose		1720		Gray Partridge		2881	
Wood Duck		1440		Ring-necked Pheasant		3091	
Green-winged Teal		1380		Spruce Grouse		2980	
American Black Duck		1330		Ruffed Grouse		3000	
Mallard		1320		Sharp-tailed Grouse		3080	
Northern Pintail		1430					
Blue-winged Teal		1400		Yellow Rail		2150	
Northern Shoveler		1420		Virginia Rail		2120	
Gadwall		1350		Sora		2140	
American Wigeon		1370		Common Moorhen		2190	
Redhead		1460		American Coot		2210	
Ring-necked Duck		1500					
Greater Scaup		1480		Semipalmated Plover		2740	
Lesser Scaup		1490		Piping Plover		2770	
Common Eider		1600		Killdeer		2730	
Black Scoter		1630		Greater Yellowlegs		2540	
Surf Scoter		1660		Solitary Sandpiper		2560	
Common Goldeneye		1510		Spotted Sandpiper		2630	
Hooded Merganser		1310		Upland Sandpiper		2610	
Common Merganser		1290		Least Sandpiper		2420	

Figure 3
Data card used for the Québec Atlas (pages 1 and 2)

the species' known breeding habitat. Reports coded "Species observed" were input into the Atlas database but were not used to produce the distribution maps, since they concerned birds considered as nonbreeders in southern Québec.

Nest building (code *CN*) was considered evidence of confirmed breeding in species other than woodpeckers. Nest building by woodpeckers (code *N*) was considered evidence of probable breeding since these birds sometimes excavate holes that they do not lay eggs in. Although nest building by wrens was originally considered evidence of probable breeding (common practice in breeding bird atlases), it was later decided that it would be classed as confirmed breeding. Therefore, all *N* codes for wren species in the database were replaced with *CN* codes.

Employment programs

During the field work portion of the project, a number of summer employees were hired by regional birdwatching clubs to assist in the atlassing effort. To ensure optimum coverage of the entire Atlas area, these employees, who were hired under 22 different employment projects, were assigned to the more remote portions of squares or areas with a scarcity of volunteer atlassers like the North Shore, Charlevoix, Abitibi–Témiscamingue, and Chibougamau regions, and part of the Eastern Townships. Summer employees, who were mainly students, were selected for their knowledge of ornithology and field identification skills.

Other data sources

In order to provide the most accurate and complete picture of the distribution of breeding birds in the province from 1984 to 1989, data from other sources were integrated into the Atlas database. These sources included the Quebec Nest Record Card Program (QNRCP), the Étude des populations des oiseaux du Québec (ÉPOQ), the Banque informatisée des oiseaux marins du Québec (BIOMQ) (a databank on seabirds in the province), and Ministère du Loisir, de la Chasse et de la Pêche du Québec (MLCP) (now Ministère de l'Environnement et de la Faune du Québec [MEF]).

Checking and processing of data

At the end of each field season, atlassers submitted their data cards to the regional co-ordinators, who checked them and sent the year's batch to the Canadian Wildlife Service. The cards were checked a second time at the Atlas office to identify reports of rare species that needed more documentation. The checking and final coding of the roughly 1000 data cards received every year at the Atlas office took approximately two months.

The data were then input into the computer and checked one last time with a data verification program to detect coding and input errors. These data were then combined with those of previous years to create a permanent file containing all the records received since the beginning of the project. This file was subsequently used to generate a separate file containing the strongest piece of breeding evidence obtained so far for each species in each square, which was updated at the end of each field season. At the end of the sixth season, this file was used to provide the data presented in the Atlas and produce the distribution maps. Another file was also kept with information on each square visited: the total number of hours of field work completed by atlassers in the square, the dates of the first and last visits, the participant number or a code indicating that the square had been covered by more than one atlasser, and the total number of species observed. This file was used to monitor the quality of coverage of the square.

Every year, these data were used to generate basic statistics on such items as the number of data cards received, number of squares visited, total number of field hours, total number of species observed, average number of field hours and species per square, most common species, and squares with the greatest number of species. Regional compilations were sent to the regional co-ordinators, who in turn passed the information on to participants in order to give them a final opportunity to validate the results of their observations.

Ensuring adequate coverage of a square

Atlassers were encouraged by both the regional co-ordinators and the *Guide for participants* to visit all the habitat types found in their squares. Participants were asked to complete at least 16 hours of field work to detect most of the breeding species present. The field work was to be done mainly in June and July, the breeding season for most species in southern Québec. Atlassers were also asked to visit their squares early in the morning when birds are most active. June is the easiest month to find and identify birds by their song, while July is the easiest month to obtain breeding evidence by spotting adult birds carrying food for their young. In addition, the nestlings and fledglings are often noisy, allowing them to be detected in or near the nest. Atlassers were also asked to spend some time visiting their squares outside the main breeding season to concentrate on early and late nesting species, and to do some field work in the evening and at night to increase their chances of finding crepuscular and nocturnal species.

For a square to be considered adequately covered, a minimum number of hours had to have been spent on field work and a minimum number of species had to have been observed. The number of hours required was assessed during the first field season by correlating the number of hours spent in the field with the number of species reported. Although the expected number of species varied by region, it was decided that at least 16 hours of field work were required to cover a square adequately in southern Québec. The number of species observed increases rapidly during the first few hours, then more slowly as time goes on until finally new species are found only at lengthy intervals. Therefore, continuing to survey a square becomes unproductive after a certain number of hours.

To establish the minimum goals for coverage in terms of the number of species observed, estimates of the potential number of breeding species present in each region were made at the beginning of the project. To do this, southern Québec was divided into broad geographical regions and a group of ornithologists who were very familiar with Québec avifauna drew up a list of all the species, including the rare ones, that could potentially breed in each region. This list was used for all the squares in a region, in full knowledge that no 10 km × 10 km unit could contain all the habitats found in the entire geographical region and therefore support all the species of birds in that region. Consequently, squares were considered to be adequately covered if half the number of species on the regional list were found. In the Montérégie region, for example, the number of breeding bird species was estimated to be 160. Therefore, a minimum of 80 species had to be reported in each square in the region during at least 16 hours of atlassing for the square to be considered adequately covered.

Gilles Falardeau

Final coverage

A total of 932 participants completed and submitted 9041 data cards containing 202 521 bird records during the six years of the Atlas. In all, 67 587 hours of field work were expended, although this figure was really higher considering the time spent travelling back and forth from the squares, casual observations made during other activities, and hours in the field that were not reported. In addition, many other people whose names were not recorded on the data cards assisted atlassers in the field or passed on their observations. The atlassing effort was spread fairly evenly over the six years of the project, except during the last year (1989), when the emphasis was on completing the coverage of less accessible regions.

Atlassing effort

	1984	1985	1986	1987	1988	1989	Total
Number of atlassers	215	420	322	255	263	164	932[1]
Hours of field work	12 201	14 020	11 469	10 565	12 862	6 470	67 587
Number of cards received	1 085	1 568	1 638	1 696	1 808	1 246	9 041
Number of records	39 880	37 325	34 291	31 534	39 958	19 533	202 521
Average number of field hours per atlasser: 72.5							
Average number of data cards per atlasser: 9.7							
Average number of field hours per card: 7.5							
Average number of species per card: 22.4							

1 The total does not correspond to the total of the individual columns for each year because many atlassers took part in more than one field season and contributed cards in more than one year.

Variability of coverage

Contrary to many breeding bird atlas projects, the Québec Atlas area could not be covered systematically and atlassing effort varied from region to region (Figure 1). In general, the number of hours of field work declined on a gradient from south to north. Almost total coverage was achieved in the southernmost part of Québec, particularly in the St. Lawrence and Ottawa valleys, the Eastern Townships, and the Laurentians north of Montréal. However, other regions were covered much less systematically, either because they were difficult to access or had only a small number of local birders, including northern Abitibi, the area between the Abitibi–Témiscamingue and Lac-Saint-Jean regions, the area northeast of Lac-Saint-Jean, the inland portion of the North Shore, part of Anticosti Island, and the interior of the Gaspé Peninsula. To counter this variability in coverage and obtain a more accurate picture of species' distribution in southern Québec, a statistical method was developed, which is discussed in the section "Frequency of occurrence by ecological region and mapping of relative abundance."

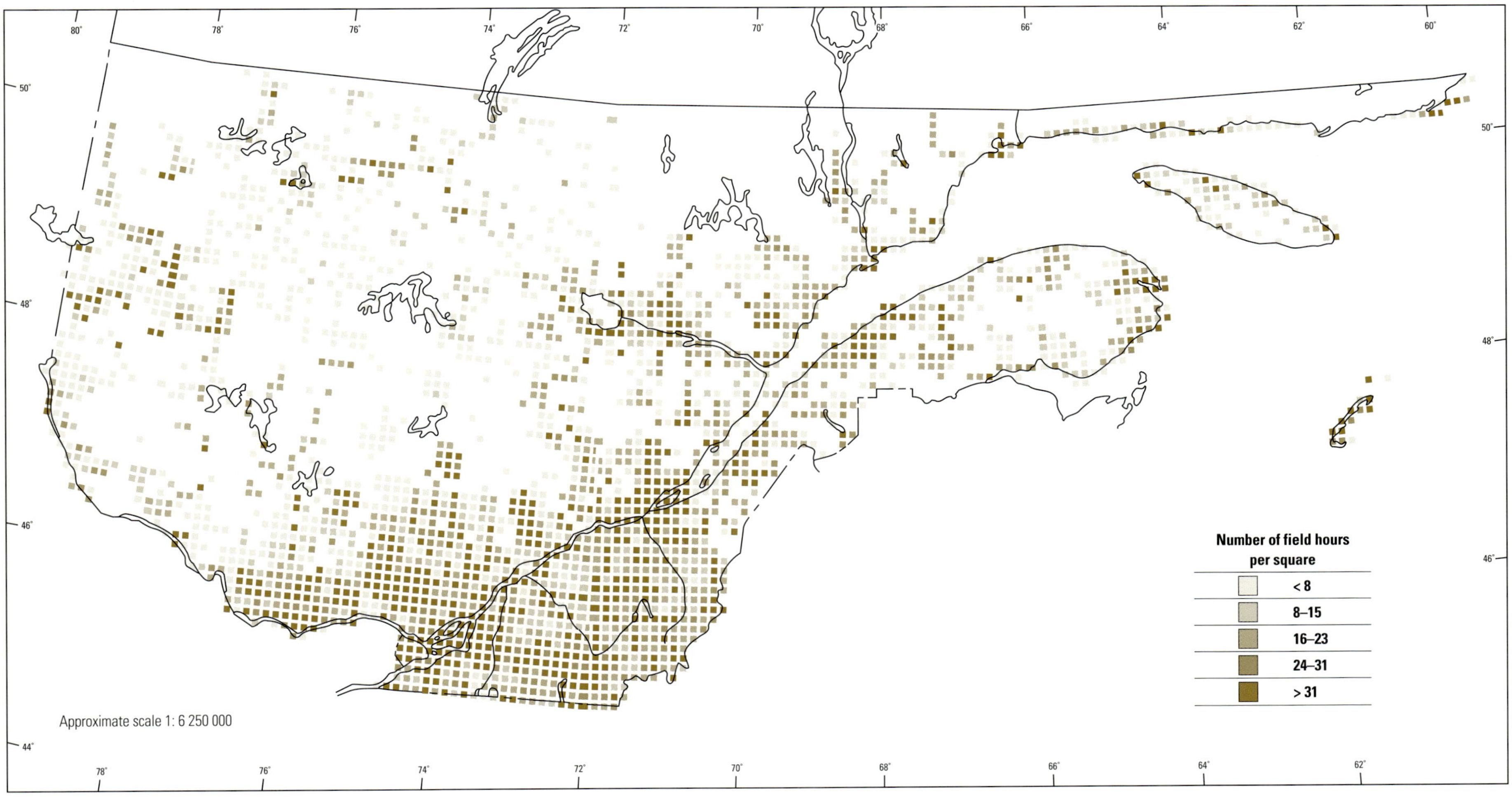

Figure 1
Number of field hours spent in each Atlas square surveyed

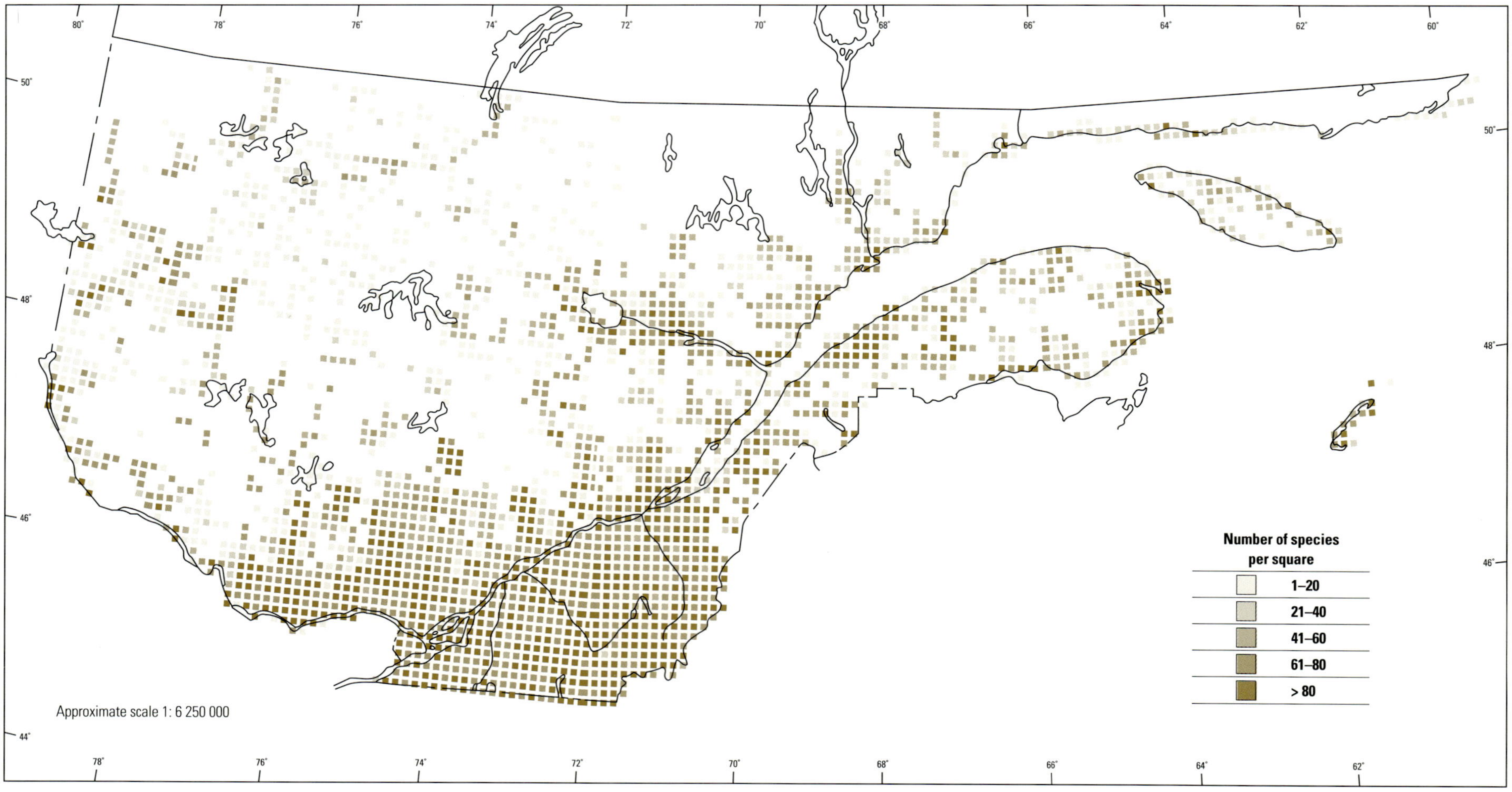

Figure 2
Number of species reported in each Atlas square surveyed

The second map shows the number of species reported in each square (Figure 2). As in the previous map, the number decreases from south to north.

Difference in species coverage

Not all species are equally easy to detect. The species that are easiest to observe during the breeding season tend to be those that are abundant, conspicuous, active during the daytime, and easy to identify both by sight and ear. The easier a species is to spot, the more truly representative will be its distribution map; the harder a species is to spot, the less likely it is to be reported in a square where it is present. Some species' habits make them more difficult to find. The owls are probably the best example; they are active at night and breed early in the season, when atlassers were less likely to be in the field. Species like the Least Bittern and the Yellow Rail are difficult to observe because they are very shy and frequent habitats that are difficult to access. Finally, species such as Cooper's Hawk are difficult to identify. These species are probably underrepresented to some extent on Atlas distribution maps.

Gilles Falardeau

Overview of results

The Atlas survey produced a number of very interesting results which are discussed in greater detail in the species accounts; only the highlights are given here. A total of 232 species were confirmed as breeders in southern Québec between 1984 and 1989. In addition, 14 species were recorded as probable breeders—Eurasian Wigeon, Barrow's Goldeneye, Willow Ptarmigan[1], Yellow Rail[1], Willet, Yellow-billed Cuckoo[1], Snowy Owl, Red-bellied Woodpecker, Carolina Wren[1], Bohemian Waxwing, Blue-winged Warbler, American Tree Sparrow, Henslow's Sparrow[1], and Western Meadowlark[1]—and 7 as possible breeders—Sandhill Crane, Short-billed Dowitcher, Barn Owl[1], Tufted Titmouse, Prairie Warbler, Louisiana Waterthrush, and Yellow-headed Blackbird.

New breeding confirmations

Five species were recorded as breeders for the first time during the Atlas survey: Great Egret, Brant, Turkey Vulture, White-winged Tern (which mated with a Black Tern), and Great Gray Owl. While the records for the Great Egret, Turkey Vulture, and Great Gray Owl were not unexpected, the records for the Brant and, above all, the White-winged Tern, were very surprising. Brewster's Warbler, a hybrid resulting from the mating of a Blue-winged Warbler and a Golden-winged Warbler, was also found breeding in Québec for the first time. This is quite remarkable since there is no confirmed evidence that the Blue-winged Warbler breeds in the province. In addition, the first breeding record for the Monk Parakeet, an exotic species, was obtained during the Atlas period. This species is discussed in the chapter "Other Species of Record in Québec" since it is usually found here only in captivity. Another discovery worthy of note is the breeding of the Greater Snow Goose in 1990 on Battures aux Loups Marins in the St. Lawrence estuary, which is roughly 2500 km south of its usual breeding range.

Rare species reported in a single square

Rare species found in only a single square were: Brant, Willow Ptarmigan, Willet, Short-billed Dowitcher, White-winged Tern, Barn Owl, Red-bellied Woodpecker, Prairie Warbler, Louisiana Waterthrush, American Tree Sparrow, and Yellow-headed Blackbird.

Most common species

The 30 species found in the greatest number of squares were:

Species	Number of squares (out of 2464)	Species	Number of squares (out of 2464)
American Robin	1843	Black-capped Chickadee	1477
White-throated Sparrow	1832	Common Grackle	1474
Tree Swallow	1781	Chipping Sparrow	1465
Northern Flicker	1684	Least Flycatcher	1403
Cedar Waxwing	1625	Alder Flycatcher	1395
Common Yellowthroat	1620	Hermit Thrush	1391
Yellow-rumped Warbler	1539	Winter Wren	1389
Spotted Sandpiper	1529	Magnolia Warbler	1385
American Redstart	1516	Belted Kingfisher	1379
Red-eyed Vireo	1512	Ovenbird	1349
American Crow	1510	Dark-eyed Junco	1344
Red-winged Blackbird	1496	Nashville Warbler	1339
Song Sparrow	1493	Swainson's Thrush	1336
Barn Swallow	1488	Ruby-crowned Kinglet	1297
Purple Finch	1481	Veery	1267

Species richness

Some parts of southern Québec are richer in breeding birds than others. The 12 squares supporting the greatest number of species, together with the associated number of field hours, were:

UTM Co-ordinate	Location	Total number of species	Number of hours
19 360 5210	Cap Tourmente	141	266
18 730 5010	Katevale	141	167
19 270 5100	Victoriaville	132	139
19 530 5360	Rimouski	132	108
19 330 5190	Beauport	129	176
18 600 5050	Saint-Vincent-de-Paul (Île Jésus)	128	345
18 570 4980	Herdman–Rockburn	128	263
18 560 5030	Hudson	127	396
18 420 5130	Maniwaki	126	73
18 710 5000	Saint-Benoît-du-Lac	124	318
18 470 5060	Lac la Blanche	124	194
19 350 5030	Lake Mégantic	124	38

Gilles Falardeau

1 Species for which breeding records in southern Québec had been obtained before the Atlas period.

Frequency of occurrence by ecological region and mapping of relative abundance

It was clear from the outset of the Atlas project that about 1500 of the 5261 10 km × 10 km squares in southern Québec were not accessible by conventional road and that it would be impossible to cover the entire area. Although the inhabited portion of southern Québec was thoroughly atlassed, portions of the northern part of the Atlas area were not visited because they were too remote, were not accessible by road, or had no resident birders (see Figure 1, page 75). In addition, the proportion of squares visited for fewer than five field hours was higher in the northern part of the Atlas area than in the southern part (see Figure 2). This disparity clearly biases distribution maps for a number of species, particularly those (such as Swainson's Thrush) that are distributed widely in the boreal forest (Figure 3) in that such species appear to be less common in the north than they really are. Consequently, a statistical method was developed to correct this bias and facilitate correct interpretation of distribution maps.

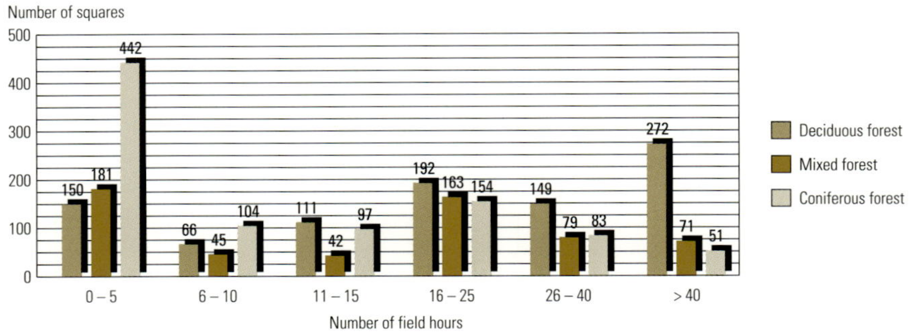

Figure 2
Atlassing effort by major forest zone

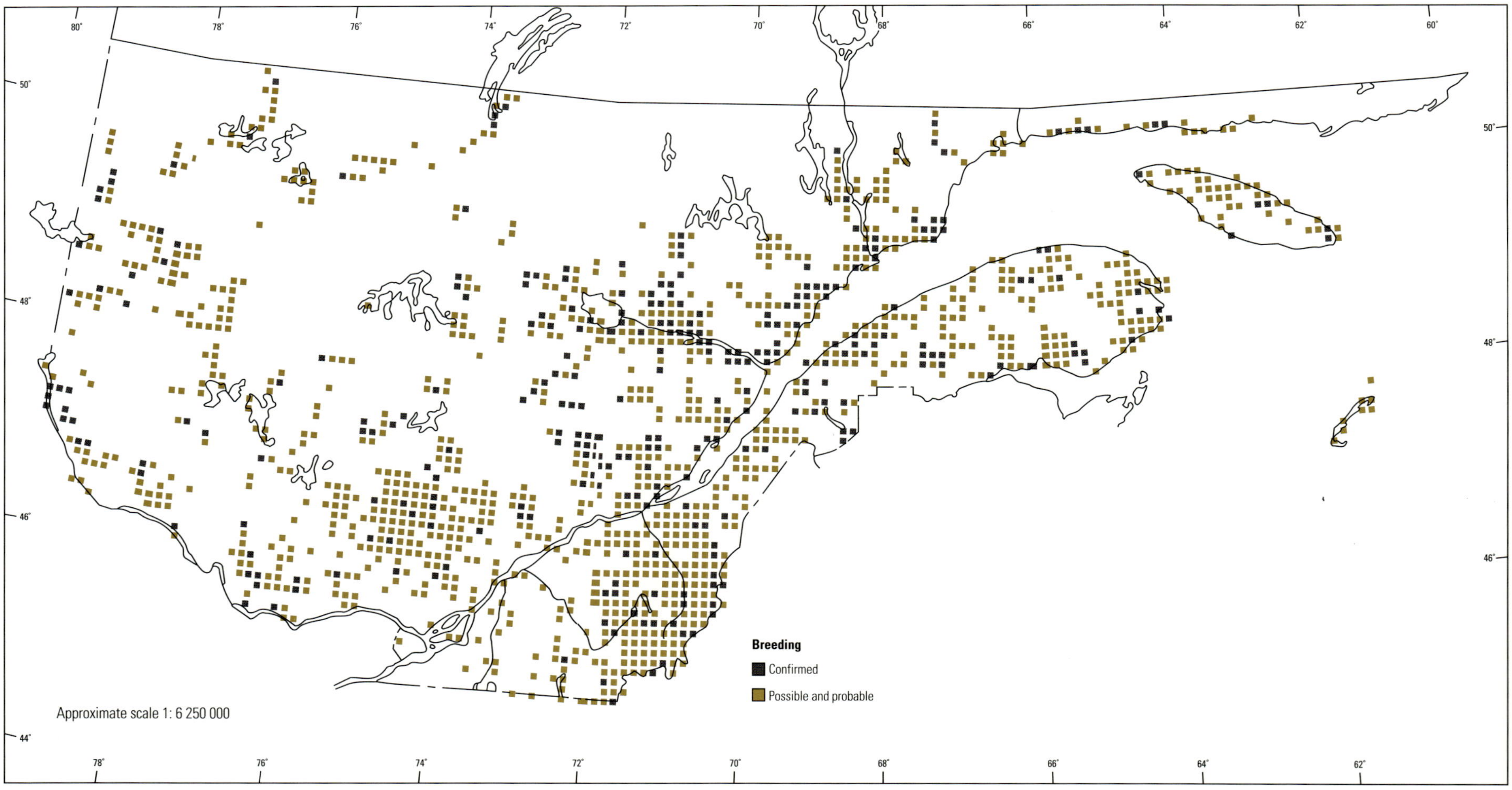

Figure 3
Distribution map for Swainson's Thrush

The concept underlying the method

The method is fairly simple: to compensate for the unevenness of atlassing effort, frequency of species occurrence is compared among ecological regions of southern Québec. The goal is not to estimate the probability of observing a particular species in a given square, but rather, using estimated relative frequencies of occurrence in the various ecological regions, to identify those southern Québec regions in which each species is most common, and hence determine its preferred regions.

Since vegetation is a key aspect of nesting habitat, southern Québec was subdivided into ecological regions, also referred to as forest regions in the Atlas, using the scheme described by Thibault (1985, 1988) and later simplified by NOVE (1990) (Figure 4). Mapping was further simplified by incorporating a few small patches of non-conforming forest vegetation (at most 14 squares) into the surrounding ecological region. For example, the Mont-Tremblant Recreation Park, which formed a small patch dominated by balsam fir and yellow birch, was included in the surrounding Sugar maple-yellow birch-beech ecological region (see Thibault 1985 or 1988).

Furthermore, the extensive forest islands in the mountainous regions of the Laurentide Wildlife Reserve and the Gaspé Peninsula belong to the same ecological region as the Middle North Shore; while these patches are fairly large and clearly distinguishable from neighbouring ecological regions, they are too small to be distinct entities. Grouping these patches into the same ecological region can bias the results, however: frequent (or infrequent) occurrence of a species in one patch will have an upward (or downward) influence on results for the other two.

The boundaries selected generally follow those of the ecological regions. It also occasionally proved useful to retain some of Thibault's (1985) subdivisions within some ecological regions. For example, by splitting the large Balsam fir-white birch ecological region into eastern and western sections using a north-south line through Lake Saint-Jean, habitat preferences for several bird species such as the Blackpoll Warbler, Fox Sparrow, and Pine Grosbeak could be described more accurately. Regions to the east of Lake Saint-Jean are more mountainous and generally receive more precipitation than do those to the west; in addition, unlike the western section, jack pine stands are practically non-existent in the eastern section.

Although they belong to the same ecological region (White spruce-balsam fir and balsam fir-white spruce), Anticosti Island and the Magdalen Islands had to be considered separately. Despite sharing the same forest type, these islands support very different avian communities. This phenomenon can be explained in part by the scarcity of large wooded areas and the ubiquitousness of the marine environment on the Magdalen Islands. Since they contain very few squares, however, the Magdalen Islands had to be excluded from the analysis. Finally, the Lower North Shore (Region 18) was excluded from the analysis because of insufficient coverage—not enough squares could be visited.

Figure 4 presents the modified map of ecological regions used to identify the preferred regions of bird species. The map reflects a compromise between a desire to use a fine-enough geographic breakdown to detect slight differences in bird distribution and the need to have enough squares in each region to obtain valid estimates of frequency of occurrence.

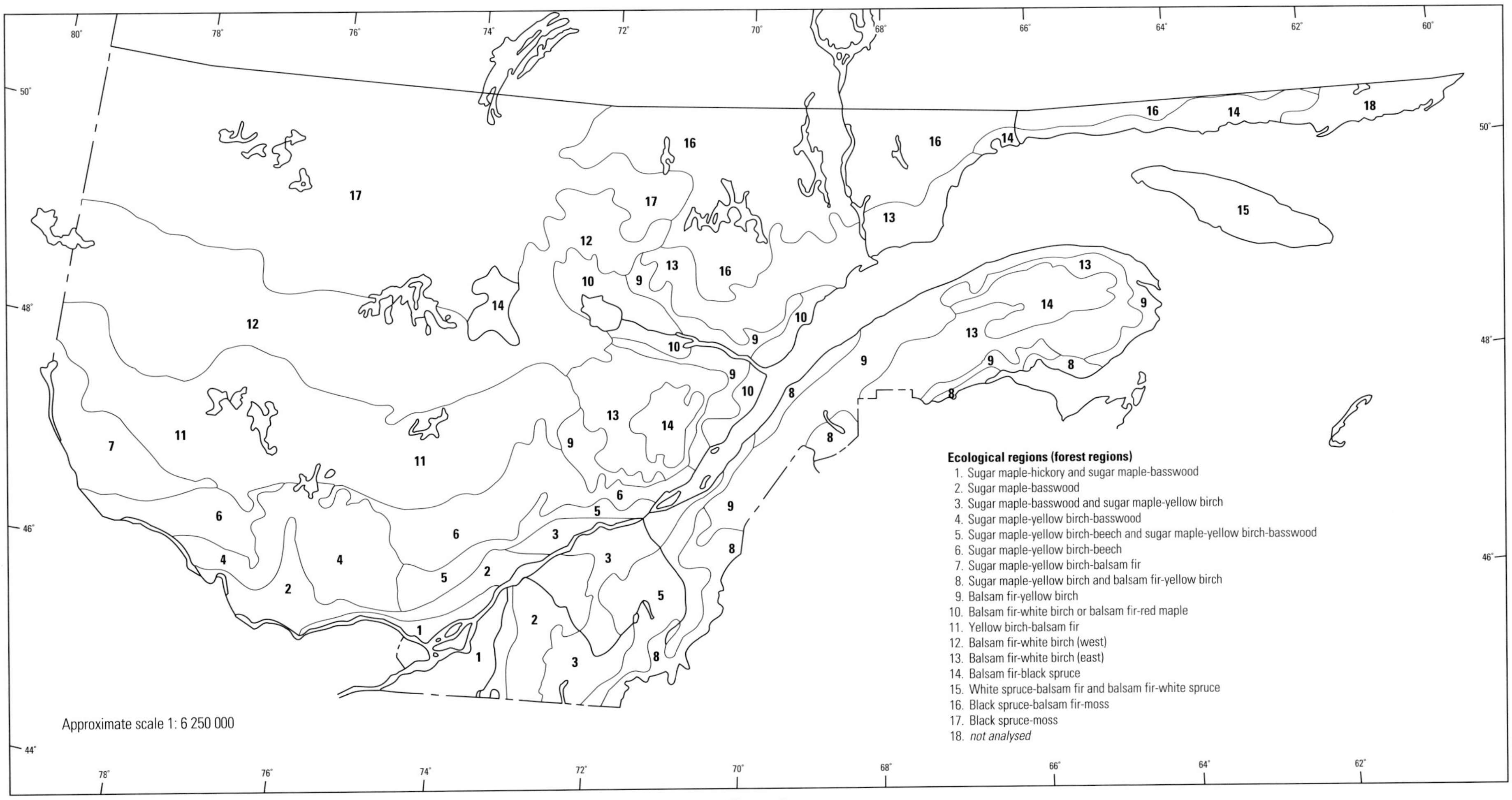

Figure 4
Map of ecological regions
Adapted from Thibault (1985, 1988).

While each ecological region is characterized by at least one climax community (Grandtner 1966), local edaphic factors or natural or man-made perturbations may produce several other types of communities there as well. The Balsam fir-white birch ecological region, for example, may contain maple, oak, basswood, yellow birch, elm, ash, or pine stands. In addition, the landscape of some Québec regions (such as the Central St. Lawrence Lowland) is no longer dominated by forest because of large-scale clearing for agriculture. In interpreting the results, therefore, the reader must always bear in mind that ecological regions as identified do not describe uniform habitats; it would be incorrect, for example, to automatically associate a given species with the sugar maple-basswood forest community simply because it is common in the Sugar maple-basswood ecological region. Rather, these regions should be viewed as "broad ecological regions" having certain climatic and biological characteristics that affect the distribution of various bird species.

A number of conditions apply to the statistical analysis of distribution maps. First, the absence of a species from a particular square must be a certainty, or nearly so, and not the result of inadequate coverage, which would underestimate the occurrence of the species in a given region. To satisfy this condition, only adequately covered squares were included in the analysis, thereby eliminating the problem of inter-regional variations in atlassing effort and making possible valid comparisons between ecological regions. The method used to select squares for inclusion in the analysis is presented later in this section. Frequency of occurrence and confidence interval calculations are based on the principle of random sampling. This principle could be adequately respected in inhabited regions in the south where all squares were easily accessible by road and had the same probability of coverage. Farther north, however, squares located close to inhabited areas (cities, villages, summer cottage country) and accessible by road or logging road had a greater probability of coverage than those that were uninhabited or difficult to access. The second condition requires that, if the sample cannot be completely random, it must at least be representative of all the squares in the entire ecological region. The squares that were adequately covered, hence included in the analysis, are shown in Figure 5; it can be seen that squares are fairly well distributed within regions.

In addition, analysis was restricted to forest birds and those whose distribution does not necessarily depend on existing forest vegetation but may be influenced by the same physical factors—for example, number of degree-days, precipitation, relief, and soil—that determine vegetation characteristics. Therefore, all species whose presence depended on a factor that could influence their frequency of occurrence in different parts of the same ecological region were generally excluded from the analysis. This is the case for open-country and wetland birds and those distributed only along the St. Lawrence. For example, in the Balsam fir-white birch (west) ecological region, one would expect the Killdeer to be much more common in Abitibi, which is largely open, than in the rest of the ecological region, which is extensively forested. The frequency of occurrence of this open-country species, therefore, would be proportional to the ratio of open-country squares to forested squares and may not be representative for the whole region. Rare species reported in fewer than 50 squares throughout the entire survey period were also excluded. Table 1 lists the species included in the analysis. It can be verified that the second condition (representativeness of the sample) is met, since the frequency of occurrence of these species should be fairly uniform throughout regions, hence less vulnerable to bias due to square distribution.

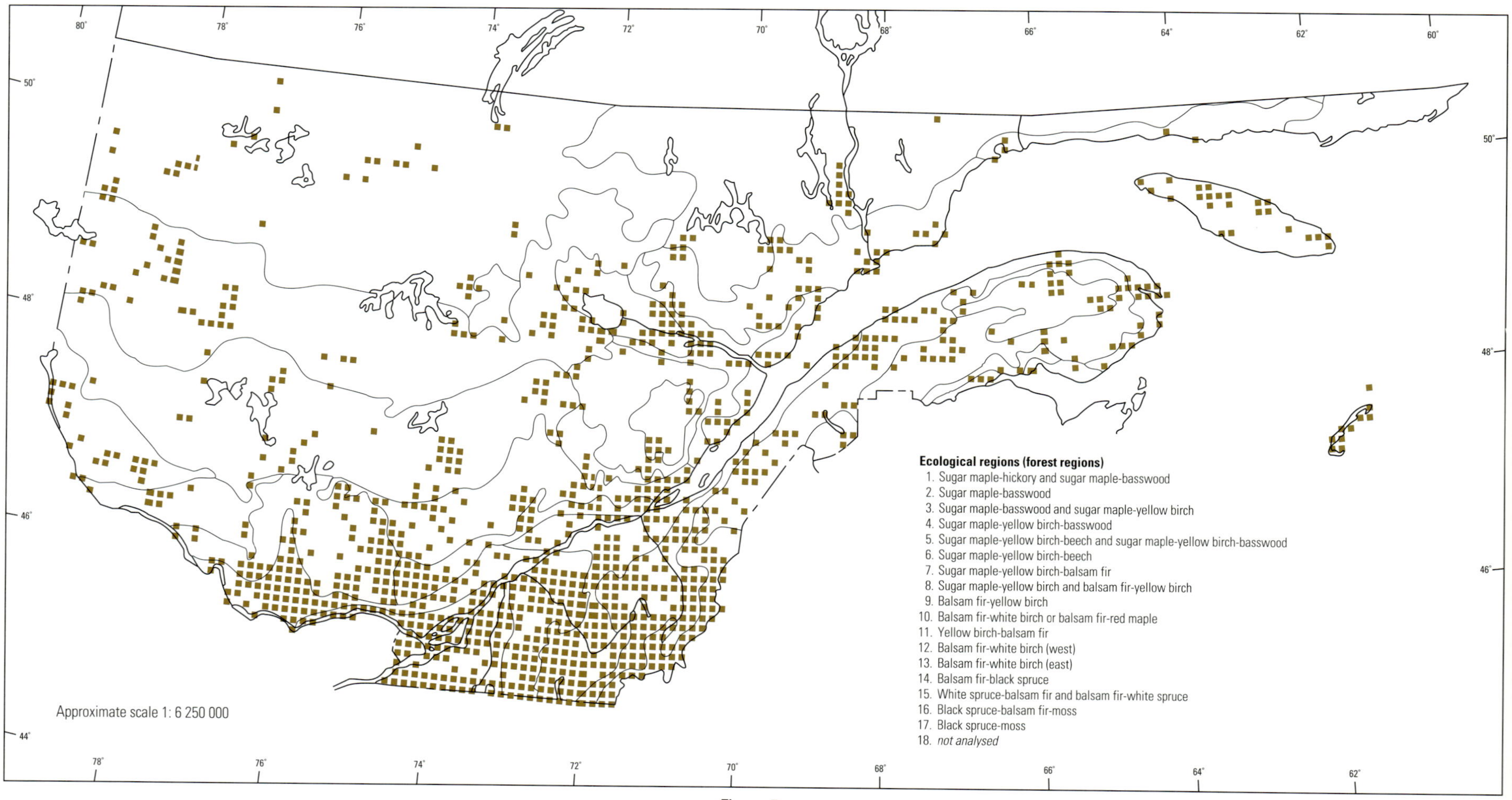

Figure 5
Location of adequately covered squares included in the analysis

Methodology

First, a threshold value had to be established for the number of observed species beyond which a square was considered to have been adequately covered. When atlassing birds in a given sector, a good birder quickly records several species in the initial hours in the field. Over time, fewer and fewer new species are observed; ultimately, the number of observed species tends to level off. This upper bound was the threshold to be identified.

Table 1
List of species included in the analysis

Common Loon	Olive-sided Flycatcher	Magnolia Warbler
American Bittern	Eastern Wood-Pewee	Cape May Warbler
Green Heron	Yellow-bellied Flycatcher	Black-throated Blue Warbler
Canada Goose	Alder Flycatcher	Yellow-rumped Warbler
Wood Duck	Willow Flycatcher	Black-throated Green Warbler
American Black Duck	Least Flycatcher	Blackburnian Warbler
Ring-necked Duck	Eastern Phoebe	Bay-breasted Warbler
Common Goldeneye	Great Crested Flycatcher	Blackpoll Warbler
Hooded Merganser	Tree Swallow	Black-and-white Warbler
Common Merganser	Gray Jay	American Redstart
Red-breasted Merganser	Blue Jay	Ovenbird
Sharp-shinned Hawk	American Crow	Northern Waterthrush
Northern Goshawk	Common Raven	Mourning Warbler
Red-shouldered Hawk	Black-capped Chickadee	Common Yellowthroat
Broad-winged Hawk	Boreal Chickadee	Wilson's Warbler
American Kestrel	Red-breasted Nuthatch	Canada Warbler
Merlin	White-breasted Nuthatch	Scarlet Tanager
Spruce Grouse	Brown Creeper	Northern Cardinal
Ruffed Grouse	House Wren	Rose-breasted Grosbeak
Greater Yellowlegs	Winter Wren	Indigo Bunting
Solitary Sandpiper	Golden-crowned Kinglet	Rufous-sided Towhee
Spotted Sandpiper	Ruby-crowned Kinglet	Chipping Sparrow
American Woodcock	Veery	Fox Sparrow
Black-billed Cuckoo	Swainson's Thrush	Song Sparrow
Great Horned Owl	Hermit Thrush	Lincoln's Sparrow
Barred Owl	Wood Thrush	Swamp Sparrow
Northern Saw-whet Owl	American Robin	White-throated Sparrow
Common Nighthawk	Gray Catbird	Dark-eyed Junco
Whip-poor-will	Brown Thrasher	Rusty Blackbird
Chimney Swift	Cedar Waxwing	Common Grackle
Ruby-throated Hummingbird	Solitary Vireo	Northern Oriole
Belted Kingfisher	Warbling Vireo	Pine Grosbeak
Yellow-bellied Sapsucker	Philadelphia Vireo	Purple Finch
Downy Woodpecker	Red-eyed Vireo	White-winged Crossbill
Hairy Woodpecker	Tennessee Warbler	Pine Siskin
Three-toed Woodpecker	Nashville Warbler	American Goldfinch
Black-backed Woodpecker	Northern Parula	Evening Grosbeak
Northern Flicker	Yellow Warbler	
Pileated Woodpecker	Chestnut-sided Warbler	

A threshold was established for each ecological region using Atlas data collected by some 150 experienced observers. Assuming that observations from these data cards were independent, the relationship between the number of observed species and field hours for each data card was modelled using non-linear regression, as follows:

$$Y_i = \beta_0 \times (1 - e^{-\beta_1 x_i}) + \varepsilon_i$$

where Y_i = is the number of species observed on data card i;

x_i = is the number of hours of observation for card i;

e = is a constant (= 2.718...);

ε = is a random error term for card i.

The parameter β_0 represents the asymptote toward which the curve tends. Regression models were fitted similarly for all ecological regions, using the least squares method. Figure 6 illustrates the regression curve obtained for the Sugar maple-basswood and sugar maple-yellow birch ecological region. Table 2 lists the estimates of β_0 for each region, as well as some other statistics. Note that this approach may eliminate some well atlassed (but less rich) squares in which the habitat is highly homogeneous because they support fewer species than squares with more diverse habitat.

Table 2
Thresholds for identifying adequately covered squares

Ecological regions using (Forest regions)	Estimate of β_0	Number of squares adequately covered using threshold = β_0	Selected threshold (lower limit of confidence interval for β_0)	Number of squares adequately covered selected threshold
1. Sugar maple-hickory and sugar maple-basswood	52.1 ± 2.7	89	49.4	97
2. Sugar maple-basswood	57.8 ± 2.1	141	55.7	153
3. Sugar maple-basswood and sugar maple-yellow birch	60.4 ± 2.8	90	57.6	101
4. Sugar maple-yellow birch-basswood	66.2 ± 4.0	49	62.2	56
5. Sugar maple-yellow birch-beech and sugar maple-yellow birch-basswood	60.6 ± 2.2	106	58.4	113
6. Sugar maple-yellow birch-beech	63.0 ± 3.0	67	60.0	71
7. Sugar maple-yellow birch-balsam fir	70.3 ± 4.6	11	65.7	19
8. Sugar maple-yellow birch and balsam fir-yellow birch	67.7 ± 2.9	39	64.8	47
9. Balsam fir-yellow birch	63.7 ± 2.2	76	61.5	87
10. Balsam fir-white birch or balsam fir-red maple	57.6 ± 2.7	35	54.9	46
11. Yellow birch-balsam fir	62.1 ± 2.9	42	59.2	52
12. Balsam fir-white birch (west)	54.4 ± 3.7	49	50.7	59
13. Balsam fir-white birch (east)	55.7 ± 1.9	69	53.8	76
14. Balsam fir-black spruce	50.9 ± 4.1	25	46.8	28
15. White spruce-balsam fir and balsam fir-white spruce	41.2 ± 3.5	14	37.7	22
16. Black spruce-balsam fir-moss	55.5 ± 8.2	6	47.3	18
17. Black spruce-moss	46.4 ± 2.7	24	43.7	32

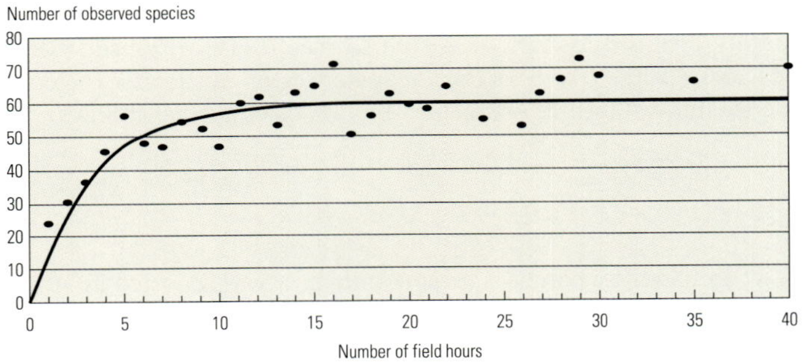

Figure 6
Non-linear regression of number of observed species as a function of time, in the Sugar maple-basswood and sugar maple-yellow birch ecological region

Use of β_0 as a criterion or threshold to determine whether or not a square was adequately covered did not admit enough such squares in all regions (Table 2). Consequently, squares were selected using the lower bound of the 95% confidence interval on the estimate of β_0. Table 2 presents the number of adequately covered squares by ecological region, after applying this new criterion.

Absolute and relative frequencies of occurrence of selected species were calculated for each region, from observations made in squares considered to be adequately covered. For example, Table 3 presents results for Swainson's Thrush. "Percentage of squares in which reported" indicates the ratio of the number of adequately covered squares in which the species was reported (squares in which reported) to the number of "adequately covered squares" in the region. The 95% confidence interval for the estimate of percentage of squares in which the species was reported is calculated using a Normal approximation (Cochran 1977:57):

$$p \pm [\, t\sqrt{1-f}\,\sqrt{pq/(n-1)} + 1/2n \,]$$

where p = the proportion of adequately covered squares in which the species was reported;

q = the proportion of adequately covered squares in which the species was not reported $(1-p)$;

t = critical value from the Normal distribution for the given confidence interval ($t = 1.96$ for a 95% confidence interval);

f = the ratio between the number of squares sampled (included in the analysis) and the number of squares in the statistical population (the total number of squares in the region, including those that were not visited);

n = the number of squares in the sample.

The confidence interval indicates the precision of the estimate of percentage of squares in which sightings were reported: the narrower the interval, the more accurate the estimate of the percentage. Confidence intervals tend to be wider in regions with few adequately covered squares (regions 7 and 16). Results must be interpreted with caution, therefore, when confidence intervals are wide compared to estimated percentages. Importantly, a number of species were withdrawn from the analysis because the estimate for this percentage was too imprecise.

"Relative abundance" indicates the ratio of the percentage of adequately covered squares in which the species was reported in a particular region to the same percentage in the region where the species is most common. An example of this calculation is provided here as an illustration.

Swainson's Thrush was reported in 58 of the adequately covered squares of region 3, Sugar maple-basswood and sugar maple-yellow birch (Table 3). The statistical population of this region consists of 145 squares, including 101 that were considered adequately covered. Swainson's Thrush was therefore observed in $(58 \div 101) \times 100 = 57.4\%$ of the adequately covered squares; the confidence interval for this estimate is ±5.8%, calculated as follows:

$$\pm [\, 1.96\sqrt{1-(101 \div 145)}\,\sqrt{(0.574 \times 0.426) \div (101-1)} + (1 \div (2 \times 101)) \,]$$

$$\pm (1.96 \times 0.551 \times 0.049 + 0.005) = \pm 0.058 = \pm 5.8\%$$

In regions where it is most common, it was reported in 100% of the squares. Therefore, its relative abundance is:

$$57.4 \div 100.0 = 0.574 = 57.4\%$$

Table 3

Frequency of occurrence of Swainson's Thrush by ecological region

Ecological regions (Forest regions)	Statistical population[1]	Adequately covered squares	Squares in which reported	% of squares in which reported	Relative abundance (%)
1. Sugar maple-hickory and sugar maple-basswood	121	97	15	15.5 ± 3.7	15.5
2. Sugar maple-basswood	198	153	46	30.1 ± 3.8	30.1
3. Sugar maple-basswood and sugar maple-yellow birch	145	101	58	57.4 ± 5.8	57.4
4. Sugar maple-yellow birch-basswood	154	56	38	67.9 ± 10.7	67.9
5. Sugar maple-yellow birch-beech and sugar maple-yellow birch-basswood	152	113	95	84.1 ± 3.9	84.1
6. Sugar maple-yellow birch-beech	253	71	64	90.1 ± 6.6	90.1
7. Sugar maple-yellow birch-balsam fir	105	19	17	89.5 ± 15.5	89.5
8. Sugar maple-yellow birch and balsam fir-yellow birch	140	47	47	100.0 ± 1.1	100.0
9. Balsam fir-yellow birch	336	87	85	97.7 ± 3.3	97.7
10. Balsam fir-white birch or balsam fir-red maple	118	46	46	100.0 ± 1.1	100.0
11. Yellow birch-balsam fir	531	52	51	98.1 ± 4.5	98.1
12. Balsam fir-white birch (west)	623	59	59	100.0 ± 0.8	100.0
13. Balsam fir-white birch (east)	439	76	75	98.7 ± 3.0	98.7
14. Balsam fir-black spruce	208	28	28	100.0 ± 1.8	100.0
15. White spruce-balsam fir and balsam fir-white spruce	114	22	22	100.0 ± 2.3	100.0
16. Black spruce-balsam fir-moss	503	18	18	100.0 ± 2.8	100.0
17. Black spruce-moss	1030	32	30	93.7 ± 10.0	93.7

1 The statistical population is the total number of squares in a given region, whether or not they were visited.

If Swainson's Thrush had been reported in only 95% of the squares in the region in which it occurs most frequently (instead of 100%), its relative abundance in the Sugar maple-basswood and sugar maple-yellow birch region would have been $57.4 \div 95 = 0.604 = 60.4\%$ and its relative abundance in the region in which it was most common would have been $95 \div 95 = 100\%$. Therefore, relative abundance is always 100% in the region where the species is most common. In addition, when a species is observed in all (100%) of the squares of at least one ecological region (as was the case with Swainson's Thrush), the relative abundance is identical to the percentage of squares in which the species was reported.

A table detailing the results of this analysis is included on the map page for each species. Appendix I presents a comprehensive table giving the frequency of occurrence for almost all species reported during the Atlas survey period in each ecological region.

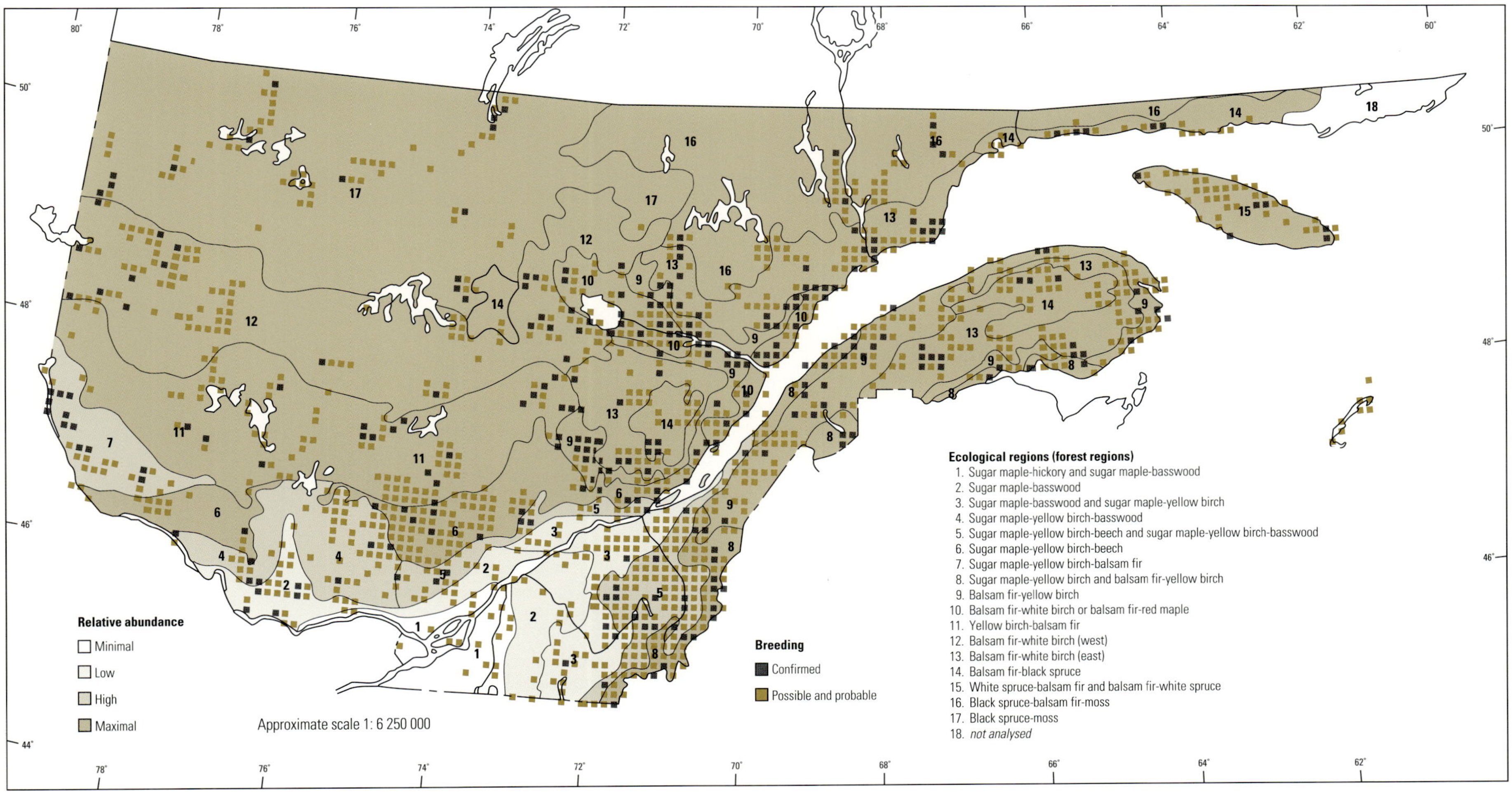

Figure 7
Distribution and relative abundance of Swainson's Trush

Relative abundance for each ecological region is illustrated on the species distribution maps by white and three tones of ochre, corresponding to the following percentage ranges: 90–100%, maximal relative abundance (dark ochre); 60–89.9%, high relative abundance (medium ochre); 30–59.9%, low relative abundance (light ochre); and 0–29.9%, minimal relative abundance (white). The main reason for presenting relative abundance instead of percentage of squares in which reported was to highlight those regions in southern Québec in which each species was most common (these regions are always dark ochre). The disadvantage of this approach, however, is that interspecies comparisons may not be made by referring to colours on the maps. Such comparisons may be made, however, on the basis of the percentage of squares in which the species was reported, by consulting the summary table accompanying the distribution map for each species included in the analysis, or by consulting the comprehensive table in Appendix I.

Figure 7 presents the final result from the approach described above, by superimposing relative abundance by region for Swainson's Thrush on the distribution map obtained from Atlas data. The figure clearly illustrates that this species is very common in the boreal forest; this result is not apparent from the raw distribution map (Figure 3).

Gilles Falardeau

Breeding chronology

Most species accounts include a chronology chart showing the duration of the various stages of the species' breeding cycle in Québec and tracking the species' occurrence throughout the year (Figure 1). In the chart, each month is divided into four parts: the 1st to the 7th day of the month, the 8th to the 14th, the 15th to the 21st, and the 22nd to the last day of the month. The top line in the chart, "Laying and incubation," shows the period when eggs can be found in the nest. The second line, "Young in nest," shows the period when young birds are in the nest and the third line, "Dependence of young out of nest," shows the period when young that have left the nest but are still dependent on their parents can be seen. Finally, the bottom line, "Seasonal occurrence," provides a general idea of the species' abundance at various times of the year.

Chronology

How to read the chart

The horizontal lines in the chart show the duration of the various stages in the breeding cycle (laying and incubation, young in nest, dependence of young out of nest), from the earliest reported or extrapolated date to the latest reported or extrapolated date throughout Québec, regardless of latitude. In general, observations of a species are not distributed uniformly throughout the period shown on the chart but tend to be concentrated in the middle, reflecting the tendency of most individuals to engage in the behaviour at that time. Furthermore, for species with a widespread distribution in the province, these dates vary considerably with latitude, with the difference ranging from a few days to a few weeks.

The black portion of the line for each stage, based on observed data, shows the interval between the earliest and latest dates the behaviour was actually observed. In some cases, this interval was extended by extrapolating from the observed data and from the literature, which is shown in ochre. (The principles behind these calculations are discussed later in the section.) For the line representing "Dependence of young out of nest," the arrow symbol is used when the period of dependence may be considerably longer, as in swans and geese, for example. The arrow signifies that the reader should refer to the species profile for further information. It is also used when this period is not precisely defined—for example, "until autumn," as in the case of the Great Egret or, "until winter," as in the case of the Gray Partridge. Lastly, where the species profile gives the period of dependence as "several weeks," this was interpreted for the purposes of the chronology chart as three weeks.

The period of dependence out of the nest is defined as the elapsed time during which the young depend wholly or in part on the parent birds for some of their needs such as food and protection, regardless if they are capable or not of flight. In altricial, semi-altricial, and semiprecocial species, this period is taken to end when the parents cease feeding their young. In precocial species, where the young are either dependent on their parents for food for only a very short time (rails) or not at all (ducks, gallinaceous birds, shorebirds, etc.), this period is taken to end when the brood is finally abandoned by both parents. The "Young in nest" period is taken to end when the young first leave the nest voluntarily—even if they remain in the branches nearby as in the case of herons and raptors, on the ground as in the case of gulls and terns, or if they return to the nest to be fed or brooded at night. In species with precocial young which leave the nest less than two days after hatching and often on the very same day, the period when young may be observed in the nest is not shown since it is so short.

Sources of data

The data for the chronology charts came mainly from the Québec Nest Record Card Program (QNRCP) administered by the Canadian Museum of Nature in Ottawa. This database contains slightly over 21 000 records, dating from 1955 to 1989. Data on nesting were also taken from 5194 records dating from 1955 until the end of 1989 in the ÉPOQ database, which contains approximately 150 000 checklists. In addition, 1123 Atlas records from 1984 to 1989 were also included, as was information from egg collections at the Canadian Museum of Nature, the Florida State Museum in Gainesville, and the Carnegie Museum of Natural History in Pittsburgh. (All these museums have egg collections from Québec.) This information was further supplemented with data from various reference works and from the personal notes and observations of a number of contributors, whose names are listed at the end of the section.

The total number of data items used to produce the chart for each species is represented by the value *n*, with each brood or clutch being considered as a single data item. For colonial species, *n* represents the number of information sources (QNRCP records, ÉPOQ checklists, study results, personal communications, etc.).

Data that could not be dated exactly or matched with one of the three stages in the breeding cycle (laying and incubation, young in nest, dependence of young out of nest) were not used. Therefore, a record of a nest observed sometime in June but not precisely dated was rejected, because it could not be assigned to one of the four periods in the month. Furthermore, breeding evidence such as adults transporting food was not used, since it was not known whether the food was for young in or out of the nest. The only Atlas records used for the chronology chart were observations of downy young or recently fledged young incapable of sustained flight (*JE* code), since the other types of breeding evidence could not be matched with any of the three stages. In addition, unusual records were rejected in many cases for lack of adequate documentation. ÉPOQ data were also validated to eliminate coding and identification errors. In short, only reliable data were used to draw up the breeding chronologies.

Extrapolated data

Once the data has been compiled, certain extrapolations could be made to extend the stages in the breeding cycle using actual known dates. For example, if the hatching date of a clutch is known, the dates for laying, nest leaving, first flight, and dependence of young can be extrapolated as long as the incubation period, nestling period, age at first flight, and duration of the period the young are dependent are known. Known laying dates or dates of first flight can also be used to calculate the dates of the various stages in the breeding cycle. Furthermore, age estimates of a chick or brood can also be used in some cases. Plumage-based classification systems have been developed for various groups of birds allowing the age of chicks to be determined fairly accurately, such as Gollop and Marshall's (1954) scheme for ducks. Data for related species were used to extrapolate the information in species for which such a system has not been developed.

The following example illustrates how information can be used for extrapolation purposes. On 13 May, a Black Tern nest containing 3 eggs is found. Hatching date is not known, consequently this nest is used only to illustrate actual dates and not for extrapolation. A second nest, also containing 3 eggs, is found on 15 May; one of the eggs in this clutch is in the process of hatching. This clutch, therefore, will be used for extrapolation purposes. We already know that, in this species, incubation lasts 19–22 days and that the young leave the nest 2–3 days after hatching and can fly at the age of

18–28 days, and become independent 13 days afterwards. Therefore, we can posit that the egg in the process of hatching was laid 19–22 days earlier; to be conservative, we will use 19 days since this does not take us so far back in time. This means that our egg (the first egg) was laid on 26 April; this is well before the first actual record of a nest with eggs we have for the species (13 May). This allows us to extend earlier the laying and incubation period by roughly a fortnight (see Figure 1). To continue our calculations, since our egg hatched on 15 May, the chick left the nest two or three days later, on 18 May. This is roughly three weeks earlier than the earliest record of young out of the nest.

To continue with our example, the latest record of eggs in the nest for the species is 15 July, roughly one week later than the latest record of young in the nest. Therefore, we can use this information to extend our chronology. Since incubation in the species begins when the first egg is laid and the female lays one egg a day, hatching is asynchronous—in other words, the second egg in the clutch will hatch a day later than the first egg, and the third egg, two days later. Although we do not know the exact hatching dates for this late clutch reported in the nest on 15 July, at a conservative estimate we can assume that the first egg hatched the same day. Therefore, we can extrapolate that the first chick to hatch left the nest two or three days later, on 17 July, and that the last chick hatched on 17 July, remained in the nest until at least 19 July, and made its first flight 18–28 days later or, assuming the earliest date (18 days), on 4 August. Lastly, the end of the period of dependence of the young out of the nest is 13 days from the date of the first flight, or 17 August. Figure 1 shows the final results of our calculations.

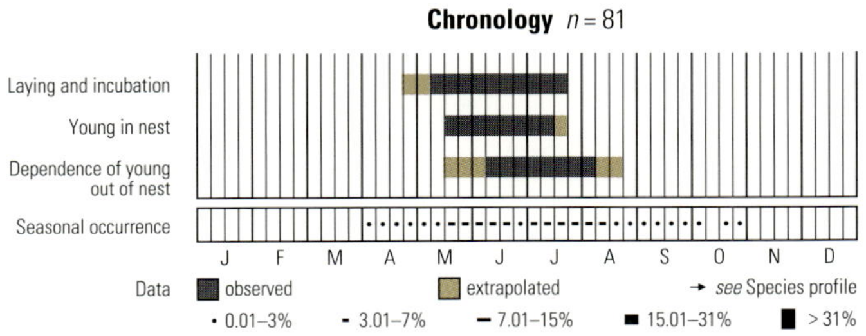

Figure 1

Breeding chronology and seasonal occurrence of the Black Tern in Québec

Special cases

Data from the species profiles were used almost exclusively to perform the calculations. However, in a few special cases where the species profile did not provide enough information, other data had to be used.

For example, when the age at first flight of an altricial or semialtricial species was not known, we assumed that it was at least equal to the number of days the young spend in the nest, which was indicated in the species profile by the phrase "according to nestling period." When no data at all were available, we used data for the most closely related species. For example, for the incubation period of the Surf Scoter (25–33 days), we used the extreme values recorded for the other two scoter species in Québec and, for the incubation period of the Cape May Warbler (10–15 days), we used the extreme values recorded for all other warbler species. For the Iceland Gull, we assumed that the nestling period was 2–3 days, which corresponds to the period observed in most other gull species.

The nestling period could not be determined for several species of warblers, including the Tennessee Warbler, Cape May Warbler, and Blackburnian Warbler. We used a period of 7–14 days, which corresponds to the extreme values recorded for all the other warbler species. The nestling period of 9–10 days used for Le Conte's Sparrow is that observed in other *Ammodramus* sparrows.

For some precocial and semiprecocial species, age at first flight was extrapolated from data of closely related species: 18–25 days for the Solitary Sandpiper (based on data for the Lesser Yellowlegs and Greater Yellowlegs); 16–20 days for Wilson's Phalarope (based on data for the Red-necked Phalarope and Red Phalarope); 35–42 days for Bonaparte's Gull (based on data for the Common Black-headed Gull); and 35–40 days for the Iceland Gull (based on data for the Ring-billed Gull).

When it was not known how long the young remained dependent, age at first flight was used to calculate the end of the period of dependence. We took this approach even if the period of dependence in the species ended before first flight, as in ducks and sandpipers. Although the young are independent at this time, they cannot fly and are still very vulnerable.

It is important to make a distinction between observed data and extrapolated data. Although the age at first flight of a very late clutch may be extrapolated, there is no assurance that the young will survive until they can fly. Indeed, this is the case for late clutches in the northern part of a species' range as they are more exposed to the inclement weather that occurs at the end of the summer and early autumn.

Seasonal occurrence

The line for seasonal occurrence in the chart gives an idea of the species' frequency of observation throughout the year in Québec as well as the periods when migrant breeders and transient species are migrating. The calculations were based on approximately 150 000 checklists in the ÉPOQ database dating from 1955 to the end of 1989. Seasonal occurrence was calculated by dividing the number of ÉPOQ checklists with the species checked off by the total number of checklists received during the period in question. This figure was then multiplied by 100 to express the result as a percentage, classed in the following ranges:

0.01–3% 3.01–7% 7.01–15% 15.01–31% >31%

Data on seasonal occurrence were also obtained from the sources listed at the end of this section. Data compiled and updated by the authors of the Québec regional report for the publication *American Birds* was also used (M. Gosselin, R. Yank, and N. David, unpubl. data), involving mainly records of rare species, earliest arrival or latest departure dates, and all winter records for migratory species. In the case of some northern species, there were periods during the breeding season for which no data could be given because of the small number of records from the north. For these species, a dot was used to replace the missing data as long as it fell within the species' breeding season as shown in the breeding chronology chart.

Although the aim in illustrating seasonal occurrence was to give a general picture of a species' presence at various times of the year in Québec as a whole, the data reflect mainly the species' frequency of observation in southern Québec, particularly the inhabited regions where most of the records came from. In other words, these data should be interpreted as the frequency with which the species was observed by birders rather than seen as a true picture of the species' frequency of observation. The number of records depends on many factors such as how easy it is to detect and identify the species, the methods used by birders, and the type of habitats they visit. For example, the abundance curve for many species, particularly passerines, drops in the middle of summer, not because the species is less abundant but because, beginning in July, it sings much less and is more difficult to spot. The Black-capped Chickadee is another example; contrary to what the chart would seem to show, numbers do not actually increase in winter. However, in winter, the birds travel in flocks and visit feeders in large numbers, making them more conspicuous. Readers should also note that it is not always possible to distinguish on the chart between species that winter regularly in Québec and species that only winter sporadically here: a continuous series of dots in the winter months may merely mean that a single individual or a single group of individuals overwintered on one occasion. Therefore, readers should refer to the species account, which usually indicates in the introduction whether or not the species is a regular permanent resident in Québec. Lastly, readers should also note that, as for the breeding period, there is usually a discrepancy between the time a species arrives in the southernmost part of the province and in more northerly regions.

Gilles Falardeau

REFERENCES USED TO PRODUCE BREEDING CHRONOLOGY CHARTS

The following individuals and organizations provided unpublished data, unpublished observations, personal communications or personal notes:

Yves Aubry (CWS); Jean Bédard (Laval U.); Daniel Bordage (CWS); Gilbert Bouchard; Pierre Brousseau (CWS); Jacques Cayouette (Laval U.); Gilles Chapdelaine (CWS); Léo-Guy de Repentigny (CWS); Jean-Luc DesGranges (CWS); Serge Dulac; Gilles Falardeau (CWS); Sylvie Gagnon (Ducks Unlimited, Rouyn-Noranda); René Gaudet; Jean Gauthier (CWS); Yves Gauthier; Pierre Gingras; GREBE; Michel Julien; François Lambert; Jean Lapointe; Pierre Laporte (CWS); Robert Lebrun; Claire Lépine; Mabel McIntosh; François Morneau; Alain Morrier; Pierre Mousseau; Henri Ouellet (CMN); Réginald Ouellet (MEF); Sylvain Paradis (Parks Canada); Austin Reed (CWS); Michel Robert (CWS); Line Rochefort (Laval U.); Jean Rodrigue (CWS); Jacques Rosa (CWS); Gilles Seutin; Daniel St-Hilaire (MEF); Serge Tanguay; Stéphane Verd; Jacqueline Vincent (CWS); Richard Yank; the staff of La Mauricie National Park; MEF specimen collection.

Other sources consulted include:

Alliston (1979a); Ball (1954); Bateman (1953); Bédard (1969b); Benoit et al. (1993b); Bent (1938); BIOMQ; Boulva and Browne (1964); Cairns (1981); Chabot and Barette (1989); Chabot et al. (1982); Chapdelaine et al. (1985); Cons. G. G. – GREBE (1990a,c, 1991a, 1992a,g,h,j,t,v, 1993a,c,g); D'Amours (1989); Dionne and St-Georges (1993); Ducas (1971); Fradette (1992); Gaboriault (1961); Gaston et al. (1985); Gilbert (1981); Gosselin (1977a); Harper (1958); Harris (1989); Heyland et al. (1970); Hildebrand (1950); Hudon (1978); le Bel (1990); Lévesque and McNeil (1985); McNicoll et al. (1991); Mousley (1924c, 1931c, 1937b); Olpinski (1986); Pilon et al. (1983a); Reed et al. (1994); Robert and Laporte (1991a); Rochefort (1983); Tardif and Morneau (1991); Terrill (1931a); Todd (1963).

World breeding range

This map illustrates each species' world breeding range, and is particularly useful in understanding the distribution of circumpolar, circumboreal, holarctic, and Nearctic species. The map also shows at a glance whether a species is at the northern or southern limits of its range in Québec. However, because of the map's small scale, the information provided is very general.

The following sources were consulted in drawing up the map:

AOU (1983, 1985, 1987, 1989, 1991, 1993); Benkman (1992); Blake (1977); Boag and Schroeder (1992); Bull and Duncan (1993); Butler (1992); Cadman et al. (1987); Campbell et al. (1990a,b); Cannings (1993); Confer (1992a); Cramp (1985, 1988, 1992, 1994a,b); Cramp and Perrins (1993); Cramp and Simmons (1977, 1980, 1983); Dement'ev (1951a,b,c, 1952, 1954a,b); Derrickson and Breitwisch (1992); Dunning (1982); Eaton (1992); Ellison (1992); Erskine (1992); Gauthier (1993a); Gibbs et al. (1992a,b); Godfrey (1986); Gratto-Trevor (1992); Haig (1992); Harrison (1982); Harrison (1983); Hayman et al. (1986); Hill (1993); Johnston (1992); Long (1981); Lowther (1993); Lowther and Cink (1992); Madge and Burn (1988); Marti (1992); W. A. Montevecchi (unpubl. data); Montevecchi and Tuck (1987); National Geographic Society (1987); Naugler (1993); Newton (1973); Palmer (1962, 1976, 1988); Parmelee (1992); Payne (1992); Peterson (1980, 1984); Robbins et al. (1980); Ryder (1993); Smith (1993); Sodhi et al. (1993); Strickland and Ouellet (1993); Tacha et al. (1992); Wheelwright and Rising (1993).

Abundance maps for waterfowl and Common Loon in southern Québec

A helicopter-based survey method was tested between 1985 and 1989 over an 100 000 km² area in the Abitibi region of Québec (Bordage 1987, 1988a,b). Beginning in 1990, as part of the Black Duck Joint Venture, the method was applied to the study of American Black Duck populations in boreal habitat south of 50°30' N. The primary objective of this census program was to verify trends in the American Black Duck breeding population, in response to a 30-year decline shown in winter Atlantic seaboard surveys by the United States Fish and Wildlife Service (Rogers and Patterson 1984; Rusch et al. 1989). At the same time, the census program provided an opportunity to collect breeding data on several other waterfowl species and map their general distribution in southern Québec.

Data used to illustrate waterfowl population densities (and consequently their abundance in southern Québec) were provided by surveys conducted in 1990, 1991, and 1992 in 83 squares, each 10 km × 10 km and situated in the southwest corner and centre of a 100 km × 100 km block (according to the Universal Transverse Mercator [UTM] grid system). Data were collected in May, during the breeding season for American Black Duck. Squares were surveyed by three teams, each composed of three individuals, as follows: a navigator-compiler sitting in the front of the helicopter (Bell 206L) indicated the flight plan to the pilot and noted, on a transparency film placed over a topographic map of the square (1:50 000), all observations dictated by two observers sitting on opposite sides of the rear of the cabin. The helicopter was equipped with "bubble" type rear windows to maximize the observers' field of vision.

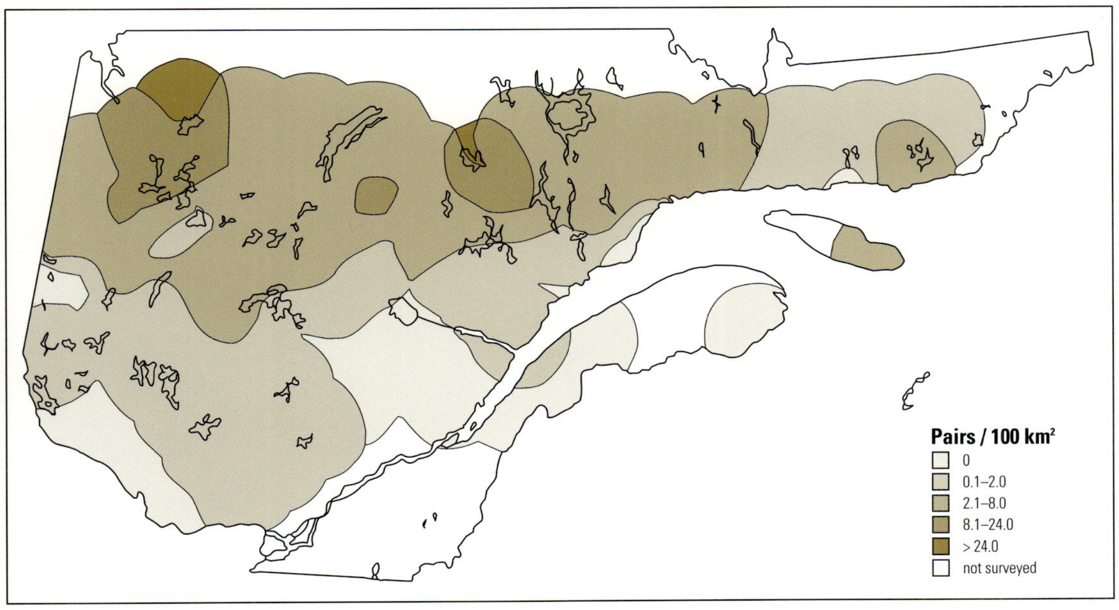

Figure 1
Canada Goose abundance in southern Québec.
Aerial surveys conducted in May 1990, 1991, and 1992
(D. Bordage, CWS, unpubl. data).

The itinerary within each square called for overflight of shores of all bodies of water, at a distance not exceeding 100 m. Flight speed varied from 50 to 100 km/hr; the helicopter flew at an altitude of between 5 and 50 m, depending on the complexity of the aquatic habitats and the topography. Flights generally occurred during good weather, when there was no precipitation and the wind was under 30 km/hr. Although surveying was done during the day, the hour after sunrise and hour before sunset were avoided because of the poor quality of the light (yellow blinding light and long shadows along shores). All 83 squares were surveyed in this way once yearly, over a period of about three weeks.

While all observations were noted during survey missions, only breeding pairs were included in the analysis. A breeding pair was defined as two birds of the same species (whether or not of opposite sex) observed together; single birds (male or female) were assumed to be mated and were treated as a pair. Data were compiled in a computerized data base and geo-referenced using the UTM cartography system.

Abundance maps were then constructed from these data, to illustrate the distribution of observations in southern Québec (Figure 1). To simplify presentation on the maps, observations have been grouped into five spatial density classes reflecting the number of pairs per 100 km². For each species, the mean number of pairs observed in each square over the 3 years of surveys was assigned to the co-ordinates of the centre of the square. Since these were discrete, systematically distributed data, interpolation was necessary to provide a continuous representation of the various bird density data. This operation was carried out using SPANS (Spatial Analysis System), available from TYDAC Technologies. Interpolation was based on circular influence zones of radius 75 km. As the centres of the squares were approximately 70 km from one another, the interpolation took account of densities observed in (as many as four) neighbouring squares. Therefore, the final abundance map is less dependent on extreme observations; in addition, density classes fit together more harmoniously on the map. If a particular influence zone does not include another square, the affected zone is represented by a circle with a 75-km radius, whose density class corresponds to the observed mean in the single square surveyed. This effect is illustrated by the distribution of species for Anticosti Island and the Gaspé.

The survey was timed to coincide with the breeding season for American Black Duck. Therefore, abundance maps are probably more accurate for early-breeding waterfowl species.

Daniel Bordage and Marcelle Grenier

THE BIRD FAMILIES OF QUÉBEC

TAXONOMY OF THE BIRDS OF QUÉBEC

SINCE TIME IMMEMORIAL, PEOPLE HAVE TRIED TO CATEGORIZE THE LIFE forms around them, and birds, of course, have not escaped their attention. A classification system may be based on many varied criteria. For instance, birds can be grouped by colour, way of life, morphology, or any other characteristic, though this can result in birds known to be different and not at all closely related being classed together. In Québec, for example, in his book *Charmants voisins*, Mélançon (1940) groups birds according to whether they run, climb, perch, hover, or swim. Similarly, Surprenant (1993), in a guide to wetland birds, groups species by size. Although these types of classifications have their own value and may be useful, they are rarely employed in modern ornithological publications. Biologists usually classify birds, other animals, and plants on the basis of phylogenetic criteria, that is, criteria that can distinguish between and group the most closely related and the most genetically similar forms. The base unit of such classifications is usually species. In biology, a species is defined as a group of individuals belonging to interbreeding populations that do not breed with other groups.

Taxonomy, the scientific discipline that is concerned with the classification of life forms, owes a great deal to the Swedish naturalist Carl Linnaeus (1707–1778), the originator of the binomial nomenclature still used today. His system consists in giving each species a two-part latinized name and grouping species in a genus on the basis of common characteristics, genera in an order, and orders in a class. Taxonomy has made enormous progress since the time of Linnaeus, and in particular since the second half of the 19th century, when the theory of evolution really gained recognition as a result of Charles Darwin's work on natural selection (Mayr and Ashlock 1991). It is now thought that all forms of life on Earth have a common ancestor, and that the approximately 10 000 species of birds (9672, according to Sibley and Monroe 1990) that now inhabit it are descended from a reptilian ancestor, probably a small dinosaur that lived in the Jurassic, at least 150 million years ago (Feduccia 1980; Sibley and Ahlquist 1990). The discovery of fossils such as those of *Archeopterix*, *Hesperornis*, and *Ichthyornis*, all of them toothed birds that lived 70 to 150 million years ago (del Hoyo et al. 1992), is the most eloquent proof of this origin and descent.

Bird evolution took place through gradual differentiations spread out over generations, over tens of millions of years. Since Darwin's time, taxonomists have been classifying birds according to evolutionary criteria, that is, on the basis of characters through which the true degree of relationship between species can be assessed. They proceed on the assumption that related birds, that is, those having a recent common ancestor, have more traits in common than do distantly related birds. They therefore group species into genera, then families and so on, as Linnaeus did. In today's system each group constitutes, theoretically, a set whose representatives are all descended from the same ancestor. The classification is thus somewhat more natural; it expresses as faithfully as possible the relationship that exists between the various groups composing it. Nonetheless, within this so-called natural classification, only at the species level can biological entities be percieved with the senses; higher categories (genus, family, and so on) are purely theoretical, conventionally accepted entities, which are nevertheless very convenient for illustrating the evolution of various lines of birds.

Until quite recently, anatomy and morphology were the main criteria taxonomists used to assess the relationship between birds. Although these characters are useful, they are sometimes deceptive, since groups of birds belonging to distantly related lines sometimes independently acquire the same trait, especially when the birds live in similar ecological conditions. In biology, this phenomenon is known as convergent evolution. Among Québec birds, for example, the Chimney Swift and swallows are similar at first glance, but the swifts are more closely related to the hummingbirds than to the swallows, which are more closely related to the kinglets than to the swifts. Conversely, divergent evolution may make closely related birds look very different. Thus, despite their dissimilar appearances, the Evening Grosbeak and White-winged Crossbill are near relatives. This is also true of the Chimney Swift and Ruby-throated Hummingbird. Convergent and divergent evolution may therefore be a serious problem in traditional taxonomy.

Modern taxonomy is now based not only on anatomy and morphology, but also on ethological, ecological, biogeographic, and even biochemical criteria. In fact, the last few years have seen rapid advances in molecular biology, and birds are now compared on the basis of similarities in their enzymes, proteins, and DNA (Sibley and Ahlquist 1990; del Hoyo et al. 1992). DNA testing is a highly revolutionary method, since DNA, the building block of heredity, is definitely the most reliable criterion of comparison for assessing the degree of kinship between living beings. Another advantage of this type of biochemical analysis is that it prevents faulty conclusions being drawn on the basis of deceptive appearances resulting from convergent and divergent evolution (Sibley and Ahlquist 1990).

The table that follows lists in taxonomic order the 293 birds treated in this Atlas, grouped by taxon, from order down to species. All birds are members of the class Aves (Latin for birds). The members of each taxon are listed in a natural sequence, in which, theoretically, birds with the most primitive characteristics come first, and near relatives are closest together. So, according to this list, the Canada Warbler is more closely related to Wilson's Warbler than to the Nashville Warbler, which belongs to a genus (*Vermivora*) having more primitive characteristics than the genus *Wilsonia* to which the Canada Warbler belongs. At another level of taxonomy, the list shows that the Gaviiformes are closer to the Podicipediformes than to the Passeriformes, the order of birds with the most highly advanced (derived) characteristics.

There are several taxonomic lists in ornithology. The Atlas list follows that of the American Ornithologists' Union (1983, 1985, 1987, 1989, 1991, 1993) for the birds of North America. The work of Sibley and Ahlquist (1990), however, which is based on the study of genetic material (DNA), has produced a list of birds of the world (Sibley and Monroe 1990) very different from this more traditional one. For example, according to Sibley and Monroe's (1990) classification, the Galliformes would head the list of birds of Québec as the order with the most primitive characteristics, followed by the Anseriformes, Piciformes, and Coraciiformes. These authors propose many more modifications (see David 1991; Salzman 1993), some of which will probably be built into future classifications, as further research corroborates that of Sibley and his colleagues.

Each of the birds on the taxonomic list is designated by its scientific name and its common name. The scientific name consists of two latinized words; the first, the genus, always takes a capital; the second, the species, always takes a lowercase letter. The common names (the English designations used by North American ornithologists) and the scientific names are those from the AOU Check-list (1983) and its supplements.

The names of families, orders, and other taxa are latinized. Under each family (except for the Emberizidae which is discussed under the subfamilies), we have shown in parentheses the number of breeding species in the world, in North America, in the continental United States, in Canada, and in Québec. The number of species of birds recognized around the world varies depending on the author and advances in ornithology. The number of species grouped in various families may also differ according to the arrangement of the various taxonomic lists. As a result, the figures in parentheses should not be considered fixed or definitive. The number of breeding species in the world was calculated on the basis of Sibley and Monroe's list (1990), while the figures for North America, the continental U.S., and Canada are taken from DeSante and Pyle (1986). The Québec figures indicate the number of species confirmed as breeding in Québec.

MICHEL ROBERT

Taxonomic List of Birds Described in Atlas

Class	Order	Suborder	Superfamily	Family (Wo, NA, US, Ca, Qc)*	Subfamily	Tribe	Scientific Name Genus Species	Common Name
Aves								
	Gaviiformes			*Gaviidae*			*Gavia stellata*	Red-throated Loon
				(5,5,5,4,3)			*Gavia pacifica*	Pacific Loon
							Gavia immer	Common Loon
	Podicipediformes			*Podicipedidae*			*Podilymbus podiceps*	Pied-billed Grebe
				(21,7,7,6,3)			*Podiceps auritus*	Horned Grebe
							Podiceps grisegena	Red-necked Grebe
	Procellariiformes			*Hydrobatidae*			*Oceanodroma leucorhoa*	Leach's Storm-Petrel
				(21,4,4,2,1)				
	Pelecaniformes	*Pelecani*		*Sulidae*			*Morus bassanus*	Northern Gannet
				(9,2,1,1,1)				
				Phalacrocoracidae			*Phalacrocorax carbo*	Great Cormorant
				(38,6,6,4,2)			*Phalacrocorax auritus*	Double-crested Cormorant
	Ciconiiformes	*Ardeae*		*Ardeidae*		*Botaurini*	*Botaurus lentiginosus*	American Bittern
				(65,12,12,7,6)			*Ixobrychus exilis*	Least Bittern
						Ardeini	*Ardea herodias*	Great Blue Heron
							Casmerodius albus	Great Egret
							Butorides virescens	Green Heron
						Nycticoracini	*Nycticorax nycticorax*	Black-crowned Night-Heron
	Anseriformes	*Anseres*		*Anatidae*	*Anserinae*	*Cygnini*	*Cygnus columbianus*	Tundra Swan
				(157,43,43,37,32)			*Cygnus olor*	Mute Swan
						Anserini	*Chen caerulescens*	Snow Goose
							Branta bernicla	Brant
							Branta canadensis	Canada Goose
					Anatinae	*Cairinini*	*Aix sponsa*	Wood Duck
						Anatini	*Anas crecca*	Green-winged Teal
							Anas rubripes	American Black Duck
							Anas platyrhynchos	Mallard
							Anas acuta	Northern Pintail
							Anas discors	Blue-winged Teal
							Anas clypeata	Northern Shoveler
							Anas strepera	Gadwall
							Anas penelope	Eurasian Wigeon +
							Anas americana	American Wigeon
						Aythyini	*Aythya valisineria*	Canvasback
							Aythya americana	Redhead
							Aythya collaris	Ring-necked Duck
							Aythya marila	Greater Scaup
							Aythya affinis	Lesser Scaup
						Mergini	*Somateria mollissima*	Common Eider
							Somateria spectabilis	King Eider
							Camptorhynchus labradorius	Labrador Duck †
							Histrionicus histrionicus	Harlequin Duck
							Clangula hyemalis	Oldsquaw
							Melanitta nigra	Black Scoter
							Melanitta perspicillata	Surf Scoter
							Melanitta fusca	White-winged Scoter
							Bucephala clangula	Common Goldeneye
							Bucephala islandica	Barrow's Goldeneye +
							Bucephala albeola	Bufflehead
							Lophodytes cucullatus	Hooded Merganser
							Mergus merganser	Common Merganser
							Mergus serrator	Red-breasted Merganser
						Oxyurini	*Oxyura jamaicensis*	Ruddy Duck
	Falconiformes	*Cathartae*	*Cathartoidea*	*Cathartidae*			*Cathartes aura*	Turkey Vulture
				(7,3,3,1,1)				
		Accipitres	*Accipitroidea*	*Accipitridae*	*Pandioninae*		*Pandion haliaetus*	Osprey
				(239,25,25,13,11)	*Accipitrinae*		*Haliaeetus leucocephalus*	Bald Eagle
							Circus cyaneus	Northern Harrier
							Accipiter striatus	Sharp-shinned Hawk
							Accipiter cooperii	Cooper's Hawk
							Accipiter gentilis	Northern Goshawk
							Buteo lineatus	Red-shouldered Hawk
							Buteo platypterus	Broad-winged Hawk
							Buteo jamaicensis	Red-tailed Hawk
							Buteo lagopus	Rough-legged Hawk
							Aquila chrysaetos	Golden Eagle
		Falcones		*Falconidae*		*Falconini*	*Falco sparverius*	American Kestrel
				(63,7,7,5,4)			*Falco columbarius*	Merlin
							Falco peregrinus	Peregrine Falcon
							Falco rusticolus	Gyrfalcon
	Galliformes		*Phasianoidea*	*Phasianidae*	*Phasianinae*	*Perdicini*	*Perdix perdix*	Gray Partridge
				(214,21,21,16,10)			*Alectoris chukar*	Chukar
							Coturnix coturnix	Common Quail
						Phasianini	*Phasianus colchicus*	Ring-necked Pheasant
					Tetraoninae		*Dendragapus canadensis*	Spruce Grouse
							Lagopus lagopus	Willow Ptarmigan
							Lagopus mutus	Rock Ptarmigan
							Bonasa umbellus	Ruffed Grouse
							Tympanuchus phasianellus	Sharp-tailed Grouse
					Meleagridinae		*Meleagris gallopavo*	Wild Turkey
					Odontophorinae		*Colinus virginianus*	Northern Bobwhite +

Class	Order	Suborder	Superfamily	Family (Wo, NA, US, Ca, Qc)*	Subfamily	Tribe	Scientific Name Genus Species	Common Name
	Gruiformes			*Rallidae* (142,10,10,6,5)	*Rallinae*		*Coturnicops noveboracensis*	Yellow Rail
							Rallus limicola	Virginia Rail
							Porzana carolina	Sora
							Gallinula chloropus	Common Moorhen
							Fulica americana	American Coot
				Gruidae (15,2,2,2,1)	*Gruinae*		*Grus canadensis*	Sandhill Crane
	Charadriiformes	*Charadrii*		*Charadriidae* (67,11,11,7,3)	*Charadriinae*		*Charadrius semipalmatus*	Semipalmated Plover
							Charadrius melodus	Piping Plover
							Charadrius vociferus	Killdeer
		Scolopaci	*Scolopacoidea*	*Scolopacidae* (88,42,40,32,13)	*Scolopacinae*	*Tringini*	*Tringa melanoleuca*	Greater Yellowlegs
							Tringa flavipes	Lesser Yellowlegs
							Tringa solitaria	Solitary Sandpiper
							Catoptrophorus semipalmatus	Willet +
							Actitis macularia	Spotted Sandpiper
						Numeniini	*Bartramia longicauda*	Upland Sandpiper
						Limosini	*Limosa fedoa*	Marbled Godwit +
						Calidridini	*Calidris pusilla*	Semipalmated Sandpiper
							Calidris minutilla	Least Sandpiper
						Limnodromini	*Limnodromus griseus*	Short-billed Dowitcher
						Gallinaginini	*Gallinago gallinago*	Common Snipe
						Scolopacini	*Scolopax minor*	American Woodcock
					Phalaropodinae		*Phalaropus tricolor*	Wilson's Phalarope
							Phalaropus lobatus	Red-necked Phalarope
							Phalaropus fulicaria	Red Phalarope
		Lari		*Laridae* (106,38,34,28,18)	*Stercorariinae*		*Stercorarius pomarinus*	Pomarine Jaeger
							Stercorarius parasiticus	Parasitic Jaeger
							Stercorarius longicaudus	Long-tailed Jaeger
					Larinae		*Larus minutus*	Little Gull
							Larus ridibundus	Common Black-headed Gull
							Larus philadelphia	Bonaparte's Gull
							Larus delawarensis	Ring-billed Gull
							Larus argentatus	Herring Gull
							Larus glaucoides	Iceland Gull
							Larus hyperboreus	Glaucous Gull
							Larus marinus	Great Black-backed Gull
							Rissa tridactyla	Black-legged Kittiwake
					Sterninae		*Sterna caspia*	Caspian Tern
							Sterna dougallii	Roseate Tern
							Sterna hirundo	Common Tern
							Sterna paradisaea	Arctic Tern
							Chlidonias leucopterus	White-winged Tern
							Chlidonias niger	Black Tern
		Alcae		*Alcidae* (23,19,13,13,5)		*Alcini*	*Uria aalge*	Common Murre
							Uria lomvia	Thick-billed Murre
							Alca torda	Razorbill
							Pinguinus impennis	Great Auk †
						Cepphini	*Cepphus grylle*	Black Guillemot
						Fraterculini	*Fratercula arctica*	Atlantic Puffin
	Columbiformes	*Columbae*		*Columbidae* (310,13,13,4,3)			*Columba livia*	Rock Dove
							Streptopelia risoria	Ringed Turtle-Dove
							Zenaida macroura	Mourning Dove
							Ectopistes migratorius	Passenger Pigeon †
	Psittaciformes			*Psittacidae* (358,5,5,1,1)	*Arinae*		*Myiopsitta monachus*	Monk Parakeet
	Cuculiformes			*Cuculidae* (142,6,6,2,2)	*Coccyzinae*		*Coccyzus erythropthalmus*	Black-billed Cuckoo
							Coccyzus americanus	Yellow-billed Cuckoo
	Strigiformes			*Tytonidae* (17,1,1,1,0)			*Tyto alba*	Barn Owl +
				Strigidae (161,18,18,15,10)			*Otus asio*	Eastern Screech-Owl
							Bubo virginianus	Great Horned Owl
							Nyctea scandiaca	Snowy Owl
							Surnia ulula	Northern Hawk Owl
							Strix varia	Barred Owl
							Strix nebulosa	Great Gray Owl
							Asio otus	Long-eared Owl
							Asio flammeus	Short-eared Owl
							Aegolius funereus	Boreal Owl
							Aegolius acadicus	Northern Saw-whet Owl

† Extinct species, not counted in number of species in family
* Wo: Number of breeding species in world
NA: Number of breeding species in North America
US: Number of breeding species in United States, including Alaska (not including Hawaii)
Ca: Number of breeding species in Canada
Qc: Number of breeding species in Québec
** Hybrid of Golden-winged Warbler and Blue-winged Warbler, not counted in number of species in subfamily Parulinae
\+ Breeding not confirmed in Québec

Class	Order	Suborder	Superfamily	Family (Wo, NA, US, Ca, Qc)*	Subfamily	Tribe	Scientific Name Genus Species	Common Name
	Caprimulgiformes			*Caprimulgidae* (76,8,8,4,2)	*Chordeilinae* *Caprimulginae*		*Chordeiles minor* *Caprimulgus vociferus*	Common Nighthawk Whip-poor-will
	Apodiformes			*Apodidae* (99,4,4,4,1)	*Chaeturinae*		*Chaetura pelagica*	Chimney Swift
				Trochilidae (319,15,15,5,1)			*Archilochus colubris*	Ruby-throated Hummingbird
	Coraciiformes	*Alcedines*	*Alcedinoidea*	*Alcedinidae* (94,3,3,1,1)	*Cerylinae*		*Ceryle alcyon*	Belted Kingfisher
	Piciformes	*Pici*		*Picidae* (215,22,22,14,8)	*Picinae*		*Melanerpes erythrocephalus* *Melanerpes carolinus* *Sphyrapicus varius* *Picoides pubescens* *Picoides villosus* *Picoides tridactylus* *Picoides arcticus* *Colaptes auratus* *Dryocopus pileatus*	Red-headed Woodpecker Red-bellied Woodpecker + Yellow-bellied Sapsucker Downy Woodpecker Hairy Woodpecker Three-toed Woodpecker Black-backed Woodpecker Northern Flicker Pileated Woodpecker
	Passeriformes	*Tyranni*	*Tyrannoidea*	*Tyrannidae* (416,34,34,16,9)	*Fluvicolinae* *Tyranninae*		*Contopus borealis* *Contopus virens* *Empidonax flaviventris* *Empidonax virescens* *Empidonax alnorum* *Empidonax traillii* *Empidonax minimus* *Sayornis phoebe* *Myiarchus crinitus* *Tyrannus tyrannus*	Olive-sided Flycatcher Eastern Wood-Pewee Yellow-bellied Flycatcher Acadian Flycatcher + Alder Flycatcher Willow Flycatcher Least Flycatcher Eastern Phoebe Great Crested Flycatcher Eastern Kingbird
		Passeres		*Alaudidae* (91,2,2,2,1)			*Eremophila alpestris*	Horned Lark
				Hirundinidae (89,8,8,7,6)	*Hirundininae*		*Progne subis* *Tachycineta bicolor* *Stelgidopteryx serripennis* *Riparia riparia* *Hirundo pyrrhonota* *Hirundo rustica*	Purple Martin Tree Swallow Northern Rough-winged Swallow Bank Swallow Cliff Swallow Barn Swallow
				Corvidae (117,16,16,8,4)			*Perisoreus canadensis* *Cyanocitta cristata* *Corvus brachyrhynchos* *Corvus corax*	Gray Jay Blue Jay American Crow Common Raven
				Paridae (53,10,10,6,2)			*Parus atricapillus* *Parus hudsonicus* *Parus bicolor*	Black-capped Chickadee Boreal Chickadee Tufted Titmouse +
				Sittidae (28,4,4,3,2)	*Sittinae*		*Sitta canadensis* *Sitta carolinensis*	Red-breasted Nuthatch White-breasted Nuthatch
				Certhiidae (7,1,1,1,1)	*Certhiinae*		*Certhia americana*	Brown Creeper
				Troglodytidae (75,9,9,8,5)			*Thryothorus ludovicianus* *Troglodytes aedon* *Troglodytes troglodytes* *Cistothorus platensis* *Cistothorus palustris*	Carolina Wren House Wren Winter Wren Sedge Wren Marsh Wren
				Muscicapidae (1492,22,22,16,12)	*Sylviinae* *Turdinae*	*Sylviini* *Polioptilini*	*Regulus satrapa* *Regulus calendula* *Polioptila caerulea* *Oenanthe oenanthe* *Sialia sialis* *Catharus fuscescens* *Catharus minimus* *Catharus bicknelli* *Catharus ustulatus* *Catharus guttatus* *Hylocichla mustelina* *Turdus migratorius*	Golden-crowned Kinglet Ruby-crowned Kinglet Blue-gray Gnatcatcher Northern Wheatear Eastern Bluebird Veery Gray-cheeked Thrush Bicknell's Thrush Swainson's Thrush Hermit Thrush Wood Thrush American Robin
				Mimidae (34,10,10,4,3)			*Dumetella carolinensis* *Mimus polyglottos* *Toxostoma rufum*	Gray Catbird Northern Mockingbird Brown Thrasher
				Motacillidae (65,6,6,3,1)			*Anthus rubescens*	American Pipit
				Bombycillidae (8,2,2,2,1)			*Bombycilla garrulus* *Bombycilla cedrorum*	Bohemian Waxwing + Cedar Waxwing
				Laniidae (87,2,2,2,2)	*Laniinae*		*Lanius excubitor* *Lanius ludovicianus*	Northern Shrike Loggerhead Shrike
				Sturnidae (114,3,2,2,1)	*Sturninae*		*Sturnus vulgaris*	European Starling
				Vireonidae (51,11,11,7,5)	*Vireoninae*		*Vireo solitarius* *Vireo flavifrons* *Vireo gilvus* *Vireo philadelphicus* *Vireo olivaceus*	Solitary Vireo Yellow-throated Vireo Warbling Vireo Philadelphia Vireo Red-eyed Vireo

Class	Order	Suborder	Superfamily	Family (Wo, NA, US, Ca, Qc)*	Subfamily	Tribe	Scientific Name Genus Species	Common Name
				Emberizidae	*Parulinae*		*Vermivora pinus*	Blue-winged Warbler +
					(115,55,55,37,27)		*Vermivora chrysoptera X pinus*	Brewster's Warbler **
							Vermivora chrysoptera	Golden-winged Warbler
							Vermivora peregrina	Tennessee Warbler
							Vermivora celata	Orange-crowned Warbler
							Vermivora ruficapilla	Nashville Warbler
							Parula americana	Northern Parula
							Dendroica petechia	Yellow Warbler
							Dendroica pensilvanica	Chestnut-sided Warbler
							Dendroica magnolia	Magnolia Warbler
							Dendroica tigrina	Cape May Warbler
							Dendroica caerulescens	Black-throated Blue Warbler
							Dendroica coronata	Yellow-rumped Warbler
							Dendroica virens	Black-throated Green Warbler
							Dendroica fusca	Blackburnian Warbler
							Dendroica pinus	Pine Warbler
							Dendroica discolor	Prairie Warbler +
							Dendroica palmarum	Palm Warbler
							Dendroica castanea	Bay-breasted Warbler
							Dendroica striata	Blackpoll Warbler
							Dendroica cerulea	Cerulean Warbler
							Mniotilta varia	Black-and-white Warbler
							Setophaga ruticilla	American Redstart
							Protonotaria citrea	Prothonotary Warbler +
							Seiurus aurocapillus	Ovenbird
							Seiurus noveboracensis	Northern Waterthrush
							Seiurus motacilla	Louisiana Waterthrush +
							Oporornis agilis	Connecticut Warbler
							Oporornis philadelphia	Mourning Warbler
							Geothlypis trichas	Common Yellowthroat
							Wilsonia pusilla	Wilson's Warbler
							Wilsonia canadensis	Canada Warbler
					Thraupinae (247,4,4,2,1)	*Thraupini*	*Piranga olivacea*	Scarlet Tanager
					Cardinalinae		*Cardinalis cardinalis*	Northern Cardinal
					(42,10,10,6,3)		*Pheucticus ludovicianus*	Rose-breasted Grosbeak
							Passerina cyanea	Indigo Bunting
					Emberizinae		*Pipilo erythrophthalmus*	Rufous-sided Towhee
					(300,46,45,29,20)		*Spizella arborea*	American Tree Sparrow
							Spizella passerina	Chipping Sparrow
							Spizella pallida	Clay-colored Sparrow
							Spizella pusilla	Field Sparrow
							Pooecetes gramineus	Vesper Sparrow
							Passerculus sandwichensis	Savannah Sparrow
							Ammodramus savannarum	Grasshopper Sparrow
							Ammodramus henslowii	Henslow's Sparrow
							Ammodramus leconteii	Le Conte's Sparrow
							Ammodramus caudacutus	Sharp-tailed Sparrow
							Passerella iliaca	Fox Sparrow
							Melospiza melodia	Song Sparrow
							Melospiza lincolnii	Lincoln's Sparrow
							Melospiza georgiana	Swamp Sparrow
							Zonotrichia albicollis	White-throated Sparrow
							Zonotrichia leucophrys	White-crowned Sparrow
							Junco hyemalis	Dark-eyed Junco
							Calcarius lapponicus	Lapland Longspur
							Plectrophenax nivalis	Snow Bunting
					Icterinae	*Dolichonychini*	*Dolichonyx oryzivorus*	Bobolink
					(97,20,20,11,8)	*Agelaiini*	*Agelaius phoeniceus*	Red-winged Blackbird
							Sturnella magna	Eastern Meadowlark
							Sturnella neglecta	Western Meadowlark
							Xanthocephalus xanthocephalus	Yellow-headed Blackbird +
							Euphagus carolinus	Rusty Blackbird
							Quiscalus quiscula	Common Grackle
							Molothrus ater	Brown-headed Cowbird
						Icterini	*Icterus spurius*	Orchard Oriole +
							Icterus galbula	Northern Oriole
				Fringillidae	*Carduelinae*		*Pinicola enucleator*	Pine Grosbeak
				(139,15,15,12,10)			*Carpodacus purpureus*	Purple Finch
							Carpodacus mexicanus	House Finch
							Loxia curvirostra	Red Crossbill
							Loxia leucoptera	White-winged Crossbill
							Carduelis flammea	Common Redpoll
							Carduelis hornemanni	Hoary Redpoll
							Carduelis pinus	Pine Siskin
							Carduelis tristis	American Goldfinch
							Coccothraustes vespertinus	Evening Grosbeak
				Passeridae (46,2,2,1,1)			*Passer domesticus*	House Sparrow

† Extinct species, not counted in number of species in family
* Wo: Number of breeding species in world
NA: Number of breeding species in North America
US: Number of breeding species in United States, including Alaska (not including Hawaii)
Ca: Number of breeding species in Canada
Qc: Number of breeding species in Québec
** Hybrid of Golden-winged Warbler and Blue-winged Warbler, not counted in number of species in subfamily Parulinae
\+ Breeding not confirmed in Québec

GAVIIDAE

THE ORDER GAVIIFORMES, AMONG THE OLDEST AND MOST PRIMITIVE OF living birds, comprises only one family (Gaviidae), which in turn has only one genus represented by five fairly homogeneous species. The Gaviidae, or loons, are widely distributed throughout almost the entire Northern Hemisphere, and three species breed in Québec: the Common Loon, Pacific Loon, and Red-throated Loon (in many ways – morphologically, behaviourally, and ecologically – the most distinct from the other species).

Loons are large (53-91 cm), narrow-winged aquatic birds that spend most of their life cycle on open water. Although they share certain physiological and behavioural characteristics with the grebes (Podicipedidae), the similarity between the two families is superficial and due more to convergent evolution than true kinship. Despite their different origins, loons, like grebes, are wonderfully adapted to their aquatic way of life. For instance, their bones are more solid with many fewer air spaces than those of birds that spend most of their lives in the air. They are also able to regulate buoyancy by adjusting the volume of air trapped among their feathers or in their air sacs, expelling it when they want to dive.

The legs are set far back on the body and the three front toes are webbed, while the fourth and smallest hind toe, the hallux, is elevated. The tarsi are laterally flattened to reduce resistance and surface contact with the water. Underwater, loons propel themselves with their feet, using their wings for balance and steering. Loons are excellent swimmers and divers, skills essential to their feeding technique of chasing fish – the main component of their diet – underwater, generally at depths of 2-20 m. In some cases, they may also eat frogs, crustaceans, molluscs, or aquatic insects. Since loons locate their prey by sight, they prefer lakes with relatively clear water.

All these adaptations to life in the water have made loons awkward on land, although, unlike grebes, they generally come ashore to nest. Soon after hatching, most species of loons lead their chicks to water, where they swim with them on their backs. The Red-throated Loon, however, raises its young on land and is known to travel relatively long distances overland. Loons are monogamous and evidence suggests they mate for life. Males tend to be larger than females, but otherwise the sexes look alike. Incubation and care of the young are shared by both parents. During the breeding season, the large species defend a small lake or a part of a big lake, but the smaller Red-throated Loon sometimes nests colonially or on floating clumps of vegetation, defending only the immediate nest vicinity. Loons generally prefer to nest in areas undisturbed by humans and return to the same nest site year after year. Their calls, especially the ones they use to defend their territory, carry farther than that of any other bird.

Only the Red-throated Loon can take off from or alight onto land. Takeoff and landing in the other four species are laborious and spectacular. Despite these apparent difficulties, loons can fly long distances during migration or in post-breeding dispersal to find ice-free areas where they can spend the winter.

With its greater ease in taking to the air, the Red-throated Loon usually forages for food outside its breeding territory in deeper lakes or in coastal waters, while the other species feed mostly on their breeding lakes.

LINE CHOINIÈRE AND ROBERT ALVO

Austin and Singer 1990; Cramp and Simmons 1977; del Hoyo et al. 1992; Godfrey 1986; Harrison, C.J.O. 1978; National Geographic Society 1987; Pettingill 1985; Robbins et al. 1966.

Red-throated Loon. *Yves Aubry.*

Pacific Loon. *Austin Reed.*

Common Loon. *Austin Reed.*

PODICIPEDIDAE

THE PODICIPEDIDAE, OR GREBE FAMILY, IS THE ONLY FAMILY IN THE ORDER Podicipediformes. It is a fairly homogeneous group that, from an evolutionary point of view, became distinct from other groups of birds fairly early, which explains why it is unrelated to any other living family of birds. There are approximately 20 species with worldwide distribution, three of them breeding in Québec. The Pied-billed Grebe is the most common, while the Horned and Red-necked Grebes breed only very locally in the province. In Québec, grebes are seen mainly during migration, south of the 50th parallel, and usually in greater numbers in fall than in spring.

Grebes spend their entire lives on water and are said to be the most perfectly adapted to water of all birds. One such adaptation is their dense, lustrous plumage composed of over 20 000 feathers, which trap a layer of air that insulates them from the cold and keeps them buoyant. To dive, they increase their specific gravity by expelling the air between the feathers and contracting their chest muscles so as to compress the air in their air sacs. Grebe bones, like those of several other diving birds, have fewer air spaces than those of most other species of the same size that are better adapted for flying.

The four toes are lobate, but not fully webbed. Other birds have lobed toes, but only grebes have a stiff horny flap fringing each toe. To provide propulsion when paddling, the lobes are positioned for maximum contact with the water. On the return stroke, when the foot is brought forward, the feet pivot 90° to reduce the area of contact so that the lobes slice through the water. This decreases resistance and thus the energy required for swimming. Other species of aquatic birds with lobed toes do not have to rotate their feet because the lobes are flexible and fold up against the toes instantly on the return stroke. The legs are set far back on the body and have laterally flattened tarsi, as in loons. Grebes propel themselves underwater with simultaneous strokes of the feet, which are manoeuvrable and serve as a rudder in diving as in flight. They can remain submerged for 10 to 40 seconds, reaching underwater speeds as a high as 2 m/s. They are awkward in flight, however, and they need to taxi a long way on the surface of the water before takeoff, so when alarmed, they dive rather than fly.

In grebes that feed mainly on fish, the bill is usually long and pointed, while it is short and conical in species that, like the Pied-billed Grebe, feed on arthropods. Grebes have a fairly long, straight neck and compact body. The tail, reduced to a downy tuft, is practically non-existent. Several species have tufts or crests on the head that they display during their courtship dances and ritualized ceremonies, which are accompanied by a variety of cries. Among these mainly monogamous birds, the female is almost as active as the male in the spectacular displays that take place throughout the breeding season. Both sexes are active parents, sharing the tasks associated with breeding – nest building, incubation, care of the young – and

Pied-billed Grebe. *Paul Perreault.*

Horned Grebe. *François Shaffer.*

Red-necked Grebe. *Yves Aubry.*

then the pair splits up. In some cases, the pair bond is broken after hatching, and the parents divide the brood between them. Crèching has also been observed. Grebes nest solitarily or in small colonies, in relatively shallow fresh or sometimes brackish water with lush emergent vegetation. Some species may also nest on slow-moving rivers or in sheltered bays along the coast. Although highly variable, the number of chicks per nest that survive is generally rather low. Nonetheless, females may produce several replacement clutches to make up for losses due to flooding or predation.

Neck outstretched a little lower than the plane of the body and feet trailing, grebes fly with rapid wing beats. Some species, especially those breeding at high latitudes, are migratory, others are altitudinal migrants or nomadic, such as during periods of heavy rains. In good climates with an abundant food supply, grebes practically never fly and are sedentary, remaining in the same locality all year long.

Grebes, both young and adult, have the unique habit of swallowing feathers, which may amount to between 10 and 60% of their stomach contents. It is thought that the feather balls enable the birds to regurgitate pellets containing the bones of prey. Some studies suggest, rather, that the feathers act as a plug to keep bones in the stomach until they are digested by an acid solution found nowhere else in the digestive tract. The pellets then contain more undigested material, such as chitin (a component of insect, arachnid, and crustacean shells), than fish bones.

Line Choinière

Austin and Singer 1990; David 1980a; del Hoyo et al. 1992; Fjeldså 1973; Godfrey 1986; Gosselin 1992b; Munro 1941b; Simmons 1956; Storer 1961, 1971; Terres 1980; Turner 1989.

HYDROBATIDAE

THE HYDROBATIDAE, OR STORM-PETREL FAMILY, COMPRISES APPROXIMATELY 20 species of small seabirds (15-25 cm long) that range over the world's oceans, spending their entire lives at sea, except when nesting. Their habit of flitting low over the water with their legs pattering the surface has earned them their Latin name, which means "to tread water," and their English name "petrel," after St. Peter, who walked on the water with Jesus' help.

Hydrobatids have darkish upperparts, often with white rump patches, and underparts of variable colour. Their dense waterproof plumage covers a layer of down, and their feet are webbed. The bill is short, black, and hooked, with the external tubular nostrils fused together. They have two well-developed salt glands near the eyes to excrete surplus salt from their bodies.

Storm-petrels usually breed on islands, but also sometimes on mainland cliffs or even in inland deserts, in the colony where they were born. The single egg is laid in the crevice of a rock or cliff, or in a ground burrow. Each parent takes a turn incubating the egg for a few days at a time while the other goes off to feed. When the chick hatches it is covered with long, thick down. The parents feed it by regurgitating an oily mixture of marine invertebrates such as shrimp and squid. The adult diet consists of fish, crustaceans, and other marine animals.

Two species of this family are found in Québec: Leach's Storm-Petrel, which breeds in the Gulf of St. Lawrence, and Wilson's Storm-Petrel, a summer visitor that breeds only in the Southern Hemisphere.

ROBERT ALVO

Bourne 1985; Cramp and Simmons 1977; del Hoyo et al. 1992; Gill 1990; Godfrey 1986; Harrison, C.J.O. 1978; Harrison 1983; Stiles and Skutch 1989; Terres 1980; Tuck 1978.

Leach's Storm-Petrel. *Yves Aubry.*

SULIDAE

Northern Gannet. *Denis Faucher.*

THE FAMILY SULIDAE (GANNETS AND BOOBIES) NUMBERS NINE SPECIES FOUND all over the world, except in the Arctic and Antarctic. Only a single representative, the Northern Gannet, breeds in Québec, in the Gulf of St. Lawrence.

Gannets and boobies are large seabirds (64-100 cm long) that plummet from great heights, head first into the sea after fish and squid, which they locate by sight. These plunge-divers have cigar-shaped bodies, long, narrow wings, a tapered tail, and short, strong legs with feet webbed between all four toes (totipalmate). The stout conical bill has serrated cutting edges specialized for catching fish. The sealed external nostrils are adapted for diving. The plumage is waterproof, and well-developed air sacs help cushion the impact when the bird strikes the water.

Species vary in the colour of the bill, feet, and the facial skin around the eyes. During the breeding season, the sexes of some species differ strikingly in the colours of the iris, bill, facial skin, and throat pouch, while in other species the sexes look identical. The females are generally larger than the males.

All sulids are territorial and lay 1 to 4 eggs. They incubate the eggs and brood the chicks with their feet. Both parents tend the young, which are almost naked at birth, but covered with thick white down two or three weeks later. Between the 10th and 16th week, the down is completely replaced by feathers. Although gannets and boobies are very loyal to their nest site and mate, pairs sometimes separate.

Outside the breeding season, groups of gannets roost at sea, while boobies roost mostly on land. Gannets live in northern and southern temperate latitudes, while boobies are tropical. Boobies gain their name from their fearlessness of humans.

ROBERT ALVO AND DIANE DAUPHIN

del Hoyo et al. 1992; Gill 1990; Godfrey 1986; Harrison, C.J.O. 1978; National Geographic Society 1987; Nelson 1978a, b; Stiles and Skutch 1989; Terres 1980; Tuck 1978.

PHALACROCORACIDAE

CORMORANTS ARE LONG-NECKED BIRDS OF MEDIUM TO LARGE SIZE (45-100 cm long) that skim low over the water. They are excellent swimmers with their strong, short legs and large totipalmate feet. As in loons, the legs are set far back on the body, also an aid in underwater swimming. One of the consequences of their adaptation for diving is wettable plumage, which forces them to spread their wings to dry, a habit that makes them easy to identify from a distance.

In most species the sexes are similar, with shiny, mostly black plumage; males, however, tend to be larger. Bare patches of skin around the eyes are often brightly coloured, and include a small distensible throat pouch. The slender bill is strongly hooked at the tip. Lacking external nostrils, adults must breathe through their mouths. Cormorants are generally silent except in the breeding season.

Most cormorants prefer salt water, although some species favour fresh water and others have no preference. Predominantly fish eaters (although they also take crustaceans and amphibians), they are pursuit-divers, chasing their prey underwater after diving from the surface. Bones, scales, and other indigestible parts of prey are regurgitated as a pellet about once a day. Interestingly, in some countries, especially in Asia, fishermen use cormorants to catch fish.

Cormorants are typically colonial breeders, often nesting in mixed-species colonies. They defend their small nest territories vigorously. The pair bond generally lasts for only one breeding season. Nests are often bulky structures on the ground, on cliff ledges, in trees, or even on artificial platforms, including wrecks washed up on shore. Both parents incubate the eggs with the webs of the feet.

Curiously, the family includes a species that cannot fly, the Galapagos Flightless Cormorant. Another interesting fact is that the guano (droppings) of the South American Guanay is sold as fertilizer.

Of the roughly 40 species of cormorants found worldwide, only two occur in Québec. The Great Cormorant colonizes the islands of the Gulf of St. Lawrence, while the Double-crested Cormorant breeds in the St. Lawrence estuary and Gulf.

ROBERT ALVO

Austin and Singer 1990; del Hoyo et al. 1992; Gill 1990; Godfrey 1986; Harrison, C.J.O. 1978; Storer 1960; Terres 1980.

Double-crested Cormorant. *Jean Giroux.*

Great Cormorant, adult (left) and young. *Yves Aubry.*

ARDEIDAE

HERONS, EGRETS, AND BITTERNS ARE WADING BIRDS OF MEDIUM TO LARGE size, with long legs, necks, and bills for hunting prey in shallow water. Ardeids can range greatly in size from the Least Bittern in North America (approximately 30 cm tall) to the Goliath Heron in Africa (approximately 130 cm tall). They live on the margins of wetlands, bays, rivers, lakes, and salt marshes. With their long, slender bill, they catch – and swallow whole – mostly fish, but also frogs and mice. They fly with their legs trailing and their long necks tucked back on the shoulders in an S shape. The bill, lores, and legs often change colour with the seasons. The sexes look identical. Bitterns have shorter legs than herons and egrets. The cryptic coloration of bitterns and their habit of "freezing" in an upright posture when alarmed provide excellent camouflage among the dense cattails, bulrushes, reeds, and sedges of their marsh habitat.

Some species have ornamental plumes on the head, lower neck, breast, and back. These usually appear before the onset of breeding and are important during pairing and courtship. Historically, the plumes of some species were much sought after for millinery, and the plume trade caused a sharp decline in the populations of some species.

A typical feature of this family is the presence of continuously growing powder-down on the flanks. This down disintegrates into a powder used for preening, especially to remove fish slime and oil or any other dirt. The powder is applied and spread throughout the feathers with the pectinated (comblike edge) claw on the third toe.

Many species breed in mixed colonies. Some are colonial in some areas, but solitary in others. The nest is usually built by the female with material presented ceremoniously by the male.

In several heron species, siblings are often highly competitive. Larger chicks may kill the smaller ones or monopolize most of the food brought by the parents. This sibling rivalry is a function of the availability of food (limited in some years) and asynchronous hatching (the first-born young are generally larger and stronger).

Herons, egrets, and bitterns occur worldwide, except in the arctic and subarctic regions of North America, northern Asia, Antarctica, and on some oceanic islands. Twelve species have been reported in Québec; only six breed here, however, one of which, the Great Egret, has invaded the province recently. Ardeids are rare in the northernmost part of Québec.

ROBERT ALVO

Austin and Singer 1990; del Hoyo et al. 1992; Gill 1990; Godfrey 1986; Hancock and Kushlan 1989; Harrison, C.J.O. 1978; Terres 1980; Van Tyne and Berger 1966.

Least Bittern. *Robertson Graham, Parks Canada.*

American Bittern. *André Cyr.*

Great Blue Heron. *Denis Faucher.*

Great Egret. *Michel Robert.*

Green Heron. *Denis Faucher.*

Black-crowned Night-Heron. *Denis Faucher.*

ANATIDAE

The family Anatidae (ducks, geese, and swans) comprises nearly 160 species distributed all over the world, except in Antarctica, and is the best represented family in Québec, with 32 breeding species. Anatids – often called waterfowl (North America) or wildfowl (Europe) – have had close ties with humans since ancient times and were among the first species domesticated. Some wild species are the highly prized targets of hunters in fall for sport, and of birders in spring for enjoyment of their nuptial plumage.

This family is highly diverse. Around the world, there are three subfamilies: the monotypic Anseranatinae, whose sole species, the Magpie Goose, is found in Australia and New Guinea. The other two subfamilies, the Anserinae and Anatinae, are represented in Québec by seven different tribes. The subfamily Anserinae is made up of the geese (tribe Anserini) and swans (tribe Cygnini). The much more diverse subfamily Anatinae consists of the perching ducks (tribe Cairinini), dabbling ducks (tribe Anatini), pochards or freshwater diving ducks (tribe Aythyini), sea ducks and mergansers (tribe Mergini), and stifftails (tribe Oxyurini). The classification of anatids is complex and is still the subject of debate, mainly because of the genetic similarities among several species and the many cases of hybridization reported in both nature and captivity.

Anatids are birds of medium to large size, varying in length between 30 and 180 cm, and weighing between 0.2 and 23 kg. They are well adapted to life on the water and are all similar in general shape with a wide body, either rounded or elongated. The legs are short, with the three front toes webbed and the hind toe, or hallux, elevated. The neck is fairly long, and the rather small head has a broad, flat bill with a rounded tip. The structure of the bill, the musculature of the head, and the length of the neck vary according to the diet and feeding habits of the various species. The wings are adapted to long-distance flying and can support considerable weight. When they moult after breeding, anatids are flightless for a short time, since they lose all their flight feathers at once. Most species have simple unmusical vocalizations. Because of their aquatic habits, anatids spend a great deal of time preening. Several times a day, they use their bill or head to smooth their feathers, coating them with an oily, waxy substance secreted by the uropygial (preen) gland at the base of the tail to keep them flexible and totally waterproof. The salt glands, more highly developed in marine species, extract salt from body fluids.

Most anatids are migratory and generally associated with wetlands. They are gregarious, congregating in their dozens, hundreds, or thousands on wintering grounds and on staging areas along the migration route. Some species are colonial breeders, while others disperse to breed alone. Unlike most other species of birds, anatids pair on the wintering grounds, so they can spend more time feeding during spring staging and building up the fat reserves needed for migration or breeding.

Egg production in anatids is extremely energy-intensive. The weight of the eggs in a clutch usually represents 20-30% of the female's weight and, in the Ruddy Duck, may be as high as 100%. Anatids normally lay one egg a day, and in most cases, the female does not start incubating until she has finished laying, so hatching is synchronous. Parental care is limited, since the precocial young hatch at an advanced stage of development and are practically independent shortly after hatching. The chicks may actually begin to swim as soon as their down has dried. Like many other precocial birds, newly hatched chicks learn to recognize their parents through imprinting. This process is so strong that if, during the critical period, they are experimentally placed in continuous contact with a human being, they will look on that person as their parent and follow him or her around everywhere, as they would instinctively follow their true parents to feeding sites. Although ducks are not obligate parasites, some species may lay eggs in the nests of conspecifics or of other species; this type of behaviour is especially common in the Redhead and Ruddy Duck.

In addition to the general characters mentioned above, each subfamily, tribe, or species has its own particular traits. The geese and swans of the first subfamily, the Anserinae, have plain plumage that is almost all-white, grey, or brown. Sexual dimorphism is not very marked; the sexes look alike, but the male is usually larger and his courtship behaviour, not very elaborate. Pairs mate for life and family ties are very strong, since the young remain with their parents for an entire year, or sometimes longer if breeding is unsuccessful the following year. Geese and swans are very sociable, and migrating flocks typically fly in V formation, honking non-stop to maintain group cohesion. They moult just once a year. Most species in this subfamily are herbivores, and staging and wintering flocks are increasingly to be found grazing in ploughed fields as well as their natural habitats. In most species the chicks' diet is supplemented with insects or aquatic invertebrates.

Tundra Swan. *Denis Faucher.*

Brant. *Austin Reed.*

Canada Goose. *Denis Faucher.*

Mute Swan. *Denis Faucher.*

Snow Goose, adult and gosling. *Austin Reed.*

Members of the subfamily Anatinae, the ducks, are smaller than the Anserinae. They may be quite colourful, and several species, especially those living in the Northern Hemisphere, exhibit marked sexual dimorphism, the male having brighter plumage, which he displays during courtship. The female is cryptically coloured and rather shy and retiring so as not to draw the attention of predators when she is incubating the eggs and rearing her chicks alone. Family ties are weak; the pair bond lasts for a single season and usually dissolves at the start of incubation. Ducks moult twice a year, once completely and the second time losing only the body feathers. The males of several species have nuptial plumage and eclipse plumage, which appears after the breeding season. This subfamily is the most diverse and can be divided into five tribes.

The perching ducks of the first tribe, Cairinini, represented in Québec by only a single breeding species, the Wood Duck, are some of the most brightly coloured ducks. These hole nesters with sharp claws for perching are generally associated with southerly woodland habitats, at least in Québec. They are sometimes classed with the dabbling ducks of the tribe Anatini.

Wood Duck, male. *André Cyr.*

The dabbling ducks of the second tribe, Anatini, have a varied diet. Some feed on aquatic plants, while others prefer seeds and invertebrates. They like stretches of shallow water where they can filter-feed at the surface or upend themselves to take food from the shoals. They can live on small ponds since they can bound straight into the air off the water with a flap of their wings. Most members of this tribe have a speculum (or "mirror"), an iridescent patch of colour on the back of the wing that helps keep members of the same species together. It is a trait more developed in species that migrate in a group than in those that migrate alone or over short distances.

Blue-winged Teal, male. *Denis Faucher.*

Green-winged Teal, male. *Jean Giroux.*

American Wigeon, male (left) and female. *Pierre Dupuis.*

Eurasian Wigeon, male (foreground). *Pierre Dupuis.*

Northern Shoveler, male. *Yves Aubry.*

Northern Pintail, male. *Denis Faucher.*

American Black Duck, female and ducklings. *Denis Faucher.*

American Black Duck x Mallard, hybrid, male. *Pierre Dupuis.*

Gadwall, female (left) and male. *Yves Aubry.*

Mallard, male. *Denis Faucher.*

The pochards, or diving ducks (Aythyini), usually live in fresh water, but may also frequent salt or brackish water, especially on their wintering grounds. Their wider legs, with lobed hallux, are set far back on the body, making it easier for these ducks to dive, but harder for them to walk. In addition, because of their heavy body weight relative to wing-surface area, becoming airborne involves a takeoff run over the surface of the water. Most species are rather quiet, even silent, except during the breeding season. The male's head plumage is iridescent, and unlike dabbling ducks, the secondaries generally have a white pattern. Courtship displays in this group are very similar. Diving ducks do not form crèches *per se*, but families may join together after the breeding period. They feed mainly on vegetable matter, but their diet may include aquatic invertebrates.

Canvasback, males. *Denis Faucher.*

Redhead, male. *Daniel Jauvin.*

Lesser Scaup, male. *André Cyr.*

Ring-necked Duck, male. *Denis Faucher.*

Greater Scaup, female (left) and males. *Austin Reed.*

The sea ducks and mergansers (Mergini) are the tribe with the most diverse species. It is hard to describe their general appearance because they are so dissimilar. They are all excellent divers and most have some of the characteristics of freshwater diving ducks. The plumage of sea ducks is generally black and white in the males, brown in the females. It is easy to recognize them flying in small flocks low over the water in single file. Each species has typical elaborate courtship behaviour. In several species, the males leave the females incubating the eggs and migrate considerable distances to moult in groups. During the rearing period, some species may form crèches. Sea ducks generally feed on molluscs and crustaceans, occasionally taking small fish.

Although they are part of the same tribe, mergansers (or sawbills) are in some respects different from the sea ducks, especially as far as their outward appearance is concerned. The bill is probably the feature that sets them most apart from other ducks and is the reason for their formerly being classified as a distinct tribe. But according to many authors, there is no valid criterion for classifying mergansers and sea ducks in separate tribes, the goldeneyes to some extent the taxonomic link between the sea ducks and the mergansers. Well adapted to chasing fish underwater, mergansers have an elongated body and long, narrow bill with saw-toothed edges. Some species favour fresh water, while others prefer salt or brackish water. Generally speaking, these birds look for woodland habitats on their breeding grounds and nest in cavities most of the time.

Common Eider, female. *Claude Nadeau.*

Common Eider, male. *Austin Reed.*

King Eider, female. *Pierre Lamothe.*

King Eider, male. *Pierre Lamothe.*

Labrador Duck, male. *Richard Sloss, American Museum of Natural History, no. 734023.*

Harlequin Duck, female (left) and male. *Serge Brodeur.*

Surf Scoter, male. *André Cyr.*

Black Scoter, male (right) and female. *Denis Faucher.*

Oldsquaw, male (right) and females. *Yves Aubry.*

Bufflehead, males (right) and female. *André Cyr.*

Common Merganser, female and ducklings. *Gaétan Duquette.*

Red-breasted Merganser, male. *André Cyr.*

Barrow's Goldeneye, male. *Christian Marcotte.*

White-winged Scoter, male (right) and female. *Pierre Lamothe.*

Hooded Merganser, male. *Denis Faucher.*

Hooded Merganser, female. *Denis Faucher.*

Common Goldeneye, female (left) and male. *Austin Reed.*

The Ruddy Duck is the only representative of the last tribe of the Anatinae, the stifftails (Oxyurini), that breeds in Québec. Stifftails are quite different from other species of anatids and are considered the most specialized for aquatic life, sharing certain similarities with the grebes in some ways. Their feet are very wide, the legs set far back on the body, and the wings rather short. When stifftails dive, they slip smoothly under the water without creating ripples. These relatively small ducks usually frequent fresh water in southern areas. They eat the seeds, stalks, and leaves of aquatic plants, and insects and invertebrates during the breeding season.

Line Choinière

Batt et al. 1992; Bellrose 1980; Cramp and Simmons 1977; Delacour 1975; del Hoyo et al. 1992; Faaborg 1988; Johnsgard 1965, 1975b; Palmer 1976; Scott 1985; Terres 1980.

Ruddy Duck, male. *Luc Leclerc.*

CATHARTIDAE

The cathartids, or New World vultures, range from southern Canada to South America. These large, dark birds (63-132 cm long) with long, broad wings are adept at soaring and can often be seen gliding and circling overhead as they ride the thermals over vast distances in their search for food.

The New World vultures appear to be the most distantly related family of the order Falconiformes. They share very few features with the other members of the order except the Old World vultures of the family Accipitridae, namely an unfeathered head except for sparse hairlike bristles and a scavenging way of life. All other traits justify the cathartids' position in a separate family. The nasal septum, which separates the nostrils, is perforated. The olfactory chamber is large, and in some genera, particularly the Turkey Vulture, the highly developed sense of smell assists in locating carrion, even from high in the air. Cathartids feed almost exclusively on carrion, which they tear apart with their strong, hooked beaks. The trachea is unspecialized, lacking a syrinx and its intrinsic muscles, and as a result they are voiceless except for hissing and grunting. Cathartids cannot close the foot, and their long toes with only slightly curved claws are relatively weak, better suited for walking than grasping. The elevated hind toe is small and non-functional. New World vultures also have a vestigial wing claw and a distinctive sequence of secondary and tail feather moult. Recent DNA data show the Cathartidae are more closely related to the storks, family Ciconiidae, than to the diurnal birds of prey.

In Québec, the Cathartidae are represented principally by the Turkey Vulture, which breeds in the southernmost regions of the province, and to a much lesser extent by the Black Vulture, which is only a rare visitor.

Derin Henderson

Brown and Amadon 1968; Rea 1983; Sibley and Ahlquist 1990.

Turkey Vulture. *Denis Faucher.*

ACCIPITRIDAE

THIS IS THE LARGEST AND MOST VARIED FAMILY OF DIURNAL BIRDS OF PREY. It is divided into two subfamilies: Accipitrinae – which in North America includes the accipiters, buteos, harriers, eagles, and kites – and the monotypic Pandioninae, with its single species, the Osprey. A number of authors actually consider the Pandioninae to be a family, the Pandionidae. All these groups of raptors are found in Québec except kites.

The subfamily Accipitrinae as a whole has a worldwide distribution, except Antarctica. Its members vary considerably in size (25-114 cm), shape, and way of life, but they are remarkably similar in morphology. They all have the typical raptorial hooked beak with the fleshy membranous covering on the base of the upper mandible known as the cere. Most have strong feet armed with sharp, curved talons and conspicuous bumps on the soles of their feet for grasping, but the size and shape of their toes vary with the type of prey hunted. Their characteristic stern, fierce look is a result of the supraorbital ridge, a bony projection that acts as a shield over the eyes.

All species in the family except the Old World vultures show reversed sexual size dimorphism: the females are larger than the males. Over 20 hypotheses, not necessarily mutually exclusive, have been advanced to explain this reversed sexual dimorphism in birds of prey. One set postulates that the size difference evolved to allow a pair to exploit a wider ecological niche, since the smaller bird hunts smaller prey than the bigger one. Other explanations centre rather on reproductive behaviour or territorial defence. For example, one theory is based on the fact that raptors are predators capable of attacking large birds. Since the male is probably more aggressive than the female, at least during pair formation, he may represent a threat to her, and she can probably use her larger size to ward off any attacks by her mate, as well as better defend the territory and young. Although there is a wealth of literature on the subject, the question of reversed sexual dimorphism has not yet been satisfactorily explained.

The accipiters are small to medium-sized mainly forest hawks with short, rounded wings and long tails that make them highly manoeuvrable when chasing prey through the trees, bushes, and underbrush. Their diet consists chiefly of birds, but they also hunt small mammals and occasionally reptiles. They have long tarsi and long, slender toes. There are more changes in eye colour with increasing age in the larger accipiters than in any other large bird of prey. Accipiters have a distinctive flight, alternating three or more rapid wing beats with periods of gliding.

Three species of accipiters breed in Québec. The Sharp-shinned Hawk is found throughout southern Québec, Cooper's Hawk occurs only very locally in the southern part of the province, and the Northern Goshawk breeds throughout the province south of the tree line.

The buteos are soaring hawks of medium to large size, characterized by robust bodies, long, broad wings, and relatively short to medium-length fan-shaped tails. They inhabit woodlands or open country, hunting mammals, reptiles, and frogs taken on the ground, with birds forming only a minor part of their diet.

Four species of buteos breed in Québec. The Broad-winged Hawk is widely distributed throughout southern Québec, the Red-shouldered Hawk breeds only locally in the extreme south of the province, the Red-tailed Hawk ranges throughout most of southern Québec up to Great Whale River, while the Rough-legged Hawk breeds almost exclusively in the arctic and subarctic regions of the province.

Harriers (genus *Circus*) are slender, medium-sized raptors with long tails and legs, and long, narrow, rounded wings. They have a distinctive hunting flight, flying low and quartering back and forth over fields, marshes, and other open country. In gliding flight, their wings are held in a V above the back. They chiefly hunt rodents, but will also feed on other small mammals, some birds, frogs, snakes, and occasionally carrion. Due to their large ear apertures and almost owl-like facial discs, they have the most highly developed sense of hearing in the family and can locate their prey by sound alone. There is also marked sexual dimorphism in coloration among harriers.

In Québec, as in the rest of North America, this genus is represented by a single species, the Northern Harrier, which breeds throughout southern Québec and north possibly as far as the Hudson Bay coast.

The last group in this subfamily, the eagles, are large raptors with large, rounded wings and a broad tail, resembling the smaller buteos in flight silhouette. Active flight consists of slow wing beats, and although they are capable of soaring and gliding over vast distances, they spend much of their time perching. Some species prey on mammals and birds, while others are primarily fish eaters. They also scavenge on carrion and sometimes resort to food piracy.

Two species of eagles occur in Québec. The Bald Eagle is typically associated with water and breeds very locally in the southern and central part of the province as far north as the Labrador border to the east and the James Bay region to the west. The Golden Eagle breeds very locally in primarily mountainous and hilly regions, such as the Gaspé Peninsula and the Great Whale River area.

The subfamily Pandioninae consists of a single species, the Osprey, which is found in all tropical and temperate regions of the world, although it is more widely distributed in the Northern Hemisphere.

Ospreys are large birds (about 61 cm long) with a small kitelike head, long legs, very long, narrow wings, and a relatively short tail. They have a strongly hooked beak similar to the other diurnal birds of prey, and the females are larger than the males, but they lack a supraorbital ridge, or bony projection, above the eye. They are anatomically different from the other Falconiformes, largely due to their specialized adaptations for catching and gripping slippery fish. Their heavy legs are bare and entirely covered with projecting scales.

Osprey, adults and nestlings. *Alain Côté.*

Golden Eagle. *Gaétan Duquette.*

Northern Harrier, female. *Paul Perreault.*

The outer toe is reversible, as in owls, and the pads on the undersurface of their feet are equipped with sharp spicules. They also have extremely long toes and very curved, sharp talons. They hunt by diving from the air and either snatch the fish from the water's surface or, more frequently, plunge spectacularly feet first, submerging completely for a few seconds. As a result, their plumage is stiff and closely packed, and the feathers on their thighs are extremely short, dense, and compact. When soaring and gliding, Ospreys have a distinctive flight silhouette in which their long, slender wings are angled in a gull-like shape, with wrists cocked forward and wing tips pointed down and back. Unlike most diurnal raptors, their eggs are neither tinted greenish nor yellowish on the inside of the shell.

In Québec, the Osprey's breeding range extends throughout the southern and central areas of the province close to water up to the Great Whale area.

Derin Henderson

Amadon 1975; Clark and Wheeler 1987; Godfrey 1986; Grossman and Hamlet 1964; Newton 1979; Snyder and Wiley 1976.

Cooper's Hawk. *Jim N. Flynn.*

Sharp-shinned Hawk. *Jeanne Lehoux.*

Red-tailed Hawk. *Yves Aubry.*

Bald Eagle. *Denis Faucher.*

Northern Goshawk. *Denis Faucher.*

Red-shouldered Hawk. *Denis Faucher.*

Rough-legged Hawk. *André Cyr.*

Broad-winged Hawk. *Paul Perreault.*

FALCONIDAE

In North America, this family consists of the falcons and caracaras, but only four members of the genus *Falco*, or the true falcons, are found breeding in Québec: the American Kestrel, Merlin, Peregrine Falcon, and Gyrfalcon.

Falcons are small to medium-sized (15-63 cm long) diurnal birds of prey with long, pointed wings and a medium to long tail. Birds of open habitats, they are virtually cosmopolitan, breeding on all continents except Antarctica and some oceanic islands. Some of the swiftest and most spectacular fliers among birds, they fly with quick powerful wing strokes and are capable of diving, or "stooping," at great speeds, while some *Falco* falcons actively hover. They usually kill their quarry, principally birds in the case of the larger falcons, in full flight, either dispatching their prey with a single blow or seizing it with their feet and bringing it to the ground. They may also hunt animals on the ground or occasionally in the water. Kestrel-type *Falco* species are adapted for catching non-avian prey; they hover and scan the ground for small mammals or catch insects on the wing. Typical of most raptors, falcons have a strong, hooked beak and sharp, curved talons designed for grasping and holding their quarry. Their vision is extremely acute, with lateral and binocular foveae (areas of high visual acuity) in the retina of each eye. They have a fleshy cere at the base of the beak and an eye ring, and while a few differ in coloration between the sexes, they all exhibit reversed sexual size dimorphism: females are larger than males.

Falcons have a prominent, central bony tubercle in each nostril surrounded by almost completely ossified nasal bones. The sternum, or breastbone, has a deep keel projecting well forward to accommodate large, powerful flight muscle attachments. The neck is short and the body strong and rigid due to important skeletal modifications, including fused thoracic vertebrae. They have an extra pair of bones attached to the base of the pygostyle, or tailbone, to house the powerful tail depressor muscles. These accessory bones apparently help to transmit and concentrate the force of the muscles in the tail to the pygostyle and tail feathers, thereby relieving pressure on the tail when the birds brake, shift, and manoeuvre quickly at great speeds. Their wings have moderately stiff to very stiff primaries. Many have a dark malar, or mustachial, stripe under each eye, which may possibly reduce the glare from the sun when they are hunting over open country. All *Falco* falcons and most other members of the family have a toothlike projection, or tomial tooth, on each side of the cutting edge of the upper mandible, with a corresponding pair of notches on the lower mandible, features also found in the shrikes. Using their fairly short but strong beak, jaw muscles, and tomial teeth, falcons are able to kill birds and mammals larger than themselves by biting and severing the neck vertebrae.

Other distinguishing features of falcons include a distinctive sequence of moult of the primaries, which starts with the fourth and continues on either side of it. Their eggs also produce a characteristic reddish-yellow translucence when held up to the light. Falcons are not nest builders, instead using a natural cavity, ledge, or nest of another bird. They also exhibit two characteristic behaviours when disturbed or during sexual activity: bobbing their heads and pumping their tails up and down.

The Merlin and Peregrine Falcon range throughout the province, while the American Kestrel breeds predominantly in southern Québec, and the Gyrfalcon nests only in the arctic and subarctic regions of the province.

Derin Henderson

Cade 1982; Clark and Wheeler 1987; Cramp and Simmons 1980.

American Kestrel. *Daniel Jauvin.*

Merlin. *Paul Perreault.*

Gyrfalcon, white phase, adult and nestlings. *David M. Bird.*

Gyrfalcon, dark phase. *David M. Bird.*

Peregrine Falcon, immature. *Denis Faucher.*

Peregrine Falcon. *Bruno Blais.*

PHASIANIDAE

WITH OVER 200 SPECIES, THIS FAMILY OF FOWL-LIKE BIRDS IS THE LARGEST and most diverse of the order Galliformes, ranging across temperate and tropical regions almost all over the world.

Despite large variations between individual species, phasianids are closely related and share numerous characteristics. They are predominantly ground-dwelling birds, heavily built, with fairly short, rounded wings. Their short bills have deep bases and the upper mandible overhangs the tip of the lower. Their thick legs have four large toes equipped with short, blunt claws designed for scratching. The hallux is usually shorter and elevated. Most have long necks and relatively small heads. They are primarily vegetarians, although some supplement their diet with invertebrates. They have well-developed crops for temporary food storage, as well as muscular gizzards adapted for grinding hard food material with the aid of grit. Although they typically feed on the ground, many roost in trees at night. All but a few species are ground nesters, and their nests are usually shallow scrapes, unlined or sparsely lined with plant material, often concealed in ground vegetation.

In North America, the family can be divided into four subfamilies: Phasianinae (including partridges and pheasants); Tetraoninae (grouse and ptarmigan); Meleagridinae (turkeys); and Odontophorinae (quails).

All species of the subfamily Phasianinae found in North America have been introduced from their native lands in Europe, the Middle East, and Asia. In North America, they frequent mainly farmland, scrub, grassland, and woodland borders. They are small to large (14-235 cm long), stocky birds with unfeathered nostrils, legs, and toes. Some species have one or more prominent pointed spurs on the back of the tarsi, used when fighting for dominance. Tails vary from short in most partridges to very long and ornate in most pheasants. Some species have areas of bare skin on the head, around the eye, and on the throat. These birds are able to propel themselves almost vertically up in the air to escape danger and are capable of fast and powerful flight but unable to sustain it for long, and thus are primarily sedentary. In most partridges the sexes look alike and are cryptically coloured, and they have a monogamous mating system. In all but one genus of pheasants there is a high degree of sexual dimorphism in both size and plumage. The larger males are boldly patterned with often iridescent reds, blues, greens, and whites. Many have elaborate, graduated or variably vaulted tails, which they spread fanlike over their back. Pheasants are usually polygynous, and the males perform lengthy and spectacular courtship displays.

Only two species in this subfamily breed in Québec. Both the Gray Partridge and Ring-necked Pheasant have an extremely local distribution in the St. Lawrence Lowlands; the Ringed-necked Pheasant maintained primarily by repeated introductions.

The grouse and ptarmigan, subfamily Tetraoninae, inhabit the boreal forests as well as prairies, alpine meadows, brushlands, and tundra. They are plump, fowl-like birds of medium to large size (30-90 cm) with short, curved, strong bills. They differ from the other subfamilies in having nostrils hidden by feathers and completely or partially feathered and unspurred legs. Their toes are either feathered, as in ptarmigan, or unfeathered with a fringe of horny, comblike pectinations along the sides of each toe in winter.

In the breeding season, most males have engorged combs over the eyes. Some species have a patch of erectile elongated feathers on the sides of their neck, while in others the patch is bare with distensible skin pouches that are temporarily inflated during displays. Males are often larger than females, and although in some species the sexes are similarly coloured, in others they are distinct. While most males are polygynous or promiscuous, ptarmigan are generally monogamous. In tetraonids males are territorial; in approximately half of the species they occupy solitary territories while in the remaining species they assemble in display areas known as leks.

Gray Partridge. *Denis Faucher.*

Grouse and ptarmigan are capable of a quick takeoff and may glide for long distances, but they cannot sustain prolonged flight. As a result, they are relatively sedentary, moving only short distances between summer and winter ranges; some ptarmigan, however, are often semimigratory and may travel far in search of food. Tetraonids are primarily vegetarian, eating buds and shoots, but they may consume invertebrates during the spring and summer. Although they usually feed on the ground, some species roost and feed in trees on buds and fruits.

Five species in this subfamily breed in Québec. The Spruce Grouse is found in woodland habitats throughout the province, the Willow Ptarmigan and Rock Ptarmigan breed in the tundra and taiga of northern and central Québec, the Ruffed Grouse nests in the hardwood and mixed-wood areas of southern Québec, while the Sharp-tailed Grouse occupies open country in the western part of the province.

Sharp-tailed Grouse. *Pierre Lamothe.*

Northern Bobwhite, female (right) and male.
Denis Faucher.

Spruce Grouse, male. *Yves Aubry.*

The subfamily Meleagridinae comprises only two species of turkey, inhabiting woodland, mixed open forests, and prairie brushland in North and Central America to Guatemala. Only the Wild Turkey breeds in Québec, along the U.S. border. Turkeys are large (89-122 cm), stocky, powerful birds with strong legs, rudimentary webbing between the toes, and a wide, fan-shaped tail. The small head and neck are bare and brightly coloured. The broad body feathers are square tipped. The males have darker and more iridescent plumage than the females, especially on the tail, and their legs have spurs, which are very small in females. Turkeys are polygynous, and in the breeding season the male has a thick swelling of stored fat on the chest that sustains him during the extremely energetic and elaborate courtship displays. Outside the breeding season, these birds are gregarious, forming flocks that forage together by day and roost together at night in trees. They feed on the ground in fields or open forests, eating mainly plant foods such as fruits, seeds, waste grain, and succulent vegetation, but they will also consume invertebrates. Although they generally walk or run, they can fly strongly for short distances. They are usually non-migratory, except for altitudinal migrations in mountainous areas, but they wander extensively in search of food.

The quail of the subfamily Odontophorinae are native to the Americas, ranging from southern Canada to Argentina, but are not represented in Québec. They frequent open grassland, cultivated areas, fields, and woodlands, and a few species are even found in desert scrubland. They are rotund birds of small to medium size (18-36 cm) with short tails, stocky necks, and rounded heads. A distinguishing feature of this subfamily is the bill, which has a slightly serrated cutting edge on the lower mandible. Some species have a conspicuous forward-pointing crest or plume on the top of their head. Although both males and females are similarly patterned in colours of black, brown, white, rufous, and grey, males tend to be more boldly marked. These gregarious, non-migratory birds are monogamous, and the family group, or covey, remains together outside the breeding season, sometimes amalgamating with other coveys to form larger flocks.

Willow Ptarmigan, female.
Yves Aubry.

Willow Ptarmigan, male. *Pierre Lamothe.*

Ruffed Grouse, male. *Paul Perreault.*

Rock Ptarmigan, male.
Austin Reed.

Chukar. *Pierre Lamothe.*

Many phasianids, such as pheasant, partridge, and grouse, are hunted as game birds. Some species have even been domesticated, notably the Red Junglefowl (which originated in southern Asia and is the ancestor of our domestic chickens) and the Wild Turkey. Today, these two types of fowl are commercially farmed and are of great economic importance. The family also includes species kept in captivity solely for their aesthetic qualities, of which peacocks are a prime example.

Derin Henderson

Delacour and Ridley 1985; Gill 1990; Godfrey 1986; Hamerstrom et al. 1985; Harrison, C.J.O. 1978; Johnsgard 1986; Terres 1980; Van Tyne and Berger 1966.

Wild Turkey, male. *Denis Faucher.*

Ring-necked Pheasant, male. *Gaétan Duquette.*

RALLIDAE

THE RALLIDAE (INCLUDING COOTS, GALLINULES, RAILS, AND THE SORA) ARE SMALL TO medium-sized (13-51 cm), ground-dwelling birds that live in marshes or other aquatic habitats. Most are dark or soft cryptic greys, browns, buffs, or greens. Few are brightly coloured, but gallinules and coots often have a colourful horny frontal shield on the forehead. The sexes usually look identical.

Most members of the family have a laterally compressed body adapted for running through low, thick vegetation, giving rise to the expression "thin as a rail." They have short, rounded wings and very short tails, and strong feet of varying length. Their long front toes help them walk over mud and grasses. Some species have lobed feet. They are usually omnivorous, although some species are exclusively herbivorous.

Most species, especially the rails, are hard to spot, but can be heard at twilight and at night. When approached, they usually prefer to run through the grass rather than fly. Their flight is weak and slow, and the bird soon drops back down into cover. The young and their parents may roost each night on different platforms or in nests built above water or wet ground. Most members of the family in North America are game birds.

This family of more than 140 species is represented worldwide except in polar regions and deserts. Seven species have been reported in Québec, but only five of them breed in the province, mostly in the southern half.

ROBERT ALVO

Austin and Singer 1990; Cramp and Simmons 1980; Godfrey 1986; Pettingill 1970; Ripley 1976; Terres 1980.

Yellow Rail, male. *Michel Robert.*

Yellow Rail, female. *Pierre Laporte.*

Yellow Rail, one-day-old chick. *Pierre Laporte.*

American Coot. *Paul Perreault.*

Sora. *Gaétan Duquette.*

Common Moorhen. *Denis Faucher.*

Virginia Rail. *Paul Perreault.*

GRUIDAE

THERE ARE ABOUT 15 SPECIES OF CRANES IN TWO SUBFAMILIES: GRUINAE and Balearicinae. Distribution of this relatively homogeneous family, which may be 40 million years old, is virtually worldwide, except in South America, Malaysia, Oceania, New Zealand, and Antarctica. Species diversity is greatest in eastern Asia and in northeastern Siberia. Only a single species, the Sandhill Crane, breeds in Québec, locally in the north of the province.

Cranes are long-necked, long-legged, large birds, up to 2 m tall. Their broad wings, spanning over 2 m, make them powerful fliers. The 10 or 11 primaries are usually black (the 11th primary, when present, is smaller), while the rest of the plumage is white, grey, brown, or blue. Several species have a red, orange, or black patch on the head. The short tail is often hidden by the relatively long tertials. The feet are anisodactylous and the backward-pointing hallux is elevated. Sexual dimorphism is not very marked, although the male is usually slightly larger than the female.

All members of the family are monogamous, and in some species pairs bond for life. Both sexes generally share the tasks associated with breeding, from building the nest to raising the young, which in migratory species remain with their parents for an entire year. Most cranes produce only one or two chicks a year and do not reach sexual maturity until the age of 2-7 years.

During migration, cranes commonly fly in a typical V formation, and, weather permitting, at very high altitudes. Migratory species tend to return to the same breeding ground year after year, the young learning the route from their parents. Cranes may also fly considerable distances each day between their feeding grounds and their roosts. They prefer open areas – bogs, marshes, stubble fields, savannas, and tundra – whether for feeding, roosting, or nesting. Their diet is highly diversified, consisting of various plant parts, insects, worms, molluscs, small mammals, reptiles, and even small birds.

Graceful both in the air and on the ground, cranes perform a highly fascinating courtship display. Facing each other, the male and female leap over 2 m into the air, flapping their wings and performing a resonant, penetrating duet. Some researchers feel that the special structure of the trachea explains the peculiar song of some cranes, in which the trachea, which is coiled up in the sternal cavity, is longer than the bird itself.

Breeding pairs of cranes are often too widely dispersed to interact with other pairs, making it hard to determine whether pairs really establish a territory during the breeding season. Although the Sandhill Crane seems to defend a territory to which it returns year after year, the Whooping Crane is one of the only species recognized as truly territorial. Cranes are solitary in the breeding season, but during the rest of the year they are usually gregarious.

Most crane species – except for the Sandhill Crane and the Common Crane, which still have fairly large populations – have greatly declined in numbers, and almost half are now considered vulnerable.

LINE CHOINIÈRE AND SIMON BARRETTE

Archibald 1985; Austin and Singer 1990; Collar 1985; Cramp and Simmons 1980; Faaborg 1988; Godfrey 1986; Harrison, C.J.O. 1978; Tacha et al. 1992; Terres 1980.

Sandhill Crane. *Denis Faucher.*

CHARADRIIDAE

THE FAMILY CHARADRIIDAE (PLOVERS AND LAPWINGS) COMPRISES NEARLY 70 species found worldwide except in Antarctica. Only three species, all of the genus *Charadrius* (true plovers), breed in Québec, while two representatives of the genus *Pluvialis*, the Black-bellied Plover and Lesser Golden-Plover, can be seen on migration. Species of the genus *Vanellus* (lapwings), which along with the *Charadrius* account for more than 80% of the members of the family, are found mainly in Africa and are only rare visitors to North America. There are other genera around the world, but most are monotypic.

The members of this family are small to medium-sized shorebirds (15-40 cm long) with a plump body, thick, short neck, and large eyes. They are usually distinguished from other Charadriiformes by their short, pigeonlike bill with a hardened swelling at the tip of the upper mandible and a narrower middle section. They have medium to long legs and are fleet of foot, often darting forward in quick spurts and then stopping abruptly. In most forms, the relatively short three front toes have only basal webbing and the hallux is absent; when present, it is small or vestigial and non-functional. They have a relatively short, rounded tail, and most North American plovers have long, pointed wings, whereas the lapwings have broad, rounded wings.

The Charadriidae are all strong, swift fliers, and many, especially those that breed at high latitudes, are long-distance migrants, covering up to 10 000 km between their wintering and breeding grounds. On migration they may make short stops to feed, day or night, but when their route takes them across oceans, they need to build up enough fat reserves for the trip before they start. Most species are fairly gregarious outside the breeding season, and on migration flocks of several thousand birds commonly occur.

Although most plovers are conspicuous in flight, with their black and white markings or white stripe on the wings, on the ground, particularly at rest, their plumage has a visually disruptive pattern, allowing them to blend in easily with their normal surroundings. In most plovers, the males are usually slightly larger than the females.

Unlike other Charadriiformes, most plovers forage in damp areas offering feeding opportunities fairly close to the surface. They avoid areas where wave action drives their prey to burrow deeper underground. With their large eyes and lack of sensitive nerve endings at the bill tip, plovers rely much more on sight than on touch when foraging, unlike sandpipers with their highly sensitive tactile bills. Plovers have a characteristic feeding method: they take a few quick steps, then pause before seizing their prey with their bill. They are occasionally seen stamping on the ground before pecking. Their diet is varied, but consists chiefly of terrestrial and aquatic invertebrates such as insects and crustaceans.

Semipalmated Plover. *Yves Aubry.*

Piping Plover. *Daniel Jauvin.*

Killdeer, adult and young.
Denis Faucher.

Most lapwings and plovers are monogamous, and pairs defend a small territory, varying in size mainly according to the type of habitat. In polyandrous species, the female has the more colourful breeding plumage, while the male incubates the eggs. Polygyny and polyandry have also been observed in the Mountain Plover. Typical nests are simple scrapes or hollows in the ground, usually unlined or sparsely lined with pebbles or plant material, and the eggs are cryptically coloured. Many species will lead potential predators away from the nest or young by feigning injury with a "broken-wing" distraction display. Despite these efforts, the nest predation rate is fairly high, and some species will produce up to three replacement clutches before bringing their eggs to term. Plovers occupy a wide variety of habitats, nesting on sandy or pebbly beaches, near salt water or fresh water, in marshes, or on grasslands, ploughed fields, uplands, or tundra.

Three representatives of this family breed in Québec. The Killdeer breeds mostly in southern Québec, while the Semipalmated Plover nests in northern Québec south to James Bay, Schefferville, on the north shore of the Gulf of St. Lawrence, Anticosti Island, and the Magdalen Islands. The Piping Plover is on the endangered list; it now nests only in the Magdalen Islands, although it used to breed on the Lower North Shore and in the Gaspé.

Derin Henderson

Cramp and Simmons 1983; Evans 1985; Gooders 1975; Harrison, C.J.O. 1978; Thomas 1985a.

SCOLOPACIDAE

THIS IS THE LARGEST FAMILY OF WADERS, WITH CLOSE TO 90 SPECIES around the world. They are the most widespread family in the world, found everywhere except Antarctica. Although they are collectively known as shorebirds, some – most notably the American Woodcock – are upland birds and rarely frequent the shore. Over 40 species breed in North America, 11 of them in Québec. This family can be divided into two subfamilies: Scolopacinae (sandpipers and their allies) and Phalaropodinae (phalaropes).

Greater Yellowlegs. *Denis Faucher.*

Most species of Scolopacinae breed in the Northern Hemisphere, ranging well into the arctic and subarctic regions; only a few species are adapted to tropical breeding. They nest in all types of wetlands and grasslands, primarily on tundra, and in boreal and temperate zones. Most are highly migratory, and some long-distance travellers may winter at the opposite end of the world. The Scolopacinae are birds of small to medium size (13-63 cm long) with relatively long, either pointed or rounded wings. Some have a short, stocky neck, but in most it is long and slender. Most have long, slim legs and all have three fairly long front toes and a short hind toe, except for the Sanderling, in which the hallux is absent or rudimentary. Bill length and shape are adapted largely to feeding behaviour and thus vary considerably from one species to another. In most species, the bill is much longer than the head. Some have a straight, narrow bill for pecking at surface prey, which they detect by sight, while others have an extremely long bill that may be straight, recurved, decurved, or even spatula-shaped. They feed mainly by probing in the soft substrate, and as a result the fairly soft, pliable bill tip is equipped with sensitive tactile sensory endings. They prey mainly on invertebrates, but plant foods may account for a good part of their diet. The eyes are placed well to the sides of their head to afford a wide field of view; this is most highly developed in woodcock, which has a 360° field of view, a complete circle.

Most are highly gregarious except during the breeding season, when solitary pairs become aggressive and territorial. Almost all of the members of this subfamily are cryptically coloured above and pale below, with distinctive breeding, non-breeding, and juvenile plumage. The sexes are alike. Each species has a characteristic combination of wing bars and tail pattern visible in flight, which aids in field identification; along with their similar but distinctive calls, this helps to maintain contact between members of a flock, which can number many thousand strong. Sandpipers generally nest on the ground, and their eggs are usually well concealed in a shallow scrape or clump of grass, but several species lay their eggs in old songbird nests in trees and bushes. Although most members of the subfamily Scolopacinae are monogamous, some species are polygynous, polyandrous, or promiscuous.

The phalaropes, subfamily Phalaropodinae, breed in the Northern Hemisphere, winter in the tropics or the Southern Hemisphere, and nest beside shallow waters, ponds, and lakes near marshy and grassy areas. The subfamily is made up of three species, Wilson's Phalarope, the Red-necked Phalarope, and Red Phalarope, all of which breed in Québec. The last two are truly oceanic shorebirds, spending much of their lives outside the breeding season on the ocean. These medium-sized shorebirds (18-25 cm in length) have relatively long necks, long pointed wings, and a moderately long tail tapering at the end. The bill is straight and needlelike in Wilson's Phalarope and the Red-necked Phalarope, and thicker and flattened in the Red Phalarope. They all possess large nasal glands, known as salt glands, which remove the excess salt from the blood. They are well adapted to aquatic life, having laterally flattened tarsi, and their three long toes are lobed and partially webbed at the base. These adaptations reduce underwater drag to aid swimming, and their dense, waterproof plumage traps a layer of air so that they float, almost corklike, high in the water. Phalaropes feed by spinning like a top, dabbling at the water's surface with their bills as they rotate. This may be a technique for stirring up invertebrates from the shallow bottom, causing them to rise to the surface. Along shorelines, they forage in shallow water and may even drop their heads below the surface of the water while swimming in search of food, primarily insects and large oceanic plankton.

Solitary Sandpiper. *Yves Aubry.*

According to current knowledge, the males and females of most species show little or no attachment to the nest site. Phalaropes can be polyandrous; the females are larger and more brightly coloured in their breeding plumage, and they take the initiative in courtship, while the males incubate the eggs and care for the young.

DERIN HENDERSON

Godfrey 1986; Gooders 1975; Oring and Lank 1984; Thomas 1985b.

Willet. *Jacques Brisson.*

Marbled Godwit. *Jacques Brisson.*

Spotted Sandpiper. *Paul Perreault.*

Lesser Yellowlegs. *Yves Aubry.*

Red-necked Phalarope, female. *Yves Aubry.*

Least Sandpiper. *François Shaffer.*

Upland Sandpiper. *Denis Faucher.*

Semipalmated Sandpiper. *Denis Faucher.*

Short-billed Dowitcher. *Ross Harris.*

Common Snipe. *Paul Perreault.*

American Woodcock. *Denis Faucher.*

Red Phalarope. *Jacques Brisson.*

Wilson's Phalarope, male (right), female (centre) and immature. *Christian Marcotte.*

LARIDAE

THERE ARE OVER 100 SPECIES OF LARIDS, MANY OF WHICH HEAD INLAND for the breeding season although they live along the coast and at sea the rest of the year. The family, which ranges over all continents and all oceans, can be divided into four large subfamilies: Stercorariinae (skuas and jaegers), Larinae (gulls), Sterninae (terns), and Rynchopinae (skimmers). Only the first three subfamilies are represented in Québec, and the skimmers live mainly in tropical and subtropical regions.

Larids are usually white, with the backs and tops of the wings varying from grey to black. The colour combinations distinctive to each species or subfamily vary widely, however. For example, one major group of gulls has a white head and the other major group has generally a black head, while most terns have only a black band or cap on the top of the head. Members of the larid family range in size from medium to large (20-81 cm long). They have fairly long, pointed wings and webbed feet. In skuas and jaegers the tail is usually long and slender, in gulls it is usually square, and in terns it is forked.

The shape of the bill varies from one subfamily to another, making it easy to recognize the feeding techniques and diet of each. Skuas and jaegers have a very hooked bill and a cere over the nostrils. As in raptors, their bill is designed to tear prey. On the other hand, their webbed feet and claws are not well adapted to handling food, and so they catch and kill prey with their bills. During the breeding season, skuas and jaegers feed on small mammals and birds, among other things, while throughout the rest of the year, they are pirates, stealing prey from other birds such as gulls, terns, ducks, and sandpipers, a behaviour termed kleptoparasitism. Gulls are opportunistic feeders and have the most varied diet in the larid family. Their bill is much less hooked than that of the skuas and jaegers, and they capture prey on the surface of the water or on land. They eat an impressive variety of food; when food is scarce, however, they tend to concentrate on a limited number of types of prey. Terns fish by diving from high in the air, but not to great depths; the size of their catch depends on the size of their long, straight, pointed bill. The skimmers have a very unusual manner of feeding; they skim the surface of the water, trawling with the lower mandible in the water to catch fish.

Larids use bill and leg colour (black, yellow, red, pink, or orange) as important cues to recognizing members of their own and other species. Moreover, in several species of gull, the red spot on the lower mandible of adults appears to be a target for the young, which will stimulate the regurgitation of food by pecking at this exact spot.

The sexes are alike in most of the family, although in several species of gulls the male is slightly larger than the female. Pairs often stay together for life, sharing the incubation and rearing of the semi-precocial young. In the first few years, the young have brown plumage unlike that of the adults.

Pomarine Jaeger. *Pierre Lamothe.*

Little Gull. *Yves Aubry.*

Parasitic Jaeger. *Yves Aubry.*

Most species of larids are migratory, although the distance travelled varies considerably with species; the Arctic Tern migrates from practically one pole to the other, while some species of gulls travel such short distances that it would be more accurate to speak of post-breeding dispersal rather than migration. The Herring Gull and Great Black-backed Gull are permanent residents in Québec.

Most larids are colonial, but skuas, jaegers, and Bonaparte's Gulls sometimes nest solitarily. Although breeding behaviour and strategies vary widely from species to species, generally speaking, members of this family nest in relatively accessible colonies and defend the nest against predators. The eggs and chicks usually blend in well with their surroundings.

Larid population trends are regularly monitored along the St. Lawrence. The Ring-billed Gull has been increasing in numbers for about 20 years, while the Roseate Tern is threatened, and the Caspian Tern is endangered.

Pierre Brousseau and Line Choinière

Cramp and Simmons 1983; del Hoyo et al. 1992; Godfrey 1986; Lack 1968; Löfgren 1984; Paquin 1993a; Storer 1971; Terres 1980.

Common Black-headed Gull. *Christian Marcotte.*

Ring-billed Gull. *Denis Faucher.*

Long-tailed Jaeger. *Yves Aubry.*

Bonaparte's Gull. *Gaétan Duquette.*

Iceland Gull, immature. *Paul Perreault.*

Herring Gull. *Denis Faucher.*

Glaucous Gull. *Pierre Lamothe.*

Black Tern. *Paul Perreault.*

Black-legged Kittiwake. *Marcel Giroux.*

Caspian Tern. *Pierre Mineau.*

Great Black-backed Gull. *Denis Faucher.*

Roseate Tern. *Daniel Jauvin.*

Common Tern. *Denis Faucher.*

White-winged Tern. *Michel Savard.*

Arctic Tern. *Yves Aubry.*

ALCIDAE

THE SOME 20 SPECIES OF ALCIDS (AUKS AND ALLIES) ARE CONFINED TO THE Northern Hemisphere and are associated with the Arctic, Atlantic, and Pacific oceans. Five species breed in Québec, and the Great Auk, now extinct, probably used to breed here.

These seabirds range in size from small to large (14-76 cm long). They have a chunky body with a short neck, short tail, and a bill that varies greatly in shape depending on the species. Their short legs set far back on the body are used to steer underwater, but give them their upright stance and awkward gait on land. The hind toe is rudimentary or absent, while the three front toes are fully webbed, with sturdy claws.

Alcids are remarkable swimmers and divers, and some species can dive to depths of over 100 m. Their short, pointed wings with their quick, regular beat help them swim underwater as well as fly. Taking to the air is more problematic, however.

They have very dense, waterproof plumage. Most have a black back and white underparts, although two Pacific species have cryptic coloration. The sexes are alike, but the female is generally slightly smaller than the male.

Alcids live in cold salt water and go ashore solely to breed. Most species nest in dense colonies, usually in association with other species of alcids or seabirds, on steep coastal cliffs, rock stacks offshore, coastal islands, or, more rarely, dozens of kilometres inland. Eggs are usually laid on the bare rock of narrow ledges, in crannies, rock tumbles, or in burrows, but one species, the Marbled Murrelet, nests in big trees. Like many other seabirds, most alcids lay a single egg, but some species, such as the Black Guillemot, lay two.

The males in the family are monogamous and the same pairs return year after year to the same breeding site. All alcids are territorial, defending the vicinity of the nest. Both parents feed the young. In some species, such as the Black Guillemot and Atlantic Puffin, the young are able to fly and are independent as soon as they leave the nest. Other species, however, like the Razorbill, Common Murre, and Thick-billed Murre, nest in less sheltered sites, and the young leave the nest before they can fly; one of the parents accompanies them to sea for a while.

Atlantic Puffin. *Yves Aubry.*

Thick-billed Murre. *Gilles Chapdelaine.*

Common Murre. *Claude Nadeau.*

The diet of most species consists chiefly of fish, but also includes marine invertebrates such as molluscs and crustaceans. Alcids catch fish by chasing them underwater, usually far from shore.

DIANE DAUPHIN

Austin and Singer 1990; Bédard 1985; Birkhead 1985; Birkhead and Harris 1985; Bradstreet and Brown 1985; Cramp 1985; Gaston 1985; Godfrey 1986; Harris and Birkhead 1985a; Harrison, C.J.O. 1978; Storer and Birkhead 1985; Terres 1980.

Great Auk. *Steven Holt/VIREO.*

Black Guillemot. *Yves Aubry.*

Razorbill.
Denis Faucher.

COLUMBIDAE

Ringed Turtle-Dove. *Christian Marcotte.*

THERE ARE OVER 300 SPECIES OF PIGEONS AND DOVES RANGING VIRTUALLY all over the world, except in the polar regions. Two species breed regularly in Québec: the native Mourning Dove and the Rock Dove, which was introduced. Two other species, the Ringed Turtle-Dove and White-winged Dove, are rare visitors. The now extinct Passenger Pigeon was also a member of this family.

Members of this family are plump birds with small heads. The rather short bill is constricted in the middle and has a fleshy cere at its base. The legs are short and the feet have three toes pointing forward and one functional hind toe. Both sexes look alike and have dense, soft plumage. Their feathers are easily detached, possibly so that they can escape predators more easily. Columbids are small to large birds (15-84 cm long). Pigeons are generally larger than doves, although there is no trenchantly different taxonomic distinction between them.

Most members of this family are strong, fast fliers with stiff wing feathers that make a whistling sound at takeoff. On the ground, they exhibit a characteristic head-bobbing gait. Their low cooing is generally accompanied by a swelling of the throat.

The nest is usually a simple, fragile platform in which the female lays 1 to 3 eggs. Both parents incubate the eggs and feed the young. The nestlings are fed by regurgitation; in the first few days they receive pigeon's milk, a curdy substance secreted by the crop of the adults; after that, they are gradually fed more and more predigested food.

Pigeons and doves eat mainly seeds, fruit, and some insects and worms. Most species have a large, very muscular gizzard which they use to grind hard foods with the aid of ingested grit. The back of the palate is soft, so these birds can swallow long drafts of water without having to tip their head back, which is unusual in birds.

About 30 species of columbids now living in several parts of the world have been introduced. Many representatives of this family are kept in captivity as pets, in zoos, or in aviaries.

DIANE DAUPHIN

Austin and Singer 1990; Cramp 1985; Gill 1990; Godfrey 1986; Harrison, C.J.O. 1978; Long 1981; Peterson 1984; Pettingill 1970; Terres 1980.

Rock Dove. *Jean Giroux.*

Mourning Dove. *Denis Faucher.*

Passenger Pigeon, male (foreground) and female. *Paul Laliberté, Laval University Collection.*

PSITTACIDAE

THERE ARE OVER 350 SPECIES OF PARROTS AND THEIR ALLIES DISTRIBUTED on all continents, except Europe. Species diversity is greatest in South America, especially in Brazil and the southern Pacific.

Members of this family, well known as pets, are easily recognized by their short, thick bill. Their thick, muscular tongue has a large number of taste buds and tactile papillae. The hooked upper mandible is attached to the skull by a hinge that makes the bill very mobile. The head is sturdy and the skull relatively short. The legs are strong, with yoked-toed (zygodactylous) feet, making them agile climbers and good at grasping objects. Most species also use their bills to climb trees.

These birds vary greatly in size (8-101 cm long). They are known for their brilliant plumage, which may be green, blue, red, or yellow, but there are a few rare species with drabber dress. In Australia and Asia, it is usually possible to tell males and females apart by their plumage, but sexual dimorphism is much less marked in South America and Africa. Most species are monogamous and pairs bond for long periods, even for life. They are usually gregarious birds that flock together in noisy groups, screeching and squawking. Several species kept in captivity are good mimics, although no species has yet been observed imitating the song of another bird species in nature. Some species live to be 30, 50, perhaps even 100 years old.

Parrots and their relatives are very common all over the Southern Hemisphere, especially in the tropics, but some species may live fairly far north. The records of occurrence in Québec are undoubtedly of birds escaped from captivity, since it is highly unlikely that any of these species could withstand the Québec climate and establish themselves so far north.

SIMON BARRETTE AND LINE CHOINIÈRE

Forshaw 1977; Grzimek 1984b; Serpell 1985; Terres 1980.

Monk Parakeet. *Denis Faucher.*

CUCULIDAE

THIS FAMILY CONSISTS OF OVER 140 SPECIES OF CUCKOOS, ANIS, roadrunners, and coucals distributed throughout the tropical and temperate zones of the world. It is particularly well represented in Europe and Africa.

These birds vary in size from small to fairly large (15-90 cm long). They have 10 primaries, which may be pointed or rounded, and the bill, usually fairly long, is downcurved in some species. Most species have drab grey or brown plumage. The skin around the eyes is often unfeathered and may be coloured. The feet are yoke-toed (zygodactylous), and the tail varies in length from long to very long. Most members of the family are insectivorous and feed principally on hairy caterpillars, which most other birds will not take. The tree dwellers have short legs, while the ground dwellers usually have longer legs. There is rarely any sexual dimorphism among these birds, but the young look different from the parents.

Depending on the taxonomic reference work consulted, representatives of this family may be placed in several subfamilies or distinct families. They are relatively heterogeneous morphologically, and their breeding behaviour varies with species or group of species.

The Old and New World cuckoos are known for their brood parasitism and various associated adaptive characteristics. The females of some European cuckoos, for example, have evolved egg mimicry, laying eggs similar in colour and size to those of the host species. The eggshell is thicker, and thus tougher, than that of most other bird species. In addition, the young cuckoo, whose loud cry may sound like that of the host species, is sometimes able to push other nestlings out of the nest and monopolize the food brought by the foster parents. The eggs require a fairly short incubation period and the young grow quickly.

The breeding behaviour of the anis of South America is also singular. Anis are unlike other members of the family in that, within the same population, some birds build their own nests, while others form social units of several monogamous pairs that build communal nests. Female Groove-billed Anis deliberately throw the eggs of other females out of the nest before laying their own first egg. Laying order corresponds to rank in the social hierarchy, or peck order, with the least dominant female laying her first egg first and the most dominant laying her first egg last. As egg rejection stops once the dominant female begins laying, it is in the subordinate females' interest to lay their last eggs after the dominant female has begun laying. The subordinate females have developed behaviours that enable them to do this by laying more eggs than dominant females and prolonging the interval between eggs laid. All the individuals in the unit, males and females alike, help incubate, with the dominant female's mate doing the lion's share. Unlike most other members of the family which are fairly solitary, anis are usually gregarious throughout the year.

The cuckoo family has colonized a wide variety of habitats: dry and wet woodlands, open and semi-open areas. One of the best-known representatives of this family, the Greater Roadrunner (familiar from cartoons), lives in desert areas ranging from the western United States down to Central America. These birds spread their back feathers in the morning sun to expose the black pigment of their skin, so they can absorb as much solar energy as possible.

Black-billed Cuckoo. *Gaétan Duquette.*

Far from illustrating the astonishing diversity of the family, the only two representatives that breed in Québec, the Yellow-billed Cuckoo and Black-billed Cuckoo, are fairly similar in appearance and behaviour. Both build rather fragile-looking nests about a metre off the ground, and may occasionally parasitize the nests of other cuckoos or other species.

LINE CHOINIÈRE

Austin and Singer 1990; Collias and Collias 1984; Ehrlich et al. 1988; Faabord 1988; Gill 1990; Grzimek 1984; Harrison, C.J.O. 1978; Skutch 1976; Terres 1980; Vehrencamp 1977.

Yellow-billed Cuckoo. *André Cyr.*

TYTONIDAE

THE FAMILY TYTONIDAE CONSISTS OF THE BARN AND GRASS OWLS (*TYTO* SPP.) and bay owls (*Phodilus* spp.); along with the family Strigidae (typical owls) it makes up the order Strigiformes. Tytonids are distributed worldwide except in the polar regions, New Zealand, and most oceanic islands, although only the Barn Owl (*Tyto alba*) is found in North America.

Mainly nocturnal or crepuscular predators, these owls have many traits in common with the typical owls of the family Strigidae. All owls have forward-facing eyes with a high number of light-sensitive rods, or cells, in the retina, and, like humans, they can blink with their upper eyelids, which other birds cannot do. The hooked beak is deflected sharply downwards to keep the already limited field of view clear, and the soft cere at the base of the beak is covered with bristles projecting laterally from the base. They also have a feathered tarsus, sharp, curved talons, and a reversible outer toe.

Tytonids have the dense, soft plumage, large, rounded wings, and silent flight characteristic of owls. Their silence on the wing derives from the fringes of the primaries, or flight feathers, with their serrated leading edges, which break up the flow of air over the wing and eliminate the vortex noise created when air flows over a smooth surface; this allows them to hear other sounds in flight without warning potential prey of their approach.

These medium-sized owls (33-53 cm long) differ from the typical owls of the family Strigidae in a number of structural characteristics. Their large round head does not have ear tufts, and they have a distinctive heart-shaped, rather than round, facial disc. Their dark eyes are comparatively small, and their longer beak (mostly concealed by feathers) has long nasal grooves. Tytonids have a longer, narrower skull and a breastbone fused with the wishbone; the two indentations in the breastbone characteristic of most owls are either very small or absent. Tytonids have longer legs than most strigids; contrary to the strigids, their inner and middle toes are nearly equal in length, and the talon of the middle toe has a comblike serrated edge. They have a rather square tail, while it is usually rounded in strigids.

Tytonids have a characteristic wavering and very buoyant flight, often dangling their legs as they slowly quarter the ground in search of rodents. They show little sexual dimorphism. Tytonids, unlike strigids, produce a chemically unique wax in the tail gland which they use for preening.

These solitary birds do not build a nest, but use deep crevices, hollow trees, burrows, lofts, or belfries. Both parents incubate and tend the young. When food is scarce, only the older chicks survive. Pairs seem to mate for life.

These owls appear to rely largely on their very acute sense of hearing for orientation and hunting. Research has shown that the Barn Owl can hunt in total darkness thanks to its special ear structure. The ear openings are long, but have an extremely small orifice covered by a large mobile flap, or operculum. The ear openings and flaps are asymmetrical, with those on the right set about 15% higher, permitting excellent directional hearing. By moving their ear flaps, they can alter the size and shape of the ear openings to make sound reception different for each ear. The feathers of the facial disc are sound-transmitting, channelling the sound frequencies towards the ear openings. Tytonids are particularly sensitive to high frequencies above the normal range of human hearing and can accurately pinpoint sound sources to within 2, both horizontally and vertically. They have the largest number of auditory neurons in the brain among owls, allowing them to detect and memorize extremely small differences in frequency.

Tytonids feed on small mammals as well as birds, frogs, insects, and fish. Like other owls, they regurgitate the undigested parts of their prey (fur, feathers, bones, etc.) in the form of pellets. Unlike Falconiformes, owls do not have a crop, the thin-walled extension of the esophagus used to store food.

The Barn Owl is the only representative of this family found in North America. It has often been reported in Québec, but has not been positively confirmed as breeding in the province.

DERIN HENDERSON

Bunn et al. 1982; Burton 1984; Gill 1990; Godfrey 1986; Harrison, C.J.O. 1978; Hirons 1985; Terres 1980; Van Tyne and Berger 1966; Voous 1989.

Barn Owl.
Denis Gingras.

STRIGIDAE

The family Strigidae, also known as the typical owls, comprises all the owls except for the Barn Owl and its allies. This family is found throughout the world except in Antarctica and on some oceanic islands. Birds of this order (Strigiformes, the nocturnal raptors) number far fewer species and are less diverse in form and general habits than the diurnal raptors of the order Falconiformes.

These owls are small to large birds (14-76 cm long) with a big round head that is flattened in front, a broad facial disc of radiating feathers, and large, forward-facing eyes. Some have ear tufts (elongated feathers above their eyes), which can be raised or lowered. This feature has no connection with their hearing, and its precise function is still not clear. They have a short tail and rounded wings. Females are usually larger than males, but this difference is not as pronounced as in some members of the Falconiformes. They all have a strongly hooked, short, downcurved beak with a fleshy cere at its base, usually partly covered by rictal bristles. Their toes are armed with sharp, curved talons and the outer toe is reversible, so they can perch or grip with two toes forward and two back. The legs – and in a number of species the toes – are feathered, possibly to aid in thermoregulation and/or to protect against bites from prey. Unlike owls of the family Tytonidae, the inner toe of the Strigidae is shorter than the unserrated middle toe, and the breastbone is not fused to the wishbone. Their plumage is soft and sombre in colour to blend in with their surroundings. Although some are active during the day, most owls are nocturnal or crepuscular hunters. They feed mainly on rodents and other small mammals, also taking birds, large insects, frogs, reptiles, crustaceans, and fish, but rarely carrion. They usually swallow their prey whole or in large chunks, regurgitating the indigestible parts, such as fur, feathers, bones, teeth, and beaks, as compact pellets. The size and shape of the pellets is often characteristic of a particular species, and their contents reveal what the owl has been eating.

Owls have evolved extensive and unique adaptations to their nocturnal and predatory way of life. They have highly developed vision and their large eyes are modified for hunting prey in poor light. The retina is packed with rods, or tubular photosensitive cells, which enable the pupil to dilate more. They have a large, highly convex lens, and the thickened cornea acts as an extra lens. The frontally placed eyes provide good binocular vision, which aids in depth perception, but the field of vision is reduced, since the eyes are fixed in their sockets and virtually immovable. To compensate for this limitation, they have remarkably mobile necks, enabling them to turn their heads through at least 270. Their reflexes are also so fast that when they turn their heads back, it often seems that they have spun their heads all the way around. Their highly acute night vision does not impede their ability to see or hunt during the day, since their pupils can readily adjust to changes in light.

Their extremely acute hearing is also an aid to night hunting. Most species have very large, vertical ear openings on the side of the head. These slits are hidden by folds of skin edged with feathers, which the bird can move to adjust the size of the opening. In addition, the facial disc typical of these birds acts as a parabolic reflector, collecting and focussing sounds towards the ear at the back of the head. Their large skull size means that the ear openings are far enough apart that the owls can pinpoint where a sound is coming from, because the ear closest to the sound actually hears it a fraction of a second earlier. Some of the genera in this family, in common with the barn owls of the family Tytonidae, have asymmetrical ear openings, allowing them to localize a sound vertically by detecting the difference in intensity of the sound received by each ear. The owl's inner ear is large, and the auditory region of the brain has many more nerve cells than in other comparably sized birds. Although the range of perceived sound frequencies is more limited than in humans, they are very sensitive to high-frequency sounds, such as the rustling of leaves and the high-pitched squeaks of rodents.

Snowy Owl. *Denis Faucher.*

As an integral component of their way of life, complementing their keen senses of sight and hearing, owls can fly in nearly complete silence. This is a consequence of the particular structure of their primaries, which break up and ease the flow of air over them, eliminating the customary noise made by air flowing over a smooth surface. The surface area of their wings is also large relative to the weight of the bird, facilitating their gliding flight with slow wing beats. Some fish-eating African and Asian species of this family lack these adaptations as their prey cannot hear them approaching.

Eastern Screech-Owl. *Denis Faucher.*

Strigids are solitary birds, and most of them are very territorial. Usually only the female incubates, but in some species incubation is shared by the male. Both parents tend the young; when food is scarce, only the older chicks survive.

Ten species of strigids breed in Québec: the Eastern Screech-Owl nests only in the southernmost part of the province; the Barred Owl, Long-eared Owl, and Northern Saw-whet Owl are found in southern Québec; and the Boreal Owl breeds in central Québec. The Great Horned Owl occurs in woodlands throughout the province, all the way up to the tree line. The Northern Hawk Owl is associated mainly with the boreal forest, but is also found in the southern part of the province, at high altitudes or in some bogs. The breeding range of the Short-eared Owl extends slightly farther north than that of these two species to include Ungava Bay. A single species, the Snowy Owl, nests on the arctic tundra of Québec. Another, the Great Gray Owl, has recently been confirmed as breeding in Québec, around James Bay and in Abitibi.

DERIN HENDERSON

Burton 1984; Cramp 1985; Gill 1990; Godfrey 1986; Harrison, C.J.O. 1978; Hirons 1985; Peterson 1969; Terres 1980; Van Tyne and Berger 1966.

Barred Owl. *Denis Faucher.*

Short-eared Owl. *Paul Perreault.*

Great Gray Owl. *Jean Giroux.*

Northern Saw-whet Owl. *Alain Côté.*

Boreal Owl. *Gaétan Duquette.*

Northern Hawk Owl. *Denis Faucher.*

Long-eared Owl. *André Cyr.*

Great Horned Owl. *Denis Faucher.*

CAPRIMULGIDAE

THERE ARE CLOSE TO 80 SPECIES OF GOATSUCKERS (NIGHTJARS AND nighthawks) distributed worldwide, except in New Zealand and most oceanic islands.

These small to medium-sized birds (about 18-30 cm long) have sleek, owl-like feathers. Their brown, buff, grey, white, and black plumage camouflages them well at rest. In several species, the sexes are distinguished by white markings on the wings or tail.

Goatsuckers have long wings, variously pointed according to species. Some New World species have falconlike wings. The males in some forms develop very long primaries during the breeding season. In other species, the tail feathers grow very long.

Caprimulgids have very short legs, and in some species the tarsi are feathered. The feet are small and weak, with three toes pointing forward and one backward. The middle claw is pectinated and is probably used to preen the head and neck feathers.

Goatsuckers have a rather flat head, short neck, and large eyes. The small, weak bill opens very wide into an enormous gape that may extend under the ears, an adaptation to their feeding technique of capturing insects in flight. In nightjars, the vibrissae, or rictal bristles, surrounding the mouth that funnel insects into the buccal cavity are probably a further such adaptation. Goatsuckers also have strong jaw muscles that snap the bill shut like a mousetrap as soon as insects touch the palate. Moths, beetles, and grasshoppers make up a good part of their diet.

Nightjars and nighthawks, as their names suggest, are usually nocturnal or crepuscular. They spend the day on the ground or in trees; perched lengthways on branches to enhance their camouflage, they are hard to spot but can be located thanks to their call or nightly singing.

Goatsuckers nest in isolated pairs. All species appear to be monogamous and are thought to be territorial. They normally nest on the ground but also use flat roofs. They typically lay one or two eggs, and generally both parents will share in caring for the young. The young are fed insects, which the parents bring to them in their bills. At least one species of goatsucker – the Common Poorwill, which is found mainly in the western United States – hibernates (in the same rock crevice every winter). Hibernation is a rare habit among birds.

Two species of goatsuckers breed in Québec: the Common Nighthawk is widespread south of the 53rd parallel, and the Whip-poor-will is found mainly in southern Québec. A third species, the Chuck-will's-widow, visits the province rarely and is considered accidental here.

DIANE DAUPHIN

Austin and Singer 1990; Cramp 1985; Godfrey 1986; Harrison, C.J.O. 1978; Terres 1980.

Common Nighthawk, female. *Jean Giroux.*

Whip-poor-will, female. *André Cyr.*

APODIDAE

There are about 100 species of swifts throughout the world, but only one, the Chimney Swift, breeds in Québec, and only in the southern part of the province.

Swifts are some of the best adapted birds to life on the wing. They are real acrobats and can fly at great speeds. Foraging, drinking, bathing, courting, gathering of nest materials, and generally copulation are all accomplished in flight. Some European species even spend the night soaring at great altitudes. Swifts superficially resemble swallows on the wing, but they use frequent short glides on stiff bowed wings.

Apodidae have stiff, dense feathers coloured somberly in black, blue-black or blackish-brown, and some have patches of white. The feet, and even the toes in some species, are feathered. The sexes look identical.

The swifts' aerodynamic shape reflects their aerial way of life. They have a streamlined body, short neck, and long, pointed backward-curving wings. In these fairly small birds (9-25 cm long), the secondaries are short, while the primaries are long and stiff. Since they rarely perch in trees or on the ground, their feet are not well developed. The tarsi are actually very short and the feet are tiny; the Latin name *Apodidae* means "footless." With their very pointed, highly curved claws and reversible hind toe, these birds can cling to vertical surfaces. Among swifts of the subfamily Chaeturinae, the shafts of the tail feathers extend beyond the vanes into a row of spines, forming a stiff tail that acts as a brace.

Members of this family have a tiny slightly decurved bill and a mouth that opens into a huge gape to sweep up small insects in flight. Swifts will look for clouds of insects, then quarter back and forth through them with the gape wide open. Some species become torpid during periods of cold weather when flying insects are not available.

Apodidae nest in cliff crevices, down chimney shafts, in hollow trees, inside buildings, or in old swallow nests. Most species glue nest materials together and to the surface with saliva that hardens; their salivary glands are very large during the nesting season. One species, the Palm Swift, found in Africa and Asia, sticks its tiny nest to the underside of a palm leaf and glues the eggs to the nest. The nests of some species of the genus *Collocalia* are actually built entirely of saliva. These are extremely valuable in southern Asia, where they are used to make bird's nest soup, sometimes wrongly called swallow's nest soup.

Both parents incubate and tend the young, ferrying food for them in a pouch under the tongue.

Most swifts are gregarious, nesting in colonies of varying sizes and often feeding in large numbers where food is abundant. Most also roost gregariously in protected spots like rock crevices, caves, chimneys, or hollow trees.

Robert Alvo and Diane Dauphin

Austin and Singer 1990; Cramp 1985; Godfrey 1986; Harrison 1978; Harrison, C.J.O. 1978; Stiles and Skutch 1989; Terres 1980; Van Tyne and Berger 1966.

Chimney Swift, immature. *Daniel Jauvin.*

TROCHILIDAE

THIS NEW WORLD FAMILY IS MADE UP OF OVER 300 SPECIES OF hummingbirds, with the greatest diversity of species found in tropical areas, particularly Ecuador and northern Brazil. The family is usually classed with the Apodidae (swifts) in the order Apodiformes, but because of their many unique features, members may also be classified separately, in the order Trochiliformes. Some taxonomists break up the family Trochilidae into two subfamilies: Phaethornithinae, which are better adapted to feeding on insects, and Trochilinae, or typical hummingbirds.

Most hummingbirds have several features that distinguish them from other families of birds. Most of them are tiny, the tiniest birds in the world: their size, including tail, ranges from 6 cm for the Bee Hummingbird of Cuba to 20 cm for the Chilean Giant Hummingbird found in the Andes; their weight averages 3-6 g, but may range between 2 and 20 g. Their long, thin bill may be straight or curved, most often downward, and in many species has evolved to fit the shape of the flowers on which they feed. The tongue is extensile, and, in some species, has silky little threads at the tip forming a sort of brush used to gather nectar. The arrangement of bones and muscles controlling the tongue is similar to that of woodpeckers. Hummingbirds' feet are so small that they cannot walk; their toes, however, are big and strong enough for perching.

In hummingbirds, the bones of the arm and forearm are small and the shoulders are very flexible. The primaries are long, the chest muscles highly developed, and the wing surface large in relation to the bird's weight. All these attributes make hummingbirds excellent fliers. The wings may beat anywhere from 8 to 80 times per second, depending on the size of the bird and the manoeuvre being executed. In flight, hummingbirds can fly up, down, backwards, and hover, an agility which is very useful in feeding.

Over time, the close association between hummingbirds and flowers has resulted in what is known as coevolution. In this case, coevolution has come about because of positive interactions, with hummingbirds helping pollinate flowers and flowers providing hummingbirds with their main source of food. Research has shown that there are many species of ornithophilous plants (pollinated by birds) having a corolla shaped exactly to match the shape of the bill of a certain species of hummingbird. In addition, some plants produce nectar only at dawn or at dusk, when hummingbirds are most active in their search for food.

Ruby-throated Hummingbird, male. *André Cyr.*

Ruby-throated Hummingbird, female. *Paul Perreault.*

Hummingbirds probably have one of the highest basal metabolic rates of any warm-blooded vertebrates; their heart may beat up to 1260 times per minute. Because of their small size, hummingbirds heat up quickly, but cool down just as quickly, since their exposed surface area is large in relation to their volume. To compensate for their relatively high energy expenditure and low capacity to store reserves, hummingbirds may become torpid, a fairly rare phenomenon in birds, but one that has also been observed in goatsuckers and swifts. When conditions are favourable, hummingbirds can maintain their body temperature, even at night. On the other hand, when food is scarce or the weather is poor, they reduce their oxygen consumption; their breathing becomes irregular, interspersed with phases of apnea. In this way, provided temperatures do not drop below a certain level, they can maintain their body temperature at a few degrees above ambient temperature. Despite this size-related metabolic constraint, by including insects in their diet, hummingbirds take in supplementary protein and fat so they can store up enough fat reserves for migration. Species that breed in the temperate zones usually migrate south, travelling long distances. Hummingbirds in mountainous areas, in the highlands of Peru or Costa Rica, are altitudinal migrants: they breed at high elevations and return to lower elevations in winter. Not all species are migratory, however. Some are sedentary, while others are nomadic, never settling anywhere and wandering about, depending on food availability.

Most hummingbirds have bright, glittering plumage; the male often has an iridescent patch on the back or throat. The pair bond is ephemeral, pairs usually staying together only for fertilization. The female builds the nest, incubates the eggs, and takes care of the young by herself. The male does not help with any of these activities, but expends a great deal of energy attracting females for fertilization, which takes place outside the territory defended during courtship.

The Ruby-throated Hummingbird, the only species found in Québec, is widely distributed throughout the southern part of the province. It is the only representative of the family to breed so far north, except for the Rufous Hummingbird found in the western part of the continent, on the south coast of Alaska.

LINE CHOINIÈRE

Austin and Singer 1990; Berger and Johansen 1989; Cleave 1989; Faaborg 1988; Gill 1990; Godfrey 1986; Harrison, C.J.O. 1978; Johnsgard 1983b; Snow 1985; Storer 1971; Terres 1980.

ALCEDINIDAE

Around the world, there are more than 90 species of kingfishers, which are divided into three subfamilies: Alcedininae, Daceloninae, and Cerylinae. Kingfishers are distributed mainly in the Old World, with the greatest diversity of species found in Southeast Asia. Of all the species that have colonized the New World, only the Belted Kingfisher breeds in Québec.

Members of this family are stocky birds, varying in length from 10 to 46 cm. They have a relatively large head and a long, straight, strong, pointed bill. The neck is short and the rounded wings have 10 primaries. The tail, which in some species is short and square, is long in others and used as a rudder in flight. Kingfishers have fairly short legs and small, weak feet with syndactylous (fused) toes. Their colourful plumage is usually green or blue with red, brown, or white, and many species have an erectile crest. The male and female are usually similar in appearance, although in some species, the female is slightly more colourful.

The New World species in this family are all fish eaters and forage for prey from branches at the water's edge. The Belted Kingfisher has a large number of droplets of red oil in its eyes, which help reduce chromatic aberration and glare on the water's surface, and are thought to be an adaptation for locating underwater prey from high up. When the kingfisher spies a small fish, it dives and snares it in its bill, propelling itself out of the water with a flap of its wings. It then usually returns to its perch for the kill, beating the fish against a branch until it stops resisting. Kingfishers are not all fish eaters, however. Several Old World species live in rain forests, or on savannas, where they feed on small mammals, reptiles, amphibians, or insects.

Most species in this family are territorial and defend their territory energetically with vocalizations and aerial displays. All are cavity nesters; most are monogamous and may mate for life. Both parents raise the young, and in some multiple-brooded species the young of the first brood help raise those of the next. Unlike most other groups of birds, kingfishers do not keep their nests clean. Most species are sedentary, but those in temperate regions are migratory.

Line Choinière

Austin and Singer 1980; Fry 1985; Godfrey 1986; Martin 1985a; Parrish et al. 1981; Pettingill 1970; Terres 1980.

Belted Kingfisher, female. *André Cyr.*

PICIDAE

SOME 200 SPECIES OF WOODPECKERS, PICULETS, AND WRYNECKS (FAMILY Picidae) are distributed virtually all over the world, except in New Zealand, New Guinea, Madagascar, Australia, the polar regions, and on most oceanic islands. All have an extremely extensile tongue, yoke-toed (zygodactylous) feet, and fairly rounded wings with 10 primaries. Of the three kinds of picids, only the true woodpeckers are found in Québec, or anywhere else in Canada.

Woodpeckers are small to medium-sized birds (9-58 cm long) with black and white, green, brown, or grey plumage. The head plumage of the sexes is usually different. Individual Downy Woodpeckers and Hairy Woodpeckers can even be identified by the variations in the black and white pattern and red patch on their heads. The flight of woodpeckers is usually strong and undulating, but slow.

Pileated Woodpecker, female. *Christian Marcotte.*

Although woodpeckers are usually found in woodlands, they sometimes frequent more open country. They have evolved specialized characters for climbing tree trunks and extracting insects from the wood or bark. Like creepers, they have very stiff, narrow tail feathers that brace them; other birds, such as nuthatches, rely on their feet alone.

Woodpeckers have short legs and strong, sharp, curved claws to cling to tree bark. Most species have four long, strong toes, although some have no hallux. The outer toe, unlike that of passerines, points backward. Most have two forward- and two backward-pointing toes, which may be an adaptation enabling them to climb trees more easily.

Woodpeckers have a straight, hard, pointed bill that they use to drill holes in wood. Their very muscular neck, extremely sturdy skull, and shock-absorbing cartilage structure at the base of the bill are further adaptations for pecking trees very hard. The short, stiff feathers around the nostrils also seem to protect them from sawdust produced by pecking.

One of the most remarkable adaptations in woodpeckers is the very extensile tongue, with which they extract insects from deep holes. The Northern Flicker, for example, can extend its tongue about 5 cm beyond the tip of its bill, and the Green Woodpecker, almost 10 cm. The bones supporting the tongue are highly developed in woodpeckers: they start at the back of the tongue and curl up around the back of the skull to the top of the head, where they connect to the front of the skull. The muscles enveloping these bones, which move them, are also highly developed.

In some ant-eating species, such as the Northern Flicker, the tongue not only has barbs at the tip but is covered with very thick saliva to trap insects. The saliva is also alkaline, which may be an adaptation to neutralize the formic acid secreted by ants. In other species, such as the Yellow-bellied Sapsucker, the brushlike tip of the tongue has silky points to sweep in the sap on which they feed. Others have a tongue with a hard, pointed tip for spearing insects and extracting them from deep holes. Although woodpeckers are largely insectivorous, they may eat seeds and fruits outside the breeding season.

Woodpeckers nest in cavities that they hollow out of living or dead trees; even stumps, telephone poles, or fence posts can be used for nesting. Although a few species are colonial, most nest in isolated pairs. Both adults share incubation, but it seems to be the male woodpecker that stays on the nest at night. Both parents feed the young. In some species, food is carried in the bill, while in others it is regurgitated. The young follow the parents for some time after leaving the nest.

Woodpeckers' voices are loud and harsh. They also communicate by drumming, in some species probably to define their territory.

Most woodpeckers are permanent residents, and only a few species, such as the Northern Flicker and Yellow-bellied Sapsucker, migrate south for the winter. In some other species, such as the Three-toed Woodpecker and Black-backed Woodpecker, part of the population makes irregular irruptions southward to winter in areas where food is more abundant. Resident species wintering in Québec survive on larvae and pupae hidden in the wood and bark inaccessible to most other birds.

Eight woodpecker species breed in Québec. They range throughout most of the province, as far north as the tree line. Another species, the Red-bellied Woodpecker, is a rare visitor here.

DIANE DAUPHIN

Austin and Singer 1990; Blume 1985; Cloutier and Houle 1987; Cramp 1985; Godfrey 1986; Harrison, C.J.O. 1978; Kilham 1983; Pettingill 1970; Storer 1971; Terres 1980; Welty 1975.

Red-bellied Woodpecker, male. *Denis Faucher.*

Red-headed Woodpecker. *Denis Faucher.*

Hairy Woodpecker, male. *Jean Giroux.*

Three-toed Woodpecker, male (left) and female. *Gilles Falardeau.*

Downy Woodpecker, male. *Denis Faucher.*

Yellow-bellied Sapsucker, female. *Denis Faucher.*

Northern Flicker, male. *Denis Faucher.*

Black-backed Woodpecker. *Gaétan Duquette.*

TYRANNIDAE

THERE ARE MORE THAN 400 SPECIES IN THE FAMILY OF NEW WORLD, OR tyrant, flycatchers, including kingbirds, phoebes, flycatchers, tyrannulets, and elaenias. As their name would suggest, they are confined to the New World, with the centre of diversity in the tropics.

These small to medium-sized birds (9-40 cm long) generally have drab brown, grey, or olive green plumage, which is usually identical in both sexes and does not vary seasonally. North American flycatchers, especially those of the genus *Empidonax*, lack conspicuous field marks so that the best, if not the only, way of identifying them is often by their song.

Flycatchers are insectivorous, but some species also eat fruits or vertebrates such as small lizards, frogs, or mice. Most North American species watch for their prey from a well-exposed perch and catch them on the wing; wasps, bees, and flying ants make up a good part of the diet of these broad-billed species. Some Central and South American species, however, glean insects directly on foliage – they have a slender bill. In several species, mainly those that take their prey in flight, the bill is surrounded by stiff bristles that probably protect the bird's eyes as it forages.

Flycatchers have a poorly developed song. They most often nest on the branch of a tree or shrub, but some *Myiarchus* forms build their nest in a tree cavity, while *Muscisaxicola* species nest in a hole in the ground. In general, only the females incubate, but the males help build the nest and rear the young. All species that breed in temperate regions are migratory.

In Québec, there are only nine breeding species, although another six are accidentals.

DIANE DAUPHIN

Austin and Singer 1990; Conover and Miller 1980; Cramp 1988; Godfrey 1986; Harrison, C.J.O. 1978; Terres 1980.

Alder Flycatcher, adult and nestlings. *Jeanne Lehoux.*

Willow Flycatcher. *Jim N. Flynn.*

Great Crested Flycatcher. *Jeanne Lehoux.*

Olive-sided Flycatcher. *Rob Curtis/VIREO.*

Yellow-bellied Flycatcher. *A. and E. Morris/VIREO.*

Least Flycatcher, adult and nestlings. *Christian Marcotte.*

Eastern Kingbird. *Paul Perreault.*

Eastern Phoebe, adult and nestlings. *André Cyr.*

Acadian Flycatcher. *Robertson Graham, Parks Canada.*

Eastern Wood-Pewee. *André Cyr.*

ALAUDIDAE

THIS FAMILY COMPRISES MORE THAN 90 SPECIES OF LARKS, MOST OF THEM originating in the Old World. Although they are distributed almost all over the world, they are particularly well represented in Africa, where all genera and almost 80% of species are found. Larks utilize a wide variety of open habitats, such as tundra, desert, fields, barrens, and beaches. Only two species occur in North America: the native Horned Lark and the Eurasian Skylark, which was introduced early in the century on Vancouver Island.

Larks are small to medium in size (13-23 cm long). They have long, pointed wings with 10 primaries, the 10th one reduced in adults. Their fairly long legs generally have a long, straight, sharp claw on the hind toe, an adaption to their terrestrial way of life. The shape of the bill varies with species as a function of diet – broad in the seed eaters, slender in the primarily insectivorous species, and in between in those with a mixed diet.

Plumage is cryptic, brownish to greyish, and lighter below. Desert-dwelling species are characterized by solid colours, while field dwellers are streaked. Their plumage blends in so well with the ground that in some species, the coloration of different races varies by region. The sexes look alike, but the male is often larger.

Flight is strong and undulating. On the ground, larks usually walk or run rather than hop. They may perch on vegetation to sing, but rarely to feed. Several species have a very musical song, often given in full flight.

Larks are monogamous. They nest on the ground, the cryptic coloration of both adults and eggs helping to reduce nest predation. Although most species are solitary nesters, at least three form loose colonies. Incubation is mostly by the female, but both parents feed the young, which leave the nest well before they are able to fly.

The family includes both migratory and sedentary species; several desert-dwelling species are nomadic.

There is only one representative of this family in Québec, the Horned Lark, which breeds both in the agricultural heartland and in the northern part of the province.

SERGE GAUTHIER AND DIANE DAUPHIN

Austin and Singer 1990; Cramp 1988; Godfrey 1986; Green 1985; Harrison, C.J.O. 1978; Lack 1968; Terres 1980.

Horned Lark. *Roger Larose.*

HIRUNDINIDAE

THE SWALLOW FAMILY IS FOUND ALL OVER THE WORLD, EXCEPT IN THE polar regions and on some oceanic islands. It comprises close to 90 species, more than 20 of them found exclusively in the New World. About 30 species are confined to Africa, where the family most likely originated. Of the six species that breed in Québec, only the Bank Swallow and Barn Swallow colonized both the Old and the New World.

Swallows share several anatomical and behavioural characteristics with swifts, but the superficial resemblance between the two families is due more to convergent evolution than true kinship. The families are members of fairly distantly related orders: swifts are of the order Apodiformes and swallows are of the order Passeriformes. The Hirundinidae are really a group unto themselves. Even placing them within the order Passeriformes was problematic because of the differences between them and other families in the same order that are much more alike.

Swallows are relatively small birds, 12-23 cm long and weighing 10-55 g. Their shape is very aerodynamic and they make use of thermals to save energy while flying. They have slim bodies, short necks, and fairly long, forked tails. The long, pointed wings have nine primaries, the outermost one reduced. Swallows have short legs, and although they can perch easily, they are awkward on the ground. The bill looks tiny at first sight, but it opens very wide into a gape that sometimes extends up to the eyes; it is designed for catching insects, of course, but also for carrying mouthfuls of mud, which is often the main building material for the nest. Swallows build several different types of nest; some species nest in natural cavities or abandoned woodpecker nests, others in burrows excavated out of banks or taluses, or in mud nests of varying size and shape.

All swallows are insectivorous and feed almost exclusively on the wing in fairly open country. Only the Tree Swallow may occasionally eat berries when insects are hard to find. Tropical swallows are sometimes sedentary; in temperate zones, however, all species are migratory and their return is often heralded as a sign of spring. Some years, the first arrivals, popularly called "scouts," may even reach the breeding grounds too early, meeting with cold and a shortage of insects.

In most species, the male and female look alike. The back is usually darker than the underparts in swallows. Although their coloration varies from blue, black, green, and iridescent mauve on the back to white or orange below, some species are completely grey or black. There is no seasonal difference in plumage. The moult, which sometimes begins on the breeding grounds, mainly occurs on the wintering quarters.

Both parents build the nest, and although mainly the female incubates, in some species the male helps. Both parents tend the young. Some species, mostly those that nest solitarily in burrows, are territorial, but the majority of species are colonial, with colonies ranging in size from a few individuals to over 1000 pairs.

LINE CHOINIÈRE

Bryant 1985; Faaborg 1988; Ginn and Melville 1983; Godfrey 1986; Grzimek 1984c; Harrison, C.J.O. 1978; Lunk 1962; Terres 1980; Turner 1989.

Purple Martin. *Roger Larose.*

Purple Martin, male (left) and female. *André Cyr.*

Bank Swallow. *Denis Faucher.*

Tree Swallow. *Jean Giroux.*

Northern Rough-winged Swallow. *Paul Perreault.*

Cliff Swallow. *Paul Perreault.*

Barn Swallow. *Denis Faucher.*

CORVIDAE

Common Raven. *Denis Faucher.*

THERE ARE OVER 100 SPECIES OF CORVIDS FOUND THROUGHOUT THE world, but only four, the American Crow, Common Raven, Blue Jay, and Gray Jay, breed in Québec. This family is distributed worldwide except in the polar regions, southern South America, New Zealand, and most isolated islands. Corvids live in a wide variety of habitats: forests, farmland, grasslands, deserts, steppes, and tundra.

These medium- to large-sized birds (20-71 cm long) include the largest of all passerines, the crows and ravens of the genus *Corvus*. Corvids are robustly built, with a powerful tapered bill, external nostrils covered with vibrissae, and strong legs. Tail length varies with species, with jays generally having a longer tail than ravens or crows. Ravens and crows are dark, sometimes all-black, while jays are more colourful. The sexes are alike, and the young of most species resemble the adults. On the ground, ravens and crows walk, while jays hop. All the members of this family are easily recognized by their raucous calls and gregarious, aggressive, attention-getting behaviour. Some species, like the American Crow, are good mimics.

Corvids are highly adaptable, as can be seen by their diet and feeding behaviour. These birds have the habit of burying food or hiding it in holes in trees or in any other suitable spot. Some species of *Corvus*, especially crows, gather their food and store it in an antelingual pouch formed by an extension of the buccal cavity, while jays store their food in the esophagus until they take it to their caches. The corvids' sturdy, generalized bill allows them to exploit a wide variety of foods. Most species are omnivorous, eating fruits, seeds, insects, and small vertebrates; often they also eat eggs and chicks of other bird species, or even carrion.

These birds are not always popular around the world. For a long time, black crows and ravens were seen as good or bad omens, depending on the time and place. Even today, prejudices against them abound. Although farmers sometimes look on them as pests, they are very useful as eaters of carrion and insects.

JOSÉE TARDIF

Eigelis and Nekrasov 1967; Farrand 1983; Godfrey 1986; Goodwin 1976; Holyoak 1990; Kilham 1989; Terres 1980.

Blue Jay. *Jean Giroux.*

American Crow. *Denis Faucher.*

Gray Jay. *Denis Faucher.*

PARIDAE

Black-capped Chickadee. *Denis Faucher.*

THE TITMOUSE FAMILY COMPRISES SOME 50 SPECIES THAT HAVE COLONIZED wooded areas all over the world, except in South America and Australia. Two of the three species found in Québec, the Black-capped Chickadee and Boreal Chickadee, are widely distributed all over the province, except the far northern tundra. The Black-capped Chickadee is typical of mixed and deciduous forests, while the Boreal Chickadee is usually found in coniferous forests. The Tufted Titmouse is only a casual visitor in southern Québec.

Although most titmice are capable of migrating long distances, they are rather sedentary birds. Despite their fairly small size (9-20 cm long), they are physiologically and behaviourally adapted to remaining in northern latitudes even in winter. The bill, although short and pointed, is sturdy enough for them to switch from a summer diet of insects to eating more seeds in winter.

Titmice nest and sleep in cavities. Interestingly, their roosting holes are smaller than their nesting holes, with less free space between the bird and the walls of the cavity, which may help reduce heat loss. Unlike most of the larger seed eaters, some species of titmice cannot store up enough energy during the day to maintain their basal metabolism and fight off the cold all night without feeding. This is especially true in winter, when the nights are colder and the daylight hours that can be spent feeding are short. Nonetheless, to counteract the effects of cold and lack of food, some species are able to conserve energy by lowering their body temperature and entering a phase of controlled hypothermia. Some species, like the Siberian Tit, dig burrows in the snow to insulate themselves against the cold.

Titmouse plumage is often denser and thicker than that of most other birds of the same size. The single moult begins after breeding and continues until fall, leaving them with insulating plumage (worn feathers lose their effectiveness against the cold). Like several other types of birds, titmice can puff up their feathers to increase the layer of trapped air and thus improve insulation.

Members of this family are very nimble on their rather short, sturdy legs. They can often be seen, upside down, carefully checking conifer branches for prey. They have short, rounded wings with 10 primaries. Usually black, white, grey, brown, or yellow, several species have white or pale cheeks and a dark crown. The male and female usually look alike.

These birds are curious and quite tame. Several species are a common sight at feeding stations and may even eat out of the hand. Except during the breeding season, titmice flock with conspecifics or other passerines.

LINE CHOINIÈRE

Austin and Singer 1990; Gill 1990; Godfrey 1986; Gosselin 1992a; Perrins 1985; Smith 1991; Terres 1980.

Boreal Chickadee. *Roger Larose.*

Tufted Titmouse. *Jeanne Lehoux.*

SITTIDAE

AROUND THE WORLD, THERE ARE ABOUT 25 SPECIES OF NUTHATCHES, found mainly in central Asia, and in smaller numbers in Eurasia and North America. Of the four species found in North America, only two occur in Québec: the Red-breasted Nuthatch, which favours coniferous and mixed forests, and the White-breasted Nuthatch, which prefers deciduous forests.

Nuthatches are stocky little birds (9-19 cm long) with short, square tails. The blue-grey of the upperparts and rather long wings contrasts with their black or brown crown. Below they vary from white to red, and several species have a dark eye-stripe. The short, thin bill may be straight or slightly downcurved. The tarsi are short and the four toes have strong, laterally compressed claws. The male and female are alike, although the female is generally slightly darker.

Nuthatches are mainly arboreal, hitching up and down tree trunks and branches, often headfirst. They are among the rare climbing birds that forage for food in this way. This gives them a different perspective on their habitat, and lets them find food left behind in crevices by woodpeckers and creepers.

Red-breasted Nuthatch, male. *Jean Giroux.*

White-breasted Nuthatch, male. *Paul Perreault.*

Unlike woodpeckers and creepers, nuthatches do not use their tail for support on tree trunks – their agility is attributable to their specially adapted feet. They have long, strong, sturdy toes; the hind claw on the hallux is longer and more curved than the others, enabling them to hop headfirst down tree trunks. On the wing, their undulating flight is very similar to that of woodpeckers.

Rather solitary in summer, after breeding, nuthatches often flock with chickadees, creepers, and kinglets to feed. Nuthatches are chiefly insectivorous, but they can also eat seeds with their strong bill. Many species prefer coniferous forests, where they find cone seeds, and in winter, they are often seen at feeders in wooded areas. They often hide away seeds in tree bark crevices for storage. Although rather sedentary, some species, including the Red-breasted Nuthatch, may migrate south when the winter is harsh or pine cones are scarce.

Nuthatches are monogamous and nest in cavities, often smeared with mud or resin around the entrance hole. Some species hollow a nest out of the soft wood of decaying trees, while others use abandoned cavities that they line with moss, leaves, feathers, or fur. They have even been reported to use nest boxes. The female usually incubates alone, but the male helps tend the young. The pair often defends its territory all year round.

CLAUDE NADEAU

Austin and Singer 1990; Cayouette 1978b; Godfrey 1986; Gosselin 1993; Harrison, C.J.O. 1978; Löhrl 1985; Terres 1980; Wallace and Mahan 1975.

CERTHIIDAE

CREEPERS ARE SMALL (UP TO ABOUT 15 CM LONG), TREE-DWELLING BIRDS OF North America, Eurasia, and Africa. They derive their name from their habit of spiralling their way, generally up, the trunks of trees in search of food; once they reach the top, they flit off to another tree, alighting near the base, and start all over again. Brown and black above with white spots and bars, the sexes look alike. The slender, pointed, downcurved bill is used to probe for food in bark crevices. The legs are short but the toes are long, with very long, sharp, curved claws for clinging to the bark.

Like woodpeckers, species of the genus *Certhia* have a stiff tail for bracing against tree trunks. The two central tail feathers are so essential that they are not moulted until the other tail feathers have grown back in; the tail can then still be used as a prop, even without them.

Creepers are insectivorous, taking insects (all life stages), as well as other invertebrates hidden in the bark. Larger, more readily available insects, like caterpillars, are also eaten regularly when they are in good supply.

Creepers generally hide their nest underneath a piece of loose bark, although they may also use cavities in trees. The female does most of the incubating, but both sexes care for the young.

The sole representative of this family in North America is the Brown Creeper. In Central and South America, the creepers' ecological niche is filled by the very similar woodcreepers (Dendrocolaptidae).

DIANE DAUPHIN AND ROBERT ALVO

Austin and Singer 1990; Godfrey 1986; Harrison, C.J.O. 1978; Terres 1980.

Brown Creeper. *Gaétan Duquette.*

TROGLODYTIDAE

WRENS ARE CHUNKY LITTLE BIRDS (10-23 CM LONG) WITH SLENDER, somewhat downcurved, extremely sharp-tipped bills well adapted for catching invertebrates (mainly spiders and insects) hidden in small crevices. The tail is often short, uptilted, and finely barred. They have strong legs and toes with long claws. Clad in shades of brown, rufous, chestnut, buff, grey, black, and white, often barred, streaked, or spotted, the sexes look identical.

Some species are polygamous. Of the five species breeding in Québec, only the Carolina Wren is strictly monogamous. All the others – the Sedge Wren, House Wren, Winter Wren, and Marsh Wren – are polygynous to varying degrees. Of these, the Marsh Wren shows the strongest tendency toward polygyny; the male builds a number of nests in an attempt to attract several females to his territory. Polygynous species devote more time than monogamous species to singing, nest building, and various other types of courtship behaviour. The males of polygynous species leave the female to finish the nest and incubate the eggs, but during the chick-rearing period when the possibility of mating with other females is limited, they will often help feed the young.

Winter Wren. *Michel Sokolyk.*

House Wren. *Paul Perreault.*

All wrens take shelter at night; some species use one of the nests built by the male, while others build special nests which, depending on the species, may or may not resemble the nest used for breeding.

Wrens sing loudly and defend their territories vigorously. Most of them have dry scolding rattles. They are extremely quick and active. Most live close to the ground and skulk under low scrubby cover. Birders can sometimes lure them into view by making squeaky noises. The various species of wrens frequent a wide variety of habitats: rain forests, marshes, rock tumbles, and semidesert, urban or agricultural areas. All of the nearly 80 species of wrens inhabit the New World and the majority are tropical. Only a single species, the Winter Wren, has spread to the Old World.

The Québec wrens are restricted to the southern part of the province, and the Winter Wren is the only one that winters here occasionally.

ROBERT ALVO AND JOSÉE TARDIF

Austin and Singer 1990; Cramp 1988; Garson 1985; Gill 1990; Harrison, C.J.O. 1978; Pettingill 1970; Terres 1980.

Carolina Wren. *Alain Gagnon.*

Marsh Wren. *Paul Perreault.*

Sedge Wren. *Gaétan Duquette.*

MUSCICAPIDAE

THE FAMILY MUSCICAPIDAE IS AN ASSEMBLAGE OF NEARLY 1500 SPECIES of songbirds formerly classified in several different families. It has its greatest development in the Old World, but two subfamilies, the Sylviinae and Turdinae, are found in Québec. The Sylviinae is represented here by two kinglets and a gnatcatcher, and the Turdinae by thrushes, bluebirds, and one species of wheatear that breeds only very locally in the northern part of the province. All members of this family have 10 primaries and are mainly insectivorous in summer, frugivorous in winter.

The subfamily Sylviinae is found mainly in Africa, Eurasia, and Australia. Kinglets and gnatcatchers, along with hummingbirds, are the smallest birds that breed in North America. These tree dwellers feed chiefly on insects sometimes caught on the wing. The male and female look alike. Unlike the Turdinae, the young have unspotted plumage similar to that of adults. Kinglets are the smallest birds that winter in Québec, sheltering in conifers and feeding on insects, eggs, and larvae in the needles and bark.

The subfamily Turdinae is distributed worldwide, except Antarctica and some oceanic islands. Representatives in Québec may be ground or tree dwellers. The Turdinae are small to medium-sized birds (11-33 cm long) with booted (undivided, unscaled) tarsi; the outermost primary is reduced. All these birds are good songsters, but the syrinx of species with a very musical song is more highly developed. In Turdinae and some other species of Muscicapidae, the syrinx is connected to both bronchi; it is made up of vibrating membranes and two distinct series of muscles that control their position, so the birds can produce two sounds at different frequencies simultaneously. The male and female may be alike or unalike, and the young always have spotted plumage. Most species are migratory and moult once a year.

LINE CHOINIÈRE

Austin and Singer 1990; Godfrey 1986; Gosselin 1991; National Geographic Society 1987; Pettingill 1985; Terres 1980.

Ruby-crowned Kinglet. *Denis Faucher.*

Blue-gray Gnatcatcher. *Denis Faucher.*

Golden-crowned Kinglet, female. *Christian Marcotte.*

Northern Wheatear, male. *André Cyr.*

Swainson's Thrush. *Roger Larose.*

Eastern Bluebird, male. *Paul Perreault.*

Veery, adult and nestlings. *Paul Perreault.*

Hermit Thrush. *Christian Marcotte.*

Wood Thrush. *Robertson Graham, Parks Canada.*

Gray-cheeked Thrush. *André Cyr.*

American Robin. *Denis Faucher.*

Bicknell's Thrush, adult and nestlings. *Arthur A. Allen.*

MIMIDAE

The family Mimidae comprises some 30 species of mockingbirds, thrashers, and allies, all confined to the New World. Mockingbirds and their relatives are distributed throughout most of the Americas as well as neighbouring islands, although the centre of abundance is in Central America. A dozen or so species breed in North America, including three in Québec: the Brown Thrasher, Gray Catbird, and Northern Mockingbird.

Mockingbirds have long been thought to be related to both thrushes (subfamily Turdinae) and wrens (family Troglodytidae). A recent genetic study has demonstrated, however, that the family Mimidae is more closely related to the Sturnidae (starlings), with which it shares the ability to mimic the songs of other birds. The Northern Mockingbird and other species of the genus *Mimus* are among the best mimics. But even though they can copy the song of other species, mockingbirds also have their own song: they may repeat the same vocal sequence up to four times before going on to the next. Interestingly, like the starlings, they keep on learning new sounds throughout their lives, while in most other species of birds the learning period lasts only from a few months to less than two years.

Mimids are medium-sized birds (20-30 cm long) with fairly uniform plumage in shades of grey, brown, rufous, or white. Some species have a triangular black mark around the eyes, which are sometimes yellow. There is no marked sexual dimorphism or seasonal difference in plumage. Since these birds are mainly ground dwellers, their legs are more powerfully developed than the fairly short, rounded wings. The tail is fairly long. The curve and length of the bill vary with feeding habits – species with curved bills probe the soft earth or ground litter, while those with pointed bills extract their food from crevices in tree bark. Some species eat insects, invertebrates, and seeds, while others feed on fruit.

Both sexes build the nest, often near the ground, in dense trees or bushes. Incubation is mainly by the female, but the male sometimes helps. In all species, both parents feed the nestlings. During the nesting season, most species are solitary and territorial. Mimids defend their territory aggressively, sometimes even injuring intruders.

Line Choinière

Austin and Singer 1990; Gill 1990; Godfrey 1986; Gosselin 1990b; Harrison, C.J.O. 1978; Miller and Gulledge 1985; Sibley and Ahlquist 1984.

Northern Mockingbird. *Denis Faucher.*

Brown Thrasher. *Jeanne Lehoux.*

Gray Catbird. *Denis Faucher.*

MOTACILLIDAE

THIS FAMILY COMPRISES OVER 60 SPECIES OF PIPITS AND WAGTAILS DISTRIButed almost all over the world, particularly the Old World. In North America, it is represented by fewer than 10 species.

Pipits and wagtails are small to medium-sized birds (13-23 cm long) with slender bodies, rather long, tapered bills, and long, thin legs. Their toes are fairly long and the hind toe has an elongated claw. They have pointed wings with nine primaries and long proximal secondaries that are often the same length as the largest primaries. The tail is fairly long, especially in wagtails; it usually also has a white or yellow edge that is quite visible in flight. In pipits, the plumage is brownish, striped, or spotted, and the sexes are alike. Wagtails have black, white, or yellow plumage with distinct markings and exhibit sexual dimorphism. Interestingly, in the genus *Motacilla* there is considerable geographic variation in the coloration of some species; in particular, male Yellow Wagtails in breeding plumage may have a yellow, blue, grey, black, or white head, depending on the region.

These ground dwellers inhabit open country such as tundra, grassland, desert, and farmland. They take insects on the ground as they walk or run, but they never hop. Most species wag their tails up and down. Representatives of this family, especially the wagtails, have a powerful, undulating flight.

Wagtails and pipits are gregarious, except during the breeding season. They nest mainly on the ground; in wagtails, both sexes share incubation, while in pipits, only the female incubates, although the male helps feed the young.

The family has only a single representative in Québec, the American Pipit, which breeds mainly in the northern part of the province.

DIANE DAUPHIN

Austin and Singer 1990; Godfrey 1986; Harrison, C.J.O. 1978; Terres 1980; Wood 1985.

American Pipit. *Denis Faucher.*

BOMBYCILLIDAE

THIS FAMILY CONSISTS OF FEWER THAN 10 SPECIES DISTRIBUTED throughout the Old and New Worlds – the waxwings, silky-flycatchers, Phainopepla, and Palmchat. Only waxwings are found in Québec or anywhere else in Canada, and they are the only members of the family associated with the subarctic and temperate zones of the Northern Hemisphere. Waxwings and their relatives all have dark plumage with a crest on the head. They have a fairly short, wide-based bill, and their short, sturdy legs have long claws.

There are three species of waxwings in the world. Two of them are found in Québec: the Bohemian Waxwing, which probably breeds in the province but is seen mainly in winter, and the Cedar Waxwing, which breeds throughout southern Québec, except on Anticosti Island. These fairly small birds (15-19 cm long) have thick, silky fawn-brown or grey plumage, with a black throat and mask, and a yellow stripe across the end of the tail. The fairly pointed wings have 10 primaries. Some of the secondaries have red waxlike appendages at the tips, giving them their name. These may be lacking in some birds, however, regardless of age or sex. The sexes are alike.

Waxwings have a strong bill with a wide gape that allows them to eat large food items. Essentially frugivorous tree dwellers, they are some of the rare songbirds capable of living solely on fruit. Their diet has resulted in a number of special physiological characteristics. For example, they have a large liver, which neutralizes the poisons in some fruits, and a short digestive tract, so that they can quickly and easily digest the fruit they eat. Waxwings are known as big eaters; given the low nutrient value of fruit, they have to eat the equivalent of two to three times their own weight in fruit and berries each day, and most of the time, they swallow them whole. At certain times of year, they also feed on insects, which they catch in the manner of flycatchers. Bombycillids are among the rare frugivorous birds that even feed fruit to their basically insectivorous nestlings. This may explain why, in Québec, waxwings nest fairly late, at a time when most plants are bearing fruit.

Bohemian Waxwing. *Jean Giroux.*

These gregarious birds have a poorly developed song and defend only the immediate vicinity of the nest.

JOSÉE TARDIF

Austin and Singer 1990; Cramp 1988; Godfrey 1986; Gosselin 1990a; Harrison, C.J.O. 1978; Terres 1980.

Cedar Waxwing. *André Cyr.*

LANIIDAE

The shrike family comprises close to 90 species, only two of which, the Northern Shrike and Loggerhead Shrike, occur in North America and breed in Québec. The Northern Shrike is a summer resident in northern Québec, wintering in the southern part. The Loggerhead Shrike's breeding range is restricted to southern Québec, where it is a summer resident. These two shrikes are similar, and it is often hard to tell them apart. The other members of the family are distributed all over the Old World, mainly in Africa and Asia.

Shrikes are small to medium-sized birds (15-35 cm long) with a strong, hooked, raptorlike bill. Several species have a toothlike notch on either side of the upper mandible. Like falcons, they use their bills to break the necks of some types of prey. Their plumage is drab – usually grey, brown, or rufous. Some species, particularly those of the genus *Lanius*, have a black mask.

Shrikes are birds of open spaces or scrub, such as open woodlands, savannas, fallow fields, and deserts. Often solitary and rather aggressive, they watch for prey from treetops or other vantage points. Shrikes are the only passerines that are true carnivores; they hunt small birds, small mammals, and other small vertebrates. A good part of their diet may also consist of grasshoppers and beetles. Characteristic of shrikes is a foraging behaviour that has earned them the common name of "butcher bird": they store food by impaling their prey on thorns or hanging it from the fork of a branch; lacking the powerful talons of raptors, they also use these techniques to tear apart larger prey which they cannot immobilize.

The Loggerhead Shrike, one of the two shrike species found in North America, is listed as endangered in Québec.

Josée Tardif

Austin and Singer 1990; Cade 1967; Robert 1989b, 1991; Robert et al. 1991; Telfer 1993; Terres 1980.

Northern Shrike. *Yves Aubry.*

Loggerhead Shrike. *Jim N. Flynn.*

STURNIDAE

THE LARGE OLD WORLD FAMILY OF STARLINGS AND ALLIES HAS ONLY THREE representatives in continental North America: the European Starling, Crested Myna, and Hill Myna, all of them introduced. While the Crested Myna has remained a localized resident of Vancouver and Vancouver Island, BC, the European Starling spread rapidly and now breeds from coast to coast. The Hill Myna breeds locally in southern Florida, but is uncommon.

This family is native to Africa (home to probably the most beautiful specimens), Eurasia, much of Oceania, and northern Australia. Through introductions, some species have become established in North America, New Zealand, southern Australia, and on many tropical islands.

Starlings are small to medium-sized passerines (16-45 cm long). They have a fairly long, straight, stout bill and strong, stocky legs. The nostrils are not covered by frontal bristles, and many family members have short, square tails. Species that live in dry, open habitats, such as the European Starling, have relatively long, pointed wings, whereas the forest-dwelling forms tend to have broader, more rounded wings. All have 10 primaries on each wing, with the outermost one reduced and often lancet-shaped. Most have glossy iridescent plumage. Although the sexes are generally similar, in some, the males are brighter than the females. Many tropical species are arboreal, but others, such as those of the genus *Sturnus*, spend much of their time on the ground and walk rather than hop. They have developed specializations for this way of life, including powerful legs and long toes equipped with sharp claws. Most starlings are omnivorous, feeding on fruit, seeds, and insects, with some species even taking pollen and nectar. They are generally gregarious, nesting in colonies, feeding in flocks, and roosting communally at night. Some may form huge flocks or roost with other species. Most *Sturnus* starlings will feed in farmland, often in close association with domestic animals. The majority of species in this family nest in holes, using a wide variety of cavities and building bulky nests within the chamber.

Starlings are recognized for their ability to mimic the song of other species of birds, as well as a whole range of mechanical and human sounds that they hear around them. Their song, although usually not very musical, includes a wide variety of sounds. Unlike most other birds, which learn to sing within a set period, starlings continue to learn sounds throughout their lives, so that the older they are, the wider their repertoire. Because of the diverse habitat they occupy, individuals can be recognized by their characteristic song.

The European Starling, like other species of the genera *Sturnus*, *Acridotheres*, and *Creatophora*, has a unique foraging technique termed "open-bill probing": the muscles of the jaw which raise the upper mandible, the protractor muscles, are well developed and are used to force the bill open. It can insert its closed bill into the soil surface or between blades of grass and push apart the substrate as it pries open its bill, exposing hidden invertebrates and seeds. The front part of its skull is relatively narrow, and as the bill opens, the eyes move forward towards each other, allowing it to rely on binocular vision without having to tilt its head to either side to see directly in front. This foraging method enables starlings to continue feeding on seeds and on insects lying dormant in vegetation during the winter months, when other insectivores have been forced to migrate south in search of food.

In Québec, the European Starling breeds throughout most of the province northeast to Ungava Bay, but not in the extreme northwest above the tree line.

DERIN HENDERSON

Beecher 1978; Dean 1985; Feare 1984; Gosselin 1989b; Orians 1985a.

European Starling. *Jean Giroux.*

VIREONIDAE

THE NEW WORLD FAMILY OF VIREOS IS REPRESENTED BY ABOUT 50 SPECIES, most of them in the subfamily Vireoninae; the other two subfamilies, Vireolaniinae and Cyclarhinae, comprise fewer than 10 species. Vireos range from central Canada south to Argentina, including the Caribbean. Of the 11 species present in North America, five breed in Québec.

Vireos are small birds, 10-18 cm long and weighing 9-24 g. Plainly dressed in greenish, grey, or brown shades above, and often yellow, grey, or white below, the sexes usually look alike, with no seasonal difference in plumage. The bill is short, and in some species that eat large insects, the upper mandible is hooked at the tip. The nostrils and part of the head of all vireos are covered with bristlelike feathers. The wings, which may be pointed or rounded, have 10 primaries, the outermost vestigial. The legs are short and usually strong.

All members of the family are arboreal, frequenting both trees and shrubs. Since it is not uncommon for several species of vireos to live sympatrically, they must share the use of the habitat to avoid competing for food (resource partitioning); that is why some species live in treetops, others midway up, and others lower down. Most species take insects, catching them on the wing or gleaning them on branches and leaves, but when the migration season approaches, the temperate species also eat berries. On their wintering grounds, some species may even become frugivorous.

Most species are territorial and some tropical species defend their territory year round. Once the territory is established, it is usually the female that builds the nest, but in several species, both parents incubate the eggs. In all species, both sexes tend the young. Generally speaking, most species of vireos are relatively common in suitable habitat.

SERGE GAUTHIER

Austin and Singer 1990; Godfrey 1986; Terres 1980.

Solitary Vireo. *Jean Giroux.*

Yellow-throated Vireo. *Paul Perreault.*

Warbling Vireo. *Jim N. Flynn.*

Red-eyed Vireo, adult and nestling. *Paul Perreault.*

Philadelphia Vireo. *Paul Perreault.*

EMBERIZIDAE

THIS LARGE FAMILY IS AN ASSEMBLAGE OF APPROXIMATELY 800 SPECIES, most native to the New World. They are small to medium-sized, variously coloured birds with wings having nine primaries. Many species are migratory, breeding in Canada and the United States and wintering in the tropics.

In Québec, this highly diversified family comprises five main subfamilies: Parulinae (wood-warblers), Thraupinae (tanagers), Cardinalinae (cardinals, grosbeaks, and allies), Emberizinae (emberizines), and Icterinae (icterines). There are also two other monotypic subfamilies, one in Central and South America, and the other confined to South America.

Austin and Singer 1990; Gill 1990; Terres 1980.

Parulinae

THIS SUBFAMILY OF THE EMBERIZIDAE COMPRISES OVER 100 SPECIES OF wood-warblers confined to the New World, from Alaska to Labrador, the Caribbean, and down to South America. Wood-warblers are well represented in Québec, with about 30 breeding species widely distributed throughout the province, except the Ungava Peninsula.

Wood-warblers are small (10-19 cm long), very active birds, primarily arboreal though some species spend much of their lives close to or on the ground. Many are brightly coloured, commonly with patterns of yellow, but also red, orange, blue, brown, black, or white. Ground dwellers, such as the Ovenbird, Northern Waterthrush, and Louisiana Waterthrush, have darker, brownish plumage with no sexual dimorphism. In breeding plumage, the males of many temperate species, however, are brighter than the females, although sexual dimorphism is not always very marked. The sharpest dimorphism is seen in the Black-throated Blue Warbler. After the late-summer moult, males of bright-plumaged species become fairly drab and can be hard to identify. Tropical warblers show no sexual dimorphism or seasonal variation in plumage, even in the most colourful species.

These birds have pointed wings with nine primaries; some have a square tail, while in others it is rounded. The bill is usually narrow and pointed, although it is broader and flatter in species such as the American Redstart, which feeds chiefly on insects caught on the wing.

Wood-warblers are almost exclusively insectivorous; the majority are foliage gleaners, primarily foraging in trees. Some species, however, such as the Common Yellowthroat, Ovenbird, Northern Waterthrush, and Louisiana Waterthrush, prefer to forage in the shrub layer or on the ground. Resource partitioning is observed, with the Cape May Warbler and the Blackburnian Warbler, for example, favouring the top of the tree, while the Yellow-rumped Warbler feeds mainly lower down. Different species also prefer different parts of the branch: for example, small *Vermivora* and *Parula* warblers chiefly feed out at the tips of branches, while others, like the Bay-breasted Warbler, forage on large branches close to the trunk. One species, the Black-and-white Warbler, gleans directly on the tree trunk, in the manner of nuthatches.

Nashville Warbler, male. *Denis Faucher.*

Generally speaking, wood-warblers nest in trees and shrubs, but not necessarily at foraging height. Some species, such as the Ovenbird, Northern Waterthrush, Louisiana Waterthrush, Tennessee Warbler, and Common Yellowthroat, build their nest on the ground. Warblers are usually monogamous; in migratory species, the male defends the pair's nesting and feeding territory. Some species also defend territories on the wintering grounds. Tropical species appear to be territorial year round and both sexes defend the territory.

The female incubates the eggs alone, but the male usually helps feed the young. The adults carry food for the young in their bills and continue to feed the chicks even after they leave the nest. Then the family usually splits up, with each parent tending part of the brood.

Pine Warbler. *Christian Marcotte.*

Mourning Warbler, male. *Roger Larose.*

Prothonotary Warbler, male. *Jacques Brisson.*

Most wood-warblers have a fairly unmusical, yet often lively, song. In species such as the Cape May Warbler and the Bay-breasted Warbler, it is so high-pitched as to be inaudible to some people.

Temperate species are migratory, and most have distinct breeding and wintering grounds. The vast majority of wood-warblers overwinter in Mexico, Central America, and South America, except for the Yellow-rumped Warbler, which winters as far north as southern Canada, sometimes even in Québec. This species thus remains near its breeding grounds, which explains why it is the first to arrive here in numbers in spring. The Blackpoll Warbler, with its rather northerly breeding grounds, probably migrates farthest south, and it is therefore one of the last warblers to return to its breeding grounds in spring.

Diane Dauphin

Austin and Singer 1990; Godfrey 1986; Griscom and Sprunt 1979; Harrison, C.J.O. 1978; Harrison 1984; MacArthur 1958; Morse 1989a; Terres 1980.

Orange-crowned Warbler. *D. and M. Zimmerman/VIREO.*

Ovenbird. *Christian Marcotte.*

Northern Parula, male.
Jim N. Flynn.

Bay-breasted Warbler, male.
Robertson Graham, Parks Canada.

Brewster's Warbler (hybrid), male. *Robertson Graham, Parks Canada.*

Golden-winged Warbler, male. *B. Schorre/VIREO.*

Blue-winged Warbler, male. *Robertson Graham, Parks Canada.*

Magnolia Warbler, male. *Paul Perreault.*

Cape May Warbler, male. *Jean Giroux.*

Yellow-rumped Warbler, male. *Roger Larose.*

Yellow Warbler, male and nestlings. *Paul Perreault.*

Chestnut-sided Warbler, male. *Robertson Graham, Parks Canada.*

Palm Warbler. *André Desrochers.*

Black-throated Blue Warbler, male. *André Cyr.*

Connecticut Warbler, male. *Doug Wechsler/VIREO.*

Cerulean Warbler, male. *Denis Faucher.*

Blackburnian Warbler, male. *André Desrochers.*

Black-and-white Warbler, male. *André Cyr.*

Black-throated Green Warbler, male. *Roger Larose.*

Louisiana Waterthrush. *André Cyr.*

American Redstart, male. *Michel Sokolyk.*

Northern Waterthrush. *Robertson Graham, Parks Canada.*

Blackpoll Warbler, male. *Robertson Graham, Parks Canada.*

Tennessee Warbler, female. *Robertson Graham, Parks Canada.*

Wilson's Warbler, male. *Yves Aubry.*

Canada Warbler, male. *Roger Larose.*

Common Yellowthroat, male. *Michel Sokolyk.*

Thraupinae

THERE ARE MORE THAN 250 SPECIES IN THIS SUBFAMILY OF THE FAMILY Emberizidae, confined to the New World and ranging from southern Alaska and Canada to the West Indies and Argentina; two thirds of all species are found only in South America, however. The best known and most numerous members of this subfamily are the tanagers, but other species, such as dacnises, chlorophonias, and euphonias are also members. The Thraupinae is represented in Canada by three species, but only one, the Scarlet Tanager, breeds in Québec, in the southern part of the province.

Scarlet Tanager, male. *André Cyr.*

Thraupinae are small to medium-sized birds (7-30 cm long) with a much reduced outer primary in addition to the nine primaries normally found in the family Emberizidae. They are some of the most brightly coloured birds, although some species are drab. There is very little seasonal difference in coloration in most species, except in those that breed in temperate latitudes. About half of all species exhibit sexual dimorphism in plumage.

Fruits and insects make up the bulk of the diet of these birds. Almost a quarter of species also supplement their diet with nectar or flowers, and some subsist almost exclusively on nectar for part of the year. Some species also regularly eat leaves and buds. Bills vary with feeding habits, with some species having a bill almost as sturdy as a cardinal's, and others, a bill almost as slender as a warbler's. Some also have a notched mandible for grasping hard-shelled insects more easily. In all species the bill is surrounded by rictal bristles.

These birds like clearings and fairly open woods, but not dense forests. They prefer sunny microhabitats, which is why they are often seen on treetops.

They are generally monogamous. Some species mate for life, which may explain why a good many of them do not sing or have a poorly developed song. During nesting, only the female incubates, but both parents share in tending the young.

DIANE DAUPHIN AND SERGE GAUTHIER

Austin and Singer 1990; Godfrey 1986; Isler and Isler 1987; Skutch 1985c; Terres 1980.

Cardinalinae

THE SUBFAMILY CARDINALINAE IS ONE OF THE SMALLER GROUPS OF THE family Emberizidae, with around 40 species of cardinals, grosbeaks, and allies. Confined to the Americas and certain adjacent islands, with the greatest speciation shown in tropical America, they inhabit woodlands of the temperate zones, tropical rain forests, thickets, arid shrublands, plantations, gardens, and fields.

These small (11-24 cm long), plump-bodied birds have longish tails and rounded wings. Their typical seed-eater's bill is strong, short, and conical; Cardinalinae are seed crunchers, while the Emberizinae, with their more slender bill, are seed peelers. Cardinals and their relatives have powerful jaws, and in most species the cutting edge is angled at the base of the lower mandible for cracking the hardest shells. They use their tongue to hold the seed against a plate in the roof of the mouth, crack the shell, then, rotating the seed, slip off the coat with the edge of the bill. They are predominantly vegetarian, eating seeds, grains, and berries, but they also feed on a wide variety of insects.

Cardinals are good songsters, and most species show marked sexual dimorphism; the males' plumage is brightly coloured with reds, yellows, or blues, while the females have more muted coloration. Most members of the family are territorial and solitary during the breeding season but fairly gregarious the rest of the year. Pairs are usually monogamous, and the males are involved in rearing the young.

Northern Cardinal, female. *Denis Faucher.*

Indigo Bunting, male. *Paul Perreault.*

Three species in this subfamily breed in Québec. The Rose-breasted Grosbeak and Indigo Bunting are summer residents in southern Québec, while the Northern Cardinal is a permanent resident in the southwestern part of the province.

DERIN HENDERSON

Austin and Singer 1990; Harrison, C.J.O. 1978; Palmer 1985; Rising and Rising 1982; Skutch 1985a; Terres 1980; Welty 1975; Ziswiler and Farner 1972.

Rose-breasted Grosbeak, male and nestlings. *Paul Perreault.*

Northern Cardinal, male. *Denis Faucher.*

Emberizinae

American Tree Sparrow. *Jeanne Lehoux.*

THIS SUBFAMILY OF THE FAMILY EMBERIZIDAE NUMBERS SOME 300 SPECIES around the world. Emberizines are found in all climates, from sea level up to very high altitudes, and virtually from pole to pole. Species diversity is much greater in the New World than in the Old, however. About 20 species, comprising a variety of sparrows, the Rufous-sided Towhee, Dark-eyed Junco, Snow Bunting, and Lapland Longspur, breed in Québec. The range of some species is restricted to the southernmost part of the province.

The subfamily Emberizinae is a fairly homogeneous group of rather small, stocky birds (9-21 cm long). They are usually dressed in dull brown, grey, or olive green, but some are showier because of the black and white contrasts in their plumage. Some species exhibit sexual dimorphism.

These birds have medium-length legs; the well-developed toes are adapted to scratching the ground. The tail is fairly long, and the wings, with nine primaries, are long and pointed in migratory species, but rather short and rounded in sedentary forms. The bill is short, conical, and pointed, with an angled cutting edge. Some *Emberiza* species have a horny structure on the palate to remove seed coats.

Emberizines feed mainly on seeds taken from the ground, but insects also account for a good part of their diet. In a number of species, arthropods are the young's staple, if not exclusive, diet.

Most species are territorial in the breeding season. Some are very musical, while others have a poorly developed song. Pairs are usually monogamous. The nest is most often built low in a tree or bush, but sometimes it is placed right on the ground. The males of some species help incubate the eggs, and almost all help feed the young.

Emberizines are mainly birds of open habitats, but some species prefer thin woodland and forest edges, shunning tall, dense vegetation, however. They are usually gregarious outside the breeding season, forming large flocks, although some species are rarely seen in groups of more than two or three birds.

SERGE GAUTHIER AND DIANE DAUPHIN

Austin and Singer 1990; Ehrlich et al. 1988; Godfrey 1986; Grzimek 1984c; Harrison, C.J.O. 1978; Pry-Jones 1985.

Henslow's Sparrow. *Robertson Graham, Parks Canada.*

Rufous-sided Towhee, male. *Roger Larose.*

White-crowned Sparrow. *Denis Faucher.*

Lapland Longspur, male and nestlings. *Yves Aubry.*

Dark-eyed Junco, male. *Paul Perreault.*

Snow Bunting, male. *Denis Faucher.*

Grasshopper Sparrow. *Paul Perreault.*

Clay-colored Sparrow. *Jeanne Lehoux.*

Field Sparrow. *Michel Sokolyk.*

Savannah Sparrow. *Denis Faucher.*

Vesper Sparrow. *Paul Perreault.*

Chipping Sparrow. *Denis Faucher.*

Fox Sparrow. *Roger Larose.*

Sharp-tailed Sparrow. *Denis Faucher.*

Le Conte's Sparrow. *Yves Aubry.*

Song Sparrow. *André Cyr.*

White-throated Sparrow. *Denis Faucher.*

Lincoln's Sparrow. *Roger Larose.*

Swamp Sparrow. *Michel Sokolyk.*

Icterinae

THIS SUBFAMILY OF THE FAMILY EMBERIZIDAE NUMBERS CLOSE TO 100 species distributed throughout the tropical and temperate zones of the New World. The majority of species are tropical, including the cassiques, oropendolas, and troupials, which are absent from the temperate zones. Eight species have been confirmed as breeding in Québec, while some others, notably the Orchard Oriole and Yellow-headed Blackbird, are fairly regular visitors in the southern part of the province. Commonly known as blackbirds, this subfamily include not only species with dark plumage, such as grackles, cowbirds, and blackbirds, but also some more colourful birds with more musical songs, such as orioles, the Bobolink, and meadowlarks.

Icterines are small to medium in size, ranging in length from 15 to 23 cm. In most non-monogamous species, the male is bigger and more colourful than the female. The characteristic conical bill is fairly long and pointed. They have nine primaries, and the tail is fairly long. Most species moult only once a year, although the brighter parts of the plumage, which play a part in courtship, may undergo a second, partial moult. Icterines exhibit a great variety of social systems: several species are monogamous, while others are polygynous, polyandrous, or promiscuous. This subfamily also includes the only obligate parasite in North America, the Brown-headed Cowbird, which never builds a nest and takes absolutely no part in rearing its young.

Common Grackle, male. *Jean Giroux.*

Red-winged Blackbird, male. *Jean Giroux.*

Thanks to their great adaptiveness, icterines have colonized virtually every type of habitat except the tundra. In Québec, they are chiefly associated with open and semi-open man-made environments, frequenting urban habitats and farmland as well as marshes. Ever since European settlement of the Americas, deforestation has been conducive to the establishment of these birds. Until recently, many populations were flourishing, so that today they are often the most numerous birds in their habitats. The Red-winged Blackbird, for example, is known to be one of the most abundant species breeding in North America. Icterines are highly gregarious, and outside the breeding season, thousands-strong feeding flocks of blackbirds, grackles, and cowbirds may do serious damage to ploughed fields. In the evening, they head back from their feeding grounds to congregate in communal roosts.

These birds show great curiosity and considerable learning ability. But aside from these two characteristics, it seems to be their special feeding method of open-bill probing that has enabled some of them to occupy a large ecological niche that they share only with some corvids and starlings. The muscles controlling the mandibles are developed and arranged such that the bill, unlike that of most other species, can open against great resistance. To feed, they insert the closed bill into the ground or inside a stalk, then open it and seize the prey caught in the space between the two mandibles. Blackbirds and some South American relatives use this method to forage in the thick carpet of wet leaves that covers the ground at certain times of the year. In other species such as the grackles, however, the muscles for closing the bill are also highly developed and they have a horny structure on the palate. These two additional features enable them to eat harder foods than most other species in the subfamily. The diet of Icterinae is highly varied; they can be said to eat almost anything. They are usually insectivorous during the breeding season, however, and seed eaters the rest of the year.

Northern Oriole, male. *Jeanne Lehoux.*

Bobolink, male. *Paul Perreault.*

Yellow-headed Blackbird, males (left) and female. *Paul Perreault.*

Orchard Oriole, male. *Denis Faucher.*

Icterines almost always have some iridescent black feathers. The coloration of some species, however, such as the Bobolink and the meadowlarks, helps them blend in more with their surroundings. The function of the black plumage of these birds is still an open question. Extremely colourful plumage is generally believed to be a selection criterion during courtship, while plumage that acts as camouflage helps make the bird less visible to predators; yet, totally black plumage seems to fulfil neither of these functions, at least not in North America, where most species are birds of open habitats. The black feathers may enable the bird to absorb more sunshine, but this is only true in some very specific environmental conditions. Above a certain wind speed, a white bird can absorb more sunshine than a black bird, because the light rays penetrate white more deeply. And if thermoregulation predominated over all other functions of plumage, why would blackbirds be just as common in northern regions as in the deserts of the Southern Hemisphere? In short, despite many hypotheses, the adaptive value of black plumage is still an open question.

Line Choinière

Blake and Orians 1985; Orians 1980, 1985a, b; Serventy 1971; Terres 1980; Weatherhead et al. 1980.

Brown-headed Cowbird, male. *Jean Giroux.*

Western Meadowlark. *Daniel Jauvin.*

Rusty Blackbird. *Normand David.*

Eastern Meadowlark. *Denis Faucher.*

FRINGILLIDAE

THE FAMILY FRINGILLIDAE COMPRISES CLOSE TO 140 SPECIES OF FINCHES distributed almost all over the world, although most are native to the Old World.

The shape of the bill varies with species, and the cutting edge, unlike that of some Emberizidae, is not angled. Fringillids are rather small birds. Most species exhibit sexual dimorphism in plumage. During breeding, only the female incubates, and usually both parents tend the young.

Only the subfamily Carduelinae is represented in Québec.

Austin and Singer 1990; Gill 1990; Godfrey 1986.

Carduelinae

THIS SUBFAMILY OF THE FAMILY FRINGILLIDAE IS FOUND IN MANY PARTS of the world. In Québec, 10 species occur regularly.

Cardueline finches are fairly small (10-25 cm long) and often coloured red or yellow. They have a highly developed song and even sing in full, undulating flight. The well-known singing canaries sold in pet shops are part of this subfamily.

With their strong skull, sturdy conical bill, powerful jaw muscles, and well-developed gizzard, these birds are adapted to peel the hard-shelled seeds that constitute the bulk of their diet, although they also eat buds, berries, and insects.

The bill varies greatly in size and shape: thick and wide in grosbeaks, it is short in goldfinches, siskins, and finches, while the tips of the mandibles overlap in the aptly named crossbills. Cardueline finches also have a notch on the upper mandible that they use to crack seeds before peeling them with their tongue.

These finches breed in isolated pairs or small groups with nests a few metres apart. They defend only the immediate vicinity of the nest, and individuals may feed in flocks during the breeding season as at other times of the year. They are usually monogamous; the female incubates the eggs alone, but both parents care for the young, feeding the chicks by regurgitation from the adult's crop. Their diet may be made up of seeds and insects, or just seeds, which is rare among birds. The Pine Grosbeak has a special adaptation: during the breeding season, it develops two pouches in the floor of its mouth, enabling it to carry extra food for the young. The redpolls and crossbills have an "esophageal diverticulum," a bilobed pocket in the esophagus, in which they can accumulate stores of food to help them get through the cold nights.

In birds, the breeding season coincides with an abundance of food; as the type of seeds they like varies from one species to another, the breeding periods of different cardueline finches also vary. The American Goldfinch, which mainly eats thistles, breeds late in the season, and crossbills breed at different times of the year, depending on the availability of conifer seeds.

Pine Grosbeak, male. *Paul Perreault.*

American Goldfinch, male. *Paul Perreault.*

Purple Finch, male. *Denis Faucher.*

House Finch, male. *André Desrochers.*

Pine Siskin. *Denis Faucher.*

White-winged Crossbill, male. *Paul Perreault.*

Hoary Redpoll, male. *Jean Giroux.*

Evening Grosbeak, male. *André Cyr.*

Red Crossbill, male. *Jean Paquin.*

Common Redpoll, male. *Denis Faucher.*

Some winters see irruptions of members of this subfamily when the trees on the northern wintering grounds fail to produce enough seeds. These food shortages over large areas force part of the bird population to invade other areas where food is more plentiful. How far north they travel in spring, and thus where they opt to breed, also partly depend on the food supply along the way. Irruptions of large mixed-species flocks of crossbills, redpolls, grosbeaks, Pine Siskins, and Purple Finches occur in Québec.

Cardueline finches frequent a variety of habitats, such as woodlands and more open areas. Some species have also successfully colonized man-made environments, often nesting in city parks, gardens, or farmland. This makes them one of the most familiar groups of birds, especially since some species regularly visit feeding stations. Members of this subfamily are very common throughout Québec; some more northern species are irregular visitors in southern Québec.

Diane Dauphin and Serge Gauthier

Austin and Singer 1990; Bock and Lepthien 1976b; Ehrlich et al. 1988; Godfrey 1986; Harrison, C.J.O. 1978; Lack 1968; Newton 1985a, b; Terres 1980; Welty 1975; Ziswiler and Farner 1972.

PASSERIDAE

THIS IS AN OLD WORLD FAMILY WITH NO SPECIES NATIVE TO THE Americas. In Québec, as in the rest of Canada, only a single species is found: the House Sparrow, which nests throughout the southern part of the province. In fact, since its introduction more or less all over the world, the House Sparrow has expanded its range to cover at least two thirds of the globe.

Passerids are small, stocky birds (10-18 cm long) with fairly uniform coloration, mainly browns and greys. They resemble the sparrows of the subfamily Emberizinae, but their legs are shorter and their bills, sturdier. Their song is poorly developed and their calls are harsh and unmusical.

These seed eaters are fairly gregarious. During the breeding season, both sexes build the nest, incubate the eggs, and raise the young. Species in this family are not migratory.

DIANE DAUPHIN

Austin and Singer 1990; Colston 1990; Godfrey 1986; Long 1981; National Geographic Society 1987.

House Sparrow, male. *Paul Perreault.*

THE BREEDING BIRDS OF SOUTHERN QUÉBEC

Red-throated Loon

Huart à gorge rousse

Plongeon catmarin

Gavia stellata

Ghislain Caron

THE VEHEMENCE AND VARIABILITY OF THE HARSH, DISSONANT, AND VOLUBLE cries of the Red-throated Loon contribute to the aura of wildness in the northern coastal areas where this smallest of the five loon species breeds. It can become airborne from water with less of a running start than the other loons and can even take off from or alight on land. The slightly uptilted appearance of the bill, even in chicks, and the adult's red throat in breeding plumage are diagnostic.

The Red-throated Loon is holarctic and breeds farther north than the other loons. It is common in Québec (Robert 1989b), breeding mostly north of the Atlas area.

Habitat and behaviour

The Red-throated Loon nests on small ponds and occasionally on lakes, both near the coast and far inland in treeless and forested country (Palmer 1962; Dickson 1992). It favours shallow pools with well-vegetated banks, promontories, and islets for nesting (Cramp and Simmons 1977); breeding pools also have areas of open water, several metres of open ground surrounding the nest site (K. A. Blanchard, QLF/Atlantic Center, unpubl. data), and a muddy substrate for the chicks to hide in (Davis 1972). On the Middle and Lower North Shore, its preferred nesting sites are islands ranging from 10 to 580 ha in size that contain 4 to 30 ponds. It chooses 0.01–1.46 ha ponds, preferring small ones (0.01–0.15 ha) located on smaller, offshore islands exposed to heavy winds that provide lift (Davis 1972). Mainland sites are an exception, with pairs opting for larger ponds averaging 0.69 ha. Two important factors in the selection of mainland sites are avoidance of mammalian predators and proximity to the sea, where adults secure most of their food as well as food for the young (K. A. Blanchard, QLF/Atlantic Center, unpubl. data).

Pairs are monogamous, forming probably lifelong bonds. The ritualized courtship ceremony is often performed on the sea close to shore and adjacent to the nesting pond. Both sexes build the nest, which is a heap of moss or aquatic vegetation, or a scrape or flattened hummock with no material added, usually within 0.5 m of the water (Cramp and Simmons 1977).

In the north, the timing of egg laying is directly related to ice-out in the nesting ponds (Dickson 1993). Both parents incubate the eggs and feed and brood the young. Incubation takes at least 24 days (Dickson 1993), and has been reported to last up to 27 days (K. A. Blanchard, QLF/Atlantic Center, unpubl. data), and even as much as 31 days (Douglas and Reimchen 1988b). Initially the young are brooded at the nest and later at any suitable spot on the shore. Adults rarely carry the young on their backs.

Except during the first few days after hatching, when the chicks may be fed crustaceans and aquatic insects (Johnson and Johnson 1935; Cramp and Simmons 1977), parents apparently almost never forage in the nesting pond (Bergman and Derksen 1977), even when the pond contains fish (Reimchen and Douglas 1984). Subsequently, the chicks are fed small fish that can be eaten whole, and by 28 days of age their diet is the same as that of adults (Cramp and Simmons 1977). On the North Shore this includes capelin, sand lance, and gunnel (Johnson and Johnson 1935). The parents ferry food to the chicks from the sea, rivers, or lakes, returning to the natal pond with one fish at a time held crosswise in the bill (Norberg and Norberg 1976). The young are often left alone—in one study for 42% of the observed time (Petersen 1973)—during adult foraging trips, which can range from 14 to 124 minutes for the round trip (Eriksson et al. 1990). The parents bring 9–14 fish to the young per day (Reimchen and Douglas 1985). In Alaska, the first chick to hatch was given priority for food (Davis 1972), but this was not the case in Sweden (Eriksson et al. 1990).

Chicks fledge at 38 to 52 days (Palmer 1962; Bundy 1976; Cramp and Simmons 1977; Douglas and Reimchen 1988b; Dickson 1993; K. A. Blanchard, QLF/Atlantic Center, unpubl. data). At fledging, the young fly to a lake, river, or to the sea, where they may still be fed by their parents as late as autumn (Reimchen and Douglas 1984).

Notes on territory

The territory is usually a pond ranging from 0.002 ha (20 m²) to 5 ha in size (Merrie 1977), which is used for display, copulation, nesting, and rearing (Palmer 1962). The same territory and even the same nest are often used from year to year, at least sometimes by the same pair. Usually only one pair nests per pond, but larger ponds may have several pairs. Pairs may also nest in island colonies within a few metres of each other (Palmer 1962), particularly when mammalian predators are a serious threat on the mainland (Bundy 1976). Families may move by land to a nearby pond if disturbed by humans (Bergman and Derksen 1977) or predators (Davis 1972).

Parents make regular foraging trips to lakes, rivers, or the sea, usually within 8 km of the pond (Merrie 1977; Douglas and Reimchen 1988a; Eriksson et al. 1990), but feeding areas apparently are not defended (Cramp and Simmons 1977).

In Scotland, each pair required about 2 km² of shallow water (marine and freshwater combined) within 8 km of the nesting area (Merrie 1977). Breeding density in Alaska was 7 pairs/km² (Schamel and Tracy 1985), while

The Red-throated Loon nests mainly along the edges of ponds and lakes in northern coastal areas close to the sea, the source of its food supply. Sainte-Marie Islands, Lower North Shore. *July 1991, Pierre Brousseau.*

Red-throated Loon

Breeding distribution

1984–1989

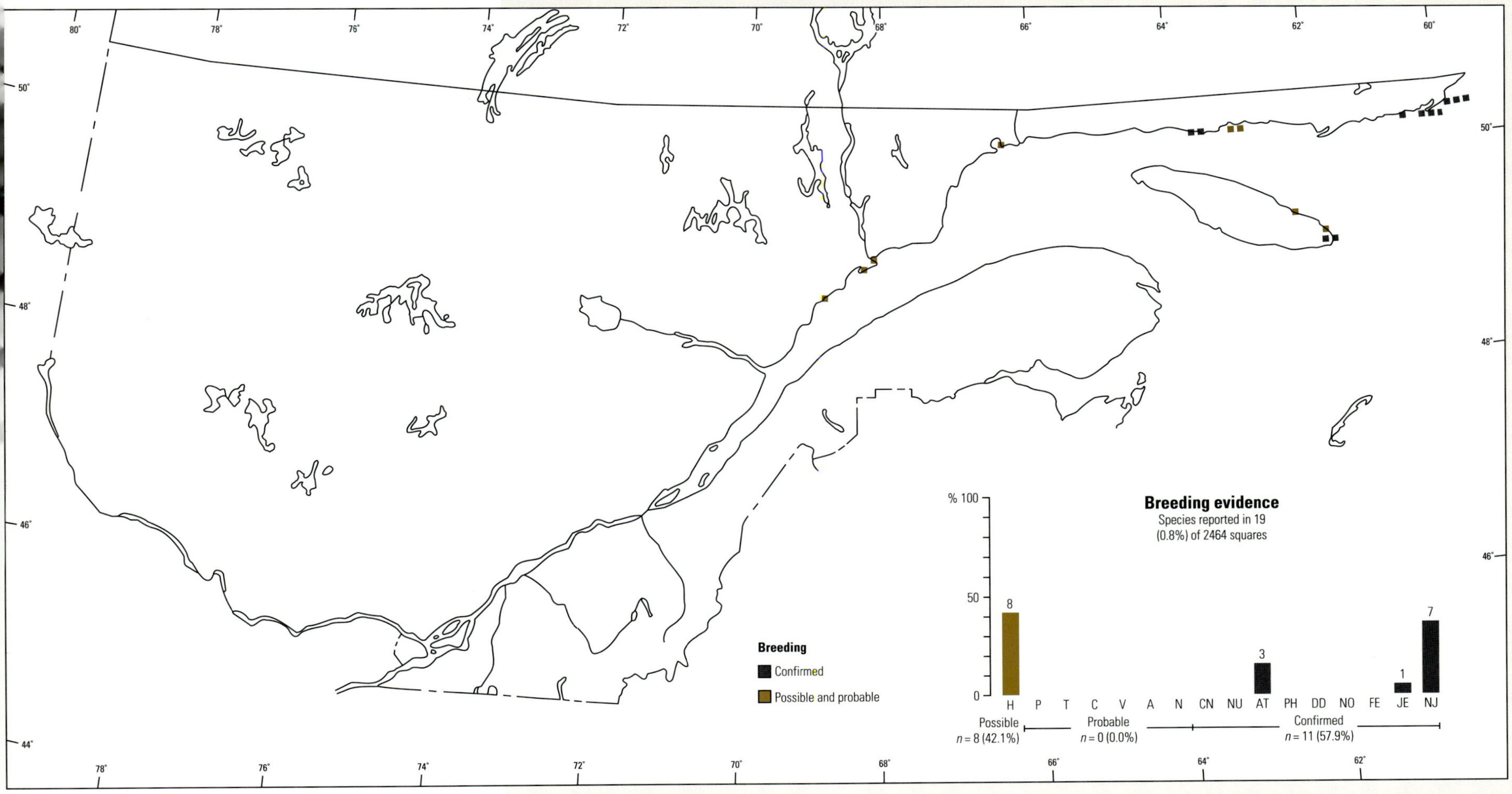

Species profile

Clutch size	usually 2 eggs (1–3)[1,2,4,6,7]
Laying interval	1 egg per 1–2 days[2,3,4,5]
Start of incubation	with first egg[2,3,4,5,7,8]
Incubation period	usually 27 days (24–31)[1,2,3,5,6,7,8]
Maturity at hatching	precocial
Care of young	by female and male[2,3,4]
Nestling period	less than 24 hours[2]
Age at first flight	38–52 days[2,3,4,5,11]
Dependence of young	probably until after first flight[2,3,4]
Broods per year	1[2,3,4,6,7,8]
Breeding age	2–3 years[1,2,4]
Mating system	monogamy[2,4,6]
Duration of pair bond	probably for life[2]
Total length	61–69 cm[8]
Wing span	106.7–114.3 cm[9]
Weight	1551 g (1150–1980)[10]
Longevity record	23 years 8 months[2]

References:
1. del Hoyo et al. 1992
2. Cramp and Simmons 1977
3. Bundy 1976
4. Palmer 1962
5. Douglas and Reimchen 1988b
6. Ehrlich et al. 1988
7. Harrison 1978
8. Godfrey 1986
9. Terres 1980
10. Dunning 1984
11. Dickson 1993

Chronology *n* = 72

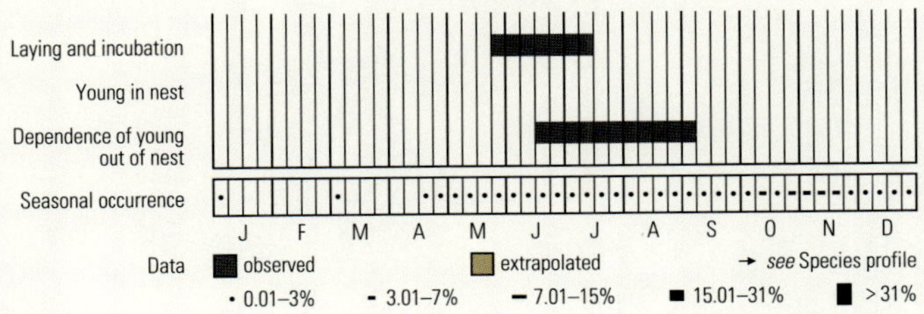

World breeding range

Notes on territory (continued)

in British Columbia it was only 0.079 pair/km² (Douglas and Reimchen 1988a). Densities in northern Yukon ranged from 0.6 to 1.8 pairs/km² and averaged 1.1 pairs/km² (Dickson 1993). No information on breeding density is available for Québec.

Favourite nest sites are on fairly bare ground on the shores of small freshwater, often shallow lakes near the coast. Blacklands Island, Lower North Shore. *June 1981, Pierre Brousseau.*

A tundra-nesting Red-throated Loon incubating its eggs on an islet not far from the shore. Bathurst Island, Northwest Territories. *August 1972, Pierre Lamothe.*

The nest is a mound of vegetation generally placed at the water's edge. Blacklands Island, Lower North Shore. *June 1981, Pierre Brousseau.*

Most individuals of pre-breeding age remain on salt water (Palmer 1962) and are not often seen in the breeding areas during the pre-nesting and nesting periods (Bergman and Derksen 1977), where they can be distinguished from breeders by their habit of occupying ponds only overnight (Douglas and Reimchen 1988a).

Distribution

Red-throated Loons were atlassed in only 19 squares, with breeding confirmed in 11 of them. All confirmations were on the Middle and Lower North Shore, between Havre-Saint-Pierre and the Sainte-Marie Islands, and near the eastern tip of Anticosti Island. There are reports of summer sightings on the Magdalen Islands and the Gaspé Peninsula (David 1980a) where the species may very well breed (Todd 1963; Fradette 1992). However, atlassers did not observe it at either location.

North of the Atlas area, the Red-throated Loon breeds along the coasts of the Ungava Peninsula (Godfrey 1986). Recent records include two broods near Lake Bienville (Cons. G. G. – GREBE 1990a), pairs with one chick each in mid-July 1992 and 1993 on two small lakes (0–5 ha) near Lake Choret south of Lake Bienville (Bergeron et al. 1993; R. Bergeron, Dryade, and A. Reed, CWS, pers. comm.), a pair north of the Eastmain River (Tardif and Morneau 1991), and a pair demonstrating nesting behaviour near Lake Chabouillé southeast of Waskaganish (Cons. G. G. – GREBE 1992j). Many individuals, including several pairs, were observed at different locations in the Great Whale and Little Whale river watersheds (Cons. G. G. – GREBE 1990a; Dryade 1992), and near the Laforge Reservoir (Bergeron et al. 1993; R. Bergeron, Dryade, and A. Reed, CWS, pers. comm.).

The North American breeding population of Red-throated Loons overwinters mainly at sea along the east coast from Newfoundland to the Gulf of Mexico and along the west coast from Alaska to Mexico (Godfrey 1986). The highest density of wintering birds is found in southern North Carolina near Cape Fear (Root 1988), and often several hundred individuals can be seen off Cape May, New Jersey (Terres 1980). One of the best areas to observe Red-throated Loons in Québec is along the Gaspé Peninsula, where they are seen frequently during migration (Girard 1988).

History and trends

The breeding population along the Middle and Lower North Shore apparently has not changed appreciably since 1925 (BIOMQ; QLF/Atlantic Center, unpubl. data). On Anticosti Island, the species was at least fairly common in the late 1800s and early 1900s, and it bred near Port-Menier (Ouellet 1969), but Atlas data suggest that it is now restricted to the east side of the island. Some Red-throated Loons are taken by people living on the Hudson Bay and James Bay coasts (James Bay and Northern Quebec Native Harvesting Research Committee 1982).

In Great Britain, Red-throated Loons are subject to pressure from peat cutting, agricultural reclamation, hydro-electric dams, and oil spills, and to disturbance from fishermen, birders, photographers, and egg collectors (Sharrock 1976; Merrie 1979; Booth 1982). Some of these problems could be limiting factors along the North Shore, while hydro-electric dams and associated hydrological changes could be a problem in the Hudson Bay region. A number of individuals become entangled in fishing nets and drown along the North Shore (K. A. Blanchard, pers. obs.).

Kathleen A. Blanchard and Robert Alvo

Common Loon

Huart à collier

Plongeon huard

Gavia immer

For many North American canoeists and cottagers, the Common Loon is the symbol of the northern lake country. Its black-and-white breeding plumage is unmistakable, as are its vocalizations. It is a large, heavy bird with legs placed far back on the body, making it adept underwater but awkward on land. Its bones contain fewer air spaces than those of most other birds, making it better adapted to life in the water than in the air, and it compresses its feathers before diving to further increase its specific gravity (Schorger 1947). This common migrant breeder is found throughout Québec (Godfrey 1986).

This islet, at some distance from the shores of Lake Tantaré, has been home to a pair of Common Loons for several years. Tantaré Ecological Reserve, near Québec City. *Summer 1990, Ministère de l'Environnement du Québec.*

Habitat and behaviour

Common Loons breed on lakes at least 5 ha in size (McIntyre 1988) with extensive areas of fairly deep open water (Cramp and Simmons 1977). Pairs probably mate for life and use the same territories every year (McIntyre 1974; Strong and Bissonette 1987; Belant and Anderson 1991a). Sexes are identical in plumage, but males tend to be larger (Olson and Marshall 1952). One pair may occupy a whole lake, although lakes larger than 50 ha are often partitioned by two or more pairs and may include extensive neutral areas that are not defended (Olson and Marshall 1952). On the other hand, one pair may defend two lakes (Belant 1991), particularly if the nesting lake is very small. Croskery (1988) has shown that if a pair is removed from its territory, another pair may come and take it over, demonstrating that there may be a floater population of adults that fail to breed for lack of vacant territories.

Nests are placed at the water's edge, usually next to deep water, to allow the incubating bird an easy escape in case of danger. Nest sites can be located on small islands, in marshy bays, or on the mainland (Olson and Marshall 1952; Alvo 1981), or even up to 800 m from the main body of the lake along inlet or outlet streams (Strong and Bissonette 1989). Both parents build the nest, which may be a shallow depression with no lining on flat ground, a simple nest of aquatic vegetation, or a large mound of aquatic vegetation in shallow water. Muskrat houses and natural mats of floating vegetation are also used. The same nest bowl or the same general area is often used in successive years (Strong and Bissonette 1987). Usually two eggs are laid, but sometimes only one, and rarely three (Henderson 1924; Vermeer 1973). Rare four-egg nests may involve two females (Peck and James 1983; Zicus et al. 1983).

Both parents incubate and care for the young. Eggs are laid 2 to 3 days apart, but hatch one day apart (McIntyre 1975). Shells and egg membrane sacs left on or next to the nest can be used to determine the year of the nesting attempt (Alvo and Prior 1986) and nest success (Alvo 1985). The downy chicks leave the nest usually within a day of hatching, often riding on the parents' backs during the first three weeks to keep warm, conserve energy, and avoid predation by large fish (Solman 1945), turtles (McIntyre 1975), and birds (Croskery 1991).

Chick mortality is highest in the first two weeks; after that, they have a very low mortality rate (Olson and Marshall 1952). They can catch some fish for themselves at 1–2 weeks, but are dependent on the parents for most of their food for 8 weeks (Barr 1973). The parents feed the chicks by diving and surfacing with a food item in the bill; they then present it to the chick, which reaches for and takes it. The chicks' initial diet is composed of invertebrates, small fish, and vegetation, changing to a diet of mostly fish within a few weeks (Alvo and Berrill 1992). They depend on food from the natal lake, and parents only rarely bring fish from neighbouring lakes (Parker 1985). Chicks can fly at about 11 weeks and may leave the lake soon afterward (McIntyre 1975). Loons start breeding in their second or third year (Olson and Marshall 1952; Cramp and Simmons 1977). Based on banding data, the maximum age recorded for this species is 7 years 10 months (Kennard 1975), but McIntyre (1988) estimated that loons may live for 25 to 30 years.

Common Loons dive from the surface and pursue prey under water, using not only their legs but also their wings for quick directional changes (Townsend 1924). They are primarily fish eaters (Barr 1973), but also take invertebrates, amphibians, aquatic vegetation (Madsen 1957), and sometimes young waterfowl (Brooks 1941). They can breed on fishless lakes that contain these other sources of food (Munro 1945).

Common Loons are extremely shy on lakes with little human activity, especially during incubation and brood rearing. Adults and chicks often retreat to a part of the lake not visible to the observer, or they dive repeatedly, surfacing only partially and briefly.

Ghislain Caron

Notes on territory

Breeding territories are 5–80 ha or more, whereas winter feeding territories are 4–8 ha (McIntyre 1978). Young chicks are raised in a shallow, vegetated, sheltered part of the territory (Belant and Anderson 1991b).

Territorial reactions to loons and other intruders include yodelling, bill-dipping, neck-raising, circle-dancing, splash-diving, standing upright, and rushing with wings flapping over the water (Sjölander and Ågren 1972).

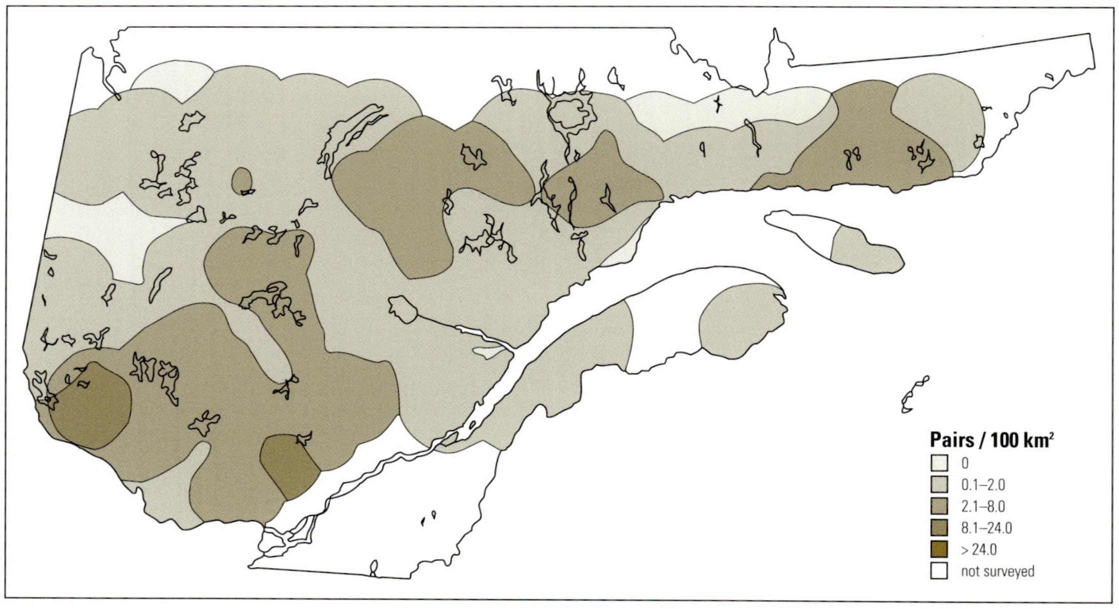

Abundance of Common Loons in southern Québec. Aerial surveys conducted in May 1990, 1991, and 1992 *(D. Bordage, CWS, unpubl. data).*

Common Loon

Breeding distribution
1984–1989

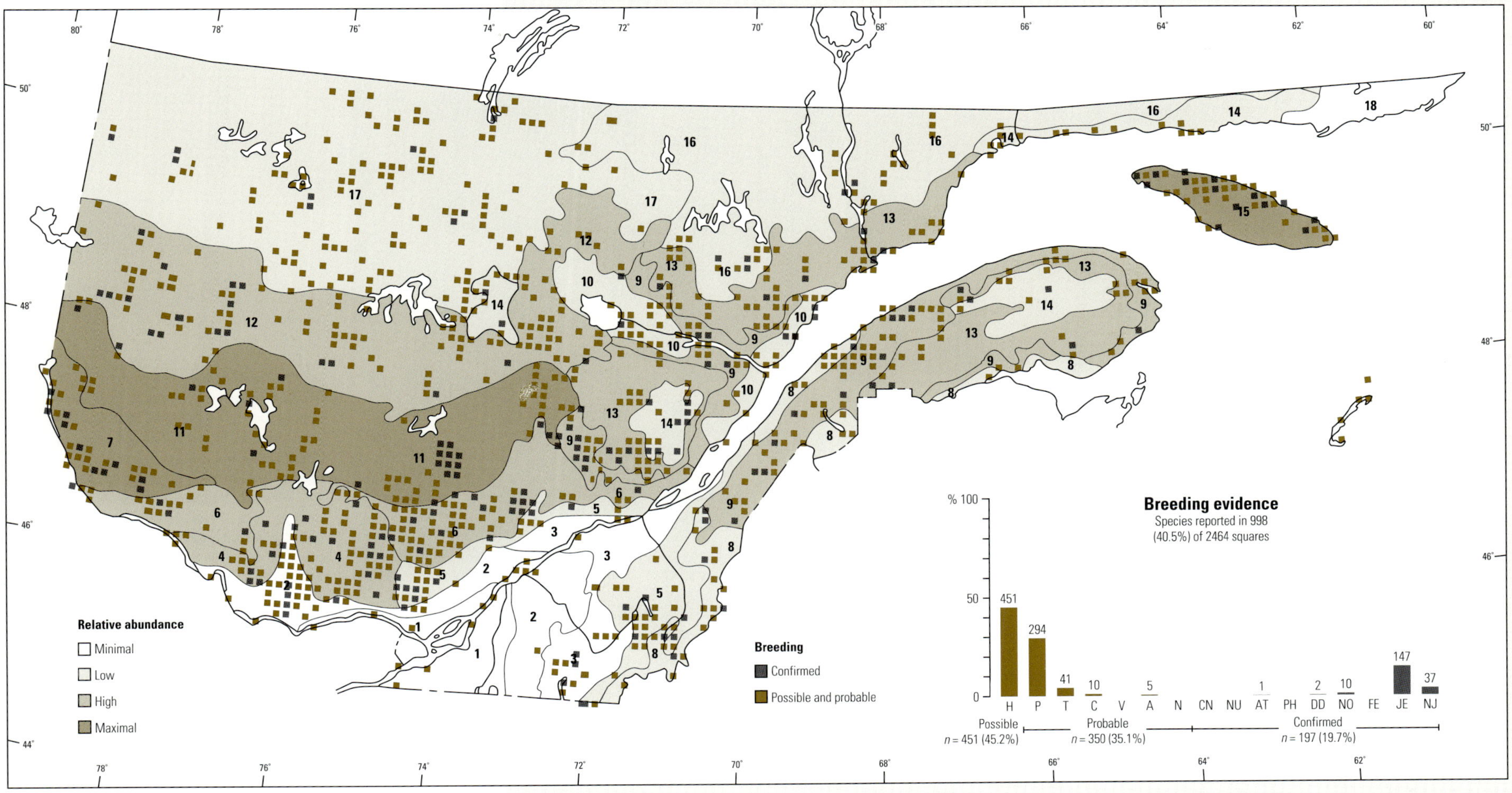

Species profile

Clutch size	usually 2 eggs (1–3)[1,2,3,4]
Laying interval	1 egg per 2–3 days[1,2]
Start of incubation	with first egg[2,3]
Incubation period	26–31 days[1,2,3,4]
Maturity at hatching	precocial
Care of young	by female and male[1,2,3,4]
Nestling period	less than 48 hours[1,3]
Age at first flight	10–12 weeks[1,2,3,4]
Dependence of young	until ≈ 8 weeks old[8]
Broods per year	1[1,2,3]
Breeding age	2–3 years[2,3]
Mating system	monogamy[2]
Duration of pair bond	probably for life[2]
Total length	71–89 cm[5]
Wing span	147.3 cm[4]
Weight	4134 g (3600–4500)[6]
Longevity record	7 years 10 months[7]

References:
1. McIntyre 1975
2. Cramp and Simmons 1977
3. Olson and Marshall 1952
4. Terres 1980
5. Godfrey 1986
6. Dunning 1984
7. Kennard 1975
8. Barr 1973

World breeding range

Chronology $n = 213$

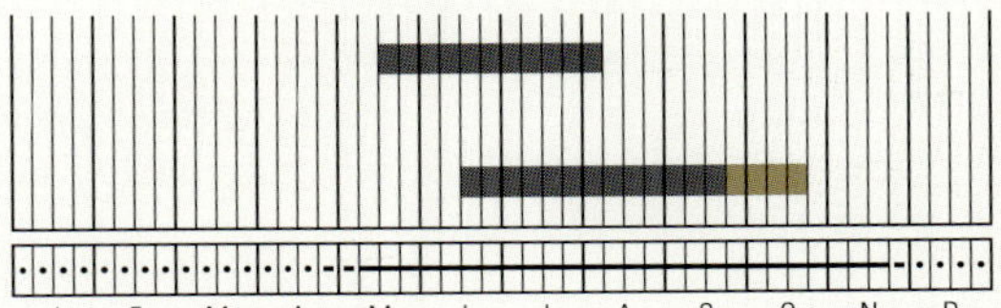

Frequency of occurrence by ecological region

Ecological regions** (forest regions)	Adequately covered squares	Squares in which reported	% of squares in which reported	Relative abundance (%)*
1. Sugar maple-hickory and sugar maple-basswood	97	7	7.2 ± 2.8	7.2
2. Sugar maple-basswood	153	39	25.5 ± 3.6	25.5
3. Sugar maple-basswood and sugar maple-yellow birch	101	18	17.8 ± 4.6	17.8
4. Sugar maple-yellow birch-basswood	56	40	71.4 ± 10.4	71.4
5. Sugar maple-yellow birch-beech and sugar maple-yellow birch-basswood	113	39	34.5 ± 4.9	34.5
6. Sugar maple-yellow birch-beech	71	51	71.8 ± 9.6	71.8
7. Sugar maple-yellow birch-balsam fir	19	18	94.7 ± 12.0	94.7
8. Sugar maple-yellow birch and balsam fir-yellow birch	47	[illegible]	[illegible]	[illegible]
9. Balsam fir-yellow birch	[illegible]	[illegible]	[illegible]	[illegible]
10. Balsam fir-white birch or balsam fir-red maple	[illegible]	[illegible]	[illegible]	[illegible]
11. Yellow birch-balsam fir	[illegible]	[illegible]	[illegible]	[illegible]
12. Balsam fir-white birch (west)	[illegible]	[illegible]	[illegible]	[illegible]
13. Balsam fir-white birch (east)	[illegible]	[illegible]	[illegible]	[illegible]
14. Balsam fir-black spruce	[illegible]	[illegible]	[illegible]	[illegible]
15. White spruce-balsam fir and balsam fir-white spruce	22	[illegible]	[illegible]	[illegible]
16. Black spruce-balsam fir-moss	18	10	[illegible]	[illegible]
17. Black spruce-moss	32	18	56.2 ± 18.8	56.2
18.			*not analysed*	

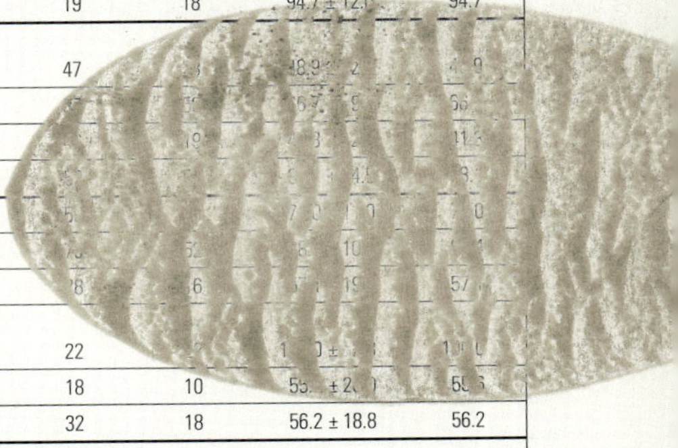

* minimal: 0–29.9%; low: 30–59.9%; high: 60–89.9%; maximal: 90–100%

** Deciduous forest: 1–7; mixed forest: 8–11; coniferous forest: 12–17

To have peace and quiet to incubate its eggs, this Common Loon has built its nest in the undergrowth on an island on Lake Malbaie almost 1 m from the water. Here, it is preparing to return to the nest. Breeding success is easily compromised by disturbance during incubation; birds are often forced to leave the nest when fishermen, in the belief that loons are a sign of good fishing, come too close to the nest site. Laurentide Wildlife Reserve, north of Québec City. *June 1994, Jack Hughes.*

The "possible" category accounted for 45.2% of the squares in which the species was atlassed. Some cases involved individuals seen, while others involved loons heard vocalizing in their breeding habitat. Three of their vocalizations (wail, tremolo, and yodel) are unmistakable, but even the soft hoots uttered between family members can signal their presence to an observer stealthily approaching a lake. Their call carries over great distances and they can often be heard calling to other loons on neighbouring lakes (Barlow 1979). Loons that are accustomed to the presence of boats and cottages are usually easy to spot, even if they have young.

Reports of young accounted for 74.6% (147/197) of the "confirmed" squares. Because of the long pre-fledging period and the fact that loon chicks only rarely make overland trips to other lakes (McIntyre 1988), virtually all chicks had probably hatched on the lakes on which they were observed.

Concentrations of up to hundreds of individuals are found on large lakes in summer (Nero 1972) and fall (Campbell 1970). On the Atlantic coast, migrants occur on the continental shelf up to 170 km from the coast (Powers and Cherry 1983). Individuals winter along the Atlantic and Pacific coasts (Godfrey 1986), usually within 100 km of land (Haney 1990). Loons are flightless in late winter when all the flight feathers are moulted simultaneously (Godfrey 1986).

Distribution

Common Loons are widely distributed in Québec in the Atlas area, but are mainly found in the Laurentians, the Abitibi Upland, and Anticosti Island, all areas where there are many large bodies of water. They also occur in the Lower St. Lawrence and around the Gaspé Peninsula in significant numbers. Recent aerial surveys conducted in May 1990, 1991, and 1992 by the Canadian Wildlife Service and the then Ministère du Loisir, de la Chasse et de la Pêche (now the Ministère de l'Environnement et de la Faune) have confirmed the Laurentians and Abitibi as significant areas where pair density varies from 2–16 pairs/100 km^2 (D. Bordage, CWS, unpubl. data). These surveys also highlight other loon areas such as north of Lac-Saint-Jean and the North Shore. The only areas where there is probably little or no breeding, as indicated on the map, are the heavily agricultural St. Lawrence Plain, with its high level of human disturbance and low lake density, and along the Ottawa River. It is interesting to note that, whereas the raw data point to the Gatineau region as a major area, statistical analysis shows that records are not as concentrated as they may seem given the extent of the statistical area under consideration. Reports of Common Loons in some squares in the upper St. Lawrence River suggest that a few may breed in that area, and they probably nest on Lake Saint-Pierre (SOCQ 1988). However, they can breed no farther downstream than about Trois-Rivières because of tidal fluctuations.

The map suggests a gradual northward decrease in frequency of occurrence, starting north of the heavily populated area of southern Québec. This is consistent with the findings of DesGranges and Laporte (1979), who estimated densities of breeding pairs to be about 10 times greater south of the 50th parallel in the Laurentians and Appalachians than north of the 50th parallel in central and northern Québec.

Common Loons are somewhat locally distributed in northern Québec (Todd 1963), possibly avoiding areas with mostly brown-water or fishless lakes (DesGranges 1989). Harper (1958) felt that the species was unusually scarce in the Ungava Peninsula.

History and trends

The Common Loon has been eliminated from many southern parts of its former North American breeding range (Titus and VanDruff 1981), due in part to shooting (Bent 1919), but mostly to heavy lake

A pair of loons nested for several years among the emergent vegetation, on a hummock at the edge of this shallow, grass-fringed bay inaccessible to pleasure boaters. Petit Lac Saint-Germains, north of the Saguenay River. *July 1991, Jean Gauthier.*

Note the entrance and exit routes to this nest, which has been in use for several years and is built on a hummock a few metres from the shore. Petit Lac Saint-Germains. *July 1991, Jean Gauthier.*

The brushy shores of a lake make possible nest sites for the Common Loon. Lake Guilbeault, Portneuf Wildlife Reserve, northwest of Québec City. *June 1982, Jean-Luc DesGranges.*

use by humans, which adversely affects loons (Ream 1976; McIntyre 1988). Fewer than 200 pairs remain in New York State (Peterson 1988e), while in Vermont, from 1977 to 1983, only 7–19 pairs attempted to nest each year, of which 50–71% were successful, producing 11–15 chicks annually (Fichtel 1985a).

Common Loons used to breed in Ontario's heavily populated Carolinian Forest zone but do so only rarely now (Dunn 1987a). They have probably been eliminated from some lakes in southern Québec, for example, the lakes on Mount Saint-Hilaire, Mount Saint-Bruno, and Mount Yamaska (Ouellet 1974a), and they will probably be eliminated from some others as lake use by humans increases. Loons are especially vulnerable to water level changes, which can flood the nest or leave it stranded on land. Although very few clear lakes have reached pH 5.0 or lower in Québec (Kelso et al. 1986), acid rain could pose a problem within the next few decades by increasing lake acidity and eventually reducing or eliminating food resources for loons (McNicol et al. 1987; Alvo et al. 1988; Parker 1988; DesGranges 1989).

The creation of large reservoirs could adversely affect Common Loons through increased mercury levels (Barr 1986). Mercury seems to have caused the massive 1983 die-off in Florida (McIntyre 1988), and general water pollution on the wintering grounds poses a continual threat.

About 50 000 Common Loons nest in the Canadian Shield in Québec (DesGranges 1989). Loons are hunted by the Natives in northern Québec, where an estimated 4591 Common, Red-throated, and Pacific loons (probably mostly Common Loons) were taken annually between 1974 and 1978 at 19 sites (DesGranges and Laporte 1979).

Robert Alvo

At nest exchange, Common Loons tend to rearrange the eggs before settling down to sit. Lake Malbaie, Laurentide Wildlife Reserve, north of Québec City. *June 1994, Austin Reed.*

Pied-billed Grebe

Grèbe à bec bigarré

Podilymbus podiceps

Jean-Marc Giroux

The Pied-billed Grebe is a small, stocky bird with a short neck, large head, and chickenlike bill. At first glance it resembles a small, tailless duck, but in the hand its partially webbed, lobed toes identify it as a grebe.

Grebes are a family of diving birds with no living relatives; they may, however, have once been related to loons (Terres 1980). The Pied-billed Grebe is the brownest of the grebes, and one of the few that does not show a sharply defined white wing patch in flight. It is also the most widespread and least gregarious (Terres 1980). The Pied-billed Grebe nests in much of southern Québec, although not in large numbers. It migrates to its breeding grounds as soon as the ponds and lakes there become ice-free, and remains through autumn until freeze-up. It then migrates to its wintering grounds in the southern United States (Bent 1919). Very secretive, under threat it can expel air from its body and feathers to lower its specific gravity, sinking like a submarine, with only its head or bill and nostrils out of the water, concealed among cattails or other emergent vegetation.

Habitat and behaviour

The breeding habitat of the Pied-billed Grebe consists of shallow lakes, ponds, marshes, and slow-moving streams with ample emergent and bottom growth of plants (Cramp and Simmons 1977). Occasionally, it will nest in saltwater bays with weak tidal influence (Palmer 1962). Open water is required for the birds to become airborne, and vegetation (type not important) is needed for nest building and concealment (Palmer 1962; DeGraaf et al. 1980). Pied-billed Grebes are solitary nesters, a single pair usually occupying ponds up to 4 ha in size (Palmer 1962). They are monogamous, and each pair defends a small area, approximately 50 m, around the nest (Glover 1953). They often arrive on the breeding grounds already paired (Kibbe 1985j), where the considerable courtship activity probably has the effect of reinforcing pair bonds (Kibbe 1985j). Courtship involves both vocal and visual displays. The characteristic call of the males starts with a complex disharmony that defies description, ending with a series of loud, resonant notes sounding like *pow-pow-pow* (Palmer 1962). This loud call, which can be heard as far away as 1.6 km (Taverner 1934), is mostly heard during the breeding season. It can be easily mistaken for the call of the Common Moorhen (Kibbe 1985j), which also nests in southern Québec (Godfrey 1986). The Pied-billed Grebe has several displays, including male-female chases, circling displays, the male carrying small pieces of vegetation in its bill, and bill touching between the two sexes (Palmer 1962).

The nest is built by both sexes and takes 3–7 days to complete. It is constructed of decaying emergent and submerged vegetation (Forbes 1983), sometimes starting from the bottom in shallow water. More often, it is built as a floating structure attached to dead or growing vegetation (Palmer 1962), usually well concealed in cattails, bulrushes, sedges, and burreeds, and located close to open water, allowing the birds to approach and leave under water (Palmer 1962). Nests are found in water ranging from a few centimetres to more than 2 m in depth (Peck and James 1983; Campbell et al. 1990a). Nest loss is highest during the egg laying period because adults are less vigilant at this time than during incubation (Forbes et al. 1989).

The female usually lays 4–7 eggs per clutch. Incubation lasts about 23 days and is shared by both sexes (Forbes and Ankney 1988). If the incubating grebe leaves the nest and is not replaced by the mate, it will cover the eggs with vegetation (Palmer 1962). The chicks are precocial. Immediately after hatching, they take to the water, where they follow their parents or ride on their backs (Terres 1980; Godfrey 1986). The chicks' plumage has a complex pattern of black and white stripes, interspersed with reddish-brown spots (Palmer 1962). The parents often dive with the chicks on their backs, and both sexes will carry and feed the young (Palmer 1962). Frequently the parents divide up the brood, and the two groups will forage separately during the day. They remain in communication with each other through individual calls and duets (Forbes and Ankney 1987). The young chicks are fed insects such as dragonfly nymphs (Palmer 1962). As they grow older, their diet becomes similar to that of the adults: small fish, snails, small frogs, tadpoles, aquatic worms, leeches, and aquatic insects (Bent 1919). Co-operative rearing has been observed, either with the young from first broods feeding siblings from a second brood, or with adults other than the parents helping to feed the chicks (Cramp and Simmons 1977; Forbes 1987).

In spite of their secretive habits during the breeding season, Pied-billed Grebes were reported in 181 squares, in 82 of which breeding was confirmed (45.3%), with records of an actual nest accounting for 14 squares (17%), and records of downy young or recently fledged young representing 65 squares (79%).

Notes on territory

The Pied-billed Grebe defends an area used for nesting, courtship, and perhaps some feeding, sometimes on ponds as small as 0.2 ha (Palmer 1962). Often one pair may occupy a pond up to 4 ha, or several pairs may nest in marshy areas of larger ponds, with their territories separated by vegetation (Munro 1941b). In three pothole marshes in Iowa, the area defended was an arc of about 50 m radius around the nest. Home range was about twice the size of

The Pied-billed Grebe's breeding habitat must have emergent vegetation to conceal the nest and provide cover under threat. Baie-du-Febvre, south shore of Lake Saint-Pierre. *July 1991, Daniel Jauvin.*

Pied-billed Grebe

Breeding distribution

1984–1989

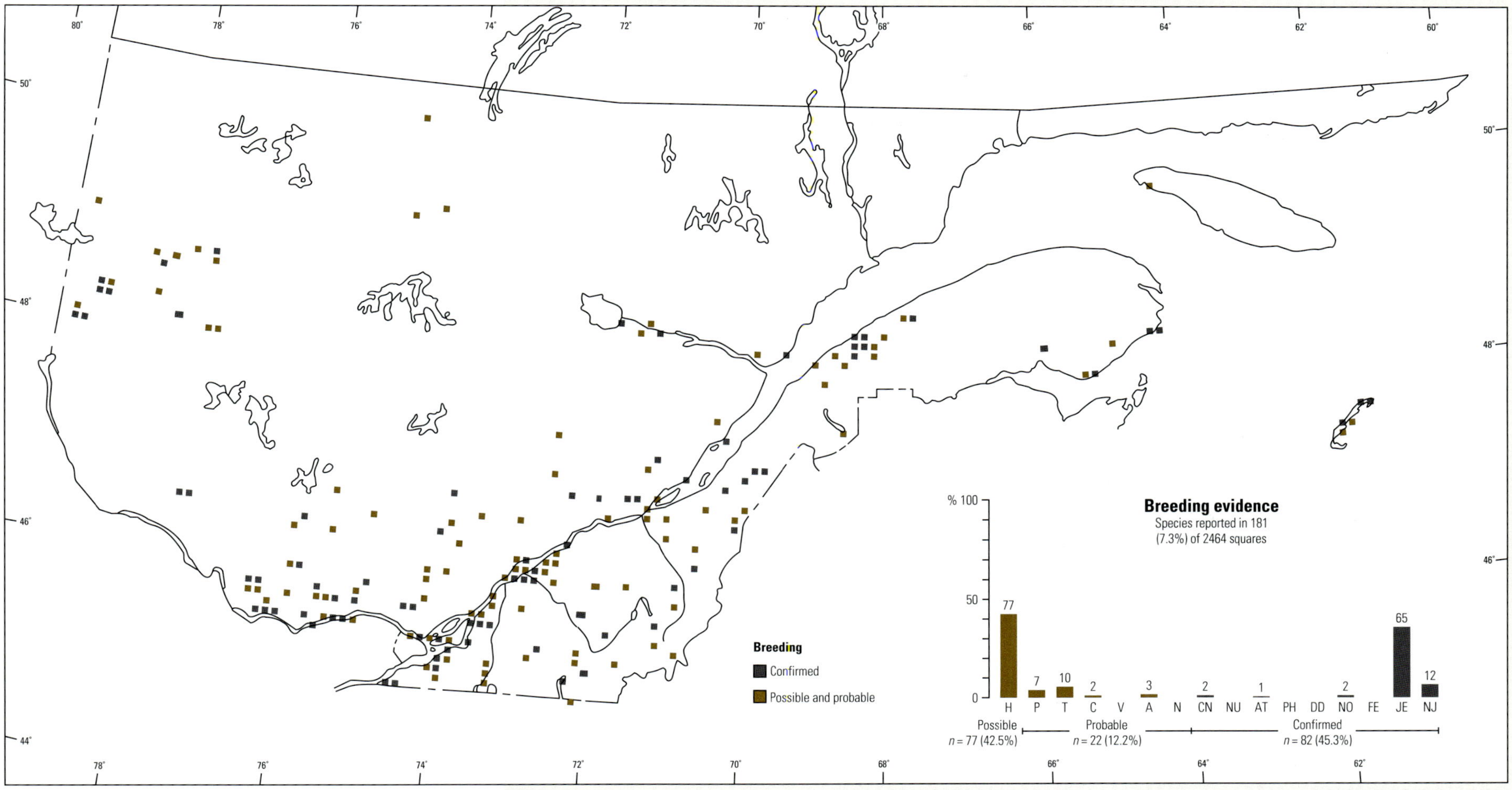

Species profile

Clutch size	usually 4–7 eggs (2–10)[1,2,3,4,6]
Laying interval	1 egg per day[2,6]
Start of incubation	with second egg[6]
Incubation period	21–27 days[1,2,4,5,6]
Maturity at hatching	precocial
Care of young	by female and male[2]
Nestling period	less than 48 hours[2,4,5]
Age at first flight	35 days[1,3]
Dependence of young	probably until 3 weeks old or first flight[2,6,9]
Broods per year	1–2[2,6]
Breeding age	1 year[1]
Mating system	monogamy[2]
Duration of pair bond	sometimes more than 1 season[2]
Total length	30.5–38.1 cm[4]
Wing span	58.4 cm[7]
Weight	442 g (343–561)[8]
Longevity record	?

References:

1. del Hoyo et al. 1992
2. Palmer 1962
3. Forbes and Ankney 1987
4. Godfrey 1986
5. Bent 1919
6. Forbes 1983
7. Terres 1980
8. Dunning 1984
9. Kirby 1976

Chronology $n = 164$

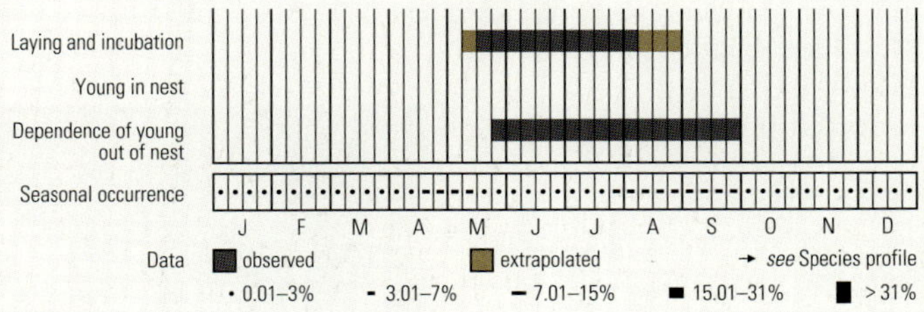

World breeding range

The Pied-billed Grebe favours shallow open water where it can feed and where the chicks can learn to fly. Nuns' Island, Verdun. *July 1991, Pierre Bannon.*

Notes on territory (continued)

nesting territory (Glover 1953). Nest density found in a study in Nova Scotia was 0.73 nest/ha, with average nearest-nest distances of 37.7 m and 53.6 m, depending on vegetation type (Forbes 1983). A study in Louisiana found 1.32 nests/ha, and an average inter-nest distance of 55 m (Chabreck 1963). The male and sometimes the female defend the area (Palmer 1962; Forbes 1983). Interspecific aggression has been reported against American Coots, Eared Grebes, and several species of ducks (Palmer 1962; Kirby 1976). Typically, the bird will drive the intruder off by diving and striking the feet of the victim while underwater.

Distribution

The Pied-billed Grebe breeds widely throughout North America, across Canada, and south through Central America, the West Indies, and South America to Chile and southern Argentina (Godfrey 1986). In Québec, it breeds primarily in the southern part of the province. Breeding was confirmed in Abitibi, the Ottawa Valley, down the St. Lawrence as far as Mont-Joli, along the Saguenay River, in the Gaspé Peninsula, in the Montréal area, the Eastern Townships, and the Mauricie region. Breeding was also confirmed on the Magdalen Islands. On Anticosti Island, Pied-billed Grebes were reported in one square where they were recorded as a possible or probable breeder.

In June 1991, an individual was sighted near Lake Soscumica, roughly 50 km north of Matagami, pointing to possible breeding in this area (Cons. G. G. – GREBE 1992j).

History and trends

Wintle (1896) listed the Pied-billed Grebe as a common "transient visitant" in the Montréal region at the end of the 1800s, commenting that a few probably bred there. A century later, Bannon (1991a) gave it a similar status as an uncommon migrant breeder, seen annually but infrequently. The species appears to have held its own in the Montréal region during the last 100 years.

Trends for the North American population of the Pied-billed Grebe are difficult to determine. Populations on some major breeding grounds have declined considerably. On Swan Lake, near Vernon, British Columbia, the local breeding population declined from 60 birds in 1932 to 4 in 1965 and to 0 in 1978 (Cannings et al. 1987). As well, *The Atlas of Breeding Birds of Vermont*, completed in 1981, reported that few breeding pairs were found in the Champlain Valley, which contains considerable prime breeding habitat for this species. For the whole state, the Pied-billed Grebe was classified as

Usually the nest is an anchored, floating structure made out of aquatic plants. Lake Georges, near Amos. *July 1986, Yves Aubry.*

Before leaving the nest, the Pied-billed Grebe generally covers the eggs with aquatic plants and other plant material. Île Brion Ecological Reserve, Magdalen Islands. *June 1994, Alain Richard.*

Two young grebes in the nest. Île Brion Ecological Reserve, Magdalen Islands. *June 1994, Alain Richard.*

rare (Kibbe 1985j). This is in contrast to the end of the 19th century when it was considered a fairly common summer resident (Perkins and Howe 1901 *in* Kibbe 1985j). The reasons for the decline and scarcity in both the Champlain Valley and Swan Lake areas are unknown (Kibbe 1985j; Campbell et al. 1990a). There has also been a decline in numbers in recent years in New York State. According to Gibbs and Melvin (1992b) and Connor (1988d), wetland loss and degradation due variously to draining and filling, pollution, and silting are the main pressures on the Pied-billed Grebe in North America. The Pied-billed Grebe's fortune seems to be brighter in Québec, where its breeding range may be expanding into the eastern part of the province. The Atlas confirmed it as breeding in the Gaspé, where Godfrey (1986) reported it as absent. It is also a recent addition to the Magdalen Islands, where it was first sighted in 1954 (Gaboriault 1961) and first confirmed as breeding in 1972 (McNeil et al. 1973).

RICHARD C. COTTER AND MICHAEL SPENCER

Horned Grebe

Grèbe cornu

Grèbe esclavon
Podiceps auritus

THE YELLOW-TIPPED BILL, BUFFY TUFTS OR "HORNS," AND CHESTNUT neck are the field marks of the Horned Grebe in breeding plumage. Like other species of grebes, this curious bird has the unusual habit of consuming feathers, which play a part in its digestive system (Storer 1969). It can control its buoyancy and, when alarmed, will submerge, submarine-like, until just its head, like a periscope, shows above the water.

Present only on the Magdalen Islands, the Québec population of the Horned Grebe is a disjunct population, isolated from the main population located in the West. Because the Québec breeding population is so small, the species is considered vulnerable in this province (Robert 1989b). Most Québec sightings occur during migration, as a small portion of the western population passes through Québec, stopping along the St. Lawrence River, the estuary, and occasionally in other parts of the province. The spring migration peaks around the end of April. The heavier fall migration is more of an event, extending from early September to early November (David 1980a), as the migrating birds, to the best of our knowledge, head from their breeding grounds west of Ontario to their wintering grounds on the Atlantic coast.

Habitat and behaviour

The earliest arrival dates for Horned Grebes on the Magdalen Islands are during the second week of May, with migration probably continuing throughout the whole month. Absent from their coastal winter quarters in spring, they appear on the ponds only after the ice has melted. Sightings are of pairs, single birds, and the occasional small flock. Pairs return fairly faithfully to the same breeding sites from year to year (Ferguson 1981).

The birds use permanent freshwater or brackish ponds as their breeding habitat. These may be heavily vegetated but must nonetheless have areas of open water. Sometimes the ponds are located along a road, even within 30 m (pers. obs.). Nests are located in emergent vegetation where water depth varies between 20 and 140 cm (Faaborg 1976; Sugden 1977; Ferguson and Sealy 1983), ensuring protection from wave action and land predators. Horned Grebes nest as solitary pairs on the smaller ponds (up to 0.1 ha) on the Magdalen Islands, whereas the Étang de l'Est system supports several pairs. This extensive pond is shared by Horned and Pied-billed grebes.

Courtship displays include head-shaking, weed ceremonies, and a penguin dance, during which the birds assume an upright posture in the water. Both sexes build the nest and share in incubating and caring for the young (Ferguson and Sealy 1983). The nest is a platform of aquatic plants, either floating or built up from the bottom, and is attached to emergent rushes, cattails, or ericaceous shrubs. It is constructed of plant debris, mainly reeds mixed with mud. While one adult sits on the eggs, the other adds new material to the nest structure. The incubating bird sometimes covers the eggs with mud and plant debris before leaving the nest to feed.

The pair will usually renest if the first attempt is unsuccessful (Ferguson and Sealy 1983). Because of successive (asynchronous) hatching, the chicks often differ in size, and the age difference in some broods may be as much as 10 days. The black-and-white striped chicks often ride on their parents' backs; they can easily go unnoticed, as often only the head shows from beneath the plumage of the adult bird. When the chicks are 8–9 days old, they no longer seek the safety of the adults' backs by day, and at 12–13 days old they begin capturing their own prey. They can fend for themselves by the age of 19–21 days. During this time, their plumage begins to turn grey and they can be left unattended. The first flight is made as early as their seventh week (Ferguson and Sealy 1983). They dive or seek cover when fleeing from danger.

Jean-Marc Giroux

Notes on territory

The Horned Grebe is aggressively territorial. The male will even drive away other species, such as Ruddy Ducks, regardless of sex, from the stretch of water it defends (Storer 1969; Otto 1988).

In a study in North Dakota, nests were separated by at least 45 m of water or a barrier of land or vegetation. The average size of pond or portion of a pond defended by a pair was 0.9 ha, with the minimum being 0.1 ha. (Faaborg 1976). In a similar study in Manitoba, a pair would defend an average of 1.2 ha of water, with the minimum being 0.1 ha. The territory is used for feeding, mating, and nesting (Ferguson and Sealy 1983).

The Magdalen Islands is the only place the Horned Grebe is known to breed in North America east of the Great Lakes. The many ponds in the Pointe de l'Est National Wildlife Area, seen here beyond Clarke Bay (foreground), support part of the population. *Summer 1990, Michel Papageorges, Canadian Wildlife Service.*

Horned Grebe

Breeding distribution
1984–1989

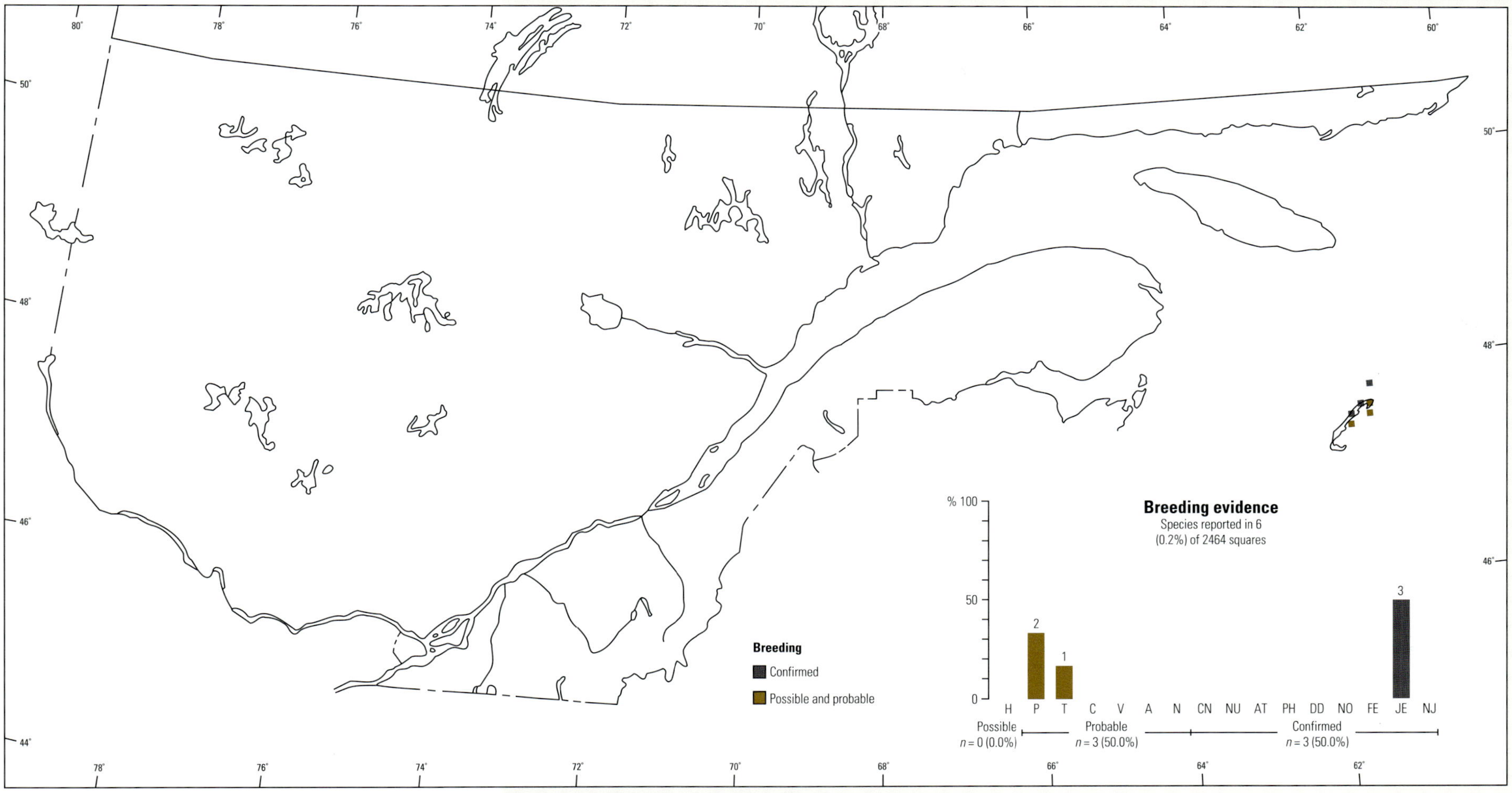

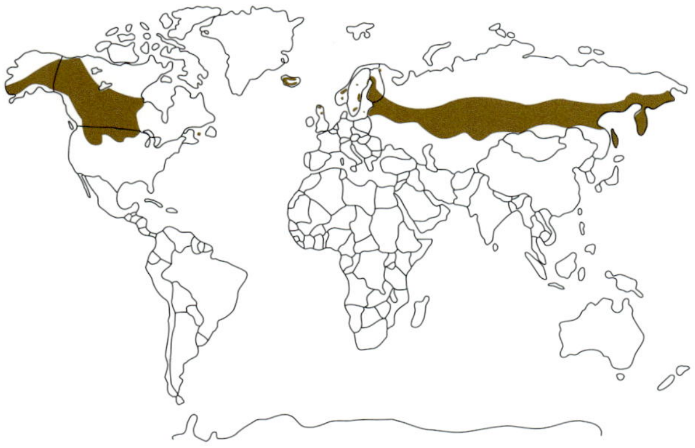

World breeding range

Species profile

Clutch size	usually 4–5 eggs (1–7)[1,4]
Laying interval	1 egg per 1–3 days[2,4]
Start of incubation	with first egg[3]
Incubation period	22–25 days[1,3]
Maturity at hatching	precocial
Care of young	by female and male[2,3,4,5]
Nestling period	less than 72 hours[2]
Age at first flight	45–60 days[1,2,3]
Dependence of young	until 19–21 days old[2]
Broods per year	1[3,4]
Breeding age	1–2 years[1,6]
Mating system	monogamy[2,3]
Duration of pair bond	more than 1 season[9]
Total length	31.5–38.1 cm[5]
Wing span	61 cm[4]
Weight	453 g (327–528)[7]
Longevity record	5 years 2 months[8]

References:
1. del Hoyo et al. 1992
2. Ferguson and Sealy 1983
3. Ehrlich et al. 1988
4. Terres 1980
5. Godfrey 1986
6. Palmer 1962
7. Dunning 1984
8. Clapp et al. 1982
9. Ferguson 1981

Chronology *n* = 30

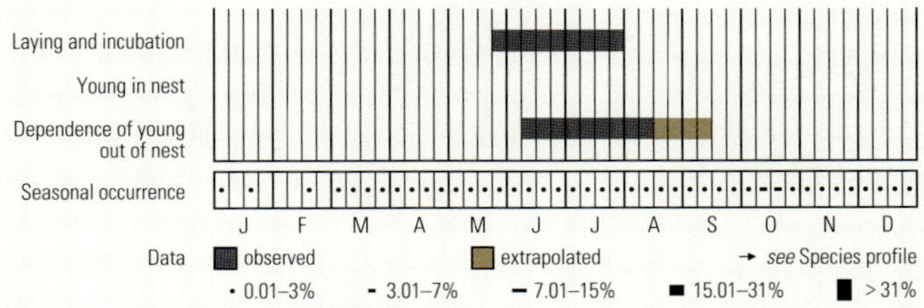

Newly hatched, this Horned Grebe chick is about to leave the nest to join its parents on the water. Its striped plumage enables it to blend in with its environment. Dune du Nord, Magdalen Islands. *July 1996, Pierre Laporte.*

The Horned Grebe, classed as a vulnerable species, lives on freshwater ponds that have some open water. Its nest is concealed in emergent aquatic vegetation. Havre-aux-Maisons, Magdalen Islands, *June 1991, Desneige Perreault.*

Although Horned Grebes conceal themselves in emergent vegetation, they can be readily spotted on the open water of their habitat. It can be hard to distinguish between the young of Horned and Pied-billed grebes when both species are nesting on the same pond.

On the Magdalen Islands, Horned Grebes usually gather on the large ponds (especially on Étang de l'Est) after the breeding period and before the fall migration, which begins at the end of August.

Distribution

No breeding evidence for the Horned Grebe was found during the Atlas years 1984–89 except on the Magdalen Islands, the species' only breeding grounds in eastern North America. In fact, no other breeding evidence has been reported in Québec since the mid-1960s. Breeding was confirmed in 3 of the 14 squares on the islands, and probable reports in 3 other squares point to nesting in almost all of the northern part of the islands, including Brion Island. This is supported by reports from before and after the Atlas period (Fradette 1992). An abandoned nest containing two eggs was reported at Baie du Havre aux Basques in the southern part of the islands in 1992 (Shaffer 1992).

Normally, when it leaves the nest, the Horned Grebe partially covers its eggs with plant material, while the Pied-billed Grebe covers them completely. Magdalen Islands. *June 1994, Pierre Laporte.*

The smaller ponds on the Magdalen Islands are generally only occupied by one pair, whereas several pairs are found on the larger Étang de l'Est. Pointe de l'Est National Wildlife Area. *June 1982, Jacques Poitras.*

The adult birds carry their young on their backs until the chicks are 10 days old. Here the striped head of a chick can be seen above the plumage of the adult bird on the right. Havre-aux-Maisons, Magdalen Islands. *June 1991, François Shaffer.*

Horned Grebe nest among bulrushes. Dune du Nord, Magdalen Islands. *June 1994, François Shaffer.*

History and trends

There is no evidence in any of the early ornithological records from Ontario, New Brunswick, or Québec (the Magdalen Islands excepted) that Horned Grebes ever bred in eastern Canada in numbers comparable with those in the West (Taverner 1922; Palmer 1962; Godfrey 1986; Eagles 1987b). There are a few historical breeding reports for Québec: on Anticosti Island in 1919, and probably in 1922 (Ouellet 1969), on Lake Sainte-Anne on the North Shore in 1959 (Ouellet and Ouellet 1963), and on Lake Perceval on the Valcartier military base in 1960 and 1964 (QNRCP; ÉPOQ). The southwestern Ungava sighting reported by Turner (Bent 1919; Pough 1951) was rejected by Todd (1963).

It appears that the breeding range of the North American subspecies *Podiceps auritus cornutus* has contracted since European settlement (Palmer 1962). The Icelandic and Eurasian subspecies *Podiceps auritus auritus* appears to have met with a similar fate but seems to be regaining lost ground in some Eurasian regions (Cramp and Simmons 1977). The species was once hunted for its feathers, which were used to trim hats (Terres 1980).

Although breeding pairs of Horned Grebes have been reported on the Magdalen Islands since the end of the 19th century (Young 1897), it is only in recent years that the size of the population has been estimated. Three surveys conducted in 1990–92 of most of the suitable habitat indicated that the local population numbered between 15 and 20 breeding pairs (Richard 1991; Fradette 1992; Shaffer 1992). In 1990, 41 Horned Grebes gathered on Étang de l'Est at the end of the breeding period. A similar count was arrived at in 1956 (Hagar 1956). The little information that we have on population trends would seem to indicate that its numbers have always been in this range. Moreover, given the area of available habitat and the number of sites, the Magdalens could not support many more breeding pairs.

PIERRE FRADETTE

Red-necked Grebe

Grèbe jougris

Podiceps grisegena

WITH ITS ELEGANT PROFILE AND HANDSOME PLUMAGE, THE RED-NECKED Grebe is a striking bird. It is easy to spot and identify on open water, especially early in the breeding season when it puts up a cacophony with its noisy vocalizations.

The Red-necked Grebe is common in central and western Canada and has recently been added to the list of breeding birds in Québec (Chabot et al. 1982). It is classed as a transient throughout most of its range in Québec, except in Abitibi–Témiscamingue, where it is becoming an increasingly less local, rare breeder. Elsewhere, it is seen mainly in autumn, along the St. Lawrence, in transit to its wintering grounds on the Atlantic coast (David 1980a).

Habitat and behaviour

During the breeding period, the Red-necked Grebe is found on shallow lakes, ponds, and sheltered bays edged with emergent vegetation such as cattails and bulrushes, both in plains and wooded habitat. Pairs may nest singly or colonially (Godfrey 1986). Both sexes build the nest from a mixture of fresh and decaying aquatic vegetation (Bent 1919; Chabot et al. 1982), and the pair can be seen diving for and ferrying new nest material to the nest site. The nest is set on a bed of marsh plants, often floating and attached to surrounding vegetation (Godfrey 1986). Red-necked Grebes, like wrens, may build several nests before settling on one (Chamberlin 1977).

Both parents share incubation and vigorously defend the nest, sometimes even attacking intruders with their bills (pers. obs.). The incubating bird, however, will attempt to flee unobserved, after covering the 4 or 5 eggs with plant matter. Both adults tend the young, often transporting them on their backs, even when diving underwater. Sometimes the young hide under the wings of the parent birds, their presence revealed only by the fluffed-out plumage of the adults.

Red-necked Grebes feed on crustaceans, small fish, tadpoles, and aquatic insects. Like many other members of the family Podicipedidae, they consume feathers, which play a part in their digestive system (Simmons 1956; Palmer 1962). According to one study, their preferred prey in British Columbia was yellow perch, followed by pumpkinseed and bass (Ohanjanian 1989). Age of first breeding is 2 years, but younger, non-breeding birds in adult plumage may be observed on breeding grounds early in the summer (Palmer 1962).

Distribution

The seven squares in which the Red-necked Grebe was reported are scattered throughout Abitibi and somewhat to the east, about 100 km west of Chibougamau. This area is not easily accessible so it is highly probable that the species is more widespread in this region than is indicated on the map.

There is no clear distribution pattern for this species. In Ontario, it is found in disparate, geographically dispersed locations, and would appear to be restricted to specific, traditional breeding areas (Armstrong 1987). In Québec, the organic pollution of certain lakes in the Rouyn-Noranda region, where breeding has been confirmed, has created suitable nesting habitat, as eutrophic lakes are rich in aquatic vegetation and food.

In June 1990, a Red-necked Grebe was sighted about 5 km west of Lake Soscumica, about 50 km north of Matagami, and this report was treated as possible breeding evidence, as the bird was observed in suitable nesting habitat (Cons. G. G. – GREBE 1992j).

History and trends

The first two known nesting attempts in Québec, on Lake Pelletier near Rouyn-Noranda, in 1980 and 1981, were unsuccessful (Chabot et al. 1982). The first record of successful breeding in Québec dates from 29 June 1982 when a pair and four young were seen foraging on Lake Pelletier. Since then, several nests and broods of Red-necked Grebes have been reported in the Rouyn-Noranda region. Some of the ponds used, notwithstanding protected areas bordered with cattails, are subject to disturbance from boating.

Recent breeding records for the species in Abitibi, records of probable breeding west of Chibougamau, and the discovery in the early 1980s of nesting sites in east-central Ontario, near the Québec border, indicate an eastward expansion of its breeding range (Armstrong 1987). Droughts on the Prairies over the last 10 years may have prompted part of the Red-necked Grebe population in that area to establish itself farther east.

A study in Manitoba showed that ingestion of organochlorine residues by Red-necked Grebes reduced reproductive success. A considerable percentage of the nests studied contained inviable or thin-shelled eggs. The residues had probably been absorbed on their wintering grounds or during migration (De Smet 1987a).

JEAN LAPOINTE

Jean-Marc Giroux

Notes on territory

Pairs usually nest alone but sometimes in groups. The defended territory is generally a 70–120 m stretch of shoreline, and nests are rarely less than 10 m apart (Palmer 1962). A study in British Columbia showed pair density on one pond to be 0.4/ha. Some of the adults, however, were feeding on a neighbouring lake (Ohanjanian 1989). A nesting pair will drive away other Red-necked Grebes of either sex from the territory. Mating and nesting both occur on the territory, which is occupied by the pair from the onset of incubation until the end of the chick-rearing period (Palmer 1962).

Red-necked Grebes are often aggressive toward other species of aquatic birds. In the West they will sometimes attack Buffleheads and Barrow's Goldeneyes (Donaghey 1975; Savard 1986a).

The Red-necked Grebe, which has recently begun nesting in Abitibi–Témiscamingue, prefers shallow lakes and ponds edged with emergent vegetation. Some eutrophic lakes with their ample supply of aquatic vegetation and food make suitable nesting sites. Rouyn-Noranda, Abitibi. *July 1986, Yves Aubry.*

Red-necked Grebe

Breeding distribution
1984–1989

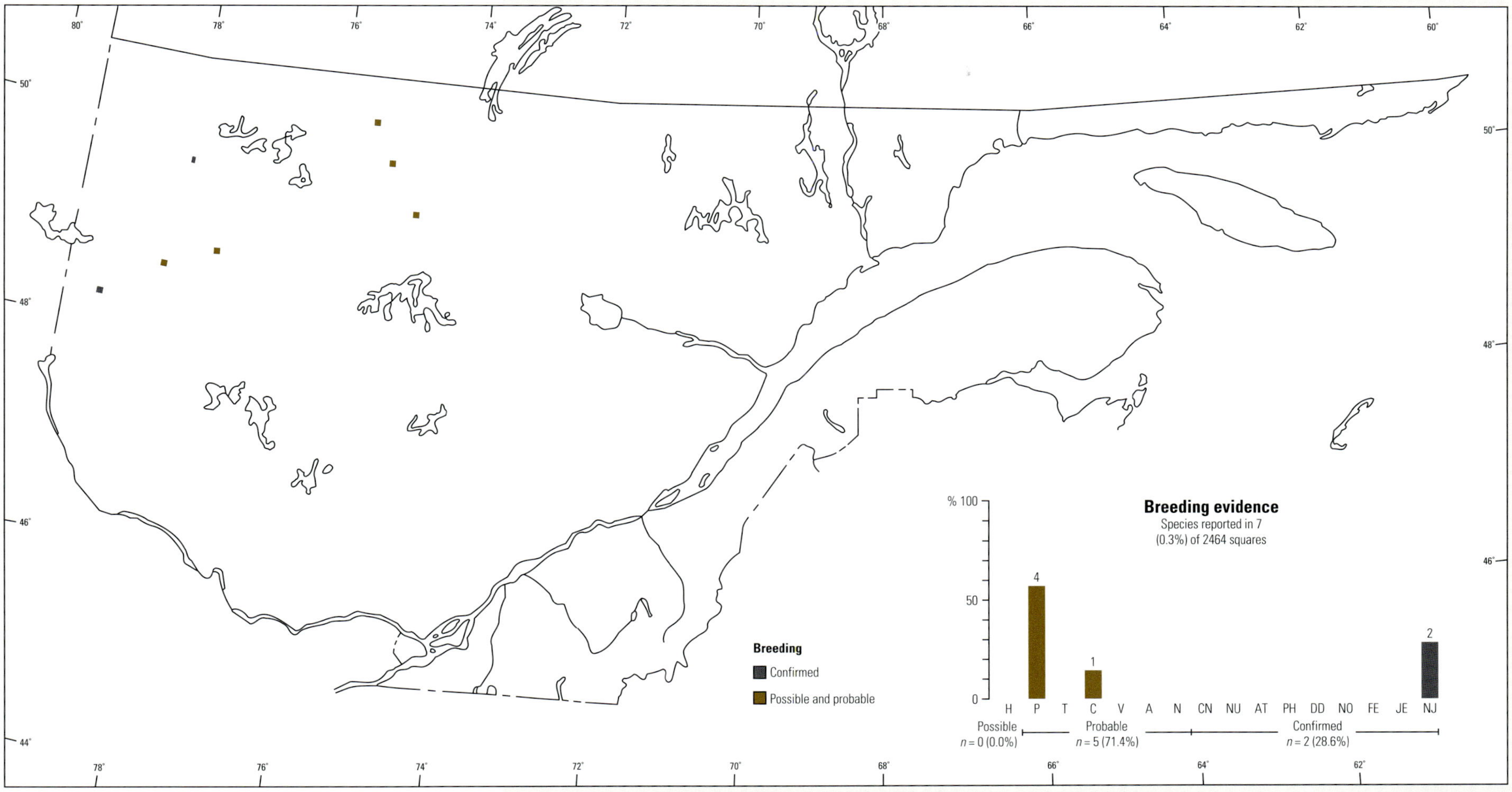

World breeding range

Species profile

Clutch size	usually 4–5 eggs (2–8)[1,2,3,4]
Laying interval	1 egg per day[4]
Start of incubation	with first egg[4]
Incubation period	usually 23 days (20–25)[1,2,3,4]
Maturity at hatching	precocial
Care of young	by female and male[2,3,4]
Nestling period	≈ 24 hours[7]
Age at first flight	probably more than 72 days[4]
Dependence of young	?
Broods per year	1, sometimes 2[1,2,4]
Breeding age	2 years[3,4]
Mating system	monogamy[2,3]
Duration of pair bond	probably 1 season[4]
Total length	46–52 cm[1]
Wing span	76.2–81.3 cm[5]
Weight	1023 g (745–1270)[3]
Longevity record	4 years 10 months[6]

References:
1. Godfrey 1986
2. Ehrlich et al. 1988
3. Palmer 1962
4. Cramp and Simmons 1977
5. Terres 1980
6. Klimkiewicz and Futcher 1989
7. Chamberlin 1977

Chronology *n* = 16

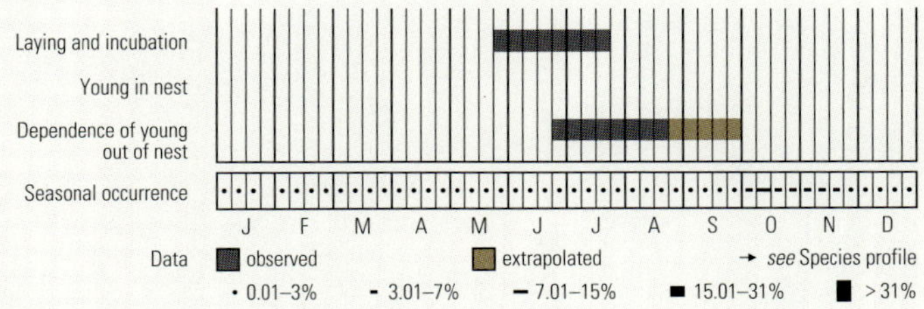

The pair builds the nest of aquatic plants, sometimes constructing several before choosing one in which to nest. Lake Rouyn, Rouyn-Noranda, Abitibi. *July 1986, Yves Aubry.*

Leach's Storm-Petrel

Pétrel cul-blanc

Océanite cul-blanc
Oceanodroma leucorhoa

THE FEEDING AND BREEDING HABITS OF LEACH'S STORM-PETREL MEAN THAT it is seldom seen, and thus sightings are much prized by birders. It is, nevertheless, the most abundant seabird in the western Atlantic (Sklepkovych and Montevecchi 1989). In Québec, it is found in the oceanographic zones of the Gulf of St. Lawrence, especially along the North Shore past Anticosti Island and around the Magdalen Shelf (Brown 1986).

A pelagic species, it comes ashore only to nest. Like most other Procellariiformes, its breeding habits keep it ashore much longer than other seabirds such as puffins and murres. It winters in the convergence zones in the tropical Atlantic off east Africa and Brazil (Brown 1986).

Habitat and behaviour

Breeding birds return to the gulf in early May. Homing experiments have shown that this petrel is very attached to its colony. Breeding birds from New Brunswick released in England covered the 4800 km separating them from their natal colony in 13.7 days, averaging 350 km per day (Cramp and Simmons 1977).

According to the scant information available, it seems that only the male excavates the nest burrow, which is used year after year (Gross 1935). The single egg (8–10 g) is incubated alternately by the male and female for 38–46 days.

Interestingly, as with the chicks of most pelagic species (shearwaters, albatrosses, etc.), young petrels can go without food for up to three days and sometimes longer, with no harmful effects on their health or development. The adults force-feed their young planktonic crustaceans, small fish, and droplets of oil or fat skimmed off the surface of the water while on the wing, often far from the colony. This particular foraging method is one of the characteristics that distinguishes Leach's Storm-Petrel from Wilson's Storm-Petrel. Rearing the chick takes 63–70 days, after which the young petrel is completely independent.

Although petrels are amazingly manoeuvrable in flight when feeding out over the ocean, their lack of speed makes them very vulnerable inshore. Thus, they return to the colony only after sunset to avoid predation by gulls. Unfortunately, this strategy does not protect them from land predators such as red foxes and weasels. On Baccalieu Island, Newfoundland, it is estimated that foxes take about 31 000 birds per year of an estimated population of 3 430 000 pairs,

Jean-Marc Giroux

Notes on territory

The petrel's use of burrows for nesting has tempered many traits usually associated with territoriality, to the point that it is viewed as one that exhibits little or no

The Île Brion Ecological Reserve, with its low, thick vegetation, harbours a colony of Leach's Storm-Petrels. Magdalen Islands. *July 1982, Yves Aubry.*

Leach's Storm-Petrel

Breeding distribution

1984–1989

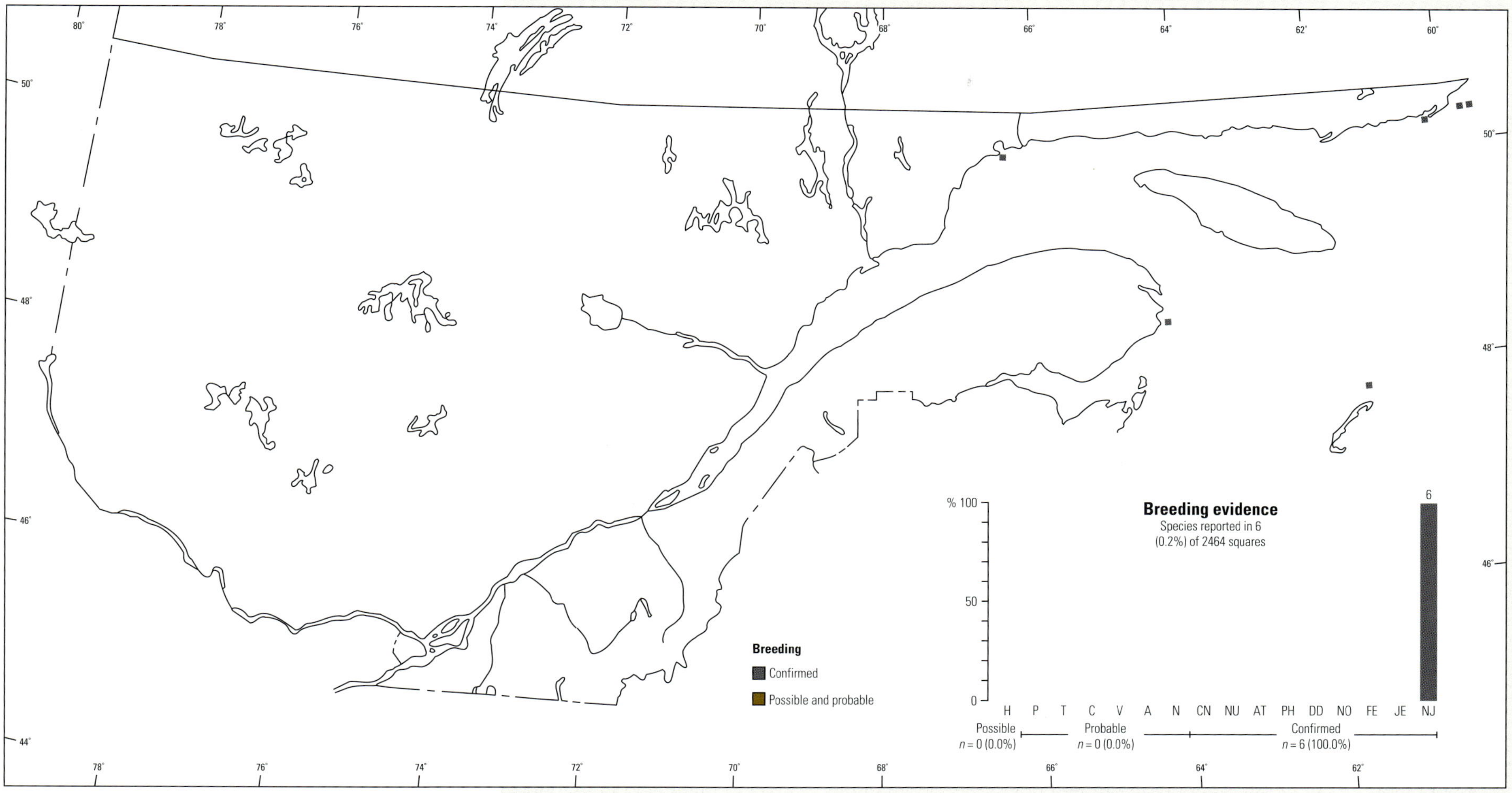

Species profile

Clutch size	1 egg[1,5]
Laying interval	–
Start of incubation	after laying of egg[1]
Incubation period	38–46 days[1,5,6]
Maturity at hatching	semi-altricial
Care of young	by female and male[1]
Nestling period	63–70 days[1]
Age at first flight	63–70 days[1]
Dependence of young	until first flight[1]
Broods per year	1[1]
Breeding age	5 years[1]
Mating system	monogamy[1,5]
Duration of pair bond	for life[1]
Total length	19–23 cm[2]
Wing span	≈ 48 cm[4]
Weight	M 43.5 g (41.8–45.6) / F 45.3 g (41.2–52.5)[6]
Longevity record	24 years[3,4]

References:
1. Cramp and Simmons 1977
2. Godfrey 1986
3. Kennard 1975
4. Terres 1980
5. Ehrlich et al. 1988
6. Palmer 1962

Chronology $n = 16$

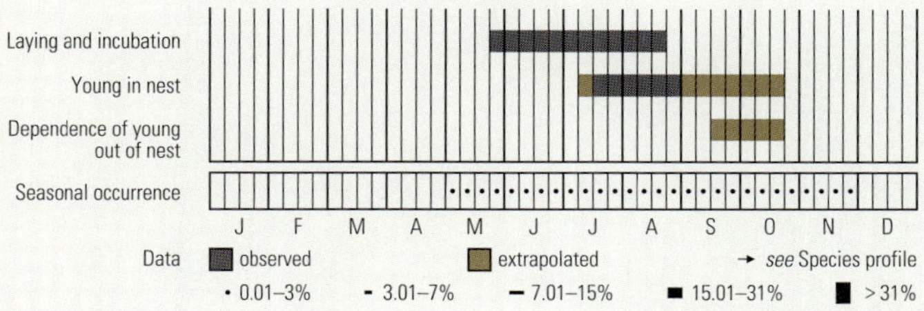

World breeding range

Notes on territory (continued)

territorial behaviour (Cramp and Simmons 1977). Petrels do defend their burrows, but conflict between breeding pairs is very limited except when the burrow is being dug. Few people have been lucky enough to observe this nocturnal activity.

Burrow density varies according to habitat, soil type, and, to a lesser degree, colony size. Even in a very small colony, burrows tend to be grouped together. On Baccalieu Island, Newfoundland, Sklepkovych and Montevecchi (1989) estimated that density varied from 0.05 to over 4 burrows/m² depending on the type of habitat studied. No estimates exist for Corossol Island in Québec because of the small size of the colonies.

A colony of 750 pairs of Leach's Storm-Petrels lives on Corossol Island. Burrows can be found under stands of stunted conifers as well as in open areas. Sept Îles Bay. *June 1992, Jean-Luc DesGranges.*

The male digs the burrow by himself right in the top soil. The burrow will be reused every year. Corossol Island, Sept Îles Bay. *June 1991, Daniel Bordage.*

amounting to a loss of 0.05% (Sklepkovych and Montevecchi 1989). In Québec, this type of predation is not so well documented, but it does occur and, depending on the colony, is attributed to foxes or weasels (Chapdelaine 1979a; Brousseau and Chapdelaine 1983; Fradette 1992).

Distribution

Locating a petrel colony requires knowledge, perseverance, and luck as the observer has to search the ground very carefully to find the burrows of this wholly nocturnal species. The petrel does have a characteristic musky odour (Chapdelaine and Brousseau 1992b), but it is only noticeable quite close to the burrow's opening early in the season.

Most colonies in Québec were located over 75 years ago: Brion Island and Bird Rocks in 1881 (Brewster 1884), and Bonaventure Island in 1914 (Demille 1926). The colony on the Forillon Peninsula in the Gaspé no longer exists (Ball 1938). On the North Shore, Bryant (1861) found a colony on Île aux Goélands, near Gethsémani, but it has never been revisited since. Other concentrations were located on the Perroquets Islands in 1923 (Lewis 1925a) and on Cliff Island in 1963 (Bédard 1969b). These two stations are now part of the Îles Sainte-Marie Migratory Bird Sanctuary on the Lower North Shore. Nettleship and Lock (1973) recorded the species on an island in the Baie-des-Loups Migratory Bird Sanctuary, some 50 km from the Sainte-Marie Islands. In 1985, the Canadian Wildlife Service also recorded the species on Corossol Island, in Sept Îles Bay (Brousseau and Chapdelaine 1987).

On Corossol Island, burrows are found both under dense stands of stunted spruce and fir as well as in open areas dominated by raspberry tangles and ferns. Most burrows are found in the second type of habitat, probably because of the deep top soil. On Bonaventure Island, burrows are dug in grassy meadows, while on Brion Island they can be found in meadows as well as in the coniferous forest (Y. Aubry, pers. comm.).

There are probably many more petrel colonies in Québec than are indicated by Atlas data. The North Shore, with its hundreds of islands, still needs to be explored, but access is difficult. Although no nests were found, Verrill (1862) and Schmitt (1904) considered at the time that the species was common around Anticosti Island. Merriam (1882) observed the species off the coast of Mingan. It was only recently that petrels were again sighted in this area (S. Paradis, pers. comm.).

History and trends

The six colonies surveyed total at least 900 pairs, of which 750 are on Corossol Island. The estimates for the Baie des Loups, Sainte-Marie Islands, and Bonaventure Island bird sanctuaries are 104, 28, and 21 pairs, respectively (BIOMQ). The data for the Magdalen Islands is too rough to make an estimate (Fradette 1992).

There is simply not enough hard data to postulate a population trend for Leach's Storm-Petrel in Québec. However, numbers can be expected to increase because all the known colonies are now protected (although this is no guarantee), either in migratory bird sanctuaries under the federal Migratory Birds Convention Act (S.C. 1917, c. 18), or in ecological reserves such as Brion Island under the provincial Ecological Reserves Regulations (Order-in-Council 1274-88, 24 August 1988; R.26, R.1.01).

Nevertheless, the Québec population is tiny compared with the Saint-Pierre and Miquelon and Newfoundland populations, which have an estimated 4 690 000 breeding pairs. In addition, the Newfoundland population has been growing for several decades (Sklepkovych and Montevecchi 1989).

Pierre Brousseau

Northern Gannet

Fou de Bassan

Morus bassanus

THIS MAGNIFICENT WHITE BIRD WITH BLACK-TIPPED WINGS COULD EASILY be the bird emblem of the Gaspé Peninsula. In the summer Northern Gannets are a familiar sight all around the coast of this peninsula washed by Chaleur Bay, the Gulf of St. Lawrence, and the St. Lawrence estuary. But despite their omnipresence, there are only three gannet colonies in the Gulf of St. Lawrence.

Habitat and behaviour

On their return from their wintering grounds off the coasts of Florida and the Gulf of Mexico, Northern Gannets are relatively easy to spot out at sea in the estuary and the Gulf of St. Lawrence from March to late October. They nest on the cliffs and top plateaus of certain islands, such as Bonaventure Island and Bird Rocks. The nest is made of mud compacted with grass, seaweed, and flotsam such as pieces of fishing net. The male brings the most material to the nest (Nelson 1978a), which, over time, builds up into a mound well off the ground.

Age of first breeding is about 5 years, and a single egg is laid that is incubated about 44 days. A particular trait of these birds, both male and female, is that they incubate the egg with their highly vascularized webbed feet. On Bonaventure Island, the first eggs are laid as early as the last week of April and hatch in early June. Both parents rear the nestling, which takes from 82 to 99 days (Poulin 1968). Consequently, fledging does not occur until the second week in September, and young can often be seen at the nest until October.

Adults usually feed fish to their young, such as Atlantic mackerel, Atlantic herring, capelin, and sometimes sand lance (Taylor and Nettleship 1974). The Northern Gannets in the Gulf of St. Lawrence consume over 8000 tonnes of Atlantic mackerel annually (Cairns et al. 1991), foraging up to 100 km in search of prey. Watching them fish is a spectacular sight. When they spot a shoal of fish, they circle it from a height of about 10 m, sometimes more, over the surface of the water. Then, with their spearlike bills down-pointed, they plunge toward the fish, sometimes diving as deep as 12 to 15 m (Nelson 1978a).

On Bonaventure Island, which is relatively accessible, a wide array of nesting behaviours can be observed at close range. These include mutual fencing, the greeting ceremony in which they knock their bills together, as well as frequent copulations which occur even after the egg has been laid. Non-breeders, their white plumage mottled with black, can be seen on the rock tumbles at the foot of the cliffs. These are the prospectors (3-, 4- or even 5-year-old birds) that are learning how to live in the colony in what are called "clubs." After this learning period they will one day settle on the cliff face or cliff top to breed.

Jean-Marc Giroux

Notes on territory

Both male and female defend the territory, protecting an area limited to the nest site and enough living space for the three family members. Nest density is a relatively stable parameter, regardless of colony size and location. On Bonaventure Island, the average inter-nest distance is 80 cm, varying from 61 to 99 cm (Poulin 1968).

Because gannets live at such close quarters, a gannet colony provides a never-ending spectacle of agonistic behaviour, generally mild in intensity. The most violent clashes take place between young birds fighting for a vacant site. When a site is occupied, intruders are easily driven away by the site-holder (Nelson 1978a).

There were 689 pairs on North Bird Rock in 1989. Magdalen Islands. *July 1984, Gilles Chapdelaine.*

Bird Rocks, including North Bird Rock in the background, harboured a total of 7640 pairs in 1989. Numbers are expected to decline because of the erosion eating away at the two islands. Magdalen Islands. *July 1983, Yves Aubry.*

Northern Gannet

Breeding distribution
1984–1989

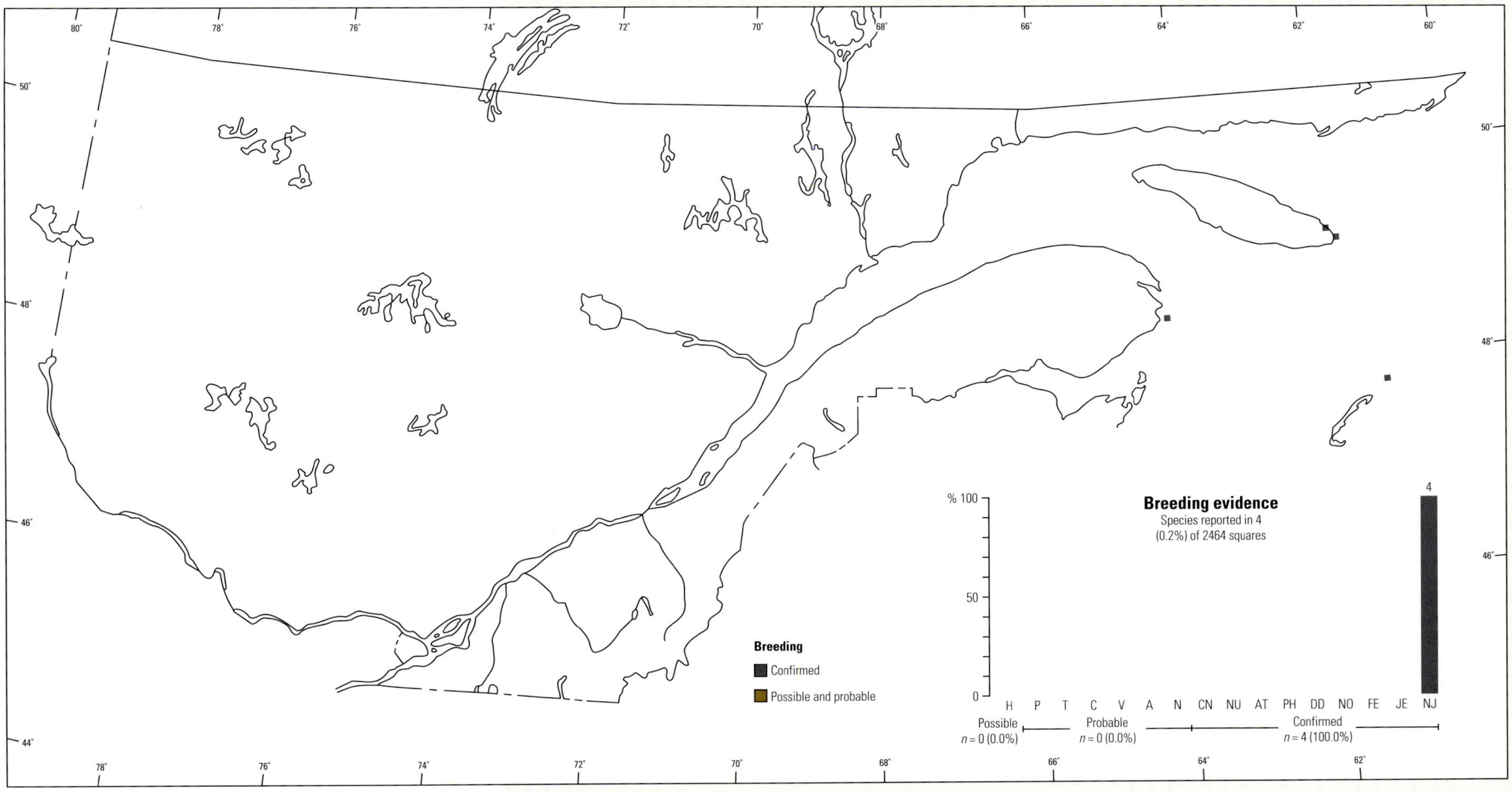

Species profile

Clutch size	1 egg, rarely 2[1,2]
Laying interval	–
Start of incubation	after laying of egg[1]
Incubation period	44 days (42–46)[1,2]
Maturity at hatching	altricial
Care of young	by female and male[1]
Nestling period	82–99 days[2]
Age at first flight	82–99 days[2]
Dependence of young	until first flight[1]
Broods per year	1[1,2]
Breeding age	5 years[1]
Mating system	monogamy[1]
Duration of pair bond	for life[1]
Total length	88–102 cm[3]
Wing span	167.6–182.8 cm[4]
Weight	M 2932 g (2470–3470) / F 3067 g (2570–3610)[6]
Longevity record	20 years 5 months[5]

References:
1. Nelson 1978a
2. Poulin 1968
3. Godfrey 1986
4. Terres 1980
5. Klimkiewicz and Futcher 1989
6. Dunning 1984

Chronology $n = 21$

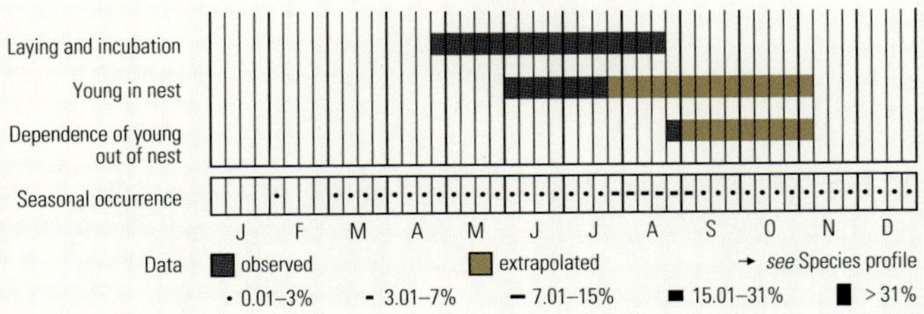

World breeding range

The three gannet colonies in the Gulf of St. Lawrence occupy cliffs and bare top plateaus, such as those on Bonaventure Island. Île Bonaventure and Rocher Percé Conservation Park. *May 1984, Gilles Chapdelaine.*

A typical gannet colony on a cliff on Great Bird Rock. Magdalen Islands. *July 1983, Yves Aubry.*

Distribution

The Northern Gannet population in the Gulf of St. Lawrence numbers approximately 31 920 pairs in three colonies: Bonaventure Island (*c.* 24 125 pairs), Bird Rocks (*c.* 7640 pairs), and the northeastern tip of Anticosti Island (*c.* 155 pairs) (BIOMQ). The Atlas map indicates breeding in four squares, two of which are on the northeastern tip of Anticosti Island, suggesting that there are two colonies there when there is actually only one: the two squares are contiguous, with a common boundary running through what is generally considered a single colony.

Historical data show that in the last century the species also bred on Île aux Perroquets in the Mingan Archipelago (Bryant 1861; Comeau 1909). But this colony disappeared before the turn of the century because of poaching. The populations in the Gulf of St. Lawrence represent 70% of the total gannet population in North America. Besides the three colonies in the gulf, there are three others on the Atlantic coast of Newfoundland (Nettleship and Chapdelaine 1988).

There are 24 125 pairs of Northern Gannets on Bonaventure Island, mainly on the eastern cliffs. *July 1984, Gilles Chapdelaine.*

The nest, which grows higher every year as nest material accumulates, consists of mud compacted with grass, seaweed, and flotsam. The regular spacing between the nests shows the small size of the territories. Bonaventure Island. *May 1994, Gilles Chapdelaine.*

Northern Gannet population trends on Bonaventure Island, 1887–1989 (BIOMQ)

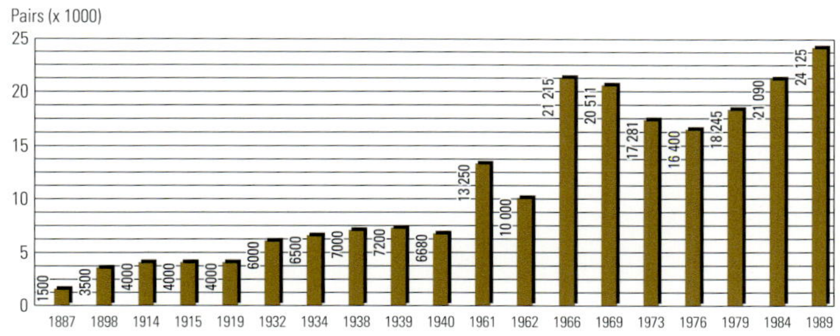

History and trends

The gannet colony on Bird Rocks is first mentioned in Hakluyt's account of Jacques Cartier's voyage to Canada in 1534. In Cartier's time there were three islands in the group, and he says: "these ilands were as full of birds, as any medow is of grasse, which there do make their nestes: and in the greatest of them, there was a great and infinite number of those that we call margaulx, that are white, and bigger than any geese" (R. Hackluyt *in* Fisher and Vevers 1943).

He was obviously referring to the Northern Gannet. Today there are only two islands there, Great Bird Rock and North Bird Rock, gradually shrinking in size due to erosion and the friability of the substrate; but they are still inhabited by *margaulx*. Fisher and Vevers (1943, 1944) estimated the total population of Bird Rocks in Cartier's time at 125 000 pairs, but that is merely speculation. Later, other navigators and naturalists visited this area. J. J. Audubon, who came in 1833, described how the fishermen used the birds as bait for cod fishing. In 1860, Bryant (1861) estimated there were 75 000 pairs of gannets breeding on Bird Rocks. The shrinking surface area of the islands and large-scale killing by fishermen contributed to this decline. Then, with the construction of a lighthouse on Great Bird Rock in 1869, the birds gradually abandoned the island's plateau for the cliffs. Persecution was commonplace at the time, extending even to egging, according to Brewster (1884). In 1919, the Canadian government declared Bird Rocks a migratory bird sanctuary. Nonetheless, the population barely exceeded 2500 birds in 1934 (Gross 1937). Since then, however, it has rebounded and today there are 6951 pairs on Great Bird Rock and 689 pairs on North Bird Rock (BIOMQ).

Despite the persecution to which the Northern Gannet was subject before this century, the world population has grown at an annual rate of 3% since 1900 (Nelson 1978a). The Bonaventure Island colony showed fairly similar growth, as evidenced by the population estimates obtained between 1887 and 1989 (see figure). However, growth slowed during the period 1969-1976, which saw hatching success fall to between 36 and 40%, with a concomitant negative effect on net productivity (only 29%). The presence of large amounts of DDT and its breakdown products in eggs was responsible for the decline. Net productivity has to be at least 67% to sustain a Northern Gannet population. In 1976, net productivity slightly exceeded this level, stabilizing at around 75% between 1979 and 1989 (Chapdelaine et al. 1987). The population is now growing again at an annual rate of 3% and should exceed 30 000 pairs by the turn of the century.

GILLES CHAPDELAINE

Great Cormorant

Grand Cormoran

Phalacrocorax carbo

PRESENT ONLY IN THE GULF OF ST. LAWRENCE, THE GREAT CORMORANT can be distinguished from the Double-crested Cormorant not so much by its plumage and size, as by its feeding habits. It clearly prefers to fish in the calm of the open ocean than in the busier river estuaries and coastal areas favoured by its cousin.

Its acknowledged prowess at fishing has earned it a special relationship with man on several continents throughout history: in 17th century England, at the time of Charles I, it played a key role in a very popular type of sport fishing, and in Asia down through the ages it has always helped fishermen earn their living. In America, and in Québec, it has had no known role of this type.

Known rather in Québec for its coloniality, the Great Cormorant winters regularly in Chaleur Bay (Ouellet 1969; Chapdelaine 1977) and occasionally on the Lower North Shore (David 1980a). However, it is not known what proportion of the population these colonies represent.

Habitat and behaviour

The Great Cormorant colonizes the steep slopes of islands on the North Shore, occupying the tops of escarpments or suitable wider ledges of cliffs up to 40 m high. Small clusters of nests are generally scattered over the face of the cliff; it is rare, even exceptional, to see dozens of nests together, as is common in a colony of Double-crested Cormorants. On Anticosti Island, the Great Cormorant nests in association with the Double-crested Cormorant on the upper decks of a shipwreck (BIOMQ).

The nest, which is reused every year, can measure up to 1 m in diameter at its base and 30–40 cm high. Both sexes build the nest, generally with the male supplying the material and the female adding it to the structure (Cramp and Simmons 1977).

The breeding biology of the Great Cormorant was studied on the Magdalen Islands (Pilon et al. 1983a). A clutch of 3–5 eggs is laid starting in the first week of May, at intervals of one egg every 2–3 days. Incubation starts as soon as the first egg is laid. Asynchronous hatching begins in early June after 28 days of incubation. Adults share the task of raising the young, feeding them groundfish, such as winter flounder and cunner, or surface fish, such as sand lance and rainbow smelt (Burton 1978; Pilon et al. 1983b). Young cormorants fledge at about 50 days, but remain dependent on their parents until the 11th or 12th week. During this time, they develop their flying and fishing skills, returning to the nest every evening like their parents (Palmer 1962). Fortunately for it in the long run, the Great Cormorant's choice of prey has no adverse effect on commercial fish stocks.

Breeding success varies from year to year; the Magdalen Islands study assessed it at 2 fledged chicks per pair in 1978, a slightly better outcome than the 1.2 observed in Nova Scotia by Lock and Ross (1973). Nothing is known about the juveniles until they reach sexual maturity between 4 and 5 years of age.

Marie-France Lelièvre

Notes on territory

The Great Cormorant's territory is limited to the nest and immediate surroundings. It is used for courting, pair formation, and raising the young until at least 4 weeks of age (Cramp and Simmons 1977).

The female sometimes helps defend the nest, but it is usually the male's responsibility. The species' defensive behaviour suggests that this is not a dominant reproductive trait. The breeding habitat of cliffs and escarpments does not lend itself to conflict.

In 1833, Audubon was the first to visit the Great Cormorant colony on Îles de Ouapitagone du Large. Today, this island is home to both Great and Double-crested cormorants. Lower North Shore. *July 1990, Pierre Brousseau.*

Great Cormorant

Breeding distribution
1984–1989

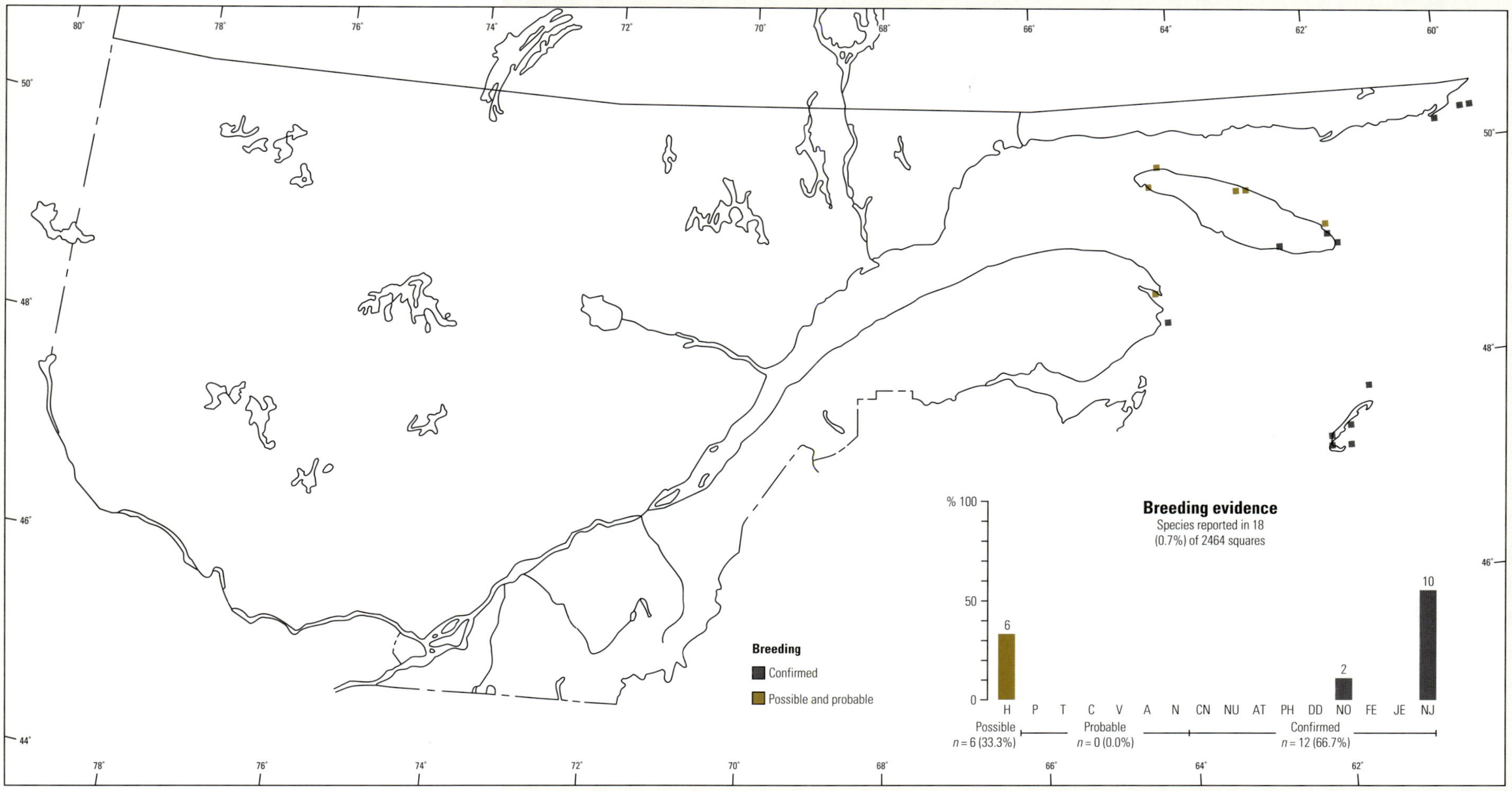

Species profile

Clutch size	usually 3–5 eggs (2–6)[1,3]
Laying interval	1 egg per 2–3 days[1]
Start of incubation	with first egg[1]
Incubation period	28–31 days[1]
Maturity at hatching	altricial
Care of young	by female and male[1]
Nestling period	≈ 50 days[1]
Age at first flight	≈ 50 days[1]
Dependence of young	until 11–12 weeks old[6]
Broods per year	1[1,6]
Breeding age	4–5 years[1]
Mating system	monogamy[1]
Duration of pair bond	usually 1 season[1]
Total length	86–101 cm[2]
Wing span	≈ 150 cm[4]
Weight	M 2283 g (1975–2687) / F 1936 g (1673–2174)[1]
Longevity record	14 years 4 months[5]

References:
1. Cramp and Simmons 1977
2. Godfrey 1986
3. Pilon et al. 1983a
4. Terres 1980
5. Clapp et al. 1982
6. Palmer 1962

Chronology $n = 24$

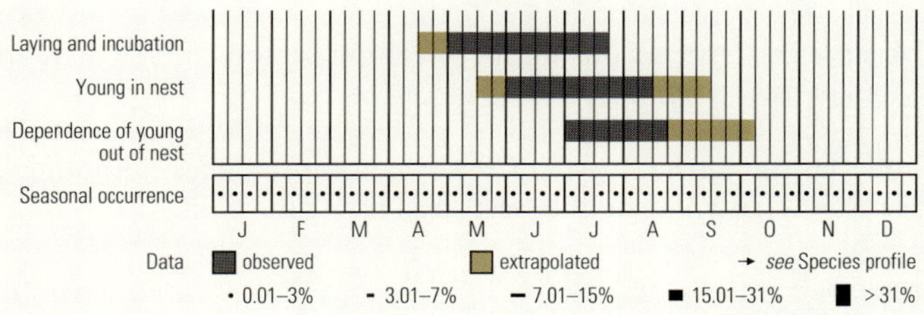

World breeding range

A recently established colony, visible from Pic de l'Aurore, scattered over the Percé Rock cliffs. It is easy to locate part of the colony by the droppings that bleach the west end of the cliff. Percé. *July 1992, Yves Aubry.*

Distribution

The main Great Cormorant colonies are located on Brion Island, Shag Island, and Île d'Entrée in the Magdalen group, and on the Lower North Shore in the Harrington Harbour area. On Anticosti Island, Atlas data show three colonies, including one at Dauphiné Point. Anticosti Island used to have about 15 colonies (Ouellet 1969; Erskine 1972c; BIOMQ) but it is not known if they are still active. The Percé Rock colony is the most accessible for birders to observe this species during the breeding season. In all, there are thought to be some 30 colonies in Québec.

Elsewhere in the Gulf of St. Lawrence there are scattered colonies along the coasts of the Atlantic provinces, although in Newfoundland only on the west coast facing the gulf.

History and trends

According to the historical data on population estimates in Québec (see table), there has never been a large number of breeding pairs. In 1990, the breeding population was estimated at about 1000 pairs, excluding the probable active colonies on Anticosti Island. Brown (1986) assessed the population in eastern Canada at 2500 pairs.

In the Magdalen Islands, the number of breeding pairs more than doubled between 1976 and 1988, from 300 to 764 pairs, and a new site was colonized on Île aux Goélands (Fradette 1992; BIOMQ). The most noticeable increases occurred at the Cap Noir, Shag Island, and Le Gros Cap colonies, while the total number of breeding birds remained stable on Brion Island and Île d'Entrée (BIOMQ).

In 1940, Lewis (1941a) estimated that of the North American population (2172 pairs), over half, 1210 pairs, nested on Anticosti Island in 15 colonies. On the basis of the scant information available for the period from 1960 to 1992 (Erskine 1972c; BIOMQ) the proportion of Great Cormorants now breeding on Anticosti Island is considerably lower, although no figure can be put on it because recent data are not complete. There are thought to be several active colonies on the island, all situated on the north shore except for the one at Dauphiné Point, which is on a shipwreck. An overflight of the colony in 1989 confirmed that the species was breeding there, but no population estimate was made.

The Canadian Wildlife Service has detailed information for the Lower North Shore (BIOMQ) going back to the 1920s. As shown in the table on population trends on the Lower North Shore, Great Cormorant colonies flourished for a time, but for two decades now the situation in the Îles Sainte-Marie Migratory Bird Sanctuary has been deteriorating and is cause for concern (Brousseau and Chapdelaine 1990).

Great Cormorant population estimates in Québec (1940–90)

Colonies by region	1940[6]	1963[7]	1974	1976	1977	1979	1982	1988	1989	1990
Gaspé										
Percé Rock			18p[1]			16p[1]			35p[1]	
Magdalen Islands										
Île d'Entrée				350i[2]	279i[3]					84p[5]
Cap Noir				40i[2]	46i[3]			124p[5]		69p[5]
Le Corps-Mort				75i[2]	48p[3]					58p[5]
Le Gros Cap										58i[5]
Shag Island				400i[2]	150p[3]					247p[5]
Brion Island				161i[2]	110i[4]					116p[1]
Île aux Goélands								151p[5]		163p[5]
North Shore										
Îles Sainte-Marie Bird Sanctuary					107p[1]		67p[1]	43p[1]		
Île du Lac					+[1]		+[1]			139p[1]
Île de Ouapitagone				+[1]					35p[1]	
Anticosti Island										
Dauphiné Point									+[1]	
Pointe de l'Est	92i									
Falaise aux Goélands	180i	125i		10i[1]					+[1]	
Cape Sandtop	18i									
Innommée Bay	210i								+[1]	
Pointe du Renard	10i	135i								
Table Head	100i			18i[1]						
Cape Tunnel	16i									
Joseph Point	34i	17i								
Pointe à la Batterie	154i									
Harvey Point	24i	75i								
Pointe de la Tour	32i									
Cap de l'Ours	32i									
Guy Point	82i	120i		44i[1]						
Cape Vauréal	30i									
Cape Observation	154i			200i[1]						

p: pairs i: individuals +: numbers not available 1 BIOMQ 2 Mousseau et al. 1976 3 Burton 1978 4 Mousseau 1984c 5 MLCP, unpubl. data 6 Lewis 1941a 7 Ouellet 1969

The upper deck of this shipwreck, off Dauphiné Point on Anticosti Island, harbours the only Great Cormorant colony on the southeastern shore of the island. *July 1989, Pierre Brousseau.*

The Île du Lac colony is probably the best known and one of the oldest on the Lower North Shore; the birds were already on the escarpments when Audubon visited the area in 1833 (Todd 1963). The 1990 survey shows an encouraging turnaround for both this colony and the one on Île de Ouapitagone. The other colonies on the Lower North Shore are situated in the Îles Sainte-Marie Migratory Bird Sanctuary and have been protected since 1925, when this sanctuary was established.

Cormorants wintering on cliff ledges on the shore of Chaleur Bay. Saint-Godefroi, Gaspé. *December 1977, Jean-Luc Desgranges.*

Great Cormorant population trends (number of pairs) in colonies on the Lower North Shore (BIOMQ)

Year	Îles Sainte-Marie Bird Sanctuary	Île de Ouapitagone	Île du Lac
1860			4500 i
1923			11
1925			22
1926			33
1927			32
1928			48
1929			49
1930	4		63
1931			70
1932			97
1933			112
1934			102
1935	30	12	95
1936		25	73
1937		48	58
1938		29	98
1939		13	97
1940	86	6	140
1945	180	45	
1947	173		85
1950	245		
1955	339		
1960	277		
1962	184	29	63
1965	182		
1972	220		
1976	173		
1977	107		+
1982	67		+
1988	43		
1990		35	139

i: individuals +: numbers not available

Despite good breeding success (2 chicks per pair) and the fact that the breeding areas in most of the colonies are not very accessible, the Great Cormorant population is not experiencing significant growth, unlike that of the Double-crested Cormorant. It is difficult to explain this relative stability as little is known about this species except its breeding biology.

Pierre Brousseau

Double-crested Cormorant

Cormoran à aigrettes

Phalacrocorax auritus

The Double-crested Cormorant can be seen almost anywhere along the St. Lawrence River standing on a stump or rock with its wings spread open. During the late 19th and early 20th centuries, this fish-eating migrant was persecuted throughout its North American range until it completely abandoned many colonies (Lewis 1929), but protection has allowed it to reoccupy many of its former island colonies in the St. Lawrence River (DesGranges et al. 1984).

Habitat and behaviour

Cormorants ride low in the water, with the tail submerged and the bill tilted up at an angle, and dive from the water surface, propelling themselves underwater using their legs. Their diet is composed of a variety of fish (Mendall 1936), including commercial species, as found in the Magdalen Islands (Burton 1978). Recent stomach content analyses found that the Double-crested Cormorant feeds primarily on capelin and gunnel in the St. Lawrence estuary, representing 37% and 22%, respectively, of the diet in terms of volume consumed; in the gulf, however, sand lance was the most important prey, representing 63% of the diet by volume (Rail et al. 1996)

Double-crested Cormorants fly singly or in small groups very low over the water. Even though they have an oil gland which, through preening, presumably allows them to render their plumage somewhat water-resistant (Terres 1980), they spend hours with their wings held open to dry in the sun and wind, possibly to reduce heat loss by rapidly drying the wet plumage (Hennemann 1984).

Double-crested Cormorants are colonial breeders on freshwater lakes and rivers, estuaries, and coastal saltwater, often nesting on islands. Nests are placed on the ledges and flat tops of cliffs, as well as in trees, and may be in mixed colonies with Great Cormorants, Great Blue Herons, Black-crowned Night-Herons, and gulls. They often use old nests of Great Blue Herons and may even take over active heron nests if the occupants spend too much time away or the eggs are depredated. Tree-nesting cormorants produce a build-up of excrement at the base of the tree that eventually kills it by disrupting the ionic balance of the soil (Dusi 1977). When the tree dies and the branches fall the heron nests are destroyed because they build their nests out in the branches. Cormorants, however, place their nests close to the trunk and can continue to nest in the dead tree for years after the herons are gone. In this way heronries are often transformed into cormorant rookeries over the years (DesGranges 1980a).

Marie-France Lelièvre

Notes on territory

The territory consists of the nest and an adjoining perch used for copulation, display, and feeding the young. The pair defends it with intimidation displays, and closely guards it to prevent other

Tree dieback is a sign of an established colony of tree-nesting Double-crested Cormorants as the heavy nitrogen content of their droppings kills off the trees supporting the nests. The white lines mark the limits of the colony. Corossol Island, Sept Îles Bay. *June 1988, Pierre Brousseau.*

Double-crested Cormorant

Breeding distribution

1984–1989

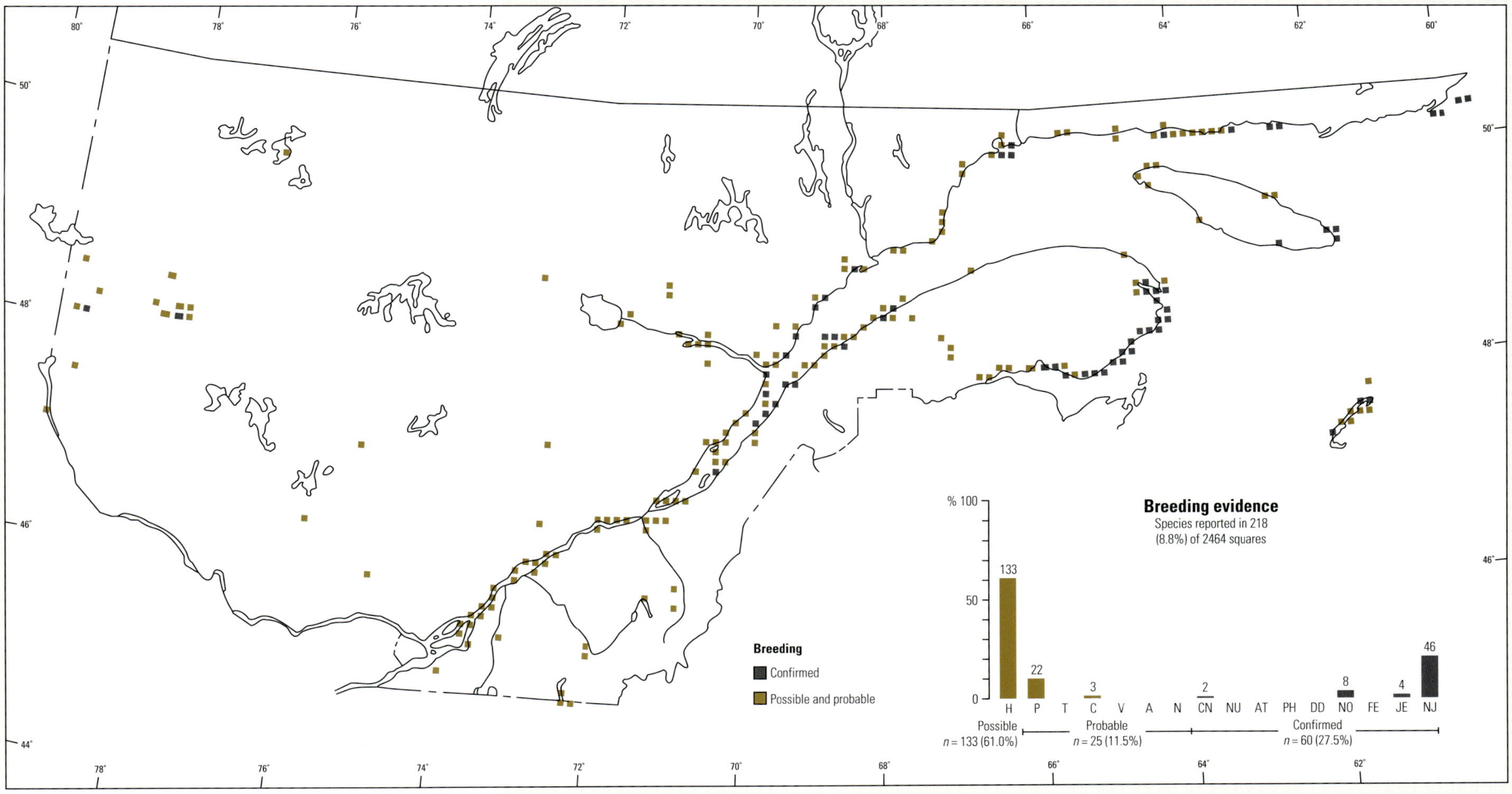

Species profile

Clutch size	usually 3–4 eggs (2–7)(1,2,3,4,5,8,9)
Laying interval	?
Start of incubation	with third egg(2,5)
Incubation period	24–29 days(1,2,3,4)
Maturity at hatching	altricial
Care of young	by female and male(1,2)
Nestling period	3–5 weeks(5)
Age at first flight	5–6 weeks(5)
Dependence of young	until 10 weeks old(2,3,5)
Broods per year	1(1,2,5)
Breeding age	2–3 years(5)
Mating system	monogamy(1)
Duration of pair bond	1 or more seasons(5)
Total length	73–89 cm(4)
Wing span	137.2 cm(3)
Weight	M 1818 g (?) / F 1540 g (?)(6)
Longevity record	23 years(7)

References:
1. Ehrlich et al. 1988
2. Harrison 1978
3. Terres 1980
4. Godfrey 1986
5. Palmer 1962
6. Dunning 1984
7. Kennard 1975
8. Pilon et al. 1983a
9. DesGranges 1982

Chronology *n* = 59

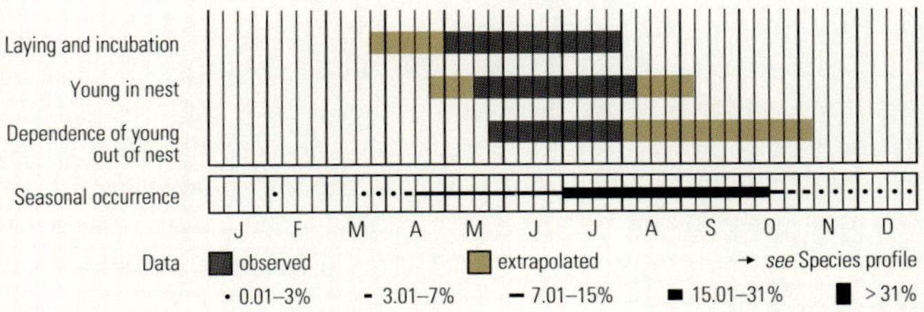

World breeding range

Notes on territory (continued)

pairs from usurping it or removing nest material.

Birds prefer the centre of established colonies, occupying them first in spring. About 25% of the individuals at colonies are 1-2 year old pre-breeders which generally remain at the periphery of the colony (Palmer 1962).

An established cormorant rookery in a stand of stunted fir, called krummholz, on Île de l'Est, Sainte-Marie Islands, Lower North Shore. *August 1992, Pierre Brousseau.*

Nesting on the ground, either in flat areas or on cliffs, may have been the original type of nesting. Arboreal nesting may be a secondary type, adopted for safety against predators such as foxes and minks or as a response to human disturbance. The clumsiness of these birds, with their webbed feet, supports this view (Lewis 1929). Yet, although perching is not natural to the Double-crested Cormorant, it is not a natural ground dweller either.

The territory consists of the nest plus an adjoining perching spot for displays, copulation, and feeding the young (Palmer 1962). Old nests are often rebuilt (Terres 1980), while new ones take the pair about 4 days to construct. Once the foundation is built, it is primarily the male's job to bring nest material to the female. Both parents incubate. The most successful pairs are the more experienced ones that nest early in the season in the centre of the colony (McNeil and Léger 1987). Hatching success is apparently not related to whether the pair nested in a tree or on the ground (Cleary 1977). Eggs are often found outside the nests, possibly having been ejected by nonparental birds; the significance of this behaviour is unclear, however (Van Scheik and Lewin 1986).

The eggs hatch over a period of 2 to 7 days (Palmer 1962), which, in the St. Lawrence estuary population, leaves the fourth chick at such a disadvantage that it often starves within two weeks (DesGranges 1982). In St. Lawrence estuary, in nests with 3 or 4 chicks, the younger ones may grow more slowly than the older ones (Cleary 1977); in contrast, younger chicks in the Magdalen Islands do not seem to be at a disadvantage (Léger and McNeil 1987).

At 8–9 days after hatching, the altricial young can sit up. At 3–4 weeks, young hatched from ground nests often leave the nest, forming bands that wander through the colony doing much social visiting. Their intrusions may be accepted or they may be gently nudged along by an adult, never viciously. At 5 weeks, bands can number 6 to 40 young which still all rush back to their own nests when their parents return to the colony with food to regurgitate to them (Palmer 1962). Parents normally feed only their own young, despite begging by strangers.

Young hatched from nests on the sloping sides of cliffs remain in the nest 4 to 7 days longer than those from nests on flat ground (Mendall 1936). Chicks in tree nests or on small ledges will remain in or near them until they are about 6 weeks old. Some of those in trees make short excursions into neighbouring branches at 4–6 weeks (Lewis 1929).

Juveniles first fly (usually to water) at about 6 weeks old, but can take off from water only at 7 weeks. They become fully independent at 10 weeks, but some return to the nesting site to roost (Lewis 1929).

Double-crested Cormorants were recorded in 218 squares, with breeding confirmed in only 27.5% and accounted possible or probable in the remaining 72.5%. Breeding is so difficult to confirm because, apart from the fact that this colonial breeder concentrates in a few locations, many breeding birds feed far from their nesting colonies (Weseloh 1987a). So the best way to confirm nesting is to locate the colonies.

Double-crested Cormorants nesting with Great Cormorants on this wreck off Dauphiné Point, Anticosti Island. *August 1984, Pierre Brousseau.*

Spread-wing posture typical of Double-crested Cormorants when drying their feathers. Florida, United States *January 1991, Denis Faucher.*

Cormorants at their bulky nests of twigs on the flat ledges of cliffs on Cliff Island. Sainte-Marie Islands, Lower North Shore. *June 1982. Pierre Brousseau.*

Distribution

The major portion of Québec's population of Double-crested Cormorants breeds along the St. Lawrence River, in the estuary, along the North Shore, on the Gaspé Peninsula, Anticosti Island, and the Magdalen Islands. On the Magdalens the species was first observed in 1949, after which its numbers increased to a high of about 3000 pairs in 1988 (Fradette 1992). A colony of some 15 pairs at Val-d'Or and 2 pairs at La Tuque represent the only inland colonies known in the province. At Lake Saint-Pierre a colony of 50 individuals was discovered in the 1950s but later disappeared (David 1980a). Four nests were discovered there in 1991 (Bannon 1991c). The species also breeds along the southeastern side of James Bay, where a colony on Way Rock off Boatswain Bay increased from 34 nests in 1912 to about 200 in 1942 (Todd 1963). However, this colony may have disappeared (Palmer 1962). About 150 individuals were seen on the McNab Rocks in Rupert Bay in 1990, and about 10 individuals were seen on some islands in Boatswain Bay, but breeding was not confirmed (Cons. G. G. – GREBE 1992j).

Double-crested Cormorants are reported every summer in areas where they apparently do not breed, such as the upper St. Lawrence River, the Lachine Rapids, and inland from the St. Lawrence River at Sainte-Famille-d'Aumond, near Maniwaki. Double-crested Cormorants sometimes start nesting after occupying a new site for a few years (Weseloh 1987a). At Young Island in Vermont in 1982, one bird incubated while 27 others looked on; the following year there were 34 nests and 108 adults (Kibbe and Laughlin 1985).

History and trends

During the last century, Double-crested Cormorants were persecuted heavily because they were considered competitors to fishermen. They were also hunted for their eggs and meat. They declined until about 1925 in eastern North America, including along the St. Lawrence River (Todd 1963). Since then they have reoccupied much of their former breeding range, with the Great Lakes population spreading eastward toward Québec and the St. Lawrence population spreading westward (Palmer 1962). The species has only recently colonized some parts of New York State (Arbib 1988b).

Known colonies of Double-crested Cormorants in Québec (BIOMQ)

Locality	Number of colonies	Number of breeding pairs	Year surveyed
St. Lawrence estuary	19	9550	1990
North Shore	14	6152	1988
Gaspé Peninsula[1]	28	4677	1989
Magdalen Islands	3	1841	1990
Anticosti Island	1	150	1976
Val-d'Or	1	15	1988
Lake Saint-Pierre	1	4	1991
La Tuque	1	2	1988
Total	**68**	**22 391**	

1 Refers to the small, scattered colonies on the cliffs and islands along the shore of the Gaspé Peninsula.

Cormorant numbers in Québec have been increasing over at least the past 20 years. More than 22 000 pairs are now estimated to breed in the province in 68 colonies (see table), whereas in 1979 only about 12 000 pairs nested in 43 colonies (DesGranges et al. 1984). A control program was launched to reduce the breeding population in the St. Lawrence estuary from about 17 000 pairs to about 10 000 by 1992 because of the damage that cormorants do to underlying vegetation, including nesting cover for Common Eiders. This provincial government-sponsored program involves shooting adults, destroying nests, and spraying eggs with an oily substance that asphyxiates the embryos (DesGranges and Reed 1981; Bédard 1988).

Along the North Shore and in other areas such as Forillon National Park or elsewhere outside Québec, efforts are being made to protect cormorant colonies. Where colonies are managed for tourists, they should only be visited late in the nesting cycle when the young are half grown, because nest abandonment is less likely to occur then than earlier in the cycle (Ellison and Cleary 1978).

ROBERT ALVO

American Bittern

Butor d'Amérique

Botaurus lentiginosus

The tall aquatic grasses and cattails in these channels afford ideal nesting sites for the American Bittern. Sainte-Foy. *June 1963, Gaston Le Page.*

Clodin Roy

THE AMERICAN BITTERN IS AN EXPERT IN THE ART OF CAMOUFLAGE. It conceals itself in the tall grass of its wetland habitat by "freezing," stretching its neck and bill skyward to reveal the brown streaks on a yellowish-white belly that blend in with the surrounding vegetation. In a breeze, it will even sway in this posture, in imitation of reeds swaying in the wind. This camouflage affords considerable protection from predators (Vesall 1940). The American Bittern is a ventriloquist, and its booming call sounds like the suction of a hand pump or a stake being driven into a bog. Its guttural call carries for as much as a kilometre (Terres 1980), making it easy to find, despite its inconspicuousness.

A migrant breeder in southern Québec (David 1980a), this American Bittern is sedentary in the warm climes in the southern part of its range (Palmer 1962; Hancock and Kushlan 1989). This bittern, which is fairly abundant in southern Québec, arrives here after nocturnal migration (del Hoyo et al. 1992), from April through mid-May. The fall migration lasts until November and may take it as far as Panama and, exceptionally, South America. It has also been sighted in Cuba, Bermuda, and the Virgin Islands (AOU 1983).

Habitat and behaviour

The American Bittern lives in freshwater and salt marshes, swamps and wet meadows, alder and willow thickets, and sometimes even dry fields (Godfrey 1986; del Hoyo et al. 1992). Outside the breeding season, it is seen in coastal marshes and mangroves in the southern part of its American range (Hancock and Kushlan 1989).

This solitary, shy bird is active mainly between dusk and dawn, although it sometimes feeds during the day (del Hoyo et al. 1992). In common with the night-herons, it has the least varied diet of the ardeids (Kushlan 1976). When fishing, it stands almost motionless or stalks slowly and stealthily. It does on occasion walk quickly (Stone 1965) or run through the tall grass, gleaning insects from the vegetation or catching them in the air (Dudones 1983).

Active pursuit and circle flights (Hancock and Kushlan 1989) precede mating, which occurs on the breeding territory, but not necessarily at the nest, as is the case with the other ardeids (Gibbs et al. 1992b). The white breeding tufts, normally concealed by the scapulars and carpal joint of the folded wing, are exposed during displays on the ground (Johnsgard 1980).

The male calls throughout the breeding period to keep other males off his territory and maintain contact with his mate. Bitterns are monogamous, although the male may be polygynous. The female builds the nest, concealing it in a tall dense clump of vegetation. It is built of dry cattails, reeds, or rushes directly on the ground, or on a low platform over water or mud (Vesall 1940; Palmer 1962). The same nest may be used year after year (Palmer 1962). The female is extremely cautious when leaving or returning to the nest, always entering and exiting by the same paths, which serve as clues to locating the nest. The female incubates the eggs and tends the young. She is a poor housekeeper, neglecting to discard the shells of hatched young or their feces (Vesall 1940). Incubation begins as soon as the first egg is laid, resulting in asychronous hatching. The young are fed mainly insects and small vertebrates that are regurgitated by the female.

American Bitterns sometimes nest along shrubby shorelines. L'Isle-Verte, northeast of Rivière-du-Loup. *Summer 1991, Jean Bachand.*

Notes on territory

Once established on its territory, the male defends it throughout the nesting season. When another male arrives, the two birds move swiftly back and forth in a crouching position, displaying tufts of white feathers on either side of the upper breast. Other defensive behaviour consists of circle flights, similar to those of the Great Blue Heron, with feet dangling and white feathers expanded (Hancock and Kushlan 1989).

Studies variously reported densities of 0.4 pair/10 ha in Pennsylvania (Leberman 1992a), 4 singing males/10 ha in

American Bittern

Breeding distribution
1984–1989

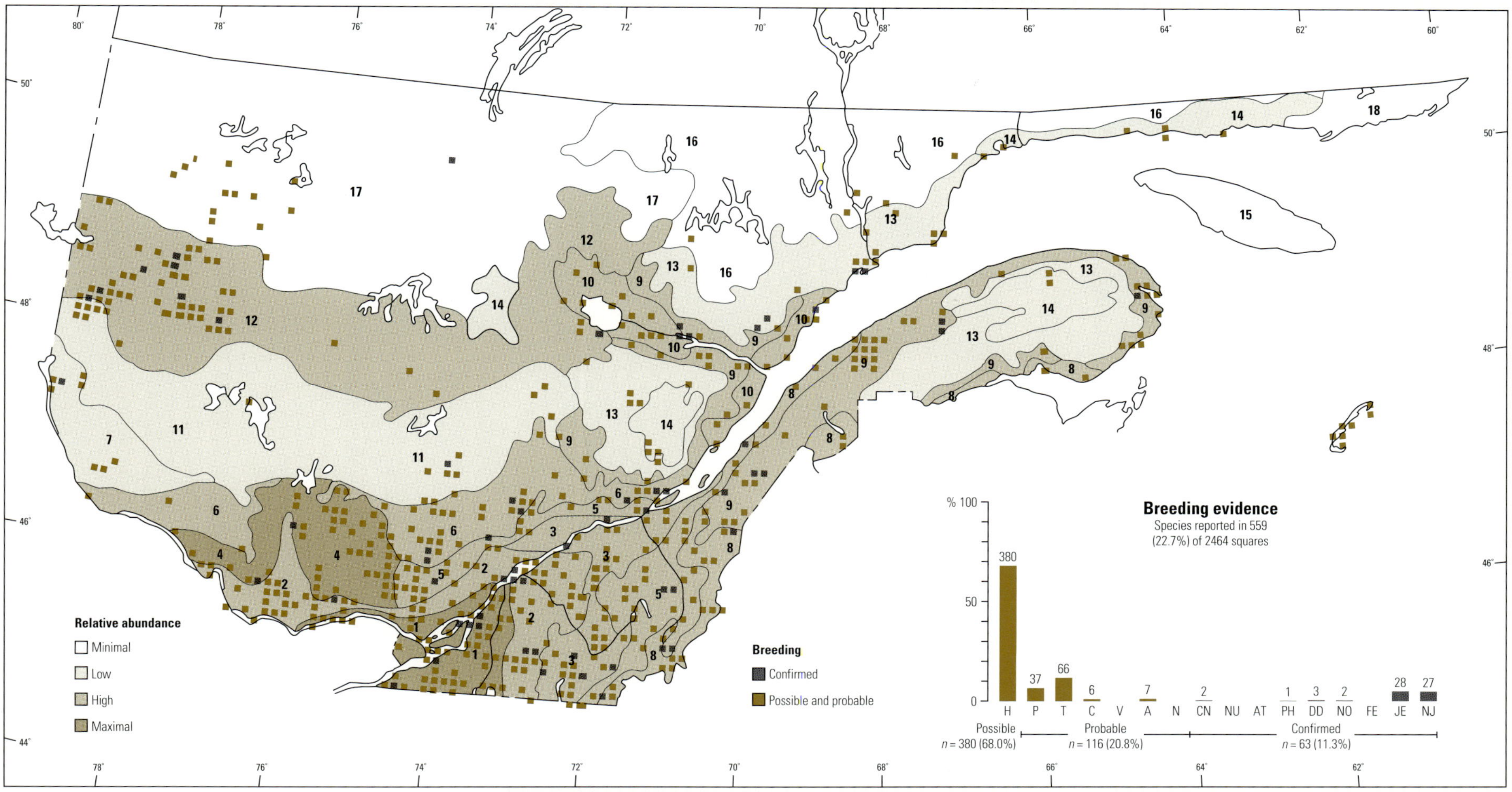

Species profile

Clutch size	usually 3–5 eggs (2–7)[2,3,4,5,9]
Laying interval	1 egg per day[3]
Start of incubation	with first egg[5,6]
Incubation period	24–29 days[2,3,4,5]
Maturity at hatching	semi-altricial
Care of young	by female[3,9]
Nestling period	7–14 days[3,4,5,9]
Age at first flight	? *see* text
Dependence of young	probably until first flight[9], *see* text
Broods per year	probably 1[6,9]
Breeding age	probably 1 year[9]
Mating system	monogamy, polygyny possible[9]
Duration of pair bond	probably during mating only[9]
Total length	61–86 cm[7]
Wing span	127 cm[3]
Weight	706 g (520–1072)[8]
Longevity record	8 years 4 months[1]

References:
1. Clapp et al. 1982
2. Ehrlich et al. 1988
3. Terres 1980
4. Harrison 1975
5. Hancock and Kushlan 1989
6. Palmer 1962
7. Godfrey 1986
8. Dunning 1984
9. Gibbs et al. 1992b

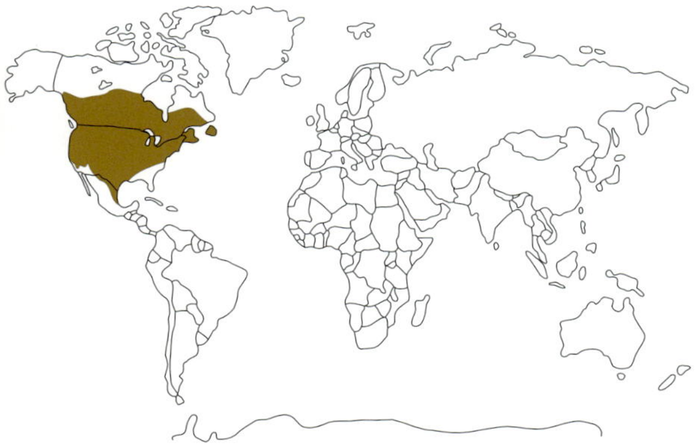

World breeding range

Chronology $n = 61$

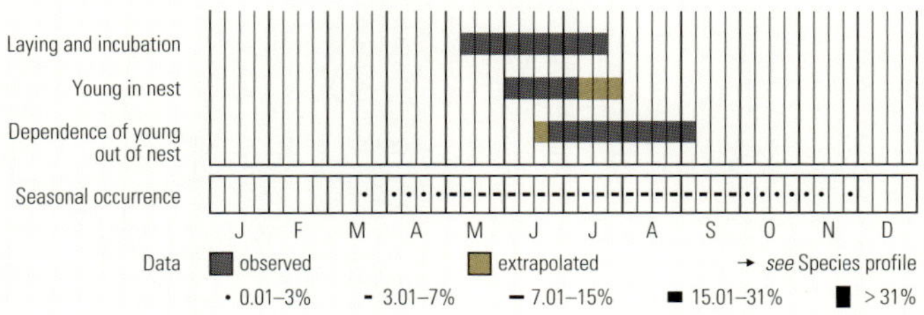

Frequency of occurrence by ecological region

Ecological regions** (forest regions)	Adequately covered squares	Squares in which reported	% of squares in which reported	Relative abundance (%)*
1. Sugar maple-hickory and sugar maple-basswood	97	52	53.6 ± 5.0	100.0
2. Sugar maple-basswood	153	68	44.4 ± 4.1	82.9
3. Sugar maple-basswood and sugar maple-yellow birch	101	37	36.6 ± 5.7	68.3
4. Sugar maple-yellow birch-basswood	56	30	53.6 ± 11.4	99.9
5. Sugar maple-yellow birch-beech and sugar maple-yellow birch-basswood	113	43	38.1 ± 5.0	71.0
6. Sugar maple-yellow birch-beech	71	33	46.5 ± 10.6	86.7
7. Sugar maple-yellow birch-balsam fir	19	4	21.1 ± 19.7	39.3
8. Sugar maple-yellow birch and balsam fir-yellow birch	47	21	44.7 ± 12.8	83.3
9. Balsam fir-yellow birch	87	37	42.5 ± 9.6	79.3
10. Balsam fir-white birch or balsam fir-red maple	46	18	39.1 ± 12.2	73.0
11. Yellow birch-balsam fir	52	13	25.0 ± 12.2	46.6
12. Balsam fir-white birch (west)	59	28	47.5 ± 13.1	88.5
13. Balsam fir-white birch (east)	76	17	22.4 ± 9.2	41.7
14. Balsam fir-black spruce	28	5	17.9 ± 15.2	33.3
15. White spruce-balsam fir and balsam fir-white spruce	22	0	0	0
16. Black spruce-balsam fir-moss	18	2	11.1 ± 17.4	20.7
17. Black spruce-moss	32	4	12.5 ± 13.0	23.3
18.			*not analysed*	

* minimal: 0–29.9%; low: 30–59.9%; high: 60–89.9%; maximal: 90–100%

** Deciduous forest: 1–7; mixed forest: 8–11; coniferous forest: 12–17

This American Bittern is incubating its eggs in a field. Hatching is asynchronous, explaining the single chick. Saint-Elphège, northwest of Drummondville. *July 1991, Jeanne Lehoux.*

Notes on territory (continued)

Wisconsin (Manci and Rusch 1988), and 0.26 singing male/10 ha in Maine (Gibbs et al. 1992b). Internest distances, with no corresponding territorial measurements, were also given: 3 nests less than 50 m apart around a half-hectare pond (Provost 1947), 2 nests 50 m apart (Vesall 1940), and 2 nests only 20 m apart (Middleton 1949).

They leave the nest at the age of 1–2 weeks, but remain in the vicinity of the nest until they are 2–4 weeks old, with the parent birds continuing to bring them food (Gibbs et al. 1992b). Age at first flight is not known, but it may be similar to that of its cousin, the Eurasian Bittern, in which fledging occurs at about 50 days (Cramp and Simmons 1977). It is not likely that the young are still dependent after their first flight (Gibbs et al. 1992b).

The reclusiveness of the American Bittern during its breeding cycle is reflected in the low percentage of reports of confirmed breeding—barely 11.3% of the squares (63/559) in which it was recorded.

Distribution

The American Bittern nests throughout southern Québec, in the interior and along large waterways. Breeding was confirmed in Abitibi, the Ottawa Valley, the Lanaudière, Bois-Francs, Montréal, Québec City, Eastern Townships, and Saguenay–Lac-Saint-Jean regions, on the North Shore, and in the Gaspé, but not on the Magdalen Islands or on Anticosti Island, where known breeding records date from the end of the 19th century and beginning of the 20th century (Verrill 1862; Fradette 1992).

The northern limit of its range is poorly known, but one bird was heard in Kuujjuaq in Ungava Bay in 1993 (ÉPOQ), and a nest was reported at the mouth of La Grande Rivière (Todd 1963). The American Bittern is considered casual inland north of the 50th parallel. In the southern James Bay region, it is present at least as far north as the Eastmain River, in areas where clay surface deposits occur, these being more productive than the oligotrophic lakes of the Lake Plateau, which have little shoreline habitat (Bordage and Aubry 1982; Bordage 1983; Cons. G. G. – GREBE 1992j). In this region, it is found mainly in beaver ponds, lakes under 20 ha in size, and streams bordered by tall grasses or areas of emergent vegetation. The American Bittern has also been observed using beaver ponds in Maine (Gibbs et al. 1991a). It has also been recorded south of the Great Whale River and near Coats River, only 20 km from the Little Whale River (Cons. G. G. – GREBE 1992u,x). Here again, the distribution pattern corresponds with areas of clay surface deposits.

The bittern stalks its prey through the marsh vegetation. This adult bird has taken a small fish in a marsh on the Saint-Charles River. Lac-Saint-Charles, near Québec City. *June 1993, Robert Morin.*

When alarmed, this wading bird will freeze in a reedlike pose, bill pointing skyward, even when it does not blend in with its surroundings. Maskinongé, north of Lake Saint-Pierre. *June 1982, Yves Aubry.*

An American Bittern on its nest in the shrubby margins of a pond. Sainte-Dorothée, Laval. *June 1942, Henry Mousley, Laval University Collection.*

History and trends

At the turn of the century, before its breeding habitat was drained for agriculture, industry, and urban development (Ouellet 1974a), the American Bittern was considered common throughout Canada (Dionne 1883; Chamberlain 1887; Wintle 1896).

The American Bittern figured on the Blue List between 1976 and 1986, and is listed as of special concern in Pennsylvania (Gill 1985). Its numbers have declined everywhere, especially in the central part of its range in America, because of the loss of wetland habitat (Ehrlich et al. 1988). In Québec, between 1950 and 1978, 3643 ha of fresh, brackish, and salt water wetland—suitable bittern breeding sites—disappeared, eliminating 6.6% of the 55 385 ha of wetlands along the St. Lawrence, Ottawa, and Richelieu rivers (Champagne and Melançon 1985; Lands Directorate 1986). In the Maritime provinces, the draining and filling of freshwater marshes considerably reduced the bittern's habitat in the past, but it appears that the impoundment of marshes for waterfowl management in recent years has somewhat offset previous losses (Erskine 1992). Needless to say, any wetland conservation effort will benefit bitterns and help ensure their long-term survival.

A recent study (DesGranges 1989) on the American Bittern's sensitivity to the acidity of wetland habitat in the Canadian Shield showed that the species actively seeks out productive alkaline or neutral lakes ($pH \geq 7.0$), such as those found in the Appalachians, or low-acid lakes in the middle and upper Laurentians, generally with large well-vegetated margins of shallow water prolific in aquatic vegetation and wildlife such as fish and amphibians.

Although the effects of acid precipitation on wetlands are still not fully understood, the repercussions on the American Bittern are fairly predictable. Wetland acidification, which can be caused by the spring melt, changes the wildlife productivity of a given habitat. It adversely affects the breeding performance of many fish species and populations of amphibians and aquatic invertebrates (Pierce 1985; Gascon and Planas 1986). As it affects the availability of food, acidification could cause major changes in the distribution of aquatic bird populations (DesGranges 1989).

Gaétan Duquette

Least Bittern

Petit Butor

Petit Blongios

Ixobrychus exilis

THE LEAST BITTERN IS THE SMALLEST OF THE ARDEIDS THAT BREED IN Québec. It is also the most difficult to locate because of its secretive behaviour and protective coloration—the dark crown and mantle and buffy wing patches that help it to blend in with its surroundings. It rarely leaves the cover of the marsh vegetation, except for the occasional short flight, quickly dropping back down out of sight. At the least sign of danger, it will make itself inconspicuous, adopting the same freezing posture as the American Bittern, neck stretched up and bill pointing skyward (Allen 1915). The bird is an infrequent caller; the male's soft, low cooing notes are most often heard in the early morning or in the evening.

In Québec, the Least Bittern is a scarce migrant breeder, only seen in the southern part of the province. It is usually present from early May to early September, although it has been observed in October. It winters from the southern United States to Panama and Colombia (AOU 1983; Root 1988).

Dense reedbeds afford excellent nest cover for the Least Bittern, now considered vulnerable in Québec. Saint-Étienne-de-Beauharnois, south of Montréal. *June 1991, Gaétan Duquette.*

Habitat and behaviour

The Least Bittern is found mainly in freshwater cattail marshes. It also nests near ponds and along slow-moving water in the dense cover of bulrushes, sedges, and reeds (Weller 1961).

Unlike other ardeids, it is considered a solitary nester. Although most evidence supports this view, population density under certain specific conditions, linked probably to the abundance of food, can be rather high during the breeding period (Kushlan 1973). Little is known about the breeding biology of this species. There is some evidence, however, to indicate that songs and calls play a large part in courtship rituals (Hancock and Kushlan 1984).

The nest, often built in year-old vegetation, is commonly located in a thicket, several metres from open water (Peck and James 1983). It is a rather flimsy platform of cattail stalks and leaves, both green and dry (Weller 1961), between 5 and 122 cm above water level, with water depth varying between a few centimetres to nearly a metre (Peck and James 1983). It appears that the nest is mainly built by the male, but both parents may add nest material after laying has begun (Weller 1961). At the onset of incubation, and more often during hatching, the male and female will peck at the bottom of the nest (Weller 1961; Aniskowicz 1981) to make small openings to aerate the nest and dispose of insects, feces of the young, and food scraps (Weller 1961). Both male and female incubate the eggs and feed the young by regurgitation (Gibbs et al. 1992a).

The Least Bittern was recorded in 1.6% (40/2464) of Atlas squares surveyed, and breeding was confirmed in 25% (10/40) of these. Six records of confirmed breeding were based on nests containing eggs or young and 3 were of recently fledged young. The low number of sightings is due to the elusiveness of the species and the small number of breeding pairs in Québec (Robert 1989b).

Distribution

The Atlas distribution map shows the Least Bittern to be confined to southern Québec. The species was mainly observed in the southern Ottawa Valley area, along the St. Lawrence, and south of the river to the area around Lake Saint-Pierre. Breeding was confirmed in the Ottawa Valley at Gatineau, Thurso, and Plaisance, and in the Montréal area at Dundee and on the Boucherville Islands. Farther east, breeding was also confirmed at Sainte-Rose-de-Watford and Cap-Tourmente, the latter record being the most northerly to date. The Least Bittern probably breeds around Lake Saint-Pierre, Lake Boivin in the Granby area, and in the Katevale marsh in the Eastern Townships.

Outside this area, the Least Bittern is considered a casual visitor, the same status given by David (1980a). Elsewhere in Canada, it breeds in southern Manitoba and southern Ontario, and probably in southwestern New Brunswick and northern Nova Scotia (Erskine 1992).

History and trends

The Least Bittern has always been considered rare in Québec and has probably always bred very locally (Dionne 1906; Ouellet 1974a; David 1980a; Sandilands and Campbell 1987b). Paucity of data, however, makes it difficult to measure fluctuations in abundance and distribution with any accuracy. Generally, the species seems to be in decline in many areas of North America (Kibbe 1985f; Sandilands and Campbell 1987b; Woodliffe 1987b; Smith, G. A. 1988a). The Least Bittern has been on the Blue List since 1979 and is even considered to have been extirpated from certain areas (Tate 1986). In Québec, it is currently classified as vulnerable (Robert 1989b). Habitat loss is generally considered to be the main cause of population decline (Kibbe 1985f; Sandilands and Campbell 1987b; Smith, G. A. 1988a; Robert 1989b; Gibbs and Melvin 1992a). The filling and draining of marshes for agriculture and urban development have considerably reduced the extent and number of marshes in densely populated areas such as the Central St. Lawrence Lowland (Lands Directorate 1986). Environmental pollution and pleasure boating may also have contributed to the declining numbers (Palmer 1962; Sandilands and Campbell 1987b). Moreover, a study on the rivers of southern Louisiana found that bioaccumulation of certain insecticide residues in organisms such as frogs, snails, and small fish, which are the mainstay of the Least Bittern's diet, resulted in a rather high concentration of contaminants in eggs (Causey and Graves 1969).

PHILIPPE FRAGNIER

Clodin Roy

Notes on territory

Territorial displays by Least Bitterns are rare. The sighting of a male in defensive posture in front of a conspecific, as reported by Davidson (1944), and the uniform distribution of nests within certain study areas in the southern United States appear, however, to indicate some form of territoriality (Weller 1961; Hands et al. 1989a). In Iowa, a greater incidence of Least Bitterns in marshes 5–11 ha and over suggests that the birds are less attracted by smaller marshes, however suitable they may appear as breeding sites (Brown and Dinsmore 1986).

Kushlan (1973) gives the results of different studies carried out in the United States. Nest density varied between 0.6 and 18 nests/10 ha; densities of 51 and 185 nests/10 ha were also reported but are considered exceptional.

A 4-week-old chick. Saint-Lambert, near Montréal. *July 1933, Henry Mousley, Laval University Collection.*

Least Bittern

Breeding distribution
1984–1989

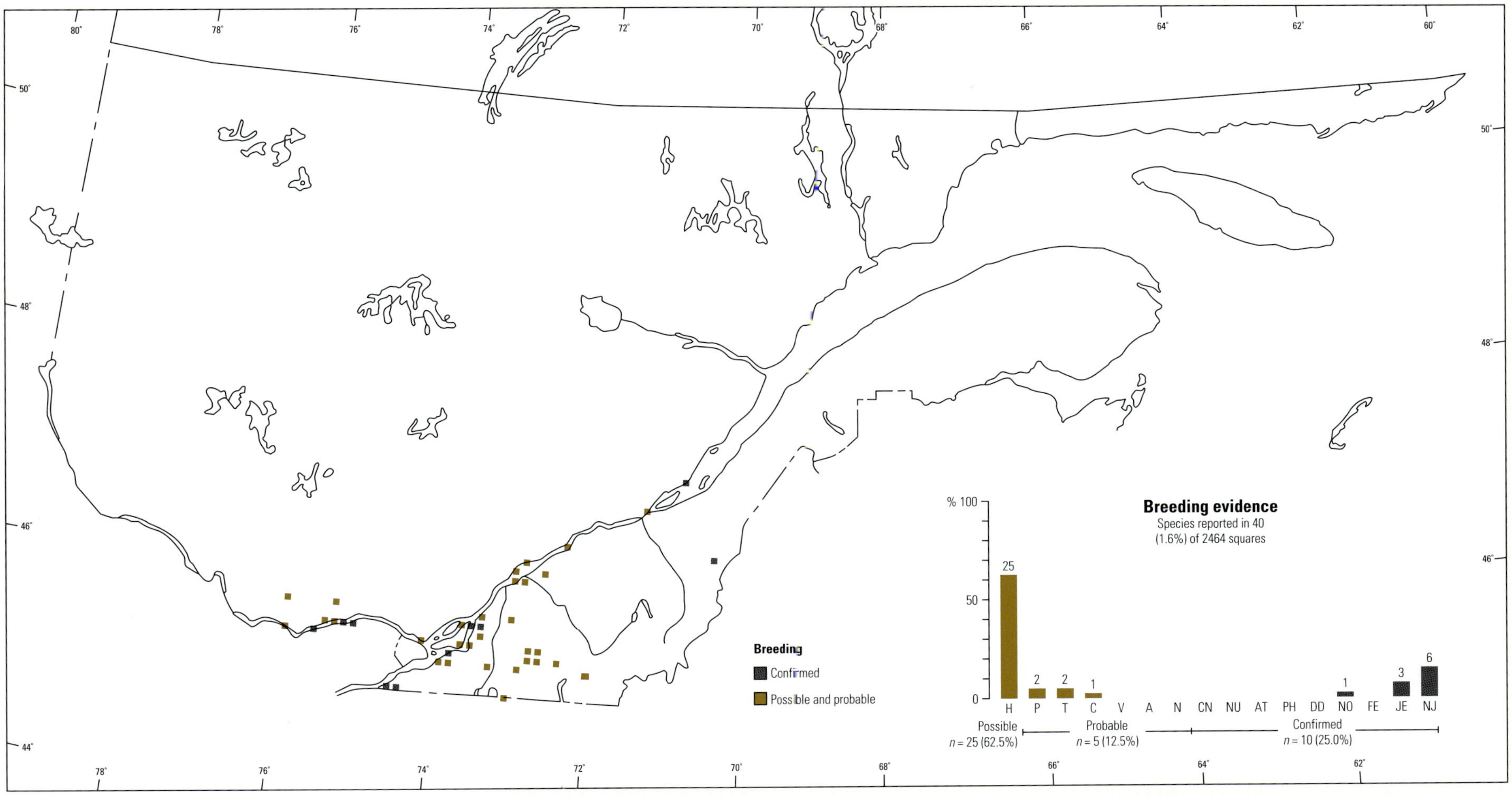

Species profile

Clutch size	usually 4–5 eggs (2–7)[2,3,4,6,7]
Laying interval	usually 1 egg per day[1,2,5,6,7]
Start of incubation	with first or second egg[6]
Incubation period	17–20 days[6,7]
Maturity at hatching	semi-altricial
Care of young	by female and male[7]
Nestling period	usually 13–15 days (6–15)[5,7]
Age at first flight	?
Dependence of young	until at least 26 days old[5]
Broods per year	1, sometimes 2[2,4,5,6,7]
Breeding age	?
Mating system	probably monogamy[5]
Duration of pair bond	probably 1 season[5]
Total length	28.0–35.5 cm[3]
Wing span	40.6–45.7 cm[4]
Weight	? g (42.5–113.4)[5]
Longevity record	?

References:
1. Kibbe 1985f
2. Weller 1961
3. Godfrey 1986
4. Terres 1980
5. Palmer 1962
6. Harrison 1978
7. Gibbs et al. 1992a

Chronology $n = 12$

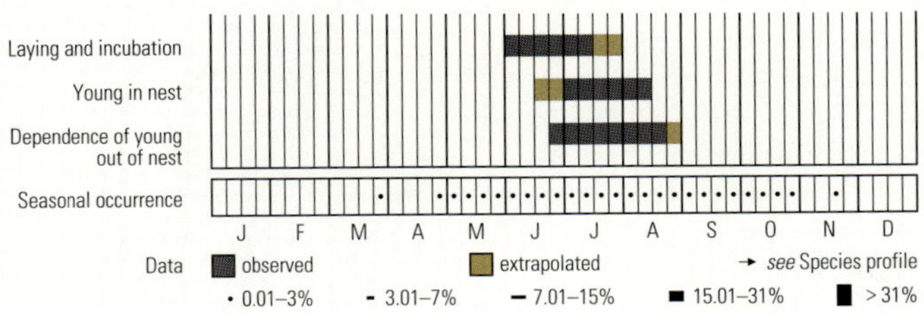

World breeding range

The Least Bittern, which breeds locally in southwestern Québec, may nest along sluggish streams edged with cattails. These serve both as camouflage and nest material. Saint-Jacques River, La Prairie, near Montréal. *June 1991, Gaétan Duquette.*

Great Blue Heron

Grand Héron

Ardea herodias

Clodin Roy

THE GREAT BLUE HERON IS QUÉBEC'S LARGEST HERON AND THE MOST widely distributed ardeid in North America (Butler 1992). It is not unusual to see this large wader during the breeding season foraging in shallow salt or fresh water along the St. Lawrence and along rivers in populated areas or near remote lakes right up into the boreal forest. It is particularly abundant around broad low-lying stretches of water with plenty of fish (DesGranges and Darveau 1985, 1988). Its size (overall length up to 1.3 m, standing height about 60 cm), its predominantly blue, grey, and white coloration, and especially its long neck, folded back in flight (Butler 1992), readily distinguish the Great Blue Heron from any other fish-eating birds in Québec, such as the Black-crowned Night-Heron, Double-crested Cormorant, or Osprey. It often feeds alongside these species and may even share with them its nest colony, or heronry (DesGranges 1980a).

Exceptionally, solitary individuals may be spotted in winter near stretches of open water in the southernmost part of the province (Gosselin et al. 1990b; Cyr and Larivée 1995). The great majority, however, disperse after breeding, heading initially in all directions (Butler 1992), then south as far as the northern parts of South America (Palmer 1962; Hancock and Kushlan 1989).

Habitat and behaviour

Heronries are nearly always located near feeding grounds (Gibbs 1991; Benoit 1991; Benoit et al. 1993a; Butler 1995), in wooded areas hard for people and ground predators to penetrate (Werschkul et al. 1976), and where the birds have a commanding view. Herons habitually seek out wooded islands, shoreline woodlands that are flooded in the spring run-off, and dead trees in beaver ponds (DesGranges 1979). Heronries are always an impressive sight, with their many huge nests (0.5–1.2 m in diameter) made of interwoven dry branches, generally built quite close to each other in the tops of trees, mostly hardwoods (DesGranges 1979). The comings and goings, displays, and quarrelling of the adult birds, accompanied by raucous cries, are a spectacle to behold (Meyerriecks 1960; DesGranges 1981a; Butler 1992), as is the sight of so many perennial nests high in the trees, all the huge birds busy night and day (McNeil et al. 1993), tending their insatiable offspring.

Generally, the Great Blue Heron is monogamous, though the pair bond lasts only a single season (Simpson et al. 1987). Clutch size is from 3 to 7 eggs, the number increasing with the latitude of the nest site (Palmer 1962; Hancock and Kushlan 1989), though 4 or 5 is the usual clutch size for Québec breeding pairs. After about a month of incubation by both adults and 10 or so weeks of diligent parenting, 2 or 3 young fledge (DesGranges 1979). At the peak of its growth spurt, a healthy chick can consume up to six meals per 24-hour period, the diet consisting, for adult and young alike, of small fishes less than 25 cm long (Mock et al. 1987a), though sometimes longer (Willard 1977), occasionally supplemented by frogs or large aquatic insects (DesGranges 1981b; Butler 1992; McNeil et al. 1993). If food is in short supply and the combined fishing success of the two parents is not enough to satisfy the growing appetite of their brood, only the stronger chicks survive. Malnourished heron chicks, usually the youngest, gradually weaken, to the point where they fall out of the nest, buffeted by their more vigorous siblings in their determination to monopolize the food brought to the nest. The parents stand impassively by as these fratricidal battles proceed and show not the slightest concern for chicks that are sick (Mock et al. 1987a).

Distribution

The Great Blue Heron is a migrant breeder usually found in Québec from March until early December. Although breeding pairs were reported throughout southern Québec (225 Atlas squares), nevertheless the majority of the 500 or so heronries in Québec are located in the hardwood forests of the southwest of the province (DesGranges and Laporte 1983; Bélanger and Tremblay 1989; Desrosiers 1993; CWS and Hydro-Québec, unpubl. data). These colonies vary in size from year to year, generally comprising several dozen pairs, though there are 18 known colonies over 50 pairs-strong, mainly along marshy stretches of the riverine lakes, the estuary, and the gulf. A heronry in Lake Saint-Pierre (a riverine lake) is thought to be the largest known colony of this species in the world, with over 1000 breeding pairs. The typical number of nests drops to about 10 per colony in the Ottawa Valley, even though this region has a third of Québec's known heronries. At the northern limit of the breeding range (beyond the area covered by the Atlas), in the James Bay lowlands as far as the 53rd parallel (Lake Kapwakami), colonies are smaller, many of them (40% of the 45 heronries in the region) having fewer than 4 breeding pairs (Bordage and Aubry 1982; Gosselin et al. 1991b; Cons. G. G. – GREBE 1992f). Occasionally (12% of heron nesting sites in Québec), solitary breeding pairs

Notes on territory

Only males defend territories, which are usually limited to the immediate vicinity of their own nests (DeGraaf et al. 1980; Hancock and Kushlan 1989). On their return in spring, herons head for their colony, sometimes even for their old nests (Simpson et al. 1987), where they engage in courtship displays. This is when most aggression is shown between rival males, both in the colony and on the feeding grounds (Hancock and

In southwestern Québec, most known heronries are in deciduous stands, like this one on Dickerson Island. Lake Saint-François, southwest of Montréal. *April 1984, Jean-Luc DesGranges.*

Great Blue Heron

Breeding distribution
1984–1989

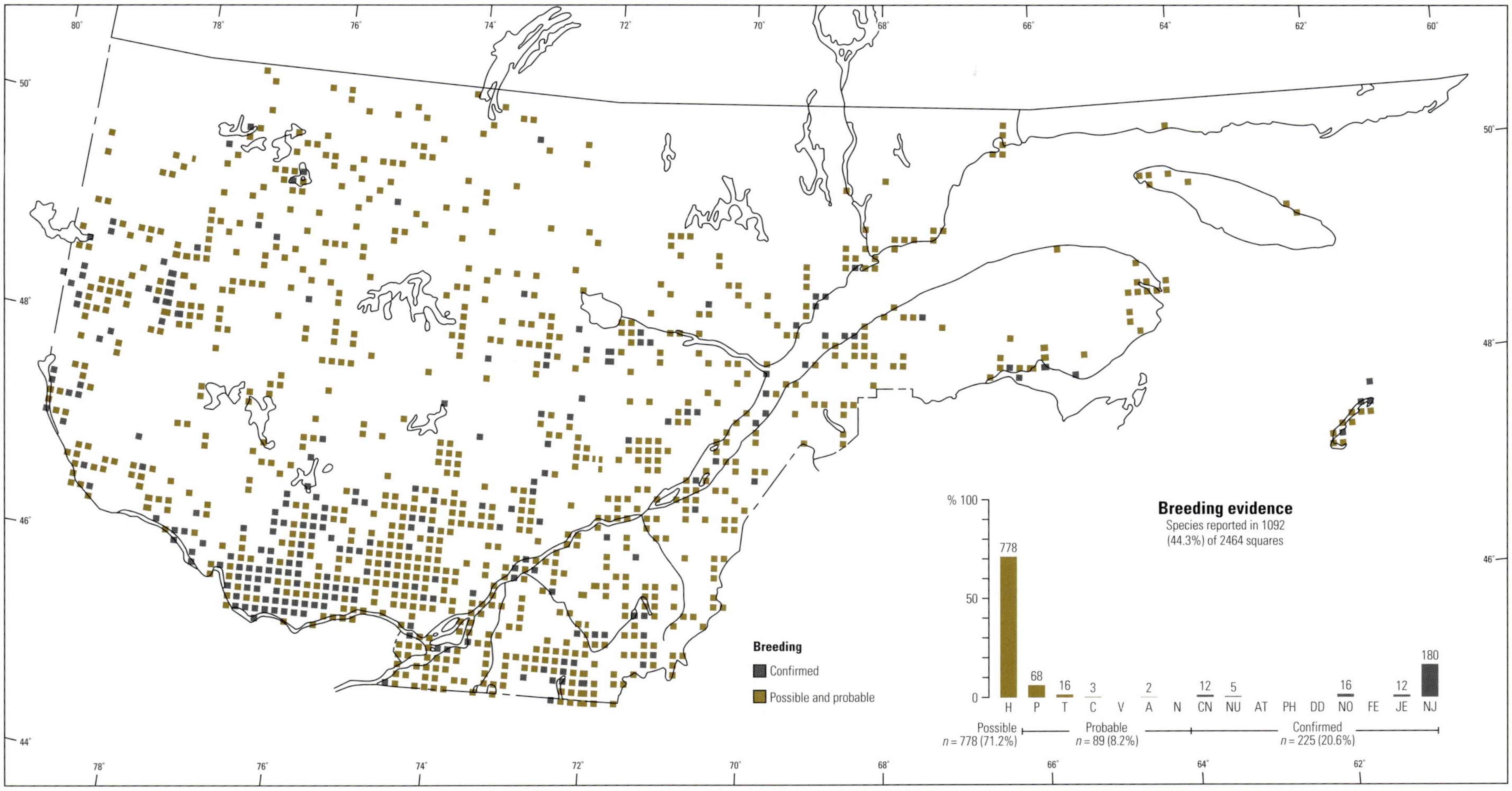

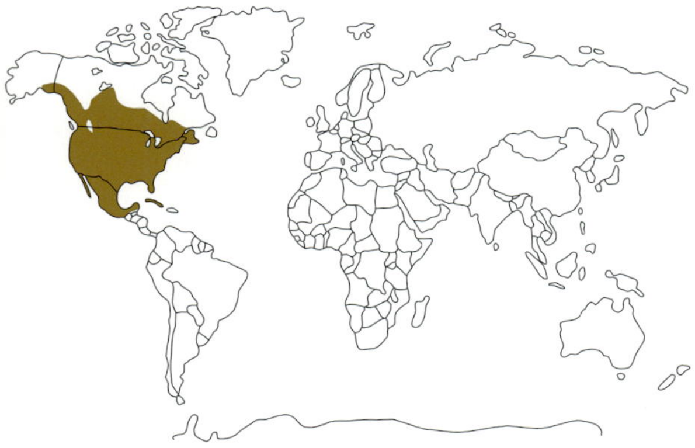

World breeding range

Species profile

Clutch size	usually 4–5 eggs (3–7)[1,4,5]
Laying interval	1 egg per 2–3 days[7]
Start of incubation	with first egg[7]
Incubation period	usually 27 days (25–30)[1,2,3,4,5,7]
Maturity at hatching	semi-altricial
Care of young	by female and male[1,2,4]
Nestling period	usually 81 days (64–91)[7]
Age at first flight	usually 60 days (51–81)[5,7]
Dependence of young	until 3 weeks after first flight[7]
Broods per year	1[1,3,5]
Breeding age	usually 2 years[5]
Mating system	monogamy[1,5,7]
Duration of pair bond	1 season[5,7]
Total length	97–137 cm[7]
Wing span	183.0–213.4 cm[4,5,8]
Weight	M 2576 g (?) / F 2204 g (?)[6]
Longevity record	23 years 3 months[9]

References:
1. Harrison 1978
2. Ehrlich et al. 1988
3. Godfrey 1986
4. Terres 1980
5. Palmer 1962
6. Dunning 1984
7. Butler 1992
8. National Geographic Society 1987
9. Clapp et al. 1982

Chronology $n = 215$

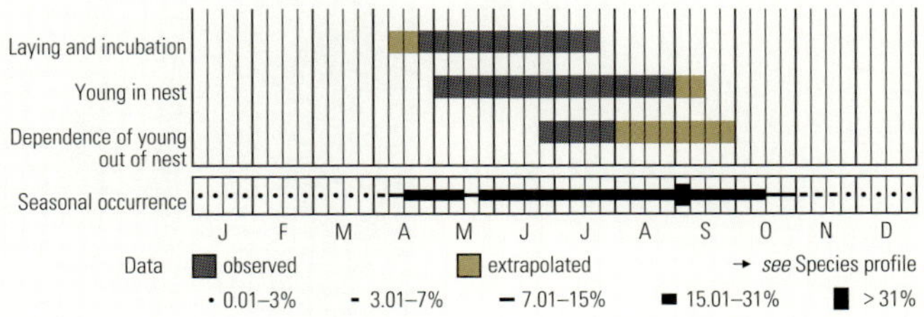

Inland colonies are generally smaller but more numerous than those on the St. Lawrence, as is the case with this heronry located in a pine stand on Oster Island in Lake Timiskaming. *July 1986, Yves Aubry.*

are found. The current late summer Great Blue Heron population of Québec is estimated at some 25 000 individuals (CWS, MLCP, and Hydro-Québec, unpubl. data). Since there are so many of these large birds frequenting open areas, it is hardly surprising that the species was reported in nearly half of all the squares atlassed. Breeding behaviour accounted for 20% of these records. In fact, the Great Blue Heron is the easiest wader to observe throughout southern Québec, as far east as the Mingan Islands, Anticosti Island, and the Magdalen Islands, and certainly, of all the ardeids, its heronries are the easiest to find.

History and trends

This species was doubtless already common in the southern part of the province at the very beginning of the colonial period, as can be seen from Champlain's reports (Laverdière *in* Ouellet 1974a). At the turn of the century, the species was common around both Montréal and Québec City, especially in the autumn (Wintle 1896; Dionne 1906), though it was rare in the lower estuary (Tanguay 1964). Over the last 25 years, population data from Québec (DesGranges and Laporte 1983; Cyr and Larivée 1993b), Ontario (Dunn 1987b), the Maritimes (Erskine 1992), and bordering U.S. states (Norse and Fichtel 1985b; Peterson 1988a) all point to a relatively stable population. In Québec, though some traditional nesting sites have been abandoned, the Great Blue Heron is known to have expanded its breeding range, and there has been a slight increase in relative colony size (CWS, MLCP, and Hydro-Québec, unpubl. data).

Even when undisturbed by humans or predators, Great Blue Herons periodically abandon their colonies and settle elsewhere. In fact, their very presence eventually renders sites unsuitable for breeding. Over the three years that she studied a heronry in a riverside woodland in the Montréal area, S. David (pers. comm.) noted that 6% of the trees had fallen and that use of other trees had dropped by 30% (10% per year) because nests had collapsed. The birds' droppings, falling on foliage or on the ground, block the normal processes of photosynthesis and transpiration. In a study of a coniferous site, Julin (1986) noted abnormal loss of needles and degradation of the soil, leading to deterioration of the nest trees. Within a few years, a general decline occurs in the forest supporting a heronry (Kerns and Howe 1967; Weseloh and Brown 1971; Wiese 1978; Dusi 1979). Short-lived trees growing in shallow soil are more susceptible to root damage and blowdown than are longer-lived species (Scharf 1989).

Notes on territory (continued)

Kushlan 1989; Butler 1992). Once pairs have formed and laying has begun, territorial behaviour drops off, and the adults go about their breeding and feeding activities in peace. Fishing is usually done well within a 30-km radius of the colony (Butler 1992; Benoit et al. 1993a).

After several years of use, the site of a Great Blue Heron colony deteriorates. Also, broken branches and fallen nests gradually force the birds to colonize a new site nearby. Le Gros Pèlerin, St. Lawrence estuary. *June 1977, Jean-Luc DesGranges.*

A pair on their nest in the top of a spruce tree. The huge nest consists of a tangle of dry branches interwoven with those of the tree. Île aux Basques, St. Lawrence estuary. *July 1968, Ronald Lepage.*

On La Grande Île, site of the largest colony in Québec, three nestlings await the return of their parents with food. Berthier Islands, west of Lake Saint-Pierre. *June 1981, Daniel Jauvin.*

There therefore has to be more suitable habitat for nesting nearby; preserving an existing heronry is not enough. When looking for an alternative nesting site, herons usually choose the same type of tree, whether deciduous or coniferous, in which to build. Often, the site chosen is near their former feeding grounds. The new colony will have a better chance of success if, in addition to being near feeding grounds and safe from disturbance (Forbes et al. 1985), it is located in an extensive habitat, with trees of uniform height, density, and species.

The Great Blue Heron is a hardy bird that can live for more than 15 years. Once it reaches adulthood, it has very few enemies, except for alligators on its wintering grounds, a few hunters, and some fish farm operators (CWS, unpubl. data), who certainly kill them from time to time. As with other species, though, mortality is very high in the early stages of life: 69% in the first year, 36% in the second, and 22% thereafter (Henny 1972). Many eggs and young fall prey to crows, ravens, gulls, raptors, raccoons, and even black bears (P. Dupuy, pers. comm.), especially where the adults are forced to leave the nest unattended because of a disturbance in the colony (Bent 1926; DesGranges 1980a; Butler 1992). In places, pesticides and contaminants may have caused deaths (Ohlendorf et al. 1981), but the data so far tend to indicate that, on the whole, Québec Great Blue Heron populations have probably not suffered on this account (Laporte 1982; DesGranges and Laporte 1983). The steady shrinking of wetland areas in the southwesternmost part of the province (Lands Directorate 1986), ongoing clearing of the few remaining wooded islands there, and increasingly frequent disturbance during breeding are much more serious threats to the maintenance of populations in this region, which is home to the largest heronries in Québec.

Jean-Luc DesGranges

Great Egret

Grande Aigrette

*Casmerodius albus**

Ghislain Caron

THE CONSPICUOUS GREAT EGRET, EASILY RECOGNIZED BY ITS ALL-WHITE plumage, is the most cosmopolitan of the heron family, breeding in all parts of the world. In the late 1970s, it was a casual visitor to Québec, with 75 records of occurrence (David 1980b). In 1982, the Great Egret was observed throughout the summer in the Dundee area in southwestern Québec, raising the possibility that it might be nesting there (Yank 1982). But it was not until the first year of atlassing, in 1984, that three nests were discovered on Dickerson Island, near the junction of the Québec, Ontario, and New York State borders (Bannon 1984). Since then, many active nests have been observed yearly, with more than a dozen individuals seen each spring. They arrive on Dickerson Island, the most northerly breeding colony in eastern North America, from their wintering sites around the third week of April.

Habitat and behaviour

The Great Egret can be found in a variety of habitats, from marshes, wet meadows, swamps, river margins, lake shorelines to drainage ditches and pastures in agricultural areas; it usually nests nearby in trees or bushes (Palmer 1962; Hancock and Kushlan 1984).

The Great Egret is colonial, and breeds on islands with other colonial heron species such as the Great Blue Heron, Black-crowned Night-Heron, and Green Heron (Palmer 1962; Krebs 1978; Ehrlich et al. 1988). In mixed colonies, the largest species nest in the canopy, while the smaller species nest lower down. Great Egrets and Great Blue Herons tend to nest in open or exposed areas, very often in the top of the vegetation, with the Great Blue Heron nesting highest when present. When the Great Blue Heron is absent, however, the Great Egret's nest is the highest. Likewise, smaller egrets and herons nest higher in the vegetation when no larger species is present, suggesting selective advantages to nesting high (Adams et al. 1973; Burger 1978, 1979; Spendelow et al. 1989). In 1992, the Great Egrets on Dickerson Island nested in the open on top of tall hawthorn thickets (J.-L. DesGranges, pers. comm.).

Prior to pair formation, the male may reclaim its former nest or build a new one. During this time, the male's territorial display, with the bill pointed skyward and all the long aigrettes erected, makes the egret very conspicuous, except when high winds and low temperatures inhibit these displays (Palmer 1962; Mock 1978). For the first 4–7 days after pair formation, the pair bond is fragile, and about 10–20% end in "divorce" (Mock 1978).

Incubation of the 3–5 eggs by both parents lasts about 25 days, and the eggs generally hatch 1–2 days apart (Mock 1985). Both parents feed the young egrets; their diet consists mainly of fish, but may include insects, amphibians, small mammals, and some birds (Hancock and Kushlan 1984). When food is abundant, all the young survive. When food is scarce, however, the smaller late-hatched chicks are often killed by their older siblings (siblicide) or else starve

Notes on territory

Great Egrets usually nest in colonies, often with 4 or 5 nests per tree, and sometimes up to 50. The nests are sometimes touching each other. After an initial period

* The scientific name of the Great Egret, *Casmerodius albus*, has recently been changed to *Ardea alba* (AOU 1995).

Dickerson Island is the most northerly colony of Great Egrets in eastern North America, with almost a dozen nesting at the top of tall hawthorns. A Great Egret beside two nests. Lake Saint-François, southwest of Montréal. *April 1984, Jean-Luc DesGranges.*

Great Egret

Breeding distribution
1984–1989

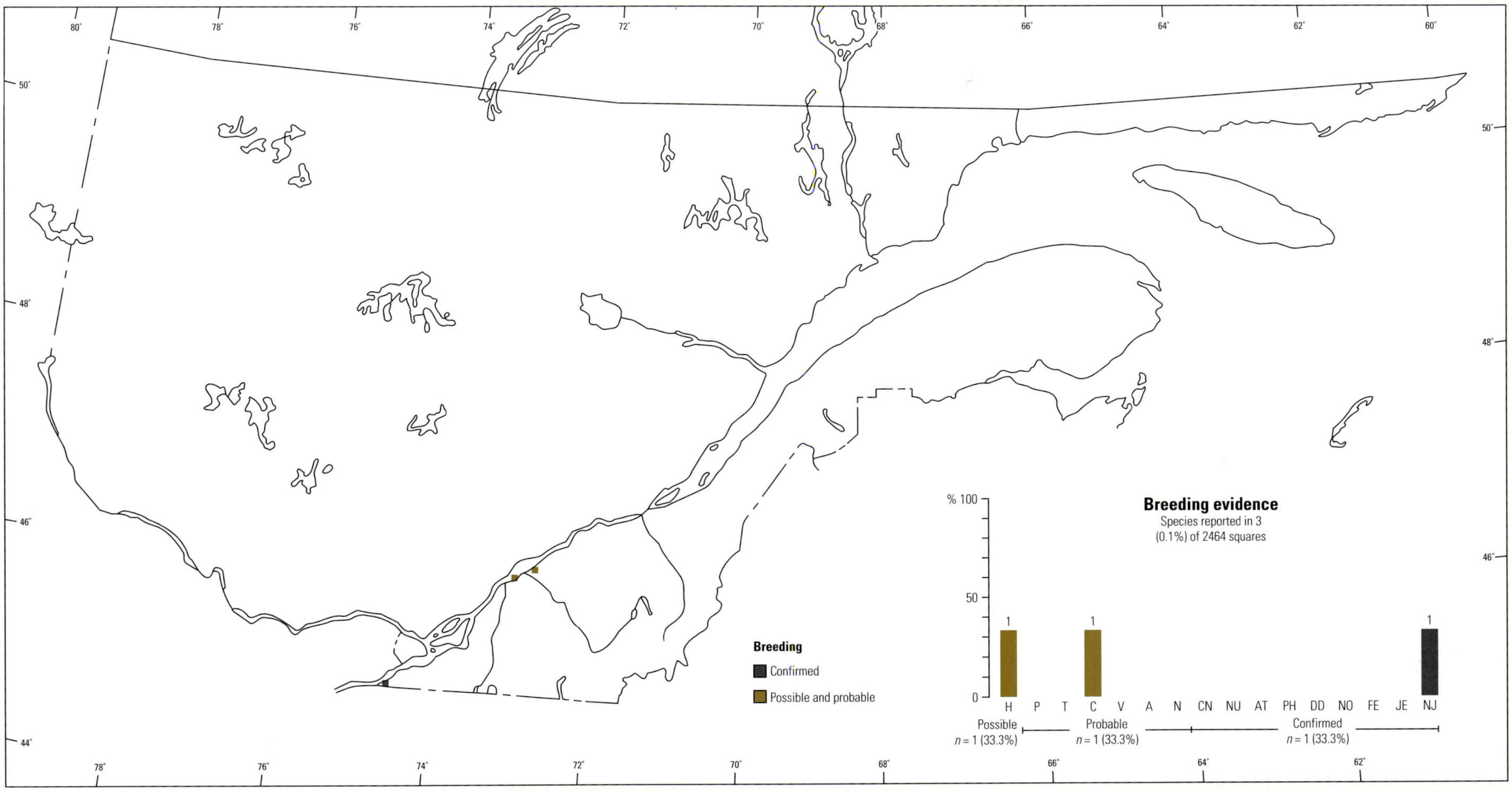

Species profile

Clutch size	usually 3–5 eggs (1–6)[1,4,5,6,8]
Laying interval	1 egg per 2 days[5]
Start of incubation	with third egg, sporadic before[5]
Incubation period	23–26 days[1,5,6,7,8]
Maturity at hatching	semi-altricial
Care of young	by female and male[1,5]
Nestling period	≈ 20 days[4,5,7]
Age at first flight	42–49 days[1,5,6]
Dependence of young	until autumn[5]
Broods per year	1[1,5]
Breeding age	probably 2 years[5]
Mating system	monogamy[1,5]
Duration of pair bond	probably 1 season[5]
Total length	88–107 cm[3]
Wing span	139.7 cm[6]
Weight	M 935 g (?) / F 812 g (?)[9]
Longevity record	22 years 10 months[2]

References:
1. Ehrlich et al. 1988
2. Kennard 1975
3. Godfrey 1986
4. Harrison 1978
5. Cramp and Simmons 1977
6. Terres 1980
7. Hancock and Kushlan 1989
8. del Hoyo et al. 1992
9. Dunning 1984

Chronology n = 5

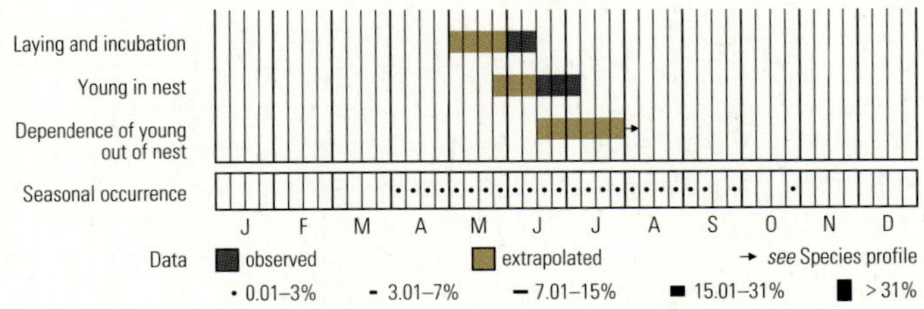

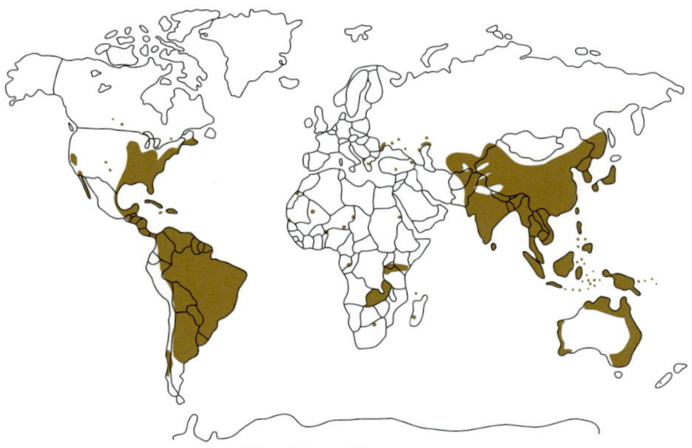

World breeding range

because they cannot compete for food (Mock et al. 1987a). Surviving young egrets wander from the nest at 3 weeks of age and can fly at 6 weeks.

Between 1987 and 1990, 3 to 5 active nests were observed on Dickerson Island. In 1985, 6 nests produced an average of 1.33 chicks/nest (Bannon 1985b), while in 1990, 3 nests produced an average of 2.67 chicks. Since the end of the Atlas survey period, the number of Great Egrets nesting on Dickerson Island has substantially increased. In 1991, an extensive ground survey by boat and an aerial survey revealed that the number of successful nests had increased to 11, producing an average of 3.09 chicks. The number of nests remained stable in 1992 and 1993. It is highly possible that the number of birds nesting during the Atlas survey period was higher than found, as the egrets were nesting in the interior of the island; they are very difficult to observe by boat and impossible to see from land.

Distribution

The Great Egret was reported in 3 Atlas squares, with breeding confirmed in one square. Individuals have also been observed elsewhere in Québec, such as on the Sorel Islands, at Cacouna, Rimouski, Chaleur Bay, along the North Shore (ÉPOQ), and even on the Magdalen Islands (Fradette 1992). These scattered records may be due to extensive post-breeding dispersal prior to fall migration as well as to the egret's tendency to wander widely throughout the year (Peck 1987a). Therefore, these records of occurrence are probably of wandering, non-breeding individuals. Spring sightings of Great Egrets are often in close proximity to active Great Blue Heron colonies.

In spring, breeding egrets concentrate around the Dickerson Island area and feed in surrounding marshes. They are often seen feeding in the Dundee region, and also within the nearby Cooper Marsh Conservation Area in Lancaster, Ontario. In Ontario and New York, all nestings of egrets occurred at established heronries, indicating that the presence of active heronries may be a factor in inducing the egret to nest (Peck 1987a; Peterson, D. M. 1988). Particularly noteworthy are the yearly sightings of Great Egrets near the La Grande Île (Berthier) heronry (the largest heronry in Québec, located in the Lake Saint-Pierre region) and, in 1990, the observation of a pair of egrets in courtship display there; this could indicate an interest in this colony as a future nesting location (Aubry et al. 1990b). The presence of established heronries comprising Great Blue Herons and Black-crowned Night-Herons may induce Great Egrets to expand their breeding range in Québec. Birders should therefore look carefully at colonies of those two species for nesting Great Egrets (Scharf 1991).

History and trends

In 1896, the Great Egret was considered a rare fall "transient visitant" in the Montréal region, with several specimens taken at Beauharnois and at Île aux Noix (Wintle 1896). In 1882 and in 1937 respectively, the species was recorded in the Godbout River and near Harrington Harbour, but Todd (1963) concluded that those individuals had outrun their normal northern limits during spring migration.

In the early 1900s, the plume trade was a lucrative business in the United States, with egret plumes fetching $32 an ounce in 1903, and later commanding $80 an ounce. Egrets were taken at the start of the breeding season, when their long aigrettes were at their best. Since this almost decimated the population, with the low ebb occurring in 1902–03, the egret was adopted as the symbol of the bird preservation movement (Palmer 1962; Ehrlich et al. 1988).

Notes on territory (continued)

of flying around the colony area, individual males begin to defend potential nest sites vigorously. The male establishes a breeding territory that is used for pair formation, courtship, copulation, and nesting. The territory is later reduced to a 1 m² area defended

Heronries seem to have a particular attraction for Great Egrets as nesting sites. This heronry on La Grande Île in the Berthier Islands could become the next breeding site as egrets have been sighted there for some years now. West of Lake Saint-Pierre. *June 1991, Jean-Luc DesGranges.*

aggressively by the pair against conspecifics (Cramp and Simmons 1977).

Small groups of Great Egrets may roost occasionally away from the colony near feeding grounds (Cramp and Simmons 1977). Some individuals feed solitarily and defend a feeding territory by chasing conspecifics away, while others feed in groups and display less aggression (Wiggins 1991).

The Great Egret has established itself in the centre of Dickerson Island. Other species such as the Great Blue Heron, Black-crowned Night-Heron, and Green Heron also nest there. Lake Saint-François, southwest of Montréal. *June 1992, Sharon David.*

With legal protection assured under the Migratory Birds Convention Act, passed in 1917, the North American population began to recover. Great Egret numbers have been increasing, but have fluctuated since the 1930s as a result of habitat loss, human disturbance, and eggshell thinning due to build-up of organochlorine residues in bird tissues in the late 1960s and early 1970s (Chapman 1932; Palmer 1962; Pratt 1972; Thompson 1978). The present northward range expansion has outstripped the historical northern limit of New Jersey (Ogden 1978). Some factors affecting egret populations, colony success, and expansion are loss of nesting and feeding habitat due to marsh drainage, development, and drought, human disturbance within colonies, predation (Pratt and Winkler 1985), and occasionally, natural disasters (Shepherd et al. 1991).

SHARON DAVID

Great Egrets feeding in the marshes impounded for waterfowl management by Ducks Unlimited in the Lac Saint-François National Wildlife Area. Southwest of Montréal. *July 1988, Jean-Luc DesGranges.*

Green Heron

Héron vert

Butorides virescens

ANYONE WHO HAS SPENT SOME TIME ON QUIET PONDS FRINGED WITH TREES and shrubs in southern Québec will recognize the Green Heron's harsh squawks, often rendered as *skeow*. Smaller than a crow, the Green Heron is one of the most diminutive members of the heron family in North America. It is a beautiful bird when seen at close range, with its yellow legs, rich chestnut neck, and dark green crown and mantle. When alarmed, it raises its handsome crest and jerks its tail. The male and female are identical, although the male is slightly larger. Today the Green Heron is considered as a fairly common migrant breeder in southern Québec, the same status given by David (1980a).

Habitat and behaviour

The Green Heron uses a wide range of wetland habitats, both saltwater and freshwater (Palmer 1962). In Québec, it is found mainly in freshwater habitats, near ponds and sluggish streams, where tangled thickets provide cover. Alder thickets on flooded land may also be used (Peck and James 1983; Godfrey 1986).

The species is generally a solitary nester, but may nest in small heronries with conspecifics or other ardeids (Meyerriecks 1960; Link 1980; Hiemenz 1981; Peck and James 1983; Kaiser and Reid 1987). The nest, a flimsy structure of sticks and twigs and usually located 3–6 m from the ground or sometimes directly on the ground, may be used for several years in a row (Ouellet 1974a; Godfrey 1986). Green Herons are monogamous, and both male and female participate in building the nest, incubating the eggs, and feeding the chicks. Courtship behaviour, consisting of short flights and grunts, clucks, and other cries performed on the nesting territory, occurs over a period of a few days (Palmer 1962). Before pair formation, the male alone defends his territory against conspecifics; after the pair bond is established, the female also helps to defend the nest, mainly against predators such as raccoons, grackles, and crows (Palmer 1962; Hancock and Kushlan 1984). The adults brood the newly hatched chicks almost constantly during the first few days and feed them by regurgitating food directly into their open mouths or by dropping food into the nest. At the age of 1 week, the chicks are already agile climbers, using their feet, wings, and bills to hoist themselves up the tree to escape danger. They may also regurgitate the contents of their crop on intruders (Bent 1926; Palmer 1962).

Clodin Roy

Notes on territory

Before pair formation, the male defends the territory against conspecifics of both sexes. After copulation, the female assists in

The Green Heron frequents the sheltered banks of sluggish streams and rivers, sometimes hunting from a perch overhanging the water. Lac Saint-François National Wildlife Area, southwest of Montréal. *September 1974, Léo-Guy de Repentigny.*

Green Heron

Breeding distribution
1984–1989

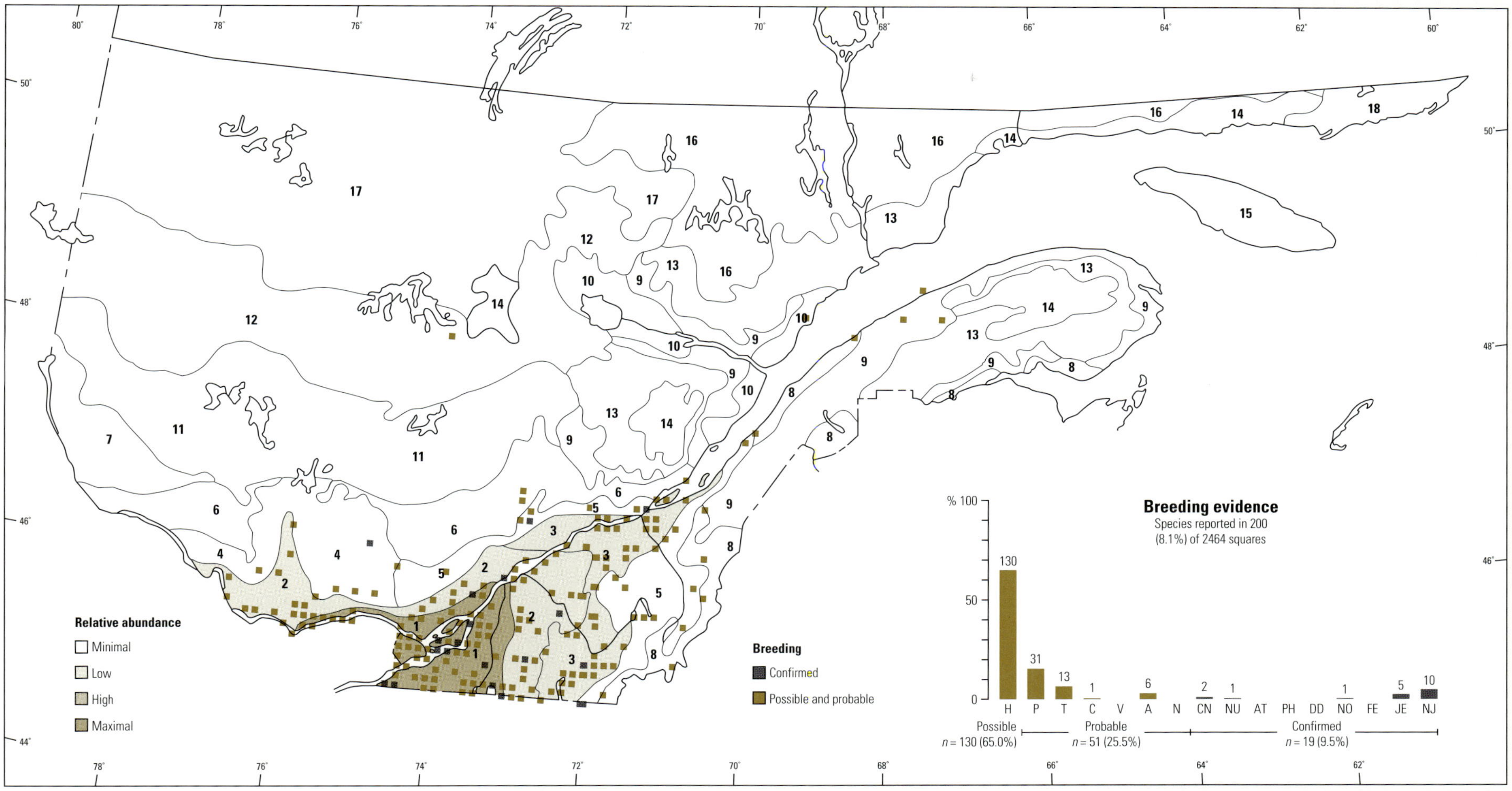

Species profile

Clutch size	usually 4–5 eggs (3–6)[1,3,5,7,9]
Laying interval	1 egg per 1–2 days[4]
Start of incubation	usually with second egg[13]
Incubation period	usually 19–21 days (17–25)[2,5,7,8]
Maturity at hatching	semi-altricial
Care of young	by female and male[4,5,8]
Nestling period	16–17 days[2,4,8,9]
Age at first flight	21–23 days[4,8]
Dependence of young	probably until 30–35 days old[8]
Broods per year	1, rarely 2[2]
Breeding age	probably 2 years, sometimes 1[13]
Mating system	monogamy[3]
Duration of pair bond	1 season[8]
Total length	40.6–55.8 cm[11]
Wing span	66 cm[6]
Weight	212–241 g (?)[12,13]
Longevity record	7 years 11 months[10]

References:
1. Peck and James 1983
2. DeGraaf and Rudis 1987
3. Link 1980
4. Kaiser and Reid 1987
5. Bent 1926
6. Terres 1980
7. Hancock and Kushlan 1984
8. Palmer 1962
9. Bull 1974
10. Clapp et al. 1982
11. Godfrey 1986
12. Dunning 1984
13. Davis and Kushlan 1994

Chronology $n = 16$

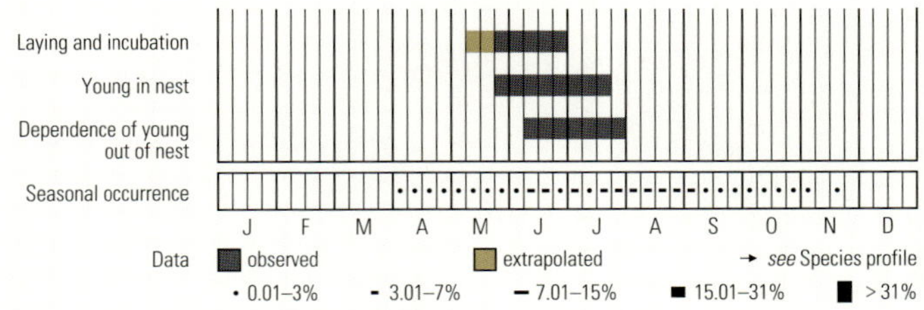

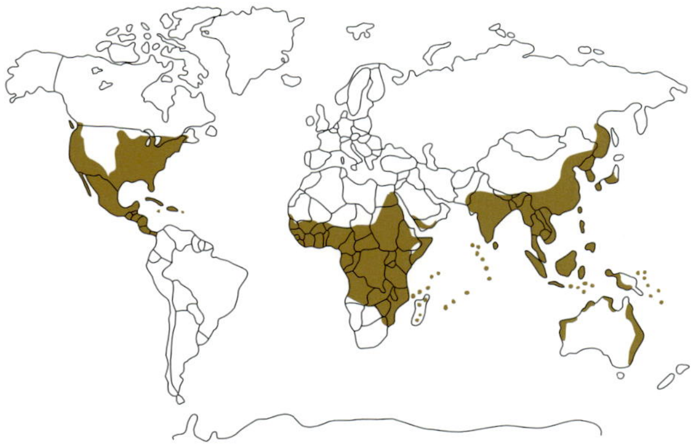

World breeding range

Frequency of occurrence by ecological region

Ecological regions** (forest regions)	Adequately covered squares	Squares in which reported	% of squares in which reported	Relative abundance (%)*
1. Sugar maple-hickory and sugar maple-basswood	97	57	58.8 ± 4.9	100.0
2. Sugar maple-basswood	153	53	34.6 ± 3.9	58.9
3. Sugar maple-basswood and sugar maple-yellow birch	101	31	30.7 ± 5.5	52.2
4. Sugar maple-yellow birch-basswood	56	6	10.7 ± 7.4	18.2
5. Sugar maple-yellow birch-beech and sugar maple-yellow birch-basswood	113	15	13.3 ± 3.6	22.6
6. Sugar maple-yellow birch-beech	71	7	9.9 ± 6.6	16.8
7. Sugar maple-yellow birch-balsam fir	19	0	0	0
8. Sugar maple-yellow birch and balsam fir-yellow birch	47	5	10.6 ± 8.3	18.1
9. Balsam fir-yellow birch	87	2	2.3 ± 3.3	3.9
10. Balsam fir-white birch or balsam fir-red maple	46	0	0	0
11. Yellow birch-balsam fir	52	0	0	0
12. Balsam fir-white birch (west)	59	1	1.7 ± 4.0	2.9
13. Balsam fir-white birch (east)	76	0	0	0
14. Balsam fir-black spruce	28	0	0	0
15. White spruce-balsam fir and balsam fir-white spruce	22	0	0	0
16. Black spruce-balsam fir-moss	18	0	0	0
17. Black spruce-moss	32	0	0	0
18.			*not analysed*	

* minimal: 0–29.9%; low: 30–59.9%; high: 60–89.9%; maximal: 90–100%

** Deciduous forest: 1–7; mixed forest: 8–11; coniferous forest: 12–17

Streit Pond, a calm pond bordered by trees and shrubs, is an ideal nesting and foraging site for this heron. Philipsburg, south of Saint-Jean-sur-Richelieu. *July 1991, Pierre Bannon.*

Notes on territory (continued)

defending the territory, particularly in fighting off predators like crows and grackles. The size of the territory depends on the habitat and the stage in the breeding cycle. Before copulation, the area defended may measure about 20 m in diameter, but afterward, may shrink to a radius of 1 or 2 m around the nest. Although the species is usually a solitary nester, it may nest in small colonies of a dozen or so pairs (Meyerriecks 1960). According to Kaiser and Reid (1987), the degree of sociability shown during nesting may be linked to the spatial distribution of food sources. The Green Heron sometimes also defends a feeding territory (Meyerriecks 1960).

Hedgerows along this pond offer good nesting habitat. Sainte-Foy. *September 1991, Yves Aubry.*

The Green Heron's nest is a fairly rudimentary platform of interwoven twigs set in a tree. The first breeding record for the Green Heron in Québec was reported by Henry Mousley at Sainte-Rose, Laval in 1942. *May 1942, Henry Mousley, Laval University Collection.*

The Green Heron feeds on a variety of small fish, amphibians, reptiles, crustaceans, molluscs, and aquatic and terrestrial insects (Palmer 1962), usually foraging in the morning and late afternoon (Bent 1926). It fishes by standing motionless, waiting to pounce on its prey as it comes by, or by stalking slowly through shallow water (Kushlan 1976). It may also place an object on the surface of the water as bait to attract fish (Higuchi 1988).

The birds tend to disperse after the breeding period, when they can be seen outside their usual range (Hancock and Kushlan 1984).

The Green Heron was recorded in only 8.1% (200/2464) of Atlas squares surveyed, and breeding was confirmed in only 9.5% (19/200) of them, mainly through reports of nests. Its retiring behaviour and cryptic coloration make it difficult to spot while it is hunting or near the nest. It is most easily identified in flight, thanks to its distinctive silhouette.

Distribution

In Québec, the Green Heron is found mainly along the wooded shores of rivers and streams in the Central St. Lawrence Lowland. It has also been observed in the Laurentians, Ottawa Valley, and the foothills of the Appalachians; here again lakes, rivers, and streams fringed with trees and shrubs seem to provide suitable nesting habitat, whereas oligotrophic bodies of water do not appear to furnish the diversity and abundance of prey it requires to establish itself.

History and trends

According to Palmer (1962), the Green Heron has expanded its breeding range northward. Whereas early this century Macoun and Macoun (1909) described it as a scarce summer resident in Montréal, today it is considered fairly common there. In midcentury, Mousley (1945a) found that sightings were increasing, "with a gradual infiltration of the species...taking place round the district of Montreal." Godfrey (1986) noted that its breeding range had expanded along the St. Lawrence River between 1966 and 1986. Given the bird's reclusive behaviour, however, and the growing popularity of birdwatching in the last few decades, the Green Heron may have been more common in Québec at the beginning of the century than ornithologists at the time thought (Goodwin 1987b).

Potential limiting factors include the draining of wetlands and diversion of streams for agriculture, as well as the clearing of trees and shrubs along ditches and banks. Another potential problem is pleasure boating.

A study conducted on rivers in the state of Missouri showed that pleasure boating is detrimental to the timid Green Heron, which is easily alarmed by passing motor boats and canoes (Kaiser and Fritzell 1984).

Jean-Denis Gauvreau and Simon Barrette

Black-crowned Night-Heron

Bihoreau à couronne noire

Bihoreau gris

Nycticorax nycticorax

Clodin Roy

Whether breeding in isolated pairs or in single-species colonies, the Black-crowned Night-Heron usually makes its nest under foliage. This night-heron's nest contains one nestling. Île aux Hérons, near Montréal. *Spring 1982, Pierre Mousseau.*

THE BLACK-CROWNED NIGHT-HERON IS UNUSUAL IN THAT IT IS ACTIVE mainly at dusk and during the night. A distinctive, stocky, short-legged heron, it is often heard uttering its characteristic *kwok*, which has earned it the nickname "squawk." In and around their colonies the birds are very noisy, making it easy to associate vocalizations with corresponding behaviour (Hancock and Kushlan 1984).

Adults look alike although the female is somewhat smaller. Immature birds are brownish with orange eyes (Godfrey 1986; Hancock and Kushlan 1984). Although Québec is at the northernmost limit of its North American range, this colonial species is fairly common in the southern part of the province from April to October. In August, the birds begin flocking in various southerly regions of the province to migrate to their winter quarters on the eastern seaboard of the United States, the Gulf of Mexico coast, in the Bahamas, Jamaica, and Cuba (AOU 1983; L'Arrivée and Blokpoel 1990).

Habitat and behaviour

In Québec, the Black-crowned Night-Heron nests along large stretches of water, in deciduous, mixed, or coniferous stands, depending on the area. In the southwestern part of the province, it uses deciduous stands of hackberry, basswood, slippery elm, and silver maple, whereas in the Gulf of St. Lawrence and the estuary it uses trees such as white birch, balsam fir, black spruce, and sometimes even large pines, as in a plantation in Baie-Saint-Paul. Nest height is anywhere from 3 to 20 m, but in some places, such as Îles aux Loups Marins or Le Pilier de Pierre in the St. Lawrence estuary, colonies have become established in low bushes, where nests are only a few centimetres from the ground (Reed 1973).

In Québec, more nests are found in deciduous trees than in conifers (Mousseau 1983, 1985), no doubt because these trees are larger. Hancock and Kushlan (1984) reported that as many as 30 nests were counted in one tree in Turkmenistan (formerly the Turkmen Soviet Socialist Republic).

Black-crowned Night-Herons are gregarious throughout the year but secretive during the breeding season. In southern Québec, adults arrive on the breeding grounds in early April. Pairs then form, and they begin building nests or repairing old platforms. A nest can in fact be reused and enlarged by different birds from year to year (Hancock and Kushlan 1984). Courtship displays and mating take place in the colony near the nest. In their handsome plumage with long ornamental head feathers (shorter in the female), pink feet and black lores at height of breeding, the adults enact a variety of behaviour rituals: hunched or threatening postures, swaying from one foot to another, clappering bills, and calling (Hancock and Kushlan 1984). In general, age of first breeding is 2 or 3 years, but some birds may reproduce as yearlings (Custer and Davis 1982).

The Black-crowned Night-Heron nests in single pairs or in colonies, sometimes in the company of other ardeids such as the Great Blue Heron, as on Île aux Hérons in the Lachine Rapids near Montréal and La Grande Île in Lake Saint-Pierre. When these species nest together, their respective niches can easily be distinguished: the

The night-heron builds its nest under the crown in both deciduous and coniferous trees. Brandypot Islands, St. Lawrence estuary. *July 1993, Jean Gauthier.*

Notes on territory

Nests in mixed-species colonies in fairly large trees are often densely clustered high in the canopy. Each pair fiercely defends its territory, as far as the sitting bird can reach with its bill. Newly fledged young, however, wander freely about the colony without evoking the hostility of adults (Cramp and Simmons 1977). Yearlings and mature birds will often return to the same colony (L'Arrivée and Blokpoel 1990). Mature birds may fly up to 20 km or more from the colony to forage and may establish their own feeding areas. Later in the season, juvenile birds will sometimes accompany them (Kibbe 1985a). In the Mediterranean, it has been found that a colony of 100 pairs requires 5587 ha for foraging (del Hoyo et al. 1992). However, it is difficult to make accurate, comparable estimates, given that the abundance of prey and the distance to foraging grounds vary from one region to another.

Outside the breeding period, birds roost in flocks during the day and disperse at dusk to forage. Individual birds defend their perches in the trees (Cramp and Simmons 1977).

Black-crowned Night-Heron

Breeding distribution

1984–1989

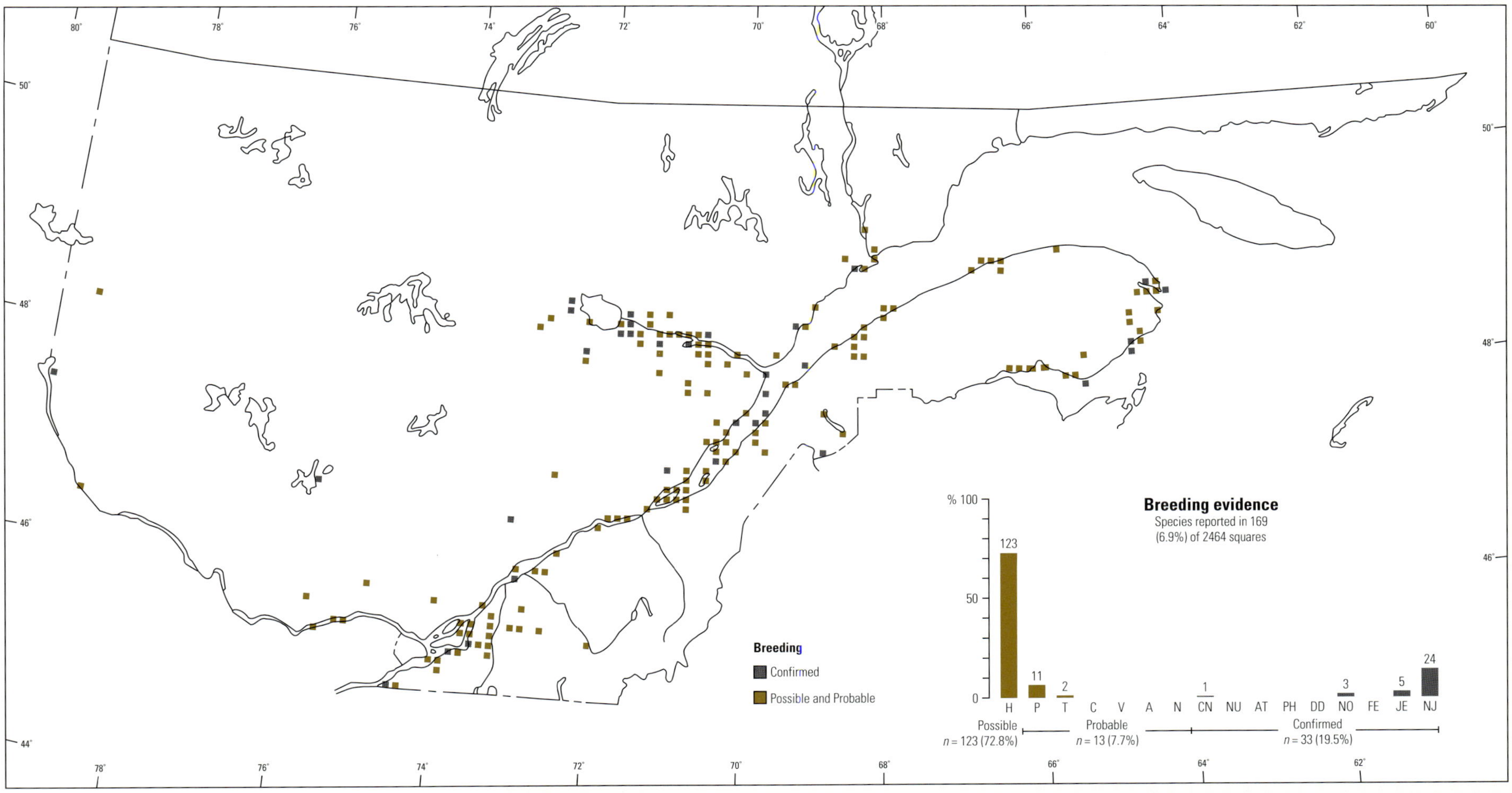

Species profile

Clutch size	usually 3–5 eggs (1–8)[5,9,12]
Laying interval	1 egg per 2 days[4,5,8]
Start of incubation	with first egg[2,4,5]
Incubation period	21–26 days[2,3,5,6,8]
Maturity at hatching	semi-altricial
Care of young	by female and male[1,4,5,8]
Nestling period	20–34 days[5,13]
Age at first flight	40–50 days[1,4,5,6,9,11,12]
Dependence of young	until a few days after first flight[5]
Broods per year	1[1,3]
Breeding age	2–3 years, sometimes 1[8,9,10]
Mating system	monogamy[1]
Duration of pair bond	probably 1 season[12]
Total length	58–71 cm[3]
Wing span	114.3 cm[6]
Weight	M ? g (785–1014) / F ? g (727–884)[11]
Longevity record	21 years 1 month[7]

References:

1. Ehrlich et al. 1988
2. Peck and James 1983
3. Godfrey 1986
4. Harrison 1978
5. Cramp and Simmons 1977
6. Terres 1980
7. Clapp et al. 1982
8. Hancock and Kushlan 1984
9. del Hoyo et al. 1992
10. Custer and Davis 1982
11. Palmer 1962
12. Davis 1993
13. Tremblay and Ellison 1979

Chronology $n = 75$

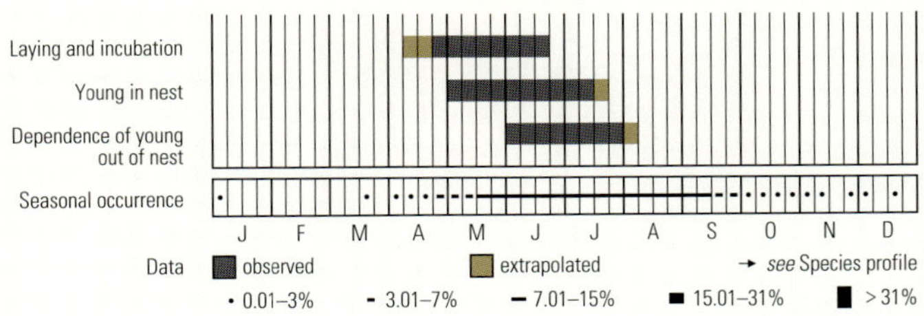

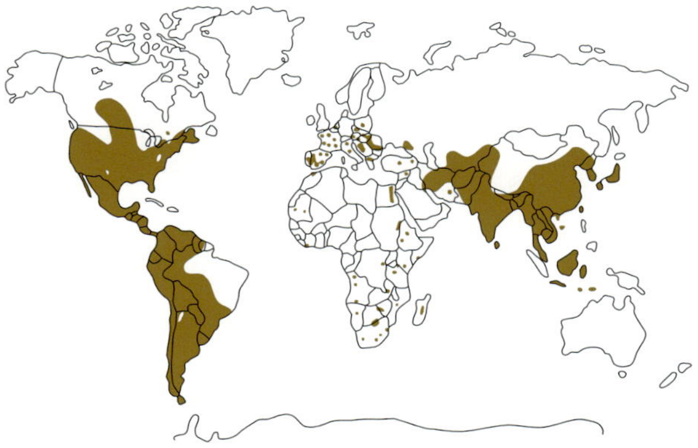

World breeding range

A new colony of night-herons on Deslauriers Island where 28 pairs were counted. The nests were built in a reedbed on dieback vegetation. Varennes Islands, near Montréal. *May 1994, Pierre Brousseau.*

Black-crowned Night-Heron takes cover deep in the tree crown, while the larger Great Blue Heron prefers exposed or peripheral forks.

Some birds lay their first eggs during the second week of April, 3 days after the first copulation and 4 or 5 days after the pair bond has been formed (Palmer 1962). The eggs are laid at 2-day intervals, and hatching is asynchronous, which is often to the advantage of the first-hatched (Custer and Peterson 1991). Both adults defend the nest, incubate the eggs, and tend the young. At the age of 2–3 weeks, the young begin venturing out of the nest along the branches. In southern Québec, first flights occur at the end of the first week of July, when the young are around 40 days old (43 days on Île aux Hérons; Mousseau 1983), and the last young herons leave the colony at the end of July. Timing, however, varies with latitude, being later farther north. As soon as they fledge, the young birds accompany the adults to the foraging grounds.

Breeding success was estimated for various colonies in Québec—Le Gros Pèlerin and Île Brûlée (Kamouraska) in the St. Lawrence estuary (Tremblay and Ellison 1980a), and Île aux Hérons, near Montréal (Mousseau 1983). In the Île aux Hérons colony, on the average 3.9 eggs were laid per nest, 3 hatched, and 2.3 young herons reached the age of 28 days (the age at which the young at Île aux Hérons were ready to venture forth on the branches). In the estuary, productivity was determined at 2 or 3 young/nest on average, depending on the island (Tremblay and Ellison 1980a).

A study on movement in the Île aux Hérons colony showed that Black-crowned Night-Herons travel between the nest site and feeding grounds mainly at dawn and dusk. When they leave Île aux Hérons, they head mainly to Lake Saint-Louis and Île Sainte-Hélène, across from Montréal (Dimension Environnement 1985b). During the chick-rearing period, night-herons are on the feeding grounds both night and day. Still-fishing, often alone but sometimes in groups, is one of their fishing techniques, as are also hovering and diving (Hancock and Kushlan 1984). Most often, they resort to foot probing and foot stirring, slowly probing the mud with their toes to detect their prey or stirring the water with their feet to dislodge prey items. This behaviour is characteristic of heron species with colourful toes (del Hoyo et al. 1992).

The Black-crowned Night-Heron is an opportunistic predator that is well adapted to its habitat. Its varied diet is made up of fish, amphibians, insects, and even the young of colonial breeding species such as gulls, terns, and other herons (Collins 1970; Wolford and Boag 1971b; Shealer and Kress 1991). In Québec, particularly in the St. Lawrence estuary, young night-herons are mainly fish eaters. Their diet includes, in decreasing order, threespine stickleback, Atlantic tomcod, and marine invertebrates (in equal proportions), some insects, mammals, other species of fish, and, very rarely, birds and amphibians (Tremblay and Ellison 1980b).

When the young leave the colony but before they undertake their true migration, adults, and young in particular, explore far afield (800 km and more) in all directions. Roosts and colonies then become centres of dispersal from which the birds explore new foraging or nest sites (del Hoyo et al. 1992). Nevertheless, analysis by L'Arrivée and Blokpoel (1990) of data gathered from marked birds in Ontario and Québec showed that the species not only has a clear attachment to its region or place of origin but also has a taste for exploring neighbouring areas, as the majority of recoveries were found within a 250-km radius.

Distribution

The Black-crowned Night-Heron is widely distributed throughout the Old and New worlds. In Canada, its range is restricted to the south and extends from Saskatchewan to Nova Scotia (Godfrey 1986). Breeding has not been confirmed in Québec north of the 49th parallel. The species nests mainly on islands and shores of large expanses of water such as the St. Lawrence River and the gulf, the Saguenay River, and Lake Saint-Jean. No colonies have been reported beyond Pointe-des-Monts on the North Shore, but there are some colonies on the shores of the Gaspé Peninsula, notably at the eastern tip, and along Chaleur Bay. The largest and greatest number of colonies are found on islands in the Montréal area and the St. Lawrence estuary, the most notable being those on Île aux Hérons (350 breeding pairs in 1988; L.-M. Soyez, pers. comm.) and Brandypot Islands (*c.* 450 nests in September 1989; Société Duvetnor 1990).

A few colonies and isolated nests are found in the interior, at the Cabonga Reservoir, Lake Timiskaming, and Lake Wayagamac, near La Tuque (78 nests; Laplante 1979). As the map shows, possible and probable breeding reports were received from practically all along the St. Lawrence. These observations were probably of birds from nearby colonies, as the species is known to travel extensively to forage (Mousseau and Beaumont 1982) or find nesting sites (L'Arrivée and Blokpoel 1990; del Hoyo et al. 1992). These may also have been non-breeding or post-breeding individuals. Such birds regularly frequent colonies of other species or gather several kilometres from colonies at diurnal roosts (Goodwin 1987a), which may be taken for colonies. Non-breeding birds have also been sighted as far north as Lake Mistassini (Todd 1963).

History and trends

According to L'Arrivée and Blokpoel (1990), the species has expanded its overall range over the last 30 years. Data are insufficient, however, especially outside the St. Lawrence system, to evaluate population trends of night-herons in Québec. In several areas, colonies have been established since at least the beginning of the century, as in the Montréal, Québec City, Kamouraska, and

A Black-crowned Night-Heron foraging in the pools of a cord-grass marsh on L'Isle-Verte. Northeast of Rivière-du-Loup. *Summer, circa 1978, Mario Laverdière.*

About 350 pairs of night-herons nested on Île aux Hérons in 1988. A view of part of the colony. **Montréal.** *Spring 1982, Pierre Mousseau.*

Baie-Comeau regions (Dionne 1906). In 1896, Wintle referred to a presumed colony at the Lachine Rapids, which may well be the one on Île aux Hérons. This colony is on the increase, having grown from 150 nests in 1970 (Ouellet 1974a), to 300 in 1982 (Mousseau 1983), to 350 nests in 1988 (L.-M. Soyez, pers. comm.).

Some colonies move after a few decades, one reason being habitat destruction from the collapse of nest trees killed by heron droppings (Erskine 1992).

Because of this mobility, it is difficult to find new colonies, just as it is difficult to census nests in active ones since these, for the most part, are located in dense woods. This means that any attempt at establishing abundance estimates for the species in Québec is a hazardous undertaking.

Some colonies that were once flourishing have disappeared or declined considerably. For example, on Île aux Basques, the number of nests decreased from 100 in 1930, to 35 in 1955, to 1 in 1976, and subsequently to 0, although sharp pulses were noted throughout this entire period (Darveau 1993). A similar situation occurred on Île Blanche, where the population declined from 105 nests in the early 1970s (Reed 1973), to 34 pairs in 1976 (BIOMQ), to 0 in 1986, after the island was developed (Bédard et al. 1987). On Île Brûlée and Le Gros Pèlerin, where 537 and 380 nests respectively were counted in 1975 (Tremblay and Ellison 1979), only a dozen or so remained in 1990 (BIOMQ).

The nest, composed of branches, may be reused and enlarged over the years by different pairs. Newport, southwest of Percé. *Spring 1983, Pierre Mousseau.*

The reverse phenomenon has also occurred: the colony on Île aux Loups Marins increased from a few nests in the 1970s (Reed 1973) to 135 in 1990 (BIOMQ), and the Brandypot Islands colony grew from some 30 nests in 1976 to about 450 in 1989 (Société Duvetnor 1990).

Compilation of data from Reed (1973), BIOMQ, and others gives about 1100 nests in the St. Lawrence estuary in the mid-1970s, compared with a scant 700 in 1989 and 1990 (pooled data). But when the Montréal area and the several recent small colonies are included, there is no indication of a decline in population.

The eastern seaboard population, including Florida but excluding the Maritimes, has been estimated at 21 954 birds in 144 colonies (Spendelow and Patton 1988); the Québec population comprises less than 10% of this number in over 30 stations.

In Québec, studies of the colonies on Les Pèlerins, Île Brûlée, and Île aux Hérons (Tremblay and Ellison 1980a; Mousseau 1983) have shown that eggshell thinning due to high concentrations of toxic chemical residues has not affected reproductive success, as it has in the Great Lakes and along the eastern seaboard of the United States (Price 1976; Ohlendorf et al. 1978).

According to Tremblay and Ellison (1979), human intrusion in colonies during the breeding period causes desertion of newly built nests, facilitates predation, and frightens young birds into falling out of the nest. They maintain that repeated visits flush adults from the colonies, leaving the nests vulnerable to predation by the American Crow. The ever-increasing population of Ring-billed Gulls poses a threat to Black-crowned Night-Heron numbers, to the point that this species could become the night-heron's major predator, as is the case in Alberta (Wolford and Boag 1971a).

The proximity of ponds and wetlands—mudflats along with freshwater and salt marshes are their preferred foraging sites—has a considerable effect on colony size over the years. Over recent decades, many of these habitats have disappeared, especially along the St. Lawrence (Lands Directorate 1986). Moreover, human disturbances such as logging and recreational activities are major causes of colony displacement and desertion. For example, after trees were cleared to build a marina, in an area 10 m from the Newport Point colony in the Gaspé, a 35% decrease was noted in one year, with 6% of the pairs moving to a new colony nearby (Mousseau 1985). By the spring of 1985, the colony, which was 130 nests-strong in 1983, had been completely abandoned.

Pierre Mousseau

Brant

Bernache cravant

Branta bernicla

Louis Ruelland

In southern Québec, this small goose is most frequently seen along the coasts of the estuary and Gulf of St. Lawrence during spring migration while en route to its breeding grounds in the eastern Canadian Arctic. Throughout most of their lives, Brant generally rarely venture great distances from salt or brackish coastlines, mainly because they rely heavily on a few species of marine plants for food. Unlike most other migratory waterfowl that stop off in the St. Lawrence Valley during both spring and fall migrations, Brant are seen mainly in spring because they generally overfly southern Québec on their fall migration. For the same reason, few sport hunters in Québec have the opportunity to take Brant, but they are a popular quarry for American hunters on the east coast. The James Bay Cree hunt Brant during both migrations, but Canada Geese are of much greater importance to their subsistence harvest (James Bay and Northern Quebec Native Harvesting Research Committee 1982; Reed 1991).

Habitat and behaviour

Because of their small body size and large wings, Brant are powerful and agile fliers. Their power and endurance are put to good use in their lengthy overland and oversea migrations. The fall flight from James Bay to the New Jersey coast covers 1400 km, generally non-stop, probably in about 24 hours. Black Brant of the Pacific coast fly directly from Alaska to Baja California, a distance of more than 4400 km over open sea in about 60 h (Dau 1992). Their agility and power is also evident in their aggressive aerial attacks on avian predators such as gulls, ravens, and jaegers that approach their nests in the summer.

No other geese rely as heavily on such a restricted range of wild plant species for food during most of the year. Indeed, at many staging and wintering locations a single species, eelgrass, may constitute more than 90% of the diet. Marine green algae (*Ulva* spp. and *Enteromorpha* spp.) may also be of regional or seasonal importance. The breeding range of Brant is beyond the range of these marine plants, however, so in summer they graze mainly on short grasses and sedges in Arctic salt marshes and wet meadows; at impoverished sites, mosses are also eaten (Lewis 1931; Cottam et al. 1944; Barry 1956; Smith et al. 1985; A. Reed, CWS, unpubl. data).

Brant breed in the Arctic, usually building their nests near the coastline on low-lying tundra, on hummocky islets in ponds, or on gravel islands in braided streams. In some of the more favourable areas, nests may be clustered in colonies of up to several hundred pairs, but many nest in small, loose groupings or individually. Like other geese, Brant mate for life, and most pairs return to the same nesting area every year. Nesting territories are established and are defended aggressively against other Brant and against avian predators. Clutch size is typically 2–6 eggs, generally averaging 3–4 eggs (Barry 1962; Abraham and Ankney 1986). Laying usually begins around 15–20 June and hatching occurs about mid-July. The eggs and young goslings are vulnerable to predation by arctic foxes, jaegers, Herring Gulls, and Common Ravens, but adult Brant vigorously defend their young (Barry 1956, 1962). Arctic foxes and Snowy Owls occasionally take adult Brant. Goslings are led from the nest at about 1 day old and are continuously accompanied by their parents, not only until flight age but throughout fall migration and often through part of their first spring migration (Reed 1993).

Distribution

Brant are circumpolar in distribution. Taxonomists recognize three subspecies: *Branta bernicla bernicla*, which is found in western Siberia–Europe, *Branta bernicla nigricans* or Black Brant, which is found in eastern Siberia and western North America, and *Branta bernicla hrota*, which occurs in eastern North America, Greenland, and western Europe. Those that migrate through Québec are of the latter subspecies, which is readily distinguished from the other two by the light grey coloration of its belly plumage. This light-bellied subspecies, *Branta bernicla hrota*, can be further broken down into three distinct populations or stocks, two of which breed in North America. One of the North American populations breeds in the eastern Canadian High Arctic and in Greenland and migrates across Greenland and Iceland to winter in Ireland; the other nests in the eastern Canadian Low and Mid Arctic—mainly on Southampton Island and the coasts and islands of Foxe Basin—and migrates through Québec via James Bay to winter along the Atlantic seaboard of the United States between Massachusetts and North Carolina (Boyd 1979; Boyd and Maltby 1979; Owen 1980; Reed et al. 1980, 1987b; Boyd et al. 1988).

Notes on territory

The size of the area defended by male Brant around the nest varies from a few square metres in the case of colonially nesting individuals to several thousand square metres for dispersed nesters. In dispersed nesting situations, the

Nesting habitat of Brant near Cape Dominion in the Arctic. Great Plain of the Koukdjuak, Baffin Island, Northwest Territories. *July 1989, Austin Reed.*

Brant

Breeding distribution
1984–1989

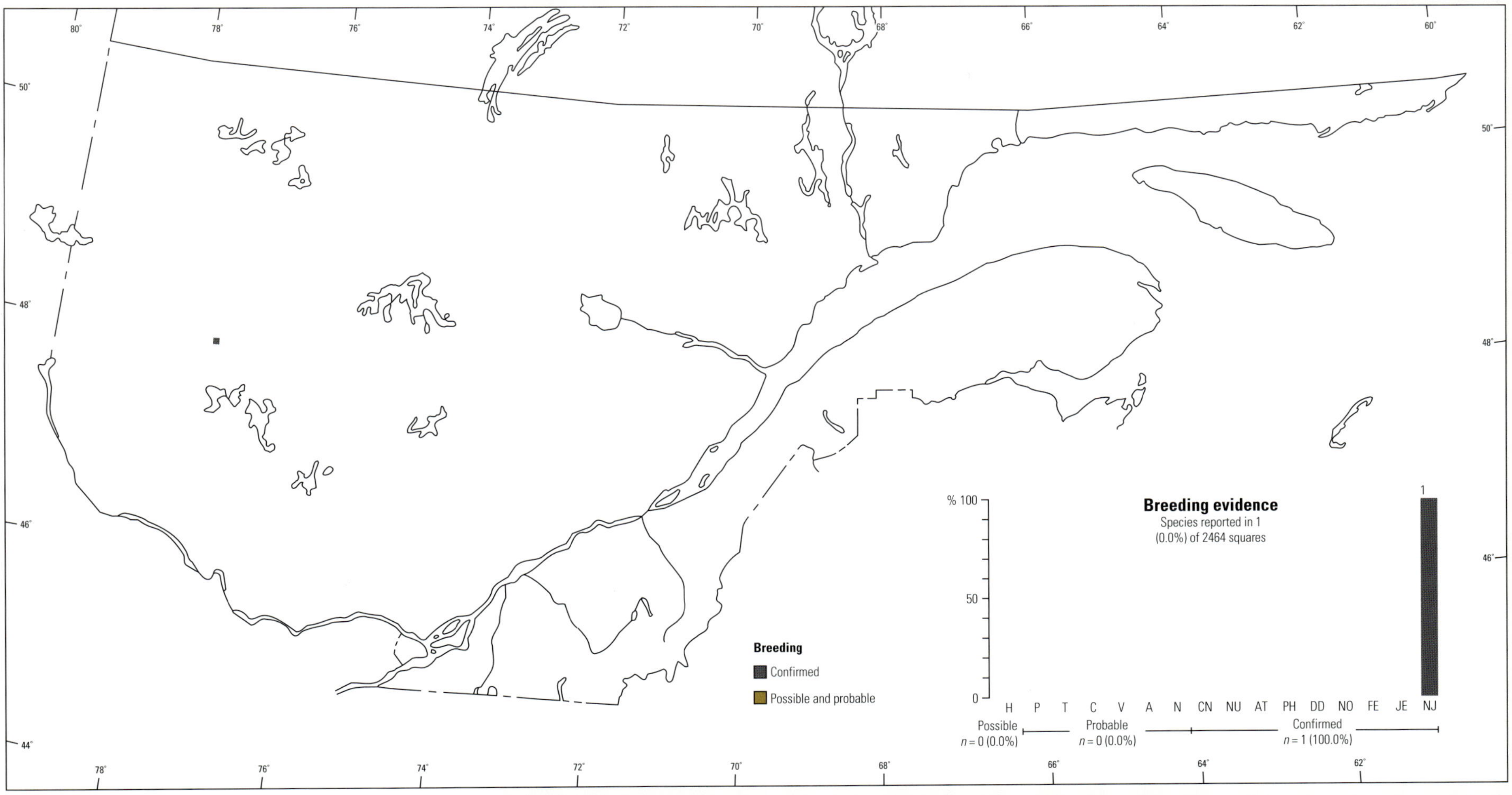

Species profile

Clutch size	usually 3–4 eggs (2–6)[5,10,11]
Laying interval	1 egg per day[3,4,5]
Start of incubation	with last egg[4,5]
Incubation period	usually 23–24 days (22–26)[1,2,5,6]
Maturity at hatching	precocial
Care of young	by female and male[1,2,3,4,5]
Nestling period	less than 24 hours[1]
Age at first flight	45–50 days[2,3,5]
Dependence of young	until ≈ 1 year old[1,4,10]
Broods per year	1[1,4]
Breeding age	2–3 years[3,4,7]
Mating system	monogamy[2,3,4]
Duration of pair bond	for life[4,5]
Total length	M 56–66 cm / F 56–61 cm[5]
Wing span	110–120 cm[4,7]
Weight	M 1670 g (1200–2010) / F 1510 g (1125–1890)[9]
Longevity record	21 years 7 months[8]

References:
1. Harrison 1978
2. Ehrlich et al. 1988
3. Johnsgard 1975b
4. Cramp and Simmons 1977
5. Bellrose 1976
6. Godfrey 1986
7. del Hoyo et al. 1992
8. Klimkiewicz and Futcher 1989
9. Reed et al. 1992b
10. Reed 1993
11. Abraham and Ankney 1986

Chronology *n* = 1

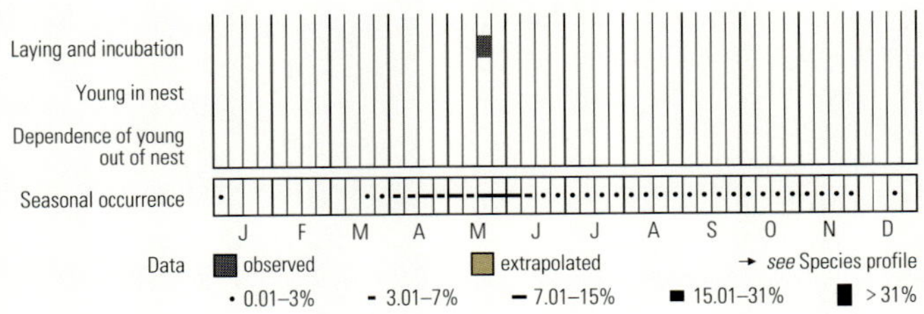

World breeding range

During the brood-rearing period, families gather on the tundra. Baffin Island, Northwest Territories. *August 1984, Austin Reed.*

Notes on territory (continued)

males may claim exclusive use of entire lakes or stretches of river more than 1 km long. The male defends its territory against other breeding pairs of Brant, as well as against potential predators, with defence generally involving aerial pursuit. Both male and female may participate in attacks against intruders at the nest. The territory always includes the nest site, but may or may not include feeding areas. In spring, males may also defend small territories around their mates as they move around foraging, which allows the female more time to feed because of

Two areas of Québec are of considerable importance to the eastern North American population of Brant: the maritime coasts of the estuary and Gulf of St. Lawrence, mainly east of Kamouraska, which serve as a major staging area for the majority of the population during spring migration (Erskine 1988), and the coast of James Bay, which is used extensively during both spring and fall migrations (Curtis and Allen 1976; Reed 1977b; Bellrose 1980; Vangilder et al. 1986; Reed et al. 1992b). A third area, centred around the confluence of the St. Lawrence and Ottawa rivers, including Lake Saint-Louis, Lake Saint-François, Lac des Deux Montagnes, and parts of eastern Lake Ontario and northern Lake Champlain, is also used by smaller numbers during migration, but its overall importance has not been adequately assessed. This third area is evidence that two migration routes are used by northward migrating Brant in spring (Erskine 1988). One route heads northward to the headwaters of the Hudson River, overland to the confluence of the St. Lawrence and Ottawa rivers, northward along the Ottawa River watershed, and then to James Bay, while the other (of greater importance) follows a coastal route northward through the Maritime provinces and into the estuary and Gulf of St. Lawrence, and from there to James Bay, probably along the Saguenay River and the watersheds emptying into James Bay. These present routes are somewhat different from those decribed earlier by Lewis (1937), which shows a certain adaptability of the species to changing conditions.

The population that migrates through Québec nests principally near the coasts of Foxe Basin, including Southampton and Baffin islands. An unusual breeding record was reported near Val-d'Or, Abitibi, on 18 May 1984 by J.-G. Lavigne (pers. comm.). An apparently wild individual nested under the eaves of a woodshed less than 10 m from Lake Villebon; the clutch of 4 eggs, typical for the species (Barry 1956, 1962; Abraham and Ankney 1986), was subsequently destroyed by a predator. Regular spring sightings of flocks in northwestern Québec (Abitibi) and northeastern Ontario indicate that this location is on one of the spring migration routes to James Bay (Erskine 1988). Brant of the eastern Arctic do not usually nest until mid-June (Barry 1956). The early date of egg laying in the Abitibi nest suggests that this individual experienced early ovarian development while on migration and remained behind to nest. There are also a few isolated records of breeding by Brant in southern Ontario (Lumsden 1987a).

Île aux Pommes, with its grassy vegetation, is sometimes used by feeding Brant during spring migration. St. Lawrence estuary. *May 1974, Austin Reed.*

reduced disturbance from other Brant. Both parents also defend small moving feeding territories around their broods for the same reason, and this form of parental care may continue for almost a full year (Barry 1956; Bellrose 1980; Reed 1993; A. Reed, CWS, unpubl. data).

An eelgrass meadow at high tide. This submerged aquatic plant is the mainstay of the Brant's diet. James Bay region. *September 1991, Austin Reed.*

History and trends

Eelgrass is so vital to Brant that they are vulnerable to starvation should the plants be affected by disease or habitat modification or should severe cold freeze over the submerged plants. Two catastrophes of this nature have struck Brant populations since 1930. In the early 1930s, a disease caused an almost complete die-off of many eelgrass beds in eastern North America, including those of the estuary and Gulf of St. Lawrence, resulting in a dramatic reduction in Brant numbers (Tremblay and Gaudry 1936; Lewis and Cottam 1954). A gradual but perhaps incomplete recovery occurred as they adapted to alternative foods such as sea-lettuce, and as eelgrass made a partial comeback. In the winter of 1976-77, record low temperatures caused most marine feeding areas to freeze over, resulting in massive starvation: from an initial 1976 fall population of 115 000, only about 50 000 could be accounted for in surveys in February 1977, and 35 000 in May (Reed 1977b; Rogers 1979). Between 1989 and 1992, the wintering population on the eastern seaboard was high, fluctuating between 135 000 and 185 000 individuals, with an average of 151 000, but it dropped to approximately 100 000 in January 1993 after a poor breeding year in 1992 (Atlantic Flyway mid-winter survey reports, unpubl. data). Eelgrass is especially abundant along the Québec coast of James Bay and constitutes an important resource for the entire population of Brant that winters in eastern North America. Although studies conducted to date show no apparent decline in eelgrass abundance since hydro-electric development began in 1987 (Reed et al. 1992b; Lalumière et al. 1992), careful monitoring is called for to detect any changes in abundance which might occur over the long term as a result of modified water characteristics (salinity, temperature, sediment load, etc.).

AUSTIN REED

In the St. Lawrence estuary, Brant can replenish their energy reserves in intertidal cordgrass (tall standing vegetation) marshes where they feed on algae and especially eelgrass which lies flattened on the mud at low tide (in the middle of the photo). Pointe-au-Père, near Rimouski. *September 1980, Jean Gauthier.*

Canada Goose

Bernache du Canada

Branta canadensis

René Roy

THE LONG, IRREGULAR V-FORMATIONS OF NORTHWARD MIGRATING FLOCKS of Canada Geese are an exciting and welcome sign of spring. Their distinctive black neck, head, and crown, white cheek patch, large body size, and, most of all, their clear, resonant, honking call help to make the Canada Goose one of the most readily identifiable birds in North America.

The Canada Goose has the largest size variations among breeding populations of any bird, with body length ranging from 55 to 110 cm and body weight from 1.6 to 5.2 kg (Madge and Burn 1988). Consequently, there are 11 recognized races of the Canada Goose (Bellrose 1976), with two races historically breeding in Québec: the Atlantic race, *Branta canadensis canadensis*, and the interior race, *Branta canadensis interior* (Reed and Joly 1977). In the last few years, giant Canada Geese, *Branta canadensis maxima,* have been invading southern Québec through the Ottawa and St. Lawrence river valleys (P. Dupuis, CWS, unpubl. data). This subspecies was introduced in the 1930s in the northeastern United States from releases of decoy flocks and purposeful stocking (Hindman and Ferrigno 1990). During migration, small Canada Geese, probably of the races *Branta canadensis parvipes* and *Branta canadensis hutchinsii*, which breed in northwestern Hudson Bay and on Baffin Island, are regularly taken by sport hunters (P. Dupuis, CWS National Harvest Survey files) and by native hunters in James Bay (Prevett et al. 1983; A. Reed, CWS, unpubl. data). Canada Geese breeding in Québec winter in the central and eastern United States, though there are a few records of birds wintering in southern Québec (David 1980a), mostly in the Montréal region (Bannon 1991a).

The Canada Goose is of major economic and cultural importance to northern Québec natives, especially the James Bay Cree (Reed 1991). Geese represent about a quarter of the wild meat eaten by the Cree, and entire Cree families assemble in outlying camps to harvest the geese during the so-called Goose Break in spring and fall. Local schools are closed to allow the children to take part in these traditional gatherings of great social and spiritual importance. The subsistence harvest by Québec Cree hunters averaged 64 000 Canada Geese annually during the 1970s, or 13% of the entire hunting kill throughout the Atlantic Flyway.

Canada Geese nest in the many bogs and fens of Québec's boreal forest, generally in open habitat. Laforge Reservoir, south of Lake Bienville, north-central Québec. *August 1992, Austin Reed.*

Habitat and behaviour

Nesting habitat of the Canada Goose varies considerably, ranging from ponds, lakes, and rivers, to marshes, wooded swamps, meadows, small islands, and cliffs (Bellrose 1976; Ehrlich et al. 1988). In Québec, the Canada Goose nests primarily in two distinct habitat types: bog-fen habitat in the boreal forest, and arctic tundra (Johnsgard 1975b). On the tundra near Arviat (Eskimo Point), Northwest Territories, favoured nesting sites were hummocky sphagnum-covered islets 0.5–5 m in diameter and only 0.5 m above summer water level (MacInnes 1962), a preference noted by many authors (Keast 1973; Raveling and Lumsden 1977; Lamothe 1982; Reese et al. 1987; Cons. G. G. – GREBE 1990a; Tardif and Morneau 1991). This type of habitat is also found in northern Québec. In bogs and fens around Lake Bienville and Laforge Reservoir in northern Québec, Canada Geese nest mostly on small islets (less than 4 m in diameter) situated in ponds, and on strips of vegetation in structured fens (Tardif and Reed 1991; Hughes et al. 1994c). Giant Canada Geese favour marshes for nesting, especially those with stands of bulrushes or cattails, and will often nest on muskrat houses (Hanson 1965; Palmer 1976). The nest is a depression with an inner lining of down and an outer lining of dry grass, forbs, moss, twigs, and various plant material (Godfrey 1986; Ehrlich et al. 1988).

Canada Geese are monogamous and start forming lifelong pair bonds at around 2 or 3 years of age (Craighead and Stockstad 1964; Cooper 1978; MacInnes and Dunn 1988a,b; Hardy and Tacha 1989). Clutch sizes of 4–6 eggs are typical, with an average of 4.6 and 5.3 eggs for populations of interior and giant races, respectively (Bellrose 1976).

Notes on territory

Canada Geese establish exclusive territories for nesting. The territory is defended, most vigorously by the male, against intruding pairs. Most territorial conflicts are resolved by posturing and threat rather than by overt fighting. Size and other characteristics of the territory vary considerably with subspecies and habitat. The territory always includes the nest site, but may or may not include feeding and brood-rearing areas. In the boreal forest in Québec, Canada Goose nesting territories are widely dispersed, with most pairs claiming exclusive use of a small pond surrounded by bog or fen (Bellrose 1976; A. Reed, CWS, unpubl. data). The size of the defended area varies with the individual male's level of

Abundance of Canada Geese in southern Québec. Aerial surveys conducted in May 1990, 1991, and 1992 *(D. Bordage, CWS, unpubl. data).*

Canada Goose

Breeding distribution
1984–1989

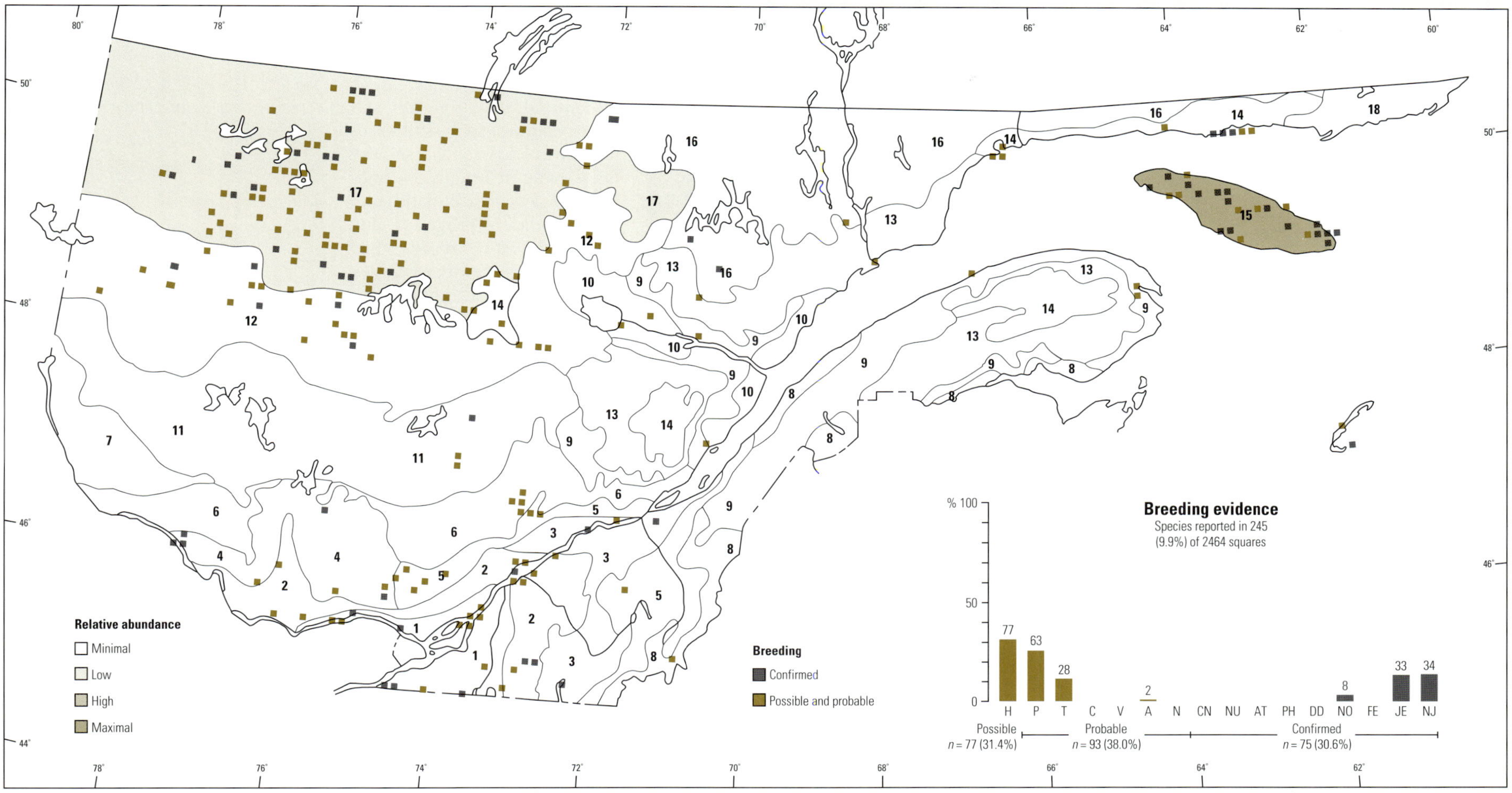

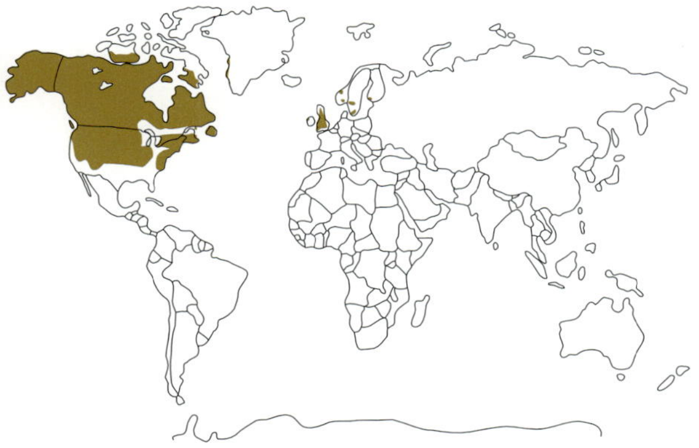

World breeding range

Species profile

Clutch size	*m** usually 5–6 eggs[1] / *c* usually 5–6 (4–10)[2] / (*i* usually 4–5, more than 8 = dump nest)[1]
Laying interval	1 egg per 1–3 days[1]
Start of incubation	with last egg[4,5]
Incubation period	*m* usually 26–28 days (25–28)[1] / *c* usually 28–30 days[2] / *i* ?
Maturity at hatching	precocial
Care of young	by female and male[1]
Nestling period	less than 24 hours[1]
Age at first flight	*m* 71–73 days[1] / *c* minimum 42 days[2] / *i* ≈ 63 days[1]
Dependence of young	until ≈ 1 year old[1]
Broods per year	1[1]
Breeding age	2–3 years[3]
Mating system	monogamy[1]
Duration of pair bond	for life[4,5]
Total length	*m* ? / *c* ? / (*i* M 92.0 cm / F 85.1 cm)[1]
Wing span	?
Weight	*m* M 5670 g (?) / F 5035 g (?)[1] *c* M 3992 g (?) / F 3447 g (?)[1] *i* M 4173 g (?) / F 3492 g (?)[1]
Longevity record	23 years 7 months[6]

* *m* = *B. c. maxima*
c = *B. c. canadensis*
i = *B. c. interior*

References:
1. Bellrose 1976
2. Bent 1925
3. del Hoyo et al. 1992
4. Palmer 1976
5. Cooper 1978
6. Clapp et al. 1982

Chronology *n* = 253

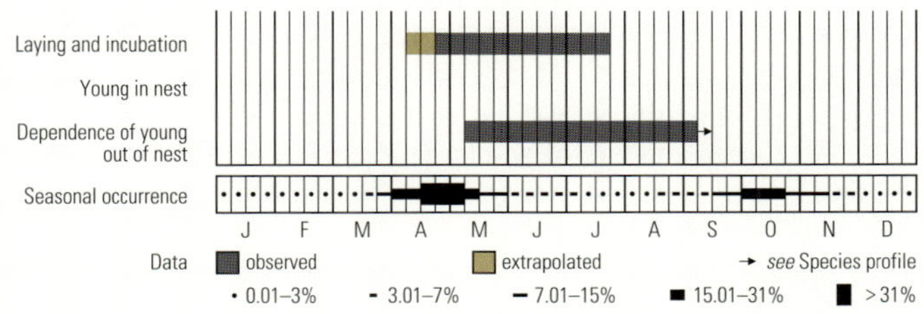

Frequency of occurrence by ecological region

Ecological regions** (forest regions)	Adequately covered squares	Squares in which reported	% of squares in which reported	Relative abundance (%)*
1. Sugar maple-hickory and sugar maple-basswood	97	14	14.4 ± 3.6	26.5
2. Sugar maple-basswood	153	12	7.8 ± 2.4	14.4
3. Sugar maple-basswood and sugar maple-yellow birch	101	3	3.0 ± 2.3	5.4
4. Sugar maple-yellow birch-basswood	56	6	10.7 ± 7.4	19.6
5. Sugar maple-yellow birch-beech and sugar maple-yellow birch-basswood	113	6	5.3 ± 2.5	9.7
6. Sugar maple-yellow birch-beech	71	5	7.0 ± 5.8	12.9
7. Sugar maple-yellow birch-balsam fir	19	0	0	0
8. Sugar maple-yellow birch and balsam fir-yellow birch	47	0	0	0
9. Balsam fir-yellow birch	87	3	3.4 ± 3.9	6.3
10. Balsam fir-white birch or balsam fir-red maple	46	2	4.3 ± 5.7	8.0
11. Yellow birch-balsam fir	52	2	3.8 ± 6.0	7.1
12. Balsam fir-white birch (west)	59	5	8.5 ± 7.7	15.5
13. Balsam fir-white birch (east)	76	2	2.6 ± 4.0	4.8
14. Balsam fir-black spruce	28	4	14.3 ± 14.1	26.2
15. White spruce-balsam fir and balsam fir-white spruce	22	12	54.5 ± 21.4	100.0
16. Black spruce-balsam fir-moss	18	1	5.6 ± 13.5	10.2
17. Black spruce-moss	32	6	18.8 ± 15.1	34.4
18.	*not analysed*			

* minimal: 0–29.9%; low: 30–59.9%; high: 60–89.9%; maximal: 90–100%

** Deciduous forest: 1–7; mixed forest: 8–11; coniferous forest: 12–17

The Atlantic race nests in large numbers in the wet meadows of Anticosti Island. *April 1992, Réginald Ouellet.*

Generally, clutch size decreases as latitude increases (Owen 1980; Dunn and MacInnes 1987; MacInnes and Dunn 1988a,b). While the female incubates, the male remains close to the nest, guarding against intruders and predators. For most races, the incubation period ranges from 25 to 30 days (Bellrose 1976).

The precocial goslings leave the nest accompanied by both parents within 24 hours after hatching. Often the goslings are first led to a nearby marshy wetland, but later they may use a larger rearing area which includes areas of open water (Palmer 1976). Near the Laforge Reservoir, young broods used mainly small boggy ponds, but made increasing use of lakes and rivers as they got older; individual brood home ranges varied from 1.7 to 14.0 km² (Hughes et al. 1994c). On some rivers, the family group may drift downstream many kilometres from their nesting area to favourable feeding meadows (Bellrose 1976). Studies in Manitoba and Québec found that families move less than 10 km (Didiuk 1980; Hickey and Bider 1981; Lamothe 1982). Occasional mixing of several broods may occur, resulting in large groups of adults, subadults, and goslings (Bellrose 1976).

As a consequence of the large variation in body size among the races, there is a wide range in the age at which goslings fledge (Bellrose 1976), with first flights usually later in the larger races. In a study of a population of the interior race breeding along James Bay, goslings were observed to fledge at about 63 days (Hanson 1965).

As is common with most species of geese, families of most races of Canada Goose remain together until they arrive on the breeding grounds the following spring. Then the yearlings, which do not breed, leave or are evicted from the nesting area by the parents at the onset of breeding (Raveling 1969, 1970; Austin 1990; Sedinger and Raveling 1990).

Canada Geese are essentially herbivores. On the breeding grounds they feed mostly on the shoots, roots, and seeds of locally abundant grasses, sedges, and other herbaceous plants, as well as on berries (Polunin and Eklund 1953; Craven 1984; Prevett et al. 1985; Reed et al. 1990; Hughes et al. 1994c). Along migration routes and on wintering grounds, Canada Geese feed heavily on waste grain, in particular corn (Bellrose 1976; Reed 1976; Reed et al. 1977; Cazelais 1992; Bergeron 1992).

The Canada Goose was reported in 9.9% (245/2464) of Atlas squares surveyed. In spite of the large size of both the nest and the incubating female, nests are often very difficult to locate because of the cryptic coloration of the incubating goose and the concealing vegetative cover in which they are hidden. Breeding was confirmed in 30.6% of the squares in which the species was reported, with all confirmations involving observations of either nests (42 squares) or broods (33 squares).

Distribution

The Canada Goose breeds across northern North America and occasionally in western Greenland. During the past three centuries, successful introductions of Canada Geese have been made in Britain, Ireland, Norway, Sweden, Finland, and New Zealand (Madge and Burn 1988). In North America, they breed locally in northern Alaska, across most of Canada as far north as northern Baffin Island, and in the United States as far south as northeastern California, Utah, to Kansas in the west, and to Maine in the east.

Two races are indigenous as breeders in Québec. The interior race nests throughout northern and central Québec, with highest densities occurring in tundra areas on the Ungava Peninsula and lower densities in the black spruce forests extending south to Lac-Saint-Jean and Rouyn-Noranda (Addy and Heyland 1968; Gillespie and Wetmore 1974; Malecki and Trost 1990); the southern portion of this breeding range extends into the northwest portion of the Atlas area. The Atlantic race nests principally in Labrador, Newfoundland, and eastern Québec; low densities are found along the north shore of the Gulf of St. Lawrence, but there is an important concentration on Anticosti Island. In recent years, Canada Geese have been found breeding in several locations in southwestern Québec; these are believed to have originated from expanding populations of introduced stocks of large-sized geese (*maxima* and other geese that have interbred with other large races) that are well established in Ontario and the northeastern United States. Groups of non-breeding geese seen in summer along the Gaspé Peninsula, in the St. Lawrence estuary, along the Saguenay River, in Lakes Saint-Jean and Saint-Pierre, in the Gouin Reservoir, and in the Rouyn-Noranda region may also be from these introduced populations.

The Atlantic race migrates south along the Atlantic coast and winters from Newfoundland to New Jersey (Palmer 1976; Bellrose 1976; Owen 1980). The interior race migrates in a broad front from the east coast of Hudson Bay and James Bay south through eastern Ontario and southern Québec. These birds winter primarily in the eastern United States, particularly from New York to South Carolina (Bellrose 1976; Lamothe 1982; Menkens and Malecki 1991). It is not known where the Canada Geese that have recently settled in southwestern Québec winter, but banded geese from the Ottawa region have been observed wintering in urban parks in cities in the eastern

The Canada Goose often builds its nest not far from water in a hollow lined with down and a variety of plant material. North of Schefferville. *July 1981, Jean-Luc DesGranges.*

Notes on territory (continued)

aggressiveness, habitat type, breeding density, and stage in the reproductive cycle (Collias and Jahn 1959; Ewaschuk and Boag 1972; Bellrose 1976; Cooper 1978). Generally, the territory is more extensive at the start of incubation, growing smaller over time until hatching.

Another form of territorial behaviour consists of protecting a small moving area around the brood (both parents participate in its defence) or around the adult female before and during nesting (protection afforded by the male) (Cooper 1978). This allows the young geese or the nesting female increased feeding opportunities by preventing disturbance by other geese.

In Québec, the largest densities of Canada Geese appear to occur along the coasts of Ungava Bay and Hudson Bay (Malecki and Trost

One of six Canada Goose nests in a ***Phragmites*** **reedbed on Deslauriers Island, near Montréal. This is one of the few known breeding sites in the southernmost part of the province. Varennes Islands.** *May 1994, Pierre Brousseau.*

1990; Hindman and Ferrigno 1990). In these regions, Malecki and Trost (1990) estimated average breeding density at 1.63 pairs/km². In the area around Great Whale River, Lamothe (1982) found density ranged between 0.49 and 0.8 nest/km².

Aerial surveys conducted in southern Québec in May 1990, 1991, and 1992 show that the greatest concentrations of Canada Geese are to be found north of Lake Evans (southeast of James Bay) and north of Lake Manouane (east of Lake Mistassini). Density was estimated at slightly more than 0.24 pair/km² (D. Bordage and M. Grenier, CWS, unpubl. data).

United States (P. Dupuis, CWS, pers. obs.). In southern Québec, peak fall migration occurs during the first three weeks of October, and peak spring migration at the end of April through the beginning of May (Reed and Joly 1977; Lehoux et al. 1985).

History and trends

Few species have adapted to the settlement of the North American continent as well as the Canada Goose. Although the geese have always been a popular quarry for subsistence and sport hunters, and although most races have ceased nesting in settled areas, their numbers in eastern North America are believed to be greater today than before the arrival of European settlers. Their success in southern Canada and in the United States can be attributed largely to their ability to adapt to the transformation of the landscape from forest to farmland. Their habit of grazing nutritious grasses and gleaning cereal grains on agricultural land has probably enhanced their survival on staging and wintering grounds and allowed the population to increase. Other contributing factors include the abolition of sport hunting in spring (since 1917) and the establishment of sanctuaries, refuges, and wildlife areas in Canada and the United States.

Canada Geese respond well to captivity. Through the early part of this century, many flocks were raised in captivity to serve as decoys to attract wild geese for market hunters. The large number of captive flocks in the Lake Saint-Pierre area undoubtedly has its roots in an earlier era when the use of live decoys was legal. Many breeding subpopulations of free-flying Canada Geese have been established through the release or escape of captive birds, both in North America and abroad. Until quite recently, such introductions were proudly acclaimed as human contributions to the welfare of the species. More recently, however, many of these introduced flocks have expanded rapidly and become a nuisance in urban areas, littering public parks and golf courses with their droppings and becoming a safety risk near airports. These resident flocks may also compete with natural populations for food on the wintering grounds.

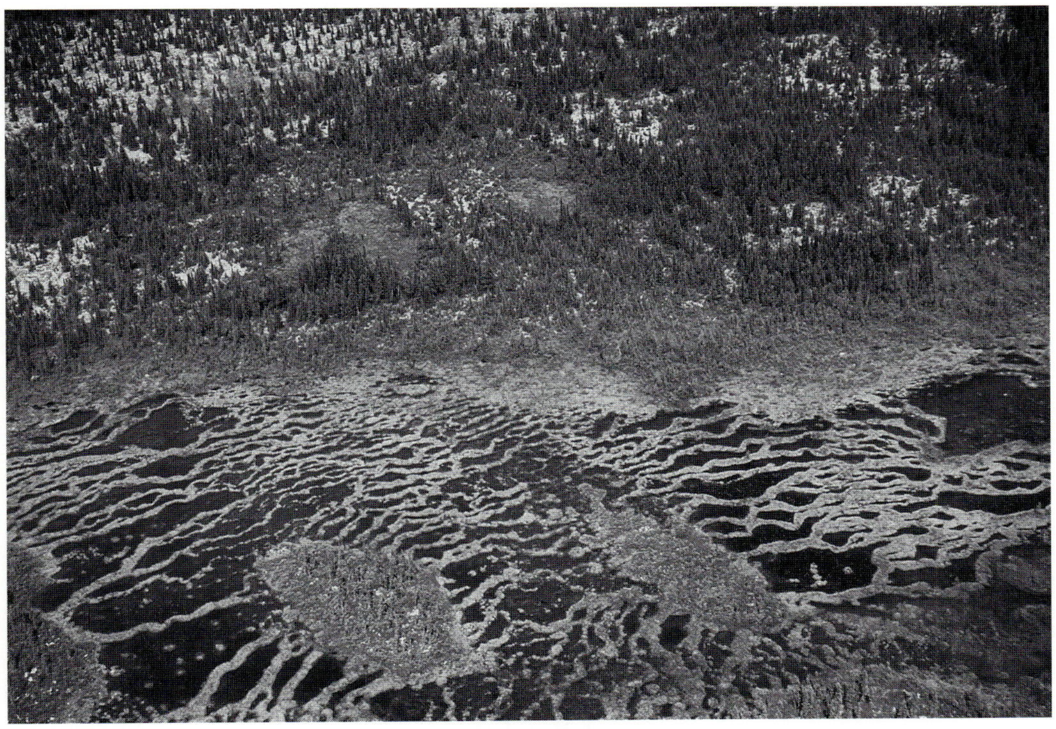

The Canada Goose nests on dry land, often on strips of vegetation and hummocks in the fens in the James Bay region. *August 1992, Austin Reed.*

Winter surveys along the eastern seaboard of the United States have been used to monitor the status of the Atlantic Flyway population since the late 1940s. The numbers recorded increased steadily from approximately 200 000 in 1948 to more than 900 000 in 1981, but then declined to about 570 000 in 1992–93 (Hindman and Ferrigno 1990; USFWS, unpubl. data). These surveys do not cover all areas of the wintering grounds and, therefore, underestimate the total population. Nevertheless, because they are conducted in the same way each year, they are considered to accurately reflect population trends. The survey includes both the wild populations that breed in northern Québec and Labrador as well as introduced flocks that reside year-round in the United States. Over the past decade while the total wintering population has been declining, the size of the introduced resident population has been increasing. Thus the decline in the wild population from northern Québec and Labrador has been greater than is indicated by the winter survey. The reasons for the recent decline are not known, but heavy hunting pressure and a series of years with poor spring weather conditions on the breeding grounds are the most likely causes. The population is not yet in danger, but the decline observed in the number of breeding pairs in northern Québec (Harvey and Bourget 1995) is of serious concern and merits closer study.

Between 1973 and 1988, the average harvest in Québec was around 39 000 birds (Appendix II). Subsequently, between 1989 and 1992, the average Canada Goose kill by sport hunters in fall was around 47 000 individuals. The trend has thus been to a higher harvest over recent years.

Richard C. Cotter, Pierre Dupuis, Josée Tardif, and Austin Reed

Wood Duck

Canard branchu

Aix sponsa

WITH ITS EXQUISITELY COLOURED PLUMAGE, THE WOOD DUCK IS UNQUESTIONABLY one of our most beautiful ducks. Conservationists also look to it as a symbol of success. The species almost disappeared at the turn of the century but has made such a successful comeback thanks to the enforcement of protective measures that it is now one of the most abundant ducks in eastern North America. In Québec, this migratory species seems to have expanded its range northward; although 20 years ago it bred only in the southern part of the province, more recently it has been confirmed as breeding as far north as Matagami, and has been observed up to the Eastmain River.

Habitat and behaviour

Wood Ducks begin pairing in early fall on their wintering grounds, at first forming short-term associations that may last anywhere from a few minutes to a few hours. As autumn wears on, these bonds become stronger (Fredrickson 1990). In late March or early April, the Wood Duck heads back to Québec (Bouvier 1974a; Chapdelaine 1979b; David 1980a), and soon commences laying (Johnsgard 1978). The female chooses the nest site, and incubates the eggs and raises the young alone. As in other species of ducks, the drake deserts the hen towards the end of incubation and leaves the breeding area (Fredrickson 1990; Kirby 1990).

Of all the ducks, the Wood Duck is the species best adapted to wetlands in woodland environments, especially those where the adjacent forest cover is deciduous (Haramis 1990). It nests in wooded areas with shallow water and mature trees that contain cavities that can be used for nesting (Johnsgard 1978; Godfrey 1986). After hatching, broods move to bodies of water with dense plant cover where they can hide; they prefer flooded areas containing trees, shrubs, and beds of emergent vegetation interspersed with small areas of open water (Prince 1991). Such sites often include deadfalls projecting into the water and aquatic vegetation (Cottrell et al. 1990). Flooded woodlands, lakes, ponds, streams, slow-moving rivers, and forest swamps all provide suitable breeding habitat (Peck and James 1983; DeGraaf and Rudis 1987; Prince 1991; Hartman 1992); beaver ponds are particularly favoured (Ducks Unlimited 1991). In southern Québec, the Wood Duck likes flooded hardwood forests, where species such as red maple, silver maple, black ash, and white elm grow (Chapdelaine 1979b).

Unlike most other species of ducks, the Wood Duck does not build its nest on the ground, but in a cavity in a tree (Bent 1923). These cavities are usually made by woodpeckers, particularly the Pileated Woodpecker, although artificial nest boxes may also be used (Johnsgard 1978; Godfrey 1986). Prefered trees are older deciduous trees measuring 35–95 cm in diameter at breast height (dbh), although occasionally conifers are used. In Ontario, most nests found were in dead or dying trees (Peck and James 1983). Cavities are usually located at a height of 6–15 m, but may be anywhere between 0.6 and 19.5 m above the ground; the cavity is 15–150 cm deep, and the diameter of the opening is 9–30 cm (DeGraaf and Rudis 1987; Prince 1991). The bottom of the cavity is lined with wood chips and the female's down (Harrison 1975).

In some places cavities are hard to come by, and Wood Ducks may have to compete not just among each other or with other duck species, such as the Hooded Merganser or Common Goldeneye, but also with the European Starling, Eastern Screech-Owl, American Kestrel, squirrels, or even bees (Johnsgard 1978). Clutch size is generally 12 eggs, which are incubated by the female (Bellrose 1976). First-time breeders lay fewer eggs than do older ones (Heusmann 1975). A few cases of double-brooding have been reported (Bellrose 1976; Fredrickson and Hansen 1983; Kennamer and Hepp 1987); the Wood Duck is the only species of waterfowl in North America known to do so. Double-brooding is more frequent in the south, because of the longer breeding season (Haramis 1990). Shortage of nest sites increases competition among females and usually results in reduced reproductive success (Jones and Leopold 1967; Haramis and Thompson 1985). Females tend to return to breed in their natal area, tolerating within their territory female siblings or offspring (Bellrose 1976). Wood Ducks frequently engage in nest parasitism, that is, laying eggs in the same nest as another female, not necessarily of the same species (Morse and Wight 1969; Clawson et al. 1979; Heusmann et al. 1980). The exact causes of this behaviour are still unclear (Rohwer and Heusmann 1991). Indeed, several different species of cavity-nesting ducks may all lay eggs in the same nest (Bellrose 1976), and, in southern Québec, nests containing eggs of the Wood Duck, Hooded Merganser, and Common Goldeneye have been found (Bouvier 1974a).

The hatchlings remain in the nest for about the first day (Bellrose 1976), then with the hen calling and urging them on, they use their small claws to pull themselves up the cavity wall and then jump out of the nest (Godfrey 1986). The ducklings' mortality rate is usually quite high in the first few days (McGilvrey 1969; Ball et al. 1975; Cottrell et al. 1990). The hen remains with the ducklings for 30–60 days (Fredrickson and Hansen 1983; Kennamer and Hepp 1987), but the bond between them seems weaker than in the other dabbling ducks. It is not unusual for the hen to abandon her brood early or for broods to be observed moving about alone, failing to respond to their mother's calls (Kirby 1990). Abandoned broods of various ages tend to flock together (Johnsgard 1978).

The Wood Duck forages in shallow water, along the shores of lakes and rivers, or on the forest floor (DeGraaf and Rudis 1987). It eats chiefly fruits and seeds of terrestrial and aquatic plants, as well as their stalks and leaves (Hocutt and Dimmick 1971; Johnsgard 1978). In contrast, up to 70% of the ducklings' diet may consist of insects (Hocutt and Dimmick 1971).

The male starts to moult shortly after leaving the female, while the female does not moult until later in summer, after leaving the brood. Both sexes are flightless for about 3 weeks (Kirby 1990); in autumn, they complete their moult of body and flight feathers, thus reacquiring their nuptial plumage.

Wary and retiring during incubation and brood rearing, the Wood Duck is easily overlooked (Biro 1987; Hartman 1992), making it all the more difficult to confirm breeding. Furthermore, nest cavities are generally located high up in trees, making nests (other than nest boxes that are being monitored) hard to find.

Distribution

The Wood Duck is a fairly common breeder in southern Québec, specifically in the Ottawa Valley, Laurentians, Eastern Townships, Lanaudière, Mauricie–Bois-Francs, and Montérégie regions. Since the species is associated with mature deciduous stands, it becomes progressively less abundant towards the north and at higher elevations, as conifers become dominant. Analysis of data from aerial surveys conducted by the Canadian Wildlife Service in May 1990, 1991, and 1992 leads to the same conclusion (D. Bordage, CWS, unpubl. data). The species has also been reported in

René Roy

Notes on territory

The females show strong nest site tenacity (Ehrlich et al. 1988). Similarly, yearling females have a strong tendency to nest within a kilometre or two of the place where they hatched (Hepp et al. 1989; Havera and Kirby 1990). The Wood Duck does not seem to defend a territory around the nest; at the very most, the drake may prevent other males from approaching his mate for a short period before laying commences (DeGraaf and Rudis 1987). During the brood-rearing period, broods travel distances of up to 5 km (Johnsgard 1978). In Tennessee, the average distance travelled by broods was 1.1 km, and their home range encompassed 46 ha of water and 21 ha of shoreline on average (Cottrell et al. 1990).

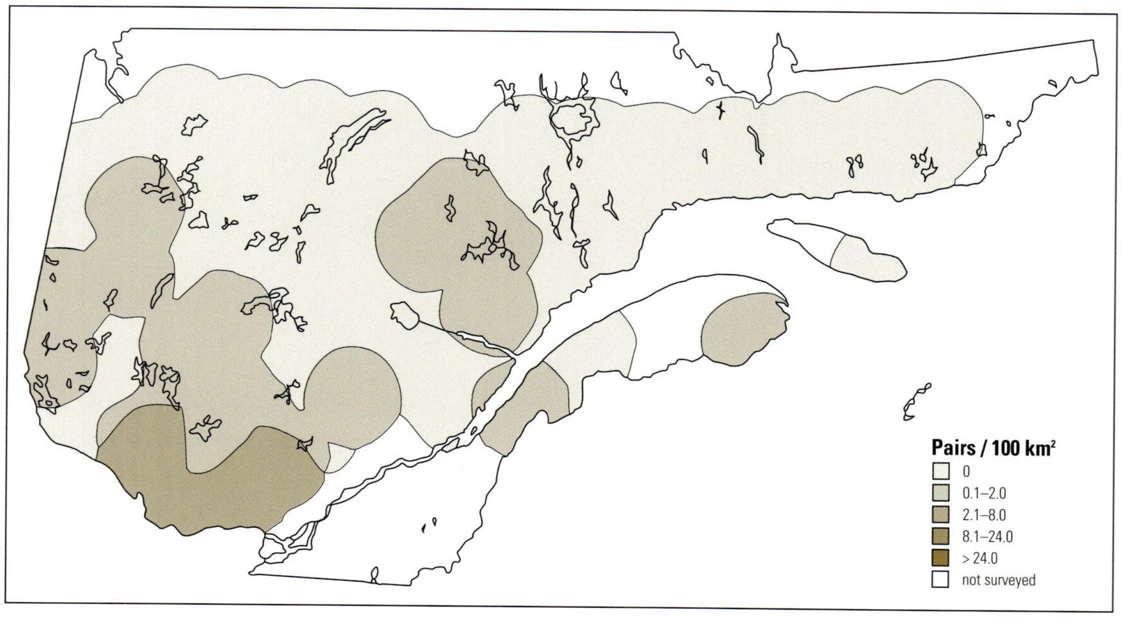

Abundance of Wood Ducks in southern Québec. Aerial surveys conducted in May 1990, 1991, and 1992 *(D. Bordage, CWS, unpubl. data).*

Wood Duck

Breeding distribution 1984–1989

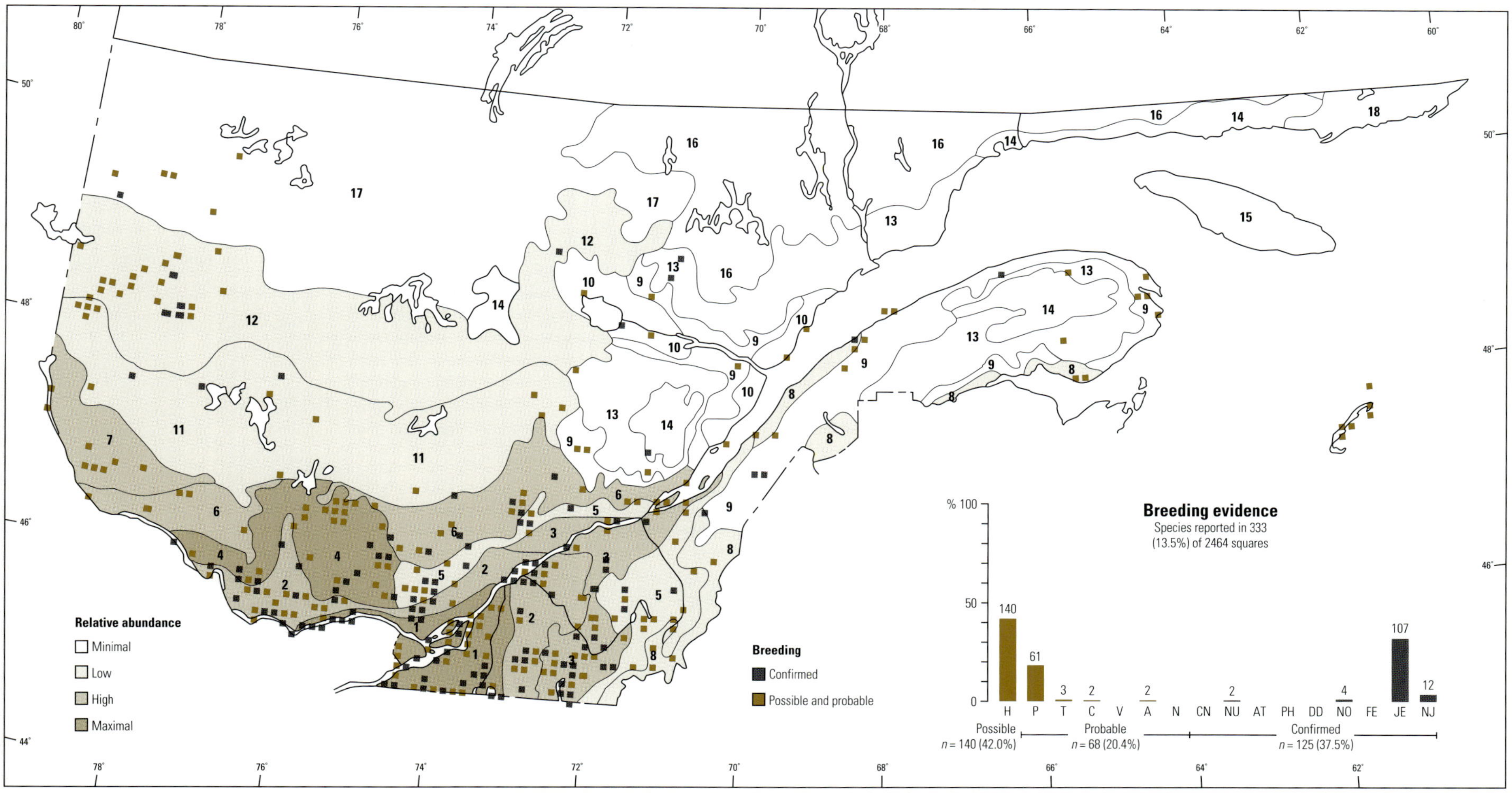

World breeding range

Species profile

Clutch size	usually 12 eggs (10–15), more than 15 = dump nest[3]
Laying interval	1 egg per day[2]
Start of incubation	with last egg[1]
Incubation period	27–37 days[2,4,5,6]
Maturity at hatching	precocial
Care of young	by female[7]
Nestling period	less than 48 hours[6]
Age at first flight	8–10 weeks[3]
Dependence of young	until ≈ 30–60 days old[3,9,10,11]
Broods per year	1, sometimes 2[6,10,11]
Breeding age	1 year[1]
Mating system	monogamy[7]
Duration of pair bond	1 season[1]
Total length	43–53 cm[4]
Wing span	71.1–76.2 cm[5]
Weight	M 681 g (?–907) / F 635 g (?–907)[1]
Longevity record	22 years 6 months[8]

References:
1. Palmer 1976
2. Leopold 1951
3. Bellrose 1976
4. Godfrey 1986
5. Terres 1980
6. DeGraaf and Rudis 1987
7. Ehrlich et al. 1988
8. Klimkiewicz and Futcher 1989
9. Ball et al. 1975
10. Fredrickson and Hansen 1983
11. Kennamer and Hepp 1987

Chronology $n = 133$

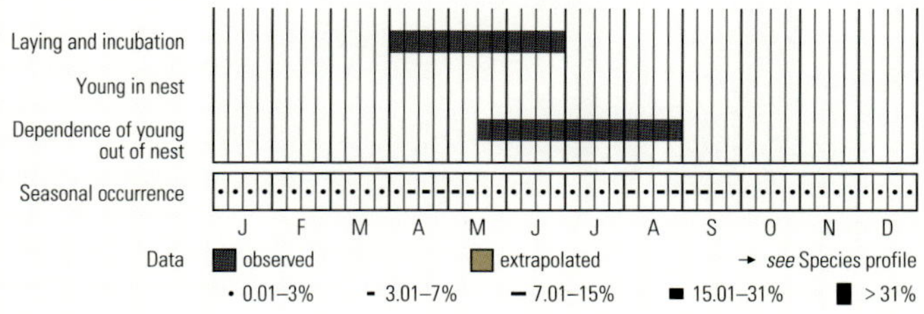

Frequency of occurrence by ecological region

Ecological regions** (forest regions)	Adequately covered squares	Squares in which reported	% of squares in which reported	Relative abundance (%)*
1. Sugar maple-hickory and sugar maple-basswood	97	44	45.4 ± 5.0	100.0
2. Sugar maple-basswood	153	56	36.6 ± 4.0	80.7
3. Sugar maple-basswood and sugar maple-yellow birch	101	33	32.7 ± 5.6	72.0
4. Sugar maple-yellow birch-basswood	56	25	44.6 ± 11.4	98.4
5. Sugar maple-yellow birch-beech and sugar maple-yellow birch-basswood	113	29	25.7 ± 4.5	56.6
6. Sugar maple-yellow birch-beech	71	25	35.2 ± 10.2	77.6
7. Sugar maple-yellow birch-balsam fir	19	6	31.6 ± 22.1	69.6
8. Sugar maple-yellow birch and balsam fir-yellow birch	47	7	14.9 ± 9.4	32.8
9. Balsam fir-yellow birch	87	9	10.3 ± 6.1	22.8
10. Balsam fir-white birch or balsam fir-red maple	46	4	8.7 ± 7.5	19.2
11. Yellow birch-balsam fir	52	11	21.2 ± 11.6	46.6
12. Balsam fir-white birch (west)	59	13	22.0 ± 11.0	48.6
13. Balsam fir-white birch (east)	76	6	7.9 ± 6.2	17.4
14. Balsam fir-black spruce	28	0	0	0
15. White spruce-balsam fir and balsam fir-white spruce	22	0	0	0
16. Black spruce-balsam fir-moss	18	0	0	0
17. Black spruce-moss	32	2	6.2 ± 10.0	13.8
18.		*not analysed*		

* minimal: 0–29.9%; low: 30–59.9%; high: 60–89.9%; maximal: 90–100%

** Deciduous forest: 1–7; mixed forest: 8–11; coniferous forest: 12–17

the Magdalen Islands, and although breeding was not confirmed during atlassing, the Wood Duck is considered to be an irregular breeder there (Fradette 1992).

In western Québec, the Wood Duck's range extends farther north than the Atlas data indicate. During recent surveys in the James Bay region, the species was found at several locations west of Waswanipi and south of Lake Evans (Cons. G. G. – GREBE 1992a). Pairs were also observed west of Rupert Bay and a brood was seen a few kilometres south of Matagami. Lastly, there are two isolated reports—an individual was sighted near the Pontax River, about 50 km west of Nemiscau, and a Wood Duck was heard in late June 1990 on the Eastmain River, north of Lake Mistassini, which is north of the 52nd parallel (Tardif and Morneau 1991).

History and trends

The Wood Duck was very abundant when the first European settlers arrived, but by the turn of the century local populations had become decimated by uncontrolled hunting and excessive habitat pressure (Taverner 1922; Bent 1923; Bellrose 1990; Ducks Unlimited 1991). The species did not receive full protection until 1918, when both the United States and Canada had enacted the protective measures stipulated in the 1916 Migratory Birds Convention (Dennis 1990). In Canada hunting was banned completely until September 1950 (Migratory Birds Regulations 1950).

At the same time, both public and private organizations initiated extensive nest box programs (Leopold 1951; Allen et al. 1990; Soulliere 1990; Ducks Unlimited 1991). It is estimated that there are currently over 100 000 of these boxes in North America (Ducks Unlimited 1991), and of those in the northeastern United States, approximately 44% are used by Wood Ducks (Soulliere 1990). According to Ducks Unlimited (1991), 300 000 ducklings hatch each year in these boxes. The number and quality of habitats have also increased in recent decades. Some forests cut down in the last century have regrown, and current logging practices protect the Wood Duck's habitat better, since strips of woodland are preserved around bodies of water (Ducks Unlimited 1991). The beaver has also made a strong comeback this century and is responsible for creating much favourable habitat for the Wood Duck (Soulliere 1990; Ducks Unlimited 1991). The effect of these combined factors has been spectacular. By 1950, the increase in the Wood Duck population made it possible to reopen the hunt in Québec to a limited degree. The population continued to expand through 1966–79 (Robbins et al. 1986). The Wood Duck has rebounded so well that it is now considered the most abundant species of duck in the eastern United States.

Notes on territory (continued)

According to aerial surveys conducted in May 1990, 1991, and 1992, the highest densities of Wood Ducks found in Québec, between 2.1 and 8.0 pairs/100 km^2, were in the Ottawa Valley in the southwestern part of the province (D. Bordage and M. Grenier, CWS, unpubl. data).

Trees and snags in this swamp have many hidden cavities where Wood Ducks can nest, while the tangles of shrubs, deadfalls, and herbaceous vegetation provide good cover for broods. Massey Road, Rockburn, south of Montréal. *June 1991, Pierre Bannon.*

A nest box like this one on a mature deciduous tree in a silver maple grove near a river is a potential nest site for the Wood Duck. Saint-Joachim-de-Courval, north of Drummondville. *June 1991, Daniel Jauvin.*

The banks of this slow-moving stream, which runs through a deciduous forest, have lush aquatic vegetation, providing excellent nesting habitat for the Wood Duck. Saint-Jean Creek, Châteauguay. *July 1991, Pierre Bannon.*

Habitat loss in the southern part of the province has been offset by the proliferation of beaver ponds since the 1940s (Hutchinson 1985; R. Lafond, MEF, pers. comm.), and especially since the 1960s (Dennis 1990). Nest boxes have also enabled the Wood Duck to make better use of its preferred habitats. In 1979, this bird was found only along a few rivers in southern Québec (Chapdelaine 1979b). Since then its range has expanded noticeably northward. Chapdelaine (1979b) estimated the number of breeding pairs at between 2000 and 10 000 in 1979, while Dennis (1990) recently proposed a figure of 4000 based on the data available, specifying that this was probably an underestimate.

DANIEL ST-HILAIRE AND ALAIN MORRIER

Green-winged Teal

Sarcelle à ailes vertes

Sarcelle d'hiver

Anas crecca

THE GREEN-WINGED TEAL IS NOT ONLY THE SMALLEST, BUT ALSO ONE OF the fastest fliers, among our ducks. It is not easy to identify on the wing. Its most reliable field characters are its swift wingbeats, iridescent green speculum with white borders, and white belly (revealed as it twists and turns agilely in flight). In the hand, this teal is about the size of a pigeon. The male is resplendent in his nuptial plumage, his rust-coloured head adorned with an iridescent green face patch sweeping back from the eye. The female is a drab mottled brown, with no outstanding field marks. The species' behaviour during the breeding season is poorly understood, mainly because nests are so hard to find.

The Green-winged Teal frequents small streams and beaver ponds, where it wades in the shallow water, filtering the mud in search of seeds and invertebrates. Near Castagnier, Abitibi. *May 1991, Yves Aubry.*

Until 1973, the Green-winged Teal (*Anas carolinensis*), which is restricted to North America, and its Eurasian counterpart the Common Teal (*Anas crecca*) were classified as two different species. They are now considered to be two races of the same species (AOU 1973).

The Green-winged Teal occurs throughout the Northern Hemisphere and is widely distributed in Québec. Despite its widespread distribution and its prominent rank in the waterfowl harvest, this dabbling duck has been largely neglected by North American waterfowl biologists, resulting in a paucity of scientific information on the species. This is due partly to the fact that on the Prairies, where most of the waterfowl research has been carried out, the Green-winged Teal is less abundant than other species of ducks. Furthermore, it occurs in low densities (despite its widespread distribution), which makes it difficult to observe and census in the field. Although the species breeds in Québec, it is easiest to observe during the spring and fall migrations when it stages along the St. Lawrence River.

Habitat and behaviour

Pair formation occurs mainly on the wintering grounds and during spring migration (Palmer 1976), when several males can be seen swimming around a female, displaying to her and whistling loudly (McKinney 1965) to attract her attention and intimidate rivals (McKinney et al. 1990). Although the Green-winged Teal is monogamous, paired males sometimes attempt to copulate with females who are also already paired (McKinney and Stolen 1982), leading to much aggressive behaviour among males. However, when no females are present, males are usually quite peaceable (Cramp and Simmons 1977). Nests are generally well spaced, although they may sometimes be found as little as only 1 m apart (Cramp and Simmons 1977).

In migration, the Green-winged Teal is found in shallow inland and brackish coastal waters with abundant floating and emergent vegetation, as well as in cultivated fields. For nesting, it prefers grassy areas or sedge or rush meadows with bushes, shrubs, or scattered trees. The nest is usually located on a dry, upland site, generally 40–100 m from a source of freshwater such as a small lake, pond, marsh, swamp, stream or river (Bellrose 1976; Palmer 1976). The female makes a shallow scrape in the ground, and lines it with dry grasses, leaves, feathers, and down. The nest is very well hidden in the dense grass, and is often located under a shrub or log (Peck and James 1983).

Typical clutch size is 8–12 eggs, ranging in colour from whitish cream to pale olive buff. The female incubates the eggs for 21–23 days. Whenever she leaves the nest, she covers her eggs with a thick layer of down and other lining materials so that they are well concealed and the nest blends in with its surroundings (Munro 1949).

Hatching is synchronous since incubation only begins after laying is complete (Cramp and Simmons 1977). At the onset of incubation, the drakes abandon their mates and gather in large marshes and lakes to moult (Bellrose 1976), leaving the females to incubate the eggs and raise the young alone. When the ducklings are only a few

Jean-Pierre Vallée

Notes on territory

During the breeding season, hostile encounters between males seem to be linked more to the protection of females from attacks by other males than to territorial defence (Cramp and Simmons 1977).

In Québec, Bordage (1987, 1988a,b), in an effort to determine the size of the species' home range, found the maximum distance between pairs' locations to be 1.8–3.3 km during aerial surveys.

In southern Québec, surveys by the Canadian Wildlife Service in May 1990, 1991, and 1992 revealed that the highest densities (4.1–16 pairs/km^2) occurred in the area stretching from the northwestern part of southern Québec and the Abitibi–Témiscamingue

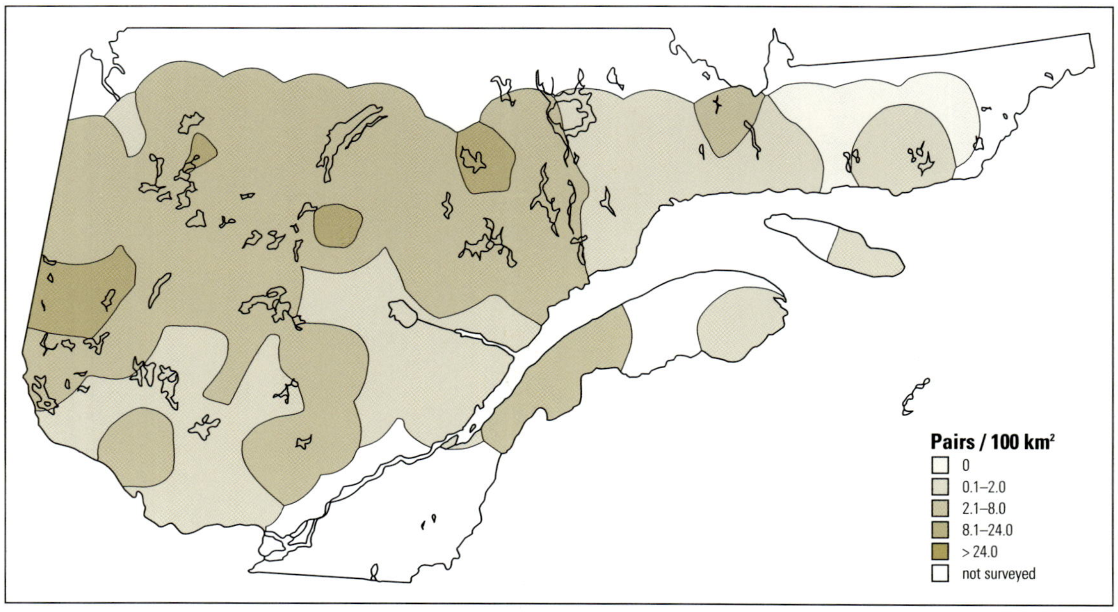

Abundance of Green-winged Teals in southern Québec. Aerial surveys conducted in May 1990, 1991, and 1992 *(D. Bordage, CWS, unpubl. data).*

Green-winged Teal

Breeding distribution

1984–1989

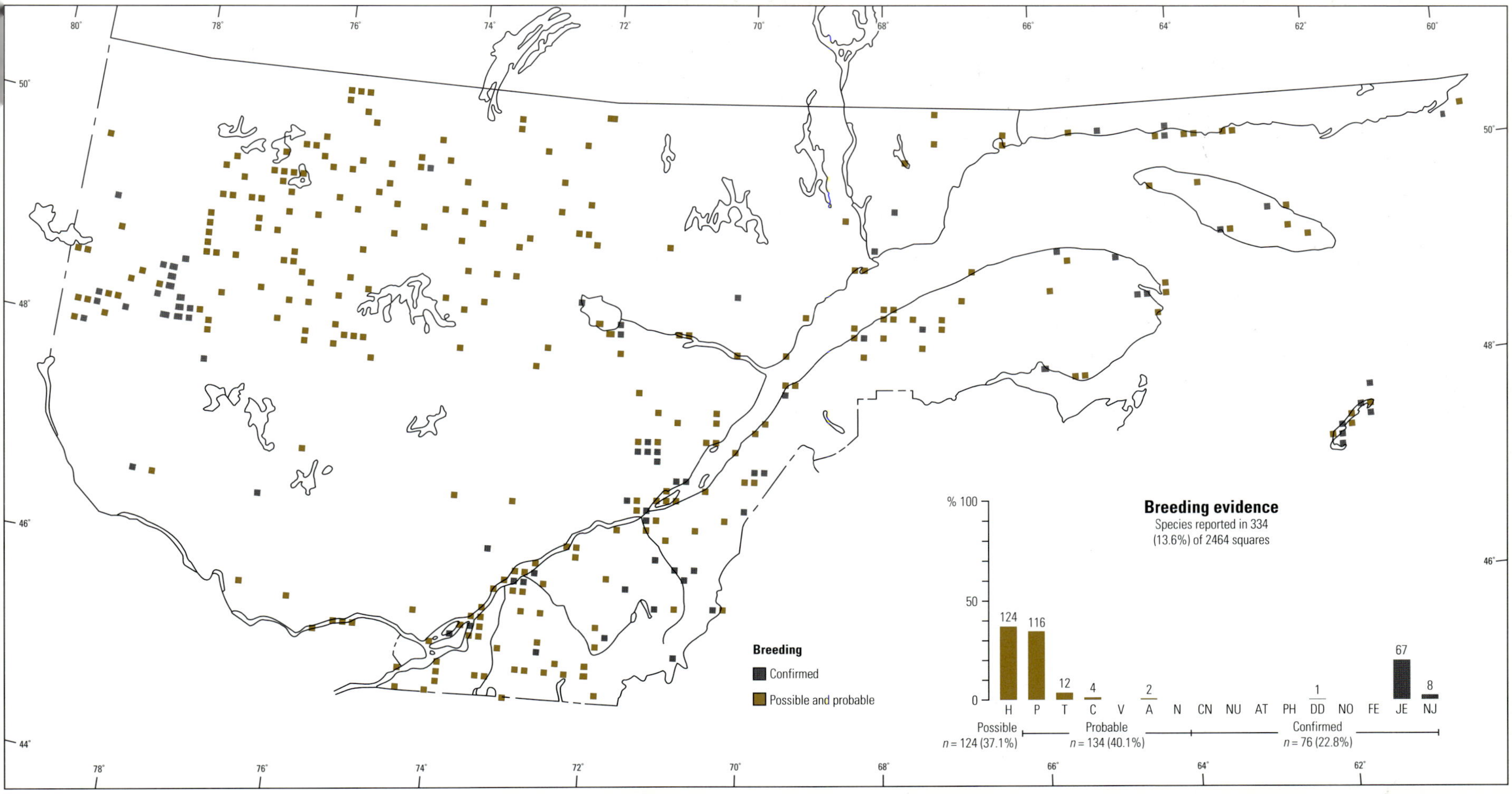

Species profile

Clutch size	usually 8–12 (5–18)[1,2,4,5,6,7,8,9,10,11,12]
Laying interval	?
Start of incubation	with last egg[5,10]
Incubation period	usually 21–23 days (20–25)[1,4,5,6,10,11,12]
Maturity at hatching	precocial
Care of young	by female[5,7]
Nestling period	less than 24 hours[2,4]
Age at first flight	34–44 days[1,2,6,10,11,12]
Dependence of young	≈ until fledging[5]
Broods per year	1[5]
Breeding age	1 year[1,2,4,5]
Mating system	monogamy[5]
Duration of pair bond	1 season[5]
Total length	M 34.8–39.4 cm / F 33.0–36.8 cm[2]
Wing span	50.8–63.5 cm[6]
Weight	M 322 g (226–500) / F 308 g (181–500)[2]
Longevity record	20 years 3 months[3]

References:
1. Palmer 1976
2. Bellrose 1976
3. Kennard 1975
4. DeGraaf and Rudis 1987
5. Cramp and Simmons 1977
6. Terres 1980
7. Bent 1923
8. Peck and James 1983
9. Godfrey 1986
10. Harrison 1978
11. Johnsgard 1975b
12. Ehrlich et al. 1988

Chronology *n* = 73

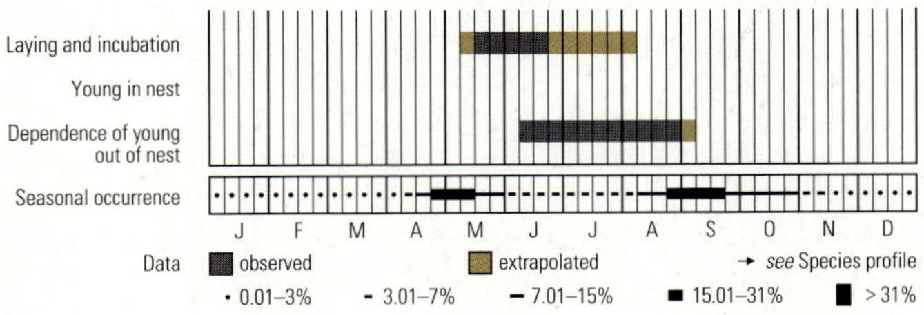

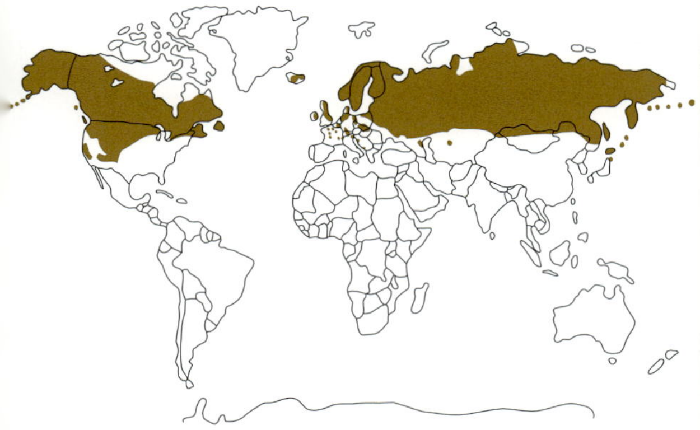

World breeding range

hours old, their mother leads them to the closest body of water, where they feed on their own (Bent 1923; Bellrose 1976; Harrison 1978). According to Bellrose (1976), average brood size ranges from 7 in class I (downy stage), to 5.7 in class II (young whose body, or contour, feathers are in the process of replacing the downy plumage), to 5.4 in class III (young with fully feathered juvenile plumage but incapable of flight); the decrease is one of the smallest observed in ducks. According to Munro (1949), the high rate of duckling survival in the species may be linked in part to the very protective behaviour of the female, who reacts energetically to intruders: on the water, she flies or runs over the surface beating her wings and quacking loudly to attract the predator's attention away from the young so as to give them a chance to hide in the nearby weeds; on land, she drags one or both of her wings in a broken-wing distraction display (Munro 1949; Johnsgard 1975b). Excited females may also join together to defend their broods (Munro 1949).

Most Green-winged Teals breed as yearlings. They produce only one brood a year but may lay a replacement clutch if the first clutch is lost early in the season (Cramp and Simmons 1977).

The species prefers to forage along muddy shores, although it also feeds in the shallow parts of marshes, ponds, slow-moving streams and rivers, and flooded farmland. It uses a wide range of foraging techniques including wading slowly through shallow water as it filters the mud with its bill and dabbling up-ended with its body half immersed, its feet kicking in the air and its bill probing the bottom. The Green-winged Teal's diet is composed mainly of the seeds of aquatic plants (sedges, smartweed, pondweed, etc.), small crustaceans and molluscs, and aquatic insects; in spring and summer, animal matter predominates (Mabbott 1920; Bellrose 1976; Cramp and Simmons 1977; Terres 1980). This teal may feed extensively at night, particularly when it is disturbed during the day (Bent 1923; Tamisier 1972, 1976; Palmer 1976).

The Green-winged Teal is one of the most agile fliers among the ducks; flocks twist and turn expertly like shorebirds. It is equally at home on land and in the air. On the water, it is an expert diver and can swim underwater for a long distance to escape danger (Bent 1923). Its main predators (of nests) are skunks and crows (Johnsgard 1975b; Bellrose 1976).

Outside the breeding season, the Green-winged Teal is gregarious. It often flocks with other species of ducks, although it appears to migrate mainly with those of its own kind, travelling in compact groups of 30–40, although larger flocks of hundreds or even thousands of individuals are sometimes seen (Bellrose 1976; Cramp and Simmons 1977).

The species was atlassed in 13.6% (334/2464) of the squares surveyed and breeding was confirmed in 22.8% (76/334) of these, mainly from reports of family groups. Nests, which were only reported in 8 squares, are very difficult to find since they are well hidden in the vegetation and the female usually does not flush until the last minute.

Distribution

In Canada, the Green-winged Teal nests in the Yukon, the Mackenzie and Keewatin districts of the Northwest Territories, and in all 10 provinces. In the United States, it breeds in Alaska, the Great Lakes region, and Maine, and in the West from North Dakota to Washington State south to California and New Mexico (National Geographic Society 1987).

In Québec, the Atlas survey confirmed the species as a breeder on the Gaspé Peninsula, where Godfrey (1986) had placed a question mark on his distribution map. In addition, the data revealed two main breeding areas in southern Québec. The first extends from the northwestern portion of southern Québec and the Abitibi region in the west to the Saguenay–Lac-Saint-Jean region in the east, while the second, farther south, consists of the Central St. Lawrence Lowland and the Appalachians. The species breeds in smaller numbers between these two areas, as confirmed by the results of aerial surveys carried out by the Canadian Wildlife Service in May 1990, 1991, and 1992, shown on the abundance map (D. Bordage and M. Grenier, CWS, unpubl. data).

Aerial surveys conducted by the Canadian Wildlife Service in the boreal forests of southern Québec from 1985 to 1987 to obtain data on the American Black Duck also provided invaluable information on the Green-winged Teal (which up to then had not been studied to any extent in Québec), confirming its presence in this region (Bordage 1987, 1988a,b). Subsequent surveys in May 1990, 1991, and 1992, which covered most of southern Québec (except for the St. Lawrence Valley, the Appalachians west of Québec City, and a few other areas), revealed that the species is particularly abundant from the northwestern part of southern Québec and the Abitibi–Témiscamingue region to the Manicouagan Reservoir, including the area northeast of Lake Saint-Jean. Densities of 4–16 pairs/100 km² were recorded in these regions (D. Bordage and M. Grenier, CWS, unpubl. data), showing that the species is more common in Québec than previously thought. An earlier study by Moisan et al. (1967) describing the species' breeding range from

In the forest, the species is found in a variety of wetland habitats, particularly small ponds. It conceals its nest in dense vegetation, under a shrub or a fallen log near the water. Decelles Reservoir, Témiscamingue region. *May 1991, Réginald Ouellet.*

Notes on territory (continued)

region in the west to the Manicouagan Reservoir in the east, including the region northeast of Lake Saint-Jean (D. Bordage and M. Grenier, CWS, unpubl. data).

This duck also nests in the drier, grassy parts of marshes, like the one shown here near the Natashquan River on the Lower North Shore. The nest is usually placed on the dry ground in a clump of tall vegetation, where it is well hidden. *August 1977, Pierre Lamothe.*

This dabbling duck is a common summer resident on the Magdalen Islands, where it is found most often near lagoons and marshy ponds. Grosse Île Bay. *August 1983, Yves Aubry.*

the survey data available at the time had concluded that densities were generally greater in western Canada than in Ontario and Québec: while densities of only 0.1–1.5 pairs/100 km^2 were reported in Québec, densities as high as 33–39 pairs/100 km^2 were obtained in the West, in the Athabasca River delta in northern Alberta and Great Slave Lake in the Mackenzie District of the Northwest Territories. The data in the Moisan et al. (1967) study come from aerial surveys using fixed-wing aircraft, while the 1990–92 surveys, co-ordinated by D. Bordage of the Canadian Wildlife Service, used helicopters, which had become more readily available by this time. Helicopters are better suited to Québec's dense forests than fixed-wing aircraft, which fly too fast even at minimum speeds and are better adapted to the more open areas in the West. Furthermore, the Green-winged Teal is difficult to observe on its breeding grounds, because of its small size and preference for areas with abundant emergent vegetation (Bellrose 1976). Therefore, the disparity in densities in the two surveys is probably due more to the use of different survey aircraft than to an actual increase in numbers in Québec.

North of the Atlas area, breeding has been confirmed as far north as the Nastapoka River (Bider and Lamothe 1982; Bordage 1985b; Cons. G. G. – GREBE 1990a; Tardif and Morneau 1991) and Kuujjuaq (Godfrey 1986).

In the fall, the most important staging area for the Green-winged Teal is the St. Lawrence estuary, particularly the Cap Tourmente area and the islands off Montmagny (Lehoux et al. 1985). Some flocks number almost 10 000 individuals (Lehoux 1978).

Bellrose (1976) stated that probably 90% of the Green-wings found along the Atlantic Flyway originate from breeding grounds in Ontario, Québec, Labrador, and the Atlantic provinces. The species winters in southern Alaska, British Columbia, southern Ontario, Nova Scotia, Newfoundland, most of the United States, and Mexico (AOU 1983; Godfrey 1986). Only a few overwintering individuals have been reported in Québec.

History and trends

Moisan et al. (1967) published an exhaustive account on the population dynamics of this species, based on the survey data available at the time, including the results of a continent-wide banding program in which 7000 of 106 000 banded individuals were recovered. Indeed, the banding station at Baie-Johan-Beetz in Québec, which has operated almost every year since 1947, has banded more Green-winged Teals than any other station in eastern North America. Before the Moisan et al. (1967) study, there were no published data on annual harvests in Canada. The following quantitative data on population dynamics are taken from this study, and Canadian Wildlife Service harvest data for 1973–92 (Appendix II).

On the basis of the annual autumn sport harvest, age ratios (immature:adult) in the harvests and the percentage of banding recoveries reported by hunters, Moisan et al. (1967) estimated that, between 1956 and 1963, the total North American population at the end of summer averaged almost 8 million, made up of 3.5 million adults and 4.5 million immatures. Furthermore, taking the results of aerial surveys from the same period and correcting against visibility bias by applying a correction factor obtained from complete ground counts in scattered locations, the same authors calculated that there were 8–10 million Green-wings in North America before the opening of the hunting season each year.

Bellrose (1976) reported an increase in the North American population on its breeding grounds between 1955 and 1974. Winter surveys undertaken during the same period also showed an upward trend along all four of the species' migration corridors (Pacific, Atlantic, Mississippi, and Central flyways). Since a large proportion of the Green-winged Teal population breeds in wetlands in the boreal forest north of Canada's farm belt, the species was relatively untouched by habitat loss compared to other species of ducks that nest mainly on the Canadian Prairies (Bellrose 1976). In addition, Reynolds (1985), after analysing the results of winter waterfowl surveys in 1964–83 along the Atlantic Flyway, concluded that there was no evidence of a decline in the breeding population using this migration corridor.

However, breeding pair surveys carried out by the Canadian Wildlife Service in Ontario pointed to declining numbers in that province after 1970 (Sandilands 1987b).

Roughly one million Green-wings have been harvested every year in North America since 1950. In Québec, an average of 55 000 teals are taken every year, or about 38% of the total Canadian harvest (Appendix II). Moisan et al. (1967) calculated that, depending on the year, 15–20% of the pre-hunting season population was harvested by hunters and the mean total annual mortality rate was 63%, with a higher rate among immatures (70%) than among adults (50%). According to Moisan (1966), increased hunting pressure could significantly increase adult mortality, making it difficult to sustain stable populations.

Between 1979 and 1991, the average number of migratory game bird hunting permits sold in Canada decreased by approximately 18 000 annually; the number of permits sold in Québec has also been dropping since the 1980s. The percentage of Green-wings in the total waterfowl harvest in Québec varied considerably during the same period. Nevertheless, the Green-winged Teal still ranks among the top three game species in the province (Appendix II).

Gaston Moisan

American Black Duck

Canard noir

Anas rubripes

Pierre Dupuis

The American Black Duck is the characteristic dabbling duck of lakes, rivers, and streams in Québec's mixed and coniferous forests and of the vast marshes along the St. Lawrence and its main tributaries. The species occurs throughout southern Québec during the breeding season, though in somewhat fewer numbers than in the past, particularly in the southwest. The American Black Duck breeds exclusively in northeastern North America, south of the tundra; southern Québec is the heart of its breeding range, with a large proportion of the world's population breeding here. For the last few decades, it has co-existed with the Mallard, a genetically similar species that breeds throughout almost all of the Northern Hemisphere below the tree line.

In migration, the American Black Duck sometimes gathers in large flocks along the province's main rivers. A few thousand overwinter in some areas in Québec (Reed and Bourget 1977), although most of the birds leave in fall for the species' main wintering grounds along the coast of the northeastern United States.

Many nesting pairs are found in peatlands in the interior of L'Isle-Verte. Indeed, some hens travel several kilometres with their young to reach a brood-rearing area in the estuary and it is not unusual to see broods crossing roads and ditches on their way to the intertidal marshes. Northeast of Rivière-du-Loup. *July 1991, Jean Bachand.*

American Black Ducks are relatively inconspicuous during the nesting period. Pairs tend to be widely dispersed over the many lakes, ponds, streams, and rivers in Québec's forests, although a few concentrations can be found in the salt marshes of the St. Lawrence estuary (Barske 1968; Reed and Moisan 1971; Reed 1975b). They are highly secretive during the breeding season, aside from a few wild aerial chases indulged in when defending the nesting territory (Seymour and Titman 1978). Moreover, both sexes have almost identical sooty brown plumage, which helps to camouflage them and makes spotting and identifying breeding pairs all the more difficult.

Habitat and behaviour

The American Black Duck is one of the first birds to nest in the spring. In the St. Lawrence Valley, laying begins in mid-April, both in the Montréal area (Laperle 1974) and the St. Lawrence estuary (Reed 1970). Typical clutch size is 6–12 eggs (mean = 9.2 eggs, *n* = 258 nests; Reed 1970), laid at an interval of one a day. If the nest is destroyed by a predator, a second clutch may be laid, though rarely after mid-June. The nest is usually built on the ground, sheltered in shrubs or herbaceous growth or under the low branches of a conifer (Reed 1970). In seasonally flooded lowlands, however, the nest may be built in the crotch of a tree, up to a few metres from the ground (Wright 1954; Laperle 1974). Islands are often selected as nest sites, especially when located near a suitable brood-rearing marsh, no doubt because of the absence of terrestrial predators. Black Ducks also nest on the mainland, usually near water, but it is not unusual to find nests that are located a fair distance from water, sometimes even more than 1 km away. Incubation lasts for 28 or 29 days on average (Reed 1970). While the female is incubating, the male goes off to moult, often fairly far from the nesting area. The male's reproductive contribution is thus ended for the year, and by the next spring, he will probably have paired with another female. But even though the pair bond only lasts a few months, females remain faithful to the breeding area, returning year after year to build their nests in the same area, which may even be where they were hatched (Stotts and Davis 1960; Coulter and Miller 1968; Reed 1970).

Shortly after the eggs hatch, the female leads her offspring to a marshy brood-rearing area. The journey from the nest to the brood-rearing site, conducted when the ducklings are only 2 or 3 days old, can be a long and hazardous one. In the St. Lawrence estuary, for example, broods hatched on Île aux Pommes have about a 5-km swim to their rearing site in a salt marsh on the south shore of the estuary (Reed 1970, 1975b; Lepage 1973). Such journeys are normally made without mishap, but if adverse weather conditions or predators are encountered, losses can be considerable (Reed 1975b). Once they reach the rearing site, the broods tend to be more sedentary. In one salt marsh in the St. Lawrence

Notes on territory

The breeding territory defended by the male depends on the type of habitat. In estuarine cord-grass marshes, the male defends an area of a few hundred square metres, beginning shortly before laying and continuing until the onset of incubation, which the pair uses mainly for feeding. Pairs which have nested in the cord-grass marsh at L'Isle-Verte Bay in the St. Lawrence estuary can be divided into 3 groups: pairs that nested on dry land next to the marsh, pairs that nested near farm fields up to a few kilometres inland, and pairs that nested on islands, up to 5 km from shore. All the pairs had territories in the marsh, but those that nested near the marsh included the nest in their territories.

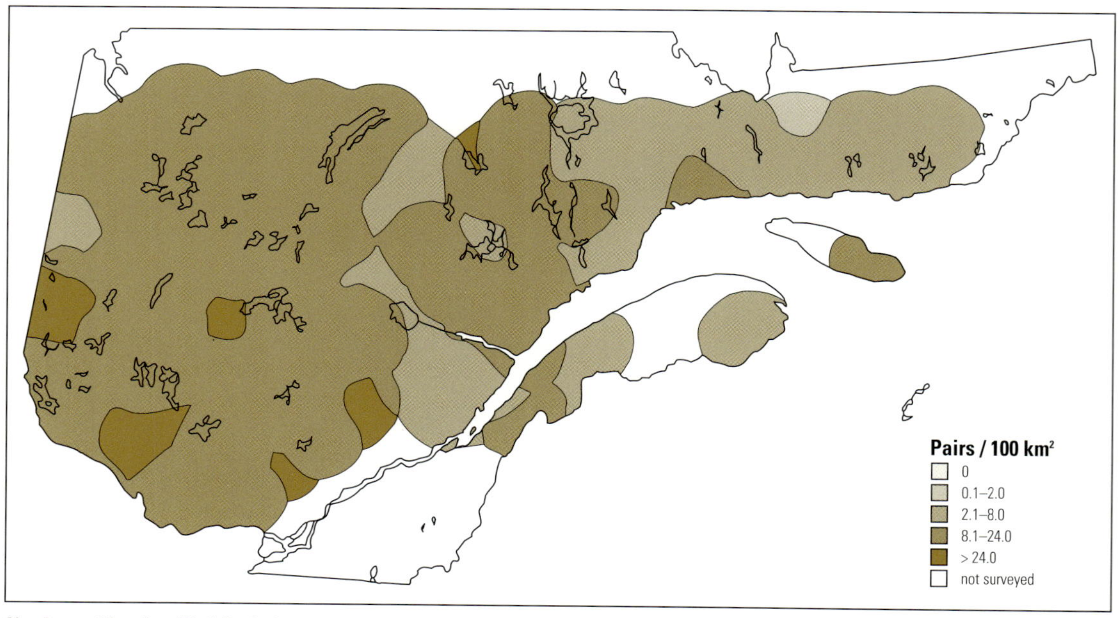

Abundance of American Black Ducks in southern Québec. Aerial surveys conducted in May 1990, 1991, and 1992 *(D. Bordage, CWS, unpubl. data).*

American Black Duck

Breeding distribution

1984–1989

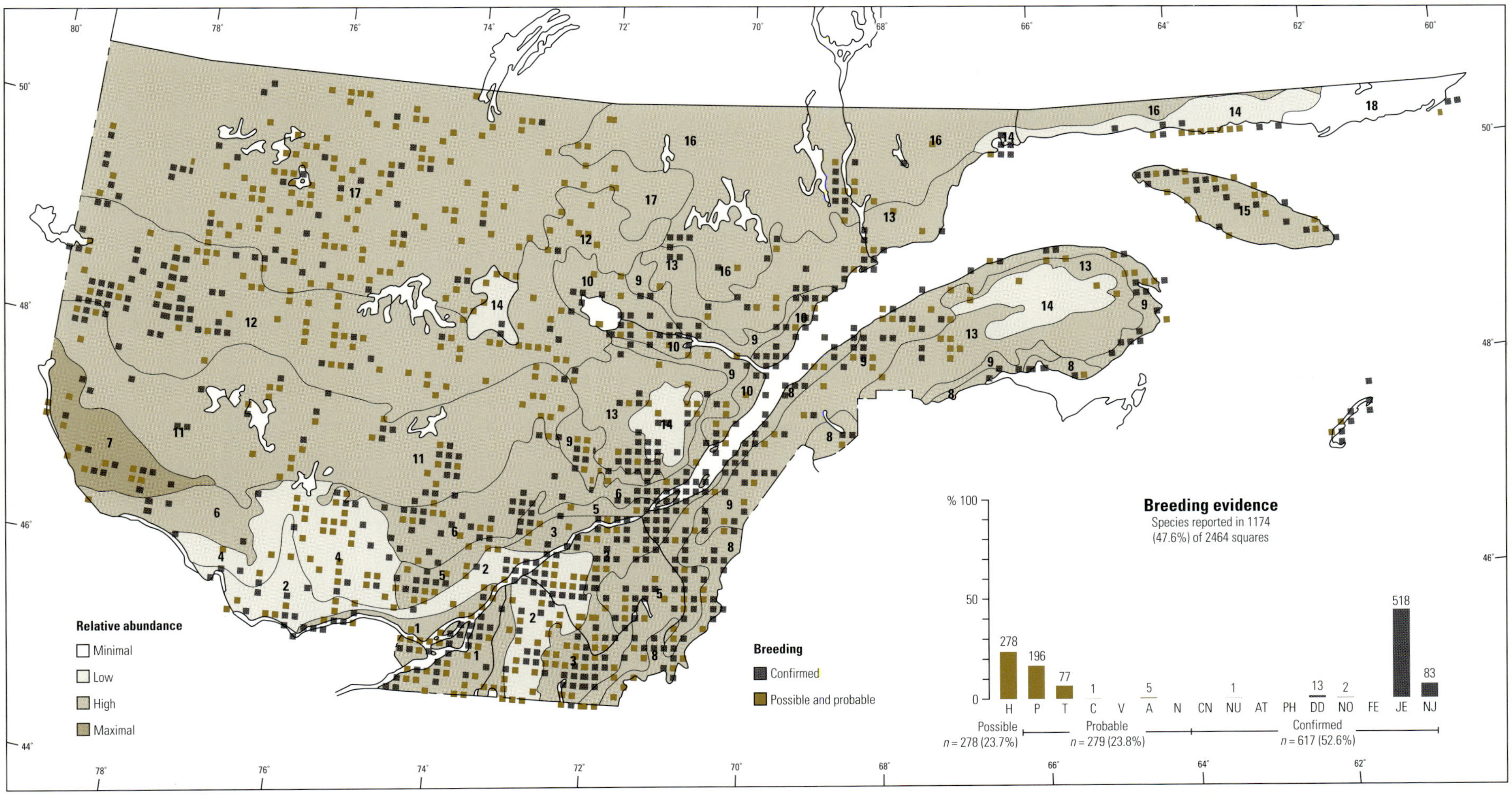

Species profile

Clutch size	usually 6–12 eggs[1,8]
Laying interval	1 egg per day[1,8]
Start of incubation	with last egg[1,8]
Incubation period	28–29 days[1,8]
Maturity at hatching	precocial
Care of young	by female[1,8]
Nestling period	less than 48 hours[1,8]
Age at first flight	58–62 days[2]
Dependence of young	until first flight[7]
Broods per year	1[1,8]
Breeding age	1 year[7]
Mating system	monogamy[7]
Duration of pair bond	1 season[7]
Total length	53.3–61.0 cm[3]
Wing span	91.4 cm[4]
Weight	M 1400 g (900–1800) / F 1100 g (900–1500)[5]
Longevity record	26 years 5 months[6]

References:
1. Reed 1970
2. Lepage 1973
3. Godfrey 1986
4. Pough 1951
5. Dunning 1984
6. Clapp et al. 1982
7. Bellrose 1976
8. A. Reed, CWS, unpubl. data

World breeding range

Chronology *n* = 1362

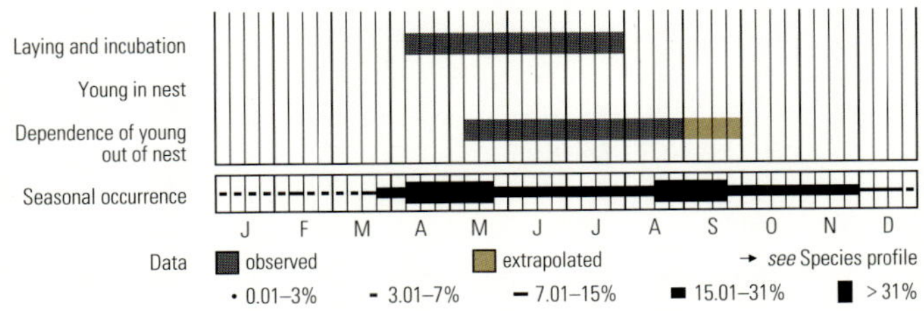

Frequency of occurrence by ecological region

Ecological regions** (forest regions)	Adequately covered squares	Squares in which reported	% of squares in which reported	Relative abundance (%)*
1. Sugar maple-hickory and sugar maple-basswood	97	59	60.8 ± 4.9	68.0
2. Sugar maple-basswood	153	79	51.6 ± 4.1	57.7
3. Sugar maple-basswood and sugar maple-yellow birch	101	64	63.4 ± 5.7	70.8
4. Sugar maple-yellow birch-basswood	56	28	50.0 ± 11.4	55.9
5. Sugar maple-yellow birch-beech and sugar maple-yellow birch-basswood	113	66	58.4 ± 5.1	65.3
6. Sugar maple-yellow birch-beech	71	49	69.0 ± 9.9	77.1
7. Sugar maple-yellow birch-balsam fir	19	17	89.5 ± 15.5	100.0
8. Sugar maple-yellow birch and balsam fir-yellow birch	47	32	68.1 ± 12.0	76.1
9. Balsam fir-yellow birch	87	53	60.9 ± 9.5	68.1
10. Balsam fir-white birch or balsam fir-red maple	46	32	69.6 ± 11.6	77.7
11. Yellow birch-balsam fir	52	39	75.0 ± 12.2	83.8
12. Balsam fir-white birch (west)	59	37	62.7 ± 12.7	70.1
13. Balsam fir-white birch (east)	76	44	57.9 ± 10.8	64.7
14. Balsam fir-black spruce	28	13	46.4 ± 19.3	51.9
15. White spruce-balsam fir and balsam fir-white spruce	22	17	77.3 ± 18.4	86.4
16. Black spruce-balsam fir-moss	18	11	61.1 ± 25.5	68.3
17. Black spruce-moss	32	21	65.6 ± 18.0	73.3
18.			*not analysed*	

* minimal: 0–29.9%; low: 30–59.9%; high: 60–89.9%; maximal: 90–100%

** Deciduous forest: 1–7; mixed forest: 8–11; coniferous forest: 12–17

This American Black Duck nest, made of grasses, plant stalks, and down, is well hidden in the surrounding growth of shrubs, grasses, and forbs. Sainte-Foy. *May 1957, Gaston Le Page.*

estuary, individual broods were found to settle along stretches of marshy shoreline ranging from 0.7 to 1.6 km in length (Reed 1970; Lepage 1973). In these areas, broods mainly frequent the upper marsh, which is dominated by salt-meadow cord-grass (L. Bélanger, CWS, unpubl. data).

Several types of habitat are used for brood-rearing, including freshwater and salt marshes, beaver ponds, and slow-moving rivers with lush emergent vegetation. These habitats have in common an abundance of invertebrates (mainly insect larvae) for food and tall, dense vegetation for cover for the ducklings. In salt marshes, broods use small tidal ponds (< 200 m²) for foraging (≈ 40% of total time budget) and the surrounding salt-meadow cord-grass for resting and cover. The young also forage on the ground for insects or snap them up in the air (L. Bélanger, CWS, unpubl. data).

In the St. Lawrence Valley, brood-rearing extends from approximately mid-May to the end of August; in the northern part of the Atlas area, it begins 1–2 weeks later but ends at basically the same time (Reed 1970; Laperle 1974). The ducklings grow rapidly, their weight increasing from about 44 g at hatching to 870 g at fledging, 58–62 days later (Lepage 1973).

Several factors limit reproductive success in the species. About 40% of nests studied around Lake Saint-Louis (Laperle 1974) and along the St. Lawrence estuary (Reed 1975b) were destroyed by predators, but in the end, since some females renested, roughly 65% of pairs produced ducklings (Reed 1975b). Numerous ducklings die during the brood-rearing period from predation, exposure, exhaustion, or disease. According to Reed (1975b), duckling mortality in the St. Lawrence estuary is around 66%; furthermore, the cumulative effects of egg and duckling loss limit productivity to less than 2 ducklings/breeding female on average. Red foxes and Peregrine Falcons have recently been identified as significant predators of ducklings in this region (L. Bélanger, CWS, unpubl. data).

Hens and their broods are still present on southern Québec's major rivers in September when the duck hunting season opens. These local populations are subjected to intensive hunting pressure (Reed and Boyd 1974), with no respite until other Black Ducks arrive from more northerly breeding sites.

The American Black Duck's diet varies widely with the season. After hatching, the young ducklings feed mainly on insect larvae and other invertebrates, but, as they grow older, consume a greater proportion of seeds of aquatic plants. These seeds continue to make up a major part of the species' diet through early autumn, although wild berries or even cereal grains may also be consumed at this time. Later in the fall, the ducks switch to a diet of mainly small molluscs and amphipods, which they continue to feed on in winter; nonbreeders congregating along the James Bay coast probably have a similar diet.

Distribution

The American Black Duck breeds throughout most of southern Québec and is the most abundant waterfowl species here, with an average density of about 15 pairs/100 km² (D. Bordage, CWS, unpubl. data). Closer examination of the Atlas map, however, shows squares where the species is absent, mainly in the St. Lawrence Plain west of Québec City, and in the interior of the Gaspé Peninsula. The gaps in distribution in the St. Lawrence Plain are no doubt linked to habitat loss from agriculture, industrialization, urbanization, and other human activities, coupled with considerable hunting pressure. On the Gaspé Peninsula, the steep terrain and fast-flowing streams and rivers do not provide the type of wetland habitat the species needs. Populations farther north are also subject to habitat disruption; acid rain may render lakes unsuitable for the American Black Duck, which avoids acidic lakes (DesGranges and Darveau 1985). Forest disturbances such as logging and forest fires may also drastically alter the landscape, but their effects on waterbirds are not fully understood. Hydro-electric megaprojects create vast flooded areas that are largely shunned by dabbling ducks (Bordage 1985a).

Recent surveys in the James Bay region (the watersheds of the Nottaway, Broadback, and Rupert rivers) yielded breeding pair and brood density estimates of 7.7 pairs/100 km² and 2.3 broods/100 km², respectively (Cons. G. G. – GREBE 1992a). Farther north, in the Eastmain River watershed, upstream from the reservoir, breeding pair density was estimated at 12.1 pairs/100 km² and brood density at 2.9 broods/100 km² (Tardif and Morneau 1991). On the northeast coast of James Bay, breeding pair densities of 4.6 pairs/100 km² and brood densities of 1.6 broods/100 km² were obtained (Benoit et al. 1993b).

Notes on territory (continued)

Pairs that nested some distance inland sometimes defended a second territory around the nest. As a general rule, pairs that nested on islands did not defend a territory around the nest. The situation is often much less complex in other habitats. In forested areas, for example, each drake defends a small lake or beaver pond, which is used both for foraging and nesting (Reed 1970).

This wide range of behaviour suggests that territoriality in the American Black Duck has two functions. The first is to allow the female to feed undisturbed during the critical period of egg formation, and the second is to ensure greater spacing between nests, which would minimize losses from terrestrial predators (Reed 1970).

View of the canals and tidal ponds in a typical cord-grass marsh used by the species for brood-rearing. Each female defends a territory encompassing one or more ponds. Kamouraska, southwest of Rivière-du-Loup. *Summer 1970, Austin Reed.*

The drake defends the territory by energetically pursuing in flight any pair that invades his domain. Occasionally, he may threaten intruders and even engage in combat on the water. Defence of the territory ends during the first half of incubation, when the male leaves for a moulting area (Reed 1970; Seymour and Titman 1978; Titman and Seymour 1981).

Studies conducted between 1989 and 1992 on the use of coastal habitats by the American Black Duck in the L'Isle-Verte region provided varying nest density estimates depending on the habitat (Bélanger et al. 1994) The lowest nest densities occurred in farmland with only 0.4 nest/100 ha. Densities in coastal meadows were nearly 6 times higher, at 2.2 nests/100 ha. The highest densities, 22.5 nests/100 ha, were in peatlands that were being, or had been, mined. Within these peatlands, areas dominated by low ericaceous shrubs and tamarack or black spruce and jack pine had the highest densities, 34.5 and 61.2 nests/100 ha, respectively. Lastly, the unexploited strips in the peatlands had the highest density, with 90.9 nests/100 ha.

History and trends

Data on population trends in the species for North America as a whole were not available until the 1950s. Although several studies had been conducted prior to this, they generally covered small areas and were too local in scope to provide a global perspective. Ornithological literature from the turn of the century mentions another race or subspecies of the American Black Duck, the Red-legged Black Duck, a "large black duck of the north;" it was often treated as a separate species, which created a certain amount of confusion about the status of the continental population at the time. Macoun and Macoun (1909) considered the American Black Duck to be "the common wild duck of Newfoundland and the Maritime Provinces...in Québec and Ontario it still breeds in diminished numbers but evidently its chief breeding grounds are towards the Atlantic coast and Hudson Bay. A few stragglers reach Manitoba..." We now know, however, that the highest breeding densities occur locally in Québec, and this in all probability has always been the case.

In the boreal forest, the species can nest in the herbaceous growth along meandering streams. Near Castagnier, north of Val-d'Or, Abitibi. *May 1991, Yves Aubry.*

The results of annual aerial waterfowl surveys conducted in midwinter since 1955, flown mainly along the eastern seaboard of the United States, point to a decline in American Black Duck populations (Rogers and Patterson 1984). The only quantitative data for southern Québec before the 1980s covering the whole area north of the St. Lawrence are from aerial surveys carried out by the United States Fish and Wildlife Service in 1955 and 1956 and between 1963 and 1966 (Kaczynski and Chamberlain 1968). These data showed a continuous decline in breeding population indices from 1963 to the end of the surveys in 1966.

Hatching is synchronous in the species, which means that all the eggs in the clutch hatch within a few hours. The ducklings' down has not yet dried, but in a few hours they will leave the nest, led by the hen. Saint-Pierre, Island of Montréal. *June 1935, Henry Mousley, Laval University Collection.*

A first symposium on management of the species was held in 1968 in Chestertown, Maryland (Barske 1968). However, in 1982, controversy erupted over Black Duck management when the Humane Society of the United States, the Maine Audubon Society, and a private citizen tried to close the Black Duck hunting season in the United States (Grandy 1983; Feierabend 1984). In 1984, the United States government reduced Black Duck harvest quotas, and in 1985 the Canadian government followed suit. The Black Duck is still one of the leading game species in eastern Canada, however. In Québec, for example, it is the most harvested species, comprising 25% of the migratory game bird harvest (Appendix II). In 1986, the North American Waterfowl Management Plan (NAWMP) set the goal of restoring American Black Duck populations to their levels in the 1970s by the year 2000. To ensure follow-up of the NAWMP objectives, the Black Duck Joint Venture (BDJV) was set up with representatives of both the United States and Canada, creating, among other things, a survey network throughout eastern North America. This joint project was implemented in 1990, its initial phase consisting of a series of annual surveys until 1994. In early 1990, a second symposium on the American Black Duck was held in Saint John, New Brunswick, to take stock of the research that had been conducted until then and to determine future research needs.

In Québec, a pilot project was carried out between 1985 and 1989 in a 100 000 km² area north of the Gouin Reservoir (Bordage 1987, 1988a,b). It revealed that the measures taken, mainly the hunting restrictions, had had a positive effect on the American Black Duck population, with numbers increasing significantly. However, recent BDJV surveys revealed a significant decrease in the Black Duck breeding population in Québec between 1990 and 1994 (D. Bordage, CWS, unpubl. data). Along with hunting pressure, habitat loss and hybridization with the Mallard are often cited as causes of the species' decline.

There is little information on current American Black Duck densities in the St. Lawrence Plain except in the Boyer River area near Québec City, where 29.0 pairs/100 km² and 27.6 broods/100 km² were reported in 1992 (Maisonneuve et al. 1993). In mixed forests, however, average densities are about 18 pairs/100 km² and, farther north, in the boreal forest, 13 pairs/100 km² (D. Bordage, CWS, unpubl. data). Although densities are locally high in southern Québec, distribution is much less uniform than in the boreal forest, as can be seen by comparing the abundance map for the species with the Atlas distribution map. Breeding habitats for the species in southern Québec, especially between the Ontario border (Cornwall) and Québec City, have undergone considerable change, particularly in agricultural and urban areas. These changes have allowed such opportunistic species as the Mallard and Gadwall, which at one time were uncommon, to colonize the region.

Daniel Bordage and Austin Reed

Mallard

Canard colvert

Anas platyrhynchos

The Mallard is one of the best known and most abundant waterfowl species in North America. Its ready association with humans makes it one of our most familiar ducks, while its large body size and the male's distinctive green head help make it one of the easiest waterfowl to identify.

The Mallard, the ancestor of all domestic duck species except the Muscovy Duck (Terres 1980), has one of the widest breeding distributions among waterfowl, ranging throughout Europe, Asia, North America (Madge and Burn 1988), and Australia (Blakers et al. 1984). Considered a scarce breeder in Québec at the turn of the century (Ouellet 1974a), the Mallard is now very common in the southern part of the province.

Pierre Dupuis

The Mallard nests in extensive marshes like these along the Ottawa River. Daragon Bay, east of Hull. *July 1988, Jacques Chabot, Ministère du Loisir, de la Chasse et de la Pêche.*

Habitat and behaviour

Mallards nest in a wide range of habitats, including marshes, grassy meadows, sloughs, ponds, lakes, and rivers throughout prairies, forested areas, and tundra regions (Johnsgard 1975a; Godfrey 1986). The nest is usually built on the ground near water (Godfrey 1986), and is well concealed in tall vegetation (25–125 cm), such as grasses, nettles, thistles, raspberry bushes, or even fallen trees (Johnsgard 1975a). In southern Québec, on small islands subject to late spring flooding, Mallards have been observed nesting up to 2 m off the ground in crotches and holes in trees (Laperle 1969). In flooded woodlands in the Montezuma National Wildlife Refuge in New York, Mallards commonly nest in sturdy snags and in the crotches of trees (Cowardin et al. 1967). Nests are made from grass, reeds, and leaves, and are lined with down (Godfrey 1986).

Mallards are seasonally monogamous, with pair formation mainly occurring on the wintering grounds and the male subsequently following the female back to her breeding area, to which she returns each year (Bellrose 1976). Upon arrival, the pair establishes a home range, encompassing a small territory which the male aggressively defends. Although males maintain firm pair bonds, promiscuity is common. Males continue to pursue females, regardless of paired status, and will attempt to mate, usually by forced copulation, with as many females as possible (Cramp and Simmons 1977). Evarts and Williams (1987) estimated that a large proportion (≥17–25%) of broods in their study area in the Prairie pothole region of southwestern Manitoba involved multiple parentage, that is, broods contained ducklings fathered by a male other than the hen's mate. Clutch size ranges from 6 to 15 eggs, but typically 7–10 eggs are laid (Bent 1923; Bellrose 1976). Feeding experiments using captive Mallards have shown that reducing the protein content

Notes on territory

Mallards do not seem to defend a territory in the classic sense; instead the female is the male's focus of defence (Raitasuo 1964). Dzubin (1955) noted that Mallards do not defend rigid areas and there is some overlap of territories with neighbouring pairs. Attacks by males are usually directed only at conspecifics, and may be launched from a drake's loafing area or from wherever he, or the pair, is feeding (Dzubin 1969). The size of the territory, to which the female goes when she is off the nest (Stokes 1979), varies with the individual, the nesting density, and the stage of the reproductive cycle. Drakes may chase intruding females or pairs for distances up to 183 m during the pre-nesting period and for at least 1 km during the early incubation

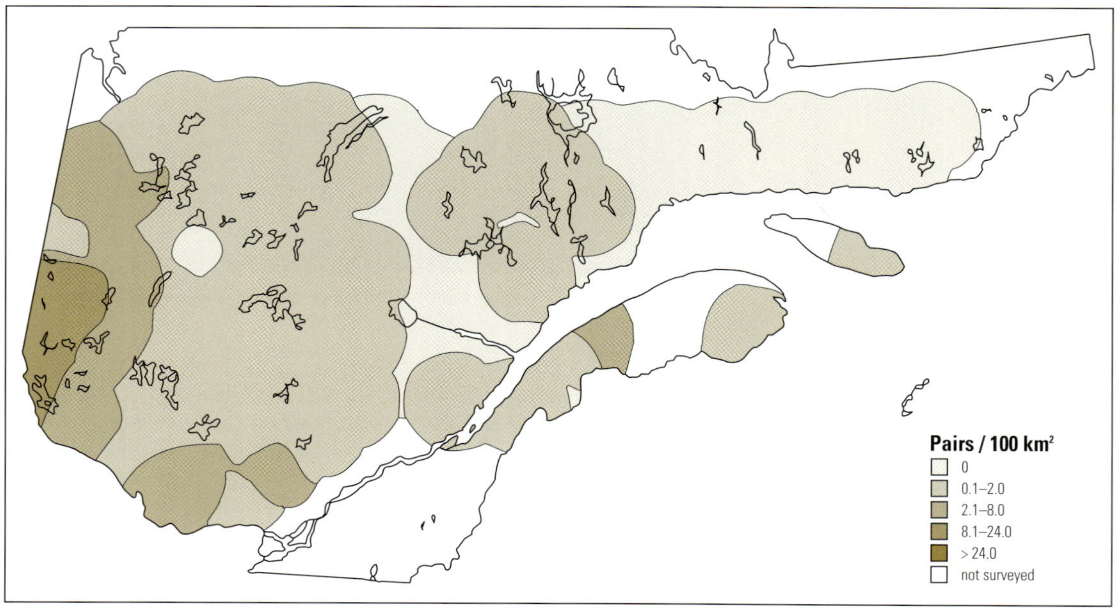

Abundance of Mallards in southern Québec. Aerial surveys conducted in May 1990, 1991, and 1992 *(D. Bordage, CWS, unpubl. data).*

Mallard

Breeding distribution
1984–1989

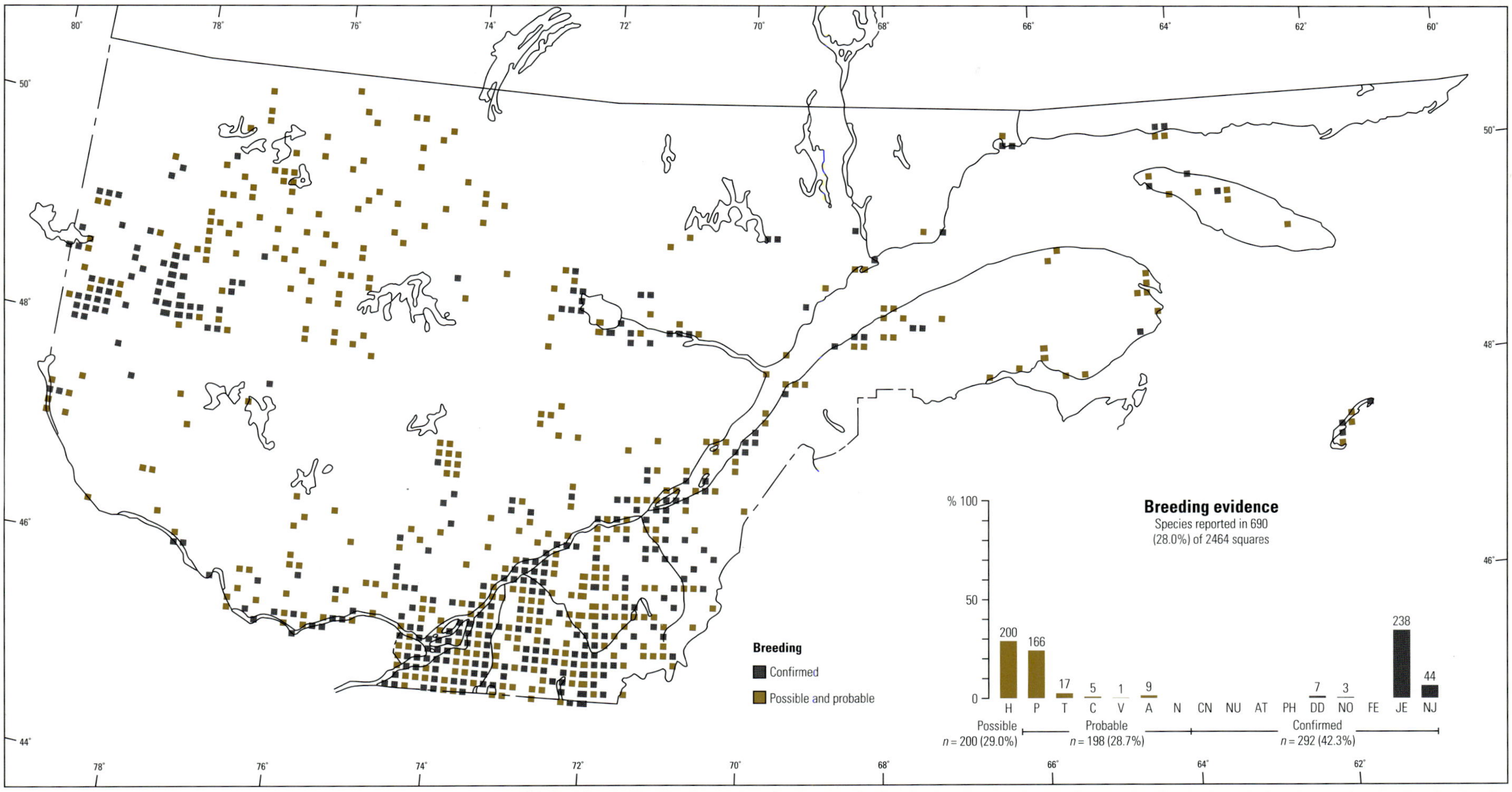

Species profile

Clutch size	usually 7–10 eggs (6–15)[1,3,6]
Laying interval	1 egg per 1–2 days[1,2]
Start of incubation	with last egg[1,3,4]
Incubation period	usually 28 days (25–31)[1,3,4]
Maturity at hatching	precocial
Care of young	by female[1,2,3]
Nestling period	less than 24 hours[1]
Age at first flight	42–60 days[2,6]
Dependence of young	usually until first flight[1,2,6]
Broods per year	1[1,2]
Breeding age	1 year[1,2]
Mating system	monogamy[1,2]
Duration of pair bond	1 season[1]
Total length	50.0–68.5 cm[4]
Wing span	78–101 cm[1]
Weight	M 1247 g (680–1724) / F 1107 g (544–1724)[6]
Longevity record	29 years 1 month[5]

References:
1. Palmer 1976
2. Johnsgard 1975a
3. Bent 1923
4. Godfrey 1986
5. Kennard 1975
6. Bellrose 1976

Chronology $n = 503$

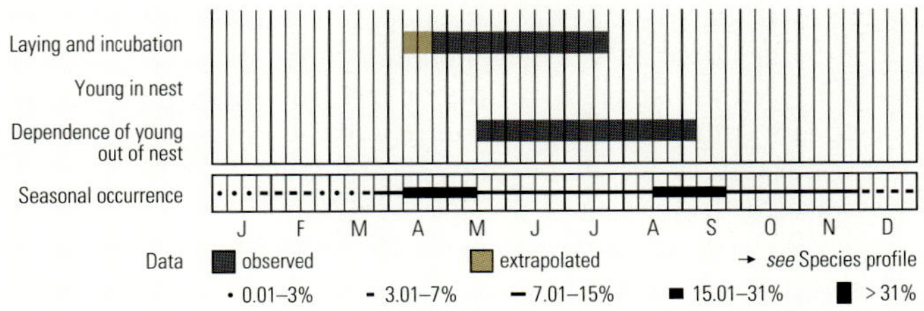

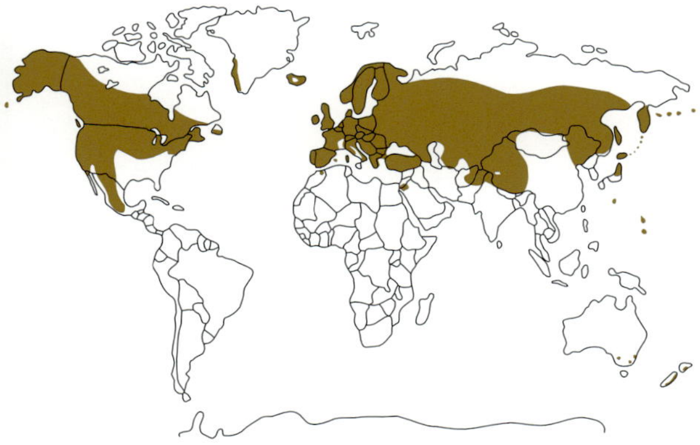

World breeding range

Ponds located in agricultural areas make excellent habitat for brood rearing. Grande Ferme sector, Cap Tourmente National Wildlife Area, northeast of Québec City. *August 1991, Yves Aubry.*

of the female's diet results in smaller egg and clutch sizes (Eldridge and Krapu 1988; Pehrsson 1991). Only the hen incubates for the 28-day incubation period. The pair bond weakens with the onset of incubation, with the male deserting the female from as early as the start of incubation until as late as the third or fourth week (Johnsgard 1975a). Soon after the ducklings hatch, the hen leads the brood to water. On Lake Saint-François, southwest of Montréal, Thompson (1974) observed that broods travelled approximately 1 km from their nest islands to other small islands with good plant cover. The hen stays with the ducklings until they fledge; age at first flight varies from 42–43 days in Alaska to 49–60 days in Delta, Manitoba (Bellrose 1976).

Like other dabbling ducks, Mallards feed mostly on vegetation, in particular the seeds and shoots of sedges, bulrushes, and aquatic vegetation, as well as grain and acorns (Bellrose 1976; Ehrlich et al. 1988). They are frequent visitors to farmland in southern Québec, but do less damage to crops there than in the Prairies (Bellrose 1976). Approximately 10% of their diet includes aquatic insects, molluscs, tadpoles, frogs, and small fish (McAtee 1918), but this can increase to nearly 90% with the onset of egg laying. For the first three weeks, ducklings forage primarily on land or the water surface for insects. Afterwards, they gradually switch from insects to seeds (Cramp and Simmons 1977).

Distribution

The Mallard breeds widely in southern Québec, and was reported in 28% (690/2464) of Atlas squares surveyed. It was particularly widespread in the Central St. Lawrence Lowland, south of the St. Lawrence River, and in the Ottawa Valley. It was also common in agricultural areas from the Gaspé to the Abitibi–Témiscamingue Clay Belt, as well as on Anticosti Island and the Magdalen Islands, and in coastal areas of the Gaspé and the North Shore. A large number of squares in the western boreal forest also had breeding Mallards.

In northern Québec, Mallard broods have been observed up to Lakes Nudant and Bienville (Savard 1977; D. Bordage, CWS, unpubl. data) while summer sightings from far north as Tasiujaq have been reported (ÉPOQ).

Mallards breeding in Québec migrate south through New England and winter in the eastern United States (Bellrose 1976). Every year, a few Mallards overwinter on ice-free stretches of water in the southern part of the province, joining small mixed groups of other dabblers (Reed and Bourget 1977; David 1980a). Spring migrants arrive in southern Québec in late March or early April (David 1980a).

History and trends

At the end of the 19th century, the Mallard nested only in central and western North America. Its range has expanded northward and eastward during the last 100 years, partly as a result of habitat modification and partly from propagation and release programs (Palmer 1976; Long 1981).

Taverner (1934) reported that at the turn of the century, Mallards were breeding throughout Canada, but mostly west of Ontario. By the 1960s, they were common in southwestern Québec, but scarce in the eastern part of the province (Heusmann 1991); hunters at that time looked upon the Mallard as quite a prize in the St. Lawrence Valley. Today the Mallard breeds throughout southeastern Québec as well as in the Atlantic provinces (Godfrey 1986; Montevecchi and Tuck 1987; Erskine 1992). Changes in land-use patterns, human disturbance, changing agricultural practices, urbanization,

Nesting sites in the Îles de la Paix National Wildlife Area are frequently flooded. When this happens during the breeding season, Mallards build their nests in low crotches in trees. Near Montréal. *April 1991, Jean-Luc DesGranges.*

Notes on territory (continued)

period (Dzubin 1957). Pairs are therefore widely dispersed, and nests well spaced, making them less susceptible to predation.

A study in Lousana, Alberta, found a very high nesting density of 36 pairs/km² (Smith 1964). In Prairie pothole habitat in South Dakota, average densities of 2.3 pairs/km² (Drewien and Springer 1969) and 2.6 pairs/km² (Duebbert 1969) have been recorded. In a study in Saskatchewan, Stoudt (1969) reported an average of 10.8 pairs/km², while on an island in Lake Champlain, Vermont, Coulter and Miller (1968) found 29 nests in an area of 2.3 ha (12.6 nests/ha).

The highest densities of Mallards reported during aerial surveys of the boreal forest in southern Québec conducted in May of 1990, 1991, and 1992 were in the Abitibi–Témiscamingue region, with a range of 8.1–24 pairs/ 100 km² (D. Bordage and M. Grenier, CWS, unpubl. data).

Mallards generally arrive on their breeding grounds already paired and immediately begin looking for a nest site. Saint-François-du-Lac, Lake Saint-Pierre. *April 1985, Daniel Jauvin.*

and intensive logging have all contributed to the Mallard's eastward expansion (Heusmann 1991).

In much of its Québec breeding range, the Mallard nests alongside the American Black Duck, with which it interbreeds. This has caused concern among wildlife biologists, who fear for the future of some Black Duck populations (Kirby et al. 1987), especially in areas where forest habitat has been subject to severe disturbance. Recent genetic studies, however, have led researchers to hypothesize that the American Black Duck is merely a colour morph of the Mallard (Ankney et al. 1986).

Mallard nest hidden in tall herbaceous vegetation. Baie-du-Febvre, south shore of Lake Saint-Pierre. *May 1981, Daniel Jauvin.*

The Mallard has been, and continues to be, one of the most sought-after game species in eastern North America. Before the Migratory Birds Convention was adopted early this century to protect and manage populations, Mallards were shot in the hundreds by market hunters in the United States (Bent 1923). In the eastern United States, population decreases of 50–90% were reported between 1880 and 1910 (Forbush 1912). Although hunting is now carefully regulated, the harvest of Mallards each fall is still very large. In Canada, from 1968 to 1982, between 1 and 2 million Mallards were taken each fall, accounting for approximately 45% of the total waterfowl harvest each year (Bartonek et al. 1984). In Québec, however, the Mallard ranks second in the harvest after the Black Duck. Between 1984 and 1992, almost 83 000 Mallards were taken each fall, accounting for 21% of the total waterfowl kill in the province (see Appendix II).

Although the North American Mallard population numbers some 6 million, it has been decreasing in recent years (USFWS and CWS 1986). The steepest declines have been recorded on the Prairies, with the major factor cited being intensive agriculture and its associated impacts: habitat destruction, increased efficiency of nest predators in residual habitat, and nest destruction caused by farm machinery (Nudds and Cole 1991).

Generally, Mallards breeding in the boreal forest are not subject to agricultural disturbances. A study comparing mean numbers of breeding pairs and broods, and the ratio of broods to pairs in a population of Mallards near Yellowknife, Northwest Territories, found no significant changes in population size and breeding success between surveys conducted in 1962–65 and 1985–88 (Nudds and Cole 1991). Although changes in hunter harvest rates affect the size of Mallard populations (Reynolds and Sauer 1991), the decline in populations on the Prairies is more the result of events on the breeding grounds that inhibit breeding success (such as habitat loss) than events that decrease survival rates at other times of the year (Nudds and Cole 1991).

Richard C. Cotter, Pierre Dupuis, and Derin Henderson

Northern Pintail

Canard pilet

Anas acuta

Ghislain Caron

THE ELEGANT NORTHERN PINTAIL IN SPRING IS AN UNFORGETTABLE SIGHT, with its sleek, streamlined profile and graceful neck. The male has a long tail and distinctive white markings extending from the breast up the neck to two thin points on either side of the head (Johnsgard 1975b; Godfrey 1986). Regionally, this dabbling duck is known by a variety of names, including *le pointu* (sharp-tail) along the Lower St. Lawrence and *le gris* (grey duck) at Lake Saint-Pierre.

An early migrant (Robert 1989b; Fredrickson and Heitmeyer 1991), the Northern Pintail is one of the first arrivals in the province in spring, and can be seen on streams, rivers, and floodlands from early April onward. During this period, large flocks congregate regularly along the freshwater course of the St. Lawrence River and the estuary (Lehoux et al. 1985; CWS and MLCP 1986).

Habitat and behaviour

Pair formation takes place around mid-December on the species' wintering grounds, so most females have found mates by January (Smith, R. I. 1968; Fredrickson and Heitmeyer 1991). Pairs arrive on the breeding grounds as soon as there is some open water, around early April.

Suitable habitats include shallow freshwater environments, such as floodlands and open marshes with low, sparse vegetation (Smith, R. I. 1970; Peck and James 1983; DeGraaf and Rudis 1987; Fredrickson and Heitmeyer 1991). The Northern Pintail also frequents agricultural habitats—pastureland, grain stubble, uncultivated fields, grainfields, and even roadsides (Bélanger 1991). According to Fredrickson and Heitmeyer (1991), the Northern Pintail, along with the Blue-winged Teal, are the dabblers that most frequently utilize farmland.

Man-made ponds are also used extensively by the species during the breeding season. Bélanger and Couture (1989) noted that, in the Lake Saint-Pierre region, Northern Pintail broods made up 64% of the broods observed in this habitat. The Northern Pintail is often one of the first species to move into newly created habitats, leading Hochbaum and Bossenmaier (1972) to describe it as a pioneering species.

Breeding begins as soon as pairs arrive on the breeding grounds. The Northern Pintail usually starts nesting earlier than other dabblers (Fredrickson and Heitmeyer 1991), and broods have been observed as early as the second week of May. The nest is a shallow scrape in the ground, lined with fine grass, plant stalks, and dead rushes. Most of the time, the Northern Pintail builds its nest under small trees or shrubs or in tall herbaceous vegetation a few metres from water, but nests are also frequently found much farther (over 90 m) from water (Bellrose 1976; Peck and James 1983).

Notes on territory

The Northern Pintail does not appear to defend a territory during the breeding season (Johnsgard 1975b). R. I. Smith (1968) noted that drakes become more aggressive during egg laying, but this

The Northern Pintail nests in open areas, like this cord-grass marsh in Gaspé Bay. Gaspé Peninsula.
April 1992, Réginald Ouellet.

Northern Pintail

Breeding distribution
1984–1989

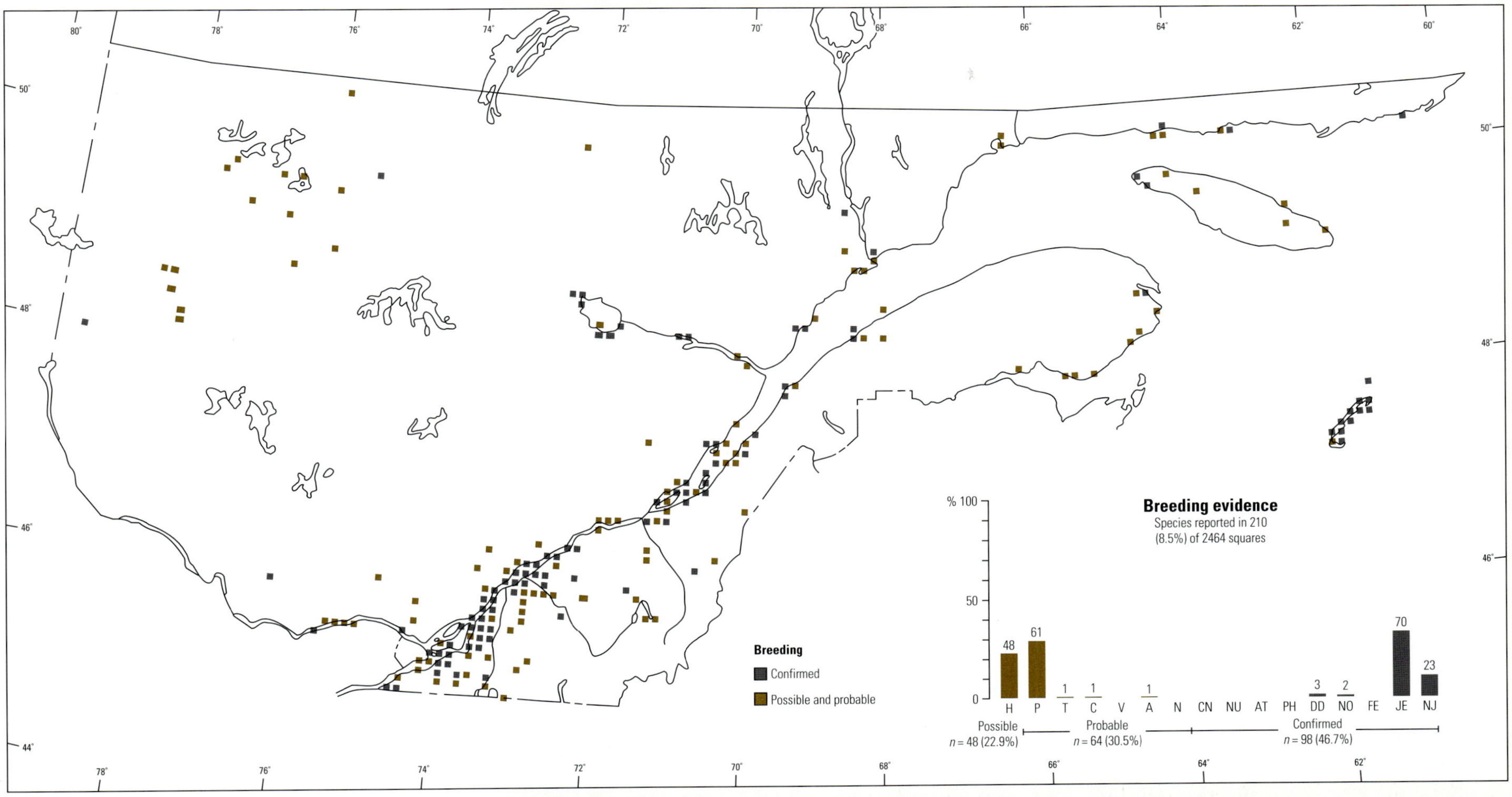

Species profile

Clutch size	usually 8 eggs (3–14)[4]
Laying interval	1 egg per day[3,10]
Start of incubation	with last egg[3,4]
Incubation period	21–25 days[1,2,5,6,10]
Maturity at hatching	precocial
Care of young	by female[6,7,8,9]
Nestling period	less than 24 hours[1]
Age at first flight	36–57 days[3,4,5,7,8]
Dependence of young	usually until first flight[10]
Broods per year	1[1,3,4]
Breeding age	1 year[1,3,4,8,9]
Mating system	monogamy[7]
Duration of pair bond	1 season[9]
Total length	M 66–76 cm / F 51–61 cm[6]
Wing span	M 92.5 cm / F 87.5 cm[10]
Weight	M 928 g (746–1111) / F 762 g (601–845)[10]
Longevity record	26 years 6 months[11]

Chronology *n* = 491

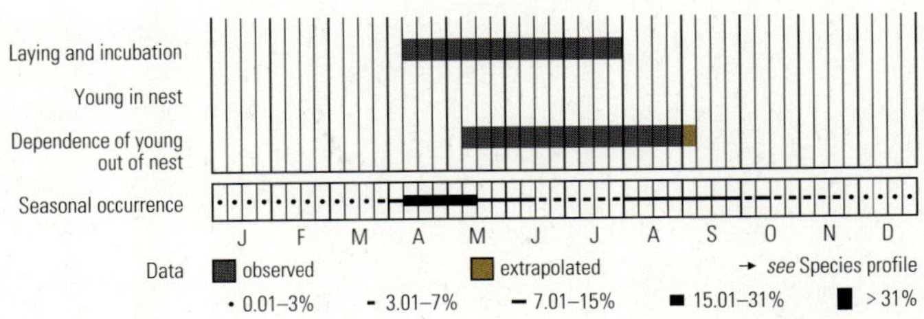

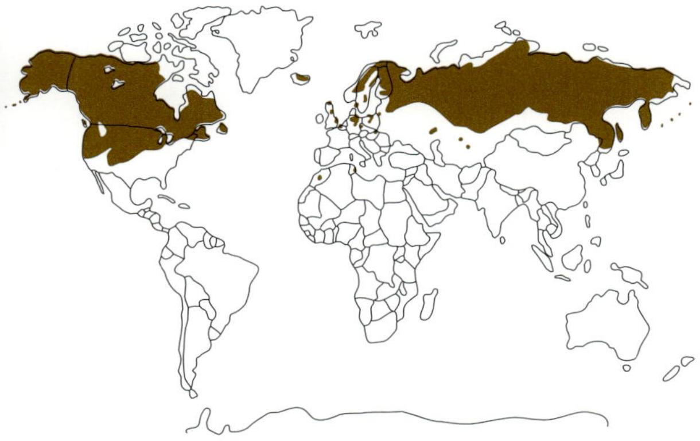

World breeding range

References:
1. DeGraaf and Rudis 1987
2. Peck and James 1983
3. Cramp and Simmons 1977
4. Bellrose 1976
5. Terres 1980
6. Godfrey 1986
7. Ehrlich et al. 1988
8. Fredrickson and Heitmeyer 1991
9. Johnsgard 1975b
10. Palmer 1976
11. Kennard 1975

This Northern Pintail nest, made of grasses, plant stalks, and down, does not have much plant cover since it is still early in the season. Varennes Islands, near Montréal. *May 1993, Luc Bélanger.*

This anatid often nests in farm areas, foraging on nearby bodies of water. Pabos, Gaspé Peninsula. *April 1992, Réginald Ouellet.*

Northern Pintail nest hidden in low growth in an uncultivated field. Lake Saint-Pierre region.
May 1991, Réginald Ouellet.

Notes on territory (continued)

was the extent of their territorial behaviour. During this time, the hen is constantly harassed by courting males.

Brood density in the Lake Saint-Pierre region was 0.68 brood/ha (27.2 broods/40 ha), according to Bélanger and Couture (1989). In Alberta, Keith (1961) reported 12 pairs/40 ha of water. In North Dakota, Stewart and Kantrud (1972) found 12 pairs/km² (4.8 pairs/40 ha), while, in South Dakota, Drewien and Springer (1969) recorded only 2 pairs/km² (0.8 pair/40 ha).

In rural environments, nests can be found out in the open, in meadows or ploughed or cultivated fields, where they are completely unprotected. Such nests are often destroyed by farm machinery (Bélanger 1991).

Typical clutch size is 8 eggs, which are incubated by the female for 21–25 days. The drakes desert the hens at the onset of incubation and flock together to moult, remaining flightless for 3–4 weeks. After moulting, the drakes begin to migrate to the wintering grounds in stages, stopping several times to feed.

After the eggs have hatched, the hen leads her offspring to a good rearing area, sometimes quite a distance away (Johnsgard 1975b; Bellrose 1976). In the Delta Marsh in Manitoba, Sowls (1955) observed that some hens and their broods travelled 0.5–0.7 km within 24 hours of hatching to reach their brood-rearing areas. The female alone tends the young, staying with them until they are able to fly. Age at first flight is 36–57 days, depending on the region (Johnsgard 1975b; Bellrose 1976; Fredrickson and Heitmeyer 1991).

The females begin their moult after raising their broods and, like the males, tend to flock together at this time. Later, in October and November, they gradually join the males on the staging areas and wintering grounds (Fredrickson and Heitmeyer 1991).

Although the Northern Pintail is considered an opportunistic omnivore (Fredrickson and Heitmeyer 1991), its diet is made up chiefly of plants and seeds. Nonetheless, in spring, before and during laying, females feed mainly on aquatic invertebrates (Krapu 1974; Krapu and Swanson 1975).

The Northern Pintail was reported in 210 (8%) of the 2464 Atlas squares surveyed, and breeding was confirmed in 46.7% of these squares.

Distribution

The Northern Pintail is not usually found in forested regions in southern Québec, but is quite common along the St. Lawrence River between Cornwall and Kamouraska. Habitats along this stretch of the river range from submerged and emergent vegetation between Cornwall and Grondines, to bulrush marshes between Grondines and La Pocatière, to cord-grass marshes between La Pocatière and Rimouski (Lehoux et al. 1985; Lehoux and Bourget 1987). Elsewhere, the Northern Pintail is less common and very local in distribution. It is also found on Anticosti Island and the Magdalen Islands.

Along the St. Lawrence River and in the Abitibi and Lac-Saint-Jean regions, the Northern Pintail is associated with farmland.

In northern Québec, the Northern Pintail is the most abundant breeding dabbler in the coastal lowlands along Hudson and James bays as well as along southern Ungava Bay (David 1980a; Bider and Lamothe 1982; Cons. G. G. – GREBE 1990a).

History and trends

At the turn of the century, Wintle (1896) considered the Northern Pintail a scarce "transient visitant" in the Montréal area. Macoun and Macoun (1915) as well as Ouellet (1974a), referred to it for that time period as an abundant transient; breeding had not as yet been confirmed, however. Subsequently, an increase in Northern Pintail numbers was documented and the species colonized the Montréal area, where breeding was reported for the first time on Nuns' Island, in June 1946 (Ouellet 1974a).

On Anticosti Island, sight and breeding records date back to the turn of the century (Ouellet 1969). On the Magdalen Islands, however, the species was not recorded until August 1949; since then, the population on the Magdalens has continued to increase and since 1981, the pintail has been the most commonly observed dabbling duck there (Fradette 1992).

Deforestation and the clearing of land for agriculture have probably encouraged the establishment of populations of Northern Pintails in Québec. However, habitat loss as a result of massive draining of marshes is a potential limiting factor on populations concentrated along the St. Lawrence River.

Reed (1978) estimated the population of Northern Pintails in southern Québec at approximately 9580 breeding pairs. Each spring, flocks of up to 10 000 birds can be observed on the floodlands around Lake Saint-Pierre, while in fall, such large flocks are more likely to be seen in the Cap-Tourmente area (Lehoux et al. 1985). The Northern Pintail is hunted for sport in southern Québec and for subsistence in the north. Since 1973, the annual sport harvest has ranged from 6000 to 40 000 birds, with the lowest figures occurring in recent years (Appendix II).

According to BBS data, the Québec population of Northern Pintails is increasing, while populations in the western part of the continent have declined considerably over the past two decades (Environment Canada 1993).

Raymond McNicoll and Josée Tardif

Blue-winged Teal

Sarcelle à ailes bleues

Anas discors

THE BLUE-WINGED TEAL CAN BE IDENTIFIED BY ITS SMALL SIZE, RAPID flight, and chalky blue forewing.

A migrant breeder in southern Québec, wintering in the southern United States, Central America, and South America (AOU 1983), the Blue-wing is one of the last dabbling ducks to return to its breeding grounds in spring and one of the first to leave in fall (Bent 1923; Bellrose 1976). The males leave the breeding grounds before the females and immatures (Bellrose 1976).

Pairs are formed on the wintering grounds and during spring migration (Bellrose 1976; Palmer 1976). It seems unlikely that females stay with the same mate for more than a year (Johnsgard 1975b).

Habitat and behaviour

The Blue-winged Teal usually nests in grass on dry ground (Lepage 1976; CWS and MLCP 1986) near shallow water: marshes, swamps, ponds, sloughs, small lakes, and slow-moving rivers (AOU 1983; Godfrey 1986). Nests can be found in grassy meadows and farm fields (Livezey 1981b) and along ditches and roadsides (Lokemoen et al. 1990). This teal is one of the species of dabblers that most often nests in agricultural areas (Bélanger 1991).

The hen chooses the nest site with her mate (Bellrose 1976). The species seems to prefer nesting in bluegrass (Bellrose 1976; Miller 1976; Livezey 1981a,b; Masse and Raymond 1988), but sometimes also selects alfalfa, grasses such as quack grass (Livezey 1981a), or sedges (Miller 1976).

This duck prefers a location with fairly tall vegetation—up to 60 cm according to Peck and James (1983)—to hide its nest. Bellrose (1976) reported that nests are usually found 20–200 m from water.

Nest building begins 3–4 days before laying (Dane 1966). According to Peck and James (1983), the nest is made of dead grass stems, straw, sedges, cattails, and bulrushes, and is lined with dead leaves, moss, feathers, and down. The hen continues to add down while she is laying (Bellrose 1976; Peck and James 1983), and continues even during incubation (Bent 1923). She also covers the eggs with down when she leaves the nest (Bent 1923; Palmer 1976).

Typical clutch size is 9–10 eggs (Bull 1985), laid at an interval of one a day (Bent 1923; Bellrose 1976). The incubation period is 23–27 days (Bellrose 1976; Peck and James 1983). The hen may, on occasion, renest if the first nest is destroyed (Palmer 1976), but the size of replacement clutches generally decreases as the season progresses (Dane 1966; Lokemoen et al. 1990). The pair bond is very strong (Bellrose 1976), and the drake does not leave the hen until the end of the second (Dzubin 1955; Oring 1964) or third week of incubation (Stewart and Titman 1980).

The young leave the nest less than 24 hours after hatching (DeGraaf and Rudis 1987) and follow the hen into a marshy area where they can find food and cover (Lepage 1976). At this age, their diet consists chiefly of aquatic invertebrates. The family remains together until the young are able to fly (Bent 1923; Oring 1964), at 35–44 days old (Gollop and Marshall 1954). The female is very protective, and when danger threatens, she makes every effort to drive off or lure away the predator or intruder, allowing the ducklings to take refuge in the tall grass (Bent 1923). By 14 months of age, the yearlings have acquired their full adult plumage (Bent 1923).

There is some difference of opinion over territoriality in the species. Stewart and Titman (1980) noted that drakes defend small, well-defined territories that contain nesting, feeding, and resting sites. According to these authors, territorial boundaries remain stable from the onset of nesting until the third week of incubation, and neighbouring territories do not overlap. Dzubin (1955) observed that the male defends a territory within its home range against conspecifics. Contrary to Stewart and Titman (1980), however, Dzubin (1955) considered the boundaries of the territory to be neither fixed nor well defined, but variable over time, and described the territory as "ameboid," in shape. McHenry (1971) concluded that the Blue-winged Teal drake defends a small space around the hen, rather than a specific area. In short, the territorial behaviour of the Blue-winged Teal seems to vary from place to place.

According to Lokemoen et al. (1990), the Blue-winged Teal has little tendency to return to nest in the same area from year to year, unlike other ducks such as the Mallard and Gadwall. One year, this fairly opportunistic bird may be abundant at a site where it was absent the year before because of poor conditions (Lokemoen et al. 1990).

Ghislain Caron

Notes on territory

Biologists are divided on the extend of territoriality in the Blue-winged Teal. Stewart and Titman (1980) reported that this species defends a territory of approximately 0.69 ha, with non-overlapping boundaries that are stable from the onset of nesting until the third week of

This shallow pond surrounded by tall aquatic grasses is ideal habitat for the Blue-winged Teal. Cap Tourmente National Wildlife Area, northeast of Québec City. *August 1991, Yves Aubry.*

Blue-winged Teal

Breeding distribution

1984–1989

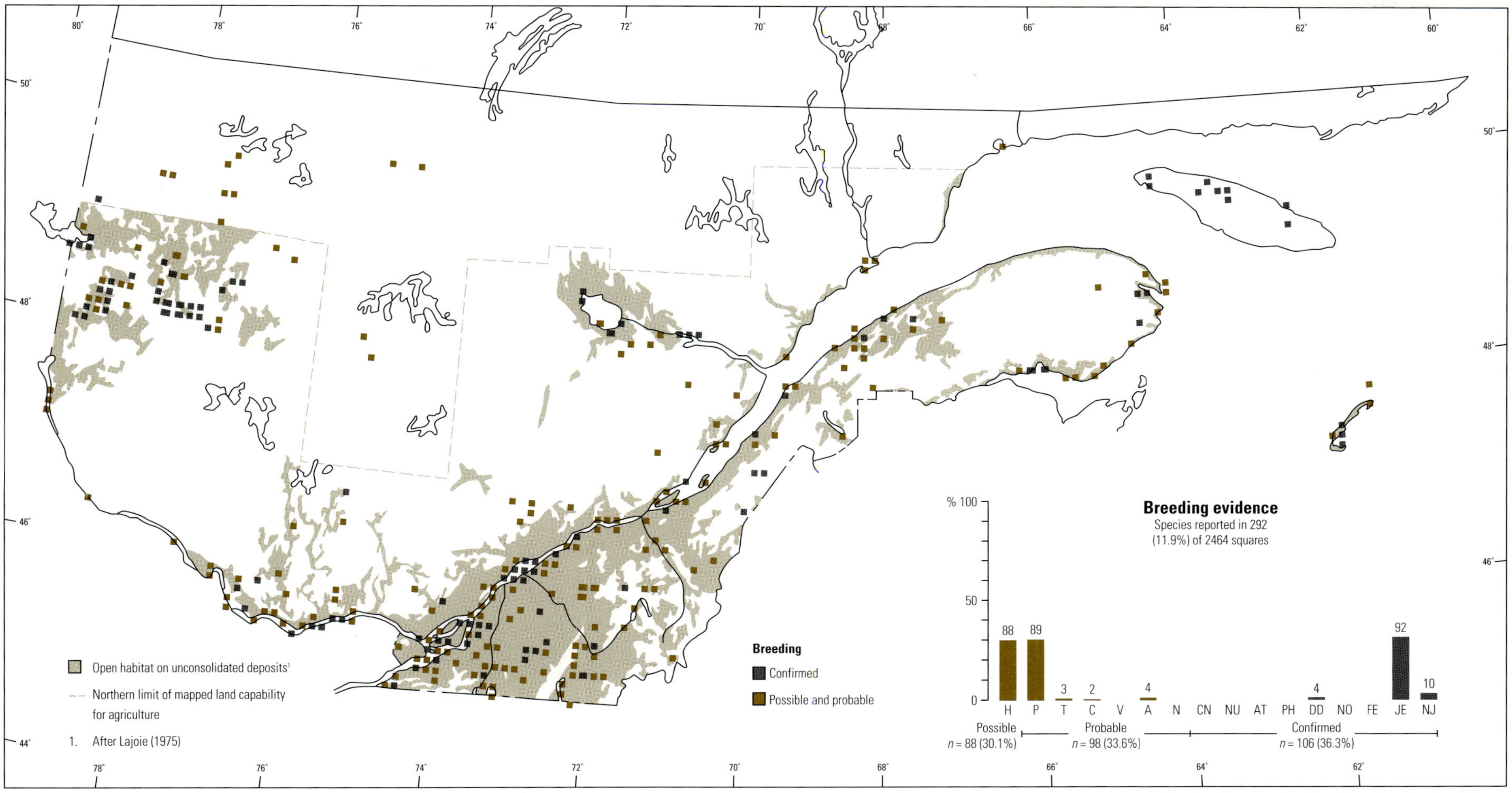

Species profile

Clutch size	usually 9–10 eggs (6–15)[1,10]
Laying interval	1 egg per day[1,9]
Start of incubation	with last egg[9]
Incubation period	23–27 days[1,12]
Maturity at hatching	precocial
Care of young	by female[5]
Nestling period	less than 24 hours[7]
Age at first flight	35–44 days[6]
Dependence of young	until sometime before first flight[13]
Broods per year	1[7]
Breeding age	1 year[7]
Mating system	monogamy[11]
Duration of pair bond	1 season[8]
Total length	35.5–40.6 cm[3]
Wing span	61.0–78.7 cm[2]
Weight	M 463 g (320–540) / F 376 g (230–500)[1]
Longevity record	22 years 4 months[4]

References:
1. Bellrose 1976
2. Terres 1980
3. Godfrey 1986
4. Klimkiewicz and Futcher 1989
5. Bent 1923
6. Gollop and Marshall 1954
7. DeGraaf and Rudis 1987
8. Oring 1964
9. Harrison 1978
10. Bull 1985
11. Ehrlich et al. 1988
12. Peck and James 1983
13. Palmer 1976

Chronology *n* = 233

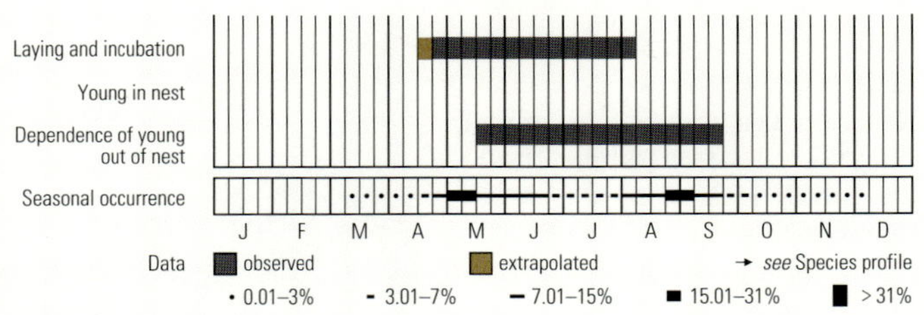

World breeding range

The species' diet changes over the breeding season. When the Blue-wing arrives on the breeding grounds, invertebrates make up 45% of its diet, rising to 80–95% when nesting begins (Dirschl 1969; Swanson et al. 1974). By the end of the summer, the Blue-winged Teal is eating mainly plant materials again (Dirschl 1969).

In its search for food, the Blue-winged Teal is constantly faced with variations in invertebrate populations, not only from pond to pond, but also depending on the point in the breeding season. But the species is well adapted to the situation: since it is not very selective in its choice of prey, it eats whatever is most readily available (Swanson et al. 1974; Swanson and Meyer 1977; DuBowy 1985). Given the morphology of its bill, the Blue-winged Teal is limited in its ability to utilize small crustaceans. The wide spacing of the lamellae on its bill permits it to ingest mainly invertebrates measuring 1.5 mm or more in diameter, such as water fleas (*Daphnia pulex*), copepods, and ostracods (Swanson et al. 1974). The species also eats snails, the larvae and adults of various insects, chironomids, amphipods, seeds, and other items (Swanson et al. 1974; Swanson and Meyer 1977).

Blue-winged Teal nest sheltered in tall grass and generously lined with down from the hen's breast. Saint-François-du-Lac, Lake Saint-Pierre. *May 1980, Daniel Jauvin.*

Distribution

The Blue-winged Teal breeds throughout Québec south of the 50th parallel, and is particularly well established in the agricultural areas of the province (USFWS and CWS 1986).

During the nesting period, the Blue-winged Teal is fairly common along the St. Lawrence River between Montréal and Lake Saint-Pierre (Lepage 1975). This stretch includes a number of islands where nesting waterfowl have had high reproductive success (CWS and MLCP 1986), which is not surprising since 43% of the total area of these islands is used for farming (Bélanger 1991). Moreover, Lake Saint-Pierre has some of the best brood-rearing marshes in Québec (CWS and MLCP 1986).

Lake Saint-François, the Ottawa River, and to a lesser extent, the upper Richelieu River, all offer an abundance of brood-rearing areas (CWS and MLCP 1986).

In the St. Lawrence estuary, the Blue-winged Teal is dispersed along the strip of farmland on the south shore. In the Gulf of St. Lawrence, it breeds along the Gaspé coast and in the Magdalen Islands, where there are small inland and coastal marshes close to fields.

Concentrations of the species are also found in the lowlands of the Abitibi and Lac-Saint-Jean regions, where there are many productive wetlands (CWS and MLCP 1986). The Atlas survey confirmed the species as a breeder on Anticosti Island, where it had previously been considered a non-breeder (Ouellet 1969; David 1980a).

In the northern part of the province, breeding has been confirmed near Matagami (Cons. G. G. – GREBE 1992a) and near Kakassituq Point on the east coast of James Bay (Yank et al. 1990b), with the northernmost

Lake Georges, near Amos, is close to agricultural areas where the species nests. The hens leads their broods along the canals and ditches to the lake. Abitibi region. *July 1991, André-Guy Bernier.*

The Blue-winged Teal raises its ducklings in these marshes created by Ducks Unlimited on Department of National Defence property at Baie-du-Febvre. South shore of Lake Saint-Pierre. *September 1988, Daniel Jauvin.*

Notes on territory (continued)

incubation. Dzubin (1955), on the other hand, stated that pairs use non-exclusive home ranges of approximately 100 ha, and that within these areas there are nesting territories with boundaries that may change over time.

Yet other biologists maintain that the Blue-winged Teal is not really territorial because the drake does not defend a well-defined area, but simply a small space around the hen (McHenry 1971).

In southern Québec, during surveys in May 1990, 1991, and 1992, the highest densities were found northeast of Matagami and around Rouyn–Noranda and Val-d'Or, where between 1.1 and 2 pairs/100 km² were observed (D. Bordage, CWS, unpubl. data).

reports from Manitounuk Sound (Cons. G. G. – GREBE 1990a). Generally speaking, however, the species seems to be very rare above the 49th parallel (Cons. G. G. – GREBE 1990a, 1992a; Benoit et al. 1991, 1993b, 1994; Bordage 1991; Tardif and Morneau 1991).

History and trends

At the turn of the century, the Blue-winged Teal was common in the province during migration (Dionne 1906; Macoun and Macoun 1915), and a few individuals probably bred in the Montréal area (Wintle 1896; Macoun and Macoun 1915). The BBS results for 1965–79 show an increase in the Blue-winged Teal population in the St. Lawrence lowlands during this period, and no significant decline elsewhere on the continent (Robbins et al. 1986). More recently, a major decrease has been recorded in North America: the breeding population dropped from 5.3 million to 3.8 million individuals between 1970 and 1985 (USFWS and CSW 1986). In Québec, ÉPOQ data show a marked decline in the frequency of occurrence from 1970 to 1989 (Cyr and Larivée 1993b). Similarly, data collected by the Canadian Wildlife Service also reveal a significant decline in harvests in the province begining in the late 1970s, as shown in the hunting table in Appendix II. The Blue-winged Teal hunt has been restricted in Québec since 1991.

Like other species of waterfowl, the Blue-winged Teal has experienced considerable changes in its habitat during the 20th century. Between 1945 and 1976, but mainly between 1945 and 1960, 3642 ha of wetlands between Cornwall and Matane were destroyed or damaged (CWS and MLCP 1986). This transformation of the landscape—as a result mainly of intensive farming, particularly the destruction of sloughs and the straightening of streams as part of a farmland drainage program in the St. Lawrence Valley—is forcing ducks like the Blue-winged Teal to make greater use of cultivated fields (Bélanger 1991). Other factors contributing to habitat loss are, in decreasing order of importance, the filling in of wetlands, residential construction, land clearing, and the installation of service and industrial infrastructures (CWS and MLCP 1986).

Furthermore, current farming practices are lowering reproductive success, since they reduce plant cover, which is needed for nesting, and result in increased exposure to cattle and farm machinery, which kill females and break eggs (Masse and Raymond 1988; Bélanger 1991). Haying delays the growth of the next year's vegetation, which may lead to greater nest predation (Martz 1967), sometimes a major cause of mortality in this duck (Bellrose 1976; Livezey 1981b). Some of the Blue-wing's principal predators are the American Crow, striped skunk, raccoon, red fox, and red squirrel (Bellrose 1976; Weller 1979; Livezey 1981b).

If Blue-winged Teal populations are to increase or at least remain stable in the future, specific measures to preserve and, if possible, improve the quality of existing habitats must be taken in order to reduce the damage caused by farming (USFWS and CWS 1986). According to Bélanger (1991), conservation of the species must take into account its use of farmland. Even though there are lower densities there than in natural habitats, these vast man-made habitats support large numbers of birds. However, increasing the productivity of cultivated land would require the use of integrated farm and waterfowl management techniques, such as preserving strips of vegetation, rotating crops, leaving fields fallow, postponing haying until 1 July, and decreasing the number of animals at pasture (Bélanger 1991). The creation of artificial islands from dredge spoil could also provide suitable nesting habitats for the species. Finally, strict controls on the number of Blue-winged Teal killed by hunters each year should also be maintained (Environment Canada 1990).

Réjean Benoit and Diane Dauphin

Northern Shoveler

Canard souchet

Anas clypeata

Ghislain Caron

THE NORTHERN SHOVELER'S LONG SHOVEL-SHAPED BILL NEVER FAILS TO provoke a chuckle from birders seeing it for the first time. This trait distinguishes the shoveler from all other ducks encountered in Québec and, in flight, makes its wings look as if they are set far back on the body. The highly colourful male somewhat resembles the Mallard, but it is not nearly as sought after for the table (Bellrose 1980). David (1980a) classified the Northern Shoveler as an uncommon migrant breeder in southern Québec.

Habitat and behaviour

During the breeding season, the Northern Shoveler frequents lakes, marshes, and shallow ponds in open areas. These bodies of water often have muddy bottoms and teem with aquatic invertebrates. Abundant aquatic vegetation and dry grassy banks with some open areas are also important (Palmer 1976).

Pair formation begins in November on the wintering grounds and may continue throughout spring (Johnsgard 1975b; Palmer 1976). The male fiercely defends a territory of 1–3 ha, and in late April or early May, can often be seen chasing away conspecifics, both pairs and individuals, that have intruded on his territory (Seymour 1974a; Titman and Seymour 1981). The female arrives on the breeding grounds with no stored energy reserves, accumulating them instead on the territory defended by the male (Afton 1979). Furthermore, because she is small and eats mainly a diet of animal matter during the breeding season, the hen probably cannot store up enough fat to carry her through the entire incubation period (Afton 1979; Ankney and Afton 1988). The Northern Shoveler is therefore more territorial than most other ducks, because the female relies heavily on an abundant food resource near the nest, where she can feed undisturbed both before and during incubation (McKinney 1973). The drake continues to defend the territory for up to 21 days after the hen has finished laying, while in most other ducks, the male abandons his mate shortly after the last egg has been laid (Afton 1979).

The nest is generally located in an open area away from any trees, and usually 25–70 m from water, although some have been found as far as 1.6 km from the nearest pond. It is a rudimentary mat of grasses and down, concealed in some grass cover (Bellrose 1980). The fact that the grass is generally less than 30 cm tall (Johnsgard 1978) may have made it easier to find nests during atlassing and 22% (11/50) of confirmed breeding records involved the discovery of nests.

The drake usually leaves the hen toward the end of the incubation period. In June, he starts to lose his nuptial plumage. In July, he heads for large bodies of water to moult with mostly other male shovelers, and is unable to fly for a few weeks (Bellrose 1980).

Notes on territory

The pair's home range during the breeding season is relatively small, measuring 6–36 ha (Poston 1969). The drake is highly territorial within this area—particularly in the centre (1–3 ha), where the birds feed and rest—and drives away all conspecifics, both male

The Northern Shoveler nests on the edges of ponds and marshes on the Varennes Islands, near Montréal. *June 1991, Luc Bélanger.*

Northern Shoveler

Breeding distribution

1984–1989

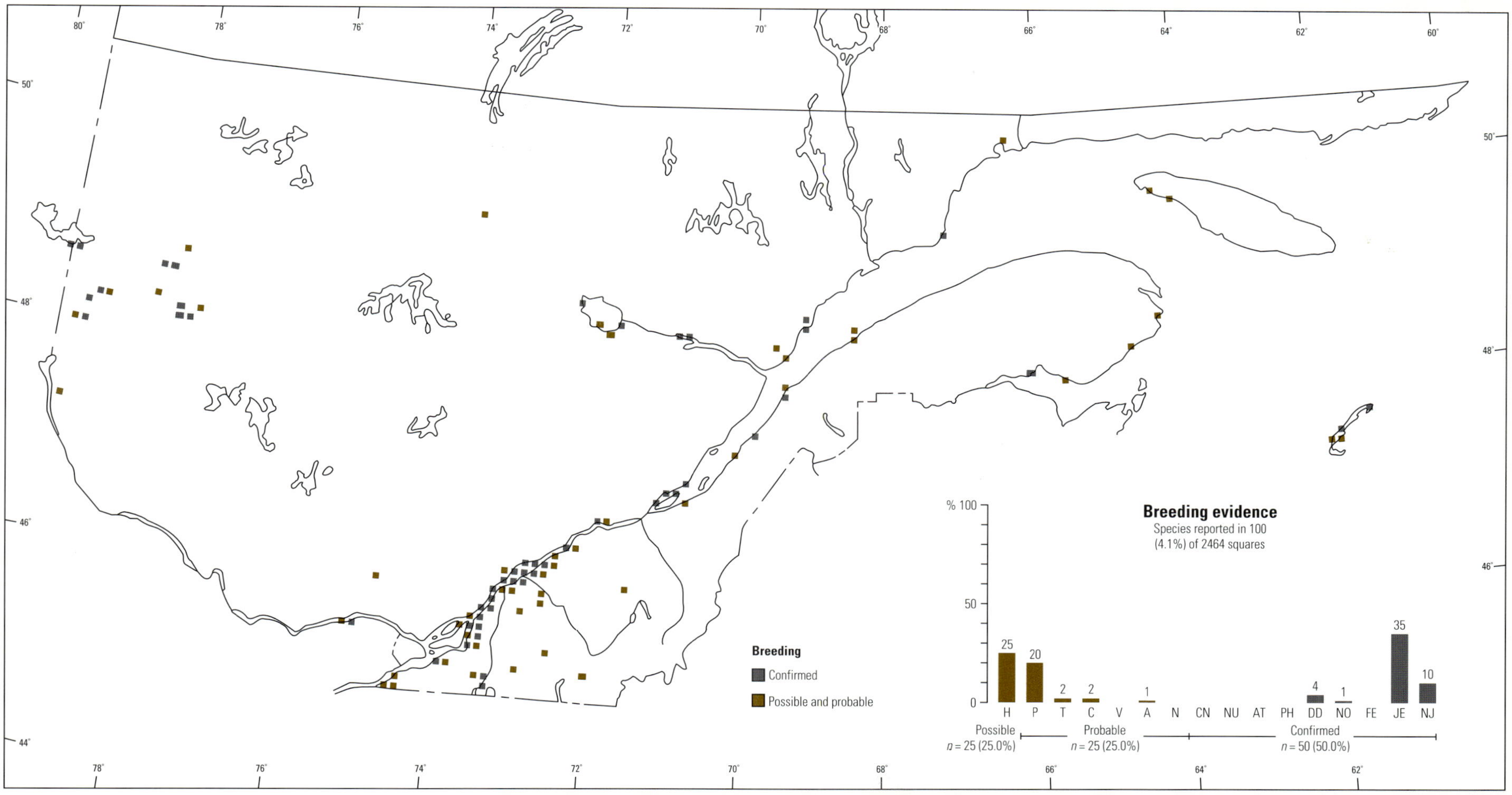

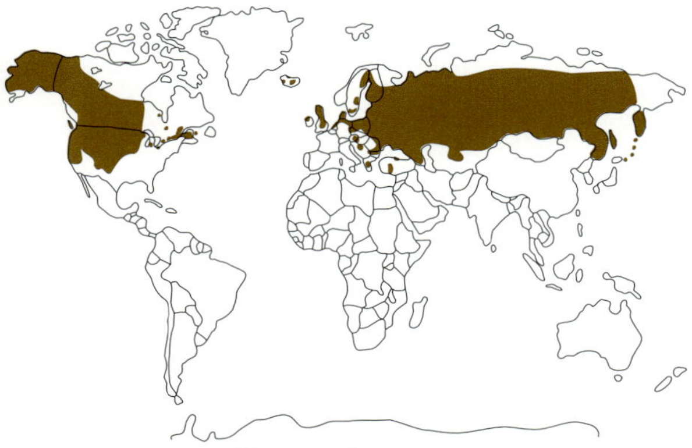

World breeding range

Species profile

Clutch size	usually 9 eggs (5–14)[3]
Laying interval	1 egg per day[3,6]
Start of incubation	with last egg[4]
Incubation period	usually 22–23 days (21–28)[2,3,4,6]
Maturity at hatching	precocial
Care of young	by female[3,4,5,6]
Nestling period	less than 24 hours[10]
Age at first flight	36–66 days[2,3,4,5,6]
Dependence of young	until first flight[2,4]
Broods per year	1[3]
Breeding age	1 year[2,3,4]
Mating system	monogamy[3,4,5,6]
Duration of pair bond	1 season[2]
Total length	43–53 cm[8]
Wing span	68.6–83.8 cm[1]
Weight	M 636 g (?–908) / F 590 g (?–726)[9]
Longevity record	16 years 7 months[7]

References:
1. Terres 1980
2. Palmer 1976
3. Bellrose 1980
4. Cramp and Simmons 1977
5. Ehrlich et al. 1988
6. Johnsgard 1978
7. Clapp et al. 1982
8. Godfrey 1986
9. Dunning 1984
10. DeGraaf and Rudis 1987

Chronology $n = 95$

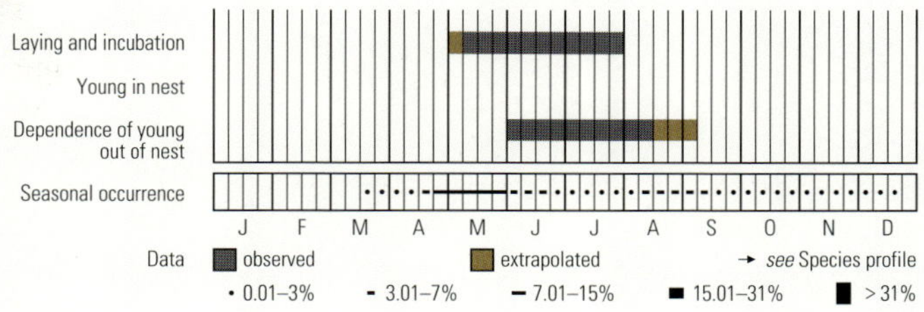

Once the eggs have hatched, the hen leads the ducklings to the closest water. The highly mobile brood may travel several hundred metres overland to reach a pond (Bellrose 1980). Broods are easy to spot, and were seen in 35% (35/100) of squares in which the species was reported and accounted for 70% (35/50) of all records of confirmed breeding.

Most individuals begin breeding as yearlings. Northern Shoveler hens, like most other female ducks, tend to breed in the same area every year (Rohwer and Anderson 1988). Hens may renest the same season if their first attempt fails (Bellrose 1980). Even though nests are sometimes destroyed by skunks, crows, or gulls, hatching success—between 42% and 90% depending on the region—is fairly high (Bellrose 1980).

During the breeding season, the Northern Shoveler's diet consists mainly of tiny invertebrates (chiefly water fleas and midges), which it filters through the comblike lamellae on its broad bill. Outside the breeding season, aquatic plants are also consumed (Palmer 1976; DuBowy 1985), and the diet is then composed mainly of molluscs, aquatic insects, the seeds of sedges—including those of the genus *Cyperus* (umbrella-sedges)—naiads, grasses, and lesser duckweed (Palmer 1976). Mornings and evenings, the shoveler can often be seen paddling placidly along on calm lakes and ponds, skimming the surface of the water with its large bill (Bellrose 1980).

Distribution

The Northern Shoveler winters on both coasts of North America, in the southern United States, Mexico, and the West Indies. Its breeding range extends from Alaska to Nova Scotia, dipping south to the American Midwest. It is also found in Europe and Asia (Palmer 1976).

In Québec, the Northern Shoveler occurs mainly along the St. Lawrence River between Montréal and Trois-Rivières. East of Trois-Rivières, it tends to be more local in distribution, although it occurs downstream as far as Sept-Îles on the north shore and Rimouski on the south shore. Small numbers breed in the Abitibi region and individuals have been seen in the Saguenay, Lac-Saint-Jean, Magdalen Islands, southeastern Gaspé, Montérégie, and Ottawa Valley regions. The species has also been reported in the Eastern Townships, and on Anticosti Island, although breeding has not been confirmed there. The Atlas results extend its breeding range slightly eastward of that described by David (1980a) and Godfrey (1986), although these authors did report that the shoveler bred in the Magdalen Islands.

In the Abitibi region, the extensive areas of farmland scattered with marshes provide favourable breeding habitat for the species. Along the St. Lawrence the large, muddy, marshy bays in the freshwater section offer particularly good breeding habitat, as does the estuary to a lesser extent.

Notes on territory (continued)

and female (Seymour 1974b; Titman and Seymour 1981). He is particularly vigilant in chasing away conspecifics that land within the territory, but may also pursue birds overflying it (Seymour 1974a; Titman and Seymour 1981).

A Manitoba study found that territories were roughly 3 ha prior to laying but decreased to slightly less than 1 ha during laying and incubation, and were no longer defended at all toward the end of incubation (Seymour 1974a). Territories are better defined than those of most other ducks and overlap very little (McKinney 1970).

This settling pond surrounded by a strip of dense aquatic plants provides good breeding habitat for the Northern Shoveler. The hen nests in the dry grass around the pond, then leads her brood to the water. Baie-du-Febvre, south shore of Lake Saint-Pierre. *July 1991, Daniel Jauvin.*

This duck prefers ponds teeming with invertebrates. It uses its large lamellate bill to skim the surface of the water and to filter the mud, and also eats seeds and aquatic vegetation. Varennes Islands, near Montréal. *August 1992, Luc Bélanger.*

The Northern Shoveler may also breed farther north, in the watersheds of the Nottaway, Broadback, and Rupert rivers (Cons. G. G. – GREBE 1992c). In addition, the species has been reported in the area around the Eastmain River (Tardif and Morneau 1991), near Pointe Louis XIV (Benoit et al. 1993b), and in the watershed of the Great Whale River (St-Louis 1978; ÉPOQ).

History and trends

The Northern Shoveler is a fairly recent breeder in Québec (Robert 1989b). Macoun and Macoun (1909) classified it as a summer visitor in eastern Canada (Ontario, Québec, and the Maritimes), while Taverner (1922) described it as rare around the Great Lakes and farther east. The first breeding record in the Montréal area dates from 1967 (Ouellet 1974a).

Deforestation and agricultural development have created favourable habitat for the Northern Shoveler, thus undoubtedly helping the species expand its breeding range eastward. Since settlement, the draining of wetlands in the Prairie provinces, the heart of the shoveler's range, may also have prompted it to move east in search of new breeding habitat (Johnson and Grier 1988). The creation of sewage lagoons, which teem with aquatic invertebrates, also seems to have been a boon to the species (Sandilands 1987c; Bélanger and Couture 1988; Erskine 1992).

Although the North American population fluctuates a great deal, it has increased overall during the last few decades. In the Atlantic Flyway, however, where the eastern population traditionally winters, a slight downward trend has been observed (Bellrose 1980). The Northern Shoveler has never been heavily hunted as a game bird in Québec. Statistics show that there has been a slight drop in the numbers killed in the province over the last 20 years (Appendix II; Boersma 1990).

RODGER D. TITMAN AND SIMON BARRETTE

Gadwall

Canard chipeau

Anas strepera

Ghislain Caron

THE HOLARCTIC GADWALL HAS BEEN EXPANDING ITS BREEDING RANGE since midcentury and has only just recently become a regular breeder in southern Québec (Cantin et al. 1976). This medium-sized grey dabbling duck is concentrated mainly along the St. Lawrence. In flight, the white inner secondaires, white belly and, in the male, black upper- and under-tail coverts are diagnostic.

Habitat and behaviour

The Gadwall usually begins breeding at one year of age. Oring (1969), however, observed that yearling males are less likely to breed than older birds. The species is usually monogamous, but polygamy has been reported, especially when breeding pair density is high. The Gadwall, normally gregarious and often associated with the Northern Pintail and the American Wigeon (Bent 1923), ceases to be sociable when nesting time approaches (Gates 1962).

By the time the birds arrive in Québec in the spring, pairs have usually formed and are easy to spot as they forage along the edges of shallow marshes. The Gadwall has also taken to using sewage lagoons during the breeding season.

Laying occurs usually in late May or June, making this species a late nester in comparison with other dabblers. Pairs arrive on the breeding grounds almost a month before the onset of laying (Oring 1969; Dwyer 1974), but wait until the plant cover is high and dense before nesting (Johnsgard 1975b). According to a number of studies (Gates 1962; Lokemoen et al. 1990), over 60% of females return to the nest site used the previous year. The hen chooses the nest site, often selecting a small island, some of which may support several dozen nests. One reason the Gadwall prefers nesting on islands is that this type of habitat provides strong protection against terrestrial predators and they often support nesting colonies of gulls or terns (Cramp and Simmons 1977). The nest, concealed in dense grasses and forbs, is built on dry ground, usually less than 50 m from water (Oring 1969). The female makes a shallow scrape, lines it with leaves, grasses, and dry twigs gathered nearby, and gradually adds down as laying and incubation progress (Palmer 1976).

Typical clutch size is around 10 eggs, creamy white in colour, laid at an interval of one a day (Gates 1962). Dump nests may contain up to 20 eggs from several different females (Bellrose 1976). Incubation, by the female alone, begins when the last egg has been laid, and takes about 26 days (Cantin et al. 1976). If the clutch is destroyed, most females will renest (Gates 1962). The hen flushes readily from the nest when approached. Drakes desert their mates about halfway through laying (Gates 1962; Duebbert 1966), flocking together in large marshes to moult.

Notes on territory

The pair occupies a 14–35 ha breeding home range around the previous year's nest site; this area includes one or more activity centres (resting or feeding ponds), with the drake defending a moving

Originally from western North America, the Gadwall established its first stronghold in Québec on the islands in the St. Lawrence River near Montréal. Several islets in the Contrecoeur Islands were created from dredge spoil from the Seaway. Îles de Contrecoeur National Wildlife Area, northeast of Montréal. *October 1993, Pierre Normand.*

Gadwall

Breeding distribution

1984–1989

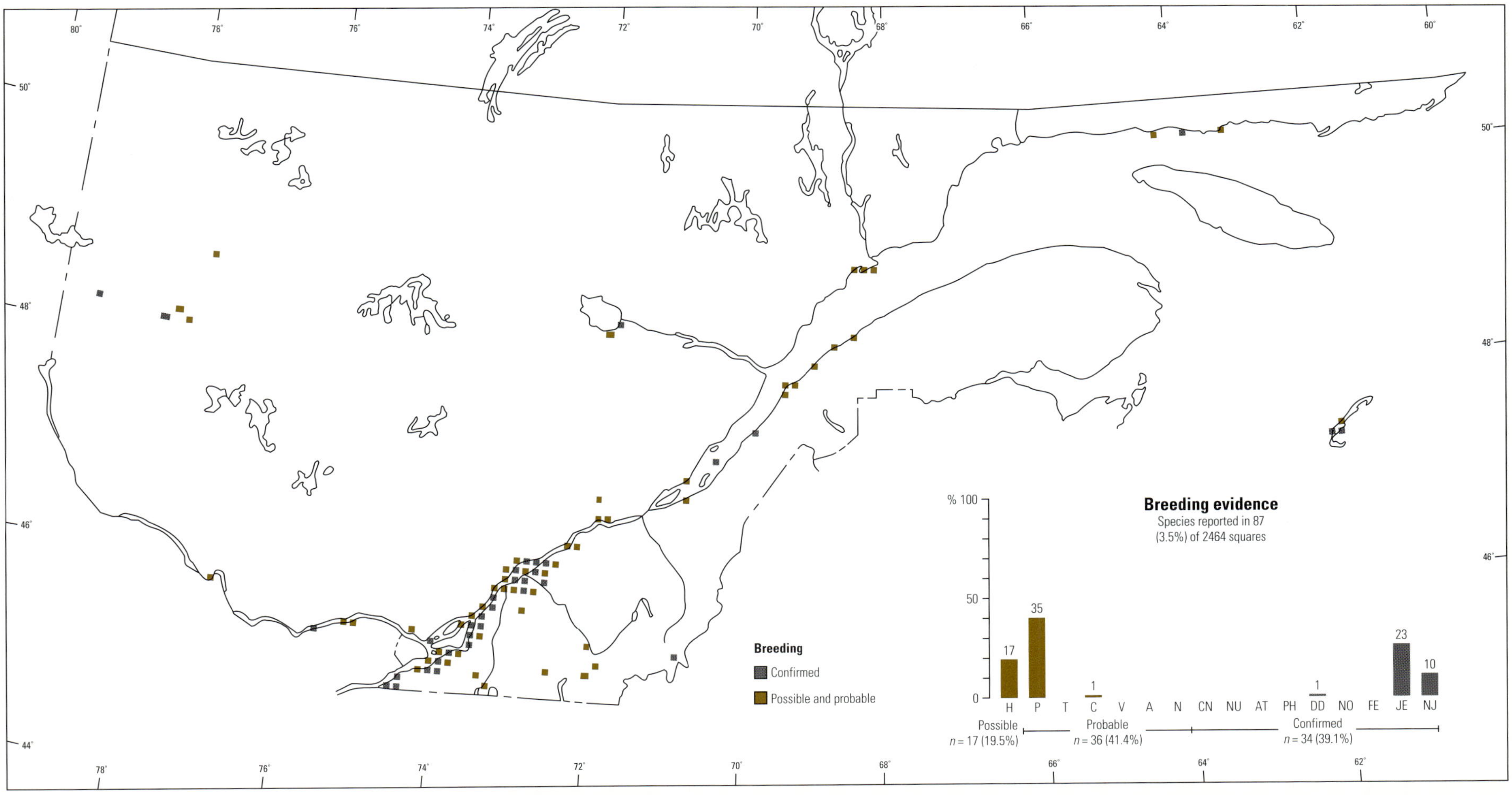

Species profile

Clutch size	usually 8–11 eggs [5–13], more than 13 = dump nest[5]
Laying interval	1 egg per day[1]
Start of incubation	with last egg[3]
Incubation period	23–27 days[2,7]
Maturity at hatching	precocial
Care of young	by female[4]
Nestling period	≈ 24 hours[4,6]
Age at first flight	45–50 days[3,4]
Dependence of young	until first flight[3]
Broods per year	1[3,4,5]
Breeding age	1 year[3]
Mating system	monogamy, sometimes polygamy[3,4]
Duration of pair bond	1 season[3,4]
Total length	46.0–55.8 cm[8]
Wing span	78.7–91.4 cm[9]
Weight	M 966 g (726–1043) / F 835 g (635–1043)[5]
Longevity record	19 years 6 months[10]

References:

1. Gates 1962
2. Oring 1969
3. Cramp and Simmons 1977
4. Palmer 1976
5. Bellrose 1976
6. Johnsgard 1975b
7. Peck and James 1983
8. Godfrey 1986
9. Terres 1980
10. Klimkiewicz and Futcher 1989

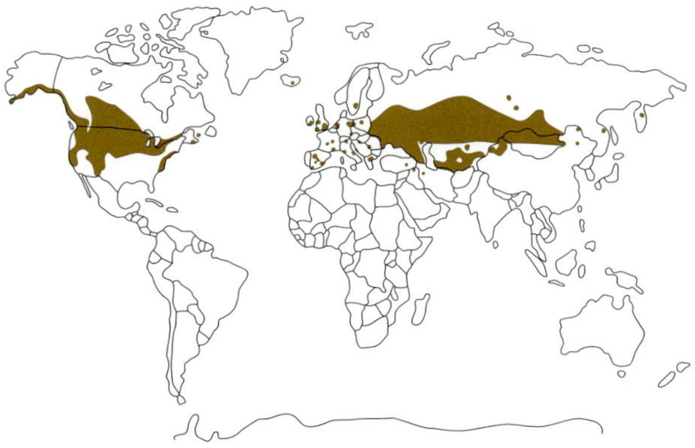

World breeding range

Chronology n = 199

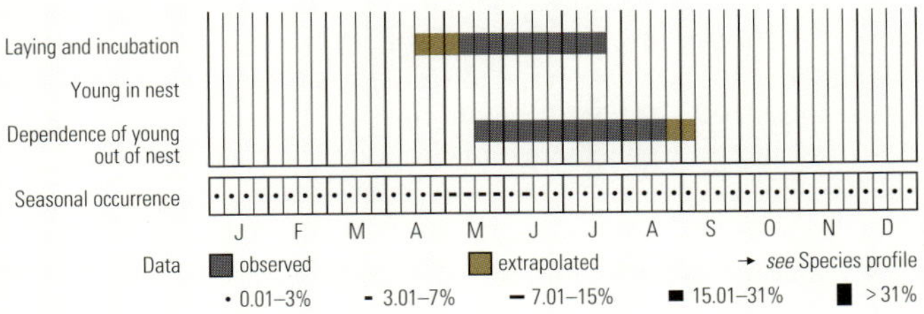

Saint-Ours Island, created from dredge spoil, once had the highest density of nesting Gadwalls in the province; in the 1970s, approximately 150 nests were found there in one season. The habitat has since changed, and nest density is now lower. Îles de Contrecoeur National Wildlife Area, northeast of Montréal. *September 1993, Luc Bélanger and Renée Langevin.*

Notes on territory (continued)

territory around the hen (Gates 1962). Dwyer (1974) believes that the drake, by chasing off other males that are trying to court the hen, eliminates competition for the food source at the activity centre and allows the hen to spend more time selecting the protein-rich food that she needs for laying. Sometimes nest density is exceptionally high for a dabbling duck: on some small islands, nests may be less than 30 cm apart. Under these circumstances, territoriality seems to be absent, or limited to protecting the hen; Duebbert (1966) believes that this behaviour enables the species to colonize islands more easily.

Since nests are built on islands and are well protected from predators by dense vegetation, hatching success is generally high (Cantin et al. 1976). Hatching of the precocial young usually takes place between mid-June and late July. Within a few hours of hatching, the hen quietly leads her brood to large beds of aquatic vegetation (Cantin et al. 1976). When breeding density is high, several broods may merge, otherwise they tend to remain separate (Palmer 1976). Age at first flight is 45–50 days (Palmer 1976; Cramp and Simmons 1977), and the females may begin to moult before the ducklings are independent.

Foraging on the surface of the water or no more than 15 cm below it, the Gadwall is the most vegetarian of our ducks. In winter, it feeds almost exclusively on the roots, seeds, and stems of aquatic plants (Paulus 1982). In summer, however, the proportion of animal food consumed—mainly aquatic insects and their larvae, but also alevins, worms, and molluscs—gradually increases until it makes up approximately 70% of the species diet during the laying period (Mabbott 1920; Serie and Swanson 1976). The young also have a diet consisting mainly of animal matter.

Gadwalls arrive on their breeding grounds between early April and early May, in small flocks consisting mainly of pairs (Cantin et al. 1976). Most leave Québec by the end of September and large flocks can be seen assembling prior to migration from late August until mid-September. Gadwalls most commonly migrate at night in small flocks, flying quickly in a long line (Bellrose 1976; Cramp and Simmons 1977).

In terms of breeding success, the Gadwall is one of the most productive of all the dabbling ducks. There are two main factors responsible for this: first, it begins nesting late in the season, when the plant cover better conceals the nest and predators are less active, and, second, it usually nests in high densities on islands, where it is protected from terrestrial predators. Furthermore, late laying probably also reduces losses due to spring flooding and ploughing (Bellrose 1976; Cantin et al. 1976; Crabtree et al. 1989).

Although Gadwalls have many predators, including striped skunks, minks, raccoons, red foxes, hawks, and corvids, the toll they take varies depending on the habitat.

Distribution

Found locally in Québec, the Gadwall was reported in only 3.5% (87/2464) of Atlas squares surveyed, and breeding was confirmed in only 39.1% (34/87) of these. It breeds primarily in the St. Lawrence Valley, from Valleyfield to Trois-Rivières, and is particularly abundant in the Îles de Contrecoeur National Wildlife Area, where close to half of all nesting ducks are Gadwalls (Cantin et al. 1976; Giroux et al. 1992).

Atlas data confirmed that the breeding range of the Gadwall has expanded to include the Abitibi region, Lac-Saint-Jean, the Ottawa River, St. Lawrence estuary, Middle North Shore, and the Magdalen Islands. The species has been sighted as far north as the Great Whale River watershed (St-Louis 1979).

This anatid nests late in the season to avoid spring flooding, and uses small islands to protect itself from predators. Like many other species of ducks, the drake abandons the hen during laying and goes off to moult. Îles de Contrecoeur National Wildlife Area, northeast of Montréal. *May 1990, Luc Bélanger.*

Pairs wait for plant cover to grow high and dense before nesting. Here, a hen broods her eggs in a well concealed nest on La Grande Île. Berthier Islands, west of Lake Saint-Pierre. *July 1981, Daniel Jauvin.*

In migration, the Gadwall is seen mostly in southwestern Québec, mainly on Lake Saint-François, in the Montréal area, and around the islands between Montréal and Lake Saint-Pierre, including the Sorel Islands (Cantin et al. 1976). According to Bellrose (1976), during fall migration about three quarters of the North American population fly to Louisiana, with the majority of birds overwintering there. Some Gadwalls winter in Mexico, the Bahamas, Cuba, along the eastern seaboard from Maryland to Florida, in the southern Great Lakes, and from southern British Columbia to California in the west (Bellrose 1976; AOU 1983). The birds tend to concentrate in migratory bird refuges during the winter (Root 1988). In Québec, only two cases of successful overwintering have been recorded: at Lake Memphrémagog in 1975–76 (David 1980a), and on the Lachine Rapids in 1990–91 (R. Yank, pers. comm.).

This Gadwall nest was found under dense cover on La Grande Île. Berthier Islands, west of Lake Saint-Pierre. *June 1981, Daniel Jauvin.*

History and trends

Early this century, the Gadwall's breeding range was limited to western—and even more precisely—southwestern North America. In 1939, an isolated breeding population was discovered in North Carolina and Delaware, almost 2000 km east of the species' traditional breeding grounds (Henny and Holgersen 1974). It has gradually expanded its range since then.

The creation of impoundments in refuges and wildlife areas in the United States and the release of birds raised in captivity there have undoubtedly helped the species increase in number and become established in eastern North America (Henny and Holgersen 1974).

At the turn of the century, Wintle (1896) and Macoun and Macoun (1909) classified the Gadwall as a rare "transient visitant" in Québec. It probably arrived southwest of Montréal sometime between 1962 and 1967, continuing its eastward expansion thereafter (Cantin et al. 1976). The species has been breeding in Québec for only a very short time. The first nest was discovered in 1968 near Dundee, on Lake Saint-François (Reed 1969 *in* Cantin et al. 1976). The creation of islands from dredge spoil from the construction of the St. Lawrence Seaway has also undoubtedly helped the Gadwall become established here (Cantin et al. 1976). Ducklings raised in captivity from eggs collected from the Contrecoeur Islands were released on Prince Edward Island and in Nova Scotia in 1985 (McAlpine et al. 1988a); the following year, pairs returned and bred successfully.

Gadwalls began to show up in hunting records in Québec in 1969 (Cantin et al. 1976). The number taken each year increased gradually until 1980, and then more or less stabilized. The species now represents between 1–2% of ducks killed during the hunting season (Appendix II). In the areas around Varennes, Sorel, and Berthierville, the Gadwall may, however, represent half of all birds shot by hunters in fall (CWS, unpubl. data).

Jean-François Giroux and Jean-François Rail

American Wigeon

Canard siffleur d'Amérique

Canard d'Amérique

Anas americana

THE DRAKES' SOFT *WID-WID-WIDGEON* WHISTLE EVOKE THIS SPECIES' commonly known name in North America: wigeon. The drake's shining white crown, which is visible at great distances, is also diagnostic. The American Wigeon is unique among dabbling ducks in that it feeds chiefly on the leaves and stems of aquatic plants. It sometimes obtains its food by pirating food from diving ducks, which dive down deep for these plants (Bellrose 1980).

The American Wigeon is a migrant breeder in Québec, found mainly in agricultural areas: the St. Lawrence Valley upstream from Trois-Rivières and the Abitibi and Lac-Saint-Jean regions.

Habitat and behaviour

American Wigeons nest mainly on shallow lakes and ponds with limited emergent vegetation, generally selecting bodies of water fringed with dry grassy meadows although they sometimes use ponds in wooded or brushy habitats (Johnsgard 1975b). Wigeons prefer permanent rather than temporary bodies of water (Palmer 1976), with shallow edges (R. D. Titman, McGill U., unpubl. data), and with submergent vegetation that grows close to the surface (Johnsgard 1978).

Males make up approximately 60% of the population (Bellrose 1980). Pair formation begins in November on the wintering grounds, and may continue until early spring. The grass-lined nest is built in a slight depression in the ground, anywhere from a few metres to several hundred metres from water. It is generally in a dry place, and is well hidden in the surrounding vegetation (rushes or sedges), although it may also be located under a tree or shrub (Palmer 1976).

Drakes leave the females during the first two weeks of incubation, sometimes travelling long distances to reach their moulting areas; on larger bodies of water, aggregations may be quite sizable. The moulting birds remain flightless for about two weeks. Females undergo their moult a little later in the season, near the nesting area (Palmer 1976; Bellrose 1980).

American Wigeon broods tend to be more sedentary than those of other ducks, and do not venture more than a few hundred metres from the hatching site. The female accompanies the brood until the young are almost fully grown (Bellrose 1980). The ducklings feed chiefly on aquatic invertebrates when they are small, but eat an increasing amount of plant material as they grow (Palmer 1976). Unlike most dabbling ducks, the female and her ducklings head for open water when disturbed, making them easy to spot.

Distribution

The American Wigeon breeds from Alaska to Newfoundland, and in the western states. It winters along both coasts of the United States as well as in Mexico, Central America, and the West Indies (Palmer 1976), with most of the Québec population probably overwintering along the eastern United States seaboard (Bellrose 1980).

In Québec, wigeons breed mainly along the St. Lawrence River from the Ontario border to Trois-Rivières; here, the large shallow bays teeming with aquatic vegetation and grassy islands provide ideal nesting habitat. The species is more thinly distributed east of Trois-Rivières. It also breeds in the Abitibi region, where the broad expanses of farmland dotted with marshes and islands of woodland are similar to the aspen parklands in the West (prairie-forest transition zone) where the species reaches its greatest abundance. The species was also atlassed in the Magdalen Islands, around Lac-Saint-Jean, and in a few isolated areas elsewhere in the province.

This pattern of distribution is corroborated by the results of aerial surveys conducted in May 1990, 1991, and 1992 (D. Bordage, CWS, unpubl. data), which confirm the species abundance in the Abitibi. In addition, the chapter by D. Lehoux (pp. 1207–1213) on the importance of the St. Lawrence to waterfowl during the breeding season highlights the importance to the species of the portion of the river upstream of Trois-Rivières. An estimated 450 pairs, nests or broods can be found in this area: 17 on Lake Saint-Pierre, 150 in the Montréal–Sorel section, 31 on Lake Saint-Louis, and 247 on Lake Saint-François.

History and trends

Traditionally, the American Wigeon's breeding range was limited mainly to the central and western part of the continent (Palmer 1976), and at the turn of the century, this duck was considered a rare transient in Québec, Nova Scotia, and New Brunswick (Macoun and Macoun 1909). The first breeding record in the province was obtained in 1950 on the coast of James Bay (Todd 1963); in the Montréal area, now the species' centre of abundance in Québec (Lehoux et al. 1985), the first record dates only from 1965 (Ouellet 1974a).

The species expanded its breeding range in the province considerably from the 1960s to the 1980s (Godfrey 1986). Although in the 1960s it nested in only a few isolated areas along James Bay and in the Abitibi, Ottawa Valley, and Lake Saint-Pierre regions, by the 1980s, it could be seen throughout southern Québec, from James Bay to the Lower St. Lawrence.

Breeding pairs have also been reported farther east, in the Magdalen Islands (Fradette 1992), Newfoundland (Goudie 1985), and various locations in New Brunswick, Nova Scotia, and Prince Edward Island (Erskine 1992), showing the extent of this eastward expansion. Reasons cited for the expansion include habitat loss on the Prairies from the draining of wetlands (Johnson and Grier 1988), in conjunction with the opening up of suitable breeding habitat in the East through the clearing of land for farming, and an increase in the number of sewage lagoons, which many waterbirds find attractive (Sandilands 1987a).

Although the North American population has fluctuated a great deal in the last few decades, surveys conducted on the breeding and wintering grounds show that populations declined slightly in the 1960s and 1970s (Bellrose 1980). The species accounts for only about 1% of the total waterfowl harvest in Québec (Appendix II), with the main hunt taking place in the marshes of southwestern Québec (Lehoux et al. 1985).

RODGER D. TITMAN AND SIMON BARRETTE

Jean-Pierre Vallée

Notes on territory

The American Wigeon does not seem to defend a well-defined territory. Nevertheless, breeding pairs are very intolerant of conspecifics around the nest, and may chase them off (Palmer 1976). A density of 6 breeding birds/km² was observed in the Mackenzie Delta in the Northwest Territories (Bellrose 1980). In the Abitibi region of Québec, aerial surveys conducted in May 1990, 1991, and 1992 showed densities of 0.02 pair/km² (D. Bordage, CWS, unpubl. data).

The American Wigeon likes to nest in areas where the emergent vegetation is not too dense, like this shallow freshwater marsh at Contrecoeur, near Montréal. *June 1992, Jean Rodrigue.*

American Wigeon

Breeding distribution

1984–1989

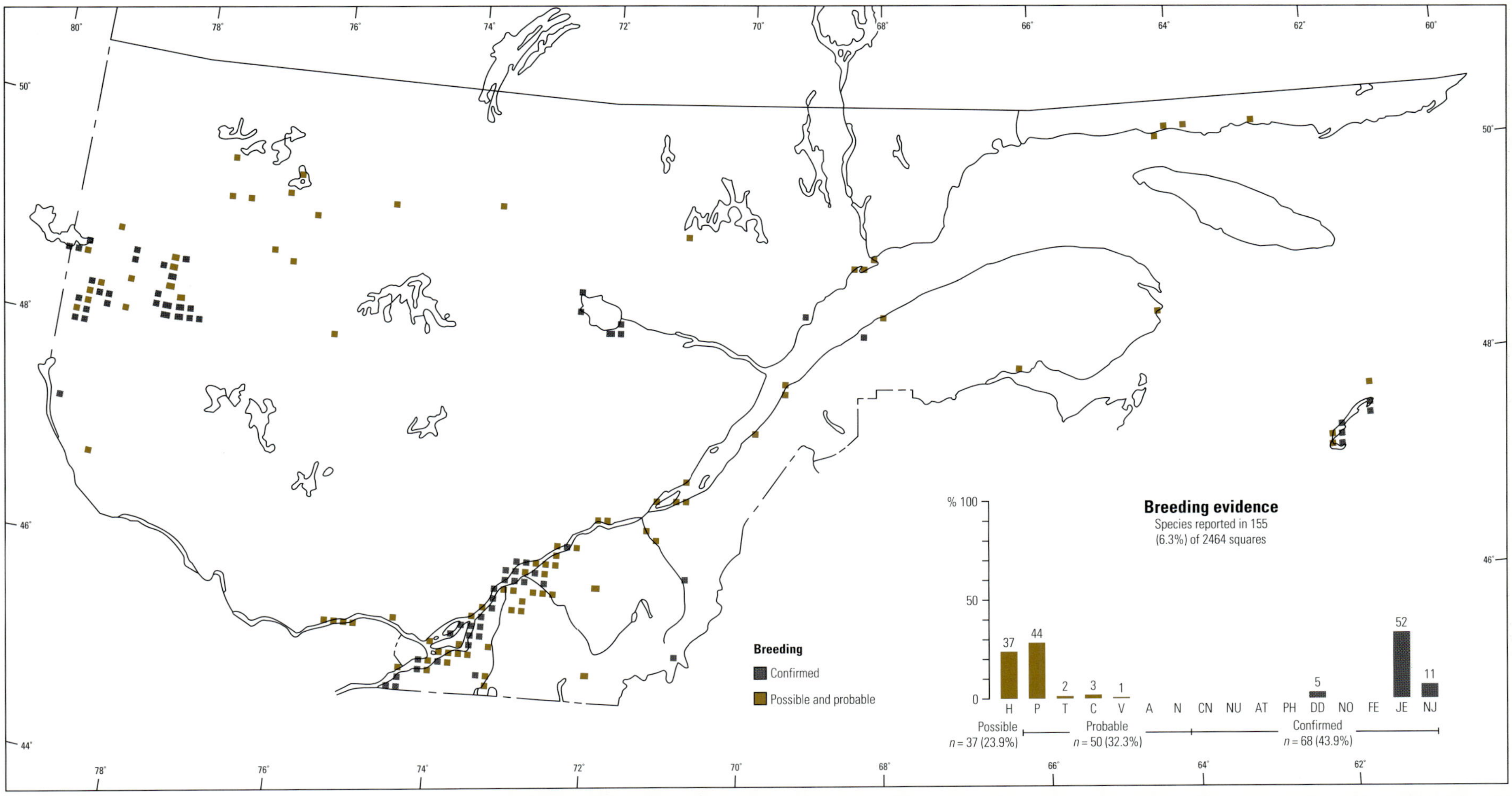

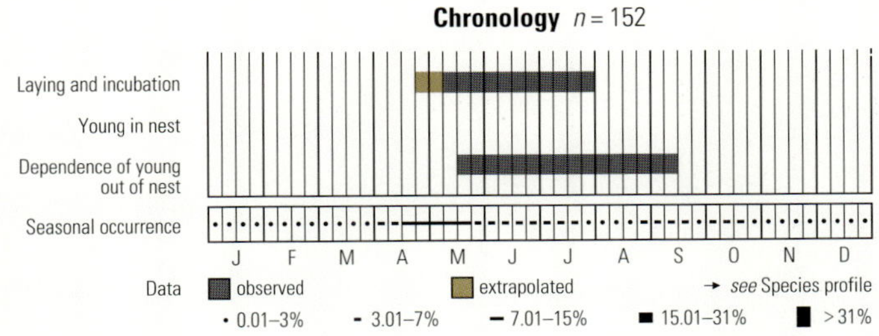

Species profile

Clutch size	usually 7–10 eggs (6–12)[1,2,3,4]
Laying interval	1 egg per day[8]
Start of incubation	with last egg[8]
Incubation period	23–35 days[1,2,3,4,8]
Maturity at hatching	precocial
Care of young	by female[1,2,3]
Nestling period	less than 24 hours[9]
Age at first flight	37–48 days[1]
Dependence of young	until first flight or a few days before[1]
Broods per year	1[1]
Breeding age	1 year[1,2]
Mating system	monogamy[1,2]
Duration of pair bond	1 season[1,2]
Total length	46–56 cm[5]
Wing span	76.2–88.9 cm[4]
Weight	M 792 g (635–1036) / F 719 g (512–872)[6]
Longevity record	21 years 4 months[7]

References:
1. Bellrose 1976
2. Palmer 1976
3. Ehrlich et al. 1988
4. Terres 1980
5. Godfrey 1986
6. Dunning 1984
7. Kennard 1975
8. Harrison 1978
9. DeGraaf and Rudis 1987

World breeding range

The herbaceous vegetation along an impoundment in this coastal marsh at Cap Tourmente provides excellent cover for the nest, which is built in a dry spot some distance from the shore. *August 1991, Yves Aubry.*

Redhead

Morillon à tête rouge

Fuligule à tête rouge

Aythya americana

Carole Bérubé

THE REDHEAD OFTEN LAYS ITS EGGS IN THE NESTS OF OTHER DUCKS, AS evidenced by the easily recognizable bright yellow ducklings seen in the broods of other species. The brick red head of the male in nuptial plumage is quite visible in the field. Noisier during the breeding season than other diving duck drakes, the Redhead has a call very much like a cat's meow (Bellrose 1980).

The Redhead is a relative newcomer to the province list of breeding birds and the main Québec breeding population at Lake Saint-François has probably only been established since the late 1950s (Alliston 1979a).

Habitat and behaviour

In the Redhead, the pairing process is particularly long in comparison with that of most other species of pochards, as is the duration of the pair bond (Weller 1967). Pair formation begins on the wintering grounds, significantly earlier than in other pochards—perhaps in January, or even earlier—while nesting begins later (Weller 1965b, 1967; Bellrose 1980). Redheads tend to return to the same nesting areas every year; Alliston (1979a) estimated that close to 70% of breeding females are philopatric.

The Redhead nests in dense emergent vegetation along the margins of permanent wetlands (Johnson and Grier 1988), such as lakes and large freshwater marshes. The nest is generally a floating structure in shallow water. Although it is sometimes located on dry land or on an island, it is rarely far from water (Bellrose 1980). The Redhead is very selective about the plant cover in which it builds its nest: bulrush beds are its first choice, followed by cattails, and then sedges (Bellrose 1980). The female uses the dry stalks of these plants to build a large, sturdy basket-shaped nest, usually with one or more access ramps (Low 1940), often bending the surrounding vegetation down to form a roof. Material is added to the nest during laying, and to keep the eggs dry when the water level rises following heavy rain. Nests are easily identifiable, since they are lined with the species' distinctive white down (Klett et al. 1986).

Redheads may breed as yearlings. Typical clutch size is 9–11 eggs, laid one day apart, although occasional days are skipped (Palmer 1976). Actual clutch size is difficult to determine, since the species is notorious for its parasitic habits. The female may lay all her eggs in her own nest or all in the nest of another duck (whether a Redhead or another species), or lay some in her own nest and the rest in other nests. Her nest may also be parasitized by other Redheads. The Mallard, Northern Pintail, Gadwall, Northern Shoveler, Ruddy Duck, and other species of ducks—and even marsh birds like the

Notes on territory

Pairs are gregarious, even during the breeding season: nesting pairs are very tolerant of one another, and nests are sometimes so close together that the species could be taken for colonial (Weller 1959).

The Redhead breeds exclusively in the southwestern part of the province, along the St. Lawrence. Lake Saint-François, with its marshes rich in aquatic vegetation is the main nesting site in Québec. Lac Saint-François National Wildlife Area, southwest of Montréal. *September 1991, Léo-Guy de Repentigny.*

Redhead

Breeding distribution
1984–1989

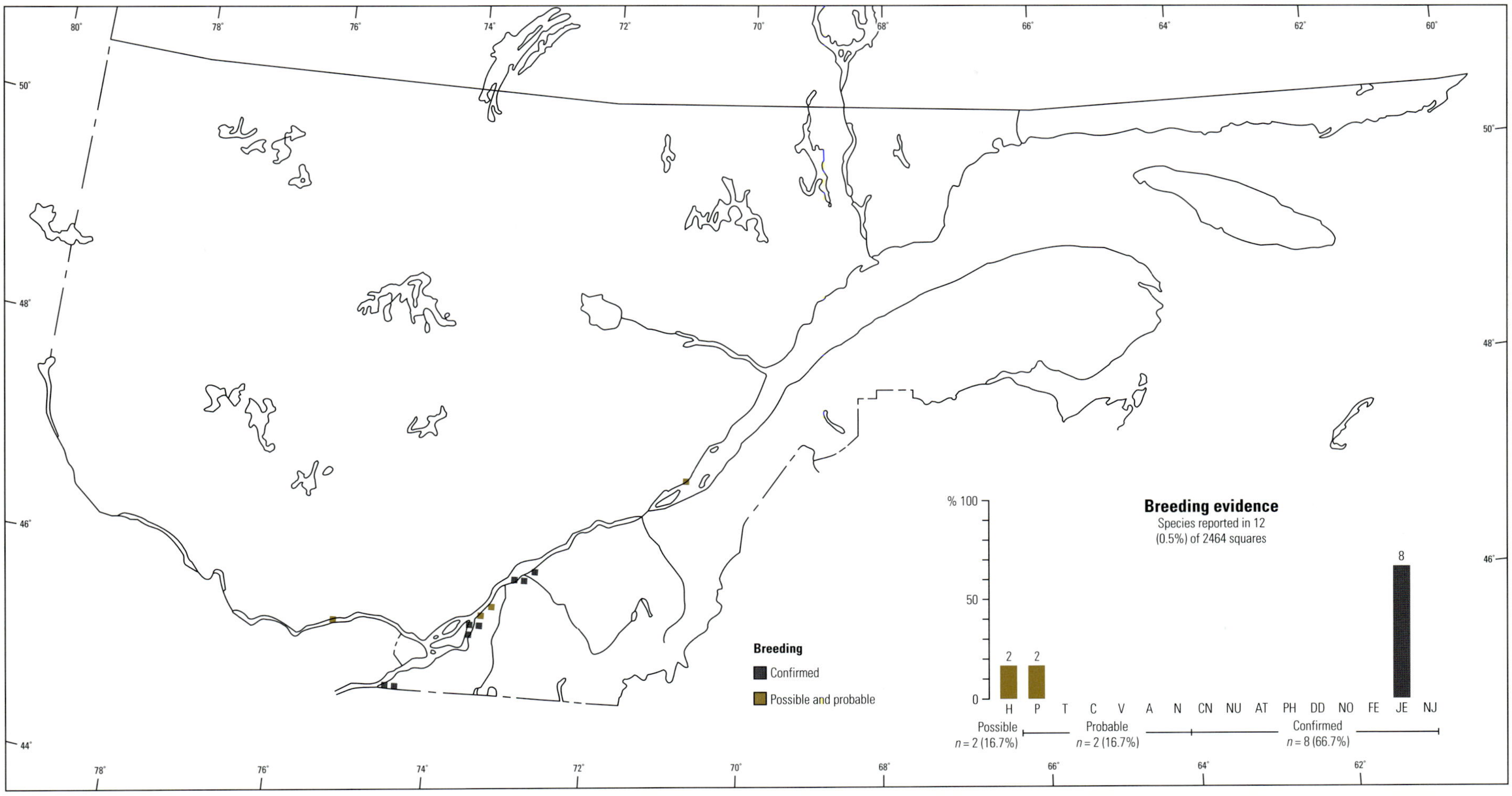

Species profile

Clutch size	usually 9–11 eggs (7–19), may use dump nest[5,7]
Laying interval	1 egg per day[5]
Start of incubation	within 48 hours of last egg[5]
Incubation period	22–28 days[4,7,8]
Maturity at hatching	precocial
Care of young	by female[9,10]
Nestling period	3–30 hours[4,5]
Age at first flight	usually 60–65 days (56–84)[4,5,7]
Dependence of young	until 3–8 weeks old, sometimes up to 12 weeks[5,7]
Broods per year	1[9,10]
Breeding age	1 year[7]
Mating system	monogamy[5,6]
Duration of pair bond	1 season[4,5,6]
Total length	43–58 cm[3]
Wing span	73.7–88.9 cm[2]
Weight	M 1107 g (952–1452) / F 971 g (680–1315)[4]
Longevity record	21 years 5 months[1]

References:
1. Clapp et al. 1982
2. Terres 1980
3. Godfrey 1986
4. Bellrose 1980
5. Palmer 1976
6. Weller 1965b
7. Alliston 1979a
8. Bent 1923
9. Ehrlich et al. 1988
10. Harrison 1978

Chronology $n = 257$

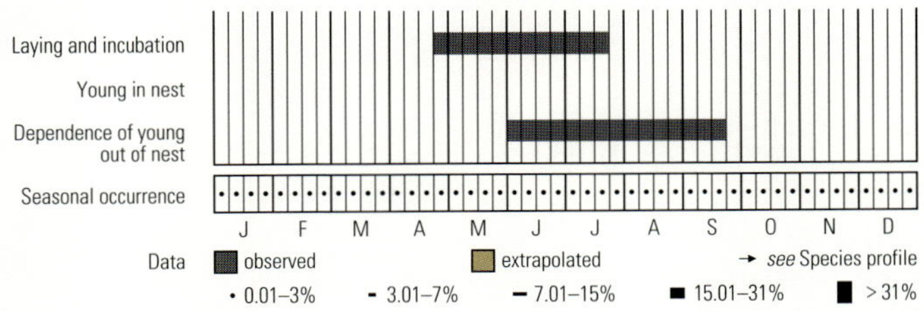

World breeding range

American Bittern and Sora—, may also play host to the parasitic Redhead (Weller 1959). On the Prairies, where the Redhead and the Canvasback are sympatric and densities are high, Canvasbacks are the most frequent victims of Redheads, since both species use similar nesting habitat (Bellrose 1980).

At the onset of incubation, the drakes desert the hens and go off to moult. Incubation begins less than 48 hours after the hen has finished laying and continues for 24–28 days (Palmer 1976; Bellrose 1980). Hatching is relatively synchronous with the whole clutch hatching within the space of 5–10 hours. Within 30 hours of hatching, the precocial chicks leave the nest and are led by the hen to a brood-rearing pond (Lokemoen 1966; Palmer 1976). In the western part of the Redhead's range, hens may abandon their broods when the ducklings are 3–5 weeks old (Palmer 1976); in Québec, however, they seem to stay with their ducklings until they are 7–8 weeks old, and sometimes until they are capable of flight, at around 8–12 weeks (9 on average) (Alliston 1979a). In the Lake Saint-François population, laying usually begins in mid-May, but may start as early as the end of April (Gervais et al. 1983).

The young that hatch from parasitic eggs are raised by the host females. Since several females may lay in the same nest, broods comprised mainly or exclusively of Redhead ducklings but led by a Mallard hen can be seen. This phenomenon, which may occur on an extremely large scale (Bouffard 1983; Joyner 1983), has been documented in breeding populations on Prairie marshes in particular. Olson (1964 *in* Bellrose 1980) calculated that about half of all Redheads on the Manitoba prairies are raised by Canvasbacks. In Québec, however, Alliston (1979b) reported that parasitism by the Redhead is rare.

There is some evidence that the Redhead is not very adept at this parasitic behaviour. According to Weller (1959), for example, 44–77% of parasitic eggs are laid in nests where the host has already begun incubation: these "late" eggs do not hatch, since the host stops incubating after her own eggs have hatched. The Redhead can also be its own worst enemy in some respects, because it has a strong tendency to abandon its nest if it has been parasitized (Bellrose 1980). Moreover, the hen takes a long time (close to 4 minutes) to lay an egg, which sometimes provokes an attack by the host female; this often results in eggs being broken and the nest being deserted (Weller 1959), thus reducing breeding success in the host species (Giroux 1981; Talent et al. 1981). Furthermore, parasitism, when taken to extremes, is counterproductive: nests have been found—deserted ones, obviously—containing 30–87 Redhead eggs (Weller 1959).

In Nevada, it has been noted that the Redhead hen expends a great deal of her energy reserves (46%), drawn from her fat stores, during laying, and somewhat less (30%) during incubation (Noyes and Jarvis 1985). As she has less energy reserves for incubation, she may abandon her eggs or not incubate them sufficiently. In addition, nests are often abandoned due to egg predation and variations in water level. In short, hatching success in the species—and thus productivity—are very low. During drought years, a large part of the population may not even attempt to breed at all (Weller 1959).

Although the Redhead is a diving duck, it forages mainly in marshes less than 2 m deep. In fact, it feeds in water so shallow that it rarely dives, foraging more in the manner of a dabbling duck (Siegfried 1976b; Bellrose 1980). The Redhead also differs from other divers by its diet, 90% of which is composed of the vegetative parts and seeds of aquatic plants, while the other divers tend to consume more invertebrates (Bergman 1973; Bellrose 1980; Bailey and Titman 1984). Even the ducklings eat mainly plants, which is surprising, given their high protein requirements to fuel their growth (Jarvis and Noyes 1986). Females, however, increase their intake of invertebrates during laying to meet the high energy and protein demands associated with egg production (Noyes and Jarvis 1985).

Distribution

The Redhead's breeding range is mainly in the western part of the continent, taking in western Canada and the western and north-central United States. It also breeds in southern Alaska, and very locally throughout northeastern North America.

In Québec, the Redhead breeds mainly in the extreme southern part of the province, in the St. Lawrence Valley. Populations are found on Lake Saint-François, near the Ontario border; along the St. Lawrence River downstream from the Island of Montréal (Boucherville, Île aux Fermiers); and on Lake Saint-Pierre. It has also been confirmed as breeding at Saint-Gédeon in the Lac-Saint-Jean region, where a female and 9 ducklings were discovered in 1993 (Club des ornithologues amateurs du Saguenay–Lac-Saint-Jean 1993b). The Dundee area, on the shores of Lake Saint-François, is a particularly good place to observe the species. Elsewhere, the Redhead is rather rare during the breeding season, although during atlassing it was reported around Cap Tourmente, downstream from Québec City. The most northerly record comes from Lake Desaulniers, near the La Grande-2 Dam (Côté 1980).

Notes on territory (continued)

Redheads do not seem to be affected by the presence of other species breeding nearby (Low 1940).

The Redhead generally builds its floating nest in bulrushes or cattails (background) in shallow water. Île aux Fermiers, Varennes Islands, near Montréal. *August 1991, Gaétan Duquette.*

The winding canals created by Ducks Unlimited in the Lac Saint-François National Wildlife Area are fringed with abundant emergent vegetation that provides excellent nesting cover. Southwest of Montréal. *April 1991, Jean-Luc DesGranges.*

Between the end of the breeding season and the start of fall migration, the drakes and young often move north and eastward to congregate on lakes (Weller 1964). In migration, Redheads, unlike other pochards, can be seen alone, in pairs, or in small tight groups of 5–15 individuals (Bellrose 1980). They can occasionally be observed in the southern part of the province during spring migration. Redheads have also been sighted around Rupert Bay in both the spring and fall of 1991 (Cons. G. G. – GREBE 1992q,r). In winter, 80% of the population is concentrated along the Gulf of Mexico, although Redheads are also found on both the Atlantic and Pacific coasts and locally inland in the central and southern United States. (Weller 1964; Root 1988).

History and trends

The Redhead only began breeding in Québec fairly recently. The first unconfirmed reports of individuals breeding at Lake Saint-François date from the 1940s; breeding was suspected in the late 1950s, but it was not confirmed until 1961 (Alliston 1979a).

The origin of the Québec breeding population is uncertain. In 1952 and between 1957 and 1963, a total of 3933 Redheads were introduced into New York State, to try to establish a breeding population: 269 of these releases were in the Wilson Hill Game Management Area, only 40 km from Lake Saint-François. Birds banded and released in New York were recaptured later at Lake Saint-François (Alliston 1979a).

There is, however, another hypothesis regarding the origin of this population. The Redhead's centre of abundance is on the Prairies, and a small population has been established in Ontario since the turn of the century. Between 1958 and 1961, there was a severe drought on the Prairies, which forced the population there to move to the edge of its breeding range. It was at this time that the species reached the Cornwall area on the Québec-Ontario border, and perhaps Lake Saint-François as well.

Both factors—the New York State releases (Alliston 1979a) and the natural expansion of the species' breeding range—may have helped the population at Lake Saint-François to become established.

Continent-wide, the Redhead is generally agreed to be at risk: population levels fluctuate rapidly and sometimes unpredictably (Weller 1959; Bellrose 1980). Annual mortality in the species is high, up to 80% for young in their first year, and 40% for adults (Bellrose 1980). The Redhead was designated a species of Special Concern in 1982 because of the declining numbers in the eastern United States (Tate and Tate 1982), undoubtedly due largely to hunting and habitat loss. Readily attracted to decoys, curious and gregarious, Redheads are easy to hunt (Palmer 1976), particularly immature birds (Bellrose 1980). In Québec, the species is marginal as a game bird, and only a few individuals are killed each fall when they mingle in with other diving ducks staging on the large riverine lakes in the southwestern part of the province. The shallow marshes that the species depends on during the nesting season are threatened by draining and filling for farming. Wetland conservation and management seem to be essential to sustaining Redhead populations (Bergman 1973; Woodin and Swanson 1989).

The Québec Redhead population is all the more vulnerable to local environmental disturbances, as it is concentrated at only a few sites.

François Shaffer and Jean-François Rail

Ring-necked Duck

Morillon à collier

Fuligule à collier
Aythya collaris

The Ring-necked Duck is a small diving duck that can be distinguished from the other pochards by the white ring around its bill. It gets its name from the chestnut collar around the base of the neck, which is hard to see from a distance. The Ring-necked Duck is a regular migrant breeder in Québec.

Hunters on Lake Saint-Pierre have dubbed this bird the "sooty bay duck," which shows how observant they are. They noted that, during migration, this "sooty" duck, as they call all pochards, frequents marshy bays, while the other pochards preferred to stay in open water. They also observed that this duck has several traits in common with dabbling ducks. The Ring-necked Duck is able to spring directly from the water like dabblers, while other pochards must run over the surface of the water to take off. The Ring-neck is also less gregarious and congregates in smaller flocks. When danger threatens, the hen hides her ducklings in the vegetation as dabblers do, instead of leading them out to open water the way most divers do. Despite these similarities to dabbling ducks, the Ring-necked Duck is a true diving duck.

At first, the nest is just a simple scrape but, as laying progresses, the hen adds a little vegetation from around the nest as well as down which she plucks from her belly and breast, thereby producing an incubation patch. Typical clutch size is 9 eggs, laid one day apart, which are incubated by the female for 25–29 days. Nests containing eggs or young were found in 20 Atlas squares. Near the end of incubation, or sometimes at the onset of hatching, the drake deserts the hen and joins other drakes to moult; in males, the post-nuptial moult begins even before they have abandoned their mates (Johnsgard 1975b). Hatching is relatively synchronous. The ducklings are brooded for 12–20 hours, then leave the nest in the company of the hen (Bellrose 1976). Breeding was confirmed in 185 (34.0%) of the 544 squares in which the species was reported. In 161 squares, confirmations were based on sightings of downy chicks.

Notes on territory

Breeding pairs are not aggressive; once nest sites have been selected, there appears to be mutual respect for them by neighbouring pairs. According to Mendall (1958), pairs may nest as close as 1.5–1.8 m apart; the species' small home range (1 breeding pair/2.4–9.3 ha, depending on habitat) and lack of aggressiveness toward conspecifics allow such a high nest density.

In Québec's boreal forest, breeding densities were recently estimated to be as high as 24 pairs/100 km² (D. Bordage and M. Grenier, CWS, unpubl. data).

In the boreal forest, the Ring-necked Duck looks for lakes with boggy, shrubby margins. Lake Malin, Laurentide Wildlife Reserve, north of Québec City. *June 1991, Réginald Ouellet.*

Habitat and behaviour

The Ring-necked Duck is chiefly associated with regions where conifers predominate. Although it frequents several types of habitat, it is particularly fond of shallow fresh water: marshes, bays, rivers, and lakes with marshy shorelines. Unlike other diving ducks, it shuns salt water and is even more rarely seen way out on large expanses of open water (Godfrey 1986).

In spring, Ring-necks head for their breeding grounds in small flocks of 6–30 birds (Mendall 1958). There, the pairs disperse to nest along the margins of open water, in bogs, fens, ponds, lakes, beaver ponds, and marshes.

According to Mendall (1958), there is little evidence of territoriality in the species. The drake seems to protect the hen much more than a specific area, and this behaviour is limited to immediately before or during nest site selection.

The hen, sometimes accompanied by the drake, chooses the nest site, preferably hidden in low vegetation on the edge of a lake or pond, on floating islands, or among hummocks and clumps of bushes in marshes (Mendall 1958). About 7–10 days elapse between site selection and the onset of laying (Johnsgard 1975b).

Abundance of Ring-necked Ducks in southern Québec. Aerial surveys conducted in May 1990, 1991, and 1992 *(D. Bordage, CWS, unpubl. data).*

Ring-necked Duck

Breeding distribution

1984–1989

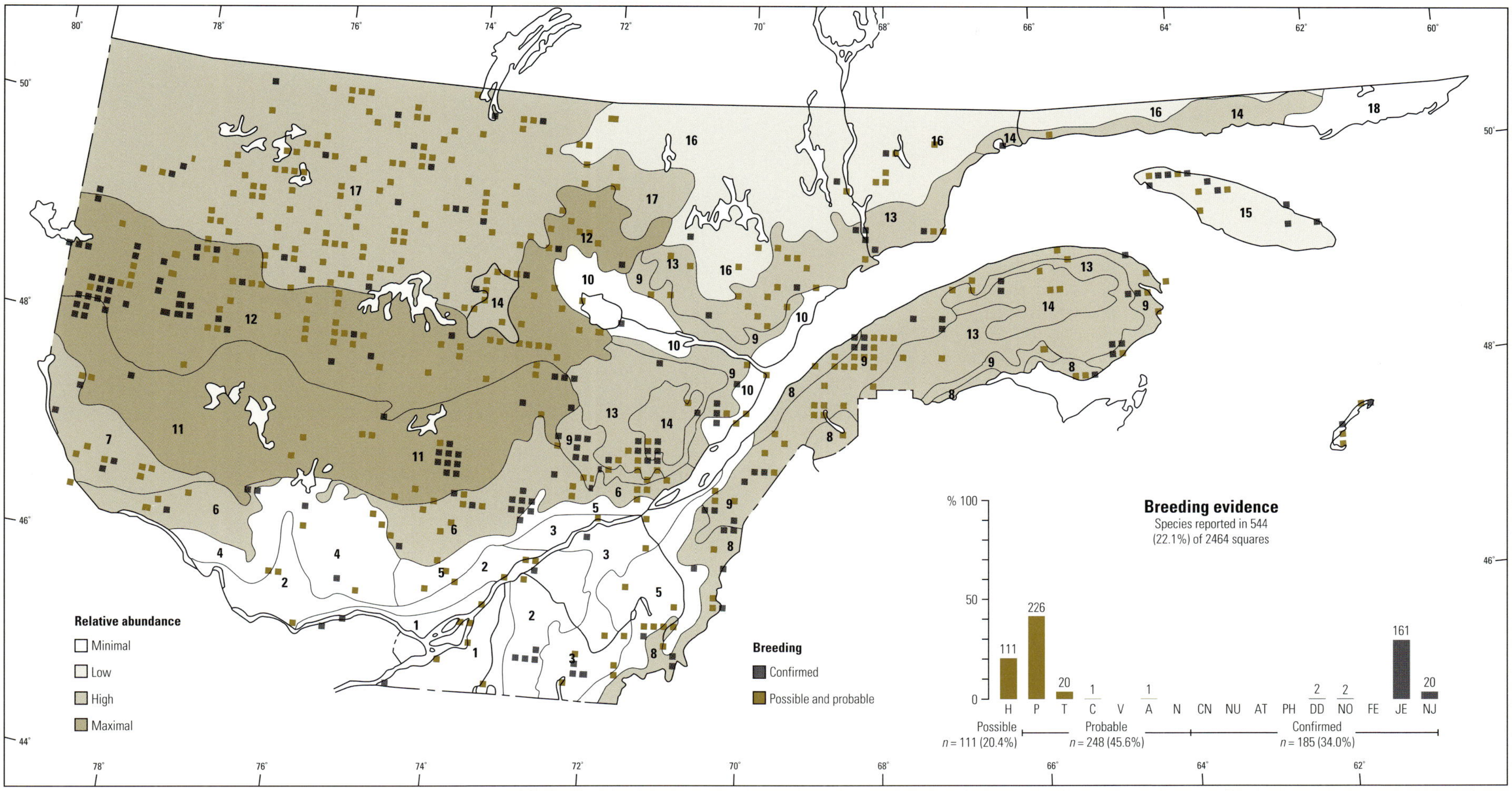

Species profile

Clutch size	usually 9 eggs (5–14)[2]
Laying interval	1 egg per day[1,2]
Start of incubation	with last egg[1,2]
Incubation period	25–29 days[1]
Maturity at hatching	precocial
Care of young	by female[1]
Nestling period	less than 24 hours[1,3]
Age at first flight	49–56 days[1,3]
Dependence of young	usually until first flight[1]
Broods per year	1[1]
Breeding age	1 year[3]
Mating system	monogamy[1]
Duration of pair bond	1 season[1]
Total length	38–46 cm[6]
Wing span	61.0–76.2 cm[5]
Weight	M 744 g (590–910) / F 671 g (544–910)[2]
Longevity record	20 years 5 months[4]

References:
1. Mendall 1958
2. Bellrose 1976
3. Johnsgard 1975b
4. Klimkiewicz and Futcher 1989
5. Terres 1980
6. Godfrey 1986

World breeding range

Chronology $n = 167$

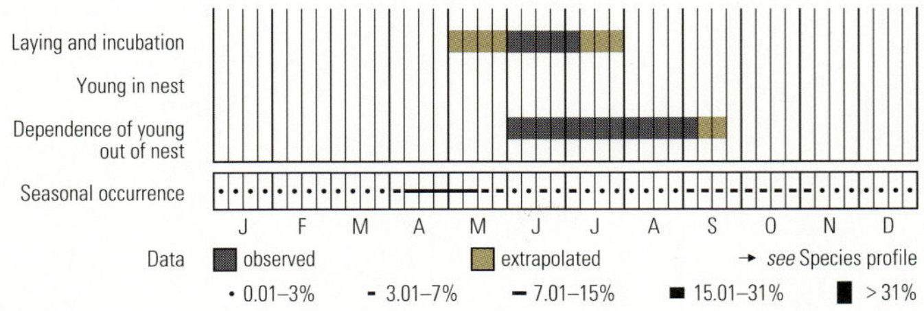

Frequency of occurrence by ecological region

Ecological regions** (forest regions)	Adequately covered squares	Squares in which reported	% of squares in which reported	Relative abundance (%)*
1. Sugar maple-hickory and sugar maple-basswood	97	8	8.2 ± 3.0	17.2
2. Sugar maple-basswood	153	11	7.2 ± 2.3	15.0
3. Sugar maple-basswood and sugar maple-yellow birch	101	10	9.9 ± 3.7	20.6
4. Sugar maple-yellow birch-basswood	56	6	10.7 ± 7.4	22.3
5. Sugar maple-yellow birch-beech and sugar maple-yellow birch-basswood	113	13	11.5 ± 3.4	23.9
6. Sugar maple-yellow birch-beech	71	26	36.6 ± 10.3	76.2
7. Sugar maple-yellow birch-balsam fir	19	6	31.6 ± 22.1	65.7
8. Sugar maple-yellow birch and balsam fir-yellow birch	47	16	34.0 ± 12.2	70.8
9. Balsam fir-yellow birch	87	35	40.2 ± 9.5	83.7
10. Balsam fir-white birch or balsam fir-red maple	46	4	8.7 ± 7.5	18.1
11. Yellow birch-balsam fir	52	25	48.1 ± 14.0	100.0
12. Balsam fir-white birch (west)	59	26	44.1 ± 13.0	91.7
13. Balsam fir-white birch (east)	76	29	38.2 ± 10.7	79.4
14. Balsam fir-black spruce	28	10	35.7 ± 18.6	74.3
15. White spruce-balsam fir and balsam fir-white spruce	22	6	27.3 ± 19.4	56.7
16. Black spruce-balsam fir-moss	18	4	22.2 ± 22.2	46.2
17. Black spruce-moss	32	12	37.5 ± 18.3	78.0
18.			*not analysed*	

* minimal: 0–29.9%; low: 30–59.9%; high: 60–89.9%; maximal: 90–100%

** Deciduous forest: 1–7; mixed forest: 8–11; coniferous forest: 12–17

This duck prefers regions where conifers dominate. It nests along a strip of shoreline where ericaceous plants are predominant, near shallow water. Martin-Valin ZEC, north of the Saguenay River. *August 1991, Jean Gauthier.*

Broods forage on the edges of lakes, small streams, and ponds, mainly for invertebrates, small crustaceans, molluscs, seeds, and aquatic vegetation (Mendall 1958; Hohman 1985; McAuley and Longcore 1988). The Ring-necked Duck can dive to depths of 12 m, although it usually finds its food at less than 1.5 m. The hen abandons her ducklings when they are able to fly (49–56 days) and goes off to moult (Mendall 1958; Johnsgard 1975b).

Distribution

The Ring-necked Duck winters on the Pacific coast from southern British Columbia to Guatemala, and on the Atlantic coast from Massachusetts to the West Indies (Godfrey 1986). In early spring, it heads north to its breeding grounds, which encompass all the Canadian provinces (Johnsgard 1975b) and the southwestern portion of the Mackenzie District, Northwest Territories (Godfrey 1986).

In southern Québec, Ring-necks are found almost everywhere south of the 50th parallel, particularly in regions of mixed and coniferous forest. The Abitibi lowlands, which supports a large beaver population, provides particularly favourable habitat; Ring-necks are less common, however, in the agricultural areas of Lac-Saint-Jean and southern Québec and, because of the limited number of lakes and ponds, the more mountainous sections of the Gaspé.

Data obtained from aerial surveys conducted by the Canadian Wildlife Service in May 1990, 1991, and 1992 provide a more precise picture of the species' abundance in certain parts of southern Québec (D. Bordage, CWS, unpubl. data).

The Ring-necked Duck has also been confirmed as a breeder north of the Atlas area, in the Rupert, Pontax, Opinaca, and Eastmain watersheds (Bordage and Aubry 1983; Bordage 1985a; Tardif and Morneau 1991; Cons. G. G. – GREBE 1992a; QNRCP). Breeding has also been confirmed near the James Bay coast (Benoit et al. 1993b), around the Great Whale and Little Whale rivers, and just a few kilometres south of the Nastapoka River (Cons. G. G. – GREBE 1992t).

On the Upper North Shore, the Ring-necked Duck frequents small lakes like these. Port-aux-Quilles, southwest of Tadoussac. *October 1993, Pierre Normand.*

The Ring-necked Duck also frequents very slow-moving rivers like the Rivière du Sud-Ouest, which has broad lake-like stretches. The nearby farming operations do not seem to affect its nesting. Near Saint-Fabien, southwest of Rimouski. *October 1993. Pierre Normand.*

The nest, which starts out as a simple scrape hidden in low vegetation, is lined with plant material and down as laying progresses. Lake Bleury, Laurentide Wildlife Reserve, north of Québec City. *June 1982, Jean-Luc DesGranges.*

History and trends

At the turn of the century, Wintle (1896) and Dionne (1906) classified the Ring-necked Duck as a rare transient in southern Québec. The first confirmed breeding record in the province is from 1943 (Ouellet 1974a). The evidence suggest that, before 1930, the Ring-necked Duck was rare in northeastern North America; during the next two decades it expanded its breeding range considerably in this part of the continent (Mendall 1958). Since then, the species has spread rapidly through the wetlands of the boreal forest. The marked increase in beaver populations since the 1940s (R. Lafond, MEF, pers. comm.) has no doubt helped create suitable habitat for the Ring-necked Duck and contributed to its expansion.

Since it feeds by grubbing at the bottom of ponds, marshes, and lakes, where hunters' spent shot accumulates, the Ring-necked Duck is highly vulnerable to lead poisoning. In Québec, Lemay et al. (1989) noted that 21% of Ring-necked Ducks examined had ingested at least one lead pellet. A study of wings turned in by hunters also showed that lead concentrations in bone were much higher in this species than in the American Black Duck or Mallard (Dickson and Scheuhammer 1993).

The Ring-necked Duck is highly prized by sport hunters, who bag some 20 000 birds each year in Québec (Appendix II).

Michel Lepage and Marie-Renée Doyon

Greater Scaup

Grand Morillon

Fuligule milouinan

Aythya marila

Guy Michaud

THE BEST TIME TO SEE THE GREATER SCAUP IN QUÉBEC IS IN SPRING AND fall as it stages on lakes in the southern part of the province on its way between its breeding and wintering grounds. In North America, most individuals breed in Alaska and Yukon and winter on the Atlantic coast (Bellrose 1980). In Québec, breeding has been confirmed in a few parts of Ungava and on the James and Hudson bay coasts (Manning and MacPherson 1952; Todd 1963; Cons. G. G. – GREBE 1990a; Benoit et al. 1993b). Robert (1989b) classifies the Greater Scaup as a migrant breeder, nesting locally in central and northern Québec. There is, however, an isolated breeding population in the Magdalen Islands and probably another on Anticosti Island.

The Greater Scaup looks very much like the Lesser Scaup, but the male in nuptial plumage has a rounder head, with a greenish rather than violet gloss, and whiter flanks. Determining the distribution and abundance of the Greater Scaup in northern Québec is highly problematic, given the difficulty of separating Greater and Lesser scaup in the field and the remoteness of the species' breeding grounds.

Habitat and behaviour

Aside from plumage, there are a number of other characteristics that distinguish the Greater Scaup from the Lesser Scaup. The Greater breeds farther north and its distribution is much wider, indeed, almost circumpolar. The Greater uses different migration routes, its wintering grounds are more northerly, and it is more of a saltwater duck than the Lesser (Bent 1923). The Greater Scaup often flies in compact, expertly co-ordinated flocks and takes wing and gains altitude easily. It is much more mobile and agile in flight than the other diving ducks. The Greater Scaup is an excellent swimmer underwater; it uses its feet alone to propel itself, diving to depths of over 6 m (Cramp and Simmons 1977) and remaining underwater for up to a minute (Ehrlich et al. 1988).

Despite its skill as a diver, the Greater Scaup prefers to forage in shallow water (generally less than 1.5 m) for aquatic plants, the seeds of herbaceous plants, and animal matter—molluscs (oysters, mussels, scallops), aquatic insects, small crustaceans, tadpoles, small fish, and fish roe. The Greater Scaup is an opportunist; its diet varies widely depending on the season, habitat, and food available. For example, in spring migration along the Pacific coast, it seems to follow the spawning herring, eating the roe (Munro 1941a).

The Greater Scaup nests primarily on the edges of taiga and tundra lakes and ponds, particularly those with islands covered with herbaceous vegetation. Suitable nesting habitat is fairly open, with a shallow body of water with grassy banks nearby. Nests can be clustered, especially on islands; the species thus appears to form loose colonies (Ehrlich et al. 1988), often in association with gulls and terns. Indeed, hatching success seems to be enhanced in larid colonies: egg

Notes on territory

Currently, there is no evidence that the Greater Scaup defends a territory during the breeding season (Johnsgard 1975b). On the contrary, Bengtson (1970) states that the species tends to flock together in nesting areas but he could not

An isolated population of Greater Scaup, consisting of a few pairs, nests along the edges of the lagoon at Baie du Havre aux Basques. Magdalen Islands. *June 1991, Desneige Perreault.*

Greater Scaup

Breeding distribution

1984–1989

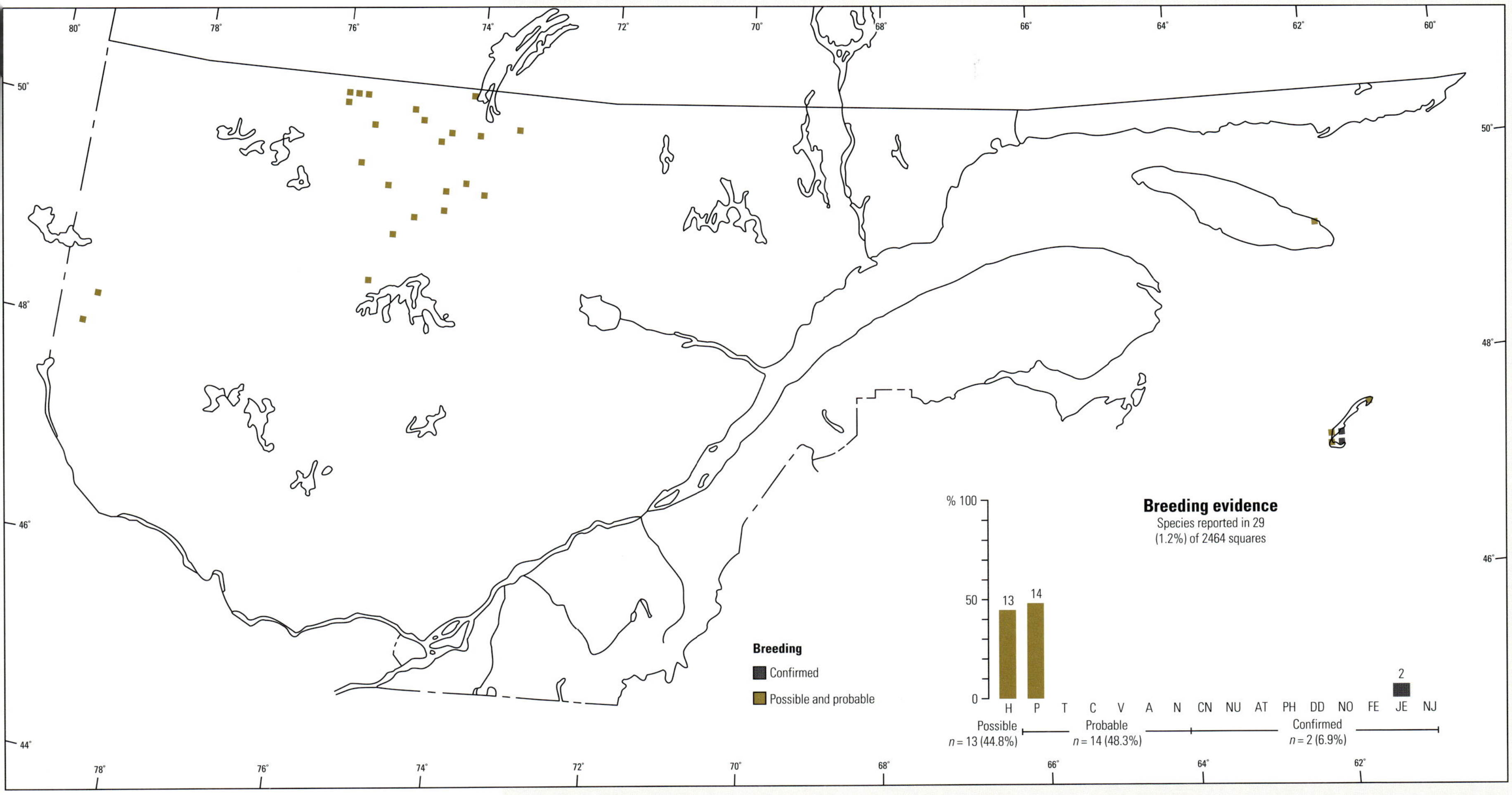

Species profile

Clutch size	usually 8–9 eggs (6–11), more than 11 = dump nest[2]
Laying interval	1 egg per day[4]
Start of incubation	with last egg[4]
Incubation period	23–28 days[1,2,3,4]
Maturity at hatching	precocial
Care of young	by female[4,5]
Nestling period	probably less than 48 hours
Age at first flight	35–45 days[4,5]
Dependence of young	until first flight or a few days before[3,4]
Broods per year	1[4,5]
Breeding age	1–2 years[1,2,4]
Mating system	monogamy[5]
Duration of pair bond	1 season[1]
Total length	43–53 cm[6]
Wing span	76.2–87.6 cm[7]
Weight	M 1052 g (862–1361) / F 975 g (726–1270)[2]
Longevity record	11–13 years[7]

References:
1. Johnsgard 1975b
2. Bellrose 1980
3. Palmer 1976
4. Cramp and Simmons 1977
5. Ehrlich et al. 1988
6. Godfrey 1986
7. Terres 1980

Chronology *n* = 78

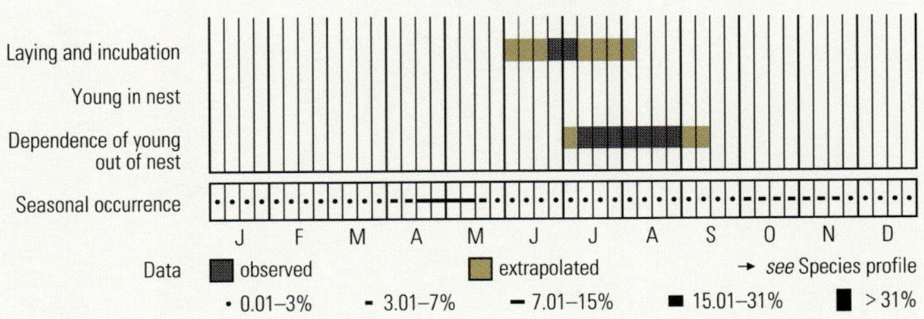

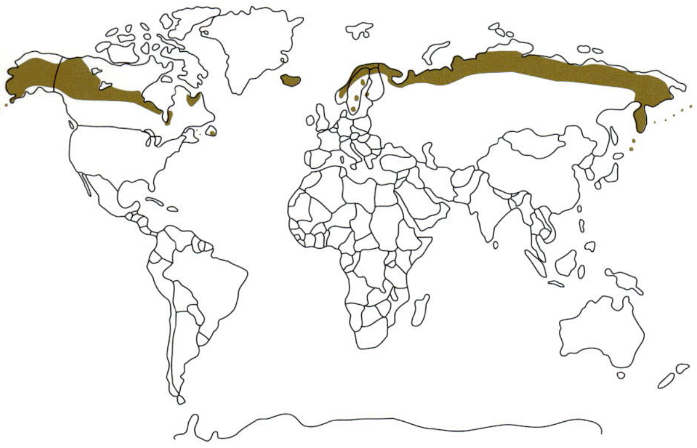

World breeding range

predation by crows and ravens is reduced since the larids chase them off, although the larids themselves may prey heavily on the ducklings (Johnsgard 1975b).

The nest, generally built in a shallow depression near the water, is slightly raised and hidden among clumps of grass, reeds, or brush. It is lined with fine grass and the female's dark down. Typical clutch size is 8–9 eggs, laid a day apart. The hen incubates the eggs alone, beginning once the clutch is complete. Females occasionally lay eggs in the nests of conspecifics (Ehrlich et al. 1988). Incubation lasts 23–28 days, and hatching is synchronous. Hens generally produce one brood a year, but may renest if the first clutch is destroyed. Ducklings feed more on the surface than adults, and their diet probably consists of a greater proportion of insects, larvae, and seeds (Cramp and Simmons 1977). The female stays with her ducklings longer than do other pochards hens, usually moulting at the nest site. Several broods (up to 30 ducklings) may be under the protection of a single hen (Cramp and Simmons 1977). The drake does not help with incubating the eggs or rearing the young.

Pair formation occurs in spring (between February and April), and the pair bond is usually maintained until sometime during incubation. Huge flocks of males then gather on large lakes on the breeding grounds to moult. They are flightless for 3–4 weeks. In September, when they can fly again, they begin slowly migrating back to large inland lakes or the ocean, then southward. They sometimes linger on lakes along the way and tend to slow down when they reach salt water. Males winter farther north than females and non-breeders (Palmer 1976). Yearling non-breeders occasionally remain on the wintering grounds during the breeding season (Madge and Burn 1988).

In migration, the main staging areas are along the St. Lawrence, particularly the riverine lakes (Lakes Saint-François, Saint-Louis, and Saint-Pierre) as well as Lac des Deux Montagnes and all along the estuary, the Ottawa River, and some inland lakes. The timing of migration is fairly similar to that of the Lesser Scaup. The largest flocks of Greater Scaup are seen from early April to mid-May and in October and November (Lehoux et al. 1985). In winter, the species mainly frequents salt water on the Pacific and Atlantic coasts, both on the open ocean and in bays, but is also often seen on large inland lakes, particularly Lake Erie and Lake Ontario (Godfrey 1986; Root 1988). There are a few winter records in southern Québec and the species is considered to winter irregularly in the region (David 1980a; Godfrey 1986).

Notes on territory (continued)

determine whether this behaviour is social or is related to particular features of the habitat. During his studies in Iceland, Bengtson (1970) recorded an average density of 273 nests/km² (islands and mainland), and 331 nests/km² on islands. On Great Slave Lake, in the Northwest Territories, nests scarcely 2 m apart have been found (Weller et al. 1969).

In southern Québec, the highest pair densities (1–2 pairs/100 km²) were found around Lake

The Greater Scaup prefers to nest on small islands, where it is protected from predators, rather than on shore. One of its haunts during the breeding season is this lake near Rivière au Phoque, southeast of Point Louis XIV. James Bay region. *July 1994, Réjean Benoit.*

Manouane, east of Lake Mistassini, during surveys conducted by the Canadian Wildlife Service between 1990 and 1992 (D. Bordage, CWS, unpubl. data).

Distribution

The Greater Scaup is very rare in southern Québec during the breeding season. It was reported in only 1.2% (29/2464) of the squares atlassed and the only evidence of breeding was obtained in the Magdalen Islands, where broods were sighted twice. There is a small breeding population of just a few pairs in the Magdalens, which has been known to exist for over a century (Job 1902; Bent 1923; Fradette 1992). The 8 July 1989 record from Anticosti Island—as well as several authenticated pre-Atlas records during the breeding season (Todd 1963; Ouellet 1969)—suggest that the species also breeds there.

The scattered reports from the Abitibi region and southwest of Lake Mistassini, however, probably involve pairs or individuals on their way north. Pairs observed in May or June in the boreal forest cannot be considered breeders, given the late nesting habit of this northerly breeder (James 1987a). In the Abitibi, on the other hand, if a pair is seen several times in June and July, it could be nesting. The cluster of records from southwest of Lake Mistassini is essentially the result of better coverage of this region, owing to aerial surveys of waterfowl conducted by the Canadian Wildlife Service in May between 1985 and 1989.

Pairs of Greater Scaup, the most northerly of our pochards, can be observed on this lake each May. This type of lake is similar to those found farther north within the species' usual breeding range. Northwest of Chibougamau. *June 1988, Réginald Ouellet.*

In Québec, the species has a northerly distribution, in the taiga and on the tundra (James 1987a). According to Bellrose (1980), the highest breeding densities are found in the forest-tundra at the tree line. North of the Atlas area, several broods have been reported on the northeast coast of James Bay (Benoit et al. 1993b) and in the watersheds of the Great Whale, Little Whale, and Nastapoka rivers (Cons. G. G. – GREBE 1990a), corroborating the findings of earlier investigators who reported the Greater Scaup as common there (Savard 1977; St-Louis 1978; Bider and Lamothe 1982). During the nesting period, the species is found mainly on the smaller lakes in this region (Cons. G. G. – GREBE 1990a). Farther north, pairs and single birds are regularly seen during the summer on small lakes around Ungava Bay and east of Hudson Bay (Bateman 1953; Todd 1963; Lamothe 1978; B. Harvey, USFWS, and D. Bordage, CWS, pers. comm.). Excluding the Magdalen Islands and Anticosti Island, the southern boundary of the species' breeding range is probably around La Grande Rivière, where the Greater Scaup is considered rare in summer (Boucher and Lalonde 1980; Côté 1980, 1981). In the area around the Nottaway, Broadback, and Rupert rivers (Cons. G. G. – GREBE 1992a,b) and around the Eastmain River (Tardif and Morneau 1991), the species is mainly seen during spring migration.

History and trends

In the mid-1970s, the breeding population in Canada and Alaska was estimated to be about 750 000 birds (Bellrose 1980), and across North America, Greater and Lesser scaup numbers seem to have remained stable until quite recently. According to Bellrose (1980), this stability is due to the species' northerly breeding range; in the North, droughts are generally less severe than on the Prairies, where several other duck species nest and where populations have been declining for several years. Nonetheless, a slight downward trend in Greater and Lesser scaup populations was noted between 1981 and 1990 (Environment Canada 1990).

Québec and Ontario have the highest hunter kills in Canada. In southern Québec, the seasons for Greater and Lesser scaup are identical. The Greater Scaup harvest has been decreasing slowly but steadily for the last 20 years, however (Appendix II), largely because the birds have moved from their traditional staging areas to other stretches of the St. Lawrence less accessible to hunters. Curiously, the Greater Scaup seems to be less affected by hunting than the Lesser Scaup, although it is more abundant in staging areas in southern Québec in the fall (Lehoux et al. 1985). There is a native subsistence hunt but its extent is unknown. So far, hunting pressure on the Greater Scaup does not seem to have had a noticeable effect on populations. Other factors, such as habitat loss or modification in the main breeding areas, could affect the North American population of the species. Breeding success may be affected by poor weather some years. Finally, the Greater Scaup's habit of gathering in large concentrations on its wintering grounds makes it particularly vulnerable to disturbances such as oil spills (Cramp and Simmons 1977).

Réjean Benoit and Jean-François Rail

Lesser Scaup

Petit Morillon

Petit Fuligule

Aythya affinis

It takes a practised eye to distinguish between the Greater Scaup and the Lesser Scaup. The head of the Lesser drake has a violet gloss visible only in good light, and its crown is somewhat more angular than in the Greater Scaup. On the water, the hens of the two species are also difficult to tell apart. On flying birds, the bold white stripe on the wing is a more reliable field mark: it extends onto the primaries in the Greater Scaup, but is restricted to the secondaries in the Lesser.

Although Robert (1989b) classifies the Lesser Scaup as a migrant breeder with a limited distribution in central and northern Québec, it also breeds very locally in southern Québec.

Guy Michaud

A few Lesser Scaup breed in southern Québec; Lake Saint-François' extensive marshes and numerous islets are particularly attractive to the species. Lac Saint-François National Wildlife Area, southwest of Montréal. *September 1991, Léo-Guy de Repentigny.*

The best time to see the Lesser Scaup is during spring and fall migration, when it can be found on large bodies of water such as the St. Lawrence River in mixed flocks of several thousands birds with the Greater Scaup (Lehoux et al. 1985).

Habitat and behaviour

In western Canada, the Lesser Scaup generally nests on the margins of lakes and ponds less than 2 ha in size and at least 3 m deep, in open or sparsely forested habitats. Most often, nests are on low-lying islands or shores of lakes and ponds where rushes, bulrushes, and sedges grow (Palmer 1976; Johnsgard 1978; Godfrey 1986). In Québec, along the St. Lawrence River, it breeds in some of the large marshes and swamps, while in the Abitibi region, it is found along the shores of lakes and ponds. An abundant supply of food—aquatic plants and especially aquatic invertebrates—is crucial (Johnsgard 1975b; Palmer 1976).

Pair formation occurs quite late, usually beginning in March and ending in April or May (Johnsgard 1975b). Generally silent, the drake makes a variety of raucous calls and whistles during courtship (Palmer 1976), which takes place mainly on the surface of the water but is interspersed with dives and aerial pursuits (Johnsgard 1965).

The pair is not very territorial and occupies a poorly defined nesting area; several nests may be found close together. Unlike the drake, the hen tends to return to breed near the site where she hatched (Trauger 1971; Palmer 1976).

The nest, in a shallow depression in the ground, is lined with plant stems and down and is well concealed in the tall grass. Water level and habitat seem to affect how far it is from water (Bellrose 1980). Despite the

Notes on territory

The Lesser Scaup is not very territorial and sometimes nests in loose colonies (Palmer 1976). Pairs will usually keep to themselves, but will tolerate the presence of conspecific pairs nearby. Nonetheless, the drake will defend the hen when other drakes get too close (Gehrman 1951). In a study in the Northwest Territories, Trauger (1971) observed that pairs separated early in the morning, but tended to flock together in late afternoon. This scaup sometimes nests on islands; Vermeer (1970) reported densities of 0.6–9.6 nests/ha on islands in some Alberta and Saskatchewan lakes.

In southern Québec, the highest densities have been observed in the 200-km stretch from

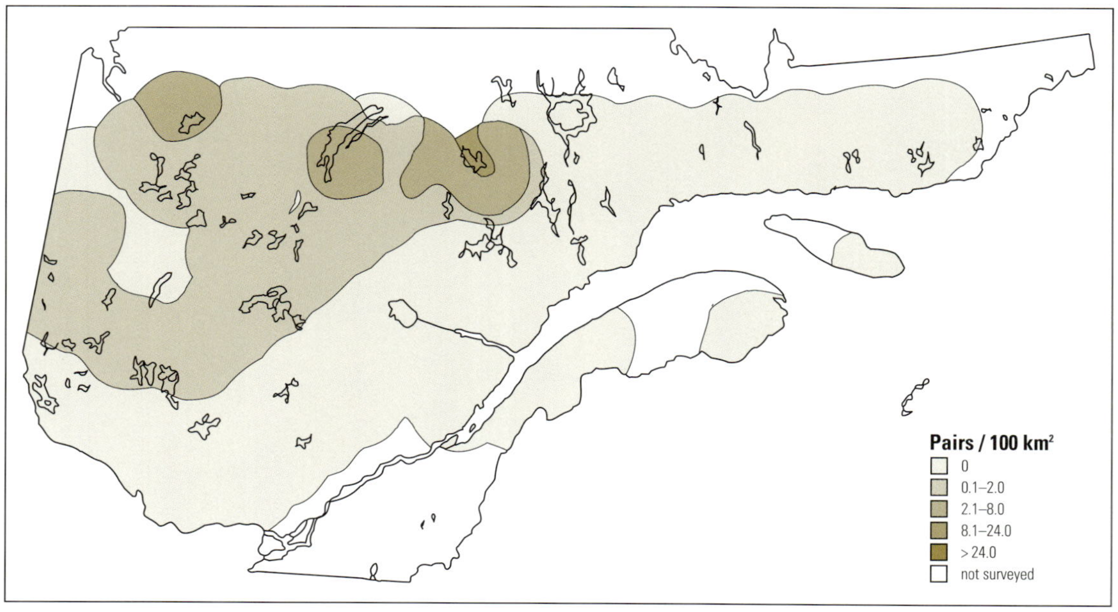

Abundance of Lesser Scaups in southern Québec. Aerial surveys conducted in May 1990, 1991, and 1992 *(D. Bordage, CWS, unpubl. data).*

Lesser Scaup

Breeding distribution
1984–1989

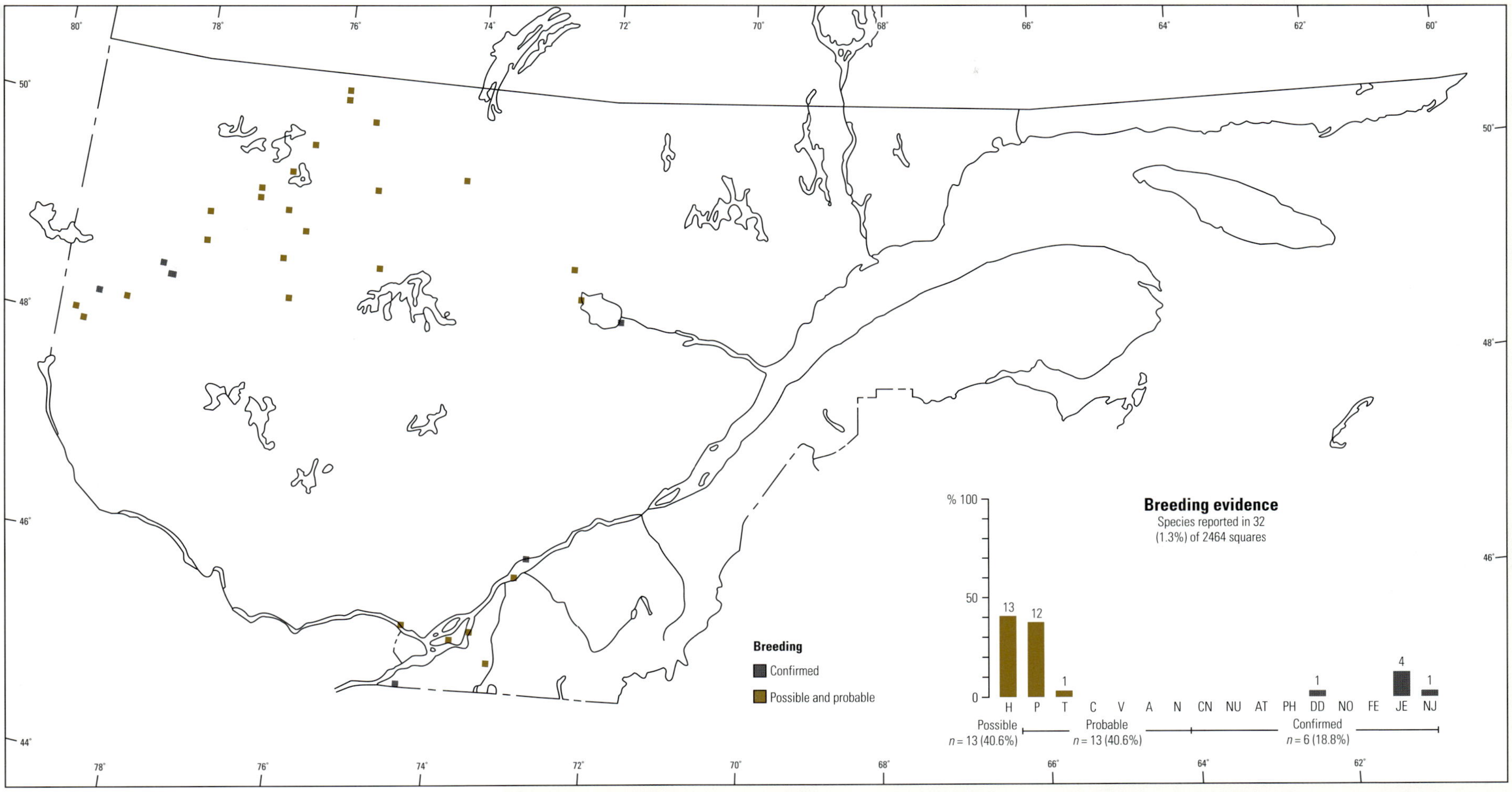

Species profile

Clutch size	usually 9 eggs (8–11)[2]
Laying interval	1 egg per day[1]
Start of incubation	with last egg[7]
Incubation period	21–28 days[1,2,4,5]
Maturity at hatching	precocial
Care of young	by female[1,2]
Nestling period	probably less than 48 hours
Age at first flight	45–54 days[1,2]
Dependence of young	until 4–5 weeks old[2]
Broods per year	1–2[2]
Breeding age	1–2 years[1,2]
Mating system	monogamy[1,2,5]
Duration of pair bond	1 season[1,2]
Total length	38.0–45.5 cm[3]
Wing span	61.0–83.4 cm[4]
Weight	M 826 g (499–1089) / F 748 g (454–953)[2]
Longevity record	18 years 4 months[6]

Chronology *n* = 50

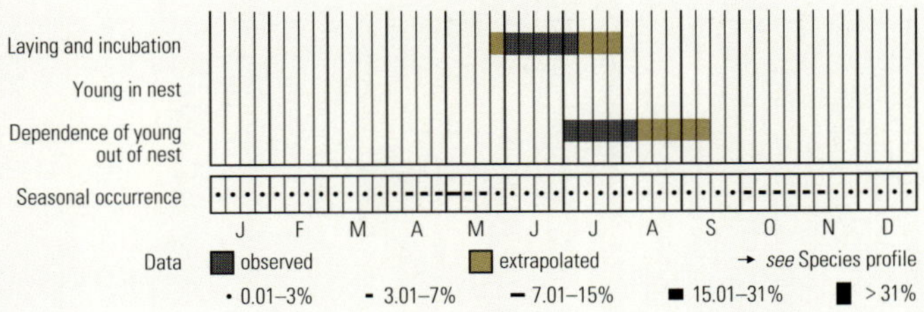

World breeding range

References:
1. Palmer 1976
2. Bellrose 1980
3. Godfrey 1986
4. Terres 1980
5. Ehrlich et al. 1988
6. Clapp et al. 1982
7. Afton 1984

late onset of breeding, double brooding may sometimes be attempted (Johnsgard 1978; Bellrose 1980). The drake does not help with either incubation or rearing the young, but abandons the hen early in the incubation period. Nest predators are primarily the striped skunk and, to a lesser extent, the American Crow (Hines 1977; Bellrose 1980).

The ducklings leave the nest shortly after hatching. Occasionally, two or more broods will merge under the joint care of more than one hen (Afton 1983). Hens often desert the ducklings when they are still in the downy stage, but most broods are abandoned at the age of 4–5 weeks (Bellrose 1980). Breeding success is quite variable, but as in the Greater Scaup, it is fairly low in comparison with that of other duck species. Productivity seems to be affected by water level variations (Palmer 1976).

The Lesser Scaup is an expert diver, reaching depths of 6 m and remaining underwater for up to a minute (Johnsgard 1975b). According to Palmer (1976), adults feed mainly on the seeds of aquatic plants, vegetable matter, and animal foods such as molluscs and aquatic insects, while the young eat mainly insects. In Saskatchewan, adults were found to consume mostly invertebrates during the nesting period (May, June, and July) (Dirschl 1969).

Of all the ducks, Lesser Scaup populations appear to have the largest proportion of males, 60–80% depending on the population (Bellrose 1980). Non-breeding males wander around the breeding grounds for a time before gradually migrating south sometime during the summer (Bellrose 1980). Many individuals do not breed until they are 2 years old, and during their first year can also be seen on the breeding grounds. During the first half of June, they congregate on lakes for the moult, where they are soon joined by drakes that have abandoned their incubating mates (Johnsgard 1975b; Palmer 1976).

Distribution

The Lesser Scaup has strictly a New World distribution, breeding from Alaska to Québec, and in the American Midwest. The Québec population probably winters on the eastern seaboard of the United States south to Florida (Bellrose 1980).

Most Atlas records were scattered across the Abitibi to Lac-Saint-Jean. A few birds were seen along the St. Lawrence River near Montréal, from Lake Saint-François to Lake Saint-Pierre, with one report from the Richelieu River near Montréal. Breeding was confirmed in the Lac Saint-François National Wildlife Area, near Yamachiche on Lake Saint-Pierre, and at Saint-Gédéon in the Lac-Saint-Jean region, as well as at Rouyn-Noranda, Figuery, and near Lake Beauchamp in the Abitibi region.

The Abitibi records need some explanation: the presence of non-breeders may have inflated the number of reports of possible breeding and birds seen in some squares in May were probably on their way to their breeding grounds in northern Québec.

Since field identification of the species is difficult during surveys, trying to define its breeding range is also problematic. It is very easy to confuse Lessers and Greaters at any time of year, and hens are particularly hard to tell apart during the breeding season and moulting period. As a result, many surveys and waterfowl studies make no distinction between the two species and conflate them into a single species. However, data from a number of studies show that

Notes on territory (continued)

Lake Mistassini in the west to Lake Manouane in the east. In this region, densities of 8–16 pairs/100 km² have been recorded (D. Bordage and M. Grenier, CWS, unpubl. data).

Along the St. Lawrence, the Lesser Scaup also nests on Île des Barques in Lake Saint-Pierre, selecting tall beds of grass that will conceal the nest well. *October 1993, Pierre Normand.*

A clutch of Lesser Scaup hatching. Hatching is synchronous and, as soon as the ducklings' down has dried, the hen will lead them to water. Chip Lake, Alberta. *June 1970, Pierre Lamothe.*

the Lesser's distribution extends north of the Atlas area. Breeding has been confirmed on the northeast coast of James Bay, near Kakassituq Point, where 1 brood was seen in 1990 and 6 broods in 1991 (Benoit et al. 1991, 1992). Sightings of 4 broods near Lake Denys on 21 July 1990, 2 broods near Lake Paimpont on 27 July 1990, and a single brood near Lake Amichinatwayach on 4 August 1990 have allowed breeding to be confirmed in the watersheds of the Great Whale and Little Whale rivers (F. Morneau, GREBE, unpubl. data). The Lesser Scaup has also been reported during the breeding season at Kuujjuarapik (ÉPOQ), at Lake Desaulniers (Côté 1981), around the Eastmain and Opinaca rivers (Bordage and Aubry 1983), and on several lakes near the Opinaca River, east of Highway 109 (D. Bordage, CWS, unpubl. data). Farther east, Lesser Scaup have been observed in May around Manic-5 (ÉPOQ).

History and trends

The Lesser Scaup has only recently expanded into eastern Canada. Early in the century, Macoun and Macoun (1909) classified this scaup as a visitant in Ontario, Québec, and New Brunswick, maintaining that the handful of breeding records more properly involved Greater Scaup. Taverner (1922) considered the Lesser Scaup exclusively a western species although he said it occasionally bred as far east as the Great Lakes. Breeding was confirmed for the first time in Québec in 1950, at Moar Bay on the east coast of James Bay (Manning and MacPherson 1952). In southern Québec, the first nests were discovered by G. W. Alliston in the Lac Saint-François National Wildlife Area in 1969 and 1970 (Ouellet 1974a), and on Île du Moine in Lake Saint-Pierre in 1972 (COQ 1977).

Continent-wide, the population is generally more stable than that of most other waterfowl species. Part of the population breeds in the wetlands of the boreal forest, where water levels tend to be more uniform than in the prairie marshes where most duck species nest, which tend to be more vulnerable to drought (Bellrose 1980).

The Lesser Scaup frequently ingests spent lead shot while feeding (Lapointe 1986; Havera et al. 1992). During a study conducted at Lake Rouyn, Abitibi, one bird was found with 96 pellets in its gizzard, a record for the species (Lapointe 1986). Birds ingesting pellets may eventually die from lead poisoning (Anderson 1975; Stout and Cornwell 1976; Bellrose 1980). Poisoning rates are nonetheless generally lower in diving ducks than in other ducks because of their diet, which consists mainly of animal food (Sanderson and Bellrose 1986).

In Québec, the Lesser Scaup hunt takes place mainly along the St. Lawrence River and its principal tributaries upstream of Sainte-Anne-de-la-Pocatière, producing an annual harvest of just over 20 000 birds in recent years (Appendix II).

SIMON BARRETTE AND RODGER D. TITMAN

Common Eider

Eider à duvet

Somateria mollissima

Pierre Dupuis

ABUNDANT JUST ABOUT EVERYWHERE IN ITS CIRCUMPOLAR BREEDING RANGE, the Common Eider nests in particularly high densities in its island colonies in the St. Lawrence estuary. The Common Eider is the largest of Québec's breeding ducks, although its weight varies with sex, subspecies, and time of year. With the drake's elegant black and white plumage and the species' intriguing habit of raising its young in crèches, the Common Eider is a noteworthy inhabitant of our coastal areas. To the casual observer, the eider provides a panoply of sights and sounds: on May evenings, the air is filled with the ghostly cooing of the drakes as they gather by the hundreds in the sheltered bays of the estuary; in the summer, crèches of dozens of peeping ducklings ply the rockweed in search of food; and, in fall, migratory flocks of hundreds of males, and later females, stretch out in long, low lines over the sea. From the species' name comes the french word *édredon*, itself taken from the German *Ederdaun*, for a comforter filled with down collected from eider nests.

Common Eiders in the St. Lawrence estuary nest on certain islands but rear their broods elsewhere, along the shoreline of other islands or the mainland. Females nesting on Bicquette Island (foreground) lead their broods to Île du Bic (middle distance) and the mainland (background). *October 1993, Pierre Normand.*

Habitat and behaviour

By the time the St. Lawrence estuary population begins its northward migration from its wintering grounds along the coast of New England (Reed and Erskine 1986), many pairs have already been formed (Gorman 1970). Adult eiders can be seen in the estuary beginning in late March (David 1980a; Larivée 1993), although the whereabouts of the immature birds is not known. On reaching the estuary, they spread out along the shores of the mainland and around the main breeding islands (Reed 1975a,c; Bédard et al. 1986b; Chapdelaine et al. 1986c), to wait until their nest sites are snow-free. The females feast on blue mussels, herring roe, and other food (Cantin et al. 1974) to store up fat reserves for nesting, while their mates keep other displaying males at bay; consequently, the females end up gaining weight and the males losing weight at this time (Gorman 1970). Pairs begin to move onto their nesting islands at the end of April.

The Common Eider nests on islands with low vegetation (grasses or shrubs) or forested with balsam fir and carpeted with moss or conifer needles (Milne and Reed 1974). On wooded islands, nests are usually built in the undergrowth under the forest canopy, while colonies of Herring Gulls or Great Black-backed Gulls occupy the grassy rim. Small colonies of Double-crested Cormorants, Black Guillemots, or Razorbills may also be found nearby.

The female chooses an existing depression in the ground or makes a small scrape, lining it with down from her breast. Laying peaks around 25 May (van Dijk 1986). Typical clutch size is 3–5 eggs (Guignion 1967; Milne and Reed 1974; Munro and Bédard 1977a; van Dijk 1986), which are laid a day apart (Watson et al. 1993); Common Eider eggs are characteristically large. Very early in the incubation period, the female lines the nest with a thick layer of down particularly around the rim, which she uses to keep the eggs warm and to cover them when she leaves, adding a little more as incubation progresses (Guignion 1967).

At the onset of laying, the drakes abandon the hens and raft with other males off certain islands. Some of the drakes then migrate farther east to invertebrate-rich coastal shoals, often located at river mouths, to moult (Gauthier and Bédard 1976; Bédard et al. 1986b), where they can feed peacefully, undisturbed by terrestrial predators, while they are flightless (Gauthier and Bédard 1976). Most males breeding in the estuary, however, leave the area before they moult for an unknown destination, although a few flightless individuals can be found off Anticosti Island and at the mouth of Chaleur Bay (Gauthier and Bédard 1976).

During incubation, the female does not feed and only leaves the nest occasionally to drink; consequently, by the end of the approximately 26-day-long period, some females have lost close to 50% of their total weight (Gorman and Milne 1972). Hatching peaks around 20 June (Gauthier and Bédard 1976; van Dijk 1986), with all ducklings in the same brood hatching within less than 18 hours. The female's calls and the peeping of the unhatched ducklings apparently contribute to the high level of synchronization. The down of the newly hatched young dries quickly and the ducklings soon become very active around the nest. Roughly 24 hours after the first egg has hatched, the hens lead the ducklings to water (Munro and Bédard 1977a).

As broods make their way along the paths through the vegetation, they converge, begin to mix and form crèches. By the time the ducklings have reached the water, single broods represent only about a third of the groups, the rest consisting of crèches of 2–7 broods. At this stage, a certain hierarchy has already been established among the females accompanying the young. The "broody" females call repeatedly to the ducklings, remain on the alert for danger and, with their aggressive behaviour, relegate the other hens to "associate" status at the edge of the crèche. "Visiting" females—failed nesters that gather in flocks by the shore—often approach the crèches, but are energetically repelled by the ducklings' mothers (Munro and Bédard 1977a).

The trek from the nest to the shore and across the inlet or bay to open water takes only a few hours but is a very critical period for the ducklings. Near the shore, the ubiquitous gulls frequently overfly the groups, sometimes swooping down on a crèche and trying to snatch away a duckling. The females respond with loud alarm calls, prompting the ducklings to take refuge under their outspread tail

Notes on territory

Although Common Eiders use the same nest sites year after year (Reed 1975c; Gauthier and Bédard 1976), they do not defend a territory strictly speaking. During courtship, the drake defends a space around the hen, which, among other things, allows her to build up fat reserves for laying and incubation. When the female is on the nest incubating the eggs, she appears to defend the area within the range of her bill. In brood-rearing areas, crèches or broods adopt a specific part of the shoreline such as a bay as their home range, sometimes for more than 60 days (Gauthier and Bédard 1976). Bédard et al. (1986b) recorded average densities

Bicquette Island alone is home to over 40% of the Common Eider population in the estuary. Most individuals nest in the island's forests, while large numbers of gulls nest at the forest edge, in grassy patches and along the rocky shore. The gulls gather on the rocks waiting for the broods to leave the island, when they snatch up large numbers of ducklings. *May 1973, Jean Munro.*

Common Eider

Breeding distribution 1984–1989

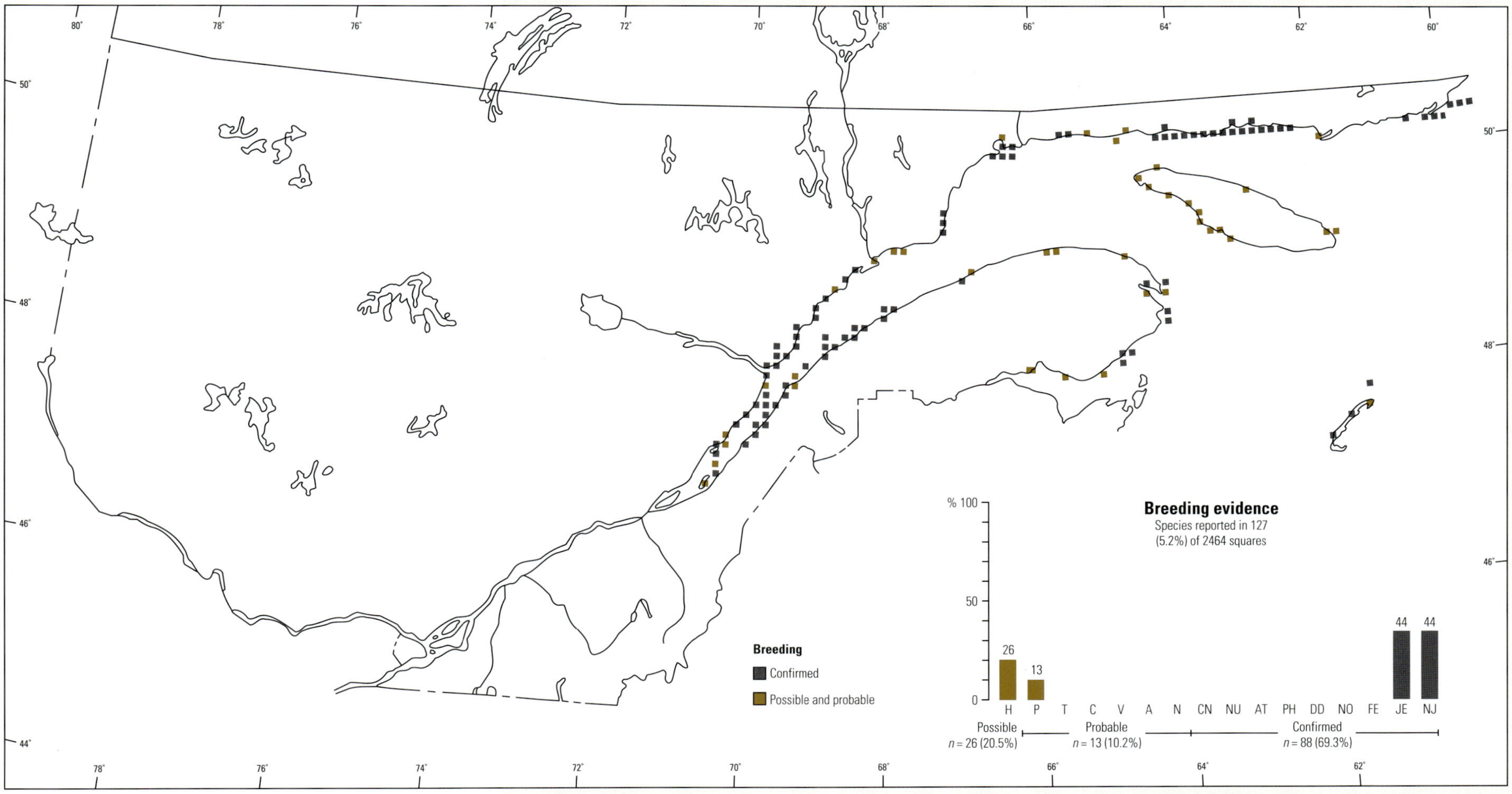

World breeding range

Species profile

Clutch size	usually 3–5, more than 7 = dump nest[1,8,9,12,13]
Laying interval	1 egg per day[2,3,5]
Start of incubation	usually with third egg[1,2,8,9]
Incubation period	26 days (23–30)[8,10]
Maturity at hatching	precocial
Care of young	by female[1,2,3]
Nestling period	less than 24 hours[1,3,8]
Age at first flight	60–75 days[2,3,9]
Dependence of young	until 40–60 days old[3,11]
Broods per year	1[3]
Breeding age	2–3 years[1,2,3,7]
Mating system	monogamy[2,3]
Duration of pair bond	usually 1 season[1,2,3,7]
Total length	53–71 cm[4]
Wing span	M 81.3–109.2 cm / F 76.2–106.7 cm[2]
Weight	M 2218 g (1384–2800) / F 2315 g (1965–2875)[3]
Longevity record	21 years 4 months[6]

References:
1. Bellrose 1980
2. Palmer 1976
3. Cramp and Simmons 1977
4. Godfrey 1986
5. Watson et al. 1993
6. Klimkiewicz and Futcher 1989
7. Johnsgard 1975b
8. Guignion 1967
9. Cooch 1965
10. J. Munro, Fisheries and Oceans Canada, unpubl. data
11. Bédard and Munro 1977
12. Milne and Reed 1974
13. van Dijk 1986

Chronology $n = 325$

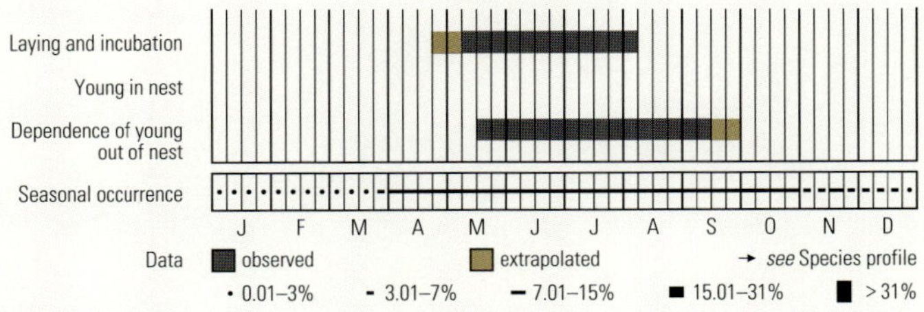

Hens almost always cover the eggs with down when they leave the nest. A portion of this down can be harvested to make parkas, sleeping bags, and comforters. Eiderdown is renowned for its lightness, insulating qualities, and loft. Île aux Basques, St. Lawrence estuary. *June 1982, Daniel Jauvin.*

and wings. These crises often lead to the formation of new crèches, since, under attack, adjacent broods tend to merge spontaneously (Munro and Bédard 1977b). Ducklings defended by more than one female generally have a better chance of survival, particularly when the crèche is attacked twice in succession by lone gulls. On the other hand, large crèches attract the attention of predators more readily; an initial attack that scatters the ducklings may result in the destruction of the entire crèche if followed by a group attack (Munro and Bédard 1977b).

The crossing to the coastal brood-rearing areas, where predation is much less intense, takes 1–12 hours (Gauthier and Bédard 1976). Once the crèches reach the shore, they spend the next few days moving to their rearing areas in bays or other protected sections of the coastline, generally within a 60-km radius of the nesting island, foraging all along the way (Gauthier and Bédard 1976). By this time, the family links, which grow progressively stronger as time passes, have become firmly established (Munro and Bédard 1977a), and for the next eight weeks or so, most of these crèches and broods will remain fairly stable, frequenting the same section of coastline (Gauthier and Bédard 1976; Bédard and Munro 1977).

Broods feed most actively at low tide and on the rising tide, foraging for amphipods and periwinkles found in abundance in the rockweed (*Fucus* spp. and *Ascophyllum* spp.) (Cantin et al. 1974; Minot 1980). Although fewer than 10% of adult females are involved in brood rearing by mid-July (Bédard et al. 1986b), most of the females not tending young remain in the coastal brood-rearing areas (Gauthier and Bédard 1976).

Between the middle and end of August, when they are 8–10 weeks the old, young have reached a size and plumage resembling that of the adult females. At this time, the family bonds gradually disintegrate, and large flocks of females and young are formed. These flocks leave the rearing sites around the end of August to join

Crèche made up of 2 females and 17 ducklings. A crèche is a group of broods cared for by one or more females. In crèches attended by more than two females, a hierarchy is established, with the broodiest female dominating and the other females remaining on the periphery. Bicquette Island, St. Lawrence estuary. *June 1973, Jean Munro.*

the males still present in the moulting areas (Gauthier and Bédard 1976). At the end of September, flocks of males begin migrating to their wintering grounds, and can be seen flying low over the sea in long lines. Immatures and females follow in October (Gauthier et al. 1976). By November, only a very small proportion of the summer population remains in the estuary (Gauthier and Bédard 1976).

Distribution

Three subspecies of the Common Eider occur in eastern North America: the Hudson Bay race, *Somateria mollissima sedentaria*, the Northern race, *Somateria mollissima borealis*, and the American race, *Somateria mollissima dresseri*, all of which breed along Québec's seacoasts. The Hudson Bay Eider breeds exclusively in Hudson Bay, with approximately 20 800 breeding pairs, or 85% of the total population, along its east coast (Abraham and Finney 1986; Chapdelaine et al. 1986b; Nakashima and Murray 1988). The Northern Eider breeds on the northern coasts of Québec and Labrador, the eastern Arctic Archipelago, and the west coast of Greenland; the heart of the Canadian population's breeding range is Hudson Strait, particularly the south side of the strait and Ungava Bay, where approximately 48 700 pairs, or 53% of the total population, can be found (Abraham and Finney 1986; Chapdelaine et al. 1986b). The American Eider breeds from Labrador to New England, and Québec is home to approximately 32% of the breeding population. Approximately 25 400 pairs, or three quarters of the southern Québec breeding population, are found in the estuary (Chapdelaine and Brousseau 1992a), while 10 000 pairs breed on the north side of the gulf (Chapdelaine and Brousseau 1991), 6000 pairs on the southern coast of Labrador, 200 pairs in Newfoundland, 8000 pairs on Nova Scotia's Atlantic coast, 7000 pairs in the Bay of Fundy (Erskine and Smith 1986; Lock 1986), and 22 000 pairs on the coast of Maine (Mendall 1968).

Crèches and broods seek shorelines with sheltered coves and bays, like this section of coast near Cap à l'Orignal. They spend several weeks in the rearing areas, foraging along the rugged, rockweed-strewn shores for invertebrates, which are plentiful there. Bic Conservation Park, southwest of Rimouski. *October 1993, Pierre Normand.*

Notes on territory (continued)

of 40 ducklings/km of shoreline at coastal sampling stations. At some stations, like the mouth of the Tartigou River, concentrations of approximately 2500 moulting birds (325 birds/km of shoreline) were observed in 1972–73.

In the St. Lawrence estuary, 88% of the nests can be found on a few small islands: Bicquette Island (west of Rimouski), Île Blanche, Brandypot Islands (off Rivière-du-Loup), Île aux Fraises, and Île aux Pommes (off L'Isle-Verte). In 1994, an exceptionally large number of nests (9000) were found on Bicquette Island, corresponding to a density of slightly over 1000 nests/ha (Bédard and Nadeau 1994). Recent surveys suggest that the population in the estuary has expanded to the northeast (Ragueneau), the north (Îlet aux Alouettes at the mouth of the Saguenay River), and the southwest (Battures aux Loups Marins) (Bédard and Nadeau 1994). On the North Shore, the species nests in single pairs or small colonies (fewer than 1500 nests: Chapdelaine and Brousseau 1992a).

In the estuary, Common Eiders nest mainly on small islands, but raise their young on the shores of larger islands and the mainland (Bédard et al. 1986b). During the brood-rearing period, more broods can be found on the south shore of the estuary than on the north, no doubt because it is closer to the nesting islands and because of the effect of currents on the broods as they leave the islands (Gauthier and Bédard 1976; Bédard et al. 1986b). The distribution of broods along the estuary shoreline is determined mainly by the nature of the substrate: the species requires a rocky shoreline, where periwinkles and gammarids are plentiful. The degree of protection from the elements provided by the coastline seems to play a

The Common Eider also nests on grassy islands. Here, too, the nests are well hidden in the vegetation and are usually found in groups. Gulls, on the other hand, nest at the edge of the vegetation and in more exposed areas. Île aux Pommes, St. Lawrence estuary. *June 1973, Jean Gauthier.*

In northern Québec, most birds nest in small, scattered colonies on the thousands of rocky islands along the coast. Gyrfalcon Islands, Northwest Territories and Québec. *August 1980, Gilles Chapdelaine.*

secondary role (Bédard et al. 1986b). The tendency for broods to be reared east of their nest sites may also be associated with the greater abundance downstream of blue mussels (Bédard et al. 1986b), believed to be the preferred food of females accompanying broods (Pethon 1967; Milne 1968).

The profusion of this bivalve downstream also explains the moulting aggregations of both sexes found on the south shore from Sainte-Luce-sur-Mer to just west of Matane and on the north shore near Baie-Sainte-Catherine and Pointe-à-Boisvert. Winter concentrations also occur in open water off Anticosti Island and Mingan, on the Middle North Shore, where there are plentiful supplies of blue mussels and other marine invertebrates (Bourget et al. 1986; Guillemette 1991; Guillemette et al. 1993).

Most of the North Shore and estuary population winters along the coast of Maine and Massachusetts, where roughly 180 000 birds gather at this time (Bourget et al. 1986). To reach their wintering grounds, most of the estuary birds migrate along a coastal route that takes them around the Gaspé Peninsula and New Brunswick and across the Chignecto Isthmus, a distance of about 2250 km (Reed 1975c). However, part of the population takes a direct overland route, travelling up the estuary to Montmagny and then making a spectacular high-altitude flight of 640 km over the Appalachians (Gauthier et al. 1976). A sizable number of eiders (150 000) winter off Mingan on the Middle North Shore and Heath and Sud-Ouest points on Anticosti Island (Bourget et al. 1986); more than half are Northern Eiders from Labrador and Hudson Strait, while the remainder are American Eiders from the North Shore and southern Labrador (Reed et al. 1986). The rest of the northern Québec population, roughly 150 000 eiders, spends the winter on the east coast of Newfoundland (Reed and Erskine 1986). The Hudson Bay Eider is believed to overwinter in polynyas in Hudson Bay, off the northwest coast and the Belcher Islands. However, the information available is sketchy and some individuals may winter in Hudson Strait or areas farther east (Reed and Erskine 1986).

History and trends

The species' colony-nesting behaviour, habit of rearing its young near the shore, and gregariousness have facilitated exploitation by humans over the centuries. A flourishing plume and eiderdown trade in the 18th and 19th centuries collapsed when populations fell drastically (Blanchard 1984). Nevertheless, local exploitation continued, particularly egging and spring shooting, and numbers were very low at the turn of the century. The adoption of the Migratory Birds Convention, signed in 1916, helped populations to recover somewhat. Intensive exploitation continued in the isolated Middle and Lower North Shore communities, however, keeping populations at low levels for many years. The increased number of motor boats and renewed poaching after World War II decimated populations even further (Cairns 1978a).

In the St. Lawrence estuary, government agencies and wildlife conservation organizations have laboured to establish bird sanctuaries and maintain a small, local industry in eider down collection. On the Middle and Lower North Shore, numbers have increased substantially since 1980 (Chapdelaine and Brousseau 1991) due to public information and education programs, the efforts of conservation officers (Blanchard 1994), and the creation of the Mingan Archipelago National Park Reserve. During the past 10 years, the estuary population has increased by some 20%, while the Middle and Lower North Shore population has almost doubled (Chapdelaine et al. 1986b; Chapdelaine and Brousseau 1992a).

Every year in eastern Canada, hunters kill roughly 500 000 Common Eiders, making it the most harvested of the sea ducks (Wendt and Silieff 1986). On the Atlantic seaboard, eider hunting has become more popular over the last 40 years, both along migration routes and on the wintering grounds (Krohn et al. 1992). Although there is no evidence that hunting is harming eider populations (Krohn et al. 1992), harvest rates and trends in breeding populations must be closely monitored to ensure that population levels are maintained. There is some concern about the potential effects of hunting on certain colonies in the estuary (Bédard and Nadeau 1994). Although the annual total harvest in the province fluctuated considerably between 1973 and 1992 (1700–9000 birds), the overall trend was downward (Appendix II), probably due in large part to the decreased number of hunters, particularly since the early 1980s.

Predation is another limiting factor on eider populations. Foxes undoubtedly do the most damage, particularly when they become marooned on islands in spring following ice break-up and are forced to subsist primarily on ground-nesting colonial birds. The harvesting of rockweed (1150 tonnes were collected in 1994) on the coast near Cacouna and between Rimouski and Métis-sur-Mer farther east—the estuary's main brood-rearing areas—seems to be having a negligible effect on populations, at least for the time being (Bédard and Nadeau 1994). Were this industry to expand, however, its impact would have to be examined more carefully, particularly since rockweed supports the organisms that young eiders need for growth.

The growing popularity of pleasure boating also poses a major threat to the species because of the disturbance caused by visits to nest sites and coastal brood-rearing areas; this facilitates gull predation, particularly during the first two weeks after the ducklings leave the nest. The greatest potential danger to the species, however, is no doubt an oil spill in the estuary—which is a major commercial shipping lane—because of the large numbers of birds present, the estuary's relatively narrow width, and tidal action.

Jean Munro

Harlequin Duck

Canard arlequin

Arlequin plongeur

Histrionicus histrionicus

THE HARLEQUIN DUCK IS NAMED FOR THE HARLEQUIN OF TRADITIONAL Italian comedy and pantomine, with his costume of multicoloured triangular patches. Males in breeding plumage have a similarly bizarre arrangement of blue-grey, chestnut, brown, black, and white stripes and patches.

In Québec, this handsome duck is a migrant breeder, although some individuals occasionally winter along the St. Lawrence. It is currently classified as an endangered species in eastern Canada.

Habitat and behaviour

The Harlequin Duck is the only duck in the Northern Hemisphere that breeds along fast-flowing rivers and streams. It is often found in forested mountainous regions, where it searches for its prey in the clear, turbulent waters of rapids. Nesting rivers seem to be selected mainly for their food resources, rather than the type of habitat present (Bengtson 1966). In Iceland, Harlequins feed on the larvae and pupae of black flies (Simuliidae) and other insects found along the rocky bottom of highly oxygenated rivers (Bengtson and Ulfstrand 1971; Bengtson 1972) and this is probably also the case in Québec. The ducks feed by diving into the current with open wings. Maintaining their balance with their wings, they walk on the rocky bottom against the current, often resurfacing in the very spot where they dived, or even upstream (Pool 1961; Bengtson 1966).

The Harlequin Duck, scarce in Québec, nests along turbulent, clear, and highly oxygenated rivers. Here, in the Sainte-Anne River, Harlequins feed on the larvae and pupae of insects that attach themselves to rocks in the rapids. Gaspé Conservation Park. *July 1991, Yves Aubry.*

Age at first breeding is 2 years. The adults often nest in the same area year after year, arriving on their breeding grounds at the end of April or in May (Bengtson 1966, 1972; Cons. G. G. – GREBE 1993a,b). First-year birds do not usually return to their natal rivers but remain at sea during the summer (Bengtson 1966, 1972). Therefore, some of the Harlequins found in summer around Bonaventure Island and in the Port-Daniel area, among other places, are probably non-breeders. Immature males can be told from adult males by their drabber plumage, particularly the scapulars (Cramp and Simmons 1977).

Harlequin Ducks are thought to build their nests on the ground, although nests have been found in snags (Cassirer and Groves 1992) and in cavities in rocky cliffs (Cassirer et al. 1993). Nests are often located on islands, usually within 5 m of water, and are well concealed under tufts of herbaceous vegetation, bushes, woody debris, or other cover. The hen builds the nest, incubates the eggs, and raises the ducklings. Drakes desert their mates at the onset of incubation (generally during the last two weeks of June), joining non-breeding birds at sea to moult. Broods linger on the rivers until the young are able to fly and then head to the seacoast at the end of summer, usually in September (Bengtson 1966, 1972; Gudmundsson 1971; Bellrose 1976; Palmer 1976).

Outside the breeding season, Harlequins are found generally on salt water, frequenting surf-pounded rocky coasts and islands that provide the abundance of crustaceans and molluscs the birds require at this time of year (Cramp and Simmons 1977; Goudie and Ankney 1986).

There are very few breeding records in Québec of the Harlequin Duck. Not only is it scarce, but also it breeds remote, relatively inaccessible areas, which is probably why it was reported in only 0.6% (16/2464) of Atlas squares. Nests are difficult to find, and the 2 confirmed breeding records involved sightings of broods. In fact, no Harlequin Duck nests have ever been found in Québec. Some of the records of probable and possible breeding most likely originated from moulting or staging areas where adults and immatures congreagate—as in

During the breeding season, pairs of Harlequin Ducks frequent certain rapids in the Little Whale River. Hudson Bay region. *October 1981, Pierre Lamothe.*

Pierre Dupuis

Notes on territory

The Harlequin Duck is highly gregarious, except during the breeding season (Cramp and Simmons 1977). Nesting pairs are generally well dispersed, although some rivers with particularly suitable habitat (the River Laxá in Iceland) may support densities of nearly 10 pairs/kilometre of shoreline, with several pairs sometimes nesting on the same island. Harlequins do not defend a territory, but often limit their activity to a specific section of their nesting river. Drakes, however, defend a small area around their mate (Bengtson 1966, 1972).

Harlequin Duck

Breeding distribution

1984–1989

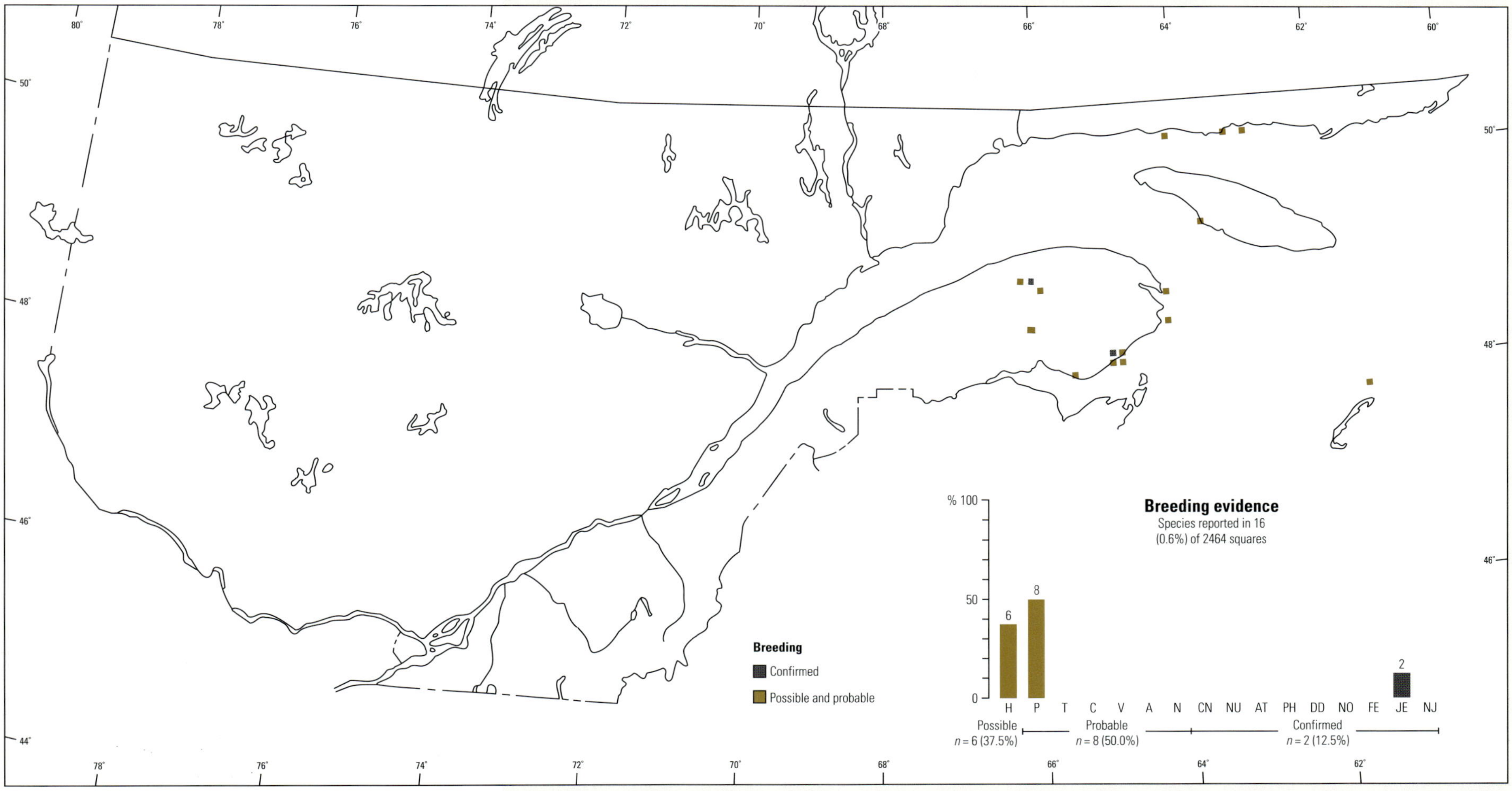

World breeding range

Species profile

Clutch size	usually 6 eggs (3–7)[5]
Laying interval	1 egg per 2–4 days[3]
Start of incubation	with last egg[1]
Incubation period	27–29 days[1,3]
Maturity at hatching	precocial
Care of young	by female[1,2]
Nestling period	probably less than 48 hours
Age at first flight	60–70 days[1]
Dependence of young	until after first flight, sometimes before[1]
Broods per year	1[1,2]
Breeding age	2 years[1,5]
Mating system	monogamy[1,3]
Duration of pair bond	1 season[1,3]
Total length	38–45 cm[4]
Wing span	63–69 cm[1]
Weight	M 653 g (590–726) / F 553 g (500–590)[5]
Longevity record	?

References:
1. Cramp and Simmons 1977
2. Bengtson 1972
3. Bengtson 1966
4. Godfrey 1986
5. Bellrose 1976

Chronology *n* = 50

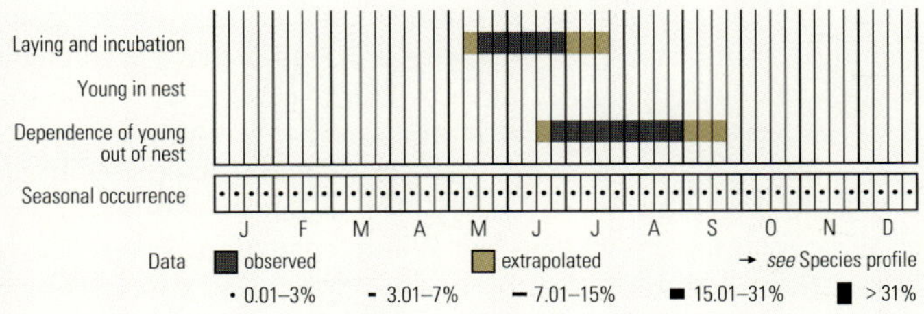

This Harlequin Duck nest on a small ledge on a rock face is well protected under the small overhang and fir tree. During incubation the female would cover the nest with down whenever she left to forage in order to protect the eggs from the cold and to camouflage them against predators. Sainte-Anne River, Gaspé Peninsula. *June 1996, Serge Brodeur.*

Harlequin Ducks are often seen resting on rocks in the water, like this male in breeding plumage on the Little Whale River. Hudson Bay region. *June 1992, Serge Brodeur.*

the squares on the Gaspé Peninsula that take in the Forillon Peninsula, Bonaventure Island, and Port-Daniel—rather than from the breeding areas.

Distribution

Most of the breeding records for southern Québec come from the Gaspé Peninsula, where 2 confirmed reports were obtained for the Atlas. To date, about 20 breeding records have been collected from half a dozen rivers on the peninsula, including the Port-Daniel, Bonaventure, Madeleine, Hall, Sainte-Anne, and Cascapédia (le Bel 1990; Robert 1990). The Gaspé Peninsula is also where the largest flocks are regularly observed in summer, around Bonaventure Island and in the Port-Daniel area. As many as 204 birds have been counted there in one day in late summer (A. Bourget, pers. comm.), which suggests that Gaspé rivers support several dozen breeding pairs.

Although there is no proof, the Harlequin Duck probably also breeds on the North Shore (Godfrey 1986). A female and brood were seen in September 1993 on the Wabouchagamou River, on the Lower North Shore (D'Astous 1994). The species was also atlassed in the Mingan area. Harlequins have regularly been sighted in summer along the North Shore since the turn of the century, in places such as Sept-Îles and the Kegaska River (Todd 1963), and near La Romaine (D'Astous 1994). There are also summer records from Anticosti Island, from the Sud-Ouest Point area (Gosselin et al. 1988a; Robert 1990). Although the species may breed on Anticosti Island rivers, it is equally possible that birds seen around the island were from the Gaspé or the North Shore—immature birds, for example, that hatched the previous year on the Gaspé Peninsula or the North Shore that went to the shores of Anticosti Island strictly to feed. The Atlas record just north of the Magdalen Islands was not of a breeding bird; Harlequins do not breed on the islands since there is no suitable breeding habitat there (Fradette 1992).

Elsewhere in Québec, the Harlequin Duck breeds in the watersheds of Ungava, Hudson, and James bays, where roughly 20 sightings of females with broods have been recorded since the turn of the century, some near the coast and others a few hundred kilometres inland (Todd 1963; Savard 1977; St-Louis 1978; David 1983; David and Gosselin 1983; Robert 1990; Cons. G. G. – GREBE 1990a, 1993a,b; ÉPOQ). This vast, remote area must certainly support a large portion of the eastern North American population. In June 1992, a total of 377 birds, 270 of which were paired, were counted on the larger rapids of some of the rivers in the Great Whale hydro-electric project study area (Cons. G. G. – GREBE 1993b), including the Great Whale, Little Whale, and Nastapoka rivers and their tributaries and the rivers that empty into Lake Guillaume Delisle. Furthermore, some of the birds that breed in the Hudson Bay drainage basin may winter in offshore polynyas near the Belcher Islands, for example. This is only speculation, however.

In eastern North America, the Harlequin Duck also breeds in Labrador and Newfoundland, and possibly on Baffin Island and in northern New Brunswick (Godfrey 1986; Goudie 1989b; Erskine 1992; W. A. Montevecchi, pers. comm.). The species is much more widespread in western North America (AOU 1983).

First nesting site of the Harlequin Duck ever found in Québec. The nest was under a small tree near the top of a rock face along the headwaters of the Port-Daniel River. Gaspé Peninsula. *June 1996, Serge Brodeur.*

Surprising as it may seem, the Harlequin Duck is able to feed in rapids like these on the Great Whale River. Hudson Bay region. *August 1989, François Morneau.*

In summer, flocks of Harlequin Ducks often forage along the reefs around Bonaventure Island, at the foot of the Port-Daniel lighthouse, and in the mouths of some rivers on Anticosti Island. Île Bonaventure and Rocher Percé Conservation Park. *July 1991, Yves Aubry.*

Female and her four ducklings foraging at the bottom of these rapids on the Port-Daniel River. Gaspé Peninsula. *June 1996, Serge Brodeur.*

History and trends

Information on the historical status of the Harlequin Duck in Québec is scarce and often anecdotal. Few ornithologists have visited the remote areas where the species breeds. Some accounts (Dionne 1883, 1906; Comeau 1909; Todd 1963) suggest that numbers were higher in the past, but there is no evidence of any significant decline in the breeding population in Québec over the past century (Robert 1990), not withstanding Goudie's (1989a,b) contention that the Harlequin Duck used to be more abundant in winter in the estuary and the Gulf of St. Lawrence.

Surveys of the species' eastern wintering grounds along the rocky coasts of the North Atlantic have indicated that the winter population in eastern North America is only about 1000 birds (Vickery 1988). More recent surveys in northwestern Québec during migration and the breeding season suggest, however, that the population may exceed 1000 birds (Cons. G. G. – GREBE 1993b). In winter, more than three quarters of the population is concentrated in a few areas in Maine, Nova Scotia, and Newfoundland (Vickery 1988; Goudie 1989a).

Goudie (1989a,b) maintains that the eastern population has declined sharply since the last century, particularly in the Atlantic provinces, estimating that Harlequin Ducks in eastern North America once numbered between 5000 and 10 000. This is why the Harlequin Duck, or at least its eastern population, has been included on the list of endangered species in Canada (Goudie 1989b). According to the information available, Québec is home to the bulk of the eastern Canadian breeding population. For this reason, the hunting of Harlequin Ducks has been prohibited in Québec since the fall of 1990, to protect the species in its few known fall staging sites in the Gulf of St. Lawrence.

MICHEL ROBERT

Surf Scoter

Macreuse à front blanc

Melanitta perspicillata

Pierre Dupuis

Of the three scoters, the Surf Scoter, a strictly North American species, is by far the one most often sighted in Québec during migration, both on salt and fresh water. The limits of its breeding range have only become better known in the last 20 years. The Surf Scoter, which is mainly a transient in southern Québec, is now acknowledged to breed over a large portion of central and northern Québec.

Beginning in August, scoters gather in a number of areas along the St. Lawrence estuary (Lehoux et al. 1985; J. Bédard, Laval U., and J. Gauthier, CWS, unpubl. data). During migration, they form immense rafts but tend to be rather discrete. Surf Scoters are not much sought after by hunters and are known in Québec as "deaf ducks" because they do not react to the sound of gunfire.

The male Surf Scoter does not defend a fixed territory but a variable moving space around the female. Similar defence of the female may be observed also within flocks. Confrontations between pairs are common on the breeding grounds, involving both females and males; sometimes the male is the instigator, sometimes the female. Level of aggression varies between pairs and between males and females. The causes of this variation are still unknown (Nero 1963; R. Ouellet, pers. obs.; A. Morrier, CWS, unpubl. data).

The Surf Scoter breeds mainly north of the 50th parallel. The many lakes around the Laforge Reservoir, located in the taiga of central Québec, provide excellent nesting habitat for the species. South of Lake Bienville. *August 1992, Austin Reed.*

Notes on territory

The antagonistic behaviour observed in the species in early spring is associated with mate defence and does not seem to be site-specific, so that a fixed territory per se is not defended.

Furthermore, observations made over a period of time at Lake Malbaie in 1994 indicate that aggressive behaviour in brood-rearing females varies from individual to individual: some females will drive off young that attempt to approach the brood or join it, while others will accept the intruders without any apparent aggression. Records of interspecific interactions include a scoter and her brood being violently driven off by a female goldeneye, accompanied by her brood, and a female scoter repelling a female American Black Duck.

Habitat and behaviour

After wintering off Georgia and South Carolina, tens of thousands of scoters begin migrating north, reaching the St. Lawrence estuary, Upper North Shore, and Chaleur Bay in early April, where they remain until break-up on the inland lakes, which can sometimes be as late as mid-May (Lehoux et al. 1985). Then, already paired, they head overland to their breeding grounds in the north.

Although some birds arrive in the Eastmain and Opinaca river watersheds and the Great Whale River region as early as mid-May, the main flights do not reach the area until late May and early June (Cons. G. G. – GREBE 1992u; D. Bordage, CWS, unpubl. data). Most of these birds go directly to their nesting grounds, though some wait on open water on large rivers and lakes until their breeding habitats are ice-free (Savard 1977; Cons. G. G. – GREBE 1992u).

Although courtship behaviour on the wintering grounds has been documented by Myres (1959a,b), behaviour on the breeding grounds is much less well known.

Abundance of Surf Scoters in southern Québec. Aerial surveys conducted in May 1990, 1991, and 1992 *(D. Bordage, CWS, unpubl. data).*

Surf Scoter

Breeding distribution 1984–1989

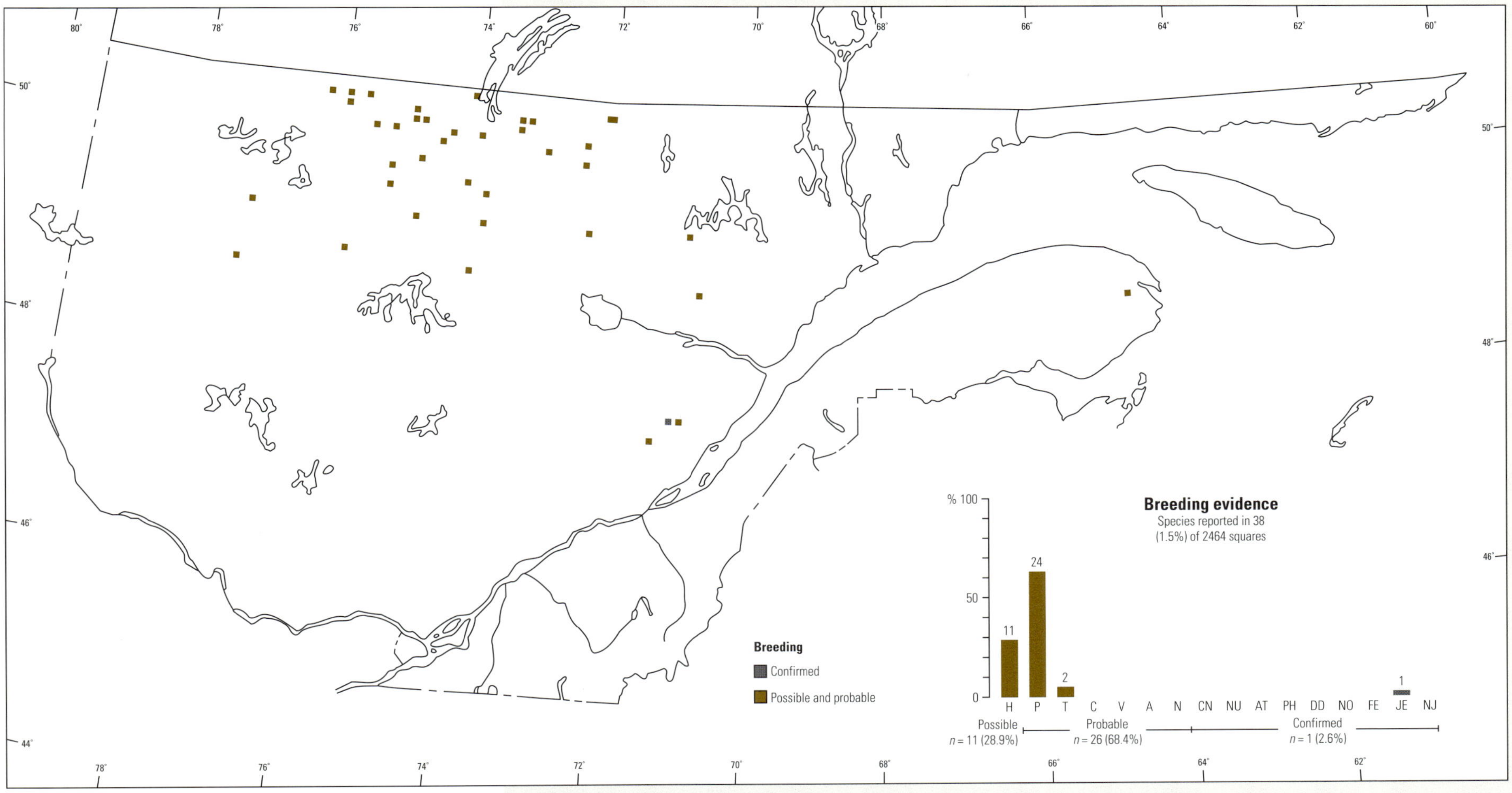

Species profile

Clutch size	usually 6–8 eggs (5–8)[2,3,4,5,8]
Laying interval	?
Start of incubation	?
Incubation period	?
Maturity at hatching	precocial
Care of young	by female[4]
Nestling period	less than 24 hours[3]
Age at first flight	≈ 60 days[8]
Dependence of young	probably until sometime before first flight[3]
Broods per year	1[4]
Breeding age	probably 2–3 years[2,7]
Mating system	monogamy[4]
Duration of pair bond	?
Total length	48–58 cm[6]
Wing span	76.2–91.4 cm[5]
Weight	M 998 g (630–1134) / F 907 g (680–997)[1]
Longevity record	?

References:
1. Bellrose 1976
2. Johnsgard 1975b
3. Palmer 1976
4. Ehrlich et al. 1988
5. Terres 1980
6. Godfrey 1986
7. Cramp and Simmons 1977
8. A. Reed and A. Morrier, CWS, unpubl. data

World breeding range

Chronology *n* = 55

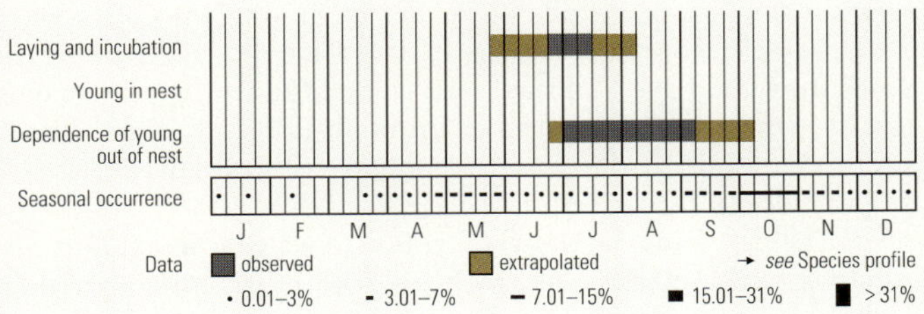

This Surf Scoter nest containing 6 eggs is built on the ground in a tangle of deadfalls. As incubation progresses, the female adds down to the nest until a thick blanket is formed. Lake Malbaie, Laurentide Wildlife Reserve, north of Québec City. *May 1994, Austin Reed.*

Paired males can also be seen making short flights past a female and then skidding to a stop in front of her (Nero 1963; A. Reed, pers. comm.). During these flying-away displays, a distinctive whistling sound is made by the wings, which are held upright as the bird alights. After this, the female joins the male. Additional forms of courtship behaviour have been described by Johnsgard (1965). As a general rule, Surf Scoters nest peaceably in solitary pairs on small lakes.

The nest is built on the ground, sometimes a considerable distance from water, and is well hidden under a log, the low branch of a conifer or in shrubby or herbaceous growth (Bellrose 1976; Palmer 1976; Cons. G. G. – GREBE 1992t; A. Reed and A. Morrier, CWS, unpubl. data). A study on Lake Malbaie, in the Laurentide Wildlife Reserve, 95 km north of Québec City, has shed new light on some aspects of the Surf Scoter's breeding behaviour (Reed et al. 1994; A. Reed and A. Morrier, CWS, unpubl. data). Most of the 10 nests found were under a fallen log or the low branches of a spruce. All were located on islands and were made of twigs, bark, conifer needles, and moss, and lined with feathers and down from the female's breast. During incubation, the female always covers the nest with down when she leaves, to conceal the eggs and keep them warm.

In the Great Whale River region, most females begin incubating between the first and third weeks of June (Cons. G. G. – GREBE 1990a, 1992w; Savard and Lamothe 1991), the first clutches hatching in early July, and hatching peaking between the second week and end of July (Cons. G. G. – GREBE 1990a; Savard and Lamothe 1991). These dates are in keeping with findings farther south in the Eastmain River (Bordage 1985a; Tardif and Morneau 1991) and Pontax River (Cons. G. G. – GREBE 1992a) regions. Farther south still, at Lake Malbaie, in 1993 and 1994, most eggs hatched during the last week of June and the first week of July, but in 1994, hatching occurred as late as early August (Reed et al. 1994; A. Reed and A. Morrier, CWS, unpubl. data).

The pair bond dissolves shortly after the onset of incubation, and the male heads to the moulting grounds (Savard and Lamothe 1991). In populations breeding near the coast and inland in the James Bay and Great Whale River regions, moulting males congregate mainly between the coastal islands or farther offshore (Cons. G. G. – GREBE 1990a, 1992p,w; Savard and Lamothe 1991; Benoit et al. 1991, 1992, 1993b). Males from the Lake Malbaie region may head to the St. Lawrence estuary, where thousands gather to moult (J. Bédard, Laval U., unpubl. data). Moulting birds have also been sighted in the Gulf of St. Lawrence (R. Ouellet, pers. obs.).

The hen leads the precocial young to water several hours after they hatch. As in the White-winged Scoter, crèching sometimes occurs (Kehoe 1986; Savard and Lamothe 1991). Crèches have been observed on several occasions on Lake Malbaie, some consisting of 30 or more ducklings just a few days old (Reed et al. 1994; A. Reed and A. Morrier, CWS, unpubl. data). At this southerly site, age at first flight is approximately 60 days.

In the Great Whale River region, broods occupy habitats similar to those used by Black Scoters (Bider and Lamothe 1982), mainly ponds and small lakes 1–5 m deep and smaller than 30 ha. Broods have also been sighted on lakes of about 100 ha (Bider and Lamothe 1982; Cons. G. G. – GREBE 1990a,b, 1993e; Dryade 1993). Very large, deep lakes do not appear to be used (Savard and Lamothe 1991). The Surf Scoter prefers lakes with clear, mineral-poor water that is alkaline or slightly acidic (Cons. G. G. – GREBE 1993e), and a rocky bottom (Savard 1977; Cons. G. G. – GREBE 1990b; Savard and Lamothe 1991). Often these lakes have no fish (Savard 1988c) but support a richer benthic population than neighbouring lakes (Bider and Lamothe 1982; Cons. G. G. – GREBE 1990b), with aquatic vegetation either sparse or absent (Cons. G. G. – GREBE 1990b, 1993e). The handful of broods that were reported in the Eastmain and Pontax river regions were found in this type of habitat (Tardif and Morneau 1991; Cons. G. G. – GREBE 1992a). Near the northeast coast of James Bay, however, broods were usually seen on lakes containing aquatic vegetation (Benoit et al. 1993b).

In the Lake Amichinatwayach area at the headwaters of the Little Whale River, ducklings feed mainly on insects and molluscs, with plant matter also accounting for a large portion of their stomach contents (Cons. G. G. – GREBE 1993e).

Only the down is visible in this nest, which is built on the ground amid a tangle of dead branches and other plant debris, under the shelter of a fallen tree. Surf Scoter nests are generally located in the undergrowth, often some distance from the water. Lake Malbaie, Laurentide Wildlife Reserve, north of Québec City. *July 1994, Jack Hughes.*

During the moulting period, Surf Scoters on the east coast of James and Hudson bays feed mainly on molluscs, particularly blue mussels (Cons. G. G. – GREBE 1990d; Benoit et al. 1993b). Birds taken during fall migration in 1987 and 1988 on the St. Lawrence between Lake Saint-Louis and Lake Saint-Pierre were found to have similar stomach contents (plants, molluscs, crustaceans), although barley seeds were also found in some cases (R. Ouellet, MEF, unpubl. data). Scoters collected in the fall of 1991 between Lake Saint-François and Montmagny were also found to have a diet of molluscs, including zebra mussels (Guillemette et al. 1994). On their wintering grounds on both the Atlantic and Pacific coasts, Surf Scoters also feed mainly on mussels—the most sought after being the blue mussel—along with gastropods, crustaceans, and sometimes barnacles and herring roe (Stott and Olson 1973; Vermeer and Levings 1977; Vermeer 1981; Vermeer and Bourne 1984; Goudie and Ankney 1986).

Surf Scoters often dive synchronously when foraging in large flocks. These synchronous dives may be related to feeding efficiency (Schenkeveld and Ydenberg 1985; Beauchamp 1992) but may also serve some social purpose (J. Bédard, pers. comm.).

The low number of Atlas breeding records (38) was due to the northerly distribution of the species as well as the fact that nests are very difficult to find. The best way to confirm breeding, therefore, is to find broods, which means that atlassers must be in the field late in summer, as Surf Scoters nest later than many other waterfowl species.

The Surf Scoter nests on the shores of shallow lakes and ponds. Pairs have been sighted on Lake Malbaie since 1988 despite the fact that it is fairly far south. Laurentide Wildlife Reserve, north of Québec City. *July 1993, Yves Aubry.*

Notes on territory (continued)

Recent studies in northern Québec, particularly in the Great Whale and Little Whale river regions, reported densities of 10 pairs/100 km² and 10 broods/100 km² in some areas, the average being approximately 2 pairs and 2 broods/100 km² (Cons. G. G. – GREBE 1990a; D. Bordage, CWS, unpubl. data). South of 51° N, the highest pair density —13 pairs/100 km² (probable breeders)—was recorded in a square near Lake Manouane, north of Lac-Saint-Jean, although the average density for this vast area is about 0.3 pair/100 km² (D. Bordage, CWS, unpubl. data).

Distribution

Unlike the other two scoter species, which have circumpolar distributions, the Surf Scoter breeds exclusively in North America. Relatively little is known about its distribution in Québec, since some areas have not been surveyed or were surveyed too early in spring to see broods.

Broods have been sighted from the coast of James and Hudson bays to the Smallwood Reservoir in Labrador, and from Lake Minto south to the Pontax River region (Bider and Lamothe 1982; Bordage 1985a; Godfrey 1986; Savard and Lamothe 1991; Tardif and Morneau 1991; Cons. G. G. – GREBE 1992t; Benoit et al. 1993b; Dryade 1993; D. Bordage, CWS, unpubl. data).

During aerial surveys conducted by the Canadian Wildlife Service, isolated pairs were sighted in suitable habitat in the southern two thirds of Labrador (Gillespie and Wetmore 1974; Goudie and Whitman 1987), on the Lower North Shore, and in various locations between 49°00' and 51°30' N in 1990–92 (D. Bordage, CWS, unpubl. data). Therefore, it can safely be said that the species' breeding range also takes in these areas.

Other data indicate, however, that considerable numbers of breeding birds occur in such places as the watersheds of the Great Whale and Little Whale rivers (Savard and Lamothe 1991; Cons. G. G. – GREBE 1992t).

Between the 49th parallel and the north shore of the St. Lawrence, pairs have been sighted in and around the Laurentide Wildlife Reserve since the early 1970s. These records were all in late June or early July on shallow lakes. Several were in late June at the northern end of Lake Jacques Cartier, where 1–3 pairs were sighted at different occasions between 1972 and 1982 (R. Ouellet, pers. obs.). Later in July, single females were seen for several minutes at the end of the day paddling around, stretching their wings and diving before moving quietly off. In 1986, a pair was sighted on 29 June on Lac Nouvel in the Jacques Cartier Conservation Park and in 1994, during an aerial survey over the Laurentide Wildlife Reserve, pairs were seen on 9 lakes and on the Malbaie River, also in the Lake Malbaie area (A. Morrier, CWS, unpubl. data).

Breeding was first confirmed in this area in 1978 on Lac des Neiges, 3 km south of Lake Malbaie, where a female and several young were sighted (G. Trencia, pers. comm.). The species has been known to breed on Lake Malbaie since 1988 and appears to do so regularly (Gosselin et al. 1988b; D. Banville, pers. comm.). In 1993, a surprising 59 females and 139 young were counted in this area (Reed et al. 1994). In 1994, in the same region, broods were seen on Lakes Malbaie, des Neiges, Carroll, and Nouvel. In this vast, mountainous area, which is almost entirely protected by virtue of the conservation parks and the wildlife reserve, all sightings of Surf Scoters occurred on lakes at elevations of 600–900 m.

Two other sightings worthy of mention are a brood of 8 ducklings on 17 July 1979 on Lake Rouvray, south of the Pipmuacan Reservoir (Breton 1980), and some juveniles on 27 August 1925 at Île de Ouapitagone, on the Lower North Shore (Lewis 1927).

On the south shore of the St. Lawrence, a pair was sighted for the first time in early July 1969 at Grand Lac Macpès, near Rimouski. In 1971, at Lake Témiscouata, 2 pairs were observed over a 10-day period between late June and early July; later, only the males were seen (R. Ouellet, pers. obs.). Farther east, a pair was seen at Lake Fronsac, southwest of the town of Gaspé, on 24 June 1985.

Within hours of hatching, the female leads her brood to water. Shallow lakes are generally used for brood-rearing since broods can forage over the entire surface and not just along the shoreline. Near the Laforge Reservoir, south of Lake Bienville, north-central Québec. *July 1994, Austin Reed.*

History and trends

At the turn of the century, most ornithological expeditions to northern Québec were limited to the coasts. In mid-summer, large rafts of Surf Scoters, often composed of thousands of males, would be seen along the James and Hudson bay coasts and the North Shore, and observers surmised that fairly large numbers of females must be nesting somewhere in the interior, leaving a question mark over the species' actual breeding range for many years. Until the 1960s, very little was known about the species, and there were only a few breeding records from the coast of James Bay and the interior of the Labrador Peninsula south of Kuujjuaq (Todd 1963).

Winter surveys and harvesting data indicate a decline in scoter numbers along the Atlantic coast since the 1960s (Kehoe 1993). Moreover, the Québec Surf Scoter harvest, like that of the Black Scoter, has fluctuated considerably over the years, with the largest harvests recorded during the 1970s and early 1980s (Appendix II).

A study in Québec of waterfowl gizzards collected in the fall of 1987 and 1988 indicated that Surf Scoters ingest lead shot (Lemay et al. 1989), which may result in lead poisoning. However, the lead accumulates mainly in the bird's skeletal system, after passing through the liver and kidneys, posing little risk to human health.

Réginald Ouellet and Daniel Bordage

Common Goldeneye

Garrot à œil d'or

Bucephala clangula

Pierre Dupuis

THE COMMON GOLDENEYE IS KNOWN FOR ITS SPECTACULAR COURTSHIP displays as well as for the whistling sound made by its wings in flight. This duck is a swift flier, achieving speeds of 72 km/h (Ryden and Kallander 1964 *in* Palmer 1976). When it swims, it pumps its head back and forth, especially when disturbed. Like many cavity-nesting ducks, it readily accepts artificial nest boxes, which humans have provided for centuries. In 1753, Linnaeus reported that it was the custom of the Karelians and Lapps of northern Finland to install nest boxes in trees to attract goldeneyes (Rever and Miller 1973). The use of nest boxes is very common in some countries; there are an estimated 75 000 or more in Sweden (Dennis and Dow 1984).

Common Goldeneyes nest in tree cavities near water. Martin-Valin ZEC, north of the Saguenay River.
August 1991, Jean Gauthier.

Habitat and behaviour

The drake's courtship displays, performed in the open water among the ice floes during migration in early spring, are a fascinating sight. As with many other species of diving ducks, males outnumber females, which intensifies the competition among the males to attract a mate (Carter 1958; Savard 1990). These displays are among the most complex and spectacular of any North American duck. For example, in one display, the drake raises his breast, draws his head back then thrusts it sharply forward, neck outstretched and bill pointing upward, all the while uttering a characteristic harsh, two-syllable note.

As the ice begins to melt on the small inland lakes, pairs disperse through the boreal forest. The diet of the Common Goldeneye varies according to season and habitat, but is mainly composed of molluscs, crustaceans, and the larvae of aquatic insects (Cottam 1939; Bellrose 1976; Eriksson 1976; Palmer 1976), which the birds procure by diving (Cramp and Simmons 1977).

Common Goldeneyes usually nest in tree cavities, most often near water but sometimes over 1 km away. Occasionally, unusual sites like chimneys or church steeples are chosen (Carter 1958). In 1940, in the former USSR in an area that was clearcut to build a reservoir, 21 of the 53 goldeneye nests in the study area were found in atypical locations: on the ground, between roots, in hollow logs, in stumps and snags, in log piles, and in haystacks (Dement'ev and Gladkov 1952). In natural habitat, the type of cavity and especially the diameter of the opening seem to be the main factors in nest site selection (Prince 1968). In general, the entrance is 10–20 cm wide and slightly longer, but can be as small as 5 cm wide and 11 cm long (Carter 1958; Prince 1968); Prince (1968) found the average cavity depth to be 46 cm. Several factors may limit breeding success, including egg predation by martens, black bears, and red squirrels, as well as competition for nest cavities with European Starlings (Bouvier 1974a; Eriksson 1979; Dow and Fredga 1983; Allen et al. 1990). Prolonged periods of rain may also be a cause of nest failure and desertion (Dement'ev and Gladkov 1952; Palmer 1976).

Finding a suitable nest cavity appears to be crucial to breeding success. Females tend to use the same site year after year and those whose nests have been destroyed by predators change their nest site more often (Johnson 1967; Dow and Fredga 1983; Eadie 1989), which lowers breeding success (Dow and Fredga 1983). Some natural cavities are used for 30 years in succession (Dement'ev and Gladkov 1952). When a hollow elm tree in the Gaspé was felled by wind, it was found to contain the remains of 28 female Common Goldeneyes. The inside of the cavity was 50 cm in diameter and approximately 6 m deep with an opening of about 13 cm in diameter; the birds had apparently been trapped while inspecting the cavity, which from the outside appeared to be ideal (Mershon 1927 *in* Palmer 1976). Female goldeneyes do not nest until the age of 2 (Palmer 1976). During the incubation period and for a short time after the eggs hatch, non-breeding females and possibly failed breeders will prospect for nest cavities for the following year in flocks of 2 to 7. During this time, they often give a characteristic prospecting call, a rapid *cuk cuk cuk* (Eadie and Gauthier 1985; Phillips 1986).

Notes on territory

In the spring, the male defends a well-defined section of water (Savard, J.-P. L. 1984), excluding all other goldeneyes, except his mate, from this territory. Two males will often face off along an imaginary line marking the boundary, swimming side by side with necks outstretched. Sometimes the ritual ends in violent combat. The female also establishes a territory for her brood, from which she excludes other goldeneyes, usually a different territory from the one defended by the male earlier in the season. In winter, some males appear to defend an area around their mate, but this behaviour has not been confirmed as being truly territorial.

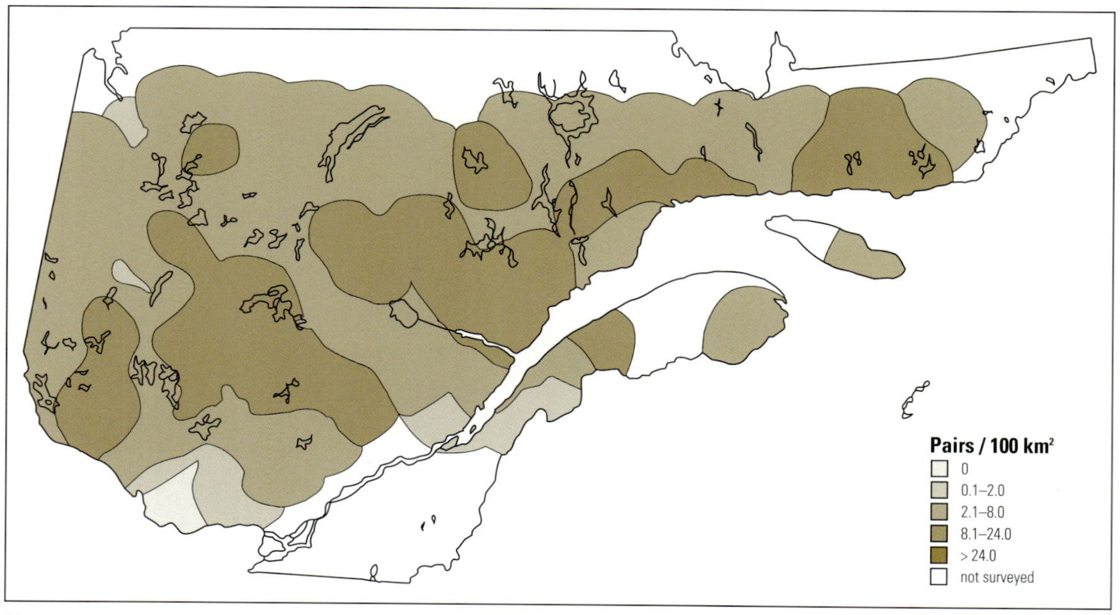

Abundance of Common Goldeneyes in southern Québec. Aerial surveys conducted in May 1990, 1991, and 1992 *(D. Bordage, CWS, unpubl. data).*

Common Goldeneye

Breeding distribution

1984–1989

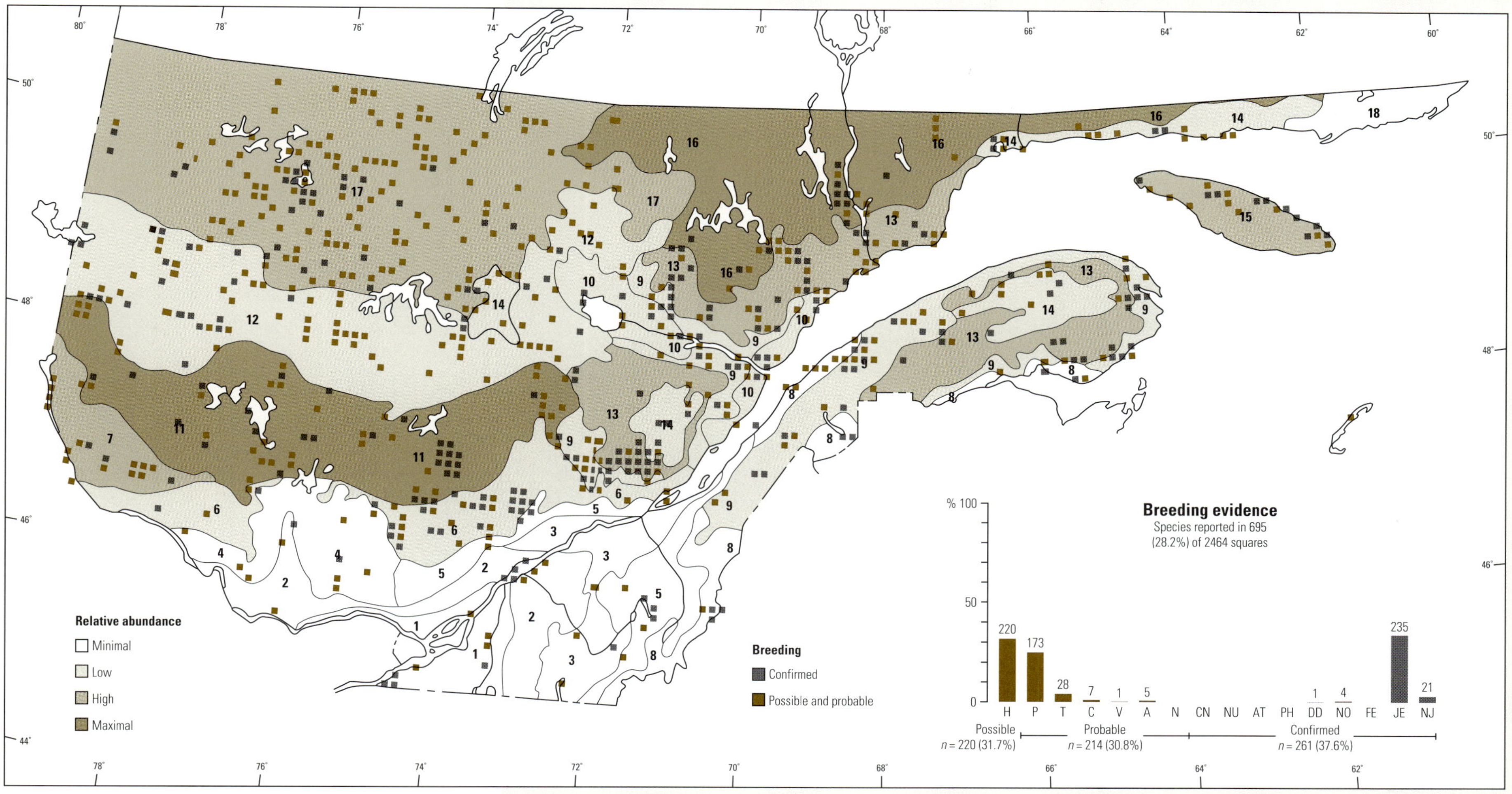

Species profile

Clutch size	usually 8–10 eggs (5–15), more than 15 = dump nest[1,2,3,4,5]
Laying interval	1 egg per 1–2 days[6]
Start of incubation	with last egg[2]
Incubation period	usually 30 days (28–32)[6]
Maturity at hatching	precocial
Care of young	by female[3]
Nestling period	24–37 hours[2]
Age at first flight	56–66 days[2,3,4,5]
Dependence of young	until ≈ 50 days old[2]
Broods per year	1[2,3]
Breeding age	2 years[4]
Mating system	monogamy[3]
Duration of pair bond	?
Total length	40.5–51.0 cm[1]
Wing span	63.5–81.3 cm[5]
Weight	M 1075 g (900–1451) / F 789 g (726–1270)[6]
Longevity record	14 years 3 months[7]

References:
1. Godfrey 1986
2. Cramp and Simmons 1977
3. Ehrlich et al. 1988
4. Palmer 1976
5. Terres 1980
6. Bellrose 1976
7. Clapp et al. 1982

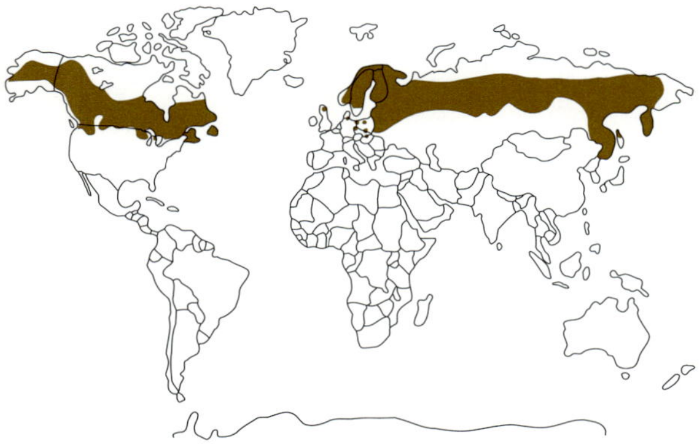
World breeding range

Chronology $n = 279$

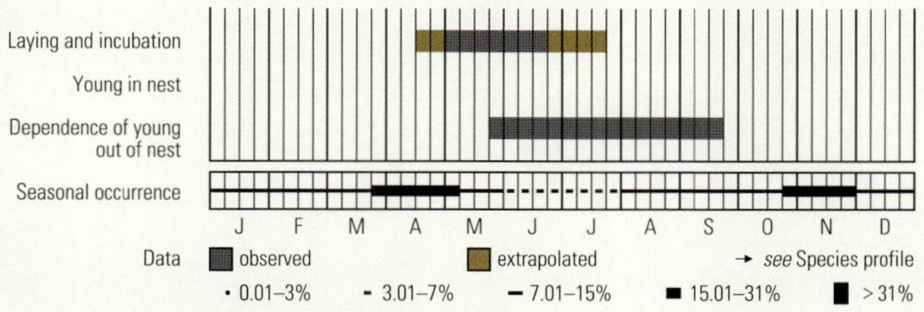

Frequency of occurrence by ecological region

Ecological regions** (forest regions)	Adequately covered squares	Squares in which reported	% of squares in which reported	Relative abundance (%)*
1. Sugar maple-hickory and sugar maple-basswood	97	8	8.2 ± 3.0	9.7
2. Sugar maple-basswood	153	10	6.5 ± 2.2	7.7
3. Sugar maple-basswood and sugar maple-yellow birch	101	5	5.0 ± 2.8	5.9
4. Sugar maple-yellow birch-basswood	56	8	14.3 ± 8.3	16.9
5. Sugar maple-yellow birch-beech and sugar maple-yellow birch-basswood	113	6	5.3 ± 2.5	6.3
6. Sugar maple-yellow birch-beech	71	30	42.3 ± 10.5	49.9
7. Sugar maple-yellow birch-balsam fir	19	12	63.2 ± 22.8	74.6
8. Sugar maple-yellow birch and balsam fir-yellow birch	47	11	23.4 ± 11.0	27.7
9. Balsam fir-yellow birch	87	38	43.7 ± 9.6	51.6
10. Balsam fir-white birch or balsam fir-red maple	46	17	37.0 ± 12.1	43.7
11. Yellow birch-balsam fir	52	44	84.6 ± 10.4	100.0
12. Balsam fir-white birch (west)	59	27	45.8 ± 13.0	54.1
13. Balsam fir-white birch (east)	76	45	59.2 ± 10.8	70.0
14. Balsam fir-black spruce	28	13	46.4 ± 19.3	54.9
15. White spruce-balsam fir and balsam fir-white spruce	22	14	63.6 ± 20.8	75.2
16. Black spruce-balsam fir-moss	18	15	83.3 ± 20.2	98.5
17. Black spruce-moss	32	20	62.5 ± 18.3	73.9
18.			*not analysed*	

* minimal: 0–29.9%; low: 30–59.9%; high: 60–89.9%; maximal: 90–100%

** Deciduous forest: 1–7; mixed forest: 8–11; coniferous forest: 12–17

Notes on territory (continued)

In Québec, there are no studies on interspecific territoriality among the three Bucephala species. In western Canada, in areas where the three species are sympatric, they maintain exclusive, non-overlapping territories. Barrow's and Common goldeneyes are dominant over Buffleheads (Savard, J.-P. L. 1984). Territorial defence is thought to procure feeding advantages for the female and the young.

The Common Goldeneye nests mainly in the boreal forest, on the margins of lakes rich in invertebrates. Chibougamau. *June 1988, Réginald Ouellet.*

The laying interval is greater than in most other ducks. Typical clutch size is 8–10 eggs, laid at a rate of 1 egg every 1 or 2 days, which corresponds to 1 egg every 1.4 days on average (Bellrose 1976). Laying takes about 13 days. Once the first egg has been laid, the female starts accumulating down in the cavity until there is enough to cover the eggs when she leaves the nest (Palmer 1976), which she does 3–6 times a day, for an average of 79 minutes for each trip (Sirén 1952). Body down clinging to the cavity entrance is often a good indication of current or recent occupation. Incubation lasts about 30 days (Bellrose 1976). Unlike many ducks, Common Goldeneyes rarely renest after their first clutch has been lost (Dow and Fredga 1983; Zicus and Hennes 1990). Intraspecific parasitism is common in this species and seems to be more frequent after a hard winter or late spring (Dow and Fredga 1984). Dump nests are generally unsuccessful (Rever and Miller 1973), although a female may successfully incubate the top layer of eggs (Dement'ev and Gladkov 1952). Mixed clutches are also reported with Hooded Mergansers, Wood Ducks, Barrow's Goldeneyes, and Buffleheads (Bouvier 1974a; Eadie 1989). These eggs may be successfully incubated (Bouvier 1974a), but the success rate is usually lower (Allen et al. 1990).

The female incubates the eggs alone. Sometime during the first few weeks of incubation, the male ceases defending his feeding territory (Cramp and Simmons 1977; Savard, J.-P. L. 1984), and leaves the area to join other males on the moulting grounds, which are often located well to the north of the nesting area (Bellrose 1976; Palmer 1976).

Between 24 and 37 hours after they hatch, usually in the morning, the ducklings jump from the nest (Dement'ev and Gladkov 1952; Sirén 1952; Cramp and Simmons 1977) and follow the female to a brood-rearing site which may be over 1 km away. The species is easiest to spot at this time. Thanks to their diet rich in aquatic invertebrates, the ducklings are able to fly 56–66 days after hatching (Gibbs 1962; Cramp and Simmons 1977).

When migrating, goldeneyes frequent mainly shallow bays where they feed during the day in small flocks, which combine in dense rafts at dusk, generally well away from shore (Palmer 1976).

Where suitable nest cavities are lacking, goldeneyes may use a stump, snag, or nest box. Martin-Valin ZEC, north of the Saguenay River. *August 1991, Jean Gauthier.*

The female leads her brood to the nearest body of water 24–37 hours after they hatch. She frequents the vegetation along the shoreline which provides cover for the young to feed and rest. Rivière du Sud-Ouest, near Saint-Simon, southwest of Rimouski. *July 1964, Austin Reed.*

Distribution

The Common Goldeneye occurs virtually throughout the boreal forest belt of the Northern Hemisphere (Bellrose 1976; Godfrey 1986). It breeds all over Québec, except on the tundra of the Ungava Peninsula, which does not provide suitable nesting habitat. However, it does breed on the taiga, which is surprising given the scarcity of trees of sufficient diameter except in certain enclaves and valleys. The Common Goldeneye nests as far north as the Little Whale River watershed (Cons. G. G. – GREBE 1990a) and has even been sighted beyond the 58th parallel at Tasiujaq, Kuuraapik River, and Inukjuak (ÉPOQ).

The Common Goldeneye was atlassed as a breeder throughout almost all of southern Québec, frequenting mainly regions with extensive old-growth forests where it can find nest cavities, and for the most part avoiding the St. Lawrence Plain, Bois-Francs, and Beauce regions, where there are fewer forests. It also seems to be less abundant between the Abitibi region and Lake Saint-Jean and in the Lower St. Lawrence region, areas that were logged heavily between 1974 and 1988 (Fragnier 1992). However, numbers may have been underestimated in the area between the Abitibi and Lake Saint-Jean, due to poor coverage of the mature forests around the Gouin Reservoir. Canadian Wildlife Service aerial surveys have shown the area around the Gouin Reservoir to be important for the species. Other centres of abundance in southern Québec identified in these surveys include the Témiscamingue region, the area northeast of the Saguenay River, and the North Shore. The highest pair density (31 pairs/100 km^2) was recorded near Lakes Manouane and Péribonka, northeast of Lake Saint-Jean. Mean density in boreal and mixed forests is around 8 pairs/100 km^2 (D. Bordage, CWS, unpubl. data).

In spring and fall, Common Goldeneyes are usually found in small flocks, though huge rafts do occur, chiefly on the large riverine lakes of the upper St. Lawrence and in the lower estuary (Lehoux et al. 1985), where the species can be seen from August onward (J. Gauthier, CWS, unpubl. data).

Although several thousand goldeneyes overwinter in the estuary, most of the Québec population winters farther south, mainly along the Atlantic seaboard. Winter sightings occur primarily along the north shore of the estuary, where several thousand birds can often be seen (Lehoux et al. 1985; Savard 1990). The species also frequents ice-free areas in the upper St. Lawrence, the highest numbers occurring on Lakes Saint-François and Saint-Louis (Reed and Bourget 1977). In these areas, the birds form dense rafts just before sundown (Reed 1971).

History and trends

In the 19th century, the Common Goldeneye was well known to ornithologists, who reported its presence throughout most of Québec (Macoun and Macoun 1915; Todd 1963). Surveys by Québec's Ministry of Recreation, Fish and Game and the Canadian Wildlife Service between 1985 and 1989 of a 100 000-km^2 area from the Abitibi region to Lake Saint-Jean showed the breeding population to be stable (Bordage 1987, 1988a,b), as did a similar survey by the CWS, begun in 1990 and covering all of the mixed and boreal forests in southern Québec (CWS 1992; D. Bordage, CWS, unpubl. data). Midwinter aerial surveys of waterfowl along the U.S. eastern seaboard, however, show a dramatic and unexplained decline in the winter population index (Steiner 1984; Serie 1987, 1989, 1991, 1992, 1993). The number of birds recorded, which averaged about 78 000 between 1954 and 1969, despite major fluctuations, dropped from 50 000 in 1970 to less than 25 000 in 1992 and to less than 10 000 in 1993. Data on sport harvests in Québec also indicate a downward trend in populations over the last 20 years. The estimated kill declined from a maximum of 50 000 birds in 1974 to annual kills of around 10 000 in recent years (Appendix II).

Because the Common Goldeneye needs mature trees with nest cavities, it is particularly vulnerable to forest disturbances such as clearcutting and forest fires. Goldeneyes seem to be more abundant in forests that have not been logged and the intensity and age of the cut also seem to have a significant effect on numbers (Tailleux and Marchand 1987). According to Dement'ev and Gladkov (1952), the Common Goldeneye's range in Europe has shrunk as a result of the destruction of forests in the south of the former USSR and western Europe.

DANIEL BORDAGE

Barrow's Goldeneye

Garrot de Barrow

Garrot d'Islande

Bucephala islandica

BARROW'S GOLDENEYE HAS A FAIRLY DISTINCTIVE DISTRIBUTION, somewhat similar to that of the Harlequin Duck. Worldwide, there are three disjunct populations: the largest one, which breeds and winters west of the Rockies and accounts for over 90% of total numbers, and two other small residual populations that breed and winter in Iceland and in Québec (Hasbrouck 1944; Gardarsson 1978; Savard 1987a). The species' winter distribution in Québec is only partially known and appears to be rather limited (Reed and Bourget 1977; Savard 1990). Its breeding range in the province is still largely a mystery.

Some pairs also defend a winter territory (Savard 1988a). Not content to chase away other Barrow's Goldeneyes from the territory, territorial birds also exclude several other species of ducks, especially other *Bucephala* ducks (Savard 1982a; Savard, J.-P. L. 1984). In both summer and winter, aggressiveness is more intense toward diving ducks than toward dabblers, suggesting that the purpose of territoriality is to secure a feeding site (Savard and Smith 1987).

Pierre Dupuis

During aerial surveys conducted in Québec in May 1990, 1991, and 1992, pairs were seen on small headwater lakes such as this one, which is still iced over in this photo. Some of these lakes contain no fish; indeed, in British Columbia, Barrow's Goldeneyes prefer to nest on alkaline lakes that contain no fish, but are very productive in invertebrates. Near Manic-5, Upper North Shore. *May 1990, Daniel Bordage.*

Notes on territory

Barrow's Goldeneye is undoubtedly the most territorial of North American ducks. In spring, the drake defends a well-defined territory on a breeding pond (Savard 1982a, 1988a; Einarsson 1990). Territories in British Columbia averaged 0.7 ha (0.5 ha, excluding water over 3 m deep, which the ducks do not use very much), but ranged from 0.05 to 1.85 ha, depending on the pond and the breeding density (Savard 1986a, 1988a). The male will chase all other Bucephala ducks, except his mate, out of the territory; making either an underwater attack, or pursuing the intruder along the surface of the water or in the air. The territory is even defended when the female is away.

Habitat and behaviour

Territoriality in birds is usually associated with passerines, not ducks. Nonetheless, territorial behaviour is highly developed in Barrow's Goldeneye (Savard 1988b; Einarsson 1990): in spring, on the breeding grounds, the drake defends a territory on a pond from which he excludes all other *Bucephala* species (Common Goldeneyes, Buffleheads), thus ensuring that the hen has an exclusive feeding site. Once the eggs have hatched, the hen becomes very aggressive and defends a territory herself, generally not the same as the first, for her brood's exclusive use.

Confrontations between broods sometimes degenerate into violent battles between females, occasionally even leading to the death of ducklings. More frequently, however, a partial or total amalgamation of the broods occurs, especially when the ducklings are of similar age and have mixed together during the fight (Savard 1987b; Einarsson 1988). When amalgamation is complete, one of the hens ends up with all the ducklings, and the other, after sometimes fruitless efforts to recover her own young, goes off and joins the small flocks of females without young.

Abundance of Barrow's Goldeneyes in southern Québec. Aerial surveys conducted in May 1990, 1991, and 1992 *(D. Bordage, CWS, unpubl. data).*

Barrow's Goldeneye

Breeding distribution

1984–1989

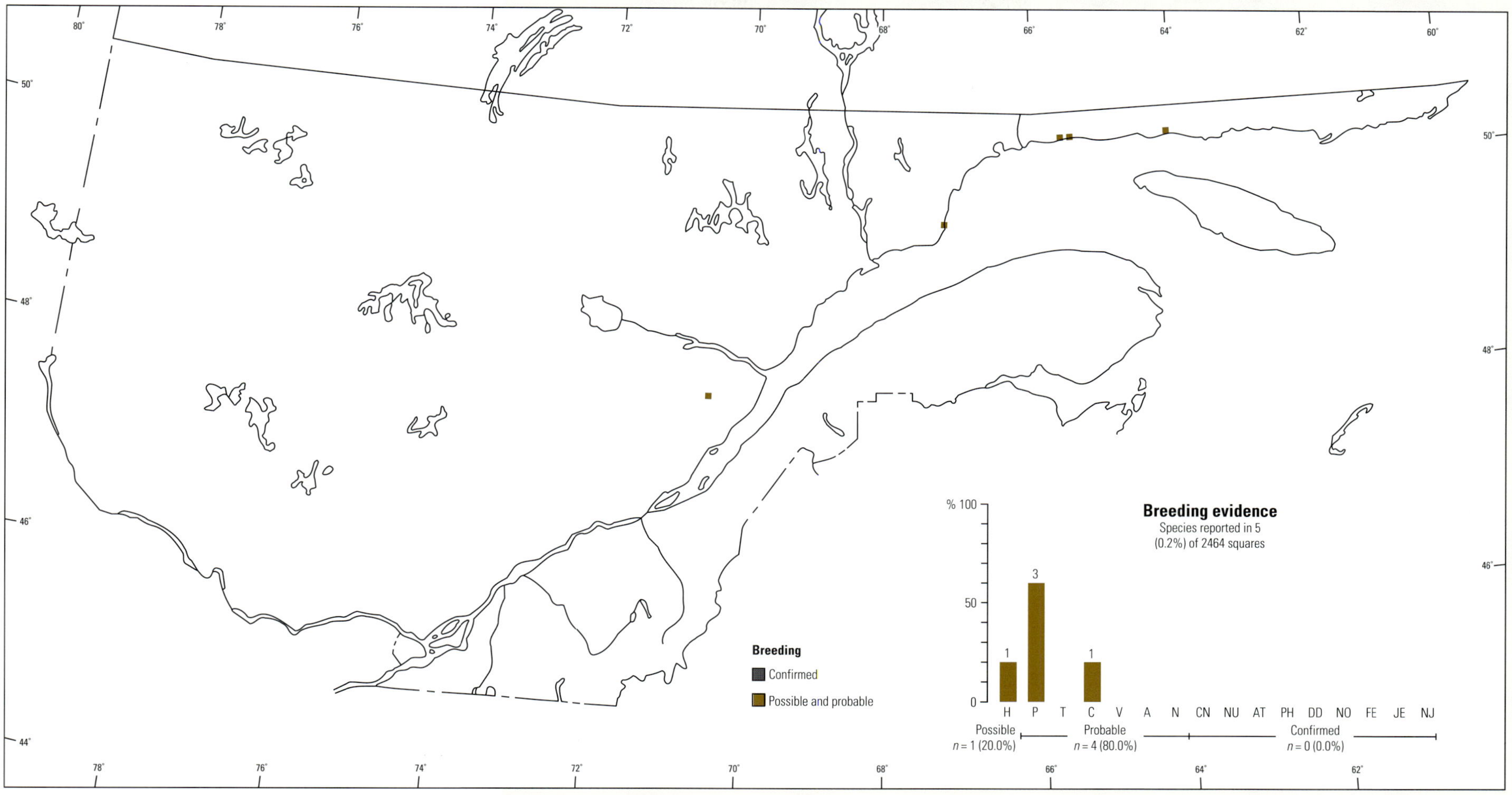

Species profile

Clutch size	usually 8–11 eggs (4–15)[3,4]
Laying interval	usually 1 egg per 1.5 days (1–4)[1]
Start of incubation	usually with last egg[1,2]
Incubation period	28–30 days[3]
Maturity at hatching	precocial
Care of young	by female[3]
Nestling period	24–36 hours[1]
Age at first flight	≈ 56 days[2]
Dependence of young	until ≈ 6 weeks old[1]
Broods per year	1[3]
Breeding age	2 years[1,2,3]
Mating system	monogamy, rarely polygyny[3]
Duration of pair bond	more than 1 season[7]
Total length	41–52 cm[4]
Wing span	≈ 76 cm[5]
Weight	M 1162 g (1077–1191) / F 840 g (822–850)[2]
Longevity record	15 years 4 months[6]

References:
1. Bellrose 1976
2. Palmer 1976
3. Cramp and Simmons 1977
4. Godfrey 1986
5. Terres 1980
6. Clapp et al. 1982
7. Savard 1985

Chronology *n* = 0

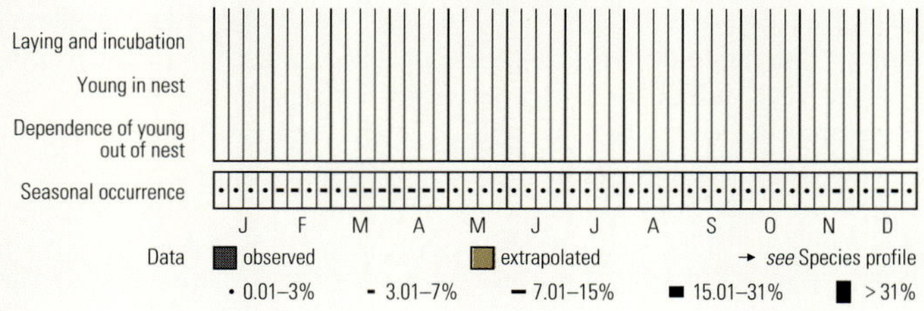

World breeding range

Barrow's Goldeneyes seek ponds and small lakes with wooded margins that have hollow trees with nesting cavities. Once the eggs have hatched, the female leads her brood to the water. Riske Creek, British Columbia. *August 1980, Jean-Pierre L. Savard.*

Apart from the findings of winter surveys, little is known about the behaviour and ecology of Barrow's Goldeneye in Québec (Savard 1990). Recent studies in Iceland and British Columbia (Einarsson 1985, 1988; Savard 1986a), however, have revealed many fascinating aspects of the species' behaviour. Intraspecific parasitism, or the laying of eggs by a female in the nest of another female of the same species, is frequent in Barrow's Goldeneye (Eadie 1989); this behaviour also occurs in the Common Goldeneye and Bufflehead. In extreme cases, especially when there is a shortage of nest sites, nests containing up to 30 eggs have been found (J.-P. L. Savard, CWS, unpubl. data). These nests are usually abandoned, although occasionally a female may hatch up to 25 eggs (pers. obs.). Polygyny, in which a male mates with several females, has been documented in the species, but appears to be rare (Savard 1986b). Forced copulation, or rape, a common phenomenon in dabbling ducks, has never been observed among *Bucephala* ducks. On the other hand, pairs copulate throughout the year, which helps to maintain the pair bond. But the most important discovery has undoubtedly been the fidelity of pairs from year to year, a phenomenon hitherto undocumented in species in which the male does not help raise the young. It appears that after a separation of close to four months, the male and female pair up again on their wintering grounds, where they return year after year (Savard 1985).

Barrow's Goldeneyes nest in natural or artificial cavities, preferring ones higher off the ground. In Iceland, Barrows use holes in lava rock cliffs (Bengtson 1970), while in British Columbia they use mainly tree cavities (Savard 1982b, 1988b). In North America, this diving duck has also been reported to occasionally use crows' nests (Edwards 1953; Sugden 1963), marmot burrows (Munro 1935), and holes in cliffs (Harris et al. 1954). The nest, usually located outside the territory, is not defended by the pair, and may be over 3 km from water (Munro 1939; Savard 1988b). The species breeds successfully in nest boxes, which have been used extensively to manage populations in the West (Savard 1988b). The hen tends to use the same nest site year after year, unless she loses her clutch. In British Columbia, egg predators include red squirrels, black bears, and martens (Savard et al. 1991).

Age at first breeding is 2 years. Females are philopatric, returning to breed in the same place where they hatched (Savard and Eadie 1989). Yearling females explore the area looking for nest sites for the following year (Eadie and Gauthier 1985), sometimes travelling in groups of two or three, visiting various tree cavities, and often vocalizing noisily. When nest boxes are provided, there may be a delay of a year before they are used and the first birds to use the nest boxes are often young females, which usually lay fewer eggs. In the first two years after nest boxes were installed in British Columbia, mean clutch size was 8 eggs, but increased to 12 in the next two years (Savard 1982b, 1988b). Yearling males have also been observed on the breeding grounds, but they are probably not as philopatric as the females. They travel in small flocks with unmated adult males.

The males leave the nesting areas before the eggs hatch, probably around mid-June in Québec, and head off to their moulting areas, where they will remain flightless for close to 3 weeks and acquire a plumage similar to the female's. It is not known where they moult in Québec. Brood size can vary enormously, owing to parasitism, egg predation, and brood amalgamation. As in most duck species, the young are particularly vulnerable during the first week after hatching (Savard et al. 1991). During the brood-rearing period, females will chase broods of other species of ducks off their territories. High densities of Barrows broods have even been associated with an increase in mortality in scaups (Einarsson 1988) and Buffleheads (Savard et al. 1991).

Hybridization between Barrow's and Common goldeneyes is fairly common. The male offspring of these crosses are easily recognized by their plumage, which is intermediate between the two species.

Notes on territory (continued)

After the eggs hatch, the hen becomes very aggressive and will establish a territory on the brood-rearing pond, driving away all other Bucephala *ducks (Savard 1986a, 1988a). Confrontations between territorial females often end in an exchange of ducklings and occasionally in the death of ducklings. The brood-rearing territory is usually not the same as the male's territory and is larger (0.9 ha on average in British Columbia) (Savard 1986a).*

On their wintering grounds on salt water, some pairs maintain a territory similar in size to the spring territory, which is defended mainly by the male (Savard 1986a, 1988a).

In defending these winter and summer territories, the male and female are aggressive toward not only conspecifics, but also toward several other duck species, particularly Common Goldeneyes and

This goldeneye nests in natural or artificial cavities, preferably well off the ground and near water. Riske Creek, British Columbia. *October 1982, Jean-Pierre L. Savard.*

Buffleheads. They tend to be more tolerant of dabbling ducks than of diving ducks. Territoriality in Barrow's Goldeneyes is probably associated with the foraging advantages afforded to the female and young (Savard 1988a; Einarsson 1988, 1990).

Several hundred Barrow's Goldeneyes regularly winter on Baie des Rochers, east of Saint-Siméon. *February 1988, Jean-Pierre L. Savard.*

Distribution

The breeding range of Barrow's Goldeneye in Québec is still a mystery. The species was reported in only 5 (0.2%) of the 2464 Atlas squares surveyed, and breeding was not confirmed. Four of these records were from the Upper and Middle North Shore. As early as the turn of the century, Dionne (1906) and Comeau (1923a) considered this duck species to be common in summer near Godbout. More evidence of the species' abundance on the Upper North Shore comes from the *Répertoire des appelants du Québec*, a catalogue of decoys from Québec; all the Barrow's Goldeneye decoys were made on the Upper North Shore (P. Dupuis and A. Bourget, CWS, unpubl. data).

Recent surveys (1990, 1991, and 1992) conducted during the second half of May by the Canadian Wildlife Service found pairs present at several inland locations. Several pairs were observed in the back country of the Upper and Middle North Shore, where the highest densities were recorded, 2–4 pairs/100 km^2 (D. Bordage, CWS, unpubl. data). In 1986, pairs were also seen on small lakes at elevations over 500 m in the Laurentians in Charlevoix County (P. Dupuis, CWS, unpubl. data).

In southern Québec, all pairs observed during aerial surveys were on small lakes (less than 10 ha), and were never in the company of Common Goldeneyes. Several types of lakes are used by Barrow's Goldeneyes. For example, a pair was sighted in 1990 on a small pond close to an access road to the Manicouagan hydro-electric complex, which had probably been created when the road was built. Most pairs, however, are found on high elevations on long, narrow headwater lakes, in other words, in a valley-type landscape.

In British Columbia, the species prefers alkaline lakes, which usually have no fish but have abundant invertebrate populations (Savard 1982a; Savard, J.-P. L. 1984). In Iceland, Barrow's Goldeneyes nest mainly on a single river and a volcanic lake, with exceptionally high productivity (Bengtson 1971b; Einarsson 1988). Barrow's Goldeneyes may also favour similarly rich environments in Québec.

The species has been observed at various locations north of the Atlas area, notably in the Ungava region (Koksoak and Arnaud rivers and Rivière à la Baleine; Todd 1963; Lamothe 1978) and in the lower part of the Little Whale River watershed, where it may breed (F. Morneau, pers. comm.).

Contrary to what most field guides say, from November to June, female Barrow's Goldeneyes in Québec can be easily identified by their bright orange bills. British Columbia females have similarly coloured bills, whereas in Iceland the orange is less extensive (A. Einarsson, pers. comm.). At the onset of incubation, however, the female's bill becomes blackish, and it then becomes extremely difficult to distinguish her from the Common Goldeneye female, which may partially explain the dearth of information on the breeding ecology of the species in Québec. Adult females' orange-coloured bills reappear in fall, while first-year females retain their blackish bills all winter.

Brood in a nest box. To leave the nest, the young will climb up the left-hand wall using the specially carved slits, and jump to the ground to join their mother. Riske Creek, British Columbia. *June 1983, Jean-Pierre L. Savard.*

History and trends

Because of the lack of historical data, it is impossible to determine the current status of the population. The Icelandic population is relatively small, with approximately 2000 goldeneyes (Gardarsson 1978). The Québec population probably numbers somewhere between 2000 and 4000 birds (Reed and Bourget 1977; Savard 1990), but it is not known how accurate this estimate is. Winter waterfowl surveys in the St. Lawrence estuary have not revealed any significant population trends, but suggest that the species has a very local distribution in winter (Savard 1990).

There are several threats to the Québec population: oil spills in fall and winter; overhunting caused by the accessibility of autumn staging areas, the difficulty of measuring, and thus controlling, harvests of scarce and very local species (the structure of national surveys does not allow for this), and the lack of information on the current status of the population in Québec (Savard 1987a); logging, which may reduce the number of suitable nesting sites; coastal development for residential or other purposes, which could modify the species' wintering grounds; and finally, the lack of information on summer and winter distribution, which prevents potential threats from being identified.

Jean-Pierre L. Savard

Bufflehead

Petit Garrot

Bucephala albeola

The Bufflehead, North America's smallest diving duck, nests exclusively in tree cavities and artificial nest boxes. It is readily identified by its small size, diving habit, and, in the male, the bold black and white plumage. Like other *Bucephala* ducks, the Bufflehead is very territorial when breeding (Donaghey 1975; Gauthier 1993a).

In migration, it is seen in small flocks throughout Québec on salt water along the coast and on freshwater ponds, lakes, and rivers, including the St. Lawrence itself. The species winters very locally in Québec, and breeding has only been confirmed in the province since the early 1980s (David 1983).

Pierre Dupuis

Several pairs of Buffleheads occupy this lake, which is surrounded by conifers and heaths and located near the Gouin Reservoir. The mature stands nearby have many snags and old trees with woodpecker holes providing suitable nest sites for the species. *May 1991, Yves Aubry.*

Habitat and behaviour

During the breeding season, the Bufflehead is primarily found on small lakes and ponds and occasionally on rivers and marshes (Campbell et al. 1990a). Preferred lakes and ponds are shallow (Erskine 1972b; Gauthier 1988) and usually somewhat alkaline (Savard 1982a; Savard, J.-P. L. 1984), with only sparse floating and emergent vegetation; such sites contain few if any fish but support large invertebrate communities (Savard 1982a). The Bufflehead typically avoids bodies of water with abundant aquatic vegetation (Palmer 1976) and with populations of pike, a predatory fish (Erskine 1972b).

In British Columbia, 80% of the ponds used by breeding pairs were less than 8 ha in size; some of the ponds chosen were surrounded by a small fringe of vegetation (Gauthier 1987b). On the northeast coast of James Bay, Benoit et al. (1993b) reported broods on lakes of less than 5 ha. The Bufflehead feeds on open water, often far from shore, and although it may forage at the edge of emergent vegetation, it rarely dives into emergent or submerged vegetation.

Pairs are first formed on the wintering grounds, when the birds are in their second year (Erskine 1972b; Eadie and Gauthier 1985), and apparently stay together for a few years, although they separate during incubation and brood rearing (Gauthier 1987a, 1993a). Courtship displays by the male include head movements and short flights that show off his bright pink webbed feet. Males frequently pursue each other on and under the water as well as in flight (Gauthier 1993a).

At the beginning of the breeding season, the male defends a territory on a body of water near the nest, sometimes the same territory as the preceding year (Gauthier 1987a). The male excludes all other Buffleheads from this territory (Savard, J.-P. L. 1984), which may cover an entire pond or, on a larger body of water, a section of water along several hundred metres of shoreline (Munro 1942; Gauthier 1987b).

The nest is in a natural tree cavity or a nest box in the female's natal area (Gauthier 1990). Hens normally prospect for sites the summer before the breeding season: small flocks of female yearlings and failed breeders will visit potential sites, whether occupied or not, and the following spring, the hens will bring their mates back to the site selected (Eadie and Gauthier 1985). Buffleheads are small enough to be able to use Northern Flicker holes, which are the preferred nest site in western Canada. They also use Pileated Woodpecker holes and artificial nest boxes (Erskine 1972b; Campbell et al. 1990a; Gauthier 1993a). However, they avoid open-top cavities (Gauthier 1993a). In general, the nest is located outside, but within close range of, the territory; the closer the nest is to water, the more likely it is to be adjacent to the territory (Gauthier 1987b). On rare occasions, two females may simultaneously occupy two holes in the same tree (Erskine 1972b). Nests are generally located 20–75 m from water, though sometimes as far away as 350 m (Erskine 1972b). Tufts of down clinging to the entrance hole are often a sign of occupation (Munro 1942).

Little lining material is used for the nest, aside from the down added by the female during incubation. Often the same cavity is used for two consecutive years, especially if nesting was successful the first year; however, nest failure usually prompts the female to move to another cavity (Gauthier 1988). Female Buffleheads rarely renest, but will do so if the clutch is destroyed before it was completed (Gauthier 1988, 1989). Egg predators include black bears, squirrels, martens, and fishers (Gauthier 1988; Campbell et al. 1990a; Savard et al. 1991). Brood parasitism is rare in the species (Erskine 1972b, 1990).

Notes on territory

Like the Common Goldeneye and Barrow's Goldeneye, the Bufflehead is highly territorial. The male defends a territory for the pair, usually located near the nest. The female later defends a brood territory, which is generally a different one from the first. Territorial defence is directed at both conspecific individuals and pairs (intraspecific competition) and those of other species, mainly diving ducks (interspecific competition) (Savard 1982a). The two goldeneyes are generally dominant over Buffleheads in territorial conflicts (Savard 1982a; Savard, J.-P. L. 1984; Gauthier 1993a).

In British Columbia, when the pair arrives on the nesting grounds, the male defends a territory that includes a section

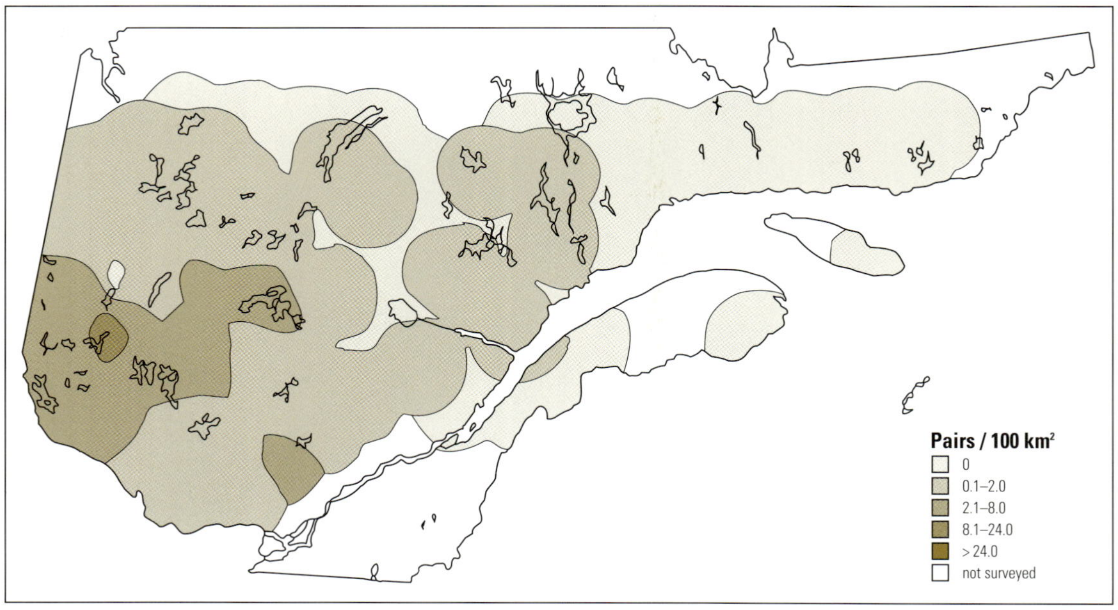

Abundance of Buffleheads in southern Québec. Aerial surveys conducted in May 1990, 1991, and 1992 *(D. Bordage, CWS, unpubl. data).*

Bufflehead

Breeding distribution

1984–1989

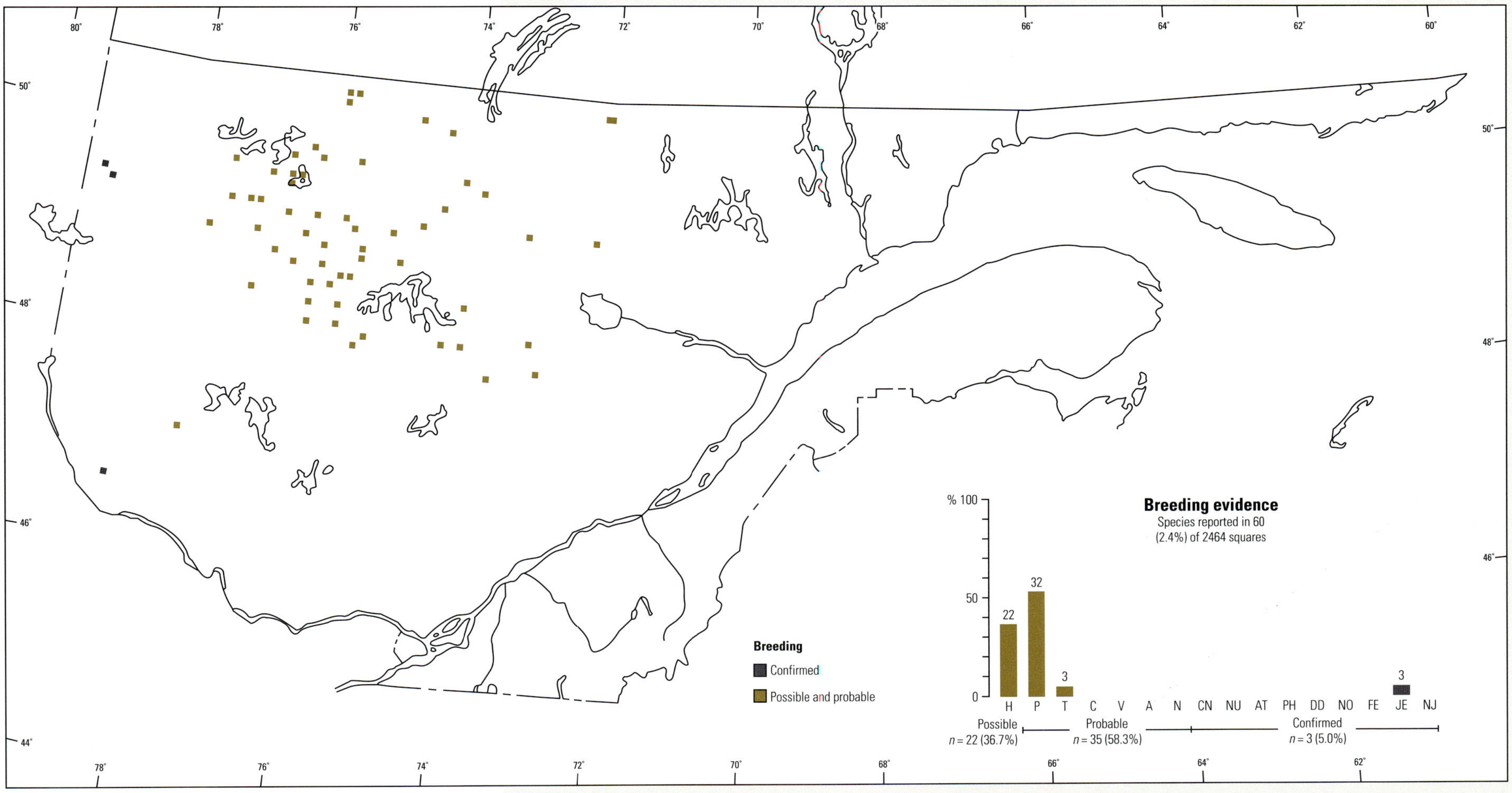

Species profile

Clutch size	usually 6–11 eggs (4–12)[1,2]
Laying interval	1 egg per 1–3 days[2]
Start of incubation	usually with last egg[2]
Incubation period	usually ≈ 30 days (28–33)[2]
Maturity at hatching	precocial
Care of young	by female[1,2,5]
Nestling period	24–36 hours[1,2]
Age at first flight	usually 50–55 days (45–55)[1,2]
Dependence of young	usually until 5–6 weeks old (3–6)[2]
Broods per year	1[1,2,6]
Breeding age	2 years[1,2]
Mating system	monogamy, rarely polygyny[2]
Duration of pair bond	sometimes more than 1 season[2]
Total length	M 35–40 cm / F 32–35 cm[2]
Wing span	50.8–61.0 cm[3]
Weight	M 473 g (424–551) / F 334 g (297–374)[4]
Longevity record	14 years 11 months[2]

References:
1. Erskine 1972b
2. Gauthier 1993a
3. Terres 1980
4. Dunning 1984
5. Gauthier 1989
6. Palmer 1976

Chronology $n = 26$

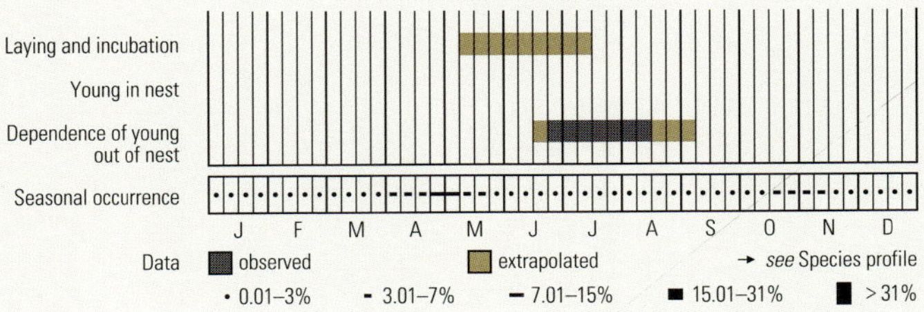

World breeding range

In 1987, a brood of Buffleheads resided on this beaver pond, which was partially covered with aquatic vegetation. The following year, the beaver dam—a temporary structure at the best of times—was destroyed and the Buffleheads left the area. South of Lake Deyglun, Témiscamingue. *September 1987, Yves Aubry.*

Although broods have been sighted in Québec, a Bufflehead nest has never been found. This biologist in British Columbia is measuring the height of a cavity in a pine snag occupied by Buffleheads. This same type of nest site might well be found in Québec. Riske Creek, British Columbia. *June 1983, Jean-Pierre L. Savard.*

The hen incubates the eggs. The drake, if present, joins the hen once she arrives on the pond after leaving the nest to forage (Munro 1942). The drake abandons the territory during incubation to assemble with other drakes for the moult.

Shortly after the young hatch, they jump out of the nest and are led by the female to a nearby pond. Early on the female broods the ducklings at night and in cold weather; she rarely leaves them unattended during the first month (Palmer 1976). She also establishes a brood territory for the young, generally in a different area and larger than that defended earlier by the male (Gauthier 1987c), excluding all individuals or broods that enter this area. Fights between two females with broods usually result in the mixing of the broods, with the evicted female often being forced to abandon a few ducklings or her entire brood (Savard 1987b).

The ducklings' diving abilities take a few days to develop, and at first, they must depend on food available on or near the surface of the water (Savard et al. 1991). Such food is scarce in rain and cold weather, reducing the ducklings' chances of survival (Sjöberg and Danell 1982; Gauthier 1987c; Savard et al. 1991). The young feed mainly on aquatic insects and their larvae and occasionally on amphipods and plant matter (Gauthier 1987c). Insects also make up a major part of the summer diet of the adult birds (Gauthier 1993a), which also feed on molluscs and, to a lesser extent, on small fish and fish roe (Munro 1942).

The hen begins to leave her brood unattended from time to time when the ducklings are at the midpoint of their growth period, at about 25 days of age (Palmer 1976), and usually abandons them altogether before they fledge, to assemble for the moult (Gauthier 1993a).

In winter, Buffleheads frequent shallow brackish water and salt water, preferring more sheltered areas than Barrow's and Common goldeneyes (Erskine 1972b). Males tend to be faithful to wintering sites after their second winter. The pair bond is probably renewed on the wintering grounds or during the moult (Gauthier 1987a).

The diet of birds wintering on salt water consists mainly of crustaceans and molluscs, which they forage for in calm shallow water with a sandy, silty, gravelly, or rocky bottom (Vermeer 1982).

Notes on territory (continued)

of shoreline. This territory is generally adjacent to the nest and shifts if the female changes nests. It serves as a kind of buffer zone to keep other Buffleheads from the nest site (Gauthier and Smith 1987). The average size of a pair's territory was estimated at 0.57 ha in British Columbia and 0.56 ha in Alberta (Gauthier 1987b). When several pairs use the same pond, territories are spaced out along the shoreline, with the central area often unoccupied (Gauthier 1987b). The male drives away other Buffleheads, not to protect the food supply, but rather to allow the hen to feed without being harassed by other males (Savard 1982a; Gauthier 1987b). The female is actually the most important resource to protect, and a male that loses a hen ceases to defend a territory (Gauthier 1993a).

A female with young will fiercely defend a brood territory, established in the first few days after hatching, to ensure an adequate food supply for the ducklings. In high-density areas in British Columbia, broods that are late in hatching occupy areas farther from the nest than broods hatching earlier. In the same region, brood territories averaged around 0.52 ha, with an intensively used area of 0.33 ha (Gauthier 1993a). When a body of water is occupied by several broods, violent conflicts may occur, occasionally resulting in the death or abandonment of ducklings. The dominant female may acquire the ducklings that the evicted female was not able to recover (Savard 1982a, 1987b).

Territoriality in the species appears to limit breeding pair density, especially in areas where the species is abundant. The absence of adequate nest

Pairs like this one in Vancouver, British Columbia, are formed on the wintering grounds. The female returns to her natal area to nest accompanied by the male. *March 1982, Jean-Pierre L. Savard.*

holes may also be a limiting factor (Gauthier 1993a).

In southern Québec, the highest densities of Buffleheads were recorded in the Abitibi–Témiscamingue region, especially around the Decelles Reservoir, where densities of 8–16 pairs/100 km² were found (D. Bordage and M. Grenier, CWS, unpubl. data).

Distribution

The Bufflehead is an exclusively North American species for which evidence of breeding in Québec is scant. When broods are present, they are relatively easy to spot, since they tend to swim off when disturbed rather than to seek cover in the vegetation (Savard et al. 1991).

The first broods of Buffleheads were sighted in Québec in 1982 and 1983, on a small pond south of Radisson in the James Bay region (David 1983; Bordage 1984). Over 20 broods have been reported in the province since then, including several during the Atlas project (Yank et al. 1987b; Gosselin et al. 1988b). In the Témiscamingue area in 1987, a brood of 12 ducklings was reported on a beaver pond of less than 1 ha, a tributary of Lake Deyglun. The pond, located in a mixed forest, had aquatic vegetation over less than 20% of its surface but dried up the following year when the beaver dam was destroyed. In 1988, 3 other broods were sighted in 2 squares in the northwestern Abitibi region: a brood of 5 ducklings on a small pond (barely 1 ha) surrounded by a strip of ericaceous vegetation and located in coniferous forest; an unattended brood of 3 ducklings about 30 days old on a 1.5-ha pond in a black spruce forest; and a brood of 6 ducklings on a small lake of nearly 4 ha. All these ponds and lakes had sparse aquatic vegetation. There are many similar ponds and lakes in the Abitibi and Témiscamingue regions that were not atlassed, so breeding numbers may very well be higher than indicated by the data.

Many possible and probable breeding records were also obtained in the northwest part of the Atlas area, most involving sightings of pairs and single birds during helicopter surveys at the beginning of the breeding season, during the third and fourth weeks of May. Although these squares contain suitable breeding habitat, the records may involve transient birds on their way to nesting grounds farther north.

Broods seen outside the Atlas period or area include 3 reported near the Opinaca River in 1984 (Bordage 1985b). Additional broods were found during surveys in northern Québec between 1989 and 1993: 2 on the west shore of Rupert Bay; 1 east of Nemiscau, near Rupert River; 1 near Lake Cawachagamite, south of the Eastmain River; 3 in 1990 and 4 in 1992 on the east coast of James Bay, between Attikuan Point and Rivière au Castor; 3 in 1993 in the Roggan River area; and finally, 2 farther north in the Lake Fagnant region near Kuujjuarapik (Cons. G. G. – GREBE 1990a, 1992a; Yank et al. 1990b; Benoit et al. 1991, 1993b; Tardif and Morneau 1991; R. Benoit, pers. comm.). These breeding records suggest that the Bufflehead's breeding range extends farther northeast than previously thought, as far east as the western side of Lake Saint-Jean and as far north as the Kuujjuarapik area. Other sightings include a pair observed in May 1977 on Lake Bienville (St-Louis 1978) and several single adults seen during the summer of 1989 in the Great Whale River watershed (Cons. G. G. – GREBE 1990a). As Gauthier (1993a) pointed out, the species is restricted to forested areas, and relies on natural cavities excavated by woodpeckers, especially the Northern Flicker.

In spring, Buffleheads gradually move northward as lakes and ponds become ice-free (Erskine 1972b), and nesting is delayed in years of late ice-out. Between the arrival of the pair on the nest site and the departure of the young for the fall migration, the Bufflehead requires an ice-free period of 120 days (Palmer 1976; Savard et al. 1991).

In southern Québec, the Bufflehead is seen in spring between 10 April and 25 May, but most often in early May. The fall migration usually takes place between 15 October and early December, and peaks at the end of October and the beginning of November (ÉPOQ, unpubl. data). Buffleheads generally winter on the Atlantic and Pacific coasts but are also found inland on water bodies in the United States and northern Mexico (Bellrose 1976; Root 1988). A small population ranging between 40 and 69 birds winters regularly at Baie-Sainte-Catherine, near Tadoussac (Reed and Bourget 1977; Savard 1990).

History and trends

The species is much more common in western Canada than in the East (Bellrose 1976). For a long time, the species was considered a transient in Québec (Wintle 1896; Macoun and Macoun 1915; Ouellet 1974a; David 1980a; Godfrey 1986), and breeding has only been confirmed in recent years, despite aerial surveys conducted early in the breeding season during the 1960s (Erskine 1972b).

Wintle (1896) noted that Buffleheads were being sold in markets in the Montréal region at the turn of the century, and to this day they are hunted in Québec in the fall. However, the species makes up only 1% of the total harvest in the province (Appendix II), most being taken in the southwest. The species tends to be unwary and decoys readily, often flying within range of hunters, who also prize it for its aesthetic qualities (Munro 1942). Bent (1925) reported that it was common on the New England coast and described it as an easy duck to shoot. According to Gauthier (1993a), the Bufflehead is one of the few ducks whose numbers have been increasing steadily since the mid-1950s.

Any limiting factors on Northern Flicker and Pileated Woodpecker populations during the breeding season may have repercussions on the Bufflehead, which breeds exclusively in wooded areas and relies on these species to provide nest cavities. Logging also has a direct effect on the Bufflehead, due to the ensuing loss of trees with nest cavities (Tate and Tate 1982). Forestry practices may also influence the type of tree used and the distance between the nest tree and the body of water (Gauthier 1993a). In western Canada, Buffleheads use large poplars and aspens growing near water (Gauthier 1993a). Cavities generally last longer in coniferous than in deciduous trees (Erskine 1978a). The Bufflehead's main competitor for nesting cavities is the European Starling, but it also vies with Northern Flickers, both species of goldeneyes, Hooded Mergansers, Wood Ducks, and one mammal, the northern flying squirrel (Bellrose 1976; Peterson and Gauthier 1985; Gauthier 1988, 1993a). Buffleheads will readily nest in suitable nest boxes (Gauthier 1988; Savard 1988b), and this management technique may help increase density in areas where natural cavities are lacking (Gauthier 1988). In areas where nest sites are plentiful, nest boxes neither increase nor decrease density (Gauthier and Smith 1987), and in areas where Buffleheads are numerous, the presence of nest boxes has little impact, and the main limiting factor on density is territorial behaviour (Gauthier 1993a).

Oil spills or changes to the tidal flats around Baie-Sainte-Catherine, the only known wintering site in Québec, could jeopardize the continuing use of this area in the future.

Yves Aubry

Hooded Merganser

Bec-scie couronné

Harle couronné
Lophodytes cucullatus

René Roy

The Hooded Merganser is the smallest of our three mergansers, and the only one found exclusively in North America. There are two disjunct breeding populations: a western one and an eastern one.

With the large white triangular patch on his crest, the male looks something like the male Bufflehead, but his long, narrow, serrated bill and reddish-brown flanks make it easy to distinguish between the two. The female can be differentiated from other female mergansers by her small size and darker coloration. David (1980a) lists the Hooded Merganser as a rare migrant breeder and uncommon in migration in southern Québec; more recent data, however, show that the species may be locally common.

Habitat and behaviour

Hooded Mergansers breed mainly on small lakes, beaver ponds, peatland ponds, rivers, swamps, and streams in or near woodland (Godfrey 1986; DeGraaf and Rudis 1987; Tardif and Morneau 1991; Cons. G. G. – GREBE 1992a). In some regions, they also frequent clear, gravel-bottomed waterways with flooded shorelines and standing drowned trees (Johnsgard 1975b; Palmer 1976).

Hooded Mergansers nest in woodpecker holes in trees or stumps, generally 6–10.5 m up. Like goldeneyes, Buffleheads, and Wood Ducks, Hooded Mergansers also use nest boxes. This duck seems to prefer cavities or nest boxes near water (Morse et al. 1969).

Age at first breeding is 2 years. The pair bond lasts only a few weeks, and the male abandons the female at the onset of incubation (Morse et al. 1969). Pair formation usually takes place over the winter, but some pairs may not form until late spring; during this time, one or more males can be seen parading around a female, raising their crests (Johnsgard 1961).

The female incubates the eggs and raises the young alone. Nest attentiveness during incubation is high and females are seldom observed on the water at this time (Bouvier 1974a). About 24 hours after hatching, the ducklings jump out of the nest and, guided by the hen, head for the nearest water (McGilvrey 1966). A secluded area is preferred for brood rearing and, in southern Québec, Hooded Merganser broods seem to favour beaver ponds or quiet bays along rivers (J. Bouvier, pers. obs.). In Wisconsin, however, broods were found most often on swift-flowing, cobble-bottomed rivers (Kitchen and Hunt 1969). The females abandon the young when they are about 5 weeks old, before they can fly (Beard 1964).

Frequently a female will lay her eggs in the nest of another Hooded Merganser, a Wood Duck or a Common Goldeneye, and the converse also occurs (Hansen et al. 1973; Bouvier 1974a; Zicus 1990). In some cases, the eggs are laid after the adoptive female has begun incubation. This is why all the eggs may not hatch at the same time or some may never hatch, since some eggs may be infertile or underdeveloped (Bouvier 1974a). Incubation of mixed broods usually proceeds normally, however, and most of the adopted eggs hatch (Bouvier 1974a). A few mixed broods older than 5 weeks have been observed around Harrington, in the Ottawa Valley (J. Bouvier, pers. obs.).

The Hooded Merganser forages chiefly by diving for crayfish, aquatic insects, and a variety of small fish, such as yellow perch, bass, and sunfish (Bellrose 1980).

This merganser is easily overlooked because of its secretive behaviour and habit of nesting on secluded woodland ponds. Although broods may be hard to detect as they swim along the winding, wooded shores of their habitat, sightings of broods still made up 89% of all confirmed breeding records for the Atlas.

Distribution

The Hooded Merganser is found throughout the central and western portions of southern Québec, but is scarcer northeast of the Saguenay River. It is absent from the Lower North Shore, but there are a few isolated records from the Gaspé, Anticosti Island, and the Magdalen Islands. Aerial surveys conducted in May 1990, 1991, and 1992 show that the species' centre of abundance in Québec is in the southwestern part of the province, particularly the Ottawa Valley, the Laurentians, and part of the Mauricie–Bois-Francs region; outside this area, densities are much lower (D. Bordage, CWS, unpubl. data).

The Hooded Merganser is considered an uncommon transient in the Gaspé (Girard 1988). In the following regions, where it is a migrant breeder, it is considered irregular on the Magdalen Islands (Fradette 1992), common in the central and southern part of the Ottawa Valley (Club des ornithologues de l'Outaouais 1985), and abundant a little farther east, in the Rivière Rouge Valley (Bouvier 1974b).

Hooded Mergansers were recorded in 13.8% (339/2464) of Atlas squares surveyed. Given the species' secretive behaviour and often relatively inaccessible habitat, it is probably more widely distributed than the Atlas map indicates.

The northern limit of the Hooded Merganser's breeding range is poorly known. According to Godfrey (1986), the species breeds up to Waskaganish, on the south end of James Bay. There are sight records in the breeding season from the Nottaway, Broadback, and Rupert river watersheds, the east coast of James Bay, and the Great Whale River area (Savard 1977; St-Louis 1978, 1979; Bordage and Aubry 1982, 1983; Bordage 1984, 1985b; Tardif and Morneau 1991; Cons. G. G. – GREBE 1992a), complemented by recent breeding records on the basis of reports of broods from the Nottaway, Broadback, Rupert, Eastmain, and Great Whale river regions (St-Louis 1978; Fraser and St-Louis 1979; Cons. G. G. – GREBE 1990a, 1992a; Tardif and Morneau 1991). In July, thousands of Hooded Mergansers fly north to moult, as far as the northern limits of the boreal forest, including lakes along the east side of James Bay and in the Great Whale River watershed (Cons. G. G. – GREBE 1990a; Benoit et al. 1993b).

Notes on territory

In Québec, the highest breeding densities (8–16 pairs/100 km²) have been recorded north of the Ottawa River (D. Bordage and M. Grenier, CWS, unpubl. data). This part of southwestern Québec is characterized by deciduous and mixed forests and some old-growth stands.

Densities of 1.9 nests/km² were reported in Minnesota (Zicus 1990). Brood densities of 1.33 broods/km of river were found in Michigan, with the highest densities recorded on rivers with thickly wooded banks; in this region, population density was found to be correlated more with the availability of food than with that of nest sites (Kitchen and Hunt 1969).

Abundance of Hooded Mergansers in southern Québec. Aerial surveys conducted in May 1990, 1991, and 1992 *(D. Bordage, CWS, unpubl. data).*

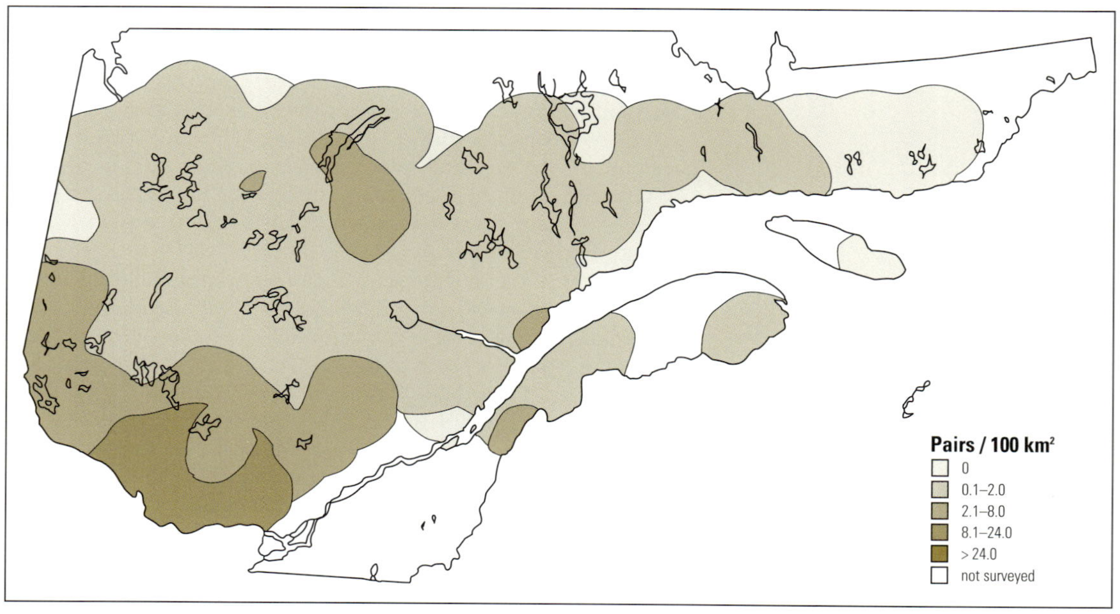

Hooded Merganser

Breeding distribution
1984–1989

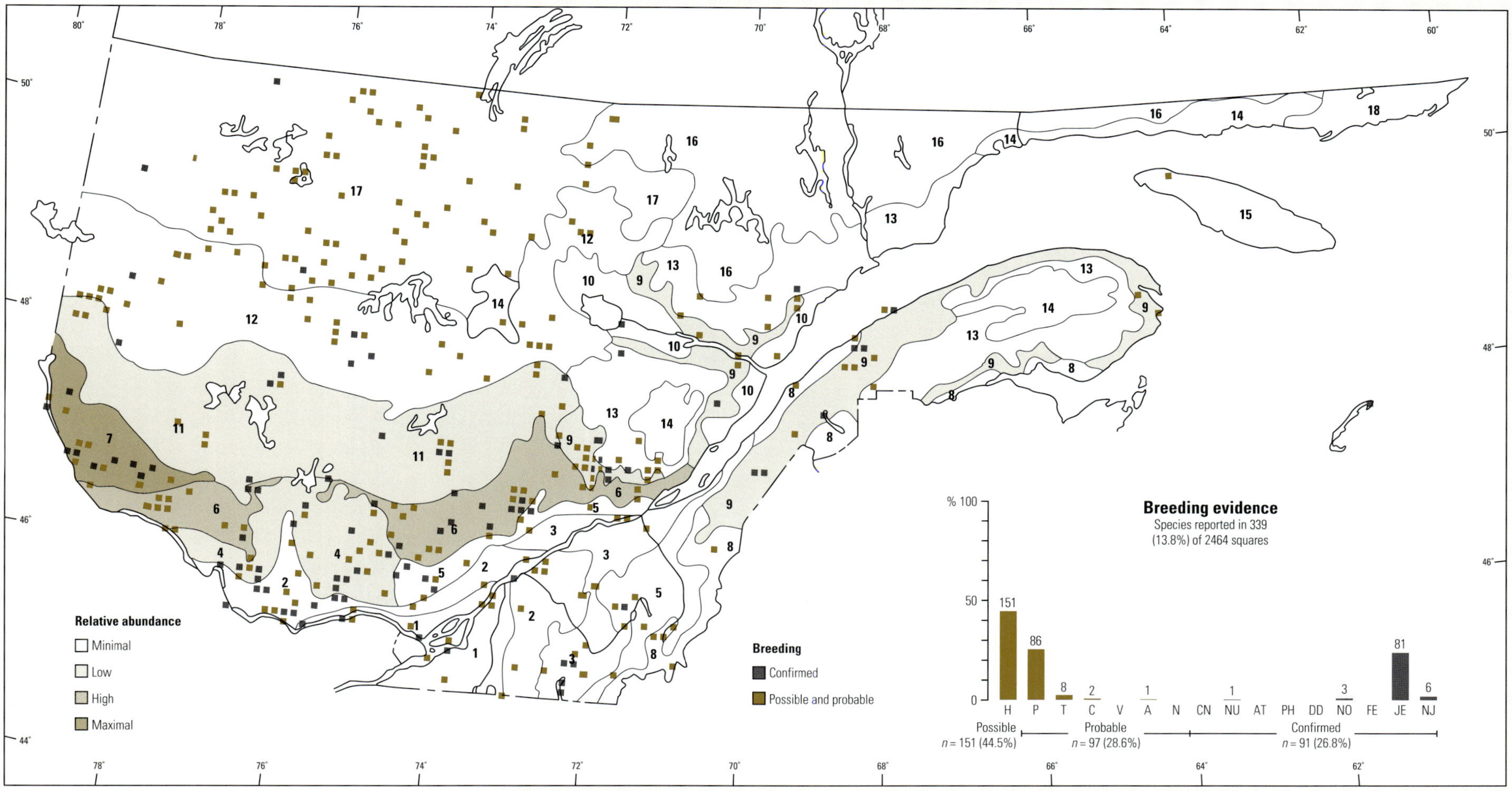

Species profile

Clutch size	usually 11 eggs (7–13), more than 13 = dump nest[3]
Laying interval	1 egg per 1–3 days[3,8]
Start of incubation	?
Incubation period	usually 29–33 days (26–41)[5,6,9]
Maturity at hatching	precocial
Care of young	by female[1,4]
Nestling period	≈ 24 hours[1,2]
Age at first flight	≈ 71 days[1,3,4,5]
Dependence of young	until ≈ 5 weeks old[1]
Broods per year	1[1]
Breeding age	2 years[1,3]
Mating system	monogamy[3,4]
Duration of pair bond	1 season[8]
Total length	43–58 cm[6]
Wing span	61.0–67.3 cm[5]
Weight	M ? g (600–950) / F ? g (450–675)[5]
Longevity record	11 years 3 months[7]

References:
1. Palmer 1976
2. Bouvier 1974a
3. Bellrose 1980
4. Ehrlich et al. 1988
5. Terres 1980
6. Godfrey 1986
7. Clapp et al. 1982
8. Morse et al. 1969
9. Dugger et al. 1994

World breeding range

Chronology $n = 80$

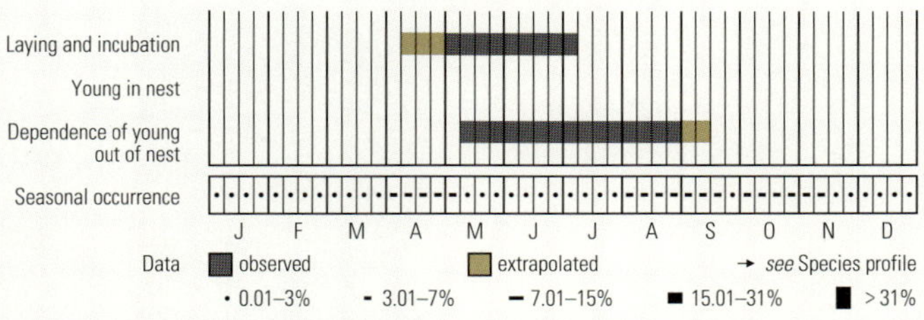

Frequency of occurrence by ecological region

Ecological regions** (forest regions)	Adequately covered squares	Squares in which reported	% of squares in which reported	Relative abundance (%)*
1. Sugar maple-hickory and sugar maple-basswood	97	12	12.4 ± 3.4	18.1
2. Sugar maple-basswood	153	29	19.0 ± 3.3	27.7
3. Sugar maple-basswood and sugar maple-yellow birch	101	11	10.9 ± 3.9	15.9
4. Sugar maple-yellow birch-basswood	56	22	39.3 ± 11.2	57.4
5. Sugar maple-yellow birch-beech and sugar maple-yellow birch-basswood	113	13	11.5 ± 3.4	16.8
6. Sugar maple-yellow birch-beech	71	32	45.1 ± 10.6	65.9
7. Sugar maple-yellow birch-balsam fir	19	13	68.4 ± 22.1	100.0
8. Sugar maple-yellow birch and balsam fir-yellow birch	47	6	12.8 ± 8.9	18.7
9. Balsam fir-yellow birch	87	18	20.7 ± 7.9	30.2
10. Balsam fir-white birch or balsam fir-red maple	46	2	4.3 ± 5.7	6.4
11. Yellow birch-balsam fir	52	21	40.4 ± 13.8	59.0
12. Balsam fir-white birch (west)	59	7	11.9 ± 8.8	17.3
13. Balsam fir-white birch (east)	76	8	10.5 ± 7.0	15.4
14. Balsam fir-black spruce	28	1	3.6 ± 8.3	5.2
15. White spruce-balsam fir and balsam fir-white spruce	22	1	4.5 ± 10.3	6.6
16. Black spruce-balsam fir-moss	18	0	0	0
17. Black spruce-moss	32	2	6.2 ± 10.0	9.1
18.			*not analysed*	

* minimal: 0–29.9%; low: 30–59.9%; high: 60–89.9%; maximal: 90–100%

** Deciduous forest: 1–7; mixed forest: 8–11; coniferous forest: 12–17

Beaver ponds teeming with small fish and invertebrates provide ideal habitat, particularly if they are cluttered with logs and other debris, which provide cover for broods. Forillon National Park, Gaspé. *July 1992, Yves Aubry.*

Ponds, small lakes, and slow-moving streams and rivers are all likely habitats for a Hooded Merganser pair. Trécesson, Abitibi. *August 1991, André-Guy Bernier.*

At Streit Pond in Philipsburg, nest boxes and natural cavities in tree trunks and stumps provide the Hooded Merganser with a good selection of nest sites. South of Saint-Jean-sur-Richelieu. *July 1991, Pierre Bannon.*

History and trends

It is difficult to compare the species' current and past distribution given the paucity of information on the subject. In the early part of the century, the Hooded Merganser had still not been confirmed as a breeder in Québec (Macoun and Macoun 1909), which may have been due solely to ornithologists' unfamiliarity with its distribution, since it was observed during spring and fall migrations in the province (Dionne 1906; Macoun and Macoun 1909). Moreover, on the basis of information passed on to him, Wintle (1896) reported that the species had nested at Lac des Deux Montagnes, near Montréal.

The Hooded Merganser has probably suffered from logging and silvicultural operations, which often result in the removal of the dead trees it nests in. Water pollution, which decreases the clarity of lakes and rivers, thus making prey harder to find, has probably also had a negative effect on populations. According to Palmer (1976), unrestricted hunting from about the 1850s to early this century is the primary cause of its local extinction and general scarcity. Nonetheless, a recovery seems to have been under way since the 1930s, partly because of nest box programs. The species is hunted in Québec, and between 1988 and 1992, Hooded Mergansers accounted for 1.3–3.8% of bags reported (Appendix II), which corresponds to an average annual harvest of 9810 individuals.

Jacques Bouvier and Simon Barrette

Common Merganser

Grand Bec-scie

Grand Harle

Mergus merganser

The Common Merganser is not only the largest of the three mergansers that breed in Québec, but also one of the largest ducks in North America. Much of the interest in this species stems from fears that it may feed extensively on young trout and salmon, species highly prized by anglers. It is common and widespread in Québec. Although a portion of the Québec population overwinters here, most winter farther south, in the United States or Mexico.

Habitat and behaviour

Common Mergansers are one of the earliest spring migrants among the waterfowl, following the edge of the thaw north in late winter, but sometimes retreating during cold spells (Palmer 1976).

Pairs usually form in winter and during spring migration (Cramp and Simmons 1977). In this monogamous species, the males and females are often seen together in the weeks immediately preceding laying. Common Mergansers generally breed in forested areas near fresh water, frequenting the wooded shores of lakes, streams, rivers, inlets, and beaver ponds, but avoiding isolated mountain lakes. This species needs clear water and small fish for its ducklings.

Females may search for nest sites together, and in Switzerland one tree contained 4 occupied nest cavities (Palmer 1976). Nest sites are usually at the water's edge, but may be up to 1 km away (Cramp and Simmons 1977). A natural cavity or an old woodpecker hole is generally used; Peck and James (1983) reported nests up to 17 m high in Ontario. Neither tree species nor cavity height is important (Palmer 1976). The same cavity may be used for several years in a row (Bellrose 1976). Nests are sometimes in crevices, on the ledges of rocky cliffs, in abandoned buildings, or even on the ground (Peck and James 1983). When not enough nest sites are available, several females may lay in a "dump nest," which is then abandoned; dump nests may contain up to 36 eggs (Palmer 1976). Females may also lay eggs in the nest of another female Common Merganser, behaviour known as intraspecific nest parasitism (Young and Titman 1988).

The typical clutch of 8–12 eggs (Peck and James 1983; Godfrey 1986) is incubated by the female upon completion of the clutch (Cramp and Simmons 1977). The young hatch in 28–35 days (Palmer 1976; Godfrey 1986) and stay in the nest for only a day or two before jumping down to the ground from the nest (Bent 1923; Bellrose 1976; Palmer 1976). Over the next few weeks, the broods move slowly downstream until they reach a larger expanse of water such as a lake or estuary. By the time they reach these wider waters, which may be as far as 80 km from their birthplace, the young can fly (Erskine 1972d; Bellrose 1976; Palmer 1976). During the journey downstream, broods often merge, probably as a consequence of aggression among females; in the resulting confusion, the two broods may become muddled, leading to a partial exchange of young. Sometimes, a clutch may be completely absorbed by the other, leaving one female bereft of all her offspring. Brood amalgamation is facilitated by the weak female-duckling bond and by incomplete imprinting of the young on their siblings, especially in larger broods (Gottlieb 1963, 1968; Hess 1973; Savard 1987b).

René Roy

Common Mergansers often nest in cavities in trees along the shores of lakes and rivers. East of the Chute-des-Passes Dam, north of Chicoutimi. *May 1991, André Bourget.*

Notes on territory

Common Mergansers do not seem to be territorial (Palmer 1976). They nest singly or in loose groups (del Hoyo et al. 1992), and sometimes three or four females may occupy cavities in the same tree (Cramp and Simmons 1977). During the nesting period, a pair may use a 3–5 km stretch of river which overlaps areas used by adjacent pairs, though the portion used most intensively, which contains loafing and roosting rocks and is probably near the nest, is rarely encroached on (Palmer 1976). Nests were reported to be about 5 km apart along a river in California (Foreman 1976).

Aerial surveys of southern Québec yielded estimated pair densities of 6.8 pairs/100 km^2 in 1990, 8.5 pairs/100 km^2 in 1991, and

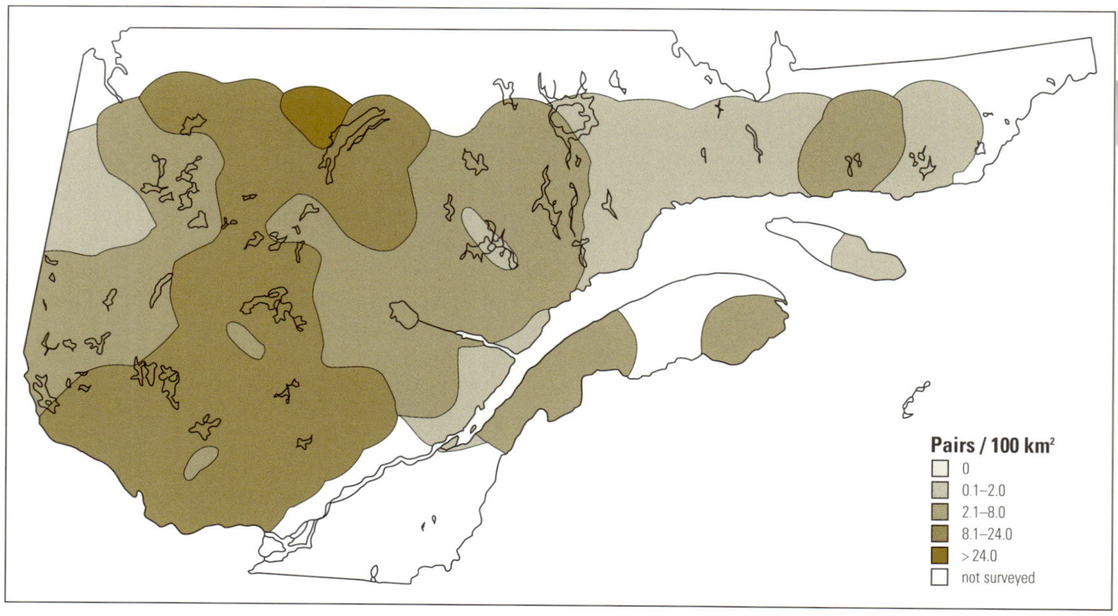

Abundance of Common Mergansers in southern Québec. Aerial surveys conducted in May 1990, 1991, and 1992 *(D. Bordage, CWS, unpubl. data).*

Common Merganser

Breeding distribution

1984–1989

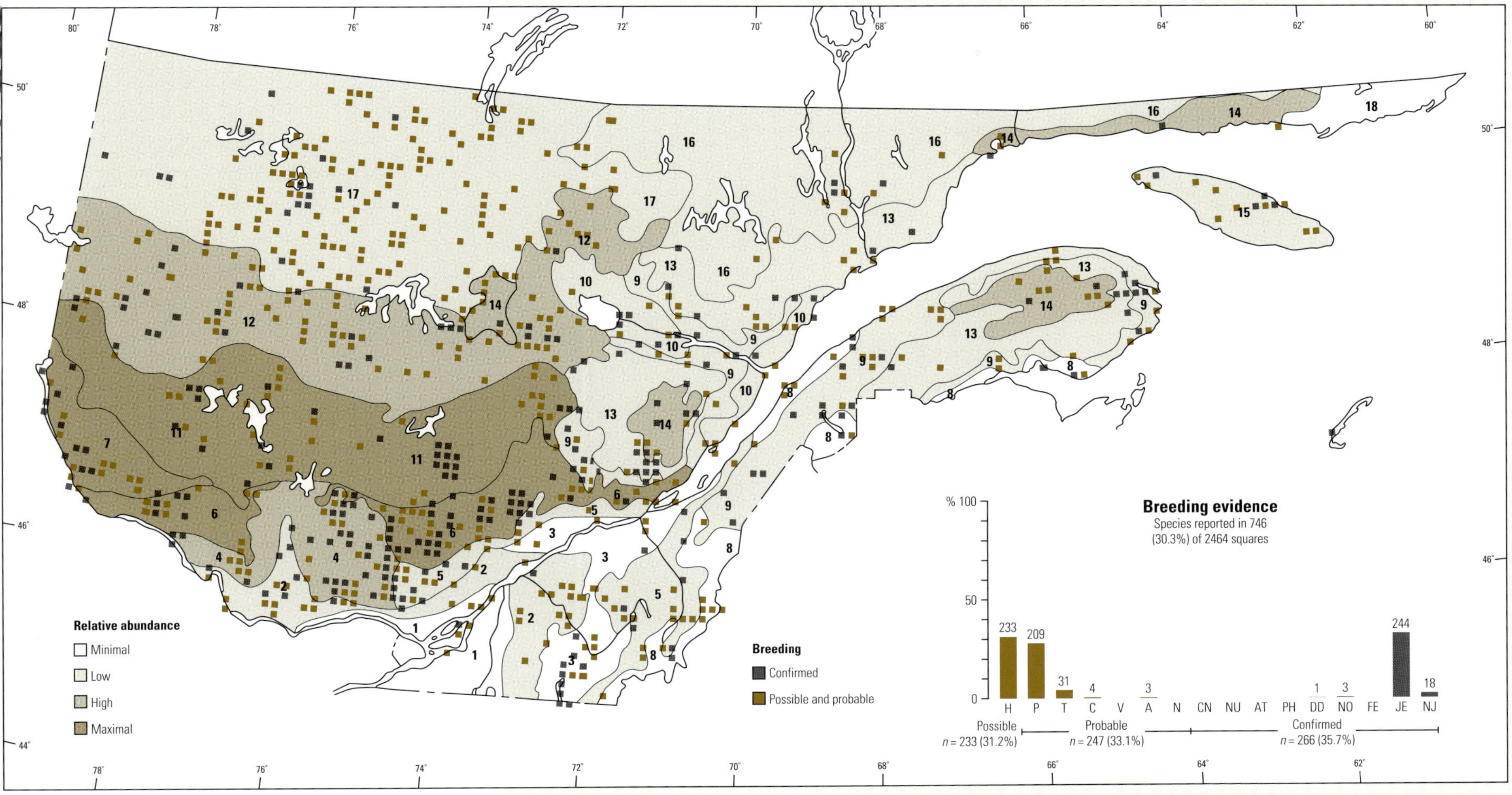

Species profile

Clutch size	usually 8–12 eggs (6–17), may use dump nest[4,6,7,11]
Laying interval	1 egg per 1–2 days[3,12]
Start of incubation	with last egg[2]
Incubation period	28–35 days[2,3,4,6,10,12]
Maturity at hatching	precocial
Care of young	by female[2,3]
Nestling period	1–2 days[3,5]
Age at first flight	60–70 days[1,2,3,5,10]
Dependence of young	until ≈ first flight[2]
Broods per year	1[2]
Breeding age	2 years[2,3,6]
Mating system	monogamy[2]
Duration of pair bond	1 season[2]
Total length	53.0–68.5 cm[7]
Wing span	82.0–99.1 cm[1,2]
Weight	M 1709 g (1528–2054) / F 1232 g (1050–1362)[9]
Longevity record	13 years 5 months[8]

References:

1. Terres 1980
2. Cramp and Simmons 1977
3. Palmer 1976
4. Peck and James 1983
5. Bellrose 1976
6. del Hoyo et al. 1992
7. Godfrey 1986
8. Clapp et al. 1982
9. Dunning 1984
10. Erskine 1972d
11. Bent 1923
12. Ehrlich et al. 1988

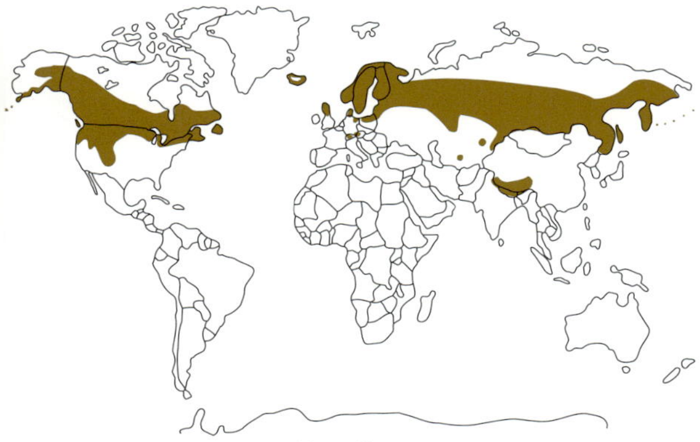

World breeding range

Chronology $n = 242$

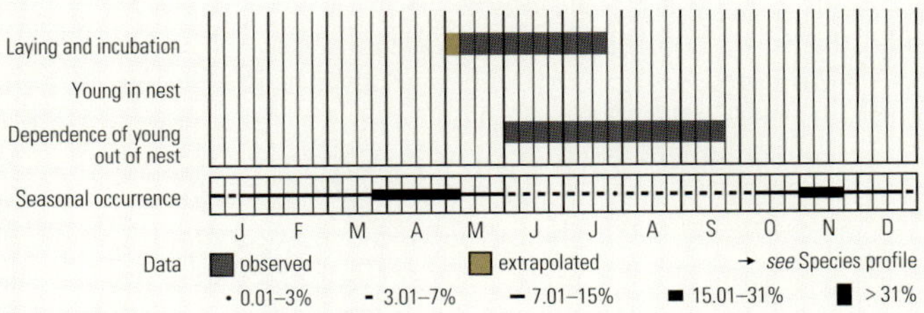

Frequency of occurrence by ecological region

Ecological regions** (forest regions)	Adequately covered squares	Squares in which reported	% of squares in which reported	Relative abundance (%)*
1. Sugar maple-hickory and sugar maple-basswood	97	6	6.2 ± 2.7	8.0
2. Sugar maple-basswood	153	37	24.2 ± 3.6	31.4
3. Sugar maple-basswood and sugar maple-yellow birch	101	20	19.8 ± 4.8	25.7
4. Sugar maple-yellow birch-basswood	56	38	67.9 ± 10.7	88.2
5. Sugar maple-yellow birch-beech and sugar maple-yellow birch-basswood	113	36	31.9 ± 4.8	41.4
6. Sugar maple-yellow birch-beech	71	51	71.8 ± 9.6	93.4
7. Sugar maple-yellow birch-balsam fir	19	14	73.7 ± 21.0	95.8
8. Sugar maple-yellow birch and balsam fir-yellow birch	47	9	19.1 ± 10.3	24.9
9. Balsam fir-yellow birch	87	36	41.4 ± 9.5	53.8
10. Balsam fir-white birch or balsam fir-red maple	46	12	26.1 ± 11.1	33.9
11. Yellow birch-balsam fir	52	40	76.9 ± 11.9	100.0
12. Balsam fir-white birch (west)	59	31	52.5 ± 13.1	68.3
13. Balsam fir-white birch (east)	76	28	36.8 ± 10.6	47.9
14. Balsam fir-black spruce	28	14	50.0 ± 19.3	65.0
15. White spruce-balsam fir and balsam fir-white spruce	22	10	45.5 ± 21.4	59.1
16. Black spruce-balsam fir-moss	18	6	33.3 ± 24.8	43.3
17. Black spruce-moss	32	10	31.3 ± 17.6	40.6
18.			*not analysed*	

* minimal: 0–29.9%; low: 30–59.9%; high: 60–89.9%; maximal: 90–100%

** Deciduous forest: 1–7; mixed forest: 8–11; coniferous forest: 12–17

Notes on territory (continued)

6.6 pairs/100 km² in 1992 (D. Bordage, CWS, unpubl. data); the highest densities (over 24 pairs/100 km²) occur northwest of Lake Mistassini (D. Bordage and M. Grenier, CWS, unpubl. data).

This species seeks out clear rivers rich in aquatic life, where it can find the fish, crustaceans, frogs, and insects needed to sustain its young. Matapédia River, Gaspé. *August 1991, Yves Aubry.*

The hen alone cares for the young, the drake having abandoned her during incubation. She sometimes carries a few ducklings on her back. Preflight young often "run" along the water to escape danger, but if they are startled at close range, they may hide under a bank or even in vegetation ashore while the hen dives or flaps away along the surface to distract the enemy's attention (Palmer 1976). The young remain dependent until they are able to fly (60–70 days), although some hens abandon their ducklings before they fledge and go off to moult (Cramp and Simmons 1977). The downy young can actually survive without parental care (Bellrose 1976).

Common Mergansers are generally fish eaters, but also consume insects, crustaceans, annelids, molluscs, frogs, small mammals, birds, and plants (del Hoyo et al. 1992). Common Mergansers forage with the head submerged, either catching their prey directly or diving after it (Sjöberg 1988). Their dives usually last no more than 30 seconds (Cramp and Simmons 1977). They prefer to feed in water less than 4 m deep. Chicks eat mainly aquatic insects at first, switching gradually to a diet of mostly fish after 10–12 days, and fish is the staple of their diet thereafter (Cramp and Simmons 1977). Young mergansers take most of their prey directly from the surface (Foreman 1976).

Distribution

In Canada, Common Mergansers breed in all 10 provinces, the Yukon Territory, and the Mackenzie District of the Northwest Territories. In the United States, they breed in Alaska, the Northeast, the Great Lakes, and in the West as far south as New Mexico and California (National Geographic Society 1987).

Common Mergansers were reported in 30.3% (746/2464) of Atlas squares, and breeding was confirmed in all parts of southern Québec, including Anticosti Island and the Magdalen Islands. The Common Merganser tends to be more common in the western half of southern Québec, according to both the Atlas data and the CWS surveys in May 1990, 1991, and 1992 (D. Bordage and M. Grenier, CWS, unpubl. data). Atlas data also show that the species is less common in the more densely populated areas of southernmost Québec.

Aerial surveys north of the 50th parallel also show mergansers (three species combined) to be more common in the western part of the province (CWS and MLCP 1986). Bordage (CWS, unpubl. data) found the highest densities in the Lake Mistassini area. Common Mergansers have also been confirmed as breeding in the watersheds east of James Bay (Tardif and Morneau 1991; Cons. G. G. – GREBE 1992a) and southeast of Hudson Bay, as far north as the Nastapoka River or roughly the 57th parallel (Cons. G. G. – GREBE 1990a).

As Bellrose (1976) points out, in the Northern Hemisphere, Common Mergansers generally breed farther south than, but not as far north as, Red-breasted Mergansers. The Common Merganser is, nevertheless, not an abundant breeder in southern Québec, and according to Peck and James (1983), it rarely breeds in southern Ontario.

Common Mergansers winter throughout their breeding range wherever open water persists (Norse and Fichtel 1985a). In Québec, the species occurs regularly in winter in the southern part of the province as far north as Lake Saint-Jean (David 1980a; Godfrey 1986), but less often in the Laurentians, and the Témiscamingue and Gaspé regions (David 1980a). It is common on the St. Lawrence River (Girard 1988; Bannon 1991b)—especially in the Montréal area (Reed and Bourget 1977), in particular on Lake Saint-François (Lehoux et al. 1985)—and on the Ottawa River (Club des ornithologues de l'Outaouais 1985). Some individuals, however, winter away from these large waterways, in the Eastern Townships, for example (Boily et al. 1988). An estimated 5000–8000 individuals of all three merganser species can be found on the St. Lawrence in winter and in migration (Lehoux et al. 1985).

History and trends

Early this century, Macoun and Macoun (1909) described the Common Merganser as common in summer in Québec, but according to Ouellet (1974a), the population has declined since then. The species has long been accused of consuming large quantities of fish from lakes and rivers where sport fishing is practised, and the resulting human persecution has driven it from some parts of Québec. The main cause of the species' disappearance in many parts of its breeding grounds in southern Québec, however, has been the clearing of forests for agriculture. In New Brunswick, the use of DDT to control spruce budworm, together with human persecution, decimated populations between 1950 and 1979 (Erskine 1992). According to Palmer (1976), in North America, the Common Merganser has been eliminated from some of its southern breeding range by logging and cottage development, with the consequent destruction of nest trees and shoreline vegetation, and by disturbance from power boats; these factors have probably also affected populations in Québec.

The logging of shorelines in potential breeding habitat prevents Common Mergansers from nesting and makes it difficult for them to maintain population levels. Near Lake Chibougamau. *September 1986, Jean Beaubien and Guy Simard, Canadian Forest Service.*

Some Common Mergansers winter on ice-free stretches of water in southern Québec. Migrants start returning to their breeding grounds in late March. Pairing usually takes place on the wintering grounds, but since there are more males than females, some drakes fail to find a mate. Salaberry-de-Valleyfield, southwest of Montréal. *March 1973, Léo-Guy de Repentigny.*

Bellrose (1976) estimated the North American population to be around 641 000 individuals, with twice as many males as females. During the 1980s, an estimated 163 000 mergansers (all three species combined) bred in Québec (CWS and MLCP 1986).

As the flesh of this species is not very palatable, the Common Merganser is not much sought after by hunters. From 1988 to 1992, an average of 8930 birds were shot annually, representing about 2% of the total annual duck harvest in the province (Appendix II).

Recent BBS data show no significant downward trend in the Common Merganser population in Québec between 1980 and 1989. Nonetheless, the acidification of lakes and rivers due to acid rain may alter the availability of food resources, thus forcing the species to abandon certain waters and ultimately affecting its distribution (DesGranges 1989). The increase in surface waters resulting from the creation of artificial reservoirs has provided the species with new winter habitat (Bellrose 1976), and some 5000 individuals winter near the Cornwall dam in Ontario, not far from the Québec border (Lehoux et al. 1985). In the James Bay and Hudson Bay watersheds, large hydro-electric projects may extend the autumn staging period by providing large reservoirs of open water that remain ice-free later into the season.

Robert Alvo

Red-breasted Merganser

Bec-scie à poitrine rousse

Harle huppé

Mergus serrator

René Roy

THE RED-BREASTED MERGANSER'S LATIN NAME *SERRATOR* REFERS TO THE backward-pointing serrations on the cutting edges of its bill, an adaptation that enables this fish-eating diving duck to hold on to its prey. It is one of the fastest fliers of the ducks, having been clocked at speeds up to 130 km/hour (Terres 1980:357). Though it is less well adapted for walking, it comes ashore more often and nests farther from the water than most other ducks (Cramp and Simmons 1977).

This species is more closely associated with marine habitats at all times of year than the Common Merganser. Nevertheless, it breeds in freshwater environments throughout Québec, though in much smaller numbers than the Common Merganser. Most individuals overwintering in Québec are found on the ice-free waters of the estuary and Gulf of St. Lawrence and Chaleur Bay (Reed and Bourget 1977; David 1980a; Godfrey 1986; Savard 1990).

Elsewhere in North America, the species winters chiefly along the Atlantic, Pacific, and Gulf of Mexico coasts. Winter populations are also found in the mountain states of the western United States, on the Great Lakes and on larger bodies of water between the Great Lakes and the Gulf of Mexico (Bellrose 1976).

Habitat and behaviour

Red-breasted Mergansers breed along bays, lagoons, estuaries, lakes, and rivers. According to a Norwegian study, they generally use lakes smaller than 15 ha only if these are part of a river system connecting them to other lakes (Råd 1980).

Pair formation usually takes place before the birds arrive on their breeding grounds (Bellrose 1976; Palmer 1976). Nesting does not begin until several weeks after their arrival. This is the only period during the breeding season when males and females are seen together, because males desert the females soon after the onset of incubation to moult (Palmer 1976).

Unlike Common and Hooded mergansers, the Red-breasted nests on the ground. A wide variety of habitats are used: marshes, small rocky or wooded islands, and secluded sections of shoreline (Palmer 1976; Peck and James 1983). Nests are never very far from water (Cramp and Simmons 1977), generally within 23 m (Townsend 1923c), although some have been found as far as 70 m away in Ontario (Peck and James 1983). Nests are not easy to find, often being hidden under low branches (fir, spruce, or willow) or fallen logs, between and under boulders, in cavities, or among low shoreline vegetation (Townsend 1923c; Bellrose 1976; Palmer 1976). In Denmark, nests are often built under driftwood washed ashore during winter, thus remaining protected and dry though only a few metres from the water (Kortegaard 1968). The female sometimes has one or more paths, or even a tunnel, through the plant growth from the nest to the water (Palmer 1976).

This duck nests in a variety of habitats, but the ideal breeding habitat is a small island with low woody growth or other low overhead shelter such as conifer limbs and nearby gravel bars or rocks that are almost completely submerged, where the male has his station early in the season and the brood later rests and preens (Palmer 1976).

Nests are usually made of grasses, leaves, and other plant materials taken from the immediate vicinity of the nest, and are lined with feathers and down, the quantity increasing as laying progresses (Peck and James 1983). This species may nest among Common Eiders or other ducks, or in gull or tern colonies (Peck and James 1983; Young 1985; Erskine 1992). In North America, typical clutch size is 8 eggs, which are incubated by the female for about 30 days (Bellrose 1976; Cramp and Simmons 1977). Nests with more than 14 eggs are probably the product of more than one female (Palmer 1976; Cramp and Simmons 1977). Red-breasted Mergansers sometimes parasitize other duck species (Pelzl 1971).

Notes on territory

Red-breasted Mergansers form loose colonies in areas of high density, with many pairs sometimes nesting on the same island. Though not truly territorial, they may defend a small area around the nest. Red-breasteds are gregarious, often gathering at the same place when feeding or when not engaged in incubation duties. In some areas, they roost communally at night, on an island (Nilsson 1965), the shore, or a sandbank. In others, they collect at a gathering place in the evening, but at night fly off to sea to roost, returning at dawn (Cramp and Simmons 1977).

Spring aerial surveys conducted south of the 51st parallel by the Canadian Wildlife Service and the Québec Ministry of Recreation, Fish and Game between 1985 and 1992 yielded mean densities of 0.02–0.43 pair/100 km² (Bordage 1988b; D. Bordage, pers. comm.).

In southern Québec, this diving duck breeds in coastal areas. The nest is built on the ground, well hidden in herbaceous vegetation or dense thickets. Watshishou Migratory Bird Sanctuary, west of Natashquan, Middle North Shore. *June 1989, Sylvain Paradis.*

Red-breasted Merganser

Breeding distribution

1984–1989

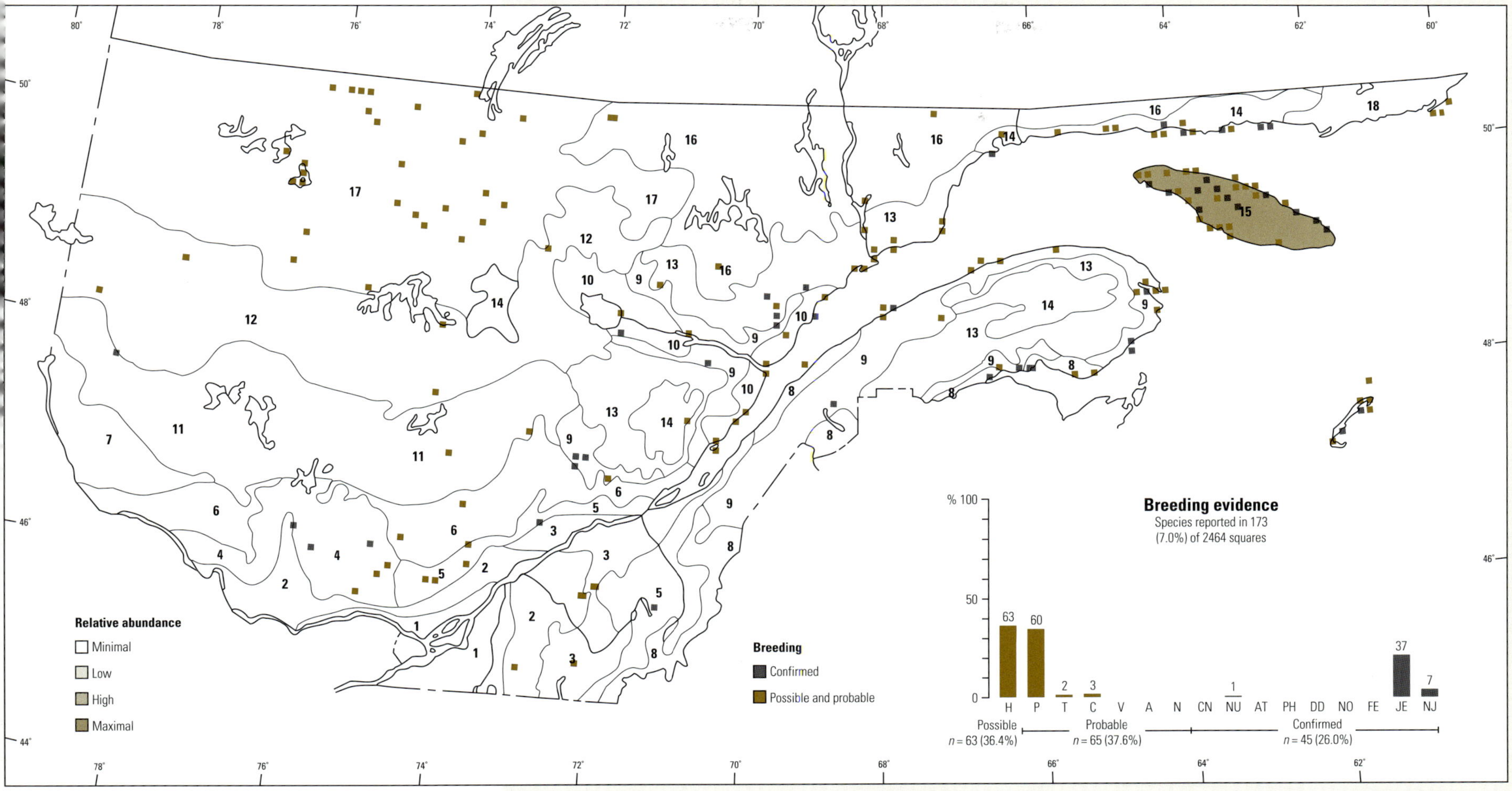

World breeding range

Species profile

Clutch size	usually 7–10 eggs (5–14), more than 14 = dump nest[5,7,9]
Laying interval	1 egg per 2 days[2,5,9]
Start of incubation	with last egg[4,5,9]
Incubation period	usually 31–32 days (28–35)[1,2,5,8,9]
Maturity at hatching	precocial
Care of young	by female[5,6,9]
Nestling period	less than 24 hours[9]
Age at first flight	59–69 days[3,4,5,6,7,8,9]
Dependence of young	until ≈ 50 days old[5]
Broods per year	1[4,5,6]
Breeding age	2 years[7,8]
Mating system	monogamy, rarely polygyny or polyandry[5,9]
Duration of pair bond	1 season[5,9]
Total length	51–64 cm[1]
Wing span	79–89 cm[3]
Weight	M 708 g (635–820) / F 689 g (590–772)[7]
Longevity record	9 years 4 months[5]

References:
1. Godfrey 1986
2. Kortegaard 1968
3. Terres 1980
4. Harrison 1978
5. Cramp and Simmons 1977
6. Ehrlich et al. 1988
7. Bellrose 1976
8. del Hoyo et al. 1992
9. Palmer 1976

Chronology $n = 152$

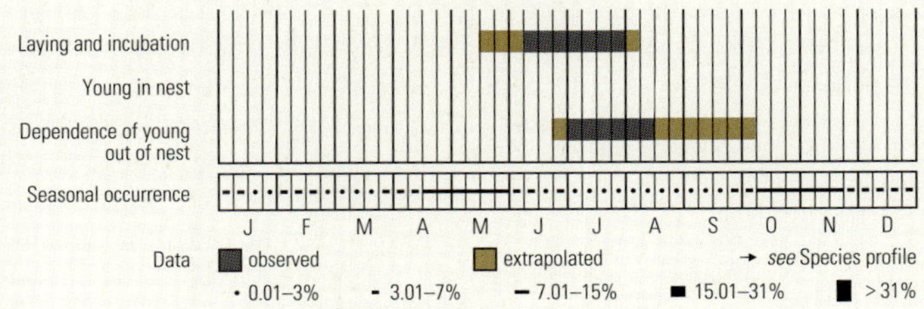

Frequency of occurrence by ecological region

Ecological regions** (forest regions)	Adequately covered squares	Squares in which reported	% of squares in which reported	Relative abundance (%)*
1. Sugar maple-hickory and sugar maple-basswood	97	0	0	0
2. Sugar maple-basswood	153	4	2.6 ± 1.5	3.8
3. Sugar maple-basswood and sugar maple-yellow birch	101	1	1.0 ± 1.6	1.5
4. Sugar maple-yellow birch-basswood	56	4	7.1 ± 6.3	10.5
5. Sugar maple-yellow birch-beech and sugar maple-yellow birch-basswood	113	4	3.5 ± 2.2	5.2
6. Sugar maple-yellow birch-beech	71	4	5.6 ± 5.3	8.3
7. Sugar maple-yellow birch-balsam fir	19	0	0	0
8. Sugar maple-yellow birch and balsam fir-yellow birch	47	2	4.3 ± 5.8	6.2
9. Balsam fir-yellow birch	87	15	17.2 ± 7.4	25.3
10. Balsam fir-white birch or balsam fir-red maple	46	7	15.2 ± 9.3	22.3
11. Yellow birch-balsam fir	52	1	1.9 ± 4.5	2.8
12. Balsam fir-white birch (west)	59	1	1.7 ± 4.0	2.5
13. Balsam fir-white birch (east)	76	11	14.5 ± 7.9	21.2
14. Balsam fir-black spruce	28	4	14.3 ± 14.1	21.0
15. White spruce-balsam fir and balsam fir-white spruce	22	15	68.2 ± 20.2	100.0
16. Black spruce-balsam fir-moss	18	1	5.6 ± 13.5	8.1
17. Black spruce-moss	32	1	3.1 ± 7.6	4.6
18.			*not analysed*	

* minimal: 0–29.9%; low: 30–59.9%; high: 60–89.9%; maximal: 90–100%

** Deciduous forest: 1–7; mixed forest: 8–11; coniferous forest: 12–17

Sitting females wait until intruders are about 4 m away before flushing, either flying up or escaping through the undergrowth (Kortegaard 1968). Hatching is synchronous, and the young remain in the nest less than 24 hours, afterward following the female closely, some even climbing on her back (Palmer 1976). The female broods the ducklings at night when they are small (Cramp and Simmons 1977) and protects them from gulls, though some may fall prey to pike (Palmer 1976; Braun et al. 1980). When a predator approaches, the ducklings do not dive but skitter away as fast as they can, often led by the female toward deep water or to a sheltered spot along the shoreline; only when danger is imminent do the young attempt to escape by diving or crouching motionless on the shore (Råd 1980). It is not uncommon for several broods to merge and for one female to have custody of several dozen young, not necessarily of the same age (Palmer 1976; Cramp and Simmons 1977).

The Red-breasted Merganser is a diving duck that feeds primarily on various species of fish, which it locates by immersing its head partly or totally underwater. Its prey are generally less than 10 cm long. In addition to fish, this merganser takes crustaceans (shrimp, crayfish), molluscs, worms, insects, and plant matter (leaves, roots, and seeds) (Palmer 1976; Cramp and Simmons 1977). When very young, the ducklings feed mainly on aquatic insects (Palmer 1976), but very soon fish become central in the diet (Råd 1980). In western Norway, the species' presence during the breeding season is strongly correlated with that of the threespine stickleback, the only small fish species that is easily accessible to ducklings in shallow water (Råd 1980). There is a recognized predator-prey relationship between some fish-eating ducks and the threespine stickleback in North America (Scott and Crossman 1973), and broods of Red-breasted Mergansers have been observed on lakes along the northeast coast of James Bay, where various stickleback species are known to occur (R. Benoit, pers. comm.).

The Red-breasted Merganser was reported in 7.0% (173/2464) of Atlas squares surveyed, and breeding was confirmed in 26.0% (45/173) of these, on the basis of records of young off the nest (37/45) and reports of nests (8/45). However, Red-breasted Mergansers can be confused with Common Mergansers, particularly when young, females, or males in eclipse plumage are involved (Kaufman 1990b).

Distribution

The Red-breasted Merganser has a holarctic distribution (Palmer 1976; Cramp and Simmons 1977). In North America, it breeds throughout Canada, except for the western Prairies and most of British Columbia. It also breeds in the Great Lakes states and northern New England, and in small isolated populations in Montana and Wyoming (Bellrose 1976).

This shore-dwelling merganser also turns up in tern colonies, like this one at Carleton, in the Gaspé. *June 1991, Pierre Brousseau.*

Although Red-breasted Mergansers are more often associated with coastal habitats than Common Mergansers, Red-breasteds breed on both coastal and inland waters in southern Québec. According to Harper (1958) and Todd (1963), the species' breeding range is not limited to coastal areas, but covers all of Québec. South of the St. Lawrence River and west of Québec City, however, the species is considered accidental; during atlassing, breeding was confirmed around Lakes Saint-François and Aylmer in the southern reaches of the Chaudière–Appalaches region. Anticosti Island was the source of the largest number of sightings. Aerial surveys carried out by the Canadian Wildlife Service in 1985–92 in co-operation with the Québec Ministry of Recreation, Fish and Game revealed that the species was present in the northern part of southern Québec and along the gulf and estuary (D. Bordage, CWS, unpubl. data).

North of the Atlas area, the Red-breasted Merganser has been reported along the coasts of James Bay, Hudson Bay, and Ungava Bay, and inland on lakes in various watersheds all the way up to the northwestern tip of the Ungava Peninsula (Cons. G. G. – GREBE 1990a; Tardif and Morneau 1991; Benoit et al. 1993b, 1994; D. Bordage and B. Harvey, unpubl. data). Gaston et al. (1985) found a nest in Digges Sound on Pitsulak City Island (Northwest Territories), near the village of Ivujivik.

In Québec, as in Ontario (Peck and James 1983), the Red-breasted Merganser appears to be much less common than the Common Merganser. North of the boreal forest, however, the Red-breasted tends to replace the Common Merganser as latitude increases (Erskine 1977).

In spring, 70% of the migratory merganser population can be found in the St. Lawrence estuary and along the Gaspé coast (Lehoux et al. 1985).

A few Red-breasted Mergansers nest along large inland lakes or on islands in the lakes. East of Lake Mistassini. *May 1991, André Bourget.*

History and trends

After the arrival of Europeans in North America, the Red-breasted Merganser population went into a decline as a result of habitat loss, predation by humans and their domestic animals, and persecution carried out in the name of preserving fish stocks (Erskine 1992).

According to a number of commentators, both in the last century and our own, the species was once very common on Anticosti Island and the Magdalen Islands (Verrill 1862; Townsend 1923c; Ouellet 1969). The population on the Magdalens has probably declined in the course of the 20th century (Fradette 1992). In the Montréal area, the Red-breasted Merganser was considered a common "transient visitant" in the 19th century (Wintle 1896; Ouellet 1974a), and Macoun and Macoun (1909) reported that the species frequently bred in the Thousand Islands section of the St. Lawrence just west of Québec, though only a few breeding pairs were left by the early 20th century.

Current threats to this species include water pollution and habitat loss on its breeding, staging, and wintering grounds (Ouellet 1974a).

About 1.5 million mergansers (all three species combined) breed in North America (USFWS and CWS 1986), including about 163 000 in Québec (CWS and MLCP 1986).

The number of Red-breasted Mergansers taken by Québec sport hunters has not varied much since 1973, generally ranging between 3000 and 4000 annually. Since the Red-breasted inhabits relatively remote areas, which have far fewer hunters than the St. Lawrence Valley, it appears only half as often as the Common Merganser in the Canadian Wildlife Service's bag surveys, and since it is not considered good eating, the species accounts for barely 1% of all ducks taken annually in Québec (Appendix II).

ROBERT ALVO AND ANDRÉ BOURGET

Flock of Red-breasted Mergansers, mainly males, on Étang de l'Est, Magdalen Islands. The males leave their mates shortly after the start of incubation to moult. *July 1982, Robert Michaud.*

Ruddy Duck

Canard roux

Érismature rousse

Oxyura jamaicensis

Luc Leclerc

THIS LITTLE DUCK, WHICH OFTEN COCKS ITS TAIL UP LIKE A WREN, HAS only appeared recently in Québec and is still rare and very local here, although it is abundant on the Prairies. The male in breeding plumage is instantly recognizable with his white cheek patches and blue bill, while the brownish female can be identified by the dark line across her light cheeks. Ruddy Ducks seem to be able to take flight only with a great deal of effort, pattering over the surface of the water for a long distance before they are airborne. Once aloft, they can be identified by the rapid beating of their short wings. Like grebes, these ducks can slowly sink below the surface without a ripple.

The Ruddy Duck is a migrant breeder in Québec, arriving on its breeding grounds here between mid-April and mid-May and leaving again in fall, between mid-September and mid-October. It overwinters in southern British Columbia south along the Pacific coast to California and along the Atlantic seaboard from Massachusetts down to Florida, as well as in the Gulf states, Mexico, and Central America (AOU 1983).

Habitat and behaviour

In summer, Ruddy Ducks are found on shallow freshwater marshes, lakes, and ponds with margins of emergent vegetation, fairly stable water levels, and areas of open water for displaying and feeding (Johnsgard 1975b; Palmer 1976). In Québec, Ruddy Ducks can also be seen feeding on municipal sewage lagoons. During migration, Ruddies prefer larger expanses of fresh or brackish water (Palmer 1976).

The drake is often noisy during mating and laying. In one of the best-known displays, the bubble display, he suddenly and forcefully strikes his swollen throat with his bill, while flicking his tail back and forth and raising his head feathers so that they stand up like two little crests. He then stretches his neck forward to utter a two-syllabled *raa-anh* (Palmer 1976).

Pairs form in spring, usually near the nesting area. The duration of the pair bond varies; some drakes desert the females during incubation, while others have been seen accompanying their mates and broods (Oring 1964; Bellrose 1976; Siegfried 1977). The hen usually builds her nest in dense emergent vegetation—cattails, bulrushes, or reeds, for example—where the water is 25–30 cm deep. The nest is generally a floating platform anchored in the surrounding vegetation, but can also be on top of a mound of dense vegetation or even a muskrat house (Palmer 1976). When the opportunity arises, the hen uses canals through the vegetation made by muskrats to slip quietly to the nest (Johnsgard 1975b). Hens will occasionally lay their eggs in the nest of another Ruddy Duck or other species, such as a Mallard, Northern Pintail, Ring-necked Duck, or American Bittern (Joyner 1976; Siegfried 1976a; Peck and James 1983).

Although the hen may renest if her first clutch is lost, ordinarily she lays only one clutch per season (Bellrose 1976). The ducklings leave the nest within 12–24 hours of hatching, prompted by their mother's incessant calls (Siegfried 1977). The young then form into a compact group, and are often heard calling softly. When danger threatens, the female lifts her tail, exposing her white under-tail coverts, and leads the brood to safety. The white patch under the tail seems to be a signal that stimulates the ducklings to follow their

Notes on territory

According to a Utah study (Joyner 1969), pairs establish a territory once they have selected a nest site, and will then chase off all other Ruddies of both sexes, although drakes do become less territorial toward the end of the incubation period. An area with a radius of approximately 3 m is defended around the nest, and the distance

This sewage lagoon was one of the Ruddy Duck's first nesting sites in Québec. Baie-du-Febvre, south shore of Lake Saint-Pierre. *June 1991, Daniel Jauvin.*

Ruddy Duck

Breeding distribution

1984–1989

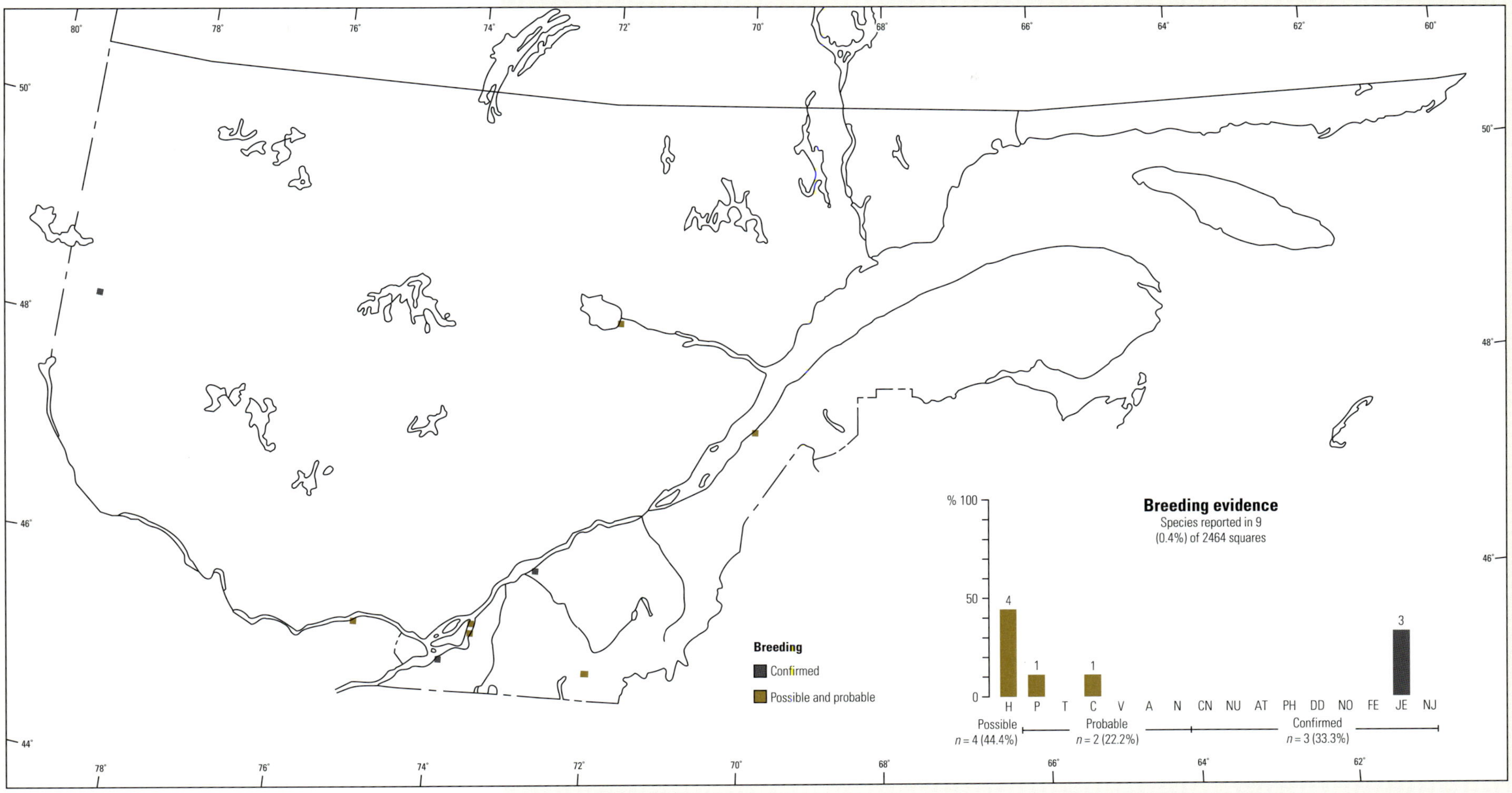

Species profile

Clutch size	usually 8 eggs (5–15), may use dump nest[1,2,3,4,5]
Laying interval	1 egg per day[1,2,5]
Start of incubation	with last egg[1,2]
Incubation period	23–26 days[1,2,3,4,7]
Maturity at hatching	precocial
Care of young	by female, sometimes by both female and male[11]
Nestling period	≈ 24 hours or less[2,10]
Age at first flight	6–7 weeks[1,2,4,5,6,10]
Dependence of young	until 4–5 weeks old[10,11]
Broods per year	1[2,3,4,5]
Breeding age	usually 2 years (1–2)[1,7,11]
Mating system	monogamy[2,4]
Duration of pair bond	1 season[2]
Total length	35.5–43.0 cm[9]
Wing span	53.3–61.0 cm[6]
Weight	M 544 g (272–635) / F 540 g (454–635)[1]
Longevity record	13 years 7 months[8]

References:
1. Bellrose 1976
2. Palmer 1976
3. Harrison 1978
4. Ehrlich et al. 1988
5. Siegfried 1976a
6. Terres 1980
7. del Hoyo et al. 1992
8. Clapp et al. 1982
9. Godfrey 1986
10. Siegfried 1977
11. Cramp and Simmons 1977

World breeding range

Chronology *n* = 9

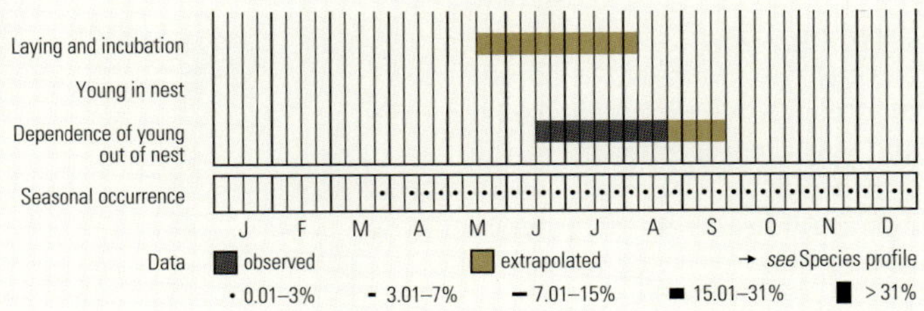

Notes on territory (continued)

between territories ranges from 7 to 90 m, although in favourable conditions territories could be less than 7 m apart. The pattern of emergent vegetation on the water seems to influence the spacing of territories (Joyner 1969).

The marshes created at Saint-Étienne-de-Beauharnois are teeming with invertebrates, which make up the bulk of the species' diet. South of Montréal. *July 1991, Pierre Bannon.*

mother (Siegfried 1977). Broods usually stay on the same body of water: most hatch, are raised, and fledge on the same pond. Ducklings are usually abandoned at around 4–5 weeks after hatching, before they can fly, and remain together at least until they fledge at about 7 weeks (Siegfried 1977). The drakes moult in midsummer on expanses of shallow water, alone or in small flocks (Palmer 1976).

According to a Manitoba study, Ruddy Ducks feed chiefly on animal food during the breeding season: the ducklings' basic fare is similar to the adults' diet, consisting mainly of larval and pupal midges found in the soft mud on the bottoms of ponds. The seeds of aquatic plants (bulrush, sedge, duckweed, dock) make up only a small percentage of the diet (Siegfried 1973). Another study conducted in North Dakota by Woodin and Swanson (1989) confirms the importance of chironomids, which were found to account for about three quarters of the Ruddy Duck's food during the breeding season. The authors believe that the species nests late so that laying will coincide with a flush of chironomids, guaranteeing the hen, which cannot build up large fat reserves because of her small size, an abundant source of food. Laying in Québec does not begin until at least mid-May.

Though rare in Québec, the Ruddy Duck is often found at popular birding hot spots, where it does not go unnoticed. The colourful drake is easy to identify, although the hen may be confused with the Black Scoter hen. No nest has been found in the province and the 3 confirmed Atlas records were all based on sightings of broods. Nests are well concealed and difficult to find, while a brood dabbling out in the middle of a pond in the company of an adult is easy to spot and identify.

This duck prefers ponds with dense emergent aquatic vegetation. The Ruddy Duck has been nesting on these impoundments on National Defence land for the last few years. Nicolet Migratory Bird Sanctuary, south shore of Lake Saint-Pierre. *June 1991, Daniel Jauvin.*

Ruddy Duck hen leading her brood on a pond at Baie-du-Febvre. South shore of Lake Saint-Pierre. *July 1981, Daniel Jauvin.*

Distribution

Ruddy Ducks were reported in 0.4% (9/2464) of Atlas squares. Five records come from the Central St. Lawrence Lowland and the Ottawa Valley, and the other 4 from the Eastern Townships, the Lower St. Lawrence, Lac-Saint-Jean, and Abitibi regions.

This duck has nested every year since 1981 at Baie-du-Febvre on Lake Saint-Pierre. In 1982, 2 broods were seen near the village settling pond, 1 in June and 1 in August. Since 1989, 1–3 broods have been sighted every year at the same pond or near other impoundments. In 1988, another brood was discovered on an impoundment at Saint-Étienne-de-Beauharnois, and another one on Lake Pelletier (Rouyn–Noranda) in the Abitibi region. In 1992, a hen and several ducklings were seen in the Lac Saint-François National Wildlife Area (A. Pelletier, pers. comm.). The species has also been spotted several times in summer at Saint-Gédéon, on Lake Saint-Jean, and on the sewage lagoon at Saint-Pascal, in the Lower St. Lawrence region.

Sewage lagoons are clearly important to the species in Québec. Their appeal is probably due to the fact that they are rich in plant life and aquatic invertebrates and are similar, both in size and other physical characteristics, to the species' breeding habitat in the West (Champagne and Surprenant 1981). It should also be noted that sewage lagoons are being watched increasingly closely by birders.

History and trends

The Ruddy Duck used to be confined to the western half of the continent. The clearing of land that accompanied European settlement in the East, as well as habitat loss due to the draining of wetlands and drought in the Prairies, probably caused it to expand its breeding range eastward. Although the species had been observed in Québec early in the century (Macoun and Macoun 1909), the first breeding record dates only from 18 September 1951, when an unfledged duckling was confiscated from a hunter on Lake Saint-Pierre (CMN, Specimen No. 39046). A pair was next seen in early June 1980, on a sewage lagoon in Napierville. More recently, the spread in the use of sewage ponds and lagoons by small municipalities has certainly encouraged the species to breed in several new locations in Québec.

The North American population has declined considerably since European settlement, primarily because of habitat loss and hunting (Palmer 1976). Since the species sometimes nests late, juveniles may not be able to fly by the time the hunting season opens, making them particularly vulnerable. Between 1984 and 1992, Ruddy Ducks were killed occasionally by sport hunters in Québec (Appendix II). In order to help the small breeding population become well established in Québec, restrictions on hunting Ruddies were imposed in the fall of 1993.

Daniel Jauvin

Turkey Vulture

Urubu à tête rouge

Cathartes aura

The first known nest site of the Turkey Vulture in Québec was spotted on a cliff next to deciduous trees on Rigaud Mountain in 1986. West of Montréal. *June 1991, Pierre Bannon.*

Ghislain Caron

Notes on territory

From his observation of a nest site on a hillside in the Ohio Valley, Coles (1938) had the impression that the Turkey Vulture was defending territory. Work and Wool (1942), however, observed several birds in adult plumage near nest sites without noting any sign of territorial behaviour. Although they are generally solitary nesters, Turkey Vultures may nest close together if there is a large food supply nearby.

THE TURKEY VULTURE HAS BEEN SEEN OVER QUÉBEC SINCE THE LATE 1970s. David (1980a) considered it a casual visitor at the beginning of the 1980s, but recently Chabot and Barrette (1989) have proposed that it be classed as an uncommon migrant breeder. In the southernmost part of the province, it is easily recognized at a distance on a sunny day by its black plumage and the open V formed by its long wings. At closer range, the adult's greyish primaries and secondaries and naked red head become visible.

Habitat and behaviour

When scanning for food, Turkey Vultures soar at differing altitudes over a variety of habitats. In Québec, they are found mainly in agricultural areas and adjoining deciduous and mixed forests. They seem to have a preference for valleys and cliffs, which create the thermals needed for their long soaring flights. In general, Turkey Vultures are found singly or in groups of two or three, but dozens of them also congregate at communal roosts in early morning or late afternoon, usually in the branches of large trees. These roosts may be used for only a few days or for an entire season; others are used for years at a time (Palmer 1988).

Several birds take part in the courtship display, which precedes pair formation. During this display, the birds perform a kind of dance, hopping on the ground in a clearing with wings open (Loftin and Tyson 1965; Palmer 1988).

After selecting the nest site, the pair roosts nearby, sometimes for several days, before egg laying begins. This behaviour can help observers locate nests (Jackson 1983).

The Turkey Vulture nests on rocky cliffs, in talus, or on small islands, in marshes, and in the underbrush of deciduous and mixed forests. It generally lays 2 eggs on the ground in a shaded clearing, in a crevice, cave, hollow tree or stump, under dense plant cover, and

In 1987, 2 chicks were found in a cavity in this boulder pile on Lafontaine Island. Ottawa River, northwest of Hull. *July 1987, Jacques Chabot.*

Turkey Vulture

Breeding distribution 1984–1989

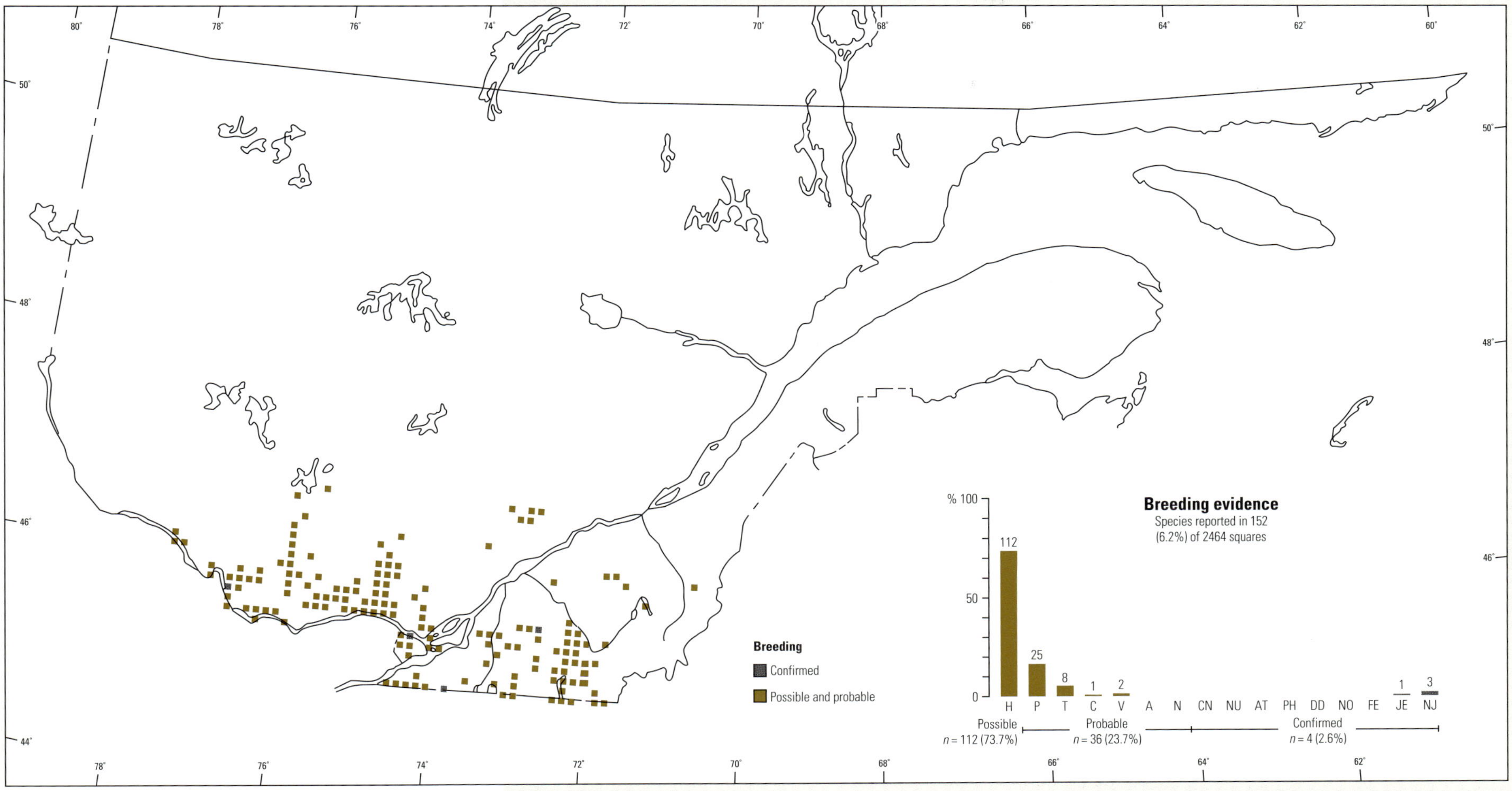

World breeding range

Species profile

Clutch size	2 eggs (1–4)[1,3,4,12]
Laying interval	1 egg per 1–3 days[2]
Start of incubation	?
Incubation period	28–41 days[5,6,7,12]
Maturity at hatching	semi-altricial
Care of young	by female and male[6,7]
Nestling period	at least 55 days[1]
Age at first flight	usually ≈ 9 weeks[2]
Dependence of young	probably until several months after first flight[12]
Broods per year	1[2]
Breeding age	probably several years[2]
Mating system	monogamy[6]
Duration of pair bond	probably for life[2]
Total length	67–81 cm[9]
Wing span	172.7–182.9 cm[8]
Weight	1467 g (?)[10]
Longevity record	16 years 10 months[11]

References:
1. Peck and James 1983
2. Palmer 1988
3. Buhnerkempe and Westemeier 1984
4. Schaefer 1978
5. Wilbur 1983
6. Ehrlich et al. 1988
7. Harrison 1978
8. Terres 1980
9. Godfrey 1986
10. Dunning 1984
11. Klimkiewicz and Futcher 1989
12. Jackson 1983

Chronology $n = 6$

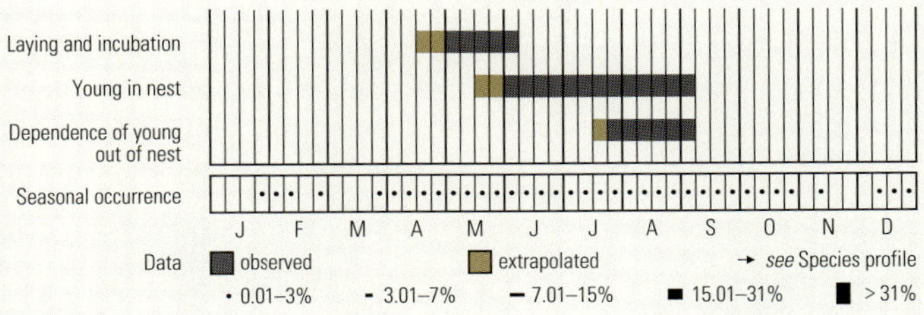

sometimes in old, abandoned buildings. It is not a colonial nester although as many as 12 nests/0.4 ha have been found in Alberta (Buhnerkempe and Westemeier 1984; Palmer 1988; Rabenold and Decker 1989).

Both parents incubate the eggs and tend the young, which they feed by regurgitation (Harrison 1978; F. Morneau, pers. comm.). Adult birds are not very aggressive toward intruders. The Turkey Vulture is faithful to its nest site and may use it for several years in a row if not disturbed (Palmer 1988).

Turkey Vultures have been visiting the Cap Tourmente National Wildlife Area for a number of years. They are often sighted near this cliff and in the eastern section of the wildlife area. Northeast of Québec City. *August 1991, Yves Aubry.*

Turkey Vulture eggs laid on the ground. The nest, sheltered by boulders, was found in a forest. Mont-Saint-Bruno Conservation Park, east of Montréal. *May 1990, François Morneau.*

The Turkey Vulture locates its food by sight or by its highly developed sense of smell. Birds have been captured in traps baited simply with carrion scent. Experiments have shown that the species can detect camouflaged food, so it is able to find food even in wooded areas (Houston 1986; Palmer 1988). A study in Ontario showed that its diet consists mainly of carcasses of small wild mammals, such as woodchucks and even shrews, and sometimes domestic animals (Prior 1990). It often feeds on dead animals found along roads and near farms. According to another Ontario study, roosts do not appear to be used as information centres for the location of food sources, contrary to what some researchers have reported (Prior and Weatherhead 1991a,b).

Sometimes Turkey Vultures will perch with their wings outspread in the sun, in the manner of cormorants, to dry their feathers and get warm. It is known that Turkey Vultures lower their body temperature a few degrees overnight, reducing energy consumption (Clark and Ohmart 1985).

Turkey Vultures can be seen and identified at a distance. It is difficult, however, to confirm breeding in a given area. Prolonged observation does not necessarily help in locating a nest, as the adult birds never carry food in their claws; nor do they carry nesting material, as they do not build nests. They are also secretive near their nest site (Coles 1938; Palmer 1988; F. Morneau, pers. comm.), and the nests are generally well concealed. Breeding was confirmed in only 4 squares and the "pairs" that were sighted in some squares may simply have been unmated birds drawn to the same site by a source of food (Coles 1938).

Notes on territory (continued)

According to Palmer (1988), the abundance of food in a region may affect the spacing of nest sites. Forbush and May (1939) estimated breeding density of Turkey Vultures in the southern part of their range to be 1 bird/km², whereas Stewart and Robbins (1958) estimated it to be 0.75 pair/km² in a Maryland habitat made up of forests, abandoned fields, and farmland.

Turkey Vulture chicks in the nest. Mount Saint-Bruno. *July 1990, François Morneau.*

Distribution

The Turkey Vulture is mainly found in the Ottawa Valley, Montérégie, and Eastern Townships; it has also been reported several times in the Mauricie region. Farmland edged by forests, as well as valleys, cliffs, and areas around main highways are suitable for the species, and the availability of these habitats no doubt explains its present distribution. The species is a newcomes to Québec, however. It colonized New England and Ontario between the 1920s and the 1960s (Bagg and Parker 1951, 1953; Martin, N. L. 1985; Cadman 1987c; Meade 1988f) before reaching southern Québec, where sight records have been accumulating since the late 1970s (Chabot and Barrette 1989). Increasingly regular sightings in the eastern part of

For several years, Turkey Vultures have been roosting on Mount Saint-Hilaire, where there are many taluses and crevices suitable for nesting. *July 1991, Pierre Bannon.*

the province indicate that it is continuing to spread. Sightings of 2 birds summering around a large talus mass of fallen rock in the Cap Tourmente National Wildlife Area in 1990 and 1991 are evidence of probable breeding that would extend the provincial range into the northeast.

The first report of breeding in Québec came from Rigaud Mountain in 1986. The nest was on a cliff beside a deciduous forest (G. Huot, pers. comm.). Two other sites were discovered in 1987, one on Lafontaine Island, on the Ottawa River in Pontiac County, and the other in Saint-Valérien-de-Milton, Shefford County. These 2 nests were found among large boulders in talus near a thinned-out mixed forest (Chabot and Barrette 1989). Finally, 2 flightless young were sighted in 1989 in a steep gorge in Covey Hill, Huntingdon County (P. Bannon, pers. comm.). After the Atlas period, another nest was found in 1990 in a moderately thinned deciduous forest at Mount Saint-Bruno (F. Morneau, pers. comm.). This nest produced 2 young in 1990 and 2 more in 1991. It was used again in 1992, but the eggs were destroyed, apparently after being disturbed.

The Turkey Vulture may not nest as extensively in Québec as is shown on the distribution map. The species is, however, easily spotted and can range up to 70 km from its nest site or roost (Coleman and Fraser 1989).

For some years now, large numbers of Turkey Vultures —sometimes upwards of 100—can be seen between July and September in Saint-Cléophas, near Joliette, being drawn by chicken carcasses discarded in fields by poultry breeders (Gingras 1993).

Potential nest sites for this cathartid are plentiful in the Mount Owl's Head area. Many birds are seen over the mountains and nearby farmlands. Southwest of Sherbrooke. *July 1964, Gaston Le Page.*

History and trends

Since the turn of the century, the Turkey Vulture's range has expanded remarkably into northeastern North America and its numbers have increased dramatically. The first Québec sighting of the bird was in Godbout in 1894 (Comeau 1909). Up until the early 1970s, sightings were sporadic, then increasingly frequent (Chabot and Barrette 1989). Atlas data show an expanded breeding range compared with that given by Godfrey (1986), who classed the bird as an uncommon visitor to the province; the range appears to extend all the way to the Maritimes, where several summer sightings have been reported, although breeding has not been confirmed (Erskine 1992).

Several factors have no doubt contributed to the spread of the Turkey Vulture. Among them are a warming trend in the northeastern North American climate, the clearing of land for agriculture, the large number of animals killed by traffic, and habitat deterioration in the southern part of this vulture's range, which has prompted birds to go farther north to breed (Bagg and Parker 1951, 1953; Wilbur 1983).

The increase in the Turkey Vulture population over the last few years and the northward expansion of its range indicate that the species will probably become more common and widespread in Québec.

SIMON BARRETTE

Osprey

Balbuzard

Balbuzard pêcheur
Pandion haliaetus

Ghislain Caron

Areas near large lakes and reservoirs are excellent breeding sites for the Osprey; these large water bodies abound with the fish on which it feeds. Grand Lac Victoria, near the Dozois Reservoir. *May 1991, Yves Aubry.*

Closely associated with water, the Osprey is nonetheless one of the most widespread species in the world. Except for polar regions where water remains frozen too much of the year, and a few remote oceanic islands, Ospreys are found on all continents, almost all coastlines, and along large lakes, rivers, and streams (Poole 1989). As a general rule, this migrant breeder stays in Québec from March to the end of November. Very occasionally isolated birds can be spotted in winter beside open water in the southern part of the province (Cyr and Larivée 1995; ÉPOQ).

The Osprey is a large, powerful raptor, the female birds weighing as much as 1.8 kg (Schaadt 1989). Its anatomy somewhat resembles that of a small eagle, but for the narrow wings which are sharply bent at the wrist in flight. Because of the unique structure of its feet and claws, it has been classified into a distinct subfamily, Pandioninae, of which it is the sole member. The Osprey is unlike other birds of prey in that its four toes are of equal length, terminating in long, sharp, curved talons. The outer toe is reversible, and the soles of the feet are covered with short, stiff spicules, which help in grasping slippery prey (Palmer 1988). The Osprey feeds on fish, which it captures live, hence its English colloquial name, fish hawk, and former French name *Aigle-pêcheur*. It is a spectacular sight when diving for prey. Its white breast gleaming against a brownish-black mantle, wings partially closed and talons thrust forward, the bird hits the water in a shower of spray to almost disappear below the surface; when it reappears a few seconds later, it usually has a fish, head pointed forward, clasped in its talons.

Habitat and behaviour

The opportunistic Osprey generally feeds on the most common fish species found in shallow waters (Prévost 1977; Poole 1989; DesGranges et al. 1994), and most of its prey are of little economic value. In Québec, in the Gulf of St. Lawrence and estuary, it takes tomcod and flounder, whereas in the rivers and freshwater lakes of the boreal forest, it takes mainly midsized pike, suckers, and walleye, and sometimes trout from aquaculture ponds (MLCP, unpubl. data). On rare occasions, blue and freshwater mussels have been found beneath Osprey nests, which suggests that molluscs may at times be consumed (Poole 1989; DesGranges et al. 1994). Ospreys have also been seen attacking young birds in a heronry near Montréal (M. Léveillé, pers. comm.), but the species generally lives peaceably alongside colonies of Great Blue Herons (pers. obs.).

The massive nest, generally about 1–1.5 m in diameter and 50–200 cm in height (Géroudet 1984; DesGranges et al. 1994), is perennial (Hagan and Walters 1990). Usually near water, it is built atop a tall tree, live or dead, that is stout enough to support the enormous mass of dry sticks that the male accumulates over the years (Géroudet 1984). He will also add all sorts of other materials, such as stakes, rope, strips of cloth, plastic, and even caribou antlers (Poole 1989; DesGranges et al. 1994). In Québec, the Osprey most often takes up residence in a healthy spruce or pine with its tip broken by the weight of snow and ice, presenting a natural nest platform. Where tall trees are scarce, as

A pair of Ospreys nesting atop a conifer on the Roggan River. James Bay region. *June 1991, Jean-Luc DesGranges.*

Notes on territory

The nest, or eyrie, is the focal point of the birds' social life, so important that when established on a solid rock pinnacle it may be used for over 100 years (Bent 1937; Thibault and Bouvet 1983). Breeding performance is often better in eyries that have a long history of use (Thibault and Patrimonio 1991).

Most pairs return to the same nest every year, while others use eyries 20–30 m away from one year to another (Terrasse and Terrasse 1977). In certain unusual conditions, the

Osprey

Breeding distribution
1984–1989

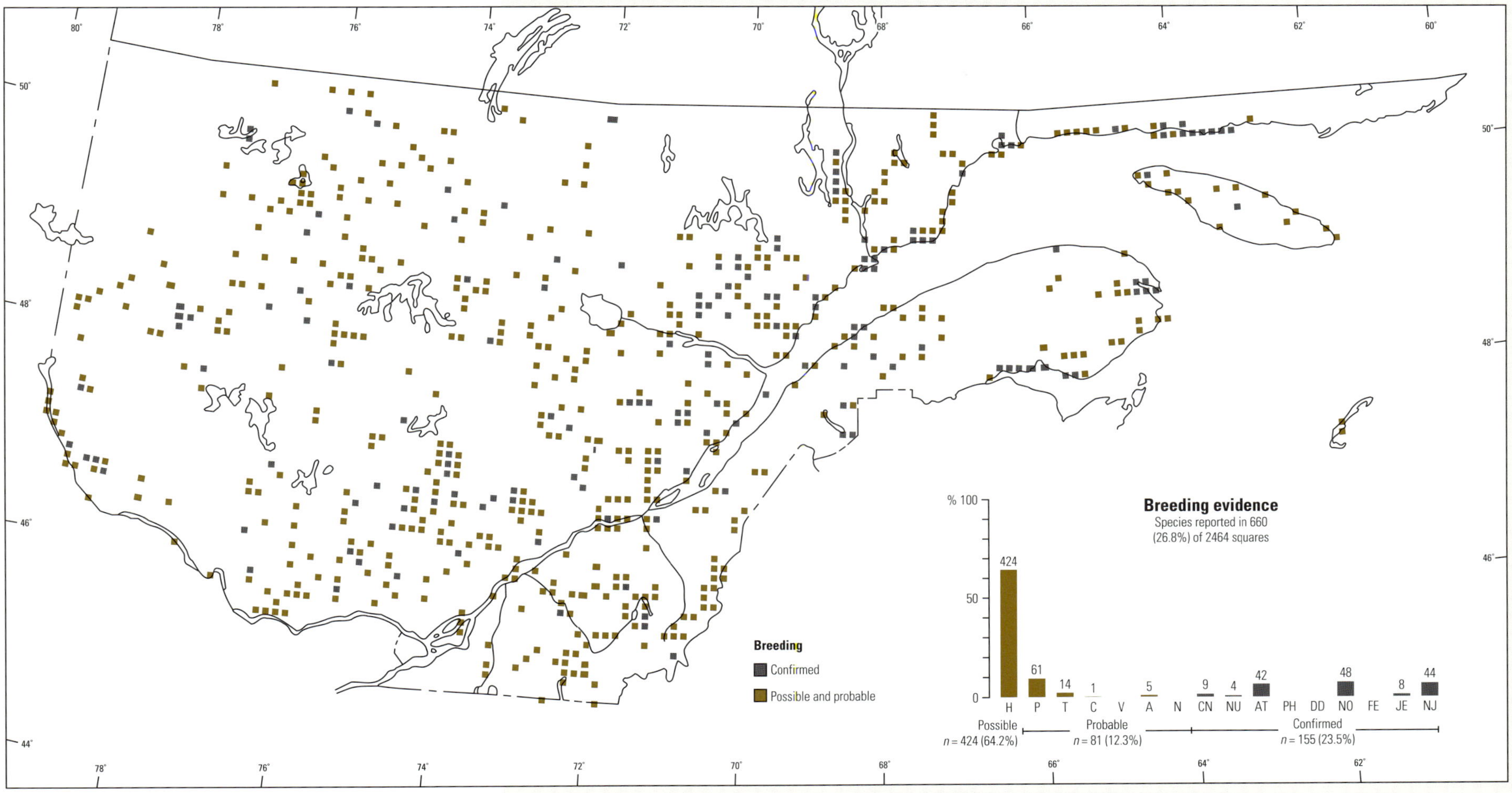

Species profile

Clutch size	usually 3 eggs (2–4)[1,2,6,7,8]
Laying interval	1 egg per 1–3 days[7]
Start of incubation	with first egg[1,2,3,7]
Incubation period	35–43 days[7]
Maturity at hatching	semi-altricial
Care of young	by female and male[3,7], *see* text
Nestling period	50–55 days[7]
Age at first flight	usually ≈ 53 days (49–57)[3,7]
Dependence of young	until ≈ 10–20 days after leaving nest[7]
Broods per year	1[3,7,8]
Breeding age	3 years[1,3]
Mating system	monogamy, sometimes polygyny[1,3,7]
Duration of pair bond	1 season[3], possibly for life[7]
Total length	43–62 cm[1,2,4]
Wing span	135–183 cm[1,2,4]
Weight	M 1428 g (1120–1740) / F 1627 g (1208–2050)[3]
Longevity record	32 years[5]

References:
1. Palmer 1988
2. Terres 1980
3. Cramp and Simmons 1980
4. Godfrey 1986
5. Kennard 1975
6. Ehrlich et al. 1988
7. Poole 1989
8. DesGranges et al. 1994

Chronology *n* = 173

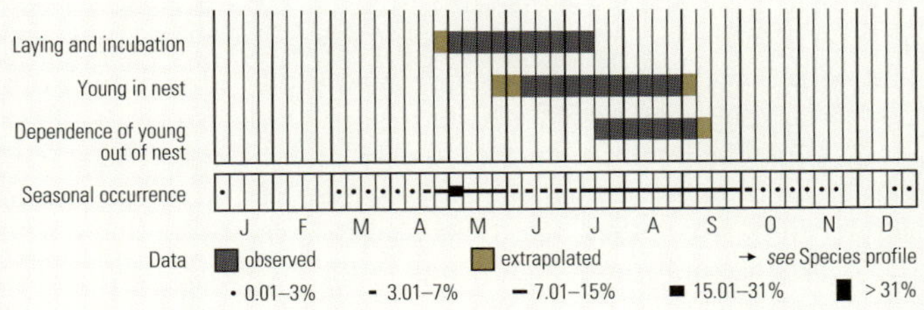

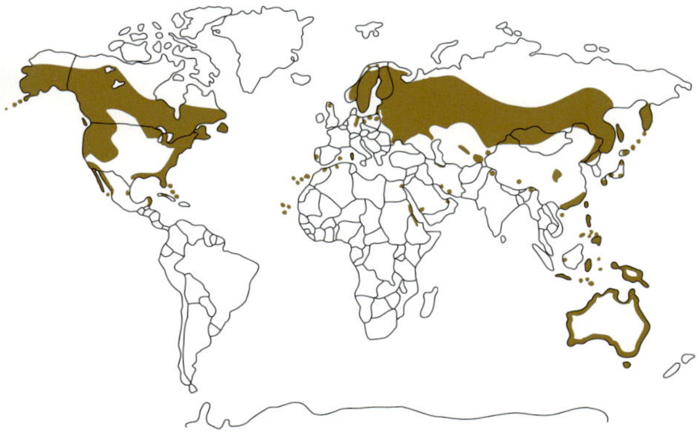

World breeding range

Over the years, the male adds dry sticks and other material to the nest, like this one built on a duck blind. Stater Pond, near Saint-Adrien-d'Irlande, west of Thetford Mines. *July 1990, Jean-Luc DesGranges.*

in areas that have been devastated by forest fires, the Osprey may nest on huge erratic boulders jutting up from shallow water, on a rock pinnacle out of reach of land predators, or sometimes even on man-made structures such as electrical transmission towers, utility poles, duck blinds, or an abandoned teepee (DesGranges 1978b; Aubry and Bordage 1982; Poole 1989; DesGranges et al. 1994).

A breeding pair in Québec will usually produce 3 eggs. Often, after 7 or 8 weeks of assiduous parental care, only 2 of the young will fledge (DesGranges et al. 1994). The youngest bird, being smaller and less aggressive, may starve before it can learn to fly (Poole 1989; Forbes 1991). When the parents are not disturbed by humans, predation on eggs and young by crows, ravens, owls, gulls, and raccoons is rare (Poole 1989). The female does almost all of the incubating and is fed by the male, which is the sole provider for the family during the chicks' first month of life. At the height of their growth period, they may consume upwards of 6 fish per day.

After fledging, the young continue to depend on the parents for food for 10–20 days, although some manage to capture prey on their own 2 or 3 days after leaving the nest (Poole 1989). The daily food requirement of an adult is 300 g, and the food consumption of a pair and 3 young for the breeding period is estimated at 300 kg, or about 1200 fish weighing 250 g each (Géroudet 1984). This is consistent with Poole's (1989) estimate of 6 medium-sized fish, measuring 25–35 cm, delivered daily to the nest by the male.

Pairs likely remain faithful to each other from one year to the nest as both males and females show strong site fidelity to the nest and territory where they raise their young (Poole 1989).

Because of its long flights high in the air and its habit of hovering to spot its prey, the Osprey rarely goes unnoticed. That is why the species was sighted in more than a quarter of the Atlas squares and at least a third of reports were of birds probably engaged in breeding activities. The Osprey, along with the American Kestrel, Broad-winged Hawk, and Northern Harrier, is one of the easiest raptors to observe in southern Québec. Its nest is also one of the easiest to locate: nests were found in 4% of the squares atlassed.

Adult and young in the nest. When food is plentiful, 2 young normally fledge from a nest. Lac à la Truite, near Saint-Adrien-d'Irlande, west of Thetford Mines. *August 1985, Alain Côté.*

Notes on territory (continued)

density of active nests is so high that there appears to be a colony (Flemming et al. 1991).

In May, the male joins the female, which normally arrives at the eyrie first (Green 1976). Nest preparation marks the beginning of bond formation; this is followed by aerial displays and proffering of food by the male (Poole 1989). Copulation takes place on the nest. The male's territorial instinct and desire for sexual exclusivity are strongest during the egg-laying period. After that, he becomes more tolerant of other Ospreys near his eyrie and at preferred hunting sites (Jamieson et al. 1982).

Distribution

Breeding pairs have been reported in almost all parts of southern Québec. To date, over 300 nests have been recorded south of 50°30' N, or about 1-active nest/1000 km^2, mostly in the Laurentian Shield and the coastal regions of the estuary, Anticosti Island, and the Gulf of St. Lawrence (CWS, unpubl. data). In the lower estuary, the birds occasionally form loose colonies, which may number up to 12 active nests in an area of approximately 10 km^2, as has been found on the Manicouagan Peninsula (CWS, unpubl. data). Nest density may be as high as 4 nests/100 km of shoreline along the large reservoirs in the western part of the upper Laurentians (DesGranges et al. 1994).

Nesting is only sporadic in the agricultural St. Lawrence Plain and in the Appalachians, where large bodies of clear water are few and far between. There are only a few records from the Eastern Townships (L. M. Terrill, unpubl. data), Bois-Francs, South Shore, and Témiscouata regions (Tanguay 1964). Breeding was not confirmed in the Magdalen Islands during the Atlas period, but Fradette (1992) cited 3 breeding records, from 1897, 1954, and 1955.

Incomplete coverage makes it difficult to assess Osprey abundance in the northern boreal forest. The species is known to be rather common, however, north of the Atlas area and almost up to the limit of contin-

This opportunistic, piscivorous member of the Accipitridae family chooses a nest site that overlooks its surroundings. The young birds can be seen in this nest built on an electrical transmission tower. Manicouagan Peninsula, Upper North Shore. *August 1989, Jean-Luc DesGranges.*

This erratic boulder is a stable nest site that the Osprey can use for years to come. Near La Grande-4, James Bay region. *June 1991, Jean-Luc DesGranges.*

Generally speaking, the female seldom leaves the eyrie during the nesting period. When not fishing, the male also spends a good deal of time there, usually perched on a nearby tree, where he normally roosts at night (Terrasse and Terrasse 1977; Jamieson et al. 1982). After they leave the nest, the young keep coming back, sometimes for as long as 3 weeks, to be fed by their parents (Poole 1989). Often 2-year-old subadult birds, which make up 5–10% of the population, may occupy and repair a nest within 100 km of where they were hatched, although they do not lay any eggs (Henny and Van Velzen 1972).

uous forest, where summers are still long enough to allow the successful rearing of their young before the onset of winter; 504 nests were reported, about 1.5 active nests/1000 km^2. In Québec, there are probably about 1200 breeding pairs around James Bay and Hudson Bay (Cons. G. G. – GREBE 1989, 1992k; DesGranges et al. 1994); the La Grande complex reservoirs, with densities of 1.5–5 nests/100 km of shoreline, appear to be especially suitable habitat for Ospreys (DesGranges et al. 1994).

History and trends

Ouellet (1974a) reviewed early ornithological literature and concluded that Ospreys were at one time more abundant in migration in southwestern Québec. He believes that some pairs may have nested in the more inaccessible parts of this region, such as Île aux Hérons in the Lachine Rapids. Populations in Ontario (Weir 1987c), the Maritimes (Erskine 1992), and those American states bordering Québec (Fichtel 1985d; Carroll 1988a) have also shown disturbing declines, leading the species to be included on the Blue List between 1971 and 1981 (Tate 1981).

The layers of this Osprey nest, which contains 3 eggs, indicate it has been used for many years. La Grande-4, James Bay region. *June 1989, Jean-Luc DesGranges.*

Like many other birds prey, Ospreys throughout the world suffered serious declines in breeding success in the 1950s and 1960s. The then widespread use of organochlorine pesticides, particularly DDT, caused eggshells to thin and consequently crack under the weight of the incubating female (Poole 1989). Now that these chemicals have been banned almost everywhere, populations have been recovering steadily in their remaining breeding habitat (Poole 1989; Titus and Fuller 1990).

This fish-eating raptor may also build its bulky nest of sticks on a rock standing out in a river. La Grande Rivière, James Bay region. *June 1991, Jean-Luc DesGranges.*

In 1990, a pair of Ospreys built a nest on a power line at the La Grande hydro-electric complex not yet in use. James Bay region. *August 1990, Jean Rodrigue.*

The Osprey's long life expectancy of 15–25 years probably helped it to survive the years of heavy pollution, despite low productivity; one banded bird lived to be 35 (Poole 1989). Its astonishing ability to live alongside humans and use artificial nest platforms put up for it where sound natural supports are lacking has contributed to the recovery of its populations. The creation of reservoirs and artificial ponds teeming with fish has been another positive factor, although fish farmers are authorized to shoot birds that raid their ponds too often (MLCP, unpubl. data).

JEAN-LUC DESGRANGES

Bald Eagle

Pygargue à tête blanche

Haliaeetus leucocephalus

Ghislain Caron

WITH A WINGSPREAD OF JUST OVER 2 M AND STRIKING WHITE HEAD AND tail attached to a chocolate brown body, an adult Bald Eagle cannot be mistaken for any other bird. Few will ever forget the moment of seeing such a majestic bird.

Bald Eagles were once quite plentiful throughout North America, but now in Québec this species is classified as threatened, with probably fewer than 40 breeding pairs in the province (Robert 1989b). In the United States, this is the only bird with its own protection bill, the National Emblem Act of 1940. Had the Bald Eagle lost out to the Wild Turkey as America's choice for its national bird in 1782, perhaps there would be less concern for its survival today. As matters stand, breeding and wintering surveys, captive-breeding and release programs, public-awareness efforts, habitat preservation, rehabilitation of sick, orphaned, and injured eagles, and well-funded research have pushed the cost of Bald Eagle conservation well into the tens, if not hundreds, of millions of dollars.

Usually seen near water, Bald Eagles soar and glide with their wings held horizontally, and in British Columbia and Alaska they sometimes roost in assemblages of up to 1000 birds near productive food sources such as salmon runs. Bald Eagles generally take 5 years to acquire fully adult plumage, and the immatures, either dark brown overall or in the mottled transition stage, may be confused with the Golden Eagle. Once heard, however, the weak, high-pitched scream of a Bald Eagle is an identifying characteristic and is easily recognized again.

Habitat and behaviour

The Bald Eagle usually nests near or along undisturbed seacoasts, inland lakes, or rivers (Brownell and Oldham 1983). Since this large raptor is primarily a fish eater, the majority of nests are located within 200 m of fresh or salt water where fish are more abundant or reliable, such as beside rapids or fast-flowing streams that are ice-free early in the year (Gerrard and Bortolotti 1988). Most Bald Eagles nest in trees, typically in old-growth stands and usually in the tallest tree. They prefer to build the nest beneath the top of the crown in a live tree for protective cover. The most successful nests are in trees with protective canopies, as the young are sheltered from the elements while the parents still have adequate aerial access (Hancock 1970). A wide variety of tree species are used, including elm, hickory, cypress, various firs, and poplar, depending on the region. In Québec, as in much of northeastern North America and across the boreal forest, various species of pines are used. Where suitable trees are not available, such as on treeless Amchitka Island in Alaska, nests are placed on rock pinnacles, steep ridges, hillsides, or on the face of steep cliffs (Sherrod et al. 1976).

Rapid chases or pursuit flights with the occasional "undulating skydance" help advertise the pair's nesting territory and secure their pair bond. Although the spectacular display is seldom witnessed, Bald Eagles are renowned for their cartwheel flight in which the two eagles gain altitude, lock toes, and fall earthward, rotating and then separating just before hitting the ground (Palmer 1988). Bald Eagles normally begin breeding at 5 years of age and are believed to form a lifelong, monogamous pair. If one of the pair dies or disappears, however, the remaining bird will take a new mate.

While both sexes help to build the nest, the female collects and arranges most of the branches and assorted debris, sometimes slamming her body into limbs to help break them off. Construction can take as little as 4 days, but nest maintenance continues throughout the breeding season (Herrick 1932). Bald Eagles build the largest nests of any bird in North America; if there is already a nest on the chosen spot the eagles may construct a new one on top of it (Stalmaster 1987). Nests that have been occupied for 50–60 consecutive years have been known to reach up to 6 m in height and weigh as much as one or two tonnes (Broley 1947)! Most nests, however, are much smaller, typically 1.5–1.8 m in diameter and 0.7–1.2 m in height (Stalmaster 1987). A refurbished nest does not always indicate occupation, as a pair will sometimes build and decorate more than one nest in its breeding territory (Grier 1969). Occasionally, a pair will inexplicably abandon its nesting effort.

Clutches vary in size from 1 to 3 eggs, but most have 2 eggs, laid 2–4 days apart (Herrick 1932). The incubation period averages 35 days and both sexes participate, although the female does the larger share during the day (Gerrard et al. 1979). Bald Eagles often lay a replacement clutch upon the loss of the first. The young hatch asynchronously, and sometimes the elder kills and even devours the second-hatched chick, although this fratricidal behaviour is seen less often in Bald Eagles than in Golden Eagles (Palmer 1988). It usually occurs in times of food shortages while the young are small, between 3 and 8 weeks of age. In the competition for food, the older and larger chick may win; the smaller nestmate is then pecked or trampled to death, or dies of sickness or exposure. Both chicks frequently survive, however, and in years of plentiful food, some nests have even produced 3 eaglets.

For the first 2 weeks after hatching, at least one adult remains in almost constant attendance at the nest, brooding the semi-altricial chicks. Food, primarily fish, is brought by either parent, but for the first month the male does most of the hunting (Stalmaster 1987). The parents feed the eaglets by tearing the prey item into pieces and then dangling it in front of the hungry young. The eaglets begin to feed themselves at 6–7 weeks and fledge at 10–12 weeks. They remain dependent on their parents for food while mastering their flying skills. The adults bring food to where the young are perched, but fledglings also meet the parents at feeding stations or else scavenge on dead prey (Stalmaster 1987). Between 4.5 and 13 weeks after fledging, the eaglets begin to break the family ties and leave the nesting area (Stalmaster 1987), but occasionally, the family may remain together throughout the winter (Sherrod et al. 1976; Palmer 1988). In a Saskatchewan study, only 20% of young survived to 3 years of age (Gerrard et al. 1978).

Despite their bulk and size, Bald Eagles are remarkably agile fliers, as well as skilled, versatile hunters. Although they generally favour fish, both live and dead, they are opportunistic foragers and will also eat birds, primarily the aquatic species, as well as mammals such as rabbits, hares, rats, and sea otter pups (Stalmaster 1987). Eagles will feed on prey of any size, from small rodents to good-sized fish, and will even dine on beached whales. Their hunting methods are as

With its many islands, the huge Cabonga Reservoir is a favourite Bald Eagle site. It has been seen nesting in the canopy of a tall pine in the Richard Island area, on the left in the photo. Lake Andou, Cabonga Reservoir. *May 1991, Yves Aubry.*

Notes on territory

Breeding Bald Eagles have a large home range made up of three components: the nest site, the surrounding territory that is defended against other eagles, and a large unattended area known as the hunting range. The hunting range is used for foraging and often overlaps those of adjacent pairs. The size of the hunting range can be difficult to determine, and has been found to vary widely, from a minimum of 10–15 km² (Gerrard et al. 1980) to over 64 km² (Haywood and Ohmart 1983).

Both members of the breeding pair will defend the area around the nest, most forcibly against

Bald Eagle

Breeding distribution
1984–1989

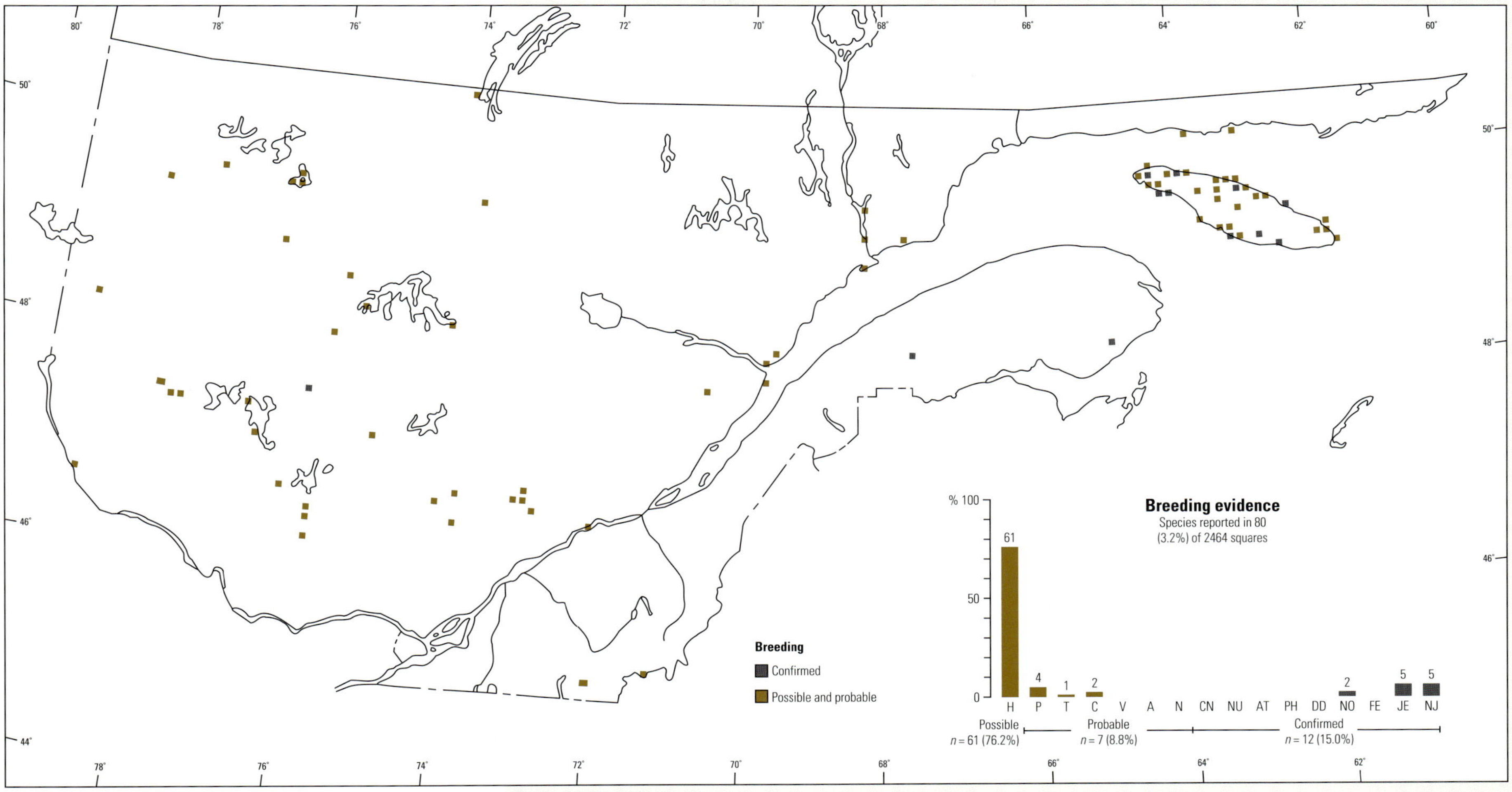

World breeding range

Species profile

Clutch size	usually 2 eggs (1–3)[1,2,3,4]
Laying interval	1 egg per 2–4 days[2,3,4]
Start of incubation	with first egg[1,3]
Incubation period	34–46 days[1,2,3,4]
Maturity at hatching	semi-altricial
Care of young	by female and male[1,2,4]
Nestling period	70–88 days[1,2,3,4,7]
Age at first flight	usually 70–88 days; M ≈ 78 days / F ≈ 82 days[1,2,3,4,8]
Dependence of young	until 4.5–13 weeks after first flight, sometimes until winter[1,3]
Broods per year	1[1,3]
Breeding age	5 years, sometimes 4[1,4,7]
Mating system	monogamy[1,2,4]
Duration of pair bond	usually for life[1,2,3,4]
Total length	M 76–86 cm / F 89–94 cm[6]
Wing span	M 175–210 cm / F 198–225 cm[1]
Weight	4300 g (2000–6200)[4]
Longevity record	21 years 11 months[5]

References:
1. Palmer 1988
2. Gerrard and Bortolotti 1988
3. Stalmaster 1987
4. Johnsgard 1990
5. Klimkiewicz and Futcher 1989
6. Godfrey 1986
7. Brownell and Oldham 1983
8. Stocek 1992

Chronology $n = 19$

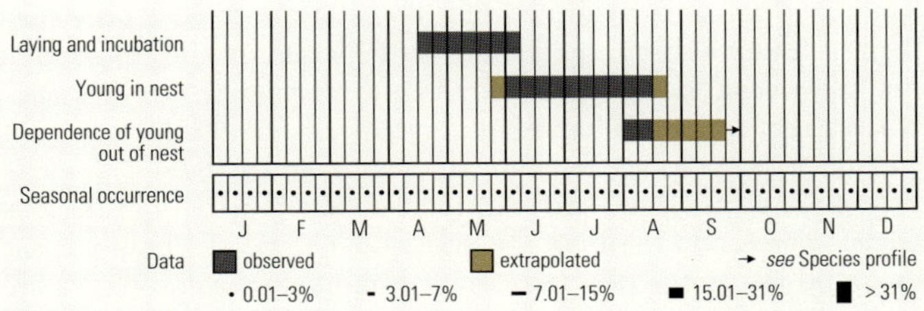

In 1991, a pair of eagles nested in a spruce tree on the banks of the Opinaca Reservoir. In 1990, the nest was used by an Osprey, which occupied it again in 1992. James Bay region. *June 1989, Jean-Luc DesGranges.*

diverse as their diet. They may hunt and kill their prey themselves, by soaring above an area then swooping down on their quarry, striking it with their talons, or by skimming the surface of the water for fish. Rarely will eagles dive into the water the way Ospreys do. They may hunt co-operatively, walking through low brush to flush rabbits for waiting eagles. Sometimes Bald Eagles will wade into water up to their breasts to drag out large spawning fish. Also, they will often resort to stealing prey from conspecifics or other species. They are known to harass and pirate kills from other predators such as Ospreys and gulls, or even sea otters and foxes. In some areas and most commonly in winter, scavenging on carrion is the preferred foraging method (Sherrod et al. 1976). Besides fish, Bald Eagles will feed on the carcasses of large mammals such as deer and moose. Each winter some birds gather on the ice of the Gulf of St. Lawrence to feed on the placentas of seals that have just pupped (W. Hoek, pers. comm.).

Although Bald Eagles are not common in Québec, their large size and conspicuous adult plumage tend to make them stand out. Individuals are often spotted soaring on thermals and updrafts, or being chased by crows and other birds. Locating nests, however is far more difficult. Not all mated pairs breed each year (Gerrard and Bortolotti 1988), so the sighting of a pair does not necessarily mean there is a nest nearby. Incubating birds lie flat on the nest and are not easily seen, but parents will often display aggressive behaviour when a nest containing young is approached. The adults may vocalize and fly above or even directly at human intruders. Once the young are a month or two old, there is much activity at the nest. Eaglets perch on the rim of the nest, and a circle of whitewash may be found below the nest tree. The young often squabble over food and prepare for eventual flight by jumping and flapping their wings at the nest. For the first few weeks after fledging, the young are often awkward in flight and will frequently perch near the nest, awaiting a food delivery (Stalmaster 1987). Young in the nest or recently fledged were the most frequently used breeding evidence codes.

Distribution

The Bald Eagle is confined almost exclusively to North America, breeding locally from northwestern Alaska across central Canada south to the southern United States and Baja California (Godfrey 1986). Populations from the northern part of the breeding range in areas where the water freezes over in winter are primarily migratory, but a few birds may remain in the north in winter. Farther south, this species generally winters throughout the breeding range but more frequently along the coast from southern Alaska and western Canada southward (Johnsgard 1990). Surveys of wintering Bald Eagles in the contiguous United States from 1979 through 1982 reported the highest concentrations along the Mississippi and Missouri rivers and their tributaries, as well as in California, Oregon, and Washington State (Millsap 1986).

The two main pockets of breeding pairs in southern Québec, as evidenced by the Atlas findings, are a relatively dense population on Anticosti Island and a population of similar size spread out over the western corner of Québec south of James Bay and northeast of the Ottawa–Hull region, particularly on the shores of large lakes and reservoirs. The species was reported in 3.2% of Atlas squares surveyed (80/2464) and breeding was confirmed in only 15% of squares (12/80).

Scattered sightings were also reported at the mouth of the Saguenay, along the North Shore, and in the Eastern Townships close to the U.S. border. These observations do not necessarily imply nestings nearby; the birds seen could be either members of the floater population of non-breeding adults or wandering immatures. In Florida and other southern breeding areas, reproduction takes place in the winter and then the immatures move north after fledging, reaching southeastern Canada in April or May, where they remain throughout the summer (Broley 1947).

With a fairly large home range, Bald Eagle pairs are probably more widely distributed throughout Québec than the Atlas map illustrates. Some nest along the Québec–Labrador border (Wetmore and Gillespie 1976) and an adult was seen as far north as the Koksoak River just south of Kuujjuaq (A. Grist, pers. comm.). A nest containing 2 eggs, but which produced only 1 chick, was found at the Opinaca Reservoir in the summer of 1991 (J.-L. DesGranges and J. Rodrigue, pers. comm.). In a 1990–91 survey of the Nottaway, Broadback, and Rupert river watersheds in the southern James Bay region, researchers reported 4 adult Bald Eagles sighted along the Rupert River, 1 along the Nottaway River, 3 at Lake Waswanipi, 1 at Lake Chevrillon, and 2 along other rivers and lakes, for a total of 11 Bald Eagles including 3 probable pairs in the region (Cons. G. G. – GREBE 1992l). Other sightings near James Bay have been made along the Opinaca River near Lake Guillaume Delisle on Hudson Bay (ÉPOQ), and in the Great Whale River watershed (Cons. G. G. – GREBE 1990c). An immature was also reported near the mouth of the La Grande Rivière in June 1992 (Benoit et al. 1993b).

Eagles may build their nests right at the top of a mature tree, as here on Anticosti Island, home to a sizable resident population of eagles. Bec-scie River. *August 1991, Gaétan Laprise.*

Notes on territory (continued)

conspecifics and Ospreys, but once incubation has begun they are often intolerant of other avian intruders, such as other raptors, crows, gulls, and other large birds (Herrick 1932; Stalmaster 1987). Usually, nests are defended with a minimum of fighting, and the sight of one of the parents perched on a prominent tree or circling overhead is most often sufficient warning to others to stay away (Gerrard and Bortolotti 1988).

The defended breeding territory varies greatly in size, depending on prey abundance, suitable nest sites, and population density. It may be nearly circular, oval, cone-shaped, or linear (Stalmaster 1987). In the northern regions of the breeding range, the size of breeding territories ranged from 0.11 km² to 12.7 km² (Hensel and Troyer 1964; Robards and King 1966; Whitfield et al. 1974; Robards and Hodges 1977; Leighton et al. 1979). Under suitable conditions, the

Eagle nest, high in the canopy of a prominent pine tree. Dozois Reservoir. *June 1990, Jean-Luc DesGranges.*

mean distance between occupied nests appears to be about 1–3 km. Where breeding populations are dense, however, shorter distances are tolerated. On Amchitka Island, Alaska, for example, the closest active nests were only 137 m apart (Sherrod et al. 1976).

Some Bald Eagles winter in southern Québec, mainly on Anticosti Island, along the North Shore, and in the southwestern Laurentians (David 1980a). The areas along the Gulf of St. Lawrence and the western tip of Anticosti Island remain ice-free throughout most of the year and are the wintering grounds for a large population of aquatic birds, which provide a continuous source of food for the eagles (Guillemette and Guillemette 1983). Bald Eagles will also remain near hydroelectric dams, such as the Daniel Johnson Dam, where the water is clear of ice throughout the winter (ÉPOQ). Carrion sustains those inland, and wintering eagles will feed on carcasses of white-tailed deer that have succumbed to starvation or illness.

The factors limiting the Bald Eagle's distribution in Québec include the presence of large mature trees for nesting, lakes well stocked with fish, and areas where human disturbance is at a minimum. Anticosti Island remains ideal Bald Eagle habitat because of its large, mature trees next to the shore coupled with near-pristine wilderness, unlike the Gaspé which, despite its numerous salmon rivers, has seen far more human activity. A study of human activity in the Chesapeake Bay area found that because eagles do not seem to adapt to living alongside humans, shoreline is irretrievably lost as eagle habitat once it is developed (Buehler et al. 1991).

History and trends

Bald Eagles were once widespread throughout North America, nesting on both coasts and probably along every large lake and major river in the interior, including the St. Lawrence. Precise numbers are difficult to determine because it is only in the last few decades that concerted efforts to census the population have been made, but some researchers suspect that there were once between a quarter of a million and half a million Bald Eagles on the continent when Europeans first arrived. The species is now extirpated in many of its former breeding areas, and the population is only a fraction of what it once was (Gerrard and Bortolotti 1988).

The first major decline probably began in the mid- to late 1800s, starting in eastern North America and progressing westward, coinciding with the spread of European settlement (Swenson 1983). As settlers cleared the forests for farms and ranches, they cut down the large old-growth trees used for nesting and roosting, irreparably altering the Bald Eagle's breeding habitat. In Québec, large-scale clearcutting operations began in the Ottawa Valley in 1806, expanding to the Laurentians by 1850, when massive pine trees were felled and floated down the St. Lawrence for export to England (Dunn 1975).

Changes in the landscape and human population expansion also had a detrimental effect on the abundance of the eagle's prey species, especially important for survival in the winter (Gerrard and Bortolotti 1988). Eagles, admired and respected by the First Nations of North America, were viewed by many settlers as vermin and a threat to livestock. Little distinction was made between Bald and Golden eagles, and both species were hunted for bounties and trophies and killed unintentionally by traps set for wolves, coyotes, or bobcats, often using poisoned bait. By the early 1900s, sightings of Bald Eagles in eastern North America were becoming rare (Taverner 1919).

A second, more rapid and drastic decline began in the 1950s and lasted until the early 1970s, caused primarily by pesticide contamination, particularly DDT and other organochlorines (Grier 1972; Sprunt et al. 1973). DDE, a metabolite of DDT, affected reproduction by causing thin-shelled eggs and dead or deformed embryos. Their position at the top of a relatively long food chain meant that fish-eating predators such as the Bald Eagle were seriously affected, due to the biomagnification of toxic substances at each step in the food chain. In Canada, populations of Bald Eagles were less affected by pesticide poisoning, except in some regions such as New Brunswick (Stocek 1979) and southern Ontario (Bortolotti 1987). Other hazards include heavy-metal poisoning, mainly from ingesting lead shotgun pellets in waterfowl wounded by hunters and mercury-contaminated fish, continued habitat destruction, electrocution by high-voltage power lines, and human disturbance at or near the nest (Hamerstrom et al. 1975; Craig et al. 1990; Elliott et al. 1992).

Since the banning of DDT in Canada and the United States in the early 1970s, followed by dieldrin in 1974 and endrin in 1984 in the United States, there has been a steady increase in the population in certain areas, such as northwestern Ontario (Grier 1982) and around Chesapeake Bay (Cline 1986). Captive-breeding and rehabilitation programs, combined with the preservation of important breeding habitat, have contributed to the slow recovery (Simons et al. 1988; Martell et al. 1991). The Bald Eagle population in North America is currently estimated at 70 000–100 000 birds, of which two thirds are found in Alaska and coastal British Columbia (Stalmaster 1987).

Atlas results show the Bald Eagle to be an uncommon permanent resident in southern Québec. Some of the eagles observed in southern Québec originated from the population in the New York–New England area, where the release of translocated or captive-bred eaglets has led to their partial recovery (Bautz and Nye 1987). According to Ouellet (1974a), historically, this species nested in the Montréal area, along the St. Lawrence and Richelieu rivers, before human pressure became too great. Since the habitat is somewhat similar throughout Ontario and Québec, it would not be surprising to learn that Québec, like Ontario, once was home to a sizable number of Bald Eagle breeding pairs (Baillie and Harrington 1936).

David M. Bird and Derin Henderson

Northern Harrier

Busard Saint-Martin

Circus cyaneus

Jean-Pierre Vallée

WITH ITS CONSPICUOUS WHITE RUMP PATCH AND CHARACTERISTIC HABIT OF quartering low over open countryside, the Northern Harrier, formerly known as the Marsh Hawk, is easy to identify as it searches for its prey. As it glides, it rocks from side to side in a leisurely fashion, holding its wings in a shallow V similar to the Turkey Vulture. Adult males have bluish-grey upperparts with pale grey to whitish underparts. In contrast, the upperparts of females are dark brown with light brown to cream-coloured underparts. Immatures resemble adult females, but are darker with rusty brown underparts. Unlike the other members of the order Falconiformes, harriers of the genus *Circus* have facial discs similar to those of owls, enabling them to locate their prey by sound (Rice 1982).

The Northern Harrier is a fairly common migrant breeder in Québec. It winters along coastal and river marshes from southern Canada to northern South America, and on rare occasions may be seen in southern Québec during the winter (Yank et al. 1989a).

Wetlands like these along the Rivière du Sud attract the Northern Harrier in the breeding season. This accipiter cruises low over the vegetation in search of small mammals and birds, reptiles, and amphibians. Tributary of the Richelieu River. *June 1991, Pierre Bannon.*

Habitat and behaviour

The Northern Harrier is a bird of open country and wetlands, nesting in fresh- and saltwater marshes, peatlands, wet meadows, fields, pastures, grasslands, and even cultivated land, such as hayfields. Northward into the boreal forest it inhabits wetlands, muskeg, clearcuts, and recent burns (Palmer 1988; Serrentino 1992). Harriers that have reared young successfully tend to return to the same general area in the following years, with adult males arriving up to 10 days earlier than the females (Brown and Amadon 1968; Hamerstrom 1969).

Although in some parts of its range, including Québec, voles are the staple prey item, the species also feeds on other small mammals, small and medium-sized birds, frogs, snakes (in Québec, family Colubridae), and sometimes carrion (Clark and Ward 1974; Palmer 1988).

The courtship flight of the male is a spectacular sight, consisting of a series of dives from a near stall at a height of 30 m and resembling a series of U-shaped loops. At the top of each leg of the U, the male may somersault before beginning a new dive. Females have been observed to join in (Bent 1937; Brown and Amadon 1968). Although a few may breed at 1 year of age, most do not nest until they are at least 2 years old (Hamerstrom 1969; Palmer 1988). While many pairs are monogamous, some males may attract up to 5 mates in a season (Simmons et al. 1986a). A study of Northern Harriers in New Brunswick found that the incidence of polygyny increased in years of abundant food supply, notably voles, yet females with polygynous mates raised relatively fewer young than their monogamous counterparts because the male was not able to provide each female with sufficient food for the duration of the breeding season (Simmons et al. 1986b). Females in polygynous pairings do not all have the same status, however, and the male provides them with food accordingly, so that the lowest-ranking females may suffer food shortages as the season wears on and the family's demands grow. Simmons (1989) concluded that these lower-status females may have been duped into choosing the male based on the large amounts of food he provided during courtship, a performance that he could not sustain as the season progressed.

After pairing, both adults inspect the chosen territory for a nest site. They will normally use the same general area year after year, but rarely the identical spot (Brown and Amadon 1968). The Northern Harrier is one of the few raptors to build its nest on the ground, typically in the shelter of grassy, weedy, or shrubby vegetation such as cattails, grasses, sedges, willows, and bushes (Hamerstrom and Konepy 1981; Simmons and Smith 1985; Christiansen and Reinert 1990). Occasionally, the nest may be built close to the ground among the branches of a shrub or bush (Palmer 1988) or may take the form of a deep platform in shallow water (Evans 1983). The nest is made with sticks, aquatic plants, and grasses. Since this species exploits sporadic food sources, it has a tendency to nest in loose colonies, even when opportunities for widely spaced nesting sites are plentiful, but nest density in the colony is related to food abundance. In years when the rodent population is high, some nests may be only a few hundred metres apart (Cramp and

Notes on territory

The home range of the Northern Harrier is made up of a large area for hunting and a smaller, defended one for nesting. Home ranges are usually established at the onset of incubation and often overlap those of neighbouring pairs. In Michigan, estimated size ranged from 0.98 to 10.1 km² (Craighead and Craighead 1956). The nest site is defended primarily by the female, although the male may come to her aid (Hecht 1951; Hamerstrom 1969). The defended area may extend to a radius of 300 m around the nest (Brown and Amadon 1968).

Even when there is plenty of suitable nesting habitat available, harriers often form loose colonies, and pairs have been found nesting within 36–366 m of each other

The Northern Harrier is widespread in the agricultural areas of southern Québec and also breeds on extensive cleared lands, as in this case near La Morandière, Abitibi. *May 1991, Yves Aubry.*

Northern Harrier

Breeding distribution 1984–1989

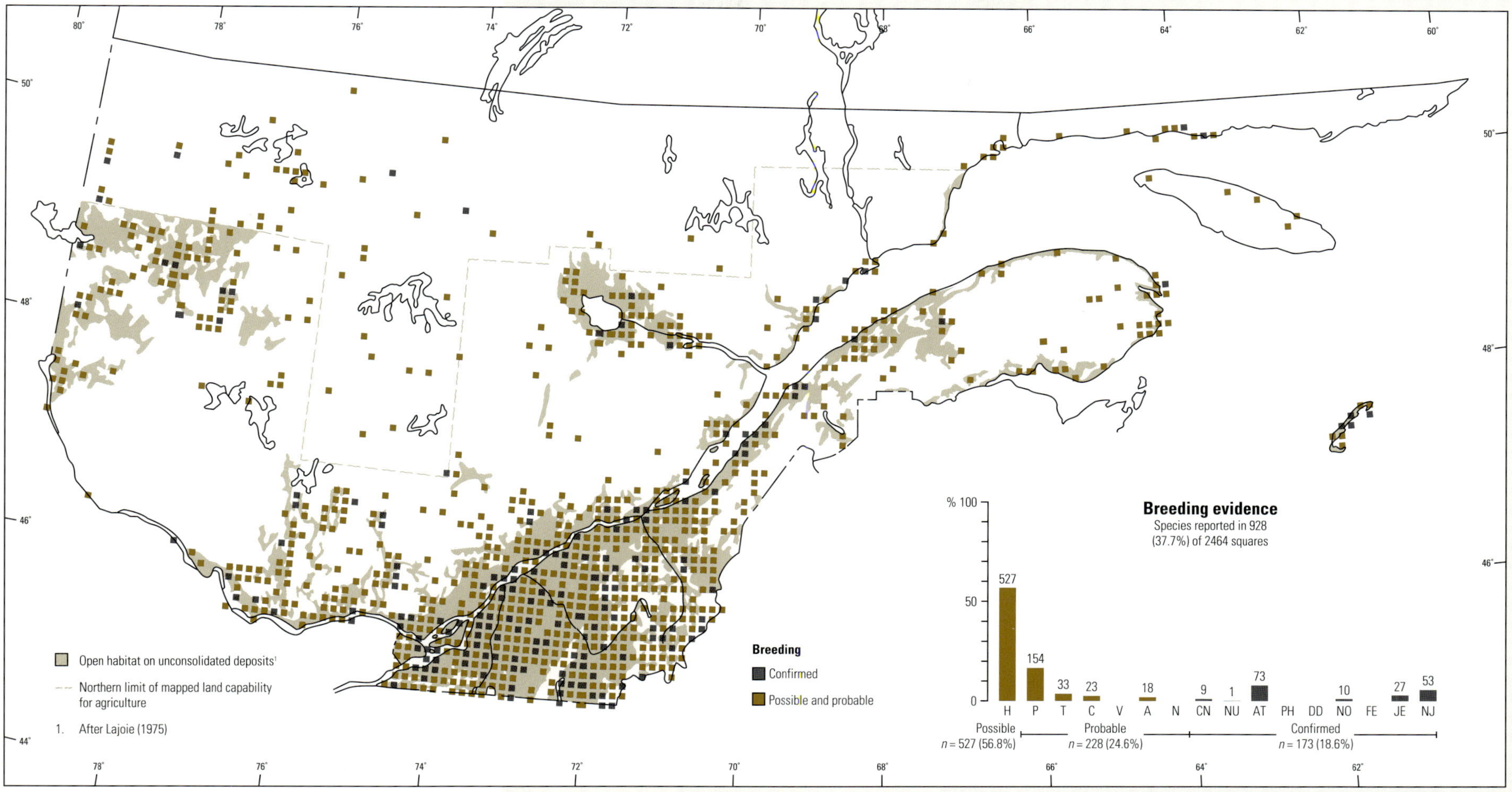

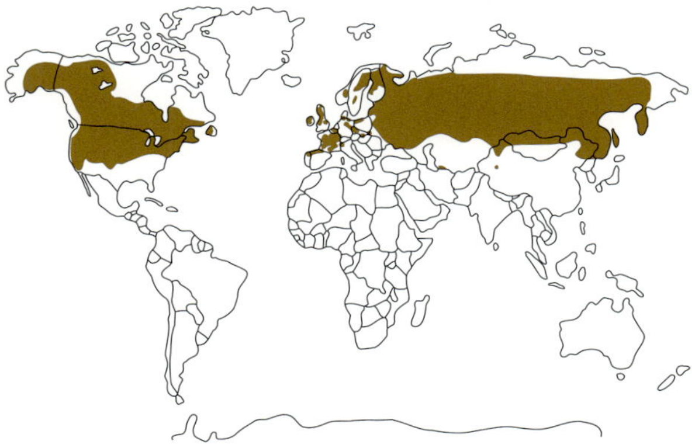

World breeding range

Species profile

Clutch size	usually 4–6 eggs (2–12)[2,3,4,5,6]
Laying interval	usually 1 egg per 2 days(1–8)[6]
Start of incubation	with second egg[1,2,4,6]
Incubation period	30–32 days[6]
Maturity at hatching	semi-altricial
Care of young	by female and male[1,4,5,6], *see* text
Nestling period	usually ≈ 15 days (7–21)[1,2,4,5,8,9]
Age at first flight	M 31–34 days / F 35–38 days[6]
Dependence of young	until ≈ 8 weeks old[1,4,6]
Broods per year	1[2,4,5,6]
Breeding age	2–3 years, sometimes 1[4,6]
Mating system	monogamy, sometimes polygyny[2,4,5]
Duration of pair bond	more than 1 season, sometimes for life[6]
Total length	M 45–51 cm / F 48–61 cm[3]
Wing span	100–120 cm[4]
Weight	M 369.6 g (297.0–469.0) / F 546.0 g (466.0–752.0)[6]
Longevity record	16 years 5 months[7]

References:
1. Bent 1937
2. Harrison 1978
3. Godfrey 1986
4. Cramp and Simmons 1980
5. Brown and Amadon 1968
6. Palmer 1988
7. Clapp et al. 1982
8. Hecht 1951
9. Bannerman 1956

Chronology $n = 90$

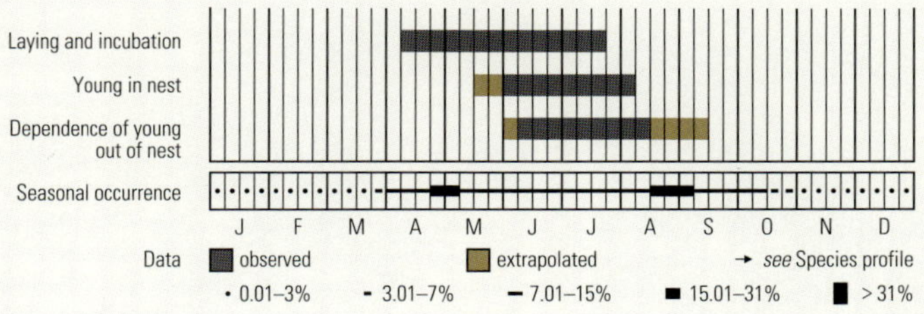

Female harriers continue to add dry twigs to their nests, even after the young have hatched. Boischatel, northeast of Québec City. *June 1986, Paul Perreault.*

Simmons 1980). During years of low vole abundance, the nests are often widely dispersed (Brown 1976). A comprehensive 20-year study (1959–78) on the effects of vole density on Northern Harrier populations in central Wisconsin showed a strong synchronization between the number of harrier nests and vole numbers (Hamerstrom 1979).

Incubation and subsequent feeding of the semi-altricial young are by the female alone, who relies on her mate to provision her at or near the nest. After the young hatch, the male transfers the food to the female in midair above or near the nest. Within 1–3 weeks, the young usually leave the nest and hide in the surrounding vegetation several metres or more from the nest (Hecht 1951; Bannerman 1956). Brood survival is thereby enhanced, since a predator may find one chick but is unlikely to find them all. In the first week or two after fledging, the young remain near the nest, returning to it to feed on the prey caught by their parents until they have gradually mastered hunting for themselves (Palmer 1988).

Outside the breeding season, harriers typically hunt and roost alone, although in some parts of the United States, the Northern Harrier uses communal roosts from late October until March in areas where voles and other prey are abundant (Palmer 1988).

The hunting habits and aerial courtship acrobatics of this species make it easy to spot throughout the breeding season. During the day, a harrier spends about 40% of its time on the wing, quartering open areas searching for prey (Root 1988). Its courtship displays take place in the air over these areas, and because of its large size it is easily seen from a distance. Territorial and courtship behaviour was recorded in 6% (56/928) of Atlas squares in which the species was reported. Nests are relatively easy to find, since they are usually built on the ground and can be located by watching for an aerial food exchange from male to female and then noting the spot to which the female returns. Reports of adults carrying food for the young accounted for the bulk of breeding confirmations (42.2%; 73/173).

Notes on territory (continued)

(Errington 1930; Munoff 1963; Clark 1972). In California, a study found that usually only females defended their winter territories, often evicting larger raptors that would attempt to steal the harriers' freshly caught prey (Temeles 1990).

The Northern Harrier nests on the ground in grassy or bushy vegetation. Sainte-Foy. *May 1957, Gaston Le Page.*

Two young harriers fighting over a meal. There is competition among the stronger chicks, while the weaker ones, usually the youngest, are sidelined. When food is scarce, the weak are unlikely to survive. Boischatel, northeast of Québec City. *June 1986, Paul Perreault.*

Female Northern Harrier tearing up prey to feed to her young. Boischatel, northeast of Québec City. *June 1986, Paul Perreault.*

Young harrier attempting to swallow a small rodent, which its older siblings will no doubt try to take away from it. Boischatel, northeast of Québec City. *June 1986, Paul Perreault.*

Distribution

The Northern Harrier is found almost everywhere in Québec where there are open areas. Its breeding range in southern Québec extends from the Ottawa Valley along the north and south shores of the St. Lawrence. It is particularly abundant in the Eastern Townships and the Beauce, likely because of the predominance of farmland and wetlands in these regions. It also nests in the Abitibi and Témiscamingue regions, along the Saguenay and around Lake Saint-Jean, along the North Shore, in the Gaspé, and on the Magdalen Islands. Few nesting sites were recorded in the Laurentians, a relatively mountainous region largely covered with mixed or coniferous forest. Northern Harriers tend to avoid areas of unbroken forest, except in migration (Palmer 1988).

The breeding range of this species extends north of the upper limits of the Atlas area. Northern Harriers are fairly common in the Eastmain Lowland, and one nest was found in this region in 1990, along the Rupert River, near Lake Nemiscau (Cons. G. G. – GREBE 1992m). As well, this species has been seen along a narrow strip of the Hudson Bay coast from the Great Whale River to north of Lake Guillaume Delisle (Todd 1963; Cons. G. G. – GREBE 1990c). It probably also nests there.

History and trends

Before European settlement, this species was probably rare in southern Québec, except near marshes, swamps, and peatlands and along the major river systems, because of the extensive forests throughout the region. This bird of the open countryside most likely increased in numbers and expanded its range with the clearing of forests. In the early 1900s, the Northern Harrier was sighted regularly in and around Montréal and Québec City, and was considered one of Québec's most common raptors (Dionne 1906), but the species subsequently suffered a significant decline over much of its range in North America. The sharpest drop began in the early 1950s (Palmer 1988). The decline has been attributed to many factors, including the widespread use of DDT and other agricultural pesticides that began in the mid-1940s (Gauthreaux 1971; Anderson and Hickey 1972), intensive farming practices, and the loss of suitable habitat, especially the draining and filling in of marshes (Cadman 1991). Yet by the mid-1970s the population seemed to have stabilized in southern Québec and Ontario, except for a marked decline in the Ottawa area (Fyfe 1976).

Although Atlas results indicate that the Northern Harrier is still a familiar sight in Québec, BBS results for the years 1980–89, based on observations along roads in the southern part of the province, suggest a downward trend. It is difficult to reach firm conclusions on harrier populations, however, since their numbers fluctuate with the abundance of voles, their preferred prey. Nevertheless, habitat loss is an ongoing concern, especially near major urban centres. Abandoned fields are prime breeding habitat, but are often targeted for residential and commercial development. Drainage of wetlands for farming and urban development was extensive along the St. Lawrence prior to 1965 and continues at a steady pace today (Lands Directorate 1986). These factors may have a detrimental effect on local populations over the long term.

Derin Henderson

Sharp-shinned Hawk

Épervier brun

Accipiter striatus

Guy Michaud

The Sharp-shinned Hawk is the smallest member of the subfamily Accipitrinae that breeds in Québec. Although it is our most abundant accipiter, it is difficult to spot, usually keeping a low profile in the foliage under the forest canopy. Under ideal circumstances, the Sharp-shin can be distinguished from the other accipiters by its smaller size and square-tipped or slightly notched tail, but observers must usually rely on a number of features to identify it correctly (Kaufman 1990a).

In Québec, a few birds regularly winter near feeding stations, which provide them with a steady source of prey (Dunn and Tessaglia 1991). The Sharp-shinned Hawk's winter range extends from the southern part of Newfoundland, the Maritimes, and southern Ontario and Québec, to Central America (Godfrey 1986). In spring, migrants arrive at the same time as the first large flocks of migratory passerines, on which they depend for food.

Habitat and behaviour

In the breeding season, this hawk generally frequents dense coniferous and mixed forests (Platt 1976b; Clarke 1982; Peck and James 1983), preferring young stands 25–50 years old with a thick canopy (Platt 1976b; Reynolds et al. 1982). In the Central St. Lawrence Lowland it inhabits dense woodlands bordered by brushy meadows, but in the hinterland it favours dense forests with clearings (Erskine 1977). The Sharp-shinned Hawk requires openings for hunting, using snags and stumps, both in woodlands and in clearings, as perches to pluck and dismember its prey.

The female's body weight is up to 1.7 times that of the male's, the most extreme case of sexual size dimorphism of any North American bird (Mueller and Berger 1970; Mueller et al. 1981). Adult plumage is identical in both sexes and is not fully acquired by immatures until the summer or early autumn of their second year. Adult birds are a dark slate blue above, contrasting with a white and reddish-brown barred breast and abdomen. Like those of many other diurnal birds of prey, Sharp-shin yearlings are brownish above and whitish below, with heavy brown streaking on the underparts (Mueller et al. 1979; Clark and Wheeler 1987). Immatures have yellow eyes, which become orangish and finally ruby red at 2 years of age (Grossman and Hamlet 1988).

When pairs arrive on their breeding grounds, they perform a courtship display consisting of slow rhythmic flapping, undulating flight, and precipitous dives. Mate feeding is also common. Pairs communicate with short, high-pitched call notes given in a series *kik-kik-kik*, resembling a woodpecker's alarm call. The male defends a territory extending up to 37.5 m from the nest, but hunts outside this area (Palmer 1988), sometimes venturing as far as 1.5 km from the nest in search of prey (Platt 1973; Mueller et al. 1981). Nest sites are characterized by a dense canopy of foliage and a high density of trees (Platt 1976b; Reynolds et al. 1982).

When birds return to the same nesting territory they used the year before, they generally do not occupy their old nest, but build a new one less than 100 m away, although they may occasionally refurbish an old one (Reynolds and Wight 1978). The nest, a platform 40–65 cm in diameter and 3–20 cm deep (Clarke 1982; Flood and Bortolotti 1984), is generally built 4–23 m up in a dense conifer, against the trunk on a horizontal branch (Reynolds et al. 1982; Peck and James 1983). It is made of sticks and twigs and lined with finer twigs and pieces of bark. Both sexes gather material for the nest, but the female does most of the building (Palmer 1988). The eggs are laid about 2 days apart. Incubation begins when the last egg has been laid and continues for 30–32 days (Brown and Amadon 1968; Platt 1976b). The male provides food for his mate and offspring from the early stages in the breeding cycle until the young are around 15 days old. After this, the female begins hunting to help supply food for the brood (Snyder and Wiley 1976).

As with other raptor nestlings, the males appear to develop faster than the females. For example, age of first flight is 21–30 days for males, and 24–33 days for females (Flood and Bortolotti 1984; Delannoy and Cruz 1988). The young are dependent on their parents until they are 7–10 weeks old. They appear to need this long period of parental care so that they can learn to hunt (Delannoy and Cruz 1988; Palmer 1988). The Sharp-shinned Hawk is an expert at sneak attacks. It generally hunts from an elevated perch, taking advantage of any concealment found on the way—whether bushes, stumps, or fencerows—to swoop down on its quarry. Any prey that tries to escape is pursued on the ground with a series of leaps, made possible by the Sharp-shin's long legs. Avian prey, mainly passerines and small shorebirds, accounts for close to 95% of its diet (Storer 1966; Snyder and Wiley 1976; Duncan 1980). Males generally take prey no larger than a thrush, whereas females can handle a robin (Storer 1966). Depending on the time of year, the Sharp-shinned Hawk also feeds on small mammals such as shrews and voles, as well as the occasional grasshopper, cricket, beetle, butterfly, or moth (Duncan 1980).

In the Central St. Lawrence Lowland, the Sharp-shinned Hawk nests in dense young woodlands with a well-developed canopy, hunting in nearby openings such as fields and clearings. Cap Tourmente National Wildlife Area, northeast of Québec City. *August 1991, Jean Gauthier.*

Notes on territory

Pairs rarely return to the same nesting territory 2 years in a row, but when they do, they usually build a new nest within 100 m of an existing one (Reynolds and Wight 1978). In New York State, males were found to defend a territory of approximately 37.5 m around the nest (Palmer 1988), but hunted elsewhere, sometimes as far as 1.5 km away (Platt 1973; Mueller et al. 1981).

In Wyoming, the home ranges used by the species were 69–132 ha in size (Craighead and Craighead 1956). In Oregon, local population densities of 3.6 pairs/100 km² were observed, with active nests spaced 1.8–6.0 km apart (Reynolds and Wight 1978).

Sharp-shinned Hawk

Breeding distribution

1984–1989

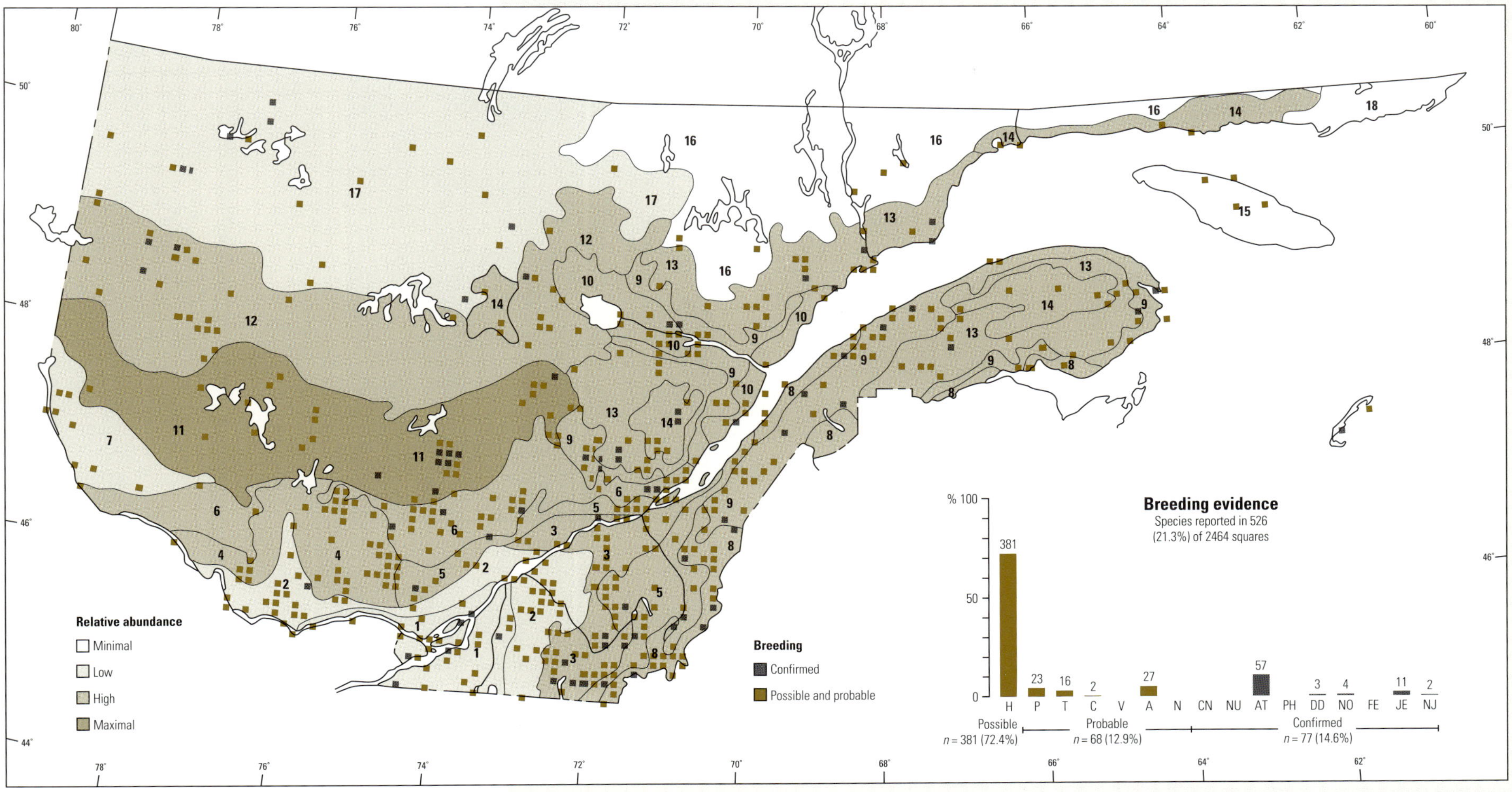

Species profile

Clutch size	usually 3–5 eggs (1–8)(1,2,3,4)
Laying interval	1 egg per 2 days(1)
Start of incubation	with last egg(3)
Incubation period	usually 30–32 days (30–35)(2,3,4)
Maturity at hatching	semi-altricial
Care of young	by female and male(3), *see* text
Nestling period	21–27 days(1,2,3)
Age at first flight	M 21–30 days / F 24–33 days(2,3)
Dependence of young	until ≈ 49–72 days old(3,4)
Broods per year	1(2,3)
Breeding age	1–2 years(1,2,4)
Mating system	monogamy(2)
Duration of pair bond	probably more than 1 season(4)
Total length	M 25–30 cm / F 30–36 cm(7)
Wing span	50.8–68.6 cm(6)
Weight	M 103 g (82–125) / F 174 g (120–210)(2)
Longevity record	9 years 10 months(5)

References:
1. Bent 1937
2. Flood and Bortolotti 1984
3. Delannoy and Cruz 1988
4. Palmer 1988
5. Clapp et al. 1982
6. Mueller et al. 1979
7. Godfrey 1986

World breeding range

Chronology *n* = 20

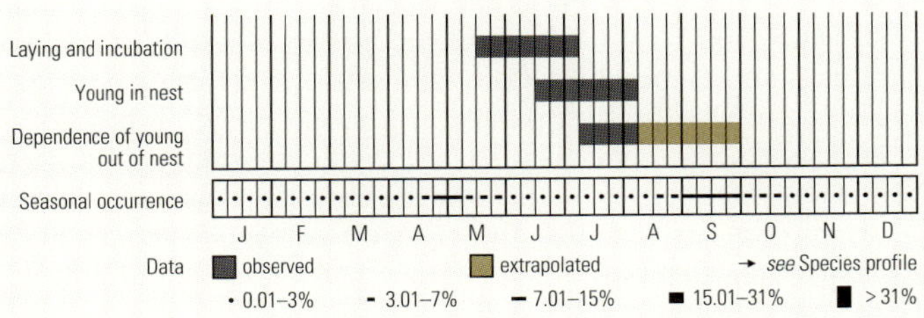

Frequency of occurrence by ecological region

Ecological regions** (forest regions)	Adequately covered squares	Squares in which reported	% of squares in which reported	Relative abundance (%)*
1. Sugar maple-hickory and sugar maple-basswood	97	24	24.7 ± 4.4	44.4
2. Sugar maple-basswood	153	50	32.7 ± 3.9	58.6
3. Sugar maple-basswood and sugar maple-yellow birch	101	44	43.6 ± 5.8	78.1
4. Sugar maple-yellow birch-basswood	56	27	48.2 ± 11.4	86.5
5. Sugar maple-yellow birch-beech and sugar maple-yellow birch-basswood	113	39	34.5 ± 4.9	61.9
6. Sugar maple-yellow birch-beech	71	34	47.9 ± 10.6	85.9
7. Sugar maple-yellow birch-balsam fir	19	5	26.3 ± 21.0	47.2
8. Sugar maple-yellow birch and balsam fir-yellow birch	47	22	46.8 ± 12.8	83.9
9. Balsam fir-yellow birch	87	41	47.1 ± 9.7	84.5
10. Balsam fir-white birch or balsam fir-red maple	46	16	34.8 ± 12.0	62.4
11. Yellow birch-balsam fir	52	29	55.8 ± 13.9	100.0
12. Balsam fir-white birch (west)	59	22	37.3 ± 12.7	66.9
13. Balsam fir-white birch (east)	76	31	40.8 ± 10.8	73.1
14. Balsam fir-black spruce	28	12	42.9 ± 19.2	76.8
15. White spruce-balsam fir and balsam fir-white spruce	22	3	13.6 ± 15.5	24.5
16. Black spruce-balsam fir-moss	18	3	16.7 ± 20.2	29.9
17. Black spruce-moss	32	10	31.3 ± 17.6	56.0
18.			*not analysed*	

* minimal: 0–29.9%; low: 30–59.9%; high: 60–89.9%; maximal: 90–100%

** Deciduous forest: 1–7; mixed forest: 8–11; coniferous forest: 12–17

In the hinterland, the species seeks dense mixed or coniferous forests with clearings or openings created by water. Jacques Cartier River, Laurentide Wildlife Reserve, north of Québec City. *June 1991, Réginald Ouellet.*

The Sharp-shinned Hawk is particularly evident during migration, when it generally flies low, at 550–1000 m above ground level (Kerlinger et al. 1985b). The fall migration begins with the juveniles in August, followed by the adults in September through October. In the St. Lawrence Valley, migration usually peaks in mid-September.

Of Québec's three accipiters, the Sharp-shinned Hawk was atlassed in the greatest number of squares. Nests were discovered in only 6 squares, however, accounting for 7.8% (6/77) of the confirmations of breeding obtained. Most records of probable or possible breeding were based on reports of individuals in suitable nesting habitat. Breeding is very difficult to confirm in this species.

Distribution

The Sharp-shinned Hawk breeds in all of the Canadian provinces and in most of the United States. In Québec, it nests in all regions south of the 51st parallel. Although breeding has not yet been confirmed on Anticosti Island, the Sharp-shinned Hawk has been seen there regularly in summer. The Atlas data show that the Sharp-shin is associated with coniferous, mixed, and deciduous forests zones, where there is a variety of open habitat for hunting. It is less frequent in the Central St. Lawrence Lowland and the neighbouring regions upstream of Trois-Rivières. This hawk was reported, though more sporadically, in the Magdalen Islands and on the Island of Montréal, with several records confirming breeding in both areas. Nests occupied by an adult were seen in Châteauguay and Dundee, and a juvenile bird was reported on Mount Saint-Hilaire. Nests containing eggs have been found near Mount Saint-Hilaire (1916), and at Chambly (1924), La Prairie (1927), and Sainte-Dorothée (1951, 1952) (Ouellet 1974a).

The Sharp-shinned Hawk was observed sporadically in the black spruce-moss forests that are found mainly north of the 49th parallel. Reports of probable breeding have come from as far north as the Rupert, Eastmain, and Opinaca rivers, and the species has also been sighted near the Great Whale River (Cons. G. G. – GREBE 1990c, 1992m; ÉPOQ). The Atlas record of confirmed breeding from Lake Ouescapis is the northernmost record for Québec.

History and trends

It is very difficult to determine the status of Sharp-shinned Hawk populations in North America. Most of the current data on populations come from the counts taken during spring and fall migration, particularly at the Hawk Mountain Sanctuary in Pennsylvania, where large numbers of these little hawks are seen every year (Nagy 1977). According to Robbins (1974) and Nagy (1977), the species declined in number between 1947 and 1971, particularly in the northeastern United States. The causes of the decline have not been ascertained, but many authors attribute it to pesticide contamination, habitat loss, and hunting, particularly in areas where the birds gather in migration (Snyder et al. 1973; Nagy 1977; Flood and Bortolotti 1984).

Adult feeding 5 chicks in a conifer in Sainte-Majorique, near Drummondville. *July 1991, Jeanne Lehoux.*

Five young Sharp-shinned Hawks in the nest; the smaller males leave the nest earlier than the females. The young are dependent on their parents until the age of 7–10 weeks. Sainte-Majorique, near Drummondville. *July 1993, Jeanne Lehoux.*

Birds of prey were not protected by all 50 American states until 1972 (Millsap 1987). Populations appear to have increased since the early 1970s (Nagy 1977; Heintzelman 1982), but this could be due in part to the growing popularity of hawk watching (Tate and Tate 1982). According to Flood and Bortolotti (1984), there are tens of thousands of Sharp-shinned Hawks in North America, and the species is neither rare nor threatened with extinction. The Sharp-shinned Hawk has been on the Blue List since 1972 (Tate 1986), however. In Québec, there has been legislation partially protecting diurnal birds of prey since 1887, but raptors were not fully protected until 1992 (see Cazelais 1994).

The Sharp-shinned Hawk is frequently seen in southern Québec during the breeding season. It is the most common accipiter, and one of the most common raptors, in the province. Population trends are difficult to determine, however, since our knowledge of the species' breeding biology and population density is very limited.

Before European settlement, the Sharp-shin was probably uncommon in the Central St. Lawrence Lowland. According to Champlain and other early explorers, this region was covered almost entirely by forests (Ouellet 1974a), and there must have been fewer of the forest edges and openings required by the species than there are today. Hall (1862) and Wintle (1896) considered the Sharp-shinned Hawk to be a common "transient visitant" in the Montréal region, and Wintle even suspected that the species nested there. At the turn of the century, W. J. Brown discovered more than 200 nests in and around Montréal over the course of several years and knew of 50 localities in the province where the species nested (Bent 1937). This information suggests that populations have probably declined since the turn of the century, which would be in keeping with the general trend in North America.

The Sharp-shinned Hawk used to be considered one of the major predators of domestic fowl, and was hunted in Québec and elsewhere by chicken farmers, who regarded the hawks as harmful vermin (Bent 1937). Today, the Sharp-shin still has a bad reputation as a killer of songbirds, and programs to educate the public on the role of raptors in the ecosystem are needed more than ever. This hawk's colourful eggs used to be prized collectors' items (Bent 1937), another reason why it may have suffered in the past.

Many Sharp-shinned Hawks die when they strike plate-glass doors, picture windows and other reflective surfaces on buildings, particularly when the image reflected is the surrounding bushes and trees. Sharp-shins account for a third of the raptor specimens brought into Québec's Ministry of Environment and Wildlife, and most have died as a result of such collisions (R. Ouellet, MEF, unpubl. data).

Réginald Ouellet and Manon Bombardier

The nest consists of a platform of dry sticks. At the turn of the century, the species suffered because its colourful eggs were prized by collectors. Ahuntsic, Montréal. *May 1942, Henry Mousley, Laval University Collection.*

Cooper's Hawk

Épervier de Cooper

Accipiter cooperii

Guy Michaud

Cooper's Hawk, easily confused with its smaller cousin, the Sharp-shinned Hawk, and the immature Northern Goshawk, is a rare summer resident in southern Québec, seen in small numbers at established hawk migration sites such as those at Valleyfield, Senneville, and south of Montréal (Barnhurst and McIntosh 1993). It is currently classified as vulnerable, with probably fewer than 60 pairs breeding exclusively in southwestern Québec (Robert 1989b).

Using its long tail and shortish rounded wings to great effect, this diurnal raptor hunts low to the ground, often exploiting thick cover and taking its prey by surprise. Its preferred diet is small birds, including the occasional nestling (Nelson 1968), but it often takes birds as large as chickens, thus earning its unfortunate colloquial name of chicken hawk. It commonly hunts mammals such as chipmunks, squirrels, and cottontails (Meng 1959; Bielefeldt et al. 1992), as well as a few reptiles, amphibians, and insects. The species occasionally over-winters in the southwesternmost part of the province and is often sighted taking birds at feeders.

Habitat and behaviour

Because of its food preferences, Cooper's Hawk frequents lightly to heavily wooded areas with an abundance of prey. The forest cover makes the species hard to observe. Solitary Cooper's Hawk nests are found in deciduous (oak, beech, maple, etc.), coniferous (normally Eastern white pine, but also Eastern hemlock and fir), or mixed woodlots, 7–20 m up in a large tree, often fairly close to water. Nests may be built of sticks and twigs in a crotch or on top of an old crow or squirrel nest, typically 70 cm in diameter and 15–20 cm deep, lined with bark and occasionally dressed with pine needles. Either sex may choose the nest site, but the male does most of the building (Palmer 1988). Often more than one nest is built each year before incubation begins (Rosenfield et al. 1991). A pair may nest in the same woodlot for several years in succession, but then may desert the area in favour of another (Bent 1937). On rare occasions, this hawk may nest in woodlots in suburban or urban areas (Murphy et al. 1988; Palmer 1988).

This stick nest in the top of a maple in a hardwood stand belongs to a Cooper's Hawk. Laval. *July 1991, Pierre Bannon.*

Notes on territory

The territory required for successful nesting tends to vary in size with prey density and geographic region. The actual nesting territory is small, with the nest at its centre, and may be defended against intruders to a distance of 100 m or so (Palmer 1988). According to Rosenfield (1990), most breeding activities take place within an area 800 m in diameter around the nest. The hunting territory is much larger and overlaps the nesting area only a little or not at all. In western Oregon, Reynolds and Wight (1978) reported a density of 1 nest/2321 ha (mean distance of 5.0 km between nests) in one year, 1 nest/1857 ha (mean distance 5.5 km) the next year, and in eastern Oregon 1 nest/2200 ha (mean distance 3.5 km). A study of 6 males and 1 female in Wisconsin showed that the males nested on average 12 km from their birthplace, and the female, 14.5 km from hers (Rosenfield and Bielefeldt 1992).

Cooper's Hawk

Breeding distribution

1984–1989

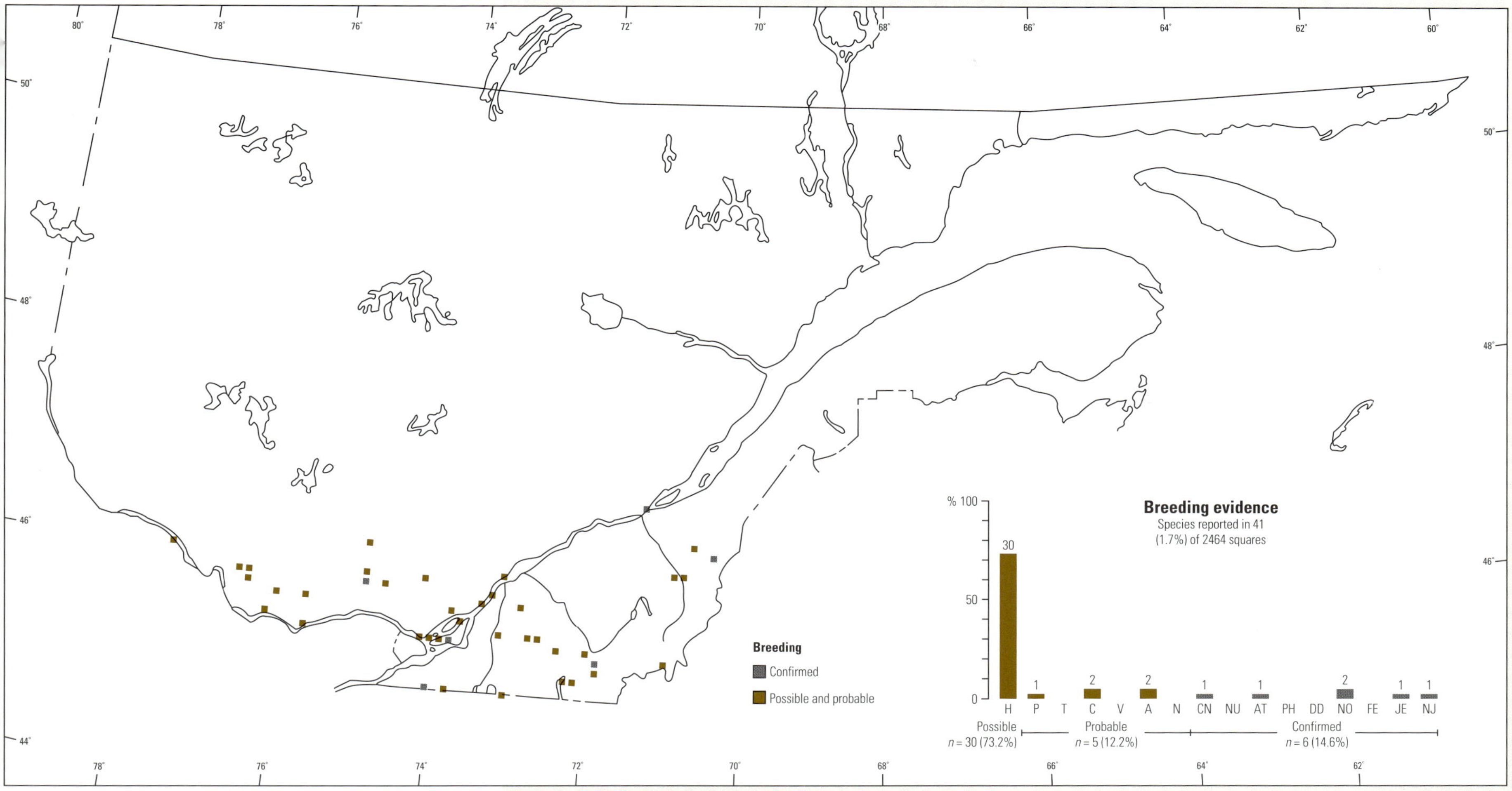

Species profile

Clutch size	usually 4–5 eggs (2–6)[1,2,3,4,5]
Laying interval	1 egg per 1–3 days[1,2,3,5,6]
Start of incubation	with third egg[5]
Incubation period	usually 32–36 days (30–39)[2,4,5]
Maturity at hatching	semi-altricial
Care of young	by female and male[7], *see* text
Nestling period	M usually 27–30 days / F usually 34 days[5,11]
Age at first flight	30–35 days[2,5]
Dependence of young	until 38–53 days after first flight[2,5]
Broods per year	1[1,3]
Breeding age	usually 2 years (1–2)[1,2,5]
Mating system	monogamy[3,5]
Duration of pair bond	probably more than 1 season, possibly for life[1,3,5]
Total length	M 35–46 cm / F 43–51 cm[4]
Wing span	68.6–91.4 cm[9]
Weight	M 349 g (297–380) / F 529 g (460–588)[10]
Longevity record	12 years[8]

References:

1. Penak 1981
2. Newton 1979
3. Bent 1937
4. Godfrey 1986
5. Palmer 1988
6. Harrison 1978
7. Ehrlich et al. 1988
8. Rosenfield and Bielefeldt 1993
9. Terres 1980
10. Dunning 1984
11. Johnsgard 1990

Chronology *n* = 5

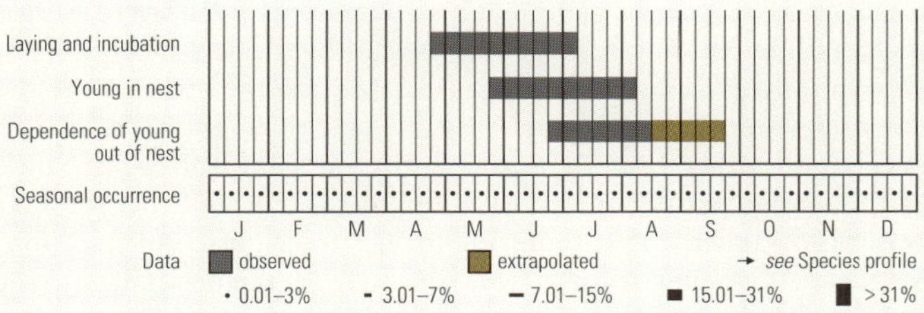

World breeding range

Courtship behaviour occurs in late March or April. The birds fly together or separately in circles with slow, jerky wingbeats, batting the air much like a Short-eared Owl. This is interspersed with folding of the wings close to the body and a brief glide, often with an upward movement, in the manner of a flicker. This hawk vocalizes infrequently, except in close proximity to the nest and at dawn, when both sexes call to each other (Rosenfield et al. 1991).

Usually, 4 or 5 eggs are laid between the end of April and mid-June, and the young hatch asynchronously in about 30–39 days (Reynolds and Wight 1978; Palmer 1988). The female does most of the incubating, but the male may take her place for short periods (Jones 1979). The male does all the hunting during the incubation period and for the first 3 weeks after the young have hatched, and the food is usually delivered in an aerial food pass near the nest (Johnsgard 1990). Results of a study in Arizona and New Mexico suggest that parents always hunt at maximal capacity to feed their young, regardless of the size of the brood. Clutch size seems to be related to the quality of the particular habitat of the breeding pair, with the largest clutches being laid in areas of high prey abundance (Snyder and Snyder 1973).

By 27–30 days of age, the male chicks are ready to leave the nest, while the larger females do not leave before about 34 days (Johnsgard 1990). When the young first leave the nest, they usually clamber out onto the branches nearby, frequently returning to the nest for food and rest. Usually, the young are flying at about 35 days and begin exploring their surroundings, while slowly learning to hunt for themselves. For the first 10 days or so after fledging, the parents continue to bring food to the nest, but soon after that the hungry young will often intercept the adults in flight, forcing them to deposit the prey on the nearest platform (Palmer 1988). The young may remain in the vicinity of the nest, relying on their parents for food, for up to 53 days after fledging (Reynolds and Wight 1978).

This rare bird is extremely difficult to identify and was recorded in only 1.7% (41/2464) of Atlas squares surveyed. It hunts either in the dense cover of wooded areas or near forest edges, but its large size may help identify it from a distance during courtship flights. Nests are even harder to find. This species nests relatively early in the breeding season, and eggs may be laid by the end of April. Nests are typically placed in the cover of a dense canopy (Palmer 1988), and the adults are quiet and secretive at the nest throughout the incubation period. Atlas records show breeding was confirmed only 6 times in the 6-year period; the most frequently used confirmation code was of adults leaving or entering an occupied nest. An additional 5 probable breeding records were reported, on the basis of courtship activity or agitated behaviour near a probable nest site. The clue to a brooding female is often merely a glimpse of tail feathers extending over the lip of a nest. In defence of a nest with young, the adults become noisy and aggressive, giving a *cuk cuk cuk* call similar to that of the Pileated Woodpecker, and they will often strike intruders from the rear. As with many accipiters and falcons, a telltale sign of the bird's presence is a scattering of feathers on and beneath a favoured plucking perch (which may be an old nest). Once the young leave the nest, their begging calls may also attract attention.

Cooper's Hawk is known to have nested in this pine stand. Gowan Road, Rockburn, south of Montréal. *May 1991, Pierre Bannon.*

Distribution

Cooper's Hawk breeds throughout southern Canada from British Columbia, across the Prairie provinces with the exception of Saskatchewan, to southern Ontario and southernmost Québec; it breeds rarely in the Maritimes. It is found almost everywhere in the United States except southern Florida, and occurs in northwestern Mexico (Godfrey 1986). In Québec, the Atlas records came largely from maple-dominated deciduous forests in the Montréal area, the Ottawa and St. Lawrence river valleys, the Eastern Townships, the Beauce, and the lower Laurentians. A few Cooper's Hawks were sighted in the mixed forests of the Appalachian region. Only 6 breeding reports were confirmed, from Sainte-Foy, Sainte-Rose-de-Watford in the Beauce, Sherbrooke, the Island of Montréal, the southern Montérégie, and the Ottawa Valley. Due to the difficulty of identifying this species and locating nests, it is probably underrepresented on the Atlas map, but possible and probable breeding reports give a good idea of its breeding range.

Pre-Atlas records have been reported during the breeding season outside the breeding range, with records coming from as far north as Baie-Comeau (1961 and 1984) on the North Shore; Val-Jalbert (1960), Lake Corbeau (1970), Falardeau (1975), and Saint-Ambroise (1979) in the Saguenay–Lac-Saint-Jean region; Île aux Basques (1974) and Saint-Mathieu near Rimouski on the Lower St. Lawrence, and Sainte-Anne-des-Monts (1957 and 1976) and Forillon National Park (1975) on the Gaspé Peninsula (ÉPOQ). Due to the difficulty of accurately identifying the species, these sightings may or may not reflect an expansion of its range (David 1978a), and breeding in these areas remains to be confirmed.

For several years, this peatland has been used as a hunting territory by a Cooper's Hawk nesting in an adjoining woodlot. Sainte-Foy. *September 1991, Yves Aubry.*

History and trends

According to historical records, Cooper's Hawk may never have been a common breeding bird in Québec. Dionne (1906) described the species as rare as both a transient and a resident around Québec City and Montréal. Macoun and Macoun (1909) listed it as "sparingly distributed through the whole St. Lawrence Valley," though common in the Restigouche Valley. Throughout much of its North American breeding range, however, the species was considered a common breeder (Bent 1937; Robbins 1991), but since the early 1900s its population has been in decline. The migration records at Hawk Mountain, Pennsylvania, show a downward trend after 1937, with particularly sharp declines after 1947 (Snyder et al. 1973). Cooper's Hawk was on the Blue List from 1972 until 1982, then in response to increasing numbers in several parts of the continent, including the East, it was downgraded to a Species of Special Concern in 1982 (Tate and Tate 1982); it was blue-listed again in 1986, however (Tate 1986). Fyfe (1976) classified the population as slightly declining or stable for southern Québec and Ontario, with low to medium relative abundance.

Three young Cooper's Hawks about 3 weeks old. Sainte-Rose, Laval. *August 1943, Henry Mousley, Laval University Collection.*

The decline in the early part of the century was due to heavy hunting pressure (Henny and Wight 1972), and this trend continued into the 1960s (Erskine 1992). This hawk had a much-exaggerated reputation as the chicken hawk, "so cordially hated by poultry farmers" (Bent 1937).

Since the late 1940s, the species appears to have been adversely affected by accumulated organochlorine residues in the tissues of its prey, mainly migratory insectivorous birds (Snyder et al. 1973; Penak 1981). According to Henny and Wight (1972), organochlorines may have been responsible for the 25% reduction in productivity in the northeastern United States and southeastern Canada for the years 1949–67, and Snyder et al. (1973) demonstrated a significant correlation between breeding failure and concentrations of DDE, a metabolite of DDT, in its eggs. Since restrictions were imposed on the use of organochlorine pesticides in Canada and the United States in the 1970s, concentrations detected in raptor eggs have generally declined. DDE residues in eggs of Cooper's Hawks from southern Ontario remained at relatively high levels between 1980 and 1988, however, and the persistent nature of these contaminants in the environment, as well as continued pesticide use on their wintering grounds in the southern United States, may be contributing factors (Noble and Elliott 1990). Further limiting factors are the sparseness of the remaining mature forest and human harassment of the birds on their breeding grounds.

Bob Barnhurst, Mabel W. McIntosh, and Derin Henderson

Northern Goshawk

Autour des palombes

Accipiter gentilis

Guy Michaud

THE NORTHERN GOSHAWK IS A RAPTOR THAT INVARIABLY EXCITES interest. To protect her brood, the female will unhesitatingly launch ferocious, repeated attacks on any potential predator, even a human being. Woe to the intruder, for the Northern Goshawk, along with the Great Horned Owl, Snowy Owl, and a few other species, is one of the few birds both bold enough and strong enough to inflict serious injury.

This denizen of the northern forests is the most sedentary of our three accipiters. The Northern Goshawk may even overwinter north of the Atlas area, preying mainly on ptarmigan and lemmings (Todd 1963), but every few years when hare and grouse populations crash, it irregularly invades southern Québec and the northern United States at the onset of winter. That is when the bird is most often seen, sometimes even in city parks.

Habitat and behaviour

In Québec, the Northern Goshawk is an early nester. In the southern part of the province, pairs return to the previous year's nesting grounds in early March or, less often, in late February. Goshawks nest in all types of forests, seeming, however, to prefer mature stands with a dense canopy and open understorey (Crocker-Bedford and Chaney 1988). According to Speiser and Bosakowski (1989), in eastern North America they prefer deciduous forests, building their nests mainly in birch or beech.

The onset of breeding is marked by noisy vocalizations and spectacular courtship displays by the male and female. A pair may reuse the same nest each year, use an alternate nest within their territory, or build a new one (Reynolds and Wight 1978; Speiser and Bosakowski 1991). On Mount Saint-Bruno, one pair—or its successors—built 10 different nests in 10 years in an area of about 20 ha (F. Morneau, pers. comm.). Most of the nest building is done by the female. Twigs and leaves are added throughout the nesting period (Schnell 1958), to a diameter of up to 0.9 m and depth of 0.5 m (Palmer 1988).

Notes on territory

The pair is generally faithful to its nesting territory, which may contain more than one old nest and, as in Oregon, cover 8–10 ha (Reynolds and Wight 1978). A nest may be used up to 8 years in a row (Speiser and Bosakowski 1991). In Oregon, 4 nests in a 117 km² forest were 5.6 km apart, yielding a density of approximately 0.036 pair/km² (1 pair/2750 ha), which is rather high compared with densities found in Alaska by McGowan in 1971 (0.021 pair/km²; 1 pair/4869 ha) and 1972 (0.024 pair/km²; 1 pair/4142 ha) (Reynolds and Wight 1978).

The size of the goshawk's home range in central Sweden varies according to the abundance of prey; in a woodland with large populations of pheasants, for example, it covers 20 km², compared with 57 km² in the boreal forest, where prey is scarcer (Kenward and Widén 1989). In Wyoming, one adult lived permanently within a range estimated at 2.1 km² (Craighead and Craighead 1956).

Built under leafy cover, in the main crotch of an aspen, this nest was home to a Great Horned Owl in 1981 and a Northern Goshawk in 1985. Saint-Lazare, west of Montréal. *June 1981, Pierre Bannon.*

Northern Goshawk

Breeding distribution 1984–1989

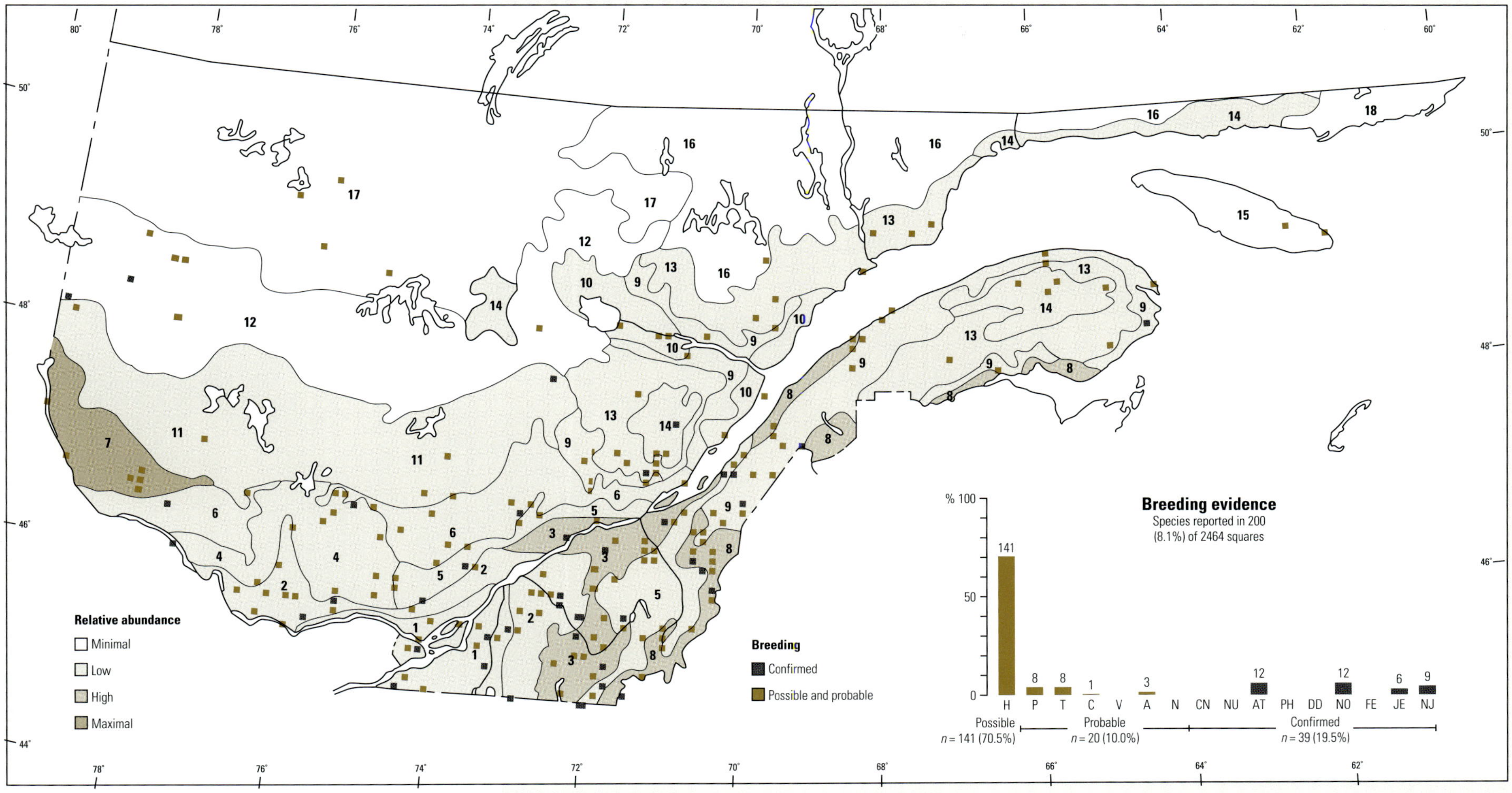

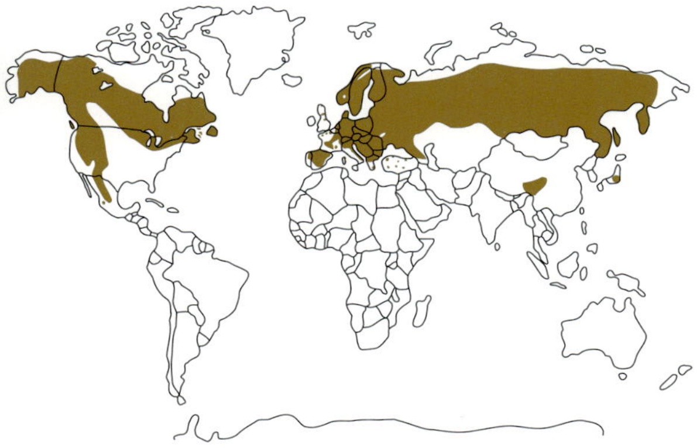

World breeding range

Species profile

Clutch size	usually 2–4 eggs (1–5)[3,4]
Laying interval	1 egg per 2–3 days[1,3,8]
Start of incubation	probably with last or penultimate egg, sporadically after second egg(9)
Incubation period	30–38 days[2,3,4]
Maturity at hatching	semi-altricial
Care of young	by female and male[1,3,4], *see* text
Nestling period	usually 35 days (34–49)[1,2,4,8]
Age at first flight	M usually 35–36 days / F usually 40–42 days[8]
Dependence of young	until ≈ 70 days old[8]
Broods per year	1[3]
Breeding age	usually 2 years (1–2)[2,3]
Mating system	monogamy[8]
Duration of pair bond	probably for life[3]
Total length	51–66 cm[5]
Wing span	48.2–68.6 cm[4]
Weight	M 912 g (735–1099) / F 1137 g (845–1364)[6]
Longevity record	13 years[7]

References:
1. Schnell 1958
2. Reynolds and Wight 1978
3. Palmer 1988
4. Terres 1980
5. Godfrey 1986
6. Dunning 1984
7. Klimkiewicz and Futcher 1989
8. Cramp and Simmons 1980
9. Anonymous 1990

Chronology $n = 34$

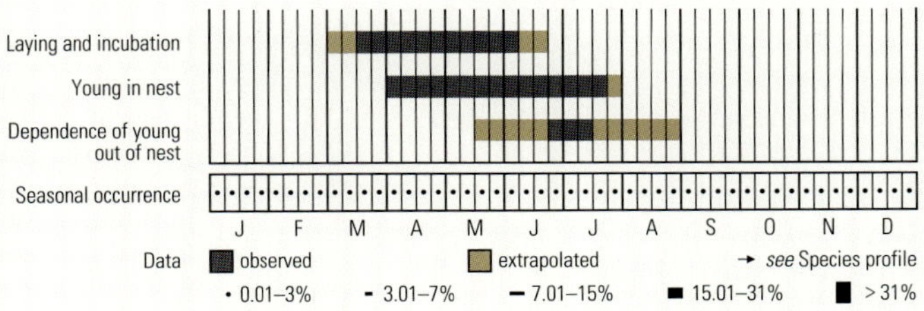

Frequency of occurrence by ecological region

Ecological regions** (forest regions)	Adequately covered squares	Squares in which reported	% of squares in which reported	Relative abundance (%)*
1. Sugar maple-hickory and sugar maple-basswood	97	14	14.4 ± 3.6	45.7
2. Sugar maple-basswood	153	23	15.0 ± 3.0	47.6
3. Sugar maple-basswood and sugar maple-yellow birch	101	21	20.8 ± 4.9	65.8
4. Sugar maple-yellow birch-basswood	56	10	17.9 ± 9.0	56.5
5. Sugar maple-yellow birch-beech and sugar maple-yellow birch-basswood	113	14	12.4 ± 3.5	39.2
6. Sugar maple-yellow birch-beech	71	13	18.3 ± 8.4	58.0
7. Sugar maple-yellow birch-balsam fir	19	6	31.6 ± 22.1	100.0
8. Sugar maple-yellow birch and balsam fir-yellow birch	47	13	27.7 ± 11.6	87.6
9. Balsam fir-yellow birch	87	14	16.1 ± 7.3	51.0
10. Balsam fir-white birch or balsam fir-red maple	46	5	10.9 ± 8.2	34.4
11. Yellow birch-balsam fir	52	6	11.5 ± 9.3	36.5
12. Balsam fir-white birch (west)	59	5	8.5 ± 7.7	26.8
13. Balsam fir-white birch (east)	76	10	13.2 ± 7.6	41.7
14. Balsam fir-black spruce	28	5	17.9 ± 15.2	56.5
15. White spruce-balsam fir and balsam fir-white spruce	22	1	4.5 ± 10.3	14.4
16. Black spruce-balsam fir-moss	18	0	0	0
17. Black spruce-moss	32	1	3.1 ± 7.6	9.9
18.			*not analysed*	

* minimal: 0–29.9%; low: 30–59.9%; high: 60–89.9%; maximal: 90–100%

** Deciduous forest: 1–7; mixed forest: 8–11; coniferous forest: 12–17

Adult and young on nest. The structure of this nest in the crotch of a beech is plain to see. Mont-Saint-Bruno Conservation Park, east of Montréal. *June 1979, François Morneau.*

Built of sticks of various sizes and lined with bark, the nest is usually placed in a high crotch of a tree, 7–20 m from the ground (Speiser and Bosakowski 1989). The nest is easy to locate early in the season, especially before sunrise, when the adults are performing their noisy territorial displays (Weir 1987b). Once incubation begins, the pair becomes quieter and chances of spotting a goshawk after late April are much slimmer, unless the nest is approached; intrusion within 100 m of the nest generally provokes an attack by one of the parents (Speiser and Bosakowski 1991). The female incubates and protects the nest and young. The small percentage (1.6%) of Atlas squares in which breeding was confirmed is no doubt a reflection of the species' secretive behaviour and low population density.

Incubation lasts 30–38 days (Reynolds and Wight 1978; Terres 1980; Palmer 1988), during which time the male provides food for the female (Schnell 1958). He typically deposits the prey at a nearby plucking perch, and stands over the eggs while she feeds (Quinton 1994). Hatching occurs at the time of year when prey is most abundant. The female feeds the young on prey delivered by the male, mostly young thrushes, corvids, and woodpeckers (Schnell 1958; Reynolds and Meslow 1984), but also squirrels (Quinton 1994). About 3 1/2 weeks after hatching, the female begins to hunt, remaining near the nest (Palmer 1988), whereas the male ranges farther afield. The young stay in the nest until they are 34–49 days old; age of first flight is 35–42 days, but fledglings return to the nest until the age of about 70 days to be fed by the adults (Schnell 1958; Reynolds and Wight 1978; Cramp and Simmons 1980; Terres 1980).

Distribution

The Northern Goshawk is found in every province and territory in Canada. It breeds throughout Québec, except on the Ungava Peninsula and the Magdalen Islands (David 1980a). Its distribution in southern Québec confirms that it prefers areas with hardwood growth, as Speiser and Bosakowski (1989) observed. Témiscamingue is certainly a major breeding area, thanks to the large stands of mature hardwoods there, although its importance has been overestimated due to poor coverage. Numbers decrease as deciduous growth gives way to coniferous, as in the large forest areas dominated by fir and spruce. We can assume, as did Weir (1987b) in Ontario, that the paucity of sightings in the boreal forest is an indication of lower densities. This is also true of the Magdalen Islands, where the species has, however, been observed on several occasions (Fradette 1992), and of Anticosti, where the small number of sightings lend credence to Schmitt (1904) and Dionne's (1920) assertions that the species was rare on the island.

The Northern Goshawk's range north of the 50th parallel extends to around Ungava Bay, a few kilometres above the tree line; sightings there are rare to uncommon, and mainly in the forest enclaves of the Rivière aux Feuilles Valley (Gariepy 1980) and at Forks (Gabrielson and Wright 1951). Records indicate that the goshawk ranges widely in this northern area but is not common anywhere (Todd 1963). Breeding has been confirmed in several spots along the James and Hudson bay coasts, around the Great Whale River (Todd 1963; Cons. G. G. – GREBE 1990c), and around the Eastmain River (F. Morneau, pers. comm.). The Northern Goshawk is considered to be a probable breeder in the area of the Nottaway, Broadback, and Rupert rivers (Cons. G. G. – GREBE 1992m). The only records from the North Shore are from Baie-Johan-Beetz (Townsend and Bent 1910) and the Bonne-Espérance region (Todd 1963).

The goshawk nests in mature stands in the mixed forest around Lake Matane. Matane Wildlife Reserve. *July 1991, Yves Aubry.*

The greatest number of records south of the 47th parallel are from the Lake Timiskaming, Eastern Townships, Beauce, and Québec City regions. Ouellet (1974a) reported that between late September and mid-December, the species is sometimes seen more often in the Montréal area. He attributes such irregular migratory movements to a decreased abundance of small mammals and of Ruffed and Spruce grouse within its range. Goshawks are often drawn to Montréal city parks and suburbs when food is scarce, and are at times seen hunting pigeons, even chasing them into dovecotes.

History and trends

The Northern Goshawk has always been considered rare in Québec, as throughout the rest of its range. Most 19th century authors (Couper 1881; Caulfield 1890a; Wintle 1896) considered the bird to be a resident or migrant, but were never able to confirm its presence during the breeding period. It must, however, have been more abundant than it is today. The massive felling of mature stands of trees and clearcutting may have had a negative effect on the overall population. On the other hand, the gradual reversion of farmland to woodland over the last 10 years in all agricultural areas outside the Greater Montréal area might prove favourable to the species. Forest regeneration in New York and Vermont appears to be behind the remarkable resurgence of the population since the 1950s (Bull 1985; Ellison 1985q; Eaton 1988k). Present data do not, however, allow us to conclude that a similar expansion has occurred in southern Québec.

At the time of European colonization, the goshawk was considered a pest, mainly because of its predation on domestic fowl. Although it is now protected by provincial law, its audacious attacks make it highly vulnerable to reprisals in populated areas. For example, during the Atlas period, a pair of goshawks was shot by some young all-terrain vehicle drivers who had been repeatedly attacked by birds defending their nest territory while they were driving by.

Because it often hunts on the ground, the Northern Goshawk is the raptor most frequently caught in rabbit snares (R. Ouellet, pers. comm.). It is not known what impact this additional threat might have on goshawk populations.

On the other hand, the species appears to have been little affected by pesticide use, as its preferred prey—hares, squirrels, grouse, ptarmigan, and crows—feeds mainly on plant matter and is less liable to accumulate high concentrations of contaminants (Bednarz et al. 1990).

PIERRE BANNON, FRANÇOIS MORNEAU, AND MANON BOMBARDIER

Red-shouldered Hawk

Buse à épaulettes

Buteo lineatus

Ghislain Caron

The Red-shouldered Hawk, ironically, is not likely to be identified by the characteristic featured in its name. While a reddish shoulder patch is visible on this bird perched at eye level, an observer is more likely to identify it by its piercing territorial call, a call often imitated by Blue Jays (James 1988). In flight, the crescent-shaped, translucent "windows" in the wrist area of the wings combined with a black tail crossed by several narrow white bars, (as opposed to the wider bars half in number in the smaller Broad-winged Hawk), are diagnostic. As indicated by David (1980a), the Red-shouldered Hawk is still listed as a migrant breeder in southern Québec, occurring irregularly in winter in the Central St. Lawrence Lowland.

Habitat and behaviour

The home range of breeding Red-shouldered Hawks is generally situated in mature deciduous woodlands, especially flood-plain forests or mature lowland forests with open water and clearings, swamps, and wooded marshy margins nearby (Bednarz and Dinsmore 1981; Titus and Mosher 1981; Palmer 1988; Bloom et al. 1993). In southwestern Québec, Red-shoulders nest in mature deciduous forests where beech and sugar maple predominate (Morris and Lemon 1983). In Ontario, mature woodlands dominated by maple, beech, and hemlock are frequently used for nesting (Sharp and Campbell 1982).

The nest is typically made of sticks and lined with bark strips, leaves, lichens, green twigs, and down (Godfrey 1986). It is usually placed 14–19 m up in the main fork of a sturdy deciduous tree (Bednarz and Dinsmore 1982; Morris et al. 1982), though on rare occasions it may be in a conifer. Bent (1937) suggested that nest construction took 4–5 weeks, but Palmer (1988) argued that one can be built in a few days. A nest may be refurbished and reused for 2 or 3 years in a row, or the pair may alternate nests within the territory. On Mount Saint-Bruno, a nest was used for 7 years out of 8, and 6 years in succession; nests are commonly reused there for more than 2 years (F. Morneau, pers. comm.). Occasionally an old crow, Cooper's Hawk, or squirrel nest is used as a foundation (Bent 1937). Old nests are generally lined with fresh sprigs of greenery, often hemlock, pine, or cedar, but even entire nightshade or violet plants have been used. Pair bonds may last for life, especially over the southern range, where this hawk is a permanent resident. Pairs have a strong attachment to their breeding territory. In Québec, on Mount Saint-Bruno, one territory had been used for 21 years, and probably longer (F. Morneau, pers. comm.). In Massachusetts, Bent (1937) recorded one territory as being occupied continuously by either the same pair or succeeding generations for at least 42 years, and another for 45 years.

Normally a quiet, shy species, this hawk is most easily detected in the spring, when it becomes noisy and conspicuous. One or both sexes can be sighted circling over the breeding territory, swinging near and then away from the other, with wings and tails fanned, accompanied by repeated loud cries of *kee-ah, kee-ah*. One might climb high and then dive toward the other (Bent 1937). Males defend the territory from conspecifics in an aggressive display of aerial attacks and talon locking.

Notes on territory

In Michigan, hunting ranges averaged 1.4 km in diameter, were circular in shape, and centred on the nest tree (Craighead and Craighead 1956). In suitable habitat, nesting densities averaged 1.2 pairs/km² in Maryland (Brown and Amadon 1968), and in central Massachusetts ranged from 1 pair/0.9 km² to 1 pair/3.0 km² (mean density was 1 pair/1.7 km²) (Portnoy and Dodge 1979). Red-shouldered Hawks appear to be more tolerant of other nesting raptors than of conspecifics. In a study in New York State, the mean inter-nest distance for Broad-winged and Red-shouldered hawks was 877 m, whereas the mean inter-nest distance for neighbouring Red-shouldered Hawks was 1271 m; furthermore, in the same study an active Barred Owl nest was found only 120 m from a Red-shouldered Hawk's nest (Crocoll and Parker 1989). On Mount Saint-Bruno, Barred Owl nests are often as close as 50 m to active Red-shouldered Hawks' nests (F. Morneau, pers. comm.).

The Red-shouldered Hawk's preferred nesting habitat is mature hardwood forest, with fields nearby for hunting. Rigaud Mountain, west of Montréal. *June 1991, Pierre Bannon.*

Red-shouldered Hawk

Breeding distribution
1984–1989

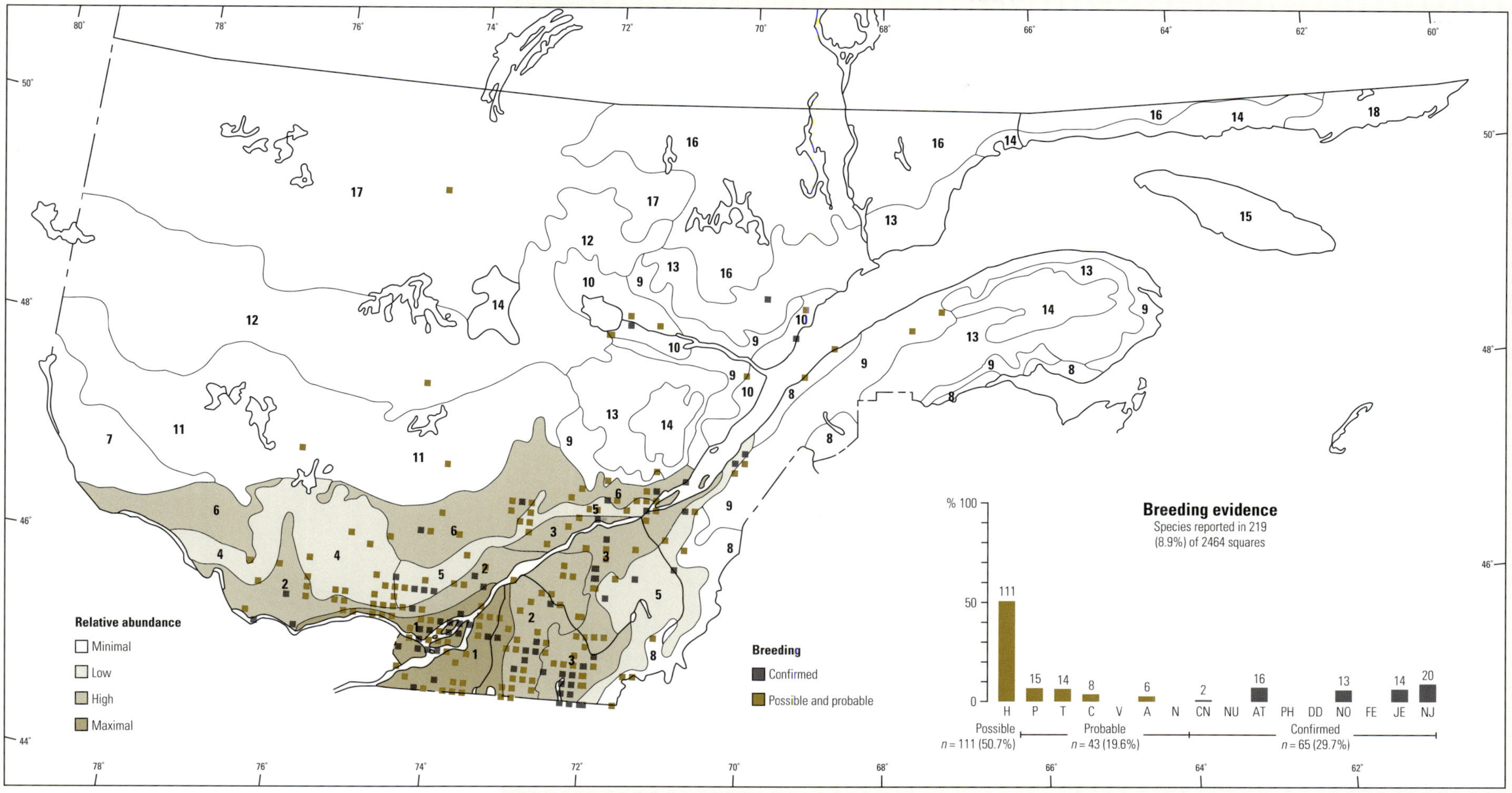

World breeding range

Species profile

Clutch size	usually 3–4 eggs (1–6)[1,2,4,5,6,7]
Laying interval	1 egg per 2–3 days[1,3,5,7]
Start of incubation	before last egg[3]
Incubation period	28–35 days[2,3,5,6,7]
Maturity at hatching	semi-altricial
Care of young	by female and male[3,4], *see* text
Nestling period	32–45 days[3,4,5,6,8,9]
Age at first flight	≈ 45 days[7]
Dependence of young	until ≈ 8–10 weeks after first flight[2,7]
Broods per year	1[3,5,7]
Breeding age	usually 2 years (1–2)[2,7]
Mating system	probably monogamy[3,7]
Duration of pair bond	probably for life[3,7]
Total length	M 45–58 cm / F 48–61 cm[1]
Wing span	M 90–112 cm / F 99–127 cm[7]
Weight	M 543.5 g (486–582) / F 670.3 g (593–774)[9]
Longevity record	20 years[7]

References:
1. Godfrey 1986
2. Newton 1979
3. Bent 1937
4. Portnoy and Dodge 1979
5. Brown and Amadon 1968
6. Crocoll and Parker 1989
7. Palmer 1988
8. DeGraaf and Rudis 1987
9. Crocoll 1994

Chronology $n = 155$

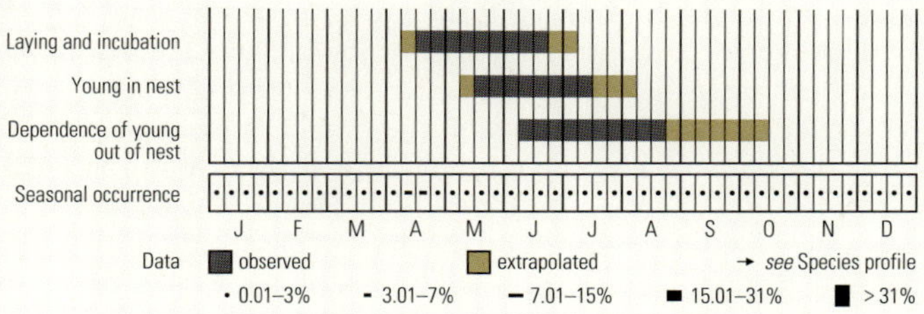

Frequency of occurrence by ecological region

Ecological regions** (forest regions)	Adequately covered squares	Squares in which reported	% of squares in which reported	Relative abundance (%)*
1. Sugar maple-hickory and sugar maple-basswood	97	44	45.4 ± 5.0	100.0
2. Sugar maple-basswood	153	46	30.1 ± 3.8	66.3
3. Sugar maple-basswood and sugar maple-yellow birch	101	34	33.7 ± 5.6	74.2
4. Sugar maple-yellow birch-basswood	56	14	25.0 ± 10.0	55.1
5. Sugar maple-yellow birch-beech and sugar maple-yellow birch-basswood	113	23	20.4 ± 4.2	44.9
6. Sugar maple-yellow birch-beech	71	20	28.2 ± 9.6	62.1
7. Sugar maple-yellow birch-balsam fir	19	0	0	0
8. Sugar maple-yellow birch and balsam fir-yellow birch	47	3	6.4 ± 6.8	14.1
9. Balsam fir-yellow birch	87	0	0	0
10. Balsam fir-white birch or balsam fir-red maple	46	3	6.5 ± 6.7	14.4
11. Yellow birch-balsam fir	52	2	3.8 ± 6.0	8.5
12. Balsam fir-white birch (west)	59	0	0	0
13. Balsam fir-white birch (east)	76	1	1.3 ± 3.0	2.9
14. Balsam fir-black spruce	28	0	0	0
15. White spruce-balsam fir and balsam fir-white spruce	22	0	0	0
16. Black spruce-balsam fir-moss	18	0	0	0
17. Black spruce-moss	32	0	0	0
18.			*not analysed*	

* minimal: 0–29.9%; low: 30–59.9%; high: 60–89.9%; maximal: 90–100%

** Deciduous forest: 1–7; mixed forest: 8–11; coniferous forest: 12–17

The usual clutch size in northern populations is 3 or 4 eggs, but nests containing 1–6 eggs have been reported (Craighead and Craighead 1956; Crocoll and Parker 1989). Eggs are laid at intervals of 2–3 days, and incubation lasts 28–35 days, with the young hatching asynchronously (Crocoll and Parker 1989). The female does most of the incubating, but males participate in some pairs. The male brings food to the female at the nest or to a perch nearby, where she will join him. Some birds will defend the nest aggressively, even striking an intruder, but most will fly away from the nest when a person approaches. If they sense a continuing danger in the nesting area, they may desert the nest. For the first 2 weeks or so after hatching, the male does most of the hunting, passing the food to the female, which then tears the prey into small pieces to feed to the young (Lyon 1978; Portnoy and Dodge 1979). At about 18 days of age, the young begin to feed themselves, and the female spends more and more time away from the nest helping the male hunt for the family. Age at first flight is about 45 days, but the young may return to the nest to feed and to roost at night for some time after, remaining dependent on their parents for food and protection for up to 8–10 weeks.

Although there are regional and seasonal differences in diet, the Red-shouldered Hawk tends to specialize in cold-blooded vertebrates, such as toads, frogs, and snakes, and small mammals like chipmunks, moles, and shrews. They will also prey upon small birds, including nestlings, the occasional small fish, and insects such as crickets and grasshoppers (Bent 1937; Penak 1979). They hunt under the forest canopy or over open terrain that is either marshy or near water, and catch their quarry by using a perch-and-wait technique, or by conducting direct searching flights, then attacking by surprise. This hawk can carry surprisingly heavy prey for short distances, sometimes even matching its own weight.

Red-shouldered Hawks were reported in only 8.9% (219/2464) of Atlas squares surveyed, and breeding was confirmed in just 65 squares. Although their nests may be fairly easy to spot and distinguish from other hawk nests because they are always covered with hemlock branches or other green sprigs of forest greenery, these hawks become quiet and secretive once incubation has begun. A nest with eggs or young was the most frequently used breeding evidence code; locating the source of the begging calls of hungry young can often help pinpoint active nests. Young that are nearly fully fledged may climb out onto the branches by day (Bent 1937), becoming more visible. Reports of recently fledged young accounted for 21.5% (14/65) of squares in which breeding was confirmed. Breeding was also confirmed on the basis of observations of an adult carrying food for the young, or adults entering or leaving nests.

Distribution

The range of this species is quite restricted due to its apparent inability to adapt to habitats other than mature hardwood forest, which is becoming rare in Québec. While a few birds were seen east of Matane on the south shore of the St. Lawrence, in the Saguenay–Lac-Saint-Jean region, and up to around Forestville on the North Shore, the vast majority of the population was found in the Central St. Lawrence Lowland, the Eastern Townships, the Beauce, and the Ottawa Valley, coinciding largely with the Québec distribution of maple-hickory-basswood forest, and to a lesser extent with yellow birch and beech forest. Red-shouldered Hawks may be found nesting wherever small pockets of mixed hardwood forest still exist. There are small local populations in the Hudson–Rigaud area, as well as on Mount Saint-Bruno and Mount Saint-Hilaire.

History and trends

In the early part of this century, the Red-shouldered Hawk was regarded as a common hawk in southern Québec (Dionne 1906), and according to Taverner (1919) was "probably the most common hawk in eastern Canada." A general decline was apparent beginning in the 1950s, however. W. Brown's (1971) analysis of nearly 10 000 Christmas Bird Count reports in the United States in the *Audubon Field Notes* for 1950–69 showed a decrease in the Red-shouldered Hawk's winter populations in every state except California and West Virginia. The species was on the Blue List from 1972 to 1986 (Tate 1986). In Canada, COSEWIC classified the Red-shouldered Hawk as a vulnerable species in 1988 (Robert 1989b), and in the same year in the United States, it was listed as endangered, threatened, or a Species of Special Concern in 17 states (Johnsgard 1990).

Accumulation of organochlorine pesticides was suspected as the cause of reproductive failure. PCBs and dieldrin were considered as possibilities in certain regions (Henny et al. 1973), but alterations in this hawk's habitat appear to be the main cause (Bryant 1986). In earlier years, the conversion of large tracts of forest to farmland favoured the Red-tailed Hawk, a more aggressive species that prefers more open areas (Craighead and Craighead 1956; Bednarz and Dinsmore 1981, 1982). According to Bryant (1986), Red-tailed Hawks cannot dislodge Red-shouldered Hawks from dense, mature forests, however, though Red-tails, being larger and more aggressive, seem to have the upper hand in interspecies competition in woodlands disturbed by human

Red-shouldered Hawk's nest against the trunk of a yellow birch. Lachute, west of Montréal. *May 1943, Henry Mousley, Laval University Collection.*

The Red-shouldered Hawk often nests in hardwood thickets or mixed stands bordering lakes or rivers and wetlands where it hunts. Lac au Chien, Saint-Raymond, west of Québec City.
August 1959, Gaston Le Page.

activity (clearing, draining, logging, and so on). More recently, Montréal's growing urban sprawl, coupled with the development of forests and wetlands for housing and commercial projects, has caused the Red-shoulder's breeding range to shrink steadily every year. On a smaller but not insignificant scale, hardwoods are still being felled for firewood and furniture.

Red-shouldered Hawks will adapt to nesting in mature trees close to human habitation if they are not disturbed (Morris and Lemon 1983). Other than a few local population studies (for example, Morris et al. 1982; Morris and Lemon 1983), little is known of this hawk's needs and population trends in Québec. According to Peterson and Crocoll (1992), the best way to provide suitable habitat for the Red-shouldered Hawk is by creating openings within extensive forest areas through selective cutting.

DAVID M. BIRD AND DERIN HENDERSON

Broad-winged Hawk

Petite Buse

Buteo platypterus

Ghislain Caron

The Broad-winged Hawk, a crow-sized inhabitant of southern Québec forests, is rarely seen in open country. The field marks of the bird in adult plumage are a reddish breast and two broad dark bands on a short tail. It can be identified by its high plaintive whistled notes given often as it soars over the forest. Although secretive, it is commonly seen during the breeding season.

In early September, it migrates to Central and South America in aggregations that sometimes number in the hundreds. In Québec, areas where one can observe kettles include the Morgan Arboretum in Sainte-Anne-de-Bellevue, on Mount Royal, and along the Richelieu River (Barnhurst and McIntosh 1976). In mid-September in Texas, up to 100 000 birds have been seen in a single day (Lasley and Sexton 1993).

Habitat and behaviour

The adult birds arrive in Québec around mid-March. Egg laying begins towards the end of April when the trees begin to leaf out, hindering nest detection. At this time, the birds can be heard calling back and forth to each other as they perform their courtship displays above the treetops.

The Broad-winged Hawk nests in trees, mainly in fairly dense deciduous or mixed forest (278 trees/ha) near clearings, which serve as hunting grounds (Titus and Mosher 1981). The species seems to prefer habitat that is wet and areas located near water (Burns 1911; Matray 1974; Titus and Mosher 1981). Keran (1978) reported nests in Minnesota and Wisconsin to be at a mean distance of 143 m (n = 29) from a wet area. The bird may occupy an abandoned nest of a crow, squirrel, or other hawk, to which it adds new material (Bent 1937). The trees used most frequently are white birch, sugar maple, white oak, and yellow birch (Armstrong and Euler 1983; Titus and Mosher 1987). This hawk rarely nests in trees with trunks less than 19 cm in diameter. The nest is most often located in the main fork of the tree, usually supported by three or four branches (Titus and Mosher 1987) at a height of 4.5–18.5 m (Peck and James 1983).

The nest is built of sticks and green foliage from a variety of coniferous and deciduous trees such as sugar maple, yellow birch, Eastern hemlock, cedar, and some spruces (Matray 1974; Peck and James 1983). Nest materials are gathered by both adults within a radius of 25 m of the nest (Lyons et al. 1986). The female lines the nest with hardwood bark, and continues to do so throughout the incubation period (Matray 1974).

Notes on territory

The Broad-winged Hawk sometimes returns to its previous year's nesting territory, but does not occupy the same nest. Rosenfield (1984) reported that of the Broad-wings that returned to the same

The Broad-winged Hawk, common in southern Québec, prefers to nest in relatively dense deciduous and mixed forests. Sainte-Foy. *September 1991, Yves Aubry.*

Broad-winged Hawk

Breeding distribution
1984–1989

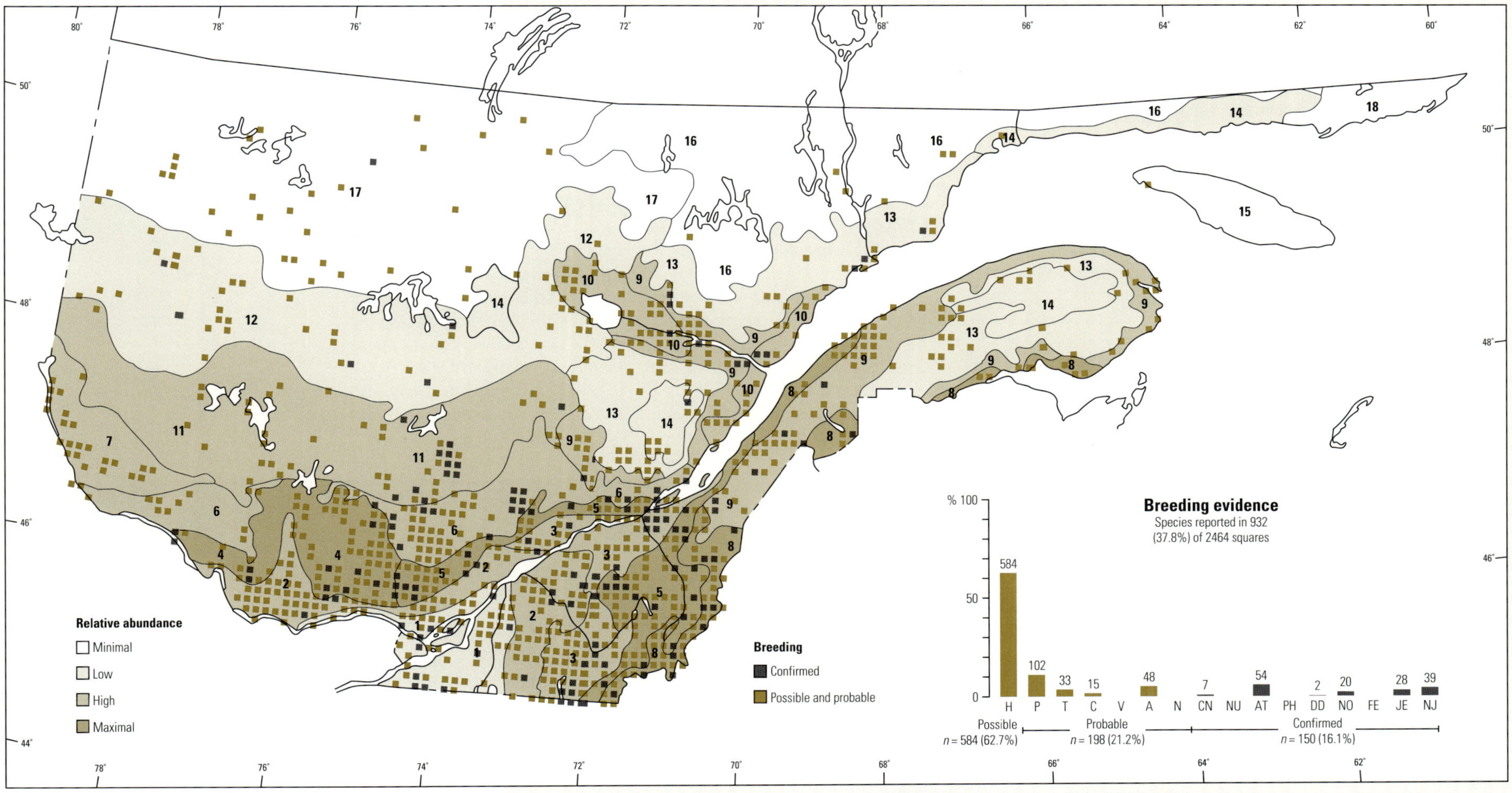

Species profile

Clutch size	usually 2–3 eggs (1–5)[2,3]
Laying interval	1 egg per 2 days, sometimes 1 per day[3]
Start of incubation	with first egg[3]
Incubation period	28–31 days (28–38)[1,3]
Maturity at hatching	semi-altricial
Care of young	by female and male[1,3], *see* text
Nestling period	≈ 29–31 days[1,3]
Age at first flight	usually 35–36 days (30–42)[1,3,4,6]
Dependence of young	until 50–56 days old[1]
Broods per year	1[9]
Breeding age	usually 2 years (1–2)[3]
Mating system	probably monogamy[4]
Duration of pair bond	probably for life[1,2,3]
Total length	M 34–42 cm / F 38–47 cm[5]
Wing span	81.3–99.1 cm[8]
Weight	M 357 g (310–400) / F 440 g (389–559)[3]
Longevity record	18 years 4 months[7]

References:
1. Matray 1974
2. Bent 1937
3. Palmer 1988
4. Ehrlich et al. 1988
5. Godfrey 1986
6. Crocoll and Parker 1989
7. Klimkiewicz and Futcher 1989
8. Terres 1980
9. DeGraaf and Rudis 1987

Chronology *n* = 62

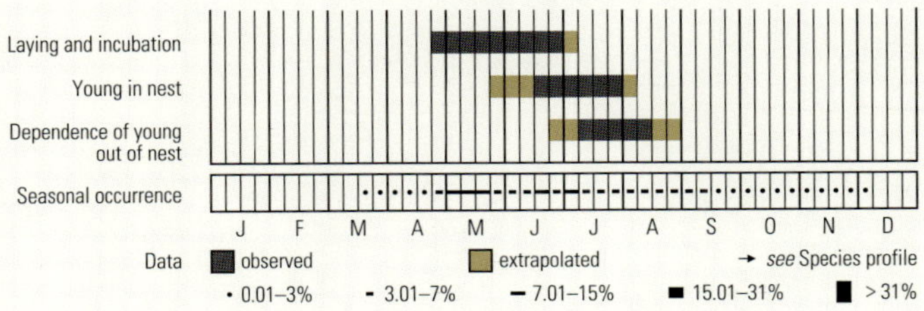

Frequency of occurrence by ecological region

Ecological regions** (forest regions)	Adequately covered squares	Squares in which reported	% of squares in which reported	Relative abundance (%)*
1. Sugar maple-hickory and sugar maple-basswood	97	50	51.5 ± 5.0	57.7
2. Sugar maple-basswood	153	100	65.4 ± 3.9	73.2
3. Sugar maple-basswood and sugar maple-yellow birch	101	80	79.2 ± 4.9	88.7
4. Sugar maple-yellow birch-basswood	56	50	89.3 ± 7.4	100.0
5. Sugar maple-yellow birch-beech and sugar maple-yellow birch-basswood	113	93	82.3 ± 4.0	92.2
6. Sugar maple-yellow birch-beech	71	57	80.3 ± 8.6	89.9
7. Sugar maple-yellow birch-balsam fir	19	15	78.9 ± 19.7	88.4
8. Sugar maple-yellow birch and balsam fir-yellow birch	47	41	87.2 ± 8.9	97.7
9. Balsam fir-yellow birch	87	59	67.8 ± 9.1	76.0
10. Balsam fir-white birch or balsam fir-red maple	46	35	76.1 ± 10.8	85.2
11. Yellow birch-balsam fir	52	36	69.2 ± 13.0	77.5
12. Balsam fir-white birch (west)	59	27	45.8 ± 13.0	51.3
13. Balsam fir-white birch (east)	76	34	44.7 ± 10.9	50.1
14. Balsam fir-black spruce	28	10	35.7 ± 18.6	40.0
15. White spruce-balsam fir and balsam fir-white spruce	22	1	4.5 ± 10.3	5.1
16. Black spruce-balsam fir-moss	18	3	16.7 ± 20.2	18.7
17. Black spruce-moss	32	7	21.9 ± 15.9	24.5
18.			*not analysed*	

* minimal: 0–29.9%; low: 30–59.9%; high: 60–89.9%; maximal: 90–100%

** Deciduous forest: 1–7; mixed forest: 8–11; coniferous forest: 12–17

The male delivers food to the female during egg laying, incubation, and the first two weeks of brood rearing; the transfer of food is often made off the nest, though usually close by. This buteo feeds mainly on amphibians, small mammals such as chipmunks, shrews, or voles, and young passerines (Fitch 1974; Mosher and Matray 1974; Rosenfield et al. 1984).

The female shows little fear when disturbed at the nest, often allowing a close approach before taking flight. Unlike other raptors, she is relatively unaggressive toward intruders. Breeding was difficult to confirm for the Atlas (6.1%, 150/2464), undoubtedly because of the birds' secretive habits during the incubation period. The species, however, is very conspicuous when soaring over open country in search of prey. Moreover, its call, a high, rather melancholy, two-syllable whistle not unlike the call of the Eastern Wood-Pewee, is unmistakable.

The female broods and feeds the young, which rely on her to tear up prey until they are 2 weeks of age. At 5 or 6 weeks old, they fledge from the nest and are able to feed themselves. They depend on their parents until they are 50–56 days old (Matray 1974).

Distribution

According to the Atlas map, the breeding range of the Broad-winged Hawk covers all of southern Québec, except for the Magdalen Islands and east of Sept-Îles on the Middle North Shore, where earlier calculations overestimated its occurrence. The species is mainly found in deciduous and mixed forests in the southern and central parts of the province. It is regularly seen south of the 49th parallel, from the Abitibi–Témiscamingue region to Baie-Comeau. Farther east it is quite scarce, Godbout being the eastern limit of its breeding range in this part of Québec.

Scattered sightings have been made north of the 49th parallel; the species occurs locally near Lakes Waswanipi and Matagami, where it appears to be associated mainly with large trembling aspen stands. The most northerly record of possible breeding was made south of Lake Evans (50°30' N) (Cons. G. G. – GREBE 1992m). These records extend the breeding range somewhat beyond the 49th parallel, the northern limit set by David (1980a). The species may also occur north of the 51st parallel in the scattered boreal poplar groves there.

Notes on territory (continued)

nesting area, 60% nested within 250 m of the old nest. The territory defended is used for nesting and foraging (Keran 1978), and covers approximately 171–200 ha (Fitch 1958; Elderkin 1989).

This buteo often builds its nest in a yellow birch. Duchesnay Forestry Station, west of Québec City.
June 1992, Robert Alvo.

Two downy young waiting to be fed in their nest in the fork of a beech tree. In addition to the chicks, fresh hardwood and coniferous boughs are evidence of recent nest occupation. Plessisville, northeast of Victoriaville. *July 1987, Alain Côté.*

History and trends

Broad-winged Hawk populations in Canada and the northeastern United States seem to have remained relatively stable since the early 1930s, although migratory bird counts between 1934 and 1987 showed a slight drop. Bednarz et al. (1990) and Titus and Fuller (1990) have shown that at known concentration points like Hawk Mountain, Pennsylvania—where thousands of birds may funnel through daily—the number of migrating birds declined slightly during this period. According to Bednarz et al. (1990), the decline may be attributable to a long-term climate change in these regions that is affecting the thermals on which this hawk depends in migration.

Some local populations in Québec may have declined slightly since the end of the 19th century. Hall (1862) and Wintle (1896) considered the species common around Montréal, but Ouellet (1974a) believes that numbers have since declined, probably through loss of habitat to agricultural development. The Broad-wing's situation appears nonetheless to have remained stable in the southern part of the province since the early 1970s (Cyr and Larivée 1993a). The increasing number of abandoned fields over the last 20 years no doubt favours its long-term establishment in places where farming is on the wane, that is, in agricultural areas outside the Greater Montréal region.

Daniel Banville and Manon Bombardier

Red-tailed Hawk

Buse à queue rousse

Buteo jamaicensis

Ghislain Caron

BY FAR THE MOST WIDESPREAD AND BEST-KNOWN BUTEO IN NORTH America, the Red-tailed Hawk breeds almost everywhere in Québec, except on the tundra, the Lower North Shore, the Magdalen Islands, and Anticosti Island. It is probably one of the first large birds of prey that a beginning birdwatcher sees in Québec. In the southern part of the province, small numbers of Red-tails can be sighted during migration and the breeding season.

It winters north into the southernmost part of the province, and a few birds also winter in the Maritimes, southern Ontario, and southwestern British Columbia (Godfrey 1986).

Habitat and behaviour

The Red-tailed Hawk prefers to nest in tall mature trees in open deciduous and coniferous woodland, usually edging a clearing or field where it can hunt. This type of habitat is often found alongside farms, large clearcuts, railway tracks, or highways, or on hillsides (Gates 1972; Bednarz and Dinsmore 1982; Speiser and Bosakowski 1988). Nests have also been reported in isolated trees in open country (Toland 1990), on a rocky cliff (Andrle 1969), and on an electrical transmission tower (Peck and James 1983). Ease of access to the nest site seems to be very important; nests are generally placed near the trunk, in the topmost crotch of the tallest tree (Toland 1990). Nests are found 4.6–27 m above the ground, but generally they are 12–18 m up. On cliffs, they may be as high as 80 m (Peck and James 1983). Preferred nest sites are found in deciduous forests—elm, oak, ash, maple, or beech—but conifers, such as pines, are used on occasion (Petersen, L. R. 1979; Peck and James 1983; Toland 1990). The Red-tailed Hawk often shares its habitat with the Great Horned Owl, sometimes nesting within 100 m of it (Toland 1990). According to L. R. Petersen (1979), the species can coexist because their daily activity cycles differ, the hawk hunting by day and the owl by night. Some researchers, however, have recorded instances of predation on the eggs and young of Red-tailed Hawks by Great Horned Owls (Bent 1937; Craighead and Craighead 1956). In Québec, remains of a mature Red-tailed Hawk and an immature bird were found in Great Horned Owl nests on Mount Saint-Bruno (F. Morneau, pers. comm.).

The Red-tailed Hawk can be seen during the breeding season, conspicuously perched as it waits for prey to appear in open fields or along woodland edges (Orde and Harrell 1977). At times, a patient observer may also see it hopping about, pouncing on crickets or grasshoppers on the ground (Bent 1937). Red-tails normally feed on rodents, especially mice, shrews, and voles (Preston 1990), and other small mammals such as snowshoe hares (Adamcik et al. 1979), but they are nonetheless opportunistic hunters, and will sometimes take birds, reptiles or amphibians (Palmer 1988).

Notes on territory

Pairs generally return each spring to the same territories defended the previous year. Territory size varies according to prey abundance and the availability of suitable nesting sites (Fitch et al. 1946).

The Red-tailed Hawk hunts in forest cutovers, but nests in tall trees along woodland edges and in patches of forest. North of Lac du Gros Ours, Gouin Reservoir. *May 1991, Yves Aubry.*

Red-tailed Hawk

Breeding distribution
1984–1989

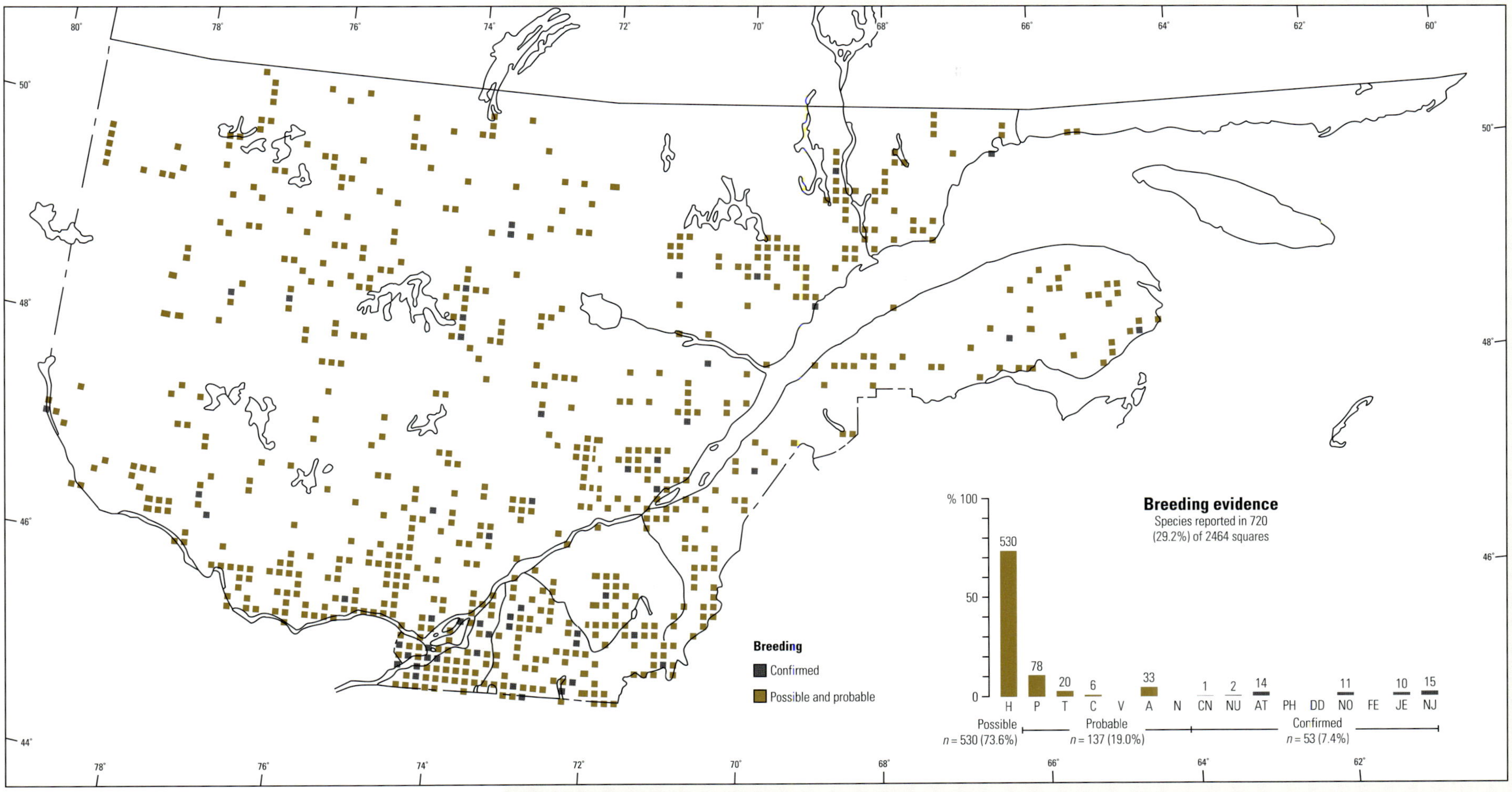

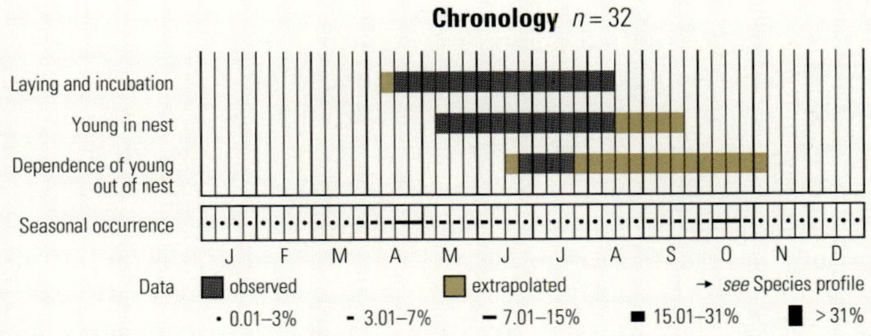

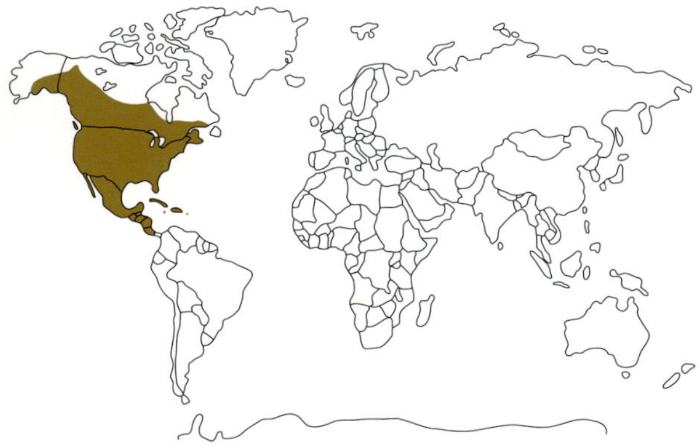
World breeding range

Species profile

Clutch size	usually 2–3 eggs (1–5)[2,3,4,8]
Laying interval	usually 1 egg per 2 days (1–3)[2]
Start of incubation	before last egg[2], with first egg[8]
Incubation period	usually 30–34 days (28–34)[2,4]
Maturity at hatching	semi-altricial
Care of young	by female and male[1,2], *see* text
Nestling period	usually 45–46 days (42–46)[1,2,3,4]
Age at first flight	usually 60–63 days (53–63)[1,2,3]
Dependence of young	until ≈ 13–15 weeks old[1,2,8]
Broods per year	1[2,7]
Breeding age	2 years[2]
Mating system	monogamy[2]
Duration of pair bond	usually for life[2]
Total length	M 48–56 cm / F 53–61 cm[5]
Wing span	116.8–147.3 cm[4]
Weight	M ? g (1028–1143) / F ? g (1371–1600)[4]
Longevity record	21 years 6 months[6]

References:
1. Petersen, L. R. 1979
2. Palmer 1988
3. Fitch et al. 1946
4. Terres 1980
5. Godfrey 1986
6. Klimkiewicz and Futcher 1989
7. DeGraaf and Rudis 1987
8. Preston and Beane 1993

This raptor likes to nest in small farm woodlots surrounded by agricultural land, and preys on rodents in nearby fields. Saint-Étienne-de-Beauharnois, south of Montréal. *July 1991, Pierre Bannon.*

Notes on territory (continued)

The female's activities are concentrated around the nest, whereas it was estimated in Wisconsin that the male hunts over an area of 163 ha. When the young are 5 weeks old, the female extends her activities to a larger area of 117 ha. The parents rarely hunt less than 100 m from one another (Petersen, L. R. 1979).

In Alaska, local population densities may reach 2.1 pairs/100 km² (Lowe 1978). Farther south, in Michigan, Craighead and Craighead (1956) recorded a density of 5.2 pairs/100 km² in 1948. In the same location in 1966, the population had increased to 11 pairs, giving a density of 11.5 pairs/100 km² (Wallace 1969).

The Red-tailed Hawk was reported in 29.2% (720/2464) of Atlas squares surveyed, although breeding was confirmed in only 7.4% of these squares. Most confirmations (52.8%; 28/53) were based on reports of nests found, and about a quarter (26.4%; 14/53) were based on reports of adults carrying food. When the young are in the nest, the aggressive behaviour of the pair, especially the female, can aid in locating the nest. At the approach of an intruder, the adult birds will fly in wide circles over the nest site, calling repeatedly; they will even dive to within 10–40 m (Santana et al. 1986; Andersen 1990). The bulky nests are often conspicuous early in the season, before the trees come into leaf. The outer diameter of the structure may be 60–90 cm, and the inner cup 5–25 cm deep (Peck and James 1983). During atlassing, Red-tailed Hawk nests proved to be easier to find than those of the Broad-winged Hawk.

When a pair arrives in spring it begins its fascinating breeding display, which consists of a kind of undulating aerial dance performed by both members; they rise upon flapping wings to great heights, then plummet with wings partially closed. The male will fly over his mate, sometimes clutching her back in his talons. The contact lasts a few seconds, and then they resume their circling (Conner 1974a).

Each year the pair may reuse an old nest on its territory or build a new one. Both adults spend 4–7 days constructing a platform (Petersen, L. R. 1979), which consists mainly of dead and fresh branches and twigs 6–40 mm in diameter (Peck and James 1983). They line the nest with bark and sprigs, adding more greenery to the structure during incubation and rearing of the young (Petersen, L. R. 1979; Palmer 1988).

Red-tailed Hawks in southern Québec typically lay 2 eggs, but 5 of the 6 nests found in the Great Whale River watershed in 1990, when rodent populations were apparently high, contained 3 or 4 eggs (Cons. G. G. – GREBE 1990c). In southern Québec, incubation begins around mid-April and lasts 30–34 days; both parents sit on the eggs, but the female does most of the incubating. The male sometimes brings prey to the female during this period, but most often she hunts for her own food. When the eggs have hatched, the male provides both the female and chicks with food. The female remains on the nest until the young are 5 weeks old, then she resumes hunting (Petersen, L. R. 1979).

The young leave the nest at around 6–7 weeks of age, remaining in the vicinity for about 18–25 days (Johnson 1973). After the 9th week, they leave the nest site and follow the adults to the hunting grounds. They continue to be fed by the parents until they can successfully hunt on their own, at about 13–15 weeks of age (Petersen, L. R. 1979; Preston and Beane 1993). They leave the parental territory 4–21 weeks later (Johnson 1973; Petersen L. R. 1979).

Distribution

The map shows that the Red-tailed Hawk is distributed throughout almost all of southern Québec, although it is somewhat more common in the boreal forest. It is absent, however, from the Magdalen Islands and Anticosti Island. The most easterly Atlas record on the North Shore was from the village of Mingan.

Above 49° N, the Red-tailed Hawk hunts in old burns, forest cutovers, and peatlands, often covering extensive areas. It is especially common in the Rupert, Nottaway, and Broadback river watersheds where 10 nests were discovered in 1990 and 1991 (Cons. G. G. – GREBE 1992m). Nests have also been found in the Great Whale River watershed, where the species is often observed. The most northerly sighting is from the Little Whale River, about 20 km from the mouth. In 1990, 3 nests were discovered 30 km southeast of this area (Cons. G. G. – GREBE 1990c). The breeding range of the Red-tailed Hawk extends to the northernmost limit of the boreal forest, as established by Payette (1975), and is north of Godfrey's (1986) limit.

The platform of sticks on which the Red-tailed Hawk nests may be found on the edge of a black spruce stand. This buteo, the most widespread in Québec, also dwells in large coniferous forests. Grands-Jardins Conservation Park, northwest of Baie-Saint-Paul. *June 1991, Réginald Ouellet.*

A large nest of sticks used by the Red-tailed Hawk in 1974, the Northern Goshawk in 1976, and the Great Horned Owl in 1980. Mont-Saint-Bruno Conservation Park, east of Montréal. *March 1980, François Morneau.*

History and trends

Red-tailed Hawk numbers seem to have increased substantially in several parts of North America over the last 70 years. In Ontario, for example, its status changed from rare at the beginning of the century to common from the 1930s on (Weir 1987d). The situation is similar in the northeastern United States; a notable increase in numbers in recent decades has been reported in New York State (Smith, G. A. 1988b). Moreover, counts of Red-tails in migration through White Marsh, Maryland, remained stable throughout the 1950s and early 1960s, whereas those of other migratory raptor species declined (Braun et al. 1977). BBS data show a slight increase in Red-tailed Hawk numbers between 1966 and 1987 in the northeastern United States (Robbins et al. 1986). Data gathered by Bednarz et al. (1990) during migration at Hawk Mountain, Pennsylvania, however, indicate the opposite trend, as fewer sightings were made between 1971 and 1986. They attributed this decline to conversion of pastureland into housing developments and intensive agricultural practices in the northeastern United States.

Little is known about the status of Red-tailed Hawk populations in Québec. However, the species is probably more abundant around Montréal today than it was in the 19th century, when Wintle (1896) considered it scarce there. Deforestation, abandoned fields, forest edge habitat, and highway construction all provide the Red-tailed Hawk with suitable hunting and breeding areas.

PIERRE CHAGNON AND MANON BOMBARDIER

Golden Eagle

Aigle royal

Aquila chrysaetos

Carole Bérubé

DESPITE ITS RESEMBLANCE TO THE BALD EAGLE, THE GOLDEN EAGLE IS more closely related to the buteos and accipiters (Sibley and Ahlquist 1990). Most of the birds of this species that nest in Québec leave the province in autumn to spend the winter in the eastern United States, although a small contingent winters here. This majestic raptor is rarely observed in winter or summer and is seen more often during the spring and fall migrations, when it delights birdwatchers.

Habitat and behaviour

The Golden Eagle is a bird of wide-open spaces. In the breeding season, it inhabits vast areas of wilderness with cliffs for nesting and open expanses such as peatlands, marshes, and tundra for hunting. It is most often found in mountainous regions deeply cut by steep rocky canyons and valleys (De Smet 1987b). The Golden Eagle feeds on a variety of animals, mostly small mammals such as rodents and hares, and avian prey such as grouse. Carrion is also an important part of its diet (Snow 1973; Olendorff 1976). In northern Québec, particularly in the watersheds of the rivers flowing into central Hudson Bay, avian prey—Canada Geese, ducks, ravens, gulls, and raptors—make up most of its diet, although it also feeds on mammals such as arctic ground squirrels, hares, and foxes, and the remains of caribou and seals (S. Brodeur, pers. comm.).

The Gaspé Peninsula has a number of high ridges as well as extensive wilderness providing suitable nesting habitat for the species. Golden Eagles can be seen soaring over the valleys near Mount Albert and over alpine tundra. Gaspé Conservation Park. *July 1991, Yves Aubry.*

In eastern North America, the Golden Eagle almost always builds its eyrie on a cliff, although it may nest in a tall tree on rare occasions (Spofford 1971; Bull 1974; Morneau et al. 1994). The species is monogamous and the pair bond lasts for many years, if not for life. Both parents participate in nest construction, although the female does most of the work. The pair may either build a new nest or build on top of a previous year's nest. In spring, the pair often refurbishes two or three old nests before selecting one for use (Snow 1973; Cramp and Simmons 1980). The same eyrie may be occupied several years in a row or irregularly over the years (McGahan 1966, 1968). With time, some cliff nests may reach a size of 3 m in diameter and 1.5 m in depth (Brown and Amadon 1968).

The male helps incubate the eggs but, here again, the female does most of the work. Hatching is asynchronous. Both parents care for the eaglets, the male doing most of the hunting and the female feeding and protecting the young (Collopy 1984). When the chicks are very young, the female tears off bits of flesh for them, but by the end of the first month, prey is left in the nest for the eaglets to eat on their own. Generally, only one chick survives to fledging, because the firstborn harasses the younger chicks (Brown and Amadon 1968; Cramp and Simmons 1980). Competition ceases when the eaglets have become fully feathered. In parts of its range where the species is resident, in Scotland, for example, the eaglets are dependent on their parents for 90–100 days after they leave the nest. This period of dependence may be shorter in areas where the species is migratory (Cramp and Simmons 1980).

Very little data were collected on the Golden Eagle during atlassing. The species is not only scarce but wary and inhabits remote, virtually inaccessible areas. Furthermore, it is easily confused with the immature Bald Eagle. The presence of an adult bird does not necessarily mean that there is a nest nearby, since adults travel long distances to hunt. By the same token, an observer may be near a nest without seeing an adult. The Golden Eagle was reported in only 0.6% (14/2464) of the squares visited and nesting was confirmed in only one square.

Distribution

Most of the Atlas breeding records come from the Gaspé Peninsula. A nest was observed near Lake Matane, a site known to birders for several decades (Baillie 1955). In addition, in 1991, a partially feathered nestling was found near the La Grande Rivière on the Gaspé Peninsula, then rehabilitated and released in the same area (P. Fradette, pers. comm.). Other parts of the Gaspé with suitable habitat include Mount Jacques Cartier, Mount Albert, and Percé, where the many seabird colonies attract Golden Eagles in summer (Brodeur and Morneau 1991). The peninsula's vast mountainous areas, most of which are inaccessible, probably support several pairs of eagles, thus making it the species' stronghold in southern Québec (David and Gosselin 1977a).

The area around Lake Guillaume Delisle is home to several breeding pairs, which patrol the valleys and other open areas in search of prey. This region has many rocky cliffs that provide ideal nesting habitat. Hudson Bay region. *July 1990, Serge Brodeur.*

Notes on territory

Pairs often occupy the same home ranges year after year. Home ranges usually cover 40–100 km², or sometimes even several hundred square kilometres (Snow 1973; Cramp and Simmons 1980). The birds do not defend their entire range, but only a small area around the nest. In ideal habitat, active nests may be found only a few kilometres apart (McGahan 1966, 1968; Cramp and Simmons 1980). Near the coast of Hudson Bay, where many pairs breed, active nests are usually 12–20 km apart (S. Brodeur, pers. comm.).

Golden Eagle

Breeding distribution
1984–1989

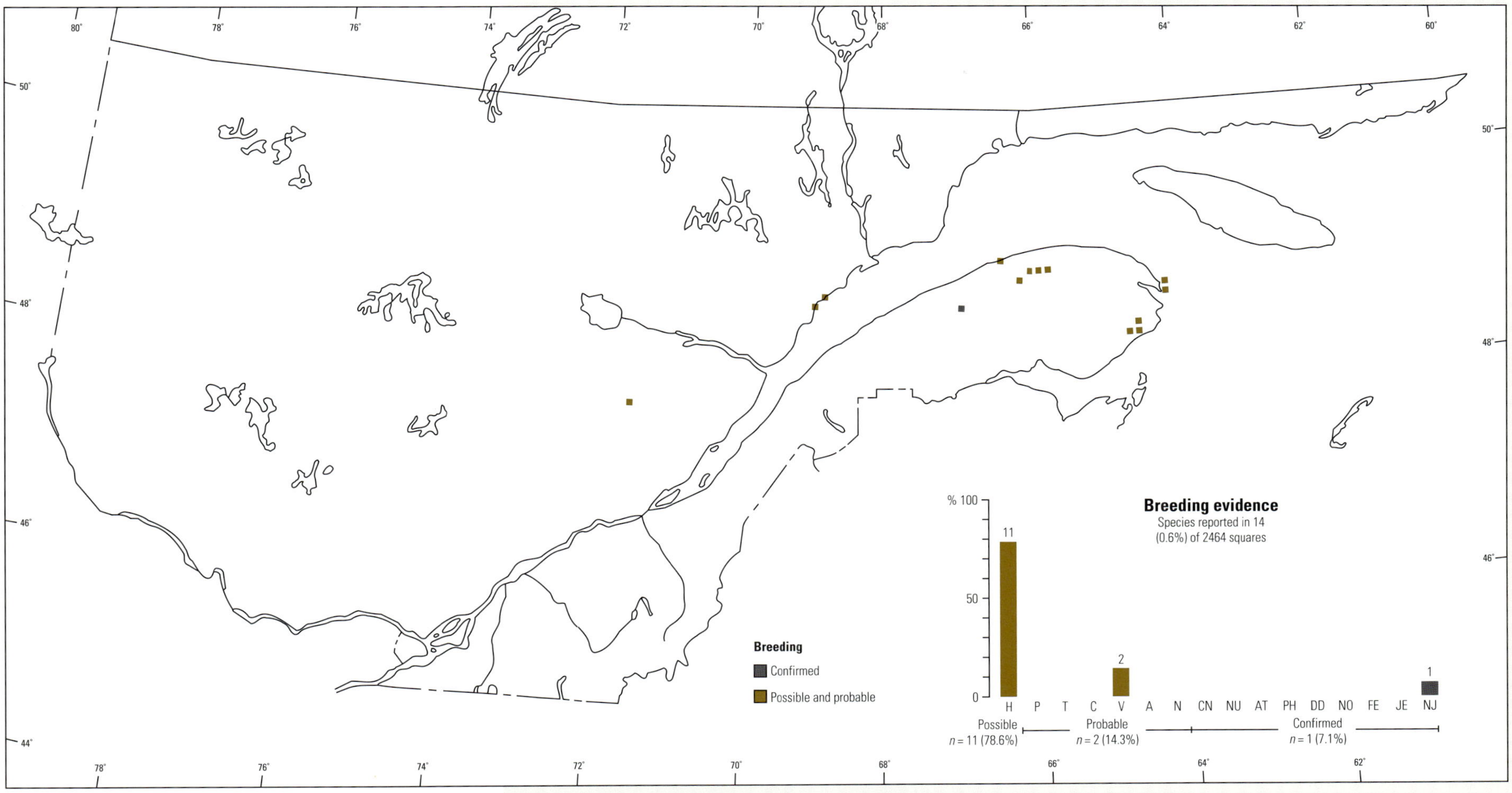

Species profile

Clutch size	usually 2 eggs (1–4)[1,4]
Laying interval	1 egg per 3–4 days[1]
Start of incubation	with first egg[1,4]
Incubation period	43–45 days per egg[1]
Maturity at hatching	semi-altricial
Care of young	by female and male[1], *see* text
Nestling period	usually 65–70 days (65–80)[1]
Age at first flight	usually 65–70 days (65–80)[1]
Dependence of young	until 90–100 days after first flight[1]
Broods per year	1[1]
Breeding age	3–4 years[1]
Mating system	monogamy[1,4]
Duration of pair bond	usually for life[1,4]
Total length	M 76–89 cm / F 89–104 cm[3]
Wing span	187–230 cm[3]
Weight	M ? g (3550–4400) / F ? g (4050–5720)[2]
Longevity record	30 years[2]

References:
1. Cramp and Simmons 1980
2. Terres 1980
3. Godfrey 1986
4. Palmer 1988

Chronology *n* = 25

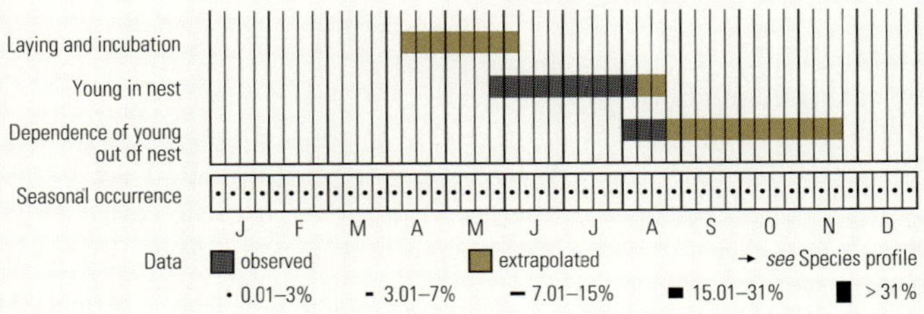

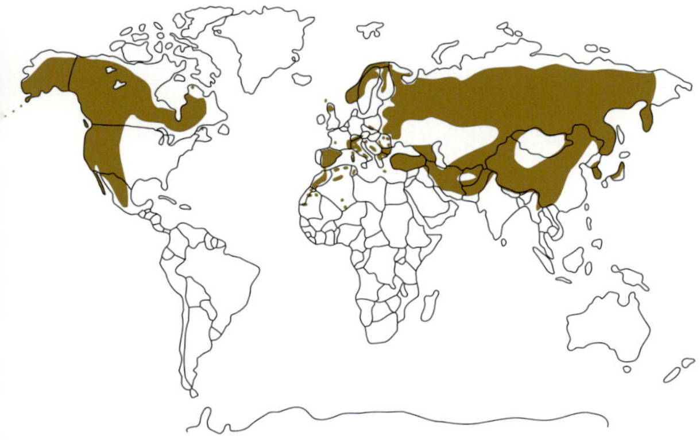

World breeding range

The cuestas along the Little Whale River are ideal nesting sites for this majestic raptor. Hudson Bay region. *June 1992, Serge Brodeur.*

A pair of Golden Eagles has nested on a cliff overlooking Lake Matane for several years. Matane Wildlife Reserve. *July 1991, Yves Aubry.*

The Golden Eagle was also atlassed on the North Shore and in the Laurentide Wildlife Reserve. Before the 1980s, there was no proof of Golden Eagles breeding on the North Shore, though there were several indications that the species did in fact breed there (Comeau 1909; Todd 1963; Cayouette 1977). Since then, however, 6 records of confirmed breeding have been obtained from along rivers to the north of Sept-Îles (Brodeur and Morneau 1991; GREBE 1994). The species may also nest in the hilly areas north of Québec City and in the Charlevoix region. One or possibly two adults carrying a hare, and another with a Spruce Grouse were observed in the Laurentide Wildlife Reserve in July of 1986 and 1987. There have also been reports of nests found near Baie-Saint-Paul and La Malbaie (Provancher 1870). These places all have cliffs and vast areas of wilderness that appear to provide suitable habitat for the species. Elsewhere in southern Québec, a few birds may possibly breed in the Ottawa Valley, the Laurentians, and the Mauricie, as well as on Anticosti Island (Brodeur and Morneau 1991). So far, however, the Gaspé Peninsula and the Middle North Shore are the only places in southern Québec where irrefutable evidence of breeding has been obtained.

This large raptor also breeds in northern Québec, both inland and near the coast. The vast area above the 54th parallel is no doubt home to the greatest number of Golden Eagles in eastern North America. At least 3 active nests were found in the area around the Caniapiscau and George rivers, and nearly 10 nests have been found around Ungava Bay over the last 40 years (Spofford 1959, 1971; Todd 1963; Brodeur and Morneau 1991). Aerial surveys of the watersheds of central Hudson Bay carried out since 1990 have revealed the presence of 20 breeding pairs; 15 of these pairs were found within a strip several dozen kilometres wide along the coast of Hudson Bay between the mouths of the Great Whale and Nastapoka rivers, including 8 pairs near Lake Guillaume Delisle. Another 3 pairs were spotted on the middle stretch of the Great Whale River (Morneau et al. 1994), and there have been reports suggesting that other pairs may be nesting in the area (F. Morneau, pers. comm.). According to Brodeur and Morneau (1991), who studied the Golden Eagle's status in the province, the Québec population probably numbers 100 pairs or so, most of them in northern Québec.

In the states and provinces bordering Québec, the Golden Eagle has been reported as nesting in New Hampshire and New York, but breeding has only been confirmed in Maine (Lee 1990). A few pairs also nest in northern Ontario and perhaps in the Maritimes (De Smet and James 1987; Erskine 1992). The species is much more common in western North America (Snow 1973).

Young Golden Eagle in a nest, which has been built on a ledge on a rocky cliff overlooking the Great Whale River. Generally, only one of the brood survives to fledging. Hudson Bay region. *July 1990, Serge Brodeur.*

Golden Eagle nest with an eaglet. The large mass of sticks shows that the nest has been used repeatedly over the years. Near the Great Whale River, Hudson Bay region. *July 1990, François Morneau.*

Adult Golden Eagle soaring over the Caribou River. This photograph gives an idea of how the eagle sees its habitat, a bird's-eye view of earth and sky. Near Lac à l'Eau Claire, Hudson Bay region. *June 1992, Serge Brodeur.*

History and trends

According to what little is known, Golden Eagles have never been very abundant in Québec. Godfrey (1986) stated that the species no longer occupies much of its former breeding range in eastern Canada. Here in Québec, however, the species still appears to occupy its traditional breeding grounds (Brodeur and Morneau 1991). Since the Québec birds inhabit remote areas where little field work is done even today, we do not have enough information to determine whether the Québec population has declined since the last century.

The number of migrating Golden Eagles observed at Hawk Mountain Sanctuary, Pennsylvania—a large proportion of which probably come from Québec—dropped sharply during the 1950s and 1960s (Spofford 1971; Bednarz et al. 1990). Because of this decline and the poor breeding success of some eagles in the Appalachians, Spofford (1971) concluded that birds in the East face serious problems, including pesticides and other contaminants, poisoning and accidental capture by trappers, and shooting by poachers, as well as habitat loss and disturbance caused by human activity (Spofford 1971; De Smet 1987b). Opinions are divided, however, on recent population trends in eastern North America. Some authors (De Smet 1987b; De Smet and James 1987) maintain that the species increased in number during the 1980s, while others, such as Bednarz et al. (1990), who recently analysed the annual counts at Hawk Mountain, are less optimistic. They found that, while the number of immature birds has increased since the 1960s and 1970s, the number of adults has not. They attribute this to the fact that the decline of the Golden Eagle in the East may not be caused solely by the use of the insecticide DDT —which was banned in Canada and in the United States in the early 1970s—but by other factors as well.

In Québec, problems affecting the species include accidental ingestion of poison meant to control predators, poaching, and trapping; every year a few birds are accidentally caught in snares and leghold traps. The Golden Eagle is classified as a vulnerable species in the province (Robert 1989b; Brodeur and Morneau 1991).

MICHEL ROBERT

American Kestrel

Crécerelle d'Amérique

Falco sparverius

Ghislain Caron

MANY A BIRDWATCHER HAS DELIGHTED IN THE SIGHT OF A TINY AMERICAN Kestrel hovering over a fallow field in search of grasshoppers or an unwary mouse. The hovering behaviour is a sure giveaway for this species, as is the bobbing tail of one perched on a utility wire along a roadside or at the top of a tree in a field. Being smaller and less stocky than a Merlin and sitting more erect than passerine species of similar size, the kestrel is difficult to mistake for anything else. In flight, its small size and typical falcon silhouette, with longish tail and pointed wings, are distinguishing. Close up, the male with his slate blue wings, black moustachial markings, and rufous back and tail is more brilliantly coloured than the drab female with her reddish-brown plumage barred with black.

American Kestrels are migrant breeders in southern Québec and are still considered, as earlier reported by David (1980a), locally casual to uncommon in winter. Besides raiding backyard feeders to prey upon small birds, such as House Sparrows, they are clever enough to seek warmth by perching in steam rising from chimneys (D. M. Bird, McGill U., unpubl. data).

Habitat and behaviour

The American Kestrel is an ecologically versatile species, breeding from the northern alpine zones to the western deserts of North America. Kestrels nest in a variety of open habitats, their two main requirements being open ground for hunting and cavities for nesting. They are found near fields, meadows, prairies, woodland edges, patches created by fires and clearcutting, farmland, as well as in suburban and urban areas (Johnsgard 1990). Kestrels, like other falcons, build no nest, but instead scrape a shallow depression in the substrate of the chosen nesting site (Palmer 1988). Commonly used sites include natural cavities in live and dead trees, woodpecker holes, ledges on cliffs or buildings, old stick nests of other birds, nest boxes of suitable size or burrows (Godfrey 1986; Village 1990). In Ontario, cavity openings were usually 5–20.5 cm in diameter (Peck and James 1983). Their opportunistic nesting habits, combined with their broad diet, have helped them colonize an extensive variety of habitats, including those modified by humans.

During the winter, the American Kestrel is found in the southern parts of British Columbia, Ontario, Québec, and Nova Scotia, as well as in the United States, Mexico, and Central America (AOU 1983), often in sex-segregated populations (Meyer and Balgooyen 1987). It generally returns to its breeding grounds in Québec by March. Non-migrating birds have been seen entering nesting cavities in the Montréal area as early as the first week in March (D. M. Bird, McGill U., unpubl. data). Males generally arrive a few days ahead, but may also arrive at the same time with the females (Palmer 1988). Nest-site fidelity and pair bonding are strong, and the same cavity may by reused in successive years (Craighead and Craighead 1956; Balgooyen 1976).

For the first few weeks after their return to their breeding grounds, males delineate and defend a small territory, chasing away any intruding males as well as other species of birds. One male inhabiting a Montréal skyscraper was seen perched in front of a release cage scolding its contents, a brood of soon-to-be released Peregrine Falcons! While rarer, direct contact in the form of talon locking with other kestrels may occur.

Notes on territory

Territorial defence in American Kestrels is variable. At Sagehen Creek, California, they were observed defending the nest site and the home range against other kestrels (Balgooyen 1976). In certain residential areas of California, Cade (1955a) found that during the nesting period, only the immediate vicinity of the nest was defended. After the breeding season, however, resident pairs defended a hunting territory against other, often migratory, kestrels.

Some pairs appear to tolerate conspecifics nesting at close range. In a Utah study, 5 nests were located within 61 m of at least 1 other nest, and 2 were only 12 m apart. No aggression was observed between neighbours, despite overlapping defended areas (Smith et al. 1972). Sherman (1913) found House Sparrows nesting in a nest box within 1 m of an incubating female American Kestrel.

A hole in a dead tree on the edge of a field or forest clearing may make a suitable nest site for the American Kestrel. Sainte-Foy. *September 1991, Yves Aubry.*

American Kestrel

Breeding distribution
1984–1989

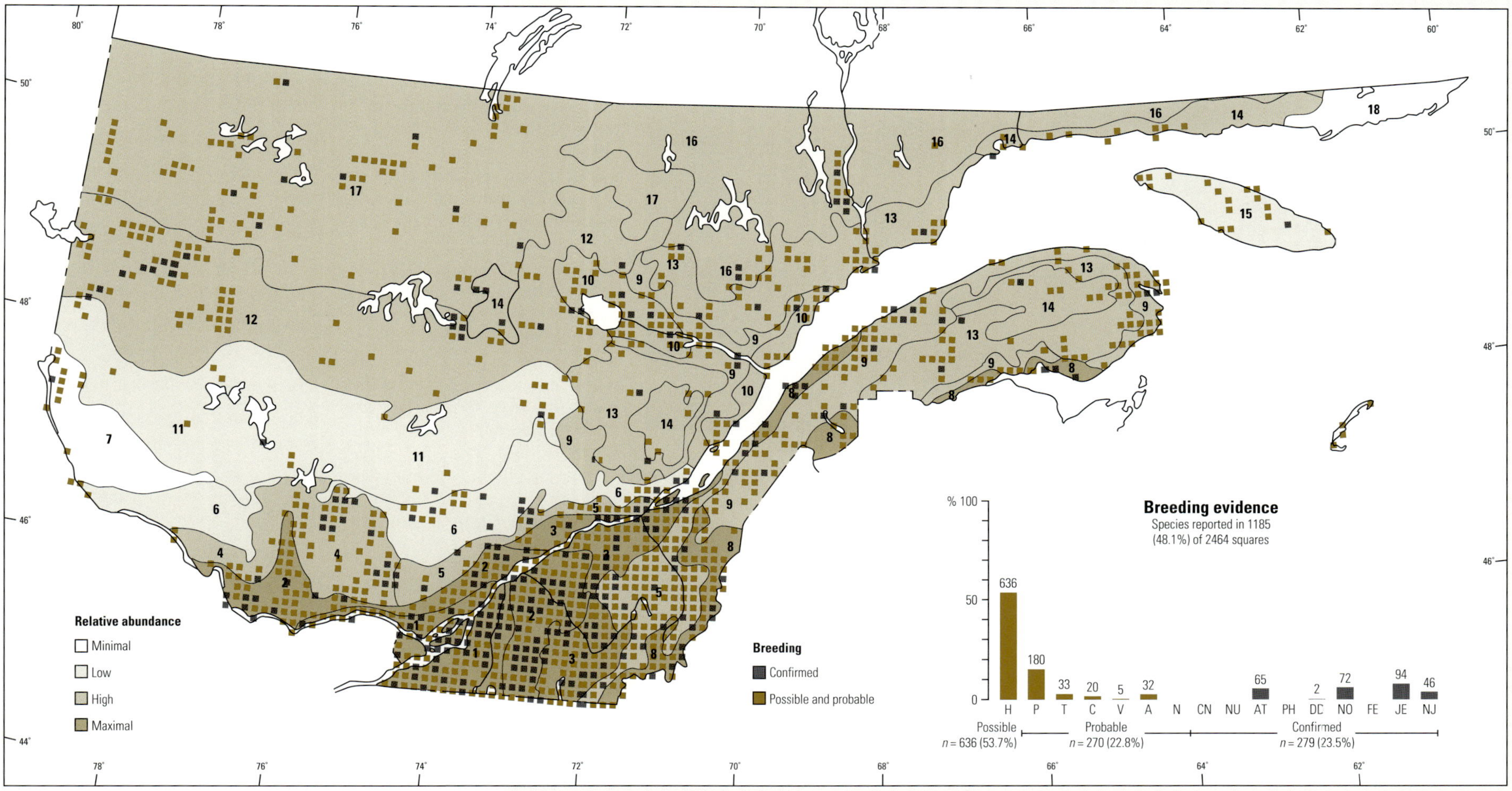

Species profile

Clutch size	usually 4–5 eggs (2–7)[2,5,7,8]
Laying interval	usually 1 egg per 2 days (1–3)[2,5,7]
Start of incubation	with third, fourth or fifth egg[2,7]
Incubation period	29–31 days[7,8]
Maturity at hatching	semi-altricial
Care of young	by female and male[2,4,6,7,8], *see* text
Nestling period	usually 29–31 days (27–35)[1,2,3,7]
Age at first flight	usually 29–31 days (27–35)[1,2,3,7]
Dependence of young	until 14–28 days or more after leaving nest[2,3]
Broods per year	1, rarely 2[7,8,9]
Breeding age	1 year[2,9]
Mating system	monogamy[2,3,7,8]
Duration of pair bond	probably for life[2,3]
Total length	M 111 g (?) / F 120 g (?)[12]
Wing span	M 51–56 cm / F 57–61 cm[2]
Weight	M 111 g (?) / F 120 g (?)[12]
Longevity record	11 years 7 months[10]

References:
1. Sherman 1913
2. Palmer 1988
3. Cade 1955a
4. Smith et al. 1972
5. Brown and Amadon 1968
6. Willoughby and Cade 1964
7. Johnsgard 1990
8. Ehrlich et al. 1988
9. Newton 1979
10. Clapp et al. 1982
11. Godfrey 1986
12. Dunning 1984

Chronology *n* = 138

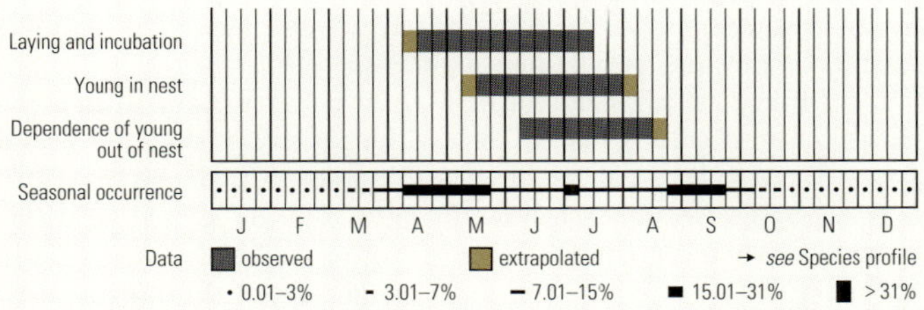

Frequency of occurrence by ecological region

Ecological regions** (forest regions)	Adequately covered squares	Squares in which reported	% of squares in which reported	Relative abundance (%)*
1. Sugar maple-hickory and sugar maple-basswood	97	89	91.8 ± 3.0	97.5
2. Sugar maple-basswood	153	138	90.2 ± 2.6	95.9
3. Sugar maple-basswood and sugar maple-yellow birch	101	95	94.1 ± 3.0	100.0
4. Sugar maple-yellow birch-basswood	56	38	67.9 ± 10.7	72.1
5. Sugar maple-yellow birch-beech and sugar maple-yellow birch-basswood	113	90	79.6 ± 4.2	84.7
6. Sugar maple-yellow birch-beech	71	32	45.1 ± 10.6	47.9
7. Sugar maple-yellow birch-balsam fir	19	3	15.8 ± 17.9	16.8
8. Sugar maple-yellow birch and balsam fir-yellow birch	47	40	85.1 ± 9.4	90.5
9. Balsam fir-yellow birch	87	64	73.6 ± 8.6	78.2
10. Balsam fir-white birch or balsam fir-red maple	46	34	73.9 ± 11.1	78.6
11. Yellow birch-balsam fir	52	16	30.8 ± 13.0	32.7
12. Balsam fir-white birch (west)	59	42	71.2 ± 11.9	75.7
13. Balsam fir-white birch (east)	76	47	61.8 ± 10.7	65.7
14. Balsam fir-black spruce	28	17	60.7 ± 18.9	64.5
15. White spruce-balsam fir and balsam fir-white spruce	22	11	50.0 ± 21.5	53.2
16. Black spruce-balsam fir-moss	18	12	66.7 ± 24.8	70.9
17. Black spruce-moss	32	26	81.2 ± 15.1	86.4
18.			*not analysed*	

* minimal: 0–29.9%; low: 30–59.9%; high: 60–89.9%; maximal: 90–100%

** Deciduous forest: 1–7; mixed forest: 8–11; coniferous forest: 12–17

The American Kestrel can often be seen hovering over one spot in a field or other open area. Cap Tourmente National Wildlife Area, northeast of Québec City. *August 1991, Léo-Guy de Repentigny.*

When females first arrive on the breeding grounds, they initially use extensive areas outside of any one male's territory, and promiscuous mating is fairly common among both sexes before a monogamous pair bond is finally formed (Fast and Barnes 1950; Cade 1955a). Old pair bonds may be re-established or new ones formed. Aerial displays, hunting in pairs, various vocalizations, nest-site inspections, copulation, and courtship feeding are important components of pair bonding (Willoughby and Cade 1964); males may also curtsey or bow when soliciting a female (Mueller 1971). Nest-site inspections are initiated by the male; he is then followed by the female, which may enter the nest periodically throughout the day and roost in it at night (Balgooyen 1976). Competition for nest sites is a factor limiting American Kestrel distribution (Hamerstrom et al. 1973). Kestrels are apparently dominant over woodpeckers and most other cavity-nesting birds using similar-sized cavities, as well as mammals such as chipmunks and squirrels (Balgooyen 1976).

Typical clutch size is 4 or 5 eggs, but may vary from 2 to 7 (Brown and Amadon 1968; Palmer 1988), with eggs being laid 24–72 hours apart, or normally on alternate days (Balgooyen 1976). If the first clutch fails, renesting frequently follows (Bowman and Bird 1985). Incubation is by both sexes, with males accounting for up to 20% of the total time spent on the nest. While the male usually feeds the female at the nest, she may sometimes capture her own prey or feed from a food cache. The incubation period lasts 29–31 days, and the young hatch asynchronously. The male supplies all the food for the first 7–10 days, while the female broods the hatchlings (Smith et al. 1972). The young grow rapidly, attaining adult weight at about 17 days, and generally fledge at 29–31 days. The fledgling survival rate is usually high—88–100%—thanks to choice of nest site and vigorous nest defence (Cade 1982; Palmer 1988). After fledging, the young remain dependent on the parents for food for 12–24 days and, if prey is abundant, may remain in the vicinity of the nest (Kellner 1990). Some may then join other fledglings in wandering groups until late summer (Lett and Bird 1987), while there is evidence that others migrate as a family group (Cade 1955a; Kellner 1990).

American Kestrels prey upon a wide variety of animals, chiefly birds (European Starlings, Horned Larks, sparrows and other small passerines), small mammals (mice and voles), insects (grasshoppers, beetles, cicadas, etc.), and the occasional reptile. Birds and mammals are taken through the early part of the breeding season, with the more easily caught insects becoming increasingly important as the season progresses and they become more readily available. In typical raptor fashion, kestrels capture their prey with their talons, but hunting methods vary from place to place and according to the type of prey taken. They may dive upon their prey from a perch or from hovering flight, and they will even occasionally forage on the ground.

While there definitely exists a floater population of unmated kestrels (Bowman and Bird 1986), it is reasonable to assume that in most of the squares where possible (636/1185) or probable (270/1185) breeding was reported, pairs were in fact nesting.

This small, colourful bird is readily recognized by most birdwatchers and was the most easily located raptor in Québec during the Atlas survey period: it was reported in 48.1% (1185/2464) of squares surveyed. It is frequently seen perching conspicuously near the tops of trees, on fence posts, telephone poles, and power lines, or hovering over open areas. Nests are readily spotted from a distance, as they are often in dead trees beside open countryside or near human dwellings. The entrance may be found by watching an adult, usually the male, carrying food into the nest; this was the breeding evidence code used in 23.2% (65/279) of all cases. A further 42.3% (118/279) of breeding confirmations involved observations of adults leaving or entering the nest site, or a nest with eggs or young. After fledging, the young remain with the adults for several weeks, and their noisy whining as they beg for food often reveals their presence. The most frequently used confirmation code was the one for recently fledged young. As family bonds loosen, kestrels can be seen hunting in social groups of as many as 20 juveniles and adults (Cade 1955a; Varland and Klaas 1991).

Notes on territory (continued)

Relative to larger North American falcons, American Kestrels are able to maintain fairly high population densities and small home ranges. Johnsgard (1990) speculated that this is due to the kestrel's small size and ability to feed on both insects and vertebrates. A study on the western end of the Island of Montréal, Île Perrot, and in eastern Vaudreuil County estimated mean territory size to be 0.24 km², with mean breeding density of 1.7 pairs/km² (Bowman and Bird 1986).

In studies in Michigan, Illinois, and Utah, estimated mean home range diameters were 2.42 km, 2.3 km, and 0.82 km, with population densities of 0.16 pair/km², 0.11 pair/km², and 1.98 pairs/km², respectively (Craighead and Craighead 1956; Enderson 1960; Smith et al. 1972). These large variations appear to be related to prey abundance and nest-site availability.

This raptor frequents a variety of open habitats, among them pastures. Logging and land clearance have played a big part in boosting the species' population. Brownlee Road, Hemmingford, south of Montréal. *July 1991, Pierre Bannon.*

After 50 nest boxes were put up in 1968–72 in a study area in central Wisconsin, the number of kestrels increased from only 3 nesting pairs over a period of more than 20 years to as many as 12 fledged broods per year (Hamerstrom et al. 1973).

Distribution

The American Kestrel is exclusive to the Americas, breeding throughout much of Canada and central Alaska from the tree line south through North, Central, and South America to Tierra del Fuego (Godfrey 1986). In Québec, this species is the most widely distributed raptor. Frequency of occurrence was highest in the St. Lawrence and Ottawa valleys, the Eastern Townships, and the Appalachians. The species was also recorded in several squares in the Gaspé, along the Upper North Shore, on Anticosti Island, and in the Saguenay–Lac-Saint-Jean and the Abitibi–Témiscamingue regions. It was also reported in the Magdalen Islands, though confirmation of breeding there was obtained only in 1990, after the end of the Atlas survey period (Fradette 1992).

This is similar to the range cited by Godfrey (1986) and corresponds to the predominantly agricultural regions of Québec. Although they usually prefer fields and other open areas, such as clearings, roads, and backwoods cutovers, these opportunistic birds also nest in towns and cities. A sizable population breeds in Montréal; cavities in trees and in skyscrapers appear to be equally acceptable. Lack of sightings in the Laurentians and neighbouring areas may have been due to the inaccessibility of some areas as well as the lack of suitable open habitat in this region. Kestrels can likely be found breeding in virtually any cleared area in southern Québec, as long as there are suitable cavities and adequate prey in the form of mice, insects, and small birds.

Farther north, the kestrel becomes scarce, and pairs are extremely localized. Godfrey (1986) cites rare records from Lake Mistassini and the lower Moisie River. The kestrel has also been reported at James Bay and on the Eastmain and Opinaca rivers (ÉPOQ) and on the Little Whale River (Cons. G. G. – GREBE 1990c). Nests have been found in the area of the Nottaway, Broadback, and Rupert rivers (Cons. G. G. – GREBE 1992m), at La Grande-2 (F. Morneau, pers. comm.), and near the Kuujjuaq airport (P. May, pers. comm.).

History and trends

In the early 1900s, Dionne (1906) listed the American Kestrel as fairly common during spring and fall migration around Québec City; Wintle (1896) regarded it as scarce in the late 19th century in the Montréal area. Widespread deforestation and land clearing seem to have been favourable to the species (Ouellet 1974a), increasing the availability of nesting sites and enlarging the hunting areas where it can find its rodent and insect prey. Since the last century, kestrel numbers have increased across the continent.

However, like most other species of the order Falconiformes, the American Kestrel is susceptible to environmental pollutants (Wiemeyer and Sparling 1991). Eggshell thinning was demonstrated in laboratory-raised kestrels when organochlorine contaminants were added to their food (Porter and Wiemeyer 1969). A study of kestrels around Ithaca, New York, in 1969–72 showed eggshells averaging 10% thinner than in the pre-DDT period (Lincer 1975). Anderson and Hickey (1972) analysed eggs collected from wild birds and found an average 3% decrease in the American Kestrel's shell index, but no population declines were noted. Noble and Elliott (1990) found that kestrel eggs analysed prior to 1973 contained less than the critical level of 10 mg/kg DDE. However, in the late 1980s, mean levels in egg samples from Ontario ranged from 0.44 to 10.8 mg/kg DDE, with the highest levels coming from central Ontario, but levels of all other contaminants were low. Due to declining numbers in several U.S. states, the kestrel was placed on the Blue List from 1972 to 1981 (Tate 1981).

The negative impact of exposure to agricultural chemicals appears to have been alleviated by the increase in suitable open habitat and the greater availability of nest boxes (Noble and Elliott 1990). BBS figures for 1966–89 show a population increase for southern Québec.

As long as there is an abundance of nest sites and food—the two main limiting factors—we will likely continue to see a rise in kestrel numbers, as well as a range expansion as more openings are created in the forest cover.

DAVID M. BIRD
AND DERIN HENDERSON

An adult at the entrance to its nest in an old dead tree. Woodpecker holes can be seen on the trunk. Sherbrooke. *June 1979, Daniel Jauvin.*

Merlin

Faucon émerillon

Falco columbarius

Ghislain Caron

At first glance, the Merlin in flight might be mistaken for a member of the pigeon family. This speedy, agile flier maintains steady, vigorous flapping when pursuing its favourite prey, small birds.

The less stocky and more conspicuous American Kestrel is more likely to hover when foraging and to bob its tail when perched than the Merlin. Sharp-shinned Hawks in hot pursuit share the Merlin's swift flight, but have more rounded wings and a longer tail. Merlins often fly low, only 2–3 m from the ground, and perch on rocks, trees, and even on the ground. They are seen more often during migration, when they fly at greater altitude, than any other time of the year. They also fly high while defending a territory, courting, and preparing to stoop on aerial prey (Palmer 1988).

This species is sexually dimorphic: the slaty-backed male is easily distinguished from the larger, brownish-backed female.

The Merlin is considered a common migrant breeder in the province (Robert 1989b) and casual in winter in southern Québec (David and Gosselin 1984). It winters mainly from southern British Columbia and the southern United States to South America and the West Indies (Godfrey 1986).

Habitat and behaviour

In the breeding period, Merlins like broken forest or grasslands with wooded areas; they generally nest in trees and hunt over open ground. They are therefore found in a variety of habitat: near lakeshores and peatlands in the boreal forest, near natural or artificial openings, on woodland edges, or on rolling plains, prairies, and along tree-lined river systems (Trimble 1975; Palmer 1988). They have recently been found nesting in urban and residential areas, including the cities of Saskatoon, Edmonton, and Regina (Oliphant and Haug 1985; Palmer 1988). They commonly use old corvid nests, especially those of crows or ravens in coniferous or deciduous trees, often near water, but may use larger unoccupied nests such as those of Ospreys (Y. Aubry, pers. comm.). A study in Montana found that Merlins prefer old nests in trees surrounded by taller trees for maximum nest concealment; these nest sites are in areas of low tree density to ensure easy access and greater visibility (Sieg and Becker 1990). They may also nest on cliff and roof ledges, in tree cavities, or they may make a scrape, scratched out of the substrate with their feet and formed by turning movements of the body (Cramp and Simmons 1980). Nests may be lined with twigs or feathers, and sometimes adorned with new greenery.

The Merlin often nests high up in a tall conifer on the margin of a mixed or coniferous stand, near a river or open area. Sainte-Anne River, Gaspé Conservation Park. *July 1991, Yves Aubry.*

Notes on territory

The size of the home range appears to vary with population density and prey abundance. In Saskatoon, where breeding densities are high, hunting ranges during the nesting period varied from 2.2 to 64.3 km² for males and from 0.6 to 17.5 km² for females, overlapping those of neighbouring pairs by up to 77% (Sodhi and Oliphant 1992). In Montana, home ranges varied from 12.6 to 28.1 km², with much overlap over the hunting areas, although not generally near the nest (Becker and Sieg 1987).

Merlins defend the area around the nest, but not their hunting areas (Sodhi et al. 1993). They are notoriously noisy and aggressive when defending their nest sites, and males have been known to fly out to scold trespassers as far away as 1.5 km (Fox 1964; Trimble 1975). The female leaves the nest only when intruders are close, and she can be highly aggressive even toward humans (Craighead and Craighead 1940). While other bird species usually avoid nesting near Merlins, some have been recorded breeding only 30m away (Newton et al. 1978), possibly enjoying greater protection against predators.

Merlin

Breeding distribution 1984–1989

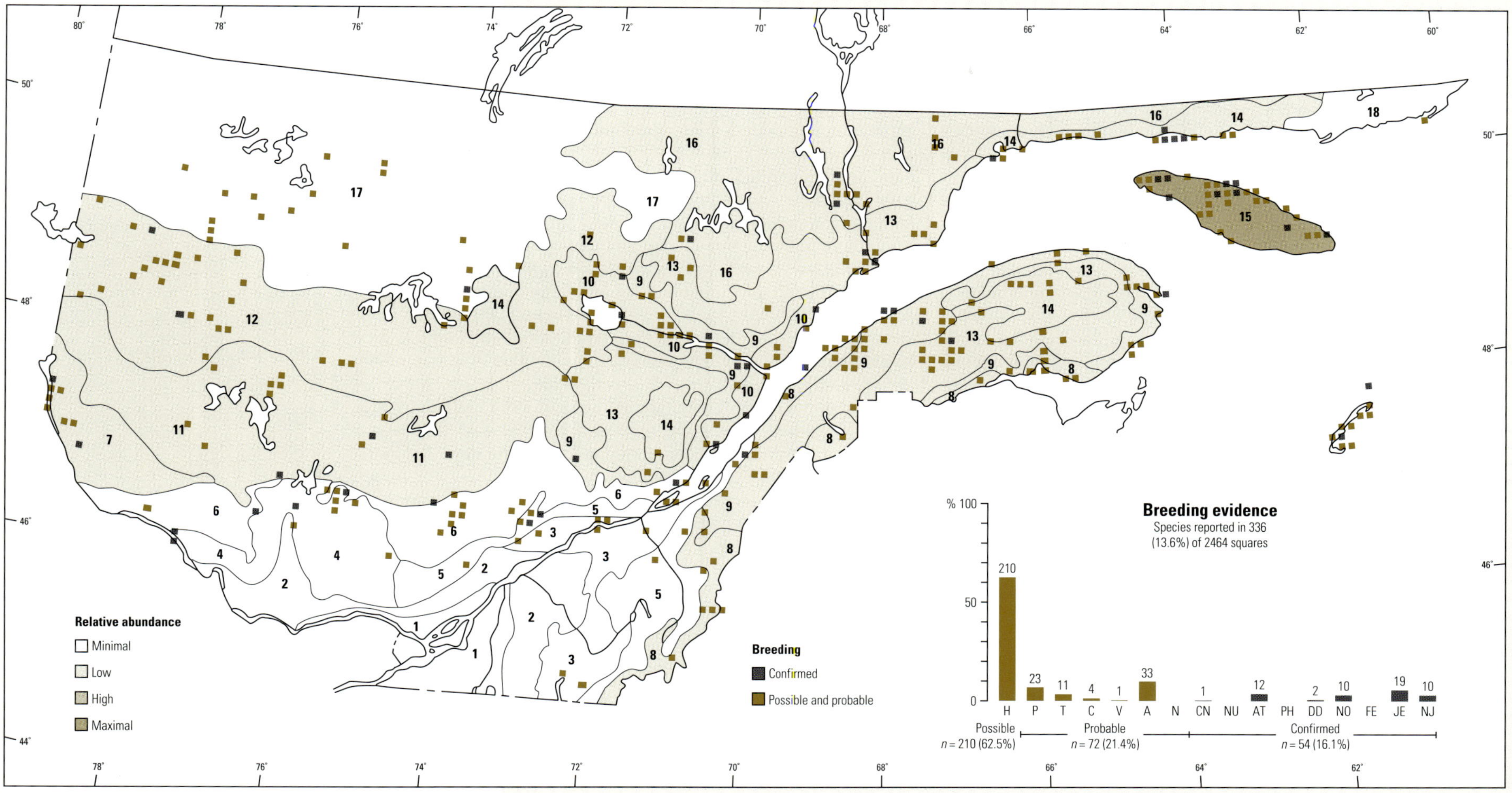

Species profile

Clutch size	usually 4–5 eggs (2–7)[1,2,3,4]
Laying interval	usually 1 egg per 2 days[1,2,3,4,12]
Start of incubation	with penultimate egg, sporadically after second egg[12]
Incubation period	≈ 30 days (28–32)[1,2,4,12]
Maturity at hatching	semi-altricial
Care of young	by female and male[2,3,4,11,12], *see* text
Nestling period	usually 25–27 days (25–35)[1,2,4]
Age at first flight	usually 29–34 days[4,12]
Dependence of young	until 1–4 weeks after leaving nest[12]
Broods per year	1[1,2,4,12]
Breeding age	usually 2 years (1–2)[1,2,4,10,12]
Mating system	monogamy, rarely polygyny[1,2,4,7,12]
Duration of pair bond	sometimes more than 1 season[1,2,9]
Total length	M 25.0–27.0 cm / F 30.5–33.0 cm[5]
Wing span	59.7–67.3 cm[8]
Weight	M 186.5 g (150.0–215.0) / F 231.5 g (187.0–255.0)[3]
Longevity record	10 years 7 months[6]

References:
1. Trimble 1975
2. Cramp and Simmons 1980
3. Brown and Amadon 1968
4. Palmer 1988
5. Godfrey 1986
6. Clapp et al. 1982
7. Sodhi 1989
8. Terres 1980
9. Warkentin et al. 1991
10. Ellis and Okill 1990
11. Sodhi et al. 1992
12. Sodhi et al. 1993

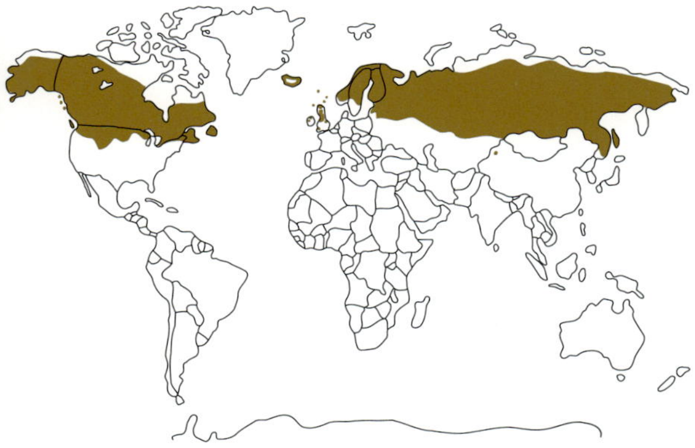

World breeding range

Chronology *n* = 26

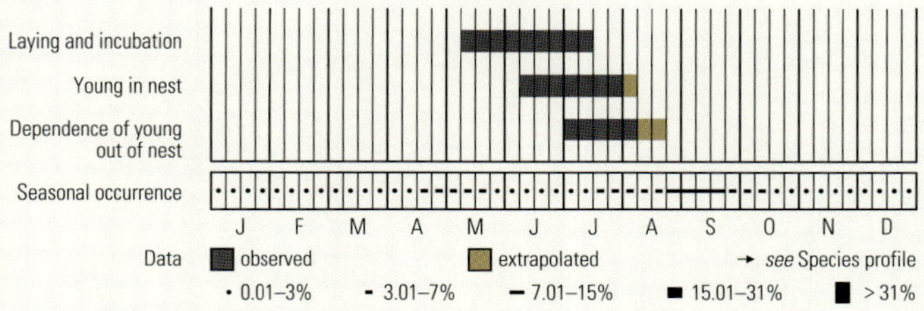

Frequency of occurrence by ecological region

Ecological regions** (forest regions)	Adequately covered squares	Squares in which reported	% of squares in which reported	Relative abundance (%)*
1. Sugar maple-hickory and sugar maple-basswood	97	0	0	0
2. Sugar maple-basswood	153	1	0.7 ± 0.9	0.8
3. Sugar maple-basswood and sugar maple-yellow birch	101	6	5.9 ± 3.0	7.7
4. Sugar maple-yellow birch-basswood	56	7	12.5 ± 7.9	16.2
5. Sugar maple-yellow birch-beech and sugar maple-yellow birch-basswood	113	9	8.0 ± 3.0	10.3
6. Sugar maple-yellow birch-beech	71	13	18.3 ± 8.4	23.7
7. Sugar maple-yellow birch-balsam fir	19	6	31.6 ± 22.1	40.9
8. Sugar maple-yellow birch and balsam fir-yellow birch	47	12	25.5 ± 11.3	33.0
9. Balsam fir-yellow birch	87	33	37.9 ± 9.4	49.1
10. Balsam fir-white birch or balsam fir-red maple	46	21	45.7 ± 12.5	59.1
11. Yellow birch-balsam fir	52	13	25.0 ± 12.2	32.4
12. Balsam fir-white birch (west)	59	25	42.4 ± 12.9	54.8
13. Balsam fir-white birch (east)	76	28	36.8 ± 10.6	47.7
14. Balsam fir-black spruce	28	10	35.7 ± 18.6	46.2
15. White spruce-balsam fir and balsam fir-white spruce	22	17	77.3 ± 18.4	100.0
16. Black spruce-balsam fir-moss	18	8	44.4 ± 26.0	57.5
17. Black spruce-moss	32	3	9.4 ± 11.7	12.1
18.			*not analysed*	

* minimal: 0–29.9%; low: 30–59.9%; high: 60–89.9%; maximal: 90–100%

** Deciduous forest: 1–7; mixed forest: 8–11; coniferous forest: 12–17

Merlins return to the same general breeding area each year, but tend to use alternative nest sites (Brown and Amadon 1968). Some birds—mostly males—show a higher level of site fidelity from year to year, however (Warkentin et al. 1991). Males arrive on the breeding territory up to a month before females (Brown and Amadon 1968; Trimble 1975). Pair bonds rarely outlast the breeding season. Even among partly non-migratory populations, such as those in Saskatoon, a strong pair bond is rarely maintained through the winter. Thus most pair bonds are newly established each year in both migratory and resident populations. Occasionally, however, a male retains the same breeding territory throughout the winter or returns to it in the spring to pair again with his old mate if she has returned to the same area (Warkentin et al. 1991). Most are monogamous, but polygynous males have been reported (Palmer 1988). They have an extensive repertoire of courtship displays, including power flying and power diving, flying in circles or figure eights with shallow wingbeats, and high circling and soaring (Feldsine and Oliphant 1985). Courtship feeding may occur air-to-air, air-to-perch with either sex perched, or with both birds perched, and is often accompanied by vocalizations (Johnsgard 1990).

Typical clutch size is 4 or 5 eggs laid at 2-day intervals, although the final egg may be laid after a slightly longer time (Johnsgard 1990). Incubation usually lasts about 30 days and is performed primarily by the female, but some observers have found the male sitting on the nest for about one third of daylight hours (Temple 1972; Newton et al. 1978). Incubation begins before the last egg is laid (Sodhi et al. 1993). The male provides the female with food, either bringing it to the nest or passing it in an aerial transfer. The hunting male is not aided by the female until the nestlings are about a week old. The young leave the nest at 25–27 days, but remain in the vicinity, and within a week they are able to take extended glides. They remain dependent on their parents for food as they hone their hunting and flying skills and achieve complete independence about 1–4 weeks after fledging, when they can catch birds (Brown and Amadon 1968; Oliphant 1974). The young migrate south 1–6 months after leaving the nest (Brown and Amadon 1968).

As adults, Merlins prey almost exclusively on birds, mostly small to medium-sized passerines, but they may take birds as large as teal and ptarmigan (Bent 1938). They also hunt mammals such as bats, squirrels, mice, and voles, and occasionally snakes, toads, frogs, and large insects such as dragonflies (Johnsgard 1990). Usually launching surprise attacks from a perch or flying low in search of prey, Merlins flush their quarry and then pursue it in a rapid and direct tail chase, snatching it in their talons. On one occasion 6 were observed hunting alongside a moving train, apparently taking advantage of the fright induced among some small birds by the noise and speed of the train (Kenyon 1942). High-speed stoops from great heights have also been observed (Brown and Amadon 1968; Palmer 1988), but the conspicuous nature of this method leaves Merlins vulnerable to kleptoparasitism by other raptors (Buchanan 1988a). When feeding, Merlins may sometimes cache the uneaten remains of their prey in trees or on the ground, to be retrieved later in the day (Oliphant and Thompson 1976; Palmer 1988).

Adults are noisy around the nest, when the male is delivering food to the female and when she is feeding the young, and therefore breeding pairs can be located from a distance. This behaviour, along with the presence of young or eggs in the nest, was used to confirm breeding. Once the young have fledged, they remain near the nest, learning from their parents how to hunt and practising their flying skills, so they are conspicuous at this vulnerable stage. The presence of recently fledged young was the breeding evidence code used most often.

Notes on territory (continued)

Reported nesting densities have generally been low, probably because of low populations (Trimble 1975). A population on the Minnesota–Ontario border was considered relatively concentrated, with nests 3.2 km apart (Craighead and Craighead 1940). The highest ever reported nesting density was in Saskatoon in 1989, at 0.25 pair/km², with nearest neighbours 161–4669 m away (Sodhi et al. 1992).

This falcon is found in broad tracts of second-growth forest with a few remaining stumps that serve as perches. North of Dieppe Hills, north of La Sarre, Abitibi. *June 1991, Desneige Perreault.*

In openings in the boreal forest, like those around the headwaters of the Jacques Cartier River, the Merlin targets small birds, its favourite prey. Laurentide Wildlife Reserve, north of Québec City. *June 1991, Réginald Ouellet.*

A general decline in the North American population was noted in the 1950s and 1960s, when Merlin reproductive success was affected by DDT and other organochlorines in the environment (De Smet 1984). This migratory raptor is at the top of the food chain and feeds on migratory birds, making it highly susceptible to the accumulated effects of pesticide residues in its breeding and wintering areas. Of the four subspecies in Canada, the most seriously affected was *Falco columbarius richardsonii*. Its range is restricted to the prairie-parkland of the northern Great Plains of the United States and the Canadian Prairie provinces, covering the most contaminated of all North American habitats (Trimble 1975; Noble and Elliott 1990). Agricultural practices in that region have also severely reduced the availability of suitable habitat and may continue to be the main limiting factor affecting Merlins on the Prairies (Oliphant 1985).

In eastern Canada, the subspecies *Falco columbarius columbarius* also suffered lowered productivity. Studies showed that between 1950 to 1969, hatching success fell to 48%, down from the pre-1950 level of almost 91%. Meanwhile, northern birds maintained a hatching success of 89% (Fox 1971). Even in regions like Newfoundland, where essentially no pesticides were used, pesticide residues in breeding birds were found to be high enough to cause eggshell thinning (Temple 1972).

In response to restrictions on the use of DDT, dieldrin, and other toxic pesticides, Merlin population levels appear to be on the rebound (Buchanan 1988b). The species was removed from the American Blue List after 1981, on which it had figured since the list's inception in 1972, but was still classified as a Species of Special Concern in 1982 and 1986 by the National Audubon Society (Tate and Tate 1982; Tate 1986). In Québec and southern Ontario, Fyfe (1976) described the Merlin's abundance as rare to medium, with a declining or stable population. In Québec, BBS data for 1966–89 showed a population increase, but numbers have not yet returned to their pre-1950s levels, and levels of toxic chemical residues in eggs remain medium to high (Fyfe 1977; De Smet 1984). Organochlorine use is still increasing on a worldwide scale, notably in Central and South America, where many Merlins, as well as the avian prey of all North American Merlins, still winter.

Distribution

In North America, the Merlin breeds from the tree line down to northern United States (National Geographic Society 1987).

The species was atlassed throughout southern Québec. Frequency of occurrence was highest on Anticosti Island and the Magdalens, where there are spruce stands with broad open tracts, and lowest in the deciduous forest region and the northwestern sector of southern Québec.

Merlins were reported in only 13.6% (336/2464) of squares surveyed; 210 reports were simply of spring and summer sightings of birds in suitable breeding habitat.

Subsisting mainly on small birds and able to nest in a variety of habitats, Merlins have adapted to breeding virtually throughout Québec, even as far north as Baie aux Feuilles in the Ungava area, where they nest on cliffs near small pockets of spruce trees (D. M. Bird, pers. obs.) and along Hudson Bay (Cons. G. G. – GREBE 1990c).

History and trends

The Merlin has been the least studied of the five species of falcons in Canada (Oliphant 1985), and historical records are lacking. Taverner (1919) believed it to be "nowhere a common bird" in eastern Canada. Dionne (1906) noted that it was fairly common during spring and fall migration in the Québec City region, but scarce around Montréal. Todd (1963) listed a few documented sightings from northern Québec during the summer months, including La Longue Pointe (then Long Point) on the east coast of James Bay in 1860 and 1926, near the Great Whale River in 1930 and 1935, and on a cliff face bordering Lake Guillaume Delisle along the Hudson Bay coast in 1915. Todd noted that most sightings were made in August and September, when the birds were presumably migrating southward and were at times fairly common. Based on his own records, he believed that the greatest number of Merlins were to be found in the extensive marshy areas along the south shore of James Bay, especially around Partridge Creeks, some 50 km west of the Québec–Ontario border (Todd 1963).

DAVID M. BIRD AND DERIN HENDERSON

Peregrine Falcon

Faucon pèlerin

Falco peregrinus

Ghislain Caron

FEW PEOPLE CAN FORGET SEEING THEIR VERY FIRST PEREGRINE FALCON, and many will agree that witnessing this spectacular bird in full stoop is an experience without equal. While records vary, it is safe to say that a stooping Peregrine can reach a speed of 200 km/h. Once prized by kings for falconry, the Peregrine Falcon heads the list as the world's most popular bird of prey (Bird and Brown 1991). Its endangered status and the efforts to restore its numbers in the 1970s, 1980s, and 1990s have only served to heighten its stature in the eyes of the general public.

Roughly crow-sized, the Peregrine Falcon is smaller than a male Gyrfalcon and larger than a female Merlin. The Peregrine can be confused with the Gyrfalcon at a distance, but at close range, it appears to have a blackish hood and it has a pronounced moustachial stripe below the eyes. Adult Peregrines are slate blue-grey with a cream or whitish breast. First-year birds are quite different, with chocolate brown plumage.

At Cap Tourmente, Peregrine Falcons nest with Common Ravens on a bare cliff. Twenty young falcons were released over 3 successive seasons, from 1982 to 1984, in this national wildlife area. Northeast of Québec City. *August 1991, Yves Aubry.*

Aside from the non-migratory subspecies inhabiting Canada's west coast, *Falco peregrinus pealei*, Peregrine Falcons in North America are members of either of the subspecies *Falco peregrinus anatum* or *Falco peregrinus tundrius*. The main distinguishing characteristics of the *anatum* birds are their slightly larger size, salmon or pink breast with more extensive barring (as opposed to plain creamy-white in *tundrius*), and more distinctive black malar markings (White 1968; White and Boyce 1988).

In Canada, the *tundrius* subspecies was listed as threatened until 1992, when it was downgraded to vulnerable. Recent surveys point to a considerable comeback in its populations (Bird and Weaver 1988; Bromley 1991; D. M. Bird, McGill U., and M. Lepage, MEF, unpubl. data). The *anatum* subspecies is listed nationally as endangered, but will probably be downgraded shortly.

Habitat and behaviour

Nesting sites of the Peregrine Falcon are extremely varied, ranging from trees or the ground to city skyscrapers and large bridges (Cade and Bird 1990), but cliffs, especially those associated with a body of water, have always been the staple choice. Peregrine pairs traditionally use the same nesting territory (Court et al. 1988; Blais 1992, 1993), but are known to use alternative cliffs or sites from year to year (Bird and Weaver 1988). In Ungava Bay, they will share nesting cliffs with Gyrfalcons and Rough-legged Hawks, but seldom with Common Ravens, as they sometimes do in southern Québec.

Males generally arrive on the territory first. Peregrines defending their territories display vigorous intraspecific aggression, as well as hostility toward other raptors and ravens. During courtship, the male engages in power dives, perches in prominent locations, and offers food to the female. While he entices the female to explore several nesting locations, she ultimately chooses the site. Generally, 3–4 eggs are laid in a hollowed-out scrape in loose gravel or earth on a ledge, anywhere from early April to early June, depending on latitude. Renesting is common if the first clutch is lost, but this is less likely in northern populations (Bird et al. 1990).

The male hunts for the pair, while the female does most of the incubating over a 28–35 day period. The young are nest-bound generally for 35–42 days, and they are still dependent on the parents for up to about 5 weeks after fledging.

Other than humans, the main enemies of Peregrines and their eggs and young are raccoons, foxes, mink, corvid species, Great Horned Owls, Red-tailed Hawks, and Northern Goshawks.

Peregrines were reported in only 1.3% (33/2464) of squares atlassed, and breeding was confirmed in only 10 squares. In 7 other squares some activity indicated probable breeding. Single Peregrines are known to inhabit a nesting territory without producing eggs or young, however (Bird and Aubry 1982;

For several years, Peregrine Falcons have again been nesting on cliff ledges on Mount Saint-Hilaire. *July 1991, Pierre Bannon.*

Notes on territory

Peregrine Falcons defend a large territory where breeding activities take place, at the heart of which is the nesting cliff. Defensive behaviour varies in intensity and type with the intruder's distance from the nest and the stage in the breeding cycle, but a minimum distance of 100 m is vigorously defended at all times (Cade 1960). Intraspecific aggression declines in frequency with distance, although attacks may be launched up to 1.6 km from the nest. In hunting territories shared by more than one pair, agonistic behaviour may be associated with some perches. Dominance is probably the factor that controls the use of perches, however, as not all individuals are on the hunting

Peregrine Falcon

Breeding distribution 1984–1989

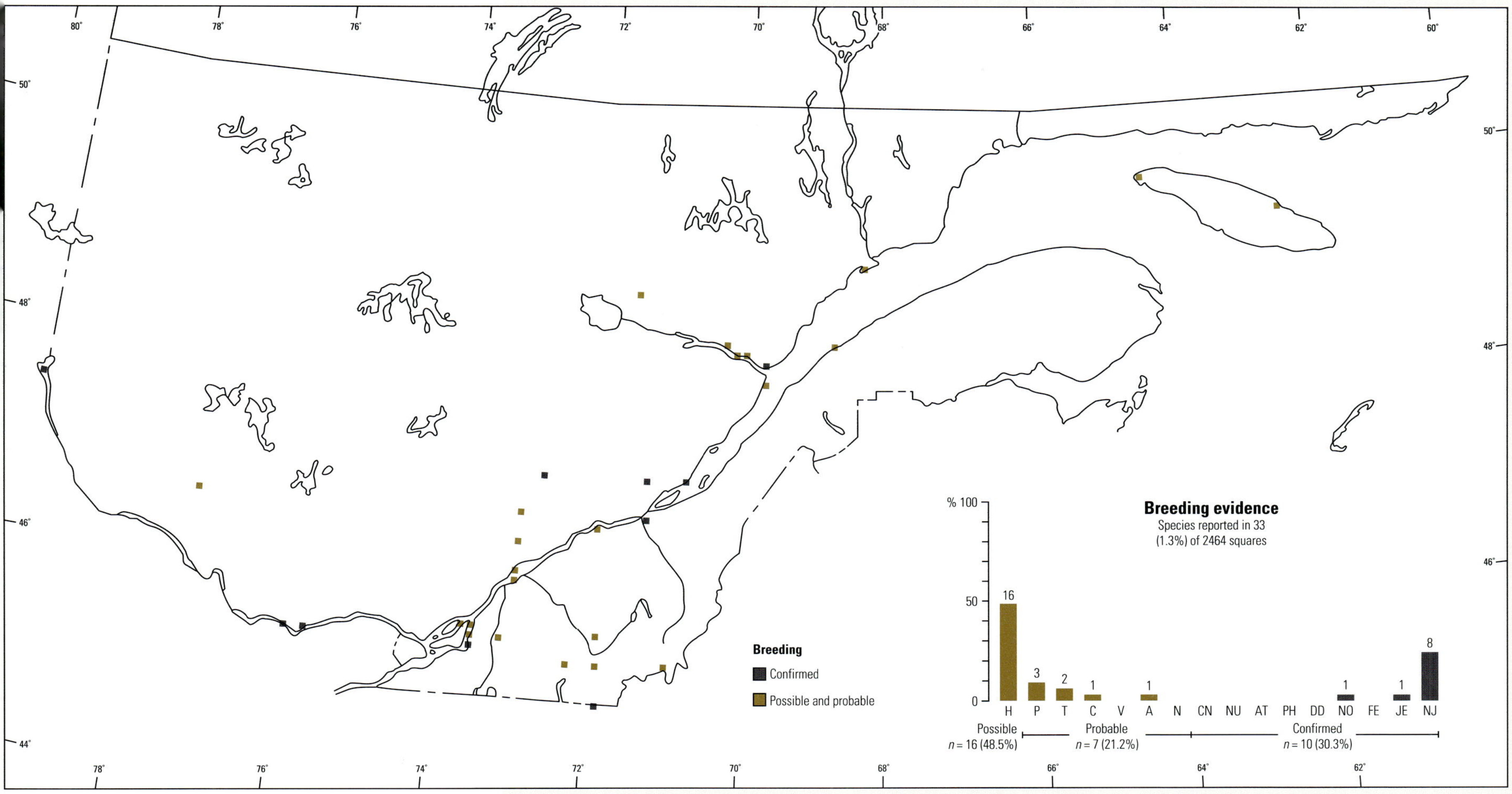

Chronology $n = 43$

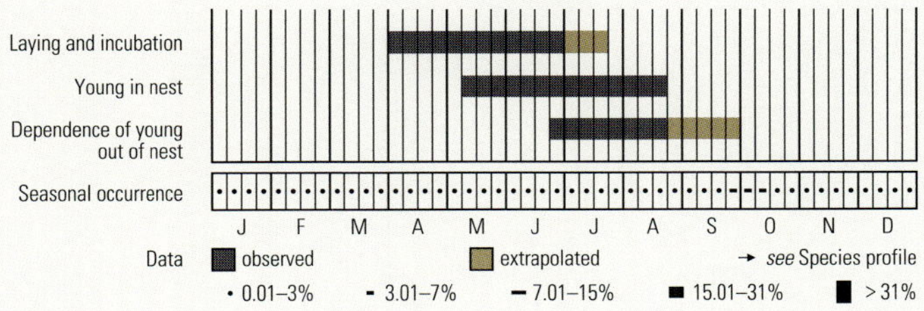

Species profile

Clutch size	usually 3–4 eggs (2–5)[1,2,3,4]
Laying interval	usually 1 egg per 2 days (2–3)[1,2,4,10]
Start of incubation	with last or penultimate egg[10]
Incubation period	28–35 days[1,2,3,4]
Maturity at hatching	semi-altricial
Care of young	by female and male[1,2,3], *see* text
Nestling period	usually 35–42 days (33–49)[1,3,4,9,10]
Age at first flight	M usually 39–46 days / F usually 41–49 days[9]
Dependence of young	until at least 30 days after first flight[1,4,10]
Broods per year	1[1,10]
Breeding age	usually 2 years (2–4)[1,2,9,10]
Mating system	monogamy, rarely polygyny[1,9]
Duration of pair bond	usually for life[9]
Total length	M 38–46 cm / F 46–54 cm[7]
Wing span	109.2–116.8 cm[6]
Weight	M 611 g (?) / F 952 g (?)[5]
Longevity record	12 years 3 months[8] / 18 years[11], *see* text

References:
1. Cade 1960
2. Cade 1982
3. Bent 1938
4. Johnsgard 1990
5. Dunning 1984
6. Terres 1980
7. Godfrey 1986
8. Clapp et al. 1982
9. Palmer 1988
10. Cramp and Simmons 1980
11. Hall 1970

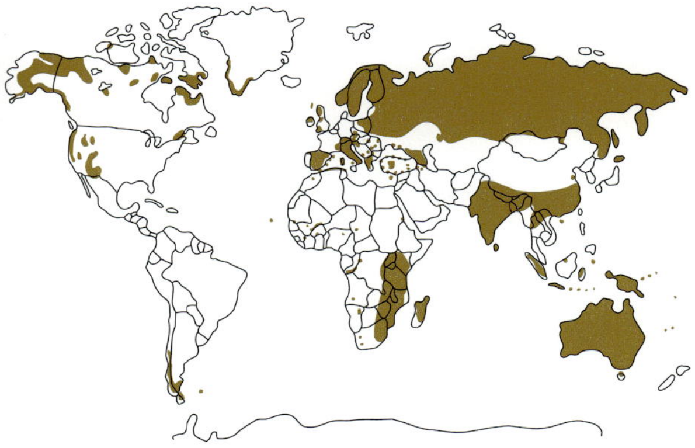

World breeding range

This humus-lined nest on a cliff ledge contains an egg and 3 hatchlings. At the end of each baby falcon's upper mandible can be seen the egg tooth, which it used to break open its shell; this will disappear within the next few days. Rankin Inlet, Northwest Territories. *July 1980, Yves Aubry.*

Blais 1992, 1993). On occasion, a pair will form but not breed, perhaps due to sexual immaturity. The 10 reports of confirmed breeding are in keeping with the results of intensive annual surveys conducted in southern Québec.

Peregrine Falcons prey primarily on birds (Ratcliffe 1980), although there are some indications that mammals play a larger role than expected in the diet of tundra birds (Court et al. 1988). Arctic birds eat mainly shorebirds and passerines, while coastal Peregrines depend on alcid and waterfowl populations. In forested areas, larger passerines like Northern Flickers and Blue Jays constitute much of the diet. Urban-dwelling falcons are quite variable in their tastes, some subsisting mainly on pigeons and others on migratory birds (Cade and Bird 1990). In southern Québec, 5 nests yielded the remains of over 25 prey species ranging from Ring-billed Gulls to small passerines; their dietary mainstays are Rock Doves, sandpipers, and Blue Jays (Blais 1992).

The Peregrine's preferred hunting technique is to circle above and behind its victim, then stoop at high speed. It sometimes launches a surprise attack from a perch on a cliff. Although the success rate of individual falcons varies, a breeding male inhabiting a cliff in the Eastern Townships captured 1 prey for every 3 attempts (Bird and Aubry 1982).

Distribution

The breeding range of the *tundrius* subspecies in Québec includes the entire northern coast, particularly Ungava Bay and the nearby Gyrfalcon Islands. Inland breeding pairs have been reported on the mainland, but are not very common (Bird and Weaver 1988).

The *anatum* subspecies occupies a wide breeding range throughout Québec south of the tree line, but at a much lower density than *tundrius*. Breeding pairs are active on the east coast of Hudson Bay near the Great Whale River, along the St. Lawrence, in the Eastern Townships, and in the Ottawa Valley. In the Saguenay River mouth and valley, there are a great many suitable cliffs, but their inaccessibility makes them difficult to survey. Whether *anatum* birds exist in any numbers in Québec's interior is not known, largely due to the lack of surveys.

The boundary between the ranges of the two subspecies is not clearly delineated and they appear to interbreed and produce intermediate forms, at least in the southern Ungava Bay region (D. M. Bird, McGill U., unpubl. data).

The most northerly record for Peregrine Falcons comes from Ivujivik, at the northern tip of the Ungava Peninsula.

By and large, birds that breed farther north migrate longer distances than southerly breeders (Hickey 1969), and *tundrius* birds fly as far south as Argentina. Captive-bred and released falcons, on the other hand, generally move only short distances. In 1983, for example, at least 4 Peregrines wintered in Montréal (R. Galbraith, unpubl. data). Hall (1970) reported 2 banding returns in 1943, one from Georgia and the other from New York City, of sibling first-year birds born in Montréal. Interestingly, their mother nested successively with 3 males, from 1936 to 1952, on the Sun Life Building in Montréal.

Notes on territory (continued)

grounds at the same time. In areas with strong concentrations of Peregrines, inter-nest distance is around 4.8 km but can vary greatly; in some places it can be as much as 8.6 km. It is rare, however, to find Peregrines nesting less than 0.5 km apart (Ratcliffe 1962).

It is not known what governs Peregrine density in suitable areas. According to Ratcliffe (1962), territorial behaviour limiting nest density is related to the food supply, whereas for Hickey and Anderson (1969), the size of the non-breeding population is an important factor.

Peregrine Falcons have often nested on the cliffs of Le Pinacle, overlooking Lake Lyster, and have regularly been seen stooping on birds flying over the lake. South of Sherbrooke. *May 1979, Yves Aubry.*

History and trends

While Peregrine Falcons are found on every continent except Antarctica, they have never been considered a common bird (Kiff 1988). By the 1960s, Peregrine populations were deemed to be in a serious state of decline in eastern North America (Hickey 1969). A pattern of thin-shelled eggs and aberrant reproductive behaviour became evident in all North American populations, including those in northern Québec (Berger et al. 1970; Cade and Fyfe 1970). It has been widely accepted that DDT was behind the population crash (Ratcliffe 1980).

Ungava Bay populations appear to have recovered fully (Bird and Weaver 1988), solely as a result of the ban on the use of organochlorines, and without the intensive management program undertaken throughout the rest of North America. Indeed, thousands of captive-bred Peregrines have been released in many states and all provinces except British Columbia, Prince Edward Island, and Newfoundland. These birds now constitute the vast majority of the existing breeding population in southern Canada and the eastern and midwestern United States.

Due to the multiplicity of habitats available in Québec—cliffs, lakes, and rivers—it is likely that other pairs have yet to be discovered. Aside from the healthy, stable numbers of Peregrines in Ungava Bay (Bird and Weaver 1988), another small population has been found on the east coast of Hudson Bay (Albright and Doidge 1985; Cons. G. G. – GREBE 1990c).

Adult on the nest brooding a hatchling and an egg on a cliff on Anderson Island. Hudson Bay. *July 1990, Serge Brodeur.*

The goal of the *Anatum* Peregrine Falcon Recovery Plan for southern Québec and the Atlantic provinces is to achieve a minimum of 10 pairs producing a mean of 15 young annually for the 5-year period between 1993 and 1997 (CWS 1988). Québec's southern population is well on its way to recovery in this range, for the number of breeding pairs is increasing, as can be seen in the table below. A few sites where pairs used to breed—on Bonaventure Island and at Cape Bon Ami in the Gaspé, for example—have not been occupied recently.

Known nesting of Peregrine Falcons in southern Québec

Year	Territorial pairs (a)	Sites occupied by singles (b)	Successful pairs with fledged young (c)	Fledged young (d)	Young/ territorial pair (d÷a)	Young/ successful pair (d÷c)
1976	0		0	0	0.00	0.00
1977	0		0	0	0.00	0.00
1978	0		0	0	0.00	0.00
1979	0		0	0	0.00	0.00
1980	1		1	2	2.00	2.00
1981	0		0	0	0.00	0.00
1982	1		0	0	0.00	0.00
1983	1		1	1	1.00	1.00
1984	3		2	5	1.67	2.50
1985	0	1	0	0	0.00	0.00
1986	3		2	3	1.00	1.50
1987	6	1	3	7	1.17	2.33
1988	10		5	11	1.10	2.20
1989	8	3	4	9	1.13	2.25
1990	11*	1	8	17	1.55	2.13
1991	11	1	6	20	1.82	3.33
1992	12	0	6	14	1.16	2.33
1993	9	3	5	10	1.11	2.00
1994	11	3	9	25	2.27	2.78
1995	13	2	10	26	2.00	2.60

* Excludes the site found on the North Shore of the Gulf of St. Lawrence during the 1990 survey.

As for current limiting factors, some Peregrines are still lost each year to indiscriminate shooting and persecution by people who regard them as pests that prey on wild game or valuable domestic birds like homing pigeons. Aside from existing and possibly new chemical threats such as organophosphates, the main threat to the Peregrine Falcon remains exploding human populations leading to habitat degradation. This can take the form of decreased prey populations through habitat loss, or disruption of breeding activities by rock climbers or other visitors to the cliffs.

The trend to urban living in this species is still on the upswing (Cade and Bird 1990). Peregrines now breed in at least 25 North American cities, including Montréal and Québec City. For example, New York City is now home to 10 pairs! As part of this trend, Peregrines are also taking to nesting on bridges. This is unfortunate, as many fledglings drown when they cannot find safe places to land (Cade and Bird 1990).

Despite all of the above, it is obvious that the spectacular Peregrine Falcon is quite a flexible species, and unless humans decide otherwise, it should be with us for some time to come.

David M. Bird, Pierre Laporte, and Michel Lepage

Gray Partridge

Perdrix grise

Perdix perdix

Jean-Marc Giroux

NATIVE TO EURASIA AND INTRODUCED INTO QUÉBEC AROUND 1940, the Gray Partridge is now a fairly common permanent resident in agricultural areas in the southern part of the province.

Despite its size, comparable to that of a pigeon, and its fairly bright coloration, the Gray Partridge is often overlooked during the breeding season because of its secretive behaviour. As a result, few observers have been able to confirm its status as a breeder.

Habitat and behaviour

The Gray Partridge is a bird of open country. It inhabits mainly farmland, but may also find suitable food and cover in urban parks like the Montréal Botanical Garden, where until the 1990s it had nested consistently for over 20 years. During the breeding season, both in its native habitats in Europe and Asia (Cramp and Simmons 1980) and in its recently adopted range in Québec (Chabot et al. 1980), the Gray Partridge prefers meadows where the grass is not much higher than its head and where there are both hedgerows for cover and open areas nearby.

The Gray Partridge is gregarious for 7 or 8 months of the year, forming coveys of 6–25 birds. Each covey is made up of a nucleus of one or more families, sometimes accompanied by non-breeders. During the breeding season, however, the Gray Partridge is monogamous and territorial (McCabe and Hawkins 1946; Cramp and Simmons 1980). Extensive courtship displays, often initiated by the female, occur from pair formation to the onset of laying (Cramp and Simmons 1980). The nest is a scrape in dry ground dug by the female and lined with twigs (McCabe and Hawkins 1946). Typical clutch size is 10–20 eggs, laid a day or two apart (Cramp and Simmons 1980). Usually only one clutch is laid in a season, but if the first clutch is lost early in the season, a second clutch of 9 or 10 eggs may be laid. During laying, the female covers the unincubated eggs with grass to help camouflage the nest against predators. She incubates the eggs herself for a period of 23–25 days. The chicks leave the nest a few hours after hatching and are generally able to feed themselves right away, although the adults lead them to food. Like other gallinaceous birds, the young develop flight feathers at an early age and can fly short distances by the time they are around 2 weeks old (Cramp and Simmons 1980).

Nesting success is fairly low. McCabe and Hawkins (1946) estimated that 13% of the eggs in clutches are infertile, and 68% of clutches are destroyed, mainly by farm machinery (78%) and predation (22%). In Québec, the chief predators include the domestic cat, red fox, striped skunk, raccoon, long-tailed weasel, American Crow, Great Horned Owl, Snowy Owl, and Northern Harrier (Chabot et al. 1979).

The Gray Partridge does not scratch the ground for its food, but rather pecks at it. The adult birds feed on grain, clover, weeds, and green grasses, while the chicks eat mainly insects (Potts 1972; Cramp and Simmons 1980; Erpelding et al. 1987).

The species was reported in only 4% (98/2464) of Atlas squares surveyed, which corresponds to slightly fewer than a third of the expected number in the Central St. Lawrence Lowland. Most reports were of pairs in suitable nesting habitat (38/98), either seen from a road or flushed by observers walking through a field. Chicks and nests were found on only a few occasions (25/98).

Distribution

Extensive releases of the Gray Partridge in the 1920s and since promoted natural population increases, allowing the species to colonize vast portions of North America. The gaps in current distribution reflect in part the original patterns of introduction (Johnsgard 1973). In Québec, the species occurs mainly in the Central St. Lawrence Lowland and surrounding areas.

According to McAtee, a small population was already established in 1929 in northwestern New York State and expanding over the border into Huntingdon County in southwestern Québec (Chabot et al. 1980). Since then, the species has gradually extended its range in Québec, assisted by major release programs in the 1940s and up until 1978 (Rand 1945; Chabot et al. 1980). By the late 1970s, the Gray Partridge had gained a firm foothold in the farmlands of the Ottawa and St. Lawrence valleys, from the American border to Portneuf and Lotbinière counties (Chabot et al. 1980). By 1989, the species had reached Bellechasse County and, by 1991, Beauport on the north shore and Saint-Denis (Kamouraska) on the south shore.

History and trends

In its native Europe, the Gray Partridge originally inhabited steppes and grassy mountain ridges. Although it still occupies this habitat, it has also colonized lowland areas cleared for farming during the Middle Ages (Cramp and Simmons 1980). The first attempt to introduce the species into North America was in New Jersey at the end of the 18th century. Like several subsequent attempts in other locations, it was a failure. The first successful release occurred in 1908, when several birds from Hungary were released near Calgary. This initial population, augmented by later releases, spread through southwestern Canada and over the border into the Pacific Northwest (Bent 1932; Long 1981; Root 1988). In Québec, the species was first reported in the winter of 1942–43 (Ouellet 1974a). The Québec population is most likely descended from birds released in New York and Ontario (Gosselin 1977b).

In North America, the Gray Partridge has not only colonized the vast farmlands of the Great Plains but has also become established on the open grassy ridges of the western part of the continent. In Washington State, it is found at elevations as high as 1600 m (Terres 1980). In New York (Peterson 1988g) and Vermont (Kibbe 1985e), the Gray Partridge has colonized only the St. Lawrence River and Lake Champlain regions. Therefore, the Central St. Lawrence Lowland could be considered a northeastern outlier of the species' main breeding range. Although it is very unlikely that the Gray Partridge will extend its range beyond this area in the near future, it should continue to expand northeastward along the St. Lawrence to colonize the remaining farming country. According to Root (1988), it does not appear that low temperatures and snow cover directly limit distribution, but rather the lack of natural and artificial grasslands providing suitable habitat.

Generally speaking, then, this newcomer is now well established as a breeder in Québec and the population is still expanding, but the agricultural use of herbicides, fungicides, and insecticides could be a limiting factor on its breeding success (Rands 1986).

MARCEL DARVEAU

Notes on territory

Starting in July, Gray Partridges form coveys that stay together until the winter months. Coveys may have overlapping home ranges, but try to stay away from each other. Pairs are solitary the rest of the year. During the first two weeks after pair formation, the pair does not stake out a fixed territory and tolerates the presence of conspecifics as long as they do not come too close. Then, the pair defends a territory for copulating, nesting, and foraging until the eggs hatch. Families may move several hundred metres from the nest in the first few days after hatching (Cramp and Simmons 1980).

In South Dakota, pairs had a home range of 9.7–17 ha, while the mean home range for coveys was 96 ha (Kibbe 1985e). In winter and spring in New York State, home ranges varied between 82–672 ha in size (Church and Porter 1990). One of the densest populations in the world (1 bird/4 ha) was in Ontario during the 1950s and 1960s (Lumsden 1987b).

The Gray Partridge, a newcomer to the province, inhabits the farm country of the Central St. Lawrence Lowland. This cornfield in Neuville, west of Québec City, provides food and cover for this member of the pheasant family. *August 1991, Marcel Darveau.*

Gray Partridge

Breeding distribution
1984–1989

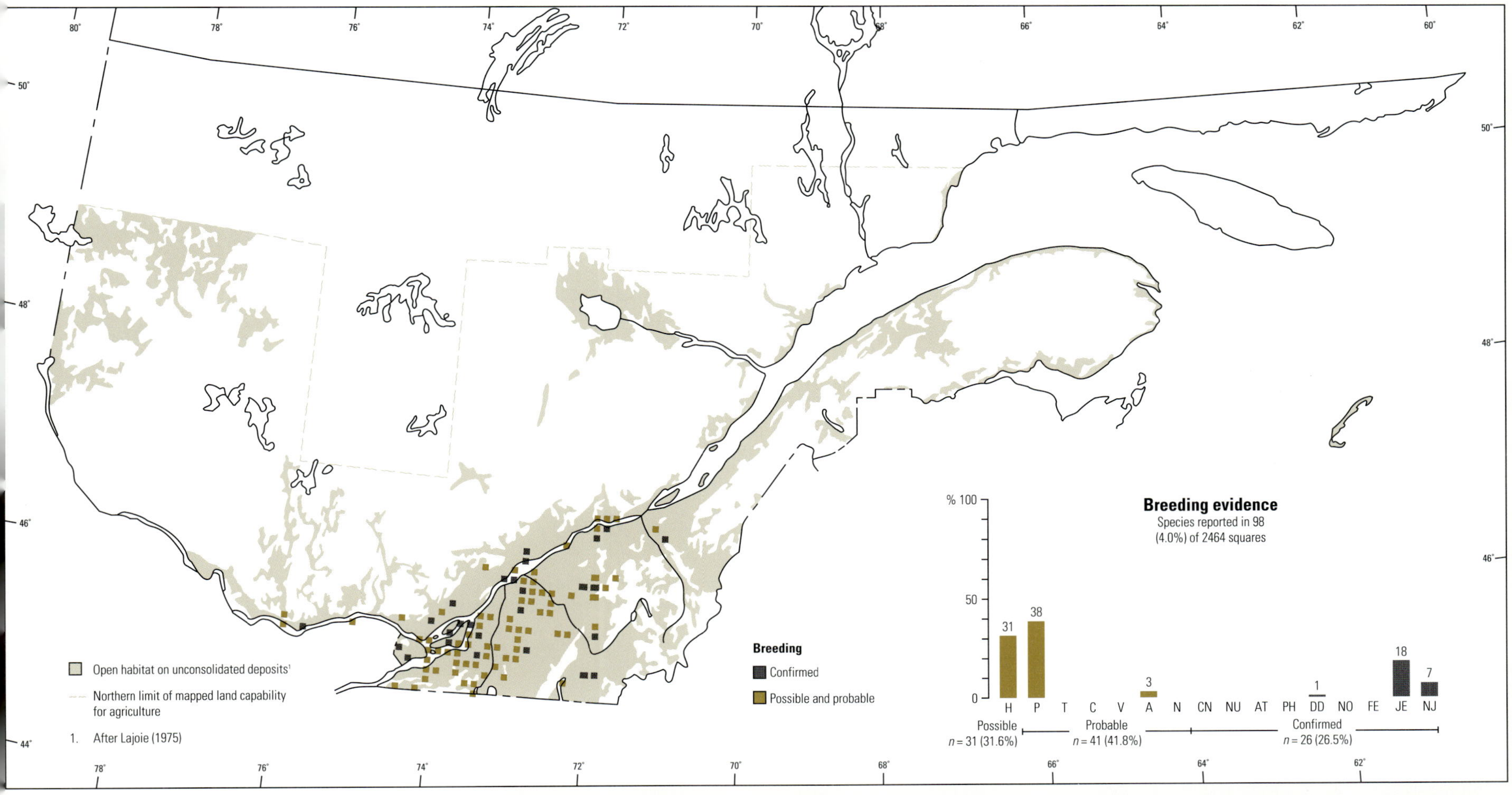

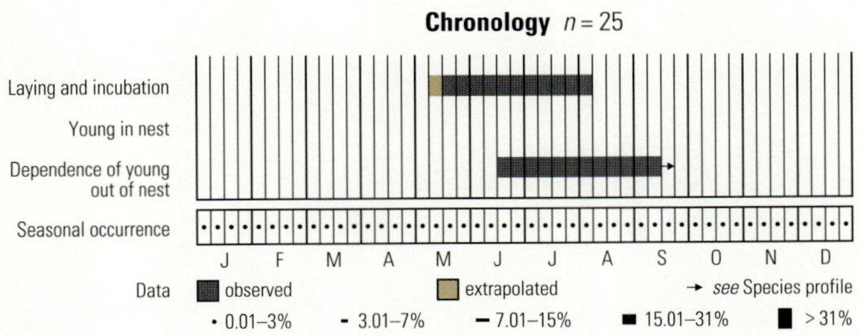

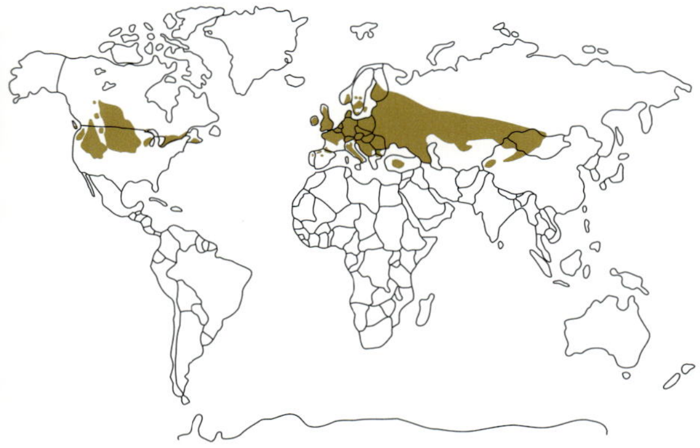

World breeding range

Species profile

Clutch size	usually 10–20 eggs (4–29)[1]
Laying interval	1 egg per 1–2 days[1]
Start of incubation	with last egg[1]
Incubation period	23–25 days[1]
Maturity at hatching	precocial
Care of young	by female and male[1]
Nestling period	less than 24 hours[1]
Age at first flight	≈ 13–16 days[1,2,3]
Dependence of young	until winter[1]
Broods per year	1[1]
Breeding age	1 year[1]
Mating system	monogamy[1]
Duration of pair bond	usually more than 1 season[1], sometimes for life[7]
Total length	30.5–33.0 cm[4]
Wing span	45.7–55.9 cm[5]
Weight	M 398 g (?–454) / F 381 g (?–434)[6]
Longevity record	5 years 2 months[1]

References:
1. Cramp and Simmons 1980
2. Ehrlich et al. 1988
3. Harrison 1978
4. Godfrey 1986
5. Terres 1980
6. Dunning 1984
7. Carroll 1993

Ring-necked Pheasant

Faisan de chasse

Faisan de Colchide

Phasianus colchicus

IMAGINE THE SURPRISE OF THE FIRST PERSON TO STUMBLE UPON THIS exotic-looking bird in the fields of Québec! The male Ring-necked Pheasant, resplendent with his iridescent green-blue feathers along the sides of his head, rich bronze plumage, bright red wattles and long, tapering, barred tail, is unlike any other wild bird in Canada. The female, dull by comparison, is a mottled brown with a long, tapering tail, and does not have the tarsal spurs of the male.

A native of Eurasia, the North American Ring-necked Pheasant is a hybrid of several introduced subspecies (Godfrey 1986). This bird is a sedentary breeder in the Central St. Lawrence Lowland, where it is seen only occasionally (David and Gosselin 1984; Club des ornithologues de l'Outaouais 1985; Bannon 1991a; Lepage 1993); birds sighted have usually been released that season. With the exception of the fluctuating resident population on Mount Royal in Montréal from the 1940s to 1982, pheasants have never really established a self-sustaining population in Québec.

Habitat and behaviour

Pheasants are birds of farmland, grassland, orchards and parks in residential areas (Peck and James 1983). The ideal habitat has grassland or cropland cover for nesting and feeding, close to dense cattail marshes or shrubland for shelter in winter. In Canada, natural populations sustained by wild breeding birds are found only in the deciduous forest region of southern Ontario (Peck and James 1983), and one of the earliest signs of spring there is the call of the courting male pheasant resounding throughout the open countryside.

Beginning in February or March, the cock establishes a territory, or crowing area, usually encompassing open field with some adjacent woods or brush for escape cover. This area, of indefinite size and boundary, is defended against trespassing males, and battling cocks will occasionally fight to the death (Dale 1956). Within the area, there may be several prominent sites where the territorial cock rises high on his feet and emits a loud, harsh call. This performance is usually combined with vigorous wing drumming, and serves both to warn rivals of his presence and to attract females (Terres 1980).

When a hen wanders onto his territory, the cock fluffs out his breast feathers, erects his ear tufts and extends his scarlet wattles. He then drops his head, fans his tail, and scrapes the wing closer to the female along the ground as he struts in semicircles around her. The male may also "tidbit" the female, offering her choice bits of food (Cramp and Simmons 1980). Pheasants are polygynous, and males with harems of as many as 18 hens have been reported (Dale 1956), but the average is 3 or 4 (Terres 1980).

Jean-Marc Giroux

Notes on territory

The exact boundaries of the male's territory are indefinite and appear to shift as the season progresses (Dale 1956). Territory size depends on population density, food supply, and habitat structure (Cramp and Simmons 1980). It may be over 40.5 ha in poor habitat and less than 0.8 ha in optimum habitat

There was a pheasant population living in the thickets of the Mount Royal cemeteries for some 40 years until 1982, sustained by the local abundance of berries and by handouts of grain. Montréal. *March 1991, Yves Aubry.*

Ring-necked Pheasant

Breeding distribution

1984–1989

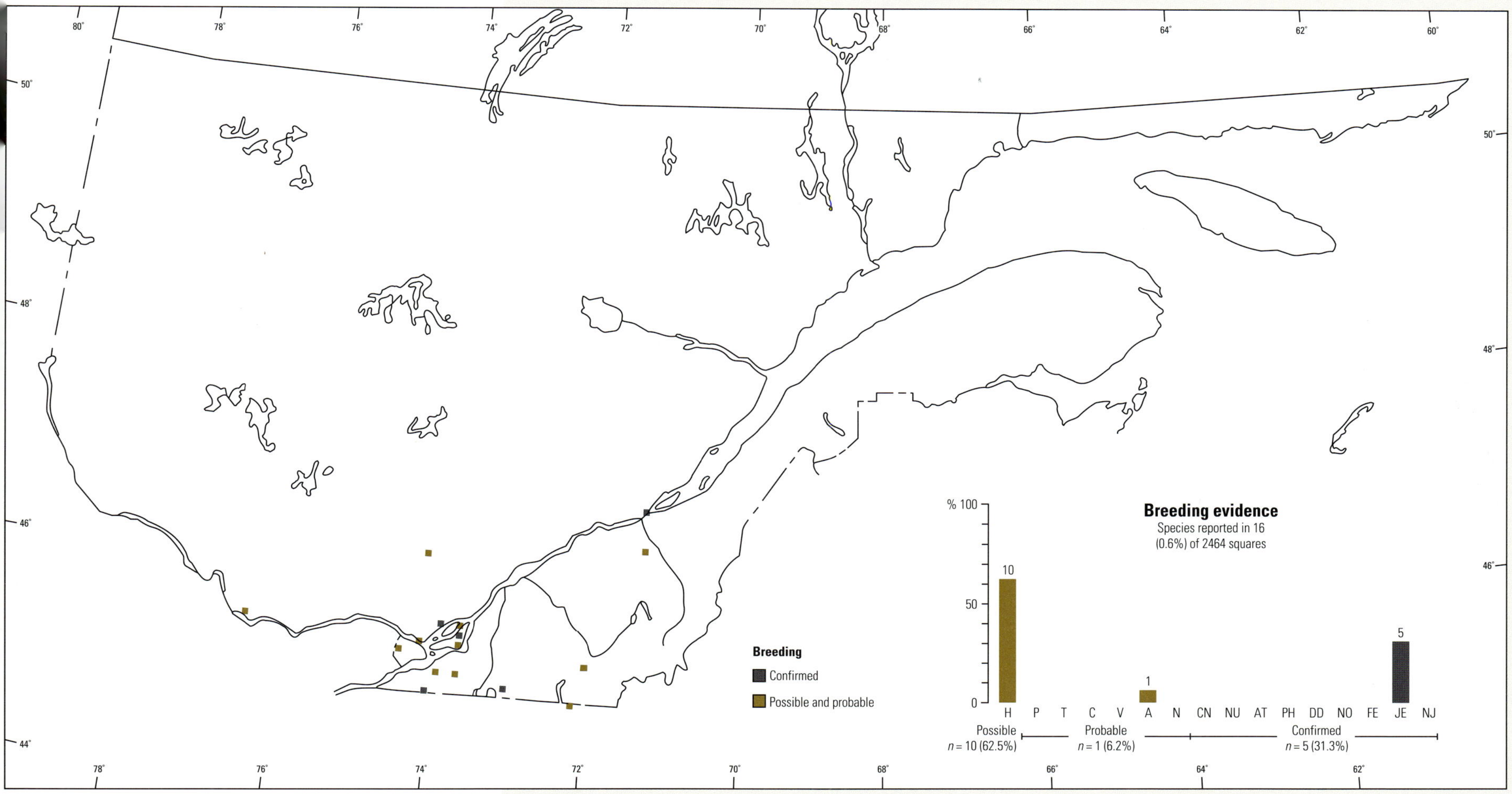

Species profile

Clutch size	usually 8–15 eggs (2–18)[1,2,3,4]
Laying interval	1 egg per day[1,3,4]
Start of incubation	with last egg[1,4]
Incubation period	23–25 days[1,2,3,4]
Maturity at hatching	precocial
Care of young	by female[1,3,4,8]
Nestling period	less than 24 hours[1,3,4]
Age at first flight	≈ 12 days[4]
Dependence of young	until 70–80 days old[4]
Broods per year	1[2,4]
Breeding age	1 year[2,4]
Mating system	polygyny[3,4,8]
Duration of pair bond	until laying[3,4]
Total length	M 82.0–91.0 cm / F 45.5–57.0 cm[7]
Wing span	81.3 cm[5]
Weight	M 1317 g (?–1861) / F 953 g (?–1453)[6]
Longevity record	?

References:
1. Johnsgard 1975a
2. Godfrey 1986
3. Dale 1956
4. Cramp and Simmons 1980
5. Terres 1980
6. Dunning 1984
7. Godfrey 1986
8. Göransson et al. 1990

Chronology $n = 11$

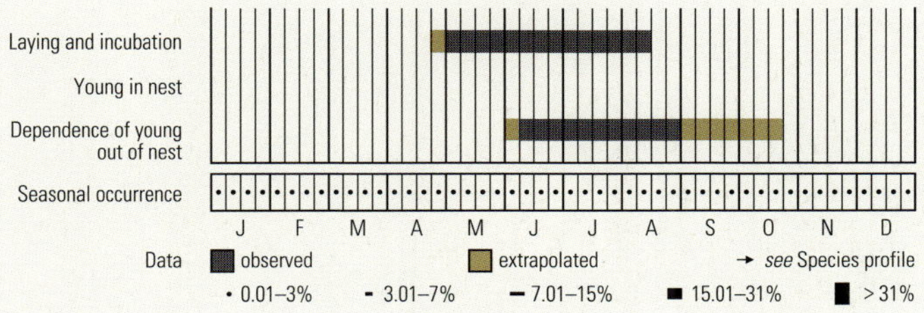

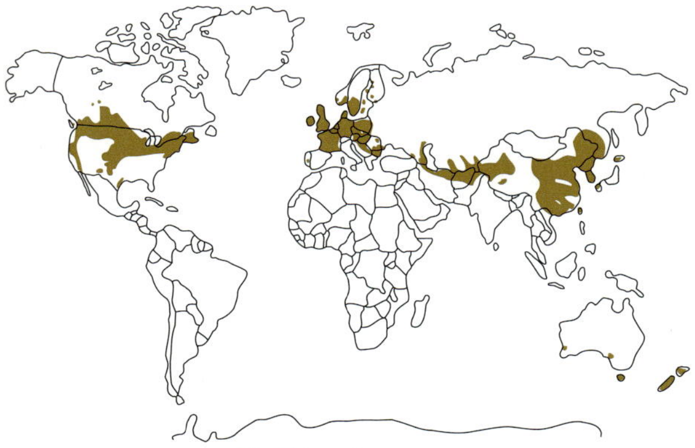

World breeding range

Females usually establish nest sites within the male's defended territory in the cover of fields of grass, weeds, or grain, under bushes, or alongside hedges, fencerows, edges of marshes, or in ditches (Peck and James 1983; Godfrey 1986). After copulation, the male usually plays no further role in the reproductive effort except to defend the territory. Nest building, incubation, and then later the care of the chicks are the exclusive role of the female, but the male may occasionally accompany the female and brood when the young are at least 6 weeks old (Bent 1932).

Nests are built on the ground in a scooped-out or natural shallow depression and lined with leaves and grass (Godfrey 1986). The female's age, experience, and body condition, as well as the habitat, all have an influence on her first egg-laying date (Hill and Robertson 1988b). Increasing day length, however, is the main factor determining the onset of laying, which is consistent at each location from year to year, despite annual variations in spring weather (Buss et al. 1951). The date when females first start to incubate depends on the temperature, and if the spring days remain cool, the first few eggs may be dropped at random or laid in dump nests. Females will only become broody with the advent of warmer weather. The temperature in the three weeks before egg laying is a crucial factor for predicting pheasant productivity and abundance. Above-average temperatures during this time are necessary for an annual increase in the population (CWS 1973). Two hens may use the same nest, or occasionally a female may lay her eggs in the nests of ducks, grouse, or partridge. There is only one brood each year. Like other ground nesters, however, the pheasant is exposed to many hazards, such as predation and nest disturbances through haying or flooding, and therefore only about a third of hens are successful in their first nesting attempt of the season (Dale 1956). If the first clutch is lost, this persistent renester will make up to two more renesting efforts in the season. Hens will also attempt to lure predators away from the brood by feigning a broken wing or performing some other distraction display.

The newly hatched young are precocial and leave the nest within a few hours. Led by the hen, which shows them where to find food but never feeds them, they subsist almost entirely on protein-rich insects for their first three weeks (Dale 1956). Adult pheasants have a varied diet. Although they feed primarily on plant matter —seeds and grains such as corn and wheat, as well as acorns and wild berries—they may also eat mice and snails (Terres 1980). Along with their food, they will ingest grit or pebbles, which help grind food in the gizzard. With the onset of winter, pheasants gather into coveys of 30–40 birds, usually segregated by sex, and take shelter in cattail marshes, along hedgerows, in shrubland, or in woodlots. As adequate food becomes scarce, pheasants will come readily into farm feedlots and feeding stations.

Low winter survival rates mean that most pheasants sighted in Québec are from seasonal releases, which take place mainly in the fall, because the birds are intended for hunting. This fact, combined with the extremely local distribution of the species, makes it difficult to atlas, and it was reported in only 16 of the 2464 squares surveyed. The female's camouflage and secretive nesting habits make it even more difficult to confirm nesting.

Notes on territory (continued)

(Dale 1956). Females form loose-knit harems within the male's territory, but will often change territory and visit those of several different males. Some females do not attempt to nest each year, and those that nest do not always do so in the territory of the most frequently visited male (Göransson et al. 1990). The male continues to defend his territory until the last female leaves her nest, whereupon he abandons it (Taber 1949).

Each year in southern Québec, a few pheasants released for sport hunting withstand the rigours of the winter and survive to nest, often around farms. Dewittville, southwest of Montréal. *July 1991, Gaétan Duquette.*

The Ring-necked Pheasant nests on the ground in a sheltered spot in fields, along fences and hedgerows, or in ditches. Philipsburg, south of Saint-Jean-sur-Richelieu. *July 1991, Pierre Bannon.*

Distribution

The long winters and less than ideal conditions have greatly restricted the range of the pheasant in southern Québec and prevented its permanent establishment. The few places pheasants were sighted or breeding was confirmed were either on farmland in the Eastern Townships and the Ottawa Valley, or in and around Montréal and Québec City, where people probably feed the birds throughout the winter. The extremely local distribution of the species is no doubt influenced by the sites of game farms, from which some captive-bred birds occasionally escape or are released as part of an introduction program. There is also the possibility that birds sighted close to the southern borders of Québec have come from nearby parts of Ontario or New York.

Even in ideal habitat, soil fertility restricts the abundance of pheasants, as they appear to be dependent on calcareous soils to meet their calcium needs (Dale 1956). While the cold itself does not bother them, deep snow makes foraging for food difficult, and they do not usually survive the winter in areas that receive more than 125 cm of snow annually unless most of their food is supplied by humans.

History and trends

The first pheasants in North America were released in 1733 on Governor's Island, New York, but the birds quickly disappeared. The first recorded nesting attempts occurred in 1790 and then 1800 in New England, but once again the birds did not survive. The first successful release took place in Oregon, where a population became established from an initial group of 28 birds introduced in 1881 (Dale 1956).

Ring-necked Pheasants were first released in Québec by private game farms in the early 1900s for sport hunting, and some birds may have moved into the province from the United States (Anonymous 1915, 1923). The provincial Ministry of Fish and Game operated a game farm in Magog and released birds regularly in the 1950s and early 1960s in Kamouraska County, around Québec City, and on Île d'Orléans (R. Ouellet, pers. comm.). In many parts of the province, pheasants have been released from game farms for sport hunting, yet due to the harsh winter conditions, few survive to breed the following spring (Gosselin 1977b). Even when hand-reared pheasants do survive their first winter, studies in Ireland and the United States have reported lower breeding success in these birds than in wild populations. Hand-reared males were less able to establish territories, and hand-reared females were more vulnerable to predation than their wild counterparts (Anderson 1964; Hill and Robertson 1988a).

The cemeteries on Mount Royal in Montréal were also home to a pheasant population from the 1940s to 1982, with the highest number—143—recorded in 1949. The numbers fell to between 12 and 75 in the 1950s and then reached another peak of 105 in 1972 (PQSPB Christmas Bird Counts). It is believed that these pheasants originated from private releases in the area beginning in 1941, and this local population found many advantages in living on the mountain: food supplied by humans during the winter, fruit from the shrubs planted around the cemeteries, lack of predators except for the occasional skunk or raccoon, and absence of competition for their prime nesting sites (Gosselin 1975b). They disappeared after 1982, either due to hunting by humans or raptors, or because the good Samaritans who had fed them were no longer supplementing their diet.

Derin Henderson

Spruce Grouse

Tétras du Canada

Dendragapus canadensis

A TYPICAL SPECIES OF THE BOREAL FOREST, THE SPRUCE GROUSE, DESPITE being a popular game bird, is one of the most trusting of birds. As a consequence, it has disappeared from most of the southern parts of its range (Godfrey 1986). According to the Atlas data, however, this sedentary breeder is still common in the north, particularly in the Black spruce-moss forest region.

The Spruce Grouse is also called "fool hen" because of its extreme tameness, and "black partridge" because of its dark flesh and the male's black plumage.

Habitat and behaviour

The Spruce Grouse is found in climax coniferous forests, but also in peatlands (Pietz and Tester 1982). It occupies different habitats depending on the season. In Québec, it prefers stands dominated by black spruce and balsam fir with a dense, well-developed shrub layer covering much of the ground (Lemay 1989; Turcotte et al. 1993, 1994). Early in the breeding season, it selects cover that is a little less dense for courtship displays, later switching during nesting and brood rearing to denser terrain, which affords safe cover for the female and her chicks, as well as providing males cover and protection from predators when moulting (Lemay 1989). In summer, it can also be seen on the edges of burned-over areas as well as in open habitats such as clearings and blueberry barrens (Godfrey 1986), where it feeds on berries, mushrooms, insects, and herbaceous plants. In winter, it mainly haunts stands of balsam fir, black spruce, red spruce, or jack pine, where it feeds on the needles (Naylor and Bendell 1989). In Québec, white spruce and tamarack appear to be important components of its diet (Lemay 1989; Turcotte et al. 1993, 1994).

During the breeding season, the male uses various displays to scare off intruders or to attract females. When a hen approaches, he beats his wings very quickly, producing a sound like muffled drumming, and makes low-pitched calls. He may also perch on a fallen log or stump, his wings drooping, or strut slowly with his tail fanned and neck feathers spread, reminiscent of a cobra (Robinson 1980). His bright crimson eye combs also play a role in attracting a mate.

In the courtship flight of the subspecies found in western Canada, the male rises almost vertically from the ground up to a low tree branch, then produces a loud double clap by striking his wings together behind his back as he drops back down to the ground. This behaviour is exhibited early in the breeding season and is very seldom observed (Boag and Schroeder 1992).

The male Spruce Grouse mates with several females, and is aggressive towards conspecifics during the breeding season. He defends only a small portion of his home range, the part that is used as a centre for breeding activities, such as courtship displays and vocalizations (Boag and Schroeder 1992).

The rudimentary nest is built on the ground, usually in a slight depression and almost always with some overhead cover. For instance, it may be under a low conifer bough, stump, or pile of dead branches. It is lined with conifer needles, fine grasses, lichens, and sometimes a few feathers (Boag and Schroeder 1992). During incubation the hen sits very still on the nest, and with her cryptic coloration, blends right into the background. When an intruder approaches, she waits until the last moment to flush. When alarmed, she often makes a low-pitched guttural *kwerrr* (Johnsgard 1973).

The female alone raises the brood, defending it vigorously against intruders (Johnsgard 1973). The young are dependent on the hen for approximately 12 weeks. When the chicks are about a week old they are able to fly short distances, and will perch on a tree branch when danger threatens. The family group begins to disperse around the end of August, and in September, most of the juveniles go off, either on their own or in small groups (Schroeder and Boag 1985). Some studies show that during this period, juvenile females travel farther (3 km) than do males of the same age (1–2 km) (Herzog and Keppie 1980; Schroeder 1986). Adults stay in their territories all year-round, even in winter (Robinson 1980).

Jean-Marc Giroux

Notes on territory

Male territoriality in spring does not seem to serve as a population-regulation mechanism. Territories appear to be used only for mating; females forage, nest, and raise their broods elsewhere. Most females nest in areas between male territories (Ellison 1971).

The size of the home range appears to vary according to sex and breeding status, and also differs from individual to individual (Turcotte et al. 1994). In a study of radio-marked birds in the Abitibi–Témiscamingue region, Turcotte et al. (1994) observed that from May to October males and broodless females occupied home ranges of 32 ha and 22 ha respectively on average, while females with broods had home ranges with a mean size of 75 ha. The same study also revealed that mean densities dropped after logging from an original 0.25 male/10 ha to fewer than 0.09 male/10 ha. Densities of 0.21 male/10 ha were found in riparian strips left intact during logging operations.

In the Kenai National Moose Range on the Kenai Peninsula, Alaska, annual home ranges were estimated to be 100 ha for territorial males, and 100–150 ha for breeding females, which travelled

The Spruce Grouse, renowned for its tameness, is a permanent resident of black spruce stands—its preferred habitat. Caribou Lake, Decelles Reservoir, Témiscamingue. *May 1991, Réginald Ouellet.*

Spruce Grouse

Breeding distribution

1984–1989

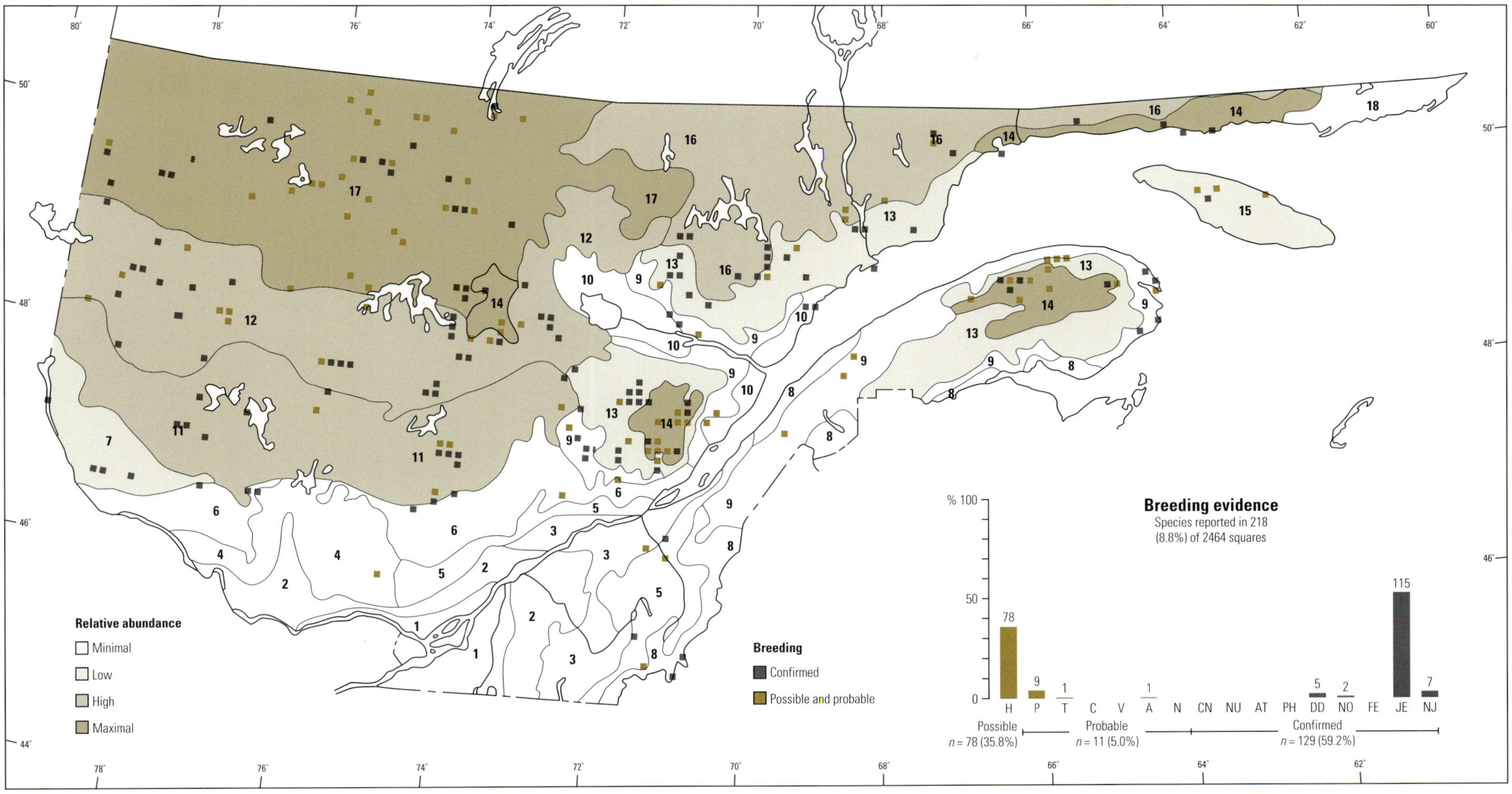

Species profile

Clutch size	usually 6–8 eggs (4–10)[2,3,4,5,6,7]
Laying interval	1 egg per 1.4 days[13]
Start of incubation	with last egg[8]
Incubation period	21–24 days[8]
Maturity at hatching	precocial
Care of young	by female[3,8]
Nestling period	≈ 8 hours[8]
Age at first flight	usually 10 days (6–12)[1,2,3,8]
Dependence of young	until 70–100 days old[8]
Broods per year	1[1,2,3,5]
Breeding age	1 year[5,8]
Mating system	polygyny[8,9], probably promiscuity[11,12]
Duration of pair bond	during mating only[11,12]
Total length	38–43 cm[4]
Wing span	?
Weight	M 492 g (?) / F 456 g (?)[10]
Longevity record	at least 13 years[8]

References:

1. Ehrlich et al. 1988
2. Terres 1980
3. Harrison 1978
4. Godfrey 1986
5. DeGraaf and Rudis 1987
6. Johnsgard 1973
7. Ferron and Lemay 1987
8. Boag and Schroeder 1992
9. Bergerud and Gratson 1988
10. Dunning 1984
11. Ellison 1971
12. Ellison 1973
13. McCourt et al. 1973

World breeding range

Chronology $n = 93$

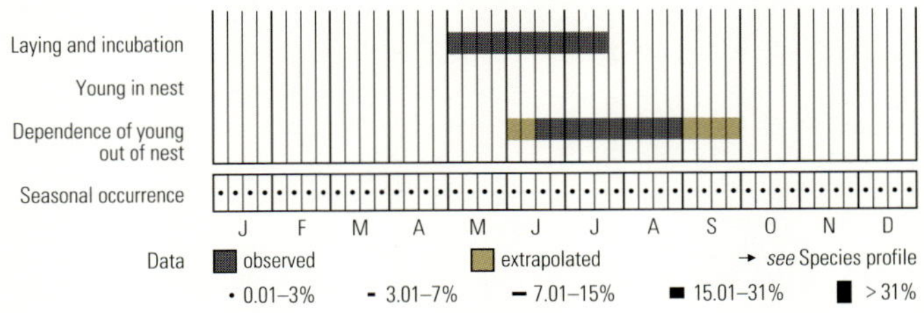

Frequency of occurrence by ecological region

Ecological regions** (forest regions)	Adequately covered squares	Squares in which reported	% of squares in which reported	Relative abundance (%)*
1. Sugar maple-hickory and sugar maple-basswood	97	0	0	0
2. Sugar maple-basswood	153	0	0	0
3. Sugar maple-basswood and sugar maple-yellow birch	101	2	2.0 ± 2.0	4.6
4. Sugar maple-yellow birch-basswood	56	1	1.8 ± 3.7	4.2
5. Sugar maple-yellow birch-beech and sugar maple-yellow birch-basswood	113	2	1.8 ± 1.7	4.1
6. Sugar maple-yellow birch-beech	71	3	4.2 ± 4.7	9.9
7. Sugar maple-yellow birch-balsam fir	19	4	21.1 ± 19.7	49.1
8. Sugar maple-yellow birch and balsam fir-yellow birch	47	1	2.1 ± 4.5	5.0
9. Balsam fir-yellow birch	87	10	11.5 ± 6.4	26.8
10. Balsam fir-white birch or balsam fir-red maple	46	3	6.5 ± 6.7	15.2
11. Yellow birch-balsam fir	52	17	32.7 ± 13.2	76.3
12. Balsam fir-white birch (west)	59	19	32.2 ± 12.3	75.1
13. Balsam fir-white birch (east)	76	11	14.5 ± 7.9	33.8
14. Balsam fir-black spruce	28	12	42.9 ± 19.2	100.0
15. White spruce-balsam fir and balsam fir-white spruce	22	3	13.6 ± 15.5	31.8
16. Black spruce-balsam fir-moss	18	5	27.8 ± 23.7	64.8
17. Black spruce-moss	32	13	40.6 ± 18.6	94.8
18.			*not analysed*	

* minimal: 0–29.9%; low: 30–59.9%; high: 60–89.9%; maximal: 90–100%

** Deciduous forest: 1–7; mixed forest: 8–11; coniferous forest: 12–17

The Spruce Grouse seeks openings and clearings in the boreal forest, where it forages for berries, mushrooms, and other foods. Decelles Reservoir, Témiscamingue. *May 1991, Réginald Ouellet.*

Notes on territory (continued)

longer distances with their broods (Ellison 1973). *In spring, in the same part of Alaska, males occupy territories ranging from 1.2 to 8.5 ha in size, with densities of 0.27–0.39 male/10 ha* (Ellison 1971).

In New Brunswick, population densities of 0.4–1.15 birds/10 ha have been reported (Keppie 1987).

Breeding was confirmed in 59.2% (129/218) of Atlas squares in which Spruce Grouse were reported. Confirmation was based on nests found in 9 squares, but in the great majority of cases —93% (120/129)—breeding was confirmed by sighting a female either with young or feigning a broken wing.

Unlike the Ruffed Grouse, the Spruce Grouse is very tame. It does not flush easily, making it more difficult to spot.

The Spruce Grouse's main terrestrial predators are red foxes, lynxes, and American martens, which attack both the birds and their eggs. Its chief avian predator is the Northern Goshawk, although Great Horned Owls may also take a bird once in a while (Keppie and Herzog 1978; pers. obs.).

In 1974, the first known case of hybridization between a Spruce Grouse and Ruffed Grouse was reported from the Saint-Maurice Wildlife Reserve (Ouellet 1974b).

Distribution

The Spruce Grouse is a permanent resident of coniferous forests all the way up to the tree line (Godfrey 1986). The species can be found in suitable habitat from Alaska to Newfoundland, and south to the northern United States (AOU 1983). There are two subspecies in Québec: *Dendragapus canadensis canadensis* in the northern and central part of the province, and *Dendragapus canadensis canace* in southern Québec as far north as Lake Saint-Jean (Godfrey 1986). Recent literature lumps these two races together as *Dendragapus canadensis canadensis* (Boag and Schroeder 1992).

The Spruce Grouse makes its rudimentary nest on the ground, usually under an overhanging conifer bough, tree trunk or pile of dead branches in stands like this balsam fir-white birch forest. Laurentide Wildlife Reserve, north of Québec City. *August 1991, Marcel Darveau.*

A hen taken by surprise in the boreal forest on the La Romaine Indian Reserve. Middle North Shore. *June 1980, Serge Jauvin.*

Although the Spruce Grouse was observed locally throughout the Atlas area, it was reported in only 8.8% of the 2464 squares surveyed. According to the distribution map, it is most common north of the 48th parallel and in the mountainous regions of the central Gaspé Peninsula and the Laurentide Wildlife Reserve, in the Black spruce-moss and Balsam fir-black spruce forest regions. It is also found, though less frequently, in the mixed and deciduous forest of the Appalachians, in their sizable patches of coniferous woodland. In the Central St. Lawrence Lowland, a brood was observed near a peatland southeast of Saint-Valérien-de-Milton in the late spring of 1993 (F. Bourret, pers. comm.).

The Spruce Grouse was introduced into Anticosti Island in 1985 and 1986. The first year, 219 juvenile and adult grouse were released at two sites, followed by 124 more birds the next year (Ferron and Lemay 1987). Since its introduction, the species has been confirmed as a breeder on the island and is presently expanding into the eastern part of the island (A. Gingras, pers. comm.).

History and trends

Before European settlement of North America, the Spruce Grouse was undoubtedly more common in the southern part of its breeding range than it is now. Nevertheless, Ouellet (1974a) doubts that the Spruce Grouse ever inhabited the Montréal and Montérégie regions, since the forests in these areas have probably never met its habitat requirements. Small populations may have lived in the peatlands in southern Québec, however, even near Montréal.

North of Québec's deciduous, maple-dominated forests—in Abitibi and Lac-Saint-Jean regions, for example—habitat was probably lost with the development of agriculture. Furthermore, in the first half of this century, successive Québec governments, in an effort to develop remote areas of the province during difficult economic times and to feed the hungry population, encouraged settlers and loggers to take advantage of the easy pickings provided by the Spruce Grouse, which can literally be killed with a stick (Martin 1990). The destruction of balsam fir stands by the spruce budworm has also been detrimental to the species, as has the harvesting of timber after these epidemics which accelerated the loss of already-fragile habitats in many places.

Since the turn of the century, intensive logging, particularly of softwoods, has no doubt also brought about steep declines in Spruce Grouse populations in some areas. According to Turcotte et al. (1994), when coniferous forests are logged acceptable densities of Spruce Grouse can only be maintained if large areas at least 150 ha of forest are left uncut, if cutovers, or clearcuts, are not too large, and if measures are taken to promote regeneration in conifers. On the positive side, regenerating cutovers should help to make up for habitat lost to logging. In addition, most of the Spruce Grouse's range is located north of commercially viable forests, so there is reason to believe that populations will remain within the limits of normal fluctuations for many years to come.

Populations of this game bird have only been monitored on a very local level. According to a study carried out in the Rimouski Wildlife Reserve (Bégin et al. 1988), numbers vary from one year to the next, mainly in response to harsh summer weather conditions, such as cold temperatures and heavy rainfall. The same is true of populations of Ruffed Grouse. Most of the grouse hunted in the Laurentide Wildlife Reserve are Spruce Grouse (Fortin 1985).

Réginald Ouellet

Willow Ptarmigan

Lagopède des saules

Lagopus lagopus

René Roy

The circumpolar Willow Ptarmigan can be identified by its fully feathered feet and whitish plumage, hence its nickname, "white grouse." In southern Québec, it is mainly seen during cyclical winter invasions into the Abitibi, upper Laurentians, North Shore, and northern Lac-Saint-Jean regions.

In southern Québec, the species' preferred breeding habitat, the tundra, is found only on the Lower North Shore below 50°30' N. A pair was observed in the breeding season on an island in this region, suggesting that the Willow Ptarmigan probably nests there.

Habitat and behaviour

In the Arctic, the Willow Ptarmigan nests on the tundra in areas with low growing shrubs, particularly alder, willow, and birch (Johnsgard 1973). In winter, it moves to sheltered valleys, where it frequents shrub thickets along lake and river margins and forest openings (Godfrey 1986). In northern Ontario, Thomas (1984) found that 68% of its winter diet consists of willow buds and twigs. Willow is also predominant in its winter diet in northern Québec: in a study carried out near La Grande Rivière and Great Whale River, willow accounted for 70.9% of the crop contents of the birds examined (Cons. G. G. – GREBE 1991c). In spring, the Willow Ptarmigan's diet consists mainly of buds and twigs of willow as well as fruit and stems of *Vaccinium* spp., whereas in summer it forages primarily on the leaves and berries of a small number of woody and herbaceous plants (Cotter 1992).

Around the end of April or beginning of May, males begin to stake out their territories in fairly open habitat with shrubby vegetation, where the herbaceous plants are lower than the ptarmigan's eye level. They defend this territory vigorously until the female and her brood desert the nest. The territorial males' ritual flight-song displays can be observed most often at dawn and dusk. A large proportion of males (91%) and females (69%) return to the same territories several years in a row (Schieck and Hannon 1989). A courtship display consisting of various postures and movements precedes copulation. The first eggs are laid 7–16 days after copulation (Martin and Hannon 1988).

Clutch size is typically 7–10 eggs, but can range from 5 to 14 and, according to Bergerud (1970), the earlier the breeding season begins, the more eggs are laid. The nest is a slight depression in the ground lined with dead vegetation and feathers. It is usually located near the centre of the territory and under cover of a shrub (Bergerud and Huxter 1969). The female lays one clutch a season and incubates the eggs herself, generally for around 20 days. The male, usually monogamous, remains with the female during laying and incubation and part of the brood-rearing period. The male defends both his mate and the eggs against intruders (Bergerud and Mossop 1984; Hannon 1984; Martin and Hannon 1988).

Notes on territory

In spring, males occupy territories ranging in size from 2.8 to 8.1 ha (Ogilvie 1976), near the centre of which the females makes their nests (Bergerud and Huxter 1969). The males vigorously defend the area required for nesting (Bergerud

The Willow Ptarmigan lives on the tundra, a terrain with patches of shrubby vegetation and many open areas. Lake Bienville, north-central Québec. *July 1976, Pierre Lamothe.*

Willow Ptarmigan

Breeding distribution
1984–1989

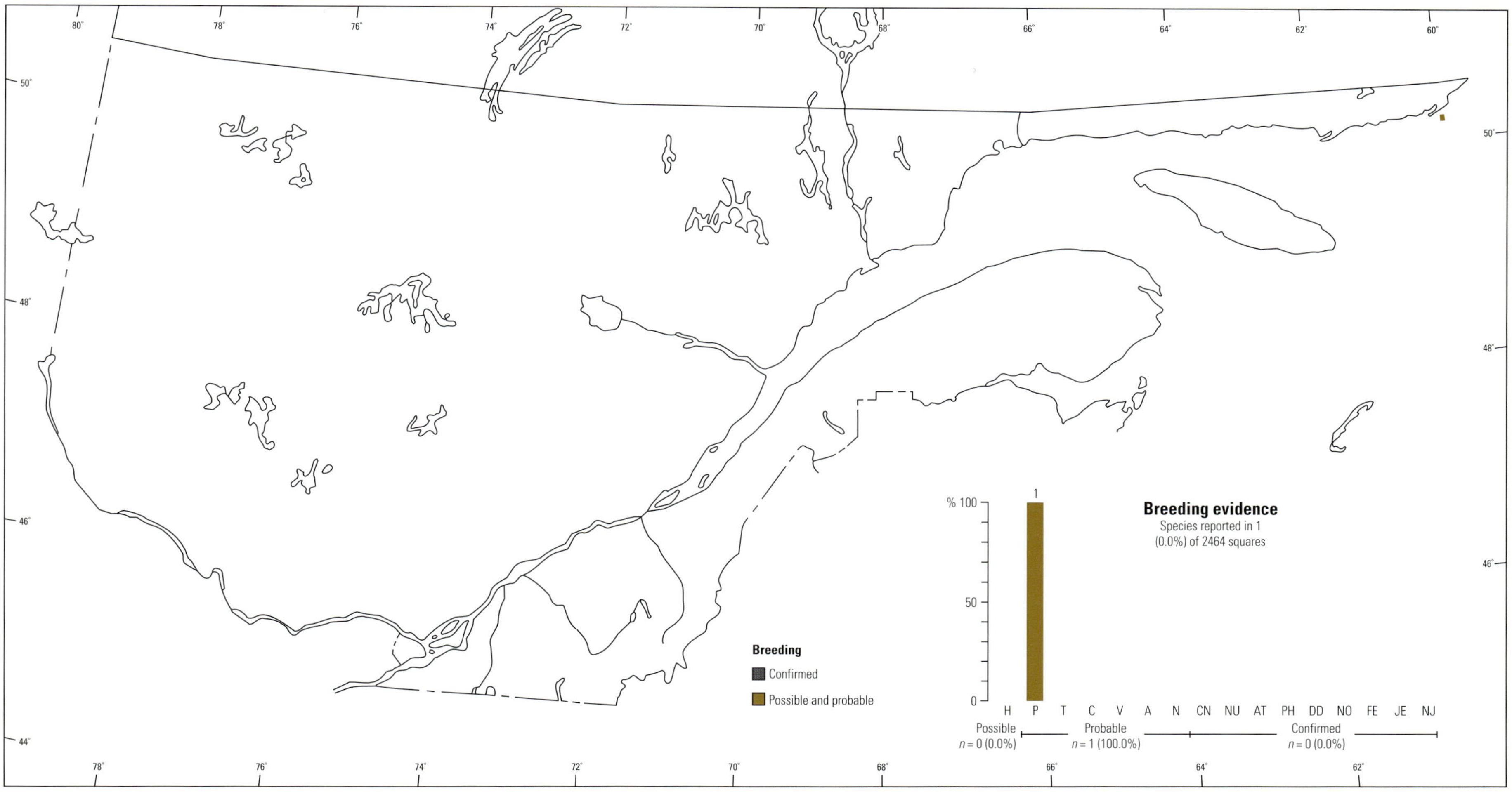

Species profile

Clutch size	usually 7–10 eggs (5–14)[1,3,5,7,10]
Laying interval	1 egg per 1–2 days[8]
Start of incubation	with last or penultimate egg[2,8]
Incubation period	19–26 days[1,2,3,5,6]
Maturity at hatching	precocial
Care of young	by female and male[2,3,4,5,8]
Nestling period	less than 24 hours[2]
Age at first flight	12–13 days[2,8]
Dependence of young	usually until 6–10 weeks old[3,8]
Broods per year	1[2,8]
Breeding age	1 year[8]
Mating system	monogamy, sometimes polygyny[3]
Duration of pair bond	sometimes more than 1 season[8]
Total length	35.5–43.0 cm[1]
Wing span	55–66 cm[8]
Weight	M 601 g (?) / F 516 g (?)[9]
Longevity record	?

References:
1. Godfrey 1986
2. Harrison 1978
3. Ehrlich et al. 1988
4. Terres 1980
5. Johnsgard 1973
6. Bergerud and Huxter 1969
7. Milonoff 1991
8. Cramp and Simmons 1980
9. Dunning 1984
10. Hannon and Barry 1986

Chronology $n = 35$

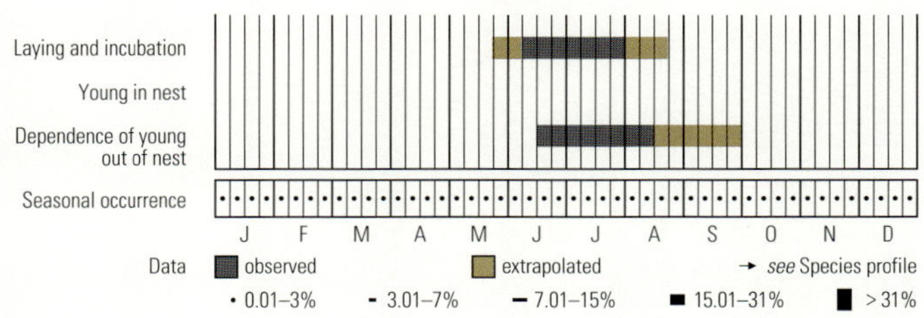

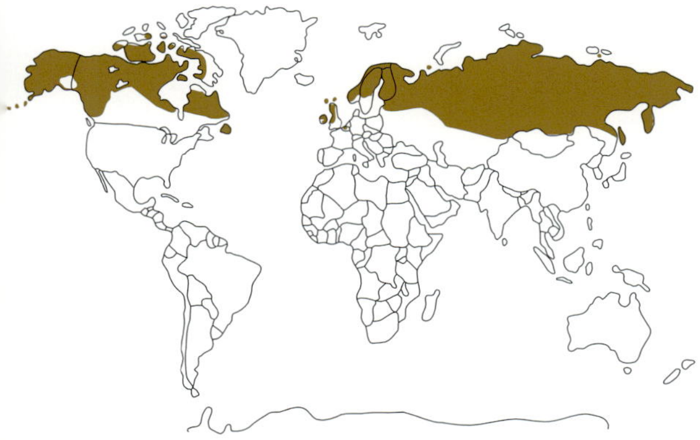

World breeding range

A day-old chick that has just left the nest. Churchill, Manitoba. *June 1980, Yves Aubry.*

The newborn chicks leave the nest a few hours after hatching in search of invertebrates and berries. For the first week, the chicks are brooded regularly by the hen and remain close to her when feeding and travelling. They grow quickly and make their first attempts at flight at 12 or 13 days and, by 20 days, have begun to fly well. The chicks do not develop a taste for willow until they are 2 months old; their bills are not strong enough to snip the twigs and their gizzards cannot grind this food until they are at least 10 weeks old.

By mid-September, the young birds begin to grow their winter plumage, becoming completely white by mid-November except for their tail feathers, which remain black. The adults, too, moult their brownish summer plumage and acquire their white winter plumage, becoming indistinguishable from the immature birds. Thanks to these changes in plumage, the Willow Ptarmigan is well camouflaged in all seasons. Both the Willow and Rock ptarmigan have black tails, and the best way to distinguish between the two species is the Willow Ptarmigan's larger size and heavier bill (Johnsgard 1973). Easily visible in winter, Rock Ptarmigan males have a thin black stripe through the eye which is not present in the Willow Ptarmigan.

Distribution

In the breeding season, the Willow Ptarmigan inhabits damp shrubby areas on the taiga and tundra of central and northern Québec.

During atlassing, a pair was seen between 18 and 29 June 1988 in the Ouapitagone Archipelago, the only extensive area of tundra habitat on the Lower North Shore. As a result, the species was classified as a probable breeder here, at the southernmost limit of its breeding range. A series of other reports suggests that the Willow Ptarmigan may nest in alpine tundra elsewhere in south central Québec. The species was reported on 8 July 1971 in the Otish Mountains north of Lake Mistassini (H. Ouellet, pers. comm.); 2 birds were seen on 10 July 1978 on Montagne Blanche, 250 km north of Port-Cartier (COQ 1978); and in May 1991, several birds, including a pair in the process of moulting into their summer breeding plumage, were observed on Mount Veyrier (Groulx Mountains), east of the Manicouagan Reservoir (M. Duchesne and S. Côté, pers. comm.)

Todd (1963) also reported that the Willow Ptarmigan bred sparingly in suitable habitat on the Lower North Shore. In 1917, Olaus J. Murie collected specimens in breeding plumage in alpine tundra on a hilltop along the Sainte-Marguerite River, close to the northern boundary of the Atlas area.

The Willow Ptarmigan generally winters within its breeding range, but sometimes travels great distances south of the tree line, occasionally down to the 50th parallel, which is the southern limit of its winter range. The species has, however, been observed on rare occasions in the Magdalen Islands (Fradette 1992), lower Laurentians, and Central St. Lawrence Lowland (David 1980a). Early in this century, Macoun and Macoun (1909) reported that the species was a transient visitor in winter around Montréal, but common in the eastern part of the province, north of the St. Lawrence.

Notes on territory (continued)

and Mossop 1984). A study of Willow Ptarmigan in the Chilkat Pass, British Columbia, shows that major population variations do occur. The study found a density of 57 territorial males/km² (5.7/10 ha) in 1971, compared with 16 males/km² (1.6/10 ha) 5 years later, and 55 males/km² (5.5/10 ha) in 1979 (Mossop 1988).

On the taiga, the Willow Ptarmigan requires bushy areas for nesting. Bay of Many Islands, James Bay. *August 1990, Yves Aubry.*

Male in breeding plumage on a rock overlooking a shrub thicket. Lake Bérard, Ungava Bay region. *June 1977, Pierre Lamothe.*

History and trends

Several studies have demonstrated that the Willow Ptarmigan undergoes major cyclical fluctuations in abundance that vary in duration and amplitude from region to region. Williams (1954) noted a regular cycle of 9 or 10 years in North American populations, while Gosselin et al. (1990a) reported a cycle of 8–12 years in Québec, depending on the region. The reasons for the cycle are still unknown, but researchers have noted that when populations are higher, the birds are distributed more widely. This phenomenon would help explain the presence of pairs south of the usual breeding range. Movements may exceed 1000 km in winter, resembling true migrations. Occasional winter incursions are still observed in the Abitibi, Saguenay–Lac-Saint-Jean, and North Shore regions, but involve fewer and fewer birds. The small groups of Willow Ptarmigan, consisting of several dozen birds at the most, seen recently near Lake Saint-Jean and Mingan cannot be compared with the thousands of birds bagged by hunters at Beauport in December 1648 or, as Comeau (1954) related, with the tens of thousands of birds that were taken between Godbout and Mingan around the turn of the century (60 000 in 1885; 30 000 in 1895; 14 000 in 1904). As these figures show, ptarmigan populations are often decimated by hunters when their winter movements take them into inhabited areas.

Owing to the poorly documented cyclical population changes in the species, it is difficult to estimate the population in Québec and its variations.

Pierre Lamothe and Marie-Renée Doyon

Ruffed Grouse

Gélinotte huppée

Bonasa umbellus

René Roy

GENERALLY A RETIRING, UNASSUMING BIRD, THE RUFFED GROUSE sometimes behaves in unexpected ways. It blends easily into its surroundings thanks to its cryptic coloration, and when an intruder approaches, often waits until the last minute to flush noisily out of the undergrowth. In spring, males produce a muffled drumming noise with their wings that makes them easier to locate. The birds also leave visual signs of their presence: droppings on drumming logs, scattered feathers near dust baths, and "forms," or roosting holes in the snow. The grouse is a highly prized game bird, challenging hunters with its wily ways and excellent camouflage. The Ruffed Grouse is a sedentary breeder in Québec.

Habitat and behaviour

The Ruffed Grouse mainly inhabits deciduous and mixed forests, where it favours wood edges, openings, and alder- and willow-bordered ravines and stream edges, though in winter it will take shelter in coniferous woods in bad weather (Godfrey 1986; Thompson and Fritzell 1988; Gullion 1990). It also frequents areas where felling, burning, and natural disturbances to the forest cover create habitats that meet its needs for food and shelter (Gullion 1984).

Sedentary and somewhat solitary by nature, the Ruffed Grouse usually spends its entire life in a home range of less than 100 ha (Archibald 1975; Gudlin and Dimmick 1984; Woolf et al. 1984; Thompson and Fritzell 1989), so a bird found in a square probably breeds there.

Except in winter, this grouse generally stays on the ground, taking flight only when necessary to flee or to perch in a tree. It flies in short, quick bursts.

Male Ruffed Grouse will mate with several females. In the mating period, the male drums from a fallen tree or similar perch to establish dominance over other males and attract females. Drumming begins after the snow melts and reaches a peak in late April or early May (Gullion 1967). Drumming is also performed at other times of the year, particularly in fall by resident males and by juveniles trying to oust established males from their territories (Johnsgard 1989). Drumming by males was the most frequent courtship behaviour reported for the Atlas.

When a female approaches, the male struts, fanning his tail and erecting his ruff. After copulation, the female makes a nest on the ground at the base of a tree, stump, rock, or bush. The nest is a hollow in the ground lined with leaves and a few feathers (Bump et al. 1947). In Québec, the 10 or so eggs are laid over a period of about 15 days, and the incubation period is generally 23 or 24 days. A second clutch may be laid if the first is destroyed early in the incubation period. The male plays no part in either incubation or rearing the young. The chicks leave the nest less than 24 hours after hatching, watched over by the female and can feed on their own immediately. At 10 days, the chicks are able to fly short distances to escape predators or other danger. To defend the young from intruders, the hen distracts their attention by feigning a broken wing and fluttering along the ground all the while making whining noises.

In winter, Ruffed Grouse feed mainly on birch and poplar buds; poplar, maple, and staghorn sumac seeds; viburnum, mountain-ash, and holly berries; cherries; and American beechnuts (Philippe 1945). In winter, three or four grouse can often be seen feeding together in a tree (Doerr et al. 1974). From fall until spring, the buds of the trembling aspen seem to be a favourite food, while in summer, fruits and mushrooms play an important role in their diet. The young birds are fond of insects and slugs (Vézina 1975).

The easiest way for atlassers to confirm breeding was sighting broods out in the open, crossing logging foads, for example. A female feigning a broken wing is also a sure sign that a brood is nearby. Records of young accounted for 50.0% (598/1197) of squares in which the species was reported and 80.9% (598/739) of squares where breeding was confirmed.

Notes on territory

The home range of the Ruffed Grouse is generally less than 100 ha (Archibald 1975; Gudlin and Dimmick 1984; Woolf et al. 1984; Thompson and Fritzell 1989). Within this range, males defend a breeding territory of up to 6.7 ha in spring (Archibald 1975), while females defend only the immediate area around the nest (Maxson 1978). In Minnesota, densities as high as 2.5 males/km² (0.25/10 ha) were found in a coniferous forest and as high as 25 males/km² (2.5/10 ha) in a poplar stand (Gullion 1990).

Sugar maple-beech stands, like this one on the south side of Rigaud Mountain, are one of the nesting habitats used by the Ruffed Grouse. West of Montréal. *June 1991, Pierre Bannon.*

Ruffed Grouse

Breeding distribution
1984–1989

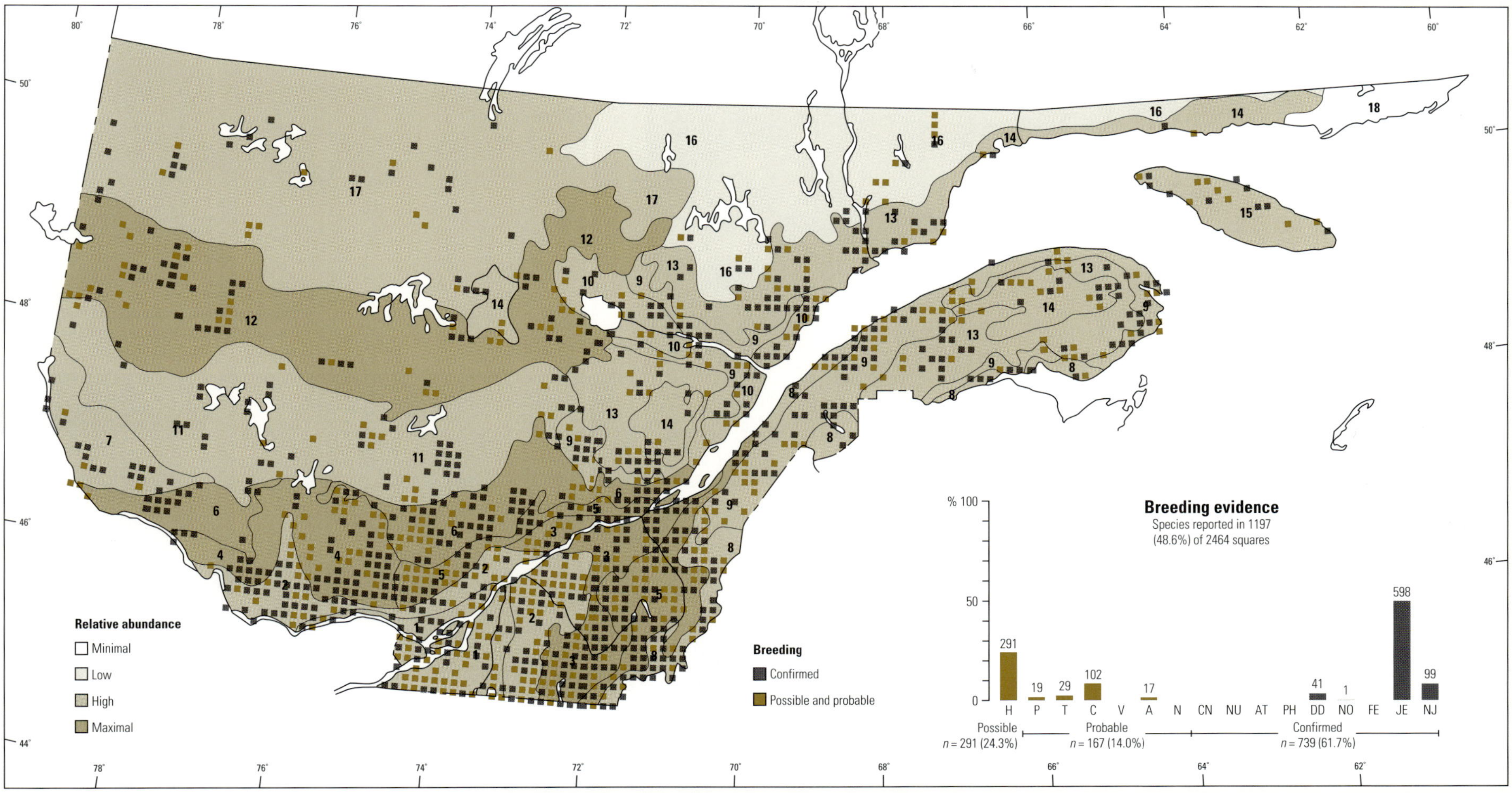

Species profile

Clutch size	usually 9–14 eggs (9–19)[2,3,4,7]
Laying interval	2 eggs per 3 days[2,4,7,12]
Start of incubation	with last egg[7]
Incubation period	usually 23–24 days (21–28)[2,4,8,13]
Maturity at hatching	precocial
Care of young	by female[3,4,12]
Nestling period	less than 24 hours[3,4]
Age at first flight	10–12 days[13]
Dependence of young	until 84–105 days old[13]
Broods per year	1[2,3]
Breeding age	1 year[2]
Mating system	promiscuity[8,10,11], polygyny[3]
Duration of pair bond	a few hours[11]
Total length	40.5–48.0 cm[6]
Wing span	55.9–63.5 cm[8]
Weight	M 621 g (?) / F 532 g (?)[9]
Longevity record	8 years 6 months[5]

References:
1. Johnsgard 1989
2. Bump et al. 1947
3. Bergerud and Gratson 1988
4. Bent 1932
5. Klimkiewicz and Futcher 1989
6. Godfrey 1986
7. Johnsgard and Maxson 1989
8. Terres 1980
9. Dunning 1984
10. Ehrlich et al. 1988
11. Johnsgard 1973
12. Harrison 1978
13. Johnsgard et al. 1989

Chronology $n = 540$

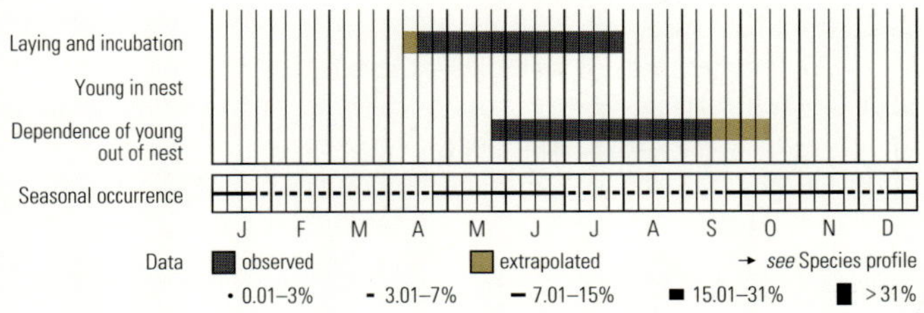

World breeding range

Frequency of occurrence by ecological region

Ecological regions** (forest regions)	Adequately covered squares	Squares in which reported	% of squares in which reported	Relative abundance (%)*
1. Sugar maple-hickory and sugar maple-basswood	97	72	74.2 ± 4.4	82.3
2. Sugar maple-basswood	153	120	78.4 ± 3.4	87.0
3. Sugar maple-basswood and sugar maple-yellow birch	101	87	86.1 ± 4.2	95.6
4. Sugar maple-yellow birch-basswood	56	48	85.7 ± 8.3	95.1
5. Sugar maple-yellow birch-beech and sugar maple-yellow birch-basswood	113	100	88.5 ± 3.4	98.2
6. Sugar maple-yellow birch-beech	71	64	90.1 ± 6.6	100.0
7. Sugar maple-yellow birch-balsam fir	19	13	68.4 ± 22.1	75.9
8. Sugar maple-yellow birch and balsam fir-yellow birch	47	35	74.5 ± 11.3	82.6
9. Balsam fir-yellow birch	87	70	80.5 ± 7.8	89.3
10. Balsam fir-white birch or balsam fir-red maple	46	31	67.4 ± 11.8	74.8
11. Yellow birch-balsam fir	52	41	78.8 ± 11.6	87.5
12. Balsam fir-white birch (west)	59	48	81.4 ± 10.4	90.3
13. Balsam fir-white birch (east)	76	56	73.7 ± 9.7	81.7
14. Balsam fir-black spruce	28	21	75.0 ± 17.0	83.2
15. White spruce-balsam fir and balsam fir-white spruce	22	14	63.6 ± 20.8	70.6
16. Black spruce-balsam fir-moss	18	7	38.9 ± 25.5	43.1
17. Black spruce-moss	32	18	56.2 ± 18.8	62.4
18.			*not analysed*	

* minimal: 0–29.9%; low: 30–59.9%; high: 60–89.9%; maximal: 90–100%

** Deciduous forest: 1–7; mixed forest: 8–11; coniferous forest: 12–17

This mossy log is a potential drumming perch for the male in spring to attract females and in fall to inform other males that he occupies the territory. Decelles Reservoir, Témiscamingue. *May 1991, Réginald Ouellet.*

This Ruffed Grouse nest, made of dead leaves and containing 10 eggs, is located at the foot of a moss-covered spruce. North of Lake Fortier, Charlesbourg. *June 1966, Gaston Le Page.*

Distribution

The Ruffed Grouse is widespread in southern Québec and was reported in 48.6% (1197/2464) of Atlas squares surveyed. It was reported everywhere except the Magdalen Islands where, according to Fradette (1992), attempts to introduce it have been unsuccessful. As well, on the Lower North Shore the species was found in only a few squares.

Because the Ruffed Grouse can fly only relatively short distances, it has not been able to colonize offshore islands (Bump et al. 1947). It was introduced into Anticosti Island early this century (Alain 1988) and, in 1990, 1991, and 1992, on Île aux Lièvres, near Rivière-du-Loup. Breeding has been confirmed in both places and, by the autumn of 1992, the population on Île aux Lièvres had swelled from the original 40 birds released to approximately 250 (Doyon 1992). In 1993, 12 females and 16 juveniles were released at Notre-Dame-des-Sept-Douleurs on Île Verte by the Société de conservation de la baie de L'Isle-Verte (J. Bachand, pers. comm.). The species is not found on any of the other small islands in the St. Lawrence.

The Ruffed Grouse's favourite habitats are found all over the Atlas area. The species is most abundant in the mixed woodland along the St. Lawrence River, the area south of the St. Lawrence from the U.S. border to the Gaspé Peninsula, the Ottawa Valley, and the Abitibi, Saguenay, and Lac-Saint-Jean regions (Alain 1988). The regions with the highest frequency of occurrence on the map correspond fairly closely to areas of white birch and poplar, young deciduous stands, young and mature mixed stands, and forests that have been recently logged (MTFQ 1977), which include the eastern part of the Abitibi and the regions surrounding the Central St. Lawrence Lowland. However, the insignificant differences between ecological regions shown in the table suggest that the species is found virtually everywhere below the 49th parallel. It is slightly less abundant above 49° N and on Anticosti Island, where conifers predominate.

Distribution of the species is patchier north of the Atlas area. Birds have been reported near Rupert River (Todd 1963), Lake Mistassini and Lake Albanel (Godfrey 1949), and around the Eastmain River, which lies at approximately 52° N (Tremblay and Belles-Isles 1991). The northernmost record comes from near Schefferville (Brower 1969), around 54° N.

Two subspecies occur in Québec: *Bonasa umbellus obscura*, which is found north of the 48th parallel, and *Bonasa umbellus togata*, which is found south of this latitude and on the Gaspé Peninsula and Anticosti Island (Alain 1988).

The new growth of shrubs and saplings after logging creates favourable habitat for the Ruffed Grouse, which finds food and shelter in these disturbed areas. Montmorency forest, Laurentide Wildlife Reserve, north of Québec City. *May 1991, Réginald Ouellet.*

This member of the subfamily Tetraoninae forages for food along the forest edge. Ottawa Valley. *October 1969, Pierre Lamothe.*

History and trends

The Ruffed Grouse's abundance and distribution do not appear to have changed significantly since the last century (Macoun and Macoun 1909; Taverner 1922). BBS data also confirm that the species' abundance has been relatively stable since 1966.

Ruffed Grouse densities fluctuate on a cycle of 8–10 years in most regions. These variations appear to be strongly related to the number of chicks produced and the number of juveniles in the population in fall (Bump et al. 1947). According to Keith and Rusch (1988), predation on juveniles is the major factor in determining population abundance cycles. These authors also believe that the predation pressure on grouse, particularly by the Northern Goshawk and Great Horned Owl, varies with cyclical fluctuations in snowshoe hare populations.

Hunting does not appear to be a significant factor in mortality and, according to an Alberta study, 80% of population losses between fall and spring were caused by predation (Rusch and Keith 1971). Juveniles are more vulnerable to hunting than adults because of their greater mobility; after the breeding season, young birds are often driven away from established territories held by adults, to areas along paths and roads where they are more likely to be found by hunters (DeStefano and Rusch 1986).

Second growth of birches and poplars after forest fires and logging in coniferous regions favours the establishment of the species. Today's more sophisticated logging practices—particularly strip cutting, limited clearcuts, and the preservation of strips along bodies of water and roads—benefit the species by promoting the growth of fruiting shrubs and increasing edge habitat. However, the replanting of logged areas with softwood species, which tends to create homogeneous stands of conifers, may be a limiting factor in some areas (Alain 1988).

Marie-Renée Doyon

Sharp-tailed Grouse

Gélinotte à queue fine

Tétras à queue fine

Tympanuchus phasianellus

René Roy

THE SHARP-TAILED GROUSE IS THE ONLY MEMBER OF THE SUBFAMILY Tetraoninae in Québec to use leks, or arenas, for mating. The males gather on these leks, which are usually found in a sparsely vegetated spot, to perform elaborate dances and vie for the favours of females. Each lek, used in some cases for generations, may be occupied by a variable number of males, although a dozen or so is the average, each of which occupies a territory of a few metres in diameter. Their cooing and cackling calls, which intensify at dawn and dusk, can be heard over long distances. The displays reach a peak in late April or early May, but sometimes occur in autumn as well (Edminster 1954; Johnsgard 1973).

The Sharp-tailed Grouse is today still considered a rare resident breeder in the northern portion of southern Québec, as reported earlier by David (1980a).

Habitat and behaviour

There are six recognized subspecies, but only one, *Tympanuchus phasianellus phasianellus,* is found in Québec. This subspecies inhabits treeless peatlands, burns, and regenerating cutovers, and occasionally abandoned pastureland in the northern and western portions of southern Québec. During the mating period, territorial males gather in openings to form leks. The most prized territory is the centre of the lek and it is probably occupied by the dominant male (Rippin and Boag 1974; Moyles and Boag 1981). The younger males occupy peripheral territories (De Vos 1979) but may move into the central territories when dominant males die. According to De Vos (1979), sometimes the same territories are occupied by the same males two or three years in a row. The territories are established in fall or early winter. Some young males are not able to defend a territory in their first year (Rippin and Boag 1974).

As mating displays reach maximum intensity, the female begins to build a nest for laying and incubating her eggs in low vegetation 0.5–1.5 km from the lek. The nest, a shallow depression in the ground lined with grasses, leaves, and a few feathers, is usually located in a clump of grass or under a bush or stump. Copulation takes place on the lek, the females tending to mate with males occupying territories close to the centre (Tsuji et al. 1992). Shortly after the females stop visiting the lek so they can attend to nesting, the males begin to disperse singly or in small groups. The female leaves the nest in the morning and evening to feed. She may attempt to renest if the first clutch is lost (Edminster 1954; Johnsgard 1973).

As soon as they hatch at the end of June, the female leads the chicks away from the nest into the surrounding open area to feed on grasshoppers, spiders, ants, and weevils. Chicks also like young shoots and blueberries (Grange 1948). Broods are fairly sedentary and generally remain within 500 m radius of the nest. By the time they are 10 days old, chicks are able to fly short distances to escape predators. They become fully independent at 6–8 weeks and reach adult size by mid-September or early October (Edminster 1954). The adults' diet consists of about 10% insects, such as grasshoppers and beetles, and 90% plant material, including the berries of *Vaccinium* spp. and mountain-ash, the buds and leaves of willow and birch, and seeds (Bent 1932).

Notes on territory

During the mating period, territorial males gather in clearings to form leks (Evans 1969). Leks, which measure on the average 450 m², are located on elevated sites 0.5–1.5 km from nesting habitat (Hamerstrom and Hamerstrom 1951).

Sharp-tailed Grouse families sometimes visit abandoned fields in the Abitibi. Trécesson, northwest of Amos, Abitibi. *August 1991, André-Guy Bernier.*

Sharp-tailed Grouse

Breeding distribution
1984–1989

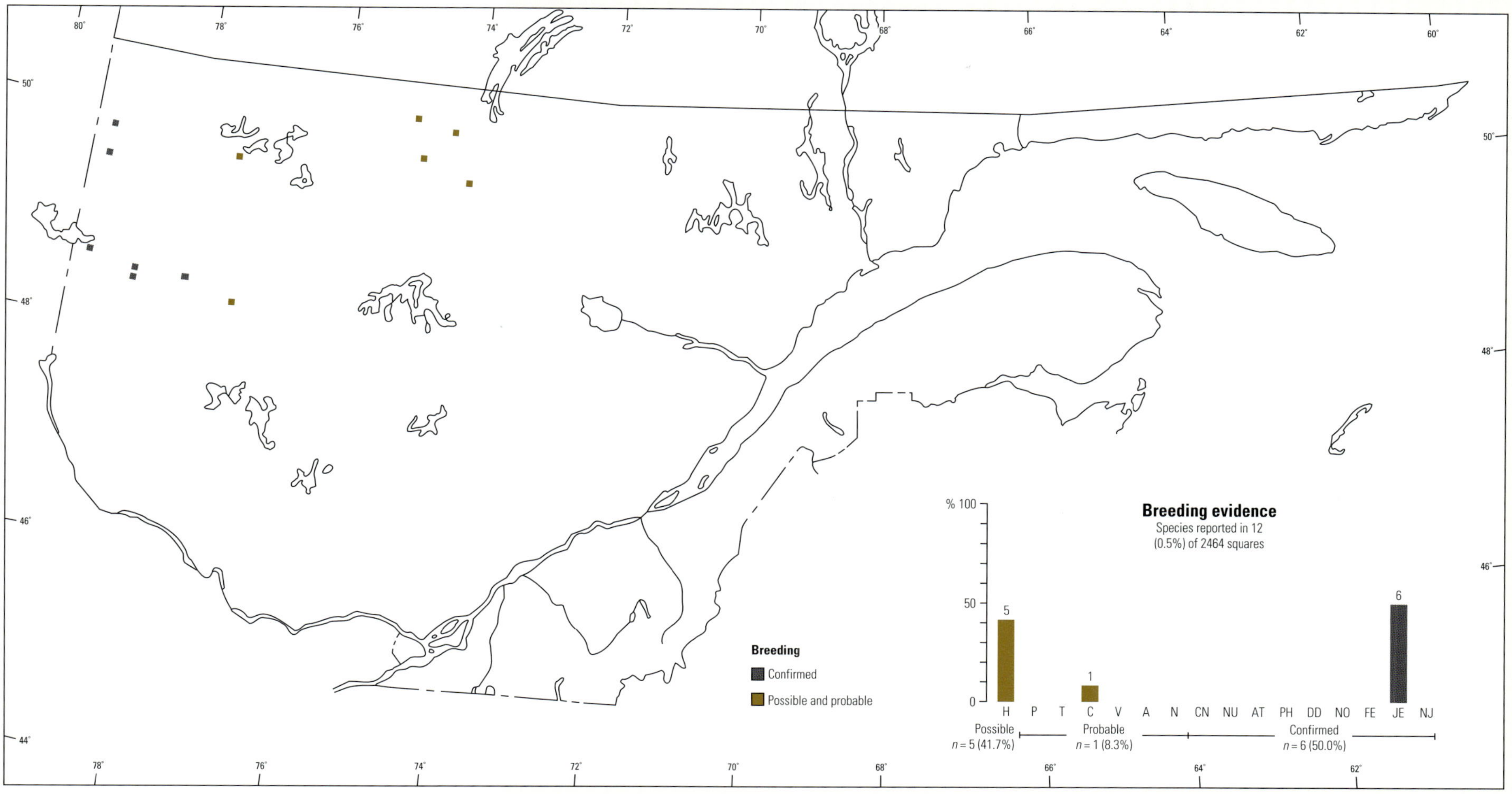

Species profile

Clutch size	usually 12–13 eggs (5–17)[1,2,3]
Laying interval	1 egg per day[2,8]
Start of incubation	with last egg[2,8]
Incubation period	23–24 days[1,2,8]
Maturity at hatching	precocial
Care of young	by female[1,2,8]
Nestling period	less than 24 hours[1,2]
Age at first flight	usually 10 days (10–14)[1,2,8]
Dependence of young	until 6–8 weeks old[2,8]
Broods per year	1[8,10]
Breeding age	1 year[2]
Mating system	promiscuity[10], polygyny[3]
Duration of pair bond	during mating only[9]
Total length	41.5–47.0 cm[4]
Wing span	68.6 cm[6]
Weight	M 953 g (?–1090) / F 817 g (?–999)[7]
Longevity record	6 years 4 months[5]

References:
1. Terres 1980
2. Johnsgard 1973
3. Bergerud and Gratson 1988
4. Godfrey 1986
5. Clapp et al. 1982
6. Pough 1951
7. Dunning 1984
8. Harrison 1978
9. Tsuji et al. 1992
10. Ehrlich et al. 1988

Chronology *n* = 8

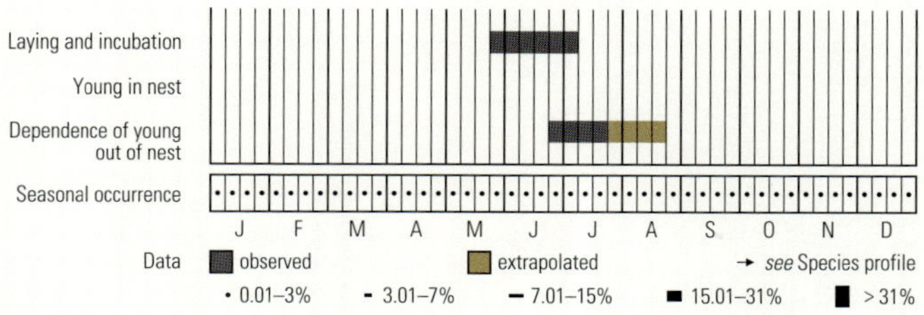

World breeding range

This regenerating cutover is typical of the nesting habitat used by the Sharp-tailed Grouse. The nest is always fairly close to the lek, the opening where males gather to perform their courtship dances and vocalizations. South of Matagami, Abitibi. *August 1993, Robert Alvo.*

Although nests may be depredated by skunks, crows, and foxes, and the occasional Northern Goshawk may make off with a chick or adult, sport hunting appears to be a more important factor than predation in species mortality (Edminster 1954). Ammann (1963) found that the number of males per lek was reduced by close to half in areas where hunting was allowed.

In years when populations are high, Sharp-tailed Grouse sometimes undertake a fall migration. The birds are gregarious in autumn and winter and gather in flocks of 10–30 or more. They roost on the ground in the open or, in harsher climates, in dense thickets or even snow burrows. When in danger, this ground-dwelling species runs away or takes flight (Edminster 1954; Johnsgard 1973).

Notes on territory (continued)

Each male defends a territory of approximately 50 m² (Baydack 1988). The number of territorial males on a lek varies with population density. In Nebraska, there was, on average, 9 or 10 males on a lek (Johnsgard and Wood 1968). In Saskatchewan, the number of males per lek was related to the amount of vegetation cover (Pepper 1972). In Montana, however, Swenson (1985) did not find a relationship between number of males per lek and habitat type.

A lek, where 2 males are dancing to compete for the favours of the females. Near St. Paul, Alberta. *April 1971, Pierre Lamothe.*

In open black spruce forests, clearings with low vegetation cover provide ideal habitat for nesting. Near Dieppe Hills, north of La Sarre, Abitibi. *June 1991, Desneige Perreault.*

The 6 confirmed Atlas breeding records were based on sightings of chicks, which are easy to spot in openings along back roads and logging roads. The Sharp-tailed Grouse was reported in only 0.5% of squares surveyed (12/2464). In flight, it can be easily distinguished from other gallinaceous birds in the province by its pointed tail, white outer tail feathers, V-marked breast, and size, and there is little likelihood of its being confused with other species.

Distribution

The Sharp-tailed Grouse is found from the American Midwest to Alaska in the north and Québec in the east. Leks have been discovered 75 km south of Poste-de-la-Baleine (P. Lamothe, pers. comm.). The handful of Atlas records came from throughout Abitibi and farther north, to around Matagami and Lake Chibougamau. Most authors consider the eastern limit of the species' range in Québec to be approximately 200 km east of the Ontario border and James Bay (Aldrich 1963; Johnsgard 1973; Godfrey 1986). Reports of Sharp-tails a few kilometres north of Lake Mistassini (Groupe Boréal 1992) and near Chapais and Lake Chibougamau extend this range approximately 400 km farther east, matching the boundary proposed by Edminster (1954). The northernmost Atlas breeding records are from north of Matagami. Farther east, breeding evidence was found at Lake Caopatina (J. Lapointe, pers. comm.). Reports from Chapais and 410 km north of Matagami during the winter of 1989–90 (Gosselin et al. 1990b) and near the La Grande Rivière and northwest of Lake Sakami during the winter of 1992–93 (R. Gingras, pers. comm.) further document the species' winter range.

The distribution of the species varies and is difficult to determine. Populations tend to fluctuate in size with changes in habitat caused by forest fires, farming, and road construction (Sexton 1979; Baydack and Hein 1987). Disturbances that affect habitat where leks are located limit the Sharp-tail's breeding potential. In addition, the subspecies found in Québec is partially migratory (Edminster 1954). The fact that the species inhabits remote areas makes it difficult to define its precise range and explains the patchy distribution on the Atlas map.

In Québec, this grouse is mainly found in the James Bay coastal plain and the southernmost part of Hudson Bay, a region characterized by peatlands carpeted with sphagnum and sedges and a scattering of spruce and tamarack. In the short term, the open, brushy habitat created by logging and fire may very well encourage the Sharp-tailed Grouse to colonize areas farther east (Edminster 1954). However, the species does not appear to be able to become permanently established very far from its traditional breeding range.

History and trends

Cyclical fluctuations in the population are punctuated by sporadic irruptions east and south of the Sharp-tail's normal range, like those that occurred during the autumns of 1865, 1896, and 1932 (Snyder 1951; Cayouette 1978a; Lumsden 1987f). Since the last century, the species has been observed outside the breeding season in such unusual locations as Saint-Raphaël, Maniwaki, Rigaud, and the Lake Saint-Jean area (Macoun and Macoun 1909; Todd 1963; David 1980a).

Unlike some subspecies in the United States and southwestern Canada, the population around southern Hudson Bay does not seem to have declined significantly (Edminster 1954). Suitable habitat is being lost to new reservoirs and hydro-electric projects, however. On the other hand, repeated forest fires, which maintain open habitat, benefit the species. On the whole, the Sharp-tailed Grouse's current status in the province does not seem to have changed substantially from its historical situation.

Simon Barrette

Wild Turkey

Dindon sauvage

Meleagris gallopavo

Ghislain Caron

BECAUSE OF ITS LARGE SIZE AND THE LOUD GOBBLING OF THE MALES IN spring, the Wild Turkey is one of Québec's most spectacular birds. This sedentary species has only recently become established in the southern part of the province; the first sight records date from 1976, and the first breeding record from 1984 (Yank and Aubry 1985b). For this reason, and because the bird is extremely wary, few birdwatchers here have had the good fortune to observe this heavyweight of North American birds.

Habitat and behaviour

Wild Turkeys are generally associated with mature hardwood forest. In Québec, the species is mainly concentrated along the Canada–United States border south of Montréal, where the forest cover is predominantly sugar maple mixed with other deciduous species, such as basswood, beech, hickory, ironwood, and ash (Robert 1988). Mature trees, beech in particular, as well as the bushes growing in clearings and along the forest edge, afford a plentiful supply of seeds, berries, and leaves. The rest of the turkey's diet consists mainly of insects, especially grasshoppers, but includes other invertebrates and small vertebrates, such as amphibians (Hewitt 1967). Invertebrates are the mainstay of the young turkey's diet for the first few weeks (Healy 1985). Turkeys spend their days on the ground, scratching in the leaf litter for food. At night they roost in mature trees, often near water, and many choose large conifers, especially in winter (Schorger 1966; Kilpatrick et al. 1988).

Wild Turkey habitat in Québec, and in many parts of North America, is not entirely forested. In southwestern Québec, the only part of the province inhabited by Wild Turkeys, nearly half the area where the species is found is pastureland, cornfields, or hayfields. Wild Turkeys spend much of the summer, fall, and winter in these open environments where both adults and young can find shelter and food in good supply (Vander Haegen et al. 1989; Kurzejeski and Lewis 1990).

The breeding season begins early in spring. In April, the males start attracting the females with their characteristic gobbling which carries for more than 1.5 km (Schorger 1966). The males are polygynous and will try to court several females, usually 4 or 5, by strutting about while puffing out their iridescent plumage, drooping their wings, and fanning their tails. The hen builds the nest, incubates the eggs, and rears the young. The nest is a leaf-lined depression in the ground, often under a fallen tree or a bush, near a clearing. At night, the poults are brooded on the ground until they are 2 or 3 weeks of age and able to fly up into the trees to roost (Schorger 1966). Several hens and their young may flock together in the summer, but in late fall, males and females separate, forming distinct groups until spring (Hewitt 1967).

Notes on territory

Wild Turkeys captured by cannon nets and fitted with radio transmitters can be monitored to measure the size of their home ranges. A study in Missouri (Kurzejeski and Lewis 1990) showed that seasonal movements vary considerably

In 1984 and 1988, a pair of Wild Turkeys nested under an Eastern white-cedar in this woodland on the edge of a cornfield. Montée Glass, Saint-Bernard-de-Lacolle, south of Saint-Jean-sur-Richelieu. *July 1991, Pierre Bannon.*

Wild Turkey

Breeding distribution

1984–1989

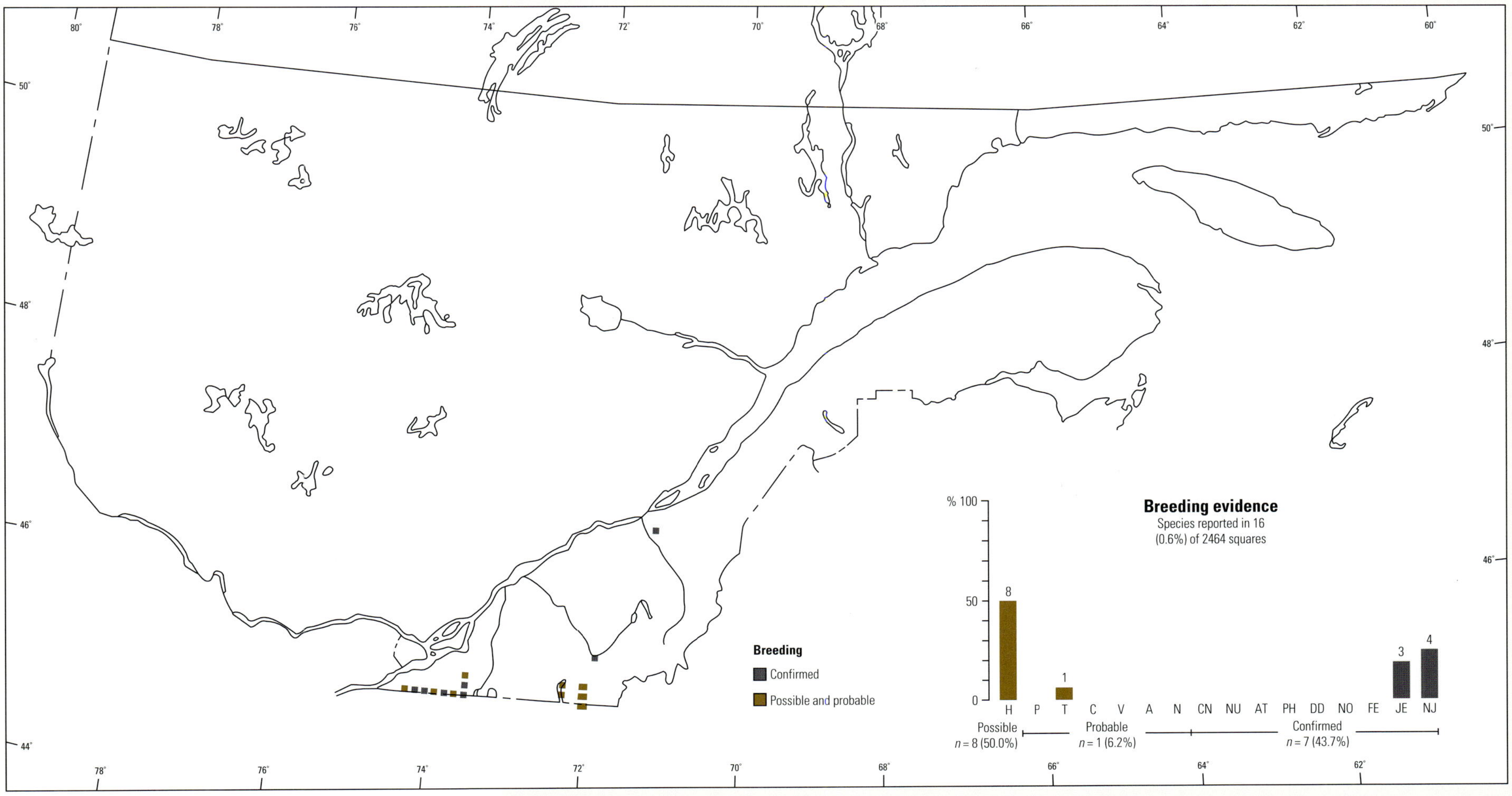

Species profile

Clutch size	usually 10–12 eggs (7–18)[1,6,7,8,12]
Laying interval	usually 1 egg per day[1,11]
Start of incubation	with last egg[1]
Incubation period	usually 28 days (25–31)[1,2,3,4,5,11]
Maturity at hatching	precocial
Care of young	by female[1,2,3,4,11]
Nestling period	≈ 24 hours[1]
Age at first flight	usually 9–14 days[11]
Dependence of young	until end of autumn or beginning of winter[1,11]
Broods per year	1[1,2,3,4,11]
Breeding age	1 year[1,6,7,11]
Mating system	polygyny[1,2,4,11]
Duration of pair bond	during mating only[1]
Total length	91–122 cm[10]
Wing span	122–152 cm[5]
Weight	M 6100–10 800 g (?) / F 3200–6500 g (?)[11]
Longevity record	12 years 6 months[9]

References:
1. Schorger 1966
2. Ehrlich et al. 1988
3. Harrison 1978
4. Harrison 1975
5. Terres 1980
6. Porter 1983
7. Lutz and Crawford 1987
8. Vander Haegen et al. 1988
9. Kennard 1975
10. Godfrey 1986
11. Eaton 1992
12. Still and Baumann 1990

Chronology *n* = 6

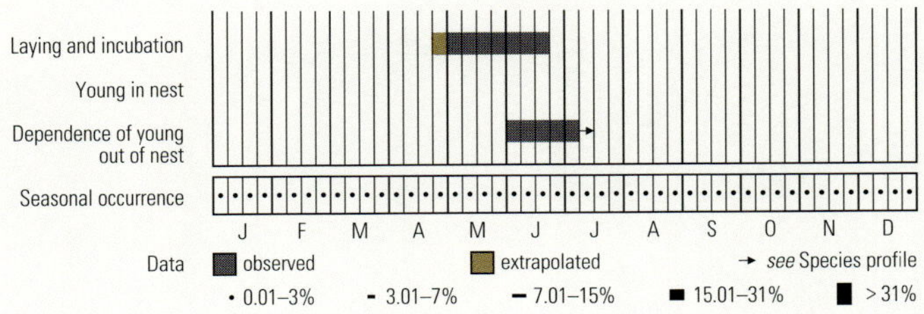

World breeding range

Being so wary and few in number, Wild Turkeys were rarely sighted by atlassers, but nesting was confirmed in over 40% of the squares in which they were reported. Four nests were discovered, even though nests of this species are generally difficult to locate. The other confirmations were based on reports of poults with hens; such sightings are more frequent in summer when flocks regularly venture out into the open.

Distribution

In Québec, Wild Turkeys are confined mainly to two distinct regions in the southern part of the province: in Huntingdon County, along the New York border, and near Magog, just north of the Vermont border. In all likelihood, there are a fair number of birds in Huntingdon County, as over 50 sightings were reported between 1982 and 1988 (Robert 1988) and many birds are still being seen there regularly. In that region, breeding was confirmed in 5 squares and 3 nests were found. The mix of deciduous forest and agricultural land that extends over the Adirondack foothills between Hemmingford and the Châteauguay River seems particularly suitable for this species.

Several reports have come from around Magog since 1982, east of Lake Memphrémagog, particularly from the Georgeville area. A nest was found near Sherbrooke in the spring of 1984 (Yank and Aubry 1985b). Breeding was also confirmed in 1988 at Saint-Lambert-de-Lévis, south of Québec City, where a nest containing 4 eggs was found in a pine plantation beside a field, but no Wild Turkeys have been seen or heard in the area since then. For this reason, and because Saint-Lambert-de-Lévis is so far from where Wild Turkeys are regularly seen, it is felt that this record may concern birds that had escaped from captivity or been released.

In the spring of 1978, 21 turkeys raised in captivity were released in the woods in Saint-Fabien-de-Panet, Montmagny County (Beaudoin 1978). The experiment initially proved successful, as turkey hens accompanied by young were seen in the area in the early 1980s (Beaudoin 1984). In the end, however, the Saint-Fabien-de-Panet experiment failed, as did similar attempts in the United States and Ontario a few decades ago (Alison 1976; Eaton 1988m). Birds from pen-reared stock do not fare well in the wild. Wild Turkeys have also recently been seen in Philipsburg, Saint-Armand-Station, and Dunham; these sites are situated between the two regions where the Wild Turkey is regularly sighted.

Notes on territory (continued)

from one bird to another and seem to be linked mainly to the availability of food. The home ranges of the 12 turkeys monitored in that study varied between 365 and 2300 ha (mean of 780 ha). It appears that these birds can cover long distances—almost 5 km, according to Kurzejeski and Lewis (1990)—in order to winter where food supplies are most accessible.

Most of Québec's Wild Turkey population is concentrated south of Montréal, along the United States border. The species generally nests on the edge of mature deciduous forest beside fields, as in Hemmingford in 1986. Fisher Road, south of Montréal. *July 1991, Pierre Bannon.*

History and trends

Contrary to popular belief, the turkeys seen in southern Québec are not domestic or barnyard birds that have been released into the wild; they are wild birds that have survived in the eastern half of North America for thousands of years. Domestic turkeys are, in fact, mostly descended from wild birds that were first domesticated by the indigenous peoples of the southern United States and Mexico, about 2500 years ago. Spanish explorers introduced these birds into Europe, where they soon became common on farms (Schorger 1966; Hewitt 1967).

Over the past 50 years, the Wild Turkey has made a remarkable comeback in the northeastern United States. Although abundant before the arrival of the Europeans, this large gallinaceous species gradually disappeared in the last century as the result of intensive clearing of hardwood forests and excessive hunting (Hewitt 1967). However, with the reversion of farmland to forest and programs for reintroducing the species using wild-caught, not domesticated, stock, the Wild Turkey is now more widespread today than it was in precolonial times (Mosby 1973; Kennamer and Kennamer 1990).

According to Tolfrey (1845), Wild Turkeys were at one time found in southern Québec. In a long discussion extolling the culinary merits of the Wild Turkey over the barnyard fowl, he stated: "the wild turkey is rarely met with in the Province of Lower Canada, although they have been seen and shot by the settlers on their first arrival to clear their land."

We can almost certainly take Tolfrey at his word—though he himself may never have observed them in Lower Canada—since Wild Turkeys are known to have inhabited southern Ontario in colonial times (Alison 1976). In any case, the species is present in southern Québec today, although the birds sighted here since the beginning of the 1980s originated in neighbouring states. In all probability, the birds in Huntingdon County and Magog come from populations in New York and Vermont, where some 65 000 and 15 000 turkeys respectively were censused in the late 1980s (Kennamer and Kennamer 1990). The species is well established in several areas only a few kilometres south of the Québec border (Gosselin 1977b; Laughlin 1985c; Eaton 1988m). Forest regeneration and the ever-increasing prevalence of corn growing have certainly helped the species gain a foothold in the province (Kurzejeski and Lewis 1990).

The Wild Turkey is currently considered vulnerable in Québec (Robert 1989b). In response to stocking of domestic turkeys for hunting, the provincial government opened a turkey hunting season in 1984. Five years later, in 1989, it decided to prohibit turkey hunting in the province's four southernmost hunting zones, after confirmation that Wild Turkeys from indigenous American stock had become established there (Robert 1988, 1989a). It also prohibited the release of domestic turkeys in these zones, with a view to preserving the integrity of the wild stock (R.S., c. C-61.1). It is thus to be hoped that Wild Turkeys will continue to spread throughout the province. Québec's harsh winters and the barrier presented by the St. Lawrence River valley make it rather unlikely, however, that the Wild Turkey will expand its range northward to any considerable degree.

Michel Robert and Pierre Bannon

Yellow Rail

Râle jaune

Coturnicops noveboracensis

Clodin Roy

THE YELLOW RAIL IS ONE OF THE MOST DIFFICULT BIRDS TO SPOT IN Québec, and to many birders, this migrant breeder is more myth than reality.

This small marsh bird resembles a week-old chick. Its size, buffy plumage with black markings, very short tail, light eyebrow, and small bill are also reminiscent of a quail, hence its scientific name of *Coturnicops*, meaning "that which looks like a quail." From the tip of its bill to the tip of its tail, it is scarcely larger than a House Sparrow, making it one of the smallest rails in the world.

At Île aux Grues, the Yellow Rail occupies huge areas of herbaceous vegetation at the upper level of the brackish marshes, which are flooded by the St. Lawrence only with the highest tides each year. Some of the marsh hay in which the species nests is harvested or burned. St. Lawrence estuary. *June 1993, Michel Robert.*

Habitat and behaviour

Like most other rails, the Yellow Rail is mainly a wetland bird. Unlike other rails in Québec, however, it does not inhabit cattail and bur-reed marshes, but prefers habitats with dense, fairly low herbaceous vegetation, where there is very little or almost no water, generally 7–15 cm. It can therefore be found in damp fields and meadows, on the flood plains of rivers or streams (wet meadows), in the herbaceous vegetation of peatlands, or at the upper levels of estuarine and salt marshes. In addition, while the Sora and the Virginia Rail, the other two rails that breed in Québec, are regularly found in small wetlands, the Yellow Rail usually prefers marshes large enough to support several pairs (Walkinshaw 1939, 1991a; Terrill 1943b; Gibbs et al. 1991b; Semenchuk 1992). In the nesting season, it is often associated with marshes dominated by sedges, including bulrushes and spike-rushes, and grasses, including fescue, cord-grass, and barnyard-grass (Peabody 1922; Fuller 1938; Devitt 1939; Walkinshaw 1939; Stalheim 1974; Bookhout and Stenzel 1987; Gibbs et al. 1991b).

The best descriptions of Yellow Rail nest sites in Québec have come from the Canadian Wildlife Service's recent work in the marsh between Île aux Grues and Île aux Oies, and at Cacouna (Robert 1993). In the first area, the species nests in the herbaceous vegetation of the upper levels of the vast marshes (brackish) dominated by grasses and sedges, where the most common plant species are red fescue, slimstem reedgrass, freshwater cord-grass, scaly sedge, moniliform sedge, aquatic sedge, Baltic rush, single-scaled spike-rush, purple loosestrife, Canada burnet, simple aster, and blue-joint. These communities are flooded only by certain high tides. In the salt (cord-grass) marshes of the St. Lawrence estuary, and specifically at the Port of Gros-Cacouna, the Yellow Rail is also found at the upper-level, where the most common plants are Mackenzie's sedge, scaly sedge, salt-marsh bulrush, black-grass, Baltic rush, single-scaled spike-rush, foxtail barley, redtop, red fescue, sweet grass, and woolly sedge (M. Robert and P. Laporte, CWS, unpubl. data).

Both birds share the first stage of nest building, making a cup in the vegetation, but the hen finishes off the selected nest by herself, and continues to add vegetation to it throughout incubation and brooding (Stalheim 1974). The cup-shaped nest is made of dead plants, and is usually covered by surrounding vegetation, especially dry remains of previous years, making it hard to find (Maltby 1915; Peabody 1922; Stenzel 1982). Built just a few centimetres above water or on the ground, the nest has an outside diameter of 10–15 cm and an inside diameter of 6–9 cm (Maltby 1915; Devitt 1939; Walkinshaw 1939; Lane 1962; Elliot and Morrison 1979; Stenzel 1982; M. Robert and P. Laporte, CWS, unpubl. data). The species sometimes uses more than one nest. For example, one of the females studied by Stalheim (1974) had built two nests; the first was for laying and incubating the eggs, the second for brooding the chicks. Similarly, Stenzel (1982) reported that a female brooded her offspring in a year-old nest, not the one in which she had laid her eggs.

In the broad expanse of marsh between Île aux Oies and Île aux Grues, more than 15 males were heard calling over the summer of 1993. This is the largest known nesting area in southern Québec. St. Lawrence estuary. *July 1993, Michel Robert.*

Notes on territory

In Michigan, a radio telemetry study conducted during the breeding season showed that mated males occupied an average area of about 8 ha, with a range of 5.8–10.5 ha. Mated females moved around much less: they stayed in an area of 1 or 2 ha before incubation and less than 0.5 ha during incubation (Bookhout and Stenzel 1987). In Québec, a similar study conducted in 1993 tracked the movements of Yellow Rails at Île aux Grues and Cacouna. The 3 males monitored at Île aux Grues had home ranges of 8.6, 13.7, and 19.8 ha, while

Yellow Rail

Breeding distribution
1984–1989

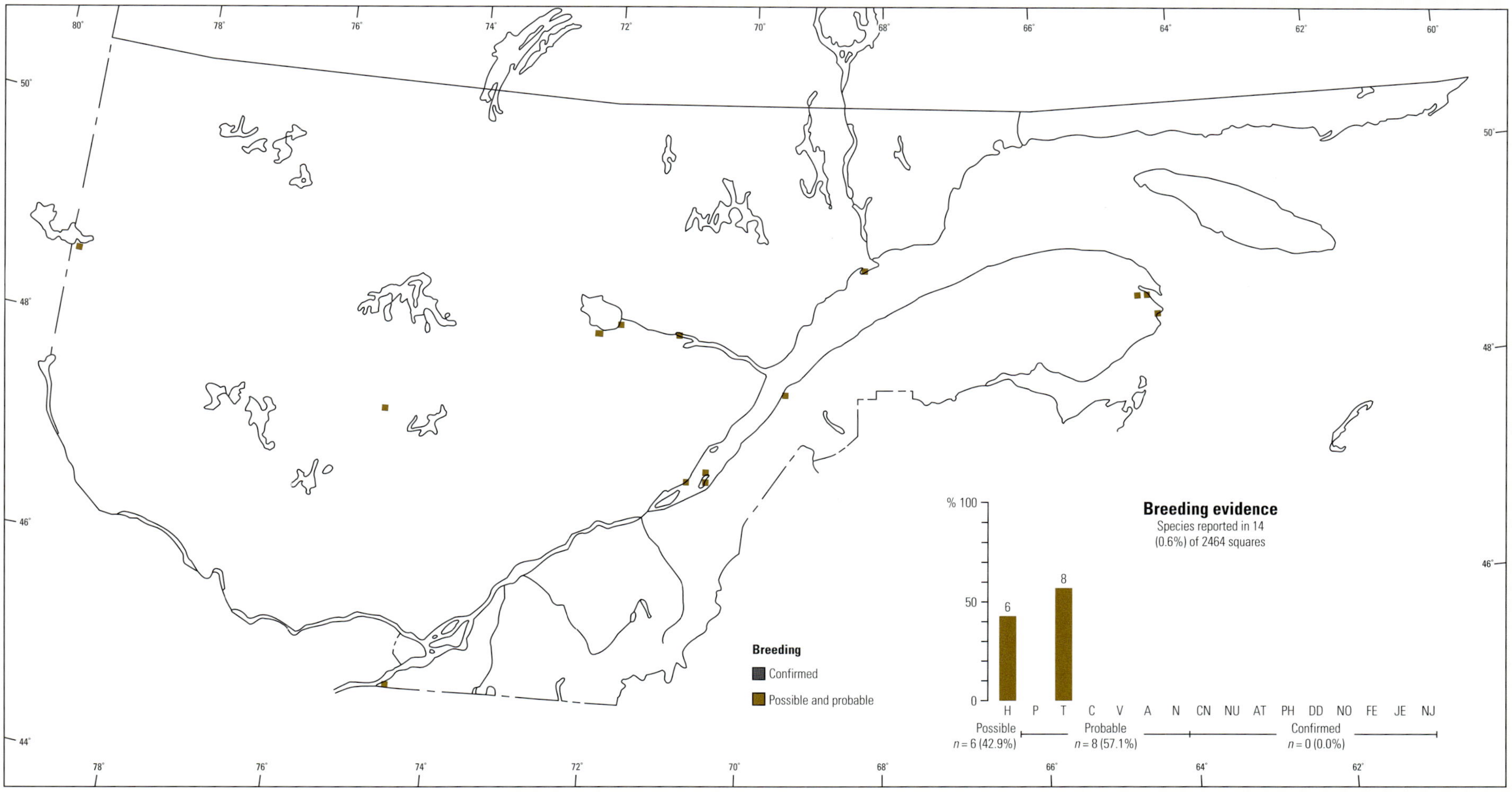

Species profile

Clutch size	usually 7–10 eggs (3–10)[4,6,7,8,9]
Laying interval	1 egg per day[6]
Start of incubation	with last egg[1,7]
Incubation period	17–18 days[4,6,7,8]
Maturity at hatching	precocial
Care of young	by female, and possibly also by male[1,8]
Nestling period	less than 48 hours[1,9]
Age at first flight	35 days[1]
Dependence of young	until ≈ 3 weeks old[1]
Broods per year	1[1]
Breeding age	?
Mating system	monogamy, sometimes polygyny[1,8]
Duration of pair bond	1 season[1]
Total length	15–19 cm[2]
Wing span	25.4–33.0 cm[3]
Weight	60.1 g (53.0–71.0)[5,9]
Longevity record	?

References:
1. Stalheim 1974
2. Godfrey 1986
3. Terres 1980
4. Elliot and Morrison 1979
5. Robert 1993
6. Peabody 1922
7. Terrill 1943b
8. Stenzel 1982
9. M. Robert and P. Laporte, CWS, unpubl. data

Chronology $n = 6$

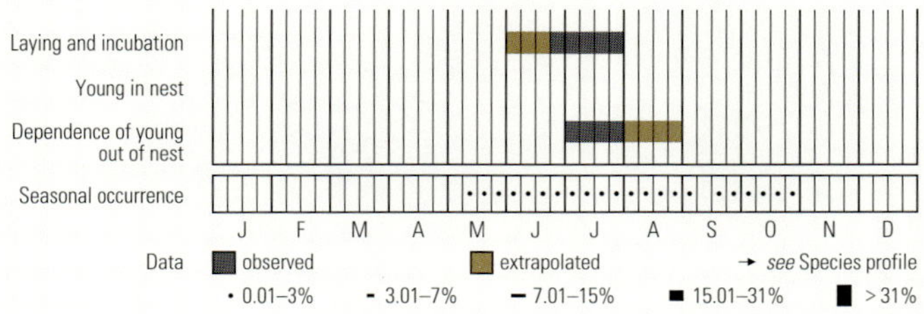

World breeding range

Both the male and female participate in the preliminary stages of building the nest, which the female continues to work on alone throughout the incubation period. The female lays a large number of eggs for a bird this size. Île aux Grues, St. Lawrence estuary. *July 1994, Pierre Laporte.*

The female's bill is darker than that of the male. She arches grasses down over the nest so that it becomes almost impossible to detect. Île aux Grues. *June 1994, Michel Robert.*

The Yellow Rail is one of the most seldom seen birds in North America (Savaloja 1981). Its small size and predilection for dense marsh cover help explain why birders have a hard time spotting it, but there are other, more important reasons why this little rail often eludes them. The Yellow Rail is expert in slipping through and hiding in marsh grasses. It is also very difficult to flush a rail, especially during the day (Peabody 1922; Walkinshaw 1939; Stalheim 1974; Stenzel 1982; Robert 1993). Ripley (1977) noted that the Yellow Rail, like other species of the genus *Coturnicops*, often remains motionless in the vegetation instead of running away. Its buffy and black plumage blends extremely well into the background of its marsh habitat, both when it is immobile and when it is moving about (Walkinshaw 1939). The Yellow Rail almost never flies during daylight hours (Stalheim 1974); when it does so, it is often because it is being chased or wants to avoid being trodden upon (Morris 1905; Swales 1912; Meredith 1935; Walkinshaw 1939; Stenzel 1982). On the other hand, it is much easier to flush one at night (Stalheim 1974; Stenzel 1982; Robert 1993).

The call of the Yellow Rail is of particular interest, since it is quite peculiar and is often the only way that birders have of detecting the bird. It is a rapid series of monotonous *ticks*, sounding like two stones tapping together. There is typically a series of five *ticks* interrupted by a short pause, *tick-tick, tick-tick-tick*, at a usual frequency of one series per second (Fryer 1937; Stalheim 1974). Only the male makes this sound (Stalheim 1974). The fact that the bird can be heard much more often at night than during the day adds to the mystery of the Yellow Rail and its call (Peabody 1922; Bart et al. 1984; Robert 1993). Once night has fallen, it can be heard for hours on end, because it calls almost non-stop, often continuing until first light (Robert 1993). According to Stalheim (1974), birds living where there is dense vegetation are usually heard more at night than during the

Notes on territory (continued)

2 at Cacouna had home ranges of 1.6 and 4.6 ha (M. Robert and P. Laporte, CWS, unpubl. data). Perhaps the Cacouna rails had smaller ranges than those at Île aux Grues because the marsh at Cacouna is small and there were a large number of rails there in 1993. In Cacouna, as in Michigan (Bookhout and Stenzel 1987), the home ranges studied overlapped (Robert 1993).

Hatching is synchronous. Two chicks with still-wet down and a third coming out of the egg, no more than 30 mm long. Île aux Grues. *July 1994, Pierre Laporte.*

Clutch size is 7–10 eggs, laid a day apart (Maltby 1915; Peabody 1922; Walkinshaw 1939; Stenzel 1982). Incubation, carried out by the female alone, usually begins when laying is complete and continues for 17 or 18 days (Terrill 1943b; Stalheim 1974; Elliot and Morrison 1979; Stenzel 1982). Once incubation has begun, the female remains on the nest almost all the time. Hatching is synchronous, and within a few hours, the young rails can already stand up. Two days after hatching, the entire brood follows the hen around outside the nest. The chicks have a small claw, or wing spur, on each wing that they use to grip onto vegetation as they roam around and when they climb back into the nest. They begin to feed themselves when they are about 5 days old, are no longer brooded three weeks after hatching, and are fledged by 35 days (Stalheim 1974, 1975).

According to Stalheim (1974, 1975), who studied the species in captivity, the male does not tend the young and stops associating with the female once incubation begins. This would make the Yellow Rail quite different from other members of the family Rallidae studied so far, since in most of them both parents incubate the eggs and tend the young (Stalheim 1974). Yet Stenzel (1982), who worked in a natural setting, observed Yellow Rail young in the company of both parents, as well as a male with chicks. He also saw a male with his mate at hatching time. The male may therefore quite possibly help take care of the young.

In the hours after hatching, the hen calls the chicks—little balls of black down—to follow her. Île aux Grues. *July 1994, Pierre Laporte.*

In the vast coastal marshes of Cabbage Willows Bay, the Yellow Rail likes areas where cottongrass and buckbean grow. Northwestern Rupert Bay. *August 1991, Michel Robert.*

day, probably because their communication system is not very dependent on vision, acoustic conditions are better at night, and the chances of attracting predators are also lessened. This does not necessarily mean, as some authors have concluded (Maltby 1915; Peabody 1922), however, that the Yellow Rail is a nocturnal bird. The work of Stalheim (1974, 1975) has shown that on the contrary, the Yellow Rail is much more active during the day than at night. For example, it does not feed at night, and its sexual and agonistic behaviour is much more frequent in the daytime.

The Yellow Rail was reported in only 0.6% (14/2464) of Atlas squares. The secretive behaviour of the species during the day and atlassers' unfamiliarity with it are doubtless responsible for this low figure and explain why breeding could not be confirmed. The only evidence of breeding reported by atlassers was the presence of the species in its habitat and territorial behaviour, due to the fact that the Yellow Rail is far more often heard than seen.

Distribution

The Yellow Rail's eastern North American breeding range is very poorly known. In Ontario, the species breeds along the Hudson Bay and James Bay coasts. Elsewhere in the province, as well as in the northeastern United States and in the Maritimes, it is very rarely observed in the breeding season (Prescott 1987b; Gibbs et al. 1991b; Erskine 1992).

This little rail, heard mostly at night, has also been observed in the marshes of the York River estuary. Gaspé. *July 1991, Yves Aubry.*

Available data indicate that in Québec, the rail is generally associated with the upper level of some of the marshes along the Saguenay River, Lake Saint-Jean, St. Lawrence River, and James Bay, in either freshwater, brackish, or salt marshes. It is also possible that some birds breed inland, however, this needs to be confirmed. Until quite recently, breeding had been confirmed only at Coin-du-Banc, on the Gaspé Peninsula, where 2 nests containing eggs were found, the first in 1940, the other in 1946 (Terrill 1943b; L. M. Terrill, unpubl. data). Then, in 1994, 4 nests were found in the marsh between Île aux Grues and Île aux Oies (M. Robert and P. Laporte, CWS, unpubl. data).

The Yellow Rail was atlassed in the Gaspé (Coin-du-Banc, York River, Saint-Jean River), on the North Shore (Pointe-Lebel), in the Saguenay–Lac-Saint-Jean region (Saint-Fulgence, Chambord, Saint-Gédéon), in the St. Lawrence estuary (Cap Tourmente, Île aux Grues, Île aux Oies, Cacouna), and at Lake Saint-François (Dundee). Other Atlas reports come from Lake Abitibi (Antoine Creek) and central Québec (Bazin River). Of all these areas, the most suitable, that is, the most regularly inhabited, are those on the Gaspé Peninsula, the Saint-Fulgence marsh, Cacouna, Île aux Grues, Île aux Oies, the Cap Tourmente National Wildlife Area, and the Lac Saint-François National Wildlife Area.

Outside the Atlas survey period, the Yellow Rail has been reported elsewhere in southern Québec during the breeding season. The main areas likely to be home to breeding birds are Île du Moine, at the southwest end of Lake Saint-Pierre, the Magdalen Islands, and some of the marshes along the Lower St. Lawrence and North Shore (Meredith 1935; COQ 1980a; David 1980a; Gosselin et al. 1991b; Fradette 1992). In all likelihood, the species also nests in the vast coastal marshes of James Bay: 5 rails were heard in Cabbage Willows Bay on 17 June 1991 (Gosselin et al. 1991b; Cons. G. G. – GREBE 1992j). The coastal marshes of James Bay probably support a large number of birds since, as in Ontario (Prescott 1987b), the species seems to be locally common. There have also been a number of summer reports from southern Rupert Bay up to Chisasibi (Todd 1963; Cons. G. G. – GREBE 1992j).

History and trends

There is not enough information on the Yellow Rail to determine whether its current status is similar to its historical status. The first sightings in Québec during the breeding period were at James Bay and date from the end of the last century (Todd 1963). Other summer reports have been made throughout the 20th century: along the Lower St. Lawrence, birds were seen at Kamouraska in 1922 and at Sainte-Anne-de-la-Pocatière in 1930 (Meredith 1935). The Yellow Rail was later observed at Coin-du-Banc, on the Gaspé Peninsula, in 1939, and at Saint-Fulgence, in the Saguenay, in 1964 (Terrill 1943b; Browne 1967). The species still inhabits most of these areas today (Gosselin et al. 1991b).

Like other wetland birds, the Yellow Rail may have suffered from the draining of its favourite habitats, particularly along the St. Lawrence. Wetland loss along the St. Lawrence, including high marsh plant communities, has been significant in the last few decades. The draining of wetlands for agricultural purposes accounted for approximately 40% of the marshes that disappeared between 1950 and 1978 along the St. Lawrence (Bouchard and Millet 1993). Dike building in the St. Lawrence estuary also reduced the upper levels of salt marshes by several metres (Gauthier et al. 1980; Couillard and Grondin 1986), with the consequent loss of a number of potential Yellow Rail habitats. For example, the intertidal marshes around Kamouraska, which stretch over approximately 26 km between Saint-Denis Point and the Rivière des Caps on the south shore of the St. Lawrence, decreased in area from 9.33 km^2 in 1930 to 3.91 km^2 in 1986. The main changes to the marshes have been caused by the building of dikes to recover land for farming (Pelletier et al. 1990).

Given its rarity and the fragility of its habitat, the Yellow Rail is classified as vulnerable in Québec (Robert 1989b).

Michel Robert

Virginia Rail

Râle de Virginie

Rallus limicola

Clodin Roy

LIKE OTHER RAILS, THE VIRGINIA RAIL HAS A LATERALLY COMPRESSED BODY allowing it to slip through dense marsh vegetation. It hides so well, it is practically impossible to see. And like other birds in the family Rallidae, it uses its strong legs and long slender front toes to walk on mud and floating plants. When disturbed, it will run away rather than fly.

The Virginia Rail is an uncommon migrant breeder in southern Québec. It may be locally common in suitable habitat.

Habitat and behaviour

With its elusive behaviour and cryptic coloration, the Virginia Rail is heard far more often than seen. This sometimes noisy bird makes a variety of strange calls and grunts night and day. Although the Virginia Rail is not as territorial as the Sora (Kaufmann 1989), it will respond aggressively to recordings or imitations of its own calls and songs or those of the Sora, and can sometimes even be lured into view (Glahn 1974; Johnson and Dinsmore 1986a).

The Virginia Rail nests in well-hidden spots in freshwater marshes of just about any size where cattails, sedges, and rushes grow (Zimmerman 1980; DeGraaf and Rudis 1987). Other nesting habitat includes brackish marshes, the marshy edges of lakes and rivers, ditches, and wet meadows. There is no significant difference between the nesting habitat of this species and that of the Sora (Johnson and Dinsmore 1986b). In fact, the two species are commonly found in the same marshes, where their territories are often adjacent and may even overlap. The Virginia Rail is more abundant in cattail marshes, however.

The female builds the nest, sometimes assisted by the male, which brings nesting materials. It is a loosely woven basket of plant matter, usually attached to a clump of marsh vegetation. Nest heights range from water level to 61 cm, but are most often 10–23 cm above the water (Peck and James 1983). The Virginia Rail places its nest on or above the ground more often than the Sora and, when it nests over water, chooses shallower areas (Townsend 1926; Peck and James 1983). The nest is often canopied with arched leaves of surrounding plants and sometimes a small entrance ramp is also constructed. Nests are well hidden and difficult to get to, and therefore very hard to find. Nests were used to confirm breeding in only 7 of the 144 squares in which the species was recorded.

Typical clutch size is 7–12 eggs. Incubation begins 1–5 days before laying is complete (Kaufmann 1989) and lasts 18–20 days; both parents take part (Ehrlich et al. 1988). Two or three days after the eggs hatch, the male builds another nest nearby for brooding the hatchlings. The precocial young can run and swim shortly after hatching, but do not leave the nest until they are 3 or 4 days old, unless the nest is disturbed (Kaufmann 1987). During the first few weeks of life, they stay close to their parents and return to the nest often to be brooded by the male or female. The daytime brooding periods gradually shorten, and cease altogether when the chicks are 2 weeks old, but night-time brooding continues for longer. The chicks do not travel together like ducklings; although they usually follow a parent, they frequently wander off on the family territory by themselves and return to the nest on their own.

The chicks are able to feed themselves when they are a few days old, but the parents continue to bring them food for more than a week. Around the fourth week, the male may begin to act aggressively toward juveniles still begging for food, to encourage them to become independent (Kaufmann 1989). Both the young and their parents consume mainly aquatic invertebrates—insects and their larvae, crayfish, and snails, for example—which they capture by probing the water and mud with their bills. The adults also eat seeds and aquatic plants, particularly duckweed (Horak 1970).

Most confirmations of breeding were based on observations of young that had left the nest but were still incapable of sustained flight. Other types of breeding evidence were obtained much less often.

Notes on territory

The Virginia Rail's territory is not as well defined as that of the Sora, and is defended mainly against conspecifics (Kaufmann 1989).

In Iowa, Johnson and Dinsmore (1985) estimated that the mean size of the species' home range was 0.18 ha. Berger (1951) found 5 nests in a small marsh of approximately 0.2 ha in

The Virginia Rail uses the cattails in this marsh to anchor and hide its cup-shaped nest. Sainte-Foy. *June 1963, Gaston Le Page.*

Virginia Rail

Breeding distribution

1984–1989

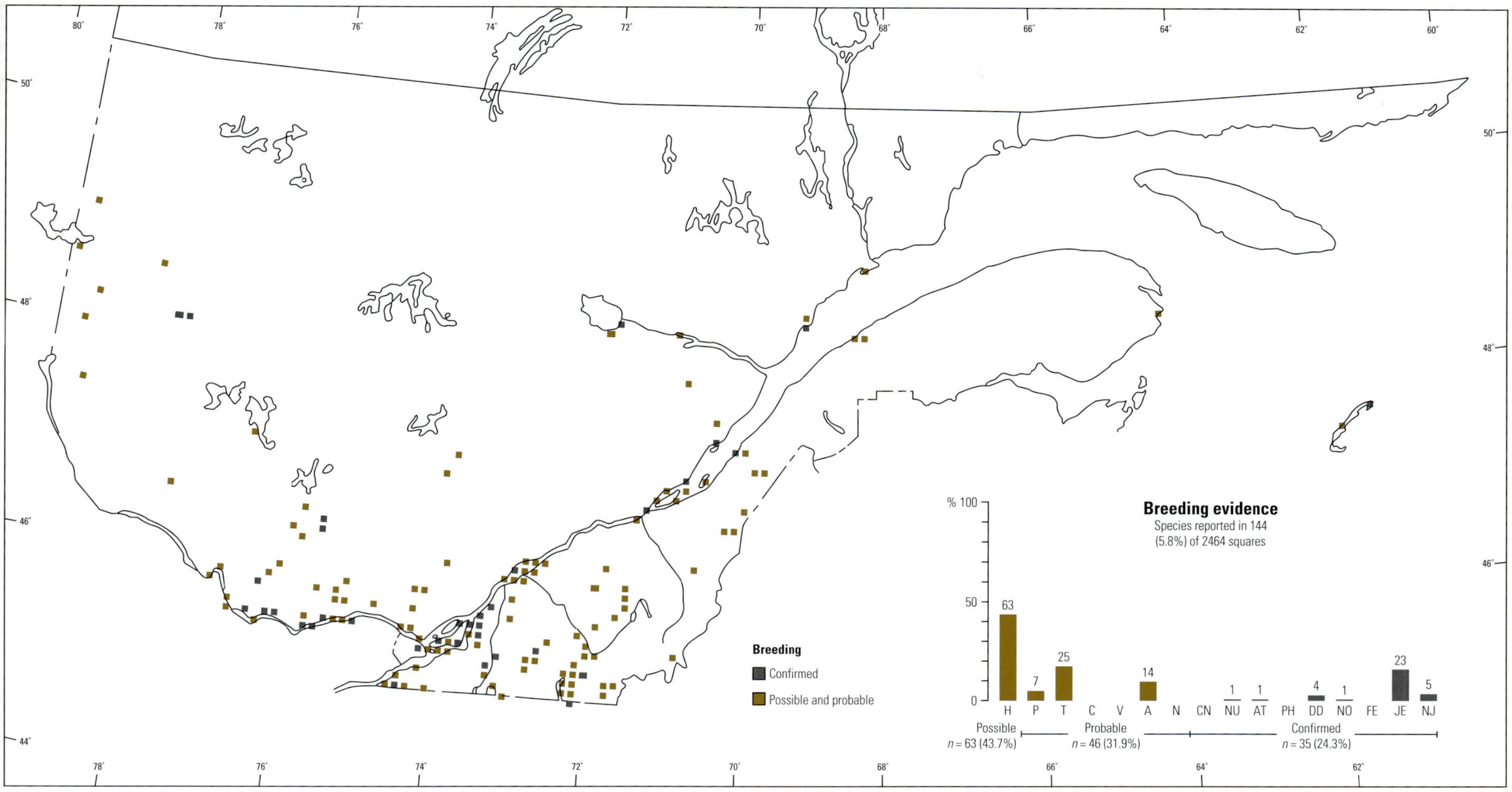

World breeding range

Species profile

Clutch size	usually 7–12 eggs (4–13)[1,4]
Laying interval	usually 1 egg per day[1]
Start of incubation	before last egg[1]
Incubation period	18–20 days[2,3,4]
Maturity at hatching	precocial
Care of young	by female and male[1,4,5]
Nestling period	usually 3–4 days[5]
Age at first flight	25–28 days[4,5]
Dependence of young	until 15–42 days old[9]
Broods per year	1, rarely 2[3,4]
Breeding age	?
Mating system	monogamy[4]
Duration of pair bond	?
Total length	23.0–25.5 cm[6]
Wing span	33.0–36.8 cm[7]
Weight	M 89.0 g (64.0–120.0) / F 74.9 g (67.0–79.6)[8]
Longevity record	?

References:
1. Kaufmann 1989
2. Walkinshaw 1957a
3. DeGraaf and Rudis 1987
4. Ehrlich et al. 1988
5. Kaufmann 1987
6. Godfrey 1986
7. Terres 1980
8. Dunning 1984
9. Johnson and Dinsmore 1985

Chronology $n = 60$

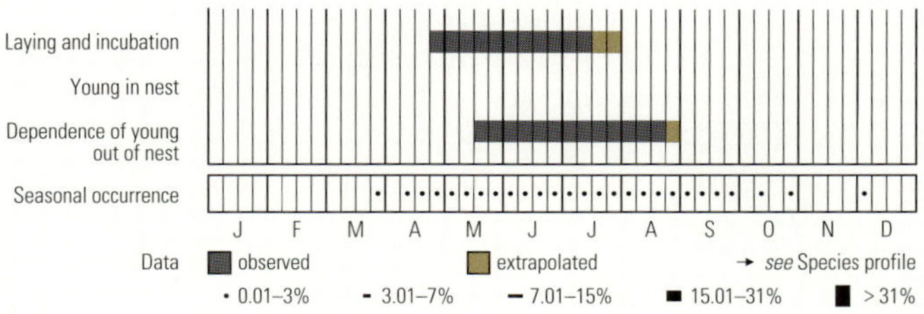

Notes on territory (continued)

Michigan, and Glahn (1974) reported densities ranging from 1.4 to 2.3 pairs/ha in Colorado.

Virginia Rail in herbaceous vegetation. Louiseville, north shore of Lake Saint-Pierre. *May 1981, Daniel Jauvin.*

Distribution

The Virginia Rail's distribution in Québec corresponds closely to the extensive marshland and with suitable nesting habitat found mainly in the southern part of the Atlas area. It is therefore not surprising that the species occurs most often in the Central St. Lawrence Lowland and along the St. Lawrence and Ottawa rivers. It was also reported as far north as La Sarre (Abitibi), at Saint-Fulgence, Métabetchouan, and Saint-Gédéon (Saguenay–Lac-Saint-Jean), at Pointe-à-Boisvert and Pointe-aux-Outardes (Upper North Shore), near Rimouski (Lower St. Lawrence), at Barachois (Gaspé Peninsula), and in the Magdalen Islands. Atlassers did not find it on Anticosti Island, and neither Ouellet (1969), David (1980a), nor Godfrey (1986) reported its occurrence there. The Atlas distribution is somewhat north of the range shown in Godfrey (1986) and David (1980a), and given its elusive nature, the Virginia Rail is probably more widespread than the Atlas map suggests.

The wetland habitat these rails depend on is often located next to pastureland and the nests may be trampled by livestock. Île aux Fermiers, Varennes Islands, near Montréal. *August 1991, Gaétan Duquette.*

Elusive by nature, this rail nests in marshes in southern Québec, where it feeds on seeds, invertebrates, and aquatic plants. Hull. *September 1991, Gilles Falardeau.*

History and trends

According to Macoun and Macoun (1915), the Virginia Rail was a common summer resident in the western part of the province early in the century. Dionne (1906) described it as uncommon on the shores of the St. Lawrence near Québec City, while Wintle (1896) listed it as a common summer resident in the Montréal area. Wintle also tells of a hunter that managed to bag 32 Virginia Rails in one day on the Verchères Islands with his pointer dog in the spring of 1890. Today, the Virginia Rail is still a game bird, but is not hunted as extensively as it used to be, so hunting does not seem to pose a threat to the species.

Since the turn of the century, the species has faced a more insidious threat, however: the destruction of its breeding habitat. According to Dryade (1981a), 2625 ha of wetlands along the Cornwall–Grondines stretch of the St. Lawrence were destroyed between 1945 and 1976. In Ouellet's (1974a) view, the Virginia Rail was much less common around Montréal in the late 1960s than in the early 1950s because so many marshes had been drained and filled.

To maintain Québec populations at their current levels, extensive wetland aeras must be protected. Wetland management practices that create nesting habitat for the rails must also be encouraged, particularly if they are compatible with the management of waterfowl habitat (Johnson and Dinsmore 1986b).

Gilles Falardeau

Incubating Virginia Rail. Grass has been arched over the nest to provide better camouflage. Saint-Pierre, Island of Montréal. *May 1936, Henry Mousley, Laval University Collection.*

Sora

Râle de Caroline

Marouette de Caroline
Porzana carolina

SELDOM SEEN, THE SORA MOST OFTEN REVEALS ITS PRESENCE THROUGH its vocalizations, which are heard at any hour of the day or night. Although in its behaviour and habits it resembles the Virginia Rail, its plumage and appearance are quite different. Anyone who succeeds in detecting it among the cattails will discover a dark, chicken-like bird with a short yellow bill that stands out against the black patch on the face and throat.

The Sora breeds locally in southern Québec, where it may, in places, be common in its habitat. It is one of the rail species that makes the longest migrations, wintering from the southern United States, through Central America and the West Indies south to Peru (AOU 1983). This comes as a surprise to anyone who has flushed a Sora from a marsh and observed its laboured flight.

Habitat and behaviour

Soras nest in approximately the same habitat as Virginia Rails, that is, in freshwater, or sometimes brackish marshes with tall, dense beds of cattails, sedges, and bulrushes. They also nest in marshy areas along lakes and rivers and in ditches (Peck and James 1983). Although there is little segregation in their habitat, Soras are more common than Virginia Rails in areas of sparser growth (Johnson and Dinsmore 1986b). Being less selective, they are also more apt to be found in smaller marshes, ditches, and wet meadows.

The breeding biology of the two species is very similar too. Assisted by the male, which may bring nesting material (Kaufmann 1989), the female Sora weaves a cup of plant matter, which she normally anchors to the marsh vegetation, no more than 45 cm, and usually less than 15 cm, above the water's surface. She generally places it in deeper water (8–120 cm) than does the Virginia Rail (Peck and James 1983). She often makes an arch of the surrounding plants to hide the nest and builds an entrance ramp.

Typical clutch size is 10–12 eggs, and sometimes as many as 16. Both parents take part in incubation, which begins between the laying of the second and eighth eggs. Hatching is more asynchronous than in the Virginia Rail; the interval between the hatching of the first and last eggs may be anywhere from 1 to 17 days (Kaufmann 1989). This means that one parent must sometimes tend the young birds while the other continues to incubate the unhatched eggs (Walkinshaw 1940). Two or three days after the eggs begin to hatch, the male builds another nest close to the first; the chicks often go there to be brooded by one of the parents in the first few weeks.

The young are able to walk and swim for short distances when they are only a day old. They do not, however, normally leave the nest until they are 3 or 4 days old, unless the nest is disturbed (Kaufmann 1987). The chicks form a loosely knit brood and tend to scatter over the family territory, rather than follow one of the parents. Both parents feed the young for 2–3 weeks (Kaufmann 1989). The chicks have a tuft of orange down at the base of the lower mandible that elicits a feeding response from the adult birds. Their begging behaviour is more elaborate than that of the young Virginia Rails (Kaufmann 1983). According to Johnson and Dinsmore (1985), young Soras remain with their parents until the age of 16–32 days.

Like the other rails, the Sora is hard to spot, as it is always concealed by vegetation. It answers the call of the Virginia Rail, but responds more readily to calls of other Soras (Johnson and Dinsmore 1986a). It is more territorial than the Virginia Rail and defends a clearly defined territory, from which it generally excludes all other species of rails (Kaufmann 1989).

The diet of the adult birds is composed mainly of the seeds of plants found in their habitat—sedges, foxtail, smartweed, duckweed—and, to a lesser degree, aquatic insects (Horak 1970). The chicks, however, eat only aquatic invertebrates. Kaufmann (1989) suggested that because the Sora has a smaller bill than the Virginia Rail, it is less efficient at capturing invertebrates, and so must protect this essential food source for its young by defending the territory from other species of rails. In the fall, the Sora feeds on wild rice, which in the freshwater marshes of the United States, may constitute up to 94% of its diet (Webster 1964).

Breeding was mostly confirmed for the Atlas through observation of young birds (17 out of 27 squares). Nests were found in only 9 squares, which is an indication of how difficult they are to discover. It is not easy to study the Sora's habits, given its furtive behaviour and inaccessible nature of its habitat.

Clodin Roy

Notes on territory

The Sora defends a well-defined territory against other Soras and Virginia Rails (Kaufmann 1989). The species exhibits a wide range of aggressive territorial defence behaviour (Kaufmann 1983).

Johnson and Dinsmore (1985) estimated the home range to be 0.19 ha (±0.02 ha) in Iowa. Berger (1951) reported 4 nests in a small marsh of about 0.2 ha in

The Sora nests not just in dense vegetation, but in marshy areas with sparser growth. Its small size and long toes enable it to slip easily through the cover of emergent plants. Lake Saint-Pierre. *May 1991, Réginald Ouellet.*

Sora

Breeding distribution
1984–1989

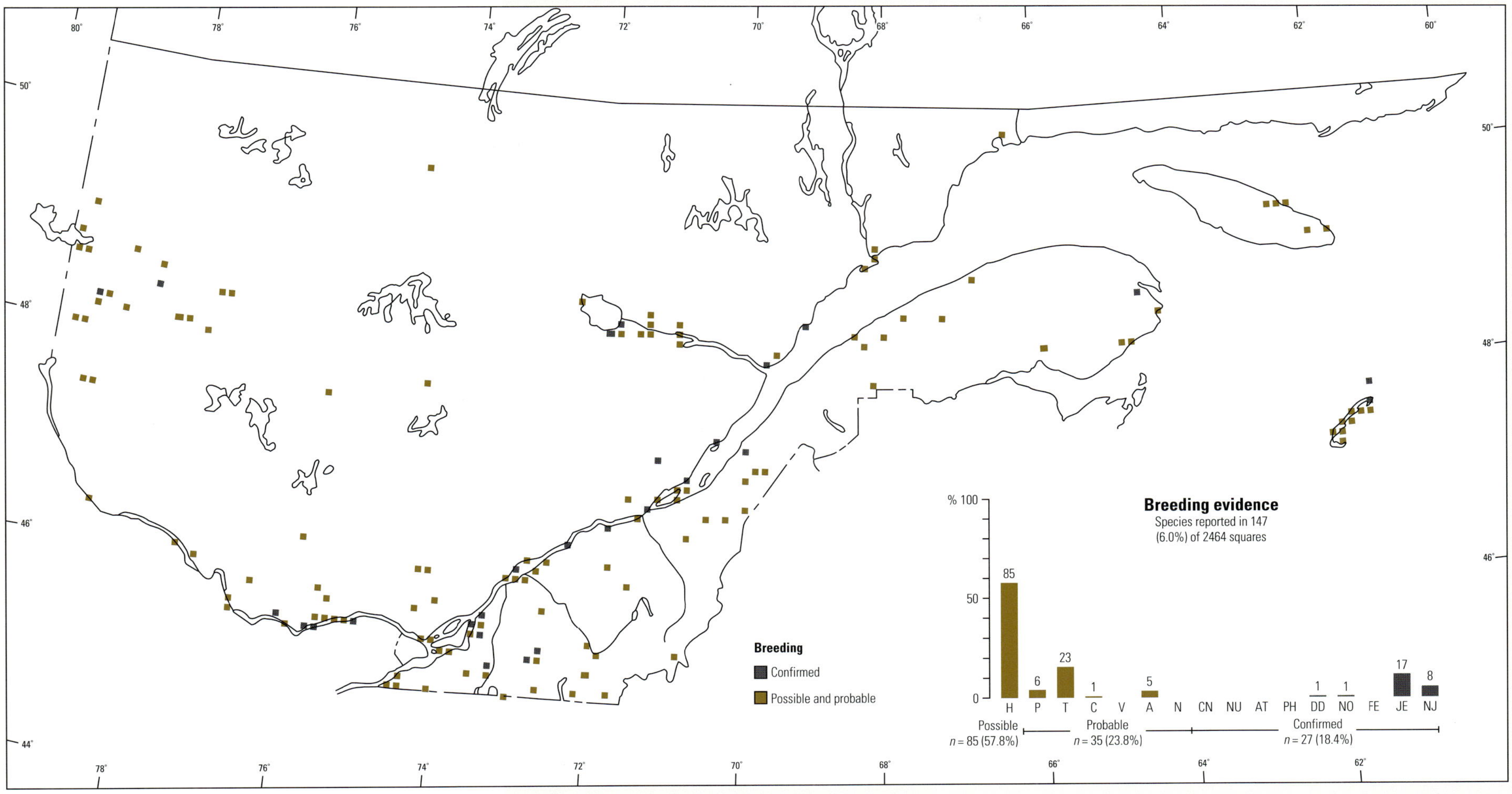

Species profile

Clutch size	usually 10–12 eggs (6–16)[1,2,3,4,5]
Laying interval	usually 1 egg per day[1]
Start of incubation	between the second and eighth egg[1]
Incubation period	16–20 days[2,3,4,5,6]
Maturity at hatching	precocial
Care of young	by female and male[1,4,7]
Nestling period	usually 3–4 days[7]
Age at first flight	21–28 days[5,7]
Dependence of young	until 16–32 days old[11]
Broods per year	1, rarely 2[3,5]
Breeding age	?
Mating system	monogamy[5]
Duration of pair bond	?
Total length	20–30 cm[8]
Wing span	30.5–36.8 cm[9]
Weight	74.6 g (?)[10]
Longevity record	?

References:
1. Kaufmann 1989
2. Walkinshaw 1940
3. Lowther 1977
4. DeGraaf and Rudis 1987
5. Ehrlich et al. 1988
6. Walkinshaw 1957a
7. Kaufmann 1987
8. Godfrey 1986
9. Terres 1980
10. Dunning 1984
11. Johnson and Dinsmore 1985

Chronology *n* = 43

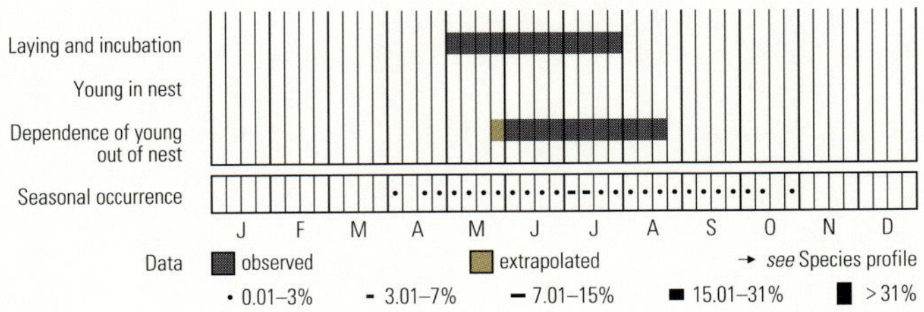

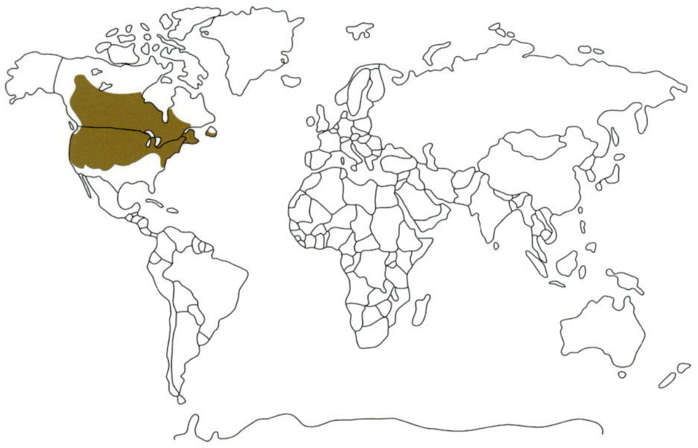

World breeding range

Both Virginia Rails and Soras inhabit the Lac Saint-François National Wildlife Area. The swampy part where there are shrubs such as those seen in the background is more likely to be home to the Sora. Southwest of Montréal. *July 1991, Gaétan Duquette.*

Notes on territory (continued)

Michigan, and Glahn (1974) reported a density of 1.0 pair/ha in Colorado. In Iowa, Tanner and Hendrickson (1956) found 26 nests in an area of about 32 ha.

Distribution

The Sora was recorded in almost the same number of squares as the Virginia Rail, but as shown in the table, the Sora has a somewhat more northerly distribution. Its ability to adapt to a greater variety of habitats enables it to occupy smaller marshes and areas with fewer cattails. The Sora was reported more frequently than the Virginia Rail in the Abitibi, Saguenay, and Gaspé regions, as well as on Anticosti Island and the Magdalen Islands. It was atlassed in 5 squares on Anticosti, where it had been seen before (Ouellet 1969; David 1980a). There is only one breeding record from this area, however: in early July 1976, a young bird was seen with an adult (Bergeron 1977). Todd (1963) cites records farther east, on the North Shore—at the mouth of the Natashquan River and at Harrington Harbour—and at Red Bay, Labrador, along with a further report from Sandwich Bay on the east coast of Labrador. Being so difficult to detect, the Sora is probably more widespread than the Atlas map indicates.

Frequency of occurrence of the Virginia Rail and Sora north and south of the 47th parallel

	South of 47th parallel	North of 47th parallel	Total
Virginia Rail	114	30	144
Sora	72	75	147

The Sora also nests in marshes on the James Bay coast. At Rupert Bay, several birds were heard on Cabbage Willows Bay on 2 July 1943, and a pair was collected at Waskaganish on 29 or 30 June 1914 (Todd 1963). The female had an almost fully developed egg, ready to be laid, in her oviduct. Todd (1963) saw a bird at Fort George on 25 and 26 July 1941, while Côté (1980) reported the species at Lake Desaulniers, near La Grande-2, in the summer of 1976. In June 1991, several Soras were heard on a lake in the Lake Soscumica area (50°10' N, 77°10' W) and around Cabbage Willows Bay (Cons. G. G. – GREBE 1992j). The species was sighted farther north on 16 June 1991, in a peatland 3 km west of the outlet of Lake Fressel

Nest containing 12 eggs. Though small in size, the Sora lays a large clutch. The nest is a woven cup securely anchored to the cattails above the surface of the water. Sainte-Foy. *July 1966, Gaston Le Page.*

Newly hatched Sora chick. Saint-Lambert, near Montréal. *May 1936, Henry Mousley, Laval University Collection.*

(55°31' N, 75°22' W), when a bird answered a recording of its call (Cons. G. G. – GREBE 1993c). According to Todd (1963), ecological conditions on the margins of most of the lakes north of Abitibi and Lac-Saint-Jean do not seem to be favourable to the species. The northernmost Atlas sighting, near Chibougamau, marks the northern limit of the Sora's range in this part of the province, which is in line with David's (1980a) description. There appears to be some discontinuity in the range between the southern Québec and James Bay coastal populations.

History and trends

Wintle (1896) listed the Sora as a common summer resident in the Montréal area. Dionne (1906) reported that it was fairly common in summer, adding that it was mainly seen in autumn on the beaches of Île d'Orléans and the north shore of the St. Lawrence, from Québec City to Saint-Joachim. Godfrey and Wilk (1948) described it as fairly common in the cattail marshes on the north side of Lake Saint-Jean.

According to DeGraaf and Rudis (1987), Sora populations have been declining throughout northeastern North America since the turn of the century. Ouellet (1974a) reported a major drop in the Montréal area throughout the 1950s and 1960s, subsequent to the loss of numerous marshes and swamps. Dryade (1981a) documented the considerable habitat loss that occurred along the St. Lawrence between 1945 and 1976, noting that 84% of wetland losses took place in the first half of the period, between 1945 and 1960.

According to Bent (1926), the Sora used to be a popular game bird, at least in the United States. Over a century ago, it was not uncommon for a good hunter to bag more than 100 rails a day in some places in the northeastern states. Birds shot nowadays are probably killed by accident.

Hunting could still have an indirect effect on the Sora. In two Maryland studies, lead pellets were found in 12% and 7% of gizzards analysed (Artmann and Martin 1975; Stendell et al. 1980). Up to 28 pellets were found in one gizzard, and lead concentrations in the livers of some birds were at levels considered fatal for waterfowl. The fall harvest of wild rice in the United States could have a local impact on the Sora, through the combined effects of disturbance and loss of food (Fannucchi et al. 1986).

The key to maintaining Sora and Virginia Rail populations in Québec and elsewhere lies in the preservation and management of wetland habitat.

Gilles Falardeau

This heavily overgrown marsh in Cazaville is typical of the Sora's preferred breeding habitat. In this type of area, the species is more often heard than seen. Southwest of Montréal. *July 1991, Gaétan Duquette.*

Common Moorhen

Poule-d'eau

Gallinule poule-d'eau
Gallinula chloropus

The Common Moorhen is aptly named, since it feeds by pecking at the ground like a chicken. Its long unwebbed toes allow it to walk surefootedly on top of the floating vegetation in the slow-moving water of its habitat. It is equally at home in the water itself, pumping its head with each stroke as it swims. When disturbed, it hides quickly in the vegetation. The Common Moorhen has dark slate grey plumage, with a white line along its flanks and a white patch under its tail. Its red frontal shield and yellow-tipped red bill give it a clownish look. The sexes are identical.

The Common Moorhen is a fairly common migrant breeder in southernmost Québec (David 1980a). It is rare and local north and east of Lake Saint-Pierre, however. There are only two known winter records for the species in southern Québec, one from the winter of 1986–87 and the other between January and March 1988, both from Léry, near Châteauguay.

In eastern North America, the Common Moorhen winters primarily from South Carolina south through the Gulf states, and in the western part of the continent from California and Arizona south to South America (AOU 1983; Godfrey 1986; National Geographical Society 1987).

Habitat and behaviour

In the breeding season, this member of the rail family is found in different types of wetland habitat, mainly marshes and ponds of various depths and sizes, but seems to prefer deep marshes with lush submergent and emergent vegetation, particularly cattails (Brackney and Bookhout 1982). The dead stalks of cattails remain throughout the winter, providing the Common Moorhen with protective cover and nesting materials early in the season (Krauth 1972). The same habitats are used for brood rearing. The water at the nest site is usually 20–91 cm deep (Fredrickson 1971), which helps reduce the risk of predation (Brackney and Bookhout 1982).

Courtship displays and pair formation begin during migration, which in Québec extends from the end of April to mid-May. One unusual characteristic of the species is that the female selects a mate from a number of candidates before territories are established, possibly because the pair's breeding success depends on the male's ability to provide adequate care for the young (Petrie 1986). The pair defends its territory in the marsh by calling repeatedly and expelling intruders aggressively (Cramp and Simmons 1980). Courtship displays, performed by both the male and female, include bill dipping, bowing, nibbling, and chasing. Solicitation and copulation take place on solid ground or on a specially constructed platform, but never on the water, as is the case with most other aquatic species (Wood 1974).

In Québec, nesting begins around mid-May. The nest, a platform of aquatic vegetation gathered from the immediate surroundings, is constructed by both adults. It is generally located in emergent growth in a marsh or pond with channels and small areas of open water. It may also be built in a small clump of terrestrial vegetation or on a solid structure such as a fallen log in water (Cramp and Simmons 1980). Well hidden, the nest often has an access ramp, or runway (Wood 1974). Roosting and brooding platforms may be found nearby.

Typical clutch size is 5–9 eggs; clutches of more than 13 eggs are usually the result of brood parasitism, where more than one female lays in the same nest (Cramp and Simmons 1980). Both sexes incubate the eggs and the change over between shifts may be marked by the presentation of additional plant material for the nest (Johnsgard 1975a). Incubation generally lasts 19–22 days. When more than one clutch is produced in a season, hatching is asynchronous in the second and later clutches.

This system creates developmental differences and competition for food among chicks of the same brood. According to Leonard et al. (1988), the adults discourage the older and larger siblings from begging for food to reduce competition. When the older chicks —usually 10–30 days old—come to the parents for food, the parents rebuff them by grabbing them by the head and shaking them roughly. The older the chicks get, the more independent they are forced to become.

The precocial young leave the nest one or two days after hatching. Some may follow their parents, but during the first two weeks of life, they return periodically to a second platform where they are fed and brooded by their parents. At 25 days of age, they can feed themselves, but the parents continue to bring them food until the age of 45 days. The chicks' diet consists of insects and plant matter (Wood 1974). The downy chicks are blackish with a white beard, and a reddish bill with a yellow tip. Age at first flight is 6–9 weeks, and full independence comes at around the 10th week.

Clodin Roy

Notes on territory

Territorial defence, carried out mainly by the male, centres on a nesting site that provides adequate cover against predators. The territory consists of a section of open water edged by vegetation that provides cover.

The most important factor in a male's ability to defend his territory, and consequently, in determing the size of the territory, appears to be his weight: in other words, heavier males can maintain larger

The Common Moorhen inhabits lentic habitats, like this pond with abundant floating vegetation. Sainte-Martine, south of Montréal. *July 1991, Gaétan Duquette.*

Common Moorhen

Breeding distribution

1984–1989

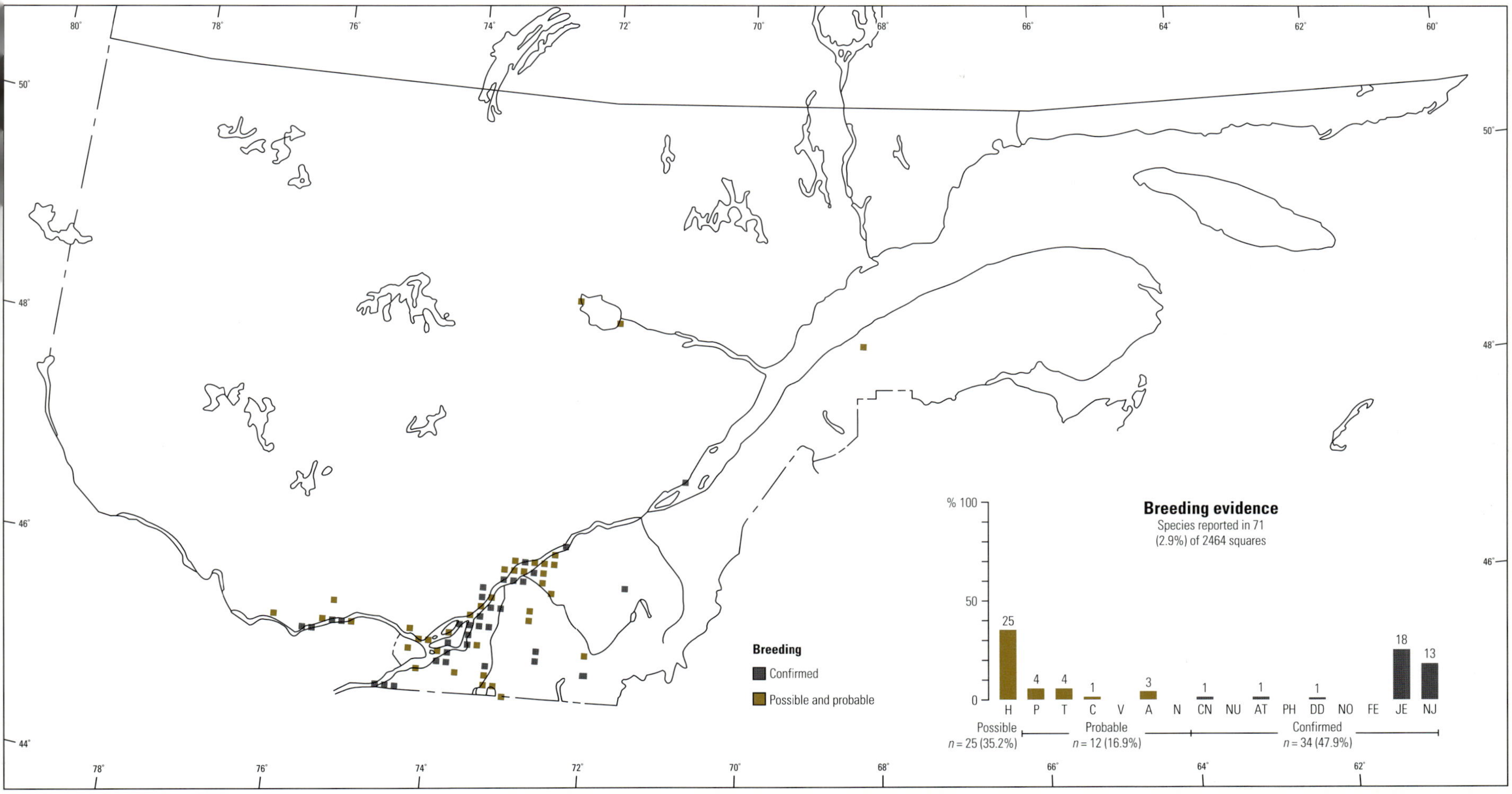

Species profile

Clutch size	usually 5–9 eggs (2–13)[1,2,3]
Laying interval	1 egg per day[1]
Start of incubation	first clutch: with last egg; second and third clutches: when ≈ half the eggs have been laid[1]
Incubation period	19–22 days[1,2,3]
Maturity at hatching	precocial
Care of young	by female and male[1,2,3,4]
Nestling period	usually less than 24 hours[1]
Age at first flight	usually 60–65 days (40–65)[1,2]
Dependence of young	until 72 days old (55–99)[1,3]
Broods per year	usually 1 (1–3)[1,2,3]
Breeding age	1 year[1,3]
Mating system	monogamy[3,4]
Duration of pair bond	1 season[3,4]
Total length	30.5–37.5 cm[7]
Wing span	50.8–58.4 cm[6]
Weight	334 g (?)[8]
Longevity record	10 years 6 months[5]

References:
1. Wood 1974
2. Ehrlich et al. 1988
3. Cramp and Simmons 1980
4. Petrie 1986
5. Clapp et al. 1982
6. Terres 1980
7. Godfrey 1986
8. Dunning 1984

Chronology $n = 147$

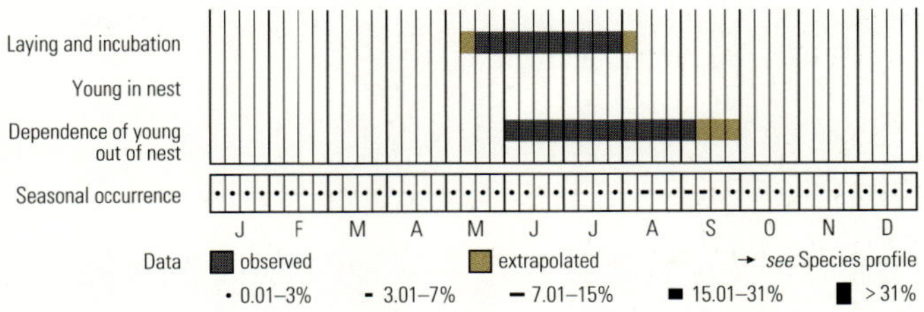

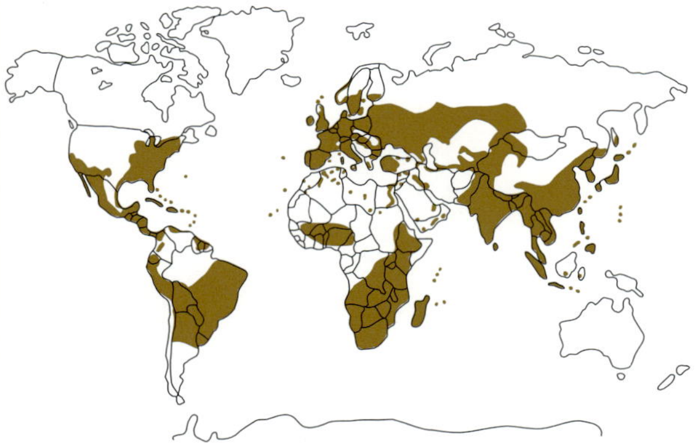

World breeding range

This moorhen's nest is a platform made of dead stems of cattails and other water plants. Lake Saint-Pierre. *June 1989, Jean-Luc DesGranges.*

Notes on territory (continued)

territories and gain access to a greater number of suitable nesting sites (Petrie 1984).

Despite the species' retiring nature, breeding was fairly easy to confirm. The very high percentage of breeding confirmations compared with other species in the family Rallidae—47.9% (37/71) of squares in which the species was reported—is attributable to the fact that the Common Moorhen tends to forage out in the open. The possibility that a second clutch may be laid later in the season also improves the chances of finding young. Most breeding confirmations were reports of chicks off the nest or on open water.

Distribution

According to the Atlas data, the Common Moorhen's distribution in Québec corresponds fairly closely to that described by Godfrey (1986). Since the species prefers fairly deep marshes with submergent and emergent vegetation, it is concentrated in the Central St. Lawrence Lowland, where this type of habitat is abundant. The greatest concentrations appear to be in the lentic freshwater stretches of the St. Lawrence River system, including the Ottawa and Richelieu rivers, Lakes Saint-Louis, Saint-François, and Saint-Pierre, and Lac des Deux Montagnes. Here, the slow currents (Gauthier 1977), numerous bays and coves, alluvial shoreline deposits, and long growing season create a wide variety of wetland habitats (Lands Directorate 1986). In addition, breeding was confirmed at Cap-Tourmente. Common Moorhens were also observed in suitable habitat, though breeding evidence was not obtained, at Saint-Gédéon and on the Ticouapé River in the Lac-Saint-Jean region and at Sainte-Blandine, near Rimouski.

The extensive reed beds along this channel may camouflage the nests of several pairs of Common Moorhens. Saint-Étienne-de-Beauharnois, south of Montréal. *July 1991, Gaétan Duquette.*

This marsh, colonized by arrowheads and other broad-leaved plants, provides ideal nesting habitat for the Common Moorhen. Châteauguay River, near Montréal. *July 1991, Pierre Bannon.*

Freshwater marshes are a defining element in the moorhen's distribution (Root 1988). Accordingly, the Common Moorhen is virtually absent north of the 48th parallel; this includes the Gaspé Peninsula, where the mean elevation is over 400 m above sea level and very little habitat of this type is found. Early in the century, Schmitt (1904) characterized the species as fairly rare in summer on Anticosti Island; this status does not appear to have changed, as atlassers did not observe the species there. Québec is at the northern limit of the moorhen's North American breeding range.

History and trends

Wintle, writing in 1896, listed the species as a common summer resident in Montréal and even more plentiful in fall. Although Wintle did not provide any specific information on nesting, the Common Moorhen appears to have been a well-established breeder in the region at the turn of the century. More recently, Ouellet (1974a) noted that breeding records seem to have declined and have become more local and ascribed the decline to habitat destruction, although its specific effects cannot be pinpointed. The Common Moorhen is still hunted throughout North America, but the number killed is negligible, at least in Québec (A. Bourget, CWS, pers. comm.), partly because it is not considered a particularly tasty table bird, so hunting is not a serious threat to the species.

This local decline must be attributed instead to habitat loss. Destruction of the province's lentic habitats has been concentrated mainly in the stretch of the St. Lawrence between Cornwall and Trois-Rivières, where the human impact on the environment has increased over the last few decades. Between 1945 and 1976, a total of 2625 ha of wetlands were lost in this region, 84% prior to 1961. Close to 50% of the losses resulted from the filling in and clearing of wetlands, while agriculture (cropland and pastureland) accounted for about 33%, and residential construction for 14%. Needless to say, further wetland destruction has occurred since 1978. A recent study showed a loss of 6.7% of the wetland areas in the Montréal region between 1966 and 1981 (Champagne and Melançon 1985). Given the continuing development expected, wetlands are in more trouble than ever (Lands Directorate 1986).

GAÉTAN DUQUETTE
AND MANON BOMBARDIER

Moulting adult. Its large toes help it walk on floating vegetation. Baie-du-Febvre, south shore of Lake Saint-Pierre. *August 1981, Daniel Jauvin.*

American Coot

Foulque d'Amérique

Fulica americana

Clodin Roy

There is something endearingly awkward about the American Coot taking off from a marsh, splattering across the water on its lobed feet and flapping furiously until it is finally airborne. In flight, its long legs trail inelegantly behind its short tail, but it swims with ease, pumping its head like a chicken with each stroke. The coot can dive to depths of 7.6 m and remain underwater for up to 16 seconds (Terres 1980), bobbing to the surface like a cork.

Gregarious outside the breeding season, in migration in the United States, the American Coot is often found amid large flocks of ducks, where it can be identified by its slate-coloured plumage and whitish bill. It winters in southern British Columbia, the United States from the southeast coast of Alaska to Maryland, Central America, and the West Indies (AOU 1983). In the United States, the greatest winter concentrations occur in New Mexico and California (Root 1988). In Québec, the American Coot is a rare, local migrant breeder; the only known winter record for the province came from Philipsburg on 18 January 1975.

Habitat and behaviour

The American Coot is monogamous and a solitary nester, and under optimal habitat conditions can be found in high breeding densities (Lang 1990). The male is highly territorial and chases off ducks and other coots with a cacophony of cackles, clucks, creaks, grunts, snorts, and whistles. Double brooding occurs in some regions (Hill, W. L. 1988). During the breeding period, the species is generally associated with freshwater marshes and ponds of all sizes, but may also nest in the strip of emergent vegetation along lakeshores. In marshes, the nest is usually placed approximately 1.5–30.5 m from open water, where it forages, and 3.5–400 m from the shoreline (Peck and James 1983). The coot does not seem to prefer any specific type of vegetation, although it is often found in marshes dominated by cattails, sedges, bulrushes, and common reeds. According to Sutherland and Maher (1987), the density and height of the emergent vegetation used to provide cover for the nest are key criteria in the selection of a nest site. Coots sometimes delay nesting until the vegetation is high enough to support the nest and be used as nesting material (Fredrickson 1970).

Both the male and female build the nest, one transporting the nesting materials and the other doing the actual construction. The nest is most often made of the leaves and stalks of cattails, umbrella-sedges, and bulrushes and then lined with reeds and grasses. It is often anchored to emergent vegetation (Peck and James 1983). The pair builds several nesting structures, one for laying and incubation and the others for copulation and brood rearing (Gullion 1954). Both sexes share the task of incubating the eggs for 21 to 27 days, with the male more often taking over at night. When the chicks hatch, the parents eat the shells or push them out of the nest. The downy chicks are blackish with bristlelike orange down on the throat and shoulders. Since hatching is asynchronous, the female feeds and tends to the first-hatched chicks and guides them away from the nest, while the male continues to incubate the unhatched eggs.

Notes on territory

The territory is established and defended mainly by the male after pair formation and is used for copulation, nesting, brood rearing, and foraging.

Territories ranged from 0.20 to 0.50 ha in California (Gullion 1953).

During brood rearing, the density and height of emergent marsh vegetation is crucial because of its role in providing cover. Saint-Étienne-de-Beauharnois, south of Montréal. *July 1991, Gaétan Duquette.*

American Coot

Breeding distribution

1984–1989

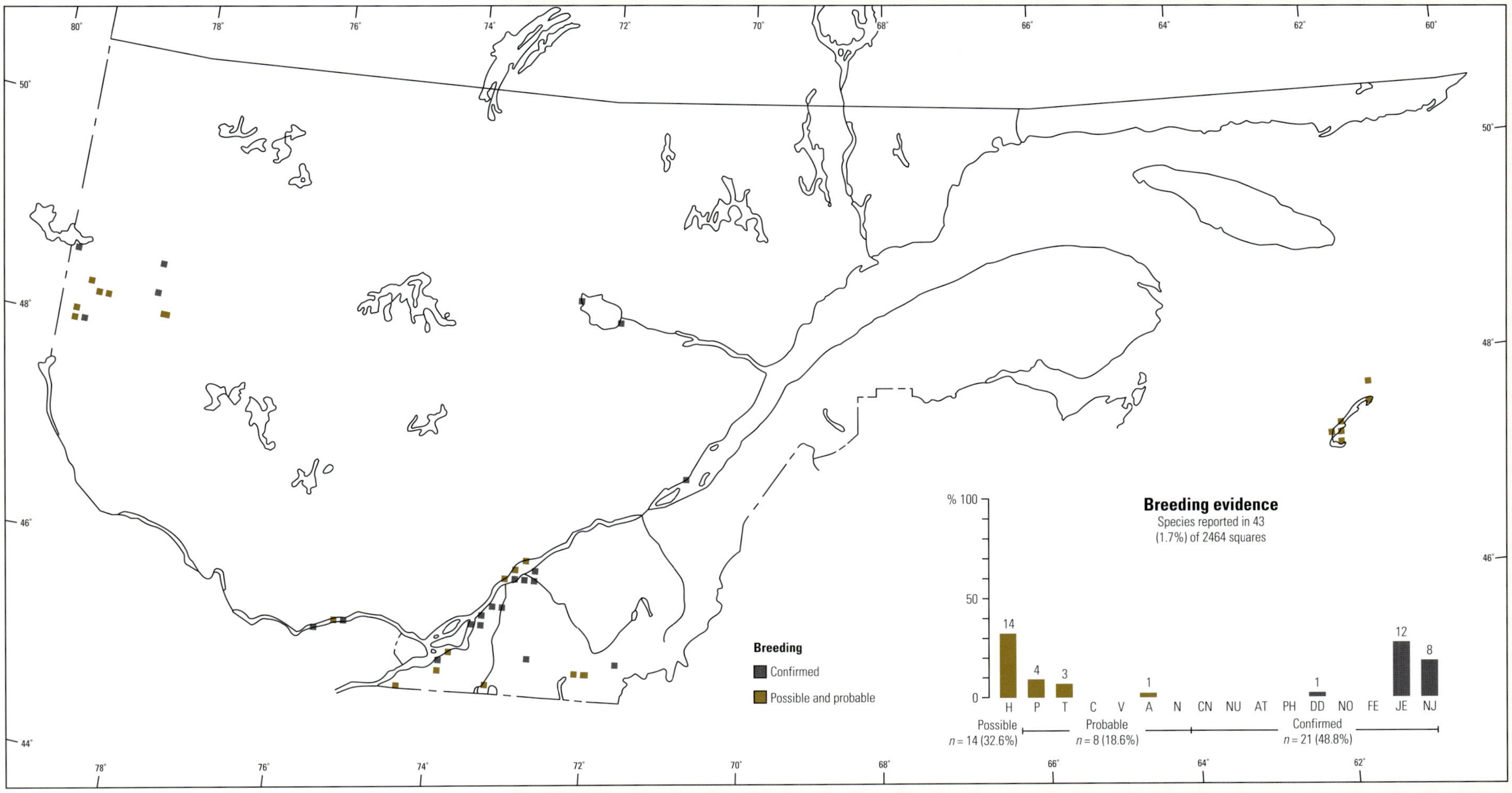

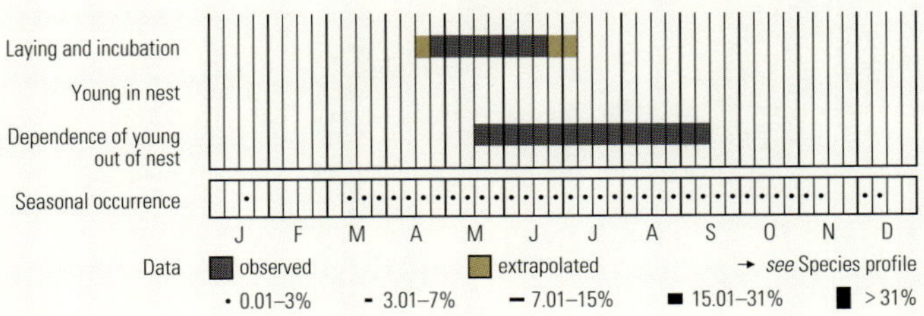

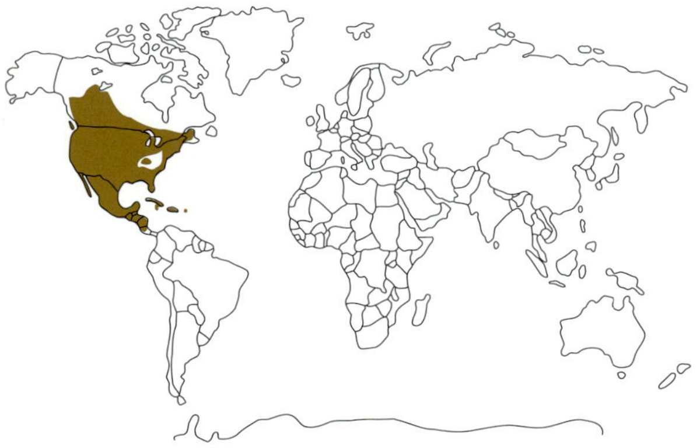

World breeding range

Species profile

Clutch size	usually 8–12 eggs (2–12)[1,3]
Laying interval	1 egg per day[1,2,4]
Start of incubation	first clutch: with second or last egg; second clutch: with first or second egg[2]
Incubation period	usually 23 days (21–27)[1,2,3,4]
Maturity at hatching	precocial
Care of young	by female and male[1,2,3]
Nestling period	usually less than 24 hours[1,4]
Age at first flight	75 days[2]
Dependence of young	until 80 days old[2]
Broods per year	usually 1 (1–2)[2,3,10]
Breeding age	probably 1 year[2,9]
Mating system	monogamy[2,3]
Duration of pair bond	probably for life[2]
Total length	33.0–40.5 cm[5]
Wing span	58.4–71.1 cm[6]
Weight	M 724 g (576–848) / F 560 g (427–628)[8]
Longevity record	22 years 4 months[7]

References:
1. Bent 1926
2. Gullion 1954
3. Ehrlich et al. 1988
4. Fredrickson 1970
5. Godfrey 1986
6. Terres 1980
7. Klimkiewicz and Futcher 1989
8. Dunning 1984
9. DeGraaf and Rudis 1987
10. Hill, W. L. 1988

The chicks follow the hen about, begging for food (Fredrickson 1970). For the first few days of life, they eat almost nothing but insects, but their diet becomes increasingly vegetarian as their digestive systems develop and become able to digest fibrous matter (Jones 1940).

When all the eggs have hatched, the male helps the female feed and rear the chicks. At 30 days of age, the chicks forage on their own, but still remain near the nest. They feed by pecking at the water's surface or diving for food (Desrochers and Ankney 1986). The young make their first flight at the age of 10 or 11 weeks (Gullion 1954) and acquire adult plumage at around 1 year (Bent 1926).

The American Coot is hard to find in Québec in the breeding season, as it is rare and breeds very locally in the province. In addition, its marsh habitat is not easily accessible to birders. It is more often heard than seen, but care must be taken in identifying it by ear, because its wide-ranging vocal repertoire is similar to that of the Common Moorhen. The American Coot was recorded in only 1.7% (43/2464) of the Atlas squares surveyed. Since nests are usually well hidden and may be difficult to distinguish from those of the Common Moorhen, breeding was most often confirmed by observations of chicks that had left the nest and were out on open water.

Distribution

In Canada, the American Coot's range is more northerly than that of the Common Moorhen. The species even nests as far north as Yellowknife, at approximately 63° N (Godfrey 1986). Nevertheless, in Québec and elsewhere it is sensitive to habitat changes, particularly variations in water level. Populations may fluctuate from year to year, and the species may not always return to the same place every year. For example, in the Montréal area, Île aux Fermiers near Verchères would appear to provide ideal nesting habitat for the species. Despite the fact that coots were seen on the island in the breeding season every year from 1983 to 1990, they only nested there twice in that time, in 1985 and 1990.

The American Coot's low frequency of occurrence in southern Québec can probably be attributed to the scarcity of suitable habitat. It nests in marshes and ponds with emergent vegetation and access to open water. This kind of habitat is found mostly in the southwestern Central St. Lawrence Lowland west of Québec City, particularly in the riverine lakes—Lakes Saint-François, Saint-Louis, and Saint-Pierre—and La Prairie Basin. Several factors contribute to the formation of vast wetlands in these aquatic systems, including the type of deposits, a weak current, and shallow water (Couillard and

Notes on territory (continued)

In Iowa, maximum nest density is reached in marsh habitat with equal areas of open water and emergent vegetation (Weller and Fredrickson 1973).

Nests are usually spaced at least 40 m apart (Sugden 1979). Intraspecific aggression by the pair reaches a peak before laying, and interspecific aggression peaks during brood rearing (Ryan and Dinsmore 1979).

The American Coot, which is fairly rare in the province, is found on marshes and ponds of various sizes. Saint-Étienne-de-Beauharnois, south of Montréal. *July 1991, Gaétan Duquette.*

This settling pond in Baie-du-Febvre provides excellent habitat for waterbirds like the coot. It is often seen in the company of Common Moorhens and other species. South shore of Lake Saint-Pierre. *July 1991, Daniel Jauvin.*

Grondin 1986; SOCQ 1988). Most of the squares in which breeding was confirmed were in these areas. Breeding was also confirmed at Cap Tourmente near Québec City, at Lake Saint-Jean, and in the Abitibi. Confirmation of nesting in these last two regions is fairly recent and neither was mentioned in accounts from the early 1980s (David 1980a; Godfrey 1986).

The American Coot was also reported in several squares in the Magdalen Islands, although atlassers did not confirm breeding there. The only evidence of breeding from the Magdalens was an observation of a family group on Île du Cap aux Meules on 9 August 1982. There are also a few pre-Atlas records from the Lower St. Lawrence, the Gaspé, and the North Shore (Todd 1963; David 1980a). The species was not atlassed on Anticosti Island, where it was considered a rare summer visitor at the turn of the century (Schmitt 1904) and casual in the late 1960s (Ouellet 1969).

History and trends

The American Coot was once one of the most abundant species of waterbirds in North America (Forbush 1912). The population seems to have declined around 1910, probably due to overhunting and a significant loss of wetlands in its breeding range (Fredrickson 1977). In the early and mid-1960s, it was also greatly affected by droughts in the United States (Eaton 1988b).

The American Coot is not a very popular game species, and between 1989 and 1991, fewer than 5000 birds were shot annually in Québec (Lévesque et al. 1993).

Wintle (1896) listed the species as scarce around Montréal, adding that a few probably bred on Lake Saint-Pierre. Lloyd (1923) considered the American Coot a common summer resident in the Ottawa Valley, while Dionne (1906) said it was not uncommon near Québec City. Mousley (1916a) did not mention the species in his paper on the birds of Hatley. According to Ouellet (1974a), the status of the American Coot has not changed a great deal in the Montréal area since the beginning of the century. This point is difficult to confirm, since there are not enough quantitative data to assess populations trends, even on a short-term basis.

Gaétan Duquette and Manon Bombardier

Semipalmated Plover

Pluvier semipalmé

Charadrius semipalmatus

Ghislain Caron

THE SEMIPALMATED PLOVER IS AN EASILY IDENTIFIABLE SMALL SHOREBIRD with a prominent single black breastband, a black-tipped yellow bill, and orange-yellow legs. As its name indicates, this species has partially webbed toes, an uncommon feature in shorebirds (Hayman et al. 1986). On the breeding grounds, these birds are easy to spot, as one or both parents anxiously fly about at the approach of an intruder, giving out a piping *chee-wee, chur-wee* or *tyoo-eep* call, rising on the second note (Terres 1980).

In Québec, this plover is associated with the North Shore and coastal and inland regions from southern James Bay to the northern tip of the Ungava Peninsula during the breeding season (Godfrey 1986). Migrating in spring by both day and night (Townsend 1929b), this species arrives here in mid-May. It is a common fall transient in southern Québec along the shores of lakes and rivers, including the St. Lawrence, migrating through in a broad front (Townsend 1929b; Cramp and Simmons 1983; Bourget 1990). The Semipalmated Plover winters on the Atlantic coast from South Carolina to Florida, on the Pacific coast from Washington to California, and along both coasts of South America as far south as central Chile and Argentina. It is also found in the Galápagos Islands and in Trinidad and Tobago (AOU 1983; Root 1988). Non-breeders do not migrate and remain in these latitudes all summer (Townsend 1929b).

Habitat and behaviour

On the Québec side of James Bay and northward, the Semipalmated Plover nests on sand or gravel bars in rivers, high on rocky beaches (R. Benoit, pers. comm.), along coastal beaches, and inland on the heath-lichen tundra of the Little Whale and Great Whale river watersheds (Cons. G. G. – GREBE 1993c). In Ontario, it nests on the shores of ponds, lakes, and rivers (Peck and James 1983). On the tundra it favours sparsely vegetated raised beach ridges. It has also adapted to populated areas in the north by nesting on gravel margins of roads, abandoned airstrips, and other gravel surfaces (Johnsgard 1981; Hussell 1987c).

Semipalmated Plovers are monogamous, with the pair bond forming after the birds arrive on the breeding grounds. The males perform an elaborate aerial display: a slow, buoyant flight during which the bird alternates between a chattering *too-wat-wat-watwatwat* vocalization that increases in tempo, and a prolonged, liquid trill (Colston and Burton 1988). In displays on the ground, the male and female face off and run at each other like fighting cocks or strut side by side, chattering (Townsend 1929b). Pairs may nest in loose colonies in suitable habitat (Terres 1980). The nest, which may be lined with a little plant material, is typically a small depression in sand or gravel, or sometimes in moss and dead seaweed, usually in areas with sparse vegetation (Townsend 1929b; Johnsgard 1981; Godfrey 1986; Fradette 1992). The male builds the nest alone, but both sexes incubate the eggs for 23–25 days (Ehrlich et al. 1988). The precocial chicks leave the nest almost immediately after hatching and are reared by both parents, fledging at about 3 weeks of age (Terres 1980; Johnsgard 1981).

While the Piping Plover's plumage is the colour of dry sand, that of the Semipalmated is the shade of wet sand. When the bird is motionless, its outline gives it a particular camouflage, mimicking an empty, washed up bivalve shell (Townsend 1929b).

The diet of this species consists of molluscs, crustaceans, worms, and insects (Terres 1980; Ehrlich et al. 1988). It usually forages at low tide on muddy, sandy, or shingle beaches (Johnsgard 1981; Michaud and Ferron 1986), feeding in typical plover fashion, that is, running about on the beach or mudflat with head up, dabbing suddenly at the ground. It stands motionless for much of the time while feeding, as it locates its prey mainly by sight (Terres 1980; Michaud and Ferron 1986). While stationary, this plover sometimes uses its feet to bring prey to the surface (Johnsgard 1981; Colston and Burton 1988).

Notes on territory

Both parents defend the territory, the size of which is unknown. When threatened, a male in charge of a brood seems to respond less nervously than a female; the female keeps up her protests for some time after the intruder has left the territory (Townsend 1929b). When disturbed near its nest, the Semipalmated Plover tries to distract intruders by feigning injury (Hussell 1987c).

On the Middle and Lower North Shore, the Semipalmated Plover nests on gravel beaches along the coast and on islands. Sainte-Marie Islands, Lower North Shore. *July 1991, Pierre Brousseau.*

Semipalmated Plover

Breeding distribution

1984–1989

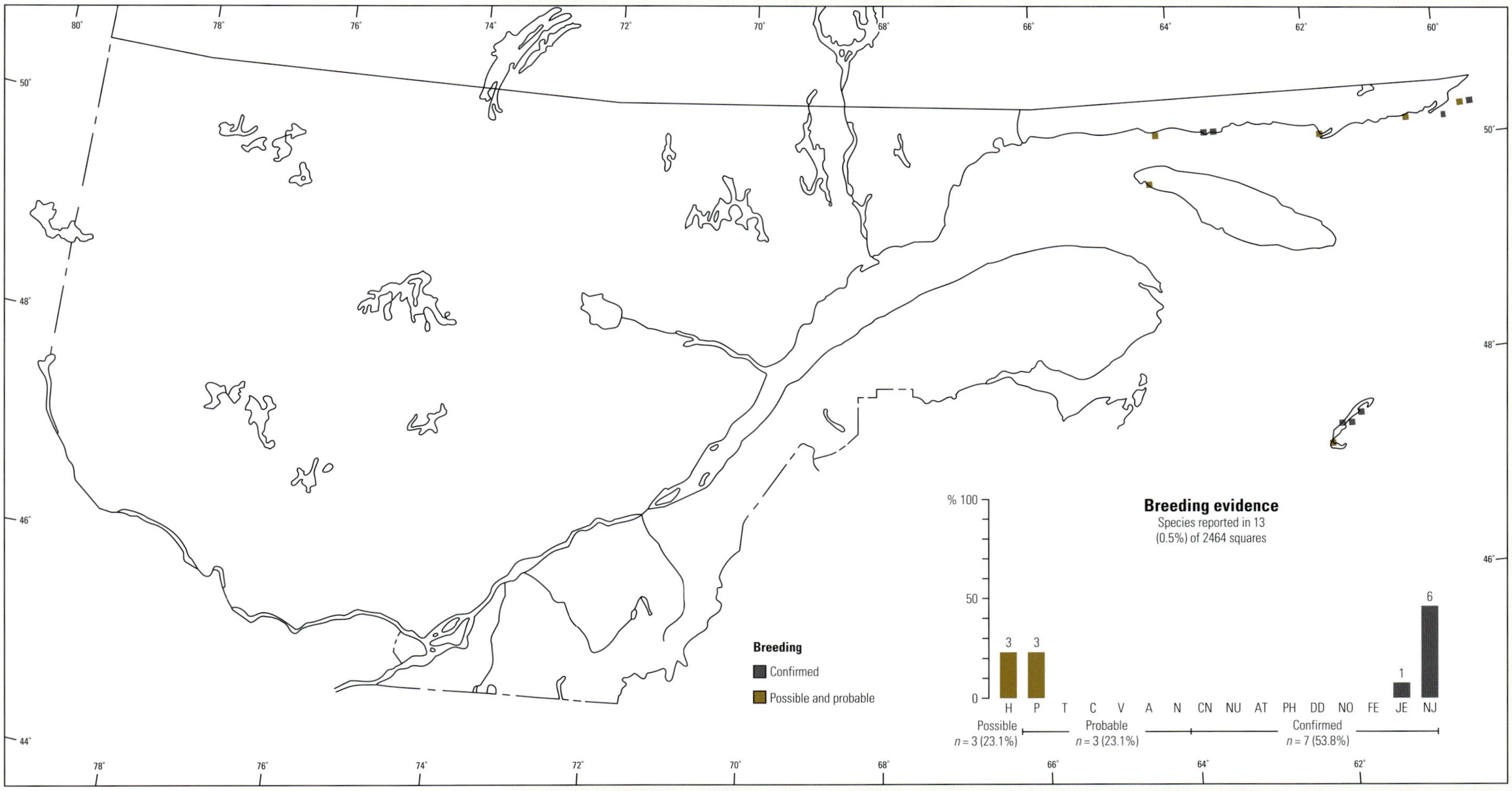

Species profile

Clutch size	usually 4 eggs (3–4)[1,2,4]
Laying interval	?
Start of incubation	probably with last or penultimate egg[5]
Incubation period	23–25 days[1,2]
Maturity at hatching	precocial
Care of young	by female and male[2]
Nestling period	less than 24 hours[1,4]
Age at first flight	probably 22–31 days[5]
Dependence of young	?
Broods per year	1[2]
Breeding age	?
Mating system	monogamy[2,4]
Duration of pair bond	?
Total length	16.5–19.5 cm[7]
Wing span	43–52 cm[3]
Weight	M 47.4 g (37.6–57.4) / F 46.1 g (39.2–56.5)[8]
Longevity record	8 years 2 months[6]

References:

1. Terres 1980
2. Ehrlich et al. 1988
3. Cramp and Simmons 1983
4. Bent 1929
5. Johnsgard 1981
6. Clapp et al. 1982
7. Godfrey 1986
8. Dunning 1984

Chronology *n* = 37

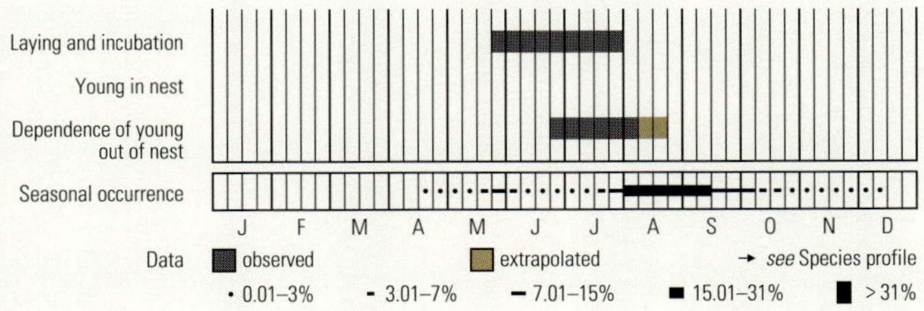

World breeding range

Semipalmated Plover's nest of twigs and scraps of dry vegetation. Shore of the Bay of Many Islands, James Bay. *June 1990, Austin Reed.*

The Semipalmated Plover is easy to spot when its nest or brood is approached, as it becomes highly agitated, bobbing its head up and down and uttering an alarm call (Hussell 1987c). In spite of this very conspicuous behaviour, the nest and eggs are so well camouflaged that they are difficult to find. Observers working in pairs can make finding nests easier: one observer leaves the territory while the other stands still until the bird returns and settles back on its nest.

Distribution

The Semipalmated Plover breeds primarily in the Low Arctic and taiga of Canada and Alaska. In northern Canada, it is found from Yukon to Labrador. In Québec, it breeds both on the tundra and in the taiga from Hudson Strait to Schefferville and Lake Bienville (Macoun 1900; Manning 1949; Savard 1977). It is found along the coasts of Hudson Bay and James Bay (Manning 1952), and recent surveys have confirmed breeding on the northeast shore of James Bay and the adjacent islands, where 21 nests were discovered in 1991 and 39 in 1992 (Benoit et al. 1993b). The species also breeds on the Middle and Lower North Shore, Anticosti Island, and the Magdalen Islands (Godfrey 1986; Fradette 1992).

The Semipalmated Plover's breeding range, therefore, lies largely north of the Atlas area. The species was reported in only 13 of the 2464 Atlas squares surveyed. In 6 of the 7 squares where breeding was confirmed, nests containing eggs or young were reported, and in the remaining square, young were seen off the nest. Breeding was confirmed only in the Magdalen Islands and along the North Shore of the Gulf of St. Lawrence between Mingan and Saint-Augustin. On Anticosti Island, the Semipalmated Plover was reported in only one square, at the western tip of the island, in suitable breeding habitat.

Semipalmated Plovers arrive on the breeding grounds around the third week of May, reaching the northern limits two or three weeks later (Cramp and Simmons 1983; Erskine 1992; Fradette 1992). In the fall migration, the adults leave the breeding areas before the juveniles. Thus, in the Magdalen Islands, the greatest concentrations of adults are found between 31 July and 30 August, whereas the juveniles peak between 1 September and 10 October (McNeil and Burton 1973; Mousseau et al. 1976; Fradette 1992). During fall migration is the best time for observing large flocks of Semipalmated Plovers in the St. Lawrence estuary and gulf (Maisonneuve 1982; Maisonneuve et al. 1990; Bourget 1990, 1991). Fall migration takes the birds to the Caribbean and northeastern South America where they overwinter (McNeil and Burton 1977). Records of banded birds picked up in the Gulf of St. Lawrence show that individuals tend to follow the same migration route each year (McNeil and Burton 1973).

The Semipalmated Plover also nests in low vegetation. Island in the Bay of Many Islands, James Bay. *June 1990, Yves Aubry.*

Semipalmated Plover's nest on a patch of gravel at Lake Pau, near the main Caniapiscau dam. The eggs are laid in a shallow depression lined with scraps of plant matter. North-central Québec. *July 1981, Yves Aubry.*

Raised beach ridge, here on an island in the Bay of Many Islands, provides suitable breeding habitat for this shorebird. James Bay. *June 1990, Yves Aubry.*

History and trends

As well as many other shorebird species, populations of Semipalmated Plovers in eastern North America were drastically reduced in the last decade of the 19th century by market hunters. Populations began to recover after 1916 with the signing of the Migratory Birds Convention. By 1925, populations had rebounded to the point that flocks of several hundred were commonly observed in migration (Townsend 1929b), and since then, numerous authors have reported populations to be stable.

No changes have been detected in this species' breeding range in recent years (Hussell 1987c) and the Maritimes are considered its southern limit (Erskine 1992). The recent discovery of 2 Semipalmated Plover nests in Oregon (Ivey et al. 1988; Ivey and Baars 1990), however, has generated some interest.

Although this shorebird is in no immediate danger (Johnsgard 1981), there are fears for its survival, particularly due to habitat loss, mainly on its wintering grounds, but also on its breeding grounds as well (Senner and Howe 1984). The Semipalmated Plover is thought to be at less risk than the Piping Plover, since the gravel beaches where it nests are subject to much less recreational traffic (Erskine 1992). As the Semipalmated Plover population is relatively stable, the slightest change, whether caused by natural or artificial factors, should be easily detected (Erskine 1992).

Richard C. Cotter and Pierre Lafontaine

Piping Plover

Pluvier siffleur

Charadrius melodus

In Québec, elsewhere in Canada, and in the Great Lakes region of the United States, the Piping Plover is listed as endangered (Haig 1985; Sidle 1985; Robert 1989b). In other parts of the United States, it is classified as threatened.

Habitat and behaviour

This small sand-coloured shorebird is well camouflaged on the beaches it haunts. It is confined, at least in eastern North America, to broad, gently sloping beaches, especially those strewn with pebbles and shells (Cairns 1982; Burger 1987; Shaffer and Pineau 1987; Prindiville Gaines and Ryan 1988).

The very rudimentary nest is no more than a shallow depression in the sand. The eggs, typically numbering 4, are often laid on white shell fragments. Nests are also sometimes located in dune vegetation. Incubation is shared by both adults and lasts about 28 days. The eggs are therefore exposed to the risk of predation for quite some time. As in a number of other shorebird species, the adults vigorously defend both nest and young, luring predators away with cries and feigned injuries. The Piping Plover's main predators are foxes, raccoons, skunks, ravens, crows, and several species of gulls. In the Magdalen Islands, avian predators and red foxes are a real threat to this plover.

The young are precocial and nidifugous, leaving the nest within a few hours of hatching. They begin feeding on their own right away, dining on marine worms, crustaceans, molluscs, and insects (Tyler 1929a; Cairns 1977), including many species of flies and beetles, chiefly rove beetles and snout beetles (Shaffer and Laporte 1994).

The young fledge at about 25 days. Until then, at least one of the adults stays with them. Soon after the brood has been reared, the Piping Plover heads for its wintering grounds. In Québec, the birds have started leaving their nesting areas in the Magdalen Islands by mid-July (Shaffer and Laporte 1992a).

This migratory plover overwinters along the east coast of North America from North Carolina to Florida and along the Gulf of Mexico (Haig and Oring 1988a), as well as on some Caribbean islands (Haig and Plissner 1993). Monitoring of banded birds has shown that the journey between the Magdalen Islands and the west coast of Florida takes about a month in either direction (Shaffer and Laporte 1992a).

The birds tend to return to broadly the same areas each year (Wiens and Cuthbert 1988), but pairs may not always stay together. Sometimes, too, birds whose nests have been destroyed form new pairs within the same breeding season (Haig and Oring 1988b).

The Piping Plover's behaviour is as unobtrusive as its plumage, making it difficult to find. Though breeding was confirmed in 11 of the 12 squares in which the species was reported, this was not because nests are easily found, but rather because there was a research program on the Piping Plover in progress at the time (Shaffer and Laporte 1992a).

Ghislain Caron

Distribution

The world Piping Plover population was calculated at 5482 birds in 1991 (Haig and Plissner 1993). In 1994, 100 individuals (50 pairs) nested in Québec, in the Magdalen Islands (Roy 1994). At present, the Magdalens are the only home to a population of this species in Québec. It has been known to nest in the Gaspé and along the Lower North Shore, yet a full census of potential sites in these two regions taken in 1988 found not a single Piping Plover (Demers and Laporte 1988). When all known potential sites throughout the species' North American breeding and wintering ranges were visited as part of the 1991 International Piping Plover Census, the species was found nowhere in Québec but the Magdalen Islands (Laporte and Shaffer 1994).

Only one Atlas record came from the Lower North Shore, where a family was sighted at Chevery in 1986. The species was not seen in the Gaspé, an area it had abandoned several years earlier and where the last breeding record dates from the summer of 1973 at Carleton (Robert 1989b; Laporte and Shaffer 1994).

Notes on territory

Throughout the breeding season, the Piping Plover defends a territory that varies in size from 0.05 to 0.8 ha (Cairns 1982). Both mates help defend it, though the male is the more active. In defending its territory, the bird adopts a horizontal posture, with the neck stretched forward and the nape and back feathers raised. It chases intruders both running and flying.

The long beaches of the Magdalen Islands are the only remaining home of the Piping Plover in the province. La Martinique beach. *September 1982, Yves Aubry.*

Piping Plover

Breeding distribution
1984–1989

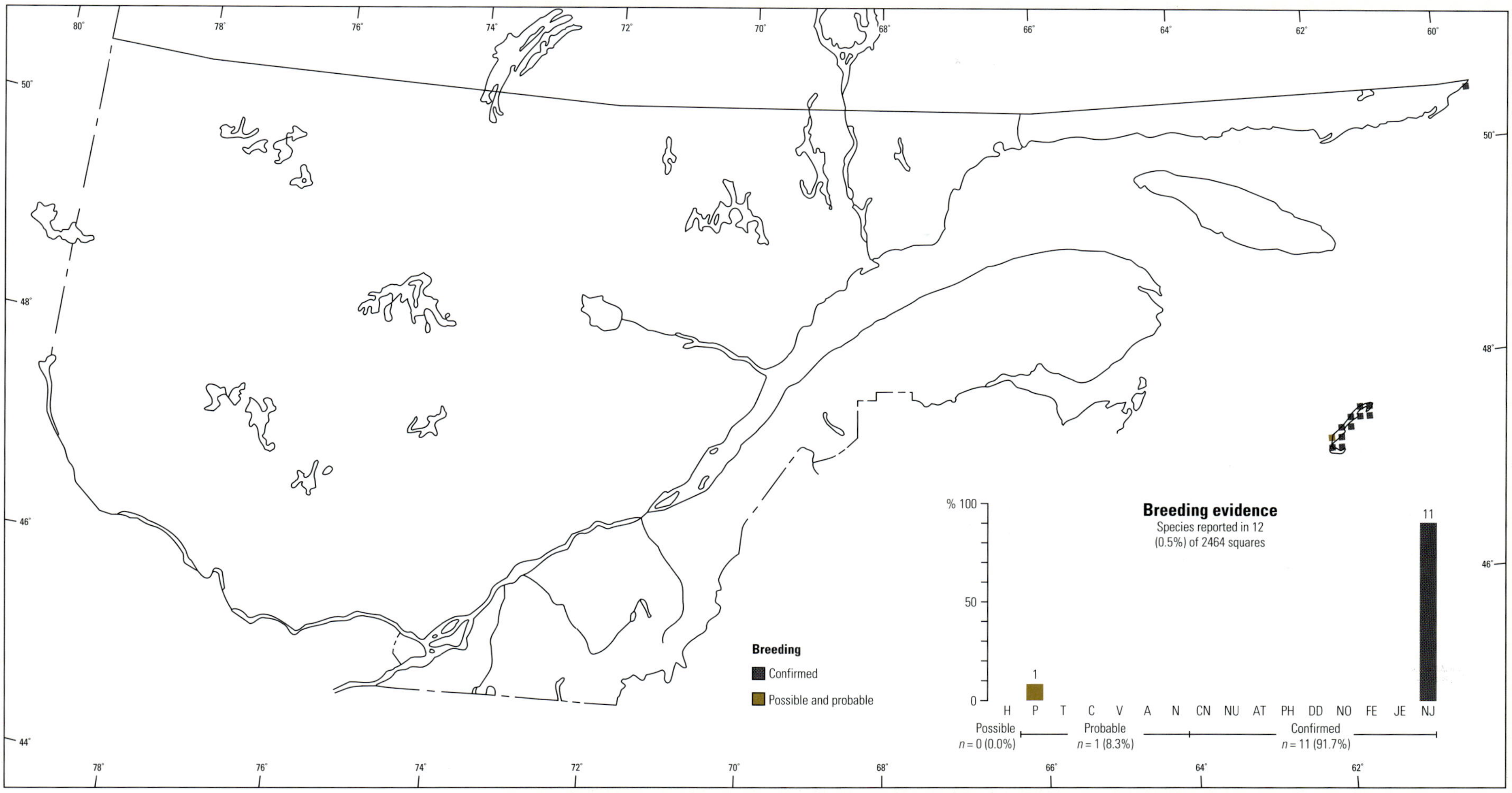

Species profile

Clutch size	usually 4 eggs[1,8,10]
Laying interval	1 egg per 2 days[2]
Start of incubation	with last egg, sporadically after third egg[9]
Incubation period	≈ 28 days[2,7,8,9,10]
Maturity at hatching	precocial
Care of young	by female and male[8]
Nestling period	a few hours[1,8,10]
Age at first flight	21–35 days[1,8]
Dependence of young	until 21–25 days old, sometimes until autumn[8]
Broods per year	1[8,10]
Breeding age	1 year[8,10]
Mating system	monogamy, sometimes polyandry[8,11]
Duration of pair bond	usually 1 season[7,10]
Total length	15.0–19.5 cm[6]
Wing span	35.6–38.7 cm[3]
Weight	55.2 g (46.4–63.7)[4]
Longevity record	14 years[5]

References:
1. Shaffer and Laporte 1992a
2. Wilcox 1959
3. Terres 1980
4. Dunning 1984
5. Clapp et al. 1982
6. Godfrey 1986
7. Haig and Oring 1988b
8. Haig 1992
9. Cairns 1977
10. Goossen 1989
11. Ehrlich et al. 1988

World breeding range

Chronology *n* = 155

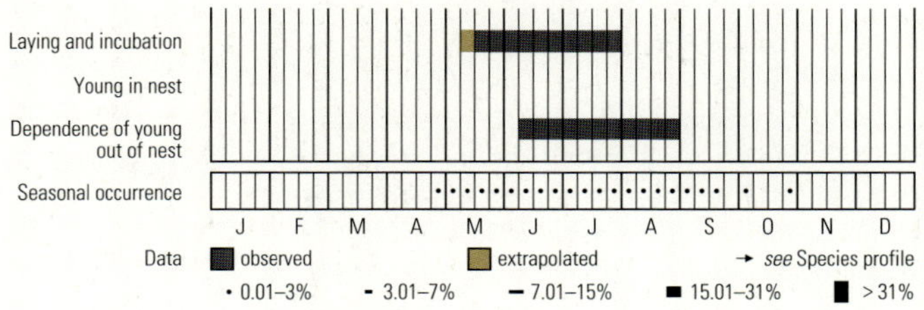

This plover prefers beaches strewn with shells and pebbles. In the lower right corner, a nest with 4 eggs can be seen. Magdalen Islands. *July 1983, Yves Aubry.*

Notes on territory (continued)

Once territories are established, the birds face off at the boundaries, standing erect, and stretching the neck to emphasize the band across the breast. They utter a series of increasingly rapid calls, finishing on a guttural note.

There is some evidence to suggest that the species also defends a feeding territory, either distinct from or included in the breeding territory, both on the breeding grounds (Cairns 1977) and on the wintering grounds (Johnson and Baldassarre 1988).

Piping Plover pairs are found on almost all the beaches of the Magdalens, though overall they occupy only some 5% of 130 km of beach (Shaffer and Laporte 1989). They also nest on two small artificial islets in Grande Entrée Lagoon. The islets were created from dredge spoil between 1980 and 1982, and in 1983 Piping Plovers were sighted there.

History and trends

There are not enough historical data to determine the precise scale of the decline of the Québec Piping Plover population, and specifically that of the Magdalen Islands. We do know that the first record of the species in Canada was made by Audubon, the famous ornithologist. He reported that on 11 June 1833, when he was visiting the Magdalens, several pairs of this plover were incubating (Audubon 1840), but we have no idea of the size of the islands' population at the time.

Many subsequent visitors to the Magdalens between 1873 and 1958 regarded the Piping Plover as a common breeder (Fradette 1992; Laporte and Shaffer 1994). Up to 1979, population estimates ranged from 50 to 70 pairs (McNeil et al. 1973; Mousseau et al. 1976). Two surveys, in 1979 and 1983, produced estimates of 30 and 20 pairs (Cairns and McLaren 1980; David 1983), which were subsequently revised to 26 and 23 pairs respectively. A series of surveys have shown that between 1987 and 1994, the population fluctuated between 35 and 50 pairs (Shaffer and Laporte 1992a,b; CWS, unpubl. data).

Number of Piping Plover breeding pairs in the Magdalen Islands, 1979–94

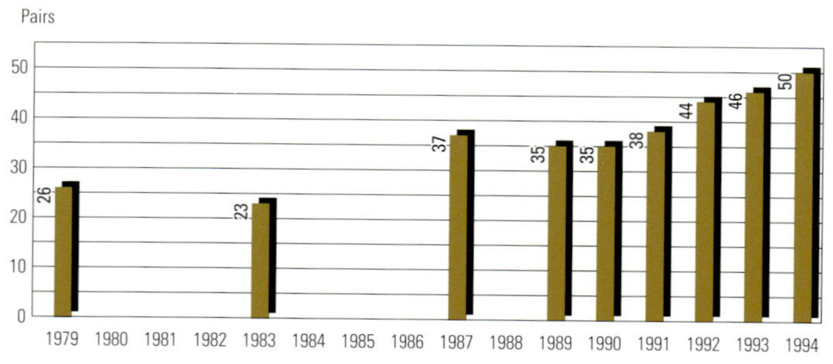

1979: Estimate based on data from W. Cairns
1983: Estimate based on data from Y. Aubry
1987–94: Full counts

The population thus appears to have declined over the last few decades. It must be pointed out, however, that prior to 1979, mere estimates were made, not accurate counts. Whatever the case, it is highly likely that the population has shrunk, as it has everywhere else in North America, though the extent of the drop is hard to gauge. Systematic censuses since 1987 in the Magdalen Islands suggest the beginning of a recovery for the population there, as can be seen from the figure.

The Piping Plover used to be seen regularly in the Gaspé and along the Lower North Shore, though in small numbers (Laporte and Shaffer 1994). These areas are at the northernmost edge of its breeding range on the east coast of North America, and because the beaches there are narrow and broken, their potential as habitat is limited.

There are many reasons for the decline of the Piping Plover in North America. Some of them are natural—storms, tides, and predation—but since the late 1940s, the world Piping Plover population has declined markedly, as a result of greater human pressure on beaches (Haig 1985). It has been shown by Flemming et al. (1988) that higher use of beaches indirectly increases the mortality rate of young plovers, as they spend

The Piping Plover's nest is a depression in the sand among the pebbles. The eggs are often laid on bleached shell fragments, where their cryptic colouring and positioning make them hard to see. Magdalen Islands. *July 1983, Yves Aubry.*

With its sand-coloured plumage, this incubating adult blends right into the background. Magdalen Islands. *July 1982, Yves Aubry.*

more time anticipating danger than searching for food. In addition, larger populations of predators have led to greater losses of nests and young (Haig 1985). In some regions, the Prairies for example, water level regulation has resulted in habitat loss and nest destruction. Disturbance on wintering grounds, notably by beachfront residential development, has eliminated much habitat and contributed to the decline of the species (Haig 1985).

In the course of a 1987–89 study conducted in the Magdalen Islands, nearly 10% of nests were destroyed each year by human activity (Shaffer and Laporte 1989), the worst being without doubt the use of motorized vehicles on beaches. In 1991 and 1992, no nests were lost in this way, thanks to protective measures and a public-awareness program. There is therefore room for hope that the situation will improve if such measures are maintained.

François Shaffer and Pierre Laporte

One of the causes of the Piping Plover's decline is human disturbance, which forces the young to spend more time on the lookout for danger than feeding, thus reducing their chances of survival. Motorized vehicles on beaches also destroy many eggs and young birds. Magdalen Islands. *June 1983, Yves Aubry.*

Killdeer

Pluvier kildir

Charadrius vociferus

Ghislain Caron

THE KILLDEER IS THE ONLY SHOREBIRD IN QUÉBEC THAT HAS TWO DARK breastbands. This plover is the earliest North American shorebird to arrive on its breeding grounds and is recognized by its shrill *kill-dee* call uttered both on the ground and in flight at the approach of danger. Its well-known broken-wing display features its bright orange-brown rump and upper tail coverts.

Killdeers are very widely distributed and are the best known of the North American shorebirds. While most other shorebirds are long-distance migrants, the Killdeer is a permanent resident throughout much of the southern part of its range, from northern South America to the central United States, as far as southern New England on the Atlantic coast and southern British Columbia on the Pacific (AOU 1983; Root 1988). In Québec, however, it is a common migrant breeder.

Habitat and behaviour

Killdeers breed in open habitat with little or no vegetation, such as grazed meadows, mowed pastures, ploughed fields, golf courses, roadside ditches, and similar disturbed areas (Godfrey 1986; Colwell and Oring 1990). Nests are often far from water. Killdeers sometimes nest on the roofs of buildings (Pickwell 1925; COQ 1978).

The nest is a shallow scrape in soft ground. Nest sites have little or no vegetation, but small stones and small plant stems are sometimes used as nesting materials (Johnsgard 1981).

Killdeers are usually monogamous, and pairs stay together for more than one breeding season, although males have a greater tendency to return to the same breeding sites than females (Lenington and Mace 1975). The male builds the nest and helps incubate the eggs and care for the young. Parents brood their chicks, guard against enemies, and warn of danger, but never feed the young (Davis 1943; Brunton 1988). Although the parents are capable of caring for the brood singly, both participate to provide greater protection against predators and prevent chicks from straying from the family (Lenington 1980).

A pair of Killdeers may raise 2 broods in a single season; double brooding is not restricted to the southern part of its breeding range, having been observed in Michigan (Nickell 1943) and Ontario (Peck and James 1983). The second clutch is laid before the first brood has fledged (Lenington 1980). Sometimes the female abandons the male and their offspring, and then nests again with a new mate in the same breeding season (Brunton 1988).

Killdeers may be hard to spot in some circumstances because of their disruptive coloration of alternating bands of white and black that break up the body outline against the background (Ehrlich et al. 1988), but their open habitat facilitates detection. They are conspicuous and noisy, often calling loudly when an intruder enters their territory, so they were rarely missed by atlassers.

The Killdeer was confirmed as a breeder in 60.3% (711/1179) of the squares in which it was reported, demonstrating that it is relatively easy to verify whether the birds are nesting where they are discovered. Reports of young accounted for 36.6% (260/711) of breeding confirmations, while false brooding, injury feigning and other distraction displays (Brunton 1986) or nest discoveries were the evidence confirming breeding in 32.2% (229/711) and 31.2% (222/711) of squares respectively. Distraction displays near the end of the breeding season can sometimes be misleading because Killdeers may continue to perform them after they have moved considerable distances from the nesting areas (Nol 1987).

Notes on territory

Killdeers defend territories by vocalizing every few seconds for hours as they stand in a conspicuous spot. They also run and fly short distances, or circle high over an area a kilometre or more in diameter, uttering their kill-dee call. On the ground, they often perform nest-scraping displays while trilling. They allow other Killdeers on the territory when the appearance of an intruder stimulates collective defensive behaviour (Johnsgard 1981). In winter, they forage in loose groups, but roost in large flocks (Stiles and Skutch 1989).

The Killdeer nests in open areas like this vacant lot with its very sparse vegetation. La Prairie, near Montréal. *July 1991, Gaétan Duquette.*

Killdeer

Breeding distribution
1984–1989

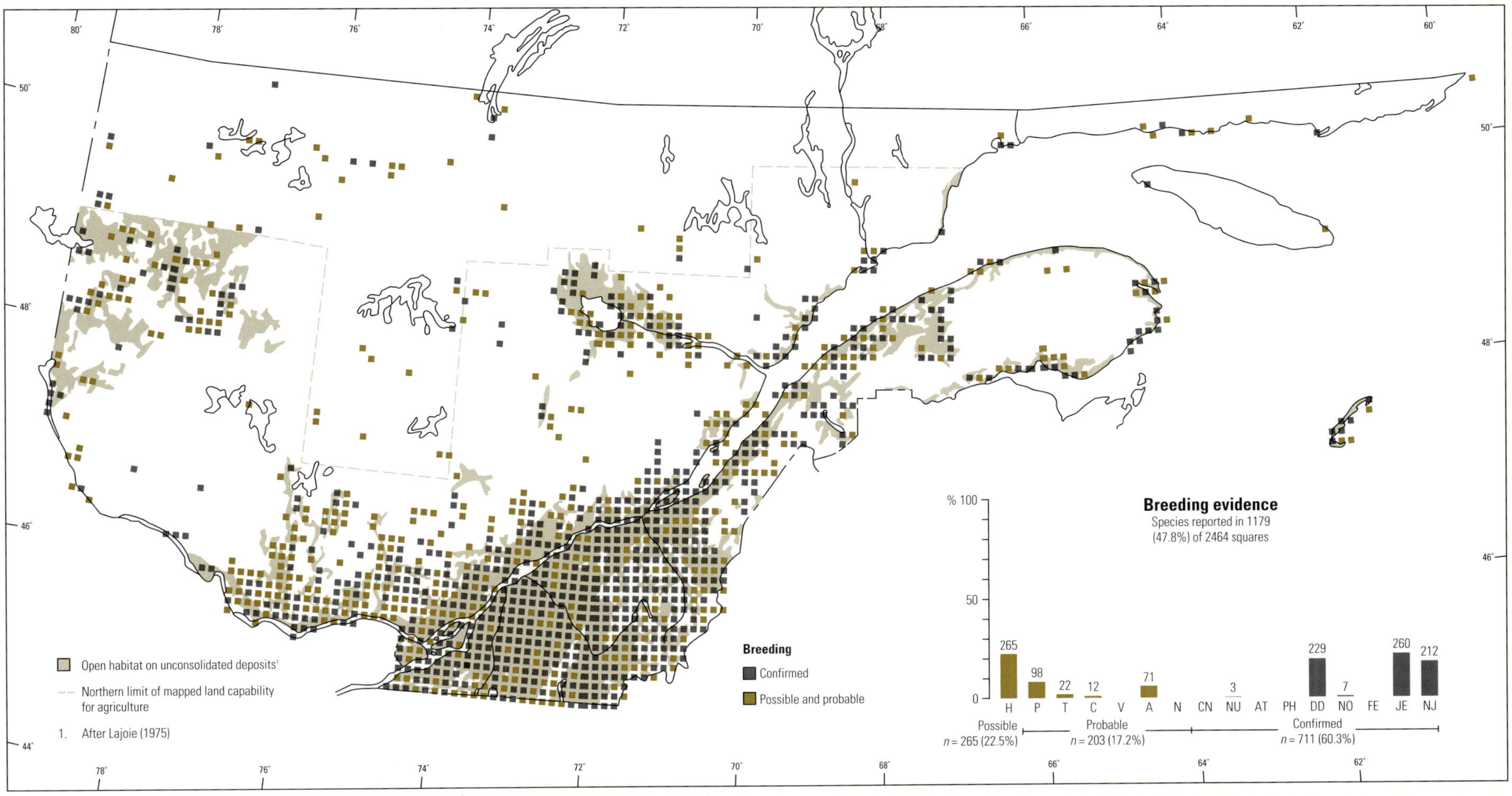

Chronology *n* = 640

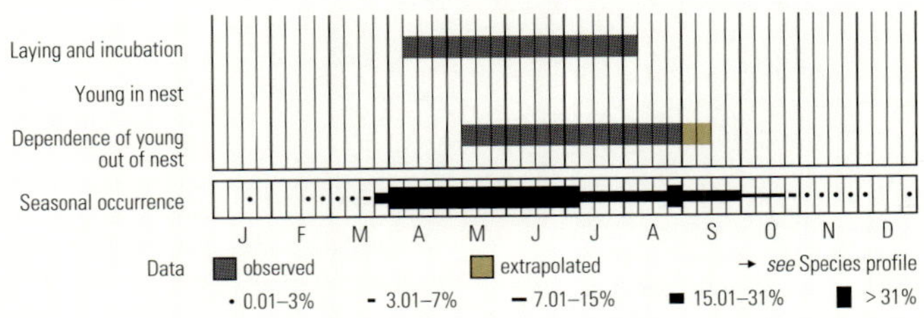

Species profile

Clutch size	usually 4 eggs (3–5)[1,2,3,5]
Laying interval	1 egg per 2–4 days[1,7]
Start of incubation	with last egg[5]
Incubation period	24–29 days[1,2,3,5]
Maturity at hatching	precocial
Care of young	by female and male[4,8]
Nestling period	less than 24 hours[1,3,6,7]
Age at first flight	25–40 days[1,4,6]
Dependence of young	until 39–43 days old, sometimes before[1,5]
Broods per year	1–2[1,2,3,5,6]
Breeding age	1 year[3]
Mating system	monogamy, rarely polyandry[4,8]
Duration of pair bond	usually more than 1 season[9]
Total length	23.0–28.5 cm[10]
Wing span	48.3–53.3 cm[2]
Weight	M 92.1 g (83.9–109.0) / F 101.0 g (87.7–121.0)[12]
Longevity record	10 years 11 months[11]

References:
1. Johnsgard 1981
2. Terres 1980
3. DeGraaf and Rudis 1987
4. Ehrlich et al. 1988
5. Nickell 1943
6. Harrison 1978
7. Davis 1943
8. Brunton 1988
9. Lenington and Mace 1975
10. Godfrey 1986
11. Clapp et al. 1982
12. Dunning 1984

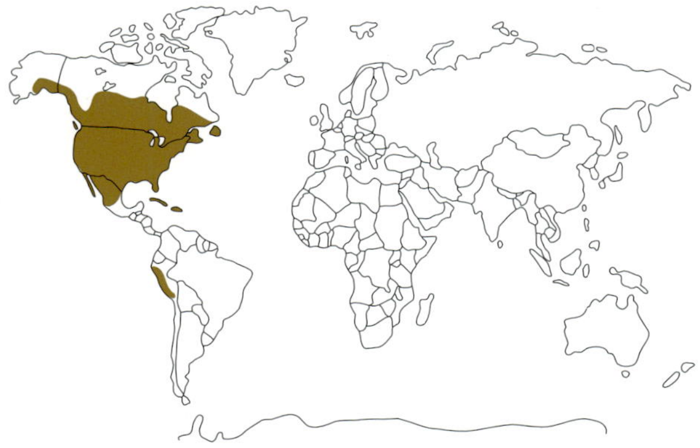

World breeding range

Killdeers are largely insect eaters, often preying upon many of the worst pests, like grasshoppers, locusts, beetles, caterpillars, and ticks (Townsend 1929a), but they sometimes eat frogs, minnows, and possibly other vertebrates (Schardien and Jackson 1982). They forage in the usual plover manner of quickly running forward, stopping, and suddenly seizing their prey from the surface, rather than probing for food (Johnsgard 1981).

This plover nests in disturbed areas such as ploughed fields and roadsides. A nearby stream will provide a source of food. Sainte-Martine, south of Montréal. *July 1991, Gaétan Duquette.*

Distribution

Killdeers breed wherever suitable open areas are available. They are widespread in southern Québec, having been reported in 47.8% of the squares surveyed. The Atlas map suggests that their occurrence declines above 47° N, except in open areas around settlements.

They were recorded in only 2 squares on Anticosti Island and were absent from many squares on the central Gaspé Peninsula, along the North Shore of the St. Lawrence, and in the central part of the Atlas area. The large forested tracts in many of these areas probably explain, at least in part, why Killdeers were not found there. Other atlasses show similar distribution patterns (Ellison 1985p; Nol 1987; Eaton 1988i; Santner 1992b).

Some birds, particularly during the laying period, leave the nest quietly and may be inconspicuous, so atlassers may have missed the species in some squares.

The Killdeer's very rudimentary nest is no more than a shallow scrape in the ground, in which its eggs blend into the surroundings. Sainte-Foy. *May 1954, Gaston Le Page.*

This clutch of 4 eggs, watched over by an adult, is on clay farmland soil. Saint-François-du-Lac, Lake Saint-Pierre. *June 1988, Daniel Jauvin.*

History and trends

Killdeers had been considered common in New England in the mid-1800s, but in the late 1800s and early 1900s, many were shot by market hunters there during spring and fall migrations, and around Montréal as well (Wintle 1896). Populations declined to such a degree that the birds became extremely rare in many parts of their range. The Killdeer was considered a scarce migrant in Vermont in the 1890s and early 1900s (Ellison 1985p). Wintle (1896) listed it as a common summer resident, but to Dionne (1906), it was a rare summer resident in Québec City. It became numerous by about 1921 after the Migratory Birds Convention Act (S.C. 1917, c. 18) outlawed hunting (Townsend 1929a).

Young Killdeer exploring the area around the nest it has just left. Baie-du-Febvre, south shore of Lake Saint-Pierre. *May 1981, Daniel Jauvin.*

Its range and numbers in eastern Canada have expanded recently (Godfrey 1986). It was first found breeding in New Brunswick in the 1940s, in Nova Scotia and Prince Edward Island in the 1960s (Erskine 1992), in Newfoundland in 1968, and in the Magdalen Islands in 1969 (Strauch 1971).

In Québec, its known range has recently expanded northward, documented by a series of breeding records: an adult with a downy chick at Laforge Reservoir in June 1977 (QNRCP), a family group at Blanc-Sablon in 1980 (ÉPOQ), a nest at Gagnon and sightings of birds in the breeding season at Caniapiscau and at the La Grande-3 airport in 1982 (Gosselin and David 1982), a nest at Attikuan Point on James Bay in 1989 (Aubry et al. 1989), and a nest at Poste-de-la-Baleine in 1990 (Yank et al. 1990b). In 1991, breeding was confirmed by the discovery of 2 birds and a nest with 4 eggs at the camp at km 105 (50°22' N, 76°58' W) in the Nottaway, Broadback, and Rupert rivers region (Cons. G. G. – GREBE 1992g). Nonetheless, its numbers in southern Québec have apparently decreased over the past 10–20 years (BBS).

While the increase in Killdeers this century has been partly due to the clearing of land for agriculture (Todd 1963; Eaton 1988i), the recent decrease in some parts of Québec, particularly in the south, may be a result of urban development and the reversion of farmland to forest. From 1966 to 1986, roughly 24 000 ha of prime agricultural land were urbanized in Québec (Government of Canada 1991).

Claire Gauvreau and Robert Alvo

Greater Yellowlegs

Grand Chevalier

Tringa melanoleuca

Clodin Roy

The Greater Yellowlegs is a tall, elegant shorebird with grey and white plumage, long bright yellow legs, and a loud flight call consisting of three to five notes. It can only be confused with the Lesser Yellowlegs, which is similar in plumage but 30% smaller, and whose flight call has only one or two notes (Terres 1980).

In the breeding season, Greater Yellowlegs are locally common in the boreal forest throughout Canada and Alaska (Terres 1980). In Québec, they can be found nesting chiefly across the central part of the province in wooded muskeg (Godfrey 1986). They can be seen migrating through southern Québec in spring, from late April to late May, but are more often seen on their way back from the breeding grounds, any time between mid-July and mid-November (Mousseau et al. 1976; ÉPOQ). This species winters along both the Atlantic and Pacific coasts, from Long Island, New York, south through the West Indies and from southern British Columbia through the southern tip of South America (Godfrey 1986; Buchanan 1988c; Root 1988; Morrison and Ross 1989).

Habitat and behaviour

Soon after the birds arrive on the breeding grounds in spring, they form pairs, then quickly disperse to find nest sites (Godfrey 1986). Preferred nesting habitat of the Greater Yellowlegs is wooded muskeg (fens and bogs) in the boreal forest and taiga. Two nests were recently discovered in northern Québec, one in a bog with pools close to a black spruce-moss stand (P. Lafontaine, pers. obs.) and the other in a relatively dry horizontal fen with herbaceous cover (Cons. G. G. – GREBE 1993c). In Ontario, one nest was discovered in a tamarack swamp and another in a partly wooded peatland (Nash and Dick 1981).

In the Great Whale River watershed, Greater Yellowlegs favour peatlands having non-filiform pools less than 0.01 ha in area (Cons. G. G. – GREBE 1993c), whereas in the Nottaway, Broadback, and Rupert river watersheds, the species is drawn mainly to bogs with pools (Cons. G. G. – GREBE 1992g). Although large expanses of grassy shoreline vegetation with small shallow pools are analogous to peatlands with pools, the Greater Yellowlegs seen in these former areas were simply foraging and exhibited no breeding behaviour (Cons. G. G. – GREBE 1992g). In peatlands, the birds can be seen feeding in shallow water.

Early in the breeding season, Greater Yellowlegs perform display flights that last 10–15 minutes, during which they give a rolling *toowhee, toowhee* call (Johnsgard 1981).

This species, probably monogamous (Ehrlich et al. 1988), generally nests close to water, selecting sites that are partly concealed by shrubs, fallen logs, or tall grasses (Johnsgard 1981; Godfrey 1986). An area of several square kilometres may, therefore, have only a few widely spaced nesting pairs (Johnsgard 1981). The nest is a simple depression or scrape in the ground lined with grass, moss, or leaves (Majka et al. 1976).

Little is known of the breeding biology of this species. It is known, however, that the female typically lays a single clutch of 4 eggs (Ehrlich et al. 1988) and that incubation and brood rearing are performed by both parents (Terres 1980; Johnsgard 1981). The precocial chicks have well-developed legs at hatching, and can follow their parents and find their own food upon leaving the nest (Bent 1927; Johnsgard 1981). The chicks grow rapidly, and by 3 weeks are competent fliers (Ehrlich et al. 1988).

In migration, Greater Yellowlegs feed on small crustaceans, small fish, worms, and aquatic insects, as well as various terrestrial insects, such as ants, flies, and grasshoppers (Johnsgard 1981). Greater Yellowlegs normally forage in shallow water along the edges of

Notes on territory

Since the Greater Yellowlegs pursues intruders for several hundred metres and is found in peatlands measuring at least 5 ha and preferably over 10 ha (Cons. G. G. – GREBE 1993c), it seems probable that it defends a breeding territory of a dozen hectares or more. Estimated breeding densities in Ontario are usually 10 pairs/100 km² or less, but may be as high as 1000 pairs along some sections of Hudson Bay (James 1987b). In the Great Whale River watershed in 1990, the density of breeding pairs was greater in peatlands with pools than in wooded heathland or shoreline scrub (Cons. G. G. – GREBE 1991a).

Greater Yellowlegs, either solitary or in small groups, have been reported to defend feeding territories on their wintering grounds in Argentina (Myers and Myers 1979).

A pair of Greater Yellowlegs has built its nest in this extensive peatland with pools some 50 km east of Lake Poncheville. Northeast of Matagami. *June 1991, Pierre Bannon.*

Greater Yellowlegs

Breeding distribution

1984–1989

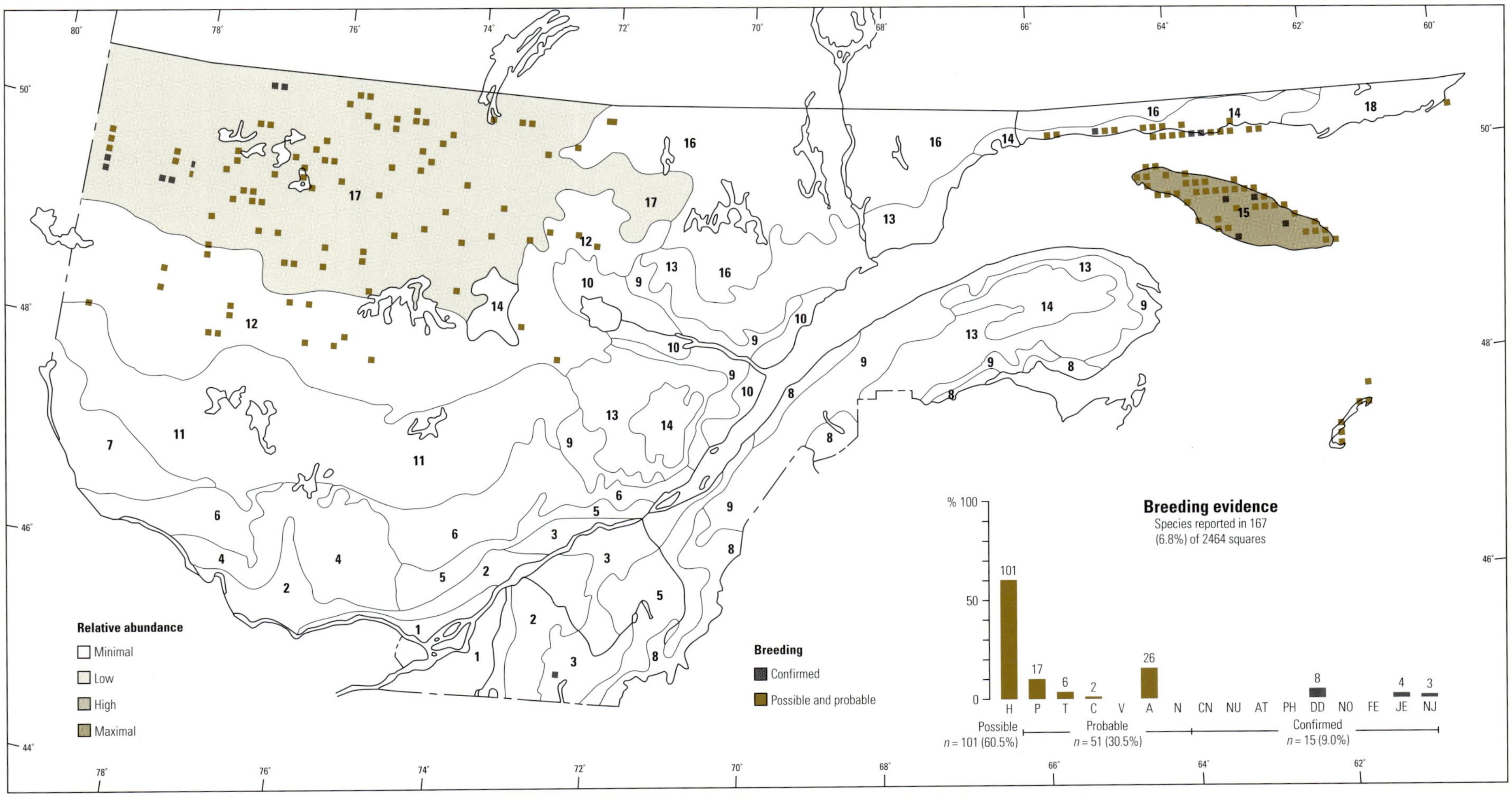

Species profile

Clutch size	usually 4 eggs[1,5]
Laying interval	?
Start of incubation	?
Incubation period	23 days[1,5]
Maturity at hatching	precocial
Care of young	by female and male[1]
Nestling period	less than 48 hours[1,5]
Age at first flight	18–20 days[1]
Dependence of young	?
Broods per year	1[1]
Breeding age	?
Mating system	probably monogamy[1]
Duration of pair bond	?
Total length	32–38 cm[2]
Wing span	58.4–66.0 cm[3]
Weight	171 g (124–224)[4]
Longevity record	?

References:
1. Ehrlich et al. 1988
2. Godfrey 1986
3. Terres 1980
4. Dunning 1984
5. Bent 1927

World breeding range

Chronology $n = 10$

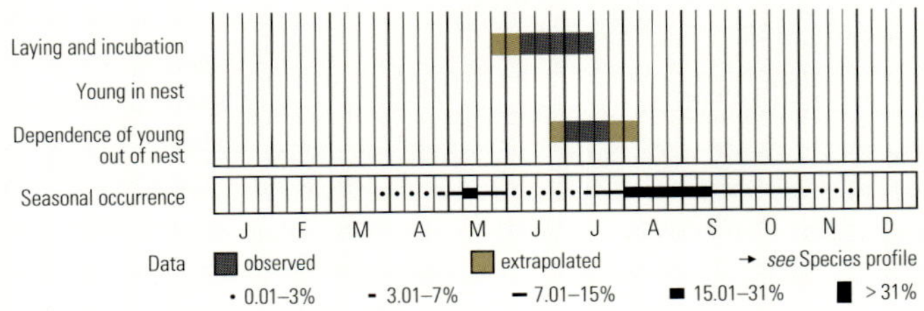

Frequency of occurrence by ecological region

Ecological regions** (forest regions)	Adequately covered squares	Squares in which reported	% of squares in which reported	Relative abundance (%)*
1. Sugar maple-hickory and sugar maple-basswood	97	0	0	0
2. Sugar maple-basswood	153	0	0	0
3. Sugar maple-basswood and sugar maple-yellow birch	101	1	1.0 ± 1.6	1.0
4. Sugar maple-yellow birch-basswood	56	0	0	0
5. Sugar maple-yellow birch-beech and sugar maple-yellow birch-basswood	113	0	0	0
6. Sugar maple-yellow birch-beech	71	0	0	0
7. Sugar maple-yellow birch-balsam fir	19	0	0	0
8. Sugar maple-yellow birch and balsam fir-yellow birch	47	0	0	0
9. Balsam fir-yellow birch	87	0	0	0
10. Balsam fir-white birch or balsam fir-red maple	46	0	0	0
11. Yellow birch-balsam fir	52	1	1.9 ± 4.5	2.0
12. Balsam fir-white birch (west)	59	6	10.2 ± 8.2	10.7
13. Balsam fir-white birch (east)	76	0	0	0
14. Balsam fir-black spruce	28	2	7.1 ± 10.8	7.5
15. White spruce-balsam fir and balsam fir-white spruce	22	21	95.5 ± 10.3	100.0
16. Black spruce-balsam fir-moss	18	0	0	0
17. Black spruce-moss	32	11	34.4 ± 18.0	36.0
18.			*not analysed*	

* minimal: 0–29.9%; low: 30–59.9%; high: 60–89.9%; maximal: 90–100%

** Deciduous forest: 1–7; mixed forest: 8–11; coniferous forest: 12–17

A peatland with pools where Greater Yellowlegs nest on Anticosti Island; the pools are a source of food. West of Port-Menier. *August 1991, Gaétan Laprise.*

ponds and lakes, probing the mud for insects with their long bills, although they sometimes fly low over the water, snatching prey from the surface or snaring small fish with their bills (Johnsgard 1981; Ehrlich et al. 1988). They are regularly seen on tidal flats (Mousseau et al. 1976; Hayes and Fox 1991).

Unlike most shorebirds, which head for the coast after breeding (Connors et al. 1979), the Greater Yellowlegs uses a variety of wetland habitats, both freshwater and salt, in migration (Burger et al. 1977; Martini et al. 1980; Burger 1984).

Greater Yellowlegs were recorded in 6.8% (167/2464) of Atlas squares surveyed, but breeding was confirmed in only 9% (15/167) of these. Breeding is difficult to confirm because nests and broods are usually well concealed by vegetation in places difficult for observers to access. Furthermore, parents often greet intruders with loud and persistent alarm notes at great distances from their nests or broods (Bent 1927; Majka et al. 1976). Similarly, in Ontario, breeding was confirmed in only 10% of squares in which the species was recorded (James 1987b). Most confirmations of breeding were based on reports of distraction displays by one or both parents. If a Greater Yellowlegs flushes and perches in a nearby tree at the approach of an intruder, the chances are very high that a nest is in the vicinity (Todd 1963). It is hard to interpret Greater Yellowlegs sightings, however, as the species exhibits loud, agitated behaviour in habitats other than those where it breeds (Erskine 1992), and many non-breeders are found among the nesting pairs on breeding grounds (Todd 1963; Erskine 1992).

Distribution

The Greater Yellowlegs breeds only in North America, nesting in the boreal forest from southern Alaska and northwestern British Columbia east across central Canada to Labrador and Newfoundland, and south into Nova Scotia (Erskine 1992). In Québec, as in Ontario (James 1987b), it breeds primarily in the northern boreal forest and southern taiga, between latitudes 48° and 58° N. South of this area, the species nests locally where suitable nesting habitat is available (Godfrey 1986). In the Maritimes, the Greater Yellowlegs is a confirmed breeder on Cape Breton Island and east-central mainland Nova Scotia (Erskine 1992).

Most Atlas records came from two areas. One extends east from the Ontario border to Lake Saint-Jean and north from the Gouin Reservoir to Lake Mistassini. This encompasses the Black spruce-moss forest region. South of this region, the birds are seen chiefly in migration. The second area where the species was found breeding is along the Middle North Shore between Moisie and Havre-Saint-Pierre, as well as on Anticosti Island and the Magdalen Islands. The paucity of observations between the two areas may be due to problems of accessibility, and hence less thorough surveying, rather than to actual absence of the species.

The nest is a shallow depression in the ground lined with moss, twigs, and leaves. Northeast of Matagami. *June 1991, Pierre Bannon.*

This shorebird also nests in peatlands with extensive herbaceous cover and a patchwork of ponds and small lakes. Havre-Saint-Pierre, Middle North Shore. *July 1987, Christian Marcotte.*

In summer, Greater Yellowlegs were sighted along much of the St. Lawrence River between Montréal and Moisie, as well as in the Eastern Townships, Chaudière–Appalaches, Gaspé, and Saguenay regions. None of these individual birds or small flocks exhibited any behaviour indicative of breeding, however, nor were any observed in what might be considered suitable breeding habitat. These birds, therefore, were probably either non-breeders or failed breeders that had begun their southward migration early. Birds observed after the middle of July may have been early nesters that had completed nesting and were heading south again (Cramp and Simmons 1983).

One Atlas breeding record from south of the boreal forest was unexpected. On 18 July 1987, a pair with 3 chicks, still too young to fly, was observed by a small pond in Knowlton, in the Eastern Townships (S. Tanguay, pers. comm.). This is the southernmost nesting record for the species.

Since 1989, 2 more nests have been discovered: one on James Bay and the other in the Great Whale River watershed. In one case, the incubating female flushed when the intruder was within 5 m. It flew up and perched on a stunted spruce overhanging its nest, then flew a few dozen metres away, calling (P. Lafontaine, pers. obs.).

In autumn, Greater Yellowlegs migrate in flocks in a broad front through the interior of North America, including southern Québec (Bourget 1990). Adults migrate first, with males preceding females, followed by juveniles about a month later (McNeil and Cadieux 1972; Mousseau et al. 1976; Cramp and Simmons 1983). In the Magdalen Islands, adults begin arriving in mid-July, with the largest concentrations passing through in August. The peak passage of juveniles, in contrast, occurs in September (Mousseau et al. 1976). In the fall migration, Greater Yellowlegs are seen mainly along the shores of the gulf and also along the St. Lawrence (Maisonneuve et al. 1990).

Spring migration is less prolonged, taking place between the end of April and the end of May; Greater Yellowlegs are seen in the northern United States in April and reach the breeding grounds in Canada by mid-May (Terres 1980; Cramp and Simmons 1983). They migrate mostly through the interior of the continent and up the Atlantic coast (Cramp and Simmons 1983).

History and trends

Population trends of the Greater Yellowlegs, like those of many northern-nesting shorebirds, are difficult to ascertain. The breeding distribution of the species was more extensive at the end of the 19th century than it is now. It probably used to nest in marshes throughout forested areas in central Canada as far south as the northern United States. Evidence of the southern limits of breeding includes a nest found by Dr. E. W. Nelson (1877 *in* Bent 1927) in an Illinois marsh in 1876. Less than 50 years later, however, a contraction in the breeding range of the species was noted by American ornithologist William Brewster. He stated that the Greater Yellowlegs had been driven from many of its former breeding areas by several factors, chief among them being habitat loss resulting from the draining and filling in of marshes (Brewster 1925 *in* Bent 1927). Hunting of this species, which was legal until 1926 (Migratory Birds Regulations 1927), probably did much to reduce its breeding range (Root 1988). It was hunted for sport or food, or because it disrupted the hunting of other species. In the Maritimes, its abundance does not seem to have changed since early colonial times (Erskine 1992).

There is no evidence from Québec that the Greater Yellowlegs has ever nested south of its present range. At the turn of the century, Wintle (1896) listed the Greater Yellowlegs as a common "transient visitant" in Montréal, as did Bannon (1991a) almost 100 years later.

Recently, Johnsgard (1981) stated that although there are no quantitative data on the Greater Yellowlegs, this species is far less common than the Lesser Yellowlegs. Since Greater Yellowlegs nest in sparsely populated regions of North America, they are probably not in any immediate danger in their breeding range, unless the creation of reservoirs for hydro-electric power plants appreciably reduces their habitat. Hunting on the winter range in Latin America, habitat loss, and environmental contamination on both breeding and winter ranges all pose potentially more serious long-term problems for the species (Senner and Howe 1984; Erskine 1992).

Richard C. Cotter and Pierre Lafontaine

Solitary Sandpiper

Chevalier solitaire

Tringa solitaria

Clodin Roy

IMAGINE A SHOREBIRD THAT LAYS ITS EGGS IN A ROBIN'S NEST 6-m UP IN A conifer, right in the middle of the boreal forest. Odd, you might say—and yet the Solitary Sandpiper does just that. Very little is known about its nesting habits except that it uses the nests of passerines. The Solitary Sandpiper is a widespread migrant breeder in Québec. It is commonest in the boreal forest as far north as the tree line, though it is never abundant.

As its name indicates, the species is not gregarious, keeping to itself even in migration. Outside the breeding period, it is retiring rather than timid. It is difficult to observe, because it usually ignores intruders. When it does react, it takes flight with a metallic *peet weet weet*, usually landing near its original location. Only slightly larger than the Spotted Sandpiper, it can be distinguished in flight by its dark brown wings and rump and the white sides of its tail barred with black. When it lands, it often holds its wings up for a moment and bobs nervously in typical sandpiper fashion, but less so than the Spotted Sandpiper (Johnsgard 1981).

Habitat and behaviour

In the breeding season, the Solitary Sandpiper is found mainly on wet, open terrain in the boreal forest. It likes swamps, small lakes, ponds, and river edges in both coniferous forest and mountainous regions south of the tree line (Johnsgard 1981; Godfrey 1986). When you come across a beaver pond, a Solitary Sandpiper is often nearby (Henderson 1923). The species appears to be more common in areas with small rather than large ponds. The Rusty Blackbird may also be found nesting in this habitat (Bent 1929), particularly when the pond is a little larger.

The species' nesting behaviour has long been a mystery, and our knowledge of it is still rudimentary. Courtship displays consist of ritual postures and calls executed in a specific sequence leading to mating (Oring 1973). In 1903, when E. Thompson found Solitary Sandpiper eggs in an old American Robin's nest in a tamarack (Street 1923), ornithologists realized that the species, unlike other shorebirds, not only nests in trees, but typically uses the abandoned nests of passerines, including the Rusty

Notes on territory

Pairs are widely dispersed and their territories, which are defended mainly by the male, are fairly large.

A wet area in the boreal forest, like this standing water on the edge of a logging road in a black spruce forest, attracts the Solitary Sandpiper. Caribou Lake, Decelles Reservoir, Témiscamingue.
May 1991, Réginald Ouellet.

Solitary Sandpiper

Breeding distribution 1984–1989

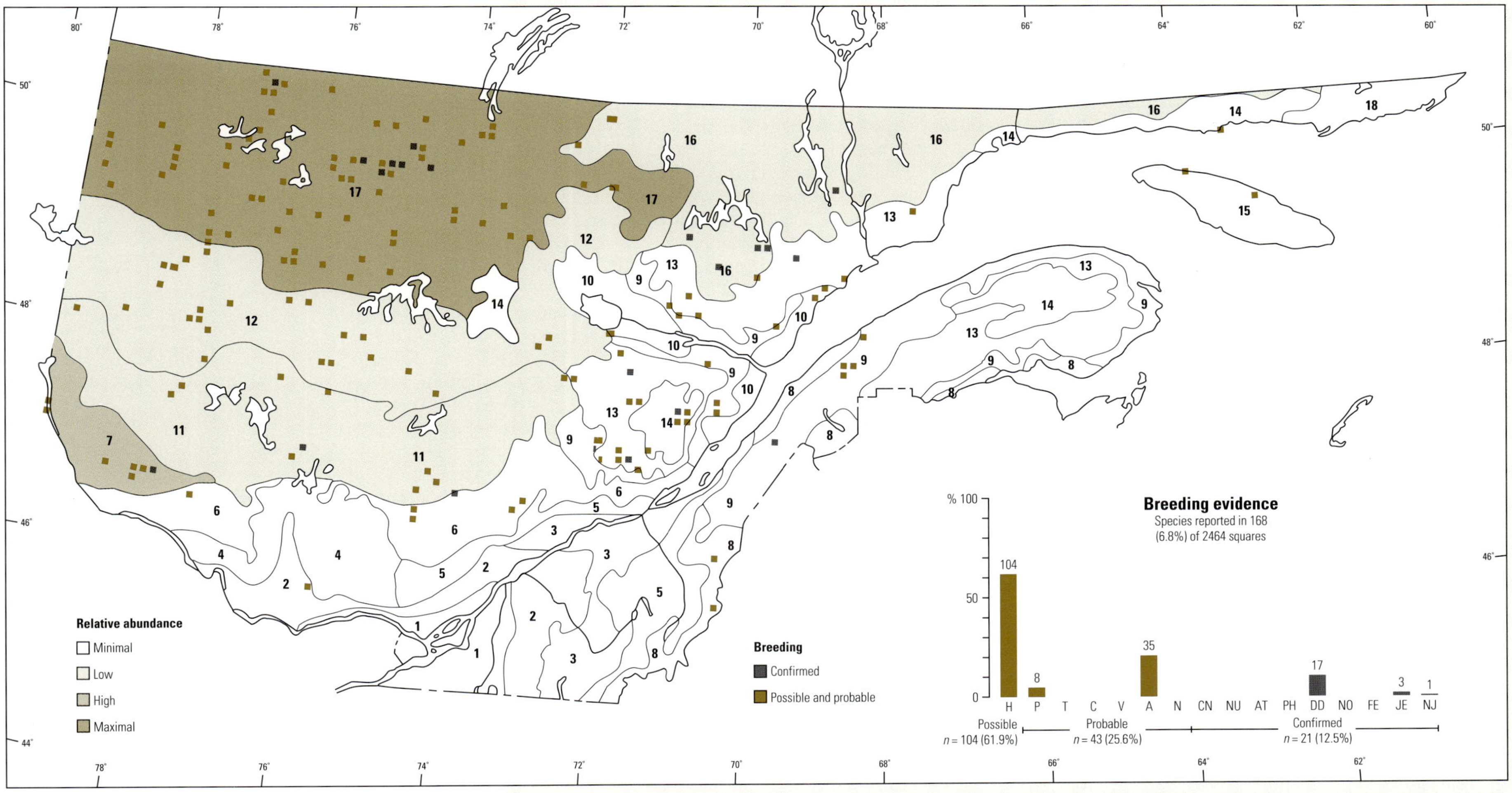

Species profile

Clutch size	usually 4 eggs (3–5)[1,2,3,4,8]
Laying interval	≈ 1 egg per day[7]
Start of incubation	?
Incubation period	23–24 days[4]
Maturity at hatching	precocial
Care of young	?
Nestling period	probably leaves nest soon after hatching[2,5]
Age at first flight	?
Dependence of young	?
Broods per year	1[2]
Breeding age	?
Mating system	monogamy[4]
Duration of pair bond	?
Total length	19–23 cm[3]
Wing span	38.1–43.2 cm[5]
Weight	51.2 g (37.9–76.5)[6]
Longevity record	?

References:
1. Peck and James 1983
2. Harrison 1978
3. Godfrey 1986
4. Ehrlich et al. 1988
5. Terres 1980
6. Dunning 1984
7. Johnsgard 1981
8. Bent 1929

World breeding range

Chronology $n = 9$

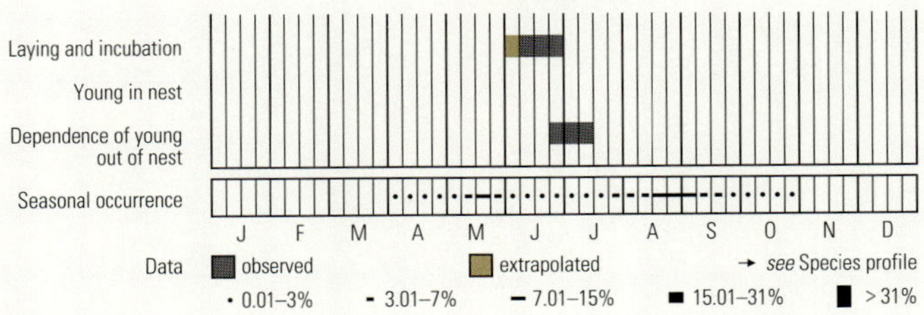

Frequency of occurrence by ecological region

Ecological regions** (forest regions)	Adequately covered squares	Squares in which reported	% of squares in which reported	Relative abundance (%)*
1. Sugar maple-hickory and sugar maple-basswood	97	0	0	0
2. Sugar maple-basswood	153	1	0.7 ± 0.9	1.3
3. Sugar maple-basswood and sugar maple-yellow birch	101	0	0	0
4. Sugar maple-yellow birch-basswood	56	0	0	0
5. Sugar maple-yellow birch-beech and sugar maple-yellow birch-basswood	113	0	0	0
6. Sugar maple-yellow birch-beech	71	4	5.6 ± 5.3	11.3
7. Sugar maple-yellow birch-balsam fir	19	7	36.8 ± 22.8	73.7
8. Sugar maple-yellow birch and balsam fir-yellow birch	47	3	6.4 ± 6.8	12.8
9. Balsam fir-yellow birch	87	5	5.7 ± 4.8	11.5
10. Balsam fir-white birch or balsam fir-red maple	46	2	4.3 ± 5.7	8.7
11. Yellow birch-balsam fir	52	9	17.3 ± 10.8	34.6
12. Balsam fir-white birch (west)	59	10	16.9 ± 10.0	33.9
13. Balsam fir-white birch (east)	76	6	7.9 ± 6.2	15.8
14. Balsam fir-black spruce	28	1	3.6 ± 8.3	7.1
15. White spruce-balsam fir and balsam fir-white spruce	22	1	4.5 ± 10.3	9.1
16. Black spruce-balsam fir-moss	18	4	22.2 ± 22.2	44.4
17. Black spruce-moss	32	16	50.0 ± 18.9	100.0
18.			*not analysed*	

* minimal: 0–29.9%; low: 30–59.9%; high: 60–89.9%; maximal: 90–100%

** Deciduous forest: 1–7; mixed forest: 8–11; coniferous forest: 12–17

Blackbird, Common Grackle, Cedar Waxwing, Bohemian Waxwing, Eastern Kingbird, and Gray Jay. The pair selects an old nest together (Ehrlich et al. 1988) or, occasionally, a newly constructed one, located 1–12 m up, usually in a conifer, but sometimes in a deciduous tree (Oring 1973). The species is monogamous and often returns to the nest used the year before; the female adds material inside as needed (Ehrlich et al. 1988).

Only females have a brood patch, which suggests that they do most, if not all, the incubating of the 4 eggs (Ehrlich et al. 1988). Incubation and brood rearing have not been studied in the field, and their duration remains unknown, but by placing eggs in an incubator, Oring (1973) calculated that the incubation period was 23 or 24 days long. The precocial, nidifugous young probably leave the nest by jumping out of the tree shortly after hatching (Harrison 1978), then forage for food on their own (Ehrlich et al. 1988). The parents, which raise only one brood a year, possibly accompany the chicks for a few weeks, as is the case with other sandpipers.

This sandpiper forages in stagnant ponds fringed with vegetation, ditches, wet meadows, and damp woods. A favourite method of feeding is to wade slowly in shallow water, stirring the bottom delicately with one foot so as not to disturb it too much. This brings up aquatic insects, their larvae, and small crustaceans, which the Solitary Sandpiper then captures (Johnsgard 1981). It also catches insects out of the air and supplements its diet with spiders, grasshoppers, and even small frogs (Bent 1929).

The Solitary Sandpiper was recorded in only 6.8% (168/2464) of Atlas squares surveyed, and nesting was confirmed in only 12.5% (21/168) of them. In the breeding season, the Solitary Sandpiper can be noisy and very agitated. It is often seen at the top of a tree or sitting conspicuously on a perch, calling loudly. Both the male and female may produce songs and calls (Oring 1968). In all, 20.8% (35/168) of Atlas records were based on agitated behaviour. Most confirmations of breeding were based on reports of distraction displays. Only one nest was discovered. The young are also well camouflaged and were seen only 3 times.

Solitary Sandpipers usually migrate singly, probably at night (Johnsgard 1981). In fall, they are often found in more open habitat, such as meadows and salt marshes also frequented by Lesser Yellowlegs (Bent 1929). In migration, the Solitary Sandpiper seeks out small marshy ponds, where dragonfly and damselfly larvae constitute a large part of its diet (Savard, M. 1984b). It can also be seen in places where a shorebird is unexpected, such as on the muddy bank of an urban stream, on the edge of a river in deep woods, or near a mud puddle or pond in a field (Cramp and Simmons 1983). The Solitary Sandpiper winters from southern California, Florida, and the Gulf of Mexico south to Argentina (Godfrey 1986). It appears to be solitary in winter, too: rarely are more than 3 birds seen together at this time of year, with 1.3 being the mean (Hayes and Fox 1991).

In the breeding season, the Solitary Sandpiper inhabits open areas in the boreal forest. Unusual for a shorebird, this member of the family Scolopacidae often lays its eggs in an old passerine nest in an evergreen. Near Matagami. *July 1987, Yves Aubry.*

Notes on territory (continued)

In Alberta's Crimson Lake Provincial Park, one pair's territory was calculated to be 50 ha (Oring 1973). The Solitary Sandpiper does not seem to defend its territory aggressively against humans the way other Tringa sandpipers do, and observers can reportedly get as close as 12 m to nesting birds before they will fly away (Bent 1929). A 1990 study in the Great Whale and Little Whale river watersheds revealed that the species was 3 times more abundant in peatlands with pools than in black spruce-moss stands (Cons. G. G. – GREBE 1991a).

The species is encountered in abandoned beaver ponds with residual wet areas criss-crossed by stumps, branches, and snags, like this pond in the Abitibi region. Aiguebelle Conservation Park, northeast of Rouyn-Noranda. *June 1991, Desneige Perreault.*

Distribution

In Canada, the Solitary Sandpiper breeds throughout the boreal forest from Yukon to Labrador. In Québec, atlassers reported it mostly in coniferous parts of the Canadian Shield, where peatlands, lakes, beaver ponds, and other wetland habitats abound. A particularly high number of records came from the area between Chibougamau and the Abitibi region, where breeding was confirmed, as it was in the Témiscamingue area. Farther south, breeding seems to be less frequent, but it was confirmed in the Mauricie, the southern Laurentians, the Rivière-Blanche ZEC north of the Portneuf Wildlife Reserve, the Laurentide Wildlife Reserve, Charlevoix, the area northeast of the Saguenay region, and even, for the first time, south of the St. Lawrence River, near Saint-Bruno-de-Kamouraska. Evidence of probable and possible breeding was obtained in the Beauce, along the Lower St. Lawrence near Rimouski, on the North Shore to just east of Havre-Saint-Pierre, and on Anticosti Island, where the species had already been sighted in summer (Ouellet 1969). These records suggest a range that extends considerably farther south than that given by Godfrey (1986), who showed the southern limit of the breeding range as a line between Témiscamingue and southwestern Labrador.

North of the Atlas area, Godfrey (1986) shows the species occurring up to the mouth of the La Grande Rivière and up to northern Labrador in the east. This puts the northern boundary of its distribution at roughly the tree line. Recent surveys support this hypothesis, but extend the northwestern boundary somewhat, since the species has been reported 100 km north of the Great Whale River, on the Hudson Bay coast (Cons. G. G. – GREBE 1991a).

History and trends

This sandpiper's retiring, solitary nature and its rather remote northern habitat make it extremely difficult to evaluate its historical status and current population trends. We know only that it was observed in migration in southern Québec at the turn of the century, as it is today. Wintle (1896) listed it as a scarce "transient visitant" in the Montréal area, while Mousley (1916a) and Lloyd (1923) described it as fairly common in migration in the Eastern Townships and Ottawa Valley.

JACQUELINE VINCENT AND JEAN-FRANÇOIS RAIL

Spotted Sandpiper

Chevalier branlequeue

Chevalier grivelé
Actitis macularia

Clodin Roy

THE SPOTTED SANDPIPER IS PROBABLY THE BEST-KNOWN, MOST WIDELY distributed and most abundant New World sandpiper, breeding commonly throughout Québec except on the Ungava Peninsula. It exhibits polyandry (Hayman et al. 1986), a mating system documented in less than 1% of the world's bird species studied, in which a female has more than one mate in a single breeding season (Maxson and Oring 1980).

The Spotted Sandpiper bobs its tail up and down almost continuously, beginning soon after hatching (Tyler 1929b). This migrant breeder is easily identified in its adult breeding plumage by the roundish black spots on the white underparts combined with a slightly metallic lustre to the greyish-brown upperparts (Godfrey 1986). It flutters from shore out over water with its wings stiffly downcurved and vibrating like a taut wire (Terres 1980).

Habitat and behaviour

Spotted Sandpipers breed in an extremely diverse array of habitats, ranging from sea level to the tree line in both alpine and subarctic regions. Some of the many natural habitats used for breeding are sandy beaches, dunes, clearings, woodland edges, herbaceous lake shores, and riverbanks. Habitats associated with human disturbance include gravel pits, fields, roadsides, and ditches. In northern Québec, the species is also seen in peatlands with pools (Cons. G. G. – GREBE 1991a). Pairs nest solitarily or in loose colonies.

Nests are on the ground and usually concealed. The cuplike depressions, or hollows, may be found almost anywhere and may be used in successive years (Mousley 1937a). They are usually located in or near vegetation, but may also be under rocks, sticks, branches, roots, or driftwood (Johnsgard 1981; Peck and James 1983).

Spotted Sandpiper females are often polyandrous, and may mate with up to 4 males in a season, while males provide most of the parental care (Oring et al. 1983). Females tend to be slightly larger than males, showing larger and blacker spots in their breeding plumage, although first-year females may resemble males. Females precede males to the breeding grounds (Oring and Lank 1982), and they take the initiative in courtship. Females begin prospecting for new males after laying the third egg of the 4-egg clutch. Incubation is shared by both sexes in monogamous pairs. A polyandrous female helps incubate her last clutch, but only the males incubate her other clutches (Hays 1972). When a female acquires a new mate, she usually deserts the previous one (Oring and Reed 1991). An exceptional female was observed simultaneously bonded to two different males (Oring and Maxson 1978).

Notes on territory

The female defends a territory against other females. The male may help defend the territory against other males, but with the arrival of a second male and the need to incubate eggs, the first

The Spotted Sandpiper occupies a wide variety of habitats, such as these grassy margins for nesting and riverbanks for feeding. Lake Belle Fontaine, Laurentide Wildlife Reserve, north of Québec City.
June 1991, Réginald Ouellet.

Spotted Sandpiper

Breeding distribution

1984–1989

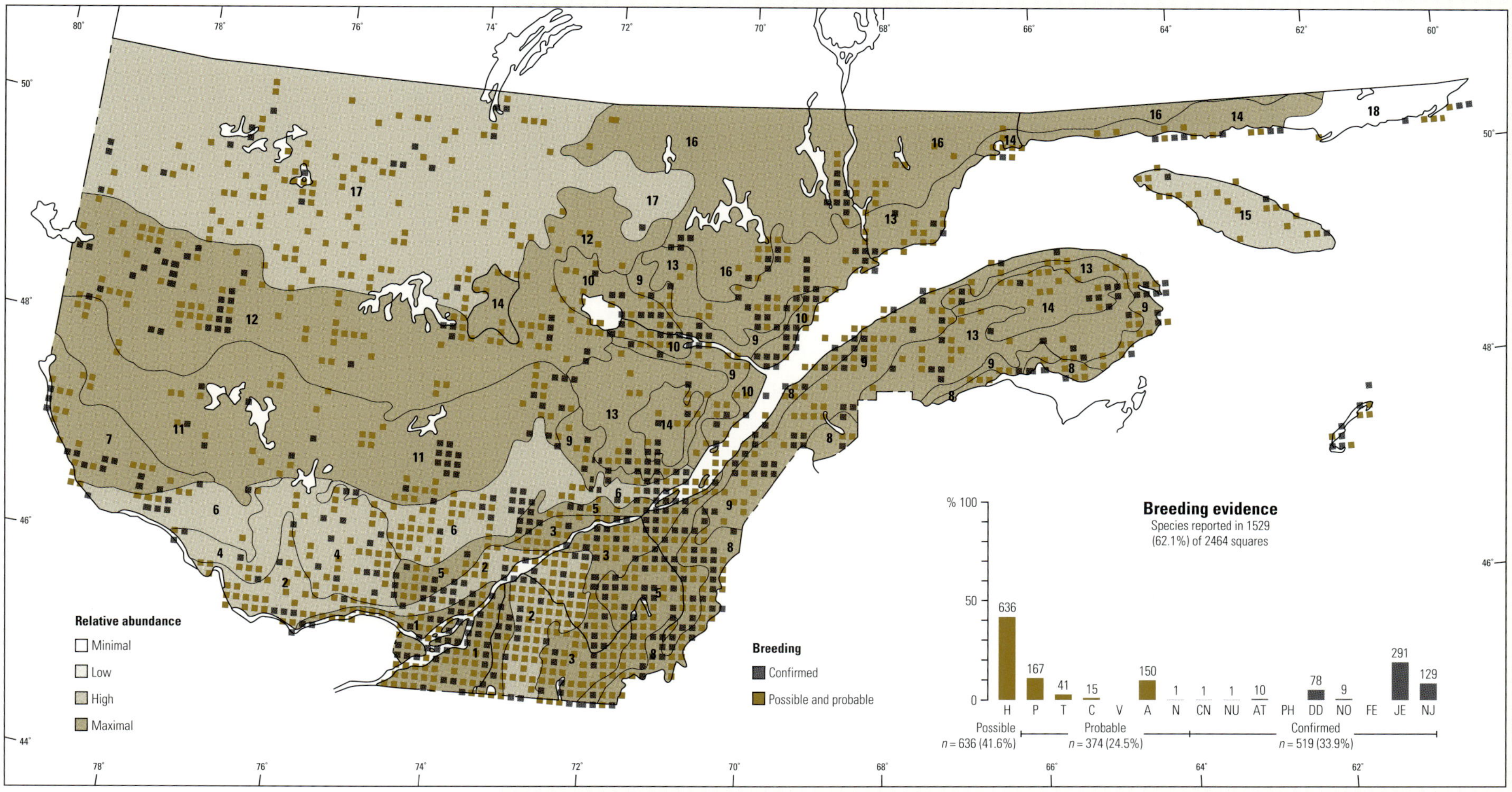

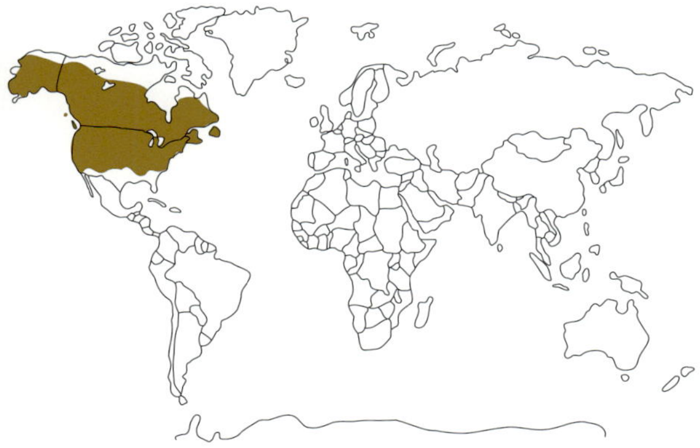

World breeding range

Species profile

Clutch size	usually 4 eggs (1–5)[1,3,5,10]
Laying interval	1 egg per day, sometimes 1 per 2 days[1]
Start of incubation	with last egg[1]
Incubation period	20–24 days[1,3,5,6]
Maturity at hatching	precocial
Care of young	mainly by male[1,11,12]
Nestling period	5–6 hours[1,5]
Age at first flight	usually 18–21 days (13–21)[1,3,4,5,6]
Dependence of young	until first flight[1]
Broods per year	usually 2–3 (1–4)[1,3]
Breeding age	1–2 years[1,2]
Mating system	monogamy or polyandry depending on habitat[1,11,12,13]
Duration of pair bond	1 season or less[1]
Total length	17.5–20.5 cm[8]
Wing span	33.0–35.6 cm[4]
Weight	40.4 g (29.4–59.8)[7]
Longevity record	12 years[9]

References:

1. Cramp and Simmons 1983
2. Oring et al. 1983
3. Ehrlich et al. 1988
4. Terres 1980
5. Harrison 1978
6. Johnsgard 1981
7. Dunning 1984
8. Godfrey 1986
9. Clapp et al. 1982
10. Peck and James 1983
11. Oring and Knudson 1972
12. Hays 1972
13. Oring and Maxson 1978

Chronology $n = 475$

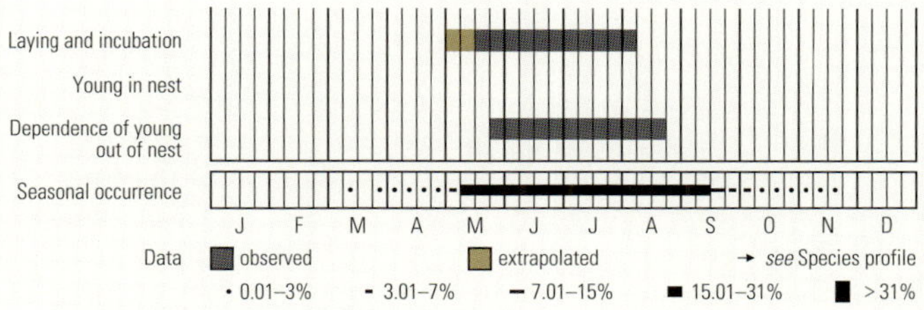

Frequency of occurrence by ecological region

Ecological regions** (forest regions)	Adequately covered squares	Squares in which reported	% of squares in which reported	Relative abundance (%)*
1. Sugar maple-hickory and sugar maple-basswood	97	92	94.8 ± 2.5	99.1
2. Sugar maple-basswood	153	121	79.1 ± 3.4	82.6
3. Sugar maple-basswood and sugar maple-yellow birch	101	92	91.1 ± 3.6	95.1
4. Sugar maple-yellow birch-basswood	56	42	75.0 ± 10.0	78.3
5. Sugar maple-yellow birch-beech and sugar maple-yellow birch-basswood	113	102	90.3 ± 3.2	94.3
6. Sugar maple-yellow birch-beech	71	60	84.5 ± 7.9	88.3
7. Sugar maple-yellow birch-balsam fir	19	17	89.5 ± 15.5	93.5
8. Sugar maple-yellow birch and balsam fir-yellow birch	47	45	95.7 ± 5.8	100.0
9. Balsam fir-yellow birch	87	77	88.5 ± 6.4	92.4
10. Balsam fir-white birch or balsam fir-red maple	46	42	91.3 ± 7.5	95.4
11. Yellow birch-balsam fir	52	47	90.4 ± 8.6	94.4
12. Balsam fir-white birch (west)	59	51	86.4 ± 9.2	90.3
13. Balsam fir-white birch (east)	76	66	86.8 ± 7.6	90.7
14. Balsam fir-black spruce	28	25	89.3 ± 12.6	93.3
15. White spruce-balsam fir and balsam fir-white spruce	22	13	59.1 ± 21.2	61.7
16. Black spruce-balsam fir-moss	18	17	94.4 ± 13.5	98.6
17. Black spruce-moss	32	27	84.4 ± 14.1	88.1
18.	*not analysed*			

* minimal: 0–29.9%; low: 30–59.9%; high: 60–89.9%; maximal: 90–100%

** Deciduous forest: 1–7; mixed forest: 8–11; coniferous forest: 12–17

This sandpiper forages along rocky and driftwood-strewn shorelines as well as open, sandy ones, in search of aquatic organisms. Lake Belle Fontaine, Laurentide Wildlife Reserve, north of Québec City. *June 1991, Réginald Ouellet.*

Notes on territory (continued)

male's territorial behaviour decreases and may become confined to his portion of the female's territory (Oring and Maxson 1978). The female may move to a new territory, however (Oring and Reed 1991). Newly independent young may also defend a territory (Cramp and Simmons 1983).

In Michigan, in habitat consisting of dry meadows, rocky banks, and sandy beaches, the Spotted Sandpiper nests in densities of up to 6 pairs/ha (Miller and Miller 1948).

The precocial young leave the nest 5–6 hours after hatching, or sooner if disturbed, and they normally do not return (Merilees 1967). Chicks generally feed themselves, although the male may feed them for the first 4 or 5 days (MacDonald 1947 *in* Cramp and Simmons 1983). They can swim and dive to escape danger, but also rely on suddenly lying motionless on the ground, where they are very difficult to see (Terres 1980). The male accompanies the chicks until they become independent at 17–21 days. During the pre-fledging period, the male acts chiefly as a sentinel, often perched on a post or tall shrub (Cramp and Simmons 1983). In monogamous pairs the female may also attend the young.

Birds typically forage solitarily, and work the water margins rather than wading. At times they will stalk a winged insect by stretching the entire body parallel to the ground and slowly moving up on it. They eat a vast diversity of aquatic and terrestrial insects and occasionally small fish (Johnsgard 1981).

The Spotted Sandpiper hides its nest in low herbaceous vegetation. This depression contains a clutch of 4 eggs. Saint-Raphaël, southwest of Montmagny. *May 1985, Ronald Lepage.*

Spotted Sandpiper hatchlings still in the nest. Saint-François-du-Lac, Lake Saint-Pierre. *June 1988, Daniel Jauvin.*

Distribution

Spotted Sandpipers were recorded in 62% (1529/2464) of Atlas squares surveyed. Atlassing revealed that the species is widespread and breeds throughout the Atlas area. This is probably also the case in the northern part of its Québec range, where it ventures beyond the tree line only exceptionally (Todd 1963). The Spotted Sandpiper was reported in the vicinity of the Eastmain River in June 1993 (D. Bordage, CWS, unpubl. data), around the site of the planned Laforge-1 Hydro-electric Station project (Dryade 1992), and in the watersheds of the Great Whale, Little Whale, and Nastapoka rivers (Cons. G. G. – GREBE 1991a, 1993c).

Its widespread distribution is not surprising, given its highly diverse breeding habitats. In Québec, it appears to be common not only in low-lying, but also in mountainous areas, contrary to what the atlasses of New York (Sibley 1988d), Vermont (Ellison 1985v), and Pennsylvania (Brauning 1992d) indicate.

History and trends

There is no evidence to suggest that the status of the Spotted Sandpiper has changed significantly in the past century (Bryant 1987; Sibley 1988d). Many species of shorebirds declined drastically due to hunting before the Migratory Birds Convention Act (S.C. 1917, c. 18) was passed in Canada, but Spotted Sandpipers suffered less, thanks to their solitary foraging habits, which made them less valuable to hunters seeking to bag several birds with a single shot (Tyler 1929b). On their wintering grounds, which extend from the southern United States to northern Argentina, they are found in virtually every aquatic habitat from rain puddles to seashore (Stiles and Skutch 1989). They migrate on broad fronts all across North America, favouring inland freshwater habitats (Cramp and Simmons 1983), and should therefore not be particularly susceptible to habitat changes during migration. They seem to be doing well even in the acid-stressed parts of southwestern Nova Scotia (Erskine 1992).

Jean-Denis Gauvreau and Robert Alvo

Upland Sandpiper

Maubèche des champs

Bartramia longicauda

Clodin Roy

Perched on a fence post near a hayfield, the Upland Sandpiper watches carefully for intruders. Unlike other sandpipers in Québec, it is seldom found on shorelines or wetlands, preferring, as its name indicates, open, grassy uplands. When it alights, it often extends its wings straight up in a characteristic way before folding them (Ailes 1976; Godfrey 1986).

Upland Sandpiper populations have fluctuated considerably in North America over the last 200 years. With the clearing of forests in the eastern part of the continent in the 18th and 19th centuries, numbers grew significantly as the agricultural landscape expanded (Taverner 1922; Bent 1929). Then, in the middle of the last century, the Upland Sandpiper became a prize game bird, and intense hunting pressure caused populations to drop spectacularly. After 1916, numbers stabilized with the signing of the Migratory Birds Convention, as a result of which the Upland Sandpiper was removed from the list of game species (Bent 1929; Osborne and Peterson 1984). The fate of the Upland Sandpiper still remains uncertain today, but for other reasons. Now the species is threatened by urbanization and agricultural practices (Osborne and Peterson 1984), and has been added to the Blue List (Tate 1986).

This migrant breeder arrives in Québec from its far-off wintering grounds on the Argentine pampas (White 1988) in early May or sometimes April (David 1980a). In late summer, when it heads south again, small flocks of about a dozen birds can be seen on the grassy verges bordering airports and highways, and in open fields. Most of the population has departed by late August, although a few stragglers may remain until October (David 1980a).

Habitat and behaviour

In the breeding season, the Upland Sandpiper makes use of a variety of open, flat habitats, like hayfields, pastures, and land growing up in short grasses, as is found around airports (Peck and James 1983; Carter 1992) or along highways. It often returns to the same nesting area year after year. The species generally avoids old fields and pastures that are reverting to shrub (Eastman 1991). In the western part of the continent, it may be found in peatlands (Godfrey 1986).

In Québec, the breeding season begins in early May and continues until mid-August. Most pairs have already formed before they arrive on the breeding grounds (Ailes 1976). Once there, they establish their nest sites and advertise their presence by a loud, musical whistling (Bent 1929).

The nest, built on the ground and hidden in tall grass, is a grass-lined scrape (Harrison 1978). Upland Sandpipers nest solitary or in small loose colonies (Hayman et al. 1986). A single clutch is produced each year, and the eggs, usually numbering 4 (Higgins and Kirsch 1975; Peck and James 1983), are incubated by both parents for 21–28 days. The precocial young leave the nest within 24 hours. The male usually tends the brood, but the female sometimes helps (Ailes 1976). In southernmost Québec, the young start to leave the nest in the first week of June, just before the haying season begins.

The species was reported in 12.6% (311/2464) of Atlas squares surveyed and breeding was confirmed in 20.9% (65/311) of them. Nests are difficult to find, as the Atlas data show: only 4 were reported. Except during the incubation period, the bird itself is fairly easy to detect, since its habitat is open and accessible, and it often perches on fence posts. When threatened, the parents actively defend the nest and young, and may feign a broken wing, the way Killdeers do.

Distribution

The Upland Sandpiper's range in Québec is co-extensive with agricultural areas and the species is still commonest south of the 46th parallel, as David (1980a) reported.

The Upland Sandpiper's stronghold in Québec comprises the St. Lawrence Plain and Ottawa Valley, as well as the agricultural areas of the Eastern Townships. Below Québec City, the lowlands are only a narrow strip along the St. Lawrence, and the Upland Sandpiper is encountered at a few isolated sites along the Lower St. Lawrence, such as fields on the edge of the Mont-Joli Airport. In the Gaspé, where the species is scarce, it has been seen at Bonaventure and at Sainte-Thérèse-de-Gaspé on Chaleur Bay. Comeau (1923a) sighted it in migration at Godbout in 1883, 1887, and 1906.

The Upland Sandpiper has also been confirmed as a breeder in the Upper Saguenay region and in the Lake Saint-Jean Lowland, where it occurs very locally and in small numbers (Club des ornithologues amateurs du Saguenay–Lac-Saint-Jean 1993b; G. Savard, pers. comm.). This region is actually an interrupted extension of the St. Lawrence Lowlands (Marie-Victorin 1964). Only one case of breeding was reported farther north, at Amos in the Clay Belt of the Abitibi, on 19 July 1989. North of the Atlas area, the species was sighted on 18 June 1981 near the Eastmain River (ÉPOQ).

History and trends

Common on the Prairies, in Québec the Upland Sandpiper has undoubtedly been a prime beneficiary of the clearing of land for agriculture. At one time, it was also hunted ruthlessly (Bent 1929).

At the end of the last century, Wintle (1896) listed the species as a scarce "transient visitant" in the Montréal area. According to Terrill (1951), the Upland Sandpiper was an uncommon summer resident on the Island of Montréal, with a few birds nesting in Côte-Saint-Luc. According to Ouellet (1974a), the species became more abundant in the Monteregian Hills between 1935 and 1965.

Unknown in the Saguenay–Lac-Saint-Jean region before 1975, the Upland Sandpiper is now a well-established breeder there (Cormier and Breton 1984).

More recently, Breeding Bird Surveys conducted between 1965 and 1979 indicate an increase in Upland Sandpiper populations in eastern North America, including the Central St. Lawrence Lowland (Robbins et al. 1986). Nonetheless, the species is likely declining near big cities as a result of residential and industrial development. The continuing abandonment of family farms and intensive corn cultivation in the southern part of the province have also been contributing to the disappearance of suitable breeding habitat for the Upland Sandpiper in Québec.

Richard Yank and Louis Breton

Notes on territory

Soon after they arrive on the breeding grounds, Upland Sandpipers establish themselves and make their presence known by loud, musical whistling (Bent 1929).

On the Wisconsin plains, 2 pairs of Upland Sandpipers had territories of 8.1 and 12.1 ha in area (DeGraaf and Rudis 1987). In Illinois, surveys conducted between 1963 and 1984 recorded a density of 1.1 nests/100 ha (Buhnerkempe and Westemeier 1988). In North Dakota, ideal habitat can accommodate up to 8 pairs/100 ha (Osborne and Peterson 1984). A density of 15 territorial males/100 ha has nonetheless been estimated in Ottawa, in a field with scattered trees (Erskine 1971a).

In Québec, the Upland Sandpiper, often seen perched on a fence post, is associated with agricultural areas, but it has also been reported in peatlands. Dewittville, southwest of Montréal. *July 1991, Gaétan Duquette.*

Upland Sandpiper

Breeding distribution 1984–1989

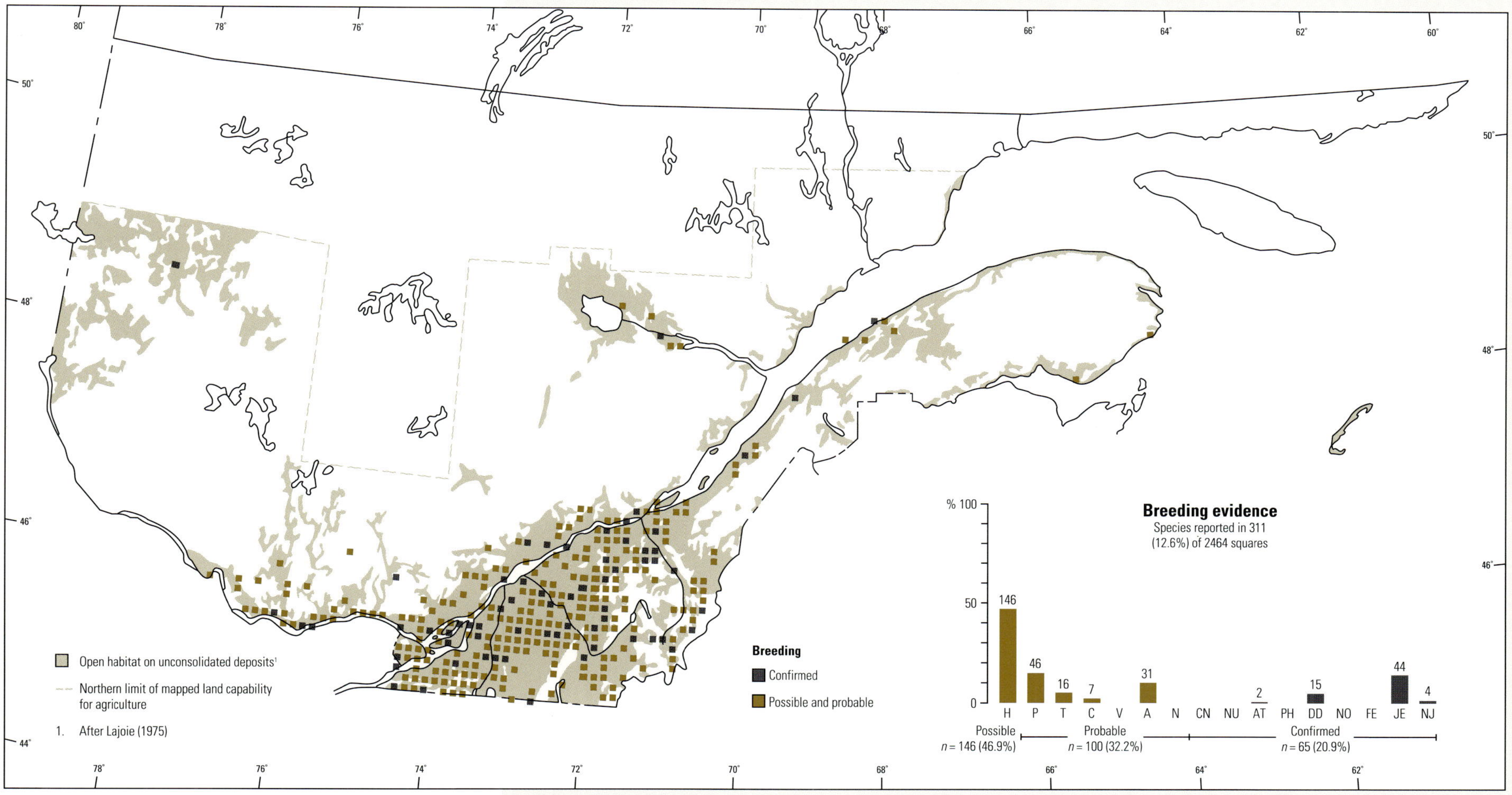

Species profile

Clutch size	usually 4 eggs (2–5)[1,2,3,4,5,8,9]
Laying interval	1 egg per 1–2 days[2,10]
Start of incubation	?
Incubation period	usually 24 days (21–28)[1,2,4,5,8,10]
Maturity at hatching	precocial
Care of young	by male, sometimes by both male and female[2]
Nestling period	less than 24 hours[1,2,8]
Age at first flight	≈ 30–34 days[2,5,8]
Dependence of young	until ≈ 32–34 days old[10]
Broods per year	1[2,3,8]
Breeding age	1 year[8]
Mating system	monogamy[2]
Duration of pair bond	usually 1 season[2]
Total length	28–32 cm[4]
Wing span	43.2–50.8 cm[5]
Weight	M 149 g (143–153) / F 190 g (165–218)[6]
Longevity record	8 years 5 months[7]

References:
1. Higgins and Kirsch 1975
2. Ailes 1976
3. Harrison 1978
4. Godfrey 1986
5. Terres 1980
6. Dunning 1984
7. Kennard 1975
8. DeGraaf and Rudis 1987
9. Peck and James 1983
10. Johnsgard 1981

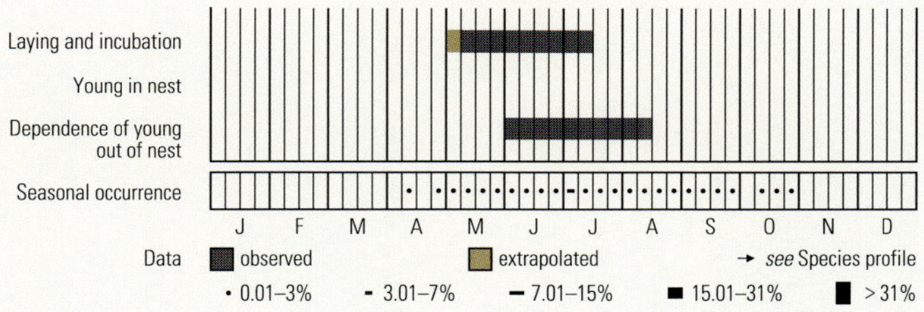

World breeding range

In 1989, an Upland Sandpiper nested in this field in Amos, a first in the Abitibi region; the species usually nests farther south in the province. *July 1991, André-Guy Bernier.*

Least Sandpiper

Bécasseau minuscule

Calidris minutilla

The Least Sandpiper is the smallest of the world's 214 shorebird species. It is slightly smaller than the other members of its genus, commonly called peeps, and is the only North American peep with greenish-yellow rather than black legs (Hayman et al. 1986). In flight, its call is a prolonged or repeated *scree-ee-ee*, similar to that of the Semipalmated Sandpiper, except for the long *ee* (Townsend 1927a).

The Least Sandpiper is a common migrant breeder in north-central Québec. In southern Québec, it breeds only in the Magdalen Islands and on Anticosti Island. As a transient, it is common along the St. Lawrence, especially on the shores of the gulf (Maisonneuve et al. 1990), but rare inland. Least Sandpipers winter along the Atlantic coast from North Carolina south through Mexico and Central America to central Brazil, and along the Pacific coast from Oregon south through California and Central America to central Chile (Cramp and Simmons 1983).

Habitat and behaviour

The Least Sandpiper is widely distributed, yet nests in similar habitat throughout its range (Miller 1983a). In the Arctic, it favours wet tundra with a mixture of tussock grasses and heath and scattered mossy hummocks (Johnsgard 1981). It is a regular breeder on the salt marshes of the Magdalen Islands (Moore 1912; Fradette 1992). In the boreal forest, it nests in peatlands and muskeg, and on Sable Island, Nova Scotia, it nests in the lush vegetation bordering small ponds (Miller 1983a).

Least Sandpipers are monogamous, and because males aggressively defend their nesting territories, breeding pairs are usually evenly spaced (Miller 1979). In the early part of the breeding season, males perform a display flight in which they rise to a height of 15–20 m and then alternate between glides and bouts of rapid fluttering. Flights last about 3 minutes, during which the male repeats a simple, monotonous call (Miller 1979, 1986).

Both sexes build the nest, typically a small depression in the ground on grass, moss, or a sedge hummock, and usually lined with dry leaves, grass, or moss (Godfrey 1986). Clutch size is normally 4 eggs, with the female laying only one clutch, though sometimes she will lay a second if the first clutch is lost (Miller 1983a). Incubation takes 19–23 days (Jehl and Smith 1970); the male incubates primarily during the day and the female at night (Miller 1985).

The chicks are precocial and leave the nest within 24 hours once the entire clutch has hatched (Miller 1983a). Their well-developed legs enable them to follow their parents and find their own food as soon as they leave the nest (Moore 1912). Chicks are strong fliers by 14–16 days of age, and though the female normally abandons the chicks within the first week or two after hatching, the male usually stays with them longer, up to 27 days (mean=20) (Miller 1985).

The diet of Least Sandpipers consists mostly of aquatic invertebrates, such as amphipods, midges and mosquitoes (including larvae and pupae), and small beetles, but also includes a few seeds (Townsend 1927a). In the breeding season, these sociable birds are often seen on mudflats with other shorebirds, like Semipalmated Sandpipers, plovers, Red Knots, and Dunlins. Least Sandpipers, however, usually remain on the drier, algae-covered mud of salt marshes and tidal flats in estuaries, where they forage by pecking for insect larvae and crustaceans (Baker and Baker 1973).

The primary breeding range of the Least Sandpiper in Québec lies north of the Atlas area, so it is not surprising that the species was reported in only 0.7% (18/2464) of the squares surveyed. It was listed as a probable breeder in only 2 and as a confirmed breeder in 1, based on a report of a bird performing distraction displays, behaviour usually indicative of a nest nearby.

Clodin Roy

Notes on territory

There is little detailed literature on territoriality in the Least Sandpiper, and the following account is drawn primarily from research carried out on Sable Island, Nova Scotia, by Miller (1979).

Sexually mature males establish, and defend against other males, territories that include the eventual nest site, but only occasionally areas used for foraging and brood rearing. These latter activities usually take place in undefended communal areas either close by or up to 2 km from the nest.

Males attract females to their territories using display flights over an area 200–400 m in diameter, and after pairing, the nest is built within this area, often near the periphery. In the study, the airspace where the display flights took place was sometimes used by a succession of unpaired males, resulting in some "overlay."

The closest nests on Sable Island were 100 m apart, and breeding densities were found to be highest when nests were surrounded by rich foraging areas. In a marsh on La Grosse Île in the Magdalen Islands, Moore (1912) found

Probably because of the prevailing maritime climate, the Least Sandpiper is found in the salt marshes of the Magdalen Islands, even though this is far south of its main breeding range. Étang-des-Caps, Magdalen Islands. *July 1983, Yves Aubry.*

Least Sandpiper

Breeding distribution
1984–1989

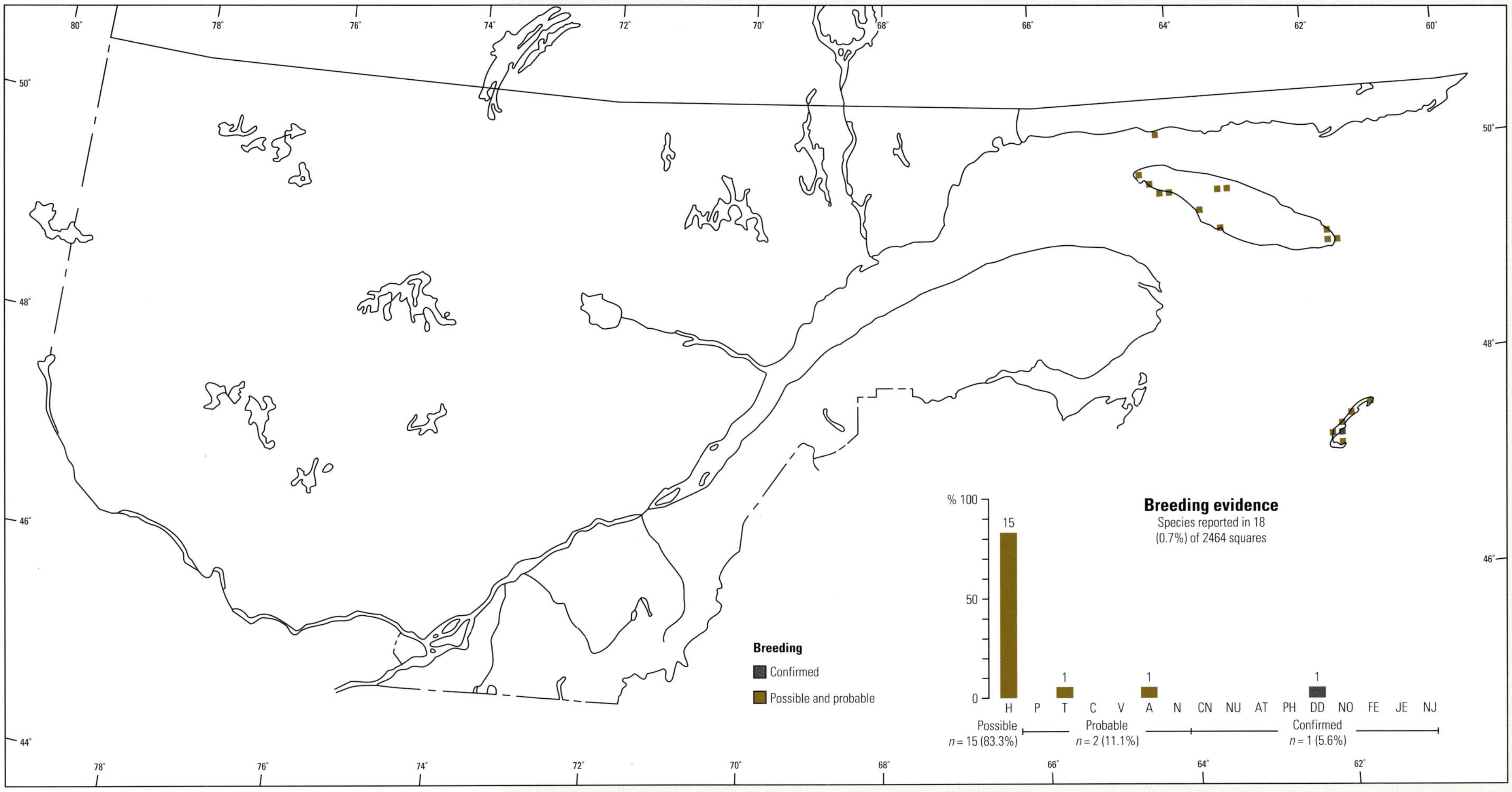

World breeding range

Species profile

Clutch size	usually 4 eggs[1,3,5,6]
Laying interval	1 egg per 1–3 days[6]
Start of incubation	usually with last egg[4,5,6]
Incubation period	19–23 days[1,6]
Maturity at hatching	precocial
Care of young	mainly by male[1,4,11]
Nestling period	≈ 24 hours[4,6]
Age at first flight	?
Dependence of young	until 14–27 days old[5]
Broods per year	1[1,5,6]
Breeding age	1 year[11]
Mating system	monogamy[5]
Duration of pair bond	sometimes more than 1 season[5]
Total length	12.5–17.0 cm[8]
Wing span	27.9–30.5 cm[2,7]
Weight	20.9 g (18.0–24.0)[10]
Longevity record	16 years[9]

References:
1. Johnsgard 1981
2. Cramp and Simmons 1983
3. Townsend 1927a
4. Moore 1912
5. Miller 1985
6. Miller 1983a
7. Terres 1980
8. Godfrey 1986
9. Klimkiewicz and Futcher 1989
10. Dunning 1984
11. Cooper 1994

Chronology n = 63

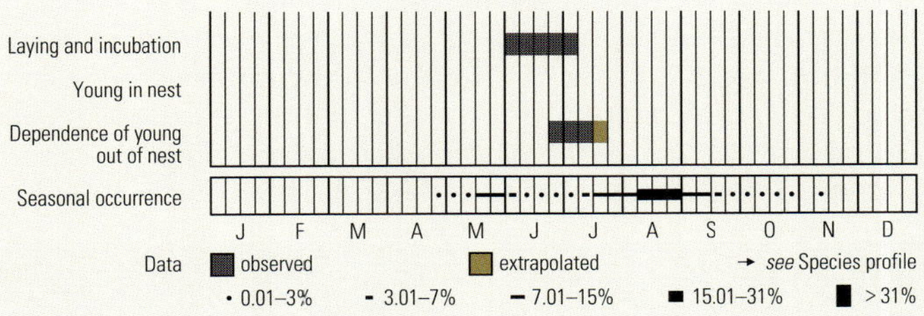

The grassy peatlands of Anticosti Island are one of this sandpiper's few nesting sites in southern Québec. West of Port-Menier. *August 1991, Gaétan Laprise.*

Notes on territory (continued)

5 nests within an area of 91 m in diameter. In the Great Whale and Little Whale river watersheds of northern Québec in 1990, this species was four times more abundant in peatlands with pools than in peatlands without pools (Cons. G. G. – GREBE 1991a).

Distribution

The Least Sandpiper breeds only in North America, ranging from southern Alaska to the Labrador coast and, very locally, as far south as Massachusetts along the east coast (Anderson 1980; Godfrey 1986). In northern Québec, the Least Sandpiper breeds in the boreal forest from James Bay and Hudson Bay east to Labrador between approximately 52° and 58° N (Godfrey 1986). It breeds in both the Great Whale and Little Whale river watersheds (Cons. G. G. – GREBE 1993c). All but one Atlas report of Least Sandpipers came from either the Magdalen Islands or Anticosti Island. The exception originated from near Longue-Pointe-de-Mingan on the Middle North Shore. The single confirmation of breeding came from the Magdalen Islands, considerably south of the main breeding range in northern Québec. These islands, located in the Gulf of St. Lawrence, enjoy a maritime climate (Fradette 1992), and the boggy habitat preferred by the Least Sandpiper for nesting is available there.

The Least Sandpiper migrates in a broad front through the interior of North America (Cramp and Simmons 1983), and is locally abundant in many areas of southern Québec during spring and fall migration (Maisonneuve et al. 1990). Birds nesting farther north pass through the Magdalen Islands during the last two weeks of May. In autumn, the greatest numbers of adults migrate through the islands between 15 July and 10 August, whereas juveniles pass through between 10 and 25 August (Mousseau et al. 1976). At Rupert Bay and Boatswain Bay, the Least Sandpiper can be seen from 20 July to 28 August (Cons. G. G. – GREBE 1992s). They migrate through the Montréal area in late May on their way north, and most of July, August, and September on their way south (Brousseau 1981).

Least Sandpiper nest with 4 eggs. The knife is shown to give the scale of the nest. East of Lake Bienville, north-central Québec. *June 1991, Serge Brodeur.*

This shorebird's nest is a shallow depression in low vegetation. The single egg suggests that laying has just begun. Étang-des-Caps, Magdalen Islands. *July 1983, Yves Aubry.*

Around Lake Bienville in north-central Québec, there are many peatlands where the Least Sandpiper nests. *July 1978, Jean-Pierre L. Savard.*

Adult on the nest, well camouflaged by the surrounding vegetation. East of Lake Bienville, north-central Québec. *June 1991, Serge Brodeur.*

History and trends

It is difficult to ascertain the historical breeding range of the Least Sandpiper, but the distribution described by Townsend (1927a) almost 70 years ago is similar to its current distribution, as described by Godfrey (1986). The earliest published record of the species breeding in Québec is probably a nest found by Audubon in 1833 at Natashquan, on the Lower North Shore (Todd 1963). Since then, this species has regularly been found breeding on Anticosti and the Magdalen Islands (Macoun and Macoun 1909; Fradette 1992).

Populations of many small shorebird species, probably including the Least Sandpiper, were drastically reduced between 1890 and 1916 by market hunters, but following enactment of the Migratory Birds Convention, numbers of these species quickly rebounded to their former levels in Canada and the United States (Townsend 1927a). Although the Least Sandpiper is now locally common or abundant through most of its breeding range (Johnsgard 1981), there is still cause for concern for this and all shorebird species in North America (Morrison and Ross 1989). Threats include hunting pressure on the wintering grounds in South America as well as environmental contamination and habitat loss and degradation in both South and North America (Senner and Howe 1984).

Richard C. Cotter and Derin Henderson

Short-billed Dowitcher

Bécasseau roux

Bécassin roux
Limnodromus griseus

Clodin Roy

THE SHORT-BILLED DOWITCHER IS A MEDIUM-SIZED SHOREBIRD THAT NESTS in the wetland areas of the boreal forest-tundra transition zone. Until the 1950s, this and the morphologically similar Long-billed Dowitcher were considered a single species (Pitelka 1950; Hayman et al. 1986). In flight, the conspicuous white patch on the back, showing as a slit between the tertials, and the rather plump outline distinguish dowitchers from other shorebirds (Pitelka 1950; Hayman et al. 1986). Although it is very difficult to differentiate the two dowitcher species by their plumage, Short-billed Dowitchers have a diagnostic call in flight, a mellow, rapid *tu-tu-tu* (Jehl 1963; Kaufman 1990a). In Québec, the Short-billed Dowitcher is a migrant breeder, whereas the Long-billed Dowitcher is a rare transient (David 1980a).

Habitat and behaviour

The Short-billed Dowitcher nests in the northern boreal forest in wet sedge and grassy meadows and peatlands with low vegetation (Godfrey 1986). In north-central Québec, in the vast Lake Plateau region around Lake Bienville, the Short-billed Dowitcher is found in areas with over 5% peatland cover. These extensive peatlands may be as small as 13.4 ha, but most are over 36 ha. Water is an important part of the species habitat; they particularly favour large pools bordered by herbaceous plants (Cons. G. G. – GREBE 1993c). In the region around Churchill, Manitoba, these birds nest in hummocky sedge marshes close to trees. Such areas are often patchily distributed in the boreal forest (Johnsgard 1981).

The nest of the Short-billed Dowitcher is usually a small depression lined with fine twigs, grasses, or leaves and situated on or near a hummock or clump of moss (Godfrey 1986). Nests are usually well concealed, often completely surrounded by water, making them difficult to locate (Harris 1989; Cons. G. G. – GREBE 1993c). As suitable nesting sites may be patchy in a peatland, many pairs may nest in one area (Johnsgard 1981). At Schefferville, birds are usually on the breeding grounds from mid-May until at least mid-August, with egg laying beginning in late May (Harris 1989).

Short-billed Dowitchers are monogamous, with pairs forming a strong bond during the incubation period (Johnsgard 1981). Early in the breeding season, the male performs a hovering flight song over the territory (Johnsgard 1981; Harris 1989). The song structure and element types are amazingly uniform across the species' range (Miller et al. 1983). Nest defence is variable. In some cases, both sexes will actively defend the breeding territory against intruders with short flights, calls, and other conspicuous behaviour. Neighbouring pairs may even join together in an attempt to lure an intruder away, making it very difficult to locate nests (Johnsgard 1981). Other individuals may sit quietly on the nest, flushing only when nearly stepped upon (Harris 1989).

Typical clutch size is 4 eggs, and both the male and female incubate. The precocial chicks hatch after 21 days and are soon mobile (Jehl and Hussell 1966). The fledging period is approximately 3–4 weeks. The male provides most, if not all, of the care to the brood (Jehl and Smith 1970; Hayman et al. 1986), with the female often leaving the breeding grounds soon after hatching (Hayman et al. 1986).

Short-billed Dowitchers forage in shallow water, rapidly probing the mud for aquatic insects, marine polychaetes, crustaceans, molluscs, seeds of aquatic plants, and eelgrass roots (Bannerman 1961). On the breeding grounds, the larvae and pupae of flies are a major part of the birds' diet (Johnsgard 1981). Although this dowitcher has been described as rarely aggressive toward conspecifics while foraging in flocks (Recher and Recher 1969), patchy food resources can elicit agonistic behaviour (Mallory and Schneider 1979).

Notes on territory

There are few detailed studies of territoriality in this species, but nest spacing appears to be a function of the availability of suitable nest sites (Johnsgard 1981).

T.-E. Randall believed solitary nesting was the most common system near Banff, Alberta (Bannerman 1961), but one year in a grassy marsh at the end of a shallow lake, he found 4 nests 50 m or more apart; a year later 3 were found in a clearing measuring about 18 m by 7.6 m.

In a study inland from Hudson Bay, in Québec, mean density in peatlands near Lakes Élizabeth and Kakupis was 1.4 pairs/100 ha in 1991 (Cons. G. G. – GREBE 1993c). Around Lakes Bienville and Amichinatwayach in 1990, the species was somewhat more abundant in peatlands with pools than in the ecotones between coniferous scrub and the peatlands (Cons. G. G. – GREBE 1991c).

A few Short-billed Dowitcher pairs nest in the vicinity of Schefferville, where they are found in openings or wetland areas in the boreal forest. *July 1980, Ross Harris.*

Short-billed Dowitcher

Breeding distribution

1984–1989

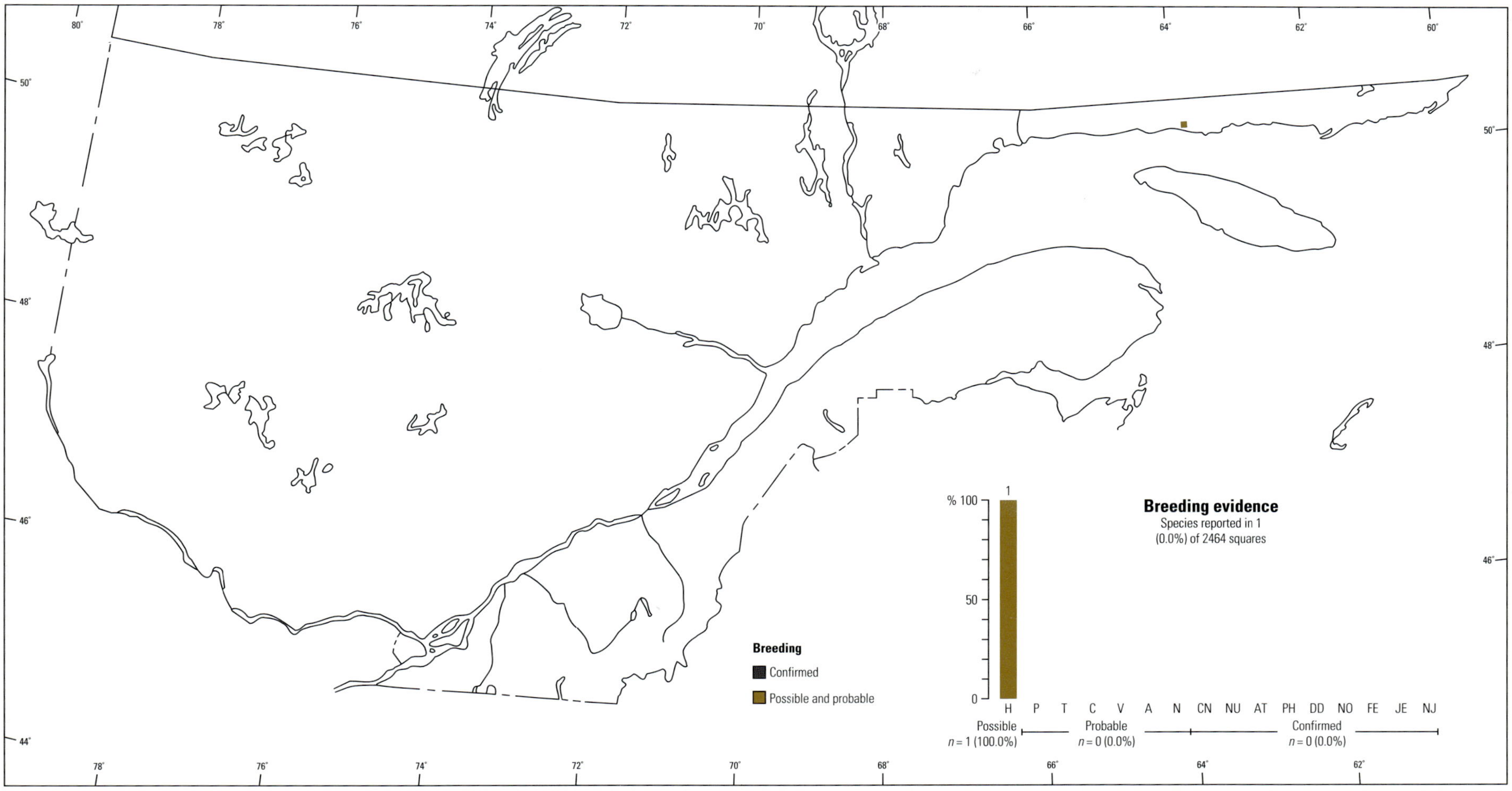

Species profile

Clutch size	usually 4 eggs (3–5)[1,2,3]
Laying interval	probably 1 egg per day[9]
Start of incubation	probably with last egg[9]
Incubation period	21 days[1,2]
Maturity at hatching	precocial
Care of young	mainly by male[1,2,4]
Nestling period	less than 24 hours[9]
Age at first flight	≈ 3–4 weeks[10]
Dependence of young	until ≈ 21 days old[9]
Broods per year	probably 1[9]
Breeding age	probably 2 years or more[9]
Mating system	monogamy[1,2,3,9]
Duration of pair bond	usually 1 season[9]
Total length	26.5–30.5 cm[8]
Wing span	48.3 cm[5]
Weight	M 111.0 g (73.0–152.0) / F 116.0 g (82.5–154.0)[6]
Longevity record	13 years 3 months[7]

References:
1. Johnsgard 1981
2. Ehrlich et al. 1988
3. Bannerman 1961
4. Cramp and Simmons 1983
5. Pough 1951
6. Dunning 1984
7. Klimkiewicz and Futcher 1989
8. Godfrey 1986
9. Harris 1989
10. Jehl and Smith 1970

Chronology *n* = 18

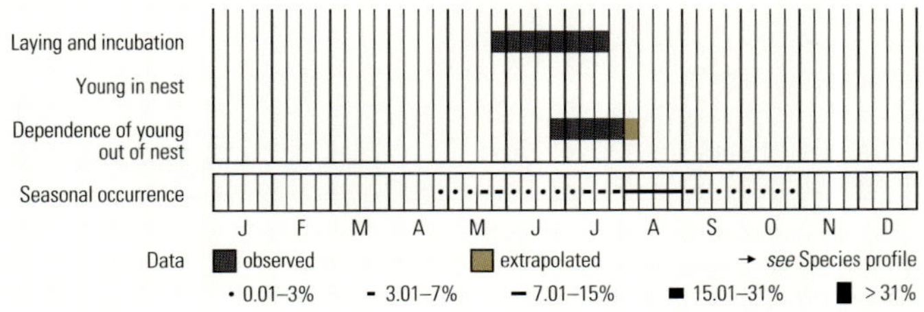

World breeding range

This brooding adult blends in with the surrounding vegetation. Schefferville. *July 1978, Ross Harris.*

Distribution

There are three subspecies of the Short-billed Dowitcher, each breeding in disjunct regions in North America. The eastern population, *Limnodromus griseus griseus*, breeds across north-central Québec and western Labrador from approximately 52° N up to Ungava Bay in the north, and from James Bay and southeastern Hudson Bay in the west to central Labrador in the east (Godfrey 1986). Its range is probably co-extensive with the peatlands throughout this region (Todd 1963). The central population, *Limnodromus griseus hendersoni*, nests primarily in the boreal forest in the western provinces, whereas the western population, *Limnodromus griseus caurinus*, nests in similar habitat in northwestern British Columbia and southern Alaska (Pitelka 1950; Godfrey 1986). In northern Québec, the Short-billed Dowitcher has been confirmed as breeding in and around Lakes Bienville, Élizabeth, and Kakupis (Cons. G. G. – GREBE 1993c).

The eastern population migrates to and from breeding grounds along the Atlantic coast, passing through the Gulf of St. Lawrence (McNeil and Burton 1977; Hayman et al. 1986). Outside the breeding season, Short-billed Dowitchers have a strong habitat preference for coastal areas, in particular mud- and sandflats in sheltered bays and estuaries, or borders of shallow pools on salt marshes

Short-billed Dowitchers arrive in the Schefferville area by about mid-May, when there is still snow on the ground; as the thaw progresses, meltwater feeds pools around which herbaceous vegetation then grows, providing suitable cover. *June 1978, Ross Harris.*

The Short-billed Dowitcher seeks out large ponds with shallow, grassy margins. Schefferville. *August 1978, Ross Harris.*

(Cramp and Simmons 1983). They are often seen among small groups of plovers and other sandpipers (Bent 1929). In fall migration, they are common in many areas along the estuary and Gulf of St. Lawrence (Bourget 1990), particularly in the Magdalen Islands (Mousseau et al. 1976; Fradette 1992). They are occasionally seen in and around Montréal in autumn, but rarely in spring (Ouellet 1974a; Bannon 1991a). In the Magdalen Islands, failed breeders arrive as early as 29 June, but the largest concentrations of birds stop over on their southward migration between 20 July and 10 August for adults and 10 and 25 August for juveniles (Mousseau et al. 1976; Fradette 1992). Short-billed Dowitchers migrate in three waves in the fall: the adult females migrate first, then the adult males, and finally, the juveniles migrate as a group (Jehl 1963).

The eastern population winters along the Atlantic coast from North Carolina south to Florida, in the Caribbean, and south to the coast of Brazil (Pitelka 1950; Hayman et al. 1986). The largest concentrations of birds occur in the northern part of South America (Morrison and Ross 1989).

The Short-billed Dowitcher was reported in only one Atlas square, on the North Shore of the St. Lawrence near Mingan: between 13 and 20 July 1986, an adult was sighted calling as it flew toward some peatlands. The date, type of habitat, location, and behaviour suggested the possibility of nesting in the area, but breeding was not confirmed.

History and trends

The Short-billed Dowitcher has a wide breeding range across North America, most of it lying in sparsely populated regions. Consequently, it is very difficult to ascertain the historical or current population of this dowitcher on the breeding grounds (Johnsgard 1981). Although the species is now protected from hunting in North America, as are most shorebirds, this was not always the case. This shorebird was a common fall game bird along the eastern seaboard of the United States, and before regulations were implemented early this century, populations were severely depleted (Bent 1927; Bannerman 1961). In South America, hunting of shorebirds continues today (Senner and Howe 1984). This factor, but even more, continuing habitat disturbance on wintering, breeding, and migration grounds, points to the need for a hemispheric management and conservation plan for all shorebirds (Morrison 1983; Morrison et al. 1991).

Richard C. Cotter

In northern Québec, this shorebird is found in wet meadows with low vegetation. Schefferville. *August 1978, Ross Harris.*

Common Snipe

Bécassine des marais

Gallinago gallinago

Clodin Roy

THE COMMON SNIPE'S SPECTACULAR AERIAL DISPLAY TAKES THE FORM OF a series of ascents and plunges, accompanied by mournful sounds that seem to come out of nowhere. This species has a worldwide distribution and is well known, partly because of its long-standing popularity with hunters. It used to be much more abundant, but overhunting decimated its populations at the turn of the century. Nevertheless, this migrant breeder is common almost everywhere in Québec south of the tree line.

The dark, spotted plumage of this medium-sized shorebird lets it blend in easily against its wetland habitat as it methodically treads the soft ground, probing for food with its bill. Once you see the snipe, however, its very long, straight bill is highly noticeable.

The Common Snipe is not nearly as stocky as the American Woodcock, a relative that inhabits woodlands, but like the woodcock, the snipe has its eyes set far back on the head, giving it excellent peripheral vision (Terres 1980). Although there are many subspecies of the Common Snipe throughout the world, *Gallinago gallinago delicata* is the only one found in North America.

Habitat and behaviour

The Common Snipe seems to migrate northward as the ice thaws and the meltwaters collects in puddles. The males arrive in Québec beginning around mid-April. They may return to their territories of the year before, which they mark out with their display flight, flying about 100 m high (Terres 1980) and making short plunges with tail feathers spread. The air passing through the outer tail feathers produces a loud noise that is modulated by the beating of the wings, which controls the way the air arrives at the tail. The resulting *who-who-who-who* carries for a kilometre. This display becomes especially frequent after the females arrive, 10–14 days after the males. Territories are fiercely defended against displaying birds, whereas those wishing to feed there are tolerated (Tuck 1972). This territorial behaviour lasts until hatching, after which point other males looking to breed may take over the territory.

The Common Snipe is associated with peatlands, freshwater marshes, and wet or flooded fields and pastures; it prefers waterlogged soils rich in organic matter, and the shallow edges of marshes are its favourite habitat in the nesting period. Various species of moss, sedge, grass, and small shrubs provide the low plant cover and substrate favourable to the snipe. It tends to avoid areas where the vegetation is too high and dense (Fogarty et al. 1980).

The species is monogamous, but the female is somewhat promiscuous early in the season. Later she gradually confines herself to the territory of a single male, which then follows her closely: the pair remains together from the time the female chooses the nesting site until the eggs have hatched (Tuck 1972).

Notes on territory

As soon as he arrives in spring, the male selects and defends a territory (Tuck 1972; Fogarty et al. 1980), which may cover as much as 10 ha at the start of the breeding season, but shrinks bit by bit during the incubation period (Tuck 1972; DeGraaf and Rudis 1987).

The territory is generally defended in the airspace above it, through display flights and pursuit flights to intimidate intruders. The male gradually stops defending the nest site once the eggs have hatched.

Tuck (1972) assessed the density of breeding pairs in the peatlands of Newfoundland at over 1.7 pairs/10 ha in the breeding season.

The Common Snipe feeds along the shallow edges of tidal pools in salt marshes. Baie de L'Isle-Verte National Wildlife Area, Lower St. Lawrence. *July 1993, Michel Robert.*

The Common Snipe is encountered mainly in open wetland habitats edged with willows and alders. Here, an almost dried-out marsh invaded by sensitive ferns and horsetail. Baie-du-Febvre, south shore of Lake Saint-Pierre. *May 1991, Réginald Ouellet.*

Common Snipe

Breeding distribution
1984–1989

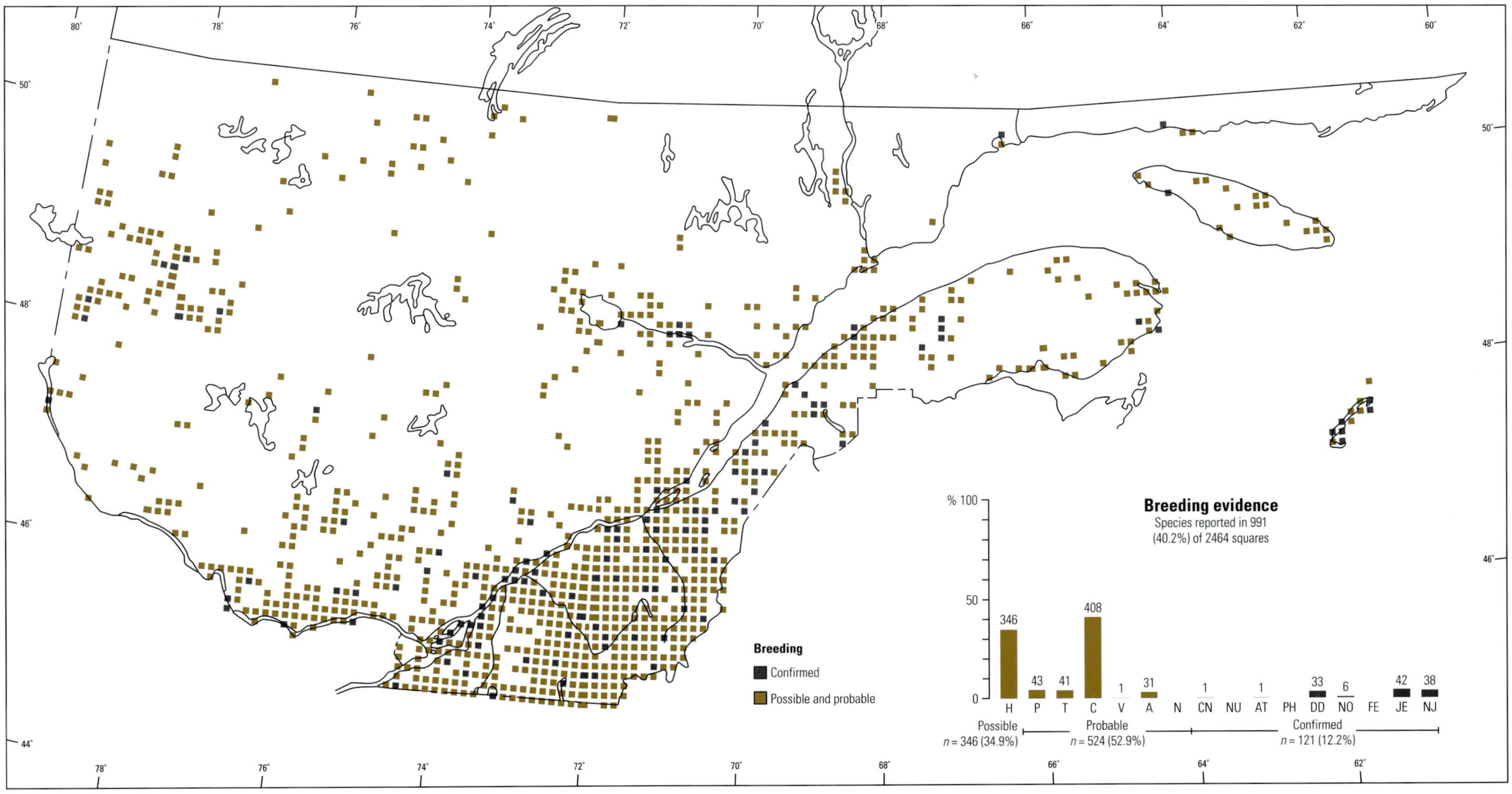

Species profile

Clutch size	usually 4 eggs (3–5)[1,3,6,7,8]
Laying interval	1 egg per 2 days[10]
Start of incubation	with third egg[5,8]
Incubation period	17–20 days[1,5,6,8,9]
Maturity at hatching	precocial
Care of young	by female and male[8]
Nestling period	less than 24 hours[7,8]
Age at first flight	17–20 days[5,6,8]
Dependence of young	until 17–20 days old[1]
Broods per year	1[7]
Breeding age	1 or 2 years[7,8]
Mating system	monogamy[1,8]
Duration of pair bond	1 season[1]
Total length	26–30 cm[9]
Wing span	44.4–50.8 cm[3]
Weight	M 128 g (?–156) / F 116 g (?–156)[4]
Longevity record	9 years 3 months[2]

References:
1. Ehrlich et al. 1988
2. Kennard 1975
3. Terres 1980
4. Dunning 1984
5. Johnsgard 1981
6. Fogarty et al. 1980
7. DeGraaf and Rudis 1987
8. Tuck 1972
9. Godfrey 1986
10. Mason and Macdonald 1976

Chronology n = 144

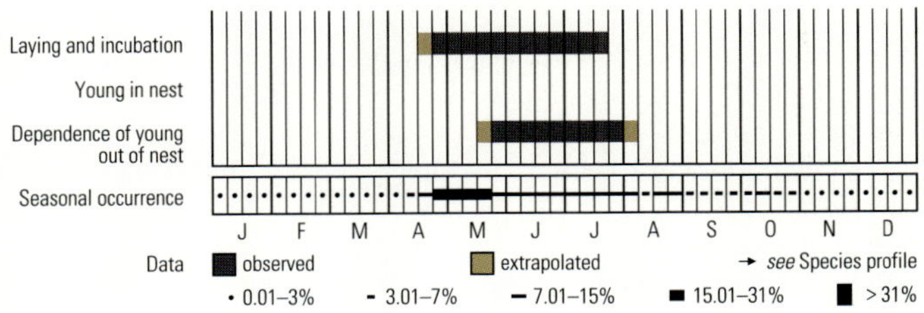

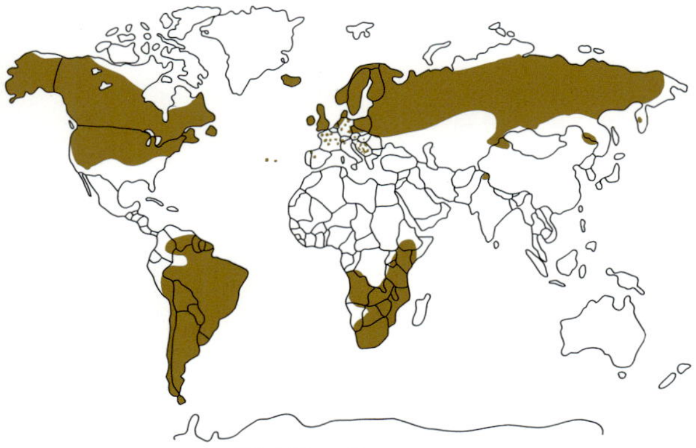

World breeding range

Notes on territory (continued)

In the Great Whale River area, during surveys in June and July 1990, close to three times as many pairs of Common Snipes were observed in peatlands with pools and on rock outcrops than in peatlands without pools and in the shoreline ecotone of coniferous woodland lakes (Cons. G. G. – GREBE 1991a).

Type of vacant lot on city outskirts used by the Common Snipe. Grasses and forbs dominate here, but the progressive invasion of shrubs will eventually make the site uninhabitable by the species. Saint-Pierre, Island of Montréal. *May 1938, Henry Mousley, Laval University Collection.*

The female digs a slight depression in the ground, in a location drier than the rest of the habitat (Johnsgard 1981), where the vegetation is low. She then lines the nest with fine grass and sometimes moss, often covering it with a canopy made of woven dry grass. She typically lays 4 eggs and incubates them for 17–20 days; all hatch within 4–8 hours (Tuck 1972).

An hour after hatching, the downy young are dry and ready to leave the nest. The pairs then divides up the brood; the male takes the first one or two chicks (Tuck 1972), but the female usually tends to take care of more offspring (Tuck 1972; Terres 1980). The pair bond is thus dissolved, with each adult rearing its chicks separately. For the first few days, the chicks are fed by a parent, but then they begin to feed themselves in adult fashion, remaining with the parent bird until they are 17–20 days old, at which time they begin to make sustained flights (Ehrlich et al. 1988). A single brood is produced each year, and individuals attain sexual maturity at 1 or 2 years (Tuck 1972).

The Common Snipe feeds mainly on larval and adult insects and earthworms, but also on small crustaceans, other small animals, and plant debris—in short, anything within reach when it plunges its very sensitive bill deep into the damp soil of its habitat or along a shore. The upper mandible is flexible and prehensile, enabling the bird to catch prey underground by opening just the tip of the bill without withdrawing it from the ground. When it feeds, the snipe moves very slowly in a semicircle, probing the ground at various locations. It drinks copious amounts of water, and regurgitates pellets of indigestible parts of food (Ehrlich et al. 1988).

The eggs, chicks, and adults are cryptically coloured, providing camouflage and protection against predators. Scarcely an hour after hatching, the young are able to hide motionless in the vegetation (Cramp and Simmons 1983). This means that to find them, you have to walk directly toward the nest or the young, or get close enough so that an adult attempts to distract you. The low percentage of confirmed breeding records (12.2%) is reflective of the fact of how difficult it is to confirm breeding in this species. Moreover, the wetland haunts of the snipe are often hard to get to.

On the other hand, the highly visible display flight alone accounted for 41.2% (408/991) of Atlas records. It may be executed by either sex for various reasons, such as establishing a territory, attracting a female, or even intimidating predators, and is a sign of probable breeding. The snipe is active mainly at dusk, and that is when it most frequently performs the display flight. The bird can be heard at any time during the day, however, and even on a clear night, but especially early in the season; after that the species is less conspicuous.

When it feels threatened, the snipe freezes, then takes off at the last minute, flying in a zigzag, and emitting a rasping *scaip* (Peterson 1980). It often sits on a stump, fence post, or other perch.

Less gregarious than other shorebirds, the Common Snipe is occasionally found in large numbers in wetlands, including the bulrush and cord-grass marshes along the St. Lawrence River. In fall, snipes migrate mainly at night, flying in tight formation (Terres 1980). Some

The female builds a nest of grass in a dry location; she often arches over and interweaves the surrounding grass to hide the nest even better. Sainte-Foy. *May 1966, Gaston Le Page.*

Common Snipe on its eggs in a field at Saint-François-du-Lac. Both parents share incubation. Lake Saint-Pierre. *June 1981, Daniel Jauvin.*

stragglers do not begin migrating until November. The species may exceptionally be seen in winter in some parts of southern Québec, where there are a few well-established wintering areas (David 1980a; Aubry et al. 1991).

The Common Snipe winters primarily in the United States along the Atlantic and Pacific coasts, as well as in the southern states south through Central America to southern South America (AOU 1983).

Common Snipe nest on a clump of vegetation in a wetland area. Ahuntsic, Montréal. *May 1943, Henry Mousley, Laval University Collection.*

Distribution

The Common Snipe is a holarctic species; it breeds mainly within the limits of the boreal forest, preferably in peatlands (Cramp and Simmons 1983), and its range extends up onto the tundra (Terres 1980).

The Atlas map shows that the snipe prefers open areas, where it is much easier to detect. In forested areas, its distribution is discontinuous; openings are few, and the bird is harder to spot. North of the St. Lawrence, it has been seen in cleared or cultivated areas, especially in the Saguenay–Lac-Saint-Jean region, near Chibougamau, north of the Ottawa Valley, and in the Abitibi. The species is common on Anticosti Island (Ouellet 1969) and in the Magdalen Islands (Fradette 1992). In northern Québec, it occurs up to the tree line (Todd 1963; Godfrey 1986).

North of the Atlas area, breeding has been confirmed around the Nottaway, Broadback, and Rupert rivers, with a report of a nest containing 4 eggs in Cabbage Willows Bay (Cons. G. G. – GREBE 1992g). Farther north, around the Great Whale River, the Common Snipe is considered a probable breeder (Cons. G. G. – GREBE 1993c).

History and trends

In the past, the Common Snipe was hunted in great numbers, at least locally. Between 1867 and 1887, a single Louisiana hunter killed a total of 69 087 snipes (Bent 1927; Tuck 1972). In the early part of the century, on the shores of the St. Lawrence near Québec City, two hunters, without reaching a comparable scale, each bagged several hundred snipes a year (Tuck 1972). Nonetheless, various historical accounts may give the impression that the snipe hunt was more extensive than it actually was because the terms "snipe shooting" or "snipe hunting" often referred to the hunting of shorebirds in general (Mackey 1945).

Prior to 1917, the species was hunted as much in migration as on its wintering grounds. Since then, the hunt has been regulated and restricted to a single period of the year. In Québec, a daily bag limit of 25 birds was set in 1920, and this was reduced to 8 in 1944. In the United States, the snipe hunt was banned between 1941 and 1953; it was then allowed again, with a daily bag limit (Tuck 1972). But the tradition of the hunt had been broken, and since it was reopened in 1954, fewer hunters have been taking part.

Still, it is estimated that close to 900 000 snipes are taken each year in North America. It is thought, however, that the total population may be greater than 5 million birds (Fogarty et al. 1980). In Canada, the hunt seems to be most intensive in Québec; between 1986 and 1990, 15 000–25 000 birds were shot each year, or more than 40% of all snipes taken in the country (Boersma 1990; Legris and Lévesque 1991a).

North American population trends cannot be discerned on the basis of the most recent studies (Johnsgard 1981; Robbins et al. 1986). A decline has been noted in Common Snipe populations in several countries of the world, however, probably due to habitat loss as a result of wetlands being drained both for agriculture and for urban expansion (Cramp and Simmons 1983). The North American snipe population is not immune to such a threat.

Jean-Denis Gauvreau and Jean-François Rail

American Woodcock

Bécasse d'Amérique

Scolopax minor

Clodin Roy

THE AMERICAN WOODCOCK IS A FASCINATING BIRD—IT TESTS THE SKILL OF hunters and its spectacular courtship display and strange appearance arouse curiosity. The 6–7 cm long bill with its soft, membranous covering (Pettingill 1936) is particularly noteworthy. The woodcock uses the flexible, tactile tip of its bill to probe the ground for prey; once detected, it opens only the tip of the upper mandible to seize it (Sheldon 1971; Owen et al. 1977; Terres 1980; Baird 1987). Its eyes are also unusual: set far back on the head, they are less likely to come into contact with the ground while the bird is feeding (Cobb 1959), and their large size and position give the woodcock 360° vision for excellent predator detection (Peterson 1969). It has evolved so that the brain is tipped back in the skull, placing the ears in front of the eye sockets (Cobb 1959; Baird 1987).

The woodcock nest is a depression in the ground lined with a few leaves and blades of grass. Lake Saint-Pierre area. *May 1991, Réginald Ouellet.*

With its short, rounded wings, the woodcock explodes into flight and manoeuvres very easily through densely wooded areas (Owen et al. 1977). It cannot fly for long periods, however (Bérubé and Couture 1986), and in migration, it travels in stages, never more than a short distance at a time (Baird 1987). The outer three primaries, stiff and narrower than the others, produce a whistling sound when the bird bursts into flight from the ground (Terres 1980) and during its courtship flights (Sheldon 1971).

The male and female have identical plumage, but the female is larger (Terres 1980). Their dead-leaf coloration is very effective camouflage when they are crouching on the forest floor (Sheldon 1971; Owen et al. 1977).

The American Woodcock is a crepuscular bird (Caldwell and Lindzey 1974). It belongs to the family Scolopacidae, but unlike most other shorebirds it also inhabits woodlands (Sheldon 1971; Owen et al. 1977; Gregg 1984).

The woodcock, a migrant breeder in Québec, winters in the southeastern United States, including eastern Oklahoma, southern Missouri, Tennessee, and Virginia, as well as the Atlantic coast all the way to Florida, the Gulf of Mexico, and eastern Texas (AOU 1983).

Habitat and behaviour

The American Woodcock is one of the first birds to return to its breeding grounds in spring, often even when there is still snow on the ground (Owen et al. 1977; McAuley et al. 1990). Each year, Québec woodcock hunters report seeing some in March, especially near Montréal (P. Dupuis, CWS, unpubl. data), although arrival dates may vary from year to year, depending on the weather (R. Couture, UQTR, pers. comm.). The males reach the breeding grounds before the females and immediately begin to establish their territories (Peppard 1963; Owen et al. 1977).

The American Woodcock needs a variety of habitats for its full life cycle (Gutzwiller et al. 1982). Its breeding grounds therefore include singing grounds, day cover, nesting cover, rearing habitat, roosting habitat, and autumn habitat. Should any of these habitats be lacking, the density or productivity of breeding populations may be affected (Gutzwiller et al. 1982).

The woodcock seems to have three main requirements of its various habitats: damp ground, since it feeds chiefly on earthworms (Bérubé and Couture 1986), shrubby escape cover for protection from predators (Rabe 1977), and watering places that it probably uses for bathing and preening (Bérubé and Couture 1986).

The American Woodcock can generally be considered an edge species (Bérubé and Couture 1986; Dupuis 1990). Along the gradient of plant succession from field to mature forest, it is found in habitats at early seral stages. Such habitat includes abandoned fields, alder and trembling aspen thickets, and second-growth (15–30-year-old) hardwood stands with scattered openings (Bérubé and Couture 1986; Couture 1990). It uses these various habitats for different purposes (Couture 1990). These are areas evolving unevenly through time; for example, a farmer's field takes longer to revert to mature forest than does a cutover, making it habitable by the woodcock longer (Couture 1990). The American Woodcock is hardly ever encountered in mature forests (Terres 1980; Bérubé and Couture 1986; Couture 1990).

Old abandoned fields seem to be the best woodcock habitat (Coulter and Baird 1982; Owen and Galbraith 1989). This habitat usually has a mix of stands that can be used over the course of the breeding cycle as well as in fall (Owen et al. 1977).

Singing grounds are sites where males display and copulation takes place (Redmond and Keppie 1988). They are found in the earliest seral stages (Sheldon 1971; Kinsley et al. 1982; Couture 1990), where the scrub is neither too dense, so that it does not interfere with the courtship display, nor too low, so it does not increase the risk of predation. Wishart and Bider (1976) reported that on the singing grounds they studied in Québec, there is rarely more than 30% of scrub cover higher than 0.9 m, and never more than 50%, while trees taller than 4.6 m never account for more than 8% of the plant cover.

In short, openings surrounded by scrub, like fields and clearings in young woods, are the preferred settings for courtship (Sheldon 1971; Wishart and Bider 1976; Owen et al. 1977; Gutzwiller et al. 1982). Size of clearings seems to be quite variable; the opening must be large enough and the nearest fringing vegetation low enough for the bird to be able to perform its courtship flight easily (Sheldon 1971; Wishart and Bider 1976).

The female looks for a wooded area, here partly open, with dense, tall grass to make her nest. Sainte-Foy. *September 1991, Yves Aubry.*

Notes on territory

The territory defended by the male American Woodcock is used solely for courtship and copulation (Sheldon 1971; Owen et al. 1977; Bérubé and Couture 1986; Redmond and Keppie 1988). The female does not defend the area around the nest (DeGraaf and Rudis 1987).

Males are loyal to their territories and often occupy the same ones throughout the courting season,

American Woodcock

Breeding distribution

1984–1989

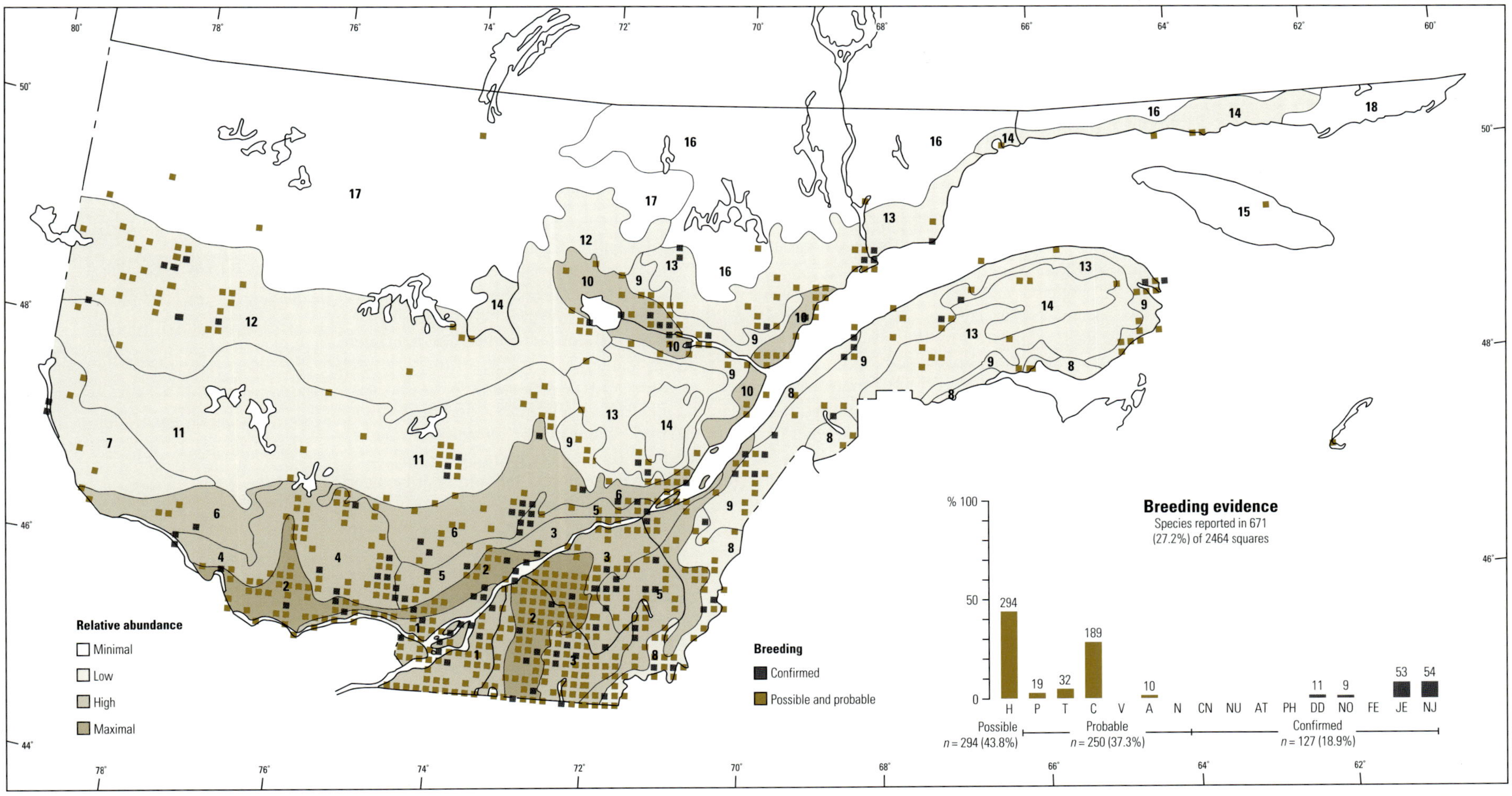

Species profile

Clutch size	usually 4 eggs (3–5)[1,4,8,12,13]
Laying interval	1 egg per day[7,8]
Start of incubation	with last egg[7,9,19]
Incubation period	19–21 days[1,4,7,10,12]
Maturity at hatching	precocial
Care of young	by female[3,9,15,16]
Nestling period	less than 24 hours[17]
Age at first flight	14–15 days[3,4,7,18]
Dependence of young	until 31–38 days old[11]
Broods per year	1[3,8,12]
Breeding age	1 year[12]
Mating system	promiscuity[7,8], polygyny[12]
Duration of pair bond	probably during mating only[8]
Total length	25.5–30.5 cm[1]
Wing span	46 cm[2]
Weight	M 176 g (?–222) / F 219 g (?–278)[5]
Longevity record	11 years 4 months[6]

References:
1. Godfrey 1986
2. Terres 1980
3. Ehrlich et al. 1988
4. Harrison 1978
5. Dunning 1984
6. Klimkiewicz and Futcher 1989
7. Johnsgard 1981
8. Sheldon 1971
9. Pettingill 1936
10. Bent 1927
11. Horton and Causey 1982
12. DeGraaf and Rudis 1987
13. Owen et al. 1977
14. Bérubé and Couture 1986
15. Dwyer et al. 1982
16. Redmond and Keppie 1988
17. Peppard 1963
18. Sepik et al. 1981
19. Keppie and Whiting 1994

World breeding range

Chronology *n* = 148

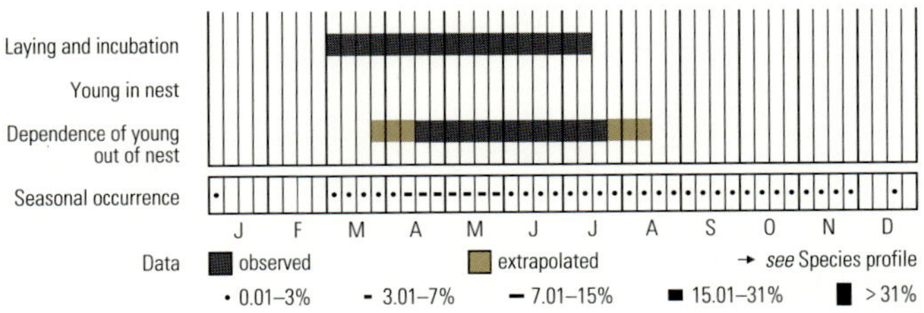

Frequency of occurrence by ecological region

Ecological regions** (forest regions)	Adequately covered squares	Squares in which reported	% of squares in which reported	Relative abundance (%)*
1. Sugar maple-hickory and sugar maple-basswood	97	65	67.0 ± 4.7	89.9
2. Sugar maple-basswood	153	114	74.5 ± 3.6	100.0
3. Sugar maple-basswood and sugar maple-yellow birch	101	65	64.4 ± 5.7	86.4
4. Sugar maple-yellow birch-basswood	56	33	58.9 ± 11.3	79.1
5. Sugar maple-yellow birch-beech and sugar maple-yellow birch-basswood	113	61	54.0 ± 5.1	72.4
6. Sugar maple-yellow birch-beech	71	35	49.3 ± 10.6	66.2
7. Sugar maple-yellow birch-balsam fir	19	5	26.3 ± 21.0	35.3
8. Sugar maple-yellow birch and balsam fir-yellow birch	47	20	42.6 ± 12.7	57.1
9. Balsam fir-yellow birch	87	37	42.5 ± 9.6	57.1
10. Balsam fir-white birch or balsam fir-red maple	46	26	56.5 ± 12.4	75.9
11. Yellow birch-balsam fir	52	19	36.5 ± 13.5	49.0
12. Balsam fir-white birch (west)	59	26	44.1 ± 13.0	59.1
13. Balsam fir-white birch (east)	76	27	35.5 ± 10.5	47.7
14. Balsam fir-black spruce	28	7	25.0 ± 17.0	33.6
15. White spruce-balsam fir and balsam fir-white spruce	22	1	4.5 ± 10.3	6.1
16. Black spruce-balsam fir-moss	18	2	11.1 ± 17.4	14.9
17. Black spruce-moss	32	3	9.4 ± 11.7	12.6
18.		*not analysed*		

* minimal: 0–29.9%; low: 30–59.9%; high: 60–89.9%; maximal: 90–100%

** Deciduous forest: 1–7; mixed forest: 8–11; coniferous forest: 12–17

The woodcock's day cover is generally not far from the male's courting ground (Sheldon 1971; Wishart and Bider 1976). Kinsley et al. (1982) reported that there is generally less than 100 m between the two types of habitat, while Sheldon (1971) estimated the distance at less than 180 m in most cases. In comparison with singing grounds, day cover is in much less open areas, either in alder thickets or regenerating hardwood and mixed stands (Owen et al. 1977). Alder thickets are chiefly used in fall (Wenstrom 1974). Woodlands are inhabited in summer, and are used as day cover by males and nesting and rearing cover by females (Couture and Bourgeois 1974; Wishart and Bider 1976; Rabe 1977).

Nesting and rearing habitats seem similar at first glance, but closer examination shows notable differences (Bourgeois 1977). Nest sites have taller, thicker herbaceous vegetation, a sparser shrub layer, and drier ground than the rearing hatitat (Bérubé and Couture 1986). In spring, females select well-drained, snow-free sites, where the nest is at little risk of flooding and the clutch is concealed in the herbaceous vegetation (Bourgeois 1977; Bérubé and Couture 1986). Once the eggs have hatched, the females lead their young to damp ground with plenty of earthworms where the sparse grass makes it easy for the family to move around (Sheldon 1971; Bérubé and Couture 1986). The shrub layer in such places is dense enough to protect the birds against avian predators, while not so thick that it prevents them from fleeing danger (Wenstrom 1974; Bourgeois 1977; Rabe 1977). Bérubé and Couture (1986) also reported the presence of thickets that could provide additional protection to the broods on the rearing grounds they studied in Québec.

Roosting sites where woodcocks flock at night are open areas which often serve as courting grounds in the spring (Sheldon 1971; Wishart and Bider 1976; Owen et al. 1977). The woodcocks go there at dusk from their day cover, generally flying, but sometimes walking (Dunford and Owen 1973; Owen and Morgan 1975). Despite Sheldon's (1961) claim that woodcocks visit nocturnal sites to feed, it seems that they actually use them solely for roosting (Wishart and Bider 1976). It has been demonstrated that woodcocks are not very active in the nocturnal habitats (Dunford and Owen 1973) and that they feed before going there (Krohn 1970, 1971).

Dunford and Owen (1973) further determined that adults travel a mean distance of 170 m between their day and night habitats, as compared with 330 m for young birds. This difference may explain the lower survival rate among the young, given their greater exposure to predators, especially raptors, during these trips (Derleth and Sepik 1990).

The American Woodcock's autumn habitats are varied. It is found most often in alder thickets, but can also be seen in poplar or trembling aspen stands, in young hardwood forests on hillsides, and near abandoned farms (Baird 1987).

The courtship display of the American Woodcock has two phases: a flight display and a ground display. The flight display begins with a low-level (1.5–3 m) flight over a distance of about 15 m (Sheldon 1971), followed by a vertical ascent, during which the male spirals up to a height of 60–90 m, producing a whistling sound with his wings (Peppard 1963; Sheldon 1971; Terres 1980; Baird 1987). The descent has been described as a spiral to the ground (Peppard 1963; Baird 1987) or as a zigzag flight like a fluttering leaf (Sheldon 1971; Terres 1980). As he descends, the male emits a musical chirping (Peppard 1963; Sheldon 1971). A complete flight display lasts about a minute (Peppard 1963; Owen et al. 1977). As soon as he lands on his territory, the bird begins his ground display. He pivots, giving a nasal *peent* call in several directions (Peppard 1963; Sheldon 1971; Baird 1987). This display lasts several minutes, after which the bird begins another flight display (Peppard 1963; Sheldon 1971). The male continues his courtship behaviour over a 30–60 minute period, at dawn and dusk, repeating the flight and ground displays 10–20 times (Peppard 1963; Sheldon 1971; Owen et al. 1977; Redmond and Keppie 1988). The duration of the display depends on the weather. If it is cloudy, rainy or windy, the flight phase is shorter and the ground phase is longer. When it is cold, the whole display is shortened (R. Couture, UQTR, pers. comm.).

The male performs the courtship display to attract a female and to mate with her, and to defend his territory against other males (Peppard 1963). According to Duke (1966), courtship behaviour peaks around the hatching period. The male woodcock does not defend his mate against other males (Redmond and Keppie 1988).

The American Woodcock's nest is a depression in the ground sparsely lined with leaves and grass (Peck and James 1983; Godfrey 1986). The female builds it by herself (Johnsgard 1981), generally at the foot of a tree or shrub (Sheldon 1971; Johnsgard 1981; Peck and James 1983), less than 10 m from the edge of the wood (Couture and Bourgeois 1974; Gregg 1974; Owen et al. 1977; Coon et al. 1982; Bérubé and Couture 1986). Typical clutch size is 4 eggs (Bent 1927; Pettingill 1936; Sheldon 1971; Peck and James 1983; Godfrey 1986), incubated by the female alone (Pettingill 1936; Godfrey 1986) for about 20 days (Bent 1927; Johnsgard 1981; Godfrey 1986).

It is very hard to spot and flush an incubating hen from her nest. When danger approaches, she tends to freeze and rely on the camouflage of her plumage (Sheldon 1971). This tactic is so effective that pointing dogs are invaluable to both fall hunters and field biologists who want to capture the bird (Ammann 1981). On the other hand, once the young have left the nest, the female reacts to the arrival of an intruder by performing a distraction display, feigning injury. Feet dangling and tail down, she flies up to draw the enemy away from the chicks, while giving a call that makes them freeze where they are (Bent 1927; Sheldon 1971; Baird 1987).

The American Woodcock male mates with several females (Pettingill 1936; Sheldon 1971; Johnsgard 1981), and plays no part in tending the young (Pettingill 1936; Dwyer et al. 1982; Redmond and Keppie 1988), which leave the nest scarcely a few hours after hatching (Peppard 1963). Although they start probing the ground themselves by 3 days of age (Sheldon 1971; Johnsgard 1981), the hen goes on bringing them earthworms for a long time (Gregg 1984). The chicks grow very quickly (Sheldon 1971) and are able to fly short distances from 2 weeks on (Johnsgard 1981; Sepik et al. 1981). By 4 weeks old, they have almost reached their full size and can fly quite well (Sheldon 1971; Johnsgard 1981; Sepik et al. 1981).

As they grow older, the young forage farther and farther away from the female (Sheldon 1971), and when they are 31–38 days old, they begin leaving her, becoming fully independent in another 6 weeks (Sheldon 1971; Horton and Causey 1982). Like the females, they then frequent moister habitats like alder thickets (Wenstrom 1974). Dwyer et al. (1982) observed that in years when the weather is poor, more young males than females survive to the age of independence. This may be due to the larger size and more rapid growth of females, which put them at a disadvantage in times of environmental stress.

Notes on territory (continued)

and even from one year to the next (Bourgeois and Couture 1979; Redmond and Keppie 1988). Couture and Bourgeois (1974) reported that the area defended is often larger than the opening used as the display site. They also observed that a male may occupy more than one courting ground. According to Sheldon (1971), in the breeding season, woodcocks use areas of several square kilometres for their various activities.

Densities of 1.5–2 courting grounds/10 ha have been observed in Québec (Bourgeois and Couture 1979), 2–2.5 courting grounds/10 ha in Massachusetts (Sheldon 1971), and 2 nests/10 ha in Pennsylvania (Coon et al. 1982). Stewart and Robbins (in DeGraaf and Rudis 1987) reported densities in Maryland of 1.4 territorial males/10 ha on an abandoned farm and 0.4 territorial male/10 ha in a woodlot after a fire and cutting.

Woodcock chicks waiting to leave the nest. Greenfield Park, near Montréal. *May 1932, Henry Mousley, Laval University Collection.*

For his courting ground, the male seeks an open space without trees, such as an abandoned field, where the scrub is not so dense that it will get in the way of his displays, but is dense enough to protect against predators. Baie-du-Febvre, south shore of Lake Saint-Pierre. *May 1991, Réginald Ouellet.*

American Woodcock on her nest in the underbrush. Sainte-Dorothée, Laval. *April 1942, Henry Mousley, Laval University Collection.*

The diet of the American Woodcock consists chiefly of earthworms, but may also include insects like flies and beetles (Sperry 1940; Krohn 1970; DeGraaf and Rudis 1987). According to Reynolds (1977a), the main species of earthworm eaten by the woodcock in Québec are the pasture worm, octagonal-tail worm, red marsh worm, and Québec worm.

Distribution

The American Woodcock occupies the agricultural heartland of southern Québec. It is most common in and around the Montérégie and in the Ottawa Valley, partly because of the abundance of farmland and the presence of young hardwood and mixed wood areas there. On the periphery of this primary centre of distribution, significant numbers are still to be found, particularly in the southern Ottawa Valley, the Laurentians, Lanaudière, and the Mauricie all the way to the Baie-Comeau area, and the Saguenay–Lac-Saint-Jean lowlands. South of the St. Lawrence, towards the east, similar concentrations are found in the Bois-Francs, Eastern Townships, and Chaudière–Appalaches regions, all the way to La Pocatière. Outside of this large secondary zone, the species is scarcer, although significant concentrations can be observed all the way to near Rimouski and in the Matapédia Valley (R. Couture, UQTR, pers. comm.).

Atlas breeding records extend to the northern Abitibi, which has plentiful early second-growth habitat as a result of logging and farm abandonment. The species was even reported around Matagami and Chibougamau. The woodcock was atlassed on the North Shore as far as Havre-Saint-Pierre, although breeding was confirmed only on the Upper North Shore. Recent CWS surveys (P. Dupuis, in prep.), however, indicate spring singing activities along the Middle North Shore and in some regenerating clearcuts as far north as Lake Mistassini and around Joutel in the northern Abitibi. The northernmost reports in eastern Québec come from Manic-5 and Lake Gaudreault on the Lower North Shore (ÉPOQ). The northern limit of the range is not well defined, but it is probably around James Bay and southern Labrador (Owen et al. 1977).

Several factors limit the northward expansion of the American Woodcock's range. They have chiefly to do with the characteristics of the soil and vegetation, which influence the abundance of earthworms, the staple item in the bird's diet.

In the North, the acidic soil of the boreal forest explains the scarcity or complete absence of earthworms (Reynolds et al. 1977; Owen and Galbraith 1989). On the other hand, earthworms are particularly abundant in the agricultural heartland of the St. Lawrence Plain, where the soil is rich in organic matter. For example, undisturbed mull contains 400–600 worms/m^2, while the more acidic podzols of the hinterland have lower concentrations, which may vary from 0 to 50 worms/m^2 (F. Pagé, MAPAQ, pers. comm.). Other factors, such as permafrost, limit earthworm populations.

Farther south, in the Abitibi, the heavy clayey soil from glacial lakes drains poorly and stays cold for a large part of the year; very few worms are found there, therefore, except in a few enclaves that have their own microclimates. This combination of factors may explain why the American Woodcock is not very abundant in the Abitibi despite the fact that farmers and loggers have been cutting openings in the forest for decades.

The American Woodcock was atlassed on Anticosti Island, but sightings there are apparently rare (Ouellet 1969). An Atlas record was the first breeding confirmation of the woodcock in the Magdalen Islands, at Étang-des-Caps.

History and trends

At the turn of the century, Wintle (1896) listed the woodcock as scarce in the Montréal area, breeding only in a few localities, but more plentiful during the fall migration. At about the same time, according to Dionne (1906), woodcocks were not uncommon in the Québec City area and there was a certain market in them in the fall. Macoun and Macoun (1915) described the American Woodcock as common in the southern part of the province.

More recently, data from the 1966–89 Breeding Bird Survey and the ÉPOQ database for the 1970–89 breeding seasons show no significant change in Québec's woodcock population (Cyr and Larivée 1993a). The eastern continental population, however, declined at an annual rate of 1.9% between 1968 and 1992 (Straw 1992).

The American Woodcock is a very popular game bird in Québec. It is estimated that between 1989 and 1991, a mean of 4664 hunters bagged 33 367 birds each year (Lévesque et al. 1993). Given the strong hunting pressure on the species (Couture and Bourgeois 1974, 1977; Hounsell 1991), it would be advisable to take measures to increase the amount of suitable habitat and encourage woodcock production (Gélinas 1990), especially in and around urban areas, where breeding habitat is constantly shrinking (Couture 1990). In fact, on some woodcock survey routes in southern Québec, not a single woodcock was reported in recent years, due to heavy urban development (P. Dupuis, CWS, unpubl. data).

In addition to natural succession and ageing of woodlands, many types of human activity can lead to habitat loss, including urbanization and recovery of idle farmland for agriculture or conifer plantations (Owen et al. 1977; Bérubé and Couture 1986; Couture 1990; USFWS 1990). Furthermore, Dwyer et al. (1983) demonstrated a correlation between the population decline observed during the North American survey and the loss of suitable woodcock habitat along northeastern U.S. roads. There are a number of measures that should be taken to restore woodcock habitat, including selective cutting to slow the ageing of forests, the maintenance of wildland around the edges of forestry and agricultural operations, and preservation of greenspace in urban development plans (Sepik et al. 1981; Gélinas 1990).

Diane Dauphin and Pierre Dupuis

Wilson's Phalarope

Phalarope de Wilson

Phalaropus tricolor

The female Wilson's Phalarope is larger and more colourful than the male, and it is the male that incubates and rears the young.

Wilson's Phalarope is a migrant breeder that returns to Québec in spring, usually in early May. The species is expanding its range in the province (Robert 1989b), where it breeds locally beginning in late May or early June. The females leave again in July, whereas the males and juveniles wait until the end of August or early September to migrate. Stragglers may remain until early October. Unlike other phalaropes, Wilson's does not winter at sea, but in the Peruvian highlands and on the inland and coastal waters of Chile, Paraguay, Uruguay, and Argentina (Burger and Howe 1975).

Habitat and behaviour

Wilson's Phalarope is found in wetlands such as ponds, short-grass marshes, and wet meadows. It sometimes feeds locally on sewage lagoons, which teem with invertebrates.

It is crucial for the females to arrive early on the breeding grounds to select their mates (Reynolds et al. 1986). The females take the initiative, performing their displays beginning in mid-May. The females are about 35% larger than the males (Thomas 1990), and their plumage is a much more colourful black and chestnut. They do not defend a territory per se, but once they have chosen a mate, they exhibit aggressive behaviour (Höhn 1967), defending the area immediately around him, and letting no other female approach (Terres 1980). Monogamy is the general rule (Ehrlich et al. 1988), but polyandry has recently been confirmed (Colwell 1986), although for a long time this behaviour was considered sexually promiscuity (Höhn 1967; Johns 1969).

The female is the first to go off looking for a nesting area, but it is the male that selects the site and builds the nest (Howe 1975). Nest dispersion in Wilson's Phalarope suggests that the selection of the nest site is strongly influenced by the presence of conspecifics (Colwell and Oring 1990). The nest is usually located in a field, near water. Though Wilson's Phalarope generally seeks the protection of dense vegetation against predators, for safety's sake, it sometimes prefers the proximity of a more aggressive species. That is why, in western Canada, it has been seen nesting in tern colonies (Colwell and Oring 1990).

The nest is a small hollow in the ground lined with dry grass. Ordinarily the female lays 1 egg a day for 4 days, and the male, which incubates the eggs and rears the young alone, begins to incubate as soon as the third egg has been laid (Johnsgard 1981). The male sometimes rests during incubation and is then often harassed by unpaired females; as he returns to the nest to incubate again, he performs a distinctive flight display, showing the females that he is paired (Colwell and Oring 1988a). At the start of incubation, the female helps the male chase off intruders, but she then leaves to moult. This is followed by her migrating onwards to the winter grounds (Höhn 1967).

The young are precocial and can swim an hour after hatching. They follow the male and feed themselves in the marsh or in the tall grass along the shoreline until they can fly. The young moult very rapidly into a plumage similar to that of the adults' winter plumage. Wilson's Phalarope can be distinguished from other phalaropes by its less-developed salt gland, which nonetheless does not keep it away from saltwater environments (Mahoney and Jehl 1985).

Phalaropes are the only shorebirds that regularly feed while swimming. They spin on the water, creating whirlpools that stir up particles of food and insect larvae, making them easier to see (Höhn 1967). As Johns (1969) demonstrated with young raised in captivity, this behaviour is innate.

Wilson's Phalarope was reported in 22 Atlas squares, and breeding was confirmed in 10 of those; half of the confirmations involved birds feigning injury. As phalaropes nest in groups in open areas, they are fairly easy to spot. An intruder in a field where there is a colony will often be mobbed, and birds will perform distraction displays. The nests themselves, sometimes hidden in clumps of vegetation, are rarely found.

Distribution

Wilson's Phalarope is a native of western North America. In Canada, it is found in small marshes, from central British Columbia to Manitoba. Expanding eastward since the early 1970s, it has been confirmed as a local breeder in southern Ontario and Québec. The first Wilson's Phalarope nest in New Brunswick was found in 1985, in the St. John River Valley (McAlpine et al. 1988b).

Because its diet limits it to extremely fertile ponds, and because it is a colonial nester, this phalarope is very locally distributed. Wilson's Phalarope was confirmed as a breeder for the Atlas in the Montréal area (Pointe-aux-Trembles, Boucherville Islands, and Île aux Fermiers), around Lake Saint-Pierre (Île du Moine and Baie-du-Febvre), and east of Québec City (Cap Tourmente, Île aux Grues, Île aux Oies, La Pocatière, and Saint-Germain, near Kamouraska). It was atlassed at other locations in the Montréal area, and also near Landrienne (Abitibi), at Saint-Gédéon (on Lake Saint-Jean), Cacouna, L'Isle-Verte, and Trois-Pistoles (Lower St. Lawrence), as well as at Saint-Paul-du-Nord and Pointe-Lebel (Upper North Shore).

Clodin Roy

Notes on territory

Wilson's Phalarope does not defend a territory per se. Early in the season, the females arrive first and are more numerous than the males; later, the proportion of males increases, although the ratio of males to females is never totally balanced (Höhn 1967).

During pair formation, females compete directly for the favours of males (Reynolds et al. 1986). Moreover, Höhn (1967), Johns (1969), and Reynolds et al. (1986) reported that the female may defend the area immediately around her mate against other females during the period leading up to incubation.

Wilson's Phalarope seeks out wet meadows to nest. Île du Moine, Sorel Islands, west of Lake Saint-Pierre. *July 1991. Daniel Jauvin.*

Wilson's Phalarope

Breeding distribution

1984–1989

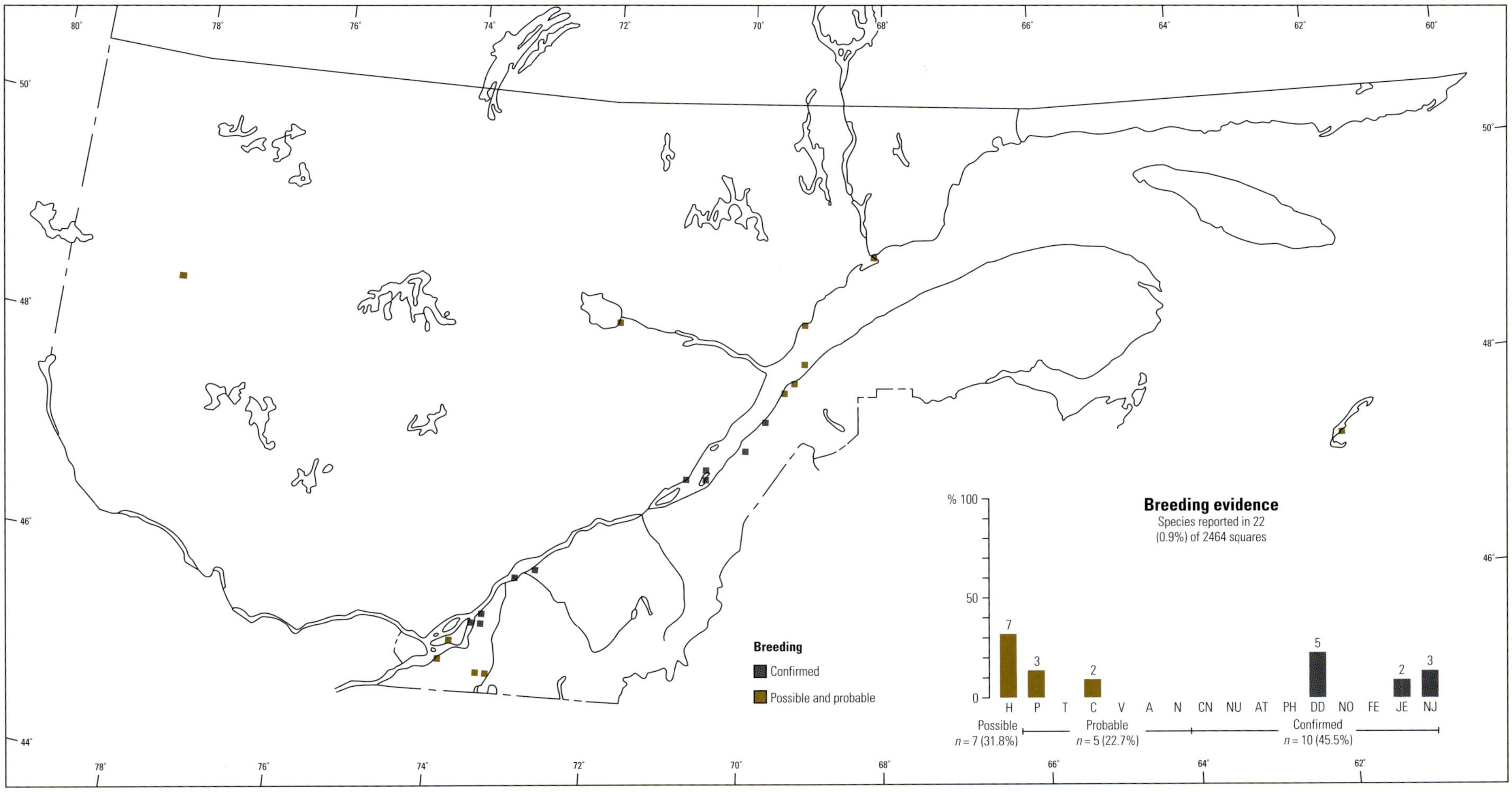

Species profile	
Clutch size	usually 4 eggs (3–4)[3,4,5,6]
Laying interval	≈ 1 egg per day[5,6,9]
Start of incubation	with third egg[7,9]
Incubation period	16–27 days[3,4,5,6,9]
Maturity at hatching	precocial
Care of young	by male[4,6,8]
Nestling period	less than 24 hours[10]
Age at first flight	?
Dependence of young	?
Broods per year	1, sometimes 2[4]
Breeding age	1 year[7]
Mating system	monogamy, sometimes polyandry[4,6,8]
Duration of pair bond	until ≈ start of incubation[5,6]
Total length	21.0–25.5 cm[2]
Wing span	36.8–40.6 cm[1]
Weight	M 51.8 g (40–62) / F 68.1 g (55–85)[6]
Longevity record	?

References:
1. Terres 1980
2. Godfrey 1986
3. Harrison 1978
4. Ehrlich et al. 1988
5. Höhn 1967
6. Colwell and Oring 1988b
7. Colwell and Jehl 1994
8. Colwell 1986
9. Johnsgard 1981
10. Thomas 1990

Chronology *n* = 15

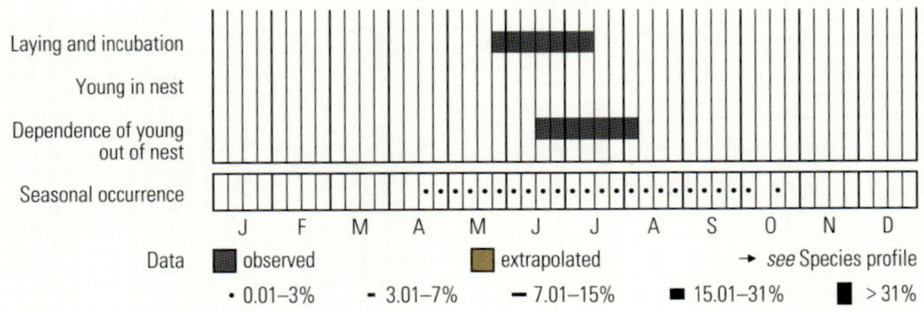

In suitable habitat like Île aux Fermiers, several Wilson's Phalaropes nest in the herbaceous vegetation close to ponds and bays. Varennes Islands, near Montréal. *August 1991, Gaétan Duquette.*

Notes on territory (continued)

The pair bond dissolves completely once laying has ended; the male is left to take care of the nest (Howe 1975). During incubation, if a female harasses him while he is resting, the male performs a distinctive flight display to indicate that he already has a mate (Colwell and Oring 1988a).

Howe (1975) reported having once seen a male attempting to get rid of a conspecific pair that was about 3 m from a nest he was working on. This observation supports the hypothesis put forward by Colwell and Oring (1990) that Wilson's Phalarope avoids nesting too close to others of the same species.

On the Québec side of James Bay, the first records go back to 7 July 1943, when 10 birds were spotted in Boatswain Bay (Todd 1963). In 1976, 3 birds were sighted at James Bay (Gosselin 1977a); more recently, on 11 and 13 August 1991, a flock of 12 birds, including at least 4 juveniles, was seen there (Gosselin et al. 1991b), feeding in pools on the upper marsh at Boatswain Bay (M. Robert and A. Morrier, pers. comm.).

History and trends

The first record of Wilson's Phalarope in Québec dates from August 1870, in Montréal (Whiteaves 1870). A specimen was then taken on Anticosti Island around 1900, but it was deemed an accidental (Lewis 1924). Ten phalaropes were seen in Boatswain Bay, on James Bay, on 7 July 1943 (Todd 1963). Later, the bird began to be sighted regularly at Île du Moine, near Sorel (PQSPB 1952), and at La Prairie (PQSPB 1958, 1959).

Reports of 2 females on Île du Moine, in May and June 1970, and 4 juveniles in July, point to probable breeding there (COQ 1971). On 5 May 1974, the species was recorded in a new area, Nuns' Island, where on 23 June of the same year, a nest containing 4 eggs was found (Steeves and Holohan 1975). In the course of that summer, 11 adults and 10 juveniles were seen on Île du Moine. In 1976, Wilson's Phalarope was observed more or less all over Québec: at La Pocatière, Pointe-au-Père, Chandler, Jonquière, Waskaganish, and a number of localities near Montréal (David and Gosselin 1976a,b, 1977a). Another nest was discovered in June of the same year on Île du Moine (Jauvin 1981).

Pair in vegetation during nesting period. Sex roles are reversed in phalaropes; the female courts the male, and leaves the incubation and rearing of the young up to him. Baie-du-Febvre, south shore of Lake Saint-Pierre. *May 1981, Daniel Jauvin.*

Wilson's Phalarope nest built by the male and containing 4 eggs. Baie-du-Febvre, south shore of Lake Saint-Pierre. *June 1981, Daniel Jauvin.*

This range expansion appears to be related to a greater utilization of available suitable habitat and a shift eastward and northward prompted by Prairie drought over several years (Bousfield et al. 1986). In the Maritimes, it has been observed that waterfowl habitat management also helps Wilson's Phalarope (Erskine 1992). Since the 1960s, the species has been sighted increasingly frequently in Europe, especially in August (Cramp and Simmons 1983). It seems that Wilson's Phalarope may be becoming a holarctic species (Gosselin 1977a).

Despite its eastward expansion, the species is not yet firmly established in its new areas, where its numbers remain low and birds nesting in colonies are more vulnerable to habitat disturbance and loss. No Wilson's Phalaropes were atlassed on Nuns' Island, although this was one of the first sites occupied by the species in the province. Their absence is doubtless attributable to habitat loss resulting from landfill and residential development there. Furthermore, nesting areas often visited by birders, such as Île du Moine, deserve some degree of protection.

Daniel Jauvin
and Pierre Lafontaine

Female Wilson's Phalarope moving through the aquatic vegetation in search of food. The female's plumage is more colourful than the male's. Baie-du-Febvre, south shore of Lake Saint-Pierre. *June 1982, Daniel Jauvin.*

Little Gull

Mouette pygmée

Larus minutus

Clodin Roy

THE LITTLE GULL, WHICH IS AS SMALL AS A MOURNING DOVE, IS THE smallest of the *Larus* gulls, and is dwarfed by its giant cousins the Great Black-backed and Glaucous gulls, which are over 10 times its weight.

Originally from Eurasia, the Little Gull has been breeding in North America only since 1962, when 3 nests were found near Oshawa, Ontario (Scott 1963). In Québec, it is now considered a rare visitant and an accidental breeder.

Habitat and behaviour

In Europe and North America, the Little Gull usually breeds in small colonies, in emergent vegetation found along broad stretches of fresh water. It nests on the edges of rivers and lakes and along coasts, wherever there is lush aquatic vegetation, shallow water, and a muddy substrate (Cramp and Simmons 1983), sometimes in association with Black Terns, Common Terns, and other larids. The nest, which closely resembles the Black Tern's, may be built on the ground in dense vegetation close to the water, or even in emergent vegetation at water level, as was the nest found in 1982 in the Lachine Rapids near Montréal. It was located in a pile of dead reeds in the water, protected from the current by only a few stones and some emergent plants. Generally, the nest begins as a shallow depression in the ground, and plant materials are added as time goes by. Both parents build the nest, incubate the eggs, and feed and care for the young (Veen 1980; Cramp and Simmons 1983). During the breeding season, mainly insects and their larvae and other invertebrates are consumed, supplemented by the occasional small fish, though fish probably plays a greater role in the diet during the rest of the year. Little Gulls capture their prey on the wing, or in typical tern fashion, flying low over the water and periodically dipping to snatch prey from the surface, often without interrupting their flight (Veen 1980; Cramp and Simmons 1983).

Like most other larids, Little Gulls are easy to spot during the breeding season because of their conspicuous behaviour. The pair that nested on a tiny island in the Lachine Rapids every year from 1982 to 1987 provided ornithologists with a unique opportunity to observe different types of breeding behaviour in the species, including aerial displays, copulation, antagonistic behaviour toward intruders, and the carrying of nesting materials. The most spectacular of these is the pursuit flight, described by Cramp and Simmons (1983), in which one of the members of the pair pursues and overtakes the other until they are directly over one another, the upper bird flying with neck extended and head held upward, giving frequent "long-call." The behaviour observed most frequently at the Lachine Rapids, however, was both members of the pair carrying nesting materials, in this case, plant stalks. This was seen not only at the beginning of the season but throughout the entire incubation period (Bannon 1983), thus providing confirmation of breeding at this site.

Distribution

Although breeding was confirmed only at the Lachine Rapids during the Atlas survey period, the Little Gull was reported in other parts of the Montréal area, at Île aux Fermiers and Pointe-aux-Trembles, but these may very well have been sightings of the same birds. Three adults also summered at Baie-du-Febvre, south of Lake Saint-Pierre, in 1989. Other sight records were obtained at Les Escoumins on the Upper North Shore, where the species is seen annually, L'Anse-Saint-Jean on the Saguenay River (28 July 1985), Aylmer on the Ottawa River (2 June–2 August 1986), and Île du Moine near Sorel (7 June 1986) (Yank and Aubry 1985a, 1986a; Aubry et al. 1989).

The Little Gull occupies only a small fraction of its potential breeding habitat. At the present time, it is found mainly along the upper St. Lawrence, which appears to be a natural extension of its range in the Great Lakes region (Johnson and Adams 1977), and even here a great deal of suitable habitat, particularly around Lakes Saint-François and Saint-Pierre, remains unoccupied. The Little Gull may also breed in small numbers around the large man-made reservoirs located farther north, but for the time being there is no evidence to support this. The Ontario Atlas, however, documented breeding at several sites in the coastal lowlands of James Bay and Hudson Bay (Carpentier 1986; Weseloh 1987c).

History and trends

Although the Little Gull was first confirmed as a breeder in North America over 30 years ago (Scott 1963), it is still quite rare today. Outside Québec, breeding has been reported in the Great Lakes region (in Wisconsin, Michigan, and Ontario) and on the James Bay and Hudson Bay coasts (in Ontario and Manitoba) (Weseloh 1987c).

The Little Gull was first reported in Québec in 1949 on the Magdalen Islands (Gaboriault 1961) but 20 years elapsed before the second sighting was made, at Beauharnois. Since then, this gull has been seen every year along the St. Lawrence, particularly in August, September, and October (ÉPOQ), and in 1971, a pair was observed on the Sorel Islands performing courtship displays. Breeding was not confirmed until 1982, however, when a chick was spotted at the Lachine Rapids (Bannon 1983). The species frequented this site from 1982 to 1987 but has not been seen there since.

The future of the Little Gull in Québec remains uncertain. The number of birds sighted annually appears to have peaked in the 1980s, despite an increase in the number of competent observers and the seeming abundance of suitable breeding habitat. Breeding success appears to be poor in Québec as no young are known to have fledged at the Lachine Rapids, despite annual nesting attempts between 1982 and 1987. Low breeding success has also been reported in Ontario (Scott 1963; Tozer and Richards 1974) and in The Netherlands (Veen 1980). One reason is thought to be a high rate of predation by muskrats, mink, and gulls. Sudden fluctuations in water level during the incubation period might also have been responsible for the numerous nest failures at the Lachine Rapids.

PIERRE BANNON AND MICHEL ROBERT

Notes on territory

Little Gulls are normally gregarious and dozens or even hundreds can often be seen loafing or feeding together peaceably without defending territories or showing any signs of aggressive behaviour. The gulls do defend a small territory when nesting. In Europe, Little Gulls, like most larids, have been found nesting very close together, generally 1–6 m apart. Adults also defend an area of about 30 m² around the young once they have left the nest and have begun moving about the colony (Veen 1980; Cramp and Simmons 1983).

Between 1982 and 1987, a pair of Little Gulls nested every year in the plant debris next to the tiny islands in the Lachine Rapids. The Little Gull is an accidental breeder in Québec. LaSalle, near Montréal. *June 1991, Pierre Bannon.*

Little Gull

Breeding distribution 1984–1989

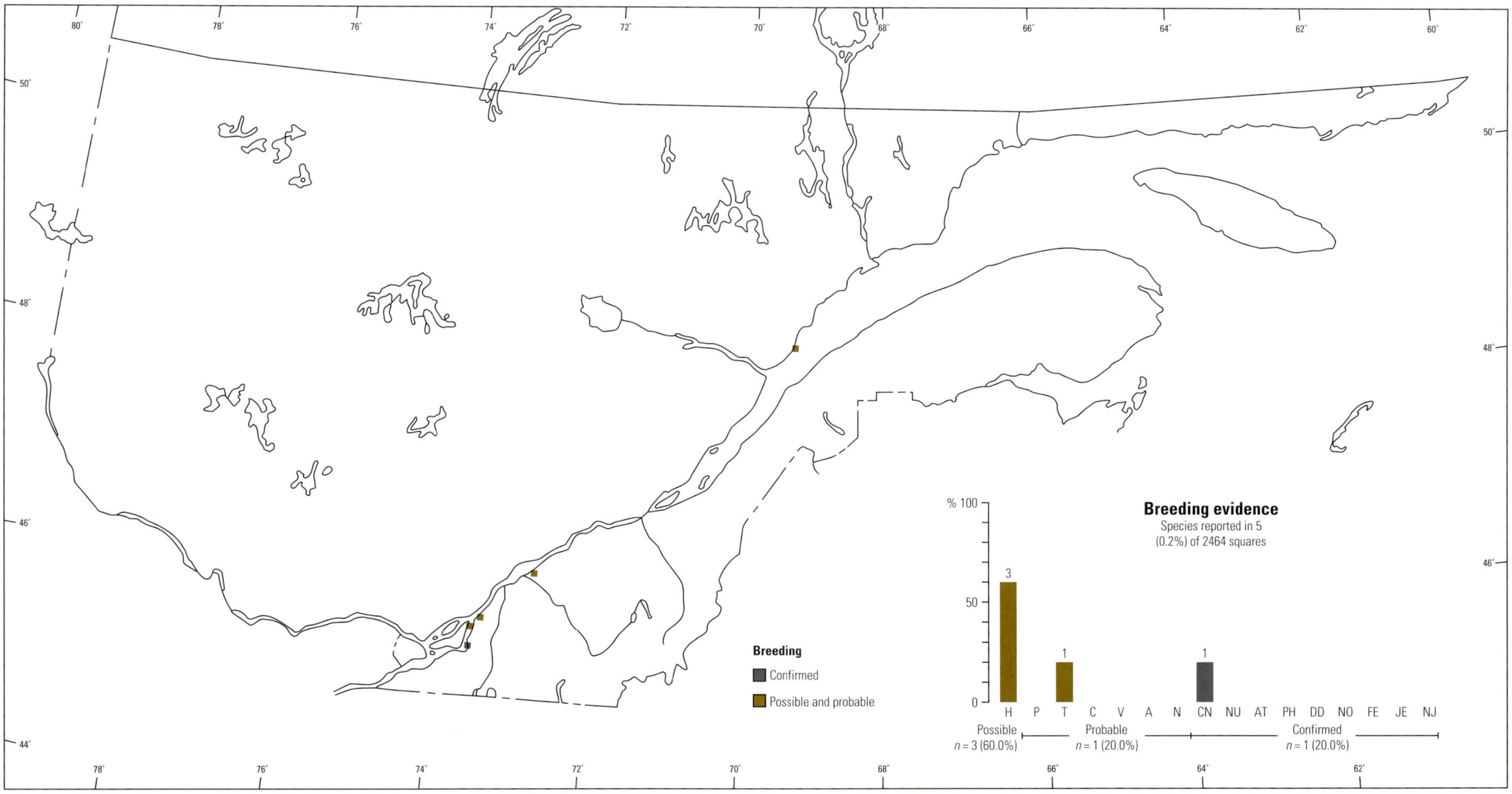

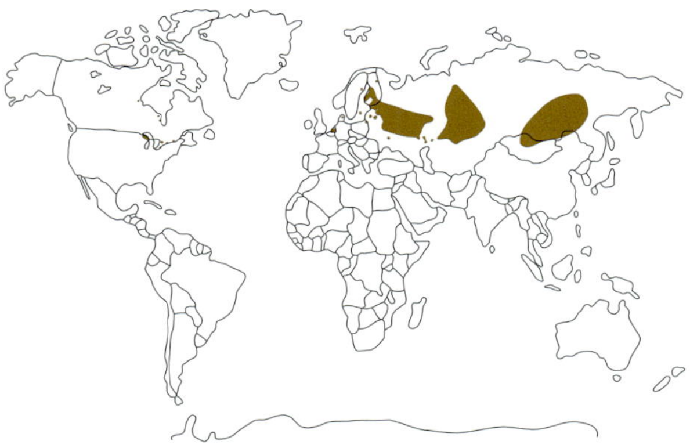

World breeding range

Species profile

Clutch size	usually 2–3 eggs (1–5)[1]
Laying interval	1 egg per 1–2 days[1,3]
Start of incubation	with first egg[1,3]
Incubation period	21–25 days[1,3]
Maturity at hatching	semiprecocial
Care of young	by female and male[1,3]
Nestling period	≈ 7 days[3]
Age at first flight	21–24 days[1,3]
Dependence of young	until shortly after first flight[1]
Broods per year	1[1]
Breeding age	2–3 years[1,3]
Mating system	monogamy[1,2]
Duration of pair bond	?
Total length	25.4–27.0 cm[1]
Wing span	75–80 cm[1]
Weight	120 g (103–140)[4]
Longevity record	?

References:
1. Cramp and Simmons 1983
2. Ehrlich et al. 1988
3. Veen 1980
4. Dunning 1984

Chronology $n = 4$

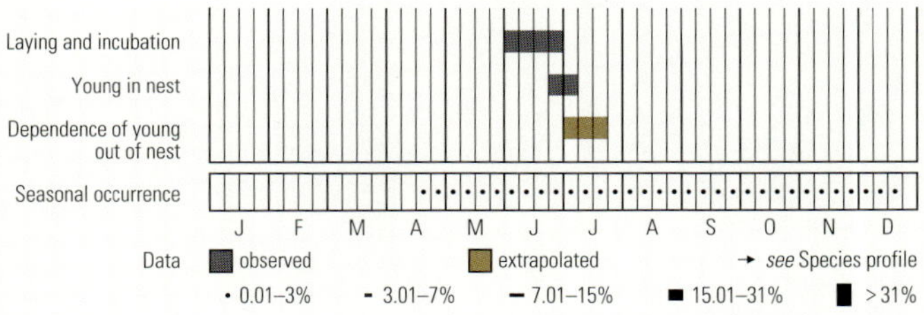

Little Gull nest in the emergent vegetation. LaSalle, near Montréal. *June 1982, Pierre Bannon.*

Common Black-headed Gull*

Mouette rieuse

Larus ridibundus

Clodin Roy

VERY ABUNDANT IN EUROPE AND IN THE BRITISH ISLES, THE COMMON Black-headed Gull has been expanding its range for over half a century, first to Iceland, then to Greenland, and recently to the east coast of North America. This larid has even reached the Gulf of St. Lawrence, colonizing the Magdalen Islands in 1981 (Aubry 1984). The Common Black-headed Gull looks a great deal like Bonaparte's Gull, but is smaller and adults in breeding plumage have dark brown rather than black heads and slightly longer, dark red bills.

Habitat and behaviour

The Common Black-headed Gull, though basically colonial, sometimes breeds in small isolated groups or near colonies of other gulls or terns. In Europe, this gull inhabits mainly lowland areas, particularly coastal marshes. It can be found along the edges of shallow marshes, as well as estuaries, and it also frequents large inland lakes and dry sites situated near water like sand dunes (Hanssen 1982; Cramp and Simmons 1983; Montevecchi et al. 1987). Pairs may even nest at high elevations (900 m), as Cochet (1980) observed in the Haute-Loire (France).

In the Magdalen Islands, breeders arrive in mid-April and start building their nests in early June (Fradette 1992). The species prefers areas of fairly dense vegetation (Isenmann 1977), particularly small islands, for its platform-shape nest, which is made of plant materials. A nest found on an islet in Baie du Havre aux Basques was made of dried vegetation, including beachgrass, and was situated in dense silverweed (Aubry 1984). In Massachusetts, where the species bred for the first time in 1984, the nest seen was also made of beachgrass (Holt et al. 1986).

The Common Black-headed Gull is an omnivore and consumes a wide variety of foods, including insects, worms, fish, plants, and often small fruits such as cherries, as well as refuse of all sorts (Isenmann 1977). Like Ring-billed Gulls, Common Black-headed Gulls visit dumps and follow tractors during ploughing to feast on earthworms (Isenmann 1977; Cramp and Simmons 1983), which may make up as much as 50% of total food mass during the breeding season (Cramp and Simmons 1983).

The Common Black-headed Gull is basically monogamous (Tinbergen 1956), although males can often be seen in colonies attempting to copulate with an already mated female (Tinbergen 1956; Van Rhijn and Groothuis 1985a). In captivity, polygyny and even pairs of males are common, but according to Van Rhijn and Groothuis (1985a,b), this type of behaviour has never been observed in the wild, although they have seen a male in a colony court two females at the same time.

Notes on territory

Studies of territoriality in the Common Black-headed Gull have only been conducted in Europe. Territorial behaviour may occur

* The name of the Common Black-headed Gull has recently been changed to Black-headed Gull (AOU 1995).

In 1988, a pair of Common Black-headed Gulls nested in this Ring-billed Gull colony on Île à Calculot in the Mingan Archipelago National Park Reserve. Middle North Shore. *June 1988, Sylvain Paradis.*

Common Black-headed Gull

Breeding distribution
1984–1989

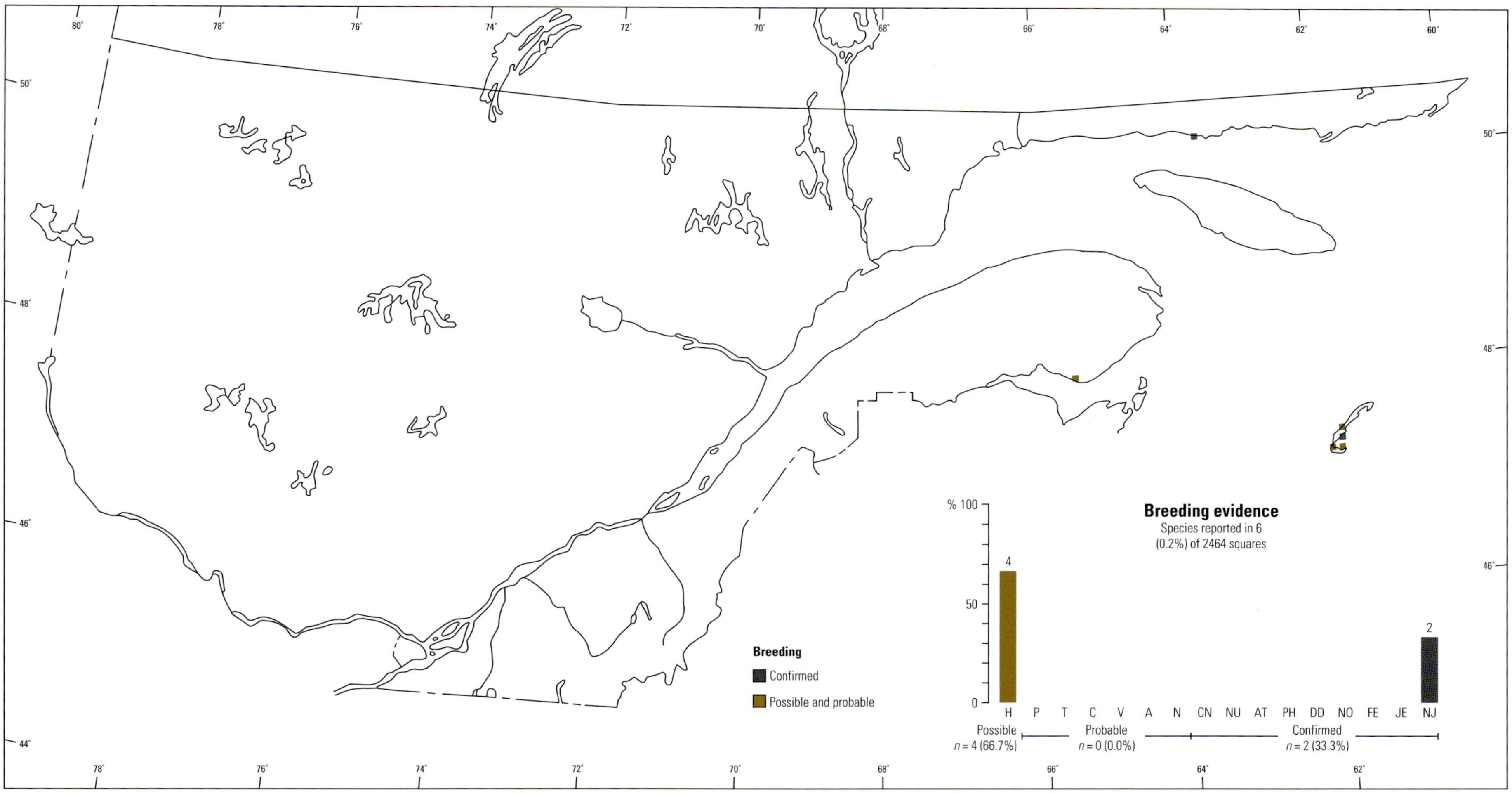

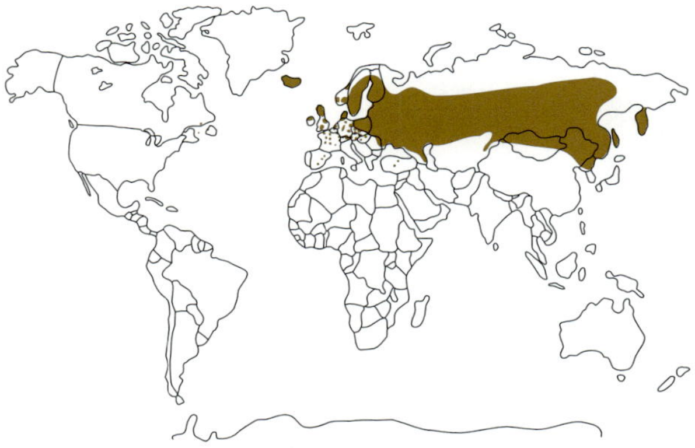

World breeding range

Species profile

Clutch size	usually 2–3 eggs (1–4)[1,2,3]
Laying interval	1 egg per 2 days[3]
Start of incubation	with first, second or last egg[3]
Incubation period	20–26 days[1,2,3]
Maturity at hatching	semiprecocial
Care of young	by female and male[1,3]
Nestling period	≈ 10 days[3]
Age at first flight	35–42 days[1,2,3]
Dependence of young	until shortly after first flight[3]
Broods per year	1[3]
Breeding age	usually 2–3 years (1–4)[1,3]
Mating system	monogamy[1,3]
Duration of pair bond	sometimes more than 1 season[3]
Total length	35.5–38.0 cm[4]
Wing span	100–110 cm[3]
Weight	284 g (195–327)[5]
Longevity record	32 years 1 month[3]

References:
1. Ehrlich et al. 1988
2. Terres 1980
3. Cramp and Simmons 1983
4. Godfrey 1986
5. Dunning 1984

Chronology $n = 4$

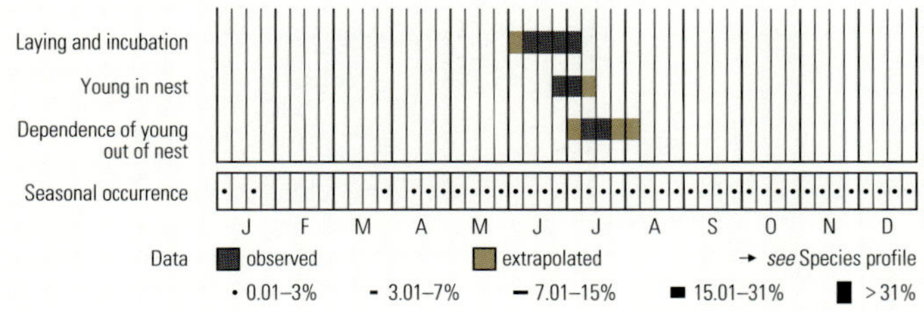

Gregarious year-round, the Common Black-headed Gull noisily defends a small territory around the nest (Tinbergen 1961). Individuals with adjoining territories seem to recognize each other, and no bird ever intrudes into the middle of a neighbour's territory, even if the other bird is away (Patterson 1965). Breeding and territorial behaviour in the species is highly complex and has been described in detail by Moynihan (1955) and Tinbergen (1956, 1961); the species adopts similarly complex behavioural strategies to detect and defend themselves against predators (Kruuk 1964; Allainé 1991). Ritualized behaviour governs territorial encounters and the pair's movements, and limits displays of aggressiveness, which become much more intense during nesting.

The nesting territory may be selected by both members of the pair (Patterson 1965) or by the male alone (Cramp and Simmons 1983). The male initiates nest building and gathers most of the plant materials required. In a colony in the former Czechoslovakia, nest building began an average of 9 days before the onset of laying (Drozdek 1991). Typical clutch size is 2–3 eggs, though 1–4 may be laid (Cramp and Simmons 1983; Drozdek 1991), which are incubated by both parents for 20–26 days. Both parents also tend the young. When the chicks are about 10 days old, they leave the nesting territory in the company of their parents (Tinbergen 1956), and in another 10 days, begin to feed themselves. Near the end of the fledgling period, the adults begin leaving the young and the colony at night to roost together, which is fairly unusual in birds, although they come back at the least sign of alarm. The young can fly at the age of 35 days. The Common Black-headed Gull is often aloft at night, unlike the larger larids (Cramp and Simmons 1983). In the Magdalen Islands, family groups break up between late August and late October, but flocks of up to 30 birds can sometimes be seen in November on Baie du Havre aux Basques (Fradette 1992).

Distribution

In Québec, the species was reported for the first time in October 1951 at Havre-Saint-Pierre. Twenty-one years later, in August 1972, it was sighted in a tern colony on the Magdalen Islands (McNeil et al. 1973). According to Aubry (1984), it may have started nesting in the Magdalens in 1981. Up to 20 birds have been reported on the islands during the breeding season (Fradette 1992). In August 1988, a pair established themselves in a Ring-billed Gull colony on an island in the Mingan Archipelago, off Havre-Saint-Pierre on the North Shore. Moreover, an immature yearling was sighted in July 1989 on Anticosti Island, north of Port-Menier, and during the same month, 2 other birds were seen near Blanc-Sablon, on the Lower North Shore (ÉPOQ).

Outside the breeding season, individuals have been reported along almost the entire length of the St. Lawrence. In winter, the species is found on Lake Ontario and Lake Erie, among other places (Root 1988). However, most birds seen in winter are probably not local birds, but from Iceland (Root 1988).

Notes on territory (continued)

in three contexts: during pair formation, during nesting, and sometimes during the winter (Tinbergen 1956).

In spring, males defend a territory of around 12 m² at the periphery of, or at some distance from, the colony—on a small island, for example (Moynihan 1955). The male may defend several territories simultaneously, which are used for attracting a female. Pair formation ensues, and within the colony the pair then establishes a territory around the nest, which is defended against all intruders. In denser colonies in Europe, territories may be just a few centimetres

A small colony of Common Black-headed Gulls has become established in Baie du Havre aux Basques. The species has probably been nesting there since 1981. Magdalen Islands. *June 1991, Desneige Perreault.*

Common Black-headed Gull nest made of plant stalks, grass, and dried seaweed, and concealed amid tufts of vegetation. Île à Calculot, Mingan Archipelago National Park Reserve, Middle North Shore. *June 1988, Sylvain Paradis.*

larger than the nest, or up to several metres in diameter (Tinbergen 1956). The average diameter is, however, 1–2 m (Patterson 1965; Cramp and Simmons 1983; Drozdek 1991). Approximately 10 days after the young hatch, they leave the territory with their parents and roam around the colony or its outskirts; the adults then defend a moving territory of 2 m or more around the chicks (Tinbergen 1956). Food gathering takes place outside these territories and the colony. In winter, some birds defend an individual resting territory, which may be up to several metres in diameter (Tinbergen 1956) and is usually near a feeding area.

History and trends

Originally from Europe, the Common Black-headed Gull has reportedly been wintering in North America for several decades (Erskine 1963), with the first record dating back to 1921 (Tufts 1986). In Iceland, it bred for the first time in 1911 and since then, the population on the island has expanded remarkably, reaching 10 000 breeding pairs in 1975 (Cramp and Simmons 1983). The species then continued to expand westward, breeding in Greenland for the first time in 1969, and in Newfoundland in 1977 (Finch 1978). Since then, reports of breeding have continued to increase.

In the United States, the Common Black-headed Gull breeds in Massachusetts (Holt et al. 1986) and Maine, where a nest was found in 1986 in the Petit Manan Island National Wildlife Refuge (Drennan et al. 1987). In eastern Canada, a number of confirmed or probable breeding records have been obtained in Newfoundland and Labrador (Montevecchi et al. 1987) and the species probably also breeds in Nova Scotia (Lock 1991). The most westerly report comes from Little Galloo Island, on the far eastern end of Lake Ontario, where an apparent Common Black-headed Gull–Ring-billed Gull hybrid nested in 1982 (Weseloh and Mineau 1986).

The species seems to be constantly expanding, both in terms of numbers and range, and should continue to do so in the years to come. Particular care should be taken to watch for it on the outskirts of Common Tern and Ring-billed Gull colonies. Further expansion of the species is to be expected and the Common Black-headed Gull should eventually colonize various sites along the North Shore, the Gaspé Peninsula, and the St. Lawrence.

Jacqueline Vincent

Bonaparte's Gull

Mouette de Bonaparte

Larus philadelphia

Clodin Roy

BONAPARTE'S GULLS DO NOT NEST ON THE GROUND LIKE MOST OTHER larids, but instead nest in trees, in the vast, silent expanses of the boreal forest. This gull generally breeds far from inhabited areas, where there is an abundance of bogs, fens, and ponds. In Québec, it is a migrant breeder, confined mainly to the James Bay coastal plain and the Lake Saint-Jean watershed. In southern Québec, the best time to see Bonaparte's Gull is from August to November, when flocks of migrating birds, sometimes numbering in the hundreds, stage along its waterways (David 1980a). The species' breeding biology has been studied very little and a number of aspects are still unknown.

Habitat and behaviour

Bonaparte's Gulls breed in isolated pairs or in small colonies; colonies in Alberta, Ontario, and Québec were found to contain 4–20 individuals (Salt and Wilk 1966; Blokpoel 1987; Cons. G. G. – GREBE 1992h).

During the breeding season, Bonaparte's Gulls can be seen near peatlands, small boggy lakes edged with clumps of conifers, and larger lakes and slow-moving rivers (Salt and Wilk 1966; Godfrey 1986; Cons. G. G. – GREBE 1992h). In the Nottaway, Broadback, and Rupert rivers region, it seems to seek out productive lakes and ponds, especially ponds within a complex of peatlands with pools (Cons. G. G. – GREBE 1992h). Nests found in this area were generally in small, isolated stands of spruce in the middle of a peatland, near a pond, or on the edge of an open forest near a small boggy lake. In the Lake Saint-Jean Lowland, nests have been seen on a small wooded island and a riverbank.

Although it is much more unusual, nests have also been found on a 1.2-m high stump (QNRCP) and on the ground. Ground nests were found at Lake Moquachéa, northeast of Matagami, and in the Magdalen Islands, where a pair nested in a colony of Common Black-headed Gulls and terns (Yank et al. 1990b). The nest is generally 1.2–7 m up in a conifer, with most nests in Québec being at a height of about 3 m (Cons. G. G. – GREBE 1992h). Both adults build the nest, which is made of twigs, moss, and lichen (Harrison 1978; Godfrey 1986; Cons. G. G. – GREBE 1992h; QNRCP).

Typical clutch size is 3 eggs, which are incubated for about 24 days (Jehl and Smith 1970; Godfrey 1986), by both parents, though apparently the female does the lion's share (Harrison 1978). The parent that is not sitting keeps watch near the nest, often perching in a treetop (Salt and Wilk 1966; Cons. G. G. – GREBE 1992h). The parents will dive-bomb potential predators, or perch a few metres away, screaming the species' characteristically nasal call. Sometimes adults join together to drive an intruder away from the colony (Bent 1921). The young leave the nest about a week after hatching (Harrison 1978), although they are not yet able to fly at this time.

It is fairly difficult to find nests from the ground, as they are often well hidden among the branches of a conifer. Nonetheless, it is easy to tell if you are in a pair's breeding territory, since the species is noisy and the pugnacious parents will attack anyone who comes too close to the nest or the young (Jehl and Smith 1970). Another excellent indicator is spotting a bird on guard, perched in a treetop; sometimes the nest can then be located by examining the surrounding conifers closely or by waiting and observing one of the parents alight on the nest (Blokpoel 1987). The main problem in spotting the species is that nest sites are isolated and hard to get to (Blokpoel 1987). Bonaparte's Gull was reported in 1% (27/2464) of the Atlas squares, and breeding was confirmed in 18% of these (5/27), by spotting nests and young incapable of sustained flight.

Bonaparte's Gull feeds mainly on insects and other invertebrates, but may also catch small fish (Baltz and Morejohn 1977; Braune and Gaskin 1982a,b; Braune 1987a; Burger 1988). A very agile and graceful flier, it catches prey on the wing, snatches them from the surface of the water, or plunges its head underwater to grab them. It may even occasionally make full dives to catch fish. Often it floats on the water and takes insects or other invertebrates near the surface (Braune and Gaskin 1982a; Braune 1987a; Burger 1988). During the breeding season, Bonaparte's Gulls forage on lakes and boggy ponds or rivers (Cons. G. G. – GREBE 1992h), and also glean insects from heaps of driftwood and logs along riverbanks or from floating log booms. When the flying ants emerge, usually in August and September, gulls often flock together to prey on the ants, which they catch on the wing or pick from the surface of the water.

Notes on territory

Little data is available on this gull's territory and home range. In Québec, mating has been observed on several occasions in late May, but it was not possible to determine whether the pairs were on their breeding territories at the time. Bonaparte's Gulls sometimes nest in colonies with contiguous territories. Adults and young leave the nest a few days after hatching, but it is not known how far they

Bonaparte's Gull breeds in Québec's boreal forest, especially near peatlands, ponds, and lakes. James Bay region. *June 1982, Vincent Létourneau.*

Bonaparte's Gull

Breeding distribution
1984–1989

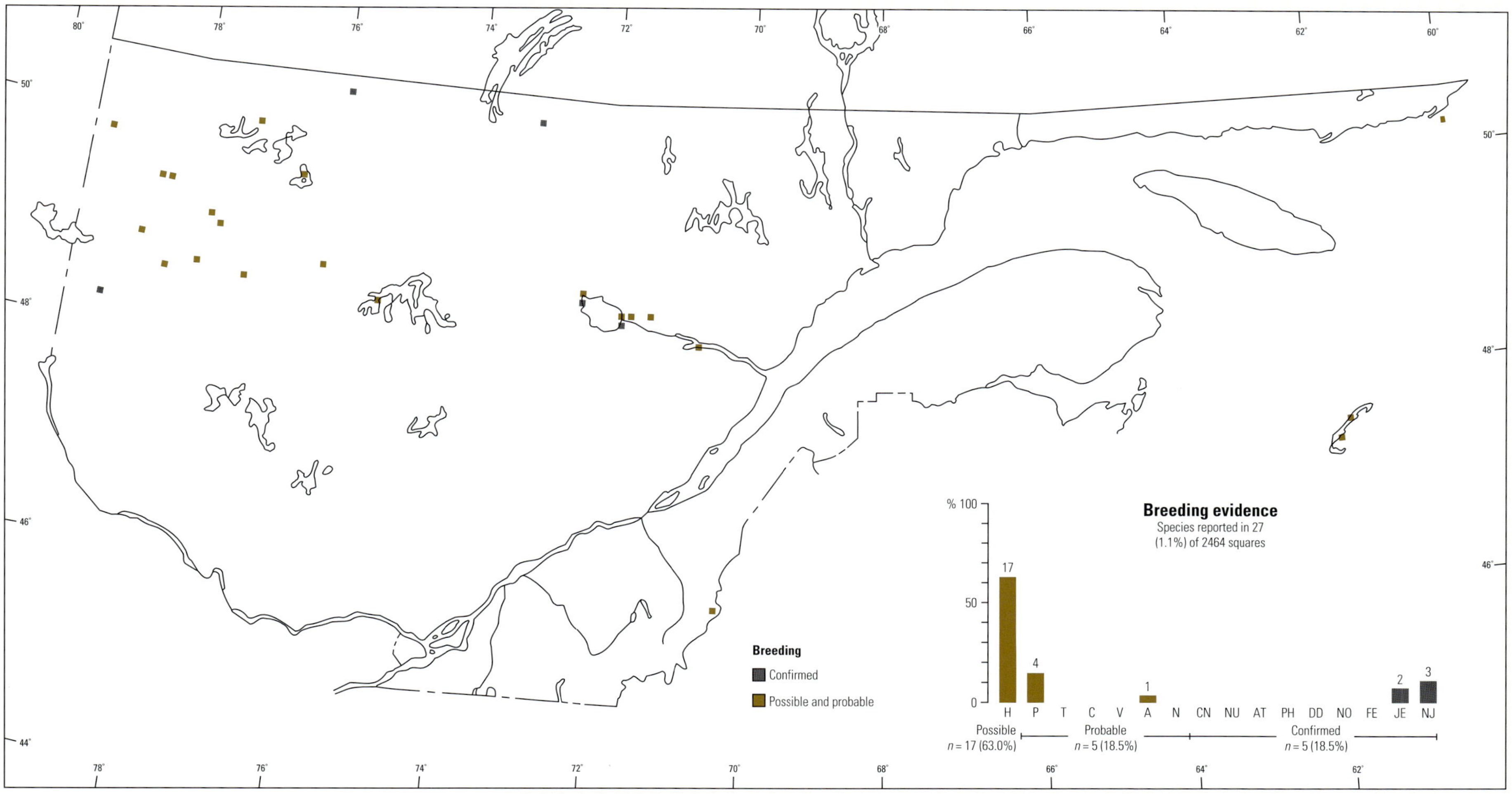

World breeding range

Species profile

Clutch size	usually 3 eggs (2–4)[3,4,5,6]
Laying interval	?
Start of incubation	?
Incubation period	24 days[5,6,7]
Maturity at hatching	semiprecocial
Care of young	by female and male[2,6]
Nestling period	≈ 7 days[2]
Age at first flight	?
Dependence of young	?
Broods per year	1[2]
Breeding age	?
Mating system	monogamy[6]
Duration of pair bond	?
Total length	30.3–37.0 cm[7]
Wing span	90–100 cm[1]
Weight	M 189 g (182–202) / F 170 g (?)[1]
Longevity record	?

References:
1. Cramp and Simmons 1983
2. Harrison 1978
3. Salt and Wilk 1966
4. Bent 1921
5. Terres 1980
6. Ehrlich et al. 1988
7. Godfrey 1986

Chronology $n = 26$

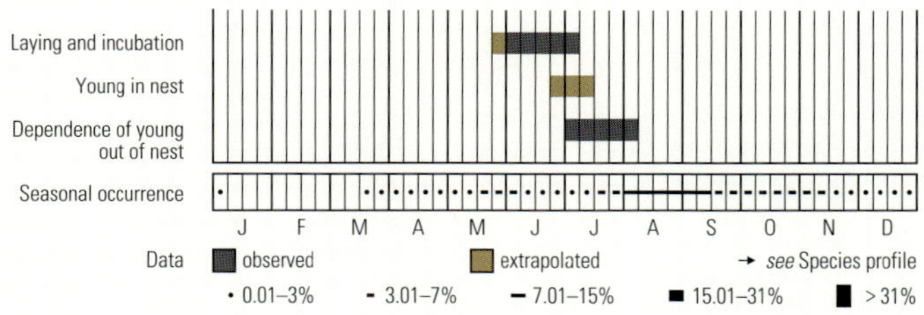

Notes on territory (continued)

travel. In the James Bay region, the birds appear to travel several kilometres from the nest territory sometimes to forage near rivers or lakes (Cons. G. G. – GREBE 1992h).

Well concealed in the branches of a spruce, the nest is made of twigs, plant stalks, moss, and lichen. It usually contains 3 eggs, as seen here. Churchill, Manitoba. *August 1980, Yves Aubry.*

All summer, flocks of Bonaparte's Gulls can be seen feeding on certains mudflats along the St. Lawrence River, such as the one at Les Escoumins, treading the mud to extract small invertebrates. Staging and wintering birds sometimes frequent saltwater habitats (Braune and Gaskin 1982a,b).

Distribution

The Nearctic Bonaparte's Gull (Harrison 1985; Grant 1986) breeds mainly in boreal and subarctic regions of central and western Canada and Alaska. Although breeding records were rare in Québec until recently (Braune 1989), pre-Atlas and Atlas records clearly show that the species is present in a large part of the province.

In western Québec, this gull is found from Rouyn-Noranda to the Opinaca Reservoir (Bordage and Aubry 1983; QNRCP), and in the east, from Lake Saint-Jean (where nests have been found at Racine Point and Saint-Gédéon) to Lake Mistassini. East of Lake Mistassini, a brood has been sighted on Lake Fleuricourt (D. Bordage, CWS, unpubl. data).

Bonaparte's Gull seems to have a very local distribution in the Nottaway, Broadback, and Rupert rivers region. In a study covering the area from Lebel-sur-Quévillon in the south, to the Pontax River in the north, the Ontario border in the west, and Lake Mistassini in the east, the species was regularly observed only in the western part, between the Ontario border and Nemiscau (Cons. G. G. – GREBE 1992h). A total of 10 nests and 2 broods, as well as numerous adults, were sighted in the area, which is located in the James Bay coastal plain, with its abundance of ponds, peatlands with pools, small lakes, and boggy ponds. The species was virtually absent from the eastern part, except for sightings north of Lake Mistassini, in a region of mostly oligotrophic lakes (Cons. G. G. – GREBE 1992h), and at Lake Moquachéa, around 100 km east of Lake Soscumica, where a single nest was found (R. Ouellet, MLCP, pers. comm.). Lake Moquachéa is shallow and bordered by lush vegetation, making it similar to lakes on the James Bay coastal plain.

There are several isolated breeding records from other areas, including one from the Magdalen Islands in 1990 (Fradette 1992). South of the St. Lawrence, an adult was observed on 9 July 1985 on a lake in the Jaro ZEC, in the Appalachians; although it may have been an early transient, the possibility that it was breeding cannot be excluded, since the species has reportedly nested east of this region in northwestern Maine (Braune 1989).

Beginning in late July, pre-migratory aggregations, often numbering in the hundreds, may be observed in the upper estuary and in Chaleur Bay. According to Braune (1987b, 1989), Québec breeding birds and most of those breeding in Ontario migrate through Québec on their way to New Brunswick and the Atlantic seaboard.

Unlike most larids, which usually nest on the ground, this gull generally builds its nest in a conifer, about 3 m up. James Bay region. *July 1982, Vincent Létourneau.*

In the James Bay region, Bonaparte's Gull nests in peatlands interspersed with small stands of spruce and ponds fringed with herbaceous vegetation. *June 1991, Vincent Létourneau.*

This route is probably also taken by some birds from the Prairies and the Northwest Territories, though most birds from western Canada head toward the Pacific coast or the Great Lakes in winter. The major waterways—particularly the Saguenay and St. Lawrence rivers—seem to be the preferred migration routes (Braune 1989). In addition, flocks of several hundred Bonaparte's Gulls have been seen along the Richelieu River during fall migration (M. Robert, pers. comm.).

This gull winters on the Atlantic and Pacific coasts, from Maine to Cuba and Haiti, and from British Columbia to Mexico (Harrison 1985; Braune 1989). Some gulls also winter inland, near large expanses of water such as the Great Lakes and some large rivers in the southern United States (Root 1988).

History and trends

A century ago, it was not unusual to observe this species along the St. Lawrence during migration (Dionne 1906) and it was also considered a fairly common summer resident on Anticosti Island and at Godbout at the time (Todd 1963; Ouellet 1969). According to Harper (1958), however, the birds sighted in the gulf were most likely breeding west of Hudson Bay. Todd (1963) described this gull as frequent in southern James Bay in the early part of this century and commented on records suggesting breeding along the inland waterways there or (more likely) insome of the lakes which they drain. In Québec, no nests were found until 1977, in the Lac-Saint-Jean and James Bay regions (QNRCP).

Data gathered during the last decade have documented Bonaparte's Gull as regular in many regions of Québec during the breeding season and migration. Recently, surveys in the area between the Ontario border, Lake Mistassini, Lebel-sur-Quévillon, and the Pontax River, produced an estimate of close to 2000 breeders in that region (Cons. G. G. – GREBE 1992h).

In Ontario, it is believed that the ecological requirements of Bonaparte's Gull have not changed during the last century and that there have been no major changes in its distribution in recent years (Blokpoel 1987). This may perhaps be also true for Québec, but not enough is known yet about the species' abundance and distribution here, either during breeding or migration, for any conclusions to be drawn.

The total population in North America has been estimated at slightly under 175 000 birds, which is not very large (Braune 1989). Hydro-electric projects and intensive logging operations in the northern part of the province, as well as excessive development pressure on some rivers must be monitored closely. The flooding of many small lakes and peatlands for hydro-electric projects, as well as the disappearance of the surrounding forest, could affect the breeding population of Bonaparte's Gull.

MICHEL SAVARD AND ALAIN MORRIER

Ring-billed Gull

Goéland à bec cerclé

Larus delawarensis

TODAY, THE RING-BILLED GULL IS THE MOST COMMON GULL IN THE farming and urban areas of southern Québec. However, it was not so long ago that Bent (1921) wrote, "The ring-billed gull yields readily to persecution, is easily driven away from its breeding grounds, and seems to prefer to breed in remote unsettled regions, far from the haunts of man." This statement demonstrates the extent to which the species has been able to adapt to the changes that have occurred in North America over the last 50 years or so.

In Québec, the Ring-billed Gull is an abundant but local breeder, which winters sporadically in small numbers in the southwestern part of the province (Bannon 1991b). Most birds overwinter on the Atlantic coast, from New England to the Gulf of Mexico (Blokpoel and Tessier 1986). Ring-bills migrate by day. In the fall, part of the population travels up the St. Lawrence to the Great Lakes, and in spring returns by the same route, mingling with other gulls.

Habitat and behaviour

Migrants begin to arrive in the Montréal region in early March, and a few days later in regions farther east and north. The species establishes its colonies on a variety of sites and substrates. Islands partially or completely covered with herbaceous vegetation are favoured, but the birds also nest on peninsulas, artificial sites created from dredge spoil or landfill, industrial yards, and sometimes even the roofs of buildings (Blokpoel and Smith 1988; Paquin 1993b).

In the Contrecoeur and Varennes islands, nest density in the Northern Pintail and Gadwall was higher on islands occupied by Ring-billed Gulls (Bélanger 1989). It is thought that the territorial defence mounted by the gulls against terrestrial predators improves the ducks' breeding success.

The breeding cycle is fairly rapid, with laying commencing in mid-April. Typical clutch size is 3 eggs, which are incubated by both parents. The cup-shaped nest, built in a scrape, is composed of twigs, sticks, and various other types of material. On one of the islands off Montmagny, DesGranges and Angers (1981) found that, of 686 nests examined, 16 (2.3%) contained 5 or 6 eggs. Some investigators have attributed these larger clutches to shared nesting and incubation by two females whose mates were lost at the beginning of the season (Ryder 1978; Ryder and Somppi 1979). An apparent dump nest of 23 eggs that was found by DesGranges (1981c) was explained as resulting either from overpopulation and/or an imbalance in the sex ratio in the colony.

The young develop rapidly and generally become independent at around the age of 40–45 days (Ryder 1993), although in Québec, this may occur as early as the 35th day (CWS, unpubl. data). The chicks are fed a variety of foods by both parents, including fish, insects, earthworms, garbage, and small mammals. On Lake Ontario, Haymes and Blokpoel (1978) found chick stomachs to contain mainly fish and insects while, at the Île de la Couvée colony near Montréal, Lagrenade and Mousseau (1981) reported a more varied diet, consisting of 39% insects, 20% garbage, 24% fish, and 6% earthworms.

In 1992, studies comparing the Ring-billed Gull's diet at two colonies, one in a farming area and the other in urban habitat, showed that only the proportions of food taken differed to any notable extent. On Saint-Ours Island, the diet was found to consist of 51% commercial waste (dead poultry chicks and viscera), 7% domestic waste, 26% insects, 6% earthworms, and 5% small mammals (Lefebvre and Giroux 1993). In Québec City, the proportions were as follows: 42% garbage, 22% insects, 23% earthworms, and 7% fish (Brousseau et al. 1996).

Young birds can be seen outside the colonies by early July, but most do not leave the surrounding area until the middle of the month, at which time they disperse. Banding studies have shown the birds to be far-ranging in their dispersal. Some birds seen in the St. Lawrence Valley originated from colonies in the Great Lakes (Blokpoel and Tessier 1986) and New York State (CWS, unpubl. data). Young gulls from the Québec City colony tend to disperse to the Lower St. Lawrence and the Gaspé during the first month (CWS, unpubl. data).

An analysis of the results of a dozen or so studies in the Great Lakes region shows productivity to be 1.0–2.2 young/nest. In Québec, breeding success was estimated to be 1.04 young/nest at Île de la Couvée (Mousseau and Lagrenade 1979), 1.67 young/nest at Duval Island in the Contrecoeur Islands (Mousseau and Lagrenade 1980), and 1.1 young/nest at the Québec City colony (CWS, unpubl. data).

Distribution

In Canada, the Ring-billed Gull breeds from southern British Columbia, through northern Alberta and Saskatchewan and central Manitoba, to Newfoundland. It also occurs in northern New Brunswick but is absent from Nova Scotia (Godfrey 1986). In Québec, it is restricted to open habitat in agricultural and urban areas during the breeding season. In its quest for food, the Ring-billed Gull forages in fields, cities, the shores of lakes and rivers (including the St. Lawrence itself), and garbage dumps. Nesting is confined to a limited number of rather large colonies, usually in open areas near water.

Since the species is colonial, breeding was fairly easy to confirm by finding nests with eggs, or young incapable of sustained flight. Some young birds, however, may have been reported in a square adjacent to the one where the colony was located. The Atlas map gives a fairly accurate picture of the species' distribution. Since the Atlas survey period, small colonies have also been reported in Trois-Rivières (1629 pairs in 1994) and at Stater Pond (25 pairs in 1992) near Thetford Mines (BIOMQ).

René Roy

Notes on territory

Once at the colony, both members of the pair defend a small territory that increases from about 1 m² to 2 m² during incubation, and to 4 m² after hatching (Ryder 1993).

The presence of vegetation appears to reduce territorial conflicts among Herring Gulls, allowing for higher breeding densities (Sootes 1975). According to Mousseau and Lagrenade (1980),

The explosive growth in the Ring-billed Gull population has led to the colonization of many new sites, both in agricultural and urban habitat. A large colony has become established in this pile of waste wood from a paper mill. Enclosures were set up to study the productivity of the species. Québec City. *May 1992, Pierre Brousseau.*

Ring-billed Gull

Breeding distribution
1984–1989

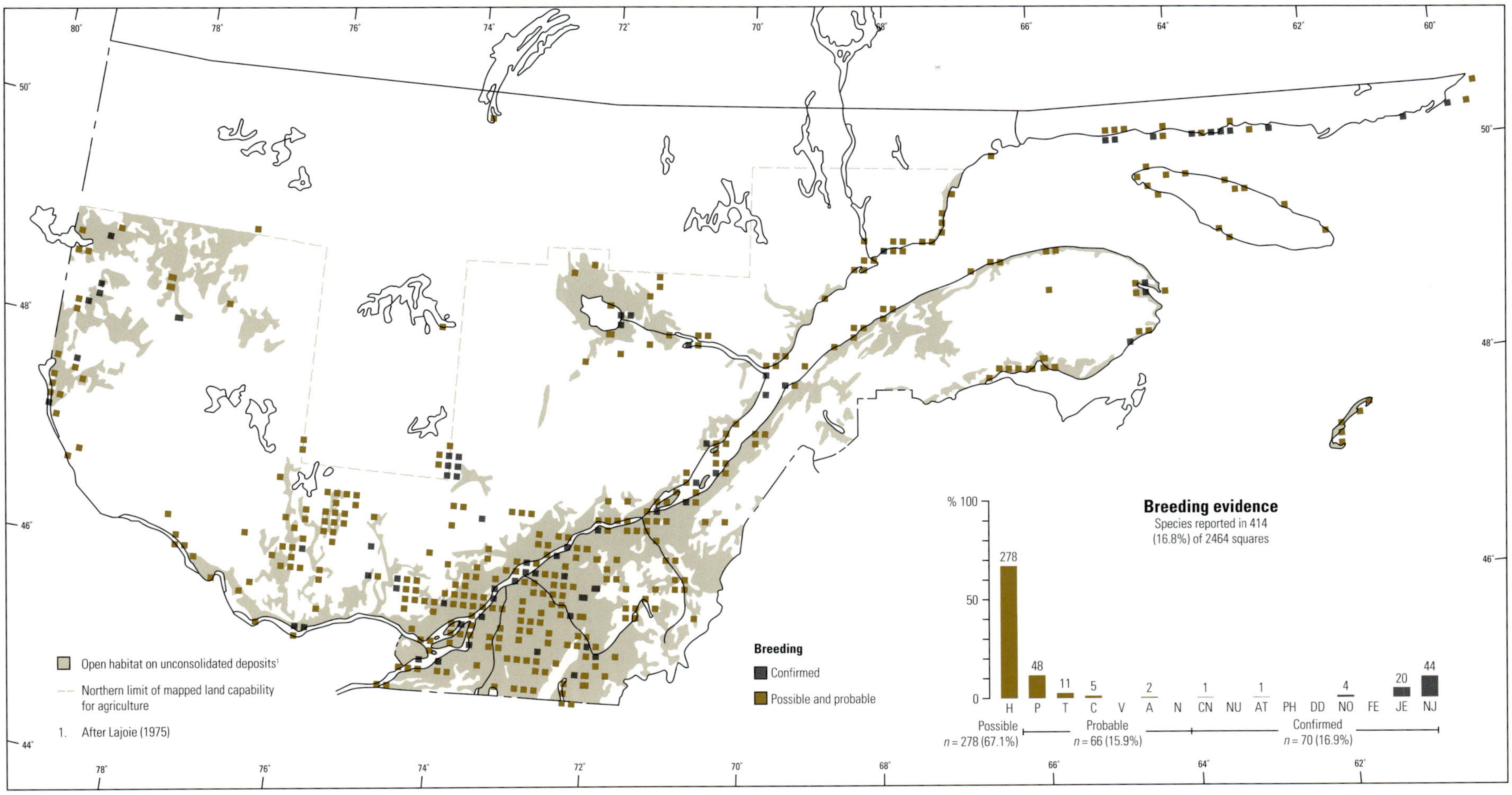

Species profile

Clutch size	usually 3 eggs (2–4)[1,2,4,5]
Laying interval	1 egg per 1–3 days[4]
Start of incubation	with last egg, sporadic before[4]
Incubation period	usually 26–27 days (23–28)[3,4]
Maturity at hatching	semiprecocial
Care of young	by female and male[1,4]
Nestling period	2–3 days[1]
Age at first flight	35–40 days[1]
Dependence of young	until first flight or 40–45 days old[4]
Broods per year	1[1,4,5]
Breeding age	usually 3 years (2–4)[2,4]
Mating system	usually monogamy[4]
Duration of pair bond	sometimes more than 1 season[4]
Total length	45.5–51.0 cm[3]
Wing span	121.9 cm[2]
Weight	M 566 g (495–650) / F 471 g (375–590)[4]
Longevity record	31 years 9 months[4]

References:
1. CWS, unpubl. data
2. Terres 1980
3. Godfrey 1986
4. Ryder 1993
5. Harrison 1978

Chronology *n* = 50

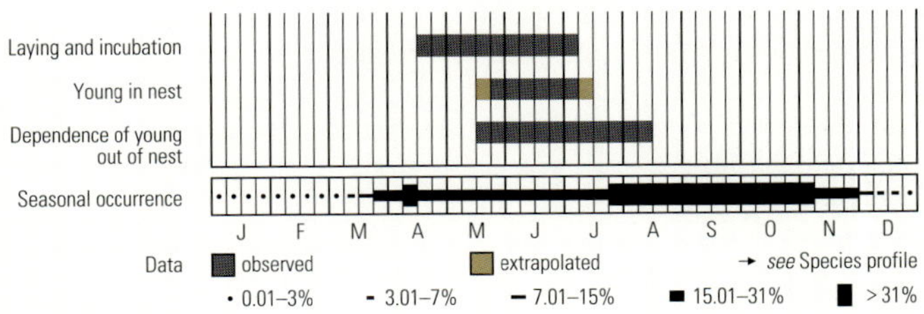

World breeding range

After a few days, the young leave the nest and move into the surrounding area. Interestingly, the presence of groups of young birds does not provoke conflict among the adults. Québec City. *June 1992, Yves Aubry.*

Notes on territory (continued)

this might also be true for Ring-billed Gulls and would explain the high densities, 0.8–1.3 nests/m², observed in three colonies in southwestern Québec. Field work by Blokpoel and Tessier (1986) in the Toronto area lends additional support to this hypothesis, with vegetation often being replaced by obstacles such as wood, rock, and concrete, although the type of obstacle does not appear to make any significant difference.

Colonies are found all along the St. Lawrence from Lake Saint-François to the Lower North Shore. There are also colonies along the major tributaries of the St. Lawrence, in the Ottawa Valley, and in the Saguenay–Lac-Saint-Jean region, as well as in the Eastern Townships, Abitibi, Mauricie, and Mont-Laurier regions. There are also a few colonies in the Mingan region, and on the Gaspé Peninsula, where breeding was confirmed at Saint-Omer, Carleton, Chandler, and Sandy Beach near the town of Gaspé (Brousseau and Chapdelaine 1989; Guillemette 1993). Non-breeders were also sighted regularly in the Lower St. Lawrence region. To date, breeding has still not been confirmed on Anticosti Island (Ouellet 1969; BIOMQ). The first breeding record from the Magdalen Islands was obtained in 1993 (Shaffer and Laporte 1993).

North of the Atlas area, there are several confirmed pre-Atlas records in the southern part of James Bay. Nests were found on an island located a dozen kilometres north of the mouth of the Eastmain River, on 30 June 1926, and on Barboteau Rock in Rupert Bay on 4 July 1941 (Todd 1963). Although located near the Québec mainland, these islands belong to the Northwest Territories. Todd (1963) reported that he had never observed the species north of the Boatswain Bay area, but added that in July 1926, George M. Sutton had several sightings of Ring-billed Gulls along the coast up to and beyond the 53rd parallel. The most northerly sighting to date is from Povungnituk, where 9 gulls were seen on 22 August 1986 (ÉPOQ). Some birds were also seen farther south in the Kuujjuarapik region on 9 August 1935 (Todd 1963) and on 18 and 25 July 1949 (Savile 1950).

History and trends

In the middle of the 19th century, the Ring-billed Gull had two disjunct breeding populations: a western one on the Prairies extending east to the Great Lakes (Huron and Michigan) (Blokpoel and Tessier 1986) and an eastern one consisting of isolated colonies on the east coast of James Bay, on the Lower North Shore of the St. Lawrence, and in Labrador (Todd 1963).

The first breeding record in the east was obtained in 1833 by J. J. Audubon, who discovered a colony of 200 pairs on Île du Petit Mécatina on the Lower North Shore (Todd 1963). Near the end of the last century, Wintle (1896) described the Ring-billed Gull as a scarce "transient visitant" in the Montréal area. It was still classified as a common transient in the 1930s (Mousseau 1984a).

In 1953, the species was found nesting on Moffat Island (now Île Notre-Dame), but the colony was destroyed during the construction of Expo 67. The species then colonized several nearby sites, finally establishing itself between 1967 and 1971 on Île de la Couvée, an island created from dredge spoil during the construction of the St. Lawrence Seaway. It quickly expanded, establishing colonies along the Ottawa River, the upper St. Lawrence, and the estuary (the Québec City area and the group of islands off Montmagny), finally reaching

Both adults build the nest using a variety of materials, ranging from plastic to seaweed. This eggshell has been pipped by a chick that will hatch the next day. Québec City. *June 1992, Pierre Brousseau.*

This man-made island, with its automated lighthouse, has been colonized by the opportunistic Ring-billed Gull. Lake Saint-François, Ontario side. *July 1989, Jean-Luc DesGranges.*

Baie-Comeau in 1973. Inland areas were colonized much later, with small colonies appearing on Lake Saint-Jean, near Mont-Laurier and in the Eastern Townships at the end of the 1970s and the early 1980s (Mousseau 1984a).

During the 6-year Atlas period, the species expanded its range toward the northwest, founding colonies near urban areas in the Abitibi region such as Val-d'Or and Rouyn-Noranda. Farther south, a new colony of 350 pairs appeared at Lake Saint-Pierre (D. Jauvin, pers. comm.). Another colony was discovered near La Tuque after atlassing was completed.

In 1979, the Ring-billed Gull population along the St. Lawrence was estimated at 36 401 pairs, as shown in the table below. Of this number, 28 236 (77.6%) were found in the four colonies in the Montréal region, including the Île de la Couvée colony with its 15 300 pairs (Mousseau 1984a). Twelve years later, in 1991, another colony was established at the foot of the Beauharnois Dam, bringing the count for the region to 94 000 pairs, which corresponds to an annual growth rate of about 7.5%: of these pairs, 47 201 were found at Deslauriers Island, and 28 114 at Île de la Couvée. By 1994, the five colonies contained a total of 95 713 pairs, which suggests that population levels may have more or less stabilized since the 1991 survey (CWS, unpubl. data).

Number of Ring-billed Gulls nesting along the upper St. Lawrence and estuary, 1978–94 (BIOMQ)

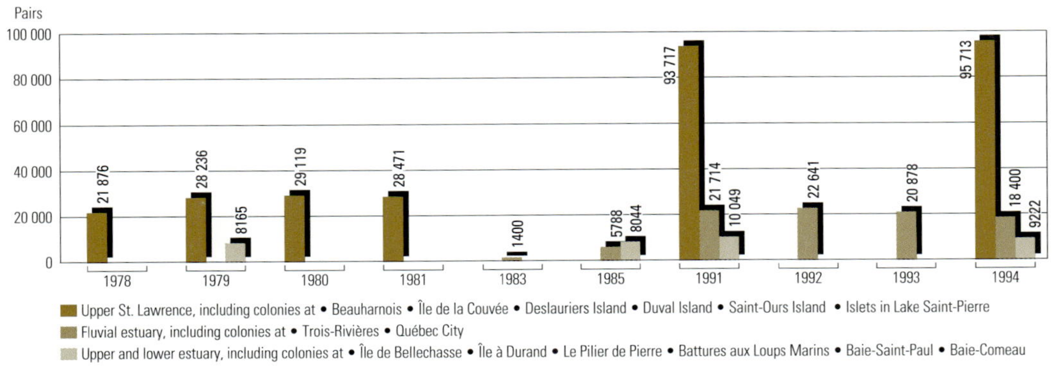

In 1981, there were some 1200 pairs in the Ottawa Valley, divided among four colonies, mostly in the Hull region (Chabot 1981). In the Eastern Townships, the species failed to gain a foothold on islands in the Saint-François River because of water-level fluctuations but did manage to establish permanent colonies at the Choinière Reservoir in 1979 and at Bromptonville in 1985–86. The Bromptonville colony has continued to increase, and in 1991 numbered 516 pairs, whereas the colony in the Yamaska Recreation Park disappeared in 1990 (BIOMQ).

In the group of islands off Montmagny, the species has been nesting on Île à Durand and Le Pilier de Pierre since at least 1971, and on Île de Bellechasse since 1979. The Québec City colony was first described by Perreault (1983), who counted 1400 pairs. The colony expanded rapidly, from 5788 pairs in 1985 to 22 641 pairs in 1992, as shown in the table. The population in 1994 was estimated at 16 771 pairs (CWS, unpubl. data).

The Baie-Comeau colony has been in existence since 1973, but the latest surveys show that numbers have stabilized at about 2500 pairs (BIOMQ). It is not known why the species has never become firmly established in the Lower St. Lawrence region. In the Gaspé, approximately 20 nests were found at Chandler and Sandy Beach Point in 1989 (Chapdelaine and Brousseau 1992b), but in 1993 close to 670 pairs were found in three colonies: Saint-Omer Bank, Carleton, and Chandler (Guillemette 1993). A year later, the Carleton and Saint-Omer colonies had 1548 pairs (Guillemette 1994).

The Ring-billed Gull became established in the Mingan region between 1940 and 1945, and the first nests were found in 1947 (Gabrielson 1952). Since then, the birds have been continually moving their breeding sites, in some years forming a single colony and in other years, several colonies (Chapdelaine and Bourget 1981; Brousseau and Chapdelaine 1990; S. Paradis, pers. comm.). A census in 1990 yielded an estimated 4016 pairs in three colonies (Grenier and Kavanagh 1993). Censuses of the Middle and Lower North Shore bird sanctuaries show colonies to be continually shifting within and outside the boundaries of these sites. This behaviour, which is typical of the region, was first described in 1884 by Frazar (Lewis 1941b). In 1940, Lewis (1941b) estimated the population in this region to be 1769 pairs, divided among six colonies. Since then, there has been no complete survey of the area to serve as a basis of comparison. The 13th census of the North Shore sanctuaries in 1988 yielded estimates of 533 birds (Chapdelaine and Brousseau 1991).

In 1993, Shaffer and Laporte (1993) reported the first breeding evidence in the Magdalen Islands, on a small island near Grande-Entrée.

The population explosion of the Ring-billed Gull in Québec is part of an overall North American trend (Blokpoel and Scharf 1991). In Québec, the species has been increasing its numbers steadily for 20 years, with the population in 1991 estimated at 125 700 pairs (BIOMQ). Between 1979 and 1991, the overall annual growth rate in the colonies along the upper St. Lawrence and estuary was 10.4%, which is slightly lower than the 12.3–13.3% recorded during the same period in Ontario's lower Great Lakes region (between the St. Clair River and Cornwall) (Blokpoel and Tessier 1991).

It is now known that a combination of factors have been responsible for this extraordinary increase, including the availability of new food sources, the creation of new nesting habitat, and factors intrinsic to the species itself, such as its prolonged life expectancy and opportunistic behaviour. Nonetheless, the 1994 census of the colonies along the St. Lawrence found a slight decrease (1.7%) in numbers since 1991, with the 1994 population estimated at 123 500 pairs. The reasons for this decrease, which has been observed in most of the large colonies in the upper St. Lawrence and fluvial estuary, are not known. We can only conclude that the dynamics of this population are still not fully understood or that monitoring has not been done at a small enough scale to have predicted such small changes.

Despite the fluctuations in its populations, the Ring-billed Gull has become increasingly common in the urban environment, as it continues to adapt its behaviour to human activity. It has expanded not only its breeding range, but also its feeding areas and the extent of its post-breeding dispersal. Most affected by this expansion are urban and, to a lesser extent, farming areas.

Pierre Brousseau

Herring Gull

Goéland argenté

Larus argentatus

René Roy

THE HERRING GULL IS UNDOUBTEDLY THE MOST COMMON GULL IN Québec. It breeds along the coastline, including the shores of the St. Lawrence, and on inland lakes, from the agricultural and forest regions of southern and central Québec to the tundra of the Ungava Peninsula. This is the gull that inevitably figures in vacation photos taken by the sea. It is a familiar fixture in our ports, perching on rooftops and scavenging food from fishermen, a bird superbly adapted to the human environment.

In fall, part of the population migrates up the St. Lawrence to the Great Lakes (Moisan and Poulin 1967), and then south to the species' wintering grounds in the United States, Mexico, the West Indies, and even Panama (Godfrey 1986). Thousands of birds winter in Québec, mainly around Tadoussac and Les Escoumins and in the gulf.

Herring Gulls often nest in isolated pairs on the outskirts of Ring-billed Gull colonies. Québec City. *May 1992, Pierre Brousseau.*

Habitat and behaviour

The Herring Gull breeds in colonies or in solitary pairs on islands, sandbars, coastal marshes, islets in lakes, and cliffs overlooking the sea. On wooded islands, it nests on the ground in grassy, rocky, or sandy areas, but rarely in the underbrush. Sometimes nests are built in conifers; although this type of nesting does not seem to be prevalent, some 400 nests were found in conifers on the Mingan Islands (Grenier and Kavanagh 1993). At the La Grande-2 Reservoir, in the James Bay region, an adult was observed incubating eggs in an old Osprey nest (J. Rodrigue, pers. comm.). During the last century, on Grand Manan Island in the Bay of Fundy, some Herring Gulls began nesting in fir trees to escape persecution by fishermen, but returned to the ground when harassment ceased (Townsend 1921b).

The pair bond is permanent, although divorce is common (Cramp and Simmons 1983). Pairs begin arriving on the breeding grounds in mid-March, taking up territories in the colony but delaying nest building and laying until the snow melts. Late arrivals are usually inexperienced birds which, although they do not nest, begin defending a territory in the year before their first breeding season, when they are about 4 years old. Such territories are most often found on the outskirts of the colony (Chabrzyk and Coulson 1976). In the following weeks, pairs establish their territories and build their nests. West of Québec City, the onset of laying is in late April or the first week of May, while farther east the process is delayed by a few days or even a few weeks (2–3 weeks on the Middle and Lower North Shore).

The nest is a bowl-like structure about 15–20 cm in diameter and 5–8 cm in depth, set in a mound of plant materials, usually of terrestrial rather than marine origin, as well as feathers and other assorted objects (Cramp and Simmons 1983). It is very similar to the nest of the Great Black-backed Gull.

Incubation, which usually lasts 28–30 days, is performed by both parents (Cramp and Simmons 1983). Hatching is asynchronous, and the semiprecocial young leave the nest 2–3 days after hatching, and seek shelter from the rain, sun, and predators under vegetation, rocks, or other objects. Early on in brood rearing, one adult is always present to attend the chicks and defend the territory. Later, attendance becomes less constant and the chicks may be left alone for several hours. The young become fully independent at roughly 40 days old, around mid-July in the upper St. Lawrence and estuary, and in early August in the gulf.

The Herring Gull's diet is highly varied, reflecting the seasonal availability of food resources. It forages in a number of different habitats, even though some individuals seem to be specialized to a certain extent (Davis 1975). Herring Gulls feed both inshore and offshore; they follow boats in all seasons and visit fish processing plants and garbage dumps. Birds collected in colonies in the Montmagny region were found to have a diet of molluscs, crustaceans, insects, fish (minnows, common carp, walleye), and fish roe (DesGranges and Angers 1981), while, in the gulf, gulls fed mainly on capelin, sand lance, and young chicks, including those of its own species (CWS, unpubl. data).

The Herring Gull is a consummate predator, scavenger, and food pirate. It has a number of feeding techniques, one being its well-known habit of dropping sea urchins and molluscs. This it does usually at low tide and in high winds, carrying the prey in its bill 15 m or so aloft and breaking it by dropping it on the rocks below. This is a highly effective technique, but success seems to depend on how experienced the individual is (Ingolfsson and Estrella 1978). The same technique is used on other small prey such as young ducklings, which are then swallowed (J. Gauthier, pers. comm.).

The Herring Gull was reported in 32% of Atlas squares (790/2464). As the species often nests in colonies in open areas, breeding was confirmed in nearly a third of the squares in which it was reported (257/790), mainly from records of nests and young incapable of sustained flight.

Notes on territory

Territory size can vary considerably depending on the type of terrain, ranging anywhere from about 15 m² to five times this size. Over the course of the breeding season, the territory will change in size: it is smallest during incubation and largest during brood rearing (Pierotti and Good 1994).

Younger birds are known to nest later than older, more experienced birds and tend to nest in low-density areas on the outskirts of the colony, moving to a more densely populated area after a few years (Chabrzyk and Coulson 1976).

On the grassy summit of Percé Rock, a colony of Herring Gulls, estimated at about 250 pairs in 1989, nests alongside Double-crested Cormorants and Great Black-backed Gulls. Île Bonaventure and Rocher Percé Conservation Park. *Summer 1991, Alain Desrosiers, Ministère du Loisir, de la Chasse et de la Pêche.*

Herring Gull

Breeding distribution
1984–1989

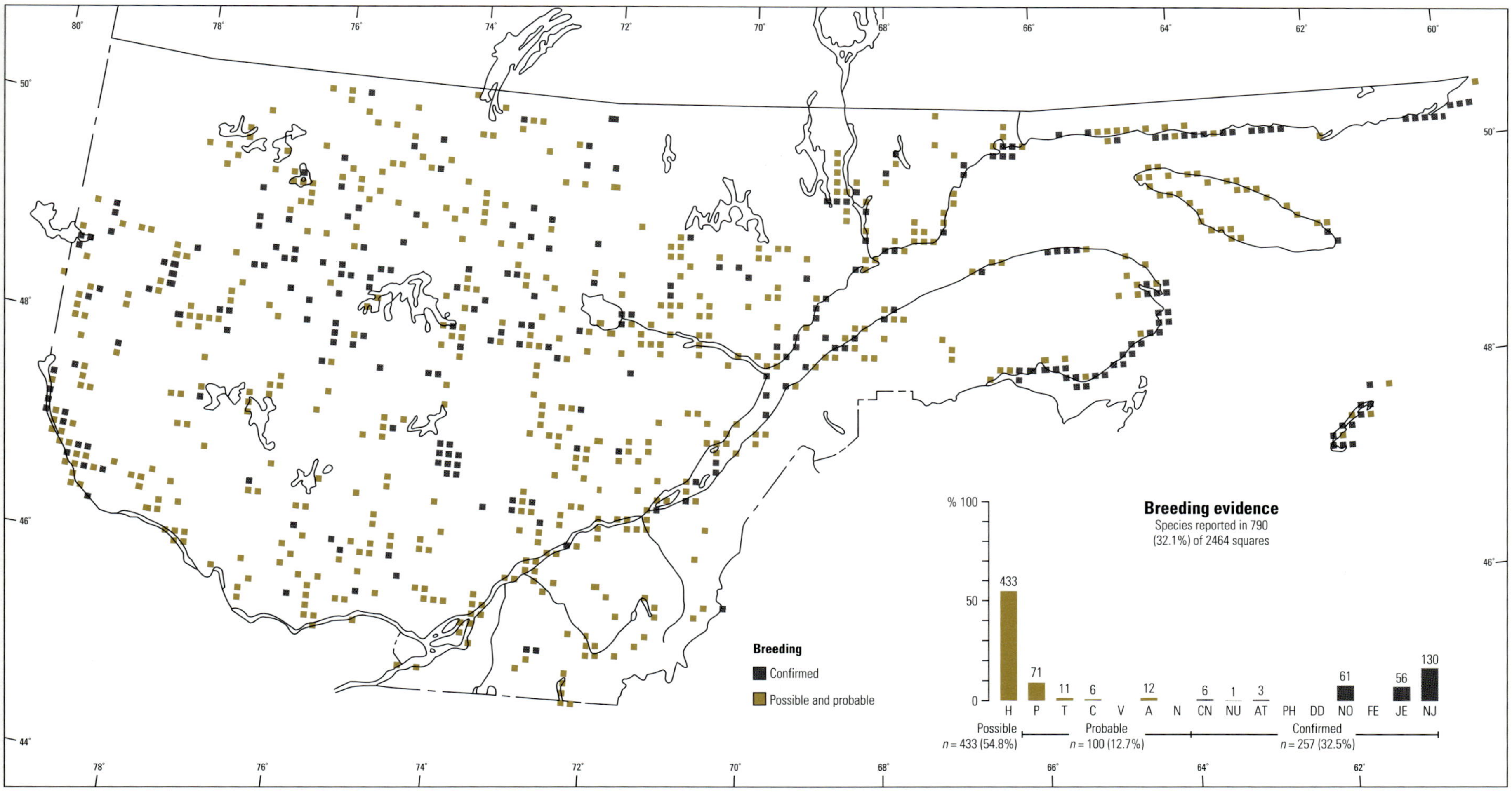

Species profile

Clutch size	usually 3 eggs (2–4)[1]
Laying interval	1 egg per 1–2 days[1]
Start of incubation	with last egg, sporadic before[1], with second egg, sporadic before[5]
Incubation period	usually 28–30 days (26–32)[1,5]
Maturity at hatching	semiprecocial
Care of young	by female and male[1]
Nestling period	2–3 days[1], less than 24 hours[5]
Age at first flight	35–50 days[1,5]
Dependence of young	usually until shortly after first flight[1], until 12–15 weeks old, sometimes up to 6 months after first flight[5]
Broods per year	1[1]
Breeding age	usually 4–5 years (3–7)[1]
Mating system	monogamy, rarely polygyny[1]
Duration of pair bond	for life[1]
Total length	58.5–66.0 cm[2]
Wing span	137.2 cm[4]
Weight	M 1226 g (755–1495) / F 1044 g (717–1385)[3]
Longevity record	31 years 11 months[1]

References:
1. Cramp and Simmons 1983
2. Godfrey 1986
3. Dunning 1984
4. Terres 1980
5. Pierotti and Good 1994

Chronology $n = 290$

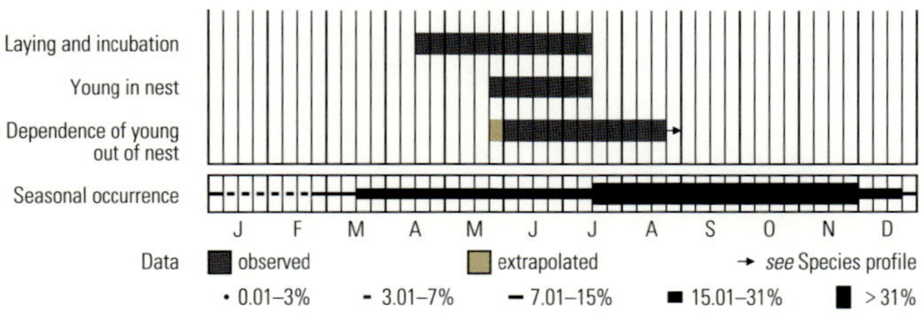

World breeding range

Plant stalks, grass, and feathers were used to construct this nest, which was found among debris and rocks on the shore of Le Gros Pèlerin. Îles de l'estuaire National Wildlife Area. *June 1956, Gaston Le Page.*

Distribution

Herring Gulls breed throughout southern Québec, but the species' centre of abundance is the St. Lawrence River east of Québec City. It also nests in solitary pairs or in small colonies around Montréal and in the Eastern Townships, on the outskirts of Ring-billed Gull colonies. In the estuary, most islands and islets have mixed colonies of anywhere from a few pairs to several thousand birds, nesting along with Great Black-backed Gulls and Common Eiders. On Île aux Pommes, for example, there are 1800 pairs (BIOMQ).

On the Gaspé coast, between New Richmond and the Forillon Peninsula, isolated pairs nest on small ledges on the cliffs. There are also colonies on some islands such as Les Îlots at Newport, Île Plate, and Percé Rock, and on the sandbars off Sandy Beach (Gaspé Bay) and Carleton. The absence of breeding records from the Gaspé interior can be explained by the scarcity of lakes in that region.

On the North Shore, most confirmed records were from three main areas: Sept Îles Bay, the Mingan region, and the area around the Sainte-Marie Islands, near Harrington Harbour. Had it not been for the inaccessibility of certain squares between the sanctuaries that were censused, the species would most likely have been found nesting along the entire North Shore, as it does on the Magdalen Islands, for example. It also breeds on the eastern tip of, and probably elsewhere on, Anticosti Island.

The breeding range of the Herring Gull also takes in the mixture of forest and farmland found in the Abitibi region of northwestern Québec, all of the boreal forest, and the tundra of the Ungava Peninsula (Todd 1963; Gaston et al. 1985; Godfrey 1986). The species was reported in June 1993 in the Ivujivik region (D. Bordage, CWS, pers. comm.). In the North, this gull generally nests in isolated pairs, occasionally in colonies, but very rarely in mixed species colonies. Nests are generally located on rocky islands or islets in the middle of large lakes over 10 ha in area.

History and trends

The data that have been compiled on the Herring Gull since the turn of the century are fragmented, local, and often qualitative, rather than quantitative, making it difficult to draw any conclusions on the status of the Québec population. Early ornithologists who visited the estuary and the gulf usually commented only on occurrence; any estimates given were of the total number of individual birds (Brewster 1884; Lucas 1891; Coote 1916; Townsend 1920; Lewis 1922; Ball 1938). These early accounts show that the Herring Gull, like most seabirds, was subject to considerable persecution in those days, causing a decline in numbers (Blanchard 1984).

With the signing of the Migratory Birds Convention in 1916 by Canada and the United States, and the creation of a network of migratory bird sanctuaries, particularly on the Middle and Lower North Shore, Herring Gull populations gradually stabilized and then began to prosper and expand. Censuses conducted by the Canadian Wildlife Service since that time document population trends by region.

The census program in the 9 sanctuaries on the Middle and Lower North Shore is a good example of population monitoring efforts. The sanctuaries, located between Sept-Îles and Blanc-Sablon, have been surveyed every 5 years beginning in 1925. As shown in the figure below, Herring Gull numbers in the sanctuaries increased steadily between 1925 and 1977, and then more than doubled between 1977 and 1988, from 10 089 to 22 409 birds (Brousseau and Chapdelaine 1990), which corresponds to an annual growth rate of 7.2%. Much of this growth took place in the Corossol Island and Saint-Augustin sanctuaries, where 11 296 and 6214 birds were counted respectively in 1988, accounting for 78% of the total Herring Gull population in the North Shore sanctuaries (Chapdelaine and Brousseau 1991). The fourteenth census, conducted in 1993, revealed a 70% decrease in the breeding population. Declines on this scale were recorded in all the sanctuaries, particularly Corossol Island, where the population plummeted from 11 296 to 1224 birds (Chapdelaine 1995).

Herring Gull populations in the Middle and Lower North Shore sanctuaries, 1925–93 (BIOMQ)

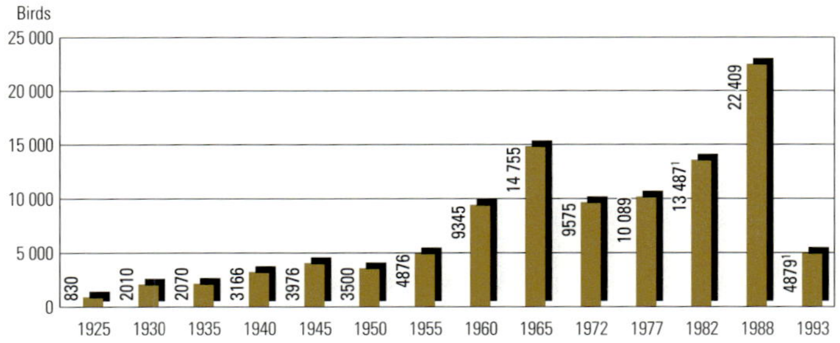

1 The Saint-Augustin sanctuary was not surveyed in 1982 and 1993.

Herring Gull preying on a young Black-legged Kittiwake. Corossol Island, Sept Îles Bay. *July 1985, Gilles Chapdelaine.*

Notes on territory (continued)

Nest density within colonies is influenced by a number of factors including vegetation, topography, colony size, and disturbance. Therefore, estimates are not very useful, unless made regularly within the same colony.

Herring Gulls often nest in solitary pairs on rocks and islets in inland lakes. On wooded islands, nests are built in a grassy, open area. Dozois Reservoir. *May 1991, Yves Aubry.*

Data from the estuary are difficult to interpret, because the various colonies were not always censused in the same year and all 48 colonies were only surveyed simultaneously once, in 1990, as the figure below shows, populations on some of the islands between Montmagny and Trois-Pistoles fell by roughly 46% over a 23-year period, a significant decrease, since these sites represent 47% of the total breeding population (10 274 pairs) in the upper and lower estuary (Chapdelaine and Brousseau 1992a). This decrease is particularly troubling since the corresponding Great Black-backed Gull populations have shown a sharp increase.

Herring Gull populations in selected colonies in the upper and lower estuary, 1967–90 (BIOMQ)

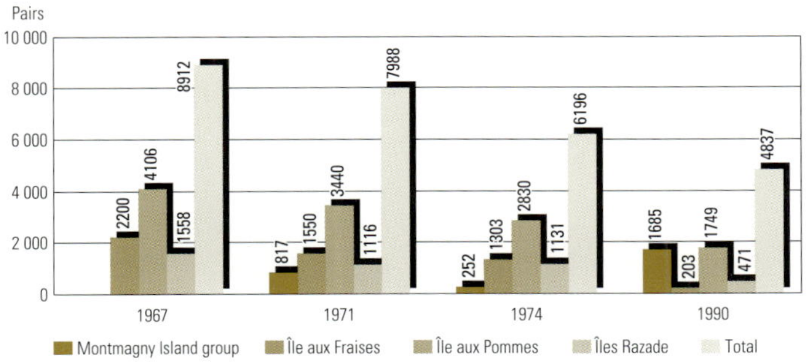

Populations along the Gaspé coast have also declined according to censuses conducted between 1979 and 1989, which show an annual decrease of 1.9% (Chapdelaine and Brousseau 1992b). During the same period, Great Black-backed Gull populations grew annually by 24.4%.

Herring Gull populations in the Gaspé and Magdalen Islands

	Gaspé		Magdalen Islands	
	1979	1989	1977	1990
Population	9792 pairs	8200 pairs	2313 birds	3070 birds
No. of colonies	23	27	8	13

Breeding populations on the Magdalen Islands increased slightly (Chapdelaine and Brousseau 1992a; Fradette 1992), although different survey methods were used on the islands.

In the province as a whole, the Herring Gull population has decreased, despite significant increases in places, such as the Montmagny and Minganie regions. Studies in Québec, elsewhere in Canada, and in Europe attribute recent increases in the populations of most seabirds that feed underwater to an abundance of small forage fish such as sand lance and capelin (Furness and Barrett 1985, 1991; Chapdelaine and Brousseau 1991). Therefore, in this context, it is difficult to explain the decreased number of Herring Gulls—which feed on many of the same prey—given that this gull is much more of a generalist than the Razorbill or Common Murre.

The most recent information from the east coast of Newfoundland, however, indicates that seabirds such as the Herring Gull and the Black-legged Kittiwake, which feed near the coast and on the surface of the water, have had abnormally low levels of breeding success over the past few years (Neuman et al. 1992). Montevecchi et al. (1992) attribute this to the cooling of the surface water, such that the breeding period in these species is now out of phase with the abundance and availability of fish, especially capelin.

Research by the Canadian Wildlife Service in the Middle and Lower North Shore sanctuaries in 1993 and on Bonaventure Island in 1994, found low productivity in the nesting species present, possibly linked to the same phenomenon. A full 46% of the nests censused on 4 and 5 June 1993 contained no eggs (CWS, unpubl. data).

In New England, Herring Gull populations have increased rapidly since the 1940s (Kadlec and Drury 1968) and are still expanding, but at a much slower rate. In Europe, annual increases of 13% were recorded until the mid-1970s, but since then numbers have dropped considerably in England (Lloyd et al. 1991) and Scandinavia (Evans 1984).

The paucity of information from the areas between the North Shore sanctuaries makes it difficult to estimate Québec's total breeding population. Recent data suggest that as many as 35 000 pairs breed along the St. Lawrence (BIOMQ). This does not include birds nesting inland in the regions of farmland and forest, populations on which we have very little information.

Pierre Brousseau

Three downy Herring Gull chicks. As soon as they are able to move about, the young gulls begin exploring the area around the nest. Forillon National Park, Gaspé. *June 1979, Daniel Jauvin.*

Great Black-backed Gull

Goéland à manteau noir

Goéland marin

Larus marinus

THE GREAT BLACK-BACKED GULL IS A TYPICAL SEABIRD. A FORMIDABLE predator, and sometimes a carrion eater, scavenger, and food pirate, it dominates other species of gulls on both breeding and feeding grounds. It may even chase away a Bald Eagle and steal its food or perch (Lien 1975). This gull breeds in colonies in the coastal areas of eastern North America, from northern Labrador south to Virginia, including the Gulf of St. Lawrence and estuary. Its breeding range extends inland to the Great Lakes, where it is most often found nesting in isolated pairs in mixed gull colonies (Blokpoel and Scharf 1991). In North America, the Great Black-back winters on the Atlantic coast from Newfoundland south to Florida and inland along the St. Lawrence and on the Great Lakes (AOU 1983).

Habitat and behaviour

The Great Black-back's breeding biology is quite similar to that of the Herring Gull, but in mixed colonies, Great Black-back eggs almost always hatch first. Pairs nest on islands, islets, and promontories, and sometimes on cliffs, but always in places where the vegetation does not obstruct their view of the surroundings, or with a commanding lookout, such as rocky mounds.

Clodin Roy

This larid feeds mainly at sea but nests mostly on coastal islands—in Québec, in the estuary and gulf. Île aux Basques, St. Lawrence estuary. *May 1972, Ronald Lepage.*

Notes on territory

In a Great Black-backed Gull colony in Ireland, Hudson (1982) observed that the diet of breeding birds varied with nest density. Birds nesting colonially fed on fish and crabs, while isolated pairs ate mainly larids and alcids taken from the colony.

According to Butler and Trivelpiece (1981), nest density does not influence egg production or hatching success in the species, but isolated pairs succeed in raising more young because of fewer

When it nests with other species of gulls, the Great Black-backed Gull generally occupies the sites with the best vantage points, such as grassy areas and rocky outcrops on hillocks and promontories. Corossol Island, Sept Îles Bay. *July 1985, Gilles Chapdelaine.*

Great Black-backed Gull

Breeding distribution

1984–1989

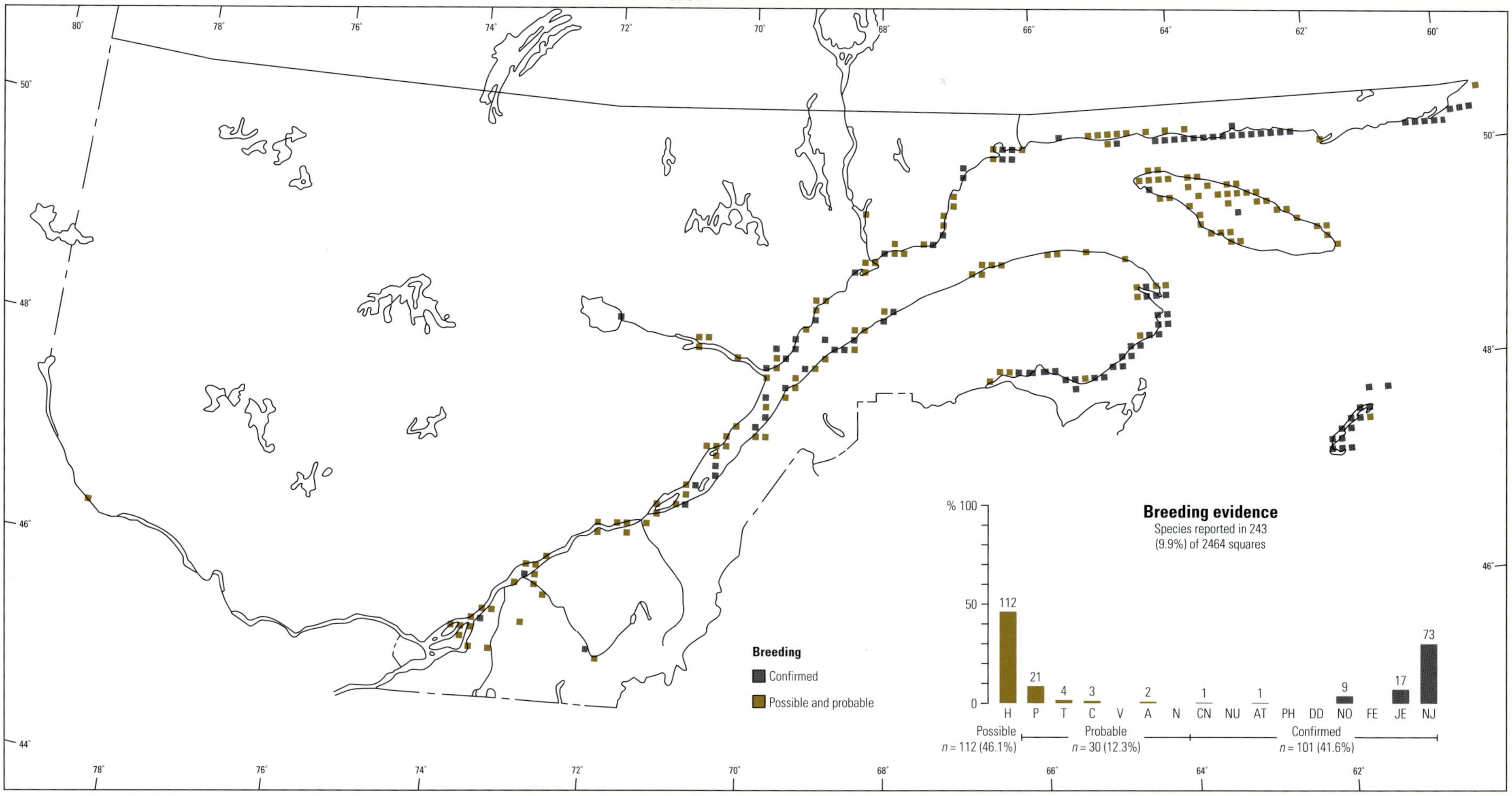

Species profile

Clutch size	usually 3 eggs (2–4)[1]
Laying interval	1 egg per 1–2 days[1]
Start of incubation	with last egg, sporadic before[1]
Incubation period	27–28 days[1]
Maturity at hatching	semiprecocial
Care of young	by female and male[1]
Nestling period	1–2 days[1]
Age at first flight	56–63 days[1]
Dependence of young	until first flight[1]
Broods per year	1[1]
Breeding age	usually 4–5 years[1]
Mating system	monogamy[1]
Duration of pair bond	probably more than 1 season[1]
Total length	71.0–78.5 cm[2]
Wing span	152.4–167.6 cm[3]
Weight	M 1829 g (1380–2272) / F 1488 g (1033–2085)[4]
Longevity record	23 years 3 months[5]

References:
1. Cramp and Simmons 1983
2. Godfrey 1986
3. Terres 1980
4. Dunning 1984
5. Klimkiewicz and Futcher 1989

Chronology $n = 60$

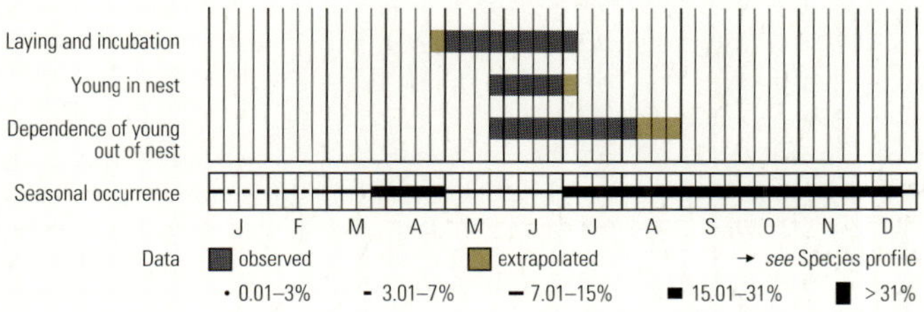

World breeding range

Notes on territory (continued)

interactions with their neighbours.

Great Black-backed Gulls are outnumbered by Herring Gulls in all but two of the mixed colonies along the St. Lawrence. The two exceptions are Île aux Fraises, where the Great Black-back population is twice as large as the Herring Gull one (476 pairs as opposed to 202), and Îlet aux Alouettes, near Sainte-Catherine Bay, where the species are evenly matched (409 pairs of each) (BIOMQ). The reasons for this are unknown.

A pair has been nesting for several years on the stack on the left, on Niapiskau Island. This advantageous location provides protection from disturbances, peace and quiet, and a commanding view of the surroundings. Mingan Archipelago National Park Reserve, Middle North Shore. *August 1980, Daniel Bordage.*

On wooded islands like those in the St. Lawrence estuary while several other colonial birds prefer forest cover, this gull nests mainly in grassy areas along the shore that afford a good view of its immediate environment and the sea beyond. This strategic location enables the gulls to spot potential prey, such as broods of ducklings that must make their way through the grass to get to the water.

The diet of the Great Black-backed Gull consists mainly of fish, invertebrates, and other items associated with the sea or the intertidal zones, and food procured at dumps and fish plants, supplemented by wild berries (black crowberries, blueberries) and birds, in proportions that vary regionally and seasonally (Cramp and Simmons 1983).

The nest, usually on the ground, is made of plant stalks, dried seaweed, and feathers. Île Blanche, Îles de l'estuaire National Wildlife Area, near Rivière-du-Loup. *May 1976, Pierre Brousseau.*

A pair of Great Black-backed Gulls near their nest, high up on Blacklands Island. Baie des Loups Migratory Bird Sanctuary, Lower North Shore. *July 1981, Pierre Brousseau.*

In the St. Lawrence estuary, the Great Black-back is a major predator on ducklings, both of Common Eiders and other species. This big gull may, despite the vigorous defence mounted by the females, seize a duckling and swallow it whole, even in flight. Bicquette Island. *June 1972, M. Durand.*

In high-density seabird colonies, the Great Black-backed Gull frequently preys on the young of other species (Herring Gulls, ducks, murres) and sometimes even of its own (Butler and Trivelpiece 1981). On Corossol Island, in Sept Îles Bay, in a mixed-species colony of 5648 pairs of Herring Gulls and 402 pairs of Great Black-backed Gulls (Brousseau and Chapdelaine 1990), predation was most intense on chicks a few days to 3 weeks old, regardless of species (CWS, unpubl. data). It is not known whether the Great Black-back's apparent predilection for chicks of this age was simple opportunism or if it resulted from a change in the abundance and availability of fish or some other food item.

On the Lower North Shore, Great Black-backs occasionally attack Atlantic Puffins in flight, on the water, or as they emerge from their burrows. They are also a major predator of Common Eider eggs and ducklings. Great Black-back predation has a demonstrated influence on the breeding success of the Common Eider during laying and incubation in the colonies (Bourget 1973a; Milne and Reed 1974; Gerell 1985), during brood rearing of the ducklings (Milne and Reed 1974; Munro and Bédard 1977b; Mendenhall and Milne 1985; van Dijk 1986), and in the aftermath of a human disturbance (Ahlund and Gotmark 1989).

In high-density colonies, Great Black-backs may attack their immediate neighbours, like this young Herring Gull. Corossol Island, Sept Îles Bay. *July 1985, Pierre Brousseau.*

Distribution

In the upper St. Lawrence, the species breeds locally on Lake Saint-Pierre and on the Contrecoeur, Varennes, and Boucherville islands. It does not usually nest farther inland than this, and only 2 confirmed breeding records were obtained in inland locations: one at Bromptonville on the Saint-François River in the Eastern Townships, and the other on a rocky islet near Saint-Gédéon on Lake Saint-Jean. In these areas, the Great Black-backed Gull nests alone or in the company of Ring-billed Gulls.

At Québec City and on several islands in the estuary, the species nests in mixed colonies with Herring Gulls and sometimes Ring-billed Gulls. On Chaleur Bay, solitary pairs breed on the cliffs all along the coast up to the Forillon Peninsula, and colonies can be seen on the islands and in some lagoons (Carleton, Paspébiac, Malbaie) (Brousseau and Chapdelaine 1989). About 15 colonies have been found on the Magdalen Islands (Fradette 1992). From Sept-Îles eastward, breeding reports are rather patchy, which is probably due more to the region's remoteness and the lack of observers rather than to any real absence of the species. Curiously, in the gulf, the species is found inland, a phenomenon noted as early as the turn of the century on Anticosti Island (Brewster *in* Ouellet 1969), but probably more recent in the Havre-Saint-Pierre area (S. Paradis, pers. comm.). Inland nests are usually located on lakes and ponds, on an islet covered with vegetation, to provide protection from predators.

In northern Québec, Godfrey (1986) reported that the Great Black-backed Gull nested on the coast of Ungava Bay (Rivière aux Feuilles). In 1990, 2 nests were discovered near Kakassituq Point (Benoit et al. 1991), and in 1991, 3 nests were found on islands off the northeast coast of James Bay (Benoit et al. 1992), confirming breeding in this region.

History and trends

Québec populations have increased noticeably in recent years. In the Magdalen Islands, the gulls made significant gains between 1976 and 1990, and the population was estimated at roughly 2400 birds in 1990 (Chapdelaine and Brousseau 1992a). The most recent surveys of the North Shore sanctuaries indicate that the population increase that began in the 1950s there is still continuing (Brousseau and Chapdelaine 1990; Chapdelaine 1995). The most spectacular increase has been on the Gaspé Peninsula, where, between 1979 and 1989, the number of breeding pairs rose from 187 to 1337, which represents an annual growth rate of 24% (Chapdelaine and Brousseau 1992b). In the estuary, a strong upward trend has also been observed, though not to the extent seen on the Gaspé. The trend in the estuary seems more pronounced, however, because of the corresponding decrease in the Herring Gull population there; the ratio of Great Black-backed Gulls to Herring Gulls strengthened from 1:13 in 1971 to 1:7 in 1990 (BIOMQ).

Population of Great Black-backed Gulls in the Middle and Lower North Shore sanctuaries, 1925–93 (BIOMQ)

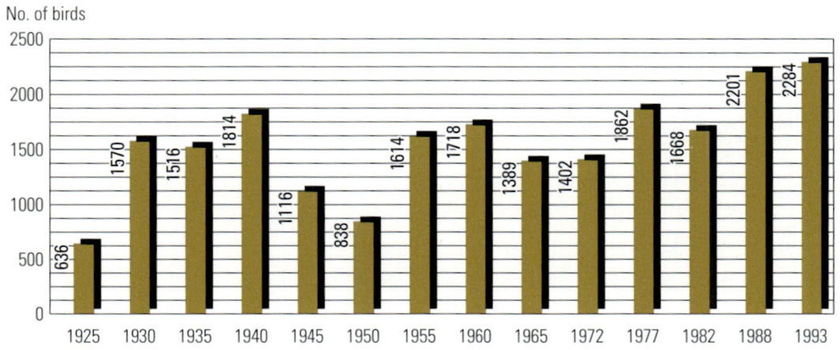

Based on the most recent data, the present population in the Atlas area is about 10 000 pairs, which indicates quite a clear increase. It is difficult, however, to predict whether this trend will continue; if it does, some seabird communities along the St. Lawrence could be affected.

PIERRE BROUSSEAU

Black-legged Kittiwake

Mouette tridactyle

Rissa tridactyla

Louis Ruelland

THE BLACK-LEGGED KITTIWAKE, A SKILFUL FLIER, IS BASICALLY PELAGIC IN winter and coastal during the breeding season. In fall, it rarely ventures farther inland than the upper estuary of the St. Lawrence, except in fierce storms.

Of the many kittiwake nests on Corossol Island, only a few are as accessible as these. The extent of the whitewash on the rocks suggests that these birds have been occupying the colony for several weeks. Sept Îles Bay. *July 1985, Pierre Brousseau.*

Habitat and behaviour

The Black-legged Kittiwake nests on the narrow ledges of cliffs or rocky escarpments. The nest, made of seaweed and plant debris cemented with mud to the ledge, is built by both the male and female, which take turns fetching nest materials and forming the cup (Danchin 1987a). Sometimes material is stolen from a neighbour's unguarded nest.

Nest spacing is determined to a large extent by cliff topography. At favourable sites, it is not uncommon to find several dozen nests only 15–30 cm apart. Nesting in high densities on narrow ledges has led to a number of unique behavioural adaptations in the species, including elaborate nests, non-cryptic young, and relatively undeveloped anti-predator behaviour.

Kittiwakes return to the gulf in March, when they frequent the ice-free areas near the shore, but usually do not venture near the cliffs. Regular occupation of the ledges begins a few weeks later, along with the first signs of territoriality, nest construction, and courtship behaviour. Immature birds usually return to the natal colony at 3 years of age. During the year preceding their first nesting attempt, some birds may form pairs and occupy and defend territories in the colony, but do not produce any eggs. Sexual maturity is attained at the age of 4 or 5 years. Pairs are monogamous (Coulson 1972).

At Forillon National Park, the onset of laying is in early May (Quenneville 1990). Clutch size is 1–3 eggs, which are incubated for 27 days on average. Hatching begins during the first week of June. Both parents attend and feed the young, which become independent at around 40 days of age. The young are fed by regurgitation at regular intervals throughout the day, but feeding is stepped up around dawn; their diet consists mainly of sand lance, capelin, and invertebrates. A study conducted on Corossol Island in Sept Îles Bay in 1985 found average clutch size to be 1.9 eggs, while productivity was only 0.62 fledgling/nest (CWS, unpubl. data), which is below the 0.88 threshold value required to maintain a stable population (Chapdelaine and Brousseau 1989). The poor results in this colony are due largely to nestling predation by Herring Gulls, which accounts for roughly 60% of losses. This situation is likely peculiar to this colony, because in most other large colonies, the cliffs are high and have narrow ledges that the gulls cannot reach.

After the young have fledged, at about 40 days of age, the adults disperse throughout the gulf and the estuary (CWS, unpubl. data), with both parents going their separate ways for 3 to 8 months. The pair bond is renewed when the birds return to the colony area. Divorce is not uncommon, however, especially among inexperienced birds. Colour marking of young and adult kittiwakes on Corossol Island showed that part of this population moves to the upper estuary around Les Escoumins and Tadoussac, where they are found in August and September. It is not known, however, whether birds from Anticosti Island and other colonies in the gulf also use this area. One young kittiwake was sighted at Red Bay, Labrador, some 800 km east of its natal colony, barely two weeks after fledging (P. Brousseau, CWS, unpubl. data).

Notes on territory

Territoriality in the Black-legged Kittiwake is limited to defence of the nest site. The level of aggression shown depends on the situation. Some antagonistic behaviour observed in the species, such as "mock fighting" and "jabbing," is very similar to that seen in other larids (Cramp and Simmons 1983). Disputes peak during nest building and laying, and taper off during incubation.

Distribution

According to the Atlas data, the Black-legged Kittiwake breeds mainly in the gulf, where most of the known colonies are located. A comparison of the Atlas data with the results obtained a dozen years earlier by Brown et al. (1975) reveals that the species has not only expanded its breeding range, but increased the number of known colonies substantially, from 11 to 46.

Young kittiwakes in their nests, on a cliff on Corossol Island. On the North Shore, hatching begins in early June. Sept Îles Bay. *July 1991, Daniel Bordage.*

Black-legged Kittiwake

Breeding distribution

1984–1989

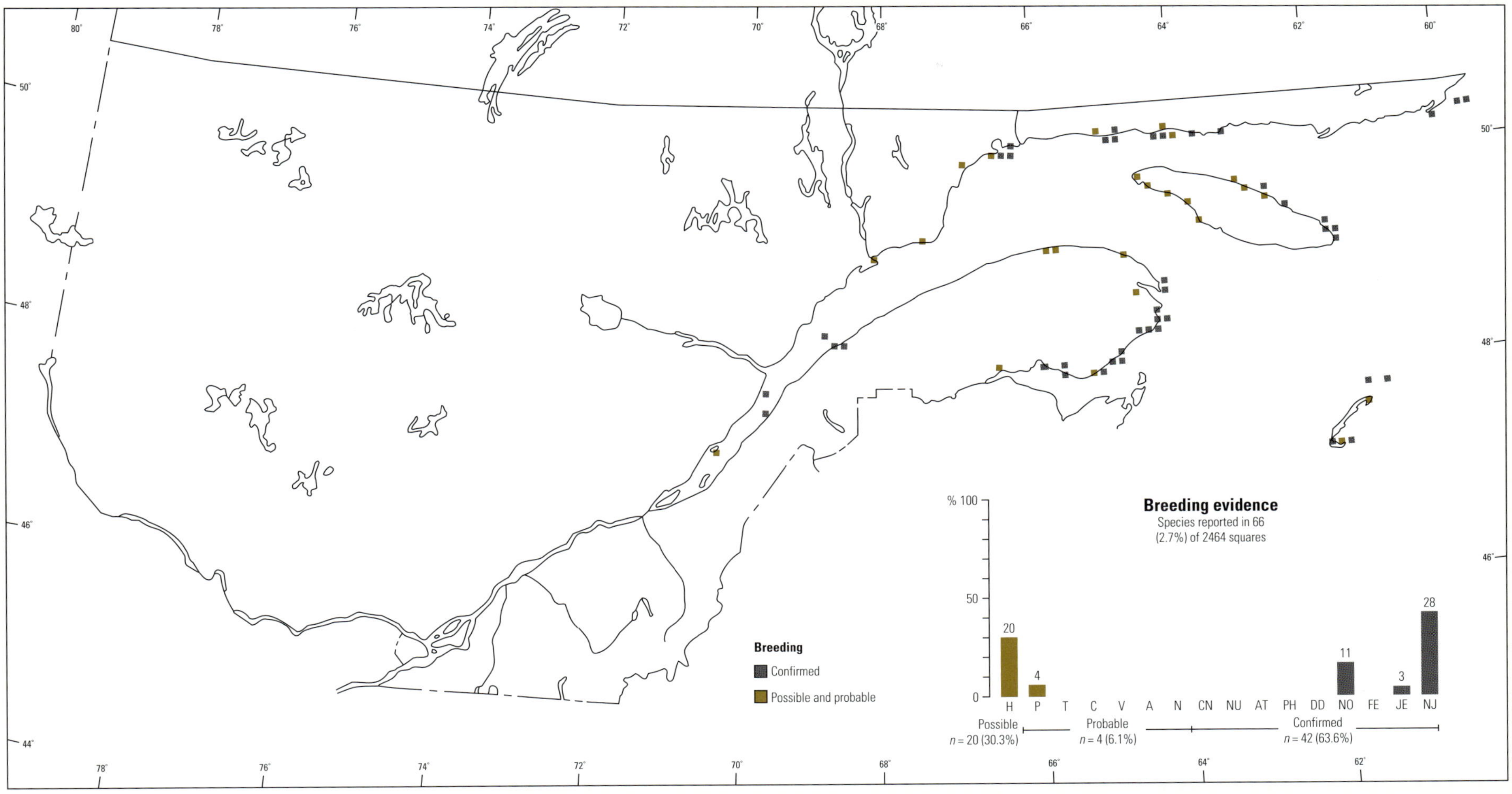

Species profile

Clutch size	usually 2 eggs (1–3)[1,5]
Laying interval	1 egg per 2 days[1]
Start of incubation	with last egg[1]
Incubation period	usually 27 days (25–32)[1]
Maturity at hatching	semiprecocial
Care of young	by female and male[1]
Nestling period	usually ≈ 42–43 days (33–58)[1,5,6]
Age at first flight	30–36 days[6]
Dependence of young	until first flight or shortly afterward[1]
Broods per year	1[1,5]
Breeding age	usually 4–5 years (3–8)[1], M 4–5 years / F 3–4 years[5]
Mating system	monogamy[1]
Duration of pair bond	usually for life[1]
Total length	40–46 cm[2]
Wing span	91.4 cm[4]
Weight	M 421 g (305–512) / F 393 g (305–525)[3]
Longevity record	21 years[1]

References:
1. Cramp and Simmons 1983
2. Godfrey 1986
3. Dunning 1984
4. Terres 1980
5. Baird 1994
6. Maunder and Threlfall 1972

Chronology *n* = 19

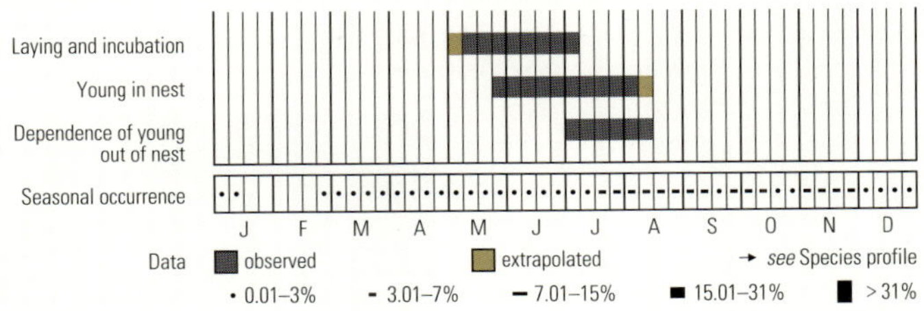

World breeding range

Kittiwakes have recently begun nesting on the cliffs of Île du Fantôme. The sedimentary rock here is constantly being eroded by the elements and is quite fragile. Mingan Archipelago National Park Reserve, Middle North Shore. *June 1991, Chantal Lepire.*

Notes on territory (continued)

Birds returning to the colony sometimes exhibit what could be considered preventive behaviour. Before landing, an individual will utter several different calls (Danchin 1987b), to establish acoustic communication with its mate, offspring, and neighbours, and to warn them of its arrival. Often, this will prompt the other parent to move or even leave the nest altogether, thus reducing the potential for conflict.

Population of Black-legged Kittiwakes in the Middle and Lower North Shore sanctuaries, 1940–1993 (BIOMQ)

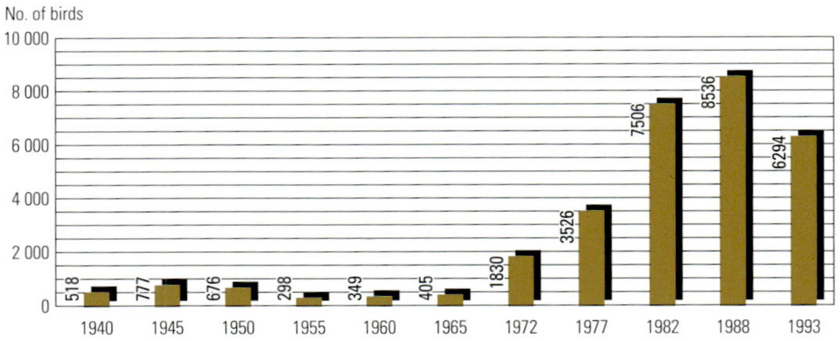

The species has also recently colonized a number of islands in the upper and lower estuary. In the early 1980s, the Sept Îles Bay colonies marked the western limit of the kittiwake's breeding range in the gulf. In 1984, however, the species became established on Bicquette Island and then went on to colonize the Brandypot Islands and Les Pèlerins (near Rivière-du-Loup), and finally reaching Laval Island (Forestville) in 1990 (Chapdelaine and Brousseau 1989; BIOMQ). The number of colonies in Chaleur Bay has also increased substantially, especially between Saint-Godefroi and Newport. A similar trend was noted in the Mingans, where the species quickly colonized the many small cliffs on islands such as Île du Fantôme and Île Nue de Mingan (Grenier 1989).

History and trends

Kittiwake populations have changed considerably over the past few decades, both in Europe and eastern Canada (Brown and Nettleship 1984b; Barrett 1985; Coulson 1985; Chapdelaine and Brousseau 1989). Chapdelaine and Brousseau (1989) compared numbers in 1974 and 1985, based on data from Brown et al. (1975) and the BIOMQ database. In 1974, the Québec population consisted of 43 000 pairs, divided among 11 colonies. By 1985, the population had risen to 71 390 pairs divided among 26 colonies, with 2 colonies —Bonaventure Island with 23 544 pairs and Falaise aux Goélands on Anticosti Island with 22 840 pairs—accounting for almost 65% of the total

The nest is fashioned out of seaweed and plant debris and held together with mud. Corossol Island, Sept Îles Bay. *July 1985, Pierre Brousseau and Gilles Chapdelaine.*

The Laval Island colony, across from Forestville. Upper North Shore. *May 1992, Jean Rodrigue.*

population. This upward trend was noted everywhere in Québec where the species breeds, but the annual rate of increase was found to be higher in small colonies (13–35%) than in the larger ones (1–4%). Similar results were reported in colonies in England (Coulson 1983). The lower rate of increase in large colonies in Québec could be due to the saturation of the available nest sites.

This upward trend continued between 1985 and 1991. As of 1991, there were 83 957 pairs in 46 colonies, which represents an increase of 20% in 6 years (see table).

Estimated Black-legged Kittiwake population in Québec, 1985–91[1]

Region	No. of pairs	No. of colonies
Upper and lower estuary	906	7
Gaspé Peninsula	43 423	15
Magdalen Islands	8 309	6
North Shore	31 319	18
Total	**83 957**	**46**

1 These estimates were obtained from surveys carried out from 1985 to 1991, with the exception of one colony on Anticosti Island that was surveyed in 1983.

Several authors (Brown and Nettleship 1984a; Bradstreet and Brown 1985; Chapdelaine et al. 1985; Chapdelaine and Brousseau 1992b) have linked the increases to an abundance of small prey fish, such as capelin and sand lance; these stocks have apparently benefited from the overfishing of the commercial species that are their predators. This appears to be a plausible explanation, but needs to be confirmed by long-term studies on breeding biology and diet to determine the relations between the various factors affecting the dynamics of these populations.

Pierre Brousseau

Falaise aux Goélands on Anticosti Island, where over 20 000 pairs of Black-legged Kittiwakes nest. The wider ledges are occupied by Northern Gannets and the narrower ones by kittiwakes. *July 1985, Pierre Brousseau.*

Caspian Tern

Sterne caspienne

Sterna caspia

Christiane Girard

The Caspian Tern is somewhat larger than the Ring-billed Gull. Its large, thick red bill, heavy wingbeats, rather direct flight, and raucous voice make it relatively easy to identify. Therefore, it is not likely to be missed by birders eager to add a rare Québec species to their lists. Although in Québec it breeds only on the Lower North Shore, over the last 10 years there have been quite a few sightings during migration and in summer elsewhere in southern Québec.

Habitat and behaviour

The Caspian Tern is primarily colonial, though it sometimes nests in isolated pairs in colonies of Common and Arctic terns, as well as in Ring-billed Gull colonies (Shugart et al. 1978). The rudimentary nest consist simply of pebbles and plant stalks, often on a sandy substrate; typical clutch size is 1–3, or sometimes 4, eggs (O'Donoghue and Gowanlack 1919; Evans et al. 1970; Shugart et al. 1978). Caspian Terns first breed at 3 years of age or later (Ludwig 1965). Incubation lasts for about 21 days, and the young fledge at 28–35 days of age, depending on the conditions present (Hayward 1935). This tern feeds on fish that it captures by plunge-diving, but it also feeds on the surface, as gulls do. There is no information available on the type of prey captured by the Caspian Terns nesting in the Île à la Brume Migratory Bird Sanctuary on the Lower North Shore. Alewife and smelts are apparently the main foods of young Caspian Terns on Lake Michigan (Shugart et al. 1978). These species are also present in freshwater and saltwater sections of the St. Lawrence River (Scott and Crossman 1978). Ludwig (1965) found yellow perch and rock bass in the stomach contents of birds captured on the Great Lakes.

Caspian Terns on Île à la Brume in 1925, when the colony consisted of 60 birds. Lower North Shore.
1925, P. A. Taverner, reproduced by permission of the Canadian Museum of Nature, Ottawa, No. 63766.

Since Caspian Terns often nest on the same sites as Herring and Ring-billed gulls, they regularly interact with these species. According to Shugart et al. (1978), in territorial conflicts, Herring Gulls are usually dominant over Caspian Terns, but the terns have the upper hand in conflicts with Ring-billed Gulls.

In the Eastern Headland colony in Lake Ontario, near Toronto, hatching success of 77% (the mean over a 2-year study period), fledging success of 92%, and a net productivity of 71% were reported (Fetterolf and Blokpoel 1983). Egg and chick loss occur for a variety of reasons, which depend on the site. For example, Shugart et al. (1978) cited flooding, human activity around colonies, and predation by Herring and Ring-billed gulls as the main causes of mortality in eggs and young on Lake Michigan. Fetterolf and Blokpoel (1983), however, identified raccoons and striped skunks as the main predators at the sites they studied. Analysis of banding return data from birds banded on the Great Lakes show that after the breeding season, the birds disperse along the Atlantic seaboard (especially the southern states) and the north coast of the Gulf of Mexico, south to the West Indies and Colombia (Ludwig 1965).

Distribution

The Caspian Tern is found on all continents except South America and Antarctica (Cramp 1985). In Canada, it nests locally in the southern Mackenzie District (Great Slave Lake region) of the Northwest Territories, in northeastern Alberta, and central Saskatchewan (Godfrey 1986). There are also well-established colonies on islands in Georgian Bay and Lake Ontario in southern Ontario (Blokpoel 1977).

Currently, the only known nesting site in Québec is the Île à la Brume Migratory Bird Sanctuary on the Lower North Shore. In 1988, in a census of the Middle and Lower North Shore sanctuaries, 15 adults were found on Île à la Brume (Chapdelaine and Brousseau 1991), but in 1993 the species was no longer nesting there (Chapdelaine 1995). Even though the terns did not return to nest at Île à la Brume in 1994, roughly two dozen birds were sighted on a nearby island, but nesting was not confirmed (A. Roswell, CWS, pers. comm.). In 1995, Caspian Terns were back nesting on Île à la Brume, with a census of the island yielding 14 active nests (pers. obs.).

Several recent sightings raise questions on the status of the species in the upper St. Lawrence west of Trois-Rivières. At the end of July and in August 1989, up to 4 birds were seen frequenting the beach at Oka (Yank et al. 1990a; A. Lagrenade, pers. comm.); on 8–10 September, on this same beach, a fledged juvenile bird begging for food from an adult was also reported (Yank et al. 1990a). In 1990, between 23 June and 14 July, up to 8 birds were sighted in LaSalle and Notre-Dame-de-Pierreville (Gosselin et al. 1991b), and a pair was seen mating at the second location on 11 June 1992 (Yank and Aubry 1992a). On Saint-Ours Island, near Contrecoeur, a pair frequented a Ring-billed Gull colony during the summer of 1991. Two adults accompanied by a fledgling were seen on 7 August of the same year (Aubry et al. 1992). It should be noted, however, that young birds may be fed by an adult for 6–9 months after they fledge (Cramp 1985).

The species has also been sighted in summer in the Lower St. Lawrence region, on Chaleur Bay, and on the Magdalen Islands.

History and trends

All the historical data on breeding population trends in Québec come from the Île à la Brume Migratory Bird Sanctuary, the only known nesting site of the Caspian Tern in the province. In 1884, Frazar (Todd 1963) found about 200 nests on Île à la Brume. The first census of the Middle and Lower North Shore sanctuaries in 1925 reported a decline in the population, with only 60 adults (Lewis 1925b). Subsequently, between 1925 and 1988, there were many fluctuations in the population but the overall trend was downward, as shown in the accompanying figure.

Gilles Chapdelaine

Caspian Tern populations in the Île à la Brume Migratory Bird Sanctuary, Lower North Shore, 1925-95 (BIOMQ)

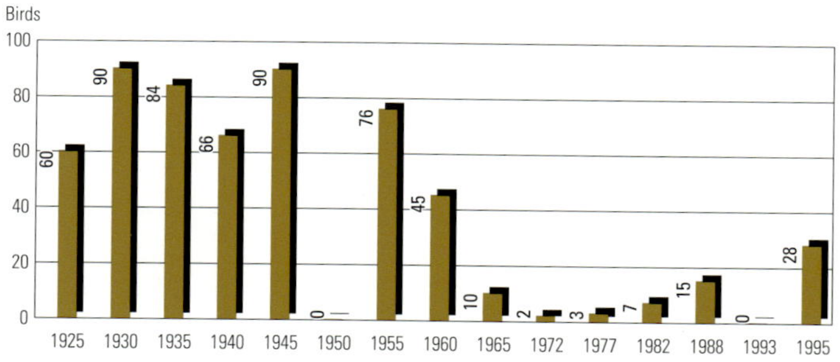

Notes on territory

Caspian Terns generally breed in dense colonies, although sometimes isolated pairs or small groups of 2 or 3 pairs can be found (Terres 1980; Cramp 1985). Colonies may become quite large. On the Great Lakes, Ludwig (1965) reported colonies ranging from 35 to 500 pairs and minimum inter-nest distances of 0.53 m.

Pairs defend a roughly circular area around their nests. In the Great Lakes, Ludwig (1965) found territories of 1–1.4 m². The territory is used for courtship displays, mating, nesting, and brood rearing, and is defended for up to 30 days after the young hatch (Cramp 1985).

First-time breeders often choose colonies other than the natal colony but, once established, show strong fidelity to the nest site (Cramp 1985).

Caspian Tern

Breeding distribution

1984–1989

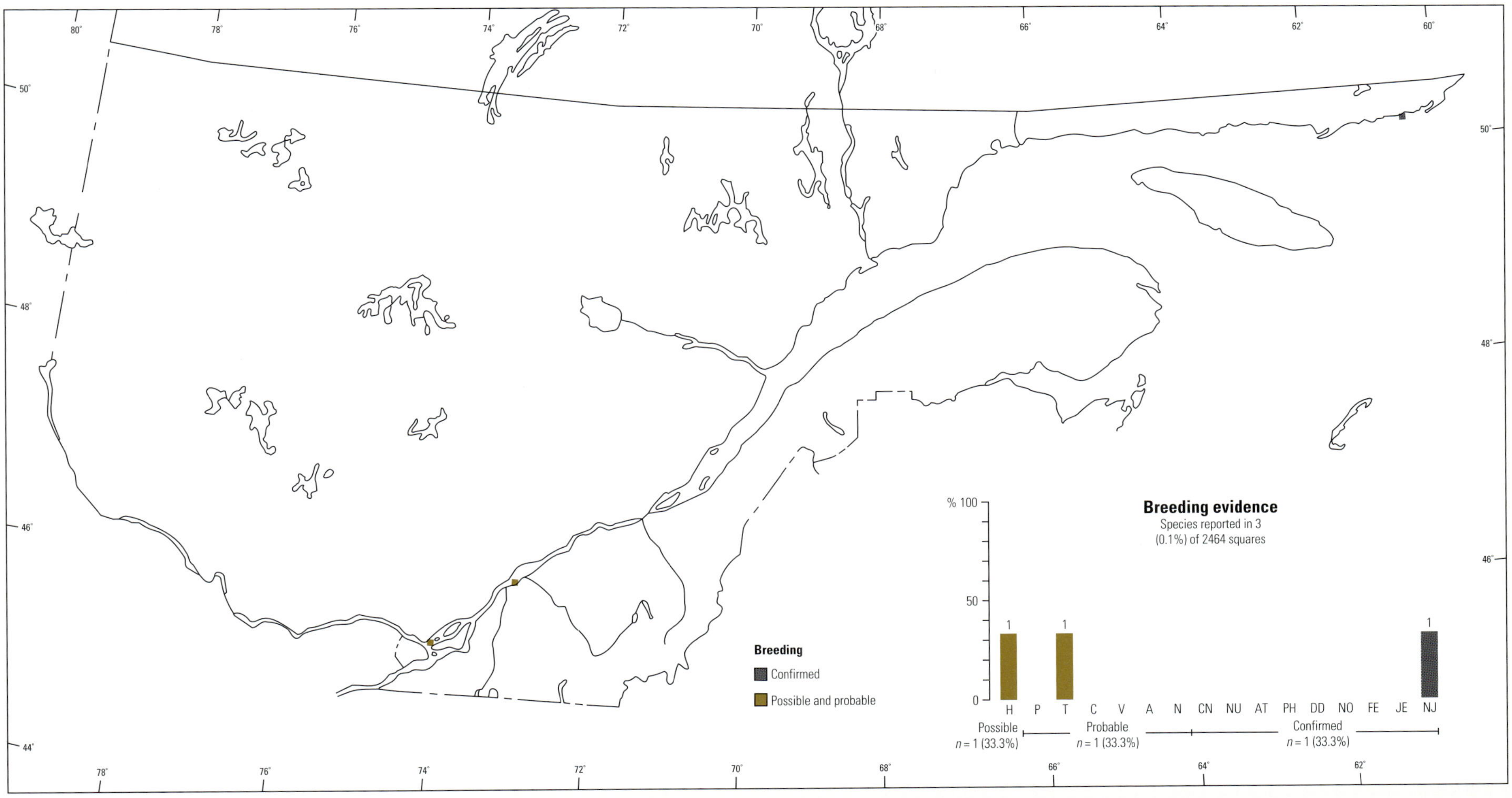

Species profile

Clutch size	usually 1–3 eggs (1–4)[2,3,4,9]
Laying interval	1 egg per day[12]
Start of incubation	with first egg[12]
Incubation period	usually 20–22 days (20–27)[1,2,3,5,6]
Maturity at hatching	semiprecocial
Care of young	by female and male[6,12]
Nestling period	usually 1–7 days[12]
Age at first flight	usually 30–35 days (28–35)[5,8,12]
Dependence of young	usually until 6–8 months (6–9) after first flight[12]
Broods per year	1[1,12]
Breeding age	3 years or more[2,12]
Mating system	monogamy[6,12]
Duration of pair bond	more than 1 season[9,12]
Total length	48.0–58.4 cm[6]
Wing span	127.0–139.7 cm[10]
Weight	661 g (604–709)[11]
Longevity record	26 years 2 months[7]

References:
1. O'Donoghue and Gowanlack 1919
2. Evans et al. 1970
3. Shugart et al. 1978
4. Ludwig 1965
5. Hayward 1935
6. Godfrey 1986
7. Clapp et al. 1982
8. Morris and Quinn 1979
9. Ehrlich et al. 1988
10. Terres 1980
11. Dunning 1984
12. Cramp 1985

Chronology *n* = 2

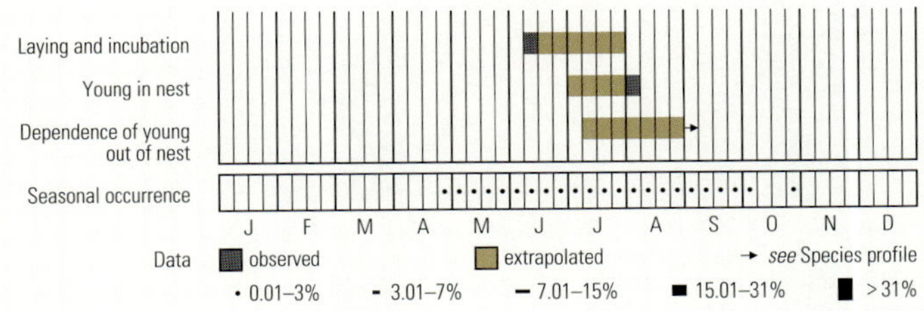

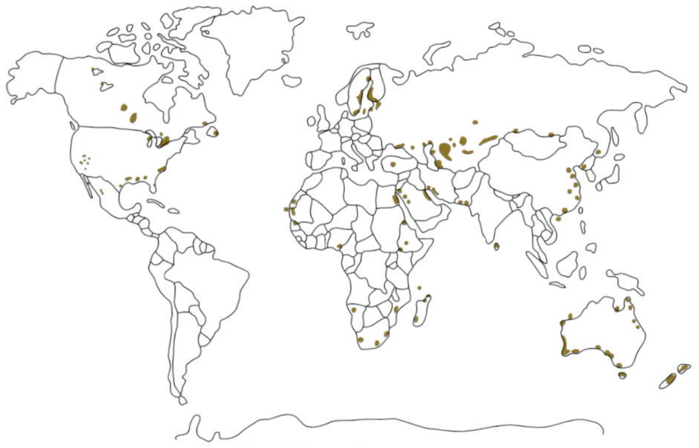

World breeding range

Built in a depression in the bare ground, this nest is made of plant stalks, dead leaves, lichens, and small pebbles. The Île à la Brume Migratory Bird Sanctuary is the only known nesting site of Caspian Terns in Québec. Lower North Shore. *June 1976, Canadian Wildlife Service.*

Roseate Tern

Sterne de Dougall

Sterna dougallii

Christiane Girard

The Roseate Tern, which has been on Canada's list of threatened birds since 1986 (Kirkham and Nettleship 1986), is not very well known in Québec because it is rare and only occurs in the Magdalen Islands. Every year, in May, a few birds arrive here from the West Indies or the northern coast of South America to breed in mixed colonies with Common Terns and Arctic Terns, which vastly outnumber the Roseates (Nisbet 1984; Fradette 1992). Although Roseate Terns are difficult to spot in these colonies, their distinctive call is invaluable in clinching identification.

Habitat and behaviour

The Roseate Tern is a saltwater specialist. In the northeastern United States, it nests mainly on small coastal islands and occasionally on pebble beaches (Nisbet 1981). In the Magdalen Islands, breeding pairs build their nests on islets. As a rule, the Roseate Tern goes to greater lengths to hide its nest under tufts of vegetation than does the Common Tern (Gochfeld and Burger 1987; Nisbet 1989). The rudimentary nest is placed on the ground and generally contains 1 or 2 eggs. As in other terns, both parents share incubation and brood-rearing duties (Nisbet 1981).

In Québec, Roseate Terns breed only in the Magdalen Islands. Here, in the foreground is a nesting area on Île aux Cochons, near Havre-aux-Maisons, used by the species. *June 1993, Pierre Laporte.*

Distribution

The Magdalen Islands represents the current northern limit of the Roseate Tern's breeding range in North America. The species has never bred elsewhere in Québec, although it has used different sites in the islands over the years. On the Magdalens, Roseates were first reported on the islets off Pointe-aux-Loups, then at the tip of Dune du Sud, and finally, on Baie du Havre aux Basques (Hardy 1979b; Robert 1989b). At present, the species frequents three terneries, one on Île aux Cochons in the Havre-aux-Maisons Lagoon, one on the islets off Pointe-aux-Loups, and one on an artificial islet off Grande-Entrée, although these colonies are not necessarily used every year (Fradette 1992; Shaffer 1995).

The first breeding evidence in Québec was a nest found in Baie du Havre aux Basques in 1983 (Fradette 1992). In 1987, a second nest was found on the artificial islet off Grande-Entrée, hidden under a tuft of beachgrass. Breeding was also confirmed in 1988 and 1989, when young Roseates were found on Île aux Cochons (Fradette 1992).

History and trends

In Québec, the Roseate Tern has only been recorded on the Magdalen Islands, the first time in 1972, when 13 birds were sighted (McNeil et al. 1973). In view of ornithologists' infrequent visits to the islands and the difficulty of spotting Roseate Terns among the hundreds of Arctic and Common terns, it is conceivable that this species frequented the area earlier and even bred there, but went unnoticed. Since the early 1980s, Roseates have been regularly sighted on the Magdalens during the breeding season (19 May–11 September). The numbers reported, however, have been very low. In 1988, only 6 adults and 2 chicks were counted; all the other records involved 2–4 birds (Fradette 1992). As a result, Kirkham and Nettleship (1986) set the Magdalen breeding population at fewer than 5 pairs. This estimate seems fairly accurate, given that a 1990 census of all likely breeding sites turned up only 2 pairs (Fradette 1992).

In the past 100 years or so, the Roseate Tern population of northeastern North America has suffered three major declines: at the turn of the century, between 1935 and 1952, and again during the 1970s. The plume trade and egging were the main causes of the decline in the last century (Kirkham and Nettleship 1987; Andrews et al 1989). Protective legislation subsequently allowed the population to recover to some extent. The second decline was probably caused by increased populations of Herring Gulls and Great Black-backed Gulls (Drury 1973, 1974), which compete with terns for nest sites. The most recent population decrease has been linked to the commercial trapping of birds on their wintering grounds and the continued growth of gull populations. Predation on the breeding grounds has also contributed to the species' decline (Buckley and Buckley 1981; Nisbet 1981, 1989; Kirkham and Nettleship 1987). Since 1977, populations have remained fairly stable: in the late 1980s, there were an estimated 3000 breeding pairs in the United States and about 100 in Canada. The most recent surveys, in 1991, point to a slight increase in numbers, with the total North American population estimated at over 3600 pairs (Nisbet 1992). In view of the small population, concentrated at just a few breeding sites, the Roseate Tern is currently classified as a threatened species in both the United States and Canada (Kirkham and Nettleship 1986, 1987; Andrews et al. 1989), including Québec (Robert 1989b).

François Shaffer and Michel Robert

Notes on territory

Once laying begins, pairs defend a territory of roughly 1 m around the nest (Nisbet 1981). According to Nisbet and Drury (1972), nest density in sample plots was 1–4 nests/m², although densities of 0.7 and 0.13 nest/m² have also been reported (Nisbet 1981). Given the proximity of breeding pairs, quarrelling is common. Nests are not evenly distributed within a colony, however, because the terns choose sites on the basis of the ground cover present (Burger and Gochfeld 1988). Roseate Terns do not defend a feeding area, and may travel as far as 19 km in search of food (Nisbet 1981).

Roseate Tern

Breeding distribution
1984–1989

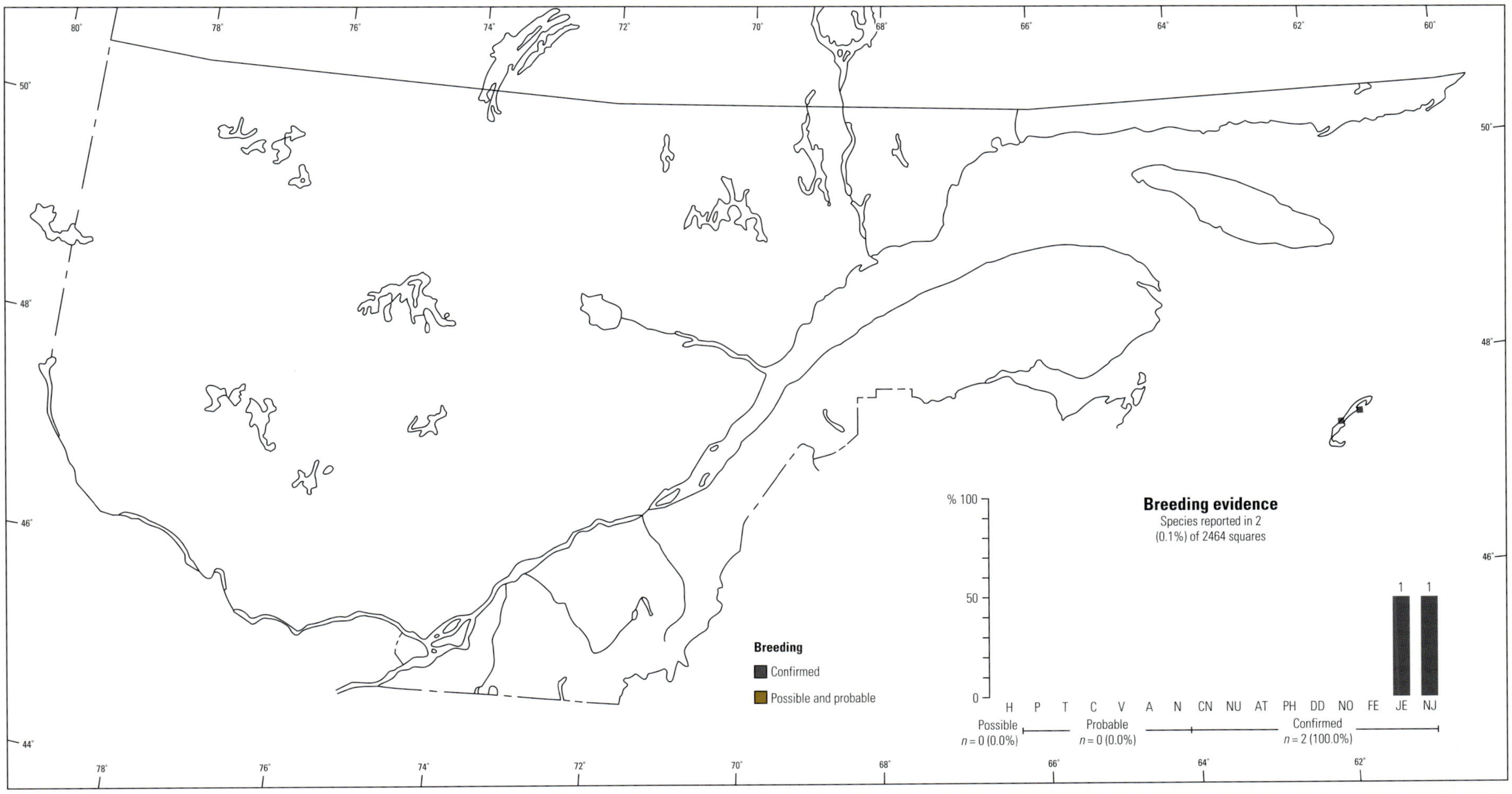

Species profile

Clutch size	usually 1–2 eggs (1–4)[3,4]
Laying interval	usually 1 egg per 3 days (2–4)[2,4]
Start of incubation	with first egg[2]
Incubation period	usually 23 days (21–31)[3,4]
Maturity at hatching	semiprecocial
Care of young	by female and male[2,4]
Nestling period	less than 72 hours[4]
Age at first flight	usually 23–24 days (21–30)[4]
Dependence of young	until at least 8 weeks old, possibly until winter[2,4]
Broods per year	1[2,4]
Breeding age	usually 3 years (2–4)[4]
Mating system	monogamy[3]
Duration of pair bond	1 season[4]
Total length	35.5–43.0 cm[6]
Wing span	72–80 cm[2]
Weight	107.6 g (92–125)[1]
Longevity record	16 years[5]

References:
1. LeCroy and Collins 1972
2. Cramp 1985
3. Kirkham and Nettleship 1986
4. Nisbet 1981
5. Nisbet 1989
6. Godfrey 1986

World breeding range

Chronology *n* = 8

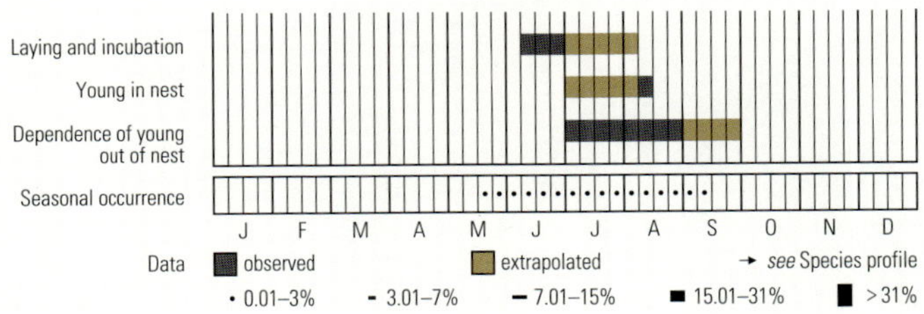

The nest of this threatened tern is made of dried grass and feathers and is hidden in the vegetation. The eggs are speckled, in contrast with those of Common and Arctic terns. This clutch of 3 eggs is unusually large. Baie du Havre aux Basques, Magdalen Islands. *June 1983, Yves Aubry.*

Common Tern

Sterne pierregarin

Sterna hirundo

Christiane Girard

With its graceful, buoyant flight, the Common Tern has been dubbed the sea swallow. Common Terns nest in colonies sometimes containing hundreds or even thousands of birds, and are found in both salt water and fresh water. In Québec, it is widespread along the upper St. Lawrence and in the gulf but tends to be more local inland. This migrant breeder returns to the southernmost part of Québec between late April and early May and to the Gaspé and the North Shore as late as mid-May. In the fall, it migrates to its wintering grounds, which extend from South Carolina and southern California to the Strait of Magellan in South America (Godfrey 1986; Root 1988). The distance travelled in migration may be considerable: a Common Tern found dead near a colony on the Magdalen Islands on 18 June 1993 had been banded on 22 April 1991 on the coast of Brazil, near the Uruguay border (F. Shaffer, CWS, unpubl. data).

Common Tern on the nest, located in low herbaceous vegetation, including American dune grass and starry false Solomon's-seal. Carleton, Charleur Bay, Gaspé. *July 1991. Eric Le Bel.*

Habitat and behaviour

The first thing one notices near a tern colony is all the noise (Chapman 1966). The adults' usual call is a descending *kierr* given as they approach their mate or young with food (usually a fish 5–10 cm long) or even as soon as they catch their prey. Alarm calls include *keeearr* and *kek-kek* (Stokes and Stokes 1989a). Common Terns are very aggressive and flock together to mob predators or intruders that threaten the colony.

Colonies are usually located near a large body of water, either a large lake, reservoir, river, or the sea. The birds generally nest on low-lying islands with sparse, low vegetation, or on sand or gravel spits, but sometimes use dredge-spoil islands, navigational structures, and jetties. Coastal colonies are larger than inland ones, which generally contain only a few pairs.

The nest of this monogamous species is a simple scrape made by the pair in sand, gravel, or shingle, often lined with plant stalks, pebbles, or other beach wrack. As the laying period approaches, the female almost never leaves the territory, relying instead on her mate to bring her food (Nisbet 1973). Typical clutch size is 2–3 (or occasionally 4) eggs, laid 1 or 2 days apart (Harrison 1978). If the first clutch is destroyed, a replacement clutch may be laid (Bent 1921), but its success will depend on the amount of time left in the breeding season. Colonies in the southernmost part of the province are obviously at an advantage because of the longer summer (Razurel 1994). Incubation begins after the laying of the first egg and is done mostly by the female (Cramp 1985). On the Mingan Islands, incubation lasts 22–23 days on average (Chapdelaine et al. 1985), but incubation periods of as long as 31 days have been reported elsewhere (Stokes and Stokes 1989a).

The young leave the nest at the age of 3–4 days but remain within their parents' territory. As they grow older, they may wander more outside the territory but return regularly to be fed by their parents until they fledge, at about 25 days of age or sometimes later (Chapdelaine et al. 1985; Cramp 1985). They are totally independent at about 2 or 3 months old, and breed for the first time at 3 or 4 years, though occasionally at 2 (Cramp 1985). Generally, immature birds only return to the colony in their second year.

Common Terns feed mainly on small fish such as sand lance, stickleback, capelin, smelt, and the fry of larger species (Chalifour 1982; Chapdelaine et al. 1985; Cramp 1985). Flocks can often be seen hovering over schools of fish. They dive headfirst from a height of about 3 m, snatching their prey from just below the surface. This diet is supplemented by small crustaceans and aquatic insects taken from the surface of the water and the occasional insect captured on the wing (Bent 1921; Cramp 1985).

Notes on territory

Territories in Common Tern colonies are no more than 1 or 2 m in diameter. The same territory is usually occupied in succeeding years by the same pair, which may remain together for several seasons (Cramp 1985).

Île de la Maison, a low-lying grassy island in Jacques Cartier Passage, is home to a Common Tern colony of several hundred pairs. Mingan Archipelago National Park Reserve, Middle North Shore. *July 1989, Michel Boulianne, Parks Canada.*

Common Tern

Breeding distribution
1984–1989

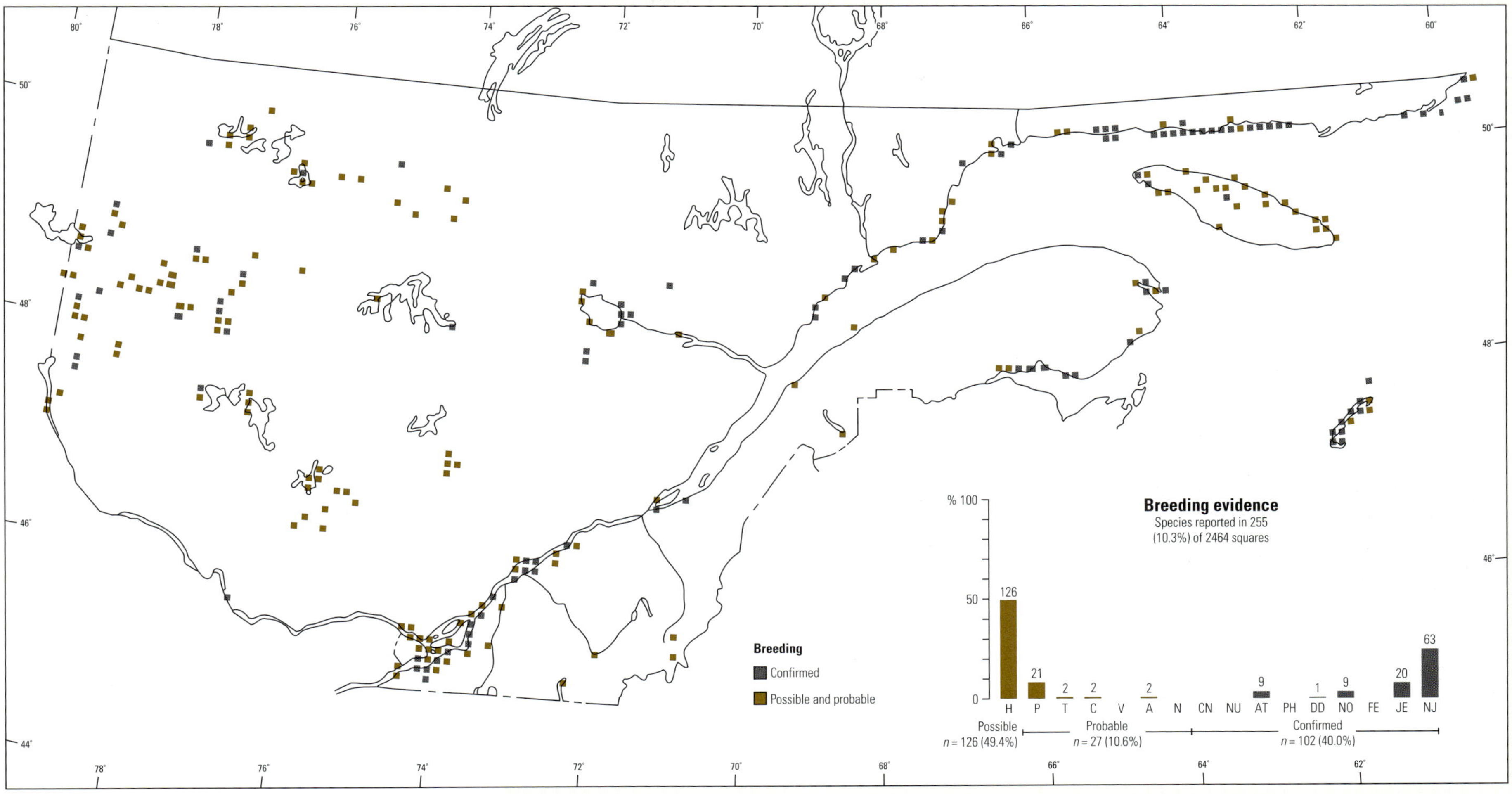

Species profile

Clutch size	usually 2–3 eggs (2–4)[1,6]
Laying interval	1 egg per 1–2 days[1,5]
Start of incubation	with first egg[1,5]
Incubation period	usually 21–27 days (21–31)[3,4]
Maturity at hatching	semiprecocial
Care of young	by female and male[1,3,5]
Nestling period	3–4 days[5]
Age at first flight	usually 25–26 days (22–33)[5]
Dependence of young	until 60–90 days old[5]
Broods per year	1[2,5]
Breeding age	usually 3–4 years, sometimes 2[5]
Mating system	monogamy[5]
Duration of pair bond	more than 1 season[5]
Total length	33.0–40.5 cm[7]
Wing span	78.7 cm[8]
Weight	120 g (103–145)[9]
Longevity record	24 years[10]

References:
1. Harrison 1978
2. Bent 1921
3. Ehrlich et al. 1988
4. Stokes and Stokes 1989a
5. Cramp 1985
6. Chapdelaine et al. 1985
7. Godfrey 1986
8. Terres 1980
9. Dunning 1984
10. Clapp et al. 1982

Chronology $n = 169$

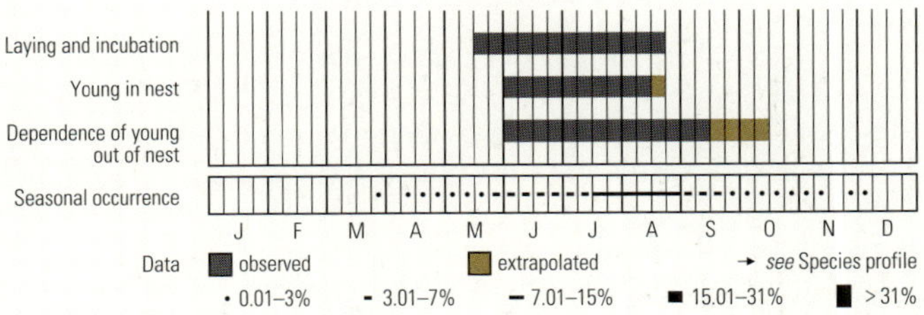

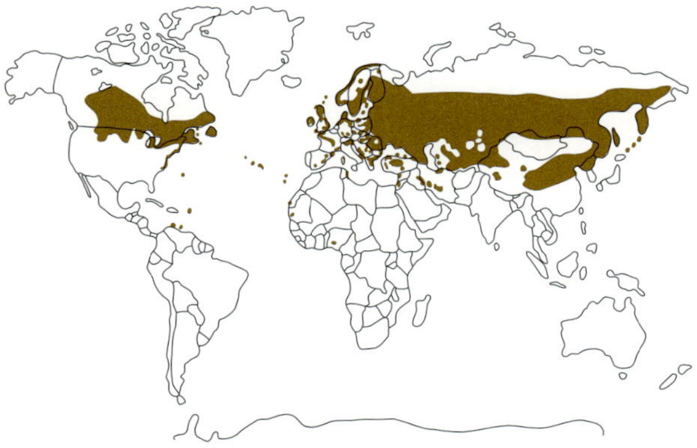

World breeding range

In the upper St. Lawrence, the species nests on artificial sites such as this rocky islet in Lake Saint-Pierre. *July 1991, Daniel Jauvin.*

Tern colonies are easy to find, as indicated by the 92 confirmed Atlas records based on discoveries of nests or of young, which generally hide in the vegetation when approached. Common Terns often breed in mixed colonies with other terns and sometimes on the edge of gull colonies, which makes breeding more difficult to confirm. Distinguishing Common Terns from Arctic Terns, for example, requires good viewing conditions and a certain amount of practice.

Distribution

The Common Tern has a worldwide distribution, frequenting almost all the world's coastal waters. It breeds in the Northern Hemisphere, particularly in the boreal zone, as well as in the north temperate zone. Less restricted to salt water than some of the other terns, it generally breeds along or near the coast, but is also found far inland, on large bodies of water. The Common Tern occurs in a wider spectrum of habitats than the other terns. It is the most common and widespread tern in Québec.

The largest Common Tern colonies are generally along the coast (Cramp 1985). In Québec, the greatest concentrations are found along the North Shore (especially in the Mingans), along Chaleur Bay in the Gaspé, on Anticosti Island, and in the Magdalen Islands (Chapdelaine and Bourget 1981; Razurel 1994). Smaller numbers are found along the upper St. Lawrence, particularly on Lakes Saint-François, Saint-Louis, and Saint-Pierre. The Atlas map shows the distribution of the main colonies and smaller colonies located in the Lac-Saint-Jean and Abitibi regions and around the large reservoirs in the northern Ottawa Valley region.

North of the Atlas area, the Common Tern breeds as far north as southern Labrador in the east and the southern tip of James Bay in the west (Godfrey 1986). In 1990 and 1991, the Common Tern was confirmed as a breeder in the Nottaway, Broadback, and Rupert rivers region, where 21 colonies containing a total of 483 birds were observed; 71% of these colonies, representing 86% of the population, were located south of 50°30' N (Cons. G. G. – GREBE 1992i). The species has also been confirmed as a breeder in the Eastmain River region, where several colonies were discovered in 1990 (Tardif and Morneau 1991).

History and trends

By the early 1900s, the Common Tern was on the verge of extinction in North America (Bent 1921; Bull 1985). The plume trade appears to have been the main cause of this decline, as tern feathers were much in demand for millinery. With the signing of the Migratory Birds Convention in 1916, the species was given full protection in North America, enabling populations to recover to their former levels (Bent 1921; Bull 1985).

More recently, numbers have declined again in some parts of the species' range, both in Europe (Cramp 1985) and in North America, where it was on the Blue List from 1978 to 1982 (Tate and Tate 1982). Locally, colonies are threatened by a number of factors,

The nest, a shallow scrape in soil or sand, is lined with plant stalks and other debris. Île Brion Ecological Reserve, Magdalen Islands. *July 1982, Yves Aubry.*

A nest in a rock pile containing 2 eggs and, nearby, a downy chick just a few hours old. **Lake Saint-Pierre.** *July 1991, Daniel Jauvin.*

both natural and human in origin. Common Tern populations in Québec have been quite unstable over the years. On the Magdalen Islands, the population on the islands in the Grande Entrée Lagoon dropped from 1750 to 450 birds between 1973 and 1976, probably due in part to egging and vandalism (Mousseau et al. 1976), although the overall population on the Magdalens has increased since 1976 (Fradette 1992). Increases were also recorded on the Mingan Islands between 1978 and 1986 (Simard 1986). In the Gaspé, Guillemette (1993) found declines of 23.9–100% in different colonies. The New Richmond colony, for example, disappeared entirely in 1993. In fact, between 1989 and 1993, the number of pairs on the entire peninsula dropped from 2407 to 838, despite an overall annual increase of 20.2% between 1979 and 1989 (Chapdelaine and Brousseau 1992b). This decline is probably due mainly to predation, followed by competition for nest sites, flooding of nesting grounds, and human disturbance, and most likely began in the summer of 1990, when breeding failure struck six colonies for more or less these reasons (Razurel 1994). Therefore, the Common Tern should be considered a sensitive species in need of protection.

One of the limiting factors on Common Tern populations is the strong competition from gulls for nest sites. Gulls not only are larger and nest earlier than terns, but their populations are increasing (Drury 1973; Stokes and Stokes 1989a). When a ternery is located next to a gull colony, any type of disturbance may have disastrous effects on the terns, as it causes them to leave their eggs and young unprotected. In an Arctic and Common tern colony in Maine, Hatch (1970) estimated losses from gull predation to be 0.48–1.2 young/pair. Common Terns are also subject to predation by nocturnal raptors such as Great Horned Owls and Short-eared Owls, which take adults or drive them from the colony; in the latter case, this compromises breeding success by allowing Black-crowned Night-Herons to prey on the young, sometimes causing considerable mortality (Bent 1921; Nisbet and Welton 1984; Shealer and Kress 1991). Shealer and Kress (1991) note that terns do not usually fear night-herons during the day, but flee from them at night because they mistake them for Great Horned Owls. Such disturbance affects the survival of eggs and young as it causes nests to be left unguarded. Other predators on the terns include terrestrial mammals such as red foxes, minks, and striped skunks (Blokpoel and Scharf 1991).

Common Terns are susceptible to bad weather, such as late snowstorms, and torrential rains and heavy seas that flood beaches. Shoreline erosion, often the result of wave action, is also a problem, as is human disturbance, whether in the form of shoreline development, recreational activities, all-terrain vehicles on beaches, or industrial and residential construction.

In Ontario, nesting rafts have been installed to encourage Common Terns to breed. The terns quickly adopted these sites, successfully defending them against gulls and other predators (Dunlop et al. 1991). Moreover, rafts are not colonized by gulls or affected by erosion, and provide protection from terrestrial predators and human disturbance.

Éric Razurel

In the vegetation, an adult tern guards its unfledged chicks. **Mingan Archipelago National Park Reserve, Middle North Shore.** *August 1983, Pierre Brousseau.*

Arctic Tern

Sterne arctique

Sterna paradisaea

Christiane Girard

THE ARCTIC TERN IS NOTABLE NOT ONLY FOR ITS STREAMLINED SILHOUETTE and elegant flight but also for the length of its migrations, which make it one of the most indefatigable avian travellers in the world. From close range, the Arctic Tern can be distinguished from the Common Tern by its shorter legs, slightly longer tail, and, during the breeding season, by its entirely blood-red bill. Arctic Terns breed in colonies, or terneries, that may contain from a few pairs to several thousand individuals. In southern Québec, the species nests along the Gulf of St. Lawrence.

Habitat and behaviour

Banding returns from birds banded on the Labrador coast and observations at sea between Newfoundland and Ireland (Austin 1928) have provided invaluable information on the migratory behaviour of populations breeding in the Gulf of St. Lawrence and in the interior of northern Québec. After the breeding season is over in August, Arctic Terns embark on their long journey to the South Atlantic, first crossing the North Atlantic to Europe, and then following the coast of Europe and West Africa to the coast of Antarctica, where they spend the winter. In spring, they make the same journey, but in reverse, covering more than 35 400 km during the round trip, a distance roughly equivalent to the circumference of the earth (Terres 1980).

Pair formation appears to occur on the breeding grounds shortly after migration (Hawksley 1957). In the Magdalen and Mingan islands, the onset of laying is around 6 June, and average clutch sizes are 2.00 and 1.98 eggs, respectively (Chalifour 1982; Chapdelaine et al. 1985). Other authors have reported somewhat smaller clutch sizes, ranging from 1.19 to 1.87 eggs (Eklund 1944; Hawksley 1957; Drury 1960; Bengtson 1971a; Evans and McNicholl 1972; Lemmetyinen 1972). These differences may be due to the birds' age and breeding experience, as well as to the abundance and quality of the food supply (Lemmetyinen 1973; Coulson and Horobin 1976). Length of incubation is on average 20–24 days (Cramp 1985), 21 days on the Mingan Islands (Chapdelaine et al. 1985). Incubation and brood-rearing duties are shared by both parents. By fledging, which occurs at the age of 23 days on average, the young birds have grown to almost their adult weight of around 103 g (Chapdelaine et al. 1985). The young terns feed on a wide range of fish in the gulf, including capelin, sand lance, and threespine stickleback, as well as marine invertebrates such as gammarids and krill shrimps (Chalifour 1982; Chapdelaine et al. 1985).

Arctic Terns inhabit a wide range of inland and marine habitats. Inland, they nest singly or in colonies on islands and islets in lakes and peatlands in the boreal forest and the taiga, where black spruce, lichens, and sphagnum moss predominate (Evans and McNicholl 1972). Arctic Terns are also found on gravel bars in large rivers. In the vast treeless areas of the Arctic, the species nests on islands,

Notes on territory

The Arctic Tern sometimes breeds in isolated pairs, but prefers nesting in colonies (Cramp 1985). On the eastern seaboard, colony size ranges between 2 and 5000 pairs (Erwin 1978). Inter-nest distances of 0.30–20 m have been variously reported (Cramp 1985).

The Arctic Tern breeds not only along the Gulf of St. Lawrence, James Bay, and Hudson Bay, but also inland in northern Québec, where it nests on islands and islets in lakes and peatlands. Lake Bienville, north-central Québec. *September 1979, Pierre Lamothe.*

Arctic Tern

Breeding distribution 1984–1989

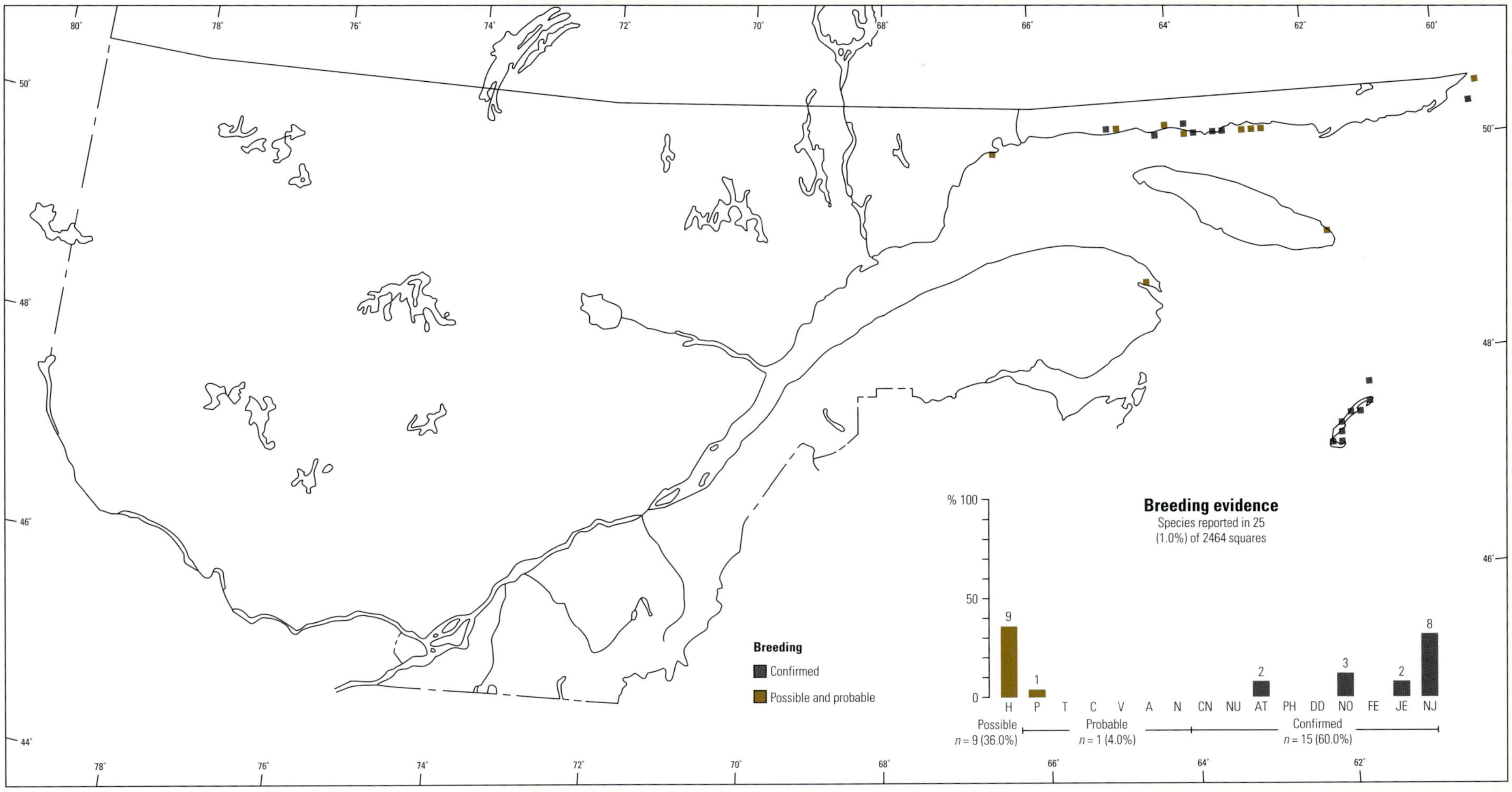

Species profile

Clutch size	usually 2 eggs (1–3)[1,2,3,5,6]
Laying interval	1 egg per 1–2 days[2]
Start of incubation	with first or second egg[2]
Incubation period	20–24 days[2,3,4]
Maturity at hatching	semiprecocial
Care of young	by female and male[2,3]
Nestling period	1–3 days[2]
Age at first flight	21–24 days[2,3,6]
Dependence of young	until at least 1–2 months after first flight[2]
Broods per year	1[6]
Breeding age	usually 3 years (1–3)[2,9]
Mating system	monogamy[2]
Duration of pair bond	more than 1 season[2]
Total length	35.5–43.0 cm[4]
Wing span	73.7–83.8 cm[5]
Weight	110 g (86–127)[8]
Longevity record	34 years[7]

References:
1. Chalifour 1982
2. Cramp 1985
3. Chapdelaine et al. 1985
4. Godfrey 1986
5. Terres 1980
6. Townsend 1921a
7. Clapp et al. 1982
8. Dunning 1984
9. Ehrlich et al. 1988

Chronology *n* = 14

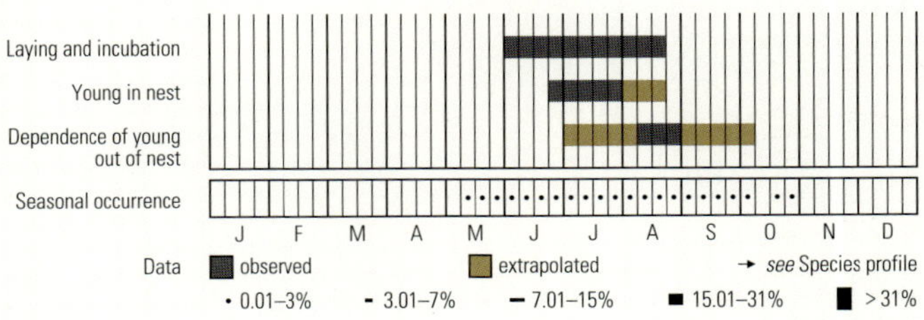

World breeding range

Young Arctic Tern hidden among lichens and heaths on the tundra around Lake Bienville. North-central Québec. *July 1976, Pierre Lamothe.*

Notes on territory (continued)

The species defends a small, roughly circular territory around the nest, which is used for the ground phases of courtship, pair formation, copulation, and nesting, and concealment and location of the young. The chicks remain near the nest for 2–3 days after hatching. The family abandons the colony 5–10 days after the young fledge (Cramp 1985).

Within a colony, birds occupying adjacent territories sometimes jointly defend a common loafing area against other individuals nesting farther away. There is strong year-to-year fidelity to the breeding site (Cramp 1985).

islets, and sandbars in the many tundra lakes and ponds. In coastal areas, it breeds on small rocky islands with sparse vegetation (mainly lichens), as well as on the spits of larger islands, habitat that is found in abundance on the islands off the Lower North Shore. The species also breeds on islands with sandy beaches dotted with clumps of American dune grass and sea beach groundsel (Chapdelaine et al. 1985). It is often found nesting in association with Common Terns, although Common Terns tend to prefer areas of slightly lusher vegetation (Fisher and Lockley 1954).

Distribution

In Québec, the Arctic Tern breeds on the Magdalen Islands and on islands off the Middle and Lower North Shore, as well as in James and Hudson bays (Todd 1963; Chapdelaine and Tremblay 1979). Colonies can also be found inland (Todd 1963; ÉPOQ; QNRCP). On the whole, the Atlas data confirm what was already known about the species' breeding range. However, the record from northeast Anticosti Island, in a square characterized by tall cliffs, is something of an anomaly in terms of the species' usual breeding habitat.

The Arctic Tern nests with the Common Tern on some islands in the Mingans, where the ground is sandy or gravelly. Wreck Island, Middle North Shore. *July 1983, Pierre Brousseau.*

An adult incubating eggs in scattered patches of sea beach groundsel on a beach on Wreck Island. Mingan Archipelago National Park Reserve, Middle North Shore. *June 1983, Pierre Brousseau.*

The Arctic Tern has a circumpolar distribution, with centres of abundance in the arctic portions of Russia, Canada, Alaska, Greenland, and Iceland (Lloyd et al. 1991). In Europe, its range extends south to Brittany in northern France and, in North America, south to Massachusetts (Croxall et al. 1984).

It is difficult to accurately estimate Arctic Tern populations because the species often nests alongside the Common Tern, a species similar in many respects. As a result, biologists surveying seabird populations usually combine the two species under *Sterna* spp., which complicates the interpretation of the results (Chapdelaine and Bourget 1981; Chapdelaine and Brousseau 1991). The largest terneries in the gulf appear to be on the Mingan Archipelago, where more than 600 pairs of Arctic Terns were tallied in 1983 (Chapdelaine et al. 1985). On the Magdalen Islands, Mousseau et al. (1976) estimated the population to be approximately 62 individuals in 1976, or around 5% of the total estimated *Sterna* spp. population (640 pairs). Recent results, including Fradette's (1992) estimate of 432 pairs in all the Magdalen Island colonies, suggest that the Arctic Tern population on the Magdalens has probably increased.

History and trends

For the reasons explained above, of all the seabird populations regularly surveyed in Québec, the status of the Arctic Tern's is the most difficult to assess. Therefore, it is impossible to confirm whether the population is declining or increasing. We can only compare the numbers reported by Mousseau et al. (1976) and Fradette (1992), which show that the population on the Magdalen Islands may have increased over the past 15 years.

Elsewhere in its breeding range—for example, in the Orkney and Shetland islands in Scotland where there are large concentrations of Arctic Terns (65 000 pairs)—the population has decreased by 50% since 1980 (Lloyd et al. 1991). For a number of years, productivity was very low there, apparently because of the scarcity of sand lance, which is harvested commercially (Monaghan et al. 1989). Nesting Arctic Terns are also vulnerable to predation by natural enemies such as arctic foxes, red foxes, ermines, and least weasels (Bird and Bird 1935), and by introduced species (Uttley et al. 1990), all of which prey on the eggs and young.

GILLES CHAPDELAINE

White-winged Tern

Guifette leucoptère

Chlidonias leucopterus

Christiane Girard

The White-winged Tern is by far the most unusual breeding bird in the Atlas. In fact, the only White-winged Tern ever seen in Québec was a female bird that, in the summers of 1985 and 1986, bred with a male Black Tern in a small Black Tern colony in the marshes on the eastern side of Lake Saint-Jean, producing hybrid young. Before then, this Eurasian migrant had only been reported a few times in Canada (in New Brunswick) and a dozen times in the eastern United States.

The White-winged Tern closely resembles the Black Tern, but in breeding plumage, can be readily distinguished by its white tail and black wing linings. Its calls given in flight and its agonistic calls at the nest are harsher and more slurred than those of the Black Tern (Glutz von Blotzheim and Bauer 1982; Cramp 1985; pers. obs.). Both are colonial species that are highly similar in behaviour, flight, feeding habits, and food preferences (Flint et al. 1984).

Habitat and behaviour

White-winged Terns are very sociable birds and, in Eurasia, colonies often contain 100–150 pairs. Although isolated pairs or small stray flocks may form mixed colonies with Black Terns (Glutz von Blotzheim and Bauer 1982), interbreeding is extremely rare and has been observed only at the edges of the species' normal ranges. The female White-winged Tern reported in Québec between 1985 and 1987 had joined a small colony of 4 or 5 pairs of Black Terns in the Saint-Gédéon marsh on the eastern side of Lake Saint-Jean. The event was closely monitored, and some of the many observations made at that time are reported here (unpubl. data).

In Eurasia, White-winged Terns nest in wetlands in the interior of the continents. These terns inhabit flooded grasslands or swampy stagnant water often bordering large rivers or freshwater lakes. Alkaline bodies of water with open areas fringed with reeds, sedges, and other aquatic plants are preferred (Cramp 1985). According to Cramp (1985), in comparison with Black Terns, White-winged Terns more often occupy unstable habitats with fluctuating water levels, and seem to prefer shallower habitats such as marshes and small ponds to larger, deeper water bodies. The nest sites selected by the mixed pair in Québec reflected these habitat preferences. In 1985, the nest was located in the Saint-Gédéon marsh on the edge of emergent vegetation, mostly common cattail and great bulrush, where the water was less than 50 cm deep; the nest was placed somewhat apart from the 4 Black Tern nests, which were better concealed. The next year, the nest was found in a dry cattail bed in the Métabetchouan marsh, 5 km south, in an area where the vegetation is interspersed with stretches of open water. The nest was less than 10 m from the secondary highway that crosses through the area.

In 1985, the nest was found a dozen or so metres from the site where the birds had been observed mating 9 days after they had arrived at the marsh. The nest was typical of both species, a floating structure constructed of plant debris and clumps of filamentous green algae anchored to the shore with standing stems and a mass of fallen dead vegetation. The structure was added to each time the birds relieved each other on the nest during incubation and brooding. The habit of adding nest materials as time goes on is characteristic of both species (Glutz von Blotzheim and Bauer 1982; Cramp 1985).

In the White-winged Tern, typical clutch size is 3 eggs, laid at 18- to 24-hour intervals (Cramp 1985) and incubated for 18–20 days (and sometimes up to 22 days). The mixed pair that nested

In 1985 and 1986, a female White-winged Tern mated with a male Black Tern from a Black Tern colony in this marsh at Saint-Gédéon. Hybrids of the two species are highly unusual, as is the occurrence of the White-winged Tern in North America. Lake Saint-Jean. *June 1985, Rémy Bouchard.*

Notes on territory

White-winged Terns are gregarious at all times, even during the breeding season (Cramp 1985). Depending on the amount of space available, colonies may contain anywhere from a few birds to several hundred pairs (Glutz von Blotzheim and Bauer 1982; Cramp 1985). A small space around the nest is defended against other White-winged Terns and is used for courtship, nesting, and resting (Cramp 1985).

The White-winged Tern of the mixed pair that nested on Lake Saint-Jean was noticeably more aggressive than the Black Terns in defending the area in and around the colony; any intruders (gulls, muskrats, and humans) were persistently mobbed.

The anti-predation strategies of the White-winged Tern are probably similar to those of the Black Tern (Glutz von Blotzheim and Bauer 1982). In Black Tern colonies, when one colony member gives an alarm call signalling the presence of an intruder or a predator, the other members will immediately congregate and attempt to intercept the intruder, and they will mob it until they succeed in driving it an acceptable distance from the colony (Cramp 1985); this behaviour was observed at the

White-winged Tern

Breeding distribution

1984–1989

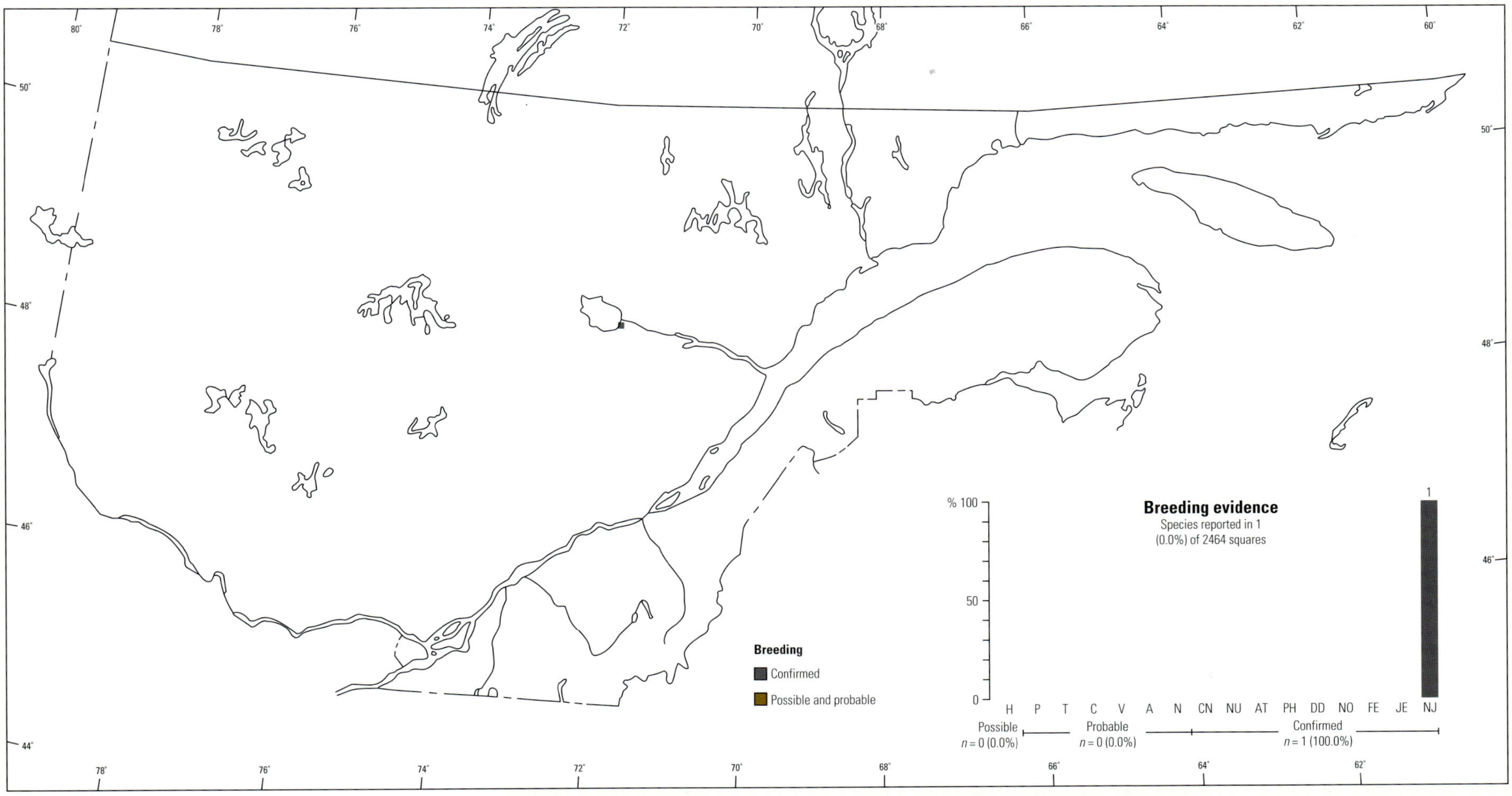

Species profile

Clutch size	usually 3 eggs (2–3)[1,2]
Laying interval	1 egg per day[2]
Start of incubation	with first or second egg[1]
Incubation period	usually 18–20 days (18–22)[2]
Maturity at hatching	semiprecocial
Care of young	by female and male[2]
Nestling period	2–3 days[1]
Age at first flight	usually 24–25 days (19–25)[2]
Dependence of young	until at least 10 days after first flight[1]
Broods per year	1[2]
Breeding age	probably not until 2 years old[2]
Mating system	monogamy[2]
Duration of pair bond	1 season[2]
Total length	23–25 cm[3]
Wing span	63–67 cm[1,2]
Weight	M 68.7 g (60.0–80.0) / F ? g (61.6–76.8)[1]
Longevity record	?

References:
1. Glutz von Blotzheim and Bauer 1982
2. Cramp 1985
3. Godfrey 1986

Note: *see* text *re* hybrids

Chronology *n* = 2

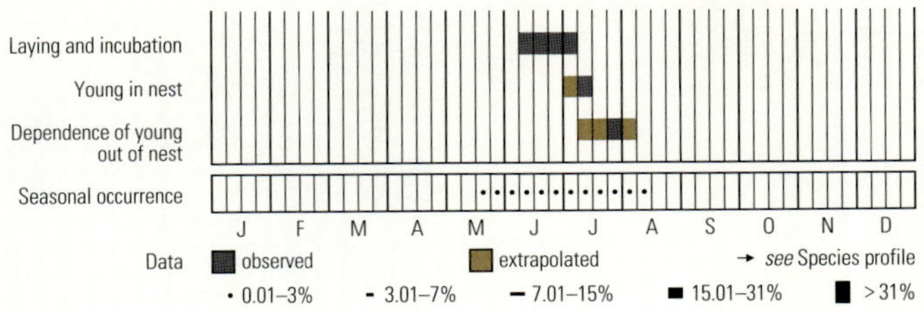

World breeding range

on Lake Saint-Jean in 1985 and 1986 produced 3-egg clutches in both years. In 1985, the male and female relieved each other on the nest every 30–45 minutes at the beginning of the incubation period; incubation lasted 21 or 22 days, slightly longer than the average for Eurasian birds.

Both parents actively participated in feeding the young. For example, one morning in 1985, when the young were 3 days old, the male and female each brought food to the chicks 37 and 30 times, respectively, in a 4-hour period. That year, the young left the nest the day after they hatched (8 July), whereas nest departure in young Black and White-winged terns normally occurs at 2–3 days old (Cramp 1985). The parents had built, for the chicks, a second floating platform among some bulrushes, about 15 m or so from the nest. The young hybrids made their first short flights at the age of 18 days. This is the minimum of the reported range for the Black Tern and is one day less than the minimum for the White-winged Tern in Eurasia. According to Cramp (1985), the age for full flight capability for both species is about 24–25 days. The female White-winged Tern left the colony on 7 August, but the male Black Tern continued to feed the young until they left the nest site on 9 August. According to Glutz von Blotzheim and Bauer (1982), juvenile White-winged Terns are often accompanied by the male but rarely by the female when they leave the colony.

The White-winged Tern observed on Lake Saint-Jean foraged much like a Black Tern; it would fly low over the water to snatch up prey from the water surface, including both aquatic insect larvae (dragonflies and beetles) and small fish, or it would swoop down, pausing momentarily to snatch its prey before flying off again. The bird also hunted insects on the wing swallow-style, skimming over the tops of the emergent vegetation or dipping down over the water and the floating mats of algae.

The female that nested in Québec participated actively in defending the colony, turning out to be much more aggressive than the Black Terns in this respect. The White-winged, with its high-pitched calls, would be the first to sound the alarm and to launch the attack on an intruder, whether a human being, Herring Gull, Ring-billed Gull, brood of ducklings, or muskrat.

In summary, a female White-winged Tern and male Black Tern were observed nesting between 30 May–7 August 1985 and 18 May–22 July 1986. The 3 hybrid young produced in 1985 were banded and remained at the marsh until 9 August. The single surviving hybrid chick of the 1986 brood was observed until 8 August. Eagerly awaited in 1987, the White-winged Tern appeared briefly at the Saint-Gédéon marsh on 26–27 May and was only seen again on 8 June. In 1985, the first Black Terns arrived at the colony 9 days earlier than the White-winged Tern and in 1986 and 1987, 6 days earlier.

White-winged Terns are seasonally monogamous; age of first breeding is 2 years (Glutz von Blotzheim and Bauer 1982).

Distribution

The White-winged Tern breeds on the temperate plains and steppes of the Palaearctic, from the Far East to eastern Central Europe (Glutz von Blotzheim and Bauer 1982; Cramp 1985). This migratory species winters in the tropics and southern temperate zones, mainly in Africa and Southeast Asia, through Malaysia and Indonesia, to Australia (Glutz von Blotzheim and Bauer 1982; Cramp 1985). In North America, the White-winged Tern is classified as accidental. Aside from the successful nestings of the mixed pair in Québec, there have only been 4 sight records in Canada (New Brunswick in 1968, 1971, and 1976; Ontario in 1991), about a dozen in the eastern United States, and one in the Lesser Antilles (Morgan and Pearce 1969; Pearce 1973; Glutz von Blotzheim and Bauer 1982; DeBenedictis 1990; M. Gosselin, pers. comm.).

Notes on territory (continued)

Saint-Gédéon marsh, on Lake Saint-Jean. Avian predators may be intercepted over 100 m from the colony, humans at a distance of over 60 m.

The nest of the White-winged–Black tern pair was located somewhat apart from those of the other colony members. Built of plant debris and placed in an area of shallow water, it contained 3 eggs. Saint-Gédéon, Lake Saint-Jean. *June 1985, Michel Savard.*

The day after they hatched, the young hybrid terns left the nest and moved to a floating platform built by the parents. In 1985, all three young fledged, as opposed to only one in 1986. Saint-Gédéon, Lake Saint-Jean. *July 1985, Rémy Bouchard.*

In Québec, the White-winged Tern observed in 1985 had joined a small colony of Black Terns that had been nesting in the Saint-Gédéon marsh since at least 1983. This is the only known Black Tern colony in the watershed of Lake Saint-Jean in recent times. In 1986, probably because of human disturbance during the laying period, the terns moved to the Métabetchouan marsh, 5 km away, but continued to forage in the Saint-Gédéon marsh. Both areas are located in the same Atlas square.

History and trends

In parts of Eastern Europe, the construction of flood-control structures and the massive draining of wetlands have driven White-winged Terns from their traditional breeding sites. Moreover, this tern's nesting habits make it highly vulnerable to severe water level fluctuations (Glutz von Blotzheim and Bauer 1982). Nevertheless, in the second half of this century, the Eurasian population has shown a strong increase and has been expanding its range westward (Glutz von Blotzheim and Bauer 1982; Harrison 1985). Since the 1930s, an increasing number of stragglers have been sighted outside the species' usual migration routes, as far as northwestern Central Europe and the British Isles. The increased number of sightings in North America since the 1960s may be part of this overall trend. Until now, however, no breeding pairs have been reported on this continent.

There have been a few reports of mixed pairs of White-winged and Black terns at the edge of the species' ranges—in northeastern Russia, in the Baltic Sea region, and in The Netherlands—although only 2 of these pairs have been studied to any extent. In Sweden (Öland Island), a female White-winged Tern and a male Black Tern attempted unsuccessfully to nest within a colony of Black Terns in 1978 and 1979 (Alexandersson 1979). Another mixed pair in The Netherlands also tried to nest in a Black Tern colony in 1979 with a similar lack of success. In this case, 2 eggs were laid and incubated but did not hatch (van Ijzendoorn 1980). In both instances, the length of the nest relief display was very irregular and sometimes abnormally long. It is thought that the clutch was abandoned because of the birds' inability to co-ordinate their nesting activities (Alexandersson 1979; van Ijzendoorn 1980).

In North America, a female specimen was collected on 5 July 1873 in a large Black Tern colony in Wisconsin (Kumlien and Hollister *in* Bent 1921). On examining the specimen, the authors found physiological signs of breeding. As no other White-winged Terns were present, the authors assumed that a mixed pair had formed but were unable to give any other information on the pair's breeding history and behaviour or whether any offspring had been produced (Bent 1921).

In England, there have been reports of juvenile terns observed in migration with mixed characters of White-winged and Black terns (Vinicombe 1980; Davis, A. H. 1982). The pair observed in Québec, however, is the only one known to have produced hybrid young that fledged and, even more remarkably, to have done this two years in a row. This pair is unique in that, in 1985, both members were able to synchronize their breeding activities: mating, nest building, incubating, raising hybrid young, and defending the colony. During that particular summer, weather and water level conditions were close to ideal. Recreational activities were controlled throughout the season so that almost no human disturbance occurred. These factors may have contributed to the pair's breeding success. In 1986, human disturbance early in the laying period caused the small colony of Black Terns to abandon the site, and they attempted, unsuccessfully, to nest again. The partial success of the mixed pair at the new site has been attributed to the highly aggressive behaviour of the female White-winged Tern in defending the territory.

Michel Savard, Rémy Bouchard, Christiane Girard, and Nicolas Tremblay

Black Tern

Guifette noire
Chlidonias niger

Christiane Girard

The Black Tern is the smallest of the terns that breed regularly in Québec and is the only one that, in breeding plumage, has a solid black head and mostly black underbody.

Its dipping, erratic flight, dark plumage, typical tern silhouette, and sharp *kik-kik-kik* call make it easy to spot and identify. This migrant breeder returns to Québec in May, after wintering along the coast of South America.

In Québec, Black Terns breed in marshes along the St. Lawrence and its main tributaries in the southwestern part of the province, among other places. Baie-du-Febvre, south shore of Lake Saint-Pierre. *June 1991, Daniel Jauvin.*

Habitat and behaviour

The Black Tern breeds in marshes and swamps, and along lakes, ponds, and rivers, generally in fresh but sometimes brackish water where the dominant vegetation consists of cattails, rushes, or sedges. For nesting, a marsh of at least 5 ha seems to be required and those over 20 ha are favoured (Brown and Dinsmore 1986). Although Black Terns sometimes nest in solitary pairs, loose colonies are usually the norm. In Ontario, Peck and James (1983) reported an average colony size of 9 nests, qualifying this figure, however, as low due to the loose nature of the aggregations and the difficulty of locating all the nests. The species is monogamous and the birds often return to nest on the site occupied by the colony in previous years. However, Black Terns may abruptly change sites when the emergent marsh vegetation is altered by drought, flooding, winter storms, or other weather conditions (Carroll 1988b). In spring, the birds arrive in small flocks to explore potential nest sites but do not alight. They settle in a common resting area away from the colony site (up to 1.6 km) and await the onset of laying. This area is also used for courtship displays and roosting. A few days before laying begins, the first pair alights on the breeding site and the others follow within a few hours; between this time and the end of incubation, male birds can often be seen proffering a fish to their mates (Baggerman et al. 1956).

The Black Tern raises only one brood a year, but may renest if the clutch is lost. Typical clutch size is 3 eggs (sometimes 2, rarely 1 or 4), laid a day apart and incubated for 20–22 days (Harrison 1978; Cramp 1985). Both members of the pair build the nest, using dead cattails and other marsh vegetation. The mass of dead vegetation forms a floating mat with a slight depression in the centre where the eggs are laid, and may either be anchored or unanchored (Davis and Ackerman 1985). The nest may also be built on a solid base such as a muskrat house, stump, half-submerged tree trunk or a small mud patch. It is often placed amid moderately dense emergent vegetation and is usually near open water averaging about 1 m deep (Bergman et al. 1970; Gerson 1988).

Both parents share incubation and brood-rearing duties. Although the semiprecocial chicks may temporarily leave the nest at 2 or 3 days old (Cramp 1985), or even earlier if danger threatens, nest exodus does not finally occurs until they are about 2 weeks old (Bent 1921; Harrison 1978; Cramp 1985). If disturbed when the young are still in the nest, the pair may build an alternate nest a little farther away (Cuthbert 1954). The young first fly at about 3–4 weeks of age, and the parents stop feeding them shortly thereafter.

Mortality during breeding is due mainly to egg loss, largely because the floating nests are often flooded, overturned, submerged, or destroyed by adverse weather conditions and changing water levels (Dunn 1979; Carroll 1988b). The chief predators of Black Tern nests include American Crows, Common Ravens, Great Horned Owls, Northern Harriers, muskrats, minks, long-tailed weasels, river otters (Gerson 1988), raccoons (Brewer 1991), and possibly Great Blue Herons (Chapman and Forbes 1984). When a human being or other potential predator approaches, the adults join together to defend the colony, screaming alarm calls and dive-bombing the intruder, sometimes even striking their target with their bills, until the danger is past. This aggressive behaviour and the conspicuous black adult plumage, which stands out against the nest, make nests and young easy to locate.

Notes on territory

Black Terns generally form loose colonies, each pair defending a territory around the nest with a radius of 3 m or less, although the boundaries are vaguely defined. The territory is defended mainly by the male (Cuthbert 1954; Baggerman et al. 1956; Cramp 1985). After fledging, the young remain with the adults, which establish a temporary feeding territory. Any terns or gulls that venture into this area (which may also be established before the young fledge) are driven away by family members (Cuthbert 1954).

Black Terns usually form small, loose colonies in moderately dense vegetation, in marshes, and along lakes, rivers, and ponds. Here, a flock is seen fluttering about over the marsh vegetation. Châteauguay. *June 1987, Paul Perreault.*

Black Tern

Breeding distribution

1984–1989

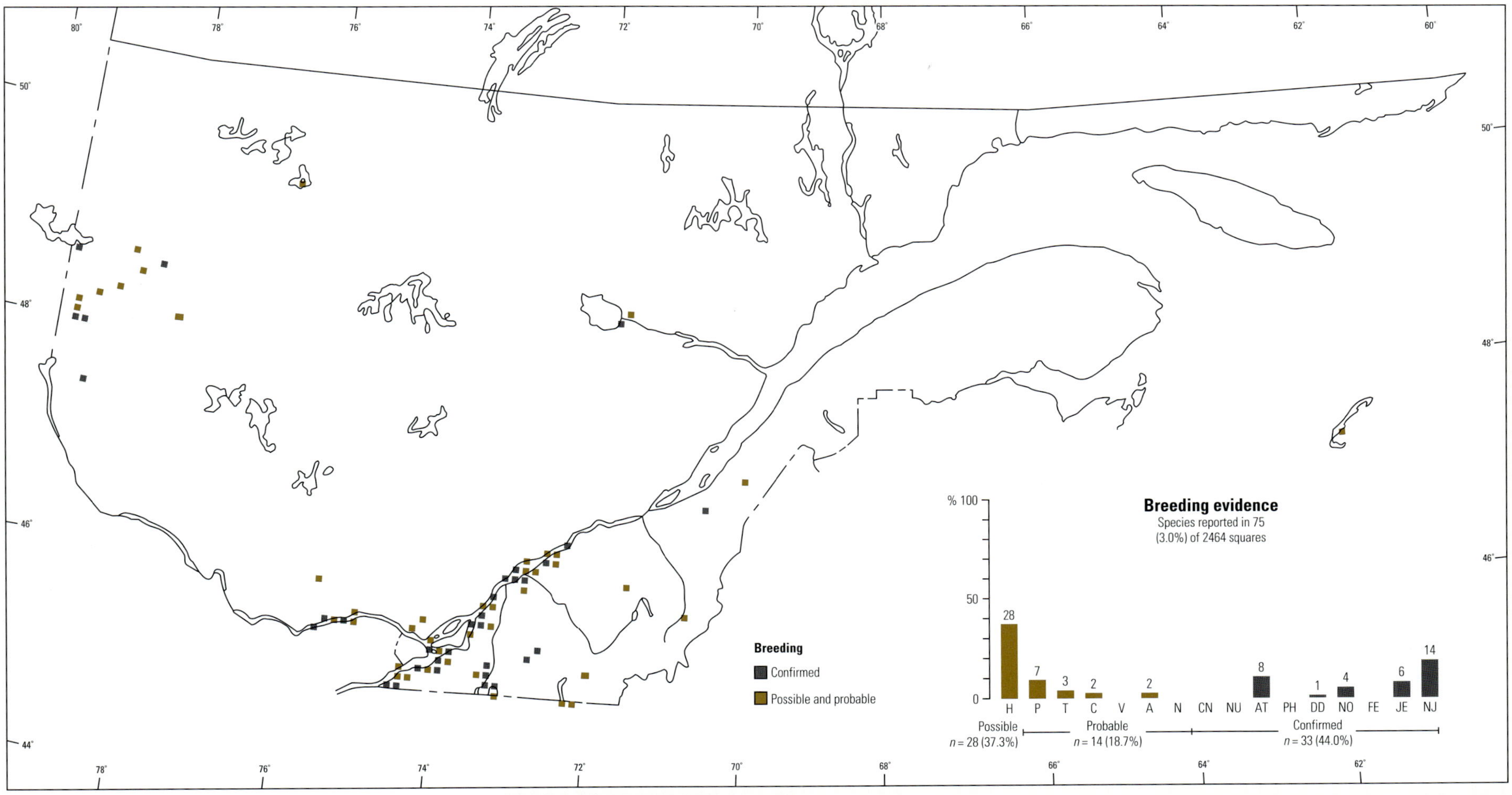

Species profile

Clutch size	usually 3 eggs (1–4)[1,6,7]
Laying interval	1 egg per day[2,6]
Start of incubation	with first egg[2]
Incubation period	19–22 days[1,3,6,7,8,12]
Maturity at hatching	semiprecocial
Care of young	by female and male[3,6]
Nestling period	2–3 days[6]
Age at first flight	usually 20–24 days (18–28)[3,4,6,7,12]
Dependence of young	until at least 13 days after first flight[6]
Broods per year	1[6]
Breeding age	probably 2 years[6]
Mating system	monogamy[6]
Duration of pair bond	?
Total length	23.0–26.5 cm[5]
Wing span	64–68 cm[6]
Weight	65.3 g (60.3–74.1)[10]
Longevity record	8 years 5 months[9]

References:
1. Bergman et al. 1970
2. Dunn 1979
3. Cuthbert 1954
4. Baggerman et al. 1956
5. Godfrey 1986
6. Cramp 1985
7. Terres 1980
8. DeGraaf and Rudis 1987
9. Clapp et al. 1982
10. Dunning 1984
11. Harrison 1978
12. Dunn and Agro 1995

Chronology *n* = 81

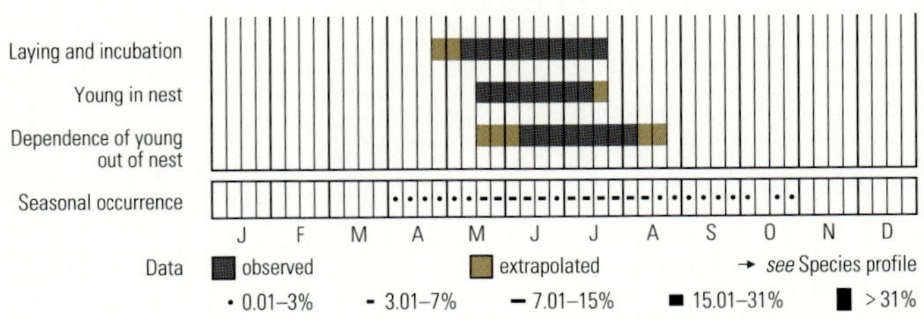

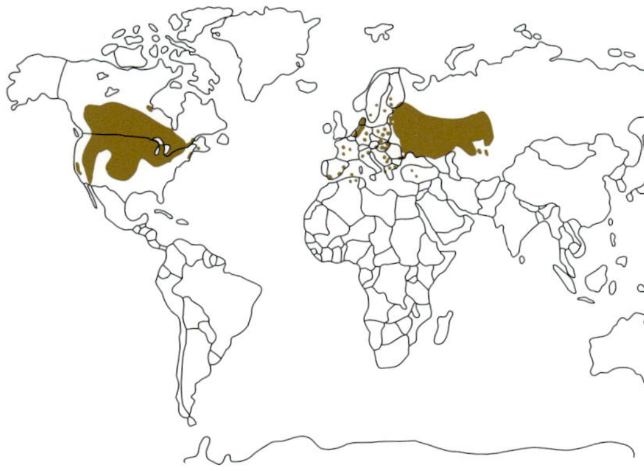

World breeding range

Black Terns feed in flocks of 2–20 individuals, flying over the open water in the marsh only a few metres above the surface. This tern's buoyant and erratic flight is somewhat like that of the nighthawk, except that it hovers frequently (Terres 1980). Black Terns feed mainly on insects, both adults and larvae, which they catch in the air, pluck from the surface of the water, or sometimes pick directly off the leaves of emergent vegetation. Unlike other terns, they rarely dive, immersing only their bills when taking prey from the water (Baggerman et al. 1956). Small fish, however, may make up a substantial portion of the young birds' diet; since fish are larger and easier to digest than insects, they probably meet at least a third of the chicks' protein requirements (Cuthbert 1954; Dunn 1979). In the marine habitats frequented during migration and in winter, Black Terns feed mainly on small fish.

Distribution

The Black Tern is a holarctic species that breeds exclusively on inland waters. In North America, its centre of abundance is in the Prairie provinces and north-central United States. In Québec, it breeds in the Central St. Lawrence Lowland, particularly along the St. Lawrence, Ottawa, and Richelieu rivers, and in the Appalachians. Breeding records were also obtained from the Abitibi and Lac-Saint-Jean regions. The species was atlassed on the Magdalen Islands, where it had already been sighted a few times during the breeding season prior to the Atlas. But since no breeding evidence has ever been obtained on the islands, the species is considered to be a rare visitor there (Fradette 1992).

To breed, Black Terns require a large marsh with fairly extensive stretches of open water. This type of habitat is not particularly abundant, occurring mainly along large bodies of water. No sightings were made, however, near the large lakes and reservoirs in south-central Québec. In all, the species was atlassed in only 3% (75/2464) of squares.

Black Tern nests are vulnerable to water-level variations and wave action (Chapman Mosher 1986; Novak 1992). Therefore, the species would find it difficult to breed in the tidal marshes of the estuary and gulf, where nests would soon be destroyed by the wind, waves, and tides.

This nest is a loose pile of dried plant stalks on a half-submerged log, although nests usually consist of a floating mat of vegetation. North of Lake Saint-Paul, Bécancour, across from Trois-Rivières. *June 1967, Gaston Le Page.*

The Black Tern favours marshes larger than 20 ha with emergent vegetation, like this one adjoining Lake Georges, near Amos. Abitibi region. *July 1991, André-Guy Bernier.*

Incubating Black Tern. Châteauguay. *June 1987, Paul Perreault.*

Nest with 1 egg and 2 downy chicks. Châteauguay. *June 1987, Paul Perreault.*

History and trends

The Black Tern was probably more abundant in Québec during the 19th century than it is now, although no specific abundance data are available from that time (Ouellet 1974a). BBS data from 1966 to 1989, however, show a decrease in the province's population. This is a continent-wide trend and the species has been on the Blue List since 1978 (Arbib 1977). Habitat loss from the draining of wetlands for agriculture and urban development is undoubtedly the main reason for the decline (Tate 1986; Gerson 1988). A similar trend, probably also attributable to habitat loss, has been observed in Europe (Cramp 1985).

The amount of apparently suitable but unoccupied habitat suggests, however, that other factors may be responsible for declining Black Tern populations. Exposure to contaminants, especially during migration and on the tern's wintering grounds along the coasts of Central and South America, could affect reproduction and survival (Gerson 1988). Similarly, the intensive use of pesticides, herbicides, and insecticides on the Prairie breeding grounds could very well reduce breeding success or at the very least jeopardize food supplies during the breeding season, when the birds feed mainly on insects (Carroll 1988b). Acid rain and increasing populations of predators such as the raccoon may also affect tern populations at this time (Brewer 1991). The species is also sensitive to bad weather, flooding, and wakes from the increasing number of pleasure and sport fishing boats on Québec's lakes and rivers. Limiting access to the main breeding sites during the nesting season could be recommended in certain cases (Leberman 1992b).

When all these factors are taken into account, it is not surprising that Gerson (1988) has proposed having the Black Tern classified as a threatened species in Canada.

Paul Messier and Jean-François Rail

Common Murre

Marmette de Troïl

Guillemot marmette

Uria aalge

THE COMMON MURRE HAS A CHOCOLATE-BROWN HEAD, NECK, AND BACK, white breast, and long, daggerlike bill. From the rear, as it is standing on the rocks, it looks like an orchestra conductor in full evening dress. It is the largest member of the alcid family, weighing 1 kg on average. These murres breed along the coast of Labrador and Newfoundland and in the Gulf of St. Lawrence, in just a few colonies, which often contain several thousand birds. Common Murres winter on the Grand Banks off Newfoundland, on Georges Bank off Nova Scotia, and farther south.

The bridled form of the Common Murre has a white eye ring and a thin white line extending backward from the eye, which should not be confused with the white line along the cutting edge of the upper mandible in the Thick-billed Murre.

Habitat and behaviour

Common Murres leave their wintering grounds in March, and by April, can be spotted in the waters near the southernmost colonies. On Bonaventure Island, the onset of laying is mid-May, but is not until the first week of June on the Sainte-Marie Islands off the Lower North Shore. The pyriform egg is thought to confer distinct advantages on this species, which nests on cliffs and lays its single egg directly on bare rock: since the egg pivots around its axis, it is less likely to roll away or fall off the rocky ledge when the parents are disturbed (Tuck 1965). Incubation, which lasts about 33 days, is shared by both the male and female. The chick remains in the nest for 18–29 days, and is fed by both parents; by the time it leaves the colony, it weighs 250 g on average (Tuck 1965). Chicks are fed mainly capelin, although Atlantic herring, rainbow smelt, sand lance, and other species of fish are also occasionally consumed.

Chicks usually leave the colony at dusk (which reduces the risk of predation by gulls and other species), accompanied by their male parent (Tuck 1965). Colony attendance by the female continues for another 2 weeks, during which she makes regular visits to the nest site; the pair will return to the same site year after year (Harris and Birkhead 1985a). Once out of the nest, the chick must learn to catch its own fish so it can build up the energy reserves it needs for its plumage to develop. This is a critical period as the chick can swim and dive but, since it does not have its flight feathers yet, it will not be capable of sustained flight for at least one or two months.

Three years later, immatures return to their natal colonies as prospectors, not to breed but to familiarize themselves with daily life at the colony and to look for a future nest site. Average age at first breeding is 5 years (Birkhead and Hudson 1977), and the pairs undoubtedly mate for life (Harris and Birkhead 1985a).

These seabirds are extremely well adapted for swimming and diving. Murres have been recovered in fishing nets set as deep as 180 m, and it is not uncommon for them to dive more than 100 m to catch fish (Piatt et al. 1984). However, this adaptation also has its drawbacks. Since their wings are shorter than those of other birds their size, murres are relatively poor fliers. To maximize lift, they often take off from a high point into the wind; on the water, they have to patter over the surface a long time before they get airborne. On the islands off the Lower North Shore, they sometimes smash into the rocks when trying to take off from the boulder scree and fissured rock formations that characterize the area. In high winds, they use their webbed feet as rudders because their tails are so short. Common Murres are superbly adapted for marine life, and like other alcids, they spend eight or nine months of the year at sea, only coming ashore to breed.

Common Murres visit islands in the Gulf of St. Lawrence only to nest. They occupy the highest ledges on rocky islands, as well as partly sheltered crevices and deep fissures in cliffs, sometimes forming very dense colonies. Cliff Island, Sainte-Marie Islands, Lower North Shore. *July 1989, David Cairns and Gilles Chapdelaine.*

Louis Ruelland

Notes on territory

The Common Murre is undoubtedly the species with the smallest territory, only 0.05 m². On flat, low-lying islands such as Funk Island off Newfoundland, breeding densities can be as high as 20 pairs/m².

Whether the colony is located on a broad flat area or a cliff, the murres often nest in direct physical contact with their neighbours, with nests less than 5 cm apart, although inter-nest distances of over 30 cm have been found in less crowded colonies (Cramp 1985).

Both parents defend the territory, which is used for courtship, mating, and for raising the nestling. If their egg moves, their territory shifts too. The parents tolerate close physical contact with incubating neighbours, but keep individuals with no egg at least 15 cm away. Common Murres nesting on the fringes of the colony may drive off intruders from adjacent neutral ground. Immature individuals may establish temporary territories in clubs (Cramp 1985).

Common Murres may also form mixed colonies with other seabirds, such as Razorbills, Black-legged Kittiwakes, and Thick-billed Murres, and will sometimes usurp the latters' nest sites (Cramp 1985).

Common Murre

Breeding distribution
1984–1989

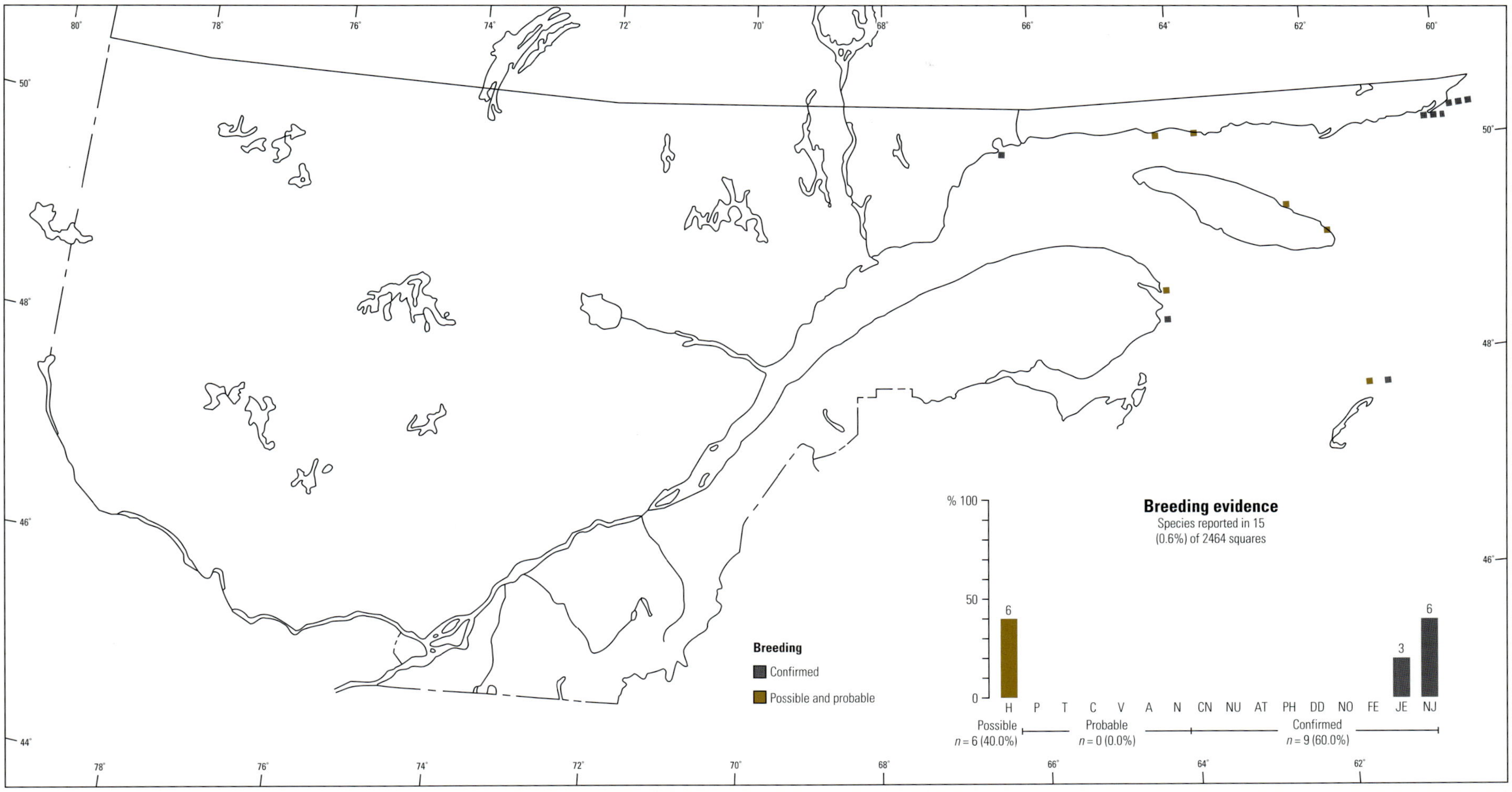

Species profile

Clutch size	1 egg[2]
Laying interval	—
Start of incubation	after laying of egg[2]
Incubation period	33 days[5]
Maturity at hatching	semiprecocial
Care of young	by female and male[5]
Nestling period	18–29 days[5]
Age at first flight	50–70 days[3]
Dependence of young	until a few weeks after leaving nest[3]
Broods per year	1[3]
Breeding age	usually 5–6 years (3–9)[1]
Mating system	monogamy[3]
Duration of pair bond	probably for life[2]
Total length	42.0–45.5 cm[7]
Wing span	76.2 cm[6]
Weight	M 1006 g (775–1202) / F 979 g (815–1187)[8]
Longevity record	26 years 5 months[4]

References:
1. Harris et al. 1994
2. Harris and Birkhead 1985a
3. Cramp 1985
4. Clapp et al. 1982
5. G. Chapdelaine, CWS, unpubl. data
6. Terres 1980
7. Godfrey 1986
8. Dunning 1984

Chronology $n = 23$

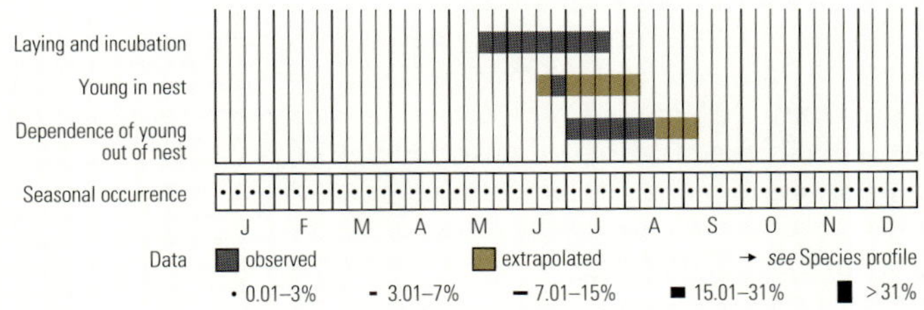

World breeding range

Bonaventure Island, with its innumerable rocky ledges, is home to over 27 000 pairs of Common Murres, which share their quarters with Black-legged Kittiwakes and Northern Gannets. Île Bonaventure and Rocher Percé Conservation Park. *August 1982, Yves Aubry.*

Common Murres generally nest in immense colonies of several thousand individuals, sometimes in the company of other seabirds, on cliffs with narrow or deep ledges, as is the case on Bonaventure Island and Bird Rocks. They also breed on flat, low-lying rocky islands, where they form huge single-species communities, somewhat like the penguin colonies of the Antarctic. Funk Island, 55 km off the northeast coast of Newfoundland, is a good example of such a colony, with over 400 000 pairs of Common Murres crowded together on a flat-top of only 10 ha (Birkhead and Nettleship 1980). In this type of habitat, the murres pratically nest on top of each other, since, incredible as it may seem, the average density on such broad, flat, rocky areas is around 20 pairs/m² (Harris and Birkhead 1985a). In Québec, there are also several small Common Murre colonies on the Sainte-Marie Islands, in partly sheltered crevices and deep fissures, where they nest in association with other alcids. However, such sites are generally cramped, with limited room for the colony to expand.

This species is rarely seen near the coast and most of the known colonies are located on islands well offshore, too far for terrestrial predators such as foxes to reach over the ice in winter. It is easy to imagine the destruction that such a predator, starving and marooned on an island by the spring break-up, could wreak on a colony of Common Murres.

Distribution

In North America, Common Murres are found along both the Pacific and Atlantic coasts, occurring in warmer waters than the Thick-billed Murre. In the eastern part of the continent, 95% of the population inhabits a small number of colonies off Newfoundland and southern Labrador (Nettleship and Evans 1985).

The only colonies in Québec are located in the Gulf of St. Lawrence. Because of the methodology used for the Atlas, the number of squares in which the species was atlassed does not correspond exactly to the number of existing colonies, since a given square may contain more than one colony. Records in 2 squares (Brion Island and Forillon National Park) where there are no actual colonies, relate not to colonies but to fledglings on the water that had obviously left their natal colony.

The current population in the gulf is estimated at just over 85 000 birds. The largest colony is on Bonaventure Island, where there are 54 000 Common Murres (Chapdelaine and Brousseau 1992a). With close to 26 000 birds, the Sainte-Marie Islands group supports the second largest population (Chapdelaine and Brousseau 1991). The rest of the population is scattered along the Middle and Lower North Shore in very small colonies. Several hundred Common Murres also nest on Bird Rocks along with Thick-billed Murres, but it is impossible to determine the proportions of the two species, since they are difficult to tell apart at a distance and these islands are quite inaccessible. Possible or probable records based on sightings of single birds or pairs in suitable habitat in squares located far from the known colonies most likely involved non-breeders.

History and trends

In the 19th century, almost all species of seabirds were commercially exploited for their meat, eggs, and feathers, and the Common Murre was no exception. Around the 1840s, over 750 000 Common Murre eggs collected from the Lower North Shore colonies were sold annually in Halifax (Johnson 1940); this figure may not reflect the actual number of pairs breeding in the area, however, since the eggs may have been from more than one laying. Johnson (1940) estimated the population on the Lower North Shore at the time to be about 375 000 pairs. This suggests that numbers have declined considerably in the last 150 years.

The single egg is laid directly on bare rock on an island or a cliff ledge. The pear-shaped egg can spin on its axis, which keeps it from rolling away and breaking. Sainte-Marie Islands, Lower North Shore. *June 1977, Pierre Dupuis.*

Common Murre populations in the Middle and Lower North Shore sanctuaries, 1925–93 (BIOMQ)

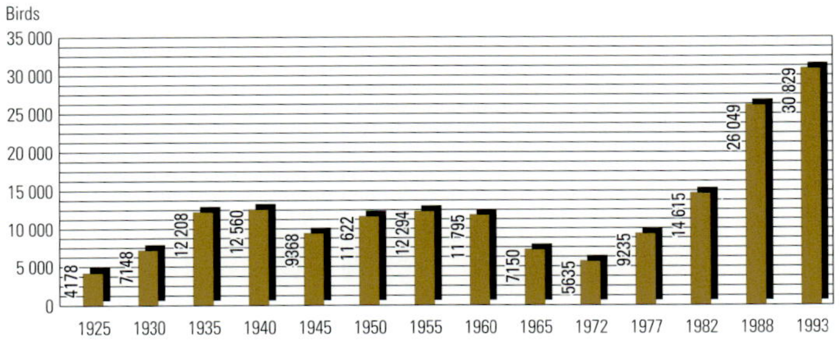

At present, although the Common Murre is still hunted on its wintering grounds off Newfoundland, the gulf population is increasing, as is the case in many seabird species (Chapdelaine and Brousseau 1992b). According to censuses carried out since 1925 in the seabird sanctuaries along the Middle and Lower North Shore, the Common Murre population has recently rebounded, reaching the highest levels recorded since the sanctuaries were created in 1925 (see figure). Nonetheless, the population is probably a far cry from the levels reported by naturalists in the 19th century.

GILLES CHAPDELAINE

The chick leaves the colony at dawn or dusk, accompanied by the male parent, remaining at sea for several years before returning to the natal colony. Bonaventure Island. *July 1979, Gilles Chapdelaine.*

Thick-billed Murre

Marmette de Brünnich

Guillemot de Brünnich
Uria lomvia

Louis Ruelland

THICK-BILLED MURRES SOMETIMES BREED IN HUGE COLONIES OF MORE than a million birds, which from afar resemble giant beehives. The Thick-billed Murre and Common Murre are so similar that, to separate them, a close study of their plumage and morphological characteristics is required. Adult Thick-billed Murres are slightly smaller than Common Murres, weighing 900 g on average (Gaston and Nettleship 1981), compared to the Common Murre's 1 kg. The best way of differentiating the two species is by examining bill shape and markings. The Thick-billed Murre's bill is, as its name implies, shorter and thicker than the Common Murre's and is slightly compressed laterally; in addition, in breeding plumage, most Thick-billed Murres have a fairly distinct white line along the cutting edge of the upper mandible. In autumn and winter, the Thick-billed also has a darker head than the Common Murre, whose head is whiter, particularly behind the eye.

Habitat and behaviour

Thick-billed Murres begin returning to their breeding grounds in March after a long winter sojourn at sea off the ice-clogged coast of Newfoundland. The birds arrive first at their southernmost colonies: Cape St. Mary's in Newfoundland and Bird Rocks in Québec. Murres begin to occupy the Bird Rocks colony around 25 March and are present in force by 3 April, according to the Bird Rocks lighthouse keepers, who did not distinguish between Common and Thick-billed murres, both of which nest there (Job 1906). More careful observations of the mixed colonies on the south coast of Newfoundland suggest that Thick-billed Murres arrive a little later than Common Murres (Tuck 1965), which is probably also true for Bird Rocks. The lighthouse keepers also reported that the murres begin to lay their eggs around 15 May (Fradette 1992).

Notes on territory

Unlike Common Murres, Thick-billed Murres breed only on narrow cliff ledges and shun large flat top surfaces (Harris and Birkhead 1985b). Nesting density is determined by the structure of the cliff. When the ledge is several metres wide, the birds nest side by side. Although density may be high on particular ledges, the overall density on the cliff may be low, depending on the configuration of the cliff face.

Adult Thick-billed Murre. Old Harry Point. Magdalen Islands. *July 1987, Daniel Jauvin.*

Breeding density is determined in part by the configuration of the cliff face. Akpatok Island, Northwest Territories. *August 1982, Gilles Chapdelaine.*

Thick-billed Murre

Breeding distribution

1984–1989

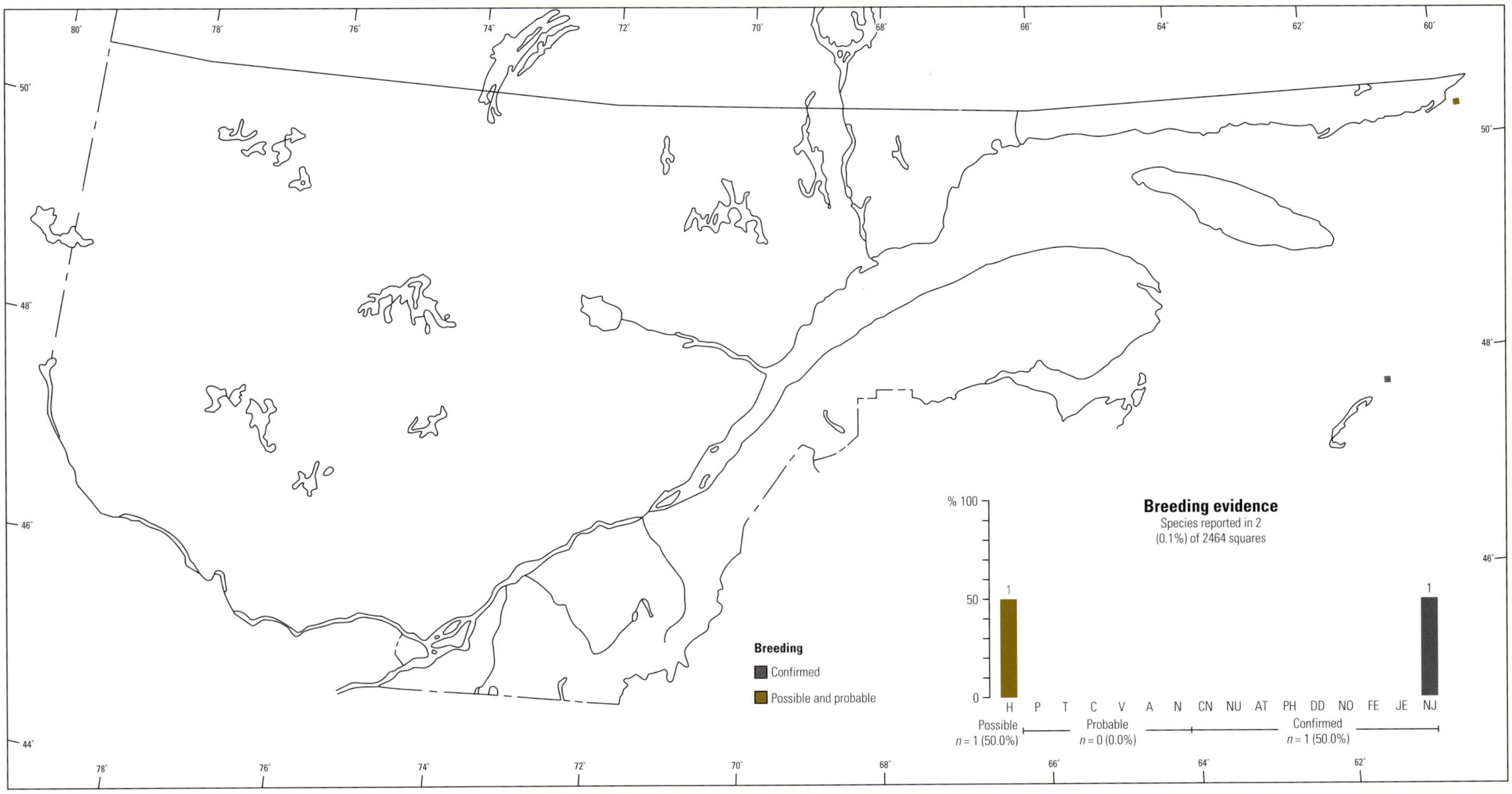

Species profile

Clutch size	1 egg[2]
Laying interval	—
Start of incubation	after laying of egg[3]
Incubation period	32 days[1]
Maturity at hatching	semiprecocial
Care of young	by female and male[2]
Nestling period	15–30 days[1]
Age at first flight	?
Dependence of young	until several weeks after leaving nest[3]
Broods per year	1[3]
Breeding age	probably 3 years or more[7]
Mating system	monogamy[3]
Duration of pair bond	for life[3]
Total length	43–48 cm[5]
Wing span	65–73 cm[3]
Weight	M 842.4–972.5 g (?) / F 821.7–968.6 g (?)[6]
Longevity record	22 years 8 months[4]

References:

1. Gaston and Nettleship 1981
2. Harris and Birkhead 1985a
3. Cramp 1985
4. Clapp et al. 1982
5. Godfrey 1986
6. Gaston et al. 1984
7. Tuck 1965

Chronology $n = 18$

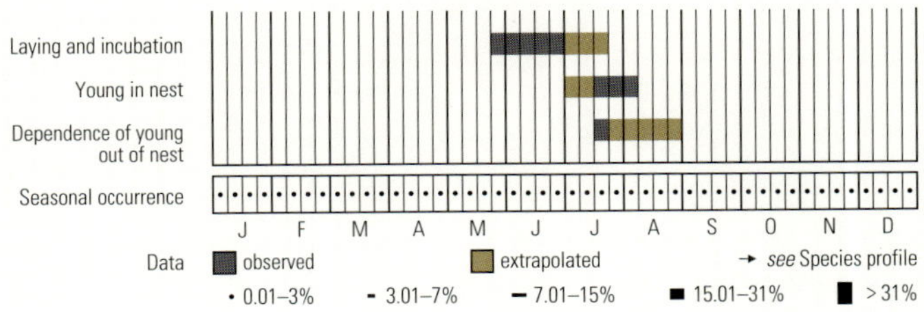

World breeding range

The steep cliffs overlooking Digges Sound, some as high as 300 m, provide excellent nesting habitat for approximately 300 000 pairs of Thick-billed Murres, which are found in huge colonies there. Cape Wolstenholme, northern tip of Ungava Peninsula. *August 1982, Gaétan Guertin.*

The Thick-billed Murre lays a single, pyriform egg, with various degrees of speckling, blotches, and scribbles, and ranging in colour from almost pure white to shades of blue, green, and brown. Unlike Common Murres, which nest both on the flat tops of rocky islands and on narrow cliff ledges, Thick-billed Murres breed exclusively on cliffs. The incubation period is around 32 days, and the chick remains in the nest for 15–30 days, with 21 days being the average. When the young birds leave the colony, they weigh about 200 g (Harris and Birkhead 1985a); this figure may vary from colony to colony, however, depending in part on the size of the colony (Gaston et al. 1983). The young do not attain their adult weight and are not capable of sustained flight until they have spent a month or two at sea (Tuck 1965; Gaston and Nettleship 1981).

No information is available on the Thick-billed Murre's diet in the Gulf of St. Lawrence, but studies carried out on the Gannet Islands, off the coast of Labrador, show that the young are fed mainly daubed shanny, followed by capelin and the occasional sand lance (Birkhead and Nettleship 1987). In winter, the adults live on fish, mainly capelin, and invertebrates (Bradstreet and Brown 1985).

Distribution

The Thick-billed Murre has a circumpolar distribution, breeding from the 46th to the 82nd parallels. In the North Atlantic, the species nests on islands with high cliffs, from the Gulf of St. Lawrence and off southeastern Newfoundland to northern Baffin Bay. It also breeds in Greenland, Iceland, northern Norway, and eastern Siberia. In the North Pacific, the species is found in the Aleutian Islands, Gulf of Alaska, and northern Japan (Nettleship and Evans 1985).

The only sizable colony in the gulf is on Bird Rocks, near the Magdalen Islands. Because of the difficulty of separating Thick-billed and Common murres and of surveying seabirds in this cliffside habitat in general, accurate population estimates are not available. Most ornithologists from 1880 to 1966, with the exception of H. Bryant, considered the Thick-billed Murre to be more abundant on Bird Rocks than the Common Murre (Fradette 1992); the estimate of 1000 pairs of Thick-billed Murres proposed by Brown et al. (1975) must be considered approximate. On the North Shore, Thick-billed Murres have been sighted on Cliff Island in the Sainte-Marie Islands by Lewis (1942), Allen (1948), Gabrielson (1952), and G. Chapdelaine (pers. obs.). These observations are all from the same cliff ledge, where many Common Murres nest, and considering the habitat, it is very probable that Thick-billed Murres breed there, too. The Atlas map provides a good picture of the species' distribution in the Gulf of St. Lawrence.

This species of murre nests on narrow cliff ledges. Akpatok Island, Ungava Bay, Northwest Territories. *July 1981, Gilles Chapdelaine.*

In the Gulf of St. Lawrence, these murres breed only on Bird Rocks, in the Magdalen Islands, in association with many other species of seabirds. *July 1983, Yves Aubry.*

The towering cliffs of Akpatok Island, in Ungava Bay, are home to between 300 000 and 400 000 pairs. Northwest Territories. *August 1982, Gilles Chapdelaine.*

Elsewhere in Québec, huge colonies of Thick-billed Murres are found much farther north, around Hudson Strait. The Digges Islands and Cape Wolstenholme colonies, on the northwest tip of the Ungava Peninsula, support over 300 000 pairs (Gaston et al. 1985, 1993), while the Akpatok Island colonies, in Ungava Bay, probably are home to between 300 000 and 400 000 pairs. Actual numbers are probably higher, however, because of the problems inherent in photographic surveys (Chapdelaine et al. 1986a).

History and trends

Thick-billed Murres are hunted in late autumn and winter off the coasts of Newfoundland and Labrador. In the past, the hunt was an important winter subsistence activity, and it still goes on extensively today. Surveys by the Canadian Wildlife Service between 1977 and 1988 show that 300 000–750 000 birds are harvested annually (Elliot et al. 1991). There is some evidence that populations in the eastern Canadian Arctic declined between 1950 and 1970 (Evans and Nettleship 1985), but it is difficult to determine the extent of this decline, since surveys of most of the large Arctic colonies have been infrequent and irregular. Some colonies have probably remained stable since 1950 (Gaston and Elliot 1991).

It is no easier determining trends in the single colony in southern Québec, on Bird Rocks, since Common Murres also nest on these islands and to date no survey differentiating the two species has been carried out.

GILLES CHAPDELAINE

Razorbill

Petit Pingouin

Alca torda

Louis Ruelland

THE RAZORBILL IS NAMED FOR THE THIN UPPER MANDIBLE OF ITS BILL, which resembles an old-fashioned cutthroat razor. Razorbills, like other alcids, are sometimes associated in the popular mind with penguins, a completely different family of birds that are limited to the Southern Hemisphere. Like penguins, alcids are seabirds and well adapted for swimming and pursuing their prey underwater. However, alcids are smaller than penguins, and, with the exception of the now-extinct Great Auk, a close relative of the Razorbill, they can fly, despite their relatively short wings.

Razorbills are migrant breeders in Québec, nesting in small island colonies scattered over the estuary and Gulf of St. Lawrence. The stocky Razorbill has dense plumage, the deep black on its back contrasting with its immaculate white breast. It has a thick, laterally compressed bill with a white line across it and a bright yellow interior. In breeding plumage, the Razorbill also has a white line from the eye to the base of the upper mandible, which gives it a distinctive look.

Habitat and behaviour

In mid-April, Razorbills arrive on their breeding grounds in the estuary, where they nest on cliffs and rocky shorelines of uninhabited islands. The species may be seen in the St. Lawrence as early as the end of March, although the island colonies on the Middle and Lower North Shore are not occupied until late April or early May. At this time, new pairs form and those from previous years are reunited (Bédard 1969b). The birds, which generally mate for life (Lloyd 1979), set about preparing their nest site, to which they return year after year (Harris and Birkhead 1985a). The nest is a rudimentary, sometimes almost non-existent structure, made of pebbles or bits of plant debris gathered nearby. It is usually placed on a thin layer of gravel in a deep rock crevice or in boulder scree overlooking the sea. Colonies generally contain only a small number of breeding pairs which share a communal social area, or club (Bédard 1969b). Razorbills often breed in mixed colonies with other alcids, such as murres and Atlantic Puffins. However, Razorbills are less tolerant than murres of contiguous nesting and do not live, as the latter do, in dense colonies. The white droppings on rocks around Razorbill colonies make nests quite easy to locate.

Notes on territory

From observations of banded birds, it has been determined that Razorbills are extremely loyal to their colonies and nesting territories (Bédard 1969b). The territory is

Some Razorbills breed on the rocky cliffs of Le Gros Pèlerin, an island in the St. Lawrence estuary.
Spring 1988, Pierre Brousseau.

Razorbill

Breeding distribution

1984–1989

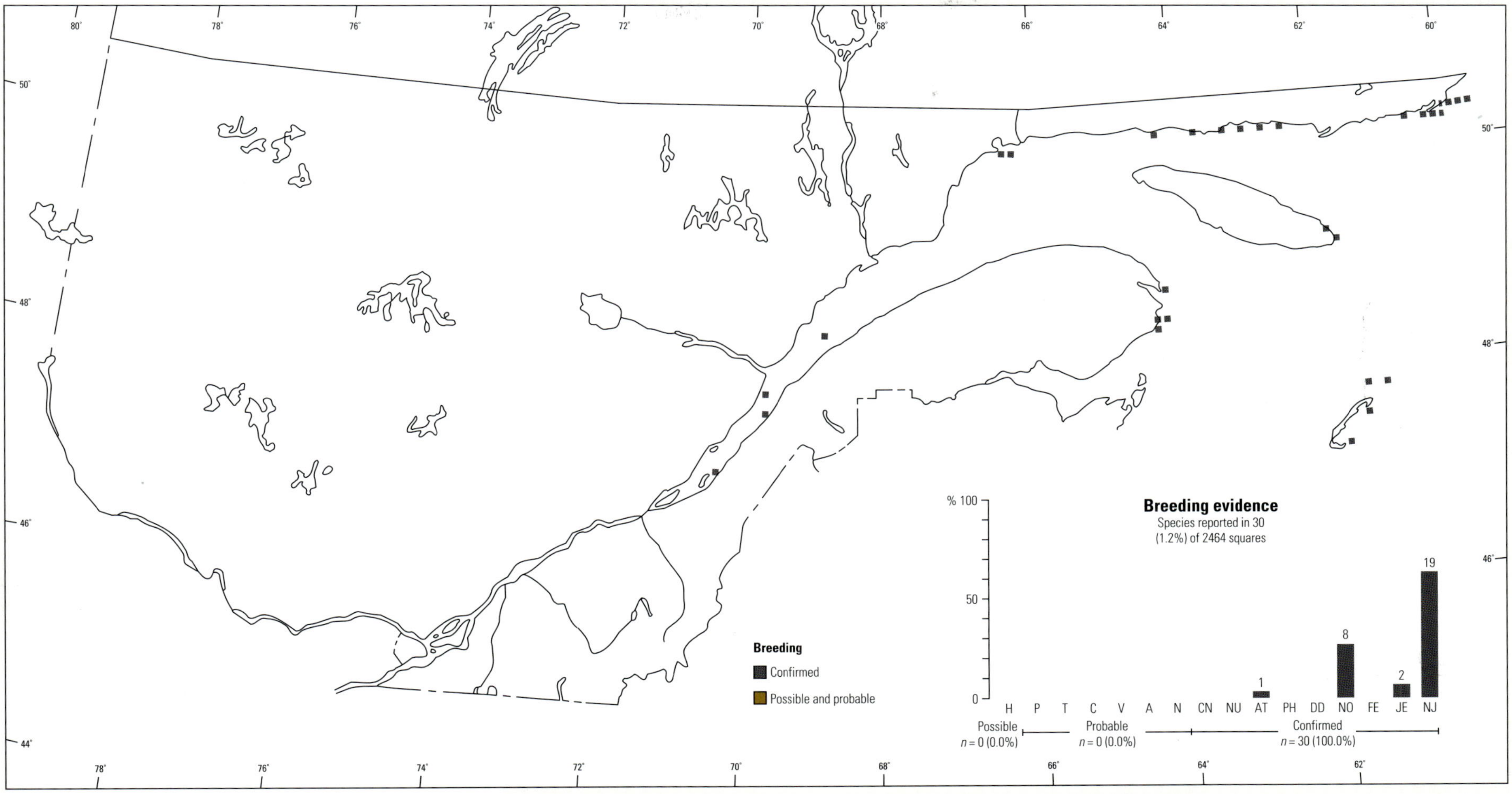

Species profile

Clutch size	1 egg[3,7]
Laying interval	–
Start of incubation	after laying of egg[1]
Incubation period	33–39 days[2]
Maturity at hatching	semiprecocial
Care of young	by female and male[1,9]
Nestling period	usually 18 days[8]
Age at first flight	?
Dependence of young	until several weeks after leaving nest[1]
Broods per year	1[9]
Breeding age	4–5 years[4,9]
Mating system	monogamy[1]
Duration of pair bond	usually for life[1,5]
Total length	40–47 cm[3]
Wing span	63.5–68.6 cm[7]
Weight	728 g (610–870)[8]
Longevity record	20 years[6]

References:
1. Harris and Birkhead 1985a
2. Bédard 1969b
3. Godfrey 1986
4. Croxall and Gaston 1988
5. Lloyd 1979
6. Hudson 1985
7. Terres 1980
8. Chapdelaine and Brousseau (in press)
9. Cramp 1985

Chronology *n* = 106

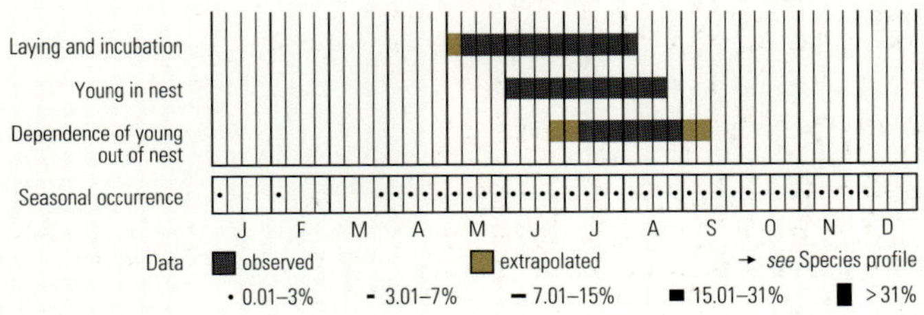

World breeding range

The nest of this alcid consists of a few bits of vegetation placed in a crevice in the rock on a soil or gravel substrate. Corossol Island, Sept Îles Bay. *June 1991, Daniel Bordage.*

On the Middle and Lower North Shore, laying occurs between late May and early June, but in the estuary, is at least 2 weeks earlier. The single, large egg is incubated by both parents for 33–39 days. If the clutch is lost early in the season, another one may be laid. The species has 2 brood patches, which suggests that in the past it produced 2-egg clutches, or that the patches are an adaptation to a particular incubation posture (Harris and Birkhead 1985a). Experienced pairs breed earlier in the season, generally lay larger eggs, and have better breeding success (Lloyd 1979). This shows the crucial role of experience in the co-ordination of breeding activities and selection of good nest sites.

On the Middle and Lower North Shore, Razorbill eggs hatch over a period of about 3 weeks in July (Chapdelaine and Brousseau, in press). The young have a greyish-black back and wings, white underparts, and a distinctive pale grey down covering their heads. The chicks are semiprecocial and are closely brooded by the parents for the first few days after they hatch, since they are not yet able to regulate their own body temperature (Harris and Birkhead 1985a). Hatching weight is about 60 g and increases to nearly 200 g at departure from the colony, which occurs at around 18 days of age. The nestlings are fed fish 3 or 4 times a day, mostly sand lance and capelin. By the time they leave the nest, Razorbill young have lost their grey down and look very much like adults, albeit a miniature version since adults weigh 720 g on average (Chapdelaine and Brousseau, in press).

Chicks leave the colony when they are still flightless. To avoid gull predation, departure usually occurs at dusk, when the male parent coaxes the chick out the nest into the water. The male will remain with the chick at sea for a few weeks (Harris and Birkhead 1985a), while the female continues to frequent the colony for a while to defend the nest site. In August, the Razorbills leave the colonies for their wintering grounds. Average age at first breeding is 4 or 5 years but, in the intervening time, the immature birds return to the colony as prospectors, to familiarize themselves with social life and explore prospective breeding sites. The life span of the Razorbill is about 20 years (Hudson 1985).

Distribution

In Québec, Razorbills breed exclusively on islands in the estuary and Gulf of St. Lawrence. They can also be sighted foraging offshore a few kilometres away from their colonies. In the estuary, a few individuals nest on Le Pilier de Bois, an islet near Saint-Jean-Port-Joli; farther east, over 2000 Razorbills nest on Les Pèlerins and Brandypot Islands off Rivière-du-Loup (BIOMQ). A few birds recently colonized Bicquette Island, but there is little room for colony expansion at this location. In 1991, 2 pairs were seen nesting on Laval

Notes on territory (continued)

a small, clearly defined area, where a breeding pair lays its single egg, incubates it, and raises the chick. Trespassing by a neighbour or intruder is not tolerated, and triggers an immediate reaction by the site holder.

Adjacent to this territory is a communal area—neutral ground frequented by breeders, prospectors, and sometimes even other species (murres). This area is used by both experienced and immature birds, and serves as a crucial information centre on breeding behaviour for the younger birds.

On the Sainte-Marie Islands, many pairs nest in the boulder scree. The field biologist in the photo gives an idea of the size of the rocks. Lower North Shore. *July 1991, Pierre Brousseau.*

Adult Razorbills loafing at the top of a cliff. Great Bird Rock, Magdalen Islands. *July 1983, Yves Aubry.*

Island, off the North Shore, east of Forestville, but only 1 pair bred successfully there (J. Rodrigue, pers. comm.). The largest colonies are located on islands in the gulf, primarily in the 8 migratory bird sanctuaries on the Middle and Lower North Shore, where over 7000 individuals breed; the largest concentrations are on the Sainte-Marie Islands and on Île aux Perroquets off Lourdes-de-Blanc-Sablon (Chapdelaine and Brousseau 1991), although the latter site is outside the Atlas area. Smaller concentrations can be found off the northeast shore of Anticosti Island, on Bonaventure Island, in Forillon National Park, and in the Magdalen Islands (Chapdelaine and Laporte 1982). A few birds have even been spotted in Digges Sound, off the northwest tip of Québec, although breeding has not been confirmed there (Gaston and Malone 1980; Gaston et al. 1985).

Razorbills winter along the New England coast, particularly off Nantucket Island, in Massachusetts, according to band recoveries (Bédard 1969b; CWS, Québec Region, unpubl. data) and sightings by birders (Nikula 1991). However, some of the gulf birds apparently spend the fall and early winter off Newfoundland (CWS, Québec Region, unpubl. data).

Two adults, one with its bill full of sand lance. Like Atlantic Puffins, Razorbills can carry more than one fish at a time, in contrast with murres, which can handle only one. Sainte-Marie Islands, Lower North Shore. *Summer 1991, Claude Nadeau.*

History and trends

As is the case with the other alcids, the best information on trends in Razorbill populations in Québec comes from the censuses carried out regularly in the Middle and Lower North Shore sanctuaries. Analysis of the census data reveals that, between 1925 and 1965, the population in the sanctuaries hovered around 10 000 individuals and even rose above 15 000 in 1955 (see figure). After 1965, however, substantial decreases were recorded, particularly in the Baie des Loups Migratory Bird Sanctuary, which once was home to over 10 000 Razorbills (Moisan and Fyfe 1967), but now supports only about 200 (Chapdelaine and Brousseau 1991). This decline is probably due to poaching, although it is not the only factor. A series of oil spills along the Atlantic coast in the 1970s, which killed a number of Razorbills, may also have played a role (Powers and Rumage 1978). Furthermore, it is becoming increasingly evident that Razorbills are being shot accidentally during the traditional winter murre hunt in Newfoundland, which is legal (CWS, Québec Region, unpubl. data).

Razorbill populations in the Middle and Lower North Shore sanctuaries, 1925–93 (BIOMQ)

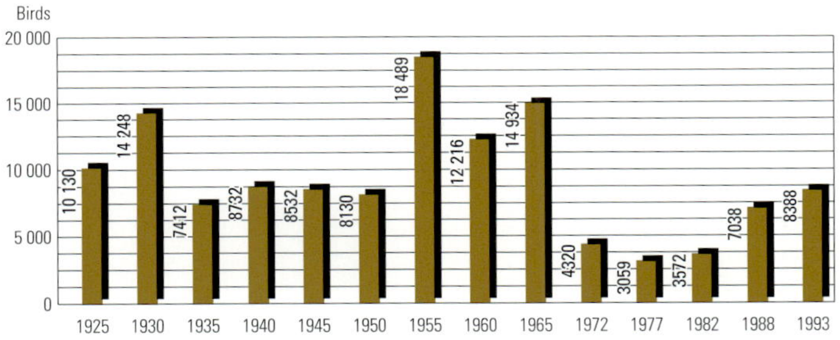

Due to limited historical data, however, it is impossible to determine whether a similar decline occurred in the gulf outside the sanctuaries during the same period. Since 1982, population increases have been observed in almost all the sanctuaries on the Middle and Lower North Shore except Baie des Loups (Chapdelaine and Brousseau 1991). This recent upturn, common to all the alcids, has also been noted on Bonaventure Island (Chapdelaine and Brousseau 1992b).

GILLES CHAPDELAINE AND JEAN-FRANÇOIS RAIL

Black Guillemot

Guillemot à miroir
Cepphus grylle

Louis Ruelland

THE BLACK GUILLEMOT IS THE MOST WIDELY DISTRIBUTED DIVING SEABIRD in southern Québec, but because it nests in small, scattered colonies, it is not as well known as gannets, puffins, and eiders, which occur in more spectacular concentrations. In summer, the guillemot's jet-black dress with white wing patches is unmistakable; only the much larger White-winged Scoter might mislead an inexperienced observer.

On land, guillemots often stand upright, their vermilion feet contrasting with their domino plumage. Their small size has earned them the nickname of "pigeon" among fishermen, but their appearance also invites anthropomorphism; guillemot is a French word meaning "Little William" (Terres 1980; Donovan and Ouellet 1993).

Habitat and behaviour

Seabirds may be divided (somewhat arbitrarily) into inshore species, which breed in many widely dispersed small colonies and produce several chicks, and offshore species, which breed in a few very large colonies and lay 1-egg clutches. Like gulls, terns, and cormorants, Black Guillemots are inshore seabirds, breeding in numerous small colonies and sometimes in solitary pairs. Guillemots nest in rocky cavities to protect themselves and their young from predation. On the mainland, they generally breed on cliffs inaccessible to mammals, but talus slopes and even artificial breakwaters are also used. On predator-free islands, Black Guillemots nest in cavities in cliffs, bedrock, or tumbled boulders that provide good drainage and sufficient shelter (Cairns 1980).

Black Guillemots are frequently seen near their breeding sites by April or early May (Cairns 1981). Sea ice does not discourage early colony attendance; the species is known as an ice-edge specialist in the arctic part of its range. During the pre-breeding period, Black Guillemots spend more and more time on shore, engaging in courtship and social displays which involve head-bobbing and raising of the wings. Fighting and copulation are also frequent.

Some Black Guillemots begin breeding at about 2 years of age, though reproduction probably starts later for most birds (Hudson 1985). Like most seabirds, Black Guillemots form long-term pair bonds, and a pair will return to the same nesting site year after year. However, some pairs may divorce, especially if reproduction fails. Both sexes contribute to the rearing of the young. Occasionally 1, but usually 2, eggs are laid on a bed of pebbles deep in a cavity, often beyond the view and reach of human observers. Laying generally occurs in May in the St. Lawrence estuary, but may be a month later on the Lower North Shore (Cairns 1981).

Both parents incubate the eggs, developing a pair of featherless brood patches on the breast for the task, which takes about 4 weeks. The newly hatched chicks are covered in dense black down. In the estuary, the nestlings' diet includes capelin, Atlantic tomcod, sand lance, and various blennies and sculpins. Principal foods on the North Shore are tomcod, sand lance, and blennies, particularly the Arctic shanny (Cairns 1981). Parents bring fish to their young crosswise in the bill, one at a time. After 5 weeks, the chicks leave the nest, usually at dusk, and make their way to the water unaccompanied by a parent, to begin an independent life. At this time, the young are close to adult size and have a mottled grey plumage resembling that of winter adults. They are flightless when they leave the nest, but probably learn to fly soon after.

In general, adults eat a mix of fish and invertebrates, in contrast to the fish-only fare they feed to their young (Bradstreet and Brown 1985). This prey spectrum suggests a benthic feeding habit, but some prey are likely taken in the water column. When ice is present, Black Guillemots congregate at the edges and feed on the abundant fauna on the under-ice surface (Bradstreet 1980; Cairns 1987). In the estuary, wintering Black Guillemots have been found to take krill shrimps, arrow worms (phylum Chaetognatha), crabs, amphipods, and small fish (Savard 1990). Diving depths range to about 40 m (Piatt and Nettleship 1985; Cairns 1987).

Notes on territory

Black Guillemots defend an area of several dozen centimetres around the nest, where almost all courtship and breeding activities are carried out. Conspecifics are chased out of this area and pairs will continue to defend it throughout the breeding season, even when the clutch has been lost or no eggs have been laid, though non-breeders are less aggressive than guillemots with eggs or young (Cramp 1985).

On the mainland, Black Guillemots prefer crannies and crevices in steep cliffs that mammals cannot reach. After foraging at sea, the birds have to fly over this road to return to their nests, sometimes startling passing motorists. Gros-Morne area, Gaspé. *June 1984, Pierre Brousseau.*

Black Guillemot

Breeding distribution

1984–1989

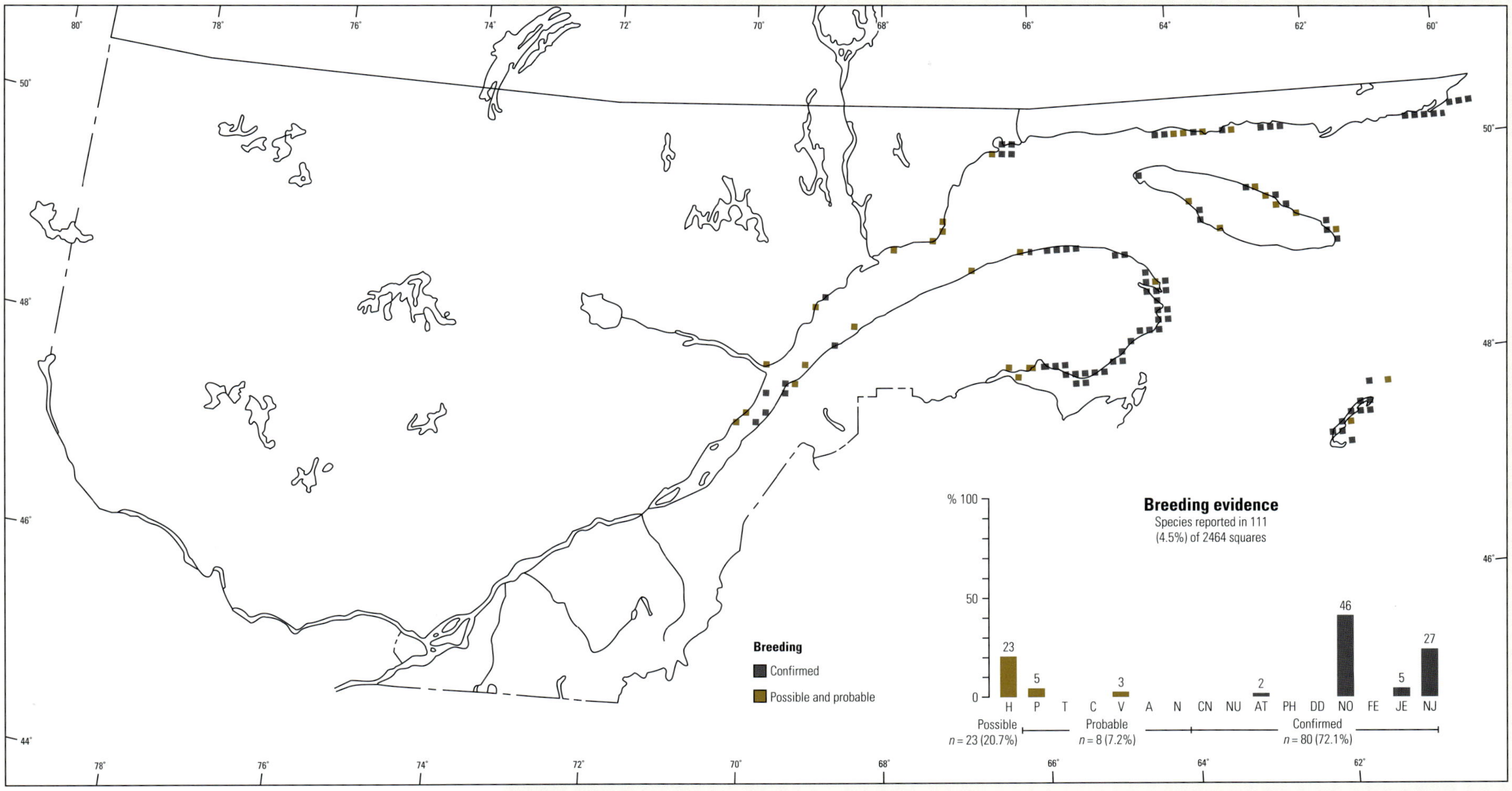

Species profile

Clutch size	2 eggs, sometimes 1[2]
Laying interval	1 egg per 3 days[2]
Start of incubation	with second egg or, when only 1 egg has been laid, up to 4–5 days later[2]
Incubation period	23–29 days[2]
Maturity at hatching	semiprecocial
Care of young	by female and male[2]
Nestling period	34–39 days[3,6]
Age at first flight	34–39 days[3,6]
Dependence of young	until first flight[2]
Broods per year	1[2]
Breeding age	usually 4 years (2–4)[5]
Mating system	monogamy[5]
Duration of pair bond	usually for life[2]
Total length	30.5–35.5 cm[4]
Wing span	52–58 cm[5]
Weight	428 g (?)[1]
Longevity record	13 years 11 months[7]

References:
1. Dunning 1984
2. Harris and Birkhead 1985a
3. Croxall and Gaston 1988
4. Godfrey 1986
5. Cramp 1985
6. Cairns 1978b
7. Clapp 1976

Chronology $n = 183$

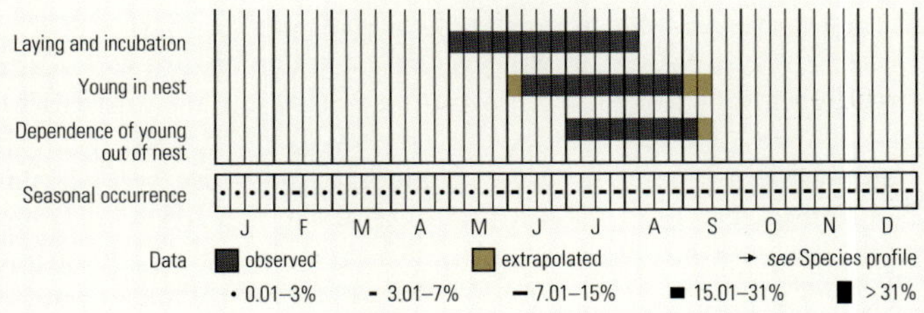

World breeding range

Guillemots nest on the rocky shores of the estuary and Gulf of St. Lawrence; in the Gaspé, the heavily layered and fractured cliffs of the Forillon Peninsula provide excellent breeding habitat. Cape Gaspé, Forillon National Park. *July 1992, Yves Aubry.*

This alcid nests in small colonies or isolated pairs, on either cliff faces or talus slopes. Manowin Island, Sept Îles Bay. *August 1993, Yves Aubry.*

Black Guillemots have a marked daily cycle. Birds are often abundant and conspicuous at the colony in the morning, but are on the fishing grounds in the afternoon and hence absent from view (Cairns 1979). Away from the colony, Black Guillemots frequent inshore waters less than 40 m deep, foraging singly or in loose groups. Birds on the water often dip their heads, apparently checking for underwater predators, or for prey or habitat in which it might be found.

Distribution

Black Guillemots have a circumpolar distribution centred in arctic waters, but they also breed along the rocky shorelines of the western Atlantic south to New England (Brown et al. 1975). In Québec, guillemots breed in large numbers in Hudson Strait and in lesser numbers in Hudson Bay. During counts taken in June and August of 1991, several hundred birds were seen on the east coast (Québec side) of Hudson Bay and on the offshore islands (Northwest Territories), mainly in and around the Nastapoka Islands and Long Island (Hydro-Québec 1993). In southern Québec, Black Guillemots are widespread breeders along the rocky shorelines of the estuary and Gulf of St. Lawrence. They nest in the estuary as far upstream as the Kamouraska Islands. The largest guillemot population in the estuary is found on Les Pèlerins and Brandypot Islands, each with several hundred breeding pairs. Smaller numbers also breed along both shores of the mainland, especially where precipitous cliffs or boulder heaps provide safe breeding habitat.

In the Gaspé, a few hundred birds at most nest in the great seabird colonies of Percé Rock and Bonaventure Island, scattered among the throngs of gannets, kittiwakes, and murres. The cliff-girt Gaspé coast, however, has innumerable guillemot-size crannies and crevices and, there, Black Guillemots make up in ubiquitous distribution what they lack in spectacular concentration. Particularly favourable breeding habitat is found around Grande-Vallée and on the Forillon Peninsula.

Guillemots breed on rocky cliffs and offshore islets throughout the Magdalen Islands, and a few birds can be found in the Great Bird Rock gannet colony. Black Guillemots are widespread on Anticosti Island, but because of the large fox population, breeding is generally limited to safe cliff habitat. The North Shore provides favourable habitat for guillemots, with its thousands of rocky islands washed by cold, productive waters; the largest concentration (over 100 breeding pairs) can be found on the Sainte-Marie Islands.

The species' winter distribution is poorly known, and it is unclear whether most southern Québec birds winter among the ice floes of the estuary and gulf or leave the area to winter in the open waters of the Atlantic. Black Guillemots may be found in the estuary throughout the winter (Savard 1990), but the colony affiliation of these birds is unknown. Black Guillemots are well adapted to sea ice, and some even winter in the High Arctic with its long polar night (Renaud and Bradstreet 1980).

An adult resting on a sandstone cliff ledge. Île Brion Ecological Reserve, Magdalen Islands. *July 1982, Yves Aubry.*

History and trends

Black Guillemots are notoriously difficult to census accurately. To find all nests in a given area requires long and arduous searching; hence, most population estimates are based on counts of birds during their early-morning assemblies at the colony. Recent estimates of guillemot populations in the Québec portion of the gulf and the estuary point to about 4400 pairs (Cairns et al. 1991). Since 1925, the Canadian Wildlife Service has surveyed seabird sanctuaries on the North Shore every 5 years. These surveys show that populations peaked at 700–800 individuals between 1940 and 1965, and then declined sharply (see figure). A survey in the Gaspé conducted in 1979 and repeated in 1989 showed an increase from 2000 to 3100 pairs (Chapdelaine and Brousseau 1992b). However, it is uncertain whether these changes are real or merely reflect the errors inherent in census methodology.

Because of legal protection and their small size, Black Guillemots are usually not a target of hunters, though some are occasionally mistaken for "small ducks" and shot. On the North Shore, nesting guillemots were a traditional delicacy, and children were often dispatched to search for "young pigeons" in the rocky crevices of offshore islands while their parents picked berries. A common subterfuge was to conceal the illegal booty in a bucket of cloudberries for transport home.

Black Guillemot populations in the Middle and Lower North Shore sanctuaries, 1925–93 (BIOMQ)

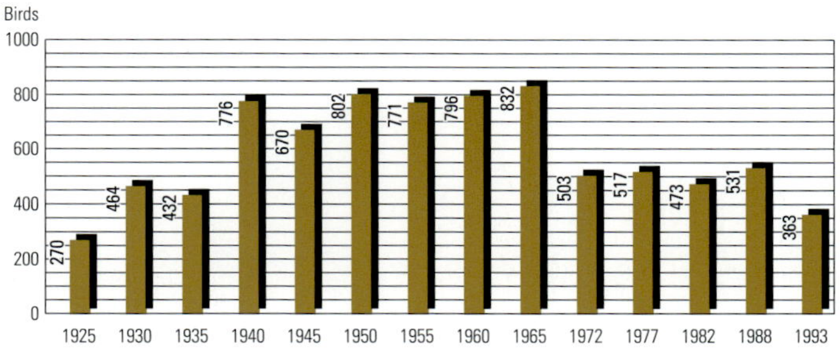

Black Guillemots are vulnerable to oil spills and chemical pollution of the seas, which now pose a greater hazard to seabirds than direct exploitation. Because they breed in widely dispersed colonies, Black Guillemots are perhaps at less risk than seabirds breeding in immense colonies, because local spills will affect only a small portion of the population. For many species of seabirds, competition with commercial fisheries may pose a threat, but this is of little concern in the case of the Black Guillemot, whose benthic prey are generally not commercially harvested.

David Cairns

Atlantic Puffin

Macareux moine

Fratercula arctica

Louis Ruelland

Of all the alcids found in the Gulf of St. Lawrence, the Atlantic Puffin would certainly be the most popular choice as a symbol for the protection of marine ecosystems.

Similar in size to the Black Guillemot, the Atlantic Puffin has a black back and white breast. It is its distinctive head that makes it so appealing, particularly its sad, expressive eyes set off by its white cheeks. The puffin's large, triangular bill—grooved and laterally compressed and gaily coloured in hues of reddish-orange, yellow, and blue—has earned it the nickname "sea parrot." In spring, the birds return to their colonies to breed, which in Québec are mainly located along the north side of the gulf. The largest concentrations in North America are found along the coasts of Newfoundland and Labrador.

Habitat and behaviour

Puffins begin to arrive at their southernmost colonies in the gulf, at Brion Island, Great Bird Rock, and Bonaventure Island, in mid-April, but do not reach the Lower North Shore colonies until early May. The Atlantic Puffin nests on offshore islands, laying its single egg either in a burrow in an earthen slope, which it digs with its feet and bill, or a crevice or cranny in a talus slope or moraine deposit. Age at first breeding is around 5 years, but sometimes occurs as early as 3. Both the male and female share incubation, which lasts 39–49 days (Harris 1984). The nestling period is 38–44 days, which is relatively long compared to that observed in Razorbills and Common Murres, at 16–23 days (Harris 1984). When the chick leaves the colony, it weighs about 250–350 g, or about 60–80% of its adult weight, is fully fledged, and is completely independent of its parents (Harris and Birkhead 1985a). Dependent chicks are fed mainly capelin and sand lance, but their diet varies considerably depending on the colony location, point in the breeding season, and the year (Harris 1984). For example, in the Lower North Shore colonies, like those on the Sainte-Marie Islands, American sand lance appears to be the main food of young nestlings, but as the season progresses, capelin replaces it as the major prey species (G. Chapdelaine, unpubl. data).

Breeding success is highest in large colonies. According to Harris (1984), at low breeding densities, puffins are more vulnerable to predation by Great Black-backed Gulls, which will even attack adults. This phenomenon has also been observed in the colonies on the Sainte-Marie Islands. In addition, adult puffins bringing fish to their young are often harassed by Herring and Great Black-backed gulls trying to steal their catch, a behaviour known as kleptoparasitism (Nettleship 1972).

Locating nests on steep talus slopes takes effort, but the contents of these nests (an egg or a nestling) are sometimes immediately visible, unlike those of burrow nests. During atlassing, the contents of most nests were identified. The main problem in observing puffins and confirming breeding is the inaccessibility of the offshore islands where they nest.

Distribution

The largest Atlantic Puffin colonies in North America are in Newfoundland and Labrador, at Witless Bay (225 000 pairs), and the Gannet Islands (41 000 pairs) (Nettleship and Evans 1985).

In Québec, puffins can be found in a number of seabird colonies in the gulf, particularly on the Lower North Shore. In the southern gulf, puffins nest on Great Bird Rock (100 individuals, Fradette 1992), Brion Island (377 individuals), and Bonaventure Island. On Bonaventure, a few pairs breed on the cliffs amid thousands of Northern Gannets and Common Murres (BIOMQ) and the best time to see puffins there is in late July and in August, when the adults are going back and forth with food for the young. A few pairs nest on Anticosti Island, probably in similar numbers as on Bonaventure Island.

Notes on territory

In puffin colonies, the burrow is used only to incubate the egg and rear the young. The burrow entrance appears to be defended mainly by the male (Harris and Birkhead 1985a; Cramp 1985). At Île aux Perroquets on the Lower North Shore, burrow densities were 0.07 burrow/m² in 1982 (Chapdelaine and Brousseau 1984). In the British Isles, density has been found to vary within a colony (0.16–0.6 burrow/m²) with densities generally higher near the cliff edge (Cramp 1985).

Steep talus slopes on offshore islands in the Gulf of St. Lawrence provide ideal nesting habitat for puffins. Blacklands Island, Baie des Loups Migratory Bird Sanctuary. *June 1981, Pierre Brousseau.*

Atlantic Puffin

Breeding distribution
1984–1989

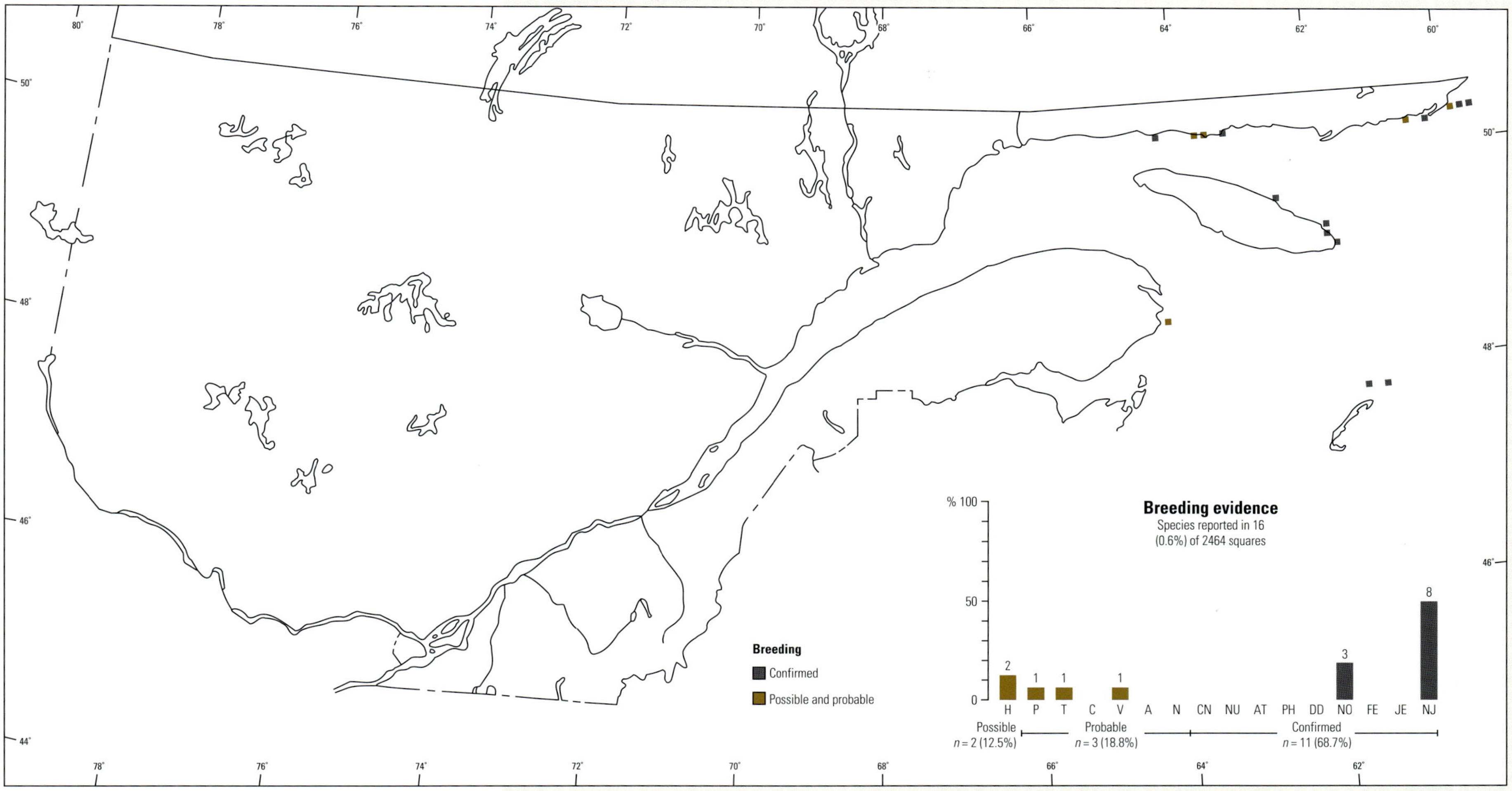

Species profile

Clutch size	1 egg[2]
Laying interval	–
Start of incubation	after laying of egg[2]
Incubation period	39–49 days[1]
Maturity at hatching	semiprecocial
Care of young	by female and male[2]
Nestling period	usually 38–41 days (34–57)[2]
Age at first flight	usually 38–41 days (34–57)[2]
Dependence of young	until first flight[2]
Broods per year	1[2]
Breeding age	usually 5 years (3–6)[1,6]
Mating system	monogamy[2]
Duration of pair bond	usually for life[2]
Total length	29.0–34.5 cm[5]
Wing span	53.3–61.0 cm[4]
Weight	381 g (?)[3]
Longevity record	21 years[4]

References:
1. Harris 1984
2. Harris and Birkhead 1985a
3. Dunning 1984
4. Terres 1980
5. Godfrey 1986
6. Cramp 1985

Chronology *n* = 13

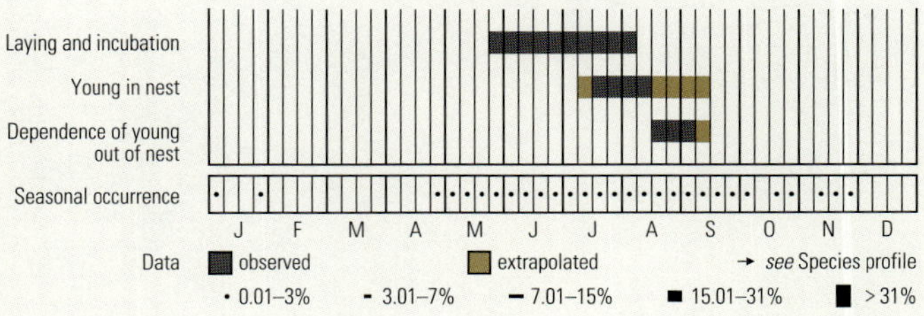

World breeding range

Puffin colonies can be found in moraine deposits on some islands off the Lower North Shore, such as Île des Loups in the Baie des Loups Migratory Bird Sanctuary. The row of whitened rocks, caused by puffin droppings, delineates this colony, where two biologists can be seen checking out burrows. *June 1977, Pierre Dupuis.*

Earthen slopes are also preferred nesting sites. Puffins dig their burrows with their feet and bills. Blacklands Island, Baie des Loups Migratory Bird Sanctuary. *June 1981, Pierre Brousseau.*

Puffins may also dig a burrow and lay their single egg in the soft sandstone under a flat rock. Great Bird Rock. *August 1983, Yves Aubry.*

In addition to serving as nest sites, cliffs are used as lookout points and gathering spots for non-breeders. Brion Island, Magdalen Islands. *July 1982, Yves Aubry.*

Atlantic Puffin populations in the Middle and Lower North Shore sanctuaries, 1925–88 (BIOMQ)

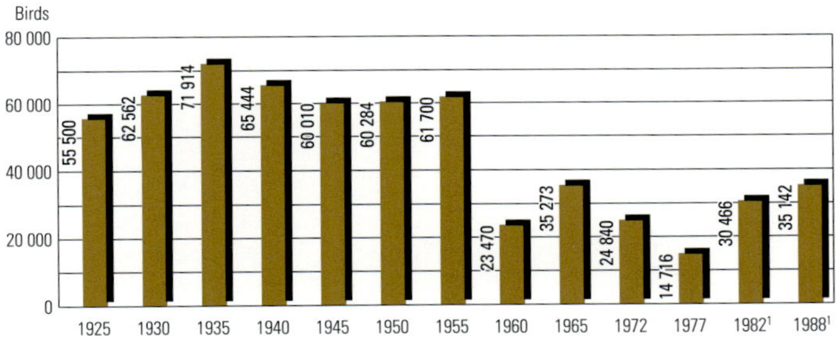

1 Including the Perroquets Islands (Sainte-Marie Islands).

In Québec, the largest puffin colonies are on the Middle and Lower North Shore. Some of the most accessible on the Middle North Shore can be found in the Mingan Islands: the largest colony in 1988 was on Île à Calculot des Betchouanes with 113 pairs (Chapdelaine and Brousseau 1991), whereas in 1993, the largest colony was on Île de la Maison with 353 pairs (Vaudry 1994). The Mingans have now become the best site to see puffins in Québec, thanks to Parks Canada's interpretation program, which has made the islands more accessible. Farther east, on the Lower North Shore, there are some large but more remote colonies: the Baie des Loups colonies, which have over 9000 birds, and the Sainte-Marie Islands colonies (including the Perroquets Islands), which have over 8000 birds (Chapdelaine and Brousseau 1991). Other large colonies in the gulf, though located outside the Atlas area, include those on the two islands in the Baie de Brador Migratory Bird Sanctuary, near Lourdes-de-Blanc-Sablon, which are home to more than 17 000 nesting puffins. Atlantic Puffins also breed in small numbers on at least two islands in Digges Sound, on the northwestern tip of the Ungava Peninsula, and the area is probably home to other breeding pairs (Gaston and Malone 1980; Gaston et al. 1985).

Very little is known about the species' winter distribution, but individuals are believed to disperse widely off the south coast of Newfoundland (Brown 1986).

History and trends

Atlantic Puffin populations were probably larger in the past than they are today, judging from the accounts—most of which are qualitative rather than quantitative—of early European explorers and 19th century naturalists, including the famed John James Audubon. However, the most reliable information comes from the surveys conducted roughly every 5 years since 1925 in the Middle and Lower North Shore seabird sanctuaries. As the figure opposite shows, from 1925 to 1955, the population stood at over 60 000 individuals, but then declined. Like several other seabird species that inhabit the area, the Atlantic Puffin was subject to heavy poaching, and this undoubtedly contributed to the decline at most of the colonies. Fortunately, over the past 15 years, some recovery has been noted, as is the case with several other alcids that breed along the Middle and Lower North Shore. These increases may be attributable to improved conservation measures and to the extensive commercial harvesting of large fish such as cod, which has boosted stocks of smaller prey species like sand lance and capelin, the favourite food of alcids. Elsewhere in the gulf, population increases have also been noted, although survey records are less exhaustive.

Gilles Chapdelaine

Rock Dove

Pigeon biset
Columba livia

A COMPANION OF MANKIND FOR THOUSANDS OF YEARS, THE ROCK DOVE IS a familiar feature of both urban and rural landscapes, and was probably the first bird to be domesticated. This Old World species has been introduced around the world, on every continent except Antarctica (Long 1981) and is sedentary both within its native range and wherever it has been introduced.

Habitat and behaviour

Rock Doves, commonly known as domestic pigeons, are gregarious birds, and feed and nest in the company of others of their own species, although the size and composition of feeding flocks may vary seasonally (Lefebvre 1985; Lévesque and McNeil 1985). These seed eaters feed on the ground in open areas, in farm fields, near barnyards and feeding troughs, in city parks, and in grain-handling and transport facilities such as ports, railway yards, grain elevators, and flour mills. Potvin et al. (1976) showed that in agricultural areas pigeons feed exclusively on plant matter (grains and weed seeds). In urban areas, however, bread and more exotic items such as popcorn and French fries given as handouts or thrown away often make up a large part of their diet.

Despite their sedentary nature, they have a highly developed sense of direction (Keeton 1977) that allows them to make spontaneous trips of varying lengths between their nests or roosts and their feeding grounds, depending on local conditions (Lévesque and McNeil 1986). People have taken advantage of this direction-finding ability for centuries and have bred a domestic race of pigeons that will travel long distances. Homing pigeons used to play a particularly important role in wartime, when they were used as messengers. Pigeon racing is still practised today, and sometimes domestic birds that have escaped or become temporarily disoriented can be found in flocks of feral pigeons (Goodwin 1977).

Rock Doves seem to use the sun to orient themselves, and also have an internal time sense to correct their courses to compensate for the movement of the sun across the sky (Matthews 1968; Wiltschko and Wiltschko 1990). On overcast days, they apparently use the earth's magnetic field to navigate (Keeton 1969). When travelling short distances, in the city, for example, they rely on landmarks (Chelazzi and Pineschi 1974). They may also depend on olfactory cues (Papi 1989).

The onset of courtship behaviour, including insistent cooing, can be observed at feeding areas, the species' favourite gathering place. Mating behaviour continues on the nesting grounds, which may be located some distance away (Goodwin 1977).

Jean-Marc Giroux

Rock Doves can usually be found nesting in groups of several pairs on ledges or roofs, and inside buildings, as well as under bridges, elevated expressways, and overpasses. They also nest in crevices in cliffs, particularly in open quarries. Any site has to provide shelter for the nest, which is a loose pile of sticks and twigs.

Both parents brood and feed the young. For the first week, the chicks are fed nothing but pigeon's milk, a substance produced by the walls of the parents' crops. Then, when the chicks are about a week old, regurgitated seeds are gradually introduced until they completely replace the milk (Johnston 1992). With time, the regurgitated material becomes less and less predigested (Goodwin 1977). The production of pigeon's milk protects squabs against reductions in the parents' food supply, thus ensuring an adequate diet for the squabs and ultimately their survival (Mondloch and Timberlake 1991).

Notes on territory

The Rock Dove is not a particularly territorial species and generally defends only the immediate vicinity of the nest against intruders, but sometimes a bird nesting on the edge of a porch roof will extend its territory to the adjacent roof area (Stokes 1979).

Johnston (1992) reported that the area defended depends on the configuration of the nest site. When the nest is on a flat surface, the pigeon's territory can be a square metre or even more.

Both parents defend the nest. Although Rock Doves usually ignore House Sparrows and European Starlings, they may become aggressive when an American Crow approaches the nest (Stokes 1979).

The Rock Dove, which is accustomed to being around people, nests inside farm buildings and feeds on spilled grain. Fraser Point, Lake Saint-François, southwest of Montréal. *May 1990, Léo-Guy de Repentigny.*

Rock Dove

Breeding distribution
1984–1989

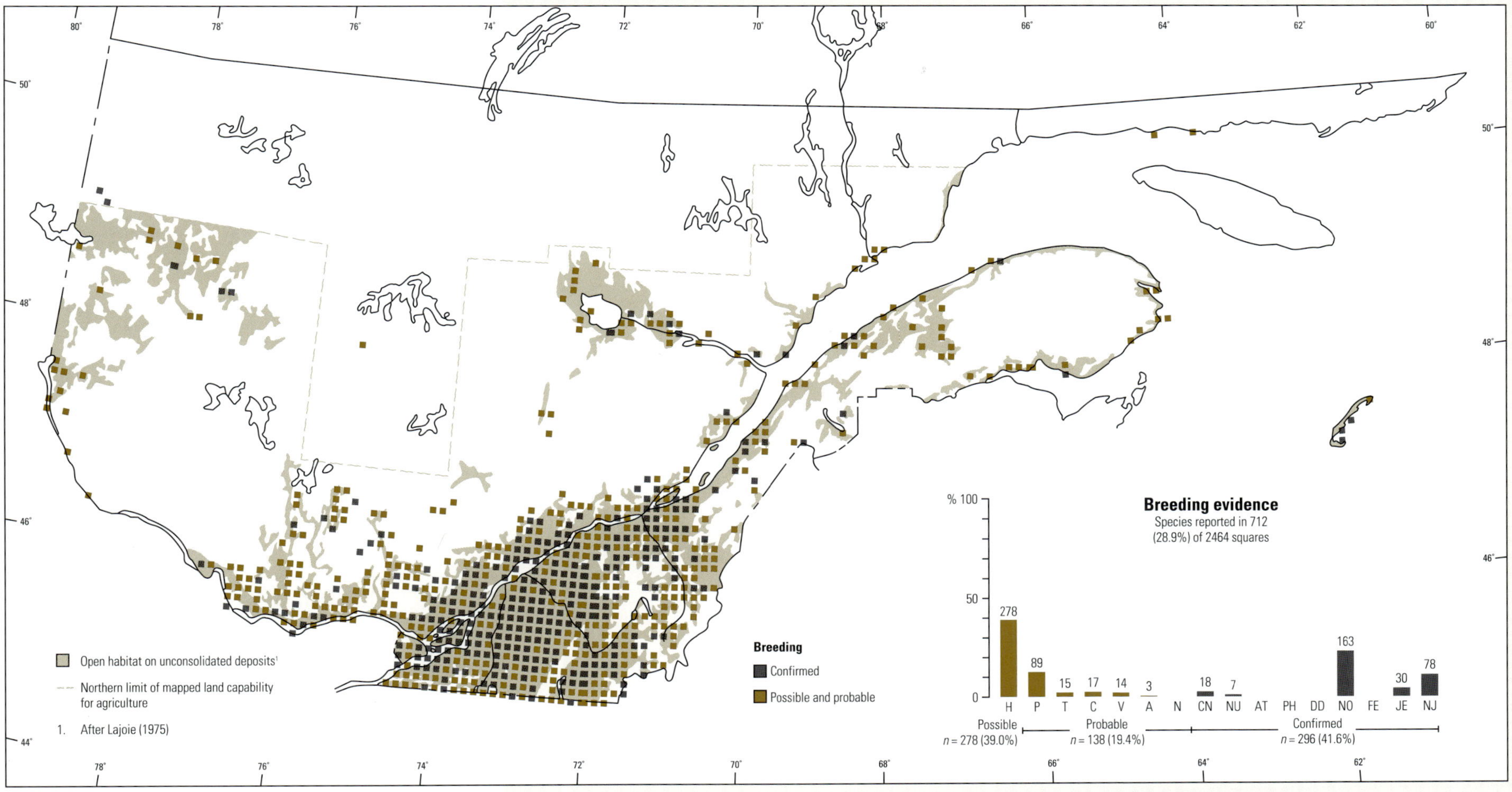

Species profile

Clutch size	usually 2 eggs (1–2)[1,2,3]
Laying interval	≈ 1 egg per 2 days[3]
Start of incubation	usually with second egg[3]
Incubation period	usually ≈ 18.5 days (16–19)[1,2,3]
Maturity at hatching	altricial
Care of young	by female and male[2,3]
Nestling period	summer 25–32 days, winter up to 45 days[3]
Age at first flight	30–32 days[3]
Dependence of young	until 30–35 days old[1]
Broods per year	usually ≈ 3, sometimes more[3,5,7]
Breeding age	6 months–1 year[2,3]
Mating system	monogamy[3]
Duration of pair bond	for life[3]
Total length	28–34 cm[5]
Wing span	63–70 cm[6]
Weight	542 g (494–616)[4]
Longevity record	6 years 2 months[3]

References:
1. Harrison 1978
2. Ehrlich et al. 1988
3. Johnston 1992
4. Dunning 1984
5. Godfrey 1986
6. Cramp 1985
7. Erskine 1976b

Chronology $n = 37$

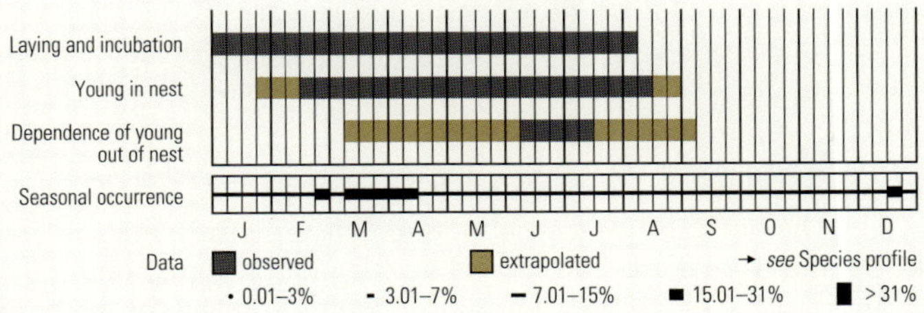

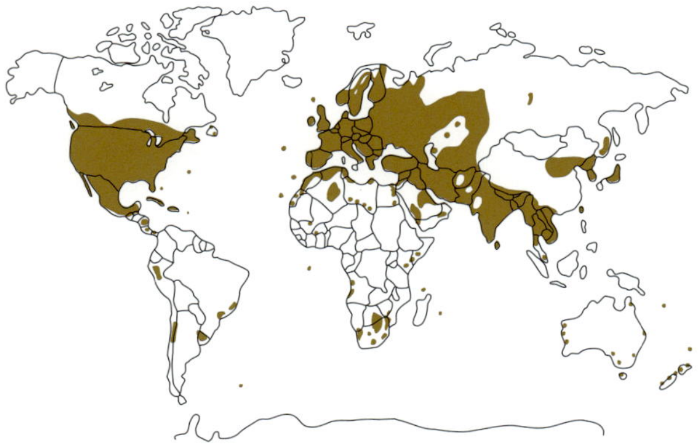

World breeding range

Several broods, three or sometimes more in Canadian cities, are produced in a breeding season (Erskine 1976b). Often the pair copulates and the male begins building a new nest before the first brood has even left the nest. The young receive parental care for only a few days after they have fledged, as the parents will soon attempt to nest again (Stokes 1979).

Although pigeons nest most actively from May to August, pairs often breed at other times of the year. In a study on Columbidae reproduction in Great Britain, Loft et al. (1966) showed that when enough food was available, approximately half the Rock Dove population studied maintained sufficient gonadal activity to breed in midwinter. Similarly, in the Port of Montréal, where the birds have access to an excellent food supply all year long in the form of various whole grains, nestlings have been observed even in January, when nighttime temperatures drop to –30°C (Lévesque and McNeil 1985). On 7 January 1993, a nest containing 2 eggs was found on Île Notre-Dame, and another containing 2 squabs approximately 10 days old was found on 18 February (Dionne and St-Georges 1993). In Alberta, McGillivray (1988) also observed a pair of Rock Doves nesting in January.

Breeding was confirmed in 41.6% (296/712) of Atlas squares in which the species was reported, mainly by records of adults near a nest.

Large concentrations of this seed-eating member of the family Columbidae are found in ports where grain is stored and transferred, like the Port of Québec at the mouth of the Saint-Charles River. *October 1993, Pierre Normand.*

Distribution

The Rock Dove's distribution in Québec corresponds to areas of human habitation. The species breeds in all regions south of the 50th parallel (David 1980a), but is most common in the southernmost part of the province, which has the most intensive urban and agricultural development, in the Montérégie, Eastern Townships, Ottawa River valley, and along the north side of the St. Lawrence down to Québec City.

Farther north, the species is found locally in the developed areas along the Gatineau and Saguenay rivers, in the Chaudière–Appalaches, Lower St. Lawrence, and Gaspé regions, and on the North Shore. The Rock Dove has also become established in remote areas like the Abitibi region and the Magdalen Islands, but is not found on Anticosti Island. The northernmost records come from Radisson and Eastmain, where single birds were seen in 1984 (Bordage 1985b).

Because of their close association with human activity and their popularity with the general public, Rock Doves rarely go unnoticed, so the Atlas map is probably an accurate reflection of the species' distribution in Québec.

History and trends

It was Marc Lescarbot, a French lawyer and author and a leading figure in the early Acadian colony, who probably first introduced the species into North America. The first release occurred in Port Royal, Nova Scotia, in 1606, and was apparently followed by several more. An illustration in the *Works of Samuel de Champlain* shows the habitation with a dovecote built in 1608, and the *Jesuit Relations* of 1646–47 mentioned a gift of 8 squabs and 6 pigeons from the

Highway infrastructures provide excellent nest sites for the Rock Dove, which can often be found nesting underneath overpasses. Québec City. *July 1994, Jean Gauthier.*

The abundance of food is at least partly responsible for the presence of the Rock Dove in towns and cities. Pigeons that are fed regularly become quite tame. Québec City. *Autumn 1968, Pierre Normand.*

Rock Dove incubating its eggs in a nest of dead leaves and twigs. Saint-Hyacinthe. *May 1985, Paul Perreault.*

governor of New France. Schorger (1952) also provides an account of the introduction of pigeons into New England in the 17th century.

The first pigeons introduced by Europeans were raised for the table (Saunders 1935). They fed in the farmyard and nested in pigeon lofts. The squabs were taken from the nest just before they fledged. As Lescarbot commented, when the birds were first introduced, the settlers had to protect them from predation, particularly by raptors, but conditions became more favourable for the species with the clearing of the forests and expanding agricultural development.

With time, pigeons that escaped from the lofts bred on ledges and other outside parts of buildings, as well as in attics, sheds, and abandoned barns. They foraged in the fields and near feeding troughs and areas where grain was handled, gradually becoming a significant semidomesticated population.

Urban development helped Rock Dove populations grow even more (Ouellet 1974a; Erskine 1980b), as tall buildings provided an open environment similar to that of cliffs (Woldow 1972). The relatively few predators in urban areas and the absence of competing native species gave pigeons, which were used to living alongside people, new opportunities to expand. They were able to exploit the rich food resources associated with grain handling and the food industry, while benefiting from the benevolence of friendly humans.

In fact, seemingly sudden jumps in Rock Dove populations generally occur when a new source of food becomes available locally. Since the species is sedentary, the possibility of further expansion in Québec is probably limited by the greater scarcity of food in winter (Lévesque and McNeil 1985).

Recent BBS data show that the Rock Dove population increased somewhat between 1966 and 1989.

Hélène Lévesque

Mourning Dove

Tourterelle triste

Zenaida macroura

THE MOURNING DOVE'S PLAINTIVE SONG, PERFORMED FROM A CONSPICUOUS perch, is a familiar sound in many parts of Québec from early spring to autumn.

Fairly abundant today in many regions of southern Québec, the Mourning Dove, like several other species, only began to breed in the province in the 20th century. The capture of a juvenile at Lanoraie in August 1922 appears to have provided the first authenticated record of breeding in the province. A pair of supposed Passenger Pigeons that nested in Oka in 1913 were most likely Mourning Doves (Mousley 1945b; Ouellet 1974a).

The migratory Mourning Dove usually returns to Québec to breed in mid-March, although part of the population remains in the province year-round (David 1980a).

Habitat and behaviour

Mourning Doves occupy many different habitats, in both urban and rural areas. They nest in open woods, groves, coniferous and deciduous woodlots, plantations, orchards, roadsides, and suburban gardens (Harrison 1975; Godfrey 1986).

With the coming of warm weather in spring, males can often be seen courting females at feeding stations as Rock Doves do. Both parents are involved in constructing the nest, the male searching for twigs, grasses, sticks, and conifer needles near the nest site and the female doing the actual building. The nest, a flimsy platform, is generally built 1–10 m up in the vertical crotch of a tree or out on a horizontal branch. It is often too fragile to withstand summer storms, so eggs and nestlings may be lost (Caron et al. 1986).

Egg laying begins once the nest has been built; typical clutch size is 2 eggs. The parents take turns incubating, the male from morning to evening and the female at night (Bent 1932; Harrison 1978). The species nests early and thus has a fairly long breeding season, allowing it to produce at least two broods a year.

Both parents feed the young, regurgitating pigeon's milk, a creamy white liquid produced in the crop, into their open bills. This is the only food the chicks receive for their first 4 days, after which regurgitated seeds are gradually introduced into their diet. By the age of 10 days, the chicks are fed almost exclusively vegetable matter (Bent 1932; Caron et al. 1986; Blockstein 1989), which is why adults are never seen carrying food to the nest.

Since Mourning Doves prefer open areas and often choose conspicuous perches, they are easy to spot, as are their nests. Breeding was confirmed in 24% (192/792) of the squares in which the species was reported.

Jean-Marc Giroux

Notes on territory

After copulation and during nest building and egg laying, the male defends a territory of slightly less than 1 ha by cooing from prominent perches in the territory.

The Mourning Dove is at home not just in farming areas, but in suburbs, particularly where there are feeders and a few large trees. Hull. *September 1991, Gilles Falardeau.*

Mourning Dove

Breeding distribution

1984–1989

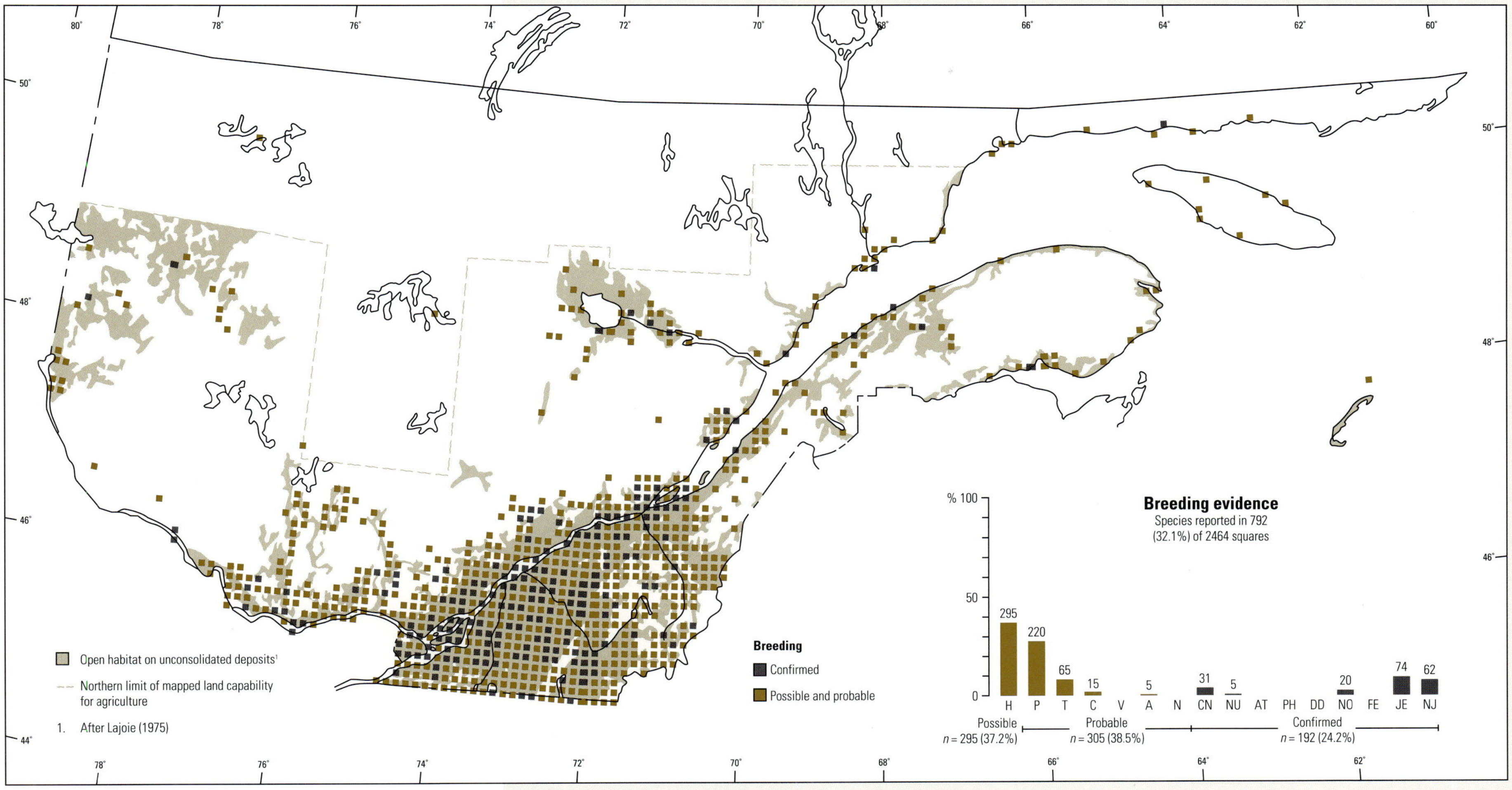

Species profile

Clutch size	2 eggs, 3–4 = dump nest[10]
Laying interval	1 egg per day[1,2]
Start of incubation	with first or second egg[10]
Incubation period	13–15 days[1,2,3,4]
Maturity at hatching	altricial
Care of young	by female and male[1,2,3,4]
Nestling period	10–15 days[4,6,7,8]
Age at first flight	10–15 days or more[12]
Dependence of young	until 15–30 days old[6,10]
Broods per year	usually 2 (1–6)[2,3,5,6,10]
Breeding age	1 year[11]
Mating system	monogamy[7]
Duration of pair bond	sometimes more than 1 season[7]
Total length	28–33 cm[3]
Wing span	43.2–48.3 cm[4]
Weight	M 130 g (110–170) / F 123 g (100–156)[10]
Longevity record	19 years 4 months[9]

References:
1. Caron et al. 1986
2. Harrison 1978
3. Godfrey 1986
4. Terres 1980
5. Peck and James 1983
6. Stokes and Stokes 1983
7. Ehrlich et al. 1988
8. Holcomb and Jaeger 1978
9. Clapp et al. 1983
10. Mirarchi and Baskett 1994
11. DeGraaf and Rudis 1987
12. according to nestling period

Chronology *n* = 170

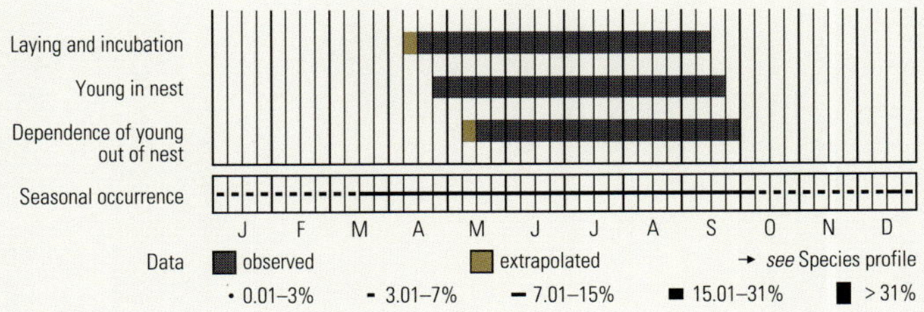

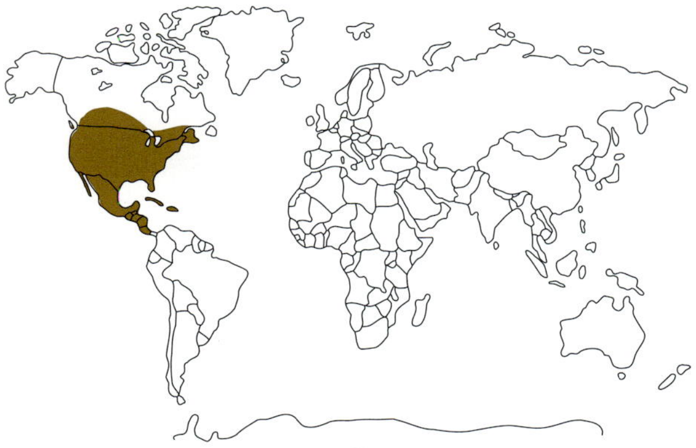

World breeding range

Adult and 2 young in the nest, a flimsy pile of grasses and twigs. Since Mourning Doves are early nesters, they produce at least two broods a year. **Montréal.** *May 1991, Pierre Bannon.*

Notes on territory (continued)

As nesting progresses, the territory gradually shrinks to a few square metres around the nest, and it is no longer defended once the young leave the nest (Stokes and Stokes 1983).

The species may also nest in loose colonies. In Ontario, 200 nests were found in a small grove of conifers (Peck and James 1983). Unmated males do not have defined territories (Stokes and Stokes 1983).

Mourning Dove incubating her eggs. This opportunistic female has built her nest between the rungs of a ladder on top of a parked truck. Shawinigan, north of Trois-Rivières. *June 1991, Michel Sokolyk.*

This seed eater has been able to overwinter in the province since 1969 and expand its range greatly thanks to bird feeders. The Mourning Dove ingests roadside gravel to help grind the seeds it eats. Île d'Orléans. *April 1984, Paul Perreault.*

Distribution

The Mourning Dove's distribution in the province roughly corresponds to the populated areas of southern Québec. The species is well established in the Ottawa and St. Lawrence valleys down to Québec City, as well as in the Montérégie region and the Eastern Townships. It also nests in the upper Laurentians and in agricultural areas in the Abitibi and Saguenay–Lac-Saint-Jean regions. On the North Shore, it can be found east to Havre-Saint-Pierre, and on the south side of the river it is found in the Lower St. Lawrence and Gaspé regions, mainly around Chaleur Bay. It has been reported on Anticosti Island as well as on the Magdalen Islands, though breeding has not been confirmed there (Fradette 1992).

History and trends

In the 19th century, Mourning Doves were seen from time to time in Québec during spring and fall migrations (Dionne 1906; Ouellet 1974a). Initially confined to the farmlands south of Montréal, the species did not begin to expand its breeding range into the province to any significant extent until the mid-1950s, exploiting the major changes to the landscape of the St. Lawrence Valley brought about by urbanization and agricultural development.

Between the mid-1950s and 1965, the species slowly spread along the St. Lawrence Valley to Québec City, particularly in the area south of the St. Lawrence River. Then, from 1966 to 1975, it began to colonize the area north of the river and the populated areas around Lake Saint-Jean. In 1976, Mourning Doves and a nest were observed near the work camp at La Grande-3 (Anonymous 1980b). An increase in the Mourning Dove population in the Eastern Townships and the Ottawa Valley, and the colonization of the Gaspé Peninsula and around Chaleur Bay also occurred at the same time. By the late 1970s, the species was found on both sides of the St. Lawrence from the Gatineau River to Baie-Comeau. In the early 1980s, it colonized the Abitibi and the upper Gatineau and Saint-Maurice river valleys, as well as the Baie-Comeau and Manicouagan River areas (Caron et al. 1986). A pair appears to have nested successfully in Kuujjuaq in 1989; 2 adults, sighted in early summer, were seen with young in midsummer (Aubry et al. 1990a). The increase in the Mourning Dove population in Québec over the last two decades is corroborated by ÉPOQ and BBS data (Falardeau and DesGranges 1991; Cyr and Larivée 1993a).

Furthermore, the number of Mourning Doves wintering in Québec has continued to grow since the early 1970s, in keeping with the overall expansion of the species' winter range in northeastern North America. The significant increase in corn acreage and the proliferation of bird feeders appear to have made it easier for the doves not only to overwinter but also to breed in Québec (Paquin 1990b). This northward expansion has entailed some problems, however, such as the frequent loss of toes or claws as a result of frostbite, making it difficult for the birds to perch and scratch the ground for food (Armstrong and Noakes 1983; Paquin 1990b).

Jean Paquin

Black-billed Cuckoo

Coulicou à bec noir

Coccyzus erythropthalmus

THIS CUCKOO DOES US A GREAT SERVICE, SINCE IT IS ONE OF THE RARE birds that eats the hairy tent caterpillars that sometimes infest deciduous or fruit trees. It consumes so many that the inside of its stomach becomes felted with a mass of hair and spines; when this mass obstructs digestion, the entire stomach lining is sloughed off and regurgitated (Forbush and May 1939). Its role in controlling these insect larvae amply compensates for the accusations of egg thief (Herrick 1910) and parasite (Bent 1940) levelled against it.

The Black-billed Cuckoo is a migrant that generally arrives in southern Québec in mid-May and departs again in late August, heading back to its wintering grounds, which extend from northern Colombia and northern Venezuela (including Trinidad) to northern Peru and central Bolivia (AOU 1983). There are some records of birds leaving as late as October. David (1980a) classifies the species as uncommon south of the 47th parallel and rare beyond it.

The Black-billed Cuckoo forages heavily on tent caterpillars, and their abundance influences the size of local cuckoo populations. These caterpillars often make their tents in deciduous trees bordering a clearing, like those in the foreground. Saint-Zéphirin, northwest of Drummondville. *June 1991, Daniel Jauvin.*

Habitat and behaviour

The habitat of the Black-billed Cuckoo is similar to that of the Yellow-billed. It seeks out the edges of forests, young woods with clearings, thickets, hawthorns, woods along the banks of rivers or swamps, and abandoned farmland growing up in trembling aspen, poplar, and birch. It seems to inhabit extensive woodlands more often than does the Yellow-billed Cuckoo, however (Pough 1949).

Before the egg laying period, the male makes a typical hollow wooden cluck, which he repeats 2–5 times, followed by a pause of extremely variable length. The bird sings both during the day and at night. One of its songs, a *kuka kuka kuka*, is similar to that of the Yellow-billed Cuckoo (Bent 1940). During nesting, the Black-billed becomes much quieter and is often more difficult to observe, since it tends to slip noiselessly through thickets. It does not move about much from branch to branch, and rarely perches out in the open. Moreover, since both species are rather unobtrusive and their plumage and some of their songs are similar, telling them apart can be quite tricky.

The nest, which is probably built by both parents (Herrick 1910), is located 1.2–1.8 m above the ground. It is well hidden in a bush or on a branch of a deciduous or coniferous tree and made of twigs roughly assembled into a platform, lined with finer materials, such as rootlets and leaves. Generally speaking, it is quite fragile, although it is bigger and better built than that of the Yellow-billed Cuckoo. Incubation, by both parents, lasts 10–14 days (Bent 1940; Spencer 1943).

Jean-Marc Giroux

Notes on territory

The Black-billed Cuckoo is very quiet during nesting. There is little specific information on density or territory size. In addition, the young leave the nest early and do not necessarily remain at the nest site until they are fully grown.

This cuckoo nests in young hardwoods along the edges of ponds or clearings. Saint-Edmond-de-Grantham, west of Drummondville. *June 1991, Daniel Jauvin.*

Black-billed Cuckoo

Breeding distribution

1984–1989

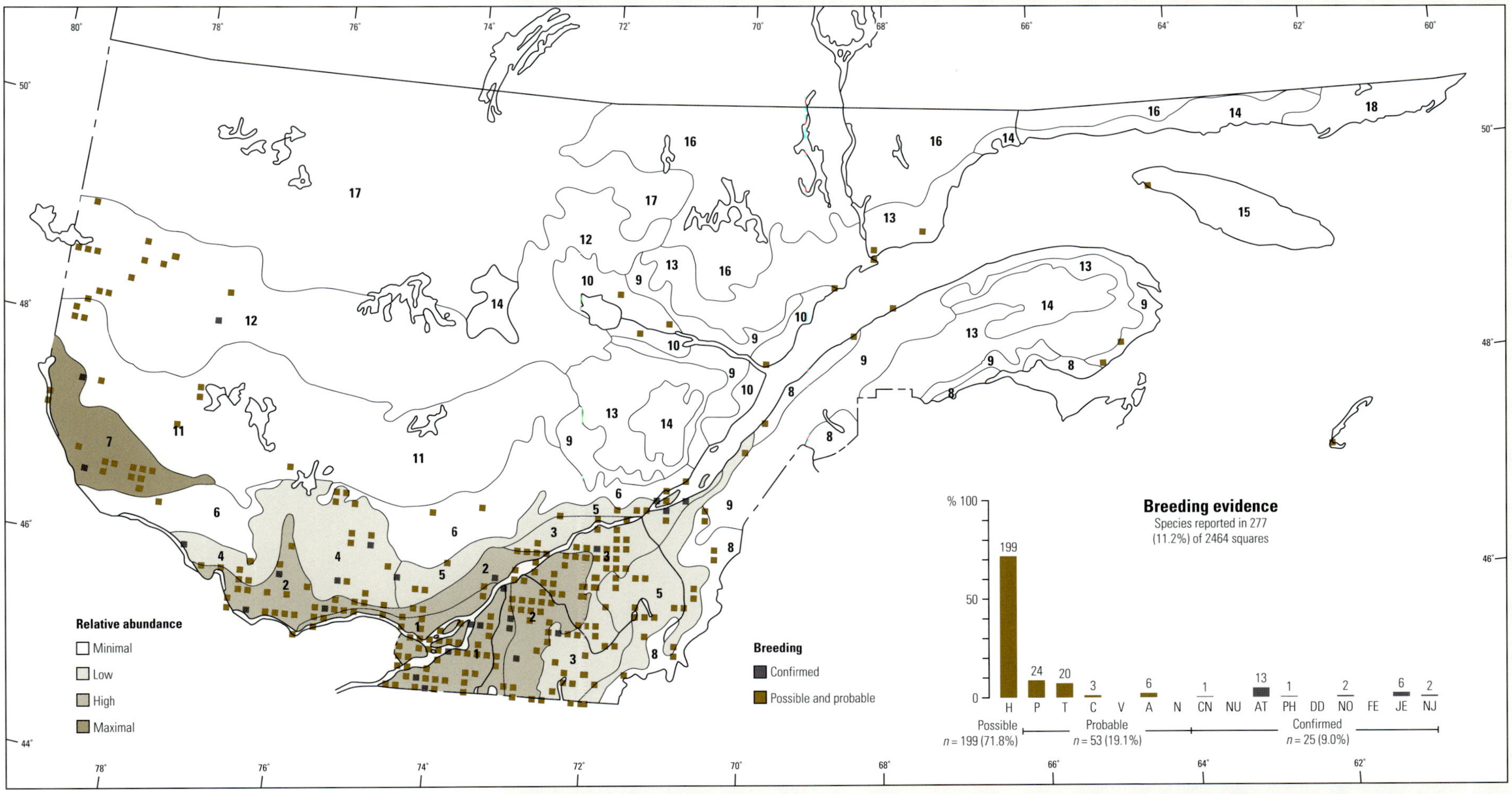

Species profile

Clutch size	usually 2–3 eggs (2–7)[1,2,3,4]
Laying interval	1 egg per 1–3 days[2,3]
Start of incubation	with first egg[2]
Incubation period	10–14 days[1,2,4]
Maturity at hatching	altricial
Care of young	by female and male[1,2,3,4]
Nestling period	6–9 days[1,2,3,4,9]
Age at first flight	≈ 21–24 days[10]
Dependence of young	?
Broods per year	1–2[3]
Breeding age	1 year[1]
Mating system	probably monogamy[4]
Duration of pair bond	?
Total length	28–32 cm[5]
Wing span	38.1–42.5 cm[8]
Weight	51.1 g (39.5–65.0)[6]
Longevity record	5 years[7]

References:
1. Bent 1940
2. Spencer 1943
3. Herrick 1910
4. Ehrlich et al. 1988
5. Godfrey 1986
6. Dunning 1984
7. Clapp et al. 1983
8. Terres 1980
9. Mousley 1931c
10. Harrison 1978

Chronology $n = 24$

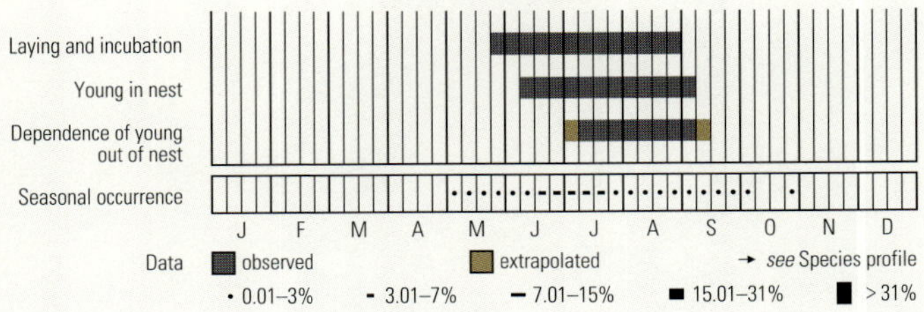

World breeding range

Frequency of occurrence by ecological region

Ecological regions** (forest regions)	Adequately covered squares	Squares in which reported	% of squares in which reported	Relative abundance (%)*
1. Sugar maple-hickory and sugar maple-basswood	97	52	53.6 ± 5.0	84.9
2. Sugar maple-basswood	153	63	41.2 ± 4.1	65.2
3. Sugar maple-basswood and sugar maple-yellow birch	101	34	33.7 ± 5.6	53.3
4. Sugar maple-yellow birch-basswood	56	12	21.4 ± 9.5	33.9
5. Sugar maple-yellow birch-beech and sugar maple-yellow birch-basswood	113	24	21.2 ± 4.3	33.6
6. Sugar maple-yellow birch-beech	71	8	11.3 ± 7.0	17.8
7. Sugar maple-yellow birch-balsam fir	19	12	63.2 ± 22.8	100.0
8. Sugar maple-yellow birch and balsam fir-yellow birch	47	5	10.6 ± 8.3	16.8
9. Balsam fir-yellow birch	87	2	2.3 ± 3.3	3.6
10. Balsam fir-white birch or balsam fir-red maple	46	3	6.5 ± 6.7	10.3
11. Yellow birch-balsam fir	52	5	9.6 ± 8.6	15.2
12. Balsam fir-white birch (west)	59	10	16.9 ± 10.0	26.8
13. Balsam fir-white birch (east)	76	4	5.3 ± 5.3	8.3
14. Balsam fir-black spruce	28	0	0	0
15. White spruce-balsam fir and balsam fir-white spruce	22	1	4.5 ± 10.3	7.2
16. Black spruce-balsam fir-moss	18	0	0	0
17. Black spruce-moss	32	0	0	0
18.			*not analysed*	

* minimal: 0–29.9%; low: 30–59.9%; high: 60–89.9%; maximal: 90–100%

** Deciduous forest: 1–7; mixed forest: 8–11; coniferous forest: 12–17

Adults and young alike eat virtually nothing but insects, mainly caterpillars, tree crickets, and katydids, which they hunt in tree foliage, and grasshoppers which they catch on the ground (Beal 1898). The parents' arrival at the nest stimulates a begging reflex in the nestlings: they open their bills, stretch their necks, and flap their wings. The prey is then deposited directly into their buccal cavities (Forbush 1927).

The young begin to climb on branches near the nest when they are about a week old. In this species, the feathers remain sheathed until they are ready to leave the nest, making the nestlings look like porcupines. At this time, the sheaths burst open and the feathers are released; it takes scarcely 6 or 7 hours for the birds to become fully feathered (Mousley 1931c).

In proportion to its weight, the young Black-billed Cuckoo is probably the strongest, hardiest altricial bird in North America (Herrick 1910), which partially explains the short nestling period (Herrick 1910; Bent 1940; Spencer 1943; Sealy 1985). When the young leave the nest at the age of 7–9 days, however, they are still flightless and are unable to sustain flight until they are about 21–24 days old (Harrison 1978; Sealy 1985). They are then able to travel long distances in search of food, possibly in the company of the adults, at least at first. This behaviour may be one of the cuckoo's ways of adapting to an environment in which food is less plentiful or more widely dispersed at the end of the breeding season. Brood mobility seems to give both the adults and young a better chance of finding a source of food, especially when food starts to become scarce at the nesting site (Sealy 1985).

A nest thickly lined with plant down containing 3 eggs in a balsam fir. Sainte-Dorothée, Laval. *June 1943, Henry Mousley, Laval University Collection.*

Young Black-billed Cuckoo whose tail-feather sheaths have not yet opened. In this species, the feathers remain sheathed until the nestlings are about to leave the nest, at which time all the feathers are released within the space of only 6 or 7 hours. Saint-Lambert, near Montréal. *September 1929, Henry Mousley, Laval University Collection.*

This bird is known for being particularly abundant during years of local caterpillar infestations (Forbush 1927). Responsiveness to these periods of exceptionally plentiful prey probably causes the female to produce an excessive number of eggs, which would partially explain why this species sometimes imitates its Eurasian cousin, the Common Cuckoo, and lays its eggs in the nests of other species (Nolan and Thomson 1975), such as the Yellow-billed Cuckoo, Gray Catbird, Eastern Wood-Pewee, Cedar Waxwing, Wood Thrush or Yellow Warbler (Bent 1940).

The Black-billed Cuckoo is heard much more often than it is seen. That is why the strongest evidence of breeding was simply possible breeding in 71.8% of Atlas squares (199/277).

Breeding was most often confirmed by reports of an adult carrying food for the young. Nests are extremely difficult to find; they were reported in only 4 of the 277 squares in which the species was recorded during the entire Atlas survey. This is chiefly because it is so quiet when nesting.

An unusually thick and sturdy Black-billed Cuckoo nest, made of small branches, to which the bird has added silky poplar catkins. Sainte-Dorothée, Laval. *June 1942, Henry Mousley, Laval University Collection.*

Distribution

The Black-billed Cuckoo breeds in the United States from Wyoming in the west to North Carolina in the east. In Canada, it breeds from east-central Alberta to Nova Scotia and Prince Edward Island (AOU 1983; Erskine 1992). It is considered very common east of Lake Nipissing, Ontario, a few kilometres from the Québec border (Lawrence 1952), and has been observed at several places in the Clay Belt, in the northwestern part of southern Québec (Smith 1957).

The Atlas map shows that the Black-billed Cuckoo's Québec range is much more northerly and extensive than that of the Yellow-billed. It inhabits mainly the St. Lawrence Plain, Eastern Townships, Ottawa Valley, Témiscamingue, and Abitibi regions. There are breeding records from Jonquière in 1978 and Saint-Fulgence in 1980 in the Saguenay (G. Savard, pers. comm.), although it was not atlassed as a breeder there. This cuckoo may also nest on the North Shore down to Godbout and along the Lower St. Lawrence as far as Métis-sur-Mer. It was atlassed in 2 squares on the Gaspé Peninsula, in one square on Anticosti Island, and in another on the Magdalen Islands.

The northeasterly expansion of the species is probably due to intensive deforestation and the subsequent creation of suitable habitat, including open areas with scattered trees and bushes, as well as reverting farmland at the sapling stage. The species winters in South America, especially the northwest (AOU 1983).

History and trends

The Black-billed Cuckoo must have been less abundant in the past, before the deforestation of southern Québec enabled it to expand its range. Its status in the Montréal area in the last century is difficult to ascertain: Hall (1862) called it scarce, while Wintle (1896) considered it a common summer resident in 1882. Nowadays it is uncommon in the southwestern part of the province and scarce elsewhere, as David (1980a) and Bannon (1991b) have reported.

BBS data indicate a continent-wide increase in abundance (Robbins et al. 1986), but the situation in Québec does not seem to reflect this general trend. BBS data for 1966–89, as well as ÉPOQ data for 1979–89 (Cyr and Larivée 1993b), show that the Québec population of the Black-billed Cuckoo is stable or rising only slightly.

Furthermore, this species' numbers vary from year to year, and seem to follow the cycles of infestations of hairy caterpillars: gypsy moth larvae, fall webworms, forest tent caterpillars, and eastern tent caterpillars (Forbush 1927; Bent 1940).

Daniel Jauvin and Manon Bombardier

Adult on nest incubating its clutch. Sainte-Dorothée, Laval. *July 1939, Henry Mousley, Laval University Collection.*

Yellow-billed Cuckoo

Coulicou à bec jaune

Coccyzus americanus

Jean-Marc Giroux

THE YELLOW-BILLED CUCKOO HAS A MORE SOUTHERLY DISTRIBUTION THAN its close relative, the Black-billed Cuckoo, with which it is easily confused.

The Yellow-billed Cuckoo is very rare in Québec. It arrives in late May, nests in June and July, and leaves at the end of September. Some stragglers have, however, been observed as late as November.

Habitat and behaviour

The habitat of the Yellow-billed Cuckoo is almost identical to that of the Black-billed, and the two species are often found together. The Yellow-billed usually shuns dense forest (Bent 1940), preferring woodlands with clearings, dense willow, alder, or poplar tangles sometimes located along waterways, brushy roadsides, and orchards (Beal 1898; Bent 1940; Terres 1980; Godfrey 1986; Anderson and Laymon 1989). In Québec, it is also found on abandoned farmland with scattered bushes, shrubs, and young trees. Laymon and Halterman (1989) have shown that in California, the species is responsive to habitat expansion.

The Yellow-billed Cuckoo's nest is even flimsier than the Black-billed's. A rough platform made of twigs, sparsely lined with vegetation slightly finer than that used by the Black-billed (Bent 1940), it is generally well concealed in the foliage of a tree or shrub, about 1–2 m up. Both parents incubate the eggs, mainly the female during the day and the male at night; incubation lasts 9–14 days (Bent 1940; Hamilton and Hamilton 1965; Potter 1980).

The timing of the onset of breeding can vary from year to year (Nolan and Thompson 1975), the species having developed the ability to synchronize breeding with periods of plentiful food. Eggs are therefore often laid at irregular intervals corresponding to the occurrence of these periods, and incubation begins as soon as the first egg is laid. As a result, hatching is asynchronous, with a large interval between hatching dates (Nolan and Thompson 1975; Potter 1980). In addition, the development of the young is remarkably quick. Hardly 16 or 17 days elapse between the start of incubation and the time the young leave the nest; Hamilton and Hamilton (1965) consider this period one of the shortest of any bird.

Although this cuckoo seems particularly adept at capitalizing on a temporary abundance of food, Potter (1980) claimed that it uses only 1 ha—half of its territory—to hunt, and that it does not necessarily take the prey that is most plentiful during the nesting period. It seems particularly fond of cicadas (Nolan and Thompson 1975; Fleischer et al. 1985) and larvae of the gypsy moth, a native of Europe that first appeared in southern Québec in 1924 (Laplante 1985) and has been responsible for major infestations of deciduous trees since 1960. It also likes other species of hairy caterpillars (Bent 1940), such as the fall webworm, forest tent caterpillar, and eastern tent caterpillar (Beal 1898).

Occasionally, the Yellow-billed Cuckoo parasitizes the nests of other birds, such as the Black-billed Cuckoo, American Robin, Wood Thrush, Gray Catbird, Cedar Waxwing, Mourning Dove, and Red-winged Blackbird (Bent 1940). This appears to be an adaptive behaviour, as a fast response to periods of exceptionally plentiful food (Nolan and Thompson 1975), such as a sudden epidemic of caterpillars, even when it has not been able to build a nest for its eggs. This behaviour is, nonetheless, much less frequent in the Yellow-billed Cuckoo than in its Eurasian cousin, the Common Cuckoo, which is an obligate parasite.

The Yellow-billed Cuckoo is fairly difficult to spot. It is a quiet bird that freezes when in danger, then quickly slips into a thicket, where it hides if anyone gets too close. Its plumage is very similar to that of the Black-billed Cuckoo. It is easier to tell the young of the two species apart, however, since the back and upper leg of young Black-billed Cuckoos are covered with white down that contrasts starkly with their blackish skin, while the down and skin of young Yellow-billed Cuckoos are a fairly uniform grey (Nolan 1975). The Yellow-billed Cuckoo can often be detected by its song. It produces a variety of vocalizations (songs and calls), some of which are impossible to distinguish from those of the Black-billed Cuckoo. The Yellow-billed Cuckoo does have its own song, however, a series of hollow clucks repeated a dozen times, rapidly at first and then deliberately, at longer intervals: *kuk-kuk-kuk-kuk-kuk-kuk-ceaow-ceaow-ceaow, ceaow, ceaow, ceaow*. The Black-billed Cuckoo never produces such a long song (Bent 1940).

Breeding was not confirmed for the Atlas. Yellow-billed Cuckoos were seen or heard in suitable habitat in 14 squares, in 4 of which breeding was recorded as probable.

Distribution

In North America, the Yellow-billed Cuckoo is 10 times as abundant as the Black-billed Cuckoo (Van Velzen and Robbins 1969). The species is common in the southeastern United States, less abundant in the northeast, and local in the west. In Canada, it is found mainly in the southernmost part of Ontario (Helleiner 1987d), and it has always been rare in the extreme southwestern part of Québec (AOU 1983; Godfrey 1986). It winters in South America (AOU 1983).

In Québec, the species is local in the Central St. Lawrence Lowland, down to east of Trois-Rivières, having been reported in the southern Ottawa Valley, the Montréal and Farnham areas, and near Drummondville, where it was sighted in 2 squares and heard in 3. All these reports originate from a long corridor along a main road, where the habitat is characterized by a regenerating forest with many openings and young trees.

History and trends

The Yellow-billed Cuckoo was probably much rarer before European settlement of North America. It took advantage of deforestation in the United States and southern Ontario early this century. Nonetheless, for about 20 years, as a result of the disappearance of hedgerows along roads, ditches, and fields, and the intensification of farming and deforestation, its numbers have dropped significantly, especially in southern Ontario (Speirs 1985). Moreover, populations, being partially dependent on the amount of available food, seem highly variable locally. In New York State, the species is most often observed during gypsy moth infestations (Eaton 1988n).

There is very little information on the Yellow-billed Cuckoo in Québec. In the last century, Hall (1862) and Wintle (1882) described it as rare in the Montréal area. As earlier indicated by David (1980a), this migrant breeder is presently considered as casual in the southernmost part of Québec, where nesting has been confirmed in the lowlands, at Beaurepaire in 1948 and at Old Chelsea in 1978, at Ulverton, near Drummondville (1956), and at Lennoxville (David 1980a; Godfrey 1986). David (1980a) classified it as a casual visitor in the central and northern parts of southern Québec. Its numbers seem to have decreased over the last 20 years or so (BBS). The species was on the Blue List between 1972 and 1981, and again in 1986 (Tate 1986). This decline may be attributable to recent efforts to control forest tent caterpillars in the province (Guérin 1987). In fall, this cuckoo may wander north or east before migrating south (Tufts 1986).

DANIEL JAUVIN

Notes on territory

Territorial defence, by the male, is limited to the nest and its immediate surroundings, in other words, the nest tree and the ground underneath its branches. Potter (1980) observed that a pair of cuckoos nesting in North Carolina foraged within a territory of approximately 1 ha.

To defend the nest site, the male seems to adopt one of two strategies. Either he heads toward the intruder, perches near it, lowers his wings and raises his tail over his back in order to distract it and drive it away from the nest, or he flies to the nest and sits on the chicks (Potter 1980).

The Yellow-billed Cuckoo has been seen and heard in this deciduous woodlot bordered by a clearing. Saint-Majorique, near Drummondville. *September 1992, Jeanne Lehoux.*

Yellow-billed Cuckoo

Breeding distribution

1984–1989

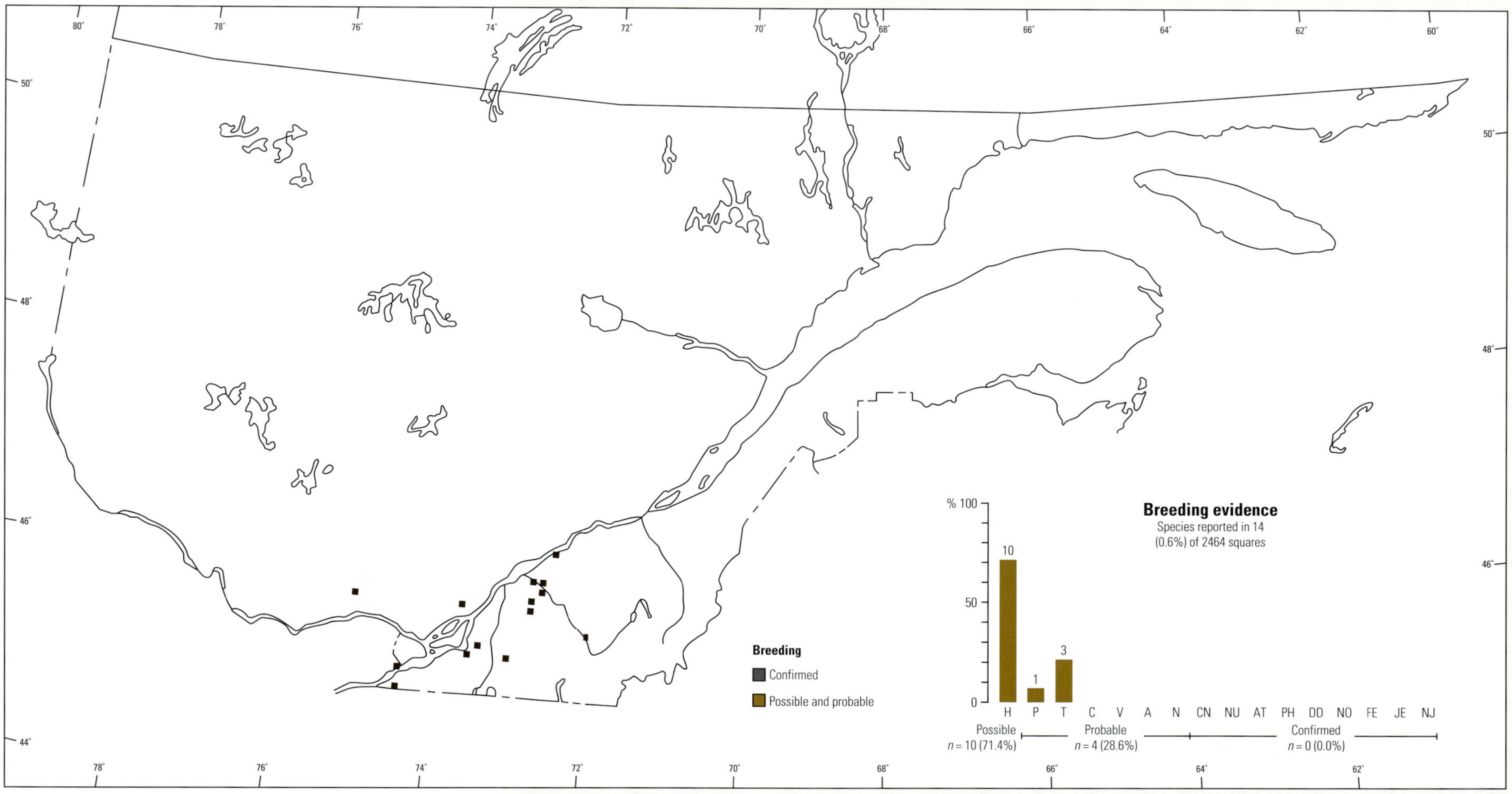

Chronology $n = 3$

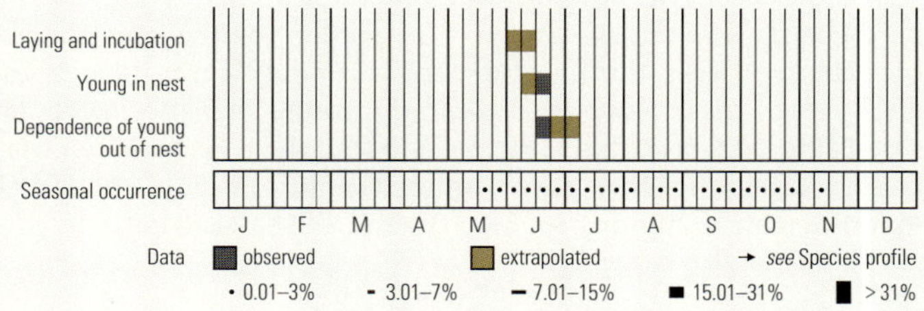

Species profile

Clutch size	usually 4 eggs (1–8)[1,3,4]
Laying interval	1 egg per 1–3 days[1,2,4]
Start of incubation	with first egg[1,4]
Incubation period	9–14 days[1,3,9]
Maturity at hatching	altricial
Care of young	by female and male[1,3]
Nestling period	7–9 days[1,2,3]
Age at first flight	≈ 21 days[3]
Dependence of young	?
Broods per year	probably 1[1,3]
Breeding age	1 year[8]
Mating system	monogamy[3]
Duration of pair bond	?
Total length	28–32 cm[5]
Wing span	39.4–43.2 cm[4]
Weight	64.0 g (50.0–84.6)[6]
Longevity record	5 years[7]

References:
1. Potter 1980
2. Preble 1957
3. Ehrlich et al. 1988
4. Terres 1980
5. Godfrey 1986
6. Dunning 1984
7. Clapp et al. 1983
8. DeGraaf and Rudis 1987
9. Bent 1940

World breeding range

The Yellow-billed Cuckoo, which shares the same haunts as the Black-billed Cuckoo, homes in on infestations of tent caterpillars in deciduous trees. Saint-Zéphirin, northwest of Drummondville. *June 1991, Daniel Jauvin.*

Barn Owl

Effraie des clochers

Tyto alba

Luc Leclerc

THE BARN OWL FIGURES ON AN EXCLUSIVE LIST OF SPECIES NESTING IN Québec that also breed in all parts of the world, except Antarctica, along with the Great Cormorant, Great Egret, Green Heron, Osprey, Peregrine Falcon, Caspian Tern, Roseate Tern, and a few introduced species. This distinction, however, rests on very slim evidence, as there are only a few sight records to even suggest the species has nested in Québec. To all intents and purposes, the Barn Owl is a casual visitor in the western St. Lawrence Valley, breeding on rare occasion in the southwesternmost part of the province.

Habitat and behaviour

Being nocturnal and generally silent, the Barn Owl often goes unnoticed, even where it is common. Unlike the Strigidae owls, this member of the family Tytonidae does not give regular territorial calls, even in the mating season (Sibley 1988a). The birds are mainly detected through observations of parents carrying food or the repeated calling of the young. The only Atlas record of Barn Owls was of 5 presumably young birds calling at night near a farmhouse.

The Barn Owl inhabits a variety of open environments at low elevations, in both urban and rural settings. Habitat selection appears to be determined primarily by the availability of nest sites and hunting grounds (Bunn et al. 1982; Marti 1992). In northeastern North America, the Barn Owl is found in agricultural areas—wet meadows, pastures, cultivated or abandoned fields, marshes—as well as around small towns (Sibley 1988a). It usually hunts by flying low to the ground, quartering over open terrain where prey is likely to be found. The many studies on the bird's diet (see Cramp 1985) indicate that it feeds on small vertebrates, particularly rodents, such as mice and voles, and insectivorous animals, such as shrews (Bunn et al. 1982; Baudvin 1983; Campbell and Campbell 1983; Marti 1988).

Nest sites are sheltered spots, sometimes in a natural cavity in a tree or cliff, but most often in a barn, unused silo, belfry, mine shaft, or other building. Among the more unusual nest sites reported are a duck blind and a drive-in movie screen (Marti 1992). Barn Owls will also occupy specially made nest boxes (Marti et al. 1979; Juillard and Beuret 1983).

The Barn Owl does not build a nest, but simply lays its eggs on the surface of the chosen spot. Incubation, by the female alone, lasts about 30 days, during which time she rarely leaves the nest and is fed by the male. She remains with the young after they hatch and shreds the prey brought by the male to feed them. When the owlets begin swallowing their prey whole, at about 15 days, the female joins the male in hunting for food (Bunn et al. 1982; Marti 1992).

Distribution

Since Québec is at the northernmost limit of the Barn Owl's range, the species is extremely rare in the province. As the bird is difficult to locate at the best of times, it is not surprising that only one Atlas record was obtained. Moreover, the birds that were seen—fledged young near North Hatley on 30 July 1987—may have hatched in the United States, as they were seen only 20 km from the border. But since young Barn Owls tend to stay near the natal site for several weeks after fledging (Cramp 1985; Marti 1992), it seems likely that these birds did indeed hatch in Québec.

There is another report of Barn Owls breeding in Québec. Some people originally from Germany, where the species is common, living near Lacolle reported seeing the eggs of a pair nesting in a barn in 1979 and 1980 (H. Miehe, pers. comm.).

There are 2 earlier reports, but they do not constitute conclusive breeding evidence. One is of a pair seen in Saint-Hubert in the summer of 1961, with no details given, and the other is of a female bird seized in July 1977 at a taxidermist's in Masson, a few kilometres from the Ontario border. An examination showed that she had laid eggs just a few weeks earlier (David 1979).

A 1931 breeding record from Berthierville (Godfrey 1966) was based on the November sighting of at least 4 birds in the church belfry. The report was not supported by any verifiable breeding evidence (David 1979), but breeding nonetheless remains a plausible explanation of why the birds were there.

Single Barn Owls were also sighted during the summer at Saint-Félix-de-Kingsey on 15 June 1936 (David 1979), Giffard on 18 and 24 June 1963 (ÉPOQ), Rigaud on 12 June 1971 (ÉPOQ), and the old Ulverton mill early in the breeding season, sometime around 1970 (J. Dion, pers. comm.).

There are 32 other sight records of the species in Québec: 9 from April and May, 19 from between August and December, and 4 for which only the year is known. Seven of the 9 spring records were obtained after 1981. The fall records can be ascribed to the post-breeding dispersal of birds that were probably from the United States since birds can wander several hundred kilometres away from the nest after breeding (Stewart 1952; Dexter 1957; Soucy 1980).

Most sightings of Barn Owls in Québec have been in the Central St. Lawrence Lowland (south of the Laurentians and west of Québec City), where the species is thought to nest from time to time. Any breeding pairs are probably migrants, as Barn Owls are unable to withstand the cold winters of northeastern North America (Campbell and Campbell 1983; Root 1988).

History and trends

It is thought that Barn Owl populations in eastern North America increased in the last century, having profited from the intensive clearing of land for agriculture. In recent decades, however, there has been some decline, apparently due to increasing urban development, draining of wetlands, and new agricultural practices. This is notably the case of certain populations in the U.S. Midwest (Campbell and Campbell 1983; Hands et al. 1989b; Marti 1989, 1992).

The Québec records do not reflect any of these apparent trends. In fact, 20 of the 42 sight records have been obtained since 1970, no doubt as a result of an improved flow of information and an increase in the number of birders.

NORMAND DAVID

Notes on territory

Barn Owls usually nest in solitary pairs, and the size of their territories depends on the abundance of food. Cramp (1985), for example, gives areas of 0.4–2.5 km^2. Moreover, Barn Owls do not always defend territorial limits. They mostly make their presence known by calling (Bunn et al. 1982). The area around the nest site, however, is fiercely defended against other Barn Owls, even when several pairs are nesting close together (Cramp 1985; Marti 1992). In New Jersey, Barn Owls typically inhabit home ranges of over 7 km^2, with a mean distance of 5.6 km between roosting sites and hunting grounds (see Marti 1992).

The Barn Owl, an inhabitant mainly of agricultural areas, is at the northern limit of its range in southern Québec. In recent years, the species has been sighted in spring and fall at Cap Tourmente, quartering the fields in search of small rodents. Cap Tourmente National Wildlife Area, northeast of Québec City.
December 1985, Yves Aubry.

Barn Owl

Breeding distribution
1984–1989

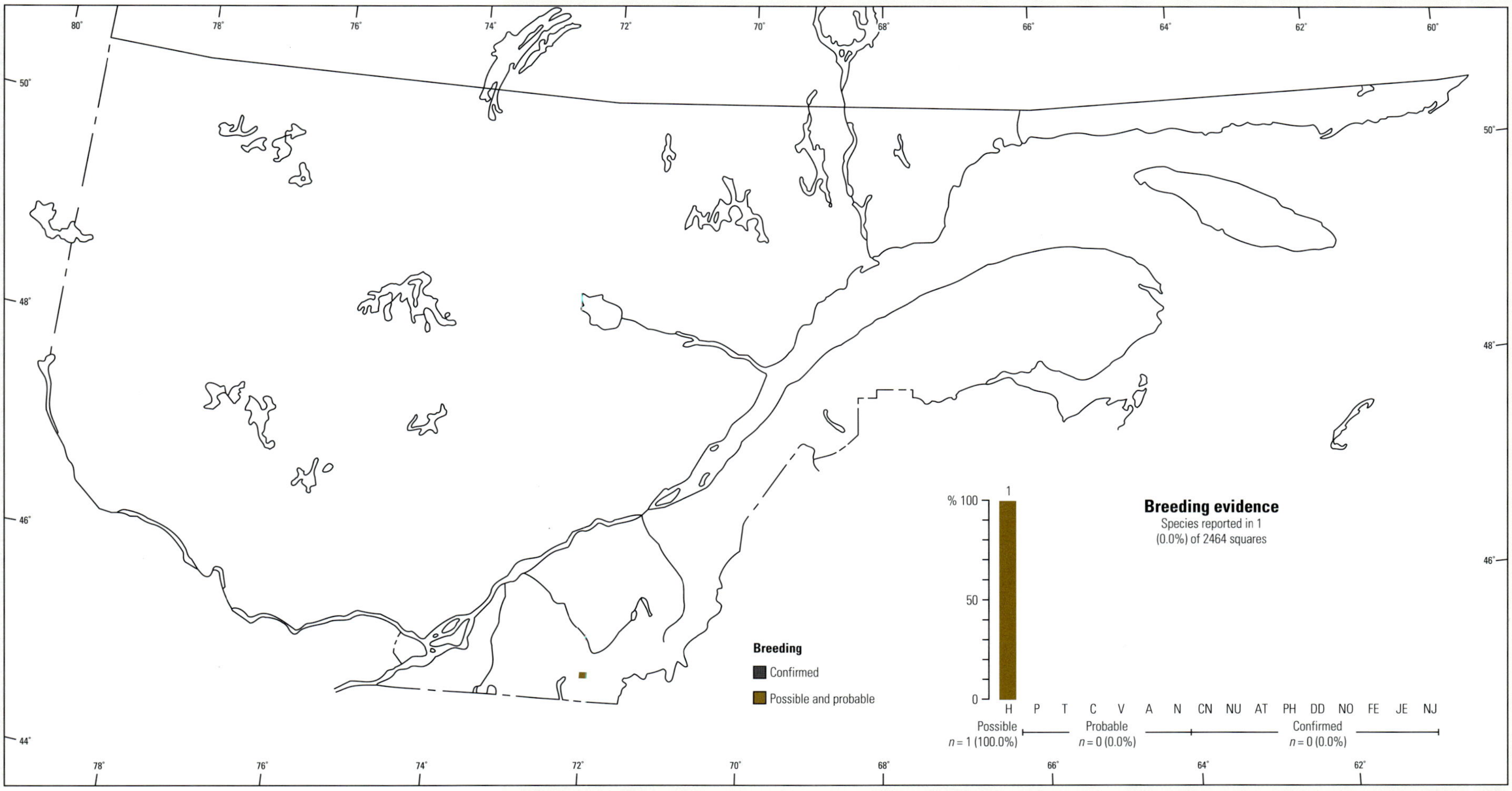

Species profile

Clutch size	usually 4–7 eggs (2–18)[1,2,3]
Laying interval	1 egg per 2–3 days[1,2,3]
Start of incubation	with first egg[1]
Incubation period	29–34 days[1,2,3]
Maturity at hatching	semi-altricial
Care of young	by female and male[1,2,3], *see* text
Nestling period	50–55 days[1]
Age at first flight	50–64 days[1]
Dependence of young	until 3–5 weeks after first flight[1,2,3]
Broods per year	1, sometimes 2[1,2,3]
Breeding age	1 year[1,3]
Mating system	monogamy, sometimes polygyny[1,2,3]
Duration of pair bond	for life[1,2]
Total length	37.0–44.5 cm[6]
Wing span	109.2–119.4 cm[5]
Weight	M 442 g (299–580) / F 490 g (382–580)[4]
Longevity record	34 years[1]

References:
1. Marti 1992
2. Bunn et al. 1982
3. Cramp 1985
4. Dunning 1984
5. Terres 1980
6. Godfrey 1986

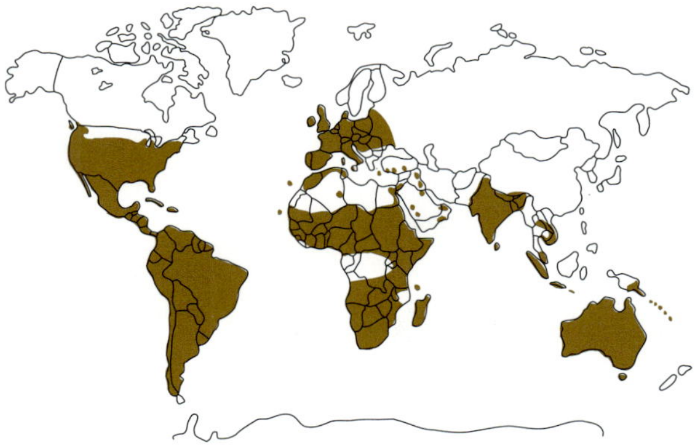

World breeding range

Chronology $n = 0$

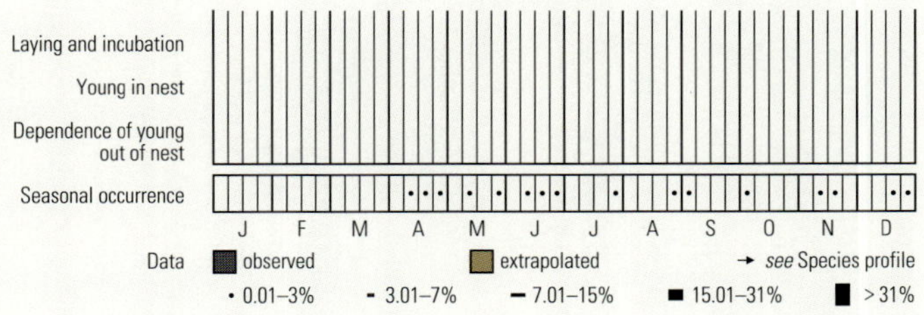

Eastern Screech-Owl

Petit-duc maculé

Otus asio

Luc Leclerc

THE EERIE SOUND OF THE EASTERN SCREECH-OWL IS MORE FAMILIAR THAN the sight of this well-camouflaged bird. Although it is a sedentary breeder in the southernmost part of the province and a casual visitor in the central and northern areas of southern Québec (David 1980b), it is rarely seen and therefore has been overlooked by early naturalists. With its round head, small ear tufts, and short, squarish tail, the Eastern Screech-Owl has the remarkable ability to blend in with the colour of the tree bark around a nesting or a roosting cavity. Although it often looks like just another branch, as it apparently dozes in a tree hole or on some sheltered sunny branch, it watches the passing scene through slanted eye slits. It is characterized by three colour phases: the grey, the less common rusty "red," and an intermediate, or chocolate brown, plumage.

Habitat and behaviour

In the breeding season, the Eastern Screech-Owl is found in a variety of open woodlands, including moderately wet habitats and those dominated by red maples, as well as edge habitats beside grainfields and meadows, in groves, orchards, parks, and shade trees in villages and suburban areas (Lynch and Smith 1984; Godfrey 1986). It can be difficult to locate by day because it often roosts in tree hollows out of sight or mimics a tree branch while dozing close to the trunk. However, males do begin advertising their presence in mid-January by their tremulous courtship song, which can be heard with increasing frequency in March and April, most often in midafternoon or early evening. Other telltale signs are the scolding of passerines, such as nuthatches, chickadees, and robins, or the splashes of whitewash on leaves or branches and the presence of small pellets at the base of trees. By early April, pairs can be heard calling to each other, but the ventriloquial quality of their sounds may make finding the birds' exact location difficult. In the Montréal area, the disappearance of this owl from its few well-known roosting sites in the spring may indicate that nesting has begun, and by mid-May, the adults can sometimes be seen standing guard near the nest cavity.

Adults form monogamous pair bonds. Then the female chooses the nest site, usually a natural cavity in a tree or an old woodpecker hole, but where these are scarce or unavailable, the owl will occupy nest boxes. The clutch usually consists of 4 or 5 eggs. The female does most or all of the incubating, which lasts 21–32 days, while the male is responsible for providing food during the incubation and brooding periods (Sherman 1911; Bent 1938). Should anyone approach the nest, the dozing male instantly becomes alert, stretching to his full height, with ear tufts pointing to the sky and yellow eyes wide open. Any perceived threat to the young will prompt him to go after the offender, swooping down on dogs, squirrels, and human intruders. The territory actually defended is very small, consisting of only the nest cavity and its immediate vicinity. Neighbouring pairs have been found nesting as close as about 45 m apart (Johnsgard 1988). Alloparenting by a third adult has been reported in Eastern Screech-Owls: in Connecticut, an unmated male that had been rehabilitated and released after an accident was observed helping a pair to provide food for their two owlets, continuing even after they had fledged (Smith and Hiestand 1990).

The young hatch asynchronously and, at dusk from late May to early July, they can be seen standing outside the nest, in increasing order of size like fluffy white organ pipes, their begging sounds and gyrating heads bringing them to the attention of the observer. Copious amounts of liquid whitewash may coat the leaves, branches, and ground near the owls' roosting and feeding area.

Notes on territory

The male defends a very small territory, just the nest cavity and its immediate vicinity. Pairs may nest as close as 45 m apart (Johnsgard 1988). In the first few weeks after the young have fledged, the whole family often roosts in the same tree at night. From the fifth week on, the juveniles become more independent and often perch at some distance from their parents, while the adults go on using the same daytime roost (Belthoff and Ritchison 1990). In central Kentucky, radio-tagged juveniles continued to roost within the bounds of their natal territory for an average of 55 days after fledging, then dispersed 0.4–16.9 km from the nest (Belthoff and Ritchison 1989).

From December to June, a female in a suburban habitat in southern Connecticut maintained a home range totalling 130.2 ha, though she used only a small portion of this area (5–16 ha) for nighttime hunting. The total area of the territory varied from month to month throughout the year; it was largest in February (100.2 ha), when the female was defending it and choosing a nest site, and in June (107.5 ha), after the nest had been destroyed by a predator.

The Eastern Screech-Owl has been nesting for several years near the gates of the Protestant cemetery on Mount Royal. Montréal. *March 1991, Yves Aubry.*

Eastern Screech-Owl

Breeding distribution

1984–1989

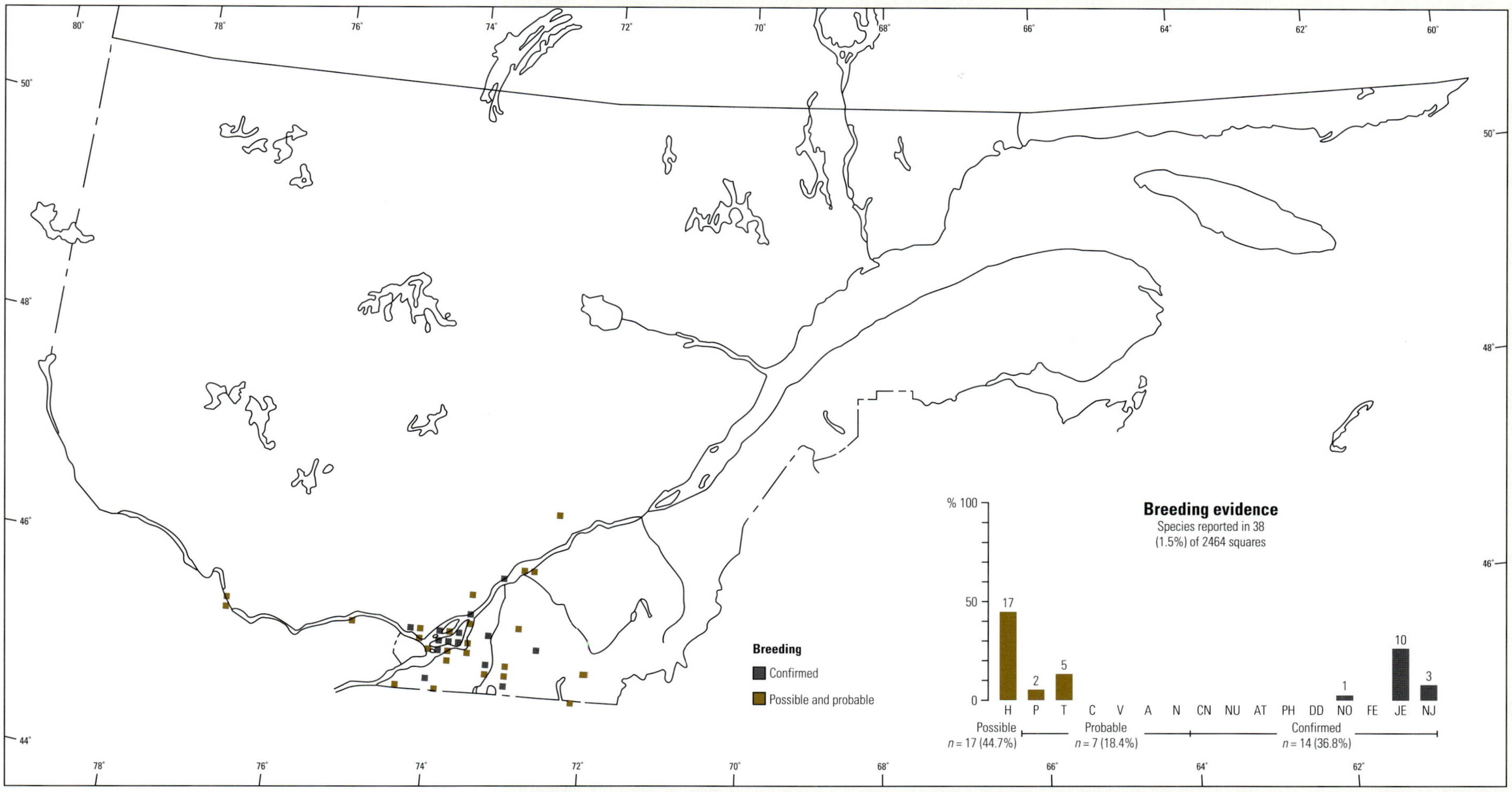

Species profile

Clutch size	usually 4–5 eggs (2–8)[1,2,3]
Laying interval	1 egg per 2–3 days[1,2]
Start of incubation	with second, third or fourth egg[1,2]
Incubation period	usually 26 days (21–32)[1,2,3]
Maturity at hatching	semi-altricial
Care of young	by female and male[1,2], *see* text
Nestling period	28–35 days[1,3]
Age at first flight	a few days after leaving nest[5]
Dependence of young	until 5–8 weeks after leaving nest[1,5]
Broods per year	1, probably 2[1,3]
Breeding age	1 year, sometimes 2[4]
Mating system	monogamy[3]
Duration of pair bond	for life[1,3]
Total length	M 18–23 cm / F 20–26 cm[1]
Wing span	M 48–57 cm / F 50–62 cm[1]
Weight	M 167 g (140–210) / F 194 g (150–235)[6]
Longevity record	13 years 6 months[3]

References:
1. Karalus and Eckert 1974
2. Bent 1938
3. Johnsgard 1988
4. DeGraaf and Rudis 1987
5. Stokes and Stokes 1989a
6. Dunning 1984

Chronology *n* = 22

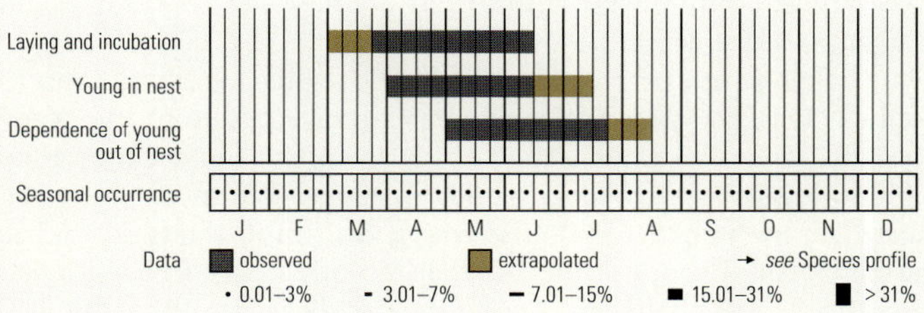

World breeding range

Orchards are ideal habitat for this small owl, a rare breeder in southern Québec. Mount Saint-Hilaire. *August 1991, Gaétan Duquette.*

Notes on territory (continued)

It was smallest in winter (8.8 ha in December) and during the breeding period in April (42.7 ha) and May (13.8 ha) (Smith and Gilbert 1984).

Craighead and Craighead (1956) estimated breeding densities at 15 pairs/100 km² in Michigan and 9 pairs/100 km² in Wyoming.

These owls feed on rodents, birds, frogs, salamanders, reptiles, large insects, and even the occasional fish (Bent 1938; Brown et al. 1986). In turn, they are vulnerable to predation by larger raptors, especially the Great Horned Owl, Barred Owl, Cooper's Hawk, and Red-tailed Hawk, as well as mammals such as raccoons (Todd 1940; Voous 1989).

By late summer and early fall, young Eastern Screech-Owls become restless and begin dispersing from their natal areas, apparently influenced by innate factors (Ritchison et al. 1992). Although they leave the nest at 28–35 days of age, most do not venture very far. In a study using radio-tagged juveniles in Kentucky, they were found to have moved only 0.5–16.9 km from the nest site 45–65 days after fledging (Belthoff and Ritchison 1989).

This cavity nester proved difficult to atlas, and was found in only 1.5% (38/2464) of Atlas squares surveyed. Due to the cryptic colouring of the adults, they elude even the most experienced ornithologists as they roost by day in a tree hole or on some sheltered sunny branch. While the young are still in the nest, they are usually overlooked, but the fact that recently fledged young remain close to the nest helps in locating the site. Because of this, the presence of recently fledged young was the most frequently found evidence confirming breeding.

Distribution

The Eastern Screech-Owl breeds eastward from southeastern Saskatchewan, across southern Manitoba, and southern Ontario to southern Québec, and possibly in the most southwesterly part of New Brunswick. It also breeds in southwestern Maine and south to northeastern Mexico and Florida (Godfrey 1986). There are casual or accidental records for Nova Scotia and Prince Edward Island (Erskine 1992).

The Atlas map illustrates a spotty distribution, primarily around Montréal and in the Eastern Townships, along the St. Lawrence Valley as far north as the Shawinigan area, and in a few squares along the Ottawa River. It is still uncertain whether the Atlas map occurately reflects the distribution of the species in Québec. For one thing, the Eastern Screech-Owl is nocturnal, whereas most atlassing was done in the daytime, and usually too late in the season to use auditory cues. Moreover, special efforts by trained teams almost every night of March 1986 and every night of April and May 1988 in the regions south of the St. Lawrence and west of the Chaudière yielded very sparse results.

In Québec, the Eastern Screech-Owl has been sighted as far west as La Sarre (1950) and Aylmer (1987); east to Saint-Eusèbe (1992) in the Lower St. Lawrence and Cap Tourmente (1972) on the north shore; and as far north as Lake Larose (1979), west of Lake Mistassini (Anonymous 1980b; David 1980a; Larivée 1993; ÉPOQ). In May 1992, a specimen was mist-netted on Île aux Lièvres near Rivière-du-Loup (Aubry and Yank 1992a). The bones of an Eastern Screech-Owl were also found in 1948 in a cave along the Saguenay Fjord, near Baie-Sainte-Catherine (Brassard 1961; Savard and Cormier 1995).

The Eastern Screech-Owl takes to nest boxes set up in deciduous woods where there are few natural cavities. Mount Royal, Montréal. *July 1991, Gaétan Duquette.*

The Eastern Screech-Owl is primarily a bird of deciduous forests. It is hard to spot as it is well camouflaged in tree cavities or against the trunk. Franklin Centre, south of Montréal. *May 1991, Pierre Bannon.*

Eastern Screech-Owl dozing in a cavity in an old hardwood. It blends into the background, escaping detection. Mount Royal, Montréal. *October 1988, Paul Perreault.*

History and trends

There are few historical records documenting the distribution of the Eastern Screech-Owl in Québec in the early part of the century, and these are confined to the St. Lawrence and Ottawa Valleys and the Eastern Townships (Terrill 1911b; Dionne 1918; Mousley 1918d, 1921c). Macoun and Macoun (1909), quoting Wintle, who probably overlooked this species as a breeder, declared the species to be a scarce winter visitant in the Montréal area. Although the species has been classified by some authors as a common permanent resident in various regions (Bull 1985; Robbins 1991), it appears to have declined in parts of its range (Temple and Temple 1976; Laughlin 1985a; Weir 1987a), and Ouellet (1974a) believes that it was more numerous in Québec in the 19th century than it is at present. In 1981, the Eastern Screech-Owl was placed on the Blue List because of falling numbers in Ontario and 4 American states (Tate 1981). It was delisted one year later, as populations appeared stable in the East, although it has remained a Species of Special Concern in certain local areas (Tate and Tate 1982). In Québec, the range may have contracted locally in the last decade or two because, even though the species used to be regularly reported around Québec City and Cap Tourmente, few new sightings have been recorded since 1972 (ÉPOQ).

Habitat alteration is believed to be a major threat to the species. According to G. K. Peck, the destruction of old orchards, which traditionally have been the favourite nesting and roosting areas of the species, and the practice of clearing dead trees in remaining orchards are limiting factors for the Eastern Screech-Owl in southern Ontario (Penak 1985). Although it is difficult to ascertain whether these factors have played a part in the lack of sightings in Québec (Larivée 1989), loss of suitable habitat might have been one of several factors. Current practice in horticulture, maple cultivation, and fruit growing, which systematically removes old and dead trees pockmarked with woodpecker holes, eliminates nest sites vital to this species, which is unable to excavate its own cavities. The spraying of fruit trees with pesticides (Temple and Temple 1976) and collisions with vehicles (Keran 1981) have also been blamed for its decline in parts of its range.

Marianne G. Ainley

Great Horned Owl

Grand-duc d'Amérique

Bubo virginianus

The Great Horned Owl, as its name implies, is one of our largest nocturnal birds of prey. It closely resembles the Long-eared Owl, but is larger and does not have vertical streaks on its underparts. When perched, it can be distinguished from other large owls found in Québec by its prominent ear tufts. This powerful predator, with its legendary hooting call, spends almost all day dozing on a branch, only becoming active when it senses prey close by. It roosts and nests in the forest, where it is often mobbed and chased by American Crows, particularly in spring.

The Great Horned Owl is found in a wide variety of habitats throughout virtually every region in all climatic zones of North and South America, from Tierra del Fuego to Alaska; it is absent from the West Indies (AOU 1983). In Québec, this sedentary breeder is the most widespread owl in the province, as well as one of the most common. An early nester, it is often forced to incubate its eggs while blanketed with snow.

Rather than build its own nest, the Great Horned Owl prefers to take over a nest of another species. This nest, in a predominantly deciduous forest, was used in 1981 and 1991 by a Great Horned Owl and by a Northern Goshawk in 1985 and 1992. Saint-Lazare, west of Montréal. *June 1991, Pierre Bannon.*

Habitat and behaviour

The Great Horned Owl is found in a wide variety of habitats, including deciduous and coniferous forests, groves, fields, meadows, stream and riverbanks, and even urban parks (Craighead and Craighead 1956; Gilmer et al. 1983). However, it is mainly associated with open habitats, woodland edges, and clearings such as those found near lakes, ponds, burns, and logged areas (McGarigal and Fraser 1984; DeGraaf and Rudis 1987).

This owl hunts mainly by employing sit and wait tactics, preying on a wide variety of birds and mammals ranging in size from shrews and small passerines to skunks and geese (Weir and Hanson 1989), but may occasionally stalk prey on the ground by turning over rocks. The Great Horned Owl also takes avian and mammalian predators, including the Red-tailed Hawk, Red-shouldered Hawk, Cooper's Hawk, Barn Owl, Eastern Screech-Owl, Barred Owl, Long-eared Owl, weasels, and mink (Bent 1938; Craighead and Craighead 1956). It is largely an opportunistic predator, taking whatever is most available at the time, so seasonal, annual, or even long-term variations in its diet may occur (Rusch et al. 1972; McInvaille and Keith 1974).

The Great Horned Owl often shares its habitat with the Red-tailed Hawk (Smith and Murphy 1982; Gilmer et al. 1983). Some authors believe they are complementary species, the hawk hunting by day and the owl by night (Bent 1938; Craighead and Craighead 1956; Houston 1975).

This large owl is very territorial. From a prominent perch, it warns off intruders with repeated hoots and snaps of its bill (Bent 1938). It is intolerant of both conspecifics and humans (Baumgartner 1939).

The Great Horned Owl becomes sexually mature at around 2 years of age (Weller 1965a). The female is significantly larger than the male, but has similar plumage. Courtship begins in December or January in southern Québec and later farther north, with the male bringing food to the female according to a well-established ritual. Other preliminaries to mating include bowing, gentle pecking, and mutual preening, punctuated by a series of hoots, bill snapping, and other noises (Bent 1938). The pair generally remains together for more than one season (Petersen, L. R. 1979).

The Great Horned Owl expends little energy in building or repairing a nest (Bent 1938; Orians and Kuhlman 1956). Instead, it takes over an abandoned nest of another bird of prey (often the Red-tailed Hawk), or of an American Crow, Great Blue Heron (Bull 1985), or even grey squirrel (Bent 1938; Bosakowski et al. 1989b). In the taiga in northern Québec, the species sometimes uses a Common Raven's cliff nest (Cons. G. G. – GREBE 1990c) or an Osprey's nest (F. Morneau, pers. comm.). Less frequently, it may also occupy an artificial nest site (Scott 1970), nest on the ground (Truslow 1966) or in the hollow top of a tree trunk (F. Morneau, pers. comm.). The nest is usually located in the crotch of a tree or on a horizontal branch (Bosakowski et al. 1989b). Generally, nests cannot be reused because they are rendered uninhabitable by the wear-and-tear inflicted by the highly active young (Bosakowski et al. 1989b). Craighead and Craighead (1956) reported that the Great Horned Owl may destroy other raptors' nests located near their own; this behaviour can probably be observed throughout the species' breeding range.

Food availability appears to be a key factor not only in the timing of the onset of breeding but also in annual variations in nesting. For example, in Saskatchewan, Houston (1987) found breeding success underwent cyclical variations that appeared to be closely linked to fluctuations in the population of its main prey, the snowshoe hare. In peak hare years, most Great Horned Owl pairs produce 4 chicks, 2.5 of which typically survive to fledging. When the hare population crashes, however, many birds migrate south (Houston 1978), and only half the resident population nests, producing as few as 1.6 chicks/brood on average (Houston 1987).

The white eggs are often laid before the snow melts, in February, March, or even January in New York and Vermont (Bent 1938; Laughlin 1985b). In southern Québec, the eggs are laid at the beginning of March and occasionally as early as February. The eggs may have to withstand intense cold, not to mention heavy snowfalls, but according to Houston (1987), are fairly resistant to cold. Incubation is mostly by the female, which is fed by the male while she is sitting.

Luc Leclerc

Notes on territory

The Great Horned Owl is highly territorial throughout the year, but particularly from late autumn or early winter until the end of the nesting period (Baumgartner 1939; Petersen, L. R. 1979).

In winter, the home range of both members of a pair is roughly the same size. In Wisconsin, L. R. Petersen (1979) found home ranges varying from 71 to 266 ha.

At the onset of laying, the male increases his home range by 26–64% and concentrates mainly on hunting. His range shrinks once the eggs have hatched (Petersen, L. R. 1979), but while the young are in the nest, he continues to hunt and will often call from a perch less than 100 m away (Rohner and Doyle 1992).

When the owlets leave the nest, the female begins hunting again, but remains close by. L. R. Petersen (1979) found that the home ranges of both parents gradually expand to 300 ha by October.

Population density was estimated to be 19 pairs/100 km² in Kansas and 10–13 pairs/100 km² in New York (Baumgartner 1939). In an agricultural area of Michigan, population density was only 6.3 pairs/100 km² in 1942, but increased to 7.3 pairs/100 km² in 1948 (Craighead and Craighead 1956).

Great Horned Owl

Breeding distribution

1984–1989

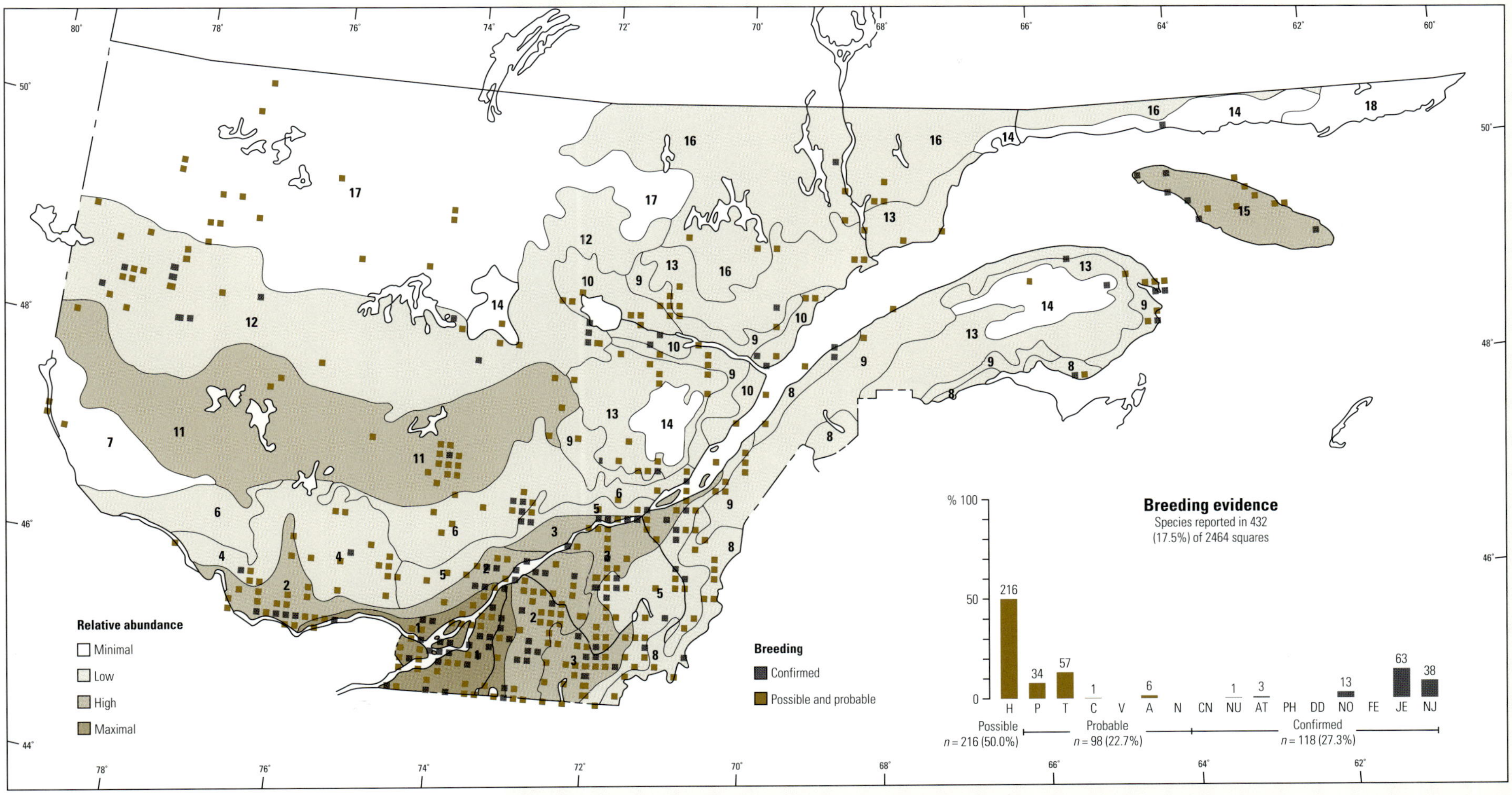

Species profile

Clutch size	usually 2–3 eggs (1–6)[2,3,4]
Laying interval	usually 1 egg per 2 days (2–4)[6]
Start of incubation	with first egg[6]
Incubation period	usually 26–30 days (26–35)[1,3,4]
Maturity at hatching	semi-altricial
Care of young	by female and male[1,2,3,4]
Nestling period	usually 28–35 days[1,3,9]
Age at first flight	35–56 days[3,6,11], 63–70 days[1,4]
Dependence of young	until 5 months old[6]
Broods per year	usually 1 (1–2)[1,2,3]
Breeding age	usually 2 years (1–2)[3,4,8]
Mating system	probably monogamy[3]
Duration of pair bond	more than 1 season[2]
Total length	M 46.0–58.5 cm / F 56.0–63.0 cm[5]
Wing span	91.4–152.4 cm[4]
Weight	M 1318 g (985–1588) / F 1769 g (1417–2503)[10]
Longevity record	20 years 7 months[7]

References:
1. Bent 1938
2. Petersen, L. R. 1979
3. Ehrlich et al. 1988
4. Terres 1980
5. Godfrey 1986
6. Stokes and Stokes 1990
7. Klimkiewicz and Futcher 1989
8. DeGraaf and Rudis 1987
9. Harrison 1978
10. Dunning 1984
11. Keith 1987

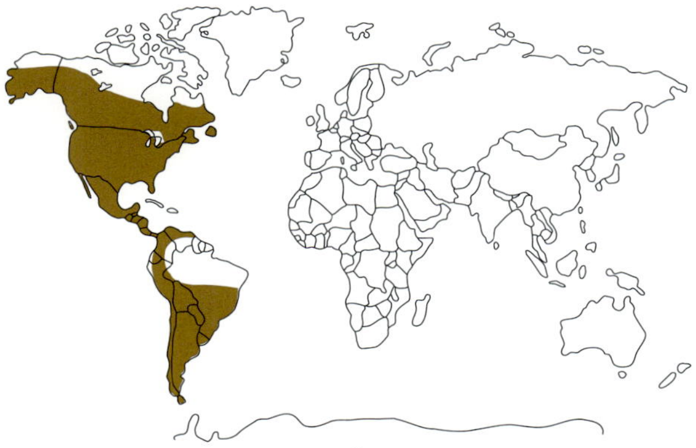

World breeding range

Chronology $n = 174$

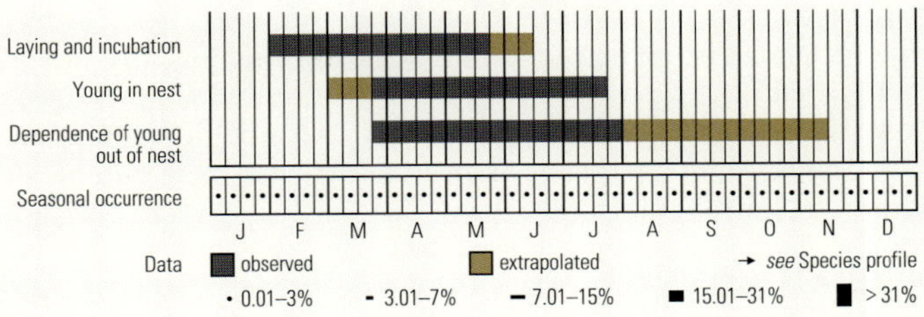

Frequency of occurrence by ecological region

Ecological regions** (forest regions)	Adequately covered squares	Squares in which reported	% of squares in which reported	Relative abundance (%)*
1. Sugar maple-hickory and sugar maple-basswood	97	53	54.6 ± 5.0	100.0
2. Sugar maple-basswood	153	62	40.5 ± 4.0	74.2
3. Sugar maple-basswood and sugar maple-yellow birch	101	47	46.5 ± 5.9	85.2
4. Sugar maple-yellow birch-basswood	56	16	28.6 ± 10.4	52.3
5. Sugar maple-yellow birch-beech and sugar maple-yellow birch-basswood	113	37	32.7 ± 4.8	59.9
6. Sugar maple-yellow birch-beech	71	16	22.5 ± 9.0	41.2
7. Sugar maple-yellow birch-balsam fir	19	3	15.8 ± 17.9	28.9
8. Sugar maple-yellow birch and balsam fir-yellow birch	47	12	25.5 ± 11.3	46.7
9. Balsam fir-yellow birch	87	15	17.2 ± 7.4	31.6
10. Balsam fir-white birch or balsam fir-red maple	46	11	23.9 ± 10.8	43.8
11. Yellow birch-balsam fir	52	18	34.6 ± 13.4	63.4
12. Balsam fir-white birch (west)	59	15	25.4 ± 11.5	46.5
13. Balsam fir-white birch (east)	76	13	17.1 ± 8.4	31.3
14. Balsam fir-black spruce	28	3	10.7 ± 12.6	19.6
15. White spruce-balsam fir and balsam fir-white spruce	22	9	40.9 ± 21.2	74.9
16. Black spruce-balsam fir-moss	18	4	22.2 ± 22.2	40.7
17. Black spruce-moss	32	3	9.4 ± 11.7	17.2
18.	*not analysed*			

* minimal: 0–29.9%; low: 30–59.9%; high: 60–89.9%; maximal: 90–100%

** Deciduous forest: 1–7; mixed forest: 8–11; coniferous forest: 12–17

Notes on territory (continued)

Laughlin (1985b) reported a pair in Vermont that occupied the same home range of 15.5 km² for 10 years in a row.

This member of the family Strigidae often nests and roosts in a grove of tall trees, sallying forth to hunt in nearby fields. Cap Tourmente National Wildlife Area, northeast of Québec City. *August 1991, Yves Aubry.*

The eggs hatch asynchronously after about a month. Covered with a thick layer of down at hatching, the noisy and ravenous owlets remain in the nest for 6 or 7 weeks. During this time, the parents defend the nest fiercely against intruders, which are attacked without hesitation. At the age of 28–35 days, the owlets may leave the nest to climb and perch on nearby branches. Both parents provide food for the young, feed them, and watch over them during this period (Terres 1980; Stokes and Stokes 1990).

By about 8 weeks of age, the young are well feathered, except for their poorly developed ear tufts, and can fly short distances (Keith 1987). At 63–70 days, they are capable of sustained flight and are able to follow their parents (Ehrlich et al. 1988; Stokes and Stokes 1990). They begin to learn basic hunting skills from their parents, but remain dependent for food until August. In Wisconsin, juveniles were observed in their parents' home range until late December (Petersen, L. R. 1979).

The Great Horned Owl was reported in 17.5% (432/2464) of Atlas squares surveyed. The breeding season is long—6 or 7 months—making it easier to find definite evidence of nesting, and breeding was confirmed in 27.3% (118/432) of the squares where the species was reported; 85.6% (101/118) of the records of confirmed breeding were reports of young out of the nest or nests containing eggs or owlets. The male's hooting during the nesting period, generally from a perch less than 100 m from the nest, may also help locate breeding pairs (Rohner and Doyle 1992).

The Great Horned Owl does not usually migrate and winters within its breeding range. Outside the breeding season, however, birds may wander extensively, often over long distances (Godfrey 1986). Adamcik and Keith (1978) reported that periodic mass movements occur every 10 years in western Canada, in tandem with cyclical crashes in the snowshoe hare population.

Two owlets in the nest, built on a protected cliff ledge. At the front of the nest is the foot of a snowshoe hare, one of the Great Horned Owl's favourite prey. La Grande-2, James Bay region. *June 1989, Jean-Luc DesGranges.*

This opportunistic predator also nests in woodlots in agricultural and semirural areas, like this stand of red oaks near Mount Saint-Hilaire. *August 1991, Yves Aubry.*

A Great Horned Owl guards its young in a nest built by a Great Blue Heron. Île aux Hérons, near Montréal. *June 1991, Jean-Luc DesGranges.*

Distribution

Two subspecies are commonly found in Québec. *Bubo virginianus virginianus*, which is medium dark with some redness in the plumage, is found throughout the southern part of the province. *Bubo virginianus heterocnemis*, which is darker, is encountered in central and northern Québec up to Kuujjuaq (Godfrey 1986). Although the western boundary of its range cannot be precisely defined due to the small number of records from the region (Todd 1963), this subspecies is thought to occur as far west as Lakes Saint-Jean and Mistassini, where it was recorded in the late 1940s (Godfrey and Wilk 1948; Godfrey 1949). A third subspecies, *Bubo virginianus subarcticus*, is occasionally found in southern Québec in winter; the eastern limit of its range is not known, but it may be distributed right to westernmost Québec, in the Eastmain Lowland (Cons. G. G. – GREBE 1992m). Its southward incursions may be synchronous with crashes in the snowshoe hare population. This subspecies is easily identified by its grey, whitish, or occasionally pure white plumage (Todd 1963).

Although the Great Horned Owl is found in all forested regions of Québec as far north as the tree line, it is not abundant anywhere. It breeds throughout the province, including the Upper and Middle North Shore as far east as Mingan, but not in the Magdalen Islands. Its highest frequency of occurrence for the Atlas was in the Montréal and Montérégie areas, the western part of the Eastern Townships, and the southern Ottawa Valley, where a patchwork of farmland and woodlots can be found.

The northernmost records in the province are from near Lake Guillaume Delisle and Kuujjuaq (Todd 1963). In addition, 2 nests were found in 1990 on cliffs in the northern part of the Great Whale River watershed (Cons. G. G. – GREBE 1990c).

History and trends

Like most birds of prey, the Great Horned Owl has been persecuted for centuries. The animosity it has traditionally inspired is reflected in nicknames such as "evil genius of the woods" common at the turn of the century. Taverner's (1919) remarks were typical of the biases held against the species at the time: "There is no danger that restriction will result in its extermination since it is perfectly able to take care of itself and there are large sections where the species can hold sway over the wilderness without molestation." Fortunately, public-education campaigns and the numerous regulations enacted in Québec and elsewhere in Canada in this century to protect birds of prey have helped to break down these prejudices (Cazelais 1994).

At the turn of the century, the Great Horned Owl was considered common in Montréal, Québec City, and along the St. Lawrence (Dionne 1906; Macoun and Macoun 1915). More recently, Ouellet (1974a) listed it as scarce in the Montérégie and Montréal areas. Although urban expansion over the last few decades has no doubt resulted in local declines, the Atlas data show that the Great Horned Owl is still extremely widespread in Québec, particularly in the southwestern part of the province.

The Great Horned Owl should continue to prosper, given its great versatility with regard to diet, nesting sites, and habitats.

Christine Langevin and Manon Bombardier

Northern Hawk Owl

Chouette épervière

Surnia ulula

Ghislain Caron

WITH ITS POINTED WINGS AND LONG TAIL, THE NORTHERN HAWK OWL IS aptly named, resembling a small accipiter hawk in both size and appearance. It flies with swift wingbeats, alternating with glides. Unlike other owls, it can hover in the air and, when perched, it flicks its tail like a kestrel. It is morphologically different from other Strigidae in many respects, including its compact rather than downy plumage, poorly developed facial disc, small eyes, and falconlike lower mandible with a distinct terminal notch. In addition, the species' ear cavities are not covered by an operculum, or ear flap, and are symmetrical rather than asymmetrical, which means that it is less adept than other owls at pinpointing the exact locations of sounds. The feathers covering its toes are almost as long, proportionally, as those of the Snowy Owl (Karalus and Eckert 1987; Voous 1989).

Despite its falconlike adaptations, the Northern Hawk Owl feeds almost exclusively on small mammals and appears to occupy a diurnal niche left vacant by other raptors (Kertell 1986).

The northern boundary of the species' breeding range corresponds roughly to the northern limit of the forest-tundra, where the arctic tundra begins. Mikkola (1983) described the Northern Hawk Owl as a species of the boreal forests of Canada, Alaska, and Eurasia; in other words, a typical species of northern coniferous forests.

Habitat and behaviour

Typical breeding habitats are open coniferous or mixed woodlands bordering open areas such as peatlands, clearings, or cutovers, where snags and clumps of trees can be found (Bent 1938; Mikkola 1983; Godfrey 1986; Jones 1987). For hunting, it seems to prefer open areas such as muskegs and burns, particularly those with stumps and snags that can be used as lookout perches. A factor in habitat selection is the risk of nest predation by animals such as mink, which frequents open riverside habitat, and marten, which prefers more heavily wooded areas (Sonerud 1985). Proximity to water (within 100 m from the nest) is also important, since a variety of prey is attracted to the water's edge, where it is more exposed and easier to capture.

The Northern Hawk Owl flies from one lookout post to another in search of prey, often selecting the tallest conifers, such as a spruce tree, as vantage point. The species is well adapted to its environment and, when perched, blends in admirably with the tops of the surrounding spruces owing to its profile and posture. The Northern Hawk Owl generally hunts using sit-and-wait tactics. From an elevated perch—the top of a spruce more than nine times out of ten—it swoops down on its prey like a shrike (Kertell 1986; Lane and Duncan 1987). According to Jones (1987), the species can spot a mouse more than 700 m away; it hunts generally within a 90–100 m radius of its nest (Simon and Simon 1980; Mikkola 1983; Jones 1987).

The Northern Hawk Owl's diet is governed by the availability of prey, mainly small mammals such as voles, mice, and shrews. It also takes squirrels, minks, frogs, fish, and occasionally birds as large as ptarmigan, which are mainly caught on the wing. It may also feed on insects (Fisher 1893; Bent 1938; Mikkola 1983; Voous 1989).

In Norway, Nybo and Sonerud (1990) demonstrated that when the ground is completely covered in snow, Northern Hawk Owls change their feeding habits radically. Since rodents in open areas tunnel under the snow to avoid detection, the owls hunt in the forest undergrowth, where they can see their prey because the canopy keeps the forest floor relatively clear of snow. This owl can adapt to the rapid changes in the environment typical of the northern boreal forest, even in the midst of breeding (Nybo and Sonerud 1990). In winter or other periods of food scarcity, the Northern Hawk Owl may pirate prey from other species, like the American Crow (Ainley 1978).

The Northern Hawk Owl generally builds its nest in the hollow top of a stub or in the crotch of a snag, 1.6–18.3 m from the ground (Bent 1938; Jones 1987; Campbell et al. 1990b). In the absence of suitable trees, the nest may be built on the ground (Terres 1980). The species is small enough to use on occasion an abandoned nest cavity of a Northern Flicker or Pileated Woodpecker (Henderson 1919, 1925) or, less often, the old nest of another diurnal raptor or American Crow. The nest is usually a roughly built structure of interlocking twigs, lined with grasses and moss. The female adds very little new material to the nest, but lines it with feathers plucked from her lower breast (Karalus and Eckert 1987).

The Northern Hawk Owl is primarily diurnal. Collins (1991) has demonstrated that its body temperature is directly related to its activity level, reaching a peak between 2:00 p.m. and 5:00 p.m. and then dropping. Little else is known about the species' breeding biology in Québec. The Northern Hawk Owl not only inhabits remote areas, where access is often difficult for biologists, but is nomadic, rarely nesting in the same place two years in a row. It is known, however, that courtship behaviour in the species consists of display flights over the treetops of the territory and wing clapping by the male as he flies among the trees (Mikkola 1983). The male becomes quite vociferous during courtship, which usually begins around mid-March or sometimes as early as February (Karalus and Eckert 1987) and may continue into mid-April, as has been observed in Finland (Johnsgard 1988). The pair establishes its territory a few weeks before the onset of nesting, which sometimes begin as late as early May (Harrison 1973 *in* Kertell 1986). The timing of egg laying varies considerably (Mikkola 1983), ranging from 16 April to 14 June in Alaska and northern Canada, and from 3 May to 20 June in Labrador and Newfoundland (Bent 1938; Karalus and Eckert 1974; Kertell 1986). Although the Atlas sample was very small ($n = 9$), it seems realistic to set egg laying and incubation in Québec between the beginning of May and mid-July.

Incubation is performed almost entirely by the female, while the male does all the hunting for the family (Kertell 1986). During incubation, the female remains on the nest, except for short periods when she leaves to receive food from the male, preen, cast pellets, or defecate. She does not always go to the male immediately when he brings food, but occasionally takes the food from a cache he has left for her (Kertell 1986; Huhtala et al. 1987). For the first two weeks after the eggs have hatched, the female remains almost constantly on the nest to brood the young, and the male brings the food directly to her there (Kertell 1986; Huhtala et al. 1987). When he is not hunting, the male defends the nest, standing sentinel in a treetop about 30 m away.

Notes on territory

The male's home range is used for mating, nesting, foraging, and brood rearing (Cramp 1985). Home ranges appear to be extensive, since individuals may roam as far as 1 km from the nest to hunt (Mikkola 1983).

In 1988, in the Martin-Valin ZEC near Saint-Fulgence, 25 km north of the Saguenay River, a male was observed for 30 hours to determine which lookout perches he used. The bird's home range, which was bordered in part by a stream, was estimated to be approximately 0.6 km² (60 ha) (J. Gauthier, CWS, unpubl. data).

In the boreal forest, the Northern Hawk Owl nests and forages in cutover areas. Water nearby means a greater number of prey. Laurentide Wildlife Reserve, north of Québec City. *June 1991, Réginald Ouellet.*

Northern Hawk Owl

Breeding distribution

1984–1989

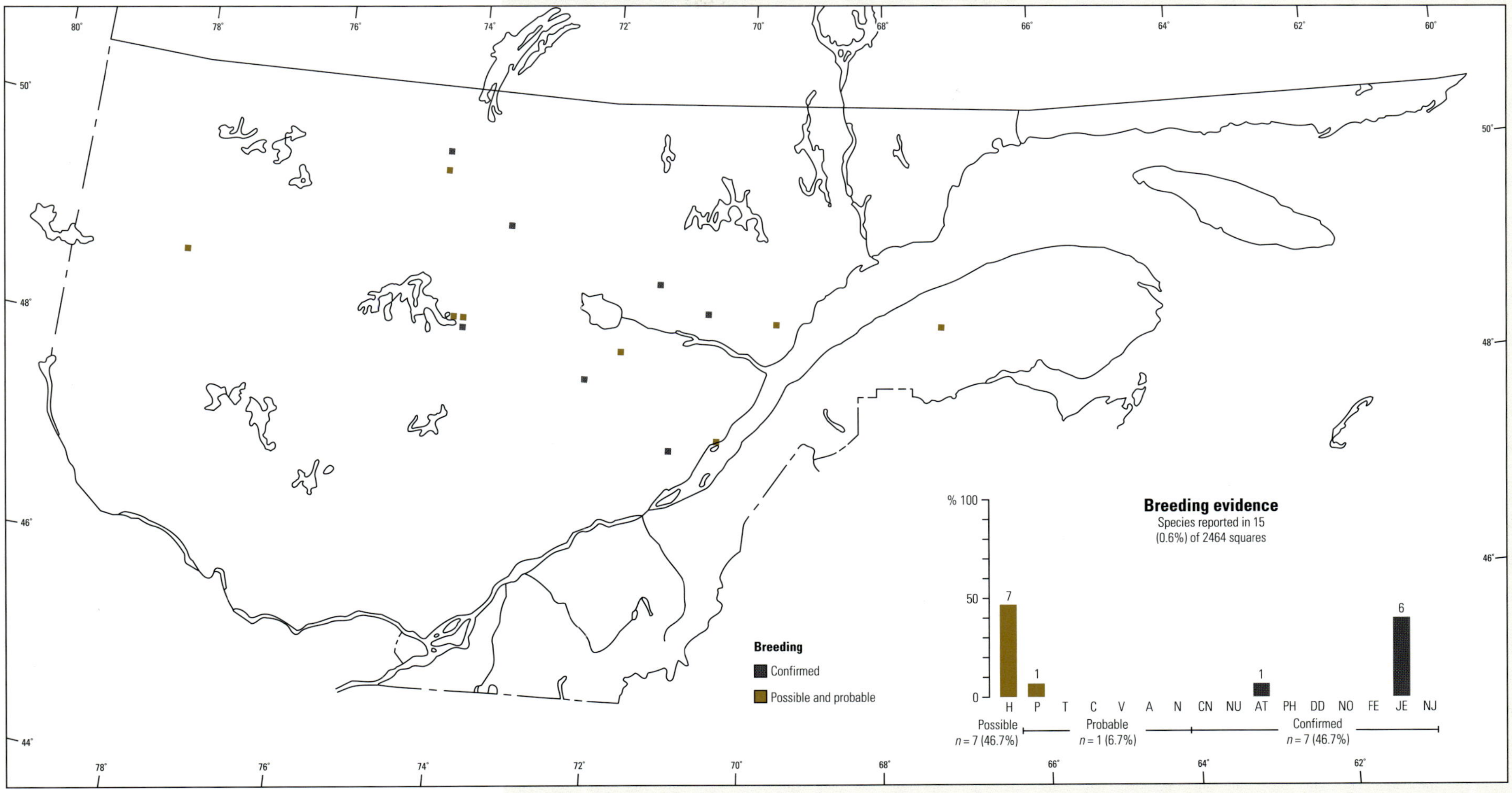

Species profile

Clutch size	usually 5–7 eggs (3–13)[1,2,3,4,11]
Laying interval	1 egg per 1–2 days[2]
Start of incubation	with first egg[1,2,3]
Incubation period	25–30 days[2,3,4,10]
Maturity at hatching	semi-altricial
Care of young	by female and male[5], *see* text
Nestling period	20–28 days[2,4]
Age at first flight	25–35 days[2,4,10]
Dependence of young	until 60–90 days old[2,10]
Broods per year	1[10]
Breeding age	1 year[2]
Mating system	monogamy, sometimes polygyny[2,9]
Duration of pair bond	until at least following spring[1,2,3]
Total length	36.5–43.0 cm[8]
Wing span	83.8 cm[7]
Weight	M 299 g (273–326) / F 345 g (306–392)[6]
Longevity record	10 years[3]

References:

1. Bent 1938
2. Johnsgard 1988
3. Karalus and Eckert 1987
4. Kertell 1986
5. Ehrlich et al. 1988
6. Dunning 1984
7. Terres 1980
8. Godfrey 1986
9. Sonerud et al. 1987
10. Cramp 1985
11. Mikkola 1983

Chronology *n* = 9

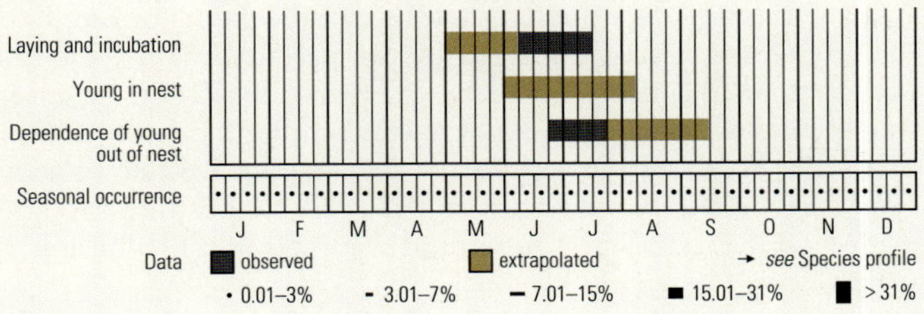

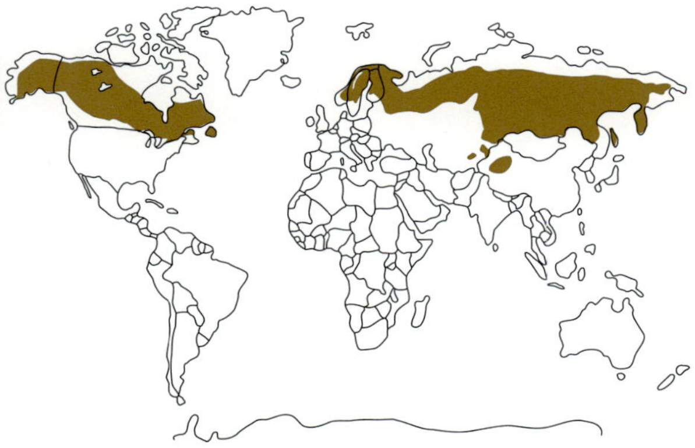

World breeding range

In 1989, a pair of Northern Hawk Owls nested in the hollow top of a snag in this second-growth stand. Martin-Valin ZEC, north of the Saguenay River. *August 1991, Jean Gauthier.*

When the nestlings are about 2 weeks old, the male again begins to bring prey to a perch near the nest, where the female goes to receive it (Kertell 1986). From this time until the end of the nestling period, the female spends more and more time perching outside the nest (Kertell 1986; Huhtala et al. 1987).

Brood rearing requires an enormous amount of energy from the parents. For their first 18 days, the owlets grow at an astonishing rate of 9 g/day. Their growth rate then slows somewhat, but by the time the owlets are 3 or 4 weeks old, they have reached 78% of their adult weight (Huhtala et al. 1987).

As Kertell (1986) and Huhtala et al. (1987) explain, the species achieves this high rate of development under the often unpredictable conditions in northern habitats by adopting a strategy of temporarily caching excess food near the nest. This allows the numerous nestlings not only to be fed almost continually but also to survive during periods of scarcity. According to Kertell (1986) and Karalus and Eckert (1987), the low juvenile mortality rates support this hypothesis, although much more information is needed on the subject.

The young leave the nest at around the age of 3 or 4, or sometimes 5, weeks, when their wings are only 60% of adult size and they are not yet capable of sustained flight (Huhtala et al. 1987). Since hatching is asynchronous, a week may pass between the time the first and last nestlings leave the nest (Huhtala et al. 1987). Once out of the nest, the owlets spend their time jumping from branch to branch, walking on the ground, and climbing the trees around the nest: for the first 10 days of this period, the female remains perched near her offspring and relies on the male to bring food. The male perches approximately 100 m from his mate when he is not hunting (Kertell 1986). Around the age of 4–5 weeks, the young begin to fly short distances, venturing up to 30 m from the nest. Then, the roles of the sexes change drastically, with the male perching near the young and feeding them, while the female goes off for long periods, presumably to hunt (Kertell 1986). Interestingly, males tend to give small prey whole to the young, while females go on shredding it for the owlets.

Until they are around 2 months old, the young remain in the vicinity of the nest and are fed by the parents, mainly the male (Johnsgard 1988). During this time, observers should not try to look into the nest even if the young have left, because the parents, particularly the male, are very aggressive and will not hesitate to attack, even striking and wounding intruders on occasion (Lane and Duncan 1987; J. Gauthier, pers. obs.). In winter, the owls seem to be less aggressive.

The young are completely independent at around 3 months old and begin to disperse (Johnsgard 1988). Some authors, including Bent (1938) and Karalus and Eckert (1987), however, have reported that family groups may remain together until the following spring.

Generally, the Northern Hawk Owl is monogamous, but like the Barn Owl (Schönfeld and Girbig 1975 *in* Sonerud et al. 1987) and the Snowy Owl (Watson 1957; Hagen 1960 *in* Sonerud et al. 1987), it may, exceptionally, be polygynous. In 1984, a male was seen in Norway bringing food to 2 females nesting in the same territory, 1050 m from one another (Sonerud et al. 1987). When this occurs, the male tends to favour the primary female, or the female with the larger brood, over the other female during the nestling stage, a phenomenon that has also been observed in passerines (Sonerud et al. 1987). Breeding success in the second brood is therefore less assured, and the nest may be abandoned altogether, as Sonerud et al. (1987) found.

The species is fairly easy to observe in the proper habitat and nests are not difficult to find once you spot one of the parents. Usually this will be the male, which during most of the breeding cycle, ranges over a much larger area than the female. If you spot a male making a kill, you can then locate the nest because he usually flies directly to the nest afterward (Jones 1987). Nests are rarely found, however, because of the species' tendency to nest in remote, inaccessible areas. The male is easier to spot at the beginning of the breeding season, when he perches in treetops, roves extensively around the territory, and is quite vociferous. In June and July, however, when atlassing was done, the species is quieter and less conspicuous. This is no doubt part of the reason why the Northern Hawk Owl was reported in only 0.6% (15/2464) of Atlas squares surveyed. Furthermore, breeding was confirmed in only 46.7% (7/15) of the squares in which the species was reported, most often from records of recently fledged young (6/15) incapable of sustained flight.

Distribution

Generally speaking, the handful of Atlas records obtained bears out the Northern Hawk Owl's known distribution north of the 47th parallel, in the northern boreal forest, with its fairly extensive tracts of conifers. Similar habitat can be found in the Appalachians, where the species was reported at Saint-Tharcisius, near Causapscal

Notes on territory (continued)

This is much smaller than the home ranges reported by Baekken et al. (1987) in Norway, which ranged between 1.4 and 8.5 km² (n = 5), but is similar to those observed by Lane and Duncan (1987) on the Manitoba–Minnesota border, which ranged from 0.20 to 1.04 km². In Norway, nests belonging to monogamous pairs were at least 600 m apart (Baekken in Sonerud et al. 1987).

The territory defended by the male is probably also quite large, since 2 nests have never been found close together (Mikkola 1983). The Northern Hawk Owl appears to be tolerant of other raptors such as the Short-eared Owl and Great Gray Owl, and nests of these species have been found within 500 m of a Northern Hawk Owl's nest (Mikkola 1983), but the

When this snag collapsed, observers were able to take a look at the Northern Hawk Owl nest inside. The entrance to the nest was on the top of the snag, which can be seen on the left. Note the accumulation of nesting materials over the years. Near the Great Whale River, Hudson Bay region. *June 1992, Serge Brodeur.*

In 1990, a pair of Northern Hawk Owls nested in this burn, with its scattered snags and conifers. Near the Great Whale River, Hudson Bay region. *June 1990, Serge Brodeur.*

Northern Hawk Owl will react to the presence of the Great Horned Owl, a predator it sees as a threat (Lane and Duncan 1987).

In Norway, breeding pair density seems to be quite low, around 4 pairs/200 km² (Hagen 1956 in Nybo and Sonerud 1990), and even lower in Sweden, 1 pair/500 km² in suitable habitat (Cramp 1985).

(Lower St. Lawrence). This lends support to an earlier observation of a territorial male in a peatland near Lake Rimouski in 1980, approximately 15 km from the city of Rimouski (R. Ouellet, pers. comm.).

The southern limit of the species' range is around the 48th parallel in Ontario (Weir 1987f) and around the 47th parallel in Québec. Farther south, suitable habitat is usually found only at higher elevations and in some peatlands. In late July of 1992, an adult feeding a juvenile was observed at Sainte-Béatrix, north of Joliette, in the Laurentian foothills on the northern edge of the St. Lawrence Plain, which is the first breeding record this far south in Québec (Aubry and Yank 1993a). In Ontario, in 1963, breeding was confirmed at roughly the same latitude, in the La Mer Bleue peatland near Ottawa, after an irruption by the species the previous fall and winter (Smith, D. A. 1970). In the Maritimes breeding bird atlas, the Northern Hawk Owl was also reported nesting at a similar latitude in New Brunswick (Erskine 1992).

The lack of Atlas records from the North Shore back country can be explained to some extent by the inaccessibility of the region, the timing of atlassing (mid-June to the end of July), and the scarcity of the species. The single summer record from this region is of a female collected at the mouth of the Natashquan River in 1928 (Todd 1963). Comeau (1909), however, noted the species there in winter. The Northern Hawk Owl probably inhabits the region, since suitable habitat is found there. The species does not appear to nest in the Magdalen Islands, although owls have occasionally been seen there (Fradette 1992), and some habitat, like stands of white spruce on the edges of fields, is suitable for the species.

The Northern Hawk Owl's distribution extends north to the tree line. According to Todd (1963), the northern limit of its range in Québec follows a line from the coast of Hudson Bay at roughly the 56th parallel, which then runs in a general northeasterly direction to southern Ungava Bay; the species is not found on the Ungava Peninsula.

Surveys in northern Québec in the summer of 1990 confirmed nesting near Nemiscau. A nest was found and 2 family groups were observed near the Great Whale River, less than 100 km from Hudson Bay (Cons. G. G. – GREBE 1990c, 1992m).

Although the Northern Hawk Owl is generally considered to be resident throughout its range, it is nomadic and roams widely within this area. In winter, sporadic movements to the southern part of its range occur, and the species can be seen in the northern United States (Mendall 1944; Lane and Duncan 1987; Root 1988) or even farther south at times. In fall and winter, it is occasionally recorded south of the boreal forest, in parks and other open habitats (Smith, D. A. 1970; Jones 1987).

History and trends

Although the Northern Hawk Owl is a permanent resident in Québec, it is very difficult to determine population trends here. Despite significant atlassing effort, this owl was reported in only 0.6% of Atlas squares surveyed, a very low frequency of occurrence compared with other species, inferring that the species is rare. Similarly, Ontario atlassers obtained only 10 records (Weir 1987f). The results also confirm Erskine's (1977) impression that the species occurs only in very low densities. The Northern Hawk Owl dwells far from urban areas, in regions difficult for observers to get to, and only invades the southern part of the province when prey becomes scarce in the north. Periodic irruptions have been reported since the turn of the century in various locations in southern Québec, particularly in the Montréal region. Several authors considered the species common in Montréal in winter at the turn of the century (Vennor 1860; Hall 1862; Caulfield 1890a), while others reported it as scarce (D'Urban 1857; Wintle 1896). The situation appears to have been similar on Anticosti Island, where Schmitt (1904) described the Northern Hawk Owl as fairly common in autumn, winter, and spring, while Dionne (1920), not long afterward, reported that it was no longer seen.

These sporadic fall and winter invasions of some regions, governed by food availability, demonstrate the essentially nomadic character of the Northern Hawk Owl (Mikkola 1983). Hagen (1969) showed that prey scarcity may be the cause of poor breeding success among birds of prey like the Northern Hawk Owl; some pairs may not even breed at all in poor prey years. Conversely, productivity is high when rodents are plentiful. In a large area like a province or country, rodent populations fluctuate regionally and reproductive success in some raptors rises and falls with rodent populations (Hagen 1969; Mikkola 1983). Although data on the species' movements in North America are scarce, observations in Eurasia suggest that population increases in a region resulting from good breeding success are followed in autumn by the pronounced movement of many individuals away from the breeding range, but not necessarily to the south (Hagen 1969; Mikkola 1983; Cramp 1985; Johnsgard 1988).

When prey populations remain at normal densities, Northern Hawk Owls generally tend to winter on their northern breeding grounds. In years when prey is unusually abundant in spring, however, breeding success will be high and many juveniles will be produced; if prey is scarce the following autumn, many of these juveniles will irrupt southward; such irruptions may consist of over 80% juveniles (Johnsgard 1988). This high percentage of juveniles is not typical, however, and the birds in normal irruptions are mainly adult females, followed by juvenile males and, at much lower levels, by adult males and juvenile females; this composition is believed to result from the effects of competition for breeding territories (Johnsgard 1988). In addition, when prey is abundant in spring on the southern wintering grounds, some birds may remain there to breed (Mikkola 1983). This occurred in Québec in 1992 (Yank and Aubry 1992b), in Ontario in 1963 (Smith, D. A. 1970), in southern Manitoba and northern Minnesota in 1987 (Lane and Duncan 1987), and in Europe in 1978 (Mikkola 1983).

MANON BOMBARDIER AND JEAN GAUTHIER

Barred Owl

Chouette rayée

Strix varia

René Roy

THE BARRED OWL IS ONE OF THE MOST COMMON OF THE NOCTURNAL typical owls, family Strigidae, in North America and was the second commonest owl atlassed. It is a permanent resident of the forests of southern Canada, from British Columbia to Nova Scotia (Godfrey 1986), and is associated with mature woodlands throughout its range. Because it is highly vocal, with a vast repertoire of calls, squeaks, and whistles, it is easy to determine if the Barred Owl is in the neighbourhood.

The field marks that differentiate it from other typical owls are its blackish eyes, barred breast, and striped abdomen. It has no ear tufts on its rounded head, and in size and habits it greatly resembles the Spotted Owl, which, however, mainly inhabits the southwestern part of North America (Voous 1989). Barred Owls are sedentary birds that are strongly attached to their nest sites (Bent 1938).

Habitat and behaviour

Typical breeding habitat is mature woodland, where it can find suitable nest cavities, containing deciduous and mixed stands (Erskine 1977). According to Elderkin (1987), the availability of appropriate nesting sites rather than the age of the forest is the more important limiting factor in habitat selection. Its association with wet areas has been widely noted (Bent 1938; Soucy 1976; Bosakowski et al. 1987), and the presence of marshes and swamps in the vicinity of its preferred woodlands does appear to afford a certain degree of protection against human intrusion, as well as providing a greater abundance and variety of prey (Bosakowski et al. 1987). Other authors have, however, found the Barred Owl in dry upland forests in some parts of its range (Nicholls and Warner 1972; Fuller 1979).

The Barred Owl often shares its habitat with the Great Horned Owl, which also prefers mature stands, but unlike the Great Horned Owl, it hunts by pouncing on its prey and so does not require such wide open spaces (McGarigal and Fraser 1984). In the southern part of its range, the Barred Owl is also associated with the Red-shouldered Hawk, with which it shares similar food and habitat requirements. It is not uncommon to find the nests of these two species close together in the same wood (Bent 1938). Although the Barred Owl preys mainly on small mammals, especially mice, voles, rats, and shrews (Devereux and Mosher 1984), it is opportunistic in its diet and will occasionally feed on birds and insects. The relative importance of certain types of prey in its diet appears to depend on latitude (Elderkin 1987).

The frequent hooting of the Barred Owl makes it easy to locate, even during the day, and it readily responds to recordings or imitations of its call, especially between the beginning of courtship and the onset of laying (Fuller and Mosher 1981). If you sit quietly and imitate the call, you can coax the bird to perch nearby, and it will carry on a dialogue, sometimes for several minutes. Its hooting is more varied and higher than that of the Great Horned Owl. A study based on recorded calls showed that the bird has six distinct types of vocalizations (McGarigal and Fraser 1984). The most frequently heard hoot is commonly transcribed as *Who cooks for you? who cooks for you-all!* (Terres 1980).

Notes on territory

The pair defends its territory all year round, generally passively, by calling, but intraspecific confrontations may occur in the breeding season (Elderkin 1987). At the onset of winter, the male expands his territory as prey becomes less abundant, or may even desert it, leaving its defence to the female (Dunstan and Sample 1972).

Pair density appears to be relatively low throughout much of North America. For example, only 0.3 and 0.7 pair/10 km² were reported in Michigan and New Jersey respectively (Craighead and Craighead 1956; Bosakowski et al. 1987). In southern Québec, however, considerably higher densities have been recorded. At Mount Saint-Bruno between 1991 and 1995, an average density of 4.5 pairs/10 km² was reported, and in 1995 at Mount Yamaska, densities of 2.7–4.7 pairs/10 km² were obtained (F. Morneau, pers. comm.). These two mountains, both forested, are surrounded by farmland.

The Barred Owl inhabits mature deciduous and mixed forests in southern Québec. Mount Saint-Hilaire, with its century-old stands of oak and sugar maple, is one of its breeding areas. *August 1991, Yves Aubry.*

Barred Owl

Breeding distribution

1984–1989

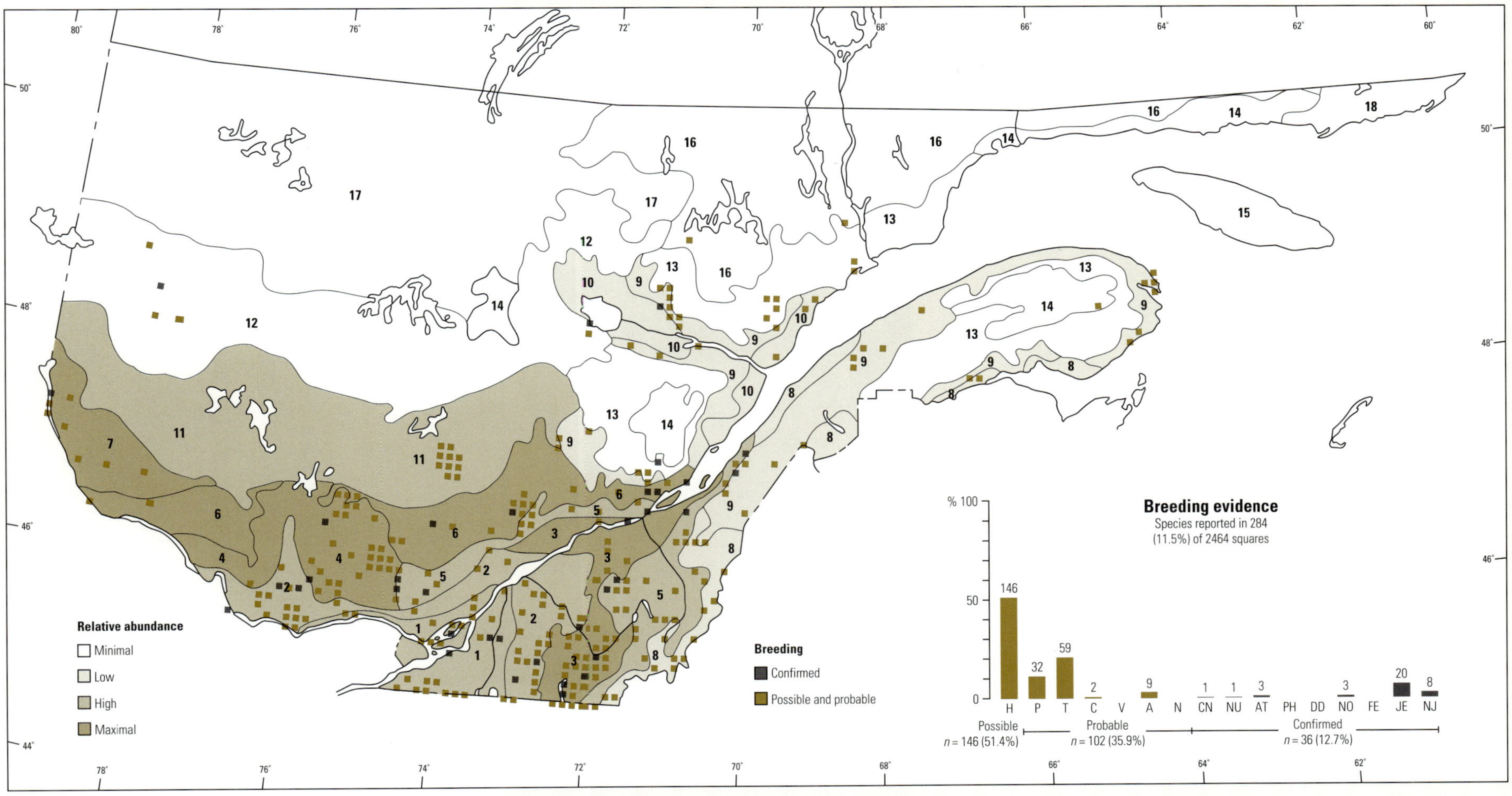

Species profile

Clutch size	usually 2 eggs (1–4)[1,2,3]
Laying interval	1 egg per 1–3 days[1]
Start of incubation	with second egg[1]
Incubation period	28–33 days[1]
Maturity at hatching	semi-altricial
Care of young	by female and male[1,2,4], *see* text
Nestling period	4–5 weeks[1,2,9]
Age at first flight	6–9 weeks[2,4,9]
Dependence of young	until end of summer or beginning of autumn[8,10]
Broods per year	usually 1 (1–3)[1]
Breeding age	?
Mating system	monogamy[1]
Duration of pair bond	for life[1]
Total length	45.5–58.0 cm[3]
Wing span	101.6–127.0 cm[7]
Weight	M 632 g (468–774) / F 801 g (610–1051)[5]
Longevity record	18 years 2 months[6]

References:
1. Elderkin 1987
2. Bent 1938
3. Godfrey 1986
4. Dunstan and Sample 1972
5. Dunning 1984
6. Klimkiewicz and Futcher 1989
7. Terres 1980
8. Stokes and Stokes 1990
9. Voous 1989
10. Devereux and Mosher 1984

Chronology $n = 39$

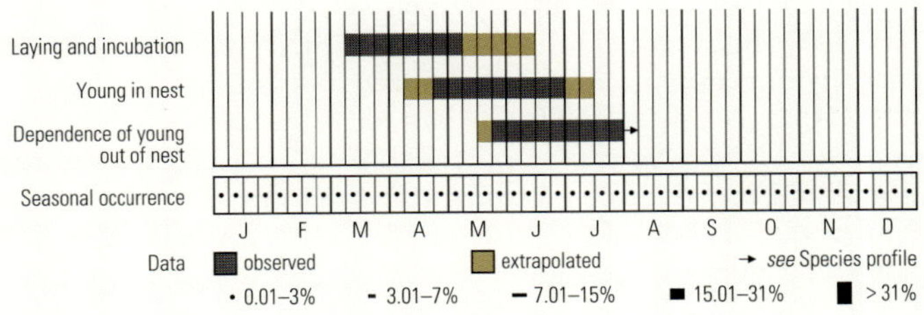

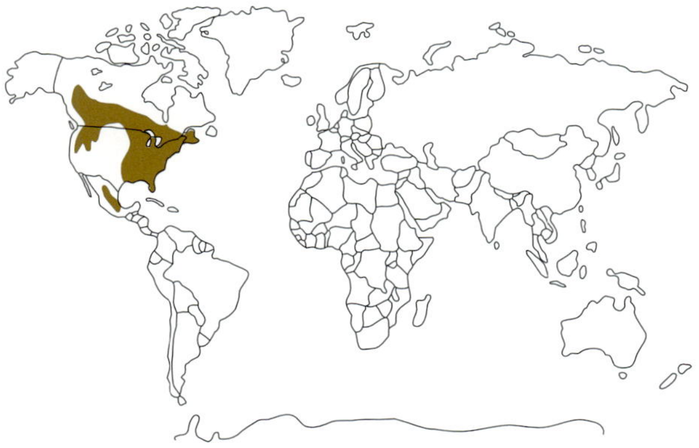

World breeding range

Frequency of occurrence by ecological region

Ecological regions** (forest regions)	Adequately covered squares	Squares in which reported	% of squares in which reported	Relative abundance (%)*
1. Sugar maple-hickory and sugar maple-basswood	97	25	25.8 ± 4.4	68.7
2. Sugar maple-basswood	153	44	28.8 ± 3.8	76.7
3. Sugar maple-basswood and sugar maple-yellow birch	101	36	35.6 ± 5.7	95.0
4. Sugar maple-yellow birch-basswood	56	21	37.5 ± 11.1	100.0
5. Sugar maple-yellow birch-beech and sugar maple-yellow birch-basswood	113	36	31.9 ± 4.8	85.0
6. Sugar maple-yellow birch-beech	71	24	33.8 ± 10.1	90.1
7. Sugar maple-yellow birch-balsam fir	19	7	36.8 ± 22.8	98.2
8. Sugar maple-yellow birch and balsam fir-yellow birch	47	8	17.0 ± 9.9	45.4
9. Balsam fir-yellow birch	87	19	21.8 ± 8.1	58.2
10. Balsam fir-white birch or balsam fir-red maple	46	6	13.0 ± 8.8	34.8
11. Yellow birch-balsam fir	52	12	23.1 ± 11.9	61.5
12. Balsam fir-white birch (west)	59	4	6.8 ± 7.0	18.1
13. Balsam fir-white birch (east)	76	7	9.2 ± 6.6	24.6
14. Balsam fir-black spruce	28	1	3.6 ± 8.3	9.5
15. White spruce-balsam fir and balsam fir-white spruce	22	0	0	0
16. Black spruce-balsam fir-moss	18	1	5.6 ± 13.5	14.8
17. Black spruce-moss	32	0	0	0
18.			*not analysed*	

* minimal: 0–29.9%; low: 30–59.9%; high: 60–89.9%; maximal: 90–100%

** Deciduous forest: 1–7; mixed forest: 8–11; coniferous forest: 12–17

Barred Owls nest in natural cavities, but also in abandoned nests of other raptors such as the Red-shouldered Hawk and Northern Goshawk (Elderkin 1987). They may also take over old squirrel or crow nests, to which they add scarcely any improvements (Bent 1938). They also use nest boxes (Johnson 1980; Johnson and Fullen 1984), the recommended diameter for the opening being 20 cm (Hamerstrom 1972). Down clinging to the tree or scattered around the vicinity of a nest are a useful indicator that the nest is active (Bent 1938). Incubation begins after the second egg is laid (Elderkin 1987), so hatching is asynchronous. Incubation is performed almost exclusively by the female for somewhere between 28 and 33 days (Bent 1938; Elderkin 1987). For the first few weeks, the nestlings are covered with a thick white down. After one week, their eyes begin to open, and in the fourth week, they begin to venture outside the nest. Age at first flight is around 6 weeks, but the young continue to be fed by the parents until they are 4 months old (Bent 1938).

The pair occupies its territory throughout the year; it is therefore crucial for the young birds to acquire their own hunting territories if they are to survive their first winter. The birds defend their territories through their vocalizations, but scuffles between Barred Owls do occur, particularly in the breeding season (Elderkin 1987).

The Barred Owl was reported in 11.5% (284/2464) of Atlas squares surveyed. The birds are easily located by their courtship calling, especially in March and April, but are secretive when on the nest. Whitewash and pellets are not always found around the nest site, so nests can be difficult to find (Dunstan and Sample 1972). Later in the season, the juveniles follow their parents on the hunting grounds, often begging noisily for food. These calls may help locate the family group, so breeding is most often confirmed through reports of young; 55.6% (20/36) of breeding confirmations were of downy or flightless young, whereas only 22.2% (8/36) were of nests containing eggs or young.

Distribution

In Canada, Barred Owls range from British Columbia to the Maritimes. In the United States, they are found in the northwest and the eastern half, and they also occur on the central plateau of Mexico (AOU 1983). In eastern North America, the Barred Owl is mainly associated with mature deciduous and mixed forests near wet areas.

Atlas records show that the Barred Owl is distributed throughout southern Québec south of the 49th parallel, except for Anticosti Island and the Magdalen Islands, for which there is only a single known record (Rowdon 1969; Fradette 1992). The northern limit of its range coincides more or less with the northern limit of the mixed forest. It is found in the Abitibi–Témiscamingue region in the west, and east to Manicouagan Point and the Gaspé. It may, however, occur as far north as Godbout, as Comeau (*in* Merriam 1882) classified it as a "tolerably common resident" there in the last century. According to David (1980a) and Godfrey (1986), its breeding range extends to 50° N, excluding, of course, Anticosti Island and the Magdalen Islands. Two recent reports are from 1990, somewhat north of the Atlas area, around Nemiscau and Lake Evans (Cons. G. G. – GREBE 1992m). In 1917, the Barred Owl was seen at Lac aux Cèdres, on the 52nd parallel, near the Labrador border (Todd 1963).

In some ways, the distribution map for the Barred Owl in southern Québec resembles that of the Red-shouldered Hawk. Both species are common in the Ottawa Valley and the Eastern Townships, where there are a number of wetlands (Couillard and Grondin 1986). The Barred Owl, however, has a wider distribution, being found in the Témiscamingue, Lac-Saint-Jean, and Gaspé regions.

The Barred Owl may nest in a natural cavity or in an abandoned nest in a sugar maple stand, one of its preferred breeding habitats. Cap Tourmente National Wildlife Area, northeast of Québec City.
August 1991, Gilles Falardeau.

A swampy area close to the woodland in which it nests provides the Barred Owl with a variety of prey species. Forillon National Park, Gaspé. *August 1990, Yves Aubry.*

Young Barred Owl, still partially covered with down, perching in a beech tree. Sherbrooke. *May 1981, Vincent Létourneau.*

History and trends

Although common in eastern North America, the Barred Owl has probably suffered from extensive felling of old-growth forests. Reforestation in several parts of its range has produced even-aged stands that do not seem to offer enough natural cavities for nesting. The species is also highly sensitive to human activity (Bosakowski et al. 1987) and, according to a New Jersey study, tends to avoid urban centres (Sutton and Sutton 1985). According to Bosakowski et al. (1987), wilderness is essential to maintaining populations of Barred Owls, and they recommend protecting continuous old-growth forest from human disturbance as a means of providing optimal habitat for the species.

Yet the expansion of the Barred Owl's range over the last 40 or 50 years and its increasing numbers in the western part of its North American range are remarkable (Grant 1966; Taylor and Forsman 1976; Harrington-Tweit et al. 1979). Expansion has been so rapid that the Barred Owl is now the most common of the large owls in certain parts of southwestern British Columbia, where it had been absent before 1953 (Munro 1956). The reasons for the expansion are not fully understood (Forsman and Bull 1987), but it appears that the Barred Owl is benefiting from the forest regeneration that has occurred since the intensive logging early in the century (Hamer et al. 1989).

The Barred Owl is currently considered a sedentary breeder, occasional in the Montréal area and common in the Laurentians (Bannon 1991a). According to Ouellet (1974a), it was apparently more abundant at one time, its numbers having been reduced in the Montréal area probably as a result of deforestation. At the turn of the century, it was common in the fall around Québec City (Dionne 1906) and was a common permanent resident of Montréal (Wintle 1896). Because of the paucity of data, it is not possible to ascertain population trends or give abundance estimates in the province. Fyfe (1976) suggests that numbers increased in southern Québec in the mid-1970s, and BBS data also show a slight increase in the province for the 1966–89 period, no doubt because forests were no longer being cleared for farming as of the middle of this century.

CHRISTIAN GAGNON AND MANON BOMBARDIER

Great Gray Owl

Chouette lapone

Strix nebulosa

René Roy

THE GREAT GRAY OWL CAN BE EASILY IDENTIFIED BY ITS LARGE SIZE, prominent facial discs, grey coloration, yellow eyes, lack of ear tufts, and black spot dividing a narrow white throat patch. These field marks allow it to be readily distinguished from its close relative, the Barred Owl, although in flight it can be mistaken for either the Barred Owl or the Great Horned Owl.

The Great Gray Owl inhabits coniferous forests in boreal and mountainous regions of North America and Eurasia (Bull and Duncan 1993). Data gathered during the Atlas period provided the first confirmation of breeding in Québec. The handful of records for this species suggests that it is scarce in the southern part of the province in the breeding season, but it can be seen in the settled regions of southern Québec during its spectacular and sporadic winter irruptions. The last major one was in the winter of 1991–92, when approximately 60 birds were reported (Bull and Duncan 1993). The Great Gray Owl is usually seen at twilight, and occasionally during the day, high in a treetop or on another perch in the open or on the forest edge, scanning for prey.

Habitat and behaviour

Within its breeding range, the Great Gray Owl hunts in a wide variety of habitats. In Manitoba, it prefers open areas and wetlands, such as tamarack bogs and treed muskeg (Servos 1987), while in Alaska, it frequents willow-fringed meadows along rivers and streams (Osborne 1987). In the Cordilleras of western North America, it favours open, parklike mature coniferous forests (Franklin 1988; Bull et al. 1988a), but in Finland, it prefers marshes, clearings, and abandoned fields (Hildén and Solonen 1987). In the Abitibi region, it has been observed on several occasions in agricultural areas and on old farmland reverting to forest (Y. Aubry, pers. comm.). Since the Great Gray Owl employs mainly sit-and-wait hunting tactics (Bull and Duncan 1993), hunting grounds must have perches such as dead trees, a sparse or absent shrub layer, and a highly developed herbaceous layer (Osborne 1987; Servos 1987).

For breeding, pairs appear to seek not just suitable nest sites, but good hunting grounds and abundant prey (Servos 1987; Bull et al. 1988a). The Great Gray Owl is not very choosy with regard to nest sites: nests have been found in mixed forests, spruce stands, ash stands, and peatlands (Nero 1980; Osborne 1987; Spreyer 1987; Bull et al. 1988a). The stands chosen, however, are usually mature, with a fairly closed canopy (Bull et al. 1988a). Leaning trees may also be sought, since they allow young to escape terrestrial predators in the few days after leaving the nest, when they are still unable to fly (Bull et al. 1988b; Franklin 1988).

Like most other nocturnal birds of prey, the Great Gray Owl does not build a nest but appropriates one built by another species, such as the Northern Goshawk, Red-tailed Hawk, or Common Raven (Henderson 1915; Nero 1980; Hildén and Solonen 1987; Bull et al. 1988a). Nests are usually located in a poplar or tamarack, and occasionally on a stump, at the top of a decaying hollow tree, or even on the ground; the Great Gray will also use artificial nest platforms built for it (Nero 1980; Bohm 1985, 1988; Hildén and Solonen 1987; Osborne 1987; Franklin 1988). Some authors believe that the bird's ready use of these platforms indicates that the number of appropriate nests available may be a limiting factor on breeding (Bull and Duncan 1993). The only nest ever found in Québec was an abandoned Osprey nest at the top of a black spruce snag (Cons. G. G. – GREBE 1992n). Nests are often used for two breeding seasons in a row and sometimes longer (Bull et al. 1988a; Franklin 1988).

The pair may inspect one or more potential nests up to 25 days before the onset of egg laying (Nero 1980; Osborne 1987; Franklin 1988), which is anytime between the end of March and May in southern Manitoba and central Alberta, and in April or May in Alaska and southern James Bay (Henderson 1915; Nero 1980; Osborne 1987; Cons. G. G. – GREBE 1992n). The onset of breeding appears to be linked to snow depth at the end of winter, and delays of approximately 5 weeks have been observed in Idaho and Alaska when deep snow has accumulated (Osborne 1987; Franklin 1988). Too much snow makes it difficult for the birds to catch prey and therefore for the female to reach breeding condition.

Although some birds may reproduce as yearlings, age at first breeding is typically 3 (Bull and Duncan 1993). The Great Gray Owl is monogamous, but polygamy may occur occasionally (Hildén and Solonen 1987). Generally, the pair bond is maintained for only one breeding season (Korpimäki 1986).

Notes on territory

The Great Gray Owl occupies a limited home range. In a year-long study of 5 males in Oregon, the mean size was 4.5 km² (1.3–6.5 km²) (Bull et al. 1988b), and near Grand Teton, Wyoming, the mean home range during the breeding season was 2.6 km² (Quinlan et al. 1990). The home range appears to be small in Manitoba, as well, at least in the breeding season (Servos 1987).

The Great Gray Owl hunts in open areas, such as this field at an airport. Nearby stands of jack pine provide cover. La Sarre, Abitibi. *June 1991, Desneige Perreault.*

Great Gray Owl

Breeding distribution
1984–1989

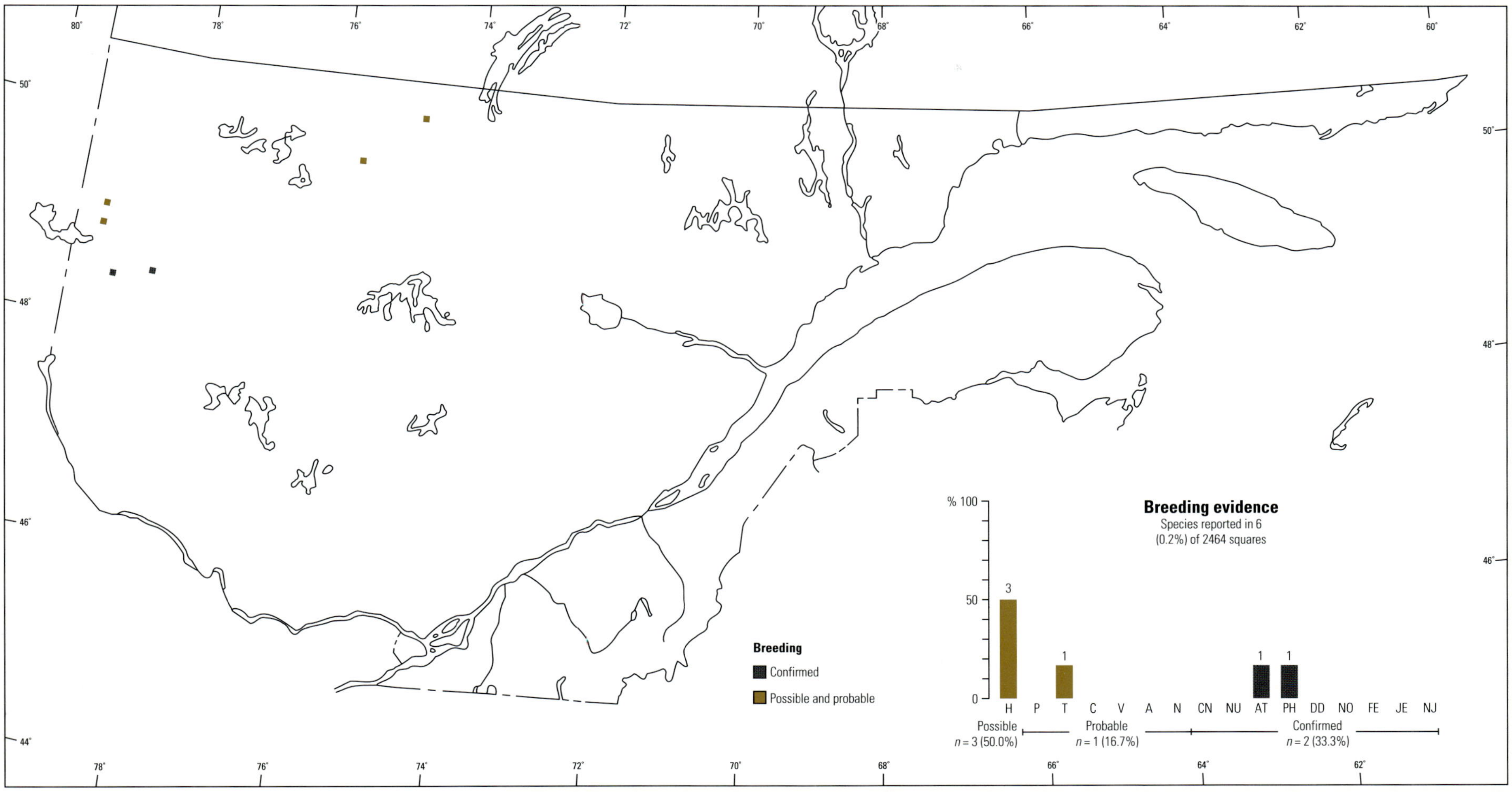

Species profile

Clutch size	usually 2–6 eggs (1–9)[1,2,3,4]
Laying interval	usually 1 egg per 1–4 days (1–12)[1,2,4]
Start of incubation	with first egg[1,2]
Incubation period	usually 28–30 days[2]
Maturity at hatching	semi-altricial
Care of young	by female and male[1,2,3], *see* text
Nestling period	usually 3–4 weeks[1,2,4]
Age at first flight	usually 4–6 weeks[1]
Dependence of young	until end of summer[1]
Broods per year	1[1,2,3]
Breeding age	usually 3 years (1–3)[1,2]
Mating system	monogamy, polygyny possible[1,2,3]
Duration of pair bond	usually 1 season, sometimes more than 1 season, sometimes for life[1]
Total length	61–84 cm[1]
Wing span	134–158 cm[2]
Weight	M 890.5 g (825.0–1050.0) / F 1267.0 g (1025.0–1700.0)[1]
Longevity record	13 years[1]

References:
1. Bull and Duncan 1993
2. Cramp 1985
3. Ehrlich et al. 1988
4. Harrison 1978

Chronology *n* = 1

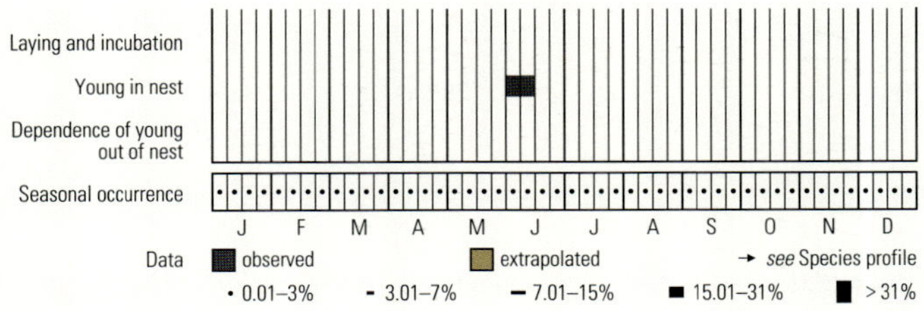

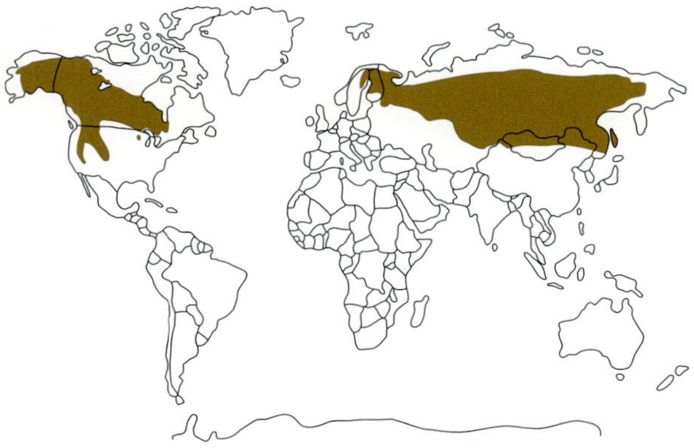

World breeding range

In the boreal forest, the species seeks open areas like this peatland, where snags provide hunting perches. The only known nest in Québec is located nearby. South of Lake Lucie, near Matagami. *May 1991, Serge Brodeur.*

The largest North American member of the family Strigidae usually appropriates the abandoned nest of another bird of prey or corvid. In Québec, the only Great Gray Owl nest found up to now was built by an Osprey. South of Lake Lucie, near Matagami. *May 1991, Serge Brodeur.*

The eggs are ovoid and chalky white in colour (Bull and Duncan 1993). Although clutches may contain 2–5 eggs, sets of 3 or 4 are most common (Terres 1980; Franklin 1988). In Finland, mean clutch size is 4.4 (Mikkola 1981 *in* Korpimäki 1986), while in Idaho, means of 3.0 were reported in 1981 and 3.3 in 1982 (Franklin 1988). Clutches of up to 9 eggs have also been reported (Blair 1962), and laying interval is normally 1–4 days but may be as much as 12 days for later eggs of large clutches (Cramp 1985).

The female begins incubating as soon as she has laid her first egg. In Idaho and Wyoming, mean length of incubation was determined to be 29.7 days (Franklin 1988). The male brings food to the female while she is on the nest and generally remains close by during the day, leaving only to hunt. He announces his return with a series of hoots (Nero 1980).

The owlets are fed by the female until they leave the nest (Bull and Duncan 1993), and afterward she remains in the vicinity until they can get around on their own. The male does most of the hunting during this period, travelling an average of 0.62 km from the nest in search of food, according to an Oregon study (Bull et al. 1988a); the maximum distance was 3.2 km (Bull and Duncan 1993).

The young leave the nest around the age of 3 or 4 weeks, before they are able to fly; mean age of departure was 28.5 days in Idaho and Wyoming (Franklin 1988) and 26 days in Oregon (Bull et al. 1989a). Once out of the nest, the owlets clamber over the branches and up the trunks of nearby leaning trees (Franklin 1988). They are able to fly a week or two after they leave the nest, but still depend on both parents for food for several more weeks, after which they are fed by the male only. The young are able to hunt alone at the age of 3 months and are fully independent by the end of the summer (Bull and Duncan 1993).

The Great Gray Owl's diet consists mainly of small mammals. Voles are its most common prey in Sweden, Finland, and Alaska (Mikkola and Sulkava 1970; Osborne 1987; Bull et al. 1989b). This owl also feeds fairly often on passerines and shrews, and occasionally takes small mustelids, rodents as large as muskrats, amphibians, and larger birds such as grouse and the Boreal Owl (Mikkola and Sulkava 1970). There are at least 2 reports of cannibalism in the species, the first involving 3 owlets captured from one nest (Mikkola and Sulkava 1970) and the second, an adult feeding on either an adult or immature bird. However, in the second case, it is not certain whether the first bird actually killed the other or whether it died of disease or starvation (Fisher 1975). In the Ottawa Valley, Great Gray Owls have been observed in winter feeding on meadow voles and the occasional squirrel (Brunton and Pittaway 1971).

The Great Gray Owl is mainly active at dusk and at night (Servos 1987). The species is able to survive the long winters characteristic of its breeding range thanks to a number of adaptations, including its size and thick plumage, which increase its ability to withstand the cold, and its asymmetrical ears and facial discs, which enable it to locate rodents under the snow. One of its hunting

The Great Gray Owl nests in both mixed and pure conifer stands in the boreal forest. Features sought are rivers and streams with clear banks and suitable perches. South of Lake Lucie, near Matagami. *May 1991, Serge Brodeur.*

There is a post-Atlas breeding record of an adult and 2 downy young in an old Osprey nest on 2 June 1990 from the James Bay lowlands (50°18' N, 78°26' W), approximately 10 km south of Lake Lucie and 60 km northwest of Matagami (Cons. G. G. – GREBE 1992n). In 1990 and 1991, the species was observed another 17 times in this region, which extends up to the southern part of the Nottaway, Broadback, Rupert, and Harricanaw river watersheds, with the northernmost record made at the latitude of the Rupert River.

Great Gray Owls have also been reported in summer at Île aux Basques, in the Charlevoix, in the Gaspé, on the North Shore, in the Lac-Saint-Jean, and in other localities as far north as Fort Mackenzie in northern Québec (Todd 1963; Hardy 1977; Anonymous 1979, 1980a; Godfrey 1986). A bird sighted is not necessarily nesting nearby, however; it may be an immature or a non-breeding adult. Given the species' elusiveness and the difficulties of surveying in the Abitibi and southern James Bay regions, the breeding range may be more extensive than the current data suggest.

Notes on territory (continued)

In Manitoba, Minnesota, Oregon, and Alaska, population densities, which were studied mainly in areas less than 100 km², ranged between 0.73 and 17.2 pairs/10 km² (Osborne 1987; Servos 1987; Spreyer 1987; Bull et al. 1988b). In Sweden, 2 nests were found less than 100 m apart (Bull et al. 1988b), and in Manitoba, less than 200 m apart (Servos 1987). The high population densities found locally and the extensive overlapping of home ranges suggest that the species does not defend a territory or that the territory is limited to the area around the nest (Bull et al. 1988b).

techniques is to plunge into the snow with talons thrust forward (Nero 1980) to capture prey under more than 20 cm of snow or even under a hard crust. A bird plunging through a crust that was hard enough to support an 80 kg man has even been reported (Norberg 1987). Great Gray Owls leave characteristic plunge-holes in the snow, a sure sign that they are in the area.

In Manitoba, the Great Horned Owl is the main predator on both adult and juvenile Great Gray Owls, but many birds taken are likely already weak with hunger. In addition, adults and juveniles may fall prey to fishers (Duncan 1987), and juveniles may be taken by Northern Goshawks and black bears (Nero 1980; Bull and Duncan 1993).

Great Gray Owls are usually sedentary in areas of abundant food resources, but when the vole population crashes in a region, some may disperse, wandering as far as 684 km from their breeding grounds (Duncan 1987). This behaviour is probably responsible for the irruptions observed in southern Canada and the northeastern United States (Eckert 1978; Nero 1980), but appears to be less pronounced in the Cordilleras (Bull et al. 1988a).

Great Gray Owls are quite bold and easily approached, but difficult to spot despite their large size (Nero 1980). The best time to observe them is during southern irruptions, when many owls invade settled areas and can be seen in urban parks, woodlots, and even right downtown. The extent of these irruptions seems to vary (Hardy 1977). Many of the birds are emaciated and die of starvation.

The Great Gray Owl was reported in only 0.2% (6/2464) of Atlas squares surveyed. Breeding was confirmed in just 2 of the 6 squares, while evidence of probable breeding was recorded in 1 and possible breeding in the 3 others. The Great Gray's retiring nature and the early start to the breeding season make it difficult to confirm nesting even when there are birds in an area, but the low number of records suggest that the species is scarce in southern Québec.

Distribution

All Great Gray Owl Atlas records came from the Abitibi region. Breeding was confirmed at Sainte-Gertrude-Manneville on 14 April 1988, when a female with a brood patch was hit by a car; after treatment, she was released in the same area (A.-G. Bernier, pers. comm.). The second confirmation came from Destor, near Mont-Brun, where a bird carrying food was sighted on 23 June 1989. The species was also reported near Chibougamau.

In winter, Great Gray Owls have been seen throughout almost all of southern Québec. For example, during the major irruption in the winter of 1978–79, a total of about 70 were observed at Saint-Gédéon, Saint-Félicien, Lévis, Duchesnay Forestry Station, Cap-Tourmente, Senneville, Hudson, and other places. In several places, groups of up to 6 birds were seen, as is common during irruptions (Hardy 1979a).

Factors like farming, logging, and forest fires are liable to affect the Great Gray Owl's distribution. In the Abitibi region, abandoned farmland may be prime hunting habitat for the species. The logging of mature forests has no doubt reduced suitable breeding and hunting habitats, and second-growth forests are too dense for hunting. Although small forest fires probably have little impact, fires that devastate large expanses deprive the Great Gray Owl of breeding habitat for many years.

History and trends

The Great Gray Owl is more prevalent in North America than has been believed, at least in the Cordilleras (Bull et al. 1988b). Nevertheless, the species is considered rare throughout its breeding range (Nero 1980). In Québec, breeding has been confirmed only recently. Because of the small number of records, neither the size of the population nor current population trends can be determined.

Some of the birds seen in southern Québec in winter probably come from farther north. The 250 or more owls sighted during the 1983–84 irruption give a general idea of minimum abundance, since most birds probably escape observation.

Unfortunately, in Manitoba, collisions with cars are a major factor in mortality some winters (Nero and Copland 1981).

François Morneau

Long-eared Owl

Hibou moyen-duc

Asio otus

Ghislain Caron

THE LONG-EARED OWL SLIPS SILENTLY THROUGH THE NIGHT AS IT HUNTS, ranging far and wide across the open countryside. Before dawn, it quietly and unobtrusively perches in dense cover and remains just as elusive to birders by day as it does by night. Records of the species are therefore sparse, and it is classified as a rare migrant breeder in Québec (David 1980a; Bannon 1991a), yet it is probably not as rare as it seems. Up to 13 birds have been seen roosting together in winter, and larger numbers are possible. These communal roosts occur locally where vole densities are high, usually in mixed woods or thick deciduous forests in southernmost Québec, where owl watchers can finally get a close look at these beautiful and mysterious birds.

Habitat and behaviour

During the breeding season, this species usually nests and roosts at or near the edge of dense coniferous or mixed woodlands and in reforested areas, but it may also be found in hardwood stands, parks, orchards, isolated copses, and farm woodlots (Johnsgard 1988; Voous 1989). This nocturnal hunter specializes in voles (Smeenk 1972; Korpimäki 1992), but also preys on mice, shrews, and the occasional small bird for variety (Holt and Childs 1991). Although the Long-eared Owl has a restricted diet, it has taken some unusual prey in coniferous forests in southern interior British Columbia and northeastern Oregon: red squirrels, bushy-tailed wood rats, and northern pocket gophers (Hooper and Nyhof 1986; Bull et al. 1989c). The species typically hunts by quartering over open countryside (Johnsgard 1988), but it may also hunt from a perch at the forest edge, in open areas, and on occasion in woodlands (Bull et al. 1989c).

The nest site is usually chosen by the female (Voous 1989). They normally nest solitarily, though occasionally they are found in small colonies of two, three, or more. These owls never build their own nests, but occupy old crow, squirrel, or raptor nests. Where there are no suitable old nests, Long-eared Owls may nest on the ground (Bent 1938; PQSPB 1952; Voous 1989); L. M. Terrill once found a ground nest in the Laurentians (Bent 1938). Occasionally, they nest in tree cavities, lining the inner nest with downy feathers from their plumage (Craighead and Craighead 1956). They may nest in the same wood for several years running, but experience has shown that nesting for only a year or two is more typical, due to the cyclical nature of prey populations (Korpimäki 1992).

Not much is known about courtship, but vocalizations are frequent and consist of a series of soft hoots; a dovelike *hoo-hoo-hoo* is typical, or catlike mewing, foxlike cries, grouselike whistles, puppylike barks, and Screech-Owl-like whistles. Other vocalizations include moans, shrieks, and twitters (Bent 1938). These are sometimes given in the daytime, when the owl is disturbed by an intruder approaching its nest. The only indication of a breeding female may be the sight of the long ears protruding well above the edge of the nest, looking something like a rabbit in a tree. The ground beneath the male's roosting tree or trees will have a scattering of regurgitated pellets and copious amounts of white droppings, commonly referred to as whitewash.

Typical clutch size is 4 or 5 eggs, laid early in spring when snow may still be on the ground. The female incubates them for about 28 days, while the male does all the hunting until the young are at least 2 weeks old (Craig et al. 1988). During the incubation period the birds are very secretive, the female remaining on the nest and the male at his roost within several hundred metres of the nest, where his cryptic coloration and upright stance allow him to blend in with the trunk or to mimic a branch.

The young hatch asynchronously, and while in the nest, they give begging calls that sound like rusty gate hinges or squeaky wheels. The adults have been observed ingesting the feces of nestlings, possibly to remove all telltale signs that might attract predators, though this has never been proven (Murphy 1992). After 2–5 weeks, they clamber onto branches near the nest (Craig et al. 1988; Holt 1992a), often sitting side by side and rotating their heads to get a better look at intruders. The adults are noisy at this time and often try to distract intruders by feigning injury, adopting threatening postures, hissing, clacking their bills, hooting cries of alarm, or even attacking (Bent 1938). Once they are 2 weeks old, the owlets may react aggressively to the appearance of a strange brood near the nest. They begin by calling and then, aided by the female, will even attack the intruders (Voronetsky 1987). Age at first flight is approximately 30–35 days old (Marks 1986; Holt 1992a).

Notes on territory

Unlike most other owls, the Long-eared Owl has no territorial hunting grounds, and nesting territories are usually small, so one patch of forest may support more than 2 pairs/km² (Voous 1989). In 1980 in Saint-Basile-le-Grand, near Montréal, 6 nests were found within a 3 km² area. All nests were in small woodlots surrounded by abandoned fields. Two nests were approximately 200 m apart, and the maximum distance

The Long-eared Owl hides at edge of a mixed forest in the daytime and comes out to hunt nearby as night falls. Sainte-Foy. *September 1991, Yves Aubry.*

Long-eared Owl

Breeding distribution

1984–1989

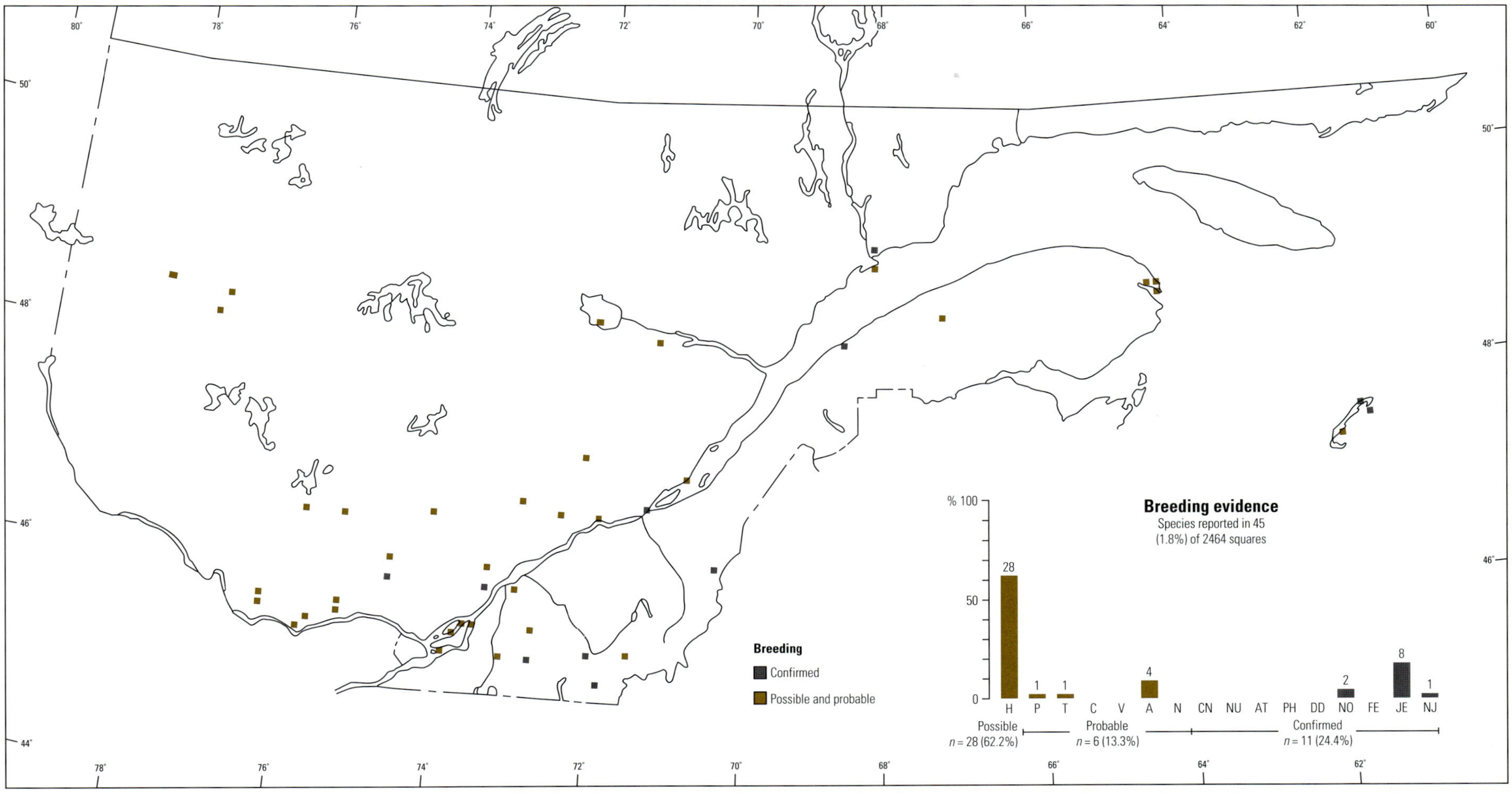

Species profile

Clutch size	usually 4–5 eggs (3–8)[4,5,6]
Laying interval	1 egg per 2 days[4,5,8]
Start of incubation	with first egg[4,5,8]
Incubation period	usually 26–28 days (25–30)[5,6,8]
Maturity at hatching	semi-altricial
Care of young	by female and male[2,4,5], *see* text
Nestling period	usually 21–24 days (16–35)[1,4,5,6,8]
Age at first flight	30–35 days[1,3,5]
Dependence of young	until 55–77 days old[4,5,6,8,9]
Broods per year	1[2,3,4]
Breeding age	1 year[3,5]
Mating system	monogamy[5,6,8]
Duration of pair bond	1 season[5,9], more than 1 season[6,8]
Total length	33–41 cm[7]
Wing span	95 cm[5]
Weight	M 250 g (220–280) / F 300 g (250–370)[5]
Longevity record	27 years 9 months[5]

References:

1. Holt 1992a
2. Voous 1988
3. Marks 1986
4. Bent 1938
5. Mebs 1989
6. Ehrlich et al. 1988
7. Godfrey 1986
8. Cramp 1985
9. Marks et al. 1994

Chronology *n* = 26

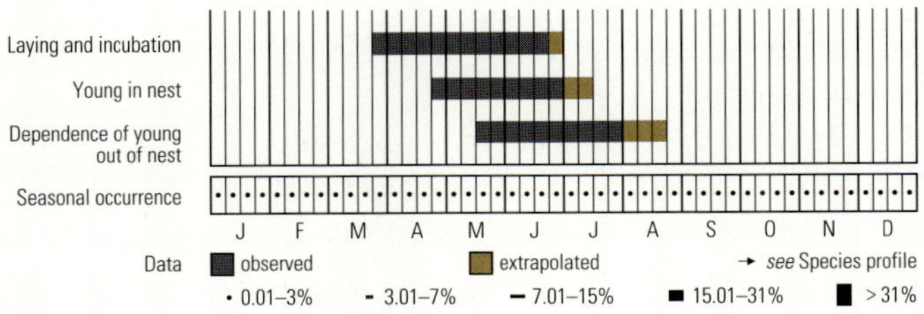

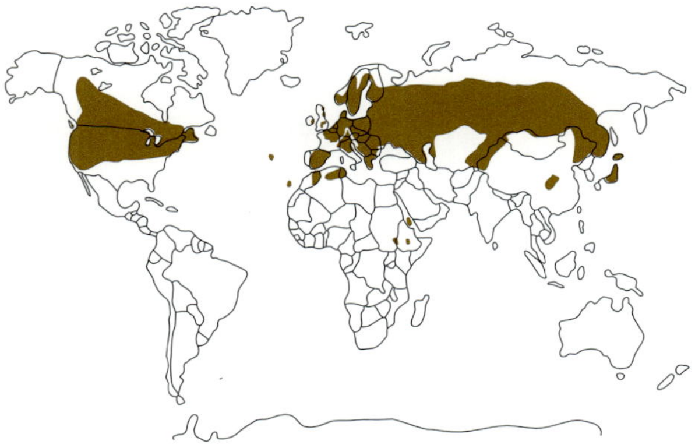

World breeding range

This shy owl uses old corvid or raptor nests and may settle in spruce plantations like this one on Île Perrot, near Montréal. *August 1991, Gaétan Duquette.*

The Long-eared Owl is apparently the most preyed upon owl of the Northern Hemisphere (Voous 1989), falling victim to Great Horned Owls, its chief predator in North America, and to Tawny Owls and Eagle Owls in Europe, as well as to Northern Goshawks, Peregrine Falcons, and mammals such as raccoons and foxes (Mikkola 1983; Marks 1986).

This is the most nocturnal of all owls in North America, and its secretive nesting habits, cryptic coloration, and preference for roosting near tree trunks or in dense cover by day make this species one of the most difficult to atlas in the breeding season. It was recorded in only 45 of the 2464 squares surveyed, with breeding confirmed in only 11. The young's habit of branching while still dependent on their parents for food, coupled with their begging calls, aids in locating nests, and this was the evidence that most frequently confirmed breeding.

Distribution

The Long-eared Owl breeds in the Northern Temperate Zone right around the world, including Eurasia and eastern Africa. In Canada, it breeds throughout most of the Prairie provinces and Ontario, southern Québec, and the Atlantic provinces except Cape Breton Island and Newfoundland (Godfrey 1986). It also breeds through much of the central and northern United States. Most Atlas records came from the Ottawa and St. Lawrence valleys, the Eastern Townships, and the Laurentians, with a few from the Gaspé, Magdalen Islands, North Shore, Lac-Saint-Jean, and Abitibi regions. Breeding has been confirmed by other sources at Chambord (1990) and Jonquière (1987–91) (G. Savard, pers. comm.).

Due to the difficulty in locating this shy and inconspicuous owl, it is probably more widespread in Québec than the Atlas map shows. The extensive coniferous forests and reforested areas of southern Québec should provide ideal nesting habitat for the species. The widely scattered sightings suggest that it probably went undetected in many of the squares in between, wherever suitable habitat exists.

The breeding range of the Long-eared Owl in Québec might extend north to the tree line, as the species has been reported near the Great Whale River and Lake Guillaume Delisle (Todd 1963).

Notes on territory (continued)

between the other nests was about 1.5 km (F. Morneau, unpubl. data).

In the Snake River Birds of Prey Area in southwestern Idaho, nesting densities along the Snake River and its tributaries were 0.42 pair/km² in 1980 and 0.28 pair/km² in 1981. The distance between adjacent owl nests ranged from 14 to 19 080 m, with a mean of 1480 m (Marks 1986). Sometimes 3 or 4 pairs nest close together and will co-operate in nest defence, with members of 2 or 3 pairs performing distraction displays near the same nest (Marks 1985).

Adult on its large nest of twigs at Saint-Basile-le-Grand. Note the owl's ear tufts, facial disc, and horizontal incubating position. East of Montréal. *May 1980, François Morneau.*

A dense conifer stand such as this one may conceal the nest of this nocturnal hunter, while the neighbouring fields serve as a hunting territory. Gould, northeast of Sherbrooke. *August 1991, Gaétan Duquette.*

Five downy Long-eared owlets, each at a different stage of growth. Sainte-Dorothée, Laval. *June 1943, Henry Mousley, Laval University Collection.*

History and trends

Long-eared Owls may have benefited from the clearing of forests by European settlers, although historical accounts are lacking. Any gains they may have made were apparently offset, however, by losses at the hands of trophy and bounty hunters (Wetmore 1935).

The Long-eared Owl has probably often eluded detection by ornithologists even in the early part of this century. Dionne (1906) wrote that it was not common either as a breeder around Québec City or in migration near Montréal. Macoun's (1900) records showed that it was not common anywhere in Canada, but most authors agreed with his assessment that "this species is so retiring in its habits that its absence from a district is more apparent than real."

Although accurate abundance estimates for this elusive owl are difficult to obtain in Québec, results of one study analysing Christmas Bird Counts from 1956 to 1986 in New Jersey indicated a decline in numbers at traditional winter roosts and suggested that nesting activity may also be decreasing (Bosakowski et al. 1989a). Several limiting factors were postulated, including loss of suitable habitat, competition with other raptors, declining vole habitat, and the increased use of chemical poisons to control rodents. Whether or not these factors have affected Long-eared Owl populations in Québec is still a matter of conjecture, as we know very little about the species in the province.

In northern Québec, the population has probably undergone little change, since the peatlands and other open and wetland habitats have remained relatively free of human disturbance. In the south, however, populations may be declining due to the agricultural practices of the last 40 years, involving the removal of trees from field and lot boundaries. On the other hand, coniferous plantations may favour the species.

Derin Henderson and Bob Barnhurst

Short-eared Owl

Hibou des marais

Asio flammeus

Ghislain Caron

THE SHORT-EARED OWL IS FOUND ON ALL CONTINENTS EXCEPT AUSTRALIA and Antarctica, making it one of the most cosmopolitan members of the Strigidae (Burton 1984). Thanks to its diurnal, and particularly crepuscular, habits, it is one of the few North American owls that can be regularly seen in action. The Short-ear is not a forest dweller. Its hunting technique is similar to the Northern Harrier's, and like the harrier, it hunts in open country, meadows, and marshes. With easy, buoyant wingbeats, it courses low over the ground, hovers in midair for a few seconds, and then pounces on its prey, usually a small rodent. It can sometimes be seen perched on a pile of dirt or a fence post, scanning the ground for prey. Although it nests in a number of regions in Québec, it is not really common anywhere. The Short-eared Owl winters irregularly in the southern Central St. Lawrence Lowland (David 1980a).

The Short-eared Owl nests in open areas like this field on Île aux Fermiers. Varennes Islands, near Montréal. *August 1991, Gaétan Duquette.*

Habitat and behaviour

In North America, the Short-eared Owl is usually associated with wide open spaces. In the winter, it is encountered in hayfields, pastures, and freshwater and salt marshes. In summer, it nests in open grassland, upper levels of marshes, scrub, peatlands, and arctic tundra. It also nests in meadows and stubblefields in agricultural areas (Clark 1975), as well as on golf courses and fields near airports (Peck and James 1983; Campbell et al. 1990b).

The Short-eared Owl winters from southern Canada to Central America. In spring, part of the population leaves its wintering grounds to breed in northern Canada, although some individuals may remain to nest farther south, generally when prey density on the wintering grounds remains high through the spring (Cadman and Page 1993). The Short-ear's diet consists of small mammals, over 95% voles (Clark 1975). In the northern part of its range, lemmings, which undergo significant cyclical fluctuations in population (Elton 1965), are also taken (Pitelka et al. 1955).

Owing to its nomadic habits—the Short-eared Owl may travel long distances on its wintering and breeding grounds in search of prey—individuals are unlikely to return to the same nesting territory two years in a row (Cramp 1985).

Courtship flights, which may extend over a period of several days before pair formation, are quite spectacular (Townsend 1938b). Typically occurring in April or May, they are performed mainly by the male while the female looks on from a nearby perch. The male performs short dives and claps his wings below his body, and the female typically responds with a barking call. Unlike most Strigidae, the Short-eared Owl has a fairly limited vocal repertoire and a weak voice. This is probably linked to its preference for open habitat, where sound carries well, or at least better than in the forest, and to its crepuscular and diurnal habits; since individuals can communicate visually, vocalizations are less important (Clark 1975).

Although the male selects the nesting territory, the female is believed to choose the actual nest site (Holt 1992b). The Short-eared Owl is one of the few Strigidae that builds its own nest. It makes a small scrape in the ground on a fairly dry site, usually under the protective cover of tall grasses (Clark 1975). The nest is lined with twigs, grass, and thin branches, usually gathered by the female but on rare occasions by the male.

The male usually hunts while the female lays and incubates the eggs and raises the young, but, when prey density is very low, the female may also hunt for short periods (Lockie 1955). Clutch size is quite variable, and depends on the female's condition, which depends, in turn, on the abundance of prey.

Since Short-ears are ground nesters, the owlets are vulnerable to predation, and this appears to influence both their rate of development, which is remarkably rapid, and the duration of the rearing period, which is fairly short. By the time they are 10 days old, the owlets often weigh 10 times their original hatching weight. They leave the nest before they are able to fly, at 12–18 days of age. They then wander about, exploring the nearby vegetation, pouncing on anything that moves, and gorging themselves on insects; this appears to be how they learn to hunt and detect prey movements (Clark 1975). One study has shown that the young may wander 50–200 m from the nest at this time (Hölzinger et al. 1973 in Cramp 1985).

Notes on territory

In the breeding season, the male defends a territory that varies in size according to prey abundance. According to Lockie (1955), the territory may be as small as 16 ha in good years and as large as 137 ha in years of scarce food supply; other authors report territories ranging from 15 ha to close to 250 ha (Clark 1975; Cramp 1985; Village 1987; Holt 1992b). Short-ears generally hunt inside the territory, except when prey is scarce, when a pair may occupy as much as 450 ha. Nests are rarely less than 145–300 m apart (Cramp 1985).

This crepuscular hunter patrols vast salt marshes in search of small land mammals. L'Isle-Verte, northeast of Rivière-du-Loup. *July 1993, Michel Robert.*

Short-eared Owl

Breeding distribution

1984–1989

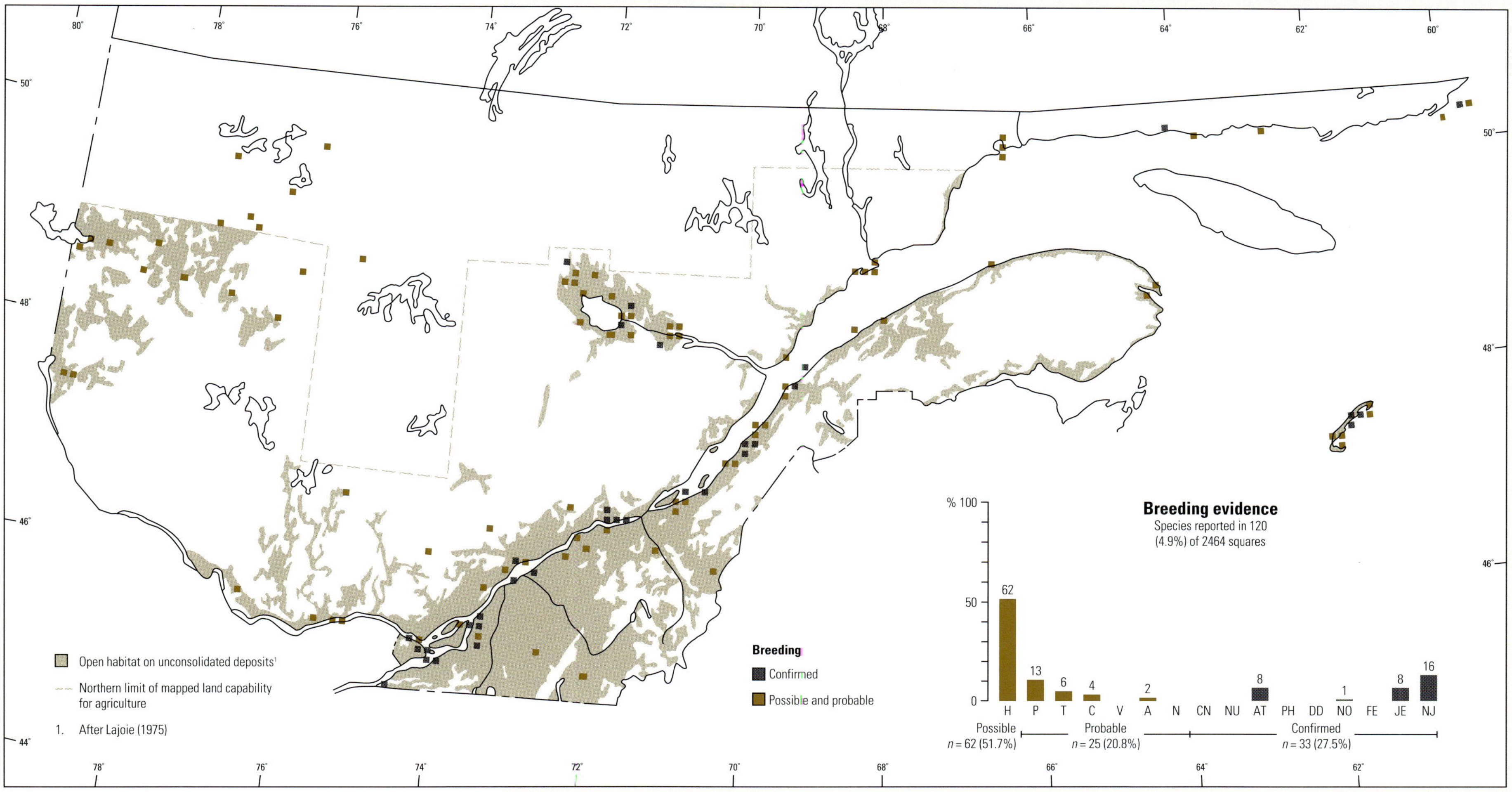

Species profile

Clutch size	usually 4–7 eggs (4–14)[1,2,3,4]
Laying interval	1 egg per 1–2 days[1]
Start of incubation	with first egg[1]
Incubation period	usually 24–29 days (21–29)[1,2,4]
Maturity at hatching	semi-altricial
Care of young	by female and male[1,3,4], *see* text
Nestling period	12–18 days[1,3,9]
Age at first flight	usually 24–27 days (24–42)[1,4,7,8]
Dependence of young	until a few weeks after first flight[1]
Broods per year	1, 2 possible[11]
Breeding age	1 year[1]
Mating system	monogamy, sometimes polygyny[1,10]
Duration of pair bond	1 season[1]
Total length	33–43 cm[6]
Wing span	96.5–111.8 cm[2]
Weight	M 315 g (206–368) / F 378 g (284–475)[5]
Longevity record	12 years 9 months[1]

References

1. Cramp 1985
2. Terres 1980
3. Clark 1975
4. Ehrlich et al. 1988
5. Dunning 1984
6. Godfrey 1986
7. DeGraaf and Rudis 1987
8. Harrison 1978
9. Holt et al. 1992
10. Tate 1992
11. Holt and Leasure 1993

Chronology $n = 25$

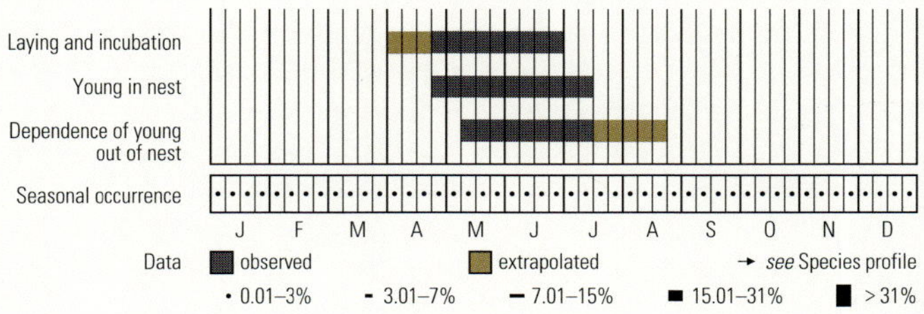

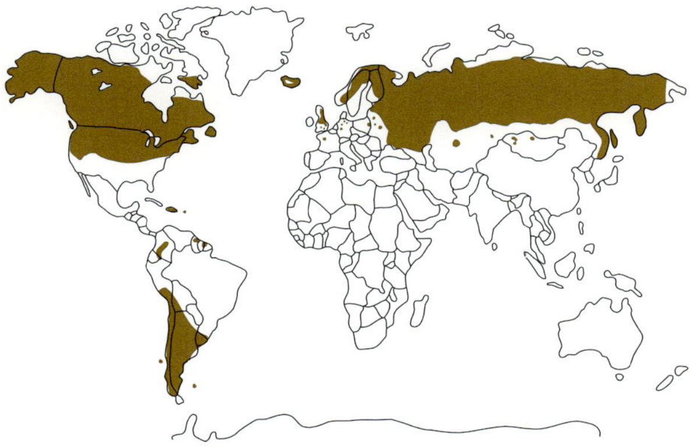

World breeding range

Short-eared owlets in the nest in the grass. There may be as many as 9 in a brood. Near Rigaud, west of Montréal. *May 1981, Daniel Jauvin.*

Once the young have left the nest, the female begins to hunt again. When prey is abundant, excess kills may be cached in tunnels 12–15 cm long dug 7–10 cm from the nest (Ingram 1959); this practice helps to reduce the risk of siblicidal attacks (Voous 1989). Age at first flight is anywhere between 24 and 42 days, but the young continue to beg for food for several weeks, until they become completely independent (Clark 1975; Harrison 1978; Cramp 1985; DeGraaf and Rudis 1987; Ehrlich et al. 1988).

Although the species nests in open areas, breeding is more difficult to confirm than might be expected. The noisy behaviour of the young at the nest site and the male's aggressiveness toward intruders, particularly when the young are still in the nest (Clark 1975), may help in locating the nest, however. Short-eared Owls were observed in 4.9% (120/2464) of the squares surveyed. Breeding was confirmed in 17 squares by finding a nest, in 8 squares by observing downy young or young incapable of sustained flight, and in 8 squares by sighting adults carrying food. In June and early July, especially, the owlets can be seen perched with their parents, so they are easier to spot then.

Distribution

In southern Québec, the Short-eared Owl shuns forested areas, and is instead associated with marshy or sandy riparian habitats, large peatlands, and agricultural areas. Its range in Québec is limited mainly to the St. Lawrence Plain and the Saguenay–Lac-Saint-Jean region, where most of the records of confirmed breeding were obtained, although nesting was also confirmed in the Magdalen Islands and on the North Shore east to the Sainte-Marie Islands. The species has been seen as far up as Blanc-Sablon in the past and is known to nest in Labrador (Todd 1963; ÉPOQ).

In summer, the Short-eared Owl is also encountered in the Abitibi–Témiscamingue region, with its abundant farmlands and open areas, suggesting that it nests there as well.

The Short-eared Owl has been reported at several locations farther north, including the Kuujjuaq area, where a nest was found in 1882 (Todd 1963). It has been seen fairly often on the shores of James Bay and Hudson Bay as far north as Inukjuak (Todd 1963; ÉPOQ). Recently, in 1989 and 1990, it was sighted several times in the Great Whale and Little Whale river watersheds, and farther north, along the downstream stretch of the Nastapoka River (Cons. G. G. – GREBE 1990c). It is found mainly in shoreline marshes, salt marshes, and peatlands in these areas. Farther south, a downy chick was seen with 2 adults on 28 June 1990 several dozen kilometres southeast of Lake Evans (Cons. G. G. – GREBE 1992m). Another sighting was reported on 10 August 1981 near Lake Maricourt (ÉPOQ). In short, during the nesting season the species is found all over Québec, except on the Ungava Peninsula. Despite the lack of sightings on the peninsula, according to Todd (1963) it probably also breeds there because it is known to nest on the tundra and few observers have visited the region.

History and trends

The Short-eared Owl's status in North America is very difficult to determine. Since the species is nomadic, high densities of nesting birds may be temporarily observed in areas with large rodent populations (Clark 1975; Korpimäki and Norrdahl 1991). Several researchers have reported a decline in Short-eared Owl populations in North America and the species was on the Blue List from 1976 to 1986 (Tate 1986). A number of eastern and midwestern states have classified this owl as endangered or threatened (Marti and Marks 1989). Recently, its numbers appear to have declined sharply in the north-eastern United States, particularly in Massachusetts, where urban development of coastal meadows has reduced the breeding population to fewer than 25 pairs, and in New Jersey and New York (Holt 1986).

The species may be found in large peatlands like this one, where a pair nested in 1990. Northwest of Lake Soscumica, near the Nottaway River, James Bay region. *June 1990, Serge Brodeur.*

Notes on territory (continued)

In winter, particularly in years of abundant prey, the Short-eared Owl usually defends a hunting territory with a mean size of 6.3 ha (Clark 1975; Cramp 1985).

Peatland at L'Isle-Verte where the species nests. Northeast of Rivière-du-Loup. *July 1991, Jean Bachand.*

Young Short-ear crouching in the vegetation. The owlets wander around the nest at a young age. Northwest of Lake Soscumica, near the Nottaway River, James Bay region. *June 1990, Serge Brodeur.*

Since the species is a ground nester, it depends on plant cover, so the removal of vegetation makes nesting birds very vulnerable to predation. Breeding success is estimated to be around 40% (Cramp 1985; Holt 1992b). According to Clark (1975), the draining of wetlands, changes in farming methods—mechanization and replacement of hay crops by intensive cereal monocultures—and industrial development have resulted in the partial or total destruction of many habitats suitable for the species.

In Canada, in particular, this owl consumes large quantities of voles in meadows and fields next to airports (Baker and Brooks 1981), and is therefore at risk of colliding with aircraft (Blokpoel 1976). Pesticide contamination in the species has not been studied extensively, although several authors have reported residues in eggs and some birds (Peakall and Kemp 1980; Henny et al. 1984). The ingestion of insecticide-contaminated prey may also cause death in the species (Mendelssohn and Paz 1977).

Short-eared Owl populations have declined in southern Ontario (Weir 1987e), but not enough data are available to determine whether the same trend is found in Québec.

This owl appears to have been fairly common in the province at the turn of the century, according to several contemporary accounts. Macoun and Macoun (1909) reported that the species was found in the Mingan Islands and the Magdalen Islands around this time, and listed it as common in the Montréal area. Wintle (1896) considered it a common "transient visitant" in Montréal, where it was sometimes observed in small flocks in autumn. Todd (1963) noted several breeding records from this time from the Sainte-Marie Islands along the Lower North Shore and from southern James Bay. Godfrey and Wilk (1948) listed it as a common summer resident in the Lake Saint-Jean area and, in *Birds of Eastern Canada*, Taverner (1919) reported that large numbers of these birds were killed in the hunting season.

The number of Short-eared Owls in Québec may have declined over the last few decades because of the loss of wetland habitat, particularly along the St. Lawrence, where close to 3643 ha of wetlands disappeared between 1950 and 1978 (Lands Directorate 1986). According to Ouellet (1974a), the Montréal area population has decreased since the early 1900s, both in the breeding season and at other times of the year. Although the clearing of forests for farming at one time may have created new habitat for the Short-eared Owl, particularly in the Abitibi–Témiscamingue and Lac-Saint-Jean regions, peat mining activities may have negative effects on the species, as may the reversion of farmland to forest taking place everywhere except around Montréal and in the Montérégie (MAPAQ).

Luc Bélanger and Manon Bombardier

Boreal Owl

Nyctale boréale

Nyctale de Tengmalm
Aegolius funereus

Louis Ruelland

The small, nocturnal Boreal Owl is a denizen of coniferous forests throughout the Northern Hemisphere. It is known for its irregular winter incursions south into second-growth and deciduous forests in eastern North America. With its chocolate brown plumage, it closely resembles the Northern Saw-whet Owl, the distinguishing field marks being its slightly larger size, yellowish beak, and black-framed facial disc, which gives it a look of constant surprise. It does not have ear tufts.

Habitat and behaviour

As its name suggests, the Boreal Owl is found mainly in the northern coniferous forest, but it also occurs in mixed forests, usually composed of spruce, pine, birch, and poplar (Clark et al. 1987).

Throughout a large part of North America, its distribution overlaps that of the Northern Saw-whet Owl, as is the case in Québec, notably in the Laurentide Wildlife Reserve. Palmer (1986) attempted to determine whether interspecific competition for habitat exists when the two species are sympatric, occupying the same area. He found that, in Colorado, the Boreal Owl avoids the dense, continuous pine forests and poplar stands that are the favoured habitat of the saw-whet. The Boreal Owl also occupies sites with high densities of voles, its preferred prey, and lower densities of mice, the preferred prey of saw-whets. The same study reported that the Boreal prefers habitats with a thinner deciduous cover, taller trees, and a denser herbaceous but thinner shrub layer in the understorey than the saw-whet. Both species use tree cavities of similar size but those selected by the saw-whet may have smaller openings (Hayward 1983). Hayward and Garton (1984) noted that in winter the Boreal Owl appears to avoid using cavities during the day, preferring to roost on a conifer branch in denser thickets than those selected by the saw-whet.

In Québec, the Boreal Owl begins nesting in late April or early May. In Finland, the onset of breeding may be as early as the end of February in years of scant snow cover and abundant prey. Early laying usually produces more eggs and young and seems to increase the young's chances of survival, as they have more time to learn to hunt and build up energy reserves for the coming winter (Korpimäki 1987). Prey abundance also appears to promote polygyny in certain Finnish and Swedish populations (Solheim 1983; Carlsson and Hörnfeldt 1989). Territorial defence, carried out by the male, appears to be limited to the immediate vicinity of the cavity (Solheim 1983), and there is no defended hunting ground (Hayward et al. 1987). Beginning in mid-February, the male advertises his presence and control over potential nesting sites by his *staccato* song, consisting of a series of monotonic hoots (Palmer 1987).

The sedentary male remains in his home range year round. He gives a *staccato* vocalization to attract a female, showing her several nesting cavities, from which she selects one (Cramp 1985). The male usually stops singing a few days after copulation (Bondrup-Nielsen 1984). The most persistent nocturnal callers are males that have not succeeded in attracting mates (Voous 1989). The song can be heard beginning at sundown, and more often in the evening than in the morning (Palmer 1987).

The Boreal Owl nests both in conifers, such as spruce and pine (Ryder et al. 1987), and in deciduous trees, such as trembling aspen (Bondrup-Nielsen 1976; Peck and James 1983). The nest cavity is usually found at a mean height of 5.27 m (Hayward and Garton 1983; Ryder et al. 1987).

Notes on territory

In Colorado, the home range of the Boreal Owl outside the breeding season is about 1130 ha. In the breeding season, however, the species only uses about 296 ha of this area (Palmer 1986). Boreal Owls in Ontario defended a breeding territory within the home range about 185 m in diameter. This estimate is based on maximum distances between perches from which the birds were heard calling (Bondrup-Nielsen 1978).

Cramp (1985), however, reported that the male will only defend the area around the nest cavity and then only when it is occupied by the female.

There are very little data on population density of the Boreal Owl in North America, but it is probably very low. Cramp (1985) reported that in Europe it varies greatly both year to year and geographically, even when nest boxes are provided.

The Boreal Owl nests in tree cavities in the white spruce stands of the Magdalen Islands. Fatima.
June 1991, Desneige Perreault.

Boreal Owl

Breeding distribution
1984–1989

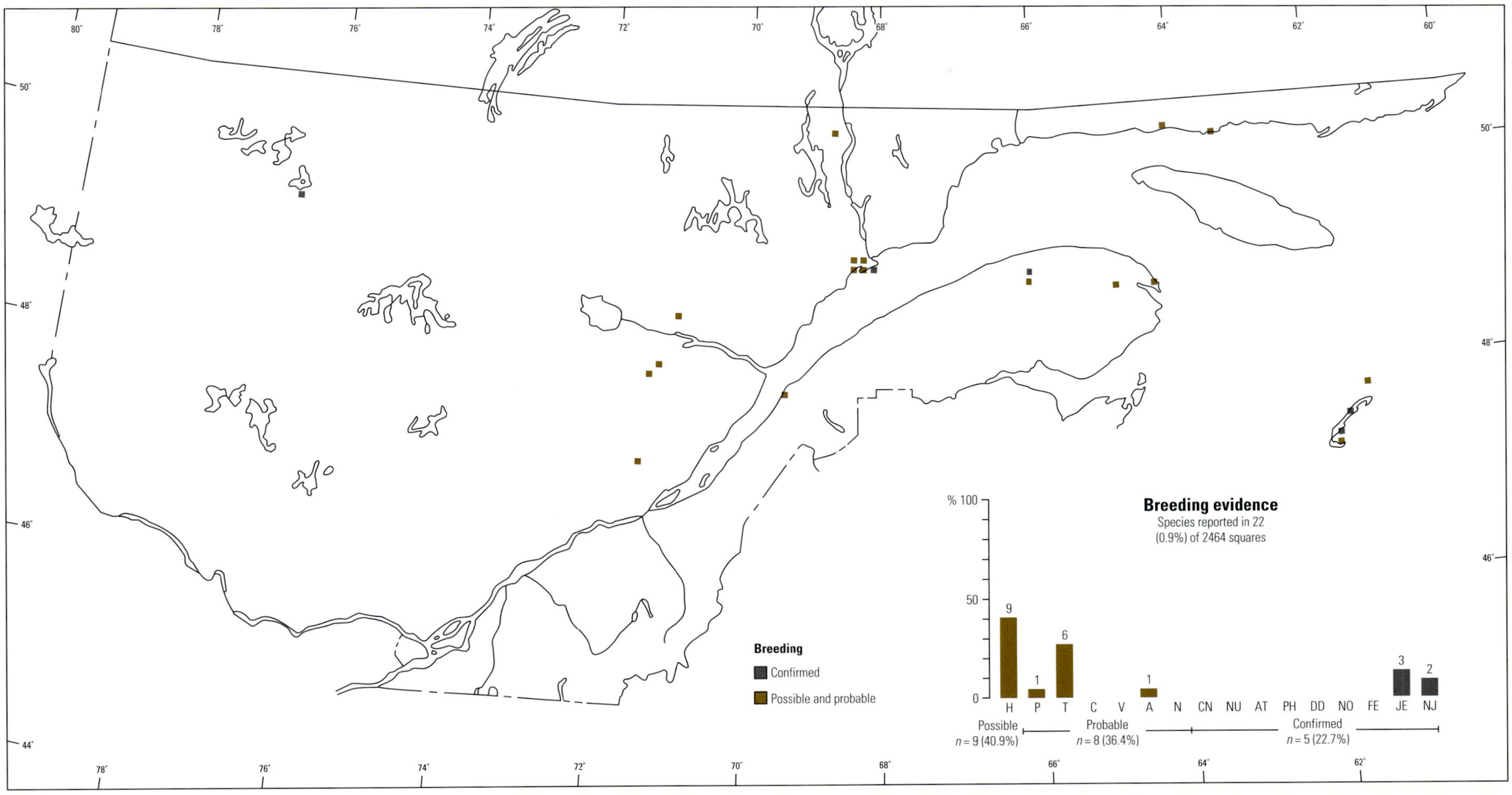

Species profile

Clutch size	usually 4–6 eggs (3–10)[1,2,3]
Laying interval	1 egg per 1–2 days[2,6]
Start of incubation	with first or second egg[1,2,6]
Incubation period	usually 29 days (25–32)[2,3,6,9]
Maturity at hatching	semi-altricial
Care of young	by female and male[4,8], *see* text
Nestling period	usually 30–32 days (28–36)[2,6,9]
Age at first flight	28–36 days[9]
Dependence of young	until 3–6 weeks after first flight[6,9]
Broods per year	1 or more[6,9]
Breeding age	1 year[2,6]
Mating system	monogamy, sometimes polygyny or polyandry [2,6,9,11]
Duration of pair bond	1 season, rarely 2[6,10]
Total length	21.5–26.5 cm[5]
Wing span	54–62 cm[6]
Weight	M 101 g (90–113) / F 167 g (126–194)[7]
Longevity record	?

References:
1. Townsend 1938a
2. Johnsgard 1988
3. Terres 1980
4. Bondrup-Nielsen 1976
5. Godfrey 1986
6. Cramp 1985
7. Dunning 1984
8. Ehrlich et al. 1988
9. Hayward and Hayward 1993
10. Korpimäki 1981
11. Korpimäki 1989

Chronology $n = 7$

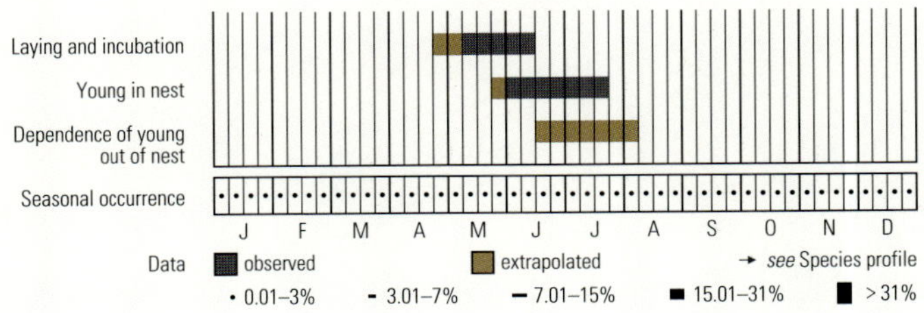

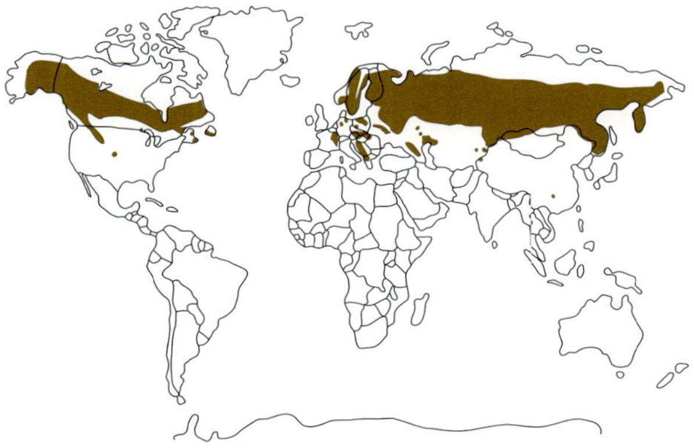

World breeding range

Incubation begins after the first or second egg is laid, leading to asynchronous hatching (Johnsgard 1988). Incubation is carried out almost exclusively by the female and typically lasts 4 weeks. During the egg laying, incubation, and brooding periods, the male brings prey to the female, and food often accumulates in the nest. Prey is kept for 1–9 days and consumed in the order in which it was brought; food may also be cached (Cramp 1985). The young can fly at about 4–5 weeks, but continue to be fed by the parents for another 3–6 weeks (Cramp 1985; Hayward and Hayward 1993). They reach sexual maturity in less than a year (Korpimäki 1981).

The Boreal Owl hunts essentially at night, but sometimes during the day as well, to keep the young owls fed (Cramp 1985). A small and agile flyer, the Boreal can twist and turn easily in the woods. It also hunts along forest edges and riverbanks. It usually hunts from a perch, locating its prey before pouncing directly on it (Cramp 1985). Voles are its principal food, but it also preys on shrews, mice, birds, and sometimes frogs and insects, when these are plentiful in its habitat (Townsend 1938a; Catling 1972; Burton 1973; Cramp 1985).

Generally considered non-migratory, the Boreal Owl does occasionally migrate, irrupting southward in winter. Boreal Owl populations are dependent on prey abundance and must adapt to fluctuations in small mammal populations. During periods of plenty, owl populations may increase, but some birds may be forced to abandon their territories when the food supply is depleted. Females and young appear to be more inclined to move than males (Cramp 1985; Palmer 1986). According to Mebs (in Catling 1972), the birds that migrate southward in winter probably do not breed the following year, even though they may be sexually mature. Boreal Owls that appear in the northern United States are often starving and many perish (Heintzelman 1979).

Various factors may have caused atlassers to miss the Boreal Owl, for it was reported in just 22 of the 2464 squares surveyed. The species is heard only early in the season, beginning in February, when few ornithologists are in the field in the owl's breeding range. It nests most often in woodpecker holes, usually those of the Pileated Woodpecker or Northern Flicker (Townsend 1938a; Bondrup-Nielsen 1976), which makes it difficult to find. Moreover, it is active mainly at night, and chances of finding it during the day are slim, unless all potential nest sites are investigated.

The Boreal Owl is found in the upper slopes and plateaus along the Jacques Cartier Valley. Jacques Cartier Conservation Park, north of Québec City. *September 1981, Jean Beaubien and Guy Simard, Canadian Forest Service.*

This owl is associated with the northern coniferous forest and prefers habitat where small rodents are plentiful. Mount Albert, Gaspé Conservation Park. *June 1976, Alain Desrosiers, Ministère du Loisir, de la Chasse et de la Pêche.*

Distribution

The Boreal Owl is probably more widespread than is generally believed. The limits of its breeding range are still poorly defined, but Atlas records show that it breeds in some areas south of the 50th parallel. Young birds were reported at Miquelon (north of Abitibi) and in the Magdalen Islands. Nests with eggs or young were found at Pointe-Lebel (Middle North Shore) and in the Gaspé Conservation Park, south of Marsoui. Evidence of possible and probable breeding was obtained in squares on the Gaspé Peninsula, in the Saguenay–Lac-Saint-Jean region, and along the Jacques Cartier Valley in the Jacques Cartier Conservation Park. Both the Boreal and Northern Saw-whet owls were recently reported in the breeding season at Lakes Minishtuk and Nixon in the northern part of the Laurentide Wildlife Reserve (B. Dumont and G. Savard, pers. comm.). There is every indication that the breeding range of the Boreal Owl in Québec is more extensive than reported by Godfrey (1986), who limited it to areas located between 50° and 54° N and to the Magdalen Islands.

The Boreal Owl is rarely seen in northern Québec, and there are only a few reports from areas beyond the 50th parallel (Todd 1963). Recently, 2 birds were heard early in the morning, about 10 km south of the Rupert River, near Lake Nemiscau (Cons. G. G. – GREBE 1992m). Another bird was sighted about 15–20 km south of the Little Whale River (55°45 N); this is the northernmost record for the species in Québec (Cons. G. G. – GREBE 1990c). It should be noted that these 3 birds were heard on the edges of a forest, a lake, and a river, all of which are suitable habitat for the small mammals upon which the Boreal Owl preys.

History and trends

Until the 1960s, the Boreal Owl was considered a rare migrant visitor south of Canada. The first breeding record for the species in the United States, from Minnesota, dates only from 1978 (Eckert and Savaloja 1979). Since then, there have been several other reports from Minnesota, as well as from the Rocky Mountains and down to northern Mexico (Ryder et al. 1987). This spate of records, however, is probably the result of increased human penetration into the bird's habitat (Palmer and Ryder 1984).

The Boreal Owl has been known to occur in Québec since the 19th century, when a few birds and a nest containing eggs were discovered in 1878 in the Magdalen Islands. According to Cory (*in* Fradette 1992), the species was common at that time on Île du Havre aux Maisons.

In southern Québec, at the turn of the century, the species was sighted only in winter on Anticosti Island, in Québec City, and in Montréal (Wintle 1896; Schmitt 1904; Dionne 1906). The forest fragmentation that has occurred as a result of European settlement, and the concomitant creation of edge habitat suitable for the small mammals upon which the Boreal Owl preys have no doubt favoured its occurrence in winter in the inhabited areas of southern Québec. Analysis of ÉPOQ records indicates that most sightings are made between October and March.

Boreal Owls depend a great deal on woodpeckers for nesting cavities, and these are generally in short supply, so loss of nest sites is a threat to the species. For this reason, the mature and overmature stands favoured by woodpeckers should be conserved.

Alain Desrosiers and Manon Bombardier

Northern Saw-whet Owl

Petite Nyctale

Aegolius acadicus

The Northern Saw-whet Owl is a species that often goes undetected, and is probably more common than is generally believed. Being secretive and nocturnal in its habits, the bird is difficult to find. About the size of a European Starling, and weighing about 90 g—barely half the weight of an Eastern Screech-Owl—the saw-whet is the smallest owl in eastern Canada.

It occurs only in North America, where it is a widely distributed permanent resident. Part of the population in the northernmost part of its range, including Québec, is migratory, however.

The snags in this hardwood-forested swamp afford suitable nest cavities for the little saw-whet. Rockburn, south of Montréal. *June 1991, Pierre Bannon.*

Habitat and behaviour

Most voluble between late winter, when they begin migrating, and early May (Bent 1938), Northern Saw-whet Owls are best found on mild, still evenings, when courting males utter their advertising song: a short, endlessly repeated monotonic whistle. This is also the best time of year to make population counts. Early in the season, the male will respond to an imitation of its song, but once mated it becomes quieter. The saw-whet is very tame and will often tolerate the approach of humans as it dozes on its roost during the day. When netted during banding operations, however, it may well snap, especially if the rictal bristles are touched (Schaeffer 1973). Some territorial males may respond aggressively to the song of another male saw-whet on their breeding territories and will sometimes even attack, but without inflicting injury.

Saw-whets generally roost in dense groves of conifers, about 1.5–4 m from the ground (Bent 1938; Swengel and Swengel 1992). The birds are found in various types of habitat just before the breeding season. In southern Canada, the species may nest in any type of dense, mature forest, but clearly prefers wet woodlands (DeGraaf et al. 1980). It is also found in dense alder thickets and cedar and tamarack swamps (Clark et al. 1987), as well as in river valleys and around lakes.

Northern Saw-whet Owls typically nest in abandoned woodpecker holes, particularly those of Northern Flickers and Hairy Woodpeckers, which are usually at a height of 6–18 m in tree trunks greater than 30.5 cm in diameter (Thomas et al. 1979). Some nests have been found in natural cavities and in nest boxes set up for Northern Flickers (Bent 1938) and Wood Ducks (Follen and Haug 1981). The female apparently does not bring new material to the nest, but simply lays her eggs on the wood chips generally found in such holes (Bent 1938).

Louis Ruelland

Notes on territory

The male stakes out a territory when he arrives on the breeding grounds through his vocal activity (Bent 1938). The size of the home range varies according to habitat and prey density. In British Columbia, Cannings (1987) estimated it to be 114 ha in a deciduous forest where prey density was high, and 159 ha in a coniferous forest where prey density was lower.

In deciduous forests, male birds hunt only in the core of the home range, about 27 ha in area, whereas males in coniferous habitat use most of their range (Cannings 1987).

In the forests of the central and southwestern United States, the maximum density recorded has been 12–13 birds/km² (Johnson et al. 1979; Swengel and Swengel 1987). Density in the nesting period, however, is probably no greater than 1 pair/km² (Cannings 1993).

The Northern Saw-whet Owl nests in deserted Northern Flicker and Pileated Woodpecker holes in this mature stand in Bernierville. East of Victoriaville. *April 1985, Alain Côté.*

Northern Saw-whet Owl

Breeding distribution

1984–1989

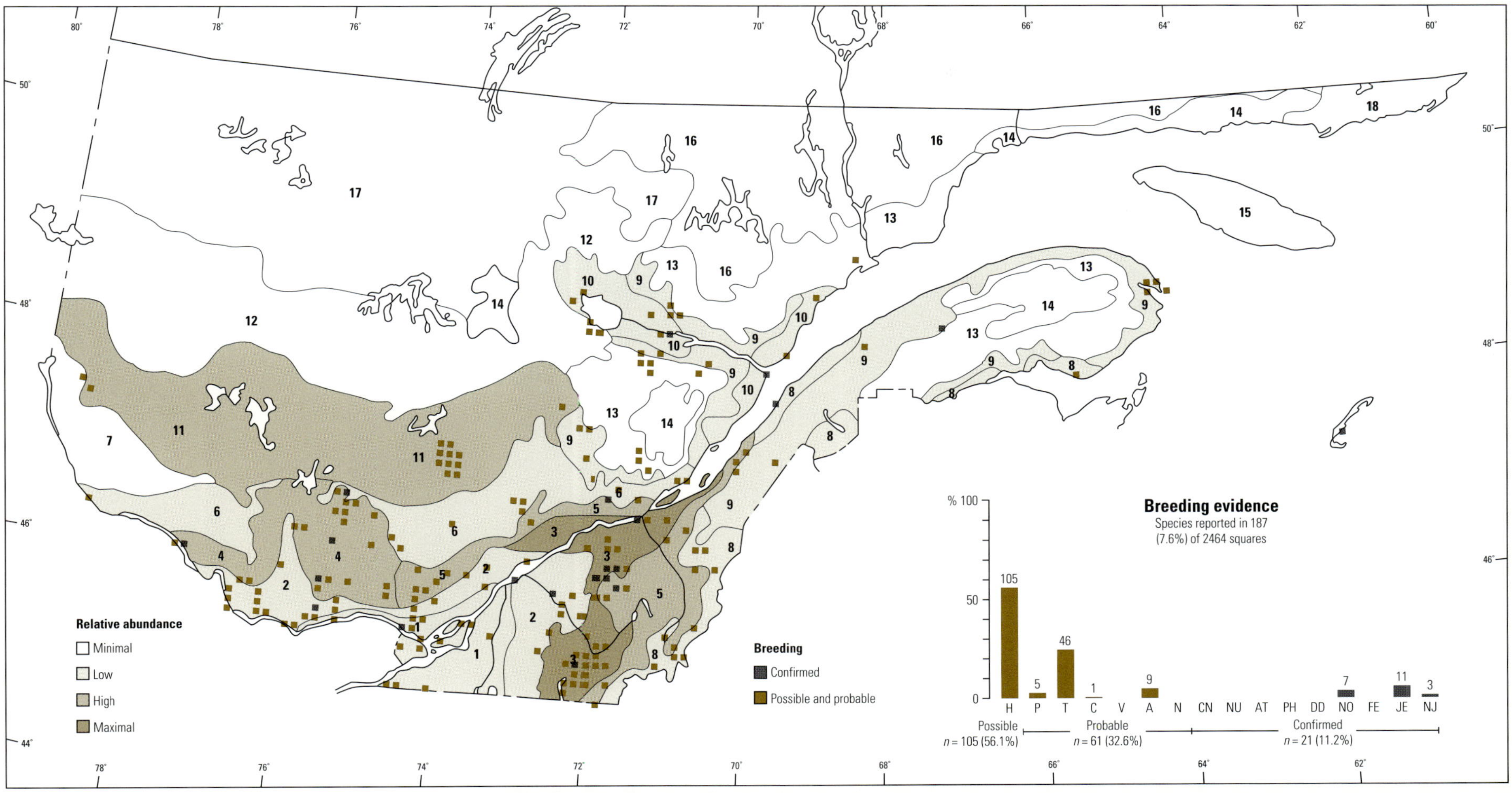

World breeding range

Species profile

Clutch size	usually 5–6 eggs (4–7)[1,2,3,4,9]
Laying interval	usually 1 egg per 2 days[9]
Start of incubation	with first egg, sometimes with second[1,3,9]
Incubation period	usually 27–29 days (21–29)[1,2,3,4,9]
Maturity at hatching	semi-altricial
Care of young	by female and male[9], *see* text
Nestling period	27–36 days[1,2,3,4]
Age at first flight	27–36 days[1,2,3]
Dependence of young	until at least 4 weeks (4–8) after leaving nest[9]
Broods per year	1, 2 possible[2,9]
Breeding age	1 year[9]
Mating system	monogamy, sometimes polygyny, polyandry possible[9]
Duration of pair bond	1 season[9]
Total length	18.0–21.5 cm[8]
Wing span	43.2–50.8 cm[6]
Weight	M 74.9 g (54.0–96.0) / F 90.8 g (65.0–124.0)[7]
Longevity record	7 years 5 months[5]

References:
1. Cannings 1987
2. Terrill 1931a
3. Bent 1938
4. Ehrlich et al. 1988
5. Klimkiewicz and Futcher 1989
6. Terres 1980
7. Dunning 1984
8. Godfrey 1986
9. Cannings 1993

Chronology *n* = 7

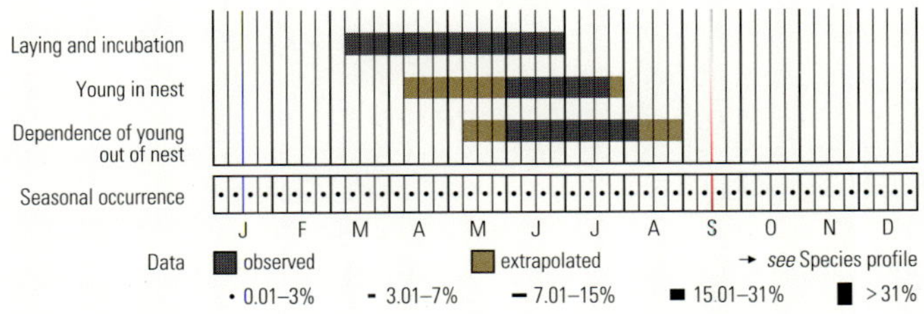

Frequency of occurrence by ecological region

Ecological regions** (forest regions)	Adequately covered squares	Squares in which reported	% of squares in which reported	Relative abundance (%)*
1. Sugar maple-hickory and sugar maple-basswood	97	13	13.4 ± 3.6	45.1
2. Sugar maple-basswood	153	27	17.6 ± 3.2	59.4
3. Sugar maple-basswood and sugar maple-yellow birch	101	30	29.7 ± 5.4	100.0
4. Sugar maple-yellow birch-basswood	56	12	21.4 ± 9.5	72.1
5. Sugar maple-yellow birch-beech and sugar maple-yellow birch-basswood	113	21	18.6 ± 4.1	62.6
6. Sugar maple-yellow birch-beech	71	12	16.9 ± 8.2	56.9
7. Sugar maple-yellow birch-balsam fir	19	0	0	0
8. Sugar maple-yellow birch and balsam fir-yellow birch	47	5	10.6 ± 8.3	35.8
9. Balsam fir-yellow birch	87	12	13.8 ± 6.8	46.4
10. Balsam fir-white birch or balsam fir-red maple	46	8	17.4 ± 9.7	58.6
11. Yellow birch-balsam fir	52	12	23.1 ± 11.9	77.7
12. Balsam fir-white birch (west)	59	2	3.4 ± 5.3	11.4
13. Balsam fir-white birch (east)	76	1	1.3 ± 3.0	4.4
14. Balsam fir-black spruce	28	0	0	0
15. White spruce-balsam fir and balsam fir-white spruce	22	0	0	0
16. Black spruce-balsam fir-moss	18	0	0	0
17. Black spruce-moss	32	0	0	0
18.			*not analysed*	

* minimal: 0–29.9%; low: 30–59.9%; high: 60–89.9%; maximal: 90–100%

** Deciduous forest: 1–7; mixed forest: 8–11; coniferous forest: 12–17

Saw-whets favour riparian woodlands in the southern part of the province, where mature deciduous trees provide hollows and cavities for nesting. Rivière du Sud, a tributary of the Richelieu River. *August 1974, Léo-Guy de Repentigny.*

The female does almost all of the incubating, beginning after the first or second egg is laid and usually sitting for 27–29 days (Bent 1938; Cannings 1987). Hatching is asynchronous, and as a result, the young of the same brood are different sizes.

Aside from a few brief moments at dusk, the female does not leave the cavity until the young are able to maintain their body temperature. The male catches prey for his mate and young. The nestlings begin to open their eyes at 8 or 9 days old and fledge between the 27th and 36th days (Terrill 1931a; Cannings 1987). They are then fed sporadically until they learn how to hunt. Like the Boreal Owl, the Northern Saw-whet preys on small mammals, mainly deer mice, voles, and shrews (Swengel and Swengel 1992), and only rarely feeds on insects (Voous 1989) or small birds (Cannings 1993).

Like most nocturnal species, the Northern Saw-whet Owl is difficult to confirm as a breeding bird. In its nesting and hunting activities it is considered to be strictly nocturnal. The nest is placed in a cavity, well out of sight. Because the bird remains immobile on its roost, it is difficult to spot during the day. Egg laying may begin in early March, and by the time atlassing began in most years, the birds had become secretive. Nonetheless, some data were obtained in March 1986 and in April and May 1988. Breeding was confirmed in only 0.9% (21/2464) of squares surveyed. Breeding confirmation was based on reports of young incapable of sustained flight in 11 squares, and on nests found in the 10 other squares.

Distribution

Atlas results support David (1980a) and Godfrey's (1986) descriptions of the Northern Saw-whet Owl's distribution in southern Québec. It appears to nest in all regions south of the 49th parallel, except in the Abitibi and on Anticosti Island, where Schmitt (1904) reported that it was quite rare all year long at the beginning of the century. There are a few sight records from the Témiscamingue region, but no breeding records. The breeding record from one square in the Magdalen Islands is worth noting, as saw-whets rarely nest in the islands, where the Boreal Owl is a resident species (Fradette 1992). Northern Saw-whet Owls are found fairly frequently in riparian woodlands in the southern part of the province, especially in predominately deciduous stands, but they appear to avoid the highly urbanized parts of the Central St. Lawrence Lowland. The largest numbers were seen in the Eastern Townships, mainly around the Saint-François River and Lake Mégantic, and in the central and northern Ottawa Valley, where wet woodlands abound.

The low frequency of occurrence in the northern part of the Atlas area may be due to the bird's reclusive nature and the fact that most atlassing there was done after the peak period of vocal activity. Because of the lack of clearings and open understorey, the owl's preferred prey may be scarce in these northern regions, making them less suitable for breeding. The sighting by Todd (1963) at Fort George in 1930 appears to be the most northerly record for western Québec.

It is easy to see why the Northern Saw-whet Owl often escapes detection. Princeville, northeast of Victoriaville. *April 1987, Alain Côté.*

Owlet, probably ready to leave the nest. Princeville, northeast of Victoriaville. *May 1987, Alain Côté.*

Saw-whet nestling peeping out of a Wood Duck nest box. Saint-Joachim-de-Courval, north of Drummondville. *July 1988, Daniel Jauvin.*

History and trends

There is very little information, either historical or recent, on Northern Saw-whet Owl populations in Canada or the rest of North America. According to Mills (1987d), it is difficult to make abundance estimates because populations fluctuate from one year to the next. Irregular southward incursions resulting from a scarcity of prey in the northern part of the range may be the cause of these annual fluctuations. A study on the fall migration of Northern Saw-whet Owls at Prince Edward Point, Ontario, showed that the annual fluctuations observed between 1976 and 1979 were due to fluctuations in the size of the juvenile population, as the number of adults remained stable (Weir et al. 1980). The authors maintain that these fluctuations may be linked to the breeding success of the species north of this stopover. Estimates of saw-whet abundance based on counts of calling birds in Colorado yielded widely varying results from year to year (Palmer 1987). These differences were attributed to fluctuations in the abundance of the birds' preferred prey, and the fact that they may be quieter in years when prey is scarce because they are less likely to breed successfully.

The status of the species in North America has long been a subject of debate. Early accounts of its movements over the continent suggested that the Northern Saw-whet Owl was a permanent resident in the regions where it was found; if it did migrate it was only for short distances, and only at the northern and southern limits of its range. Similarly, Fisher (1893) maintained that the species was not migratory and attributed its fall and winter movements to the search for food. It was not until 1911 that the first evidence of saw-whet migration was reported (Taverner and Swales 1911). We now know that the northern populations migrate south regularly, and we know some of the concentration points that they funnel through. One of them is Prince Edward Point, where a total of 3595 saw-whets were banded in fall between 1957 and 1983, most between the end of September and 20 October (Sprague and Weir 1984). Migratory movement has also been noted in Québec, where up to 50 saw-whets a year have been seen in the western part of the Island of Montréal (Barnhurst 1979). Analysis of Christmas Bird Count data shows that Northern Saw-whet Owls may winter in areas between the 35th and 50th parallels (Root 1988). The Québec literature indicates that a certain proportion of the population resides year-round in the southern part of the province (David 1980a; Bannon 1991a).

In the late 19th century, Caulfield (1890a) considered the species to be resident in Québec and Wintle (1896) listed it as a permanent resident in Montréal. Ouellet (1974a) maintained that subsequent intensive felling of boreal forests and the collection of specimens may have affected abundance and precipitated the species' decline. Preservation of sites with dead trees and snags would certainly be beneficial (Evans and Conner 1979), as would education of the general public regarding the role saw-whets play in controlling rodent populations (Gross 1992c).

Alain Côté and Manon Bombardier

Common Nighthawk

Engoulevent d'Amérique

Chordeiles minor

Christiane Girard

THE COMMON NIGHTHAWK, WITH ITS PIERCING, NASAL *PEENT* CALLS, aerial acrobatics, and often urban habitat, is likely to arouse the curiosity of people who usually pay little attention to birds. On summer evenings, males of this crepuscular species can be observed performing their swooping dives high in the sky over their nests, which are often located on the flat gravel roofs of buildings.

With a mouth so large that it opens back under the ears, the nighthawk is a voracious eater of insects, hence its nickname "mosquito hawk." The Common Nighthawk can be distinguished from other Caprimulgidae, particularly the Whip-poor-will, by its slightly forked tail and white wing patches, which stand out against its dark plumage.

The Common Nighthawk winters from northern South America to central Argentina (AOU 1983). Flocks ranging from several dozen to sometimes several hundred birds can be seen over lakes and ponds in spring migration (Terres 1980; Kibbe 1985d). The Common Nighthawk arrives in Québec to nest in May and leaves in September or October. Its breeding range extends from the southern Northwest Territories and Yukon to northern Central America (AOU 1983).

Habitat and behaviour

Common Nighthawks nest in open areas with little or no vegetation. Outside urban areas, they select woodland openings and clearings, burns, rock outcrops, and gravel and sand beaches. They may also use old fields, cultivated fields, or cutovers, but generally prefer bare ground (Terres 1980). When buildings with flat gravel roofs appeared around the middle of the 19th century, they also adopted them as nest sites. In a study in Madison, Wisconsin, nests were found on buildings 5–15 m high with flat roofs surrounded by parapets or low walls (Gramza 1967). A British Columbia study found that nighthawks preferred nesting in natural habitats even when gravel roofs were available and, in fact, all the nests observed were on the ground (Brigham 1989). Nests may also be located on stumps, fence posts, and rails up to 2 m from the ground (Terres 1980).

Unlike the other Caprimulgidae subfamily, the nightjars, nighthawks lack the long bristles on the sides of their mouths (Cramp 1985), but this does not prevent them from visually locating and capturing huge quantities of insects in the air, often in virtual darkness. Two migrating birds examined in Virginia on 4 September 1971 had feasted so prodigiously on flying ants (Formicinae) that their stomach contents made up close to 25% of their body weight, the caloric value of the meal representing 3.7 to 3.8 times the birds' daily standard metabolism (Blem 1972). In comparison, passerines' stomach contents usually represent roughly 1.5% of their body weight (Stevenson 1933 *in* Blem 1972).

Nighthawks also occasionally fly low over the ground to capture grasshoppers and moths, or snatch emerging insects by swooping low over water (Boyce 1980). They eat roughly the same foods as certain bats, and often forage in the same places. Prey is selected according to ease of detection rather than availability. A study comparing the foraging habits of Common Nighthawks and big brown bats showed that neither species ate chironomids, even though they constituted at least 60% of available prey, but instead chose larger prey (Brigham 1990). The two species use different foraging strategies, however, and their foraging times are determined by different factors (Shields and Bildstein 1979; Aldridge and Brigham 1991). With Common Nighthawks, foraging time appears to be limited by the duration of twilight, and tends to be short and relatively constant throughout the summer, which is not the case with bats (Aldridge and Brigham 1991). Nighthawks may occasionally travel up to 12 km to forage (Brigham 1989).

Common Nighthawks drink on the wing like swallows by skimming the water with their bills. They may even bathe by flying through heavy rain (McGowan and Woolfenden 1986). Since their feet are not adapted for perching, they usually roost lengthwise, sometimes crosswise, on a horizontal limb (Wintle 1896). Unlike Whip-poor-wills, Common Nighthawks are sometimes active during the day, particularly on cloudy days or when disturbed (Gross 1940).

Flight is an integral part of their existence: food and water are obtained solely on the wing, and courtship and mate selection are carried out in flight, as well (Caccamise 1974). Nighthawks may soar for long periods, often riding the thermals and updrafts common at most nesting sites, particularly in urban areas (Caccamise 1974). It is not known to what extent the behaviour of nighthawks in cities is influenced by the canyon effect, which may produce strong air currents in an area, depending on the number and configuration of tall buildings.

Females reach sexual maturity at one year of age. They arrive on the breeding grounds a day to a week ahead of the males (Rust 1947; Mills 1986) and appear to be responsible for choosing the nest site (Gross 1940; Terres 1980). Gross (1940) reported that one banded female returned to the same nest site four years in a row.

The male performs courtship displays at dawn or dusk, when he can be seen flying in wide circles above a nest site occupied by a female, uttering his characteristic *peent* calls. Every so often, he goes into a nosedive and, at the bottom of the dive, turns abruptly upward in front of the motionless female (Gross 1940). A characteristic hollow booming sound, like the one made by blowing across an empty bottle, is produced by the vibration of his primaries when he turns upward; this sound can be heard only near the nest site (Weller 1958). After these preliminaries, the male alights near the female, wagging his fanned tail from side to side and rocking his body. Sometimes, he puffs out his white throat and utters a few guttural notes that are a very important part of courtship (Gross 1940). The male's aerial acrobatics continue throughout the nesting period, but after the young have hatched, he spends more time procuring food for his family (Gross 1940).

Notes on territory

The male stakes out his territory with display flights and chases off males from adjacent territories when they trespass, while the female incubates the eggs (Weller 1958; Sutherland 1963).

In Detroit, the territories of 13 pairs of Common Nighthawks studied ranged in size from 4.14 to 22.8 ha, with a mean of 10.4—2.25 ha was flat rooftop (Armstrong 1965).

Several pairs may sometimes nest close together. In Idaho, Rust (1947) reported 4 nests in an 8.1-ha pasture, with the closest nests only 47.5 m apart.

This area, which was razed by a forest fire, has the openings and bare rocks preferred by the Common Nighthawk. Near Blanc Reservoir, northwest of La Tuque. *May 1991, Yves Aubry.*

Common Nighthawk

Breeding distribution 1984–1989

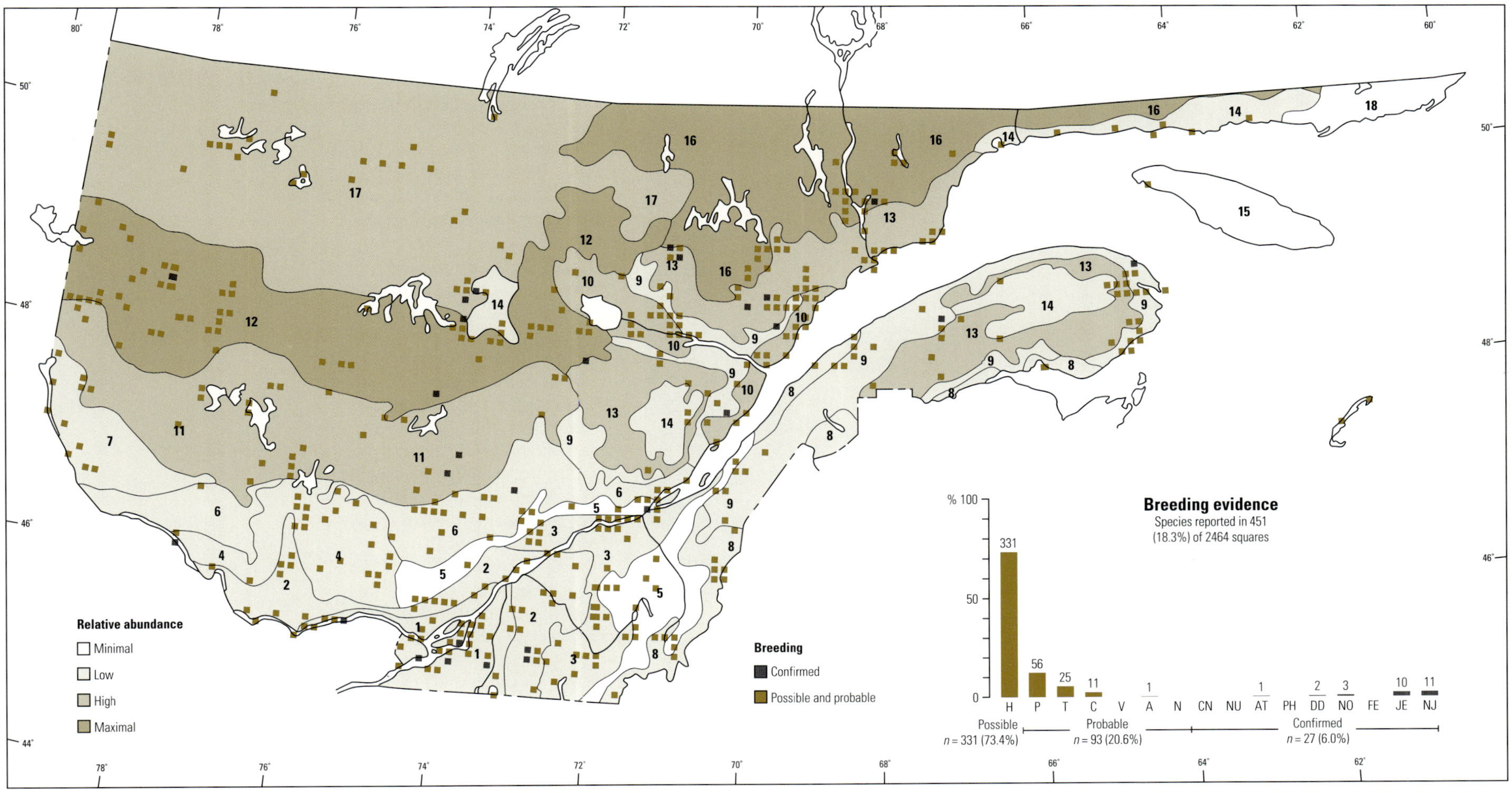

Species profile

Clutch size	usually 2 eggs (1–2)[1,2,3,6]
Laying interval	1 egg per day[3]
Start of incubation	with first egg[3]
Incubation period	19–20 days[1,2,3,6]
Maturity at hatching	semiprecocial
Care of young	by female and male[6]
Nestling period	≈ 3 days[9]
Age at first flight	21–23 days[1,2,3]
Dependence of young	until ≈ 30 days old[3]
Broods per year	1–2[2,10]
Breeding age	1 year[2]
Mating system	probably monogamy[8]
Duration of pair bond	?
Total length	21.0–25.5 cm[7]
Wing span	53.3–61.0 cm[1]
Weight	M 60 g (?) / F 64 g (?)[4]
Longevity record	8 years[5]

References:
1. Terres 1980
2. DeGraaf and Rudis 1987
3. Harrison 1978
4. Dunning 1984
5. Clapp et al. 1983
6. Ehrlich et al. 1988
7. Godfrey 1986
8. Harrison, C. J. O. 1978
9. Gross 1940
10. Weller 1958

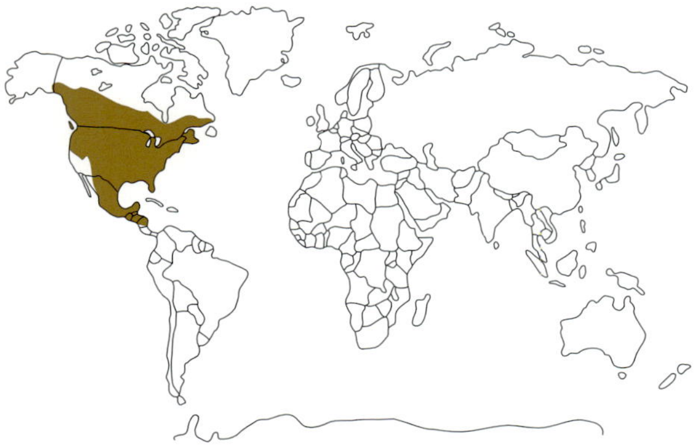

World breeding range

Chronology $n = 40$

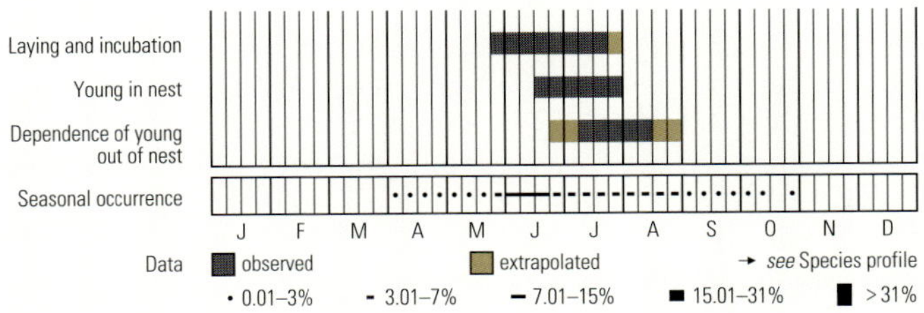

Frequency of occurrence by ecological region

Ecological regions** (forest regions)	Adequately covered squares	Squares in which reported	% of squares in which reported	Relative abundance (%)*
1. Sugar maple-hickory and sugar maple-basswood	97	28	28.9 ± 4.6	47.3
2. Sugar maple-basswood	153	34	22.2 ± 3.5	36.4
3. Sugar maple-basswood and sugar maple-yellow birch	101	21	20.8 ± 4.9	34.1
4. Sugar maple-yellow birch-basswood	56	15	26.8 ± 10.2	43.9
5. Sugar maple-yellow birch-beech and sugar maple-yellow birch-basswood	113	19	16.8 ± 4.0	27.6
6. Sugar maple-yellow birch-beech	71	19	26.8 ± 9.5	43.9
7. Sugar maple-yellow birch-balsam fir	19	5	26.3 ± 21.0	43.1
8. Sugar maple-yellow birch and balsam fir-yellow birch	47	11	23.4 ± 11.0	38.4
9. Balsam fir-yellow birch	87	21	24.1 ± 8.4	39.6
10. Balsam fir-white birch or balsam fir-red maple	46	22	47.8 ± 12.5	78.4
11. Yellow birch-balsam fir	52	22	42.3 ± 13.8	69.3
12. Balsam fir-white birch (west)	59	36	61.0 ± 12.8	100.0
13. Balsam fir-white birch (east)	76	35	46.1 ± 10.9	75.5
14. Balsam fir-black spruce	28	7	25.0 ± 17.0	41.0
15. White spruce-balsam fir and balsam fir-white spruce	22	1	4.5 ± 10.3	7.4
16. Black spruce-balsam fir-moss	18	10	55.6 ± 26.0	91.0
17. Black spruce-moss	32	12	37.5 ± 18.3	61.5
18.		not analysed		

* minimal: 0–29.9%; low: 30–59.9%; high: 60–89.9%; maximal: 90–100%

** Deciduous forest: 1–7; mixed forest: 8–11; coniferous forest: 12–17

Female incubating her eggs, which she has laid directly on the ground of a clearcut. Note how well she blends into the background. Gouin Reservoir. *July 1990, Alain Gagnon.*

In the breeding season, the female may react to the approach of a predator or human intruder with a graduated diversionary response in three stages (Gramza 1967; Hiam and Sutherland 1979). First, she flushes from the nest, her tail fanned out and pointing downward. Then, landing a short distance away, she crouches on the ground, trembling. Last, she opens her mouth, holds her wings outstretched, and hisses, glaring at the intruder. The flushing distance depends on the point in the incubation period; the closer it is to hatching, the closer an intruder can get. This strategy of waiting until the last minute to betray the location of the nest may result in fewer nests being destroyed by predators (Gramza 1967).

At 3 weeks of age, the young begin to flutter about for short distances, and by 25–30 days, they can already catch a few insects (Weller 1958). At this point, the female returns only occasionally to the nest to feed the young, but continues to keep a sharp eye on them until they are capable of sustained flight (Rust 1947).

The male usually defends a large territory. In Detroit, for example, territories were estimated to be around 10 ha (Armstrong 1965). In some habitats, a paucity of nesting sites may lead pairs to nest in close proximity (Warren 1888). According to Gross (1940), a good nest site, rather than gregariousness, accounts for the closeness of nests. The male defends his territory fiercely, immediately chasing away any other male that flies over it (Sutherland 1963). He usually roosts near the nest site (Weller 1958), in a tree or on the ground in a treeless area.

Notes on territory (continued)

A group of 16 nests on the Florida coast were an average of 73 m apart (Sutherland 1963).

Unlike other caprimulgids, the Common Nighthawk does not appear to synchronize its reproductive cycle with the lunar cycle, probably because the species is partially diurnal (Mills 1986). Therefore, it does not prolong its foraging period to take advantage of a full moon (Brigham 1991).

No nest is built and the eggs are laid directly on the ground on sand, gravel, or rock. On clear days, the female incubates the eggs with her back to the sun to protect her head from solar radiation and prevent her body temperature from rising (Weller 1958), but she has not been observed turning in any particular direction on cloudy days. During a heat wave, when the roof temperature may reach 43°C, her steadfast presence on the nest keeps the eggs 15°C cooler than the ambient temperature; at night, incubation keeps the eggs 6°C warmer (Weller 1958).

It was formerly believed that nighthawks could carry their eggs in their mouths when their clutches were endangered (Gross 1940), but this has never been proven and it appears to be a myth. The rare cases when members of this family have been seen transporting eggs are most likely accidental. One of the explanations proposed by Jackson (1985) is that the albumen from the hatching eggs may cause the shell to adhere to the plumage of the incubating hen, so that when she takes flight, she appears to be carrying an egg.

On the same day they hatch, the downy chicks can already run short distances in response to their mother's calls. Soon afterward, they are able to roam easily around their territory to seek shade from the midday sun under nearby vegetation or structures on roofs (Weller 1958). Incubation is by the female alone (Gross 1940; Weller 1958; Brigham 1989); after they hatch, she broods the nestlings assiduously for another 19 days or so (Gramza 1967). While she is on the nest, she and the chicks are fed by the male (Brigham 1989), but if she lays a second clutch, the male takes care of the young alone. The female may leave the eggs or chicks for a few minutes a day to feed (Weller 1958). When the male approaches the nest with food, he announces his arrival with a special call; the young respond by peeping and running over to him to be fed (Weller 1958; Sutherland 1963).

Common Nighthawk roosting on sandy ground. This member of the family Caprimulgidae spends most of its day resting, and is active mainly in the evenings and early mornings. Villemontel, west of Amos, Abitibi. *June 1979, André Cyr.*

Nighthawks lay their eggs right on the ground, in short vegetation, on bare soil, or on sandy or gravel-strewn areas, frequently gravel-covered flat roofs. Here, two nestlings huddle on the roof of a house in Québec City. *June 1957, Gaston Le Page.*

For a variety of reasons, it is difficult to confirm breeding of the Common Nighthawk, as reflected by the fact that atlassers found only 21 nests. For one thing, nighthawks do not build actual nests, and for another, they are never seen carrying food, since they swallow their prey and regurgitate it into the mouths of their offspring. Other factors accounting for the poor confirmation rate are the size and variety of nighthawk habitats; the cryptic coloration of the eggs, chicks, and female; the female's tendency to sit tight, remaining perfectly still on the nest; the species' crepuscular habits; and, finally, the difficulty of getting at nests on roofs (Kibbe 1985d).

Many records were obtained in urban areas. Insects are easier to see under the artificial lighting of cities and towns, and nighthawks are attracted by the prospect of easier meals (Shields and Bildstein 1979). Brigham (1991) maintains, however, that the Common Nighthawk prefers to nest in natural habitat. In uninhabited areas, it is most often observed in mixed and coniferous forests, particularly in places where the most intensive logging is being done (Fragnier 1992).

Egg and downy chick in ground litter on the forest floor. The eggs, usually numbering 2, hatch asynchronously. The nestling's marbled down blends in perfectly with the surrounding lichens and twigs. Gouin Reservoir. *July 1989, Alain Gagnon.*

Distribution

The Common Nighthawk was atlassed all over southern Québec, particularly in areas where the forests are dominated by conifers. Both in the Gaspé and the rest of the province, twice as many nighthawks were sighted north of the boundary between the deciduous forest and mixed and coniferous forests, which corresponds to the isotherm for a mean annual temperature of 2.5°C.

Common Nighthawks are observed less often in high-elevation fir and spruce forests, like those found on the mountaintops of the Laurentide Wildlife Reserve and the Gaspé Peninsula. It is not known whether the insect populations on which the species feeds in the skies above these forests are affected by the high winds and colder temperatures prevalent there.

No Common Nighthawks were recorded on the Lower North Shore, and they were spotted in just one square on Anticosti Island, although Ouellet (1969) reported that he and several other observers had found it there. There are several records from the Magdalen Islands, but breeding has not yet been confirmed there (Fradette 1992).

The northern limit of the Common Nighthawk's Québec breeding range is beyond the Atlas area. Todd (1963) recorded the species at Fort George on the James Bay coast, and it has also been sighted farther east, at La Grande-2 and Gagnon (ÉPOQ). The Common Nighthawk probably does not breed above 53° N.

Although the Common Nighthawk has a wide distribution, it is not really common anywhere within its range (Erskine 1992). A low number of breeding confirmations and poor abundance figures were also reported in the breeding bird atlasses of the Maritimes (Erskine 1992), Maine (Adamus 1987), and Ontario (Mills 1987b).

History and trends

The Common Nighthawk seems to have capitalized on human modifications to ecosystems. In the early days of European settlement, the razing of forests for cropland and pastureland provided the species with ideal nesting habitat (Kibbe 1985d). In the second half of the 19th century, nighthawks began to nest on flat gravel roofs, which may account in part for the increase in populations around the same time. In Ohio and Pennsylvania, nighthawks now nest almost exclusively on artificial structures (Peterjohn and Rice 1991; Brauning 1992b). Since the turn of the century, forest-fire prevention, extensive logging and the subsequent natural regrowth of forests, the reversion of abandoned farmland to forest, and the intensive cultivation of the remaining arable land have in turn played a part in modifying the regional distribution of the species.

Traditionally, the Common Nighthawk has been considered common and even abundant in its breeding range as a whole, including northeastern North America. More recently, however, observers have noted a significant decline in populations (Goodwin and Rosche 1970, 1974; Sibley 1988b). The species was on the Blue List from 1975 to 1986 (Tate 1986), and is listed as a Species of Special Concern in New York State (Sibley 1988b). According to ÉPOQ (1970–89) and BBS data (1969–89), the size of the Québec breeding population has been declining (Cyr and Larivée 1993a).

The decline in Common Nighthawk populations is closely tied to the decrease in available prey. In urban areas, prey abundance has probably been lowered by intensified human activity and associated pollution, as well as postwar insect-control programs. Intensive spraying against defoliator insects appears to be responsible for the decline in prey in our forests, although some forestry practices still create extensive habitat for the species (Erskine 1992).

Benoît Limoges

Whip-poor-will

Engoulevent bois-pourri

Caprimulgus vociferus

Christiane Girard

THE WHIP-POOR-WILL WAS NOT RECOGNIZED AS A SEPARATE SPECIES UNTIL the early 19th century. Before then, people mistook its song for that of its commoner and more widespread relative, the Common Nighthawk (Lawrence 1974).

Like the Common Nighthawk, the Whip-poor-will only appears on the scene at dusk, when most other birds are settling in for the night. During the day, it sleeps crouched on a branch or flat rock, protected from predators by its cryptic coloration. This migrant breeder returns to Québec from its wintering grounds at the end of April to nest.

Habitat and behaviour

The Whip-poor-will's breeding habitat is dry woodland of various types and ages, generally stands with many openings and clearings, or next to cultivated fields or shrubby pastures. Whip-poor-wills particularly favour young pine, oak, and beech forests and shun mountainous areas (Peck and James 1983; DeGraaf and Rudis 1987).

In the northern parts of its range, the first male Whip-poor-wills start to arrive in late April. Just after the sun sets, he announces his presence with his ringing song, repeated tirelessly at short intervals, commonly up to 100 times. The song has been transcribed in a variety of ways, including *whip poor-weel*, and *purple-rib* (Peterson 1980). The Whip-poor-will is most vocal in the first two hours after sunset (Lawrence 1974).

When the female arrives, male's behaviour changes markedly. When she is nearby, his singing accelerates dramatically until it is almost continuous and then dissolves into protracted cooing. When the female reaches his side, the cooing changes to soft, guttural, spun-out *cooorah-cooorah* notes (Lawrence 1974).

The female chooses a nesting site in or near a thicket. A generous layer of dead leaves or pine needles appears to be an important factor in her choice. No nest is built, although the incubating parent may cause a slight depression in the litter around the eggs. The female incubates the eggs for 19 or 20 days, the male relieving her only when she leaves the nest to feed. Both parents defend the nest. When an intruder approaches, the female is slow to flush, relying instead on her cryptic coloration for protection. When she goes, she drags her wings and spreads her tail, feigning injury. Atlassers observed a distraction display at Racine Point on Lake Saint-Jean.

According to Mills (1986), the Whip-poor-will's breeding cycle is synchronized with the lunar cycle. The two weeks after hatching seem to be the most critical for nestling survival, since at that age the young cannot fly and must be closely attended by their parents. Mills believes that the parents synchronize this crucial period in the young's development with the full moon, when it is easier for the parents to hunt for insects. The following dark half of the lunar cycle appears to coincide with the period when the young mature, and they become independent during the next bright phase, at the age of 5–6 weeks (Mills 1986).

The Whip-poor-will lives almost entirely on insects, mainly large moths, which it specializes in taking on the wing. Unlike the Common Nighthawk, which prefers foraging in continuous aerial pursuit of its prey (Brigham and Fenton 1991), the Whip-poor-will also hawks insects, generally from the ground. It can also be seen foraging for beetles and other insects on the ground around stumps and rotting wood (Tyler 1940b). The Whip-poor-will is more active on nights with a full moon (Mills 1986).

The species was reported in just 6.8% (168/2464) of Atlas squares, probably because it is only active for brief periods at dawn and dusk. Breeding was confirmed in only 6 of the 168 squares in which Whip-poor-wills were reported. The difficulty of finding nests, the female's cryptic coloration and reluctance to flush at the approach of an intruder, and the fact that adults are never seen carrying food but feed the young by regurgitation no doubt made it more difficult to confirm breeding. Accordingly, 90% (109/168 + 43/168) of Atlas records were reports of birds seen in suitable habitat or heard singing.

Notes on territory

The size of the territory defended by the Whip-poor-will generally depends on the characteristics of the forest in which it is found. In Kansas, home ranges were estimated to be 2.8, 6, and 11 ha in elm, oak, and walnut forests, respectively (Fitch 1958).

In Maryland, density of territorial males has been estimated at 0.35 male/10 ha (Stewart and Robbins 1958).

For a number of years, the Whip-poor-will has been seen in open areas like vegetable gardens and fields between the town and dunes of Tadoussac. *July 1991, Yves Aubry.*

Whip-poor-will

Breeding distribution

1984–1989

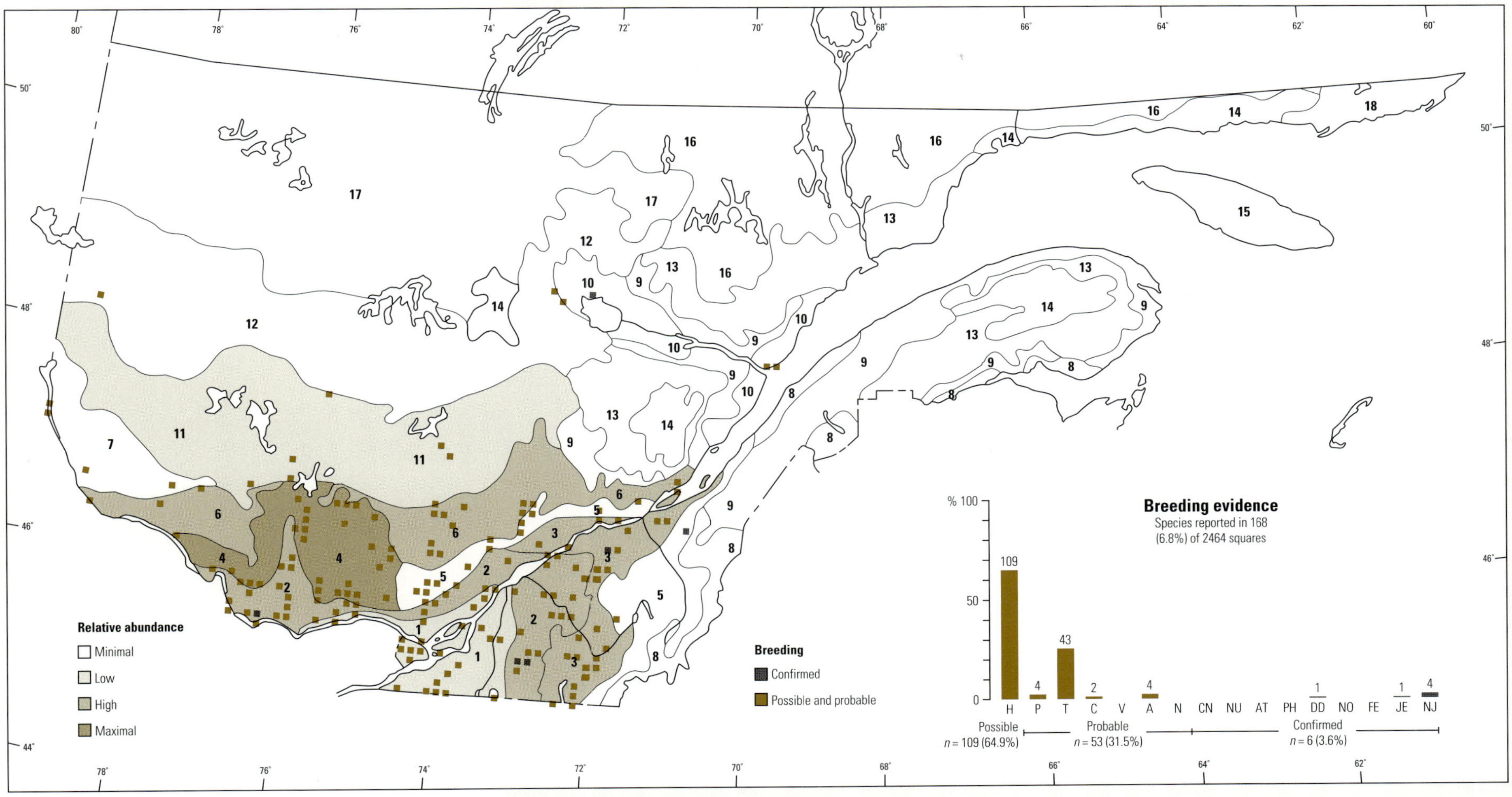

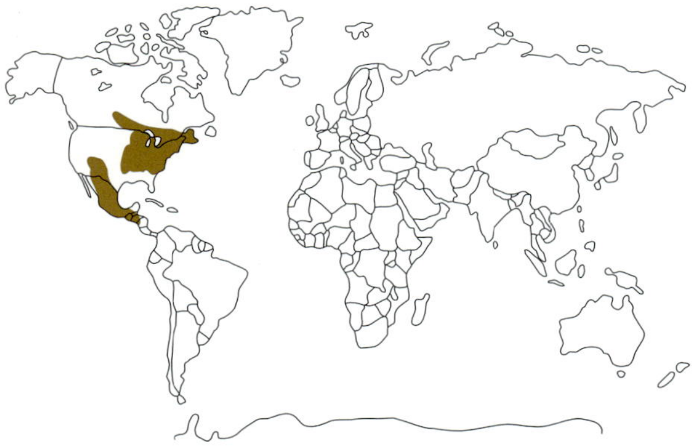

Species profile

Clutch size	usually 2 eggs (1–2)[1,2,3,4,9]
Laying interval	1 egg per day[1]
Start of incubation	with first egg[1]
Incubation period	19–20 days[1,2,3,4]
Maturity at hatching	semiprecocial
Care of young	mainly by female[1]
Nestling period	a few days[1,2]
Age at first flight	20 days[4,9]
Dependence of young	until 5–6 weeks old[10]
Broods per year	1, sometimes 2[3,4]
Breeding age	1 year[3]
Mating system	monogamy[5,9]
Duration of pair bond	?
Total length	23.0–25.5 cm[7]
Wing span	40.6–49.5 cm[4]
Weight	M 55.3 g (43.0–63.7) / F 50.6 g (44.5–61.2)[8]
Longevity record	4 years[6]

References:
1. Tyler 1940b
2. Lawrence 1974
3. DeGraaf and Rudis 1987
4. Terres 1980
5. Harrison, C. J. O. 1978
6. Clapp et al. 1983
7. Godfrey 1986
8. Dunning 1984
9. Ehrlich et al. 1988
10. Mills 1986

Chronology $n = 11$

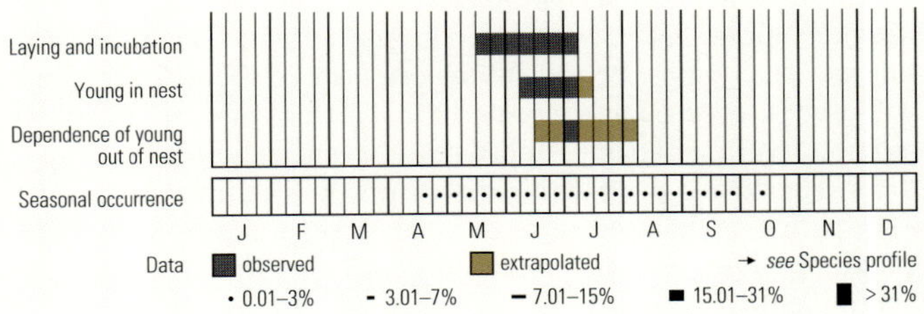

Frequency of occurrence by ecological region

Ecological regions** (forest regions)	Adequately covered squares	Squares in which reported	% of squares in which reported	Relative abundance (%)*
1. Sugar maple-hickory and sugar maple-basswood	97	19	19.6 ± 4.1	54.8
2. Sugar maple-basswood	153	45	29.4 ± 3.8	82.4
3. Sugar maple-basswood and sugar maple-yellow birch	101	22	21.8 ± 5.0	61.0
4. Sugar maple-yellow birch-basswood	56	20	35.7 ± 11.0	100.0
5. Sugar maple-yellow birch-beech and sugar maple-yellow birch-basswood	113	12	10.6 ± 3.3	29.7
6. Sugar maple-yellow birch-beech	71	16	22.5 ± 9.0	63.1
7. Sugar maple-yellow birch-balsam fir	19	2	10.5 ± 15.5	29.5
8. Sugar maple-yellow birch and balsam fir-yellow birch	47	0	0	0
9. Balsam fir-yellow birch	87	0	0	0
10. Balsam fir-white birch or balsam fir-red maple	46	2	4.3 ± 5.7	12.2
11. Yellow birch-balsam fir	52	6	11.5 ± 9.3	32.3
12. Balsam fir-white birch (west)	59	2	3.4 ± 5.3	9.5
13. Balsam fir-white birch (east)	76	0	0	0
14. Balsam fir-black spruce	28	0	0	0
15. White spruce-balsam fir and balsam fir-white spruce	22	0	0	0
16. Black spruce-balsam fir-moss	18	0	0	0
17. Black spruce-moss	32	0	0	0
18.		*not analysed*		

* minimal: 0–29.9%; low: 30–59.9%; high: 60–89.9%; maximal: 90–100%

** Deciduous forest: 1–7; mixed forest: 8–11; coniferous forest: 12–17

Closely associated with deciduous and mixed forests in southern Québec, the Whip-poor-will also nests in young pine forests with a thick layer of needles. Pin-Rigide Ecological Reserve, Saint-Chrysostome, south of Montréal. *May 1991, Pierre Bannon.*

Distribution

Whip-poor-wills breed from central Saskatchewan to Nova Scotia and in the eastern half of the United States, as well as in the southwestern states down to Central America (AOU 1983; Godfrey 1986). In Québec, they appear to be closely associated with the deciduous and mixed forest. Most of the Atlas breeding records come from the Eastern Townships, the Central St. Lawrence Lowland, and the southern part of the Laurentians. Nests containing eggs or young were found near Wyman (Ottawa Valley), Milton, Granby (Eastern Townships), and Val-Alain (Lotbinière), young incapable of sustained flight were observed near Saint-Lazare (Bellechasse), and an adult performing a distraction display confirmed breeding at Racine Point on Lake Saint-Jean. The species was also observed outside the Atlas survey period at Vauvert, near Péribonka. Whip-poor-wills seem to nest fairly regularly in La Doré and, in 1979, were reported nesting in Saint-Ambroise (G. Savard, pers. comm.).

The Whip-poor-will's distribution is patchy north of the 47th parallel, where it seems to be associated mainly with areas of deciduous forest. It was observed near Lake Timiskaming, where mature hardwood forests and open country are found together, and at the Rouyn–Noranda cemetery, to which it seems to return every year (J. Lapointe, pers. comm.). The northernmost records in the province are from the Lake Saint-Jean Lowland. Over the past several years, singing males have been heard a number of times in the dunes along the St. Lawrence River at Tadoussac. The particularly mild microclimate, as well as the presence of deciduous and mixed woodland and vegetable gardens in the area provide excellent nesting and foraging habitat for the species. Tadoussac is the eastern limit of the Whip-poor-will's range in the province. On the south shore of the St. Lawrence, it is seldom found east of Montmagny, and Tanguay (1965) considered it rare in this area.

The Whip-poor-will winters from central and eastern South Carolina to Panama, including the Gulf of Mexico coast, southern Texas, northern Mexico, and Central America (AOU 1983).

Female Whip-poor-will near her eggs, which she lays directly on the leaf litter. Saint-Lambert, near Montréal. *June 1933, Henry Mousley, Laval University Collection.*

Two 9-day-old Whip-poor-wills. The cryptic coloration of the adults and young allows them to blend into their surroundings. Saint-Lambert, near Montréal. *June 1933, Henry Mousley. Laval University Collection.*

Young Whip-poor-will, approximately 21 days old, sleeping on a rock. Montréal. *July 1940, Henry Mousley, Laval University Collection.*

History and trends

The status of the Whip-poor-will in Québec does not appear to have changed substantially since the turn of the century, when Macoun and Macoun (1909) reported that only small numbers visited Québec in summer and that they were scarce in the Montréal area. In the early 1970s, Ouellet (1974a) still concurred, listing the few known nesting sites around Greater Montréal. More recently, Bannon (1991b) described it as rare in the Montréal region and occasional in the Laurentians.

In Ontario, its breeding range appears to have shrunk somewhat since the last century. Habitat changes due to forest clearing for extensive agriculture in some areas and to the maturation of forests in others seem to be the main causes of local disappearances and declines of Whip-poor-wills in various parts of that province (Mills 1987f). Not enough data are available to determine whether a similar range contraction has occurred in Québec, but fewer and fewer birds were observed in Québec between 1980 and 1989 (BBS).

The Whip-poor-will was on the Blue List in eastern North America in 1980-81 (Tate 1981) and again in 1986, and on the list of Species of Special Concern in 1982 (Tate 1986). Shrinking populations of large moths, due to logging and the use of insecticides, are probably responsible for this decline (Robbins et al. 1986).

LAVAL ROY AND MANON BOMBARDIER

Chimney Swift

Martinet ramoneur

Chaetura pelagica

Christiane Girard

At first glance, the Chimney Swift may be mistaken for a swallow, but a closer look shows its wings to be longer, narrower, and more tapered. They are also very curved. Jerky, hurried wingbeats result in its typical erratic flight.

The Chimney Swift is a fairly common migrant breeder in southern Québec. Its zigzagging flight against the summer sky is a familiar sight in city and country alike.

Habitat and behaviour

The Chimney Swift requires a dark, sheltered spot in which to make its nest (Fisher 1958). Prior to European settlement of North America, it probably nested only in hollow trees and caves (Dionne 1906; Bull 1974) but has since taken advantage of artificial habitat that offers similar conditions.

Today, unused chimneys, barns, air shafts, silos, and other structures are the most common nest sites (Tyler 1940a; Fisher 1958), but in Québec, as elsewhere, some swifts continue to nest in hollow trees (QNRCP).

Chimney Swifts often nest in colonies, although usually only one pair will nest in a small chimney (Terres 1980). Colonies are highly variable in size. In 1985, some 125 nesting pairs were found using a chimney in Sainte-Anne-de-la-Pocatière (COQ 1985), and several hundred pairs nested on the McGill University campus in Montréal in the late 1960s (Ouellet 1974a). During the nesting period and migration, huge flocks of swifts can be seen entering a large chimney just after sundown, rising again the next morning in a steady stream like a column of smoke (Zammuto and Franks 1981).

Peak arrival in southern Québec is around mid-May, with departure in August (David 1980a). Cold weather may prove disastrous to this small, insectivorous species. For example, 109 birds were found dead at the bottom of a chimney in Sainte-Anne-de-la-Pocatière on 23 May 1990, having apparently expired from the cold and snow (Aubry et al. 1990b). The location of the species' wintering grounds had long been a mystery, until 1944, when swifts were discovered in the upper Amazon watershed in South America (Tufts 1961; AOU 1983).

The nest is built of small dead twigs that the birds snap off trees while in flight. Swifts glue their nests to the inside of a chimney or other vertical wall with their own glutinous saliva. Nests in chimneys are usually attached several metres below the opening. Egg laying begins once the nest is half finished. Both parents build the nest, incubate the eggs, and feed the young (Tyler 1940a; Fisher 1958). Sometimes one or two non-breeding swifts, rarely more, will join up with a breeding pair for the nesting season, sometimes helping to incubate the eggs and feed the young. These are usually immature or unmated birds and they may or may not be related to the pair (Dexter 1974, 1981, 1992).

Notes on territory

The Chimney Swift does not appear to be territorial either when breeding or migrating. In fact, swifts are most often found in flocks. Some nest sites may be used by only one pair, while others may house a great number of birds. In such situations,

After spending the day feeding in the air, the Chimney Swifts return to their nests in the many chimneys so characteristic of the old town. Québec City. *May 1923, National Air Photo Library, Natural Resources Canada.*

Chimney Swift

Breeding distribution
1984–1989

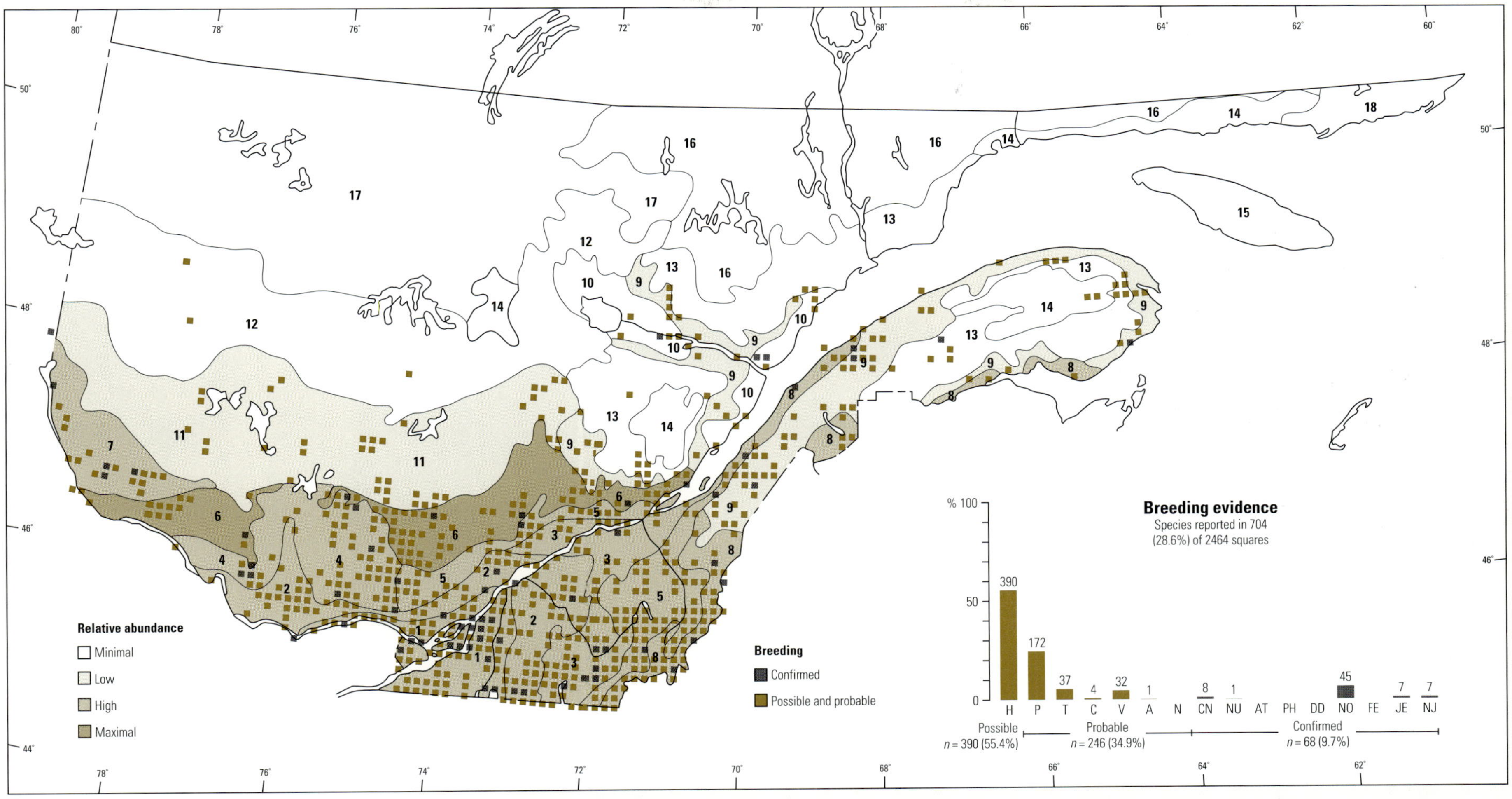

Species profile

Clutch size	usually 4–5 eggs (2–7)[2,6,7,8]
Laying interval	1 egg per 2 days[4]
Start of incubation	before last egg[4]
Incubation period	19–21 days[1,2,6]
Maturity at hatching	altricial
Care of young	by female and male[6,7]
Nestling period	14–19 days[3,5]
Age at first flight	28–30 days[2,6]
Dependence of young	until ≈ 1 week after first flight[3]
Broods per year	1[3,4]
Breeding age	1 year[3,7]
Mating system	?
Duration of pair bond	sometimes more than 1 season[9]
Total length	12.0–14.4 cm[10]
Wing span	30.5–32.4 cm[1,10]
Weight	22.8 g (16.1–33.5)[1]
Longevity record	14 years[11]

References:
1. Terres 1980
2. Tyler 1940a
3. Stokes 1989
4. Harrison 1975
5. Harrison 1978
6. Fisher 1958
7. Dexter 1981
8. Peck and James 1983
9. Dexter 1992
10. Godfrey 1986
11. Clapp et al. 1983

World breeding range

Chronology $n = 21$

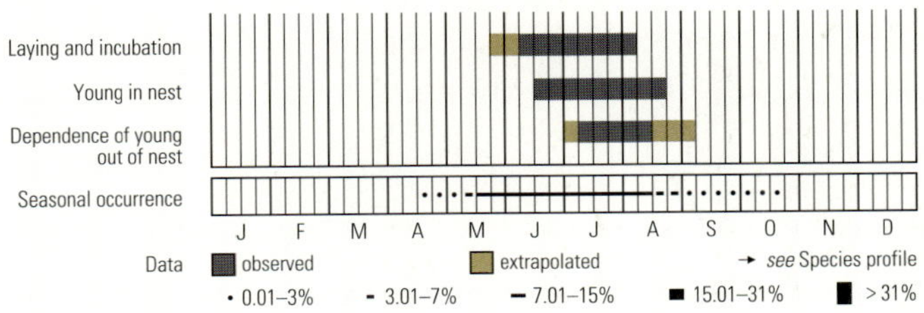

Frequency of occurrence by ecological region

Ecological regions** (forest regions)	Adequately covered squares	Squares in which reported	% of squares in which reported	Relative abundance (%)*
1. Sugar maple-hickory and sugar maple-basswood	97	76	78.4 ± 4.2	86.9
2. Sugar maple-basswood	153	89	58.2 ± 4.1	64.5
3. Sugar maple-basswood and sugar maple-yellow birch	101	64	63.4 ± 5.7	70.3
4. Sugar maple-yellow birch-basswood	56	43	76.8 ± 9.8	85.2
5. Sugar maple-yellow birch-beech and sugar maple-yellow birch-basswood	113	85	75.2 ± 4.5	83.4
6. Sugar maple-yellow birch-beech	71	64	90.1 ± 6.6	100.0
7. Sugar maple-yellow birch-balsam fir	19	13	68.4 ± 22.1	75.9
8. Sugar maple-yellow birch and balsam fir-yellow birch	47	32	68.1 ± 12.0	75.5
9. Balsam fir-yellow birch	87	41	47.1 ± 9.7	52.3
10. Balsam fir-white birch or balsam fir-red maple	46	12	26.1 ± 11.1	28.9
11. Yellow birch-balsam fir	52	23	44.2 ± 13.9	49.1
12. Balsam fir-white birch (west)	59	1	1.7 ± 4.0	1.9
13. Balsam fir-white birch (east)	76	20	26.3 ± 9.7	29.2
14. Balsam fir-black spruce	28	3	10.7 ± 12.6	11.9
15. White spruce-balsam fir and balsam fir-white spruce	22	0	0	0
16. Black spruce-balsam fir-moss	18	0	0	0
17. Black spruce-moss	32	0	0	0
18.			*not analysed*	

* minimal: 0–29.9%; low: 30–59.9%; high: 60–89.9%; maximal: 90–100%

** Deciduous forest: 1–7; mixed forest: 8–11; coniferous forest: 12–17

The Chimney Swift often nests in unused chimneys, like the one in the Maison des Français in the Cap Tourmente National Wildlife Area. Northeast of Québec City. *August 1991, Léo-Guy de Repentigny.*

Spotting the Chimney Swift is easy, as it spends most of its time in the air, foraging on the wing, chittering characteristically all the while. Confirming it as a breeding bird, however, is another matter. A lot of patient observation is needed to spot a bird entering a chimney and leaving a few minutes later. In fact, breeding was confirmed in only 9.7% (68/704) of Atlas squares in which it was reported, mostly on the basis of reports of swifts entering or leaving the nest site. In most squares, it was not possible to strengthen the breeding evidence beyond possible or probable breeding.

Distribution

The breeding range of the Chimney Swift covers almost all of the eastern half of the United States and southern Canada from Saskatchewan east to Nova Scotia (Robbins et al. 1966). In Québec, it breeds mainly south of the 49th parallel (Godfrey 1986). Atlas records show it to be found mainly in the populated areas in the southern parts of the province, probably because of the availability of nest sites. There is a marked decline in occurrence north of the mixed forest region corresponding with the isotherm for a mean annual temperature of 2.5°C. This is probably more than a mere coincidence, for temperature is known to play an important role in the development and hatching of insect populations, upon which the Chimney Swift depends.

The most northerly Atlas records are from Saint-Maurice-de-Dalquier in the Abitibi, the La Mothe Reservoir in the Saguenay–Lac-Saint-Jean region, and Forestville on the Upper North Shore. All of these regions were already known to be within the Chimney Swift's breeding range (Todd 1963).

There are pre-Atlas sight records of swifts during the breeding season from the Magdalen Islands (Fradette 1992) and Anticosti Island (Ouellet 1969), but no breeding records, however, and no Atlas records either. The situation is the same on the Middle and Lower North Shore, where the species was reported at Matamec and Harrington Harbour in the summertime (Todd 1963). The Chimney Swift also turns up as an accidental visitor in regions well north of the known breeding range. One was seen in Digges Sound, near the northwestern tip of Québec, in August 1980 (Gaston et al. 1985).

Notes on territory (continued)

swifts probably defend only the immediate vicinity of the nest (Stokes 1989), as do other similar species (Cramp 1985). Zammuto and Franks (1978) reported the case of a chimney that contained only one active nest, but served as a nocturnal roost for 40 or so other birds that were not nesting there. Some of them spent the night only 2 m from the nest, which prompted the young to beg them for food.

Two young Chimney Swifts that have survived a fall from a chimney. The young of the species reflexively cling to perpendicular surfaces to maintain an upright position. Saint-François-du-Lac, Lake Saint-Pierre. *July 1993, Daniel Jauvin.*

History and trends

By making use of chimneys as nest sites, the Chimney Swift has capitalized on European colonization of North America. It also nests in hollow trees, far from human settlement, but it has long been associated with human habitation (Fisher 1958).

Accounts by Dionne (1906) and Todd (1963) indicate that its distribution in Québec appears to have remained unchanged since the beginning of the century. Although comparison is difficult, swifts do appear to have been more abundant in the last century (Dionne 1906), despite the fact that there are more chimneys nowadays than there were then (Ouellet 1974a). Recent BBS and ÉPOQ data seem to indicate a downward trend from the early 1970s until the late 1980s (Dunn 1991; Cyr and Larivée 1993a). The species may have been affected by the loss of farm buildings due to abandonment of farms, by aerial spraying of pesticides, or by a recent decline in the number of chimneys as more and more homes are heated electrically (Erskine 1992; Mulvihill 1992b), but there is no conclusive evidence of any of these possibilities.

GUY LEMIEUX AND MICHEL ROBERT

Ruby-throated Hummingbird

Colibri à gorge rubis

Archilochus colubris

Christiane Girard

THE RUBY-THROATED HUMMINGBIRD, THE SMALLEST BREEDING BIRD IN Québec, is a regular visitor to flower gardens. Several of the distinguishing characteristics of this common migrant breeder (David 1980a) have to do with its diminutive size. Its rapid wingbeats —20–80 strokes a second, depending on sex and type of flight —enable it to hover, as well as fly backward, forward, or straight up and down with ease (Terres 1980; Ehrlich et al. 1988). Sometimes it is mistaken for a big bumble bee or a sphinx moth. Thanks to its flying ability, it can feed on the energy-rich nectar of flowers and catch insects on the wing. At night, when the temperature goes down and it cannot feed, its body temparature drops and it becomes torpid, reducing its heart rate and slowing down its vital functions (Terres 1980).

Habitat and behaviour

Hummingbirds show a preference for open disturbed habitat where they can find flowers as a source of nectar—gardens and orchards or clearings and edges of mixed and hardwood forests (Peck and James 1983; Godfrey 1986).

The species is attracted to tubular blossoms, such as those of honeysuckle, columbine, touch-me-not, monarda, and lobelia, as well as milkweed and the flowers of various ericaceous shrubs (Martin et al. 1951). Hummingbirds readily visit sugar-water feeders, and are sometimes seen drinking sap from rows of holes drilled by Yellow-bellied Sapsuckers in poplar, birch, and maple trees. Like sapsuckers, they also feed on the tiny insects drawn to the sap (Rushmore 1969; Godfrey 1986). According to Miller et al. (1985), hummingbirds are able to memorize the sites of the best food sources, first, by location and, second, by colour. An interesting study by Bertin (1982) demonstrated a link between the Ruby-throat and the plants it pollinates, such as spotted touch-me-not. The spatial distribution of the flowers it feeds upon makes it become territorial, driving other Ruby-throats as well as butterflies and bumble bees away from the area. When the food supply is very plentiful, as in a flowering hedge or garden, it may tolerate the presence of other birds.

During courtship, the male adopts a distinctive behaviour pattern when it encounters other males or females. It performs a series of swoops and dives in a pendulum arc, all the while uttering its grating squeaks and buzzing its wings. It may swoop from heights of 12 m. Other acrobatics in the form of repeated vertical and horizontal movements are linked with courtship behaviour, as well. Some of the precopulation behaviour also occurs as part of the intimidation displays between males and females, directed particularly toward rival birds attempting to use the same feeding grounds (Terres 1980; Stokes and Stokes 1989a).

Notes on territory

As soon as he arrives on the breeding grounds, the male establishes a territory of about 0.1 ha and defends it against intruding males of his own species. He will also drive away passerines, bees, and butterflies. Territorial behaviour appears to have two functions: courtship and defence of a food resource, the latter purpose causing the territory holder to drive out females as well as males.

The female defends a small area around the nest, and may defend a feeding territory as well (Stokes and Stokes 1989a).

In forest habitat, Ruby-throated Hummingbirds require clearings with an abundance of nectar-producing flowers. They also catch insects on the wing. Sainte-Foy. *September 1991, Yves Aubry.*

This diminutive migrant favours open areas where blossoms can provide it with nectar. It is attracted to both wild and garden flowers, in town and country alike. Saint-Hubert, near Montréal. *August 1991, Gaétan Duquette.*

Ruby-throated Hummingbird

Breeding distribution

1984–1989

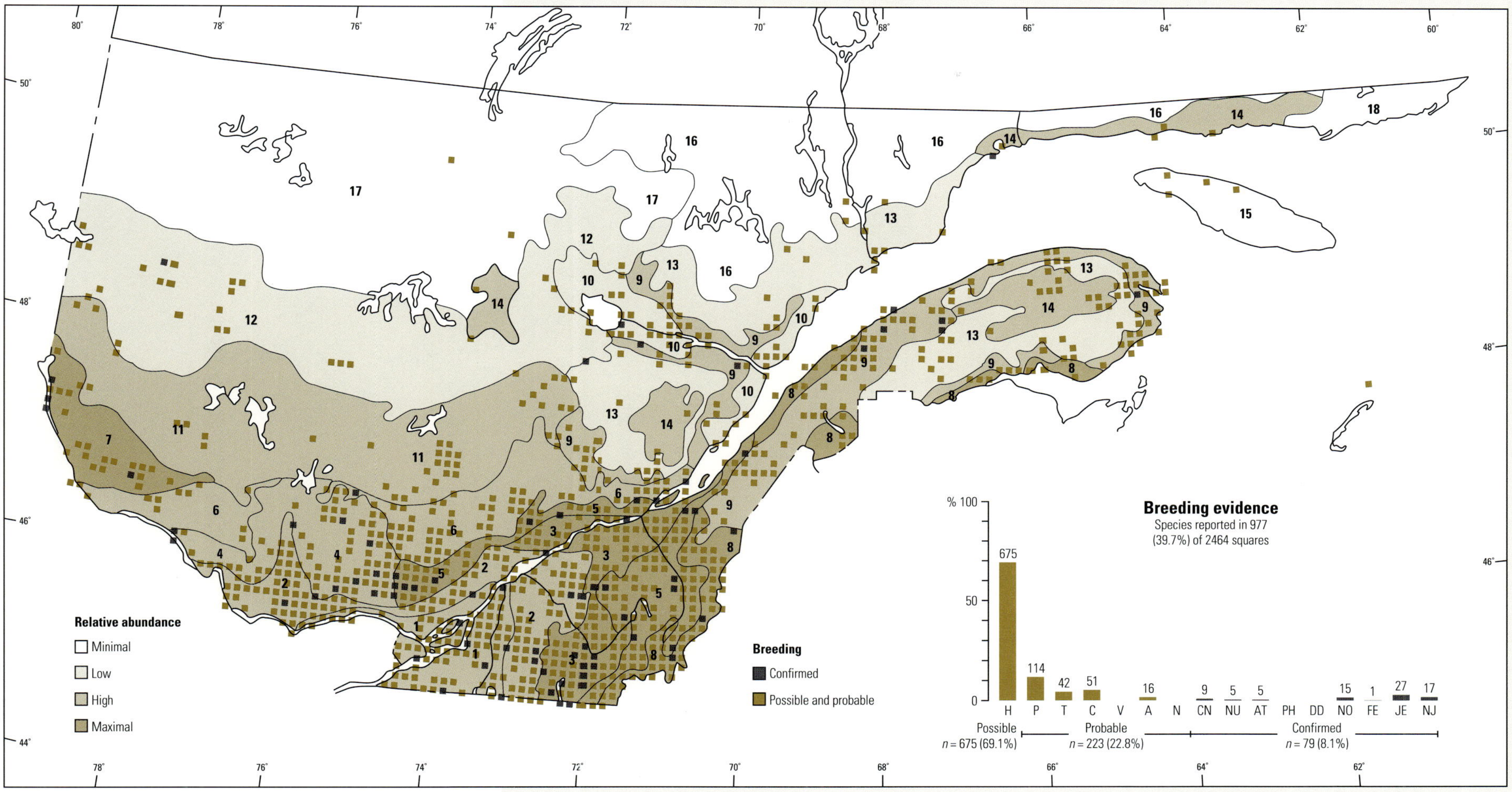

Species profile

Clutch size	2 eggs[1,2,3,4,5]
Laying interval	1 egg per 2–3 days[1,5]
Start of incubation	with second egg[5]
Incubation period	usually 16 days (11–16)[1,2,3,4,5]
Maturity at hatching	altricial
Care of young	by female[1,2,3,5]
Nestling period	14–28 days[1,2,3,5]
Age at first flight	14–28 days[1,3,5]
Dependence of young	until ≈ 1 day after leaving nest[1]
Broods per year	1–2[9]
Breeding age	?
Mating system	promiscuity[2,11], polygyny[10]
Duration of pair bond	during mating only[11]
Total length	7.5–9.4 cm[7]
Wing span	10.2–12.0 cm[4]
Weight	M 3.0 g (2.4–4.1) / F 3.3 g (2.7–4.8)[8]
Longevity record	9 years[6]

References:
1. Allen 1930
2. Ehrlich et al. 1988
3. Harrison 1975
4. Terres 1980
5. Johnsgard 1983b
6. Klimkiewicz and Futcher 1989
7. Godfrey 1986
8. Dunning 1984
9. Tyler 1940c
10. DeGraaf and Rudis 1987
11. Stokes and Stokes 1989a

Chronology $n = 30$

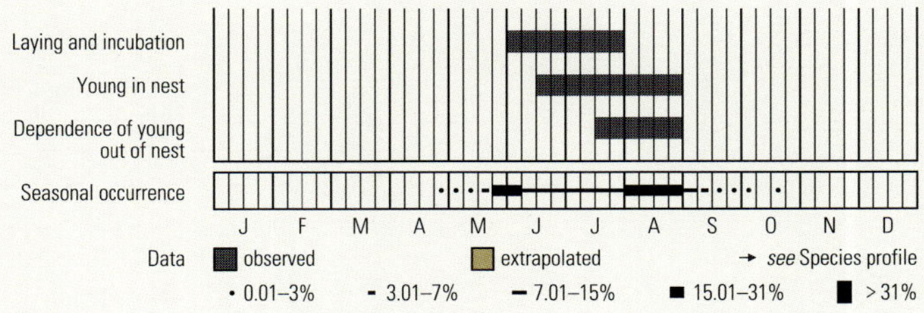

Frequency of occurrence by ecological region

Ecological regions** (forest regions)	Adequately covered squares	Squares in which reported	% of squares in which reported	Relative abundance (%)*
1. Sugar maple-hickory and sugar maple-basswood	97	76	78.4 ± 4.2	82.7
2. Sugar maple-basswood	153	121	79.1 ± 3.4	83.5
3. Sugar maple-basswood and sugar maple-yellow birch	101	88	87.1 ± 4.1	92.0
4. Sugar maple-yellow birch-basswood	56	47	83.9 ± 8.6	88.6
5. Sugar maple-yellow birch-beech and sugar maple-yellow birch-basswood	113	98	86.7 ± 3.6	91.5
6. Sugar maple-yellow birch-beech	71	57	80.3 ± 8.6	84.7
7. Sugar maple-yellow birch-balsam fir	19	18	94.7 ± 12.0	100.0
8. Sugar maple-yellow birch and balsam fir-yellow birch	47	43	91.5 ± 7.6	96.6
9. Balsam fir-yellow birch	87	70	80.5 ± 7.8	84.9
10. Balsam fir-white birch or balsam fir-red maple	46	26	56.5 ± 12.4	59.7
11. Yellow birch-balsam fir	52	33	63.5 ± 13.5	67.0
12. Balsam fir-white birch (west)	59	22	37.3 ± 12.7	39.4
13. Balsam fir-white birch (east)	76	41	53.9 ± 10.9	56.9
14. Balsam fir-black spruce	28	17	60.7 ± 18.9	64.1
15. White spruce-balsam fir and balsam fir-white spruce	22	3	13.6 ± 15.5	14.4
16. Black spruce-balsam fir-moss	18	1	5.6 ± 13.5	5.9
17. Black spruce-moss	32	1	3.1 ± 7.6	3.3
18.			*not analysed*	

* minimal: 0–29.9%; low: 30–59.9%; high: 60–89.9%; maximal: 90–100%

** Deciduous forest: 1–7; mixed forest: 8–11; coniferous forest: 12–17

The female alone builds the nest, often not far from water, and defends a small area around it. The nest is made of bud scales and plant down held together by spider silk (McKenzie 1991), and is decorated on the outside with bits of lichen (Harrison 1978). These flexible materials allow the nest to stretch as the young grow. It is placed on a tree branch at a height ranging from 1.2 to 15 m, but most often 3–6 m up (Peck and James 1983). The nest is easily identified, being scarcely larger in circumference than a dollar, with an outer diameter of 2.5–4.4 cm (Harrison 1975). It contains 2 eggs about the size of peas.

The promiscuous male only stays with the female during the mating period, then goes off to mate again, leaving the female to incubate alone. The female guards the nest, driving away other birds, even those as large as robins or as aggressive as House Wrens. She feeds the nestlings usually by regurgitation (Tyler 1940c). The young open their eyes after about a week and leave the nest at about 19 days of age (Harrison 1978). The female appears to return to the same nesting area each year (Terres 1980).

Hummingbirds are easy to observe and identify, but breeding evidence is difficult to obtain. This explains the great difference between the number of records of possible breeding (69.1%; 675/977) and those of probable and confirmed breeding (30.9%; 302/977). Birds at feeders and non-breeding males are easily found, but it is far more difficult to locate the nest or see females returning to the nest carrying food to feed their young. Displaying males probably accounted for the majority of observations of courtship behaviour.

In Québec, hummingbirds begin to include insects in their diet in September, which may explain their quasi disappearance from their usual haunts then. Between January and March, on the wintering grounds, insects comprise the bulk of their diet (Wolf 1970).

The Ruby-throated Hummingbird builds her nest on the branch of a bush or tree, as in this apple tree. The nest is made of plant down, held together by spider silk, and is decorated with bits of lichen. These 2 eggs are about 13 mm long. Glen Roy, Ontario. *July 1941, Henry Mousley, Laval University Collection.*

With her long bill and extensible tongue, this female Ruby-throated Hummingbird sips nectar from a purple foxglove. Hummingbirds are attracted by the vivid, contrasting colours of garden flowers. Lac-Saint-Charles, near Québec City. *August 1991, Robert Morin.*

In migration, Ruby-throated Hummingbirds fly non-stop across the Gulf of Mexico. In preparation for the crossing, they accumulate the equivalent of their own weight (3 g) in fat reserves. Lasiewski (1962) estimated that a hummingbird needs 2 g of fat to make a non-stop flight of 1000 km in 26 hours at 40 km/h—more than enough for the 650-km Gulf crossing (Gill 1990). This hummingbird winters from the southernmost United States down through Mexico to Costa Rica (Godfrey 1986).

Distribution

The Ruby-throated Hummingbird, the only hummingbird found in the eastern half of the continent, ranges throughout all of southern Québec below 49° N, between Abitibi and Lac-Saint-Jean. It also occurs farther east, on the North Shore, out to Havre-Saint-Pierre. It is seen all over the Gaspé, and there are a few sight records from Anticosti Island and the Magdalen Islands, not mentioned in Godfrey's (1986) distribution description. Breeding was not

Hummingbird nest in a cedar, containing two 11-day-old young, whose bills can be seen protruding above the rim. Glen Roy, Ontario. *July 1941, Henry Mousley, Laval University Collection.*

confirmed on either Anticosti Island or the Magdalen Islands for the Atlas however, and the species is uncommon there. According to both Ouellet (1969) and Fradette (1992), it is a rare visitor in both places.

A study by Miller and Nero (1983) showed that the northern limit of the Ruby-throat's range may be dependent on its association with the Yellow-bellied Sapsucker. Where nectar is scarce, Ruby-throats can sustain themselves with sap from sapsucker wells. This commensal relationship would allow the hummingbird to occupy a variety of habitat.

The Ruby-throat's Atlas distribution corresponds largely to the deciduous and mixed forest zones, corroborating Bertin's (1982) earlier range description. The Balsam fir-white birch forest region appears to represent the northern limit of the species' current range.

History and trends

The clearing of forests for agriculture and urban development since European settlement has probably benefited the Ruby-throated Hummingbird.

It is not possible to assess from the literature at the end of the last century whether any major changes in distribution or abundance have occurred in Québec since that time. Dionne (1906) reported that it was fairly common in summer in Québec City, where he had found several nests. It was also common in Montréal and accidental on Anticosti Island.

More recently, ÉPOQ records show a decline in numbers between 1970 and 1989, and BBS data for 1966–89 also recorded a decline, though not a significant one (Cyr and Larivée 1993a). The increasing popularity of flower gardens could prove beneficial to the species (Cyr 1988; Dion 1988).

André Cyr

The nest is made of flexible material so it can expand as the young grow. These 2 young Ruby-throats, 14 days old, will soon make their first flight. Glen Roy, Ontario. *July 1941, Henry Mousley, Laval University Collection.*

Belted Kingfisher

Martin-pêcheur d'Amérique

Ceryle alcyon

Christiane Girard

THE BELTED KINGFISHER, THE ONLY MEMBER OF ITS FAMILY IN QUÉBEC, has some very intriguing characteristics. Its eyes contain droplets of a red oil that reduces chromatic aberration and glare on water, making it easier for the bird to locate its prey below the surface (Parrish et al. 1981; Martin, G. R. 1985). Its circulatory system has an extra functioning artery near the aorta that distinguishes it from many other birds (Glenny 1939). It is also one of the rare Québec species in which the female is more colourful than the male. Interestingly, too, this bird lays many eggs, 8–10 on average in Québec (P. Laporte, CWS, unpubl. data), even though its nest is well protected from predators.

Habitat and behaviour

The Belted Kingfisher lives near water, either salt or fresh: lakes, ponds, rivers, brooks, marshlands, fish hatcheries, lagoons, and along the banks of the St. Lawrence. This versatility makes populations less vulnerable to wetland degradation. In the spring, as soon as the lakes and rivers begin to thaw, it returns to its breeding grounds, where its rattling call makes it easy to locate. The nest, burrowed into a cutaway bank, is often found along a stream, but may be far from the stretch of water where it hunts for food. It may also nest in sand pits or road embankments, or sometimes in the midst of a colony of Bank Swallows, where the nest is easy to spot because the opening is twice as big as a swallow's and rounder. The Belted Kingfisher has syndactyl feet—its third and fourth toes are joined for almost their entire length—and the grooves they typically make at the entrance indicate an active nest. Because it does not always forage in the same area that it nests in, the kingfisher defends two types of territories: a breeding territory and a feeding territory (Mousley 1938; Davis, W. J. 1982).

Usually solitary, kingfishers form pairs only for the breeding season. Courtship feeding of the female by the male can often be observed (Hamel 1975).

The nest site is chosen before the breeding territory is clearly established (Davis 1980). Both birds take turns building the nest, alternating every 5 or 10 minutes (P. Laporte, pers. obs.). One begins to dig with its bill, then moves out of the way. The other goes into the hole, kicking back with its feet to scrape out the sand piled up at the entrance by its mate, and then continues to tunnel. The tunnel is 1–2 m long and widens into a chamber where the eggs are laid directly on the bare sand.

Both parents incubate the eggs (Davis 1980). The altricial young are born naked, with their eyes closed, and completely helpless. While they are very young, their digestive tracts secrete highly acidic juices that enable them to completely digest the food brought by the parents. Once the chicks are fully feathered, their digestive juices lose some of their acidity and they regurgitate undigested fish bones and scales (White 1939). Bleached and cleaned by the digestive process, the debris builds up into a magnificent silvery cup at the end of the tunnel, so the chamber is clean most of the nestling phase. The chicks spend a relatively long time in the nest and only when they are about 27 days old does the decrease in the amount of food brought by the parents encourage the young to leave their home. The role of the female in raising the young now gradually diminishes (Davis 1980). The first flight is critical, because the birds will not return to the nest. The young stay with an adult, often the male, on the feeding territory for several weeks. The adult brings them food at first, before teaching them the proper fishing techniques. This is a crucial period for their survival.

Distribution

The Belted Kingfisher breeds from Alaska to Newfoundland, including southern Ontario and Québec. It can also be found in the northern and central United States, from the Atlantic to the Pacific. The Atlas map shows that the species is a common breeder all over southern Québec, wherever there is water nearby. It ranges beyond the 50th parallel: birds have been observed as far north as Schefferville in 1975, Maricourt Camp in 1980, Grand Lac Delay in 1981, Great Whale River in 1989, and Lake Bienville in 1990 (ÉPOQ; Cons. G. G. – GREBE 1991a). There are breeding records from the Eastmain River (QNRCP) and the Great Whale River, where a nest was found in July 1991 (L. Bélanger, pers. comm.).

Notes on territory

In the breeding season, the Belted Kingfisher has two kinds of territories: a breeding territoriy, which the male and female defend aggressively, and a feeding territory. In suitable habitat, both types of territories may be found in the same area, and this is frequently the case when pairs feed in streams and their nests are located nearby. On the other hand, the two territories may be quite far apart. For example, if pairs feed in lakes or ponds and

Kingfisher nest entrances, like this one in a cutaway bank, are characterized by grooves from the feet of the kingfisher as it passes in and out of the nest. These grooves occasionally give the nest entrance a keyhole shape. Saint-Roch-de-Mékinac, north of Trois-Rivières. *May 1978, Pierre Laporte.*

Belted Kingfisher

Breeding distribution

1984–1989

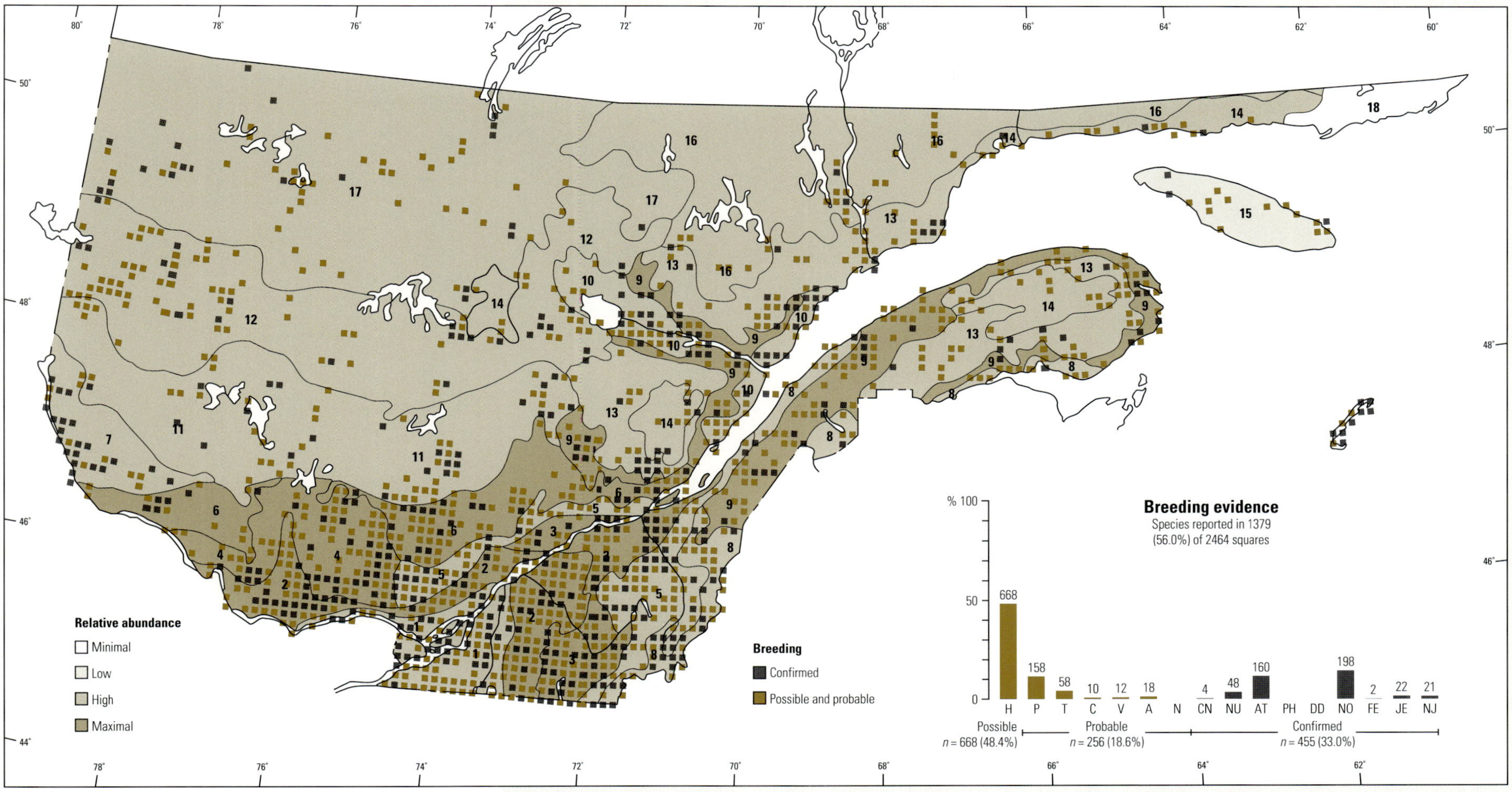

Species profile

Clutch size	5–10 eggs, more than 10 probably = dump nest[2,7,9,12]
Laying interval	1 egg per day[12]
Start of incubation	with last egg[12]
Incubation period	22–24 days[1,2,3,11,12]
Maturity at hatching	altricial
Care of young	by female and male[2]
Nestling period	23–35 days[1,2,3,7,8,12]
Age at first flight	≈ 28 days or on leaving nest[8,12]
Dependence of young	until ≈ 3 weeks after leaving nest[12]
Broods per year	1[7,11]
Breeding age	1 year[10]
Mating system	monogamy[9]
Duration of pair bond	1 season[2]
Total length	28.0–37.5 cm[4]
Wing span	47–52 cm[5]
Weight	148 g (125–215)[6]
Longevity record	?

References:
1. Harrison 1978
2. Ehrlich et al. 1988
3. Terres 1980
4. Godfrey 1986
5. Cramp 1985
6. Dunning 1984
7. P. Laporte, CWS, unpubl. data
8. Bent 1940
9. Davis 1980
10. DeGraaf and Rudis 1987
11. Harrison 1975
12. Hamas 1994

Chronology $n = 121$

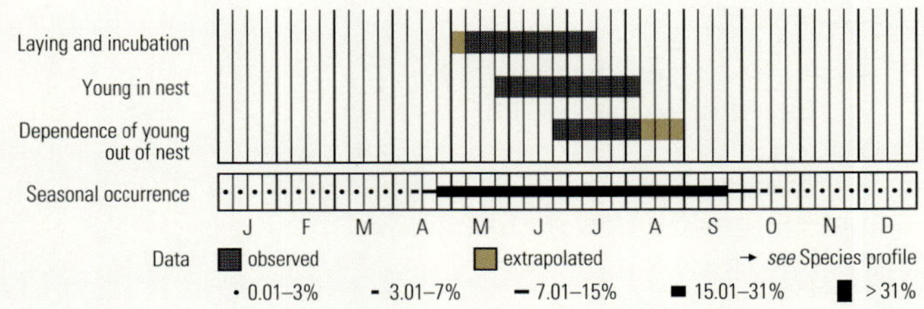

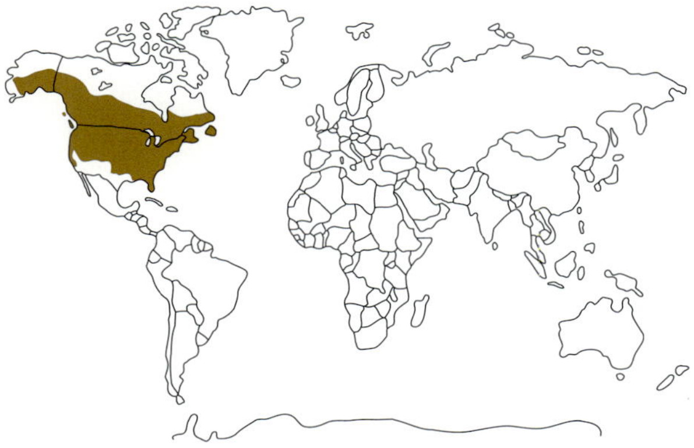

World breeding range

Frequency of occurrence by ecological region

Ecological regions** (forest regions)	Adequately covered squares	Squares in which reported	% of squares in which reported	Relative abundance (%)*
1. Sugar maple-hickory and sugar maple-basswood	97	85	87.6 ± 3.4	89.2
2. Sugar maple-basswood	153	136	88.9 ± 2.7	90.5
3. Sugar maple-basswood and sugar maple-yellow birch	101	94	93.1 ± 3.2	94.8
4. Sugar maple-yellow birch-basswood	56	55	98.2 ± 3.7	100.0
5. Sugar maple-yellow birch-beech and sugar maple-yellow birch-basswood	113	89	78.8 ± 4.3	80.2
6. Sugar maple-yellow birch-beech	71	67	94.4 ± 5.3	96.1
7. Sugar maple-yellow birch-balsam fir	19	16	84.2 ± 17.9	85.7
8. Sugar maple-yellow birch and balsam fir-yellow birch	47	38	80.9 ± 10.3	82.3
9. Balsam fir-yellow birch	87	81	93.1 ± 5.2	94.8
10. Balsam fir-white birch or balsam fir-red maple	46	38	82.6 ± 9.7	84.1
11. Yellow birch-balsam fir	52	45	86.5 ± 9.9	88.1
12. Balsam fir-white birch (west)	59	49	83.1 ± 10.0	84.6
13. Balsam fir-white birch (east)	76	57	75.0 ± 9.6	76.4
14. Balsam fir-black spruce	28	22	78.6 ± 16.2	80.0
15. White spruce-balsam fir and balsam fir-white spruce	22	11	50.0 ± 21.5	50.9
16. Black spruce-balsam fir-moss	18	15	83.3 ± 20.2	84.8
17. Black spruce-moss	32	20	62.5 ± 18.3	63.6
18.			not analysed	

* minimal: 0–29.9%; low: 30–59.9%; high: 60–89.9%; maximal: 90–100%

** Deciduous forest: 1–7; mixed forest: 8–11; coniferous forest: 12–17

A pair of Belted Kingfishers has nested near the top of this working sand quarry. Shawinigan, north of Trois-Rivières. *June 1978, Pierre Laporte.*

The Belted Kingfisher is partially migratory, with only the northern populations, including those in Québec, truly migrating. They generally leave when ice-up starts in the fall and, according to the sparse banding data available, winter in the southern United States. But kingfishers have been known to overwinter at places in Québec where lakes and rivers do not freeze and they can fish throughout the winter.

History and trends

According to Québec BBS data for 1970–89, there were no significant variations in the Belted Kingfisher population (Cyr and Larivée 1993b). This finding corroborates continent-wide BBS data for 1965–79 (Robbins et al. 1986), but it is difficult to assess population trends in the species. It does not seem likely that its numbers have diminished. In fact, certain types of development have created new breeding and feeding habitat. Construction of the road network in the Canadian Shield has produced earth banks, which are potential nesting sites. The proliferation of fish hatcheries has also provided the kingfisher with new sources of food.

Notes on territory (continued)

there are no suitable banks nearby, their nests may be up to 8 km from the feeding grounds (Cornwell 1963; Ellison 1985z). In such cases, pairs still attempt to maintain exclusive use of the resources on their feeding grounds (Cornwell 1963).

According to Davis (1980), if a feeding territory is along a stream, its length is inversely proportional to the abundance of food. Davis's study, conducted in Ohio during the breeding season, found that a streamside territory that combines feeding and breeding grounds can vary in length from 800 to 1300 m (Davis, W. J. 1982). Davis also reported that outside the breeding season, the feeding ground stretches for 300–500 m. Once the young are

The Belted Kingfisher fishes in a wide variety of aquatic habitats, though it prefers streams with riffles less than 15 cm deep. It perches on branches overhanging streams, then dives to catch its prey. It may nest far from water if it cannot find banks of soft sand or earth close by. Martin-Valin ZEC, north of the Saguenay River. *August 1991, Jean Gauthier.*

able to fish for themselves, they are driven from the feeding territory. The parent birds then become solitary, each holding its own individual feeding territory, which it defends (Bent 1940; Davis, W. J. 1982; Ehrlich et al. 1988).

A density of 1 pair/4.7 km² (0.021 pair/10 ha) *has been recorded in Minnesota* (Cornwell 1963).

Three young Belted Kingfishers resting after their first flight. Sainte-Dorothée, Laval. *July 1939, Henry Mousley, Laval University Collection.*

The Belted Kingfisher burrows a tunnel leading to a chamber where it lays its eggs. An adult watching over its eggs. Sainte-Dorothée, Laval. *June 1940, Henry Mousley, Laval University Collection.*

Although the Belted Kingfisher is a fish eater, it does not seem to have suffered from DDT contamination in the food chain, probably because of low accumulations of residues in the small prey it eats (P. Laporte, CWS, unpubl. data).

Although the species is migratory, it is not protected by the Migratory Birds Convention Act because of its feeding habits. The Belted Kingfisher feeds mainly on small fish and crayfish and is often regarded prejudicially as a pest. It is, however, protected under the provincial Act respecting the conservation and development of wildlife. Furthermore, many of the studies done on the Belted Kingfisher's prey have been unable to demonstrate any truly harmful effects on commercially valuable fish stocks (White 1953; Elson 1962).

PIERRE LAPORTE

Red-headed Woodpecker

Pic à tête rouge

Melanerpes erythrocephalus

Ghislain Caron

WITH ITS SHOWY BEHAVIOUR AND BRIGHTLY COLOURED PLUMAGE, THE Red-headed Woodpecker is easily spotted and puts most other woodpecker species in the shade. This scarce migrant breeder usually arrives in April and leaves in November, although it may occasionally winter here (David 1980a).

Red-heads are primarily opportunistic feeders. They feast on grasshoppers in August, then in the fall, they turn their attention to fruits, relying in the winter months on nuts and acorns (Beal 1911). During the breeding season, they can often be seen darting out from roadside perches to take insects on the wing (Williams and Batzli 1979), a habit which frequently leads to fatal collisions with vehicles (Bull 1985). They occasionally raid the nests of Cliff Swallows, Eastern Bluebirds, and other species (Dionne 1906; Bent 1939). These highly ingenious birds cache food in cavities drilled in tree trunks, which they then seal with damp wood chips that harden as they dry (Hay 1887; Kilham 1983).

Habitat and behaviour

Red-headed Woodpeckers seek out clumps of trees and bushes, hedgerows in farmland, and the marshy shores of lakes or waterways (Short 1982). Look for them also in burntlands or open woods, another preferred breeding habitat (Godfrey 1986). The female chooses the nest site, usually a dead tree, utility pole, or fence post, but the male does most of the drilling and excavating of the cavity (Kilham 1983; Bull 1985). The white elm seems to be their tree of choice in Ontario (Peck and James 1983); the ravages of Dutch elm disease have probably done much to make this tree one of its favourites in Québec, as well.

Excavation of the nest takes 12–14 days. Incubation of the 4 or 5 eggs lasts 12 or 13 days, and the young remain in the nest for about another 31 days (Short 1982). The male and female share incubating and brood-rearing duties equally. As their plumage is alike, the sexes are hard to distinguish (Kilham 1983). After leaving the nest, the young stay nearby for some time and continue to be fed by the adults, though they may be chased away if the pair renests. If this happens, the same cavity may be used (Short 1982).

The Red-headed Woodpecker was reported in only 26 squares during the entire Atlas survey period, and breeding was confirmed in only 8, a confirmation rate comparable to that reported by Spahn (1988) for New York State. The low frequency of occurrence clearly highlights how scarce the Red-headed Woodpecker is in Québec because the species is not easily missed, as it is very flamboyant and noisy, inhabits relatively open areas, and nests in dead trees.

Distribution

The Red-headed Woodpecker breeds throughout southern Canada from Saskatchewan to New Brunswick and in most parts of the United States (Godfrey 1986). It winters south of the border, concentrating especially in the Midwest and the Mississippi Valley (Bock and Lepthien 1975). Some observers have noted that its numbers on certain wintering grounds are proportional to the forest nut crop (Beal 1911). When it winters here, it relies largely on feeders and sometimes on corncribs for sustenance.

In Québec, it breeds chiefly in the Central St. Lawrence Lowland. Most Atlas records are from the Ottawa Valley, Montréal area, Montérégie, and Eastern Townships. It was also reported in the Berthier and Sorel islands (twice), Lanaudière, and Mauricie. The easternmost Atlas record is from the Québec City area, although there is a pre-Atlas record from farther east, at Saint-Pierre-de-Montmagny, in 1982 (Gosselin and David 1982). There are also pre-Atlas records of non-breeders from the Gaspé, the North Shore, and Lac-Saint-Jean regions (David 1980a), and a post-Atlas record from the summer of 1992 of a nest found on Île Ronde in the Berthier Islands (S. Dulac, pers. comm.).

Godfrey (1986) is of the opinion that the range of the Red-headed Woodpecker in Canada has shrunk somewhat; continuing habitat degradation may be partly to blame (Ouellet 1974a; Peck and James 1983).

History and trends

In New York State, this woodpecker was once more common in suitable habitat than it is now (Bull 1985; Spahn 1988). Declines have also been reported for Ontario (Peck and James 1983) and Québec (Ouellet 1974a). Even at the turn of the century, Wintle (1896) considered the Red-headed Woodpecker scarce in the Montréal area, and it is now classified as threatened in Québec (Robert 1989b). A number of authors blame the European Starling, which competes strongly for nest cavities (Bent 1939; Ouellet 1974a; Short 1979; Bull 1985), and delayed nesting due to usurpation by the starling may prolong the breeding season. In any case, the Red-head nests later than any other woodpecker: young have been observed still in the nest as late as the second week of September.

A number of factors have probably contributed to the species' decline. In the United States in the last century, the Red-headed Woodpecker was hunted, sometimes intensively, for its depredations on orchards (Bent 1939). Its habit of chiselling out its nest in utility poles caused significant damage each year, so that people were even hired to dislodge the birds (McAtee 1911). The toxic effects of the creosote wood preservative were probably responsible for the deaths of many young birds, as well (Rumsey 1970). The current forestry practice of culling the dead or diseased trees favoured by this species for nesting is also to blame (Woodliffe 1987d).

There is still hope for recovery, however. In fact, BBS data point to a slight upturn in the northeastern United States (Robbins et al. 1986).

In Québec, however, the species may not rebound in the foreseeable future. In and around Greater Montréal, more and more hedgerows are disappearing, taking with them the last remaining large trees. In the few surviving forests and mature stands, mostly on private land, dead and diseased trees are likewise being felled, and each year, many older hardwood stands are cut to meet the growing demand for lumber.

SERGE LEMIEUX

Notes on territory

The Red-headed Woodpecker is territorial, defending mainly the vicinity of its nest. Little is known about the size of its territory in the breeding season (Short 1982), but a winter study in Maryland by Kilham (1958b) found resident woodpeckers using an area of 0.1–0.2 ha, while Moskovits (1978) reported a mean of 0.97 ha in Florida. Territory size appears to vary considerably depending on habitat type (Short 1982).

Around Sarnia, Ontario, density in a mixed forest dominated by deciduous trees was 2.0 males/10 ha (Erskine 1980a). Density in shrubby thickets was estimated to be 5.5 birds/10 ha in Illinois (Graber and Graber 1963).

The Red-headed Woodpecker nests in dead trees in thickets and wooded margins. Luskville, near Hull. *July 1991, Daniel St-Hilaire.*

Red-headed Woodpecker

Breeding distribution

1984–1989

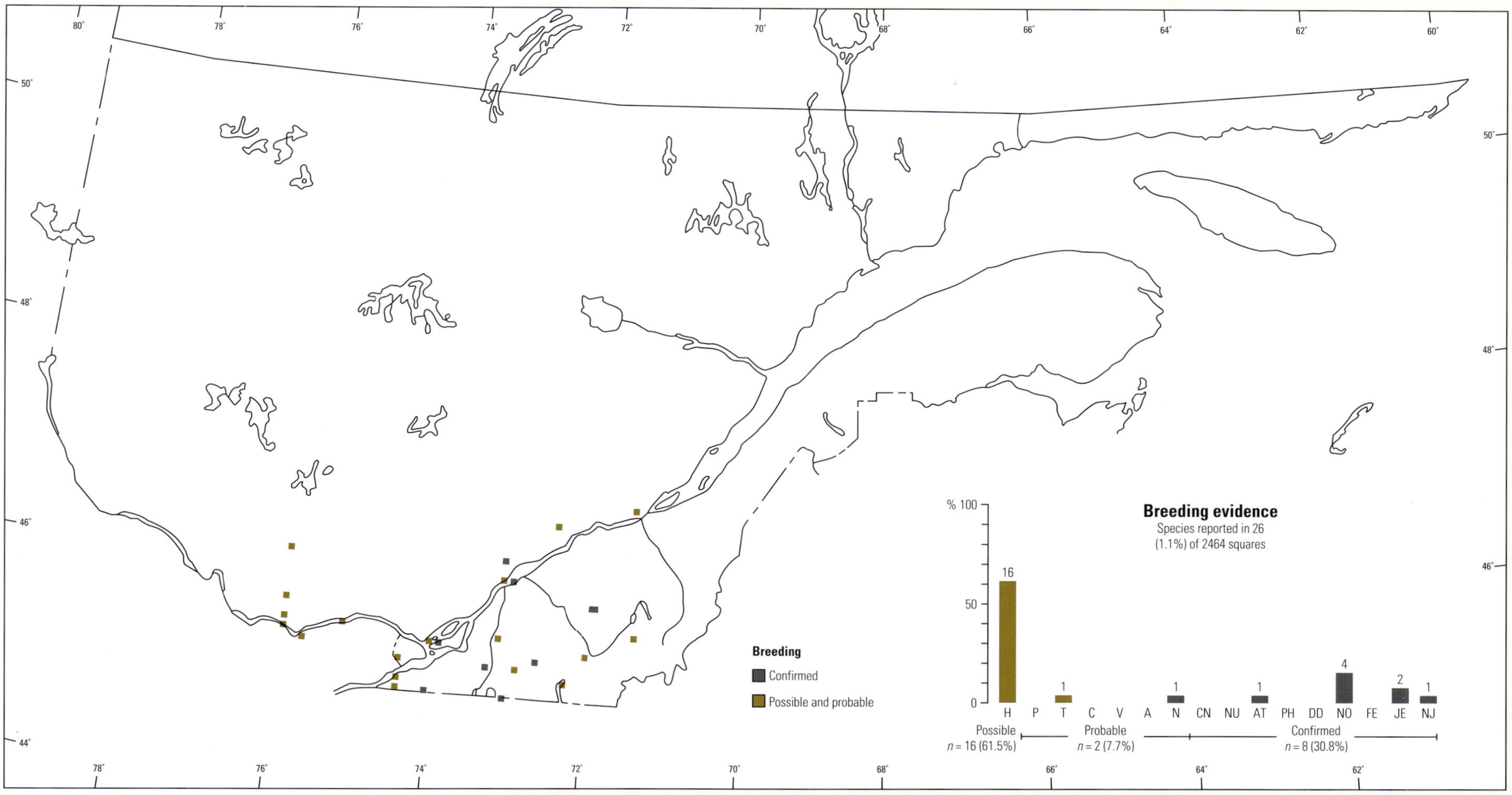

Species profile	
Clutch size	usually 4–5 eggs (3–10)[1,2,3,4]
Laying interval	1 egg per day[2,4]
Start of incubation	before last egg[2,4]
Incubation period	12–14 days[1,2,3,4]
Maturity at hatching	altricial
Care of young	by female and male[2,3,4]
Nestling period	27–31 days[2,3,4]
Age at first flight	27–31 days or more[9]
Dependence of young	?
Broods per year	1–2[1,2,3,4]
Breeding age	1 year[8]
Mating system	monogamy[3]
Duration of pair bond	?
Total length	22–25 cm[5]
Wing span	40.6–45.7 cm[1]
Weight	71.6 g (56.1–90.5)[6]
Longevity record	9 years 11 months[7]

References:
1. Terres 1980
2. Harrison 1978
3. Ehrlich et al. 1988
4. Short 1982
5. Godfrey 1986
6. Dunning 1984
7. Clapp et al. 1983
8. DeGraaf and Rudis 1987
9. according to nestling period

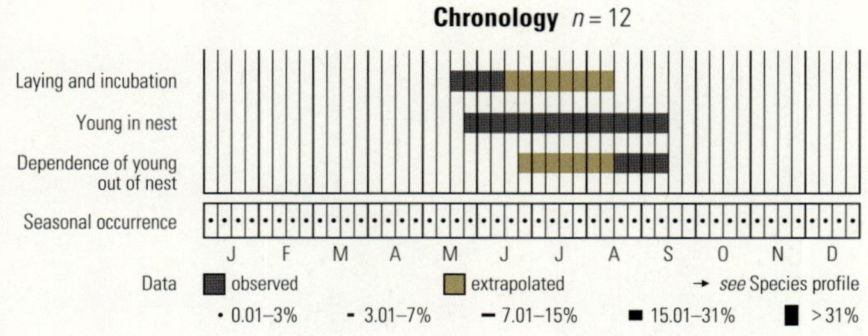

World breeding range

Swamp along Massey Road, Rockburn, where a Red-head nested in 1984. South of Montréal. *June 1991, Pierre Bannon.*

Yellow-bellied Sapsucker

Pic maculé

Sphyrapicus varius

Ghislain Caron

The Yellow-bellied Sapsucker has a highly specific diet, its brushlike tongue being unsuited to the usual woodpecker feeding method. It drills a series of holes a few millimetres in diameter, or "wells," to get to the inner bark of the tree and dine on the sap and the insects it attracts, hence its name. The neat lines of small holes are unmistakable.

Like the Northern Flicker, the sapsucker sometimes drums on telephone poles, roofs, and metal structures, using their resonance to make its presence known.

This common migrant breeder is abundant in the southern part of Québec from late March to early November. In spring the males arrive about a week ahead of the females. Their appearance marks the end of the sugaring-off season for Québec maple syrup producers.

Habitat and behaviour

In addition to numerous insects and some small mammals (Foster and Tate 1966), commensal feeding by up to 35 bird species has been observed on the sap oozing from sapsucker wells (Skutch 1985b). The Ruby-throated Hummingbird, for example, frequently comes to drink there (Tyler 1939). Surprisingly, the sapsucker occasionally eats wasps attracted by the sap, which are deemed inedible by most birds (Lawrence 1967).

Sap is not the mainstay of this woodpecker's diet, however. Beal (1911) and Lawrence (1955) reported that it is almost evenly balanced between plant matter—sap, cambium, small berries—and insects, and varies seasonally. On arrival in our part of the world in April, the bird rarely takes sap, but consumes ants, foraging on the ground for them like the flicker. In summer, it hawks beetles, dragonflies, and moths flycatcher-fashion. In late summer, it also looks for berries. Its feeding habits change again once the young have fledged: from late July until they leave in October or November, both adults and young live on sap found in the nesting territory (Lawrence 1967).

The sapsucker nests in mature deciduous or mixed woodlands, especially those where deciduous trees predominate, close to small clearings or sources of water. More rarely, it chooses coniferous forests, preferring dry areas (DeGraaf et al. 1980; Elderkin 1989). In Québec, it is the commonest woodpecker in sugar maple stands (DesGranges et al. 1987), and it is sometimes found in open country, along roadsides, and in orchards.

Notes on territory

The male generally returns each spring to his breeding territory of the previous year, but does not use the same cavity (Kilham 1983).

The Yellow-bellied Sapsucker is associated with mature deciduous woodlands, like this sugar maple-yellow birch forest. Sainte-Foy. *September 1991, Yves Aubry.*

Yellow-bellied Sapsucker

Breeding distribution

1984–1989

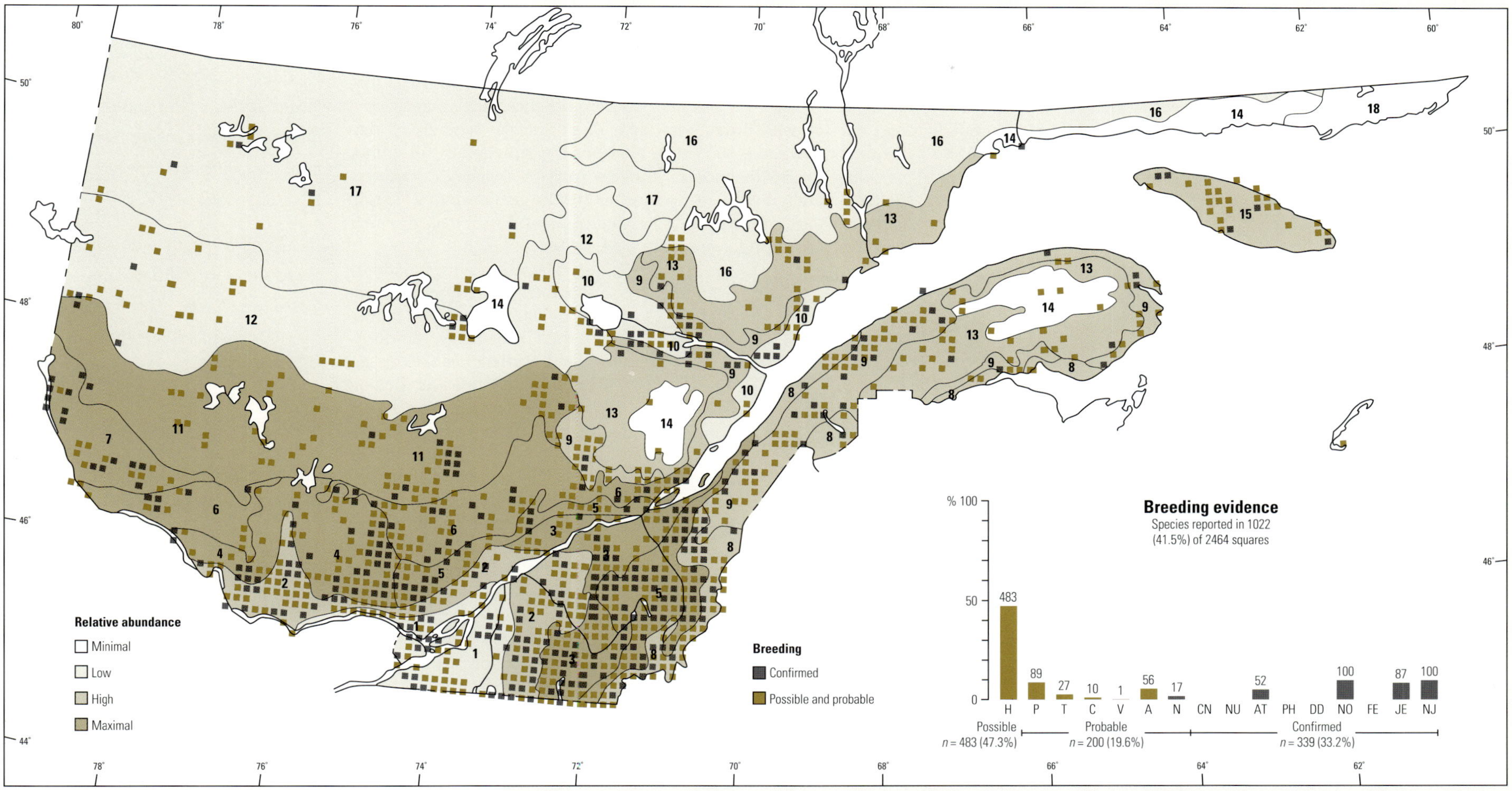

Species profile

Clutch size	usually 5–6 eggs (3–7)[1,2,3,4]
Laying interval	?
Start of incubation	?
Incubation period	12–14 days[2,3,4]
Maturity at hatching	altricial
Care of young	by female and male[1,2,3,4]
Nestling period	25–29 days[1,3,4]
Age at first flight	25–29 days or more[9]
Dependence of young	until 1–2 weeks after leaving nest[3,4,10]
Broods per year	1[4,8]
Breeding age	1 year[8]
Mating system	monogamy[4]
Duration of pair bond	?
Total length	19.5–22.0 cm[5]
Wing span	35.6–40.6 cm[7]
Weight	50.3 g (40.7–62.2)[11]
Longevity record	6 years 8 months[6]

References:
1. Kilham 1983
2. Tyler 1939
3. Short 1982
4. Ehrlich et al. 1988
5. Godfrey 1986
6. Clapp et al. 1983
7. Terres 1980
8. DeGraaf and Rudis 1987
9. according to nestling period
10. Harrison 1978
11. Dunning 1984

World breeding range

Chronology $n = 154$

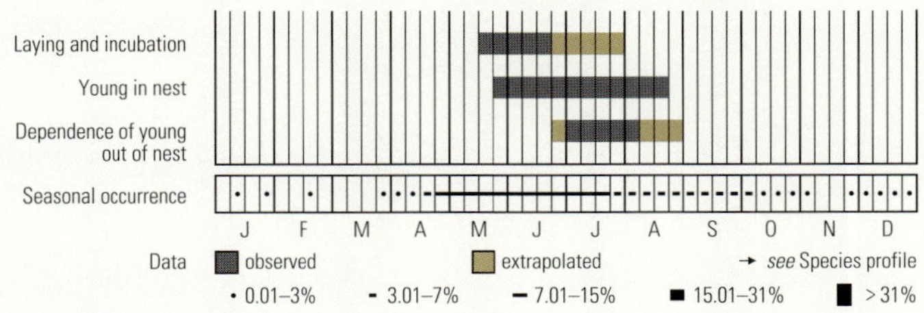

Frequency of occurrence by ecological region

Ecological regions** (forest regions)	Adequately covered squares	Squares in which reported	% of squares in which reported	Relative abundance (%)*
1. Sugar maple-hickory and sugar maple-basswood	97	51	52.6 ± 5.0	55.5
2. Sugar maple-basswood	153	123	80.4 ± 3.3	84.9
3. Sugar maple-basswood and sugar maple-yellow birch	101	91	90.1 ± 3.7	95.1
4. Sugar maple-yellow birch-basswood	56	52	92.9 ± 6.3	98.0
5. Sugar maple-yellow birch-beech and sugar maple-yellow birch-basswood	113	104	92.0 ± 3.0	97.1
6. Sugar maple-yellow birch-beech	71	64	90.1 ± 6.6	95.1
7. Sugar maple-yellow birch-balsam fir	19	18	94.7 ± 12.0	100.0
8. Sugar maple-yellow birch and balsam fir-yellow birch	47	40	85.1 ± 9.4	89.8
9. Balsam fir-yellow birch	87	70	80.5 ± 7.8	84.9
10. Balsam fir-white birch or balsam fir-red maple	46	20	43.5 ± 12.4	45.9
11. Yellow birch-balsam fir	52	47	90.4 ± 8.6	95.4
12. Balsam fir-white birch (west)	59	29	49.2 ± 13.1	51.9
13. Balsam fir-white birch (east)	76	45	59.2 ± 10.8	62.5
14. Balsam fir-black spruce	28	6	21.4 ± 16.2	22.6
15. White spruce-balsam fir and balsam fir-white spruce	22	17	77.3 ± 18.4	81.6
16. Black spruce-balsam fir-moss	18	8	44.4 ± 26.0	46.9
17. Black spruce-moss	32	12	37.5 ± 18.3	39.6
18.			*not analysed*	

* minimal: 0–29.9%; low: 30–59.9%; high: 60–89.9%; maximal: 90–100%

** Deciduous forest: 1–7; mixed forest: 8–11; coniferous forest: 12–17

Communication in this species is remarkably varied during the breeding season. It has a variety of plaintive cries and peeps often resembling the mewling of a cat (Pough 1949). It also drums early in the season as a territorial signal or to attract a mate, starting rapidly and gradually slowing to disconnected taps, rather like Morse code. The sapsucker sometimes drums on utility poles, roofs, and other structures, of metal or other material, with good resonance. The noise of the wings during pursuit flights or courtship displays reminds one of the noise made by the Common Snipe (Kilham 1962a).

The sexes co-operate in selecting a nest site, but the male plays the principal role. He usually excavates a hole himself rather than using a natural or second-hand cavity. Live trees are preferred for nesting, most often poplars (Peck and James 1983), with their thin bark and soft wood. It also likes birches, beeches, elms, maples, and trees infested with fungi (*Fomes* spp.). The presence of fungi is a sign that the heartwood is starting to rot while the sapwood is still sound, providing good protection for the nest cavity (Kilham 1983). The sapsucker generally makes its nest in tree trunks greater than 25 cm in diameter (Thomas et al. 1979) at heights of 1.7–23 m, but sometimes even uses large upright branches (Peck and James 1983).

The nest cavities made by sapsuckers are sometimes reused by Tree Swallows, Black-capped Chickadees, or Red-breasted Nuthatches (Erskine and McLaren 1972). Hall and Kelson (1959) reported seeing a flying squirrel using one in South Carolina.

The male defends the territory, but both sexes incubate, though the male's role is predominant, especially at night (Kilham 1962a). Both parents feed the young for 3–4 weeks in the nest and another 2 weeks after they fledge (Short 1982). The young then follow the parents to trees used for feeding in previous years and practise drilling holes to extract sap (Kilham 1983).

Breeding was confirmed in 33.2% (339/1022) of Atlas squares in which the species was reported. This relatively high percentage reflects the noisy behaviour of the young, whether in the nest or outside. It is easy to spook this woodpecker by imitating the call of the Barred Owl; the sapsucker immediately responds with its alarm call, a reaction that is probably even stronger when the young are just leaving the nest (Kilham 1983).

This woodpecker has an unusual feeding technique: it drills a series of small holes, often in a line, from which it takes sap and trapped insects. Lac Saint-François National Wildlife Area, southwest of Montréal. *May 1975, Léo-Guy de Repentigny.*

Notes on territory (continued)

He claims his territory by various means, including drumming and threat displays in the form of neck stretching and wing spreading (Kilham 1962a, 1983).

In a Mirabel sugar maple stand, the territory defended was estimated to be 0.7 ha (Cyr 1974). In the coniferous forests of the Blue Mountains in the western United States, the average territory was found to be 4 ha (Thomas et al. 1979).

Yellow-bellied Sapsuckers are found in densities of 2.3–3.3 territorial males/10 ha in the sugar maple stands of southern Québec (Cyr 1974; Erskine 1976a, 1984) and 2.5–6.0 males/10 ha in the fir and spruce forests of New Brunswick (Erskine 1971a).

Male Yellow-bellied Sapsucker near his nest, hollowed out of a maple tree. Charlesbourg. *June 1956, Gaston Le Page.*

Distribution

The Atlas data confirm the known distribution of the Yellow-bellied Sapsucker in southern Québec (David 1980a; Godfrey 1986). The species breeds throughout the Atlas area except for the Magdalen Islands, where it is irregular (Fradette 1992). It is much more common south of the 49th parallel, where deciduous and mixed woodlands are plentiful.

The sapsucker is less common in predominantly coniferous habitats, except on Anticosti Island, where the species is almost as common as in the southern parts of the province. It occupies primarily cutovers and burns that are regenerating, mainly in the centre and north of the island, where trembling aspen, birch, and other hardwoods offer suitable habitat.

Farther north, it has been reported a few kilometres south of the Eastmain River, near Highway 109 (52°11' N, 77°08' W) (Bordage 1984); according to Godfrey (1986), this is at the northern limit of the Yellow-bellied Sapsucker's Québec range. There have also been a few reports from Waswanipi Lake, where the species has been observed in a black spruce-moss stand, a dry peatland without pools, and in shoreline scrub (Cons. G. G – GREBE 1992o). A nest with at least 2 young was found near the Harricanaw River in 1978 (Anonymous 1980b).

Of all our woodpeckers, this one migrates farthest. It winters from the southern boundary of its breeding range in the United States into Central America and the West Indies.

History and trends

Two studies dating from the early part of the century (Beal 1911; McAtee 1911) condemned the sapsucker as being highly destructive and even suggested methods by which it might be eliminated. It stood accused of deforming trees by sapping them, thereby rendering them commercially worthless. Eventually, people came to realize that the bird tends to concentrate on a small number of trees, many of which are already diseased (Kilham 1983), and that the harm it does is minimal and is offset by its feeding on insects whose larvae are harmful to trees (Lawrence 1955). This woodpecker has also been blamed for the appearance of bumps on the trunk after it has finished feeding there. In fact, it helps prevent the development of those excrescences, which are caused by an accumulation of nutrients under the bark after a tree has been damaged (Kilham 1983).

The scant historical data on the Yellow-bellied Sapsucker do not point to any major changes in its population since the beginning of the century. ÉPOQ data give no indication of any variations in Québec breeding populations between 1970 and 1989, but do show a downward trend in the numbers observed in the province in spring and fall (Cyr and Larivée 1993a). Diamond (1986) noted that the species' winter range seems to be seriously threatened and forecast an 80% reduction in area by the year 2000 as a result of deforestation in the tropics.

Sylvain Paradis

Downy Woodpecker

Pic mineur

Picoides pubescens

Ghislain Caron

THE DOWNY WOODPECKER RETURNS FAITHFULLY TO ITS SUMMER breeding territory every year, having spent the winter wandering with chickadees and nuthatches. This tame visitor to our gardens and orchards is probably the most familiar of the woodpeckers.

This resident breeder is common here even in winter, anywhere that the mean January temperature is not below –23°C (Root 1988). It has been recorded as far north as Lake Mistassini (Macoun and Macoun 1909), though some birds in the northernmost part of the range move south in autumn or early winter (Bent 1939). Lawrence (1967) reported that in Ontario, at the same latitude as Lake Saint-Pierre (46° N), some 80% of these woodpeckers leave their summer territories in winter. It is highly likely that Québec populations behave in a similar fashion.

An adult near its nest hole in a dead tree. Cap Tourmente National Wildlife Area, northeast of Québec City. *April 1993, Robert Morin.*

Habitat and behaviour

The Downy Woodpecker is at home virtually everywhere—in orchards, city parks, or treed residential areas, as much as in deciduous or mixed forests (Short 1982; Godfrey 1986). The species seems to be most abundant where there are woodlands with very young trees and bushes (Anderson and Shugart 1974; Conner and Adkisson 1976; DeGraaf and Chadwick 1987).

As with most woodpeckers, drumming on hollow trees is a mating call that may serve much the same purpose as singing among the passerines (Kilham 1962b; Lawrence 1967). Drumming is followed by courtship displays and the selection of a nest site. Three quarters of the time, the nest cavity opening faces south or east (Lawrence 1967). In Ontario, the Downy Woodpecker nests mainly in deciduous trees, including poplar, maple, elm, and birch (Peck and James 1983).

Most often, the choice is a dead tree with crumbly, decaying wood (Lawrence 1967; Schroeder 1982; Campbell et al. 1990b). This predilection has the disadvantage of making the nest more vulnerable to predators and bad weather, which may explain why the nestling period is only 3 weeks, rather than 4, as in most other woodpecker species (Kilham 1979b).

Nest excavation takes 13–20 days. Incubation of the 4 or 5 eggs lasts 12 days, and the young stay in the nest for about 21 days. The male and female share incubation and feeding duties equally (Kilham 1979b). Once out of the nest, the young continue to depend on their parents for a further 3 weeks (Harrison 1978).

For the first few days out of the nest, the young follow their parents in search of food, climbing tree trunks and searching cracks in the bark in imitation of the adults' foraging behaviour. Their parents continue to feed them, however, bringing them such things as fat grubs, often as big as the fledglings' own heads (J. Gauthier, pers. comm.).

The Downy Woodpecker feeds mainly by penetrating the bark with its bill in search of insects; it may take insects on the bark, but rarely captures any on the wing (Jackson 1970). It forages on the branches of small trees, preferably live rather than dead (Williams 1975; Conner 1980). In winter, large numbers of Downies can sometimes be seen during epidemics of bark beetles, insects that live mainly underneath the bark of conifers (Crockett and Hansley 1978). Though it eats huge quantities of insects, it does not neglect other sources of food; in the autumn, it feeds heartily on poison-ivy berries, without any apparent discomfort, and so helps spread the plant (Beal 1911). In winter, it is sometimes found in old abandoned fields, where it likes to feed on the larvae of the goldenrod gall fly (Schlichter 1978).

Like other woodpeckers, the Downy can easily be detected by its drumming. Depending on observation conditions, however, an inexperienced birder may confuse it with the Hairy Woodpecker, of which it is a miniature replica. The nest is quite easy to find because of the comings and goings of the adults and the ceaseless cries of the young, even though, as pointed out by Lawrence (1967), Downy nestlings are quieter than those of other woodpecker species.

Distribution

The Downy Woodpecker is found throughout most of Canada and the United States. In Québec, it breeds almost everywhere south of the 52nd parallel (Godfrey 1986). It is common in the Central St. Lawrence Lowland, Ottawa Valley, Témiscamingue, Chaleur Bay, and Eastern Townships, wherever hardwoods predominate. It is a little less common in the mixed forest zone: the North Shore, the Gaspé Peninsula, Anticosti Island, and Abitibi. It becomes scarcer in the boreal forest as the proportion of conifers increases.

The Downy Woodpecker was reported in 45.8% (1129/2464) of Atlas squares, and breeding was confirmed in 25.2% (284/1129) of those. Its distribution is similar to that of the Hairy Woodpecker, although, unlike the Hairy, it is not known to breed in the Magdalen Islands, where it is classified as an irregular visitor and possible breeder (Fradette 1992). The near-total absence of hardwood on the islands for over a century partly explains why there is no breeding population there.

History and trends

The Downy Woodpecker was considered common at the turn of the century by Wintle (1896) and Dionne (1906), and still listed as common by Ouellet (1974a). Ouellet nevertheless feels that the Downy Woodpecker is less common now in the Montréal area than in the last century as a result of deforestation with its attendant loss of hardwood stands.

Today, ongoing logging and forest management mean that dead trees suitable for nesting are becoming harder to find. It seems certain that such activities are reducing this species' habitat, as is the case with the Northern Flicker (Erskine 1978b).

SERGE LEMIEUX

The Downy Woodpecker nests in both mixed and deciduous forests, preferably where there are young trees and bushes. Here, this small woodpecker has made its nest hole in one of the beeches on the left, obviously not disturbed by its proximity to humans. Mont-Saint-Bruno Conservation Park, east of Montréal. *August 1991, Gaétan Duquette.*

Notes on territory

Downies defend a territory from late winter until the end of the breeding season, using it for both nesting and mating. Both adults drum to signal their tenure of the site and will chase away conspecifics.

The size of the home range varies between 2 and 4.5 ha, but is not defended in its entirety, as neighbouring ranges may overlap. In fact, once laying has begun, the birds confine their defence to a radius of 15–30 m around the nest (Stokes and Stokes 1983).

In an Illinois forest, a density of 36 birds/40 ha was reported (DeGraaf and Rudis 1987).

Downy Woodpecker

Breeding distribution

1984–1989

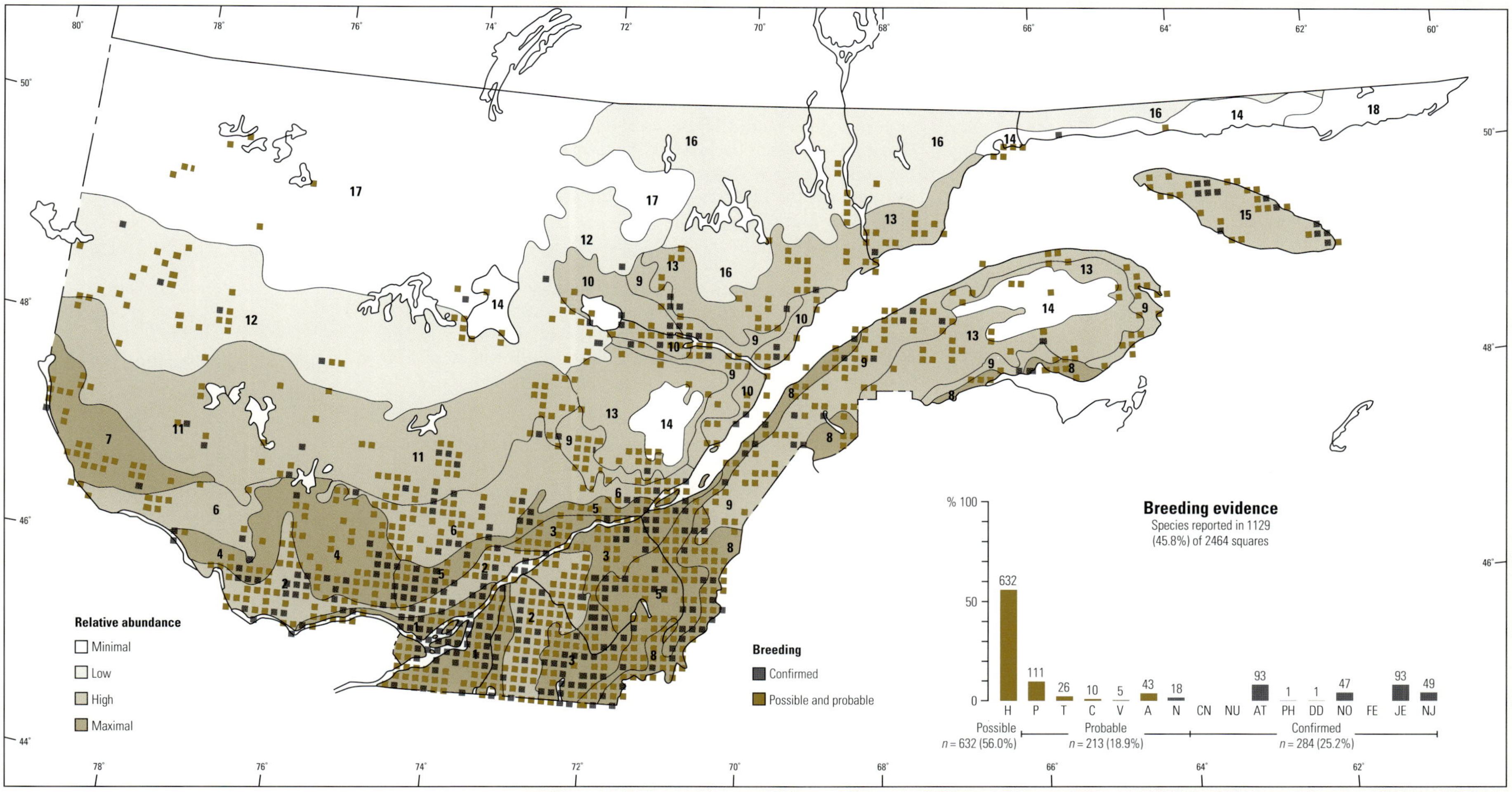

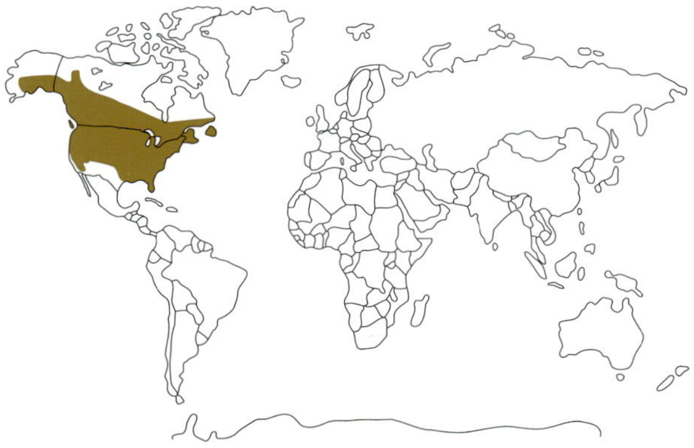

World breeding range

Species profile

Clutch size	usually 4–5 eggs (3–6)[2,3,4,6]
Laying interval	?
Start of incubation	?
Incubation period	12 days[2,3,4,6]
Maturity at hatching	altricial
Care of young	by female and male[1,2]
Nestling period	20–25 days[2,3,4]
Age at first flight	20–25 days or more[11]
Dependence of young	until ≈ 21 days after leaving nest[8,10]
Broods per year	1[3,4]
Breeding age	1 year[8]
Mating system	monogamy[4]
Duration of pair bond	?
Total length	16.0–18.5 cm[6]
Wing span	27.9–30.5 cm[7]
Weight	27.0 g (20.7–32.2)[9]
Longevity record	11 years 5 months[5]

References:
1. Kilham 1983
2. Short 1982
3. Stokes and Stokes 1983
4. Ehrlich et al. 1988
5. Clapp et al. 1983
6. Godfrey 1986
7. Terres 1980
8. DeGraaf and Rudis 1987
9. Dunning 1984
10. Harrison 1978
11. according to nestling period

Chronology *n* = 90

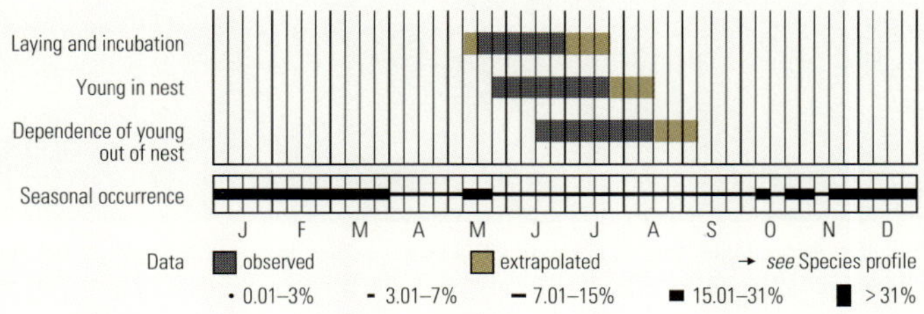

Frequency of occurrence by ecological region

Ecological regions** (forest regions)	Adequately covered squares	Squares in which reported	% of squares in which reported	Relative abundance (%)*
1. Sugar maple-hickory and sugar maple-basswood	97	90	92.8 ± 2.8	94.8
2. Sugar maple-basswood	153	125	81.7 ± 3.3	83.5
3. Sugar maple-basswood and sugar maple-yellow birch	101	90	89.1 ± 3.9	91.0
4. Sugar maple-yellow birch-basswood	56	50	89.3 ± 7.4	91.2
5. Sugar maple-yellow birch-beech and sugar maple-yellow birch-basswood	113	100	88.5 ± 3.4	90.4
6. Sugar maple-yellow birch-beech	71	57	80.3 ± 8.6	82.0
7. Sugar maple-yellow birch-balsam fir	19	18	94.7 ± 12.0	96.8
8. Sugar maple-yellow birch and balsam fir-yellow birch	47	46	97.9 ± 4.5	100.0
9. Balsam fir-yellow birch	87	72	82.8 ± 7.4	84.6
10. Balsam fir-white birch or balsam fir-red maple	46	31	67.4 ± 11.8	68.9
11. Yellow birch-balsam fir	52	43	82.7 ± 10.8	84.5
12. Balsam fir-white birch (west)	59	34	57.6 ± 12.9	58.9
13. Balsam fir-white birch (east)	76	54	71.1 ± 10.0	72.6
14. Balsam fir-black spruce	28	8	28.6 ± 17.6	29.2
15. White spruce-balsam fir and balsam fir-white spruce	22	19	86.4 ± 15.5	88.2
16. Black spruce-balsam fir-moss	18	7	38.9 ± 25.5	39.7
17. Black spruce-moss	32	8	25.0 ± 16.6	25.5
18.			*not analysed*	

* minimal: 0–29.9%; low: 30–59.9%; high: 60–89.9%; maximal: 90–100%

** Deciduous forest: 1–7; mixed forest: 8–11; coniferous forest: 12–17

Hairy Woodpecker

Pic chevelu

Picoides villosus

Ghislain Caron

AT FIRST GLANCE, THE HAIRY WOODPECKER LOOKS LIKE A BIGGER REPLICA of the Downy. The Hairy is not as tame, however, and is inclined to hide by always keeping on the opposite side of the tree from an approaching observer. It owes its name to the white feathers on its back, which tend to overlap the black edges of the wings like wispy hair (Taverner 1922).

The Hairy Woodpecker is more sedentary than the Downy, with which it usually shares its range (Lawrence 1967), though in winter it must often travel quite far to find enough food (Ouellet 1974a). It then ventures into open areas, such as city parks and orchards (Terres 1980).

Its larger size, proportionately longer bill, and lack of black spots on the white outer tail feathers are the clearest field marks distinguishing the Hairy Woodpecker from the Downy. Their similarity is not confined to their plumage; they are alike, too, in their call and their drumming, though the Hairy is louder.

The Hairy Woodpecker likes older forests with a denser canopy than does the Downy. Sainte-Foy. *September 1991, Yves Aubry.*

Habitat and behaviour

The Hairy Woodpecker seeks out mature deciduous, mixed or coniferous forests with some openings, but also takes to burns. In coniferous forests, it prefers places with at least a few deciduous trees. It is even found in wooded suburban areas (Godfrey 1986). In contrast to the Downy Woodpecker, it likes dense-canopied, more mature forests (Anderson and Shugart 1974). It rarely inhabits stands of less than 10 ha (Galli et al. 1976; Robbins et al. 1989a).

Drumming to reinforce the pair bond begins as early as late December and is often accompanied by display flights (Lawrence 1967). The male excavates a new nest cavity each year, over a period of 7–21 days (Bent 1939), preferably in a dead or rotting limb of a living tree (Lawrence 1967). Like other woodpeckers, the Hairy has a strong predilection for trees afflicted by heart rot, caused by the heartwood being infested with fungi. These trees have sound outer wood to protect the nest against predators and soft wood at the centre that is easy to excavate (Conner et al. 1976). The pair shares incubation duties, with the male taking over at night. The young remain in the nest for 28–30 days (Short 1982).

The Hairy Woodpecker's diet is nearly 80% animal (Beal 1911). It wages unrelenting war all year round on beetle grubs that bore beneath tree bark and into the wood (Lester 1980). It takes advantage of epidemics of bark beetles as well as eastern tent caterpillars to feed its young (Lawrence 1967; Lester 1980). Kilham (1983) noted that the male and female have different foraging styles: whereas the female proceeds methodically, lifting pieces of bark with short, regular, oblique pecks, the male goes after prey lodged deeper in the wood with strong perpendicular blows of the bill. When the Hairy and the Black-backed Woodpecker both feed in the same stand, the Hairy is dominant and sometimes usurps holes the Black-backed is drilling in search of grubs (Villard and Beninger 1993).

Atlas data on the Hairy Woodpecker show similarities with those on the Downy. The Hairy Woodpecker is a very early nester; it then becomes quieter and harder to locate, just at the time of year that atlassing was in full swing. After hatching, however, the species is easier to spot, as the young are very noisy and the adults much more conspicuous around the nest.

Distribution

In Québec, the Hairy Woodpecker is found everywhere south of the 52nd parallel, having been seen as far north as the region around the Nottaway, Broadback, and Rupert rivers (Cons. G. G – GREBE 1992o) and at Fermont (ÉPOQ). It is more abundant farther south, though, where there are mixed and deciduous forests. Moving north, it becomes less and less common as the deciduous trees where it usually nests (Peck and James 1983) become increasingly scarce. It adapts better to coniferous woods than does the Downy and inhabits somewhat more boreal regions. Unlike its smaller cousin, it was atlassed in the Magdalen Islands.

It breeds in wooded areas from central Alaska east across Canada and the United States and south through Central America to Panama and the Bahamas (Godfrey 1986).

History and trends

The Hairy Woodpecker is a typical bird of the deep woods, but has shown remarkable adaptability to a wide range of habitats, including second-growth forests and open areas where the Downy is customarily found. Ouellet (1974a) believes, though, that the Hairy Woodpecker was even more common in the Montréal area in the last century, before local woodlands had been largely cut down.

The Hairy Woodpecker is probably affected by forestry operations. Logging and forest management techniques like selective cutting eliminate much of its habitat in mature forests and reduce the number of dead trees that serve as its larder. On the other hand, deciduous regrowth following clearcutting or fire may partly explain why it is found so far north, within the coniferous forest belt.

In response to declines in many parts of its range in North America, the species was on the Blue List from 1975 to 1982 (Tate 1986).

Like most other woodpeckers, this one must contend with predation by red squirrels, which appear to have a particular liking for its eggs (Lawrence 1967). With its powerful bill, it can defend its nest quite successfully, however, and sometimes even drives away animals as large as raccoons (Kilham 1983). It also faces fierce competition from European Starlings for its nest holes at times (Kilham 1983).

SERGE LEMIEUX

In mature coniferous forests, the Hairy Woodpecker prefers places where there are at least a few deciduous trees, living or dead, for its nest hole. Lake Boulrisse, La Vérendrye Wildlife Reserve. *May 1991, Yves Aubry.*

Notes on territory

Hairy Woodpeckers defend small territories of about 0.1 ha, used solely for nesting, which are defended by both adults from the time they choose their nest site until the breeding season is over. The territory holders stake their claim by calling and drumming.

Feeding territories, 2.4–3.2 ha in size, are not systematically defended the way breeding territories are (Stokes 1979).

Density in a clearcut replanted with spruce in southern New Brunswick was 1.4 males/10 ha (Erskine 1984), but in a stand of various shade-tolerant deciduous species in southern Ontario was as high as 3 males/10 ha (Erskine 1972a).

Hairy Woodpecker

Breeding distribution

1984–1989

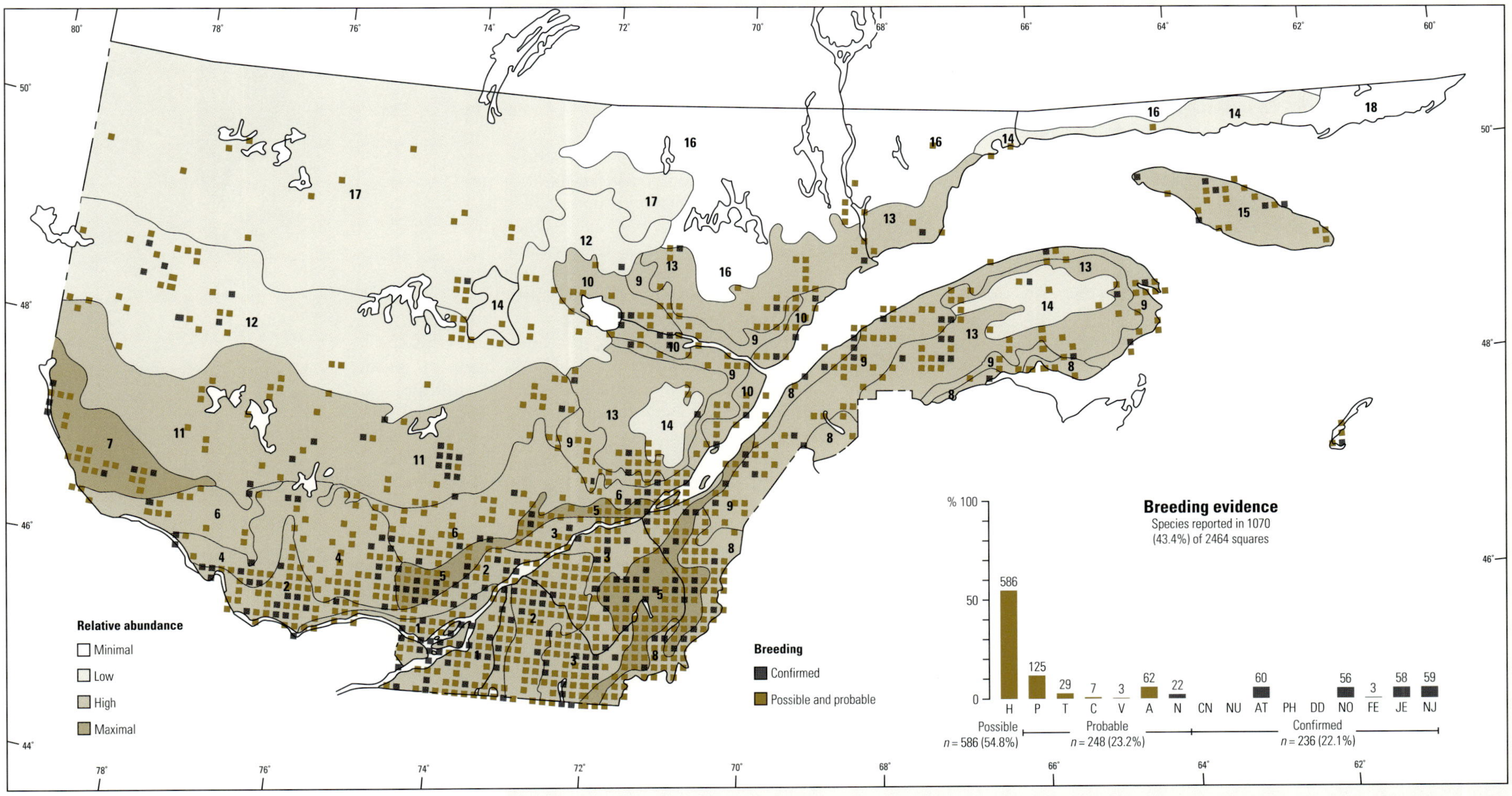

Species profile

Clutch size	usually 4 eggs (3–6)[1,4,5,6]
Laying interval	?
Start of incubation	?
Incubation period	11–15 days[2,3,5]
Maturity at hatching	altricial
Care of young	by female and male[1,2]
Nestling period	28–30 days[4,5,6]
Age at first flight	28–30 days or more[10]
Dependence of young	until ≈ 14 days after leaving nest[1]
Broods per year	1[4,6,7]
Breeding age	1 year[6]
Mating system	monogamy[2]
Duration of pair bond	?
Total length	21.5–26.5 cm[3]
Wing span	38.1–44.4 cm[4]
Weight	M 70.0 g (60.8–79.6) / F 62.5 g (59.3–65.9)[8]
Longevity record	15 years 10 months[9]

References:
1. Harrison 1978
2. Ehrlich et al. 1988
3. Godfrey 1986
4. Terres 1980
5. Short 1982
6. DeGraaf and Rudis 1987
7. Stokes 1979
8. Dunning 1984
9. Clapp et al. 1983
10. according to nestling period

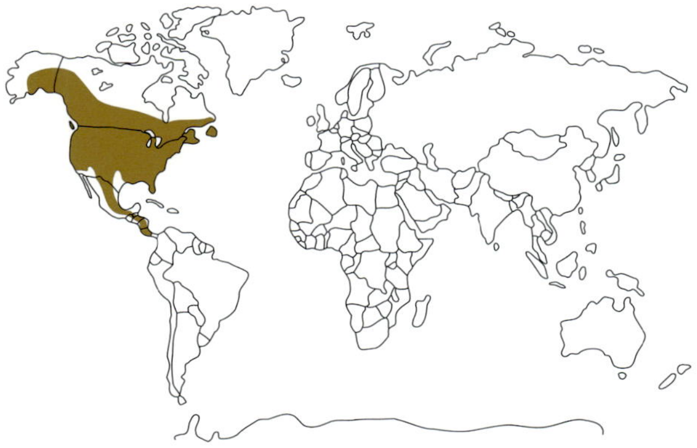

World breeding range

Chronology $n = 70$

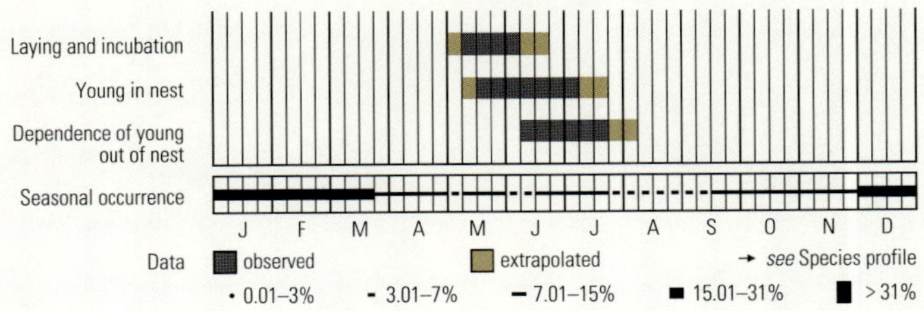

Frequency of occurrence by ecological region

Ecological regions** (forest regions)	Adequately covered squares	Squares in which reported	% of squares in which reported	Relative abundance (%)*
1. Sugar maple-hickory and sugar maple-basswood	97	79	81.4 ± 4.0	81.4
2. Sugar maple-basswood	153	126	82.4 ± 3.2	82.4
3. Sugar maple-basswood and sugar maple-yellow birch	101	86	85.1 ± 4.3	85.1
4. Sugar maple-yellow birch-basswood	56	48	85.7 ± 8.3	85.7
5. Sugar maple-yellow birch-beech and sugar maple-yellow birch-basswood	113	106	93.8 ± 2.7	93.8
6. Sugar maple-yellow birch-beech	71	54	76.1 ± 9.2	76.1
7. Sugar maple-yellow birch-balsam fir	19	19	100.0 ± 2.6	100.0
8. Sugar maple-yellow birch and balsam fir-yellow birch	47	40	85.1 ± 9.4	85.1
9. Balsam fir-yellow birch	87	68	78.2 ± 8.1	78.2
10. Balsam fir-white birch or balsam fir-red maple	46	30	65.2 ± 12.0	65.2
11. Yellow birch-balsam fir	52	43	82.7 ± 10.8	82.7
12. Balsam fir-white birch (west)	59	31	52.5 ± 13.1	52.5
13. Balsam fir-white birch (east)	76	50	65.8 ± 10.4	65.8
14. Balsam fir-black spruce	28	12	42.9 ± 19.2	42.9
15. White spruce-balsam fir and balsam fir-white spruce	22	16	72.7 ± 19.4	72.7
16. Black spruce-balsam fir-moss	18	4	22.2 ± 22.2	22.2
17. Black spruce-moss	32	12	37.5 ± 18.3	37.5
18.			*not analysed*	

* minimal: 0–29.9%; low: 30–59.9%; high: 60–89.9%; maximal: 90–100%

** Deciduous forest: 1–7; mixed forest: 8–11; coniferous forest: 12–17

Three-toed Woodpecker

Pic tridactyle

Picoides tridactylus

Ghislain Caron

BIRDERS RARELY SEE THE ELUSIVE THREE-TOED WOODPECKER IN EASTERN North America. This uncommon denizen of the remote boreal forest (Todd 1963) is resident throughout the year within its breeding range (Bent 1939). Of the nine woodpecker species breeding in Québec, it is the only one that also breeds in Eurasia (AOU 1983; Cramp 1985). It is closely associated with northern spruce forests throughout its circumboreal range (Bock and Bock 1974).

The Three-toed Woodpecker is recorded in Québec about four times as often in winter as in summer (ÉPOQ); irregular winter irruptions south of its breeding range sometimes delight observers in urban woods. These invasions may be caused by increases in wood-boring insects resulting from Dutch elm disease in the south or recent forest fires in the northern part of its range (West and Speirs 1959; Yunick 1985).

Three instead of four toes distinguish the Three-toed Woodpecker and the Black-backed Woodpecker from other North American woodpeckers; another distinguishing characteristic of the two species is a yellow crown patch in the males. Unlike the Black-backed Woodpecker, whose back is solid black, the Three-toed usually has black and white barring down the middle of its back, but there is considerable geographical variation. In western Canada, the back may be entirely white, while in the east, it may be quite black with only partial barring or a few flecks of white.

Habitat and behaviour

A typical species of spruce and fir communities of the boreal forest (Erskine 1977), the Three-toed Woodpecker is often found in coniferous areas burned by forest fires, flooded by beaver dams, or stricken by disease (Oatman 1985c). It needs dead or dying conifers because about 75% of its food consists of wood-boring larvae of beetles and lepidopterans (Bent 1939). It specializes in bark beetles and responds positively to bark beetle epidemics (Lester 1980), but takes some other insects and spiders. It also spends considerable time pecking horizontal series of holes, sometimes spanning up to a quarter the circumference of the tree trunk, to obtain sap and resin, in much the same way as the Yellow-bellied Sapsucker. It was formerly thought that this species ate wood, but according to Martin et al. (1951), the fragments of cambium and wood found in the stomachs of some specimens are ingested accidentally while extracting grubs.

The best way to find either of the two three-toed woodpeckers in any season is to listen carefully for the quiet tapping of the feeding birds, or, in spring and summer, courtship drumming, which carries a long distance. Because it is difficult to identify woodpeckers from their tapping sounds, this is only a means of finding the birds

Notes on territory

Three-toed Woodpeckers defend breeding and winter territories. Most birds probably leave their breeding territories and defend winter feeding territories elsewhere, but some overwinter in and defend part of the breeding territory.

Birds winter in territories of 4–10 ha, singly or in pairs (Cramp 1985); sharing is made possible by differences in feeding behaviour that reduce intersexual competition (Hogstad 1978). In Norway, over 3–4 days of the nestling period, pairs maintained minimum feeding areas of 11–19 ha, with the male and female foraging in different areas (Cramp 1985).

A study in the watersheds of the Great Whale and Little Whale rivers found that Three-toed Woodpecker pairs were three times as common in black spruce-moss stands as in shoreline ecotones in the coniferous forest (Cons. G. G. – GREBE 1991a).

In Ontario, Manitoba, and British Columbia, Erskine (1977) determined densities of 0.1–0.3 male/10 ha for similar habitats. Densities in Europe ranged from 0.03 to 0.15 breeding pair (or drumming male)/10 ha (Cramp 1985).

The Three-toed Woodpecker is primarily a bird of the great northern spruce forests. East of the Daniel Johnson Dam, north of Baie-Comeau. *May 1991, Daniel Bordage.*

Three-toed Woodpecker

Breeding distribution 1984–1989

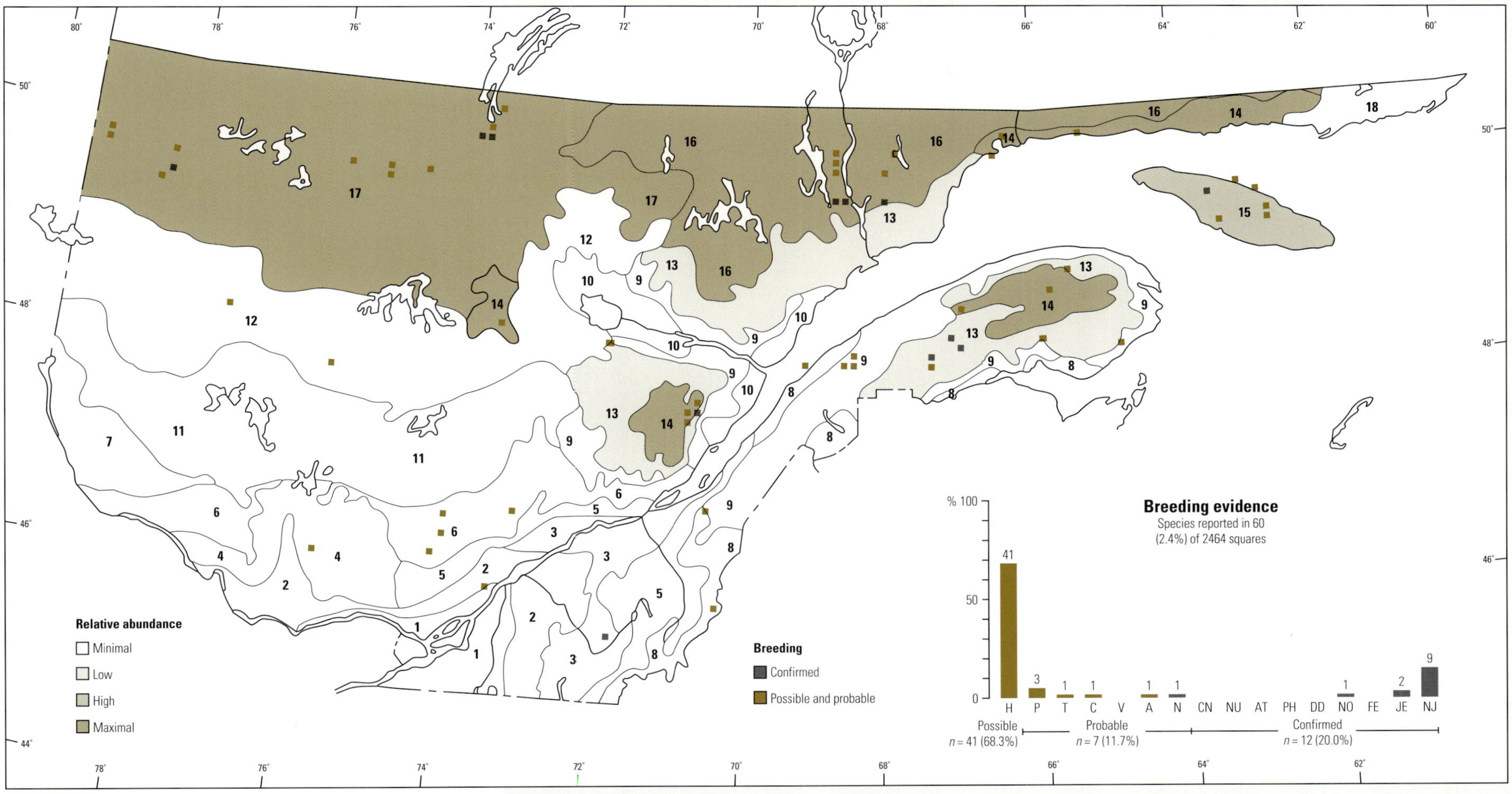

Species profile

Clutch size	usually 3–5 eggs (3–7)[1,2,3,9,10]
Laying interval	1 egg per day[1]
Start of incubation	with last egg[1]
Incubation period	11–14 days[1,2,3,4,5]
Maturity at hatching	altricial
Care of young	by female and male[2,4,5]
Nestling period	probably ≈ 22–26 days[1,6]
Age at first flight	22–26 days[1,6]
Dependence of young	until 1–2 months after leaving nest[1,6]
Broods per year	1[1,2,4]
Breeding age	?
Mating system	monogamy[1,4]
Duration of pair bond	sometimes several seasons[4]
Total length	20.3–24.5 cm[7]
Wing span	33.0–38.1 cm[3]
Weight	M 70.1 g (65.0–74.0) / F 61.2 g (57.0–66.0)[8]
Longevity record	?

References:
1. Cramp 1985
2. Harrison 1978
3. Terres 1980
4. Ehrlich et al. 1988
5. Bent 1939
6. Short 1982
7. Godfrey 1986
8. Dunning 1984
9. Peck and James 1983
10. Campbell et al. 1990b

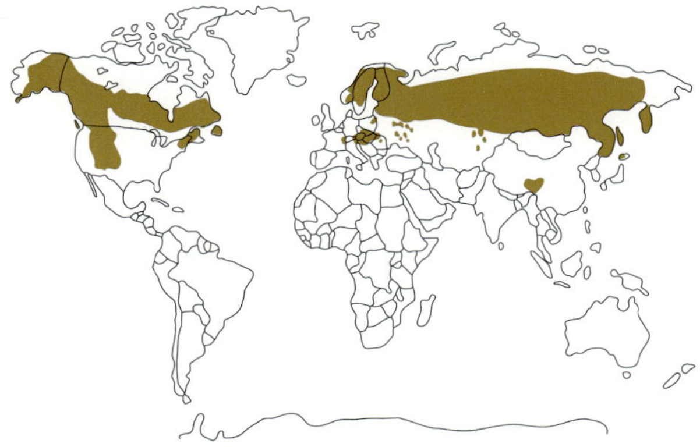

World breeding range

Chronology $n = 13$

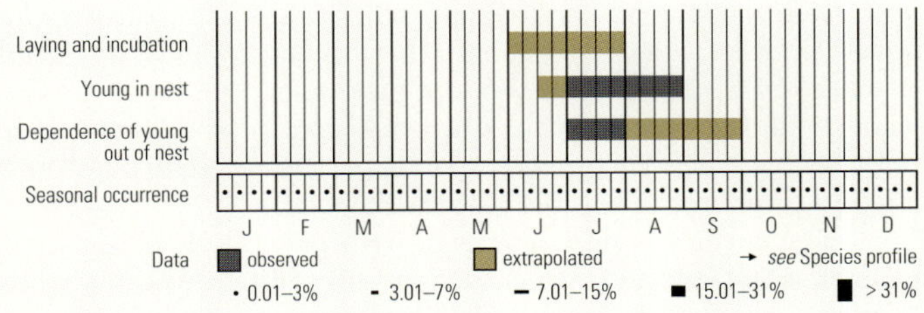

Frequency of occurrence by ecological region

Ecological regions** (forest regions)	Adequately covered squares	Squares in which reported	% of squares in which reported	Relative abundance (%)*
1. Sugar maple-hickory and sugar maple-basswood	97	0	0	0
2. Sugar maple-basswood	153	1	0.7 ± 0.9	2.9
3. Sugar maple-basswood and sugar maple-yellow birch	101	1	1.0 ± 1.6	4.5
4. Sugar maple-yellow birch-basswood	56	0	0	0
5. Sugar maple-yellow birch-beech and sugar maple-yellow birch-basswood	113	0	0	0
6. Sugar maple-yellow birch-beech	71	4	5.6 ± 5.3	25.4
7. Sugar maple-yellow birch-balsam fir	19	0	0	0
8. Sugar maple-yellow birch and balsam fir-yellow birch	47	3	6.4 ± 6.8	28.7
9. Balsam fir-yellow birch	87	2	2.3 ± 3.3	10.3
10. Balsam fir-white birch or balsam fir-red maple	46	0	0	0
11. Yellow birch-balsam fir	52	0	0	0
12. Balsam fir-white birch (west)	59	1	1.7 ± 4.0	7.6
13. Balsam fir-white birch (east)	76	7	9.2 ± 6.6	41.4
14. Balsam fir-black spruce	28	6	21.4 ± 16.2	96.4
15. White spruce-balsam fir and balsam fir-white spruce	22	3	13.6 ± 15.5	61.4
16. Black spruce-balsam fir-moss	18	4	22.2 ± 22.2	100.0
17. Black spruce-moss	32	7	21.9 ± 15.9	98.4
18.			*not analysed*	

* minimal: 0–29.9%; low: 30–59.9%; high: 60–89.9%; maximal: 90–100%

** Deciduous forest: 1–7; mixed forest: 8–11; coniferous forest: 12–17

Female Three-toed Woodpecker at the entrance to her nest, carved in a still-healthy jack pine. Eastmain River, James Bay region. *May 1982, Vincent Létourneau.*

to identify them visually. Once located, Three-toed Woodpeckers are easy to observe because they are very tame, even near the nest hole, often allowing humans to approach to within 5 m, but they frequently hide behind the trunk once they become aware of an intruder (Cramp 1985). Three-toed Woodpecker nests, like those of other woodpeckers, are easy to find when they contain young, which are noisy (Cramp 1985). This species spends more time feeding and resting on a given tree than other woodpeckers do.

The species is monogamous, with pair bonds occasionally maintained all year and even sometimes for several successive years. Nests are excavated 1–24 m up (usually 1.8–4.6 m) in stumps or in trees such as spruces, firs, tamaracks, pines, cedars, or aspens; the entrance hole diameter varies from 4.0–7.5 cm (Terres 1980; Peck and James 1983; Campbell et al. 1990b). The gourd-shaped cavity widens toward the bottom, with the lower lip of the entrance angled downward on the inside (Harrison 1975). The Three-toed Woodpecker usually selects trees infected by fungal heart rot (Lester 1980).

One clutch, generally 3–5 eggs, is laid per year (Terres 1980; Peck and James 1983; Cramp 1985; Campbell et al. 1990b). Both sexes incubate. The female roosts nightly in the nest from two days before laying until the end of the laying period, then is replaced by the male, which roosts nightly in the nest throughout incubation and until the young are 17–21 days old (Cramp 1985). The brood usually consists of 2 young, sometimes 3 and occasionally 4 (Campbell et al. 1990b). Both parents bring food, mostly beetle and moth larvae, but also other insects and sap. The young fledge at the age of 22–26 days (Short 1982; Cramp 1985). Afterward, the family is often encountered in the neighbourhood of the nest, the young calling loudly and frequently. Sometimes the young fly after the parents, which apparently divide the brood between them. When habitat is suitable and food plentiful, several pairs may nest in quite close proximity (Harrison 1975). Age at first breeding is not known.

Distribution

The Three-toed Woodpecker nests in coniferous forests (Erskine 1977). It is locally common in western North America, but rare in the East (Robbins et al. 1966). It was reported in only 2.4% (60/2464) of Atlas squares, and breeding was confirmed in 20% (12/60) of those.

Sightings are very scattered and reflect well the species' presence in southern Québec. Its distribution is closely associated with coniferous forest and it does not appear to breed in the Central St. Lawrence Lowland, where hardwoods predominate. It is seen occasionally in isolated stands, but its optimum habitat is the boreal spruce forest and the alpine coniferous forest. It is thought to nest as far north as the tree line (Todd 1963), farther than any other Québec woodpecker (Godfrey 1986). Among the northernmost sightings are those at Lake Bienville and Schefferville (ÉPOQ).

This woodpecker, which is found in the mountains of the Gaspé Peninsula, seeks out dead or dying conifers infested with wood-boring larvae, the mainstay of its diet. Mount Albert, Gaspé Conservation Park. *July 1991, Yves Aubry.*

This woodpecker can be found in spruce-moss or spruce-lichen stands in northern Québec and in mountainous areas in the southern parts of the province. La Grande-3, James Bay region. *July 1992, Réginald Ouellet.*

This species' nesting habitat is less accessible than the watersides or burns often used by the Black-backed Woodpecker (McIlveen 1987), so the Three-toed is likely more common than the small number of Atlas records would suggest. Given that it is a permanent resident, there is every reason to suspect that it breeds in many of the squares where it was merely reported during the breeding season. This woodpecker occurs only very sporadically throughout southern Québec at any time of year (A. Cyr, unpubl. data).

History and trends

Québec data are unfortunately too fragmentary to provide a clear picture of population trends for the species, although Todd (1963) considered the Three-toed Woodpecker to be much less common than the Black-backed in this province.

While abundance estimates for Ontario confirm that the Three-toed Woodpecker is generally uncommon in that province, they suggest that in favourable areas it reaches densities comparable with those of the Black-backed (McIlveen 1987). There is little to indicate significant changes in this woodpecker's Ontario distribution or abundance (McIlveen 1987).

In New England, however, its numbers may have decreased when much of the region's original spruce forest was logged in the 18th and 19th centuries (Oatman 1985c).

André Cyr and Robert Alvo

Black-backed Woodpecker

Pic à dos noir

Picoides arcticus

QUÉBEC BIRDERS TEND TO ASSOCIATE THE BLACK-BACKED WOODPECKER with the vast northern forests, yet it was atlassed as a nesting bird virtually throughout the province, except in the southern Ottawa Valley, the Montérégie, and the lowlands of the St. Lawrence River upstream from Québec City.

Some winters, this resident of the coniferous forest irrupts southward into deciduous woodland and settled areas, adjusting the length of its stay to the abundance of bark beetles, small insects that it finds under tree bark (Yunick 1985).

The Black-backed Woodpecker nests in this spruce stand that has been thinned by a budworm epidemic. It can be found in coniferous forests throughout the province. Mount Albert, Gaspé Conservation Park. *July 1991, Yves Aubry.*

Habitat and behaviour

The Black-backed Woodpecker is a typical, yet patchily distributed and uncommon resident of coniferous forests (Erskine 1977). Bock and Bock (1974) reported that in addition to living in the spruce forests favoured by the more northern Three-toed Woodpecker, the Black-backed is often found in pine and other coniferous forests at lower elevations and farther south. Much less is known of its breeding behaviour than that of other woodpecker species. Although it mainly haunts the deepest woods, it usually chooses an open area for nesting. It prefers stands that have been decimated by wind, clearcutting, or fire, or woods that have been flooded by beaver dams, as long as there are still some standing trees. The nest is a cavity excavated by the male and female—with the male doing more of the drilling and excavating (Short 1982)—0.6–4.6 m up in a living or dead tree (Harrison 1975), usually a conifer, but sometimes a deciduous tree or even a utility pole (Peck and James 1983). The species is monogamous and both parents incubate the clutch of 2–6 eggs, most commonly 4 (Koenig 1986), for about 14 days (Short 1982). The nestling period lasts about 25 days (Ehrlich et al. 1988).

Nothing is known about the nestlings' diet, but it is probably similar to that of the adults, that is, about 89% animal content —mostly beetles (68%), but also caterpillars (13%), ants (6%), and other arthropods (2%). This woodpecker is considered to be an important predator of bark beetles, which it finds by flaking off pieces of bark from trees with easily peeled or scaly bark, particularly conifers (Beal 1911; Lester 1980), but also hardwoods such as elms (Koenig 1986). It usually peels the trunk, rarely branches, so the presence of stripped trunks may be a sign of its activity.

The Black-backed Woodpecker was reported in 14.2% (350/2464) of Atlas squares. Breeding was confirmed in 23.1% (81/350) of these squares, but reported as probable in only 19.1% (67/350). A partial explanation could be the bird's unobtrusiveness before the eggs hatch, in contrast to its more conspicuous behaviour while raising the young, when parents come and go often to the nest, and the young call loudly. Seven of the 81 breeding confirmations were of adults carrying prey in their bills. Little is known about the feeding habits of the young, and our observations seem to differ from those of Bent (1939), who stated that the young are fed by regurgitation. Bent's observations may have been confined to nestlings still too young to swallow prey that had not been predigested, however.

Distribution

In Québec, the Black-backed Woodpecker is almost entirely absent from the hardwood forests in the St. Lawrence Valley, although breeding was confirmed at Terrebonne in 1982 (Gosselin and David 1982). It becomes gradually commoner as conifers become more predominant and latitude and elevation increase. It probably seems scarce in the region extending from the Abitibi to Lake Saint-Jean because most atlassing was done in the more accessible cleared areas in the Abitibi.

The Black-backed Woodpecker's range extends beyond the boundary of the Atlas area up to 55° N (Todd 1963). A male and female feeding nestlings were observed at Eastmain, close to the east coast of James Bay (ÉPOQ). A nest was also found in the Nottaway, Broadback, and Rupert rivers region (Cons. G. G. – GREBE 1992o).

Although the Black-backed Woodpecker typically occurs in fir and spruce forests, Erskine (1977) did not find it to be common anywhere in the boreal zone, which suggests that it tends to prefer the transition zone between the northern hardwood and boreal coniferous forests. Its ability to utilize pine and tamarack trees partly explains its southern range in Québec. In its search for larvae, such as those of the whitespotted sawyer (Villard and Beninger 1993), it will also visit burns and stands of dead, insect-infested trees (Bock and Bock 1974).

This woodpecker, which nests on the edge of a clearing, also forages in dense coniferous forest. Decelles Reservoir, Témiscamingue. *May 1991, Réginald Ouellet.*

History and trends

The Black-backed Woodpecker's range does not seem to have changed in Québec since the beginning of the century. Taverner (1922) associated it with the northern coniferous forest, confirming Macoun and Macoun (1915) who mentioned that the bird could be found at Hudson Bay. Todd (1963) discounted this, however, giving the northernmost record for James Bay and Hudson Bay as the mouth of the Eastmain River, where he collected a young bird on 9 July 1941. It was considered a scarce winter visitant in the Central St. Lawrence Lowland (Wintle 1896; Macoun and Macoun 1915; Taverner 1922).

The abundance of the Black-backed Woodpecker does not seem to have changed at all in the last 25 years, according to BBS data. In the long term, however, its numbers could decrease as a result of human activity, particularly forestry practices that tend to eliminate the overmature forests and decaying trees that the species prefers.

YVES GAUTHIER AND MARCEL DARVEAU

Ghislain Caron

Notes on territory

Woodpeckers usually defend their territories throughout the year (Short 1982). The Black-backed Woodpecker rules its domain fiercely and does not hesitate to drive off the Hairy Woodpecker. In fact, these birds are often in conflict because they both can be found in similar habitats (Bull et al. 1986).

The Black-backed Woodpecker is also very aggressive toward Tree Swallows, which covet nests sites left unattended, and toward all other woodpecker species that trespass on its territory (Short 1982).

Black-backed Woodpecker

Breeding distribution

1984–1989

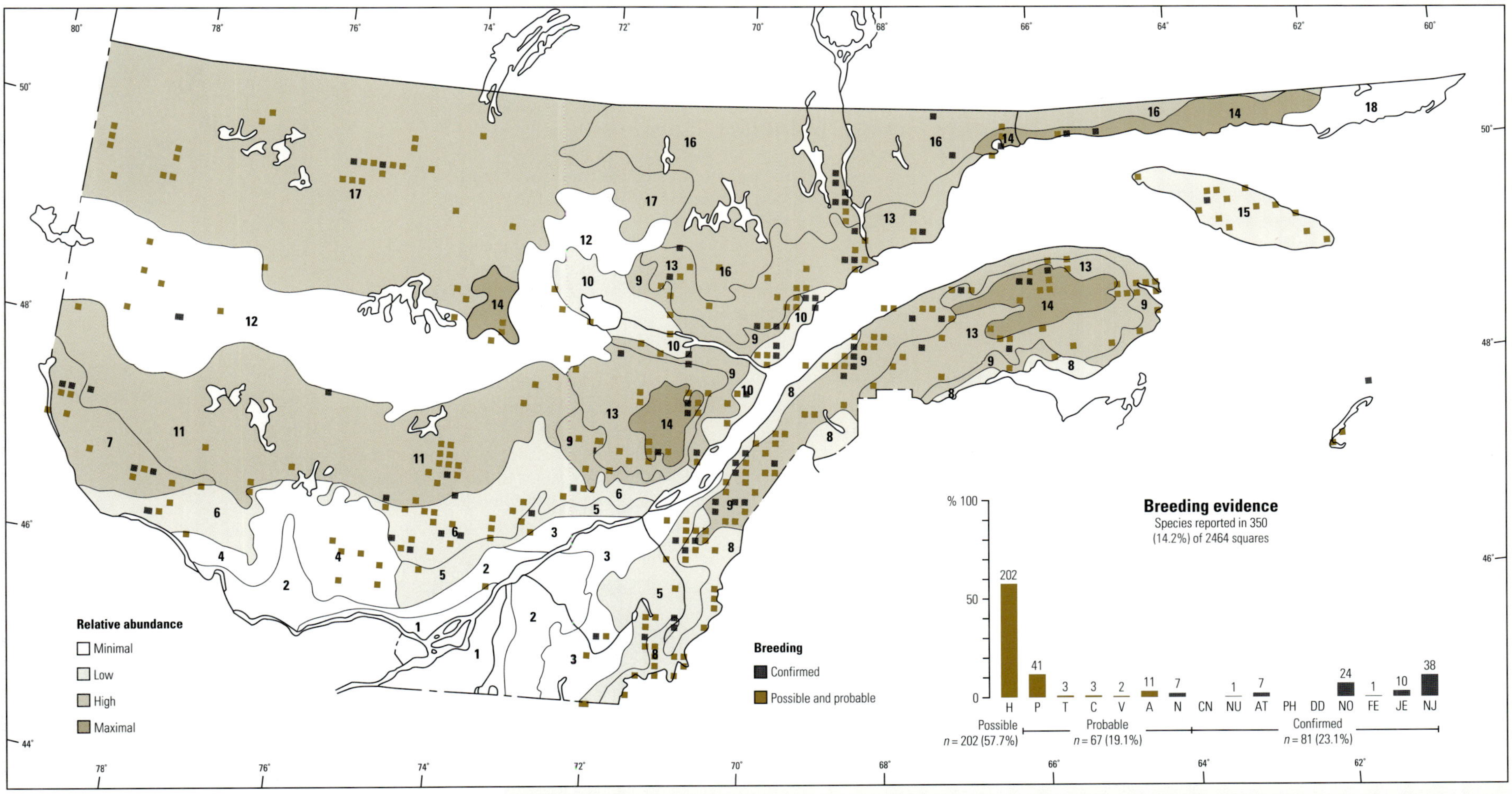

World breeding range

Species profile

Clutch size	usually 4 eggs (2–6)[1,2]
Laying interval	1 egg per day[1]
Start of incubation	?
Incubation period	probably 12–14 days[1]
Maturity at hatching	altricial
Care of young	by female and male[1,2]
Nestling period	probably ≈ 25 days[2]
Age at first flight	probably ≈ 25 days or more[7]
Dependence of young	probably until a few weeks after leaving nest[1]
Broods per year	1[2]
Breeding age	1 year[6]
Mating system	monogamy[2]
Duration of pair bond	?
Total length	23.0–25.5 cm[3]
Wing span	35.6–40.6 cm[4]
Weight	M 72.0 g (61.3–88.1) / F 66.6 g (63.4–71.3)[5]
Longevity record	?

References:
1. Short 1982
2. Ehrlich et al. 1988
3. Godfrey 1986
4. Terres 1980
5. Dunning 1984
6. DeGraaf and Rudis 1987
7. according to nestling period

Chronology $n = 62$

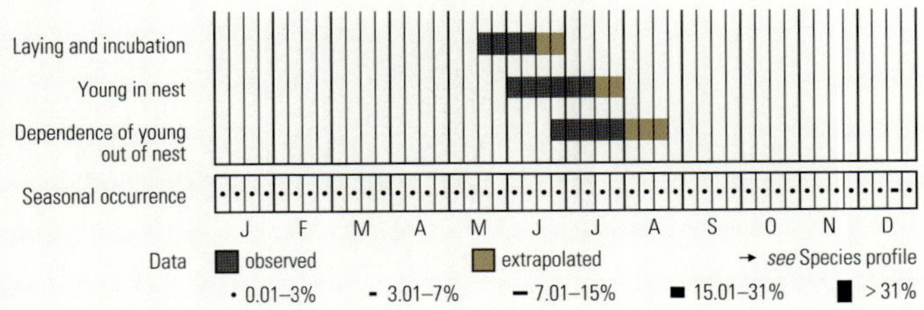

Frequency of occurrence by ecological region

Ecological regions** (forest regions)	Adequately covered squares	Squares in which reported	% of squares in which reported	Relative abundance (%)*
1. Sugar maple-hickory and sugar maple-basswood	97	0	0	0
2. Sugar maple-basswood	153	1	0.7 ± 0.9	1.0
3. Sugar maple-basswood and sugar maple-yellow birch	101	6	5.9 ± 3.0	9.2
4. Sugar maple-yellow birch-basswood	56	4	7.1 ± 6.3	11.1
5. Sugar maple-yellow birch-beech and sugar maple-yellow birch-basswood	113	22	19.5 ± 4.2	30.3
6. Sugar maple-yellow birch-beech	71	23	32.4 ± 10.0	50.4
7. Sugar maple-yellow birch-balsam fir	19	8	42.1 ± 23.3	65.5
8. Sugar maple-yellow birch and balsam fir-yellow birch	47	17	36.2 ± 12.4	56.3
9. Balsam fir-yellow birch	87	40	46.0 ± 9.6	71.5
10. Balsam fir-white birch or balsam fir-red maple	46	11	23.9 ± 10.8	37.2
11. Yellow birch-balsam fir	52	22	42.3 ± 13.8	65.8
12. Balsam fir-white birch (west)	59	9	15.3 ± 9.7	23.7
13. Balsam fir-white birch (east)	76	31	40.8 ± 10.8	63.5
14. Balsam fir-black spruce	28	18	64.3 ± 18.6	100.0
15. White spruce-balsam fir and balsam fir-white spruce	22	8	36.4 ± 20.8	56.6
16. Black spruce-balsam fir-moss	18	8	44.4 ± 26.0	69.1
17. Black spruce-moss	32	15	46.9 ± 18.9	72.9
18.			*not analysed*	

* minimal: 0–29.9%; low: 30–59.9%; high: 60–89.9%; maximal: 90–100%

** Deciduous forest: 1–7; mixed forest: 8–11; coniferous forest: 12–17

Northern Flicker

Pic flamboyant

Colaptes auratus

Ghislain Caron

APRIL MARKS THE RETURN OF THE NORTHERN FLICKER—THE ELEGANT carpenter with a voice like thunder, in the words of Québec poet Pierre Morency (1992). This migrant breeder is common in Québec and winters from southern Canada through the central and southern United States down to Florida. A number of other species, including the Bufflehead, American Kestrel, European Starling, and Eastern Bluebird, rely on the holes it excavates in trees to use for their own nest sites.

Unlike other woodpeckers, the flicker routinely perches like a passerine, and finds most of its food on the ground. Ants, which it often takes directly from the nest, constitute close to 50% of its summer diet (Beal 1911; Bent 1939). According to Test (1969), this food is the source of the yellow pigment that colours its plumage. In fall, it is attracted to cultivated fruit trees (Campbell et al. 1990b).

Habitat and behaviour

The flicker occupies a wide variety of habitats, but it generally prefers open country, such as clearings in deciduous or coniferous woodlands, burntlands, the edges of clearcuts, open agricultural land, and suburban areas where there are trees (Conner et al. 1975; Terres 1980; Godfrey 1986). It is not as fussy as other woodpeckers and will nest in a hole made in a tree trunk, utility pole, or fence post, sometimes even in a bank of sand or earth. It will occasionally move into a nest box (Short 1982; Peck and James 1983).

Excavation of a new nest cavity is hard work for the pair and takes anywhere from 5 to 28 days (Harrison 1978; Kilham 1983; J. Gauthier, pers. comm.). Sherman (1910) noted that in this philopatric species new cavities are often in the same trees used the previous year. Flickers are more prolific than other woodpeckers, generally laying 6–8 eggs, which hatch after 11 or 12 days. The parents share the tasks of incubation and brood rearing, and the young remain in the nest for about 25–28 days (Harrison 1978).

As a rule, pairs form for life. The male's piercing cry, followed by a long burst of drumming, frequently attracts other females besides his mate, leading to a contest in which each female tries to intimidate the others by shaking its head up and down or from side to side while displaying the golden undersides of its wings and tail; rivals are usually driven off (Noble 1936; Stokes 1979; Kilham 1983). The mates then face each other for their courtship dances and tap bills lightly (Lawrence 1967).

This woodpecker remains noisy all summer long, and its extroverted behaviour makes it an easy species to locate. Moreover, its white rump and the yellow flash of the wing linings make it easy to identify in flight. It is often flushed from its feeding sites along roadside ditches and shoulders in open country and forested areas.

Notes on territory

Mating, breeding, and foraging all take place on a territory defended by both adults from the time they arrive in spring until the young have become independent.

An open hardwood forest is suitable habitat for Northern Flickers, which nest in holes excavated in tree trunks. Summit Park, Westmount, near Montréal. *July 1991, Gaétan Duquette.*

Northern Flicker

Breeding distribution

1984–1989

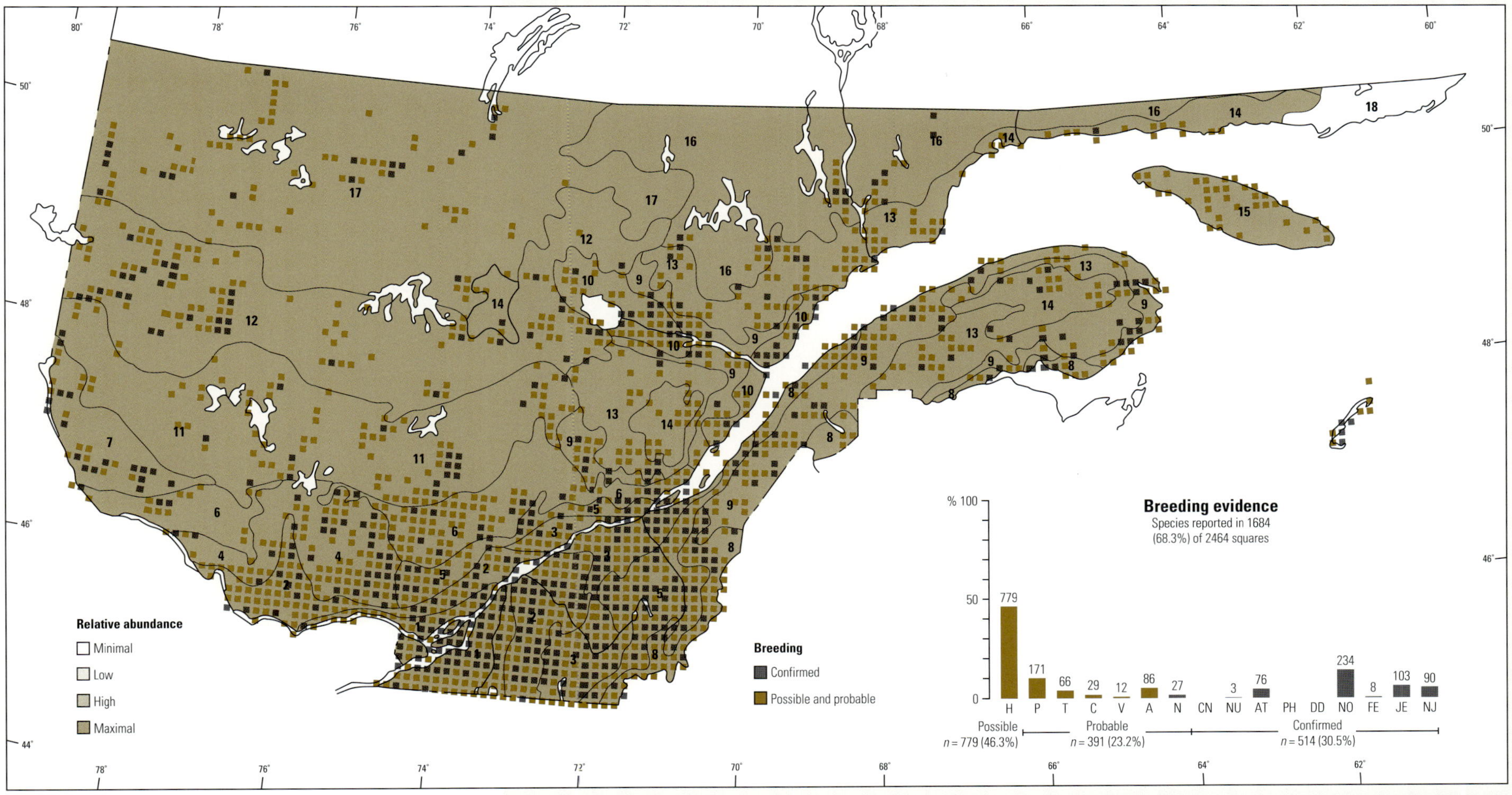

Species profile

Clutch size	usually 6–8 eggs (3–14)[1,2,4,5]
Laying interval	1 egg per day[1]
Start of incubation	?
Incubation period	11–12 days[1,3,4,5]
Maturity at hatching	altricial
Care of young	by female and male[1,2]
Nestling period	25–28 days[1,2,3,6]
Age at first flight	25–28 days or more[10]
Dependence of young	until 2–3 weeks after leaving nest[7]
Broods per year	1[1,2,5]
Breeding age	1 year[5]
Mating system	monogamy[2]
Duration of pair bond	usually for life[7]
Total length	30.5–35.0 cm[4]
Wing span	47.0–53.3 cm[3]
Weight	M 135 g (114–160) / F 129 g (106–164)[9]
Longevity record	9 years 2 months[8]

References:
1. Harrison 1978
2. Ehrlich et al. 1988
3. Terres 1980
4. Godfrey 1986
5. DeGraaf and Rudis 1987
6. Short 1982
7. Stokes 1979
8. Clapp et al. 1983
9. Dunning 1984
10. according to nestling period

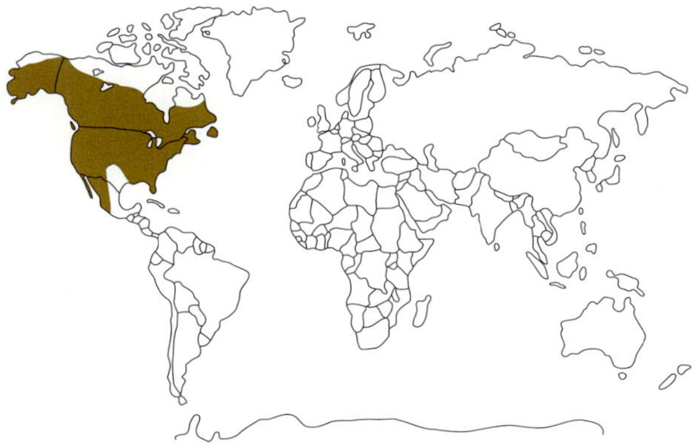

World breeding range

Chronology $n = 228$

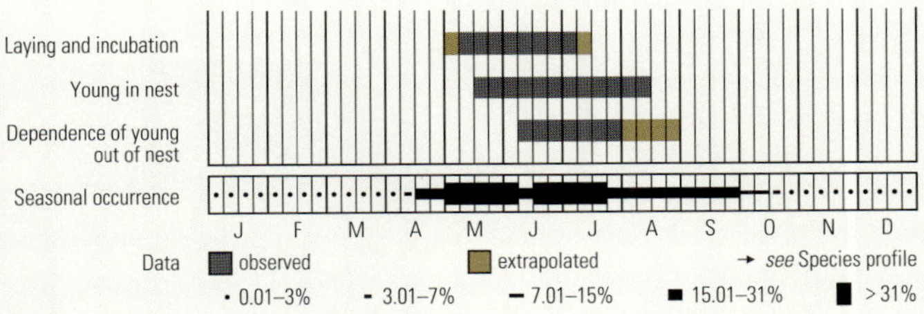

Frequency of occurrence by ecological region

Ecological regions** (forest regions)	Adequately covered squares	Squares in which reported	% of squares in which reported	Relative abundance (%)*
1. Sugar maple-hickory and sugar maple-basswood	97	97	100.0 ± 0.5	100.0
2. Sugar maple-basswood	153	153	100.0 ± 0.3	100.0
3. Sugar maple-basswood and sugar maple-yellow birch	101	100	99.0 ± 1.6	99.0
4. Sugar maple-yellow birch-basswood	56	54	96.4 ± 4.8	96.4
5. Sugar maple-yellow birch-beech and sugar maple-yellow birch-basswood	113	113	100.0 ± 0.4	100.0
6. Sugar maple-yellow birch-beech	71	71	100.0 ± 0.7	100.0
7. Sugar maple-yellow birch-balsam fir	19	19	100.0 ± 2.6	100.0
8. Sugar maple-yellow birch and balsam fir-yellow birch	47	47	100.0 ± 1.1	100.0
9. Balsam fir-yellow birch	87	85	97.7 ± 3.3	97.7
10. Balsam fir-white birch or balsam fir-red maple	46	46	100.0 ± 1.1	100.0
11. Yellow birch-balsam fir	52	52	100.0 ± 1.0	100.0
12. Balsam fir-white birch (west)	59	58	98.3 ± 4.0	98.3
13. Balsam fir-white birch (east)	76	72	94.7 ± 5.3	94.7
14. Balsam fir-black spruce	28	26	92.9 ± 10.8	92.9
15. White spruce-balsam fir and balsam fir-white spruce	22	21	95.5 ± 10.3	95.5
16. Black spruce-balsam fir-moss	18	18	100.0 ± 2.8	100.0
17. Black spruce-moss	32	32	100.0 ± 1.6	100.0
18.			*not analysed*	

* minimal: 0–29.9%; low: 30–59.9%; high: 60–89.9%; maximal: 90–100%

** Deciduous forest: 1–7; mixed forest: 8–11; coniferous forest: 12–17

Streit Pond, with its many snags, is well suited to nesting by the Northern Flicker. Philipsburg, south of Saint-Jean-sur-Richelieu. *July 1991, Pierre Bannon.*

Distribution

The Northern Flicker was reported in 68.3% (1684/2464) of Atlas squares, making it one of the commonest species atlassed. Breeding was confirmed in 30.5% (514/1684) of squares in which it was recorded. It is found throughout the Atlas area, its wide distribution in Québec doubtless due to the ubiquity of its habitats. Godfrey (1986) placed the northern limit of its breeding range at about 56° N, though Macoun and Macoun (1909) reported on a flicker collected at Ungava Bay, and more recently the species was sighted at Lake Guillaume Delisle in 1981 and at Kuujjuarapik in 1987 (ÉPOQ).

The dietary habits of this woodpecker explain why it does not winter in Québec: as a frequent ground feeder, its foraging would be impeded by snow. Thus, in winter it is generally more abundant in areas where the annual snowfall is no more than 30 cm (Root 1988).

History and trends

Flickers are especially fond of cherries in the fall, and in the United States in the last century, this often provided an excuse for wholesale slaughter of the birds, which commanded a good price on the market (Beal 1911). Yet if ornithological observations from the turn of the century (Macoun and Macoun 1909) are accurate, distribution and abundance do not appear to have been much different then from what they are today.

Notes on territory (continued)

Territory size is usually about 65 ha, but after egg laying, the birds defend only the immediate vicinity of the nest (Stokes 1979). In fact, Dennis (1969) found 2 nests less than 7 m apart, and Kilham (1973a) found 3 pairs nesting within an area of less than 0.4 ha.

In Québec, densities of 2.3 pairs/10 ha were reported in a mature maple stand at Mirabel (Cyr 1974), 1 pair/10 ha in a treed urban cemetery in Montréal (Erskine 1972a), and 0.71 pair/10 ha in a mixed forest at Saint-Michel-des-Saints (Lafontaine 1988).

In a study of the Nottaway, Broadback, and Rupert river region, the Northern Flicker was found to be twice as abundant in white birch stands as in recent burns, but even there the density was twice as high as in black spruce-moss stands (Cons. G. G. – GREBE 1992o).

Flickers are faithful to their territories and a pair has returned two years running to nest in the same hollow in this maple in a belt of trees separating farmland from a residential area. Saint-Augustin-de-Desmaures, west of Québec City. *Summer 1993, Jean Gauthier.*

Flickers also nest in trees and snags in beaver ponds. Forillon National Park, Gaspé Peninsula. *July 1992, Yves Aubry.*

Adult feeding its young at the entrance to the nest. Sherbrooke. *July 1978, Daniel Jauvin.*

Nonetheless, over the last 30 years, the Northern Flicker population seems to have declined. BBS data show a drop in southern Ontario and Québec for the period from 1968 to 1991 (Erskine et al. 1992). Erskine (1978b) attributes the decline to an increase in forestry operations, which reduces the number of dead trees suitable for nesting. Yet logging may also benefit the species by creating the open habitat that it requires. Competition from the European Starling for nest cavities and several particularly harsh winters may have contributed to the decrease, as well (Bock and Smith 1971; Robbins et al. 1986). Nonetheless, the Northern Flicker remains common in Québec.

Serge Lemieux

Pileated Woodpecker

Grand pic

Dryocopus pileatus

Ghislain Caron

The Pileated Woodpecker is by far the largest woodpecker in Québec. It is a black, crow-sized bird with a white-striped head surmounted by a conspicuous scarlet crest. The mere sight of it is a thrill.

Summer and winter, this permanent resident can be found in older forests, where it advertises its presence in various ways. It is generally quite easy to spot the characteristic rectangular holes it makes in trees infested with the insects it feeds on. It also attracts attention with its slow, muffled drumming and its voice, which is similar to that of the Northern Flicker, but deeper and much more powerful.

Habitat and behaviour

Mature mixed, deciduous, or coniferous woodlands are its habitats of choice (Hoyt 1957; Conner et al. 1975; Short 1982; Renken and Wiggers 1989). The Pileated Woodpecker prefers mature forests with numerous snags, but it may also inhabit younger forests (Mellen et al. 1992). In its home range, however, there must be big trees (often over 18 m tall) with a diameter at breast height greater than 30 cm, in which it can excavate its nest hole (Conner et al. 1975; Bull and Meslow 1977; McClelland 1979; Mellen et al. 1992). As long as it remains undisturbed, the Pileated Woodpecker may be found close to dwellings in small woods and parks where the trees are fairly mature (Godfrey 1986).

The Pileated Woodpecker mates for life and does not migrate, remaining in the same territory year after year (Kilham 1959, 1983; Short 1982). Renken and Wiggers (1989) demonstrated that the size of the territory is inversely proportional to the number of snags and dead trees, as well as the density of the canopy. Within its territory, the Pileated Woodpecker excavates two kinds of cavities: roosting holes, to which it retreats at night and in bad weather, and nesting holes (McClelland 1979).

Each spring, the pair seeks out a new nest site. Probably as a way of avoiding parasite infestations, a pair very rarely uses the same nesting cavity two years in a row (Christy 1939; McClelland 1979; Bull and Meslow 1988), but old nest holes may eventually come to serve as roosts (Kilham 1983). When nesting habitat is scarce, the pair may drill a new nest hole in the same tree each year (Hoyt 1957; Leberman 1992e). In early spring, the pair often starts several holes in different trees before making a final choice (Kilham 1983; Bull and Meslow 1988). Abandoned holes are usually completed and used for nesting in subsequent years (Hoyt 1957; Bull and Meslow 1988).

All types of mature forests are suitable breeding habitat for the Pileated Woodpecker. It is territorial year round and betrays its presence by the numerous holes, mostly rectangular, that it drills into insect-infested trees. Notre-Dame-de-Portneuf, west of Québec City. *March 1982, Christian Marcotte.*

Notes on territory

The Pileated Woodpecker is a permanent resident that is territorial year round (Kilham 1979a, 1983). All its holes and feeding sites are located within the territory.

The size of the territory established by a pair or by a lone bird depends on the structure of the habitat (Short 1982; Mannan 1984; Renken and Wiggers 1989). During the breeding season, Bull and Meslow (1977) estimated territories at around 130–140 ha in Oregon, whereas Renken and Wiggers (1989) reported areas of 53–160 ha in Missouri. Pairs with attendant young seem to defend larger territories (Mellen et al. 1992).

Density varies according to habitat; Graber et al. (1977) reported a maximum of 0.75 bird/10 ha in mature forests in Illinois, whereas Bull (1975) reported a density of 0.015 pair/10 ha in extensive conifer stands in Oregon.

Pileated Woodpecker

Breeding distribution

1984–1989

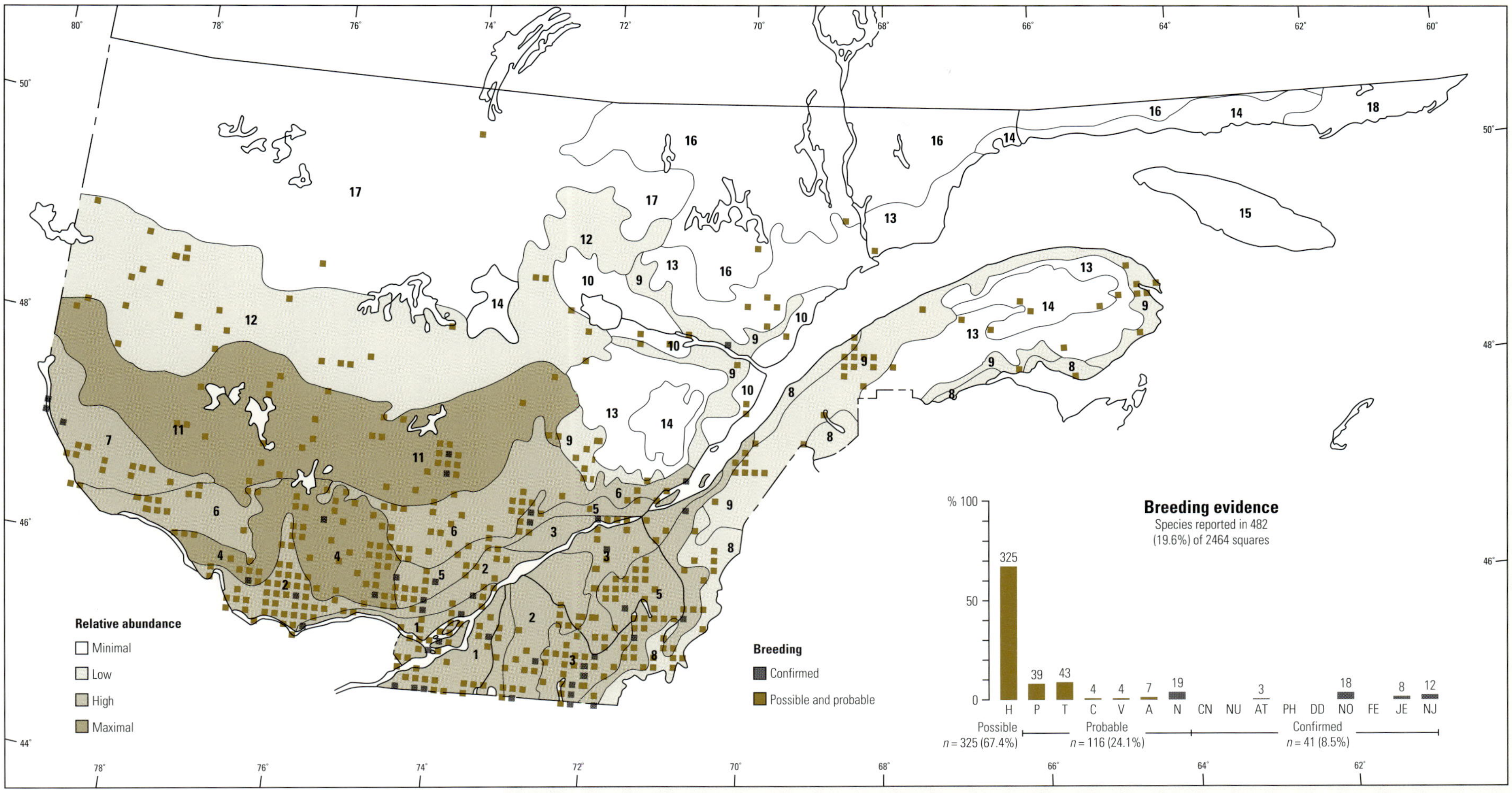

Species profile

Clutch size	usually 4 eggs (3–5)[3,5,6]
Laying interval	1 egg per day[6]
Start of incubation	often before last egg[11]
Incubation period	usually 18 days (15–18)[2,3,5,6,8]
Maturity at hatching	altricial
Care of young	by female and male[3,5,6,7]
Nestling period	24–28 days[3,6,7]
Age at first flight	24–28 days[3,6,7]
Dependence of young	probably until September[3,6]
Broods per year	1[3,6,7]
Breeding age	1 year[1,7]
Mating system	monogamy[2,6]
Duration of pair bond	usually for life[6]
Total length	40.5–49.5 cm[5]
Wing span	68.6–76.2 cm[4]
Weight	M 308 g (308–309) / F 266 g (250–284)[9]
Longevity record	9 years 11 months[10]

References:
1. DeGraaf and Rudis 1987
2. Ehrlich et al. 1988
3. Hoyt 1957
4. Terres 1980
5. Godfrey 1986
6. Short 1982
7. Bull and Meslow 1988
8. Kilham 1979a
9. Dunning 1984
10. Klimkiewicz and Futcher 1989
11. Bull and Jackson 1995

World breeding range

Chronology $n = 26$

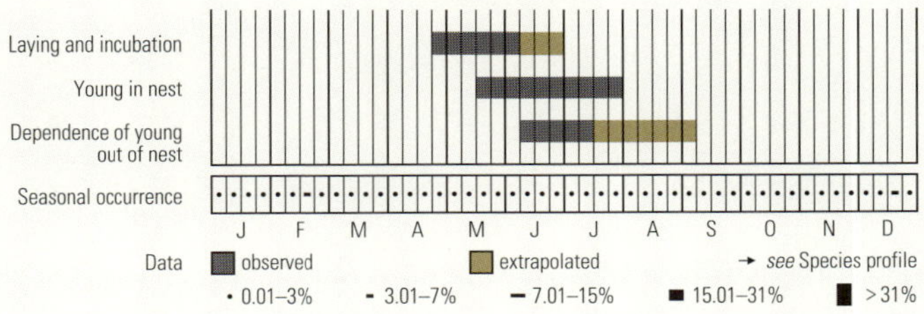

Frequency of occurrence by ecological region

Ecological regions** (forest regions)	Adequately covered squares	Squares in which reported	% of squares in which reported	Relative abundance (%)*
1. Sugar maple-hickory and sugar maple-basswood	97	47	48.5 ± 5.0	71.4
2. Sugar maple-basswood	153	73	47.7 ± 4.1	70.3
3. Sugar maple-basswood and sugar maple-yellow birch	101	47	46.5 ± 5.9	68.6
4. Sugar maple-yellow birch-basswood	56	38	67.9 ± 10.7	100.0
5. Sugar maple-yellow birch-beech and sugar maple-yellow birch-basswood	113	53	46.9 ± 5.1	69.1
6. Sugar maple-yellow birch-beech	71	43	60.6 ± 10.4	89.3
7. Sugar maple-yellow birch-balsam fir	19	11	57.9 ± 23.3	85.3
8. Sugar maple-yellow birch and balsam fir-yellow birch	47	11	23.4 ± 11.0	34.5
9. Balsam fir-yellow birch	87	21	24.1 ± 8.4	35.6
10. Balsam fir-white birch or balsam fir-red maple	46	5	10.9 ± 8.2	16.0
11. Yellow birch-balsam fir	52	35	67.3 ± 13.2	99.2
12. Balsam fir-white birch (west)	59	18	30.5 ± 12.1	45.0
13. Balsam fir-white birch (east)	76	7	9.2 ± 6.6	13.6
14. Balsam fir-black spruce	28	1	3.6 ± 8.3	5.3
15. White spruce-balsam fir and balsam fir-white spruce	22	0	0	0
16. Black spruce-balsam fir-moss	18	1	5.6 ± 13.5	8.2
17. Black spruce-moss	32	0	0	0
18.			*not analysed*	

* minimal: 0–29.9%; low: 30–59.9%; high: 60–89.9%; maximal: 90–100%

** Deciduous forest: 1–7; mixed forest: 8–11; coniferous forest: 12–17

A number of Pileated Woodpeckers come to search for insects in these dead elms killed by Dutch elm disease. Cap Tourmente National Wildlife Area, northeast of Québec City. *August 1991, Jean Gauthier.*

The nest hole is usually chiselled out of a dead tree or sometimes the desiccated top of a living tree where the heartwood has been attacked by fungi (Hoyt 1957; Conner et al. 1975). Conner et al. (1975) reported that most nests are within 50 m of water.

The pair excavates an impressive cavity with an inside diameter of some 20 cm at a height of anywhere from 4.5 to 25.5 m (Christy 1939; Hoyt 1957; Conner et al. 1975; McClelland 1979). Both the male and female do the excavating, though the male contributes much more time and effort (Kilham 1959, 1979a, 1983; Bull and Meslow 1988). When the hole is finished, no material is added, and only a few wood chips line the bottom of the nest (Hoyt 1957; Short 1982). Making the nest may take up to 30 days and work sometimes continues into the incubation period.

Clutch size is 3–5 eggs; they are incubated by both parents during the day but only by the male at night (Hoyt 1957; Short 1982; Bull and Meslow 1988). The male also broods the young at night after they hatch (Terres 1980).

Truslow (1967) recounted an interesting and probably unique occurrence: after the top of the nest tree had broken off in a storm at the level of the nest hole, the female was seen carrying the eggs in her bill to a new nest hole, one by one.

The parents feed the young by regurgitation and continue to do so for some time after they have left the nest (Hoyt 1957). Then the parents, each accompanied by one or two young, go off into the territory to forage. Each parent uses a separate area, thus increasing the pair's home range (Mellen et al. 1992). According to Hoyt (1957) and Short (1982), the young do not leave their parents' territory until sometime in the fall.

This woodpecker's year-round diet consists essentially of wood-boring insects in both the larval and adult forms (Beal 1911; Hoyt 1957; Beckwith and Bull 1985). In summer and fall, it also eats quantities of wild berries such as choke cherries (F. Hone, pers. comm.). In summer, it forages mainly on the ground, drilling into rotting stumps and logs to get at insects (Swallow et al. 1988). It has also been known to perch on anthills to feed (Beal 1911; Short 1982; Bull et al. 1986). Boreal carpenter ants are its favourite food, and it obtains them, mainly in winter, by digging holes of varying depth into infested timber, either standing or fallen (Bull and Meslow 1977; McClelland 1979; Mannan et al. 1980).

The Pileated Woodpecker was reported in 19.6% of Atlas squares. Breeding was confirmed in only 8.5% of those, whereas breeding confirmation rates for other woodpeckers were 20–34%. The bird's wariness and large territory make it hard to locate, but it is easier to find after hatching because of the peeping of the young. Moreover, as it is a generally sedentary species, unconfirmed reports probably do indicate breeding in many cases.

This woodpecker's territory includes tall trees with a diameter of at least 30 cm. Nest holes and foraging sites are all located within its territory. Cap Tourmente National Wildlife Area, northeast of Québec City. *January 1994, Stéphane Turgeon.*

Distribution

The Pileated Woodpecker is found in all forested parts of Canada and the United States. It has been reported in just about every region of southern Québec, but is most frequently associated with the mature mixed or deciduous forests of the lower half of the Atlas area. The extensive Eastern white pine and red pine stands of the Ottawa Valley and Témiscamingue are also much favoured by this species. It is much less common, however, on the North Shore and in the Gaspé, Lac-Saint-Jean, and Abitibi regions, where there are not as many mature deciduous stands. It has been sighted as far east as Anticosti Island in 1937 (Ouellet 1969) and at Port-Cartier on the North Shore in 1990 (G. Cyr, pers. comm.), but never in the Magdalen Islands. The easternmost Atlas records from the North Shore are from the Baie-Comeau area. There, in 1991, breeding was finally confirmed in the vicinity of the Outardes-3 Reservoir. Only one sighting, from Lake Mistassini, was reported above 50° N. The species is less common in the northerly boreal forests, where there are fewer stands of tall trees, but its presence in the Forestville, Outardes-3, Godbout, and Port-Cartier areas suggest that, as in Ontario, its range extends as far as the southern edge of the taiga, where good-sized trees can still be found in some valleys.

The bird winters within its breeding range, but when winters are harsh it may withdraw from the northernmost reaches (Godfrey 1986). In the mountains of Vermont, it sometimes leaves higher elevations for the shelter of the valleys (Norse 1985).

History and trends

By the first decade of this century, the Pileated Woodpecker had disappeared from many areas of northeastern North America, including southern Québec where it had once been common (Dionne 1906; Mousley 1916a; Taverner 1922). In the 19th century, its numbers were depleted by market hunting, together with logging and forest fires, which devastated its habitat. In fact, in the last century, the Pileated Woodpecker was considered a game bird, probably on account of its large size. It was hunted intensively and sold each fall in the markets of a number of towns, including Montréal (Wintle 1896; Sutton 1930).

In Ontario, in response to efforts to control forest fires, the ageing of stands, and changes to hunting regulations, the species began to reappear in the 1920s; two decades later, significant increases were reported (Dance 1987). In Québec, observers have been reporting increasing numbers since the mid-1950s (Larivée 1989). According to Christy (1939), the Pileated Woodpecker's ability to adapt to habitat affected by human activity is what enables it to recover in this way. Though the species was originally associated with extensive mature forests, the deforestation that caused it so much harm forced it to frequent smaller woods close to towns and villages.

BBS data for 1980–89 show no particular trend for the Pileated Woodpecker's Québec population.

The Pileated Woodpecker is highly beneficial to our forests, consuming huge quantities of wood-boring insects that would otherwise damage the trees. The numerous holes it makes can also be used by many other species of birds, mammals, and reptiles (Bull and Meslow 1977; Peterson and Gauthier 1985). Theirs are practically the only woodpecker holes big enough to be used by Wood Ducks, Common Goldeneyes, and Barrow's Goldeneyes, as well. Unfortunately, dead trees are still very often removed to halt the spread of wood-boring insects and the diseases that are the bane of foresters and maple syrup producers.

Benoît Limoges and Josée Tardif

Olive-sided Flycatcher

Moucherolle à côtés olive

Contopus borealis

Christiane Girard

THE OLIVE-SIDED FLYCATCHER IS A TYPICAL SPECIES OF CONIFEROUS AND mixed woodlands. Its presence is generally heralded by the familiar resounding three-syllable song, *quick, three beers*, and its habit of perching at the top of tall dead trees also facilitates detection. This migrant breeder, arriving in spring from South America, is widespread throughout Québec south of the 52nd parallel, but is not abundant anywhere. It is one of our largest flycatchers and also the one we know the least about.

The Olive-sided Flycatcher often perches clearly in view at the top of a tall snag, where it gives its characteristic call at all times of the day. From this vantage point, it searches for insects and darts out to snatch them on the wing. Mount Albert, Gaspé Conservation Park. *July 1991, Yves Aubry.*

Habitat and behaviour

The Olive-sided Flycatcher appears to be the forest-dwelling equivalent of the Eastern Kingbird. Like the latter, it frequents fairly open habitats with perches providing a commanding view of its surroundings. The availability of suitable perches appears to be a key factor in habitat selection. This species generally perches in snags, or standing dead trees, but is also seen in the bare tops or dead branches of living trees (Godfrey 1986). Like the Eastern Kingbird, the Olive-sided Flycatcher captures its prey almost exclusively on the wing (Peterson and Fichtel 1992), waiting on its perch until it detects an insect and then sallying forth to snatch it up in its bill. Its diet is composed almost entirely of insects, including bees and flying ants (Bent 1942; Terres 1980). It mainly frequents coniferous and mixed forests, most often near water (Bent 1942; Godfrey 1986). Favourite haunts include burns; the edges of logged areas, clearings, and peatlands; beaver ponds; and the wooded shores of rivers, lakes, and streams with standing dead trees (Bent 1942; Godfrey 1986). It is often found at higher elevations, particularly in southern Québec and the United States, no doubt because the coniferous habitats it favours are more prevalent there (Fichtel 1985c; Cheskey 1987; DeGraaf and Rudis 1987; Peterson 1988h; Evers 1991a).

The Olive-sided Flycatcher calls from dawn to dusk, and is often the first bird to be heard in the morning and the last one to be heard at night (Bent 1942). Nest sites are always well spaced (Bent 1942; Todd 1963). When the nest is threatened, the parents become very noisy and agitated, to the point where they may attack human intruders (Bent 1942; Terres 1980).

The well-camouflaged nest, most often built in a conifer but occasionally in a deciduous tree, is usually located well out on a horizontal branch 2–15 m up (Harrison 1975; Harrison 1978). It is made of twigs, lichen, rootlets, conifer needles, and mosses (Harrison 1978; Godfrey 1986).

Very little is known about the species' breeding biology. The female appears to incubate the eggs alone (Ehrlich et al. 1988), but both parents feed the young (Bent 1942).

The Olive-sided Flycatcher is easy to detect because of its loud, ringing call, which carries long distances, and its conspicuous perches. Breeding is difficult to confirm, however, and the best way is to spot adults carrying food or fecal sacs (Cheskey 1987).

Distribution

The Olive-sided Flycatcher was reported throughout the entire Atlas area. However, it is closely associated with conifers, whether in the coniferous forest, mixed forest or in deciduous woodland with scattered patches of conifers. Although it is widely distributed throughout southern Québec, it is less prevalent in the Montréal and Montérégie regions, where open habitats and deciduous forests predominate. It also appears to be less abundant from Abitibi to Lake Saint-Jean, but these populations may have been underestimated because many of the squares surveyed were in cleared areas rather than the forests preferred by the species. The Olive-sided Flycatcher is also uncommon in the western part of the Eastern Townships, where it is found mainly at higher elevations, but becomes more common in the eastern part of this region as well as north of the Ottawa Valley.

Although reports of confirmed breeding were rare, they were obtained in almost all regions, which suggests that the Olive-sided Flycatcher nests in most parts of southern Québec.

The species' range in western Québec has been documented as least as far north as the Eastmain and Opinaca rivers in the James Bay region (Bordage and Aubry 1982; Bordage 1985b; Tremblay and Belles-Isles 1991). In the east, the species' range extends east of the Québec border to a portion of Labrador (Godfrey 1986).

History and trends

At the turn of the century, the Olive-sided Flycatcher was generally considered scarce (Dionne 1889, 1906; Wintle 1896) or uncommon (Taverner 1922). This probably reflects the paucity of data on the species at the time rather than its actual rarity. In Ontario and Vermont, the species' distribution is not believed to have changed significantly during the past century (Fichtel 1985c; Cheskey 1987). Breeding populations, however, declined significantly in British Columbia between 1966 and 1979, and some declines probably also occurred during this period in Québec and New England (Robbins et al. 1986). The Olive-sided Flycatcher may be more abundant today in the Maritimes than during previous centuries, when the unexploited closed-canopied forests provided fewer favourable habitats for the species (Erskine 1992). This could hold true for Québec as well.

Clearcutting may reduce suitable habitat since fewer isolated trees are left standing. In general, however, logging helps to create suitable habitat such as forest edges and clearings (Fichtel 1985c; Erskine 1992). Insect epidemics and small forest fires also appear to be beneficial for the same reason. Extensive forest disturbances, however, are clearly detrimental to the species and the deforestation of the Montréal region may explain why nesting individuals are so rare in this area (Ouellet 1974a).

RENÉ SÉGUIN

This flycatcher usually builds its nest on a conifer branch in mixed or coniferous woodland, often near water. Martin-Valin ZEC, north of the Saguenay River. *August 1991, Jean Gauthier.*

Notes on territory

Although the species is considered territorial (Bent 1942), there is almost no information on this aspect of its biology. The Atlas data show that, in Québec, territories may be occupied from at least the first week of June to the end of July. In 1979, densities of 2.5 males/10 ha were observed north of Beauport, in a regenerating deciduous forest (Erskine 1984). During surveys in the area around the Nottaway, Broadback and Rupert rivers, the highest densities of nesting pairs were found in jack pine stands, shoreline scrub, and riparian ecotones in coniferous forest (Cons. G. G. – GREBE 1992a). In Ontario, local densities of up to 0.1 pair/10 ha have been recorded (Cheskey 1987).

Olive-sided Flycatcher

Breeding distribution

1984–1989

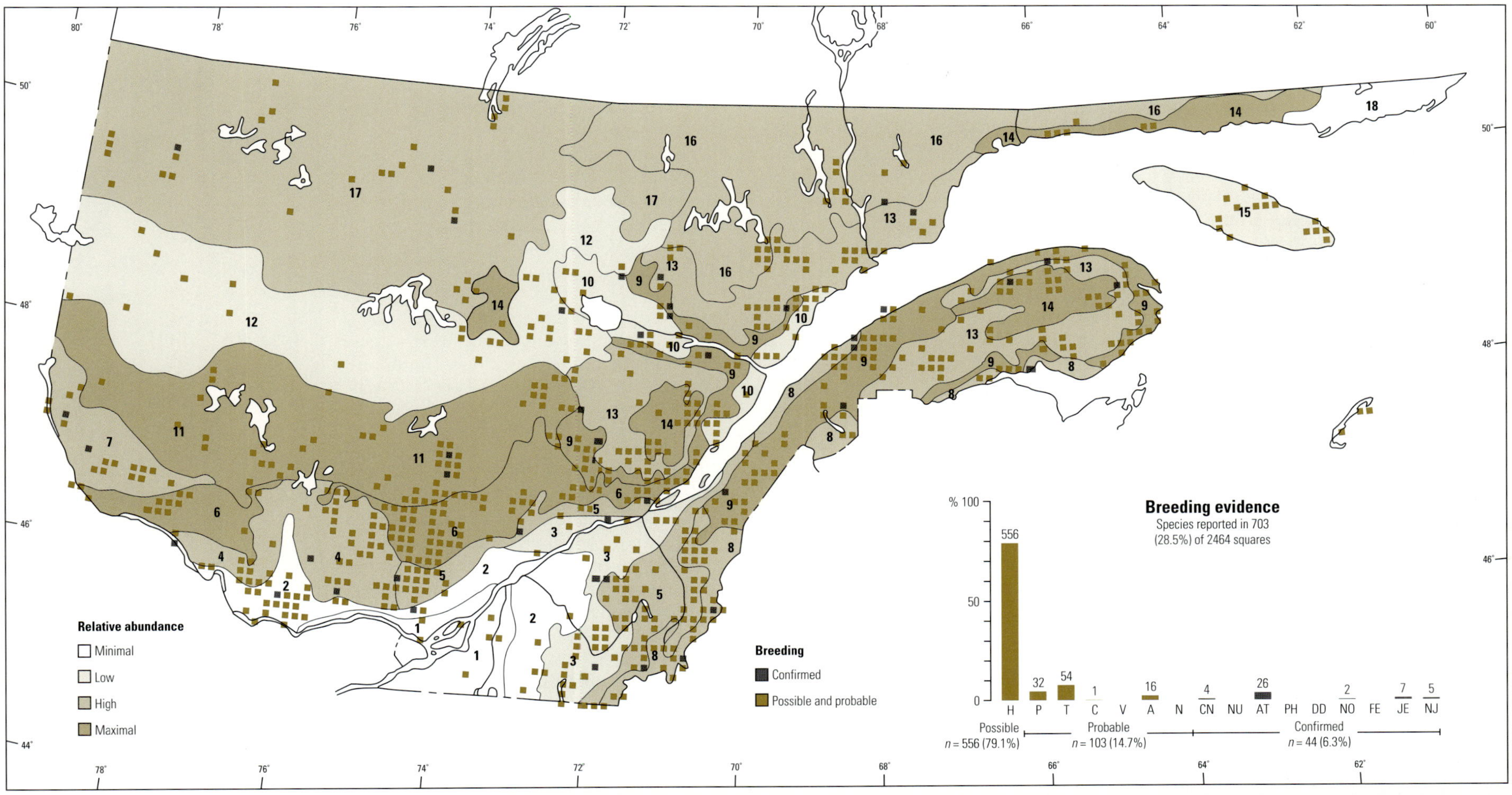

Species profile

Clutch size	usually 3–4 eggs (2–4)[1,2,3,5]
Laying interval	?
Start of incubation	?
Incubation period	14–17 days[1,2,3,5]
Maturity at hatching	altricial
Care of young	by female and male[1,5]
Nestling period	23 days[3,5]
Age at first flight	23 days or more[7]
Dependence of young	?
Broods per year	1[3]
Breeding age	1 year[3]
Mating system	monogamy[1]
Duration of pair bond	?
Total length	18.0–19.8 cm[2]
Wing span	?
Weight	32.1 g (26.7–42.2)[4]
Longevity record	7 years 1 month[6]

References:
1. Ehrlich et al. 1988
2. Godfrey 1986
3. DeGraaf and Rudis 1987
4. Dunning 1984
5. Bent 1942
6. Klimkiewicz and Futcher 1989
7. according to nestling period

Chronology $n = 6$

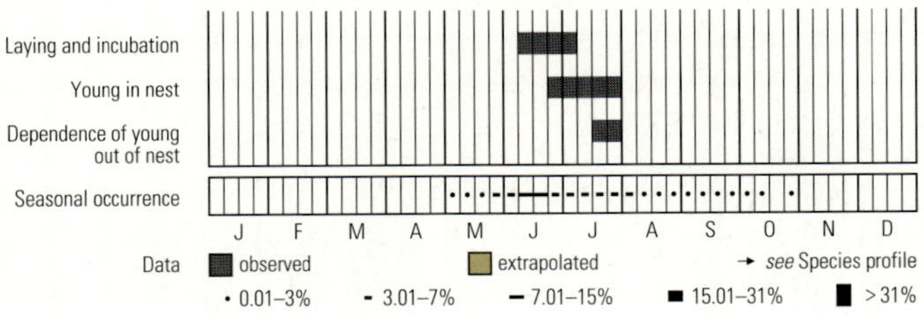

World breeding range

Frequency of occurrence by ecological region

Ecological regions** (forest regions)	Adequately covered squares	Squares in which reported	% of squares in which reported	Relative abundance (%)*
1. Sugar maple-hickory and sugar maple-basswood	97	5	5.2 ± 2.5	6.3
2. Sugar maple-basswood	153	30	19.6 ± 3.3	23.9
3. Sugar maple-basswood and sugar maple-yellow birch	101	36	35.6 ± 5.7	43.4
4. Sugar maple-yellow birch-basswood	56	33	58.9 ± 11.3	71.7
5. Sugar maple-yellow birch-beech and sugar maple-yellow birch-basswood	113	56	49.6 ± 5.1	60.3
6. Sugar maple-yellow birch-beech	71	58	81.7 ± 8.4	99.4
7. Sugar maple-yellow birch-balsam fir	19	14	73.7 ± 21.0	89.7
8. Sugar maple-yellow birch and balsam fir-yellow birch	47	27	57.4 ± 12.7	69.9
9. Balsam fir-yellow birch	87	65	74.7 ± 8.5	91.0
10. Balsam fir-white birch or balsam fir-red maple	46	19	41.3 ± 12.3	50.3
11. Yellow birch-balsam fir	52	40	76.9 ± 11.9	93.6
12. Balsam fir-white birch (west)	59	19	32.2 ± 12.3	39.2
13. Balsam fir-white birch (east)	76	54	71.1 ± 10.0	86.5
14. Balsam fir-black spruce	28	23	82.1 ± 15.2	100.0
15. White spruce-balsam fir and balsam fir-white spruce	22	10	45.5 ± 21.4	55.3
16. Black spruce-balsam fir-moss	18	12	66.7 ± 24.8	81.2
17. Black spruce-moss	32	16	50.0 ± 18.9	60.9
18.	*not analysed*			

* minimal: 0–29.9%; low: 30–59.9%; high: 60–89.9%; maximal: 90–100%

** Deciduous forest: 1–7; mixed forest: 8–11; coniferous forest: 12–17

Eastern Wood-Pewee

Pioui de l'Est

Contopus virens

Christiane Girard

THE EASTERN WOOD-PEWEE WOULD OFTEN BE OVERLOOKED WERE IT NOT for its song, which it performs from the pre-dawn hours to past sunset. Its plaintive, clearly whistled *pee-a-wee* can be heard at all times of the day, evoking its name.

Despite its co-operativeness in advertising its presence with its song, the Eastern Wood-Pewee, though common, is difficult to spot. When it returns from its wintering grounds in northern South America, this small, greenish bird perches just beneath the canopy of tall trees, where it is hidden by the already luxuriant foliage. The pewee is a late migrant and arrives on its breeding grounds when many other forest-dwelling birds have already started nesting.

The Eastern Wood-Pewee is easily confused with other flycatchers and is best identified by its song. In addition, it has white wing bars and generally remains still when perched, which allows it to be distinguished from the tail-bobbing Eastern Phoebe, which lacks wing bars.

Habitat and behaviour

Chiefly associated with openings and margins of deciduous and mixed forests (Hespenheide 1971), the Eastern Wood-Pewee nests in oaks, maples, and elms (Graber et al. 1974), or other large deciduous trees. Although several species of flycatchers may be sympatric, sometimes occupying the same area (Hespenheide 1971; Johnston 1971), they each select different microhabitats, which minimizes interspecific conflicts (Johnston 1971). Within a forest, each flycatcher occupies a different layer of vegetation and is further differentiated by its diet, the size of its prey, and its hunting techniques. The Eastern Wood-Pewee perches in dead trees more often than other flycatchers (Via 1979), foraging primarily from dead twigs on the outer periphery, generally around 11 m from the ground (Johnston 1971). Its diet is composed mainly of small insects: flies, beetles, and homopterans (Johnston 1971).

The males begin singing as soon as they arrive in spring. For the next three weeks, the number of singing males gradually increases to a peak and then declines, with individuals singing less continuously as the nesting season progresses (Smith, W. J. 1988). Pair formation occurs when the birds arrive on the breeding grounds or perhaps even during spring migration (Graber et al. 1974). The female has identical plumage to the male but is much more reclusive. She usually builds the nest at the end of a horizontal branch 2–21 m up; often a dead branch is used (Peck and James 1987). The small nest, covered with lichens bound together by strands of spider webs, can be easily confused with a knob on a dead branch and is frequently overlooked even by the most experienced birders. The eggs are incubated for approximately two weeks, and the young leave the nest 14–18 days after hatching.

Distribution

The Eastern Wood-Pewee breeds mainly in the deciduous forests of southern Québec, but is also found in mixed forests in the Appalachians (Eastern Québec Uplands) and the Chaleur Bay area. Breeding was confirmed from the Ottawa Valley to the eastern tip of the Gaspé Peninsula. The species was also observed in the Abitibi–Témiscamingue, Saguenay–Lac-Saint-Jean, and Upper North Shore regions and on the Magdalen Islands. Although the Eastern Wood-Pewee is mainly associated with deciduous forests, the northern limit of its range corresponds to the transition zone between mixed and pure coniferous forests.

Although the Eastern Wood-Pewee is common in southern Québec, it does not seem to be very abundant anywhere. Darveau et al. (1990) ranked it ninth in abundance among 30 species surveyed at different counting stations in eight sugar maple stands in the southern part of the Bois-Francs region, between Victoriaville and Plessisville; it was less common than the Least Flycatcher, Red-eyed Vireo, and Black-throated Blue Warbler.

In Canada, the species' breeding range extends roughly from southeastern Saskatchewan in the west to the Maritimes in the east; south of the border, it is found in almost the entire eastern United States from the eastern Dakotas to Texas (AOU 1983).

History and trends

At the turn of the century, the Eastern Wood-Pewee was considered a common summer resident in the Montréal and Québec City regions by Wintle (1896) and Dionne (1906) respectively. The species, which is associated with the mature deciduous forests of southern Québec, was probably adversely affected by the extensive clearing of forests for agriculture and urban development in the 19th century, which significantly reduced the total area of available habitat (Ouellet 1974a).

Populations in Québec appear to have declined slightly since 1970 (Cyr and Larivée 1993a). However, since the species is still present throughout its breeding range, the situation is not yet considered to be alarming. Webb et al. (1977), in a long-term study on avian communities in the deciduous forests of New Hampshire, observed that the Eastern Wood-Pewee does not appear to be overly affected by selective cutting. However, Forman et al. (1976) believe that significantly reducing the size of the forest islands preserved during logging operations may be detrimental to the Eastern Wood-Pewee because it usually does not frequent woodlots smaller than 2 ha.

GUY LEMIEUX AND LINE CHOINIÈRE

Notes on territory

The Eastern Wood-Pewee appears to be very intolerant of any Red-winged Blackbirds, grackles, jays, and squirrels that invade its territory (Bent 1942; Nice 1961 in Graber et al. 1974). It energetically defends a territory ranging from 0.6 ha to 1.3 ha (Fawver 1947; Calef 1953 in Graber et al. 1974), and its home range may be 4 ha or more (Odum and Kuenzler 1955).

In the Mirabel region, Cyr (1974) reported an average density of 6.1 territorial males/10 ha in sugar maple stands and 1.7 males/10 ha in Eastern hemlock stands. The average size of a pair's territory in the same region was 0.71 ha in sugar maple stands and 0.46 ha in hemlock stands. These figures are similar to McNeil's (1964 in Cyr 1974) findings, that territories averaged 0.60 ha.

The Eastern Wood-Pewee breeds mainly in mixed and deciduous forests, where it is associated with forest openings and margins. It often builds its nest at the end of a dead branch of a large deciduous tree. Gowan Road, Herdman, southwest of Montréal. *June 1991, Pierre Bannon.*

Eastern Wood-Pewee

Breeding distribution
1984–1989

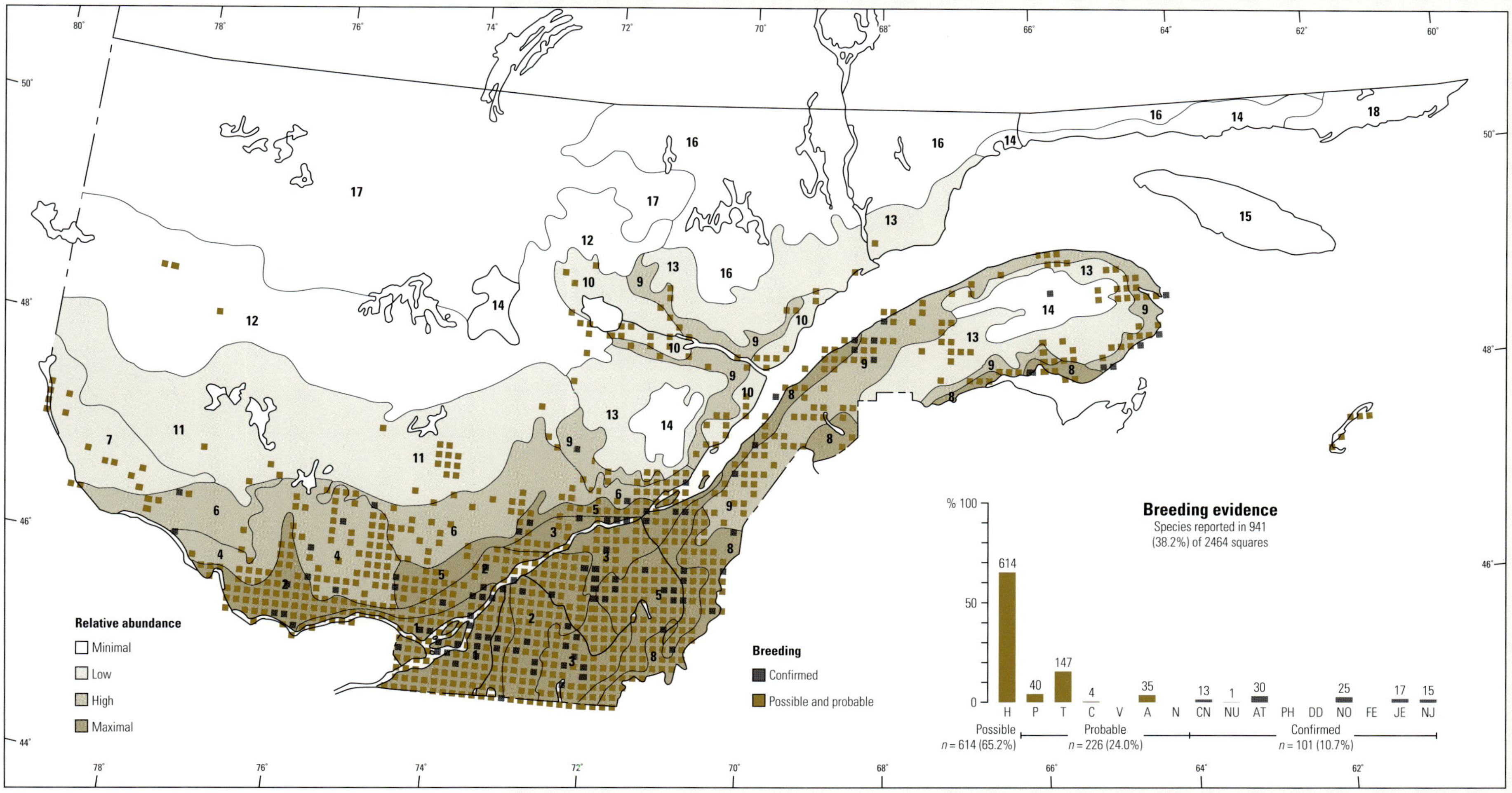

Species profile

Clutch size	usually 3 eggs (2–4)[1,2,3,5,7]
Laying interval	?
Start of incubation	?
Incubation period	12–13 days[1,3,5,7]
Maturity at hatching	altricial
Care of young	by female and male[5,7]
Nestling period	14–18 days[3,5,7]
Age at first flight	14–18 days or more[9]
Dependence of young	?
Broods per year	1, 2 possible[6,7]
Breeding age	?
Mating system	monogamy[7]
Duration of pair bond	?
Total length	15–17 cm[1,3]
Wing span	?
Weight	14.1 g (10.4–18.2)[4]
Longevity record	7 years 11 months[8]

References:
1. Godfrey 1986
2. Peck and James 1987
3. Terres 1980
4. Dunning 1984
5. Harrison 1978
6. DeGraaf and Rudis 1987
7. Ehrlich et al. 1988
8. Clapp et al. 1983
9. according to nestling period

World breeding range

Chronology n = 44

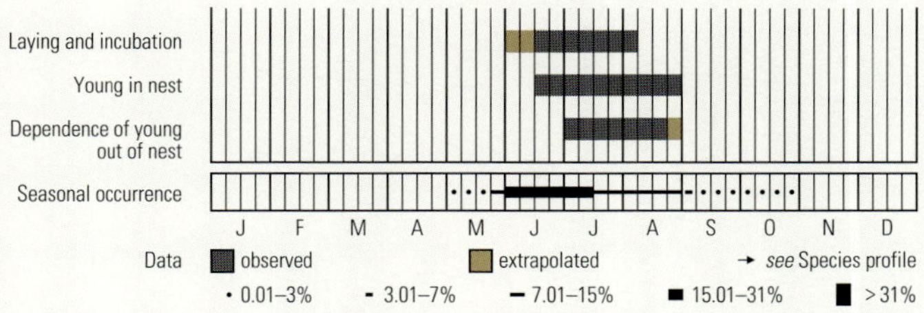

Frequency of occurrence by ecological region

Ecological regions** (forest regions)	Adequately covered squares	Squares in which reported	% of squares in which reported	Relative abundance (%)*
1. Sugar maple-hickory and sugar maple-basswood	97	97	100.0 ± 0.5	100.0
2. Sugar maple-basswood	153	147	96.1 ± 1.8	96.1
3. Sugar maple-basswood and sugar maple-yellow birch	101	96	95.0 ± 2.8	95.0
4. Sugar maple-yellow birch-basswood	56	49	87.5 ± 7.9	87.5
5. Sugar maple-yellow birch-beech and sugar maple-yellow birch-basswood	113	104	92.0 ± 3.0	92.0
6. Sugar maple-yellow birch-beech	71	52	73.2 ± 9.5	73.2
7. Sugar maple-yellow birch-balsam fir	19	10	52.6 ± 23.5	52.6
8. Sugar maple-yellow birch and balsam fir-yellow birch	47	44	93.6 ± 6.8	93.6
9. Balsam fir-yellow birch	87	61	70.1 ± 8.9	70.1
10. Balsam fir-white birch or balsam fir-red maple	46	24	52.2 ± 12.5	52.2
11. Yellow birch-balsam fir	52	18	34.6 ± 13.4	34.6
12. Balsam fir-white birch (west)	59	7	11.9 ± 8.8	11.9
13. Balsam fir-white birch (east)	76	26	34.2 ± 10.4	34.2
14. Balsam fir-black spruce	28	4	14.3 ± 14.1	14.3
15. White spruce-balsam fir and balsam fir-white spruce	22	0	0	0
16. Black spruce-balsam fir-moss	18	0	0	0
17. Black spruce-moss	32	0	0	0
18.			*not analysed*	

* minimal: 0–29.9%; low: 30–59.9%; high: 60–89.9%; maximal: 90–100%

** Deciduous forest: 1–7; mixed forest: 8–11; coniferous forest: 12–17

Yellow-bellied Flycatcher

Moucherolle à ventre jaune

Empidonax flaviventris

Christiane Girard

THE YELLOW-BELLIED FLYCATCHER IS THE EASIEST *EMPIDONAX* FLYCATCHER in Québec to identify, but one of the most difficult to find. Its yellow throat and belly distinguish it from the other flycatchers.

The Yellow-bellied Flycatcher is heard more than it is seen. Rather than trying to penetrate the shady, inaccessible habitat of this "child of the shadows" (Bent 1942), one should lure the bird into the open by imitating the call of a bird in distress (Bent 1942; Terres 1980). It is a common migrant breeder in the boreal forest, but is uncommon and local in the deciduous woodlands of the southern part of the province.

Habitat and behaviour

The Yellow-bellied Flycatcher is partial to dense, wet boreal forest, peatlands, and other damp coniferous habitat (Brunton and Crins 1975). It also frequents mixed woodland consisting of conifers, birch, and mountain-ash (Bent 1942). Such habitat, which is relatively inaccessible, is usually carpeted with a thick layer of sphagnum and other mosses, strewn with boulders and fallen trees. Although drier shady pine and spruce forests with dense shrubs are also used (Prescott 1987c), it nonetheless prefers cool humid mountainsides (DeGraaf and Rudis 1987).

Despite its preference for wet forest, the Yellow-bellied Flycatcher may also breed in thick undergrowth on mountain slopes. Mount Albert, Gaspé Conservation Park. *July 1991, Yves Aubry.*

In northern Québec, Yellow-bellied Flycatchers are found mainly in stands of intolerant hardwoods or jack pine and are less common or even absent from spruce stands and peatlands (Cons. G. G. – GREBE 1991a, 1992o).

Yellow-bellied Flycatchers perch low and catch insects flying close to the ground or in the lower part of the canopy. Studies by Beal (1912 *in* Bent 1942), who analysed 102 stomach contents, showed that 97% of the bird's diet is composed of insects, 46% of which are hymenopterans, such as flying ants, bees, and wasps. The Yellow-bellied eats more Hymenoptera than any other flycatcher and eats more spiders than most other birds. During cool, rainy periods in autumn, when insects are scarce, the Yellow-bellied Flycatcher will eat American mountain-ash berries, squeezing them in its bill to extract the pulp and then spitting out the skin (Stickney 1923).

The Yellow-bellied Flycatcher arrives in Québec late in spring. The male, singing at the top of a tree to proclaim his territory, may be difficult to observe, especially if silhouetted against the sun. More often than not it is only heard; its *killik* song resembles that of the Least Flycatcher, and its other vocalization, a descending *chu-wee* is like the Eastern Wood-Pewee's song. During the breeding period Yellow-bellied Flycatchers are not likely to be confused with wood-pewees, as the latter are found in drier habitat in deciduous woodland (DeGraaf and Rudis 1987).

The female builds a cup-shaped nest, near the ground in a fairly open area that is carpeted with sphagnum and other mosses (Walkinshaw and Henry 1957). She sometimes selects a cavity between the roots of a tree or in a spot beside a rock or hummock, or more rarely, against an uprooted tree (Brunton and Crins 1975). She begins by removing part of the moss to form a cavity and then builds what is a rather bulky nest, considering the size of the bird (Bent 1942). The nest is made of mosses, rootlets, and sedge, and is lined with fine grasses and pine needles (Walkinshaw and Henry 1957).

The female incubates alone and is not fed by the male. Walkinshaw and Henry (1957) observed that she spends about 81% of her time on the nest, for an average of about 32 minutes at a time, leaving for periods of about 9 minutes to feed. When the eggs hatch, both parents feed the young.

In southern Ontario, Hussell (1982) noted that even though Yellow-bellied Flycatchers arrive late on the nesting grounds, they leave early to fly south, having spent, at the most, 66 days on their breeding grounds.

Distribution

This flycatcher breeds mainly in the Low Subarctic and in the boreal regions of Canada and the northeastern United States. It winters in eastern Mexico and Central America (Graber et al. 1974).

The Yellow-bellied Flycatcher was reported in 24% (586/2464) of Atlas squares surveyed, and breeding was confirmed in only 8.5% (50 squares), distributed throughout most of southern Québec. There were few breeding confirmations in the agricultural areas and deciduous forest of the St. Lawrence Plain and the Ottawa Valley. The species prefers to nest in the damp coniferous and mixed forests of central Québec, the Lower St. Lawrence, and the Gaspé.

In northern Québec, possible breeding has been reported from the Great Whale River watershed (Cons. G. G. – GREBE 1991a).

History and trends

Yellow-bellied Flycatcher populations have probably not suffered much from the agricultural and urban development in the hardwood-dominated habitat of southern Québec. The species may, however, have been affected locally by drainage of the large peatlands in the St. Lawrence Plain. Dionne (1906) recorded it as being scarce around Québec City, and Wintle (1896) listed it as an uncommon summer resident in the Montréal region. The Atlas map shows a current distribution similar to that described at the beginning of the century.

Robbins et al. (1986) report an increase in breeding populations in eastern North America. BBS data for Québec for 1966–89 indicate a similar trend. Red spruce die-off, attributed to acid precipitation, may have had an impact on the species in the Adirondacks in New York State (Peterson 1988p). This, along with massive deforestation, spruce budworm epidemics, and loss of habitat on the wintering grounds (Diamond 1991), may have contributed to a decrease in the species' numbers.

BENOÎT LIMOGES

Notes on territory

The Yellow-bellied Flycatcher is a shy, elusive species that is easier to locate by ear than by sight. For this reason, little is known about its territorial behaviour. On its breeding grounds, however, its call, given from the top of a coniferous or deciduous tree, is heard regularly. Breeding territory was estimated at 1.01 ha in the swampy alder groves of the Black Sturgeon Lake region, north of Lake Superior in Ontario (Kendeigh 1947).

In a balsam fir-white birch stand on Mount Mégantic, estimated densities were 5.0 males/10 ha (Erskine 1984); 2.4 males/10 ha in a coniferous forest in Saint-Fabien-de-Panet, south of Montmagny (Erskine 1971a); and 4.7 males/10 ha in a fir-spruce forest infested with budworm at Asselin Creek in Témiscouata (Erskine 1980a). In Terra Nova National Park in Newfoundland, a density of 13.5 males/10 ha was noted in a young hardwood-dominated forest (Erskine 1984).

Yellow-bellied Flycatcher

Breeding distribution

1984–1989

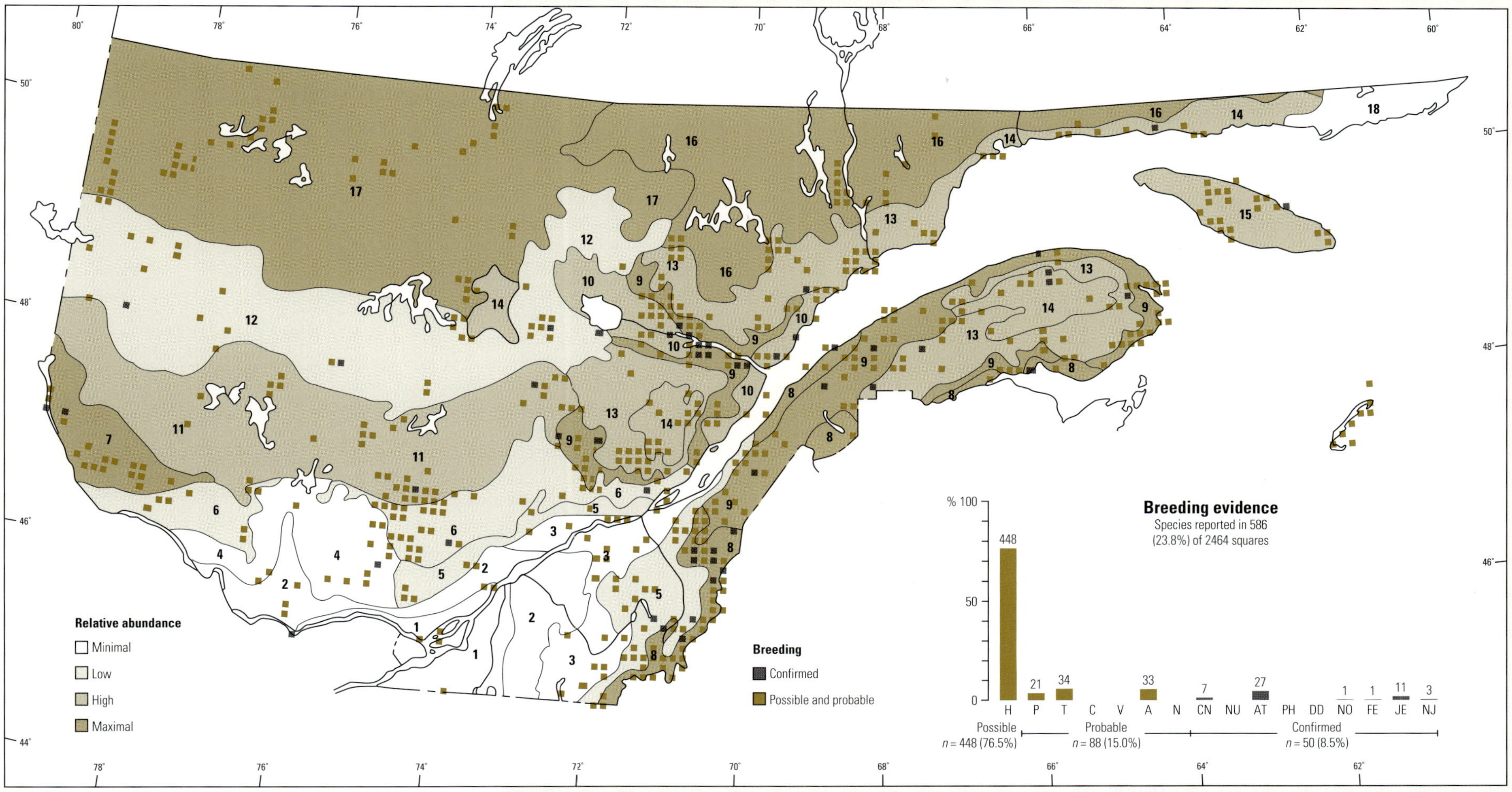

World breeding range

Species profile

Clutch size	usually 4 eggs (3–5)[1,2,3]
Laying interval	1 egg per day[3]
Start of incubation	?
Incubation period	usually 14 days (12–15)[1,2,3]
Maturity at hatching	altricial
Care of young	by female and male[1,3]
Nestling period	13–14 days[1,2,3]
Age at first flight	13–14 days or more[8]
Dependence of young	?
Broods per year	?
Breeding age	1 year[7]
Mating system	monogamy[1]
Duration of pair bond	?
Total length	12.7–14.7 cm[4]
Wing span	?
Weight	11.6 g (9.2–15.5)[6]
Longevity record	3 years 11 months[5]

References:
1. Ehrlich et al. 1988
2. Terres 1980
3. Walkinshaw and Henry 1957
4. Godfrey 1986
5. Clapp et al. 1983
6. Dunning 1984
7. DeGraaf and Rudis 1987
8. according to nestling period

Chronology $n = 11$

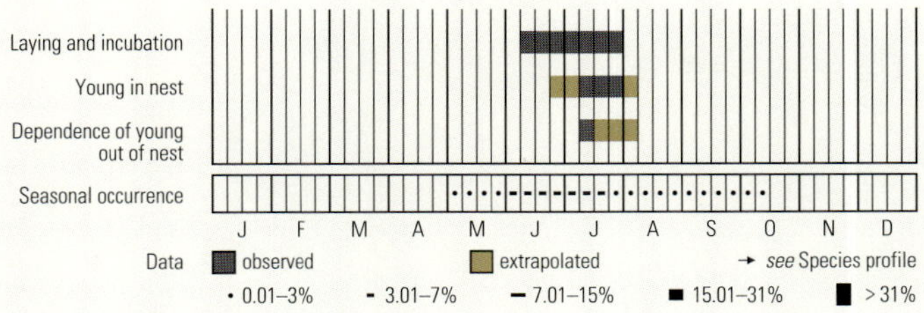

Frequency of occurrence by ecological region

Ecological regions** (forest regions)	Adequately covered squares	Squares in which reported	% of squares in which reported	Relative abundance (%)*
1. Sugar maple-hickory and sugar maple-basswood	97	4	4.1 ± 2.3	5.3
2. Sugar maple-basswood	153	7	4.6 ± 1.9	5.9
3. Sugar maple-basswood and sugar maple-yellow birch	101	23	22.8 ± 5.0	29.3
4. Sugar maple-yellow birch-basswood	56	11	19.6 ± 9.3	25.3
5. Sugar maple-yellow birch-beech and sugar maple-yellow birch-basswood	113	38	33.6 ± 4.9	43.2
6. Sugar maple-yellow birch-beech	71	29	40.8 ± 10.5	52.5
7. Sugar maple-yellow birch-balsam fir	19	14	73.7 ± 21.0	94.7
8. Sugar maple-yellow birch and balsam fir-yellow birch	47	36	76.6 ± 11.0	98.5
9. Balsam fir-yellow birch	87	64	73.6 ± 8.6	94.6
10. Balsam fir-white birch or balsam fir-red maple	46	22	47.8 ± 12.5	61.5
11. Yellow birch-balsam fir	52	26	50.0 ± 14.0	64.3
12. Balsam fir-white birch (west)	59	22	37.3 ± 12.7	47.9
13. Balsam fir-white birch (east)	76	46	60.5 ± 10.7	77.8
14. Balsam fir-black spruce	28	17	60.7 ± 18.9	78.1
15. White spruce-balsam fir and balsam fir-white spruce	22	14	63.6 ± 20.8	81.8
16. Black spruce-balsam fir-moss	18	14	77.8 ± 22.2	100.0
17. Black spruce-moss	32	23	71.9 ± 17.1	92.4
18.			*not analysed*	

* minimal: 0–29.9%; low: 30–59.9%; high: 60–89.9%; maximal: 90–100%

** Deciduous forest: 1–7; mixed forest: 8–11; coniferous forest: 12–17

Alder Flycatcher

Moucherolle des aulnes

Empidonax alnorum

Christiane Girard

The Alder Flycatcher is one of the most widespread, yet most elusive passerines in Québec. As in all *Empidonax* flycatchers, its plumage is drab, and its simple song is only heard for a short time in summer. It is a typical species of the shrubby wetlands found from the Central St. Lawrence Lowland to the boreal forest. Its winter range in South America is poorly known, although it is thought to winter generally farther south than the Willow Flycatcher (AOU 1983).

The recognition of the Alder Flycatcher as a species distinct from the Willow Flycatcher is largely the result of the work of R. C. Stein (1963). Before that, the two unrecognized taxa were treated jointly under the designation Traill's Flycatcher. Stein demonstrated that the two species differed not only by their song, which was already well known, but also by several other behavioural and ecological characteristics. The two taxa were formally recognized as distinct species by the American Ornithologists' Union in 1973 (AOU 1973). Since then, these two flycatchers have been extensively studied to establish their evolutionary (Seutin and Simon 1988a), behavioural (Prescott 1987a), and ecological (Zink and Fall 1981; Barlow and McGillivray 1983) affinities.

Habitat and behaviour

The Alder Flycatcher favours fairly wet habitats, particularly thickets and bushy shrubbery bordering lakes, rivers, streams, ponds, marshes, and peatlands, as well as damp cut-over or burned-over sites (Godfrey 1986). It is typically associated with alder, willow, elder, sumac, and viburnum (Bent 1942; DeGraaf and Rudis 1987). In the Eastmain Lowland, it is found mainly in shoreline scrub and tamarack peatlands (Cons. G. G. – GREBE 1992o).

The Alder Flycatcher builds its nest in a shrub or sapling (DeGraaf and Rudis 1987; Peck and James 1987), generally 0.6–0.9 m up (Bent 1942; Stein 1958, 1963; Peck and James 1987). The nest is made out of grass, plant stalks, plant down, moss, bark fibre, twigs, rootlets, yarn, feathers, pine needles, and spider webs and is lined with fine grasses and cattail down (Peck and James 1987). It is a rather loosely built structure with less downy material than the nest of the Willow Flycatcher (Stein 1958). Material typically dangling below the nest gives it an untidy appearance (Stein 1958; Godfrey 1986; Peck and James 1987).

Clutch size is usually 3 or 4 eggs, which are incubated for 12–14 days (DeGraaf and Rudis 1987). Both parents care for the young (Harrison 1978), which leave the nest at the age of 13–16 days (DeGraaf and Rudis 1987).

The Alder Flycatcher is thought to be monogamous (Verner and Willson 1966; Eckhardt 1976). Stein (1958) reported average territory sizes of 1.3 ha, 1.5 ha, and 3.0 ha for three populations in New York State. According to Prescott (1987a), Alder and Willow flycatchers defend exclusive territories. In Ontario, Peck and James (1987) reported cowbird parasitism in 14.9% of 47 Alder Flycatcher nests.

During the breeding season, the Alder Flycatcher is relatively difficult to observe because it spends most of its time in dense thickets where it nests and forages (Bent 1942). However, the male is often conspicuous early in the breeding season (generally the first half of June), as it energetically advertises its presence from a prominent perch, using its raspy *fee-bee-o* or *free-bee-o* song. It is heard most often at dawn and dusk, and less commonly during the day. Therefore, spotting the Alder Flycatcher greatly depends on the time of day and the time of year.

Notes on territory

Stein (1958) reported average territory sizes of 1.3 ha, 1.5 ha, and 3.0 ha in three Alder Flycatcher populations in New York State. According to Prescott (1987a), the Alder and Willow flycatchers defend exclusive territories.

In Québec's Gatineau Park, an average density of 0.65 pair/10 ha was observed in a mosaic habitat of forest and open areas (Dauphin 1985). In the area around the Nottaway, Broadback, and Rupert rivers, the highest densities were found in shoreline scrub, where the species was two to three times more abundant than in tamarack peatlands or recent burns (Cons. G. G. – GREBE 1992o). In Ontario, densities of 13.3–20.6 pairs/10 ha were estimated in regenerating coniferous forests 11–19 years after logging. In regenerating mixed forests, densities of 8.3 and 4.4 pairs/10 ha were noted 9 and 3 years after logging, respectively (Erskine 1984).

The Alder Flycatcher favours dense thickets and clumps of shrubs along water. Saint-Dominique-du-Rosaire, north of Amos, Abitibi. *August 1991, André-Guy Bernier.*

Alder Flycatcher

Breeding distribution
1984–1989

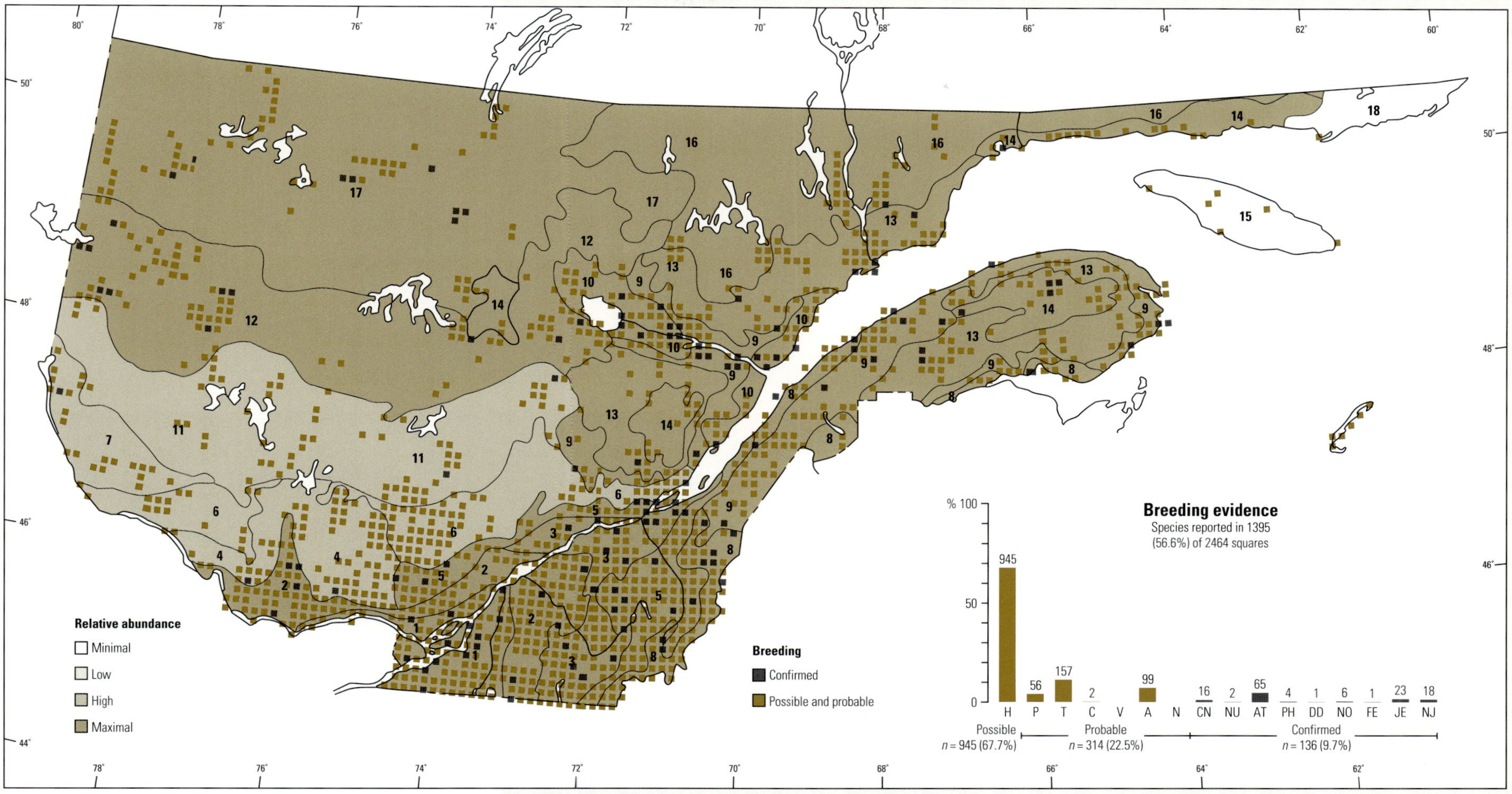

Species profile

Clutch size	usually 3–4 eggs (1–5)[1,2,4,6,7,8]
Laying interval	?
Start of incubation	?
Incubation period	12–14 days[2,3,6]
Maturity at hatching	altricial
Care of young	by female and male[7]
Nestling period	13–16 days[4,6]
Age at first flight	13–16 days or more[11]
Dependence of young	?
Broods per year	1[6]
Breeding age	1 year[6]
Mating system	probably monogamy[5,12]
Duration of pair bond	?
Total length	13–17 cm[2]
Wing span	19.7–22.9 cm[4]
Weight	M 12.7 g (10.0–14.4) / F 13.1 g (10.1–15.6)[10]
Longevity record	3 years 2 months[9]

References:
1. Bent 1942
2. Godfrey 1986
3. Stein 1958
4. Terres 1980
5. Verner and Willson 1966
6. DeGraaf and Rudis 1987
7. Harrison 1978
8. Peck and James 1987
9. Clapp et al. 1983
10. Dunning 1984
11. according to nestling period
12. Eckhardt 1976

World breeding range

Chronology *n* = 23

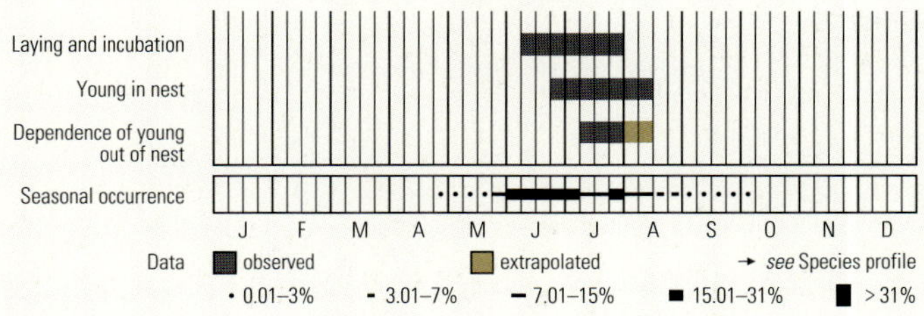

Frequency of occurrence by ecological region

Ecological regions** (forest regions)	Adequately covered squares	Squares in which reported	% of squares in which reported	Relative abundance (%)*
1. Sugar maple-hickory and sugar maple-basswood	97	86	88.7 ± 3.3	91.1
2. Sugar maple-basswood	153	138	90.2 ± 2.6	92.7
3. Sugar maple-basswood and sugar maple-yellow birch	101	96	95.0 ± 2.8	97.6
4. Sugar maple-yellow birch-basswood	56	46	82.1 ± 9.0	84.4
5. Sugar maple-yellow birch-beech and sugar maple-yellow birch-basswood	113	110	97.3 ± 2.0	100.0
6. Sugar maple-yellow birch-beech	71	58	81.7 ± 8.4	83.9
7. Sugar maple-yellow birch-balsam fir	19	15	78.9 ± 19.7	81.1
8. Sugar maple-yellow birch and balsam fir-yellow birch	47	45	95.7 ± 5.8	98.4
9. Balsam fir-yellow birch	87	80	92.0 ± 5.5	94.5
10. Balsam fir-white birch or balsam fir-red maple	46	44	95.7 ± 5.7	98.3
11. Yellow birch-balsam fir	52	43	82.7 ± 10.8	84.9
12. Balsam fir-white birch (west)	59	53	89.8 ± 8.2	92.3
13. Balsam fir-white birch (east)	76	68	89.5 ± 7.0	91.9
14. Balsam fir-black spruce	28	25	89.3 ± 12.6	91.7
15. White spruce-balsam fir and balsam fir-white spruce	22	5	22.7 ± 18.4	23.3
16. Black spruce-balsam fir-moss	18	17	94.4 ± 13.5	97.0
17. Black spruce-moss	32	31	96.9 ± 7.6	99.5
18.			*not analysed*	

* minimal: 0–29.9%; low: 30–59.9%; high: 60–89.9%; maximal: 90–100%

** Deciduous forest: 1–7; mixed forest: 8–11; coniferous forest: 12–17

This flycatcher, a typical species of shrub communities, frequents willow and alder thickets on the border of marshes. Gowan Road, Rockburn, south of Montréal. *June 1991, Pierre Bannon.*

The Alder Flycatcher was reported in 56.6% (1395/2464) of Atlas squares. Early in the season, the presence of a singing male is often the only breeding evidence obtainable. Lively chases involving two birds, usually a territorial male and an intruder, are also a fairly common sight. Breeding was recorded as probable in 22.5% (314/1395) of squares in which the species was reported, based mainly on the observation of territorial behaviour, while it was confirmed in only 9.7% (136/1395) of squares. Most breeding confirmations were based on records of adults carrying food for their young; this is seen mostly in July. Nests were found in only 2.4% (34/1395) of squares in which the species was reported.

Distribution

The Alder Flycatcher breeds in Alaska, all 10 Canadian provinces, and the northeastern United States (AOU 1983). It is found throughout southern Québec except on the Lower North Shore. According to David (1980a), it probably breeds as far north as the 54th parallel. Frequency of occurrence throughout its range in Québec is quite high, except on Anticosti Island, where it is uncommon.

Although the Alder Flycatcher's nest typically has an untidy look because of the material hanging below the cup, it is securely attached to a shrub. Sainte-Foy. *July 1966, Gaston Le Page.*

An adult Alder Flycatcher perched in a shrub. Saint-Majorique, near Drummondville. *July 1991, Jeanne Lehoux.*

History and trends

Before European settlement, the Alder Flycatcher was probably less common in southern Québec than it is today. Populations most likely began to increase when the forests of the St. Lawrence Plain were cleared for agriculture.

At the turn of the century, Dionne (1906) reported that the species was a summer resident in the province and was fairly common in the Québec City area. According to Wintle (1896), the species was scarce in the Montréal region, where it bred in Mount Royal Park. BBS data suggest that populations have increased over the last two decades in southern Québec in response to suitable habitat expansion resulting from the abandonment of farmland, a widespread trend throughout southern Québec except for the Montréal and Montérégie areas (M. Carignan, MAPAQ, unpubl. data).

There are two potential limiting factors on Alder Flycatcher populations in Québec. First, increased human activity could lead to the loss of suitable breeding habitat in the southern part of the breeding range. Second, the expansion of the Willow Flycatcher in southernmost Québec could also be detrimental. The Willow Flycatcher is currently expanding its breeding range eastward (Robbins et al. 1986) and northward (Barlow and McGillivray 1983; Prescott 1987a), and the breeding ranges of the two species now overlap. The Willow Flycatcher is more aggressive than its congener (Stein 1963; Prescott 1987a), putting the latter at a disadvantage in conflicts for nesting territories (Prescott 1987a). However, there is no indication at this time that Alder Flycatcher populations will decline significantly in the long term.

Gilles Seutin

This secretive member of the family Tyrannidae frequents a variety of habitats and likes to nest in shrubs next to wet, idle land. Cap Tourmente National Wildlife Area, northeast of Québec City. *August 1991, Yves Aubry.*

Willow Flycatcher

Moucherolle des saules

Empidonax traillii

Christiane Girard

THE WILLOW FLYCATCHER HAS BECOME AN INCREASINGLY COMMON migrant breeder in the Central St. Lawrence Lowland since the early 1980s and is still expanding its range in southern Québec. Unfortunately, the dynamics of this expansion, covering most of the northeast of the continent (Barlow and McGillivray 1983; Robbins et al. 1986), are not well understood, because this taxon has only recently been recognized as specifically distinct from the Alder Flycatcher (Stein 1963; AOU 1973). The Willow Flycatcher winters in Central America and the northern parts of South America (AOU 1983).

This typical *Empidonax* flycatcher is difficult, indeed impossible to distinguish visually from its congeners (Phillips et al. 1966; Seutin 1991b). Its song, a *fitz-bew*, with the accent on the first syllable rather than the second as in the Alder Flycatcher's *fee-bee-o*, is the best way of identifying it. It is also usually the first indication of its presence. The Willow Flycatcher is not a very vocal bird. It sings mainly at dawn and dusk but, even then, it may be difficult to hear because of the simultaneous presence in its preferred nesting habitats of more boisterous songbirds such as the Yellow Warbler, Common Yellowthroat, Red-winged Blackbird, and various sparrows.

Habitat and behaviour

The Willow Flycatcher likes relatively open habitats consisting of thickets interspersed with clearings (King 1955; Walkinshaw 1966b; Hespenheide 1971; Godfrey 1986; DeGraaf and Rudis 1987; Peck and James 1987). It will use dry or partially wet areas, but often prefers the former (King 1955; Stein 1958, 1963; Peck and James 1987; Sedgwick and Knopf 1992). In northeastern North America, including the Central St. Lawrence Lowland, it is most often found in drained wetlands and abandoned farmland. In contrast, Barlow and McGillivray (1983) reported that in Ontario it may inhabit mesic habitat, and in their view this has allowed it to expand its range in the central part of the province. The Willow Flycatcher is partial to willow, hawthorn, wild rose, and dogwood shrubs (Bent 1942; King 1955; Speirs 1985; Godfrey 1986; Taylor and Littlefield 1986; Peck and James 1987; Sanders and Flett 1989; Sedgwick and Knopf 1992).

The female is accompanied by the male when choosing the nest site (Ehrlich et al. 1988; Sedgwick and Knopf 1992). However, she builds the nest alone, which takes 5–7 days (Bent 1942; Verner and Willson 1969). The nest is a relatively compact structure, often with silky tendrils dangling from it. It incorporates a great variety of materials, both plant (grass, fibres, bark, rootlets, pine needles) and animal (hair, cobwebs), as well as paper or plastic. The inside is lined with fine grass, plant down, weed stems, hair, moss, and pine needles (Stein 1958; Peck and James 1987). The nest contains more downy material than that of the Alder Flycatcher (Stein 1958). At our latitudes, nests are generally placed 1–1.5 m up (Peck and James 1987), in the fork of a hawthorn (Peck and James 1987) or in a willow thicket (pers. obs.).

Clutch size is usually 3 or 4 eggs (Bent 1942; Terres 1980), which are incubated by the female alone for 12–15 days (Walkinshaw 1966b; Harrison 1978). The young are fed by both parents (Holcomb 1972; Ettinger and King 1980), and generally leave the nest after 11–14 days (Walkinshaw 1966b).

Although the Willow Flycatcher is considered to be monogamous (Verner and Willson 1969), polygyny has been documented in a few instances (Prescott 1986; Sedgwick and Knopf 1989). Males advertise their territories by singing from exposed perches. They will also actively chase intruders. Prescott (1987a) observed that territorial defence was directed not only at conspecifics, but also at the Alder Flycatcher. According to Prescott (1986), females defend a smaller area around the nest.

The territorial vocalization of this flycatcher is innate, whereas that of most passerines is learned (Kroodsma, D. E. 1984). Occasionally, females give the territorial song (Seutin 1987), which is exceptional in most passerines.

Brood parasitism by the Brown-headed Cowbird was reported in 5.3% of 94 nests in Michigan (Walkinshaw 1966b), 8% of 88 nests in Ohio and Nebraska (Holcomb 1972), and 26.8% of 71 nests in Ontario (Peck and James 1987).

The Willow Flycatcher was recorded in 4.6% (114/2464) of Atlas squares, though breeding was confirmed in only 12 squares.

Distribution

In North America, the Willow Flycatcher breeds across southern Canada, from British Columbia to Prince Edward Island (possibly in Nova Scotia), down to California and Georgia (AOU 1983; Erskine 1992). In southern Québec, it occurs mainly in the Central St. Lawrence Lowland, where suitable habitat is particularly abundant. It is also seen, though much less frequently, in the Ottawa Valley, Eastern Townships, and Chaudière–Appalaches region. It has also been reported in the Québec City area, the Lower St. Lawrence, and the Gaspé.

History and trends

The Willow Flycatcher was first reported in Québec near Melocheville in 1971 (David and Gosselin 1977b). Since then, the species has steadily expanded its range in the province. In 1983, the northernmost record was from Piedmont, in the Laurentians (David 1983). Then, in 1986, the species was seen at Cap-Tourmente in the Québec City area (Yank and Aubry 1986a), and in 1987 it was reported from the Gaspé Peninsula (Yank et al. 1987b, 1990b; Gosselin et al. 1991b). No other species except the House Finch (Paquin 1989) has expanded its range in Québec as far and as rapidly in historical time (Yank et al. 1987b). Not surprisingly, BBS data show a population increase between 1966 and 1989.

The first confirmed breeding record of the Willow Flycatcher in Québec dates from 22 June 1973, with the discovery of a nest at Ville Saint-Laurent on the territory of a singing male (QNRCP; M. Hendrick, pers. comm.). At that time this taxon was not yet recognized as a distinct species. Built in a hawthorn bush, the nest had pieces of plant fibre and other materials hanging from it. A second nest, containing 3 eggs, was found at Châteauguay-Station on 21 July 1985 (pers. obs.). Farther north, in the Cap Tourmente National Wildlife Area, a nest with 3 eggs was reported on 11 July 1987 (G. Lemelin, pers. comm.).

Notes on territory

The Willow Flycatcher uses its territory for breeding (Sedgwick and Knopf 1992), foraging (Frakes and Johnson 1982), and feeding its young (Prescott and Middleton 1988). Average territory size varies from 0.7–1.8 ha (Stein 1958; Walkinshaw 1966b). In Ontario scrubland, densities of 3.0–7.5 territorial males/10 ha have been reported (Clark 1991, 1992, 1993a).

Territorial defence is not only directed at conspecifics, but also at Alder Flycatchers (Prescott 1987a).

In the Central St. Lawrence Lowland, the Willow Flycatcher has expanded its breeding range with the proliferation of open shrub habitat. Alhough it prefers drier land, it will use thickets in swampy areas. La Prairie, near Montréal. *July 1991, Gaétan Duquette.*

The preferred habitat of the Willow Flycatcher in northeastern North America is, to a large degree, the product of human activity. Habitat expansion over the last half century seems to be the main reason for this species' recent advance toward the northeast. A similar explanation has been given for the range expansion of the Golden-winged Warbler (Gill 1980).

The presence of suitable habitat suggests that we can expect in the near future to see the Willow Flycatcher in greater numbers in the Lower St. Lawrence and the Gaspé Peninsula. The species may also soon be found in the Lake Saint-Jean Lowland, or along the northern reaches of the Saguenay, or in the Témiscamingue region.

GILLES SEUTIN

Willow Flycatcher

Breeding distribution
1984–1989

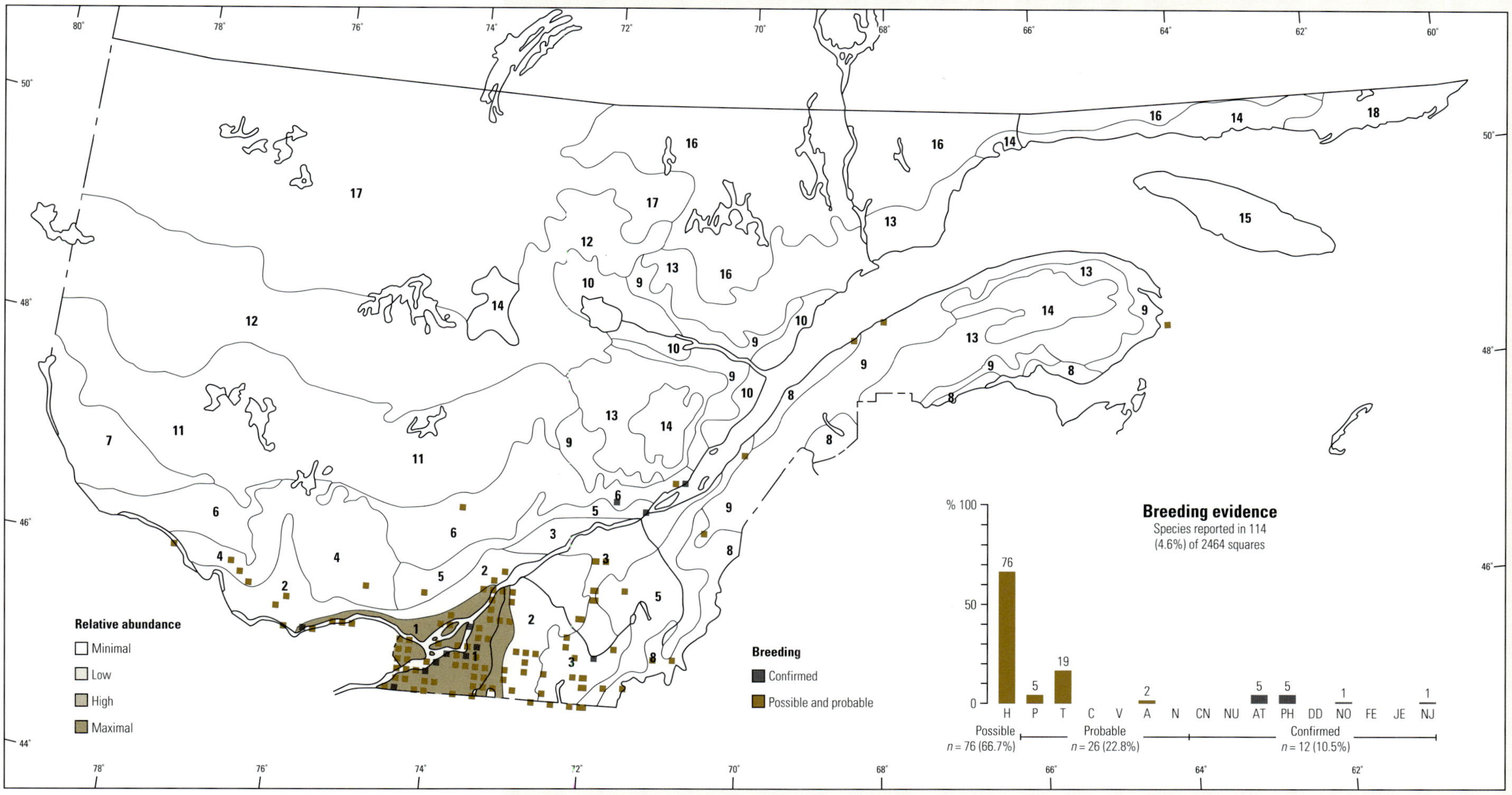

World breeding range

Species profile

Clutch size	usually 3–4 eggs (1–5)[1,2,3]
Laying interval	?
Start of incubation	?
Incubation period	12–15 days[1,2,3]
Maturity at hatching	altricial
Care of young	by female and male[1]
Nestling period	usually 11–14 days (11–19)[1,2,4,5,11]
Age at first flight	usually 11–14 days (11–19) or more[10]
Dependence of young	?
Broods per year	1[5]
Breeding age	1 year[5]
Mating system	monogamy, rarely polygyny[9,12,13]
Duration of pair bond	?
Total length	13–17 cm[6]
Wing span	?
Weight	M 13.1 g (12.0–15.7) / F 13.7 g (11.3–16.4)[7]
Longevity record	7 years[8]

References:
1. Holcomb 1972
2. Harrison 1978
3. Peck and James 1987
4. Terres 1980
5. DeGraaf and Rudis 1987
6. Godfrey 1986
7. Dunning 1984
8. Clapp et al. 1983
9. Verner and Willson 1969
10. according to nestling period
11. Walkinshaw 1966b
12. Prescott 1986
13. Sedgwick and Knopf 1989

Chronology n = 2

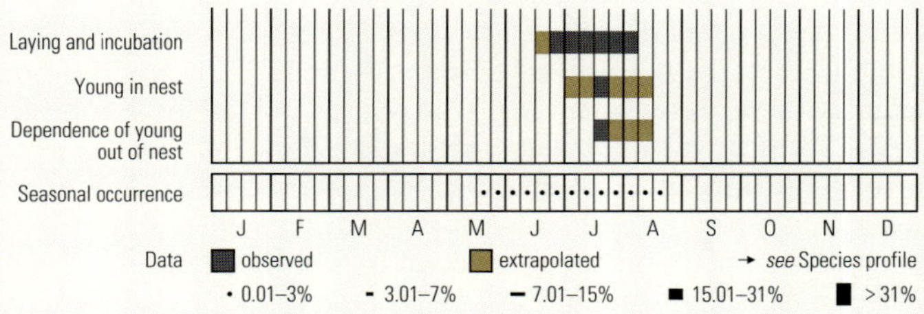

Frequency of occurrence by ecological region

Ecological regions** (forest regions)	Adequately covered squares	Squares in which reported	% of squares in which reported	Relative abundance (%)*
1. Sugar maple-hickory and sugar maple-basswood	97	51	52.6 ± 5.0	100.0
2. Sugar maple-basswood	153	16	10.5 ± 2.6	19.9
3. Sugar maple-basswood and sugar maple-yellow birch	101	15	14.9 ± 4.3	28.2
4. Sugar maple-yellow birch-basswood	56	3	5.4 ± 5.6	10.2
5. Sugar maple-yellow birch-beech and sugar maple-yellow birch-basswood	113	6	5.3 ± 2.5	10.1
6. Sugar maple-yellow birch-beech	71	4	5.6 ± 5.3	10.7
7. Sugar maple-yellow birch-balsam fir	19	0	0	0
8. Sugar maple-yellow birch and balsam fir-yellow birch	47	2	4.3 ± 5.8	8.1
9. Balsam fir-yellow birch	87	1	1.1 ± 2.5	2.2
10. Balsam fir-white birch or balsam fir-red maple	46	0	0	0
11. Yellow birch-balsam fir	52	0	0	0
12. Balsam fir-white birch (west)	59	0	0	0
13. Balsam fir-white birch (east)	76	0	0	0
14. Balsam fir-black spruce	28	0	0	0
15. White spruce-balsam fir and balsam fir-white spruce	22	0	0	0
16. Black spruce-balsam fir-moss	18	0	0	0
17. Black spruce-moss	32	0	0	0
18.		*not analysed*		

* minimal: 0–29.9%; low: 30–59.9%; high: 60–89.9%; maximal: 90–100%

** Deciduous forest: 1–7; mixed forest: 8–11; coniferous forest: 12–17

Least Flycatcher

Moucherolle tchébec

Empidonax minimus

Christiane Girard

THE LEAST FLYCATCHER IS THE SMALLEST OF THE TYRANNIDAE FOUND in Québec. In spite of its small size, it is just as aggressive as the other members of the family (Murphy 1983a). The Least Flycatcher has two distinctive traits. First of all, it tends to nest in loose colonies within which each pair has its own territory (Sherry and Holmes 1985), quite an unusual habit among forest passerines. Second, it hunts mainly by "hovering," picking its insect prey from a substrate, although it is capable of hawking on the wing like other flycatchers (Robinson and Holmes 1982). This distinctive hunting style means that its main competitor for food is the American Redstart (Sherry 1979).

The Least Flycatcher, an abundant species in hardwood and mixed forests, is nonetheless very vulnerable to maple dieback. This phenomenon particularly affects treetop foliage, opening up gaps in the canopy, and this is detrimental to the species. Appalachians. *Summer 1984, Jean Sylvain, Ministère des Ressources naturelles du Québec.*

Habitat and behaviour

This migrant breeder arrives from its wintering grounds in Mexico and Central America (Robbins et al. 1989b) in early May. It is generally considered monogamous (Holmes et al. 1979), though an incidence of polygyny has been reported (Briskie and Sealy 1987). The male establishes and defends the territory (Sherry 1979). It will use open areas with scattered trees and thin stands of timber, but prefers dense forest (Cyr 1975; Holmes and Sturges 1975). The female, which arrives within a week of the male, chooses the nest site with her mate, but builds the nest on her own (Holmes et al. 1979).

The nest, an open cup made of bark, twigs, leaf stalks, roots, and animal hair, is built in the crotch of the trunk or of a main limb of a deciduous tree or in a terminal fork. In the maple stands of the Bois-Francs region, the height of 94 nests ranged from 1 to 18 m above the ground, with an average of 8 m (Darveau et al. 1990, 1993).

The female usually lays 4 eggs (Darveau et al. 1990, 1993), more rarely 3, 5, or 6. She alone incubates the eggs for 12–14 days, generally leaving the nest only to feed, even though the male regularly feeds her while she is sitting. During incubation, it is relatively easy, more so than for other forest species, to pinpoint nests of the Least Flycatcher by locating the male, which often repeats its loud *che-bek* song near the nest, or by locating both parents through their call, usually a thin, nervous but readily identifiable *weet*. Parents are especially aggressive when defending the area around their nest against other birds or mammals, whether predatory or not. This may account for the virtual absence of nest parasitism by the Brown-headed Cowbird (Briskie and Sealy 1989a). Hatching is asynchronous over a period of about 26 hours for 3-egg clutches and about 46 hours for 5-egg clutches (Briskie and Sealy 1989b). In the Bois-Francs region, hatching peaks about 20 June, and the young are cared for in the nest for a further 12–16 days. The parents feed the young mainly on insects, chiefly butterflies and moths, flies, beetles, and hymenopterans (Darveau et al. 1990, 1993). The Least Flycatcher starts its southward migration as early as the end of July (Sealy and Biermann 1983), but some individuals stay until September (David 1980a).

The Least Flycatcher was the easiest of all the *Empidonax* flycatchers to confirm as a breeder for the Atlas. Because of its abundance, small territory, noisy behaviour, and the visibility of its nest, it is quite easy to locate and observe while it is building its nest and feeding its young.

Distribution

The Least Flycatcher is found throughout southern Québec and was reported in 57% (1403/2464) of Atlas squares, placing it among the 20 commonest species. It appears to be scarce, however, on Anticosti Island, where it was reported in only 2 squares. It is considered a possible breeder in the Magdalen Islands (Fradette 1992), though it was not seen there during the atlassing period.

The Least Flycatcher is abundant in the hardwood and mixed forests of the southern part of the province. In the maple stands of the Chaudière–Appalaches region, it is the dominant breeder, with a density of 17 pairs/10 ha (Darveau et al. 1992). It becomes less common toward the north, where conifers predominate. In northern areas, such as the Nottaway, Broadback, and Rupert river watersheds, it is found in pockets of poplar and birch (Cons. G. G. – GREBE 1992o). The northernmost sightings are from around the Eastmain River (Bordage and Aubry 1982; Tremblay and Belles-Isles 1991).

This flycatcher, which is quite easy to spot during the breeding season, prefers dense hardwood forest. Île Perrot, near Montréal. *August 1991, Gaétan Duquette.*

History and trends

A comparison of the early literature (Macoun and Macoun 1915; Taverner 1922) with current data shows that the distribution of the Least Flycatcher has not changed much over the past century.

Its numbers are bound to have declined, given the huge forested areas of southern Québec cleared for agriculture and urban development. Moreover, the Least Flycatcher is sensitive to habitat disturbance: of all birds, it is the most seriously affected by maple dieback (Darveau et al. 1990), and it has little tolerance for silvicultural practices such as thinning (DellaSala and Rabe 1987). Management of Québec maple stands for syrup production, by which American beech and yellow birch are eliminated for the most part, also seems to discourage it (Darveau et al. 1990), as this flycatcher is better suited to foraging on these tree species than on sugar maple (Robinson and Holmes 1984).

BBS data show that during the 1960s, 1970s, and 1980s, Least Flycatcher populations remained stable in Québec and throughout North America generally (Robbins et al. 1986; BBS). The species does not seem to be threatened by deforestation on its wintering grounds, where it seeks shrub cover (Diamond 1991). A limiting factor could, however, be forest fragmentation on its breeding grounds (Askins et al. 1990).

MARCEL DARVEAU

Notes on territory

The territory includes the nest and a considerable part of the feeding grounds. It is defended chiefly by the male against intruding conspecifics and often against the American Redstart. The female generally defends just an area within a 6-m radius of the nest. Territories vary between 0.012 and 0.53 ha (Kendeigh 1947; MacQueen 1950; Cyr 1974; Sherry 1979).

Densities of between 2.5 and 39.3 males/10 ha have been estimated for various maple stands in southern Québec (Erskine 1971a, 1976a, 1980a, 1984; Cyr 1974). The highest density (39.3 males/10 ha) was recorded in Saint-Fabien-de-Panet, south of Montmagny, in a sugar maple-beech-yellow birch forest (Erskine 1971a).

In the Nottaway, Broadback, and Rupert river watersheds, the Least Flycatcher appears to be twice as abundant in the ecotone between deciduous stands and bodies of water as it is in birch stands and three times as abundant as in aspen stands (Cons. G. G. – GREBE 1992o).

Least Flycatcher

Breeding distribution

1984–1989

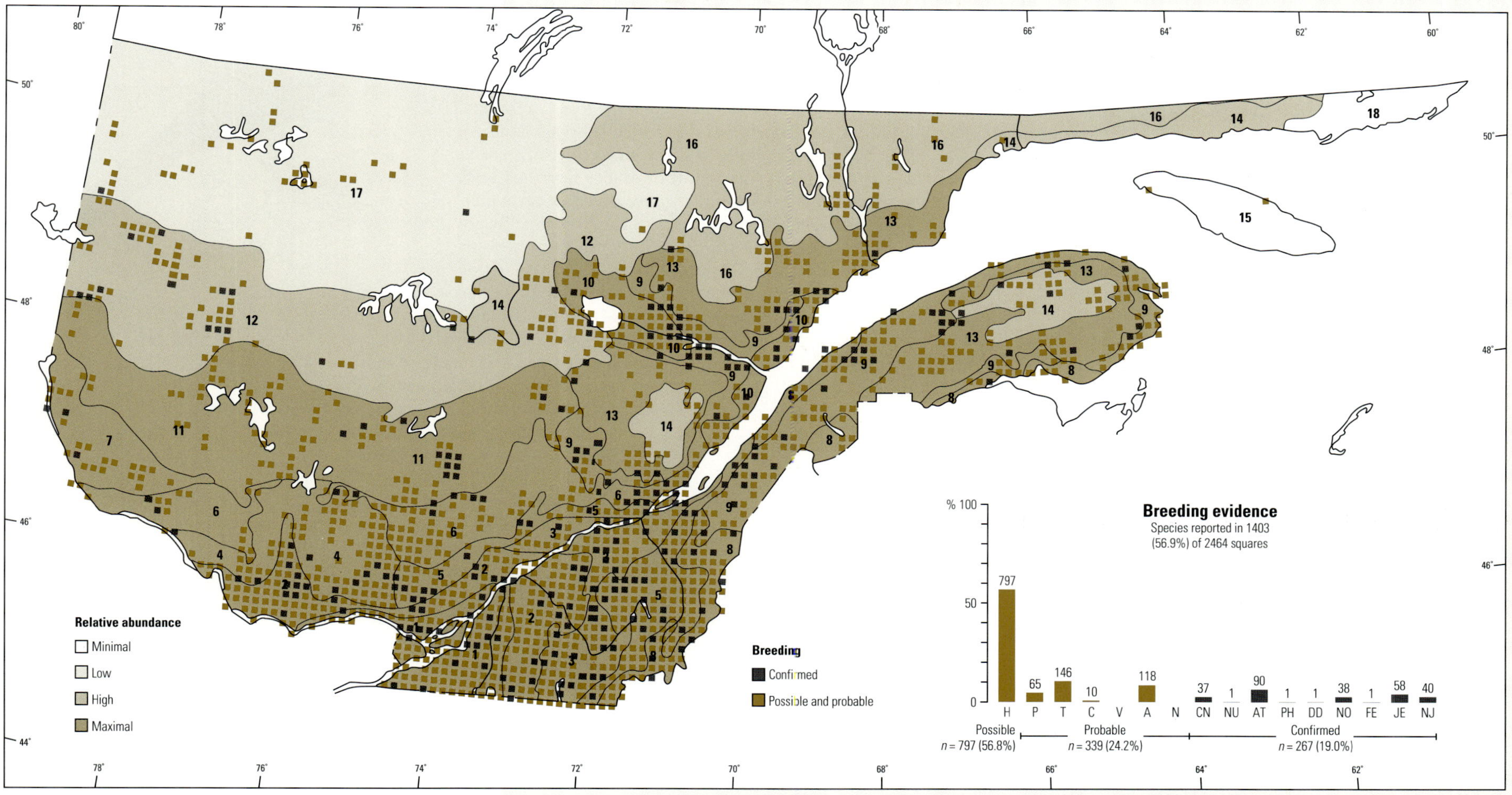

World breeding range

Species profile

Clutch size	usually 4 eggs (3–6)[1,2,3,4]
Laying interval	1 egg per day[6]
Start of incubation	with third or fourth egg[1,12]
Incubation period	12–15 days[2,3,4,5,11]
Maturity at hatching	altricial
Care of young	by female and male[1,2,3,6]
Nestling period	12–16 days[1,2,3,5,6]
Age at first flight	12–16 days or more[9]
Dependence of young	until 10–21 days after leaving nest[3,6,11]
Broods per year	1, sometimes 2[2,3,11]
Breeding age	1 year[5]
Mating system	monogamy, rarely polygyny[10]
Duration of pair bond	?
Total length	12.5–14.0 cm[4]
Wing span	19.0–21.6 cm[3]
Weight	10.3 g (8.2–14.9)[8]
Longevity record	8 years[7]

References:
1. Harrison 1978
2. Ehrlich et al. 1988
3. Terres 1980
4. Godfrey 1986
5. DeGraaf and Rudis 1987
6. MacQueen 1950
7. Klimkiewicz and Futcher 1989
8. Dunning 1984
9. according to nestling period
10. Briskie and Sealy 1987
11. Briskie 1994
12. Briskie and Sealy 1989b

Chronology *n* = 183

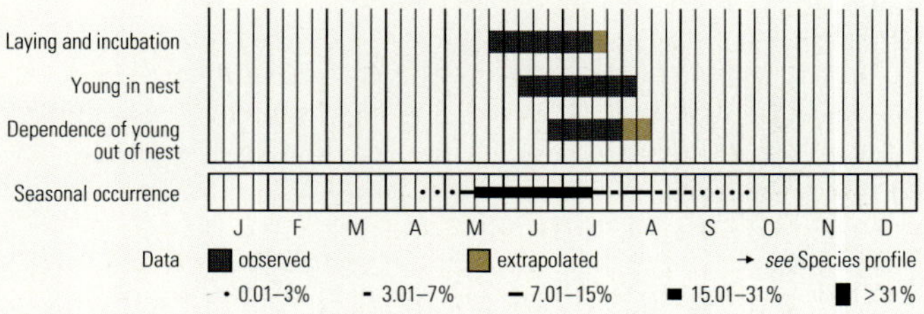

Frequency of occurrence by ecological region

Ecological regions** (forest regions)	Adequately covered squares	Squares in which reported	% of squares in which reported	Relative abundance (%)*
1. Sugar maple-hickory and sugar maple-basswood	97	93	95.9 ± 2.3	95.9
2. Sugar maple-basswood	153	149	97.4 ± 1.5	97.4
3. Sugar maple-basswood and sugar maple-yellow birch	101	96	95.0 ± 2.8	95.0
4. Sugar maple-yellow birch-basswood	56	52	92.9 ± 6.3	92.9
5. Sugar maple-yellow birch-beech and sugar maple-yellow birch-basswood	113	109	96.5 ± 2.2	96.5
6. Sugar maple-yellow birch-beech	71	64	90.1 ± 6.6	90.1
7. Sugar maple-yellow birch-balsam fir	19	19	100.0 ± 2.6	100.0
8. Sugar maple-yellow birch and balsam fir-yellow birch	47	45	95.7 ± 5.8	95.7
9. Balsam fir-yellow birch	87	86	98.9 ± 2.5	98.9
10. Balsam fir-white birch or balsam fir-red maple	46	46	100.0 ± 1.1	100.0
11. Yellow birch-balsam fir	52	49	94.2 ± 7.0	94.2
12. Balsam fir-white birch (west)	59	50	84.7 ± 9.7	84.7
13. Balsam fir-white birch (east)	76	69	90.8 ± 6.6	90.8
14. Balsam fir-black spruce	28	21	75.0 ± 17.0	75.0
15. White spruce-balsam fir and balsam fir-white spruce	22	2	9.1 ± 13.3	9.1
16. Black spruce-balsam fir-moss	18	16	88.9 ± 17.4	88.9
17. Black spruce-moss	32	17	53.1 ± 18.9	53.1
18.			*not analysed*	

* minimal: 0–29.9%; low: 30–59.9%; high: 60–89.9%; maximal: 90–100%

** Deciduous forest: 1–7; mixed forest: 8–11; coniferous forest: 12–17

Eastern Phoebe

Moucherolle phébi

Sayornis phoebe

Christiane Girard

The Eastern Phoebe is the first flycatcher to arrive in the spring, often in early April and sometimes even in March. This migrant breeder is found virtually throughout southern Québec, nesting in a large variety of habitats; it is most common, however, in the deciduous forest zone. Its nest is easy to spot, typically located in such places as the rafters of old buildings and under small bridges. The name "phoebe" comes from its song, *fee-be*, with the accent on the first syllable. About the size of a sparrow, it can be distinguished from other flycatchers of the same size mainly by its call, the lack of obvious wing bars, and its habit of wagging its tail.

Fall migration may last until late October. The Eastern Phoebe winters mainly in the southeastern United States, from Virginia and the Gulf states down to Mexico (Godfrey 1986; Root 1988).

Habitat and behaviour

The phoebe is the least discriminating of the flycatchers when it comes to habitat; it is often found in immature mixed and deciduous woodlands, and sometimes in coniferous forests as well (Peck and James 1987). Nests in remoter areas, far from human habitation, are sometimes placed in rocky ravines, sheltered by overhangs. This was originally one of the bird's breeding sites before European settlement (Tyler 1942b). It can also be found deep in the forest, although it prefers more open areas, to which it is well adapted (Via 1979). It frequents wooded suburbs, farmland, roadsides, fencerows, golf courses, marshes and swamps, and gardens. It usually breeds near water where it can feed on the insects it catches while flying over the water, and where it finds mud to build its nest. The nest, which is quite large for the size of the bird, is built almost exclusively by the female (Smith 1942; Weeks 1979; Stokes and Stokes 1989b) in 3–13 days and is located 1–3 m above the ground. Nests on horizontal surfaces such as beams and posts are cup-shaped structures made of mud, moss, roots, and grasses, whereas nests on vertical surfaces such as bridge pilings or rocky walls are semicircular, like the Barn Swallow's nest, and made mainly of mud.

The phoebe has adapted particularly well to man-made structures: old buildings, sheds, porches, and especially the underside of bridges. In Ontario, 96% of the 1349 nests found were built on man-made structures (Peck and James 1987).

Although the phoebe closely associates with humans because of its choice of nesting sites, it is not a bothersome bird. In fact, not only is it a quiet bird that catches insects all summer, it does not mind being observed and has become quite accustomed to people, even tolerating them quite close to its nest. These traits make it a very popular bird with country dwellers. Interestingly, Audubon became the pioneer of bird banding when he put the first band on an Eastern Phoebe (Audubon 1840; Tyler 1942b).

Clutch size is 2–6 (average 5) white eggs, laid a day apart (Weeks 1994). There is sometimes a period of several days (Harrison 1975) between the time the last egg is laid and incubation begins, though it usually starts on the same day (Conrad and Robertson 1992). The female alone incubates the eggs for 13–20 days (Godfrey 1986). Two broods are usually produced between early May and mid-August, and on very rare occasions, three, as has been observed in Ontario (Peck and James 1987). Incubation time seems to vary with ambient

The Eastern Phoebe often nests on man-made structures, such as on the undersides of small bridges. Alberton Road, Saint-Bernard-de-Lacolle, south of Saint-Jean-sur-Richelieu. *July 1991, Pierre Bannon.*

Notes on territory

In this monogamous species (Ehrlich et al. 1988), territories are usually widely dispersed. Territorial behaviour is particularly evident in the spring as the male establishes his territory, and when the pair starts a second clutch. Interspecific conflicts are quite rare and end quickly with the pair expelling the intruder from their area (Smith 1942; Stokes and Stokes 1989b). In Québec, in the Mirabel area, Cyr (1974) estimated territory size at 0.4 ha in a sugar maple stand, whereas McNeil (1964 in Cyr 1974) had assessed it at 1.2 ha in an unidentified forested area. In the United States, a pair's territory was estimated at 0.3 ha in wet woodland in Illinois, and at 1.3 ha and 2.9 ha for two other pairs nesting in buildings in Kansas (DeGraaf and Rudis 1987).

In sugar maple stands in Québec, density has been estimated between 1.1 and 3.3 pairs/10 ha in the Mirabel region (Cyr 1974). In the United States, densities of 4.96 pairs/10 ha in optimal habitat in Illinois, 1.75 pairs/10 ha in mixed agricultural habitat in Maryland, and 0.15 pair/10 ha in a mosaic of forest and fields in Maryland have been recorded (DeGraaf and Rudis 1987).

Eastern Phoebe

Breeding distribution
1984–1989

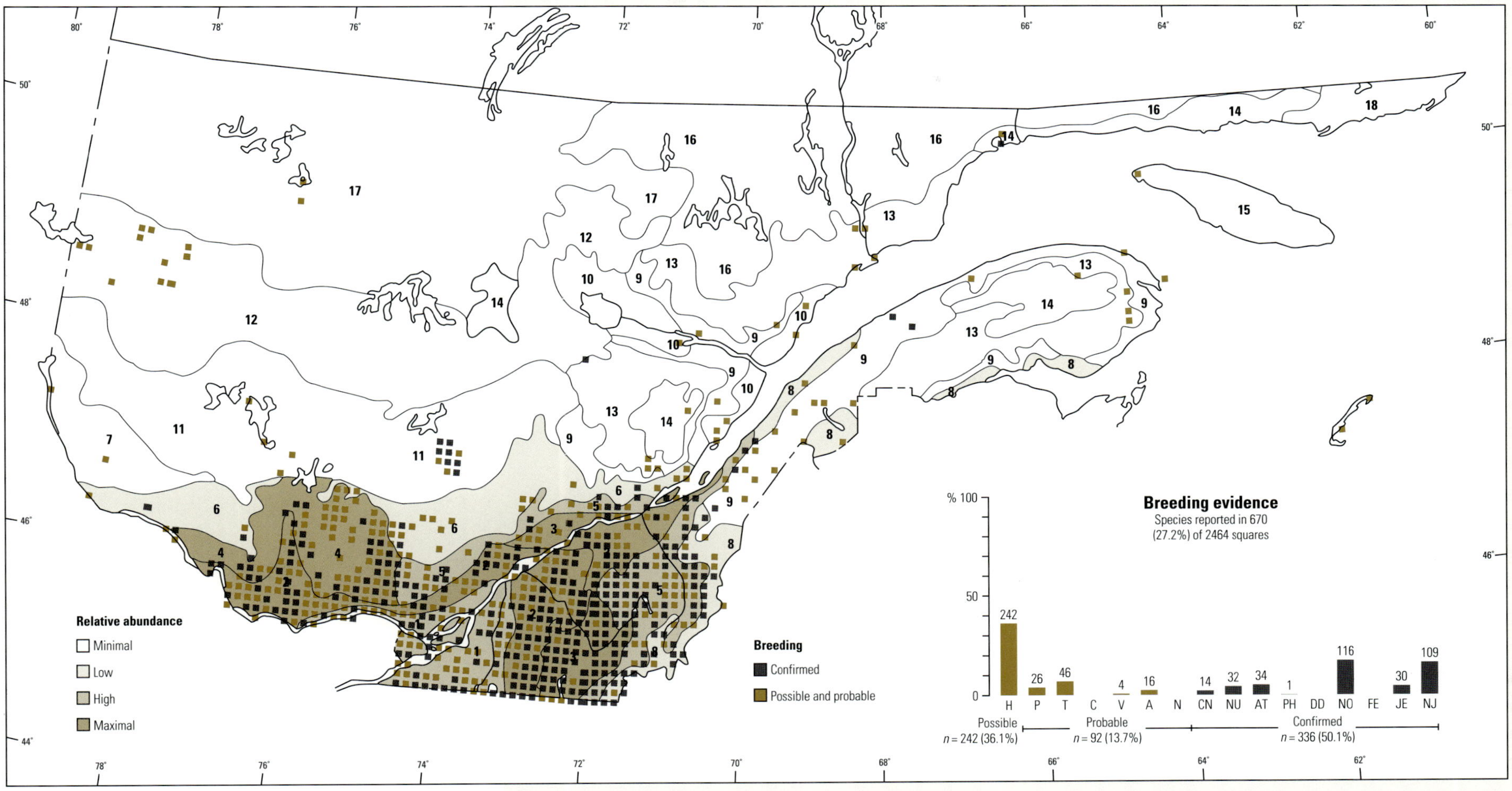

World breeding range

Species profile

Clutch size	usually 5 eggs (2–6), more than 6 = dump nest[1,2,3,11,15]
Laying interval	1 egg per day[5,10,12]
Start of incubation	with last or penultimate egg[5,12]
Incubation period	usually 16 days (13–20)[2,3,4,6,11]
Maturity at hatching	altricial
Care of young	by female and male[1,3,5]
Nestling period	usually 15–17 days (15–20)[1,3,5,12,15]
Age at first flight	usually 15–17 days (15–20) or more[14]
Dependence of young	until 2–3 weeks after leaving nest[3,5,10]
Broods per year	usually 2 (2–3)[1,2,3,7]
Breeding age	1 year[13]
Mating system	monogamy[1]
Duration of pair bond	often more than 1 season[15]
Total length	16–18 cm[6]
Wing span	?
Weight	19.8 g (11.4–24.4)[8]
Longevity record	9 years 1 month[9]

References:
1. Ehrlich et al. 1988
2. Harrison 1975
3. Harrison 1978
4. Peck and James 1987
5. Smith 1942
6. Godfrey 1986
7. Weeks 1978
8. Dunning 1984
9. Clapp et al. 1983
10. Stokes and Stokes 1989b
11. Faanes 1980
12. Conrad and Robertson 1992
13. DeGraaf and Rudis 1987
14. according to nestling period
15. Weeks 1994

Chronology $n = 291$

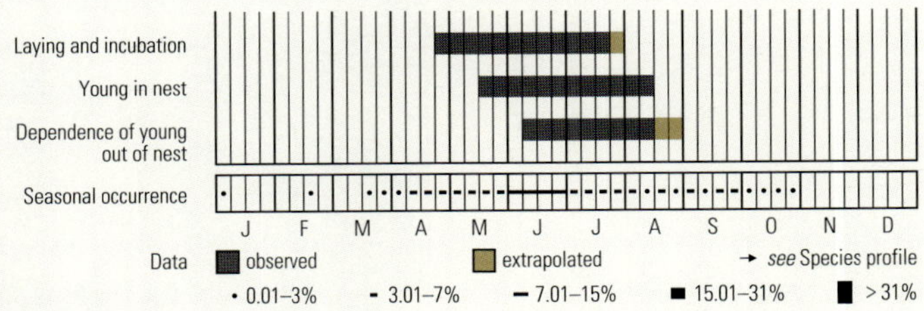

Frequency of occurrence by ecological region

Ecological regions** (forest regions)	Adequately covered squares	Squares in which reported	% of squares in which reported	Relative abundance (%)*
1. Sugar maple-hickory and sugar maple-basswood	97	72	74.2 ± 4.4	78.4
2. Sugar maple-basswood	153	139	90.8 ± 2.5	96.0
3. Sugar maple-basswood and sugar maple-yellow birch	101	88	87.1 ± 4.1	92.1
4. Sugar maple-yellow birch-basswood	56	53	94.6 ± 5.6	100.0
5. Sugar maple-yellow birch-beech and sugar maple-yellow birch-basswood	113	92	81.4 ± 4.1	86.0
6. Sugar maple-yellow birch-beech	71	29	40.8 ± 10.5	43.2
7. Sugar maple-yellow birch-balsam fir	19	2	10.5 ± 15.5	11.1
8. Sugar maple-yellow birch and balsam fir-yellow birch	47	16	34.0 ± 12.2	36.0
9. Balsam fir-yellow birch	87	9	10.3 ± 6.1	10.9
10. Balsam fir-white birch or balsam fir-red maple	46	3	6.5 ± 6.7	6.9
11. Yellow birch-balsam fir	52	13	25.0 ± 12.2	26.4
12. Balsam fir-white birch (west)	59	7	11.9 ± 8.8	12.5
13. Balsam fir-white birch (east)	76	6	7.9 ± 6.2	8.3
14. Balsam fir-black spruce	28	3	10.7 ± 12.6	11.3
15. White spruce-balsam fir and balsam fir-white spruce	22	1	4.5 ± 10.3	4.8
16. Black spruce-balsam fir-moss	18	0	0	0
17. Black spruce-moss	32	0	0	0
18.			not analysed	

* minimal: 0–29.9%; low: 30–59.9%; high: 60–89.9%; maximal: 90–100%

** Deciduous forest: 1–7; mixed forest: 8–11; coniferous forest: 12–17

Early in the spring, the Eastern Phoebe builds its nest near water, in bushes and stands of trees, where it darts from its perch to catch insects on the wing. Rivière du Sud, a tributary of the Richelieu River. *August 1974, Léo-Guy de Repentigy.*

In a sugarbush, a sugar shack often provides the Eastern Phoebe with a horizontal or vertical surface where it can build its nest. Cap Tourmente National Wildlife Area, northeast of Québec City. *August 1991, Jean Gauthier.*

This nest with four nestlings was placed on a beam in an old building. Oka, west of Montréal.
June 1987, André Cyr.

temperature and is normally shorter for the second clutch when the summer days are warmer (Skutch 1976). The young stay in the nest for 15–17 days and are initially fed by regurgitation. The parents may make up to 845 trips a day to bring them insects (Benyus 1989). After leaving the nest, the young depend on the parents for another 2–3 weeks, while they gradually learn to forage for themselves (Smith 1942; Harrison 1978; Stokes and Stokes 1989b). The adult's diet is composed of a wide variety of insects, sometimes supplemented with berries, seeds, and occasionally small fish and frogs.

The Eastern Phoebe is generally regarded as a frequent cowbird host. However, nest parasitism varies considerably from region to region. Friedmann (in Tyler 1942b) maintained that where both species were common, up to 75% of phoebe nests were parasitized. Other studies show that parasitism can range from 0.9 to 26% (Terrill 1961; Weeks 1979; Faanes 1980; Peck and James 1987). According to Weeks (1979), some habitats are less accessible to cowbirds, and only 3% of nests on man-made structures contained a cowbird egg, compared with 33% for nests in natural sites. If a cowbird lays the first egg, the phoebe usually deserts her nest to build another. Birds that nest deep in the forest, however, are likely not affected by the cowbird, which is not a forest bird.

Pairs usually return to the same territory every year and often reuse old nests after repairing them, or build another one on top (Tyler 1942b; Weeks 1978). They may even use the nest of another species, such as the Barn Swallow or American Robin (Weeks 1977; David and Parmelee 1979).

Distribution

The breeding range of this New World species extends from northeastern British Columbia to Nova Scotia down to New Mexico and northern Georgia (AOU 1983).

In Québec, the Atlas distribution map clearly illustrates the phoebe's preference for rural areas and urban outskirts, which provide it with many man-made breeding sites close to woods. It is common south of the boreal forest, from western Québec to Saint-Georges in the Beauce at longitude 70°40' W in the east. Its stronghold thus seems to be the disturbed parts of the various forest regions dominated by the sugar maple, where there are many open habitats and regenerating stands that offer numerous man-made nesting sites. The Abitibi, Saguenay, North Shore, Lower St. Lawrence, Gaspé, Magdalen Islands, and Anticosti Island produced a scattering of records. All the squares in which the species was reported have immature mixed deciduous stands. This species is much less common at higher elevations or in the mountains (Norse and Fichtel 1985c; Eaton 1988a). Two individuals sighted at Rivière au Castor (53°24' N), about 50 km south of La Grande-2, in 1979 constitute the northernmost record for this species in Québec (ÉPOQ).

History and trends

The Eastern Phoebe has obviously benefited from human settlement and the accompanying proliferation of nest sites. It was common in the Montréal region in the last century (Hall 1862; Wintle 1896), and until very recently, its status in southern Québec did not appear to have changed (Macoun and Macoun 1915; Ouellet 1974a; David 1980a). Over the last few years, however, various sources have indicated a continent-wide decline in its population. Between 1968 and 1979, Robbins et al. (1986) reported that the species was stable on the eastern half of the continent until the exceptionally cold winter of 1976–77, which seems to have ravaged the populations in the northern part of the wintering range. The species appear to be particularly vulnerable to long severe winters on its wintering grounds and late storms on its breeding grounds, because it migrates north very early in spring when winter is not yet over. The Eastern Phoebe was even blue-listed in 1980 (Tate 1986), with the qualification that while many sources reported a considerable decrease in abundance in certain parts of its range, others felt the species was stable or increasing in the same regions. Furthermore, because this insect eater frequents primarily rural areas, it is undoubtedly affected to some extent by the intensive use of pesticides in rural, agricultural, and wooded areas.

BBS data for Québec indicate that the species was stable from 1966 to 1989, although an increase was noted from 1980 to 1989. The factors that influence this flycatcher's abundance are still not very well known.

Guy Lemieux and Jean-François Rail

Great Crested Flycatcher

Tyran huppé

Myiarchus crinitus

Christiane Girard

USUALLY CONCEALED IN THE UPPER BRANCHES OF A MATURE DECIDUOUS tree, the Great Crested Flycatcher is more often heard than seen. Its distinctive call, a loud, whistled *wheep*, sometimes accompanied by a hoarse, rolling *preeet*, is usually the first indication of its presence. A bit of patient searching, however, usually brings a glimpse of the bright yellow abdomen. About the size of an Eastern Kingbird, it has a tuft of erect feathers on its forehead that has earned it the name "crested."

Essentially a migratory species, the Great Crested Flycatcher nests in mature deciduous or mixed forests in the south of the province, where it is widespread and common without being abundant. It is eclectic in its choice of habitat, food, nest sites, and nest materials. Like most of the other flycatchers, the Great Crested has a short stay in Québec, generally only from May to mid-September.

Habitat and behaviour

The species is mainly associated with mature forests dominated by tall deciduous trees. Bent (1942) considered that at one time the species was probably originally a forest-loving bird, but clearcutting and thinning reduced the forest and the number of cavities where it could nest. Consequently, it has had to adapt to more open habitat; it is seldom found in the depths of the forest. Its preference for open habitat (Robbins et al. 1989a) means that it can be seen in orchards, city parks, mature stands in farm woodlots, swampy and wet woodlands.

The Great Crested Flycatcher seems to be faithful to the nest site, and sometimes mates with the same partner for more than one season (Bent 1942). The nest is most often placed in a natural cavity or in an old woodpecker hole (Harrison 1975; Harrison 1978), usually in a live or dead deciduous tree. However, competition with the European Starling, House Wren, Eastern Bluebird, and the red squirrel over nest cavities has caused it to seek out other sites, notably such man-made structures as nest boxes, stovepipes, drainpipes, and utility poles. Nest height is also variable, ranging from 0.6 to 23 m above the ground (DeGraaf and Rudis

The Great Crested Flycatcher nests mainly in mature, hardwood-dominated forests. Its bulky nest is most often found in a natural cavity or old woodpecker hole, usually in a deciduous tree. Sainte-Foy. *September 1991, Yves Aubry.*

Notes on territory

Like its cousin, the Eastern Kingbird, the Great Crested Flycatcher is intolerant of and aggressive toward intruding birds and will vigorously defend its territory—vocally and physically—against many species of passerines including, of course, conspecifics. Unlike the kingbird, however, it does not seem to be intolerant of large predatory birds, probably because they are not able to invade its nest cavity (Bent 1942; Taylor and Kershner 1991).

In the central and northeastern United States, the Great Crested's territory ranges from 0.2 to 3.2 ha but most often from 1.5 to 2.5 ha (DeGraaf and Rudis 1987). In Québec, in a sugar maple stand and in a hemlock forest in the Mirabel region, territories defended were 1.2 ha and 1.9 ha, respectively (Cyr 1974). Breeding density also varies with habitat. In southern Québec, it ranges anywhere from 0.52 pair/10 ha, as in a mosaic of coniferous, deciduous, and mixed woodlands with open areas and wetlands in Gatineau Park (Dauphin 1985), to 7.5 males/10 ha, as recorded in both a mature sugar maple-beech stand in Rigaud and in a young balsam fir-red maple stand in Lachute (Erskine 1984).

Great Crested Flycatcher

Breeding distribution
1984–1989

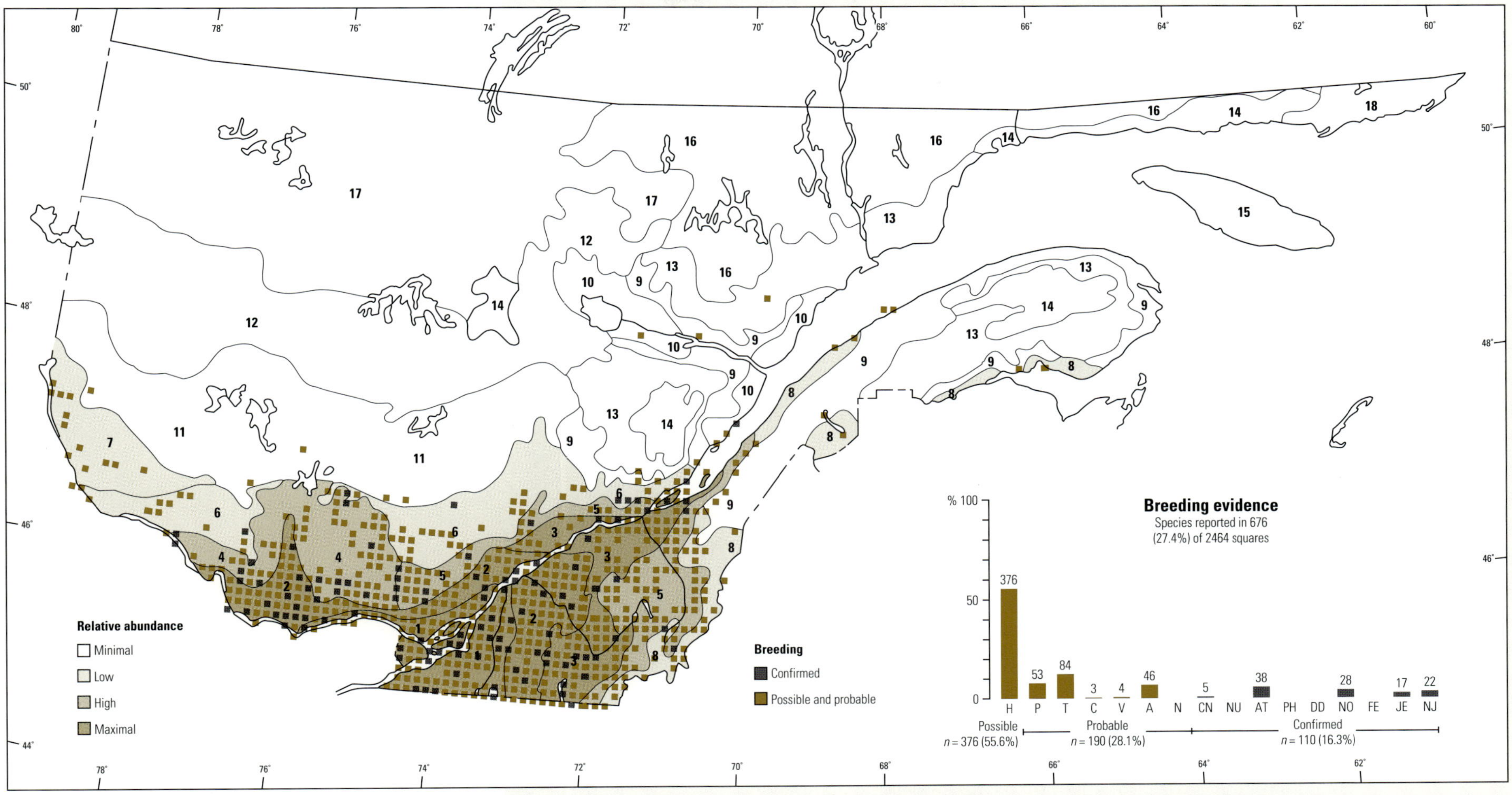

Species profile

Clutch size	usually 5 eggs (4–8)[1,2,10,11]
Laying interval	1 egg per day[1,2]
Start of incubation	with last or penultimate egg[1]
Incubation period	13–15 days[1,2,5]
Maturity at hatching	altricial
Care of young	by female and male[1,5,6]
Nestling period	13–15 days[1,5]
Age at first flight	13–15 days or more[12]
Dependence of young	until 3 weeks after first flight[1]
Broods per year	1[4]
Breeding age	1 year[4]
Mating system	monogamy[6]
Duration of pair bond	sometimes more than 1 season[2]
Total length	20.2–23.0 cm[8]
Wing span	32.4–35.6 cm[3]
Weight	33.5 g (27.2–39.6)[7]
Longevity record	13 years 11 months[9]

References:
1. Taylor and Kershner 1991
2. Bent 1942
3. Terres 1980
4. DeGraaf and Rudis 1987
5. Harrison 1978
6. Ehrlich et al. 1988
7. Dunning 1984
8. Godfrey 1986
9. Clapp et al. 1983
10. Bull 1985
11. Peck and James 1987
12. according to nestling period

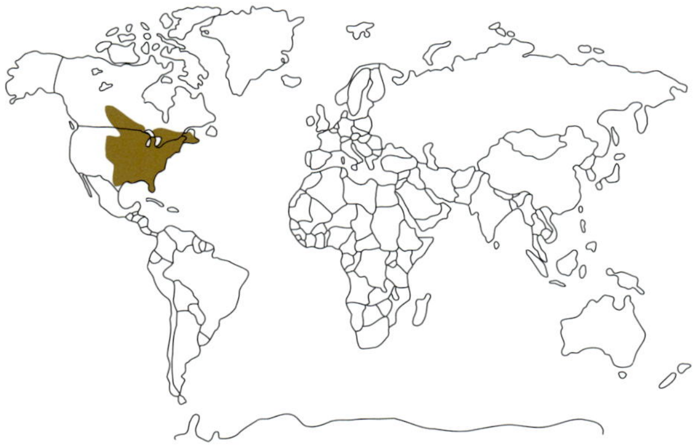

World breeding range

Chronology $n = 59$

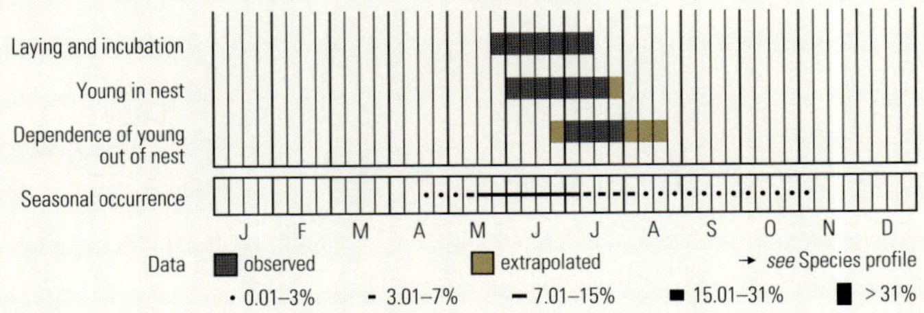

Frequency of occurrence by ecological region

Ecological regions** (forest regions)	Adequately covered squares	Squares in which reported	% of squares in which reported	Relative abundance (%)*
1. Sugar maple-hickory and sugar maple-basswood	97	94	96.9 ± 2.1	100.0
2. Sugar maple-basswood	153	147	96.1 ± 1.8	99.1
3. Sugar maple-basswood and sugar maple-yellow birch	101	92	91.1 ± 3.6	94.0
4. Sugar maple-yellow birch-basswood	56	48	85.7 ± 8.3	88.4
5. Sugar maple-yellow birch-beech and sugar maple-yellow birch-basswood	113	82	72.6 ± 4.6	74.9
6. Sugar maple-yellow birch-beech	71	41	57.7 ± 10.5	59.6
7. Sugar maple-yellow birch-balsam fir	19	10	52.6 ± 23.5	54.3
8. Sugar maple-yellow birch and balsam fir-yellow birch	47	19	40.4 ± 12.6	41.7
9. Balsam fir-yellow birch	87	8	9.2 ± 5.8	9.5
10. Balsam fir-white birch or balsam fir-red maple	46	4	8.7 ± 7.5	9.0
11. Yellow birch-balsam fir	52	3	5.8 ± 7.0	6.0
12. Balsam fir-white birch (west)	59	0	0	0
13. Balsam fir-white birch (east)	76	2	2.6 ± 4.0	2.7
14. Balsam fir-black spruce	28	0	0	0
15. White spruce-balsam fir and balsam fir-white spruce	22	0	0	0
16. Black spruce-balsam fir-moss	18	0	0	0
17. Black spruce-moss	32	0	0	0
18.			*not analysed*	

* minimal: 0–29.9%; low: 30–59.9%; high: 60–89.9%; maximal: 90–100%

** Deciduous forest: 1–7; mixed forest: 8–11; coniferous forest: 12–17

Great Crested Flycatchers are opportunistic in their selection of nest sites, as was this bird, which chose to nest in a hole in a telephone pole. The species is also found in orchards, city parks, and farm woodlots bordering on cultivated fields. Saint-Lambert, near Montréal. *July 1932, Henry Mousley, Laval University Collection.*

1987), but most often between 3 and 7 m. Great Crested Flycatchers have even been found nesting in sand dunes (Peck and James 1987). In other words, almost any type of cavity will do, provided it is large enough to hold the rather bulky nest (Bent 1942).

Although earlier accounts have stated otherwise, the female probably builds the nest without the help of the male (Taylor and Kershner 1991). A large cavity is first filled with a deep foundation of various materials: rubbish, leaves, straw, rootlets, woodchips, and bark. The nest itself is made of grasses, rope ends, feathers, bits of mammal hair and fur, and moss. Sometimes a cast snake skin, or even several, may be interwoven with the nest material, but other materials with a similar translucent appearance, such as cellophane or onion skins, are used (Harrison 1975; Harrison 1978; Terres 1980). Several studies (Bancroft 1984, 1986; Taylor and Kershner 1991) have shown the use of snake skins to be much less widespread in Canada than in the United States.

Clutch size is 4–8 eggs, most often 5 (Harrison 1975), laid a day apart (Bent 1942; Taylor and Kershner 1991). Cowbird parasitism seems to be rare (Terres 1980; Peck and James 1987). After the last or penultimate egg has been laid, the female alone incubates for 13–15 days. The nestlings remain in the nest for about 2 weeks after hatching, and are fed copiously and often, mainly by the female (Mousley 1934b; Taylor and Kershner 1991). After the young make their first flight, the family remains together, the parents continuing to protect and feed the fledglings as they gradually learn to fend for themselves (Bent 1942; Taylor and Kershner 1991).

The diet of the Great Crested Flycatcher is composed mainly of adult and larval insects: beetles, wasps and bees, flies, caterpillars and adult butterflies and moths, mosquitoes, crickets and grasshoppers, and dragonflies. It forages either by hawking its prey in flight or by gleaning on the leaves, bark, or on the ground. Spiders and berries are also part of the diet (Bent 1942; Terres 1980; Taylor and Kershner 1991).

Where natural cavities are lacking, this flycatcher will use nest boxes. It requires a wide opening and will enlarge the entrance if it is too small. Saint-Majorique, near Drummondville. *July 1991, Jeanne Lehoux.*

The Great Crested Flycatcher is difficult to spot, and breeding was confirmed in only 16.3% (110/676) of Atlas squares in which it was reported. Records of nests and of adult birds arriving at or leaving the nest accounted for half of the confirmed breeding reports, while sightings of young off the nest or parents carrying food accounted for the other half.

Distribution

The Great Crested Flycatcher breeds throughout southernmost Québec, especially in the Central St. Lawrence Lowland, Ottawa Valley, Eastern Townships, and southern Appalachians. It is confined almost exclusively to the deciduous forest and is absent from the coniferous forest. It was atlassed slightly farther north, along the St. Lawrence estuary, near Rimouski, and in the Gaspé along Chaleur Bay, in pockets of deciduous and mixed forest. The Great Crested Flycatcher is also found in the Saguenay, where it is considered accidental (G. Savard, pers. comm.).

Altitude seems to be a limiting factor on the Great Crested Flycatcher in Québec and elsewhere, as it is rare or absent in mountainous areas (Bull 1985; Levine 1988b; Ickes 1992). Although not an urban species, it now readily associates with humans, as indicated by its unusual choice of nest sites. The species winters from southern Texas and Florida to Colombia and northern Venezuela (AOU 1983; Godfrey 1986).

History and trends

A northward range expansion was noted at the beginning of the century. Macoun and Macoun (1915), for example, reported that the species had become more common in Ontario at that time. Despite subsequent loss of mature forests, Breeding Bird Surveys since 1965 in both the eastern United States (Robbins et al. 1986) and Québec (Cyr and Larivée 1993a; BBS) indicate that numbers have remained stable since 1965. Breeding Bird Surveys in Québec between 1980 and 1989, however, have shown a downward trend. This may have been caused by deforestation and selective cutting and the ensuing loss of nest cavities. On the other hand, the Great Crested Flycatcher, like several woodpecker species in whose holes it nests, may benefit from the cutting of coniferous forests, when regeneration produces hardwood growth (Bull 1985). Since it prefers fairly open habitat, the creation of openings in the continuous forest may also prove to be beneficial (Bennett 1987). It appears, however, that about 50% of its habitat on the wintering grounds may disappear by the year 2000 (Diamond 1986).

To ensure breeding success, nest boxes should be installed and snags and dead trees left standing in the bird's habitat, even though they may not always appear to be optimal nest sites (Caine and Marion 1991). According to Ickes (1992), nest boxes should be placed fairly low (1–2.5 m) to eliminate competition from starlings.

Jean-François Rail

Eastern Kingbird

Tyran tritri

Tyrannus tyrannus

Christiane Girard

THE EASTERN KINGBIRD IS WITHOUT A DOUBT THE EASIEST FLYCATCHER TO observe in Québec. It is a common, even abundant, migrant breeder throughout virtually all the settled areas of the province and is extremely visible because it frequents open areas and always perches in clear view. It is also a very noisy, active bird that aggressively harasses any intruders that come too close to its territory. A high-pitched *tzi, tzee, tzi, tzee, tzi, tzee* is its characteristic call (Terres 1980).

Habitat and behaviour

The Eastern Kingbird is hard to miss along country roads. It is often in the open, perched pertly on a fence, wire, or the top of a shrub. It likes fields and overgrown pastures, orchards, and the edges of woods and roads, as well as swamps and marshes, and the margins of ponds, lakes, and streams. The kingbird watches alertly for passing insects. When interesting prey flies by, it darts off to catch it in its bill, then usually returns to its perch. It may also hover for a short time to pick insects off leaves, and on occasion may even feed on wild berries (Morehouse and Brewer 1968). The Eastern Kingbird is in fact a generalist that can adapt its feeding habits to its surroundings, varying its diet in bad weather, for example, when there are fewer insects (Murphy 1987).

The Eastern Kingbird's nest is a rather bulky structure of twigs and grasses, lined with animal hair, rootlets, dry grasses, and plant fibres (Peck and James 1987). The female usually builds it in an isolated elm or apple tree, or low shrub, close to water (Murphy 1983a). Breeding success is higher when the nest is located halfway up the tree and between the trunk and the outer foliage in a dense cluster of leaves (Murphy 1983a). Nests may also be located in stumps or snags on the edge of lakes or swamps, at the top of hollow trees, or on fence posts. In the suburbs, the birds may even nest in rain gutters. Sometimes the kingbird chooses even more uncommon places such as street lamps, telephone poles, or old oriole, robin, or Cedar Waxwing nests (Tyler 1942a; Peck and James 1987).

Clutch size is 2–5 eggs, usually laid a day apart (Murphy 1983b, 1986b). They are white with spots and blotches of brown and often some grey. The kingbird is an uncommon cowbird host, no doubt because of its aggressiveness (Murphy 1986a), and is known for ejecting or damaging cowbird eggs (Rothstein 1975). Peck and James (1987) reported that only 1.2% of nests in Ontario were parasitized by the cowbird, and it does not figure on the list of cowbird hosts in southern Québec (Terrill 1961).

Incubation is by the female alone and takes 14–17 days. As soon as the eggs hatch, usually asynchronously (Murphy 1983b), both parents become very active, making the birds and their nest quite easy to spot. The male then has an important role in defending the nest from predators as well as feeding the young (Hayes and Robertson 1989). This passerine is single-brooded, and rarely renests if a clutch is lost (Murphy 1983a). At the end of August, the Eastern Kingbird migrates to South America, where it winters from Colombia down to northern Chile and Argentina (AOU 1983).

Notes on territory

Both the male and female are known for their fierceness in defending their territory, which can range from 5.7 to 14.2 ha in size (DeGraaf and Rudis 1987). In Illinois, Graber et al. (1974) reported densities of 0.3–2.3 individuals/10 ha, whereas in Maryland, Stewart and Robbins (1958) recorded up to 2.5 pairs/10 ha.

In the Ottawa–Carleton region, 2.2 territories/10 ha were counted in an abandoned overgrown pasture (Romanow 1993).

This insect-eating bird often nests in hedgerows, in roadside trees, or along the edge of any other open area. Cap Tourmente National Wildlife Area, northeast of Québec City. *August 1991, Jean Gauthier.*

Eastern Kingbird

Breeding distribution

1984–1989

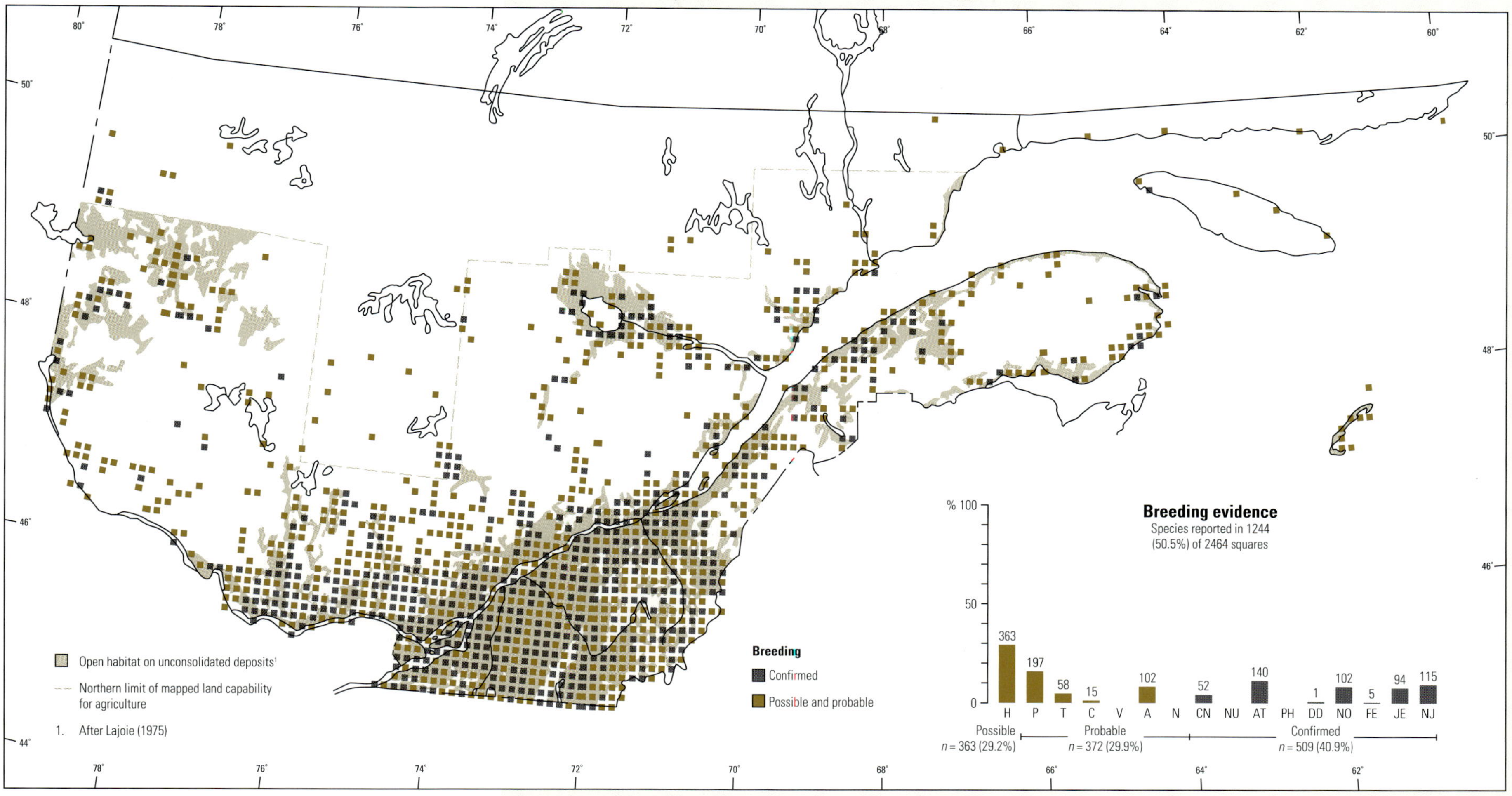

Species profile

Clutch size	usually 3–4 eggs (2–5)[1,5,6,7]
Laying interval	1 egg per day[4]
Start of incubation	before last egg[4]
Incubation period	14–17 days[4,5]
Maturity at hatching	altricial
Care of young	by female and male[10]
Nestling period	15–19 days[4,7]
Age at first flight	15–19 days or more[13]
Dependence of young	until 4–5 weeks after leaving nest[4,14]
Broods per year	1[1,3,4]
Breeding age	1 year[11]
Mating system	monogamy[1,7]
Duration of pair bond	?
Total length	19.5–22.0 cm[9]
Wing span	35.6–38.1 cm[8]
Weight	39.5 g (35.8–40.8)[2]
Longevity record	9 years 11 months[12]

References:
1. Murphy 1986b
2. Dunning 1984
3. Murphy 1983a
4. Murphy 1983b
5. Morehouse and Brewer 1968
6. Blancher and Robertson 1982
7. Ehrlich et al. 1988
8. Terres 1980
9. Godfrey 1986
10. Hayes and Robertson 1989
11. DeGraaf and Rudis 1987
12. Clapp et al. 1983
13. according to nestling period
14. Harrison 1978

Chronology $n = 308$

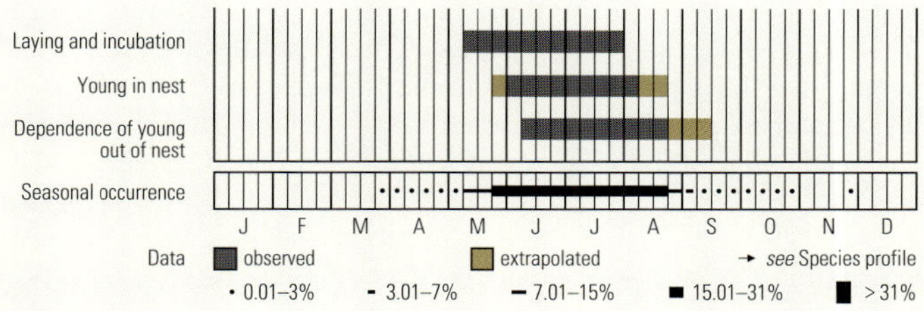

World breeding range

Unusually aggressive in defending its territory, this member of the tyrant flycatcher family lives up to that name. It does not hesitate to chase crows and birds of prey, even if these big birds are only crossing its territory at high altitudes (Stokes 1989). Intruders are chased mercilessly, with the kingbird violently diving on them from above even after they have left its territory.

There are some curious aspects to the pair's relationship. When the female arrives on the territory in the spring, the male is somewhat aggressive toward her (Stokes 1989). This may be due to the similarity in their plumage, but whatever the cause, the hostility disappears after a few days. Another form of aggressiveness occurs during nest building. In this case, the female keeps the male away from the nest site. While the female is away from the nest, the male may perch on a nearby branch at a respectful distance from the nest, but as soon as she appears, he moves farther away. After the eggs hatch, the male is finally allowed to approach the nest and help feed the nestlings (Stokes 1989).

The kingbird may build its nest in stumps or snags in a marsh. Philipsburg, south of Saint-Jean-sur-Richelieu. *May 1971, Léo-Guy de Repentigny.*

The skill required to catch insects in flight and the altricial state of the chicks at hatching probably explains the very long brood-raising period (Morehouse and Brewer 1968). The nestling phase lasts 15–19 days and the parents continue to feed the young for another 4–5 weeks (Murphy 1983b). This slow development of the young may be seen as an evolutionary adaptation that reduces the risk of starvation when cold systems or too much precipitation limit the availability of flying insects (O'Connor 1978; Murphy 1983b).

During this extended period in the nest, which can last up to 40 days, the eggs and young are exposed to predators and bad weather. Breeding success remains relatively high, however, thanks to two factors: nest-site location, featuring a well-anchored nest that is difficult for earth-bound predators to reach, and strong territoriality, expressed as aggressive behaviour toward avian predators (Murphy 1983a).

Distribution

The Eastern Kingbird breeds throughout Canada in all provinces and southwestern Northwest Territories (Godfrey 1986). Farther south, it breeds in the United States east of a line extending from Oregon to Texas, and even into northeastern Mexico (AOU 1983). In Québec, this tyrant flycatcher is distributed widely throughout agricultural areas in the south and is quite common in the agricultural areas around Abitibi, Saguenay–Lac-Saint-Jean, and the Gaspé. The Eastern Kingbird was reported in 50.5% (1244/2464) of Atlas squares and confirmed as a breeder in 509 squares. Farther north, it is much less common and is often associated with strong shoreline scrub growth. In fact, its presence in the hinterland seems limited mainly to the mixed and hardwood forests of southwestern Québec, which have been widely disturbed and are now scattered with clearings and roads. Clearings created by logging have added to the number of large natural forest clearings and marshy lake outflows, thereby encouraging the

The nest is often located in mid-branch, in an isolated, living tree close to water, as at the edge of this beaver pond. Forillon National Park, Gaspé. *July 1992, Yves Aubry.*

regeneration of the shrub vegetation preferred by the Eastern Kingbird.

Several northern sightings have been reported. Isolated individuals were observed on the east coast of James Bay (Manning and MacPherson 1952) and along the Broadback River (Todd 1963).

Three young Eastern Kingbirds ready to leave the nest located in a hawthorn. Saint-Pierre, Island of Montréal. *Henry Mousley, Laval University Collection.*

History and trends

Before intensive deforestation began in the 19th century, the Eastern Kingbird was probably confined to the boggy flats bordering rivers and lakes, and natural open areas such as burns and marshes. It was quite likely found in open areas created by beaver ponds. Since it is absent from extensive wooded areas, it is reasonable to suppose that it has become more abundant as the forests have given way to agricultural landscapes. Dionne (1906) stated that in the early 20th century it was common in its habitat throughout the province. In the Gulf of St. Lawrence area (Anticosti Island, Middle and Lower North Shore), the species does not seem to be any more widespread than in the past. Ouellet (1969) reports Schmitt's description of it as a rare summer bird on Anticosti Island early this century, which reflects its present-day status fairly closely.

Since the mid-1960s, the Eastern Kingbird population has been stable across the continent and in Québec (Robbins et al. 1986; BBS). However, the 1980–89 period shows a downward trend throughout the province (BBS), probably as a result of the considerable increase in idle land and immature woodland throughout Québec's agricultural areas in the last 20 years, except in the Montréal and Montérégie areas. It is not very likely that the species was affected when the fields were initially abandoned. However, as the idle land has slowly reverted to forest, once optimal habitat has been progressively lost, to the point that there has been a decrease in the Eastern Kingbird populations that depended on it.

Laval Roy and Line Choinière

Eastern Kingbird nest to which unusual building material—fine strips of paper—has been added. It contains 3 eggs, including a rare, unspotted one. Saint-Pierre, Island of Montréal. *June 1946, Henry Mousley, Laval University Collection.*

Horned Lark

Alouette cornue

Alouette hausse-col

Eremophila alpestris

Christiane Girard

The Horned Lark is the only native North American member of the lark family, which consists of approximately 80 species worldwide. Although it is one of the most widespread of our North American birds, in some regions it is present only very locally.

In southern Québec, the Prairie Horned Lark nests in cultivated and ploughed fields, meadows, and pastures. Baie-du-Febvre, south shore of Lake Saint-Pierre. *June 1991, Daniel Jauvin.*

The Horned Lark breeds from Alaska and the Canadian Arctic down to Mexico, nesting anywhere it finds suitable habitat. An isolated population can even be found on the high plateaus of Colombia. Of the roughly 40 subspecies in the Northern Hemisphere, 21 occur in North America (Hussell 1987a), and 2 in Québec. The Prairie Horned Lark (*Eremophila alpestris praticola*) is found in southern Québec, and the Northern Horned Lark (*Eremophila alpestris alpestris*) ranges from the southern James Bay region, the North Shore, Anticosti Island, Gaspé Peninsula, and the Magdalen Islands to extreme northern Québec (Dionne 1906; Taverner 1934; Godfrey 1986). Distinguishing field marks for the species include black sideburns, a black collar, and black horns (erectile feathers). The Northern Horned Lark is larger than the Prairie Horned Lark and has a decidedly yellow forehead, eyebrow, and throat, while these are white or much paler in the Prairie race.

The Horned Lark is considered a fairly common migrant breeder and common transient in southern Québec (David 1980a), nesting sporadically in the rest of the province. In late fall, both subspecies migrate south to the United States, although small numbers now winter irregularly in southern Québec, particularly in the Central St. Lawrence Lowland and in the southern parts of the Appalachians (Ouellet 1974a; David 1980a; Bannon 1991b). In winter and during the spring migration, Horned Larks gather in flocks of up to several thousand, descending on fields to feed on plants protruding above the snow (Todd 1940).

Habitat and behaviour

In the new year, the first flocks of Horned Larks return to southern Québec in February and congregate in fields, sometimes in the company of Lapland Longspurs (Macoun and Macoun 1915; ÉPOQ) and Snow Buntings (Dionne 1906; Macoun and Macoun 1915). Mating appears to be triggered by two consecutive days of mean temperatures over 4°C (Bent 1942). The Horned Lark nests very early, when the snow is still on the ground and the brood is exposed to the unpredictable weather of late winter (Mousley 1916b). When flocks of geese are migrating northward, adult larks feeding their young can already be seen. In Québec, the breeding season extends until mid-July. During atlassing, nests containing eggs were found over an extended period of almost four months in southern Québec; this suggests, as Dionne (1906) reported, that the species raises two broods.

The Horned Lark is found in open areas, including cultivated and ploughed fields, meadows, pastures, large grassy expanses such as airfields and golf courses, barren arctic tundra, and sand dunes and shores with sparse vegetation (Taverner 1934; Bull 1985; Godfrey 1986; Reid 1992b).

The Horned Lark's song is a rapid series of crystalline notes, a little like tinkling bells. The male sometimes sings from a perch on the ground or a fence post, but the song is more often heard during his spectacular courtship flights. In these flights, he climbs silently on the thermals to several hundred metres, then soars in wide circles for a few minutes singing his high-pitched staccato song. Abruptly folding his wings, he plummets headlong to earth, stopping a few times during his fall by straightening and beating his wings. Almost at the last second, he pulls up and lands close to his mate (Reid 1992b).

The Horned Lark is a solitary nester. The female builds the nest of grasses, plant down, feathers, and hair in a slight hollow in the ground, often next to a clod of earth or clump of grass (Terres 1980). She is very vigilant during the incubation period. When an intruder gets within 50–100 m of the nest, she leaves surreptitiously (Ellison 1985n), running or flying low over the ground to a lookout where she can watch the intruder and where she is joined by the male. She is very reluctant to go back to the nest as long as the intruder is in sight. When the nest is threatened, both birds affect a

Notes on territory

Territory size appears to vary with type of habitat. The average territory was estimated to be 0.8 ha in a burned-over grassland in Illinois, 1.5 ha in alpine tundra in Wyoming, 4.9 ha in a Kansas field, and 5.0 ha in a grainfield next to a garden in eastern New York State (Verbeek 1967; DeGraaf and Rudis 1987).

In areas with high breeding densities, the average distance between nests is roughly 142 m (Verbeek 1967). The species is usually very philopatric; one pair reportedly used the same territory for several years in a row even though road construction for urban expansion had reduced its territory to less than 0.4 ha (Reid 1992b). This strong nest-site attachment was also demonstrated in a study by

The Northern Horned Lark can be found nesting in alpine tundra on the top of Mount Albert. Gaspé Conservation Park. *July 1991, Yves Aubry.*

Horned Lark

Breeding distribution
1984–1989

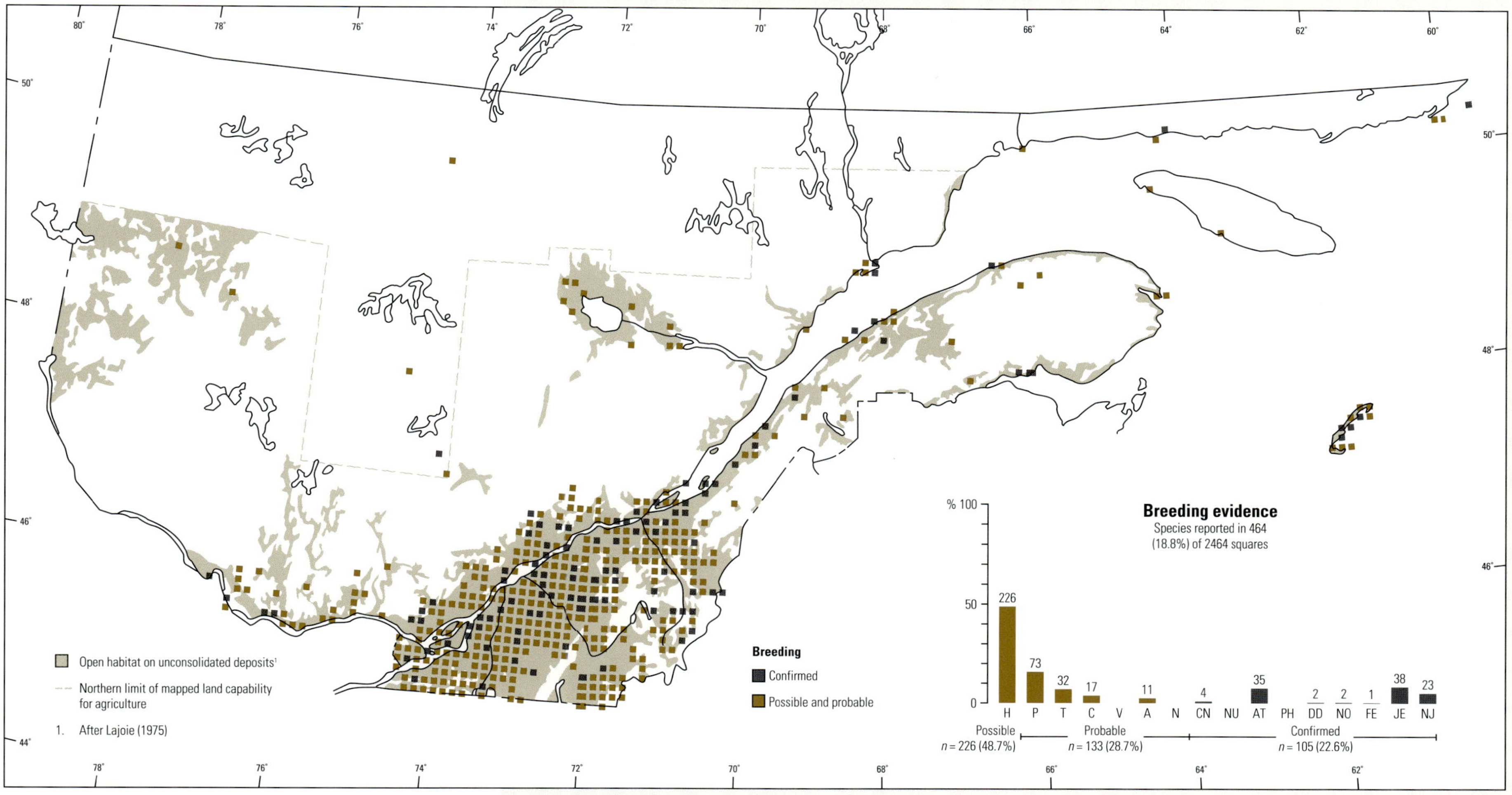

Species profile

Clutch size	usually 3–4 eggs (2–7)[1,2,4]
Laying interval	1 egg per day[1]
Start of incubation	?
Incubation period	usually 11–12 days (10–14)[1,2,4]
Maturity at hatching	altricial
Care of young	by female and male[3]
Nestling period	8–12 days[1,2,11]
Age at first flight	12–17 days[1]
Dependence of young	probably more than 1 year[10]
Broods per year	1 in the north, 2–3 in the south[2]
Breeding age	1 year[9]
Mating system	monogamy[2]
Duration of pair bond	?
Total length	17.2–20.0 cm[7]
Wing span	31.8–35.6 cm[8]
Weight	M 31.9 g (?) / F 30.8 g (?)[5]
Longevity record	7 years 11 months[6]

References:
1. Harrison 1978
2. Ehrlich et al. 1988
3. Verbeek 1967
4. Harrison 1975
5. Dunning 1984
6. Klimkiewicz and Futcher 1989
7. Godfrey 1986
8. Terres 1980
9. DeGraaf and Rudis 1987
10. Reid 1992b
11. Drury 1961

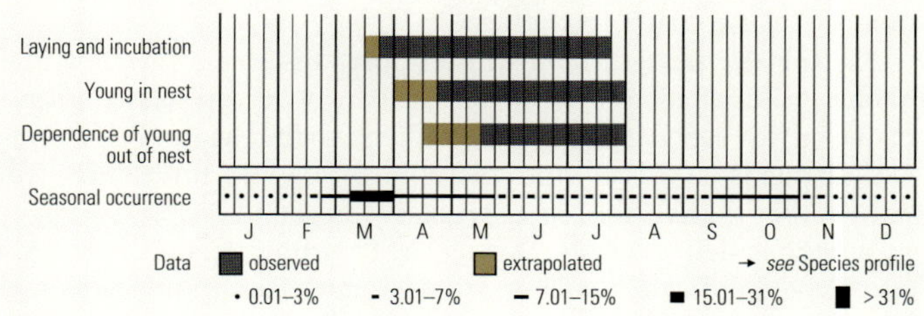

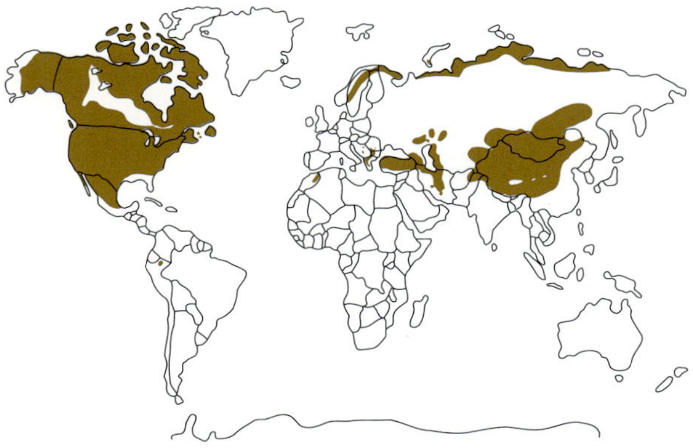

World breeding range

On the Magdalen Islands, the Horned Lark nests in dunes and other sandy habitats with sparse vegetation. Baie du Havre aux Basques. *June 1991, Desneige Perreault.*

casual unconcern, rather than displaying agitated behaviour that could attract attention. In fact, agitated behaviour or alarm calls were reported in only 8.3% (11/133) of the squares in which the species was atlassed as a probable breeder. Nests were seen in only 5.4% (25/464) of the squares in which the species was reported, a testimony to the species' inconspicuousness during nesting.

The altricial young are fed a staple diet of insects by both parents, which also look after nest sanitation (Verbeek 1967). The young leave the nest when they are around 10 days old, 3–5 days before they are able to fly. They are still fed by their parents at this time (Hussell 1987a), making it an ideal moment to confirm breeding. Beason (1984) reported a pair of Horned Larks which tended and fed a brood abandoned by another pair; the young had left the nest but were not yet able to fly. Beason attributes this behaviour to the proximity of the two pairs' territories.

The Horned Lark's brownish upperparts blend in well with its habitat, camouflaging it so effectively that once you take your eyes off a bird, for even an instant, it is difficult to find again. The Horned Lark depends on this cryptic coloration and, when threatened, runs away rather than flushes. It is primarily a ground feeder, subsisting on a wide variety of seeds. During the breeding season, however, it also feeds on insects. On alpine tundra, it forages on the snowfields for insects carried from lower altitudes by the afternoon updrafts (Norvell and Creighton 1990).

Distribution

The Atlas data show that the Horned Lark's breeding range in southern Québec corresponds roughly to areas with open habitat. The largest concentrations of Horned Larks are found mainly in the St. Lawrence Plain between Montréal and Québec City, and in the Montérégie, Bois-Francs, Eastern Townships, and Chaudière–Appalaches regions. Frequency of occurrence declines progressively farther west along the Ottawa River and toward the east in the Lower St. Lawrence region. Although nesting is often overlooked due to the early onset of breeding, the large number of observations in southern Québec may be explained by the fact that the species is double-brooded, which increases the chances of observing it and confirming breeding.

On the North Shore and Anticosti Island, the species is found mostly in open coastal habitats, while in the Abitibi and Lac-Saint-Jean regions it is associated with agricultural areas. Both types of habitats are frequented on the Gaspé Peninsula and in the Lower St. Lawrence region. In the Magdalen Islands, with its vast expanses of dunes and other types of open habitat, the number of observations and confirmed breeding records were both high.

History and trends

At the time of European settlement, the Horned Lark was much less widespread in the eastern half of North America than it is today. Only the Northern Horned Lark was present then, nesting along the coast of James Bay, Hudson Bay, and the Gulf of St. Lawrence, as well as on the tundra. This migratory subspecies was seen frequently in the southern part of the province, but was scarce in the Montréal area (Macoun and Macoun 1915). At the turn of the century, a few birds observed in spring began to remain in the Québec City region in summer (Dionne 1906; Fortin et al. 1978).

The Prairie Horned Lark traditionally nested no farther east than the Mississippi Valley and the Great Plains south of the Great Lakes (Hurtley and Franks 1976). However, it began to gradually expand eastward to occupy the suitable habitats such as cultivated fields and pasturelands that resulted from agricultural development (Eaton 1914; Ellison 1985n).

Notes on territory (continued)

Wiens et al. (1986), who showed that deliberate manipulation of vegetation structure in certain territories resulted in only minor changes in pair density and the use of space.

The Horned Lark energetically defends its territory throughout the breeding season, from the time the territory is established to when the young leave the nest. Males engage in high speed chases on the ground, sometimes involving several birds. Two males may also start a fight on the ground and continue it in the air, climbing to 20 m and higher (Verbeek 1967).

Horned Lark nest between two lettuce plants. This species often uses the same route to access and leave the nest, as is the case here. Sainte-Dorothée, Laval. *July 1916, Henry Mousley, Laval University Collection.*

Since the Horned Lark is associated with fields and meadows, large-scale farming has greatly facilitated the species' expansion and it is now the commonest bird in North Dakota (Stewart and Kantrud 1972). Densities of 2.1 pairs/10 ha were observed in a Kansas field (DeGraaf and Rudis 1987).

In northern Québec, the Horned Lark nests on the ground in tundra habitat. Rocks and mounds are used as lookouts. Near Lake Brisson, Ungava Bay region. *July 1981, Jean-Luc DesGranges.*

The first reports of Horned Larks breeding in New York State and New England date back to the last half of the 19th century (Bull 1974). Formerly, the only Horned Larks present in these regions were wintering *alpestris* birds and suspicions arose that the breeding birds and the winter birds were not the same. In 1884, the subspecies *praticola* was described for the first time by H. W. Henshaw, who differentiated it from the subspecies *alpestris* seen in the winter.

The Prairie Horned Lark was first recorded in Ontario in 1868, probably birds from Michigan, and subsequently spread through the cleared parts of southern and northeastern Ontario (Hussell 1987a).

In Montréal, the Prairie Horned Lark began to be sighted regularly in summer around the turn of the century, and was reported as nesting on the highlands above Hochelaga (Wintle 1896), incubating its eggs even before the snow melted (Dionne 1906). It was also an accidental visitant in Québec City at this time (Dionne 1906).

Taverner (1934) reported that the Prairie Horned Lark was present throughout southernmost Québec and Ontario. From then on, numerous instances of breeding throughout the St. Lawrence lowlands were reported, particularly from the Montréal region, and the species was mentioned in the PQSPB's annual reports from 1935 to 1965. During the same period, the species expanded its range into the Maritimes, where the population peaked between 1920 and 1950 (Erskine 1992).

Subsequently in Québec, ÉPOQ data show a decrease in the number of observations between 1970 and 1989 (Cyr and Larivée 1993b). However, BBS data for 1966–89 found no change.

In Vermont, the decline of agriculture and the reforestation of farmland have already drastically reduced the species' breeding range, restricting larks to the lowlands around Lake Champlain (Ellison 1985n).

In the Maritimes, Horned Lark populations have declined and even disappeared altogether from some regions in which it used to be quite common. This is probably due to the increased use of chemical insecticides in agriculture, which has greatly reduced insect populations in suitable breeding habitat, particularly potato fields. Furthermore, many open areas are reverting to forest or are used too intensively by man to provide appropriate habitat (Erskine 1992).

In south-central Ontario, the species' range has probably contracted in recent years due to the abandonment of farms and decline in acreage of pastureland (Hussell 1987a).

The abandonment of farmland, a phenomenon that began 20 years ago in southern Québec in almost every region except the Montérégie and Montréal areas, may also threaten the future of the Horned Lark here. Furthermore, it is feared that some types of intensive agriculture or reforestation may also harm populations in southern Québec by reducing the amount of habitat formerly provided by traditional farming and pasturing methods (Thibodeau et al. 1986).

Richard Pelletier

Young Horned Larks, aged 10 days old, in a nest located in corn stubble. Several factors help to hide and protect the nest: the cryptic coloration of the young and the fact that the nestlings are not brooded much during the day, when the adults visit only briefly for feedings. Montréal. *April 1942, Henry Mousley, Laval University Collection.*

Purple Martin

Hirondelle noire

Progne subis

The Purple Martin is the largest swallow found in Québec. Nowadays, it usually nests colonially in bird condominiums. This insectivorous species is very beneficial to man and has become very popular with the general public, particularly in the United States. The Purple Martin is a fairly common migrant breeder in the Central St. Lawrence Lowland (David 1980a), but is scarce east of Cap-Tourmente. The male is uniformly blue-black above and below, while the female is dark above and light below.

Habitat and behaviour

The Purple Martin catches insects on the wing, either in the air column up to 150 m above the ground (Johnston 1967), or by swooping low over bodies of water, marshes, and other open areas. It is often said that one bird can consume up to 2000 mosquitoes a day (Ducas 1971), but this is probably greatly exaggerated (Kale 1968). The Purple Martin also feeds on dragonflies, ants, wasps, beetles, grasshoppers, and other types of insects. A martin colony in a garden substantially obviates the need to use insecticides.

The Purple Martin arrives in Québec early, generally around the end of April (Ducas 1971). The adverse weather it may encounter at this time can bring about starvation and even death (Stewart 1972). Finlay (1976) has shown that insects tend to stay on the ground rather than fly at temperatures below 13°C or in heavy rain. Under these conditions, Purple Martins may not be able to feed, sometimes for long periods.

Formerly, this swallow frequented clearings and other woodland openings, where it nested in hollow trees or abandoned woodpecker holes. It also nested in cracks and crevices in rocky cliffs, as well as in hollow dead trees in coastal marshes and burnt-over areas. Today, these types of nesting habitats are still used extensively in the western part of North America (Armistead 1983), but rarely in the east (Erskine 1979). Here, the species is found mainly in cities, towns, urban parks, and farmland, most often near water or a wetland. Although nests have been observed in arc lamp reflectors in Québec (Terrill 1911b), most birds use the special multi-compartmented martin houses provided for them. The species has nested in man-made structures for centuries. Some Indian tribes in the southeastern United States held the Purple Martin in high esteem and put out hollow gourds for them to nest in (Wilson and Bonaparte 1831). Although some pairs nest singly, most Purple Martins nest in colonies of up to 200 pairs (Bull and Farrand 1977).

In spring, males aged 2 years and older usually return to the breeding grounds 2–7 days earlier than the females (Morton and Derrickson 1990). At the start of the breeding season, the males usually defend 3–12 rooms in a martin house (Brown 1979). This early arrival allows them to get a head start on later-arriving martins as well as other species such as the House Sparrow and European Starling (Brown 1978a). When the females arrive, they choose a nest site and, by extension, the male defending the site (Allen and Nice 1952). The younger males arrive at least a month later, when the females have already begun to lay and incubate the eggs (Morton et al. 1990). To breed, the younger males must take over a nesting compartment, which puts them in competition with established males, against whom they sometimes launch a veritable war of attrition (Stutchbury 1991). Sometimes, however, the older males allow the younger ones, paired or not, to move into empty cavities. This allows the older males to copulate with the females paired with the younger males.

Christiane Girard

Notes on territory

The Purple Martin's habit of nesting in colonies may, among other things, increase opportunities for extra-pair copulation, which is the only way males can increase their fecundity (Morton et al. 1990). At the start of the breeding season, males 2 years or older defend 3–12 holes in a martin house (Brown 1979); later-arriving younger males and females may attempt to take over some of these holes. By allowing the younger birds to occupy one or more holes, the adult males increase their opportunities of copulating with additional females (Morton et al. 1990). As the season progresses and more and more young males take over and defend empty holes, the territory occupied by the established males shrinks, which allows them to concentrate on defending the hole containing the nest and helping to feed the young (Brown 1979; Morton et al. 1990).

The Purple Martin, which is fairly common in open areas in the Central St. Lawrence Lowland, often nests near wetlands like this marsh at Katevale. Lake Magog, southwest of Sherbrooke.
October 1974, Vincent Létourneau.

Purple Martin

Breeding distribution

1984–1989

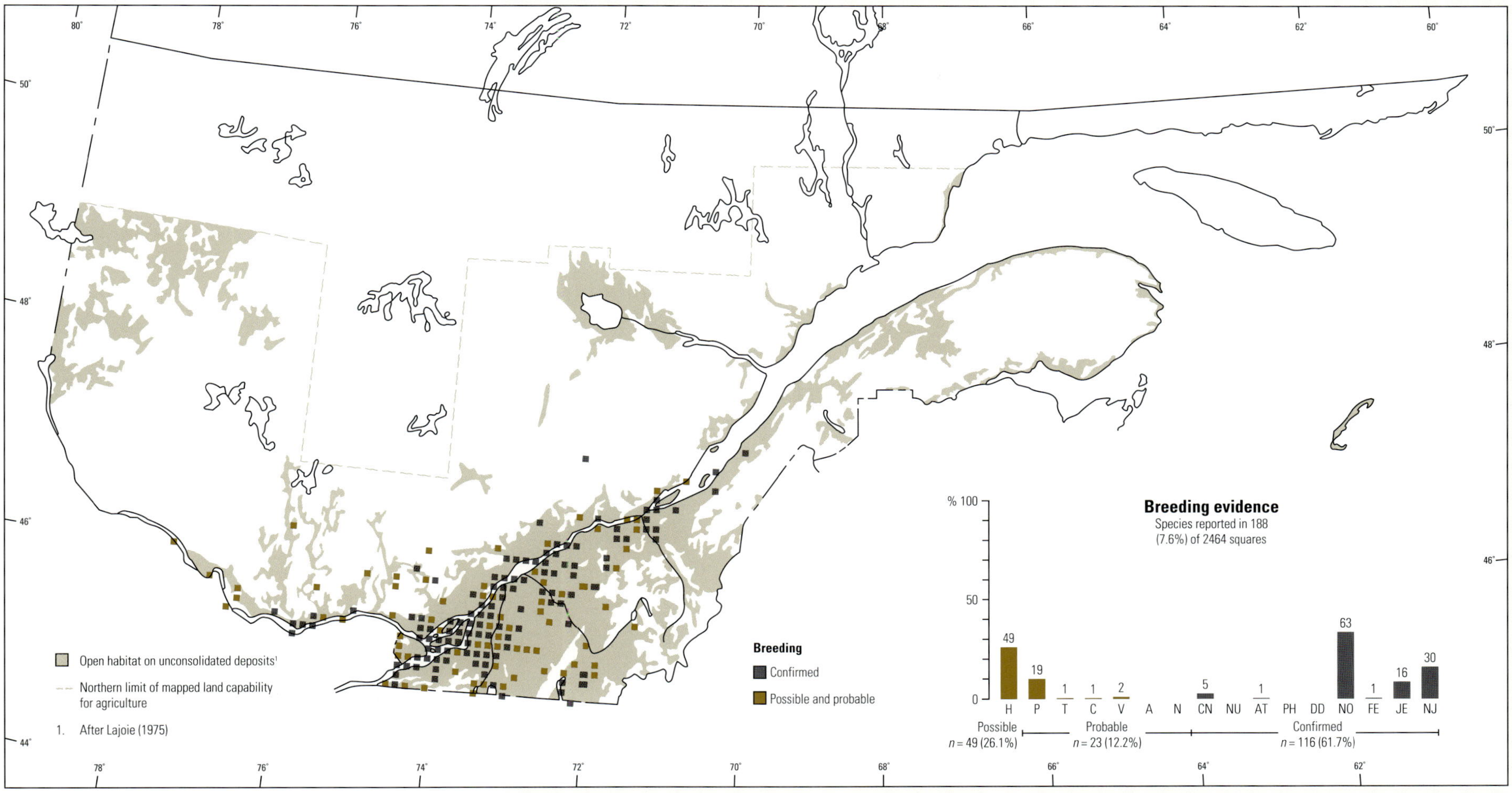

Species profile

Clutch size	usually 4–5 eggs (3–8)[1,2,3,4]
Laying interval	1 egg per day[1]
Start of incubation	with last egg[1,2]
Incubation period	15–18 days[1,2,3,4]
Maturity at hatching	altricial
Care of young	by female and male[1,2,4]
Nestling period	26–31 days[1,2,3,4]
Age at first flight	26–31 days[1,2]
Dependence of young	?
Broods per year	usually 1 (1–2)[1,2,3]
Breeding age	1 year[1,2,3]
Mating system	monogamy, sometimes polygyny[4,6]
Duration of pair bond	?
Total length	18.5–21.5 cm[8]
Wing span	39.4–42.5 cm[7]
Weight	49.4 g (?)[9]
Longevity record	13 years 9 months[5]

References:
1. Allen and Nice 1952
2. Finlay 1971
3. DeGraaf and Rudis 1987
4. Ehrlich et al. 1988
5. Clapp et al. 1983
6. Brown 1979
7. Terres 1980
8. Godfrey 1986
9. Dunning 1984

Chronology $n = 55$

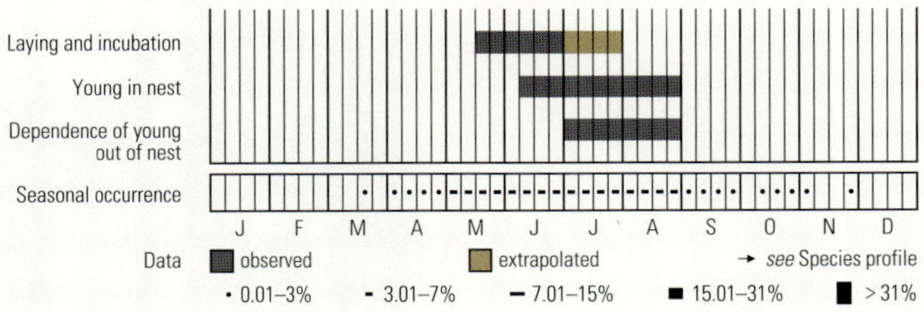

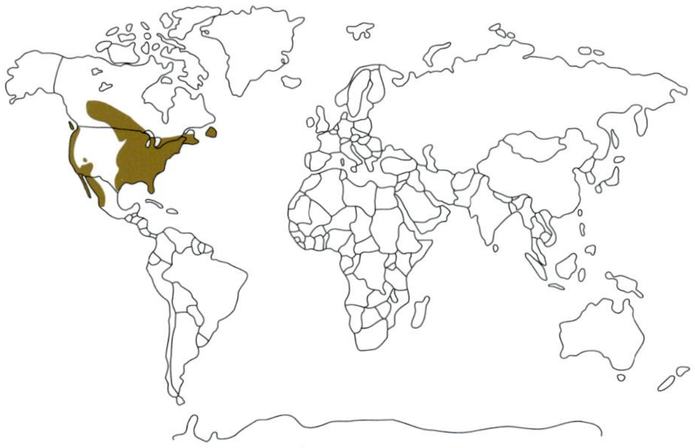

World breeding range

This swallow selects a nest site near a body of water or wetland, where it forages for insects like dragonflies and damselflies by swooping down over the surface of the water. Streit Pond, Philipsburg, south of Saint-Jean-sur-Richelieu. *July 1991, Pierre Bannon.*

Purple Martins can be observed from time to time in the martin houses installed at the Petite Ferme at Cap Tourmente. Nearby fields and marshes provide the birds with food. Cap Tourmente National Wildlife Area, northeast of Québec City. *August 1991, Yves Aubry.*

The Purple Martin nests in colonies in special multi-compartmented houses. Pierreville, east of Sorel. *June 1991, Daniel Jauvin.*

Such extra-pair copulation appears to be fairly common in the species (Morton et al. 1990). True polygyny is rare, however; in a Texas study, it was observed in only 5% of males (Brown 1979). Polygyny may occur when the number of males is reduced, usually when the early arrivals perish in a cold snap or when several females move into cavities defended by one male (Brown 1975).

Copulation occurs during the week after the female arrives. During the next 2–4 weeks, the male helps the female build the fairly rudimentary nest, the extent of male assistance depending on the pair (Sprunt 1942). The base of the nest is constructed from mud and twigs and, as a final touch before laying, is decorated with green leaves (Sprunt 1942), mainly by the male. According to Morton (1987), the bringing of green leaves by the male may represent a signal which stimulates copulation when the female is most fertile. Generally only one clutch is laid. The female does almost all of the incubation, which begins when the last egg has been laid. The reproductive cycles of the pairs in a colony are often synchronized (Ducas 1971), so that the eggs hatch at almost the same time. Both parents, the male to a lesser extent, feed the young (Allen and Nice 1952). Young Purple Martins remain in the nest longer than other swallow species (Ducas 1971). In addition, their parents defend them against harassment by other martin "raiders" (Brown 1978b). It is thought that such harassment is a strategy to encourage the young to disperse, so that they do not return to the same nest site to breed in future years, thus reducing competition (Brown and Bitterbaum 1980).

Breeding was confirmed in a high percentage of Atlas squares (61.7%; 116/188) in which the species was reported, thanks to its highly visible martin houses, often located in inhabited areas. Most of the unconfirmed breeding records probably involved individuals foraging for food or visiting nest sites.

Distribution

Southern Québec is at the northern limit of the Purple Martin's breeding range, probably because of the climate and the smaller number of nest boxes found here (Taylor 1987). The species is not widespread but rather localized, its presence being determined mainly by the proximity of bodies of water and wetlands, where prey are abundant, and by the availability of appropriate nest boxes. The Purple Martin is confined almost exclusively to the Central St. Lawrence Lowland and Ottawa Valley, with outliers in the Eastern Townships and the lower Laurentians, and is absent from mountainous regions above 300–400 m in elevation. The northernmost as well as the easternmost Atlas breeding records came from La Pocatière. The Purple Martin has also been reported from Jonquière (Kénogami), the Saguenay–Lac-Saint-Jean region (Browne 1967), Bonaventure and Forillon in the Gaspé region, the Magdalen Islands, and Godbout (in 1896) on the North Shore (David 1980a).

During migration, individuals gather in large flocks in urban areas (Robbins et al. 1966; Ouellet 1974a). The species winters in South America (AOU 1983).

History and trends

Before European settlement, the Purple Martin was probably uncommon, although some indigenous peoples in the southeastern United States encouraged it to breed (Wilson and Bonaparte 1831). The clearing of forests for farming at the beginning of the 19th century destroyed some types of breeding habitat, but was generally favourable to the species, which soon learned to use artificial structures for nesting (Sibley 1988f).

At the end of the 19th century, two exotic species were introduced into North America: the European Starling and the House Sparrow, both of which thrive in proximity to human habitation. These species gradually usurped the Purple Martin's nest sites (Eaton 1914), leading to a steep decline in the martin population. In New York State, for example, Eaton (1914) estimated that the population declined by between 30 and 50% between 1860 and 1910. In Montréal, however, Terrill (1911b) considered the Purple Martin to be the third commonest bird species at the beginning of the 20th century. More recently, Ouellet (1974a) reports that colonies are increasingly scattered, with the largest consisting of 70 individuals at most. In addition, cold waves may have a very detrimental effect on populations; Benton and Tucker (1968) estimated that mortality from the cold wave that struck New York State in the spring of 1966 was between 75 and 80%. More recently, in April 1982, the Purple Martin suffered a serious decline in the Northeast (New York, Ontario, and probably Québec), because of the very cold temperatures (Adams 1982). The species was on the Blue List from 1975 to 1981 and on the list of Species of Special Concern from 1982 to 1986 (Tate 1986).

Today, Purple Martin populations appear to be fairly stable (Sibley 1988f), which is confirmed by BBS data. This is most likely because of the dedicated efforts by martin house owners. Jackson and Tate (1974) recommended that white aluminum houses be used to discourage starlings and sparrows, which prefer dark-coloured wooden houses. In addition, to prevent other species from occupying the houses, owners are advised to wait until the first Purple Martins appear in spring before putting up their houses or allowing access. Furthermore, all compartments should be cleaned regularly to destroy ectoparasites, such as mites, that thrive in nesting materials.

Benoît Limoges and Manon Bombardier

Tree Swallow

Hirondelle bicolore

Tachycineta bicolor

The Tree Swallow is the earliest swallow to return to Québec, sometimes appearing here by the end of March. It is usually in mid-April, however, that it arrives from Florida and the Gulf of Mexico, where most of the population winters, although individuals winter as far north as Massachusetts, and others as far south as Venezuela (AOU 1983).

The male energetically defends a nest site upon his arrival. Nonetheless, it is not until the end of April that he can be seen carrying feathers in his bill with which to line the nest.

Habitat and behaviour

Although the Tree Swallow is extremely eclectic in its choice of habitat and nest site (Munro and Rounds 1985), it prefers to nest on open terrain where old or dead trees provide nest cavities. It is particularly fond of sites near water, usually 15–75 m from the water's edge. It generally nests in a cavity of a tree that is 60–90 cm in circumference and 2–9 m tall (Rendell and Robertson 1991). The inside of the cavity may be 4–25 cm in diameter and 5–92 cm deep (Peck and James 1987).

According to Rendell and Robertson (1990), the Tree Swallow, when it has a choice, avoids nesting near forest edges where the Eastern Bluebird and House Wren nest. As it is very aggressive, this may contribute to reducing the energy losses associated with competition. In addition, the Tree Swallow is primarily insectivorous, catching the insects on the wing, preferably close to the nest; it is therefore to its advantage to have open space free of obstacles around its nesting area (Rendell and Robertson 1990). In Québec, competition between the Tree Swallow and the Eastern Bluebird is a factor that must be taken into account in siting nest boxes under the Eastern Bluebird Recovery Program.

This species accepts a wide variety of nest boxes, including those with several compartments. It prefers those oriented toward the southeast, however, allowing it to benefit from the warmth of the morning sun and to be sheltered from the wind (Lumsden 1986). Nests in natural cavities vary in depth from about 30 to 50 cm, with extremes from 5 to 92 cm (Peck and James 1987). Old Cliff Swallow nests may also be used (Benton 1968) and, exceptionally, hollows in the ground (Erskine 1979). The presence of a suitable nest box seems to be a very important factor in the selection of a breeding habitat, since 84.6% of nests counted in Ontario were in artificial nest boxes (Peck and James 1987).

The males arrive on the breeding grounds in flocks of various size, usually at dawn on a mild spring day (Stocek 1986; Cohen 1987). At sunrise, they become very noisy and immediately begin to defend their territories energetically: perched near the entrance to a cavity, each male sings to make his presence known to other Tree Swallows. Some males defend two cavities (Harris 1979), usually at most 5 m apart (Dunn and Hannon 1991). This behaviour may be advantageous to these individuals, since it increases their chances of attracting more than one mate (Quinney 1983). Nonetheless, monogamy is probably the rule for this species, as the number of cavities is usually limited (Muldal et al. 1985; Dunn and Hannon 1991); approximately 5–8% of pairs are polygynous (Quinney 1983; Dunn 1989).

The females sometimes arrive on the nesting sites at the same time as the males, but they generally join them 1–7 days later (Cohen 1987). Copulation takes place right away, near the cavity (Leffelaar and Robertson 1984). At this time of year, a great many unpaired individuals, especially yearling females, explore the surroundings of the nest sites looking for a vacant cavity (Stutchbury and Robertson 1985, 1987). These young females are easily identifiable by the brown plumage that distinguishes them from their adult counterparts, which resemble the males (Cohen 1980; Hussell 1983). The paired females then begin a fierce defence of their territories, and may even go so far as to kill an intruding female; this behaviour can be observed throughout the nest-building and laying period (Robertson et al. 1986). According to Dunn and Hannon (1991), paired females vigorously defend the immediate surroundings of the cavity in order to avoid the risk of nest usurpation and parasitism.

Building the nest takes anywhere from a few days to two weeks, depending on environmental conditions, and the female does most of the work. The nest is essentially composed of dry grass, to which sometimes pine needles, stalks, twigs, roots, and leaves are added. It is sometimes lined with duck and gull feathers, gathered by the male and added to the cup during laying (Austin and Low 1932; Kuerzi 1941). This feather lining, which provides extra insulation, accelerates the development of the chicks (Winkler 1993). All material is gathered within a radius of approximately 30 m of the nest (Stocek 1970).

Christiane Girard

Notes on territory

As soon as it arrives, the male defends a breeding territory of 6–8 m around the cavity.

The Tree Swallow is eclectic in its choice of nest sites, but prefers open areas near water. Sainte-Rose-de-Watford, northeast of Saint-Georges. *June 1991, Jacqueline Vincent.*

Tree Swallow

Breeding distribution

1984–1989

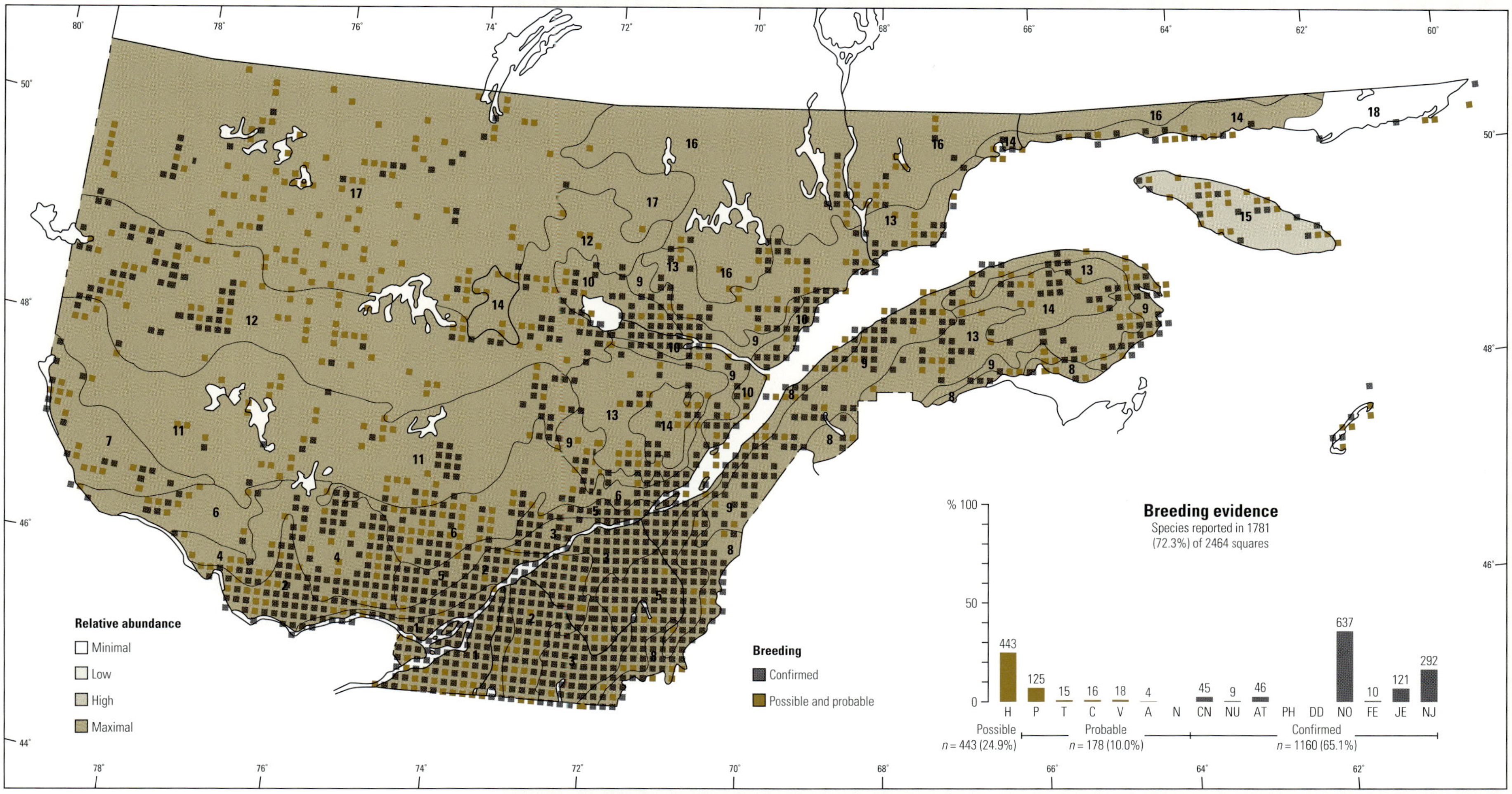

Species profile

Clutch size	usually 3–7 eggs (2–8)[1,2,3,4]
Laying interval	1 egg per day[2,4]
Start of incubation	with penultimate egg[2,4]
Incubation period	usually 13–16 days (11–19)[1,2,4]
Maturity at hatching	altricial
Care of young	by female and male[1,2,3,4]
Nestling period	usually 18–22 days (15–25)[1,4]
Age at first flight	usually 18–22 days (15–25)[1,4]
Dependence of young	until at least 3 days after leaving nest[4,8]
Broods per year	1, rarely 2[4]
Breeding age	1 year[3,4]
Mating system	monogamy, sometimes polygyny[1,4]
Duration of pair bond	1 season[1,4]
Total length	12.5–15.7 cm[5]
Wing span	30.5–33.7 cm[9]
Weight	20.1 g (15.6–25.4)[7]
Longevity record	11 years[6]

References:
1. Tyler 1942c
2. Austin and Low 1932
3. De Steven 1978
4. Robertson et al. 1992
5. Godfrey 1986
6. Clapp et al. 1983
7. Dunning 1984
8. Stokes 1989
9. Terres 1980

World breeding range

Chronology $n = 1122$

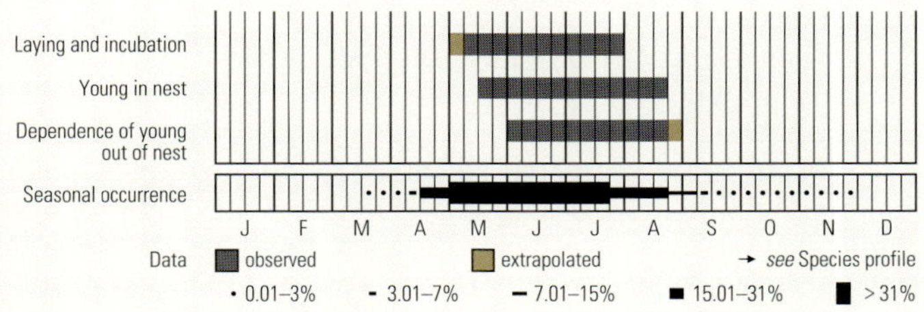

Frequency of occurrence by ecological region

Ecological regions** (forest regions)	Adequately covered squares	Squares in which reported	% of squares in which reported	Relative abundance (%)*
1. Sugar maple-hickory and sugar maple-basswood	97	97	100.0 ± 0.5	100.0
2. Sugar maple-basswood	153	152	99.3 ± 0.9	99.3
3. Sugar maple-basswood and sugar maple-yellow birch	101	101	100.0 ± 0.5	100.0
4. Sugar maple-yellow birch-basswood	56	55	98.2 ± 3.7	98.2
5. Sugar maple-yellow birch-beech and sugar maple-yellow birch-basswood	113	113	100.0 ± 0.4	100.0
6. Sugar maple-yellow birch-beech	71	69	97.2 ± 4.0	97.2
7. Sugar maple-yellow birch-balsam fir	19	18	94.7 ± 12.0	94.7
8. Sugar maple-yellow birch and balsam fir-yellow birch	47	47	100.0 ± 1.1	100.0
9. Balsam fir-yellow birch	87	86	98.9 ± 2.5	98.9
10. Balsam fir-white birch or balsam fir-red maple	46	46	100.0 ± 1.1	100.0
11. Yellow birch-balsam fir	52	50	96.2 ± 6.0	96.2
12. Balsam fir-white birch (west)	59	57	96.6 ± 5.3	96.6
13. Balsam fir-white birch (east)	76	75	98.7 ± 3.0	98.7
14. Balsam fir-black spruce	28	28	100.0 ± 1.8	100.0
15. White spruce-balsam fir and balsam fir-white spruce	22	19	86.4 ± 15.5	86.4
16. Black spruce-balsam fir-moss	18	18	100.0 ± 2.8	100.0
17. Black spruce-moss	32	29	90.6 ± 11.7	90.6
18.			*not analysed*	

* minimal: 0–29.9%; low: 30–59.9%; high: 60–89.9%; maximal: 90–100%

** Deciduous forest: 1–7; mixed forest: 8–11; coniferous forest: 12–17

Laying begins 1–3 weeks later, in early May. Incubation lasts about two weeks and is done exclusively by the female. Hatching is generally synchronous, except in poor environmental conditions (Robertson et al. 1992). When asynchronous hatching occurs, a hierarchy based on weight becomes established among the young of the same clutch, and the firstborn, heavier than those born later, has a greater chance of reaching maturity than do the others (Zach 1982).

The parents eat the fecal sacs of the young until they are 8 days old; after that the sacs become too large and must be removed from the nest. It is thought that eating them rather than carrying them away allows the female to brood with fewer interruptions and is less energy-consuming for both parents (Hurd et al. 1991). Both the male and female defend the nest against intruders during the brood-rearing period (Lombardo 1985). When the female leaves the nest looking for food, the male blocks the entrance to the cavity, and his white breast, well in evidence, warns potential intruders to keep their distance (Kuerzi 1941; Paquin 1990a). The presence of a person near the nest usually triggers typical defence behaviour in the species. The parents give alarm calls while overflying the intruder and circling overhead. Sometimes they also dive to within 0.5–2 m of the intruder (Winkler 1992).

The young are fed by both parents about every 3–6 minutes (Leffelaar and Robertson 1986; Quinney 1986). Their diet consists chiefly of aquatic insects such as mayflies and midges, but also includes a smaller proportion of terrestrial species belonging to the orders Diptera, Coleoptera, and Hymenoptera (Blancher and McNicol 1991); on occasion, they are also fed mollusc shells and crayfish fragments (Blancher et al. 1987). At around the age of 3 weeks, the parents usually stop returning to the nest (Austin and Low 1932); nestlings fledge at this time, but the parents continue to feed them for another 3 days at least (Robertson et al. 1992). After fledging, the young explore the nest sites occupied by other broods, possibly scouting for the following year (Lombardo 1987). This behaviour produces large concentrations of birds around nests toward the end of the nesting season. Adult swallows are almost exclusively insectivorous, but also eat berries when insects are scarce. The birds sometimes form large flocks around plentiful food sources (Tyler 1942c).

The territorial behaviour of the Tree Swallow, the fact that it frequents open areas, and the large numbers of individuals wandering near the nest sites all make it easy to observe the species.

In any given year, approximately 13–60% of adults, mainly males, will return the following year to the same site to nest, and two thirds of these males will reuse the same cavity (Cohen 1985). The females are less faithful to the nest site, nesting within 15–30 km of the site used the preceding year (Chapman 1955). Most yearlings will return to their natal site to nest (Butler 1988), but many will move 20–40 km to nest (Robertson et al. 1992). In fall migration, the species sometimes forms impressive flocks; in the southern part of the province, 200 000 birds were sighted between 21 and 23 September 1976 at Beauharnois, and on 10 October of the same year, a flock of an estimated one million birds was reported over the St. Lawrence River, in an area covering approximately 38 km^2 near Côte-Sainte-Catherine Park (David and Gosselin 1977a).

This swallow is a cavity nester. Here, an adult brings a billful of insects back to its nest in a natural tree cavity. Saint-Raymond, west of Québec City. *June 1968, Gaston Le Page.*

The Tree Swallow also uses nest boxes set up along fields, competing for them with other species such as the Eastern Bluebird and House Wren. Cap Tourmente National Wildlife Area, northeast of Québec City. *August 1991, Yves Aubry.*

Prolonged floods may kill a good many trees growing in marshes. Dead trees and snags can then become home to several nesting pairs. Îles de la Paix National Wildlife Area, near Montréal. *June 1987, Léo-Guy de Repentigny.*

Distribution

The Tree Swallow was reported in 72.3% (1781/2464) of Atlas squares, making it one of the most widespread breeders in Québec. It was common throughout the southern parts of the province, except on the Lower North Shore, east of Natashquan, where tundralike habitat predominates and the low numbers of trees of the appropriate size discourage its establishment.

The Tree Swallow is common in the northern part of Canada, from Alaska to Labrador and Newfoundland (Todd 1963), except in the eastern Northwest Territories and in the extreme northwestern part of Québec (Godfrey 1986). In Québec, it can range up to the 58th parallel and is occasionally seen around Kuujjuaq (ÉPOQ). Its habit of nesting in cavities enables it to find many suitable sites in the northern woodlands.

Notes on territory (continued)

The female, on the other hand, limits her territorial defence to the immediate surroundings of the nest (Robertson and Gibbs 1982). Often several pairs of Tree Swallows will live in the same multi-compartmented nest box; they may also exceptionally share a nest box with Purple Martins (Peck and James 1987). Prior to the start of incubation, this swallow may forage for insects up to 60 km from the nest site; this distance gradually decreases to 2–5 km during brood rearing (Robertson et al. 1992).

Once the nesting season is over, Tree Swallows become more gregarious, flocking together on perches. Robertson et al. (1992) reported an average distance of 10 cm between birds, but according to Grubb (1973), they may be no more than 0–2 cm apart in cold weather.

History and trends

In the past, Tree Swallow populations have declined sharply during harsh winters. The rigours of the 1939–40 winter, for example, probably caused the death of thousands of swallows wintering in the United States; the following summer, populations were much smaller than normal (Tyler 1942c).

BBS data show that Tree Swallow populations increased in North America between 1965 and 1979, particularly in the northeastern and Great Lakes states, and in eastern Canada (Robbins et al. 1986). More recently, between 1990 and 1991, an increase in numbers during the breeding season in most Canadian provinces, including Québec, was also documented (Erskine et al. 1992).

The status of the species does not seem to have changed in Québec since the 19th century, when it was generally considered common (Hall 1862; Wintle 1896; Dionne 1906). Since European settlement, the loss of natural nest sites as a result of deforestation and selective cutting of dead trees in the province, particularly in the Central St. Lawrence Lowland (Erskine 1979), probably caused an increase in the use of nest boxes in some parts of southern Québec. That is why it is important to preserve sites with dead trees and to continue to set up nest boxes to maintain Tree Swallow populations in the province.

André Dion and Manon Bombardier

An adult carrying away a nestling's fecal sac to keep the nest clean. Uneaten fecal sacs are usually dropped into bodies of water near the nest, including swimming pools in urban areas. Lac-Saint-Charles, near Québec City. *July 1993, Robert Morin.*

Northern Rough-winged Swallow

Hirondelle à ailes hérissées

Stelgidopteryx serripennis

Christiane Girard

THE GENUS *STELGIDOPTERYX* TAKES ITS NAME FROM THE GREEK WORDS *stelgido*, which means scraper or type of comb, and *pteryx*, which means wing. This is an apt description of the hook- or tooth-shaped barbs on the outer primaries, which are difficult to see but can be readily felt by passing a finger along the edge of the wing toward the tip. The exact function of these hooklets is not known, but Dingle (1942) postulated that they may be used to produce a sound during courtship flights.

Habitat and behaviour

The Northern Rough-winged Swallow breeds singly or in small colonies (Beecher et al. 1986), depending on the availability of nest sites. It often nests on the fringe of Bank Swallow colonies. In Canada, colonies generally contain only 4 pairs, except in British Columbia, where the average is 15 pairs (Erskine 1979). The species seems to be well adapted to northern conditions. According to David (1980a), it arrives in southern Québec around 15 April and leaves around 15 September. Since it does not begin nesting immediately, it probably suffers less than other swallows from inclement spring weather conditions (Lunk 1962).

During the breeding season, Rough-wings can be found by inspecting potential nest sites in banks of sand, clay or gravel, riverbanks, and cavities in rocky cliffs. The species takes advantage of human activity, nesting in man-made structures such as gutters and drainage holes in bridges. It may also nest in quarries and caves, as well as in abandoned Bank Swallow or Belted Kingfisher burrows (DeGraaf et al. 1980). In Canada, close to 75% of nests are in natural sites (Erskine 1979). Burrows, dug by both the male and female, are generally between 22 and 50 cm deep but may be as much as 180 cm deep (Dingle 1942). The height of the nesting cavity is most often between 0.3 and 15 m from the foot of the slope or bank, depending on the substrate. Since this species uses only its feet to dig, it usually selects crumbly material underneath a harder layer which serves as a roof. The entrance to a cavity excavated by this species can be distinguished from a Bank Swallow burrow by its circular shape and larger size. The Northern Rough-winged Swallow uses leaves, twigs, pine needles, and sometimes even flower petals to construct the nest, which is then lined with grass.

Apparently only the female incubates the eggs. During the 15–16 day incubation period, the male rarely enters the nest but perches nearby. Hatching is asynchronous. Both sexes feed the young. The species is almost entirely insectivorous, catching its prey mostly on the wing (Dingle 1942). The diet is comprised mainly of adult Diptera and Hymenoptera (Lunk 1962). The young fledge from the nest when they are 18–21 days old. Breeding success is generally high; in Massachusetts and Michigan respectively, Blake (1953) and Lunk (1962) reported that young were produced from 68 and 61% of the total eggs laid. Ricklefs (1972) demonstrated that productivity, defined as the ratio of immature to adult birds, increases with latitude in North America, ranging from 0.42 at the equator to 1.75 at latitude 40° N. Consequently, Ricklefs estimated that 67% of the young raised at the equator, compared with only 33% of those raised at latitude 40° N, must reach sexual maturity to maintain the population.

This species is vocal and aggressive at the nest (Lunk 1962), which makes breeding easy to confirm. Breeding was confirmed in 39.6% (84/212) of Atlas squares in which the species was recorded, mostly (71.4%, 60/84) by reports of adults entering or leaving the nest.

Distribution

This migratory species nests from southern Canada (locally) to Costa Rica (AOU 1983). Québec is therefore at the northern limit of its breeding range in eastern North America. Interestingly, the Bank Swallow is widely distributed throughout southern Québec, while the Northern Rough-winged Swallow, which often nests in the same habitat as the Bank Swallow and sometimes even uses the Bank Swallow's abandoned nests, is concentrated in the Central St. Lawrence Lowland and in the Ottawa and Gatineau river valleys. The Northern Rough-winged Swallow becomes scarcer toward the Appalachians, is occasional in the Montréal region, and rare in the lower Laurentians (Bannon 1991b). Farther north, Breton (1980) reported that, in 1967, 6 individuals were observed under a bridge over the Sainte-Marguerite River in the Saguenay region, although nesting could not be confirmed. The species is considered accidental in the Saguenay–Lac-Saint-Jean region (Savard and Cormier 1995).

The Northern Rough-winged Swallow winters mainly in Mexico and Central America as far south as Panama, but can also be found in the southern United States (AOU 1983; Root 1988). The principal factor responsible for the southward movement in winter is probably the decrease in the number of insects in the northern part of the species' breeding range (Young 1982).

History and trends

The Northern Rough-winged Swallow was scarce in the northeastern United States until the last third of the 19th century (Eaton 1914), and according to Bull (1985) was unknown to naturalists in New York State until 1870. Between 1950 and 1970, it expanded its North American range northward (Bull 1985). This trend seems to be continuing in North America as a whole according to BBS data (Robbins et al. 1986), while in Québec, no modifications appear to have occurred in the species' range between 1966 and 1989.

The increase in abundance since the turn of the century (Robbins et al. 1986) is due first to the fact that the species has adopted artificial nesting structures, which have become more common and more accessible during this time. In addition, the number and competence of observers have increased significantly since then. The first sight record of a Northern Rough-winged Swallow in Québec dates from the spring of 1948 on Île Sainte-Hélène in the Montréal region (PQSPB 1950), and the first breeding record was obtained in 1959 (Gosselin 1989a).

According to Lunk (1962), the Northern Rough-winged Swallow's ability to extend its range is limited by the availability of nest sites, particularly abandoned cavities of Bank Swallows and suitable man-made structures. However, given the Bank Swallow's much more northern range, other factors are probably responsible for the fact that the Northern Rough-winged Swallow has not been able to expand beyond southernmost Québec.

LOUIS MESSELY

Notes on territory

The Northern Rough-winged Swallow only establishes a small territory around the entrance to the nest. The male proclaims his territory by continued calling, perching, and fluttering in the vicinity, but does not attempt to defend his perches. The female also helps defend the nest until the young fledge. It is possible that the females, and probably males as well, use a site for more than one season (Lunk 1962).

Along the Little Rouge Creek, near Scarborough, Ontario, breeding density was estimated at 1.9 males/10 ha in a ravine with both deciduous and evergreen trees and open areas (Erskine 1984).

The Northern Rough-winged Swallow nests singly or in small colonies, often taking advantage of man-made structures near water. For example, it may build its nest in an opening in a concrete wall such as a drainage pipe. Saint-Jean Creek, Châteauguay. *July 1991, Gaétan Duquette.*

Northern Rough-winged Swallow

Breeding distribution
1984–1989

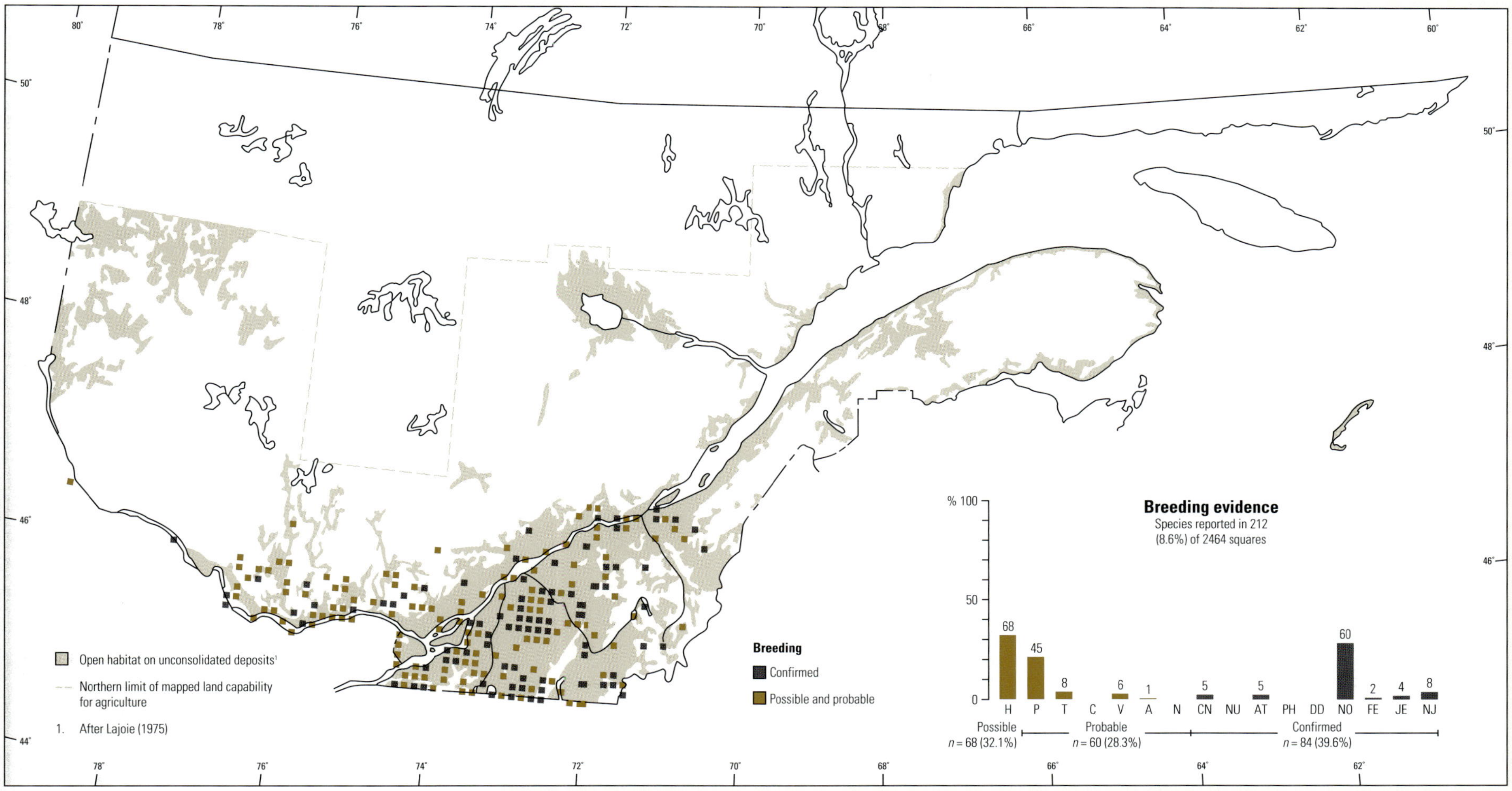

Species profile

Clutch size	usually 5–7 eggs (3–8)[1]
Laying interval	1 egg per day[1]
Start of incubation	with penultimate egg[1]
Incubation period	15–16 days[1,3]
Maturity at hatching	altricial
Care of young	by female and male[1,2]
Nestling period	18–21 days[1,2,3]
Age at first flight	18–21 days or more[8]
Dependence of young	?
Broods per year	usually 1 (1–2)[1]
Breeding age	1 year[1]
Mating system	monogamy[2]
Duration of pair bond	?
Total length	12.8–14.5 cm[5]
Wing span	29.2–31.1 cm[7]
Weight	15.9 g (10.3–18.3)[4]
Longevity record	5 years 11 months[6]

References:
1. Lunk 1962
2. Ehrlich et al. 1988
3. DeGraaf and Rudis 1987
4. Dunning 1984
5. Godfrey 1986
6. Clapp et al. 1983
7. Terres 1980
8. according to nestling period

Chronology $n = 53$

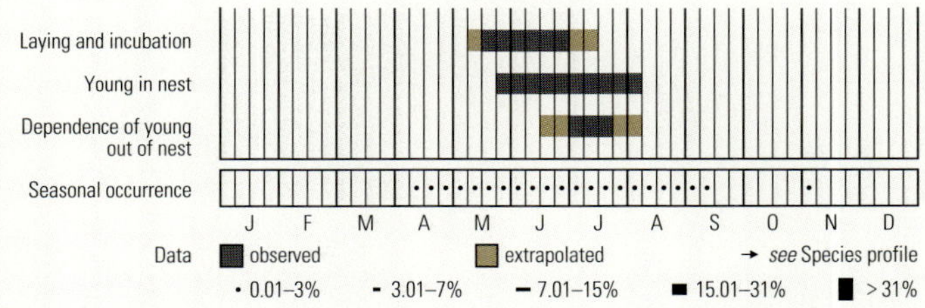

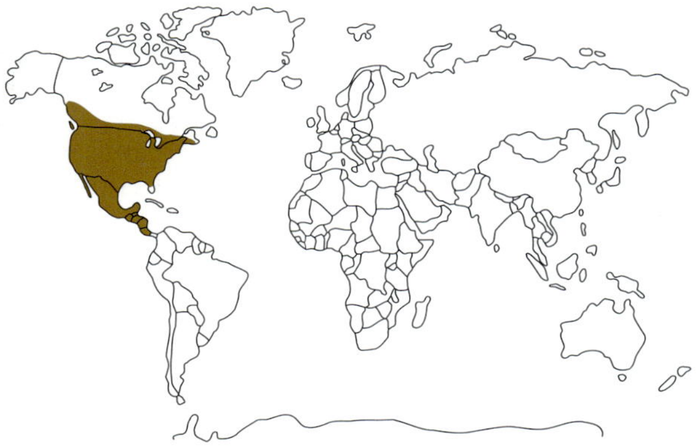

World breeding range

The species prefers to nest and forage near bodies of water bordered by fields. Lac Saint-François National Wildlife Area, southwest of Montréal. *September 1991, Léo-Guy de Repentigny.*

Bank Swallow

Hirondelle de rivage

Riparia riparia

Christiane Girard

THE BANK SWALLOW, THE SMALLEST SWALLOW IN NORTH AMERICA, usually inhabits open areas, most often near water, and nests in colonies that can range in size from several dozen to several hundred pairs.

The species has a widespread, holarctic distribution, breeding in North America and throughout Eurasia. North American populations winter in South America, mainly in Brazil.

Habitat and behaviour

The Bank Swallow is unusual in that it nests in a burrow tunnelled in a steep sand, clay, or gravel bank near fresh or salt water. It also uses cliffs and steep embankments alongside road and railway cuts, sand and gravel borrow pits, and sometimes even heaps of sand, sawdust or peat (Knapton 1988b; J.-L. DesGranges, pers. comm.). It also occupies drainage holes in concrete walls (Ouellet 1974a).

Large numbers of Bank Swallows can be seen around mid-May, as they make their way north, coursing over streams and clearings, where they find a wide variety of emerging insects to eat. Their diet consists mainly of beetles, flies, and bugs, and occasionally hymenopterans and damselflies (Stoner 1936). A study conducted by Stutchbury (1988) showed that within colonies Bank Swallows do not seem to communicate with each other about feeding sites the way Cliff Swallows do.

This swallow can be recognized by its irregular, zigzag flight, and the brown band across its white throat and breast. Birds in a colony are sometimes observed in pursuit flights shortly after their arrival on the breeding grounds. Petersen (1955) noted that these chases usually involved a male and a female and were part of the courtship display. According to Beecher and Beecher (1979), most of the time the male guards his mate as she travels, especially during the period between pair formation and the start of incubation, which would correspond to her fertile period, to prevent her from copulating with other males in the colony.

Both sexes take part in nesting activities, such as digging the burrow, building the nest, incubating the eggs, and tending the nestlings; working together in this way undoubtedly helps reinforce the pair bond (Gross 1942a). Pairs seem to prefer to burrow near the top of a bank. The firmness of the ground and lack of organic matter are also important factors in the selection of the nest site (John 1991). Height of burrows above land or water in Ontario ranged from 0.6 to 45.7 m, and their depth varied between 20 and 150 cm (Peck and James 1987). The entrance is usually oval or rounded, with a vertical diameter of approximately 2.5–5.0 cm and a horizontal diameter of 5.0–6.4 cm (Stoner 1936; Peck and James 1987; Hjertaas et al. 1988). Other cavity nesters, including the Belted Kingfisher, European Starling, and Northern Rough-winged Swallow, are often found nesting near Bank Swallow colonies (Gross 1942a; Peck and James 1987).

The nest itself is built in a wider chamber at the end of the burrow. The materials used depend on what is available, but usually include blades of grass or straw, and less often stalks, leaves, and rootlets (Stoner 1936; Peck and James 1987). Theft of nesting materials seems to be fairly common in Bank Swallow colonies and has been reported by several authors: a bird visits the cavities of conspecifics until it finds one that has been left unguarded. It goes in and comes out a few seconds later with materials that it takes back to its own nest.

Observations of Bank Swallows carrying feathers have also been reported. This swallow usually adds other birds' feathers to its nest during the incubation period, and feather fights often ensue when colony members detect a bird returning to its burrow with a feather in its bill, resulting in the feather-bearer often losing its prize to another bird (Hoogland and Sherman 1976).

Notes on territory

While excavating and building the nest, adults must defend their burrows against conspecifics, which sometimes results in fights (Hoogland and Sherman 1976). Once laying has commenced, however, conflicts cease. As the Bank Swallow is colonial, its territory is restricted to the area immediately surrounding the nest site (DeGraaf and Rudis 1987). The minimum spacing of nest holes recorded in a Wisconsin study was 10.2 cm, while most were 12.7–17.8 cm apart (Petersen 1955).

Any predator that approaches the colony is soon mobbed by most of the adults, even those whose burrows are not directly threatened (Hoogland and Sherman 1976).

The Bank Swallow usually nests in eroded banks of salt- and freshwater systems. The substrate must be easy for the bird to dig, as it is in the red sandstone cliffs of the Magdalen Islands. Île Brion Ecological Reserve. *July 1982, Yves Aubry.*

Bank Swallow

Breeding distribution

1984–1989

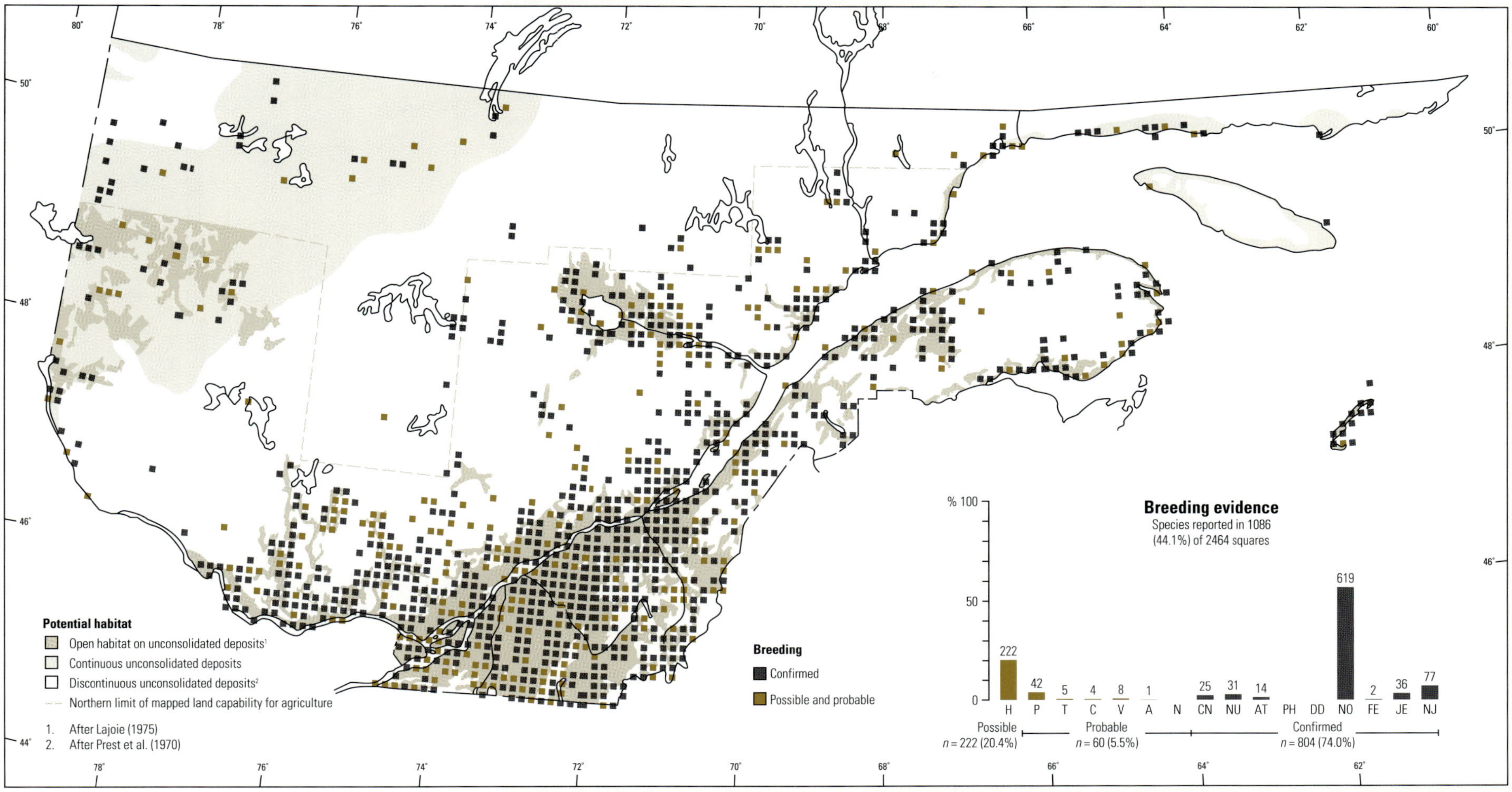

Species profile

Clutch size	usually 4–5 eggs (2–6)[1,2,3,4]
Laying interval	1 egg per day[1,2]
Start of incubation	with third, fourth or fifth egg[1]
Incubation period	13–16 days[1,2]
Maturity at hatching	altricial
Care of young	by female and male[3]
Nestling period	18–22 days[2]
Age at first flight	usually ≈ 23 days (17–23)[1,10]
Dependence of young	until 7–8 days after first flight[2]
Broods per year	1[1,2]
Breeding age	1 year[5]
Mating system	monogamy[9]
Duration of pair bond	?
Total length	12–14 cm[6]
Wing span	25.4–27.9 cm[4]
Weight	14.6 g (12.0–18.6)[8]
Longevity record	8 years[7]

References:
1. Petersen 1955
2. Stoner 1936
3. Gross 1942a
4. Terres 1980
5. DeGraaf and Rudis 1987
6. Godfrey 1986
7. Clapp et al. 1983
8. Dunning 1984
9. Ehrlich et al. 1988
10. Harrison 1978

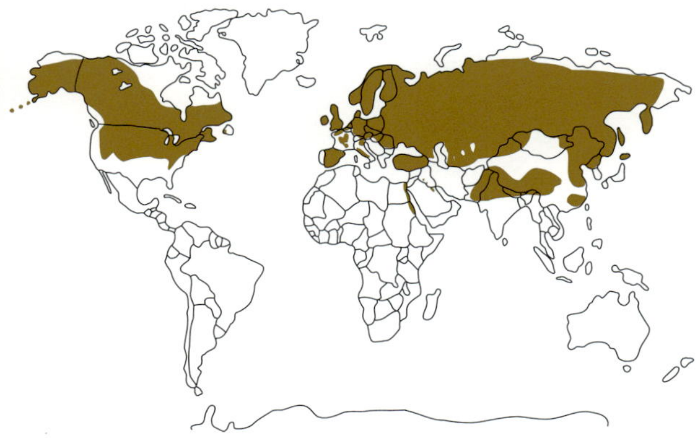

World breeding range

Chronology $n = 228$

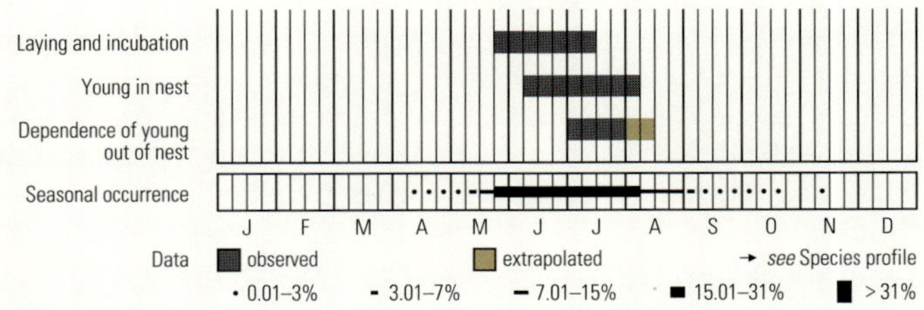

Potential Bank Swallow Habitat

Nest sites are mainly in banks with accessible unconsolidated deposits. The Atlas map shows the open areas of southern Québec (adapted from Lajoie 1975) and the distribution of unconsolidated deposits left after the last glaciation (adapted from Prest et al. 1970).

Open habitat on unconsolidated deposits

Abundant, thick, and varied in plains and valleys of marine origin (St. Lawrence and Ottawa lowlands, Chaleur Bay), unconsolidated deposits have been exposed by human activity and erosion. This process has produced open and partially open areas (such as eroded banks, sand terraces, grabens, sandpits, and cuts and fills) suitable for colonies. In the Abitibi–Témiscamingue region, unconsolidated deposits are of glacial lacustrine origin, and are generally thinner than those on the St. Lawrence Plain.

Continuous unconsolidated deposits

During atlassing, colonies were discovered on some fairly thick continuous unconsolidated deposits accessible to the species, such as in the Clay Belt that stretches from Témiscamingue to James Bay, where glaciation left deltas, raised beaches, kames, and eskers.

The areas on the Lower North Shore and Anticosti Island affected by the marine transgression of the last glaciation, and which are beyond the northern limit of land capability for agriculture mapped by Lajoie (1975), also fall into this category.

Discontinuous unconsolidated deposits

On the high plateaus and mountains of the Canadian Shield, where the bedrock is covered with a thin layer of moraine at best, thick unconsolidated deposits (outwash plains, eskers) are located in valleys. In these cases, as in those described above, steep river banks are suitable habitats for colonies. The species also uses outcroppings less than 1.5 m thick.

This sandpit at Beauport, near Québec City, is home to a colony of Bank Swallows. *July 1981, Paul Perreault.*

Laying may begin toward the end of May, but usually starts in early June. Incubation takes about 2 weeks, and the eggs usually hatch in June or early July. For the first few days, the young are brooded constantly and fed every 5 minutes. When they are about 18–22 days old, they launch themselves into the air, first hesitantly, then more confidently. For another week, they return to the nest to be fed. The colony is highly active during this period, with dozens, even hundreds of birds entering and leaving the burrows; yet the adults seem to recognize their offspring and feed only their own (Petersen 1955). Gradually the young stop returning to the nest except to roost for the night. During the day, they sometimes perch at the top of the bank and dig small holes, their first instinctive attempts at burrowing. They leave the natal site when they are 7 or 8 weeks old (Stoner 1936).

The highly visible burrows, often clustered near the top of a bank, and the considerable activity of the birds near their nests enabled atlassers to confirm breeding in 74% of squares in which the species was reported, mainly from seeing adults entering or leaving the nest, rather than eggs or nestlings, since it is hard to check on nest contents at the end of the burrow.

One unusual aspect of Bank Swallow behaviour is that they flock with other swallow species in autumn. They perch on shrubs along the water, just before dusk, then do an aerial ballet of spirals and loops, before returning to roost for the night (Gross 1942a). These gatherings begin around the end of July and continue until migration, in late August or early September.

Distribution

The Bank Swallow breeds pretty well throughout southern Québec, especially in inhabited areas. To some extent, its presence is associated with the distribution of unconsolidated deposits (such as sand or clay) that have been exposed and worn away by water action (eroded embankments) or by human activity (borrow pits). In inhabited areas, there are also totally artificial habitats like drainpipes in the riprap alongside roads, rivers, or bridges, or piles of sawdust, peat moss, or various other loose materials. Outside these areas, the species is rarer and rather local, being limited to valleys, eskers, and moraines, as are the deposits and embankments it uses, partly because in a good part of the province, the layer of soil and loose materials is rather thin. The species also nests in the crumbly red sandstone banks of the Magdalen Islands, the Gaspé Peninsula, and the North Shore, or the thin humic layer on cliff tops.

The species is even rarer and more local north of the area covered by the Atlas. A colony was reported at Fermont in July 1976 (ÉPOQ). D'Amours (1989) observed several flocks of swallows over the

Whether the banks chosen are of sand, gravel, or clay, the swallows only hollow out a nest in certain strata, so that the nest does not collapse during breeding. Martin-Valin ZEC, north of the Saguenay River. *August 1991, Jean Gauthier.*

Borrow pits like the ones that provide materials for road construction provide the species with new breeding sites in the hinterland. Martin-Valin ZEC, north of the Saguenay River. *July 1993, Jean Gauthier.*

De Pas River and a colony of about 20 nests approximately 25 km from the confluence of the De Pas and George rivers in June 1989. Todd (1963) also encountered the species north of Sept-Îles, up the Sainte-Marguerite River near its confluence with the Rivière de la Montagne Blanche, on 25 June 1917 and, a little farther north, on the northeastern arm of the Sainte-Marguerite, on 2 July 1917, at a spot he called Long Portage (51°51' N). Todd also observed it in the Ungava Peninsula about 40 km north of Kuujjuaq, on the Koksoak River, on 21 August 1917, and reported that it had nested near Rivière à la Baleine in the past. More recently in the Ungava Peninsula, it was seen at Tasiujaq, on 2 August 1982, and near the Kuuraapik River, on 1 and 7 August 1983 (ÉPOQ). Todd (1963) acknowledged only one record along the entire James Bay and Hudson Bay east coast, that of Manning and MacPherson (1952), who reported 3 birds at the mouth of the Roggan River on 13 August 1950 and found a few old nests along the river banks. Several new sightings have been made in the area since then. Bordage and Aubry (1983) found a colony on the Eastmain River in the summer of 1982, and another colony consisting of a dozen or so nests was discovered along the banks of the La Grande Rivière, near Fort George, on 24 June 1992 (G. Morissette, pers. comm.). The species has also been sighted near Lakes Élizabeth and Kakupis (Cons. G. G. – GREBE 1991a) and, farther north, at Poste-de-la-Baleine, on 30 July 1983 (ÉPOQ).

In the spring of 1992, a colony of Bank Swallows settled in a pile of granite dust left temporarily at the Old Port of Montréal. *June 1992, Pierre Brousseau.*

Adult and young Bank Swallows at the entrance to their nest. Shediac, New Brunswick. *July 1993, Robert Morin.*

History and trends

According to Robbins et al. (1986), the instability of colonies explains the large variations in annual counts and makes it hard to detect long-term continent-wide changes. Interestingly, these authors noted that between 1966 and 1979, the highest means for the species in North America documented by the BBS were obtained in Québec and New Brunswick.

Data on past abundance of this species in Québec are sparse. It used to be a very common summer resident around Québec City, and was considered common in Montréal (Wintle 1896), the Eastern Townships, and the Ottawa Valley (Mousley 1916a; Lloyd 1924); it also nested on the Gaspé Peninsula (Demille 1926; Ball 1943).

According to Cyr and Larivée (1993a), a variety of factors, such as the mining of sandpits, disturbance caused by all-terrain vehicles, the low diversity of insects as a result of pesticide use on farms, and the draining of ponds and streams, could adversely affect the species, particularly in the southern part of the province. Development of river banks and modern road-building techniques, replacing steep escarpments with gentle grassy slopes, are other factors that may reduce the number of breeding sites available to the species in the province. This notwithstanding, no marked trend has been detected in Bank Swallow numbers in southern Québec since the late 1960s (Falardeau and DesGranges 1991; Cyr and Larivée 1993a; BBS).

LOUISE CHAMPOUX AND MANON BOMBARDIER

Cliff Swallow

Hirondelle à front blanc

Hirundo pyrrhonota

Christiane Girard

A NATIVE OF THE CANADIAN WEST, THE CLIFF SWALLOW IS NOW FOUND throughout almost the entire country, from the Atlantic to the Pacific. It is not known why, but in eastern Canada, it nests mainly on artificial structures, while in the west, colonies often nest on canyon cliffs, hence its name.

There are few passerine species in which colonial nesting is more highly developed than the Cliff Swallow (Emlen 1954). Some colonies may consist of up to 2000 or 3000 pairs, although there are usually only a few hundred (Brown 1985). The species is also gregarious outside the breeding season. The Cliff Swallow is a fairly common breeder in the settled areas of southern Québec and winters in South America.

Habitat and behaviour

The Cliff Swallow begins building its nest as soon as it arrives in Québec in late April or early May. Pairs prefer to reuse nests that are in good condition, but tend to alternate between sites, occupying them every second year, probably as a way of avoiding parasitic insects in old nests (Gauthier et al. 1994). Axtel (1947) also saw the species use old robin and Eastern Phoebe nests. Before it began using man-made structures, it nested on steep cliffs and canyon walls in the mountainous parts of its range. Many Cliff Swallows still nest in areas where this habitat is available. The species quickly accepted man-made constructions and now often nests on the concrete structures of bridges and buildings, preferably in the angle where a wall and roof or two beams meet, or in a corner (Robidoux and Cyr 1989). It looks for sites with a vertical surface beneath a ledge or overhang, at least several metres above the ground or water (Emlen 1954).

Nest building in the Cliff Swallow is highly time and energy intensive (Gauthier and Thomas 1993), which explains why it usually selects sites that require the least investment. It also tries to make use of existing nests, both to provide better adhesion for its own new nest and to save time and energy. Pairs therefore prefer to build a nest that is semidetached (abutting the one beside it) or in a row (abutting those on either side), rather than isolated.

The Cliff Swallow often affixes its nest, a mud flask, to the concrete structure of a bridge. This gregarious species breeds in colonies, with nests abutting one another. Île des Soeurs Grises, Châteauguay. *July 1991, Gaétan Duquette.*

Notes on territory

The male arrives first on the breeding grounds, selects a site or territory, and aggressively defends it against all comers. One to five days later, a persistent female manages to penetrate the territory and mate with its holder. Territorial behaviour throughout the entire breeding period means that the openings of two adjacent nests are rarely less than 10 cm apart (Emlen 1952).

Cliff Swallow

Breeding distribution
1984–1989

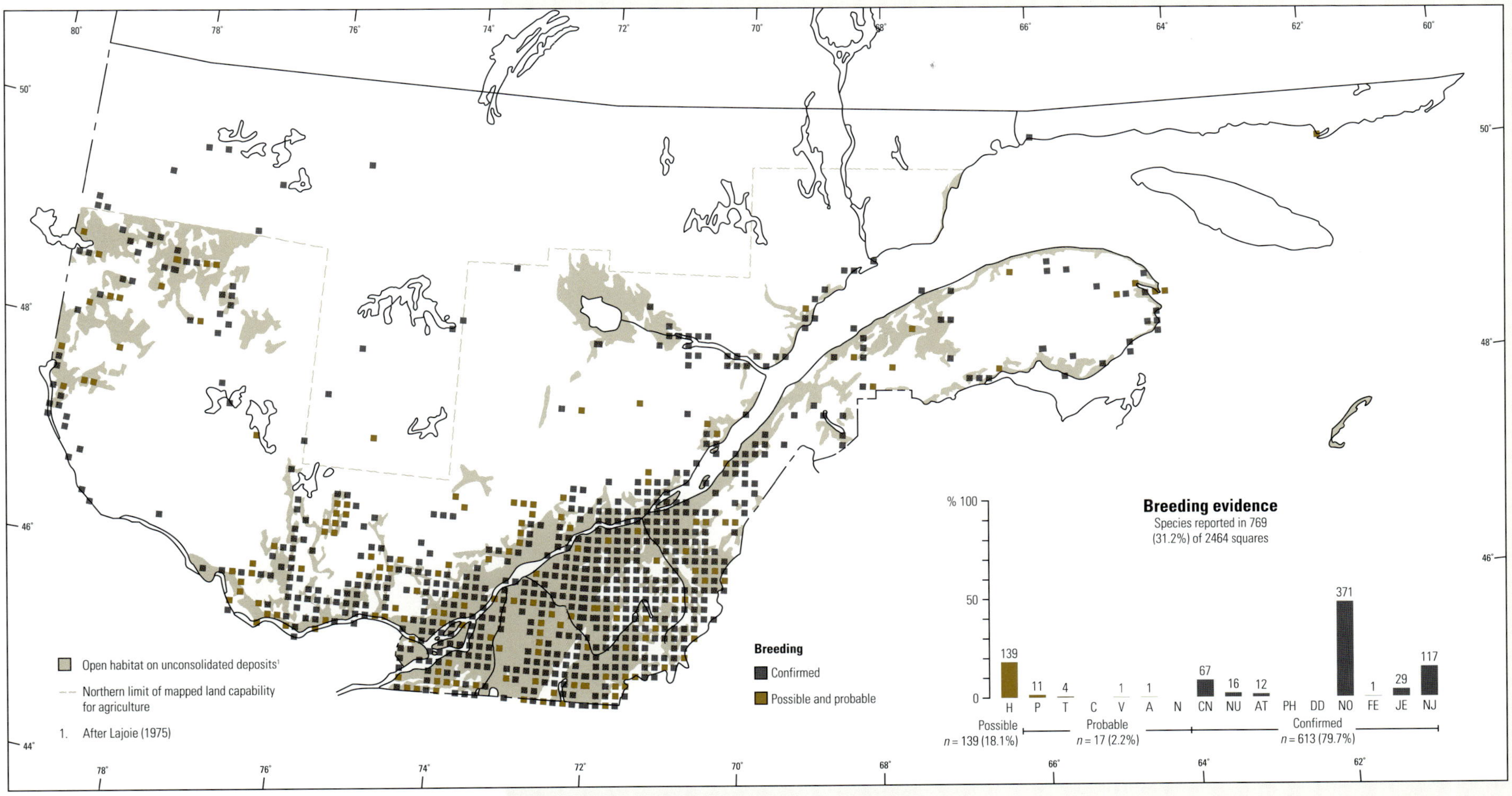

Species profile

Clutch size	usually 4–5 eggs (3–6)[1,2,3,4]
Laying interval	1 egg per day[2]
Start of incubation	with penultimate egg, sporadically after second or third egg[2]
Incubation period	usually 13–15 days (11–19)[1,2,3,4,7]
Maturity at hatching	altricial
Care of young	by female and male[1,2,4]
Nestling period	21–24 days[2,3,4]
Age at first flight	21–24 days or more[9]
Dependence of young	until 3–5 days after leaving nest[7]
Broods per year	1, rarely more[1,2,3,7]
Breeding age	1 year[8]
Mating system	monogamy, polygamy possible[1,2]
Duration of pair bond	?
Total length	12.5–15.0 cm[5]
Wing span	30.5–31.1 cm[3]
Weight	21.6 g (17.5–26.7)[6], M 23.9 g (?) / F 24.2 g (?)[7]
Longevity record	11 years[7]

References:
1. Gross 1942b
2. Samuel 1971
3. Terres 1980
4. Ehrlich et al. 1988
5. Godfrey 1986
6. Dunning 1984
7. Brown and Brown 1995
8. DeGraaf and Rudis 1987
9. according to nestling period

Chronology $n = 594$

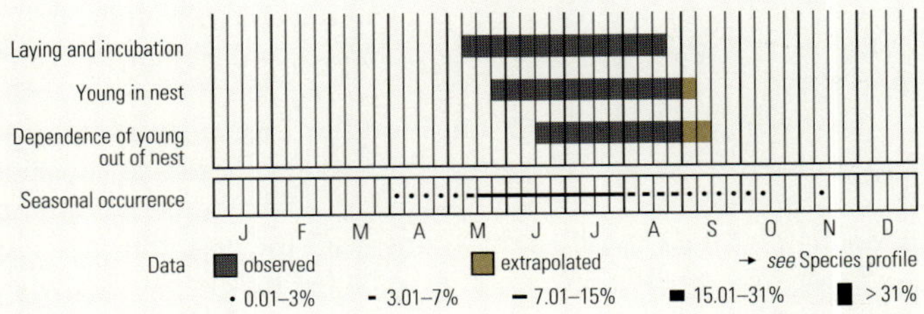

World breeding range

Cliff Swallows often place their nests under the eaves of houses or overhangs of a variety of other buildings. Cap Tourmente National Wildlife Area, northeast of Québec City. *August 1991, Léo-Guy de Repentigny.*

The Cliff Swallow probably builds one of the most elaborate and energy-consuming nests of any species, in relation to its weight. According to Gauthier et al. (1994), the weight of a typical nest is about 30 times that of an adult bird, and its construction requires between 1400 and 1800 trips to sources of mud, representing an investment of another 122–206 kJ. Its structure is also very interesting: shaped like a gourd, it is made essentially of mud pellets. In colonies, nests are often close together; they usually have a roof and often have a circular entrance tunnel with the mouth directed more or less downward (Emlen 1954). The Cliff Swallow's nest, unlike that of the Barn Swallow, has a high sand content (58.6–65%) and little organic matter, which is why it is more fragile (Kilgore and Knudsen 1977). Robidoux and Cyr (1989) have demonstrated that the importance of adhesion, cohesiveness, and stability in suspended nests dictates this swallow's choice of building materials.

Nest construction is intermittent, so that each mud pellet can dry and harden to form a firm base for further pellets (Buss 1942). The work takes 5–21 days, depending on environmental conditions: precipitation, wind, availability of materials and food, and so on (Robidoux and Cyr 1989). The mud is gathered from nearby ponds or the muddy banks of lakes and rivers. This activity is highly social, since all participants are concentrated in a limited area, and all the members of the colony engage in mud gathering at the same time, usually in the afternoon, spending most mornings on breeding and territorial defence. Pair bonds, although well established before nest construction starts, are reinforced by working together on the nest as well as by defending the territory, incubating, and feeding the young (Emlen 1954).

Most pairs produce only one brood per season, but some have two. Unlike the Barn Swallow, the Cliff Swallow begins to lay before the nest is completed. Incubation takes 13–15 days and the nestling period is 21–24 days. When the young leave the nest, they perch on a wire or tree, where they continue to be fed by the parents, which recognize their offspring by their calls (Beecher et al. 1986). Analysis of several factors, including laying and hatching dates, indicates synchrony of reproductive cycles within colonies (Emlen 1952).

Foraging is another social activity in this species, since individuals tend to leave the nest site and head for the same feeding grounds together. Flock feeding is encouraged by the exchange of information among members of a colony. Thus, less adept birds observe the more successful foragers and follow them to their feeding sites. In bad weather, when insects are less active, making them harder to find, Cliff Swallows seem to communicate vocally to locate sources of food. Passive exchanges seem to take place only at nest sites, while vocal, or active, exchanges are more frequent away from them (Brown 1988, 1991). The Cliff Swallow forages by flying over fields, marshes, or open water, following insects in the thermals (Emlen 1954). It eats mainly dipterans, homopterans, and hymenopterans (Brown 1985).

The colonial behaviour of the species seems to have disadvantages for some individuals, however. For example, the intraspecific parasitism observed in this swallow is certainly detrimental to the host, since it reduces her reproductive fitness. According to Brown and Brown (1988, 1989), there are two ways this reduction occurs. First, when a parasitic egg is laid in the host's nest during her laying period, the host reacts by laying fewer eggs. Second, when a parasitic

The nest is made of mud pellets that have dried and hardened. Here, an adult feeds one of its young. Saint-Augustin-de-Desmaures, west of Québec City. *July 1957, Gaston Le Page.*

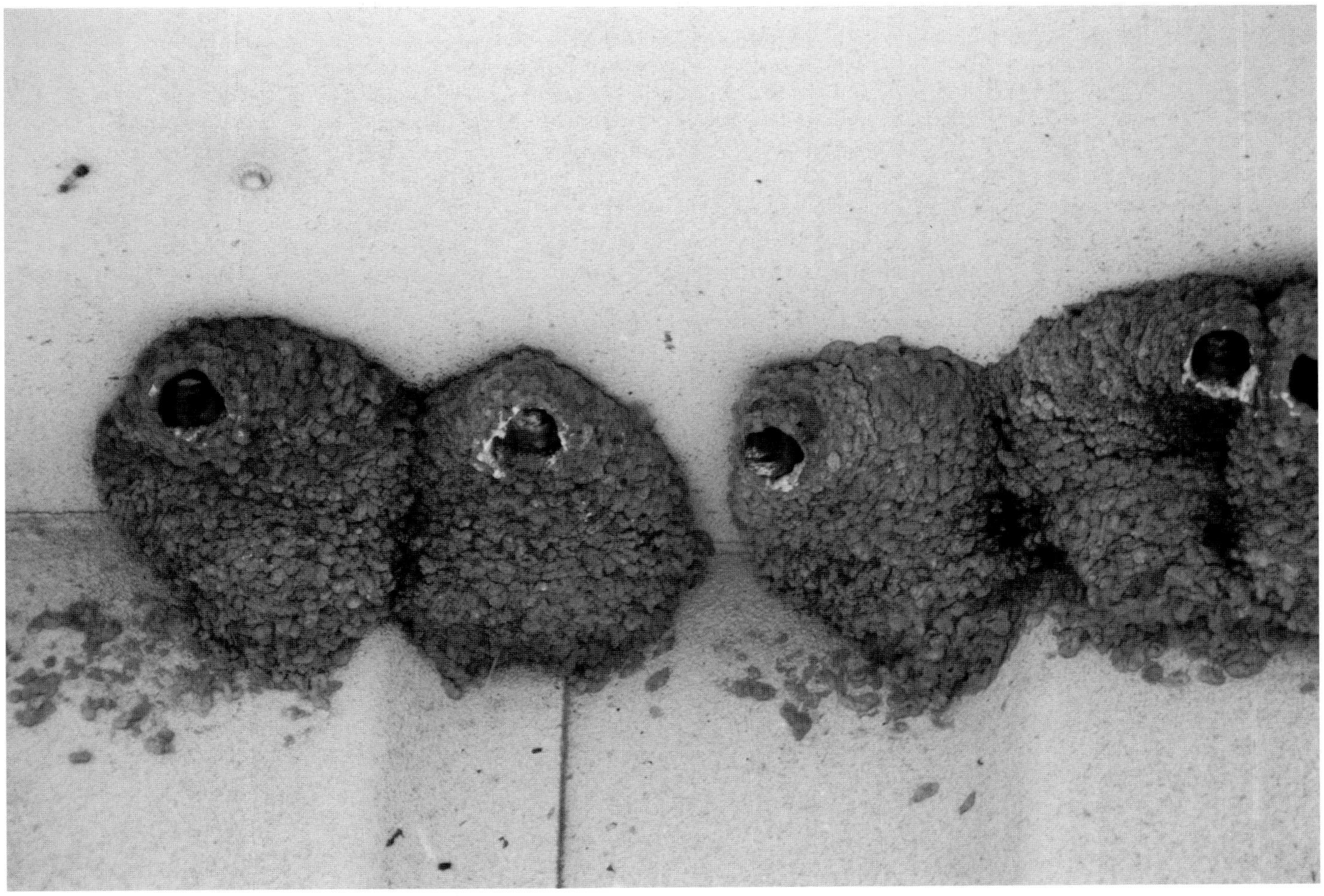

Gourd-shaped nests under the overhang of a building. The substrate must be porous enough that the swallows can attach their nests firmly. Unlike the cup-shaped nest of the Barn Swallow, that of the Cliff Swallow is almost always enclosed, with a side entrance that slopes slightly downward. Île Notre-Dame, Montréal. *July 1981, Daniel Jauvin.*

egg is deposited or transferred into the host's nest after she has finished laying, the parasitizing bird first destroys a few of the host's eggs, probably to reduce competition among nestlings.

Given its association with human structures, preference for open habitats, and colonial nature, the Cliff Swallow is easy to spot and observe, as are its breeding activities. Breeding was confirmed in close to 80% of squares in which the species was atlassed. Since the nest's structure and location render it fairly inaccessible, 60.5% (321/613) of confirmations were based on reports of adults leaving or entering nests, the contents of which could not be examined.

Distribution

During the breeding season, this swallow may be seen in the populated regions of southern Québec. Most breeding evidence for the Atlas came from the southernmost part of the province, chiefly the Eastern Townships, Beauce, Ottawa Valley, and Central St. Lawrence Lowland. The Cliff Swallow also breeds locally in Abitibi, Témiscamingue, the Saguenay, on the Upper North Shore, in the Lower St. Lawrence, and along the Gaspé coast. It is uncommon in Lac-Saint-Jean (Savard and Cormier 1995), despite the presence of suitable habitat and structures. As Todd (1963) indicated, the Gulf of St. Lawrence seems to mark the northern and eastern limits of its breeding range.

Breeding was reported for the Atlas up to the Sept-Îles area on the North Shore, with sight records, though no evidence of nesting, at Natashquan. The species was not reported on Anticosti Island, nor on the Magdalen Islands. Indeed, Fradette (1992) considers the Cliff Swallow a rare transient on the Magdalen Islands, where the only known breeding record relates to the discovery of a colony in 1971.

History and trends

The Cliff Swallow, which used to nest on cliffs, was probably rare and confined to certain spots in eastern North America prior to European settlement and the clearing of forests (Gross 1942b).

In eastern North America, and more specifically in New York State, where it bred for the first time in 1817, this swallow was abundant in the early part of the 20th century (Bull 1985). However, the increasing number of painted buildings then resulted in a drop in numbers in the eastern United States, since nests do not stick nearly as well to painted surfaces. The construction of dams and bridges, on the other hand, created many new sites and allowed the Cliff Swallow to recolonize areas it had formerly frequented, and to expand its range toward the southeastern part of the continent (Robbins et al. 1986). BBS data indicate an increase in populations in the western and southeastern United States and in Canada between 1966 and 1979.

There is insufficient information available to determine Québec Cliff Swallow population trends. According to Hall (1862) and Wintle (1896), it was common in summer around Montréal at the end of the 19th century; Wintle (1896) also reported it as nesting on the Island of Montréal. Another historical record of this swallow is that of Verrill (1862), who reported observing several hundred pairs of Cliff Swallows nesting at Cap à l'Aigle near Port-Menier in July 1861. Brewster (1884), however, believed that the colony had disappeared from there by 1881. Comeau reported a colony near Godbout in 1882 (Merriam 1882), and Hubel (1907) rated the species as fairly common at Lake Timiskaming in the summer of 1905. According to Ouellet (1974a), the species is less abundant today than it was at the end of the last century, yet BBS and ÉPOQ data (Cyr and Larivée 1993a) show no marked population trend between the late 1960s and the late 1980s.

In fact, it is very difficult to estimate the current local abundance of Cliff Swallow populations because colonies are highly mobile from year to year (S. Gauthier, pers. comm.).

Lucie Landry and Manon Bombardier

Barn Swallow

Hirondelle des granges

Hirondelle rustique

Hirundo rustica

A SYMBOL OF SUMMER, THE BARN SWALLOW IS A HOLARCTIC SPECIES WITH several subspecies established in North America, Europe, Africa, and Asia. New World populations, which belong to the subspecies *erythrogaster*, winter in South America and breed throughout North America, except Florida, from north-central Alaska to central Mexico (Godfrey 1986). In Québec, this swallow is a common migrant breeder widespread south of the 50th parallel.

Habitat and behaviour

The Barn Swallow is the only North American member of its family with a deeply forked tail and white spots on its tail feathers. The male's tail seems to play an important role in attracting a mate. During courtship, the male flies over and in front of the female, spreading his tail in a fan, and when the female visits various territories before mating, the territory holders spread their tails, while clinging to potential nesting sites (Møller 1990a).

This swallow inhabits a variety of open areas, where it hunts insects on the wing. It is most often encountered close to farms and near waterways, where it can find buildings and other structures that provide a sheltered site for its nest. Most nests are built on artificial structures: bridges, barns, and tunnels, for example. Nests inside buildings are usually on a beam, several metres up, near the ceiling. Apparently the only natural nest sites used are the undersides of rocky outcroppings on cliff tops, trees, and the inside of caves, as has been observed near Lake La Mothe in the Saguenay–Lac-Saint-Jean region (N. Tremblay, pers. obs.). The nest is anchored to surfaces of compacted sand, gravel, rock, or wood. Availability of mud to build the nest seems to be an essential criterion in the selection of a nest site (Speich et al. 1986).

In spring, the earliest arrivals generally make their appearance in the first week of May, but the main influx occurs around mid-May, with stragglers arriving in early June (Fortin et al. 1978; Gauthier and Larivée 1989). The adults, particularly the males, exhibit strong nest-site tenacity, and may keep the same mate from year to year (Shields 1984). In contrast, yearlings ready to breed for the first time generally disperse from their natal site. In 1983 at Adirondack Park, New York, Shields (1984) observed that yearlings banded as fledglings the preceding year settled a median distance of 6.4 km from their natal site. This dispersal seems to be the result of intra-sexual competition between yearlings and adults (Shields 1984).

It takes 6–8 days (Harrison 1975)—sometimes up to 12—to build a nest, depending on environmental conditions. Barn Swallows seem to prefer to repair old nests, however, which requires only a few days' work (Bent 1942). Both partners participate, but the female, working mainly in the morning, tends to do more. The nest is a deep cup made of dry mud; it is normally reused several years running. When mud is scarce, the nest is made of plant materials, usually dry grass, but sometimes also algae (Duffin 1973). The female then lines the nest with feathers. The number of feathers used seems to influence the time spent on incubation, the duration of the nestling period, and the body weight of nestlings, but not egg temperature or reproductive success (Møller 1991a).

Christiane Girard

Notes on territory

Despite the fact that the Barn Swallow is usually a solitary nester, colonial nesting is observed regularly (Speich et al. 1986). Colonies generally number 6–8 nests, on average, although concentrations of 55–63 pairs have been found in a single barn (DeGraaf and Rudis 1987).

Although both parents defend the nest, the male is more aggressive, especially after hatching (Møller 1991b). The breeding territory is larger during nest building and laying, corresponding to the female's fertile period (Møller 1990b).

In a rural environment, farm buildings offer a large number of nest sites, while the plentiful insects in pastures and fields are a good source of food. Cap Tourmente National Wildlife Area, northeast of Québec City. *August 1991, Yves Aubry.*

Barn Swallow

Breeding distribution
1984–1989

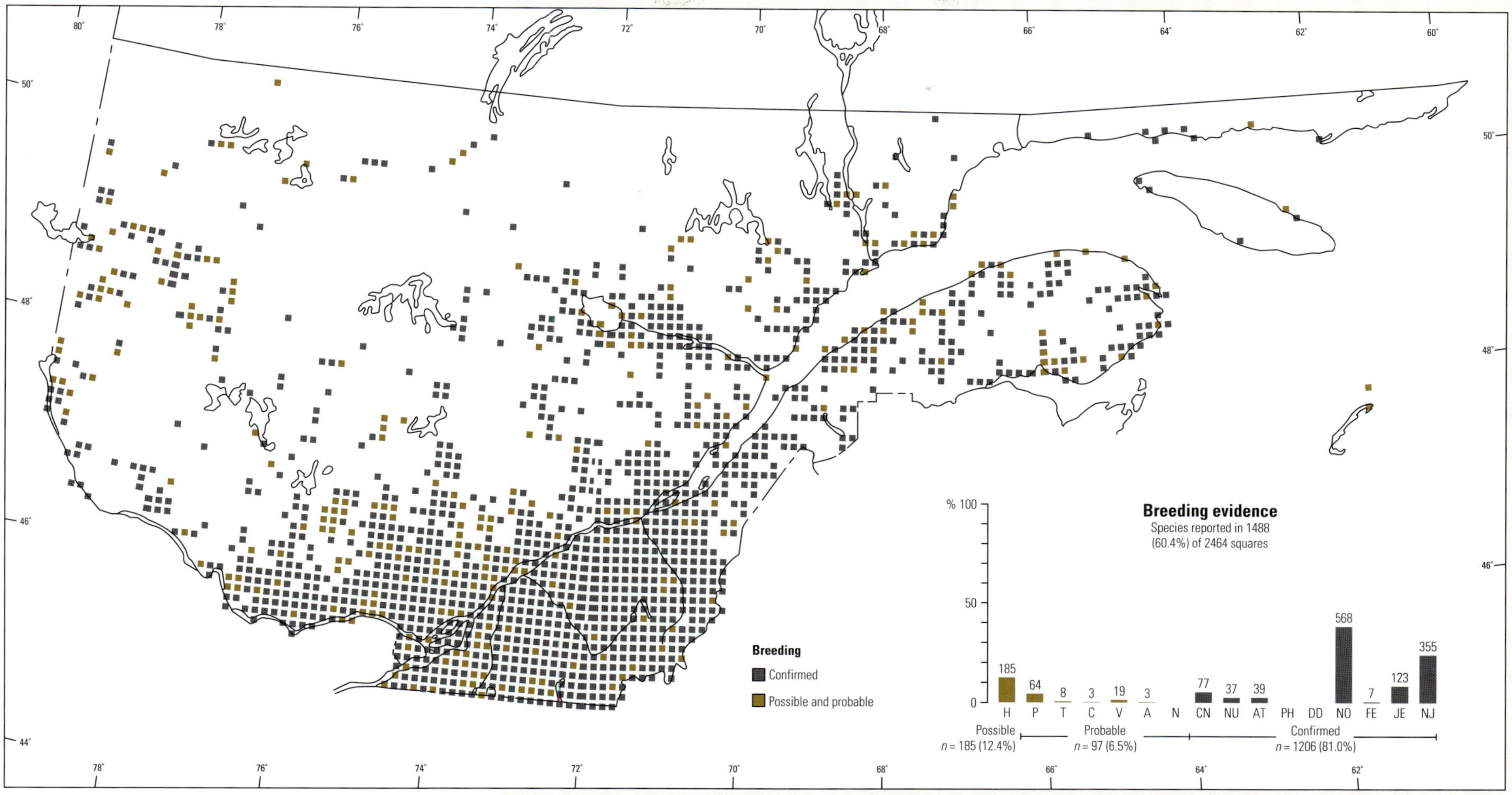

Species profile

Clutch size	usually 4–5 eggs (2–7)[1,2,3,4]
Laying interval	1 egg per day[11]
Start of incubation	with last egg, sometimes before[11]
Incubation period	usually 14–17 days (11–19)[1,2,3]
Maturity at hatching	altricial
Care of young	by female and male[1,2,4]
Nestling period	18–23 days[1,2,3,4]
Age at first flight	18–23 days or more[10]
Dependence of young	until a few days after leaving nest[1]
Broods per year	1–2[1,2,3,9]
Breeding age	F 1 year / M 1 year, sometimes 2[2]
Mating system	monogamy, sometimes polygyny[2,4]
Duration of pair bond	sometimes more than 1 season[4]
Total length	14.8–19.5 cm[5]
Wing span	31.8–34.3 cm[6]
Weight	18.6 g (13.4–23.4)[7]
Longevity record	15 years 11 months[2]

References:
1. Bent 1942
2. Turner 1989
3. Shields and Crook 1987
4. Ehrlich et al. 1988
5. Godfrey 1986
6. Kennard 1975
7. Dunning 1984
8. Terres 1980
9. DeGraaf and Rudis 1987
10. according to nestling period
11. Stokes and Stokes 1989b

Chronology $n = 881$

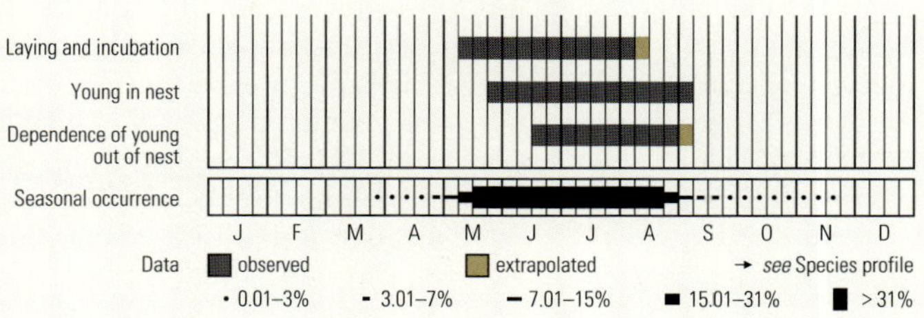

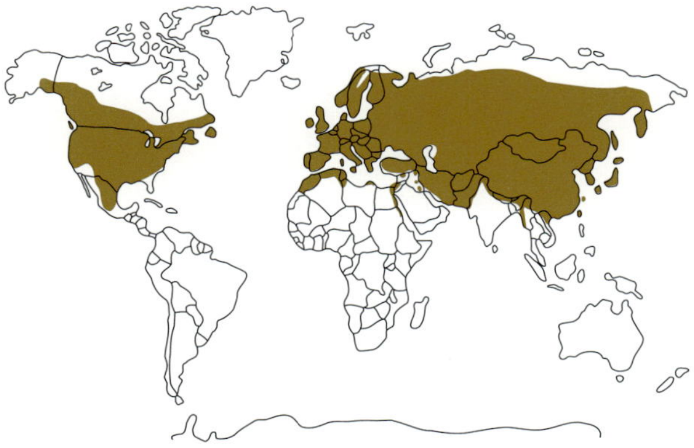

World breeding range

This swallow often makes its nest in an old building near a field or a body of water. It can slip in through the smallest opening and make itself at home. Le Bic, near Rimouski. *June 1984, Yves Aubry.*

Incubation generally lasts 14–17 days. In North America, the female does the lion's share of incubation, but she is occasionally relieved by the male (Ball 1983), especially in the early morning hours (Smith and Montgomerie 1992). In suitable habitat, pairs may nest singly or in small colonies of no more than 30 pairs (Speich et al. 1986). A study conducted in the state of New York found relatively high hatching success (about 78%), with just over half (51.1%) of the young surviving to fledging (Shields and Crook 1987). According to Shields and Crook (1987), ectoparasites, which flourish in colonies, are the principal cause of death in nestlings (65%).

The adults become noticeably aggressive when the young hatch, attacking any intruder that gets too close to the nest. Though harmless to humans, these attacks are sometimes made by a flock against winged or terrestrial predators, and are usually enough to deter them. Defensive behaviour seems to vary depending on the subspecies. According to Møller (1991b), for example, *erythrogaster* males (the North American subspecies) are more actively involved in the defence of the young than are those of the European subspecies, *Hirundo rustica rustica*. This difference is probably due to the number of extra-pair copulations, which seems higher in *Hirundo rustica rustica*; in effect, the intensity of a male's nest defence seems related to his rate of participation in the female's total number of copulations.

The young leave the nest 18–23 days after hatching, but for several days more they return to it to roost. Double-brooding seems to occur in the temperate parts of its range (Bent 1942), but this has not been confirmed in Québec. After nesting, family groups gather around hunting grounds, often forming flocks with other species of swallows. The Barn Swallow's diet is composed mainly of flies, especially house flies and also horse flies (Bent 1942).

Barn Swallow nest on the roof beam of a farm building. The nest is a cup made of mud and dry grass, often lined with feathers. Saint-François-du-Lac, Lake Saint-Pierre. *June 1980, Daniel Jauvin.*

The Barn Swallow inhabits open areas, often near water, where it catches large numbers of insects on the wing. Structures found nearby, such as this covered bridge, may be home to several breeding pairs. Routhierville, Gaspé Peninsula. *August 1991, Yves Aubry.*

Fall migration begins slowly in the early days of August. This swallow can still be seen here in September, but only occasionally. Although it has pretty well left the northeastern part of the continent by October, the Barn Swallow is known for late sightings in November, especially along the Lower St. Lawrence (Gauthier and Larivée 1989). Birds spotted then are probably young males prospecting for suitable nest sites for the following year (Davies 1990).

Breeding was confirmed in 81.0% (1206/1488) of Atlas squares in which the species was reported. The discovery of active nests or nests that had been occupied during the season, along with sightings of adults returning to or leaving the nest, accounted for 76.6% (960/1206) of breeding confirmations. The species is easy to observe, since not only does it nest close to human structures, its behaviour is conspicuous, especially near the nest.

Distribution

The Barn Swallow is abundant in southernmost Québec, especially in the rural areas that provide a large number of potential nest sites. This swallow is listed as a very common breeder in the southern and central Ottawa Valley area (Club des ornithologues de l'Outaouais 1985), as well at Lake Saint-Pierre (Société ornithologique du centre du Québec 1986). It is common in the Montréal (Bannon 1991b), Eastern Townships (Lepage 1993), and Lac-Saint-Jean (Breton 1980) regions. Gaboriault (1961) and Schmitt (1904) classed it as very scarce on the Magdalen Islands and on Anticosti Island. The Barn Swallow was atlassed as a confirmed breeder on Anticosti Island, including at Port-Menier, Baie-Sainte-Claire, and Rivière-aux-Saumons, but not on the Magdalen Islands, where the most recent breeding records date back to the early 1970s (Fradette 1992).

The Barn Swallow is found throughout southern Québec. Although it is primarily distributed below the 50th parallel, there are records from as far north as the 58th parallel: at Tasiujaq in 1984 and at Inukjuak in 1988 (ÉPOQ). It has been reported around Lake Evans, south of Nemiscau, but breeding has not been confirmed in this region. From west to east, it is found from the Abitibi–Témiscamingue region to the Magdalen Islands and on the North Shore as far as Natashquan.

History and trends

The Barn Swallow has long been considered a widespread breeder in North America (Eaton 1914; Bull 1985), yet it was probably not as common before European settlement (Eaton 1988d). Baird et al. (1905) reported that it originally nested in caves, under rocky outcroppings of cliffs, and in other similar locations. With settlement, the species gradually abandoned these natural sites for farm buildings; its numbers have increased greatly since. It also seems to have adapted to urban settings well before the Cliff Swallow, probably a full century before (De Kay 1844; Baird et al. 1905). The Barn Swallow has recently colonized Alberta, where it has followed settlement, especially farming (Erskine 1979).

According to Hall (1862) and Wintle (1882), the species was a common summer resident around Montréal in the last century. Ouellet (1974a) believes that this has not changed, although the Barn Swallow seems more abundant today because of the increased number of nest sites. Throughout Québec as a whole, the building of cottages and hunting and fishing camps in the hinterland, the building of dams and the construction of bridges for roads, not to mention the protection this species has enjoyed under the Migratory Birds Convention Act (S.C. 1917, c. 18), have undoubtedly contributed to its expansion.

According to Robbins et al. (1986), the Barn Swallow population increased between 1966 and 1979 throughout North America; the highest figures were obtained in the St. Lawrence Plain (encompassing Québec, Ontario, and New York). Data collected in southern Québec between 1970 and 1989, however, do not seem to support this trend, since they show a decrease in numbers (Cyr and Larivée 1993a). This decline is probably due to the fact that for some 15 years now, farm abandonment, with concomitant loss of farm buildings has been widespread all over southern Québec, except in the lowlands around Greater Montréal and in the Montérégie (R. Carignan, MAPAQ, unpubl. data). As well, the Barn Swallow is probably increasingly making use of resort facilities being built in the hinterland, while deserting a number of rural settings that have become less suitable, due to abandonment of farming.

LUCIE LANDRY AND MANON BOMBARDIER

Gray Jay

Geai du Canada

Mésangeai du Canada

Perisoreus canadensis

Christiane Girard

THE GRAY JAY EVOKES THE VAST BOREAL FORESTS OF QUÉBEC LIKE NO other bird. With its soft plumage and confiding ways—it is so trusting that it will eat from your hand, as many campers and loggers will attest—it is one of the most popular birds in Québec's hinterland.

Habitat and behaviour

The Gray Jay inhabits coniferous forests, particularly where spruce is found. It is monogamous and defends a permanent all-purpose territory. The average territory in La Vérendrye Wildlife Reserve was estimated at 69 ha (Strickland 1969, 1991; Strickland and Ouellet 1993).

This jay is omnivorous and feeds mainly on arthropods, berries, fungi, the eggs and nestlings of other birds, and large animal carcasses left by wolves and human beings (Ouellet 1970b; Strickland and Ouellet 1993). It can hawk for insects like a flycatcher and can even kill mice, but it cannot extract the seeds from conifer cones. It is able to survive northern winters on relatively small territories, suffering mortality rates of only 17% or less (Strickland and Ouellet 1993), thanks to its habit of caching provisions gathered in the summer and fall. With saliva from its large salivary glands, the Gray Jay works pieces of food into small sticky boli (Dow 1965), which it then hides throughout the forest, mainly under bark scales or arboreal lichens. The distribution of caches represents a compromise between the need to minimize food caching time and the need to space caches as far apart as possible to keep them from being stolen by competitors (Waite 1991). Memory appears to play an important role in the jay's retrieval of cached food (Bunch and Tomback 1986; Strickland 1991).

The Gray Jay nests very early, well before the return of most migrants, in March and April, when snow is still on the ground and it is very cold. The reason the species is able to breed when food is so scarce is no doubt its ability to access food stored the summer before. The jay's early breeding season is advantageous in that it frees pairs to concentrate on building up stores when food is most abundant, thus improving the species' ability to survive the following winter (Strickland and Ouellet 1993).

The nest is usually built in a black spruce, white spruce, or balsam fir, usually against the trunk, but sometimes well out on a horizontal branch when a tall white spruce is chosen. The nest is generally on the south side of the tree, about 1 m from the edge of the foliage, where it is well hidden and sheltered, but still receives a little sun. The nest tree is often located close to the northern edge of an opening in the forest such as a stream or peatland, where it also receives more sunlight. The nest is built of twigs interwoven with caterpillar cocoons, and is lined with feathers, particularly those of the Spruce Grouse and Ruffed Grouse, bark strips, lichens, and animal hairs (Rutter 1969). The height of 46 nests found in La Vérendrye Wildlife Reserve ranged from 2.2 to 7.7 m, with an average of 4.9 m (Strickland and Ouellet 1993).

The female sits on the nest as soon as the first egg has been laid to keep the eggs from freezing, but does not begin incubating in earnest until the clutch is complete. The clutch, which ranges from 2 to 6 eggs, but usually consists of 3 or 4, is incubated for an estimated 18.5 days (Godfrey 1986; Strickland and Ouellet 1993). The eggs all hatch the same day, which puts the nestlings on an equal footing, with no nestling developing at the expense of another. During the first 4 days after hatching, the female broods the nestlings almost continuously, while the male brings food every 45–50 minutes. After this, she plays an increasing role in feeding, eventually accompanying the male on all his foraging trips. The young leave the nest in early May, usually at 23 days of age (Strickland and Ouellet 1993).

Notes on territory

Permanent territories are used for nesting and foraging. The average size of territories has been estimated at 69 ha in La Vérendrye Wildlife Reserve and 146 ha in Algonquin Provincial Park (Strickland 1969, 1991; Strickland and Ouellet 1993). The pair chases trespassing neighbours from its territory (Rutter 1969), but is often accompanied by a third, non-breeding individual, usually one of its own offspring from the preceding season, but sometimes a non-breeding immigrant from elsewhere (Strickland 1991). In the Eastmain River region, higher densities of Gray Jays were found in mature conifer stands than in ecotones between peatlands and conifer stands (Tremblay and Belles-Isles 1991), while in the Great Whale River area, densities were higher in black spruce-moss stands than in peatlands (Cons. G. G. – GREBE 1991a). It is possible that densities may be overestimated during fall and winter, when the jays tend to follow observers, while densities in spring and summer may be underestimated, since the birds are silent and inconspicuous then.

The Gray Jay sometimes nests near openings in black spruce lichen woodlands. It often builds its nest in a tree near the forest edge, to take advantage of the sun. Grands-Jardins Conservation Park, northwest of Baie-Saint-Paul. *September 1991, Yves Aubry.*

Gray Jay

Breeding distribution
1984–1989

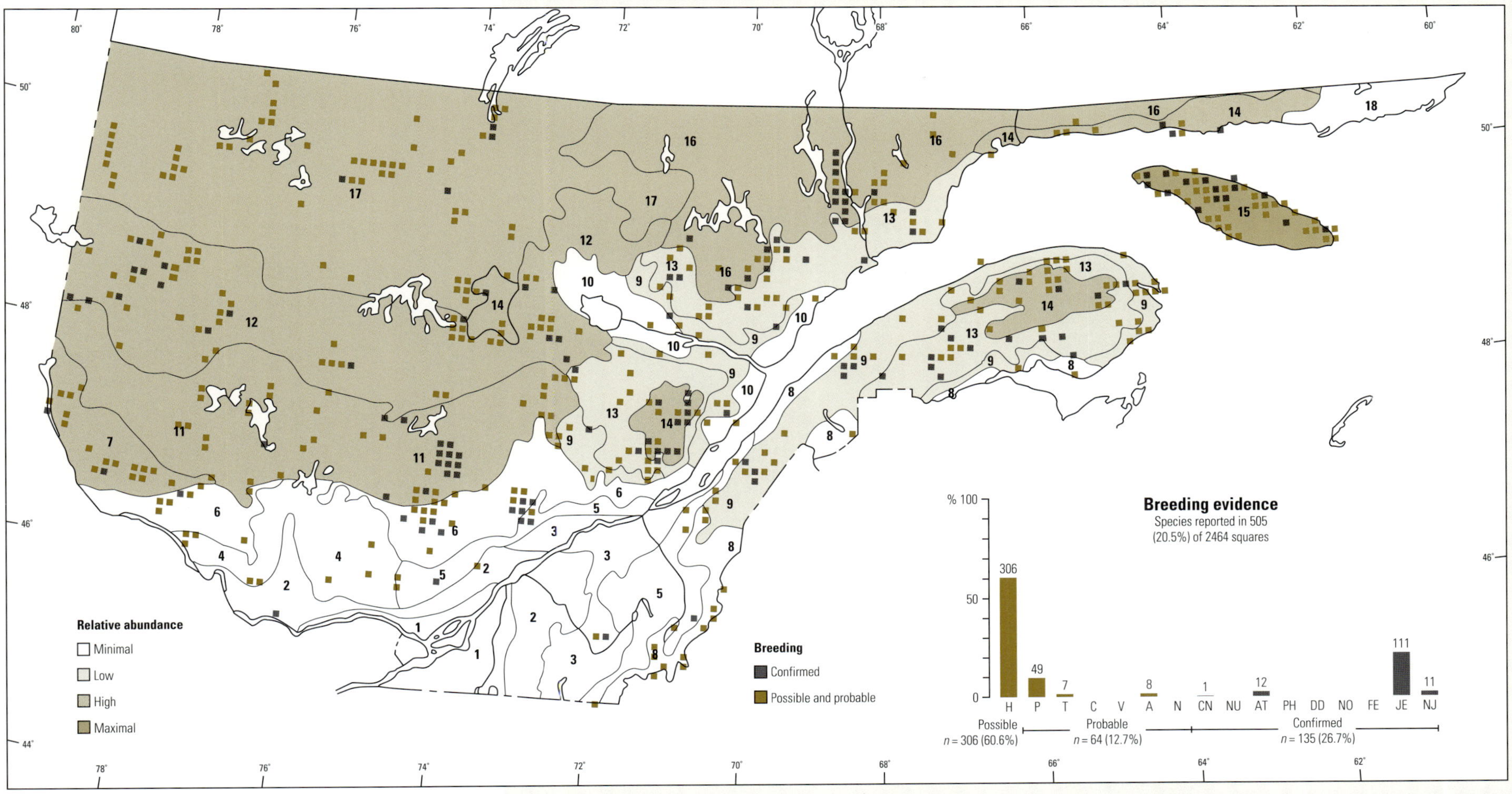

Species profile

Clutch size	usually 3–4 eggs (2–6)[1,2,3,4]
Laying interval	1 egg per day[7]
Start of incubation	with last egg, sporadic before[7]
Incubation period	16–18 days[1,2,3,4,7]
Maturity at hatching	altricial
Care of young	by female and male[1]
Nestling period	22–24 days[7]
Age at first flight	22–24 days[7]
Dependence of young	until at least 41 days old[7]
Broods per year	1[4,7]
Breeding age	1 year[4,7]
Mating system	monogamy[1,7]
Duration of pair bond	usually for life[7]
Total length	27.5–30.5 cm[3]
Wing span	40.6–44.4 cm[2]
Weight	71.1 g (64.8–73.6)[5]
Longevity record	10 years 2 months[6]

References:
1. Ehrlich et al. 1988
2. Terres 1980
3. Godfrey 1986
4. DeGraaf and Rudis 1987
5. Dunning 1984
6. Clapp et al. 1983
7. Strickland and Ouellet 1993

World breeding range

Chronology $n = 76$

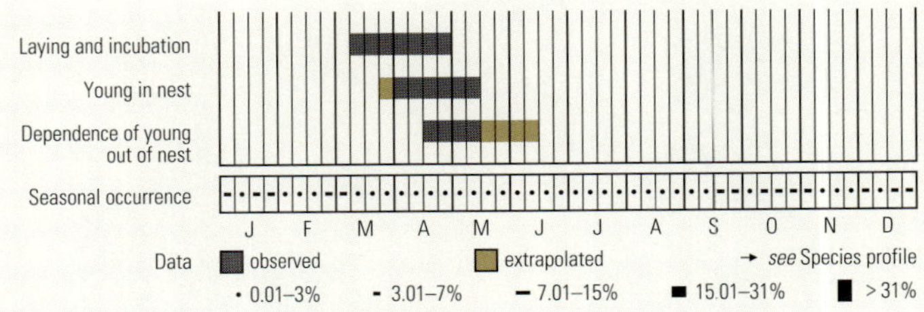

Frequency of occurrence by ecological region

Ecological regions** (forest regions)	Adequately covered squares	Squares in which reported	% of squares in which reported	Relative abundance (%)*
1. Sugar maple-hickory and sugar maple-basswood	97	0	0	0
2. Sugar maple-basswood	153	1	0.7 ± 0.9	0.7
3. Sugar maple-basswood and sugar maple-yellow birch	101	4	4.0 ± 2.6	4.0
4. Sugar maple-yellow birch-basswood	56	6	10.7 ± 7.4	10.7
5. Sugar maple-yellow birch-beech and sugar maple-yellow birch-basswood	113	9	8.0 ± 3.0	8.0
6. Sugar maple-yellow birch-beech	71	21	29.6 ± 9.8	29.6
7. Sugar maple-yellow birch-balsam fir	19	15	78.9 ± 19.7	78.9
8. Sugar maple-yellow birch and balsam fir-yellow birch	47	12	25.5 ± 11.3	25.5
9. Balsam fir-yellow birch	87	30	34.5 ± 9.2	34.5
10. Balsam fir-white birch or balsam fir-red maple	46	4	8.7 ± 7.5	8.7
11. Yellow birch-balsam fir	52	39	75.0 ± 12.2	75.0
12. Balsam fir-white birch (west)	59	38	64.4 ± 12.6	64.4
13. Balsam fir-white birch (east)	76	40	52.6 ± 10.9	52.6
14. Balsam fir-black spruce	28	23	82.1 ± 15.2	82.1
15. White spruce-balsam fir and balsam fir-white spruce	22	22	100.0 ± 2.3	100.0
16. Black spruce-balsam fir-moss	18	13	72.2 ± 23.7	72.2
17. Black spruce-moss	32	27	84.4 ± 14.1	84.4
18.			*not analysed*	

* minimal: 0–29.9%; low: 30–59.9%; high: 60–89.9%; maximal: 90–100%

** Deciduous forest: 1–7; mixed forest: 8–11; coniferous forest: 12–17

This corvid is a typical species of coniferous boreal forests. North Shore. *July 1986, Vincent Létourneau.*

Gray Jay nests are usually built on a branch against the trunk of a spruce or fir near a clearing like this one. Montmorency forest, Laurentide Wildlife Reserve, north of Québec City. *June 1991, Réginald Ouellet.*

Female Gray Jay incubating her eggs in a snowstorm on 1 April 1969. She was seen shaking the snow off her back from time to time. She also kept her wings spread more than usual. The 3 eggs hatched 3 days later and the young left the nest on 27 April. La Vérendrye Wildlife Reserve. *April 1969, Dan Strickland.*

Because of the species' early breeding season, 82% of confirmed Atlas breeding records were based on reports of juveniles. The young remain with their parents until mid-June, when intrabrood aggression results in the dominant juvenile expelling his or her siblings from the natal territory within 10 days or so (Strickland 1991). Some of the expelled birds succeed in getting themselves adopted by pairs which failed to breed that year. Therefore, the sighting of a group of three Gray Jays, even when the third bird is a juvenile, does not necessarily mean that the juvenile was produced by that pair. However, there is little risk of associating juveniles with Atlas squares where they were not hatched because most adopted juveniles probably are from nearby territories. At the beginning of the following breeding season, the surviving juveniles, whether dominant juveniles remaining with their parents or juveniles adopted by unrelated adults, are expelled by the adults (Strickland 1991).

Distribution

The Gray Jay was atlassed in all forested regions of southern Québec, particularly those dominated by coniferous and mixed forest.

The species is largely absent from the Central St. Lawrence Lowland upstream from Québec City, a primarily agricultural and urban region with very few coniferous stands except in a small number of peatlands. The species is completely absent from the Magdalen Islands.

Although the Gray Jay readily associates with humans, it is rarely seen from May to October, at least in southern Québec, perhaps because humans are not an essential source of storable food at this time. Since atlassers generally were not out in the field until early May, after the return of migrants, the species' presence was probably underestimated.

The Gray Jay's range extends well north of the Atlas area to the tree line (Godfrey 1986).

History and trends

Since the species is closely associated with coniferous forests, which are characteristic of areas with rocky, non-arable soil, agriculture has had little effect on Gray Jay populations. Although clearcutting may have temporarily reduced populations in boreal forests to some extent, fire control may have benefited the species.

At sporadic intervals, large numbers of Gray Jays leave the north in autumn for regions sometimes well south of their normal range. These movements, whose causes are unknown, are particularly apparent on Québec's North Shore, where the birds, stopped by the barrier of the Gulf of St. Lawrence, becomes concentrated along the seacoast before swinging to the southwest. On 23 August 1964, for example, Campbell (1965) estimated that more than 750 Gray Jays passed over Le Cormoran (Seal House Cove) on Moisie Bay during a three-hour period.

DAN STRICKLAND

Blue Jay

Geai bleu

Cyanocitta cristata

Christiane Girard

THE FLAMBOYANT BLUE JAY, WITH ITS BOLD PLUMAGE AND NOISY CALLS, rarely goes unnoticed. Its unmistakable bright cobalt blue coloration and prominent crest make it stand out from all other birds in Québec. Its loud *jeeah, jeeah* and its whistled *too-weedle, too-weedle*, which sounds like a squeaky clothesline, are familiar sounds in Québec woods and suburban backyards. The Blue Jay warns other birds and mammals of approaching predators with loud shrieks. It also imitates other birds, particularly the Red-shouldered Hawk (James 1988). In addition, it has a soft warbling song which is rarely heard (Tyler 1946; Godfrey 1986). The Blue Jay has a gular pouch which it uses to carry several seeds at a time (Eigelis and Nekrasov 1967; Savage 1985).

Once exclusively a bird of wilderness forests, the Blue Jay has adapted to urbanization and agriculture (Tyler 1946; Bock and Lepthien 1976a; Terres 1980) and is now often found in inhabited areas, particularly suburbs. It is a common resident breeder in most parts of southern Québec (David 1980a). As winter approaches, however, a portion of the population withdraws from the northernmost part of the breeding range in Canada and the northern United States to winter farther south within its breeding range (Hardy 1961; Stewart 1982; AOU 1983).

Habitat and behaviour

The Blue Jay frequents a wide variety of habitats. It is particularly common in mixed forest, but also inhabits coniferous and deciduous stands (DeGraaf and Rudis 1987; Peck and James 1987). Although it is found in dense forest, it prefers more open habitat (Peck and James 1987), particularly woodlots in suburban and rural areas and city parks (Goodwin 1976; DeGraaf and Rudis 1987). Because of its feeding habits, beech and oak stands are its favourite haunts (Goodwin 1976; Godfrey 1986).

As is typical of a bird with an omnivorous diet, the Blue Jay has a heavy bill useful in pecking open acorns and other hard-shelled nuts (Nero 1988). Its diet consists of three times more vegetable than animal matter, chiefly acorns, beechnuts, and corn (Martin et al. 1951; Terres 1980). Animal food includes mainly insects such as grasshoppers, caterpillars (tent caterpillars and sphinx moths), and beetles, including many harmful species such as click beetles, scarab beetles, and chrysomelids (Beal 1897; Terres 1980). Blue Jays also occasionally eat the eggs and young of other birds, as well as fishes, frogs, and mice (Martin et al. 1951; Terres

Notes on territory

The Blue Jay is territorial, although the species' territorial boundaries appear to be poorly defined and not always strictly

In mixed and deciduous forests, the Blue Jay prefers to build its nest in a conifer, often near an opening created by a body of water or clearing. Near Lake Montauban, Notre-Dame-de-Montauban, west of Québec City. *September 1966, Gaston Le Page.*

Blue Jay

Breeding distribution
1984–1989

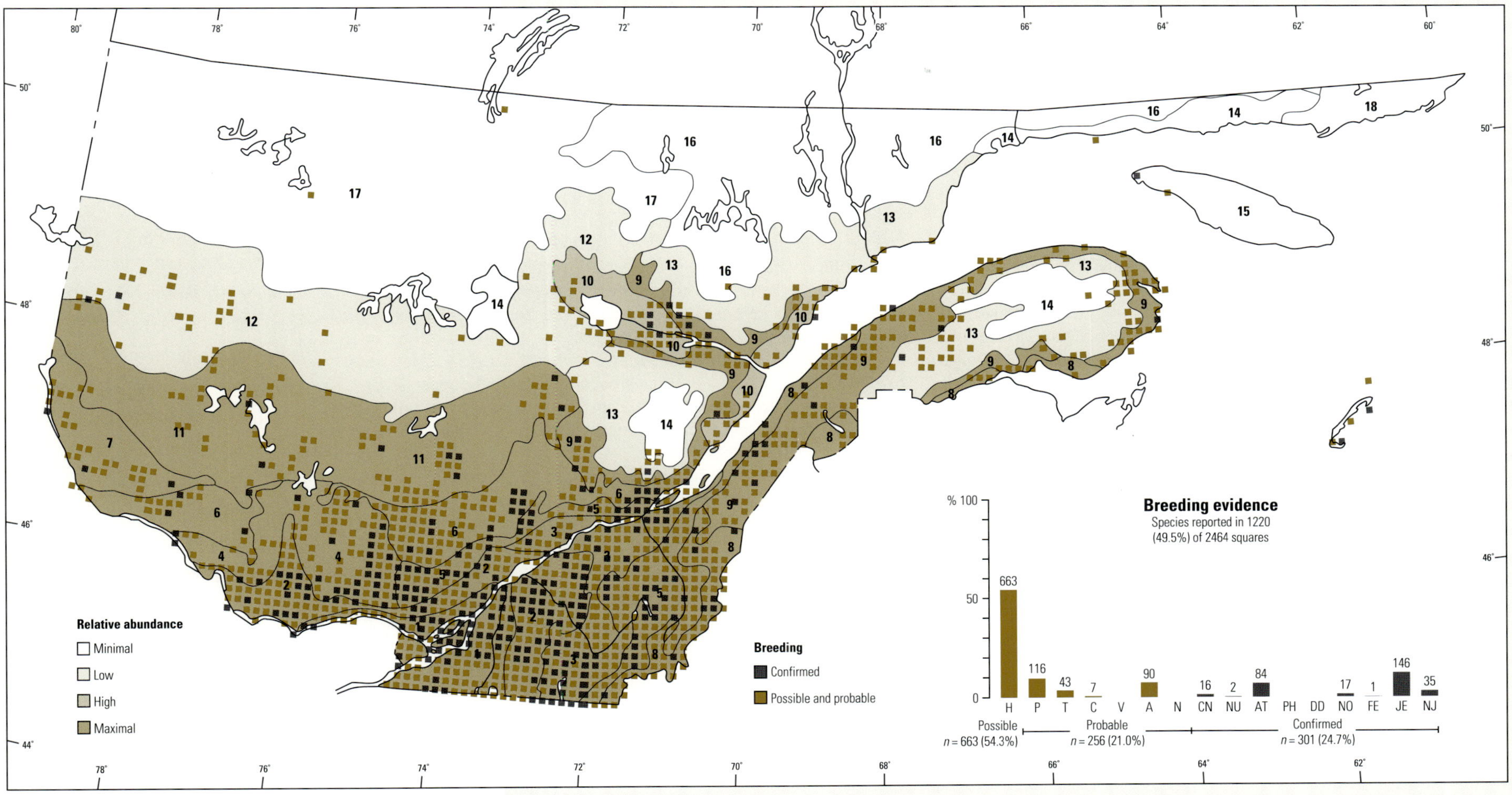

Species profile

Clutch size	usually 4–5 eggs (2–7)[1,2,3,4]
Laying interval	?
Start of incubation	probably before last egg[3]
Incubation period	usually 16–18 days (13–18)[1,2,3,4]
Maturity at hatching	altricial
Care of young	by female and male[1,2,4,11]
Nestling period	17–21 days[1,2,4,5]
Age at first flight	17–21 days or more[12]
Dependence of young	until 7–11 weeks after leaving nest, sometimes up to 4 months old[1,13]
Broods per year	usually 1 (1–3)[2,3,4,6]
Breeding age	usually 2 years (1–2)[4,6]
Mating system	monogamy[11]
Duration of pair bond	?
Total length	28.0–31.5 cm[7]
Wing span	40.0–44.4 cm[8]
Weight	86.8 g (64.1–109.0)[10]
Longevity record	18 years 4 months[9]

References:
1. Nero 1988
2. Ehrlich et al. 1988
3. Peck and James 1987
4. Goodwin 1976
5. Tyler 1946
6. DeGraaf and Rudis 1987
7. Godfrey 1986
8. Terres 1980
9. Clapp et al. 1983
10. Dunning 1984
11. Laine 1983
12. according to nestling period
13. Laskey 1958

World breeding range

Chronology $n = 104$

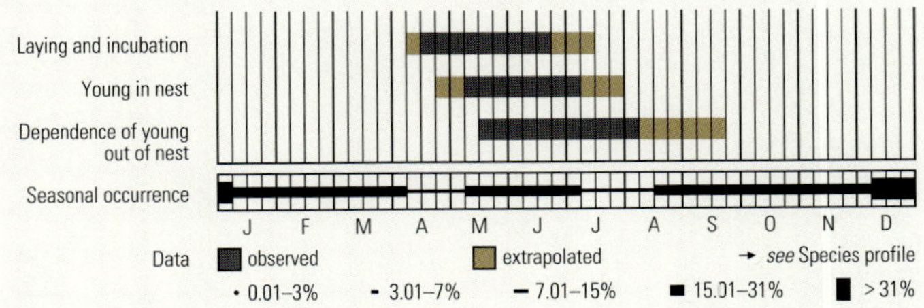

Frequency of occurrence by ecological region

Ecological regions** (forest regions)	Adequately covered squares	Squares in which reported	% of squares in which reported	Relative abundance (%)*
1. Sugar maple-hickory and sugar maple-basswood	97	95	97.9 ± 1.8	97.9
2. Sugar maple-basswood	153	151	98.7 ± 1.2	98.7
3. Sugar maple-basswood and sugar maple-yellow birch	101	101	100.0 ± 0.5	100.0
4. Sugar maple-yellow birch-basswood	56	56	100.0 ± 0.9	100.0
5. Sugar maple-yellow birch-beech and sugar maple-yellow birch-basswood	113	112	99.1 ± 1.3	99.1
6. Sugar maple-yellow birch-beech	71	71	100.0 ± 0.7	100.0
7. Sugar maple-yellow birch-balsam fir	19	19	100.0 ± 2.6	100.0
8. Sugar maple-yellow birch and balsam fir-yellow birch	47	45	95.7 ± 5.8	95.7
9. Balsam fir-yellow birch	87	86	98.9 ± 2.5	98.9
10. Balsam fir-white birch or balsam fir-red maple	46	31	67.4 ± 11.8	67.4
11. Yellow birch-balsam fir	52	48	92.3 ± 7.9	92.3
12. Balsam fir-white birch (west)	59	28	47.5 ± 13.1	47.5
13. Balsam fir-white birch (east)	76	40	52.6 ± 10.9	52.6
14. Balsam fir-black spruce	28	5	17.9 ± 15.2	17.9
15. White spruce-balsam fir and balsam fir-white spruce	22	2	9.1 ± 13.3	9.1
16. Black spruce-balsam fir-moss	18	0	0	0
17. Black spruce-moss	32	1	3.1 ± 7.6	3.1
18.	*not analysed*			

* minimal: 0–29.9%; low: 30–59.9%; high: 60–89.9%; maximal: 90–100%

** Deciduous forest: 1–7; mixed forest: 8–11; coniferous forest: 12–17

1980). At Saint-Nicolas, near Québec City, a Blue Jay was seen catching a small colubrid snake; it took it up to a tree, killed it by striking it with its bill and swallowed it whole (C. Sommelier, pers. comm.). The Blue Jay is also well known around the bird feeder, and is particularly fond of peanuts, cracked corn, sunflower seeds, and nuts (Lane 1987). It hides food in the ground, apparently as a provision for periods of scarcity, and thereby helps reforest woodlands with the acorns and beechnuts it forgets (Terres 1980; Nero 1988).

The Blue Jay is an early nester. During the first phase of courtship, small courting parties each composed of a female and 2–5 suitors are formed. The male that succeeds in intimidating and driving off his rivals then pairs with the female. The unsuccessful males form a new group or join an existing group, and the process continues until most of the adults are paired off. During the second phase of courtship, the male offers food to the female and brings her twigs he has pulled directly off a tree, which she uses to build a false nest (Hardy 1961).

The actual nest is a bulky structure which takes up to 5 days to build (Hardy 1961). In some pairs, the male and female share the work equally, while in others the female does most of the work (Hardy 1961). The nest is made of sticks and twigs, to which may be added coarse grasses, birch bark, mosses, lichens, leaves, mud, and bits of yarn, cloth, plastic, or paper. The lining usually consists of rootlets, but may also contain grasses, feathers, hair, and moss (Peck and James 1987). The nest is built 1–20 m up (Goodwin 1976; Peck and James 1987), generally in a conifer (Godfrey 1986; Peck and James 1987). Preferred nest trees include Eastern white-cedar, pine, hawthorn, spruce, hemlock, and maple (Peck and James 1987).

Incubation is usually by the female alone (Laskey 1958; Hardy 1961; Goodwin 1976; Laine 1981), probably beginning before the last egg is laid (Peck and James 1987). While she is incubating, the male brings her food (Laskey 1958; Hardy 1961). The eggs hatch in 13–18 days (Peck and James 1987), and both parents tend the nestlings (Hardy 1961). The nestlings devour hundreds of caterpillar pupae, which their parents extract from the cocoons (Laskey 1958; Nero 1988). The young leave the nest at 17–21 days (Tyler 1946), but remain with their parents for some time. The family then begins to occupy a much larger range than the original breeding territory (Hardy 1961). Three weeks after leaving the nest, the young begin searching for food on their own, but are fed occasionally by their parents for one or two months longer, sometimes until they are as old as 4 months (Laskey 1958; Nero 1988).

Most individuals occupy roughly the same home range all year. In autumn, three or four families whose home ranges overlap may gather at abundant sources of food, forming loose flocks (Hardy 1961; Racine and Thompson 1983). At the approach of winter, these flocks break up into smaller groups of 2–4 birds (Hardy 1961), most of which will nest in the same area the following season (Racine and Thompson 1983). First-year jays make up most of the migrators (Hardy 1961; Savage 1985).

When a predator approaches—whether a dog, cat, Great Horned Owl, or hawk—Blue Jays become extremely aggressive. A group of jays often encircle a raptor and drive it away with repeated loud calls (Savage 1985). Blue Jays rarely appear to be parasitized by cowbirds (Ehrlich et al. 1988).

The Blue Jay, like the American Robin and European Starling, is one of the North American species most often observed anting. In this behaviour, which is generally seen only in passerines, birds apply ants to their plumage with their bills or by rubbing themselves against an anthill. Apparently, substances secreted by ants such as formic acid have a soothing effect on the bird's skin during moulting (Potter and Hauser 1974; Potter 1985).

The Blue Jay was atlassed in almost 50% (1220/2464) of the squares surveyed and was confirmed as a breeder in 24.7% (301/1220) of the squares in which it was reported, mainly through observations of adults carrying food and of recently fledged young. The Blue Jay tends to be quiet and unobtrusive during the nesting season. However, the young are easy to spot since they are quite vociferous when begging for food and are fed by their parents for a long time after they leave the nest.

Notes on territory (continued)

defended. Aggressiveness between conspecifics is only observed between adults at the same stage of nesting. Therefore, a pair with their young still in the nest will allow adults with fledged young to enter their territory and even to land in the tree in which they are nesting (Hardy 1961). The male usually defends the territory, with the female participating only occasionally. Territories are used for nesting and feeding the young until they leave the nest (Hardy 1961). In Québec, McNeil (1964 in Cyr 1974) estimated the average Blue Jay territory to be 2.61 ha. However, Hardy (1961), in a study in the United States, reported that he had never observed territorial defence more than 60 m from the nest.

In Québec, mean densities of 0.7 and 0.8 pair/10 ha were observed in hemlock and maple stands, respectively, at Mirabel (Cyr 1974) and 1.1 pairs/10 ha in mixed and balsam fir stands in La Mauricie National Park (DesGranges 1980b). In Ontario, densities of 2.8 pairs/10 ha were recorded in abandoned pasture scrubland (Romanow 1993).

The Blue Jay may also nest in a deciduous tree on the edge of the forest. Martin-Valin ZEC, north of the Saguenay River. *August 1991, Jean Gauthier.*

In urban and semi-rural areas, this opportunistic corvid often nests in groves of trees near houses, where it takes advantage of feeders and fruit trees. Lake Saint-Augustin, near Québec City. *October 1993, Pierre Normand.*

Distribution

The Blue Jay breeds in all 10 Canadian provinces and in the United States from Montana and New Mexico in the west to the eastern seaboard (AOU 1983).

In Québec, the Blue Jay is much less common in coniferous forest than in mixed and deciduous stands. South of the St. Lawrence River, it is most common from Montérégie to the Lower St. Lawrence, and along the coast of the Gaspé Peninsula. North of the St. Lawrence, it is quite common in all regions south of the 48th parallel, except in high-elevation conifer stands on the Gaspé Peninsula and the Laurentide Wildlife Reserve. It is also fairly common in the Saguenay–Lac-Saint-Jean region.

Todd (1963) considered the species accidental on the North Shore and reports only three records east of Tadoussac. A few Atlas records were obtained on the Upper North Shore, and one on the Middle North Shore near Sheldrake. The species was also observed in 2 squares on Anticosti Island, but breeding was not confirmed. Ouellet (1969) also failed to obtain any breeding evidence on Anticosti Island. The species has been observed regularly on the Magdalen Islands for several decades (Fradette 1992), but there are only 2 confirmed breeding records, both obtained during the Atlas survey. The northernmost records in western Québec were obtained in 1991 at Cabbage Willows Bay in spring (Cons. G. G. – GREBE 1992b) and in a jack pine stand near Lake Mesgouez in summer (Cons. G. G. – GREBE 1992o).

History and trends

At the turn of the century, the Blue Jay was considered a common permanent resident in many regions of the province, including the Ottawa Valley (Lloyd 1923), Eastern Townships (Mousley 1916a), and the Québec City area (Dionne 1906). Wintle (1896) listed it as a common "transient visitant" in Montréal, but did not know of any individuals breeding there, probably because of the species' secretive behaviour during the nesting season. A few years later, Terrill (1911b) was able to confirm breeding in this region. Later in the century, the Blue Jay was described as occasional in the Gaspé (Demille 1926) and uncommon in the Lac-Saint-Jean region (Godfrey and Wilk 1948), which suggests that it may be more abundant now than it was in the past in these two areas.

More recently, BBS data for 1965–79 suggest a slight but significant decline in the breeding population in eastern North America (Robbins et al. 1986). However, a similar trend has not been observed in Québec (Cyr and Larivée 1993a; BBS).

Furthermore, Bock and Lepthien (1976a) reported that, between 1962 and 1971, the number of Blue Jays in North America migrating appears to have decreased, accompanied by an increase in the number of overwintering individuals. They believe these phenomena are linked to the increase in bird feeders.

During the last few decades, the Blue Jay appears to have extended its breeding and wintering ranges westward (Smith, K. G. 1978). This is probably due in part to habitat modification caused by urbanization, but here, too, the most important factor is probably the proliferation of feeding stations (Bock and Lepthien 1976).

Christine Langevin and Diane Dauphin

American Crow

Corneille d'Amérique

Corvus brachyrhynchos

Christiane Girard

FEW BIRDS OCCUPY SUCH A PROMINENT PLACE IN OUR COLLECTIVE IMAGINATION and popular culture as the crow. This bird, familiar to everyone, does not always have the best reputation and, being noisy and conspicuous, draws attention to itself wherever it goes. Its large size, coal-black plumage (a colour associated with darkness and death), harsh garrulous vocalizations, cunning ways and obvious intelligence in its dealings with humans have made it the source of many legends, folktales and superstitions (Goodwin 1976; Leahy 1982).

The American Crow is so closely related to the Carrion Crow, a European species, that some authors consider them a single species; another close relative, which is found in coastal areas in northwestern North America, is the appropriately named Northwestern Crow, whose status as a separate species or subspecies has yet to be determined conclusively (Goodwin 1976; AOU 1983; Godfrey 1986). Among the four recognized subspecies, *Corvus brachyrhynchos brachyrhynchos* is the crow that occurs in Québec, eastern Canada, and the northeastern United States (Good 1952; Goodwin 1976; Godfrey 1986).

Of all the passerines in the province, the American Crow is the second largest, surpassed only by the Common Raven, with which it may be confused. However, it is distinguished from the Common Raven by its higher pitched vocalizations, smaller size, less robust bill, squarish rather than wedge-shaped tail, and tendency to flap continuously in flight rather than glide. In addition, it lacks the raven's shaggy throat feathers (Goodwin 1976; Godfrey 1986). Furthermore, the American Crow readily associates with humans, while the Common Raven tends to shun civilization, inhabiting densely forested and northerly areas (Taverner 1922; Good 1952). The two species have attempted to mate in the past (Jefferson 1991).

Both sexes of the American Crow are similar although the female is sometimes smaller and has less glossy plumage (Gross 1946; Goodwin 1976). Yearlings can be identified by the brownish cast to their wings, tail, and back (Verbeek and Butler 1981; Kilham 1984a). In addition, adults have rectrices that are generally intact, producing a neat-looking tail with rounded corners, while those of yearlings are frayed and pointed at the ends, forming a squarer tail (Emlen 1936 *in* Kilham 1984a). Cases of total or partial albinism are also observed in the species (Warne 1926; Gross 1946; Good 1952), as is erythrism on occasion (Good 1952).

Northern populations are partially migratory, wintering from southernmost Canada to the southern United States and northern Mexico (Long 1981; AOU 1983). Although some individuals winter in Québec, foraging at landfills and feeding on carcasses found on roadsides and ice-free shores, most of the province's population migrates south, leaving Québec as early as mid-October and returning between the end of February and the end of March.

Habitat and behaviour

The American Crow frequents a wide range of habitats, including those created by logging, agriculture, and urban development. Principally a ground feeder, it requires open habitat for foraging and wooded areas for nesting, resting, and roosting (Good 1952; Kilham 1989). The greatest numbers are found in rural areas with a good scattering of woods, which provide the mixed habitat the species needs. The American Crow is often found in meadows, pastureland, and cultivated fields with hedges and trees, as well as orchards. It is also found in open forests, on the margins of deciduous, mixed, and coniferous stands, along the shores of lakes, rivers, marshes, and peatlands (Goodwin 1976; AOU 1983; Godfrey 1986; Peck and James 1987), and along the coast, which it frequents at all times of the year (Erskine 1992). In urban areas, the American Crow nests in parks, cemeteries, and small woodlots and forages in landfills, gardens, lawns, and vacant lots (Peck and James 1987).

Because of its varied diet and opportunistic foraging habits, the species is extremely adaptable. The American Crow feeds on a vast range of plant and animal foods, including insects, crustaceans, molluscs, and other invertebrates, amphibians, reptiles, fish, the eggs and young of other birds, small mammals, animal carcasses, food wastes, and various fruits and seeds (Forbush and May 1955; Terres 1980; Leahy 1982). Its characteristic antelingual pouch allows it to carry and store large quantities of food to feed its young or build up caches for future use (Eigelis and Nekrasov 1967; Turcek and Kelso 1968; Goodwin 1976).

The American Crow is considered a major predator of the eggs and young of other birds, particularly colony-nesting species and waterfowl (Gross 1946; Good 1952; Sugden and Beyersbergen 1986; Sullivan and Dinsmore 1990). This behaviour, which is more frequent in non-breeding individuals (Good 1952), is often denounced by hunters and has been the subject of numerous studies (Yahner and Wright 1985; Sugden and Beyersbergen 1986; Sullivan and Dinsmore 1990, 1992; Andrén 1992). The species may also attack adult birds such as Rock Doves (Duquette and Lépine 1981). Like birds of prey, crows regurgitate indigestible material such as hair and bones in the form of pellets (Good 1952).

American Crows cause considerable damage to field crops (Good 1952). In cornfields, the greatest destruction occurs in spring, when they eat the young shoots (Forbush and May 1955), and in early autumn, when they tear open the husks to eat a few kernels, exposing the ears to insects and fungus. In winter, crows consume large amounts of waste grain left in fields after the harvest. They also damage other grain crops like buckwheat, sorghum, and, in the Canadian Prairies, wheat, oats, barley, and rye, particularly in summer (Good 1952).

The crow's depredations are not always so serious, however, and to evaluate its overall economic impact, its services to mankind must also be taken into account, including the devouring of huge quantities of insects, most of them harmful to crops (Good 1952). The species is invaluable in controlling some insect infestations (Gross 1946; Good 1952).

Corvids have a large brain in comparison with other birds and are considered to be at the top of the avian pyramid in intelligence (Pasquier 1977; Terres 1980). Intelligence in the American Crow is demonstrated in a number of different ways. Crows have a great sense of play, and wild birds have often been observed manipulating objects, performing repeated aerial manoeuvres, and teasing or provoking other birds or sometimes larger animals (Good 1952; Kilham 1989; Faucher 1994). The American Crow can also use objects as tools (Pasquier 1977; Jefferson 1991; Rolando and Zunino 1992) and drop stones on intruders while in flight (Good 1952). It has a greater repertoire of calls than any other passerine, another proof of its highly developed cerebral faculties (Good 1952; Pasquier 1977). Kilham (1989) and Chamberlain and Cornwell (1971 *in* Goodwin 1976) have documented over 20 calls made by the species, each associated with a specific function; crows also imitate other species of birds and can even mimic human speech (Warne 1926; Gross 1946; Good 1952).

In urban areas, the American Crow nests in large trees in residential neighbourhoods, parks, and cemeteries and feeds in gardens, landfills, and on lawns. Mount Royal, Montréal.
September 1981, Richard Pelletier.

Notes on territory

Traditionally, the American Crow has been considered a rather non-territorial species that only exhibits territorial behaviour during the breeding season (Good 1952; DeGraaf and Rudis 1987). However, recent research has revealed that family groups defend territories throughout the year, at least in some sedentary populations in the south (Kilham 1985, 1989; Chamberlain-Auger et al. 1990). In studies in Massachusetts and Florida, territories were found to be 42.1 ha and 50 ha, respectively; Kilham (1989) even reported territories as large as 90 ha, within which the breeding pair

American Crow

Breeding distribution 1984–1989

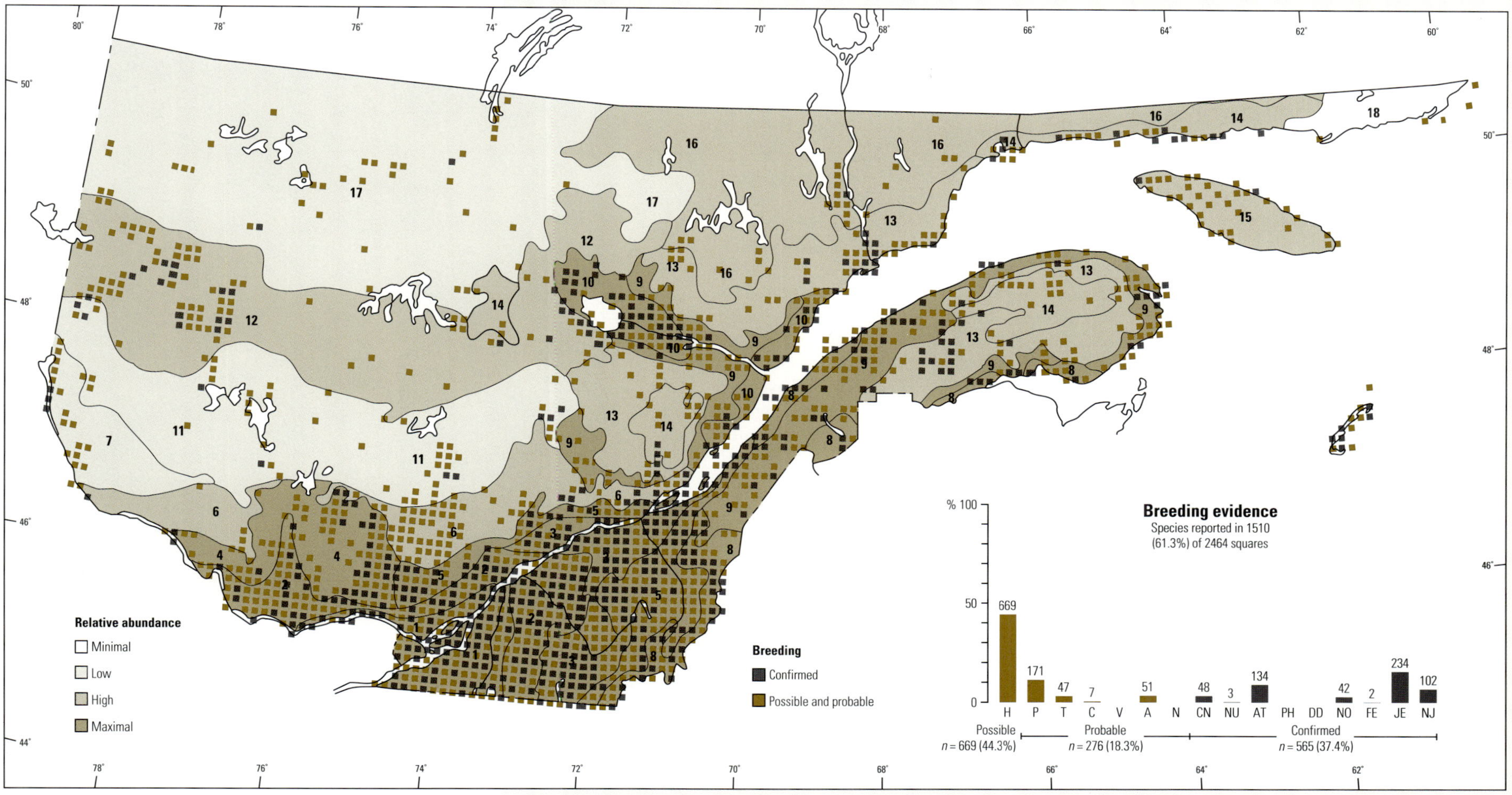

Species profile

Clutch size	usually 4–5 (3–8)[6,10]
Laying interval	1 egg per day[8]
Start of incubation	with last egg[10]
Incubation period	16–21 days[10]
Maturity at hatching	altricial
Care of young	by female and male[4,9]
Nestling period	4–5 weeks[8,9,10]
Age at first flight	28–35 days[4,7]
Dependence of young	≈ 2 weeks after departure from nest[8]
Broods per year	usually 1[6], sometimes 2[4,5,9,11]
Breeding age	2 years[5]
Mating system	monogamy[4]
Duration of pair bond	probably for life[10]
Total length	43–53 cm[3]
Wing span	83.8–101.6 cm[7]
Weight	M 458 g (?) / F 438 g (?)[2]
Longevity record	14 years 7 months[1]

References:
1. Clapp et al. 1983
2. Dunning 1984
3. Godfrey 1986
4. Ehrlich et al. 1988
5. DeGraaf and Rudis 1987
6. Peck and James 1987
7. Terres 1980
8. Stokes 1989
9. Gross 1946
10. Good 1952
11. Harrison 1975

Chronology $n = 258$

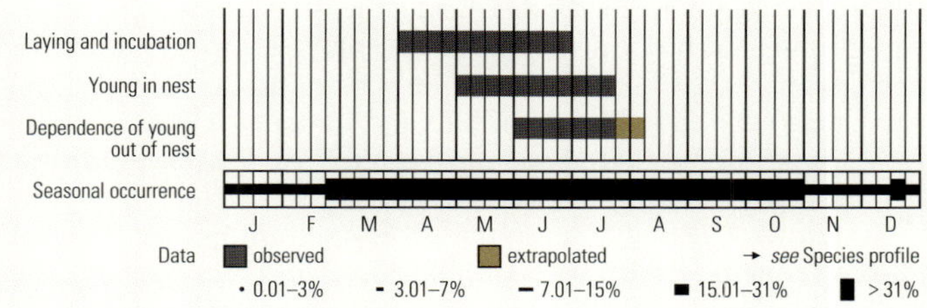

Frequency of occurrence by ecological region

Ecological regions** (forest regions)	Adequately covered squares	Squares in which reported	% of squares in which reported	Relative abundance (%)*
1. Sugar maple-hickory and sugar maple-basswood	97	97	100.0 ± 0.5	100.0
2. Sugar maple-basswood	153	151	98.7 ± 1.2	98.7
3. Sugar maple-basswood and sugar maple-yellow birch	101	101	100.0 ± 0.5	100.0
4. Sugar maple-yellow birch-basswood	56	51	91.1 ± 6.9	91.1
5. Sugar maple-yellow birch-beech and sugar maple-yellow birch-basswood	113	112	99.1 ± 1.3	99.1
6. Sugar maple-yellow birch-beech	71	61	85.9 ± 7.6	85.9
7. Sugar maple-yellow birch-balsam fir	19	8	42.1 ± 23.3	42.1
8. Sugar maple-yellow birch and balsam fir-yellow birch	47	46	97.9 ± 4.5	97.9
9. Balsam fir-yellow birch	87	85	97.7 ± 3.3	97.7
10. Balsam fir-white birch or balsam fir-red maple	46	46	100.0 ± 1.1	100.0
11. Yellow birch-balsam fir	52	31	59.6 ± 13.8	59.6
12. Balsam fir-white birch (west)	59	47	79.7 ± 10.7	79.7
13. Balsam fir-white birch (east)	76	57	75.0 ± 9.6	75.0
14. Balsam fir-black spruce	28	25	89.3 ± 12.6	89.3
15. White spruce-balsam fir and balsam fir-white spruce	22	14	63.6 ± 20.8	63.6
16. Black spruce-balsam fir-moss	18	16	88.9 ± 17.4	88.9
17. Black spruce-moss	32	18	56.2 ± 18.8	56.2
18.			*not analysed*	

* minimal: 0–29.9%; low: 30–59.9%; high: 60–89.9%; maximal: 90–100%

** Deciduous forest: 1–7; mixed forest: 8–11; coniferous forest: 12–17

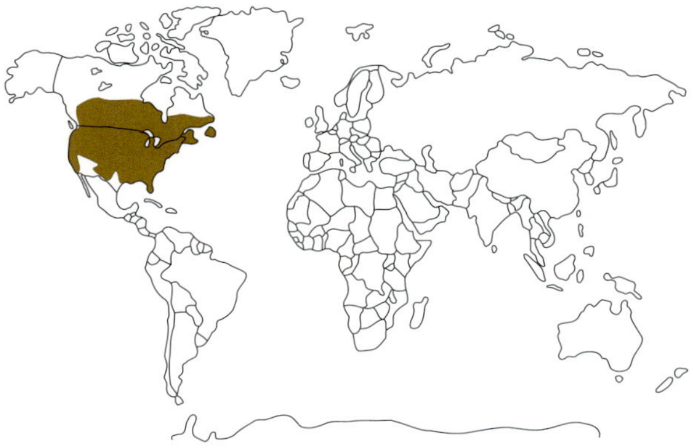

World breeding range

The American Crow builds its bulky twig nest in woodlots, in the top of a large tree. The nest is reused the next year by other species such as hawks and the Great Horned Owl. Mount Saint-Hilaire. *August 1991, Yves Aubry.*

The American Crow is a gregarious species with highly developed social behaviour (Kilham 1984a). Such behaviour can be observed in spring and autumn, when crows gather in large flocks to migrate, in winter, when they congregate in huge communal roosts, and, in summer, when non-breeding yearlings may also gather in small flocks. In California, a colony of around 60 pairs, of the subspecies *Corvus brachyrhynchos hesperis*, has been observed (Good 1952).

The American Crow's most noteworthy trait is its system of co-operative breeding, which has been described in detail by Kilham (1984a, 1989) in the subspecies *Corvus brachyrhynchos pascuus* in Florida, by Chamberlain-Auger et al. (1990) in the subspecies *Corvus brachyrhynchos brachyrhynchos* in Massachusetts, and by Verbeek and Butler (1981) in the Northwestern Crow in Alaska. This system involves the formation of small family groups, made up, for the most part, of a breeding pair and its offspring of the previous year; however, a family group consisting of 10 individuals, including the 4-year-old offspring of the breeding pair, has also been reported (Chamberlain-Auger et al. 1990). Gross (1946), in his observations of the species in New England and the Maritimes, made no mention of this system. In Ohio, however, Good (1952) reported that the previous year's offspring accompanied the parents during the following breeding season. Therefore, the system may not be present universally, but may be more frequent in sedentary populations, where family groups and territories are more stable (Kilham 1984a).

Age of first breeding is 2 years (DeGraaf et al. 1980). Therefore, yearlings can assist their parents with the tasks involved in raising the current year's brood: defending the territory and nest, building the nest, feeding the female while she is incubating the eggs, feeding the nestlings, and keeping the nest clean (Kilham 1984a, 1989; Chamberlain-Auger et al. 1990).

In spring, large flocks break up gradually as the pairs, which probably mate for life (Good 1952), reach their breeding grounds, sometimes in the company of their previous year's offspring. As in other species with a lifelong pair bond, courtship displays are rather perfunctory. In one display, the male chases the female, while performing aerial gyrations of diving and wheeling. In another display, when the two are perched side by side, the male spreads his tail and wings, puffs out his body feathers, bobs his head up and down and bows deeply several times while giving the rattle call. The pair may also caress each other with their bills and smooth each other's feathers (Gross 1946; Good 1952). Pairs often build their new nest near their old one (Good 1952).

Crows generally nest on the edges of clearings, near the margins of coniferous, deciduous or mixed forests or in rural woodlots. The nest is built in a tall, isolated tree, or occasionally in a smaller tree or large shrub in a field, hedgerow, or orchard. In urban areas, the species selects large trees in cemeteries, parks, gardens, and residential areas. Trees of various deciduous and coniferous species may be chosen, but tall pines and oaks are preferred (Gross 1946; Peck and James 1987). Nests have also been observed in such unusual locations as chimneys, posts, low bushes, and even on the ground (Gross 1946; Harrison 1975, 1978). The nest is usually placed high in the tree, in a well-concealed spot in a fork or on a large branch near the trunk; nest heights range from 0.6 to 26 m but are most often between 5.5 and 10.7 m (Peck and James 1987). The solidly built nest consists of a spacious platform of twigs covered with dry soil, on which rests a deep cup carefully lined with fine materials such as grass, plant fibres, pine needles, moss, bark strips, pieces of cloth, rootlets, leaves, hair, and feathers.

Both parents build the nest, which usually requires between 5 and 13 days to complete (Ehrlich et al. 1988). Some pairs take several weeks, giving the impression that the nest has been abandoned (Good 1952). Construction activities generally take place early in the morning, shortly after sunrise (Reaume 1987), when the birds collect nesting materials; most of the twigs for the platform are obtained by breaking them directly off trees (Reimann 1942). When building the platform, the crows place the broken ends of the twigs toward the outside of the nest, rather than in the opposite direction, as is the case with Cooper's Hawk, whose nest is similar (Reimann 1942). The new nest is usually built in the same territory and sometimes in the same tree as the previous year's nest; sometimes an old nest of the same or another species may be repaired and reused (Reimann 1942; Good 1952; Peck and James 1987). Crows may also build preliminary nests that are never completed (Goodwin 1976; Chamberlain-Auger et al. 1990). Crow nests stand up well to the weather and are often reused by other species such as Cooper's Hawk, Broad-winged Hawk, Great Horned Owl, and Long-eared Owl or by squirrels, mice, and raccoons (Reimann 1942; Good 1952).

Once the nest has been completed, activity around it ceases for 6–16 days (Kilham 1989). After this hiatus, the female starts laying her clutch of 4 or 5 eggs, producing one egg a day, usually in the morning (Peck and James 1987). Although eggs are variable in size, colour, and shape, the eggs within the same clutch are uniform (Good 1952). The eggs are incubated for 16–21 days, beginning usually after the last egg has been laid, but sometimes earlier in larger clutches (Reimann 1942; Good 1952). Gross (1946) stated that both the male and the female participate in incubating the eggs, while Good (1952), Kilham (1984a,b, 1989), and Chamberlain-Auger et al. (1990) maintain that only the female is involved. According to Goodwin (1976), the female alone incubates, while the male merely stands or crouches over the nest or sits on it for brief periods. While she is incubating, the female is fed by the male or by her year-old offspring and only leaves the nest for short periods. The American Crow is a rare cowbird host (Terres 1980; Ehrlich et al. 1988).

The young are brooded by the female for several days after they hatch and are fed by the female, the male, and any immature birds in the family group. The adults, which are quiet around the nest during the incubation and egg-laying periods, become more vociferous at this time and protest vehemently when an intruder gets too close (Good 1952). Schaefer and Dinsmore (1992) have observed an adult carrying a chick from one nest to another. Around the age of 6 days, the nestlings' skin, which is pinkish at birth, turns black, their eyes open and their feathers begin to emerge.

By the time they are almost 4 weeks old, the young birds are able to perch on the rim of the nest or even venture out on nearby branches. Between 28 and 35 days after hatching, they leave the nest and make their first flight (Reimann 1942; Gross 1946; Good 1952). Subsequently, they move farther and farther from the nest, following their parents, which continue to feed them for another 2 weeks (Good 1952). The family group remains together through the autumn and winter, and even during the following breeding season (Knopf and Knopf 1983; Ellison 1985b; Kilham 1989).

After the breeding season, family groups go off in search of new sources of food, gradually congregating into larger flocks: as time goes on, the size of these flocks increases while the number of flocks decreases. Flocks in the southern part of the species' breeding range, originally made up of resident birds, are swollen by the influx of migrants from the north (Good 1952). Migration takes place mainly during the day (Forbush and May 1955). Birds nesting in Saskatchewan and Alberta on the Canadian Prairies may fly 2000 km or more to reach their wintering grounds in Kansas, Oklahoma, and Texas (Gross 1946; Good 1952; Stokes 1989). Other northern birds may migrate

Notes on territory (continued)

defended a smaller territory from all conspecifics except their yearling offspring, which were acting as helpers (Kilham 1984a). In a Manitoba study, home ranges were found to average 260 ha (Sullivan and Dinsmore 1992).

In Saskatchewan, Ignatiuk and Clark (1991) found densities of 0.04 and 0.07 pair/10 ha in two aspen stands. In the United States, densities from 0.2 to 1.8 pairs/10 ha have been reported, varying according to region and habitat (deciduous, coniferous or mixed forests, urban woodlots, and agricultural habitats) (Reimann 1942; Braunfield 1993; Eddleman 1993; Keller 1993; Künstler 1993; Magee 1993a; Richard 1993). A density of 3.2 pairs/10 ha was found in a mixed deciduous forest in the District of Columbia (Briggs 1993).

An opportunistic species, this member of the corvid family also nests in clumps of trees in fields and pastures, foraging in the open spaces for insects, amphibians, small rodents, and seeds. Varennes, near Montréal. *August 1991, Gaétan Duquette.*

shorter distances, and populations on the Atlantic coast, for example, do not seem to travel far at all but rather establish roosts along the coast to take advantage of the wide range of food washed up on ice-free tidal shores (Gross 1946; Good 1952; Wenger 1991; Erskine 1992).

These large gatherings are probably the most striking example of the American Crow's social behaviour. The birds congregate in roosts during migration and on their wintering grounds. After spending the night in the roost, they leave at dawn along well-established flight lines, sometimes travelling as far as 80 km in search of food (Good 1952; Forbush and May 1955; Terres 1980). At the end of the day, they return to the roost in steady streams along the same flight lines, after stopping momentarily at secondary roosts (Gross 1946; Goodwin 1976; Knopf and Knopf 1983; Lapierre 1991). Individuals appear to adhere to the same itinerary and foraging areas day after day, particularly resident birds (Goodwin 1976; Stouffer and Caccamise 1991). During the roosting season, the family group still remains the basic social unit (Knopf and Knopf 1983).

This nest in a hawthorn tree between two fields is only 2.5 m from the ground. Saint-Pierre, Island of Montréal. *May 1926, Henry Mousley, Laval University Collection.*

Roosts may harbour from a few hundred to tens of thousands of birds. The largest roosts are found in the United States, in Oklahoma, northern Texas, Illinois, Indiana, and Ohio, as well as along the Atlantic coast north of the Carolinas (Good 1952). In Pennsylvania, one roost was found containing 500 000 individuals (Gross 1946). One of the greatest concentrations of crows ever observed in Québec was in Granby in the winter of 1990–91, when 20 000 birds were seen in a roost that had been used the previous year (Lapierre 1991).

Distribution

In Canada, the American Crow breeds from Newfoundland to the interior of British Columbia but is replaced on the West Coast by the Northwestern Crow. Farther south, the species is found throughout the United States except in the desert regions of the southwest (Goodwin 1976; AOU 1983; Godfrey 1986).

In Québec, the species occurs in deciduous, mixed, and coniferous forests. Although it frequents large tracts of forest with scattered openings, it is much more abundant in agricultural and densely populated regions. Large concentrations are found in the Central St. Lawrence Lowland, Ottawa Valley, Eastern Townships, and the Québec City, Montérégie, and Chaudière–Appalaches regions, as well as in open habitats in the Abitibi, Saguenay–Lac-Saint-Jean, Gaspé, and North Shore regions. On the Gaspé Peninsula and North Shore, the American Crow is practically a fixture of the coastal landscape. It is found in fewer numbers in heavily forested regions, where it is mainly associated with bodies of water, clearings, and openings created by logging and road construction.

The American Crow breeds on Anticosti Island, where it has been observed regularly in the past (Ouellet 1969), as well as on the Magdalen Islands, where it is abundant year-round (Fradette 1992).

The species is considered a resident breeder throughout the province south of the 51st parallel (David 1980a), although its breeding range extends as far north as the 52nd parallel (ÉPOQ) where, near the Eastmain River, a nest with eggs was found (Bordage 1984). It has been reported sporadically farther north, notably along the coasts of James Bay and Hudson Bay; the northernmost record is from Inukjuak (ÉPOQ).

History and trends

The American Crow is probably more widespread today in eastern North America than it was before European settlement (John 1987; Bonney 1988i; Peterjohn and Rice 1991; Wenger 1991; Erskine 1992; Gross 1992b), having capitalized on forest fragmentation resulting from logging and farming, in particular corn growing.

In Québec, the American Crow was reported in Natashquan in 1833 by Audubon (Todd 1963). At the turn of the century, Dionne (1906) described it as being abundant in summer in southern Québec; it also nested throughout the Magdalen Islands (Fradette 1992) and in the Montréal region (Wintle 1896) at that time. More recently, Ouellet (1974a) classified it as an abundant summer resident and uncommon winter resident in the Montréal region.

Because of its depredations, the American Crow has acquired a bad reputation with farmers and hunters (Gross 1946; Goodwin 1976; Skutch 1976) and is one of the most persecuted of all passerines (Root 1988). Thousands of individuals have been shot, poisoned or even dynamited for economic reasons (Terres 1980). According to Gross (1946), over 80 000 crows were killed in Texas in 1937 and 328 000 crows were killed in Illinois in 1940 when roosts were dynamited.

Such persecution has not only provoked vehement criticism and protest but, in the long run, has proven ineffective against the species, which continues to prosper despite all the forces marshalled against it. An analysis of BBS data revealed a slight increase in populations between 1966 and 1979 in North America as a whole, particularly in the East (Robbins et al. 1986). This increase was also observed in Québec between 1966 and 1989 (BBS), but ÉPOQ data from 1970 to 1989 do not reveal any significant change (Cyr and Larivée 1993a). In Québec, the American Crow is protected between 1 May and 30 June under the provincial act respecting the conservation and development of wildlife (*Loi sur la conservation et la mise en valeur de la faune*), which was enacted in 1990.

RICHARD PELLETIER

Common Raven

Grand Corbeau

Corvus corax

THE COMMON RAVEN'S LARGE SIZE, WIDE VARIETY OF VOCALIZATIONS —ranging from hoarse croaks to bell-like tones—and raptorlike behaviour have inspired many myths, not always favourable ones. For example, sheep farmers once relentlessly persecuted this largest member of the crow family because they thought it was killing their newborn lambs and diseased sheep (Forbush 1927).

To the unpractised eye, a Common Raven may be easily confused with a large American Crow. However, a number of field marks can be used to distinguish the two species. The raven is larger and has a greater wingspan. In addition, its tail is wedge-shaped, rounded, and slightly longer. In flight, the raven glides more, while crows tend to flap their wings constantly. The Common Raven does not migrate, being sedentary throughout most of its range; it is a permanent resident throughout its Québec range, where it inhabits mixed and coniferous forests, mountainous regions, seacoasts, and, more recently, urban areas.

Habitat and behaviour

The Common Raven has a holarctic distribution and occupies a wide range of habitats, usually in mountainous areas and near cliffs (AOU 1983). Given a choice, it generally prefers nesting on cliffs rather than in trees. Harlow (1922) estimated that cliff nests outnumbered tree nests by 8 to 1; more recently, Hooper (1977) found ratios of 17 to 1. In the West, elevation appears to be the environmental factor with the greatest influence on the raven's distribution (Root 1988). Ravens nesting in trees generally build their nests in the fork of a conifer 14–30 m up. In areas of dense foliage, the highest tree is often chosen (Harlow 1922). Cliff nests are usually located under an overhanging ledge and spaced well apart (Hooper 1977). Hooper et al. (1975) reported minimum distances of 2.2 km and an average distance of 4.3 km between nests on cliffs in Virginia. Shade also appears to be a key factor in nest site selection; Harlow (1922) observed that cliff nests were invariably in a dark, shady place. Hydro pylons or quarries sometimes serve as nest sites (Fradette 1992). In the north, the Common Raven and Peregrine Falcon utilize the same breeding habitat and often nest on the same cliffs (White and Cade 1971). Raven nests have occasionally been found next to Peregrine Falcon nests, for example, at Cap-Tourmente near Québec City and Lake Lyster in the Eastern Townships, where the falcon drove the pair of ravens from the site (Bird and Aubry 1982).

The Common Raven is wary of humans throughout its range. When the nest is approached, it watches carefully, leaving the nest well before the intruder arrives and not returning until the intruder has left the area. Recently, it seems to be settling more in urban areas, where it appears to have adapted to manmade habitats (Blomme 1987b; Jefferson 1989). Nevertheless, Hooper (1977) reported that it usually does not tolerate human activity within 200 m of the nest.

The Common Raven prefers mountainous areas with rocky cliffs. Mingan River, north of Havre-Saint-Pierre, Middle North Shore. *May 1991, Daniel Bordage.*

Nest sites may be used for several years in a row, but most pairs seem to alternate between two or three sites (Goodwin 1976). Like most other corvids, pairs mate for life, with the male establishing and defending the territory. Courtship begins in early February, when the pair returns to the nest area. Ravens are probably noisiest at this time (Harlow 1922).

Both sexes participate in nest building, a process that takes approximately 2½ weeks. Depending on the pair and the stage in nest construction, either the male or the female will do most of the work. The large, bulky nest may be up to 1.2 m in height and 0.6–0.9 m in diameter. It is made from sticks that the birds break off conifers and is lined with bark strips and moss (Fleming and Speich 1988). The female incubates the eggs for 18–22 days, beginning with the laying of the second egg. The male brings food to the female and gives it to her on or near the nest. Both parents feed the young. The young defecate over the rim of the nest by lifting their tails, leaving large white stains of whitewash on the nest rim or cliff ledge, a characteristic sign of their presence. The nestlings leave the nest at the age of 35–44 days, and are able to fly short distances by that time (Terres 1980). They remain near the nest for a week or more, where they are still fed by their parents; the fledglings are very noisy when their parents bring them food (Harlow 1922). Once the young are capable of sustained flight, they form large flocks which roam widely for food. According to Heinrich (1988), ravens do not forage in flocks, but disperse individually or in small groups to find food, some individuals travelling over 90 km a day to forage (Stiehl 1978). When an individual finds a carcass, it uses a special call to recruit other ravens to the feast (Heinrich 1988). Heinrich believes that this type of information sharing facilitates the discovery of widely dispersed food sources.

Christiane Girard

Notes on territory

Pairs show strong fidelity to their nesting territory, which may include several alternate nest sites (Ratcliffe 1962). The territory is defended against conspecifics and raptors (Craighead and Craighead 1956). The species does not appear, however, to use or defend a specific feeding territory (Ratcliffe 1962). In Wyoming, home ranges varied from 6.7 to 10.9 km² (Craighead and Craighead 1956).

Common Raven

Breeding distribution

1984–1989

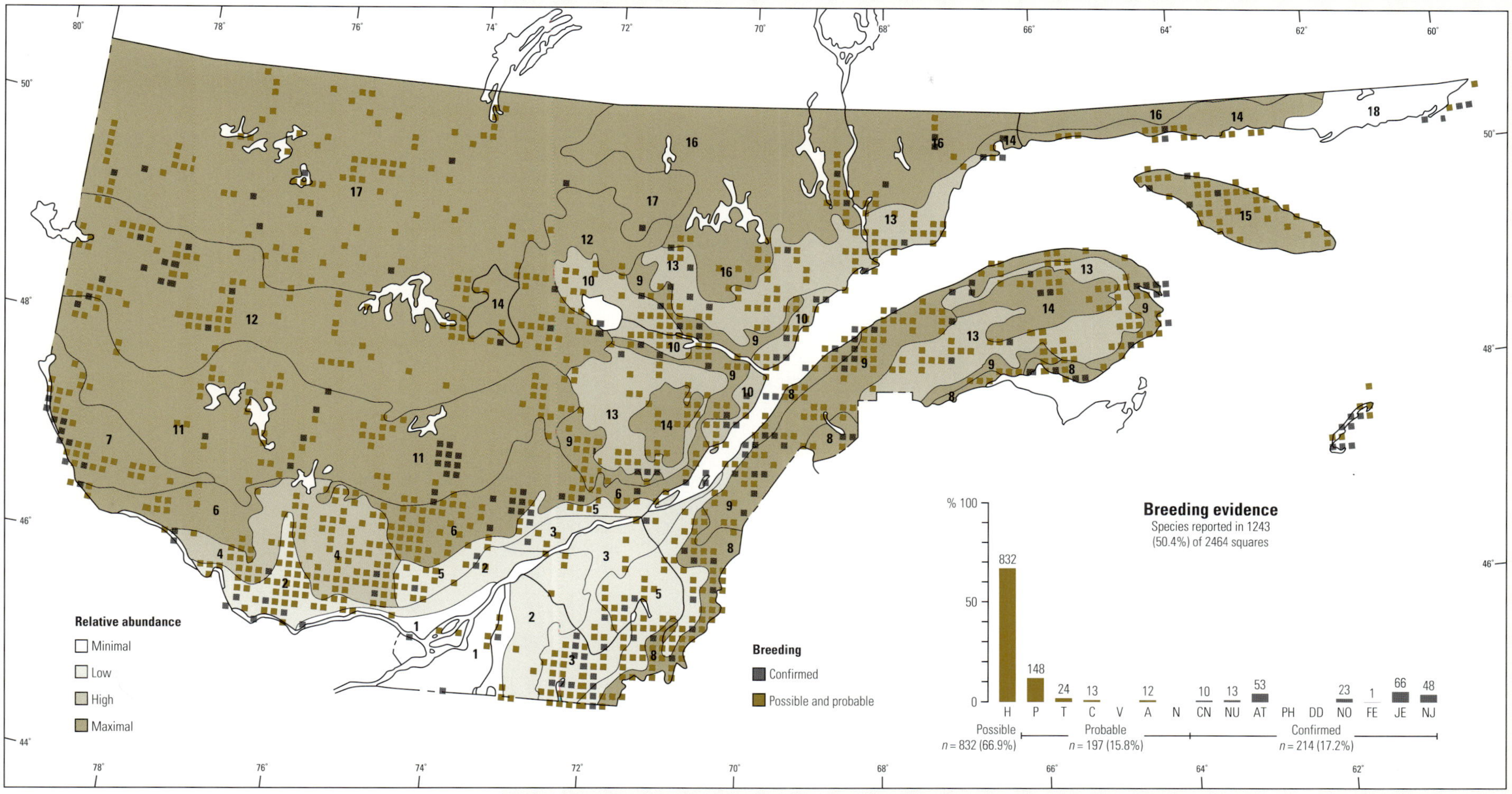

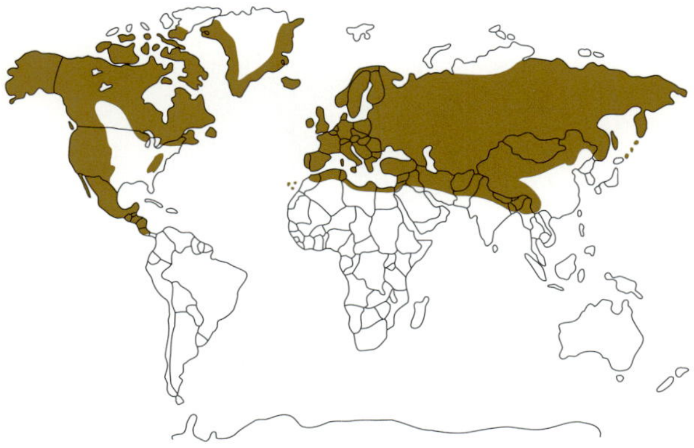
World breeding range

Species profile

Clutch size	usually 5–7 eggs (3–8)[1,2,3]
Laying interval	1 egg per day[1]
Start of incubation	with second egg[1]
Incubation period	18–22 days[1,2,3]
Maturity at hatching	altricial
Care of young	by female and male[1,2,3]
Nestling period	35–44 days[1,2,8]
Age at first flight	35–44 days[1,2,3]
Dependence of young	until 50–78 days old, possibly up to 6 months[1,2]
Broods per year	1[1,3]
Breeding age	probably 3 years or more[7]
Mating system	monogamy[3]
Duration of pair bond	for life[3]
Total length	56–67 cm[5]
Wing span	116.8–142.2 cm[3]
Weight	M 1240 g (1100–1400) / F 1158 g (1050–1300)[6]
Longevity record	13 years 4 months[4]

References:
1. Stiehl 1985
2. Goodwin 1976
3. Terres 1980
4. Clapp et al. 1983
5. Godfrey 1986
6. Dunning 1984
7. DeGraaf and Rudis 1987
8. Harrison 1978

Chronology $n = 104$

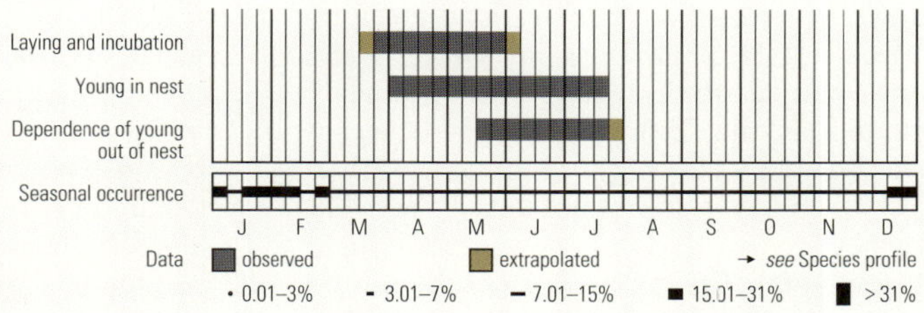

Frequency of occurrence by ecological region

Ecological regions** (forest regions)	Adequately covered squares	Squares in which reported	% of squares in which reported	Relative abundance (%)*
1. Sugar maple-hickory and sugar maple-basswood	97	11	11.3 ± 3.3	11.7
2. Sugar maple-basswood	153	51	33.3 ± 3.9	34.4
3. Sugar maple-basswood and sugar maple-yellow birch	101	50	49.5 ± 5.9	51.1
4. Sugar maple-yellow birch-basswood	56	48	85.7 ± 8.3	88.5
5. Sugar maple-yellow birch-beech and sugar maple-yellow birch-basswood	113	64	56.6 ± 5.1	58.5
6. Sugar maple-yellow birch-beech	71	62	87.3 ± 7.3	90.1
7. Sugar maple-yellow birch-balsam fir	19	18	94.7 ± 12.0	97.8
8. Sugar maple-yellow birch and balsam fir-yellow birch	47	42	89.4 ± 8.3	92.2
9. Balsam fir-yellow birch	87	83	95.4 ± 4.4	98.5
10. Balsam fir-white birch or balsam fir-red maple	46	38	82.6 ± 9.7	85.3
11. Yellow birch-balsam fir	52	46	88.5 ± 9.3	91.3
12. Balsam fir-white birch (west)	59	55	93.2 ± 7.0	96.2
13. Balsam fir-white birch (east)	76	60	78.9 ± 9.0	81.5
14. Balsam fir-black spruce	28	26	92.9 ± 10.8	95.9
15. White spruce-balsam fir and balsam fir-white spruce	22	20	90.9 ± 13.3	93.8
16. Black spruce-balsam fir-moss	18	17	94.4 ± 13.5	97.5
17. Black spruce-moss	32	31	96.9 ± 7.6	100.0
18.			*not analysed*	

* minimal: 0–29.9%; low: 30–59.9%; high: 60–89.9%; maximal: 90–100%

** Deciduous forest: 1–7; mixed forest: 8–11; coniferous forest: 12–17

The Common Raven usually seeks sheltered shady areas on cliffs to build its bulky stick nest. Mount Saint-Alban, Forillon National Park, Gaspé Peninsula. *July 1992, Yves Aubry.*

Distribution

In the Western Hemisphere, the Common Raven breeds from subarctic Alaska, arctic Canada and Greenland, south through the western United States, Mexico, and Nicaragua; in eastern North America, it is found south to the Great Lakes states and in the Appalachian Mountains. It winters mainly within its breeding range (Godfrey 1986).

In the southern part of its Québec range, the Common Raven often frequents rocky cliffs along the seacoast, where nearby seabird colonies provide food in the form of eggs and nestlings. In the northern part of its range, it is found on the tundra, particularly along seacoasts and riverbanks (Terres 1980), and in the boreal forest, where it occurs mainly in stands of spruce or other conifers (Erskine 1977). It has also been seen in many localities around the Ungava Peninsula (D. Bordage, CWS, unpubl. data; ÉPOQ).

In southern Québec, the Common Raven prefers mountainous and forested regions. Frequency of occurrence is fairly low in the Central St. Lawrence Lowland, particularly Montérégie, where the natural landscape has been greatly modified by human activity. Frequency of occurrence is higher in regions with extensive mixed and coniferous forest. In the ecological regions that lie within the deciduous forest zone, the Common Raven is also found, sometimes in great numbers, in the mountainous parts of the Laurentian and Appalachian regions.

Notes on territory (continued)

A study in Virginia in an area with many cliffs showed that mean distance between the closest active nests was approximately 5 km, and density was 0.03 pair/100 ha (Hooper et al. 1975).

In summer, the omnivorous Common Raven feeds on insects, fruit, seeds, small rodents, the eggs and nestlings of other birds, and any other animal or vegetable food it can find (Nelson 1934; Bent 1946; Harlow et al. 1975). However, in winter, such food is usually unavailable, particularly in regions with frequent heavy snowfall. In such conditions, the Common Raven associates with carnivores—wolves (Mech 1970), polar bears (Tinbergen 1958), and Gyrfalcons (Magoun 1976)—and feeds from their kills (Heinrich 1988). The Common Raven caches food throughout the year, burying it in the snow in winter (Kilham 1989). The morphology of the raven is well adapted to scavenging. Its large, heavy bill (7.5–9.5 cm long) is well suited for shearing meat from bones and pecking into fissures of bones. Furthermore, its large body mass allows it to prevail in aggressive encounters with other birds at carcasses and probably helps it to survive during periods of food shortage (Heinrich 1988).

Breeding was confirmed in 17.2% (214/1243) of Atlas squares in which the species was reported on the basis of various types of breeding evidence. The highly visible nest, the species' habit of using more than one nest site, the noisy young, and the aggressiveness of adults at the nest site, particularly during incubation (Knight 1986), no doubt facilitated confirmation. However, the incidence of confirmed breeding was slightly lower for the Common Raven than for the other corvids, probably because of the large size of its territory and its habit of nesting in remote locations.

When cliffs are not available, this corvid will nest in a tall conifer in the boreal forest, and even occasionally in a tall deciduous tree. Martin-Valin ZEC, north of the Saguenay River. *August 1991, Jean Gauthier.*

Formerly persecuted by humans, the Common Raven is increasingly found in urban habitat. Here, it is nesting on a cliff in the middle of the town of Saint-Romuald, near Québec City. *September 1994, Pierre Brousseau.*

In addition, the Common Raven appears to be drawn to the corridor along the St. Lawrence River downstream from Québec City, where seabird colonies no doubt provide an excellent source of food (eggs and nestlings) (Fradette 1992). It is also found in the Lac-Saint-Jean region, where its status appears to have changed considerably since Godfrey and Wilk's (1948) visit in the late 1940s, when no individuals were observed. Distribution is much patchier in the south, particularly in the St. Lawrence Plain upstream from Lake Saint-Pierre, where the low relief, scant forest cover, and intense human activity probably discourage it from nesting. However, the confirmed breeding records obtained in southernmost Québec—in the hilly country around Covey Hill, Rigaud, Mont-Saint-Hilaire, and, more recently, Mount Yamaska (M. A. Guertin, pers. comm.)—suggest that the Common Raven is returning to this area. Nesting was also confirmed in the Québec City area, in cliffs facing the St. Lawrence River and overlooking a residential area in the town of Saint-Romuald, and also near Charlesbourg. As these records show, the Common Raven appears able at times to adapt to the presence of humans. The sector between Québec City and Lake Saint-Pierre, with its more undulating and wooded countryside, appears to be a transition area for the species.

History and trends

The Common Raven's status in North America has fluctuated greatly from region to region since European settlement. Although the species was originally found throughout northeastern United States (Bull 1974), by the beginning of the 19th century, it had almost completely disappeared, and it has only made a comeback during the last 20 years (Oatman 1985a). The decline in the region's raven population was due largely to the destruction of large portions of forests—in Vermont, forest covered 82% of the state in 1790, but only 36% in 1880 (Garland 1977)—and the associated decline in deer populations, on whose carcasses the raven depended for food (Garland 1977; Bishop 1980). Other factors that may have been responsible for population declines across much of North America include persecution by man and a decrease in large predators—the lynx and wolf, for example—that provide carcasses for ravens to feed on in winter (Peterson 1988f).

The Common Raven's fortunes have also changed considerably in the Ottawa region since the turn of the century. In 1911, it was rare on the Ontario side of the Ottawa River, but common on the Québec side (Eifrig 1910–11), reflecting the differences in landscape and land use in the two regions since settlement. On the Québec side, north of the river lies the Canadian Shield, with its rougher terrain and largely forested land, while the Ontario side consists of a flat agricultural plain with significantly less forest cover; today, forest covers only one fifth of the traditionally wooded area (Villard 1990). Nevertheless, Common Raven populations have recently increased on the Ontario side. The species has been found there year-round since the late 1970s, and in 1987, breeding was confirmed for the first time in the Ottawa–Carleton region east of Ottawa. According to Di Labio and Martin (1989), the raven is expanding its range eastward in Ontario, an example of its ability to adapt to human habitat when natural nest sites are limited. On the Québec side, it is often observed in Gatineau Park (Villard 1990), where suitable nesting cliffs are available.

According to historical records, the Common Raven was a permanent resident in Québec at the turn of the century, but was rather rare (Dionne 1906; Macoun and Macoun 1915). Hall (1862) listed it as rare in the Montréal area, but reported that several pairs nested on cliffs on the Monteregian Hills. In Montréal, it was also occasionally seen at the city's river ice dump (Wintle 1896). Ouellet (1974a) commented that the last time the species had been seen in Montréal was probably in 1948. However, Atlas data show a return of the Common Raven, as occurred in southern Ontario, to agricultural, urban, and semi-rural areas in southwestern Québec, where the species was observed in several squares during the Atlas period. In the Magdalen Islands, its 19th century status as a breeder (Bishop 1889) does not appear to have changed (Fradette 1992).

The species' greater incidence north of the 46th parallel during the Atlas survey suggests that it has a greater affinity for northern habitats, despite any southern expansion. In North America as a whole, BBS (Robbins et al. 1986) and Christmas Bird Count (CBC) (Root 1988) data show an upward trend in the east between 1962 and 1979, particularly in New Brunswick, where the highest mean count was recorded during this period (Robbins et al. 1986).

LAVAL ROY AND MANON BOMBARDIER

Black-capped Chickadee

Mésange à tête noire

Parus atricapillus

Christiane Girard

THE INQUISITIVE, GREGARIOUS, AND BOLD BLACK-CAPPED CHICKADEE rarely goes unnoticed on a walk in the woods. It is well known, not only to the general public, but also to scientists. Over a hundred specialized articles have been published on it (Smith 1991, 1993), yet many aspects of its ecology are still poorly known. For example, is it an occasional migrant? What factors influence its choice of habitat and its abundance? In Québec, the Black-capped Chickadee is considered a resident breeder (David 1980a).

Habitat and behaviour

The Black-capped Chickadee is a species typical of deciduous or mixed woodlands. It is just as often found in large forests as in islands of trees or shrubs in urban settings. Woodlands of various age, tall thickets, and shrubbery also suit it (Godfrey 1986). The males are indistinguishable from the females by their plumage, making it difficult to study their breeding behaviour.

Most pair bonds are apparently formed in fall when the birds flock together; a few chickadees may pair in winter or spring, especially in response to the mortality of high-ranking members of the winter flock (Smith 1991). Chickadees become very quiet and unobtrusive once they have staked out their territories, and can only be detected by their simple whistling song, a plaintive *fee-bee*.

The Black-capped Chickadee excavates its own nest in a tree branch or trunk, in part of its territory where there are clearings and the trees are mature. As it needs rotting wood to make its cavity, managed woodlots with no dead trees are less suitable for nesting. It sometimes uses a nest box, however, particularly one that has been stuffed with wood shavings or similar materials. Driven by its apparent need to excavate a cavity, it usually nests in a different spot each year.

The cavity-excavation phase is a good time to look for nests. Since the trees still have little or no foliage and the chickadees frequently visit the nest site, it is fairly easy to follow them back to their cavity. In April, chickadee pairs adopt a typical routine. The male and female crisscross the territory far from the nest, looking for food, then suddenly fly straight back to the nest. They then work on the nest for a few minutes before beginning their territorial patrol all over again. In the days after they have finished excavating the nest, an observer may, with a little luck, see the male feeding the female, which flutters her wings and emits a call similar to that of the young soon after they leave the nest. Care must be taken not to disturb the parents shortly after laying, because that is when they are most likely to desert the nest.

According to observations made in Ontario, birch and poplar are the preferred species for nesting. The diameter at breast height (dbh) of trees used varies from 7.5 to 61 cm, while nest height ranges from 0 to 15 m (Peck and James 1987).

Egg laying usually takes place in early May, but the laying date varies greatly depending on the year and the age of the parents. If the first clutch is lost to predators or poor weather, the female may renest and the pair may then be feeding nestlings as late as August. A second clutch may occasionally be laid shortly after the first brood has left the nest, but usually in another cavity (Smith 1991). During the incubation period, the male often brings food to the female and it is therefore easy for an observer to confuse the incubation and brood-rearing periods.

Notes on territory

The breeding territory is established between 5 and 7 weeks before laying. Territory size is generally between 1.5 and 5.3 ha, and seems to vary inversely with territory quality, with the smallest territories being found in heavily wooded areas (Smith 1991). The size of the territory, which is at its largest during nest building, decreases as the breeding season progresses (Stefanski 1967; Glase 1973). The pair defends the territory against conspecific intruders of both sexes, and uses it for nesting, copulation, and feeding (Smith 1991).

Three Québec studies reported the following densities, expressed as the number of territorial males/10 ha, for different habitats: 2.0 in a maple stand and 2.4 in a hemlock forest, both at Mirabel (Cyr 1974); 3.8 in a white birch stand and 1.1 in both a mixed wood and in a fir stand in La Mauricie National Park (DesGranges 1980b); and 2.5 in an Eastern white pine stand at Oka (Erskine 1984).

The Black-capped Chickadee is fairly eclectic in its choice of home range, haunting all kinds of damp or dry deciduous or mixed woodlands of various age composition and expanse. Mont-Carmel, southwest of Rivière-du-Loup. *May 1991, Sylvain St-Onge.*

Black-capped Chickadee

Breeding distribution

1984–1989

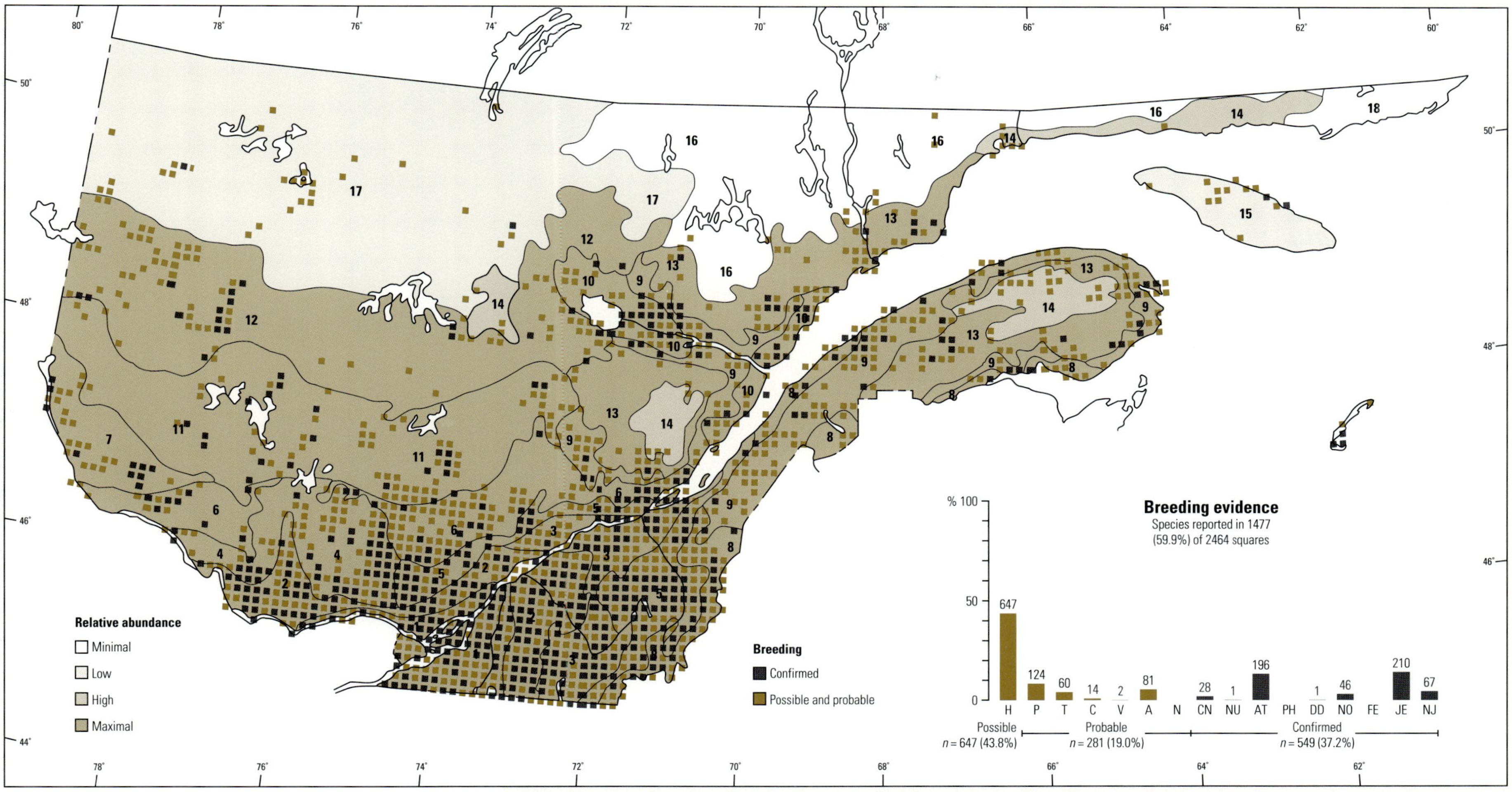

World breeding range

Species profile

Clutch size	usually 6–8 eggs (3–10)[1]
Laying interval	1 egg per day[1,7]
Start of incubation	with penultimate egg[1,7]
Incubation period	12–13 days[1,7]
Maturity at hatching	altricial
Care of young	by female and male[1,7]
Nestling period	≈ 16 days[1,7]
Age at first flight	≈ 16 days[1,7]
Dependence of young	until 14–28 days after first flight[1,7]
Broods per year	1, rarely 2[1,7]
Breeding age	1 year[1,7]
Mating system	monogamy, rarely polygyny or polyandry[1,7]
Duration of pair bond	sometimes more than 1 season[1,2,7]
Total length	12.3–14.5 cm[3]
Wing span	19.0–21.6 cm[4]
Weight	10.8 g (8.2–13.6)[5]
Longevity record	12 years 5 months[6]

References
1. Smith 1991
2. Smith, S. M. 1992
3. Godfrey 1986
4. Terres 1980
5. Dunning 1984
6. Clapp et al. 1983
7. Smith 1993

Chronology *n* = 253

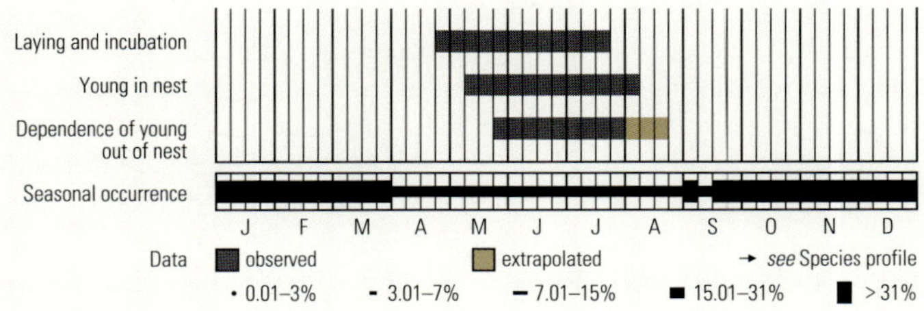

Frequency of occurrence by ecological region

Ecological regions** (forest regions)	Adequately covered squares	Squares in which reported	% of squares in which reported	Relative abundance (%)*
1. Sugar maple-hickory and sugar maple-basswood	97	97	100.0 ± 0.5	100.0
2. Sugar maple-basswood	153	151	98.7 ± 1.2	98.7
3. Sugar maple-basswood and sugar maple-yellow birch	101	101	100.0 ± 0.5	100.0
4. Sugar maple-yellow birch-basswood	56	56	100.0 ± 0.9	100.0
5. Sugar maple-yellow birch-beech and sugar maple-yellow birch-basswood	113	111	98.2 ± 1.7	98.2
6. Sugar maple-yellow birch-beech	71	70	98.6 ± 3.0	98.6
7. Sugar maple-yellow birch-balsam fir	19	19	100.0 ± 2.6	100.0
8. Sugar maple-yellow birch and balsam fir-yellow birch	47	46	97.9 ± 4.5	97.9
9. Balsam fir-yellow birch	87	86	98.9 ± 2.5	98.9
10. Balsam fir-white birch or balsam fir-red maple	46	44	95.7 ± 5.7	95.7
11. Yellow birch-balsam fir	52	52	100.0 ± 1.0	100.0
12. Balsam fir-white birch (west)	59	55	93.2 ± 7.0	93.2
13. Balsam fir-white birch (east)	76	70	92.1 ± 6.2	92.1
14. Balsam fir-black spruce	28	19	67.9 ± 18.2	67.9
15. White spruce-balsam fir and balsam fir-white spruce	22	9	40.9 ± 21.2	40.9
16. Black spruce-balsam fir-moss	18	5	27.8 ± 23.7	27.8
17. Black spruce-moss	32	16	50.0 ± 18.9	50.0
18.			*not analysed*	

* minimal: 0–29.9%; low: 30–59.9%; high: 60–89.9%; maximal: 90–100%

** Deciduous forest: 1–7; mixed forest: 8–11; coniferous forest: 12–17

Friendly and tame, the Black-capped Chickadee nests in coppices in city parks. Île Sainte-Hélène, Montréal. *October 1980, Richard Pelletier.*

Chickadees become noisy again once the fledglings leave the nest, when the cries of the young begging for food can often be heard as they wait to be fed by both parents. The young then begin their fall wanderings, which can take them several hundred metres or up to several kilometres from their natal site. Like most monogamous songbirds (Greenwood and Harvey 1982), the females tend to disperse greater distances than the males. Once this period of dispersal and mixing has begun, family groups begin to break up and winter flocks form, made up of individuals from different families.

During the breeding season, the chickadee's diet consists of 80–90% animal foods, while in winter, it is composed of equal parts animal and vegetable matter (Smith 1993).

Breeding is usually easy to confirm in this familiar species, as the 37.2% (549/1477) rate obtained during atlassing shows. Observation of the often noisy family and adults carrying food were the types of breeding evidence most often reported.

This chickadee excavates a nest cavity in the rotten wood of a branch or trunk, preferably in a hardwood tree, or it may use a nest box. Lac-Saint-Charles, near Québec City. *May 1994, Robert Morin.*

This chickadee occupies urban woodlands all year long, nesting there as well. It also visits bird feeders in residential neighbourhoods at any time of year. Cap-Rouge. *October 1993, Pierre Normand.*

Distribution

The Black-capped Chickadee's range extends from Alaska to Newfoundland and includes all the northern United States (Godfrey 1986). In Québec, its presence in 59.9% of squares covering all Atlas regions makes it one of the commonest and most widespread species in the Atlas area. Most observations were in the south, not simply because of better field coverage, but also because of the abundance of deciduous and mixed woodlands there that are optimal habitat for the species.

The only woodland areas where the Black-capped Chickadee seems to be not too common are in the boreal forest, such as the Black spruce and White spruce forest regions, and the Balsam fir-black spruce forest region, possibly as the result of competition with the Boreal Chickadee. A more accurate count of chickadee populations in various habitats is needed to attempt to explain the factors governing habitat selection by the Black-capped Chickadee.

North of the Atlas area, the species has been reported around the Nottaway, Broadback, and Rupert rivers (Cons. G. G. – GREBE 1992o), at Cabbage Willows Bay (Todd 1963), near the Eastmain River (Tremblay and Belles-Isles 1991), at the rest stop (km 381) on Highway 109 ten kilometres south of the Eastmain River (Bordage 1985b), and farther east in the province at Lake Plétipi, Gagnon, and in the Groulx Mountains (ÉPOQ). On the North Shore, it probably does not range beyond Natashquan (Todd 1963). The northern limit of its range is therefore likely to be around the 52nd parallel.

History and trends

The abundance and distribution of the Black-capped Chickadee do not seem to have changed significantly since the turn of the century, if contemporary writings (Dionne 1906; Macoun and Macoun 1909; Taverner 1922) can be relied upon. ÉPOQ data (Cyr and Larivée 1993b) for the summers 1970–89 and BBS data from 1980–89 show an increase in the population.

Many ornithologists believe that the Black-capped Chickadee has greatly benefited from the artificial provision of food in the form of the growing number of bird feeders, but two studies show that although feeding effectively improves their winter survival there is no concomitant increase in breeding density (Brittingham and Temple 1988; Desrochers et al. 1988). In other words, feeders simply seem to create a floating population of survivors that are unable to find territories in spring. Little is known about the fate of birds unsuccessful in competition for territories, but spring mortality is probably high. Nonetheless, some chickadees manage to subsist throughout the summer in territories that do not belong to them. Desrochers et al. (1988) demonstrated the existence of such birds, which quickly capitalized on any territory left free by the disappearance of a territorial bird of the same sex. Accordingly, it would be premature to conclude that chickadee populations have increased thanks to bird feeders.

While logging is probably detrimental to the species in the short term, over the medium term the deciduous regeneration in birch and poplar it promotes, creates suitable Black-capped Chickadee habitat.

André Desrochers

Boreal Chickadee

Mésange à tête brune

Parus hudsonicus

Christiane Girard

THE BOREAL CHICKADEE CAN BE DISTINGUISHED FROM ITS COUSIN THE Black-capped Chickadee by the somewhat nasal tones of its call. It is not often seen, however, and chances of observing it are best in northern coniferous forest. Birders and ornithologists alike know little about the habits of the Boreal Chickadee, which is secretive throughout the year. There is very little scientific literature on the species.

David (1980a) has classified the Boreal Chickadee as a resident breeder in southern Québec. Several authors consider the species a partial migrant, based mainly on occasional sightings of large irruptive flights in spring and autumn. Birds in autumn concentrations, however, may be juveniles looking for somewhere to overwinter, whereas spring flights may be made up of subordinate birds that have been driven from their wintering grounds by territorial birds.

This inconspicuous species breeds in relatively inaccessible habitat in the boreal forest. Nests are built in old woodpecker holes in stumps or snags. Near Caribou Lake, Decelles Reservoir, Témiscamingue. *May 1991, Réginald Ouellet.*

Habitat and behaviour

Although Boreal Chickadees are usually associated with coniferous habitat, they may also be found foraging in stands of deciduous shrubs such as alders. McLaren (1975) estimated territory size at 5 ha, a rather large area for such a small bird. However, as with the Black-capped Chickadee, this is probably a maximum corresponding to the egg-laying period, after which territory size generally decreases (Møller 1990b).

The female, assisted by the male, excavates a nest hole or uses an old woodpecker or nuthatch hole. The nest is commonly found in a stump, trunk, or branch, in either a coniferous or deciduous tree; the choice seems to be affected more by the softness of the heartwood than a preference for a particular species of tree (McLaren 1975). Nest boxes are sometimes also used (Peck and James 1987).

Breeding was confirmed in only 13.9% (97/700) of Atlas squares in which the species was reported. Nests seem to have been particularly difficult to locate, only 8 having been found. This somewhat retiring species seeks concealment in highly inaccessible areas. Sightings of males proffering food to females, which respond with wing fluttering and a soft warbling call, are a reliable means of confirming nesting early in the season (Bent 1946).

Incubation in the Boreal Chickadee is by the female alone. Once the eggs hatch, Boreals are more easily spotted. The incessant foraging trips by both parents to feed the young is the easiest breeding evidence to obtain and accounted for 56.7% of breeding confirmations. Parental activity is extremely high during this period —Pettingill counted 362 deliveries of food to a brood of Boreal Chickadees in a single day (Bent 1946)! The young are fed for some time after they leave the nest; 32% of breeding confirmations were of young birds.

Distribution

The Boreal Chickadee occurs throughout Québec wherever large expanses of coniferous forest are found. Thus it is rare in the agricultural and strictly hardwood areas in the south of the province, such as in the Central St. Lawrence Lowland and Ottawa Valley regions. Its low frequency of occurrence in the Balsam fir-white birch forest region between Abitibi and Lac-Saint-Jean is probably due to poor coverage, most atlassing having been done in the more accessible deforested regions in that area.

One lesser known aspect of the species' distribution in Québec is its regular occurrence in the Appalachians, where the forest cover is often boreal in nature down into New York State.

The species ranges well beyond the Atlas area; Todd (1963) reported it at Kuujjuaq and Godfrey (1986) at Koroc River, both located on the 58th parallel.

Density appears to be higher in the open spruce-lichen woodland, or taiga, than in the Black spruce-moss forest region farther south (Cons. G. G. – GREBE 1991a, 1992o; Tremblay and Belles-Isles 1991). It frequents several types of habitat, although population densities are higher in spruce stands.

History and trends

The species must have been more abundant in suitable habitat, such as the forested peatlands of the Central St. Lawrence Lowland, before the deforestation that came with European settlement (Ouellet 1974a). There have probably not been any major changes in the species' northern range since then (Dionne 1906; Macoun and Macoun 1909).

Populations do not seem to have fluctuated much since 1966 in those parts of the province visited by birders (Cyr and Larivée 1993b). There are some data suggestive of a local decline, however, in the southern part of its breeding range (Erskine et al. 1992; BBS).

Logging and forest fires probably do not have too much of an effect on the abundance of this species, as it does not require mature forests for breeding (Erskine 1977). Nor are its population dynamics likely to be affected by the increasing prevalence of bird feeders, as Boreal Chickadees rarely visit feeding stations. As with tits in the Scandinavian countries (Jansson et al. 1981), winter survival seems to be the main limiting factor on population size.

ANDRÉ DESROCHERS

Notes on territory

Territories do not appear to be constantly advertised vocally by their holders, and territorial limits are often trespassed by other birds. When intruders are spotted, the chickadees respond with alarm calls and aerial pursuit (McLaren 1975). McLaren (1975) recorded territories of over 5 ha in a spruce-fir stand in Algonquin Provincial Park in Ontario. However, on mountain tops in New Hampshire, the average territory was only 1.2 ha (Sabo 1980). Erskine (1980a) reported a density of 1.9 birds/10 ha in a spruce-trembling aspen stand at Big Trout Lake, in northern Ontario.

The Boreal Chickadee, although associated with large tracts of coniferous forest, also nests in more open areas. Grands-Jardins Conservation Park, northwest of Baie-Saint-Paul. *June 1991, Réginald Ouellet.*

Boreal Chickadee

Breeding distribution
1984–1989

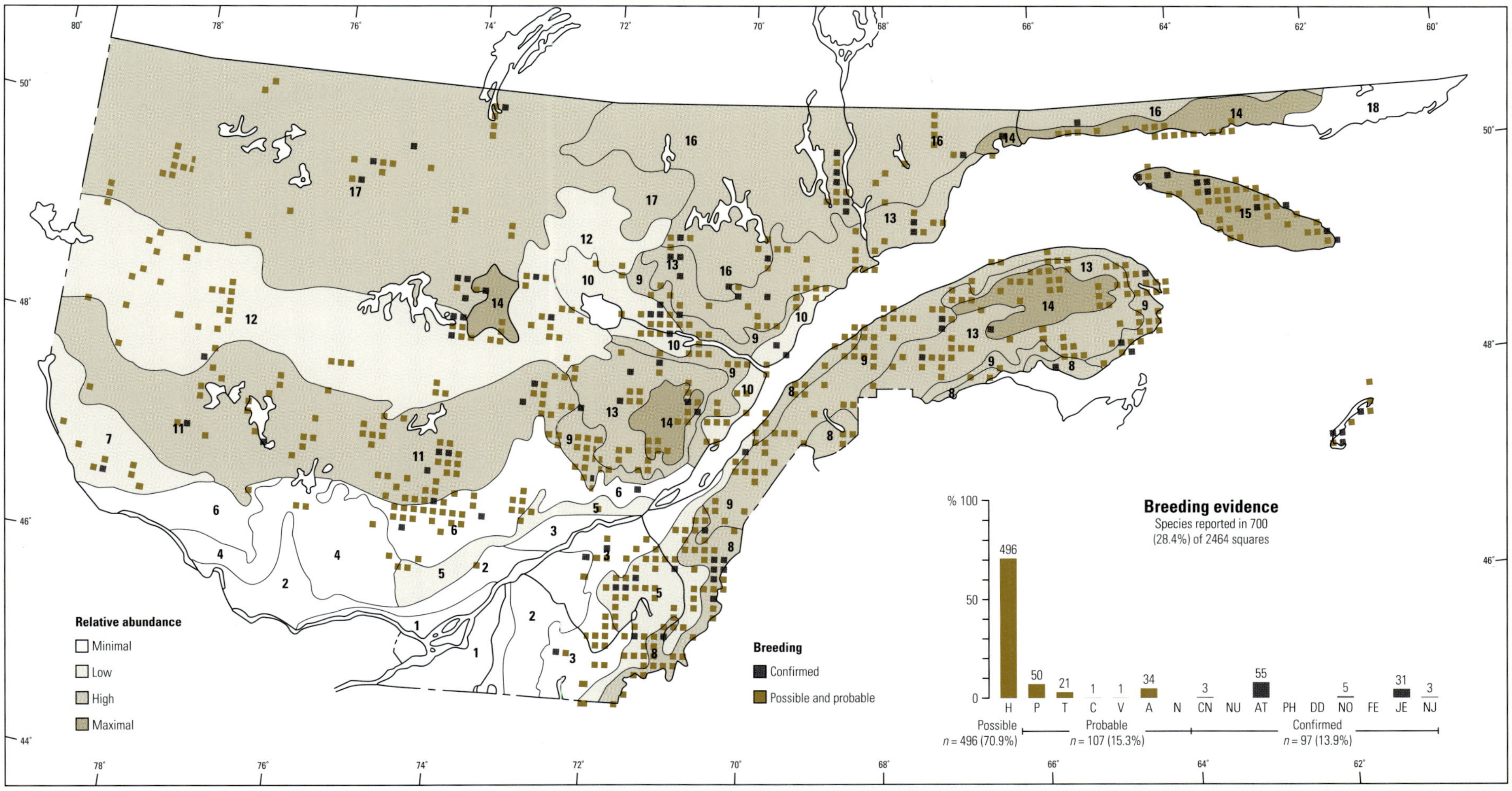

World breeding range

Species profile

Clutch size	usually 5–7 eggs (4–9)[1,2,3]
Laying interval	?
Start of incubation	?
Incubation period	usually 12–13 days (11–16)[1,2]
Maturity at hatching	altricial
Care of young	by female and male[2]
Nestling period	≈ 18 days[1,2]
Age at first flight	≈ 18 days or more[5]
Dependence of young	until 3 weeks after first flight[2]
Broods per year	1[1]
Breeding age	1 year[1]
Mating system	monogamy[2]
Duration of pair bond	probably for life[2]
Total length	12.5–14.0 cm[3]
Wing span	?
Weight	9.8 g (7.0–12.4)[6]
Longevity record	10 years 1 month[4]

References:
1. DeGraaf et al. 1980
2. Ehrlich et al. 1988
3. Godfrey 1986
4. Kennard 1975
5. according to nestling period
6. Dunning 1984

Chronology $n = 24$

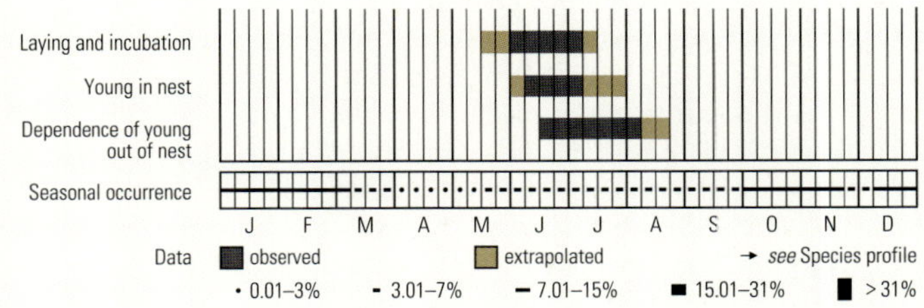

Frequency of occurrence by ecological region

Ecological regions** (forest regions)	Adequately covered squares	Squares in which reported	% of squares in which reported	Relative abundance (%)*
1. Sugar maple-hickory and sugar maple-basswood	97	0	0	0
2. Sugar maple-basswood	153	2	1.3 ± 1.2	1.3
3. Sugar maple-basswood and sugar maple-yellow birch	101	29	28.7 ± 5.4	28.7
4. Sugar maple-yellow birch-basswood	56	5	8.9 ± 6.9	8.9
5. Sugar maple-yellow birch-beech and sugar maple-yellow birch-basswood	113	40	35.4 ± 4.9	35.4
6. Sugar maple-yellow birch-beech	71	20	28.2 ± 9.6	28.2
7. Sugar maple-yellow birch-balsam fir	19	8	42.1 ± 23.3	42.1
8. Sugar maple-yellow birch and balsam fir-yellow birch	47	38	80.9 ± 10.3	80.9
9. Balsam fir-yellow birch	87	64	73.6 ± 8.6	73.6
10. Balsam fir-white birch or balsam fir-red maple	46	22	47.8 ± 12.5	47.8
11. Yellow birch-balsam fir	52	43	82.7 ± 10.8	82.7
12. Balsam fir-white birch (west)	59	33	55.9 ± 13.0	55.9
13. Balsam fir-white birch (east)	76	61	80.3 ± 8.8	80.3
14. Balsam fir-black spruce	28	26	92.9 ± 10.8	92.9
15. White spruce-balsam fir and balsam fir-white spruce	22	22	100.0 ± 2.3	100.0
16. Black spruce-balsam fir-moss	18	16	88.9 ± 17.4	88.9
17. Black spruce-moss	32	21	65.6 ± 18.0	65.6
18.			*not analysed*	

* minimal: 0–29.9%; low: 30–59.9%; high: 60–89.9%; maximal: 90–100%

** Deciduous forest: 1–7; mixed forest: 8–11; coniferous forest: 12–17

Red-breasted Nuthatch

Sittelle à poitrine rousse

Sitta canadensis

Christiane Girard

THIS VIVACIOUS LITTLE BIRD IS EASILY RECOGNIZED BY ITS BLACK CAP, reddish breast, and black eye stripe. The Red-breasted Nuthatch, slightly smaller than its white-breasted cousin, can also be distinguished by its call, a sharper, nasal, slow *ank* or *nyak*. Females and juveniles have a paler cap and underparts. The Red-breasted Nuthatch also exhibits typical nuthatch behaviour, regularly working its way up and down tree trunks and branches, headfirst.

Widely distributed throughout the province, this species is a permanent resident throughout most of its Québec breeding range. Nonetheless, populations in northern areas and at high elevations (Ehrlich et al. 1988), or in areas where the yearly cone production is insufficient to meet their needs (Bock and Lepthien 1972), migrate and winter farther south or at lower elevations. This nuthatch occurs throughout Canada up to the tree line, but prefers coniferous or mixed woodlands.

Habitat and behaviour

When food is plentiful, a pair may occupy the same territory all year round, possibly remaining together for several seasons (Tyler 1948b; Kilham 1973b). In migratory populations, however, pair bonds seem to dissolve in winter, when the species tends to form small flocks (Matthysen et al. 1992).

The monogamous male does not begin to court a female until late March. During courtship, he engages in a number of displays, including one where, with his back to the female, he lowers his wings, raises his head, and cocks up his tail, wagging it from side to side (Kilham 1973b). During the breeding season, an astonishing variety of vocalizations by both sexes can also be heard, and the male can be seen offering food to the female (Kilham 1973b).

The Red-breasted Nuthatch nests mainly in mature coniferous or mixed woodlands. Unlike the Hairy Woodpecker, this cavity nester is unable to excavate hard wood, preferring instead to dig into dead or rotten wood. It sometimes uses an old woodpecker hole or a nest box (Harestad and Keisker 1989). Nest height is 0.8–36 m though on average it is 4–5 m. The species shows no marked preference for any particular type of tree (Harestad and Keisker 1989; Cheskey 1990).

Apparently, the female chooses the nest site, and she may start to excavate several cavities at the same time. The male then attempts to encourage his mate to select a definitive site (Kilham 1973b) and helps dig. A downward-slanting passage with a mean entrance diameter of 3.8 cm leads to the cavity, which is about 10 cm deep. Inside, the nest proper consists of bark, moss, grass, feathers, or hair (Harrison 1975; Harrison 1978).

Interestingly, the Red-breasted Nuthatch has the habit of constantly smearing the cavity entrance with conifer resin. This circle of resin seems to prevent insects, especially ants, from getting in. However, it also obliges the nuthatches to enter the hole directly, without perching at the entrance, in order to avoid getting resin on themselves (Tyler 1948b).

Typical clutch size is 5 or 6 eggs (Terres 1980; Peck and James 1987), which the female alone incubates for 12 or 13 days. The nidicolous young are fed insects by the parents at frequent intervals. The female usually spends the night on the nest, which the young leave at 18–21 days of age (Tyler 1948b; Ehrlich et al. 1988; Cheskey 1990).

It is not easy to detect breeding in this little songbird because it often inhabits forests that are difficult of access, partly because they are so dense.

Adults are partial to conifer seeds, which they skilfully extract by opening the scales of the cones with their bills; they also eat nuts and a variety of other seeds, which they sometimes wedge into a bark crevice so that they can then remove the kernel. They can often be seen feeding in the treetops and at the tips of branches, searching for cones. They also rapidly scour trunks and branches for insect larvae and are skilled at flycatching. Bird feeders filled with sunflower seeds and bits of suet readily attract these birds. Red-breasted Nuthatches also have the curious habit of hiding food on tree trunks, in cracks in the bark or small holes left by woodpeckers. They usually retrieve their booty the same day (Tyler 1948b; Grubb and Waite 1987).

Notes on territory

Little is known about territoriality in this species. Apparently, the territory may in some cases be fiercely defended against several

Mature coniferous and mixed forests are the preferred haunts of the Red-breasted Nuthatch. Mount Albert, Gaspé Conservation Park. *July 1991, Yves Aubry.*

Red-breasted Nuthatch

Breeding distribution

1984–1989

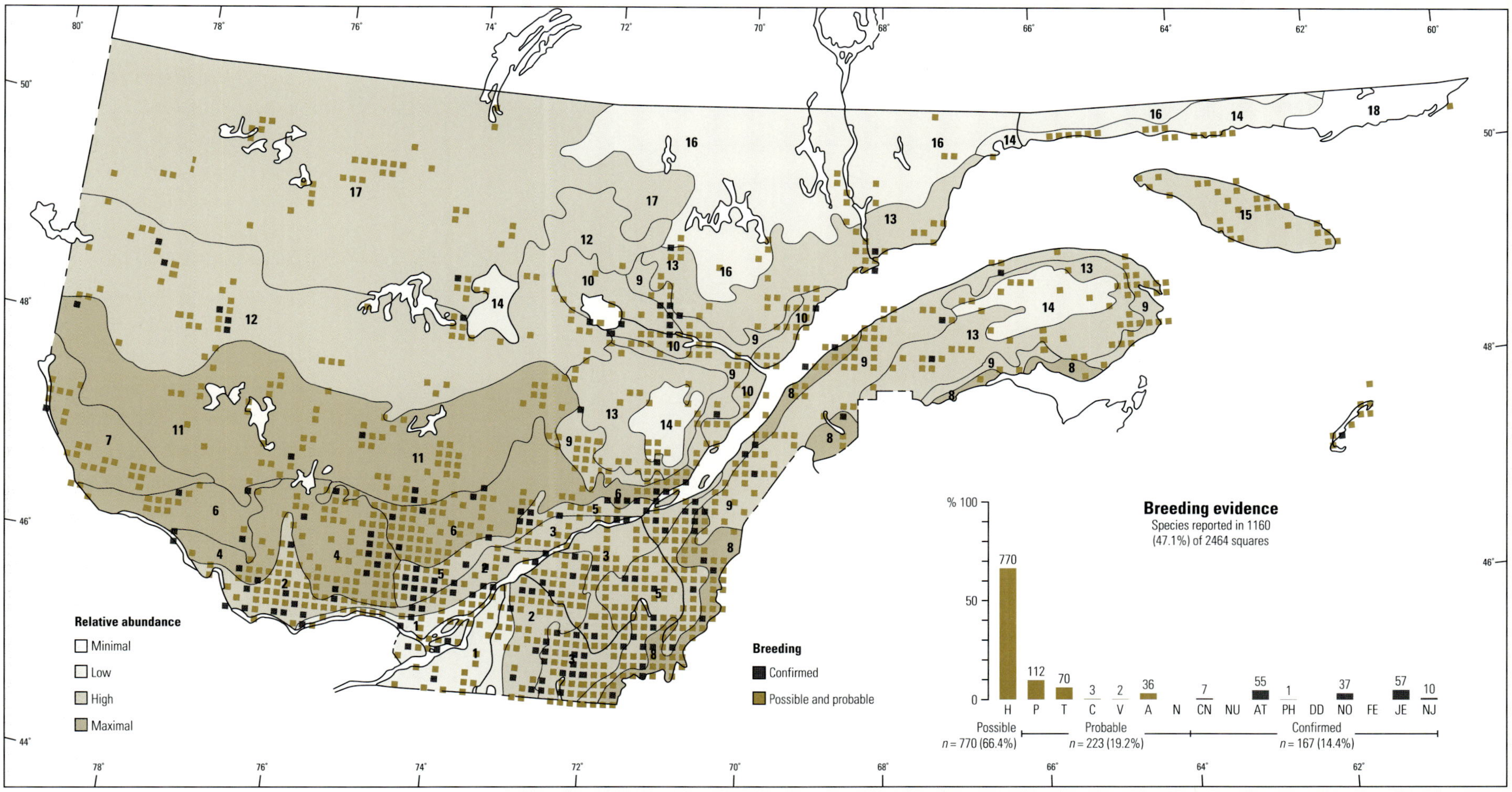

Species profile

Clutch size	usually 5–6 eggs (4–7)[1,3,4,5]
Laying interval	?
Start of incubation	?
Incubation period	12–13 days[1,6]
Maturity at hatching	altricial
Care of young	by female and male[5]
Nestling period	18–21 days[5,6,7]
Age at first flight	18–21 days or more[13]
Dependence of young	?
Broods per year	1[7,8,9]
Breeding age	1 year[8]
Mating system	monogamy[2]
Duration of pair bond	1 or more seasons[1,2]
Total length	10.2–12.4 cm[11]
Wing span	20.3–21.6 cm[3]
Weight	9.8 g (8.0–12.7)[10]
Longevity record	7 years 6 months[12]

References:
1. Tyler 1948b
2. Kilham 1973b
3. Terres 1980
4. Peck and James 1987
5. Ehrlich et al. 1988
6. Cheskey 1990
7. Harrison 1978
8. DeGraaf and Rudis 1987
9. Gosnell 1985
10. Dunning 1984
11. Godfrey 1986
12. Kennard 1975
13. according to nestling period

World breeding range

Chronology $n = 45$

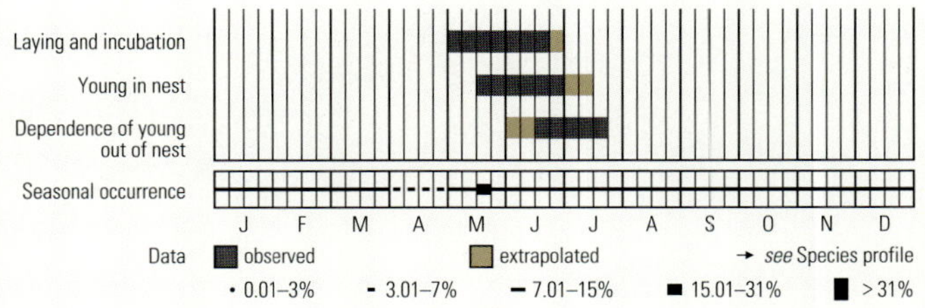

Frequency of occurrence by ecological region

Ecological regions** (forest regions)	Adequately covered squares	Squares in which reported	% of squares in which reported	Relative abundance (%)*
1. Sugar maple-hickory and sugar maple-basswood	97	46	47.4 ± 5.0	47.4
2. Sugar maple-basswood	153	117	76.5 ± 3.5	76.5
3. Sugar maple-basswood and sugar maple-yellow birch	101	89	88.1 ± 4.0	88.1
4. Sugar maple-yellow birch-basswood	56	56	100.0 ± 0.9	100.0
5. Sugar maple-yellow birch-beech and sugar maple-yellow birch-basswood	113	97	85.8 ± 3.7	85.8
6. Sugar maple-yellow birch-beech	71	66	93.0 ± 5.8	93.0
7. Sugar maple-yellow birch-balsam fir	19	19	100.0 ± 2.6	100.0
8. Sugar maple-yellow birch and balsam fir-yellow birch	47	43	91.5 ± 7.6	91.5
9. Balsam fir-yellow birch	87	73	83.9 ± 7.3	83.9
10. Balsam fir-white birch or balsam fir-red maple	46	32	69.6 ± 11.6	69.6
11. Yellow birch-balsam fir	52	48	92.3 ± 7.9	92.3
12. Balsam fir-white birch (west)	59	37	62.7 ± 12.7	62.7
13. Balsam fir-white birch (east)	76	50	65.8 ± 10.4	65.8
14. Balsam fir-black spruce	28	16	57.1 ± 19.2	57.1
15. White spruce-balsam fir and balsam fir-white spruce	22	18	81.8 ± 17.1	81.8
16. Black spruce-balsam fir-moss	18	10	55.6 ± 26.0	55.6
17. Black spruce-moss	32	20	62.5 ± 18.3	62.5
18.			not analysed	

* minimal: 0–29.9%; low: 30–59.9%; high: 60–89.9%; maximal: 90–100%

** Deciduous forest: 1–7; mixed forest: 8–11; coniferous forest: 12–17

Female Red-breasted Nuthatch feeding a nestling an insect. The adults smear the nest entrance with conifer resin to keep insects, in particular ants, out. Sainte-Foy. *June 1960, Gaston Le Page.*

Notes on territory (continued)

other species of birds that have the same feeding habits or also nest in cavities (Cheskey 1990). Territory size may also vary considerably depending on habitat, from 0.2 ha in a hemlock forest in the Mirabel area (Cyr 1974) to about 10 ha in a subalpine habitat with relatively few dead trees in New Hampshire (Sabo 1980). Territoriality in migratory individuals seems to disappear in winter (Matthysen et al. 1992). In Québec, breeding density was estimated to vary from 1.1–5.0 singing males/10 ha in coniferous woods or conifer-dominated mixed woods in La Mauricie National Park and near Cowansville (DesGranges 1980b; Erskine 1984).

The Red-breasted Nuthatch nests in the mixed woods of the Laurentide Wildlife Reserve, north of Québec City. *May 1991, Réginald Ouellet.*

In winter, conifer seeds are naturally an important component of the species' diet. Some years, when cone production is low, massive winter invasions of Red-breasted Nuthatches can be seen farther south (Bock and Lepthien 1972). These irruptions seem to occur every two years, but the factors affecting conifer productivity are not well known (Larson and Bock 1986). The winter range covers all of southern Canada and most of the United States except southern Florida and the deserts in the southwest (Root 1988).

Distribution

The Red-breasted Nuthatch breeds from southeastern Alaska and Yukon all across southern Canada to eastern Labrador and Newfoundland. Its range also covers the entire northern United States as far south as southern California in the west and North Carolina in the east.

This nuthatch is found throughout southern Québec, having been reported in 47.1% (1160/2464) of Atlas squares surveyed. At first glance, its wide southern distribution is somewhat surprising because most authors associate this nuthatch with coniferous or mixed woodlands. However, it is probably a generalist, occurring in a variety of habitats well to the south of the boreal and mixed forests. It seems particularly common in the mixed and predominantly deciduous woodlands of the Laurentians north of Montréal, and in the Ottawa Valley, where it favours scattered conifers. It is also found on the south shore of the St. Lawrence, particularly in the Appalachians. The Red-breasted Nuthatch probably becomes less abundant farther north, up to the tundra, as conifers progressively grow less densely and become smaller and stunted.

History and trends

Ouellet (1974a) believes that populations may have declined since the middle of the last century as a result of intensive logging. At the turn of this century, this nuthatch was generally considered a common "transient visitant" but rare resident in the Montréal area (Wintle 1896; Ouellet 1974a). Bannon (1991b) recently classified it as an occasional breeder. BBS data indicate an increase in the breeding population in Québec between 1966 and 1989. Its breeding range seems to be expanding eastward (Ehrlich et al. 1988); Bull (1985) commented that this nuthatch, common in the mountains of New York, has begun to frequent the lowlands, from which it had hitherto been absent. The species also seems to be more common in southern Ontario (Mills 1987e), where it appears to be taking advantage of new habitat created by maturing conifer plantations (Mills 1987e; Peterson 1988k). Thus, even far south in the mixed forest, it can be found in scattered conifers at higher elevations or in plantations, or even once in a while in ornamental spruce trees (Bull 1974).

Jean-François Rail

When it is not scouring trunks and branches for insects, this nuthatch eats seeds which it extracts from cones, either from the tops of conifers or at the tips of their branches. Martin-Valin ZEC, north of the Saguenay River. *August 1991, Jean Gauthier.*

White-breasted Nuthatch

Sittelle à poitrine blanche

Sitta carolinensis

Christiane Girard

THE WHITE-BREASTED NUTHATCH IS CLOSELY ASSOCIATED WITH THE deciduous forest zone. It is easy to spot in deciduous stands near towns and cities, and in winter it is a regular visitor to bird feeders; it is tame enough that it may even eat seeds out of your hand. The species is considered non-migratory (AOU 1983), although in winter, movements do occur in populations in the northern part of the range and at higher elevations, in the Appalachians, for example (Butts 1931; Tyler 1948c; Bull 1985).

Novice birders can soon learn to recognize the White-breasted Nuthatch by its silhouette and singular behaviour. As it works its way down a tree trunk, headfirst, it carefully inspects grooves and crevices in the bark for food. The Red-breasted Nuthatch, which is slightly smaller and has a black eyebrow, displays the same behaviour, but is mostly associated with the coniferous forest. The two nuthatches may occasionally be seen together in mixed forest or at bird feeders.

Habitat and behaviour

Pairs of this monogamous species bond for several years, perhaps even for life. All winter long, the male remains within hearing distance of his mate, but seems to pay her little attention (Tyler 1948c). Males and females roost apart, in individual cavities, which they keep clean by removing their feces. In spring, the male gives up his cavity to the female, and it becomes the future nest site (Kilham 1971b).

As early as April, the male begins to sing, almost throughout the day. His song, a series of 8–11 nasal whistled notes, is sweeter than the usual *yank, yank, yank* emitted by both sexes throughout the year. As he sings, he bows forward, often turning his back to the female (Tyler 1948c; Kilham 1972). He commonly engages in courtship feeding. The female waits motionless for her mate to line up his bill with hers before she takes the offering. According to Kilham (1972), in this ritual the female plays the part of a nestling, thereby arousing the male's paternal instinct, while at the same time getting him to practise feeding the young. The male stops feeding the female soon after the eggs have hatched (Stokes and Stokes 1989b).

The White-breasted Nuthatch inhabits primarily deciduous woodland, but can also be found in mixed stands, and very occasionally, in coniferous forest. It prefers stands of large, mature trees, and although it most often nests in a live tree, it does not shun sick or dead trees. It likes areas with open spaces nearby; territories are often located close to water, a road, or a field, or in an orchard or wooded area in a residential neighbourhood (Godfrey 1986; Peck and James 1987).

This cavity nester frequently uses knotholes in the trunk where branches have rotted and fallen off (Tyler 1948c; Stokes and Stokes 1989b). The selected tree is usually deciduous and the cavity is located 0.5–20 m up, though usually between 2 and 9 m (Peck and James 1987). Occasionally, the White-breasted Nuthatch may use an old woodpecker hole or a nest box (Harrison 1978). In other words, this species rarely digs a hole itself (DeGraaf and Rudis 1987), although it may enlarge one if necessary. The entrance to the nest cavity is at least 3 or 4 cm in diameter, although for unknown reasons, these nuthatches often prefer openings two or three times larger (Kilham 1971b; Stokes and Stokes 1989b). They may nest in the same cavity for several years (Butts 1931).

This nuthatch nests in deciduous and mixed woodland. It works its way up and down tree trunks looking for arthropods. Sainte-Foy. *September 1991, Yves Aubry.*

Notes on territory

The White-breasted Nuthatch has a relatively large territory, covering 10–20 ha (Butts 1931; Kilham 1972, 1981). It is defended all year-round, although its summer and winter boundaries do not always correspond exactly. Nests are built within or near the winter feeding territory, however (Butts 1931). Territories are usually sparsely scattered throughout suitable habitat, and confrontations between males are infrequent (Kilham 1972). These birds are fairly tolerant, especially in winter, with territories having been known to overlap and pairs even feeding at the same spot (Butts 1931).

Erskine (1984) reported densities of 2.5 territorial males/10 ha in a sugar maple-beech stand at Mount Orford and in a silver maple-black ash swamp forest at Oka. In a mature sugar maple forest at Mirabel, Cyr (1974) counted 2.3 territorial males/ 10 ha. DesGranges (1980b) reported densities of 4.3 birds/ 10 ha in a white birch stand in La Mauricie National Park.

White-breasted Nuthatch

Breeding distribution

1984–1989

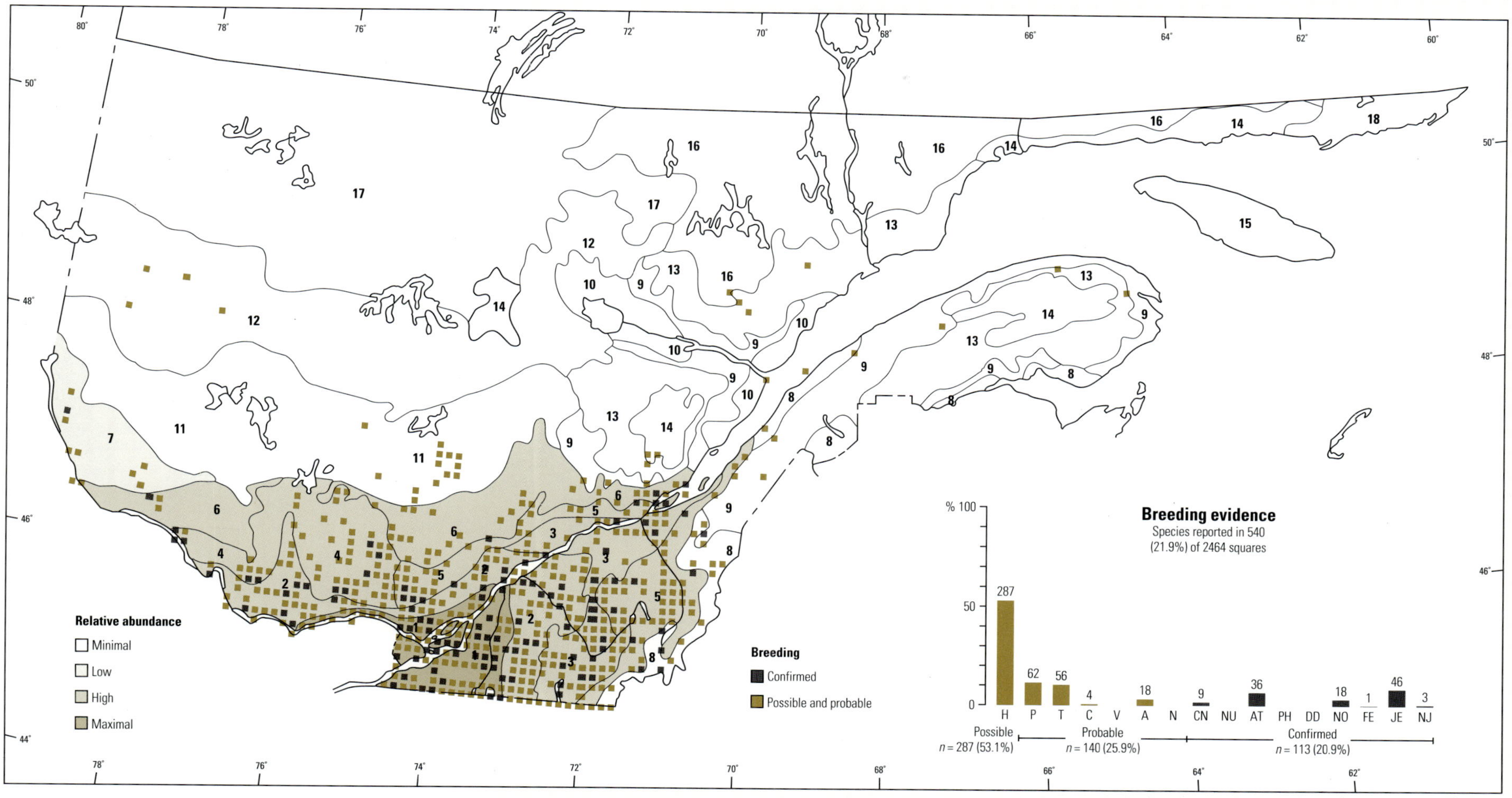

Species profile

Clutch size	usually 5–8 eggs (4–10)[3,4,5]
Laying interval	1 egg per day[1,6]
Start of incubation	with last egg[1]
Incubation period	12–14 days[1,3,7,14]
Maturity at hatching	altricial
Care of young	by female and male[7]
Nestling period	≈ 14 days[1,3,8], 26 days[14]
Age at first flight	≈ 14 days[1,8]
Dependence of young	until ≈ 14 days after first flight[1,7]
Broods per year	1[5]
Breeding age	1 year[8]
Mating system	monogamy[3]
Duration of pair bond	for life[7,12,13,14]
Total length	13.0–15.5 cm[2]
Wing span	23.5–29.2 cm[9]
Weight	21.1 g (18.5–26.7)[10]
Longevity record	9 years 10 months[11]

References:
1. Stokes and Stokes 1989b
2. Godfrey 1986
3. Ehrlich et al. 1988
4. Peck and James 1987
5. Harrison 1978
6. Kilham 1968
7. Tyler 1948c
8. DeGraaf and Rudis 1987
9. Terres 1980
10. Dunning 1984
11. Klimkiewicz et al. 1983
12. Butts 1931
13. Lack 1945
14. Pravosudov and Grubb 1993

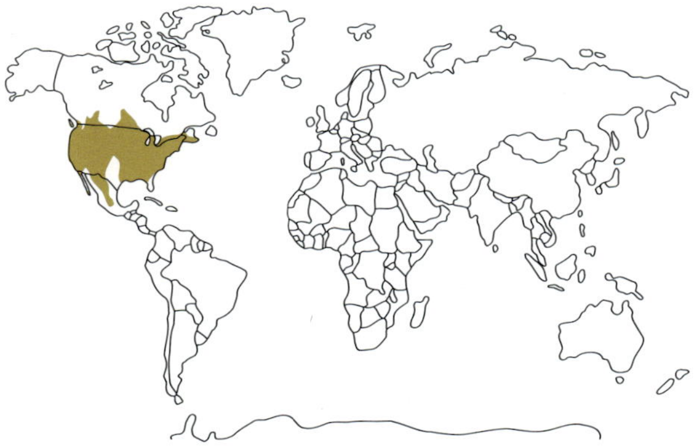

World breeding range

Chronology *n* = 37

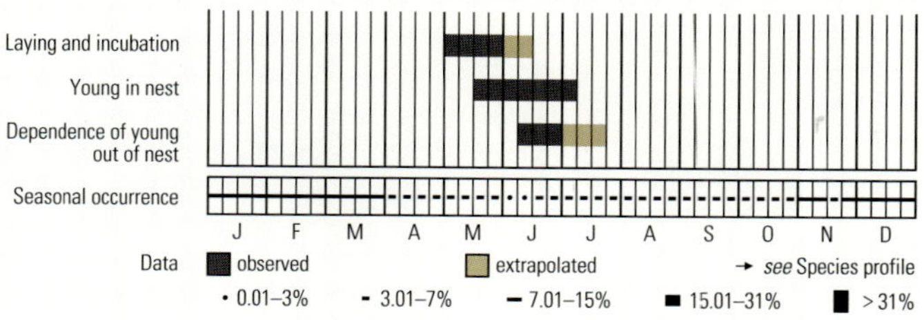

Frequency of occurrence by ecological region

Ecological regions** (forest regions)	Adequately covered squares	Squares in which reported	% of squares in which reported	Relative abundance (%)*
1. Sugar maple-hickory and sugar maple-basswood	97	84	86.6 ± 3.6	100.0
2. Sugar maple-basswood	153	116	75.8 ± 3.6	87.6
3. Sugar maple-basswood and sugar maple-yellow birch	101	67	66.3 ± 5.6	76.6
4. Sugar maple-yellow birch-basswood	56	41	73.2 ± 10.2	84.5
5. Sugar maple-yellow birch-beech and sugar maple-yellow birch-basswood	113	68	60.2 ± 5.0	69.5
6. Sugar maple-yellow birch-beech	71	42	59.2 ± 10.5	68.3
7. Sugar maple-yellow birch-balsam fir	19	7	36.8 ± 22.8	42.5
8. Sugar maple-yellow birch and balsam fir-yellow birch	47	10	21.3 ± 10.7	24.6
9. Balsam fir-yellow birch	87	4	4.6 ± 4.4	5.3
10. Balsam fir-white birch or balsam fir-red maple	46	0	0	0
11. Yellow birch-balsam fir	52	10	19.2 ± 11.2	22.2
12. Balsam fir-white birch (west)	59	3	5.1 ± 6.2	5.9
13. Balsam fir-white birch (east)	76	6	7.9 ± 6.2	9.1
14. Balsam fir-black spruce	28	1	3.6 ± 8.3	4.1
15. White spruce-balsam fir and balsam fir-white spruce	22	0	0	0
16. Black spruce-balsam fir-moss	18	0	0	0
17. Black spruce-moss	32	0	0	0
18.			*not analysed*	

* minimal: 0–29.9%; low: 30–59.9%; high: 60–89.9%; maximal: 90–100%

** Deciduous forest: 1–7; mixed forest: 8–11; coniferous forest: 12–17

The White-breasted Nuthatch usually builds its nest in a cavity in a deciduous tree, here a maple. Adult bringing an insect to its brood. Saint-Augustin-de-Desmaures, west of Québec City. *June 1965, Gaston Le Page.*

The White-breasted Nuthatch, typical of hardwood forests, nests in this maple stand on Mount Royal. Summit Park, Westmount. *October 1980, Richard Pelletier.*

The inside of the cavity is lined with wood shavings, hair, and feathers, brought in mostly by the female. Clutch size is 5–8 eggs (Ehrlich et al. 1988), laid at the rate of one a day (Stokes and Stokes 1989b). Once the clutch is complete, the female incubates the eggs for about 12 days, during which time the male feeds her frequently (Tyler 1948c). After the eggs hatch, the male brings food to the entire family, and the female helps feed the altricial young (Tyler 1948c; Stokes and Stokes 1989b). The nestlings fledge and leave the nest at 2 weeks of age, but they remain dependent on the parents for food for another 2 weeks. Families may stay together a long time, and the adults continue to feed the young occasionally, sometimes until the fall (Tyler 1948c).

During the breeding season, the White-breasted Nuthatch eats mostly insects, while in winter its diet consists primarily of nuts and seeds (Martin et al. 1951). The nestlings, however, may be fed nothing but animal food (Ehrlich et al. 1988). This nuthatch forages by moving nimbly along the trunk or large branches of a mature tree, taking caterpillars, beetles, spiders, and ants hidden in the bark. As to the plant matter in its diet, Tyler (1948c) states that it likes beechnuts and acorns, among other things, with Terres (1980) adding walnuts and corn to the list. It is also very fond of sunflower seeds and suet, and these are what attract it to bird feeders all winter long, where it often feeds in the company of chickadees, creepers, and woodpeckers. Males and females may have different diets (Grubb 1982); males exercise dominance over females by controlling access to food (Waite 1987). Nuthatches can often be seen wedging a seed or nut into a crevice in the tree bark, then using the bill as a pick to open the seed up or extract the kernel. They have the habit of carefully hiding stores of food in cracks around the nest (Waite and Grubb 1988; Petit et al. 1989; Woodrey 1991).

During nesting activities, and especially during nest building itself, birds of both sexes sometimes exhibit a very peculiar behaviour known as bill-sweeping: swaying from side to side, usually holding an insect in their bill, often for several minutes at a time, they seem to be sweeping the entrance to the nest and the area around it (Kilham 1968). One of these insect species was discovered by Kilham (1971a) to secrete an oily irritant, and he believes that nuthatches smear the entrance to their cavities with various substances in order to discourage squirrels, which often try to take them over (Kilham 1968). He also observed that White-breasted Nuthatches may sometimes put on an astonishing display when squirrels approach the nest: body and bill pointing up, wings and tail spread, they slowly sway from side to side. This performance apparently wards off intruders.

Breeding was confirmed in 20.9% of the 540 squares in which the White-breasted Nuthatch was atlassed. The species is easy to spot as it moves up and down tree trunks, and even a novice could mistake its distinctive call only for that of the Red-breasted Nuthatch. The species is an early nester and becomes very inconspicuous during breeding (Mills 1987g); the nest, often in a natural cavity fairly high up, is not always easy to find.

Distribution

The range of the White-breasted Nuthatch extends from southern Canada to southern Mexico, yet the species is absent from a large portion of the Great Plains (AOU 1983; Godfrey 1986). Its Québec range is limited mainly to the deciduous forest zone, and accordingly the bird is well represented throughout the southern part of the province west of Île d'Orléans, near Québec City. It is abundant in the Montréal area, probably because there are mature hardwoods near urban communities.

North of the deciduous forest zone, the White-breasted Nuthatch is fairly local and rare. It can occasionally be found in deciduous trees in mixed forests. Some of the most northerly reports may have come from logged areas growing up in hardwoods. There is a scattering of Atlas records from the Abitibi region, north of the Saguenay River, the Laurentians, the Appalachians, the Lower St. Lawrence, and the Gaspé Peninsula. According to Atlas data, the species does not breed in the Saguenay–Lac-Saint-Jean region, despite the fact that deciduous forest occurs there. Nor is it recognized as a breeder by the Club des ornithologues du Saguenay–Lac-Saint-Jean, which has records of observations going back even earlier (Savard and Cormier 1995). The presence of beech, oak, and walnut trees is apparently favourable to the species (Ellison 1985x; Bonney 1988g). As Bull (1985) observed in New York State, the species seems to be scarcer along the coast and in high mountainous regions.

History and trends

Neither the range nor abundance of the White-breasted Nuthatch appear to have changed much over the last century, although BBS data point to a slight increase in eastern and central populations over the course of recent decades (Robbins et al. 1986).

At the end of the last century, the White-breasted Nuthatch was considered rare in Québec except in Montréal where it was a resident and common during migration (Dionne 1883; Wintle 1896). According to Ouellet (1974a), historical data seem to indicate an appreciable increase in populations in the Montréal region since the turn of the century, despite loss of habitat through deforestation; in Ouellet's opinion, however, numbers must have been higher locally before. For the last few decades, the species has been considered a common sedentary breeder in the Montréal area (David 1980a; Bannon 1991b).

BBS data from southern Québec do not seem to show any significant trends between 1966 and 1989. The species is common in the southern parts of the province, which are dominated by deciduous stands, but remains rare farther north. Even there, however, the White-breasted Nuthatch is being reported with increasing frequency at bird feeders.

Logging is a potential limiting factor on White-breasted Nuthatch populations. Cutovers in the mixed or coniferous forest regenerating in hardwoods are favourable for this species of mature deciduous forest. The same applies to abandoned fields, which, once grown up in deciduous trees, create new habitat. A study in the Adirondacks of New York indicated that the species could tolerate certain types of partial logging in hardwood forests (Webb et al. 1977), although selective cutting involving the removal of sick or dead trees would reduce the number of cavities available and the abundance of prey. The cutting down of urban woodlots will certainly have an impact on local populations.

Alain Lanoue and Jean-François Rail

Brown Creeper

Grimpereau brun

Certhia americana

Christiane Girard

THE BROWN CREEPER SPENDS MOST OF ITS LIFE ON TREE TRUNKS, WHERE IT feeds, nests, roosts, and hides from danger. This small, shy forest dweller has the characteristic habit of gleaning small insects with the precision of a dentist while climbing in a straight or spiral course and keeping itself propped with its stiff tail. When almost at the top of the tree, it flies down to the base of a nearby tree and starts over.

This species breeds throughout southern Québec except in the easternmost portion of the North Shore (Godfrey 1986). Part of the breeding population winters in Québec (David 1980a), but the rest migrates south down to the southern United States (AOU 1983).

Habitat and behaviour

The Brown Creeper breeds in mature forests, be they coniferous, deciduous, or mixed. It favours forests at least 60 years old with snags (Zarnowitz and Manuwal 1985), where the textured and peeling bark create suitable habitat for nesting and foraging. This creeper prefers large trees because of the greater expanse they offer for foraging, which reduces the amount of energy consumed in flying from one tree to the next (Franzreb 1985). It also breeds in densely wooded swamps, spruce peatlands, and beaver ponds (Terres 1980; Peck and James 1987). In Michigan, of 20 nests, 2 were in a swamp and the others within 60 m of running water (Davis 1978). In New York State, it is found mainly in areas with at least 50% forest cover (Levine 1988a). It may require openings in the forest canopy to allow light penetration to the nest site (Davis 1978). Leck (1989) observed an adult sunbathing, probably, as in the case of various other species, either to warm itself or to rid itself of ectoparasites (Terres 1980). Recently burned forests continue to support Brown Creepers as long as the remaining standing trees provide them with suitable nesting sites (Apfelbaum and Haney 1977).

Brown Creepers are monogamous, and the sexes look identical. During courtship, various displays can be observed, such as display flights, chases, wing beating, and courtship feeding (Davis 1978).

Even in courtship the Brown Creeper is a creature of its spiralling habit. The pair engages in a game of tag flying up close around the tree twisting and turning in pursuit of each other (Tyler 1948a). The female then alights on the trunk, adopting the posture of a begging chick, and allows the male to deposit his offering into her throat (Davis 1978).

The pair selects the nest site and both sexes gather nesting materials, but the nest is built mostly by the female in 6–30 days (Harrison 1975; Davis 1978). The nest is usually placed between the detached bark and the trunk of a large dead or dying tree, 0.6–15 m up. As it is attached to the concave inner surface of the selected piece of bark, the nest comes loose if the bark is peeled away. Nests are also found in live trees, fence posts, limbs of dead trees, old woodpecker holes, or even behind a piece of tin under the roof logs of an outhouse (Terres 1980; Peck and James 1987; Ehrlich et al. 1988).

The nest is a loose cup of twigs, conifer needles, moss, and grass, and may be built up with twigs. It is lined with shredded bark and feathers. The lateral extremities of the nest are elevated, making it look like a well-filled crescent or hammock.

Notes on territory

In a hemlock stand near Mirabel, Cyr (1974) estimated density at 1 territorial male/10 ha and the territory of a pair at 0.465 ha. In Ontario, north of Lake Superior, Kendeigh (1947) estimated the average territory of a Brown Creeper pair at 1.0 ha in coniferous forest dominated by fir, black spruce, and white spruce.

In Michigan, territorial singing occurred from April until the young fledged. Territory size was 2.3–6.4 ha, and birds holding smaller territories engaged in more vocal defence than birds with larger territories. During the nestling stage, adults gathered food up to 150 m or more from the nest. During the fledgling stage, family groups observed were always within about 500 m of their former nests, sometimes in territories previously held by adjacent pairs (Davis 1978).

The Brown Creeper breeds in all types of mature forest, preferably where there are dead trees or old trees with peeling bark. Trois-Rivières. *June 1992, Michel Sokolyk.*

Brown Creeper

Breeding distribution
1984–1989

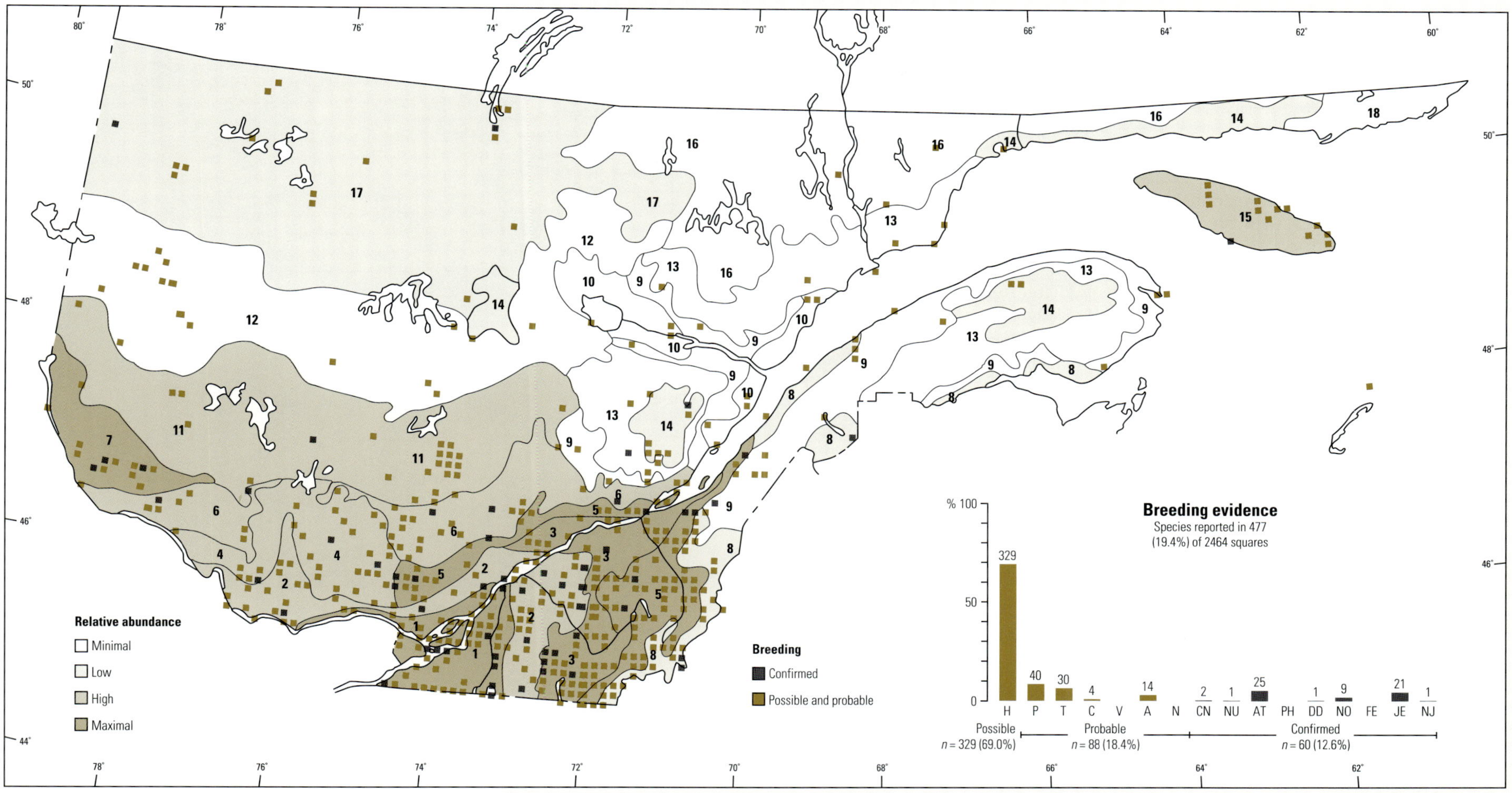

Species profile

Clutch size	usually 5–6 eggs (4–8)[2,4,5,6,8]
Laying interval	1 egg per day[6]
Start of incubation	with last egg[5,6,7]
Incubation period	14–17 days[1,3,5,6,8]
Maturity at hatching	altricial
Care of young	by female and male[3,5,6,7]
Nestling period	13–16 days[1,2,4,6,7]
Age at first flight	13–16 days or more[11]
Dependence of young	until at least 26 days old[6]
Broods per year	1–2[1]
Breeding age	1 year[1]
Mating system	monogamy[3]
Duration of pair bond	?
Total length	12.5–14.5 cm[8]
Wing span	17.8–20.3 cm[2]
Weight	8.4 g (7.2–9.9)[10]
Longevity record	4 years 4 months[9]

References:
1. DeGraaf and Rudis 1987
2. Terres 1980
3. Ehrlich et al. 1988
4. Tyler 1948a
5. Braaten 1975
6. Davis 1978
7. Harrison 1978
8. Godfrey 1986
9. Kennard 1975
10. Dunning 1984
11. according to nestling period

World breeding range

Chronology *n* = 32

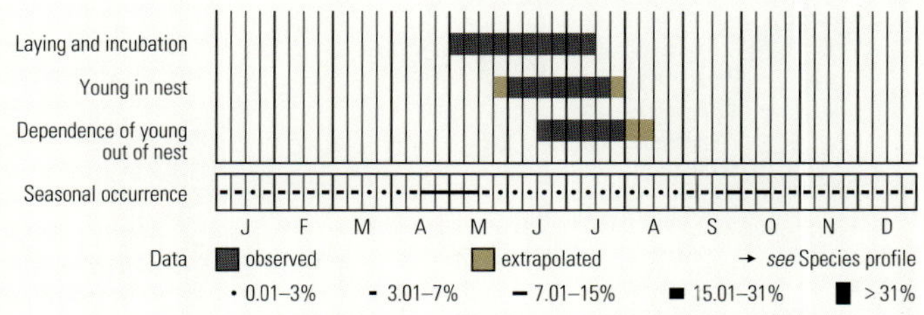

Frequency of occurrence by ecological region

Ecological regions** (forest regions)	Adequately covered squares	Squares in which reported	% of squares in which reported	Relative abundance (%)*
1. Sugar maple-hickory and sugar maple-basswood	97	55	56.7 ± 4.9	97.9
2. Sugar maple-basswood	153	65	42.5 ± 4.1	73.4
3. Sugar maple-basswood and sugar maple-yellow birch	101	53	52.5 ± 5.9	90.6
4. Sugar maple-yellow birch-basswood	56	26	46.4 ± 11.4	80.2
5. Sugar maple-yellow birch-beech and sugar maple-yellow birch-basswood	113	62	54.9 ± 5.1	94.8
6. Sugar maple-yellow birch-beech	71	26	36.6 ± 10.3	63.3
7. Sugar maple-yellow birch-balsam fir	19	11	57.9 ± 23.3	100.0
8. Sugar maple-yellow birch and balsam fir-yellow birch	47	15	31.9 ± 12.0	55.1
9. Balsam fir-yellow birch	87	13	14.9 ± 7.1	25.8
10. Balsam fir-white birch or balsam fir-red maple	46	7	15.2 ± 9.3	26.3
11. Yellow birch-balsam fir	52	23	44.2 ± 13.9	76.4
12. Balsam fir-white birch (west)	59	9	15.3 ± 9.7	26.3
13. Balsam fir-white birch (east)	76	7	9.2 ± 6.6	15.9
14. Balsam fir-black spruce	28	6	21.4 ± 16.2	37.0
15. White spruce-balsam fir and balsam fir-white spruce	22	9	40.9 ± 21.2	70.7
16. Black spruce-balsam fir-moss	18	1	5.6 ± 13.5	9.6
17. Black spruce-moss	32	10	31.3 ± 17.6	54.0
18.			*not analysed*	

* minimal: 0–29.9%; low: 30–59.9%; high: 60–89.9%; maximal: 90–100%

** Deciduous forest: 1–7; mixed forest: 8–11; coniferous forest: 12–17

The species is generally single-brooded, but renests if an early attempt fails (Davis 1978). The female incubates the 4–8 (usually 5–6) eggs for 14–17 days. Only the female broods the altricial young, but both parents feed them with small insects and other arthropods (Davis 1978; Kuitunen and Törmälä 1983). Fecal sacs are deposited on a tree trunk or eaten by the parents (Davis 1978). When they are 13–16 days old, the young creep up the tree and may take short flights to other trees. They are very difficult to find because they are small and cryptically coloured, often remaining motionless in the shadow of a horizontal branch. However, their high-pitched begging calls to their parents may facilitate detection. Older fledglings chase their parents from trunk to trunk to be fed, but once they are about 26 days old they forage for themselves more than they beg. In Michigan, cohesion of family groups was observed until mid-July. Fledglings roost in tight circles with their heads toward the centre (Davis 1978).

Relying on their cryptic coloration, adult Brown Creepers flatten themselves against the tree trunk when danger threatens, remaining still and virtually invisible for several minutes (Tyler 1948a). This is why it is easier to locate them by their short flights between trunks and by their faint notes. The little-known song (Ellison 1985e) carries up to about 120 m in mixed deciduous woods. It can be heard especially from courtship (in April) to until the young fledge (Davis 1978).

The Brown Creeper sometimes nests in a hole in a tree, but most often it builds its nest, constructed of twigs, behind a flap of loose bark. Trois-Rivières. *June 1992, Michel Sokolyk.*

A Brown Creeper's nest, hidden behind peeling bark on the trunk of a dead tree. James Bay region. *June 1991, Vincent Létourneau.*

Woodlands with dead trees are essential nesting habitat for the Brown Creeper. Near Lake Saint-Charles, north of Québec City. *October 1994, Christine LePage.*

The nest is hard to find and difficult to examine without damaging it. However, within several metres of a nest, it is possible to hear the begging cries of the young, starting when they are about 9 days old (Davis 1978). Breeding was confirmed in 60 Atlas squares, mostly on the basis of reports of parents transporting food (42%, 25/60), fledglings (35%, 21/60), and adults seen entering or leaving nests (15%, 9/60).

Distribution

The Brown Creeper, a New World species, breeds from southern Alaska all across Canada down to Nicaragua (AOU 1983). It was reported in 477 squares distributed throughout the Atlas area. A general decrease in frequency from south to north is interrupted on the map by a band of a little lower density from the Abitibi through the Lac-Saint-Jean region to the Lower St. Lawrence. Much of this band has been logged extensively in recent decades, with more than 15% of the commercial forest being removed during the period 1974–88 (Fragnier 1992). This habitat loss is all the more serious in that the Brown Creeper shuns regenerating forests, immature woodland, and open areas. Its distribution corresponds to areas supporting large dead or live trees with loosened bark, like the Central St. Lawrence Lowland which still has mature hardwood stands.

Breeding has never been confirmed on the Magdalen Islands (Fradette 1992). The species was atlassed on Brion Island, the only island in the Magdalens that has not lost its original forest cover (MENVIQ 1991).

Brown Creepers breed up to latitude 51° N (David 1980a). Single individuals seen near the Eastmain River on 5 June and 8 July 1982 (Gosselin and David 1982) and observations of birds during the breeding season at Harrington Harbour in 1982, Manic-5 in 1990, and Lake Plétipi in 1964 (ÉPOQ) suggest that the species may breed up to at least latitude 52° N. In Ontario, it breeds throughout the province in treed areas (Peck and James 1987).

The Québec population seems to undergo a southward shift after the breeding season, with numbers increasing in the southern part of the Québec breeding range during migration. This bird can even be seen in winter, as part of the breeding population overwinters here (David 1980a).

History and trends

The Brown Creeper's range has expanded southward in some parts of the United States during this century. Reforestation has allowed the Vermont population to increase and to broaden its distribution (Ellison 1985e), while Dutch elm disease probably boosted Brown Creeper populations in New York (Levine 1988a), though the benefits of the disease may be short-lived because the bark of the white elm falls off quickly (Davis 1978). However, the Brown Creeper has few competitors, given its unique foraging and nesting styles. According to Hall (1969), this may partly account for the southward expansion of its breeding range.

In the Greater Montréal area, the breeding population probably started to decrease in the mid-1800s due to intensified clearing of forests (Ouellet 1974a). In the northern part of the Québec breeding range, the population probably declined because of the large areas being logged; Brown Creepers seem to be sensitive to logging, which occurs mainly in mature forests, prime habitat for this species. Selective logging in which dead, old and sick trees, as well as stumps, are removed, further affects this species' breeding success.

The few data available for this species fail to show any discernible population trends (BBS). However, land-use decision makers might consider the creeper's rather specialized feeding and nesting habits when regulating logging, which can dramatically affect it, particularly in the north, where it needs large areas of old forest, as reported by Kuitunen and Helle (1988) in the case of the Eurasian Treecreeper. Moreover, logging may be detrimental not only by reducing available breeding habitat but also by making remaining habitat more vulnerable to heavy rain and wind, which can destroy nests by loosening or removing bark (Davis 1978). As been observed for the Eurasian Treecreeper, the installation of inverted V-shaped nesting boxes for Brown Creepers could offset the shortage of nesting sites and lead to an increase in population size (Moilanen and Kuitunen 1977).

François Shaffer and Robert Alvo

Carolina Wren

Troglodyte de Caroline

Thryothorus ludovicianus

The Carolina Wren has been known to occur in Québec only over the last three decades, and breeding in the province is very rare. Like the Northern Cardinal and Northern Mockingbird, this sedentary species, typical of the southeastern United States, has been expanding its range northward since the beginning of the 20th century. While the Northern Cardinal is now firmly established in Québec because of the proliferation of bird feeders, and while the Northern Mockingbird appears able to migrate south to escape the severe winter (David et al. 1990), the less adaptable Carolina Wren appears unable to establish itself here because of its feeding habits and sedentary nature.

Habitat and behaviour

The Carolina Wren inhabits areas with dense shrubby undergrowth, particularly in open woodland, swampy areas and beside streams and rivers, provided it can find suitable nesting cavities. It also nests in man-made habitats, such as suburban gardens and city parks with large beds of dense, thorny shrubs (Bent 1948). In the northern part of its range, and especially in Québec, the species is observed almost exclusively in such habitats, where it often benefits from artificial food sources such as feeders and fruit-bearing shrubs. These are indispensable to the bird's survival in winter, when snow cover prevents it from foraging on the ground. At feeders, the Carolina Wren eats mainly chopped nuts, suet, peanut butter, and sunflower seeds; some people maintain that raw ground beef will help it survive the winter (David and Duquette 1983). It is primarily insectivorous, however, foraging mainly on the ground, but also on leaves and bark, sometimes supplementing its diet with small vertebrates and berries (Terres 1980; Ehrlich et al. 1988).

Very early nesting can occur. In Vermont, where the climate is similar to that of southern Québec, nest building has been observed as early as 31 March (Ellison 1985g). The nest is constructed by both adults, but is completed by the female (Laskey 1948). A bulky structure of leaves, grasses, bark, twigs, and moss, usually domed and having a side entrance, it is lined with feathers, hair, and fine grasses. Normally located less than 3 m from the ground, it is difficult to find, unless a nest box is being used. It may occupy just about any natural or artificial cavity, or a nook or cranny in a man-made structure (Laskey 1948; Harrison 1975; Harrison 1978).

The Carolina Wren builds larger nests in colder areas, in the northern part of its range, and also when it builds early in the season. The species uses structures more often early in the season, and has even been reported to take shelter in an old hornet's nest on cold nights (Ramsay 1987).

Typical clutch size is 5 eggs laid a day apart, which are incubated by the female for 12–15 days (Laskey 1948; Ehrlich et al. 1988). This wren has a particularly long incubation period, and the female is a tight sitter, seldom leaving the nest (Laskey 1948). The altricial young are nest-bound for about 12–15 days and are fed by both adults (Laskey 1948; Nice and Thomas 1948). The pair may raise 2 broods a year and, in exceptional cases, 3 in the southern part of the range (Laskey 1948; Harrison 1975), in the southern United States and northern Mexico. The species is seemingly a rare cowbird host (Luther 1974; Terres 1980).

The Carolina Wren is the only member of the *Thryothorus* genus found in temperate zones. It differs from its tropical cousins in that only the male sings (Morton 1982). The species is monogamous and sedentary, mating for life, and pairs defend their territories throughout the year, perhaps even more vigorously in winter than during the breeding season (Morton and Shalter 1977). Survival appears to depend greatly on a pair's ability to defend a territory large enough to provide it with sufficient food for the winter (Strain and Mumme 1988). Since the female does not sing, she, unlike the male, cannot defend the territory alone. Therefore, her chances of survival are directly linked to those of the male, which spends a large portion of his time and energy vocalizing, the main purpose being to defend

Christiane Girard

Notes on territory

Pairs mate for life and throughout the year defend permanent territories large enough to ensure subsistence during the winter (Morton and Shalter 1977; Strain and Mumme 1988). Territory size therefore varies with habitat and availability of food in winter, when snow may prevent the birds from foraging on the ground. In Kansas, Fitch estimated the area defended by a pair nesting in a wooded ravine-pond habitat over a 4-year period at 1.6–3.7 ha. In a damp, shrubby habitat in Illinois, Brewer estimated the average territory size at 0.1 ha (DeGraaf and Rudis 1987).

The Carolina Wren, sometimes reported from the south of the province, was seen in this suburban garden in Pointe-Claire. Island of Montréal. *August 1991, Pierre Bannon.*

Carolina Wren

Breeding distribution
1984–1989

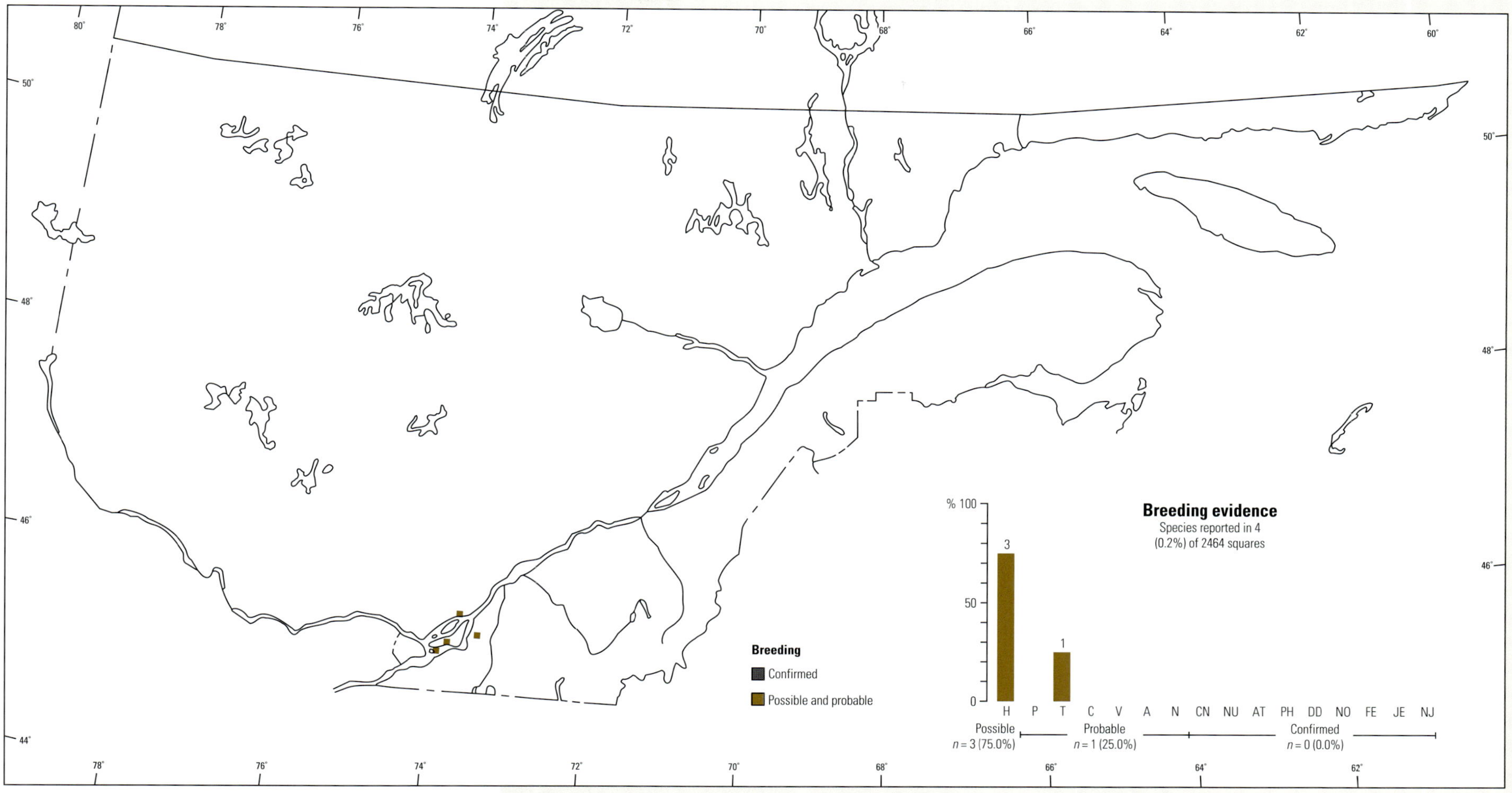

Species profile

Clutch size	usually 5–6 eggs (3–8)[2,4,6,7]
Laying interval	1 egg per day[4]
Start of incubation	?
Incubation period	12–15 days[1,4,6,7]
Maturity at hatching	altricial
Care of young	by female and male[1,4,8]
Nestling period	12–15 days[1,4,6,7]
Age at first flight	12–15 days or more[12]
Dependence of young	?
Broods per year	2[1,2,4,9,10]
Breeding age	1 year[10]
Mating system	monogamy[5]
Duration of pair bond	for life[5]
Total length	14.0–15.2 cm[6]
Wing span	17.1–19.7 cm[6]
Weight	21 g (?)[11]
Longevity record	6 years 1 month[3]

References:
1. Harrison 1978
2. Harrison 1975
3. Klimkiewicz et al. 1983
4. Laskey 1948
5. Morton and Shalter 1977
6. Terres 1980
7. Ehrlich et al. 1988
8. Nice and Thomas 1948
9. Bent 1948
10. DeGraaf and Rudis 1987
11. Dunning 1984
12. according to nestling period

World breeding range

Chronology $n = 2$

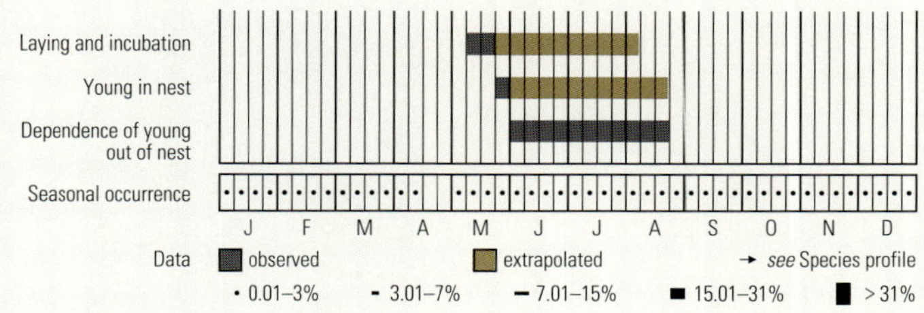

In spring 1992, a pair of Carolina Wrens nested in this shed, entering and exiting above and below the doors. The shed adjoins a house located in a residential area with many trees, bushes, and nice gardens. Sainte-Anne-de-Bellevue, Island of Montréal. *May 1992, Pierre Bannon.*

Aware of the importance of what was happening in his shed—the second confirmed nesting of the species in Québec—Mr. W. E. Sackston let the pair of Carolina Wrens nest behind this metal grill, against the wall. The more or less globe-shaped nest was made of twigs and dry grasses. At least 2 young left the nest on the morning of 4 June 1992. Sainte-Anne-de-Bellevue, Island of Montréal. *May 1992, Pierre Bannon.*

This wren, which favours dense layers of mature shrubbery, may nest in these thick clumps of vines. Nuns' Island, Verdun. *July 1991, Pierre Bannon.*

the territory rather than attract a female, as the pair mate for life (Morton and Shalter 1977; Simpson 1985). The female has the task of watching out for predators and alerting the male to imminent danger by her calls. This allows the male to spend more time and energy foraging on the ground undisturbed and to defend the territory with his singing, thereby ensuring the survival of the pair during the winter (Morton and Shalter 1977).

Carolina Wrens most often appear in Québec in the fall and early winter, almost always near a feeder. Northward dispersion from neighbouring U.S. states by young of the year is the probable explanation for these fall occurrences (Pough 1949). Most of these birds do not appear to survive the winter. Of the 83 birds observed up to the end of 1990 (see table opposite), only a dozen or so overwintered successfully. Moreover, there are only 14 summer records (April to mid-August), most of them of lone males, not a sufficient number to indicate spring migratory movement.

The male sings throughout the year, most often at the end of winter and in spring. The loud, rich musical song is highly developed, and each male may have a repertoire of an average of 32 different songs (Morton 1982). The songs' modulation varies as a function of the distance between two singing males (Simpson 1985). The song makes the bird easy to locate, but it may be confused with that of the Northern Cardinal. The Carolina Wren is a very inquisitive bird that can easily be lured by imitating squeaks of small mammals or the hooting of owls.

Atlas data show how rare the species is in Québec during the breeding season: only 4 singing males were reported during the entire Atlas survey.

Distribution

The Carolina Wren inhabits southernmost Ontario and Québec and ranges across the central states to the Atlantic seaboard and along the Gulf of Mexico in the United States (AOU 1983). Almost all records in Québec during the last 20 years have been from the southern part of the province, near urban centres: 60% from the Greater Montréal region and 40% from the Québec City, Aylmer, and Sherbrooke areas. There have been only a few reports in areas far from urban centres: the Magdalen Islands in August 1971 (Fradette 1992), Saint-Jovite in the fall of 1988 (B. Clément, pers. comm.), Petite-Rivière-Saint-François in August 1991 (Gosselin et al. 1991b), Île aux Basques and Jonquière in October 1991 (Aubry et al. 1992), Chicoutimi and Baie-Comeau during the winter of 1992–93 (Yank and Aubry 1993b), Rimouski in the fall of 1993 (Aubry and Yank 1994), and Dubuisson, near Val-d'Or, in October 1993 (Aubry and Bannon 1994c).

The Carolina Wren has been seen during the breeding season only in the Montréal and Aylmer areas, where it frequents urban woodlands and suburban gardens, often near bird feeders, which it requires to survive the winter. Temperature is the factor most closely linked to the Carolina Wren's distribution (Root 1988). The Atlas records have all come from populated areas such as Pointe-Claire, Mascouche, Boucherville, and Senneville, all well-established suburbs of Montréal.

History and trends

Since the turn of the century, the Carolina Wren has been gradually expanding its range northward (Bent 1948; Bull 1974; Richard, M. A. 1987). Population size in the northern part of the range, however, has fluctuated considerably, according to winter conditions. For example, after the two severe winters of 1976–77 and 1977–78, the breeding population declined sharply in several regions of eastern North America (Robbins et al. 1986). According to Morton (1982), mortality may be as high as 80–90% during winters when the snow cover remains for long periods. The species was even placed on the Blue List in 1980 and 1981 (Tate 1981).

The Carolina Wren had been reported only once in Québec prior to 1971, in Chambly during the winter of 1957–58 (Abbott et al. 1957). Since 1971, the species has been sighted every year except 1972 and 1978. The only known breeding records are from Chambly in 1975 (David and Gosselin 1975; Godfrey 1986), Sainte-Anne-de-Bellevue in the spring of 1992 (Aubry and Yank 1992a), and Beaconsfield in August 1992 (Aubry and Yank 1993b).

The table below gives the number of records by 5-year periods. It shows that the first nesting in Chambly, near Montréal, was preceded by an increased number of sightings during the 1971–75 period. This, however, was followed by two severe winters that depressed the small Québec population until 1985. Since then, however, the trend has reversed and the species has made a strong comeback. In 1991, there were 15 sight records, 4 of which were reported during the breeding season (Gosselin et al. 1991b). Then in 1992, the second and third breeding records for Québec were reported, at Sainte-Anne-de-Bellevue and Beaconsfield, near Montréal.

Observations of the Carolina Wren in Québec

	Total records	Summer records
Pre-1971	1	0
1971–75	16	5
1976–80	9	2
1981–85	11	2
1986–90	46	5
Total	**83**	**14**

If the present trend continues, more breeding evidence could be forthcoming in the years ahead. A few mild winters would foster temporary establishment of the Carolina Wren in southern Québec, in areas where snow cover would not impede this ground forager. A series of severe winters, however, like that of 1993–94, could provoke another decline.

Pierre Bannon and Jean-François Rail

House Wren

Troglodyte familier

Troglodytes aedon

The House Wren is a small, energetic bird that is easy to spot. Whereas most birds are inactive during the noonday heat, the House Wren sings incessantly from dawn until dusk. Its species name *aedon* means "nightingale," probably an allusion to its lively, varied song.

The House Wren is a stocky, drab bird with brown upperparts and dingy white underparts. The longish bill is slightly downcurved. Like most wrens, it cocks its tail in the air when perched or agitated. This common migrant breeder in southern Québec (David 1980a) arrives in the province during the second week in May from its winter quarters in Mexico and the southernmost United States.

Habitat and behaviour

As its name indicates, the House Wren has adapted well to the presence of humans. It is found mainly in thickets and shrubs, whether in forest habitat, city parks, or residential areas. In forest habitat, it often occupies the edges of beaver ponds or other open woodland with dead trees. It builds its nest in any sort of cavity, be it in a tree, stump, or nest box (Godfrey 1986; Lumsden 1985), preferring those found in shrubbery or open woodland (Belles-Isles and Picman 1986b; Finch 1989; Parren 1991). It sometimes uses abandoned nests of other birds and has even been known to take up residence in such odd places as a flower pot, a mail box, a metal pipe, the skull of a large mammal, farm machinery, and even a shoe (Gross 1948b).

On arrival on the breeding grounds, the male, often polygamous, defines its territory by singing and driving off other males. It then begins building nests, usually 2 or 3 but sometimes as many as 7. These enable him to attract one or more females, not only when they arrive, about one week later, but also throughout the breeding season (Kendeigh 1941; Johnson and Kermott 1991). Nests are built at varying heights and consist of an accumulation of materials, mainly twigs and dry leaves, that almost fill the selected cavity, except for the space where the eggs are to be deposited (McCabe 1965; Kennedy and White 1992). Once a female is attracted, she inspects the nests, chooses one, and completes the construction by adding the lining. She alone incubates the eggs, although the nestlings are fed by both parents (Kendeigh 1941; Gross 1948b). A pair may produce 2 broods a season (Peck and James 1987); on rare occasions the female will lay in the nests of other House Wrens (Picman and Belles-Isles 1988).

House Wrens normally return to the same breeding sites from one year to another (Drilling and Thompson 1988). The males are highly territorial and fiercely defend most of the cavities in their territory. It is not unusual among House Wrens for unmated males and females to destroy the nests and eggs or kill the young of conspecifics or of other species nesting in the same area (Belles-Isles and Picman 1986a; Quinn and Holroyd 1989). This behaviour, which seems to be more frequent when population levels are high (Kendeigh 1941), enables the birds to take possession of several nest cavities (Pribil and Picman 1991). Other birds, such as the Marsh and Sedge wrens and some mimids, exhibit similar behaviour.

The House Wren is an easy bird to find, as it sings continuously throughout the summer and is often found in populated areas. Its territory is small, which facilitates observation of its comings and goings. It is therefore not surprising that breeding was confirmed in 40.3% (152/377) of the squares in which it was reported, mostly through records of occupied nests and observations of adults entering and leaving the nest.

Christiane Girard

Notes on territory

The House Wren fiercely defends a small territory that ranges in size between 0.1 and 1.44 ha but is on average only 0.4 ha. The size is inversely proportional to population density (Kendeigh 1941). Sometimes a male will infiltrate another territory, usually one adjacent to his, thus affording him opportunities to copulate with females other than those nesting on his own territory (Johnson and Kermott 1989). A male may evict another male from his territory after song duels and pursuits, which may last for several hours. The presence of one or more females on the territory to be conquered appears to be important, as the conquering males usually mate with these females (Johnson and Kermott 1990). The density of singing males varies according to habitat; for example, it was found to be 1.8/10 ha in a small residential area with wooded vacant lots in Fort Nelson, British Columbia and 16.0/10 ha in a red oak-sugar maple forest at Long Point, Ontario (Erskine 1976a). An intermediate density of 3 males/10 ha in an open field of brome-grass and quack grass was found in Point Pelee National Park in Ontario. In Edmonton, Alberta, in a wooded ravine dominated by aspen and white spruce, density was found to be 5.2 males/10 ha (Erskine 1972a, 1980a).

Over a number of years, House Wrens have nested in nest boxes set up in areas of dense shrubbery along this road, the Allée d'Ormes at Cap Tourmente. Cap Tourmente National Wildlife Area, northeast of Québec City. *August 1991, Yves Aubry.*

House Wren

Breeding distribution
1984–1989

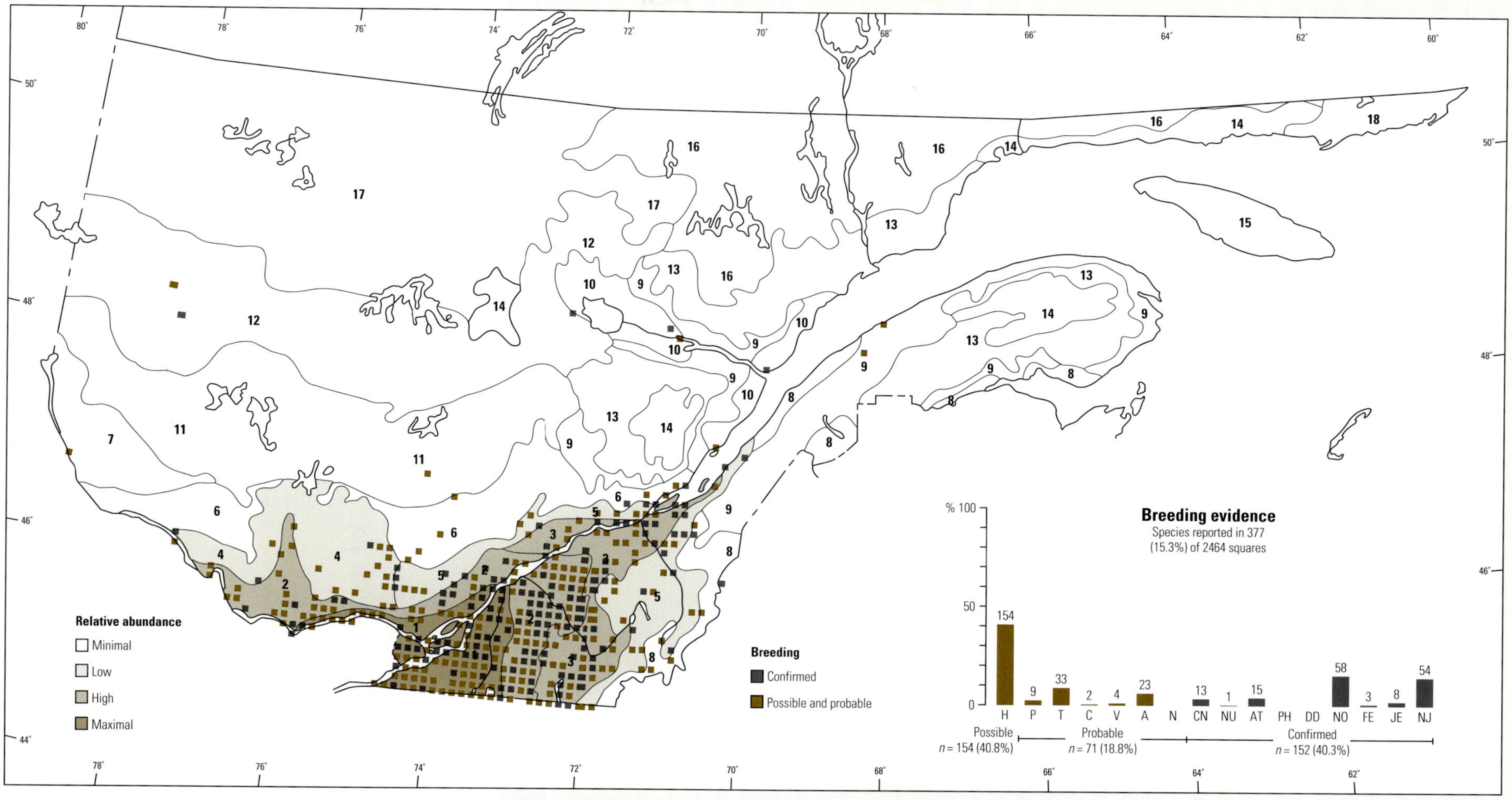

World breeding range

Species profile

Clutch size	usually 6–8 eggs (5–12)[1]
Laying interval	1 egg per day[1]
Start of incubation	with last egg[1]
Incubation period	13 days[1]
Maturity at hatching	altricial
Care of young	by female and male[1]
Nestling period	15 days (12–18)[1]
Age at first flight	15 days (12–18) or more[7]
Dependence of young	until 12–13 days after leaving nest[1,2]
Broods per year	1–2[1,2]
Breeding age	1 year[1,2]
Mating system	monogamy, often polygyny[2,6,8]
Duration of pair bond	1 season[1,2]
Total length	11.4–13.3 cm[3]
Wing span	15.2–17.8 cm[3]
Weight	M 10.3 g (9.1–11.1) / F 10.3 g (9.6–11.3)[4]
Longevity record	7 years 2 months[5]

References:
1. Gross 1948b
2. Drilling and Thompson 1991
3. Terres 1980
4. Taylor et al. 1983
5. Kennard 1975
6. Ehrlich et al. 1988
7. according to nestling period
8. Johnson and Kermott 1991

Chronology $n = 209$

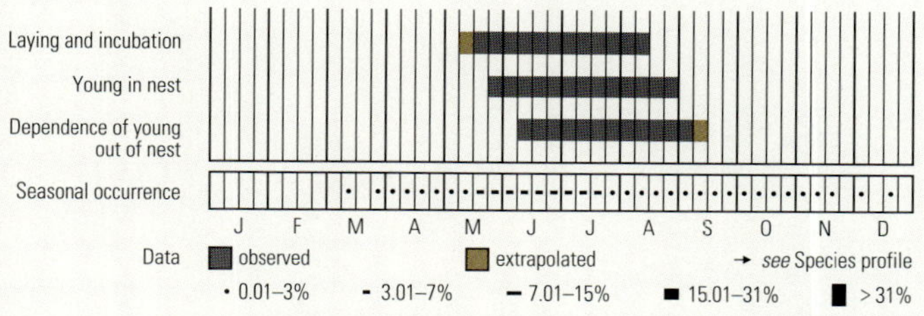

Frequency of occurrence by ecological region

Ecological regions** (forest regions)	Adequately covered squares	Squares in which reported	% of squares in which reported	Relative abundance (%)*
1. Sugar maple-hickory and sugar maple-basswood	97	82	84.5 ± 3.7	100.0
2. Sugar maple-basswood	153	109	71.2 ± 3.8	84.3
3. Sugar maple-basswood and sugar maple-yellow birch	101	58	57.4 ± 5.8	67.9
4. Sugar maple-yellow birch-basswood	56	18	32.1 ± 10.7	38.0
5. Sugar maple-yellow birch-beech and sugar maple-yellow birch-basswood	113	33	29.2 ± 4.7	34.5
6. Sugar maple-yellow birch-beech	71	11	15.5 ± 7.9	18.3
7. Sugar maple-yellow birch-balsam fir	19	1	5.3 ± 12.0	6.2
8. Sugar maple-yellow birch and balsam fir-yellow birch	47	2	4.3 ± 5.8	5.0
9. Balsam fir-yellow birch	87	2	2.3 ± 3.3	2.7
10. Balsam fir-white birch or balsam fir-red maple	46	4	8.7 ± 7.5	10.3
11. Yellow birch-balsam fir	52	2	3.8 ± 6.0	4.5
12. Balsam fir-white birch (west)	59	3	5.1 ± 6.2	6.0
13. Balsam fir-white birch (east)	76	0	0	0
14. Balsam fir-black spruce	28	0	0	0
15. White spruce-balsam fir and balsam fir-white spruce	22	0	0	0
16. Black spruce-balsam fir-moss	18	0	0	0
17. Black spruce-moss	32	0	0	0
18.	not analysed			

* minimal: 0–29.9%; low: 30–59.9%; high: 60–89.9%; maximal: 90–100%

** Deciduous forest: 1–7; mixed forest: 8–11; coniferous forest: 12–17

A House Wren with a spider in its bill, seen at the entrance to its nest cavity. Rivière du Nord, northwest of Montréal. *June 1987, Paul Perreault.*

This small, energetic passerine frequents woodland openings where dead trees afford nesting cavities. Rivière du Nord, northwest of Montréal. *May 1987, Paul Perreault.*

Distribution

The House Wren breeds across southern Canada to southern South America. In Québec, it breeds almost exclusively in the southwest, primarily in thickets and bushes in gardens and at the edge of isolated woodlots in Montérégie and the adjacent lowlands. It also nests in openings in the surrounding deciduous forest. Almost all of the squares in which the species was reported are located south of the 47th parallel, the approximate latitude of Québec City. North of this latitude, breeding was confirmed in only 4 squares: Val-d'Or in the Abitibi region, Saint-Prime in the Lac-Saint-Jean region, Saint-Honoré in the Saguenay region, and Tadoussac. There are also records from Lake La Motte, Grand-Métis, Sainte-Blandine, and Saint-Fulgence.

Generally speaking, the Québec range covers the Ottawa Valley, the Central St. Lawrence Lowland, Montérégie, the lower Laurentians, Eastern Townships, and southwestern Appalachians. The species is very localized north of this area. This differs from Godfrey's (1986) distribution map, which showed a range including all of southern Québec north to Abitibi and Lac-Saint-Jean, including the balsam fir forest regions.

Distribution of the House Wren is mostly restricted to the deciduous forest, or, more specifically, regions where extensive agricultural and urban development has provided the thicket and shrub habitat favoured by the species.

History and trends

Like all birds associated with open woodland, House Wrens are probably more common today than in earlier times. The clearing of the St. Lawrence Plain for agriculture and urban development (Ouellet 1974a) over the last few centuries has most certainly benefited the species, as has the felling of a number of forest stands north of the St. Lawrence River over the last hundred years. House Wrens also use the nest boxes commonly provided for them nowadays.

There is very little precise information on the historical status of the House Wren in Québec. In 1862, Hall noted that it was a common summer resident in the Montréal area (Ouellet 1974a), while by 1896 Wintle listed it as a scarce summer resident. According to Ouellet (1974a), House Wrens have never been very numerous in Montérégie and for some unknown reason declined alarmingly between 1961 and 1968. The Québec population has probably experienced major fluctuations since the last century, as was the case in New England and New York State (Ellison 1985o; Bonney 1988d). Some authors, including Bonney (1988d), attributed these fluctuations to weather conditions on the wintering grounds and to the establishment of the European Starling and the House Sparrow, both of which nest in cavities near human habitation.

Raymond McNicoll and Michel Robert

Winter Wren

Troglodyte des forêts
Troglodyte mignon
Troglodytes troglodytes

Christiane Girard

The Winter Wren is a tiny bird known for its exuberant song, which is particularly long and musical. Lasting for up to 9 seconds, the song is characterized by a series of varied notes and is astonishingly loud considering the size of the bird. Aside from its singing, which is often performed from a tall treetop, all of the bird's activities take place lower down in the trees.

The Winter Wren is the only member of the wren family that is not confined to the New World. It is found in the temperate and boreal regions of North America, Europe, and Asia (Cramp 1988). In Québec, it nests in coniferous and mixed forests, leaving in the fall to winter in the United States, giving the lie to its common name of "winter" wren, at least in this province (Todd 1963). On rare occasions, a few individuals will sometimes overwinter in the southwestern part of the Central St. Lawrence Lowland, however (Ouellet 1974a; David 1980a; Aubry et al. 1991).

Habitat and behaviour

The Winter Wren is a forest bird that haunts dense, moist coniferous woodlands with a ground cover of tangled bushes, upturned trees, mosses, and herbaceous plants (Bent 1948; Godfrey 1986; Peck and James 1987). It is often found near streams or lakes and in poorly drained or swampy forest. It also occurs in mature, dense deciduous or hemlock woods. Strangely enough, in Great Britain, where it is one of the most widespread species, the Wren, as it is simply known there, nests in a variety of habitat, from rocky slopes to city gardens (Armstrong 1955).

The Winter Wren normally builds its nest in a cavity on or near the ground, most often in the earthy roots of an upturned or live tree, in or under a rotted stump, or in a similar spot (Godfrey 1986). A Winter Wren nest was once discovered in the door frame of a cottage in Mont-Saint-Hilaire, in Montérégie (QNRCP). The rather bulky nest is built of mosses, twigs, and grasses and is lined with feathers and hair (Godfrey 1986). As with other wrens, the male builds several nests in an attempt to attract more than one female to his territory, though he is not always successful. The female finishes the inside of the selected nest, incubates the eggs by herself, and, in most cases, rears the nestlings alone (Armstrong 1955; Cramp 1988). The young and adults feed on various types of invertebrates, including beetles and spiders, which they find mostly on or near the ground (Bent 1948; Van Horne and Bader 1990).

Because of the type of habitat it frequents and its secretive behaviour near the nest, it is not surprising that the Winter Wren was confirmed as breeding in so few squares. The rare confirmations were made for the most part through sightings of recently fledged young and of adults carrying food, because the nest itself is difficult to find. In fact it is easier to locate the bird by its loud, ringing song than by sight.

Distribution

The Winter Wren is distributed throughout southern Québec, wherever coniferous woodlands are found. Given this predilection for forest habitat, it is less common in the agricultural regions of the southwest of the province, where it occurs mainly in mature, isolated woodlots with some conifer growth, or in peatlands. In alpine and coastal coniferous forests, as well as in other areas with extensive stands of conifers, the Winter Wren is also very common as shown in the table of frequency of occurrence by ecological region.

There are various records from north of the 50th parallel, from the Eastmain River to Lake Hélène in the James Bay region (ÉPOQ). On the James Bay coast, it is found up to at least the 54th parallel; several singing males were in fact heard north of Chisasibi (Y. Aubry and G. Falardeau, pers. comm.; see also Todd 1963).

History and trends

As Ouellet (1974a) pointed out in his study on the birds of the Monteregian Hills, it is difficult to establish whether the Winter Wren was once more abundant than it is today. If we give credence to Wintle (1896), then the species has long been relatively scarce in the southwest corner of the St. Lawrence Plain. Outside the agricultural and populated areas of southwestern Québec, its status has probably remained unchanged, although we cannot be certain.

The Winter Wren is particularly sensitive to harsh winters, which sometimes provoke major population declines in Europe and North America alike (Cramp 1988). In the southeastern United States, where a large number of Winter Wrens overwinter, abundance declined sharply after the two consecutive severe winters of 1976–77 and 1977–78 (Robbins et al. 1986). BBS and ÉPOQ data for 1969–89 do not indicate any definite population trends for the Winter Wren in Québec.

Christian Gagnon
and Michel Robert

This mixed forest, with its stream, upturned trees, and ground cover of brushwood, grasses and forbs, and mosses, constitutes ideal habitat for the Winter Wren. It inhabits the lower strata and forages for arthropods near the ground in the tangled vegetation. Duchesnay Forestry Station, west of Québec City. *June 1992, Robert Alvo.*

Notes on territory

The Winter Wren establishes a territory that is used for both nesting and foraging (Van Horne and Bader 1990). Armstrong (1956) found territories to measure between 0.8 and 1.2 ha, and Cyr (1974) estimated average Winter Wren territory size at 0.9 ha in a mature hemlock stand in the Mirabel area.

Zarnowitz and Manuwal (1985) observed up to 20.5 pairs/10 ha in a mature forest in Washington State, whereas only 1.7 pairs/ 10 ha were found in a recent cutover where stumps had been left. Erskine (1980a) reported densities of 4.5 territorial males/ 10 ha in four mixed forest habitats in Québec, two of which were dominated by fir and spruce (at Sainte-Anne-des-Monts and Lake Cascapédia) and two others by fir (at Carleton Centre and near Rivière-Nouvelle).

In the Nottaway, Broadback, and Rupert river watersheds, the relative density of Winter Wrens in stands of birch was 1½ times higher than in black spruce-moss forests and more than twice as high as in old burns and shoreline ecotones in coniferous habitat (Cons. G. G. – GREBE 1992o).

Winter Wren

Breeding distribution
1984–1989

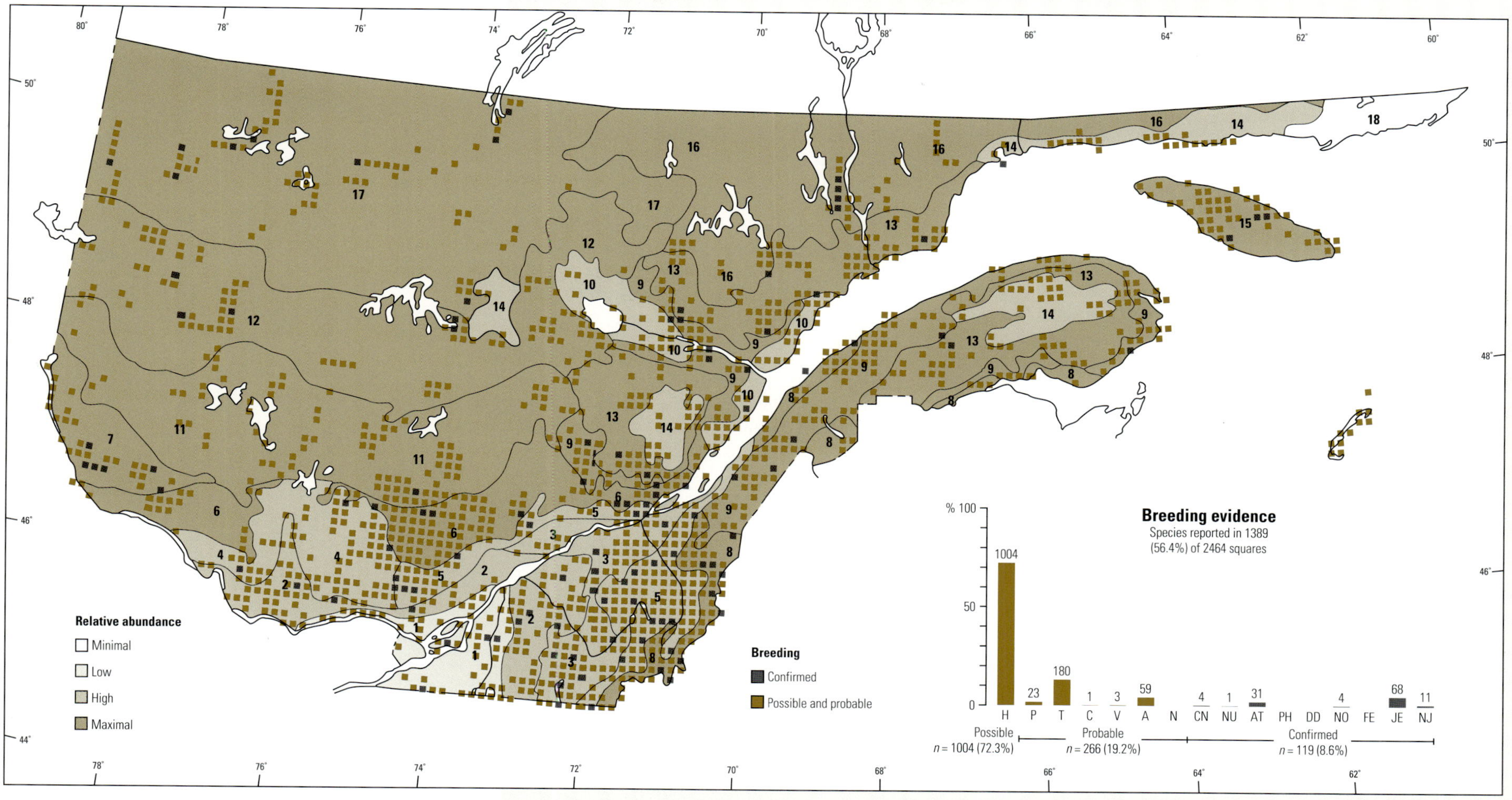

Species profile

Clutch size	usually 4–7 eggs[2,3,5]
Laying interval	1 egg per day[8]
Start of incubation	before last egg[8]
Incubation period	usually 16 days (11–20)[2,3,5,8]
Maturity at hatching	altricial
Care of young	mainly by female[1]
Nestling period	16 days (14–19)[1,8]
Age at first flight	16 days (14–19)[1,8]
Dependence of young	until 9–18 days after first flight[8]
Broods per year	2[2]
Breeding age	1 year[2]
Mating system	monogamy or polygyny depending on habitat[1,8,9]
Duration of pair bond	during mating only[8]
Total length	10.0–11.4 cm[6]
Wing span	14.0–16.5 cm[3]
Weight	8.9 g (7.5–10.5)[4]
Longevity record	4 years 1 month[7]

References:
1. Van Horne and Bader 1990
2. DeGraaf and Rudis 1987
3. Terres 1980
4. Dunning 1984
5. Ehrlich et al. 1988
6. Godfrey 1986
7. Klimkiewicz et al. 1983
8. Cramp 1988
9. Armstrong and Whitehouse 1977

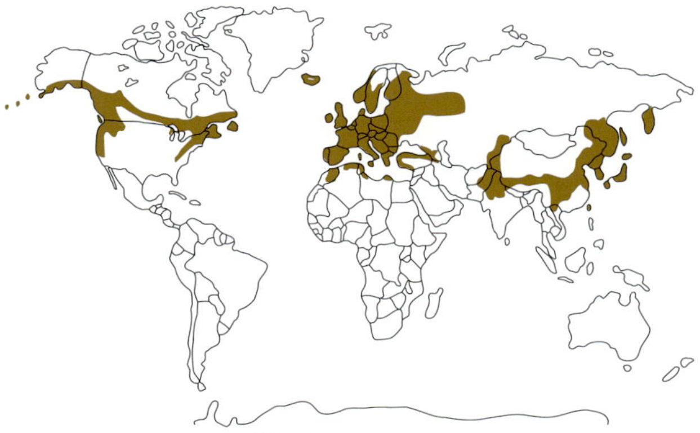

World breeding range

Chronology $n = 41$

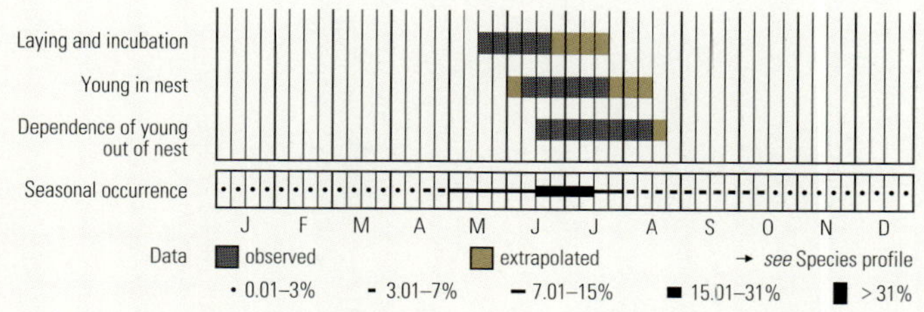

Frequency of occurrence by ecological region

Ecological regions** (forest regions)	Adequately covered squares	Squares in which reported	% of squares in which reported	Relative abundance (%)*
1. Sugar maple-hickory and sugar maple-basswood	97	48	49.5 ± 5.0	49.5
2. Sugar maple-basswood	153	92	60.1 ± 4.0	60.1
3. Sugar maple-basswood and sugar maple-yellow birch	101	85	84.2 ± 4.4	84.2
4. Sugar maple-yellow birch-basswood	56	48	85.7 ± 8.3	85.7
5. Sugar maple-yellow birch-beech and sugar maple-yellow birch-basswood	113	101	89.4 ± 3.3	89.4
6. Sugar maple-yellow birch-beech	71	67	94.4 ± 5.3	94.4
7. Sugar maple-yellow birch-balsam fir	19	18	94.7 ± 12.0	94.7
8. Sugar maple-yellow birch and balsam fir-yellow birch	47	46	97.9 ± 4.5	97.9
9. Balsam fir-yellow birch	87	83	95.4 ± 4.4	95.4
10. Balsam fir-white birch or balsam fir-red maple	46	35	76.1 ± 10.8	76.1
11. Yellow birch-balsam fir	52	52	100.0 ± 1.0	100.0
12. Balsam fir-white birch (west)	59	55	93.2 ± 7.0	93.2
13. Balsam fir-white birch (east)	76	74	97.4 ± 4.0	97.4
14. Balsam fir-black spruce	28	25	89.3 ± 12.6	89.3
15. White spruce-balsam fir and balsam fir-white spruce	22	21	95.5 ± 10.3	95.5
16. Black spruce-balsam fir-moss	18	18	100.0 ± 2.8	100.0
17. Black spruce-moss	32	32	100.0 ± 1.6	100.0
18.			*not analysed*	

* minimal: 0–29.9%; low: 30–59.9%; high: 60–89.9%; maximal: 90–100%

** Deciduous forest: 1–7; mixed forest: 8–11; coniferous forest: 12–17

Sedge Wren

Troglodyte à bec court

Cistothorus platensis

The Sedge Wren is the rarest of the wrens that breed regularly in Québec, where it is considered vulnerable (Robert 1989b).

It is also one of the smallest birds that breed here, and its elusive habits make it difficult to observe. Normally concealed by tall grasses, it prefers to flee below the cover when frightened, rather than flush. When it does flush, it flies for a short distance, only to drop back into the vegetation. A lot of patience is required to study it. Its song, a muted, dry staccato chatter, may also go undetected, although the bird is a tireless singer and will vocalize up to 20 hours daily, day and night (Walkinshaw 1935; Desrochers and Fragnier 1983).

Habitat and behaviour

The Sedge Wren favours wet sedge meadows (hence its name) with scattered alders or willows. It also frequents freshwater marshes, wet fields, and sometimes the margins of peatlands (Terrill 1922; Bent 1948). It is generally found in drier habitat than its close relative, the Marsh Wren (Walkinshaw 1935; Bent 1948), and normally shuns the cattail marshes that are the Marsh Wren's usual habitat.

Like other wrens, the male Sedge Wren will build several nests, anywhere from 5 to 13, according to Burns (1982). The female selects one of these to lay in and lines it with fine materials (Burns 1982). The nest used for egg laying and raising the young is more difficult to find than the dummy nests, as it is well hidden in vegetation near the ground or water, rarely higher than 60 cm. Spherical in shape, it is made of grasses intertwined with the surrounding vegetation (Harrison 1975). The female does all the incubating and most of the feeding (Mousley 1934a; Burns 1982). The male spends most of his time singing and building new nests, probably by way of defending his territory and attracting another mate, as male Sedge Wrens are often polygynous (Crawford 1977; Burns 1982).

The Sedge Wren nests solitarily or sometimes in small colonies that may number a few dozen singing males (Terrill 1922; Walkinshaw 1935). Whether nesting singly or in colonies, Sedge Wrens will often occupy different nest sites over the years, and for unknown reasons, habitat that is apparently suitable may not be used regularly from year to year (Burns 1982). Terrill (1943b) ascribed this to annual variations in dampness conditions at the nest sites.

The species was reported in fewer than 1% (21/2464) of Atlas squares surveyed. Breeding was confirmed in only 3 squares, through sightings of adults carrying food or nesting materials, illustrating how hard it is to find a Sedge Wren's nest.

Christiane Girard

Distribution

In Québec, the Sedge Wren is at the northern limit of its range. It occurs locally in the southern Laurentians, the Central St. Lawrence Lowland, and the southern Appalachians (David 1980a; Godfrey 1986).

Nesting was confirmed at Eardley and Schwartz in the Ottawa Valley, as well as around Lake Memphrémagog in the Eastern Townships. It was also reported near Montréal, Lake Saint-François, Lake Saint-Pierre, Missisquoi Bay, and Cap Tourmente. Given its reclusive nature and how hard it is to spot, distribution may very well be underestimated.

Notes on territory

Males fiercely defend a breeding territory of only a few hundred square metres (0.12–0.36 ha, according to Burns 1982). The boundaries of this territory may vary throughout the summer. The territory is used for courtship, nest building, and feeding (Burns 1982). Sedge Wrens often destroy the eggs of other wrens and other birds nesting nearby, thereby reducing competition (Picman and Picman 1980).

At Fesserton, Ontario, density in a sedge meadow was estimated at 3.6 males/10 ha (Erskine 1984). At Mafeking, Manitoba, where the species is more common than in Québec and Ontario, density may be as high as 29.4 males/10 ha in shrubby sedge marshes (Erskine 1976a).

One of the few sites where the Sedge Wren has been known to breed in recent years is in the tall grass of the wet meadows at Hopkins Point. Near the Lac Saint-François National Wildlife Area, southwest of Montréal. *July 1991, Gaétan Duquette.*

Sedge Wren

Breeding distribution
1984–1989

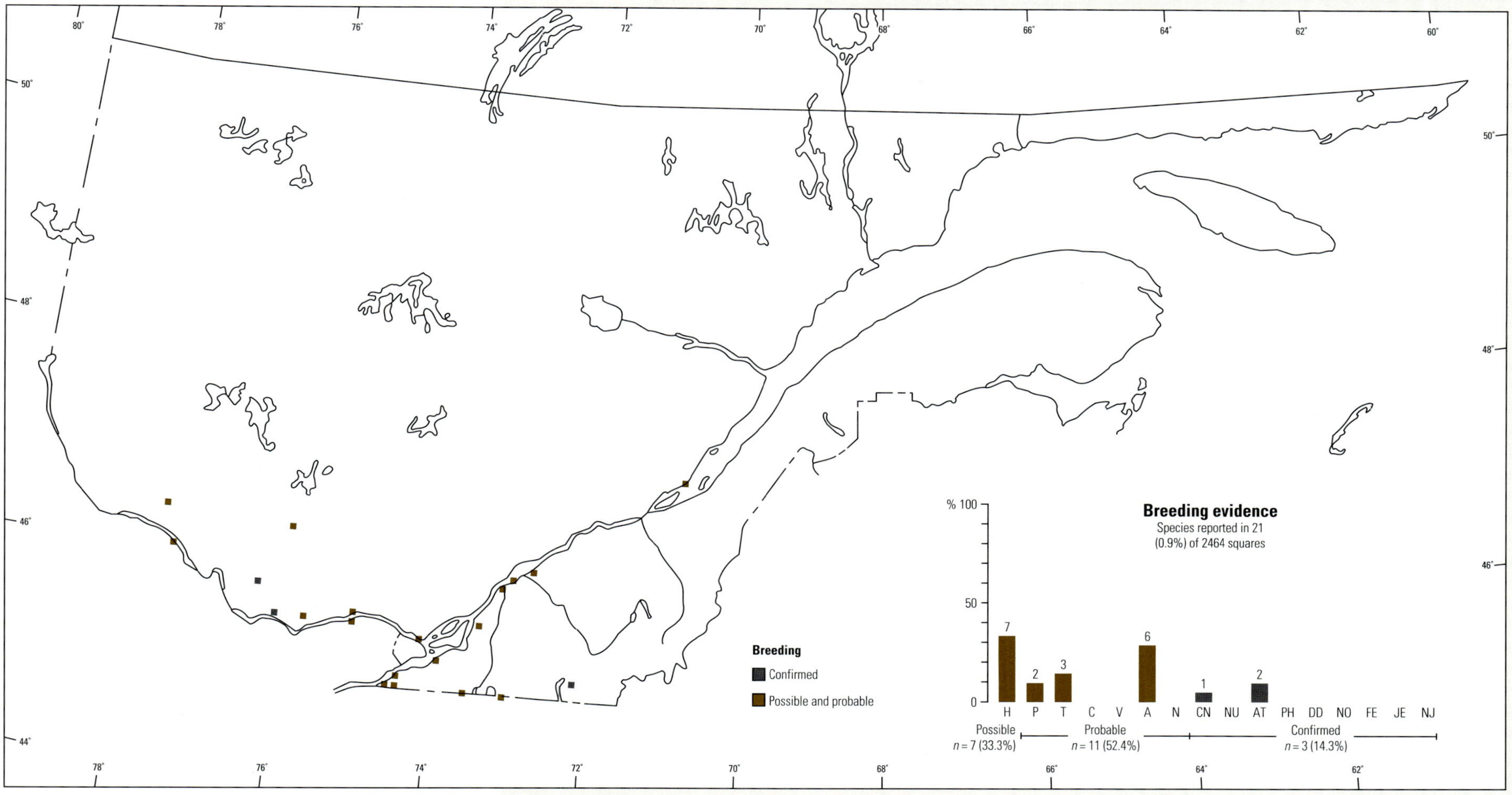

Species profile

Clutch size	usually 6–7 eggs (4–8)[1,2,3,4]
Laying interval	1 egg per day[4]
Start of incubation	before last egg[3]
Incubation period	12–14 days[2,3]
Maturity at hatching	altricial
Care of young	mainly by female[1,2,4]
Nestling period	12–14 days[1,2,4]
Age at first flight	12–14 days or more[9]
Dependence of young	?
Broods per year	usually 2 (1–3)[3,4]
Breeding age	1 year[8]
Mating system	monogamy, often polygyny[10]
Duration of pair bond	?
Total length	10.0–11.4 cm[5]
Wing span	13.3–15.2 cm[6]
Weight	9.0 g (7.2–10.3)[7]
Longevity record	?

References
1. Ehrlich et al. 1988
2. Harrison 1978
3. Harrison 1975
4. Walkinshaw 1935
5. Godfrey 1986
6. Terres 1980
7. Dunning 1984
8. DeGraaf and Rudis 1987
9. according to nestling period
10. Burns 1982

Chronology $n = 3$

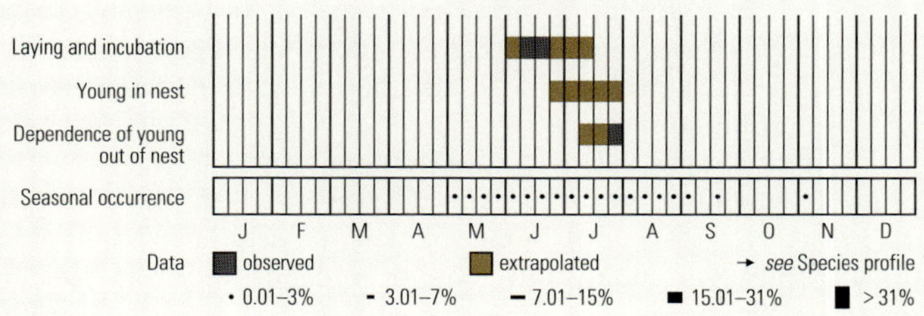

World breeding range

The nest is spherical with a side entrance. Saint-Lambert, near Montréal. *July 1933, Henry Mousley, Laval University Collection.*

The distribution map shows the few known areas in southern Québec where the species breeds regularly. The largest number of Sedge Wrens can be found in the Lac Saint-François National Wildlife Area and vicinity (Dundee, Akwesasne), where a cursory survey in July 1991 tallied 8 singing males (M. Robert, pers. obs.).

In Canada, the Sedge Wren nests on the Prairies and in southern Ontario and Québec. It winters from western Tennessee and Maryland south to Florida in the east, and from southeastern New Mexico to northeastern Mexico in the west.

History and trends

The Sedge Wren was first reported in Québec early this century, in the Eastern Townships and the Montréal area (Terrill 1922; Mousley 1934a), but as Ouellet (1974a) pointed out, it had probably been nesting in the province during the last century, but had been overlooked. Even today, little is known about its distribution and abundance in Québec, to the extent that it is not possible to say whether the species is indeed less common than it once was (Robert 1989b). In 1922, Terrill estimated that there were at least 100 pairs in and around Montréal.

The species appears to have declined in the northeastern part of the continent in recent decades (Robbins et al. 1986; Tate 1986; Gibbs and Melvin 1992c; Jalava 1993). Some attribute the decline to habitat loss, notably through draining and filling of wetlands for urban development and intensive agriculture (Gould 1987; Connor 1988e; Gibbs and Melvin 1992c; Jalava 1993). The situation may be the same in Québec, since the Sedge Wren frequents mainly the settled lowlands, where farming prevails. On the basis of the scant data available, however, we cannot seriously maintain, as does Jalava (1993), that there has been a significant decline in numbers in southern Québec over recent decades.

In 1933, Henry Mousley investigated the different nesting stages of the Sedge Wren in this old field, in Saint-Lambert, near Montréal. *June 1933, Henry Mousley, Laval University Collection.*

Seven-day old Sedge Wrens in their nest. Saint-Lambert, near Montréal. *Henry Mousley, Laval University Collection.*

This wren nests in tall grass, low down among the grasses and sedges. Ahuntsic, Montréal. *June 1941, Henry Mousley, Laval University Collection.*

The present low Sedge Wren count in the Lac Saint-François National Wildlife Area is apparently due to reversion to shrubs in some marshes there. Before the federal government acquired the area in 1971, native people regularly cleared trails by burning the vegetation. This practice has since been discontinued, and several wet meadows have been gradually invaded by speckled alder (Jean and Bouchard 1991). The reversion process is still continuing, and the Sedge Wren population has dropped by two thirds, from 28 singing males in 1983 (Desrochers and Fragnier 1983) to 9 in 1993 (M. Robert, pers. obs.).

Philippe Fragnier and Michel Robert

Marsh Wren

Troglodyte des marais

Cistothorus palustris

Christiane Girard

THE MARSH WREN IS HEARD MORE OFTEN THAN IT IS SEEN. BIRDWATCHERS who visit the cattail marshes of southwestern Québec during the summer months often hear the strong, exuberant, staccato song of this little wren. They can sometimes see it rise up above the marsh, then drift down again singing. The quietest and most patient may also see it close up, tail held erect, feet wide apart, clinging to two cattail stems.

In spring, it arrives from its wintering grounds, mainly across the southern United States and Mexico (AOU 1983).

Habitat and behaviour

This wren lives in marshes where cattails and other tall water plants, such as bulrushes, bur-reeds, and common reeds, grow. It prefers shallow water, from a few centimetres to no more than about 1 m deep, and so is found along the banks of rivers, lakes, or ponds, as well as in marshes both large and small (Bent 1948; Leonard and Picman 1987a).

The males arrive on the nesting grounds about 10 days earlier than the females and use this time to establish their territories and build several unlined nests (Welter 1935). These are built close together and seem to play a role in courting females, which prospect the territories of males (Welter 1935; Verner 1965). The nest is rather oval in shape, unlike the round nest of the Sedge Wren, and is most often built in a clump of cattails 30–100 cm above the water. It is the size of a large grapefruit and is made out of interwoven leaves of emergent plants.

The female either chooses one of the nests already built, or starts to build a new one that will, however, be mostly completed by the male. Not until she has lined the nest with feathers and small strips of dry leaves does the female commence laying. Typical clutch size is 4–6 eggs, and incubation is by the female alone, though both adults feed the

Notes on territory

Males usually defend very small breeding territories, varying in size according to the type of habitat. In Georgia, for example, territories of only 0.006 ha were recorded in marshes with tall and short grasses (Kale 1965 in DeGraaf and Rudis 1987), while in Washington State, territories averaging 0.05–0.07 ha were found in cattail and bulrush marshes (Verner 1965; Verner and Engelsen 1970). Depending on habitat, recorded densities of Marsh Wrens have ranged from 3 pairs/10 ha in an intertidal salt marsh in New York State to 3.75 pairs/10 ha in shrubby sedge habitat in Connecticut, 9 pairs/10 ha in a cattail marsh in Maryland, and even 26 pairs/10 ha in a rush marsh in Maryland (DeGraaf and Rudis 1987; Aquila 1993; Magee 1993b).

Marsh Wrens, in particular the males, have a habit of destroying eggs and nestlings—whether of their own or of other species, such as the Red-winged Blackbird—that nest in or near their territories (Picman 1977a,b).

This pond full of cattails at La Grande Ferme in the Cap Tourmente National Wildlife Area on the north shore of the St. Lawrence marks the eastern boundary of the breeding range of the species. Northeast of Québec City. *August 1991, Yves Aubry.*

Marsh Wren

Breeding distribution 1984–1989

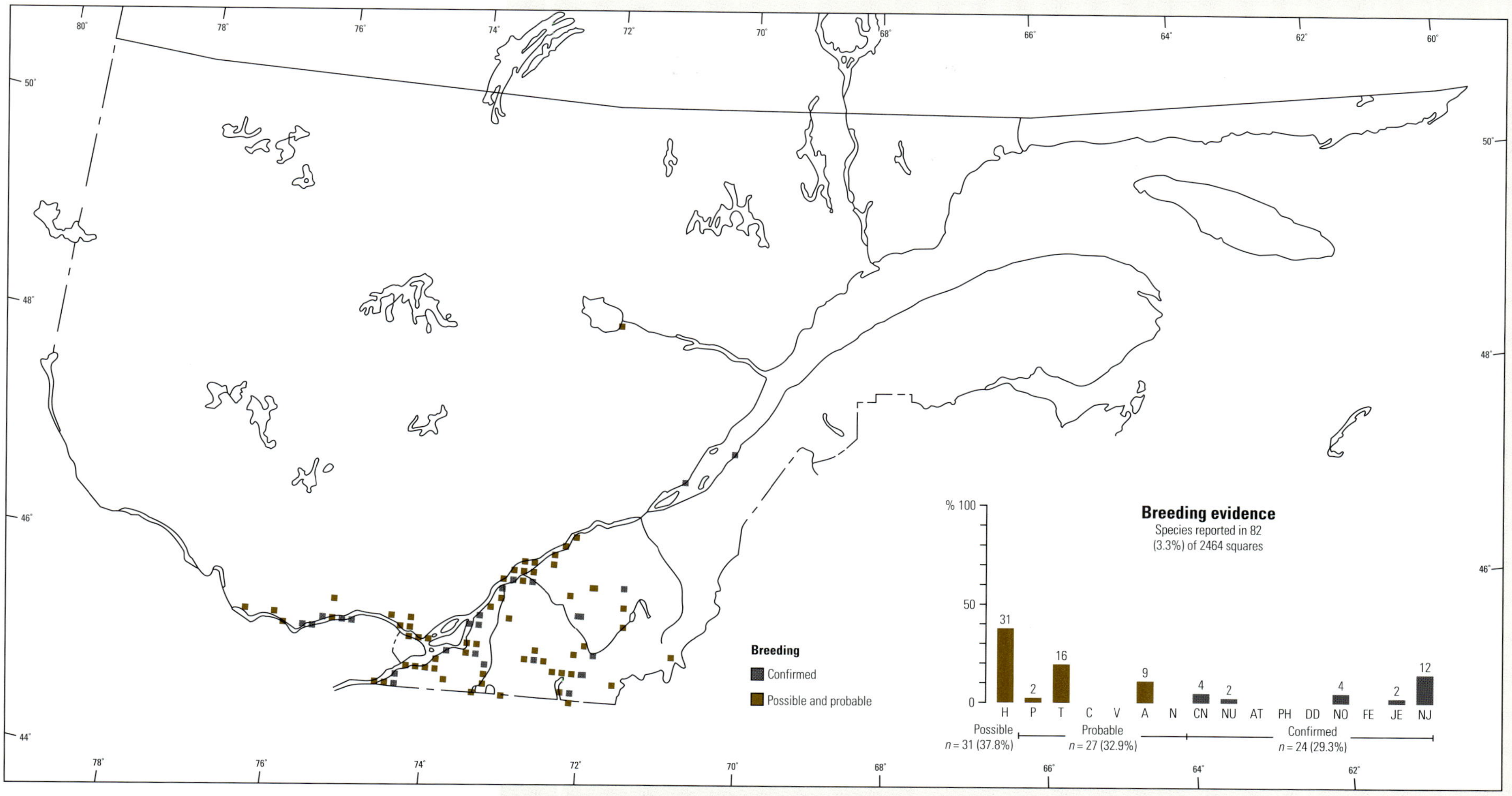

Species profile

Clutch size	usually 4–6 eggs (3–8)[1]
Laying interval	1 egg per day[1]
Start of incubation	before last egg[1]
Incubation period	14–16 days[1]
Maturity at hatching	altricial
Care of young	mainly by female[1,8]
Nestling period	usually 14 days (11–16)[1]
Age at first flight	usually 14 days (11–16) or more[11]
Dependence of young	until 12 days or more after leaving nest[1]
Broods per year	2[4]
Breeding age	1 year[7]
Mating system	monogamy, often polygyny[1,2,9,10]
Duration of pair bond	usually 1 season[1]
Total length	10.0–13.3 cm[5]
Wing span	12.7–17.8 cm[3]
Weight	M 11.9 g (10.5–13.5) / F 10.6 g (9.0–13.5)[6]
Longevity record	?

References:
1. Verner 1965
2. Leonard and Picman 1987b
3. Terres 1980
4. Ehrlich et al. 1988
5. Godfrey 1986
6. Dunning 1984
7. DeGraaf and Rudis 1987
8. Harrison 1978
9. Burns 1982
10. Metz 1991
11. according to nestling period

Chronology *n* = 36

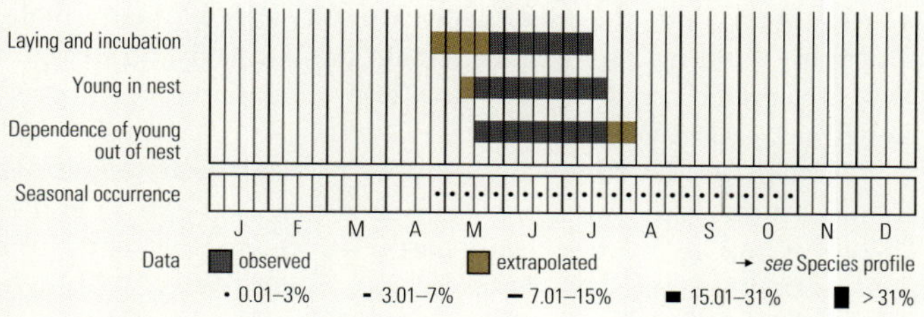

World breeding range

Along the St. Lawrence River, the Marsh Wren mainly prefers areas of large marshes dominated by cattails and ponds covered in emergent vegetation. Île aux Fermiers, Varennes Islands, near Montréal. *August 1991, Gaétan Duquette.*

Notes on territory (continued)

According to one study conducted in two consecutive summers, when wrens and blackbirds nest in close proximity to each other, it seems that their interactions lead them to use distinct nesting areas to reduce mutual competition. Wrens use the denser or thicker sections of a marsh, and blackbirds the less dense areas (Picman 1980).

young (Welter 1935; Verner 1965). Once laying begins, the male starts building new nests in another part of his territory and attempts to attract another female. Up to 50% of males manage to attract more than one female, usually two but sometimes more (Verner 1965; Leonard and Picman 1987a). This explains why the Marsh Wren sings so constantly throughout the nesting season and why so many males, apparently yearlings, sometimes remain unmated (Verner and Engelsen 1970). This wren sings virtually non-stop all summer long, even at night (Barclay et al. 1985).

Males build a surprising number of nests in a summer: the 26 males monitored by Verner (1965) each built between 11 and 32. These dummy nests are not only part of courtship, but may also be used as shelters or roosts by adults and young (Verner 1965). It has also been suggested that the chief purpose of building all these nests is to reduce the risk of predation on those used for laying eggs and brood rearing (Leonard and Picman 1987b); however, the true primary function of this behaviour remains unknown (Metz 1991).

The Marsh Wren is easy to spot, but confirming breeding is harder. Finding an active nest, one with eggs or nestlings, is often more difficult than finding dummy nests. Over a third of Atlas records (37.8%) were simply of possible breeding.

The Marsh Wren's nest is an oval sphere with a side entrance, made of woven plant material attached to surrounding vegetation. Saint-Lambert, near Montréal. *June 1933, Henry Mousley, Laval University Collection.*

Marsh Wren building a nest. Building several nests helps the male attract more than one female. Châteauguay. *July 1987, Paul Perreault.*

Marsh Wren singing from atop a cattail. The male sings constantly throughout the nesting season to court more females. Bécancour, opposite Trois-Rivières. *May 1987, Paul Perreault.*

Distribution

The southernmost part of Québec is the northern limit of the Marsh Wren's range in eastern North America. This wren is found in marshes in the southwestern part of the province, mainly along the Ottawa River and the St. Lawrence River upstream of Trois-Rivières. It is also found in the Eastern Townships and in Montérégie, as well as in a variety of other places in the Central St. Lawrence Lowland. Cap Tourmente is the most easterly location where it breeds regularly. It has been reported elsewhere during the breeding season, however. For example, in July 1988, a singing male was heard several days in a row at Saint-Gédéon in the Lac-Saint-Jean region (Gosselin et al. 1988b). Moreover, at Saint-Roch-des-Aulnaies, a dozen kilometres southwest of Sainte-Anne-de-la-Pocatière, in 1986 a bird built several nests in May and June (QNRCP). The Marsh Wren may also nest in the large marsh between Île aux Oies and Île aux Grues, as well as in the Cacouna marsh, where it has been seen for several years (M. Robert, pers. obs.).

In Québec, as elsewhere, the Marsh Wren's distribution is closely associated with that of cattail marshes. It is locally fairly common in suitable expanses of habitat in southwestern Québec.

History and trends

According to Ouellet (1974a), the Marsh Wren was probably fairly common in the last century, at least in marshes in Montérégie, although Wintle (1896) considered it fairly scarce there. Just as today, local distribution of the species must have matched marsh distribution. Over time, the gradual disappearance of wetland habitat as a result of agricultural and urban development probably had the same effect here as it did in parts of the northeastern United States (Ouellet 1974a; McCracken 1987a; Connor 1988b).

Serge Tanguay and Michel Robert

Golden-crowned Kinglet

Roitelet à couronne dorée

Regulus satrapa

ONE HAS TO LEARN TO LISTEN FOR THE GOLDEN-CROWNED KINGLET IN order to detect its presence and to be able to distinguish its frequent series of shrill *tsees* from the rallying calls of the chickadees and warblers which share its habitat.

Its high-pitched song is near our upper threshold of hearing, and some people, even in their youth, cannot hear it. This tiny, greenish-white bird, often perched high up in the trees, is not easy to spot. Weighing about 6 g, it is the smallest bird that regularly overwinters in Québec. It breeds throughout southern Québec.

The Golden-crowned Kinglet is often found in balsam fir and spruce stands, where it builds its nest in the top of a conifer, close to the trunk. Montmorency forest, Laurentide Wildlife Reserve, north of Québec City. *May 1991, Réginald Ouellet.*

Habitat and behaviour

The Golden-crowned Kinglet is basically a bird of coniferous woodland. It occurs in fir and spruce stands in the boreal forest, but can also be found in hemlock stands, tamarack peatlands, and spruce plantations in the south (Andrle 1971; Galati and Galati 1985; DeGraaf and Chadwick 1987; Derleth et al. 1989). It is also sometimes found in deciduous stands (Holmes and Sturges 1975).

The nest, always built in a conifer, usually 8–20 m up (Galati and Galati 1985; Peck and James 1987), is hard to distinguish from the Ruby-crowned Kinglet's (Harrison 1975). It is a small, oblong construction, open at the top, with the outside wall made of lichen. The female takes about 5 days to build it, usually a few centimetres from the trunk. She lays 8 or 9 eggs at the rate of one a day, generally in the morning, and incubates them alone for about 15 days. The male does not appear to feed the sitting female, although he does feed her during the courtship period (Galati and Galati 1985). It was also found that, in Minnesota, the Golden-crowned Kinglet usually renests before the first brood is reared. The male takes little part in feeding the young of the first brood until they are about 8 days old, at which time the female begins to build the second nest. The male then participates more actively in raising the first brood; these fledge at about 16–19 days of age, but are still fed for another 16 or 17 days. Meanwhile, the female lays a second clutch, generally as large as the first. The male participates little in feeding the second brood. It is not known whether this kinglet is double-brooded in Québec, but the fact that downy young have been seen in late September (M. Darveau, pers. obs.) indicates that there are at least some late nesting attempts. In their study, Galati and Galati (1985) found that of 102 eggs laid, 76 hatched, 62 of the hatchlings fledged, and at least 38 fledglings survived until they could fend for themselves.

The Golden-crowned Kinglet's diet consists mainly of insects. It catches relatively small prey, such as aphids or geometrid moth larvae (Galati and Galati 1985), which it gleans from the foliage of conifers by walking along the branches or even hanging from them (Franzreb 1984; Keast and Saunders 1991).

Finding nests of the Golden-crowned Kinglet was difficult. Atlassers reported them in fewer than 1% of the squares in which the species was reported, accounting for barely 3% of confirmed breeding records. Breeding confirmations were mostly based on observations of adults carrying food or of recently fledged young, because the adults and chicks are relatively noisy.

Distribution

The breeding range of the Golden-crowned Kinglet extends from Alaska to Newfoundland, including some northern U.S. states (AOU 1983). It was reported throughout the Atlas area, which includes all of Québec south of the 50th parallel within the range described by David (1980a). This kinglet is not only a typical bird of conifer stands in Québec's boreal forest (DesGranges 1980b; Beauchesne 1991; Darveau et al. 1995), but can also be found in mixed and deciduous stands containing some conifers. It nests particularly in the western part of southern Québec, especially in mature coniferous stands. Farther east, it is common in coniferous habitat on Anticosti Island.

The Golden-crowned Kinglet ranges north of the Atlas area. Four birds were reported on 3 July 1984 along the road between Matagami and La Grande-2, near the Eastmain and Opinaca rivers slightly north of the 52nd parallel (Bordage and Aubry 1983; Bordage 1985b). There are a few records from the same area in July 1982 and May 1990 (Bordage and Aubry 1983; ÉPOQ). A little farther south, 2 birds were seen near the Harricanaw River, near the 51st parallel, on 23 July 1982. This kinglet is a common breeder in the Nottaway, Broadback, and Rupert river watersheds (Cons. G. G. – GREBE 1992o). Farther east, 2 birds were reported at Gagnon on 11 July 1976, and 10 others at Manic-5 on 7 June 1992 (ÉPOQ).

Although the Golden-crowned Kinglet is present year-round in Québec, it is relatively rare in winter (David 1980a), and it is assumed that most of the birds winter south of the province. Peak migration occurs in late April and from late September to late October. Christmas Bird Count data show that it winters in regions such as southern Canada and the United States where the mean January temperature is above –18°C (Root 1988).

History and trends

There is little data on the distribution of the Golden-crowned Kinglet before the 1960s. Wintle (1896) merely listed it as a common "transient visitant" to Montréal. Macoun and Macoun (1915) mentioned that it nested in the Magdalen Islands, and Taverner (1922) associated it with the coniferous belt. Although none of these authors considered the Golden-crowned Kinglet a breeder in the St. Lawrence Plain or southern Québec in general, it probably was nesting, but was overlooked. However, it may not have been as widespread 100 years ago as it is today, as is the case in New York State (Andrle 1971).

Over the last 30 years, the abundance of the Golden-crowned Kinglet has fluctuated considerably because of natural disturbances and human activity. The severe winters of 1976–77 and 1977–78 were a major contributor to the sharp drop in populations observed both across the continent (Robbins et al. 1986) and in Québec, where the percentage of field trips on which the species was observed dropped from 5 to 1% (ÉPOQ).

As for human activity, massive deforestation in southern Québec, which started with European settlement, caused habitat loss, especially of peatlands containing small stands of conifers. However, reforestation of unproductive farmland with conifers has created new habitat (Andrle 1971), as has been happening for some years in the Beauce and the Eastern Townships. On the other hand, recent activity in the boreal forest, with intensified logging and the loss of mature forest, may be detrimental to this kinglet (Mannan and Meslow 1984; DeGraaf and Chadwick 1987; Johnson and Brown 1990; Beauchesne 1991).

JACQUES LARIVÉE AND MARCEL DARVEAU

Christiane Girard

Notes on territory

Males use song as their main defence in protecting their territories of 0.9–2.5 ha against conspecifics. In Maryland, Stewart and Robbins (1958) recorded a density of 8 pairs/10 ha in a spruce and hemlock peatland, whereas Andrle (1971) noted 12 pairs/10 ha in a conifer forest in the Adirondacks. In La Mauricie National Park, DesGranges (1980b) reported densities of 15 birds/10 ha in a white spruce plantation, 16 birds/10 ha in a red spruce stand, and 13 birds/10 ha in a balsam fir stand. Erskine (1984) reported a density of 5 territorial males/10 ha in a balsam fir-white birch stand in the Mount Mégantic region. In the Nottaway, Broadback, and Rupert river watersheds the relative abundance of Golden-crowned Kinglets was higher in jack pine stands than in spruce-moss stands and the other habitats surveyed (Cons. G. G. – GREBE 1992o).

Golden-crowned Kinglet

Breeding distribution

1984–1989

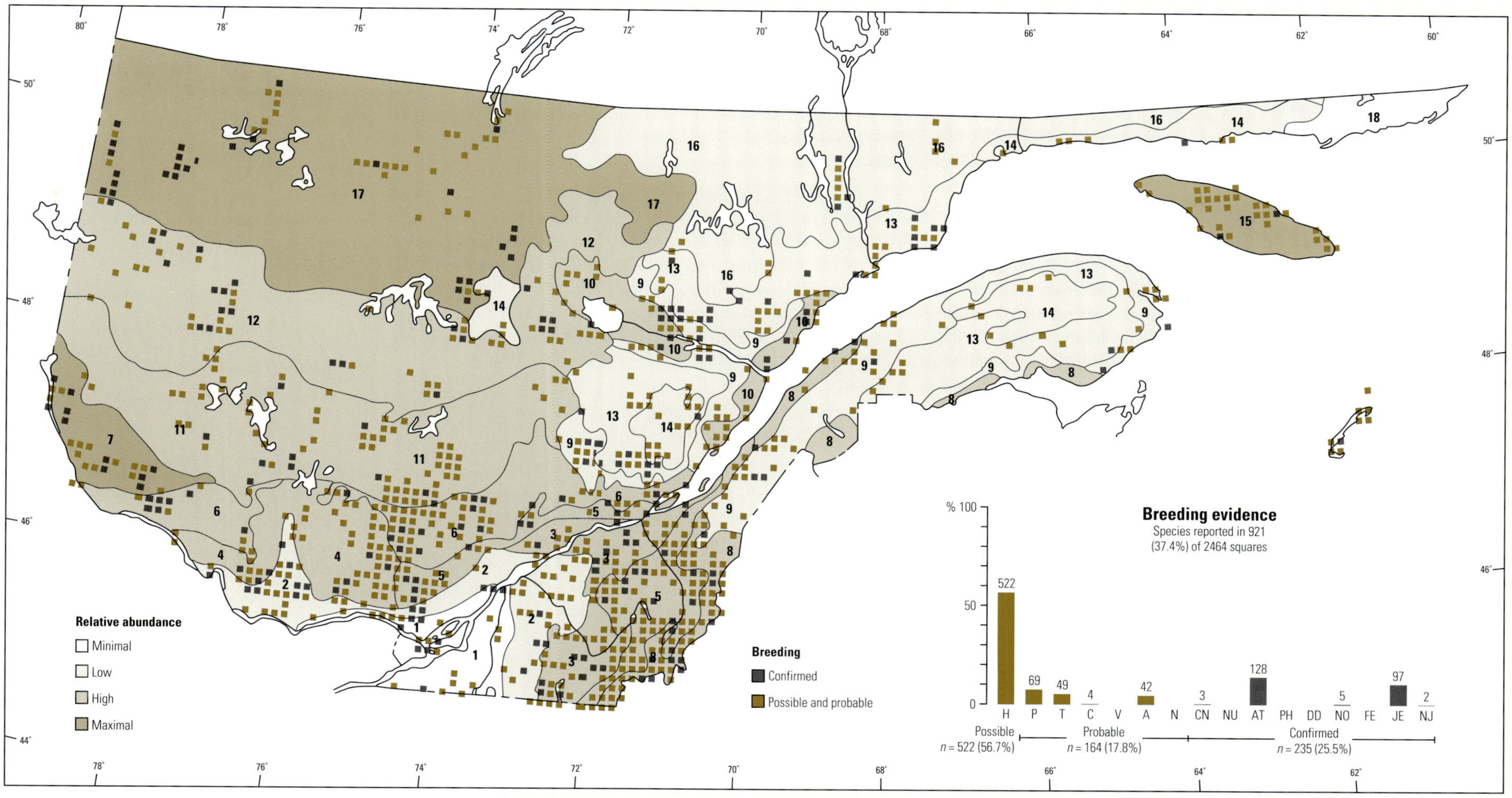

Species profile

Clutch size	usually 8–9 eggs (5–11)[1,2,3]
Laying interval	1 egg per day[3]
Start of incubation	?
Incubation period	14–15 days[1,2,3]
Maturity at hatching	altricial
Care of young	by female and male[1,2,3]
Nestling period	usually 16–19 days (14–19)[1,3]
Age at first flight	usually 16–19 days (14–19)[3]
Dependence of young	until 32–36 days old[3]
Broods per year	usually 2[1,3]
Breeding age	?
Mating system	monogamy[1]
Duration of pair bond	?
Total length	9–10 cm[4]
Wing span	16.5–17.8 cm[2]
Weight	M 6.3 g (4.9–7.7) / F 6.1 g (4.5–7.8)[5]
Longevity record	5 years 4 months[6]

References:
1. Ehrlich et al. 1988
2. Terres 1980
3. Galati and Galati 1985
4. Godfrey 1986
5. Dunning 1984
6. Klimkiewicz et al. 1983

Chronology $n = 20$

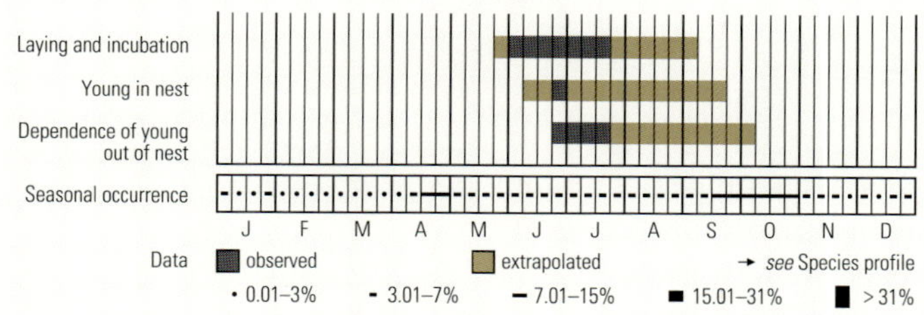

Frequency of occurrence by ecological region

Ecological regions** (forest regions)	Adequately covered squares	Squares in which reported	% of squares in which reported	Relative abundance (%)*
1. Sugar maple-hickory and sugar maple-basswood	97	21	21.6 ± 4.2	22.9
2. Sugar maple-basswood	153	68	44.4 ± 4.1	46.9
3. Sugar maple-basswood and sugar maple-yellow birch	101	70	69.3 ± 5.5	73.2
4. Sugar maple-yellow birch-basswood	56	45	80.4 ± 9.3	84.8
5. Sugar maple-yellow birch-beech and sugar maple-yellow birch-basswood	113	83	73.5 ± 4.6	77.5
6. Sugar maple-yellow birch-beech	71	50	70.4 ± 9.8	74.3
7. Sugar maple-yellow birch-balsam fir	19	18	94.7 ± 12.0	100.0
8. Sugar maple-yellow birch and balsam fir-yellow birch	47	38	80.9 ± 10.3	85.3
9. Balsam fir-yellow birch	87	44	50.6 ± 9.7	53.4
10. Balsam fir-white birch or balsam fir-red maple	46	27	58.7 ± 12.3	62.0
11. Yellow birch-balsam fir	52	42	80.8 ± 11.2	85.3
12. Balsam fir-white birch (west)	59	40	67.8 ± 12.3	71.6
13. Balsam fir-white birch (east)	76	34	44.7 ± 10.9	47.2
14. Balsam fir-black spruce	28	13	46.4 ± 9.3	49.0
15. White spruce-balsam fir and balsam fir-white spruce	22	19	86.4 ± 15.5	91.2
16. Black spruce-balsam fir-moss	18	10	55.6 ± 26.0	58.6
17. Black spruce-moss	32	28	87.5 ± 13.0	92.4
18.			*not analysed*	

* minimal: 0–29.9%; low: 30–59.9%; high: 60–89.9%; maximal: 90–100%

** Deciduous forest: 1–7; mixed forest: 8–11; coniferous forest: 12–17

Ruby-crowned Kinglet

Roitelet à couronne rubis

Regulus calendula

Christiane Girard

THIS SMALL, GREENISH-GREY BIRD ENLIVENS OUR CONIFEROUS FORESTS with its loud and oft-repeated song.

A nervous bird, the Ruby-crowned Kinglet never stays in one place for long and flutters its wings nervously even when perched. It is often seen flitting about in the branches to catch insects. Although it is a common migrant breeder in southern Québec (David 1980a), little is known about its breeding behaviour. It winters from the southern United States to Guatemala (Lepthien and Bock 1976), but some individuals overwinter in southern Québec.

Habitat and behaviour

The Ruby-crowned Kinglet is something of a habitat generalist, nesting in mixed woodlands, dense coniferous forest, and stands of scattered spruce on the edge of the tundra (Lepthien and Bock 1976; Derleth et al. 1989), and sometimes in hardwood forests (Holmes and Sturges 1975). In Québec, it has been observed in jack pine plantations (McLeod 1974) and in stands of white spruce (DesGranges 1980b) over 50 years old.

The deep, cup-shaped nest, almost spherical in shape, is built of mosses, lichens, plant fibres and down, grass, and bark strips, all held together with long hairs and spider webs. The nest is generally suspended from a branch 1.3–30 m up in a conifer (Terres 1980; Peck and James 1987). Apparently, the female of this monogamous species builds the nest and incubates the 7–9 eggs alone for about 13 or 14 days, although the male helps rear the young, which leave the nest when 14–16 days old.

The Ruby-crowned Kinglet is a rare cowbird host (Terres 1980). In Québec, Terrill (1961) found only 2 of 23 nests examined to have been parasitized. In Ontario, in an examination of 15 nests, Peck and James (1987) did not report a single case of cowbird parasitism. However, this phenomenon may occur more often than is thought, since several other cases have been reported (Friedmann 1963; Boxall 1981), though nowhere was this kinglet a common or preferred host.

Various studies conducted in summer in Ontario (Keast and Saunders 1991) and Arizona (Franzreb 1984), and in winter in Colorado (Laurenzi et al. 1982), indicate that the Ruby-crowned Kinglet forages as often in deciduous as in coniferous trees or shrubs, commonly hopping from branch to branch or fluttering about gleaning insects. These same studies also show that it may, on occasion, leave the canopy to forage on tree trunks or on the forest floor or to hawk insects. Laurenzi et al. (1982) noted that 95% of the homopterans, beetles, and flies which make up the majority of its prey were no more than 4.5 mm in length.

Although Ruby-crowned Kinglets begin arriving on their Québec breeding grounds by late April and do not leave until near the end of October, there is no evidence that they raise more than one brood a year.

Breeding was confirmed in 277 of the 2464 Atlas squares surveyed. Only 5% (14/277) of breeding confirmations were based on records of nests found, an indication of how difficult nests are to locate, even though the male becomes agitated and scolds when the nest is approached. The nest is deep enough that the female is completely hidden while incubating. In contrast, it must be easy to detect parents carrying food and recently fledged young, since these two breeding codes accounted for 90% (251/277) of breeding confirmations.

Distribution

The Ruby-crowned Kinglet breeds in northern coniferous forests from Alaska to Newfoundland. The southern limit of its range extends from New Mexico and southern California in the west to Michigan and Maine in the east (AOU 1983). This bird, with its resounding and easily recognized song, was reported in 53% (1297/2464) of Atlas squares. Breeding was confirmed throughout southern Québec reflecting the known breeding range that extends as far north as the tree line, in every part of the province except the tundra of the Ungava Peninsula (Harper 1958; Todd 1963; David 1980a; Godfrey 1986). The bird is very common throughout the Atlas area, with the exception of the St. Lawrence Plain and Ottawa Valley, particularly in Montérégie and around Montréal, where farmland abounds and mixed and coniferous woodland is scarce.

History and trends

According to Todd (1963), J. J. Audubon was the first to observe this species on its Québec breeding grounds, at Natashquan in 1833. In the early part of this century, it was already known that the Ruby-crowned Kinglet's range included the boreal forest (Macoun and Macoun 1915; Taverner 1922; Todd 1963), but it was thought to be only a transient in southern Québec (Wintle 1896; Taverner 1922). It must have been a regular breeder in the Montréal area and Montérégie during this period (Ouellet 1974a). There have been no obvious signs of change in its distribution since the turn of the century.

BBS data indicate that populations have remained stable across the continent since the late 1960s, despite a decline in the east (Robbins et al. 1986). In Québec, ÉPOQ summer data (Cyr and Larivée 1993a) and BBS data show no significant trends.

Although little historical information exists on abundance in southern Québec, clearing of woodlands for farming and urban development entailed loss of habitat. However, replanting of abandoned farmland with conifers could create favourable habitat, as has been the case in the Beauce region, where the original forest cover was deciduous.

In the boreal forest, logging is the greatest potential threat to the Ruby-crowned Kinglet. For the moment, however, it does not seem to have had a serious effect on populations of this generalist species, which actually appears to tolerate such logging practices as stand thinning and selective cutting (Beedy 1981; Mannan and Meslow 1984).

SIMON BARRETTE AND MARCEL DARVEAU

Notes on territory

Little is known about territoriality in this species. Tremblay and Belles-Isles (1991) reported densities of up to 9.5 pairs/10 ha in the area just north of Nemiscau. Erskine (1976a, 1980a) reported densities of 2.3–9.9 pairs/10 ha in the La Tuque and Sainte-Anne-des-Monts regions, in coniferous stands with shade-intolerant hardwoods. In a subalpine forest in New Hampshire, Sabo (1980) estimated territory size at 0.6 ha, whereas Kendeigh (1947) set it at 1.01 ha for the boreal forest of Ontario.

The Ruby-crowned Kinglet nests in mature spruce stands, building its nest on a branch 1–12 m high. Decelles Reservoir, Témiscamingue. *May 1991, Réginald Ouellet.*

Ruby-crowned Kinglet

Breeding distribution

1984–1989

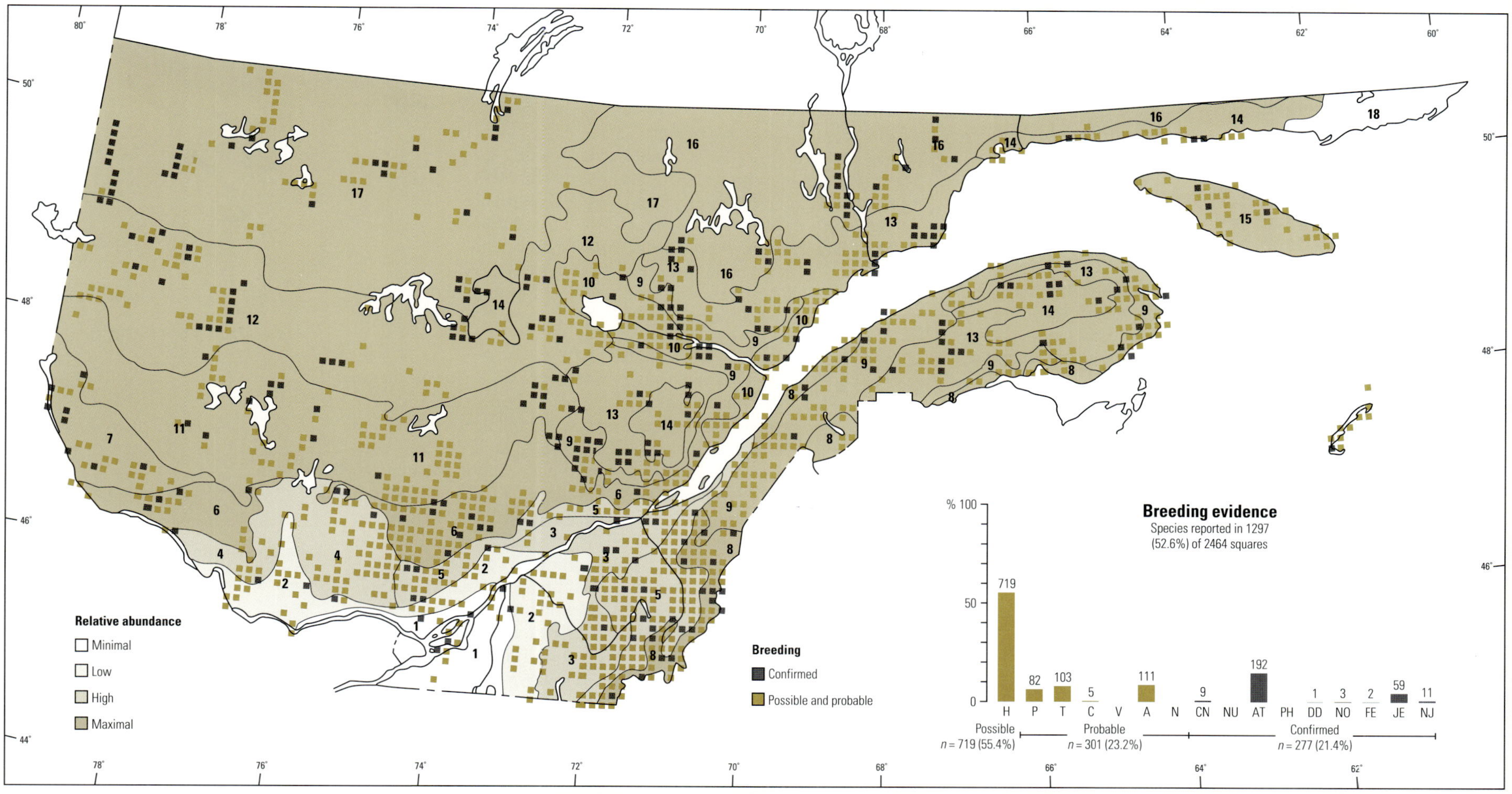

Species profile

Clutch size	usually 7–9 eggs (5–11)[1,2,3,4]
Laying interval	≈ 1 egg per day[10]
Start of incubation	before last egg possible[10]
Incubation period	usually 13–14 days (13–16)[10,11]
Maturity at hatching	altricial
Care of young	by female and male[2,4]
Nestling period	14–16 days[9]
Age at first flight	14–16 days or more[8]
Dependence of young	until at least 10 days after leaving nest[10]
Broods per year	1[3]
Breeding age	1 year[3]
Mating system	monogamy[2]
Duration of pair bond	1 season[10]
Total length	9.5–11.3 cm[5]
Wing span	17.1–19.0 cm[1]
Weight	M 6.9 g (5.0–9.7) / F 6.4 g (5.1–8.9)[6]
Longevity record	5 years 7 months[7]

References:

1. Terres 1980
2. Ehrlich et al. 1988
3. DeGraaf and Rudis 1987
4. Bent 1949
5. Godfrey 1986
6. Dunning 1984
7. Klimkiewicz et al. 1983
8. according to nestling period
9. R. Cayouette, pers. comm.
10. Ingold and Wallace 1994
11. Peck and James 1987

World breeding range

Chronology *n* = 27

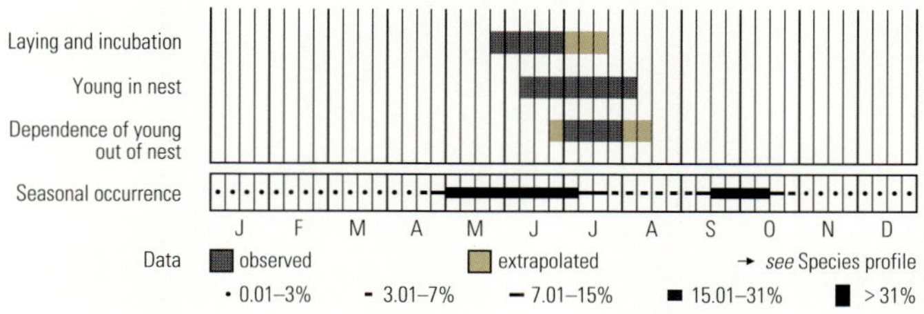

Frequency of occurrence by ecological region

Ecological regions** (forest regions)	Adequately covered squares	Squares in which reported	% of squares in which reported	Relative abundance (%)*
1. Sugar maple-hickory and sugar maple-basswood	97	14	14.4 ± 3.6	14.4
2. Sugar maple-basswood	153	62	40.5 ± 4.0	40.5
3. Sugar maple-basswood and sugar maple-yellow birch	101	64	63.4 ± 5.7	63.4
4. Sugar maple-yellow birch-basswood	56	44	78.6 ± 9.5	78.6
5. Sugar maple-yellow birch-beech and sugar maple-yellow birch-basswood	113	98	86.7 ± 3.6	86.7
6. Sugar maple-yellow birch-beech	71	64	90.1 ± 6.6	90.1
7. Sugar maple-yellow birch-balsam fir	19	19	100.0 ± 2.6	100.0
8. Sugar maple-yellow birch and balsam fir-yellow birch	47	44	93.6 ± 6.8	93.6
9. Balsam fir-yellow birch	87	87	100.0 ± 0.6	100.0
10. Balsam fir-white birch or balsam fir-red maple	46	43	93.5 ± 6.7	93.5
11. Yellow birch-balsam fir	52	51	98.1 ± 4.5	98.1
12. Balsam fir-white birch (west)	59	56	94.9 ± 6.2	94.9
13. Balsam fir-white birch (east)	76	74	97.4 ± 4.0	97.4
14. Balsam fir-black spruce	28	28	100.0 ± 1.8	100.0
15. White spruce-balsam fir and balsam fir-white spruce	22	21	95.5 ± 10.3	95.5
16. Black spruce-balsam fir-moss	18	18	100.0 ± 2.8	100.0
17. Black spruce-moss	32	32	100.0 ± 1.6	100.0
18.			*not analysed*	

* minimal: 0–29.9%; low: 30–59.9%; high: 60–89.9%; maximal: 90–100%

** Deciduous forest: 1–7; mixed forest: 8–11; coniferous forest: 12–17

Blue-gray Gnatcatcher

Gobemoucheron gris-bleu

Polioptila caerulea

THE BLUE-GRAY GNATCATCHER, A SMALL RELATIVE OF THE KINGLETS AND newcomer to southern Québec, is a rare migrant breeder here.

It is one of the few birds in Québec that use the flush-chase method of foraging: it walks along tree and shrub branches, shaking its outspread wings to flush its prey, chiefly small flying insects. It does not sing much, but regularly issues a plaintive *spee* that makes it easy to spot.

Habitat and behaviour

The Blue-gray Gnatcatcher breeds in a wide range of habitats in North America: mixed and hardwood forest, oak stands, wooded river banks, and chaparrals, for instance (Root 1967; Sedgwick 1987; Cooper et al. 1990). In Québec, the few confirmed breeders were found in a variety of wooded areas: a damp site beside a lake in a silver maple-black ash stand, in a red oak stand on a mesic site beside a lake, in a swamp white oak stand on a floodplain, and in a mature gray birch thicket in a field reverting to forest.

The nest of this territorial and monogamous species is a deep cup, narrowing slightly at the rim, covered with lichen and spider webs. It is built by both adults in the fork of a tree (Harrison 1978). In Ontario, nests were on the horizontal limbs of trees—mostly deciduous—and more rarely on vines, at heights varying between 0.6 and 26 m, but usually between 6.7 and 12 m (Peck and James 1987). The nest may also be built against the trunk. In California, Root (1967) reported that the 64 nests he had found were all located in the bottom half of the tree, no matter how tall the tree.

The Blue-gray Gnatcatcher, a newcomer to Québec, nests in this hardwood grove in Brossard, near Montréal. *August 1991, Gaétan Duquette.*

The male and female share incubation duties. Terres (1980) reported that Blue-gray Gnatcatcher nests are often parasitized by Brown-headed Cowbirds. In California, double-brooding is the norm (Root 1967), but this does not seem to be the case in the northeast (Clark 1985a; Sutherland and Gartshore 1987).

The Blue-gray Gnatcatcher eats nothing but arthropods, which it catches using mainly the flush-chase technique; more rarely, it can also be seen diving for prey on a substrate or hawking (Root 1967). Its diet consists largely of small prey (mean length of 5 mm, according to Root 1967), such as small beetles and homopterans, but may include larger insects, such as grasshoppers or caterpillars (Root 1967; Cooper et al. 1990).

Distribution

The Blue-gray Gnatcatcher breeding range extends from Mexico up to the Great Lakes; it winters from the southern U.S. states down through Central America, as well as in the western Greater Antilles and the Bahamas (AOU 1983). Breeding was confirmed for the Atlas at Hull in the Ottawa Valley, and in Montérégie, where breeding birds were reported at Brossard, Kahnawake, Mount Saint-Bruno, and at Pointe du Gouvernement (Henryville). The other Atlas records were of possible breeding only and came from Lake Bernard in the Ottawa Valley, Saint-Colomban near Saint-Jérôme, Mount Saint-Hilaire, Baie-du-Febvre, Ditchfield near Lac-Mégantic, and Cap Tourmente. All the squares in which the gnatcatcher was atlassed—except for those at Cap Tourmente and Baie-du-Febvre—lie south of the 46th parallel, and less than 100 km from Ontario or the United States, where the species is much more widespread.

One post-Atlas record was reported in 1992 of a nest at Saint-Quentin Island on the Saint-Maurice River, north of Trois-Rivières (Aubry and Yank 1992a). Breeding has not been confirmed anywhere else, although the species was observed in the spring of 1991 at Cap Tourmente, Sainte-Hedwidge near Roberval, and Beauport (Yank et al. 1991b), as well as at Mount Saint-Grégoire (P. Bannon, pers. comm.).

History and trends

In the last century, the Blue-gray Gnatcatcher apparently did not breed in southern Québec, nor in most of northeastern North America. But in the last 50 years or so, its range has extended considerably northward and now covers a good part of the New England states, southern Ontario, and the most southwesterly parts of Québec (Sutherland and Gartshore 1987). The gnatcatcher was confirmed as a breeder in Maine in 1979 (Adamus 1987), and the first Vermont record goes back only to 1954 (Clark 1985a). It is more and more often found in New York State (Meade 1988a) and in Ontario (Sutherland and Gartshore 1987), where it seems to have appeared toward the middle of the 19th century and is generally considered an uncommon breeder.

Wintle (1896) was apparently the first author to mention the gnatcatcher in Québec. In his book *The birds of Montreal*, he calls it an accidental visitant on the basis of the presumed capture of a specimen on the Island of Montréal at the end of the last century. There is practically no information on the status of the species at the turn of the century and over the next few decades. A few birds were sighted between 1954 and 1968 (Ouellet 1974a), but the first breeding record, of two adults feeding a nestling on Mount Royal, dates only from 8 July 1970 (Ouellet 1974a). The species also nested on Mount Saint-Bruno in 1978 (David and Gosselin 1978).

The Québec range of the Blue-gray Gnatcatcher can be expected to continue to expand slowly northward over the next few decades. The species could also become less rare within its current range, but it will probably remain fairly rare at the northeastern limit of its range.

MARCEL DARVEAU

Christiane Girard

Notes on territory

On arrival on the breeding grounds, the male defends a 2-ha territory, sometimes with the help of the female (Clark 1985a). The size of the territory may vary in the weeks following initial establishment depending on changes in food abundance; then, as the breeding season progresses, the Blue-gray Gnatcatcher shrinks its territory, but makes sure to include dense masses of evergreen vegetation, where it can seek shelter and settle its brood. During periods when the adults make frequent trips to the nest with either material or food for the young, the male patrols the perimeter of the territory less often, which may then be occupied by neighbouring pairs that have abandoned their own nests. In Pennsylvania, Viverette (1993) counted 0.6 gnatcatcher/10 ha in an oak-maple slope forest, Grove (1993) reported 1.0 territory/10 ha in a hardwood forest with scattered pines, and Gross (1993) recorded 1.8 territories/10 ha in a mixed oak-pine forest.

Blue-gray Gnatcatcher

Breeding distribution
1984–1989

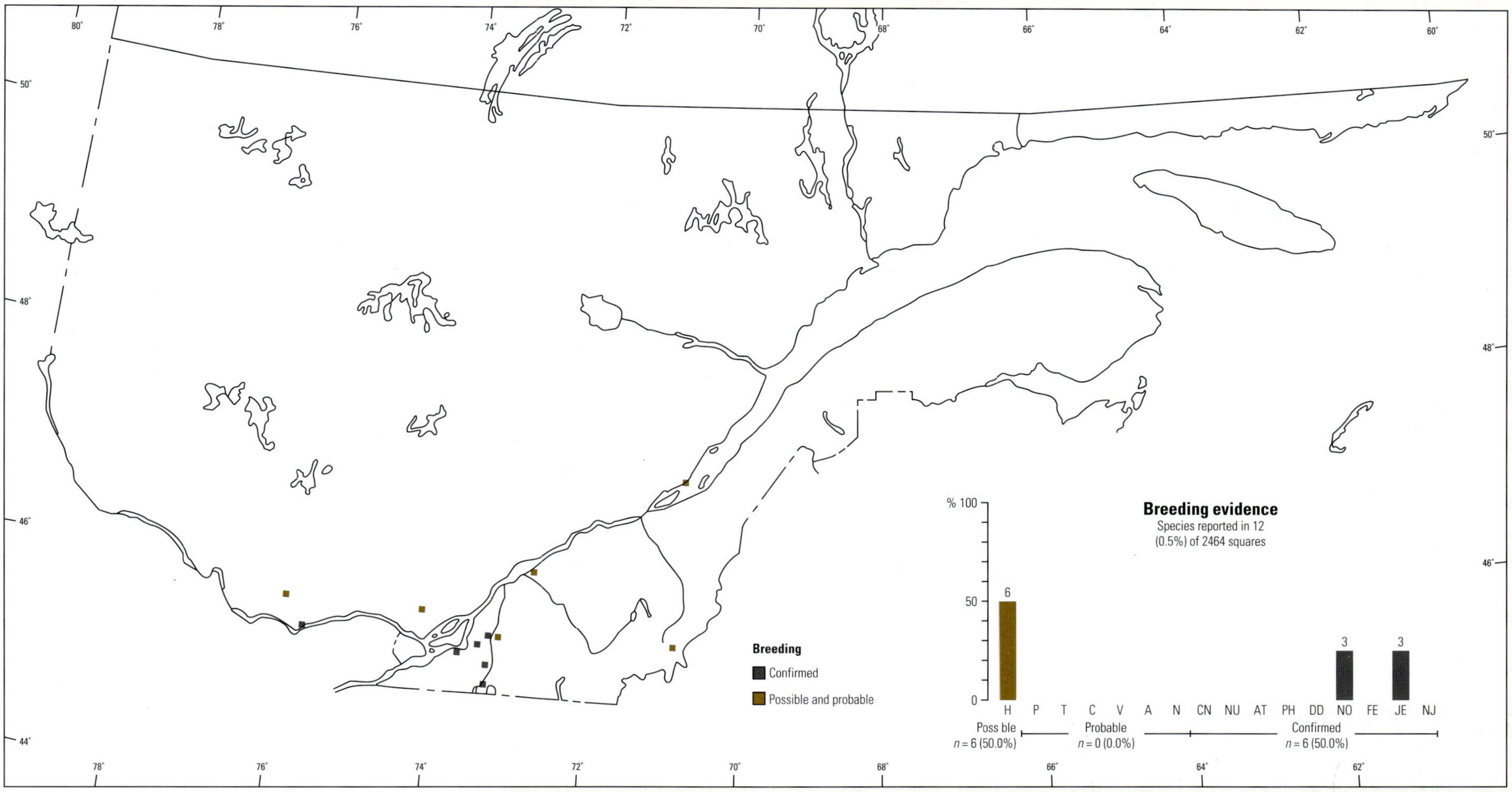

Species profile

Clutch size	usually 4–5 eggs (3–6)[1,2]
Laying interval	1 egg per day[9]
Start of incubation	with last egg[9]
Incubation period	usually 13 days (11–15)[1,2,3,9]
Maturity at hatching	altricial
Care of young	by female and male[2,3]
Nestling period	10–15 days[2,3,4,9]
Age at first flight	≈ 14–19 days[9]
Dependence of young	until ≈ 19–21 days after leaving nest[9]
Broods per year	usually 1 (1–2)[9]
Breeding age	1 year[8,9]
Mating system	monogamy[4,9]
Duration of pair bond	probably 1 season[9]
Total length	10.0–12.5 cm[5]
Wing span	14.6–16.5 cm[2]
Weight	6.0 g (4.8–8.9)[6]
Longevity record	4 years 2 months[7]

References:
1. Peck and James 1987
2. Terres 1980
3. Harrison 1978
4. Ehrlich et al. 1988
5. Godfrey 1986
6. Dunning 1984
7. Klimkiewicz et al. 1983
8. DeGraaf and Rudis 1987
9. Ellison 1992

Chronology $n = 9$

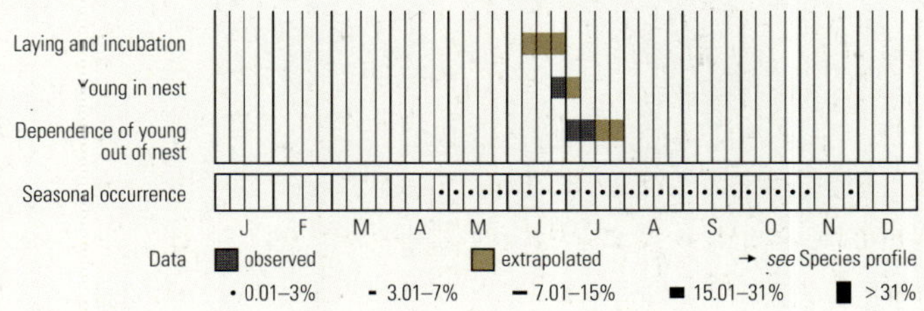

World breeding range

A Blue-gray Gnatcatcher adding the final touches to its nest. The species generally nests on a horizontal branch or in the fork of a tree. Point Pelee National Park, Ontario. *May 1986, Paul Perreault.*

Eastern Bluebird

Merle-bleu de l'Est

Sialia sialis

Christiane Girard

The Eastern Bluebird is the favourite songbird of many Québec bird lovers. This is not surprising since this colourful bird not only lives in places where it can be easily observed, but nests in birdhouses built specifically for it. Its name derives from the beautiful dark blue plumage of the upperparts of the male. The male's brick-red throat, breast, and flanks make it even more attractive. The female's coloration is similar, but much duller. The young do not show much blue, and their breast, speckled like that of young American Robins, is a reminder that they are close relatives.

Habitat and behaviour

Like the Tree Swallow, Great Crested Flycatcher, and other species, the Eastern Bluebird makes its nest in a cavity that it does not excavate itself. It may nest in a tree, fencepost, or telephone pole, and often in a woodpecker hole or a nest box (Bent 1949; Pinkowski 1976). It is found in a variety of habitats: pastures, uncultivated fields, orchards, gardens, burns, meadows, cutovers, and other open sites. Suitable nesting habitat must have nest holes as well as protective cover and foraging areas (Risley 1984). In summer, this bluebird feeds mainly on ground insects, such as grasshoppers, crickets, and beetles, which it spots from perches. Its habitat must therefore include perches such as dead trees, shrubs, posts, electric wires, tall herbs like mullein, or big rocks (Pinkowski 1977; Risley 1984) that dominate large expanses of low grass (Pinkowski 1977).

The Eastern Bluebird is one of the first migrants to return here in spring, and some birds arrive in southern Québec as early as mid-March or early April (Gosselin 1975a; David 1980a). The males arrive a few days before the females. They then stake out their territories and attempt to attract a female. Although the species is generally monogamous, studies have uncovered evidence of extra-pair copulation (Gowaty 1983; Gowaty and Karlin 1984; Karlin et al. 1990; Tucker 1990), as well as adoption of or assistance to another pair on the part of the male (Pinkowski 1975, 1978; Plissner and Gowaty 1988), which goes to show the complexity of pairing in this species.

The activities leading up to nest building itself, such as courtship displays and copulation, take place near the cavity where the eggs will be laid. The female alone is responsible for building the nest and incubating. The nestlings are fed by both parents (Hartshorne 1962; Zeleny 1976), but the female is able to raise them alone should the male disappear (Gowaty 1983). In Québec, as in Ontario, bluebirds may be double-brooded (Risley 1984), and the young of the first brood may help feed the second set of nestlings (Laskey 1939; Pinkowski 1975).

The Eastern Bluebird is a fairly easy bird to spot. It can often be seen quite plainly perching on an electric wire or fencepost and diving to the ground to capture prey, then returning to its observation post a couple of seconds later. Before even seeing it, you can often hear its sweet, musical song floating down from the perch. As Atlas findings show, it is easy to confirm breeding of the species (72.8%). Close to 70% (211/308) of confirmations involved discovery of nests, which can be explained by the fact that the species often uses nest boxes set up for it.

Notes on territory

Home ranges are highly variable in size. A study conducted in Michigan showed, for example, that bluebirds foraged in areas ranging from 4.5 to close to 40 ha (Pinkowski 1977); their commonest foraging tactic was to spot prey on the ground from a perch and dart down to seize it.

The number of potential perches for spotting prey therefore seems to influence the size of the home range. Despite considerable size differences, all the home ranges observed included a potential foraging area (containing perches) that varied fairly consistently between 3.9 and 8.4 ha (Pinkowski 1977).

A study conducted in New York State recorded densities of 1.25 pairs/10 ha in an abandoned pasture (Pitzrick 1993).

The many bluebird houses set up in open areas have helped the species rebuild its numbers.
La Ferme, near Amos, Abitibi. *July 1991, André-Guy Bernier.*

Eastern Bluebird

Breeding distribution 1984–1989

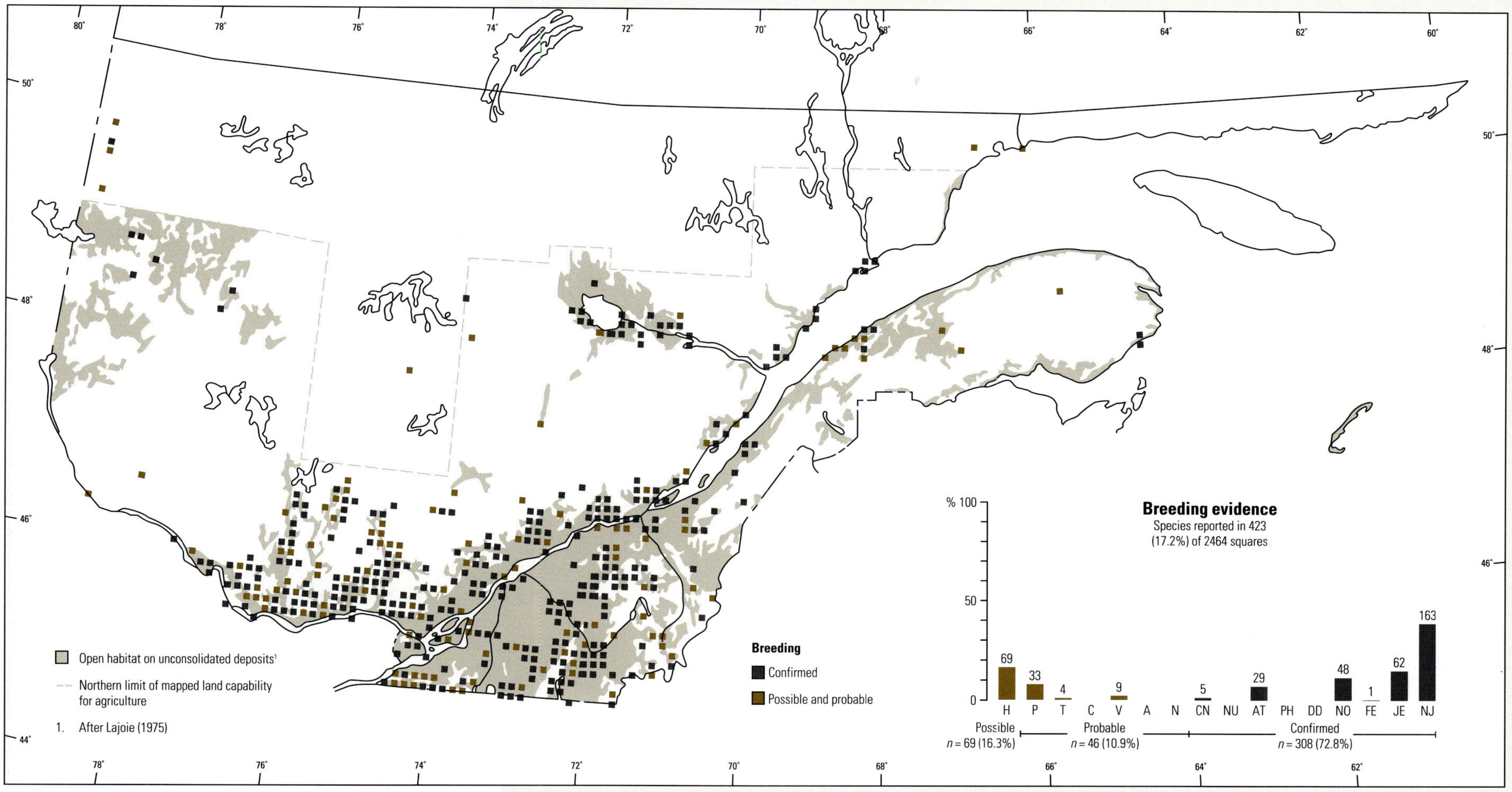

Species profile

Clutch size	usually 4–5 eggs (2–7)[1,2,3,6,7]
Laying interval	1 egg per day[1,3]
Start of incubation	with last egg[1,3]
Incubation period	usually 13–15 days (12–16)[1,3]
Maturity at hatching	altricial
Care of young	by female and male[1,2,7,8]
Nestling period	usually 17–18 days (16–19)[1,2,3,7]
Age at first flight	usually 17–18 days (16–19)[1,4]
Dependence of young	until 18–24 days after first flight[2,3]
Broods per year	usually 2 (2–3)[2,3,6,7,8]
Breeding age	1 year[10]
Mating system	monogamy, rarely polygamy[8,11,12]
Duration of pair bond	?
Total length	16.5–19.0 cm[5]
Wing span	29.2–33.7 cm[5]
Weight	M ? g (27.6–31.6) / F ? g (27.3–34.0)[5]
Longevity record	8 years[9]

References:
1. Hartshorne 1962
2. Fisley 1984
3. Zeleny 1976
4. Gowaty 1983
5. Terres 1980
6. Harrison 1975
7. Harrison 1978
8. Ehrlich et al. 1988
9. Klimkiewicz et al. 1983
10. DeGraaf and Rudis 1987
11. Gowaty and Karlin 1984
12. Tucker 1990

Chronology $n = 967$

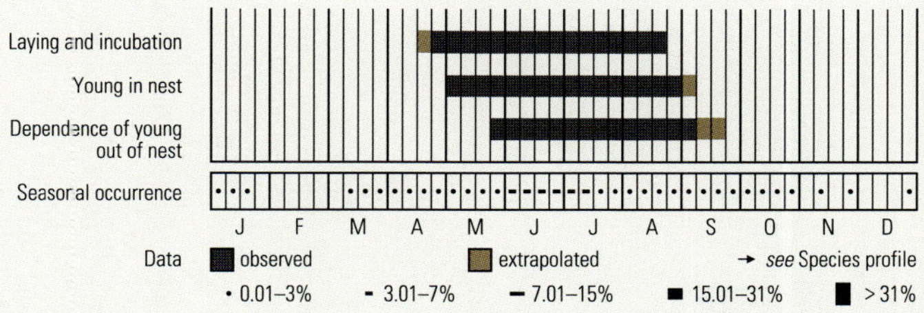

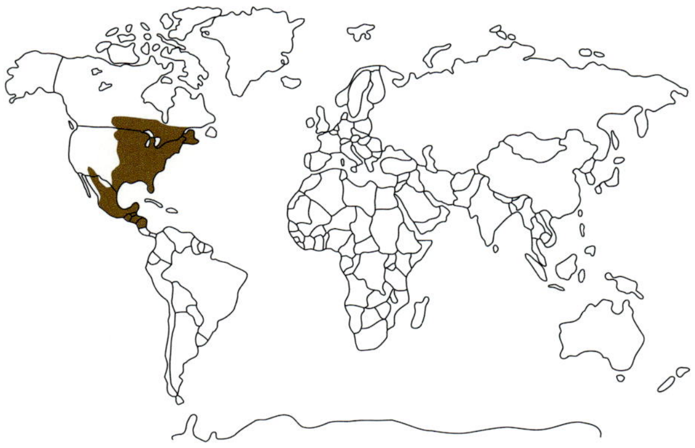
World breeding range

Distribution

In North America, the Eastern Bluebird breeds in southern Canada, from Saskatchewan to Nova Scotia; farther south, it is found from Arizona all the way to the Atlantic coast and the Gulf of Mexico, as well as in parts of Central America (Godfrey 1986).

In Québec, it breeds in all regions south of the 50th parallel, except on Anticosti Island and the Magdalen Islands, where it is a rare visitor (David 1980a; Godfrey 1986; Fradette 1992). It is more widespread and more abundant in the southernmost part of the province, especially along the edge of the Canadian Shield and the Appalachians, perhaps because of the marginal farmland and many open areas found there, but certainly also because of the large number of birdhouses that have been set up. During a survey conducted in 1991, for example, 423 clutches had been produced in the 3038 boxes surveyed, and close to half of them were in the following five counties, in decreasing order: Portneuf, Berthier, Saint-Maurice, Richelieu, and Shefford (Asselin 1992).

The most northerly confirmations of breeding were from Abitibi (Villebois, Languedoc, Villemontel), the area around the Gouin Reservoir (Wabano River), Saguenay–Lac-Saint-Jean, and the North Shore (Forestville, Baie-Comeau). Breeding birds were also observed on the far eastern tip of the Gaspé Peninsula (Pabos). In uninhabited regions where there are neither farms nor nest boxes, the Eastern Bluebird inhabits burns, cutovers, the edges of beaver dams, open swamps, and so on (Conner 1974b; Kane 1982; Gosselin et al. 1988b). It is also probably much more widespread than the map shows, particularly in squares not surveyed in the northern half of the Atlas area, since these inaccessible regions include many cutovers, burns, and other open areas favourable to the species.

Recently, the Eastern Bluebird has been seen at James Bay during the breeding season. In June 1991, birds were observed in at least three different locations in the Rupert Bay lowlands (Gosselin et al. 1991b; Cons. G. G. – GREBE 1992o). These recent reports are the most northerly for Québec. Although breeding has not yet been confirmed, it is probable that the species nests in the area: the birds seen were on burns and a peatland—open habitats suited to the bluebird. There are various sight records for the Ontario side of James Bay (Todd 1963). Moreover, it was reported in a burn in Schefferville in northern Québec, in June 1953 (Harper 1958).

Stumps with cavities in cutovers make good nest sites. Montmorency forest, Laurentide Wildlife Reserve, north of Québec City. *June 1991, Réginald Ouellet.*

Abandoned orchards provide good nest sites for the bluebird. Mount Saint-Hilaire. *August 1991, Gaétan Duquette.*

An Eastern Bluebird has made its nest in this old apple tree trunk. Frelighsburg, south of Granby. *June 1988, Douglas Graham.*

Close-up of a cavity in which a pair of bluebirds has nested. Frelighsburg, south of Granby. *June 1988, Jean-Luc DesGranges.*

History and trends

The Eastern Bluebird has undoubtedly benefited from European settlement on the St. Lawrence Plain. The clearing and expansion of pastureland and farmland that took place throughout the last century must have been beneficial to it, since these practices increased the number of potential nesting sites. As early as the turn of the century, the Eastern Bluebird was considered common in the Montréal area (Wintle 1896) and fairly common near Québec City (Dionne 1906).

Since then, bluebird populations have fluctuated significantly for a number of reasons. First, the species is highly sensitive to cold spells, ice, or snow lasting several days, either during harsh winters on the wintering grounds (especially in the United States), or during late storms on the nesting grounds (Musselman 1941; Bent 1949; Root 1988). Population declines of over 50% have regularly been reported as a result of such conditions. The population has declined this way several times since the end of the last century, most recently after the hard winters of 1957–58, 1976–77, and 1977–78 (Risley 1984). The decrease in the number of cavities suitable for nesting has also been responsible for the decline of the Eastern Bluebird in the 20th century.

The cutting of branches and felling of dead trees and changes in farming practices, such as replacing cedar fences with metal ones, have reduced the number of nest sites and increased competition among the various cavity-nesting species (Risley 1984). The Tree Swallow and House Sparrow, and to a lesser extent the House Wren and European Starling, are the species that most often compete for cavities with the Eastern Bluebird and regularly evict it from its nest (Bent 1949; Kibler 1969; Zeleny 1976; Gowaty 1984; Risley 1984). Nonetheless, no matter what some people say, in Québec at least, these birds are not a serious threat to the Eastern Bluebird, since a good many nest boxes set up for it often go unused each year. In 1991, for example, 1102 of the 3038 boxes (33%) censused by the Linnean Society of Québec (and there are probably many more) remained unoccupied. Another 1369 (45%) were occupied by the Tree Swallow, 423 (14%) by the Eastern Bluebird, 41 (1.3%) by the House Wren (Asselin 1992), and only 27 (0.9%) by the House Sparrow.

In Québec, as elsewhere in northeastern North America, the Eastern Bluebird population was probably at its lowest in the late 1970s (Risley 1984; Robbins et al. 1986). David (1980a) considered it an uncommon migrant breeder south of the 47th parallel, and rare north of that latitude. In 1984, it was designated a vulnerable species by the Committee on the Status of Endangered Wildlife in Canada (COSEWIC 1993).

Since then, the number of Eastern Bluebirds has increased a great deal, to the extent that the species is increasingly reported in the northern part of its range. The annual number of sightings by Québec birdwatchers is also on the rise (Robert 1989b; Asselin 1991, 1992). In short, this bluebird is now encountered in all the same areas where it occurred earlier (Aubry et al. 1989) and is a fairly common summer resident, at least in southern Québec. The series of relatively mild winters since 1979 on the wintering grounds, and the ability of the species to recover gradually after severe cold weather has certainly contributed to the rise in numbers.

The enthusiasm of birdwatchers has also borne fruit; the setting up of bluebird trails, that is, installing nest boxes along lines in suitable habitat, in Québec and elsewhere, has given bluebirds thousands of new breeding sites (Zeleny 1976). In Québec, several hundred volunteers maintain and monitor these trails in about a hundred locations (Asselin 1991, 1992). It is therefore to be hoped that unless we have very harsh winters, the Eastern Bluebird will continue to charm an ever-increasing number of bird lovers.

DANIEL BANVILLE AND MICHEL ROBERT

Veery

Grive fauve
Catharus fuscescens

Christiane Girard

FROM A STRICTLY MUSICAL POINT OF VIEW, THE VEERY IS MAYBE NOT THE most talented thrush, but its song is gay, communicative, and distinctive—a series of four or five downward-spiraling *whree-u*.

The Veery is a common breeder in deciduous woodland in the south of the province. It is a neotropical migrant that winters in South America east of the Andes, from northern Colombia and Venezuela to central Brazil (AOU 1983).

Habitat and behaviour

Like the Wood Thrush, the Veery is a characteristic species of deciduous woodland, but prefers a fairly open canopy and dense understorey (Noon 1981; Clark et al. 1983; Thompson and Capen 1988). Thus, it is found most often in immature stands and second-growth forest (Bertin 1977). Several investigators have noted its preference for damp habitat (Dilger 1956a; Morse 1971a; Bertin 1977). It is also found in mixed forest.

Forbush (1929) reported that it takes 6–10 days for the female to build the nest, which is usually placed on the ground or at a height of 1 m or less, though sometimes up to 5 m high (Peck and James 1987). In Ontario, nests are found somewhat more regularly in deciduous trees (maple, alder, birch) than conifers (fir, spruce). When placed on the ground, they are usually at the foot of a shrub, tree, or stump, or concealed in a grass tussock or near fallen branches or felled trees. Incubation, which lasts 10–14 days, is carried out exclusively by the female, whereas both parents care for the young (Peck and James 1987). From the first day, the parents bring small caterpillars and other small insects to the nestlings (Forbush 1929), but as the young grow, larger food items are brought. After the fifth day, the parents bring dragonflies, slugs, and, a few days later, butterflies. The Veery may raise 2 broods a year, at least in the United States (Forbush 1929; DeGraaf and Rudis 1987; Moskoff 1995b). The Veery is a fairly frequent cowbird host (Terres 1980). Peck and James (1987) reported a parasitism rate of 19% (70/368) in Veery nests in Ontario, while Terrill (1961) found cowbird eggs in 13.3% (17/128) of nests studied in southern Québec.

The Veery forages mostly on the ground and in shrubs less than 2 m tall (Holmes and Robinson 1988). It perches on rocks, logs, or low branches, often near streams, where it forages along the banks for insects. When prey is detected, it dives from its perch, taking it from the ground or foliage. It feeds mainly on beetles, flies, Lepidoptera larvae, ants, and wasps (Holmes and Robinson 1988), but it will also consume fruit and seeds (Tyler 1949; DeGraaf and Rudis 1987).

This thrush forages on the ground and in the underbrush of deciduous forest for insects, berries, and seeds. Sainte-Foy. *September 1991, Yves Aubry.*

Notes on territory

Males arrive on the breeding grounds before the females and defend a territory against conspecifics (Dilger 1956a,b). Territorial defence consists of a series of postures and behaviours that has been described by Dilger (1956b). Territory size in a sugar maple stand and a hemlock stand north of Montréal was found to be about 0.5 ha (Cyr 1974).

Density estimates have been made in various types of forest in Québec. Densities ranging between 2 and 7.5 pairs/10 ha were reported in fairly mature sugar maple stands associated with other deciduous species such as beech, yellow birch, or basswood. In hemlock, fir, and birch stands or mixed forest of variable age, density ranged between 4.9 and 10 pairs/10 ha. The highest density, 12.5 pairs/10 ha, was recorded in a young sugar maple stand associated with yellow birch (Erskine 1971a, 1972a, 1976a, 1980a, 1984; Cyr 1974; DesGranges 1980b). Densities of up to 11 pairs/10 ha, depending on the age of the stand, were reported in the deciduous forests of Nova Scotia (Morgan and Freedman 1986).

Veery

Breeding distribution
1984–1989

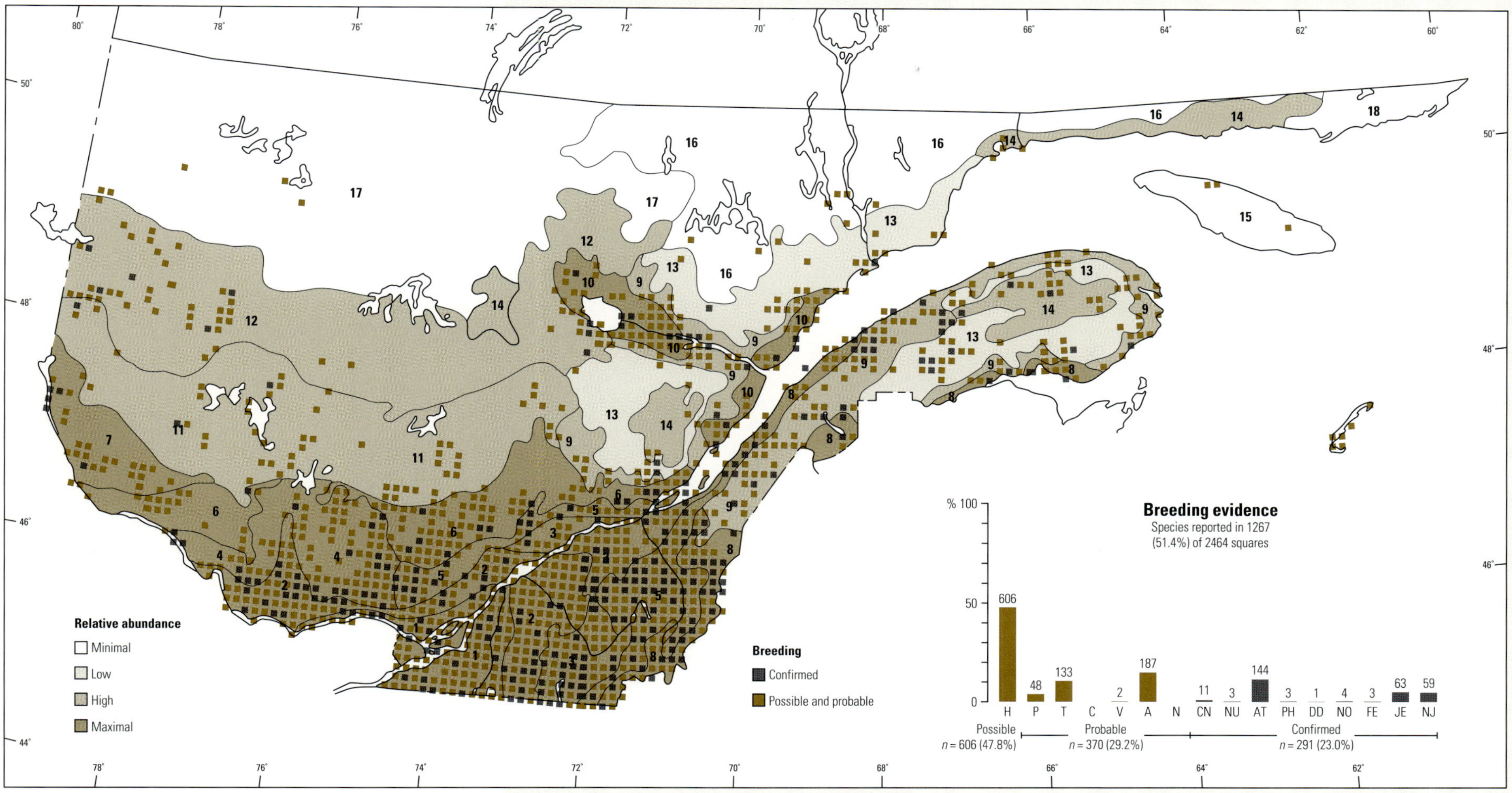

World breeding range

Species profile

Clutch size	usually 4 eggs (3–5)[1,2,3]
Laying interval	probably 1 egg per day[12]
Start of incubation	?
Incubation period	10–14 days[1,2,3,11]
Maturity at hatching	altricial
Care of young	by female and male[1,2,3]
Nestling period	10–12 days[4,10]
Age at first flight	10–12 days or more[9]
Dependence of young	until ≈ 2 weeks after leaving nest[12]
Broods per year	1, 2 possible[1,2,3,12]
Breeding age	1 year[3]
Mating system	monogamy[2]
Duration of pair bond	?
Total length	16.5–19.5 cm[5]
Wing span	28.0–31.5 cm[7]
Weight	31.2 g (26.2–41.7)[6]
Longevity record	10 years 1 month[8]

References:
1. Forbush 1929
2. Ehrlich et al. 1988
3. DeGraaf and Rudis 1987
4. Tyler 1949
5. Godfrey 1986
6. Dunning 1984
7. Cramp 1988
8. Dunning 1992
9. according to nestling period
10. Harrison 1978
11. Peck and James 1987
12. Moskoff 1995b

Chronology $n = 179$

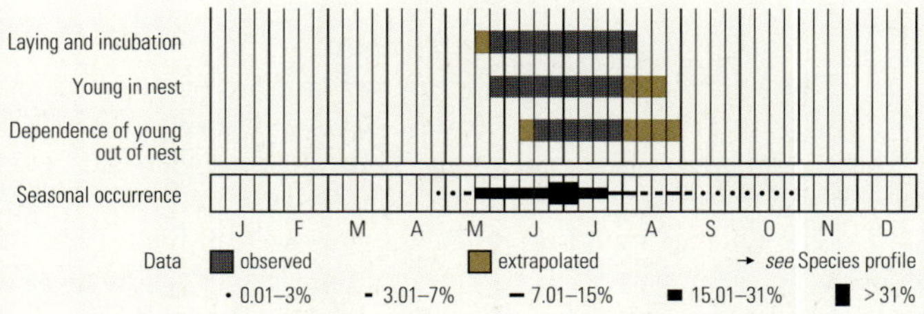

Frequency of occurrence by ecological region

Ecological regions** (forest regions)	Adequately covered squares	Squares in which reported	% of squares in which reported	Relative abundance (%)*
1. Sugar maple-hickory and sugar maple-basswood	97	95	97.9 ± 1.8	97.9
2. Sugar maple-basswood	153	153	100.0 ± 0.3	100.0
3. Sugar maple-basswood and sugar maple-yellow birch	101	101	100.0 ± 0.5	100.0
4. Sugar maple-yellow birch-basswood	56	56	100.0 ± 0.9	100.0
5. Sugar maple-yellow birch-beech and sugar maple-yellow birch-basswood	113	111	98.2 ± 1.7	98.2
6. Sugar maple-yellow birch-beech	71	67	94.4 ± 5.3	94.4
7. Sugar maple-yellow birch-balsam fir	19	19	100.0 ± 2.6	100.0
8. Sugar maple-yellow birch and balsam fir-yellow birch	47	47	100.0 ± 1.1	100.0
9. Balsam fir-yellow birch	87	78	89.7 ± 6.1	89.7
10. Balsam fir-white birch or balsam fir-red maple	46	45	97.8 ± 4.4	97.8
11. Yellow birch-balsam fir	52	41	78.8 ± 11.6	78.8
12. Balsam fir-white birch (west)	59	36	61.0 ± 12.8	61.0
13. Balsam fir-white birch (east)	76	45	59.2 ± 10.8	59.2
14. Balsam fir-black spruce	28	17	60.7 ± 18.9	60.7
15. White spruce-balsam fir and balsam fir-white spruce	22	1	4.5 ± 10.3	4.5
16. Black spruce-balsam fir-moss	18	5	27.8 ± 23.7	27.8
17. Black spruce-moss	32	3	9.4 ± 11.7	9.4
18.			*not analysed*	

* minimal: 0–29.9%; low: 30–59.9%; high: 60–89.9%; maximal: 90–100%

** Deciduous forest: 1–7; mixed forest: 8–11; coniferous forest: 12–17

The Veery favours damp habitat and is found in younger deciduous forest with a denser understorey than that used by the Wood and Hermit thrushes. Lake Leamy, Hull. *July 1991, Daniel St-Hilaire.*

The Veery is easy to detect because of its distinctive and readily identifiable song. Like most thrushes and unlike most passerines, it sings more often at dusk than in the morning (Dilger 1956a). Atlas results show Veery nests to be easier to find than those of other thrushes, even though they are well concealed. Sightings of young incapable of sustained flight accounted for a large number of confirmations, but the most frequent breeding evidence reported was of adults carrying food, which accounted for almost half of the confirmations.

Distribution

The Veery breeds in southern Canada, from British Columbia to Newfoundland. In the United States, it ranges from the Appalachians, the northeastern states, throughout the Great Lakes region and the western states to Arizona (AOU 1983).

The Québec breeding range encompasses all regions located south of the 49th parallel. It is common in the ecological regions that belong to the deciduous and mixed forest zones. It is also fairly common in Balsam fir forest regions, but frequency of occurrence drops in Spruce forest regions.

It may have expanded its breeding range, as breeding was confirmed somewhat north of the limits described by David (1980a) and Godfrey (1986), who were unaware, for example, that the species nested in Abitibi; Godfrey also doubted that it bred on the extreme eastern Gaspé Peninsula. The northward expansion of its range is probably due to deciduous regrowth over large tracts of coniferous forest that were logged out.

The species was atlassed on the north side of Anticosti Island in habitat with a substantial deciduous component, but the only evidence of possible breeding from the island is from Port-Menier where, on 24 July 1881, a pair was observed displaying agitated behaviour (Brewster 1884). Ouellet (1969) reported a few other observations from the beginning of the century and added that to his knowledge the bird had not been seen since. Bishop and Audubon found the species common in summer on the Magdalen Islands during the last century, but there is only a single record for the first half of this century, although the species has been observed regularly since the 1950s (Fradette 1992). As to the Magdalen Islands, according to Fradette (1992), both Rowdon (1969) and Godfrey (1986) list the Veery as a breeder without providing formal proof, and for this reason he still considers the species a probable breeder on the Magdalen Islands. It is likely only a matter of time before breeding is confirmed there.

A Veery with three 7-day-old young. When an adult bird alights on the edge of the nest, the young respond by opening their mouths wide to receive a beakful. Sainte-Foy. *July 1958, Gaston Le Page.*

Placed on the ground or low in shrubs or ferns, the Veery's nest is built of weed stems, grass stalks, and dry leaves, and is well camouflaged among the surrounding vegetation. Sainte-Foy. *June 1962, Gaston Le Page.*

History and trends

The Veery was common in the south of the province around the turn of the century (Wintle 1896; Dionne 1906; Macoun and Macoun 1915; Mousley 1916a; Lloyd 1924, 1944). Godfrey and Wilk (1948) considered it fairly common in mixed forest in certain areas around Lake Saint-Jean, and Tanguay (1964) listed it as a probable breeder in the area of Kamouraska, L'Islet, and Montmagny. It was rarely reported in the Gaspé (Demille 1926; Ball 1954).

Today, the Veery is still common in the south of the province. BBS (1969–89) and ÉPOQ data (1970–89) do not indicate any change in breeding populations in Québec (Cyr and Larivée 1993a). BBS data for North America, however, indicate an increase over all of the eastern part of the continent between 1965 and 1979 (Robbins et al. 1986).

The Veery may benefit from selective cutting that opens the canopy and allows the understorey to develop (Webb et al. 1977; Thompson and Capen 1988). The species does not appear to be particularly affected by clearcutting of small areas of 1–8 ha (Derleth et al. 1989), but after larger areas are logged out, it is virtually absent for about a 6-year period (Morgan and Freedman 1986).

The Veery winters in mature tropical forest and might therefore be affected by the current deforestation of tropical regions (Terborgh 1980; Askins et al. 1990; Diamond 1991). Diamond (1991) estimates that 22% of the species' winter habitat may be lost by the year 2000. BBS data for all of eastern North America show that numbers declined during the 1980s (Robbins et al. 1989b). The data show an increase over the 1966–78 period and a decrease between 1979 and 1987. Deforestation in Central and South America is cited as a possible reason for this recent decline, but fragmentation of large tracts of forest on the breeding grounds may be another limiting factor (Askins 1993). The Veery nonetheless remains a very common, familiar species in southern Québec.

Gilles Falardeau

Bicknell's Thrush

Grive de Bicknell

Catharus bicknelli

Christiane Girard

UNTIL VERY RECENTLY, BICKNELL'S THRUSH (*CATHARUS BICKNELLI*) (RIDGWAY 1882) was considered a subspecies of the Gray-cheeked Thrush (*Catharus minimus*). According to recent analyses of field and laboratory data, however, these two taxa are significantly different and, in fact, constitute two separate species (Ouellet 1993; AOU 1995).

Bicknell's Thrush is not only distinct genetically from the Gray-cheeked Thrush, but has different behavioural and external morphological characteristics. For example, its body mass and dimensions, particularly its wing length, are much smaller than those of the Gray-cheeked Thrush. It has a chestnut tail and brownish back, while the Gray-cheeked Thrush tends to be olive on both the back and tail. Its song, a reliable sign of its presence, is also diagnostic. While the two species' songs are somewhat similar, the final phrase is consistently high in Bicknell's Thrush, the frequency remaining constant or increasing slightly (from 150 to 200 kHz) rather than falling, as does the last part of the Gray-cheeked Thrush's song. Like most members of the genus *Catharus*, the Gray-cheeked Thrush and Bicknell's Thrush are very similar, and observers must carefully compare their characteristics and listen to their songs to identify them correctly.

Bicknell's Thrush is a migrant breeder in southern Québec, nesting very locally on forested mountaintops and in second-growth stands.

Habitat and behaviour

Bicknell's Thrush has traditionally been considered a habitat specialist, associated exclusively with stands of stunted conifers in mountainous and coastal regions of the eastern United States and Canada. However, recent data show that, at least in Québec and New Brunswick, it also inhabits second-growth stands regenerating after a forest fire or clearcutting. This appears to be a rather recent phenomenon. In southern Québec, the species is still found mainly in coniferous stands at elevations between 200 m above sea level and the tree line. These generally very dense and sometimes stunted stands are dominated mainly by white spruce, and less frequently by black spruce and balsam fir. Such stands are common on the mountaintops of the Gaspé Peninsula and Eastern Townships, although they can also occasionally be found at elevations as low as 100 m or so.

In contrast, the regenerating second-growth stands colonized by the species probably over the last 20 years are located consistently at elevations above 200 m. They consist mainly of dense clumps of conifers no more than 10 m high, usually dominated by balsam fir and intermixed with deciduous species typical of second-growth regeneration, including cherry, birch, and poplar.

The behaviour of Bicknell's Thrush has not been studied extensively in Québec, but data obtained in New England (Wallace 1939) are quite relevant, since the populations in the two regions are similar. Despite its fairly southerly distribution, Bicknell's Thrush is one of the last migrants to return to its breeding grounds, usually at the end of May or beginning of June. Little information is available on the fall migration, but it appears that these thrushes begin their return trip to the wintering grounds in the Caribbean in mid-August, although some may linger until the end of September or later.

Males establish their territories shortly after arrival on the breeding grounds, and actual nesting begins around the second week of June, when the melodious but modulated song can be heard, generally just before dawn and at dusk. Males may even sing late into the night, particularly on calm, windless nights. The song may also be heard during the day at unpredictable and irregular

Bicknell's Thrush is very selective in its choice of habitat, frequenting coniferous forests on some of the highest mountains in southern Québec, like Mount Mégantic. East of Sherbrooke. *August 1991, Gaétan Duquette.*

Notes on territory

Like most other passerines, the male Bicknell's Thrush defends a breeding territory and attempts to attract a female to it by engaging in several types of courtship behaviour, particularly singing (Wallace 1939).

Erskine (1984) reported densities of 2.5 males/10 ha in a stand of stunted balsam fir and white birch on Mount Mégantic, and densities of 4.9 males/10 ha in a fir-spruce stand near the Petite Rivière Cascapédia (Erskine 1980a). In the United States, Sabo (1980) recorded an average of 9 pairs/km² in the White Mountains of New Hampshire.

Bicknell's Thrush

Breeding distribution

1984–1989

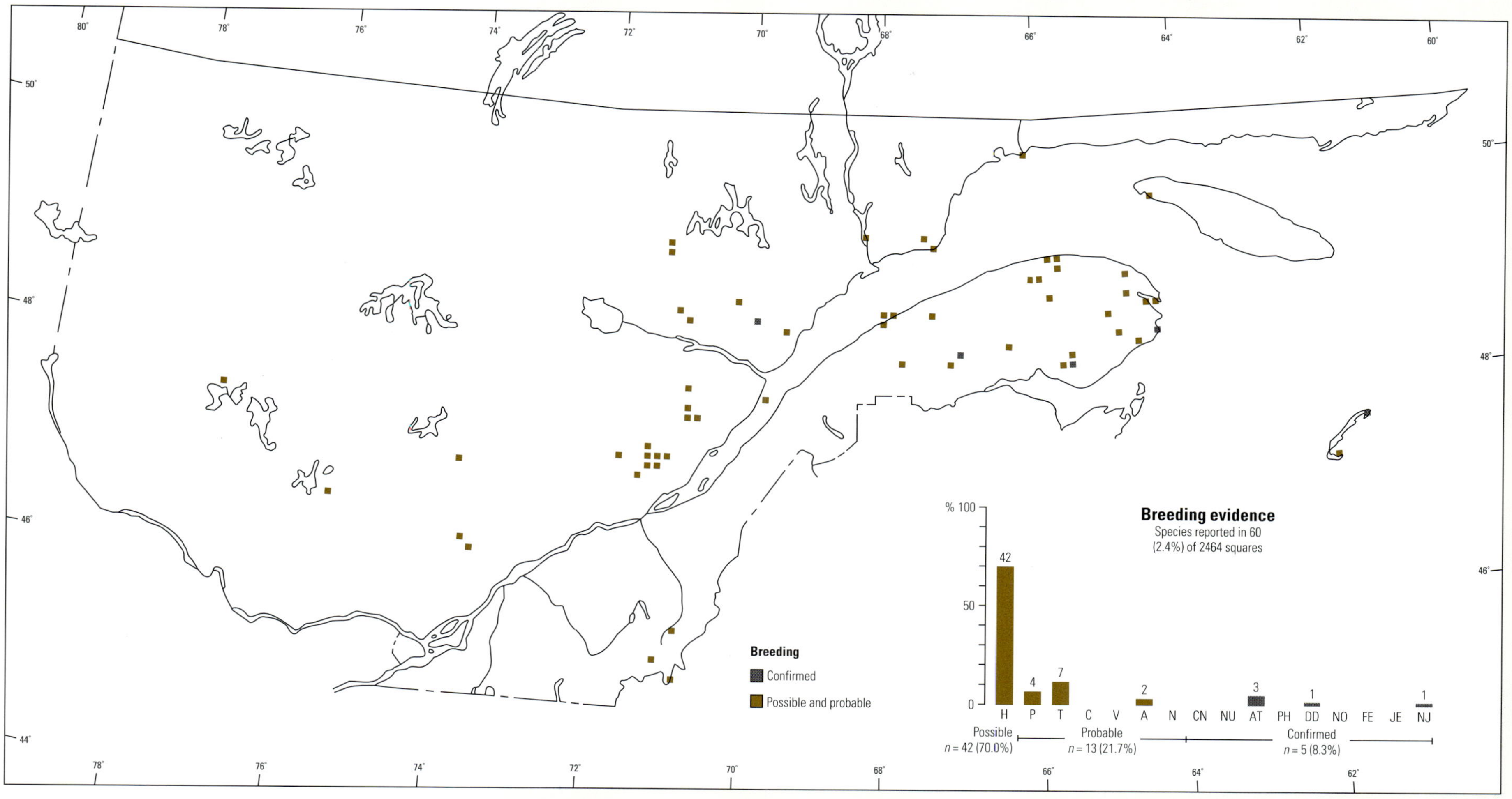

Species profile

Clutch size	usually 4 eggs (3–4)[1,2]
Laying interval	1 egg per day[1]
Start of incubation	with third egg[1]
Incubation period	13–14 days[1]
Maturity at hatching	altricial
Care of young	by female and male[1]
Nestling period	10–13 days[1]
Age at first flight	10–13 days[1]
Dependence of young	until ≈ 10 days after first flight[2]
Broods per year	usually 1 (1–2)[1]
Breeding age	1 year[2]
Mating system	probably monogamy[2]
Duration of pair bond	?
Total length	M 17.7 cm (16.8–19.0) / F 18.0 cm (17.7–18.3)[3]
Wing span	?
Weight	M 28.2 g (20.5–33.0) / F 32.0 g (28.7–36.8)[3]
Longevity record	?

References:
1. Wallace 1939
2. H. Ouellet, CMN, unpubl. data
3. Ouellet 1993

Chronology $n = 11$

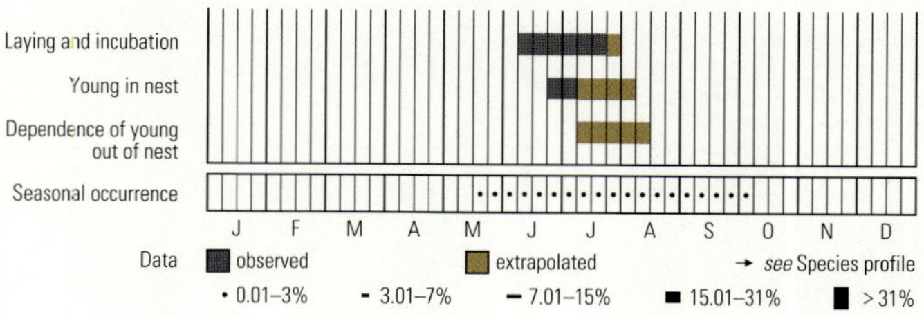

World breeding range

intervals, particularly at the beginning of the breeding season on misty or rainy days with no wind. This shy and retiring species —which could even be described as a skulker—is easily overlooked, particularly when it is not singing, when the weather conditions are unfavourable, or outside its customary singing times of dawn and dusk (Wallace 1939). Interestingly, the females also sing, and some have been heard singing at the nest while incubating, during hatching, and even after hatching (Wallace 1939).

The nest is built in either a coniferous or deciduous tree, fairly close to the ground (0.6–3.8 m). At Percé, a nest was found in a forked branch 1.8 m up, in a striped maple growing on the edge of a regenerating stand dominated by balsam fir. Average clutch size is 4 eggs, which are bluish-green, sometimes lightly spotted with brown. The female does most of the incubating, which lasts 13 or 14 days (Wallace 1939). The male's part in raising the brood varies, depending on the pair, although the female appears to do most of the feeding of the young, which are brought a variety of insects, including caterpillars, ants, wasps, beetles, and flies. The young leave the nest 10–13 days after hatching (Wallace 1939).

Although the species was atlassed in 60 squares, breeding was confirmed in only 5. The predominance of records of possible breeding (42/60) shows how difficult it is to observe the comings and goings of this very secretive species.

On the Magdalen Islands, where this member of the family Muscicapidae has been observed irregularly, Bicknell's Thrush has been found in dense stands of white spruce on the highest hills in the island chain. Butte du Vent, Île du Cap aux Meules. *June 1991, Desneige Perreault.*

Among the species' preferred nesting habitats in Québec are the high-altitude spruce forests of the Laurentians. The species builds its nest low in a conifer or deciduous tree. Montmorency forest, Laurentide Wildlife Reserve, north of Québec City. *June 1991, Réginald Ouellet.*

Bicknell's Thrush has recently colonized a new habitat: high-altitude second-growth stands regenerating after logging or a forest fire. These stands usually consist primarily of balsam fir and deciduous trees. Gaspé Peninsula. *June 1989, Henri Ouellet.*

In southern Québec, the coniferous habitats preferred by this elusive species are found only in mountainous areas, like Mount Sir Wilfrid. Mont-Laurier. *May 1991, Yves Aubry.*

Distribution

Bicknell's Thrush in southern Québec occupies a narrow range of habitat types: stunted conifer and regenerating stands at elevations above 200 m. Although the latter habitat is fairly abundant in southern Québec, the species is present only very locally in a small number of sites. According to the Atlas survey and taxonomic studies, it is confined mainly to the Gaspé Peninsula, the Laurentide Wildlife Reserve, Charlevoix County, and perhaps the region north of the Saguenay River. Bicknell's Thrush can also be observed very locally on the North Shore of the Gulf of St. Lawrence as far east as Sept-Îles, and irregularly on the Magdalen Islands. Farther west, it was reported from the Eastern Townships, around Saint-Michel-des-Saints, Mount Sir Wilfrid near Mont-Laurier, and in the La Vérendrye Wildlife Reserve. It was also reported from Montagne Noire near Saint-Donat on 18 June 1994 (P. Bannon, pers. comm.).

History and trends

It was long believed that Bicknell's Thrush had a more extensive breeding range in Québec (Todd 1963). Though this may have been the case, the number of specimens in major ornithological collections indicates the contrary. Furthermore, some of the past *bicknelli* records were probably misidentified Gray-cheeked Thrushes, which was not a great problem as long as the Gray-cheeked Thrush and Bicknell's Thrush were considered conspecific. However, since these species have now been demonstrated to be allopatric, perhaps parapatric, some records from the northern periphery of the Bicknell's range or ones that were obtained very early during migration, in mid-May, can no longer be accepted without question.

Bicknell's Thrush populations may be increasing in Québec, since the species not only continues to occupy its traditional habitats, but also appears to be able to adapt successfully to the new habitats it has recently colonized. The one exception to this trend is the Magdalen Islands. Populations must nevertheless be carefully monitored, since it seems that the most serious threat to the species is the destruction of tropical forests in its winter range in the Caribbean, which is poorly known.

Henri Ouellet

Swainson's Thrush

Grive à dos olive

Catharus ustulatus

Christiane Girard

Swainson's Thrush is a typical bird of the coniferous forest, where its crystal-clear song can often be heard in concert with that of the White-throated Sparrow.

This very common migrant breeder in the boreal forest is somewhat less common in the southernmost part of the province. Although it may linger in Québec until December, no winter observations have been reported here. The species winters from northern Mexico—occasionally from southern and coastal Texas—to Paraguay and northwestern Argentina (AOU 1983).

Habitat and behaviour

Swainson's Thrush prefers relatively mature coniferous forest dominated by fir and spruce, with a not very heavy understorey (Dilger 1956a; Clark et al. 1983). It is also often found in mixed forest, but rarely in deciduous woodland (Dilger 1956a; Noon 1981; Peck and James 1987). It can also be found in dense stands as well as in second-growth forests, in young sapling stands, and in thickets along forest edges. It is not averse to wooded peatlands and swamps, either (Godfrey 1986; Peck and James 1987). In the more mountainous regions of northeastern North America, there is altitudinal segregation between Swainson's Thrush and Bicknell's Thrush: Swainson's lives at lower altitudes, and the only overlap is at the line of demarcation between the two zones (Bent 1949; Noon 1981).

The nest, built by the female, is usually located in a sapling or shrub, and less frequently in a large tree (Peck and James 1987). The nest tree is usually a conifer, such as a spruce or fir, but it may sometimes be deciduous, for instance an alder, willow, or birch. The nest may be placed anywhere from the foot of the shrub or tree to the top, with no apparent preference (Peck and James 1987). The female alone incubates the eggs for 10–14 days. For the first few days after hatching, the female broods the young, leaving the nest only for short periods. At first, the adults feed the young by regurgitation and also bring them small insects. After a few days, they start to bring larger prey like moths, caterpillars (including spruce budworm grubs), and various other insects (Stanwood 1913). Both parents help keep the nest clean. Swainson's Thrush raises only one brood a year.

Swainson's Thrush catches its prey mainly on the ground, but more often than other thrushes it also feeds in the tops of trees and shrubs (Holmes and Robinson 1988). Its most frequent hunting technique is to examine the leaf litter from a low branch. It goes from one perch to another, hopping along the branch, and sometimes flying down to the ground and hopping around briefly before returning to another perch. When it spots an insect, it swoops down on it from its perch and catches it on the ground or in the foliage, somewhat in the manner of the Veery. It also sometimes hawks insects on the wing. It eats mainly beetles of all sorts and ants (Holmes and Robinson 1988), and also occasionally small wild berries and seeds (Bent 1949).

Like all thrushes, it sings a very pleasant song, heard mainly mornings and evenings. This makes it easy to locate and identify, although the song may be confused in poor listening conditions (wind, distance) with that of the Hermit Thrush. Breeding was confirmed in 21.9% (292/1336) of squares in which it was reported. More than half of the breeding confirmations were sightings of adults carrying food. Records of young that had recently left the nest but were still incapable of sustained flight and nests also accounted for some confirmations. Nests were found in only 44 squares, which shows how difficult they are to locate.

Swainson's Thrush is a bird of the extensive spruce forests of southern Québec. Mount Albert, Gaspé Conservation Park. *July 1991, Yves Aubry.*

Notes on territory

The male establishes and defends a territory against conspecifics (Dilger 1956a,b). Sabo (1980) determined that the mean size of Swainson's Thrush territories in the White Mountains of New Hampshire was 2.1 ha and mean density 2.2 pairs/10 ha.

In Gatineau Park, Dauphin (1985) calculated the mean density of Swainson's Thrush to be 0.5 pair/10 ha in mixed habitat that included open areas and woodlands. Lafontaine (1988) reported a density of 0.48 pair/10 ha in a mixed forest in the Saint-Michel-des-Saints area, but noted that densities may have been higher in more suitable habitat. In La Mauricie National Park, DesGranges (1980b) counted 14.0 birds/10 ha in a white spruce plantation, 7.5/10 ha in a red spruce stand, and 7.5/10 ha in a fir stand. Erskine (1984) reported densities ranging from 2.5 to 10.0 territorial males/10 ha in three balsam fir-white birch forests in the Eastern Townships—two on Mount Mégantic and another at Woburn.

Swainson's Thrush

Breeding distribution 1984–1989

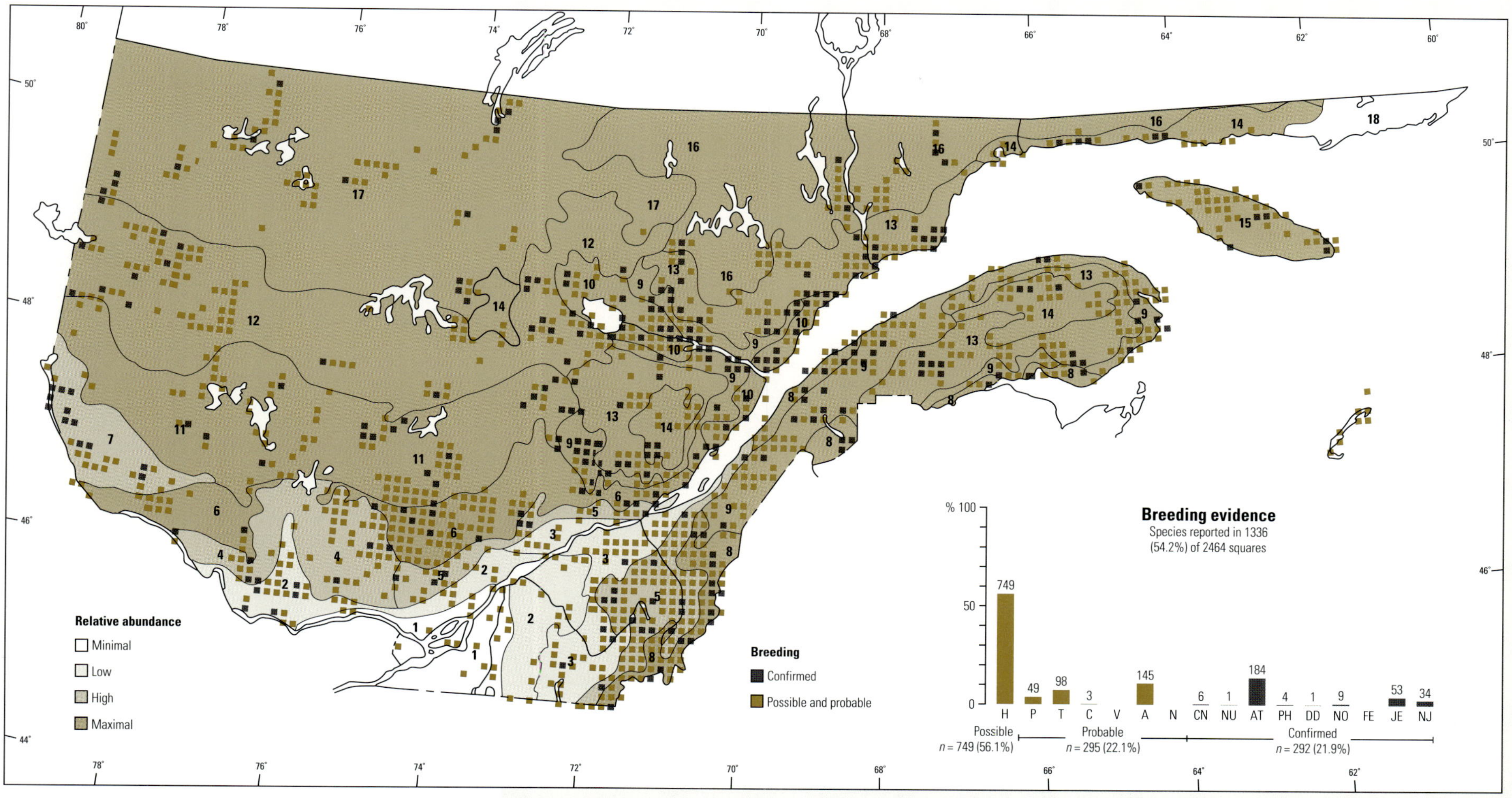

Species profile

Clutch size	usually 3–4 eggs (3–5)[1,2,3,4]
Laying interval	1 egg per day[2]
Start of incubation	with last egg[2]
Incubation period	10–14 days[1,2,3,4]
Maturity at hatching	altricial
Care of young	by female and male[3]
Nestling period	10–13 days[2,3,4]
Age at first flight	10–13 days or more[9]
Dependence of young	?
Broods per year	1[1,4]
Breeding age	1 year[4]
Mating system	monogamy[3]
Duration of pair bond	?
Total length	16.0–19.5 cm[5]
Wing span	27–30 cm[6]
Weight	30.8 g (21.9–50.7)[7]
Longevity record	10 years 1 month[8]

References:
1. Forbush 1929
2. Stanwood 1913
3. Ehrlich et al. 1988
4. DeGraaf and Rudis 1987
5. Godfrey 1986
6. Cramp 1988
7. Dunning 1984
8. Klimkiewicz et al. 1983
9. according to nestling period

World breeding range

Chronology n = 87

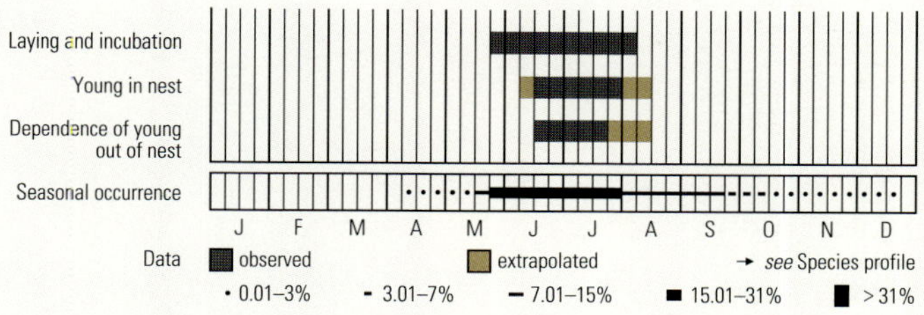

Frequency of occurrence by ecological region

Ecological regions** (forest regions)	Adequately covered squares	Squares in which reported	% of squares in which reported	Relative abundance (%)*
1. Sugar maple-hickory and sugar maple-basswood	97	15	15.5 ± 3.7	15.5
2. Sugar maple-basswood	153	46	30.1 ± 3.8	30.1
3. Sugar maple-basswood and sugar maple-yellow birch	101	58	57.4 ± 5.8	57.4
4. Sugar maple-yellow birch-basswood	56	38	67.9 ± 10.7	67.9
5. Sugar maple-yellow birch-beech and sugar maple-yellow birch-basswood	113	95	84.1 ± 3.9	84.1
6. Sugar maple-yellow birch-beech	71	64	90.1 ± 6.6	90.1
7. Sugar maple-yellow birch-balsam fir	19	17	89.5 ± 15.5	89.5
8. Sugar maple-yellow birch and balsam fir-yellow birch	47	47	100.0 ± 1.1	100.0
9. Balsam fir-yellow birch	87	85	97.7 ± 3.3	97.7
10. Balsam fir-white birch or balsam fir-red maple	46	46	100.0 ± 1.1	100.0
11. Yellow birch-balsam fir	52	51	98.1 ± 4.5	98.1
12. Balsam fir-white birch (west)	59	59	100.0 ± 0.8	100.0
13. Balsam fir-white birch (east)	76	75	98.7 ± 3.0	98.7
14. Balsam fir-black spruce	28	28	100.0 ± 1.8	100.0
15. White spruce-balsam fir and balsam fir-white spruce	22	22	100.0 ± 2.3	100.0
16. Black spruce-balsam fir-moss	18	18	100.0 ± 2.8	100.0
17. Black spruce-moss	32	30	93.7 ± 10.0	93.7
18.			*not analysed*	

* minimal: 0–29.9%; low: 30–59.9%; high: 60–89.9%; maximal: 90–100%

** Deciduous forest: 1–7; mixed forest: 8–11; coniferous forest: 12–17

Swainson's Thrush nest made of grass, twigs, and small branches, in a young spruce tree. Adult feeding its 4 young. Lake Fortier, Charlesbourg. *June 1966, Gaston Le Page.*

This thrush also feeds in shrub thickets along the edges of mixed and coniferous forests. Forillon National Park, Gaspé Peninsula. *July 1991, Yves Aubry.*

Distribution

Swainson's Thrush ranges all across Canada (Godfrey 1986). On the west coast of North America, it breeds from Alaska to southern California; in the Rocky Mountains, it is found as far south as northern Arizona and New Mexico, and, farther east, throughout southern Canada and many northern U.S. states (Robbins et al. 1980; AOU 1983). This thrush was atlassed as a breeder throughout southern Québec, except in the Montréal area, on the Magdalen Islands, and on the Lower North Shore. Both Atlas data and Bannon (1991b) suggest, however, that it is a probable breeder in the Montréal area, since some summer sightings have been made there. There is a handful of sight records from the Magdalen Islands, but no evidence of breeding has yet been found (Fradette 1992). Swainson's Thrush was not atlassed east of Johan Beetz Bay, on the Middle North Shore, probably because of the small number of squares surveyed in this fairly inaccessible region. In June 1928, Todd (1963) found it quite common at the mouths of the Natashquan and Little Mecatina rivers, where members of his expedition even discovered three nests. On this part of the North Shore, however, Swainson's Thrush must be mostly confined to valley enclaves, because on the plateaus the landscape is basically tundra. Farther into the hinterland, the species may take advantage of the fact that the tundra gives way to the boreal forest.

Swainson's Thrush is very common throughout southern Québec, except in the Central St. Lawrence Lowland upstream of Saint-Jean-Port-Joli, particularly around Montréal and in Montérégie and the Bois-Francs. The pre-dominance of farmland, combined with a scarcity of coniferous woodlots in these areas, means that there is little suitable nesting habitat for the species.

The breeding range of Swainson's Thrush extends beyond the Atlas area. Todd (1963) reported that it had been observed at several locations along the east coast of James Bay as far north as Fort George. Todd also called it common around Waskaganish and reported securing two young birds just out of the nest at Eastmain on 20 July 1912 and finding a nest with young there on 29 June 1926. He also reported finding a nest at the mouth of the Jolicoeur River on 27 June 1926, and at Ispisakas Island, just north of the Eastmain River, on 30 June 1926.

Recently, it has been reported fairly regularly around the Eastmain and Opinaca rivers, as well as at La Grande-2 (Bordage and Aubry 1982, 1983; Bordage 1984, 1985b; Tremblay and Belles-Isles 1991; ÉPOQ). Between 30 June and 3 July 1990 it was observed on the east coast of James Bay just south of Kakassituq Point, in white spruce stands and in shoreline scrub with white spruce clumps (Y. Aubry, CWS, unpubl. data). Curiously, in the watersheds of the Nottaway, Broadback, and Rupert rivers this boreal forest thrush was observed chiefly in stands of trembling aspen and birch during a study conducted in 1991 (Cons. G. G. – GREBE 1992o).

Farther east, the species was reported at Gagnon and Fermont during the summer of 1976 (ÉPOQ). It was also reported several times in the Schefferville area in June 1951 and in 1958, with the most northerly record coming from Lake Matemace (Todd 1963). It was also reported along the Sainte-Marguerite and Moisie rivers and north of the Labrador border in June and July 1917 (Todd 1963). The northern limit of the species' Québec range seems to be around the 54th parallel.

History and trends

At the end of the last century, Swainson's Thrush was a scarce "transient visitant" in the Montréal region, according to Wintle (1896), although he knew of a nest with eggs found on the Island of Montréal. Mousley (1916a) listed the species as a scarce summer resident in the Eastern Townships, where he had managed to find one nest. On the other hand, Macoun and Macoun (1915) considered it a fairly common summer resident in the eastern part of the province. Lloyd (1924) considered it a common "transient visitant" in the Ottawa–Hull area and mooted the possibility that it might breed there. In the 1940s it was listed as a common summer resident at Lake Saint-Jean (Godfrey and Wilk 1948) and around Lakes Mistassini and Albanel (Godfrey 1949). More recently, BBS and ÉPOQ data have shown no major changes in Swainson's Thrush populations in southern Québec since the late 1960s (Cyr and Larivée 1993a).

Little work has been done on the possible impact of logging on Swainson's Thrush. Webb et al. (1977) stated that it was little affected by stand thinning. On the other hand, Titterington et al. (1979) found it absent for 7–12 years after a coniferous forest in Maine had been clearcut, and associated more with older stands (more than 12 years old) where trees were over 4.5 m tall with a diameter (dbh) of 10–15 cm.

Swainson's Thrush may be affected by deforestation on its tropical wintering grounds (Terborgh 1980; Askins et al. 1990; Diamond 1991), where 26% of its winter habitat could disappear between 1985 and the year 2000 (Diamond 1991). The population of eastern North America seems to have levelled off between 1978 and 1989, after increasing between 1966 and 1978 (Robbins et al. 1989b). Deforestation of its winter habitat is thought to have put a halt to the upward trend. On the other hand, in Québec, replanting of cutovers with conifers may help the species.

Gilles Falardeau

Hermit Thrush

Grive solitaire

Catharus guttatus

Christiane Girard

The Hermit Thrush, a common migrant breeder in southern Québec, is the first thrush to return here in spring, probably because it winters much farther north than the other thrushes; it is also, incidentally, the only one known to have wintered in Québec.

The Hermit Thrush regularly winters from southern British Columbia east to southern Ontario and southern New England, and south to Guatemala and El Salvador; in the central part of the continent, northern Texas is the northern limit of its winter range (AOU 1983). According to Root (1988), it is usually absent in winter from areas where the mean January temperature is less than –4°C. There have been a dozen or so winter sightings in Québec, including a complete overwintering at Shannon (Aubry et al. 1988).

Habitat and behaviour

The Hermit Thrush is the most versatile thrush in terms of habitat (Noon 1981). It inhabits mostly mixed and coniferous forest (Dilger 1956a; Godfrey 1986; Elderkin 1989), whether damp or dry (Dilger 1956a; Peck and James 1987). Dilger (1956a) noted that it prefers the margins of lakes, clearings, burns, and second-growth areas, rather than the interior of mature stands, which is what Todd (1963) and Erskine (1977) also found in the Canadian boreal forest. In a mountainous part of Vermont, it seems to prefer mid-successional forest with a high, fairly closed canopy, and trees with a diameter at breast height of 15–38 cm (Noon 1981).

The nest, built by the female, is usually located on the ground in a natural depression, often under a small conifer and sheltered by branches that touch the ground (Gross 1949; Peck and James 1987). Occasionally, it is placed low down in a shrub, less than 2 m up. During incubation, which lasts 12 or 13 days, the female rarely leaves the nest, with her mate bringing her food on a regular basis. After hatching, both parents keep the nest clean and feed the young. For the first few days, the young are fed mainly small insect larvae or larger prey, torn up. As the young grow, prey size increases: small grasshoppers, moths, beetles, and spiders are added to their diet (Gross 1949). In the United States the Hermit Thrush may raise two, or even three, broods a year (Forbush 1929; Gross 1949). In Québec, egg dates would seem to suggest occasional double-brooding, but this has not been documented.

More than other thrushes, adults of this species feed largely on beetles of all sorts (Holmes and Robinson 1988); their diet is rounded out with caterpillars, ants, and small wild fruit (Gross 1949; Holmes and Robinson 1988). Although the Hermit Thrush feeds mainly on the ground, it is the thrush that forages most often in shrubs and small trees less than 8 m high (Holmes and Robinson 1988). It is also the thrush

Notes on territory

The male defends a territory against conspecifics (Dilger 1956a,b). Territory size was determined at 0.6 ha in a hemlock stand at Mirabel (Cyr 1974), whereas in an unspecified

More versatile than the other members of its family, the Hermit Thrush can inhabit a variety of forest types, although it prefers the edges of lakes, clearings, burns, and mid-successional areas. Decelles Reservoir, Témiscamingue. *May 1991, Réginald Ouellet.*

Hermit Thrush

Breeding distribution

1984–1989

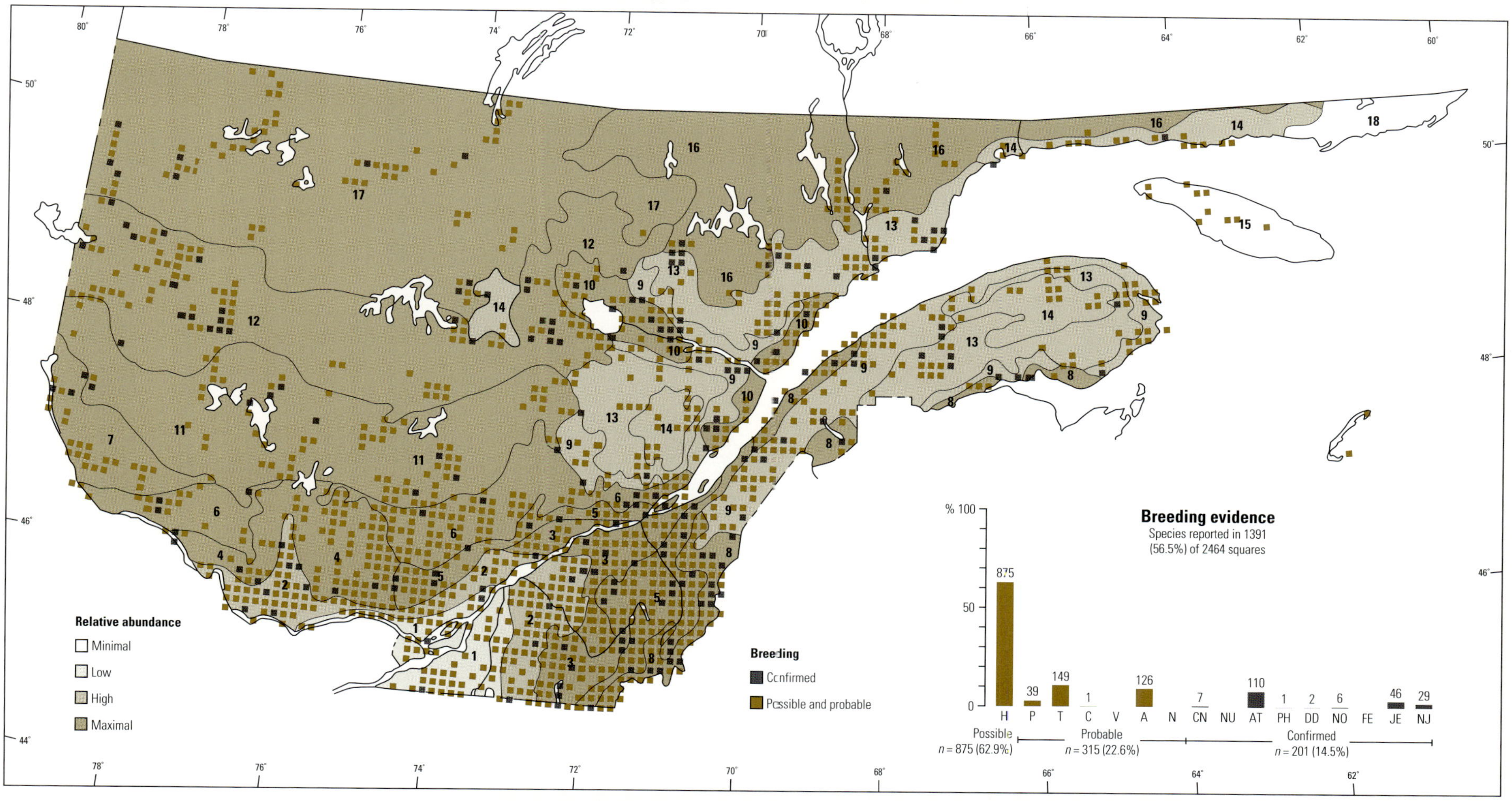

Species profile

Clutch size	usually 3–4 eggs (3–6)[1,2,3,4]
Laying interval	1 egg per day[1]
Start of incubation	with last egg[1]
Incubation period	12–13 days[1,2,3,4]
Maturity at hatching	altricial
Care of young	by female and male[1,2,4]
Nestling period	12 days[1,2,3,4]
Age at first flight	12 days or more[9]
Dependence of young	?
Broods per year	2, 3 possible[1,2,3,4]
Breeding age	1 year[3]
Mating system	monogamy[2]
Duration of pair bond	?
Total length	16–19 cm[5]
Wing span	25.0–28.5 cm[6]
Weight	31.0 g (26.6–37.4)[7]
Longevity record	8 years 8 months[8]

References:
1. Forbush 1929
2. Ehrlich et al. 1988
3. DeGraaf and Rudis 1987
4. Gross 1949
5. Godfrey 1986
6. Cramp 1988
7. Dunning 1984
8. Klimkiewicz et al. 1983
9. according to nestling period

World breeding range

Chronology $n = 70$

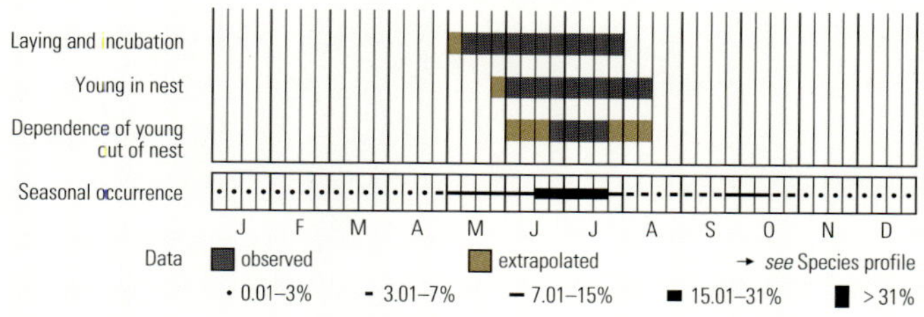

Frequency of occurrence by ecological region

Ecological regions** (forest regions)	Adequately covered squares	Squares in which reported	% of squares in which reported	Relative abundance (%)*
1. Sugar maple-hickory and sugar maple-basswood	97	51	52.6 ± 5.0	52.6
2. Sugar maple-basswood	153	130	85.0 ± 3.0	85.0
3. Sugar maple-basswood and sugar maple-yellow birch	101	93	92.1 ± 3.4	92.1
4. Sugar maple-yellow birch-basswood	56	52	92.9 ± 6.3	92.9
5. Sugar maple-yellow birch-beech and sugar maple-yellow birch-basswood	113	108	95.6 ± 2.4	95.6
6. Sugar maple-yellow birch-beech	71	68	95.8 ± 4.7	95.8
7. Sugar maple-yellow birch-balsam fir	19	18	94.7 ± 12.0	94.7
8. Sugar maple-yellow birch and balsam fir-yellow birch	47	47	100.0 ± 1.1	100.0
9. Balsam fir-yellow birch	87	73	83.9 ± 7.3	83.9
10. Balsam fir-white birch or balsam fir-red maple	46	43	93.5 ± 6.7	93.5
11. Yellow birch-balsam fir	52	50	96.2 ± 6.0	96.2
12. Balsam fir-white birch (west)	59	56	94.9 ± 6.2	94.9
13. Balsam fir-white birch (east)	76	66	86.8 ± 7.6	86.8
14. Balsam fir-black spruce	28	23	82.1 ± 15.2	82.1
15. White spruce-balsam fir and balsam fir-white spruce	22	4	18.2 ± 17.1	18.2
16. Black spruce-balsam fir-moss	18	17	94.4 ± 13.5	94.4
17. Black spruce-moss	32	32	100.0 ± 1.6	100.0
18.			*not analysed*	

* minimal: 0–29.9%; low: 30–59.9%; high: 60–89.9%; maximal: 90–100%

** Deciduous forest: 1–7; mixed forest: 8–11; coniferous forest: 12–17

Though less abundant in the Montréal area, this thrush nonetheless can be found nesting in black spruce plantations in Saint-Lazare. West of Montréal. *June 1991, Pierre Bannon.*

Notes on territory (continued)

habitat in Québec, McNeil (1964 in Cyr 1974) noted that pairs occupied average territories of 1.0 ha.

According to data from several studies, density varies according to the type and geographic location of the stand of trees. In a mixed habitat consisting of forested and open areas in Gatineau Park, Québec, density was found to be 0.37 pair/10 ha (D. Dauphin, unpubl. data). Densities of 2.2–7.5 pairs/10 ha have been recorded in stands of sugar maple-yellow birch, white birch, mixed or pure conifers, near Oka, Mirabel, Piedmont, and La Mauricie National Park (Erskine 1971a, 1976a, 1984; Cyr 1974; DesGranges 1980b). A little farther north, in a jack pine stand near La Tuque, Hermit Thrush densities between 1.5 and 2.7 pairs/10 ha were found (Erskine 1971a, 1976a). In Abitibi, in a jack pine forest west of Senneterre, Hermit Thrush density was determined to be 0.7 pair/10 ha (Erskine 1976a). Finally, in Nova Scotia, Morgan and Freedman (1986) reported densities as high as 13 pairs/10 ha, depending on the age of the deciduous stand.

that most often catches insects on the wing; however, most of the time it catches its prey from the ground or in the foliage.

The Hermit Thrush is easy to spot and identify by its song. Breeding was confirmed, however, in only 14.5% (201/1391) of Atlas squares in which it was reported, largely on the basis of records of adults carrying food, then young still incapable of sustained flight, since the nest itself was hard to find.

Distribution

In North America, the Hermit Thrush's breeding range covers most of the forested regions of Canada; it extends from Alaska to Newfoundland, south to southern California in the west and Maryland in the east (Godfrey 1986).

The Hermit Thrush was atlassed as a breeder throughout southern Québec except the Magdalen Islands and Anticosti Island, for which there were sight records but no breeding records. With these exceptions, it is also very common throughout southern Québec outside the Montréal area and Montérégie. According to Fradette (1992), it is fairly rare in summer on the Magdalen Islands, where the only known evidence of breeding dates from the last century, when Cory collected a young bird in the summer of 1878. There are a number of pre-Atlas records from Anticosti Island, where the only known breeding record is of a female, with a developing egg (without shell) in her oviduct, caught near the Sainte-Marie River on 10 June 1963 (Ouellet 1969). Todd (1963) considered the Hermit Thrush rare on the Middle and Lower North Shore. He did, however, report a few sightings in the area, the most easterly of which was made at the mouth of the Kécarpoui River, southwest of Saint-Augustin, on 22 June 1938. More recently, this thrush was also observed at Harrington Harbour on 5 and 7 May 1982, and near Brador Bay on 26 July 1981 (ÉPOQ).

In the James Bay region, this species frequents coniferous forest with openings. La Grande-3. *July 1992, Réginald Ouellet.*

Hermit Thrush nest containing 4 eggs, well hidden in the moss on the floor of a coniferous forest. La Grande-3, James Bay region. *July 1992, Réginald Ouellet.*

On the James Bay coast, the Hermit Thrush has been reported at Waskaganish, Eastmain, and Moar Bay (Todd 1963). Inland, it is common around the Eastmain and Opinaca rivers, where it is by far the most common thrush (Bordage and Aubry 1982, 1983; Bordage 1984, 1985b; Tremblay and Belles-Isles 1991), and it is one of the most common species in the Nottaway, Broadback, and Rupert river watersheds (Cons. G. G. – GREBE 1992o). It has also been recorded several times at La Grande-2, and even farther inland, it was observed at Gagnon and Fermont in July 1976 (ÉPOQ). The species has also been observed at several places along the Sainte-Marguerite River, and as far north as the upper reaches of the Carheil River (Todd 1963).

History and trends

At the turn of the century and until the 1940s, the Hermit Thrush was listed as a common summer resident in the various localities in question (Wintle 1896; Dionne 1906; Macoun and Macoun 1915; Mousley 1916a; Godfrey and Wilk 1948; Godfrey 1949). Terrill (1911b), however, reported that it had become rare on Mount Royal, in Montréal, early in the century. Ouellet (1974a) considered it less abundant than in the preceding century in the Montréal area and in Montérégie, as a result of deforestation. More recently, however, according to BBS data, Hermit Thrush populations in southern Québec grew between 1969 and 1989, while ÉPOQ data show them as stable between 1970 and 1989 (Cyr and Larivée 1993a).

The Hermit Thrush is not particularly affected by stand thinning (Webb et al. 1977), but several studies have found a decline in density after clearcutting (Freedman et al. 1981; Morgan and Freedman 1986), even in small areas (Derleth et al. 1989). According to Morgan and Freedman (1986), it takes at least 8 years of forest regeneration before Hermit Thrush densities begin to recover.

As it winters chiefly in North America, the Hermit Thrush is less affected than other thrushes by tropical deforestation. On the other hand, it sometimes suffers from the effects of particularly harsh North American winters (Robbins et al. 1986).

GILLES FALARDEAU

Wood Thrush

Grive des bois

Hylocichla mustelina

Christiane Girard

THE SOLEMN SONG OF THE WOOD THRUSH, TINGED WITH SERENITY, is often heard at dusk, echoing through shady groves of tall old maples with lush undergrowth.

This song is loud, clear, and musical, with rounder phrases than those of the other thrushes. The typically three-part song has a brief introductory motif of two or three hurried, monotonic notes. The central motif is a loud, flutelike *eeohlay*. Each bird has 2–8 versions of this that are usually variations of the main motif (Whitney 1991). The song ends on a bell-like trill. When alarmed, the bird raises its head feathers and utters a sharp *pit pit* or *quirt* (Godfrey 1986). When heard at a distance, the song might be confused with that of the Hermit Thrush, and the two species do sometimes answer each other (Erskine 1978b). Young Wood Thrushes learn their repertoire of songs from their parents, which, over the years, can lead to a certain regional uniformity or similarity in repertoire (Whitney 1992). When defending its territory, the Wood Thrush does not imitate songs of conspecifics, but answers with a song that departs as much as possible from the song of the antagonist (Whitney 1991).

Distinguishing field marks are the bright rusty head and the numerous large, distinct round spots on its breast and sides.

Nest initiation begins on arrival in early May; a territory is established that is defended throughout the breeding period, until early August. Before the leaves fall, the birds begin migrating, mostly at night, to their wintering grounds, which extend from northern Mexico to Panama and northwest Colombia (Winker et al. 1990a).

Habitat and behaviour

The preferred habitat of the Wood Thrush is mature forest, deciduous or mixed, and preferably undisturbed (Sadler 1987). Its territory must include trees at least 12 m tall which it can use as song perches (Bertin 1977). It favours cool, damp forests and is often found in bottomlands near water, but also on wooded slopes (Terres 1980; DeGraaf and Rudis 1987). It also occupies small woodlots in residential areas (Roth 1987) and has even been found in hedges adjacent to human habitation (Brackbill 1958).

Microclimate characteristics, which determine the composition and relative abundance of the vegetation, appear to be the determining factor in territory selection (Bertin 1977). In wooded areas of over 5 ha (Sadler 1987) and with greater than 70% canopy closure (Roth 1987), the most sought-after physical components are damp ground, dense shrubby undergrowth, and a thick layer of ground litter (Bertin 1977; Roth 1987; Petit et al. 1988a). The abundance of microfauna in the litter no doubt plays a key role in selection, as the Wood Thrush seldom feeds in the upper strata (Roth 1987).

Notes on territory

Territory size varies between 0.08 and 0.8 ha, depending on habitat. Within its territory, it requires a relatively undisturbed area within about 5 m of the nest, with young trees 1.1–8 cm in diameter, spaced on average about

The habitat favoured by this thrush is mature mixed or deciduous forest. Near Lake Saint-Charles, north of Québec City. *October 1981, Jean Beaubien and Guy Simard, Canadian Forest Service.*

Wood Thrush

Breeding distribution
1984–1989

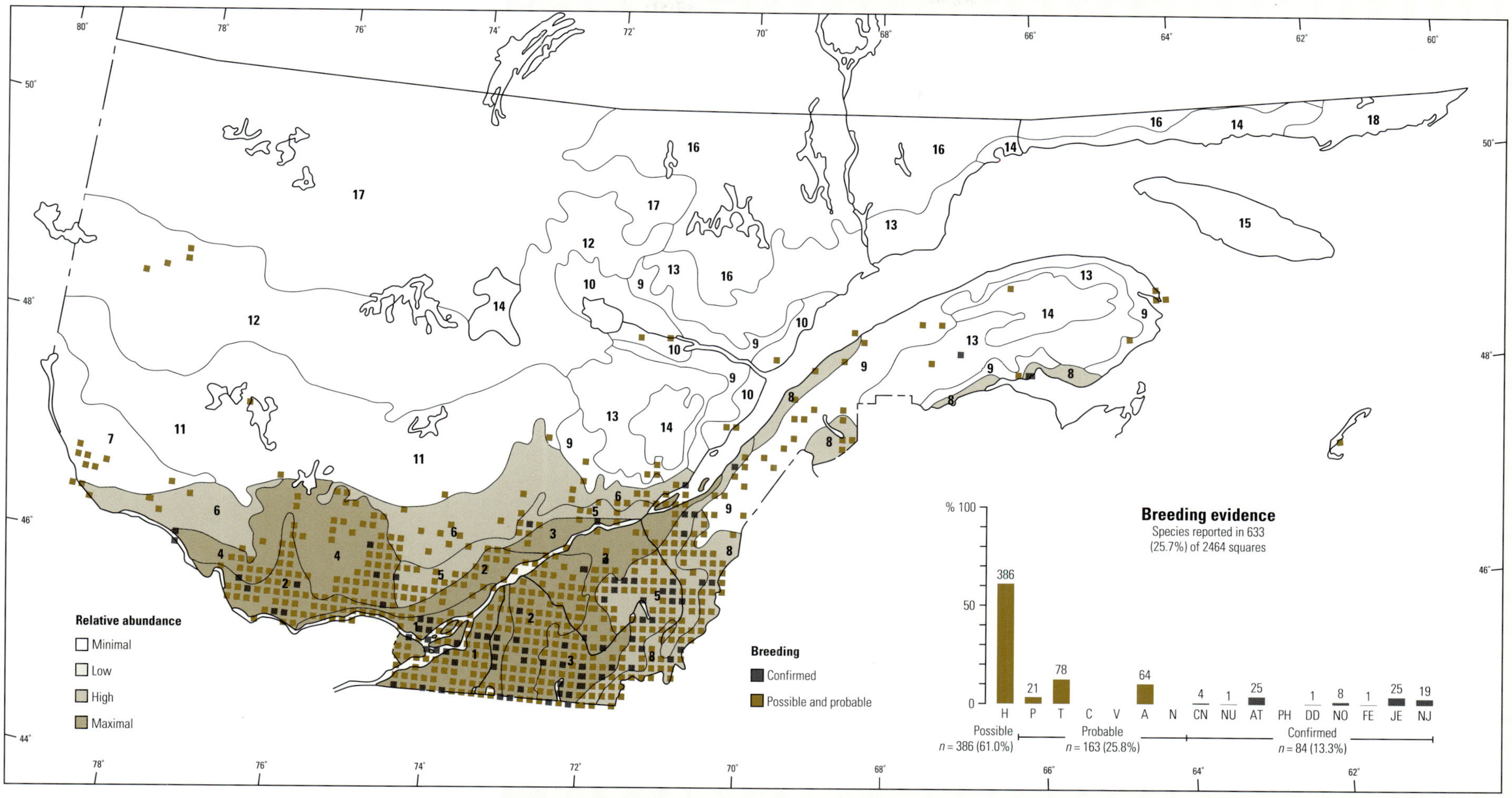

Species profile

Clutch size	usually 3–4 eggs (2–5)[1,2,3,4,6]
Laying interval	1 egg per day[6]
Start of incubation	with second or third egg[6]
Incubation period	12–14 days[1,2,4,6]
Maturity at hatching	altricial
Care of young	by female and male[4]
Nestling period	12–14 days[1,2,3,4,6]
Age at first flight	12–14 days[4,6]
Dependence of young	until ≈ 32 days old[2,4,6]
Broods per year	usually 1 (1–2)[6]
Breeding age	1 year[10]
Mating system	monogamy, rarely polygyny[2,11]
Duration of pair bond	1 season[5]
Total length	19.0–21.5 cm[8]
Wing span	33.0–35.6 cm[3]
Weight	47.4 g (39.2–57.7)[9]
Longevity record	8 years 10 months[7]

References:
1. Harriscn 1978
2. Ehrlich et al. 1988
3. Terres 1980
4. Weaver 1949
5. Petit et al. 1988a
6. Brackbill 1958
7. Kennard 1975
8. Godfrey 1986
9. Dunning 1984
10. DeGraaf and Rudis 1987
11. Johnson et al. 1991

Chronology $n = 104$

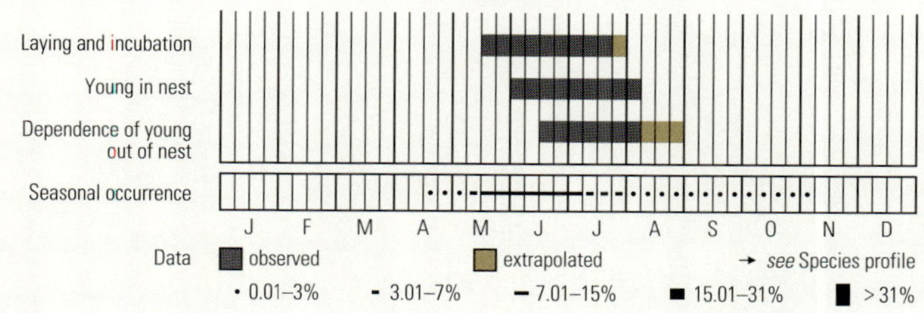

World breeding range

Frequency of occurrence by ecological region

Ecological regions** (forest regions)	Adequately covered squares	Squares in which reported	% of squares in which reported	Relative abundance (%)*
1. Sugar maple-hickory and sugar maple-basswood	97	83	85.6 ± 3.6	99.2
2. Sugar maple-basswood	153	132	86.3 ± 2.9	100.0
3. Sugar maple-basswood and sugar maple-yellow birch	101	81	80.2 ± 4.8	93.0
4. Sugar maple-yellow birch-basswood	56	46	82.1 ± 9.0	95.2
5. Sugar maple-yellow birch-beech and sugar maple-yellow birch-basswood	113	87	77.0 ± 4.4	89.2
6. Sugar maple-yellow birch-beech	71	38	53.5 ± 10.6	62.0
7. Sugar maple-yellow birch-balsam fir	19	3	15.8 ± 17.9	18.3
8. Sugar maple-yellow birch and balsam fir-yellow birch	47	36	76.6 ± 11.0	88.8
9. Balsam fir-yellow birch	87	16	18.4 ± 7.6	21.3
10. Balsam fir-white birch or balsam fir-red maple	46	5	10.9 ± 8.2	12.6
11. Yellow birch-balsam fir	52	1	1.9 ± 4.5	2.2
12. Balsam fir-white birch (west)	59	2	3.4 ± 5.3	3.9
13. Balsam fir-white birch (east)	76	4	5.3 ± 5.3	6.1
14. Balsam fir-black spruce	28	0	0	0
15. White spruce-balsam fir and balsam fir-white spruce	22	0	0	0
16. Black spruce-balsam fir-moss	18	0	0	0
17. Black spruce-moss	32	0	0	0
18.			not analysed	

* minimal: 0–29.9%; low: 30–59.9%; high: 60–89.9%; maximal: 90–100%

** Deciduous forest: 1–7; mixed forest: 8–11; coniferous forest: 12–17

The Wood Thrush uses an intermediary (mesic) habitat which sometimes overlaps the dry (xeric) habitat of the Hermit Thrush or the damp (hydric) habitat of the Veery. The territory is established early in the season, and the Wood Thrush, the dominant species in more than size, will often appropriate territory occupied the previous year by the Hermit Thrush or, less frequently, the Veery. This behaviour, indicative of a degree of interspecific competition, may account for the decrease in density of the latter two species in some parts of their breeding ranges (Morse 1971a).

The male generally returns to the same territory from year to year, arriving a few days before the female (Weaver 1949; Roth and Johnson 1989; Whitney 1992). Upon arrival, he uses the highest perches in his territory to give forth his loudest, most complete, and most varied song of the season. Starting in the second week, he sings mostly from the ground, from a low branch, a stump, or a felled tree trunk.

After the pair bond, which lasts throughout the breeding season, is formed, the female selects the nest site, where she will be relatively undisturbed in the dense understorey (Petit et al. 1988a). She builds the nest in less than 5 days (Brackbill 1958) in a tree fork or saddled on a branch. The nest is normally located 1.5–4.5 m up (Godfrey 1986), almost always in a deciduous tree, usually a young maple or an American beech (Peck and James 1987).

The nest resembles that of the American Robin, but is smaller (Brackbill 1958) and is lined with rootlets rather than dried grass (Harrison 1978). Often a piece of paper, cellophane, or white cloth will be incorporated into the structure (Weaver 1949), presumably because it serves as camouflage by breaking the nest contour.

Typical clutch size is 3 or 4 eggs, which are laid a day apart. The female begins incubating after the second or third egg is laid (Brackbill 1958). The young are brooded essentially by the female, although the male participates in feeding. If food items are too large or tough for the young birds, the female will reduce them to manageable morsels and even apportion them among the nestlings (Weaver 1949).

The young leave the nest at about 12–14 days after hatching, whereupon the parents divide the brood, each tending some of the young (Harrison 1978). The fledglings begin to forage for themselves at 21–25 days of age, although they may beg for food until they are 27 days old. They remain with the parents until they are 32 days old (Brackbill 1958). In Maryland, adults were observed to stop feeding the young of the first brood up to 2 days before the second clutch is commenced (Brackbill 1958). A new nest is build for the second clutch, generally not far from the first. Egg dates in Ontario and Québec suggest possible double-brooding (Peck and James 1987; QNRCP). Instances of polygyny among Wood Thrushes have only recently been reported (Johnson et al. 1991).

The Wood Thrush feeds on insects and berries that it finds on or near the ground (Terres 1980). Sometimes birds are seen scratching forest floor litter with their feet and turning leaves over with their bills (Weaver 1949).

The Brown-headed Cowbird often lays its eggs in Wood Thrush nests. The thrush reacts to this in several ways (Weaver 1949); according to Friedmann (1929), in most cases the eggs and young of the cowbird are tolerated, with the cowbirds often being the only survivors of the brood. Peck and James (1987) reported that 27.2% of 195 nests were parasitized in Ontario.

Notes on territory (continued)

2.45 m apart (Petit et al. 1988a). Densities reported for the Wood Thrush in Québec range from 2.2 pairs/10 ha in less frequented habitat to 5.4 pairs/10 ha in mixed forest (Erskine 1976a, 1980a, 1984).

In mature deciduous woodlands, such as this sugar maple stand, the Wood Thrush seeks areas of damp soil, dense shrubbery, and a thick layer of litter where it can forage in the organic matter on the ground. Sainte-Foy. *September 1991, Yves Aubry.*

This Wood Thrush nest, set in a crotch formed by two saplings, is made of plants stalks and twigs. Lake Timiskaming, Témiscamingue. *July 1986, Yves Aubry.*

Distribution

The Wood Thrush is found only in eastern North America, from the Gulf of Mexico to southern Canada. The Québec distribution of the species is uniform and associated with sugar maple stands in the Central St. Lawrence Lowland, lower Laurentians, Ottawa Valley, and southern Appalachians. Beyond this primary centre, it occupies habitat where deciduous regrowth has reached a certain level of maturity. These more localized populations are found toward the north, in the Abitibi region, along the Saguenay River, east of Québec City to Rimouski, in the Matapédia Valley, and along the south shore of the Gaspé Peninsula to the northeast tip. Breeding was confirmed near Causapscal and Carleton, evidence of a gradual spread to regions where apparently it had been absent until very recently (Godfrey 1986).

History and trends

Toward the end of the last century, the Wood Thrush expanded its breeding range into several regions of the northeastern United States and southeastern Canada (Weaver 1949), spreading into settled areas, and is now found in parks and isolated woodlots that have either regrown or been left standing in agricultural and suburban areas (Weaver 1949). Wood Thrush populations are currently in decline in the central regions of the continent, but have increased slightly in the northeastern United States (Robbins et al. 1986; Bonney and Burrill 1988) and southeastern Ontario (Erskine 1978b). Although it has been expanding its range in Québec, ÉPOQ data show a decline in abundance between 1970 and 1989 (Cyr and Larivée 1993a), as do BBS data for the years 1980–89.

In 1895, the Wood Thrush was rare north of Massachusetts, except around Lake Champlain in Vermont (James et al. 1984). According to Ouellet (1974a), it occurred very locally in small numbers in the Montréal area and the Eastern Townships at the beginning of the century, but by 1949 had become established in Montréal (Weaver 1949). Dionne (1906) apparently did not find it in the Québec City area at the beginning of the century. The only sightings from areas farther east are from Kamouraska in 1928 (Tanguay 1965) and Madeleine River in 1924 (Demille 1926). Godfrey (1966) set the northern limit of the species' range around Cap-Tourmente, but extended it to Rivière-du-Loup in the 1986 edition. Its breeding range appears to have expanded gradually, beginning along the south shore of the St. Lawrence, and it is now common in southernmost Québec. The Atlas distribution map shows an eastward spread in the mixed forest zone in the Sugar maple-yellow birch and Balsam fir-yellow birch forest regions. This expansion has also been noted in the Maritime provinces, where the species was rare until the 1950s. Today it occupies all of New Brunswick, being particularly prevalent in the valleys in the south of the province, less so in Nova Scotia (Erskine 1992).

The wet, tropical forest on the Wood Thrush's wintering grounds has been considerably depleted over the last several years, and if deforestation continues at the present rate, its winter habitat will be reduced by about 40% by the year 2000 (Diamond 1991). As with many other species, forest fragmentation on its breeding grounds is also a problem (Askins 1993). The species is considered vulnerable to such habitat disturbance (Winker et al. 1990a,b).

VINCENT LÉTOURNEAU AND PIERRE LAFONTAINE

American Robin

Merle d'Amérique

Turdus migratorius

The familiar song of the American Robin is often considered a harbinger of spring. The series of two- or three-syllable phrases (Godfrey 1986) can be heard all day long, but especially at dawn and dusk. It is hard to mistake this bird with its dark grey back and brick-red breast, which the first settlers called "robin redbreast" after its resemblance to the European Robin.

Found almost all over North America, the American Robin is an abundant migrant breeder in Québec (David 1980a). It lives in widely diverse habitats, often close to human habitation. Because of its abundance in semi-urban environments and its habit of running across lawns in search of food, it is easy to observe. Indeed, these traits have made possible many studies of its ecology and of the impact of extensive human disturbance of its habitat.

The American Robin is ubiquitous in residential neighbourhoods, where it runs across lawns searching for earthworms and insects. Sainte-Foy. *September 1991, Yves Aubry.*

In Québec, the American Robin occasionally overwinters in the Central St. Lawrence Lowland and the southern Appalachians (David 1980a). It has been reported in winter, though rarely, along the Lower St. Lawrence, in the Gaspé, on the North Shore, and in the Lac-Saint-Jean region (David 1980a). Fradette (1992) considered it an irregular winter visitor to the Magdalen Islands.

Habitat and behaviour

Rural and residential areas, deciduous and coniferous forests with scattered openings, burns, and clearings are all suitable habitat for this species (Eiserer 1976; Godfrey 1986). Preferred habitat has open areas near small woodlots, with trees for shelter and nesting and clearings for foraging.

Soon after they arrive in mid-April, the males establish their territories, which they defend until the end of the breeding season. The females arrive a few days later; and pair bonding ensues. Pair bonds last only one season.

The female selects the nest site within the territory. She also builds the nest, first choosing stalks and twigs to form a cup, and then covering the cup with dry grass before cementing it with a thick coating of mud (Howe 1898; Klimstra and Stieglitz 1957). The last step normally requires a rainy day, but birds can sometimes be seen making mud by dipping billfuls of dry earth into a small pond or birdbath (Eiserer 1976). The nest takes 3–6 days to build (Klimstra and Stieglitz 1957) and may be placed on a wide variety of supports: deciduous or coniferous trees, bushes, stumps, buildings, or other man-made structures. In city suburbs, nests are commonly found in quite diverse locations, such as on poles, fences, gutters, window ledges or lamp posts near human dwellings (Klimstra and Stieglitz 1957; Eiserer 1976). They may be placed directly on the ground or as high as 21 m in a tree, but usually they are found at a height of 1.4–3.0 m (Peck and James 1987).

Christiane Girard

Notes on territory

Interestingly, the American Robin always, or almost always, returns to the same breeding site year after year (Eiserer 1976). The males show greater site fidelity, usually returning to their birthplace. In this species, territorial behaviour is displayed almost exclusively during the breeding period. The male stakes out his territory as soon as he arrives in spring and usually before the female arrives. Territory size varies according to population density, but it is usually between 0.1 and 0.3 ha (Young 1951; Collins and Boyajian 1965; Eiserer 1976).

In open areas like pastures and old fields, the robin nests in shrubs as well as in groves of trees. Forillon National Park, Gaspé Peninsula. *July 1991, Yves Aubry.*

American Robin

Breeding distribution
1984–1989

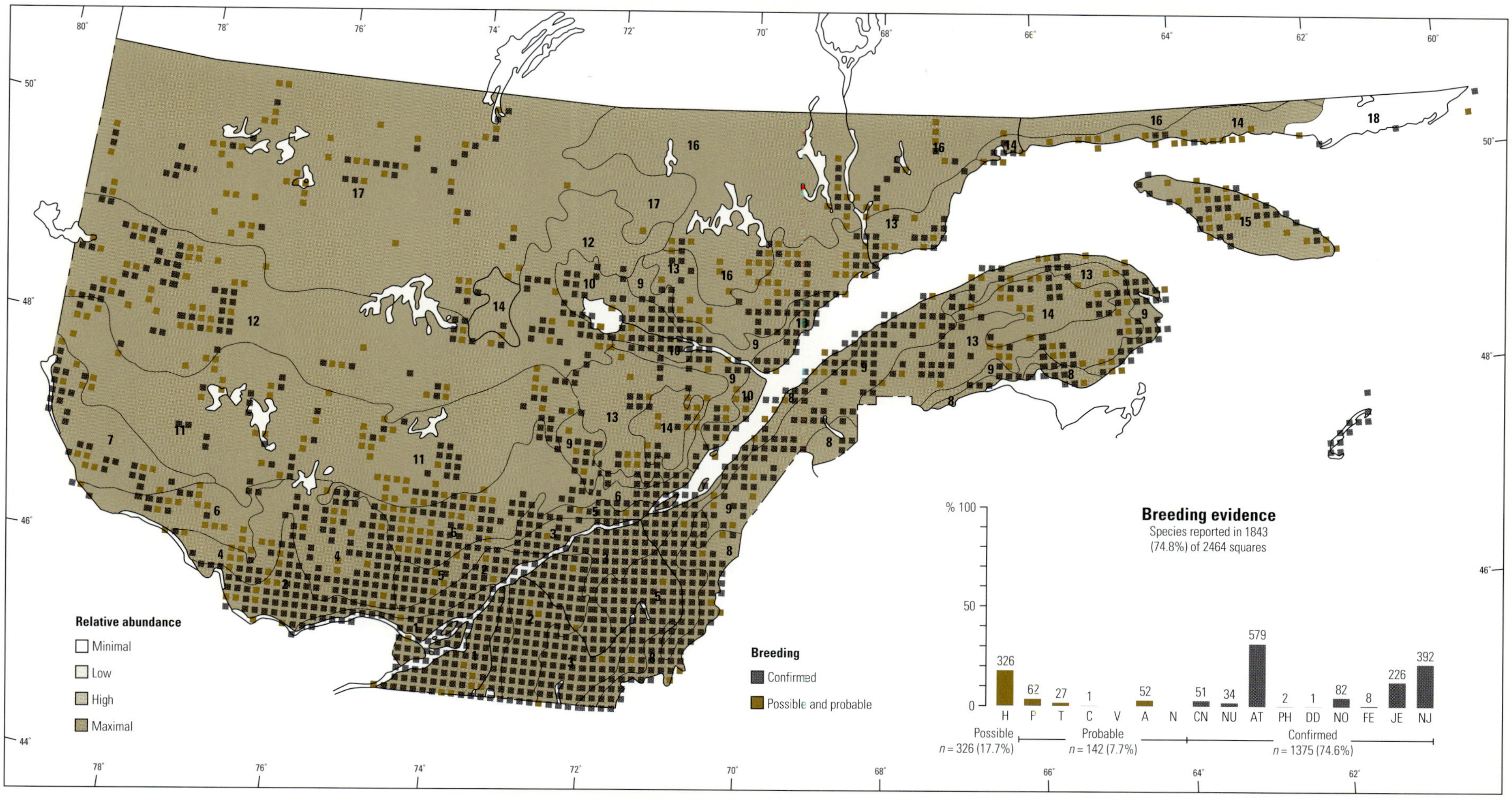

Species profile

Clutch size	usually 3–4 eggs (2–7)[3,4,7,8,9,13]
Laying interval	1 egg per day[7]
Start of incubation	with last egg[7]
Incubation period	usually 11–14 days (9–15)[3,4,7]
Maturity at hatching	altricial
Care of young	by female and male[2]
Nestling period	14–16 days[3]
Age at first flight	14–16 days[4]
Dependence of young	until 15 days after first flight (maximum)[5]
Broods per year	usually 2 (1–3)[3,4,7,8]
Breeding age	1 year[3]
Mating system	monogamy[6]
Duration of pair bond	1 season[10]
Total length	23.0–27.5 cm[11]
Wing span	36.9–41.2 cm[4]
Weight	77.3 g (63.5–103.0)[12]
Longevity record	13 years 11 months[1]

References:
1. Klimkiewicz et al. 1983
2. Ehrlich et al. 1988
3. DeGraaf and Rudis 1987
4. Terres 1980
5. Brackbill 1947
6. Gowaty and Plissner 1987
7. Peck and James 1987
8. Howard 1967
9. Klimstra and Stieglitz 1957
10. Eiserer 1976
11. Godfrey 1986
12. Dunning 1984
13. Morneau et al. 1990

World breeding range

Chronology *n* = 1747

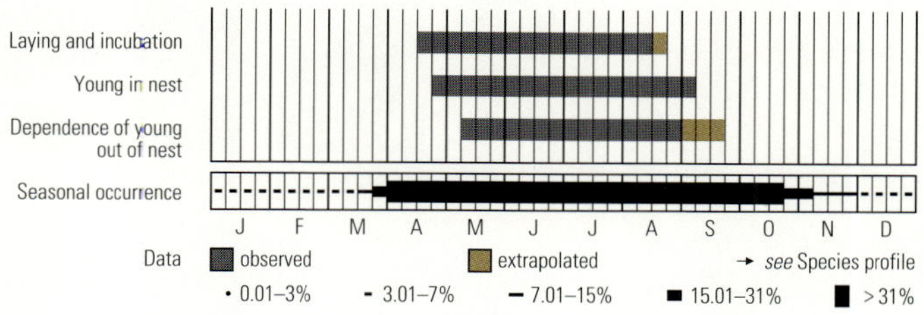

Frequency of occurrence by ecological region

Ecological regions** (forest regions)	Adequately covered squares	Squares in which reported	% of squares in which reported	Relative abundance (%)*
1. Sugar maple-hickory and sugar maple-basswood	97	97	100.0 ± 0.5	100.0
2. Sugar maple-basswood	153	152	99.3 ± 0.9	99.3
3. Sugar maple-basswood and sugar maple-yellow birch	101	101	100.0 ± 0.5	100.0
4. Sugar maple-yellow birch-basswood	56	55	98.2 ± 3.7	98.2
5. Sugar maple-yellow birch-beech and sugar maple-yellow birch-basswood	113	113	100.0 ± 0.4	100.0
6. Sugar maple-yellow birch-beech	71	71	100.0 ± 0.7	100.0
7. Sugar maple-yellow birch-balsam fir	19	19	100.0 ± 2.6	100.0
8. Sugar maple-yellow birch and balsam fir-yellow birch	47	47	100.0 ± 1.1	100.0
9. Balsam fir-yellow birch	87	87	100.0 ± 0.6	100.0
10. Balsam fir-white birch or balsam fir-red maple	46	46	100.0 ± 1.1	100.0
11. Yellow birch-balsam fir	52	52	100.0 ± 1.0	100.0
12. Balsam fir-white birch (west)	59	59	100.0 ± 0.8	100.0
13. Balsam fir-white birch (east)	76	75	98.7 ± 3.0	98.7
14. Balsam fir-black spruce	28	28	100.0 ± 1.8	100.0
15. White spruce-balsam fir and balsam fir-white spruce	22	22	100.0 ± 2.3	100.0
16. Black spruce-balsam fir-moss	18	18	100.0 ± 2.8	100.0
17. Black spruce-moss	32	31	96.9 ± 7.6	96.9
18.			*not analysed*	

* minimal: 0–29.9%; low: 30–59.9%; high: 60–89.9%; maximal: 90–100%

** Deciduous forest: 1–7; mixed forest: 8–11; coniferous forest: 12–17

Once the nest is complete, the female lays her eggs, which she alone incubates. The American Robin usually produces two broods a season, but the first clutch is often larger (Howard 1967). A Québec study reported mean clutch sizes of 3.9 and 3.2 for first and second clutches, respectively (Morneau et al. 1990). The same nest, after some refurbishing, may be used for both clutches, but usually the female builds a new nest (Howell 1942; Nickell 1957). Nest initiation occurs very early in the spring, even before the leaves are on the trees. Because of the permanent cover they provide, conifers are more often used for the first nest, while deciduous trees seem to be preferred for the second (Howell 1942; Savard and Falls 1981; Yahner 1983; Morneau et al. 1990).

Both parents defend the territory against intruders, especially conspecifics. They chase off any other robin that ventures onto their territory, whether male or female, even if they are not always sure of success. Despite this intraspecific aggressiveness, robin territories often overlap. Each pair claims the area of overlap, but often neither is able to assert dominance. The American Robin is ordinarily quite tolerant of other bird species, but it will chase Red-winged Blackbirds, Common Grackles, Blue Jays, House Sparrows, European Starlings, and a few other species out of its territory (Young 1951). The aggressiveness of adults defending the nest varies depending on the pair, but seems to increase with nest exposure (McLean et al. 1986). When a dog or cat enters its territory, the bird usually flees, but not without first sounding the alarm and hopping from perch to perch after the intruder, calling constantly. In urban areas, cats are one of the chief causes of death of young American Robins (Eiserer 1976).

The robin's diet varies throughout the season, depending on its habitat (Eiserer 1976, 1980; Wheelwright 1986, 1988). In spring, earthworms, along with larval and adult insects, usually make up the bulk of its diet (Heppner 1965; Paszkowski 1982; Wheelwright 1986, 1988). The robin then gradually eats more and more fruit as the summer gives way to autumn (Paszkowski 1982; Wheelwright 1986, 1988; White and Stiles 1991).

The nestlings are fed by both parents, first by regurgitation, then mainly on larvae or whole earthworms (Smith and Montgomerie 1991). When the young leave the nest at the age of 14–16 days (DeGraaf and Rudis 1987), the male takes charge of them, leaving the female to repair the nest or to build a new one for her second clutch (Brackbill 1947; Klimstra and Stieglitz 1957). According to Brackbill (1947), the young are dependent on their parents, particularly on the male, for about 4 weeks after hatching.

American Robin incubating its eggs. The nest, which may be reused from one year to another, is a cup made of mud mixed with grass resting on a base of twigs and leaves. Sainte-Foy. *May 1955, Gaston Le Page.*

The American Robin was the species reported in the greatest number of Atlas squares. Also, breeding was confirmed in 74.8% (1375/1843) of the squares in which it occurred. Breeding confirmations were mostly of sightings of adults carrying food, nests containing eggs or young, downy chicks, or young just out of the nest but incapable of sustained flight. The fact that the species usually produces two broods and feeds mainly out in the open undoubtedly made it easier to spot in the squares surveyed.

Distribution

The American Robin breeds throughout southern Québec, up to the tree line (Eiserer 1976; David 1980a), but does not inhabit unbroken closed forest. It is, therefore, probably less abundant in the unlogged northern forests (Eiserer 1976). Moreover, since the robin prefers to nest in fairly dry habitat (Peck and James 1987), the extensive peatlands of the boreal forest are not suitable. Rather, it occupies those areas that have been subject to human disturbance and that are near human habitation (Eiserer 1976; Savard and Falls 1981; Yahner 1983).

Notes on territory (continued)

Population density of American Robins varies depending on habitat type. In Québec, recorded densities range from 0.5 pair/10 ha in a black spruce peatland to 9.9 pairs/10 ha in a balsam fir stand with birch and trembling aspen (Cyr 1974; Erskine 1980a). In several types of forest or urban habitat, as long as conditions are neither too dense nor too damp, density is most commonly 2.2–5.4 pairs/10 ha (Erskine 1971a, 1972a, 1980a, 1984; Cyr 1974; DesGranges 1980b). In a Wisconsin arboretum, density was 9, 10, and 8 pairs/10 ha in three successive years (Young 1951). In Illinois, Graber et al. (1971) recorded a density of 33 birds/10 ha in an urban residential area, and 14 birds/10 ha in shrub habitat. Fawks (1937) counted 3.5 birds/10 ha in Illinois habitats that had been disturbed by logging.

In coniferous habitat, the robin nests along the margins of logged areas and clearings, where it forages. Montmorency forest, Laurentide Wildlife Reserve, north of Québec City. *June 1991, Réginald Ouellet.*

Most of the squares surveyed for the Atlas included the types of clearings associated with burns, roads, streams, ponds, and so on, suitable for the American Robin. Thus the probability of encountering a robin anywhere in southern Québec was high, regardless of habitat type. Frequency of occurrence was high in all habitats as a whole, varying from 96.9 to 100% throughout southern Québec.

Adult feeding a nestling an earthworm. Lac-Saint-Charles, near Québec City. *June 1993, Robert Morin.*

History and trends

North American habitats before European settlement (closed forest) were less suited to the American Robin. At that time, it must have been most abundant in forest clearings and on the edges of woods and meadows (Eiserer 1976). However, European settlement and the ensuing changes to the countryside—clearing of land for farming, urbanization, cultivation of lawns—favoured expansion of the robin's range as well as increases in its numbers.

According to Macoun and Macoun (1915), this bird was abundant all over the country early in the century. In Québec, Wintle (1896) listed the American Robin as a common summer resident in the Montréal area, as it was elsewhere in the province. In the 19th century, it was considered abundant on the Magdalen Islands as well as along the shores of the Gulf of St. Lawrence (Fradette 1992).

Starting in the 1950s, in Québec as in many U.S. states, the intensive use of pesticides to control the insect vectors of Dutch elm disease severely harmed robin populations (Wallace et al. 1961; Weller 1971; Ouellet 1974a). Later, in the 1960s, a decrease in pesticide contamination (Robbins et al. 1986) seemed to bring populations back up to their earlier levels (Ouellet 1974a). No decline was recorded between 1966 and 1979, with highest densities being found in Massachusetts and Pennsylvania (Robbins et al. 1986). Today, populations seem to be recovering well throughout North America, particularly in the central and eastern regions.

In Québec, according to studies on population trends undertaken between 1969 and 1989, American Robin populations appear stable or slightly increasing (Falardeau and DesGranges 1991; Cyr and Larivée 1993a,b).

The American Robin is especially sensitive to the pesticides used in urban areas and orchards (Wallace et al. 1961; Hunt 1969; Weller 1971; Eiserer 1976; Johnson et al. 1976; Morneau et al. 1990; Décarie et al. 1993), so intensive use could be detrimental to it (Décarie et al. 1993). So far, in southern Québec occasional spraying of insecticides on ornamental trees does not seem to have affected it, since it usually feeds on nearby unsprayed sites (Morneau et al. 1990; Décarie et al. 1993).

Josée Tardif and Alain Lanoue

Gray Catbird

Moqueur chat

Dumetella carolinensis

Christiane Girard

MANY PEOPLE STROLLING ALONG THE EDGE OF A FOREST HAVE THOUGHT they heard the meowing of a cat, when it was actually the Gray Catbird. This slate-grey bird can imitate the calls of other birds, stringing them together without interruption. Unlike its cousins, the Northern Mockingbird and the Brown Thrasher, it does not repeat its phrases. In spring, this common migrant breeder arrives here from its wintering grounds, which extend from the southeastern United States to Panama, and include the Caribbean islands (Godfrey 1986).

Habitat and behaviour

The Gray Catbird is generally found in dense vegetation, especially along woodland edges, in thickets, and in ravines. In precolonial times, it was probably confined to shrubby margins along swamps and rivers, as well as the undergrowth in lightly wooded areas. It now seems to like man-made habitats such as large gardens and forested parks, both for nesting and foraging. Forestry and farming practices have also helped create suitable habitat, including roadside shrub fringes, open fields, and abandoned farmlands overgrown with thickets and tangles (Nickell 1965), as well as deciduous saplings in some cutover areas.

The Gray Catbird usually nests in a bushy shrub or in the low branches of a tree, 1–3 m up. The rather bulky nest seems carelessly built; its shape and composition may vary depending on the plant cover and construction materials available. It is usually made of twigs, bark strips, stems, leaves, and grass (Lent 1992), with pieces of plastic and paper sometimes built into the structure or used to conceal it (Nickell 1965). The inner cup, however, is woven carefully (Gross 1948a).

In Québec, the Gray Catbird may raise two broods per year, as has been documented in Ontario (Peck and James 1987; Scott et al. 1988). When the first brood fledges successfully, the pair raises the second in the same territory (Darley et al. 1971). If, on the other hand, the first nesting attempt fails, they will settle elsewhere, up to 450 m away. Males show strong nest-site fidelity, but females much less so (Darley et al. 1977). Pairs are generally seasonally monogamous, although the same birds have occasionally been observed forming pair bonds two years in a row (Gross 1948a; Darley et al. 1977).

Incubation takes 12–15 days and is by the female alone. She then helps feed the young, but to a lesser extent than the male, especially during the first week after hatching, when she is occupied brooding the nestlings against rain, cold, and excessive heat from the sun (Johnson and Best 1982). The female is a close sitter and often does not react to an intruder approaching the nest until the last moment. In this species the nest is almost always guarded, the female being reluctant to leave it before her mate returns to keep watch nearby. This may partly explain why the catbird is seldom host to the Brown-headed Cowbird (Slack 1976). The low cowbird parasitism rate is also due to the catbird's ability to recognize parasitic eggs and immediately eject them from the nest (Rothstein 1974). The Gray Catbird is nonetheless subject to some degree of parasitism, as is evidenced by a few reports of young cowbirds in catbird nests (Woodward 1976).

Notes on territory

During the breeding season, the male vigorously defends a territory centred on the nest (Nickell 1965). His song seems to play a predominant role in establishing and defending the territory (Nickell 1965; Slack 1976).

In Ontario, reported densities are 3.4 pairs/10 ha in a tamarack slough (Goodyear 1993), 6.7 pairs/10 ha in an abandoned pasture growing up in brush (Romanow 1993), 13.7 pairs/10 ha in shrubby or woody habitat (Clark 1993b), and 39.3 pairs/10 ha in a shrubby field (Clark 1993a).

The Gray Catbird frequents dense fringes of shrubs near streams and open areas. Gowan Road, Herdman, southwest of Montréal. *June 1991, Pierre Bannon.*

Gray Catbird

Breeding distribution
1984–1989

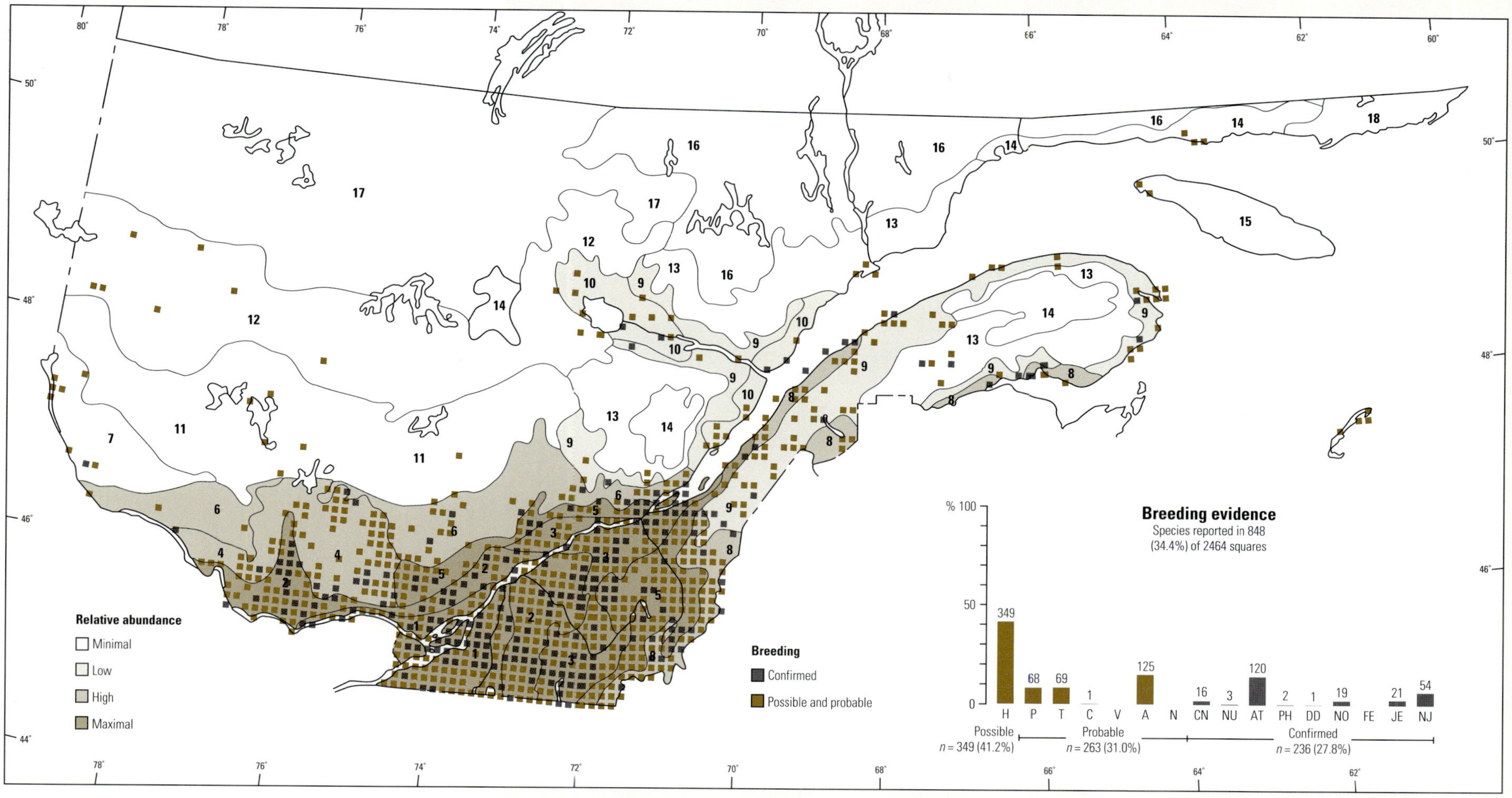

Species profile

Clutch size	usually 4 eggs (1–6)[1,2,3]
Laying interval	1 egg per day[1,3]
Start of incubation	with last egg, sometimes before[1]
Incubation period	12–15 days[1,3]
Maturity at hatching	altricial
Care of young	by female and male[4]
Nestling period	9–15 days[1,3,4,5]
Age at first flight	9–15 days or more[10]
Dependence of young	until 14 days after leaving nest[1]
Broods per year	1–2[1,2,5,11]
Breeding age	1 year[9]
Mating system	monogamy[4]
Duration of pair bond	1 season, sometimes 2[3]
Total length	21–24 cm[6]
Wing span	27.9–30.5 cm[5]
Weight	36.9 g (26.6–56.5)[7]
Longevity record	10 years 11 months[8]

References:
1. Nickell 1965
2. Scott et al. 1988
3. Gross 1948a
4. Ehrlich et al. 1988
5. Terres 1980
6. Godfrey 1986
7. Dunning 1984
8. Klimkiewicz et al. 1983
9. DeGraaf and Rudis 1987
10. according to nestling period
11. Peck and James 1987

World breeding range

Chronology $n = 179$

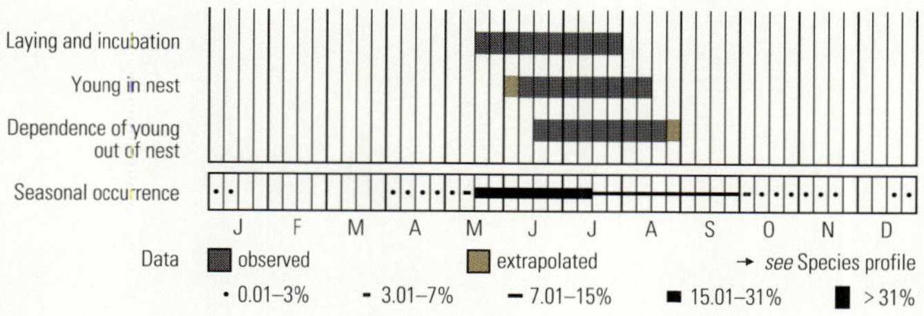

Frequency of occurrence by ecological region

Ecological regions** (forest regions)	Adequately covered squares	Squares in which reported	% of squares in which reported	Relative abundance (%)*
1. Sugar maple-hickory and sugar maple-basswood	97	93	95.9 ± 2.3	95.9
2. Sugar maple-basswood	153	146	95.4 ± 1.9	95.4
3. Sugar maple-basswood and sugar maple-yellow birch	101	101	100.0 ± 0.5	100.0
4. Sugar maple-yellow birch-basswood	56	47	83.9 ± 8.6	83.9
5. Sugar maple-yellow birch-beech and sugar maple-yellow birch-basswood	113	103	91.2 ± 3.1	91.2
6. Sugar maple-yellow birch-beech	71	46	64.8 ± 10.2	64.8
7. Sugar maple-yellow birch-balsam fir	19	5	26.3 ± 21.0	26.3
8. Sugar maple-yellow birch and balsam fir-yellow birch	47	42	89.4 ± 8.3	89.4
9. Balsam fir-yellow birch	87	43	49.4 ± 9.7	49.4
10. Balsam fir-white birch or balsam fir-red maple	46	15	32.6 ± 11.8	32.6
11. Yellow birch-balsam fir	52	7	13.5 ± 9.9	13.5
12. Balsam fir-white birch (west)	59	6	10.2 ± 8.2	10.2
13. Balsam fir-white birch (east)	76	10	13.2 ± 7.6	13.2
14. Balsam fir-black spruce	28	1	3.6 ± 8.3	3.6
15. White spruce-balsam fir and balsam fir-white spruce	22	2	9.1 ± 13.3	9.1
16. Black spruce-balsam fir-moss	18	0	0	0
17. Black spruce-moss	32	0	0	0
18.			*not analysed*	

* minimal: 0–29.9%; low: 30–59.9%; high: 60–89.9%; maximal: 90–100%

** Deciduous forest: 1–7; mixed forest: 8–11; coniferous forest: 12–17

This species also nests in leafy shrubs bordering clearings and fields. Neuville, west of Québec City. *August 1991, Marcel Darveau.*

This bulky Gray Catbird nest containing 3 eggs is made of rough materials outside and finer materials inside. Beauport, near Québec City. *May 1985, Paul Perreault.*

Under cover of a hawthorn, a Gray Catbird feeds its 3 nestlings. The rather crude nest is made of twigs, stalks, leaves, and a piece of plastic. Sainte-Foy. *June 1961, Gaston Le Page.*

The Gray Catbird's diet varies depending on the time of year. Early in the breeding season, it eats mainly insects, such as caterpillars and grasshoppers (Johnson and Best 1982), but switches to fruit as the season advances (Gross 1948a). It is quite tame, and often approaches humans for food (Gross 1948a), such as nuts and butter supplied at bird feeders (J. Vincent, pers. obs.).

The Gray Catbird was reported in 34.4% (848/2464) of Atlas squares surveyed, and breeding was confirmed in 27.8% (236/848) of those. Almost half of confirmations were based on sightings of adults carrying food.

Distribution

The Gray Catbird breeds in southern Canada from British Columbia to Nova Scotia, and in the United States in every state east of a line between Washington State and Texas (National Geographic Society 1987). Frequency of occurrence is highest in the deciduous forest regions, especially in the southernmost part of Québec, and patchy north of the 47th parallel. Occurrence in the Central St. Lawrence Lowland is presumably largely related to the scrubby habitat along the forest edges and woodlands in agricultural and urban areas, as well as some logged-over areas and fields grown up in shrubs and sapling groves.

The most northerly Atlas breeding confirmations were obtained in the Saguenay–Lac-Saint-Jean region and on the Gaspé Peninsula, corresponding fairly closely to the limit of the breeding range determined in Ontario (Curry 1987), at about the 48th parallel. Confirmations of breeding from the Matapédia Valley district also seem to form a continuum with reports from northwestern New Brunswick (Erskine 1992). In the Magdalen Islands, where it was atlassed as a probable breeder, the Gray Catbird is considered an irregular transient (Fradette 1992). The northernmost Atlas reports came from Abitibi, Anticosti Island, and the Middle North Shore.

History and trends

According to BBS data (Erskine et al. 1990), the ÉPOQ database (Cyr and Larivée 1993b), and the Migration Monitoring Program (Hussell 1981), comparison of numbers for southern Québec from the 1960s with those from the 1980s shows a downward trend. In the rest of Canada, however, populations seem to be stable (Dunn 1991).

In the 1820s, Audubon (1990) criticized overhunting of this bird in orchards, since it was liable to affect the survival of the species. Today, hunting is obviously no longer a threat to the Gray Catbird. The exact reason for the decline in abundance in Québec over the last several years is unknown, but it could have something to do with the use of pesticides in orchards and gardens. The clearing of shrubs and undergrowth in parks and on private property could also partly explain the decline, as could the elimination of natural fringes of vegetation along many waterways in the Québec lowlands, but the increase in abandoned farmland in the last 15 years or so outside the Greater Montréal area (M. Carignan, MAPAQ, unpubl. data) may be favourable to the species over the medium term.

Jacqueline Vincent and Manon Bombardier

Northern Mockingbird

Moqueur polyglotte

Mimus polyglottos

Christiane Girard

THE NORTHERN MOCKINGBIRD HOLDS A SPECIAL PLACE IN THE HEARTS of North Americans and is viewed with a mixture of affection and admiration, especially in the southern United States, where the bird is most abundant. The Northern Mockingbird owes its popularity mainly to its vocal abilities. Traditionally, this bird was said to be the finest singer in the New World, every bit a match for the Old World Nightingale. The Northern Mockingbird is an unrivalled mimic, surpassing the other Mimidae found in southern Québec and including in its medley the calls and songs of an impressive number of other bird species, as well as reproductions of the most unusual noises (Whittle 1922).

In Québec, the Northern Mockingbird is a rather rare migrant breeder (David 1980a; David et al. 1990). Some individuals regularly attempt to overwinter, but not all of them are successful.

Habitat and behaviour

The Northern Mockingbird is a bird of open country, often frequenting areas inhabited by humans. Its favourite habitat, in both Québec and New England, is lawns planted with hedges, large trees, and clumps of shrubs. This man-made landscape is remarkably uniform everywhere the bird is found—in small villages, older suburbs, cemeteries, golf courses, and around country cottages.

The nest, made of twigs, leaves, and rootlets, is built by both sexes in a conifer or deciduous shrub, usually 1–3 m up, though sometimes higher (Harrison 1978). The female alone incubates the 3–5 eggs, but both parents feed the nestlings (Breitwisch et al. 1986; Derrickson and Breitwisch 1992). In the southern part of its breeding range, it is not unusual for a pair to raise two broods a season, in most cases building a new nest for the second clutch. Double-brooding may also occur in Québec, since it has been reported in Ontario (Peck and James 1987). The species is essentially monogamous, although both males and females occasionally have more than one mate in one season (Laskey 1962; Logan 1991; Derrickson and Breitwisch 1992).

This bird is very easy to spot. It occupies its song perches —often highly visible locations such as television antennae or electrical wires—for long periods of time and stages lengthy concerts of extremely varied and powerful singing. Some individuals, usually unmated males, sing indefatigably all night long (Logan and Hyatt 1991; Derrickson and Breitwisch 1992). In addition, this mockingbird feeds near human habitation, foraging especially for terrestrial insects and wild fruit. It is thus not surprising that records of birds in suitable habitat account for nearly half of all Atlas observations and that sightings of nests or broods accounted for a large percentage of breeding confirmations.

Distribution

The Northern Mockingbird ranges discontinuously from southern Canada to southern Mexico and the West Indies (Terres 1980). In Québec, it is found mainly around urban centres and villages where suitable habitat is available. It was atlassed in 182 squares (7.4%), and breeding was confirmed in 61 (33.5%) of these. In fact, the Northern Mockingbird is found not only in highly urbanized areas such as the western part of the St. Lawrence Valley (upstream from Québec City), but also in such far-flung locations as Abitibi, Chibougamau (July 1989), around Lake Saint-Jean, the Gaspé Peninsula, and even Sept-Îles and Havre-Saint-Pierre on the North Shore. It was also atlassed on Anticosti Island, and breeding was confirmed on the Magdalen Islands. There are ÉPOQ records from near Blanc-Sablon on the Lower North Shore, and from the Opinaca River, the northernmost sighting of this species.

Analysis of breeding records since 1960 (David et al. 1990) indicates, however, that breeding pairs rarely occupy a nest site for more than two consecutive years, so that occurrence at most sites is highly irregular.

History and trends

At the turn of the century the Northern Mockingbird's range east of the Mississippi River was limited to areas south of Maryland, where it was sedentary (Coues 1903). However, nesting and sporadic overwintering had been reported in southern New England since the early 19th century (Wright 1921). This irregular occurrence north of its traditional range is probably explained by pre- and post-breeding dispersal in search of a suitable place to nest or overwinter.

Using data from the Christmas Bird Counts, Stiles (1982) demonstrated that winter populations of the Northern Mockingbird in New England have continued to increase since 1945. This is the result in large part of the greater availability of fruit-bearing shrubs, notably the multiflora rose. The planting of ornamental shrubs linked to the phenomenal expansion of suburbs in the post-war years provided the mockingbird with an abundant winter food source north of its range, enabling the species to survive and nest there in increasing numbers and thus extend its breeding range (Stiles 1982). The bird nested in New York for the first time in 1950 and, by 1970, was present throughout the state with the exception of the Catskills and Adirondacks (Bull 1974).

The Northern Mockingbird nested in Québec for the first time in 1960 and since then has been observed in the province every year. It had been recorded 14 times in Québec before breeding was confirmed (David et al. 1990), which occurred simultaneously at Como, near Vaudreuil and at Tadoussac, at the mouth of the Saguenay River, two locations about 500 km apart. The Northern Mockingbird now breeds all across southern Québec. Suitable nesting sites are available throughout this area, so distribution is fairly dispersed with no particular pattern of northward expansion within the limits of the province.

Sight records for the breeding season (June, July, and August) from 1960 to 1984 indicate that the species is increasingly reported, but that the increase in sightings is very slow, the Northern Mockingbird being, even today, rarely seen by Québec birders (David et al. 1990).

Most of the birds that nest in Québec are migratory. In all likelihood, they spend the winter in the northern United States, where they encounter competition from the resident population there. Since the species is fiercely territorial in winter, migrant birds from the north are likely at a disadvantage and succumb in the depths of the winter. An unknown number of birds therefore do not return to the places where they bred or were born the previous year. This may explain why few mockingbirds seem to return to nest in the same locations from year to year (David et al. 1990).

NORMAND DAVID

Notes on territory

Males and females defend their territories ferociously against conspecifics of the same sex throughout the year. The largest territories are those defended by pairs, followed by those defended by males and then those defended by unmated females. Breeding territories are generally 0.5–2.0 ha, with the birds occupying smaller areas the rest of the year (Derrickson and Breitwisch 1992).

Breeding densities were calculated in the United States, in habitat of varying suitability. For example, density was only 0.3 pair/10 ha in an oak woodland in California, but 4.7 pairs/10 ha in a suburban Alabama cemetery (Imhof 1993; Petersen 1993).

The Northern Mockingbird prefers urban environments. The landscaping around suburban houses provides plantings of shrubs and trees which are ideal for the species. La Prairie, near Montréal. *July 1991, Gaétan Duquette.*

Northern Mockingbird

Breeding distribution
1984–1989

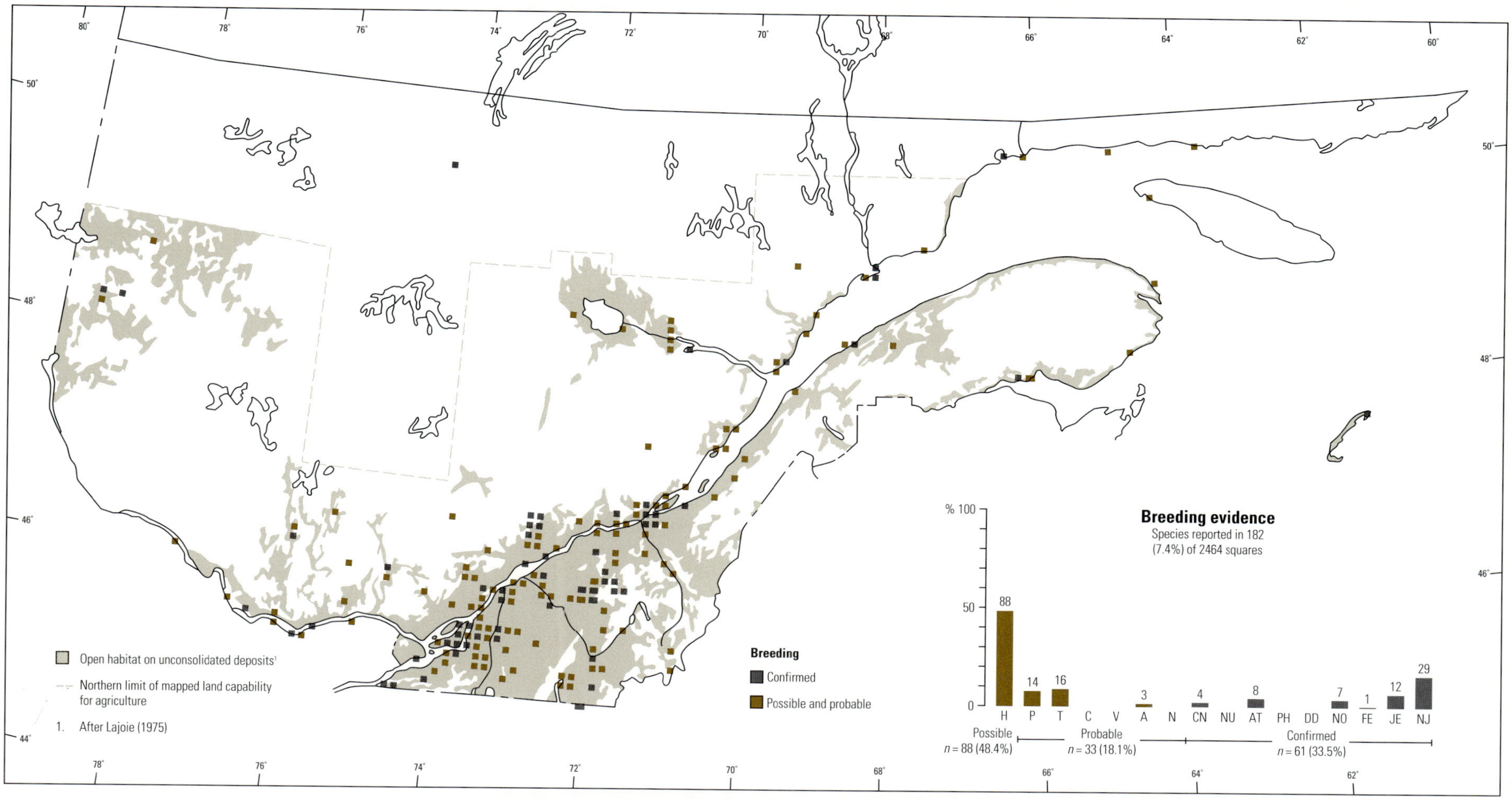

World breeding range

Species profile

Clutch size	usually 3–5 eggs (2–6)[1,4,5]
Laying interval	usually 1 egg per day[1]
Start of incubation	with last egg, sporadic before[1]
Incubation period	usually 12–13 days[1]
Maturity at hatching	altricial
Care of young	by female and male[1]
Nestling period	usually 12 days (10–15)[1]
Age at first flight	usually 20 days (18–23)[1]
Dependence of young	until 21 days after leaving nest[1]
Broods per year	usually 2–3[1]
Breeding age	1 year[1]
Mating system	monogamy, rarely polygamy[1]
Duration of pair bond	usually 1 season, sometimes for life[1]
Total length	M 22.0–25.5 cm / F 20.8–23.5 cm[1]
Wing span	33.0–38.1 cm[3]
Weight	48.5 g (36.2–55.7)[6]
Longevity record	9 years 2 months[2]

References:
1. Derrickson and Breitwisch 1992
2. Klimkiewicz et al. 1983
3. Terres 1980
4. Harrison 1975
5. Harrison 1978
6. Dunning 1984

Chronology *n* = 85

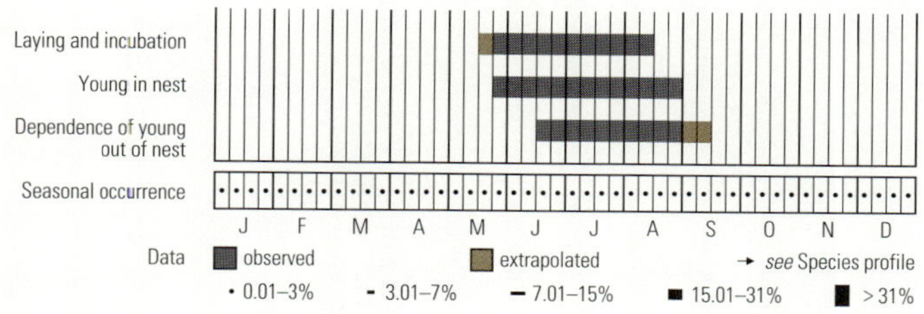

This mockingbird also nests in clumps of ornamental shrubs and in large trees in cemeteries. Mount Royal, Montréal. *July 1991. Gaétan Duquette.*

Brown Thrasher

Moqueur roux

Toxostoma rufum

Christiane Girard

The Brown Thrasher, the largest member of the family Mimidae in Québec, has a remarkable ability to mimic the songs of other birds, along with a wide variety of other sounds. The male has one of the most extensive documented song repertoire of all North American birds (Ehrlich et al. 1988), with over 1000 different song types, whereas the Gray Catbird has only a few hundred (Kroodsma and Parker 1977). The characteristic sequencing of phrases, which are usually paired, is an aid to identifying the species (Godfrey 1986).

Like most mimids, the Brown Thrasher has a long tail and rounded wings for manoeuvring in the underbrush. With its rusty brown coloration and heavily streaked breast, it resembles a thrush, which probably explains why it was formerly known as the Brown Thrush (Macoun and Macoun 1909; Taverner 1919). However, its larger size, long tail, yellow eyes, and white wing bars are reliable field marks. Juveniles can be told from adults by their eyes, which have grey instead of yellow irises (Stokes and Stokes 1989b). In Québec, the Brown Thrasher is a common migrant breeder (Robert 1989b). Most birds from northern regions such as Québec migrate toward the southeastern United States to overwinter (Godfrey 1986).

Habitat and behaviour

The males arrive on the breeding grounds a few days before the females. Early in the breeding season, the male sings loudly to attract a female and define his territory and is often observed singing from the top of a tree. The loud, musical song consists of a long series of phrases punctuated by short pauses.

The species favours underbrush and hedges in dry, well-drained habitat. It occurs in agricultural habitat (old fields, berry plantations, orchards), woodland edges, suburban and rural habitat, and sometimes near human habitation. The Brown Thrasher has a marked preference for young second-growth woods with a well-developed shrub layer (Erwin 1935; Johnston 1958; Stauffer and Best 1980, 1986; DeGraaf and Rudis 1987).

Once the pair is formed and nesting is under way, the male is more retiring (Erwin 1935; Kroodsma and Parker 1977). Both birds build the nest, which is composed mainly of twigs, stems, dead leaves, rootlets, and strips of bark (Erwin 1935; Johnston 1958; Peck and James 1987; Stokes and Stokes 1989b). The nest may be placed on the ground, in thickets, shrubbery, and occasionally in trees, in which case it is built between 0.3 and 3 m up. In Iowa, Stauffer and Best (1986) reported that the average nest height for 31 nests was 1.2 m.

Clutch size varies between 2 and 6 eggs, but is usually 4 or 5. The eggs are pale bluish-white, covered with tiny brown spots (Bent 1948; Godfrey 1986). Incubation is performed by both parents and lasts for 11–14 days (Erwin 1935; Murphy and Fleischer 1986; DeGraaf and Rudis 1987). In Kansas, Murphy and Fleischer (1986) noted that hatching is usually asynchronous. The young fledge from the nest 9–13 days after hatching, but remain dependent on the parents for a few days thereafter (Erwin 1935; Bent 1948). Defence of young against intruders during the breeding season is noisy and aggressive. In fact, agitated behaviour and alarm calls were used as probable breeding evidence in more than 37 squares. The Brown Thrasher is an uncommon cowbird host (Lloyd 1924; Peck and James 1987).

Notes on territory

Upon arrival, the male defends a nesting territory by singing and, if required, pursuing intruders (Erwin 1935; Kroodsma and Parker 1977). All of the pair's activities appear to take place within this territory (Erwin 1935). In Illinois, Graber et al. (1970) estimated that territories along woodland edges averaged 0.6 ha. A density of 1.1 males/10 ha was reported for the Mount Royal Cemetery in Montréal (Erskine 1972a). In Ontario, recorded density was 3.4 territories/10 ha in a tamarack swamp, 2.7 territories/10 ha in an area with shrubs and woodlands, and 3.0 territories/10 ha in habitat containing fields, shrubs, and small trees (Clark 1993a,b; Goodyear 1993). Graber and Graber (1963) noted different densities of Brown Thrashers in different types of agricultural habitat in Illinois: up to 47.2 birds/10 ha in shrubby habitat along fields and water courses, and 19.0 birds/10 ha in shrubbery adjacent to woodland. In second-growth habitat that had been logged, Fawks (1937) found a density of 21.5 birds/10 ha.

The Brown Thrasher is found in open habitat such as long-abandoned farmland containing stands of trees. Mont-Laurier. *May 1991, Yves Aubry.*

Brown Thrasher

Breeding distribution
1984–1989

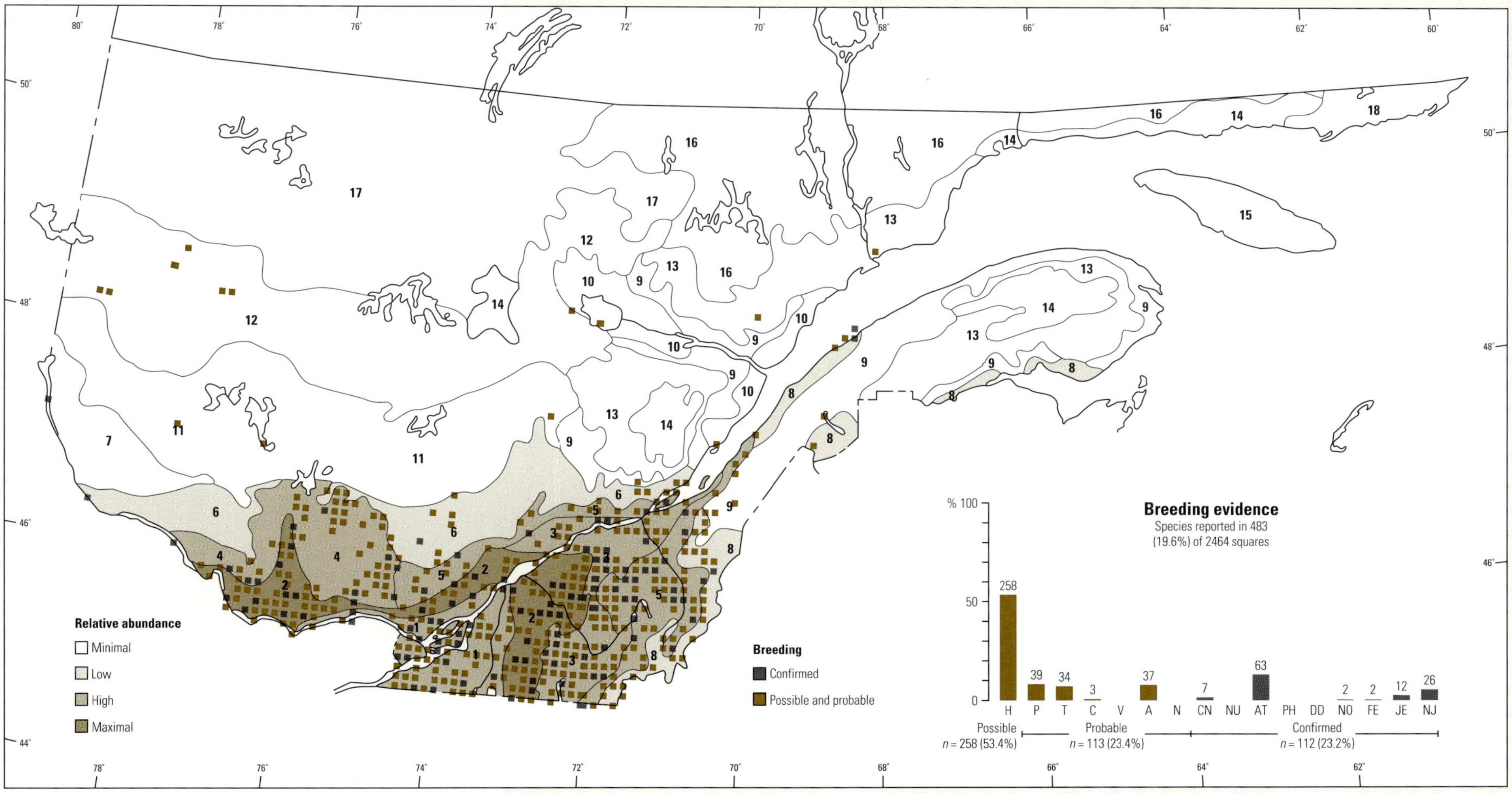

World breeding range

Species profile

Clutch size	usually 4–5 eggs (2–6)[1,2,3,6,8]
Laying interval	1 egg per day[3,7]
Start of incubation	with first egg[3]
Incubation period	11–14 days[1,3,4,5,8]
Maturity at hatching	altricial
Care of young	by female and male[1,5,8]
Nestling period	usually 11 days (9–13)[1,3,8]
Age at first flight	9–13 days[3,5]
Dependence of young	until a few days after leaving nest[1]
Broods per year	usually 1–2[1,2,3]
Breeding age	1 year[4]
Mating system	monogamy[2]
Duration of pair bond	1 season[1]
Total length	26.5–30.5 cm[6]
Wing span	31.8–35.6 cm[5]
Weight	68.8 g (57.6–89.0)[10]
Longevity record	12 years 10 months[9]

References:
1. Erwin 1935
2. Johnston 1958
3. Murphy and Fleischer 1986
4. DeGraaf and Rudis 1987
5. Terres 1980
6. Godfrey 1986
7. Stokes and Stokes 1989b
8. Bent 1948
9. Kennard 1975
10. Dunning 1984

Chronology *n* = 95

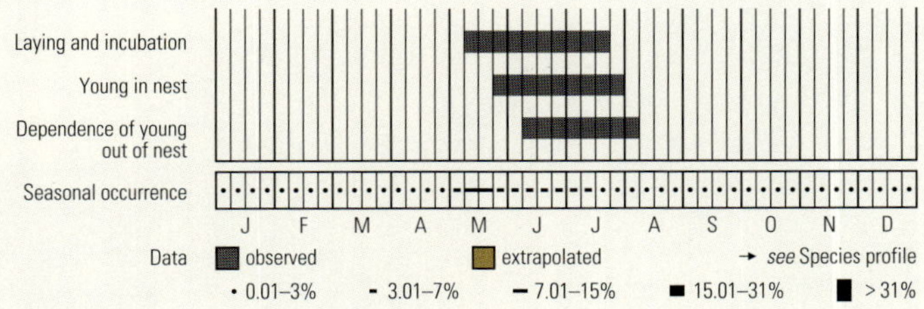

Frequency of occurrence by ecological region

Ecological regions** (forest regions)	Adequately covered squares	Squares in which reported	% of squares in which reported	Relative abundance (%)*
1. Sugar maple-hickory and sugar maple-basswood	97	66	68.0 ± 4.7	86.8
2. Sugar maple-basswood	153	120	78.4 ± 3.4	100.0
3. Sugar maple-basswood and sugar maple-yellow birch	101	71	70.3 ± 5.4	89.6
4. Sugar maple-yellow birch-basswood	56	35	62.5 ± 11.1	79.7
5. Sugar maple-yellow birch-beech and sugar maple-yellow birch-basswood	113	68	60.2 ± 5.0	76.7
6. Sugar maple-yellow birch-beech	71	18	25.4 ± 9.3	32.3
7. Sugar maple-yellow birch-balsam fir	19	1	5.3 ± 12.0	6.7
8. Sugar maple-yellow birch and balsam fir-yellow birch	47	17	36.2 ± 12.4	46.1
9. Balsam fir-yellow birch	87	3	3.4 ± 3.9	4.4
10. Balsam fir-white birch or balsam fir-red maple	46	1	2.2 ± 4.4	2.8
11. Yellow birch-balsam fir	52	3	5.8 ± 7.0	7.4
12. Balsam fir-white birch (west)	59	5	8.5 ± 7.7	10.8
13. Balsam fir-white birch (east)	76	2	2.6 ± 4.0	3.4
14. Balsam fir-black spruce	28	0	0	0
15. White spruce-balsam fir and balsam fir-white spruce	22	0	0	0
16. Black spruce-balsam fir-moss	18	0	0	0
17. Black spruce-moss	32	0	0	0
18.			*not analysed*	

* minimal: 0–29.9%; low: 30–59.9%; high: 60–89.9%; maximal: 90–100%

** Deciduous forest: 1–7; mixed forest: 8–11; coniferous forest: 12–17

The Brown Thrasher nests primarily in trees and shrubs adjacent to fields and clearings. The bird is often observed singing from the top of a snag or tree. Louiseville, north shore of Lake Saint-Pierre. *May 1991, Réginald Ouellet.*

Brown Thrashers favour young second-growth forests but also occur in fields edged with shrubbery. Cap Tourmente National Wildlife Area, northeast of Québec City. *August 1991, Jean Gauthier.*

A Brown Thrasher brooding its young under a brush pile. Nests are normally placed in shrubs. **Drummondville-Sud.** *June 1989, Jeanne Lehoux.*

After one brood has been raised, the pair may raise a second (Erwin 1935; Johnston 1958; Ehrlich et al. 1988). When time permits, the female will start building a second nest for the next brood while the male continues to tend the first one. If there is only one brood, the parents share the duties of caring for the young, often each being responsible for half the brood (Erwin 1935; Stokes and Stokes 1989b). In Kansas, as many as 2 out of 3 pairs may nest twice a year (Murphy and Fleisher 1986). Double-brooding has never been documented in Québec, but Peck and James (1987) report it in Ontario, which indicates that it may well occur here also.

The Brown Thrasher is omnivorous, feeding mainly on insects, fruit, and seeds, depending on the season (DeGraaf and Rudis 1987). Using its strong feet and bill, it often scratches and tosses leaves aside when foraging (Ehrlich et al. 1988).

The Brown Thrasher was reported in 19.6% (483/2464) of Atlas squares surveyed, and breeding was confirmed in 23.2% (112/483) of these, on the basis of 63 records of adults carrying food to the nest, 26 of nests containing eggs or young, and 12 of young that had left the nest. Reports of nest building, occupied nests, and adults carrying fecal sacs were also recorded.

The Brown Thrasher is a permanent resident in the snow-free areas of the United States. Winter occurrences of the species in Québec are extremely rare. A few instances of birds attempting to overwinter have been reported in the Montréal and Aylmer areas (winters of 1936–37, 1962–63, 1976–77, 1991–92, and 1993–94) (David 1980a; Yank and Aubry 1992b; Aubry and Bannon 1994a). Only the overwintering attempt of 1962–63 was successful.

Distribution

From southeastern Alberta, the breeding range of the Brown Thrasher extends east to New Brunswick and south to the coastal states along the Gulf of Mexico (AOU 1983). In Québec, it is commonest in open habitat in deciduous forests and rare in mixed forests. It appears to prefer shrubby and second-growth areas associated with agricultural habitat. Frequency of occurrence was highest in the Central St. Lawrence Lowland, particularly along the St. Lawrence River to Québec City, and in the Ottawa Valley. The species also proved abundant in the Bois-Francs and Beauce regions. A few sightings were reported in Abitibi and Lac-Saint-Jean, where its occurrence can probably be explained by the clearing of land for farming. A post-Atlas record in July 1990 from Alma of a Brown Thrasher feeding a young bird constituted the first breeding record for the species in the Saguenay–Lac-Saint-Jean region (Yank et al. 1990a). The species nested again in this area in 1991 (Ibarzabal and Imbeau 1992).

The species has been sighted a few times in migration on the North Shore as far east as Blanc-Sablon, as well as in the Gaspé and on the Magdalen Islands (David 1980a; Girard 1988; Fradette 1992).

History and trends

At the beginning of the century, most sightings of the Brown Thrasher in Québec were from the Montréal area (Macoun and Macoun 1909; Ouellet 1974a). According to Ouellet (1974a), however, the population in the Monteregian Hills underwent a brief decline during that period.

Over the years, continued clearing of forest and expansion of agriculture appear to have created the various types of habitat preferred by the species, including small woodlots near old fields, and the breeding range has expanded accordingly.

According to the Breeding Bird Survey (BBS), the highest concentrations of Brown Thrashers are found in the central and southern United States. The North American population appears to be stable (Robbins et al. 1986). BBS data in Québec show the species to be declining slightly, whereas analysis of ÉPOQ data for 1970–89 shows a major decline, and the Brown Thrasher has been sighted less often and in lower numbers in recent years than in the late 1960s (Cyr and Larivée 1993b).

Jean Rodrigue and Josée Tardif

American Pipit

Pipit d'Amérique

Anthus rubescens

Christiane Girard

QUÉBEC BIRDWATCHERS RARELY HAVE A CHANCE TO OBSERVE THE American Pipit except in migration as it breeds very locally in southern Québec and its nesting sites are relatively inaccessible.

The American Pipit generally appears in late April or early May, on migration to its breeding grounds. During most of May, it is seen regularly in flocks of a few dozen or even several hundred. Even in migration, the species frequents open habitat and is found, for example, in fields with very low vegetation, ploughed fields, shores, and gravel pits (Bent 1950; Godfrey 1986). It is most often observed in flight, when it gives its characteristic *tsip-it* call. It arrives on its breeding grounds in late May or early June, nesting from mid-June to early August. Between early September and late October, it migrates to its winter quarters, which extend from the southern half of the United States to Guatemala (AOU 1983; Root 1988).

The American Pipit has only recently been recognized as being a distinct species from the Water Pipit, which breeds in Europe (AOU 1989). Both birds were once considered to be races of the same species, *Anthus spinoletta*, until they were observed breeding sympatrically in southern Siberia without hybridizing.

Habitat and behaviour

The American Pipit nests on the tundra or on high mountain ridges (Verbeek 1970; Terres 1980; Godfrey 1986). In the Arctic, it frequents rocky habitat with tussocks of vegetation such as mosses, lichens, and dwarf shrubs. In the mountains, it nests above the timberline in alpine meadows, also known as alpine tundra, an environment similar to arctic tundra.

During its courtship flight, the male sings weak, tinkling notes. It flies up vertically from a rock to a height of 15–50 m and then flutters down, like a falling leaf, feet and wings outspread (Bent 1950; Verbeek 1970).

The nest is built on the ground, often in the shelter of a rock or tuft of vegetation, in areas that are free of snow early in the spring (Verbeek 1970). Many suitable nesting sites are unoccupied because they remain covered with snow too long until after territories have been established. Nest construction begins with a shallow scrape, around which a ring of nest material is placed. More material is added over 4–5 days (Verbeek 1970). The nest may have a side entrance, depending on the surrounding vegetation and the slope of the terrain (Peck and James 1987). A study in Wyoming showed that the entrance usually faces north (Verbeek 1970, 1981). Sometimes nests from previous years are refurbished and reused (Verbeek 1970).

The female alone incubates the eggs for 13–15 days. Meanwhile, the male brings her food, landing some distance from the nest where he then feeds her (Verbeek 1970; Verbeek and Hendricks 1994). She broods the young for 5 days after they hatch, and the male continues to bring food to the nest (Verbeek 1970; Hendricks 1987a). Afterward, the young are fed by both parents. Hendricks (1987a) noted that the male and female leave the nest in opposite directions to avoid competing with one another in their search for food. The

Notes on territory

When establishing its territory, the male proclaims ownership through song flights. Often it is seen walking, with wings lowered and tail upraised. It takes up this position and gives an alarm call whenever another pipit flies over the territory, and it drives away

In the Gaspé, the American Pipit is found in open habitat such as the alpine tundra zone of Mount Albert. Gaspé Conservation Park. *July 1991, Yves Aubry.*

American Pipit

Breeding distribution

1984–1989

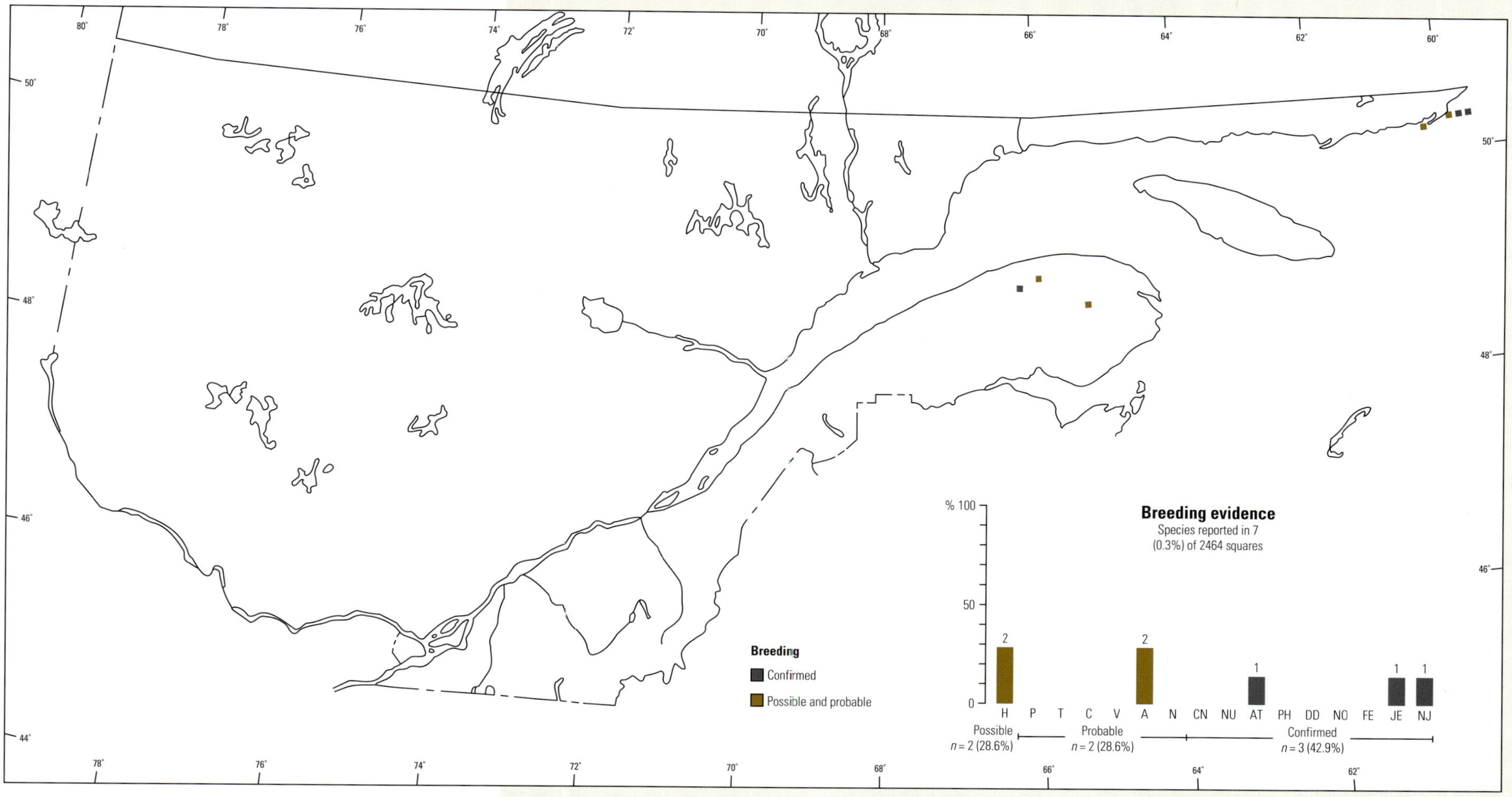

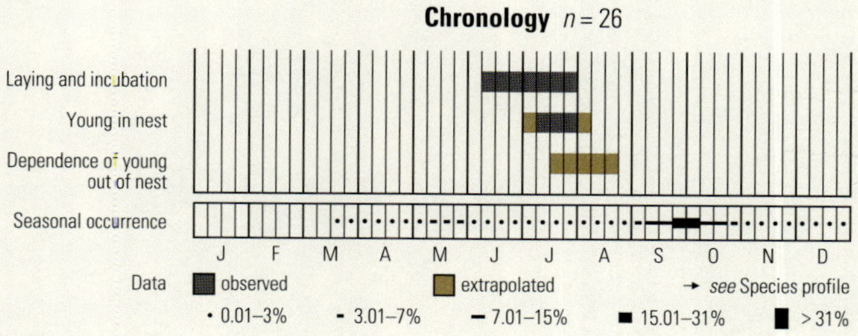

Species profile

Clutch size	usually 4–6 eggs (3–7)[1,2,3]
Laying interval	1 egg per day[1]
Start of incubation	before last egg possible[9]
Incubation period	13–15 days[9]
Maturity at hatching	altricial
Care of young	by female and male[1,3]
Nestling period	12–15 days[1,3]
Age at first flight	14–15 days[1]
Dependence of young	until ≈ 14 days after leaving nest[1,3]
Broods per year	1[1]
Breeding age	1 year[9]
Mating system	monogamy[4]
Duration of pair bond	?
Total length	15.0–17.5 cm[5]
Wing span	25.4–27.9 cm[6]
Weight	M 21.6 g (19.2–25.5) / F 20.1 g (18.6–23.2)[7]
Longevity record	5 years 1 month[8]

World breeding range

References:
1. Verbeek 1970
2. Drury 1961
3. Johnson 1933
4. Ehrlich et al. 1988
5. Godfrey 1986
6. Terres 1980
7. Dunning 1984
8. Klimkiewicz et al. 1983
9. Verbeek and Hendricks 1994

pair forages in an area that is larger than the defended territory (Hendricks 1987b). As the young grow, the parents forage closer to the nest, enabling them to bring back food more quickly and to balance their use of the resources available on the territory. The diet of both young and adult birds is composed mainly of insects taken by foraging on the ground. In general, the American Pipit forages in wetter areas with denser vegetation than those used by the Horned Lark, a species that often occupies the same habitat (Drury 1961; Verbeek 1970; Norvell and Creighton 1990). Pipits, for example, are commonly found along streams. After the young fledge, they remain near the nest for one week and become independent of their parents after a further week (Verbeek 1970).

American Pipits are not difficult to observe in their habitat. In spring, they are easy to locate because of their remarkable flight displays. The few areas in which they nest in southern Québec, however, are practically inaccessible, which is why the species was reported in only 7 squares, with breeding confirmed in only 3, over the entire Atlas period.

Distribution

The American Pipit breeds on the tundra and all along the coast, from northern Alaska to Newfoundland and western Greenland, and in the mountains in western North America south to New Mexico. Nesting in the east is sparse and restricted to the highest ridges of the Appalachians. In Québec, the American Pipit is found in the Gaspé, on the highest peaks of the Shickshock Mountains, and on the Lower North Shore. The 3 sightings from the Shickshocks were on Mount Albert and Mount Jacques Cartier, and another peak located about 20 km southeast of Murdochville. On the North Shore, the American Pipit was observed in the Baie des Loups Migratory Bird Sanctuary, on Matchiatic Island, and a few times in the Sainte-Marie Islands (Cliff Islands, Île aux Perroquets). Todd (1963) considered it a regular summer resident east of Cape Whittle to the Strait of Belle Isle and farther north on the Labrador coast. He also reported finding a nest containing 6 eggs on 21 June 1928 on Galibois Island in the Sainte-Marie Islands. The American Pipit is probably more widespread east of the Mingan region, where the many coastal islands provide suitable breeding habitat. It may also nest on other mountain peaks in the Gaspé that were not atlassed, given their relative inaccessibility. According to Fradette (1992), the species probably bred on the Magdalen Islands at the turn of the century.

The American Pipit is primarily a northern species and a common breeder on the tundra. In Québec, it nests all along the coast and on the coastal islands of James and Hudson bays, north of 53° N (Todd 1963). It is distributed along the coast of the Ungava Peninsula and the coasts of Ungava Bay and Labrador south to the Lower North Shore (Todd 1963). Among other sightings, Todd reported that it was observed and confirmed as a breeder in the interior of the Ungava Peninsula. Breeding records and summer sightings in the interior are scant, however. Near Schefferville, the American Pipit nests in the alpine zone of the mountains, above 670 m. Five nests, each containing 5 eggs, were found in this region in June 1957 (Todd 1963). A pair was also sighted near Lac de la Hutte Sauvage, near George River, on 1 July 1944 (Todd 1963), and another was reported at Caniapiscau, on 9 June 1982 (Bordage and Aubry 1983). The American Pipit was also sighted recently on the rocky outcroppings and lichen heath of the Little Whale and Great Whale river watersheds (Cons. G. G. – GREBE 1991a). During the same study, it was also sighted farther east, in the Lake Bienville area, and breeding was confirmed a little to the south of Lake Amichinatwayach.

Well camouflaged by low-growing vegetation, this pipit nest containing 6 eggs is composed mainly of twigs and grasses. Baie des Loups Migratory Bird Sanctuary, Lower North Shore. *June 1981, Pierre Brousseau.*

Notes on territory (continued)

any birds that land within its territorial boundaries (Verbeek 1970).

In Wyoming, Verbeek (1970) reported that the territories of 5 pairs ranged between 0.15 and 0.34 ha, the average being 0.18 ha. He estimated density to be about 5 pairs/10 ha. In Québec, in the Little Whale and Great Whale river watersheds, recorded densities were 3.7 breeding pairs/10 ha in lichen heath and 3.8 pairs/10 ha in rocky outcroppings (Cons. G. G. – GREBE 1991a). Densities of 0.05 pair/10 ha were obtained in the barrens of the arctic tundra (Baffin Island) and 1.4 pairs/10 ha in a subalpine growth of willow (Banff National Park, Alberta) (Erskine 1971a, 1984). Drury (1961) reported territories of about 0.56 ha on Bylot Island. Pairs forage over a home range of 3.29–10.14 ha, which extends well beyond the defended territory (Hendricks 1987b).

In the James Bay barrens, the pipit often makes its nest on the ground, in the shelter of a rock or tussock. Bay of Many Islands. *June 1990, Austin Reed.*

History and trends

Audubon in 1833 was the first to report the American Pipit on the Lower North Shore and to confirm breeding in Québec. It was Townsend (1923a), however, on climbing Mount Albert in July 1922 who first discovered a breeding population, on the higher peaks in the Gaspé.

At the turn of the century, the species was known as a common spring and fall "transient visitant" in several regions of southern Québec (Wintle 1896; Dionne 1906; Lloyd 1924). Few sightings were reported by Mousley (1918d), however, in the Hatley area of the Eastern Townships. Moreover, the species breeds in such remote areas that it is difficult to give an accurate account of population trends.

Gilles Falardeau

Cedar Waxwing

Jaseur des cèdres

Jaseur d'Amérique
Bombycilla cedrorum

Christiane Girard

THE EARLY FRENCH SETTLERS REFERRED TO CEDAR WAXWINGS AS "RÉCOLLETS" because the brown plumage and hoodlike crest brought to mind the garb of the first Canadian missionaries (Mélançon 1964).

The lisping call notes of a flock of waxwings, the sleek brown plumage, prominent crest, and tameness are some of the characteristics that draw the observer's attention. These highly gregarious birds do not sing, but communicate with each other through soft, whistling call notes (Lea 1942; Putnam 1949).

The plumage is identical in both sexes, and adults of both Cedar and Bohemian waxwings have small red appendages, similar in appearance to tiny droplets of wax, at the tips of the secondary wing feathers; these prolongations of the feather shaft have given rise to the name "waxwing" (Terres 1980). Their function remained unknown for a long time, but recently Mountjoy and Robertson (1988a) postulated that they may be indicative of age and social status. These waxy appendages are absent or fewer in number and less developed in first-year breeders, thus allowing older, more experienced birds to recognize each other more readily within a flock and form more productive pairs.

This abundant migrant breeder returns from the south in late May or early June. The wintering range extends across southern Canada from Nova Scotia to British Columbia and south to Panama in Central America, though the species is casual in the northwestern United States at this time of year (AOU 1983).

Habitat and behaviour

Cedar Waxwings display a variety of interesting behaviours, the most unusual being the exchange of fruit between mates during pair bonding and incubation (Crouch 1936; Putnam 1949). This behaviour is usually initiated by the male, which hops sideways to the female to offer a piece of fruit or a berry held in his bill. The female accepts it, hops away, then back, and returns the offering to the male, which repeats the performance. This continues until one of the birds decides to eat the fruit. Flower petals are also exchanged on occasion. A male may initiate this type of behaviour toward another male, which often responds aggressively (Crouch 1936; Putnam 1949; Stokes and Stokes 1989b).

Nest sites are often associated with fruit-bearing trees or shrubs (Crouch 1936; Lea 1942; Ellison 1985h; McPherson 1988). The timing of the breeding cycle is also influenced by the species' partiality to fruit (Putnam 1949). The Cedar Waxwing is one of the latest nesting birds in Québec. Breeding commences upon arrival on the nesting grounds, and may extend until early autumn, a prolonged time span suggestive of double-brooding. Pair bonds are formed during migration, and pairs begin looking for nest sites on arrival (Putnam 1949; Mountjoy and Robertson 1988a).

The nest is built by both male and female (Crouch 1936; Lea 1942; Putnam 1949). Cedar Waxwings often use materials taken from abandoned nests or even from active nests of conspecifics or other species (Mountjoy and Robertson 1988b). The nest is usually constructed of twigs, dry leaves, pine needles, roots, and sometimes such man-made materials as bits of yarn, string, paper, and cloth (Crouch 1936; Putnam 1949). It is placed in a bush or tree, often in the horizontal fork of a branch and frequently well out from the trunk, about 1.5–15 m up (Lea 1942; Putnam 1949; Peck and James 1987). Only the female incubates, but during this period she is fed by the male, obviating the need for her to leave the nest for prolonged periods (Crouch 1936; Lea 1942; Putnam 1949; Mountjoy and Robertson 1988a). Brood raising is timed to coincide when fruit and berries are abundant, for although the nestlings' diet is initially comprised of insects, fruit is soon added (Crouch 1936; Putnam 1949).

Notes on territory

The Cedar Waxwing is a gregarious bird that displays little territorial behaviour. During the breeding period, its territory consists for the most part of a perch and a small area around the nest. The territory, which includes the nest site, is used for mating but not for foraging (Putnam 1949). A study conducted on an island off Ohio in Lake Erie yielded areas of 0.02, 0.02, and 0.09 ha for 3 territories (Putnam 1949).

In Ontario, densities of 1.1–1.5 territories/10 ha were obtained in a tamarack slough, in a shrubby field, and in mixed habitat of woodland and shrubs (Clark 1993a,b; Goodyear 1993). Elsewhere in Ontario, a density of 2.3 territories/10 ha was obtained in a red pine-Eastern white pine forest (Erskine 1980a) and 4.4 territories/10 ha in an abandoned, overgrown pastureland (Romanow 1993).

The Cedar Waxwing nests and feeds in fruit trees and clumps of shrubs or trees in various urban and suburban habitat in southern Québec. Sainte-Foy. *September 1994, Jean Gauthier.*

Cedar Waxwing

Breeding distribution
1984–1989

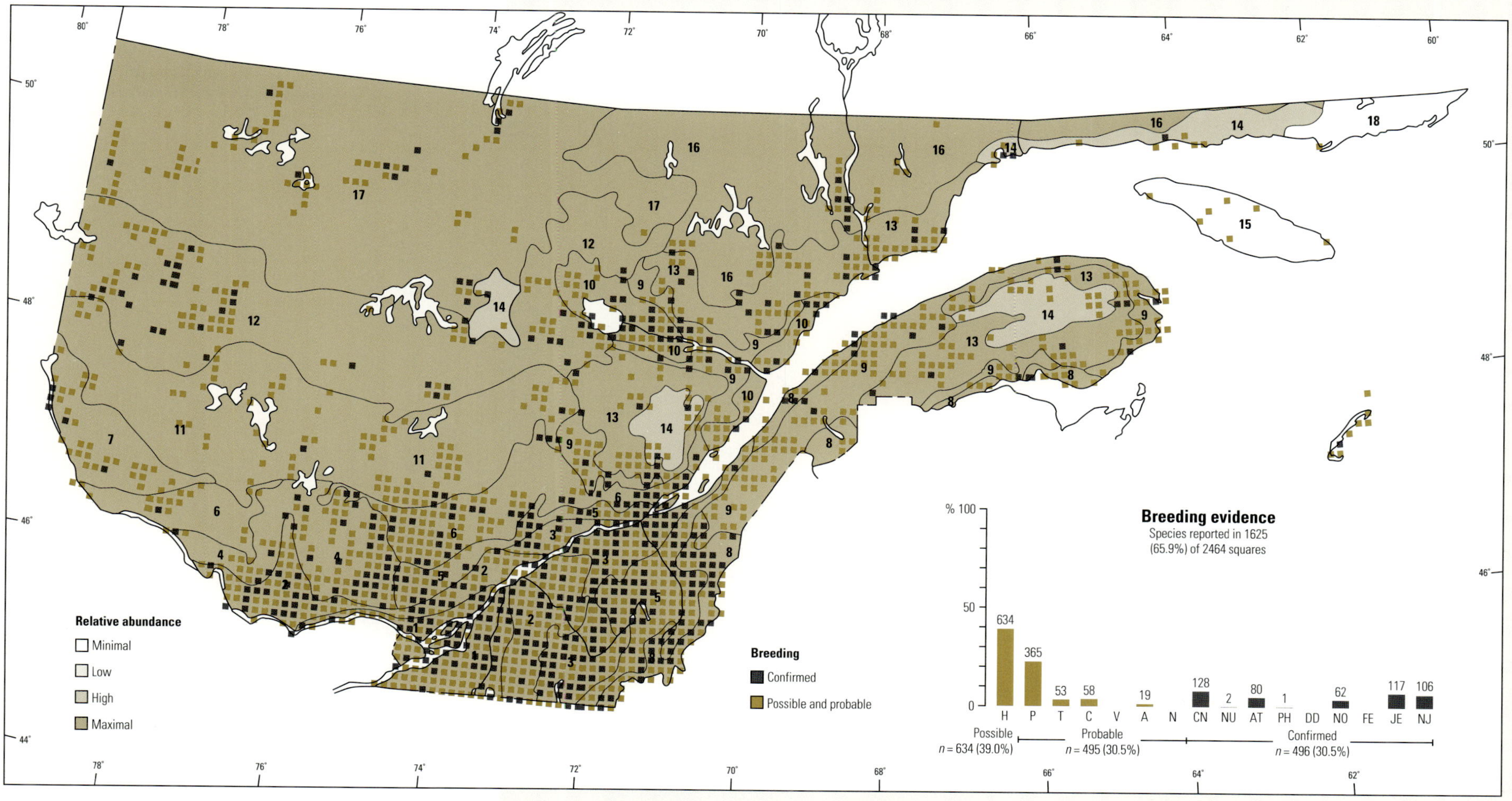

Species profile

Clutch size	usually 4–5 eggs (2–6)[1,2,3,4,6,7]
Laying interval	1 egg per day[4,5]
Start of incubation	with last egg[3,4]
Incubation period	usually 12 days (10–16)[1,2,3,4,5,6,7]
Maturity at hatching	altricial
Care of young	by female and male[3,5,6]
Nestling period	usually 16 days (12–18)[1,3,5]
Age at first flight	usually 16 days (12–18) or more[10]
Dependence of young	until a few weeks after leaving nest[3]
Broods per year	1–2[1,4,6]
Breeding age	1 year[1]
Mating system	monogamy[6]
Duration of pair bond	?
Total length	16.5–20.0 cm[2]
Wing span	27.9–31.1 cm[7]
Weight	M 30.6 g (25.5–39.6) / F 33.1 g (28.0–40.2)[9]
Longevity record	7 years[8]

References:
1. DeGraaf and Rudis 1987
2. Godfrey 1986
3. Lea 1942
4. Crouch 1936
5. Putnam 1949
6. Ehrlich et al. 1988
7. Terres 1980
8. Klimkiewicz et al. 1983
9. Dunning 1984
10. according to nestling period

Chronology $n = 353$

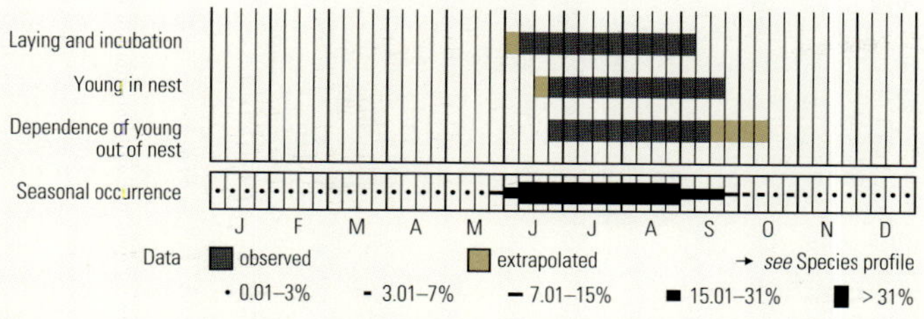

Frequency of occurrence by ecological region

Ecological regions** (forest regions)	Adequately covered squares	Squares in which reported	% of squares in which reported	Relative abundance (%)*
1. Sugar maple-hickory and sugar maple-basswood	97	97	100.0 ± 0.5	100.0
2. Sugar maple-basswood	153	153	100.0 ± 0.3	100.0
3. Sugar maple-basswood and sugar maple-yellow birch	101	101	100.0 ± 0.5	100.0
4. Sugar maple-yellow birch-basswood	56	55	98.2 ± 3.7	98.2
5. Sugar maple-yellow birch-beech and sugar maple-yellow birch-basswood	113	110	97.3 ± 2.0	97.3
6. Sugar maple-yellow birch-beech	71	71	100.0 ± 0.7	100.0
7. Sugar maple-yellow birch-balsam fir	19	19	100.0 ± 2.6	100.0
8. Sugar maple-yellow birch and balsam fir-yellow birch	47	47	100.0 ± 1.1	100.0
9. Balsam fir-yellow birch	87	85	97.7 ± 3.3	97.7
10. Balsam fir-white birch or balsam fir-red maple	46	45	97.8 ± 4.4	97.8
11. Yellow birch-balsam fir	52	51	98.1 ± 4.5	98.1
12. Balsam fir-white birch (west)	59	57	96.6 ± 5.3	96.6
13. Balsam fir-white birch (east)	76	76	100.0 ± 0.7	100.0
14. Balsam fir-black spruce	28	24	85.7 ± 14.1	85.7
15. White spruce-balsam fir and balsam fir-white spruce	22	4	18.2 ± 17.1	18.2
16. Black spruce-balsam fir-moss	18	17	94.4 ± 13.5	94.4
17. Black spruce-moss	32	31	96.9 ± 7.6	96.9
18.			*not analysed*	

* minimal: 0–29.9%; low: 30–59.9%; high: 60–89.9%; maximal: 90–100%

** Deciduous forest: 1–7; mixed forest: 8–11; coniferous forest: 12–17

World breeding range

The species is found in a variety of habitats, notably in deciduous forest edges containing native fruit-bearing trees, such as this magnificent hawthorn with its brightly coloured berries. Cap Tourmente National Wildlife Area, northeast of Québec City. *August 1991, Yves Aubry.*

The Cedar Waxwing is often parasitized by the Brown-headed Cowbird (Rothstein 1975, 1976a,b), yet studies conducted primarily in Michigan have shown the species to be intolerant of parasitism, rejecting parasitic eggs more than two times out of three, either by abandoning the nest (60% of the time), ejecting cowbird eggs from the nest (25%), or breaking them (15%) (Rothstein 1976b). Nest abandonment is the most common response to cowbird parasitism in the waxwing, but it is uncommon in other rejecter species (Rothstein 1976b). This behaviour forces the waxwing to renest and probably affects the length of the breeding period. In Ontario, Peck and James (1987) found 7.5% of 890 nests examined to have been parasitized. They consider this an underestimate as the waxwing is a known rejecter species; as Rothstein (1976a) shows, cowbird eggs are often ejected before nests are examined.

Cedar Waxwings are most often seen in small flocks. Even during the breeding period, the bird will regularly leave the nest to feed with other waxwings (Stokes and Stokes 1989b). The diet of the Cedar Waxwing consists mainly of fruit, as well as insects that are captured on the wing flycatcher-fashion (Terres 1980; Ehrlich et al. 1988; Avery et al. 1993). Like other fruit-eating species, the Cedar Waxwing plays an important ecological role (Stiles 1980; Gorchov 1988) by disseminating the seeds of fruit trees through its droppings (Stiles 1980; Gosselin 1990a).

The Cedar Waxwing frequents a variety of habitats such as urban parks, suburbs, waterside areas, open woodland, and edges of mixed and deciduous forests. Its use of open habitat makes it an easy species to observe, and it is found near residential areas as well as in uninhabited regions. The species was reported in more than 65% (1625/2464) of Atlas squares. Breeding evidence is easily obtained, and breeding was confirmed in 30.5% (496/1625) of the squares in which it was reported.

Cedar Waxwings often feed in flocks over water, darting from their perches to catch insects on the wing. Saint-Léonard-de-Portneuf. *September 1966, Gaston Le Page.*

Four young Cedar Waxwings at the nest, ready to make their first flight. In hot weather, the young birds open their mouths to regulate their body temperature. Saint-Lambert, near Montréal. *August 1930, Henry Mousley, Laval University Collection.*

A Cedar Waxwing in a pine tree, feeding a berry to one of its young. The young are fed for several days after they leave the nest, mostly on berries. Saint-Majorique, near Drummondville. *August 1992, Jeanne Lehoux.*

Distribution

Cedar Waxwings nest in all 10 Canadian provinces and in the United States as far south as California in the west and Georgia in the east (National Geographic Society 1987).

The species inhabits all regions of southern Québec, and this widespread distribution can be explained by the variety of habitat that it occupies. Frequency of occurrence is high almost everywhere. The Cedar Waxwing was the fifth commonest species atlassed. Breeding was confirmed everywhere except Anticosti Island despite numerous summer sightings. Neither Ouellet (1969) nor David (1980a) reported breeding on the island either. Breeding was confirmed for the Atlas on the Magdalen Islands and up to Havre-Saint-Pierre on the North Shore. Farther north, Cedar Waxwings have been observed as far as the James Bay region, at Lake Hélène (Levasseur and Letendre 1979).

In winter, Cedar Waxwings are sighted irregularly in southern Québec, and in some winters the species may be absent altogether (David 1980a). Winter sightings are even less frequent in Lac-Saint-Jean, in the Lower St. Lawrence, and on the Magdalen Islands (David 1980a; Fradette 1992).

History and trends

The deforestation that accompanied European settlement has undoubtedly allowed the Cedar Waxwing to expand its range. Moreover, the second-growth forest that occurs after logging greatly favours the establishment of fruit-bearing trees and shrubs such as cherry, mountain-ash, and raspberry—ideal habitat for the fruit-eating waxwing. The increasing popularity in residential area of ornamental shrubs, many of which bear fruit, also favours waxwing populations.

BBS data for 1965–79 indicate a general increase in abundance in the east, both in Canada and the United States. The species appears, however, to have declined locally in central North America (Robbins et al. 1986). ÉPOQ summer data for 1970–89 show the Québec population to have remained stable, whereas Québec BBS data for 1966–89 show an increase (Cyr and Larivée 1993b). The erratic movements of Cedar Waxwing flocks in fall and winter may explain the erratic distribution of increases and decreases observed in neighbouring regions or from year to year during the breeding season (Robbins et al. 1986).

François Shaffer and Josée Tardif

Loggerhead Shrike

Pie-grièche migratrice

Lanius ludovicianus

Christiane Girard

In the early 1990s, the Québec breeding population of the Loggerhead Shrike was estimated at barely a dozen pairs in the southern part of the province. This is the most seriously endangered species in Québec.

The Loggerhead Shrike is seriously at risk in many parts of North America. It is threatened in some midwestern and eastern U.S. states (Haas 1987; Luukkonen 1987), while in Canada, the Prairie population (*Lanius ludovicianus excubitorides*) has been designated "threatened" and the eastern population (*Lanius ludovicianus migrans*) "endangered" by the Committee on the Status of Endangered Wildlife in Canada (Cadman 1986, 1990). The species winters from the central United States to southern Mexico (Godfrey 1986).

Habitat and behaviour

In this part of its range, the Loggerhead Shrike inhabits pastures, meadows, and abandoned fields with hedgerows, thorny bushes, or conifers. As in Ontario and New York (Novak 1989; Cadman 1990), this bird often frequents marginal agricultural land characterized by a mosaic of small patches of pasture, fallow land, cultivated fields, second-growth areas, and woodland. The nest is built in an isolated tree or hedge. In the northeastern part of its range, the bird often chooses hawthorns, although it sometimes takes up residence in apple or spruce trees. The majority of the 300 or so nests found in Québec since the end of the last century were built in hawthorns (Robert and Laporte 1991a). The dense crown of these bushes affords good camouflage for the nest in the early part of the breeding season, before the leaves emerge (Porter et al. 1975; Luukkonen 1987; Gawlik and Bildstein 1990). Their thorns also enable the Loggerhead Shrike to protect its nest and impale its prey; its feet are not strong enough to grasp and manipulate large prey for any length of time, and it therefore habitually impales its prey in order to tear it into pieces for consumption (Miller 1931; Smith, S. M. 1972).

Individuals of this species migrate singly and arrive in Québec early, often by late March. Pair formation therefore takes place on the breeding grounds. Both males and females gather nesting material, but the female alone builds the nest (Miller 1931; Scott and Morrison 1990), which is usually located near the middle of the nest tree or bush (Gawlik and Bildstein 1990). The female incubates the eggs, with the male assuming the task of providing food for the female (Applegate 1977; Kridelbaugh 1983; Novak 1989). The nestlings are fed by both parents. The Loggerhead Shrike feeds mainly on insects (in particular grasshoppers and crickets), small mammals, and small songbirds (Judd 1898; Miller 1931). It spots its prey from lookouts on electrical wires, fence posts, and dead trees, where it sometimes remains motionless for long periods of time (Craig 1978; Morrison 1980). This is thus an easy bird to observe, especially since it frequents open areas and its white wing patches are highly visible in flight. The Loggerhead Shrike may, however, be confused with the Northern Shrike, which occasionally lingers in the southern part of the province until May, and with the Northern Mockingbird (Robert 1991). Be that as it may, the low number of breeding confirmations testifies clearly to the rarity of the Loggerhead Shrike in Québec, since nesting by this species is, all in all, fairly easy to confirm.

Notes on territory

The size of the home range during the breeding season is highly variable and depends mainly on such factors as the structure and composition of the vegetation, the concentration and location of food sources, the availability of nesting substrates, and local species abundance (Miller 1931). The birds sometimes inhabit areas of several dozen hectares (Scott and Morrison 1990), but most shrike pairs occupy territories smaller than 10 ha. For example, Novak (1989), who studied the Loggerhead Shrike in northern New York and in Ontario, reported a mean territory size of 7.5 ha (5.7–9.3 ha).

The Loggerhead Shrike is the most seriously endangered species in Québec. It frequents open land with scattered shrubs and trees in which it can nest. Bury, east of Sherbrooke. *July 1990, Michel Robert.*

Loggerhead Shrike

Breeding distribution

1984–1989

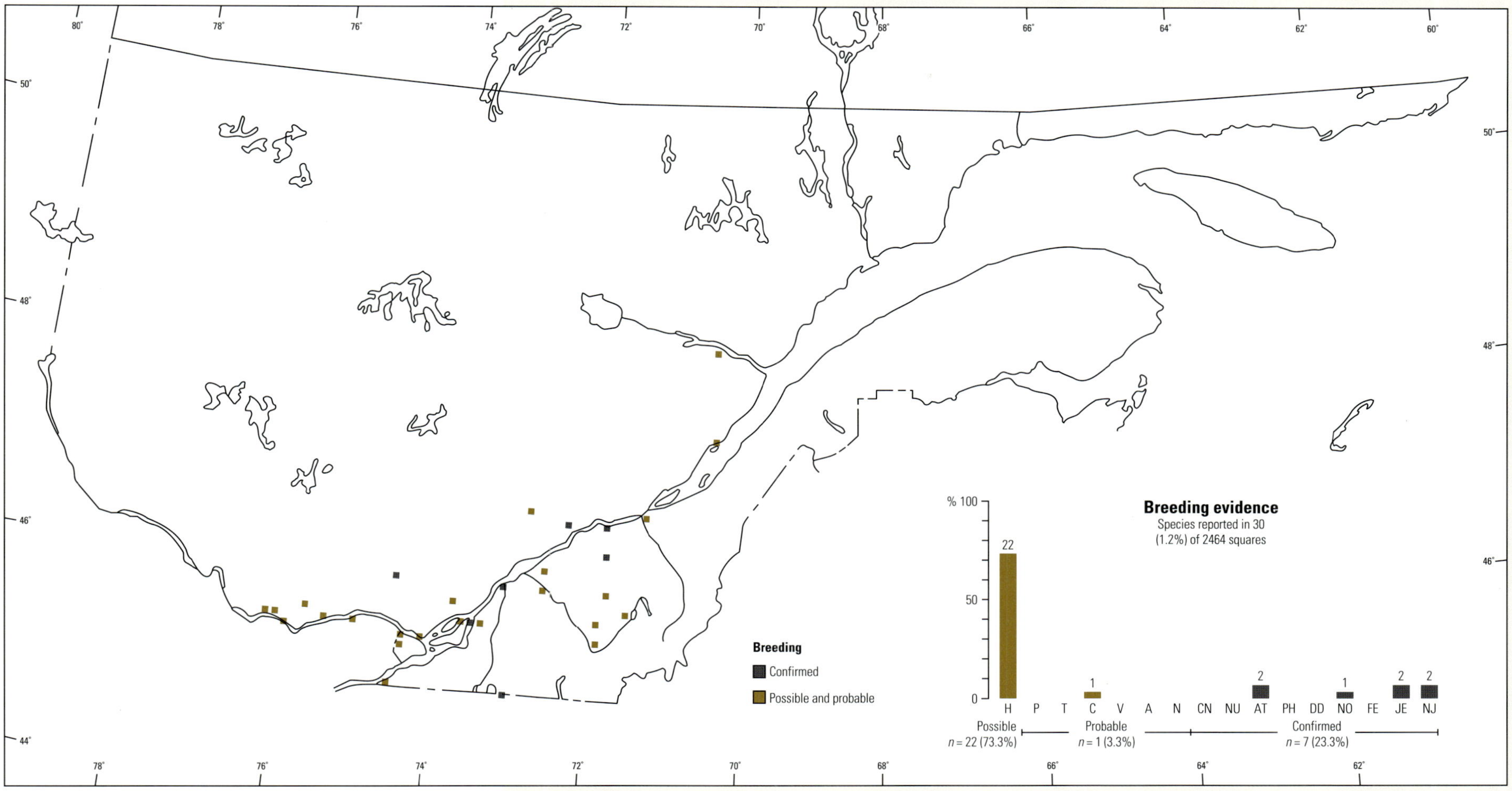

Chronology *n* = 211

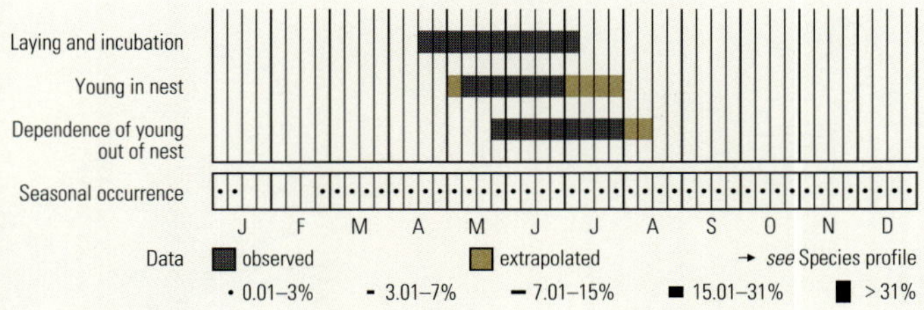

Species profile

Clutch size	usually 5–6 eggs (3–8)[1,2]
Laying interval	1 egg per day[2]
Start of incubation	with penultimate egg[2]
Incubation period	16–18 days[2,7]
Maturity at hatching	altricial
Care of young	by female and male[2]
Nestling period	16–21 days[1,2,7]
Age at first flight	16–21 days[1,2,7]
Dependence of young	until 16–25 days after first flight[2,7]
Broods per year	usually 1 (1–2)[1,2]
Breeding age	1 year[2]
Mating system	monogamy[2]
Duration of pair bond	sometimes more than 1 season[2]
Total length	22.0–24.5 cm[3]
Wing span	31.8–33.0 cm[4]
Weight	47.4 g (40.5–54.1)[6]
Longevity record	6 years 1 month[5]

References:
1. Robert and Laporte 1991a
2. Miller 1931
3. Godfrey 1986
4. Terres 1980
5. Klimkiewicz and Futcher 1989
6. Dunning 1984
7. Scott and Morrison 1990

World breeding range

This predatory passerine looks for fairly open habitat which provides lookout posts. It impales its prey on thorny shrubs or barbed wire. A pair nested in this field at Grondines from 1990 to 1992. Southwest of Québec City. *May 1990, Michel Robert.*

This shrike prefers to build its nest in the dense crown of a hawthorn—a tree with long, sturdy thorns on which to impale its prey. Batiscan, southeast of Trois-Rivières. *May 1989, Pierre Laporte.*

Two adults, accompanied by 4 young, were observed on the edge of this field in the Le Gardeur area in 1991. A pair occupied the same site the following year. Near Montréal. *July 1991, Pierre Bannon.*

Loggerhead Shrike nest with 3 nestlings in a hawthorn. Saint-Pierre, Island of Montréal. *June 1935, Henry Mousley, Laval University Collection*

These two young birds, about 7 days old, have just been banded. After being weighed and measured, they were returned to the nest. Grondines, southwest of Québec City. *July 1990, Michel Robert.*

Distribution

The breeding range of the Loggerhead Shrike extends from southern Canada to the southern United States and Mexico (Godfrey 1986). It was atlassed in about 30 squares, mainly in the Central St. Lawrence Lowland and in the Laurentian and Appalachian foothills. It was found breeding in only 7 Atlas squares between 1984 and 1989, and never more than one pair was reported in any one of those squares. Owing to the precarious status of this bird in Québec (Robert 1989b), the Canadian Wildlife Service has, since 1989, done extensive surveys of the southern part of the province in order to map its distribution. Birders have been made aware of the importance of reporting sightings. Despite these efforts, very few birds have been discovered. In 1989, the dozen sites where the species had been nesting since the early 1980s were surveyed, and no shrikes were found. Two pairs were sighted, however, that year at Notre-Dame-de-Lourdes, near Victoriaville, and Saint-Prosper, northeast of Trois-Rivières. Almost 13 000 km^2 of survey coverage in 1990, including about 50 sites where it historically bred, produced only one pair, in the Grondines area (Robert and Laporte 1991b). In 1991 and 1992, the Grondines birds were again spotted, and another pair was sighted in the Le Gardeur region, near Montréal. The Loggerhead Shrike has now disappeared from all the sites where breeding had been confirmed during atlassing.

History and trends

At the time of European settlement, the Loggerhead Shrike did not breed in Québec. The earliest observations of this species here date from the 19th century, when extensive deforestation for settlement and farming greatly expanded suitable shrike breeding habitat. It is not known precisely when the bird became established in southern Québec, but it seems fairly certain that it was nesting here in the 1860s or 1870s. The earliest breeding records date from the beginning of the next decade (Robert and Laporte 1991a,b), by which time the species was common here. Wintle (1896) listed the Loggerhead as common in Montréal, and the personal notes of contemporary ornithologists (mostly Lewis McIver Terrill) reveal that a single field trip in Montréal and the surrounding area could regularly turn up several nests (up to 5). At the turn of the century, the species was also found in the Eastern Townships and the Ottawa Valley. It then spread eastward, but although there are sight records for other parts of Québec (David 1980a), the Loggerhead Shrike has never been reported breeding outside the Ottawa Valley, the Laurentian foothills (as far as Cap Tourmente), the Eastern Townships, the St. Lawrence Plain, and the Lower St. Lawrence west of Rimouski (Robert and Laporte 1991a,b).

The decline observed in this species over recent decades has been spectacular. In Québec, the first indications date from the late 1940s, and in all probability the downward trend is still continuing. In fact, frequency of occurrence of the Loggerhead Shrike in the southern part of the province has dropped dramatically over the past 20 years (Robert 1989b). Changes in the agricultural landscape in recent decades have likely contributed to the decline in numbers in Québec (Robert and Laporte 1991b) and other parts of northeastern North America (Telfer 1992). The gradual disappearance of pastureland, the increased acreage under cultivation, the widespread growing of corn, the disappearance of windbreaks along boundaries, and the reversion to forest of marginal farmland have eliminated much suitable breeding habitat in southern Québec (Robert and Laporte 1991b; Robert et al. 1991).

Michel Robert, Pierre Laporte, and Alain Demers

European Starling

Étourneau sansonnet

Sturnus vulgaris

Christiane Girard

THE EUROPEAN STARLING IS ONE OF THE BEST-KNOWN BIRDS IN NORTH America, and with a population of almost 200 million (Cabe 1993) it is also one of the most numerous. The current size of the population is extraordinary considering that this exotic species got its start on this continent with the introduction of a hundred birds at the end of the last century to Central Park in New York City (Feare 1984).

Though many people assume that the name "starling" derives from the numerous white flecks that appear on its plumage in late summer, Choate (1973) thinks that it was so named because of the four birds on the silver coins of King Edward the Confessor. He also states that the silhouette of the birds in flight may suggest little stars. Despite its fame as a versatile vocal mimic of other songbirds (Allard 1939), the European Starling is not one of the best-liked bird species. In many regions it causes extensive damage to crops, along with Red-winged Blackbirds, and successfully outcompetes native bird species for nesting sites. Consequently, the European Starling is the bane of many birdwatchers and a pest to most farmers. Since its introduction, it has spread rapidly, such that today it breeds almost everywhere on the continent south of the tundra. The European Starling is associated primarily with urban environments and is one of the few bird species to winter within cities and towns (Feare 1984). It is a common year-round resident throughout much of southern Québec (David 1980a).

Habitat and behaviour

The European Starling is historically a temperate grassland species, requiring open areas of short grass for foraging (Feare 1984). It has proved very adaptable to human activity and today can be found in all types of open and semi-open habitats (Bonney 1988c). It breeds in cities, towns, farmlands, and woods (Feare 1984), although it is usually absent from heavily forested areas (Ehrlich et al. 1988).

European Starlings may occasionally nest in loose colonies (Ehrlich et al. 1988) and are primarily monogamous, with a few cases of polygyny having been documented (Feare 1984). They nest mostly in cavities, often using flicker holes or any convenient cavity in a building or other man-made structure (Ellison, G. F. 1985).

Starlings cannot excavate their own cavities and must therefore rely on preexisting cavities to breed. In descending order of frequency, the most common nest locations of 2518 nests in an Ontario study were: trees (including snags and stumps), nest boxes, buildings, fence posts, mail boxes, utility and flag poles, pipes, miscellaneous man-made objects, cliffs and banks, bridges and dams, and raptor nests. Nests in trees ranged from 2.4 to 9 m above ground (Peck and James 1987). Lumdsen (1980) found that nest boxes occupied by starlings in wooded areas were always located within about 500 m of cleared land.

Unlike most cavity-nesting species, European Starlings build a nest within the cavity (Ehrlich et al. 1988). Nest material depends on what is available in the immediate vicinity, but most often consists of grass, straw, fine twigs, and conifer needles (Kessel 1957). Nests are often lined with fresh green vegetation containing chemicals that act as fumigants against parasites and pathogens (Clark and Mason 1985). Both sexes participate in nest construction (Kessel 1957).

European Starlings compete with Eastern Bluebirds, Tree Swallows, House Sparrows, Purple Martins, Great Crested Flycatchers, Red-headed Woodpeckers, Northern Flickers, and American Kestrels for nest cavities (Bent 1950; Ellison, G. F. 1985; Davis et al. 1986; Curley et al. 1987; Wilmers 1987; Bonney 1988c). They have also been documented competing for nest boxes that were set up for Common Goldeneye and Wood Duck (Lumdsen 1980; Feare 1984). They will often attempt to usurp cavities by puncturing or removing resident starling eggs or those of species as large as Wood Ducks (Wilmers 1987; Lombardo et al. 1989).

Starlings may also parasitize a conspecific's nest by adding one of their own eggs to the existing clutch, and these are seldom removed by the host after clutch initiation (Stouffer et al. 1987). In New Jersey, Romagnano et al. (1990) found that one out of every four early nests (initiated in April) contained at least one parasitic egg from another starling. Hosts appeared to use countermeasures to minimize the cost of parasitism: throwing out the parasitic egg if the resident had not begun laying; beginning incubation before the clutch is completed, thus reducing nest vulnerability to parasitism; laying synchronously within the population, thus limiting opportunities for parasitism to a short period of time; and reducing clutch size to avoid raising an overcrowded brood (Power et al. 1981; Lombardo et al. 1989; Power et al. 1989; Romagnano et al. 1990; Cabe 1993).

Notes on territory

First males and then females begin showing interest in nest sites in the fall or winter, though serious nest building does not begin until the following spring (Kessel 1957). The defended breeding territory appears to be usually a 25–50 cm radius about the hole of the cavity (Kessel 1957). Starlings may build their nests near those of conspecifics or other species (Bent 1950), but within limits, and some individuals may be more tolerant than others (Kessel 1957).

The European Starling breeds in a variety of habitats, including cliffs, to which it has managed to adapt. In 1936, Stanley C. Ball first observed several pairs nesting in crevices exposed to sea spray on the south side of the Forillon Peninsula. Forillon National Park. *July 1992, Yves Aubry.*

European Starling

Breeding distribution

1984–1989

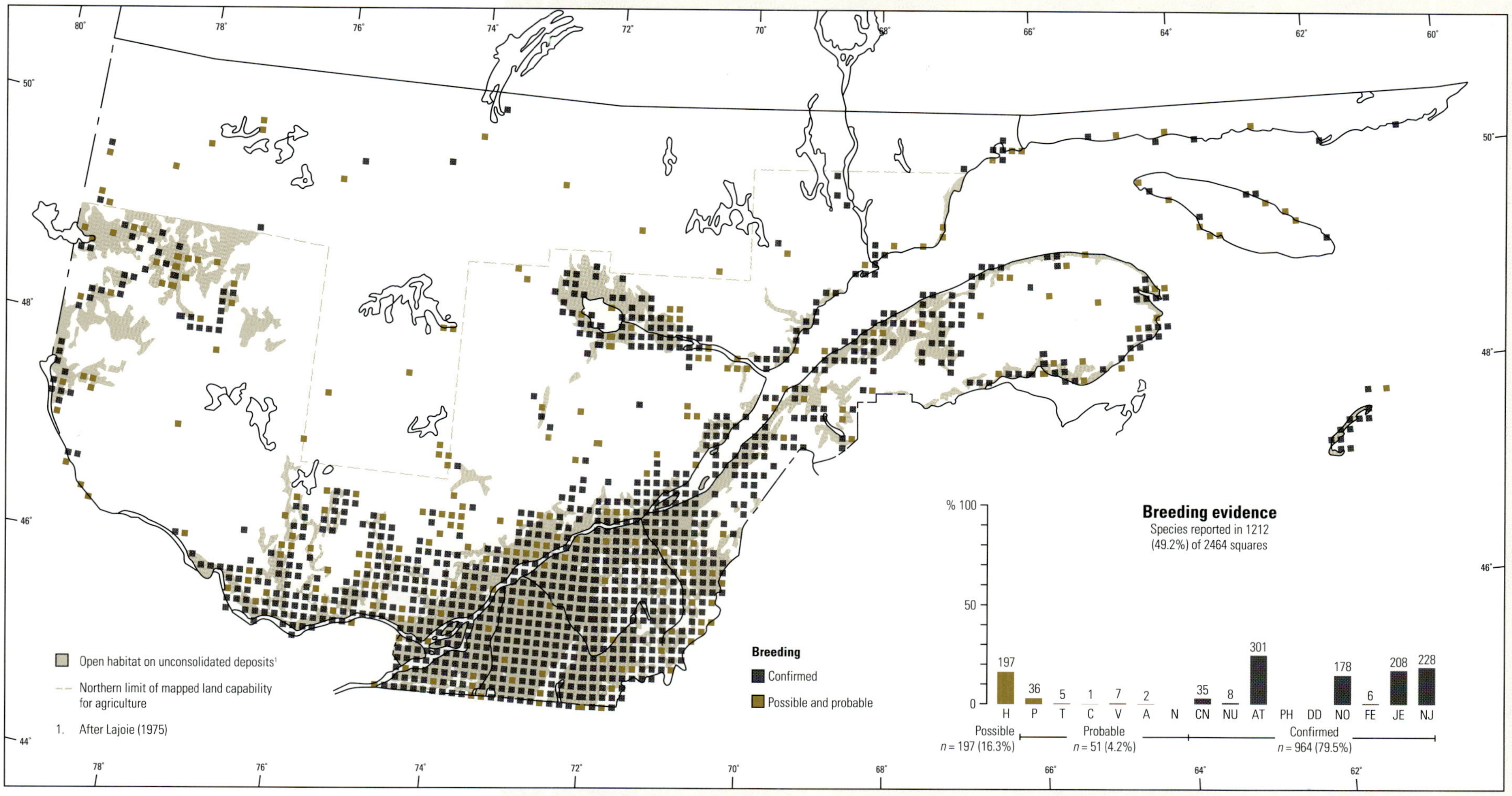

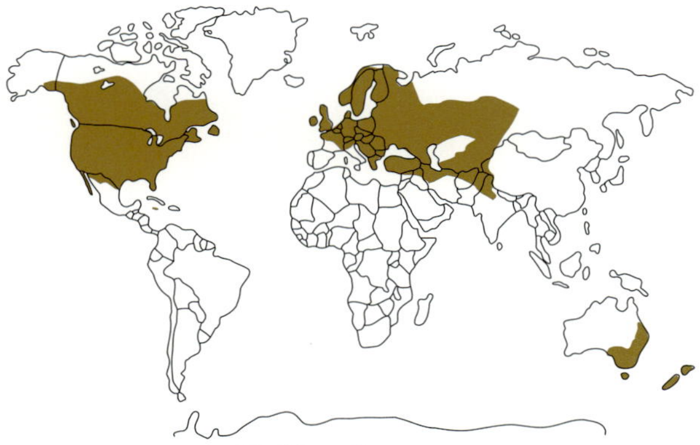

Species profile

Clutch size	usually 4–6 eggs (2–8)[1,2,3,4]
Laying interval	1 egg per day[1,5,8]
Start of incubation	with last or penultimate egg, sporadic before[1,8]
Incubation period	usually 11–12 days (11–14)[1,3,4,8]
Maturity at hatching	altricial
Care of young	by female and male[1,3,6,8]
Nestling period	14–23 days[4,6,8]
Age at first flight	usually 21 days (21–23)[8]
Dependence of young	until 4–12 days after leaving nest[1,8]
Broods per year	usually 2 (1–3)[1,2,5,6,8]
Breeding age	1 year[8]
Mating system	monogamy, sometimes polygyny[1,6,8]
Duration of pair bond	1 season[6,8]
Total length	19.0–21.5 cm[3]
Wing span	39.4 cm[7]
Weight	M 83.8 g (75.5–97.0) / F 79.4 g (75.0–89.0)[8]
Longevity record	21 years 4 months[8]

References:
1. Kessel 1957
2. Long 1981
3. Godfrey 1986
4. Bent 1950
5. Royall 1966
6. Ehrlich et al. 1988
7. Terres 1980
8. Cabe 1993

Chronology $n = 335$

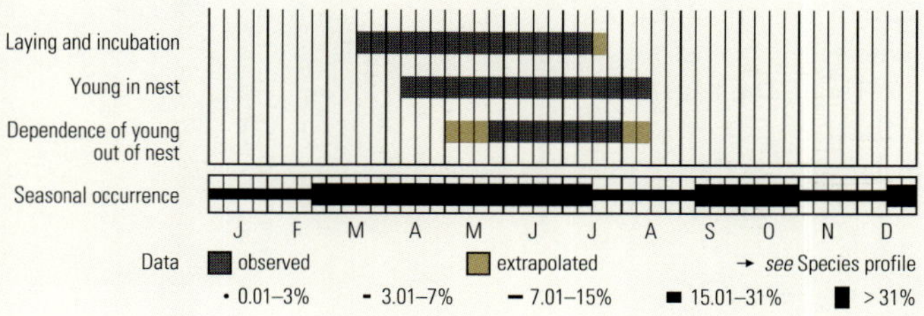

The European Starling nests in all sorts of man-made structures, from farm buildings in the country to highrises in city centres. Sainte-Foy. *June 1987, Pierre Laporte.*

Initiation of laying is highly synchronous in any given starling population, with 4–6 eggs laid in a typical clutch (Feare 1984). Incubation lasts 11–14 days (Bent 1950; Godfrey 1986). The female incubates throughout the night and for most of the day, whereas the male incubates a little during the day. During the first few days, the altricial chicks are very vulnerable to heat loss and are brooded, mostly by the female, during the night and part of the day (Feare 1984). Meanwhile, the parents feed the nestlings small caterpillars, grubs, and other insects. The chicks grow rapidly and fledge at about 21 days of age. At fledging, the young are almost as fully feathered as the parents and can fly well. After leaving the nest, they become independent in 4–12 days and then join together in flocks with other juveniles, feeding independently with one or more adult birds usually among them (Kessel 1957; Cabe 1993).

European Starlings roost communally outside the breeding season, and flocks of more than 100 000 birds have been reported. They roost most often in association with Common Grackles, but mixed flocks may also contain a small percentage of other species such as Red-winged Blackbirds, American Robins, or Mourning Doves (Caccamise and Fischl 1985). These flocks of "blackbirds" can do extensive damage to agricultural crops (Weatherhead et al. 1980; Cummings et al. 1989).

Because European Starlings breed primarily in urban and rural areas (Fischl and Caccamise 1985), where they are easily observed, the Atlas map probably gives an accurate representation of their distribution in southern Québec. Furthermore, because their nests are easy to find, breeding was confirmed in 79.5% (964/1212) of squares in which the species was reported, mostly on the basis of records of actual nests or of adults entering or leaving a nest cavity (42.1%, or 406/964), followed by observations of adults carrying food (31.2%, or 301/964).

In wooded margins of fields, marshes, or bodies of water, the starling often builds its nest in an unoccupied hole in a dead tree or in one that it has wrested from a bird of some other species. Rivière du Sud, a tributary of the Richelieu River. *August 1974, Léo-Guy de Repentigny.*

Distribution

The European Starling has one of the most extensive breeding ranges of any bird species in Canada. It breeds throughout every province, except for the northern regions of Québec and Labrador where, as in Yukon and the Northwest Territories, it breeds as far north as the tree line (Godfrey 1986).

The breeding distribution of the European Starling in southern Québec closely reflects the pattern of human habitation. It was confirmed as a breeder in the majority of squares south of the St. Lawrence River from the Ontario border east to Montmagny as well as in the Ottawa Valley and in the Saguenay–Lac-Saint-Jean region. It breeds throughout the Gaspé Peninsula, where it has adapted to local conditions by nesting in cavities in the cliffs along the shore (Ball 1945). It also nests on the Magdalen Islands, on Anticosti Island, along the North Shore as far east as La Romaine, and in inhabited areas scattered throughout the boreal forest between Lake Saint-Jean and Rouyn-Noranda. It was also found breeding at the northern limit of the Atlas area, at the Chalifour River near Mistassini.

The European Starling often usurps the nest sites of other species, like a hole carved by a flicker in a hydro pole. Fraser Road, Lac Saint-François National Wildlife Area, southwest of Montréal. *June 1990, Léo-Guy de Repentigny.*

Notes on territory (continued)

In urban areas, the species' density in the breeding season may vary considerably according to habitat. Erskine (1971a, 1972a, 1976a, 1980a) reports densities of anything from 0.5 to 24.3 males/10 ha in a number of cities, including Montréal, Toronto, Kitchener, and Waterloo. In Toronto, Savard (1978) observed the highest densities of breeding pairs (24.2/10 ha) in residential neighbourhoods of two-storey semi-detached houses with small lawns and a few large trees.

After breeding, European Starlings start to gather in flocks as early as mid-July. Here, a startled flock flees the roost where it had settled at sunset. Henry Mousley estimated the size of this flock in Saint-Lambert, near Montréal, at some 30 000 birds. A number of roosts of this kind can be found in the southern part of the province. *September 1931, Henry Mousley, Laval University Collection.*

A Red-headed Woodpecker was busy excavating its nest in this snag on Île Ronde when it was evicted by a starling. Berthier Islands, west of Lake Saint-Pierre. *June 1992, Daniel Jauvin.*

In northern Québec, a pair of starlings was observed feeding 5 young at Fort Mackenzie in July 1952 near the Caniapiscau River (Bleakney 1953). Starlings have also bred at Kuujjuaq, on Ungava Bay, in 1967 (Heyland 1970) and in 1976 (COQ 1976). They have been recorded without evidence of breeding at such places as Blanc-Sablon, Schefferville, Tasiujaq, Salluit, Povungnituk, Inukjuak, and Kuujjuarapik (Heyland 1970; Godfrey 1986; ÉPOQ).

Northern breeding populations in North America are partially migratory (Godfrey 1986). A part of the Montréal population migrates south in autumn (Ouellet 1974a). Banding records have shown that some birds migrate some winters and not others and that differing migratory tendencies may occur even among siblings (Kessel 1953). Some birds, instead of migrating, may gather close to permanent food sources in urban and agricultural areas, a common example being landfills.

History and trends

In Eurasia, the European Starling has a long historical association with man, possibly going back to the advent of agriculture (Ehrlich et al. 1988). The introduction of the European Starling into North America has been well documented. The first successful introduction was in Central Park in New York City, where 60 birds were introduced in 1890 and 40 birds the following year. These birds were released by Eugene Scheifflin in his attempt to introduce into North America all the species mentioned in Shakespeare's plays (Chapman 1925; Bent 1950). He also introduced the House Sparrow into North America (Chapman 1925).

Attempts were made to introduce the European Starling into Québec in 1875, 1889, and 1892, but all apparently failed (Long 1981). A specimen was collected in 1878 on the Labrador coast (Todd 1963); it presumably came from the attempted introduction 3 years earlier in Québec or a drift across the Atlantic from Europe (Gosselin 1977b). The first Canadian record of European Starlings derived from the original introduction in New York City was in 1914 at Niagara Falls, Ontario, and the first in Québec was a specimen collected by F. W. Salzman at Betchouane on the North Shore in April 1917 (Todd 1963). The first breeding record came 5 years later in 1922, when a pair was observed nesting in Saint-Lambert, near Montréal (Terrill 1924). A decade later, the confirmed breeding range had expanded to Charlevoix County on the north shore of the St. Lawrence, 120 km northeast of Québec City, and into the Gaspé Peninsula (Lewis 1934b).

Analysis of BBS and ÉPOQ data shows that the Québec breeding population has remained stable over the last 20 years (Cyr and Larivée 1993b). The European Starling, adaptable as it is to human activity (Cabe 1993), is secure here in Québec for the foreseeable future.

RICHARD C. COTTER, DERIN HENDERSON, AND MICHAEL SPENCER

Solitary Vireo

Viréo à tête bleue

Vireo solitarius

Christiane Girard

The Solitary Vireo, as its name indicates, nests and lives alone, isolated from others of the species. It is rarely seen in cities or near dwellings, except in migration. It owes its French name to its blue head, but the colour is not very visible except in full light.

It arrives in Québec at the end of April, several days ahead of the other vireos that visit the province. The Solitary Vireo is also the last of the vireos to head south in fall (Levine 1988c). It winters from the southeastern United States all the way to El Salvador (Kibbe 1985k).

This vireo usually builds its hanging nest in the fork of a horizontal branch in a conifer. For several years now, the species has been nesting in this conifer plantation at Saint Lazare. West of Montréal.
June 1991, Pierre Bannon.

Habitat and behaviour

The Solitary Vireo breeds in young, mature, or overmature (Elderkin 1989) mixed or coniferous forests (DeGraaf and Rudis 1987; Peck and James 1987) with openings and dense underbrush (DeGraaf et al. 1980). It is the only vireo that makes extensive use of coniferous trees for feeding and nesting (James 1987c).

Often perching in treetops, the Solitary Vireo moves slowly and carefully; its highly dignified style and white eye ring have earned it the sobriquet of "aristocrat." Its rich and varied song, which is the highest in pitch of the vireos (Bent 1950), is often confused with that of the Red-eyed Vireo (Levine 1988c). As the Solitary Vireo has no specific song perch and males rarely sing near each other, it is hard to determine the size of its territory and the exact number of pairs in an area (Bent 1950). This vireo may be heard singing until the third week of August (Bent 1950), with non-breeders singing most persistently (James 1978).

On arrival on their territories, males begin to look for suitable locations and start to build a nest, sometimes more than one (James 1978). When the females arrive, the males court them by fluffing out their yellow flank feathers, then bob and bow to them and follow them about singing (Bent 1950). The male performs a special display, associated with nest building, which plays an important role in pair formation (James 1978). Near the nest, he crouches almost horizontally, holding material in his bill, and repeats this action several times until the female enters the nest. The frequency of this behaviour decreases as the female begins to participate more in construction.

The pair thus builds the nest together, although the female does most of the work, with the male mainly providing materials. The nest of the Solitary Vireo is a basket suspended from the fork of a horizontal branch, generally in a conifer and usually at least 3 m up (Bent 1950). Strips of birch bark and lichen form the body of the nest, which is lined inside with dry grasses.

Notes on territory

The Solitary Vireo is highly territorial, on both its nesting and wintering grounds, and it is rarely seen in the company of conspecifics, even in migration (Kibbe 1985k). It has a large territory, on average 5 ha (Sabo 1980; Kibbe 1985k), within which the presence of a stream seems to be particularly important (Sabo 1980). Its territory sometimes overlaps that of the Red-eyed Vireo, with no hostility between them. This observation made Saunders (1938 in Bent 1950) sceptical about territoriality in vireos and some other birds.

In Québec, Erskine (1980a, 1984) reported densities of 2.5 males/10 ha in a pine and spruce plantation at Lachute, 4.9 males/10 ha in a stand of fir and spruce at Sainte-Anne-des-Monts, and 12.5 males/10 ha in a mature fir forest at Woburn. In La Mauricie National Park, DesGranges (1980b) reported a density of 3.2 individuals/10 ha in both a white spruce plantation and in a fir stand.

Solitary Vireo

Breeding distribution

1984–1989

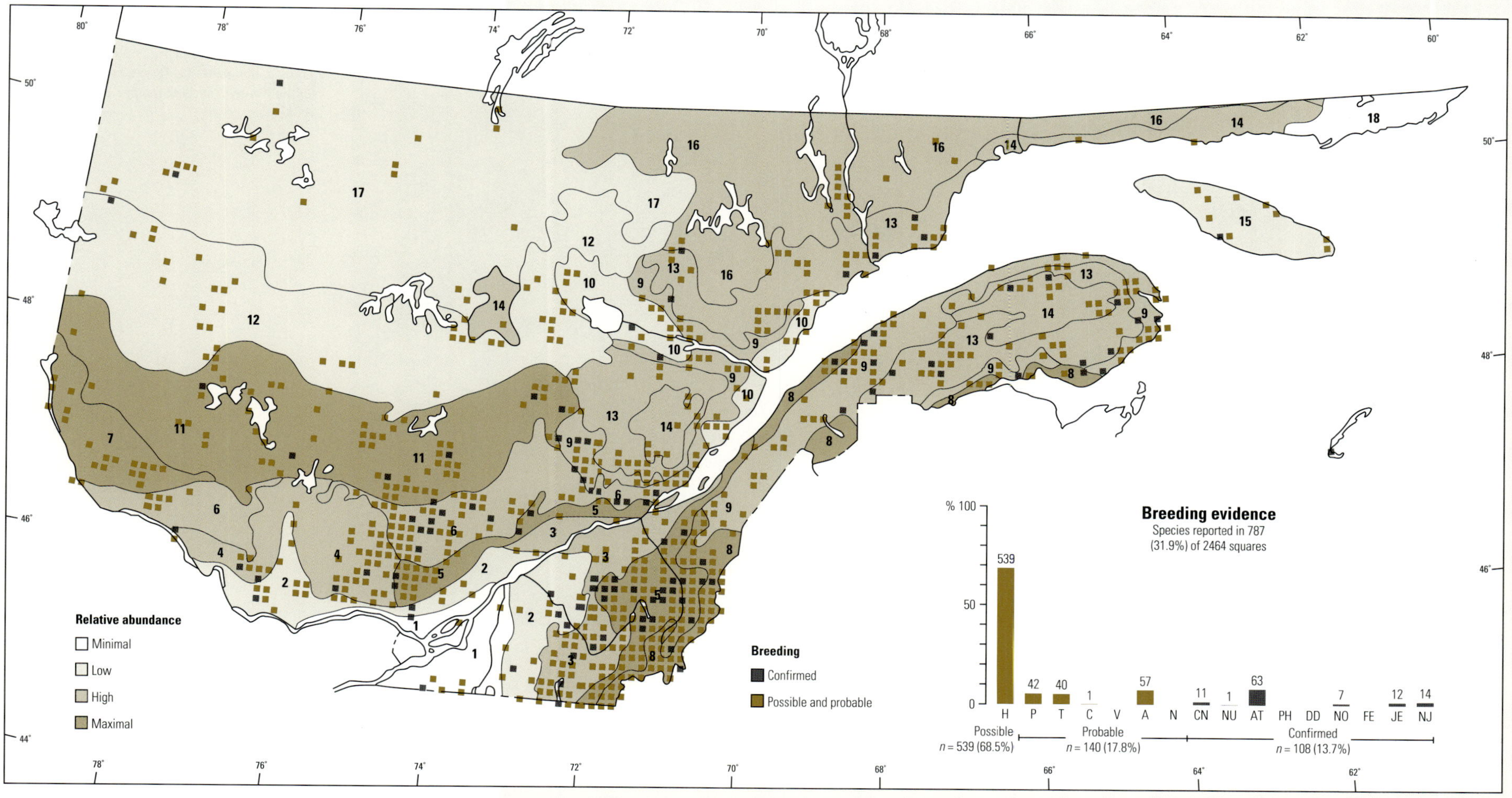

Species profile

Clutch size	usually 4 eggs (3–5)[1,2,5]
Laying interval	?
Start of incubation	?
Incubation period	10–15 days[2,4,5,7]
Maturity at hatching	altricial
Care of young	by female and male[1]
Nestling period	14–17 days[1,4]
Age at first flight	14–17 days or more[9]
Dependence of young	?
Broods per year	1, sometimes 2[4,5]
Breeding age	1 year[4]
Mating system	monogamy[1]
Duration of pair bond	1 season[10]
Total length	12.5–15.0 cm[3]
Wing span	21.0–24.8 cm[2]
Weight	16.6 g (14.1–19.3)[6]
Longevity record	7 years 5 months[8]

References:
1. Ehrlich et al. 1988
2. Terres 1980
3. Godfrey 1986
4. DeGraaf and Rudis 1987
5. Bent 1950
6. Dunning 1984
7. Peck and James 1987
8. Klimkiewicz et al. 1983
9. according to nestling period
10. James 1978

Chronology $n = 18$

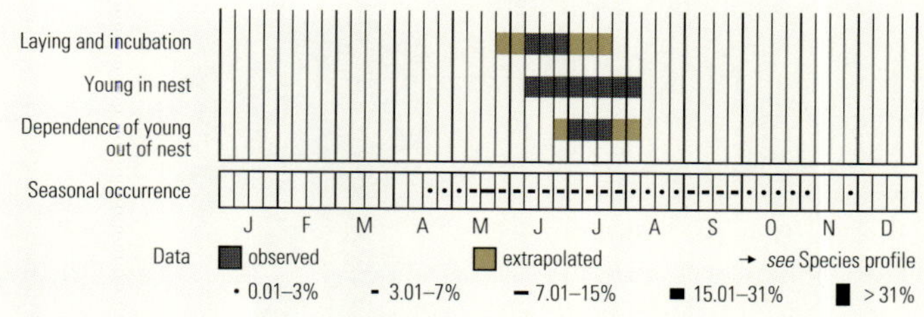

World breeding range

Frequency of occurrence by ecological region

Ecological regions** (forest regions)	Adequately covered squares	Squares in which reported	% of squares in which reported	Relative abundance (%)*
1. Sugar maple-hickory and sugar maple-basswood	97	9	9.3 ± 3.1	10.9
2. Sugar maple-basswood	153	39	25.5 ± 3.6	30
3. Sugar maple-basswood and sugar maple-yellow birch	101	62	61.4 ± 5.8	72.1
4. Sugar maple-yellow birch-basswood	56	35	62.5 ± 11.1	73.4
5. Sugar maple-yellow birch-beech and sugar maple-yellow birch-basswood	113	87	77.0 ± 4.4	90.5
6. Sugar maple-yellow birch-beech	71	52	73.2 ± 9.5	86.1
7. Sugar maple-yellow birch-balsam fir	19	15	78.9 ± 19.7	92.8
8. Sugar maple-yellow birch and balsam fir-yellow birch	47	40	85.1 ± 9.4	100.0
9. Balsam fir-yellow birch	87	61	70.1 ± 8.9	82.4
10. Balsam fir-white birch or balsam fir-red maple	46	20	43.5 ± 12.4	51.1
11. Yellow birch-balsam fir	52	42	80.8 ± 11.2	94.9
12. Balsam fir-white birch (west)	59	26	44.1 ± 13.0	51.8
13. Balsam fir-white birch (east)	76	48	63.2 ± 10.6	74.2
14. Balsam fir-black spruce	28	20	71.4 ± 17.6	83.9
15. White spruce-balsam fir and balsam fir-white spruce	22	6	27.3 ± 19.4	32.0
16. Black spruce-balsam fir-moss	18	13	72.2 ± 23.7	84.9
17. Black spruce-moss	32	15	46.9 ± 18.9	55.1
18.			*not analysed*	

* minimal: 0–29.9%; low: 30–59.9%; high: 60–89.9%; maximal: 90–100%

** Deciduous forest: 1–7; mixed forest: 8–11; coniferous forest: 12–17

The only vireo to nest so faithfully in coniferous woodlands, the Solitary Vireo prefers young, mature, or overmature spruce woods with openings. Montmorency forest, Laurentide Wildlife Reserve, north of Québec City. *May 1991, Réginald Ouellet.*

Laying may begin at the end of May. During incubation, which lasts 10–15 days, nest relief is announced by a specific call by which the incubating bird is notified by its mate. The Solitary Vireo is quite tame during this period, and can be approached easily, even touched. After hatching, however, when feeding the nestlings, it emits a nasal alarm call if disturbed (Bent 1950). The bird stops singing and becomes quiet after the eggs have hatched and does not begin to sing again until the young are independent. The pair bond dissolves when the young fledge from the nest. The young still continue to follow a parent for several days, however (James 1978). As eggs have been found in nests late in the season, it is thought that the Solitary Vireo may double-brood in some locations (Bent 1950; Kibbe 1985k), but this has not been confirmed in Québec.

The Solitary Vireo is chiefly an insect eater, but its diet is more varied on the wintering grounds (Chapin 1925). It gleans insects from the lower and intermediate parts of tree foliage (DeGraaf et al. 1980). Although it builds its nest mainly in a conifer, it can also be seen foraging for insects in deciduous trees (James 1978).

Of all the vireos, the Solitary had the lowest rate of Atlas breeding confirmations (13.7%; 108/787), but the highest rate of possible breeding (68.5%; 539/787). There are a number of reasons for the difficulty in confirming breeding in this species: its quietness on the nest, its infrequent singing, which means that it often goes unnoticed, the fewer visits by birders to its chiefly coniferous habitat, and the difficulty in locating nests in conifers.

The Solitary Vireo is also found in mixed and deciduous forests with clearings and dense underbrush, as in this open spruce forest after a spruce budworm epidemic. Mount Albert, Gaspé Conservation Park. *July 1991, Yves Aubry.*

Solitary Vireo perched on the rim of its suspended nest, to which shreds of white plastic have been added. This nest was found on a small island in Rivière Noire, at Sainte-Émélie-de-l'Énergie, north of Montréal. *July 1978, Jean Giroux.*

The Solitary Vireo is sometimes parasitized by the Brown-headed Cowbird (Terres 1980). A parasitism rate of 4.5% was found in Ontario (Peck and James 1987), while a much higher rate was found in Colorado, 34–75% (Marvil and Cruz 1989). Nests initiated early are less subject to parasitism, however (Marvil and Cruz 1989), and if a cowbird's egg has already been laid in the nest when the Solitary Vireo is about to lay, the vireo builds a second tier on top of the parasitic egg for its own clutch (Ehrlich et al. 1988). When a cowbird egg hatches, the nestling is often raised to the detriment of one or all the Solitary Vireo nestlings, and this considerably reduces the species' breeding success (Marvil and Cruz 1989).

Distribution

The Solitary Vireo was reported in 31.9% (787/2464) of Atlas squares surveyed. It is found in mixed and coniferous forests in various parts of southern Québec. It is less common in deciduous stands in the southern part of the province, and generally avoids densely inhabited areas. It was confirmed as a breeder throughout southern Québec.

North of the Atlas area, reports of Solitary Vireos have come from northeast of Lake Colomb (Cons. G. G. – GREBE 1992o), the Rupert River area downstream of Nemiscau (Todd 1963), Eastmain (Bordage 1985b), and around Great Whale River (Cons. G. G. – GREBE 1991a). It is possible, however, that as in Ontario (James 1987c), it is rarer in the Hudson Bay coastal plain, since the tree cover there is much sparser.

History and trends

The Solitary Vireo was considered rare in Québec in the late 19th and early 20th centuries (Dionne 1883, 1906; Chamberlain 1887; Macoun 1903), while some time later, it was held to be a common inhabitant of woodlands and orchards during migration (Taverner 1922). It is not surprising that it was considered rare at the time; not only is it quiet, its irregular song is sometimes similar to that of other vireos, with which it can be confused. It was the relatively recent upsurge in birdwatching that has helped clear up this confusion. Likewise, the significant increase indicated by the BBS for the Maritimes probably simply reflects greater familiarity of BBS observers with the Solitary Vireo's song (Erskine 1992).

ÉPOQ and BBS data also indicate an upward trend in Québec (Cyr and Larivée 1993a). A corresponding increase has been observed in northwestern New York State (Robbins et al. 1986) and in Pennsylvania (Brauning 1992c). In Ontario (James 1987c) and Vermont (Kibbe 1985k), however, the upward trend is not evident.

In places where a decline in Solitary Vireo populations have been noted, as in Alberta, for example, where it was common early in the century (Salt 1973), the main factor believed to be responsible is loss of suitable breeding habitat, as this species is very sensitive to logging, among other things. In New York State, on the other hand, reforestation of abandoned farmland, especially in mountainous areas, seems to be beneficial to the Solitary Vireo (Levine 1988c). In southern Québec, the recent increase in replanting farmlands with conifers could also help the species.

CHRISTINE LANGEVIN AND PIERRE LAFONTAINE

Yellow-throated Vireo

Viréo à gorge jaune

Vireo flavifrons

Christiane Girard

The song of the Yellow-throated Vireo is similar to that of the Red-eyed but is lower-pitched and has longer pauses between phrases of 2 or 3 notes. It can be distinguished from the songs of other vireos found in Québec by the somewhat hoarse, tanager-like quality of the notes. In appearance the Yellow-throat cannot be confused with any other vireo, as it is the only vireo in Canada with white wing bars and a bright yellow throat and breast. It is the rarest of the vireos nesting in Québec, with the most local distribution. Small breeding populations of a few pairs occur, however, in sites such as the Mount Orford campgrounds and the edges of Lake Hertel on Mount Saint-Hilaire, where singing males can be heard in the treetops before and even during incubation. The Yellow-throated Vireo is a migrant breeder that arrives in Québec in late April or early May from its winter quarters, which extend mainly from southern Mexico south to Panama (Godfrey 1986).

Habitat and behaviour

The Yellow-throated Vireo prefers mature, deciduous forests, often with trees well spaced and dominated by oak, maple, and hickory (Todd 1940). It is also found in moist woodlands, orchards, and large shade trees in residential areas.

The males scout for nest sites and may begin building several nests before the females arrive. The selected nest is completed after the pair bond is formed (James 1978). The brief but complex courtship display includes a nest-building ritual, whereby the male crouches in a horizontal position, bill pointing downward; this is an important behaviour as it serves to attract potential mates (James 1978).

The nest is built by both birds over a one-week period, during which time the male stays close to the female, even accompanying her while she forages. His only vocalization at this time is a special song uttered near the nest that apparently encourages the female in her nest-building activities (James 1984).

The Yellow-throated Vireo, the rarest of Québec's vireos, nests around Lake Hertel on Mount Saint-Hilaire. This tall tree contains a nest at the end of a limb, about 8 m up. *August 1991, Yves Aubry.*

Notes on territory

On arrival in spring, the males tend to wander and may sing within an area 1 km in diameter. Most activity occurs in tall trees, whence the song carries farther (James 1978). Males cease to wander when the females arrive. The selected site is defended against other passerines, but only when intruders alight in the same tree (James 1976). Williamson (1971) found the species used a territory of 4 ha in the deciduous forests of the eastern United States. DeGraaf and Rudis (1987) reported that different habitats in Maryland yielded densities ranging from 0.7 to 4.7 singing males/10 ha, whereas Gross (1993) found 0.9 territory/10 ha in a mixed forest of oak and pine in Pennsylvania.

Yellow-throated Vireo

Breeding distribution

1984–1989

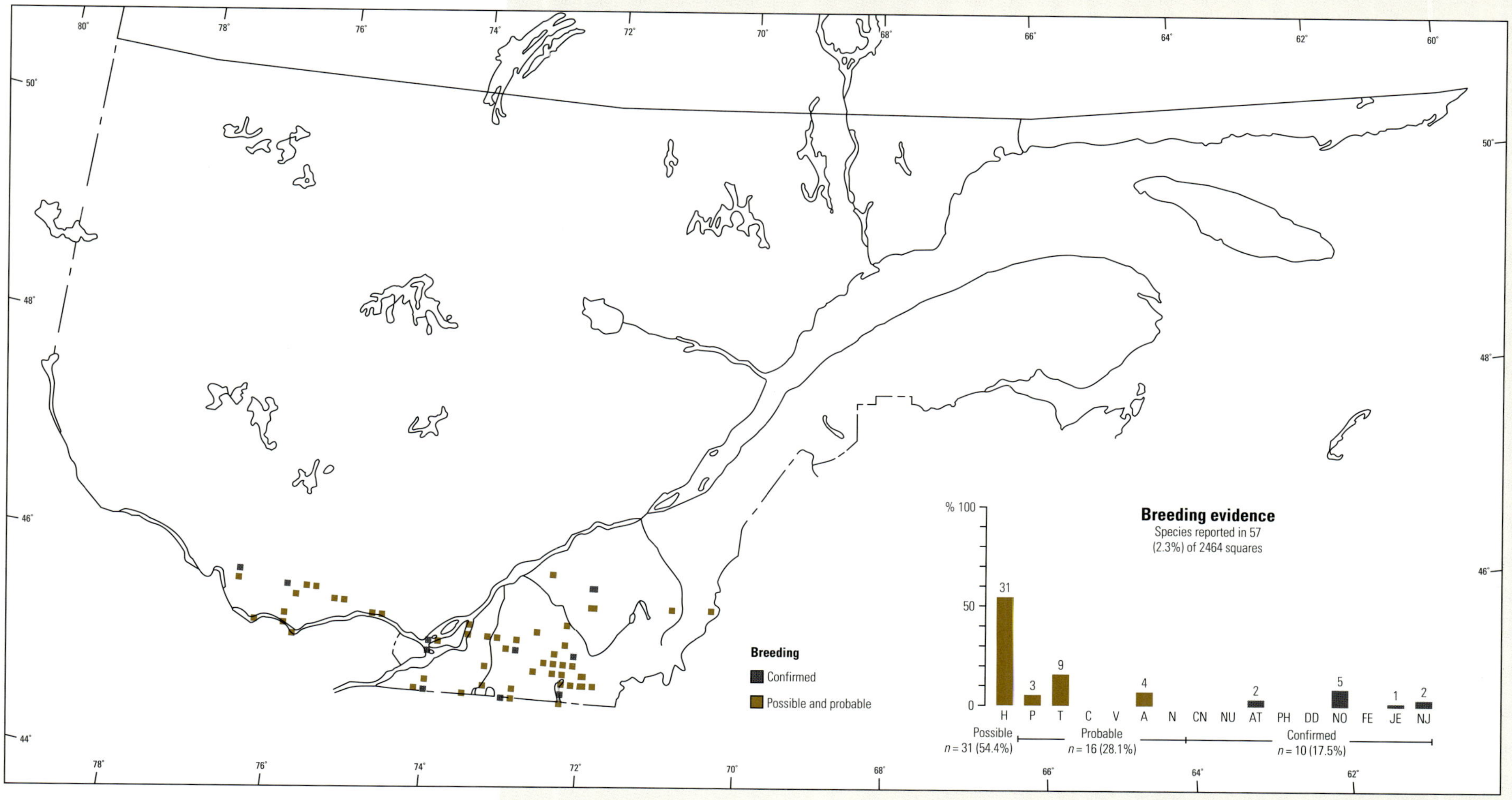

Species profile

Clutch size	usually 4 eggs (3–5)[1,2,3,4]
Laying interval	?
Start of incubation	?
Incubation period	usually 14 days (12–15)[2,3,4,10]
Maturity at hatching	altricial
Care of young	by female and male[2,5]
Nestling period	usually 14 days (14–15)[2,3,4]
Age at first flight	usually 14 days (14–15) or more[9]
Dependence of young	?
Broods per year	1[8]
Breeding age	1 year[8]
Mating system	monogamy[3]
Duration of pair bond	?
Total length	12.5–15.0 cm[1]
Wing span	24.1–25.4 cm[4]
Weight	18.0 g (15.6–21.4)[7]
Longevity record	6 years 1 month[6]

References:
1. Godfrey 1986
2. Bent 1950
3. Ehrlich et al. 1988
4. Terres 1980
5. James 1978
6. Klimkiewicz et al. 1983
7. Dunning 1984
8. DeGraaf and Rudis 1987
9. according to nestling period
10. Peck and James 1987

Chronology $n = 11$

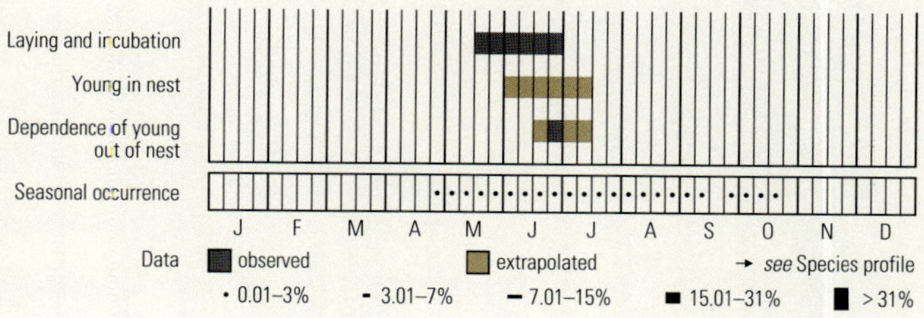

World breeding range

The Yellow-throat's is a typical vireo nest, a pensile structure with the sides strongly bound to a twig fork in a deciduous tree. It is a complex, thick-walled cup of bark strips, grasses, and rootlets, interwoven and bound together with insect silk and spider webs, lined with fine grasses on the inside and decorated on the outside with lichens and spider egg cases (Peck and James 1987). It is normally placed on a branch near the top of the tree, between the trunk and the end of a limb at a height of over 10 m (Peck and James 1987), which makes it difficult to spot.

The male resumes his energetic singing as the nest is completed, at which time copulation occurs (James 1984). Clutch size is 3–5 eggs, normally 4, which are incubated by both adults for a period of 12–15 days (Terres 1980; Peck and James 1987). The male often sings when incubating, which is an aid to locating the nest. The parents respond aggressively when the nest is approached and attempt to drive away intruders by scolding and threatening to attack.

The nestlings are fed by both parents and fledge at about 2 weeks of age. Once they have left the nest, the parents divide the brood (James 1978). Nesting occurs between mid-May and mid-July, with the nest becoming less visible as foliage develops. Atlas results clearly highlight the species' low abundance and the difficulty of confirming breeding.

Outside the breeding season, Yellow-throated Vireos do not flock together and are rarely seen in pairs. Some birds may, however, join mixed flocks of small passerines.

The Yellow-throated Vireo is a common cowbird host. It defends itself against parasitism by building a second tier to the nest over the cowbird's egg or by abandoning it altogether (Bent 1950).

A Yellow-throated Vireo building its nest in a deciduous tree. Great Smoky Mountains, Tennessee. *May 1987, André Cyr.*

The species feeds primarily on insects such as butterflies and moths, especially caterpillars, and bugs. Its diet includes only a few spiders (2%) and a little vegetable matter, such as berries and seeds (2%) (Chapin 1925). It forages for insects along branches and tree trunks, in the central or topmost portion of the tree (James 1976). Its slow, deliberate movements while foraging are often interspersed with bouts of song, which reveal its presence.

In 1993, the species nested in the tall trees of the Marcel Raymond Ecological Reserve, near the Richelieu River. Henryville, south of Saint-Jean-sur-Richelieu. *May 1991, Pierre Bannon.*

The Yellow-throated Vireo is local in southernmost Québec, nesting in tall, usually mature deciduous trees, about 10 m up. Philipsburg, south of Saint-Jean-sur-Richelieu. *July 1991, Pierre Bannon.*

Distribution

The Yellow-throated Vireo breeds very locally in the southernmost part of the province. Distribution corresponds to the Sugar maple-hickory, Sugar maple-basswood, and Sugar maple-yellow birch forest regions in the Central St. Lawrence Lowland and the Ottawa Valley. The species appears to be mostly associated with the mature forest remnants in southern Québec, notably in the Monteregian Hills and Appalachians.

History and trends

Several authors, including Ambuel and Temple (1982) in Wisconsin and James (1987e) in Ontario, have reported that Yellow-throated Vireo populations are declining.

In New York State, the situation may be somewhat different; Temple and Temple (1976) have reported a decline, whereas BBS data indicate an increase in abundance (Robbins et al. 1986). In Québec, BBS and ÉPOQ data are insufficient to yield statistically valid results. The most one can say is that the Yellow-throated Vireo is being reported much less frequently in the province by birders (Larivée 1989).

At the end of the last century, Wintle (1896) considered the Yellow-throated Vireo rare in the Montréal area. Today, few of these birds are found in the wooded areas of the region, and it remains a rare and local species wherever it occurs. Bent (1950) noted that extensive spraying of orchards and shade trees probably had a major impact on the species. The clearing and fragmentation of maple stands and mature woodlands, as well as the ravages of Dutch elm disease (transmitted by the native elm bark beetle) have probably had a similar effect (Temple and Temple 1976).

André Cyr and Derin Henderson

Warbling Vireo

Viréo mélodieux

Vireo gilvus

Christiane Girard

ON FINE SUMMER DAYS IN SOUTHERNMOST QUÉBEC, A LIMPID WARBLING IS often heard from the tops of tall deciduous trees, a song not unlike that of the Purple or House finch, but softer and more languid. Anyone who succeeds in pinpointing the source of this sound will no doubt be surprised at the drabness of the bird's plumage. This is the Warbling Vireo, so named because of the unbroken phraseology of its song, unlike the short, repetitive songs of the other vireos.

Although the Warbling Vireo is easily detected by its constant singing at any hour of the day and is common in parks and suburbs, it is a species that most people are unfamiliar with, being secretive and usually hidden in dense foliage. It is a common migrant breeder in southern Québec (David 1980a), arriving in May from its wintering grounds in El Salvador, Guatemala, and Mexico (Godfrey 1986).

The Warbling Vireo is regularly found in edge habitat in the upper canopy of tall trees, such as elm, poplar, and maple. Nuns' Island, Verdun. *July 1991, Pierre Bannon.*

Habitat and behaviour

The Warbling Vireo favours the mature elms, poplars, and maples that line roads, rivers, and streams. It is also found in clumps of deciduous trees in urban or semi-urban settings and in flooded woodlands. The general profile of its habitat appears to be more important than tree species composition (James 1976). It seems to prefer woodland edges where treed areas give way to open habitat such as lawns or a body of water (James 1976).

The nest of the Warbling Vireo is generally located 1.4–18 m up, usually near an open area (Peck and James 1987). Like other vireo nests, it is a carefully woven cup suspended from a twig fork near the end of a branch. It is built by the female alone, of hair, bark, twigs, plant down, and lichen, bound with spider webs and lined with fine grasses. Both parents incubate the eggs and tend the nestlings, which leave the nest about 2 weeks after hatching; they are fed by both adults for a few more days. Within a week of fledging, the pair bond dissolves (Howes-Jones 1985). The Warbling Vireo is essentially insectivorous, and is partial to caterpillars, although berries may be included in the diet at the end of the nesting season (Chapin 1925).

Although readily identified by its song, the species is difficult to confirm as a nester, as Atlas results indicate. It was confirmed as breeding in only 15.0% of the 532 squares in which it was reported. Most confirmations were made through sightings of adult birds carrying food or of nests or young. The young are noisy on leaving the nest and relatively easy to find (Howes-Jones 1985); this breeding evidence accounted for 13.8% (11/80) of breeding confirmations.

Distribution

The breeding range of the Warbling Vireo extends across Canada from the western provinces to southern New Brunswick and as far south as central Mexico (Godfrey 1986). In Québec, the Warbling Vireo is found mainly in the Central St. Lawrence Lowland, the Ottawa Valley, and the southern Appalachians, all areas where the patchwork of open areas and deciduous forest affords much suitable breeding habitat.

The Warbling Vireo was also atlassed in the Laurentians, Témiscamingue, Abitibi, the Lower St. Lawrence, and the Gaspé regions. The creation of open habitat through agricultural and urban development and the preservation of isolated deciduous woodlots in these areas have fostered localized establishment of the species.

The Atlas map shows a more extensive breeding range than the distribution given by Godfrey (1986), who includes only the southeastern Ottawa Valley, the St. Lawrence Valley to Québec City, and the southern Appalachians. The species was atlassed as breeding locally in other regions: the Lower St. Lawrence, Témiscamingue, and near Mont-Laurier, in the Laurentians. This divergence from the range given by Godfrey (1986), however, can probably be explained by the special coverage for the Atlas and does not necessarily mean that the species has extended its breeding range.

History and trends

The Warbling Vireo was considered scarce in the province at the turn of the century (Wintle 1896; Dionne 1906). Earlier it had probably been restricted to the wooded shores of streams. The range then expanded, favoured by deforestation for farming and urban development (Bonney 1988a).

It is generally thought that the species declined in the northeastern United States during the 1950s and 1960s, affected by pesticide spraying of shade trees. The population is considered to have recovered since then (Bull 1974), and BBS data indicate a slight increase in Warbling Vireo numbers (Robbins et al. 1986).

Although human activity has to some extent favoured population and northward range expansion, the bulk of the Québec population will no doubt continue to be restricted to the deciduous forest in the southern part of the province.

SYLVAIN PARADIS

Notes on territory

The male defends the territory mainly by singing, but often has to drive off recalcitrant conspecifics. According to Howes-Jones (1985), the female will sometimes help the male chase away other Warbling Vireos that infiltrate the territory to steal nest material. Howes-Jones also noted male territoriality in defence of the nest and the female against extra-pair copulation and predation and to prevent the theft of nest material. Moreover, by remaining near the female, the male can readily take advantage of the times when she is receptive to copulation.

In Ontario, in a forest containing both a disturbed deciduous stand and a mixed stand, a density of 2.3 territorial males/10 ha was recorded (Erskine 1971a). Counts of 1.0 territory/10 ha and 4.2 territories/10 ha were found respectively in a red ash-red oak stand (Wojnowski and Goodyear 1993) and a red oak-ironwood stand (Wojnowski 1993).

Warbling Vireo

Breeding distribution
1984–1989

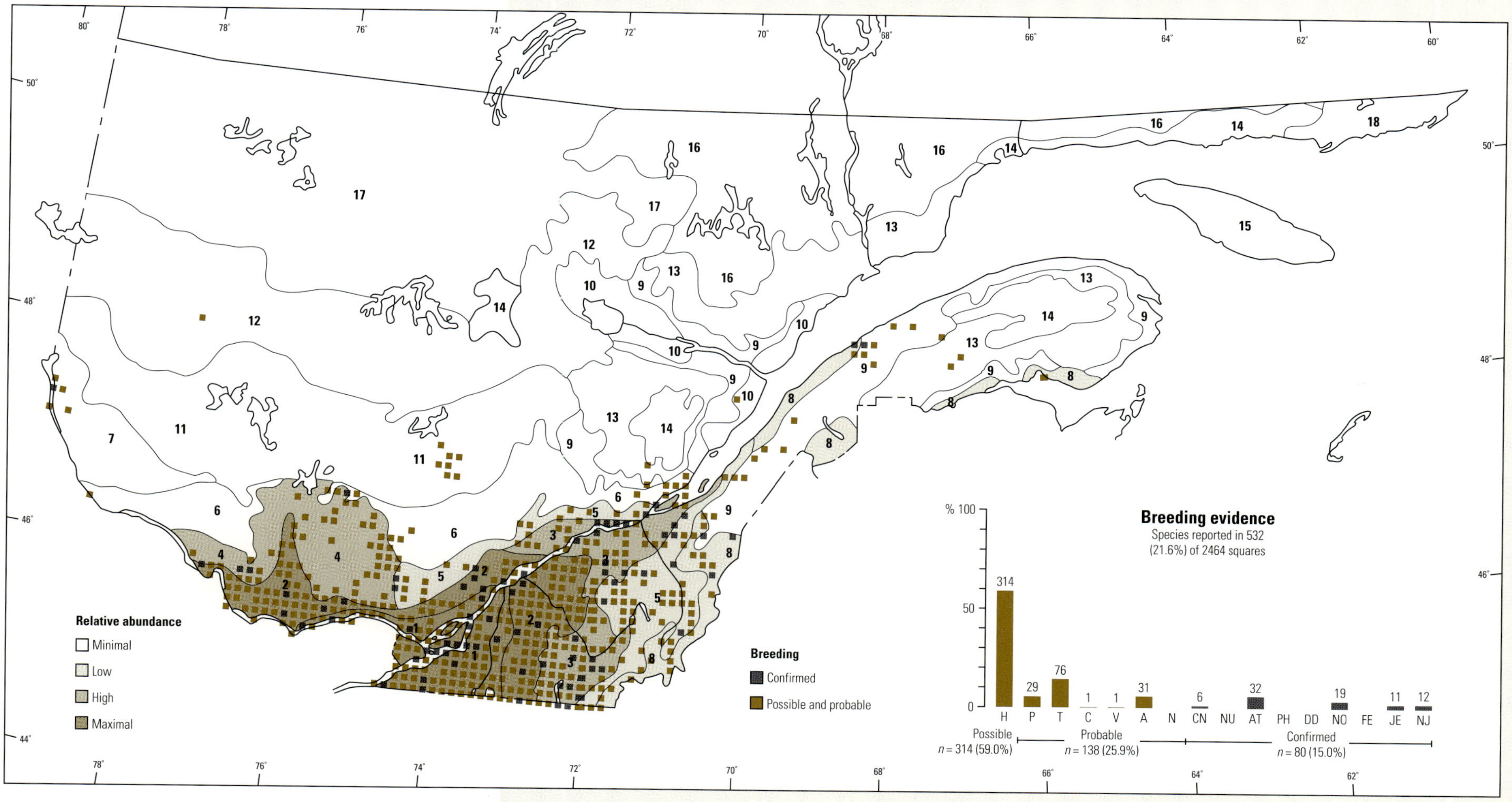

Species profile

Clutch size	usually 4 eggs (3–5)[1,2,3,4]
Laying interval	?
Start of incubation	?
Incubation period	12–14 days[1,2,3,4]
Maturity at hatching	altricial
Care of young	by female and male[2,3,4]
Nestling period	12–16 days[2,3]
Age at first flight	12–16 days or more[10]
Dependence of young	less than 1 week after leaving nest[3]
Broods per year	1[9]
Breeding age	1 year[9]
Mating system	monogamy[4]
Duration of pair bond	1 season[3]
Total length	12.5–15.0 cm[6]
Wing span	21.6–23.5 cm[5]
Weight	12.0 g (9.2–14.9)[8]
Longevity record	13 years 1 month[7]

References:
1. Peck and James 1987
2. Tyler 1950
3. Howes-Jones 1985
4. Ehrlich et al. 1988
5. Terres 1980
6. Godfrey 1986
7. Klimkiewicz et al. 1983
8. Dunning 1984
9. DeGraaf and Rudis 1987
10. according to nestling period

World breeding range

Chronology $n = 47$

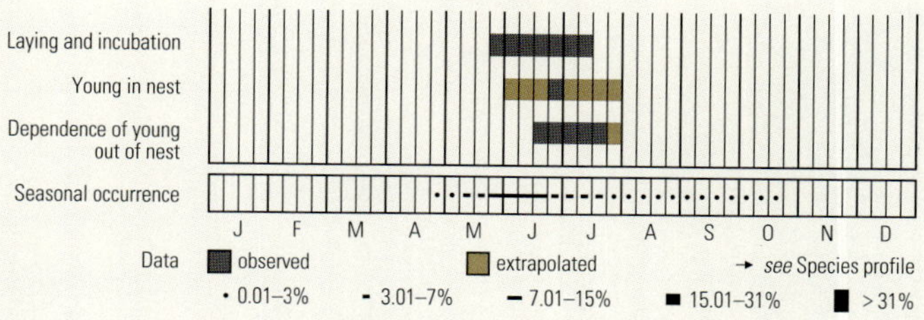

Frequency of occurrence by ecological region

Ecological regions** (forest regions)	Adequately covered squares	Squares in which reported	% of squares in which reported	Relative abundance (%)*
1. Sugar maple-hickory and sugar maple-basswood	97	90	92.8 ± 2.8	100.0
2. Sugar maple-basswood	153	135	88.2 ± 2.8	95.1
3. Sugar maple-basswood and sugar maple-yellow birch	101	74	73.3 ± 5.3	79.0
4. Sugar maple-yellow birch-basswood	56	33	58.9 ± 11.3	63.5
5. Sugar maple-yellow birch-beech and sugar maple-yellow birch-basswood	113	53	46.9 ± 5.1	50.6
6. Sugar maple-yellow birch-beech	71	17	23.9 ± 9.2	25.8
7. Sugar maple-yellow birch-balsam fir	19	4	21.1 ± 19.7	22.7
8. Sugar maple-yellow birch and balsam fir-yellow birch	47	17	36.2 ± 12.4	39.0
9. Balsam fir-yellow birch	87	11	12.6 ± 6.6	13.6
10. Balsam fir-white birch or balsam fir-red maple	46	0	0	0
11. Yellow birch-balsam fir	52	7	13.5 ± 9.9	14.5
12. Balsam fir-white birch (west)	59	0	0	0
13. Balsam fir-white birch (east)	76	3	3.9 ± 4.7	4.3
14. Balsam fir-black spruce	28	0	0	0
15. White spruce-balsam fir and balsam fir-white spruce	22	0	0	0
16. Black spruce-balsam fir-moss	18	0	0	0
17. Black spruce-moss	32	0	0	0
18.			*not analysed*	

* minimal: 0–29.9%; low: 30–59.9%; high: 60–89.9%; maximal: 90–100%

** Deciduous forest: 1–7; mixed forest: 8–11; coniferous forest: 12–17

Philadelphia Vireo

Viréo de Philadelphie

Vireo philadelphicus

Christiane Girard

FEW BIRDERS CAN BOAST THAT THEY CAN UNHESITATINGLY TELL THE difference between the songs of the Philadelphia Vireo and the much more common Red-eyed Vireo. In fact, the Philadelphia Vireo seems to be the only one that actually can; the undiscriminating Red-eyed Vireo displays the same territorial behaviour toward the Philadelphia Vireo as it does toward conspecifics (Barlow and Power 1970; Rice 1978a). The yellowish underparts and smaller size of the Philadelphia Vireo are diagnostic, however.

The Québec breeding ranges of these two vireos are sympatric, overlapping to some extent, but the Philadelphia Vireo is found more often in the boreal forest, while the Red-eyed is more common in the southern part of the province. The Philadelphia Vireo usually leaves Québec late in the summer to winter from Guatemala to Panama (AOU 1983).

Habitat and behaviour

In the breeding season, the Philadelphia Vireo is found mainly high in the canopy of deciduous or mixed forests (Peck and James 1987). It prefers poplar and birch stands with an understorey of mainly alder (Barlow and Rice 1977; Rice 1978a). Second growth, woodland edges, burns, abandoned fields grown up in small trees, and so on, are all suitable breeding habitat (DeGraaf and Rudis 1987). In one study, Robinson (1981) noted that, unlike the Red-eyed Vireo, which occupied almost the entire study area, the Philadelphia Vireo preferred sites where fallen trees had opened up the canopy. The male and female seem to arrive on the breeding grounds at the same time (Barlow and Rice 1977) and the male wastes no time in defending his territory by vigorous singing.

Given their lack of sexual dimorphism, the male and female must recognize each other by their behaviour in the breeding season. A female that enters a male's territory actually runs a high risk of being attacked: to be recognized, she must refrain from answering the male's alarm call and offer no resistance to his aggression (Barlow and Rice 1977). After a short courtship, the pair forms, and the female soon finds a suitable nest location.

The female builds the nest alone, but the male accompanies her as she looks for the thin strips of birch bark, lichen, and pieces of grass she needs. The pendant cup-shaped nest is usually suspended by its rim from the fork of a twig (Godfrey 1986), high in the canopy. Male and female alike incubate the eggs, and after hatching, they almost always forage together for their offspring (Barlow and Rice 1977). The young leave the nest about 2 weeks after hatching (Ehrlich et al. 1988).

The Philadelphia Vireo feeds on insects such as caterpillars, butterflies and moths, beetles, wasps, bees, and ants, as well as on wild berries. Most of the time, it forages for insects in the foliage, but it may also hawk its prey on the wing. This hunting technique is more effective in some types of trees, such as ash and yellow birch (Robinson and Holmes 1984).

Studies on the foraging behaviour of the Philadelphia Vireo do not provide an explanation for the bird's preference for the treetops (Robinson and Holmes 1984). Some authors have hypothesized that the Philadelphia Vireo is displaced by the more dominant Red-eyed Vireo (Robinson 1981; Robinson and Holmes 1984). When the Philadelphia Vireo is less abundant than the Red-eyed, their territories overlap considerably; however, the two species occupy different vegetational strata fairly peaceably, with the Philadelphia Vireo concentrating its activities higher up (Robinson 1981; James and Peck 1986).

Notes on territory

The male begins to defend a territory as soon as he arrives on the breeding grounds, singing and flying from branch to branch around the periphery of the defended area (Barlow and Rice 1977). He protects his mate against attacks by conspecifics and defends the resources in the territory against other species, such as the Red-eyed Vireo, with which it is often sympatric (Rice 1978a,b).

Densities of 2.0–3.5 birds/10 ha have been reported in New Hampshire (Robinson 1981). In Québec, densities of 3.2 pairs/10 ha have been noted in a fir-spruce-alder forest, 4.9 pairs/10 ha in a spruce-fir stand with cedar and birch, and 5.7 pairs/10 ha in a mature poplar forest (Erskine 1972a, 1980a). In Ontario, mean territory size was 0.31 ha (Rice 1978b).

Second-growth forests growing up in trembling aspen saplings and shrubs make good habitat for the Philadelphia Vireo. Near the Pikauba River, Laurentide Wildlife Reserve, north of Québec City. *July 1987, Réginald Ouellet.*

Philadelphia Vireo

Breeding distribution

1984–1989

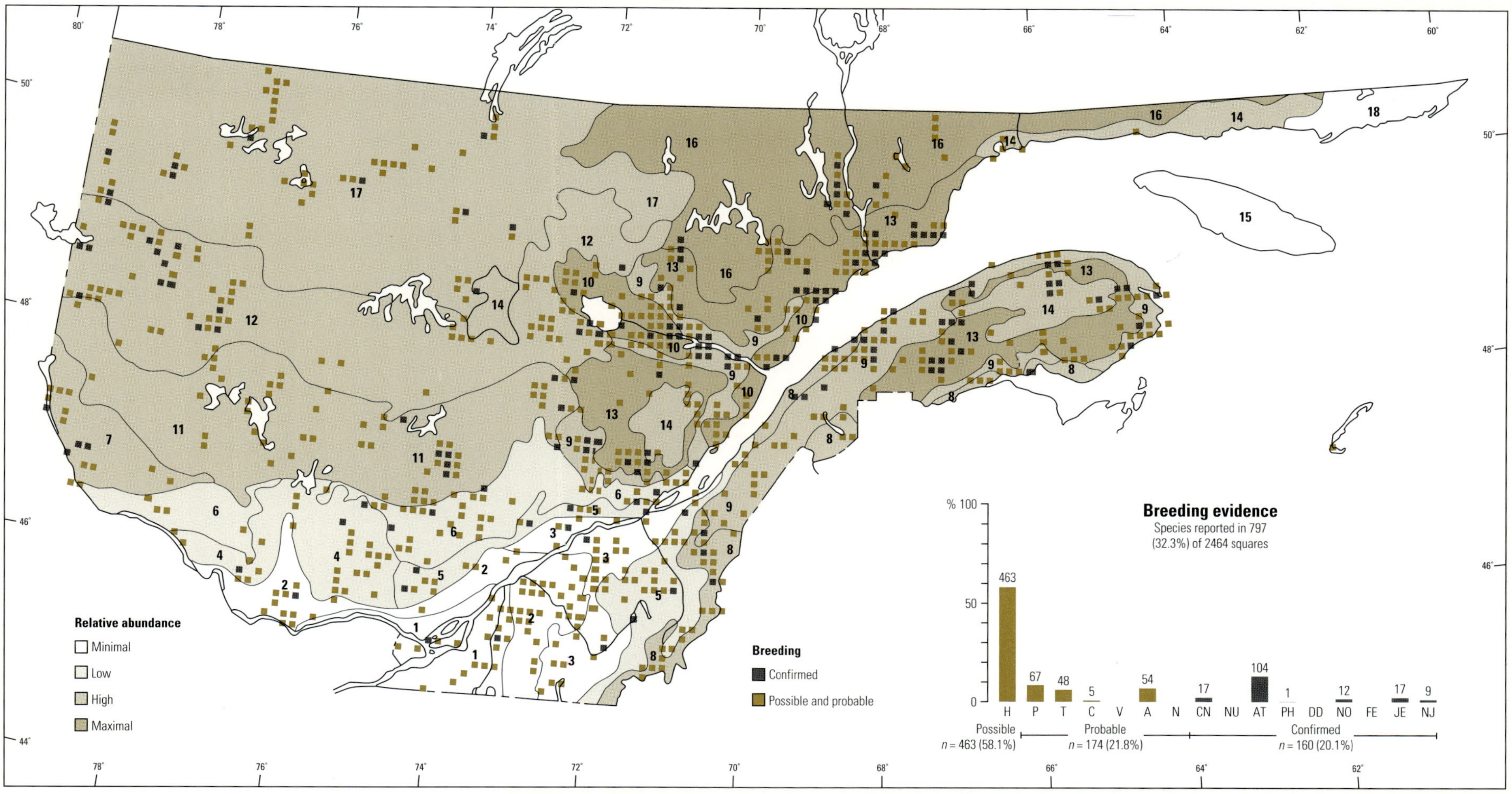

Species profile

Clutch size	usually 4 eggs (2–5)[1,2,3,4,11]
Laying interval	?
Start of incubation	probably before last egg[1,11]
Incubation period	11–14 days[1,2,3,4,11]
Maturity at hatching	altricial
Care of young	by female and male[3,5]
Nestling period	12–14 days[1,3]
Age at first flight	12–14 days or more[10]
Dependence of young	?
Broods per year	1[6]
Breeding age	1 year[6]
Mating system	monogamy[3]
Duration of pair bond	?
Total length	11.0–12.5 cm[7]
Wing span	20.3–22.9 cm[2]
Weight	12.2 g (10.3–16.1)[8]
Longevity record	8 years 10 months[9]

References:
1. Bent 1950
2. Terres 1980
3. Ehrlich et al. 1988
4. Harrison 1975
5. Barlow and Rice 1977
6. DeGraaf and Rudis 1987
7. Godfrey 1986
8. Dunning 1984
9. Klimkiewicz et al. 1983
10. according to nestling period
11. Peck and James 1987

Chronology $n = 15$

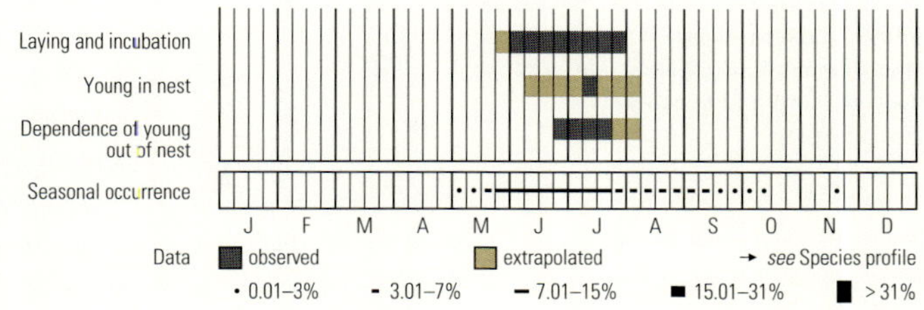

Frequency of occurrence by ecological region

Ecological regions** (forest regions)	Adequately covered squares	Squares in which reported	% of squares in which reported	Relative abundance (%)*
1. Sugar maple-hickory and sugar maple-basswood	97	18	18.6 ± 4.0	19.6
2. Sugar maple-basswood	153	43	28.1 ± 3.7	29.8
3. Sugar maple-basswood and sugar maple-yellow birch	101	27	26.7 ± 5.3	28.3
4. Sugar maple-yellow birch-basswood	56	19	33.9 ± 10.9	35.9
5. Sugar maple-yellow birch-beech and sugar maple-yellow birch-basswood	113	34	30.1 ± 4.7	31.9
6. Sugar maple-yellow birch-beech	71	36	50.7 ± 10.6	53.7
7. Sugar maple-yellow birch-balsam fir	19	11	57.9 ± 23.3	61.3
8. Sugar maple-yellow birch and balsam fir-yellow birch	47	29	61.7 ± 12.5	65.3
9. Balsam fir-yellow birch	87	70	80.5 ± 7.8	85.2
10. Balsam fir-white birch or balsam fir-red maple	46	42	91.3 ± 7.5	96.7
11. Yellow birch-balsam fir	52	39	75.0 ± 12.2	79.4
12. Balsam fir-white birch (west)	59	48	81.4 ± 10.4	86.1
13. Balsam fir-white birch (east)	76	69	90.8 ± 6.6	96.1
14. Balsam fir-black spruce	28	21	75.0 ± 17.0	79.4
15. White spruce-balsam fir and balsam fir-white spruce	22	0	0	0
16. Black spruce-balsam fir-moss	18	17	94.4 ± 13.5	100.0
17. Black spruce-moss	32	26	81.2 ± 15.1	86.0
18.			*not analysed*	

* minimal: 0–29.9%; low: 30–59.9%; high: 60–89.9%; maximal: 90–100%

** Deciduous forest: 1–7; mixed forest: 8–11; coniferous forest: 12–17

World breeding range

In the boreal forest, this vireo is closely associated with regenerating logged areas and forest edges of young hardwoods such as poplars, willows, and alders. Charlevoix. *October 1981, Jean Beaubien and Guy Simard, Canadian Forest Service.*

This vireo can be found in the tops of clumps of trembling aspen, where it usually nests. Montmorency forest, Laurentide Wildlife Reserve, north of Québec City. *May 1991, Réginald Ouellet.*

The Philadelphia Vireo frequents parts of the forest where fallen trees have created openings; it feeds higher in the canopy than the Red-eyed Vireo. La Vérendrye Wildlife Reserve. *June 1979, Jean Beaubien and Guy Simard, Canadian Forest Service.*

In summer, aggressive interactions are more frequent when the young of the two species are leaving the nest; then the Philadelphia Vireos, young and adult, use the subcanopy/sapling layer, as do the Red-eyed Vireos (Robinson 1981). Interactions between the two species are complex, however, and vary with region and phase in the breeding cycle. Their territories may overlap vertically, horizontally, or not at all, depending on the relative population density of each species, the stage of development of the brood, and food abundance (Rice 1978b; Robinson 1981).

With its greenish plumage and fondness for the treetops, the Philadelphia Vireo may sometimes be overlooked. It may also be confused with the Red-eyed Vireo because of its song. Nonetheless, the species was atlassed in 32.3% (797/2464) of squares surveyed, and breeding was confirmed in 20.1% (160/797) of those. Nests, usually high up in trees, are hard to spot. The male, a persistent singer even during incubation, may, however, make the observer's task easier by singing from the nest (Lewis 1921a). Of the 160 confirmed reports of breeding, 104 were based on sightings of adults carrying food. This behaviour is very easy to observe at this stage of breeding, since the birds forage lower down in the shrub layer, when competition is not too strong.

Distribution

The Philadelphia Vireo breeds from northeastern British Columbia to the Maritimes, perhaps even southwestern Newfoundland. The southern limit of the breeding range is the northernmost U.S. states, from North Dakota to Maine (Erskine 1992). Its Canadian range in the provinces east of the Rockies corresponds quite closely to that of the trembling aspen (James 1987f).

The Philadelphia Vireo was atlassed as a breeder throughout almost all of southern Québec. It is very widespread from west to east, from Abitibi–Témiscamingue to the Upper North Shore and the Gaspé Peninsula. It was reported on the Moisie River and at the mouth of the Magpie River on the Middle North Shore, but not on Anticosti Island, and it was atlassed in only one square on the Magdalen Islands, where breeding was not confirmed. Lewis (1941 *in* Ouellet 1969) reported seeing a bird on Anticosti in June 1940, while Fradette (1992) lists 6 sightings from the Magdalen Islands between 1956 and 1989. This vireo is less common in southernmost Québec, although it was confirmed as a breeder there. It breeds in the northern New England states, which is just about the southern limit of its breeding range (Kibbe 1985i; DeGraaf and Rudis 1987; Peterson 1988i). Its preference for trembling aspen and birch groves may explain why it breeds mostly in the boreal forest, and farther south in mountainous regions where this type of habitat is found at higher elevations, in second-growth areas (cutovers, for example), some burns, or hillsides with southern exposure. Judging by its frequency of occurrence as mapped for the Atlas, the sugar maple stands and farming areas in the St. Lawrence Plain seem to be less suitable habitat for the species.

The Québec breeding range of the Philadelphia Vireo extends beyond the Atlas area. In 1981, the species was reported on the Eastmain River (Bordage and Aubry 1982) and in June 1991, during a survey conducted in the Nottaway, Broadback, and Rupert river watersheds, in a regenerating burn (Cons. G. G. – GREBE 1992o). The northernmost breeding record, of a bird that was starting to build a nest, comes from Nemiscau (Cons. G. G. – GREBE 1992o).

History and trends

Wintle (1896) had never seen a Philadelphia Vireo although Montréal was within its geographical range and is now within the current breeding range of the species. Macoun and Macoun (1915) considered it uncommon in the Ottawa area, while Lloyd (1944) thought it quite rare. Ouellet (1974a) considered the bird a rare summer resident in the Monteregian Hills and around Montréal. In Todd's (1963) view, the Philadelphia Vireo was by no means a rare bird in the northern part of the province, but rather was often confused with the Red-eyed Vireo. ÉPOQ (1970–89) and BBS (1966–89) data reveal no significant population trends (Cyr and Larivée 1993a; BBS). The large expanses of second growth after extensive logging may favour the species, although the type of conifer reforestation now being practised may actually be detrimental; natural regrowth of mixed hardwoods is being replaced by softwood plantations that do not provide suitable habitat for the species (Robinson 1981).

MARC-ANDRÉ VILLARD

Red-eyed Vireo

Viréo aux yeux rouges

Vireo olivaceus

Christiane Girard

THE RED-EYED VIREO IS THE MOST ABUNDANT VIREO IN SOUTHERN QUÉBEC forests, and one of the commonest birds atlassed in the province. A tireless singer which may extend its breeding season to early autumn, this neotropical migrant winters in South America and arrives here toward the end of May.

Habitat and behaviour

The Red-eyed Vireo commonly nests in dense mixed and deciduous forest (Barlow and Rice 1977), as well as in open habitat such as regenerating cutovers (Titterington et al. 1979), thinned-out stands (Webb et al. 1977), and even city parks and suburbs (Terres 1980). In general, its abundance increases with the density and height of the tree cover (Crawford et al. 1981; Morgan and Freedman 1986). In the sugar maple stands of the Amiante and Bois-Francs regions, it is the second most common bird after the Least Flycatcher (Darveau et al. 1992).

Upon arrival in spring, the male establishes and defends its territory (Barlow and Rice 1977). The female arrives 3 or 4 days later and spends 4 or 5 days building the nest, a well-made cup suspended from a horizontal branch of a tree or shrub (Lawrence 1953a). Nest height varies from 1 to 25 m, with an average of 3.3 m for 23 nests in the Bois-Francs region (M. Darveau, unpubl. data), 3.2 m for 176 nests in Maryland (Williamson 1971), and 10.7 m for 31 nests in New Hampshire (Holmes 1986).

Clutch size is typically 4 eggs (Terres 1980; DeGraaf and Rudis 1987), which the female alone incubates (Barlow and Rice 1977) for 11-15 days (Peck and James 1987), during which time the male regularly brings food to the female on the nest. The young remain in the nest for 10-12 days (DeGraaf and Rudis 1987). Both parents tend the young, with the male making about 30% of trips to the nest (Holmes et al. 1979). After fledging, the young remain dependent on their parents for several more weeks (Holmes et al. 1979). In Québec, late breeders have been seen feeding young as late as September (QNRCP).

The Red-eyed Vireo usually feeds by walking along branches, either gleaning insects from the leaves or hovering to catch prey on the leaves or, less commonly, on the bark (Williamson 1971; Maurer and Whitmore 1981; Robinson and Holmes 1982). It forages at an average height of 12 m in mature stands of sugar maple, but is frequently seen at the base of shrubs and in treetops (Holmes 1986). In general, the Red-eyed Vireo forages lower down in the vegetation than the Philadelphia Vireo, probably because it is socially dominant over the Philadelphia Vireo, which is then relegated to the canopy (Robinson 1981; Robinson and Holmes 1982).

The Red-eyed Vireo nests in dense deciduous and mixed forest. When nesting alongside the Philadelphia Vireo, it concentrates its activities in the intermediate and lower strata of the stand. Gowan Road, Herdman, southwest of Montréal. *June 1991, Pierre Bannon.*

Notes on territory

The territory is used for mating, nesting, and foraging. The male defends it against conspecifics (Stokes 1979) and also against Philadelphia Vireos (Robinson 1981). The territory ranges from 0.3 to 0.8 ha in size (Kendeigh 1947; McNeil 1964 in Cyr 1974; Williamson 1971; Cyr 1974; Barlow and Rice 1977; Stokes 1979) and is defended from the time the male arrives in spring until the young leave the nest.

In the sugar maple stands of the Amiante and Bois-Francs regions, Darveau et al. (1992) recorded a mean density of 12 pairs/10 ha. Densities ranging from 7.5 to 15 pairs/10 ha were found in Québec sugar maple stands in various locations south of the 47th parallel (Erskine 1971a, 1976a, 1984; Cyr 1974; DesGranges 1980b). In a beech-maple stand at Mount Saint-Hilaire, a density of 5.7 pairs/10 ha was recorded (Erskine 1972a), and in La Mauricie National Park, density was 5.9 pairs/10 ha in a white birch stand (Erskine 1976a). Density varied from 3.2 to 10.6 pairs/10 ha in stands dominated by birch and fir at Saint-Pascal, Kamouraska County, and at Carleton and in the Rivière Nouvelle area on the Gaspé Peninsula (Erskine 1980a).

Red-eyed Vireo

Breeding distribution
1984–1989

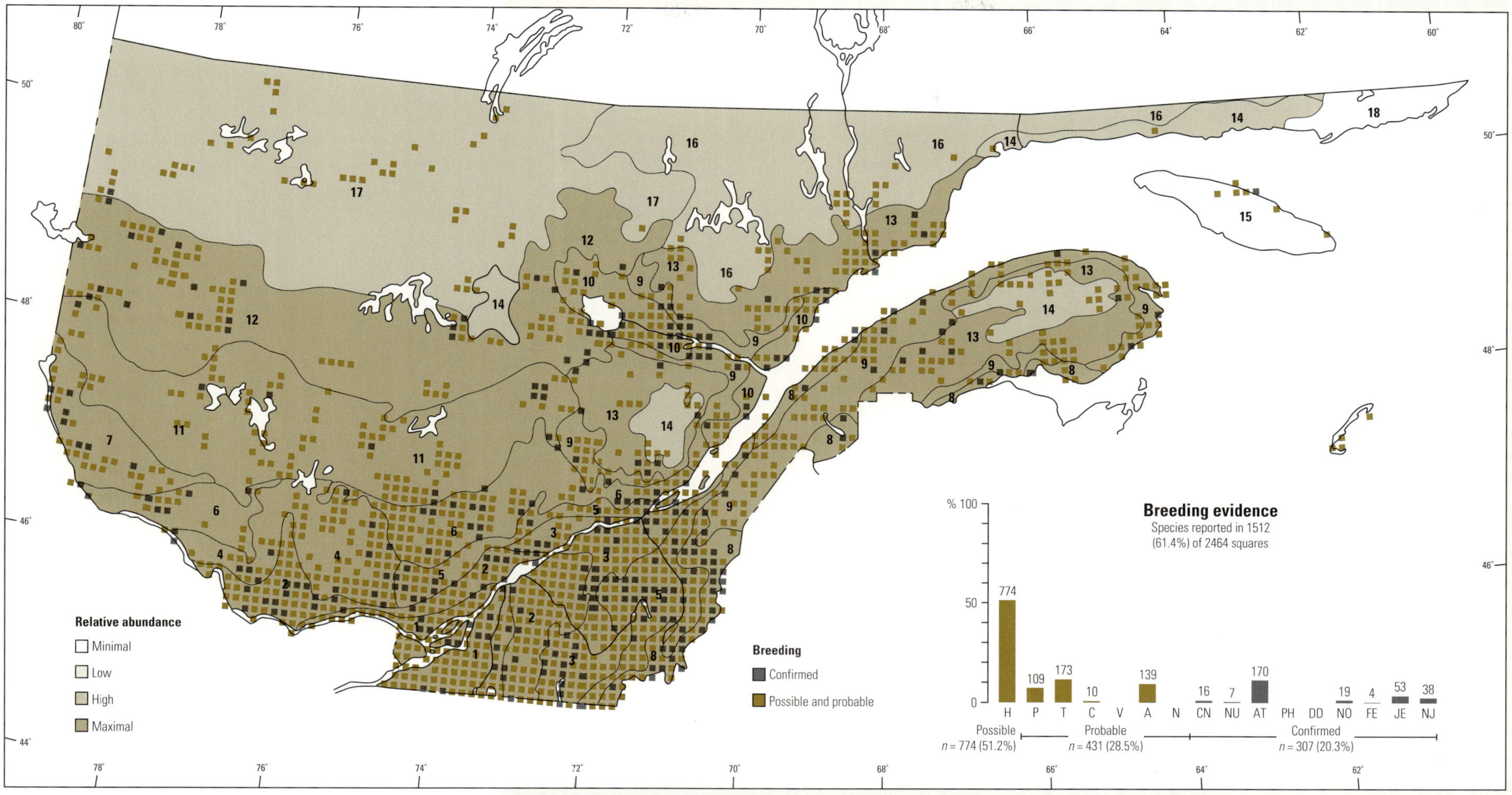

Species profile

Clutch size	usually 4 eggs (2–5)[1,2,3,4,8,10]
Laying interval	1 egg per day[5]
Start of incubation	before last egg possible[13]
Incubation period	11–15 days[1,2,3,4,5,10,11]
Maturity at hatching	altricial
Care of young	by female and male[1,2]
Nestling period	10–12 days[2,4,5,8]
Age at first flight	10–12 days or more[9]
Dependence of young	until 2–3 weeks after first flight[5,12]
Broods per year	1–2[5,8]
Breeding age	1 year[8]
Mating system	monogamy[1]
Duration of pair bond	?
Total length	13.5–16.5 cm[3]
Wing span	24.8–27.3 cm[4]
Weight	16.7 g (12.0–25.1)[6]
Longevity record	10 years[7]

References:
1. Ehrlich et al. 1988
2. Harrison 1978
3. Godfrey 1986
4. Terres 1980
5. Stokes 1979
6. Dunning 1984
7. Kennard 1975
8. DeGraaf and Rudis 1987
9. according to nestling period
10. Peck and James 1987
11. Lawrence 1953a
12. Holmes et al. 1979
13. Harrison 1975

World breeding range

Chronology *n* = 155

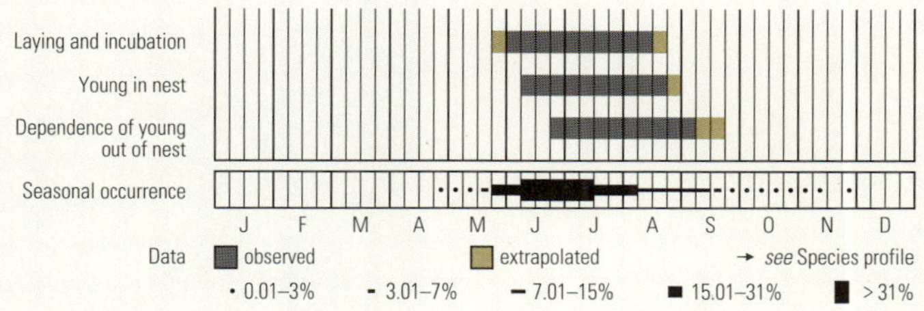

Frequency of occurrence by ecological region

Ecological regions** (forest regions)	Adequately covered squares	Squares in which reported	% of squares in which reported	Relative abundance (%)*
1. Sugar maple-hickory and sugar maple-basswood	97	95	97.9 ± 1.8	97.9
2. Sugar maple-basswood	153	151	98.7 ± 1.2	98.7
3. Sugar maple-basswood and sugar maple-yellow birch	101	100	99.0 ± 1.6	99.0
4. Sugar maple-yellow birch-basswood	56	55	98.2 ± 3.7	98.2
5. Sugar maple-yellow birch-beech and sugar maple-yellow birch-basswood	113	111	98.2 ± 1.7	98.2
6. Sugar maple-yellow birch-beech	71	71	100.0 ± 0.7	100.0
7. Sugar maple-yellow birch-balsam fir	19	18	94.7 ± 12.0	94.7
8. Sugar maple-yellow birch and balsam fir-yellow birch	47	47	100.0 ± 1.1	100.0
9. Balsam fir-yellow birch	87	81	93.1 ± 5.2	93.1
10. Balsam fir-white birch or balsam fir-red maple	46	45	97.8 ± 4.4	97.8
11. Yellow birch-balsam fir	52	50	96.2 ± 6.0	96.2
12. Balsam fir-white birch (west)	59	57	96.6 ± 5.3	96.6
13. Balsam fir-white birch (east)	76	71	93.4 ± 5.8	93.4
14. Balsam fir-black spruce	28	18	64.3 ± 18.6	64.3
15. White spruce-balsam fir and balsam fir-white spruce	22	5	22.7 ± 18.4	22.7
16. Black spruce-balsam fir-moss	18	11	61.1 ± 25.5	61.1
17. Black spruce-moss	32	25	78.1 ± 15.9	78.1
18.			*not analysed*	

* minimal: 0–29.9%; low: 30–59.9%; high: 60–89.9%; maximal: 90–100%

** Deciduous forest: 1–7; mixed forest: 8–11; coniferous forest: 12–17

The Red-eyed Vireo is also abundant in areas with relatively widely spaced trees, such as urban and suburban environments, in groves of large trees on golf courses and bordering roads, fields, and rivers. Saint-Lambert, near Montréal. *August 1991, Gaétan Duquette.*

Notes on territory (continued)

Densities of 0.7-2.7 pairs/10 ha were reported in coniferous stands at Mirabel and in La Mauricie National Park (Cyr 1974; DesGranges 1980b), and density averaged 1.92 pairs/10 ha in Gatineau Park and at La Barrière, near Saint-Michel-des-Saints in mixed habitat of forested and open areas (Dauphin 1985).

In northern coniferous habitat, this vireo is found mainly in pockets and stands of deciduous trees. Near Lake Chibougamau. *September 1986, Jean Beaubien and Guy Simard, Canadian Forest Service.*

Adult bringing a caterpillar to its young. The nest of the Red-eyed Vireo is a well-made cup suspended by its rim from a horizontal branch, and made of strips of bark, leaves, and paper from wasps' nests, bound together with animal silk and other fibres. Sillery. *July 1956, Gaston Le Page.*

During the breeding season, the Red-eyed Vireo feeds mainly on caterpillars, beetles, bugs, flies, and hymenopterans (Williamson 1971; Robinson and Holmes 1982). Chapin (1925) reported that in the fall, up to 50% of the diet of late migrants still on the breeding grounds may consist of plant matter, particularly small fleshy fruit.

The Red-eyed Vireo is easily detected by its song, which can be heard at all hours of the day, even in the afternoon, when other species are silent. However, its dull plumage makes it difficult to spot among the leaves, particularly when it is high up in the trees.

Distribution

In North America, the Red-eyed Vireo nests in the Mackenzie District of the Northwest Territories, in every Canadian province, and in the United States from the Atlantic seaboard south to Texas and to Oregon in the west (AOU 1983).

It was recorded in 61.4% (1512/2464) of Atlas squares surveyed, and breeding was confirmed in 20.3% of these (307/1512).

Breeding was confirmed in every region south of the 49th parallel, but the bird was atlassed as far north as Lake Mistassini and 20 km south of Lake Evans (James Bay). In the eastern part of the province, the northernmost sighting was at Mingan. There are even more northerly records, from as far north as the Eastmain River (ÉPOQ). Overall, the Atlas distribution matches that described by David (1980a) and Godfrey (1986).

Frequency of occurrence is highest in the deciduous and mixed forest zones. It declines, although it remains high, toward the coniferous forest zone, except in the White spruce-balsam fir and balsam fir-white spruce forest region on Anticosti Island, where frequency of occurrence is very low. The Red-eyed Vireo is found only within the strip of mixed forest along the north coast of the island (Thibault 1985). The species is considered a probable breeder on the Magdalen Islands.

History and trends

Comparison of the early literature (Macoun and Macoun 1915; Taverner 1922) with current data shows that the distribution of the Red-eyed Vireo has changed little over the past century. However, deforestation for farming has caused loss of habitat on its breeding grounds and, to some extent, on its wintering grounds (Robbins et al. 1989b; Diamond 1991; Askins 1993).

The expansion of the Brown-headed Cowbird into the northeastern United States, following the development of agriculture at the end of the last century, had a negative impact on the breeding success of the Red-eyed Vireo in the remnant woodlands (Brittingham and Temple 1983). The Red-eyed Vireo is a frequent cowbird host. Terrill (1961) estimated that 42% of 64 nests in southern Québec were parasitized, and based on data collected on 75 nests in the province Darveau (1988) set the figure at 32%. Peck and James (1987) reported parasitism in 38% of 354 nests in Ontario, and Southern (1958) reported a rate of 72% in 104 nests in Michigan. In unbroken forest, as in the Québec interior, the risk of parasitism is reduced, since the Brown-headed Cowbird does not frequent such areas.

It is difficult to assess current population trends for the Red-eyed Vireo in Québec. ÉPOQ summer data for 1970-89 show no significant trends, while BBS data for 1966-89 indicate an increase (Cyr and Larivée 1993b). Erskine et al. (1992) analysed BBS data for 1966-91 and found an increase for the Maritimes and the central areas of Québec and Ontario, but no significant trend for the southern parts of Québec and Ontario.

The Red-eyed Vireo is quite tolerant of habitat disturbance and continues to occupy stands that have been extensively thinned (Webb et al. 1977; Freedman et al. 1981) or disturbed by recreational activities (Clark et al. 1983). It recolonizes second-growth stands less than 10 years after clearcutting (Titterington et al. 1979; Darveau et al. 1982; Holmes 1990). This vireo is nonetheless sensitive to habitat fragmentation: relative abundance (per surface unit) increases with the size of stands and reaches its maximum only in areas larger than 3000 ha (Robbins et al. 1989a).

Marcel Darveau

Brewster's Warbler

Paruline de Brewster

Vermivora chrysoptera x pinus

Christiane Girard

THIS BIRD IS THE MOST COMMON MANIFESTATION OF HYBRIDIZATION between the Golden-winged and Blue-winged warblers. These two species, considered by specialists to be very closely related, interbreed regularly. Brewster's Warbler is a phenotype that shows the dominant genetic characteristics of both parent species. Subsequent generations may sometimes produce another hybrid with different plumage and expressing recessive genetic characteristics. The name "Lawrence's Warbler" has been given to this hybrid, which is a much less common phenotype than Brewster's (Pough 1949; Parkes 1983). Lawrence's Warbler has been recorded only once in Québec, on Mount Royal in Montréal during the 1978 spring migration (David and Gosselin 1978). Brewster's Warbler is a quite recent addition to the list of breeding birds in Québec.

Habitat and behaviour

In spite of the variability of its plumage, this hybrid is distinctive enough to be identifiable. Its song, on the other hand, is likely to generate confusion among birders, as it may adopt the vocal patterns of either parent. At Fitch Bay in the Eastern Townships, for instance, two individuals were observed, one singing like a Blue-winged Warbler and the other like a Golden-winged. To identify warblers in this group with certainty, it is therefore essential to see them.

Although a hybrid, Brewster's Warbler is capable of reproduction. It does not usually mate with other hybrids, though this is possible (Parkes 1983), but rather with Blue-winged or Golden-winged warblers. The offspring are hybrids as well, with varying plumage, generally either very similar to Brewster's or to one or other of the original parent species, though carrying a different genetic legacy from pure-bred birds (Pough 1949; Parkes 1951).

The breeding behaviour of both original parent species is very similar, which facilitates hybridization. In fact, the courtship displays of the Golden-winged and Blue-winged warblers are essentially the same, and pair bonding occurs soon after the adults arrive on their breeding grounds (Ficken and Ficken 1968d). The two species' preferred nesting habitats are also similar. The Golden-winged Warbler frequents clearings, abandoned fields, and margins of wet shrubby areas. The Blue-winged Warbler occupies essentially the same habitats, but also frequents more wooded areas such as forest edges and clearings. In general, it uses older succession habitats than the Golden-winged Warbler. This overlapping of habitats also favours hybridization.

Distribution

The breeding range of Brewster's Warbler in North America coincides with the range overlap of the Golden-winged and Blue-winged warblers. Brewster's Warbler was reported in 5 Atlas squares, in the Ottawa Valley, in the Montérégie region, and in the Eastern Townships, all in the southernmost part of the province, near the borders with Ontario and the United States. Three individuals were observed at Fitch Bay, Eastern Townships, in 1985, and in the Montérégie area the bird was reported in Frelighsburg. The only confirmed breeding report for Québec is from the Huntingdon area, southwest of Montréal, where in 1984 an adult was seen feeding a fledgling outside the nest (Bannon 1986). In the Ottawa Valley, this hybrid has been observed at Cantley and at Lake Leslie.

History and trends

Brewster's Warbler is a newcomer to Québec. The first recorded sighting was made during the first year of atlassing (Yank and Aubry 1984), and since then this hybrid has been reported about 10 times, always from the same regions of southern Québec. This addition to Québec's avian fauna is the result of an expansion in the breeding ranges of the Golden-winged and Blue-winged warblers. The increasingly frequent presence of these two warblers points to the probability that sightings of Brewster's Warblers will continue to grow in number over the next few years.

Historically, the two parent species were allopatric, with non-overlapping breeding ranges. For the last 175 years, in the case of the Golden-winged Warbler, and 150 years in the case of the Blue-winged Warbler, these two species have been expanding northeastward across North America (Gill 1980; Confer and Knapp 1981). Abandonment of agricultural land has allowed establishment of plant communities that first attracted the Golden-winged Warbler. Then, through succession the plant community would gradually change to the point where it was suitable for the Blue-winged Warbler. Range overlap of the two species resulted, and interspecific contact occurred, which led to the phenomenon of hybridization. Nevertheless, other conditions were essential: similar courtship rituals and feeding habits, and mutual tolerance between the two species in the same breeding areas (Gill 1980).

François Shaffer

Notes on territory

The territorial behaviour of a Brewster's Warbler is likely to be similar to that of a Blue-winged and a Golden-winged warbler, since it generally interbreeds with one or other of these species (Faxon 1917; Parkes 1983).

A Brewster's Warbler was reported nesting in this area of regeneration along Gowan Road in Rockburn in 1990. This was one of the rare reports of this hybrid's breeding in Québec. South of Montréal. *June 1991, Pierre Bannon.*

Brewster's Warbler

Breeding distribution

1984–1989

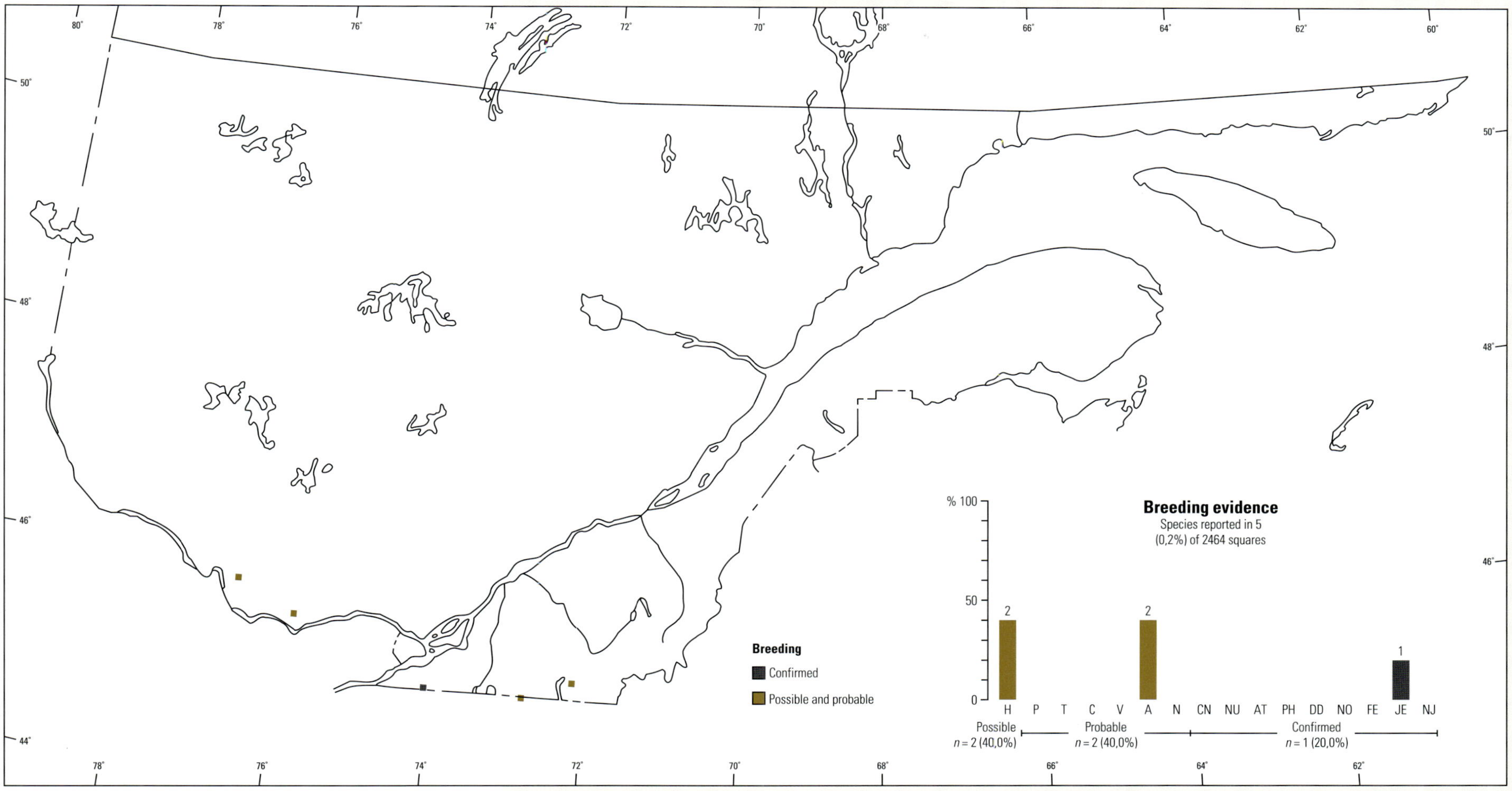

Species profile

Clutch size	4–5 eggs[1,2]
Laying interval	probably 1 egg per day[3,4,5]
Start of incubation	probably with last egg[4,5,6,9]
Incubation period	probably 10–12 days[3,4,6,9]
Maturity at hatching	altricial
Care of young	by female and male[4,5,6,9]
Nestling period	7–10 days[1]
Age at first flight	7–10 days or more[10]
Dependence of young	?
Broods per year	1[7,9]
Breeding age	probably 1 year[7,9]
Mating system	probably monogamy[4,9]
Duration of pair bond	?
Total length	10.2–12.7 cm[3]
Wing span	?
Weight	?
Longevity record	6 years[8]

References:
1. Faxon 1917
2. Peck and James 1987
3. Terres 1980
4. Ehrlich et al. 1988
5. Harrison 1978
6. Bent 1953
7. DeGraaf and Rudis 1987
8. Kennard 1975
9. extrapolated from information on Blue-winged and Golden-winged warblers
10. according to nestling period

World breeding range

Chronology *n* = 2

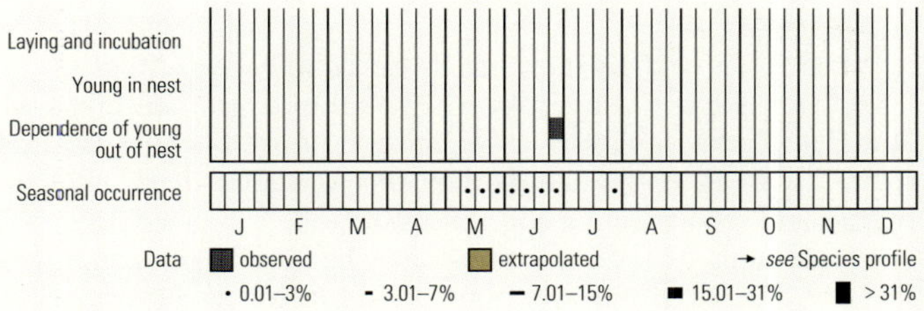

Golden-winged Warbler

Paruline à ailes dorées

Vermivora chrysoptera

Christiane Girard

THE GOLDEN-WINGED WARBLER, AS ITS NAME IMPLIES, HAS A BRIGHT yellow wing patch, the only warbler so endowed. Other obvious field marks include a black bib and ear patch, white underparts, and a habit of hanging upside down from twigs to feed in the manner of the Black-capped Chickadee, all of which makes it easy to identify. Its song—a soft, buzzy, four-syllable note usually transcribed as *bee-bz-bz-bz*—sounds more like an insect than a bird, inevitably arousing the curiosity of observers attentive to the sounds of the forest. Nevertheless, this rare and elusive bird is difficult to spot, and birders who wish to do so should concentrate on the southernmost part of the province.

In June, the Golden-winged Warbler returns to Québec from its wintering grounds in Central and South America, from the Yucatan Peninsula and Guatemala to eastern Colombia and northern Venezuela (AOU 1983).

The Golden-winged Warbler frequents clumps of bushes in clearings, forest margins, and abandoned fields in the southernmost part of the province. Gowan Road, Herdman, southwest of Montréal. *June 1991, Pierre Bannon.*

Habitat and behaviour

The Golden-winged Warbler is found in open areas with abundant herbaceous vegetation, large clumps of bushes, and relatively few trees. It shows a marked preference for dense clumps of bushes and shrubs on the edges of woodlands (Ficken and Ficken 1968a; Confer and Knapp 1981). Such habitats may be found in clearings, forest margins, the edges of beaver ponds, and brushy overgrown pastures surrounded by forest.

When the males arrive, they immediately set out to establish their territories by perching conspicuously at the top of a shrub or small tree and singing their buzzy song. They stop singing and become much more difficult to spot when the young are ready to fledge. The females are always very secretive, as shown by a survey in June 1984 in the Huntingdon area, during which 19 territorial males but only 6 females were counted (pers. obs.). Females were generally observed during the brood-raising period transporting food or displaying agitated behaviour.

The well-camouflaged nest is built on the ground and is almost impossible to find unless you can spot a female carrying material to it. Only one nest was discovered during the Atlas period and, in all, nesting was confirmed in only 5 squares, which explains the rather poor breeding confirmation rate of 16.1% (5/31).

The female builds the nest and incubates the eggs. The male assists her only in raising the brood (Broun 1979).

Distribution

The Golden-winged Warbler's breeding range in North America includes southeastern Manitoba, southern Ontario and Québec, the Great Lakes states, and central New England (National Geographic Society 1987). In Québec, Godfrey (1986) noted that the species nests at Lake Philippe in the Ottawa Valley and at Philipsburg in the Montérégie region, based on information from the 1970s. More recent data suggest that since then the Golden-winged Warbler has expanded outward from these two areas, creating 2 distinct breeding zones in the province. The expansion has been mainly along east-west lines rather than to the north. In the Ottawa Valley, the species is now found along a 200 km strip running from west to east between Rapides-des-Joachims and Plaisance, which extends no more than 50 km north of the Ottawa River. The population in the Montérégie and Eastern Townships regions is scattered between Huntingdon in the west and Sherbrooke in the east, along a 190 km corridor paralleling the U.S. border. These two corridors are characterized by regenerating deciduous forests, abandoned fields, and relatively small areas of cultivated land. In the region separating the two corridors—the western part of the Central St. Lawrence Lowland—the landscape is dominated (90%) by cultivated fields, which do not provide suitable habitat for the Golden-winged Warbler. The species was also observed at an isolated site in Mont-Tremblant Recreation Park between 1973 and 1975, and again in 1989.

History and trends

The Golden-winged Warbler has been expanding its range northward and eastward in eastern North America for the last 200 years. It was first recorded in Québec in 1957 at Cap Tourmente (Anonymous 1957) and, in 1971, was observed nesting in Philipsburg (PQSPB 1971). South of the border, its northward expansion has been well documented: it appeared in central New York State around 1920 (Confer and Knapp 1981) and reached the northern part of the state only recently (Confer 1988). In Ontario, it expanded its range 250 km northward during the first half of the 20th century (Snyder 1957), and has moved north another 100 km since 1950 (Mills 1987c).

However, the situation is completely different in the species' traditional range in the United States, where entire nesting populations have disappeared in some areas. These disappearances have very often occurred following the expansion of the Blue-winged Warbler into the area. Two explanations have been proposed for this. Gill (1980) suggests that the high rate of hybridization between the Golden-winged and Blue-winged warblers, which has resulted in hybrids such as Brewster's Warbler, slowly results in the Golden-winged Warbler's genetic introgression, or assimilation. If current trends continue, Gill thinks that the Golden-winged Warbler may become extinct in the 21st century. However, Confer and Knapp (1981) believe that the disappearance of the Golden-winged Warbler may also be due to plant succession; the Golden-winged Warbler, a habitat specialist, nests almost exclusively in abandoned farmland about 10–30 years into secondary succession, while the Blue-winged Warbler, a habitat generalist, not only nests in this early-succession habitat but also in second-growth stands as old as 70 years. Therefore, the two species may occupy the same habitat for a time but, as the habitat matures, the Golden-winged Warbler will leave the area, while the Blue-winged Warbler will continue to prosper for some time (Marsi 1980).

In Québec, where the Blue-winged Warbler is still extremely rare, interactions between the two species have been few and far between. However, the Blue-winged Warbler has recently been seen a few times in the province, and a Brewster's Warbler was reported nesting during the Atlas period (Bannon 1986). It will be interesting and indeed essential to monitor the evolution of the Golden-winged Warbler's status should the Blue-winged Warbler become established in Québec.

PIERRE BANNON

Notes on territory

Territorial relations between the closely related Golden-winged Warbler and Blue-winged Warbler and their hybrids were studied by Ficken and Ficken (1968c). They discovered that individuals use plumage rather than song to determine if other birds are the same or a different species. Therefore, males with similar plumage but different songs defend separate territories, while males with similar songs but different plumage do not. As expected, males with similar plumage and songs defend distinct territories, while males with different plumage and songs do not. Differences in face patterns play an important role in territorial behaviour (Ficken and Ficken 1968c).

In Minnesota, Coy et al. (1993) found densities of 2.0 territories/10 ha in a young stand of trembling aspen interspersed with mature pines.

Golden-winged Warbler

Breeding distribution

1984–1989

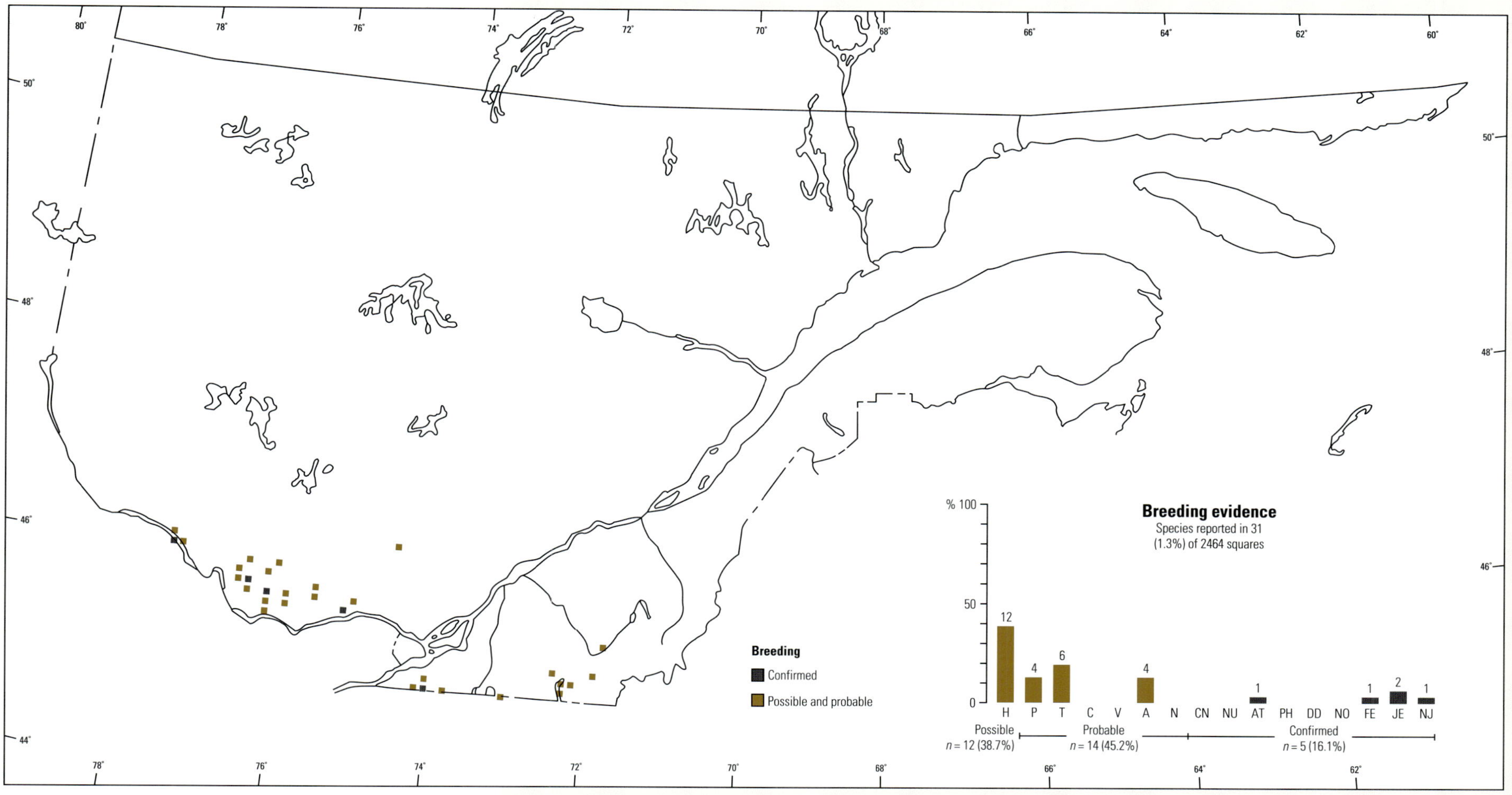

World breeding range

Species profile

Clutch size	usually 4–6 eggs (2–7)[1,2,3,4,12]
Laying interval	usually 1 egg per day[5,11]
Start of incubation	with penultimate egg[4,11]
Incubation period	usually 10–11 days (10–12)[1,2,11]
Maturity at hatching	altricial
Care of young	by female and male[6,11]
Nestling period	8–9 days[11]
Age at first flight	8–10 days[2,3,6,11]
Dependence of young	for up to 31 days after first flight[11]
Broods per year	1[10,11]
Breeding age	1 year[10,11]
Mating system	monogamy[4]
Duration of pair bond	?
Total length	12.4–13.5 cm[7]
Wing span	?
Weight	M 8.7 g (7.2–11.2) / F 8.9 g (7.5–11.8)[9]
Longevity record	7 years 11 months[8]

References:
1. Peck and James 1987
2. Broun 1979
3. Terres 1980
4. Ehrlich et al. 1988
5. Harrison 1975
6. Harrison 1978
7. Godfrey 1986
8. Klimkiewicz et al. 1983
9. Dunning 1984
10. DeGraaf and Rudis 1987
11. Confer 1992a
12. Confer 1992b

Chronology *n* = 5

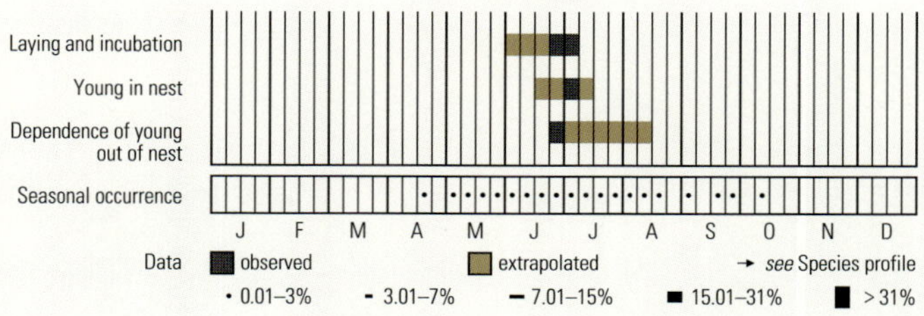

This Golden-winged Warbler's nest, one of the few ever found in Québec, is built on the ground amidst the leaf litter. Huntingdon, southwest of Montréal. *June 1984, Pierre Bannon.*

Tennessee Warbler

Paruline obscure

Vermivora peregrina

Christiane Girard

WITH ITS DRAB PLUMAGE, THE TENNESSEE WARBLER LOOKS MORE LIKE A vireo than a warbler. This very beneficial species devours large quantities of harmful insect pests, particularly the spruce budworm (Morse 1989a). The Tennessee Warbler is a common migrant breeder of the boreal forests of Québec, where it can be detected by its loud and persistent song.

On its wintering grounds, which extend from central Mexico to northern South America (Peterson 1984; Godfrey 1986), the Tennessee Warbler is gregarious and is often seen in flocks of 10–200 individuals (Ehrlich et al. 1988).

Habitat and behaviour

According to studies of Tennessee Warblers in spring migration (Quay 1985, 1989), some individuals develop a precocious physiological ability for breeding even before they reach their breeding grounds. Early copulations have been observed in southern Ontario, Missouri, and Texas between females and males already capable of producing sperm. Although it is not known what role these copulations play in fertilization and egg production, this behaviour appears to be an adaptation to the short breeding season in the northern parts of the species' breeding range (Quay 1985, 1989).

The Tennessee Warbler is mainly found in coniferous and mixed forests, although it also frequents young deciduous stands. Its preferred habitat includes stands about 20 years old with a well-developed shrub layer. The species also appears to be associated with forest openings with herbaceous vegetation, dense shrubs, and regenerating young deciduous trees. In addition, it is often found in poplar, alder, and willow thickets, and peatlands. Man-made clearings, burns, and cutovers less than 10 years old also provide suitable habitat for the species (Erskine 1984; Godfrey 1986; DeGraaf and Rudis 1987; Welsh 1987e; Ehrlich et al. 1988; Peterson 1988n).

Because of its greenish plumage, this warbler is fairly difficult to spot, but the male can usually be detected by its persistent song, which it performs from tree tops (Bowdish and Philipp 1916). The three-part song is similar to the Nashville Warbler's, but the Tennessee Warbler begins its song more slowly, as if it were having trouble getting started. Frequencies as high as 6600–9150 Hz may be reached, compared with 4000 Hz on average for most other passerines (Brand 1938), making the song inaudible to some people (Peterson 1988n). Songs may be repeated 6–9 times a minute, from dawn to midday or even later (Oatman 1985b; Welsh 1987e). The male also performs a flight song that may play a role in courtship (Ehrlich et al. 1988). The female is more difficult to observe since she remains near the ground and is difficult to flush from the nest (Bent 1953; Gunn 1979; Ehrlich et al. 1988). The nest is built on the ground and is very difficult to find.

In suitable habitat, several Tennessee Warbler nests can often be found in a small area, forming a loose colony (Bowdish and Philipp 1916; Harrison 1984; Ehrlich et al. 1988). The female generally builds the nest under a bush, in a hollow in a clump of sphagnum or other moss, or under an overhanging tussock of grass, most often in a wet place but sometimes on a dry hillside (DeGraaf et al. 1980; Godfrey 1986; Ehrlich et al. 1988). The nest is made of dried grass and lined with fine grass and hair (Bowdish and Philipp 1916; Bent 1953; Peck and James 1987). It is often well hidden even from the most astute predator, and the only way to find it is to push aside the overhanging vegetation.

The Tennessee Warbler is a common species of boreal forests, such as those found in the mountainous sections of the Laurentide Wildlife Reserve. Grands-Jardins Conservation Park. *September 1992, Yves Aubry.*

Notes on territory

Sometimes several pairs of Tennessee Warblers may nest in close proximity in suitable habitat; several nests may be found in a small area, forming a loose colony (Bowdish and Philipp 1916; Bent 1953; Harrison 1975; Ehrlich et al. 1988). The proximity of these nests suggests that only small territories are defended (Bowdish and Philipp 1916). In Ontario, in stands dominated by balsam fir and spruce, Kendeigh (1947) found 52 territories ranging between 0.16 and 0.68 ha, or 0.27 ha on average.

Pair density varies considerably with habitat type. In Québec, the greatest densities are found in predominantly coniferous forests; densities of up to 34.6 pairs/10 ha were recorded in medium-aged stands dominated by balsam fir and spruce that had been disturbed by logging operations or spruce budworm infestations (Erskine 1980a).

In Ontario, densities between 1.5 and 79.4 pairs/10 ha were recorded in coniferous and mixed forests around the 49th parallel; the greatest concentrations were found in 20-year-old forests with a dense shrub layer (Erskine 1971a, 1980a, 1984). In more mature forests in Ontario, densities of up to 61 pairs/10 ha have been reported, while densities of up to 33.5 pairs/10 ha were found in cutover areas less than 10 years into regeneration (Welsh 1987e).

Tennessee Warbler

Breeding distribution
1984–1989

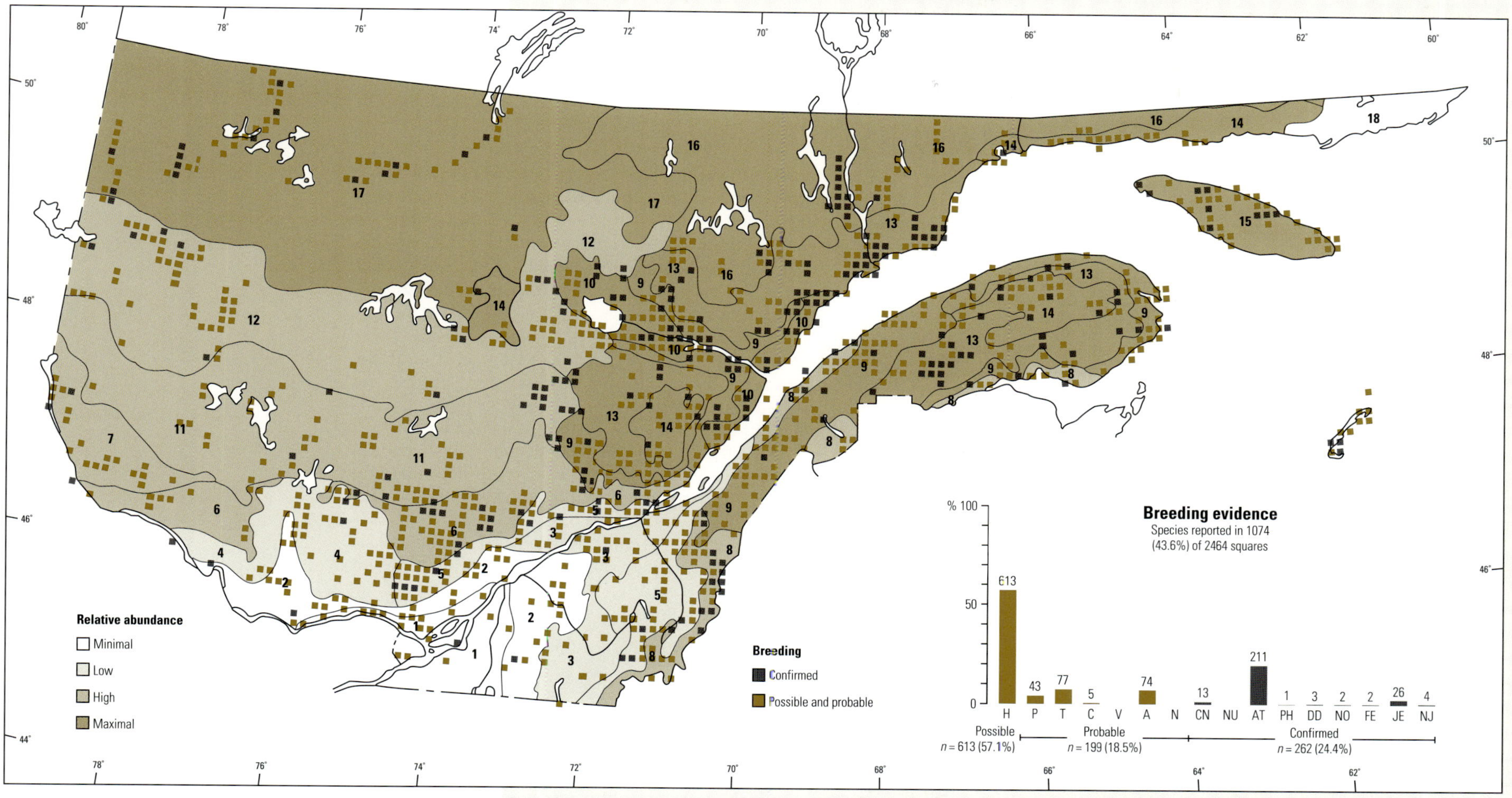

Species profile

Clutch size	usually 4–6 eggs (4–8)[1,4,6,7,9]
Laying interval	?
Start of incubation	?
Incubation period	11–12 days[1,4]
Maturity at hatching	altricial
Care of young	by female and male[1]
Nestling period	?
Age at first flight	?
Dependence of young	?
Broods per year	1[6]
Breeding age	1 year[6]
Mating system	monogamy[1]
Duration of pair bond	?
Total length	11.4–12.7 cm[2]
Wing span	18.8–21.1 cm[8]
Weight	M 10.2 g (7.3–18.4) / F 9.8 g (7.8–13.4)[3]
Longevity record	6 years 5 months[5]

References:
1. Ehrlich et al. 1988
2. Godfrey 1986
3. Dunning 1984
4. Harrison 1975
5. Klimkiewicz et al. 1983
6. DeGraaf and Rudis 1987
7. Bent 1953
8. Forbush 1929
9. Morse 1989a

Chronology $n = 31$

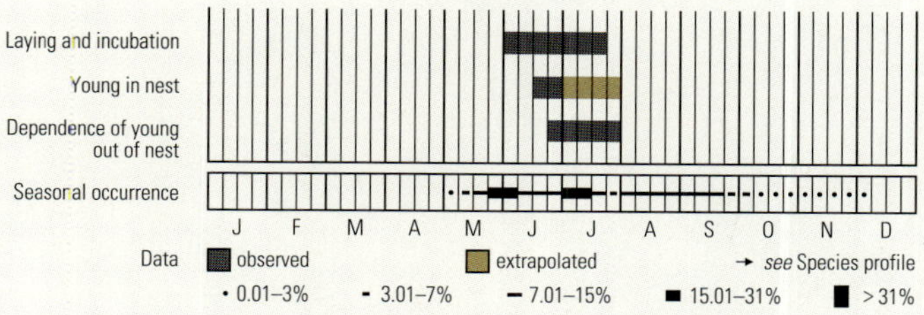

World breeding range

Frequency of occurrence by ecological region

Ecological regions** (forest regions)	Adequately covered squares	Squares in which reported	% of squares in which reported	Relative abundance (%)*
1. Sugar maple-hickory and sugar maple-basswood	97	14	14.4 ± 3.6	14.4
2. Sugar maple-basswood	153	43	28.1 ± 3.7	28.1
3. Sugar maple-basswood and sugar maple-yellow birch	101	31	30.7 ± 5.5	30.7
4. Sugar maple-yellow birch-basswood	56	31	55.4 ± 11.4	55.4
5. Sugar maple-yellow birch-beech and sugar maple-yellow birch-basswood	113	62	54.9 ± 5.1	54.9
6. Sugar maple-yellow birch-beech	71	55	77.5 ± 9.0	77.5
7. Sugar maple-yellow birch-balsam fir	19	17	89.5 ± 15.5	89.5
8. Sugar maple-yellow birch and balsam fir-yellow birch	47	34	72.3 ± 11.6	72.3
9. Balsam fir-yellow birch	87	80	92.0 ± 5.5	92.0
10. Balsam fir-white birch or balsam fir-red maple	46	44	95.7 ± 5.7	95.7
11. Yellow birch-balsam fir	52	43	82.7 ± 10.8	82.7
12. Balsam fir-white birch (west)	59	46	78.0 ± 11.0	78.0
13. Balsam fir-white birch (east)	76	74	97.4 ± 4.0	97.4
14. Balsam fir-black spruce	28	27	96.4 ± 8.3	96.4
15. White spruce-balsam fir and balsam fir-white spruce	22	21	95.5 ± 10.3	95.5
16. Black spruce-balsam fir-moss	18	18	100.0 ± 2.8	100.0
17. Black spruce-moss	32	29	90.6 ± 11.7	90.6
18.			not analysed	

* minimal: 0–29.9%; low: 30–59.9%; high: 60–89.9%; maximal: 90–100%

** Deciduous forest: 1–7; mixed forest: 8–11; coniferous forest: 12–17

A set of 4–6, and sometimes up to 8, eggs is laid (Godfrey 1986; Peck and James 1987; Morse 1989a), which is larger than most northern warblers' clutches. Clutch size may increase even more during insect infestations, mainly spruce budworm outbreaks (MacArthur 1958; Welsh 1987e; Morse 1989a). The Tennessee Warbler is adept at exploiting these outbreaks, moving to affected areas and increasing its clutch size by 50% or more (Welsh 1987e). In New Brunswick, Morris et al. (1958) found that Tennessee Warbler populations increased in budworm-infested areas while, in Québec, DesGranges and Rondeau (1995) observed that the species was significantly less abundant in balsam fir stands in which the epidemic had been allowed to run its course than in stands that had been treated with insecticides during the epidemic. In other Québec forests, predation by four warbler species, including the Tennessee Warbler, has been found to have a negligible impact on budworm numbers when the insects are abundant but to be an important factor in budworm mortality when budworm numbers are lower (Régnière and Welsh 1988).

The Tennessee Warbler is monogamous and only produces one brood a year (DeGraaf and Rudis 1987). Nest parasitism by the Brown-headed Cowbird is rare (Terres 1980; Ehrlich et al. 1988). Incubation lasts 11 or 12 days and is performed solely by the female (Harrison 1975; Ehrlich et al. 1988). However, both parents feed the young (Ehrlich et al. 1988).

During the breeding season, the species' diet consists almost entirely of insects, which it obtains by gleaning the foliage and the branches as it hops from branch to branch. Favourite foods include weevils, flies, aphids, grasshoppers, caterpillars, worms, scarab beetles, and spiders (DeGraaf et al. 1980). The Tennessee Warbler forages at heights below 12 m, but in spring can be seen hunting for insects in the catkins and flowers of tall trees. Its diet may also be supplemented by seeds and berries, particularly in autumn (Tramer and Tramer 1977; Terres 1980).

Distribution

The Tennessee Warbler breeds from northwestern Yukon to Newfoundland in the north and from northwestern Montana to New England in the south (AOU 1983). In Québec, it ranges from the extreme south to north of the Atlas area, where it is found in trembling aspen and birch stands and other habitats (Cons. G. G. – GREBE 1992o). The species was reported in 44% (1074/2464) of Atlas squares surveyed, and breeding was confirmed in 262 squares. The highest frequency of occurrence was from forested habitats dominated by fir or spruce. In summer, the Tennessee Warbler is uncommon in the Central St. Lawrence Lowland and southwestern part of the province. Unpaired males probably account for some of the individuals seen in the southernmost part of Québec, a phenomenon which has also been observed in Ontario and the neighbouring American states (Oatman 1985b; Welsh 1987e). Breeding has been confirmed on several occasions in northern New England (Dionne 1906; DeGraaf et al. 1980; Oatman 1985b).

The northernmost records come from Lake Guillaume Delisle (56°15') on the coast of Hudson Bay and from Schefferville (54°48') in the northeast (ÉPOQ).

Black spruce-white birch forest provides suitable nesting habitat for the species, which usually builds its nest on the ground in sphagnum and other mosses. Montmorency forest, Laurentide Wildlife Reserve, north of Québec City. *June 1991, Réginald Ouellet.*

The Tennessee Warbler flourishes during spruce budworm epidemics. Here, a spruce budworm caterpillar is busy eating away the needles of a sprig of fir. Gaspé Peninsula. *July 1990, Lina Breton, Ministère des Forêts.*

History and trends

According to Taverner (1922), the Tennessee Warbler was rare in almost all regions of eastern Canada at the turn of the century. This view was not shared by all observers of the time, however. For example, Macoun and Macoun (1915) considered the species common, but noted that it was often overlooked because of its drab plumage. Todd (1963) stated that it was one of the most common warblers at the turn of the century, particularly in the boreal regions (the Canadian Life-Zone), where it found optimum conditions and reached its greatest abundance, but it also extended into subarctic regions (the Hudsonian Life-Zone) in smaller numbers. On the North Shore, the Tennessee Warbler was described as fairly common at Godbout (Wintle 1896; Macoun and Macoun 1915) and at Pointe-des-Monts (Bowdish and Philipp 1916).

The Tennessee Warbler is probably much more abundant now than during the 19th century (Bull 1974). One possible explanation for this increase is that the species has taken advantage of the logging and clearing of coniferous forests (Bowdish and Philipp 1916). In addition, in the northern part of southern Québec, it has probably also benefited from spruce budworm outbreaks and the growth of successional deciduous stands after logging (Kendeigh 1947; Bent 1953; Bull 1974; Robbins et al. 1986). BBS data between 1966 and 1979 show a slight increase in populations in eastern Canada, correlated with a spruce budworm outbreak in the same area (Robbins et al. 1986).

Unfortunately, the Tennessee Warbler is one of the animal species directly affected by the spraying of insecticides. Insecticides were first used in the 1940s to control spruce budworm epidemics; they poisoned the birds rather than depriving them of the insects they fed on (Kendeigh 1947; Fowle 1972; Busby et al. 1987). Today, less potent insecticides are used, but their effects on birds, and particularly on the eggs and young, are still not known (Busby et al. 1987). Since 1987 in Québec, *Bacillus thuringiensis* has been the only product used to control spruce budworm outbreaks (Bordeleau and Auger 1992). This biological control agent appears to be less hazardous to birds.

In some regions, particularly the Middle North Shore and Gaspé Peninsula, the use of insecticides has reduced the seriousness of spruce budworm outbreaks without completely eliminating the insects (Bordeleau 1988; Bordeleau and Boulet 1989, 1990; Bordeleau and Auger 1991, 1992); this has to some extent helped to maintain Tennessee Warbler populations at fairly stable levels.

Benoît Limoges and Serge Gauthier

Orange-crowned Warbler

Paruline verdâtre

Vermivora celata

Christiane Girard

THE ORANGE-CROWNED WARBLER IS PRINCIPALLY A SPECIES OF WESTERN North America, where it is much more common and widespread than it is in Québec.

This warbler is not found in the inhabited areas of southern Québec in summer, but is observed mainly during migration in May and October, though it is still quite scarce at those times. Furthermore, because of its drab colouring, the species is easily overlooked and may be confused with the Tennessee Warbler, which is much more common during migration. Its one distinguishing field mark, the orange crown, is almost never visible, even in the male. The inconspicuousness of the crown is alluded to in the species' Latin name, *celata*, which means "hidden."

The Orange-crowned Warbler winters in the southern United States south to Guatemala (AOU 1983).

Habitat and behaviour

In summer, this warbler frequents mainly brushy open woods, the deciduous sections of mixed forests, and regenerating second-growth stands in clearings and former burns (Godfrey 1986; DeGraaf et al. 1991). The Orange-crowned Warbler is not associated with coniferous forests, even though it is limited mainly to the boreal regions of Québec, where the forests are dominated by conifers. As in Ontario, the Orange-crowned Warbler is often found in poplar stands with an understorey of shrubs such as willow or alder (Welsh and Helleiner 1987). In Québec, Todd (1963) also observed the species in an alder swamp, in alder and willow along rivers, as well as in muskeg and on a dry pine ridge, the latter according to Todd being a particularly unusual habitat for the species. During a study carried out to the southeast of the Opinaca Reservoir, this species was also sighted in stands dominated by poplar and birch (Tremblay and Belles-Isles 1991).

Very little is known about this warbler's breeding biology. The nest is generally built on the ground, but occasionally a few centimetres above the ground in a bush. It is made of coarse grasses, dead leaves, bark strips, and mosses, and is lined with fine grasses, feathers, and animal hair (Bent 1953; Terres 1980). The only nest reported from Québec was found by Todd (1963) at Waskaganish on 13 June 1943. The very rudimentary nest was located on a hillock in a peatland, in a clump of Labrador-tea. It is not known whether the male helps build the nest or incubate the eggs. However, as in other warbler species, the male helps to bring food to the nestlings (Sogge et al. 1994).

Like other *Vermivora* warblers, the Orange-crowned Warbler feeds mainly on invertebrates, which it captures by using probing and gaping motions to inspect leaves, buds, flowers, and bark crevices (Remsen et al. 1989). In the West, it also drinks the sap from woodpecker holes and visits hummingbird feeders, and on its wintering grounds, it pierces the base of flowers to drink the nectar (Ehrlich and Daily 1988; Remsen et al. 1989). A few individuals occasionally winter in the northeastern United States, where they may feed on suet or peanut butter from birdfeeders (Griscom and Sprunt 1979).

Like many other passerines, the Orange-crowned Warbler is located most often by its song, a loose two-toned trill weakening at the end. An observer unfamiliar with this song could easily confuse it with that of the Dark-eyed Junco, Chipping Sparrow, or Yellow-rumped Warbler, three species that nest at the same latitudes as the Orange-crowned Warbler. Less experienced birders may also have problems distinguishing its plumage from that of the Tennessee Warbler, which is also found in central Québec. The best way of differentiating between the two is to note the colour of the undertail coverts, which are whitish in the Tennessee Warbler and greenish yellow in the Orange-crowned Warbler.

The Orange-crowned Warbler was reported in 0.4% (11/2464) of the Atlas squares surveyed. Eight records of possible breeding and one of confirmed breeding based on a report of an adult bird carrying food were obtained. These low numbers are probably due to the fact that many squares in the northern part of the Atlas area were not surveyed exhaustively, as well as to the fact that this species, which frequents mainly the forested parts of central Québec, has a scattered distribution in the northern part of the Atlas area.

Distribution

In western North America, the Orange-crowned Warbler breeds from central Alaska to California (AOU 1983). In Canada, it is found from Yukon east to southern Labrador (Godfrey 1986).

Very few data were obtained on the Orange-crowned Warbler for the Atlas, mainly because the species does not nest in the southern part of the Atlas area. Consequently, the distribution map must be interpreted with caution because many squares in the northern part of the Atlas area were not surveyed.

The records involving birds observed in suitable nesting habitat during the breeding season all came from north of the St. Lawrence River, except the record on Anticosti Island, which was a first for the island. The southernmost record was in the Laurentide Wildlife Reserve, a first for the reserve. In the other regions where breeding records were obtained, ornithologists had already suspected that the species nested there. The only square where breeding was actually confirmed is located northeast of Matagami, in a region which produced several reports.

According to Godfrey (1986), the Orange-crowned Warbler breeds in both coastal and inland areas between the 49th and 55th parallels. Todd (1963), who sighted the species on the coast of James Bay, in the Abitibi region, and on the North Shore, maintained that the species probably breeds in central Québec from Labrador to James Bay. Todd also captured an individual at Lake Guillaume Delisle, near the shores of Hudson Bay in 1926. More recently, this warbler has been observed in the region around the Nottaway, Broadback, and Rupert rivers, particularly in the coastal plain around Rupert Bay (Cons. G. G. – GREBE 1992o), farther inland near the Eastmain and Opinaca rivers (Bordage and Aubry 1982), and near Kuujjuarapik (Cons. G. G. – GREBE 1991a). The range depicted above corresponds to the species' distribution in Ontario, where it is described as breeding almost everywhere between the 50th parallel and the tree line (Welsh and Helleiner 1987). Although the bird may be distributed over a large area in Québec, it is not found in large numbers anywhere here or in Ontario. Nevertheless, the Orange-crowned Warbler is fairly common locally in the James Bay lowlands, for example, at Waskaganish (pers. obs.)

History and trends

It was formerly thought that the Orange-crowned Warbler did not breed in eastern North America, but only passed through in migration. In 1919, for example, Taverner wrote that the species did not nest east of Manitoba and, in 1934, noted that it was rare east of the Great Lakes (Taverner 1919, 1934). Todd (1963) appears to be the first ornithologist to have obtained enough data to describe the status of the species in Québec. Therefore, the fairly recent confirmation of the presence of the Orange-crowned Warbler in central Québec must not be considered as evidence of a range expansion, but rather as a phenomenon that has only been discovered fairly recently.

The first Québec records, from the James Bay lowlands, date from the turn of the century. Most of the nesting areas described by Todd (1963) appear to be still frequented by the Orange-crowned Warbler today.

The species may benefit from forest fires and certain logging practices that modify the landscape of the boreal forest by creating large enclaves of deciduous regeneration.

MICHEL ROBERT

This warbler, which is fairly common in the lowlands around James Bay frequents shrubby areas, in both wet and dry habitats. Bay of Many Islands, James Bay. *August 1992, Michel Robert.*

Notes on territory

In Oregon, researchers surveyed a dozen clearings created by logging; the areas had been clearcut, the stumps burned, and conifer seedlings planted. Four to nine years later, in the brush and sapling successional stage, they recorded an average density of 7.7 individuals/10 ha, with a range of 2.7–15.6 individuals/10 ha (Morrison and Meslow 1983).

Orange-crowned Warbler

Breeding distribution
1984–1989

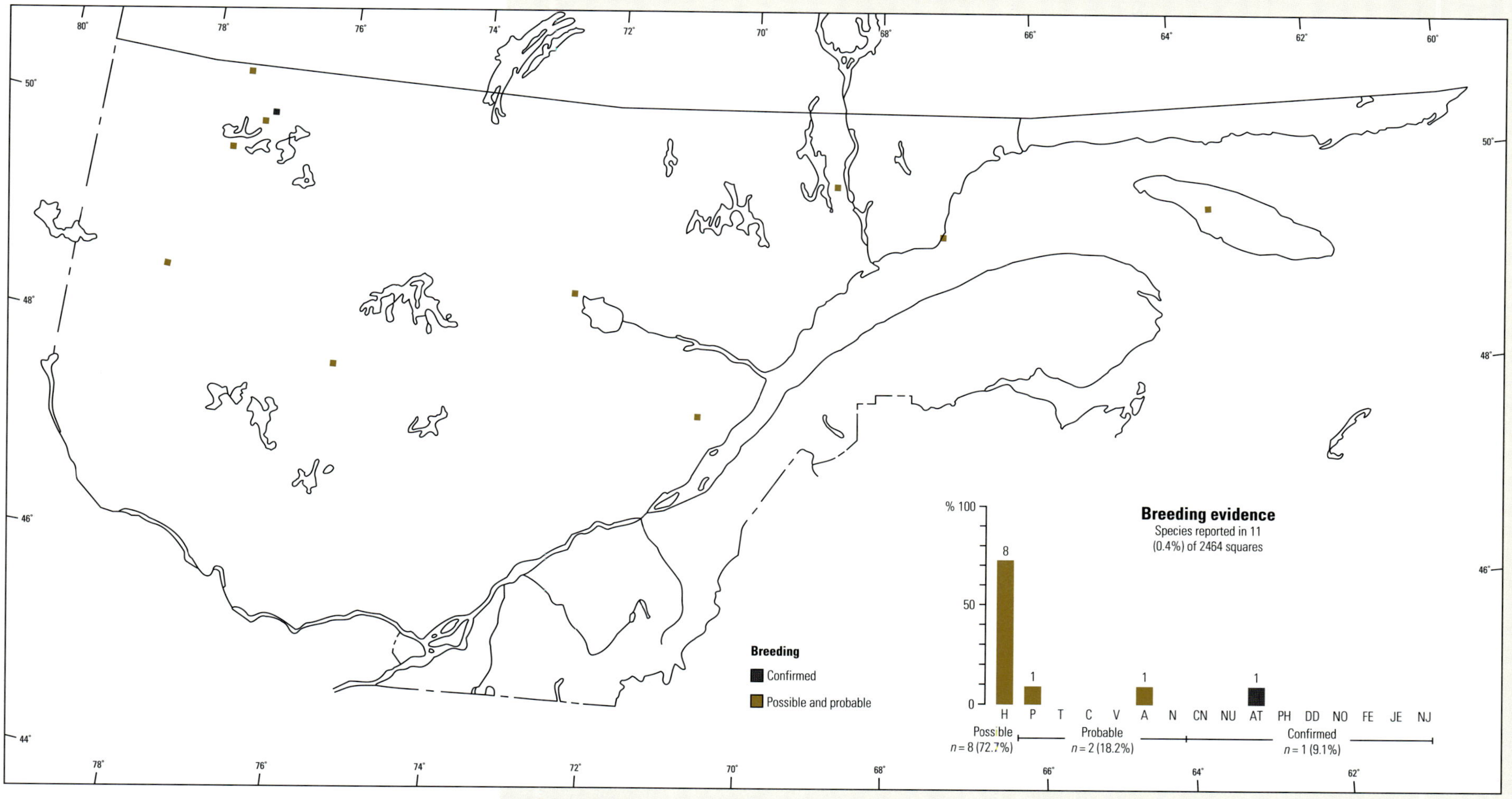

Species profile	
Clutch size	usually 4–5 eggs (3–6)[1,2]
Laying interval	1 egg per day, sometimes 1 per 2 days[7]
Start of incubation	usually with last egg[7]
Incubation period	12–14 days[1]
Maturity at hatching	altricial
Care of young	usually by female and male[7]
Nestling period	8–10 days[1]
Age at first flight	8–10 days or more[6]
Dependence of young	?
Broods per year	1, rarely 2[7]
Breeding age	?
Mating system	monogamy[1]
Duration of pair bond	?
Total length	12.2–13.4 cm[3]
Wing span	?
Weight	9.0 g (7.3–11.6)[4]
Longevity record	6 years 9 months[5]

References:
1. Ehrlich et al. 1988
2. Harrison 1978
3. Godfrey 1986
4. Dunning 1984
5. Klimkiewicz et al. 1983
6. according to nestling period
7. Sogge et al. 1994

Chronology $n = 1$

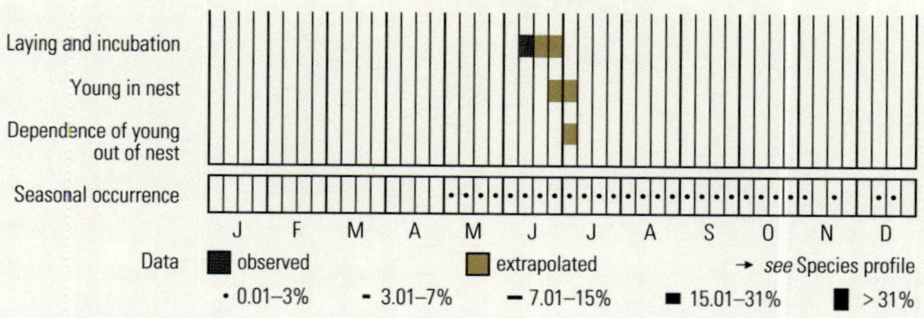

World breeding range

Nashville Warbler

Paruline à joues grises

Vermivora ruficapilla

Christiane Girard

THE NASHVILLE WARBLER IS ONE OF THE WARBLERS THAT HAS BENEFITED most from the European settlement of North America (Morse 1989a). Rare at the beginning of the 19th century, the species is now common throughout northeastern North America, where it has colonized the forest edges and regenerating second-growth forests resulting from human activity.

Of all the warblers breeding in southern Québec, the Nashville Warbler is one of the first to return in spring and one of the last to leave in fall. It may linger until late October or even November on rare occasions. The Nashville Warbler winters from southern Texas to Honduras and El Salvador (AOU 1983).

Habitat and behaviour

The Nashville Warbler is common because it can adapt to a wide range of habitats. In Québec, it is found in mixed and coniferous forest and, to a lesser extent, in deciduous forest; the same choice of habitats was observed in Ontario by Peck and James (1987). This warbler frequents clearings and young immature stands, particularly forest margins and regenerating second growth dominated by species such as trembling aspen, white birch, and balsam fir (Griscom and Sprunt 1979; Collins et al. 1982; Godfrey 1986; Peck and James 1987). In Vermont, Thompson and Capen (1988) found the species to be generally associated with dense vegetation less than 3 m tall dominated by young, shade intolerant deciduous species. In the northern part of its range, the Nashville Warbler frequents peatlands with a sprinkling of conifers such as spruce and tamarack (Griscom and Sprunt 1979; Terres 1980).

The small, cup-shaped nest is usually built in a hollow on the ground. It is made of grasses, mosses, and plant fibres, and is lined with animal hair and plant material (Harrison 1978; Godfrey 1986; Peck and James 1987). The nest is often hidden under a fern, bush, branch, in a clump of grass or under leaf litter. Typical clutch size is 4–5 eggs, which the female incubates for 11–12 days. According to Terres (1980), parasitism by the Brown-headed Cowbird is rarely a problem. After the young hatch, they remain in the nest for approximately 11 days, during which time they are fed by both parents. According to Knapton (1984a), both sexes contribute equally to feeding the young, in terms of the number, size, and type of prey. The nestlings' diet does not change substantially with time, except that the parents tend to bring larger prey items as the young mature. According to the same study, caterpillars make up 89% of the prey items brought to the nest. The adult birds feed on a variety of insects, which they capture on the trunk, branches, and leaves of trees and shrubs, working their way up from the bottom to the top (DeGraaf and Rudis 1987). The Nashville Warbler may also catch insects on the wing in the manner of a flycatcher.

Notes on territory

The Nashville Warbler defends an average territory of 1.1 ha in New Hampshire's White Mountains, where the average breeding density is 0.9 pair/10 ha (Sabo 1980). In Québec, density was found to be

The Nashville Warbler nests in dense vegetation on the margins of mixed or coniferous forests and in second-growth deciduous forests regenerating after logging operations. Montmorency forest, Laurentide Wildlife Reserve, north of Québec City. *June 1991, Réginald Ouellet.*

Nashville Warbler

Breeding distribution

1984–1989

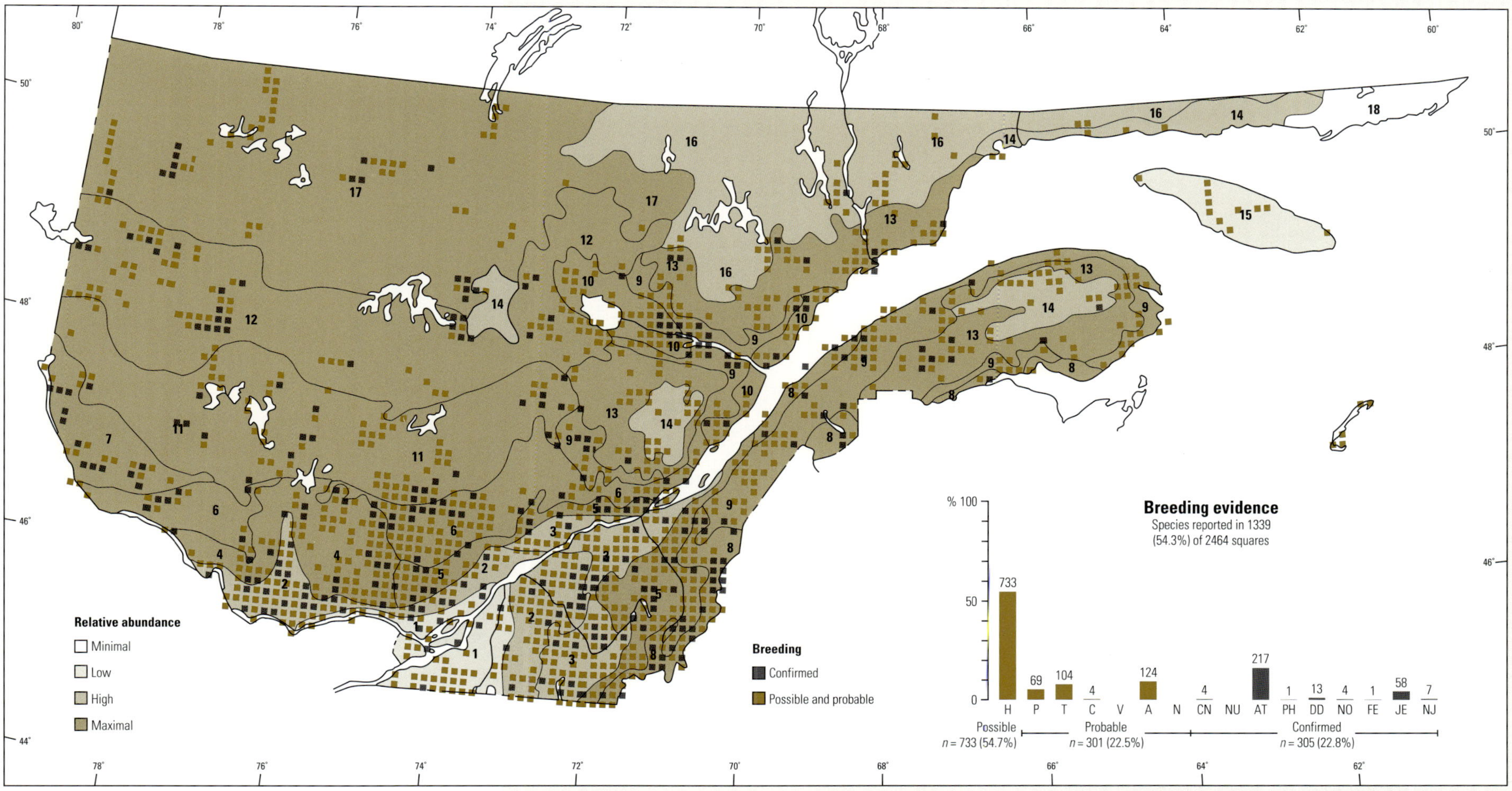

Species profile

Clutch size	usually 4–5 eggs (3–7)[2,3,6,7,8]
Laying interval	1 egg per day[8]
Start of incubation	with first egg[8]
Incubation period	10–12 days[3,6]
Maturity at hatching	altricial
Care of young	by female and male[1,3,7,8]
Nestling period	11–12 days[1,3,10]
Age at first flight	11–12 days or more[11]
Dependence of young	?
Broods per year	1[8,10]
Breeding age	1 year[10]
Mating system	monogamy[3,7]
Duration of pair bond	?
Total length	11.4–12.7 cm[4]
Wing span	?
Weight	M 8.9 g (7.0–13.9) / F 8.6 g (6.7–11.1)[5]
Longevity record	7 years 3 months[9]

References:
1. Bent 1953
2. Terres 1980
3. Ehrlich et al. 1988
4. Godfrey 1986
5. Dunning 1984
6. Peck and James 1987
7. Knapton 1984a
8. Forbush 1929
9. Klimkiewicz et al. 1983
10. DeGraaf and Rudis 1987
11. according to nestling period

Chronology $n = 38$

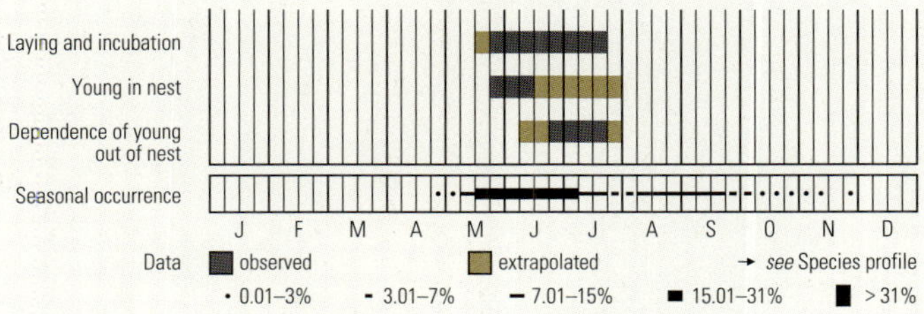

World breeding range

Frequency of occurrence by ecological region

Ecological regions** (forest regions)	Adequately covered squares	Squares in which reported	% of squares in which reported	Relative abundance (%)*
1. Sugar maple-hickory and sugar maple-basswood	97	50	51.5 ± 5.0	53.2
2. Sugar maple-basswood	153	128	83.7 ± 3.1	86.4
3. Sugar maple-basswood and sugar maple-yellow birch	101	86	85.1 ± 4.3	87.9
4. Sugar maple-yellow birch-basswood	56	52	92.9 ± 6.3	95.9
5. Sugar maple-yellow birch-beech and sugar maple-yellow birch-basswood	113	105	92.9 ± 2.8	95.9
6. Sugar maple-yellow birch-beech	71	68	95.8 ± 4.7	98.9
7. Sugar maple-yellow birch-balsam fir	19	17	89.5 ± 15.5	92.4
8. Sugar maple-yellow birch and balsam fir-yellow birch	47	45	95.7 ± 5.8	98.8
9. Balsam fir-yellow birch	87	76	87.4 ± 6.6	90.2
10. Balsam fir-white birch or balsam fir-red maple	46	44	95.7 ± 5.7	98.7
11. Yellow birch-balsam fir	52	47	90.4 ± 8.6	93.3
12. Balsam fir-white birch (west)	59	54	91.5 ± 7.7	94.5
13. Balsam fir-white birch (east)	76	68	89.5 ± 7.0	92.4
14. Balsam fir-black spruce	28	22	78.6 ± 16.2	81.1
15. White spruce-balsam fir and balsam fir-white spruce	22	7	31.8 ± 20.2	32.8
16. Black spruce-balsam fir-moss	18	15	83.3 ± 20.2	86.0
17. Black spruce-moss	32	31	96.9 ± 7.6	100.0
18.			not analysed	

* minimal: 0–29.9%; low: 30–59.9%; high: 60–89.9%; maximal: 90–100%

** Deciduous forest: 1–7; mixed forest: 8–11; coniferous forest: 12–17

Nashville Warblers almost always build their nests on the ground in the underbrush hidden by vegetation. This nest contains 4 eggs, 2 of which were laid by a Brown-headed Cowbird. Sainte-Dorothée, Laval. *June 1943, Henry Mousley, Laval University Collection.*

Notes on territory (continued)

1.1 pairs/10 ha in an area of Gatineau Park containing a mixture of open and wooded habitats (Dauphin 1985). In La Mauricie National Park, DesGranges (1980b) found densities of 4.3 birds/10 ha in a red spruce stand, 5.4 birds/10 ha in a balsam fir stand, 6.5 birds/10 ha in a white birch stand, and 10.8 birds/10 ha in an Eastern white pine plantation. Surveys in various locations in southern Québec have revealed densities of between 2.5 and 9.9 males/10 ha in mixed forests usually containing balsam fir and white birch and between 0.5 and 14.8 males/10 ha in coniferous stands (Erskine 1971a, 1972a, 1976a, 1980a, 1984).

In the area around the Nottaway, Broadback, and Rupert rivers, densities were higher in black spruce-moss stands than in all other habitats surveyed (birch forests, ecotones between burns and coniferous habitats, etc.) (Cons. G. G. – GREBE 1992o).

Forest edges and regenerating woodlands where such species as trembling aspen, white birch, and balsam fir predominate provide suitable habitat for this species. Lotbinière, southwest of Québec City. *July 1994, Jean-Pierre L. Savard.*

The Nashville Warbler's song carries well and is distinctive and easily identifiable, at least for experienced birders. However, it may be confused with one of the songs of the Yellow-rumped Warbler (Buskirk 1984). The Nashville Warbler was reported in 1339 Atlas squares, and breeding was confirmed in 22.8% of those. Nests, which are very difficult to find, were found in only 11 squares, or less than 1% of the squares where the species was observed. Breeding confirmation was most often based on reports of adults carrying food, followed by reports of recently fledged young.

Distribution

In Canada, the Nashville Warbler breeds in southern British Columbia, central Saskatchewan, southern Manitoba, Ontario and Québec, and in the Maritime provinces (Godfrey 1986) as well as in southwestern Newfoundland (Montevecchi and Tuck 1987; B. Mactavish, pers. comm.).

The Nashville Warbler was atlassed throughout southern Québec, but less frequently on Anticosti Island and in the Montréal and Montérégie regions. Anticosti Island has little suitable successional habitat; nor does Montréal or the Montérégie, where most of the land has been cleared for farming or urban development. Breeding was not confirmed on the Middle North Shore, Anticosti Island, and Magdalen Islands. Todd (1963) reported that the species was local on the North Shore, at Havre-Saint-Pierre, Baie-Johan-Beetz, Natashquan, and at the mouth of the Little Mecatina and Kécarpoui rivers. It has also been observed a number of times on Anticosti Island, and 3 males in breeding condition were collected at Baie-Sainte-Claire on 23 July 1963 (Ouellet 1969). Although the Nashville Warbler has been observed occasionally on the Magdalen Islands during the summer, no actual breeding evidence has been obtained (Fradette 1992).

The species' distribution extends slightly north of the Atlas area. Todd (1963) observed the species around the upper reaches of the Broadback River. A survey carried out in 1991 showed that it was common in the area around the Nottaway, Broadback, and Rupert rivers (Cons. G. G. – GREBE 1992o). However, the Nashville Warbler is much scarcer farther north around the Eastmain and Opinaca rivers (Bordage and Aubry 1982, 1983; Bordage 1984, 1985b).

History and trends

At the beginning of the 19th century, the Nashville Warbler was rarely seen in North America (Bent 1953). However, it became common in the middle of the century as the human population expanded and the forests were cleared extensively. Nashville Warbler populations subsequently declined according to authors such as Bent (1953) and Forbush (1929), who believed the species to be less common in the northeastern United States in the early 20th century than the late 19th century.

In Québec, data on the species is lacking that far back, although Wintle (1896) considered it a scarce "transient visitant" in the Montréal region at the turn of the century, having received reports of it nesting in swampy woods on the Island of Montréal. According to Ouellet (1974a), an observer of the time found the Nashville Warbler to be rare in the Montréal region during the 1860s, though it had been more common a few years earlier. At the beginning of the 20th century, the Nashville Warbler was considered an uncommon breeder in the Hatley area of the Eastern Townships (Mousley 1916a), and a regular summer resident in the Ottawa–Hull area (Lloyd 1924). It was also regarded as an uncommon summer resident around Québec City (Lewis 1920), but not uncommon in some parts along the Gaspé Peninsula coast (Demille 1926).

Several studies have shown that the Nashville Warbler is common in stands regenerating after clearcutting (Thompson and Capen 1988; Derleth et al. 1989). Even on its wintering grounds, this species does not frequent mature forests (Askins et al. 1990; Diamond 1991), and therefore no reduction in its winter habitat is anticipated between now and the year 2000, despite the current deforestation occurring in the tropics (Diamond 1991). The Nashville Warbler population in eastern North America remained stable from 1966 to 1987 (Robbins et al. 1989b), as well as in Québec from 1970 to 1989 (Cyr and Larivée 1993a).

GILLES FALARDEAU

Northern Parula

Paruline à collier

Parula americana

THE DAINTY NORTHERN PARULA IS THE SMALLEST OF OUR WARBLERS. It has a distinctive but unmelodious song: a thin, buzzy *zeeeeeeeee-up* that climbs the scale and ends abruptly in an explosive note that is somewhat incongruous in this retiring woodland bird (Bent 1953). The Northern Parula, which has bluish upperparts, a yellow throat and breast, and a chestnut band across the chest, has a scattered distribution in Québec and is rarely observed in large numbers. It breeds from the southern United States to southern Canada, from Manitoba in the west to Newfoundland in the east. Therefore, southern Canada, and particularly southern Québec, are at the northern limit of the Northern Parula's range.

The Northern Parula generally winters in Florida, the Bahamas, West Indies, Mexico, and Central America (AOU 1983).

Habitat and behaviour

The Northern Parula nests in mature or overmature forests. In Nova Scotia, Morgan and Freedman (1986) found higher densities of these warblers in stands 20 years and older than in stands 12 years or younger. In the northern part of its range, the species is found in coniferous habitats such as swampy, damp spruce forests, mixed forests, or deciduous forests with small stands of conifers. At these latitudes, the Northern Parula is generally associated with old man's beard, a tree lichen which it uses to build its nest. This lichen is most often found in fairly damp habitats, where it grows on the branches of conifers. In the southern part of its range in the southern United States, the Northern Parula uses another epiphyte, Spanish moss. This warbler often nests in habitats with an abundance of old man's beard, but no direct relation has been observed between the density of old man's beard and that of the Northern Parula (Morse 1989a).

In the northern part of its range, the parula must compete with the Golden-crowned Kinglet and probably the Ruby-crowned Kinglet, which also concentrate their foraging at the tips of branches. Because of this competition, the Northern Parula is usually found near the forest edge, an area not regularly frequented by kinglets. Farther south, in the United States, the Northern Parula inhabits deciduous forests where kinglets are absent, and there they forage in the forest interior rather than being restricted to the periphery (Morse 1967a, 1985, 1989a; Bull 1985; Elderkin 1989; Robbins et al. 1989a).

The species is fairly easy to observe, since the male sings regularly from the time he arrives in spring until well into July, when the young leave the nest. Males often sing from a favourite singing tree, located between 9 and 20 m from the nest (Mousley 1926; Graber and Graber 1951). The song, which is reminiscent of that of the Black-throated Blue and Cerulean warblers, may sometimes go unnoticed in the dawn chorus.

Christiane Girard

Notes on territory

On the Maine coast, Northern Parulas were found to defend territories ranging from 0.2 to 0.4 ha. The size of territories depended on the density of the species and that of other competing warblers (Morse 1989a). When the growing young required greater quantities of food, foraging extended beyond the territory originally defended (Morse 1967a). In deciduous forests in Nova Scotia, densities ranging from 2 to 9 pairs/10 ha were found in stands 20 years or older, compared with densities of 1 or 2 pairs/10 ha in stands 12 years old or less (Morgan and Freedman 1986).

The Northern Parula frequents fairly damp mature forests, whether deciduous, mixed, or coniferous. However, in mixed and deciduous forests it is found mainly in islands of conifers. Mount Albert, Gaspé Conservation Park. *July 1991, Yves Aubry.*

Northern Parula

Breeding distribution

1984–1989

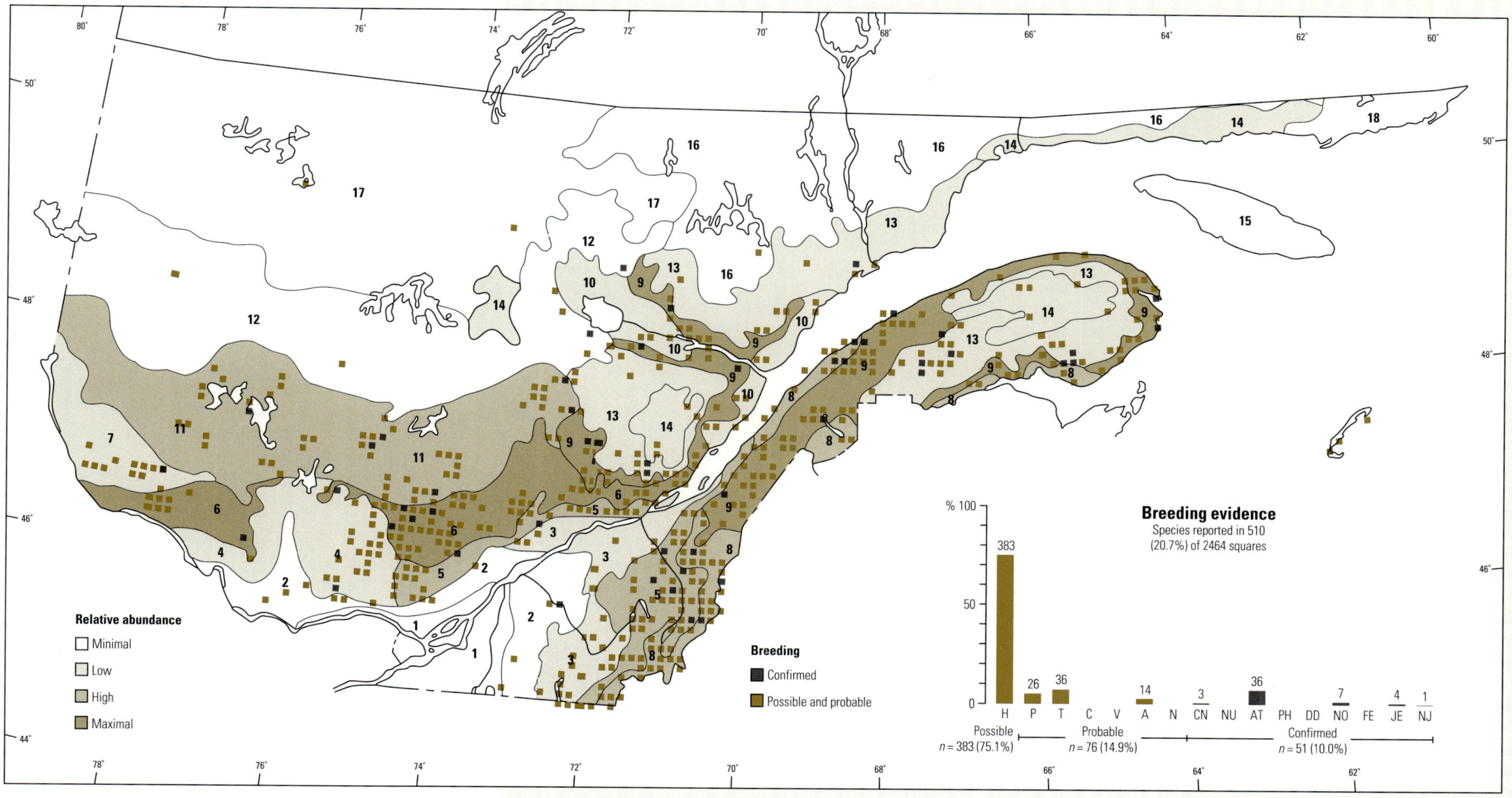

World breeding range

Species profile

Clutch size	usually 4–5 eggs (3–7)[1,2,4,10]
Laying interval	?
Start of incubation	with first egg[3]
Incubation period	12–14 days[3,4,10]
Maturity at hatching	altricial
Care of young	by female and male[3,7]
Nestling period	11 days[10]
Age at first flight	11 days or more[11]
Dependence of young	?
Broods per year	usually 1 (1–2)[6,10]
Breeding age	1 year[10]
Mating system	monogamy[7]
Duration of pair bond	?
Total length	10.9–12.5 cm[5]
Wing span	17.8–19.7 cm[4]
Weight	8.6 g (7.1–10.2)[8]
Longevity record	4 years 11 months[8]

References:
1. Bent 1953
2. Griscom and Sprunt 1957
3. Graber and Graber 1951
4. Terres 1980
5. Godfrey 1986
6. Forbush 1929
7. Ehrlich et al. 1988
8. Dunning 1984
9. Klimkiewicz et al. 1983
10. DeGraaf and Rudis 1987
11. according to nestling period

Chronology *n* = 10

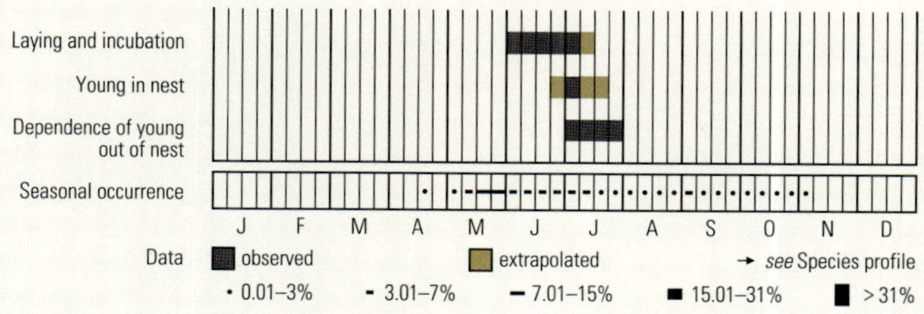

Frequency of occurrence by ecological region

Ecological regions** (forest regions)	Adequately covered squares	Squares in which reported	% of squares in which reported	Relative abundance (%)*
1. Sugar maple-hickory and sugar maple-basswood	97	1	1.0 ± 1.4	1.3
2. Sugar maple-basswood	153	10	6.5 ± 2.2	8.1
3. Sugar maple-basswood and sugar maple-yellow birch	101	28	27.7 ± 5.3	34.5
4. Sugar maple-yellow birch-basswood	56	19	33.9 ± 10.9	42.3
5. Sugar maple-yellow birch-beech and sugar maple-yellow birch-basswood	113	60	53.1 ± 5.1	66.1
6. Sugar maple-yellow birch-beech	71	57	80.3 ± 8.6	100.0
7. Sugar maple-yellow birch-balsam fir	19	8	42.1 ± 23.3	52.4
8. Sugar maple-yellow birch and balsam fir-yellow birch	47	33	70.2 ± 11.8	87.5
9. Balsam fir-yellow birch	87	66	75.9 ± 8.4	94.5
10. Balsam fir-white birch or balsam fir-red maple	46	14	30.4 ± 11.6	37.9
11. Yellow birch-balsam fir	52	35	67.3 ± 13.2	83.8
12. Balsam fir-white birch (west)	59	8	13.6 ± 9.2	16.9
13. Balsam fir-white birch (east)	76	31	40.8 ± 10.8	50.8
14. Balsam fir-black spruce	28	7	25.0 ± 17.0	31.1
15. White spruce-balsam fir and balsam fir-white spruce	22	0	0	0
16. Black spruce-balsam fir-moss	18	1	5.6 ± 13.5	6.9
17. Black spruce-moss	32	1	3.1 ± 7.6	3.9
18.			*not analysed*	

* minimal: 0–29.9%; low: 30–59.9%; high: 60–89.9%; maximal: 90–100%

** Deciduous forest: 1–7; mixed forest: 8–11; coniferous forest: 12–17

Northern Parulas are often found in mature conifers festooned with hanging old man's beard. No link has been found, however, between the density of the lichens and of the bird. Lake Fortier, Charlesbourg. *August 1966, Gaston Le Page.*

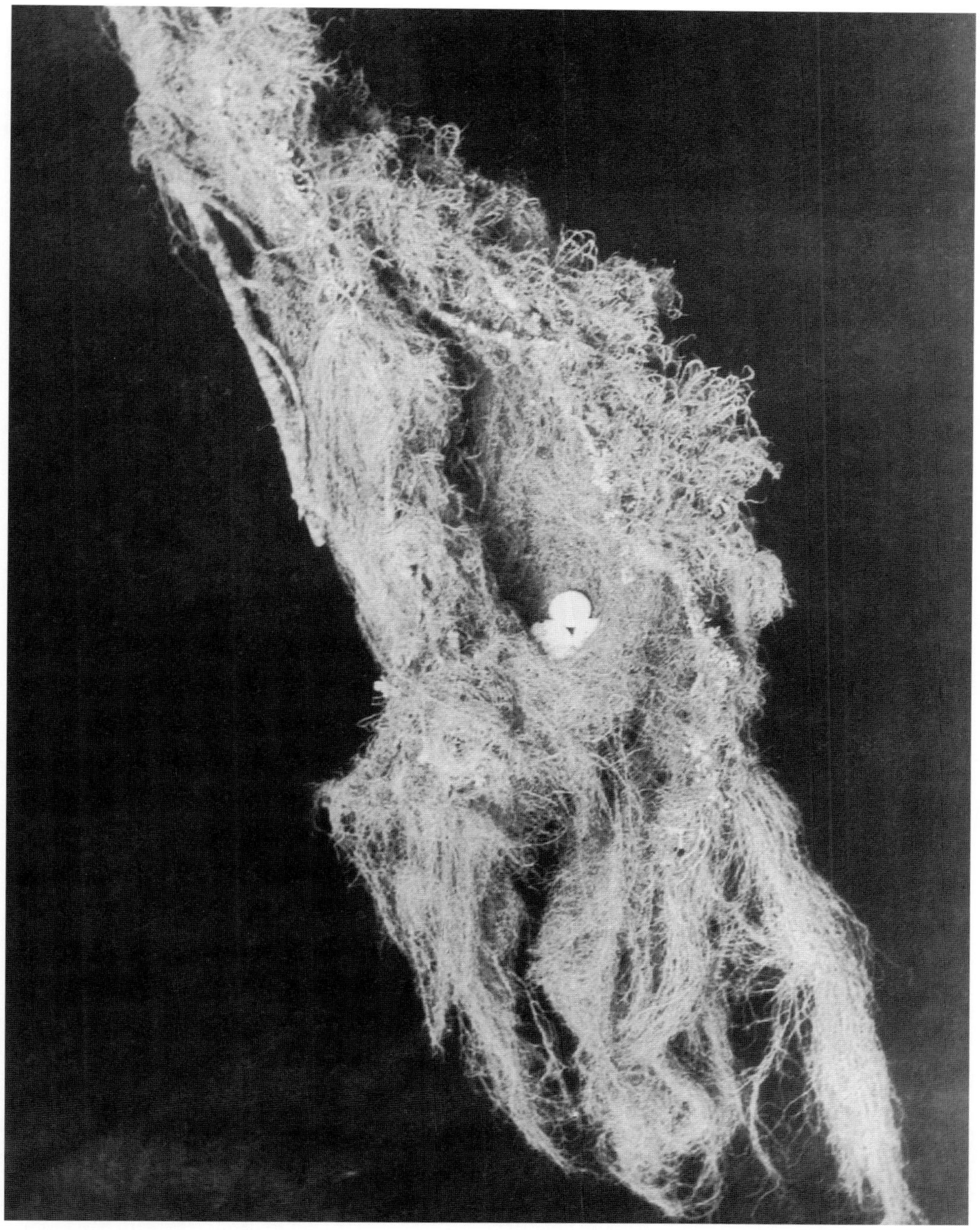

Northern Parula nest built in a clump of old man's beard suspended from the end of a conifer branch. The nest, which contains 4 eggs, was built completely by the female. Hatley, south of Sherbrooke. *June 1921, Henry Mousley, Laval University Collection.*

The nest is usually built in a dangling clump of old man's beard. When this lichen is scarce, a hanging cluster of hemlock or spruce may be used and some lichen added (Graber and Graber 1951; Morse 1967a; Godfrey 1986; Peck and James 1987). However, Bent (1953) found several nests that did not contain old man's beard. The nest is usually placed in a branch 1–12 m off the ground (Mousley 1926), and may be used two years in a row if it survives the winter (Forbush 1929). In Hatley, Québec, Mousley (1926) estimated that the female, in building the nest, made from 500 to 750 trips carrying materials, or approximately one trip every five minutes. The final structure, which is small and very well hidden, may be only 7 cm in diameter (Graber and Graber 1951) and weigh only 7 g (Bent 1953). In suitable habitat, several nests may be found fairly close together (Bent 1953).

The female does most of the incubation, during which time she is fed by the male (Forbush 1929); from time to time he relieves the female on the nest when she goes off to search for food on her own (Mousley 1924b; Forbush 1929). Both parents feed the young, which leave the nest around the age of 11 days (Graber and Graber 1951; DeGraaf and Rudis 1987). Parasitism by the Brown-headed Cowbird is rare in this species, probably because the nest is well hidden and too small to support the weight of a female cowbird (Friedmann 1963). Parasitism is further reduced by the fact that cowbirds do not often frequent the unbroken tracts of forest typical of the hinterland regions in Québec where the Northern Parula is found.

The Northern Parula gleans for insects in the mid and upper canopy in trees (Morse 1967a; Godfrey 1986). It sometimes forages along tree trunks like a nuthatch or may search for insects on the ground (Griscom and Sprunt 1957). Because of its tiny size, the Northern Parula is ideally suited to foraging on the tips of branches, allowing it to escape competition from larger warblers (Morse 1967a). It can often be seen hanging upside down from a twig like a chickadee, carefully examining every nook and cranny for insects. According to Wetmore (1916), 98% of its diet is composed of animal matter, including insects and spiders.

The Northern Parula was reported in 20.7% (510/2464) of Atlas squares surveyed. This low frequency of occurrence was no doubt due to the fact that the species is uncommon.

Distribution

Although it is primarily southern in distribution, the Northern Parula has succeeded in colonizing forests in more northerly regions (Mengel 1964). However, in northern regions it faces stiff competition, mainly from kinglets, which no doubt limits its numbers and distribution (Morse 1967a).

During the Atlas surveys the Northern Parula was reported most frequently in regions of deciduous and mixed forests. The species nests from the Ottawa Valley and Témiscamingue in the west to the Gaspé Peninsula in the east, as well as in the Saguenay–Lac-Saint-Jean region. It does not nest in the Montréal region, however, where it is considered only an occasional migrant (Bannon 1991b). On the North Shore, it nests as far east as Baie-Comeau, but was not found east of the town, despite the presence of potentially suitable nesting habitat on the Middle and Lower North Shore. Although it was reported a number of times on the Magdalen Islands and evidence of probable breeding was obtained, it was not confirmed as a breeder there. It was not observed on Anticosti Island.

History and trends

According to authors from the turn of the century, the Northern Parula was found in most regions of southern Québec at that time, although it was never very abundant. It was considered uncommon in the Québec City area (Dionne 1906), but was common in the Montréal region (Wintle 1896). In 1881, Brewster reported it at Baie-du-Renard on Anticosti Island (Ouellet 1969).

According to BBS data, North American populations appear to be increasing (Robbins et al. 1986). However, ÉPOQ data do not reveal any significant trend in either direction in Québec (Cyr and Larivée 1993a).

Given the link between the Northern Parula and old man's beard lichens (Morse 1967a), the impact of harvesting and managing fir forests must be closely monitored. Old-growth fir forests with high densities of old man's beard are very vulnerable to the spruce budworm, an insect which provides an excellent food source for the species (Morse 1989a). To protect fir forests from spruce budworm outbreaks without using pesticides, foresters currently recommend that forests not be allowed to mature to this stage and, if an outbreak is imminent, that the timber be harvested immediately. Unfortunately, both this strategy and resorting to the use of pesticides are detrimental to the Northern Parula's habitat and prey (Fowle 1972; Pearce et al. 1979). Therefore, it is difficult to reconcile efforts to control forest pests with protection of this warbler.

Another potential threat to the species in industrialized regions is old man's beard dieback caused by pollution, since lichens are sensitive to sulphur oxides. Furthermore, the species' habitat on its wintering grounds is threatened by extensive deforestation (Diamond 1991).

Guy Lemieux, Benoît Limoges, and Line Choinière

Yellow Warbler

Paruline jaune

Dendroica petechia

Christiane Girard

FROM MAY TO JULY, COUNTRYSIDE AND SUBURBS ALIKE ARE BRIGHTENED BY the clear, lively song of the Yellow Warbler, one of our most familiar birds. It is abundant in settled areas and nests as far north as the tree line (Godfrey 1986). This migrant breeder begins arriving in southern Québec in early May from its wintering grounds, which extend from Mexico to Peru and Brazil (Godfrey 1986).

Habitat and behaviour

On arriving in spring, males establish their territories in damp thickets, bordering ditches alongside roads and fields, for example, in shrubby areas along streams, lakes, peatlands, and marshes, as well as in orchards and ornamental shrubbery in gardens and parks (Godfrey 1986). The male requires a tree or other high perch from which it can sing and be seen by its rivals (Ficken and Ficken 1965b; Collins 1981).

Male Yellow Warblers appear to have two types of song (Morse 1966), a song with an accented ending, which is used to attract the female, and an unaccented song which is given during confrontation with other males. The bird repeats this song several times before resuming the first, which is the more familiar one. This warbler species is unusual in that the female will sing in instances of prolonged conflict with other females (Hobson and Sealy 1990).

The Yellow Warbler is at times polygynous (DellaSala 1986; Hobson and Sealy 1989b). The female builds the nest alone, using twigs, grasses, and plant fibres. The inside is lined with feathers, hair, and grasses. Usually placed in the fork of a shrub, 0.3–6.0 m up, the cuplike structure with an inner diameter of 4.5 cm is completed within the space of about 4 days (Schrantz 1943; Peck and James 1987). Because it is flimsy, the female builds a new nest each year.

Clutch size is 2–6 eggs (Schrantz 1943)—normally 4 or 5 (Harrison 1978; Ehrlich et al. 1988)—laid at the rate of 1 per day, in the morning. Incubation is by the female alone. Hatching begins 8–10 days after the last egg is laid, and this last one hatches after 10 to 12 days. The young are fed by both parents.

To keep the nest clean, the parents remove the excreta by ingesting the smaller fecal sacs and carrying off the larger ones.

Nine to 10 days after the first egg has hatched, the young begin to leave the nest. The chicks grow rapidly, their weight increasing from an average of 1.27 g on hatching to 8.78 g after 7 days. The last nestling to hatch grows at the same rate as the first. When the young leave the nest, they remain nearby, where they are fed by the parents. Schrantz (1943) believed that broods of 5 leave the nest earlier than those of 1 or 2, possibly because of the high level of activity in the nest.

Although the male may feed the female during incubation (Schrantz 1943; Busby and Sealy 1979; Hobson and Sealy 1989a), she does leave the nest periodically, probably to feed (Schrantz 1943). The nest at this time is unattended because the male follows her to protect her from other males and to keep her from mating with them (Hobson and Sealy 1989a). Thus the nest is vulnerable to intruders such as the Brown-headed Cowbird, which will lay in it, with consequences for breeding success. In Ottawa, the rate of nest parasitism was found to be as high as 31%, compared with 21% in one location in Manitoba (Burgham and Picman 1989; Hobson and Sealy 1989c). It is almost nil, however, when the warbler nests near a colony of Red-winged Blackbirds, because this species will not tolerate the presence of the cowbird (Sutton 1928).

The Yellow Warbler commonly frequents the shrubbery that grows along ditches, fields, and waterways. Lac Saint-François National Wildlife Area, southwest of Montréal. *September 1987, Léo-Guy de Repentigny.*

Notes on territory

The territory is used for mating, nesting, and foraging. Ranging in size from 0.2 to 1.2 ha, it is defended from the time the male arrives in spring until the end of the breeding period. The male establishes and maintains its territory by singing from a conspicuous perch and driving out male conspecifics and sometimes male Chestnut-sided Warblers (Stokes and Stokes 1983) and American Redstarts (Morse 1973). The female does not defend the territory but will defend the nest area (Stokes and Stokes 1983).

In Gatineau Park, in mixed habitat containing both forested and open areas, 0.52 pair/10 ha were found (Dauphin 1985). In southern Ontario, 2.0 males/10 ha were counted in an abandoned field, 5.8–33.0 males/10 ha in swamps, 5.4–12.5 males/10 ha in shrubby areas, 6.4–6.7 males/10 ha in deciduous woodlands, and 7.5–18.5 males/10 ha in deciduous and mixed stands either located on islands or with open habitat (Erskine 1971a, 1976a, 1984). In the Nottaway, Broadback, and Rupert rivers region, the highest relative densities were noted in the shoreline ecotones between forest habitat and water bodies, and also in

Yellow Warbler

Breeding distribution
1984–1989

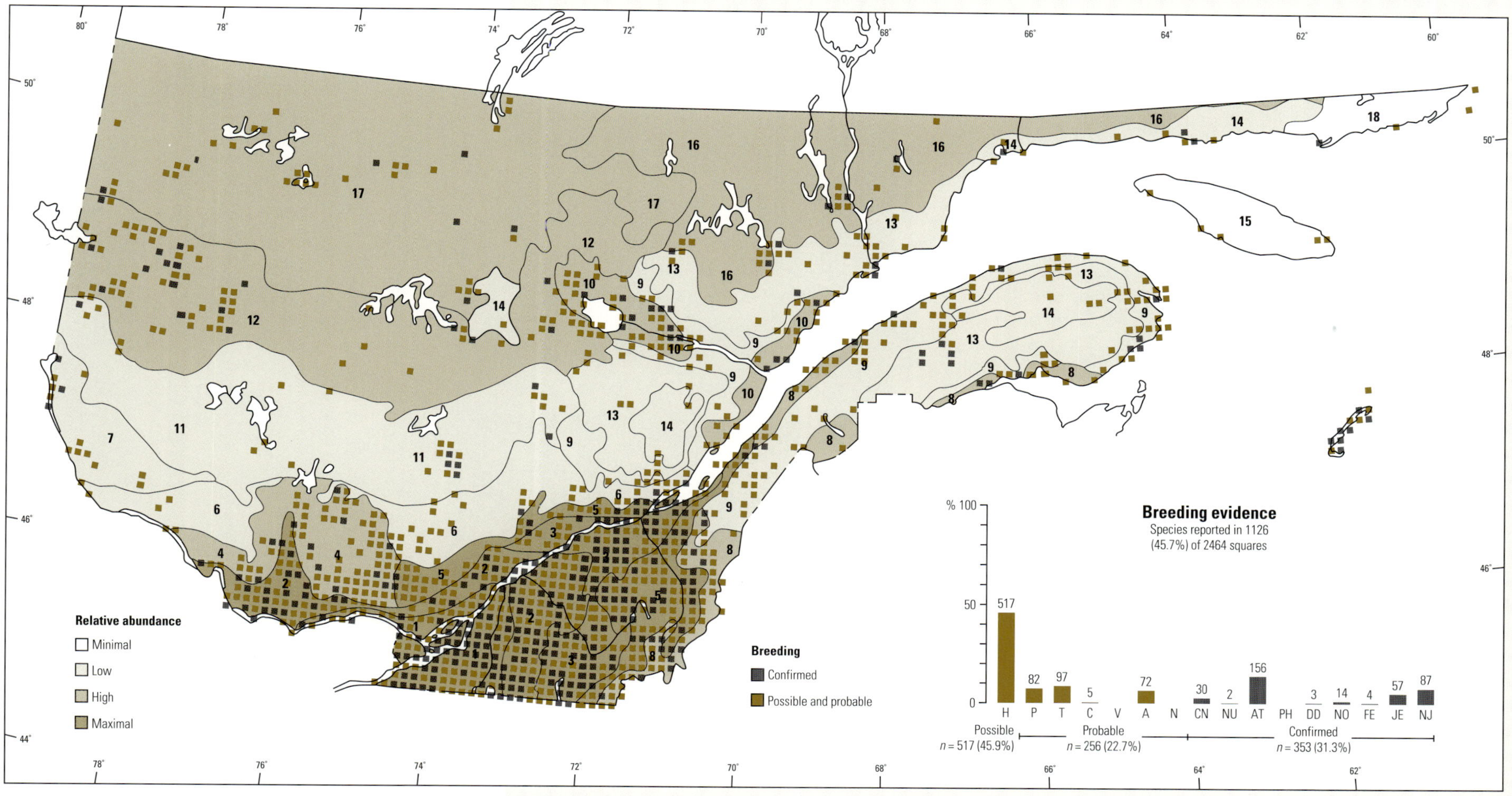

Species profile

Clutch size	usually 4–5 eggs (2–6)[1,2,3,5,10]
Laying interval	1 egg per day[3]
Start of incubation	before last egg[3]
Incubation period	10–12 days[1,2,3,5,7]
Maturity at hatching	altricial
Care of young	by female and male[2,3]
Nestling period	9–12 days[1,2,3,7]
Age at first flight	9–12 days or more[8]
Dependence of young	until ≈ 14 days after first flight[4]
Broods per year	1[4,7]
Breeding age	1 year[7]
Mating system	monogamy, sometimes polygyny[2,11,12]
Duration of pair bond	1 season[4]
Total length	12.0–13.3 cm[5]
Wing span	?
Weight	M 9.8 g (7.9–12.8) / F 9.2 g (7.4–16.0)[9]
Longevity record	8 years 11 months[6]

References:
1. Terres 1980
2. Ehrlich et al. 1988
3. Harrison 1978
4. Stokes and Stokes 1983
5. Godfrey 1986
6. Klimkiewicz et al. 1983
7. DeGraaf and Rudis 1987
8. according to nestling period
9. Dunning 1984
10. Schrantz 1943
11. DellaSala 1986
12. Hobson and Sealy 1989b

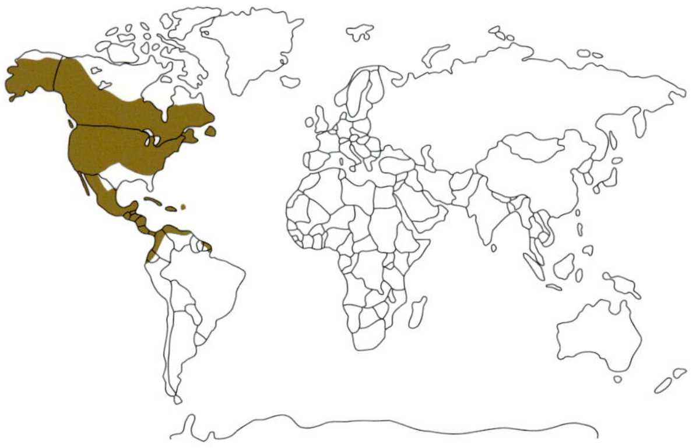

World breeding range

Chronology $n = 285$

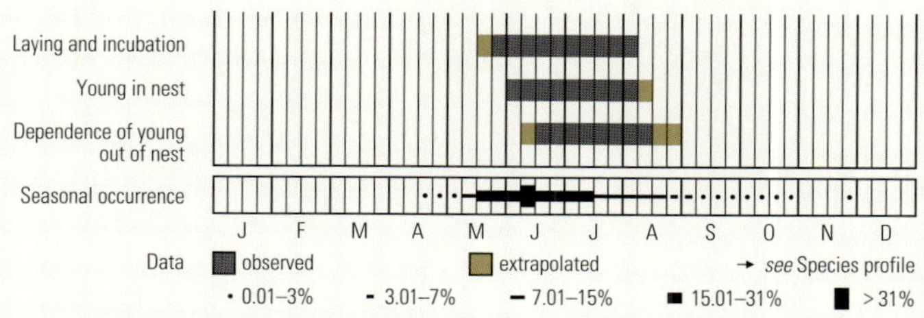

Frequency of occurrence by ecological region

Ecological regions** (forest regions)	Adequately covered squares	Squares in which reported	% of squares in which reported	Relative abundance (%)*
1. Sugar maple-hickory and sugar maple-basswood	97	95	97.9 ± 1.8	100.0
2. Sugar maple-basswood	153	146	95.4 ± 1.9	97.4
3. Sugar maple-basswood and sugar maple-yellow birch	101	97	96.0 ± 2.6	98.1
4. Sugar maple-yellow birch-basswood	56	49	87.5 ± 7.9	89.3
5. Sugar maple-yellow birch-beech and sugar maple-yellow birch-basswood	113	103	91.2 ± 3.1	93.1
6. Sugar maple-yellow birch-beech	71	41	57.7 ± 10.5	59.0
7. Sugar maple-yellow birch-balsam fir	19	9	47.4 ± 23.5	48.4
8. Sugar maple-yellow birch and balsam fir-yellow birch	47	33	70.2 ± 11.8	71.7
9. Balsam fir-yellow birch	87	51	58.6 ± 9.5	59.9
10. Balsam fir-white birch or balsam fir-red maple	46	37	80.4 ± 10.1	82.1
11. Yellow birch-balsam fir	52	24	46.2 ± 14.0	47.1
12. Balsam fir-white birch (west)	59	50	84.7 ± 9.7	86.5
13. Balsam fir-white birch (east)	76	39	51.3 ± 10.9	52.4
14. Balsam fir-black spruce	28	15	53.6 ± 19.3	54.7
15. White spruce-balsam fir and balsam fir-white spruce	22	4	18.2 ± 17.1	18.6
16. Black spruce-balsam fir-moss	18	15	83.3 ± 20.2	85.1
17. Black spruce-moss	32	23	71.9 ± 17.1	73.4
18.			*not analysed*	

* minimal: 0–29.9%; low: 30–59.9%; high: 60–89.9%; maximal: 90–100%

** Deciduous forest: 1–7; mixed forest: 8–11; coniferous forest: 12–17

To prevent the female cowbird from laying in its nest, the female places herself on it as if she were incubating and extends her wings (Burgham and Picman 1989; Hobson and Sealy 1989d). Younger females, however, simply respond by alarm-calling. If the nest is unattended, the cowbird can easily lay its egg and leave. Yellow Warblers, however, are able to distinguish cowbird eggs from their own (Sealy et al. 1989). When the nest is parasitized, the female can abandon it, build another nest over the first one or accept the foreign egg. According to Burgham and Picman (1989), she abandons the nest or builds another one when 1 or 2 (an average of 1.4) cowbird eggs have been laid in it. If, however, 3 or more foreign eggs (an average of 3.1) are present, she accepts them and tends the whole brood.

Busby and Sealy (1979), in Manitoba, found the diet of the Yellow Warbler to be comprised of invertebrates, over half of which were chironomids, which are taken as available. They noted that the female forages in the vicinity of the nest, whereas the male captures its prey in tree or shrub tops. These strategies save the pair energy and time and they are better able to tend the nest and defend the territory.

Some authors have noted that more colourful males are more aggressive in defending their territories, but devote less energy to feeding their young; less colourful males spend less time defending their territories, which are of lower quality, but expend more energy tending their young (Morse 1966; Studd and Robertson 1985).

The Yellow Warbler has to contend with a number of nest predators such as Blue Jays, squirrels, and snakes (Bent 1953). During distraction displays, the bird appears to fall from the tree and then walks away from the nest feigning injury. When the predator moves off, the warbler returns to the nest (Dionne 1906; Bonney 1988h).

The Yellow Warbler was reported in 45.7% (1126/2464) of Atlas squares surveyed, and breeding was confirmed in 31.3% (353/1126) of these squares. The species is easy to identify visually, but its song varies throughout the season and may be mistaken for that of the American Redstart or Chestnut-sided Warbler, even by experienced birdwatchers.

Yellow Warblers nest in thickets and shrubby edge habitat, preferably moist or located near water. Cap Tourmente National Wildlife Area, northeast of Québec City. *August 1991, Yves Aubry.*

Distribution

The breeding range of the Yellow Warbler covers most of North America (National Geographic Society 1987). According to Bent (1953), there are eight subspecies in North America, two of which are found in Québec. *Dendroica petechia aestiva*, which occurs in the southern part of the province, including the Québec City region and the Gaspé, has a lighter plumage. *Dendroica petechia amnicola*, which is darker and more olivaceous above, is found farther north, in the Lac-Saint-Jean region, on Anticosti Island, the North Shore, and the Magdalen Islands (Godfrey 1986).

Notes on territory (continued)

birch stands (Cons. G. G. – GREBE 1992o). In the Great Whale River area, the Yellow Warbler has been observed in shoreline scrub (Cons. G. G. – GREBE 1991a).

The scattered clumps of willow in this cattail marsh could very well contain a nest and could also be used as singing perches by the male. Commune Marsh, Châteauguay. *July 1991, Gaétan Duquette.*

Yellow Warblers are also found in the ornamental shrubs of city parks and cemeteries. Mount Royal, Montréal. *July 1991, Gaétan Duquette.*

The Yellow Warbler is widely distributed throughout southern Québec but is most abundant in the Central St. Lawrence Lowland, the Montérégie, Eastern Townships, Bois-Francs, and the southern Ottawa Valley regions, where it occupies the abundant open habitat, both natural and man-made, of farmlands, suburbs, and city parks. Here it finds suitable microhabitats in thickets and shrubby edges. Though suitable habitat is less abundant in more northerly areas, the species may occur around clearings (resulting from clearcutting or fires), lakeshores, and roadsides, in short, all regenerating sites that contain shrubs.

North of the Atlas area, the Yellow Warbler has been observed in the Great Whale and Little Whale river watersheds, mainly in riparian shrub habitat, where it is considered a probable breeder (Cons. G. G. – GREBE 1991a). It was also reported before the Atlas period as far north as Richmond Gulf, now known as Lake Guillaume Delisle (Todd 1963; ÉPOQ).

A female Yellow Warbler at the nest. The nest is placed in a fork of a shrub and is built of twigs, grasses, and plant fibres. Sainte-Foy. *June 1961, Gaston Le Page.*

History and trends

Specimens of Yellow Warblers were collected along the St. Lawrence as early as the 18th century (Browning 1989). In the 19th century, the species was noted in southern Québec (Ouellet 1974a), on Anticosti Island (Ouellet 1969), the Magdalen Islands (Fradette 1992) and, farther north, at least to the 54th parallel (Todd 1963). Atlas results and ÉPOQ data do not appear to indicate any appreciable change in distribution in Québec, although nowadays there are more sightings in northern areas.

Data collected over the last 20 years suggest that numbers increased during this period in the agricultural areas in the southern part of the province (Falardeau and DesGranges 1991). This trend has been observed throughout northeastern North America (Robbins et al. 1986; Erskine et al. 1990). The increase may be partly due to a decline in abundance of the Brown-headed Cowbird (Erskine et al. 1990; Falardeau and DesGranges 1991), which parasitizes the Yellow Warbler and may reduce breeding success (Goossen and Sealy 1982). The species may also have benefited from the abandonment and reversion of farmland, which has been widespread over the last 15 years throughout southern Québec, except in the Greater Montréal area and Montérégie (M. Carignan, MAPAQ, unpubl. data).

Alain Gauthier

Chestnut-sided Warbler

Paruline à flancs marron

Dendroica pensylvanica

THE CHESTNUT-SIDED WARBLER, WHICH IS ASSOCIATED WITH SECOND-growth habitats, is one of the species that has benefited most from the European settlement of North America. The development of agriculture and forestry created openings in the mixed and deciduous forests, allowing the species to prosper, and it still profits today from human activity in rural and forested regions. In pre-Columbian times, this warbler was probably limited to alder and willow thickets along streams and rivers and forests regenerating after a fire (Freedman et al. 1981; Elderkin 1989).

Today, the Chestnut-sided Warbler is one of Québec's most common warblers. The species is very easy to observe and identify, and is also quite tame. Its song may occasionally be mistaken for that of the Yellow Warbler, which often nests in the same habitats (Elderkin 1989), or that of the American Redstart (Ficken and Ficken 1962, 1965a). This migrant breeder winters in Central America and is one of the most common species in Costa Rica and Panama then (Bent 1953; AOU 1983).

Habitat and behaviour

The Chestnut-sided Warbler is a typical species of the second-growth habitats that develop on the margins of mixed or deciduous forests 4–10 years after logging (Morgan and Freedman 1986; Derleth et al. 1989; Elderkin 1989). It is also found in clearings, forest edges, thickets, overgrown fields, and along roadsides (Bent 1953; Collins et al. 1982). Although its habitat preferences are similar to those of the Yellow Warbler, the Chestnut-sided Warbler chooses denser, generally drier places (Peterjohn and Rice 1991).

As in many species of birds, in spring the males arrive on the breeding grounds a few days ahead of the females and immediately stake out their territories. There, they perch on the top of a bush or lower branch of a tree, tirelessly repeating their song, 5 or 6 times a minute (Kendeigh 1945). Although it often appears that the Chestnut-sided Warbler has only two songs, ornithologists have identified five song types, distinguished mainly by whether the ending is accented or unaccented. Each song type has a different significance and is used in different situations for courtship or territorial defence (Lein 1978; Kroodsma et al. 1989).

Shortly after the female chooses a mate, she begins to search for a nest site, sometimes even outside the male's original territory (Kendeigh 1945). Common supporting substrates include hazel, raspberry, meadowsweet, and viburnum bushes, as well as alder and maple saplings (Bent 1953; Peck and James 1987). The well-hidden nest is generally built less than 1 m from the ground in a tangle of vines, clump of ferns, bush, shrub, sapling, or the lowest branches of a small tree. The female fashions a compact cup of bark strips, grasses, fine weed stalks, plant fibres, and plant down, and lines it with fine grasses, rootlets, and animal hair (Godfrey 1986; Peck and James 1987).

Christiane Girard

Notes on territory

After observing 5 males that had established territories close to each other, Ficken and Ficken (1965a) estimated the average size of a territory occupied by a pair at about 0.2 ha. Kendeigh (1945) found that, at the onset of

The Chestnut-sided Warbler is a typical species of second-growth habitats in mixed and deciduous forests. Sainte-Foy. *September 1991, Yves Aubry.*

Chestnut-sided Warbler

Breeding distribution
1984–1989

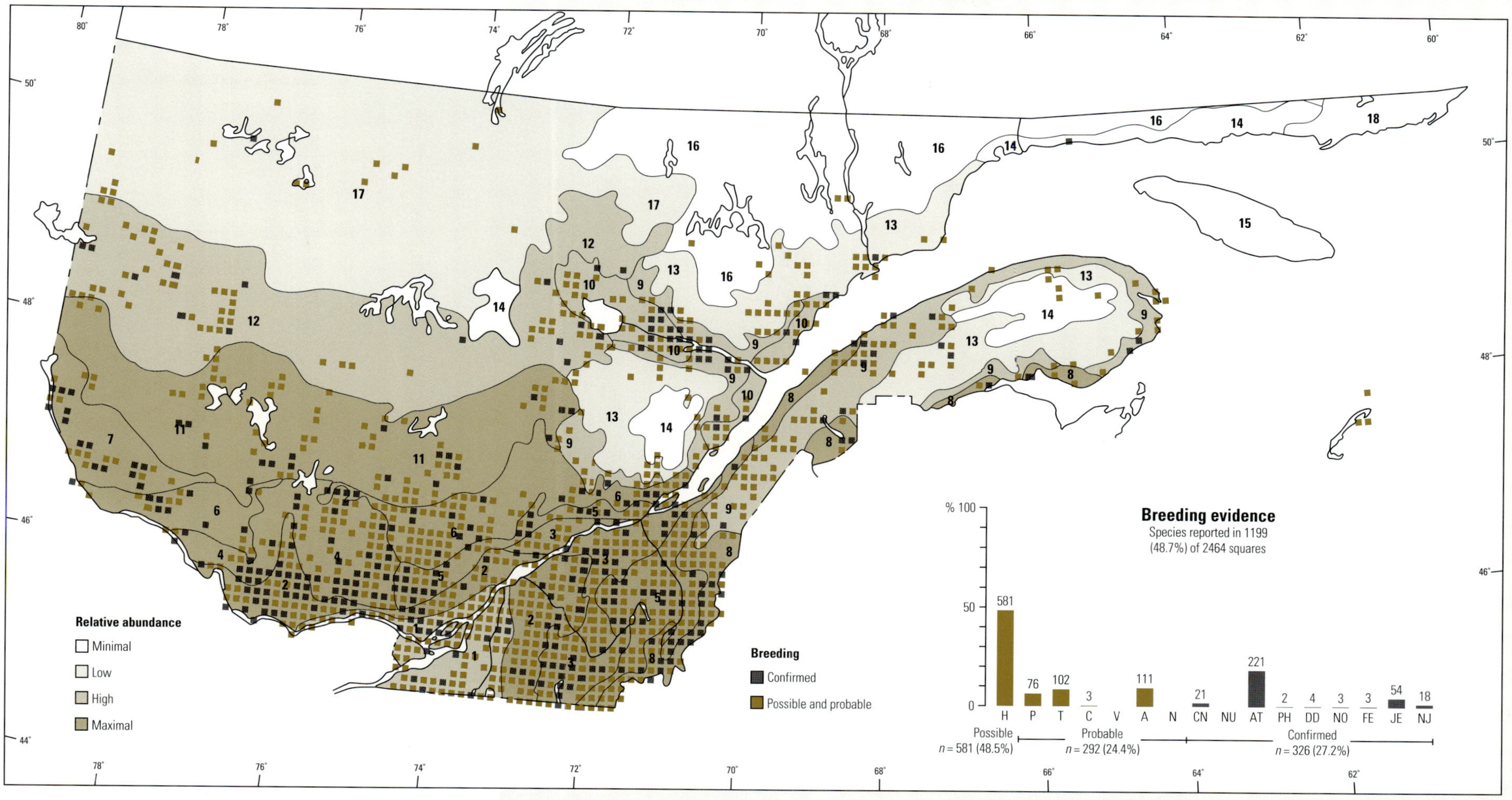

Species profile

Clutch size	usually 4 eggs (3–5)[1,2,3]
Laying interval	1 egg per day[2,4]
Start of incubation	with last egg[2,5]
Incubation period	12–13 days[1,3]
Maturity at hatching	altricial
Care of young	by female and male[1,2,3]
Nestling period	10–12 days[1,3,8]
Age at first flight	10–12 days or more[11]
Dependence of young	until ≈ 28 days after first flight[2]
Broods per year	usually 1 (1–2)[8,9]
Breeding age	1 year[8]
Mating system	monogamy[3]
Duration of pair bond	?
Total length	12.0–14.4 cm[6]
Wing span	?
Weight	M 9.8 g (8.1–13.1) / F 9.4 g (7.5–10.9)[7]
Longevity record	6 years 11 months[10]

References:
1. Bent 1953
2. Lawrence 1948
3. Ehrlich et al. 1988
4. Peck and James 1987
5. Kendeigh 1945
6. Godfrey 1986
7. Dunning 1984
8. DeGraaf and Rudis 1987
9. Kroodsma et al. 1989
10. Klimkiewicz et al. 1983
11. according to nestling period

World breeding range

Chronology $n = 66$

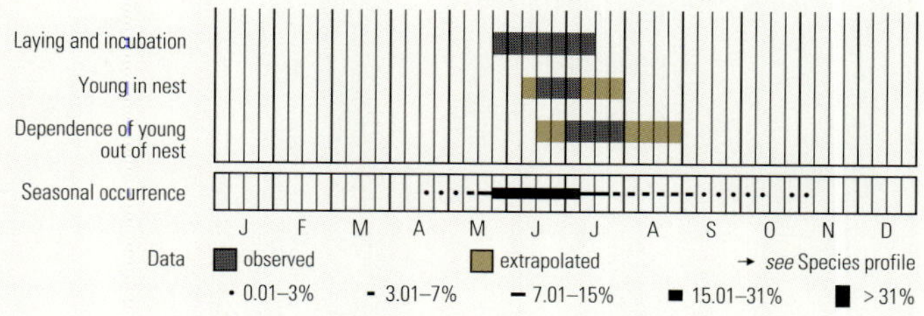

Frequency of occurrence by ecological region

Ecological regions** (forest regions)	Adequately covered squares	Squares in which reported	% of squares in which reported	Relative abundance (%)*
1. Sugar maple-hickory and sugar maple-basswood	97	81	83.5 ± 3.8	83.5
2. Sugar maple-basswood	153	144	94.1 ± 2.1	94.1
3. Sugar maple-basswood and sugar maple-yellow birch	101	96	95.0 ± 2.8	95.0
4. Sugar maple-yellow birch-basswood	56	54	96.4 ± 4.8	96.4
5. Sugar maple-yellow birch-beech and sugar maple-yellow birch-basswood	113	107	94.7 ± 2.5	94.7
6. Sugar maple-yellow birch-beech	71	67	94.4 ± 5.3	94.4
7. Sugar maple-yellow birch-balsam fir	19	19	100.0 ± 2.6	100.0
8. Sugar maple-yellow birch and balsam fir-yellow birch	47	47	100.0 ± 1.1	100.0
9. Balsam fir-yellow birch	87	73	83.9 ± 7.3	83.9
10. Balsam fir-white birch or balsam fir-red maple	46	41	89.1 ± 8.2	89.1
11. Yellow birch-balsam fir	52	51	98.1 ± 4.5	98.1
12. Balsam fir-white birch (west)	59	52	88.1 ± 8.8	88.1
13. Balsam fir-white birch (east)	76	40	52.6 ± 10.9	52.6
14. Balsam fir-black spruce	28	6	21.4 ± 16.2	21.4
15. White spruce-balsam fir and balsam fir-white spruce	22	0	0	0
16. Black spruce-balsam fir-moss	18	4	22.2 ± 22.2	22.2
17. Black spruce-moss	32	10	31.3 ± 17.6	31.3
18.			*not analysed*	

* minimal: 0–29.9%; low: 30–59.9%; high: 60–89.9%; maximal: 90–100%

** Deciduous forest: 1–7; mixed forest: 8–11; coniferous forest: 12–17

The Chestnut-sided Warbler's nest is usually built in a shrub or sapling, where it is well concealed by the leaves. This one contains two Brown-headed Cowbird eggs, which are distinctly larger than those of the warbler. Sainte-Dorothée, Laval. *Around 1940, Henry Mousley, Laval University Collection.*

Notes on territory (continued)

breeding, males defended territories ranging from 0.5 to 1 ha, but during incubation they expanded their activities to an area of 4.9 ha or more, which appeared to correspond to the species' home range. Although the male sang wherever he went in this larger area, the singing was much less vigorous and less frequent than it was inside the initial territory (Kendeigh 1945).

In Maryland, densities of 19.8 territorial males/10 ha were found in a dense second-growth forest, and of 16.8 territorial males/10 ha in an open slash area in a deciduous stand (Stewart and Robbins 1958).

In Québec, densities of 5.2 pairs/10 ha were recorded in a second-growth deciduous stand along a roadside, 4.9 pairs/10 ha in a young balsam fir stand with some birch and poplar, and 4.2 pairs/10 ha in a disturbed coniferous stand with patches of alder (Erskine 1971a, 1972a, 1980a).

This species often frequents second-growth stands 4–10 years into regeneration, shrub thickets along roadsides, fields, and clearings. Saint-Athanase, south of Rivière-du-Loup. *June 1988, Jean-Luc DesGranges.*

Once nesting has begun, the male sings a little less freely. The female alone incubates the eggs, but both parents feed the young, initially by regurgitation, and later by bringing them insects, whole or in pieces (Lawrence 1948; Bent 1953). Once the young have left the nest, the male tends to sing more often, at least in regions where two broods are raised (Kroodsma et al. 1989). Nests are often parasitized by the Brown-headed Cowbird. In Québec, Terrill (1961) reported that 29% of nests found were parasitized as against 21% recorded by Peck and James (1987) in Ontario. In response, the Chestnut-sided Warbler may destroy the cowbird egg by burying it in the bottom of the nest, and then laying her clutch (Ehrlich et al. 1988).

Like other warblers, the Chestnut-sided Warbler is insectivorous. It generally forages within 10 m of the ground, and can be seen hopping along from branch to branch, gleaning the undersides of leaves for its favourite leaf-eating insects. It can also catch insects on the wing, but seldom forages directly on the ground (Bent 1953; Greenberg 1979).

The Chestnut-sided Warbler was reported in 48.7% (1199/2464) of Atlas squares surveyed. Breeding was confirmed in 27.2% (326/1199) of squares in which the species was recorded, a fairly high rate for a warbler, mainly by reports of adults carrying food, since, as with many warblers, the nest is fairly hard to find.

Distribution

The Chestnut-sided Warbler breeds in central and southern Canada from east-central Alberta to Nova Scotia, as well as in the eastern United States as far south as the Appalachians (AOU 1983). Atlas data suggest that, in Québec, the species is found mainly south of the 49th parallel and especially below the 47th parallel, in areas where numerous openings have been created in mixed and deciduous forests by human activity. In the primarily coniferous forests farther north, the Chestnut-sided Warbler is associated with coppices of young trembling aspen or white birch that have grown up after forest fires or logging, as are found near Lakes Mistassini and Matagami, where the species has been confirmed as a breeder. Confirmed breeding records were also obtained east of Sept-Îles and on the Gaspé Peninsula. The Chestnut-sided Warbler occurs on the Magdalen Islands, where it has been observed off and on for several decades, but has not been confirmed as a breeder (Fradette 1992). It has never yet been sighted on Anticosti Island.

History and trends

At the beginning of the 19th century, ornithologists like Audubon, Nuttall, and Wilson regarded the Chestnut-sided Warbler as rare (Bent 1953). However, by the turn of the 20th century, it had become one of the most common warblers in eastern North America (Forbush 1929). Wintle (1896) listed it as a common summer resident in Montréal, nesting in Mount Royal Park. Dionne (1906) also considered the Chestnut-sided Warbler to be fairly common in southern Québec and from Québec City to Beaupré. According to Ouellet (1974a), the Chestnut-sided Warbler is probably more common in the Montréal region than it was in the past because of the increased amount of second-growth habitat available.

Although there are insufficient data to evaluate past population trends in Québec, the species is undoubtedly more common and widespread now than it was at the turn of the century. Human activities, particularly clearcutting and the razing of forests for agriculture, have increased the types of suitable breeding habitat for this species.

No major changes in abundance between 1969 and 1989 are evident from BBS and ÉPOQ data (Cyr and Larivée 1993a). However, population increases may be expected in regions where large tracts of farmland have been abandoned and invaded by trees (M. Carignan, MAPAQ, unpubl. data), a phenomenon that does not apply to the Montréal and Montérégie regions.

GUY LEMIEUX AND LINE CHOINIÈRE

Magnolia Warbler

Paruline à tête cendrée

Dendroica magnolia

Christiane Girard

THE COMMON AND VERY WIDESPREAD MAGNOLIA WARBLER, A TYPICAL species of mixed and coniferous forests throughout southern Québec, is undoubtedly one of the easiest warblers to see and hear during the breeding season, as it forages among the lower branches in trees, displaying its brilliant plumage.

The Magnolia Warbler sports a pleasing combination of colours, particularly its yellow rump and the alternating yellow and black of its underparts. Like the American Redstart, it often droops its wings and spreads its tail to show its plumage to best advantage. This allows the broad white band across the tail to be seen, a distinguishing field mark in both males and females.

This migrant breeder returns to Québec at the end of May, after spending the winter in Mexico, Central America, or the West Indies (Godfrey 1986), and has usually departed for its wintering grounds by the end of September.

Habitat and behaviour

The Magnolia Warbler inhabits mostly mixed and coniferous woodlands, but also frequents deciduous forests (Godfrey 1986; Peck and James 1987; Morse 1989a), where it seeks out islands or clumps of conifers. It is a habitat generalist and nests in stands of all ages (Morse 1985; Erskine 1992), whether disturbed or undisturbed (Morse 1989a) or in wet or dry habitats (Martin 1960).

The Magnolia Warbler's preferred breeding habitat is young forest dominated by fir and spruce (Collins 1983a; Godfrey 1986; Erskine 1992), but it also nests in mature forest with natural openings where an understorey of young conifers has sprung up (Doepker and Ozoga 1991a). Suitable nesting habitats also include forest edges, clearings, abandoned fields, burned or logged-over areas that are several years into regeneration, and young conifer plantations (Collins 1983a; Hall 1984; Godfrey 1986; Welsh 1987c; Morse 1989a).

Magnolia Warblers usually forage on the tips of branches low down in the tree (Morse 1968). However, during the breeding season, the male tends to forage higher up in the tree, near the singing perch from which he defends his territory, whereas the female limits herself to the lower levels near the nest (Morse 1968, 1989a). Magnolia Warblers feed mainly on insects such as snout beetles, ants, butterflies, moths, and homopterans, which they glean from leaves and branches or catch on the wing (Bent 1953; Morse 1968; DeGraaf and Rudis 1987).

Magnolia Warblers can be very territorial, particularly when establishing their territories and when the young are fledging. Aggressive behaviour may be directed toward other species as well as conspecifics; the Magnolia Warbler is not always dominant in these encounters, however (Morse 1976a).

During courtship, the male chases the female from branch to branch or sometimes in flight, tail and wings spread in display (Kendeigh 1945).

The nest, which is built by both the male and female (Bent 1953; Ehrlich et al. 1988), is made of grasses and twigs and lined with hair and rootlets (Godfrey 1986). The flimsy, poorly built nest is usually located near the trunk or on a horizontal branch, and is generally weakly secured to its support (Bent 1953; Harrison 1975, 1978). It is usually built in a small conifer or thick shrub, between 0.3 and 3 m from the ground (Harrison 1975, 1984; Peck and James 1987), although nests have been found as high as 10.7 m (AOU 1983).

The female alone incubates the clutch of 3–6 eggs for 11–13 days (Bent 1953). When the eggs hatch, both parents feed the young (Mousley 1924b; Kendeigh 1945; Bent 1953).

The Magnolia Warbler is a denizen of mixed and coniferous forests. In deciduous forests, it is found in islands of conifers. Near Lake Pythonga, northwest of Hull. *September 1981, Jean Beaubien and Guy Simard, Canadian Forest Service.*

Notes on territory

During the breeding season, the Magnolia Warbler defends a territory that it uses for courtship, nest building, brood raising, and foraging (Morse 1989a). To defend its territory, the male sings, engages in aerial chases and direct attacks, and displays his plumage (Morse 1980). The size and shape of the territory vary depending on the pair, the type of forest, and the year (Kendeigh 1945; Martin 1960; Morse 1976b; AOU 1983). Territories range in size from 0.02 to 1.7 ha, but most are 0.7 or 0.8 ha (Kendeigh 1945; Martin 1960; Morse 1976b, 1977, 1989a; AOU 1983; Crawford and Jennings 1989). Warblers' territories are often largest at the start of the breeding season before incubation begins, shrinking afterward (Morse 1989a).

In Québec, in fir and spruce stands containing birches and poplars Magnolia Warblers were found at densities ranging between 4.9 and 10.4 pairs/10 ha (Erskine 1971a, 1972a, 1976a, 1980a; DesGranges 1980b). In other types of stands, lower densities have been observed: 2.5 pairs/10 ha in a black spruce forest, 1.5 pairs/10 ha in a jack pine forest, and 0.2 pair/10 ha in a tamarack peatland (Erskine 1972a, 1976a). In Gatineau Park, in a mixed habitat of woodland with open areas, densities of 0.49 pair/10 ha were found (D. Dauphin, unpubl. data).

Magnolia Warbler

Breeding distribution

1984–1989

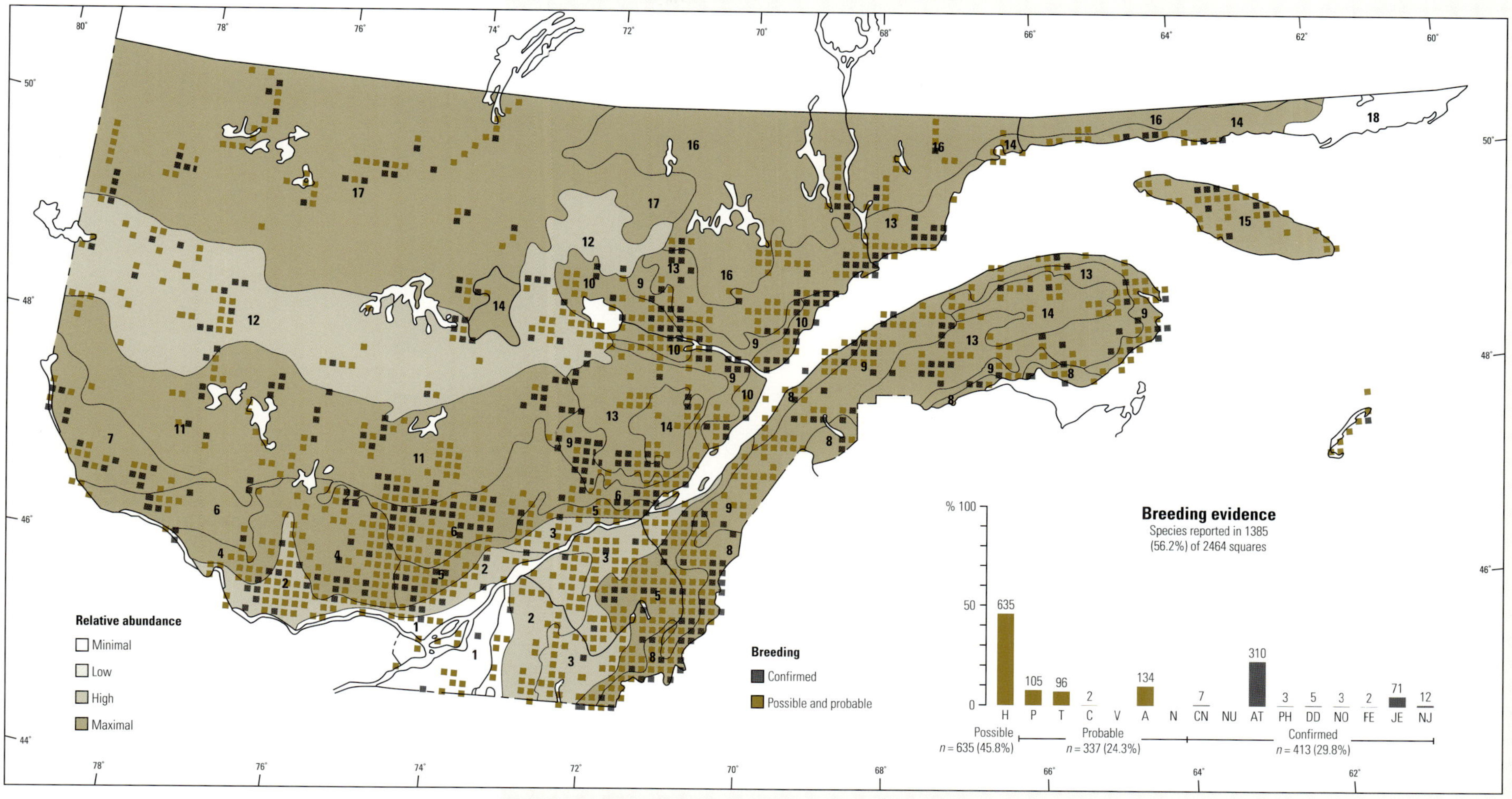

Species profile

Clutch size	usually 4 eggs (3–6)[1,3,9]
Laying interval	1 egg per day[2]
Start of incubation	with last egg, sometimes with second[12]
Incubation period	11–13 days[1,2,3,4]
Maturity at hatching	altricial
Care of young	by female and male[2,3]
Nestling period	8–10 days[2,3]
Age at first flight	8–10 days or more[11]
Dependence of young	until 25 days after leaving nest[12]
Broods per year	1, 2 possible[12]
Breeding age	1 year[10]
Mating system	monogamy[3]
Duration of pair bond	1 season[12]
Total length	10.2–13.0 cm[7]
Wing span	?
Weight	M 8.9 g (7.0–12.9) / F 8.5 g (6.6–12.6)[8]
Longevity record	6 years 11 months[6]

References:
1. Bent 1953
2. Harrison 1978
3. Ehrlich et al. 1988
4. Terres 1980
5. Kendeigh 1945
6. Kennard 1975
7. Godfrey 1986
8. Dunning 1984
9. Peck and James 1987
10. DeGraaf and Rudis 1987
11. according to nestling period
12. Hall 1994

World breeding range

Chronology $n = 79$

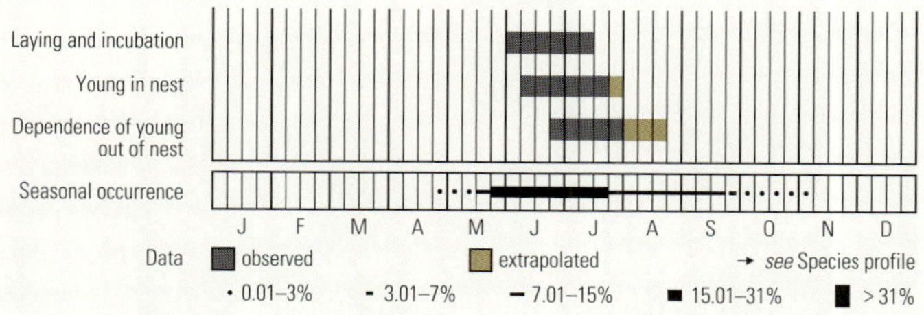

Frequency of occurrence by ecological region

Ecological regions** (forest regions)	Adequately covered squares	Squares in which reported	% of squares in which reported	Relative abundance (%)*
1. Sugar maple-hickory and sugar maple-basswood	97	20	20.6 ± 4.1	20.6
2. Sugar maple-basswood	153	92	60.1 ± 4.0	60.1
3. Sugar maple-basswood and sugar maple-yellow birch	101	67	66.3 ± 5.6	66.3
4. Sugar maple-yellow birch-basswood	56	51	91.1 ± 6.9	91.1
5. Sugar maple-yellow birch-beech and sugar maple-yellow birch-basswood	113	106	93.8 ± 2.7	93.8
6. Sugar maple-yellow birch-beech	71	69	97.2 ± 4.0	97.2
7. Sugar maple-yellow birch-balsam fir	19	19	100.0 ± 2.6	100.0
8. Sugar maple-yellow birch and balsam fir-yellow birch	47	47	100.0 ± 1.1	100.0
9. Balsam fir-yellow birch	87	87	100.0 ± 0.6	100.0
10. Balsam fir-white birch or balsam fir-red maple	46	44	95.7 ± 5.7	95.7
11. Yellow birch-balsam fir	52	52	100.0 ± 1.0	100.0
12. Balsam fir-white birch (west)	59	52	88.1 ± 8.8	88.1
13. Balsam fir-white birch (east)	76	76	100.0 ± 0.7	100.0
14. Balsam fir-black spruce	28	27	96.4 ± 8.3	96.4
15. White spruce-balsam fir and balsam fir-white spruce	22	21	95.5 ± 10.3	95.5
16. Black spruce-balsam fir-moss	18	18	100.0 ± 2.8	100.0
17. Black spruce-moss	32	32	100.0 ± 1.6	100.0
18.			*not analysed*	

* minimal: 0–29.9%; low: 30–59.9%; high: 60–89.9%; maximal: 90–100%

** Deciduous forest: 1–7; mixed forest: 8–11; coniferous forest: 12–17

Magnolia Warbler nest in a conifer, containing a Brown-headed Cowbird egg, which is larger than the host species' eggs. Saint-Ubalde, northeast of Trois-Rivières. *June 1984, Christian Marcotte.*

Young stands dominated by fir and spruce are ideal nesting habitat for this warbler. It frequents the edges of forests, cutover areas, clearings as well as regenerating second-growth forest and young conifer plantations. Laurentide Wildlife Reserve, north of Québec City. *July 1991, Guy Lemieux.*

Female Magnolia Warbler incubating her eggs in a nest in a conifer. The nest, which is usually built on a branch close to the trunk, though sometimes may be located further out on the limb, appears to be rather flimsy. Sainte-Dorothée, Laval. *June 1941, Henry Mousley, Laval University Collection.*

The Magnolia Warbler spends a great deal of time in the lower levels of the canopy, which makes it easier to detect (Welsh 1987c; Eaton 1988o; Erskine 1992). In addition, it frequents habitats that can be easily accessed by atlassers. Few nests were discovered since they are often hidden in the undergrowth. Breeding was confirmed in 29.8% (413/1385) of the squares in which the species was reported, mainly by observations of adults carrying food (310/413). Recently fledged young are easily observed (Mousley 1924b) and accounted for 71 squares in which breeding was confirmed.

Distribution

The Magnolia Warbler's range in North America extends from northeastern British Columbia to southern Newfoundland in the north and from south-central British Columbia to western Virginia in the south (AOU 1983).

In Québec, the Magnolia Warbler was confirmed as a breeder throughout the Atlas area. However, it nests mainly in mixed and coniferous forests, or in deciduous forests containing some stands of conifers. In the Montréal and Montérégie regions, the species is confined to conifer plantations as well as a few remaining islands of conifers and mixed woodlands. The species' breeding range extends north of the Atlas area as a nest was found in 1914 near Lake Evans (Todd 1963). It has also been observed in summer in various places from the southern coast of James Bay, up to 20 km north of the Eastmain River, and even as far north as the mouth of the La Grande Rivière (Todd 1963). Farther inland, it is common at least as far north as the Pontax River (Cons. G. G. – GREBE 1992o), and can be found up to the Opinaca Reservoir (Bordage and Aubry 1983; Bordage 1984). Farther east, the species is present as far north as the Churchill River in Labrador (Todd 1963).

History and trends

At the turn of the century, this warbler was considered a common summer resident in the northern United States (Eaton 1914) and was classified as common, even abundant, in some parts of eastern Canada (Macoun and Macoun 1909). Wintle (1896) listed it as a common "transient visitant" in the Montréal region. Between 1966 and 1979, populations in the northeastern United States appear to have increased (Robbins et al. 1986); in fact, in Vermont populations climbed 6% each year during this period (Kibbe 1985h). An upward trend was also noted in Québec between 1969 and 1989 (Cyr and Larivée 1993a). According to Morse (1989a) and Erskine (1992), the Magnolia Warbler is probably more abundant today than it was when the first European settlers arrived, mainly because the clearing of forests has created suitable habitat for the species.

Locally, however, Magnolia Warbler populations may suffer rapid declines. For example, in the Appalachian Mountains in Virginia, a significant decline was observed over a period of 5 years, coinciding with the maturation of the conifers and subsequent closing of the canopy, which gradually crowded out the shade-intelorant understorey and eliminated the niche for ground-dwelling birds, including the Magnolia Warbler (Hall 1984).

Alain Lanoue and Alain Morrier

Cape May Warbler

Paruline tigrée

Dendroica tigrina

Christiane Girard

THE CAPE MAY WARBLER'S SONG IS SO THIN AND HIGH-PITCHED THAT IT seems to vanish into the spires of the spruce that the bird chooses for its song perches. Populations of this species rises and falls in accordance with its primary prey, the spruce budworm, increasing during the insect's outbreaks, but declining as infestations subside. The male is easy to identify with his chestnut cheeks and yellow underparts striped with black like a tiger.

The Cape May Warbler returns to Québec in May from its wintering grounds in the West Indies and southern Florida. Although it is fairly common during migration, it is uncommon in summer in southern Québec.

Habitat and behaviour

The Cape May Warbler frequents coniferous woodland (Godfrey 1986), but may also nest in mixed forest (Morse 1978). It prefers open stands to dense ones and, when nesting in dense woods, chooses the more open spaces within them (Bent 1953; DeGraaf and Rudis 1987). Its preference appears to be for mature spruce stands (DeGraaf and Rudis 1987; Erskine 1992). On the western coast of James Bay, it is only found in coniferous forests over 50 years old (Welsh 1987b).

During the breeding season, the Cape May Warbler feeds chiefly on insects (Bent 1953), mainly the spruce budworm, which it gleans from the needles at the tips of branches near the top of the tree (MacArthur 1958). Its forages at heights of 6–24 m above the ground, most often between 15 and 21 m (Morse 1978). In rainy weather, it may feed closer to the ground, in low willows or flowers (MacArthur 1958). Unlike most warblers, it adopts a radically different foraging pattern during migration and on its wintering grounds, when it feeds mainly on the ground. Its thin, pointed bill and tubular tongue, unique among *Dendroica* warblers, allow it to puncture fruit and drink the juice (MacArthur 1958). It may even occasionally damage grape crops (Bent 1953).

The female builds the nest (Bent 1953), which is very difficult to detect since it is placed in a conifer near the trunk, usually at the top of the tree in a dense cluster of branches (MacArthur 1958), 9–20 m up (Harrison 1978; Godfrey 1986). It may also be found closer to the ground (DeGraaf and Rudis 1987). The nest is made of grasses, moss, and small twigs and lined with fine grass, hair, and feathers (Godfrey 1986).

While most warblers lay from 3 to 5 eggs (Morse 1989a), the Cape May generally produces 6 or 7 (Terres 1980; Morse 1989a); this adaptation allows the species to quickly boost its population during spruce budworm outbreaks. In years of food scarcity, clutches are smaller (MacArthur 1958; Harrison 1984). It is not known for certain which of the parents incubates the eggs and feeds the young, but since in most warbler species the female incubates the eggs and both parents feed the young (Morse 1980, 1989a), it is believed the same pattern occurs in the Cape May Warbler (Ehrlich et al. 1988).

This species is often difficult to detect because of its habit of foraging at the top of trees, the location of its nest, and its unobtrusive behaviour. The weak, high-pitched song of the male is often inaudible except at close range. The best way to spot the Cape May Warbler is to scan the tops of conifers where the species feeds; it often flies from tree to tree in search of food. Many researchers have commented on the difficulty of confirming breeding in the species (Peck and James 1987; Welsh 1987b; Peterson 1988b; Erskine 1992). According to Harrison (1984), the well-camouflaged nest, which is located high up in the tree in dense foliage, is even more difficult to find because of the female's behaviour. She does not fly directly to the nest but enters the tree from below and makes her way up from branch to branch near the trunk. She leaves the nest in the same manner, instead of flying out horizontally. The best way to confirm breeding is to observe the parents carrying food to the nest.

Distribution

In Canada, the Cape May Warbler breeds from the Mackenzie District of the Northwest Territories and northeastern British Columbia in the west to Newfoundland in the east. In the United States, it breeds in the northeastern and Great Lakes states (AOU 1983).

During atlassing, the highest frequencies of occurrence were obtained in the mixed forest zone and part of the coniferous forest zone (the Gaspé Peninsula, northeastern part of southern Québec, and alpine areas). The poor results obtained in the Black spruce-moss forest region in the northwestern part of the Atlas area can be partly attributed to the fact that the spruce budworm, the Cape May Warbler's favorite prey, seldom attacks black spruce stands (Morris 1963). The northernmost records were obtained near Matagami and Chibougamau in the west, and as far north as Havre-Saint-Pierre, on the North Shore, in the east. The Cape May Warbler breeds throughout southern Québec, including the Magdalen Islands and Anticosti Island, which until recently were not part of its breeding range (David 1980a; Godfrey 1986).

The distribution of the species as atlassed corresponds fairly well to that of the spruce budworm (Bordeleau 1988; Bordeleau and Boulet 1989, 1990). Consequently, the Atlas map must be used with caution since a more widespread spruce budworm outbreak would probably result in an expansion of the species' range, while a decline would result in a population decrease (Morse 1978, 1989a).

A typical species of coniferous forests, the Cape May Warbler nests and forages near the tops of firs and spruces. Martin-Valin ZEC, north of the Saguenay River. *August 1991, Jean Gauthier.*

History and trends

At the turn of the century, the Cape May Warbler was described as rare in eastern Canada (Taverner 1922), where it was observed mainly during migration in Québec and Ontario (Dionne 1883), particularly in the Montréal region (Wintle 1896). The first published breeding record concerned a nest found in 1916 at Lake Édouard (Merriam 1917 in Bent 1953).

The Cape May Warbler's distribution and abundance are both strongly correlated with spruce budworm numbers (Kendeigh 1947; MacArthur 1958; Morse 1978). Very abundant during spruce budworm outbreaks, the species becomes rare or even disappears altogether from certain parts of its breeding range when the infestation subsides (Kendeigh 1947; MacArthur 1958; Sanders 1970; Morse 1989a; DesGranges and Rondeau 1995). For example, high breeding densities were found in two locations in Ontario during an initial epidemic but, once the epidemic was over, the species practically disappeared altogether, only to reappear during a subsequent outbreak (Welsh 1987b). The most recent spruce budworm epidemic in Québec began in the early 1970s and peaked between 1974 and 1979. Since the 1980s, the outbreak has abated (Bordeleau 1988; Bordeleau and Boulet 1989; DesGranges and Rondeau 1995), which may affect the species' abundance and distribution in the province.

ALAIN LANOUE AND ALAIN MORRIER

Notes on territory

During the breeding season, warblers use territories for courtship, nest construction, brood rearing, and foraging (Morse 1989a). In Ontario, Kendeigh (1947) found Cape May Warbler territories ranging between 0.2 and 1 ha, or 0.5 ha on average. The species may also defend temporary feeding territories during migration (Kale 1967; Sealy 1988) and on its wintering grounds (Emlen 1973) against conspecifics and several other species (Morse 1989a).

In Québec, in stands containing balsam fir, spruce, and shade-intolerant hardwoods, densities between 7.4 and 12.4 pairs/10 ha were recorded at Bogan Creek in the Gaspé region in 1977, between 8.2 and 10.6 pairs/10 ha at Manie River, Kamouraska County, from 1977 to 1979, and 7.4 pairs/10 ha at Sainte-Anne Creek, Lac-Saint-Jean, in 1978 (Erskine 1980a). Densities of 9.9 pairs/10 ha were found in a balsam fir stand with Eastern white cedar and tamarack near Lac de l'Est, Kamouraska County, in 1977 (Erskine 1980a). On the Gaspé Peninsula, densities of 7.4 and 4.9 pairs/10 ha, respectively, were reported near Rivière Nouvelle in 1977 and Cascapédia River in 1978 in balsam fir stands interspersed with some deciduous trees (birch) and, at Cascapédia River only, spruce (Erskine 1980a). At La Mauricie National Park, a density of 6.5 pairs/10 ha was recorded in a white spruce plantation in 1973 (DesGranges 1980b).

Cape May Warbler

Breeding distribution
1984–1989

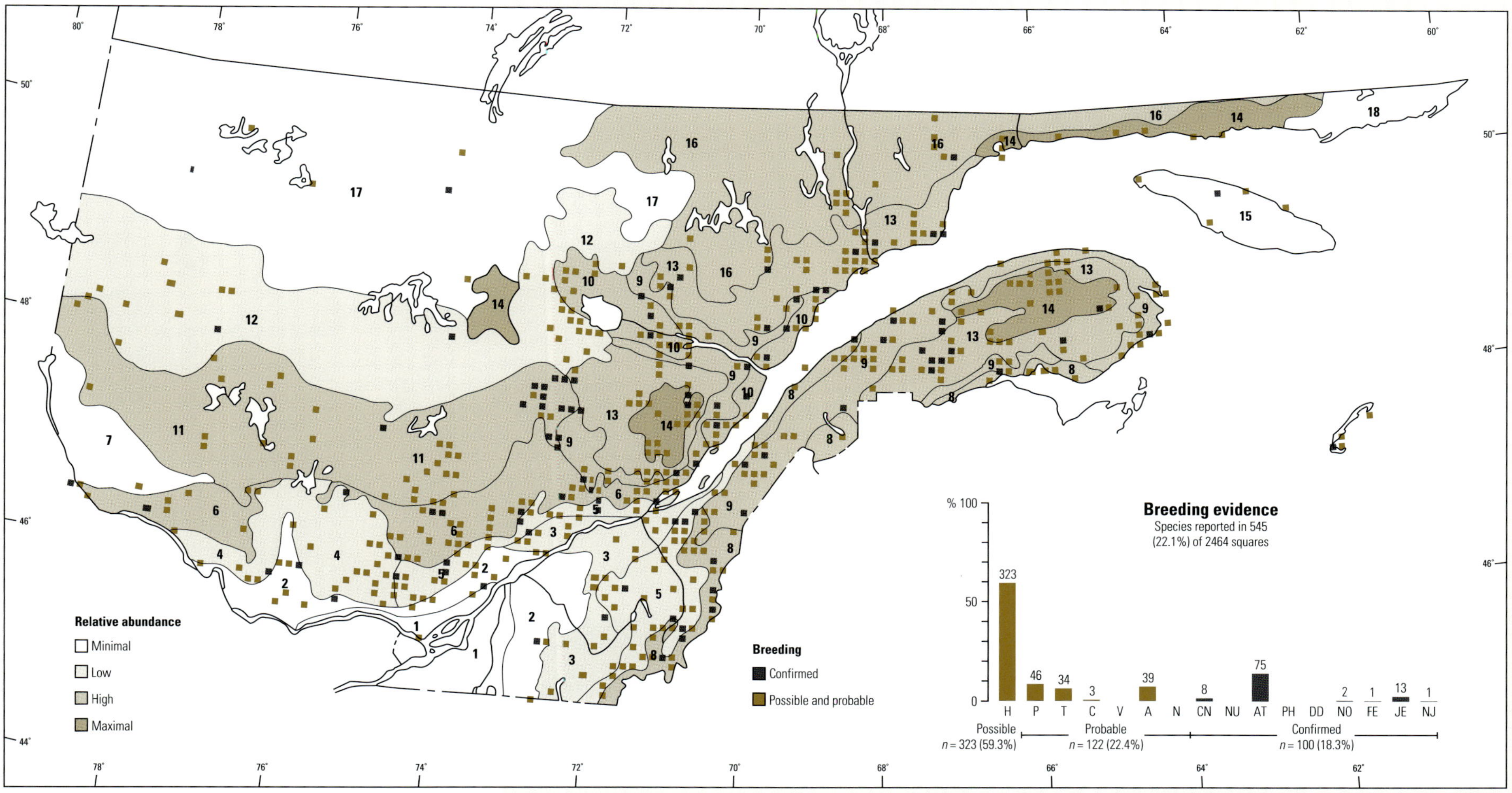

World breeding range

Species profile

Clutch size	usually 6–7 eggs (4–9)[1,2,3,4]
Laying interval	?
Start of incubation	?
Incubation period	?
Maturity at hatching	altricial
Care of young	probably by female and male[4]
Nestling period	?
Age at first flight	?
Dependence of young	?
Broods per year	1[2]
Breeding age	1 year[2]
Mating system	monogamy[4]
Duration of pair bond	?
Total length	11.8–14.0 cm[1]
Wing span	19.0–21.6 cm[3]
Weight	11.0 g (9.3–17.3)[5]
Longevity record	4 years 3 months[6]

References:
1. Godfrey 1986
2. DeGraaf and Rudis 1987
3. Terres 1980
4. Ehrlich et al. 1988
5. Dunning 1984
6. Klimkiewicz et al. 1983

Chronology $n = 9$

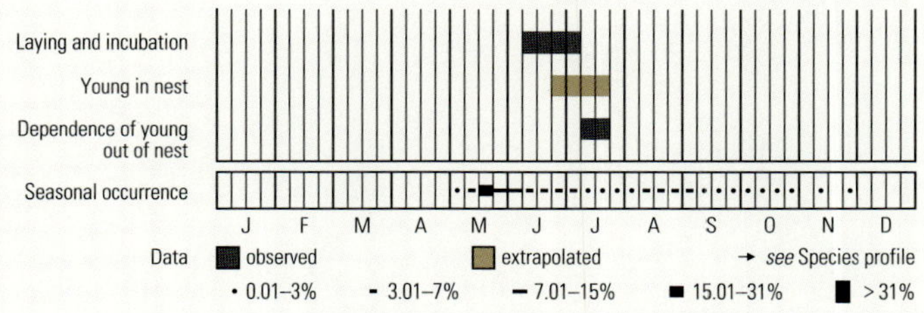

Frequency of occurrence by ecological region

Ecological regions** (forest regions)	Adequately covered squares	Squares in which reported	% of squares in which reported	Relative abundance (%)*
1. Sugar maple-hickory and sugar maple-basswood	97	1	1.0 ± 1.4	1.3
2. Sugar maple-basswood	153	17	11.1 ± 2.7	14.1
3. Sugar maple-basswood and sugar maple-yellow birch	101	26	25.7 ± 5.2	32.8
4. Sugar maple-yellow birch-basswood	56	23	41.1 ± 11.3	52.3
5. Sugar maple-yellow birch-beech and sugar maple-yellow birch-basswood	113	48	42.5 ± 5.1	54.1
6. Sugar maple-yellow birch-beech	71	44	62.0 ± 10.3	78.9
7. Sugar maple-yellow birch-balsam fir	19	1	5.3 ± 12.0	6.7
8. Sugar maple-yellow birch and balsam fir-yellow birch	47	25	53.2 ± 12.8	67.7
9. Balsam fir-yellow birch	87	53	60.9 ± 9.5	77.5
10. Balsam fir-white birch or balsam fir-red maple	46	29	63.0 ± 12.1	80.2
11. Yellow birch-balsam fir	52	36	69.2 ± 13.0	88.1
12. Balsam fir-white birch (west)	59	21	35.6 ± 12.6	45.3
13. Balsam fir-white birch (east)	76	47	61.8 ± 10.7	78.7
14. Balsam fir-black spruce	28	22	78.6 ± 16.2	100.0
15. White spruce-balsam fir and balsam fir-white spruce	22	4	18.2 ± 17.1	23.1
16. Black spruce-balsam fir-moss	18	9	50.0 ± 26.1	63.6
17. Black spruce-moss	32	2	6.2 ± 10.0	8.0
18.			*not analysed*	

* minimal: 0–29.9%; low: 30–59.9%; high: 60–89.9%; maximal: 90–100%

** Deciduous forest: 1–7; mixed forest: 8–11; coniferous forest: 12–17

Black-throated Blue Warbler

Paruline bleue à gorge noire

Paruline bleue

Dendroica caerulescens

Christiane Girard

THE BLACK-THROATED BLUE WARBLER IS A TYPICAL MEMBER OF THE insectivorous bird communities found in deciduous and mixed forests in eastern North America. Although this migrant breeder is widespread, it is not abundant anywhere.

As its name suggests, the elegantly attired male has bluish-grey upperparts and black cheeks, throat, and sides, which contrast with its sparkling white underparts. The drabber female has an olive-brown back, dull olive-yellow underparts, and a whitish stripe over her eye. The species' pocket handkerchief, or white wing spot, is less evident in the female. The Black-throated Blue Warbler has an easily identified song, a husky *zur, zur, zur, zree*.

The Black-throated Blue Warbler winters mainly in the Caribbean, Central America, and northern South America (AOU 1983; Diamond 1991). It arrives on its breeding grounds in Québec in mid-May.

Habitat and behaviour

Breeding territories are established in deciduous or mixed stands. Although the density of the canopy may vary, Black-throated Blue Warblers require a well-developed understorey and, more importantly, a dense upper shrub layer (Holmes and Sturges 1975; Thompson and Capen 1988; Darveau et al. 1992). In Nova Scotia, for example, they were found in deciduous stands at least 20 years old with a dense canopy and well-developed upper shrub layer (Morgan and Freedman 1986). They also nest in mixed forests with young spruce, fir or hemlock (Bent 1953).

The Black-throated Blue Warbler is usually monogamous, but Petit et al. (1988b) reported an instance of polygyny in which one of the two females raised two clutches successfully in a single season. Nest building and incubation are done by the female alone. Incubation of the 4 eggs lasts 12–13 days. The male feeds the female on the nest and also helps feed the nestlings, which remain in the nest for generally 8–9 days (Holmes 1994). The species is not victimized to any great extent by the Brown-headed Cowbird; in the QNRCP data base, only 2 of 21 nests were recorded as being parasitized, and in the Ontario Nest Records Scheme dataset only 3 cases of parasitism were found in 23 nests (Peck and James 1987).

The nest is a well-formed cup attached to the branch of a dense deciduous or evergreen shrub; it is generally located less than 1 m above the ground, fairly near the trunk (Harrison 1978; Peck and James 1987). In a deciduous forest in New Hampshire dominated by beech, sugar maple, and yellow birch, 60 out of 82 nests were in viburnum bushes at an average height of 50 cm (Holway 1991). However, in similar stands in the Bois-Francs region, 6 out of 7 nests were found in conifers (M. Darveau, unpubl. data).

In mixed and deciduous forests like this sugar maple-yellow birch stand, the Black-throated Blue Warbler frequents areas with a dense understorey. Ottawa Valley. *September 1986, Daniel St-Hilaire, Ministère du Loisir, de la Chasse et de la Pêche.*

Notes on territory

In spring, males establish territories generally less than 2 ha in size (Freemark and Merriam 1986; Rodenhouse and Holmes 1992). The young may remain in this territory to be fed by their parents for up to 5–10 days after leaving the nest. After the young have become completely independent, the male may defend the territory for another week or so through singing (Black 1975 in Kibbe 1985b).

In Québec, the lowest breeding densities have been found in coniferous stands: 1.7 pairs/10 ha in a hemlock stand in Mirabel, and 1.6 pairs/10 ha in a red spruce stand and 2.2 pairs/10 ha in a balsam fir stand in La Mauricie National Park (Cyr 1974; DesGranges 1980b). In the same park, a density of 6.4 pairs/10 ha was recorded in a white birch stand. The greatest variations in density were obtained in heterogenous deciduous stands, with the lowest densities —2.5 pairs/10 ha—recorded in a mature sugar maple-beech stand and the highest densities —10.8 pairs/10 ha—in a sugar maple-yellow birch stand. Several intermediate values were also obtained (Erskine 1971a, 1976a, 1984).

Black-throated Blue Warbler

Breeding distribution

1984–1989

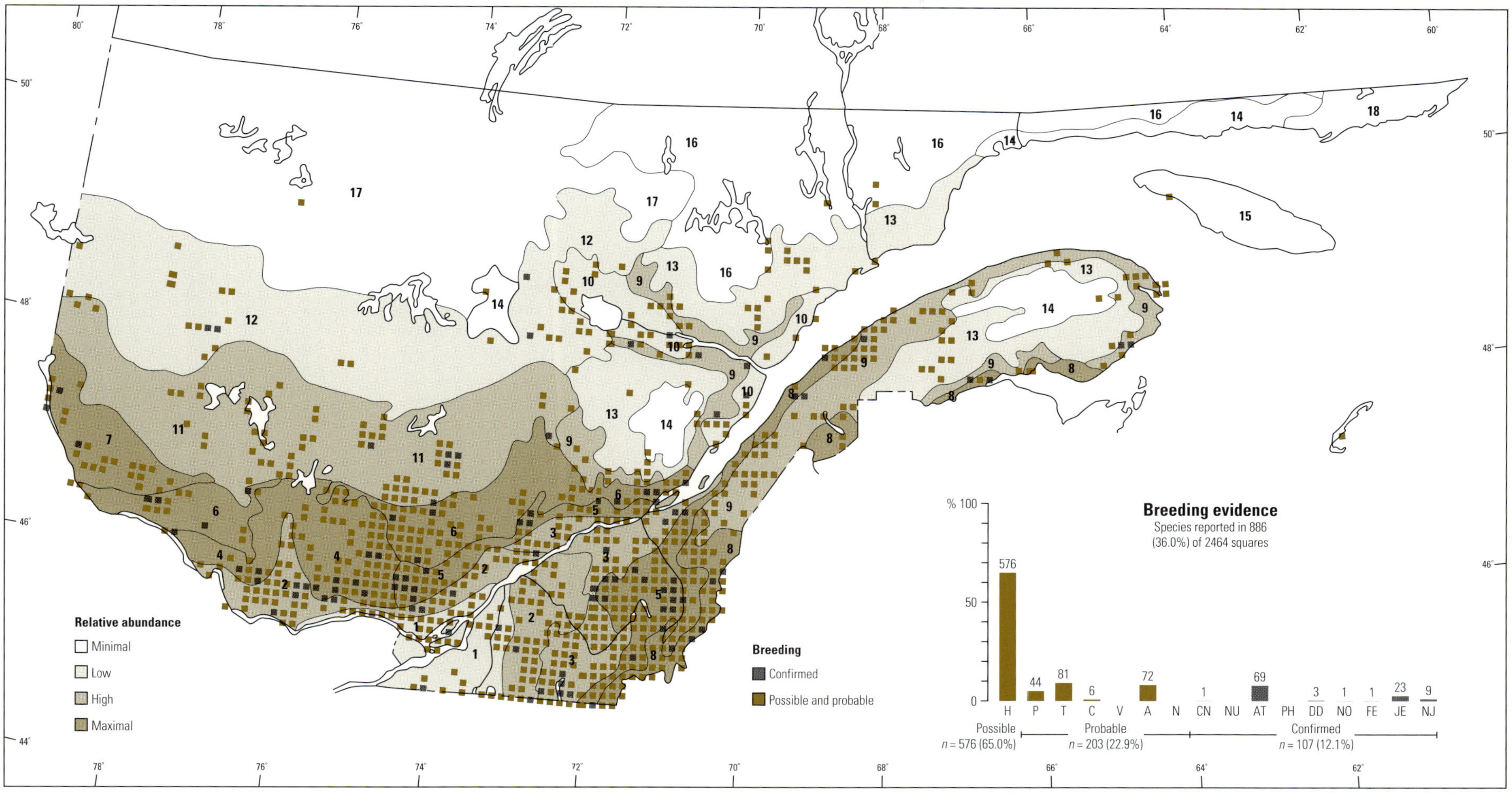

Species profile

Clutch size	usually 4 eggs (2–5)[1,3,4,5,6,11]
Laying interval	1 egg per day[7]
Start of incubation	with last or penultimate egg[7,8,11]
Incubation period	12–13 days[1,7,8]
Maturity at hatching	altricial
Care of young	by female and male[1,4,5]
Nestling period	usually 8–9 days (8–14)[4,6,7,8,11]
Age at first flight	usually 8–9 days[11]
Dependence of young	until 5–10 days after leaving nest[8], probably until 2–3 weeks after leaving nest, sometimes up to 4 weeks[11]
Broods per year	usually 1–2 (1–3)[7]
Breeding age	1 year[2]
Mating system	monogamy, sometimes polygyny[1,7]
Duration of pair bond	often more than 1 season[11]
Total length	12.0–14.0 cm[3]
Wing span	17.8–20.3 cm[5]
Weight	M 10.5 g (8.4–12.4) / F 9.8 g (8.8–12.1)[10]
Longevity record	9 years 8 months[9]

References:

1. Ehrlich et al. 1988
2. DeGraaf and Rudis 1987
3. Godfrey 1986
4. Harrison 1978
5. Terres 1980
6. Griscom and Sprunt 1979
7. Holmes et al. 1992
8. Kibbe 1985b
9. Klimkiewicz and Futcher 1989
10. Dunning 1984
11. Holmes 1994

World breeding range

Chronology $n = 26$

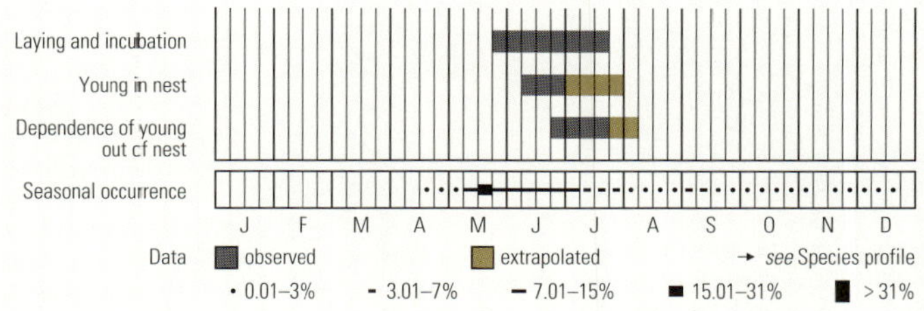

Frequency of occurrence by ecological region

Ecological regions** (forest regions)	Adequately covered squares	Squares in which reported	% of squares in which reported	Relative abundance (%)*
1. Sugar maple-hickory and sugar maple-basswood	97	40	41.2 ± 4.9	43.1
2. Sugar maple-basswood	153	103	67.3 ± 3.9	70.3
3. Sugar maple-basswood and sugar maple-yellow birch	101	87	86.1 ± 4.2	89.9
4. Sugar maple-yellow birch-basswood	56	53	94.6 ± 5.6	98.8
5. Sugar maple-yellow birch-beech and sugar maple-yellow birch-basswood	113	104	92.0 ± 3.0	96.1
6. Sugar maple-yellow birch-beech	71	68	95.8 ± 4.7	100.0
7. Sugar maple-yellow birch-balsam fir	19	17	89.5 ± 15.5	93.4
8. Sugar maple-yellow birch and balsam fir-yellow birch	47	43	91.5 ± 7.6	95.5
9. Balsam fir-yellow birch	87	53	60.9 ± 9.5	63.6
10. Balsam fir-white birch or balsam fir-red maple	46	26	56.5 ± 12.4	59.0
11. Yellow birch-balsam fir	52	37	71.2 ± 12.8	74.3
12. Balsam fir-white birch (west)	59	25	42.4 ± 12.9	44.2
13. Balsam fir-white birch (east)	76	28	36.8 ± 10.6	38.5
14. Balsam fir-black spruce	28	3	10.7 ± 12.6	11.2
15. White spruce-balsam fir and balsam fir-white spruce	22	1	4.5 ± 10.3	4.7
16. Black spruce-balsam fir-moss	18	4	22.2 ± 22.2	23.2
17. Black spruce-moss	32	0	0	0
18.			*not analysed*	

* minimal: 0–29.9%; low: 30–59.9%; high: 60–89.9%; maximal: 90–100%

** Deciduous forest: 1–7; mixed forest: 8–11; coniferous forest: 12–17

The Black-throated Blue Warbler feeds mainly on insects, most often butterflies and moths, beetles, and flies, which it captures by hovering under the foliage or by gleaning the top of the foliage directly (Robinson and Holmes 1982; Whelan 1989). The male usually feeds 2–10 m above the ground near its singing perches, while the female forages at lower levels closer to the nest (Holmes 1986).

The Black-throated Blue Warbler was reported in 36% (886/2464) of the Atlas squares surveyed, but breeding was confirmed in only 12% (107/886) of these squares, mainly by reports of recently fledged young and adults carrying food. According to Terres (1980), the adults are very tame when they are feeding the young.

Distribution

The Black-throated Blue Warbler nests in southeastern Canada, from western Ontario to Nova Scotia, and has also been observed in southeastern Manitoba. South of the border, it nests in the Great Lakes area, northeastern states, and in the Appalachians as far south as northern Georgia (AOU 1983).

In Québec, the Black-throated Blue Warbler has been recorded throughout most of the province south of the 50th parallel. However, it is more abundant in the deciduous and mixed forest zones than the coniferous forest zone farther north or in the urbanized and agricultural regions in the extreme south, where the forest cover has been modified extensively by human activity. The species occurs locally in the boreal forest, in mixed second growth or shade-intolerant hardwoods appearing after logging or fire (Erskine 1977). The lower frequency of occurrence in the Montréal and Montérégie regions is due to the predominance of open habitats created when the regions' sugar maple stands were cut down to make way for agricultural and urban development.

During the Atlas surveys, the Black-throated Blue Warbler was observed on Anticosti Island, the only time it has been recorded there since it was first reported in 1937 (Ouellet 1969). The species was not observed on the Magdalen Islands, where it is considered a rare visitant, although several sightings have been made recently (Fradette 1992).

The northernmost records were obtained by Townsend and Bent, who saw an individual at Mingan on 20 June 1909, and by Todd, who reported seeing a male near his camp on the Sainte-Marguerite River on 6 June 1917 (Todd 1963). Todd also noted that Audubon had found a dead specimen in Labrador.

History and trends

Although the historical data on Black-throated Blue Warbler populations in Québec are sketchy at best, the evidence suggests that little change has occurred. Ouellet (1974a) repeated Vennor's comment that the species was exceedingly rare in Lower Canada in the middle of the 19th century and that it nested occasionally on the outskirts of Montréal. Dionne (1906) considered it uncommon in the province at the turn of the century, while Wintle (1896) listed it as a common "transient visitant" in Montréal that probably nested there in small numbers. Without specifying the region, Macoun and Macoun (1915) stated that this warbler was common during spring migration and a large number were found throughout the summer. More recently, Ouellet (1974a) listed it as a common migrant and uncommon summer resident in the Montréal region. David (1980a) described it as a fairly common migrant breeder in the southern part of southern Québec, but as an uncommon migrant breeder farther north.

Notes on territory (continued)

Lastly, in a study of 18 sugar maple stands in southern Québec affected to varying degrees by maple dieback, the average density was 4 pairs/10 ha (Darveau et al. 1990, 1992).

The Black-throated Blue Warbler's preferred habitat includes sugar maple stands with a well-developed understorey. Cap Tourmente National Wildlife Area, northeast of Québec City. *August 1991, Jean Gauthier.*

The female builds the nest in dense shrubbery, usually less than 1 m above the ground. Lake Saint-Pierre region. *May 1991. Réginald Ouellet.*

Thus, the Black-throated Blue Warbler never seems to have been very abundant in the province, although it is observed fairly often in the southernmost parts of Québec.

BBS data indicate that, in general, Black-throated Blue Warbler populations remained stable from 1968 to 1979 (Robbins et al. 1986). BBS and ÉPOQ summer data for Québec also show stable populations between 1966 and 1989.

The disappearance of the species from some areas since the turn of the century has been noted in the breeding bird atlases of Ontario, the Maritimes, and several northeastern states (Bull 1985; Kibbe 1985b; Fraser 1987a; Peck and James 1987; Bonney 1988b; Binford 1991; Erskine 1992; Reid 1992a). The areas in question are mainly lowland regions where the forests have gradually yielded to urban and agricultural development. In Québec, similar changes to habitat have occurred in the Central St. Lawrence Lowland, but the effects on the species have been less apparent since it has traditionally been less abundant in this area.

There is little chance, therefore, that the Black-throated Blue Warbler will become more abundant in Québec in the future. On the other hand, there is little reason to fear for the species' short-term prospects, since it has a large range, where suitable habitat in the form of mountainous areas and extensive forested tracts with a dense shrub layer are plentiful (Reid 1992a). This warbler also seems to tolerate human disturbance fairly well and is not affected to any great degree by selective logging practices (Webb et al. 1977). The Black-throated Blue Warbler will return to clearcut areas 15–20 years after felling (Darveau et al. 1982; Thompson and Capen 1988) and will nest in forest stands with extensive cottage development (Clark et al. 1983). In addition, a study carried out in sugar maple stands in southern Québec, in the Bois-Francs and Amiante regions, showed that the Black-throated Blue Warbler is not affected by maple dieback and frequents stands with either a healthy or diseased canopy as long as areas of thick undergrowth are present (Darveau et al. 1990, 1992).

Marcel Darveau

Yellow-rumped Warbler

Paruline à croupion jaune

Dendroica coronata

Christiane Girard

THE YELLOW-RUMPED WARBLER IS ONE OF THE MOST WIDESPREAD AND abundant of the 27 warbler species breeding in Québec. It nests throughout the province from the U.S. border to the tree line, with the exception of the Ungava Peninsula. Its yellow rump is visible in both its breeding and fall plumages, making it easy to identify at all times of the year. Abundant and widespread during migration, it is the earliest warbler to return in the spring and the last to leave in the autumn.

The first males arrive in Québec around the end of April, followed by the females a few days later. The species is abundant everywhere you look during most of May but, by mid-June, most pairs have reached their breeding grounds.

The Yellow-rumped Warbler winters farther north than most warblers, tending to overwinter in the southern United States rather than in Central America (AOU 1983). Some individuals even attempt to spend the winter in Québec on occasion.

Habitat and behaviour

Although the Yellow-rumped Warbler is typically a boreal species, it also breeds in areas farther south where conifers are found. It prefers mature or medium-aged second-growth forest to immature forest, and coniferous or mixed woodlands to deciduous ones. It especially favours spruce or balsam fir stands on mesic sites. The Yellow-rumped Warbler can also be found in wetter habitats with black spruce or tamarack and in conifer plantations. It nests mainly in the more open parts of woodlands or close to forest edges (Ellison 1985y; Godfrey 1986; Crins 1987b; DeGraaf and Rudis 1987; Peck and James 1987).

The male's typical song is a weak, warbling, quavering trill reminiscent of the Nashville Warbler's. Songs can vary seasonally and even daily. The male is very vocal and sings frequently to defend his territory, at all hours of the day. According to Morse (1989b), the particular song sung most frequently in the evening is territorial in nature, though not necessarily sung only in response to a direct conflict with a neighbour. Songs are learned by imitation (Van Buskirk 1984).

In Ontario, Peck and James (1987) found most Yellow-rumped Warbler nests in conifers, usually 2–6 m above the ground. The loose, bulky nest, which is built by the female, is a deep cup made out of twigs, bark strips, tufts of moss, lichen, and other plant materials. It is lined with feathers in a characteristic pattern that is unique to the species: the shafts are woven or imbedded into the lining of the nest so that the tops bend inward over the cup, screening the eggs when the bird is not sitting (Bent 1953; Harrison 1975, 1984). In Ontario, Peck and James (1987) found Ruffed Grouse, Spruce Grouse, and Common Loon feathers in the nest. In Nova Scotia (Tufts in Bent 1953) and New England (DeGraaf and Rudis 1987), the species may produce two broods a year; this may also occur in Québec. The female does most of the incubating with the male incubating only on very rare occasions (Bent 1953). The female broods the young assiduously the day after they hatch, but much less regularly afterward. Both the male and the female feed the young (Bent 1953). The young leave the nest when they are 12–14 days old, and are able to fly 2 or 3 days later (Bent 1953).

The Yellow-rumped Warbler has the most varied feeding habits of all the warblers: it forages in more zones of the tree (although it concentrates its activities generally in the lower part) and has the most varied foraging patterns (MacArthur 1958). In a spruce forest in Maine, Morse (1967c) observed that the Yellow-rumped Warbler forages significantly lower in the tree than the Black-throated Green warbler, which feeds in the layer just below that used by the Blackburnian Warbler, near the top of the tree. However, there appears to be little

The Yellow-rumped Warbler is common in mature stands of black spruce in the boreal forest. It nests primarily in relatively open areas. Near Lake Chibougamau. *September 1986, Jean Beaubien and Guy Simard, Canadian Forest Service.*

Notes on territory

Yellow-rumped Warblers use their territories for nesting as well as foraging (Welty 1975). In spruce stands in Maine, territories ranged between 0.75 and 0.95 ha in size (Morse 1976b).

In southern Québec, densities of 9.9 pairs/10 ha were found in a balsam fir and spruce stand interspersed with shade-intolerant deciduous trees and containing a regenerating understorey (Erskine 1980a). Densities generally between 4.9 and 7.4 pairs/10 ha were found in stands of balsam fir with spruce or shade-intolerant deciduous trees, or both, or in stands of mixed conifer species (Erskine 1971a, 1980a, 1984). In sawfly-infested jack pine stands near La Tuque, densities ranged between 0.7 and 3.2 pairs/10 ha (Erskine 1971a, 1976a). In mixed habitats composed of woodland and openings at Gatineau Park and La Barrière, near Saint-Michel-des-Saints, the average density was 0.55 pair/10 ha (Dauphin 1985).

During surveys in the area around the Nottaway, Broadback, and Rupert rivers and in the Great Whale River region, Yellow-rumps were most abundant in black spruce-moss and black spruce-lichen forests, in the ecotones between conifer stands and bodies of water, and in the zone between conifer stands and burns (Cons. G. G. – GREBE 1991a, 1992o).

Yellow-rumped Warbler

Breeding distribution
1984–1989

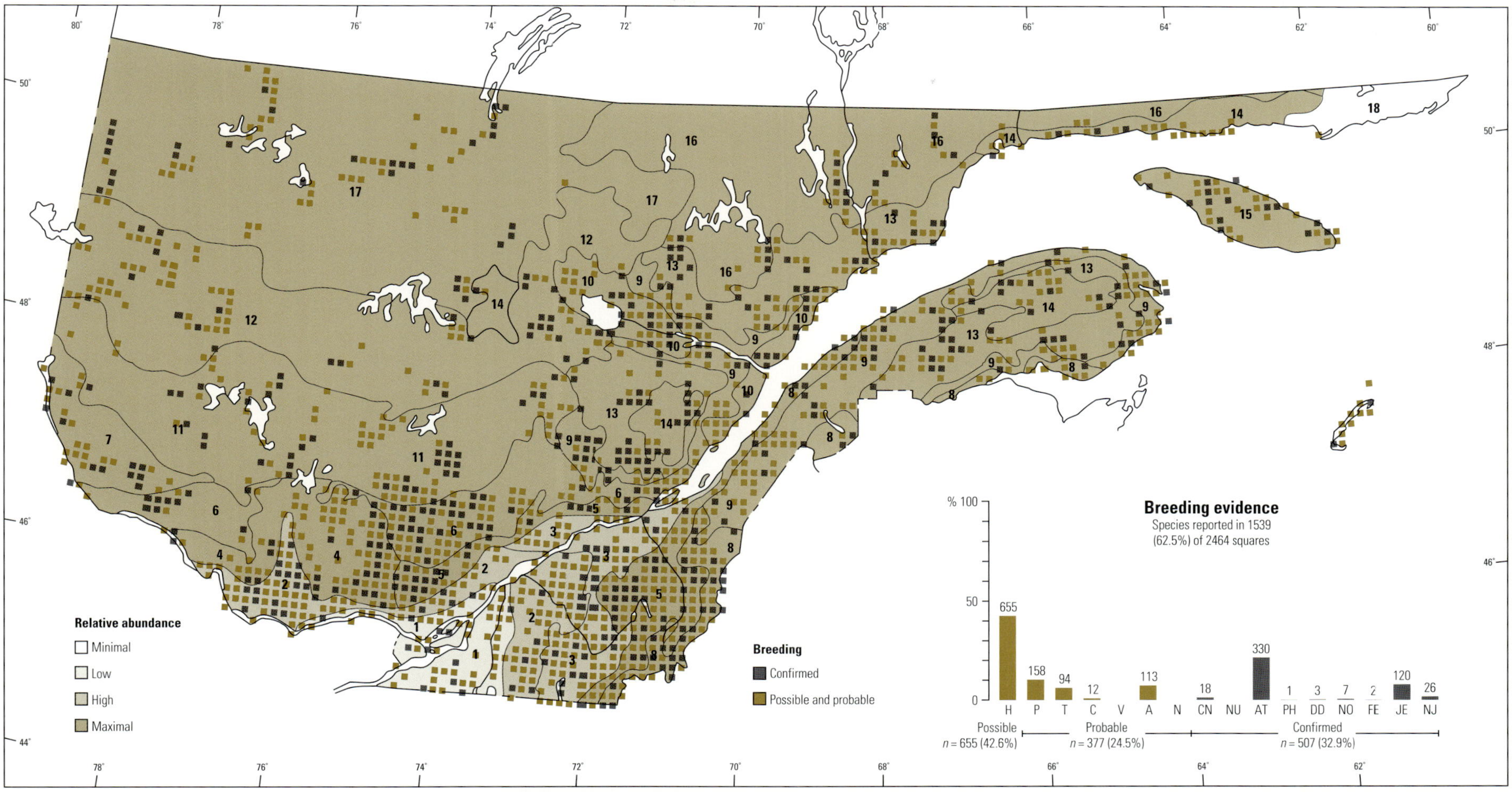

Species profile

Clutch size	usually 3–4 eggs (3–6)[2,3,5,6]
Laying interval	?
Start of incubation	?
Incubation period	11–13 days[2,3,4,6]
Maturity at hatching	altricial
Care of young	by female and male[4]
Nestling period	12–14 days[2,4]
Age at first flight	14–17 days[4]
Dependence of young	?
Broods per year	1–2[2,4]
Breeding age	1 year[2]
Mating system	monogamy[6]
Duration of pair bond	?
Total length	12.0–15.5 cm[1]
Wing span	?
Weight	M 12.9 g (10.6–16.7) / F 12.2 g (9.9–15.3)[7]
Longevity record	6 years 11 months[8]

References:
1. Godfrey 1986
2. DeGraaf and Rudis 1987
3. Peck and James 1987
4. Bent 1953
5. Terres 1980
6. Ehrlich et al. 1988
7. Dunning 1984
8. Klimkiewicz et al. 1983

World breeding range

Chronology $n = 50$

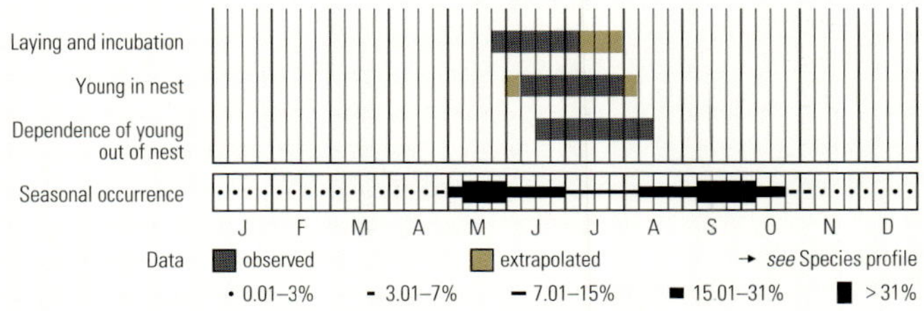

Frequency of occurrence by ecological region

Ecological regions** (forest regions)	Adequately covered squares	Squares in which reported	% of squares in which reported	Relative abundance (%)*
1. Sugar maple-hickory and sugar maple-basswood	97	46	47.4 ± 5.0	47.4
2. Sugar maple-basswood	153	130	85.0 ± 3.0	85.0
3. Sugar maple-basswood and sugar maple-yellow birch	101	88	87.1 ± 4.1	87.1
4. Sugar maple-yellow birch-basswood	56	56	100.0 ± 0.9	100.0
5. Sugar maple-yellow birch-beech and sugar maple-yellow birch-basswood	113	110	97.3 ± 2.0	97.3
6. Sugar maple-yellow birch-beech	71	70	98.6 ± 3.0	98.6
7. Sugar maple-yellow birch-balsam fir	19	19	100.0 ± 2.6	100.0
8. Sugar maple-yellow birch and balsam fir-yellow birch	47	47	100.0 ± 1.1	100.0
9. Balsam fir-yellow birch	87	86	98.9 ± 2.5	98.9
10. Balsam fir-white birch or balsam fir-red maple	46	44	95.7 ± 5.7	95.7
11. Yellow birch-balsam fir	52	52	100.0 ± 1.0	100.0
12. Balsam fir-white birch (west)	59	54	91.5 ± 7.7	91.5
13. Balsam fir-white birch (east)	76	72	94.7 ± 5.3	94.7
14. Balsam fir-black spruce	28	27	96.4 ± 8.3	96.4
15. White spruce-balsam fir and balsam fir-white spruce	22	22	100.0 ± 2.3	100.0
16. Black spruce-balsam fir-moss	18	18	100.0 ± 2.8	100.0
17. Black spruce-moss	32	32	100.0 ± 1.6	100.0
18.			*not analysed*	

* minimal: 0–29.9%; low: 30–59.9%; high: 60–89.9%; maximal: 90–100%

** Deciduous forest: 1–7; mixed forest: 8–11; coniferous forest: 12–17

Although the Yellow-rumped Warbler is a typical boreal species, it is also found farther south in mixed forests and other areas containing conifers. Ottawa Valley. *September 1986, Daniel St-Hilaire, Ministère du Loisir, de la Chasse et de la Pêche.*

competition between the Yellow-rumped and Black-throated Green warblers. On the ground, the Yellow-rump's foraging is limited almost entirely to hunting for emerging crane flies to feed to its newly hatched young (MacArthur 1958). The abundant and readily identified Yellow-rumped Warbler is one of the easiest warblers to detect when foraging, since it flits from tree to tree more frequently than most other warblers, often covering its whole territory from end to end (MacArthur 1958; Morse 1989b).

After the Common Yellowthroat, the Yellow-rumped Warbler was the warbler with the highest frequency of occurrence in the Atlas survey and was reported in 62.5% (1539/2464) of the squares surveyed. In addition, the species was one of the warblers with the greatest number of squares with confirmed breeding records. The majority of the confirmations—65.1% (330/507)—were based on reports of adults carrying food, while sightings of recently fledged young made up most of the rest.

The Yellow-rumped Warbler begins to migrate south in August but numbers peak from mid-September to when the trees are completely bare. Then, the little birds can be seen in groves of trees, shrubbery, and on the edge of fields and roadsides. They can even be spotted clinging to the walls and gutters of houses, inspecting every nook and cranny for insects, to meet their considerable energy needs during migration. The Yellow-rumped Warbler reaches its peak foraging efficiency at this time (Moore and Simm 1985). In fall, the species generally migrates at night (Terrill and Crawford 1988), lingering until conditions on the breeding grounds become unfavourable, sometimes to the extent of endangering its survival (Terrill and Ohmart 1984). The Yellow-rumped Warbler therefore has the shortest wintering period of all the warblers (Terrill and Ohmart 1984). In Québec, most individuals leave the province before mid-November, but departure dates vary from year to year depending on weather conditions.

During the breeding season, the Yellow-rumped Warbler feeds mainly on insects. It is known to devour huge quantities of spruce budworm (DesGranges and Rondeau 1995). However, during a budworm epidemic, the presence of more aggressive competitors such as the Cape May and Bay-breasted warblers may limit Yellow-rumped Warbler numbers (Morris et al. 1958; DesGranges and Rondeau 1995). On its wintering grounds, the Yellow-rumped Warbler eats mainly the seeds and berries of juniper, viburnum, sumac, and other plants (Bent 1953). When it is cold and there is heavy snow cover, this

Among the *Dendroica* warblers studied by Robert MacArthur, the Yellow-rumped Warbler uses the greatest number of feeding zones in the tree, although it concentrates mostly lower down. Laurentide Wildlife Reserve, north of Québec City. *August 1991, Marcel Darveau.*

States, this is not true in Québec with its more northerly latitude and greater abundance of conifers (Ellison 1985y; Peterson 1988q; Gross 1992a).

In recent surveys, the Yellow-rumped Warbler was found to be one of the most abundant species in the area extending from southern James Bay to Lake Guillaume Delisle, just north of the Little Whale River (Cons. G. G. – GREBE 1991a, 1992o). It was found in almost all of the habitats surveyed, but particularly both on the margins of and within conifer stands, as well as in recent and old burns.

This warbler also nests in second-growth stands and along the edges of roads and bodies of water. Grands-Jardins Conservation Park, northwest of Baie-Saint-Paul. *June 1991, Réginald Ouellet.*

warbler is forced to subsist almost entirely on the berries of wax myrtle, which is why the eastern subspecies is commonly known as the Myrtle Warbler. Wax myrtle berries are covered with a waxy substance consisting of saturated fatty acids. This substance no doubt has a high nutritional value since it is able to meet the warblers' high metabolic requirements in winter. The Yellow-rumped Warbler is one of the few animal species that can efficiently assimilate this wax, which is why it can winter much farther north than other warblers (Hausman 1927; Emlen 1966; Wiltz and Giampa 1978; Place and Stiles 1992), often close to the southern edge of its breeding range (Bent 1953; Godfrey 1986). Interestingly, the Yellow-rumped Warbler may drink the sap of trees and feed on the insects attracted to the sap (Mousley 1917i; Terres 1980).

The Yellow-rumped Warbler also helps to propagate the seeds of various plants, due in part to the fact that seeds remain in its digestive tract for a fairly long period of time, promoting germination. The seeds are spread through droppings as the warblers move about. The Yellow-rumped Warbler, along with other birds, may contribute to the rapid regeneration of Eastern redcedar in abandoned fields in the southeastern United States (Holthuijzen and Sharik 1985; Malmborg and Willson 1988).

Distribution

The species' breeding range extends from Alaska and the northwestern part of the MacKenzie District in the Northwest Territories eastward across northern Manitoba to Labrador and south to New Mexico and Mexico in the west and the northeastern states in the east (AOU 1983; National Geographic Society 1987). It is found as far north as the tree line. The Yellow-rumped Warbler breeds throughout southern Québec. Its frequency of occurrence is high everywhere except in Montérégie and a narrow band north of the Ottawa and St. Lawrence rivers as far east as Lake Saint-Pierre. These areas do not provide suitable habitat since they are primarily agricultural with few conifer stands. Although the Yellow-rumped Warbler tends to be less prevalent at low elevations in the United

History and trends

Yellow-rumped Warbler populations in eastern North America increased from 1966 to 1979 (Robbins et al. 1986). In New York State, this increase was attributed to the species' colonization of low-elevation conifer plantations from the early 1950s onward (Bull 1974). In Vermont, it has recently begun to colonize abandoned farmland grown over in conifers (Ellison 1985y). In Québec, the establishment of conifer plantations on abandoned farmland began only in the 1970s and these plantations could be one of the reasons populations here have increased. Furthermore, the planting of conifer stands in regions traditionally dominated by deciduous forests—as in the Beauce region—appears to have benefited the species. In addition, the Yellow-rumped Warbler appears to be more abundant at lower elevations than it was in the past; for example, at the turn of the century, the Yellow-rumped Warbler was considered uncommon in summer in such areas (Wintle 1896; Dionne 1906; Mousley 1916a; Lloyd 1924). More recently, Godfrey and Wilk (1948) described this warbler as surprisingly uncommon in the Lac-Saint-Jean region. However, the Atlas distribution map for the species suggests that it is now more common in southern Québec than it was in the past, and ÉPOQ summer data also indicate an increase in numbers between 1970 and 1989. It is more difficult to determine the status of populations in northern Québec, since few data are available for this region. Todd (1963) considered the Yellow-rumped one of the most common and widespread warblers in the north, which is undoubtedly still true today.

Forest fires and selective logging seem to have little effect on the species' abundance (Apfelbaum and Haney 1981; Clark et al. 1983). However, a study carried out in Québec shows that strip-cutting leads to decreased densities (G. Falardeau and J.-P. L. Savard, CWS, unpubl. data). In addition, when carried out in inhabited areas, this type of logging increases the risk of parasitism by the Brown-headed Cowbird. In southern Québec, Terrill (1961) found that out of 33 nests examined, only 6% had been parasitized, whereas in Ontario, cowbird eggs were found in 31% of the 122 Yellow-rumped Warbler nests examined (Peck and James 1987). Cowbirds tend to be scarce farther north in the boreal forest, and so pose little threat to brood survival.

VINCENT LÉTOURNEAU AND PIERRE LAFONTAINE

Black-throated Green Warbler

Paruline verte à gorge noire

Paruline à gorge noire
Dendroica virens

Christiane Girard

THE BLACK-THROATED GREEN WARBLER PROBABLY HAS THE MOST EASILY recognizable song of any warbler. Its simple song comes in two forms: a high-pitched *zee zee zee zee zoo zee* and a lazier *zoo zee zoo zoo zee*, the latter having been described by the late American ornithologist, Margaret Morse Nice, as "trees trees murmuring trees" (Harrison 1984). It is an easy song to remember and identify in the forest. Although this warbler is very active, it can be hard to detect because of its predilection for foraging in dense foliage in the crown. However, it is easy to identify, with its yellow face contrasting with its black throat and breast and its yellow-green upperparts. The female is similar but less colourful and has yellow on her throat.

In the boreal forest, the Black-throated Green Warbler is a typical species of balsam fir stands (Erskine 1977). It returns to Québec in May during the peak of the warbler migration and leaves in September. Its wintering grounds extend from southern Florida and Texas to most of Central America, including eastern and southern Mexico, Cuba, and the Bahamas (AOU 1983; Morse 1989a).

Habitat and behaviour

Often associated with Eastern white pine and Eastern hemlock (Todd 1940 *in* Santner 1992a; Ouellet 1974a), the Black-throated Green Warbler nests in mixed and coniferous forests from the southern Appalachians to the northern reaches of the boreal forest (Eaton 1914; Erskine 1977, 1992; Ellison 1985aa; Peck and James 1987; Doepker and Ozoga 1991b).

On its breeding grounds, the Black-throated Green Warbler is confined to forested habitats (Crins 1987a). Although it is found mostly in coniferous forests, it sometimes frequents deciduous stands with a scattering of conifers (Griscom and Sprunt 1979). Habitats used range from coniferous forests of Eastern white-cedar and spruce, mixed forests of yellow birch, sugar maple, American beech, balsam fir, and red spruce, to deciduous forests of maple or beech. They are also found in stands of hemlock or cedar, and conifer plantations (Collins 1983b; Crins 1987a). In the Adirondacks, the species seems to avoid stands at elevations above 1200 m (Saunders 1929a *in* Eaton 1988e), where the harsh growing conditions limit the height of trees (Able and Noon 1976 *in* Eaton 1988e). The Black-throated Green Warbler prefers a multilayered leaf canopy (Holmes and Robinson 1981 *in* Collins 1983b). Since three-dimensional structure rather than species composition appears to be the critical element in this warbler's choice of habitat (Collins 1983b), it is able to use different types of stands (Beals 1960 *in* Morse 1989a), preferring dense multilayered stands over 20 years old with a closed canopy and well-developed tall shrub layer (Morgan and Freedman 1986).

In the boreal forest, the Black-throated Green Warbler is a typical species of balsam fir stands. Port-Menier, Anticosti Island. *August 1991, Gaétan Laprise.*

Notes on territory

The Black-throated Green Warbler is generally dominant over other warblers frequenting the same habitat, particularly the Blackburnian Warbler, which occupies a niche adjacent to the Black-throated Green Warbler but forages from the tops of the same trees (MacArthur 1958; Morse 1967c, 1976b). Morse (1989a) reported an average territory size of 0.4 ha, while DeGraaf et al. (1980) found territories ranging between 0.2 and 1.0 ha, averaging 0.6 ha, in a hemlock and beech forest in New York State. Territory size appears to be a function of population density in the species (Morse 1989a). In high density populations, foliage volume and territory size appear to be closely related (Morse 1989a). Densities ranging between 2.25 and 20 pairs/10 ha have been recorded (AOU 1983). In mature stands in Nova Scotia, a density of 5.5 pairs/10 ha was obtained (Morgan and Freedman 1986), compared with 7.65 pairs/10 ha in mixed and coniferous forests in Vermont (Carpenter 1973; Metcalf 1977), 13.6 pairs/10 ha in red and Eastern white pine plantations in New York State (Klingensmith 1970), and 20.8 pairs/10 ha in white spruce stands in Maine (Morse 1976b). In Québec, densities of 7.9 pairs/10 ha were recorded in a hemlock stand near Mirabel (Cyr 1974),

Black-throated Green Warbler

Breeding distribution

1984–1989

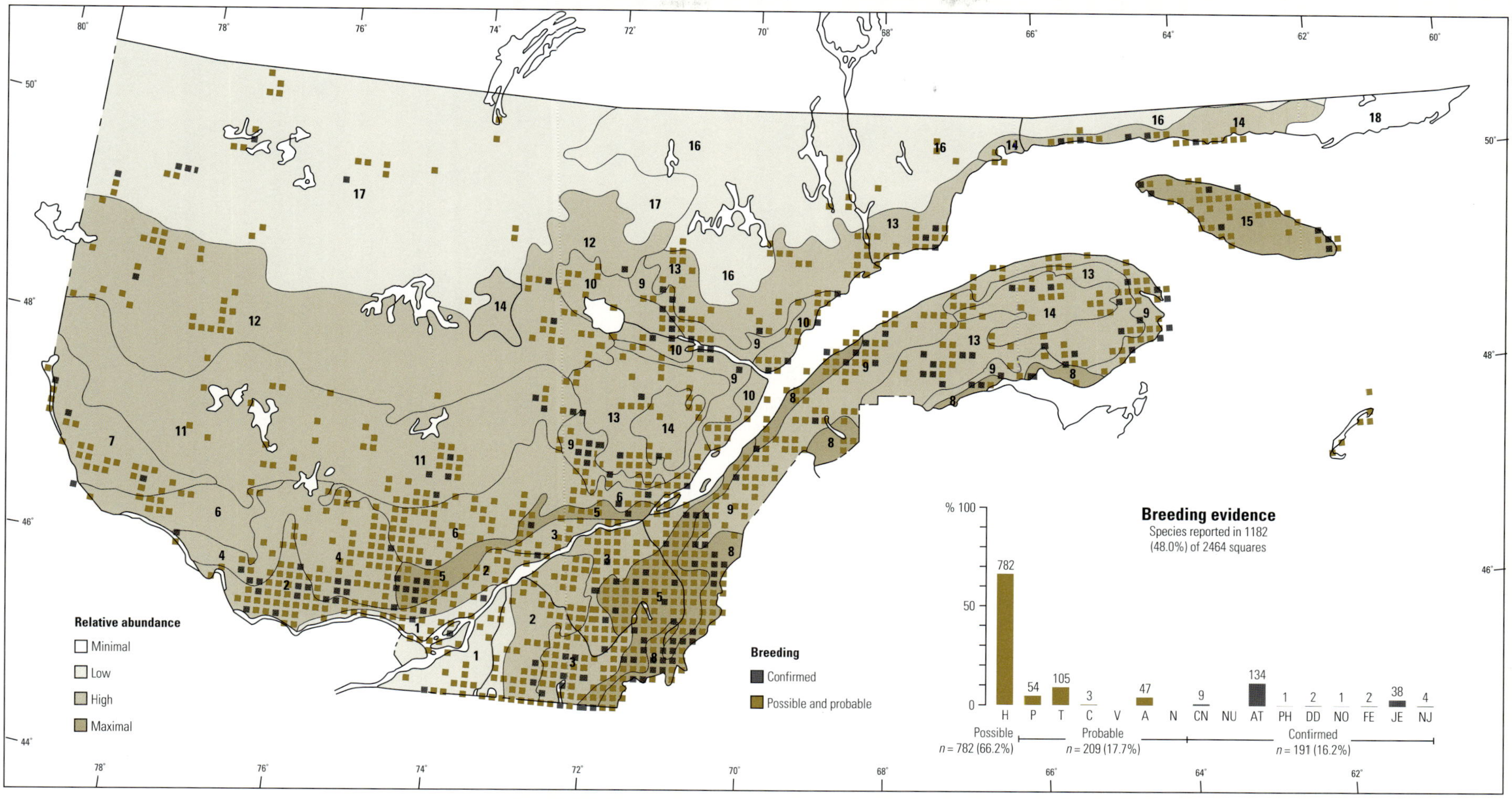

Species profile

Clutch size	4–5 eggs[2,3,4]
Laying interval	?
Start of incubation	probably with last egg[1]
Incubation period	12 days[1,2,3,4]
Maturity at hatching	altricial
Care of young	by female and male[1]
Nestling period	8–11 days[2,3,8]
Age at first flight	10–14 days[8]
Dependence of young	probably until 4 weeks after leaving nest(1)
Broods per year	1[5]
Breeding age	1 year[5]
Mating system	monogamy[3]
Duration of pair bond	1 season[8]
Total length	11.0–13.4 cm[6]
Wing span	17.8–20.3 cm[7]
Weight	M 9.4 g (8.3–10.9) / F 9.0 g (7.5–11.2)[8]
Longevity record	5 years 11 months[9]

References:
1. Pitelka 1940
2. Harrison 1978
3. Ehrlich et al. 1988
4. Harrison 1975
5. DeGraaf and Rudis 1987
6. Godfrey 1986
7. Terres 1980
8. Morse 1993
9. Kennard 1975

World breeding range

Chronology $n = 28$

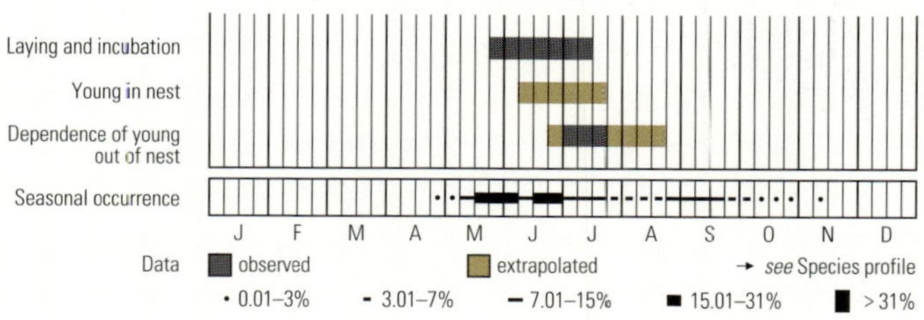

Frequency of occurrence by ecological region

Ecological regions** (forest regions)	Adequately covered squares	Squares in which reported	% of squares in which reported	Relative abundance (%)*
1. Sugar maple-hickory and sugar maple-basswood	97	36	37.1 ± 4.8	37.1
2. Sugar maple-basswood	153	108	70.6 ± 3.8	70.6
3. Sugar maple-basswood and sugar maple-yellow birch	101	89	88.1 ± 4.0	88.1
4. Sugar maple-yellow birch-basswood	56	46	82.1 ± 9.0	82.1
5. Sugar maple-yellow birch-beech and sugar maple-yellow birch-basswood	113	104	92.0 ± 3.0	92.0
6. Sugar maple-yellow birch-beech	71	56	78.9 ± 8.8	78.9
7. Sugar maple-yellow birch-balsam fir	19	16	84.2 ± 17.9	84.2
8. Sugar maple-yellow birch and balsam fir-yellow birch	47	47	100.0 ± 1.1	100.0
9. Balsam fir-yellow birch	87	78	89.7 ± 6.1	89.7
10. Balsam fir-white birch or balsam fir-red maple	46	33	71.7 ± 11.4	71.7
11. Yellow birch-balsam fir	52	40	76.9 ± 11.9	76.9
12. Balsam fir-white birch (west)	59	37	62.7 ± 12.7	62.7
13. Balsam fir-white birch (east)	76	65	85.5 ± 7.9	85.5
14. Balsam fir-black spruce	28	22	78.6 ± 16.2	78.6
15. White spruce-balsam fir and balsam fir-white spruce	22	22	100.0 ± 2.3	100.0
16. Black spruce-balsam fir-moss	18	8	44.4 ± 26.0	44.4
17. Black spruce-moss	32	17	53.1 ± 18.9	53.1
18.			*not analysed*	

* minimal: 0–29.9%; low: 30–59.9%; high: 60–89.9%; maximal: 90–100%

** Deciduous forest: 1–7; mixed forest: 8–11; coniferous forest: 12–17

In deciduous forests, this warbler frequents small pockets of conifers, such as hemlock, fir, spruce, and pine. It prefers dense multilayer stands with a closed canopy and tall shrubs. **Sainte-Foy.** *September 1991, Yves Aubry.*

During the breeding season, the best way to detect the species is by listening for its song. Like other warblers, the Black-throated Green has two different song types which it uses in different contexts (Pitelka 1940; Morse 1967c; Lemon et al. 1987; Morse 1989a). The B song type, a *zee zee zee zee zoo zee*, is sung most often at the start of the breeding season, 5 or 6 times a minute, while the male forages in the midlevels of the vegetation. This song is used to establish territorial boundaries or when a female is present. The A song type, a slower *zoo zee zoo zoo zee*, is sung about 3 times a minute from a conspicuous song perch high in the tree. This song is used for territorial defence, often during confrontations and chases with intruders in attempts to drive them off the territory (Kendeigh 1945). Aggressive behaviour is reported to be much more intense in *Dendroica* warblers than in other warbler genera (Morse 1967c). Occasionally an incomplete A song type is sung, which can be easily confused with the Black-throated Blue Warbler's song. Singing continues until at least mid-July (Crins 1987a).

Nest building is initially a joint effort but apparently the female completes the nest alone (Pitelka 1940). Conifers are preferred for nesting, but deciduous trees may also be used (Ellison 1985aa). The nest is built on a horizontal branch, near the trunk, 6–10 m above the ground, or occasionally up to 20 m or more (Harrison 1978; Ehrlich et al. 1988). The deep cup is made of thin bark strips, twigs, mosses, grasses, lichens, rootlets, and spiderwebs and is lined with hair, feathers, grasses, and rootlets. Curled, papery strips of white birch bark are woven into the exterior (Harrison 1975).

The more secretive and less visible female incubates the 4 or 5 eggs alone for the 12-day incubation period (Pitelka 1940; Harrison 1978). After the eggs hatch, the adults are easier to observe because both the male and female actively search for food for the nestlings, mainly insects and insect larvae. Most breeding confirmations (70%; 134/171) were based on reports of an adult carrying food. The male makes fewer trips to the nest with food than the female but brings more on each trip (Pitelka 1940). On approaching the nest, the male utters a wispy song or a series of soft chips to let the female know he is back. She then leaves the nest to let the male feed the young, though sometimes she remains to help him distribute the food (Pitelka 1940). The Black-throated Green Warbler is an uncommon cowbird host (Ehrlich et al. 1988).

Black-throated Green Warblers are particularly noisy when foraging, chipping incessantly, with the male singing frequently early in the breeding season (MacArthur 1958). They are more adaptable in finding and catching insects than many other insectivorous species (Petit et al. 1990). Foraging flights are generally short but frequent, with the birds typically using a rapid peering method of foraging, scanning the foliage and flying up and hovering under the branch to pick insects off the leaves (MacArthur 1958). They usually concentrate their efforts on the surface of the leaves or needles in the densely foliated exterior part of the branch on the outside of the tree (MacArthur 1958; Robinson and Holmes 1982).

Although the Black-throated Green Warbler shares the same habitat and even uses the same trees as other *Dendroica* warblers, it exploits a distinct ecological niche (Morse 1989a). This warbler concentrates its foraging and other activities, such as nest construction, in the upper middle zones of the tree (MacArthur 1958), leaving the top and lower zones and the area near the trunk to other warblers (Morse 1989a). However, its niche may partially overlap those of other species, resulting in interspecific competition (Morse 1976a, 1985). The more aggressive Black-throated Green Warbler is usually dominant over the other warblers inhabiting coniferous forests, in particular the Blackburnian Warbler, Yellow-rumped Warbler, and Northern Parula (Morse 1971b, 1976a, 1989a).

Distribution

In Canada, the Black-throated Green Warbler breeds in northern Alberta, central Saskatchewan, central and southern Manitoba, Ontario, and Québec, and the Maritimes and Newfoundland; thus it can be found throughout the southeastern part of the country. In the United States, it ranges from the Great Lakes states in the west to the northeastern states in the east, and from the Appalachian Mountains south to Georgia and Alabama (Godfrey 1986). The northern boundary of its range corresponds roughly to the northern limits of the boreal forest.

The most northerly records originate from the James Bay region, where the species has been observed in stands of shade-intolerant hardwoods around Lakes Waswanipi, Olga, and Soscumica, at the headwaters of the Nottaway River, as well as near Lake La Bardelière on the upper stretches of the Rupert River (Cons. G. G. – GREBE 1992o). Todd (1963) also observed the Black-throated Green Warbler on the Rupert River. Farther north, it has only been reported twice in the Eastmain River region (Bordage and Aubry 1982; Tremblay and Belles-Isles 1991). Farther east, its range extends to the Goose Bay region in southern Labrador (Todd 1963; Godfrey 1986). In Ontario (Crins 1987a) and the Maritimes (Erskine 1992), it occurs in coniferous and mixed forests but is absent in small woodlots (Peck and James 1987).

During the Atlas survey, the Black-throated Green Warbler was observed in all regions of southern Québec, but was reported more often in the southern part of the Atlas area, which contains mixed forests as well as deciduous forests interspersed with stands of conifers. Frequency of occurrence was greatest in the Eastern Townships, Appalachians, and Laurentians, particularly in the Sugar maple-yellow birch forest regions. Farther north, it is also present in the Balsam fir regions, where it is most often found in preferred habitats such as mixed stands of balsam fir, white spruce, and

Notes on territory (continued)

4.8 pairs/10 ha in a balsam fir stand in La Mauricie National Park (DesGranges 1980b), and between 2.5 and 9.9 pairs/10 ha in mixed forests in various other regions (Erskine 1971a, 1980a, 1984).

The Black-throated Green Warbler, which dominates any other *Dendroica* warblers present, occupies mainly the midlevels of the tree, forcing other warblers to forage higher or lower. Mount Albert, Gaspé Conservation Park. *July 1991, Yves Aubry.*

poplar, or riparian habitat of white spruce and shrubs (Erskine 1977). Its frequency decreases gradually nearing the Black spruce regions, where it uses stands of shade-intolerant hardwoods (Cons. G. G. – GREBE 1992o).

Numbers also drop considerably in the agricultural areas of the St. Lawrence Plain, where undisturbed forests are rare. In these areas, the species frequents rich, conifer-dominated sites, particularly those with hemlock, pine, fir, Eastern white-cedar, and spruce, a phenomenon which has also been observed in Ontario and the Maritimes (Erskine 1977, 1992; Crins 1987a).

Breeding has also been confirmed on Anticosti Island, where a large number of sightings have been made. The Black-throated Green Warbler is considered to be common there, particularly in regenerating conifer stands composed of fir, birch, aspen, spruce, and a well-developed shrub layer (Ouellet 1969). On the Magdalen Islands, it is considered a migrant and regular breeder, although breeding has only been confirmed once and confirmation was not obtained during atlassing (Fradette 1992). On the Magdalen Islands, the species is limited to white spruce stands.

The wide range of habitats used by the species is not surprising, since it appears to select habitat based on the availability of suitable food resources (Futuyma and Gould 1979). MacArthur (1958) and Morse (1971b) both concluded that, compared with other warblers, the Black-throated Green Warbler displays quite stereotypic behaviour in exploiting its habitat, thus limiting the ways it can use the resources available. However, data obtained by Petit et al. (1990) demonstrated that this species shows great adaptability in foraging and in catching insects, which enables it to respond more easily to variations in resource availability and to modifications in the vegetation structure. Therefore, the Black-throated Green Warbler is able to frequent stands of variable structure and composition, depending on the specific site occupied (Collins 1983b).

History and trends

At the turn of the century, the Black-throated Green Warbler was considered to be common in southern and eastern Québec and in the Maritimes (Macoun and Macoun 1909). Both Hall (1862) and Wintle (1896) described it as a scarce summer resident in the Montréal region, but Wintle added that it was common during spring migration. According to Ouellet (1974a), the species has never been numerous in summer in the Montréal region, but may have been more abundant before the turn of the century, since coniferous stands were no doubt more numerous and extensive at that time. Urban development has probably driven away the Black-throated Green Warbler from some of its former breeding grounds (Elliot 1958).

An analysis of BBS data showed little change in the size of breeding populations in the Maritimes since 1966 (Robbins et al. 1986); numbers seem to increase during spruce budworm outbreaks, but perhaps not as much as for the other spruce-woods warblers (Erskine 1992). In fact, populations sometimes decrease during an epidemic (Kendeigh 1947; Morris et al. 1958). BBS data for the eastern United States and North America as a whole up to 1979 do not show any significant population changes (Robbins et al. 1986). Maine was the only state where an increase occurred; the species appeared to reach its greatest abundance in the eastern section of the spruce-hardwood forest and in the Adirondack Mountains (Robbins et al. 1986). The species' range in New York State appears to be expanding, due to the additional habitat which has become available with the maturing of conifer plantations (Eaton 1988e). ÉPOQ data between 1970 and 1989 do not show any significant change in the breeding population for Québec (Cyr and Larivée 1993a).

It has been suggested that the Black-throated Green Warbler only frequents undisturbed forests (Clark et al. 1983). A study in deciduous woodland in Nova Scotia showed that this warbler is scarce in areas that have been strip-cut and absent from those that have been recently clearcut or thinned extensively (Freedman et al. 1981). Studies in fir-dominated coniferous forests in Québec show that the species moves into areas that have been strip-cut 5–15 years after cutting (G. Falardeau and J.-P. L. Savard, unpubl. data). However, it may be present in small numbers in the early successional stages after clearcutting, though not immediately afterward (Titterington et al. 1979). After selective cutting, populations appear to decline only when the cutting has been intensive, with a light cut having no effects (Webb et al. 1977). In a study carried out in a second-growth sugar maple-yellow birch forest, 4 areas were studied that had been clearcut and were in various stages of succession, some as old as 70 years; the Black-throated Green Warbler was found to be most common in the 70-year-old forests (Darveau et al. 1982). After clearcutting, in both mixed and coniferous forests, density decreases during the early successional stages, whereas it increases in deciduous forests (Derleth et al. 1989). In another Nova Scotia study of 23 stands, ranging from 1 to 74 years old, this warbler demonstrated a marked preference for stands around 20 years old, followed by more mature stands (Morgan and Freedman 1986). Near inhabited areas, owing to its extreme sensitivity to habitat size, the Black-throated Green Warbler may be becoming increasingly rare because of the steady decrease in the size of woodlots (Morse 1976b; Askins 1993).

Habitat changes on the wintering grounds appear to pose little threat to the species, especially given the expectation of a 12% increase in winter habitat between the years 1985 and 2000 (Diamond 1991).

Alain Lanoue and Pierre Lafontaine

Blackburnian Warbler

Paruline à gorge orangée

Dendroica fusca

Christiane Girard

EARLY IN THE MORNING IN SPRINGTIME, WHEN THE FIRST RAYS OF SUN strike the spires of the conifers, observers may be drawn to a brilliant orange flame burning amidst the dark foliage. This is the Blackburnian Warbler, whose striking orange throat has earned him the nickname "fire throat." In his characteristic breeding plumage, the male's flaming orange throat contrasts strongly with his black cheeks and upperparts.

In southern Québec, this warbler is a migrant breeder, frequenting forests with conifers. It winters from Guatemala to Bolivia (AOU 1983; Godfrey 1986).

Habitat and behaviour

The Blackburnian Warbler's song is high pitched, thin, and wiry (Godfrey 1986; Lemon et al. 1987). The species has two song types which are used in different contexts. The first, a *zip zip zip titi tseeeeee*, ends on a very high upslurred note inaudible to some ears (Peterson 1980). It is usually associated with territoriality and is sung from the top of a tall conifer (Morse 1967c). The second song type, a *teetsa teetsa teetsa teetsa zizizizi* (Peterson 1980), is sung mainly while foraging (Morse 1967c). According to Ficken and Ficken (1962), upper strata dwelling species like the Blackburnian Warbler have a high-pitched song that propagates well because it encounters less obstruction high up in the tree tops. Warblers occupying the lower strata have a lower pitched song that is an adaptation to their more acoustically cluttered environment, since the lower frequencies carry longer distances and penetrate barriers more easily.

During the breeding season, this spruce-woods warbler (Morse 1976a, 1989a) frequents forests with conifers. In Canada, it is associated with balsam fir (Erskine 1977), whereas in the United States, it is associated more with spruce and hemlock (Sherry and Holmes 1985; DeGraaf and Rudis 1987; Morse 1989a). The Blackburnian Warbler favours mature stands (Erskine 1977; Titterington et al. 1979; Doepker et al. 1992) where taller spires project above the canopy (Morse 1985). It rarely nests in forests with trees less than 18 m tall (Morse 1976b), and it avoids small islands, even when the forest cover is suitable (Morse 1971b). In the southern part of its breeding range in the United States, however, the species frequents forests with few or even no conifers (Harrison 1984).

The Blackburnian Warbler spends most of its time in the treetops (MacArthur 1958; Morse 1968). It is not known whether this is an ecological preference or due to competition from the Black-throated Green Warbler (Morse 1985), which nests and forages in the same habitat as the Blackburnian, but occupies the level just below. Naturally the two species' areas of activity overlap somewhat; in a 18-m-tall spruce, the Blackburnian Warbler will feed at a height of between 9 and 16 m (50–90% of the height of the tree), whereas the Black-throated Green Warbler will feed between 5 and 13 m (30–70% of the height of the tree) (Morse 1967c). Since the Black-throated Green Warbler

Notes on territory

The female does not participate in defending the breeding territory, except for a small area around the nest from which she excludes conspecifics of either sex as well as other warblers during laying and incubation (Kendeigh 1945).

In mixed forests, the Blackburnian Warbler is common in mature balsam fir stands, where it seeks tall spires towering over the rest of the canopy. Tantaré Ecological Reserve, near Québec City.
Summer of 1990, Ministère de l'Environnement du Québec.

Blackburnian Warbler

Breeding distribution

1984–1989

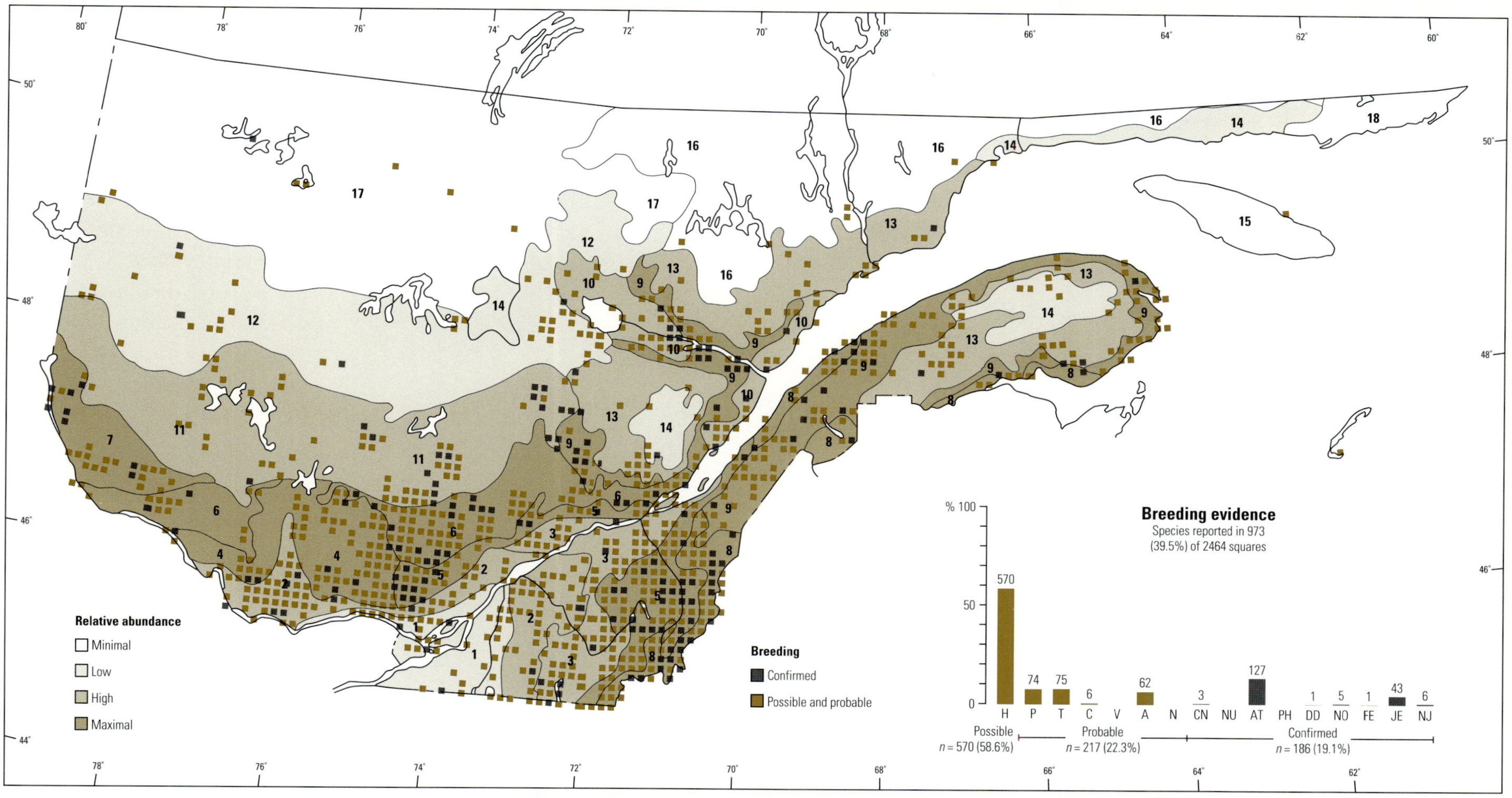

Species profile

Clutch size	usually 4 eggs (1–5)[3,5,8,9,10]
Laying interval	?
Start of incubation	before last egg[2]
Incubation period	12–13 days[2,3,6,8]
Maturity at hatching	altricial
Care of young	by female and male[2,6,10]
Nestling period	?
Age at first flight	upon leaving nest[1]
Dependence of young	?
Broods per year	1[7]
Breeding age	1 year[7]
Mating system	monogamy[4]
Duration of pair bond	?
Total length	11.4–14.0 cm[8]
Wing span	19.1–21.6 cm[9]
Weight	M 10.1 g (9.5–12.6) / F 10.3 g (8.9–11.9)[1]
Longevity record	8 years 2 months[11]

References:
1. Morse 1994
2. Kendeigh 1945
3. Peck and James 1987
4. Lemon et al. 1987
5. Bent 1953
6. Lawrence 1953b
7. DeGraaf and Rudis 1987
8. Godfrey 1986
9. Terres 1980
10. Harrison 1978
11. Klimkiewicz and Futcher 1989

Chronology $n = 24$

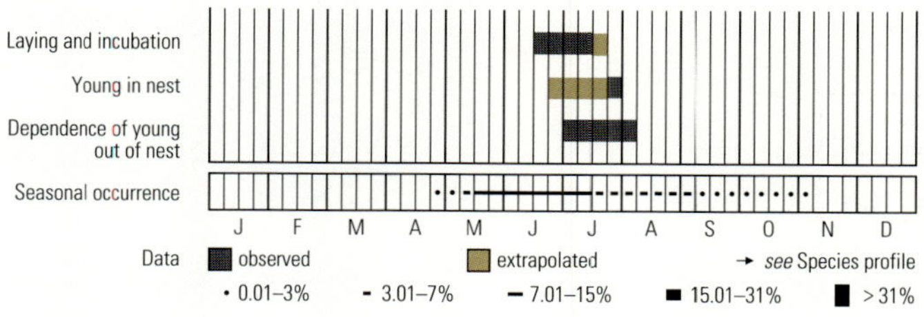

World breeding range

Frequency of occurrence by ecological region

Ecological regions** (forest regions)	Adequately covered squares	Squares in which reported	% of squares in which reported	Relative abundance (%)*
1. Sugar maple-hickory and sugar maple-basswood	97	33	34.0 ± 4.7	34.8
2. Sugar maple-basswood	153	102	66.7 ± 3.9	68.1
3. Sugar maple-basswood and sugar maple-yellow birch	101	76	75.2 ± 5.2	76.9
4. Sugar maple-yellow birch-basswood	56	51	91.1 ± 6.9	93.1
5. Sugar maple-yellow birch-beech and sugar maple-yellow birch-basswood	113	101	89.4 ± 3.3	91.3
6. Sugar maple-yellow birch-beech	71	66	93.0 ± 5.8	95.0
7. Sugar maple-yellow birch-balsam fir	19	18	94.7 ± 12.0	96.8
8. Sugar maple-yellow birch and balsam fir-yellow birch	47	46	97.3 ± 4.5	100.0
9. Balsam fir-yellow birch	87	78	89.7 ± 6.1	91.6
10. Balsam fir-white birch or balsam fir-red maple	46	36	78.3 ± 10.5	80.0
11. Yellow birch-balsam fir	52	45	86.5 ± 9.9	88.4
12. Balsam fir-white birch (west)	59	31	52.5 ± 13.1	53.7
13. Balsam fir-white birch (east)	76	47	61.8 ± 10.7	63.2
14. Balsam fir-black spruce	28	10	35.7 ± 18.6	36.5
15. White spruce-balsam fir and balsam fir-white spruce	22	1	4.5 ± 10.3	4.6
16. Black spruce-balsam fir-moss	18	4	22.2 ± 22.2	22.7
17. Black spruce-moss	32	4	12.5 ± 13.0	12.8
18.			*not analysed*	

* minimal: 0–29.9%; low: 30–59.9%; high: 60–89.9%; maximal: 90–100%

** Deciduous forest: 1–7; mixed forest: 8–11; coniferous forest: 12–17

The Blackburnian Warbler nests and feeds in the upper levels of mature conifer forests. Mount Saint-Alban, Forillon National Park, Gaspé Peninsula. *July 1991, Yves Aubry.*

Notes on territory (continued)

Territories were found to range between 0.4 and 0.5 ha in size in a spruce forest in Maine (Morse 1976b), and averaged 0.5 ha in a hemlock-beech forest in New York State (Kendeigh 1945); Elderkin (1989) reported territories of 1.1 ha. In Québec, territories were found to range between 0.6 and 1.7 ha in hemlock stands in Mirabel, with 1 ha being the average (Cyr 1974).

Also in Québec, densities of 10 and 15 pairs/10 ha were found at Woburn and Lachute in balsam fir stands with an admixture of deciduous species (Erskine 1984). Densities ranged from 3.2 to 8.2 pairs/10 ha in fir-spruce stands infested with budworms near Asselin Creek in the Abitibi region (Erskine 1980a), and densities of 7.5 pairs/10 ha were found in hemlock stands with an admixture of deciduous species at Lachute and Saint-Augustin (Erskine 1984). Densities of 5.4 pairs/10 ha, 3.2 pairs/10 ha, and 2.7 pairs/10 ha were recorded, respectively, in a red spruce stand, in a white spruce plantation, and in a mixed forest in La Mauricie National Park (DesGranges 1980b). Densities of 4.8 and 4.9 pairs/10 ha were found in balsam fir stands in La Mauricie National Park and near Lac de l'Est in Kamouraska County, respectively (Erskine 1976a, 1980a). In hemlock stands in Mirabel, an average density of 2.7 pairs/10 ha was reported (Cyr 1974), compared with 2.5 pairs/10 ha in a mixed Eastern white pine-white oak stand at Oka and in a sugar maple-yellow birch stand at Woburn (Erskine 1984).

is dominant over the Blackburnian, the latter is forced to occupy the higher levels (Morse 1976a), where the foliage on which it can forage is sparser (Morse 1967c).

It takes the female Blackburnian Warbler 3–6 days to build the nest (Harrison 1984), which is made of conifer twigs and needles, bark strips, spiderwebs, grass, lichens, and plant down. The interior is lined with fine grasses and rootlets (Peck and James 1987).

The nest is usually located on a horizontal branch near the top of a conifer (pine, hemlock, spruce, fir, or cedar), generally well out from the trunk (Harrison 1984; Peck and James 1987). Although the nest may be placed anywhere from 1.5 to 26 m above the ground, it is usually found between 6 and 15 m (Bent 1953; Peck and James 1987; Ehrlich et al. 1988).

Clutch size is generally 4 eggs (Terres 1980; Godfrey 1986), which the female incubates for 12 or 13 days (Godfrey 1986). Both parents feed the young (Lawrence 1953b; Terres 1980), mostly on caterpillars (Morse 1985).

The male defends a territory that is used for mating, nesting, and foraging (Welty 1975). Although the female does not help the male in territorial defence, she defends a small area around the nest against other Blackburnian Warblers (of either sex) as well as against other warbler species. The area is defended most energetically during laying and incubation and less actively after hatching (Kendeigh 1945). In this species, interspecific territoriality (against other warbler species) is less intense than intraspecific territoriality (MacArthur 1958).

Blackburnian Warbler nests are rarely parasitized by the Brown-headed Cowbird because of the height at which they are built (Griscom and Sprunt 1979). According to Harrison (1984), however, the small number of instances of parasitism reported is due rather to the fact that the nest is difficult to find. Out of 12 nests found in Ontario, 4 (33.3%) had been parasitized (Peck and James 1987).

Although the Blackburnian Warbler consumes large quantities of spruce budworm larvae, insect populations seem little affected by this warbler during epidemics. Top of balsam fir ravaged by the spruce budworm. Gaspé Peninsula. *July 1990, Lina Breton, Ministère des Forêts.*

Lastly, in a mixed habitat consisting of woodland with openings in Gatineau Park, a density of 0.29 pair/10 ha was obtained (Dauphin 1985).

Like other warblers, the Blackburnian Warbler is mainly insectivorous (Ficken and Ficken 1962), making only short foraging trips (0.3–0.6 m) from branch to branch, enabling it to detect small or well-hidden insects that other birds have missed (Robinson and Holmes 1982).

The Blackburnian Warbler is primarily a foliage gleaner, pecking at the vegetation underfoot as it moves along (Robinson and Holmes 1982; Holmes 1986; Morse 1989a). Occasionally, it may hover to pick up insects it has detected on a nearby substrate during its movements from branch to branch (Robinson and Holmes 1982).

The Blackburnian Warbler's diet comprises mainly forest pests (Bent 1953), notably homopterans (chiefly leaf hoppers), the larvae and adult forms of butterflies and moths, in particular the spruce budworm, and flies (Robinson and Holmes 1982). It may also eat fruit and berries when insects are scarce (Ehrlich et al. 1988).

Although the Blackburnian Warbler shares the same habitat as the Cape May, Yellow-rumped, Black-throated Green, and Bay-breasted warblers, these birds' foraging behaviour is quite different, in terms of both the techniques and the zones of the tree used. This minimizes interspecific competition, as does the fact that foraging is carried out within the breeding territory (MacArthur 1958).

Blackburnian Warbler populations, like those of the Black-throated Green Warbler, may decrease during spruce budworm epidemics because densities of Bay-breasted and Cape May warblers generally increase during such outbreaks, and these two warblers are dominant over the Blackburnian Warbler in interspecific interactions (Kendeigh 1947; Morris et al. 1958; Morse 1989a).

The Blackburnian Warbler was reported in 39.5% (973/2464) of the Atlas squares surveyed, and of those breeding was confirmed in 19.1% (136/973), mainly by reports of adults carrying food to their young.

Distribution

In Canada, the Blackburnian Warbler ranges from Newfoundland to Saskatchewan. In the United States, it nests in the Great Lakes states, the Appalachians, and along the eastern seaboard south to Georgia (AOU 1983).

The Blackburnian Warbler breeds throughout southern Québec. The likelihood of observing this species is quite high in deciduous forest with conifers—except in the Montérégie region, where extensive deforestation has occurred—as well as in mixed forest, particularly where fir is found. The species becomes less common as the percentage of conifers increases and pure coniferous stands are found, particularly those composed of spruce. Thus it is not surprising that in the northern part of the Atlas area the species is considered an uncommon migrant breeder (David 1980a). Breeding was not confirmed on Anticosti Island or the Magdalen Islands, although the species nested on the latter in midcentury (Fradette 1992). North of the Atlas area, this warbler was observed twice in the summer of 1990 in the Great Whale River region, well to the north of its known breeding range (Cons. G. G. – GREBE 1991a).

History and trends

At the turn of the century, Dionne (1906) described the Blackburnian Warbler as uncommon in the wooded areas around Québec City, whereas in the Montréal region Wintle (1896) listed it as a common spring transient but absent in fall.

ÉPOQ summer data indicate that Québec populations increased between 1970 and 1989 (Cyr and Larivée 1993b).

The Blackburnian Warbler is generally associated with undisturbed forests (Webb et al. 1977; Titterington et al. 1979; Doepker et al. 1992). Webb et al. (1977) found that a 25% loss of timber volume from thinning brought about a significant decline in populations. However, a Québec study has shown that strip cutting does not result in a significant decrease in density (G. Falardeau and J.-P. L. Savard, CWS, unpubl. data). Forest fires are detrimental to the species, since they remove the canopy required for foraging (Apfelbaum and Haney 1981). Although extensive deforestation is occurring on the Blackburnian Warbler's wintering grounds, there is insufficient information, according to Holmes and Sherry (1988), to attribute the decline in breeding populations occurring in some parts of North America to this phenomenon. In fact, as Askins (1993) emphasized, forest fragmentation within the species' breeding range may also affect breeding populations.

Alain Lanoue and Diane Dauphin

Pine Warbler

Paruline des pins

Dendroica pinus

Christiane Girard

In Québec, the best place to look for the Pine Warbler is, not unnaturally, in mature pine woods. Pines used to be fairly common in the southern part of the province, but they were systematically felled in the course of the 19th century, and today this migrant's breeding distribution broadly reflects the scattered occurrence of pine stands in southwestern Québec.

This bird used to be called the *Pine-creeping Warbler* because of its slow and methodical progress up trunks and along branches in search of insects and pine seeds (Bent 1953). Observers familiar with bird calls will recognize its song, a short trill or series of rapid notes (Bent 1953), though it may be confused with those of two other summer residents in southern Québec, the Dark-eyed Junco and Chipping Sparrow (Ellison 1985s; Godfrey 1986). This warbler also has another song similar to that of the Yellow-rumped Warbler (Peterson 1988j). Unlike most other songbirds, the Pine Warbler sings throughout the winter (Root 1988).

Habitat and behaviour

The Pine Warbler is more selective than most other warblers about its breeding habitat, nesting only in pine forests (Bent 1953; Capen 1979; Peterson 1988j). It favours stands of tall, mature pines where the trees are well spaced with a sparse understorey (Anderson and Shugart 1974; Conner et al. 1983; Evers 1991b). Although it confines itself to pines, it nests in a wide variety of pine species (Bent 1953). In Québec and Ontario, it nests chiefly in Eastern white pine and occasionally in red pine (Eagles 1987d). In the United States, it also uses pitch, loblolly, jack, shortleaf, and scrub pines (Bent 1953; Morse 1974; Conner et al. 1983). It does not need pure stands, however, as it has been observed in a Maryland forest dominated by hardwoods with only 10% pines, where it ignored the hardwoods, flying from one pine tree to another (Morse 1989a).

The female builds a cup-shaped nest 3–25 m above the ground on a horizontal branch or in a cluster of needles at the end of a branch (Griscom and Sprunt 1979). In Ontario, 5 out of 12 nests found were in Eastern white pines, 4 in red pines, 2 in jack pines, and one in a spruce tree, most being 8.5–15 m above the ground (Peck and James 1987). Nests are made of weed stems, shreds of bark, or pine needles and are lined with pine needles, hair, or feathers (Griscom and Sprunt 1979).

The Pine Warbler is monogamous. Typically, a single clutch of 4 eggs is laid (Ehrlich et al. 1988), with the female doing most, if not all, of the incubating. Incubation takes 10–13 days. Both parents, however, raise the brood (Bent 1953; Ehrlich et al. 1988). The altricial young leave the nest at about the age of 10 days (Ehrlich et al. 1988).

As nests are often placed high up in tall pines they are hard to spot, and in fact, none were found during atlassing. Breeding was confirmed in only 17% (17/101) of the squares in which the species was reported. Confirmations were based on records of adults bringing food for their young (65%), young as yet unable to fly (29%), and adults carrying nest-building material (6%).

Notes on territory

The Pine Warbler is intimately associated with pine woods. Territory size is inversely related to the proportion of pines in the forest cover (Morse 1989a).

Erskine (1971a) reported densities of 5.5–8.2 pairs/10 ha in various types of pine stands in southern Ontario. In West Virginia, in contrast, an oak-hickory forest hosted 2.2–3.2 pairs/10 ha during the breeding season (Cooper et al. 1990), while 1.2–4.2 pairs/10 ha were recorded between May and July in New Jersey pine-oak forests (Brush and Stiles 1990).

The Pine Warbler is more selective in its choice of habitat than most other warblers and is largely confined in Québec to pine stands in the southwest corner of the province. It nests in these mature Eastern white pines near the Pin-Rigide Ecological Reserve, where singing males are often heard. Saint-Chrysostome, south of Montréal. *May 1991, Pierre Bannon.*

Pine Warbler

Breeding distribution

1984–1989

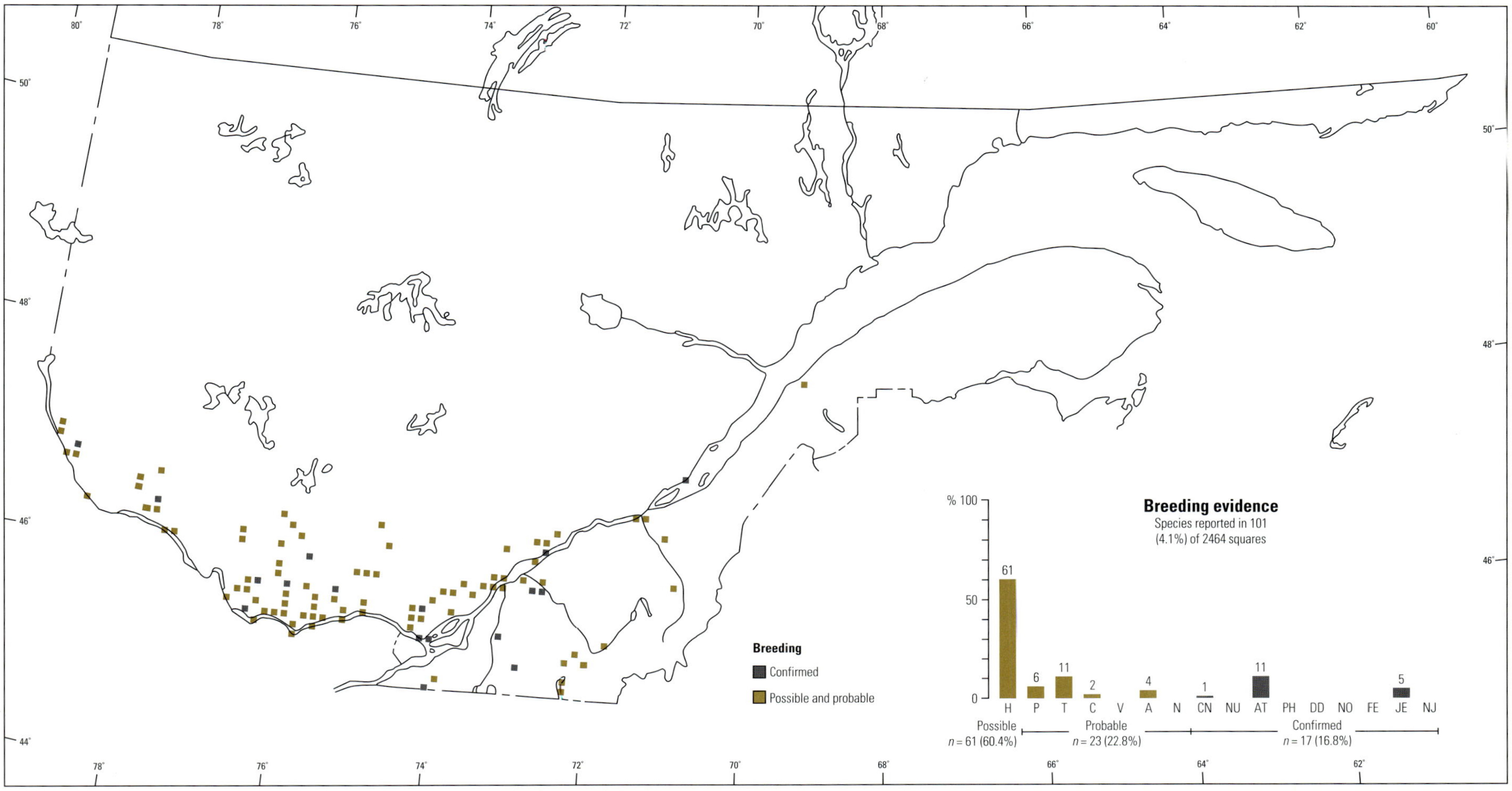

Species profile

Clutch size	usually 4 eggs (3–5)[1,3,4,5,7]
Laying interval	?
Start of incubation	?
Incubation period	probably 10–13 days[1,2]
Maturity at hatching	altricial
Care of young	by female and male[1,4,6]
Nestling period	probably ≈ 10 days[1]
Age at first flight	probably ≈ 10 days or more[10]
Dependence of young	?
Broods per year	usually 1–2[7]
Breeding age	1 year[2]
Mating system	monogamy[1]
Duration of pair bond	?
Total length	12.5–14.5 cm[3]
Wing span	?
Weight	11.9 g (9.4–15.1)[9]
Longevity record	6 years 10 months[8]

References
1. Ehrlich et al. 1988
2. DeGraaf and Rudis 1987
3. Godfrey 1986
4. Harrison 1978
5. Terres 1980
6. Chapman 1968
7. Griscom and Sprunt 1979
8. Klimkiewicz et al. 1983
9. Dunning 1984
10. according to nestling period

World breeding range

Chronology $n = 4$

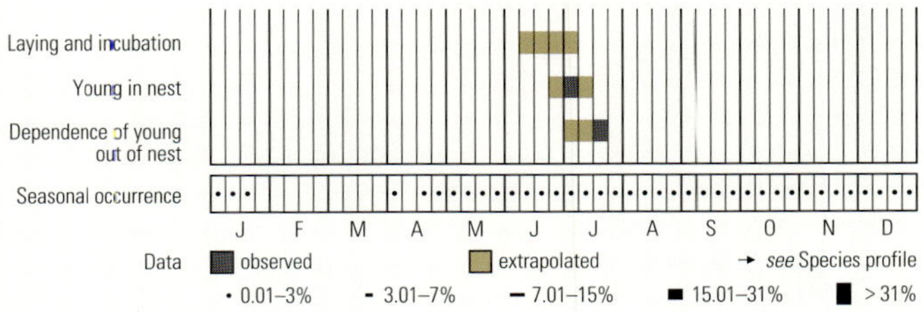

This warbler forages exclusively on pine trees. It typically builds its nest on a horizontal branch or in a cluster of needles at the end of a branch. Saint-Elphège, near Lake Saint-Pierre. *June 1991, Daniel Jauvin.*

The extensive pine stands of the Ottawa Valley and Témiscamingue, like this one on the shore of Lake Timiskaming, are crucial to the species' survival in Québec. *July 1986, Yves Aubry.*

In deciduous and mixed forests, the Pine Warbler settles in pine stands, like this one at Lake Boulrisse in the La Vérendrye Wildlife Reserve. *May 1991, Yves Aubry.*

The Pine Warbler winters in the pine forests of the southeastern United States (AOU 1983). Its wintering range extends farther north than that of most other warblers except the Yellow-rumped Warbler, which means that it may encounter quite severe winter weather conditions (Morse 1989a). A high mortality rate was recorded for this species during the very cold winter of 1976–77 (Robbins et al. 1986). The Pine Warbler is extremely aggressive in winter and attacks not just conspecifics but also other ecologically similar species such as the Brown-headed Nuthatch and Yellow-throated Warbler (Morse 1967b; Gaddis 1983).

Distribution

In North America, the Pine Warbler breeds from southern Québec to Florida, and as far west as southern Manitoba and south to eastern Texas (Griscom and Sprunt 1979). There are also breeding populations in the Bahamas and Haiti (Godfrey 1986).

In Québec, the Pine Warbler breeds only in the southwestern part of the province. It was reported in only 101 (4%) of the 2464 squares surveyed. Most of these squares were in the Ottawa Valley, probably because of the preponderance of pine stands in that region (Club des ornithologues de l'Outaouais 1985). Godfrey (1986) showed the Pine Warbler breeding throughout the Ottawa Valley to just east of Montréal; the birds were atlassed, however, in several squares between Montréal and Québec City, with even one instance of confirmed breeding as far east as Cap Tourmente, 50 km east of Québec City. It is uncertain, though, whether this warbler has always nested so far east or whether this represents a recent expansion of its breeding range. Dionne collected specimens around Québec City before 1889 (Macoun and Macoun 1909).

History and trends

Throughout the New England states, the status of the Pine Warbler varies locally from common to rare (DeGraaf and Rudis 1987). Unfortunately, human disturbance of nesting areas and the felling of pitch pines have relentlessly degraded this region's breeding habitat (Griscom and Sprunt 1979). In Québec, the sparse BBS and ÉPOQ data on the Pine Warbler are insufficient to identify population trends over the last three decades. BBS data for all of southern Ontario show no trends between 1966 and 1989 (Dunn 1991).

The fate of the Pine Warbler depends on that of mature pine stands. The history of the Pine Warbler in Québec is doubtless similar to that of Ontario: prior to the 19th century it was probably abundant, but since declined as the pine forests were cut down. As the forests have recovered to a certain extent, the Ontario Pine Warbler population has probably increased slightly. This warbler is starting to colonize 50-year-old pine plantations in the southwest of that province (Eagles 1987d), and the species has also recently started breeding in New Brunswick (Erskine 1992). There is every reason to suppose that Québec's Pine Warbler population will follow the pattern set in Ontario.

Michael Spencer and Richard C. Cotter

Palm Warbler

Paruline à couronne rousse

Dendroica palmarum

An early spring migrant, the Palm Warbler arrives in Québec with the first wave of warblers at the end of April or beginning of May. This denizen of northern peatlands can be identified by its chestnut cap and constantly twitching tail, unique among our warblers. It appears to breed only locally in most parts of southern Québec, but is more common in the northern part of its breeding range, which is remote and difficult to access. The Palm Warbler winters mainly from the southeastern United States south to the Yucatan Peninsula, as well as in the Greater Antilles and Bahamas (AOU 1983).

Habitat and behaviour

The Palm Warbler's variable trill is somewhat reminiscent of the trill of the Chipping Sparrow or Pine Warbler, but is shorter, weaker, and buzzier. At the start of the breeding season, this warbler usually sings from a conspicuous perch at the top of a spruce but, as the season progresses, sings from lower in the shrubbery. A male may have from 10 to 30 favourite song perches (Welsh 1971).

In the area around the Nottaway, Broadback, and Rupert rivers and the Opinaca Reservoir, the Palm Warbler breeds in a wide range of open habitats, generally damp and scattered with shrubs, including several types of peatlands, burns, spruce forests with moss and lichen ground cover, shade-intolerant hardwood stands, riparian shrub habitats, and the ecotones between coniferous forests and burns, peatlands, and lakes (Tremblay and Belles-Isles 1991; Cons. G. G. – GREBE 1992o). South of the boreal forest, it frequents mostly peatlands with clumps of black spruce or tamarack (Tyler 1953c; Welsh 1987d; Erskine 1992). In peatlands in Maine, the highest densities of these warblers were found in areas with a ground cover of sphagnum moss, a well-developed ericaceous shrub layer, and a 25–50% forest cover consisting of black spruce, tamarack, and Eastern white pine (Stockwell 1994). The species was also very common in peatlands with only 5–25% forest cover. In large open peatlands, the Palm Warbler nests on the edges of the nearby forest, eschewing breeding territories consisting exclusively of treeless areas (Welsh 1971, 1987d). A number of authors have recognized the importance of the wooded margin as an integral component of the Palm Warbler's territory (see Welsh 1987d).

The nest is usualy built on the ground, under a shrub or in the lower branches of a small tree (Taverner 1922; Tyler 1953c; Peck and James 1987). It is made of plant stalks, grasses, bark strips, and moss (Harrison 1978, 1984). Both parents incubate the clutch of 4 or 5 eggs for 11 or 12 days, and feed the nestlings. The young leave the nest at 10–12 days of age (Tyler 1953c; Welsh 1971; Ehrlich et al. 1988). Occasionally, a second brood may be produced (DeGraaf and Rudis 1987), even in Québec (S. Haddad and A. Desrochers, Laval U., unpubl. data).

The Palm Warbler feeds on insects, which it captures from the ground or in low shrubs (Ficken and Ficken 1962; Terres 1980), in the more open sections of its territory (Welsh 1987d).

Christiane Girard

Notes on territory

There is little information on territoriality in this species. If the Palm Warbler is like other warblers, the territory is probably used for courtship, nest building, brood rearing, and foraging (Morse 1989a). In Nova Scotia, territories ranged between 1.02 and 2.14 ha and did not appear to be defended by the female (Welsh 1971). A male was observed defending a flowering tree in Florida in March (Wunderle 1978), suggesting territorial behaviour may occasionally be observed in winter. In a tamarack peatland near Val-d'Or, breeding densities were estimated at approximately 0.5 pair/10 ha (Erskine 1972a). In a peatland with scattered black spruce near Dolbeau, in the Lac-Saint-Jean region, 4 pairs of Palm Warblers occupied territories of 1.3, 3.4, 3.6, and 13.2 ha (S. Haddad and A. Desrochers, Laval U., unpubl. data).

In the Central St. Lawrence Lowland, peatlands frequented by Palm Warblers are few and far between. Plée de Saint-Charles, a bog east of Lévis. *October 1993, Pierre Normand.*

Palm Warbler

Breeding distribution

1984–1989

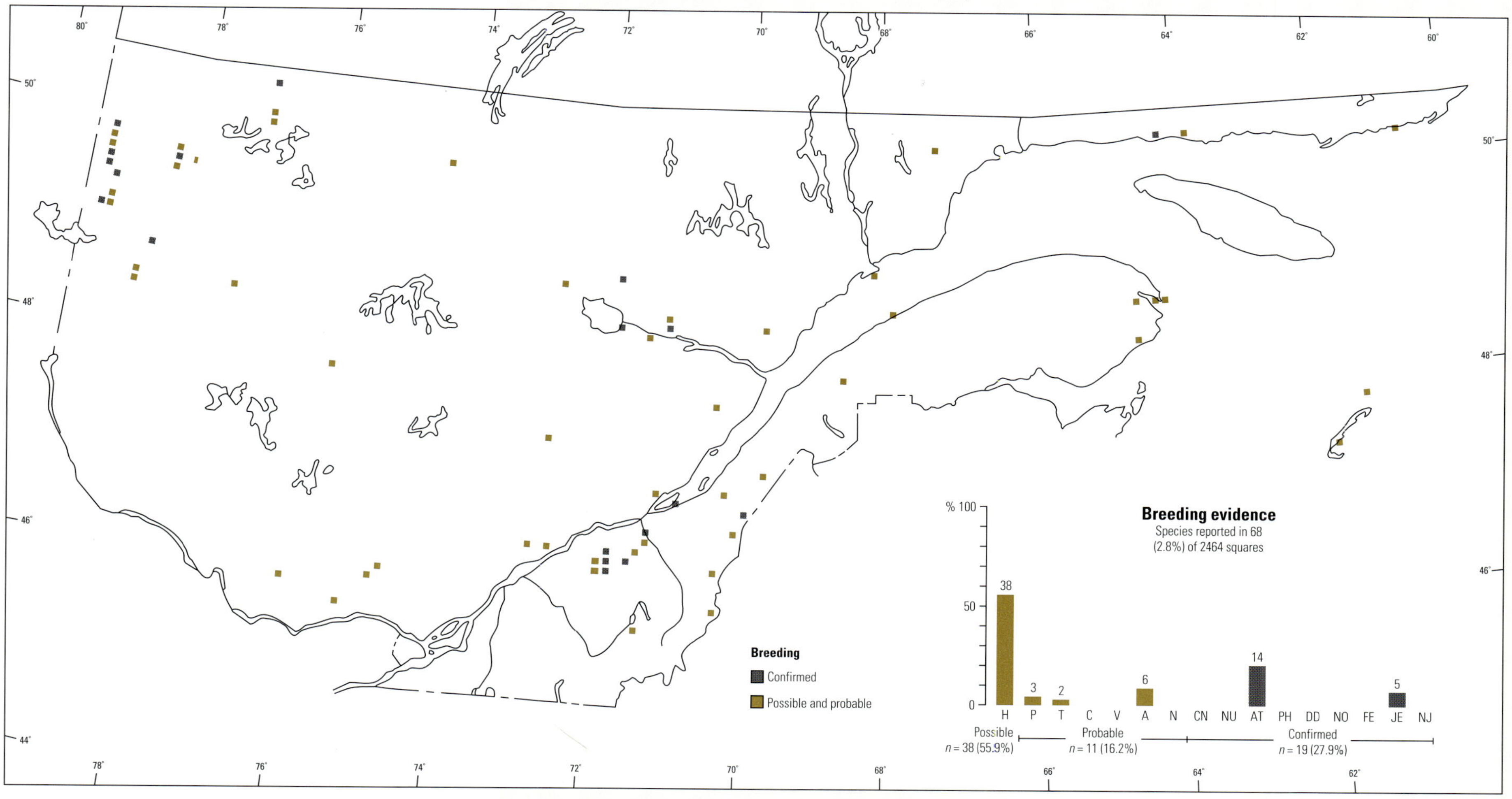

Species profile

Clutch size	usually 4–5 eggs (3–5)[1,2,3,5]
Laying interval	?
Start of incubation	?
Incubation period	11–12 days[1,2,3,5]
Maturity at hatching	altricial
Care of young	by female and male[1,4,5]
Nestling period	10–12 days[1,4,5]
Age at first flight	12–13 days[2,5]
Dependence of young	?
Broods per year	1–2[8]
Breeding age	1 year[8]
Mating system	monogamy, sometimes polygyny[1,5]
Duration of pair bond	?
Total length	12.5–14.5 cm[3]
Wing span	?
Weight	10.3 g (7.0–12.9)[7]
Longevity record	6 years 7 months[6]

References:
1. Ehrlich et al. 1988
2. Terres 1980
3. Godfrey 1986
4. Harrison 1978
5. Welsh 1971
6. Kennard 1975
7. Dunning 1984
8. DeGraaf and Rudis 1987

Chronology $n = 10$

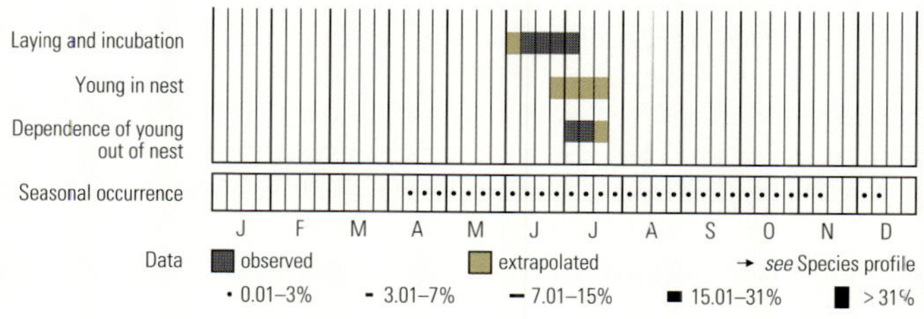

World breeding range

In open bogs, the Palm Warbler seeks small groves of trees to provide cover for its nest. Grande Plée Bleue, a bog near Pintendre, southeast of Lévis. *August 1992, Yves Aubry.*

Breeding was confirmed in 19 of the squares in which the species was recorded, mainly by reports of recently fledged young not yet capable of sustained flight, and of adults carrying food. No nests were found during the Atlas survey. The nest is difficult to find, since it is well hidden and the incubating bird leaves the nest long before an intruder gets close enough to observe it (Tyler 1953c).

Distribution

The Palm Warbler nests almost exclusively in the boreal forest (Erskine 1977). In Canada, it ranges from northern Alberta and the Northwest Territories east to Ontario, the southern half of Québec, and the Atlantic provinces (Godfrey 1986). In the United States, it is restricted to Minnesota, Wisconsin, Michigan, Maine, and New York states (AOU 1983; Carroll 1988c).

The Palm Warbler was reported in only 68 (2.8%) of Atlas squares surveyed, in fairly scattered locations. It is associated mainly with large peatlands, which are found only very locally in southern Québec. The species is probably somewhat more widespread than the Atlas map shows, however, since its habitat is often inaccessible and many observers are unfamiliar with its song. Breeding was confirmed for the Atlas in the Abitibi, Lotbinière, and Saguenay–Lac-Saint-Jean regions, on Île d'Orléans, near Lake Leverrier in the Appalachians, and near Longue-Pointe-de-Mingan on the North Shore. The species was also observed in scattered locations in southern Québec.

Information from other sources helps to clarify the species' status in regions where it was observed during the Atlas survey but not confirmed as a breeder, as well as in areas where it was not reported at all. For example, during the Atlas survey, the species was not reported in the Témiscamingue region, where it is considered a rare migrant (St-Hilaire 1991). It was atlassed in 4 squares in the Ottawa Valley, however, where traditionally it has been considered a regular migrant (Club des ornithologues de l'Outaouais 1985). It was not observed in the Montréal region, where it only breeds exceptionally (Bannon 1991b). It was observed in 2 squares in the Lower St. Lawrence region, where Gendron and Gauthier (1984) considered it an uncommon migrant and probable breeder. It also breeds irregularly in the Gaspé (Girard 1988), where it was observed in only 4 squares. Ouellet (1969) did not report the species on Anticosti Island; nor was it atlassed there, despite the fact that Anticosti Island has a number of peatlands. Although the species was sighted on the Magdalen Islands before and during the Atlas survey, breeding evidence has not yet been obtained (Fradette 1992).

The Palm Warbler's distribution extends north of the Atlas area, where it is much more common. During recent surveys southeast of James Bay, the species was confirmed as a breeder near Lakes Waswanipi and Nemiscau (Cons. G. G. – GREBE 1992o). Todd (1963) reported it breeding at Waskaganish, near Rupert Bay. It is common at least as far north as the Eastmain River area (Bordage and Aubry 1983; Tremblay and Belles-Isles 1991; Cons. G. G. – GREBE 1992o). The species has also been observed on the coast of James Bay and is thought to breed as far north as La Grande Rivière (Harper 1958; Todd 1963; Godfrey 1986). Two individuals were also seen much farther north, near the Kuuraapik River on the Ungava Peninsula on 7 August 1983 (ÉPOQ).

The nest, which is usually placed on the ground at the foot of a shrub, is made mainly of twigs and plant stalks. James Bay region. *July 1982, Vincent Létourneau.*

Although peatlands are its favourite haunt, the Palm Warbler also nests in black spruce swamps with scattered trees. La Grande-3, James Bay region. *July 1992, Réginald Ouellet.*

The Palm Warbler also frequents the wooded margins of open peatlands. Villebois, north of La Sarre, Abitibi. *June 1991, Desneige Perreault.*

History and trends

It is difficult to determine the Palm Warbler's past status in the province and whether populations have fluctuated, since a large portion of its breeding range, located in northern Québec, has been explored only recently and is still poorly known. At the turn of the century, the species appears to have been uncommon during the breeding season in southern Québec. However, the scarcity of breeding evidence and even of summer sightings at the time may be due simply to early ornithologists' poor knowledge of the species. The Palm Warbler was considered a rare "transient visitant" in the Montréal region (Wintle 1896) and Eastern Townships (Mousley 1916a) and a fairly common migrant in the Ottawa Valley (Lloyd 1924, 1944) and eastern Gaspé Peninsula (Ball 1954). Godfrey and Wilk (1948) reported that it bred locally in the Lac-Saint-Jean region, near Saint-Félicien.

In southern Québec, peatlands are fairly scarce and are being intensively exploited, which may lead to the destruction of the species' breeding habitat. If the peatlands disappear, local Palm Warbler populations in southern Québec will no doubt disappear as well.

Another threat to the species is tropical in origin. Diamond (1991) has predicted that around 50% of the species' winter habitat may be lost by the year 2000, if deforestation continues at the expected rate in these areas.

Jacques Ibarzabal and Alain Morrier

Bay-breasted Warbler

Paruline à poitrine baie

Dendroica castanea

The Bay-breasted Warbler's fortunes are closely tied to that of its main summer food, the spruce budworm. Because the species' abundance is strongly correlated with cyclical fluctuations in spruce budworm populations, it may vary widely from year to year and region to region (Morse 1978, 1980; Lehoux et al. 1981). This migrant breeder is one of the last warblers to return to Québec in spring from its wintering grounds, which extend from Panama to northern South America (AOU 1983).

The Bay-breasted Warbler frequents the midlevels of mature coniferous forests with a scattering of deciduous trees. This spatial organization allows the species to share the habitat with other warblers. Laurentide Wildlife Reserve, north of Québec City. *August 1991, Marcel Darveau.*

Habitat and behaviour

Although it is found in mixed woodlands, the Bay-breasted Warbler mainly frequents mature coniferous forests with a scattering of deciduous trees, wood edges and openings, and second-growth stands dominated by conifers (Godfrey 1986; Peck and James 1987; Darveau et al. 1995).

During the breeding season, the Bay-breasted Warbler is often found with other *Dendroica* warblers, including the Cape May, Blackburnian, Yellow-rumped, and Black-throated Green warblers. The similarities in the habitat occupied by these species were studied extensively by MacArthur (1958), who discovered that each species occupies a different niche within this habitat. According to MacArthur, the Bay-breasted Warbler usually forages in the midlevels of trees, closer to the trunk than the other species. It moves slowly and deliberately through the foliage, much like a vireo, often remaining in the same tree for a long period of time. It may even sit and wait for the budworms to descend on their threads (Morse 1978).

The cup-shaped nest is made from grasses, bark strips, and cocoons and is lined with hair and rootlets. It is usually affixed to a horizontal branch of a conifer, 2–7 m up (Harrison 1978). Generally 4 or 5 eggs are laid, but as many as 6 or 7 may be produced during years of spruce budworm outbreaks (MacArthur 1958; Morse 1989a). Only the female incubates, though the male frequently brings her food during this period. The male also helps feed the nestlings. When an intruder approaches, the female sits tight, waiting until the last minute to flush and feigning a broken wing (Mendall 1937).

Although the male sings for long periods from the same perch, he is difficult to detect because he remains in the inner part of the foliage fairly high in the tree. His weak song is high in frequency and thus inaudible to some people. The song may be confused with that of other warblers, including the Cape May Warbler, Black-and-white Warbler, and American Redstart.

The species was reported in 35.2% of Atlas squares surveyed (868/2464), with breeding confirmed in 224 squares, primarily by reports of adults carrying food (192/224), but also by observations of adults feeding recently fledged young (18/224).

Distribution

The Bay-breasted Warbler breeds in the narrow band of closed boreal forest extending across Canada from southeastern Yukon to the Maritime provinces and western Newfoundland in the north, and from northeastern British Columbia to Maine in the south, including the Great Lake states (AOU 1983; Erskine 1992). In Québec, the Bay-breasted Warbler breeds throughout the Atlas area, particularly in regions containing mixed and coniferous forests. The species appears to be less abundant in the St. Lawrence Plain and Ottawa Valley, probably because of the predominance of agricultural habitats and deciduous forests. Its distribution extends somewhat northward of the Atlas area; the Bay-breasted Warbler has been sighted near the Rupert River (Cons. G. G. – GREBE 1992o) as well as in Harrington Harbour in 1979 (ÉPOQ). According to David (1980a), the Bay-breasted Warbler nests mainly south of the 50th parallel.

The species was atlassed more frequently in areas that, according to Bordeleau and Boulet (1989, 1990), were subjected to spruce budworm outbreaks. This observation appears to be corroborated by a study in the northern New England states, which showed that the Bay-breasted Warbler mainly frequents stands where the budworm is abundant (Crawford et al. 1983). However, once the outbreak subsides, populations appear to be greater in balsam fir stands that have been treated with insecticides than in stands that have not been treated (DesGranges and Rondeau 1995).

History and trends

Most authors at the turn of the century, and just afterward, considered the Bay-breasted Warbler scarce or uncommon in Québec (Dionne 1889, 1906; Wintle 1896; Macoun and Macoun 1909). The species probably nested occasionally in the Québec City region. Abundance may have been underestimated, however, because the species is so hard to detect.

Population trends over the years are difficult to gauge because the number of individuals in a given area depends on the extent of the spruce budworm infestation. However, according to Morse (1989a), populations may have stabilized in northeastern North America over the past 45 years due to budworm control programs, which have kept budworm populations at low to medium levels, instead of letting them reach epidemic proportions.

In addition, the Bay-breasted Warbler has benefited from forest fire control programs, which allow stands to reach maturity, when they are more vulnerable to a spruce budworm attack.

René Séguin

Christiane Girard

Notes on territory

In the White Mountains of New Hampshire, the average size of breeding territories was found to be 1.5 ha (Sabo 1980).

In Québec, breeding pair density depends on the type and condition of the stand. In spruce and fir forests infested with spruce budworm, densities as high as 48.6 pairs/10 ha have been recorded (Erskine 1980a; Régnière and Welsh 1988), whereas densities in forests of the same type but with an unspecified degree of infestation ranged between 4.9 and 22.2 pairs/10 ha (Erskine 1976a, 1980a; DesGranges 1980b). In red spruce stands and white spruce plantations in La Mauricie National Park, a density of 3.2 pairs/10 ha was found (DesGranges 1980b). In a study in a mixed habitat consisting of woodland and open areas, density was 0.5 pair/10 ha (Dauphin 1985).

Bay-breasted Warbler

Breeding distribution

1984–1989

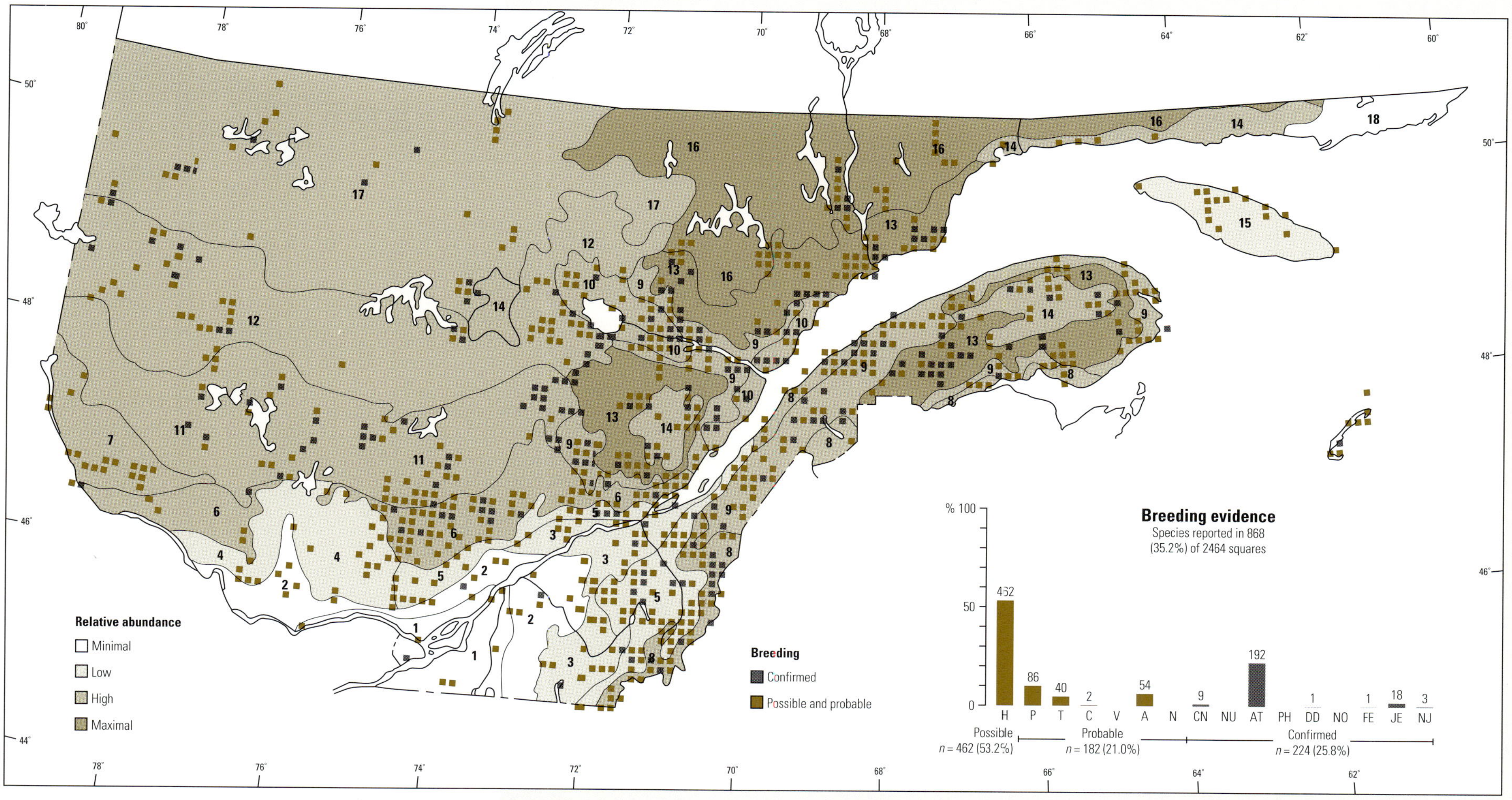

Species profile

Clutch size	usually 4–5 eggs (3–7)[2,3]
Laying interval	?
Start of incubation	before last egg[4]
Incubation period	12–13 days[1,2,3,4]
Maturity at hatching	altricial
Care of young	by female and male[3,4]
Nestling period	11–12 days[3,4]
Age at first flight	11–12 days[2]
Dependence of young	until at least 8 days after first flight[4]
Broods per year	1[5]
Breeding age	1 year[5]
Mating system	monogamy[3]
Duration of pair bond	?
Total length	12.5–15.3 cm[1]
Wing span	21.0–23.5 cm[2]
Weight	M 13.1 g (11.6–15.1) / F 12.0 g (10.7–13.6)[7]
Longevity record	3 years 5 months[6]

References:
1. Godfrey 1986
2. Terres 1980
3. Ehrlich et al. 1988
4. Mendall 1937
5. DeGraaf and Rudis 1987
6. Klimkiewicz and Futcher 1989
7. Dunning 1984

World breeding range

Chronology n = 14

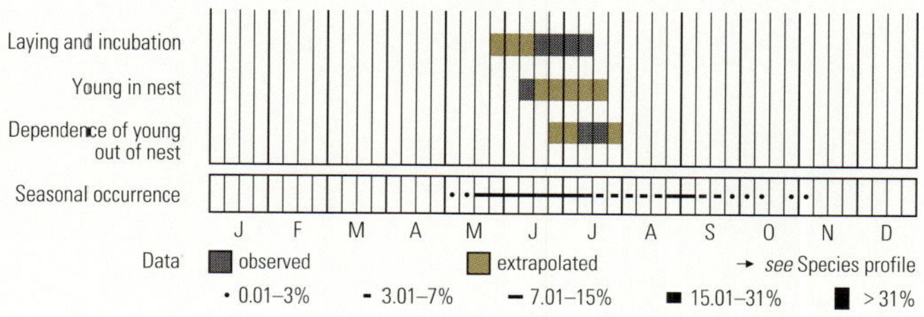

Frequency of occurrence by ecological region

Ecological regions** (forest regions)	Adequately covered squares	Squares in which reported	% of squares in which reported	Relative abundance (%)*
1. Sugar maple-hickory and sugar maple-basswood	97	9	9.3 ± 3.1	9.3
2. Sugar maple-basswood	153	22	14.4 ± 3.0	14.4
3. Sugar maple-basswood and sugar maple-yellow birch	101	31	30.7 ± 5.5	30.7
4. Sugar maple-yellow birch-basswood	56	17	30.4 ± 10.6	30.4
5. Sugar maple-yellow birch-beech and sugar maple-yellow birch-basswood	113	67	59.3 ± 5.1	59.3
6. Sugar maple-yellow birch-beech	71	51	71.8 ± 9.6	71.8
7. Sugar maple-yellow birch-balsam fir	19	14	73.7 ± 21.0	73.7
8. Sugar maple-yellow birch and balsam fir-yellow birch	47	38	80.9 ± 10.3	80.9
9. Balsam fir-yellow birch	87	73	83.9 ± 7.3	83.9
10. Balsam fir-white birch or balsam fir-red maple	46	39	84.8 ± 9.3	84.8
11. Yellow birch-balsam fir	52	40	76.9 ± 11.9	76.9
12. Balsam fir-white birch (west)	59	43	72.9 ± 11.7	72.9
13. Balsam fir-white birch (east)	76	69	90.8 ± 6.6	90.8
14. Balsam fir-black spruce	28	24	85.7 ± 14.1	85.7
15. White spruce-balsam fir and balsam fir-white spruce	22	11	50.0 ± 21.5	50.0
16. Black spruce-balsam fir-moss	18	18	100.0 ± 2.8	100.0
17. Black spruce-moss	32	24	75.0 ± 16.6	75.0
18.			not analysed	

* minimal: 0–29.9%; low: 30–59.9%; high: 60–89.9%; maximal: 90–100%

** Deciduous forest: 1–7; mixed forest: 8–11; coniferous forest: 12–17

Blackpoll Warbler

Paruline rayée

Dendroica striata

Christiane Girard

The Blackpoll Warbler has one of the northernmost breeding ranges of all the warblers, extending as far north as the tree line or, in Québec, to the Ungava Peninsula (Godfrey 1986).

The Blackpoll Warbler is easiest to observe during spring migration, when the male often perches in a conspicuous spot to sing his quite distinctive and easily recognized song. However, the song does not carry well and is so high pitched that it is inaudible to some people.

Each fall, this long-distance migrant undertakes an arduous journey of no less than 4000 km to return to its wintering grounds in South America, which extend from Colombia and Venezuela to eastern Peru, northern Argentina, and southern Brazil (AOU 1983).

Habitat and behaviour

The Blackpoll Warbler is primarily a northern breeding species that prefers open conifer woodland with old man's beard lichens growing on the trees and with a ground cover of sphagnum mosses. It also frequents stands of stunted black spruce, or krummholz, in mountainous and coastal areas (Tufts 1961; Sabo 1980; Elderkin 1989), as well as young regenerating conifer stands and the margins of burns and logged areas (Tufts 1961; Erskine 1977; Godfrey 1986). Todd (1963) found it to be also fairly common in alder swamps in the Great Whale River area in northern Québec. During recent surveys in this region, the Blackpoll Warbler was observed most often in shrubs in riparian habitat, coniferous stands on the margins of burns and lakes, peatlands, black spruce-moss forests, and recent burns (Cons. G. G. – GREBE 1991a). Farther south, around the Nottaway, Broadback, and Rupert rivers, it was found in peatlands (mainly tamarack peatlands), as well as in black spruce-moss and black spruce-lichen stands (Cons. G. G. – GREBE 1992o). In southern Québec, the species occurs mainly at higher elevations with coniferous habitats similar to those in boreal regions.

The nest, which is built by the female, is usually located against the trunk of a spruce and supported by one or two horizontal branches. Well concealed by the foliage, it is usually 0.6–4.0 m up (Peck and James 1987; Ehrlich et al. 1988), but may also be built directly on the ground in a clump of vegetation. The nest is made of grasses, twigs, lichens, fine moss, and bark, and is closely lined with plant fibres, rootlets, hairs, and feathers (Harrison 1978).

The female does all of the incubating but both parents feed the young. The young remain in the nest for 8–12 days after hatching (Harrison 1978); during this period, they are fed mainly insects. The Blackpoll Warbler sometimes captures insects on the wing like a flycatcher, but generally gleans them from conifer branches at midlevels (Gross 1953; Morse 1979). It also occasionally feeds on seeds and berries (DeGraaf and Rudis 1987).

The Blackpoll Warbler is an uncommon cowbird host as their breeding ranges overlap very little (Gross 1953). This occasionally polygynous warbler shows a strong attachment to its nesting site, to which it returns almost every year (Eliason 1986a).

In regions where both the Blackpoll Warbler and the Bay-breasted Warbler breed, they appear to exclude each other, the Blackpoll occupying mainly mountaintops and the Bay-breasted valleys and lower elevations. According to Morse (1979, 1989a), the two species cannot co-exist because they are morphologically very similar and exploit almost the same ecological niche. The Blackpoll Warbler also nests farther north and is probably better adapted than the Bay-breasted Warbler to the climatic conditions at higher elevations, which are similar to those found in boreal forests farther north (Sabo 1980).

Notes on territory

In a New Brunswick study, territory size ranged from approximately 0.5 to 1.5 ha, with 1.06 ha being the average (Eliason 1986b). In the White Mountains of New Hampshire, Sabo (1980) found that territory size decreased with elevation, while population density increased; the average territory was 0.4 ha and the average density was 15.4 pairs/10 ha. In New Hampshire, Morse (1979) found densities ranging between 1.5 and 25.3 pairs/10 ha, depending on the elevation. In Québec, densities of 7.5 pairs/10 ha were found on Mount Mégantic (Erskine 1984), and from 4.9 to 17.3 pairs/10 ha at different sites on the Gaspé Peninsula (Erskine 1980a).

The Blackpoll Warbler nests in stunted conifers, such as are found at high elevations and in coastal areas. Mount Albert, Gaspé Conservation Park. *July 1991, Yves Aubry.*

Blackpoll Warbler

Breeding distribution
1984–1989

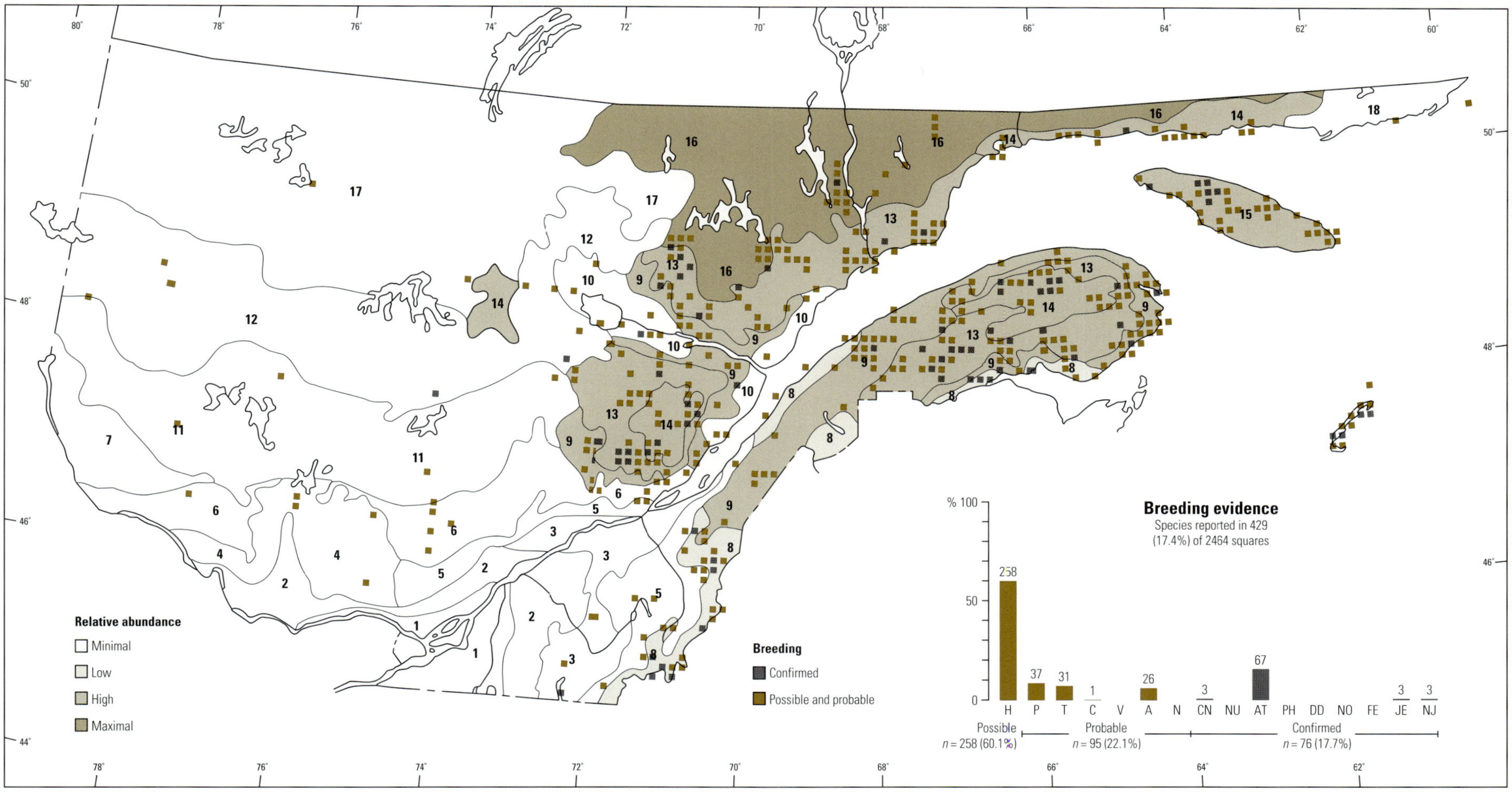

World breeding range

Species profile

Clutch size	usually 4–5 eggs (3–5)[1,2,3,4]
Laying interval	?
Start of incubation	before last egg[1]
Incubation period	11–12 days[1,2,4]
Maturity at hatching	altricial
Care of young	by female and male[1,2,4]
Nestling period	8–12 days[1,4,7]
Age at first flight	11–12 days[2]
Dependence of young	until 25 days after leaving nest[1]
Broods per year	1–2[1]
Breeding age	1 year[1]
Mating system	monogamy, sometimes polygyny[1,4]
Duration of pair bond	1 season[4]
Total length	12.5–14.5 cm[3]
Wing span	20.3–24.8 cm[2]
Weight	13.0 g (9.7–20.9)[6]
Longevity record	4 years 3 months[5]

References:
1. Eliason 1986b
2. Terres 1980
3. Godfrey 1986
4. Ehrlich et al. 1988
5. Klimkiewicz and Futcher 1989
6. Dunning 1984
7. Harrison 1978

Chronology $n = 36$

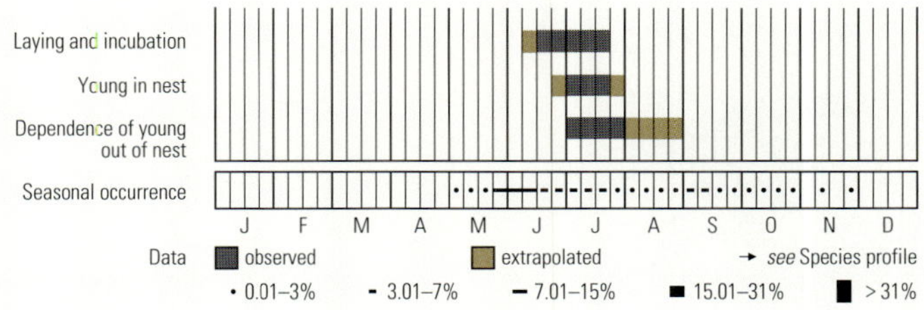

Frequency of occurrence by ecological region

Ecological regions** (forest regions)	Adequately covered squares	Squares in which reported	% of squares in which reported	Relative abundance (%)*
1. Sugar maple-hickory and sugar maple-basswood	97	0	0	0
2. Sugar maple-basswood	153	1	0.7 ± 0.9	0.7
3. Sugar maple-basswood and sugar maple-yellow birch	101	4	4.0 ± 2.6	4.0
4. Sugar maple-yellow birch-basswood	56	2	3.6 ± 4.8	3.6
5. Sugar maple-yellow birch-beech and sugar maple-yellow birch-basswood	113	9	8.0 ± 3.0	8.0
6. Sugar maple-yellow birch-beech	71	10	14.1 ± 7.6	14.1
7. Sugar maple-yellow birch-balsam fir	19	0	0	0
8. Sugar maple-yellow birch and balsam fir-yellow birch	47	20	42.6 ± 12.7	42.6
9. Balsam fir-yellow birch	87	55	63.2 ± 9.3	63.2
10. Balsam fir-white birch or balsam fir-red maple	46	11	23.9 ± 10.8	23.9
11. Yellow birch-balsam fir	52	6	11.5 ± 9.3	11.5
12. Balsam fir-white birch (west)	59	5	8.5 ± 7.7	8.5
13. Balsam fir-white birch (east)	76	63	82.9 ± 8.4	82.9
14. Balsam fir-black spruce	28	25	89.3 ± 12.6	89.3
15. White spruce-balsam fir and balsam fir-white spruce	22	17	77.3 ± 18.4	77.3
16. Black spruce-balsam fir-moss	18	18	100.0 ± 2.8	100.0
17. Black spruce-moss	32	1	3.1 ± 7.6	3.1
18.			*not analysed*	

* minimal: 0–29.9%; low: 30–59.9%; high: 60–89.9%; maximal: 90–100%

** Deciduous forest: 1–7; mixed forest: 8–11; coniferous forest: 12–17

Breeding was confirmed in only 18% of the squares in which the species was reported, and in most cases (88%) confirmation was based on observations of adults carrying food to their young. In comparison with other warbler species, very few recently fledged young were seen.

Distribution

Among warblers, the Blackpoll Warbler's breeding distribution is one of the most northerly. In Canada, it nests in the Atlantic provinces and in the northern regions of all other provinces, as well as in the Northwest Territories and Yukon (Godfrey 1986). In southern Québec, the Blackpoll Warbler is much more abundant in the east, including the Appalachians, from the Eastern Townships, to the Gaspé Peninsula, Magdalen Islands, Anticosti Island, North Shore (both on the coast and inland), and in the Laurentide Wildlife Reserve north to the Saguenay River. The higher and more undulating relief and wetter climate in this part of the province (Thibault 1985, 1988) favour the growth of the conifer stands frequented by the species. The same holds true for coastline areas and offshore islands, which often have stands of stunted spruce, making ideal habitat for this warbler.

In Québec, breeding has been confirmed as far south as the southern Eastern Townships, including Mounts Orford, Mégantic, and Owl's Head. The Blackpoll Warbler also nests south of the border, on the higher mountaintops in New York, Vermont, Massachusetts, New Hampshire, and Maine (AOU 1983; Kibbe 1985c; Peterson 1988c).

Farther north, the Blackpoll Warbler's breeding range extends to the tree line, well beyond the Atlas area. The northern boundary of its range in Québec runs from Lake Guillaume Delisle in the west to Ungava Bay in the east along the Rivière aux Feuilles (Todd 1963; Godfrey 1986). The Blackpoll Warbler is one of the commonest species in the Great Whale River region (Cons. G. G. – GREBE 1991a), but is rare in the region around the Nottaway, Broadback, and Rupert rivers (Cons. G. G. – GREBE 1992o). It is also fairly scarce in the Eastmain and Opinaca rivers region (ÉPOQ).

This northerly species also frequents black spruce-moss forests, like this one in the Laurentide Wildlife Reserve north of Québec City. *June 1991, Réginald Ouellet.*

Mature fir stand where the Blackpoll Warbler nests. Port-Menier, Anticosti Island. *August 1991, Gaétan Laprise.*

In northern regions, this warbler frequents shoreline thickets and coniferous stands, which are confined to the valleys. Koroc River, Ungava Bay. *July 1973, Henri Ouellet.*

This species is also found in the southernmost reaches of Québec on some high mountains with a wet climate and a cover of conifers draped with old man's beard, as on Mount Mégantic. East of Sherbrooke. *July 1990, Henri Ouellet.*

On the North Shore, this warbler also frequents open spruce peatlands with a ground cover of lichens. Near Grand Lac Germain, north of Sept-Îles. *Summer 1980, Jean Beaubien and Guy Simard, Canadian Forest Service.*

History and trends

In the past, the Blackpoll Warbler was seen in the Montréal, Québec City, Ottawa Valley, and Lac-Saint-Jean regions only during migration (Wintle 1896; Dionne 1906; Lloyd 1924; Godfrey and Wilk 1948), and this is still true today. Early this century it was considered a common breeder on the Gaspé Peninsula, North Shore, Anticosti Island, and Magdalen Islands (Macoun and Macoun 1909; Ball 1938). In fact, it was, and still appears to be, the most common warbler on the Magdalens (Fradette 1992), despite the extensive deforestation that has occurred there.

ÉPOQ and BBS data have come mainly from regions south of the species' breeding range and cannot be used as a reliable indicator of population trends. However, since the remote and often inaccessible regions where the species breeds have remained relatively untouched, populations have probably not undergone major changes.

Claude Nadeau

Cerulean Warbler

Paruline azurée

Dendroica cerulea

Christiane Girard

Of all our warblers, the Cerulean Warbler is the most difficult to observe and to study. In southern Québec, countless observers have scanned the canopy of tall deciduous trees for hours, with only a stiff neck to show for their efforts. Countless birders have strained their ears to hear what they thought was the Cerulean's song, only to find an American Redstart, Black-throated Blue Warbler or Blackburnian Warbler instead. The very rare Cerulean Warbler, whose population in Québec is thought to number only a few dozen pairs, is found in a few sites in the southern part of the province, where it spends most of its time hidden in the tops of tall deciduous trees.

The Cerulean Warbler winters in South America from Colombia and Venezuela south to eastern Peru and northern Bolivia (AOU 1983).

Habitat and behaviour

The Cerulean Warbler inhabits the upper tree canopy of extensive, mature, deciduous forests (Hamel 1992; McCracken 1992). In Québec, it is most often found in the Central St. Lawrence Lowland, in both damp and dry habitats. It usually selects sites with very tall trees that form a closed canopy (Lynch 1981), generally in blocks of woodland several hundred hectares in size (Robbins et al. 1989a). In Québec, it is observed most often in forests dominated by sugar maple, red oak, and American beech (Ouellet, H. 1966). The Cerulean Warbler limits itself almost exclusively to the top part of the canopy, where it flits from branch to branch in search of insects.

Observing the Cerulean Warbler in this lofty habitat is a daunting task. This sky-blue warbler inhabits the dome of an immense cathedral of dappled green, illuminated by constantly shifting shafts of sunlight and broken by patches of sky. To detect the Cerulean Warbler, observers must be familiar with the male's song, which is often the only sign of a pair's presence. The female is almost impossible to spot, as is the nest, which is built on a horizontal branch high in the treetops (7–18 m up), several metres out from the trunk (Godfrey 1986). Seen from below, the Cerulean Warbler's nest resembles that of the Eastern Wood-Pewee or the Blue-gray Gnatcatcher. The delicate, compact, rather shallow nest is made of lichens and fine plant materials interwoven with spiderwebs. The exterior is often greyish because of its covering of spiderwebs. As in many other warblers, the female builds the nest and incubates the eggs, but both sexes feed the young (Harrison 1975, 1984).

Distribution

The Cerulean Warbler nests in southern Ontario, southern Québec and the eastern and central United States as far south as Texas and Georgia (AOU 1983).

The Cerulean Warbler was reported in only 6 Atlas squares. Despite the difficulties of observing the species, confirmed breeding records were obtained in 3 of these squares: a male was seen carrying food near Rockburn on 20 July 1986; a pair was observed feeding 3 young near Lac la Blanche in the Ottawa Valley region on 14 June 1987; and a pair was seen building a nest at Philipsburg on 22 May 1989. Although the Cerulean Warbler is believed to have nested in southern Québec since the 1950s or 1960s, these 3 records are the only breeding evidence ever obtained. The species was also atlassed on Mount Saint-Bruno and Mount Saint-Hilaire, near Montréal. Mount Saint-Hilaire is the site where the species was seen most regularly during the 1960s and 1970s; however, in the early 1980s, a severe ice storm destroyed the warbler's habitat on the mountain, and since then, it has been seen only on very rare occasions.

Prior to 1984, Cerulean Warblers were also recorded during the breeding season on Mount Yamaska, Mount Rougemont, and Rigaud Mountain, as well as at Portage-du-Fort, Gatineau Park, Vale Perkins, and Lake Trois Saumons (Ouellet, H. 1966; Ouellet 1967; David 1980a; ÉPOQ). More recently, a pair was seen in Gatineau Park in July 1991 (Gosselin et al. 1991b), and a male carrying food was observed in June 1993 at Pointe du Gouvernement in Montérégie (Bannon 1993).

History and trends

The first record of the Cerulean Warbler in Québec is from Île Sainte-Hélène in 1950. It was sighted at different locations in the southern part of the province over the following three years and again in 1957, but was not recorded regularly until 1961 (Ouellet, H. 1966; Ouellet 1967). The colonization of southern Québec and other areas in the northeast is part of the eastward expansion of its breeding range. During the 19th century, it spread from southern Ontario into New York State along the corridor between Lake Ontario and Lake Erie (Bull 1974). During the 20th century, it expanded farther east, dispersing to the north and south (Connor 1988a). In southern Ontario, for example, it expanded its range to the northeast and today is found in a number of locations along the southern margin of the Canadian Shield (Eagles 1987a). The Cerulean Warblers in northwestern New York State may have come from Ontario (Connor 1988a). The Québec breeding population appears to be an offshoot of the New York or Ontario populations near the St. Lawrence River. Some of the Québec birds may also have arrived by way of the Lake Champlain and Hudson River lowlands (Ouellet, H. 1966; Ouellet 1967; Ellison 1985i).

The Québec population does not appear to have changed significantly over the last 25 years in terms of abundance or distribution and, according to H. Ouellet's data (1966, 1967), is probably roughly the same as it was in the early 1960s. The Cerulean Warbler is found today in the same places as in the 1960s, and the northward expansion observed in Ontario and New York State does not appear to be occurring in Québec, at least at the present time. Extensive blocks of mature deciduous forest are rare in the St. Lawrence and Ottawa valleys and are under threat from clearing for urban development, agriculture, and lumber and firewood production. Consequently, the Canadian Wildlife Service recently classified the Cerulean Warbler as a vulnerable species (Robert 1989b). It is also considered rare or in decline in several other regions of Canada (McCracken 1992) and the United States (Robbins et al. 1989c).

Fortunately, some of the extensive mature deciduous forests that provide breeding habitat for the species or that it frequents during the breeding season are protected to a certain extent, as in Gatineau Park, the Centre éducatif forestier de l'Outaouais, Mont-Saint-Bruno Conservation Park, Mont-Saint-Hilaire Biosphere Reserve, and the Philipsburg Migratory Bird Sanctuary. Furthermore, the less fertile land along the edges of the Canadian Shield and Appalachians is gradually reverting to forest and one day may provide a haven for the Cerulean Warbler.

Pierre Bannon and Michel Robert

Notes on territory

The Cerulean Warbler breeds locally throughout eastern North America, yet in some regions the forests frequented by the species often have a high number of nesting pairs. For example, densities of 30 pairs/10 ha were reported in Ohio and 12.4 pairs/10 ha in Ontario (McCracken 1992). Of course, not all forests providing suitable habitat harbour so many birds; in Québec, often only one singing male is found in any one area. However, some of the better-known areas where Cerulean Warblers are known to breed are home to several pairs: for example, 4 singing males were observed at Mount Yamaska (Ouellet 1974a) and 3 at the Philipsburg Migratory Bird Sanctuary.

The Cerulean Warbler nests and forages in the tops of large deciduous trees forming a closed canopy, like this mature sugar maple stand at Philipsburg, where a pair nested in 1989. South of Saint-Jean-sur-Richelieu. *July 1991, Pierre Bannon.*

Cerulean Warbler

Breeding distribution

1984–1989

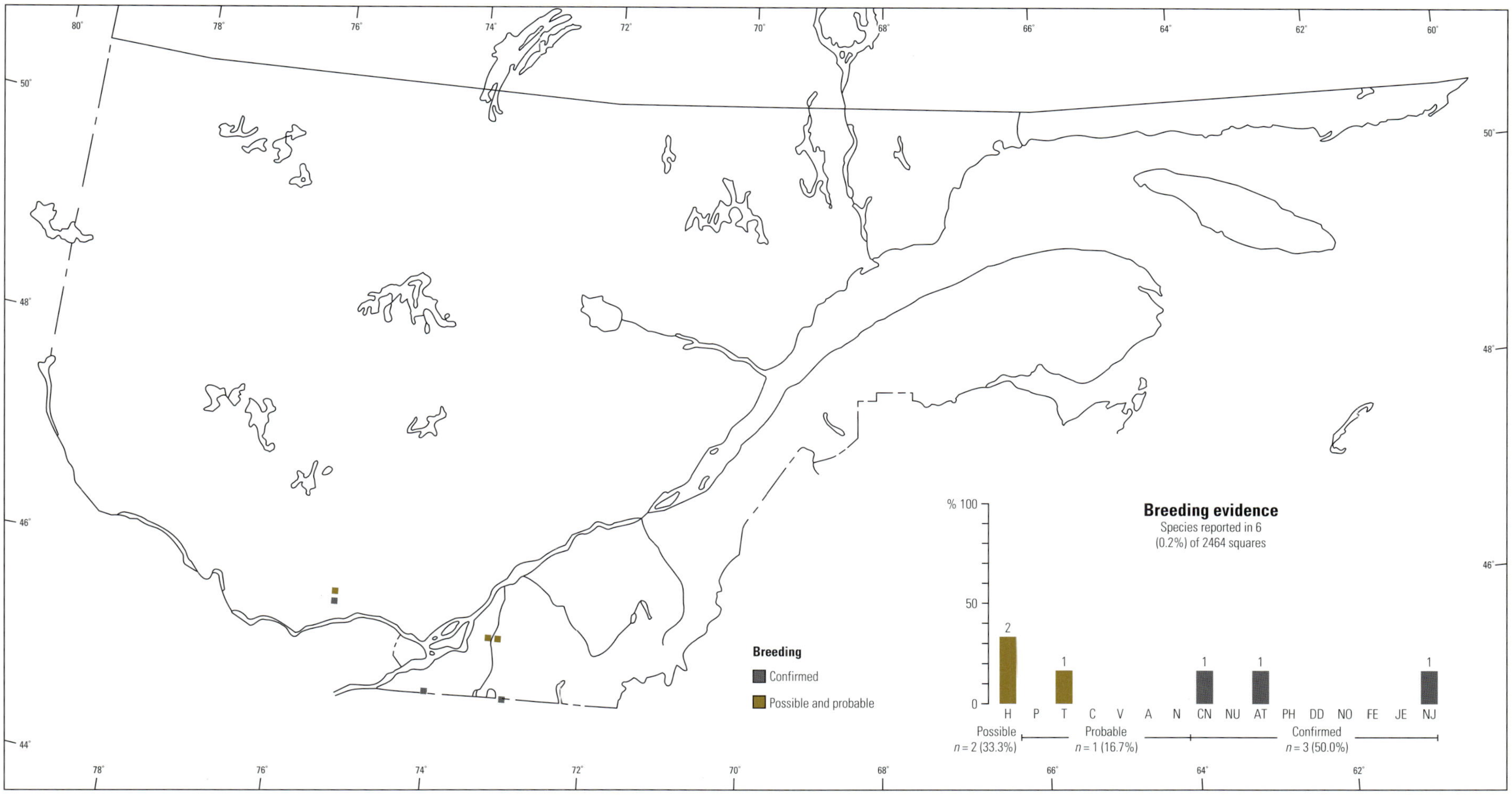

Species profile

Clutch size	usually 4 eggs (3–5)[1,2,3,5]
Laying interval	?
Start of incubation	?
Incubation period	probably 12–13 days[2,3,5]
Maturity at hatching	altricial
Care of young	by female and male[1]
Nestling period	9–10 days[5]
Age at first flight	9–10 days or more[8]
Dependence of young	?
Broods per year	1[1,5]
Breeding age	1 year[5]
Mating system	monogamy[2]
Duration of pair bond	?
Total length	10.0–12.5 cm[6]
Wing span	18.4–21.0 cm[4]
Weight	M 9.5 g (8.6–10.2) / F 9.2 g (8.4–10.3)[7]
Longevity record	?

References:
1. Harrison 1984
2. Ehrlich et al. 1988
3. Harrison 1975
4. Terres 1980
5. McCracken 1992
6. Godfrey 1986
7. Dunning 1984
8. according to nestling period

Chronology $n = 1$

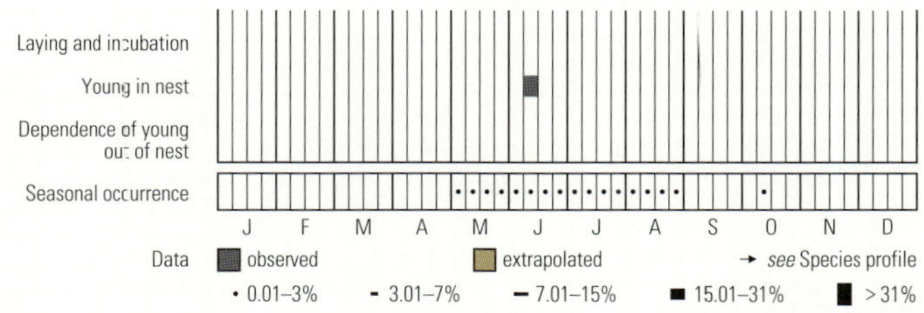

World breeding range

The Cerulean Warbler's song is heard regularly from the edge of this woodlot. It nests farther in, among the mature deciduous trees. Gowan Road, Herdman, southwest of Montréal. *June 1991, Pierre Bannon.*

Black-and-white Warbler

Paruline noir et blanc

Mniotilta varia

Christiane Girard

Unlike other warblers, the Black-and-white Warbler is a tree creeper, foraging in the manner of woodpeckers, nuthatches, and creepers by gleaning insects from trunks and branches, flitting from tree to tree, pausing now and then to search bark crevices for insects. It has a shorter tarsus and longer hind toe and claw than most warbler species, which helps it cling to bark (Eaton 1988f).

It is one of the first warblers to arrive in spring (Fortin et al. 1978; David 1980a); because it forages for insect eggs and larvae on bark, it does not need to wait until the trees are in full leaf, as do other warblers (Griscom and Sprunt 1979; Ehrlich et al. 1988). In autumn, it migrates south to its wintering grounds, which extend from the southernmost United States to Colombia and Ecuador in South America (Diamond 1986; Godfrey 1986; Root 1988).

Habitat and behaviour

Black-and-white Warblers inhabit deciduous woodlands and deciduous areas within the mixed and boreal forests. They are found in stands of varying ages, but show a preference for those growing in damp areas. They also frequent willow and alder thickets (AOU 1983; Godfrey 1986; Elderkin 1989).

The cuplike nest is built on the ground by the female, usually at the foot of a tree, shrub, or stump, or beside a rock. Sometimes, it is set up off the ground, generally no higher than 2 m, in the roots of a stump or upturned tree (Peck and James 1987). The most commonly used nesting materials are dry leaves, grass stems, strips of bark, pine needles, mosses, and rootlets. The inside is lined with finer material such as hair and fibres (Godfrey 1986; Peck and James 1987).

Clutch size is generally 5 eggs, which are white to creamy, flecked and sometimes blotched with various shades of brown (Harrison 1975; Griscom and Sprunt 1979; Kricher 1995). In southern Québec, 28.5% of the 14 nests examined by Terrill (1961) had been parasitized by the Brown-headed Cowbird, as had been 20.9% of 43 nests found in Ontario (Peck and James 1987). A nest in Michigan contained a record number of 8 cowbird eggs and only 2 warbler eggs (Harrison 1975). Double-brooding may occur, though this has not been documented (Peck and James 1987).

Notes on territory

The breeding territory is used for mating, nesting, and foraging (Welty 1975; Morse 1989a).

The Black-and-white Warbler frequents relatively wet mixed and deciduous woodlands, where it searches the bark of trunks and branches for insects. Neuville, west of Québec City. *August 1991, Marcel Darveau.*

Black-and-white Warbler

Breeding distribution

1984–1989

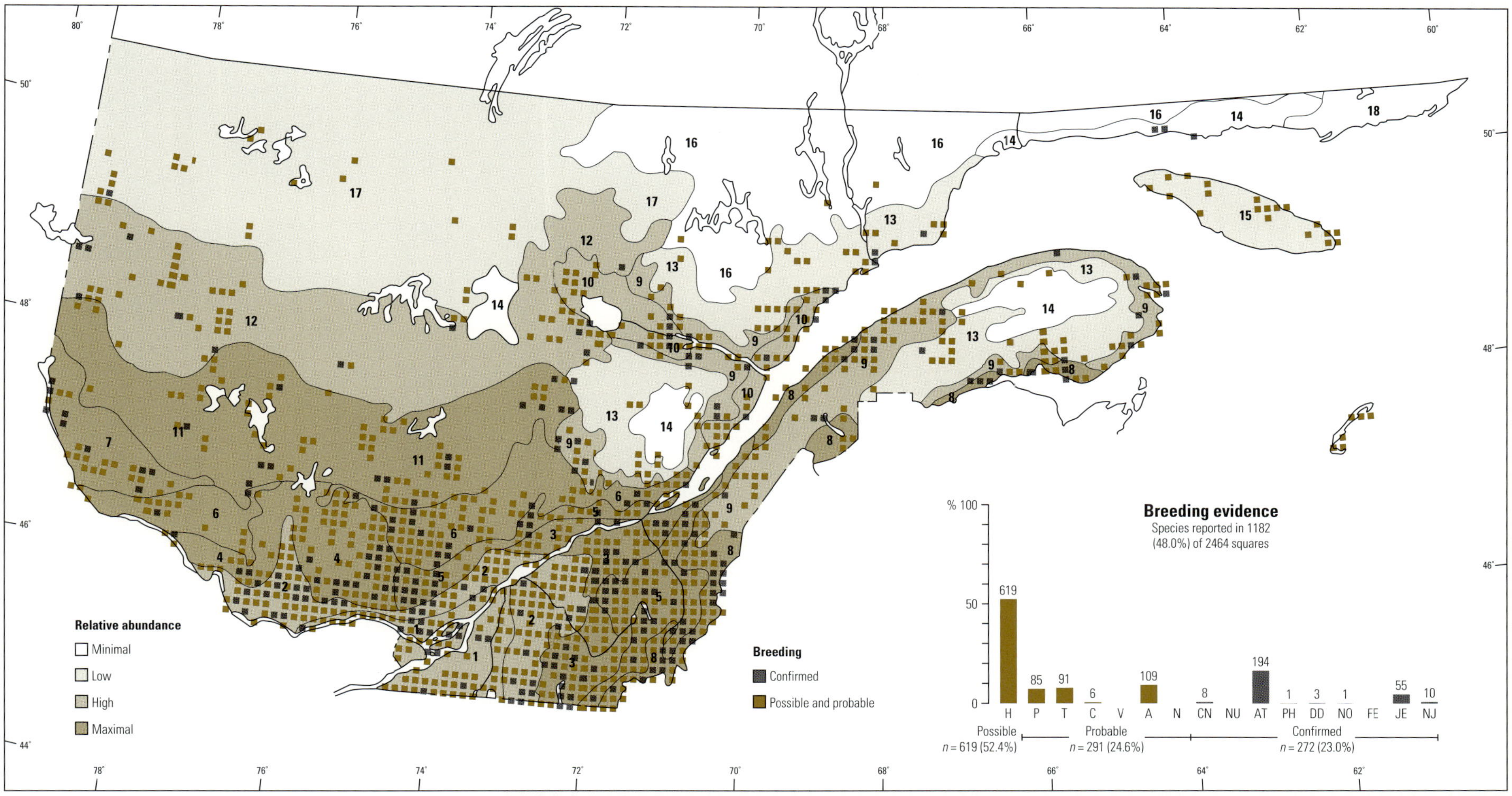

World breeding range

Species profile

Clutch size	usually 5 eggs (4–6)[1,2,3,4,5,8,11]
Laying interval	1 egg per day[11]
Start of incubation	with last or penultimate egg[11]
Incubation period	10–13 days[1,2,3]
Maturity at hatching	altricial
Care of young	by female and male[1]
Nestling period	8–12 days[1,3]
Age at first flight	8–12 days or more[9]
Dependence of young	?
Broods per year	1, 2 possible[3,8,10]
Breeding age	1 year[8]
Mating system	monogamy[1]
Duration of pair bond	?
Total length	11–14 cm[2]
Wing span	21.0–22.9 cm[5]
Weight	M 11.0 g (8.8–15.2) / F 10.6 g (9.0–12.7)[7]
Longevity record	11 years 3 months[6]

References:
1. Ehrlich et al. 1988
2. Godfrey 1986
3. Harrison 1975
4. Griscom and Sprunt 1979
5. Terres 1980
6. Kennard 1975
7. Dunning 1984
8. DeGraaf and Rudis 1987
9. according to nestling period
10. Peck and James 1987
11. Kricher 1995

Chronology $n = 31$

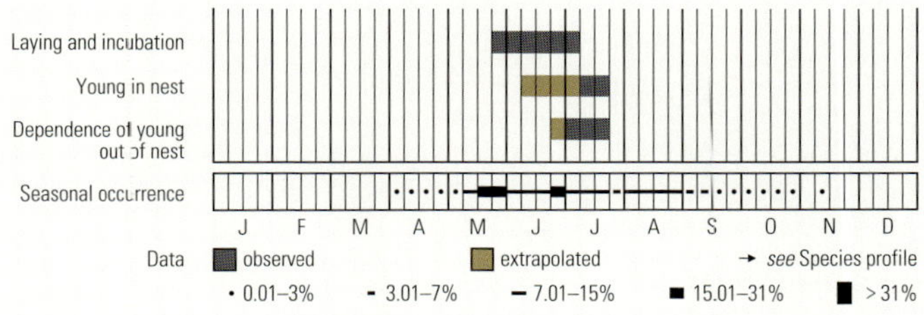

Frequency of occurrence by ecological region

Ecological regions** (forest regions)	Adequately covered squares	Squares in which reported	% of squares in which reported	Relative abundance (%)*
1. Sugar maple-hickory and sugar maple-basswood	97	73	75.3 ± 4.4	75.3
2. Sugar maple-basswood	153	137	89.5 ± 2.6	89.5
3. Sugar maple-basswood and sugar maple-yellow birch	101	94	93.1 ± 3.2	93.1
4. Sugar maple-yellow birch-basswood	56	56	100.0 ± 0.9	100.0
5. Sugar maple-yellow birch-beech and sugar maple-yellow birch-basswood	113	108	95.6 ± 2.4	95.6
6. Sugar maple-yellow birch-beech	71	67	94.4 ± 5.3	94.4
7. Sugar maple-yellow birch-balsam fir	19	19	100.0 ± 2.6	100.0
8. Sugar maple-yellow birch and balsam fir-yellow birch	47	46	97.9 ± 4.5	97.9
9. Balsam fir-yellow birch	87	67	77.0 ± 8.2	77.0
10. Balsam fir-white birch or balsam fir-red maple	46	38	82.6 ± 9.7	82.6
11. Yellow birch-balsam fir	52	50	96.2 ± 6.0	96.2
12. Balsam fir-white birch (west)	59	44	74.6 ± 11.5	74.6
13. Balsam fir-white birch (east)	76	44	57.9 ± 10.8	57.9
14. Balsam fir-black spruce	28	8	28.6 ± 17.6	28.6
15. White spruce-balsam fir and balsam fir-white spruce	22	12	54.5 ± 21.4	54.5
16. Black spruce-balsam fir-moss	18	4	22.2 ± 22.2	22.2
17. Black spruce-moss	32	13	40.6 ± 18.6	40.6
18.			*not analysed*	

* minimal: 0–29.9%; low: 30–59.9%; high: 60–89.9%; maximal: 90–100%

** Deciduous forest: 1–7; mixed forest: 8–11; coniferous forest: 12–17

A Black-and-white Warbler nest built among dead leaves at the base of a deciduous tree. Saint-Lambert, near Montréal. *Circa 1940, Henry Mousley, Laval University Collection.*

The female alone incubates for 10–13 days (Terres 1980; Godfrey 1986), but both parents feed the nestlings. The young leave the nest at 8–12 days of age (Terres 1980; Ehrlich et al. 1988).

During the breeding season, the Black-and-white Warbler is strictly insectivorous. Although most often seen foraging on the trunks of trees, it occasionally hawks insects on the wing (Ehrlich et al. 1988). It also gleans insects on the foliage in typical warbler fashion (Morse 1970).

Distribution

In North America, the Black-and-white Warbler breeds from British Columbia to Newfoundland and south to Texas and Georgia (AOU 1983; Erskine 1992).

The species was recorded in 48% (1182/2464) of Atlas squares surveyed; it was reported throughout southern Québec and confirmed as a breeder everywhere except Anticosti Island and the Magdalen Islands. Frequency of occurrence is highest in deciduous and mixed woodlands, but decreases as conifers become more dominant, as in the northern part of southern Québec, on Anticosti Island, and in the mountainous areas of the Gaspé Peninsula and the Laurentide Wildlife Reserve.

Notes on territory (continued)

In Québec, territories have been found to range in size between 0.35 and 0.63 ha (mean 0.43 ha) (McNeil 1964 in Cyr 1974). The Black-and-white Warbler tends to defend its territory until late in the summer, when other warblers have ceased to be territorial, continuing to drive out conspecifics and even other warblers (Morse 1970, 1989a). The Black-and-white Warbler will even defend temporary territories while migrating (Morse 1989a). The female sits tight on the nest at the approach of an intruder and performs a distraction display when flushed (Eaton 1988f; Ehrlich et al. 1988).

In Québec, densities of 7.5 pairs/10 ha were reported in a young balsam fir-red maple stand in Lachute (Erskine 1984), 5.3 pairs/10 ha in a mixed forest in Saint-Fabien-de-Panet (Erskine 1971a), and 5.0 pairs/10 ha in a white birch-gray birch stand in Val-Racine (Erskine 1984). In La Mauricie National Park, a white birch stand and a sugar maple-yellow birch stand yielded a count of 1.1 pairs/10 ha (DesGranges 1980b). An average density of 0.48 pair/10 ha was obtained in a mosaic of forested and open areas in Gatineau Park and at La Barrière, near Saint-Michel-des-Saints (D. Dauphin, unpubl. data).

Black-and-white Warblers are often found in saplings at the edges of damp wooded areas, regenerating clearcuts, and roadsides. Grands-Jardins Conservation Park, northwest of Baie-Saint-Paul.
May 1987, Réginald Ouellet.

Breeding was confirmed in only a few squares in the Montérégie, no doubt because of habitat loss due to intensive agricultural and urban development. The same pattern has been noted in southwestern Ontario (Cartwright 1987), the Champlain Lowlands in Vermont (Ellison 1985d), and various parts of New York State (Eaton 1988f). The species may tolerate moderate fragmentation of its habitat (Villard 1991), as indicated by its high frequency of occurrence in the other agricultural regions of Québec.

During a recent survey in the Nottaway, Broadback, and Rupert river watersheds (Cons. G. G. – GREBE 1992o), the Black-and-white Warbler was reported in stands dominated by white birch and trembling aspen, in line with Erskine's (1977) classification of the Black-and-white Warbler as a species typical of such stands in the Canadian boreal forest.

History and trends

At the turn of the century, Wintle (1896) listed the Black-and-white Warbler as a scarce summer resident, but common migrant in the Montréal area. According to Dionne (1906), it was uncommon around Québec City.

BBS data for 1969–88 show a population increase in eastern North America. ÉPOQ summer data, as well as BBS data from Québec, likewise show an upward trend for the province (Cyr and Larivée 1993b).

Considering only settled areas, however, would probably produce quite different results. In fact, rapid urban development and intensification of agriculture, particularly in the Montréal area, have undoubtedly caused a decline in breeding populations since the middle of the last century (Ouellet 1974a). The resulting forest fragmentation has increased edge habitat between woodlots and open areas. As the Black-and-white Warbler nests on or near the ground, it is highly vulnerable to predation by raccoons, striped skunks, and other mammals that move between wooded areas and farm fields or suburbs. Forest fragmentation also tends to increase the incidence of cowbird parasitism (Brittingham and Temple 1983; Robinson et al. 1993). Moreover, field work in agricultural habitat in the Ottawa region shows that density of territorial males in woodlots decreases with distance from the nearest wooded area occupied the preceding year (Villard 1991; Villard et al. 1995). All these factors combine to explain how a species such as the Black-and-white Warbler can gradually be extirpated from highly fragmented habitat, as in the Montérégie.

Marc-André Villard

American Redstart

Paruline flamboyante

Setophaga ruticilla

Christiane Girard

THE AMERICAN REDSTART RESEMBLES A FLYCATCHER WHEN DARTING OUT to catch insects flying past. Ornithologists of the 19th century placed it in the Tyrannidae family with the *Empidonax* genus (Eaton 1988c). Like flycatchers and other birds that catch insects in flight (whip-poor-wills, swallows, and swifts), the redstart has a bill fringed with small rictal bristles that widen the gape and can help the bird snatch insects out of the air.

Redstarts are sexually dimorphic in plumage; where the male is mostly black with patches of orange on wings, sides, and tail, the female is grey with yellow instead of orange patches. In their first breeding season, males resemble females and acquire adult plumage only in their second year. The American Redstart is one of 31 North American passerines that exhibit this characteristic (Rohwer et al. 1980), the function of which is still unclear, despite numerous studies in this area (Procter-Gray 1991). We do know, however, that during spring migration, males in adult plumage generally arrive a few days before yearling birds (Francis and Cooke 1986; Morris and Lemon 1988), giving them the advantage in nest site selection (Ficken and Ficken 1967; Howe 1974).

The American Redstart is found in a variety of habitat. It is a common migrant breeder throughout southern Québec and was the third most frequently atlassed warbler after the Common Yellowthroat and the Yellow-rumped Warbler. It returns to Québec around mid-May from its wintering grounds in Florida, the Bahamas, the West Indies, Mexico, Central America, and South America to Brazil (AOU 1983).

The American Redstart, one of the most widespread of the warblers, nests in deciduous woods with sapling undergrowth rather than in mature forests. Brossard, near Montréal. *August 1991, Gaétan Duquette.*

Habitat and behaviour

Rather than mature forest, the American Redstart prefers second-growth woodland or woods with a dense understorey of saplings, with or without regenerating undergrowth. It favours deciduous over mixed or coniferous stands; in fact, Robbins et al. (1989a) found that redstart numbers decrease as coniferous cover increases. Furthermore, according to Mulvihill (1992a), the American Redstart is three times more common in stands over 32 ha in size than in woodlands of less than 14 ha. Redstarts also occupy other types of habitat, such as overgrown fields, orchards, and forest edges, swamps, and roadsides (DeGraaf and Rudis 1987; Peck and James 1987).

Distribution is a function of habitat characteristics, intraspecific competition, and in particular the occurrence of the Least Flycatcher, a species that occupies a similar ecological niche. The flycatcher is dominant over the redstart, confining it to sites that the flycatcher does not occupy and potentially excluding it from certain areas within the deciduous forest (Sherry 1979; Sherry and Holmes 1988).

Notes on territory

In spring, the females arrive a few days after the males (Bent 1953; Francis and Cooke 1986). Males are highly territorial and advertise their presence by singing. They use intimidation displays and aggressive behaviour toward conspecifics in defence of their territories. As well as intraspecific competition, the redstart has to vie with other species for food and is sometimes ousted from suitable breeding habitat. The Least Flycatcher, for example, is a strong competitor and usually wins out (Bent 1953; Cruickshank 1979; Sherry 1979; Sherry and Holmes 1988). Redstarts do, however, live alongside other insectivorous species such as the Ovenbird without any apparent hostility (Bent 1953). In Québec, McNeil (1964 in Cyr 1974) reported territory sizes ranging from 0.35 to 0.93 ha (mean 0.64 ha), and Cyr (1974) found redstarts in maple stands at Mirabel with territories ranging between 0.17 and 0.61 ha (mean 0.39 ha). In New Hampshire, Sabo (1980) found the average territory size to be 0.6 ha in subalpine habitat, and Sherry (1979) found it to be 0.43 ha in deciduous woodland.

In Québec, recorded densities include 14.8 pairs/10 ha in both a disturbed red maple stand with fir and birch and in a fir-birch stand (Erskine 1980a), 12.4 pairs/10 ha in a young fir-birch stand with

American Redstart

Breeding distribution

1984–1989

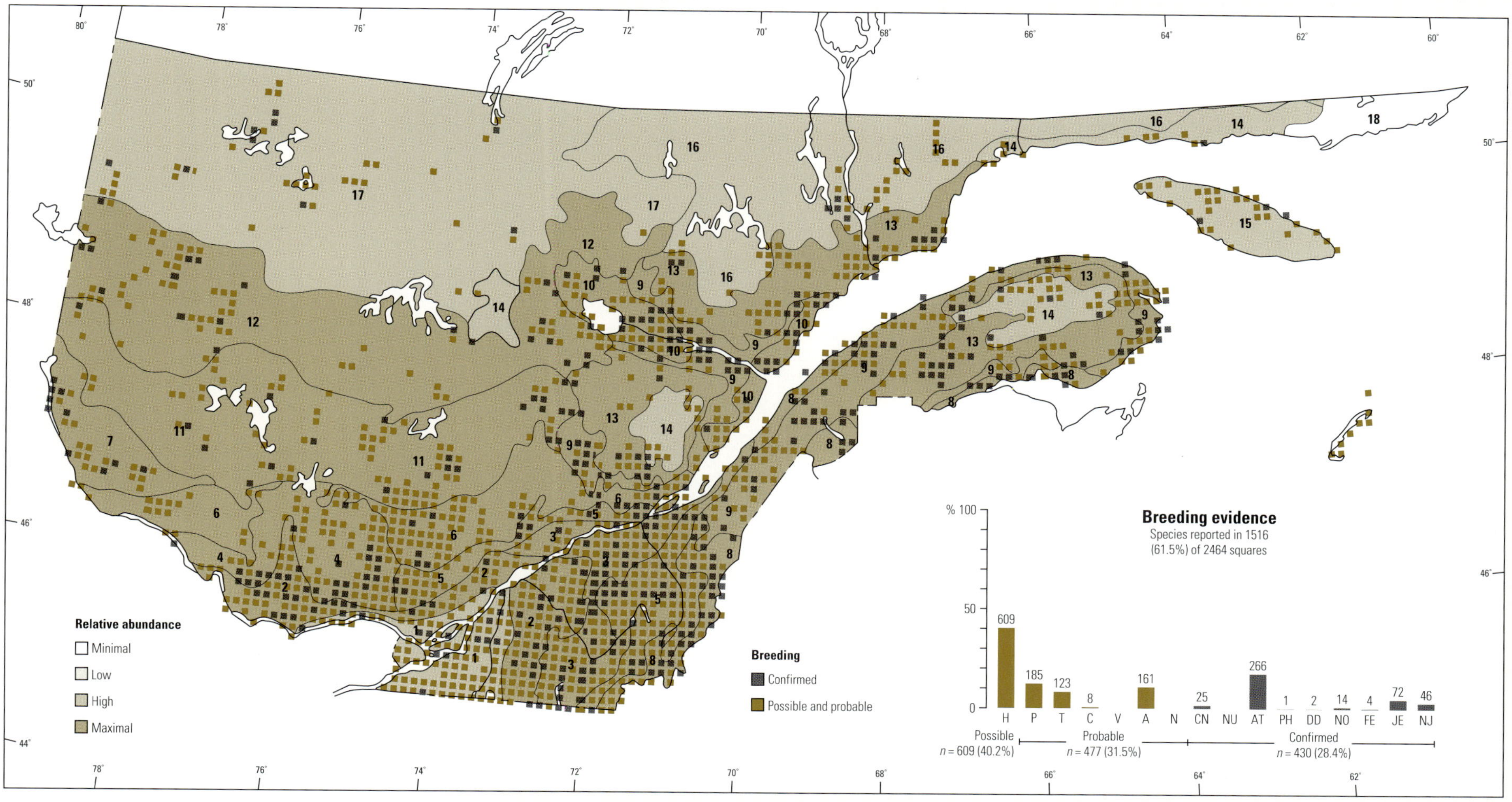

Species profile

Clutch size	usually 4 eggs (2–5)[1,4]
Laying interval	1 egg per day[4]
Start of incubation	with last egg[4]
Incubation period	10–14 days[1,2,3,4,5,6]
Maturity at hatching	altricial
Care of young	by female and male[1,3,6,11]
Nestling period	8–9 days[1,3,5,6]
Age at first flight	8–9 days or more[9]
Dependence of young	?
Broods per year	1[3,5,6]
Breeding age	1 year[5]
Mating system	monogamy, rarely polygyny[10]
Duration of pair bond	?
Total length	12.0–14.5 cm[2]
Wing span	?
Weight	M 8.5 g (7.0–12.0) / F 8.1 g (6.7–11.2)[8]
Longevity record	10 years[7]

References:
1. Ehrlich et al. 1988
2. Godfrey 1986
3. Harrison 1978
4. Peck and James 1987
5. DeGraaf and Rudis 1987
6. Bent 1953
7. Klimkiewicz et al. 1983
8. Dunning 1984
9. according to nestling period
10. Secunda and Sherry 1991
11. Boxall 1983

World breeding range

Chronology $n = 130$

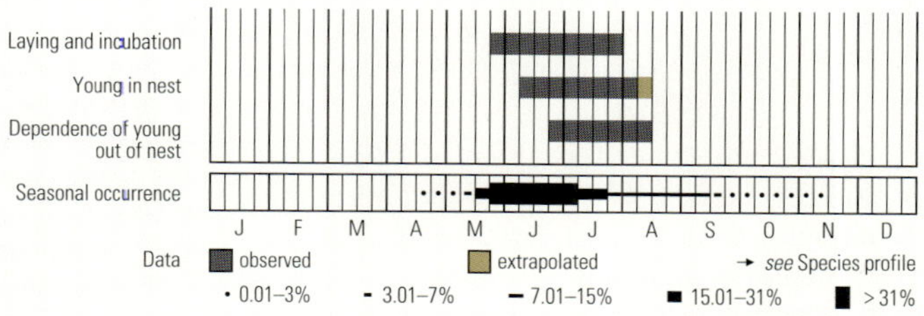

Frequency of occurrence by ecological region

Ecological regions** (forest regions)	Adequately covered squares	Squares in which reported	% of squares in which reported	Relative abundance (%)*
1. Sugar maple-hickory and sugar maple-basswood	97	86	88.7 ± 3.3	88.7
2. Sugar maple-basswood	153	143	93.5 ± 2.2	93.5
3. Sugar maple-basswood and sugar maple-yellow birch	101	98	97.0 ± 2.3	97.0
4. Sugar maple-yellow birch-basswood	56	54	96.4 ± 4.8	96.4
5. Sugar maple-yellow birch-beech and sugar maple-yellow birch-basswood	113	112	99.1 ± 1.3	99.1
6. Sugar maple-yellow birch-beech	71	68	95.8 ± 4.7	95.8
7. Sugar maple-yellow birch-balsam fir	19	19	100.0 ± 2.6	100.0
8. Sugar maple-yellow birch and balsam fir-yellow birch	47	46	97.9 ± 4.5	97.9
9. Balsam fir-yellow birch	87	85	97.7 ± 3.3	97.7
10. Balsam fir-white birch or balsam fir-red maple	46	45	97.8 ± 4.4	97.8
11. Yellow birch-balsam fir	52	50	96.2 ± 6.0	96.2
12. Balsam fir-white birch (west)	59	55	93.2 ± 7.0	93.2
13. Balsam fir-white birch (east)	76	74	97.4 ± 4.0	97.4
14. Balsam fir-black spruce	28	25	89.3 ± 12.6	89.3
15. White spruce-balsam fir and balsam fir-white spruce	22	18	81.8 ± 17.1	81.8
16. Black spruce-balsam fir-moss	18	16	88.9 ± 17.4	88.9
17. Black spruce-moss	32	20	62.5 ± 18.3	62.5
18.			*not analysed*	

* minimal: 0–29.9%; low: 30–59.9%; high: 60–89.9%; maximal: 90–100%

** Deciduous forest: 1–7; mixed forest: 8–11; coniferous forest: 12–17

Notes on territory (continued)

trembling aspen, alder, and underbrush (Erskine 1980a), 10 pairs/10 ha in both a maple-basswood stand and in a maple-beech stand (Erskine 1984), 8.7 pairs/10 ha in a beech-maple stand with small areas of swamp (Erskine 1972a), and 4.9–9.9 pairs/10 ha in fir stands with or without spruce and with an admixture of shade-intolerant hardwoods (Erskine 1980a). In the Nottaway, Broadback, and Rupert river watersheds, highest densities were obtained in birch stands and in ecotones between deciduous stands and water (Cons. G. G. – GREBE 1992o).

Redstarts are also found in dense deciduous shrubbery and second-growth edge habitat, whether in deciduous, mixed or coniferous forest. Sainte-Foy. *September 1991, Yves Aubry.*

The American Redstart is essentially monogamous, but there are a few recorded instances of a polygynous male defending 2 or 3 territories 200–500 m apart (Secunda and Sherry 1991). The female alone builds the nest, usually over a 7–10 day period (Bent 1953; Harrison 1975). Sturm (1945), however, reported that the nest may be built in only 4 days, during which time the female may make up to 700 trips to gather nest material. The nest is an open cup built in an upright fork of a young tree or shrub, preferably deciduous. Nests are found less often in mature trees (Bent 1953; Godfrey 1986; Peck and James 1987). According to Peck and James (1987), in Ontario the nest is usually built 2–6 m up. It is a compact structure of plant fibres, grasses, and rootlets, lined with fine material and often decorated with birch bark (Godfrey 1986; Ehrlich et al. 1988). Redstarts sometimes use old or abandoned nests of other species such as vireos (Ehrlich et al. 1988). Typical clutch size is 4 eggs, incubated by the female alone for a period of 10–14 days. Both parents feed the young, which leave the nest after 8 or 9 days (Bent 1953; Godfrey 1986; Peck and James 1987; Secunda and Sherry 1991).

At the approach of a predator, the redstart gives an alarm call and fans its tail, displaying the orange or yellow outer tail feathers in an attempt to distract and draw it away from the nest. In a more conspicuous demonstration of distress the bird sits tight until flushed, then feigning injury, flutters to the ground and drags itself off under a shrub, performing a broken-wing display (A. Cyr, pers. obs.).

In northern latitudes, the American Redstart lays only one clutch per season (DeGraaf and Rudis 1987).

This warbler is often parasitized by the Brown-headed Cowbird; in southern Québec and Ontario respectively, 15.8% of 145 nests (Terrill 1961) and 20% of 285 nests (Peck and James 1987) studied were found to contain cowbird eggs.

Male redstarts assist in feeding the young. The cup-shaped nest seen here was built in the branches of an elder. Sainte-Foy. *July 1962, Gaston Le Page.*

A female redstart incubating. The nest, built of grasses, strips of bark, and plant fibres, is often found in the fork of a deciduous tree, as in this young maple. Sillery. *June 1955, Gaston Le Page.*

Fewer than half of yearling males mate in their first year, but those that do are as successful as older males (Sherry and Holmes 1989). Having plumage similar to that of females contributes to the breeding success of yearling males as it allows them to infiltrate adult territories or the outskirts of their territories and to use these resources without being attacked by the highly territorial males. In confirming breeding of redstarts, the behaviour of birds in drab plumage should be taken into consideration, as these are not necessarily females. First-year males sing and defend a territory, though less vigorously than do adult males (G. Michaud and A. Cyr, U. of Sherbrooke, unpubl. data).

Redstarts feed on insects and spiders (Bent 1953; Ehrlich et al. 1988). Robinson and Holmes (1982) found that the species feeds mainly by taking insects from a substrate while "hovering." Insects are also gleaned from branches and leaves of trees. It also uses the flush-chase technique: the bird moves along the branches and twigs, fanning its tail and flicking its wings, actions which appear to flush its prey. The bird then chases after its prey as it drops from the foliage or flies off. Redstarts also hawk for insects, catching them in flight, from ground level up to heights of 25 m. Foraging behaviour changes during the day according to the availability of insects: they flush-chase and forage nearer the ground at midday than in the morning (Holmes et al. 1978).

American Redstarts were reported in 61.5% (1516/2464) of Atlas squares surveyed, and breeding was confirmed in 28.4% (430/1516) of these. The species is one of the easiest warblers to confirm as a breeder, because it is readily identified by sight and can be lured by "spishing." It also calls and sings frequently, though its repertoire of songs can be confusing to the untrained ear.

Distribution

The American Redstart breeds from Yukon, southwestern Northwest Territories, and all Canadian provinces south to the United States, from Oregon in the west to Georgia in the east (National Geographic Society 1987).

As pointed out by various authors (David 1980a; Cyr and Larivée 1995), the redstart is a fairly widespread breeder throughout southern Québec, where its status varies between common and abundant (David 1980a). Frequency of occurrence is high throughout the deciduous and mixed forest regions and throughout most coniferous forest regions, where the species occupies mainly forest edges and second-growth areas growing up in shade-intolerant hardwoods after forest fires or logging. Redstarts also nest on the Magdalen Islands (Fradette 1992), although breeding evidence was not obtained for the Atlas, and on Anticosti Island (Ouellet 1969), which is not surprising considering the amount of available habitat on the island.

History and trends

According to Dionne (1906), the American Redstart was common in summer in the province at the beginning of the century. Wintle (1896) considered it an abundant summer resident in the Montréal area, and Macoun and Macoun (1915) described it as being common in summer on the Magdalen Islands and occurring in fairly large numbers on Anticosti Island and at Mingan.

BBS data for 1968–79 show a population increase in eastern North America (Robbins et al. 1986). BBS data for Québec show an increase between 1966 and 1989 along survey routes. ÉPOQ summer data, however, do not reveal any significant trend between 1970 and 1989 (Cyr and Larivée 1993b). According to Ouellet (1974a), the American Redstart population probably began growing after the middle of the century as logged-over areas regenerated.

The American Redstart is apparently little affected by forestry operations, the major cause of disturbance within its breeding range. It will return to disturbed sites 6 years after clearcutting, attaining maximum densities after 20 years (Morgan and Freedman 1986). The species does not appear to be overly affected by strip cutting (Freedman et al. 1981) and benefits from commercial thinning (Webb et al. 1977; Johnson and Brown 1990).

Forest clearing for agriculture and urbanization, with resulting loss of breeding and wintering habitat are limiting factors on the American Redstart and other forest species (Askins 1993). Between 1985 and the year 2000 the redstart could lose 31% of its wintering grounds in Central and South America and the Caribbean (Diamond 1991).

André Cyr and Marcel Darveau

Ovenbird

Paruline couronnée

Seiurus aurocapillus

Christiane Girard

THE OVENBIRD, LIKE THE NORTHERN WATERTHRUSH, RESEMBLES THE NEW World thrushes, in both its cryptic plumage and its habit of feeding on the ground. Because of this resemblance, it was once placed in the genus *Turdus* (Harrison 1984), and until the end of the 1950s was known as the Golden-crowned Thrush (Dionne 1889; Wintle 1896). The Ovenbird advertises its presence mostly through its characteristic song, an energetic *teacher teacher teacher*. Less musical than a thrush song, it is repeated from dawn to dusk and is one of the first songs learned by the novice birder. The sweeter more musical flight song is given at twilight and at night, during the breeding season.

The Ovenbird is a common migrant breeder in southern Québec up to about the 49th parallel, though the northern limit of its Québec range is around the 51st parallel (Todd 1963; Godfrey 1986). It winters from the southern United States down to Panama and Venezuela (AOU 1983).

Habitat and behaviour

In Québec, the Ovenbird nests in several types of forest communities. It is common in deciduous woodland dominated by sugar maple, trembling aspen, or white birch, and in coniferous forest dominated by Eastern hemlock, red pine, or Eastern white pine. It also occurs in balsam fir and jack pine stands (Erskine 1977). In Ontario, it nests more often in mixed or deciduous than in coniferous forests (Peck and James 1987).

The Ovenbird is often associated with mature forest with a closed canopy and an open understorey (Erskine 1977; Crawford et al. 1981; Derleth et al. 1989). These structural aspects of the vegetation are among the criteria used by Blanchette and Larue (1993) in their index of habitat quality for the Ovenbird in Québec. The species may, however, regularly occupy second-growth cutovers, as has been noted in the southern part of the province (Drapeau 1993). After strip cutting in a conifer-dominated forest in Lotbinière County, it took 12–15 years for Ovenbird abundance to reach the levels of neighbouring uncut areas (G. Falardeau and J.-P. L. Savard, CWS, unpubl. data).

The nest appears to be built entirely by the female. It is a cup-shaped structure placed on the ground and covered with a grassy dome or concealed under branches or a fallen trunk (Peck and James 1987). It somewhat resembles an old-fashioned Dutch oven, hence the name Ovenbird (Bent 1953). The female alone incubates the 4 or 5 eggs (Peck and James 1987), for a period of about 12 days. The young normally leave the nest around the age of 8–10 days, and begin to fly within 10 days of fledging (Terres 1980; Van Horn and Donovan 1994). The Ovenbird is considered a frequent cowbird host (Terres 1980), with parasitism as high as 52% recorded in Michigan (Leberman 1992d). However, only 10% of 61 nests in

Strip cutting has a detrimental effect on local Ovenbird abundance for more than 10 years. Near Leclercville, Lotbinière. *October 1993, Pierre Normand.*

Notes on territory

When it arrives on its breeding grounds in May, the male establishes and alone defends a territory, which is used for mating, nesting, and feeding of both adults and young (Smith and Shugart 1987; Quinlan et al. 1990). Territory size ranges from 0.2 to 1.8 ha (mean 0.27 ha) (DeGraaf and Rudis 1987; Smith and Shugart 1987; Quinlan et al. 1990). Territories are apparently larger when the forest canopy is high and dense and smaller when the understorey is dense.

In Gatineau Park and at La Barrière (near Saint-Michel-des-Saints), a mosaic habitat of forest and open areas supported an average of 2.3 pairs/10 ha (Dauphin 1985). In the Mirabel area, Cyr (1974) reported densities of 10.2 territorial males/10 ha in a mature maple stand and 12.7 territorial males/10 ha in a mature hemlock stand. In a study conducted in La Mauricie National Park, DesGranges (1980b) recorded 27 birds/10 ha in a white birch stand, 8.6 birds/10 ha in a maple-yellow birch stand, 18.3 birds/10 ha in a mixed stand, 7.5 birds in a red spruce stand, 20.4 birds/10 ha in a balsam fir stand, and 5.4 birds/10 ha in a white spruce plantation. In various locations in Québec at or south of the 48th parallel, Erskine (1971a,

Ovenbird

Breeding distribution
1984–1989

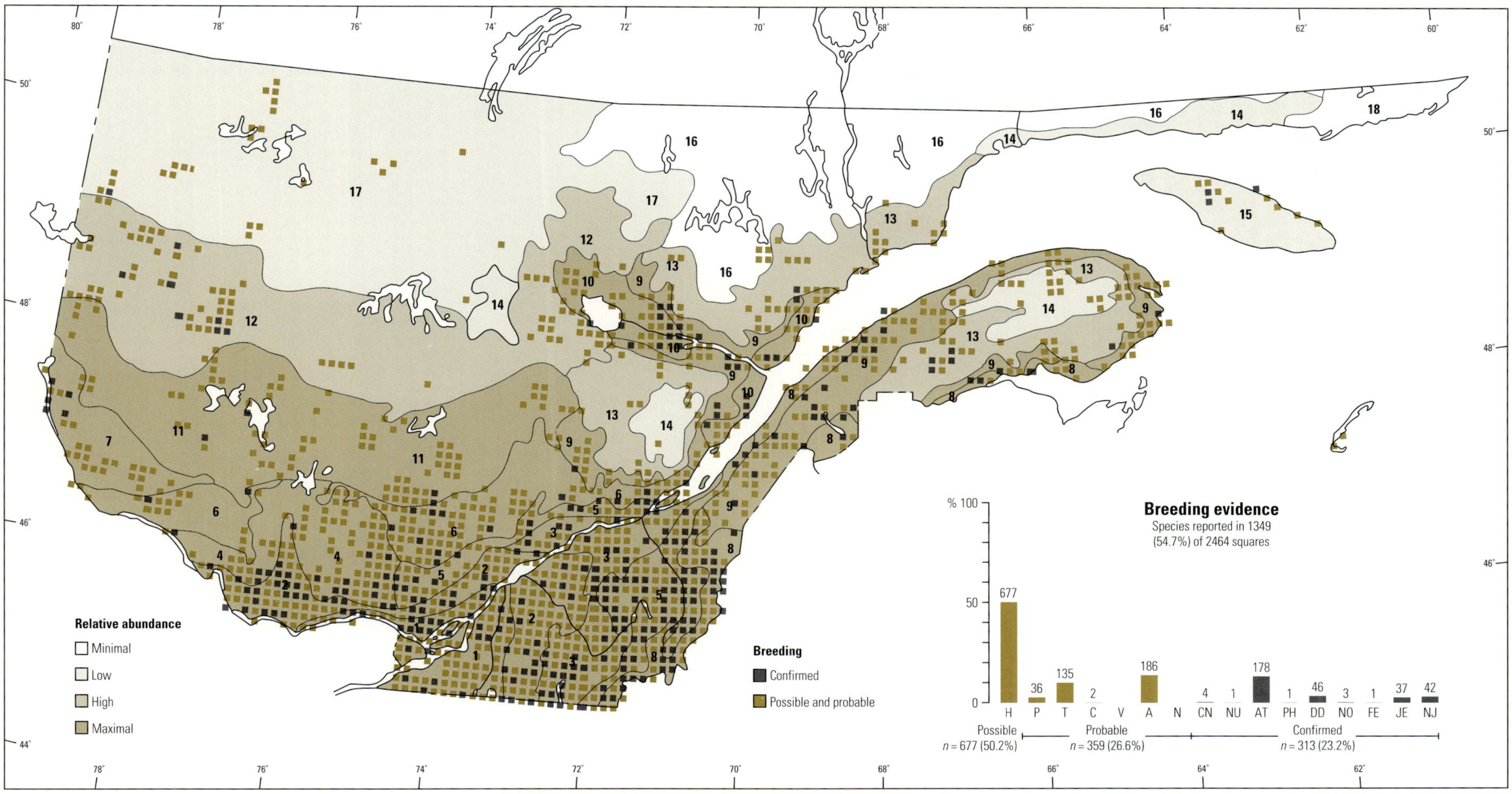

World breeding range

Species profile

Clutch size	usually 4–5 eggs (3–6)[1,3,4,6,10]
Laying interval	usually 1 egg per day[6,10]
Start of incubation	with penultimate egg[10]
Incubation period	usually 12 days (11–14)[2,4,6,10]
Maturity at hatching	altricial
Care of young	by female and male[1,4,6]
Nestling period	8–10 days[1,2,4,6]
Age at first flight	8–20 days[8,9]
Dependence of young	until 30–40 days old[6,10]
Broods per year	usually 1 (1–2)[1,3,6,9]
Breeding age	1 year[3]
Mating system	monogamy, sometimes polygamy[1,6,9,10,11]
Duration of pair bond	sometimes more than 1 season[6]
Total length	14.0–16.5 cm[2]
Wing span	22.9–25.4 cm[8]
Weight	M 22.5 g (17.8–27.5) / F 21.7 g (18.4–25.8)[5]
Longevity record	9 years[7]

References:
1. Ehrlich et al. 1988
2. Godfrey 1986
3. DeGraaf and Rudis 1987
4. Harrison 1978
5. Taylor 1972
6. Bent 1953
7. Klimkiewicz et al. 1983
8. Terres 1980
9. Van Horn and Donovan 1994
10. Hann 1937
11. Hann 1940

Chronology $n = 125$

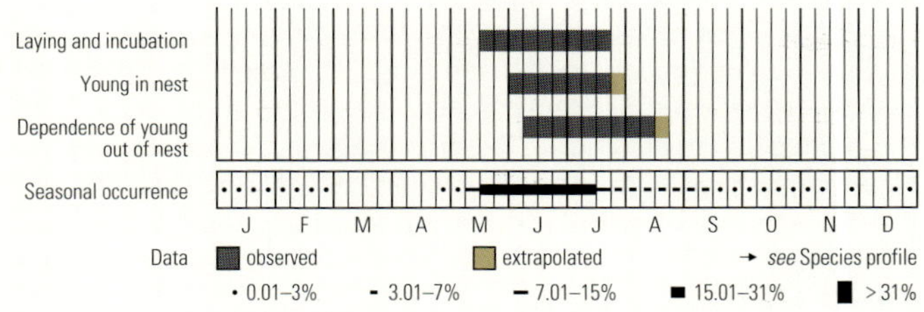

Frequency of occurrence by ecological region

Ecological regions** (forest regions)	Adequately covered squares	Squares in which reported	% of squares in which reported	Relative abundance (%)*
1. Sugar maple-hickory and sugar maple-basswood	97	90	92.8 ± 2.8	92.8
2. Sugar maple-basswood	153	147	96.1 ± 1.8	96.1
3. Sugar maple-basswood and sugar maple-yellow birch	101	100	99.0 ± 1.6	99.0
4. Sugar maple-yellow birch-basswood	56	52	92.9 ± 6.3	92.9
5. Sugar maple-yellow birch-beech and sugar maple-yellow birch-basswood	113	110	97.3 ± 2.0	97.3
6. Sugar maple-yellow birch-beech	71	71	100.0 ± 0.7	100.0
7. Sugar maple-yellow birch-balsam fir	19	19	100.0 ± 2.6	100.0
8. Sugar maple-yellow birch and balsam fir-yellow birch	47	47	100.0 ± 1.1	100.0
9. Balsam fir-yellow birch	87	83	95.4 ± 4.4	95.4
10. Balsam fir-white birch or balsam fir-red maple	46	46	100.0 ± 1.1	100.0
11. Yellow birch-balsam fir	52	50	96.2 ± 6.0	96.2
12. Balsam fir-white birch (west)	59	49	83.1 ± 10.0	83.1
13. Balsam fir-white birch (east)	76	54	71.1 ± 10.0	71.1
14. Balsam fir-black spruce	28	15	53.6 ± 19.3	53.6
15. White spruce-balsam fir and balsam fir-white spruce	22	10	45.5 ± 21.4	45.5
16. Black spruce-balsam fir-moss	18	2	11.1 ± 17.4	11.1
17. Black spruce-moss	32	15	46.9 ± 18.9	46.9
18.			*not analysed*	

* minimal: 0–29.9%; low: 30–59.9%; high: 60–89.9%; maximal: 90–100%

** Deciduous forest: 1–7; mixed forest: 8–11; coniferous forest: 12–17

Ovenbirds prefer to nest in mixed and deciduous woodlands with an almost completely closed canopy and a fairly open understorey. Sainte-Foy. *September 1991, Yves Aubry.*

Québec (Terrill 1961) and 12% of 260 nests in Ontario (Peck and James 1987) were found to have been parasitized. Generally, the species appears to be single-brooded (DeGraaf and Rudis 1987; Ehrlich et al. 1988). Zach and Falls (1975), however, did find one male that had raised two successive broods on the same territory, presumably with the same female, during a spruce budworm outbreak. They also noted a higher incidence of renesting in pairs that had lost their first clutch, in response to an increase in the budworm population.

The Ovenbird feeds mainly on invertebrates—insects, spiders, millipedes, centipedes, snails, and slugs (Zach and Falls 1979)—and is essentially a ground feeder. In second-growth hardwood stands dominated by sugar maple, American beech, and yellow birch in New Hampshire, Holmes and Robinson (1988) found that 83% of the Ovenbird's prey came from the litter, 12% from shrubs, and 5% from trees. Of all the passerine species in these woods, only the Wood Thrush, which consumed 87% of its prey from the litter, specializes more on ground feeding. The Ovenbird forages by walking slowly along the ground, gleaning prey from the surface of the litter. Less often, it forages by turning over the litter or by pouncing in flight on its prey (Smith and Shugart 1987; Holmes and Robinson 1988).

The Ovenbird was reported in 54.7% (1349/2464) of Atlas squares surveyed, and breeding was confirmed in almost a quarter (23%; 313/1349) of these. Only 26.5% of breeding confirmations were based on records of nests or downy young; 56.9% were based on sightings of adults carrying food and 14.7% on distraction displays. These figures are an accurate reflection of the characteristics of this species, in which the young are relatively quiet and nests are difficult to find. Harrison (1984) pointed out that most Ovenbird nests are discovered by accident.

An Ovenbird nest containing 5 eggs at the base of a grassy tussock. Sainte-Dorothée, Laval. *June 1941, Henry Mousley, Laval University Collection.*

Notes on territory (continued)

1972a, 1976a, 1980a, 1984) reported densities of 2.7–5.2 males/10 ha in deciduous habitat with open areas, 4.9–14.1 males/10 ha in deciduous or mixed habitats disturbed by logging, 2.5–15.0 males/10 ha in deciduous stands, and 4.9–15.0 males/10 ha in mixed stands.

In the Nottaway, Broadback, and Rupert river watersheds, the Ovenbird was reported in trembling aspen stands and in shoreline scrub; pair density was almost twelve times higher in the aspen stands than in shoreline scrub (Cons. G. G. – GREBE 1992o).

Distribution

In North America, Ovenbirds breed east of the Rockies from northeastern British Columbia and the District of Mackenzie to Newfoundland and as far south as the northern part of the states bordering the Gulf of Mexico (Godfrey 1986; National Geographic Society 1987). Breeding was confirmed virtually throughout the Atlas area south of the 49th parallel, with sporadic records beyond. On the North Shore, it was reported only as far as the Baie-Trinité area, northeast of Baie-Comeau. There are some old records from the Middle North Shore, however. Todd (1963) cites records from Mingan and the Mingan Islands in June 1921, 1922, and 1926, from Betchouane in June and September 1923, and from the mouths of the Matamec and Washicoutai rivers on 18 August 1928 and 11 June 1928, respectively. The latter record was his own report of the species based on its unmistakable song. The Ovenbird breeds on Anticosti Island, and Ouellet (1969) reported that it was seen regularly in the island's coniferous forests. Although breeding has not been confirmed on the Magdalen Islands, Fradette (1992) considers it probable.

The species is common in the deciduous and mixed forest regions, and less common in the coniferous forest. It is known to occur in fir stands and is one of the most typical species of poplar and birch stands in the boreal forest. It is rare, however, in spruce stands.

A female incubating. The dome of interlaced grass makes the nest difficult to detect. Sainte-Dorothée, Laval. *June 1941, Henry Mousley, Laval University Collection.*

An Ovenbird that has just fed its young. This nest, built of twigs, grasses, and dead leaves, has been built on the ground in a sugar maple stand. With its domed roof, it resembles a miniature Dutch oven. Saint-Augustin-de-Desmaures, west of Québec City. *July 1965, Gaston Le Page.*

North of the Atlas area, there are a few sightings from trembling aspen stands in the Nottaway, Broadback, and Rupert river watersheds (Cons. G. G. – GREBE 1992o), with the most northerly record from less than 1 km south of the Rupert River, corresponding approximately to the northern limit of the range established by Todd (1963) and Godfrey (1986).

History and trends

Comparison of current data with the rather fragmentary information from the turn of the century does not reveal any appreciable change in distribution and abundance. Wintle (1896), Dionne (1906), Macoun and Macoun (1915), and Taverner (1922) all listed the species as a common summer resident in Québec City, Montréal, and Ottawa. Macoun and Macoun (1915) reported a pair on Anticosti Island and, in more recent years, Todd (1963) set the approximate northern limit of the range at the 51st parallel.

The Ovenbird is affected by habitat disturbance, whether selective cutting of mature trees, strip cutting (Webb et al. 1977; G. Falardeau and J.-P. L. Savard, CWS, unpubl. data), or installation of power lines through the forest (Kroodsma, R. L. 1984). It takes 10–15 years for the Ovenbird to reestablish itself in second-growth habitat after logging (Titterington et al. 1979; Morgan and Freedman 1986; Drapeau 1993; G. Falardeau and J.-P. L. Savard, CWS, unpubl. data).

Forest fragmentation, however, seems to be more detrimental to the species. Galli et al. (1976) noted that in New Jersey the Ovenbird is absent from woods of less than 4 ha, and Robbins et al. (1989a) estimated that in the central states along the eastern seaboard, a woodland must be about 320 ha for Ovenbird populations to reach their maximum densities. Gibbs and Faaborg (1990) also noted that mating success depends a good deal on the size of the woodland. In Missouri, barely a quarter of the males were mated in woodlands of 9–140 ha, as opposed to three quarters of the males in forests of 500 ha or more. Forest fragmentation has been blamed for substantial decreases in Ovenbird populations in small urban, semi-urban, and rural woodlots monitored annually over a number of decades (Askins et al. 1987; Tilghman 1987; Askins et al. 1990). In Ontario, Clark et al. (1984) showed that, among other things, the Ovenbird does not tolerate resort development within its habitat. In Québec, surveys of the regional parks on the Island of Montréal found the species uncommon, and absent from some suitable habitat (Bourdages et al. 1988, 1990; Drapeau 1990).

BBS data for 1968–79 show a slight increase in abundance continent-wide (Robbins et al. 1986), followed by a decline from then to 1987 (Robbins et al. 1989b).

Pierre Drapeau and Marcel Darveau

Northern Waterthrush

Paruline des ruisseaux

Seiurus noveboracensis

Christiane Girard

Beginners may have a hard time spotting the Northern Waterthrush, since in forests they tend to look for it in the shrub layer, while it is usually on the ground, near a river or stream. This small thrush-like bird, which, despite its name, is actually a warbler, continually bobs its tail as it perches on a low branch, then gives its loud, spirited song: *twit twit twit sweet sweet sweet chew chew chew* (Peterson 1980). The Northern Waterthrush is a fairly common migrant breeder in southern Québec, arriving in early May from its wintering grounds which extend from the southeastern United States (southern Florida and occasionally the Gulf of Mexico) down through Central America to Ecuador, and southern Venezuela and Guyana (AOU 1983; Root 1988).

Bushy thickets along waterways, like this streamside alder thicket, are one of this waterthrush's favourite habitats. Montmorency forest, Laurentide Wildlife Reserve, north of Québec City. *August 1991, Marcel Darveau.*

Habitat and behaviour

This waterthrush likes the edges of ponds, lakes, or small streams, and floodplain forests (Craig 1985; Elderkin 1989), usually seeking cool, shady, damp ground with a cover of moss, ferns, and club-moss (Forbush 1929; Craig 1985) in the interior of deciduous, coniferous, or mixed forests. As a ground nester, the nature of the herb and shrub layer is more important than forest overstorey type. It is difficult to predict species abundance on the basis of a general habitat study (Elderkin 1989). Even in a large area of seemingly suitable breeding habitat, it is mainly found close to nest support structures or cover, such as the exposed roots of upended trees and hollow stumps (Wander 1985), or the underside of mossy trunks.

The usually bulky nest may also be built in a depression in the ground or hole in the river bank. Rarely more than 1 m off the ground (Peck and James 1987), it is made of mosses, leaves, grasses, twigs, pine needles, strips of inner bark, and rootlets; the inside is lined with animal hair, mosses, or fine grasses (Griscom and Sprunt 1957).

Notes on territory

During the breeding season, the Northern Waterthrush is very aggressive toward conspecifics. Even when staging for a few days during migration, it may defend

This species lives in the shade along the water's edge and may build its nest in eroded parts of the banks. Cap Tourmente National Wildlife Area, northeast of Québec City. *August 1991, Jean Gauthier.*

Northern Waterthrush

Breeding distribution

1984–1989

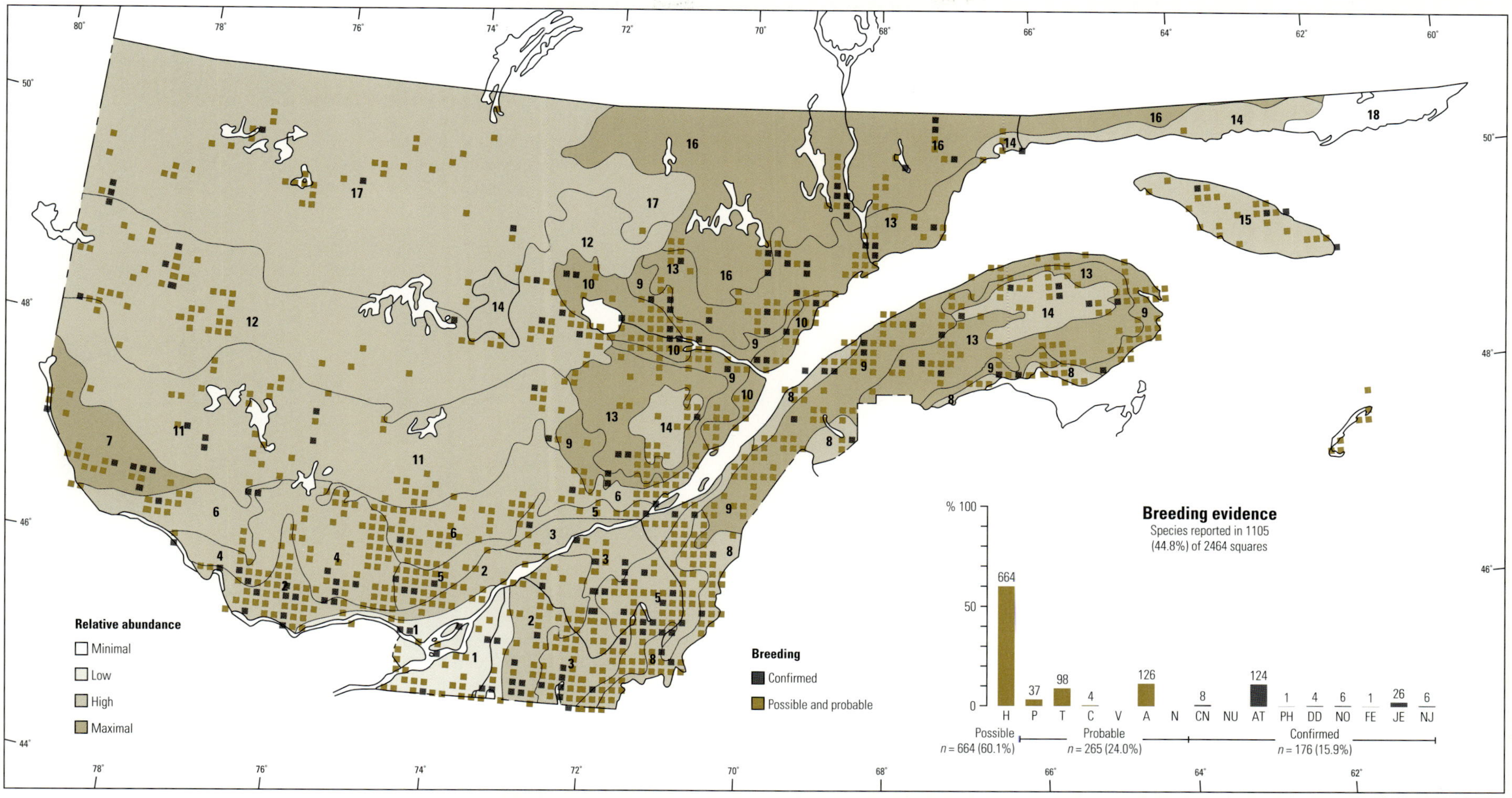

Species profile

Clutch size	usually 4–5 eggs (3–6)[3,4,7,9]
Laying interval	?
Start of incubation	probably with third egg[1]
Incubation period	13–14 days[1,4,7,8,9]
Maturity at hatching	altricial
Care of young	probably by female and male[7]
Nestling period	probably ≈ 10 days[7]
Age at first flight	probably ≈ 10 days or more[10]
Dependence of young	until at least 3 weeks old[11]
Broods per year	1[8,9]
Breeding age	1 year[8]
Mating system	monogamy[7]
Duration of pair bond	?
Total length	12.5–15.0 cm[6]
Wing span	21.6–25.5 cm[9]
Weight	17.8 g (13.8–24.4)[5]
Longevity record	8 years 11 months[2]

References:
1. Wander 1985
2. Klimkiewicz and Futcher 1989
3. Harrison 1978
4. Peck and James 1987
5. Dunning 1984
6. Godfrey 1986
7. Ehrlich et al. 1988
8. DeGraaf and Rudis 1987
9. Forbush 1929
10. according to nestling period
11. Eaton 1957

World breeding range

Chronology $n = 29$

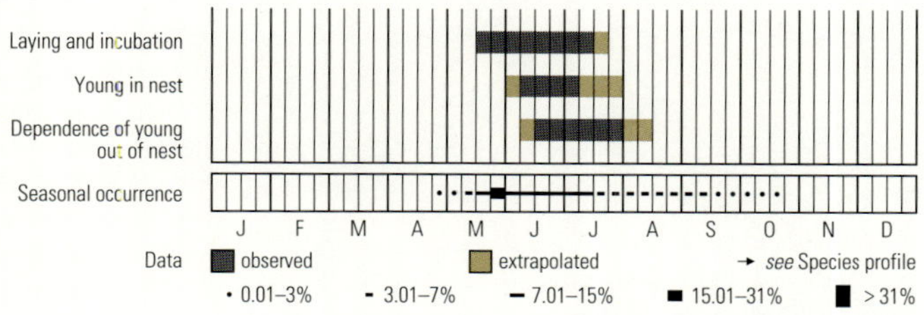

Frequency of occurrence by ecological region

Ecological regions** (forest regions)	Adequately covered squares	Squares in which reported	% of squares in which reported	Relative abundance (%)*
1. Sugar maple-hickory and sugar maple-basswood	97	38	39.2 ± 4.9	41.5
2. Sugar maple-basswood	153	90	58.8 ± 4.1	62.3
3. Sugar maple-basswood and sugar maple-yellow birch	101	59	58.4 ± 5.8	61.9
4. Sugar maple-yellow birch-basswood	56	45	80.4 ± 9.3	85.1
5. Sugar maple-yellow birch-beech and sugar maple-yellow birch-basswood	113	90	79.6 ± 4.2	84.3
6. Sugar maple-yellow birch-beech	71	59	83.1 ± 8.2	88.0
7. Sugar maple-yellow birch-balsam fir	19	17	89.5 ± 15.5	94.7
8. Sugar maple-yellow birch and balsam fir-yellow birch	47	39	83.0 ± 9.9	87.9
9. Balsam fir-yellow birch	87	81	93.1 ± 5.2	98.6
10. Balsam fir-white birch or balsam fir-red maple	46	43	93.5 ± 6.7	99.0
11. Yellow birch-balsam fir	52	32	61.5 ± 13.6	65.2
12. Balsam fir-white birch (west)	59	50	84.7 ± 9.7	89.7
13. Balsam fir-white birch (east)	76	70	92.1 ± 6.2	97.5
14. Balsam fir-black spruce	28	23	82.1 ± 15.2	87.0
15. White spruce-balsam fir and balsam fir-white spruce	22	16	72.7 ± 19.4	77.0
16. Black spruce-balsam fir-moss	18	17	94.4 ± 13.5	100.0
17. Black spruce-moss	32	22	68.7 ± 17.6	72.8
18.			*not analysed*	

* minimal: 0–29.9%; low: 30–59.9%; high: 60–89.9%; maximal: 90–100%

** Deciduous forest: 1–7; mixed forest: 8–11; coniferous forest: 12–17

The roots of upended trees hold earth and conceal many cavities in which this waterthrush often makes its nest. Cap Tourmente National Wildlife Area, northeast of Québec City. *August 1991, Jean Gauthier.*

In early spring, the Northern Waterthrush goes wading in small pools of water, turning over leaves with its bill to examine the undersides for prey. It may also consume benthic or swimming organisms. On the ground, it forages in mud or leaf litter. After leaves emerge in spring, it inspects the stalks and foliage of woody plants; it may even capture insects on the wing (Craig 1984). Although varied, its diet is composed mainly of the larvae of midges, caddisflies, and mayflies, as well as gastropods (Craig 1987). On Île aux Basques in 1975, adults were observed feeding spruce budworms, which were very plentiful there that year, to their young (COQ 1975).

Little is known about the biology of the species. It seems to produce only one brood per season (Forbush 1929), even though it arrives early and has usually finished laying by the time most other warblers begin to nest (Griscom and Sprunt 1957). The female alone incubates the eggs for 13 or 14 days (Peck and James 1987). When an intruder approaches during incubation or brooding, the female sits tight, flushing only at the last moment, sometimes performing a broken-wing distraction display similar to that of Killdeer. The young, like those of the Louisiana Waterthrush and Ovenbird, may fledge about 9 days after hatching (Wander 1985), but at 3 weeks of age, they are still occasionally fed by their parents (Eaton 1957). In southern Québec, Terrill (1961) reported that 17% of 18 nests had been parasitized by the Brown-headed Cowbird. In Ontario, 13.3% of 83 nests found were hosts to cowbird eggs (Peck and James 1987).

The species was reported in 44.8% (1105/2464) of Atlas squares surveyed, but breeding was confirmed in only 176 squares, with 70% of breeding confirmations based on records of adults carrying food to the nest.

The Northern Waterthrush forages on cool, damp, shady ground and in small streams. Forillon National Park, Gaspé Peninsula. *July 1992, Yves Aubry.*

Distribution

In North America, the Northern Waterthrush breeds from Alaska to the Atlantic provinces down into most of the northern United States (AOU 1983).

The species was atlassed as common throughout southern Québec, except in the St. Lawrence Plain around Greater Montréal; this area is primarily agricultural land where it is unlikely to breed because the courses and banks of many waterways, even the smallest, have been disturbed. The Québec range of this waterthrush extends beyond the Atlas area up to the area around Kuujjuarapik, Lake Guillaume Delisle, and Kuujjuaq (ÉPOQ), corresponding to the range up to the tree line shown by Godfrey (1986).

Notes on territory (continued)

a foraging territory from trespassers (Rappole and Warner 1976). In Connecticut, in a mesic mixedwood habitat along a stream, the average territory defended was 0.47 ha in size (Craig 1984).

In Canada, the Northern Waterthrush has been observed in several types of habitats. Its density has been estimated at 11.5 pairs/10 ha in willow scrub along a creek in Manitoba, 6.5 pairs/10 ha in a virgin balsam fir forest in Newfoundland, 6.2 pairs/10 ha in a tall bottomland willow stand in Alberta, 4.0 pairs/10 ha in a silver maple swamp in Ontario, 3.7 pairs/10 ha in a black cottonwood floodplain forest in British Columbia, 2.7 pairs/10 ha in a white spruce-tamarack forest in Manitoba, 2.2 pairs/10 ha in a mature balsam fir forest with alders and peatland in Newfoundland, and 2.0 pairs/10 ha in a riverine sedge marsh in Newfoundland (Erskine 1971a, 1976a, 1980a, 1984).

This species eats invertebrates that it takes from both the surface of the water and the beds of shallow streams, and from the moss on the rocks and banks. Forillon National Park, Gaspé Peninsula. *July 1992, Yves Aubry.*

Northern Waterthrush nest hidden among the roots of an upended tree; the two largest eggs are those of a Brown-headed Cowbird. Sainte-Dorothée, Laval. *May 1941, Henry Mousley, Laval University Collection.*

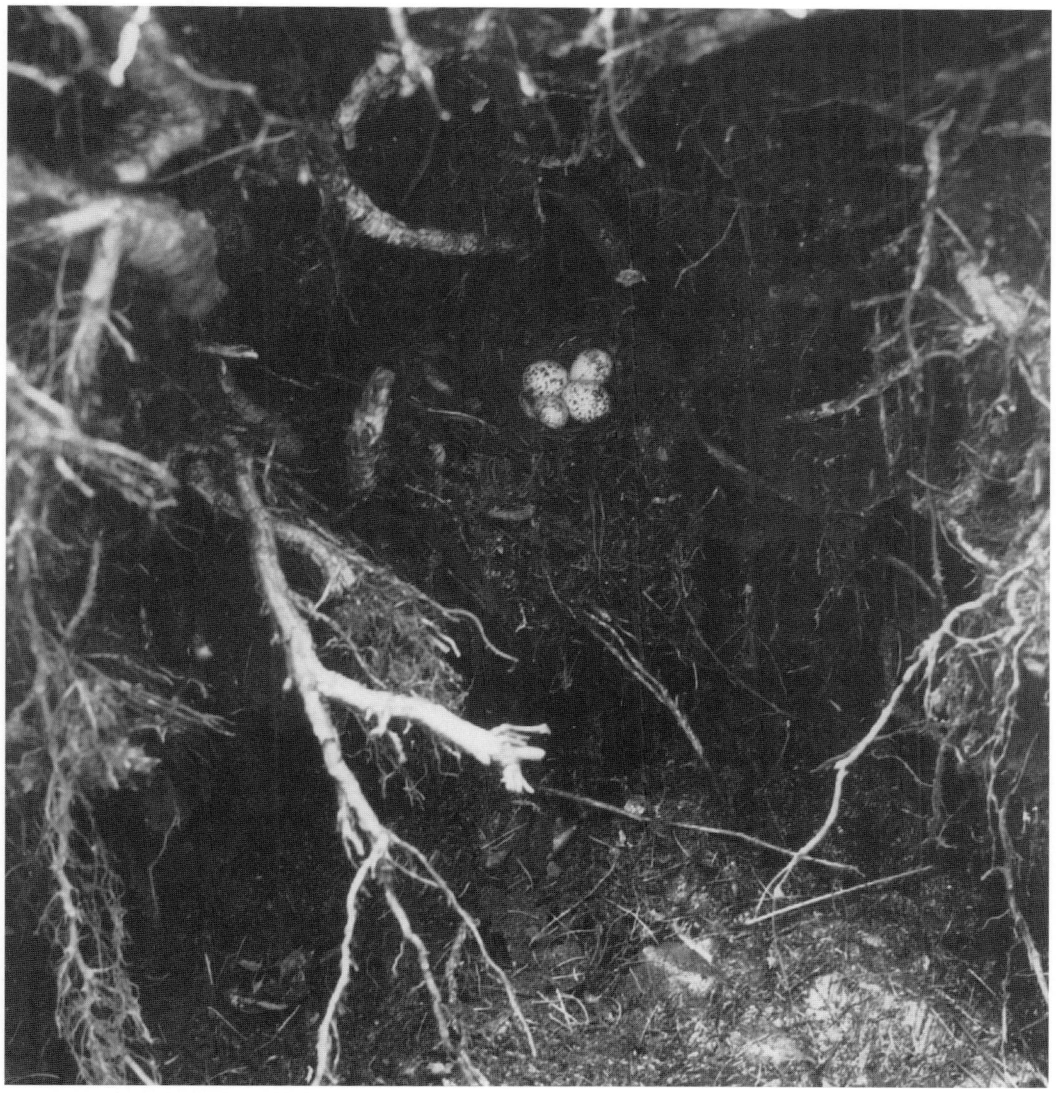

History and trends

At the turn of the century, the Northern Waterthrush was listed as scarce in the Montréal area and the rest of the province (Wintle 1896; Dionne 1906). According to Ouellet (1974a), however, the species was probably commoner around Montréal in the last century than it is now. The draining of wetlands and the straightening of small streams for farming and other purposes are most likely to blame for the decline of the species in Greater Montréal and Montérégie areas. It is also highly probable that increased cottaging on certain lakes in southern Québec has been detrimental to the species.

Numbers in the province seem to have been stable since 1969 (Cyr and Larivée 1993a; BBS), while in New Brunswick, the species' population has experienced a major increase (Robbins et al. 1986). Although the exact causes of this population growth are unknown, Elderkin (1989) says that new logging practices, which keep a riparian buffer strip of vegetation intact, have certainly helped protect suitable habitat. According to data gathered by Askins (1993), the species does not seem to be overly sensitive to the fragmentation of its North American forest habitat, although it is vulnerable to disturbance on its wintering grounds, where it is closely associated with mangrove forests and narrow strips of dense woodland along waterways, habitats that account for a small part of tropical forests and are therefore fragile and vulnerable to logging (Diamond 1991).

Guy Lemieux, Benoît Limoges, and Line Choinière

Connecticut Warbler

Paruline à gorge grise

Oporornis agilis

THE SECRETIVE AND ELUSIVE CONNECTICUT WARBLER ESCAPED ORNITHOLOGISTS' notice until 1808, when it was finally identified by Alexander Wilson (Farrand 1993). However, the first nest was not discovered until 70 years later in 1883, in Manitoba, and another 40 years elapsed before other nests were reported (Bent 1953).

The breeding biology and behaviour of this species still remain poorly known. Although evidence was found as early as 1942 suggesting that the Connecticut Warbler breeds near Rupert Bay (Todd 1963), there is only a handful of pre-Atlas records for Québec. The Atlas survey and recent studies in the southern part of the James Bay region have cleared up some of the mystery and have provided a more accurate picture of the Connecticut Warbler's breeding range, which extends farther east than was previously believed.

The Connecticut Warbler winters from northeastern Colombia to central Brazil (AOU 1983).

Habitat and behaviour

In Ontario and on the Prairies, the species nests in vast open spruce bogs or tamarack fens that have a well-developed understorey of shrubs and herbaceous vegetation, as well as in young open jack pine stands (Helleiner 1987a). It also frequents poplar woods, particularly in Alberta (Godfrey 1986). In Québec, a number of sightings have been made in jack pine stands, with fewer records from poplar stands, open spruce forests, and black spruce peatlands. In the northern part of its range, in the Nottaway, Broadback, and Rupert river watersheds, this warbler was observed mainly in peatlands, black spruce-moss forests, and the ecotone between these two habitats (Cons. G. G. – GREBE 1992o).

The Connecticut Warbler feeds on or near the ground, walking along branches to pick up insects from cracks and crevices in the bark (Terres 1980). The nest, made of mosses or grasses, is usually built on the ground at the base of a shrub, often in a small mound of moss, and is well hidden in the vegetation (Harrison 1978).

Pairs are monogamous, and the female lays 4 or 5 eggs (Ehrlich et al. 1988). Very little is known of the species' incubating and brood-raising habits or of its breeding chronology. Earliest spring arrival dates in Québec are around the second week of May, and the latest fall departure date is early October.

The Connecticut Warbler is one of the most retiring of the wood warblers (Helleiner 1987a), spending most of its time hidden on or near the ground, which makes it very difficult to spot. It can be detected by ear, however, and during the breeding season, its clear, loud, ringing song can be heard at considerable distances.

The Connecticut Warbler was reported in 21 Atlas squares, with most records involving probable or possible breeding. The species was confirmed as breeding in only 3 squares, based on records of adults carrying food or nesting materials. The nest is very difficult to find, particularly because of the species' wariness; when the female approaches the nest, she usually lands on the ground a dozen metres away and walks quietly through the underbrush, and the male appears to refrain from singing near the nest (Harrison 1978). No nests have yet been found in Québec.

Distribution

The Connecticut Warbler breeds from eastern British Columbia to western Québec, including central Ontario, the central Prairie provinces, and the northern portions of some of the Great Lakes states (AOU 1983; Godfrey 1986). Godfrey's (1986) account of the species' distribution in Québec is based on little solid data. Breeding was confirmed for the Atlas in the Abitibi region, where a pair was observed displaying agitated behaviour and the female was observed carrying food on 14 July 1988 south of Lake Dieppe; at La Doré in the Lac-Saint-Jean region on 22 June 1988 based on a report of an adult carrying nesting materials; and near Saint-Honoré, north of Chicoutimi, on 15 July 1989 based on a report of an adult carrying food. Sight records came from Témiscamingue Point, Arntfield, north of Villebois, Villemontel, Berry, Matagami, Louvicourt, between Desmaraisville and Chapais, and the Lac-Saint-Jean region.

Post-Atlas records have come from north of the Abitibi region near Waswanipi, Lac au Goéland, and Lakes Soscumica, Poncheville, and Evans (Cons. G. G. – GREBE 1992o), providing a more accurate picture of the species' breeding range, which is now known to extend almost 300 km farther east than previously thought.

Christiane Girard

History and trends

The Connecticut Warbler remained a mystery for many years, since it was rarely seen in migration and its breeding range was not explored extensively until recently, particularly in the east. As a result, its past status in Québec is difficult to ascertain. Todd (1963) observed this warbler near Waskaganish, east of Rupert Bay, in 1912, and again in 1942, when he confirmed its status as a breeder. John W. Smith believed that he had heard a Connecticut Warbler singing along the lower reaches of the Bell River in 1954 (Todd 1963). The species was also atlassed in the same area. The fact that it was seen in the same places at 30-year intervals suggests that it had been nesting there for some time.

The factors most likely to affect the Connecticut Warbler's abundance and distribution are undoubtedly logging and insecticide spraying for pest control in forestry, particularly in the southern part of its range, where forests are exploited commercially. Their impact cannot be assessed, however, due to our ignorance of the species' biology.

JACQUES IBARZABAL, GERMAIN SAVARD, AND ALAIN MORRIER

Notes on territory

There is almost no information on territoriality in this species. A survey in a tamarack bog at Seddon's Corner, Manitoba, revealed densities of 4 pairs/10 ha (Erskine 1976a). In a black spruce peatland 59 km southeast of Bissett, Manitoba, densities were estimated at 0.6 pair/10 ha (Erskine 1980a).

The Connecticut Warbler frequents poplar stands, spruce peatlands, open spruce forests, and jack pine stands. The species is very secretive and forages on the ground or on low branches. The nest, built on the ground and well concealed in the surrounding vegetation, is very difficult to find. Villemontel, Abitibi. *July 1986, Yves Aubry.*

Connecticut Warbler

Breeding distribution

1984–1989

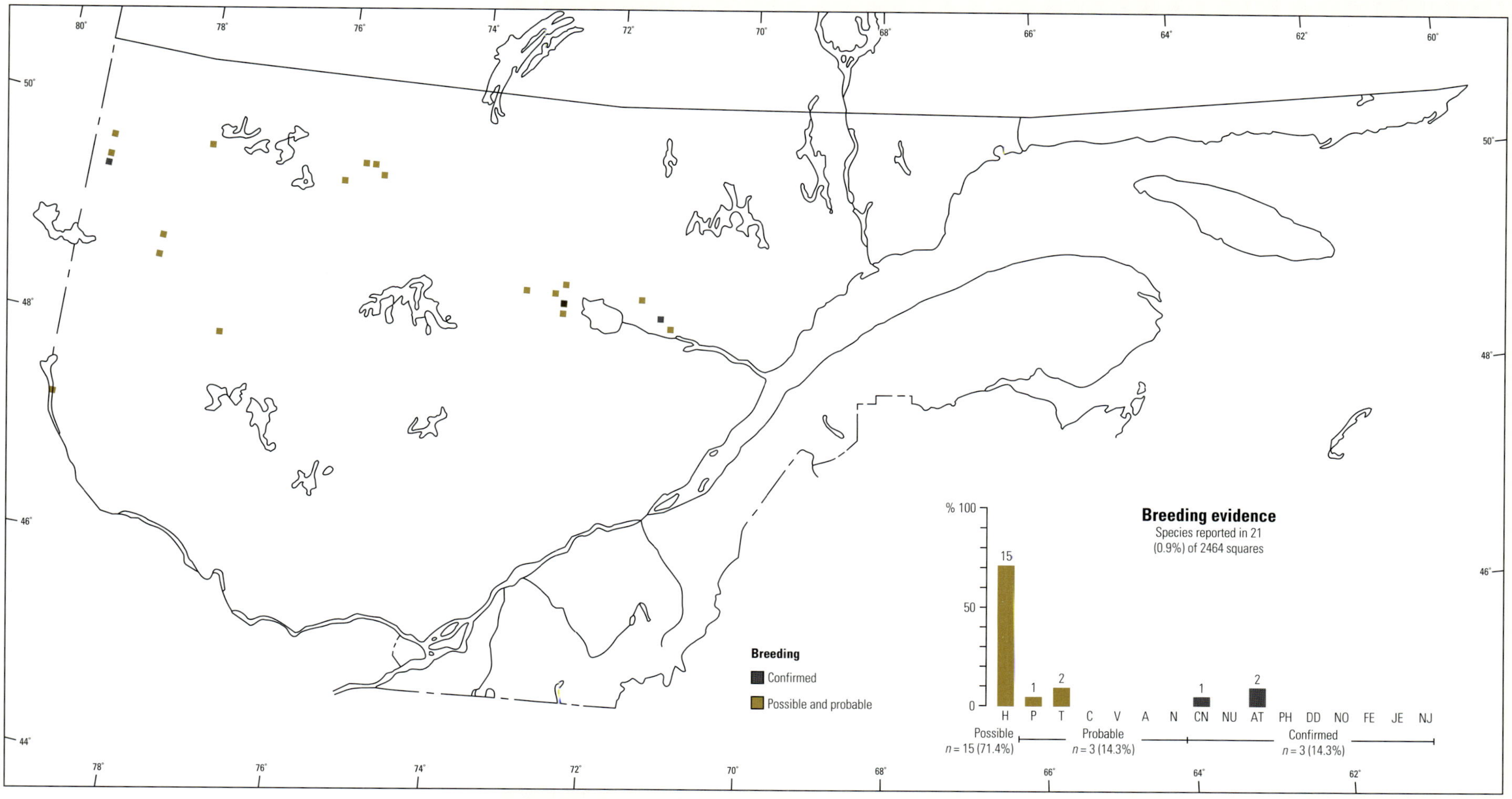

Species profile

Clutch size	usually 4–5 eggs[2,3]
Laying interval	?
Start of incubation	?
Incubation period	?
Maturity at hatching	altricial
Care of young	probably by female and male[2]
Nestling period	?
Age at first flight	?
Dependence of young	?
Broods per year	?
Breeding age	?
Mating system	monogamy[2]
Duration of pair bond	?
Total length	13.0–15.2 cm[3]
Wing span	?
Weight	15.2 g (10.7–26.8)[4]
Longevity record	4 years 3 months[1]

References:

1. Klimkiewicz and Futcher 1989
2. Ehrlich et al. 1988
3. Godfrey 1986
4. Dunning 1984

Chronology $n = 0$

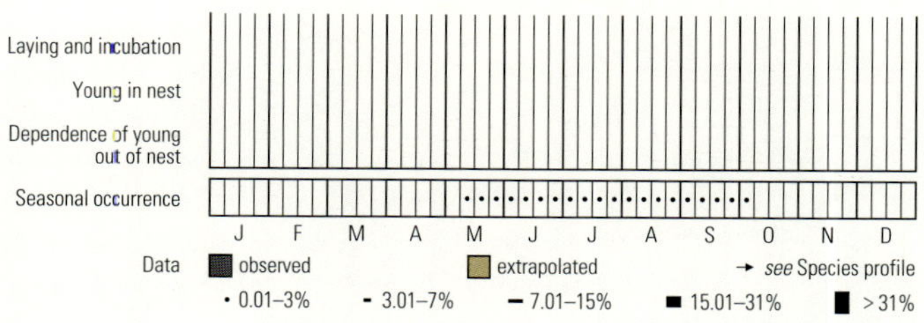

World breeding range

Mourning Warbler

Paruline triste

Oporornis philadelphia

Christiane Girard

The Mourning Warbler was given its unlikely name by Alexander Wilson in 1832, not because of its temperament, which is as lively and animated as any other warbler's, but because the black bib on its breast suggests the black crepe, a band worn then as a symbol of mourning (Bent 1953; Harrison 1984). This species' plumage is striking, its grey hood contrasting strongly with its bright yellow underparts. The Mourning Warbler is often overlooked because of its habit of skulking low to the ground in dense shrubbery. However, it can often be detected by its loud, rolling, easily recognized song, usually transcribed as *chirry, chirry, chirry, chorry, chorry*.

This late spring migrant arrives in Québec at the end of May. It breeds throughout southern Québec (David 1980a), and although it is usually found in small numbers, it may be common in suitable habitat. The Mourning Warbler winters in Central and South America (AOU 1983).

Habitat and behaviour

The Mourning Warbler can be found in deciduous, mixed, and coniferous habitat with openings and a dense shrub layer (Cox 1960; Peck and James 1987). In Québec, the species frequents second growth, wooded slopes, forest margins, ravines, and natural clearings (windfalls and burns) overgrown with raspberry thickets and brambles. It also nests in mature forests that have been thinned, for example in sugar maple stands with an understorey of brush and saplings. The Mourning Warbler particularly favours openings created by logging and is one of the dominant species in the early successional stages (Dessecker and Yahner 1987; Leberman 1992c). In northern Maine, Titterington et al. (1979) found the species particularly abundant in second-growth stands that were 3–5 years into regeneration and overgrown with brambles and fast-growing early succession trees, such as pin cherry, trembling aspen, and white birch, under 2 m high.

The female, probably assisted by the male, builds the nest on the ground (Ehrlich et al. 1988). It is often hidden at the base of a tangle of briers or clump of ferns or weeds, but may occasionally be built in a shrub less than 1 m up (Terres 1980; Godfrey 1986; Peck and James 1987). The rather bulky nest is made of dead leaves, vine stalks, coarse grass, twigs, and bark strips and is lined with rootlets, fine grasses, and hair (Godfrey 1986; Peck and James 1987). The lining materials are often black (Eaton 1914; Peck and James 1987). Typical clutch size is 4 eggs. Except locally, the species is rarely parasitized by the Brown-headed Cowbird (Bent 1953; Terres 1980; Harrison 1984; Ehrlich et al. 1988). The female alone incubates the eggs, but she is fed by the male during this period, generally while on or near the nest. Both parents tend the nestlings. The nest is often difficult to find because it is generally well concealed in dense vegetation and because the female, when disturbed, walks stealthily away from the nest rather than taking flight (Nichols 1985b). Mourning Warblers are usually rather secretive except when they are tending their young, when they become quite agitated and bold if an intruder approaches the nest. The young leave the nest at 7–9 days of age, but remain with their parents until they are 21–30 days old (Cox 1960; Harrison 1978; Ehrlich et al. 1988). Little is known about the species' diet, which seems to consist mainly of insects and spiders (Bent 1953).

Notes on territory

Both male and female defend the territory (Nichols 1985b). Territories generally range from 0.6 to 1 ha in size (Cox 1960; Nichols 1985b; DeGraaf and Rudis 1987).

Densities between 1 and 3.2 territorial males/10 ha were found in dense second growth and in sites that had been logged 4 or 5 years earlier (DeGraaf and Rudis 1987; Smith, C. R. 1988a). In a young balsam fir stand with an admixture of birch and poplar in the Rivière Nouvelle area of the Gaspé Peninsula, densities of 4.9 pairs/10 ha were recorded (Erskine 1980a). In Ontario, densities generally range from 0.02 to 10 pairs/10 ha, regardless of habitat (Fraser 1987b).

The Mourning Warbler frequents bushy areas and clearings in second-growth forest. Lake Saint-Pierre region. *May 1991, Réginald Ouellet.*

Mourning Warbler

Breeding distribution

1984–1989

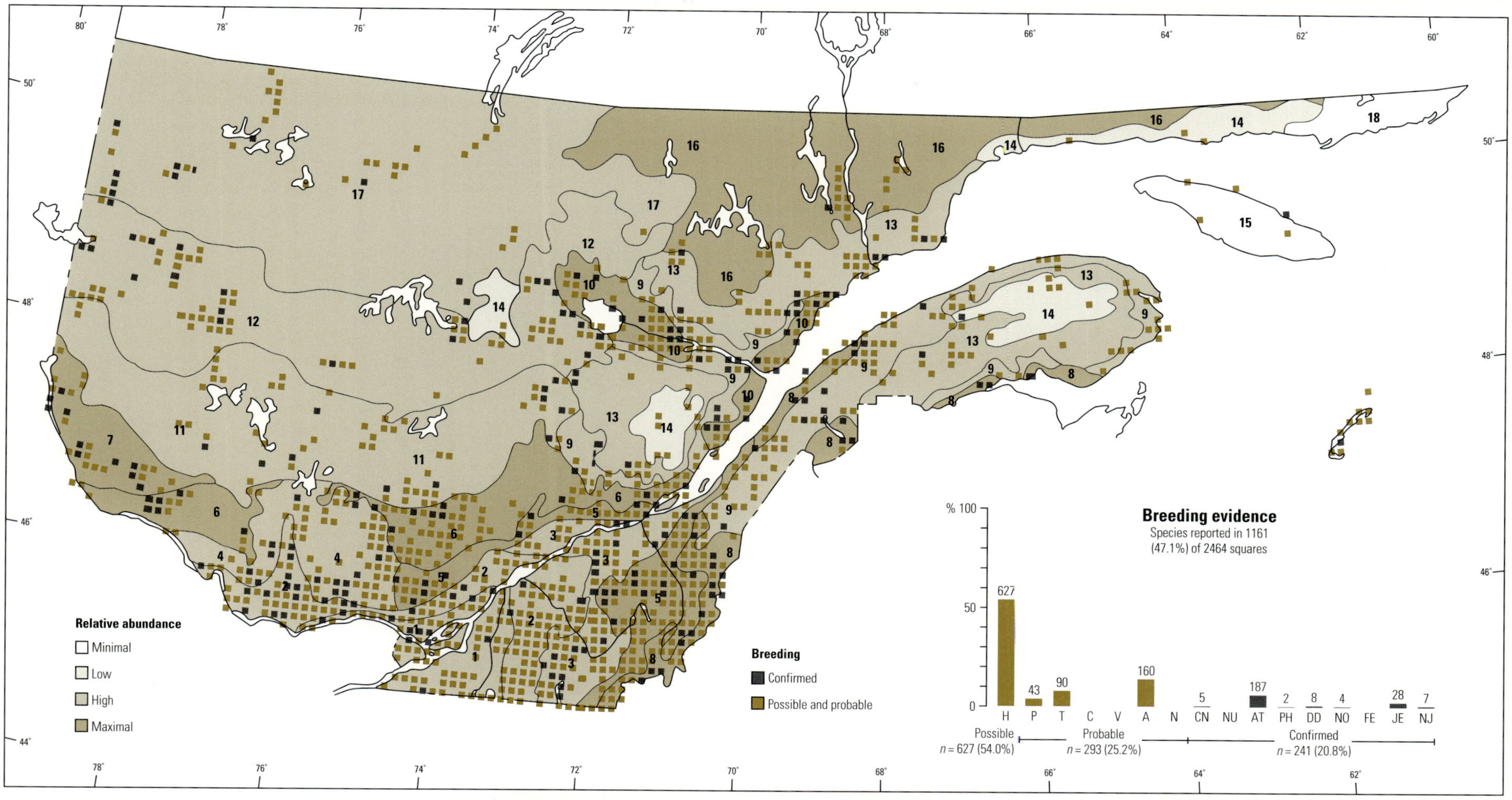

Species profile

Clutch size	usually 4 eggs (2–5)[1,2,3,5]
Laying interval	1 egg per day[9]
Start of incubation	with first egg[9]
Incubation period	12 days[1,2,3,5]
Maturity at hatching	altricial
Care of young	by female and male[1]
Nestling period	7–9 days[1,2,3,5,8]
Age at first flight	≈ 13–15 days[5]
Dependence of young	until 21–30 days old[1,5]
Broods per year	1[2]
Breeding age	1 year[2]
Mating system	probably monogamy[1]
Duration of pair bond	?
Total length	12–14 cm[4]
Wing span	?
Weight	M 13.0 g (9.6–17.9) / F 12.0 g (10.0–14.7)[6]
Longevity record	7 years 11 months[7]

References:
1. Ehrlich et al. 1988
2. DeGraaf and Rudis 1987
3. Terres 1980
4. Godfrey 1986
5. Cox 1960
6. Dunning 1984
7. Klimkiewicz et al. 1983
8. Harrison 1978
9. Pitocchelli 1993

World breeding range

Chronology $n = 22$

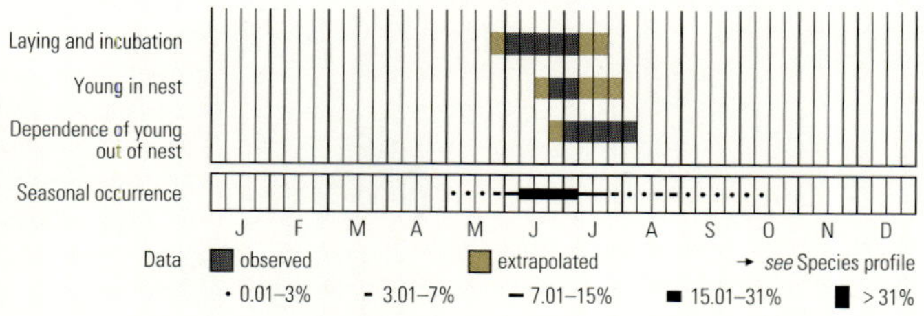

Frequency of occurrence by ecological region

Ecological regions** (forest regions)	Adequately covered squares	Squares in which reported	% of squares in which reported	Relative abundance (%)*
1. Sugar maple-hickory and sugar maple-basswood	97	67	69.1 ± 4.6	70.6
2. Sugar maple-basswood	153	127	83.0 ± 3.2	84.9
3. Sugar maple-basswood and sugar maple-yellow birch	101	81	80.2 ± 4.8	82.0
4. Sugar maple-yellow birch-basswood	56	46	82.1 ± 9.0	84.0
5. Sugar maple-yellow birch-beech and sugar maple-yellow birch-basswood	113	101	89.4 ± 3.3	91.4
6. Sugar maple-yellow birch-beech	71	63	88.7 ± 7.0	90.7
7. Sugar maple-yellow birch-balsam fir	19	18	94.7 ± 12.0	96.8
8. Sugar maple-yellow birch and balsam fir-yellow birch	47	42	89.4 ± 8.3	91.3
9. Balsam fir-yellow birch	87	69	79.3 ± 7.9	81.1
10. Balsam fir-white birch or balsam fir-red maple	46	45	97.8 ± 4.4	100.0
11. Yellow birch-balsam fir	52	40	76.9 ± 11.9	78.6
12. Balsam fir-white birch (west)	59	49	83.1 ± 10.0	84.9
13. Balsam fir-white birch (east)	76	52	68.4 ± 10.2	69.9
14. Balsam fir-black spruce	28	11	39.3 ± 18.9	40.2
15. White spruce-balsam fir and balsam fir-white spruce	22	2	9.1 ± 13.3	9.3
16. Black spruce-balsam fir-moss	18	16	88.9 ± 17.4	90.9
17. Black spruce-moss	32	25	78.1 ± 15.9	79.9
18.			*not analysed*	

* minimal: 0–29.9%; low: 30–59.9%; high: 60–89.9%; maximal: 90–100%

** Deciduous forest: 1–7; mixed forest: 8–11; coniferous forest: 12–17

The Mourning Warbler prefers early stages of succession like this cutover, which is covered with bushes, herbaceous plants, and raspberry brambles. Montmorency forest, Laurentide Wildlife Reserve, north of Québec City. *June 1991, Réginald Ouellet.*

This warbler colonizes a wide range of disturbed habitat like roadsides and power lines, as long as they have a dense shrub layer. Grands-Jardins Conservation Park, northwest of Baie-Saint-Paul. *June 1987, Réginald Ouellet.*

The Mourning Warbler's nest is usually built on the ground, often in a tangle of briers or in a clump of ferns or herbaceous plants. Sainte-Dorothée, Laval. *June 1941, Henry Mousley, Laval University Collection.*

The Mourning Warbler was reported in 47.1% (1161/2464) of Atlas squares surveyed, and breeding was confirmed in 20.8% (241/1161), mainly through records of adults carrying food.

Distribution

In Canada, the Mourning Warbler nests from central Alberta to Newfoundland and, in the United States, in New England, the Great Lake states, and the Appalachians south to Virginia (AOU 1983).

Breeding was confirmed for the Atlas throughout southern Québec except on the Middle North Shore, where there were sight records down to Havre-Saint-Pierre, however. In 1928, Todd (1963) found it fairly common near the mouth of the Little Mecatina River, although he did not obtain any local breeding evidence. More recently, the species was observed at Harrington Harbour on 14 June 1978 and on 6 June 1982 (ÉPOQ).

The Mourning Warbler is quite common throughout southern Québec except in the high-elevation Lake Marceau region (east of Gouin Reservoir), the Laurentide Wildlife Reserve, the central Gaspé Peninsula, the Middle and Lower North Shore, and on Anticosti Island. As the Atlas data show, the Mourning Warbler is mainly associated with successional habitats in both the boreal and deciduous forest, where brushy openings and forest margins are found in abundance. The disturbed habitat it prefers is prevalent throughout southern Québec.

Although this warbler nests regularly as far north as the 51st parallel according to both David (1980a) and Godfrey (1986), several records have also been obtained north of this latitude. Todd (1963) reported the species at Waskaganish in June and July 1914 and June 1942 and at Eastmain in July 1941. It was also spotted a few times near the Eastmain River in June and July 1981 (ÉPOQ; Bordage and Aubry 1982).

History and trends

The Mourning Warbler was scarce in Québec a century ago (Wintle 1896; Macoun and Macoun 1915; Lloyd 1924; Mousley 1924c). Dionne (1906) described it as accidental in Montréal and Québec City but common on the Magdalen Islands. It was also considered rare then in Ontario (Fraser 1987b), Vermont (Nichols 1985b), and New York State (Smith, C. R. 1988a), areas where the species is now widespread. As Ouellet (1974a) suggested, this warbler has undoubtedly benefited from the deforestation that has occurred over the last hundred years, which has created the second-growth habitat it prefers. It has been suggested, however, that early observers, who were not as familiar with bird songs as we are today, simply overlooked the species (Griscom and Sprunt 1979). It is difficult to determine the exact role of these two phenomena.

More recent studies have shown that Mourning Warbler populations have remained stable during the last few decades (BBS; Cyr and Larivée 1993a).

Claude Nadeau and Serge Gauthier

Common Yellowthroat

Paruline masquée

Geothlypis trichas

Christiane Girard

IN SUMMER, PARTICULARLY NEAR DAMP AREAS, THERE ARE FEW TANGLES OF vegetation in Québec, from Nuns' Island in the south to Chibougamau in the north, that do not harbour at least one pair of Common Yellowthroats. This warbler is easy to observe, since its preferred breeding habitat is abundant and accessible. Therefore, it is not surprising that the Common Yellowthroat was atlassed as the commonest breeding warbler in the province.

Both in spring and summer, the male is instantly recognizable, with his black mask, yellow throat, and yellow breast. Identifying the female and immature birds is slightly more difficult and requires some knowledge of the species' habitat and characteristic call note, a husky *tchep*.

Common Yellowthroats are very active and energetic, and become easily agitated when an intruder invades their territories. Then, they can often be heard nervously raising the alarm as they dart from branch to branch, disappearing in the undergrowth and then reappearing a few metres away to begin the performance again.

The Common Yellowthroat winters in the southern United States, the West Indies, and Central America south to Panama (AOU 1983).

Habitat and behaviour

The Common Yellowthroat is common in open, brushy habitat. It is most frequently found in damp areas, but is quite common in dry places as well (Peck and James 1987). This warbler occurs in a wide range of habitat, from marshes and overgrown fields to country roadsides and in brush along pastures, woodlots, lakes, and rivers. It spends most of its time in the lower levels of the vegetation, emerging only to sing (McLaren 1987). The Common Yellowthroat forages on the ground and in low bushes, gleaning gypsy moths, cankerworms, fall webworms, caterpillars, grasshoppers, leafhoppers, aphids, and spiders (DeGraaf and Rudis 1987).

The Common Yellowthroat usually arrives in Québec in early May and is soon much in evidence. Males can be seen emerging one by one from the shrubbery to proclaim their territories with a loud *witchity, witchity, witchity, witch* from a prominent perch. The Common Yellowthroat also has a flight song, which can be heard at any time of the day and sometimes even at night, consisting of short, sputtering notes combined with phrases from its more usual song. It is uttered as the male climbs laboriously to a height of almost 5 m from the ground, after which the bird silently drops into a nearby shrub. According to Ritchison (1991), the flight song is not territorial, but is used to warn the mate of a predator and draw the predator's attention away from the nest to the male. However, according to Kowalski (1983), since the song is performed more frequently as the breeding season progresses, it is probably also used to maintain territorial boundaries once they have been established and to discourage other males from trespassing.

Males may sometimes have more than one female on their territory (Powell and Jones 1978). The female builds the nest directly on or not far off the ground. The cup-shaped nest is a fairly large, bulky structure hidden in a tussock of tall grass or a dense shrub. The interior is usually fairly compact and neatly woven, while the exterior is more roughly fashioned (Peck and James 1987). The eggs are incubated by the female, and the nestlings leave the nest 8–10 days after hatching, when they are still unable to fly. They spend the next two or three days hopping around in the nearby bushes before making their first flight. The parents feed the young for at least two weeks after they have left the nest (Stewart 1953). Throughout the brood-raising period, both parents feed the young, but sometimes the male may

The Common Yellowthroat frequents a wide range of habitat, including shrub thickets on the shores of lakes and rivers. Jacques Cartier River. Laurentide Wildlife Reserve, north of Québec City. *June 1991, Réginald Ouellet.*

Notes on territory

As soon as Common Yellowthroats arrive in the spring, they stake out and defend a breeding territory, which is maintained until just after the commencement of moulting in early August (Stewart 1953). In Ontario, territories averaging 0.5 ha were found (McLaren 1987). In Michigan, territories of 10 monogamous males ranged between 0.3 and 0.7 ha, while a polygynous male with 2 females occupied a 1.4-ha territory (Stewart 1953). In New York State, territories in shrubby field habitat were 0.4 ha on average (DeGraaf and Rudis 1987), compared with 1.56 ha in Vermont in a habitat composed of mixed wood and overgrown fields (Ellison 1985k). In a New York State study, Kendeigh (1945) reported that once the young had left the nest, adults with their young began to roam around, and birds wanting to lay a second clutch had to adjust their territory boundaries.

In Massachusetts, the density of Common Yellowthroats in young pine stands was approximately 2.2 pairs/ha (Morimoto and Wasserman 1991a). In marshy habitat in Algonquin Park near the Québec border, Erskine (1971a) reported densities between 4.5 and 33.5 pairs/10 ha, which illustrates the high densities that can occur in good habitat. In Nova Scotia, in stands less than 12 years old,

Common Yellowthroat

Breeding distribution
1984–1989

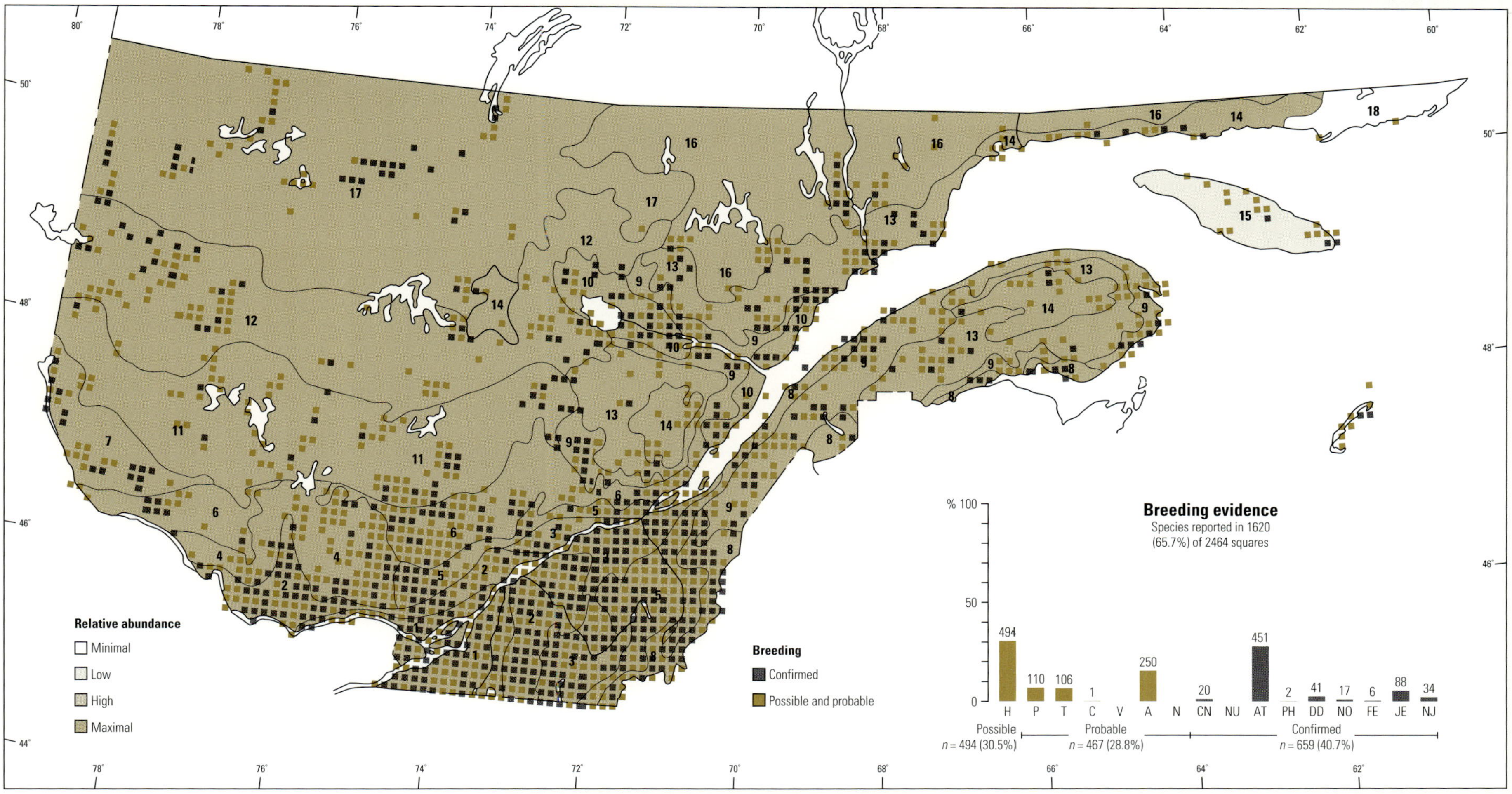

Species profile

Clutch size	usually 3–5 eggs (3–6)[4]
Laying interval	?
Start of incubation	?
Incubation period	11–13 days[1,2,3,4]
Maturity at hatching	altricial
Care of young	by female and male[4]
Nestling period	8–10 days[1,2,3,4]
Age at first flight	11–13 days[10]
Dependence of young	until 2 weeks after leaving nest[8]
Broods per year	1–2[2,4]
Breeding age	1 year[2]
Mating system	monogamy, rarely polygyny[3,4,9]
Duration of pair bond	?
Total length	12–14 cm[7]
Wing span	?
Weight	M 10.3 g (7.6–15.5) / F 9.9 g (7.6–15.3)[5]
Longevity record	9 years 11 months[6]

References:
1. Harrison 1978
2. DeGraaf and Rudis 1987
3. Terres 1980
4. Ehrlich et al. 1988
5. Dunning 1984
6. Klimkiewicz et al. 1983
7. Godfrey 1986
8. Stewart 1953
9. Powell and Jones 1978
10. Stokes 1979

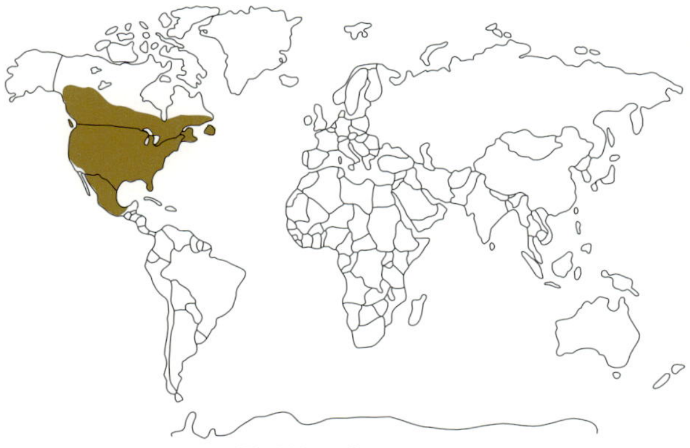

World breeding range

Chronology $n = 89$

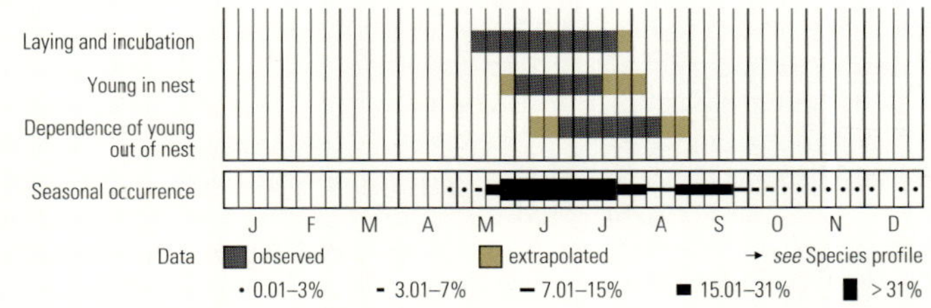

Frequency of occurrence by ecological region

Ecological regions** (forest regions)	Adequately covered squares	Squares in which reported	% of squares in which reported	Relative abundance (%)*
1. Sugar maple-hickory and sugar maple-basswood	97	97	100.0 ± 0.5	100.0
2. Sugar maple-basswood	153	151	98.7 ± 1.2	98.7
3. Sugar maple-basswood and sugar maple-yellow birch	101	101	100.0 ± 0.5	100.0
4. Sugar maple-yellow birch-basswood	56	54	96.4 ± 4.8	96.4
5. Sugar maple-yellow birch-beech and sugar maple-yellow birch-basswood	113	113	100.0 ± 0.4	100.0
6. Sugar maple-yellow birch-beech	71	71	100.0 ± 0.7	100.0
7. Sugar maple-yellow birch-balsam fir	19	19	100.0 ± 2.6	100.0
8. Sugar maple-yellow birch and balsam fir-yellow birch	47	47	100.0 ± 1.1	100.0
9. Balsam fir-yellow birch	87	85	97.7 ± 3.3	97.7
10. Balsam fir-white birch or balsam fir-red maple	46	46	100.0 ± 1.1	100.0
11. Yellow birch-balsam fir	52	52	100.0 ± 1.0	100.0
12. Balsam fir-white birch (west)	59	57	96.6 ± 5.3	96.6
13. Balsam fir-white birch (east)	76	74	97.4 ± 4.0	97.4
14. Balsam fir-black spruce	28	27	96.4 ± 8.3	96.4
15. White spruce-balsam fir and balsam fir-white spruce	22	10	45.5 ± 21.4	45.5
16. Black spruce-balsam fir-moss	18	18	100.0 ± 2.8	100.0
17. Black spruce-moss	32	31	96.9 ± 7.6	96.9
18.		*not analysed*		

* minimal: 0–29.9%; low: 30–59.9%; high: 60–89.9%; maximal: 90–100%

** Deciduous forest: 1–7; mixed forest: 8–11; coniferous forest: 12–17

The Common Yellowthroat is also found in overgrown fields and other dry, bushy habitat. La Prairie, near Montréal. *July 1991, Gaétan Duquette.*

Notes on territory (continued)

densities of 11.4 pairs/10 ha were found (Morgan and Freedman 1986). In Québec, densities of 6.7 pairs/ 10 ha were recorded in alder and willow thickets north of Duparquet in the Abitibi region (Erskine 1972a), and densities of 10.0 pairs/ 10 ha were observed in a red maple-black ash swamp forest in Dunham, in the southernmost part of the province (Erskine 1984).

be left to tend the young while the female builds a nest for a second clutch (Stokes 1989).

In the southernmost part of its range, the species regularly raises a second brood (Ehrlich et al. 1988), but this has never been reported in Québec, despite the fact that the length of the laying and incubation period would allow a second brood to be produced. The second brood is fed by the parents up to migration time.

According to Ehrlich et al. (1988), the Common Yellowthroat is one of the three most frequent cowbird hosts. Terrill (1961), however, estimated that only 7% of Common Yellowthroat nests are parasitized in southern Québec. An analysis of QNRCP nest record cards indicates that 18.3% of the 64 nests in the data set had been parasitized (M. Gosselin, pers. comm.), similar to the rate of 19% found by Peck and James (1987) in Ontario. In Michigan, over 45% of nests found had been parasitized by cowbirds, with, on average, only 0.1 fledgling/parasitized nest, compared to 1.9 fledglings/unparasitized nest (Stewart 1953). In New York State, the Common Yellowthroat is the seventh most frequent cowbird host (Arbib 1988a).

Mourning Warbler and Common Yellowthroat hybrids have been reported, proving conclusively that warblers retain the ability to produce hybrids long after genetic isolation (Bledsoe 1988; Aubry et al. 1989).

In keeping with its ubiquitous presence throughout southern Québec, the species was atlassed in 65.7% of the squares surveyed (1620/2464), and breeding was confirmed in 40.7% (659/1620) of these. Records of adults carrying food were the main evidence used to confirm breeding (451/659), while reports of recently fledged young came in a distant second (88/659).

Distribution

The Common Yellowthroat breeds throughout North America. In Canada, its range extends north to northern British Columbia and Alberta and the central portions of the prairie provinces, Ontario, Québec, and Newfoundland, and throughout the Maritimes (AOU 1983).

The female builds the nest on the ground, or a few centimetres above, in a dense bush or clump of tall grass. The rather bulky nest is made of weed stems, grasses, and leaves. Sainte-Foy. *June 1955, Gaston Le Page.*

The Common Yellowthroat is the most frequently observed warbler in Québec. It frequents the lower levels of bushy open areas, but emerges to sing from a prominent perch. Cap Tourmente National Wildlife Area, northeast of Québec City. *August 1991, Yves Aubry.*

In Québec, the Common Yellowthroat was among the 10 most atlassed species. It breeds commonly everywhere throughout southern Québec except on Anticosti Island. Although its open habitat is mainly associated with agricultural and populated regions, this species is also very common in remoter areas, including the northern part of the Atlas area, where its haunts—the shores of lakes and rivers, roadsides, peatlands, and other shrubby habitat—are found in abundance. The species' poor showing on Anticosti Island is probably due to the fact that open shrubby habitat is less common and more localized on the island than elsewhere in southern Québec.

According to Godfrey (1986), the northern boundary of the species' range in Québec corresponds roughly to the 52nd parallel. Manning and MacPherson (1952), however, found this warbler farther north: at Moar Bay on the coast of James Bay on 30 June and 14 July 1950 and at Paul Bay on 23 and 24 July 1950. More recently, surveys of the James Bay region have shown that the species occurs regularly in the Eastmain River area (Bordage and Aubry 1982, 1983; ÉPOQ), where it is the principal species in shoreline scrub (Tremblay and Belles-Isles 1991), as well as around the Opinaca River and the La Grande-2 complex (ÉPOQ). It has been reported as a possible breeder as far north as Lake Bienville (Cons. G. G. – GREBE 1991a), more than 400 km north of the species' known breeding range. Farther east, it has been seen at Lake Plétipi, northwest of the Manicouagan Reservoir, near the Labrador border, and at Gagnon (ÉPOQ).

History and trends

The Common Yellowthroat was common in Québec a century ago (Macoun and Macoun 1915) and populations in the St. Lawrence Plain have not changed substantially since the turn of the century (Ouellet 1974a). Although many suitable habitats like peatlands and marshes have disappeared, many others have been created by the clearing of forests. Numbers on the fringes of populated areas have probably been increasing for the last two centuries owing to road construction, logging, and the clearing of forests for agriculture. However, modern farming methods, which produce earlier and more frequent harvests, may have a negative impact on some local populations. In a recent Iowa study, Common Yellowthroat males were found to desert territories established early in the season when the first alfalfa crop was harvested (Frawley and Best 1991). Underground drainage systems have been used for a number of years in Québec to increase arable land, which has resulted in the disappearance of the traditional shrub-lined drainage ditches between fields, habitat used by the species.

The Common Yellowthroat is generally considered the most abundant breeding warbler in southern Québec, and the Breeding Bird Survey has shown it to be the most abundant breeding warbler in eastern North America (Robbins et al. 1986). In the eastern United States, populations have recently increased, probably due to habitat fragmentation, which has produced more edge habitat (Rolfinke et al. 1990). In the Maritimes, however, no changes in Common Yellowthroat populations were observed between 1966 and 1992 (Erskine 1992).

In Québec, no significant population trend is evident in either ÉPOQ data for 1970–89 or BBS data for 1969–89 (Cyr and Larivée 1993a). However, BBS data for 1980–89 do show a decrease. Abandonment of agricultural land is increasingly widespread in most regions of southern Québec (M. Carignan, MAPAQ, unpubl. data), and although this probably has a short-term positive effect on local populations as successional shrub communities replace open fields, the beneficial impact is lost as the scrub reverts to forest.

VINCENT LÉTOURNEAU AND PIERRE LAFONTAINE

Wilson's Warbler

Paruline à calotte noire

Wilsonia pusilla

Christiane Girard

WILSON'S WARBLER HAUNTS SHRUBBY WET AREAS SELDOM VISITED BY birders. This bundle of energy breeds throughout Québec south of the tree line, except in some of the more populated areas, but its habitat is more localized in the southern part of the province than in the north. It is apparently not abundant anywhere in eastern North America, and it is rare in southern Québec except in migration.

This species winters from southern Texas, northern Mexico, and Florida to Panama (AOU 1983; Godfrey 1986).

Habitat and behaviour

Although its breeding range is largely associated with northern coniferous forests, Wilson's Warbler does not nest in the forest itself, but in wet habitat with tall shrubs or short deciduous trees, especially alder or willow. Boggy areas with nearby tamarack or stunted spruce trees are typical. There is often a stream nearby (Helleiner 1987c).

Primarily an insectivore (Tyler 1953b), it is both a gleaner and a darting flycatcher, usually staying within 3 m of the ground (Terres 1980). It is very perky, twitching its tail up and down spasmodically. The males, and some females, are easily identified by a black crown spot during spring migration and in the breeding season.

Whether the nest is located in moss or soil at the base of a shrub or sapling, hidden in a grass tussock or in dry grass under low willow branches, or in a hollow in a bank with overhanging plants (Harrison 1978), it is usually bulky for the size of the bird (Tyler 1953b). In Ontario, the few nests found were on the ground, in hummocks or clumps of grassy weeds or moss (Peck and James 1987).

According to a California study, the female constructs the nest in about 5 days (Stewart 1973). The female may wait 1–2 days after completing the nest before laying. The male may feed her while she is incubating. After a disturbance, the female returns to the nest very quickly, even when an observer remains standing nearby. Both parents feed the young, and in California, Stewart (1973) estimated that in one 14-hour day when the 4 young were 9 days old, the two parents made 812 feeding trips to the nest.

Notes on territory

Older males arrive on the breeding grounds before females and second-year males, often returning to the same territories in

In the boreal forest, wet shrubby areas are excellent breeding habitat for Wilson's Warbler. Montmorency forest, Laurentide Wildlife Reserve, north of Québec City. *June 1991, Réginald Ouellet.*

Wilson's Warbler

Breeding distribution

1984–1989

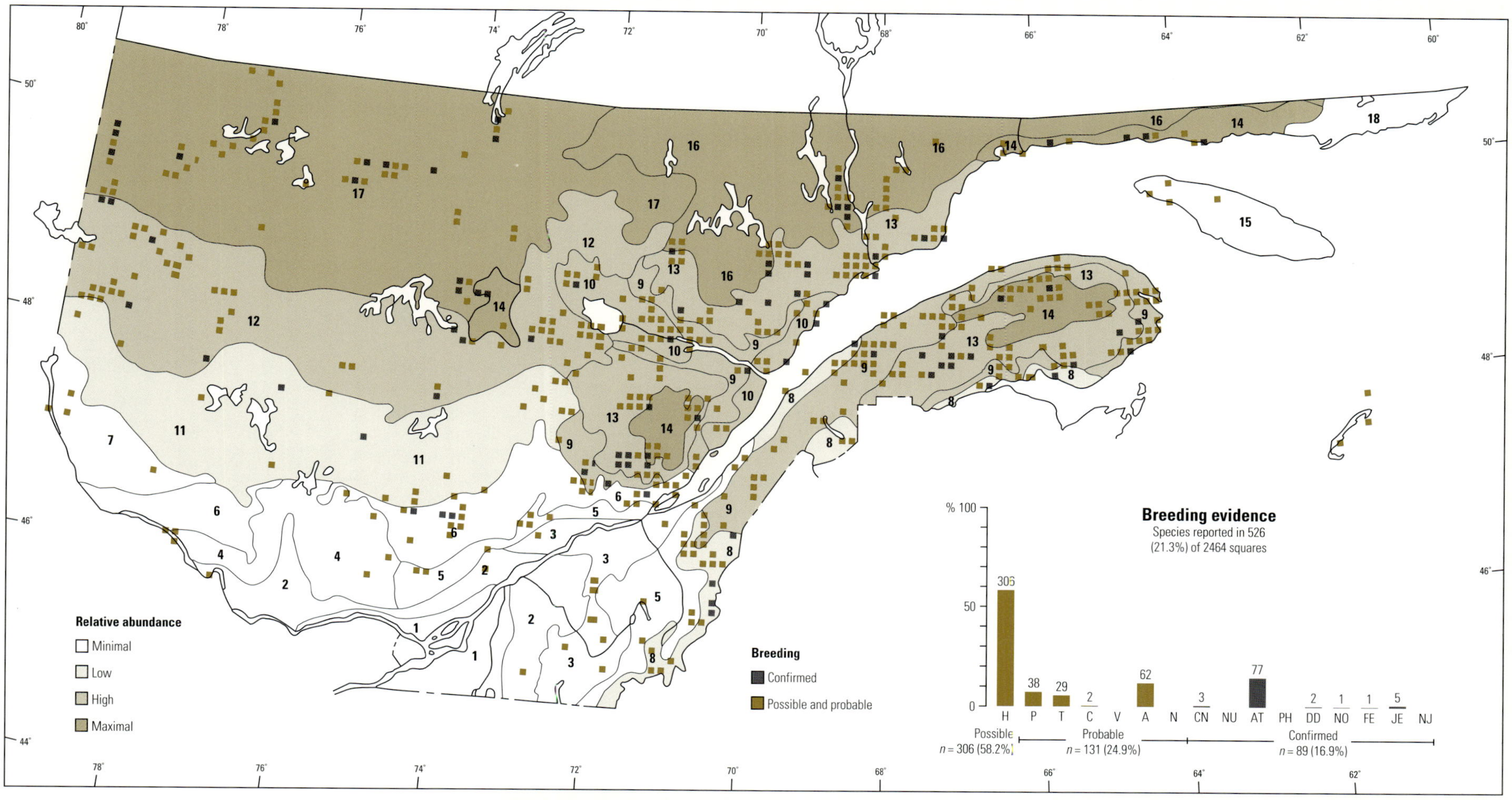

Species profile

Clutch size	usually 4–6 eggs (4–7)[1,2,4,5,6]
Laying interval	1 egg per day[3,4]
Start of incubation	with last egg[3,4,11]
Incubation period	usually 11–13 days (10–15)[2,3,4,5,6]
Maturity at hatching	altricial
Care of young	by female and male[2,11]
Nestling period	8–11 days[2,3,4,5,11]
Age at first flight	8–11 days or more[10]
Dependence of young	until ≈ 32 days old[3]
Broods per year	usually 1 (1–2)[2,3,12]
Breeding age	1 year[9]
Mating system	monogamy, sometimes polygyny[2,3]
Duration of pair bond	sometimes more than 1 season[4]
Total length	10.5–13.0 cm[5]
Wing span	17.1–17.8 cm[5]
Weight	7.7 g (6.3–10.5)[7]
Longevity record	6 years 10 months[8]

References:
1. Harrison 1984
2. Ehrlich et al. 1988
3. Stewart 1973
4. Stewart et al. 1977
5. Terres 1980
6. Godfrey 1986
7. Dunning 1984
8. Klimkiewicz et al. 1983
9. DeGraaf and Rudis 1987
10. according to nestling period
11. Harrison 1978
12. Harrison 1975

Chronology $n = 19$

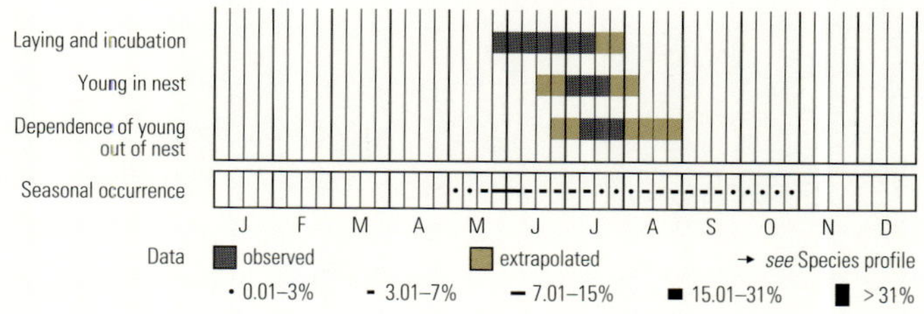

World breeding range

Frequency of occurrence by ecological region

Ecological regions** (forest regions)	Adequately covered squares	Squares in which reported	% of squares in which reported	Relative abundance (%)*
1. Sugar maple-hickory and sugar maple-basswood	97	0	0	0
2. Sugar maple-basswood	153	4	2.6 ± 1.5	2.8
3. Sugar maple-basswood and sugar maple-yellow birch	101	6	5.9 ± 3.0	6.4
4. Sugar maple-yellow birch-basswood	56	6	10.7 ± 7.4	11.5
5. Sugar maple-yellow birch-beech and sugar maple-yellow birch-basswood	113	15	13.3 ± 3.6	14.3
6. Sugar maple-yellow birch-beech	71	18	25.4 ± 9.3	27.3
7. Sugar maple-yellow birch-balsam fir	19	4	21.1 ± 19.7	22.7
8. Sugar maple-yellow birch and balsam fir-yellow birch	47	25	53.2 ± 12.8	57.3
9. Balsam fir-yellow birch	87	59	67.8 ± 9.1	73.0
10. Balsam fir-white birch or balsam fir-red maple	46	27	58.7 ± 12.3	63.2
11. Yellow birch-balsam fir	52	16	30.8 ± 13.0	33.1
12. Balsam fir-white birch (west)	59	41	69.5 ± 12.1	74.8
13. Balsam fir-white birch (east)	76	61	80.3 ± 8.8	86.4
14. Balsam fir-black spruce	28	26	92.9 ± 10.8	100.0
15. White spruce-balsam fir and balsam fir-white spruce	22	4	18.2 ± 17.1	19.6
16. Black spruce-balsam fir-moss	18	16	88.9 ± 17.4	95.7
17. Black spruce-moss	32	27	84.4 ± 14.1	90.9
18.			*not analysed*	

* minimal: 0–29.9%; low: 30–59.9%; high: 60–89.9%; maximal: 90–100%

** Deciduous forest: 1–7; mixed forest: 8–11; coniferous forest: 12–17

The young fledge in 8–11 days. By the time they are about 32 days old, they feed mostly on their own and a few days later begin leaving the parents' territory (Stewart 1973). Stewart (1973) noted that in California the peak of adult post-breeding "wandering" occurred in late June, when young had fledged, but he does not specify whether they move in groups or individually. However, during a field study near Rivière-du-Loup (Lehoux et al. 198[illegible]), 11 recently fledged young and one moulting adult were all mist-netted together on 4 July 1980, suggesting that the birds may move through the understorey in groups (J. Gauthier, CWS, unpubl. data).

In California, only 4 of 22 pairs attempted a second clutch after successfully raising a first brood, and only 2 pairs successfully fledged young in the renest (Stewart 1973); double-brooding thus seems unlikely in Québec.

Reports of adults carrying food accounted for 86.5% of Atlas squares in which breeding was confirmed. The small number of nest records is probably due to the female's habit of remaining on the nest until the observer is very close, as well as to this species' preference for habitat with poor accessibility (Helleiner 1987c).

Distribution

Wilson's Warbler breeds across the northern part of North America, from Alaska to southern California in the west to Newfoundland and Maine, including all the Canadian provinces and northern New York and Vermont (AOU 1983; Oatman 1985d; Peterson 1988o).

In Québec, the species is found throughout the Atlas area. It is uncommon in deciduous forests, but becomes more abundant as conifers predominate and as latitude and altitude increase, so that it is most numerous in the boreal spruce forest and also in the mountainous Balsam fir-black spruce forest region of the Laurentide Wildlife Reserve and the Gaspé Peninsula.

In southern Québec, the species is absent from the Montréal area, but it becomes increasingly common farther away. Observations reported in the Eastern Townships indicate that the Wilson's Warbler may be breeding in small numbers in the eastern reaches of the townships around Lake Drolet, Lake Mégantic, the Louise-Gosford ZEC, and Weedon.

Breeding was not confirmed for the Atlas on either the Magdalen Islands or Anticosti Island, although evidence of possible breeding was obtained. These results confirm those of Fradette (1992) for the Magdalens and suggest that the species may breed on Anticosti, as had been reported by Ouellet (1969).

In northern Québec, Wilson's Warbler breeds up to the tree line. Known breeding locations include Lake Guillaume Delisle on the Hudson Bay coast, and near Rivière aux Feuilles in the Ungava Bay region (Godfrey 1986; ÉPOQ).

Notes on territory (continued)

successive years. Males dispute territorial boundaries, apparently without physical contact (Stewart 1973), often singing from perches 2–3 m high (Griscom and Sprunt 1979). In California, they defend territories of 0.2–2.0 ha (Stewart et al. 1977).

In two stands of fir, spruce, and birch near Lake Cascapédia, in the Gaspé Peninsula, 4.9 males/ 10 ha were counted (Erskine 1980a). North of Duparquet, in the Abitibi region, a density of 4.5 males/ 10 ha was obtained in abandoned fields overgrown with willow and alder thickets (Erskine 1976a).

In the Nottaway, Broadback, and Rupert river watersheds, as in the Eastmain River and Great Whale River regions, the highest densities were found in shoreline scrub. Other habitats heavily frequented by this species in those three regions were burns and ecotones between burns and coniferous habitat (Tremblay and Belles-Isles 1991; Cons. G. G. – GREBE 1991a, 1992o).

In mixed and coniferous forests, this warbler is especially drawn to thickets in clearings, peatlands, and along streams and lakes; it is also fond of young deciduous stands. The female may build her nest in grass tussocks or at the base of a shrub. Portneuf Wildlife Reserve, northwest of Québec City.
July 1984, Christian Marcotte.

Clearcutting in the vast coniferous forest has a negative impact on the Wilson's Warbler in the short term, but subsequent regrowth of shrubs and young deciduous trees encourages its return to wetter areas. Near Lake Ashuapmushuan, northwest of Lake Saint-Jean. *September 1983, Jean Beaubien and Guy Simard, Canadian Forest Service.*

History and trends

The distribution of this species in Québec has probably not changed very much except in the southern part of the province, especially in the Montréal area, where habitat destruction seems to have virtually eliminated it as a breeder (Ouellet 1974a; Bannon 1991b). Ouellet (1974a) surmised that it had been common at one time around Montréal, where, according to Terrill (1922), it used to frequent alder thickets around peatlands north of the city. It was not atlassed in the Hull area, where it used to occur in large numbers in summer (Macoun and Macoun 1915), but it does breed in small numbers in the northern Ottawa Valley (Club des ornithologues de l'Outaouais 1985).

Population trends for Wilson's Warbler vary depending on the data set and scale used. Neither Breeding Bird Survey data for the period 1969–89 nor ÉPOQ summer data for 1970–89 show any clear population trends in Québec (Cyr and Larivée 1993a), though Robbins et al. (1986), on examining continent-wide BBS data, gathered between 1965 and 1979, reported an increase in the species' numbers in eastern North America. However, long-term spring migration data showed a significant decline from 1978 to 1987 in Massachusetts (Hill and Hagan 1991). In some cases, especially in Québec, such conflicting results may be related to the species' generally low numbers.

This bird is sensitive to clearcutting (Morrison 1981), but replanting with conifers in some parts of southern Québec, such as the Chaudière–Appalaches region, may be to its advantage. All this suggests that biologists and birders should keep an eye on population trends.

Robert Alvo

Canada Warbler

Paruline du Canada

Wilsonia canadensis

Christiane Girard

THE CANADA WARBLER BREEDS NOT ONLY IN CANADA BUT OVER A LARGE part of North America. The male has a blue-grey back and a bright yellow breast. The most telling field marks are the necklace of short black streaks and the yellow "spectacles" (Peterson 1980 Godfrey 1986). The song is less distinctive than the plumage and may vary from one bird to another. It is difficult to describe, consisting of a *chip* followed by abruptly tumbling musical phrases (Kendeigh 1945; Cayouette and Grondin 1977).

The Canada Warbler arrives in Québec late in spring and leaves at the end of August; it is rarely seen after the first week of October. The wintering grounds of this long-distance migrant extend from Colombia and southern Venezuela to central Peru (AOU 1983).

Habitat and behaviour

The Canada Warbler frequents fairly open mixed forest with a dense undergrowth of shrubs. Nests are often found in saplings and tall shrubs in woodland near wet areas along rivers or streams (Casebere 1979; Collins et al. 1982; Darveau et al. 1982; DeGraaf and Chadwick 1987). It also frequents midsuccessional forests (Titterington et al. 1979).

Among warblers, males usually precede the females to the nesting grounds (Morse 1989a), where they establish their territories and begin singing. Pairs are formed when the females arrive on the territories (Morse 1989a). In their study in Panama, Greenberg and Gradwohl (1980) found pairs of Canada Warblers mingling with flocks of small insectivorous birds during the spring and fall migration and postulated that if these are actually permanent monogamous associations then some birds should arrive on the breeding territories as pairs; this has never been documented, however. This study raises a number of questions regarding pair formation in Canada Warblers.

Male Canada Warblers sing to defend their territories, to attract a female during breeding displays, and to maintain the pair bond (Kendeigh 1945). The female builds the nest and incubates the eggs, of which there are usually 4. The nest is generally built on the ground or on a mossy hummock, stump, or log, in the roots of an upturned tree, or in some other opening (Taverner 1919; Terres 1980; Milosevich and Olson 1981; Peck and James 1987). It is made of plant stems, grass, leaves, and strips of bark and is lined with plant fibres, rootlets, and often hair (Taverner 1919; Godfrey 1986; Ehrlich et al. 1988). There is little information on the exact length of incubation, but Kendeigh (1945) estimated it to be about 12 days. During this time, the male brings food to the nest, even when the female is not there. When this occurs, the male hesitates and then swallows the food himself; this is an example of anticipatory feeding, behaviour prompted by imminent hatching (Harrison 1984). Both parents feed the young. The young make their first flight at about 8–10 days (Kendeigh 1945). In Ontario, cowbird parasitism has been estimated at about 20% (Peck and James 1987).

When defending its territory, the Canada Warbler is one of the most aggressive warblers, a characteristic which aids in detection. Squeaking will often lure one of the parents, which will set up a protest of loud calls, but sometimes the mere presence of an observer will suffice. Birds tend to be most agitated at the beginning of July, when the eggs have just hatched. The parent bird may also feign injury to distract the intruder away from the nest (Kendeigh 1945; Pierson 1978; Harrison 1984).

Notes on territory

The male sings loudly to maintain its territory, usually from a perch in a shrub. The female generally defends a small area around the nest site, especially during laying and incubation (Kendeigh 1945; Harrison 1984).

In Québec, recorded densities vary between 2.1 and 10.0 pairs/10 ha. Highest densities were found in young mixed forest stands or in forests with a dense understorey (Erskine 1971a, 1984; Cyr 1974). In a hemlock stand in the Mirabel region, the defended territory was estimated to be 0.4 ha (Cyr 1974). Studies by Martin (1960) in Algonquin Park in Ontario showed that the species there defends an average area of 0.2 ha. Martin also found densities ranging from 3.0 to 5.8 singing males/10 ha in different types of woodland, the highest densities being in mixed forest.

In a hemlock-beech forest in New York State, Kendeigh (1945) found that singing males defended an area of about 0.2 ha before nesting but 0.8 ha during nesting.

The Canada Warbler frequents midsuccessional forest with a dense understorey of shrubs. Mastigouche Wildlife Reserve, northwest of Trois-Rivières. *July 1991, Desneige Perreault.*

Canada Warbler

Breeding distribution

1984–1989

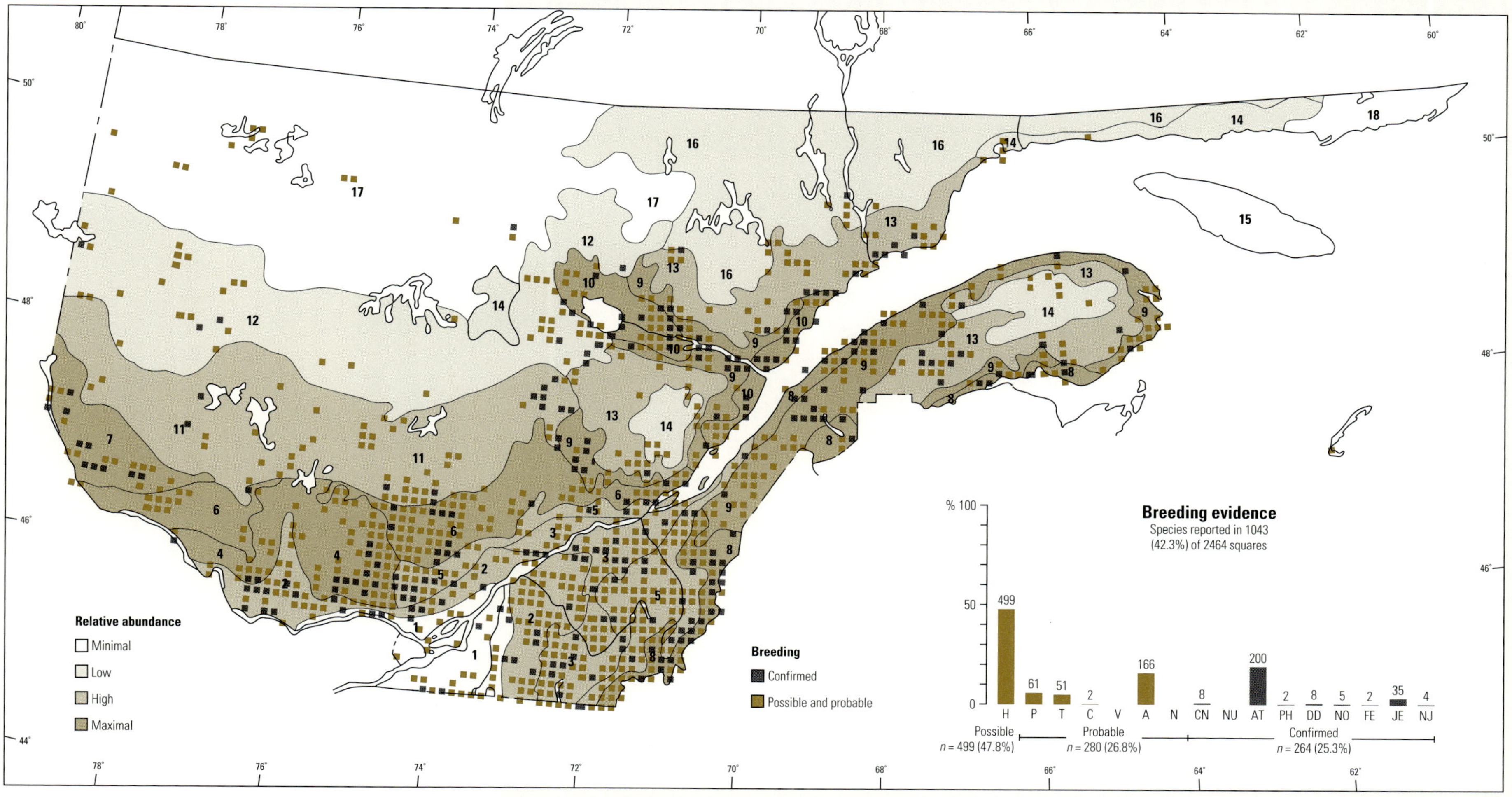

Species profile

Clutch size	usually 4 eggs (2–6)[1,2,3,4,6,7]
Laying interval	?
Start of incubation	?
Incubation period	≈ 12 days[5]
Maturity at hatching	altricial
Care of young	by female and male[2,3,5]
Nestling period	≈ 8–10 days[5]
Age at first flight	≈ 8–10 days or more[10]
Dependence of young	?
Broods per year	1[4]
Breeding age	1 year[4]
Mating system	monogamy[2]
Duration of pair bond	?
Total length	12.7–14.6 cm[1]
Wing span	?
Weight	M 10.6 g (8.7–13.5) / F 10.2 g (9.1–12.6)[9]
Longevity record	8 years[8]

References:
1. Terres 1980
2. Ehrlich et al. 1988
3. Bent 1953
4. DeGraaf and Rudis 1987
5. Kendeigh 1945
6. Godfrey 1986
7. Peck and James 1987
8. Klimkiewicz and Futcher 1989
9. Dunning 1984
10. according to nestling period

World breeding range

Chronology $n = 21$

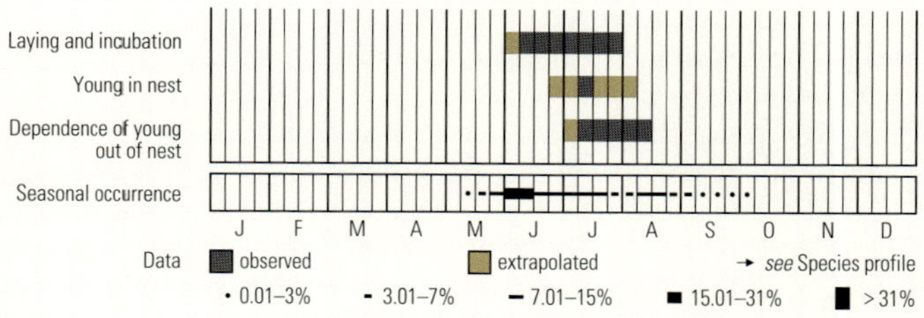

Frequency of occurrence by ecological region

Ecological regions** (forest regions)	Adequately covered squares	Squares in which reported	% of squares in which reported	Relative abundance (%)*
1. Sugar maple-hickory and sugar maple-basswood	97	29	29.9 ± 4.6	29.9
2. Sugar maple-basswood	153	99	64.7 ± 3.9	64.7
3. Sugar maple-basswood and sugar maple-yellow birch	101	84	83.2 ± 4.5	83.2
4. Sugar maple-yellow birch-basswood	56	51	91.1 ± 6.9	91.1
5. Sugar maple-yellow birch-beech and sugar maple-yellow birch-basswood	113	98	86.7 ± 3.6	86.7
6. Sugar maple-yellow birch-beech	71	66	93.0 ± 5.8	93.0
7. Sugar maple-yellow birch-balsam fir	19	19	100.0 ± 2.6	100.0
8. Sugar maple-yellow birch and balsam fir-yellow birch	47	47	100.0 ± 1.1	100.0
9. Balsam fir-yellow birch	87	82	94.3 ± 4.8	94.3
10. Balsam fir-white birch or balsam fir-red maple	46	45	97.8 ± 4.4	97.8
11. Yellow birch-balsam fir	52	36	69.2 ± 13.0	69.2
12. Balsam fir-white birch (west)	59	34	57.6 ± 12.9	57.6
13. Balsam fir-white birch (east)	76	58	76.3 ± 9.4	76.3
14. Balsam fir-black spruce	28	13	46.4 ± 19.3	46.4
15. White spruce-balsam fir and balsam fir-white spruce	22	0	0	0
16. Black spruce-balsam fir-moss	18	9	50.0 ± 26.1	50.0
17. Black spruce-moss	32	9	28.1 ± 17.1	28.1
18.			*not analysed*	

* minimal: 0–29.9%; low: 30–59.9%; high: 60–89.9%; maximal: 90–100%

** Deciduous forest: 1–7; mixed forest: 8–11; coniferous forest: 12–17

Canada Warblers also nest in large shrubs near wet areas or in cedar stands, as shown here. Cap Tourmente National Wildlife Area, northeast of Québec City. *August 1991, Jean Gauthier.*

Distribution

In the northern part of its range, the Canada Warbler breeds from east of the Rockies to the Maritimes and, in the south, from the northern Great Lakes states to New England and the Appalachians (AOU 1983). The Canada Warbler was reported in 42% (1043/2464) of the Atlas squares surveyed, with breeding confirmed in 25.3% of these. The species is primarily found in mixed, predominantly deciduous forests. Canada Warblers are less abundant in the north, where the proportion of conifers is too high, and in the extreme south numbers decrease when the proportion of deciduous trees is too high. Thus it is less common on the mountaintops in the Grands-Jardins and Gaspé conservation parks, in the Balsam fir-black spruce forest region. Frequency of occurrence is also lower in the Sugar maple-hickory and sugar maple-basswood forest region, which is heavily urbanized and agricultural with mature forest; such habitat is of less interest to the Canada Warbler, which prefers mid-succession growth (Collins et al. 1982; Darveau et al. 1982; DeGraaf and Rudis 1987).

There were a few Atlas sight records from the Matagami region; north of the Gouin Reservoir; near Havre-Saint-Pierre; and the Magdalen Islands. The Canada Warbler was not atlassed on Anticosti Island, although a pair had been sighted there in the summer of 1917, with the female carrying nesting material (Lewis 1924).

The nest of a Canada Warbler built on the ground in moss and forest litter. The female often sets the nest on a mossy hummock or in a tuft of grass under a branch or root. Saint-Hippolyte, Island of Montréal. *July 1929, Henry Mousley, Laval University Collection.*

This juvenile bird was photographed on 16 August, as it was about to make its fall migration. Canada Warblers migrate earlier than most other wood warblers. Québec City. *August 1954, Gaston Le Page.*

History and trends

At the beginning of the century, the Canada Warbler was a fairly common summer resident in the Hatley area of the Eastern Townships (Mousley 1916a) and in the Ottawa-Hull region (Macoun and Macoun 1915). Wintle (1896) considered it scarce in the Montréal area, and Dionne (1906) described it as generally uncommon in the province. Lewis (1924) reported a pair on Anticosti Island in the summer of 1917. It has also been reported from Lake Mistassini and the Matamec region (Moisie) (Godfrey 1949). Godfrey and Wilk (1948) considered it one of the commonest warblers in the Lac-Saint-Jean region, which led Todd (1963) to conclude that the Lake Mistassini and Matamec regions constituted the northern limit of the species' Québec range. According to Ouellet (1974a), the species is probably less abundant now in the Montréal region than during the last century, having been affected, like many other species, by habitat degradation.

Today, the Canada Warbler is considered a common migrant breeder and even commoner in migration (Ouellet 1974a; David 1980a; Robert 1989b). Because it inhabits second-growth areas, the species has benefited from logging. In Québec, Canada Warbler populations increased somewhat between 1980 and 1989 along BBS routes. This is probably due to reversion of abandoned farmland to forest, a fairly widespread phenomenon over the last 15 years throughout southern Québec except in the Montréal and Montérégie areas (M. Carignan, MAPAQ, unpubl. data).

Forest management and logging practices appear to have an effect on Canada Warbler population trends (Webb et al. 1977; Derleth et al. 1989). Increased densities have been noted after selective cutting, in areas that were not too heavily logged (Webb et al. 1977), but numbers decrease when small areas have been clearcut. This decrease, however, is not significant during the first years of regrowth in several types of forest (Derleth et al. 1989). Forestry practices that are detrimental to the species are those that completely eliminate the shrub layer or affect slightly wet areas. Moreover, reforestation after clearcutting produces even-aged stands with little or no understorey, unsuitable for this denizen of the dense understorey in midsuccessional forest.

Desneige Perreault and Josée Tardif

Scarlet Tanager

Tangara écarlate

Piranga olivacea

Christiane Girard

DESPITE ITS BRILLIANT COLOURING, THE SCARLET TANAGER IS OFTEN overlooked because of its elusive behaviour, and when it is detected, it is usually by its song or call.

The Scarlet Tanager is one of the few members of the subfamily Thraupinae that migrate from South America to breed in the Northern Hemisphere (Holmes et al. 1986; Holmes and Sherry 1988). The male's plumage undergoes significant seasonal changes, from scarlet with a black tail and wings during the breeding season to dull greenish and yellowish in winter, when he resembles the female (Campbell and Lack 1985). The Scarlet Tanager is a fairly common migrant breeder in the southernmost part of the province, but is scarcer elsewhere in southern Québec.

Habitat and behaviour

The Scarlet Tanager is found in deciduous and mixed forest dominated by oak, birch, American beech, and sugar maple (Shy 1984; Ellison 1985u; Holmes and Sherry 1988; Smith, C. R. 1988b). It also frequents conifers, particularly hemlock (Ellison 1985u). The species prefers forest with a tall, dense canopy, a variety of tree species, a high density of trees with trunks over 22 cm in diameter, and sparse ground cover (Shy 1984). According to Prescott (1965), it appears to favour mature deciduous forest.

The Scarlet Tanager is usually detected by its song, which is often compared to that of a robin with a sore throat. Identifications are frequently clinched by its diagnostic chip-burr call. The female also sings occasionally, and her song is similar to the male's (Prescott 1965).

The Scarlet Tanager is a prodigious consumer of insects, although it also feeds on berries. It moves quietly through the leaves and branches of the upper canopy, methodically searching for larvae and adult insects, particularly caterpillars (Prescott 1965; Holmes et al. 1986). Indeed, this mainly insectivorous species appears to be directly affected by insect abundance in its habitat; in a study in New Hampshire, Holmes et al. (1986) reported high numbers of Scarlet Tanagers during infestations of several lepidopteran species, including the saddled prominent moth, which affects mainly birch and, to a lesser extent, maple, oak, and beech. However, during a cold spring when insects were late to emerge, the same authors discovered that many tanagers had died of starvation.

During the breeding season, the Scarlet Tanager seeks tall, dense deciduous forest with many large trees and a sparse shrub layer. The female usually builds the nest 5–15 m from the ground. Philipsburg, south of Saint-Jean-sur-Richelieu. *July 1991, Pierre Bannon.*

Notes on territory

Males arrive in Québec in spring ahead of the females and lay claim to territories by singing from the tops of mature trees. Territorial boundaries remain flexible until a few days after the females arrive (Prescott 1965).

In Québec, territories ranged from 0.9 to 1.2 ha in mature hemlock and sugar maple stands in the Mirabel area (Cyr 1974), and from 0.80 to 3.42 ha in forests of unidentified composition (McNeil 1964 in Cyr 1974). Similar figures, in the range 0.8–2.4 ha, were given by Stokes and Stokes (1989b). Breeding densities varied from 1.5 to 7.5 pairs/10 ha in old or mature deciduous, coniferous, and mixed forests generally located in southwestern Québec (Erskine 1971a, 1984; Cyr 1974; DesGranges 1980b).

In Maryland, breeding densities of 3.5 territorial males/10 ha were found in mature deciduous forest and 6.5 territorial males/10 ha in virgin hardwood stands (DeGraaf and Rudis 1987).

Scarlet Tanager

Breeding distribution

1984–1989

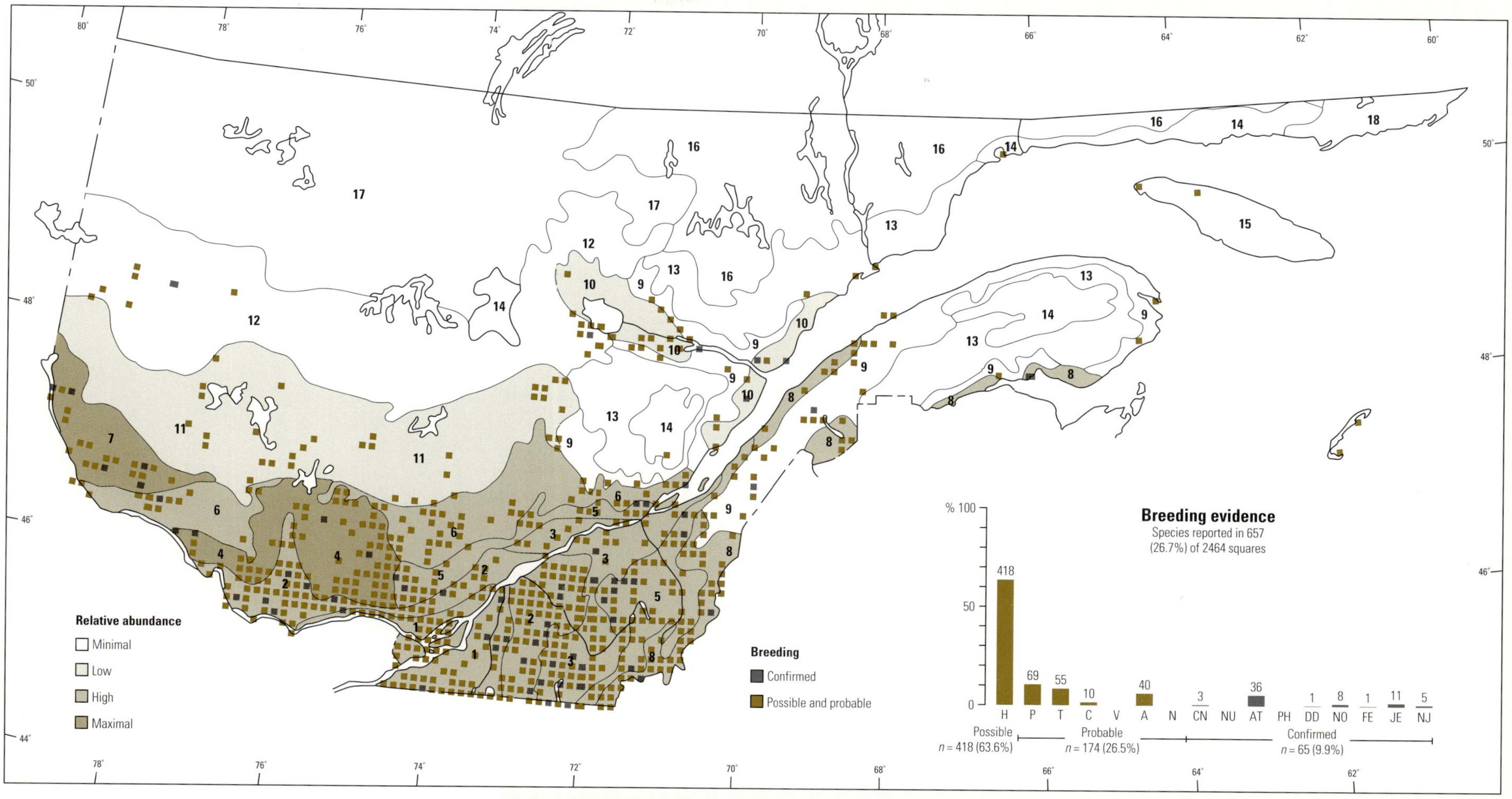

World breeding range

Species profile

Clutch size	usually 4 eggs (2–5)[1,2,3,5]
Laying interval	?
Start of incubation	probably with last egg[1]
Incubation period	12–14 days[1,2,5]
Maturity at hatching	altricial
Care of young	by female and male[1,2]
Nestling period	usually 10 days (9–15)[1,2,4]
Age at first flight	9–11 days[4]
Dependence of young	until 12–15 days after leaving nest[10]
Broods per year	usually 1[1,4,9]
Breeding age	1 year[9]
Mating system	monogamy[1,2,4]
Duration of pair bond	?
Total length	16.5–19.0 cm[5]
Wing span	27.9–30.5 cm[6]
Weight	28.6 g (17.5–35.2)[7]
Longevity record	10 years 1 month[8]

References:
1. Prescott 1965
2. Kohler 1915
3. Ellison 1985u
4. Ehrlich et al. 1988
5. Godfrey 1986
6. Terres 1980
7. Dunning 1984
8. Klimkiewicz and Futcher 1987
9. DeGraaf and Rudis 1987
10. Tyler 1958b

Chronology n = 29

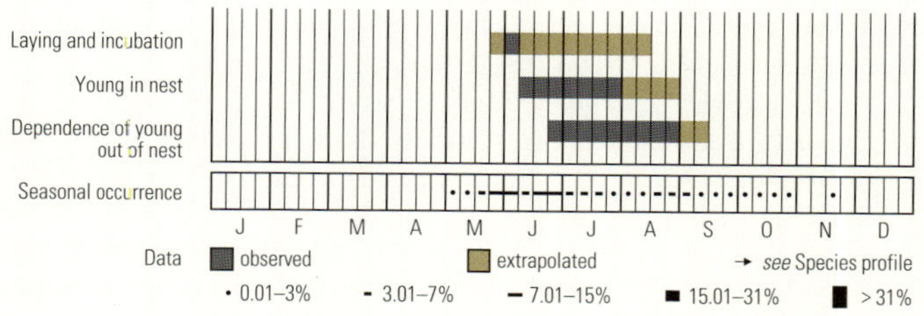

Frequency of occurrence by ecological region

Ecological regions** (forest regions)	Adequately covered squares	Squares in which reported	% of squares in which reported	Relative abundance (%)*
1. Sugar maple-hickory and sugar maple-basswood	97	62	63.9 ± 4.8	67.5
2. Sugar maple-basswood	153	124	81.0 ± 3.3	85.5
3. Sugar maple-basswood and sugar maple-yellow birch	101	67	66.3 ± 5.6	70
4. Sugar maple-yellow birch-basswood	56	50	89.3 ± 7.4	94.2
5. Sugar maple-yellow birch-beech and sugar maple-yellow birch-basswood	113	71	62.8 ± 5.0	66.3
6. Sugar maple-yellow birch-beech	71	53	74.6 ± 9.3	78.8
7. Sugar maple-yellow birch-balsam fir	19	18	94.7 ± 12.0	100.0
8. Sugar maple-yellow birch and balsam fir-yellow birch	47	34	72.3 ± 11.6	76.4
9. Balsam fir-yellow birch	87	23	26.4 ± 8.6	27.9
10. Balsam fir-white birch or balsam fir-red maple	46	15	32.6 ± 11.8	34.4
11. Yellow birch-balsam fir	52	28	53.8 ± 14.0	56.8
12. Balsam fir-white birch (west)	59	11	18.6 ± 10.4	19.7
13. Balsam fir-white birch (east)	76	3	3.9 ± 4.7	4.2
14. Balsam fir-black spruce	28	2	7.1 ± 10.8	7.5
15. White spruce-balsam fir and balsam fir-white spruce	22	2	9.1 ± 13.3	9.6
16. Black spruce-balsam fir-moss	18	0	0	0
17. Black spruce-moss	32	0	0	0
18.			*not analysed*	

* minimal: 0–29.9%; low: 30–59.9%; high: 60–89.9%; maximal: 90–100%

** Deciduous forest: 1–7; mixed forest: 8–11; coniferous forest: 12–17

The Scarlet Tanager is a typical inhabitant of mixed and deciduous forest. Extensive oak forests, like this red oak stand on Mount Saint-Hilaire, provide ideal habitat. *August 1991, Yves Aubry.*

Although males and females have the same diet, they forage in quite different ways (Holmes 1986). The female appears to feed higher in the trees than the male, at an average height of 14.8 m compared with 11.7 m for the male, and to forage more in the outer branches of the tree, while the male remains closer to the trunk (Holmes 1986). Both sexes feed mainly on the foliage by gleaning or hovering; however, the female forages by hawking from a perch more often than the male (Holmes 1986). Curiously, the male's foraging behaviour is closer to that of the Rose-breasted Grosbeak than to that of the female Scarlet Tanager (Holmes 1986).

The Scarlet Tanager winters in South America, mainly east of the Andes from Colombia to northern Bolivia (Godfrey 1986; Isler and Isler 1987; Ehrlich et al. 1988). The males return to southern Québec in spring when the trees are just coming into leaf. Indeed, visibility seems to be a key factor in territorial establishment, since the male stakes out a territory by singing from conspicuous perches in the upper canopy of mature trees. However, the boundaries of the territory are not permanently established until a few days after the arrival of the females (Prescott 1965). Good visibility in the forest also appears to facilitate pair formation (Prescott 1965).

When the females arrive, generally around a week after the males, they choose a mate, usually from among several suitors. The male's courtship displays include a characteristic posture: leaning forward toward the female and slightly drooping his outspread wings, he raises his tail so that his brilliant scarlet back can be seen. Although a similar posture is used to intimidate rivals, the male is quieter and less conspicuous during courtship, spending much of his time on branches less than 3 m up (Prescott 1965).

The female makes the final choice of nest site, but the male may also play a role in the process. The female, which builds the nest alone, usually starts work on the structure a week after she arrives. Nest construction may take as long as a week or more, or as little as a day (Prescott 1965).

The nest is a shallow, loosely woven cup of twigs, grasses, plant stalks, and rootlets. It is usually placed on a horizontal branch, closer to the end of the branch than the trunk, generally at least 5 m up, but sometimes as high as 15 m or more (Kohler 1915; Prescott 1965; Ellison 1985u; Smith, C. R. 1988b). The nest is usually well camouflaged in the foliage (Prescott 1965). Nests were found in only 2.0% (13/657) of Atlas squares in which the species was reported.

This species often breeds in mature sugar maple and beech stands. Cap Tourmente National Wildlife Area, northeast of Québec City. *August 1991, Jean Gauthier.*

The male and female generally feed in the treetops, although the latter forages higher and farther out in the tree than the male. Cap Tourmente National Wildlife Area, northeast of Québec City. *August 1991, Gilles Falardeau.*

The species is a common cowbird host (Ehrlich et al. 1988): of 36 nests found in Ontario, 7 (19.4%) had been parasitized (Peck and James 1987), and the situation is probably very similar in Québec. According to Prescott (1965), although the tanagers do their best to chase away the female cowbird, she may slip quietly into the nest while they are not watching, remove a tanager egg, and replace it with her own. Tanagers do not seem able to distinguish cowbird eggs from their own. When a nest has been parasitized with one or two cowbird eggs, a decrease in clutch size usually ensues. However, survival of the young tanagers does not seem to be compromised by the presence of the other species, and young in parasitized nests appear to be as strong and healthy as young in non-parasitized nests (Prescott 1965).

Clutch size is usually 4 eggs, which are incubated by the female for 12–14 days (Kohler 1915; Prescott 1965). Incubation probably begins as soon as the last egg has been laid (Prescott 1965). Females can be seen sitting on their nests as early as the end of May. In some pairs, the male feeds the female on the nest during incubation (Prescott 1965).

Both parents feed the young, but the female makes more trips to the nest with food as the nestlings get older. During the first 4 days after hatching, the parents usually eat the fecal sacs, but later, they dispose of them away from the nest (Prescott 1965).

The young make their first flight when 9–11 days old (Ehrlich et al. 1988). Their coloration, which is similar to the female's, allows them to hide among the leaves undetected. The female remains with the young in the vicinity of the nest for close to 2 weeks (Tyler 1958b). Breeding was confirmed in 1.7% (11/657) of Atlas squares in which the species was reported through records of recently fledged young.

Distribution

The Scarlet Tanager breeds in southern Canada from New Brunswick to Manitoba and in the United States south to southern Oklahoma and Georgia (Godfrey 1986; Erskine 1992). Confirmed breeding records in Riding Mountain National Park, Manitoba, between 1982 and 1986 suggest that the species' Canadian range may now extend as far west as western Manitoba (Walley 1989).

The species was reported in 26.7% (657/2464) of Atlas squares surveyed, particularly in regions where the forests are dominated by maple, although it is also found in mixed forest.

Most common in the southwestern part of the province, the Scarlet Tanager becomes less abundant as the deciduous forest gives way to mixed and coniferous forest. Some of the Atlas reports suggest probable breeding as far north as Sept-Îles, Anticosti Island, and the Magdalen Islands. Its occurrence there seems to be associated with a few deciduous and mixed stands, such as the strip of deciduous forest mixed with fir trees that extends about halfway along the north side of Anticosti Island from its western tip.

History and trends

The Scarlet Tanager's range does not appear to have changed a great deal since the turn of the century. Dionne (1883) described the species as being very scarce around Québec City, but much more common in Ontario and even around Montréal.

However, the species has been affected by the logging of hardwood forests, which has fragmented suitable habitats (Ouellet 1974a). In New Jersey, Galli et al. (1976) reported that the tanager's presence depended on the size of the forest, being found only in wooded areas 3 ha and larger. Despite forest fragmentation in eastern North America (Askins 1993), an analysis of BBS data for 1966–89 shows no significant downward trend, however (Robbins et al. 1986; Cyr and Larivée 1993a,b; BBS).

Alain Lanoue and Serge Gauthier

Northern Cardinal

Cardinal rouge

Cardinalis cardinalis

Christiane Girard

FAMILIAR BECAUSE OF THE BRIGHTLY COLOURED PLUMAGE OF THE MALE, the Northern Cardinal is also an excellent singer. Less well known is the interesting fact that the female, with her buffy plumage, sings just as well as the male—and pairs frequently countersing. In the breeding season, behaviour must be observed in order to determine whether the song is territorial or simply countersinging between mates.

The Northern Cardinal's breeding range in northeastern North America has been expanding for about 30 years (Bull 1985). This occasional sedentary breeder in the Montréal area (Bannon 1991b) is a newcomer to Québec's list of breeding birds. In fact, the first breeding record dates back only to 1964, when a pair was observed in summer on Mount Royal, in Montréal and a nest was then discovered (David 1976a).

Habitat and behaviour

Preferred breeding habitat is woodland edges, thickets, bushy tangles, and streamside shrubbery, as well as the shrubby borders of old fields, clearings, gardens, and parks (Peterson 1984; Godfrey 1986). It is fairly easy to spot, by both its song and the male's brilliant red plumage, and a good many Atlas records were of singing birds.

The male starts to sing in midwinter and may thus be heard even in late January and February. Its strong, varied song consists of clear whistled notes and trills, and is sung frequently in the weeks leading up to nest building.

The female sings mostly just before the onset of breeding, once the territory has been established. As she habitually countersings in response to the male, it is thought that her song serves chiefly to strengthen the pair bond rather than to establish the territory (Ritchison 1986).

In addition to countersinging, pairs adopt other distinctive behaviours during courtship. Near feeders, mate feeding can be observed, with the male taking a seed and offering it to the female. The special pose the two adopt when they are close together is harder to see: one or both birds tilt one side of the body up, giving a lopsided impression, then sway quickly from side to side (Stokes and Stokes 1983).

Notes on territory

Stokes and Stokes (1983) reported that the size of the territory used for mating, nesting, and feeding ranges from 1.2 to 4.1 ha, and is

A sedentary species and still scarce in Québec, the Northern Cardinal establishes its territory in dense thickets along the edges of old fields. Saint-Bernard-de-Lacolle, south of Saint-Jean-sur-Richelieu. *July 1991, Pierre Bannon.*

Northern Cardinal

Breeding distribution

1984–1989

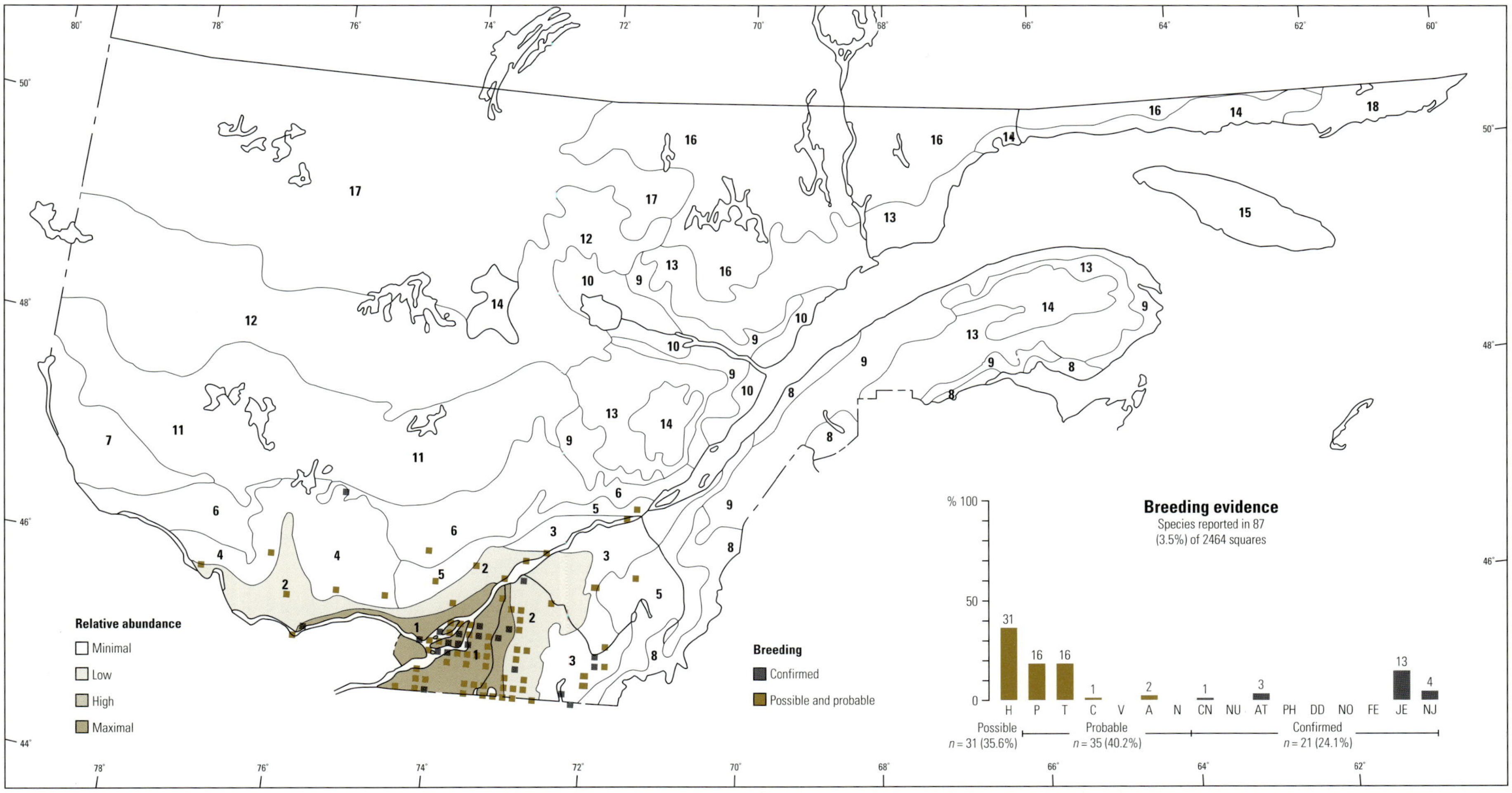

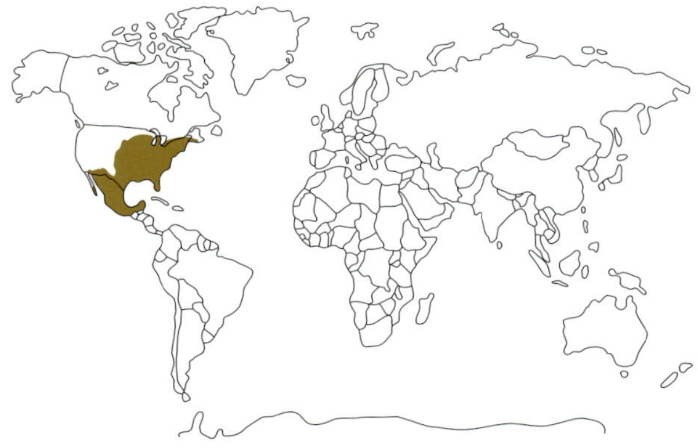

World breeding range

Species profile

Clutch size	usually 3–4 eggs (2–5)[1,2,3]
Laying interval	1 egg per day[3]
Start of incubation	with last egg[2,3]
Incubation period	11–13 days[1,2,3]
Maturity at hatching	altricial
Care of young	by female and male[5]
Nestling period	usually 9–10 days (7–11)[1,2,3]
Age at first flight	usually 9–10 days (7–11) or more[12]
Dependence of young	until ≈ 45 days old[2]
Broods per year	usually 2–3 (1–4)[1,2,3,4,11]
Breeding age	1 year[10]
Mating system	monogamy[2]
Duration of pair bond	sometimes more than 1 season[3]
Total length	19.0–23.5 cm[6]
Wing span	26.0–30.5 cm[7]
Weight	M 45.4 g (33.7–63.2) / F 43.9 g (33.6–64.9)[8]
Longevity record	15 years 9 months[9]

References:
1. Laskey 1944
2. Bent 1968
3. Stokes and Stokes 1983
4. Ehrlich et al. 1988
5. Harrison 1978
6. Godfrey 1986
7. Terres 1980
8. Dunning 1984
9. Klimkiewicz and Futcher 1987
10. DeGraaf and Rudis 1987
11. David and Gosselin 1979a
12. according to nestling period

Chronology $n = 30$

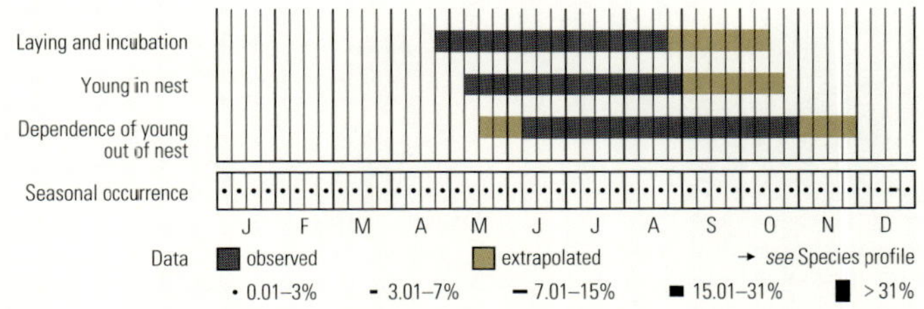

Frequency of occurrence by ecological region

Ecological regions** (forest regions)	Adequately covered squares	Squares in which reported	% of squares in which reported	Relative abundance (%)*
1. Sugar maple-hickory and sugar maple-basswood	97	42	43.3 ± 4.9	100.0
2. Sugar maple-basswood	153	21	13.7 ± 2.9	31.7
3. Sugar maple-basswood and sugar maple-yellow birch	101	9	8.9 ± 3.6	20.6
4. Sugar maple-yellow birch-basswood	56	3	5.4 ± 5.6	12.4
5. Sugar maple-yellow birch-beech and sugar maple-yellow birch-basswood	113	5	4.4 ± 2.4	10.2
6. Sugar maple-yellow birch-beech	71	1	1.4 ± 3.0	3.3
7. Sugar maple-yellow birch-balsam fir	19	0	0	0
8. Sugar maple-yellow birch and balsam fir-yellow birch	47	0	0	0
9. Balsam fir-yellow birch	87	0	0	0
10. Balsam fir-white birch or balsam fir-red maple	46	0	0	0
11. Yellow birch-balsam fir	52	0	0	0
12. Balsam fir-white birch (west)	59	0	0	0
13. Balsam fir-white birch (east)	76	0	0	0
14. Balsam fir-black spruce	28	0	0	0
15. White spruce-balsam fir and balsam fir-white spruce	22	0	0	0
16. Black spruce-balsam fir-moss	18	0	0	0
17. Black spruce-moss	32	0	0	0
18.			not analysed	

* minimal: 0–29.9%; low: 30–59.9%; high: 60–89.9%; maximal: 90–100%

** Deciduous forest: 1–7; mixed forest: 8–11; coniferous forest: 12–17

During the breeding season, both sexes modify their behaviour. Outside this period, the Northern Cardinal may sometimes be seen in small flocks (Ehrlich et al. 1988), but once territory has been established, the pair no longer tolerates the presence of conspecifics. Intruders are chased off immediately, the male pursuing other males, the female other females (Laskey 1944).

The Northern Cardinal commences nesting in the last week of April in Québec and at the end of March in Tennessee, producing up to 4 broods in the same season (Laskey 1944; Ehrlich et al. 1988). In most of the United States, however, it usually raises only 3 broods a year (Bull 1985; Robbins 1991). In Québec, triple brooding was observed in Hull in 1979 (David and Gosselin 1979a), and in Montréal in 1981 (J. Léveillé *fide* N. David). The cardinal's breeding season is therefore exceptionally long, with the last brood leaving the nest as late as October, as evidenced by the sighting of a pair feeding their young in Hudson in the last week of October in 1987 (Gosselin et al. 1988a). This long breeding season probably contributed to the success of the species as it expanded northward.

The female usually builds the nest in the densest part of a thicket, small bush, vine, or hedge, anywhere from 1.5 to 2.4 m up (Godfrey 1986). It is made of twigs, shredded bark, grass, pieces of paper, or leaves, and is lined with grass, rootlets, and hair (Godfrey 1986; Peck and James 1987). Clutch size is 2–5 eggs, but usually 3 or 4. The cardinal is a frequent cowbird host, although nests containing parasitic eggs are often abandoned (Graham 1988b). The female does most of the incubating, which takes 11–13 days (Laskey 1944; Bent 1968), though she is fed and occasionally relieved by the male. The male also feeds the nestlings for the first few days after hatching, and is helped shortly after by the female. The nestlings are fed almost entirely on insects. They generally leave the nest at the age of 9 or 10 days, but remain nearby for several days longer. The male then takes sole charge of them if the female renests (Stokes and Stokes 1983).

Although the species is essentially monogamous, Rice (1969) documented an instance of two females sharing incubation and then, with the help of the male, raising all the chicks together.

The adults' diet consists principally of plant material, although about a third of the food consumed is of animal origin, mainly insects such as beetles, cicadas, grasshoppers, ants, and dragonflies, and other invertebrates such as spiders, centipedes, and snails. Their plant food consists of fruits, wild seeds, and cultivated grains like wheat, corn, and oats (McAtee 1908). The Northern Cardinal is also a regular visitor to bird feeders, drawn by sunflower seeds, dried fruits, and millet.

The young are often seen at feeders along with their parents. The arrival of a pair accompanied by birds that resemble the female, but with a darker bill, usually indicates that the species is breeding nearby. Most confirmed records of breeding (13/21) were based on sightings of fledglings; nests were discovered in only 4 squares during atlassing.

Distribution

The cardinal's breeding range includes southern Ontario and Québec, as well as the entire central and eastern United States, south to Central America. In Québec, the Northern Cardinal is found mainly in the southwest, especially southwest of the Saint-François River. Breeding was not reported for the Atlas, however, in Lanaudière, Bois-Francs or the Québec City area, even though the species is seen there regularly in winter.

The Northern Cardinal in firmly established from Montréal south to the U.S. border and also nests in the Ottawa Valley. Breeding has even been confirmed near Mont-Laurier.

Notes on territory (continued)

defended from late winter to the end of breeding. Conspecifics often intrude and are chased away by the bird of the same sex. Dow (1969) studied the home ranges of males during the breeding season and reported means of 1.8 ha in western Tennessee and 18.8 ha in southern Ontario, at the northern limit of the species range. A study in eastern Texas found that breeding territories were larger in areas having mature trees and smaller where only shrubby vegetation was present. These smaller, better-quality territories were defended by birds with shorter and less complex songs (Conner et al. 1986).

In winter, observations in central Kentucky showed that maximum home ranges averaged 21.2 ha, and overlapped those of conspecifics by as much as 3.36 ha. Areas used on a daily basis ranged from 0.32 to 3.38 ha, with a mean of 1.17 ha (Ritchison and Omer 1990).

Southern Québec is the northern limit of its range, so the Northern Cardinal depends on bird feeders to help it survive the harsh winters. It can also be seen at feeding stations with its young in summer. Beloeil, east of Montréal. *July 1991, Gaétan Duquette.*

This species benefits from the plantings in landscaped urban areas of southern Québec. Montréal Botanical Gardens. *October 1981, Richard Pelletier.*

In Québec, the Northern Cardinal is at the northern limit of its range. The appearance of this "northern" population is a spillover from the expansion of the species' breeding range in eastern Ontario and northeastern New York State. The first major incursion into Québec (1964–69) coincided with the first reports of breeding in northeastern New York, as well as with an increase in winter reports from eastern Ontario (David 1976a), and favourable changes to habitat in southern Québec, with the gradual spread of urban woodlands and abandoned farmlands. As it prefers dense thickets and second-growth shrubs and deciduous saplings, the cardinal is perfectly at home in the rural environments, residential areas, and abandoned farmlands (Beddall 1963) of southwestern Québec.

History and trends

The first records of the Northern Cardinal in Québec date from the end of the last century and concern one bird caught in Montréal in June 1862 and another in Québec City in August 1869. It was another 73 years before a third was seen at a feeder in Hudson, in February 1942 (Ouellet 1974a). Then, 14 years later in 1956, 2 males were observed on Mount Royal and in Montréal-Ouest from January to early April (David 1976a).

The cardinal probably began to settle in Québec in 1956, as small numbers have been seen each year since then near Montréal (Ouellet 1974a; David 1976a). It was not until 1964, however, that the species was reported more widely in the Montréal area and in the Ottawa Valley and Eastern Townships. After a slight decline between 1969 and the winter of 1971–72, its numbers have increased considerably since 1972 (David 1976a). The proliferation of winter bird feeders in residential areas has greatly promoted expansion and establishment of the species (David 1976a; Bull 1985; McNicholl 1987; Meade 1988c). These ready sources of food have helped many birds survive the winter. The rigorous winter climate is a limiting factor on cardinal populations (David 1976a). The northern limit of the cardinal's range seems to correspond geographically to the mean minimum January –15°C isotherm (Root 1988). Although several decades of milder winters have helped the species expand toward the northern United States and southern Québec, the province's semi-nordic climate is still the main factor limiting the Northern Cardinal's range in Québec.

Jean Paquin

Rose-breasted Grosbeak

Cardinal à poitrine rose

Pheucticus ludovicianus

Christiane Girard

"SONGSTER IN FORMAL ATTIRE" IS PROBABLY THE MOST FITTING OF ALL the epithets that have been applied to the Rose-breasted Grosbeak. The handsome male in breeding plumage is admired as much for its flutelike, liquid song as for its striking black-and-white coloration. The crowning elegance is the triangular carmine red marking in the centre of the breast (Godfrey 1986). The streaked brown plumage of the female is not as showy. The female is unusual in that she sings like the male during the breeding season, though the song is softer and shorter (Terres 1980).

The Rose-breasted Grosbeak is a common breeder in southern Québec (David 1980a). This neotropical migrant winters from Mexico to northern South America (AOU 1983). There have been a few, rare reports of winter sightings in Québec (ÉPOQ).

Habitat and behaviour

The Rose-breasted Grosbeak prefers second-growth open forest (Peck and James 1987) and regenerating woodlands about 15 years after logging. In some places, however, it may be found 5 years after strip cutting (G. Falardeau and J.-P. L. Savard, CWS, unpubl. data). It also frequents shrubs and undergrowth in edges and clearings of mature forest, either deciduous or mixed. It is found in orchards, parks, small woodlots in rural and suburban areas, thickets in old fields, and in waterside habitat (Pough 1949; Bent 1968; Peck and James 1987).

On arrival in early May, the male chooses conspicuous song perches where it strikes up its serenade. The song is similar to that of the American Robin, but is more melodious. Lemon and Chatfield (1973) found it to be composed of 15–23 syllables. The male also sings when pursuing a female or chasing away other males. The females arrive a few days after the males, and pairs are formed shortly thereafter. Both adults defend the territory against conspecifics, each driving away intruders of its own sex (Stokes and Stokes 1983).

The nest is made of twigs, stems, rootlets, and fine grasses, all so loosely interwoven that the eggs can sometimes be seen through the bottom. The 3 or 4 eggs are greenish-blue with brown spots. The male selects the nest site and helps the female build the nest, which takes about 2–3 days (Stokes and Stokes 1983). The nest is usually placed 1.5–4.5 m up (Godfrey 1986) in a shrub or young deciduous tree, fairly near the trunk and most often in an upright fork (Peck and James 1987). Although vulnerable in his conspicuous plumage, the male participates in incubation, which lasts 11–14 days. Male Rose-breasted Grosbeaks have the curious habit of singing while incubating, sometimes even in the middle of the night (Bent 1968).

The altricial young are fed by both parents. The diet consists mainly of insects, such as Eastern tent and forest tent caterpillars, and beetles, particularly click, metallic wood-boring, and leaf beetles (McAtee 1908; Batts 1958 *in* Bent 1968). The young leave the nest at about 9–12 days of age, and the adults continue feeding them for 2–3 weeks until they become independent (Stokes and Stokes 1983). Sometimes the pair will raise a second brood. In Michigan, a female was observed initiating a second clutch while the young of the first brood were only about 2–6 days old and still in the nest (Rothstein 1973). The second nest was located about 12 m from the first, and neither parent appeared to be in charge of a particular nest, as they were seen incubating the second brood in turn.

Notes on territory

Shortly after arrival on the breeding grounds, males begin defending territories ranging in size from 0.8 to 1.2 ha (Stokes and Stokes 1983). The birds feed within this area and defend the territory by singing as they fly about to define the boundaries prior to nesting (Stokes and Stokes 1983). Once the pair bond has been formed, the female helps the male chase away intruders, each pursuing birds of its own sex (Stokes and Stokes 1983). The family uses the territory until the young have become independent (Robbins 1991), about 2–3 weeks after leaving the nest (Stokes and Stokes 1983).

Density in a maple-yellow birch stand in New Hampshire was estimated at 6.1 birds/10 ha (Holmes et al. 1986).

In Québec, density was found to range from 1.1 pairs/10 ha in a white birch stand (DesGranges 1980b) to 7.5 pairs/10 ha in a mature sugar maple-beech stand (Erskine 1984).

The Rose-breasted Grosbeak frequents the edges of mature deciduous and mixed forest as well as woodlands in urban and suburban areas. Sainte-Foy. *September 1991, Yves Aubry.*

Young Rose-breasted Grosbeaks about to make their first flight. Placed in the fork of a young tree, the nest is typically loosely woven, allowing a view of the contents through the structure. Sainte-Dorothée, Laval. *June 1940, Henry Mousley, Laval University Collection.*

rare species. There is no breeding evidence for the Magdalen Islands, even though it has been reported there every year since 1979 (Fradette 1992).

History and trends

The Rose-breasted Grosbeak is more widely distributed in Québec today than it ever was in the past. At the end of the last century, Wintle (1896) considered the species a common "transient visitant" near Montréal, which bred farther north. According to Ouellet (1974a), however, it is unlikely that the species was absent from the Montréal area in summer at that time and it must have been nesting there, at least in small numbers. Moreover, Terrill (1911b) reported finding a nest in 1897 on Mount Royal, commenting that from the turn of the century, the species had been becoming more and more common in the Montréal area, where today it is a common summer resident (Bannon 1991b). Lloyd (1924) considered the grosbeak a fairly common summer resident in the Ottawa Valley. It appears these two regions were the only places where it was common at the turn of the century. Mousley (1916a) considered it a rare breeder in the Eastern Townships, and Dionne (1906) found it uncommon around Québec City. By the early 1960s, however, it was a regular breeder in the Montmagny, L'Islet, and Kamouraska areas (Tanguay 1965). In the 1940s, it was still uncommon in the Lac-Saint-Jean region (Godfrey and Wilk 1948) and was reported for the first time in the Gaspé on 27 September 1941 (Ball 1943). Atlas data show that the Rose-breasted Grosbeak is now fairly common in both of these regions.

In the virgin forest, natural openings such as clearings and water edges, as well as old burns and areas flooded by beaver ponds were the preferred habitat of the Rose-breasted Grosbeak. Subsequent clearing for urban and agricultural development considerably altered the natural habitat in certain areas. Unlike some other forest species, the Rose-breasted Grosbeak benefited from the disturbance (Todd 1963) and began to use the new open habitat. Thus it began inhabiting forest fragments, second-growth forest, and woodland edges—all habitats that were multiplying—drawing ever closer to areas of human habitation (Todd 1940). This may explain, at least to some extent, why the breeding range of the Rose-breasted Grosbeak has expanded and its numbers have increased since the beginning of the century (Ellison 1985a; Bonney 1988f).

BBS data show a sizeable population increase for the species in eastern North America between 1965 and 1979 (Robbins et al. 1986). In Québec, on the other hand, neither BBS nor ÉPOQ data (Cyr and Larivée 1993a) for the years 1966–89 show any appreciable change in abundance. The future of the Rose-breasted Grosbeak does not look bright, however, as habitat loss from the destruction of the tropical forests on its wintering grounds may prove to be a determining factor in population trends (Terborgh 1992). Diamond (1991) predicts that between 1985 and the year 2000, the species will lose 41% of its winter habitat. It is difficult, however, to predict what impact this will have, as it is not known whether abundance of this species is affected more by changes to breeding habitat or by loss of wintering habitat.

Richard Pelletier and Diane Dauphin

Indigo Bunting

Passerin indigo

Passerina cyanea

Christiane Girard

THE MALE OF THIS SPECIES IS CERTAINLY ONE OF THE JEWELS OF QUÉBEC'S birdlife, resplendent in its breeding dress of unrelieved blue. Its sharp and strident song, which it repeats incessantly from a treetop at the edge of the woods or from its perch on a wire, is also diagnostic. The female is secretive and more difficult to detect, particularly prior to hatching (Stokes and Stokes 1983). Her drab plumage can test the skills of any birders.

The Indigo Bunting is a common migrant breeder in southern Québec, wintering mainly from Mexico to Colombia, as well as in the Greater Antilles and the Bahamas (AOU 1983).

Habitat and behaviour

The Indigo Bunting's preferred breeding habitat has fairly open canopy and dense undergrowth (Crawford et al. 1981). The leafy edges of forests, clearings, riverbanks, marshes, second growths, roadsides, abandoned farms, and power line rights-of-way are favourite sites (Godfrey 1986; Payne 1992). While it nests in shrubs and bushes, it needs trees for perching and singing (Taber and Johnston 1968; Cayouette and Grondin 1977; Godfrey 1986).

The male vigorously defends its territory, which may vary from 0.4 to 8 ha in size, depending on habitat (Payne 1992). The female generally nests in her mate's territory, but may also nest in a territory defended by another male or even in an area without a resident male (Payne and Payne 1989; Payne 1992). The pair forages mainly in the territory defended by the male, but may sometimes venture outside (Payne 1992).

In general, the Indigo Bunting is monogamous, with most males having only one female nesting in their territory. However, 10–15% of males are polygynous and have 2 (in rare cases 3 or 4) mates (Carey and Nolan 1979; Payne 1982). In addition, a fairly high rate of extra-pair copulation has been observed (Westneat 1987a,b), and 20–40% of young are raised by pairs in which the male is not the biological father, as shown by DNA analysis (Westneat 1987b).

Typically the pair stays together throughout the breeding season (Payne 1992) and may produce a second clutch (Terres 1980; DeGraaf and Rudis 1987; Payne 1992). Occasionally, the female changes territory and produces a second clutch with a new male (Payne 1992).

The female selects the location for the nest (Stokes and Stokes 1983; Payne 1992), which she alone builds (Terres 1980; Payne 1992), although she is sometimes accompanied by the male (Payne 1992). This takes 8–10 days early in the season, but sometimes as little as 2 days if the nest is built later in the summer. The female is active mainly in the early morning, but work may go on all day (Payne 1992). The cup-shaped nest is usually made in brambles (Peck and James 1987) but also in shrubs (Cayouette and Grondin 1977; Godfrey 1986), generally about 0.3–1 m up (Payne 1992). The outside is made of twigs, weeds, bark shreds, and a few leaves. The inside is lined with fine grasses and often with hair (Godfrey 1986; Payne 1992).

Incubation begins only after the last egg has been laid (Payne 1992) and generally lasts 12 or 13 days (Taber and Johnston 1968; Terres 1980; DeGraaf and Rudis 1987). The female alone incubates the eggs (Morgan 1976; Payne 1992). She is very attentive, spending close to 80% of her time on the nest (Morgan 1976) and leaving it only to forage (Payne 1992). The male shows no interest in the nest during incubation (Payne 1992).

The Indigo Bunting prefers open stands of tall trees with dense undergrowth, as well as the edges of deciduous forests and clearings. Dundee, southwest of Montréal. *July 1991, Gaétan Duquette.*

Notes on territory

Males vigorously defend their territories, which may vary from 0.4 to 8 ha in area, depending on habitat (Payne 1992). The smaller territories are found in forest edges and old fields; the larger ones are located in shrubby marshes (Payne 1992).

The female generally nests in her mate's territory, but may also settle in a territory defended by another male or in an area without a resident male (Payne and Payne 1989; Payne 1992). Most breeding males have only one female nesting in their territory, but some have 2 mates, and in rare cases, 3 or 4 (Carey and Nolan 1979; Payne 1982). The pair forages mainly within the territory, but may sometimes venture outside (Payne 1992).

In Ontario, recorded densities include 4.0 pairs/10 ha in a hackberry-choke cherry stand (Erskine 1972a); 3.5 pairs/10 ha in a site containing a wooded ravine and floodplain (Erskine 1976a); 3.2 pairs/10 ha in a hardwood stand dotted with clearings (Erskine 1971a); 3.0 pairs/10 ha in a wooded ravine in an urban setting (Erskine 1971a); 2.5 pairs/10 ha in a couch-grass field and a blue-joint and stubble field (Erskine 1972a); and 1.4 pairs/10 ha in habitat combining old fields with hackberry, blue ash, and chestnut oak (Erskine 1980a).

Indigo Bunting

Breeding distribution
1984–1989

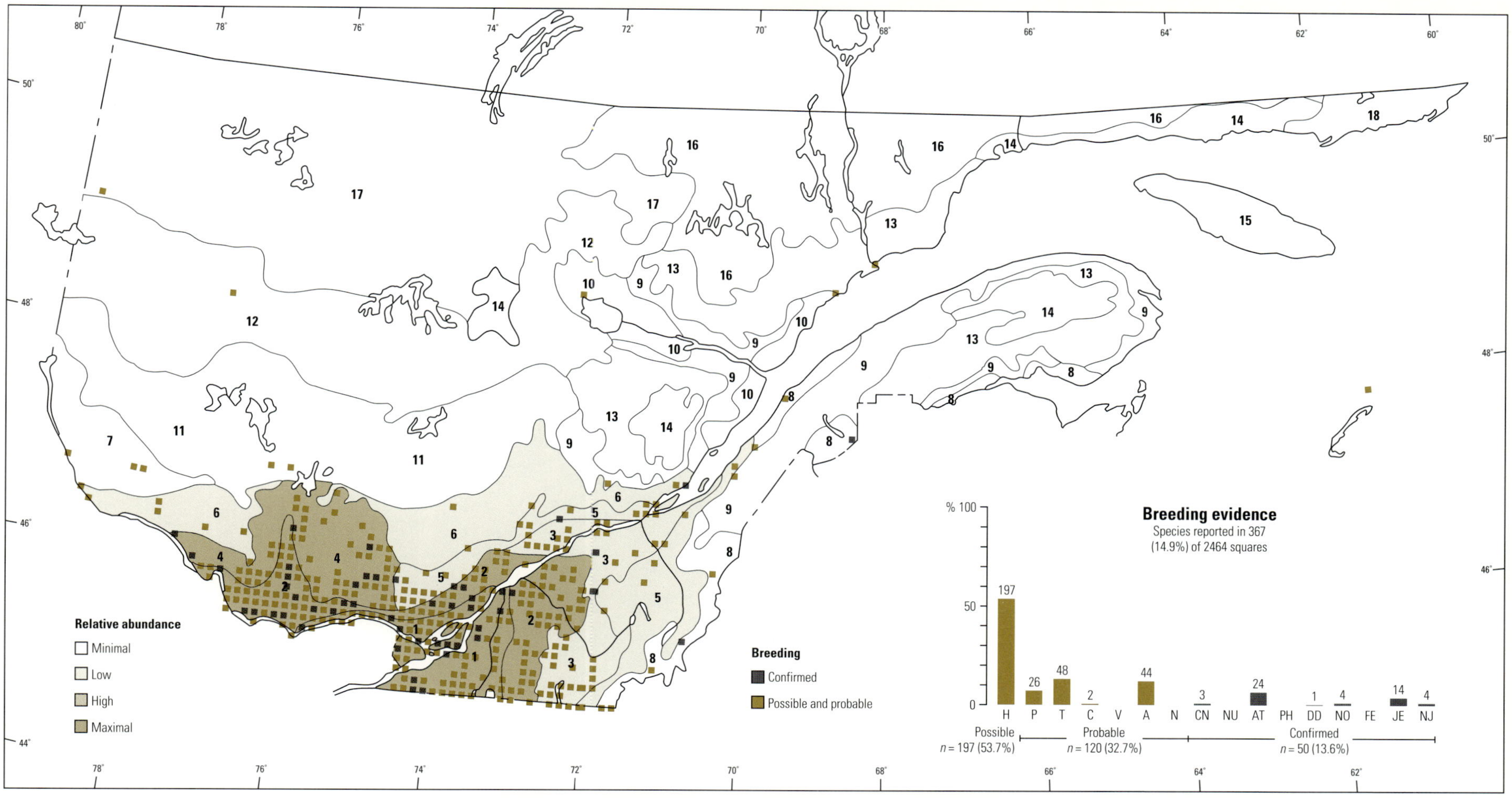

Species profile

Clutch size	usually 3–4 eggs[2]
Laying interval	1 egg per day[2]
Start of incubation	with last egg[2]
Incubation period	usually 12–13 days (11–14)[2,4,5]
Maturity at hatching	altricial
Care of young	mainly by female[2,6]
Nestling period	usually 9–12 days (8–14)[2,7]
Age at first flight	usually 9–12 days (8–14)[2,7]
Dependence of young	until ≈ 21 days after first flight[2]
Broods per year	usually 1–2 (1–3)[1,2,4]
Breeding age	1 year[2,4]
Mating system	monogamy, sometimes polygyny[2,6]
Duration of pair bond	1 season[2]
Total length	13.4–14.5 cm[3]
Wing span	20.3–22.9 cm[1]
Weight	M 14.9 g (12.3–21.4) / F 14.1 g (11.2–18.6)[8]
Longevity record	9 years 3 months[9]

References:
1. Terres 1980
2. Payne 1992
3. Godfrey 1986
4. DeGraaf and Rudis 1987
5. Stokes and Stokes 1983
6. Carey and Nolan 1979
7. Payne 1982
8. Dunning 1984
9. Klimkiewicz and Futcher 1987

World breeding range

Chronology $n = 30$

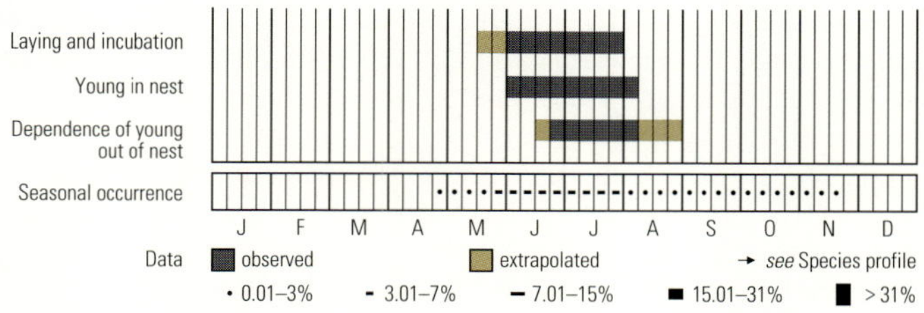

Frequency of occurrence by ecological region

Ecological regions** (forest regions)	Adequately covered squares	Squares in which reported	% of squares in which reported	Relative abundance (%)*
1. Sugar maple-hickory and sugar maple-basswood	97	71	73.2 ± 4.5	98.2
2. Sugar maple-basswood	153	114	74.5 ± 3.6	100.0
3. Sugar maple-basswood and sugar maple-yellow birch	101	36	35.6 ± 5.7	47.8
4. Sugar maple-yellow birch-basswood	56	39	69.6 ± 10.6	93.5
5. Sugar maple-yellow birch-beech and sugar maple-yellow birch-basswood	113	27	23.9 ± 4.4	32.1
6. Sugar maple-yellow birch-beech	71	16	22.5 ± 9.0	30.2
7. Sugar maple-yellow birch-balsam fir	19	3	15.8 ± 17.9	21.2
8. Sugar maple-yellow birch and balsam fir-yellow birch	47	3	6.4 ± 6.8	8.6
9. Balsam fir-yellow birch	87	0	0	0
10. Balsam fir-white birch or balsam fir-red maple	46	1	2.2 ± 4.4	2.9
11. Yellow birch-balsam fir	52	1	1.9 ± 4.5	2.6
12. Balsam fir-white birch (west)	59	1	1.7 ± 4.0	2.3
13. Balsam fir-white birch (east)	76	1	1.3 ± 3.0	1.8
14. Balsam fir-black spruce	28	0	0	0
15. White spruce-balsam fir and balsam fir-white spruce	22	0	0	0
16. Black spruce-balsam fir-moss	18	0	0	0
17. Black spruce-moss	32	1	3.1 ± 7.6	4.2
18.			not analysed	

* minimal: 0–29.9%; low: 30–59.9%; high: 60–89.9%; maximal: 90–100%

** Deciduous forest: 1–7; mixed forest: 8–11; coniferous forest: 12–17

The species sings from the tops of tall trees and nests in nearby shrubbery. Île Perrot, near Montréal. *July 1991, Gaétan Duquette.*

Hatching is synchronous, with the nestlings, covered in sparse down, usually all hatching on the same day (Payne 1992). The female eats the shells (Morgan 1976) and feeds the nestlings from the first day (Payne 1992), mainly on small insects, spiders, caterpillars, and berries (Morgan 1976; Payne 1992).

The young usually fledge at 9–12 days, but are still unable to fend for themselves (Payne 1992). They become independent around 3 weeks after leaving the nest (Payne 1992).

Participation by the male in raising the young varies, but is, in general, quite limited (Carey and Nolan 1979; Ehrlich et al. 1988; Westneat 1988a,b; Payne 1992). However, he may become more actively involved after the young birds have left the nest (Westneat 1988a; Payne 1992).

The Indigo Bunting is a frequent cowbird host (Taber and Johnston 1968; Payne 1992). The cowbird usually lays 1 egg, sometimes 2 and, in rare cases, 3 (Payne 1992). Of 30 nests studied in southern Québec, 20% had been parasitized (Terrill 1961). In Ontario, 24% of 165 nests examined had been parasitized (Peck and James 1987). When the cowbird lays its egg after the bunting has started laying, the bunting accepts the parasitic egg and incubates it along with her own. However, she abandons the nest if the cowbird egg appears before she has started to lay (Payne 1992).

The Indigo Bunting was reported in 14.9% (367/2464) of Atlas squares surveyed, and breeding was confirmed in 13.6% (50/367) of these, mainly from observations of adults carrying food for the young.

The outside of the nest is usually made of coarse material, such as stubble, while the inside is carefully lined with soft grass and hair. The Indigo Bunting is a frequent cowbird host, as this nest shows. Saint-Prosper, northeast of Trois-Rivières. *June 1984, Christian Marcotte.*

Shrubby habitat along roadsides and power line rights-of-way also suit the Indigo Bunting. **Varennes, near Montréal.** *July 1991, Gaétan Duquette.*

Distribution

The breeding range of the Indigo Bunting extends from Saskatchewan across to New Brunswick in the north down to New Mexico and Florida in the south (AOU 1983).

In Québec, the Indigo Bunting generally frequents the deciduous forest zone. Frequency of occurrence was highest in Montérégie, the Ottawa Valley, the Laurentians, and in a narrow band north of the St. Lawrence between Montréal and Lake Saint-Pierre. In these areas, the forest is fragmented, and there are many edges. North of the 47th parallel, distribution is more localized. The northernmost Atlas records come from Val-Paradis in the northwestern part of southern Québec, Saint-Méthode in the Lac-Saint-Jean region, and Pointe-Lebel on the Upper North Shore. Breeding was confirmed at Dégelis, on the Lower St. Lawrence, in 1984; this was the first breeding record for this region (Larivée 1993). The species was also atlassed on the Magdalen Islands, but breeding was not confirmed; it is considered a rare visitor to the islands (Fradette 1992). It was not reported on Anticosti Island.

The Indigo Bunting nests in thickets, often in brambles, at a height of usually less than 1 m. Notice the typical straggly exterior of the nest, built of grass and leaves in a young maple, on the rim of which two well-developed nestlings are perched. Sainte-Dorothée, Laval. *August 1942, Henry Mousley, Laval University Collection.*

History and trends

At the end of the last century, Wintle (1896) listed the Indigo Bunting as a common summer resident in the Montréal area and reported nesting on Mount Royal, while Dionne (1906) considered it fairly scarce around Québec City.

Continent-wide BBS data indicate a slight but steady increase in abundance between 1965 and 1979 (Robbins et al. 1986). In Québec, BBS data for 1966–89 show no significant trend, but ÉPOQ summer data for 1970–89 indicate that numbers have declined (Cyr and Larivée 1993b).

Farming practices involving the clearing of bushes and repeat harvesting of forage crops, urban sprawl, roadside maintenance, and the reversion of old fields to forest all limit the habitat of the Indigo Bunting. Other practices, however, including clearing for power line corridors and downhill ski trails, the abandonment of farmland, and logging, favour the species. The Indigo Bunting colonizes second-growth cutovers until crown closure, so selective cutting, which opens up the canopy, creates suitable habitat (Crawford et al. 1981).

Serge Labonté and Diane Dauphin

Rufous-sided Towhee*

Tohi à flancs roux

Pipilo erythrophthalmus

Christiane Girard

THE RUFOUS-SIDED TOWHEE IS OFTEN REVEALED BY A LOUD RUSTLING OF leaves as it forages vigorously in the litter either by stirring up the leaves with its bill or by kicking energetically backward with both feet. Its presence is also betrayed by its characteristic *to-whee* call (also transcribed as *chewink*) and its song, which consists of three or four clear, musical notes, often rendered as *drink-your-teeeee*. The species' contrasting black and robin-red plumage and its red eyes give it an exotic look. The towhee sometimes imitates the songs of other passerines such as the Carolina Wren and the Field Sparrow (Borror 1977). The female may also sing, particularly early in the morning (Dickinson 1968).

The Rufous-sided Towhee belongs to a group of species (the genera *Pipilo*, *Melozone*, *Arremon*, *Arremonops*, and *Atlapetes*, among others) found in tropical and subtropical Central and South America. Although these birds are closely related to the sparrows, they are distinguished by their generally larger size and brighter plumage. Aside from the Green-tailed Towhee, which is found only in the western United States, the Rufous-sided Towhee is the only representative of this group whose breeding range extends significantly north of the tropics.

In Québec, the Rufous-sided Towhee is a rather rare migrant breeder found only in the extreme southern part of the province. Some hardy individuals manage to winter here occasionally by taking advantage of feeding stations. The species' usual winter range extends from British Columbia to Guatemala in the west and from the central United States to the Gulf of Mexico in the east (AOU 1983).

Habitat and behaviour

The Rufous-sided Towhee is widespread throughout the United States, where it breeds in a variety of habitats ranging from California chaparral to New England pastureland and Florida pine forests. It seeks dry, brushy areas (Bull 1974), which it finds in thickets, old fields reverting to scrub, clearings, open woodland, hedgerows, and forest margins.

The female builds the nest, usually on or slightly above the ground and hidden in a bush or clump of grass. The nest may also be located in dense shrubbery or vines, up to 1.5 m above the ground. The female alone incubates the eggs and broods the young, but the male helps feed the young (Ehrlich et al. 1988), mainly on insects and occasionally fruit (Dickinson 1968). The adults' diet consists of approximately 70% vegetable matter (seeds, acorns, wild fruit); the rest is composed of animal matter, mainly beetles (particularly snout beetles), butterflies, caterpillars, bugs, ants, spiders, and slugs (McAtee 1926), which it finds on or near the ground (Morimoto and Wasserman 1991b). After the young fledge, they remain with their parents for a while before dispersing (Forbush 1929).

This species usually produces a second brood in the same territory and with the same mate (Dickinson 1968). However, since this requires a fairly long breeding season (Greenlaw 1978), birds in Québec probably have only one brood. In many regions of North America, the Rufous-sided Towhee is one of the most frequent cowbird hosts (Friedmann 1966).

Distribution

The Rufous-sided Towhee breeds from southern Mexico to southern Canada, but not in Texas (National Geographic Society 1987).

In Québec, the species is found mainly in the Ottawa Valley and the southwestern part of the St. Lawrence Plain. It has been reported as far northeast as Saint-Léon-de-Standon and Sainte-Euphémie, near Québec City. Farther west, a pair was sighted on and after 10 June 1984 on the edge of a burn near Lake Lusignan, near Saint-Michel-des-Saints (P. Lafonfaine, pers. comm.).

The species' distribution in Québec is rather patchy. It is not found in wetlands, cultivated land or urban areas, perhaps because of its preference for dry habitat, particularly pine and oak stands.

Ideal habitat is found in the pine stands that occur on the sand terraces of the Ottawa Valley, in places like Kazabazua, as well as in the young pine plantations scattered here and there in the southern part of the province. Indeed, in southern Massachusetts, the Rufous-sided Towhee is one of the most abundant birds in pitch pine stands, which have an open canopy and a very dense shrub layer (Morimoto and Wasserman 1991a). Mousseau (1984b) reports that the highest density of Rufous-sided Towhees in Québec is found in the Pin-Rigide Ecological Reserve, on a rock outcrop northeast of Saint-Chrysostome.

This towhee also frequents to a lesser extent the red oak stands that grow on the rocky soil of the Monteregian Hills and Laurentian foothills, often in association with red and Eastern white pine. The species may also be found in other dry, brushy habitat such as abandoned fields and pastures.

Interestingly, individuals found overwintering in Québec are often observed far to the north and east of the species' usual breeding range, in locations like Alma, Caplan, Chicoutimi, Jonquière, Rimouski, and Sept-Îles (ÉPOQ). These individuals are sometimes western birds, which can be distinguished from their eastern counterparts by their white wing spots. This suggests that at least some of the overwintering birds do not come from the local breeding population but rather from elsewhere in North America (ÉPOQ).

History and trends

The Rufous-sided Towhee does not appear to have occurred in Québec at the turn of the century, since there is only one record involving a pair at Cap-Rouge in 1879 (Dionne 1906). The species may have been more common in southern Québec before the province's pine forests were razed for timber during the first half of the 19th century (Ouellet, F. 1966), but this is just speculation.

In the Montréal region, the species appears to have benefited from new habitat created by forest clearing, since it has been observed regularly but sporadically since at least 1932 (Wynne-Edwards 1932). Ouellet (1974a) related that, during the 1960s, there was evidence that the species may have been expanding its range in the Montréal area, which it was also doing in northern New York State at the time (Bull 1974). However, the Québec population does not appear to have changed much since then, although a steady decrease in eastern North America as a whole was observed between 1965 and 1989 (Robbins et al. 1986; Hagan 1993).

NORMAND DAVID

The Rufous-sided Towhee prefers dry brushy habitats, like old fields reverting to scrub. Saint-Antoine-Abbé, south of Montréal. *May 1991, Pierre Bannon.*

Notes on territory

According to studies conducted by Franzblau and Collins (1980) in Arizona and Wasserman (1983) in Massachusetts, there appears to be no direct relation between the size of the territory defended by the Rufous-sided Towhee during the breeding season and abundance of food resources. In fact, neither when the quantity of prey was increased artificially inside the birds' territories nor when up to 30% of the birds' territories were made inaccessible, was any major change found in territory size. In both cases, the defended area (about 0.5 ha) remained the same. This suggests either that territoriality in the species may not be associated only with the availability of food or that territories defended may contain more food than the birds can consume (Wasserman 1983).

In southern Ontario, densities of 11.5 males/10 ha were found in a meadow, 2.0 males/10 ha in a mixed hardwood forest, 5.0 males/10 ha in a red pine plantation, and 4.0 and 9.5 males/10 ha in Eastern white-cedar and Eastern white pine stands, respectively (Erskine 1971a, 1976a, 1980a).

* The Rufous-sided Towhee (*Pipilo erythrophthalmus*) has recently been split into two species (AOU 1995), and both are encountered in Québec: the Eastern Towhee (*Pipilo erythrophthalmus*, Tohi à flancs roux) is a migrant breeder, and the Spotted Towhee (*Pipilo maculatus*, Tohi tacheté) is a visitant.

Rufous-sided Towhee

Breeding distribution
1984–1989

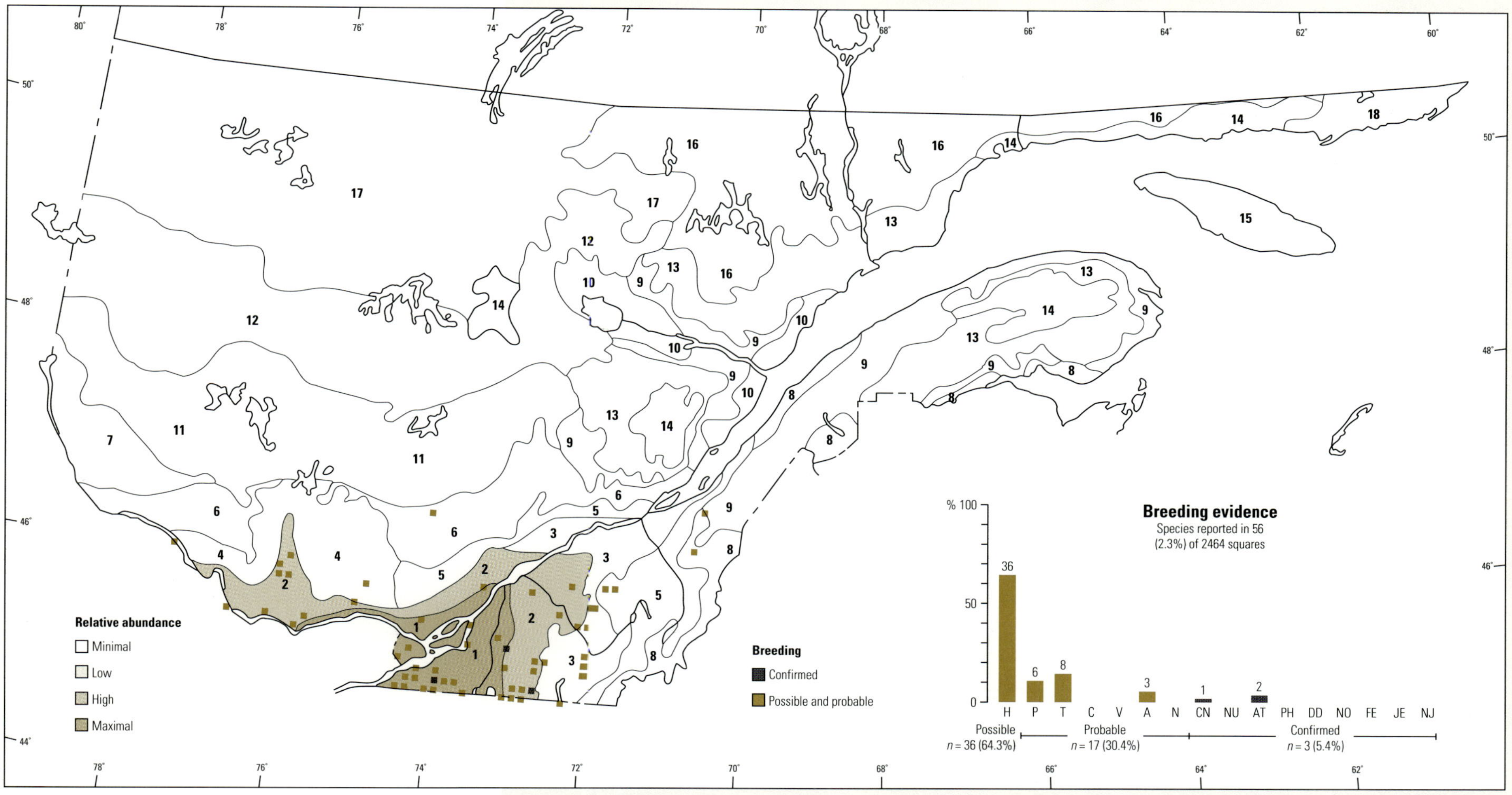

Species profile

Clutch size	usually 3–4 eggs (2–6)[1,2,3,5]
Laying interval	?
Start of incubation	with second or last egg[3]
Incubation period	12–13 days[3,4,5]
Maturity at hatching	altricial
Care of young	by female and male[4]
Nestling period	10–12 days[3,4,5]
Age at first flight	10–12 days or more[10]
Dependence of young	?
Broods per year	usually 2 (1–3)[1,3,4,9]
Breeding age	1 year[9]
Mating system	monogamy[4]
Duration of pair bond	?
Total length	19–22 cm[7]
Wing span	25.4–31.1 cm[5]
Weight	M 41.7 g (32.1–50.0) / F 39.3 g (32.1–52.3)[8]
Longevity record	12 years 3 months[6]

References:
1. Greenlaw 1978
2. Bull 1974
3. Dickinson 1968
4. Ehrlich et al. 1988
5. Terres 1980
6. Klimkiewicz and Futcher 1987
7. Godfrey 1986
8. Dunning 1984
9. DeGraaf and Rudis 1987
10. according to nestling period

World breeding range

Chronology $n = 6$

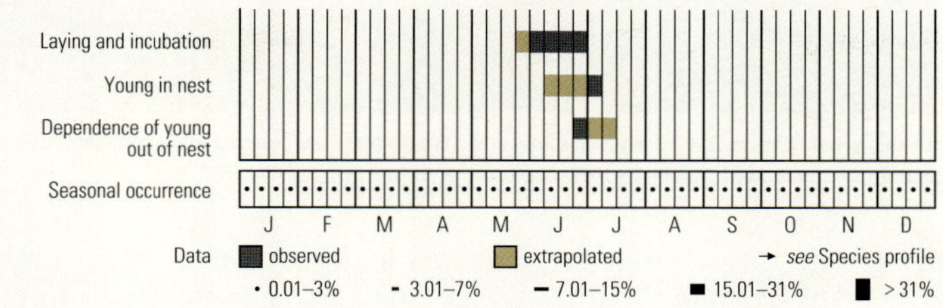

Frequency of occurrence by ecological region

Ecological regions** (forest regions)	Adequately covered squares	Squares in which reported	% of squares in which reported	Relative abundance (%)*
1. Sugar maple-hickory and sugar maple-basswood	97	19	19.6 ± 4.1	100.0
2. Sugar maple-basswood	153	25	16.3 ± 3.1	83.4
3. Sugar maple-basswood and sugar maple-yellow birch	101	4	4.0 ± 2.6	20.2
4. Sugar maple-yellow birch-basswood	56	1	1.8 ± 3.7	9.1
5. Sugar maple-yellow birch-beech and sugar maple-yellow birch-basswood	113	2	1.8 ± 1.7	9.0
6. Sugar maple-yellow birch-beech	71	0	0	0
7. Sugar maple-yellow birch-balsam fir	19	0	0	0
8. Sugar maple-yellow birch and balsam fir-yellow birch	47	2	4.3 ± 5.8	21.7
9. Balsam fir-yellow birch	87	0	0	0
10. Balsam fir-white birch or balsam fir-red maple	46	0	0	0
11. Yellow birch-balsam fir	52	0	0	0
12. Balsam fir-white birch (west)	59	0	0	0
13. Balsam fir-white birch (east)	76	0	0	0
14. Balsam fir-black spruce	28	0	0	0
15. White spruce-balsam fir and balsam fir-white spruce	22	0	0	0
16. Black spruce-balsam fir-moss	18	0	0	0
17. Black spruce-moss	32	0	0	0
18.			*not analysed*	

* minimal: 0–29.9%; low: 30–59.9%; high: 60–89.9%; maximal: 90–100%

** Deciduous forest: 1–7; mixed forest: 8–11; coniferous forest: 12–17

Chipping Sparrow

Bruant familier

Spizella passerina

Christiane Girard

THIS SPARROW USED TO LINE ITS NEST WITH HORSEHAIR, EARNING IT THE name of "hair bird." When motor vehicles replaced horses, it gradually switched to twigs and rootlets.

Starting in late April, this abundant sparrow breeds in a variety of open habitats in southern Québec. Although most of the population leaves again in October for the southern United States and Mexico, a few stragglers linger near bird feeders until November or December. A few cases of overwintering have even been reported in the southern part of the province. The sparrow's French name is very apt: because it is so easy to watch and is often found in urban areas, it is indeed one of our most familiar sparrows.

Territorial conflicts between males are frequent; they often chase conspecifics while uttering a short song. Copulation sometimes takes place immediately after the male returns to his mate (Reynolds and Knapton 1984). The male courts the female by staying with her until incubation, at which point he generally stops singing. The species is apparently polygynous on occasion (Middleton and Prescott 1989).

The female defends the immediate vicinity of the nest (Stull 1968). She lays the first egg as soon as the nest is completed and starts incubating a day before

This sparrow is especially common in urban areas where there are conifers and ornamental trees and shrubs. Sainte-Foy. *September 1994, Jean Gauthier.*

Notes on territory

The territory, varying from 0.3 to 3.1 ha, is used for feeding, mating, and nesting. It is defended by the male from the time he first arrives in spring until the young of the second brood leave the nest (Stokes and Stokes 1983; DeGraaf and Rudis 1987). The female defends the immediate vicinity of the nest (Stull 1968). Reported densities in Utah were 45.2 birds/10 ha in wet meadows bordered by pine forest, 28.7 birds/10 ha in clearcut areas with second-growth pines, and 23.7 birds/10 ha in dry meadows with a scattering of pines (Austin and Perry 1979).

Habitat and behaviour

The Chipping Sparrow forages in openings, woodland edges, and clumps of trees bordering open grassy areas (Hebrard 1978; Terres 1980; Godfrey 1986). The species seems to need at least a few conifers because it often nests in them (Perez 1982; Peck and James 1987). This sparrow has adapted to changes caused by human activity. City parks, orchards, cemeteries, golf courses, farms, fields, conifer plantations, highway and transmission line rights-of-way, are now among its breeding habitats.

As soon as the males arrive, they perch well in view and defend their territory with a long dry metallic trill repeated all day long (Stull 1968). The females arrive a week later. Once pairs have formed, they inspect various potential nest sites. In the suburbs, they often select a small conifer such as are increasingly common in landscaped yards. The pairs studied by Reynolds and Knapton (1984) in Ontario forests nested mainly in white spruce, at a mean height of 1.1 m, and usually on the southeast side of the tree.

Unlike most other sparrows, both the male and female Chipping Sparrow gather material and work together to build the nest, taking about 3 or 4 days (Reynolds and Knapton 1984). The cup-shaped nest, which is usually at the end of a horizontal branch, is made of grass stems and rootlets, and is lined with hair, including horsehair, various types of animal fur, fine grasses, plant down, and sometimes fine rootlets (Peck and James 1987). This sparrow's attraction to hair can actually be fatal: one bird was found dangling at the edge of the nest by two long hairs in which it had become entangled (Fillmore and Titman 1977).

The Chipping Sparrow likes to nest in the branches of a conifer, hedge, or dense shrubbery near dwellings, often placing the nest on the southeast side. Mount Saint-Hilaire. *April 1991, Yves Aubry.*

Chipping Sparrow

Breeding distribution

1984–1989

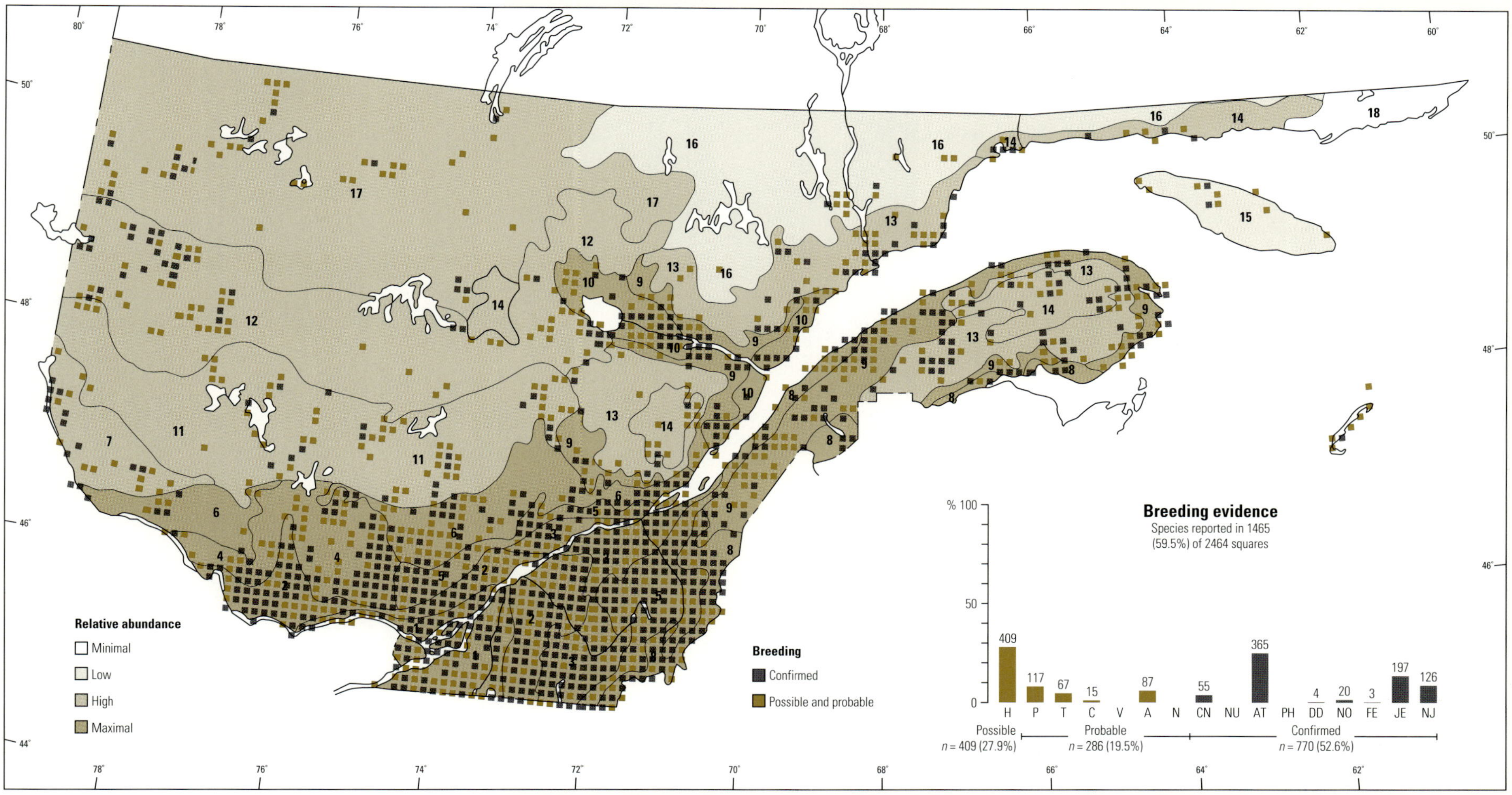

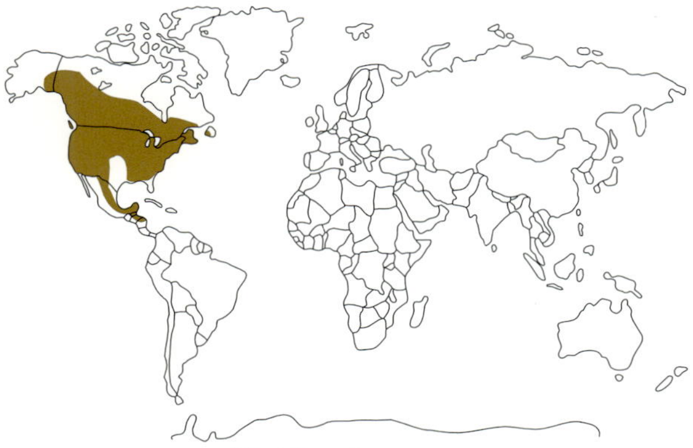

World breeding range

Species profile

Clutch size	usually 4 eggs (3–5)[1,2,3,4]
Laying interval	1 egg per day[2]
Start of incubation	before last egg[2]
Incubation period	11–14 days[1,2,3]
Maturity at hatching	altricial
Care of young	by female and male[1,2]
Nestling period	8–12 days[1,2,3]
Age at first flight	14 days[2]
Dependence of young	until 24 days old[5]
Broods per year	1–2[1,3]
Breeding age	1 year[6]
Mating system	monogamy, sometimes polygyny[8]
Duration of pair bond	?
Total length	12.5–14.5 cm[4]
Wing span	20.3–22.9 cm[3]
Weight	? g (10.3–19.7)[3]
Longevity record	9 years 9 months[7]

References:
1. Ehrlich et al. 1988
2. Harrison 1978
3. Terres 1980
4. Godfrey 1986
5. Perez 1982
6. DeGraaf and Rudis 1987
7. Klimkiewicz and Futcher 1987
8. Middleton and Prescott 1989

Chronology $n = 431$

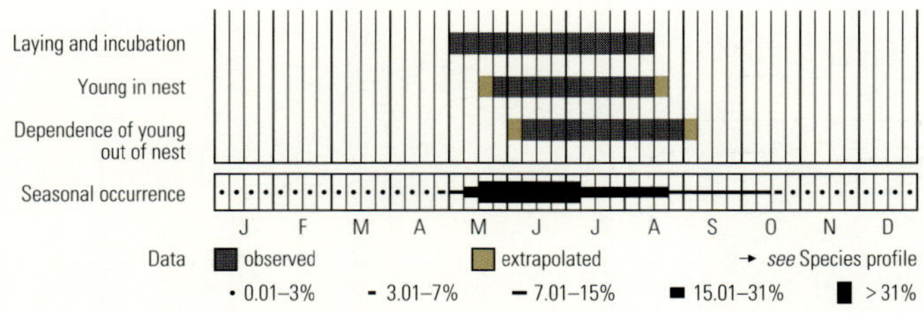

Frequency of occurrence by ecological region

Ecological regions** (forest regions)	Adequately covered squares	Squares in which reported	% of squares in which reported	Relative abundance (%)*
1. Sugar maple-hickory and sugar maple-basswood	97	94	96.9 ± 2.1	96.9
2. Sugar maple-basswood	153	153	100.0 ± 0.3	100.0
3. Sugar maple-basswood and sugar maple-yellow birch	101	101	100.0 ± 0.5	100.0
4. Sugar maple-yellow birch-basswood	56	55	98.2 ± 3.7	98.2
5. Sugar maple-yellow birch-beech and sugar maple-yellow birch-basswood	113	113	100.0 ± 0.4	100.0
6. Sugar maple-yellow birch-beech	71	64	90.1 ± 6.6	90.1
7. Sugar maple-yellow birch-balsam fir	19	16	84.2 ± 17.9	84.2
8. Sugar maple-yellow birch and balsam fir-yellow birch	47	47	100.0 ± 1.1	100.0
9. Balsam fir-yellow birch	87	87	100.0 ± 0.6	100.0
10. Balsam fir-white birch or balsam fir-red maple	46	43	93.5 ± 6.7	93.5
11. Yellow birch-balsam fir	52	45	86.5 ± 9.9	86.5
12. Balsam fir-white birch (west)	59	50	84.7 ± 9.7	84.7
13. Balsam fir-white birch (east)	76	65	85.5 ± 7.9	85.5
14. Balsam fir-black spruce	28	25	89.3 ± 12.6	89.3
15. White spruce-balsam fir and balsam fir-white spruce	22	8	36.4 ± 20.8	36.4
16. Black spruce-balsam fir-moss	18	6	33.3 ± 24.8	33.3
17. Black spruce-moss	32	25	78.1 ± 15.9	78.1
18.			*not analysed*	

* minimal: 0–29.9%; low: 30–59.9%; high: 60–89.9%; maximal: 90–100%

** Deciduous forest: 1–7; mixed forest: 8–11; coniferous forest: 12–17

laying the last egg. The eggs, usually numbering 4, are incubated for 11–14 days. The female broods the hatchlings on the nest for 4 or 5 days, during which time the male brings her food. Both parents then feed the young, which leave the nest 8–12 days after hatching. The parents continue to feed them on the branches for another 10 days or so. The Chipping Sparrow sometimes produces 2 broods in 1 year (Terres 1980).

A confirmed seed eater in winter (Allaire and Fisher 1975; Pulliam and Dunning 1987), the Chipping Sparrow includes insects like the jack pine budworm in its summer diet (Sloan and Simmons 1973; Simmons and Sloan 1974).

Nests are often parasitized by the Brown-headed Cowbird (Friedmann 1963). This sparrow's usual reaction is to abandon the nest (Graham 1988b), but it also commonly raises the young cowbird to fledging. Parasitism rates of 12% in southern Québec (Terrill 1961) and 32% in Ontario (Peck and James 1987) have been reported. In forests, however, parasitism may be virtually non-existent (Reynolds and Knapton 1984).

The Chipping Sparrow commonly frequents backyards, so it is easy to observe. Breeding is also fairly easy to confirm by observation of adults carrying food, unfledged young, or nest-building activities; furthermore, nests are quite easy to find.

Distribution

In Canada, the Chipping Sparrow breeds from Yukon and British Columbia east to southwestern Newfoundland (Godfrey 1986). It is also found in east-central and southeastern Alaska, throughout most of the coterminous United States south to Nicaragua (AOU 1983).

The Chipping Sparrow breeds throughout southern Québec. It is also very common in almost all Atlas regions, particularly in the settled southern part of the province, where urbanization and farming have resulted in open areas suitable for nesting. It is less common on Anticosti Island and in the northeastern part of the Atlas area, however, places that are not very highly developed or populated. It was atlassed as a breeder as far as the Havre-Saint-Pierre area on the North Shore, although there are pre-Atlas records from Natashquan (Todd 1963) and near Brador Bay (ÉPOQ).

North of the Atlas area, there are breeding records from Waskaganish and Eastmain and a sight record from Fort George (Todd 1963). The northernmost record comes from Lake Guillaume Delisle, on 15 June 1981 (ÉPOQ). More recently, it has also been reported at a few sites in the Rupert River watershed as well as near Lake Evans, mainly in tamarack peatlands and burns (Cons. G. G. – GREBE 1992o). Farther east, it was seen at Gagnon in May 1982 (ÉPOQ).

Brood in a spruce tree. The young instinctively beg for food as soon as a parent approaches. The male and female combine their efforts, both to build the nest and to feed the young. Québec City. *June 1955, Gaston Le Page.*

In coniferous habitat, the Chipping Sparrow is found in clearings, the edges of logged areas, and sites colonized by young conifers. Saint-Elzéar, southwest of Percé. *July 1988, Jean-Luc DesGranges.*

Notes on territory (continued)

In Québec, reported densities include 9.7 birds/10 ha in a white spruce plantation in La Mauricie National Park (DesGranges 1980b), 4 pairs/10 ha in a mixed conifer forest at Piedmont, near Saint-Sauveur-des-Monts (Erskine 1971a), and 4.9 pairs/10 ha in a young mixed stand dominated by fir adjacent to an orchard at Carleton Centre, on the Gaspé Peninsula (Erskine 1980a).

Aside from clearings, openings in the forest cover resulting from logging or the construction of transmission line rights-of-way are favourable habitats. Saint-Paul-du-Nord, northeast of Tadoussac. *October 1993, Pierre Normand.*

This sparrow nests in Christmas tree plantations—thick, dense environments reminiscent of ornamental hedges. Near Sherbrooke. *June 1990, Guy Rondeau.*

History and trends

At the turn of the century, the Chipping Sparrow was common in the settled areas of southern Québec (Dionne 1889; Macoun and Macoun 1915; Taverner 1922). The clearing of land for settlement probably favoured this species by creating suitable habitat (Stull 1968). More recently, however, the species has been declining in several U.S. states. Some authors cite competition with the House Sparrow as an explanation (Root 1988), but cowbird parasitism has certainly also played a role (Buech 1982; Graham 1988b). The Chipping Sparrow seems to need a certain minimum amount of open space in its territory and the small yards typical of new residential areas do not seem to be big enough for it (Sutton 1960).

BBS data from the agricultural areas of southern Québec for 1970–86 indicate that Chipping Sparrow populations have increased slightly (Falardeau and DesGranges 1991). On the other hand, neither Québec ÉPOQ data for the period 1970–89 (Cyr and Larivée 1993a) nor 1966–89 BBS data for the entire southern part of the province show any trend.

Since the Chipping Sparrow is common in agricultural areas, pesticides may be harmful. A Québec study has shown that, in Chipping Sparrows exposed to the organophosphates sprayed in orchards, the activity of the enzyme cholinesterase, which transmits nerve impulses, decreases (Graham and DesGranges 1993).

Desneige Perreault and Simon Barrette

Clay-colored Sparrow

Bruant des plaines

Spizella pallida

Christiane Girard

The Clay-colored Sparrow is a relative newcomer to Québec. Originally found only in western North America, it is now widespread in the Great Plains and Great Lakes regions (Knapton 1987a).

Over the last 50 years or so, this passerine has expanded its range eastward. The first Québec record dates from 1960, when an individual was observed in Philipsburg from 23 May to 5 June (Abbott et al. 1960). Breeding was not confirmed until 1975, at Saint-Vital-de-Clermont, near La Sarre in the Abitibi region, a few kilometres from the Ontario border (COQ 1975). Today, the Clay-colored Sparrow is considered a rare migrant breeder in various areas of the province.

The Clay-colored Sparrow winters from New Mexico and Texas to southern Mexico (AOU 1983), returning to Québec toward the end of May.

Habitat and behaviour

In its traditional range, the Clay-colored Sparrow breeds in a variety of open, uncultivated habitats, which are generally grass-covered and dotted with shrubs (Root 1968; Knapton 1987a). In Québec, this type of habitat is found in pastures and other areas, such as abandoned fields, roadsides, and logged areas. However, in Québec, as in New York State (Novak 1988), this sparrow appears to be associated mainly with conifer plantations, particularly pines and firs. It is usually found in young stands, although it sometimes frequents older plantations with grassy areas nearby.

Although the Clay-colored Sparrow's song is rather weak, it is very distinctive, and attentive observers will have no problem identifying the series of 2–5 insectlike buzzes, a flat, low-pitched *bzzz-bzzz-bzzz*. Spotting this rare bird is also made easier by the fact that it often sings for extended periods from a conspicuous perch and by its lack of shyness (Knapton 1987a).

According to Fox (1961), the adults perform simultaneous looping flights during courtship, touching their ventral areas together at the bottom of each loop. These flights culminate in copulation in a small shrub. The female takes 2–4 days to build the nest (Fox 1961), which resembles the Chipping Sparrow's, but is not quite so compact (Root 1968). Early in the season nests are often built in a tussock of grass at the base of a shrub or occasionally at the base of herbaceous plants. Later in the season, nests are usually placed in the low branches of a young conifer or of a small deciduous tree or shrub such as a hawthorn or serviceberry, up to 2.1 m from the ground (Root 1968). The female does most of the incubating and brooding, but is fed regularly by the male (Root 1968), which also relieves her on the nest for short periods (Fox 1961).

According to Root (1968), the Clay-colored Sparrow feeds mainly on plant material, including a wide variety of seeds. In Manitoba, Knapton (1980) found that this sparrow feeds its young invertebrates, either lepidopteran larvae, orthopteran nymphs, or various types of Odonata, depending on habitat frequented. Adults and young find most of their food in grassy areas outside the breeding territory (Knapton 1980). This behaviour, which is unique among open-country sparrows, explains why the breeding territories defended by this species are so small (from 739 to 1053 m^2 in Manitoba on average). The males defend their territories vigorously by aggressive behaviour or singing from nearby branches on adjacent territories. Unlike females, males usually return to the territory defended the year before (Knapton 1979b).

This species occasionally mates with its close relative the Chipping Sparrow, producing hybrids (Storer 1954; McIlroy 1961; Novak 1988). The breeding record from North Hatley in 1981 (Gosselin and David 1981) may be an example of such hybridization, because the Clay-colored Sparrow observed was feeding 2 young and was accompanied by a Chipping Sparrow (P. Boily, pers. comm.).

The Clay-colored Sparrow is a frequent cowbird host. Parasitized nests are often abandoned, after which the pair often makes a second attempt at nesting, usually near the first nest (Root 1968). In non-abandoned parasitized nests, it has been estimated that 38–56% of the sparrow eggs laid result in fledged young (Root 1968; Buech 1982).

Distribution

The Clay-colored Sparrow breeds in the southwestern Northwest Territories, eastern British Columbia, Alberta, Saskatchewan, southern Manitoba, the north-central United States, southern Ontario, and southern Québec (National Geographic Society 1987).

Québec is at the northeastern limit of the Clay-colored Sparrow's breeding range. Although the species is rare and has a scattered distribution, it occurs in many cultivated areas of the province. In addition, there are isolated records from the Québec City, Lanaudière, and Bois-Francs regions. The two QNRCP records involve 3 young at a nest in Saint-Placide on 8 June 1976 and one adult carrying food in Roberval on the same date.

Although the species' nests are considered easy to find (Knapton 1979a), only one was reported for the Atlas. There were 5 reports of confirmed breeding, each in a different region: the Ottawa Valley (near Maniwaki), the Bois-Francs region (near Victoriaville), the Eastern Townships (Sainte-Catherine-de-Hatley), the Saguenay region (Saint-Honoré), and near the New York border, southeast of Huntingdon. As the breeding season progresses, singing males that have been observed in suitable habitat often disappear, which suggests that they have not been able to find a mate.

Although the Clay-colored Sparrow has taken advantage of the clearing of forests and the establishment of plantations to expand its range (Gosselin 1989a), it is nonetheless confined to transitory habitat which changes over the years. It may well abandon plantations when the trees become too tall, as may have happened at Saint-Colomban, near Lachute, where the species was seen from the mid-1970s until 1989, but not since (P. Bannon, pers. comm.). Breeding was not actually confirmed at Saint-Colomban for the Atlas.

In Québec, the Clay-colored Sparrow is generally found south of the 49th parallel (ÉPOQ). However, it has been reported in northern Ontario (Godfrey 1986; Knapton 1987a), and therefore may very well turn up in the James Bay coastal plain some day.

History and trends

In western Canada, habitat loss was undoubtedly extensive when the West was first settled and vast expanses of brushy grassland were ploughed under for farming. Since the turn of the century, the species appears to have compensated for these losses by expanding northward and eastward to take advantage of the clearing of forests in other regions (Root 1968).

Prior to 1975, there were a mere 15 or so records for Québec, mainly from the southwesternmost part of the province, of singing males that in general made short incursions in spring and summer. After 1975, the number of records began to increase steadily, and a little over 70 sightings have been reported to date. The species has also been seen in several new areas, notably the Lower St. Lawrence, the North Shore, and the Gaspé Peninsula.

Though present in Québec for more than 30 years, the Clay-colored Sparrow remains a rare breeder.

Jean Paquin and Vincent Létourneau

The Clay-colored Sparrow frequents mainly open grassy habitats not under cultivation, like overgrown fields and pastures, as well as young plantations of pine or fir. Hemmingford, south of Montréal. *July 1991, Pierre Bannon.*

Notes on territory

Unlike other Spizella sparrows, Clay-colored Sparrows use their territories only for nesting, and not for foraging (Knapton 1994). Studies in the Prairies show that they defend these territories against males of the same species (Root 1968). The area defended is very small and, in a Manitoba study, was found to range from 0.07 to 0.1 ha (Knapton 1979b). In another study, in Saskatchewan, the distance between nests was 20–900 m (mean 385 m) (Fox 1961). Densities of up to 10 pairs/ha have been found in some locations (Quinlan et al. 1990).

Clay-colored Sparrow

Breeding distribution
1984–1989

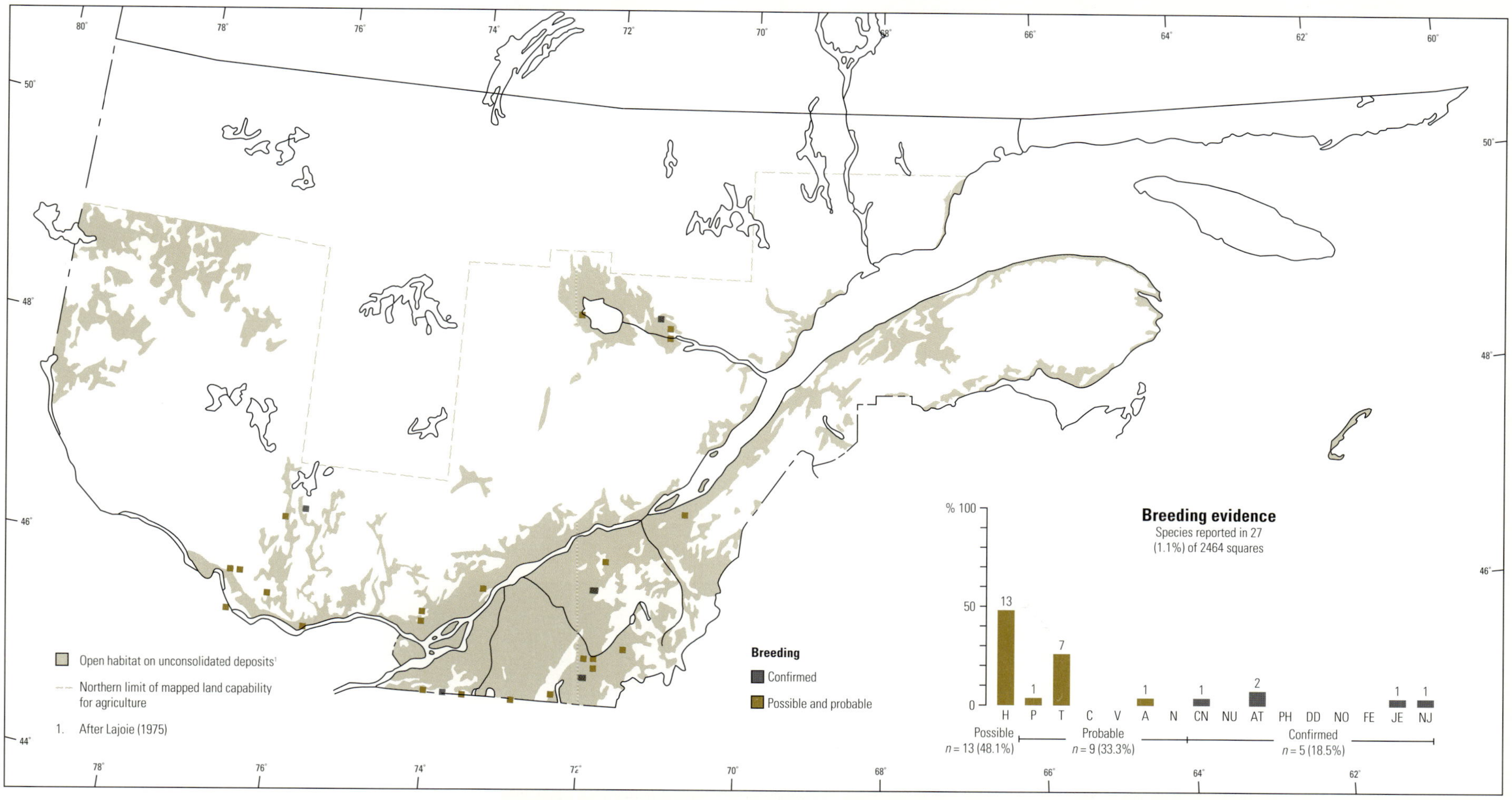

World breeding range

Species profile

Clutch size	usually 3–4 eggs (3–5)[1,3,4]
Laying interval	1 egg per day[1,2]
Start of incubation	usually with last or penultimate egg[1,2,8]
Incubation period	10–14 days[1,2,4]
Maturity at hatching	altricial
Care of young	mainly by female[1,2]
Nestling period	7–9 days[1,4]
Age at first flight	14–15 days[8]
Dependence of young	until at least 8 days after leaving nest[2]
Broods per year	1–2[2,4]
Breeding age	≈ 1 year[8]
Mating system	monogamy[4,5]
Duration of pair bond	rarely more than 1 season[5,8]
Total length	12.5–14.0 cm[6]
Wing span	?
Weight	M 11.6 g (9.8–12.5) / F 12.8 g (10.8–14.5)[2]
Longevity record	5 years 1 month[7]

References:
1. Fox 1961
2. Root 1968
3. Buech 1982
4. Ehrlich et al. 1988
5. Knapton 1979a
6. Godfrey 1986
7. Klimkiewicz and Futcher 1987
8. Knapton 1994

Chronology *n* = 3

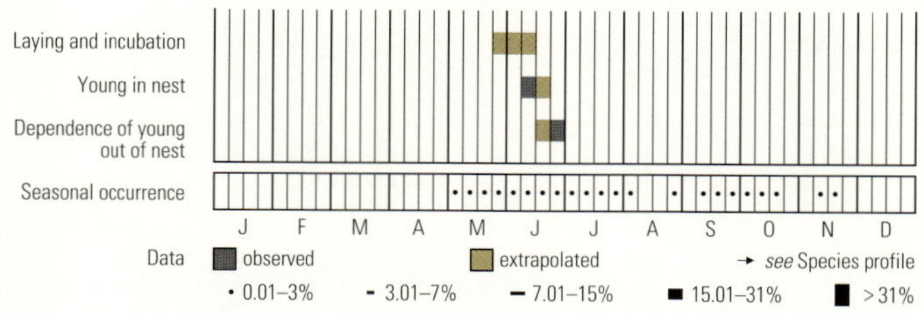

Field Sparrow

Bruant des champs

Spizella pusilla

Christiane Girard

THE FIELD SPARROW HAS A DISTINCTIVE CLEAR, SWEET, PLAINTIVE SONG, making it easy to tell apart from other emberizine species.

This sparrow returns to southern Québec in early April, or occasionally in March. The male, which arrives up to 3 weeks before the female, establishes a breeding territory upon arrival and defends it by singing plaintively for hours on end. Its song begins with two or more slow sweet whistles, then speeds up to a trill at the end (Godfrey 1986). Shortly after pair formation, the mates become inseparable and usually remain together for the rest of the breeding season (Walkinshaw 1968a).

The Field Sparrow leaves Québec toward the end of October to winter from Kansas east to Massachusetts and south to Florida and Mexico (AOU 1983). In Québec, this sparrow is fairly rare in summer and accidental in winter; scarcely 5 or 6 cases of overwintering have been recorded so far (R. Yank, unpubl. data).

Habitat and behaviour

In Québec, the Field Sparrow's breeding period runs from mid-May to mid-August. Preferred breeding habitats are open areas, where it can find bushes and shrubs in which to perch and conceal its nest (Best 1978; Evans 1978; Hebrard 1978). In Québec, the Field Sparrow is found mainly in fields, land reverting to scrub, or hedgerows bordering pastures, as well as in young pine plantations. Forest fires may therefore create new habitat for the species (Best 1979).

The habits of the Field Sparrow during the breeding period have been well documented (Walkinshaw 1968a; Best 1977b, 1978; Brooks 1980; Buech 1982; Petter et al. 1990). The male is highly territorial, with 4 or 5—usually quite visible—favourite singing perches atop shrubs or bushes (Walkinshaw 1968a), from which he defends his territorial boundaries. The male has two songs which are used in territorial advertisement or defence: a complex one sung most frequently in early morning or during conflicts with other males, and a simple song heard at other times of day (Nelson and Croner 1991).

The male usually chases off the female when she first approaches his territory. After she makes another few tries, however, he finally accepts her. The next year, if she is not the first to arrive in the same male's territory and is displaced, she will seek another mate in a nearby territory. Males generally use the same territories year after year (Walkinshaw 1968a).

Double or even triple brooding has been reported (Walkinshaw 1968a; Terres 1980; DeGraaf and Rudis 1987). At our latitudes, however, although double brooding seems possible (Peck and James 1987), it is unlikely that it nests a third time, given the short breeding season. Best (1974) also reported that in the United States the Field Sparrow is rarely able to raise a third brood, due to predation and parasitism.

Notes on territory

The breeding territory of the Field Sparrow is defended by the male and used by the family group until the young become independent.

Best (1977c) reported that in Illinois, in this sparrow's preferred habitat—fields with shrubs—nest territories are contiguous, and all available space is used. He also found that although the territory may extend into a field without shrubs, it is never limited to this type of area. Walkinshaw (1968a), however, reported that there is often open space between territories that is not defended by any male and that neighbouring territory holders may feed there at the same time with no displays of hostility.

Reported territory sizes in different locations include a range of 0.8–2.4 ha (mean 1.2 ha) in Michigan (Walkinshaw 1968a), 0.3–1.6 ha (mean 0.8 ha) in Illinois (Best 1977c), and 0.3–0.8 ha (mean 0.5 ha) in Iowa (Crooks and Hendrickson 1953).

The female alone builds the nest in the territory staked out by the male, placing it on the ground, or near it in a shrub on the edge of a field or pasture. Mount Owl's Head, southwest of Sherbrooke. *July 1964, Gaston Le Page.*

Field Sparrow

Breeding distribution

1984–1989

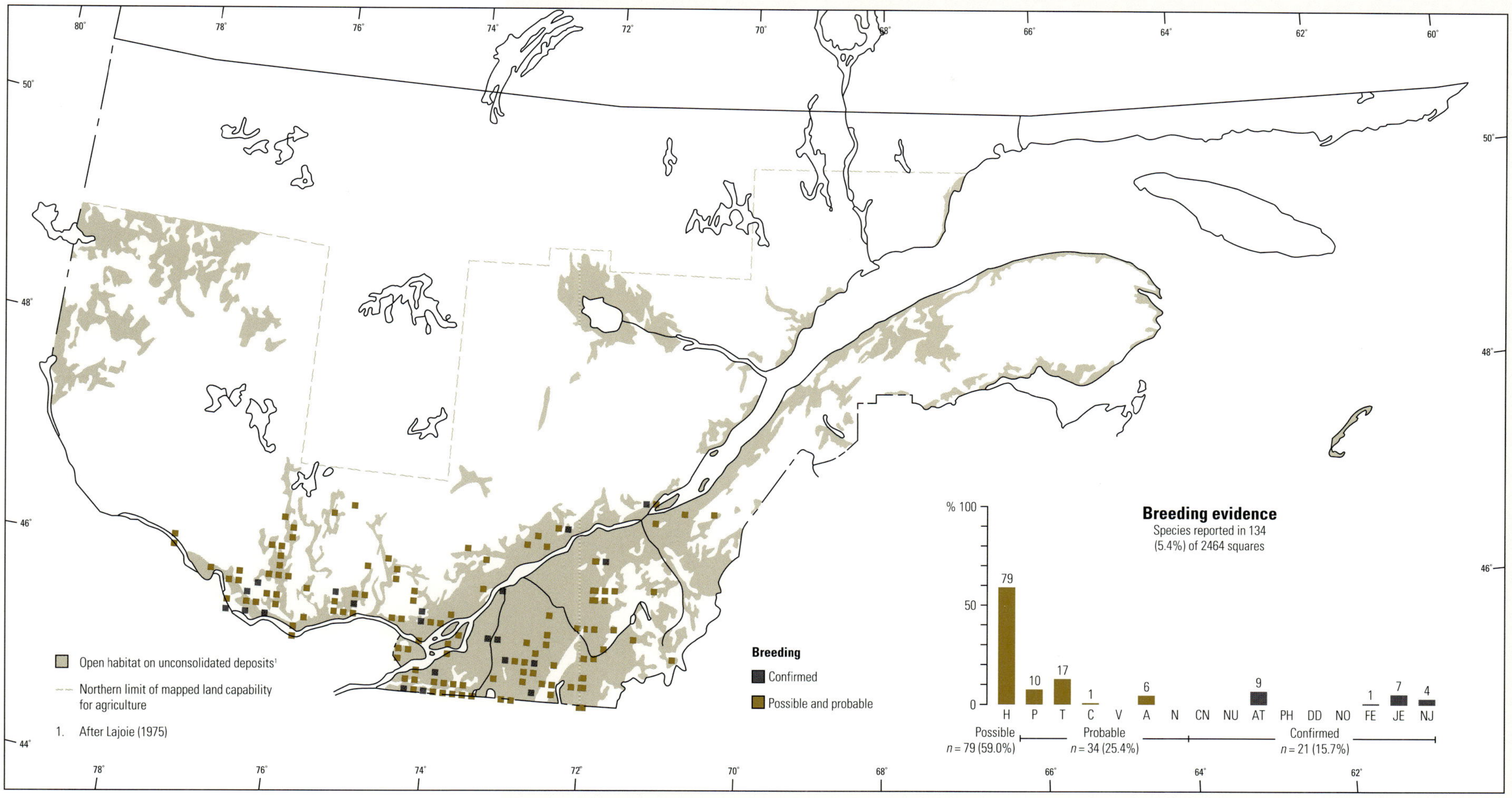

Species profile

Clutch size	usually 3–5 eggs (2–6)[1,2,3]
Laying interval	1 egg per day[1]
Start of incubation	with last egg, sometimes with penultimate egg[1,12]
Incubation period	usually 10–12 days (10–17)[1,2,3,4]
Maturity at hatching	altricial
Care of young	by female and male[1,2,4,5]
Nestling period	usually 7–8 days (7–10)[1,2,3]
Age at first flight	usually 12–13 days (12–15)[1]
Dependence of young	until 25–34 days old[1]
Broods per year	1–3[1,3,6,10]
Breeding age	1 year[3]
Mating system	monogamy[2,4,5,11]
Duration of pair bond	sometimes more than 1 season[1]
Total length	12.6–15.0 cm[7]
Wing span	19.7–21.6 cm[6]
Weight	12.5 g (10.2–16.5)[9]
Longevity record	8 years 9 months[8]

References:
1. Walkinshaw 1968a
2. Ehrlich et al. 1988
3. DeGraaf and Rudis 1987
4. Petter et al. 1990
5. Carey 1990
6. Terres 1980
7. Godfrey 1986
8. Klimkiewicz and Futcher 1989
9. Dunning 1984
10. Peck and James 1987
11. Best 1977c
12. Carey et al. 1994

Chronology *n* = 21

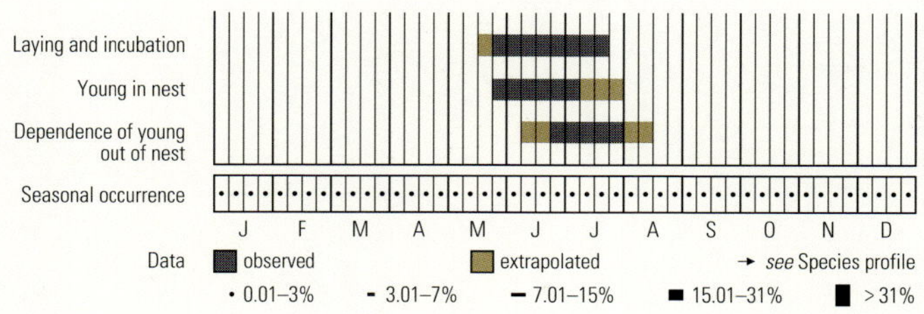

World breeding range

Several nests may be built in a single breeding season, as the female will readily renest if a clutch is lost (Best 1974). Early nests are placed on or near the ground in a tuft of grass; later ones are more often found up to 1.5 m up in a shrub or tree (Walkinshaw 1968a; DeGraaf and Rudis 1987; Peck and James 1987). The female builds the cup-shaped nest alone, of interwoven dead leaves and grass, lined with fine grasses, roots, or hair (Walkinshaw 1968a; Allaire 1972). Typical clutch size is 3–5 eggs (Ehrlich et al. 1988).

Copulation occurs early in the morning, usually during the nest-building period. The female alone incubates the eggs, usually for 10–12 days, though both parents feed the young, with feeding trips increasing as the young grow (Walkinshaw 1968a; Best 1977a; Carey 1990). The young usually leave the nest when they are 7 or 8 days old, but stay nearby, hidden on the ground or in low bushes, and are still fed by their parents. Five days later they can fly short distances, but they stay within the territory. They are independent at 25–34 days of age (Walkinshaw 1968a). Although the Field Sparrow is monogamous, the male and even the female may engage in extra-pair copulation (Petter et al. 1990).

The Field Sparrow is a frequent cowbird host (Friedmann 1963). In Ontario, for example, 18.6% of 334 nests examined were parasitized (Peck and James 1987). Although the cowbird is considered the Field Sparrow's chief natural enemy (Walkinshaw 1968a), its impact on the sparrow population seems to be minor (Buech 1982).

The Field Sparrow resembles the Chipping Sparrow and the American Tree Sparrow, but unlike them, it has a pink bill. The best way of identifying it in the breeding season, however, is by its characteristic song.

The Field Sparrow was reported in only 5.4% (134/2464) of Atlas squares surveyed, and breeding was confirmed in 15.7% (21/134) of these. This bird's nest is generally well concealed, and only 4 nests were actually found. Breeding confirmations resulted largely from observations of adults carrying food and of young still incapable of sustained flight.

This bird of open country nests mainly in shrubs along the edges of fields or in thickets on land reverting to scrub. Massey Road, Rockburn, south of Montréal. *June 1991, Pierre Bannon.*

This sparrow likes young pine plantations, where it can hide its nest and find perches from which to defend its territory. Erskin Road, Franklin, south of Montréal. *May 1991, Pierre Bannon.*

In the southern part of the province, abandoned farmland has created transition areas suited to the species. Saint-Hubert, near Montréal.
October 1993, Pierre Normand.

Distribution

The Field Sparrow breeds in Canada from southern Québec to southern Manitoba, and in the United States from Montana east to New England and south to Florida and Texas (AOU 1983). In Québec, it is a fairly common migrant breeder in the southern Ottawa Valley (Club des ornithologues de l'Outaouais 1985), occasional in the Montréal region (Bannon 1991b), and scarce in the Eastern Townships (Lepage 1993) and Laurentians (Bannon 1991b); in the Québec City region it was found breeding at Saint-Jean-Chrysostome, at Beauport (Sainte-Thérèse-de-Lisieux) (Otis et al. 1993), and at Loretteville (Atlas data).

Almost all Atlas records came from south of the 47th parallel. Displaying a marked preference for open areas, the species shuns mountainous regions and extensive forests. Accordingly, it is well established on the arable land of the Ottawa Valley, as far as about Rapides-des-Joachims, and in the Gatineau Valley. Farther east, along the north shore of the St. Lawrence River, distribution is patchy on the edge of the Laurentians and even more local east of Trois-Rivières, up to around Québec City, the easternmost breeding locality along the north shore of the St. Lawrence.

In the lowlands south of the St. Lawrence River, the species is not found in intensively farmed areas, but is present on hilly terrain near the U.S. border, at the base of the Appalachian Mountains, as well as in the Eastern Townships. The Appalachian Mountains limit its eastward distribution, but it was atlassed as far east as the Montmagny region in the lowlands.

Outside the Atlas period, records have come from as far north as Jonquière in the Saguenay region and Moisie on the North Shore, and as far east as Cape Bon Ami, at the tip of the Gaspé Peninsula, and the Magdalen Islands; all these records referred to non-breeders (David 1980a; David and Gosselin 1981b; ÉPOQ).

History and trends

The Field Sparrow probably did not move into southern Québec until the settlers had cleared much of the land. Around the turn of the century, Wintle (1896) listed it as scarce around Montréal, and Dionne (1906) stated that it was absent from the Québec City area. That is why Ouellet (1970a) considered the species to have become established only recently in the south of the province. Even though much suitable habitat has been created since then, this sparrow still has a limited distribution in southern Québec (David 1980a; Robert 1989b).

BBS data for 1966–79 indicate an increase in the Canadian population (Robbins et al. 1986). However, Québec BBS data for 1966–89 and ÉPOQ summer data for 1970–89 show no significant change (Cyr and Larivée 1993b). Eventual habitat loss to residential or commercial development, or even to reversion of abandoned farmland to forest, is probably a more serious threat to the species.

RICHARD YANK AND DIANE DAUPHIN

Vesper Sparrow

Bruant vespéral

Pooecetes gramineus

Christiane Girard

In summer, the tinkling, melodious song of the Vesper Sparrow is heard at day's end, the hour of "Vespers," whence the name (Donovan and Ouellet 1993), while its specific Latin name, *gramineus*, reflects its fondness for grass (Terres 1980). Indeed, this is a species of cultivated cropfields and short-grass areas (Simpson 1978; Godfrey 1986). The open nature of its breeding habitat and its widespread distribution during the summer months in the agricultural regions of southern Québec, along with its distinctive song and characteristic flash of white lateral tail feathers in flight (Godfrey 1986), help to make this species one of the most easily observed sparrows in Québec. Although the Vesper Sparrow was atlassed as a relatively common migrant breeder in Québec, in recent years it appears to be declining in numbers throughout North America, possibly because of recent changes in farming practices.

The Vesper Sparrow winters in the southern United States from the southern parts of California and Connecticut to southern Mexico, though not in the extreme south of Florida (AOU 1983; Godfrey 1986).

Habitat and behaviour

The Vesper Sparrow is principally a ground-dwelling species that prefers dry, short-grass fields for nesting (Bull 1974), although it can also be found in a wide variety of other similar habitats. In Ontario, in decreasing order of frequency, nests have been found in grassy fields and pastures; in cultivated grain, vegetable, and berry fields; in young open conifer plantations; along road and railway rights of way; in gravel pits; and in very open woodlands and recent burns (Peck and James 1987). Vesper Sparrows were also atlassed nesting in the short grass bordering runways at airports and in wild blueberry meadows.

Vesper Sparrows are seasonally monogamous, with individuals choosing new mates each spring. During laying and incubation the male defends a territory 0.29–3.57 ha in size against other males (Reed 1985; Perritt and Best 1989). Even though courtship, nesting, and feeding all occur on the ground, male Vesper Sparrows rarely sing from the ground (Terres 1980). Instead, they use the highest perch available, such as the tops of fence posts, tall weeds, shrubs, and trees bordering fields (Berger 1968; Rodenhouse and Best 1983).

The nest is built in a small depression in the ground and is usually well concealed under dead weed stems or surrounding plants. It consists of dry grass, weed stalks, and rootlets, and is lined with fine grass, rootlets, and horse hair (Godfrey 1986). Nests are difficult to find because they are well hidden and the parents are very secretive when they enter and leave the nest. Vesper Sparrows were reported in only 16.8% (415/2464) of Atlas squares. The male's habit of singing from a tall perch helped locate this species in a large number of squares; however, breeding was confirmed in only 16.9% (70/415) of squares in which the species was reported.

Typical clutch size is 3–5 eggs (Berger 1968). The female usually does all or most of the incubating for 11–14 days (Krueger 1981; Godfrey 1986). The altricial chicks are brooded primarily by the female, but occasionally by the male as well. Both parents share responsibility for feeding the nestlings and carrying away the fecal sacs from the nest (Berger 1968). They will often attempt to distract predators by feigning injury and fluttering along the ground away from the nest or even by attacking the intruder (Berger 1968; Krueger 1981). The young leave the nest after 9-14 days (Harrison 1978; DeGraaf and Rudis 1987), although they are still weak fliers (Berger 1968). They remain in the vicinity of the nest and are fed by the parents until they become independent at 29–35 days (Berger 1968; Harrison 1978). It is not known whether Vesper Sparrows lay more

The Vesper Sparrow is a bird of the fields and short-grass prairie. Sainte-Foy. *September 1991, Yves Aubry.*

Notes on territory

Males arrive about a week before the females and establish territories (Rodenhouse and Best 1983). This species' territories are generally larger than those of other sparrows (Berger 1968). Prior to incubation, the male delineates his territory by singing from elevated song perches at the boundary. Once incubation has begun, he sings from the field and chases off intruding conspecifics by patrolling territorial borders (Rodenhouse and Best 1983). Studies in the north-central United States reported territories ranging from 0.29 ha to 3.57 ha (Krueger 1981; Rodenhouse and Best 1983; Reed 1985, 1986; Perritt and Best 1989). The largest territories appear to have the scarcest and/or most widely dispersed food resources (Krueger 1981; Rodenhouse and Best 1983).

In southern Ontario, densities of 3.1 males/10 ha were reported in an old field, 1.2–4.7 males/10 ha in fields of grain or forage crops, 1.0 male/10 ha and 6.2 males/10 ha in two pastures, and 8.2 males/10 ha in a red pine plantation (Erskine 1971a, 1984). Vesper Sparrows appear to tolerate other species nesting in close proximity—one Vesper Sparrow nest was about 13 m from a Killdeer nest and 19 m from a Field Sparrow nest (Berger 1968).

Vesper Sparrow

Breeding distribution
1984–1989

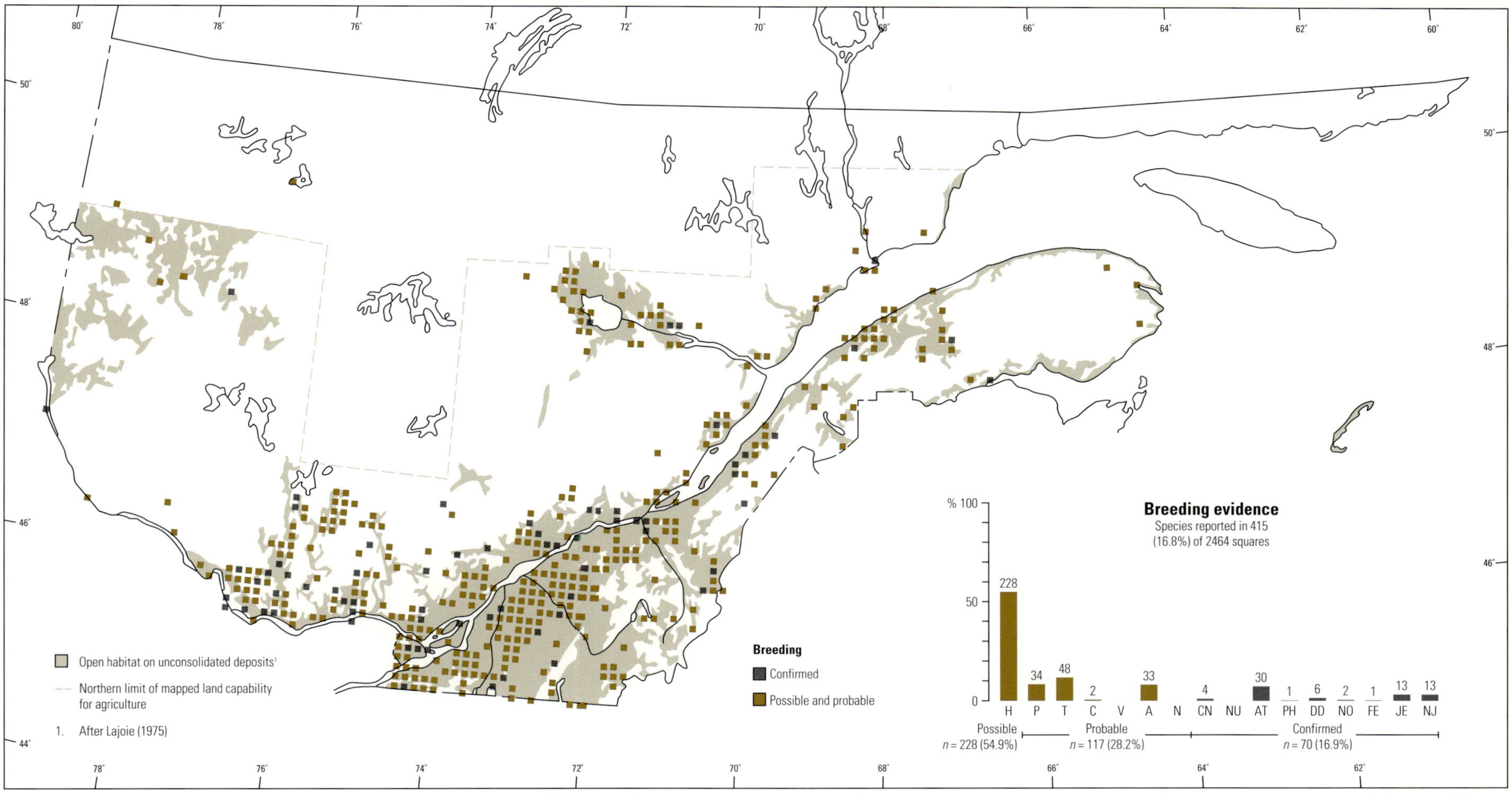

World breeding range

Species profile

Clutch size	usually 3–5 eggs (2–6)[1,2,3,4]
Laying interval	?
Start of incubation	?
Incubation period	11–14 days[1,2,3,4]
Maturity at hatching	altricial
Care of young	by female and male[1]
Nestling period	9–14 days[1,4,8]
Age at first flight	a few days after leaving nest[1]
Dependence of young	until 29–35 days old[1,4]
Broods per year	usually 2 (1–3)[1,4,8]
Breeding age	1 year[8]
Mating system	monogamy[1,2]
Duration of pair bond	1 season, *see* text
Total length	14–17 cm[3]
Wing span	25.4–27.9 cm[6]
Weight	M 26.5 g (?) / F 24.9 g (?)[7]
Longevity record	7 years 1 month[5]

References:
1. Berger 1968
2. Krueger 1981
3. Godfrey 1986
4. Harrison 1978
5. Klimkiewicz and Futcher 1989
6. Terres 1980
7. Dunning 1984
8. DeGraaf and Rudis 1987

Chronology $n = 48$

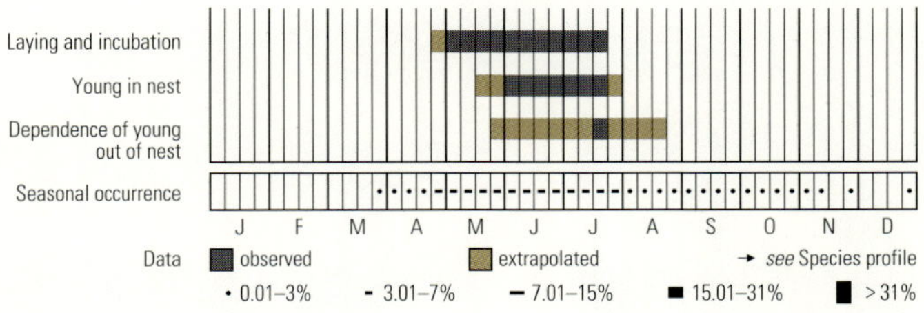

In forested areas, this sparrow may nest in the short grass alongside airport runways. Mont-Laurier. *May 1991, Yves Aubry.*

Although it is a bird of open country, this sparrow rarely sings on the ground, preferring to vocalize from a high perch such as a bush or tree. Saint-Colomban, northwest of Montréal. *June 1991, Pierre Bannon.*

In forested areas, this species may nest on the ground in blueberry patches or other dry, open situations. La Sarre Airport, Abitibi. *June 1991, Desneige Perreault.*

than one clutch a season in Québec, but double-brooding has been recorded in Michigan (Berger 1968; Krueger 1981) and New York (Bull 1974), and is thought to occur in Ontario as well (Peck and James 1987). The female will often leave the responsibility of brood rearing to her mate when she begins laying a second clutch, probably fertilized by the same mate (Berger 1968). Vesper Sparrows are frequent cowbird hosts. In Ontario, Peck and James (1987) found 11% of 442 nests parasitized. In West Virginia, it has been shown that predation by crows and other birds, snakes, and mammals can be a major cause of nestling and egg loss (Wray and Whitmore 1979; Wray et al. 1982).

Distribution

In Canada, the Vesper Sparrow breeds in the southern half of all the provinces from British Columbia to Québec and in all of Alberta and the Maritime provinces; south of the border its breeding range extends into the south-central United States (AOU 1983; National Geographic Society 1987). Although this species was probably originally associated with burned-over coniferous areas or other naturally denuded areas (Best and Rodenhouse 1984), the present distribution of the Vesper Sparrow in Québec corresponds closely with the distribution of farmland. It is widespread in the Central St. Lawrence Lowland upstream from Québec City, in the Eastern Townships, and in the lower Ottawa River watershed. It is also found around Lake Saint-Jean, in Abitibi, and in some localities in the Lower St. Lawrence region, and there are a few scattered records from the Gaspé and the Upper North Shore. Distribution in Québec is essentially restricted to the Atlas area, but it is worth noting that a nest was found recently at the airport at the La Grande-4 hydro-electric station on the La Grande Rivière (Yank et al. 1991a), which considerably expands the known breeding range of this species.

History and trends

At the end of the last century, Dionne (1889) considered this "bay-winged sparrow" to be generally uncommon in the province, though Wintle (1896) listed it as common in the Montréal area. Recent surveys have indicated that breeding populations of Vesper Sparrows are generally in decline throughout northeastern North America (Ellison 1985w; Rising 1987f; Eaton 1988p). In Québec, BBS data suggest a decline in abundance and a shrinking range between 1966 and 1989 (Falardeau and DesGranges 1991). The population decline in Québec and throughout northeastern North America appears to be a result of changing agricultural practices since 1970, including pesticide use (Falardeau and DesGranges 1991). In the last 20–30 years, hay harvesting in central and eastern Canada has changed from one cut a year to several cuts, a practice particularly common in southern Québec. Whereas hay used to be cut after the middle of July, by which time most Vesper Sparrows had completed nesting, today the first cut is often as early as the middle of June, a time when most nests still contain eggs or young. Unfortunately, the majority of these nests are destroyed during cutting, and this is believed to be a major factor behind this species' overall decline (Robbins et al. 1986; Perritt and Best 1989; Graham and DesGranges 1990, 1993).

Although the Vesper Sparrow appears to be the most adaptable of the grassland sparrows to human activity (Santner 1992d), current agricultural practices are still changing. The transitory nature of farming methods makes it difficult for Vesper Sparrows to continue to adapt (Best and Rodenhouse 1984), and therefore the current downward trend is unlikely to change for the better.

DOUGLAS GRAHAM AND RICHARD C. COTTER

Savannah Sparrow

Bruant des prés

Passerculus sandwichensis

Christiane Girard

The Savannah Sparrow, which is slightly smaller than a House Sparrow, may be mistaken for other species, particularly the Song Sparrow, with which it is sympatric to a large extent. However, its yellowish lores and eyebrow stripe and whitish crown stripe distinguish it from the latter. The Savannah Sparrow lacks any obvious sexual dimorphism, and the birds appear to use behavioural rather than visual cues to determine the sex of conspecifics (Weatherhead and Robertson 1980). It is the most common sparrow in fields throughout Canada (Godfrey 1986).

The Savannah Sparrow is a species of open areas. However, the habitat it frequents varies widely from region to region, resulting in corresponding differences in the species' morphology and plumage. A total of 17 subspecies have been identified so far in North and Central America (Wheelwright and Rising 1993); two of these are found in Québec: *Passerculus sandwichensis labradorius* in the north and *Passerculus sandwichensis savanna* in the south and on the Magdalen Islands. Interestingly, Wheelwright and Rising (1993) consider the *savanna* and *mediogriseus* subspecies a single subspecies *Passerculus sandwichensis savanna*, while Godfrey (1986) considers them two distinct subspecies.

This sparrow winters from southern New England, the south-central states, and southern British Columbia to northern Honduras and the West Indies. Occasionally it overwinters in some Canadian provinces and northern U.S. states within its breeding range (Wheelwright and Rising 1993).

Habitat and behaviour

The Savannah Sparrow frequents open habitat with a dense cover of herbaceous plants and scattered shrubs, but avoids areas with extensive tree cover (Godfrey 1986). It is found mainly in hayfields, pastures, meadows, and grassy dunes, as well as in wetlands such as saltwater and freshwater marshes, peatlands, and grassy streamsides (Terres 1980; Godfrey 1986; Peck and James 1987; Rising 1987b). It also frequents roadsides and transmission line rights-of-way. The Savannah Sparrow seems to prefer tall, dense herbaceous vegetation in wettish areas (Wiens 1973). It rarely nests in short grass like the Vesper Sparrow or in the tall, dry grassland preferred by the Grasshopper Sparrow (Bull 1985).

The nest is often difficult to find because it is generally hidden under vegetation. Usually built on the ground in a natural hollow or depression scratched in the ground (Harrison 1975; Peck and James 1987), it is made of grasses, plant stems, and sometimes moss and lined with fine grasses, rootlets, and occasionally hair (Baird 1968; Godfrey 1986; Ehrlich et al. 1988). The female takes 4 or 5 days to build the nest (Welsh 1975), and then lays her clutch of 2–6 eggs (Terres 1980; Peck and James 1987; Wheelwright and Rising 1993). In Québec, typical clutch size is 4 eggs, but may range from 2 to 5 (Bédard and Meunier 1983). Clutch size often varies from region to region; in the northernmost part of the species' range, the set is often 5 eggs, while farther south it is only 4 eggs, or even 3 in Mexico. The Savannah Sparrow is an uncommon cowbird host (Terres 1980)—in Ontario only 7.3% of 531 nests examined were found to have been parasitized (Peck and James 1987).

In Québec, egg dates range from the first week of May to the beginning of August. Both parents incubate (Baird 1968; Terres 1980; Ehrlich et al. 1988), although some authors believe that the male plays a minor role, particularly when polygynous (Welsh 1975; Bédard and Meunier 1983; Godfrey 1986). Incubation usually takes 12 or 13 days (Wheelwright and Rising 1993), and both parents tend the nestlings (Welsh 1975; Ehrlich et al. 1988). The young remain in the nest for 7–13 days (Welsh 1975; Bédard and Meunier 1983; Wheelwright and Rising 1993). Primarily granivorous on its wintering grounds, the Savannah Sparrow feeds mainly on insects and other arthropods during the breeding season, and the young's diet consists mainly of lepidopteran,

Notes on territory

The male relies mainly on song to establish a territory; although he occasionally tries to intimidate his opponents by using various types of aggressive behaviour, actual physical contact is rare (Baird 1968). The female also participates in the defence of the territory, chasing away other females that venture inside the boundaries (Welsh 1975). The size of the territory may vary from one year to another depending on the region, habitat type, and how

This is a common sparrow in agricultural areas and some types of wetlands. It is particularly abundant in hayfields, pastures, and meadows, and in the upper, drier part of marshes, like this cordgrass marsh seen here on the left side of the photo. Saint-André, southwest of Rivière-du-Loup. *October 1993, Pierre Normand.*

Savannah Sparrow

Breeding distribution

1984–1989

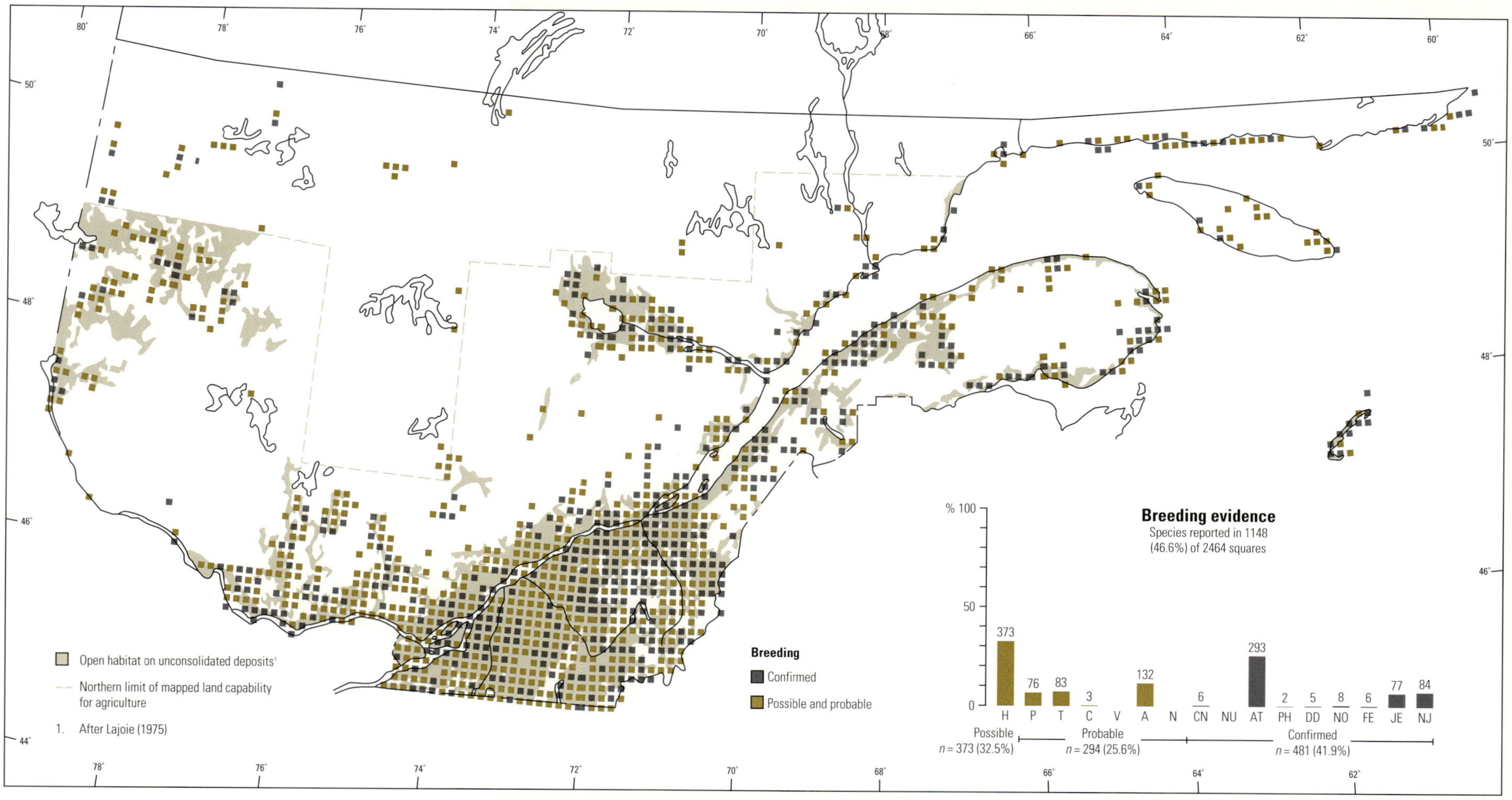

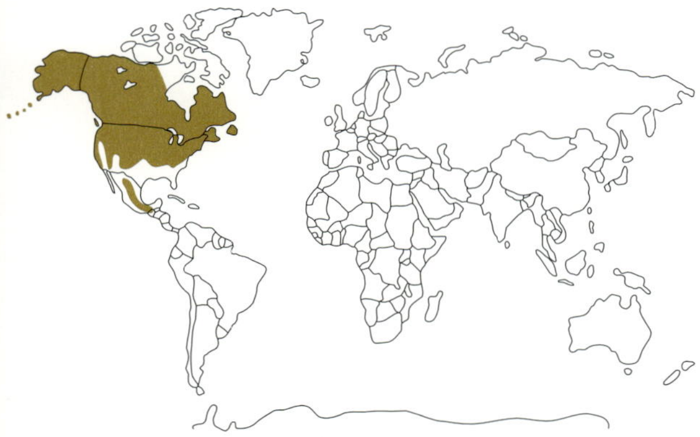

Species profile

Clutch size	usually 4 eggs (2–6)[1,2,3,4,8,9]
Laying interval	usually 1 egg per day[1]
Start of incubation	with penultimate egg, sporadic before[1]
Incubation period	usually 12–13 days (9–15)[1]
Maturity at hatching	altricial
Care of young	by female and male[1,2,5]
Nestling period	usually 9–11 days (7–13)[1,5,8]
Age at first flight	≈ 13–20 days[1]
Dependence of young	until ≈ 3 weeks old[1]
Broods per year	1–2[3,4,9]
Breeding age	1 year[1]
Mating system	monogamy, often polygyny[1,2,5]
Duration of pair bond	usually more than 1 season[1]
Total length	13.3–16.5 cm[6]
Wing span	20.3–24.1 cm[10]
Weight	M ? g (15.5–25.4)[1] / F 19.5 g (?)[11]
Longevity record	6 years 10 months[7]

References:
1. Wheelwright and Rising 1993
2. Ehrlich et al. 1988
3. DeGraaf and Rudis 1987
4. Baird 1968
5. Welsh 1975
6. Godfrey 1986
7. Klimkiewicz and Futcher 1987
8. Bédard and Meunier 1983
9. LaPointe and Bédard 1986
10. Terres 1980
11. Dunning 1984

Chronology *n* = 236

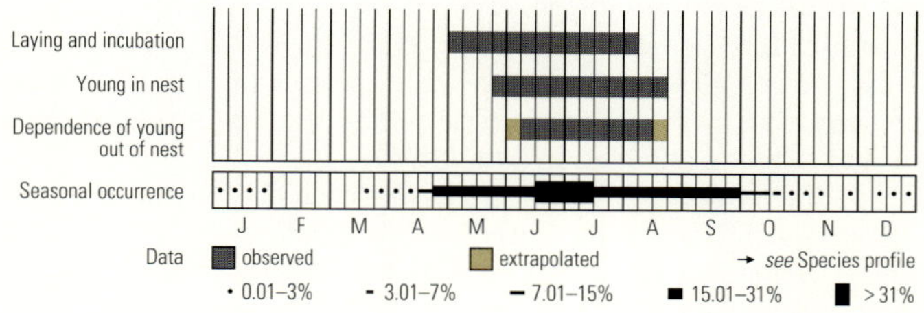

The Savannah Sparrow is the most common breeding bird in fields, meadows, and other open habitat, where it tolerates a few shrubs and fence posts that it can use as perches. Sainte-Foy. *September 1991, Yves Aubry.*

dipteran, homopteran, and orthopteran insects, as well as spiders (Baird 1968; Terres 1980; Bédard and Meunier 1983; DeGraaf and Rudis 1987; Ehrlich et al. 1988; Wheelwright and Rising 1993). Young hatched later in the season grow faster than earlier hatched young because of the warmer temperatures and greater food availability (Bédard and LaPointe 1985). In some populations double brooding may occur (Welsh 1975; Peck and James 1987), but at our latitudes usually only 1 brood is produced each year (Dixon 1978; Weatherhead and Robertson 1980). A study in southern Québec found only 9.8% of the females produced a second brood (LaPointe and Bédard 1986). Savannah Sparrows are generally monogamous, although polygyny is observed locally (Welsh 1975; Bédard and LaPointe 1984b; Ehrlich et al. 1988; Wheelwright and Rising 1993). Polygynous males do not court a second female until the first female is incubating, allowing them to help raise the first brood while the second female is incubating (Welsh 1975).

The females are very secretive and usually remain on or near the ground hidden in the vegetation (Weatherhead and Robertson 1980). The Savannah Sparrow can be difficult to detect because it often moves about under dense cover. It is said to be an excellent walker, and runs mouselike through the grass when startled (Terres 1980). Territorial males are more visible and are revealed most often by their song, which is performed from a perch on a tall weed, rock, or fence post. This sparrow sings most often in the morning and evening. The rather weak song, a *tsit-tsit-tsit tseeee-tsaay* (Peterson 1988m), can only be heard at close range. The song is used to establish and maintain a territory and to attract females; it is composed of three parts, the first part containing information allowing other birds to recognize the species, the second to identify the individual, and the third to identify regional variation (Chew 1981). The species also has four calls, associated with alarm, appeasement, defence, and feeding (Gobeil 1970).

Distribution

The Savannah Sparrow is widely distributed in North America. Its breeding range extends from Alaska, across Canada, and throughout the coterminous United States except for the southern states.

In southern Québec, the species' distribution corresponds to areas with open habitat, which is found mainly in settled agricultural areas, near rivers and streams, and along the coast in meadows and grassy dunes. The species is not found in densely forested areas. The Savannah Sparrow was reported in 46.6% (1148/2464) of Atlas squares surveyed, and breeding was confirmed in 41.9% (481/1148) of those.

The Savannah Sparrow was reported in the Ottawa Valley, St. Lawrence Plain, Appalachians, and open areas of the Abitibi–Témiscamingue region. It was also found in the Lower St. Lawrence region, around the Gaspé Peninsula, on the North Shore, on the shores of the Saguenay River, around Lake Saint-Jean, and on Anticosti Island. It is classified as a regular migrant breeder on the Magdalen Islands, where it is considered the commonest sparrow and the most abundant passerine (Fradette 1992).

North of the Atlas area, the breeding range extends to the tip of the Ungava Peninsula (Godfrey 1986). Although in this area most observations and breeding records have come from the coast (Todd 1963; Godfrey 1986; ÉPOQ), the species may occur inland as well. The Savannah Sparrow is widespread and relatively abundant in the Great Whale River watershed, where the many peatlands, sedge meadows, and shrubby areas provide suitable habitat (Cons. G. G. – GREBE 1991a).

Notes on territory (continued)

early or late it is in the breeding season (Wheelwright and Rising 1993).

On L'Isle-Verte in Québec, in habitat consisting of a cord-grass marsh and an abandoned field, Savannah Sparrows were found to return to the same site in subsequent years, even if they had not bred the year before or had lost the clutch. Territories were approximately 60 m in diameter, or roughly 0.3 ha (Bédard and LaPointe 1984a).

In Wisconsin and Nova Scotia, various researchers have recorded territories ranging in size from 0.2 to 1.9 ha in open habitat such as meadows and coastal dunes (Welsh 1975; DeGraaf and Rudis 1987).

In Québec and Ontario, the highest densities (between 4.3 and 18.5 pairs/10 ha) were found in fields and pasture, some still

This small passerine seems to prefer tall, dense vegetation in damp areas. Saint-Germain, southwest of Rivière-du-Loup. *October 1991, Réginald Ouellet.*

in use, and others gradually being colonized by trees (Erskine 1971a, 1972a, 1976a, 1980a, 1984). In tamarack peatlands and black spruce peatlands, densities were estimated at 1.5 and 1.7 pairs/10 ha, respectively (Erskine 1972a).

Along the coast, the species nests on grass-covered dunes and in other grassy areas along sandy shorelines. Baie du Havre aux Basques, Magdalen Islands. *June 1991, Desneige Perreault.*

The nest, which is well hidden in the vegetation, is constructed in a natural or scratched-out depression in the ground. Québec City. *June 1956, Gaston Le Page.*

History and trends

Before European settlement of North America, the Savannah Sparrow was probably confined mainly to wet meadows, peatlands, coastal marshes, clearings, grassy ecotones, and tundra. It has benefited from the development of agriculture (Wheelwright and Rising 1993), particularly the dairy industry, which requires extensive acreage for growing hay as a fodder crop (Peterson 1988m). At the turn of the century, the species was abundant in Québec and nested almost everywhere where grassy areas could be found, particularly on the islands and shores of the St. Lawrence, including Anticosti Island (Ouellet 1969). The Central St. Lawrence Lowland supports some of the greatest concentrations of the species in North America; forest clearing and development of agriculture at the turn of the century have undoubtedly enabled the species to extend its range (Ouellet 1974a; Robbins et al. 1986). Today, in the inhabited regions of Québec, as in other parts of North America, habitat loss could result from trends such as conversion of pasture into cornfields, abandonment and subsequent reversion to forest of farmland (M. Carignan, MAPAQ, unpubl. data), urbanization, and industrialization (Peterson 1988m; Erskine 1992). BBS data for 1966–79 suggest that the population is stable both in the East and in North America as a whole (Robbins et al. 1986). The reduced number of sightings in the Maritimes since 1966 may be linked to abandonment of farmland (Robbins et al. 1986; Erskine 1992). This phenomenon could also explain the drop in numbers observed in Québec between 1966 and 1989 (BBS) in similarly affected regions. Indeed, abandonment of agricultural land is prevalent throughout rural Québec, except in the Montréal and Montérégie regions (M. Carignan, MAPAQ, pers. comm.).

ROBERT BONNEAU

Grasshopper Sparrow

Bruant sauterelle

Ammodramus savannarum

Christiane Girard

ONE OF THE GRASSHOPPER SPARROW'S SONGS, ALMOST INAUDIBLE EVEN TO the practised ear, ends with a trill not unlike the chirring of an insect. Given also that the species is a grassland dweller, it comes quite easily by the name "grasshopper."

This elusive little bird is difficult to find in its habitat, as it generally keeps to the ground, under the cover of grass. Sometimes the male will venture out to sing, mostly in the early morning or before dusk.

This migrant breeder arrives in the province around mid-May, from its wintering grounds in the southern United States, Mexico, and Central America south to Costa Rica (AOU 1983). It has recently been listed as vulnerable in Québec (Robert 1989b).

Habitat and behaviour

The Grasshopper Sparrow is highly selective in its choice of breeding habitat, showing a marked preference for fallow fields and old pastures where soil is sandy and well drained. It is partial to areas with such grasses as timothy and quack grass, along with other plants that thrive in poor, acidic soil, such as viper's bugloss, tufted vetch, and common milkweed (Hainault and St-Hilaire 1989). Grasses growing in clumps allow the birds freedom to move about and are preferred over sod-forming grasses (Whitmore 1981). The presence of perches is another factor that influences site selection. Young pine plantations (30–50 cm tall) and fields with mullein and some small shrubs appear to fulfill this requirement. However, the species is absent from fields with more than 35% shrub cover (Johnston and Odum 1956).

Bent (1968) described the species as nesting in colonies, a reference to the fact that some suitable habitat may be occupied by several pairs nesting close together, while other apparently equally suitable spots remain unoccupied. Grasshopper Sparrows are often locally abundant; for example, population densities in West Virginia were estimated by Whitmore (1979, 1981) to be as high as 120 pairs/100 ha. A survey in southern Québec by Hainault and St-Hilaire (1989) yielded several isolated pairs but also as many as 4 pairs in one site. The nest, placed on the ground, is difficult to find. Often in a depression or at the base of a grass clump, it is concealed by a dome of overhanging grasses (Whitmore 1981; Ehrlich et al. 1988). Although opinions differ, nest building and incubation are apparently performed solely by the female. Both parents, however, feed the young. Pairs often produce 2 broods per season in the United States (Reilly 1968; Terres 1980), and probably in Ontario as well (Peck and James 1987). Nesting data are insufficient to determine whether the species is double-brooded in Québec.

In all of the 21 Atlas squares in which the species was reported, only one nest was found. This is probably due to the fact that both adults become highly secretive at the approach of an intruder. The male stops singing and hides in the grass, whereas the female sits tight and if flushed will first run through the grass before taking flight (Smith, R. L. 1968). The birds are easier to spot when the parents are feeding the young on an exposed perch. In fact, breeding was confirmed in 6 squares either by sightings of adults carrying food or of young.

Notes on territory

The male defends its territory by singing its "grasshopper song" from the tops of the highest perches in the territory, usually 50–126 m from the nest. Territories range in size generally from 0.5 to 1.3 ha, and boundaries are fiercely defended during territorial establishment, nest building, and incubation.

The Grasshopper Sparrow nests in old pastures and fallow fields with few shrubs and sandy well-drained soil. Sainte-Marthe, west of Montréal. *June 1991, Pierre Bannon.*

Grasshopper Sparrow

Breeding distribution

1984–1989

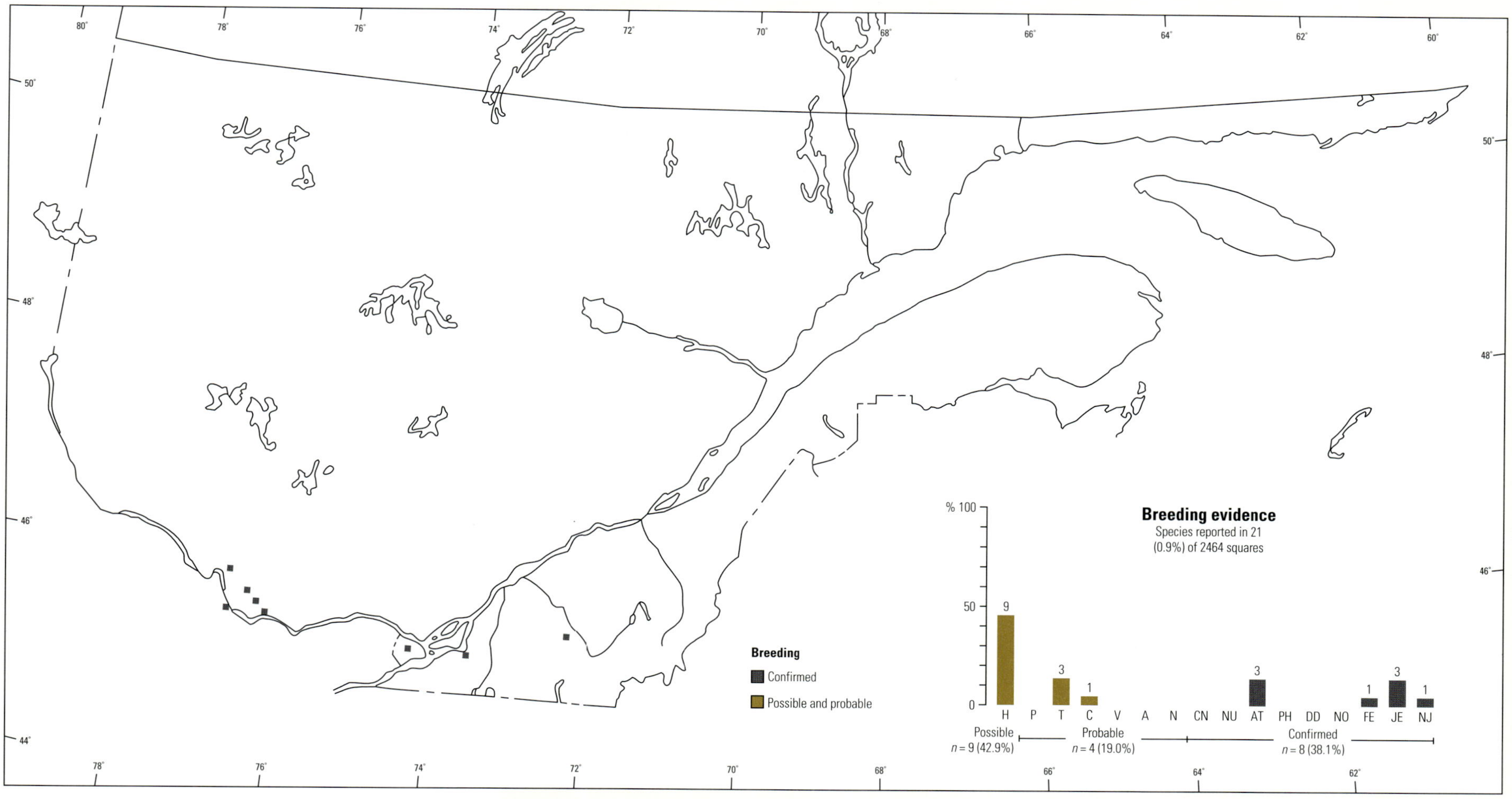

World breeding range

Species profile

Clutch size	usually 4–5 eggs (3–6)[1,2,5,6]
Laying interval	?
Start of incubation	?
Incubation period	11–12 days[2,4,5,6]
Maturity at hatching	altricial
Care of young	by female and male[2]
Nestling period	9 days[5,6]
Age at first flight	probably a few days after leaving nest[5]
Dependence of young	?
Broods per year	usually 2 (1–3)[1,2,3,5]
Breeding age	1 year[9]
Mating system	monogamy[2]
Duration of pair bond	?
Total length	12.3–13.7 cm[4]
Wing span	20.3–21.6 cm[6]
Weight	17.0 g (13.4–28.4)[8]
Longevity record	3 years 1 month[7]

References:
1. Forbush 1929
2. Ehrlich et al. 1988
3. Peck and James 1987
4. Godfrey 1986
5. Harrison 1978
6. Terres 1980
7. Klimkiewicz and Futcher 1987
8. Dunning 1984
9. DeGraaf and Rudis 1987

Chronology $n = 6$

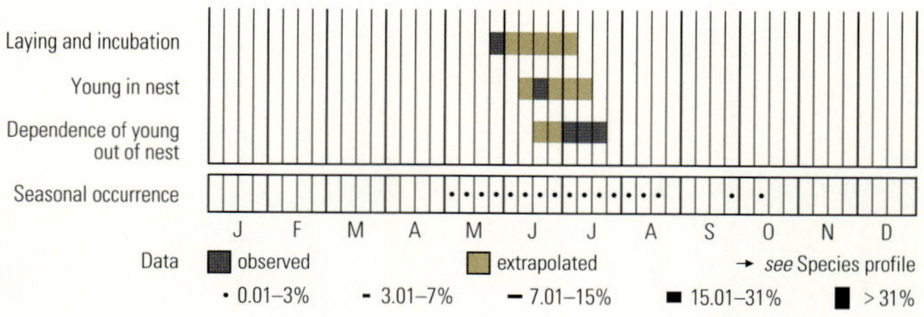

Distribution

The Grasshopper Sparrow breeds across southern Canada, excluding the Atlantic provinces, and from the western United States to the Atlantic coast down to the Gulf of Mexico (AOU 1983).

Distribution in Québec is largely confined to two regions, the southern St. Lawrence Plain and the southern Ottawa Valley. In the southern St. Lawrence Plain, around Montréal, populations of varying sizes are locally distributed at Saint-Lazare, Saint-Constant, and Sainte-Marthe, south of Rigaud Mountain, where this sparrow has been nesting every year since 1967 (Ouellet 1974a; David 1980a). Breeding was also confirmed for the Atlas near Richmond.

In the southern Ottawa Valley, west of Hull, there are several sites with 2 or 3 breeding pairs. A number of suitable breeding sites have been found in the Ottawa Valley lowlands, at the foot of the Canadian Shield escarpment (Bristol Ridge, Campbell Bay, Fort-Coulonge). Once an agricultural region, this area affords ideal nesting conditions. As many as 7 singing males/km^2 have been counted in prime habitat (St-Hilaire 1990). The species is also widely distributed in the area just south of Ottawa and locally in the agricultural areas of southern Ontario (Peck and James 1987).

The Grasshopper Sparrow has also been observed elsewhere in Québec as far north as Québec City (Otis et al. 1993).

History and trends

The Grasshopper Sparrow has benefited from deforestation and the vast tracts of land that have come under cultivation since the beginning of European settlement. It appears to have been slow to occupy this new cropland, no doubt because its original population was small and very local in distribution (Arbib 1988d). This species of the dry grasslands is a relative newcomer to Québec. The first Québec record was from Hull, and dates from 24 June 1898 (Eifrig 1910–11); the next record was obtained on 26 June 1920, when three singing males were detected near Chambly, in the Montréal region. Grasshopper Sparrows have been observed regularly in Chambly in subsequent decades (Ouellet 1974a).

The Grasshopper Sparrow has been on the Blue List since 1974. Its population is reportedly declining in New York State and Vermont as well as in several other regions in the United States (Tate 1986; Arbib 1988d; Askins 1993). Continent-wide BBS data also indicate a decline in abundance between 1965 and 1979. A notable decrease was recorded in the eastern portion of the continent (Robbins et al. 1986). These declines have been attributed to habitat changes, as abandoned farmland reverts to woodland or is used for urban development (Bull 1974; Arbib 1988d). Pesticide use and the early harvesting of hay fields, which often forces the birds to abandon nesting, have probably also been detrimental to the species (Ellison 1985m; Arbib 1988d).

Notes on territory (continued)

Once the last brood has left the nest, territories are no longer defended, but adults and young remain in the vicinity until fall migration (Smith, R. L. 1968). During a study conducted over three consecutive breeding seasons in good habitat in West Virginia, Whitmore (1979) found that densities varied considerably from one year and site to another, from 2.17 to 12.09 breeding pairs/10 ha. Breeding densities in various habitats in Ontario ranged from 3.9 to 10.2 singing males/10 ha (Erskine 1971a, 1980a, 1984).

This elusive species also frequents fields containing milkweed, on which it often perches. Sainte-Marthe, west of Montréal. *June 1991, Pierre Bannon.*

Typical Grasshopper Sparrow nesting habitat in the Ottawa Valley lowlands, where the species breeds. A large number of the birds observed in Québec inhabit this area. Near Fort-Coulonge, west of Hull. *July 1991, Daniel St-Hilaire.*

Aside from changes in farming practices, fragmentation of agricultural habitat and a reduced acreage of land under cultivation are also major factors that have contributed to the decline of this grassland specialist. In Missouri and Illinois, it was found that Grasshopper Sparrows occur less often or are even absent from habitat of less than 10 ha, and in Maine the critical size was found to be 32 ha (Askins 1993). Nests in less extensive habitat are more vulnerable to predation as well as to cowbird parasitism than those in large areas, especially when located near woodland or pasture (Askins 1993).

In the Ottawa Valley, where the greatest numbers occur in Québec, many sites where the species once nested have been altered (planting of conifers, resumption of cultivation, grazing), activities that are only tolerated by the species when exercised in moderation (Delany et al. 1985). Nevertheless, the species is colonizing new areas in the Ottawa Valley (Hainault and St-Hilaire 1989). Grasshopper Sparrows may be more threatened in the Montréal region where numbers are lower and where some breeding sites could disappear as a result of residential and industrial development or crop substitution.

Pierre Hainault

Henslow's Sparrow

Bruant de Henslow

Ammodramus henslowii

Henslow's Sparrow is an unobtrusive species that frequents a particular type of habitat—open grassy areas interspersed with tall weeds—that is often neglected by birders. The sightings in July 1985 in two locations in southern Québec were the first in the province in 15 years. Between 1943 and 1970 the species was seen only 9 times during the breeding season, and it is considered accidental in Québec (Robert 1989b). Breeding was not confirmed for the Atlas, although substantial breeding evidence was obtained in the Montréal and Ottawa Valley regions in the past.

Habitat and behaviour

In eastern North America, Henslow's Sparrow is found mainly in rural habitats, such as cultivated fields and unmowed meadows in the early stages of secondary succession (Knapton 1982; Smith, C. R. 1992). This habitat is often transitory: abandoned fields are quickly invaded by weeds and shrubs, and fallow fields are often replanted or used as pasture. In low-lying areas, this sparrow is sometimes found in wet fields colonized by bulrushes and sedges.

Suitable habitat often contains a thick ground mat of dead vegetation (Robins 1971; Knapton 1987b) and always dense, tall grassy vegetation (Knapton 1982; Smith and Smith 1992). In Kansas, Zimmerman (1988) found that males preferred to establish territories in areas with abundant standing dead vegetation. This sparrow also avoids fields where the view of the horizon is obstructed by a neighbouring woodlot or a line of trees (Peterson 1983), which is probably why it frequents mainly large fields (Smith and Smith 1992).

Males establish and maintain their territories mainly with song; territorial boundaries may shift during the season (Wiens 1969; Robins 1971). In early spring, males often sing from a clump of dead vegetation protruding above the new vegetation (Robins 1971). During the rest of the season, this quiet, drab sparrow hides all day in the tall dense grass, and can only be detected by its song, an endless repetition of the same two syllables *tsi-lik*. The song, which can barely be heard beyond 150 m, is easily drowned out by the chorus of meadowlarks, Bobolinks, and other sparrows that inhabit these areas. However, Henslow's Sparrow sings late in the evening and sometimes during the night, when the other birds are quiet.

Nest construction takes 4–6 days and is done mainly or exclusively by the female. The nest of dried stalks is placed on or just above the ground in a clump of dense vegetation. It is well camouflaged by the surrounding grasses, which sometimes form an arched roof over the nest. Only the female incubates, but the male helps her in brooding and feeding the young (Harrison 1975; Ehrlich et al. 1988). In summer, the adults' diet is over 80% animal matter, particularly ground crickets, while the young are fed exclusively insects (Hyde 1939). This sparrow usually raises 2 broods a year (Robins 1971).

Christiane Girard

Notes on territory

A Michigan study showed that males defend a territory of 0.32 ha on average: smaller territories (0.28 ha) were defended in areas where the vegetation was denser and taller, while larger territories (0.45 ha)

A very irregular breeder, Henslow's Sparrow has very specific habitat requirements: large, unmowed fields with tall dense vegetation. Huntingdon, southwest of Montréal. *August 1985, Pierre Bannon.*

Henslow's Sparrow

Breeding distribution

1984–1989

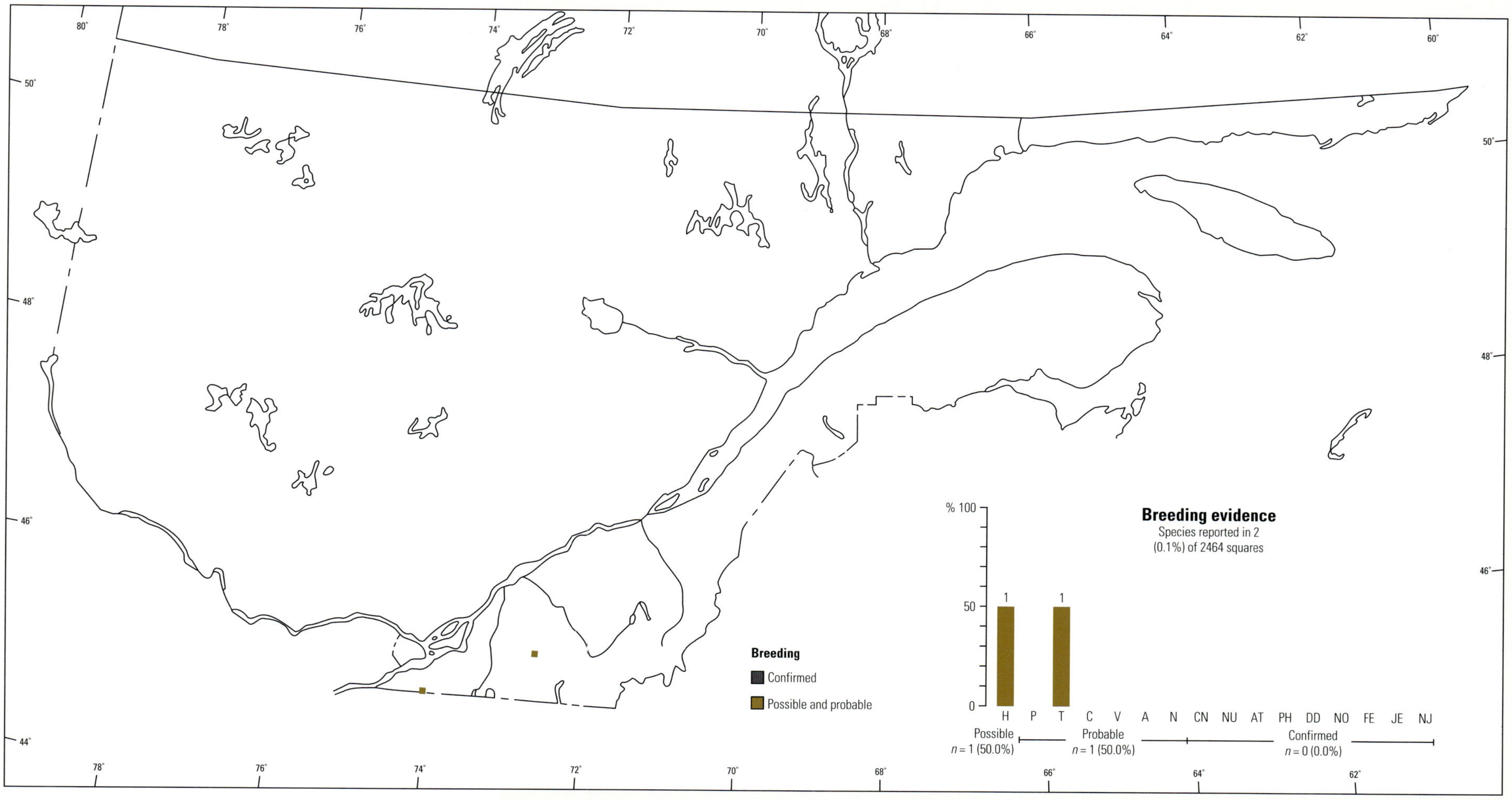

Species profile

Clutch size	3–5 eggs[1,2,3,7]
Laying interval	?
Start of incubation	with last or penultimate egg[2,6]
Incubation period	11 days[2,5,6,7,8]
Maturity at hatching	altricial
Care of young	by female and male[1,2,5,7]
Nestling period	9–10 days[2,5,7,8]
Age at first flight	9–10 days or more[11]
Dependence of young	?
Broods per year	2, sometimes 3[1,10]
Breeding age	1 year[10]
Mating system	probably monogamy[7]
Duration of pair bond	?
Total length	11.8–13.5 cm[9]
Wing span	17.1–19.0 cm[8]
Weight	13.1 g (11.1–14.9)[4]
Longevity record	?

References:
1. Robins 197
2. Graber 1968
3. Zimmerman 1988
4. Dunning 1984
5. Harrison 1978
6. Harrison 1975
7. Ehrlich et al. 1988
8. Terres 1980
9. Godfrey 1986
10. DeGraaf and Rudis 1987
11. according to nestling period

Chronology $n = 0$

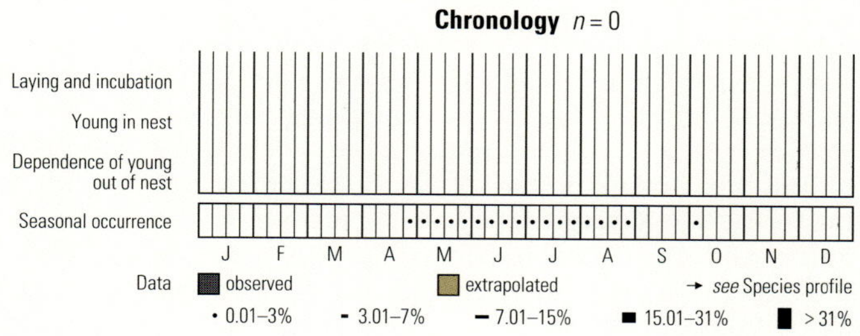

World breeding range

In 1985, two Henslow's Sparrows were observed in this field. The species is difficult to find because it spends the day hidden in the tall, dense grass, and because its song is easily drowned out by those of the Bobolink, Eastern Meadowlark, and Red-winged Blackbird. First Concession Road, Huntingdon, southwest of Montréal. *August 1985, Pierre Bannon.*

Detecting Henslow's Sparrow is complicated not just by its secretive behaviour, but by several other factors. The species is rare, and most birders in Québec are unfamiliar with it. Also, its fairly specific habitat is difficult to identify and not often frequented by birders. In addition, because of the transitory nature of this habitat, the species rarely nests in the same place for more than a few years. Finally, the nest is very difficult to find. Only 8 nests were found for the New York atlas, even though the species was reported in 348 squares (Eaton 1988h). According to Robins (1971), the male often sings right after feeding the young; birders should therefore keep an ear open during chick rearing, even though singing is less frequent then than at the start of the breeding season.

Distribution

Henslow's Sparrow breeds in the eastern United States and southern Ontario and winters in coastal areas from South Carolina to Texas (AOU 1983).

The species was atlassed in only 2 squares. One was located in Huntingdon County, in the foothills of the Adirondacks near the New York border, where 2 singing males were observed between 20 and 30 July 1985, in the upper part of a gently sloping field with dense grasses 1.2–1.5 m tall. The field, which had been planted with brome-grass, timothy, and alfalfa, had been mowed every year for at least 15 years (Bannon and Laporte 1985). In June 1989, in the same square, a male was discovered in a very similar field about 1 km away. Two males were also observed on 31 July and 1 August 1985 in a square in Yamaska Recreation Park near Granby, in a wet meadow of bulrushes and sedges.

History and trends

The first record of Henslow's Sparrow in Québec dates from May 1943, when it was observed at Notre-Dame-de-Stanbridge in the Eastern Townships. Between 1943 and 1950, it was also seen at Coteau-du-Lac in Montérégie, at Eccles Hill in the Eastern Townships, and at Côte-Saint-Luc in Montréal (David 1980a). The Côte-Saint-Luc record is particularly interesting because as many as 3 birds were observed between 31 May and 26 June 1947, including an agitated pair with food in their bills (Montgomery and Rawlings 1947). The period between 1965 and 1970 was particularly productive for sightings. Between 1965 and 1968, for example, up to 4 singing males frequented the same wet meadow in Aylmer in the Ottawa Valley for several summers (M. Brigham, pers. comm.); this suggests that the species probably nested there. Two birds were also seen in Granby in late August 1970 (David 1980a). Most of these records come from transitional sites between agricultural and forest areas, where the soil is less fertile and where poorly maintained or recently abandoned fields are more prevalent.

Henslow's Sparrow colonized southern Ontario at the end of the 19th century and gradually expanded its range toward the northeast to Ottawa and Kingston (Knapton 1984b, 1987b). The Ontario population has declined sharply during the last few decades, particularly during recent years and stood at probably fewer than 50 pairs by 1987 (Knapton 1987b). This decline in numbers and the fact that breeding is largely restricted to Ontario prompted the Committee on the Status of Endangered Wildlife in Canada (COSEWIC) to classify the species as threatened. Furthermore, an exhaustive survey in 1992 in southern Ontario revealed only one singing male. The Canadian population probably consists of only a few nesting pairs (possibly fewer than 10), and COSEWIC is expected to upgrade its status to endangered soon (Austin and Cadman 1992).

Notes on territory (continued)

were defended where the vegetation was shorter and sparser (Robins 1971). In addition, it has been documented that some individuals may occupy up to 4 different territories during a single breeding season, while other individuals will disappear during the breeding season; others will even appear and establish territories in the course of the summer. Territories are maintained by singing and do not overlap (Robins 1971). Some habitats, particularly larger ones, may support several breeding pairs (Smith, C. R. 1992).

In New York State, the species is much more widespread, particularly in the central and western portions of the state. During the 1930s, about 20 birds even colonized a site in St. Lawrence County, less than 125 km from the Québec border (Bull 1974). However, Henslow's Sparrow is currently at risk in New York State, as it is throughout its range (Smith, C. R. 1992; Askins 1993), and has been on the Blue List for the last 10 years (Tate and Tate 1982; Tate 1986; Austin and Cadman 1992).

During the first half of the 20th century, Henslow's Sparrow appeared to have benefited from extensive deforestation in the Northeast, abandonment of farmland, and agricultural practices such as periodically leaving fields fallow. However, the species' distribution has shrunk considerably during the last 30 years. Much of its potential habitat either has been, or is being, destroyed, particularly by the draining of low-lying areas and intensive cultivation (Knapton 1982, 1987b). Habitat fragmentation, the disappearance of fields larger than 10 ha—in Missouri and Illinois it is absent from fields smaller than 10 ha (Askins 1993)—, and the successive mowing of crops during a single breeding season (Smith, C. R. 1992) have also contributed to the species' decline. In Québec, this very secretive species may have been more common from 1940 to 1970 than published data show, but its hold here remains tenuous. Given current trends in the United States, it is very unlikely that Henslow's Sparrow will become a regular breeder here, at least in the short term.

Pierre Bannon

Le Conte's Sparrow

Bruant de Le Conte

Ammodramus leconteii

Le Conte's is one of the most elusive of the North American sparrows. A birdwatcher might spend a good deal of time counting singing males in a field without ever seeing a single one. The nest is so well camouflaged that by 1968, probably no more than 50 nests had ever been found in all the history of North American ornithology (Walkinshaw 1968b).

Le Conte's Sparrow is a very difficult species to observe, as it seldom perches in the open and when disturbed will flee by running along the ground. Even if it does flush, it flies for a brief moment, only to disappear in the thick grass. Its thin note could be mistaken for the squeaking of a mouse and its song for that of the Savannah Sparrow. When located, it will usually stop singing while the approaching observer is still some distance away (Walkinshaw 1968b; Sandilands and Campbell 1987a).

This rare migrant breeder is still an interesting member of our avifauna. Its field marks are the short pointed tail, bright buff-ochre eyebrow stripe and breast, and the fine chestnut streaking on the grey nape. It spends most of its time on the ground among the stubble and dense new growth (Walkinshaw 1968b), where it feeds on seeds, insects, and spiders (Terres 1980).

Habitat and behaviour

Le Conte's Sparrow is found in a variety of grassland habitat. In southern Québec, it generally frequents fairly broad, open areas thickly carpeted with dead grass and rank herbaceous growth. There is usually a scattering of low shrubs or tall herbaceous plants, such as canary-grass, used for perching (Y. Aubry, pers. obs.). These meadows may be nearly pure sedge communities or mixed communities of sedges and grasses. The species also uses certain agricultural habitats such as pasture, dense wet meadows, and fallow or recently abandoned hayfields, where shrub growth is not overly advanced. It is often observed in relatively wet habitat, but is also found in dry areas (Cooper 1984; Villard and Bracken 1989; M. Savard, pers. comm.).

Several pairs may nest in a given area. The male's singing may go unnoticed during the day, but is easily heard in the very early morning, late evening, or even late at night when other birds are less vocal (Walkinshaw 1968b; Sandilands and Campbell 1987a; M. Savard, pers. comm.).

Christiane Girard

Notes on territory

A study in North Dakota by Murray (1969) showed that the male defends a territory that is used for feeding and nesting.

Le Conte's Sparrow frequents rather wet habitat such as large marshes with a lush growth of sedges and grasses. It was observed in the upper level of this cord-grass marsh at Pointe-à-Boisvert. Northeast of Tadoussac. *October 1993, Pierre Normand.*

Le Conte's Sparrow

Breeding distribution
1984–1989

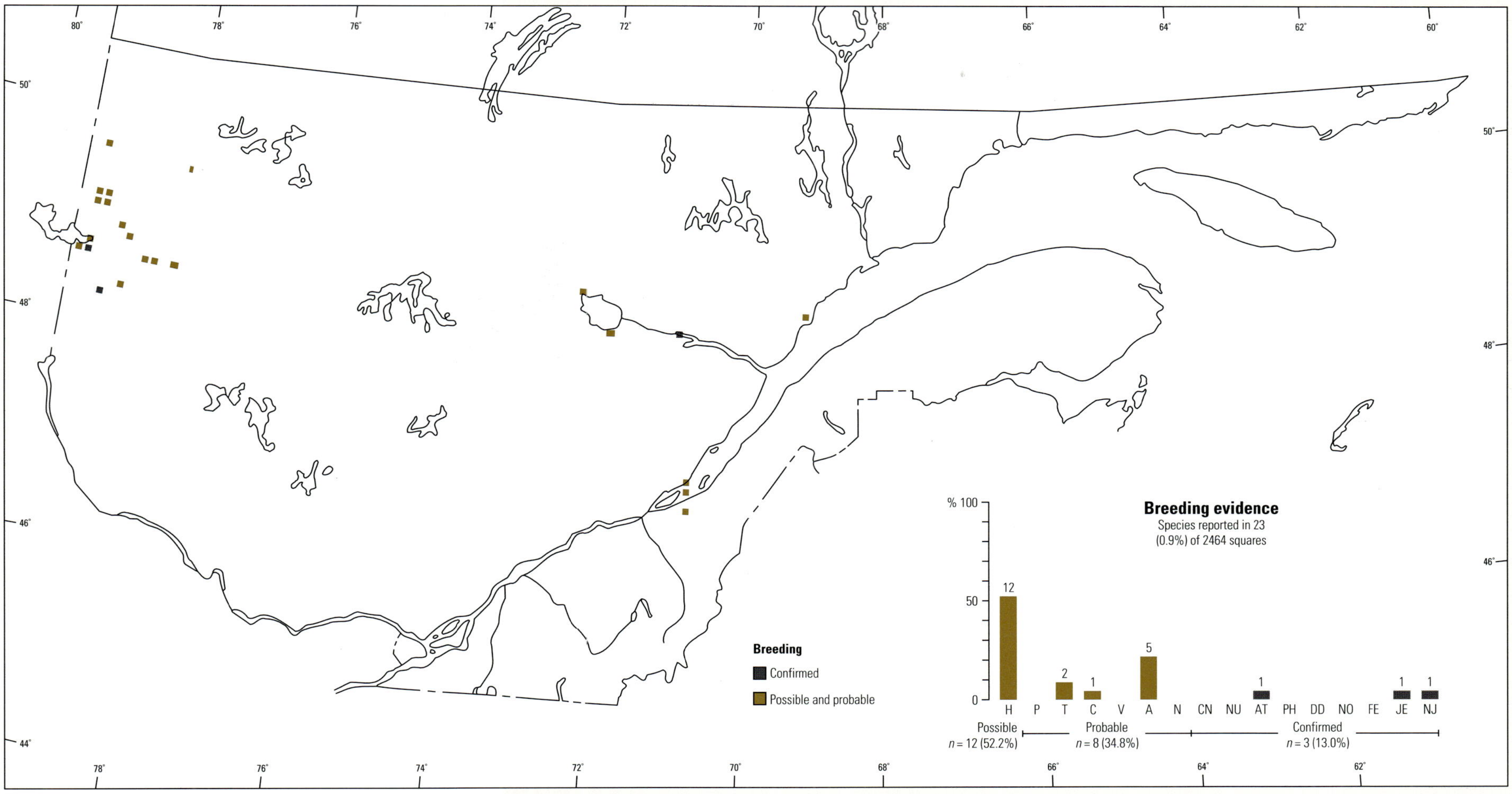

Species profile

Clutch size	usually 4 eggs (3–5)[1]
Laying interval	?
Start of incubation	?
Incubation period	11–13 days[1,4,5]
Maturity at hatching	altricial
Care of young	probably mainly by female[1,5]
Nestling period	?
Age at first flight	?
Dependence of young	?
Broods per year	?
Breeding age	?
Mating system	?
Duration of pair bond	?
Total length	11.4–13.5 cm[2]
Wing span	?
Weight	13.4 g (12.4–15.2)[3]
Longevity record	?

References:
1. Walkinshaw 1968b
2. Godfrey 1986
3. Dunning 1984
4. Harrison 1975
5. Harrison 1978

Chronology $n = 7$

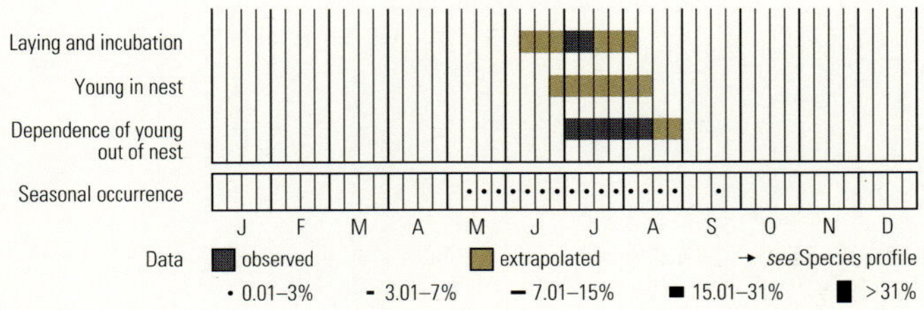

World breeding range

The nest is built on or near the ground, in a clump of dead grass. It is so well concealed in the tangled growth of dried grasses or sedges that it is extremely difficult to find. At the turn of the century, Peabody described how the bird began nest construction by interweaving strands of grass among standing stems, forming a rudimentary nest (Walkinshaw 1968b). Within this is placed the nest proper, a neat, well-rounded, deeply cupped structure composed of the very finest grasses. In most cases, it is under a fairly dense covering of dry grass.

According to Harrison (1978), the female alone incubates and possibly tends the young alone as well. Apparently, the young are fed almost entirely on insects (Walkinshaw 1968b). Little is known about the duration of the nestling period, the number of broods per season, or the rate of cowbird parasitism.

The species was reported in only 22 Atlas squares. Most records were of singing or territorial males or birds displaying agitated behaviour. Four new breeding records were obtained, however, two of which concerned the first nests ever found in Québec. These nests were found in the same square, near Rouyn-Noranda, in a field adjacent to Stadacona Pond, on 5 and 11 July 1984. Each contained 4 bluish eggs spotted with browns, mainly at the larger end. They were discovered accidentally when an adult bird (presumably a female) flushed from underfoot (J. Lapointe, pers. comm.).

Distribution

Breeding was confirmed for the Atlas at Rouyn-Noranda and Gallichan in Abitibi, and at Saint-Fulgence in the Saguenay region. Most observations were from the agricultural areas of the Abitibi and Lac-Saint-Jean regions. In Abitibi, the species was also observed in a sedge meadow bordering a large marsh and in a peatland with abundant grasses and sedges.

Since the early 1970s, several Le Conte's Sparrows have been reported from the Central St. Lawrence Lowland: in the Lac Saint-François National Wildlife Area, in May and June 1994; at Beauport, Saint-François (Île d'Orléans), Cap Tourmente, Baie-Saint-Paul, and Saint-Paul-du-Nord on the north shore; and at Saint-Raphaël, Île aux Grues, La Pocatière, Trois-Pistoles, and Pointe-au-Père on the south shore (COQ 1972, 1986, 1987, 1989, 1992; Gendron and Gauthier 1984; Yank and Aubry 1986a, 1992a, 1993a; Aubry and Bannon 1994b). Some of these observations included evidence of breeding behaviour (Saint-Paul-du-Nord, Cap Tourmente, Beauport, Trois-Pistoles, and La Pocatière). All of these summer records are suggestive of at least sporadic breeding in the St. Lawrence Plain.

This sparrow is relatively rare in Québec and until recently had been seen mostly in the Abitibi and Lac-Saint-Jean regions, in open habitat such as these abandoned fields. Amos, Abitibi. *June 1991, Desneige Perreault.*

Notes on territory (continued)

On rare occasions, territories may overlap. Boundaries appear to remain stable throughout the breeding season. Inter- and intraspecific interactions appear to be brief and subtle.

Eleven territories studied by Murray (1969) in the summers of 1965 and 1966 ranged in size from 12.65 m² (0.0012 ha) to 44.73 m², averaging 26.82 m². In Minnesota, 4 territories studied by Cooper (1984) averaged 29.54 m², with a range of 22.35–35.56 m² [territory sizes extrapolated from Murray (1969) and Cooper (1984)].

Nesting requirements of Le Conte's Sparrow include a thick matting of dead grasses, tall herbaceous growth, and a few shrubs that can serve as perches. Villemontel, Abitibi. *July 1985, Yves Aubry.*

View of the Saint-Fulgence marsh near Chicoutimi, where breeding was confirmed for the first time in Québec in 1963. *August 1991, Jean Gauthier.*

Aside from these areas, the species has also been reported from the Lake Saint-Jean Lowland and the Upper Saguenay region (M. Savard, pers. comm.), at Lachute in early August 1992 (Yank and Aubry 1992a), at Wotton in the Eastern Townships, where two birds summered in 1990 (Yank et al. 1990b), and in the Témiscamingue region at Latulipe on 8 July 1993 and at Fabre on 26 June 1994 (Aubry and Bannon 1994b; D. St-Hilaire, pers. comm.).

The nest of Le Conte's Sparrow is a rather bulky structure, built on the ground or anchored to concealing grasses. Inside is a cup woven of finer grass. Rouyn-Noranda, Abitibi. *July 1984, Jean Lapointe.*

North of the Atlas area, Le Conte's Sparrow was recorded during the breeding season for the first time in the extensive sedge community of the salt marsh at Boatswain Bay in 1941 (Todd 1963). A more recent survey conducted on 17 June 1991 in a small area of Cabbage Willows Bay yielded several singing males (Cons. G. G. – GREBE 1992o). Given the extent of the tidal meadows in this immense bay, the breeding population could be sizable. As in Saint-Fulgence, Le Conte's Sparrow shares this habitat with the Sharp-tailed Sparrow, and there may well be territorial interactions (Gill 1990).

In Canada, the species breeds from the Rocky Mountains through the Prairie provinces, and in Ontario and Québec. It occupies the north-central United States and the area west of the Great Lakes. It winters in the southeastern United States, from Florida to Texas (National Geographic Society 1987).

History and trends

The species is named for John L. Le Conte, a physician friend of John J. Audubon, and distinguished entomologist (Terres 1980). It was described in 1790 from a single specimen; 53 years later Audubon described a second bird (Walkinshaw 1968b). The first Québec sightings were made in 1935, on 21 May in Beaupré, near Québec City (Boulva and Browne 1964), and on 18 September in the southern James Bay region, near Waskaganish (Todd 1963). Breeding was confirmed for the first time in 1963 at Saint-Fulgence by the observation of a pair and a chick recently fledged from the nest and moulted into its juvenal plumage but still with a few down feathers (Boulva and Browne 1964).

Walkinshaw (1968b) maintained that the species' secretive ways have often and probably always will cause its numbers to be underestimated in areas where it is actually fairly common. The distribution as presently described for southern Québec may be an example of such underestimation. The main strongholds, the Saguenay–Lac-Saint-Jean region and especially the Abitibi region, were identified in 1963 and 1965, respectively (David 1980a).

Nowadays, there are more birdwatchers skilled in identifying elusive species and equipped with powerful binoculars and spotting scopes. This, along with the Atlas field work, has helped determine the species' status and breeding range in Québec, which will no doubt become more clearly defined as time goes on.

Vincent Létourneau and Yves Aubry

Sharp-tailed Sparrow*

Bruant à queue aiguë

Ammodramus caudacutus

Christiane Girard

IN QUÉBEC, THE SHARP-TAILED SPARROW IS FOUND MAINLY IN SALT AND brackish marshes. First reported in 1877 in the salt marshes of Kamouraska (Dionne 1883), it has since been confirmed breeding farther east, in the estuary and Gulf of St. Lawrence. More recently, it has been found breeding westward in the freshwater marshes of some islands in the St. Lawrence. A small bird with ochre-yellow facial markings and no apparent sexual dimorphism, the Sharp-tailed Sparrow is a difficult species to observe. In Québec, it is classified as a rare migrant breeder. The Sharp-tail winters along the Atlantic and Gulf of Mexico coasts (AOU 1983).

Habitat and behaviour

This late spring migrant (Gilbert 1981) is faithful to its breeding site from year to year (Post and Greenlaw 1982). Most birds arrive in early June. In the St. Lawrence estuary, the Sharp-tailed Sparrow occupies the upper level of salt marshes (Shaffer 1990), dominated by salt-meadow cord-grass and interspersed in the lower reaches with numerous tidal pools. As the pools give way to drier ground, this grass is replaced by a variety of coastal plants, including freshwater cord-grass (Couillard and Grondin 1986), a tall plant used for cover and perching by male Sharp-tails. Adjacent to salt marshes, this sparrow is sometimes found in fields of forage crops and idle fields (Gilbert 1981). At L'Isle-Verte, it also occurs in the shrub layer of the salt marsh. On Île du Moine and Île aux Fermiers (freshwater habitat), it is found in the tall grasses of wet meadows (Shaffer 1990).

On arriving, the male begins delivering its *te sheeeee* song, generally sustained, which consists of two sharp tics followed by a wheezy hiss like the sizzle of a drop of water on a hot stove. In calm weather, this can be heard up to 150 m away (Gilbert 1981). The Sharp-tailed Sparrow is a retiring, wary species of furtive, mouselike movements. Although sometimes difficult to find, it sings often, conspicuously perched on a tall herbaceous plant, a shrub, or fence. It is mostly heard in the early morning or at the end of the day; some birds sing at night and some appear even to sing only at night (Rising 1987c).

Each male confines its activities to a relatively extensive activity area of about 1.5 ha. This undefended area is frequented by several males and is therefore not a true territory (Woolfenden 1956; Gilbert 1981). Males expend all their energy in singing and patrolling their activity area in the hope of finding receptive females (Post and Greenlaw 1982). Gilbert (1981) reports that although the habitat is homogeneous, the birds exhibit cluster dispersion, concentrating their activities in a fraction of the habitat.

The male also sings in flight, rising vertically up to 10 m. Other males often join in this song flight, in a sort of collective display. Copulation usually occurs during or directly after these displays (Woolfenden 1956; Gilbert 1981). Because males and females do not form pairs and each individual mates with different partners, the species is often described as being promiscuous. However, Gilbert's (1981) work seems to indicate the existence of a hierarchical relationship among males established through direct confrontations over females. Certain males dominate the breeding grounds and copulate with a greater number of females.

Notes on territory

The Sharp-tailed Sparrow is a non-territorial species (Woolfenden 1956; Post and Greenlaw 1982). The male concentrates its activities

* The Sharp-tailed Sparrow (*Ammodramus caudacutus*) has recently been split into two species (AOU 1995). Only one, Nelson's Sharp-tailed Sparrow (*Ammodramus nelsoni*, Bruant de Nelson), occurs in Québec, where it is a migrant breeder.

In the estuary and Gulf of St. Lawrence, the Sharp-tailed Sparrow inhabits the upper level of salt marshes dominated by freshwater cord-grass. Sainte-Anne Cove, Saint-Roch-des-Aulnaies, north-east of Montmagny. *August 1987, Réginald Ouellet.*

Sharp-tailed Sparrow

Breeding distribution

1984–1989

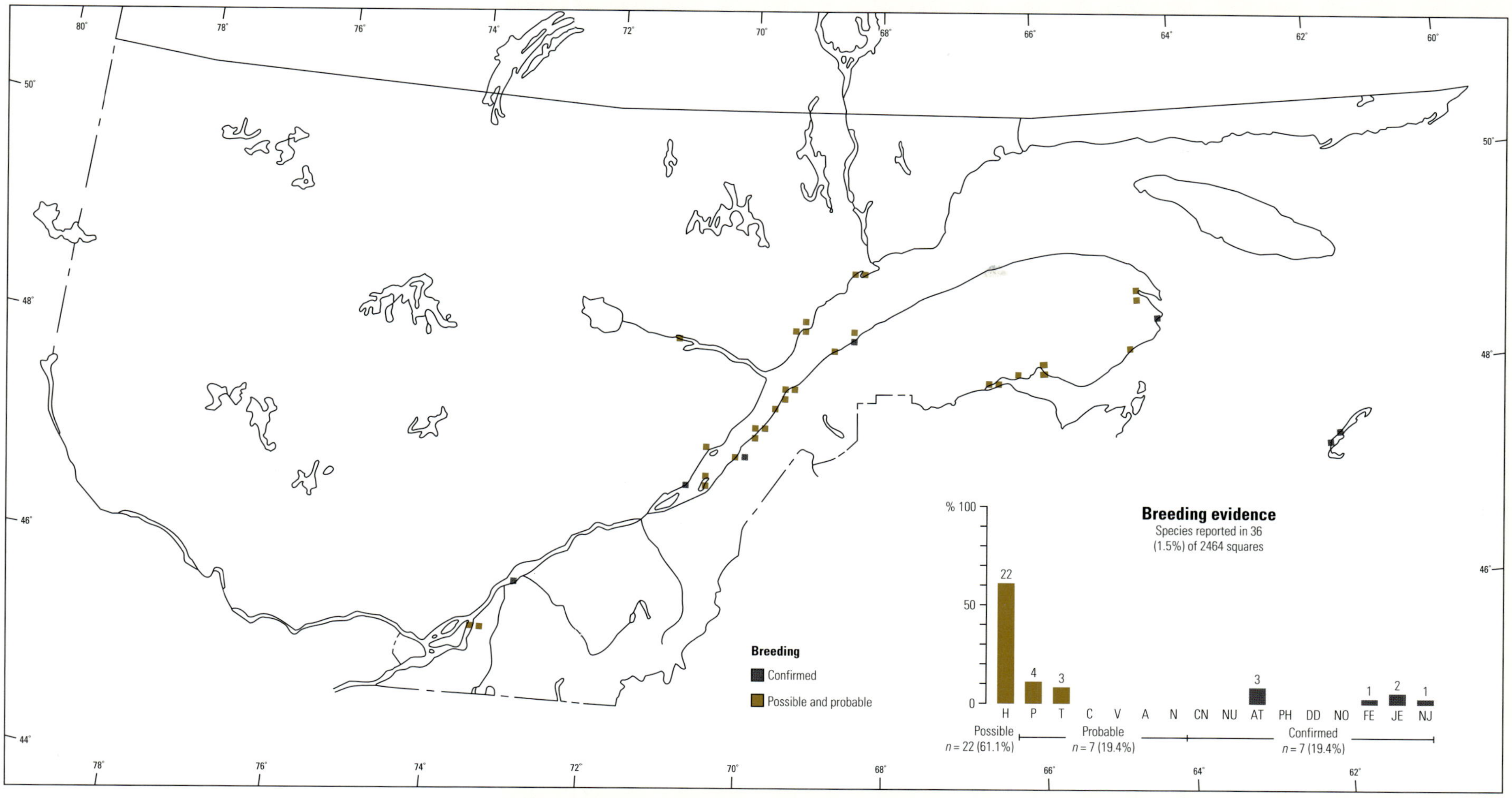

Chronology *n* = 10

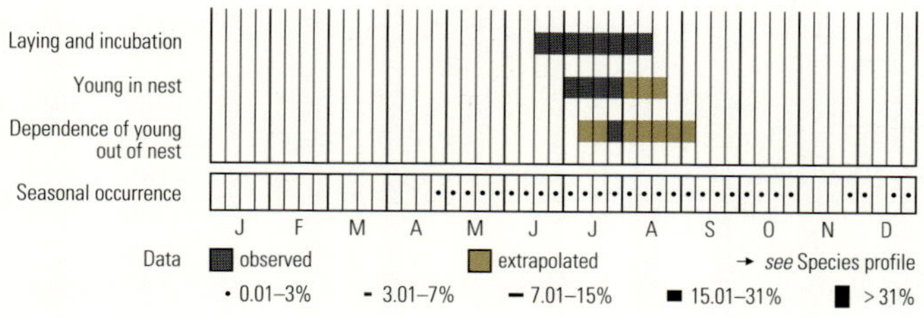

World breeding range

Species profile

Clutch size	usually 3–5 eggs (2–7)[1,2,6,11]
Laying interval	1 egg per day[11]
Start of incubation	with last or penultimate egg[11]
Incubation period	10–12 days[1,2,3,6,7,11]
Maturity at hatching	altricial
Care of young	by female[1,2,5,7]
Nestling period	8–11 days[2,3,11]
Age at first flight	11–16 days[11]
Dependence of young	until 15–20 days after leaving nest[2,5,11]
Broods per year	usually 2[1,5]
Breeding age	1 year[8]
Mating system	promiscuity[4,7]
Duration of pair bond	during mating only[7]
Total length	12.8–14.8 cm[9]
Wing span	?
Weight	M 20.7 g (18.0–25.8) / F 17.8 g (15.3–19.0)[10]
Longevity record	10 years[11]

References:
1. Ehrlich et al 1988
2. Harrison 1975
3. Terres 1980
4. Post and Greenlaw 1982
5. Woolfenden 1956
6. Shaffer 1990
7. Gilbert 1981
8. DeGraaf and Rudis 1987
9. Godfrey 1986
10. Dunning 1984
11. Greenlaw and Rising 1994

On the Gaspé Peninsula, the species nests in the grassy upper level of the estuarine marshes, as in the York River, near Wakeham. *July 1991, Yves Aubry.*

Notes on territory (continued)

in an area of about 1.5 ha (Gilbert 1981) shared with several others (Woolfenden 1956). The female uses a smaller area of 0.5 ha (Gilbert 1981) and defends the immediate vicinity of the nest (Woolfenden 1956). Stewart and Robbins (1958) counted 25 males/10 ha in Maryland marshes during the breeding season.

Sharp-tailed Sparrows breeding alongside Le Conte's Sparrows may be subject to aggression from the latter, which are highly territorial (Murray 1969).

The female frequents a relatively small area of less than 0.4 ha (Woolfenden 1956) and defends the immediate area around the nest (Post and Greenlaw 1982). Nests are constructed on the ground or are attached to stems of vegetation just above the ground (Harrison 1975). The female performs all parental care, raising an average of 4 young which may be the offspring of different fathers (Seutin and Simon 1988b). Double-brooding is common (Harrison 1975), and is assumed to occur in Québec, where Gilbert (1981) has shown that a second song period occurs between the end of July and mid-August after a roughly 2-week period of silence. Discovery of a nest containing eggs on 2 August 1978 at L'Isle-Verte supports this assumption (Gilbert 1981).

Post and Greenlaw (1982) compared the breeding strategy of the Sharp-tailed Sparrow to that of the Seaside Sparrow, a species that it closely resembles in morphology, habitat selection, and feeding habits, although the male Seaside Sparrow is territorial, monogamous, and provides parental care. They noted similar growth and survival rates among the young of both species and identical mortality rates among females. Despite its emancipation from parental duties, the male Sharp-tailed Sparrow does not have a higher survival rate than the female. The researchers concluded that the highly productive habitat used by both species makes promiscuity as effective a strategy as shared parental duties.

Post and Greenlaw (1982) also found the number of fledged young produced annually by each female Sharp-tail to be 4.03 and the annual survival rate of adults to be between 53 and 63%.

Sharp-tailed Sparrows feed on the ground among the vegetation (Hill 1968). In summer, invertebrates make up over 80% of the diet, which also includes the seeds of grasses and various other plants (Judd 1901). The bill, longer and less conical than that of other sparrows, is apparently adapted to the chiefly insectivorous diet (Woolfenden 1956).

Breeding was confirmed in 7 of the 36 Atlas squares in which the species was reported; most confirmations were based on sightings of females carrying food and on observations of young just out of the nest. Nests are difficult to find because a threatened female will flee along the ground, through the vegetation. The young can be located, however, by listening for the females' repeated distress *chip* calls and by observing females carrying food.

Recently, this emberizine sparrow has been found breeding in wet meadows associated with the freshwater marshes of some islands in the St. Lawrence River. Île du Moine, Sorel Islands, west of Lake Saint-Pierre. *July 1991, Daniel Jauvin.*

A banded female feeding her young. Nests are located near water and are placed on or slightly above the ground among supporting plant stems. L'Isle-Verte, northeast of Rivière-du-Loup. *July 1978, Jean Bédard, Société Duvetnor.*

Distribution

There are five subspecies of the Sharp-tailed Sparrow in North America (AOU 1957). Except on the Prairies, the species is associated mainly with salt and brackish marshes and is distributed along a narrow coastal strip. In Québec, there are two disjunct populations: *Ammodramus caudacutus alterus* inhabits the James Bay coast and *Ammodramus caudacutus subvirgatus* occupies the shores of the St. Lawrence River (Shaffer 1990). A recent study based on genetic fingerprinting of different populations of the five subspecies shows that they belong to two population groups: the first includes the three Canadian subspecies and the second the two subspecies that breed along the Atlantic coast as far north as southern Maine (Rising and Avise 1993).

According to Shaffer (1990), *Ammodramus caudacutus subvirgatus* is common in suitable habitat throughout its range. In Québec, this subspecies is found mainly between Montmagny and Pointe-au-Père, on the Gaspé Peninsula from Gaspé Bay to the head of Chaleur Bay, in the estuaries of certain rivers, such as the Dartmouth and York, and in Barachois-de-Malbaie. It also occurs very locally along the north shore of the St. Lawrence between Cap Tourmente and Pointe-aux-Outardes (Baie-Saint-Paul, Baie-des-Bacon, Bergeronnes, Saint-Paul-du-Nord, and Pointe-à-Boisvert), at Saint-Fulgence in the Saguenay region, and on the Magdalen Islands. The Sorel Islands, in the freshwater upper St. Lawrence, supported 2 or 3 birds in 1975, apparently for the first time (ÉPOQ). That population was recently estimated to be at most 50 individuals (Seutin and Simon 1988b). Even farther west, the species was reported from Masta Island and Île aux Fermiers, near Varennes, a further instance of recent colonization (Shaffer 1990). There were no Atlas reports from Anticosti Island, despite an earlier record from Grand Lac Salé (Shaffer 1990). During surveys conducted in 1985 and 1986 at Pointe-à-la-Croix in Chaleur Bay, 112 singing males were counted (Shaffer 1990), making this one of the largest breeding populations in southern Québec.

The *alterus* subspecies has been reported on the James Bay coast, at Waskaganish and Eastmain on the Québec side, and in a few locations on the Ontario side (Todd 1963; Rising 1987c). Observations at Cabbage Willows Bay in 1991 (Cons. G. G. – GREBE 1992o) and in the Eastmain area in 1990 (Tremblay and Belles-Isles 1991) show that the species still occurs in these regions.

History and trends

The Sharp-tailed Sparrow has been observed since 1877 in the salt marshes of Saint-Denis in Kamouraska County. It was also reported at L'Isle-Verte, Rivière-du-Loup, and in other locations along the south shore of the St. Lawrence estuary and considered an uncommon summer resident (Dionne 1883, 1906). Breeding sites have since been discovered at over 40 locations along the shores of the estuary and gulf with the Québec population estimated to be at least 1000 individuals (Shaffer 1990). Atlas data show that the St. Lawrence population of Sharp-tailed Sparrows still occurs in the great majority of sites that the species has traditionally occupied.

Habitat loss along the St. Lawrence is mainly attributable to human activity, particularly draining for agriculture and landfill for construction (Shaffer 1990). Between 1945 and 1976, agriculture, landfill, and construction of a variety of infrastructure gradually eliminated 18% of the salt marshes in the Lower St. Lawrence region (Gauthier et al. 1980; Dryade 1981b). Haying and conversion to pasture have often disturbed the birds and damaged their nests (Shaffer 1990). Although the *subvirgatus* subspecies still appears to occupy most of its traditional sites, the vulnerability of this habitat should be reassessed.

James Bay hydro-electric development projects and diking could affect some portions of the habitat of the *alterus* subspecies by altering the movement of ice. Decreased salinity levels may affect the ice cover or delay the thaw (Milko 1986; Prinsenberg 1992). Other factors could come into play, such as the extensive isostatic rebound in Rupert Bay (Champagne 1982), which alternately causes habitat to appear and disappear.

Réginald Ouellet

Fox Sparrow

Bruant fauve

Passerella iliaca

Christiane Girard

THE FOX SPARROW OWES ITS NAME TO THE OVERALL COLORATION OF ITS plumage, a combination of rusty brown, chestnut, and grey, set off by a foxy red tail, rump, and wings. Its rich colouring is equalled only by the richness of its song, which Dionne (1906) described as having the brilliance, purity of notes and fullness of tone that easily make it one of our best feathered songsters.

There are marked differences in size, colour, and shape among the many subspecies. Recent DNA analyses have indicated that these subspecies may soon be reorganized into four distinct species (Zink 1991).

The Fox Sparrow is the largest and hardiest of the sparrows that breed in Québec. The birds generally travel in small flocks, passing quickly through the southernmost part of the province on spring migration. The Gaspé Peninsula or, better still, the North Shore are the best places to observe the bird in its summer quarters, from mid-May on.

When spring is late arriving, inclement weather may entice it to linger at feeders in southernmost Québec. This is a good opportunity to hear the song and especially to observe this retiring species before it heads north to breed. Fox Sparrows very seldom overwinter in southern Québec, unless attracted by feeding stations. The species winters on the Pacific coast, in the southwestern states and in the eastern United States, and locally in southern Canada (AOU 1983; Godfrey 1986).

Habitat and behaviour

In general, the Fox Sparrow haunts low, dense tree and shrub cover, from which it rarely strays. It frequents thickets and forest edges, scrubby woodland, either coniferous or mixed, burns, and streamside willow and alder thickets (Pough 1949; Godfrey 1986) and has a penchant for rocky shores with stunted and gnarled coniferous growth that is more or less impenetrable (Terrill 1968). It is also found in cutovers regenerating in conifers (Hardy and DesGranges 1989).

Despite its reclusive ways, the Fox Sparrow does not go unnoticed. From dawn to dusk it sings its clear, ringing song from the tip of a conifer or sometimes sheltered in the shrubs. On the Magdalen Islands, one bird was heard singing non-stop from 9 a.m. to 5 p.m. Given that the song lasts 2.75 seconds, with a 10-second pause between renditions, the bird sang about 2880 times that day (COQ 1974).

Many ornithologists have been fascinated by the intensity, purity, and variety of the Fox Sparrow's song. Brewster (*in* Terrill 1968) characterized it as the typical songster of the bleak regions along the Gulf of St. Lawrence coastline, and Moore (1913) considered it to be the virtuoso of the Magdalen Islands. This song can be heard irregularly, even during the fall migration (Fradette 1992).

The Fox Sparrow can also be located by the sound it makes scratching among the leaves (Terrill 1968; Hailman 1976) as it forages for insects and seeds (Dionne 1906). This behaviour is noted most often during migration, as the bird energetically turns over the leaf litter of the underbrush (Forbush and May 1939).

Males and females return to the nesting grounds at the same time, and pairs are formed within a week of arriving. The territory is established by the male (Threlfall and Blacquiere 1982), whose singing serves to proclaim his domination far and wide and whose loud alarm note (Moore 1913) is used in defending its territory during direct confrontations with conspecifics (Martin 1980). The nest is built under foliage on the ground or low in a tree or shrub (Godfrey 1986). It is made of grass, moss, lichens, and rootlets, and also twigs if built above ground (Terrill 1968; Ehrlich et al. 1988). In general, nests built early in the season are higher than those built later on, when there is no longer snow on the ground (Philipp 1925); nests built in trees are much larger than those built on the ground (Threlfall and Blacquiere 1982).

Incubation, which lasts about 12 days, is solely by the female (Threlfall and Blacquiere 1982), and rarely begins before the last egg is laid; consequently, hatching is more or less synchronous, with all eggs hatching within 12–35 hours. The young are fed by both parents and remain in the nest barely 10 days. The adults may continue feeding them for up to 21 days after they leave the nest. Some observations suggest that a second nesting attempt may be made during the same season (Terrill 1968), but this has not yet been confirmed (Threlfall and Blacquiere 1982).

The Fox Sparrow was reported in 14.5% of Atlas squares surveyed (358/2464), and nesting was confirmed in only 6.1% (22/358) of those, mainly through sightings of adults carrying food and observations of young just off the nest. Only three nests were found.

Notes on territory

A Newfoundland study in the 1970s showed that the species occupies territories that vary in size, depending on habitat conditions, from 0.25 to 1 ha. It remains faithful to this territory from one year to another (Threlfall and Blacquiere 1982). A Québec study on Les Pèlerins, a group of islands in the St. Lawrence estuary, yielded densities of 1–4 territorial males/10 ha (Filion 1985).

The Fox Sparrow dwells in coniferous and mixed forests, where it mostly favours thickets and edges, as well as shoreline scrub. Near Saint-Gabriel-de-Gaspé, west of Percé.
November 1983, Jean Beaubien and Guy Simard, Canadian Forest Service.

Fox Sparrow

Breeding distribution
1984–1989

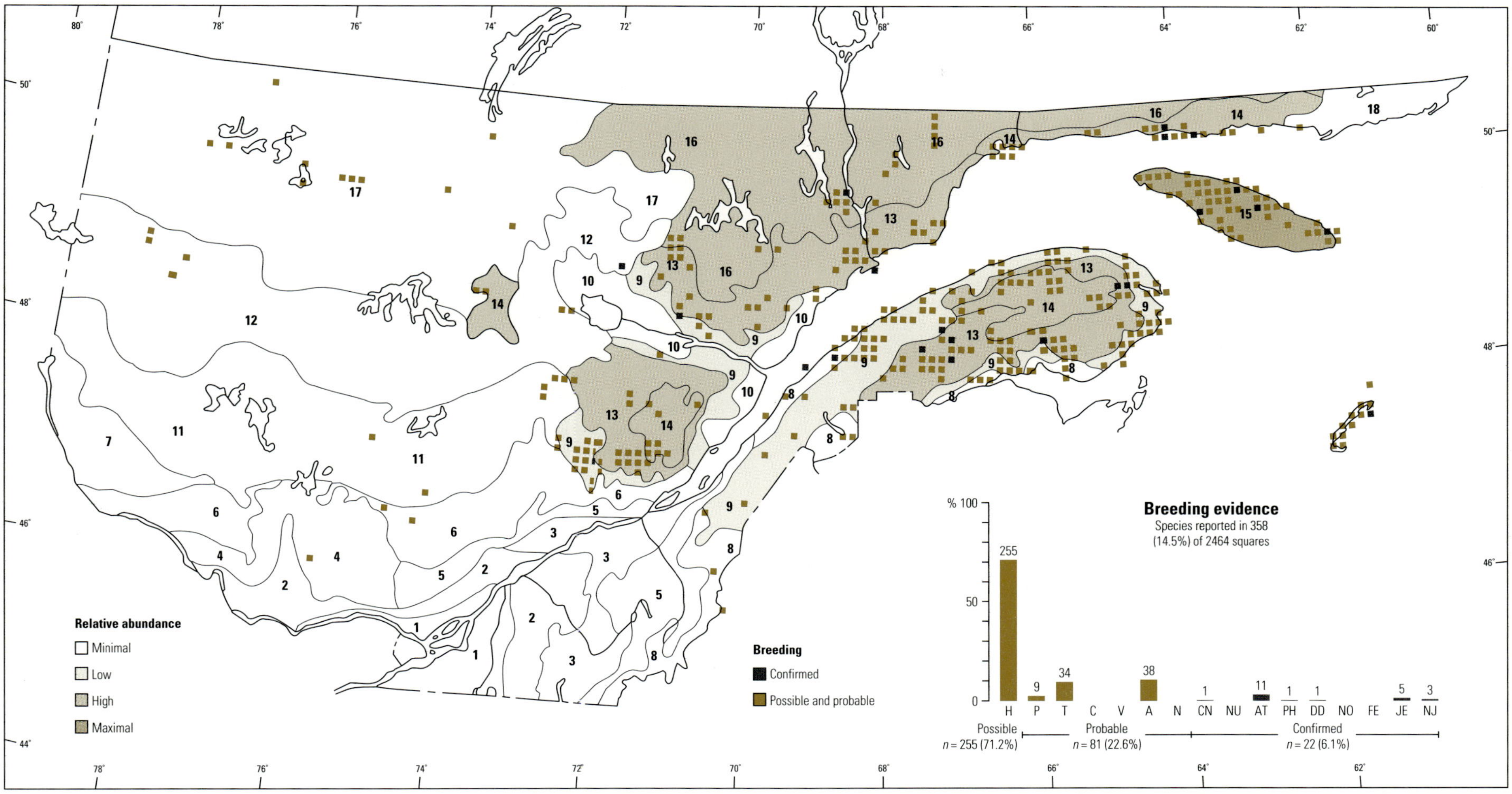

World breeding range

Species profile

Clutch size	2–5 eggs[1,2,3,4]
Laying interval	?
Start of incubation	with last egg[1]
Incubation period	12–14 days[1,3,4]
Maturity at hatching	altricial
Care of young	by female and male[1]
Nestling period	9–11 days[1,3]
Age at first flight	9–11 days[1,3]
Dependence of young	until 21 days after first flight[1]
Broods per year	1, 2 possible[1,3,4]
Breeding age	?
Mating system	probably monogamy[3]
Duration of pair bond	?
Total length	17–19 cm[2]
Wing span	26.7–29.8 cm[7]
Weight	32.3 g (21.7–42.1)[6]
Longevity record	9 years 9 months[5]

References:
1. Threlfall and Blacquiere 1982
2. Godfrey 1986
3. Ehrlich et al. 1988
4. Terrill 1968
5. Kennard 1975
6. Dunning 1984
7. Terres 1980

Chronology $n = 19$

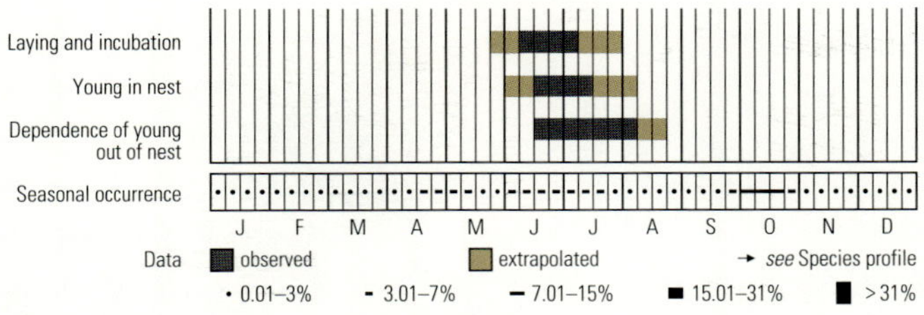

Frequency of occurrence by ecological region

Ecological regions** (forest regions)	Adequately covered squares	Squares in which reported	% of squares in which reported	Relative abundance (%)*
1. Sugar maple-hickory and sugar maple-basswood	97	0	0	0
2. Sugar maple-basswood	153	0	0	0
3. Sugar maple-basswood and sugar maple-yellow birch	101	0	0	0
4. Sugar maple-yellow birch-basswood	56	1	1.8 ± 3.7	1.8
5. Sugar maple-yellow birch-beech and sugar maple-yellow birch-basswood	113	0	0	0
6. Sugar maple-yellow birch-beech	71	2	2.8 ± 4.0	2.8
7. Sugar maple-yellow birch-balsam fir	19	0	0	0
8. Sugar maple-yellow birch and balsam fir-yellow birch	47	13	27.7 ± 11.6	27.7
9. Balsam fir-yellow birch	87	46	52.9 ± 9.7	52.9
10. Balsam fir-white birch or balsam fir-red maple	46	1	2.2 ± 4.4	2.2
11. Yellow birch-balsam fir	52	7	13.5 ± 9.9	13.5
12. Balsam fir-white birch (west)	59	6	10.2 ± 8.2	10.2
13. Balsam fir-white birch (east)	76	52	68.4 ± 10.2	68.4
14. Balsam fir-black spruce	28	22	78.6 ± 16.2	78.6
15. White spruce-balsam fir and balsam fir-white spruce	22	22	100.0 ± 2.3	100.0
16. Black spruce-balsam fir-moss	18	11	61.1 ± 25.5	61.1
17. Black spruce-moss	32	5	15.6 ± 14.1	15.6
18.	*not analysed*			

* minimal: 0–29.9%; low: 30–59.9%; high: 60–89.9%; maximal: 90–100%

** Deciduous forest: 1–7; mixed forest: 8–11; coniferous forest: 12–17

Distribution

The Fox Sparrow breeds in Alaska and in the subarctic and boreal regions of Canada east to Labrador and Newfoundland; in the south it is found in the mountains of the western United States to California (AOU 1983; Godfrey 1986).

Frequency of occurrence was highest in the eastern half of southern Québec; the Fox Sparrow is, in fact, rare west of the 73rd meridian. Breeding was reported north of the Saguenay River and Lake Saint-Jean, on the North Shore, the Gaspé Peninsula, Anticosti Island, the Magdalen Islands, Île aux Basques in the St. Lawrence estuary, and in the Laurentide Wildlife Reserve.

The eastern parts of this range are characterized by higher elevation, a damper climate, and more abundant precipitation, with forest cover consisting primarily of stands of balsam fir with yellow birch, white birch, white spruce, or black spruce. The Fox Sparrow is often found in large areas regenerating after logging or fire, as in the interior of the Gaspé Peninsula, on Anticosti Island, and in the Laurentide Wildlife Reserve. It is particularly partial to the stunted coniferous stands and the scrubby forests of the islands and shores of the Gulf of St. Lawrence and estuary (Terrill 1968; Filion 1985; Godfrey 1986). The species has often been reported from the Magdalen Islands and Anticosti Island, where it is associated with the dense coniferous growth in the interior and with stunted white spruce stands along the coast (Terrill 1968; Ouellet 1969; Fradette 1992). On the Lower North Shore, it is attracted to the characteristic alternating patches of stunted forest and open rocky areas (G. Chapdelaine, pers. comm.).

The Atlas data expand the Fox Sparrow's breeding range south of the limit established by Godfrey (1986), notably to the Abitibi region, where the species had been observed previously (Erskine 1971b), to the Saguenay–Lac-Saint-Jean region, to the Laurentide Wildlife Reserve, to the Ottawa Valley (Lake Giles), to the Laurentians (Mont-Tremblant Recreation Park), to the Lower St. Lawrence region, and to the Chaudière–Appalaches region (Sainte-Rose-de-Watford and Portage Lake).

North of the Atlas area, the breeding range extends north to the tree line (Lake Guillaume Delisle, Kuujjuaq, Koroc River), encompassing Québec's subarctic zone and part of its boreal zone. Godfrey (1986) indicated, however, that the species occurs very locally in the interior and is apparently absent from certain regions. Todd (1963) also noted its absence from vast areas between the Atlantic coast and the Hudson Bay and James Bay coasts. The most northerly record for Québec is from Aupaluk, on the west coast of Ungava Bay (ÉPOQ).

Fox Sparrows breed commonly in stunted spruce stands, where the nest is built on the ground or in the lower branches of a spruce tree. Pointe de l'Est National Wildlife Area, Magdalen Islands. *June 1991, Desneige Perreault.*

Notes on territory (continued)

Other densities have been reported elsewhere in Canada; for example, in a mountainous area of the Mackenzie District, an alder thicket yielded a density of 0.3 male/10 ha; at the other extreme, a young hemlock stand in British Columbia had an estimated density of 3.7 males/10 ha (Erskine 1976a).

The species also frequents rocky outcrops and stunted coniferous stands along the weather-beaten shores of the Gulf of St. Lawrence and estuary. Saint-Germain, southwest of Rivière-du-Loup. *October 1991, Réginald Ouellet.*

This emberizine sparrow can be found in mountainous regions, where it often nests in areas regenerating in rank, dense scrub. Mount Albert, Gaspé Conservation Park. *July 1991, Yves Aubry.*

History and trends

The profile of the Fox Sparrow population has probably changed little since John J Audubon stopped off at Natashquan in 1833. Audubon, who discovered a nest in this area, was the first ornithologist to describe the bird in its breeding habitat (Terrill 1968).

In boreal habitat, the Fox Sparrow is found in edges and clearings. It sings all day long, often from the tip of a coniferous tree. Laurentide Wildlife Reserve, north of Québec City. *August 1991, Marcel Darveau.*

At the beginning of this century, the Fox Sparrow was known to nest in large numbers on the Magdalen Islands, on Anticosti Island, and all along the North Shore of the Gulf of St. Lawrence (Dionne 1906; Gaboriault 1961; Ouellet 1969). At that time, it was a great ornithological attraction on the Magdalen Islands, where many nests were found as of 1889 (Fradette 1992). In the summer of 1956, Gaboriault (*in* Fradette 1992) counted 50 Fox Sparrows between Cap-aux-Meules and Grande-Entrée. In 1989, roughly 50 birds were found on Île de l'Est alone (Fradette 1992).

The Fox Sparrow population certainly shows no signs of decline. In fact, BBS and ÉPOQ data indicate an increase in southern Québec since the early 1970s (Cyr and Larivée 1993a).

Local fluctuations in Fox Sparrow populations are difficult to interpret. Aside from the stable habitat of the stunted maritime forest, the species occupies open habitat at a certain stage of reversion to scrub and dense coniferous forest—optimum Fox Sparrow habitat (Ouellet 1969; Hardy and DesGranges 1989). Being of an ephemeral nature, such habitat may produce local fluctuations (Terrill 1968) or short-term appearances that cannot really be construed as general population trends.

Réal Bisson and Benoît Limoges

Song Sparrow

Bruant chanteur

Melospiza melodia

THIS IRREPRESSIBLE SINGER IS FAMED FOR ITS MELODIOUS SONG, ONE OF the most well-developed and complex of all Québec's sparrows. Since the Song Sparrow frequents disturbed habitat, it has been described as the "common commensal of humans" (Greenberg 1990). A migrant breeder in Québec, it generally leaves in October and returns in early April, except for a few individuals that overwinter in the province.

Of the eight subspecies in Canada described by Godfrey (1986), only *Melospiza melodia melodia* is found in Québec.

The Song Sparrow winters throughout the United States, except in the north-central part, south to northern Mexico. Its winter range also includes a thin strip along the west coast of Canada and Alaska from the Aleutian Islands to southern British Columbia (National Geographic Society 1987).

Habitat and behaviour

The Song Sparrow is a generalist in terms of both diet and breeding habitat. This omnivorous species forages on the ground or at the base of shrubs and bushes for seeds, berries, insects, and small worms. During the breeding season, both adults and young subsist mainly on insects (Tompa 1964; Pulliam and Enders 1971). The Song Sparrow is a denizen of fairly open habitat with shrubby or bushy growth, often near water (Rising 1987d). It can be found in low, dense vegetation around ponds, lakes, rivers, and streams, in clearings, swamps, and forest margins, in shrubbery about buildings, in brushy pastures (Godfrey 1986; Rising 1987d), and even in small patches of forest or shrubbery (Yeaton and Cody 1974; Morse 1977). This highly adaptable and opportunistic species frequents both natural habitats and those modified by man (Tompa 1964; Rising 1987d; Greenberg 1989). Because of its adaptability, the Song Sparrow is widespread in North America and is one of the most plastic species, with over 30 subspecies (Miller 1956; Nolan 1968; Terres 1980).

Although this sparrow is not limited to a specific habitat, it does require certain basic elements for breeding (Marshall 1948; Tompa 1964). The vegetation has to be either dense and rather short, or tall and sparse, so that sunlight can penetrate; forest undergrowth would not be appropriate, for example. The male requires song perches overlooking his territory so that he can establish and maintain the boundaries; in some habitat, tall herbaceous plants may be used instead of shrubs (DeGraaf 1989). In addition, as a ground feeder, the species must have cover for foraging. Territories are often located near rivers, streams, ditches, marshes, or beaches. The species is not usually found in dense woods or forest, or in the middle of fields or other open areas where there are no overhanging perches or vegetation to provide cover.

The Song Sparrow is primarily monogamous, but may be polygynous in some circumstances: when a male annexes the territory of a neighbour after the latter disappears or is evicted; when a second female settles in a pair's territory (Arcese 1989a,b); or when females outnumber males (Smith et al. 1982).

Both males and females are highly territorial, but when the population is dense, males have a lower summer survival rate than females, possibly because of the strong competition among males for the best territories (Arcese et al. 1992).

Some forms of courtship behaviour in this species are quite surprising, and can be confused with nesting behaviour or aggression. One ritual, for example, is the symbolic carrying of nesting materials. In another display, pouncing, the male swoops down on the female and brushes against her, or sometimes even collides with her, as if he were defending the territory (Nice 1943).

Christiane Girard

Notes on territory

The Song Sparrow's breeding territory is generally used for mating, nesting, singing, and foraging (Nice 1943). However, on Mandarte Island in British Columbia, Tompa (1964) observed individuals foraging outside their territories.

The male establishes his territory in spring. If he succeeds in attracting a female onto the territory, she will choose the nest site. If the site is located too close to, or outside, the territorial boundaries, the male may extend the boundaries accordingly (Arcese 1989a). The territory is usually defended against individuals of the same sex, with the male expelling other males, while the female expels other females (Nice 1937; Arcese 1989a). The result is a kind of dual territoriality in which both members of the pair defend their own territories, the boundaries of which are almost exactly the same (Arcese 1989a).

Arcese (1987) observed that when the density of territorial birds was high, some males, known as floaters, lived unobtrusively on other males' territories, apparently waiting for a chance to take over the territory, either by

Song Sparrows seek fairly open areas containing clumps of shrubs, often near water. They also frequent habitat modified by man. Rivière aux Saumons, near the Lac Saint-François National Wildlife Area, southwest of Montréal. *September 1987, Léo-Guy de Repentigny.*

Song Sparrow

Breeding distribution

1984–1989

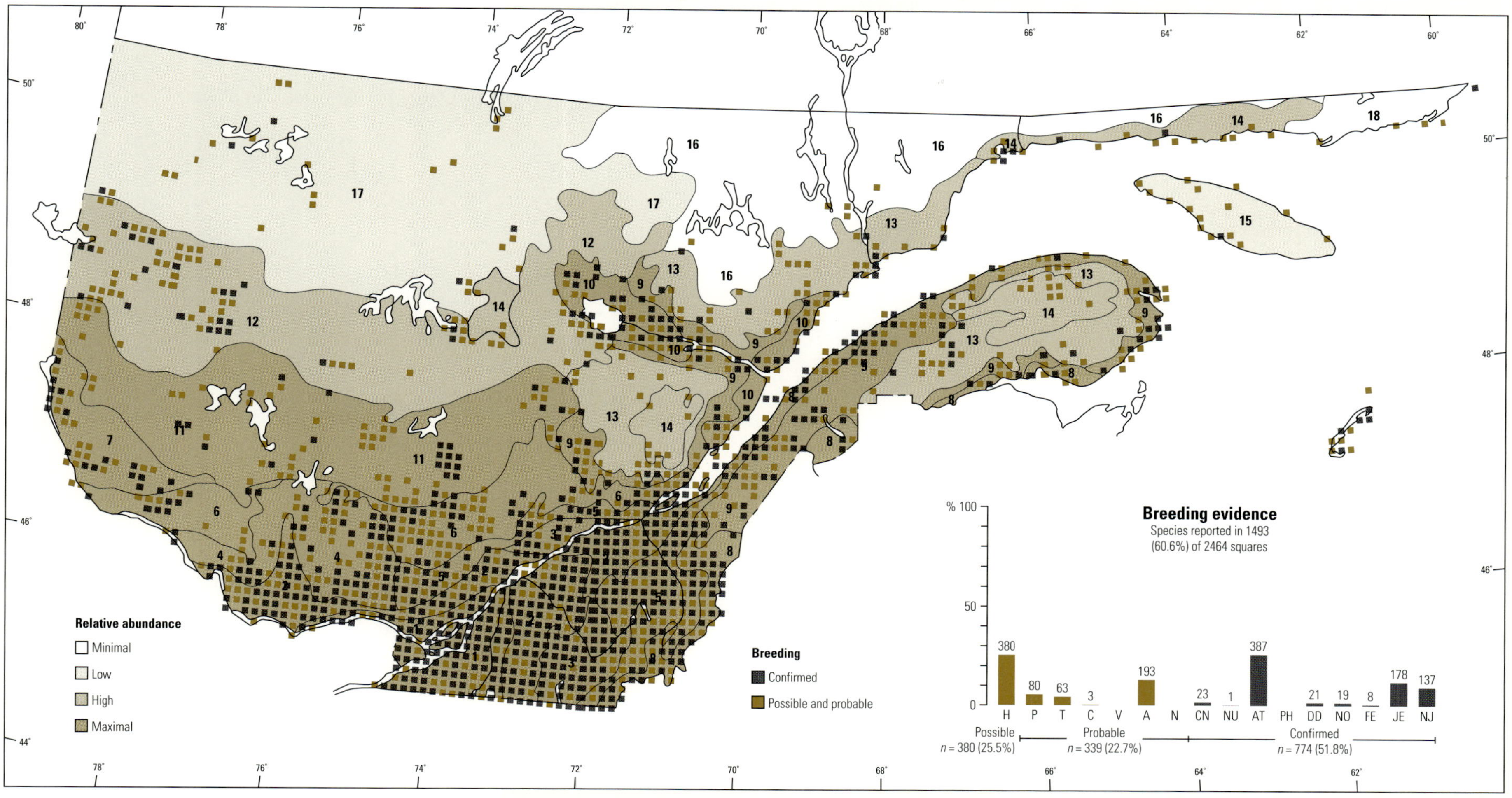

World breeding range

Species profile

Clutch size	usually 3–4 eggs (1–6)[2,3,4,5,6,12]
Laying interval	?
Start of incubation	probably with last egg[11]
Incubation period	12–14 days[3,4,5,6,7,12]
Maturity at hatching	altricial
Care of young	by female and male[1]
Nestling period	9–11 days[1,5,6]
Age at first flight	≈ 17 days[11]
Dependence of young	until 24–34 days old[11,12]
Broods per year	2 (2–4)[3,6,12]
Breeding age	1 year[8,12,13]
Mating system	monogamy, sometimes polygyny[3,8,13]
Duration of pair bond	sometimes more than 1 season[3]
Total length	15.3–17.7 cm[7]
Wing span	20.9–23.5 cm[6]
Weight	M 21.0 g (18.2–29.8) / F 20.5 g (11.9–26.1)[10]
Longevity record	11 years 4 months[9]

References:
1. Smith, J. N. 1978
2. Tompa 1964
3. Ehrlich et al. 1988
4. Nice 1937
5. Harrison 1978
6. Terres 1980
7. Godfrey 1986
8. Arcese 1989a
9. Klimkiewicz and Futcher 1987
10. Dunning 1984
11. Nolan 1968
12. Arcese et al. 1992
13. Arcese 1987

Chronology *n* = 481

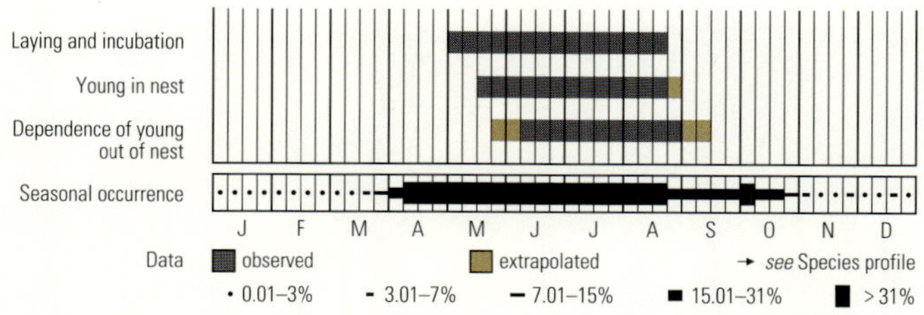

Frequency of occurrence by ecological region

Ecological regions** (forest regions)	Adequately covered squares	Squares in which reported	% of squares in which reported	Relative abundance (%)*
1. Sugar maple-hickory and sugar maple-basswood	97	97	100.0 ± 0.5	100.0
2. Sugar maple-basswood	153	153	100.0 ± 0.3	100.0
3. Sugar maple-basswood and sugar maple-yellow birch	101	101	100.0 ± 0.5	100.0
4. Sugar maple-yellow birch-basswood	56	56	100.0 ± 0.9	100.0
5. Sugar maple-yellow birch-beech and sugar maple-yellow birch-basswood	113	113	100.0 ± 0.4	100.0
6. Sugar maple-yellow birch-beech	71	69	97.2 ± 4.0	97.2
7. Sugar maple-yellow birch-balsam fir	19	19	100.0 ± 2.6	100.0
8. Sugar maple-yellow birch and balsam fir-yellow birch	47	47	100.0 ± 1.1	100.0
9. Balsam fir-yellow birch	87	84	96.6 ± 3.9	96.6
10. Balsam fir-white birch or balsam fir-red maple	46	44	95.7 ± 5.7	95.7
11. Yellow birch-balsam fir	52	50	96.2 ± 6.0	96.2
12. Balsam fir-white birch (west)	59	51	86.4 ± 9.2	86.4
13. Balsam fir-white birch (east)	76	53	69.7 ± 10.1	69.7
14. Balsam fir-black spruce	28	20	71.4 ± 17.6	71.4
15. White spruce-balsam fir and balsam fir-white spruce	22	9	40.9 ± 21.2	40.9
16. Black spruce-balsam fir-moss	18	5	27.8 ± 23.7	27.8
17. Black spruce-moss	32	16	50.0 ± 18.9	50.0
18.			*not analysed*	

* minimal: 0–29.9%; low: 30–59.9%; high: 60–89.9%; maximal: 90–100%

** Deciduous forest: 1–7; mixed forest: 8–11; coniferous forest: 12–17

In fields and second-growth habitat, this sparrow sings and patrols its territory from perches on tall herbaceous plants, shrubs, and trees. Lake Saint-Pierre region. *May 1991, Réginald Ouellet.*

The nest, built by the female of grasses, twigs, plant stalks, leaves, or bark strips, takes 4–10 days to complete (Ehrlich et al. 1988). It is usually located in herbaceous vegetation on the ground, although it may be as high as 4 m in a shrub, small conifer, or tangle of vines (Nice 1943; Tompa 1964; Nolan 1968; Ehrlich et al. 1988). The nest is usually extremely well hidden (Tompa 1964). Incubation, by the female alone (Nice 1937), takes an average of 13 days (Arcese et al. 1992).

Following hatching, both parents share more or less equally the task of feeding the nestlings, which fledge from the nest at about 9–11 days and disperse to areas of dense shrub cover up to 20 m from the nest (Smith, J. N. 1978). Both parents continue to feed the young, each parent caring exclusively for certain fledglings (Smith, J. N. 1978). When there is an even number of chicks, the brood is generally divided equally between the parents; when there is an odd number, however, the female usually takes care of the extra fledgling.

Although the young can feed on their own at 17 days, they still beg from their parents until they become independent, usually at 24–34 days (Arcese and Smith 1988; Arcese et al. 1992). They then form flocks with other juvenile birds (Arcese et al. 1992).

In British Columbia, the species may raise as many as 4 broods a year, but 2 seems to be the norm (Arcese et al. 1992). In Québec, with its shorter summer, raising more than 2 broods would be highly exceptional.

Arcese et al. (1992) found that when breeding density increased, the rate of nest failure also rose and clutch size decreased, with the number of young produced per female per year falling from 5 to 1. Food supply appears to play a key role in population regulation, though the main factor may be predation on eggs and nestlings, in part generated by cowbird parasitism.

Song allows individuals to communicate information on such things as their sex, aggressiveness, and sexual availability (Mulligan 1966). The Song Sparrow is usually detected by its complex and distinctive song rather than by its behaviour, which is generally unobtrusive; it is often noticed when it is singing from a perch or flying from one bush to another.

Males each have their own repertoire of songs, though similarities have been observed in the songs of neighbours (Harris and Lemon 1972), and individuals of the same locality are said to have a local "dialect" (Borror 1965; Harris and Lemon 1972). Consequently, the songs of Montréal birds will be somewhat similar, but differ widely from the songs of Mount Saint-Hilaire birds (Harris and Lemon 1972). Individuals seem to react more quickly to songs in their own dialect and are apparently more likely to mate with birds of their own group (Harris and Lemon 1972).

Males sometimes develop variations on the songs in their repertoire (Borror 1965; Podos et al. 1992). Females in captivity show a preference for males with very complex songs (Searcy and Marler 1981).

The range of Song Sparrow repertoires varies among individuals and among regions: 7–11 songs in the Montréal area (Harris and Lemon 1972), 5–11 in southern Ontario (Kramer et al. 1985), and as many as 20 in New York State (Podos et al. 1992). Males with large repertoires seem to be able to establish territories at a younger age and defend them longer; they also appear to be the most successful breeders (Hiebert et al. 1989; Hochachka et al. 1989).

In a confrontation between two males, the males will switch song types faster as tension mounts; the rapid succession of songs appears to be a manifestation of their aggressiveness (Kramer et al. 1985).

The female often builds her nest on the ground and hides it skilfully under vegetation. Québec City. *July 1955, Gaston Le Page.*

Notes on territory (continued)

evicting the territorial male or by taking advantage of his death; 2- or 3-year-old territorial males were more likely to chase away floaters than were younger or older males.

Territory size varies greatly and seems to depend on several factors: food availability (Arcese 1989a), habitat type (Yeaton and Cody 1974; DeGraaf 1989), the extent of the male's repertoire (Hiebert et al. 1989), and the number of adjacent territories (DeGraaf 1989). Thus, territories as small as 0.02 ha have been observed in Minnesota (Beer et al. 1956 in DeGraaf 1989) while on Mandarte Island, British Columbia, territories varied from 0.02 to 0.07 ha (Tompa 1964; Hiebert et al. 1989). Several studies in the United States report territories ranging from 0.2 to 0.6 ha (Tompa 1964; Morse 1977; DeGraaf 1989). Breeding pair density also varies according to habitat type. A density of 1.62 breeding pairs/10 ha was recorded in mixed habitat in Gatineau Park composed of forest and open areas (Dauphin 1985). In the Montréal region, a cemetery with scattered trees had 1.1 males/10 ha (Erskine 1972a). Erskine (1971a, 1972a, 1976a, 1984) reported various densities in a variety of habitats in Ontario: 1.0–6.2 males/10 ha in fields and pastures, 1.5–10.5 males/ 10 ha in fields with isolated trees or bordered by a row of trees, 2.0–7.5 males/10 ha in marshes, swamps, and flooded habitats, 8.0 males/10 ha in a habitat containing shrubs and small trees, 1.9–11.2 males/10 ha in deciduous, coniferous, and mixed woodlots with or without openings, 0.2–7.7 males/10 ha in urban areas (residential neighbourhoods, parks, cemeteries), and 19.3 males/10 ha in a transmission line right-of-way. A meadowsweet and sedge marsh near Hope, British Columbia, had 42.2 males/10 ha (Erskine 1972a).

The Song Sparrow may also build its nest in the low branches of a conifer. The female incubates the eggs, but the male helps in feeding the young. Food is found on the ground, under the cover of vegetation. Sainte-Foy. *July 1955, Gaston Le Page.*

Distribution

The Song Sparrow breeds in southern Canada from British Columbia to Nova Scotia and southwestern Newfoundland; its breeding range extends south to Mexico and the northern Gulf States (Godfrey 1986).

The species is widespread throughout southern Québec. Frequency of occurrence is highest in the heavily disturbed south, with its abundance of abandoned fields and edge habitat from forest fragmentation. The Song Sparrow is less abundant in the northern parts of southern Québec, where the landscape is dominated by coniferous forest, although it can be seen at these latitudes in clearings, along logging roads, and in other places. In several such regions, farming, logging, and mining have opened up the forest, creating suitable habitat. The species was atlassed as far as the northern boundary of the Atlas area (around 50°30'), and Bordage and Aubry (1982, 1983) have reported its presence in the Eastmain River region. According to Godfrey (1986), the species' breeding range in Québec extends to the extreme limit of the Lower North Shore in the east and to the 53rd parallel at James Bay.

History and trends

Although there is no actual proof, it would seem that the Québec Song Sparrow population was small before European settlement, when the province was almost completely forested. By the turn of the century, however, Dionne (1906) and Macoun and Macoun (1909) both described the species as abundant in southern Québec. The population probably increased as the human population grew and further modified the landscape, thus creating additional suitable habitat.

More recently, BBS data indicate an overall decline in Song Sparrow populations in North America east of the Mississippi River (Robbins et al. 1986), yet an analysis of Québec data reveals only a very slight decrease (Cyr and Larivée 1993a), or even stable numbers (Falardeau and DesGranges 1991). This is understandable, given the fact that the species' habitat has remained stable, at least in the Montréal region, where agriculture is holding its own, and where gains in improved farmland more or less offset losses (M. Carignan, MAPAQ, unpubl. data). In the rest of rural Québec, however, abandonment of agricultural land has accelerated over the last 15 years, though this trend has not been reflected in the Song Sparrow population up to now (Falardeau and DesGranges 1991).

Some trends, like deforestation and forest fragmentation, benefit the species and may allow it to extend its breeding range farther north while others, such as farming practices that reduce or eliminate hedgerows between fields or shrubbery along streams, may be detrimental.

While habitat modification seems to explain certain population trends, at least at first glance, a study on Mandarte Island in British Columbia (Arcese et al. 1992) has underlined the importance of other factors governing abundance and population stability. This study showed that severe cold waves, in conjunction with high population density, occasionally result in the demise of over 90% of the population of both young and adult birds, creating population instability. The main factor in population growth, however, appears to be the rate of juvenile recruitment into the population, which decreases as the number of adults increases. Furthermore, Arcese and Smith (1988) also observed that the availability of food during the breeding season affects breeding success and that food scarcity can limit population size.

Population declines observed in various areas of eastern North America should inspire caution in habitat management, particularly since little is known about the ecology of the subspecies *Melospiza melodia melodia* found in Québec.

Robert Bonneau

Lincoln's Sparrow

Bruant de Lincoln

Melospiza lincolnii

Christiane Girard

LINCOLN'S SPARROW WAS NAMED IN HONOUR OF THOMAS LINCOLN, THE young man who accompanied Audubon on his 1833 Labrador expedition and who bagged the first specimen at Natashquan (Todd 1963). The little-known Lincoln's Sparrow is a master of the art of concealment and, at the slightest noise, raises its crown feathers and chirps nervously before scurrying off into the shrubbery with the agility and speed of a mouse (Speirs and Speirs 1968). This migrant breeder has a distinctive and melodious song. It winters from the southern United States to Honduras and El Salvador (AOU 1983).

Habitat and behaviour

In migration, this sparrow frequents thickets, fallow land, hedges, the understorey of open woods, clearings, forest margins, and roadsides (AOU 1983; Godfrey 1986), where it forages for insects and seeds (DeGraaf and Rudis 1987). It is often overlooked in mixed flocks of sparrows, all the more so because of its furtive ways (Speirs and Speirs 1968; Godfrey 1986).

Lincoln's Sparrow begins arriving on its breeding grounds in late May. During the breeding season, it prefers rather remote habitat: peatlands, wet meadows, thickets of alder, willow, or stunted conifers along rivers, streams, and lakes, and clearings created by fire or logging (Godfrey 1986; DeGraaf and Rudis 1987; Peck and James 1987). In the James Bay and Hudson Bay regions, this sparrow is found in, or on the margins of, most open habitat, particularly damp areas (Cons. G. G. – GREBE 1991a, 1992o). It seems to require patches of brushy vegetation 1.2–2.4 m tall interspersed with openings colonized by grasses and sedges (Speirs and Speirs 1968).

The male, whose song somewhat resembles that of the House Wren or Purple Finch, is most conspicuous during the nesting period, when he defends a territory of roughly 0.4 ha. Song is the principal territorial manifestation in the species, and males rarely chase each other or fight (Speirs and Speirs 1968).

As the breeding season progresses, the adults become more agitated in their behaviour and thus easier to detect (Nichols 1985d). Although the male and female have similar plumage, they can be distinguished at this time by their behaviour. The female usually takes the initiative in courtship, inviting copulation by crouching, fluttering her wings, and uttering *dzee-dzee-dzee* calls that are very similar to those of the Song Sparrow or Swamp Sparrow (Speirs and Speirs 1968). Between or after the pair's numerous copulations, the female works on the nest while the male displays his singing ability, either from a nearby perch or sometimes in flight. The cup-shaped nest is usually built in a shallow depression in the ground, in a tussock of grass or sedge or on a carpet of moss or lichen (Terres 1980; Godfrey 1986; DeGraaf and Rudis 1987; Peck and James 1987; Ehrlich et al. 1988). It is sometimes also found in a shrub 0.2–0.3 m above the ground (Peck and James 1987). The nest is made of coarse grasses, mosses, and leaves, and is lined with fine grasses and rootlets (Godfrey 1986; Peck and James 1987). It is very difficult to find, since it is hidden by the surrounding vegetation (Speirs and Speirs 1968). Nests are rarely parasitized by the Brown-headed Cowbird (Speirs and Speirs 1968; Friedmann et al. 1977; Peck and James 1987; Ehrlich et al. 1988).

Laying typically commences at the beginning of June, and the protracted period of egg dates for nests in Ontario suggests that the Lincoln's Sparrow may produce 2 broods at our latitudes (Peck and James 1987). The 4–5 eggs are whitish or greenish-white, blotched and speckled with browns (Terres 1980; Godfrey 1986; Ehrlich et al. 1988). The female incubates for 12–14 days (Godfrey 1986; Ehrlich et al. 1988).

For nesting, Lincoln's Sparrow prefers wet habitat like peatlands. Grande Plée Bleue peatland, Pintendre, southeast of Lévis. *August 1992, Yves Aubry.*

Notes on territory

Males stake out territories almost as soon as they arrive on the breeding grounds and defend them by singing. The species may also compete for territories with the Song Sparrow, which is usually dominant; nevertheless, the territories of the two species often overlap. The size of the territory varies, but averages 0.4 ha. Individuals remain on their territories until autumn, shortly before they leave for their wintering grounds (Speirs and Speirs 1968).

In logged areas of Ontario's boreal forest, densities were 13.3 pairs/10 ha in a 13-year-old lowland cutover (alder thicket), 10.0 in a 9-year-old lowland cutover (low shrubs, conifer regeneration), 7.8 in a 5-year-old lowland cutover (low alder scrub), 6.7 in a 6-year-old lowland clearcut (grass and forbs), and 6.1 in a 3-year-old lowland cutover (forbs and bare ground) (Erskine 1984). In southern Québec, densities of 4.9 pairs/10 ha were found in a spruce-fir stand (Erskine 1980a) and 2.5 pairs/10 ha in hardwood regeneration 3 years after cutting (Erskine 1984). Densities of approximately 3 pairs/10 ha were found in a black spruce peatland in southern Québec and in a heath peatland with swamp birch in Ontario (Erskine 1972a).

Lincoln's Sparrow

Breeding distribution 1984–1989

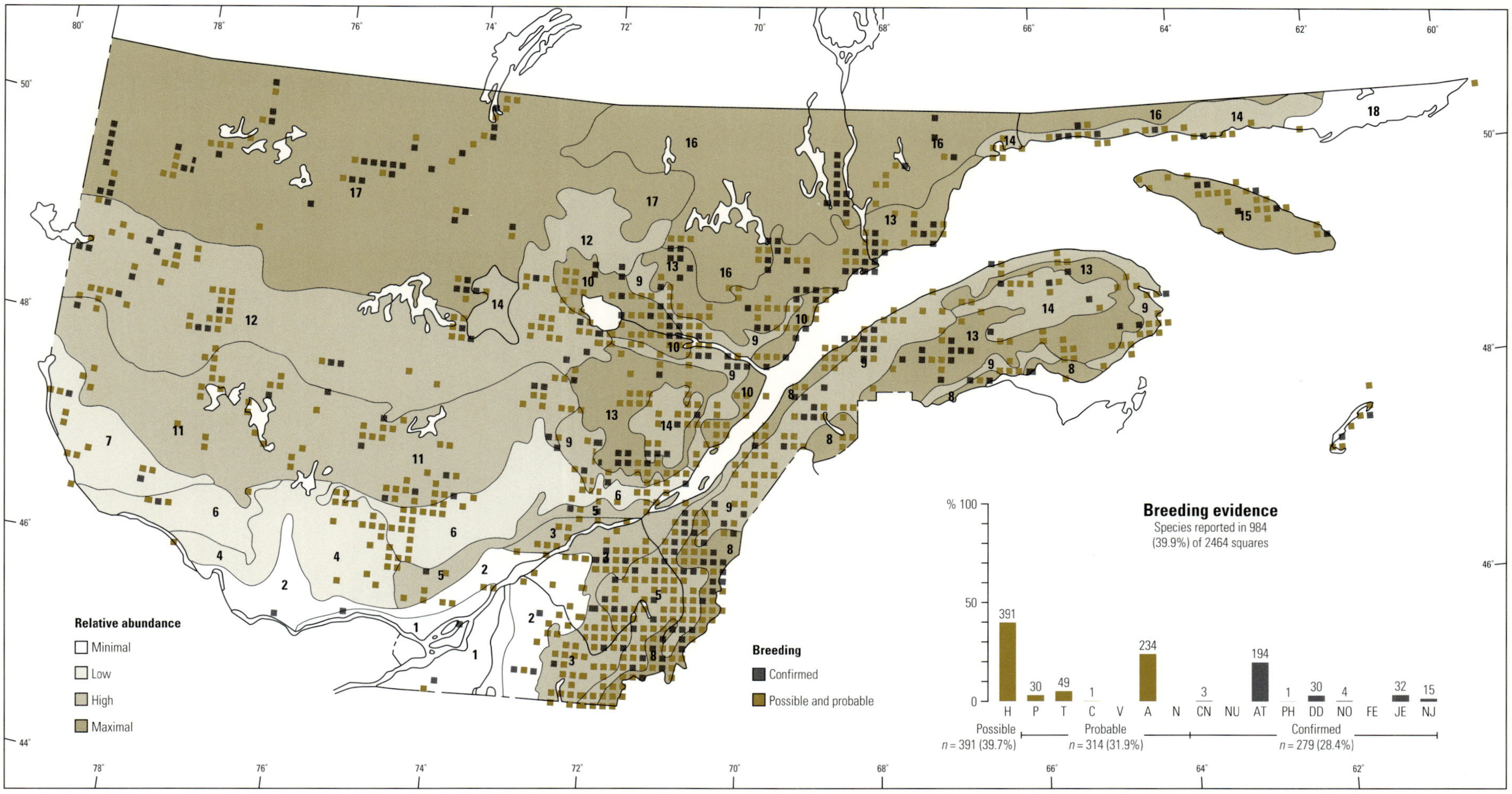

Species profile

Clutch size	usually 4–5 eggs (3–6)[1,2,3,4,5]
Laying interval	1 egg per day[6]
Start of incubation	?
Incubation period	12–14 days[1,2,3,4,5]
Maturity at hatching	altricial
Care of young	by female and male[2,6]
Nestling period	9–12 days[2,3,5,6]
Age at first flight	9–12 days or more[9]
Dependence of young	probably until at least 28 days old[6]
Broods per year	1, 2 possible[2,4]
Breeding age	1 year[3]
Mating system	monogamy[2]
Duration of pair bond	?
Total length	13.5–15.3 cm[1]
Wing span	?
Weight	17.4 g (10.4–24.0)[8]
Longevity record	7 years 7 months[7]

References:
1. Godfrey 1986
2. Ehrlich et al. 1988
3. DeGraaf and Rudis 1987
4. Peck and James 1987
5. Terres 1980
6. Speirs and Speirs 1968
7. Klimkiewicz and Futcher 1987
8. Dunning 1984
9. according to nestling period

World breeding range

Chronology $n = 36$

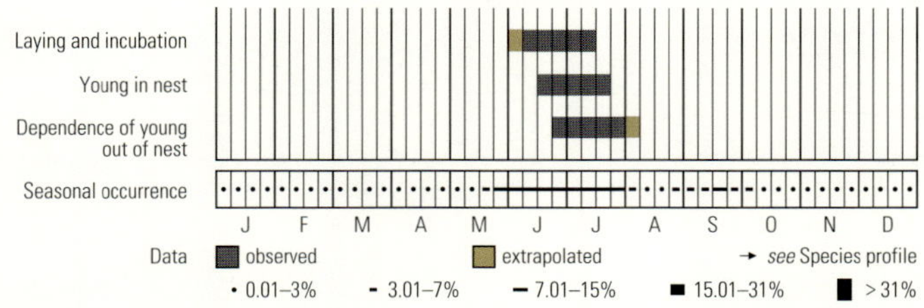

Frequency of occurrence by ecological region

Ecological regions** (forest regions)	Adequately covered squares	Squares in which reported	% of squares in which reported	Relative abundance (%)*
1. Sugar maple-hickory and sugar maple-basswood	97	3	3.1 ± 2.1	3.1
2. Sugar maple-basswood	153	27	17.6 ± 3.2	17.6
3. Sugar maple-basswood and sugar maple-yellow birch	101	68	67.3 ± 5.6	67.3
4. Sugar maple-yellow birch-basswood	56	17	30.4 ± 10.6	30.4
5. Sugar maple-yellow birch-beech and sugar maple-yellow birch-basswood	113	81	71.7 ± 4.7	71.7
6. Sugar maple-yellow birch-beech	71	33	46.5 ± 10.6	46.5
7. Sugar maple-yellow birch-balsam fir	19	9	47.4 ± 23.5	47.4
8. Sugar maple-yellow birch and balsam fir-yellow birch	47	46	97.9 ± 4.5	97.9
9. Balsam fir-yellow birch	87	71	81.6 ± 7.6	81.6
10. Balsam fir-white birch or balsam fir-red maple	46	44	95.7 ± 5.7	95.7
11. Yellow birch-balsam fir	52	40	76.9 ± 11.9	76.9
12. Balsam fir-white birch (west)	59	50	84.7 ± 9.7	84.7
13. Balsam fir-white birch (east)	76	71	93.4 ± 5.8	93.4
14. Balsam fir-black spruce	28	22	78.6 ± 16.2	78.6
15. White spruce-balsam fir and balsam fir-white spruce	22	20	90.9 ± 13.3	90.9
16. Black spruce-balsam fir-moss	18	18	100.0 ± 2.8	100.0
17. Black spruce-moss	32	32	100.0 ± 1.6	100.0
18.			*not analysed*	

* minimal: 0–29.9%; low: 30–59.9%; high: 60–89.9%; maximal: 90–100%

** Deciduous forest: 1–7; mixed forest: 8–11; coniferous forest: 12–17

In the boreal forest, Lincoln's Sparrow inhabits alder thickets in open areas, with small bodies of water nearby. Martin-Valin ZEC, north of the Saguenay River. *July 1991, Jean Gauthier.*

Notes on territory (continued)

In a tamarack peatland in southern Québec, a density of only 0.7 pair/10 ha was found (Erskine 1972a). Farther north, in the Nottaway, Broadback, and Rupert river watersheds, the greatest densities were found in tamarack peatlands, while in the Great Whale River region, the species was most abundant in shoreline scrub (Cons. G. G. – GREBE 1991a, 1992o).

The female builds her nest in a shallow depression in the ground and hides it expertly in grass or weeds. She may even nest a few metres inside the forest. Near La Grande-2 hydroelectric complex, James Bay region. *June 1976, François Morneau.*

This secretive sparrow frequents overgrown fields and clearings created by fire or logging, where it finds tussocks of grass and a little water. Trécesson, northwest of Amos, Abitibi. *August 1991, André-Guy Bernier.*

The male sings very little during incubation. The nestlings are naked and blind when they hatch and remain in the nest for about 10 days. Both parents feed the young (Speirs and Speirs 1968) on a diet exclusively of insects (Ehrlich et al. 1988). The adults subsist on a mixture of insects (over 60% in summer) and seeds (Terres 1980; DeGraaf and Rudis 1987).

When an intruder approaches the nest, the female scurries away on the ground like a mouse, fluttering her wings, jerking her tail and uttering nervous *tit* notes. She may also feign injury to distract the intruder (Speirs and Speirs 1968; Ehrlich et al. 1988).

The young remain in the vicinity of the nest until the fall migration (Speirs and Speirs 1968; Nichols 1985d), which, in Québec, takes place mainly in September and October. This is when Lincoln's Sparrows can be observed at feeders. Only 3 individuals have been reported in the province during the winter, one at Sherbrooke in 1967–68 (Boily 1978), one at Saint-Gérard-des-Laurentides in 1988–89 (Yank et al. 1989a), and one at Lévis in 1993 (Yank and Aubry 1993b).

Lincoln's Sparrow was observed in 39.9% (984/2464) of Atlas squares surveyed, and breeding was confirmed in 28.4% (279/984) of these, most often by sightings of adults carrying food.

Distribution

Lincoln's Sparrow breeds in all 10 Canadian provinces, the Yukon Territory, and the Mackenzie District of the Northwest Territories. In the United States, it is found in Alaska, the northeastern states, the Great Lakes states, and west of the Rockies (National Geographic Society 1987).

Among the three subspecies found in Canada, only one, *Melospiza lincolnii lincolnii*, occurs in Québec. This subspecies is also found in Ontario, Labrador, Newfoundland, New Brunswick, Prince Edward Island, and Nova Scotia.

Atlas data show that frequency of occurrence increases with proximity to the northern coniferous forest. Occurrence was lowest in the Central St. Lawrence Lowland and the Gatineau River Valley, undoubtedly because of the scarcity of suitable breeding habitat in these highly populated areas. Nevertheless, in July 1980 an adult carrying food was seen in the city of Longueuil (Gosselin and David 1980b).

Breeding was confirmed throughout southern Québec. Farther north, the species breeds north to Ungava Bay in the east (Todd 1963; Godfrey 1986), but in the west the northern limit of its breeding range is probably the Great Whale River region, where it is considered a probable breeder (Cons. G. G. – GREBE 1991a). Breeding was also confirmed in the area around the Nottaway, Broadback, and Rupert rivers, where Lincoln's Sparrow is common (Cons. G. G. – GREBE 1992o).

History and trends

At the turn of the century, Wintle (1896) included Lincoln's Sparrow on faith in his list as a "transient visitant" in the Montréal region, although he had never actually seen one himself. A few years later, Dionne (1906) reported the species accidental around Québec City, with a few individuals breeding on the Magdalen Islands. Taverner (1922) described the species as a rare sparrow seen only as a passing migrant in the settled sections of Canada. Accounts later in the century describe the species as being fairly common on the North Shore (Todd 1963), but uncommon around Lake Saint-Jean (Godfrey and Wilk 1948).

BBS data show a significant increase in the Lincoln's Sparrow population in eastern North America between 1966 and 1979 (Robbins et al. 1986). No significant trend was found in Québec from 1966 to 1989, but ÉPOQ summer data for 1970–89 show a marked increase in Québec (Cyr and Larivée 1993b).

Originally, suitable habitat must have been limited to burns and the edges of lakes, streams, and peatlands. More recently, Lincoln's Sparrow seems to have learned to take advantage of the shrubby areas and drainage ditches along highways, overgrown fields, logged areas, and power transmission line rights-of-way (Speirs and Speirs 1968; Nichols 1985d). Thus, human activities appear to have helped expand the species' breeding habitat.

CHRISTINE LANGEVIN

Swamp Sparrow

Bruant des marais

Melospiza georgiana

Christiane Girard

The Swamp Sparrow is an inconspicuous bird, both in habits and plumage, and but for its song it might often go unnoticed. Yet this abundant, widespread species breeds throughout southern Québec and is the commonest sparrow in wetland habitat. It is easily identified by its grey face, rusty brown back and wings, chestnut cap, outlined white throat, and pale grey breast. In wetland areas, it makes its presence known mostly by its song, a loud, melodious trill. Often it is the last bird to stop singing at the end of the day and the first to begin in the morning; sometimes it even sings at night (Wetherbee 1968). It winters mostly in the eastern United States south to the Gulf States and northern Mexico, but has occasionally been observed in winter in Québec.

Habitat and behaviour

The Swamp Sparrow breeds in wetland habitat, mainly in the tall grasses and cattails of freshwater marshes, in shoreline scrub, wet fields, riparian meadows, and the grassy, shrubby margins along bodies of water (Forbush 1929; Terres 1980; Godfrey 1986). It can also be found in herbaceous growth in the upper level of salt marshes (Greenberg and Droege 1990), as well as in bogs and fens with growths of small conifers and shrubs (Peck and James 1987), though density in this latter habitat is generally low. It avoids heavily wooded swamps (DeGraaf and Rudis 1987; Leberman 1992f). In general, the presence of surface water is the main criterion in nest site selection (Greenberg 1988). Some studies have shown a preference for tangled mixed growth of sedges, reeds, cattails, and shrubs such as willow and alder, although the species has also been reported in more homogeneous habitat, such as in pure cattail stands (Wetherbee 1968; DeGraaf and Rudis 1987).

In migration, the Swamp Sparrow frequents the same type of habitat, but it can also be seen in drier areas, such as fields, clearings, and forest edge thickets and underbrush, where it can forage for seeds (Forbush 1929; AOU 1983; Godfrey 1986).

In spring, the male is the first to arrive on the breeding grounds, where he establishes his territory by singing from a perch such as a cattail, a shrub, a tall clump of grass, or any other similar perch (Wetherbee 1968; Terres 1980). The song is a regularly repeated *weet-weet-weet-weet-weet*, very much like the trill of the Chipping Sparrow, but slower, louder, and more musical (Cayouette and Grondin 1977; Godfrey 1986). The nest, which is built by the female (Ehrlich et al. 1988), is made of coarse dry grasses and sedges, and stems of other plants; the interior is lined with fine grasses. It is placed less than 1 m (often less than 0.3 m) above the ground or water, in a shrub or suspended from rigid grass stalks or cattails. Sometimes it is built directly on the ground, concealed by low-growing vegetation (Peck and James 1987). It is well camouflaged and most often situated in a spot inaccessible not only to birdwatchers but to predators as well. At the approach of an intruder, the female slips quietly off the nest and disappears into the surrounding vegetation, which makes the nest all the more difficult to find (Rising 1987e).

Notes on territory

The male sings persistently from a conspicuous perch on a cattail, shrubby willow, or alder. On the shores of Delaware Bay and in the upstream section of the Delaware River, territories ranging between 0.04 and 0.5 ha were found in shrub—salt-meadow cord-grass associations in the salt marshes (Greenberg and Droege 1990). Pairs often nest solitarily, but sometimes fairly large aggregations are observed (Wetherbee 1968; Harrison 1975).

In a study conducted in the marshes of the Lac Saint-François National Wildlife Area, Whittam (1971) estimated average territory size at 0.15 ha. He also noted that Swamp Sparrows often left their territories to feed in fields and on undefended lawns at the edge of the study area.

Population densities have been calculated for a number of habitats in Ontario. Densities of 2.5–19.1 territorial males/10 ha were found in different types of marshes. In dense cattail stands it was found to vary between 12.7 and 18.8 males/10 ha (Erskine 1972a, 1984). The lowest density, 2.5 males/10 ha, was found in a 14-year-old

Swamp Sparrows are abundant in the broad marshes along the St. Lawrence River, especially upstream from Trois-Rivières. Saint-François Bay, Lake Saint-Pierre. *October 1993, Pierre Normand.*

Swamp Sparrow

Breeding distribution

1984–1989

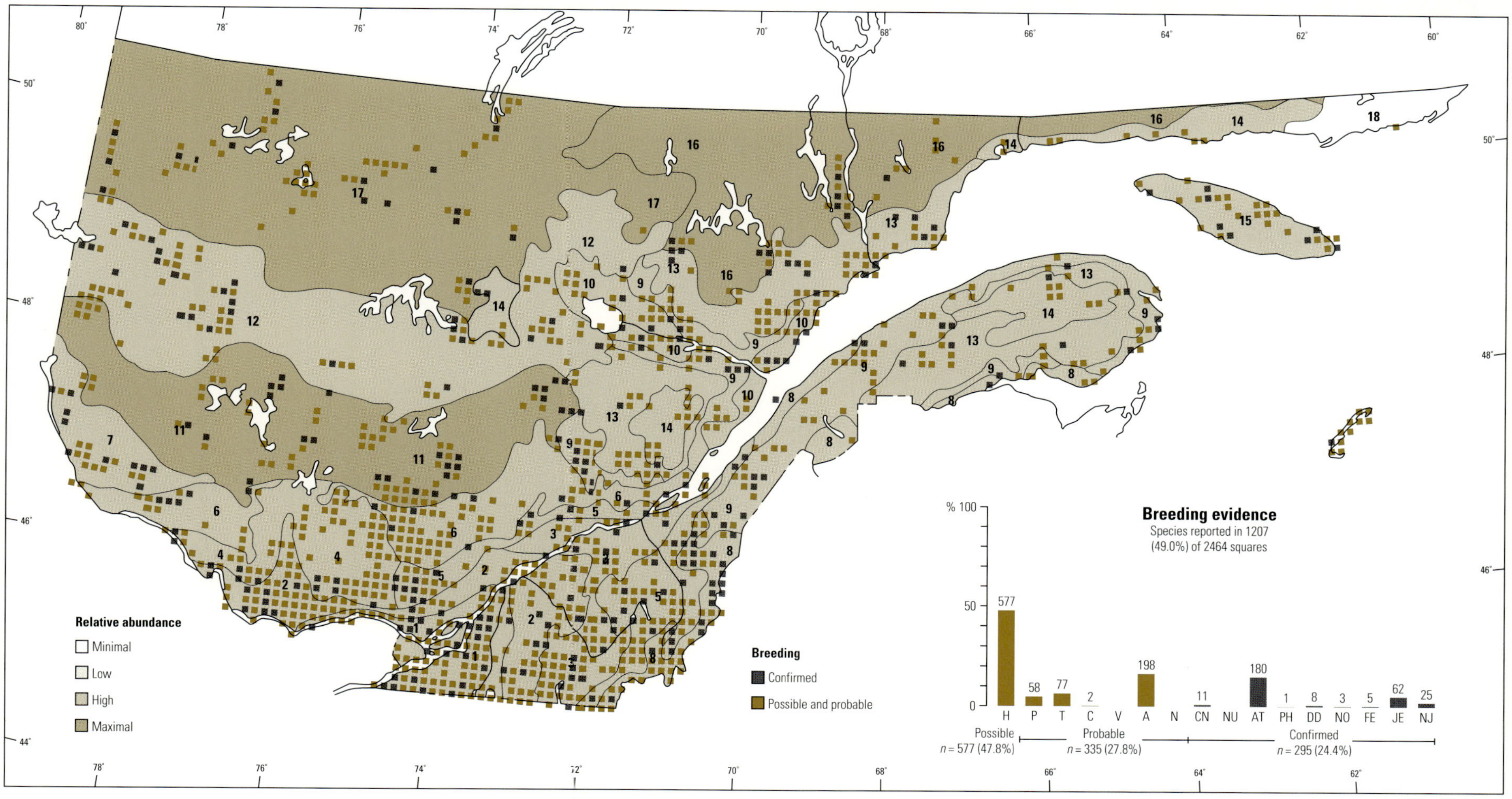

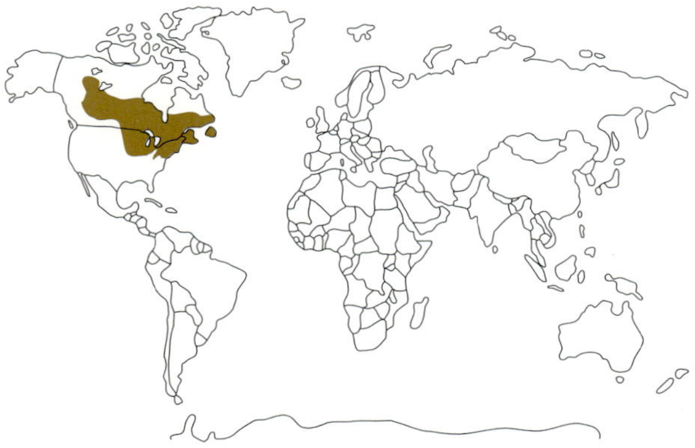

World breeding range

Species profile

Clutch size	usually 4–5 eggs (3–6)[2,3,4,5,7]
Laying interval	?
Start of incubation	?
Incubation period	12–15 days[2,3,4,5]
Maturity at hatching	altricial
Care of young	by female and male[2]
Nestling period	9–13 days[2,5,6,7]
Age at first flight	9–13 days or more[8]
Dependence of young	?
Broods per year	1–2[2,4,5]
Breeding age	1 year[5]
Mating system	monogamy[2]
Duration of pair bond	?
Total length	12.7–14.7 cm[3]
Wing span	19.0–20.3 cm[6]
Weight	17.0 g (10.9–22.2)[9]
Longevity record	6 years 4 months[1]

References:
1. Kennard 1975
2. Ehrlich et al. 1988
3. Godfrey 1986
4. Harrison 1975
5. DeGraaf and Rudis 1987
6. Terres 1980
7. Wetherbee 1968
8. according to nestling period
9. Dunning 1984

Chronology *n* = 58

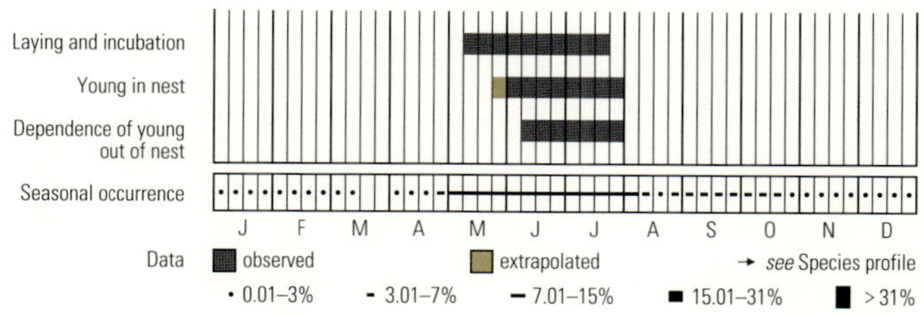

Frequency of occurrence by ecological region

Ecological regions** (forest regions)	Adequately covered squares	Squares in which reported	% of squares in which reported	Relative abundance (%)*
1. Sugar maple-hickory and sugar maple-basswood	97	82	84.5 ± 3.7	84.5
2. Sugar maple-basswood	153	115	75.2 ± 3.6	75.2
3. Sugar maple-basswood and sugar maple-yellow birch	101	68	67.3 ± 5.6	67.3
4. Sugar maple-yellow birch-basswood	56	48	85.7 ± 8.3	85.7
5. Sugar maple-yellow birch-beech and sugar maple-yellow birch-basswood	113	79	69.9 ± 4.7	69.9
6. Sugar maple-yellow birch-beech	71	59	83.1 ± 8.2	83.1
7. Sugar maple-yellow birch-balsam fir	19	17	89.5 ± 15.5	89.5
8. Sugar maple-yellow birch and balsam fir-yellow birch	47	41	87.2 ± 8.9	87.2
9. Balsam fir-yellow birch	87	60	69.0 ± 9.0	69.0
10. Balsam fir-white birch or balsam fir-red maple	46	35	76.1 ± 10.8	76.1
11. Yellow birch-balsam fir	52	50	96.2 ± 6.0	96.2
12. Balsam fir-white birch (west)	59	50	84.7 ± 9.7	84.7
13. Balsam fir-white birch (east)	76	59	77.6 ± 9.2	77.6
14. Balsam fir-black spruce	28	20	71.4 ± 17.6	71.4
15. White spruce-balsam fir and balsam fir-white spruce	22	15	68.2 ± 20.2	68.2
16. Black spruce-balsam fir-moss	18	18	100.0 ± 2.8	100.0
17. Black spruce-moss	32	29	90.6 ± 11.7	90.6
18.			*not analysed*	

* minimal: 0–29.9%; low: 30–59.9%; high: 60–89.9%; maximal: 90–100%

** Deciduous forest: 1–7; mixed forest: 8–11; coniferous forest: 12–17

Notes on territory (continued)

regenerating burn in Whitby township (Erskine 1972a); the highest, 43.9 males/10 ha, was in a meadow overgrown by sedges at Fesserton (Erskine 1984).

A typical wetland species, the Swamp Sparrow favours areas of mixed vegetation, although it may also nest in pure cattail stands. Châteauguay. *June 1991, Gaétan Duquette.*

The eggs, generally 4 or 5 in number, are usually pale bluish-green, blotched and speckled with various shades of brown (Godfrey 1986). Swamp Sparrows are often parasitized by Brown-headed Cowbirds where the two species co-exist (Harrison 1975; Arbib 1988e; Ehrlich et al. 1988). Incubation, which lasts 12–15 days, is solely by the female, though she is fed by the male throughout this time. The young are reared by both parents (Ehrlich et al. 1988). The altricial chicks hatch with their eyes closed. The pinkish skin is so transparent that the viscera and blood vessels show through the abdomen wall (Wetherbee 1968). They leave the nest at 9–13 days of age. The song is learned during the first two months after hatching (Ehrlich et al. 1988; Marler and Peters 1988). The Swamp Sparrow is the most highly insectivorous of the *Melospiza* sparrows. This is reflected in the reduced size of its skull and bill as compared with those of the more granivorous sparrows (Beecher 1951 *in* Wetherbee 1968). From the time it arrives in April and throughout the breeding season, the Swamp Sparrow feeds mainly on insects (88%), which it finds mostly on the ground and by wading through the shallow marsh water. In late summer and fall, the birds become more granivorous, with seeds comprising up to 97% of their diet (Martin et al. 1951 *in* Wetherbee 1968; DeGraaf and Rudis 1987).

Swamp Sparrows usually nest near water and are just as likely to be found in emergent vegetation in a beaver pond as in a shrub in a peatland. Gowan Road, Herdman, southwest of Montréal. *June 1991, Pierre Bannon.*

The Swamp Sparrow's nest is often suspended from rigid plant stalks, less than 1 m above the ground or water, in a relatively inaccessible area. Sainte-Foy. *June 1954, Gaston Le Page.*

Given the relative inaccessibility of its habitat, the Swamp Sparrow may at times be overlooked, all the more so because it rarely ventures into the open, even to feed (Forbush 1929; Greenberg 1989). Thus, it is difficult to study its habits during the breeding period. It was reported in 49% (1207/2464) of Atlas squares, with breeding confirmed in 24.4% (295/1207) of these, mostly by sightings of adults carrying food or of recently fledged young.

Distribution

The Swamp Sparrow is a common bird that occurs from the Mackenzie District of the Northwest Territories and eastern British Columbia to southern Labrador and Newfoundland, and throughout southern Canada except on the Prairies. It also nests in the northeastern United States (AOU 1983). It was atlassed as a breeder throughout southern Québec except on the North Shore east of Godbout, where it was nonetheless reported as far as Gethsémani, at the mouth of the Olomane River. There are pre-Atlas records locally on the Middle and Lower North Shore to Blanc-Sablon (Todd 1963; ÉPOQ). There are Atlas breeding records from Anticosti Island, where it was considered by Ouellet (1969) to be an uncommon, but evenly distributed, breeder, and from the Magdalen Islands, where Fradette (1992) considers it a regular breeder. It was observed in wetlands in all three forest zones, deciduous, mixed, and coniferous, as well as in agricultural and highly urbanized areas, where it occurs in marshes and wet fields, pond and stream edges, ditches, and flooded lowlands. In fact, the Swamp Sparrow occurs in almost all wetland habitats, regardless of size, provided there is sufficient plant cover. This is probably why it is still fairly common in the St. Lawrence Plain, despite all that region's loss of wetland habitat through drainage, stream straightening, and the reclamation of marshland for agricultural, residential, and other purposes.

Farther north, Todd (1963) reported the Swamp Sparrow on the James Bay coast as far as Paul Bay, about 23 km north of Chisasibi. Recent studies have shown the Swamp Sparrow to be relatively abundant in the Nottaway, Broadback, and Rupert river watersheds, where it occurs mostly in shoreline scrub and meadows and in tamarack peatlands (Cons. G. G. – GREBE 1992o). North of this area, it has been reported in the Great Whale and Little Whale river watersheds (Cons. G. G. – GREBE 1991a), especially in shoreline scrub near the Hudson Bay coast; the species seems to be quite rare farther east. The most northerly record is from the outlet of Lake Guillaume Delisle.

History and trends

There are few records of the Swamp Sparrow from the last century (Ouellet 1974a), probably because of the bird's reclusive ways and impenetrable habitat. Then too, it may have been poorly known at that time.

Wintle (1896), the only author who mentions it, listed it as a common summer resident, plentiful in the marshes in the Montréal region at the end of the last century, and there were several confirmed breeding reports for the area during the early part of this century (Ouellet 1974a). At this same period, it was considered a common summer resident in the Gaspé (Demille 1926) and in the southwestern Ottawa Valley, although breeding data for this region were scant (Lloyd 1924). In the Hatley area of the Eastern Townships, the species was not at all plentiful except during fall migration (Mousley 1916a). Macoun and Macoun (1915) reported that the species was seen in fairly large numbers on the Magdalen Islands, and that it was abundant on Anticosti Island and nested there. More recently, the species was described by Ouellet (1974a) as being a rare summer resident in the Montréal region, where numbers had declined sharply since the early part of the century.

There is no question that human activity has had an impact on the abundance of the Swamp Sparrow since the beginning of European settlement. In certain cases, the extensive felling of forests early in the century favoured creation of suitable breeding habitat, but extensive draining of marshes for farming and industry has certainly had the opposite effect (Wetherbee 1968; Ouellet 1974a). Moreover, rising water, whether due to natural or artificial causes, may mean disaster for the Swamp Sparrow. As the nest is often placed just above the water, even a minimal rise in water level can flood nests over a fairly large area (Wetherbee 1968).

BBS data for North America for 1965–79 (Robbins et al. 1986) and for 1966–89 in southern Québec do not suggest any change in frequency of occurrence. These data, however, are based on roadside surveys that rarely sample extensive wetland areas (Erskine 1992). In Québec, protection and restoration of wetland habitat afforded by the Eastern Habitat Joint Venture (part of the North American Waterfowl Management Plan) should promote recovery of some lost habitat and perhaps bring about an increase in Swamp Sparrow abundance, as Erskine (1992) predicts may occur in the Maritimes in future.

DANIEL BANVILLE AND SERGE GAUTHIER

White-throated Sparrow

Bruant à gorge blanche

Zonotrichia albicollis

Christiane Girard

A SLIGHT RUSTLE ON THE FOREST FLOOR IN SPRING AND AUTUMN WILL draw your attention to a flock of small brown birds. A closer look will reveal White-throated Sparrows busily turning over leaves in search of seeds and insects.

On its breeding grounds, the species is easily located by its song—perhaps one of the best-known of all bird songs. Its clear whistle has been variously interpreted as *Sweet, Sweet, Sweet Canada, Canada, Canada* and *Où es-tu, Frédéric, Frédéric, Frédéric*. The striped crown, bright white throat patch, and the yellow spot between the eye and bill are distinguishing features of adults (Godfrey 1986). White-throated Sparrows are polymorphic: crown stripes can be either tan and brown or white and black. Plumage variability in this species appears to enhance individual recognition (Watt 1983).

In Québec, this summer resident is one of the most widespread of the sparrows (David 1980a), and was second only to the American Robin in the total number of Atlas squares with sightings.

The species' wintering grounds extend from the southwestern United States and northern Mexico to the eastern seaboard (National Geographic Society 1987), with the largest concentrations found in the southeastern United States (Root 1988), although a few have overwintered as far north as southern Québec (David 1980a), where they have been noted during Christmas Bird Counts.

Habitat and behaviour

The White-throated Sparrow is primarily a woodland bird, inhabiting mostly coniferous and mixed wood forest edges (Godfrey 1986) and nesting most frequently in second-growth and semi-open coniferous or mixed forests (Peck and James 1987). In Algonquin Provincial Park in eastern Ontario, it nests in stands of balsam fir and spruce, and to a lesser extent in mixed stands of balsam fir, spruce, poplar, and white birch (Lowther and Falls 1968). Nests have also been found in deciduous woodlots, muskeg, overgrown pastureland, shrubby areas, clearcut and burned-over woodlands, swamps, and blueberry barrens (Peck and James 1987; Connor 1988f).

Males arrive on the breeding grounds 1–2 weeks ahead of females, with males showing a stronger tendency to return to the same territory each year than females (Lowther and Falls 1968; Knapton et al. 1984). Males of this species are very conspicuous during courtship and laying, singing for long bouts throughout the day and aggressively chasing any intruders from their territories. In Algonquin Provincial Park, males sing primarily from 2–4 song posts located 6–12 m up and usually in the outer branches of white spruce trees (Lowther and Falls 1968).

The well-concealed, cup-shaped nest is built of coarse grass, conifer needles, wood chips, and twigs (Terres 1980). The female builds the nest, usually on or near the ground, often at the base of a shrub or small tree at the edge of a small clearing (Lowther and Falls 1968). Occasionally, a nest may be built in the upturned roots of a fallen tree, or as high as 1 m up in a small balsam fir (Terres 1980).

Typical clutch size is 4–6 eggs, incubated by the female for 11–14 days (Lowther and Falls 1968). The altricial young are brooded mostly by the female, but are fed by both parents through to independence, which probably occurs at the age of 3–4 weeks (Stokes and Stokes 1989a). The young leave the nest when 8 or 9 days old and can fly 2–5 days later (Lowther and Falls 1968; Falls and Kopachena 1994).

Notes on territory

Soon after arrival on the breeding grounds, males begin to sing and chase intruders from their territories (Lowther and Falls 1968). Until the young become independent, all activities of the pair occur within the territory, with singing and to a lesser extent foraging taking place primarily toward the centre, close to the nest (Wasserman 1982).

In an open coniferous stand, territories ranged from 0.4 to 1.3 ha (Lainevool in Lowther and Falls 1968). In Algonquin Provincial Park, Ontario, one study reported territories ranging from 0.02 to 1.1 ha, averaging 0.2 ha (Martin 1960). In the same area and also near Port Arthur, not far from Lake Nipigon, Ontario, Kendeigh (1947) estimated territories in coniferous forests at 0.2–1.0 ha, averaging 0.46 ha. In a spruce stand in the White Mountains of New Hampshire, Sabo (1980) estimated territory size at 0.6 ha and density at 6.4 pairs/10 ha. In Gatineau Park and at La Barrière (near Saint-Michel-des-Saints) in Québec, Dauphin (1985) counted 0.9 pair/10 ha in mixed habitats of forest and open areas. Densities calculated by Cyr (1974) in the Mirabel area were 1.9 males/10 ha in a mature maple stand and 6.9 males/10 ha in a mature hemlock stand.

This sparrow frequents the edges of mixed and coniferous forests and second-growth areas. Montmorency forest, Laurentide Wildlife Reserve, north of Québec City. *June 1991, Réginald Ouellet.*

White-throated Sparrow

Breeding distribution

1984–1989

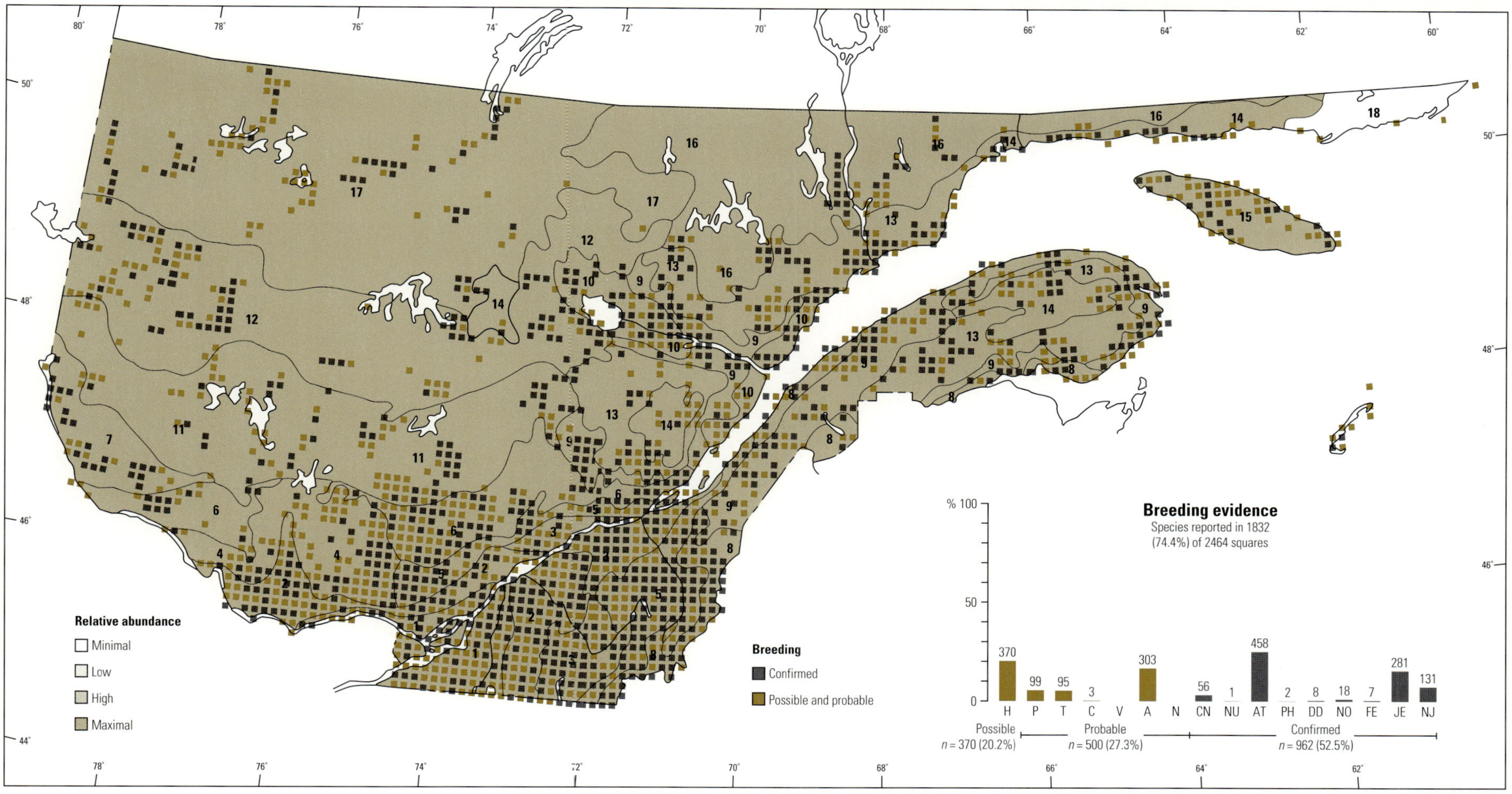

Species profile

Clutch size	usually 4–6 eggs (3–6)[1,2,3,4]
Laying interval	1 egg per day[2]
Start of incubation	with last or penultimate egg[13]
Incubation period	usually 12 days (11–14)[2,3,4,13]
Maturity at hatching	altricial
Care of young	by female and male[2,4]
Nestling period	usually 8–9 days (7–12)[1,2,4]
Age at first flight	usually 10–14 days (9–17)[2,13]
Dependence of young	until 13–20 days after leaving nest[5]
Broods per year	1–2, rarely 3[1,4,12,13]
Breeding age	1 year[9]
Mating system	monogamy, sometimes polygyny[10,11]
Duration of pair bond	probably 1 season[13]
Total length	16–18 cm[3]
Wing span	22.2–25.4 cm[7]
Weight	25.9 g (19.0–35.4)[8]
Longevity record	9 years 8 months[6]

References:
1. Knapton et al. 1984
2. Lowther and Falls 1968
3. Godfrey 1986
4. Ehrlich et al. 1988
5. Stokes and Stokes 1989a
6. Klimkiewicz and Futcher 1987
7. Terres 1980
8. Dunning 1984
9. DeGraaf and Rudis 1987
10. Knapton and Falls 1983
11. Kopachena and Falls 1993
12. Loncke and Falls 1973
13. Falls and Kopachena 1994

World breeding range

Chronology $n = 352$

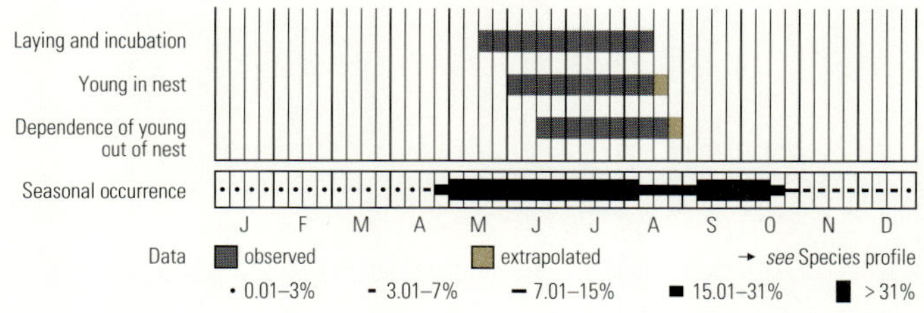

Frequency of occurrence by ecological region

Ecological regions** (forest regions)	Adequately covered squares	Squares in which reported	% of squares in which reported	Relative abundance (%)*
1. Sugar maple-hickory and sugar maple-basswood	97	95	97.9 ± 1.8	97.9
2. Sugar maple-basswood	153	152	99.3 ± 0.9	99.3
3. Sugar maple-basswood and sugar maple-yellow birch	101	101	100.0 ± 0.5	100.0
4. Sugar maple-yellow birch-basswood	56	55	98.2 ± 3.7	98.2
5. Sugar maple-yellow birch-beech and sugar maple-yellow birch-basswood	113	112	99.1 ± 1.3	99.1
6. Sugar maple-yellow birch-beech	71	71	100.0 ± 0.7	100.0
7. Sugar maple-yellow birch-balsam fir	19	19	100.0 ± 2.6	100.0
8. Sugar maple-yellow birch and balsam fir-yellow birch	47	47	100.0 ± 1.1	100.0
9. Balsam fir-yellow birch	87	87	100.0 ± 0.6	100.0
10. Balsam fir-white birch or balsam fir-red maple	46	46	100.0 ± 1.1	100.0
11. Yellow birch-balsam fir	52	52	100.0 ± 1.0	100.0
12. Balsam fir-white birch (west)	59	59	100.0 ± 0.8	100.0
13. Balsam fir-white birch (east)	76	76	100.0 ± 0.7	100.0
14. Balsam fir-black spruce	28	28	100.0 ± 1.8	100.0
15. White spruce-balsam fir and balsam fir-white spruce	22	22	100.0 ± 2.3	100.0
16. Black spruce-balsam fir-moss	18	18	100.0 ± 2.8	100.0
17. Black spruce-moss	32	32	100.0 ± 1.6	100.0
18.			*not analysed*	

* minimal: 0–29.9%; low: 30–59.9%; high: 60–89.9%; maximal: 90–100%

** Deciduous forest: 1–7; mixed forest: 8–11; coniferous forest: 12–17

In forest areas, this species also nests on the edges of clearings and wetlands. Saint-Charles, east of Lévis. *July 1991, Guy Lemieux.*

Passerine parents are usually found to feed their young more during the nestling period than during the fledgling period. In White-throated Sparrows, however, Kopachena and Falls (1993) observed that at fledging the number of trips per hour by females increased by 65%, and in males the feeding rate more than doubled. As the female brought on average more food items per trip, her parental contribution to fledglings was higher than that of the male. During both periods, the size and type of prey remained the same, but parents brought fewer items per trip to fledglings than to nestlings. The increased amount of food received by fledglings was thus less than is suggested by the increased number of trips per hour made by the parents. The young are fed animal matter only (Lowther and Falls 1968).

The White-throated Sparrow's cup-shaped nest is usually built on the ground, in most cases at the foot of a bush or conifer. Sainte-Foy. *June 1966, Gaston Le Page.*

The White-throated Sparrow forages both on the ground and in trees. It eats both animal and vegetable matter, but in summer grubs and adult insects are the major component of its diet.

Studies playing recordings of White-throated Sparrow distress calls during the breeding season showed that conspecifics gathered in the vicinity of the recordings, apparently trying to help the "victim" escape by performing distraction displays and diving upon known predators in the area, such as red squirrels and Blue Jays. Responses were strongest during the late nestling and fledgling stage (Stefanski and Falls 1972a). Members of other species also responded, but to a lesser extent (Stefanski and Falls 1972b).

Comparisons of behaviour between brightly coloured white-striped White-throated Sparrows and the duller tan-striped individuals have shown many differences between the two morphs. White-striped males sing more and are more aggressive toward other singing individuals than tan-striped males. White-striped females sing on occasion whereas tan-striped females do not (Lowther and Falls 1968). White-striped males usually occupy territories in open forest, while tan-striped males defend territories in a broader range of habitats, from open to dense (Knapton and Falls 1982). Black-and-white striped birds usually mate with tan-striped ones (Lowther 1961). White-striped males contribute less parental care than their tan-striped counterparts, and polygamy has been observed among the former (Knapton and Falls 1983).

Double-brooding has not been documented in Québec, but it has been reported in Ontario (Peck and James 1987) and New York (Bull 1985), and the wide range of egg dates for Québec suggests that double-brooding may occur here as well.

Notes on territory (continued)

In La Mauricie National Park, DesGranges (1980b) reported 2.2 individuals/10 ha in a red spruce stand, 6.5 individuals/10 ha in a white birch stand, 6.5 individuals/10 ha in a fir stand, and 7.6 individuals/10 ha in a white spruce plantation. In Saint-Michel-des-Saints, Lafontaine (1988) recorded a density of 7.4 pairs/10 ha on burnt land. Erskine (1971a, 1972a, 1976a, 1980a, 1984) reported for various places in southern Québec densities of 2.2–14.8 males/10 ha in mixed habitat (usually fir or spruce stands with intolerant hardwoods), 4.7–15.0 males/10 ha in deciduous forest, 1.0–12.4 males/10 ha in coniferous forest, 0.2 and 5.0 males/10 ha respectively in a peatland and a wooded swamp, and 4.5 males/10 ha in a field overgrown with willow and alder thickets.

In the Nottaway, Broadback, and Rupert river watersheds, White-throated Sparrows were more abundant on old or recent burns than in any other habitat surveyed

Four young White-throated Sparrows, about one week old. The nest, built on the ground, is well camouflaged and hidden among club moss. Sainte-Dorothée, Laval. *July 1940, Henry Mousley, Laval University Collection.*

(Cons. G. G. – GREBE 1992o). In the Eastmain River region, in contrast, the habitat most favoured by pairs of this species was shoreline scrub (Tremblay and Belles-Isles 1991). In the Great Whale River region, the highest densities of breeding pairs were reported in shoreline scrub, burns reverted to the shrub stage, and black spruce-moss stands (Cons. G. G. – GREBE 1991a).

The White-throated Sparrow is an uncommon cowbird host (Lowther and Falls 1968). Of 507 nests found over a period of almost 60 years in southern Québec, only 20 (4%) had been parasitized (Terrill 1961), and in Ontario, Peck and James (1987) reported 24 (6.2%) of 390 nests parasitized by the cowbird.

Nests are difficult to find because incubating females do not flush easily and because both parents are secretive when approaching the well-concealed nest (Lowther and Falls 1968).

Distribution

The White-throated Sparrow breeds only in North America, ranging in a broad band across the forested regions of Canada and the northeastern United States from southern Yukon and the Rocky Mountains in east-central British Columbia east to Newfoundland and West Virginia (Godfrey 1986; National Geographic Society 1987).

The species was widely atlassed throughout southern Québec, especially in the coniferous and mixed forests with which it is primarily associated (Godfrey 1986; Peck and James 1987); it is also quite abundant in deciduous forests and in urban and agricultural areas in the southern part of the province, since the latter contain plenty of suitable breeding habitat such as shrubbery, brush, and thickets on the edges of woods and clearings, overgrown fields, and roadsides and the shorelines of lakes and rivers (Peck and James 1987). Small mixed and deciduous woodlots, both public and private, are also common and constitute suitable habitat. On migration, these sparrows can also be seen feeding on the berries of ornamental bushes in urban areas.

In the Great Whale and Little Whale river watersheds, near the northern limit of its breeding range, the species apparently frequents shoreline scrub, burns reverted to the shrub stage, and black spruce-moss stands. It becomes less common toward the northern limit of its breeding range (Cons. G. G. – GREBE 1991a).

This sparrow ranges beyond the 55th parallel (Godfrey 1986) and has been reported as far north as the Ungava Bay region, around Tasiujaq and the Kuuraapik River (ÉPOQ).

History and trends

In southern Québec, the White-throated Sparrow probably expanded its range with the advent of European settlement, since clearing and cutting of the formerly denser, unbroken forest opened up more favourable habitat. However, Ouellet (1974a) believes that the species has declined considerably since the beginning of this century due to habitat loss from clearcutting of woodlots for urban development or farming. This bird prefers to nest in second-growth areas, so allowing cutover areas to regenerate should be to its advantage. Indeed, Freedman et al. (1981) found higher densities after a 5-year period in cutover areas that had been allowed to regenerate than in uncut areas.

BBS data for 1966–89 for Québec show that the population overall remained stable; there was an apparent decline, however, between 1980 and 1989. Declines were also recorded in the northeastern United States between 1965 and 1979 (Robbins et al. 1986).

Recent declines in certain areas may be attributable to aerial spraying of forests with insecticides used in spruce budworm control programs. A field study on the effects of fenitrothion spraying on reproduction in the White-throated Sparrow found that the main effects of spraying included disappearance of one third of adults due to death or territory abandonment, inability to defend territory, disruption of normal incubation, and clutch desertion, ultimately resulting in fewer young being produced (Busby et al. 1990). MacKinnon (1989), who studied the effects of 2,4,5-T (a herbicide used to control angiosperm regrowth) on breeding birds in a Nova Scotia forest, noted a 55% drop in the average abundance of the White-throated Sparrow in the second year after treatment.

This species is fortunate, however, in that it has a wide breeding distribution and relatively flexible nesting requirements, so this common migrant breeder should remain so and be heard throughout southern Québec for many generations to come.

MICHAEL SPENCER, RICHARD C. COTTER, AND DERIN HENDERSON

White-crowned Sparrow

Bruant à couronne blanche

Zonotrichia leucophrys

Christiane Girard

In the densely populated regions of southern Québec, the White-crowned Sparrow is seen only during spring and fall migration, when it is just passing through. Thus, our acquaintance with this elegant bird is often all too fleeting. Its white and black striped crown contrasts sharply with its grey face, neck, and breast, giving it a certain air of distinction according to Taverner (1922). Although the similar White-throated Sparrow also has a black and white crown, the White-crowned Sparrow can be distinguished by its slightly larger size, paler bill (pinkish or yellowish rather than dusky), absence of a yellow spot in front of the eye, and lack of a sharply defined white throat patch. Males and females are identical in plumage. Its song—one or more clear plaintive whistles—is diagnostic (Peterson 1980). It is not surprising that the White-crowned Sparrow was one of the first birds to attract Audubon's attention when he arrived at Natashquan in June 1833 (Todd 1963).

The White-crowned Sparrow breeds mainly in subarctic and low arctic regions, as well as in boreal, subalpine, and alpine portions of the Western Cordilleras. There are several subspecies, which are found primarily in the west; *Zonotrichia leucophrys leucophrys* occurs in Québec, Ontario, Labrador, and Newfoundland (Lewis 1975b; Godfrey 1986). The White-crowned Sparrow winters on the west coast, from southern British Columbia to central Mexico, and throughout the central and southern parts of the United States, except Florida.

Habitat and behaviour

In Québec, the White-crowned Sparrow is found in the taiga and tundra, in open lichen woodland, lichen tundra, forest-tundra, and even arctic tundra above the tree line.

On the taiga, the species is found in thickets and scrub in open areas such as burns, open spruce woodland, and clearings. Farther north, in the forest tundra, it is found in alder, dwarf willow, and scrub birch along the edges of rivers and streams and in peatlands. It also occurs in clumps of prostrate or stunted spruce on windswept coasts, as well as in mountainside shrubbery (Erskine 1977; AOU 1983; Godfrey 1986; Peck and James 1987). It finds similar conditions in some parts of the Middle and Lower North Shore.

The males arrive on the breeding grounds in May, roughly 2–3 weeks before the females (Ehrlich et al. 1988). During this time, they sing persistently. The species' melodious song, similar to the White-throated Sparrow's (Clement 1968b), consists of 1–3 clear, plaintive whistled notes followed by husky trilled whistles that drop in volume and pitch toward the end.

Because the sexes are alike, during the breeding season it is difficult to distinguish territorial disputes from courtship chases (Clement 1968b). However, the female can be seen crouching to solicit copulation by fluttering her wings while uttering a trill (Harper 1958; Clement 1968b). Once mated, the male sings far less (Blanchard 1941 in Skutch 1976).

The nest is built by the female (Harrison 1978), usually on the ground in moss, lichen, or a tussock of grass and generally well hidden at the base of a shrub, sapling, or low spruce, or it may be located near the ground in a bush or small tree (Terres 1980; Godfrey 1986).

According to Webb and King (1984), nest-site selection depends on environmental conditions like sunlight, shade, and wind. In colder regions, the female orients the nest to take maximum advantage of the sun. The nest is generally made of grasses, and often includes twigs, moss, and lichen; it is lined with fine grasses and sometimes hair or ptarmigan feathers (Clement 1968b; Harrison 1978; Peck and James 1987).

The White-crowned Sparrow is common in open spruce forest and regenerating burns in northern Québec. La Grande-3, James Bay region, *July 1992, Réginald Ouellet.*

Notes on territory

According to a study on the west coast of the United States (Lewis 1975a), the males arrive on the breeding grounds first and quickly establish their territories by singing from prominent perches and patrolling the borders. Pairs show little aggression to neighbours of the same species and may even join them in foraging. Clement (1968b) observed two adjacent territories of 0.4 and 0.8 ha.

Clement (1968b) reported a density of 2.7 pairs/10 ha in an open spruce-lichen forest in Labrador. Other densities have been found in different habitats. For example, on the western part of the Ungava Peninsula, Manning (1949 in Clement 1968b) found densities of 0.4–1.2 pairs/10 ha, with the highest densities occurring in burns. In the Great Whale River region, the White-crowned Sparrow occupies a very wide range of habitats, but is most abundant in recent burns and lichen heath (Cons. G. G. – GREBE 1991a). In northern Manitoba, a density of 1.7 pairs/10 ha was found in a white spruce-tamarack stand and in a black spruce peatland (Erskine 1971a). In British Columbia, a density of 3.7 pairs/10 ha was reported from a regenerating coastal forest 8 years after logging, part of it replanted with fir (Erskine 1980a).

White-crowned Sparrow

Breeding distribution

1984–1989

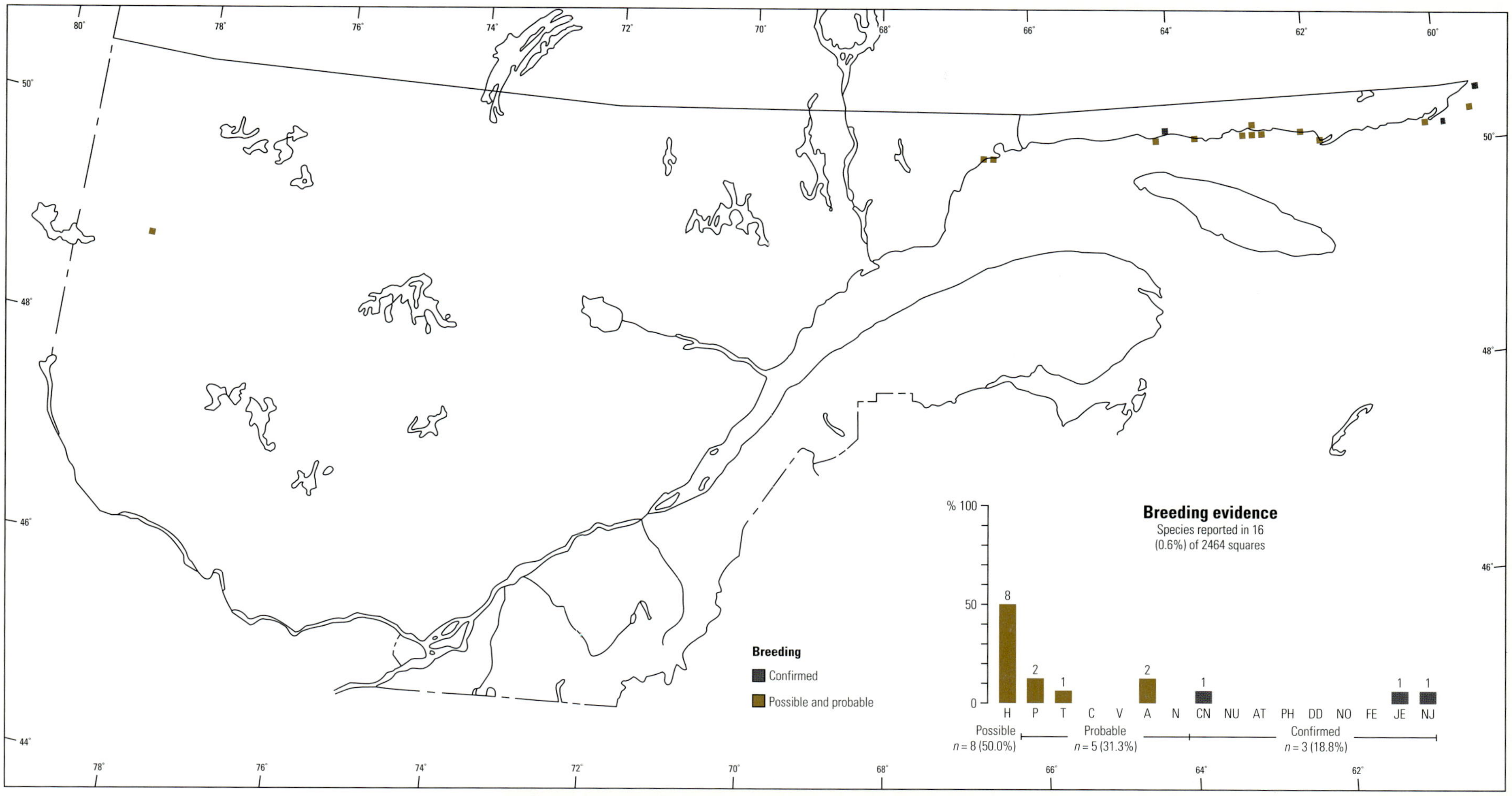

Species profile

Clutch size	usually 4 eggs (2–5)[1,3]
Laying interval	1 egg per day[1]
Start of incubation	with last egg, sometimes before[1]
Incubation period	usually 12 days (9–15)[1,2,3,4,9]
Maturity at hatching	altricial
Care of young	by female and male[1,2,3,4]
Nestling period	7–12 days[1,2]
Age at first flight	11–12 days[1]
Dependence of young	until ≈ 30 days old[1]
Broods per year	usually 1 (1–3)[4]
Breeding age	?
Mating system	monogamy, sometimes polygyny[3,4,10]
Duration of pair bond	?
Total length	16.5–19.0 cm[5]
Wing span	23.4–26.2 cm[6]
Weight	29.4 g (21.6–38.5)[7]
Longevity record	13 years 4 months[8]

References:
1. Lewis 1975a
2. King and Hubbard 1981
3. Ehrlich et al. 1988
4. Clement 1968b
5. Godfrey 1986
6. Forbush 1929
7. Dunning 1984
8. Klimkiewicz and Futcher 1989
9. Harrison 1978
10. Skutch 1976

Chronology $n = 99$

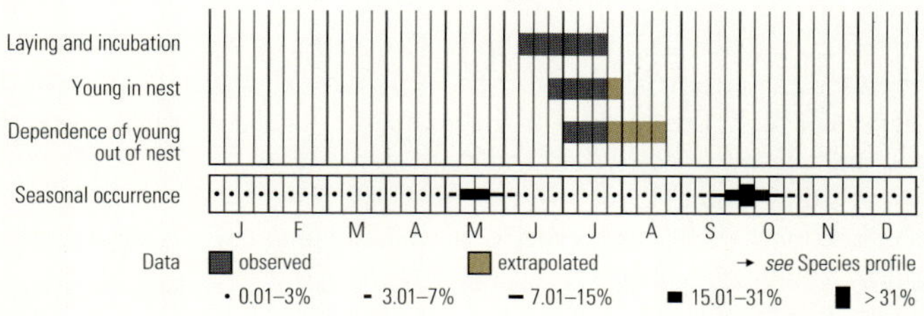

Notes on territory (continued)

The highest reported density—5.7 pairs/10 ha—was obtained in both a stunted white spruce forest and a riparian willow thicket in Manitoba (Erskine 1971a).

Stunted conifers are a typical component of the habitat used by the White-crowned Sparrow on the Lower North Shore. Sainte-Marie Islands. *July 1991, Pierre Brousseau.*

The clutch of 2–5 eggs is incubated by the female for roughly 12 days (Clement 1968b; Lewis 1975a; King and Hubbard 1981). Clement (1968b) observed that, during the first two days after hatching, the female turns over the nestlings just as she did the eggs earlier. Both parents feed the young and look after nest sanitation, although the male does a little more than the female. Fecal sacs are usually disposed of outside the nest, although the parents may sometimes eat them. Nest attentiveness and brooding by the female appears to depend on the frequency of the male's visits to the nest, since the female leaves the nest when the male arrives with food. The nestlings grow quickly and leave the nest at 7–12 days (Lewis 1975a; King and Hubbard 1981). Their parents feed them insects, seeds, and berries (Ehrlich et al. 1988) until they are around 30 days old; the young remain in the vicinity of the nest for another 20 days, and it is during this crucial period that they learn their species' songs (Baker et al. 1982; Petrinovich and Baptista 1987).

When an intruder approaches the nest, the adults repeatedly give their alarm call and may feign a broken wing if guarding fledglings. When the female is disturbed while incubating, she runs mouselike through the vegetation to elude the intruder.

The White-crowned Sparrow usually produces only one brood a year. However, Clement (1968b) reported that he had observed courtship behaviour in a pair whose young were around 7 days old on 10 July 1958 in the Schefferville region. The western races commonly raise more than one brood (Lewis 1975a). In such cases, both parents take care of the first brood until the female finishes laying the second clutch; at that time the male takes over the care of the first brood, which become independent of the female at around 17 days and of the male at around 30 days (Lewis 1975a).

The White-crowned Sparrow is primarily monogamous. However, in the west, males are occasionally polygynous; the females divide the male's territory between them and defend their own subterritory from the other females (Blanchard 1941 *in* Skutch 1976). Clement (1968b) reported a case of polygyny in Québec in which the two females laid their eggs in the same nest.

The nest is built on the ground, often in moss, and is hidden under the low branches of a tree or shrub. La Grande-3, James Bay region. *July 1992, Réginald Ouellet.*

In spring, on the breeding grounds, White-crowned Sparrows gorge on the young, green capsules of hairy-cap moss (Harper 1958). This plant appears to play an important role in the species' diet; Clement (1968b) saw one bird eat more than 120 capsules in one minute. The species also eats seeds and black flies. A little later in the season, it may also feed on willow catkins, insects, and spiders (Clement 1968b).

In the north, this sparrow also nests in the shrubby margins of lakes and rivers, as here in the Great Whale River region. Hudson Bay region. *September 1975, Pierre Lamothe.*

Distribution

The White-crowned Sparrow breeds in Alaska, Yukon, Northwest Territories, and across the northern parts of Canada's provinces as far east as Newfoundland, and in the Western Cordilleras south to southern California (AOU 1983; Godfrey 1986). In Québec, the species nests in arctic and subarctic regions from the Ungava Peninsula south to around the 52nd parallel and, in the east, as far south as the Middle North Shore. It is considered a transient elsewhere in southern Québec (David 1980a).

The White-crowned Sparrow was reported in 0.6% (16/2464) of Atlas squares surveyed, and breeding was confirmed in 18.8% (3/16) of these. All the breeding records came from the Middle and Lower North Shore; an accidental was reported in July 1985 near Lake Chicobi in the Abitibi region.

As Todd (1953 *in* Clement 1968b) explained, the White-crowned Sparrow is essentially a northern species that occurs in more southerly regions during the breeding season only when conditions similar to those in subarctic regions are found. Occurrence of this sparrow on the Middle and Lower North Shore is thus related to the presence of open forests of stunted spruce and fir with scattered shrubs (Todd 1963; Clement 1968b). The species' increased presence on the Middle and Lower North Shore from the eastern tip of the Mingan Islands eastward was observed by Lewis (Clement 1968b), who attributed it to the sudden change in the nature of the underlying bedrock and accompanying shift in the type of vegetation found. Authors such as Thibault (1983) have recognized the existence of tundra habitats in this region.

North of the Atlas area, the White-crowned Sparrow breeds throughout northern Québec as far north as Déception Bay, which is well above the tree line (Ouellet and Bider 1973). The species is transient in the Nottaway, Broadback, and Rupert river watersheds (Cons. G. G. – GREBE 1992o). It has been seen in the Groulx Mountains and near the Daniel Johnson Dam (ÉPOQ); it has also been observed a number of times in summer near the Eastmain River, La Grande Rivière, Great Whale River, on the coasts of James Bay and Hudson Bay, and as far as Kuujjuaq (Harper 1958; Todd 1963; Clement 1968b; Bordage and Aubry 1982; Cons. G. G. – GREBE 1991a).

History and trends

The status of the White-crowned Sparrow in Québec has changed little since the turn of the century, when it was reportedly an abundant summer resident in Labrador, at Kuujjuaq, and on the Ungava Peninsula south to Lake Guillaume Delisle (Macoun and Macoun 1915). Audubon observed the species at Natashquan in the 19th century and it was later observed on several occasions nesting on the Middle and Lower North Shore (Todd 1963). Dionne (1906) reported it as a transient in eastern Québec, while Wintle (1896) considered it a scarce "transient visitant" in Montréal. More recently, Ouellet (1974a) described this sparrow as a common transient in the Montréal region and commented that the species' apparent scarcity in the past may have been due to the fact that it had been overlooked or that, since then, habitat modifications had made the region more attractive for the species during migration.

BBS data are insufficient to discern trends in the Québec population. Most of BBS routes in Québec lie south of the 49th parallel (Erskine 1978b; Robbins et al. 1986). However, given the extent and inaccessibility of the species' breeding range in Québec, there has probably been no significant change in the population since the turn of the century.

Louis Breton

Dark-eyed Junco

Junco ardoisé

Junco hyemalis

Christiane Girard

THE DARK-EYED JUNCO IS ONE OF THE COMMONEST BIRDS OF MIXED AND coniferous forest. It is easy to identify with its slate grey back and white outer tail feathers, which are particularly evident in flight. This small passerine is noisy and easily agitated during the breeding season. The commotion it creates not only aids in confirming breeding, but also attracts other birds in the neighbourhood.

In Québec, the Dark-eyed Junco is a migrant breeder, with most of the population leaving in September and October for wintering grounds in southern Canada and throughout most of the United States. The southern part of its breeding range overlaps the northern part of its winter range, and it winters regularly in southern Québec (David 1980a). Juncos of different ages and sex tend to winter at different latitudes, with the young males farthest north, followed by the adult males, young females, and adult females (Nolan and Ketterson 1990). This distribution pattern can be explained at least in part by the fact that males are generally fatter and can survive longer when snow cover prevents them from feeding (Ketterson and Nolan 1976).

Habitat and behaviour

The Dark-eyed Junco breeds mainly in mixed and coniferous forest, where it favours edges, clearings, and other semi-open areas rather than the forest interior (Godfrey 1986; Peck and James 1987; Thompson and Capen 1988). It is also found in stands roughly 20 years into secondary succession after clear- or selective cutting; these provide the open habitat and dense understorey it requires (Thompson and Capen 1988). In mountainous areas, the Dark-eyed Junco can be found at elevations over 1000 m, demonstrating its ability to withstand very cold, wet conditions (Able and Noon 1976; Sabo 1980; Mulvihill 1992c).

The males arrive on the breeding grounds in April, a few days ahead of the females. They stake out their territories by singing from the tallest tree; the song is a trill reminiscent of the Chipping Sparrow's (Eaton 1968). To attract mates, the males flash their white outer tail feathers. Once the pair bond has been formed, the male usually helps the female find nesting materials, but she builds the nest alone (Eaton 1968).

The nest is usually built right on the ground, sometimes in a natural depression and sometimes in a shallow scrape when the ground is soft enough. It is often located at the edge of the forest and is usually well concealed under some form of shelter, abutting to a stump or exposed root, or in dense shrubbery or forbs (ferns, club-mosses) (Hadley 1969). It is made of a variety of materials, including bark, rootlets, twigs, moss, string, and hair (Eaton 1968).

At our latitudes, reports of Dark-eyed Juncos producing 2 broods in a year has been made in Ontario (Peck and James 1987). The clutch of 4 eggs is incubated by the female for 12 or 13 days.

The pair can often be seen hopping over the ground or scratching the dirt for insects such as ants, beetles, and leafhoppers, the nestlings' main foods (Eaton 1968; Smith and Andersen 1982). Initially, the chicks are fed by regurgitation, but are then gradually introduced to whole prey (Eaton 1968).

Rapid tarsal development allows nestlings to escape predators by running away from the nest, well before they can fly (Smith and Andersen 1982). The young leave the nest 10–12 days after hatching, but remain dependent on their parents for another 2 weeks (Wolf et al. 1988).

Notes on territory

Males generally arrive on the breeding grounds before females and stake out their territories by singing from the tallest tree (Eaton 1968). In Québec, McNeil (1964 in Cyr 1974) found the average territory defended by pairs to be 1.6 ha (1.4–1.9 ha), while in Ontario, Kendeigh (1947) reported it as 1.5 ha. In the White Mountains of New Hampshire, territories were approximately 1.1 ha (Sabo 1980).

The Dark-eyed Junco is found in mixed and coniferous forest, where it prefers forest margins and islands of conifers, as in this peatland. Plée de Saint-Charles, east of Lévis. *October 1993, Pierre Normand.*

Dark-eyed Junco

Breeding distribution
1984–1989

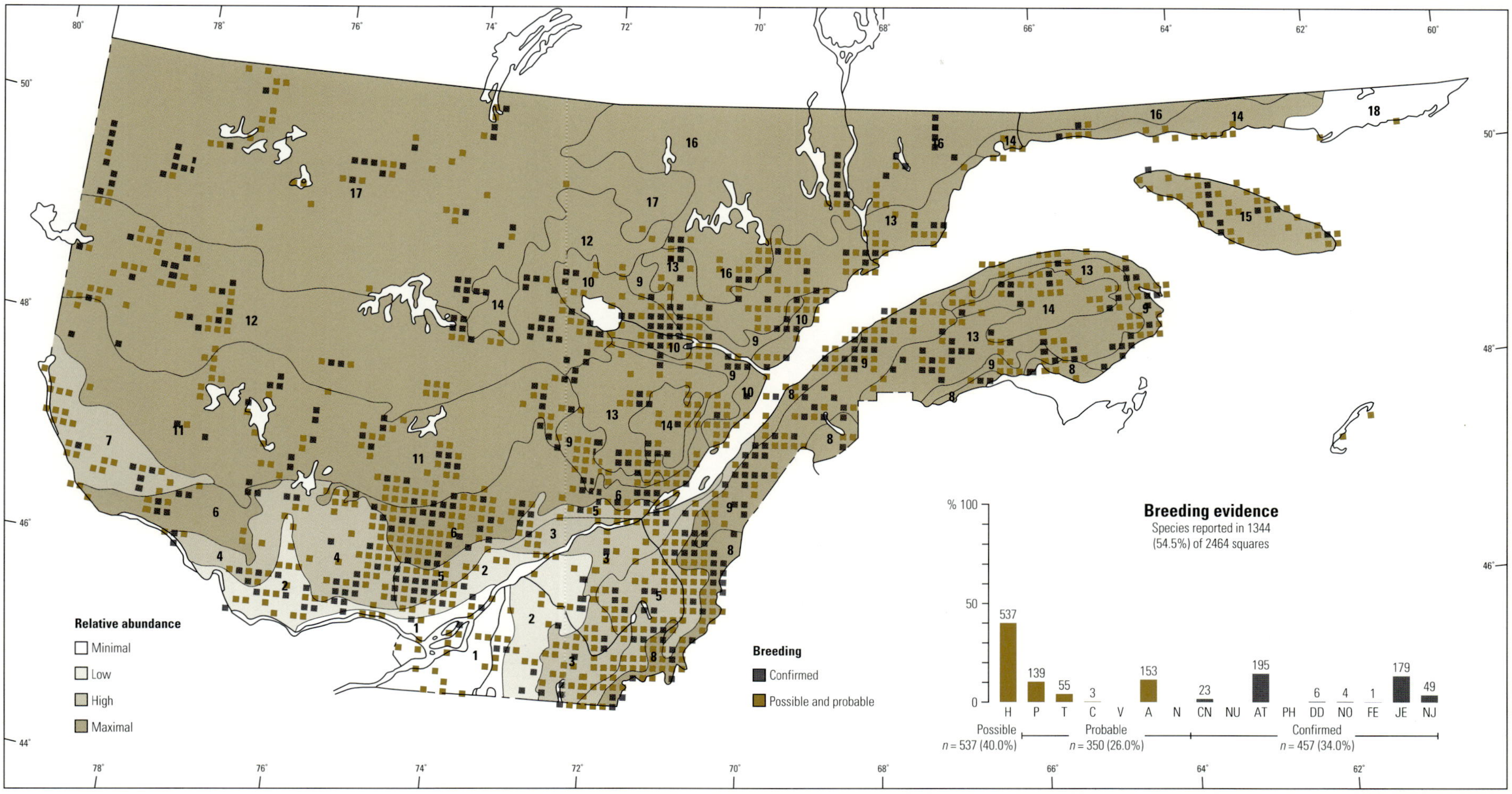

Species profile

Clutch size	usually 4 eggs (3–5)[2,3,4]
Laying interval	?
Start of incubation	?
Incubation period	usually 12–13 days (11–15)[1,2,4]
Maturity at hatching	altricial
Care of young	by female and male[1,3]
Nestling period	usually 10–12 days (9–14)[1,2,3,4]
Age at first flight	usually 10–12 days (9–14) or more[9]
Dependence of young	until ≈ 14 days after leaving nest[3]
Broods per year	usually 2 (2–4)[1,3,4]
Breeding age	1 year[8]
Mating system	monogamy[3]
Duration of pair bond	?
Total length	14.5–16.5 cm[6]
Wing span	24.1–25.4 cm[4]
Weight	M 20.4 g (14.3–26.7) / F 18.8 g (14.3–25.1)[7]
Longevity record	10 years 9 months[5]

References:
1. Eaton 1968
2. Smith and Andersen 1982
3. Wolf et al. 1988
4. Terres 1980
5. Klimkiewicz and Futcher 1987
6. Godfrey 1986
7. Dunning 1984
8. DeGraaf and Rudis 1987
9. according to nestling period

World breeding range

Chronology *n* = 162

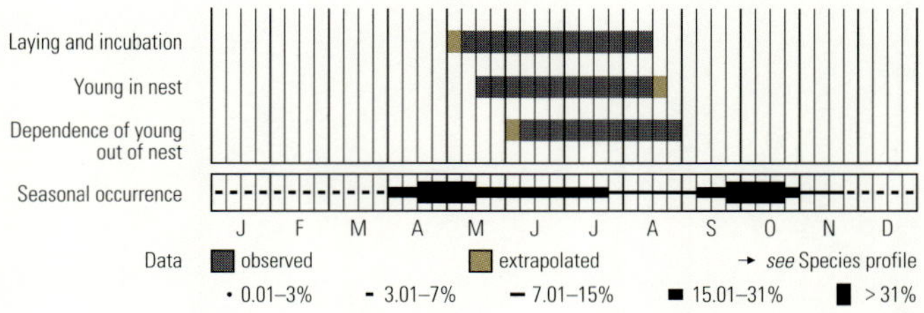

Frequency of occurrence by ecological region

Ecological regions** (forest regions)	Adequately covered squares	Squares in which reported	% of squares in which reported	Relative abundance (%)*
1. Sugar maple-hickory and sugar maple-basswood	97	25	25.8 ± 4.4	25.8
2. Sugar maple-basswood	153	58	37.9 ± 4.0	37.9
3. Sugar maple-basswood and sugar maple-yellow birch	101	65	64.4 ± 5.7	64.4
4. Sugar maple-yellow birch-basswood	56	46	82.1 ± 9.0	82.1
5. Sugar maple-yellow birch-beech and sugar maple-yellow birch-basswood	113	98	86.7 ± 3.6	86.7
6. Sugar maple-yellow birch-beech	71	67	94.4 ± 5.3	94.4
7. Sugar maple-yellow birch-balsam fir	19	17	89.5 ± 15.5	89.5
8. Sugar maple-yellow birch and balsam fir-yellow birch	47	45	95.7 ± 5.8	95.7
9. Balsam fir-yellow birch	87	86	98.9 ± 2.5	98.9
10. Balsam fir-white birch or balsam fir-red maple	46	45	97.8 ± 4.4	97.8
11. Yellow birch-balsam fir	52	51	98.1 ± 4.5	98.1
12. Balsam fir-white birch (west)	59	56	94.9 ± 6.2	94.9
13. Balsam fir-white birch (east)	76	75	98.7 ± 3.0	98.7
14. Balsam fir-black spruce	28	27	96.4 ± 8.3	96.4
15. White spruce-balsam fir and balsam fir-white spruce	22	22	100.0 ± 2.3	100.0
16. Black spruce-balsam fir-moss	18	18	100.0 ± 2.8	100.0
17. Black spruce-moss	32	32	100.0 ± 1.6	100.0
18.			*not analysed*	

* minimal: 0–29.9%; low: 30–59.9%; high: 60–89.9%; maximal: 90–100%

** Deciduous forest: 1–7; mixed forest: 8–11; coniferous forest: 12–17

In the forest, this species frequents the edges of clearings, where it spends much time foraging on the ground. Laurentide Wildlife Reserve, north of Québec City. *August 1991, Marcel Darveau.*

The nest is built on the ground, in this case in a hollow in the moss. Although the nest may be located on the edge of the forest or sometimes in an opening, it is always situated so that it is sheltered by a fallen log or other structure, or hidden in the surrounding vegetation. Montmorency forest, Laurentide Wildlife Reserve, north of Québec City. *June 1986, Paul Perreault.*

Notes on territory (continued)

In Québec, breeding pair density has been estimated at 2.5–7.4 pairs/10 ha in regenerating deciduous forest (3 years after cutting), in mixed conifer forest (mainly balsam fir and spruce), and in jack pine stands (Erskine 1971a, 1976a, 1980a, 1984). Near Val-d'Or in the Abitibi region, density was estimated at 1.5 pairs/10 ha in a black spruce stand and 0.2 pair/10 ha in a tamarack peatland (Erskine 1972a). In New Brunswick, Elderkin (1989) discovered that the Dark-eyed Junco is most abundant in stands during the first 5 years after cutting.

In coniferous habitats, for several years after logging or fire the Dark-eyed Junco can be found in second-growth areas with openings. Anticosti Island. *July 1987, Jean Beaubien and Guy Simard, Canadian Forest Service.*

Breeding was relatively easy to confirm for the Atlas; 81.8% of confirmations (374/457) were based on sightings of adults carrying food or of recently fledged young incapable of sustained flight. Only 11.6% of confirmations were based on finding a nest; because of the female's cryptic coloration, she is difficult to see on the nest. Most of the probable breeding records were reports of agitated behaviour resulting from observers getting too close to the nest.

Distribution

The Dark-eyed Junco's breeding range extends from Alaska across Canada, and in the western United States to southern California, the Great Lakes states, New England, and the Appalachians south to Georgia. The species breeds in all parts of Québec except the Ungava Peninsula (Todd 1963; Godfrey 1986).

Breeding was confirmed for the Atlas throughout southern Québec except the Magdalen Islands, though breeding evidence had been obtained on the islands prior to the Atlas project (Fradette 1992). The species is common throughout southern Québec except in the western part of the St. Lawrence Plain where, especially in Montérégie and Greater Montréal, the mainly agricultural and urban landscapes and the scarcity of conifer stands mean less suitable habitat.

Beyond the Atlas area, the Dark-eyed Junco is found as far north as the tree line (Todd 1963; Godfrey 1986). According to ÉPOQ data, the northernmost records originate from Lake Bienville, Lac à l'Eau Claire, and the Kuuraapik River near Kuujjuaq.

History and trends

At the turn of the century, according to Wintle (1896), the Dark-eyed Junco was an abundant summer resident in the Montréal area, although Terrill (1911b) stated that he had never been able to find the junco in or near Montréal during the breeding season and had to go to the Monteregian Hills or the foothills of the Laurentians or Appalachians to see it. Dionne (1906) described the species as being very common around Québec City. The Dark-eyed Junco is probably less abundant in these regions today, although reforestation of farmland, particularly in conifers, may have a positive long-term effect on populations. Farther east, on Anticosti Island, the species' status does not appear to have changed a great deal since the turn of the century, with authors both then and now describing it as regular to abundant (Verrill 1862; Brewster 1884; Schmitt 1904; Braund and McCullagh 1940; Ouellet 1969). Although it is considered a regular summer resident on the Magdalen Islands, the most recent breeding record dates from 1976 (Fradette 1992).

In Québec, BBS data suggest that the population increased between 1966 and 1989, while ÉPOQ data for 1970–89 (Cyr and Larivée 1993a) show no significant change. Since the Dark-eyed Junco is not a species of dense, mature forest, it may benefit from new practices such as the reforestation of farmland (Ellison 1985l). Newer timber harvesting methods such as selective cutting and strip cutting may also be beneficial if a mixture of age classes and a conifer component are maintained and regeneration of a dense understorey is promoted (Thompson and Capen 1988).

Sylvain Lessard and Benoît Limoges

Bobolink

Goglu

Goglu des prés
Dolichonyx oryzivorus

The Bobolink is probably one of the commonest and most easily identified birds found in Québec fields and meadows in spring and summer, enlivening the countryside with its bold plumage—black and white with a buff-gold nape—and its bubbling, tinkling song. No other bird in this family sings as energetically.

The Bobolink arrives in southern Québec around mid-May, when most passerines are already here, and leaves very early, toward the end of summer. It winters down to northern Argentina, which means it may travel more than 17 000 km annually, the longest migration trek of any North American songbird (Stokes and Stokes 1989a).

Habitat and behaviour

The Bobolink haunts fields and meadows of tall grass, clover, alfalfa, and grain (Godfrey 1986), normally showing a marked preference for hayfields and avoiding flooded habitat, dry areas distant from water (Bollinger et al. 1990), and areas with large amounts of willow, alder, or meadowsweet (Ellison 1985c).

Upon their arrival in spring, males are very conspicuous both in behaviour and song. They sing from perches on tufts of grass or at the top of a shrub and while performing aerial displays to define their territories. These they defend until the young leave the nest. The Bobolink has a two-part song, the first part consisting of several distinct notes and the second of a series of ascending warbles (Stokes and Stokes 1989a). There are two versions of the song, a longer 7-second version and a shorter version that lasts about 4 seconds (Capp 1992). Both versions seem to convey the same message, as territorial males respond to them with equal intensity (Capp 1992).

Bobolinks often renest in the same area, sometimes less than 50 m from the site occupied the previous year (Martin 1973; Bollinger and Gavin 1989; Stokes and Stokes 1989a). In Oregon and New York, it was estimated that 44–60% of adults returned to the same sites (Wittenberger 1978; Gavin and Bollinger 1988). High breeding success in previous years and high-quality nesting sites appear to be the main factors that prompt birds to return to the same area from one year to another (Gavin and Bollinger 1988; Bollinger and Gavin 1989). Bobolinks tend to situate their territories closer to woodland and forest edge habitat than do other grassland species (Wiens 1969). A certain hierarchy exists among males occupying the same field: older, dominant males establish territories in the centre of the field, while later-arriving males, usually younger, form territories on the periphery (Stokes and Stokes 1989a).

The breeding plumage of the male is unusual in that it is black below and largely white above. The yellowish-buff females are much more retiring than the males and remain hidden in the tall grasses, well camouflaged by their cryptic plumage. They arrive a few days after the males, at which time they are courted, the males pursuing them from one end of the field to the other in zigzag fashion, singing all the while. The pursuit often terminates in the vegetation, where the display continues. The males begin moulting at the end of July and gradually acquire a coloration similar to that of the females and young. Bobolinks form flocks in fields and swamps before beginning their long journey south. They migrate mostly at night and feed during the day (Stokes and Stokes 1989a).

Christiane Girard

Notes on territory

The territory is used for mating, nesting, and feeding, though males may go outside the territory to forage in neighbouring fields (Ellison 1985c; Stokes and Stokes 1989a).

When establishing their territories, males perform in aerial displays, often flying parallel to each other along a common boundary. Neighbouring Bobolinks are often seen walking or hopping side by side along common boundaries of defended areas (Stokes and Stokes 1989a). According to Stokes and Stokes (1989a), the defended area is generally about 0.5 ha in suitable habitat, but may be as much as 4.5 ha in habitat of lesser quality. Wittenberger (1980b) estimated average territory size to be 0.74 ha in fields in Oregon.

Because of the type of habitat used, distribution is usually very patchy, and average density most often ranges between 0.01 and 1 pair/10 ha, although densities as high as 10 pairs/10 ha have been recorded (Weatherhead 1987c). Densities reported by Erskine (1971a, 1972a, 1976a, 1984) for

Bobolink nests in large numbers in the St. Lawrence Plain, where farmland and meadows abound. Northeast of Berthierville, near Lake Saint-Pierre. *October 1993, Pierre Normand.*

Bobolink

Breeding distribution
1984–1989

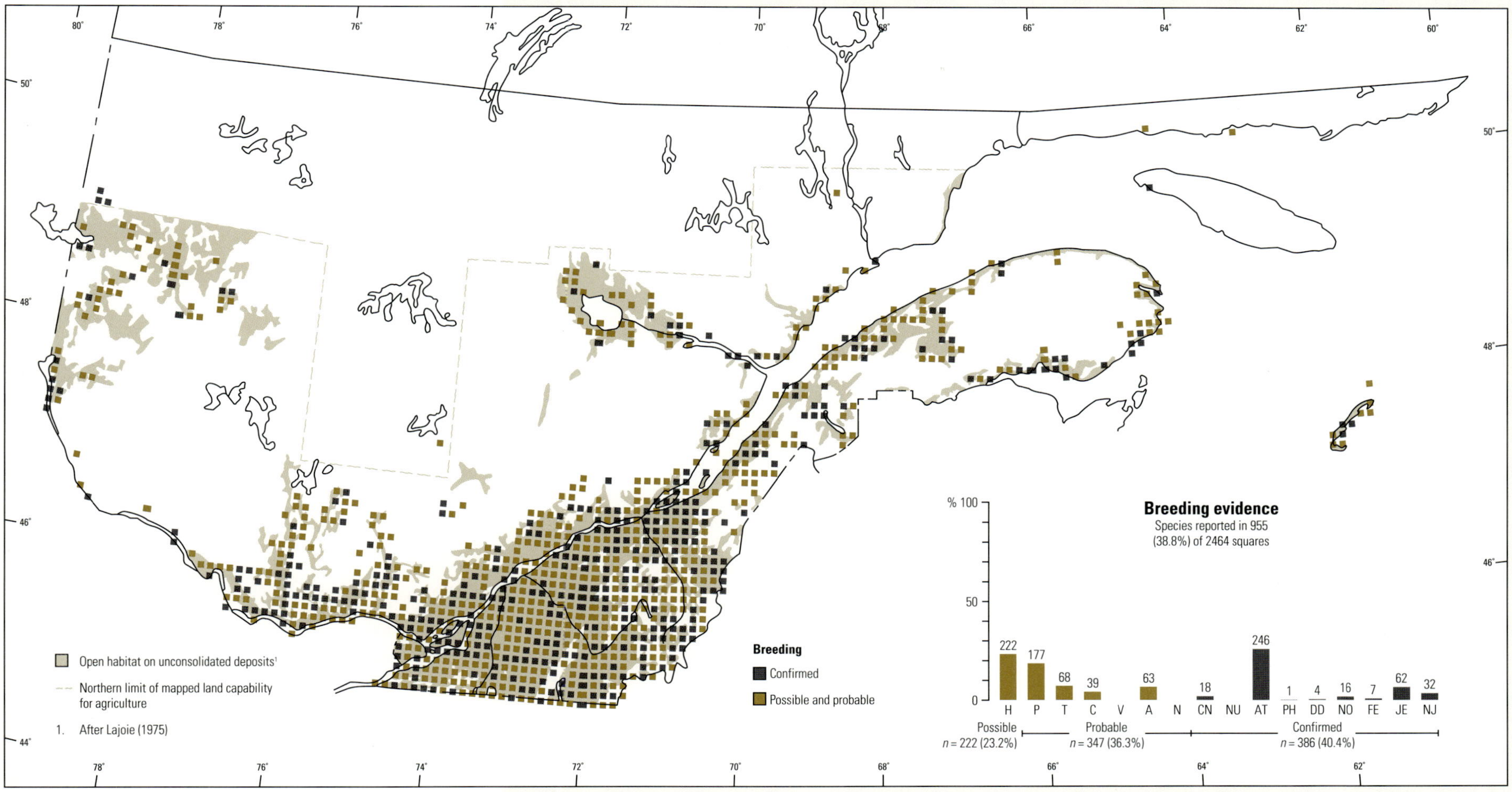

World breeding range

Species profile

Clutch size	usually 5–6 eggs (4–7)[1,2,3,5]
Laying interval	1 egg per day[3]
Start of incubation	before last egg[3]
Incubation period	10–13 days[1,2,3,5]
Maturity at hatching	altricial
Care of young	mainly by female[8,9,10,11]
Nestling period	10–14 days[3,5,13]
Age at first flight	≈ 13 days[3]
Dependence of young	until 21 days after leaving nest[3]
Broods per year	1, rarely 2[2,3,13]
Breeding age	1 year[3]
Mating system	monogamy, often polygyny[8,9,12]
Duration of pair bond	usually 1 season[8]
Total length	16.5–20.2 cm[1]
Wing span	26.0–31.8 cm[6]
Weight	M 47.0 g (28.5–56.3) / F 37.1 g (26.5–44.3)[7]
Longevity record	8 years 1 month[4]

References:
1. Godfrey 1986
2. Harrison 1975
3. Stokes and Stokes 1989a
4. Klimkiewicz and Futcher 1989
5. Ehrlich et al. 1988
6. Terres 1980
7. Dunning 1984
8. Wittenberger 1980b
9. Wittenberger 1978
10. Wittenberger 1980a
11. Wittenberger 1982
12. Wootton et al. 1986
13. Gavin 1984

Chronology $n = 55$

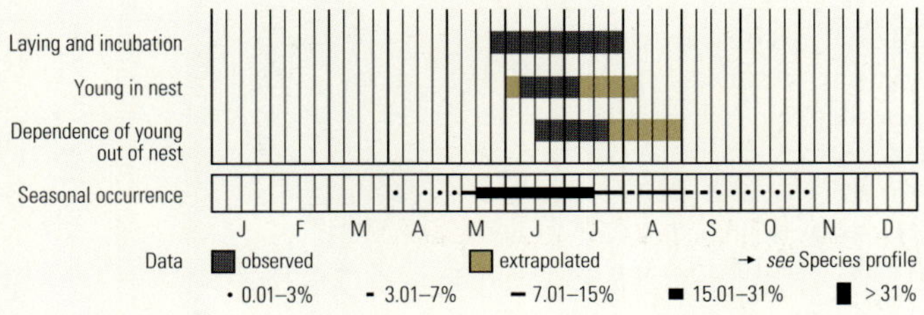

The male gives its characteristic call from atop a shrub or the tall stem of an annual plant. Bobolinks often renest in the same area from year to year. Saint-Constant, near Montréal. *July 1991, Pierre Bannon.*

Notes on territory (continued)

fields and pastures in southern Ontario—0.5–20 territorial males/10 ha—confirm this variability.

Bobolinks are highly territorial, and pursuits are frequent among males during the breeding season, which lasts from May to the end of July. It is not unusual, however, to see more than one pair sharing the same field. Like other icterines, Bobolinks are often polygynous, with the males mating with 2 or even 3 or 4 females (Wittenberger 1978; Stokes and Stokes 1989a). The quality of the territory, particularly as regards the abundance of food, appears to determine the number of females associated with one male (Orians 1969; Wittenberger 1980b). However, since food is also obtained outside the territory, other factors doubtless come into play (Wootton et al. 1986). Despite the vigilance of the territorial male, the female sometimes copulates with another male, in which case her brood may originate partly from the territorial male and partly from a competitor (Gavin and Bollinger 1985; Bollinger and Gavin 1991).

The nest is a rather flimsy, cup-shaped construction, built on the ground by the female during a period of about 2 days (Stokes and Stokes 1989a). It is placed in a slight depression, either natural or excavated by the bird, and consists of grasses and dry leaves. The inside is lined with fine grasses (Harrison 1975; Godfrey 1986). It is generally well hidden in the grass and is usually found only after a great deal of searching. When leaving the nest, the female walks through the grass for a certain distance before taking wing. She makes this same detour when returning. About 5–7 days after the pair is formed, the female begins laying her 5 or 6 eggs, at daily intervals (Wittenberger 1978; Gavin and Bollinger 1985). Incubation, by the female alone, begins before the last egg is laid and lasts for 10–13 days (Bent 1958; Harrison 1975; Gavin and Bollinger 1985; Godfrey 1986). Hatching is synchronous (Gavin and Bollinger 1985). Generally speaking, the species is single-brooded, but the birds often attempt a second nesting if the clutch is lost early in the season (Gavin 1984).

Both the male and female feed the young (Stokes and Stokes 1989a). Polygynous males will usually assist only the first female to nest on their territory (Martin 1974), although some have been seen tending more than one brood (Wittenberger 1980a, 1982). In general, the first female seems to have a higher success rate than the second, especially when food is scarce or weather conditions are less conducive to nesting (Wittenberger 1978, 1982). There have been instances

A male Bobolink seen in the background bringing a billful of insects to its five young. Both adults feed the nestlings, mostly with caterpillars and grasshoppers. Beauport, near Québec City. *July 1984, Paul Perreault.*

Five young Bobolinks soon to leave the nest, a flimsy construction of grass and leaves, built on the ground. Montréal-Ouest. *June 1924, Henry Mousley, Laval University Collection.*

of additional adults, both male and female, tending the young. These may be birds that have failed in their nesting attempt and are still physiologically motivated to feed the nestlings (Beason and Trout 1984). The exact nature of the relationship between the parents and these adults is not clear, and further study is required to define it (Beason and Trout 1984; Bollinger et al. 1986). Although this behaviour brings the birds into the territory of a competitor, such intruders are apparently tolerated because territorial behaviour diminishes after the eggs have hatched (Wittenberger *in* Beason and Trout 1984; Stokes and Stokes 1989a). At this time of the year, the diet of the young is composed mainly of lepidopteran insects (mostly caterpillars), but also includes other insects such as grasshoppers (Wittenberger 1978, 1980b, 1982). The young leave the nest at around 11 days of age (Gavin 1984) and follow the parents about the fields. As they learn to fly (they are proficient by 16 days of age) they practice foraging on their own but are still fed regularly by the parents (Stokes and Stokes 1989a). While the young are still dependent, the parents are much more active both on and off the nest and consequently more visible.

In winter, Bobolinks are granivorous and can be found feeding extensively on the pampas in South America (Wittenberger 1978). Bobolinks often live to be 5 years old (Martin 1973), and one bird was recaptured at 8 years 1 month of age (Klimkiewicz and Futcher 1989). Such longevity is surprising in such a small bird, especially, as it is often polygynous and thus has high energy expenditures. Also, its conspicuous plumage makes it vulnerable to predation, both during breeding and on its long migration (Martin 1973).

Distribution

In North America, the Bobolink breeds from southern Canada to the central United States (Godfrey 1986).

Its distribution in Québec reflects its association with agriculture. It was atlassed in many squares in the Central St. Lawrence Lowland upstream of the Saguenay River, in the southern and eastern Ottawa Valley, Eastern Townships, Bois-Francs, and the Chaudière–Appalaches region. It was also reported in the agricultural areas of Abitibi–Témiscamingue, around Lake Saint-Jean, along the Saguenay River, and in the Lower St. Lawrence region. Even in large tracts of forest, the Bobolink occupies openings created by agriculture and was observed locally in the central Laurentians and the Lanaudière region, on the Upper North Shore between Tadoussac and Baie-Comeau, and on the Gaspé Peninsula. The species was also reported on the Magdalen Islands and was sighted sporadically on the Middle North Shore, between Mingan and Natashquan, as well as around Port-Menier on Anticosti Island.

History and trends

The centre of abundance of the Bobolink must originally have been the prairies of the upper Mississippi Valley, from which it spread out as open country increased with settlement (Todd 1963).

The Bobolink population probably grew during the 17th and 18th centuries, as settlers cleared land for farming. By the turn of the century, however, the trend had reversed. Bobolink populations migrating down the eastern seaboard began feeding on rice crops, especially in the southeastern United States where Bobolinks became known as rice birds (the species name *oryzivorus* means rice eating) and were hunted relentlessly. Thus the species became a game bird and was sold in public markets and restaurants (Bent 1958).

During the last century in Québec, the Bobolink was abundant in the Montréal region (Wintle 1896). By the middle of the 20th century, it had reached areas farther north, such as the Lac-Saint-Jean region, where it was considered a fairly recent immigrant, and the North Shore (Godbout and Sept-Îles). Before this, it had long been restricted to southeastern Québec, notably the Eastern Townships, Chaudière–Appalaches, and the Lower St. Lawrence (Todd 1963). BBS data for 1965–79 showed that some northern populations in Ontario, New Brunswick, and Nova Scotia had expanded their ranges, and although expansion was not noted in this province, Québec had some of the highest densities of breeding pairs at that time (Robbins et al. 1986).

ÉPOQ summer data for 1970–89 do not show any particular trend in Bobolink populations in Québec. On the other hand, BBS data show a decrease in numbers between 1966 and 1989 (Cyr and Larivée 1993b). The Bobolink may suffer from certain agricultural practices such as early mowing of hayfields, which generally occurs when the birds are nesting and rearing their young. Few nests are spared by this practice, and as many as 94% of eggs and nestlings may be destroyed (Bollinger et al. 1990; Bollinger and Gavin 1992). Furthermore, abandonment of agricultural land, prevalent over the last 15 years throughout southern Québec (except in the Greater Montréal area and Montérégie) (M. Carignan, MAPAQ, pers. comm.) may have contributed to a decline in Bobolink numbers.

Daniel Banville and Serge Gauthier

Red-winged Blackbird

Carouge à épaulettes

Agelaius phoeniceus

Christiane Girard

THE RED-WINGED BLACKBIRD IS A COMBINATION OF SUPERLATIVES: IT IS the most abundant and widespread blackbird (Dolbeer and Stehn 1979, 1983), probably the most abundant North American bird (Dolbeer 1990), and one of the most frequently observed species in Québec. Furthermore, because of its economic impact on agriculture, it is one of the most studied species.

The male is easy to recognize with his flashing red epaulettes and jet black plumage. The Red-wing is a fierce defender of its territory and will pugnaciously challenge any intruder, be it a mink, crow, raccoon, or human. Birders who have been "buzzed" or even struck by a Red-wing when they got too close to a nest will attest to its aggressive behaviour.

The species winters from southern British Columbia eastward through the central U.S. states to southern Ontario and New England and as far south as Cuba, the Bahamas, and northern Costa Rica (AOU 1983).

Habitat and behaviour

Traditionally found only in cattail and bulrush marshes, the Red-winged Blackbird has successfully expanded its breeding habitat to include dry uplands such as cultivated fields, roadsides, and old pastures with dense vegetation of a fairly uniform height (Bent 1958; Orians 1961). The highest breeding densities are still found in marshes but, since this habitat only covers a limited area in most regions, the great majority of the population nests in other habitat (Dyer et al. 1972; Clark and Weatherhead 1986; Weatherhead 1987b; Clark et al. 1988). A study in the St. Lawrence Valley on the use of various habitats by male Red-wings has shown that they prefer to establish their territories in wetlands, hayfields, and pastures, and avoid houses, farm buildings, grain fields, and forest (Clark and Weatherhead 1987). Clark (1985d) found that, in some farming areas in southern Québec and Ontario, more Red-wings nested in mixed landscapes consisting of cornfields, hayfields, and wetlands than in habitat consisting mostly of hayfields and monocultures.

Although the proportions of plant and animal matter in the species' diet vary depending on the time of year, habitat, and the sex of the bird, some general trends can be discerned; for example, in early spring, Red-wings feed mainly on various seeds and waste grain such as oats and corn. As the breeding season progresses, consumption of insects increases, although waste grain may still play an important role in the diet of individuals nesting in agricultural habitat. Although Red-wings are territorial, they do most of their foraging in adjacent fields outside their territories. The diet of post-breeding flocks consists mainly of standing corn and oats (Bent 1958; Potvin et al. 1976; Potvin 1977; McNicol 1979; McNicol et al. 1982), although they also feed on insects, including the European corn borer and the fourspotted sap beetle (Bendell 1980; Bendell et al. 1981).

The Red-winged Blackbird is well adapted morphologically to deriving sustenance from the habitat it frequents, thus allowing it to occupy a specific ecological niche. It is not by pure chance that it is often seen perched on cattails or foraging in ditches and other wet areas. Like other icterines, its powerful pointed bill is designed more for gaping than for crushing. Red-wings can easily extract prey from a mat of wet leaves by piercing it with their sharp bills and then prying open a hole. They use the same method to pluck insects from cattail stems or remove kernels from an ear of corn still covered by the husk (Orians 1985a).

In Québec, according to Weatherhead et al. (1980), the males arrive on their breeding grounds between mid-March and early April and immediately begin to stake out their territories. The females follow 2–3 weeks later. To defend his territory, the male sings repeatedly from an exposed perch overlooking the surrounding vegetation. Early in the season, territorial defence does not involve much actual fighting or even interaction, mainly because of the limited number of competitors present. However, as the season progresses and the density of territorial males increases, aggressive behaviour becomes more frequent and intense (Nero 1956b; Beletsky and Orians 1989). Generally, yearling males do not breed and make up most of the population of floaters; they may, however, try to establish a territory, which they usually abandon shortly thereafter (Nero 1956b; Orians 1961). Successful immature males are usually those with plumage most closely resembling that of adult males; they usually establish territories in suboptimal habitat where competition with mature males is less intense (Greenwood 1985). Territorial males may occupy the same or almost the same territory year after year (Nero 1956b).

In the Red-winged Blackbird, the onset of the breeding season varies from year to year by several weeks, and these variations appear to be linked to annual variations in the quality and abundance of the food supply (Ewald and Rohwer 1982).

Notes on territory

Male Red-wings establish territories for mating and nesting. The area defended may range from 300 to 1000 m² (Nero 1956b; Stokes 1989).

Cattail marshes are the species' traditional nesting sites. Several females may build nests inside a male's territory. Sainte-Foy. *May 1966, Gaston Le Page.*

Red-winged Blackbird

Breeding distribution
1984–1989

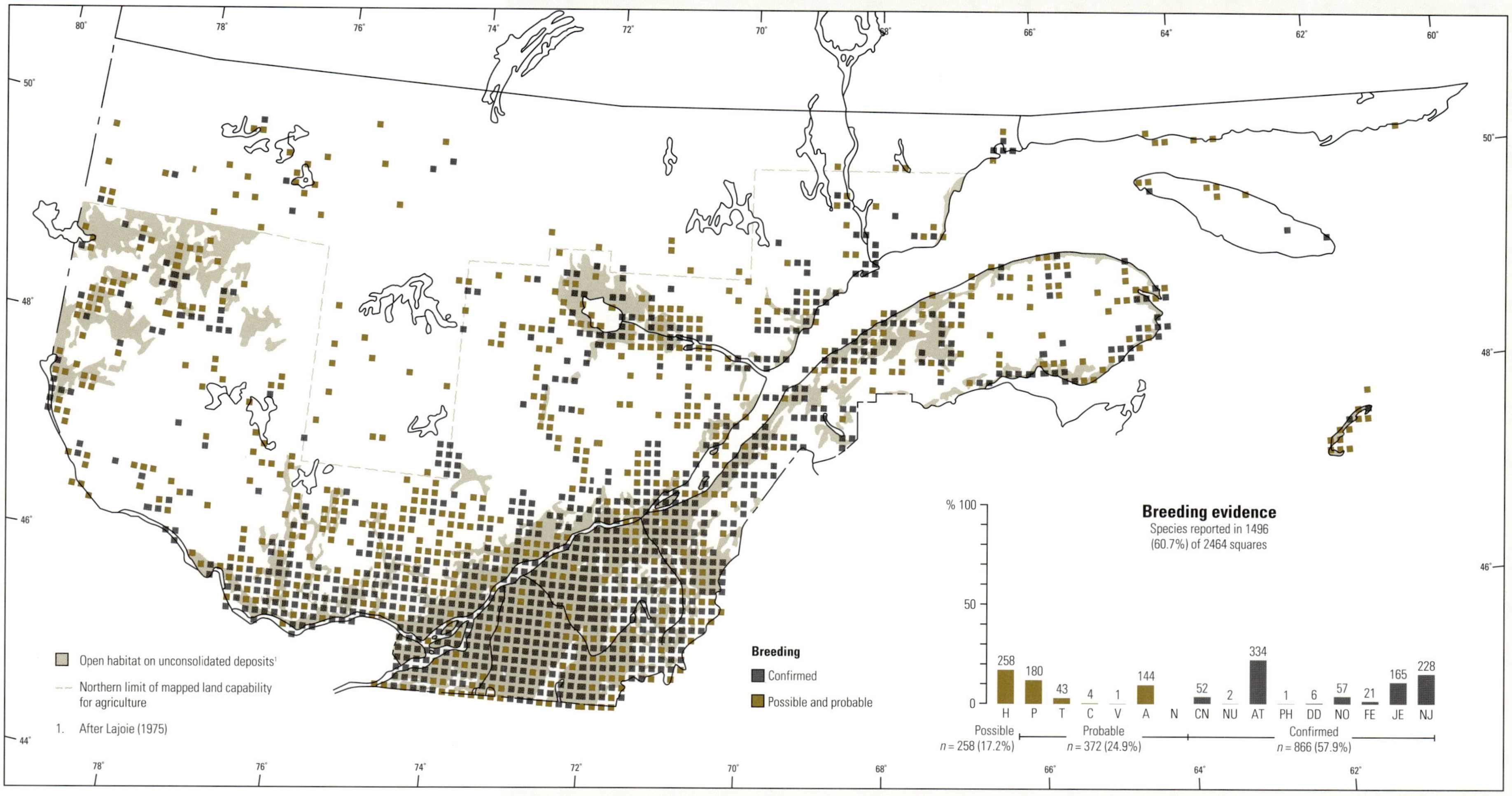

Species profile

Clutch size	usually 4 eggs (3–5)[1,2,3,5]
Laying interval	1 egg per day[11]
Start of incubation	with third egg[3]
Incubation period	10–12 days[1,2,3,5]
Maturity at hatching	altricial
Care of young	mainly by female[5]
Nestling period	11–15 days[2,5]
Age at first flight	11–15 days or more[15]
Dependence of young	until 7–10 days after leaving nest[2]
Broods per year	usually 1 (1–3)[2,3,11,12,13]
Breeding age	M usually 2 years / F 1 year[6,7]
Mating system	usually polygyny[2,5,8,14,16]
Duration of pair bond	1 season[2]
Total length	19.0–25.5 cm[1]
Wing span	30.5–36.8 cm[10]
Weight	M 63.9 g (52.9–81.1) / F 41.5 g (29.0–55.0)[9]
Longevity record	15 years 9 months[4]

References:

1. Godfrey 1986
2. Stokes 1989
3. Bent 1958
4. Klimkiewicz and Futcher 1987
5. Payne 1969
6. Beletsky and Orians 1991
7. Nero 1956b
8. Nero 1956a
9. Dunning 1984
10. Terres 1980
11. Peck and James 1987
12. DeGraaf and Rudis 1987
13. Ehrlich et al. 1988
14. Searcy and Yasukawa 1989
15. according to nestling period
16. Orians 1961

Chronology *n* = 1398

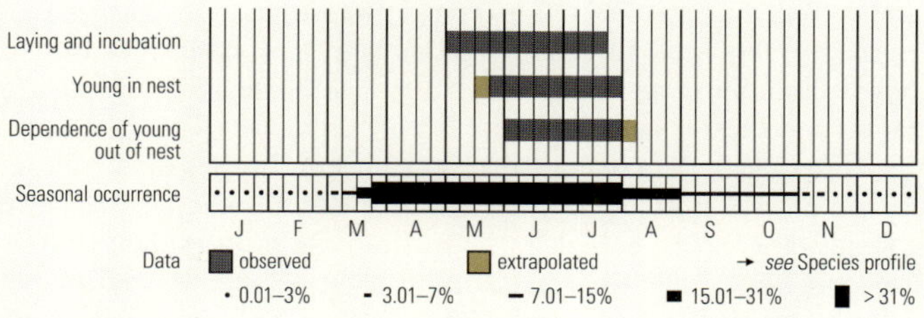

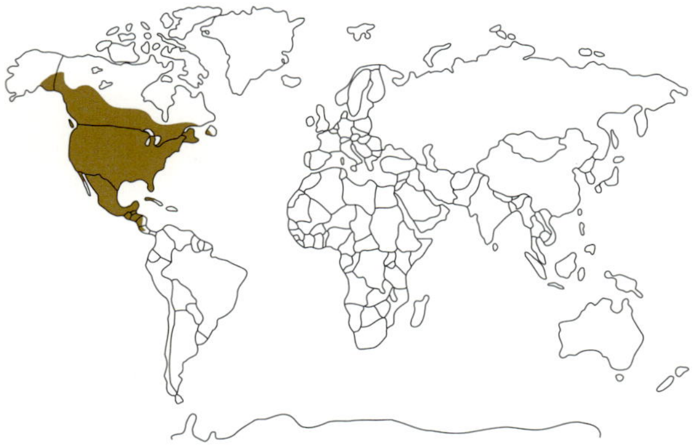

World breeding range

Female above her nest, which is well hidden in dense vegetation. She, rather than the male, feeds the young. Sainte-Foy. *June 1959, Gaston Le Page.*

The species has succeeded in expanding its nesting habitat to include cultivated fields, roadsides, and overgrown pastures. In southern Québec, agricultural areas with a combination of cornfields, hayfields, and wetlands have more breeding pairs than areas planted with monocultures or with a homogeneous landscape. Saint-Esprit, north of Montréal. *October 1994, André Bilodeau.*

Polygyny is widespread in this species, and every male has on average 2 or 3 mates, which occupy subterritories within the area defended by the male (Nero 1956a,b; Orians 1961). Therefore, several nests can often be found in one male's territory. Generally, the females occupying subterritories are at different stages in the breeding cycle (Orians 1961; Payne 1969; Stokes 1989). Finally, Gibbs et al. (1990) have shown that over 20% of the young fathered by a territorial male originate from copulations with females in other males' territories.

The nest is usually built 0.1–1.4 m up on cattail or sedge stalks (Peck and James 1987). It is often so well hidden in the vegetation that observers must rely on escape or aggressive behaviour from the female and her mate to locate it; if a female flushes from dense vegetation and flies straight up, there is a good chance that a nest is nearby (Stokes 1989). Less often, the nest is built in bushes or shrubs, and in some cases nests have even been found in apple trees or in the tops of other trees as high as 12 m from the ground (Bent 1958; Peck and James 1987). The nest is made of the leaves of marsh plants such as sedges and rushes and lined with fine grasses (Harrison 1975; Godfrey 1986). The female does all the construction work (Stokes 1989), completing the nest in 4–8 days (Bent 1958; Payne 1969). Nests are often destroyed by predators, and the female may have to build several during the season. In a study in Washington State, almost half the females that attempted to renest selected new territories, and 20% even changed marshes (Beletsky and Orians 1991).

Red-winged Blackbird nests are often parasitized by the Brown-headed Cowbird, which pushes a Red-wing egg out of the nest and lays its own. The Red-wing seems to prefer to keep the cowbird egg in the nest, because if it tries to dispose of it, it may break its own eggs (Roskaft et al. 1990).

Clutch size is 3–5 eggs, incubated by the female usually for 11 days (Bent 1958; Godfrey 1986). Females breed as yearlings (Beletsky and Orians 1991).

The female does most of the feeding of the young. The proportion of males participating in brood rearing varies widely. In some populations, this behaviour is observed very little or not at all (Payne 1969; Searcy 1979 *in* Beletsky and Orians 1990; Beletsky and Orians 1990), while in others, it is regular or even frequent (Patterson 1979 *in* Beletsky and Orians 1990; Muldal et al. 1986 *in* Beletsky and Orians 1990; Patterson 1991). Those males that do participate in feeding their young, often the oldest and most experienced males (Yasukawa et al. 1990; Patterson 1991), usually concentrate their efforts on the nestlings from the first nest in their territory (Patterson 1979 *in* Beletsky and Orians 1990; Yasukawa and Searcy 1982 *in* Beletsky and Orians 1990; Patterson 1991). Whittingham and Robertson (1993) have shown that feeding behaviour in the male may be triggered by the begging of the hungry chicks. The nestlings are fed mainly insects (McNicol 1979).

In Québec, most female Red-winged Blackbirds probably produce only one brood a year, but some may double-brood, as in Ontario (Peck and James 1987). Bent (1958) and Ehrlich et al. (1988) have reported 2 or even 3 broods a season in more southerly regions.

In late July or early August, Red-wings abandon their territories and become more secretive, hiding in marshes, where they undergo a complete moult (Bent 1958; Stokes 1989).

From this time on, the birds become gregarious and congregate in roosts at nightfall, sometimes thousands of individuals strong. In the autumn of 1979, a roost of 459 000 blackbirds was observed at Saint-Étienne-de-Beauharnois, 60–75% of them Red-wings; in the spring of 1980, a roost of 147 600 was found, 71% of them Red-wings (Weatherhead 1980b). Other large roosts were discovered at Farnham and Missisquoi in the fall of 1979 (Weatherhead 1980a). Interestingly, Red-winged Blackbirds often forage in sexually segregated flocks (Orians 1961; Orians and Christman 1968).

Males often migrate in fall in the company of other blackbirds such as grackles, cowbirds, and starlings (Bent 1958; Stokes 1989), followed by the females and immature birds a few weeks later (Stokes 1989). The species has usually disappeared altogether from the province by the end of October, although some individuals may overwinter if there is enough food.

This young Red-winged Blackbird, whose head is still covered with down, is beginning to explore the area around the nest, while two siblings are still in the nest. Saint-Pierre, Island of Montréal. *June 1935, Henry Mousley, Laval University Collection.*

Notes on territory (continued)

Males adopt a typical posture when defending their territories: arching forward, wings slightly spread to show off their red epaulettes, they perform their characteristic konk-la-ree *song, often flicking and spreading their tails at the same time. The size and red colour of the epaulettes clearly play a role in territorial defence but studies on this question have produced contradictory results (Smith, D. G. 1972; Eckert and Weatherhead 1987; Roskaft and Rohwer 1987).*

Territorial behaviour in females has not been studied extensively. According to Searcy (1988), females do not try to limit the number of other females in a male's territory and show no hesitation about nesting in a territory already occupied by several other females.

The nest, often attached to cattails or other herbaceous plants, is usually located less than 1 m above the ground. Sainte-Foy. *June 1357, Gaston Le Page.*

DesGranges (1980b) reported 4.3 individuals/10 ha in mixed forest in La Mauricie National Park. Erskine (1971a, 1972a, 1976a, 1980a, 1984) found varying densities in southern Ontario: 2.1–365.9 males/10 ha in marshes, swamps, and other wetlands; 1.0–36.4 males/10 ha in fields, pastures, grasslands, and dunes; 1.6–15.0 males/10 ha in mixed and deciduous forests; 3.5–5.5 males/10 ha in various urban habitats; and 2.8 males/10 ha in a transmission line right-of-way.

Distribution

In North America, the Red-winged Blackbird breeds from southern Alaska to central Newfoundland, and south to Mexico, Cuba, and the Bahamas (AOU 1983).

In southern Québec, breeding was confirmed in 57.9% of Atlas squares and throughout the Atlas area. Most breeding confirmations were based on reports of adults carrying food or nests containing eggs or young.

Frequency of occurrence was high in southernmost Québec, but much less so farther north and on Anticosti Island. This pattern of distribution is due to the greater prevalence of open habitat and agricultural areas in the south. In more densely forested regions, the Red-winged Blackbird frequents lake outlets, marshes, roadside and drainage ditches, areas cleared for recreation, and other open areas containing wetlands.

North of the Atlas area, the Red-winged Blackbird has been reported on several occasions during the breeding season in the region around the Nottaway, Broadback, and Rupert rivers (Cons. G. G. – GREBE 1992o). Records have also come from the Opinaca River, near James Bay and around Lake Opocopa, near Fermont (ÉPOQ).

History and trends

Although much of the original hardwood forest in eastern North America and the Great Lakes region was cleared for agriculture in the 18th century, this massive deforestation did not reach Québec until the beginning of the 19th century. These changes resulted in considerable increases in the populations of species such as the Red-winged Blackbird, Common Grackle, and Brown-headed Cowbird, which are well adapted to open habitat. However, the Red-winged Blackbird's close association with agriculture and, in particular, its depredation of corn crops, has often brought it into conflict with humans. The species' tendency to forage and roost in large flocks at certain times of the year is the main problem as far as crop damage is concerned (Weatherhead and Bider 1979); indeed, the percentage of ears in cornfields damaged by Red-wings is greater when the field is near a roost (Martin 1977). According to some estimates, the species causes annual losses of 360 000 tonnes of corn in North America every year (Dolbeer 1990).

According to Dolbeer and Stehn (1979, 1983), the greatest increases in Red-wing populations have occurred in the lowlands of eastern Ontario and southern Québec; the St. Lawrence Valley population alone grew by roughly 66% between 1966 and 1981 (Dolbeer and Stehn 1983). This increase in numbers in the farming areas of southern Québec and Ontario is undoubtedly linked to the expansion of corn acreage and the fact that the species feeds on waste grain left on the ground after the harvest (Weatherhead et al. 1985; Weatherhead 1987b). During the last few years, some corn growers in the St. Lawrence Valley have complained about increasing depredation by Red-wings (Martin 1977; Potvin 1977; Clark et al. 1982). A number of different recommendations have been made on how to reduce the Red-wing population or, at least, to make sites less attractive to the birds (Dolbeer 1990). The methods proposed include various devices to scare the birds (Martin et al. 1976; Potvin and Bergeron 1981; Dolbeer et al. 1986), the spraying of the roosts with surfactants, which reduces the birds' body temperature (Lefebvre and Seubert 1970), the use of repellents (De Grazio et al. 1971, 1972; Schafer and Brunton 1971; Potvin et al. 1978), the administration of soporifics so that individuals can be captured and eliminated (Prud'Homme-Cyr 1975), and even sterilization (Davis 1961; Potvin et al. 1982; Lacombe et al. 1986, 1990; Cyr and Lacombe 1988). However, these methods have had variable results and are often difficult to implement on a large scale.

More recently, the rising population trend seems to have reversed to some extent. An analysis of BBS data suggests that the Québec population decreased between 1980 and 1989. The decline is undoubtedly due in part to the draining of marshes, particularly cattail marshes. Although an increase in monoculture farming is not implicated directly in the decline, it may play an indirect role by bringing about the destruction and modification of habitat (hedgerows, ditches, and marshes) adjacent to cultivated fields and through the intensive use of pesticides that accompanies single-crop farming. Pesticides reduce the diversity of plant species in the fields where they are applied and in adjacent habitat through pesticide drift, as well as the diversity of the invertebrates (insects, spiders, and worms) present, which can cause significant dietary deficiencies in the birds. Blackbird control programs in the southern United States may also have contributed to the decline (Falardeau and DesGranges 1991).

BENOÎT JOBIN AND SERGE GAUTHIER

Eastern Meadowlark

Sturnelle des prés

Sturnella magna

Christiane Girard

THE LONG, WHISTLED SONG OF THE EASTERN MEADOWLARK IS HEARD upon its arrival in southern Québec—four or five melodious slurred notes which cannot be mistaken for the song of any other species.

It is easily identified by the broad black V on its bright yellow breast. In flight, it is not unlike a small partridge, with its chunky body and short wings, except for the white outer tailfeathers visible on the outspread tail. A common migrant breeder in southernmost Québec, it is often seen in fields in agricultural regions and is probably one of the most characteristic species of this type of habitat in eastern North America. Population density here, however, is lower than in the United States.

Habitat and behaviour

The Eastern Meadowlark is very similar to the somewhat paler Western Meadowlark. The two species are most easily differentiated by their songs; this difference discourages hybridization, as females appear to be responsive only to the songs of males of their respective species (Rohwer 1972). Hybridization is rare, even where the breeding ranges overlap, such as in southern Ontario and the Great Plains of the central United States, although cases have been reported in Québec. The result is reduced hatching success; hybrid young have low fledging success and are apparently sterile when mature (Welty 1975).

The Eastern Meadowlark is conspicuously active during its rather long breeding season. Males claim their territories toward the end of April, on arriving from their wintering grounds in the midwestern and eastern United States (Root 1988). They are found in fields of grass, hay, or clover, and in pastures and meadows where the herbaceous cover is dense and continuous over a relatively thick layer of dead grass (Askins 1993). They occasionally establish their territories in an orchard, wheat field, or golf course. The habitat is preferably broad, open, and fairly dry, with only a herb layer 25–50 cm high (Roseberry and Klimstra 1970; DeGraaf and Rudis 1987). A few elevated perches, such as fence posts or wires, allow the male to survey his territory (Gross 1958b). A few weeks later, the females appear, which gives rise to a variety of behaviours, including fighting between males and pursuit of females. The male displays by stretching himself upright, bill pointed skyward. He flirts his wings, spreads his tail, and puffs out his breast, showing off his bright plumage. During the breeding period, males and females adopt other postures and utter various vocalizations, apart from the main song (Gross 1958b).

The female alone builds the nest, often in a slight depression in the ground, which may be natural or excavated with her bill (Roseberry and Klimstra 1970). Several nests may be prepared simultaneously, but only one is used (Gross 1958b). The inside is lined with fine grasses, and the outside is made of grasses and weed stalks (Godfrey 1986). The nest is then covered with grass interwoven with the vegetation above the nest, forming a dome with a side entrance (Gross 1958b). It is difficult to find, and only 16.2% (25/154) of confirmed breeding records were made of nests found.

When double brooding occurs, the first clutch, in May, usually has 5 eggs and the second 4 (Gross 1958b). The female alone incubates the eggs, beginning with the laying of the last egg (Lanyon 1995). Meadowlarks are infrequent cowbird hosts. In fact, Terrill (1961) in southern Québec and Peck and James (1987) in Ontario reported parasitism in only 2% of the nests examined. Polygynous males may mate with as many as 3 females (Gross 1958b; Knapton 1988a). The male's role consists of defending the territory, as well as helping to some extent in feeding the young. He performs the greater share of tending the young when the female begins incubating the second clutch (Gross 1958b). According to Knapton (1988a), females paired with polygynous males appear to have better success than those paired with monogamous males: hatching success is higher, as is fledging success. Knapton also found that predation was the major

Notes on territory

The territory of the Eastern Meadowlark is used for nesting and foraging (Gross 1958b). Territories of 1.2–6.1 ha (Lanyon 1957) and 1.7–3.2 ha (mean 2.3 ha) were reported in Wisconsin (Wiens 1969), whereas a mean of 2.8 ha was found in New York State (Gross 1958b). Population densities of 5.25 nests/10 ha were found in pastures and 3.25 nests/10 ha in hayfields in Illinois (Roseberry and Klimstra 1970). A survey of grasslands in Wisconsin yielded a density of 3.7 territorial males/10 ha (Wiens 1969).

The Eastern Meadowlark nests in broad open spaces such as fields of wheat, grass or clover. Berthier Islands, near Lake Saint-Pierre.
October 1993, Pierre Normand.

Eastern Meadowlark

Breeding distribution
1984–1989

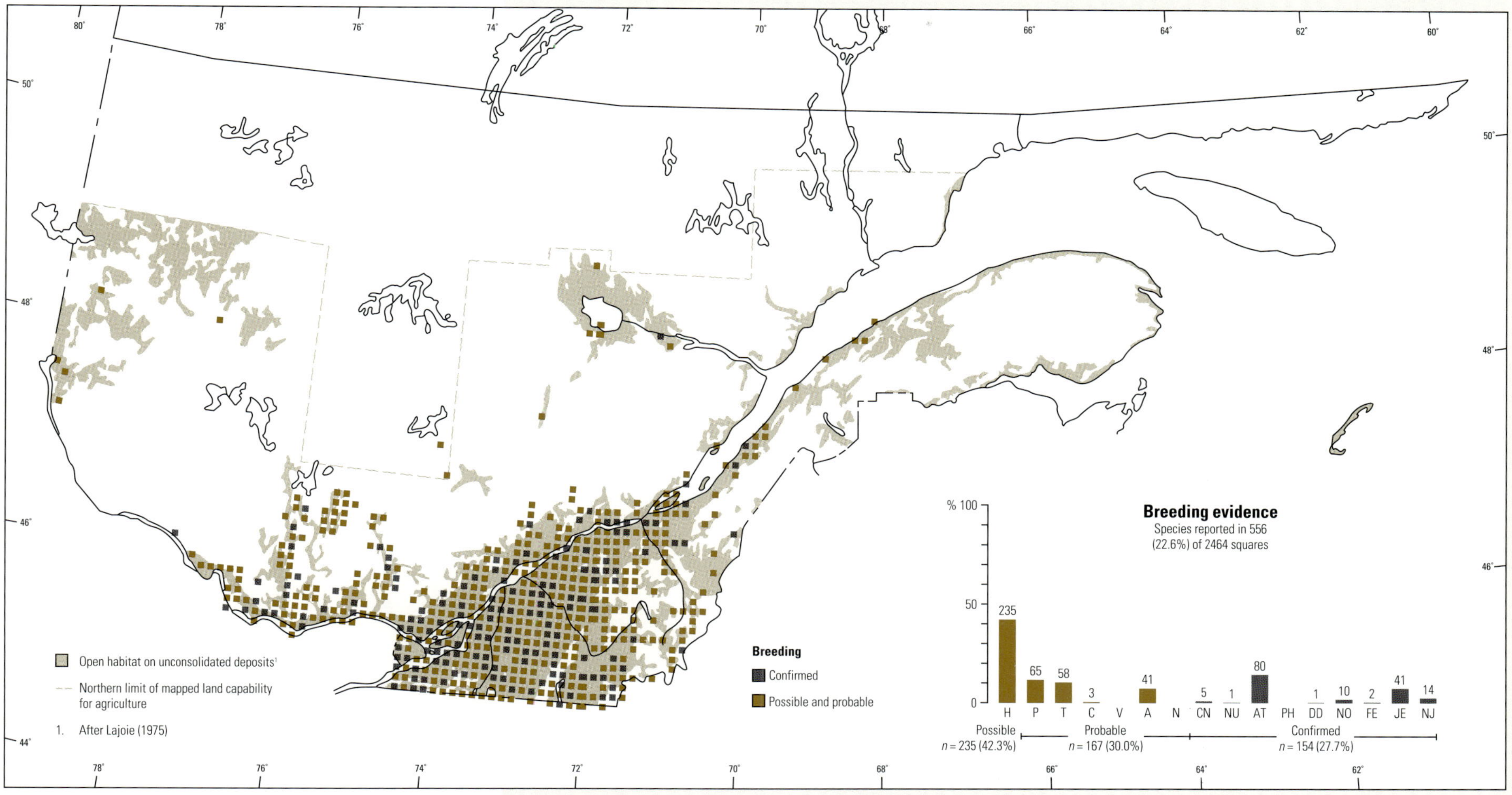

Species profile

Clutch size	usually 4–5 eggs (2–7)[1,2,3,4,5]
Laying interval	1 egg per day[11]
Start of incubation	with last egg[11]
Incubation period	usually 13–14 days (13–16)[5,6,7,9,11]
Maturity at hatching	altricial
Care of young	by female and male[5,6,7,8]
Nestling period	10–12 days[5,6,9]
Age at first flight	15–16 days[5,11]
Dependence of young	until 14 days or more after leaving nest[5]
Broods per year	2[1,5,6,9]
Breeding age	1 year[9]
Mating system	polygyny, often monogamy[1,2,5,11]
Duration of pair bond	1 or more seasons[11]
Total length	21.5–28.0 cm[7]
Wing span	34.3–43.2 cm[1]
Weight	M 123.2 g (101.5–150.0) / F 100.1 g (90.0–112.4)[11]
Longevity record	8 years 8 months[10]

References:
1. Terres 1980
2. Knapton 1988a
3. Roseberry and Klimstra 1970
4. Peck and James 1987
5. Gross 1958b
6. Harrison 1978
7. Godfrey 1986
8. Ehrlich et al. 1988
9. DeGraaf and Rudis 1987
10. Klimkiewicz and Futcher 1987
11. Lanyon 1995

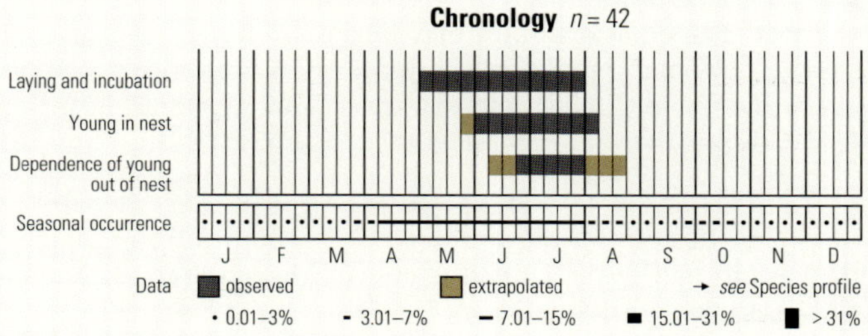

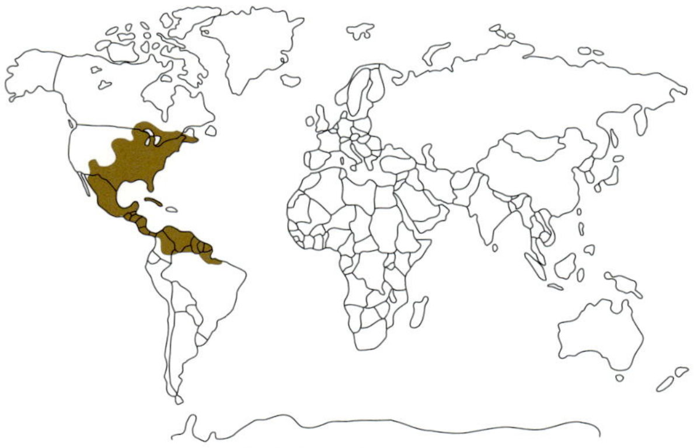

World breeding range

The meadowlark is found in fairly dry meadows with dense, continuous cover. Hemmingford, south of Montréal. *July 1991, Pierre Bannon.*

cause of mortality and that it was lower for nests of females paired with polygynous males. He suggested that polygynous males are older and more experienced than monogamous males and are thus better able to protect their offspring from predators. The female begins building the second nest 2 or 3 days after the nest is deserted by the first brood, which she continues to feed, assisted by the male (Gross 1958b).

The meadowlark feeds on or near the ground. About three quarters of the diet is composed of insects often considered pests: plant bugs, grasshoppers, scarab beetles, caterpillars, crickets, ants. The rest of the diet consists of a variety of seeds, the proportion of which is larger in spring and autumn (Terres 1980). Migratory flocks leave for their wintering grounds around late September and into October. The occasional bird may linger at a feeder or near a farm during the winter, but there is only one known instance of complete overwintering in Québec, and that was at Lac-Supérieur, near Saint-Jovite, during the winter of 1988–89 (Yank et al. 1989b).

Distribution

The Eastern Meadowlark breeds in southeastern Canada, in the eastern half of the United States, in Mexico, Central America, and northern South America (AOU 1983).

It nests in fields and meadows in the southernmost part of the province. It is not surprising, therefore, that it is closely associated with the farmlands of Québec, particularly throughout the Central St. Lawrence Lowland. North of these regions, breeding is patchier and soon becomes local and rare. There were a few scattered Atlas records from the Lower St. Lawrence, Saguenay–Lac-Saint-Jean, the Laurentians, Témiscamingue, and Abitibi regions. In the Lower St. Lawrence, breeding has since been confirmed with the finding of a nest at Rimouski (Larivée 1993). The meadowlark is a rare visitor in the Magdalen Islands, although breeding has occurred there (Fradette 1992).

A passageway to an Eastern Meadowlark nest where the young have just hatched. Strands of grass over the nest have been interwoven to form a dome. Saint-Basile-le-Grand, east of Montréal. *June 1992, François Morneau.*

Deforestation for agriculture has enabled the species to become established in certain hinterland locations such as La Tuque. North of Trois-Rivières. *August 1984, Jean Beaubien and Guy Simard, Canadian Forest Service.*

The female builds her nest on the ground. This one is partially domed and contains 7 eggs, a large clutch according to Henry Mousley. Saint-Pierre, Island of Montréal. *June 1939, Henry Mousley, Laval University Collection.*

History and trends

Originally, the Eastern Meadowlark occupied the grasslands of central North America and is one of the species that has benefited most from the deforestation that accompanied European settlement. It has extended its range considerably, both northward and eastward (Gross 1958b). Wintle (1896) and Dionne (1906) listed it as a scarce summer resident in the Montréal and Québec City regions at the turn of the century, whereas now it is common. This expansion has occurred despite losses from haying with attendant destruction of nests and young, and hunting, as the birds were mistaken for game birds (Gross 1958b). Being a ground nester, the species is prey to terrestrial predators such as skunks, dogs, and cats (Gross 1958b; Roseberry and Klimstra 1970).

Today Eastern Meadowlark numbers are declining. The species was placed on the Blue List between 1980 and 1982 because of a decline in numbers, especially in the northeast of the range (Tate 1986). Between 1969 and 1989, a decline was noted in Québec populations as well (Cyr and Larivée 1993a; BBS). Several factors are responsible for this reversal. Many nests are destroyed because of the modern farming practice of earlier and more frequent hay mowing. Also, meadowlarks do not nest in cornfields, which are becoming ever more prevalent. Its overall habitat is being fragmented, and Askins (1993) noted a relationship between habitat size and density: density increases with the size of the field. Moreover, Adamus (1987) found that the species is affected by intensive pesticide use. Currently, suitable habitat is becoming scarce, as farmland is either overexploited or abandoned and allowed to revert to scrub or reforested—not to mention the acres that are lost to highway construction and residential and industrial development (Arbib 1988c).

Jean-François Rail

Western Meadowlark

Sturnelle de l'Ouest

Sturnella neglecta

Christiane Girard

The Western Meadowlark is virtually identical to the Eastern Meadowlark. It is so hard to tell them apart that early ornithologists originally thought they were two races of the same species (Bent 1958). In the field, their songs are diagnostic. While the Eastern Meadowlark emits long clear whistles, the Western Meadowlark's song is a rich melody consisting of short flutelike notes. This difference is a sort of ecological barrier that keeps sexual interaction between the two species to a minimum in the few places where they are sympatric (Lanyon 1979). A few instances of hybridization have been reported, however.

The Western Meadowlark is considered a rare summer visitor in southern Québec, where it usually arrives in May or June, later than the Eastern Meadowlark, which appears here any time between late March and the end of April.

Habitat and behaviour

This meadowlark prefers open, grassy areas with abundant vegetation, such as hayfields, grasslands, and pastures (Bent 1958), the same types of habitat as the Eastern Meadowlark, with which Québec birders are much more familiar. Where the breeding ranges of the two species overlap and there are a good number of meadowlarks—which is not the case in Québec—the Western Meadowlark tends to prefer drier ground (Lanyon 1956; Hochstetler et al. 1991).

The male is almost always identified by his distinctive song, but the female can only be distinguished from the Eastern Meadowlark by her call. The few Western Meadowlarks seen in Québec are very likely singing males, which have a varied repertoire of 3–12 songs, depending on the individual. Horn and Falls (1991) have shown that the period during which a bird sings the same song or switches from one song to another varies depending on the situation; for example, males change songs more often when attempting to attract a female, when they lose sight of their mate, or when they are involved in territorial interactions with conspecifics.

The nests and eggs of the two species are identical. The nest is built on the ground by the female, and is generally hard to find since the male never visits it during the incubation period (Harrison 1975). Breeding is more easily confirmed by observation of adults carrying food to the young.

Given the difficulty of telling the females of the two meadowlarks apart, the possible hybridizing of the two species, and the extreme rarity of the Western Meadowlark in Québec, there is practically no information on its breeding here. Nonetheless, extensive observations in 1957 at Giffard (Beauport), a suburb of Québec City, confirmed the breeding of a hybrid pair, a male Western Meadowlark and a female Eastern Meadowlark, that produced young (Cayouette 1957; Lanyon 1966). Interbreeding is rare even though the two species share the same habitat and are closely related.

During atlassing, neither breeding nor hybridization of the Western Meadowlark was confirmed. Only 7 singing males were recorded in 6 squares, which shows how rare the species is in Québec.

Distribution

This meadowlark breeds from the Pacific coast east to Ontario and the Great Lakes states and south to central Mexico (National Geographic Society 1987). It winters from southern British Columbia to Mexico and east to western Kentucky, central Tennessee, and Alabama (AOU 1983).

The Western Meadowlark is found mainly in the Central St. Lawrence Lowland, where forest has been replaced by pastures and hayfields, whether cultivated or not. It was atlassed at Mont-Joli (twice), Rimouski, Bic, Sainte-Rita, Saint-André, and Gracefield. All these sites are along the Lower St. Lawrence, except Gracefield, which is in the Ottawa Valley. In almost every case, they were singing males that were seen for only one or two days. At Gracefield, the species was seen from 8 to 22 July 1989.

Altogether there are almost 40 sight records of the Western Meadowlark in Québec. Surprisingly, the species has been sighted at locations quite far apart. Most were in the Central St. Lawrence Lowland and the Lower St. Lawrence, as far east as Mont-Joli, about 600 km from Montréal. There is also one record from each of the following regions: Abitibi, Témiscamingue, the Laurentians, the Eastern Townships, the Saguenay, and the North Shore (David 1980a).

There is not yet any solid proof (photograph or specimen) of a pair of Western Meadowlarks breeding in Québec (Robert 1989b). There is, however, a breeding record based on observations at Verchères, where a pair is reported to have produced a chick in 1970 (Boyajian 1970). Other breeding records involve a hybrid pair at Giffard in 1957 (Cayouette 1957), a male Western Meadowlark possibly mated with a female Eastern Meadowlark at Saint-Anicet in 1975 (PQSPB 1975), and a male carrying food was observed at the same spot in 1976 (QNRCP). No link could be established, however, between the presence of a singing male in Chicoutimi from 18 May to 28 June 1980 (Blackburn 1981a,b) and a meadowlark nest (QNRCP) found in the same area.

History and trends

The Western Meadowlark formerly bred only west of the Mississippi. The bird's range has expanded eastward throughout the 20th century, especially from the north-central states (Lanyon 1956). The first reports of nesting in southern Ontario and New York State were in 1932 (Eagles 1987e) and in 1957 (Bull 1974), respectively.

In Québec, the first sight record of the Western Meadowlark dates from 1952 and breeding record (hybridization) from 1957 (Cayouette 1957). There are a dozen known records for the period 1952–69, and a dozen for each of the following decades (ÉPOQ).

Although the Western Meadowlark is seen almost every year in the southern part of the province, no stable population seems to have become established anywhere. Of course, the number of birds that visit southern Québec is very small, making encounters between males and females unlikely. Moreover, most reports of breeding from Québec seem to involve hybridization with the Eastern Meadowlark; the offspring of such pairings are quite probably sterile, which would prevent a population from becoming established (Lanyon 1979). As a result, it is highly probable that the presence of the Western Meadowlark here depends on the annual arrival of a few vagrants from the traditional breeding range of the species, the Prairies, all the more so since in Québec, individuals are usually observed late in the season.

Pierre Bannon

Notes on territory

Males chase any intruding conspecifics off their breeding territories. Horn and Falls (1991) observed 24 territorial pursuits; 42% involved males other than those from neighbouring territories, 33%, single females, and 25%, males from neighbouring territories that were themselves chasing a female. In other words, a purely territorial pursuit between two males with adjacent territories was never observed. It appears that interactions between neighbouring males are limited to visual displays along the common border between their territories.

The Western Meadowlarks studied by Schaeff and Picman (1988) occupied territories of approximately 7 ha. On uncultivated Saskatchewan prairie, density was 2 males/10 ha (Erskine 1971a).

The Western Meadowlark, a rare visitor in Québec, has the same habitat preferences as the Eastern Meadowlark, chiefly pastures and hayfields located in the southern part of the province. Boucherville, near Montréal. *July 1991, Gaétan Duquette.*

Western Meadowlark

Breeding distribution

1984–1989

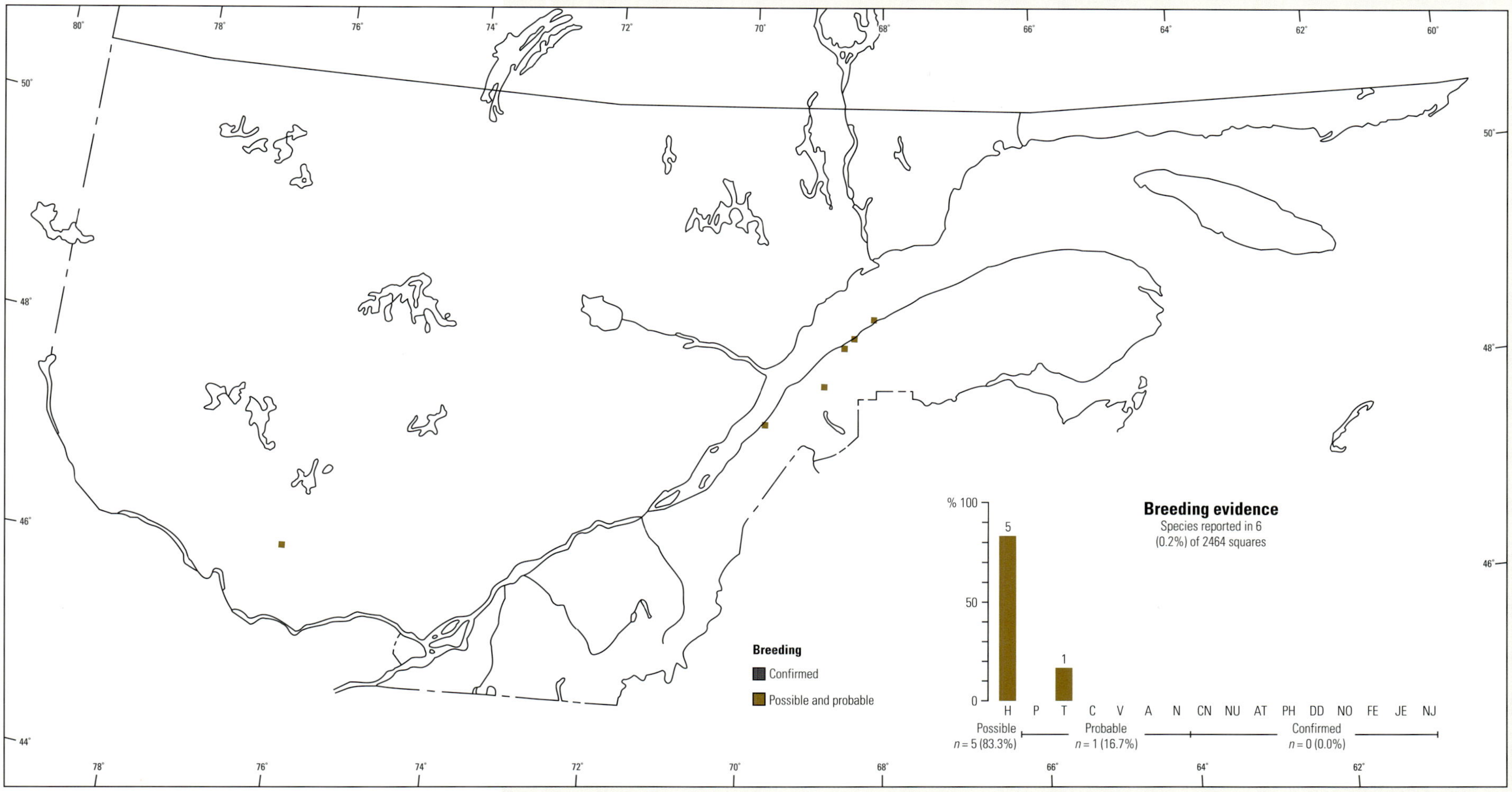

Species profile

Clutch size	usually 5 eggs (3–7)[1,2,3]
Laying interval	1 egg per day[3]
Start of incubation	with last egg[7]
Incubation period	13–15 days[1,2,3]
Maturity at hatching	altricial
Care of young	by female and male[1,2,3], mainly by female[7]
Nestling period	usually 10–12 days[1,2,3,7]
Age at first flight	15–16 days[7]
Dependence of young	until ≈ 14 days after leaving nest[1,3,7]
Broods per year	2[1,2,3]
Breeding age	1 year[7]
Mating system	polygyny, often monogamy[2,7]
Duration of pair bond	1 or more seasons[7]
Total length	21.5–28.0 cm[6]
Wing span	?
Weight	M 106.0 g (?) / F 89.4 g (?)[4]
Longevity record	6 years 6 months[5]

References:
1. Bent 1958
2. Ehrlich et al. 1988
3. Harrison 1978
4. Dunning 1984
5. Klimkiewicz and Futcher 1987
6. Godfrey 1986
7. Lanyon 1994

Chronology $n = 5$

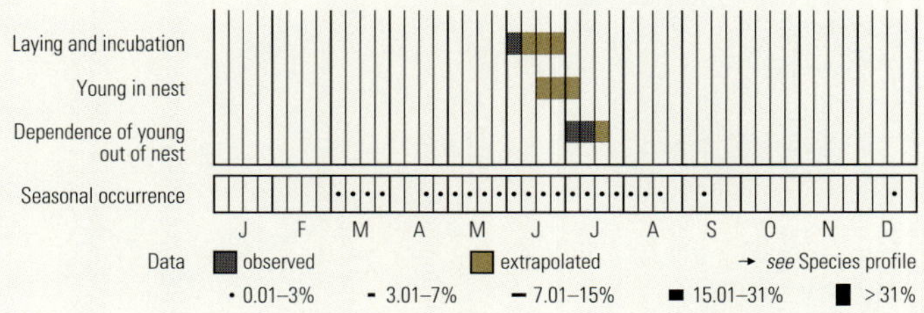

World breeding range

Rusty Blackbird

Quiscale rouilleux

Euphagus carolinus

Christiane Girard

UNLIKE OTHER BLACKBIRDS, THE RUSTY BLACKBIRD USUALLY BREEDS FAR from inhabited areas, preferring wetlands and the edges of lakes and streams in the boreal forest. Nonetheless, it is frequently observed in migration throughout Québec (David 1980a). Small numbers can be seen in fields and wetlands, often in the company of cowbirds, starlings, and various other blackbirds. In fall and winter, the Rusty's plumage has the brownish-red tinge to which it owes its name. The rest of the time, the female is slate coloured and the male is sooty black with a faint gloss.

Habitat and behaviour

During the breeding season, the Rusty Blackbird is invariably found near water. It likes peatlands, swamps, swampy forest edges, wet woods, and tall shrubbery with pools of standing water. It can also be found on partly flooded lake and stream edges, beaver ponds, and the banks of rivers and streams where willows and alders are dominant (DeGraaf et al. 1980; Godfrey 1986). In northern Québec, it can be spotted near water, where it is particularly fond of open areas with patchy shrubs and spruce (DesGranges 1989). In the southern part of its breeding range, however, including the state of New York, its distribution appears to be closely associated with beaver ponds (Peterson 1988l), and on Cape Breton Island, it also uses drier sites, such as pasture edges (Erskine 1992).

Although the species sometimes nests in small, loose colonies (Kennard 1920; Todd 1963), it tends to nest in solitary pairs (Kennard 1920; Harrison 1975; DeGraaf and Rudis 1987). According to available data, nests are usually at least 800 m apart, with 400 m being the shortest distance recorded (Kennard 1920; Bent 1958; Harrison 1975). The bird is faithful to its breeding site, returning year after year (Bent 1958; Harrison 1975).

This blackbird usually nests over or near water, often in a dense clump of young conifers, less frequently in the branches of a dead tree or a bush. The nest is built between 0.6 and 6 m off the ground, though usually less than 3 m up (Kennard 1920; Bent 1958; Godfrey 1986; Peck and James 1987). Made of twigs, grass, moss, and lichen, it may be as large as 18 cm in diameter (Harrison 1975; Godfrey 1986). The inside is a deep, well-modelled bowl made of a mudlike paste of rotting vegetation, which dries very hard and resistant. The cup is lined with long green leaves of aquatic plants that eventually turn brown with age (Kennard 1920). The nest itself may last several years, but it is not reused, and a new nest is built each spring (Bent 1958; Harrison 1975).

Notes on territory

The male starts staking out his nesting territory in early May, soon after pairs leave the large flocks of blackbirds with which they have been travelling (Kennard 1920; Bent 1958; Nichols 1985e). The

This boreal member of the subfamily Icterinae is found on forest edges near water, where alders and willows dominate. Montmorency forest, Laurentide Wildlife Reserve, north of Québec City.
May 1991, Réginald Ouellet.

Rusty Blackbird

Breeding distribution
1984–1989

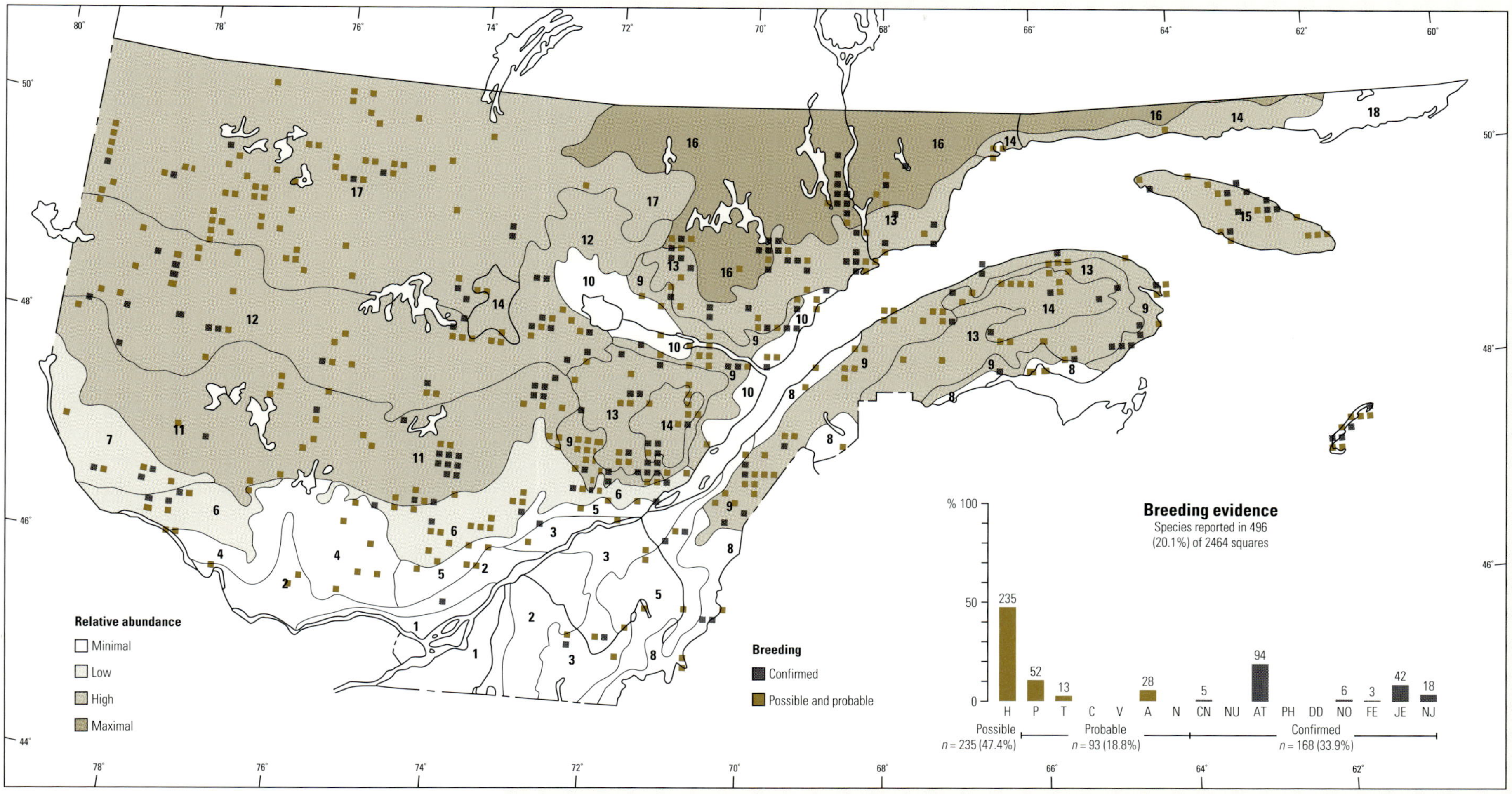

Species profile

Clutch size	usually 4–5 eggs (4–6)[1,2,3,4,9]
Laying interval	?
Start of incubation	with first egg[4,9]
Incubation period	14 days[1,2,3,4]
Maturity at hatching	altricial
Care of young	by female and male[4]
Nestling period	13 days[3,4,10]
Age at first flight	13 days or more[11]
Dependence of young	?
Broods per year	1[3,4]
Breeding age	1 year[3]
Mating system	monogamy[5]
Duration of pair bond	?
Total length	21.5–24.8 cm[1]
Wing span	33.0–38.1 cm[7]
Weight	M 64.3 g (45.9–80.4) / F 55.2 g (47.0–76.5)[6]
Longevity record	8 years 7 months[8]

References:
1. Godfrey 1986
2. Harrison 1975
3. DeGraaf and Rudis 1987
4. Nichols 1985e
5. Ehrlich et al. 1988
6. Dunning 1984
7. Terres 1980
8. Klimkiewicz and Futcher 1987
9. Harrison 1978
10. Bent 1958
11. according to nestling period

World breeding range

Chronology $n = 48$

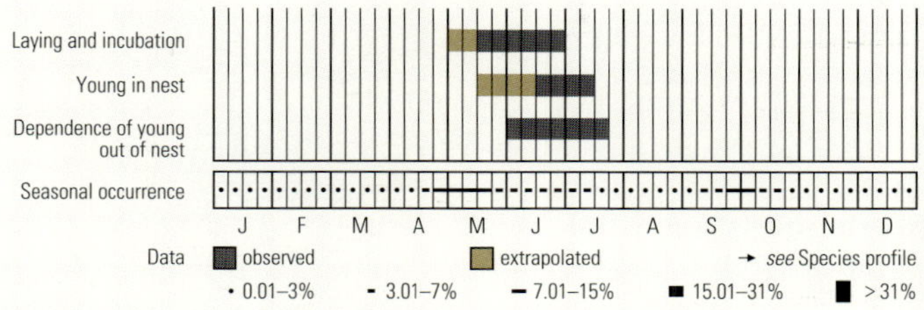

Frequency of occurrence by ecological region

Ecological regions** (forest regions)	Adequately covered squares	Squares in which reported	% of squares in which reported	Relative abundance (%)*
1. Sugar maple-hickory and sugar maple-basswood	97	0	0	0
2. Sugar maple-basswood	153	4	2.6 ± 1.5	2.9
3. Sugar maple-basswood and sugar maple-yellow birch	101	10	9.9 ± 3.7	11.1
4. Sugar maple-yellow birch-basswood	56	6	10.7 ± 7.4	12.1
5. Sugar maple-yellow birch-beech and sugar maple-yellow birch-basswood	113	11	9.7 ± 3.2	11.0
6. Sugar maple-yellow birch-beech	71	27	38.0 ± 10.3	42.8
7. Sugar maple-yellow birch-balsam fir	19	6	31.6 ± 22.1	35.5
8. Sugar maple-yellow birch and balsam fir-yellow birch	47	5	10.6 ± 8.3	12.0
9. Balsam fir-yellow birch	87	48	55.2 ± 9.6	62.1
10. Balsam fir-white birch or balsam fir-red maple	46	9	19.6 ± 10.1	22.0
11. Yellow birch-balsam fir	52	37	71.2 ± 12.8	80
12. Balsam fir-white birch (west)	59	32	54.2 ± 13.0	61.0
13. Balsam fir-white birch (east)	76	51	67.1 ± 10.3	75.5
14. Balsam fir-black spruce	28	20	71.4 ± 17.6	80.4
15. White spruce-balsam fir and balsam fir-white spruce	22	14	63.6 ± 20.8	71.6
16. Black spruce-balsam fir-moss	18	16	88.9 ± 17.4	100.0
17. Black spruce-moss	32	22	68.7 ± 17.6	77.3
18.		not analysed		

* minimal: 0–29.9%; low: 30–59.9%; high: 60–89.9%; maximal: 90–100%

** Deciduous forest: 1–7; mixed forest: 8–11; coniferous forest: 12–17

The Rusty Blackbird builds its nest in dense scrub or small conifers at the water's edge. Martin-Valin ZEC, north of the Saguenay River. *August 1991, Jean Gauthier.*

Compared with other blackbirds, the Rusty is fairly elusive during the breeding season. It can usually be detected by its song, which has been described as a split creak like a rusty hinge: *kush-a-lee* alternating with *ksh-lay* (Peterson 1980). The song is often heard at a distance, as the bird flies over its territory or watches over it from a perch.

In the breeding season, the Rusty Blackbird consumes mainly insects, while at other times its diet is composed of equal parts animal and plant matter (Bent 1958; DeGraaf et al. 1980). The female alone incubates the 4 or 5 eggs, starting when she lays the first one (Harrison 1978). During incubation, which lasts 14 days, the male defends the territory and feeds his mate regularly (Harrison 1975). Kennard (1920) reported that the nest of the Rusty Blackbird can be found fairly easily by watching the male's movements. During incubation, he sings frequently near the nest, often holding food for the female in his bill.

Notes on territory (continued)

monogamous Rusty Blackbird appears to be rather solitary. Nests have rarely been observed less than 400 m from each other, and often more than 800 m apart (Kennard 1920; Bent 1958; Harrison 1975). The birds are faithful to their nest site, returning to it year after year (Bent 1958; Harrison 1975).

On the breeding grounds, density records vary from 0.7 pair/10 ha in a black spruce peatland east of Senneterre, Québec (Erskine 1972a), to 2.0 pairs/10 ha in a white and black spruce and tamarack forest, with scattered openings, at Churchill, Manitoba (Erskine 1971a). Annual counts of birds on their wintering grounds, especially in Mississippi, have yielded densities of 0.4 bird/10 ha in an oak-jack pine forest and 1.7 birds/10 ha in an oak-hickory forest (Moore 1992).

This blackbird nests in peatlands found all over Québec. Source of the Mingan River, Middle North Shore. *August 1977, Pierre Lamothe.*

On the Lower North Shore, the Rusty Blackbird lives in wooded areas along the edges of wetlands and lakes, as well as in some valleys. North of Jacques Cartier Bay. *August 1979, Jean Beaubien and Guy Simard, Canadian Forest Service.*

In the boreal forest, pairs usually breed alone in lakeside shrubbery. North of Lake Bienville, north-central Québec. *July 1977, Pierre Lamothe.*

The young are fed frequently by both parents, which forage on open ground, especially grassy or scrubby shores. The Rusty Blackbird raises only one brood a year (Nichols 1985e; DeGraaf and Rudis 1987). It breeds fairly early (Kennard 1920) and seems to desert its territory as soon as the breeding period is over (Flood 1987b).

The Rusty was the blackbird least often atlassed in southern Québec. It was reported in only 20% (496/2464) of squares surveyed, and the vast majority of these observations were of an adult or pair in suitable breeding habitat. Breeding was confirmed in only 168 squares, distributed throughout most regions of southern Québec.

Distribution

The Rusty Blackbird's North American breeding range extends from the tree line down to southern Canada and the northeastern United States (National Geographic Society 1987).

In Québec, frequency of occurrence was highest in the black spruce forests of the North Shore hinterlands, near the Pipmuacan and Manicouagan reservoirs. The species is slightly less common throughout the mixed and coniferous forest zones.

The distribution of the Rusty Blackbird is not limited to southern Québec; the bird is actually common in the north up to the tree line. It winters in the southernmost part of Ontario and in the eastern United States. In Québec, a few birds are seen at feeders in some winters (Yank and Aubry 1985c, 1986b; Aubry et al. 1991).

History and trends

Current monitoring methods do not permit an accurate determination of the status of populations of this species. At the turn of the century, Dionne (1906), like Macoun and Macoun (1915), described the Rusty Blackbird as a common summer resident throughout almost all of central and northern Québec, as well as in the Magdalen Islands. Todd (1963) stated that it was common in northern Québec, as well.

In short, little is known about the true population trends of the Rusty Blackbird, but its future does not seem threatened either by lake acidification (DesGranges 1989) or by loss of such wetlands as the numerous peatlands found in the back country.

CLAUDE NADEAU

Common Grackle

Quiscale bronzé

Quiscalus quiscula

Christiane Girard

THE COMMON GRACKLE IS A FAMILIAR SIGHT AMONG THE FLOCKS OF "blackbirds" that often invade fields of grain crops, especially corn. Its adaptability has enabled it to take advantage of man's alteration of the environment; it nests in many kinds of habitat and is a common migrant breeder in southern Québec (David 1980a). This large, bold passerine has iridescent black plumage and yellow eyes, the female being slightly smaller and duller than the male. During the breeding season, the Common Grackle might be confused with the smaller (robin size) Rusty Blackbird, but the latter species has less of a sheen on its plumage and a shorter tail.

The grackle winters in the southern portion of its breeding range, which extends from south of the Great Lakes across to Nova Scotia and down to southeastern Mexico and southern Florida (AOU 1983). It is sometimes seen in Québec in winter (David 1980a).

Habitat and behaviour

The Common Grackle often arrives on its Québec breeding grounds while there is still snow on the ground. It chooses a variety of partially open habitat: woodlands with openings; wetlands such as marshes, ponds, swamps, and peatlands; ecotones between forest and more open environments like lakes, rivers, fields, roads, and clearcuts; abandoned fields scattered with trees; rocky cliffs; urban areas; farmland; and golf courses (AOU 1983; Peck and James 1987). This common urban and suburban species also breeds in the boreal forest where there is suitable habitat, for example around resorts. In Vermont and New York State, it is uncommon, even rare in densely or extensively wooded areas (Ellison 1985j; Peterson 1988d). In Québec, the greatest concentrations occur in the farmlands of the south (Erskine 1971c). Pairs may nest either singly or in loose colonies, with nests quite close together. Colonial grackles may mob potential nest predators, such as eagles, Ospreys, and squirrels (Gross 1958a).

Transition from typical winter flocking behaviour to pair formation in the breeding season is gradual (Maxwell 1970). Initially, the female associates with one or more males for foraging, display flights, and other activities (Maxwell 1970). After a number of flights resembling chases, the males settle on higher perches to maintain their dominance (Maxwell 1970). Keeling of the tail is an important display used by males during pair formation (Terres 1980; Hickman 1981). As the female begins to bond with a particular male, the number of males participating in flights decreases (Maxwell 1970). The pair bond is confirmed when the flights involve only the two birds and when the male initiates some of the flights and stops perching higher up to assert dominance (Maxwell 1970).

Nest construction, almost exclusively the responsibility of the female, can take from 7 straight days to 8 weeks (Maxwell 1970). Nests are usually 1.2–4.0 m above the ground, but range from 0.2 to 18 m up (Peck and James 1987). They are bulky structures of loosely woven twigs, grass, leaves, and other materials. The inside is often plastered with mud and lined with fine grass, feathers, and rootlets. Nests near human habitation may also contain such materials as paper, bits of string, and scraps of cloth (Gross 1958a; Harrison 1975; Harisson 1978; Terres 1980). The Common Grackle nests in urban as well as in agricultural areas and woodlands. In populated areas, it seeks out well hidden and inaccessible nest sites, such as tall conifers, vine-clad buildings, abandoned farm buildings, and the girders of railway bridges (Erskine 1971c). In wilderness areas, nests are built mainly in trees (especially conifers), bushes, thickets, and alder swamps, on either dry or marshy ground (Harrison 1978; Peck and James 1987; Peterson 1988d). Grackles also use stubs in marshes, beaver ponds, and grassland that is either flooded or subject to flooding. They also often take over old woodpecker holes (Peterson 1988d).

Notes on territory

In Ontario, Peck and James (1987) reported that the Common Grackle nests either singly or in loose colonies, averaging 10 nests but running as high as 200 in some cases. Within the colony, nests are anything from 0.6 m (Gross 1958a) to 4.6 m or more apart (Peck and James 1987). Both male and female defend a small territory around the nest (DeGraaf and Rudis 1987). Grackles may range as far as 1.6 km or more from the nest to forage (DeGraaf and Rudis 1987).

During the breeding season, population densities vary considerably depending on habitat. In Ontario, an estimate of 68.1 pairs/10 ha was made for a cattail marsh (Erskine 1984), 12 pairs/10 ha for grassland scattered with marshes, ponds, and trees (red oak, basswood, and white ash), and 1.4–4.9 pairs/10 ha in various open or partially open habitats (Erskine 1972a, 1976a, 1980a, 1984). In urban areas—residential neighbourhoods, parks, and suburbs—reported densities run from 1.2 to 11.9 pairs/10 ha (Erskine 1972a, 1976a, 1980a, 1984).

The Common Grackle is a bird of towns and farmlands, foraging in fields and other open areas and nesting in forest edges or thickets. Breckenridge, near Hull. *September 1991, Gilles Falardeau.*

Common Grackle

Breeding distribution
1984–1989

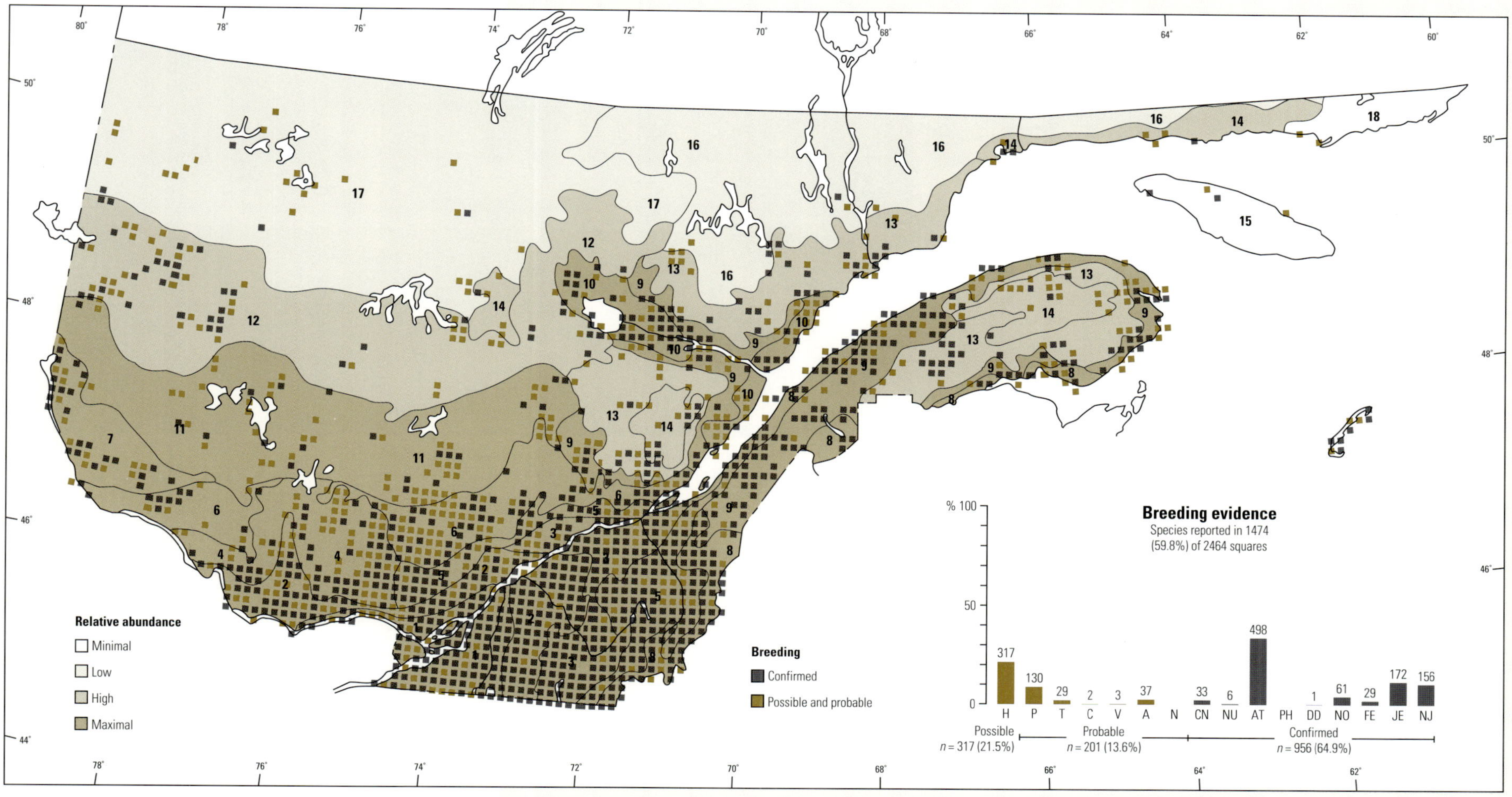

Chronology $n = 452$

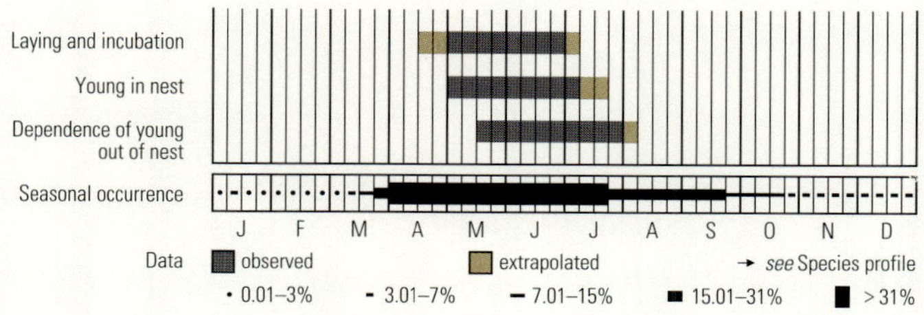

World breeding range

Species profile

Clutch size	usually 4–5 eggs (1–7)[1,2,4,9]
Laying interval	1 egg per day[2,9]
Start of incubation	before or after last egg depending on clutch size[9]
Incubation period	11–14 days[1,2,4,5]
Maturity at hatching	altricial
Care of young	by female and male[1,4,9]
Nestling period	12–15 days[4,7,8]
Age at first flight	14–20 days[4,5]
Dependence of young	probably until a few weeks after leaving nest[7]
Broods per year	usually 1 (1–2)[1,11]
Breeding age	?
Mating system	monogamy, sometimes polygyny[1,7]
Duration of pair bond	?
Total length	28–34 cm[3]
Wing span	43.2–47.0 cm[5]
Weight	M 127 g[12](105–140)[6] / F 100 g[12](95–104)[6]
Longevity record	20 years 11 months[10]

References:
1. Ehrlich et al. 1988
2. Peck and James 1987
3. Godfrey 1986
4. Harrison 1978
5. Terres 1980
6. Snyder 1937 *in* Terres 1980
7. Howe 1976
8. Maxwell and Putnam 1972
9. Howe 1978
10. Klimkiewicz and Futcher 1987
11. DeGraaf and Rudis 1987
12. Dunning 1984

Frequency of occurrence by ecological region

Ecological regions** (forest regions)	Adequately covered squares	Squares in which reported	% of squares in which reported	Relative abundance (%)*
1. Sugar maple-hickory and sugar maple-basswood	97	97	100.0 ± 0.5	100.0
2. Sugar maple-basswood	153	149	97.4 ± 1.5	97.4
3. Sugar maple-basswood and sugar maple-yellow birch	101	100	99.0 ± 1.6	99.0
4. Sugar maple-yellow birch-basswood	56	55	98.2 ± 3.7	98.2
5. Sugar maple-yellow birch-beech and sugar maple-yellow birch-basswood	113	113	100.0 ± 0.4	100.0
6. Sugar maple-yellow birch-beech	71	70	98.6 ± 3.0	98.6
7. Sugar maple-yellow birch-balsam fir	19	19	100.0 ± 2.6	100.0
8. Sugar maple-yellow birch and balsam fir-yellow birch	47	47	100.0 ± 1.1	100.0
9. Balsam fir-yellow birch	87	86	98.9 ± 2.5	98.9
10. Balsam fir-white birch or balsam fir-red maple	46	46	100.0 ± 1.1	100.0
11. Yellow birch-balsam fir	52	48	92.3 ± 7.9	92.3
12. Balsam fir-white birch (west)	59	45	76.3 ± 11.3	76.3
13. Balsam fir-white birch (east)	76	63	82.9 ± 8.4	82.9
14. Balsam fir-black spruce	28	23	82.1 ± 15.2	82.1
15. White spruce-balsam fir and balsam fir-white spruce	22	4	18.2 ± 17.1	18.2
16. Black spruce-balsam fir-moss	18	8	44.4 ± 26.0	44.4
17. Black spruce-moss	32	12	37.5 ± 18.3	37.5
18.			*not analysed*	

* minimal: 0–29.9%; low: 30–59.9%; high: 60–89.9%; maximal: 90–100%

** Deciduous forest: 1–7; mixed forest: 8–11; coniferous forest: 12–17

Typical clutch size is 4 or 5 eggs, incubated by the female for 11–14 days; meanwhile, the male protects both the female and her clutch. While the female is solely responsible for brooding the hatchlings, both parents feed the young both in the nest and after they leave it, and both male and female eat or otherwise dispose of the fecal sacs (Maxwell and Putnam 1972; Peck and James 1987; Peterson 1988d). In the first few days after hatching, the parents sometimes get rid of the fecal sacs by dropping them in flight into bodies of water; in urban areas, they may target swimming pools or vehicle windshields, whose reflective surfaces look like water. The male sometimes leaves his mate for another female, in which case the first female raises her brood alone (Peterson 1988d). Brood parasitism by the Brown-headed Cowbird seems to be rare; in Ontario, it has been estimated at barely 0.2% (Peck and James 1987).

The fledglings leave the nest at the age of 12–15 days and may move farther afield within another week, though the parents probably continue to care for them for a few more weeks. Even when adults are seen feeding juveniles as big as themselves in early July (Howe 1976), many of the young have already joined the flocks that roost at nightfall (Gross 1958a). The Common Grackle sometimes raises two broods (Ehrlich et al. 1988), but this behaviour appears to be confined to the southern parts of the breeding range. During

The Common Grackle often nests in tall conifers in built-up areas, but may also sometimes form small colonies in plantations of fir or other conifers. Baldwin Mills, south of Sherbrooke. *Summer 1991, Jean-Luc DesGranges.*

Notes on territory (continued)

During the breeding season, non-breeding individuals gather at nightfall in roosts. These roosts often include birds of other species, such as starlings, cowbirds, blackbirds, and robins, and may number in the thousands (Terres 1980). In a winter roost in Tennessee comprising some 600 000–800 000 individuals of various species (61% grackles, 35% starlings, 3% blackbirds, and 1% cowbirds), Twedt (1984) estimated grackle density at 31–41 individuals/m².

This species also nests in woodlands with clearings or in and around wetlands. In this marsh, a pair has built its nest in the rolled up wire fence in the centre of the photograph. Saint-Pierre, Island of Montréal. *May 1939, Henry Mousley, Laval University Collection.*

This species often builds its nest on the girders of bridges or in the nooks and crannies of barns and other man-made structures. Routhierville bridge, Matapédia River, Gaspé Peninsula. *August 1991, Yves Aubry.*

Close-up of the nest in the bale of wire. The bulky structure is made of loosely woven twigs and other material. Saint-Pierre, Island of Montréal. *May 1940, Henry Mousley, Laval University Collection.*

atlassing, eggs were observed in nests until the end of June, but no second broods were reported. In Ontario, Peck and James (1987) reported only replacement clutches.

Common Grackles give themselves away by their squeaky song, sounding like an unoiled wheelbarrow; their call is a loud chuck. Within their raucous colonies, they utter various kinds of vocalizations, apparently as a way of maintaining contact between individuals located quite close to each other (usually within 2 m) (Wiley 1976).

Common Grackles are often seen foraging on the ground, especially where wet, in open areas such as fields, pastures, lawns, golf courses, wetlands, and shores of ponds, lakes, and streams. This bird has a bill capable of crushing, but its morphology also allows it to open its bill against resistance, for probing the ground (Orians 1985a). Grackles are thus able to exploit a wide variety of food, both animal and vegetable: insects and other arthropods, earthworms, crayfish, small fish and frogs, small mammals (bats and mice), small birds (eggs, young, and adults), grains and seeds of all kinds, and fruit, both wild and cultivated. The relative proportions of these items in their diet vary with the seasons (Gross 1958a; Laporte 1974b; Terres 1980). They also use a variety of foraging techniques, depending on habitat: they walk on the ground with a slow, dignified gait, and occasionally run and leap high after insects. They forage in shrubs and trees, hawk in flight, probe the ground for earthworms, follow ploughs for grubs, and wade in streams (Terres 1980). The grackle is a true predator of some other bird species; Laporte (1974b) reports seeing one kill a swallow with a blow of the bill to the head.

The Common Grackle was confirmed as a breeder in 64.9% (956/1474) of the Atlas squares in which it was reported. Most breeding confirmations were obtained through observations of adults carrying food, as well as through discovery of nests and of fledglings off the nest but not yet capable of sustained flight.

Distribution

The Common Grackle breeds throughout most of North America east of the Rocky Mountains and south of the taiga. In Québec, it is very common in urban and agricultural areas in the south of the province. In the hinterland, the species is less common, though it does breed in openings around bodies of water, clearings, roads, cutovers, and resorts. It is even less common in the Black spruce forest regions. This drop in numbers with increasing latitude has also been noted in Ontario (Weatherhead 1987a). In Québec, the most northerly sightings are from the region of the Rupert, Broadback, and Nottaway rivers (Cons. G. G. – GREBE 1992o). From mid-July on, large numbers gather in the western St. Lawrence Plain to form roosting colonies with various other blackbirds. Depending on the place and date, these colonies may number from 1500 to over 100 000 individuals (Weatherhead 1980a). In this part of its range, annual losses in corn and other grain crops (oats, barley, etc.) caused by these blackbirds run into the millions of dollars (Weatherhead et al. 1980).

In winter, the Common Grackle has been reported on the North Shore, Anticosti Island, the Gaspé Peninsula, the Lower St. Lawrence, and the St. Lawrence Plain, though it is rare at that time of year (David 1980a).

History and trends

According to Macoun and Macoun (1909), the Common Grackle was indeed common in Québec at the turn of the century. In Montérégie, the clearing of forest for farming led to a considerable increase in numbers between the mid 19th and early 20th centuries, with the population stabilizing afterward (Ouellet 1974a). Though apparently absent from the Magdalen Islands at the turn of the century (Macoun and Macoun 1909), the Common Grackle was first confirmed as nesting there in 1923 and now breeds there regularly (Fradette 1992). In Québec and Ontario, increases in Common Grackle populations can probably be attributed to the availability of seed and waste grain and of stunted seed heads, especially in agricultural areas (Potvin et al. 1976; Weatherhead 1987a). In Vermont, where the grackle is one of the commonest breeding birds, historical data show no significant variation in the population's status for at least the last 130 years (Ellison 1985j).

Québec data from 1966 through 1989 show no significant shift in Common Grackle numbers (Cyr and Larivée 1993a; BBS; ÉPOQ); however, a number of harsh winters and blackbird control operations in some of the largest winter roosts in the United States (Carlson 1991) may have triggered a slow but steady population decline in the agricultural areas of southern Québec between 1970 and 1986 (Falardeau and DesGranges 1991). Though urban sprawl continues to create new grackle habitat, populations may decline in places where farmland is abandoned and fields gradually revert to forest (Peterson 1988d).

Robert Alvo

Brown-headed Cowbird

Vacher à tête brune

Molothrus ater

Christiane Girard

The Brown-headed Cowbird is widely known as a brood parasite. It never builds a nest and divests itself of all parental care by laying in the nests of other insectivorous and granivorous species. Like all parasites, it lives at the expense of its hosts, which gain no benefit whatsoever from the relationship. Although its breeding behaviour may arouse the indignation of some, the cowbird remains nonetheless a species of interesting and complex habits.

The Brown-headed Cowbird derives its name from its close association with the herds of cattle it follows, feeding on insects stirred up by grazing (Friedmann 1929; Lowther 1993). Once restricted to the Prairies of central North America, the species began expanding northward and eastward at the turn of the century, as the forest cover disappeared (Mayfield 1965).

The cowbird is a typically gregarious and abundant species in Québec. It may feed in flocks even during the breeding period (Bent 1958). In March and April and from July to October, it flocks in huge aggregations with Red-winged Blackbirds, Common Grackles, and European Starlings.

The winter range stretches from southern Ontario east to southern New Brunswick and the northeastern United States, and extends south to Florida, California, and Mexico (AOU 1983; National Geographic Society 1987). Disjunct wintering populations are found throughout the southwestern United States, in California, Arizona, and New Mexico (Root 1988). The first record of successful overwintering in Québec was in 1959–60 (David 1976b). According to David (1976b), there were still very few instances of overwintering in the mid-1970s. Although increased corn cultivation and the proliferation of cattle feedlots and feeding stations offering corn have enabled more birds to overwinter, cowbirds are the least common of the blackbirds wintering in Québec (N. David, pers. comm.).

Notes on territory

Published information about territoriality in the Brown-headed Cowbird reveals a complex and poorly understood pattern of behaviour. In populations of monogamous pairs, the female is sometimes reported as defending a territory (Dufty 1982; Darley 1983) and sometimes not (Laskey 1950; Darley 1982; Yokel 1989). In Ontario, Darley (1983) estimated average territory size to be 5 ha, whereas in New York State Dufty (1982) found it to be 20.4 ha. Elliott (1980) did not observe territorial behaviour in a population of sexually promiscuous individuals. According to Yokel (1989), female territoriality is not linked to cowbird population density.

The male occupies a definite area that he does not defend, though he does defend his mate against other males (Darley 1982; Dufty 1982). The home range of the male overlaps a considerable part or all of the territory of his mate or her home range (in the case of non-territorial females). According to Dufty (1982), the home range of the male is the same size as the territory of his mate—20.4 ha on average.

Formerly restricted to the western Prairies, the Brown-headed Cowbird followed the expansion of agriculture; in fact, it is often seen feeding on insects stirred up by grazing cattle. In the hinterland, it is mainly associated with open habitat and forest edges. It may, however, venture a few kilometres into unbroken forest to parasitize nests of other passerine species. Mont-Saint-Pierre, east of Matane, Gaspé Peninsula. *August 1958, Gaston Le Page.*

Brown-headed Cowbird

Breeding distribution

1984–1989

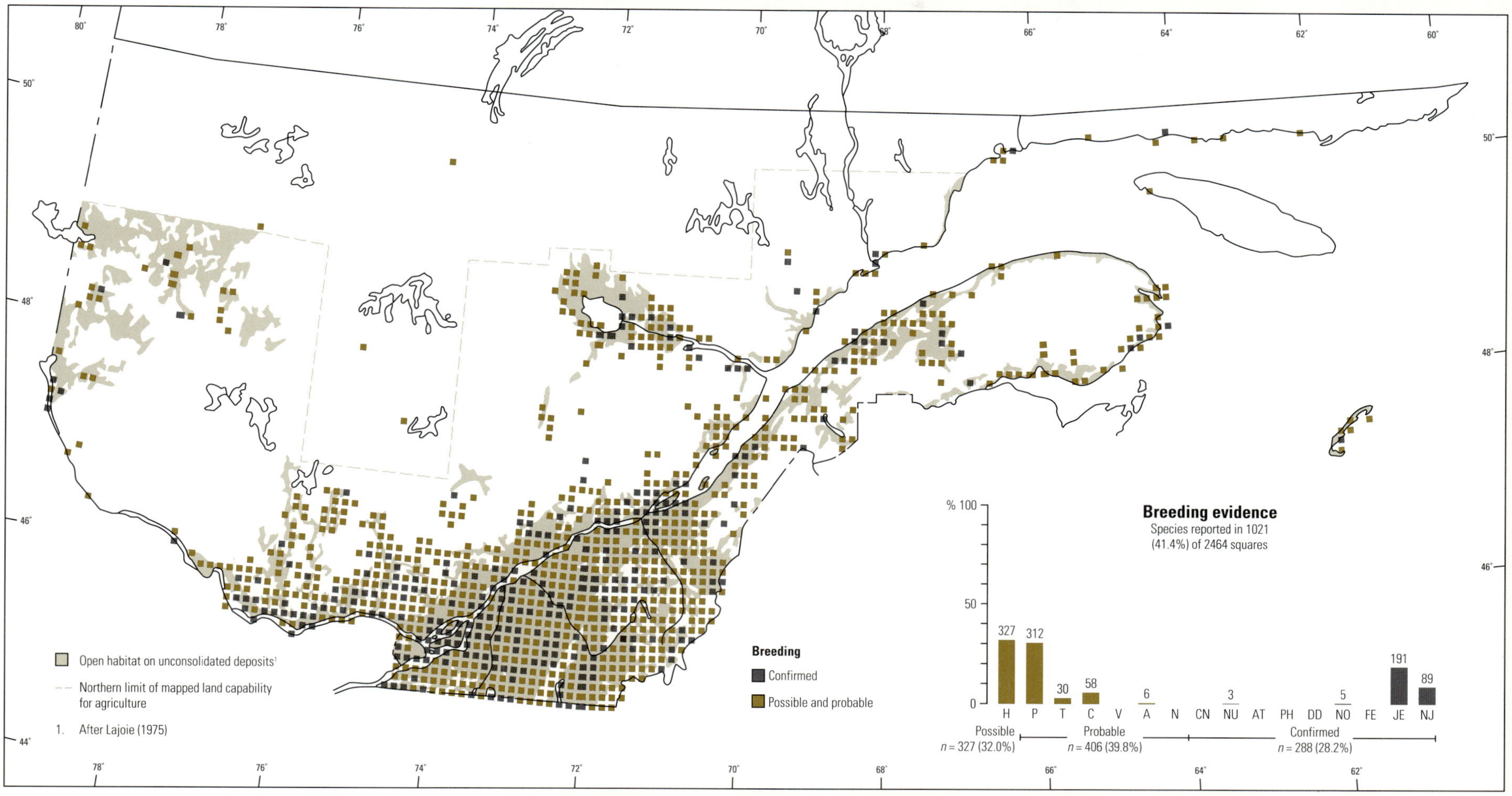

Species profile

Clutch size	up to ≈ 40 eggs[3,11,13,14,15], *see* text
Laying interval	variable, usually 1 egg per day[3,11,12]
Start of incubation	depends on host species
Incubation period	10–12 days[2,4,7,11]
Maturity at hatching	altricial
Care of young	by host species[11]
Nestling period	8–13 days[4,7,11]
Age at first flight	≈ 10–13 days[10,11]
Dependence of young	until 25–39 days old[6,11]
Broods per year	–
Breeding age	1 year[8,11]
Mating system	variable: monogamy, polygyny, polyandry, promiscuity[4,11]
Duration of pair bond	variable: mating only, 1 season, more than 1 season[11]
Total length	17–21 cm[9]
Wing span	29.8–34.9 cm[2]
Weight	M 49.0 g (32.4–58.0) / F 38.8 g (30.5–51.2)[5]
Longevity record	16 years 11 months[1]

References:
1. Klimkiewicz and Futcher 1989
2. Terres 1980
3. Jackson and Roby 1992
4. Ehrlich et al. 1988
5. Dunning 1984
6. Woodward 1983
7. Harrison 1978
8. DeGraaf and Rudis 1987
9. Godfrey 1986
10. Friedmann 1929
11. Lowther 1993
12. Scott and Ankney 1983
13. Scott and Ankney 1980
14. Payne 1965
15. Walkinshaw 1949b

Chronology *n* = 363

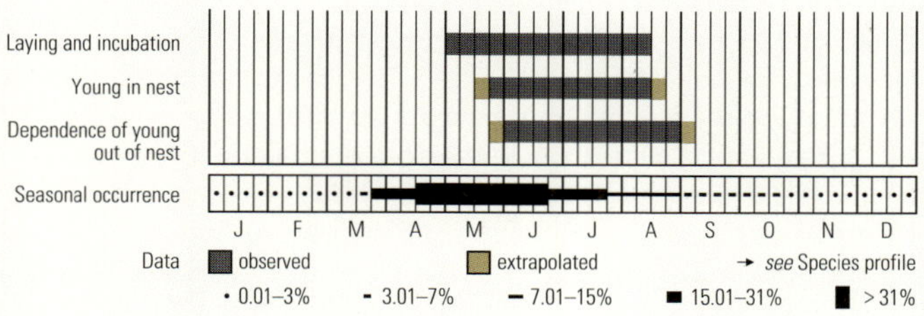

World breeding range

In agricultural areas, stands of shrubs, thickets, edges, underbrush, and woodlands all contain nests of several bird species vulnerable to parasitism by the cowbird. Baie de L'Isle-Verte National Wildlife Area, northeast of Rivière-du-Loup. *August 1991, Léo-Guy de Repentigny.*

Habitat and behaviour

The cowbird is the only obligate brood parasite, breeding exclusively through brood parasitism, in North America (Brittingham and Temple 1983). It frequents open habitat such as pastures, fields, and urban lawns (Middleton 1988b). Although forest habitat is not well suited to its feeding requirements, forest fragmentation has enabled the species to penetrate woodlands, where it can parasitize nests and at the same time feed along the edges (Brittingham and Temple 1983). In fact, it may lay its eggs up to 4 km from open habitat (Darveau 1988).

Despite its dull plumage, the cowbird is easily recognized and observed. On arriving in Québec at the end of March, it perches conspicuously to deliver its distinctive song (Bent 1958). The males arrive before the females (Friedmann 1929). The mating system varies throughout the range, and depending on the sex ratio of the local population and the density of host species nests, cowbirds may be monogamous, polygynous, polyandrous or promiscuous (Ankney and Scott 1982; Teather and Robertson 1986; Yokel 1986). In general, in forest habitat where nest density of host species is high, cowbirds are monogamous (Dufty 1982). In prairie habitat, where nest density is low, breeding cowbirds do not form any sort of lasting bond and are promiscuous (Elliott 1980).

Since the cowbird has only recently become established in this part of the world, the species it parasitizes are extremely vulnerable, as few have yet developed defence mechanisms (Rothstein 1975). Some hosts do, however, exhibit certain reactions, such as deserting the nest, constructing a floor over the eggs, breaking the parasitic egg, or ejecting it from the nest (Rothstein 1975; Graham 1988b).

In southern Québec, Terrill (1961) reported that some 40 species are parasitized by the cowbird. In North America, more than 220 species have been documented as cowbird hosts, of which only 144 have successfully reared young cowbirds (Lowther 1993). The female cowbird is not very discriminating when choosing a host, which results in highly variable survival rates. Although young cowbirds are altricial and are mainly insectivorous, the female may lay in nests of precocial (Hatch 1971) or strictly granivorous species (Middleton 1991). In such cases, the young cowbird's chances of survival are low to nil, given the incompatibility of host and parasite.

In southern Québec, Terrill (1961) noted that the Red-eyed Vireo was the most common cowbird host, with the Black-throated Blue, Mourning, Chestnut-sided, and Black-and-white warblers, as well as the Eastern Phoebe, also frequent victims. The parasitism rate in populations of some passerines may be greater than 50% (Elliott 1978; Zimmerman 1983).

The female cowbird normally only lays one egg per nest, but a nest may be parasitized by more than one female cowbird. This Common Yellowthroat nest contains 3 cowbird eggs, which means that the warbler will have its work cut out for it, unless it decides to abandon the nest. Sainte-Foy. *July 1959, Gaston Le Page.*

Notes on territory (continued)

In Ontario, Darley (1982) found that males have home ranges of 0.4–25 ha (mean 7.9 ha), which are generally larger than those of all of their mates; polygynous males had home ranges that overlapped those of their mates. Teather and Robertson (1986) also noted that in Ontario the home range of the male is greater than that of its mate: they reported areas of 5.5 ha for the male and 3.5 ha for the female of a monogamous pair, 4.7 ha for the male and 2.5 and 3.2 ha for two females of a polygynous male, and 4.9 ha for the male and 2.0, 2.6, and 3.1 ha for the three females of another polygynous male.

In Ontario, Yokel (1989) recorded densities of 3.0 males/10 ha and 1.8 females/10 ha. Stewart and Kantrud (1972) noted densities of 6 pairs/10 ha in North Dakota.

Clutch size, by the usual definition—*x* number of eggs laid one or more days apart—is not applicable to the cowbird. The cowbird's laying cycle is similar to that of a chicken that is not allowed to complete its clutch, a phenomenon that has not been observed in any other wild passerines (Scott and Ankney 1983; Jackson and Roby 1992). The cowbird generally lays a set of 1–7 eggs at a rate of about one a day, with intervening pauses of a few days (typically 2) (Scott and Ankney 1983). One female in captivity reportedly laid one egg a day for 32 consecutive days (Jackson and Roby 1992). Scott and Ankney (1980) reported that in southern Ontario a female may lay 40 or so eggs in one breeding season.

Thus it is not surprising that the Brown-headed Cowbird is considered responsible for decreased abundance of certain passerines (Mayfield 1977; Goldwasser et al. 1980; Clark and Robertson 1981; Finch 1983). The offspring of the host birds are generally smaller than the young cowbirds, and their chances of survival in the company of the cowbird are low. Breeding success in the host may be affected in several ways. The female cowbird may remove the eggs or young of the host from the nest, or she may eat the eggs (though not the young) (Earley 1991; Scott et al. 1992). Usually, she lays only one egg per nest. The incubation period of the cowbird is usually shorter than that of the host species, so that the young cowbird is often the first to hatch. It maintains its advantage by its rapid growth rate, which it owes to vociferous begging for food from its foster parents. Young cowbirds always receive more food than the young of the host (Woodward 1983; Lowther 1993) and may even accidentally crowd the other nestlings out of the nest (Bent 1958).

This Common Yellowthroat nest contains two nestlings; the larger is a 3-day-old cowbird and the smaller a Common Yellowthroat. Being dominant because of its size, the young cowbird will receive most of the food and thus will develop more quickly. Smaller nestlings are often crowded out of the nest. Snowdon, Montréal. *June 1931, Henry Mousley, Laval University Collection.*

Although there are many reasons for considering the cowbird a nuisance, its impact on the population dynamics of host species is not easy to assess. The fragmentation of forest habitat that has most assuredly been the cause of the cowbird's invasion has at the same time reduced habitat availability and increased the vulnerability to nest predation of several passerines (Yahner and Delong 1992).

The Brown-headed Cowbird was reported in 41.4% (1021/2464) of Atlas squares surveyed, which gives an indication of its visibility and high rate of occurrence in most of the inhabited regions of southern Québec. Most breeding confirmations were based on records of newly fledged young or discovery of an egg or young cowbird in the nest of a host species.

Distribution

The Brown-headed Cowbird is found only in North America. It breeds in the Mackenzie District of the Northwest Territories as well as in all 10 Canadian provinces and is fairly widespread throughout the United States and Mexico (National Geographic Society 1987). The cowbird nests in almost all of the settled regions of Québec. Its distribution coincides with the occurrence of open habitat, especially that created by agriculture. Breeding evidence was obtained all the way to the Havre-Saint-Pierre area and in the Magdalen Islands. Sight records but no breeding confirmations were obtained around Chibougamau, in the northernmost regions of southern Québec, and on Anticosti Island, where Ouellet (1969) and Godfrey (1986) described it as breeding.

There is a scattering of records from north of the Atlas area. Savile (1950) reported two birds that were observed a few times at Kuujjuarapik in June 1949, and Todd (1963) reported a young specimen taken at the same location on 15 August 1935. More recent sightings include a bird observed in the La Grande Rivière area in 1982 (Côté 1982) and one seen on 29 April 1991 near Cabbage Willows Bay (Cons. G. G. – GREBE 1992o).

History and trends

The Brown-headed Cowbird is thought to have been derived from South America ancestors which spread through Mexico to the central plains of North America by following roving herds of wild cattle (Bent 1958). Gradually it extended its range as European settlement opened the forest (Bent 1958; Lowther 1993). During the 19th century, the species became established in the Great Lakes region and in the northeastern United States, then spread into the southeastern states and Canada (Robbins et al. 1986). By the beginning of the 20th century, the cowbird had reached Québec (Lowther 1993). Westward expansion occurred in a similar fashion during the 19th century (Lowther 1993).

BBS data for Canada indicate an increasing population trend between 1966 and 1979 (Robbins et al. 1986). In Québec, however, a decline in numbers was recorded between the late 1960s and late 1980s (Falardeau and DesGranges 1991; Cyr and Larivée 1993b). The decline in the United States has been attributed mainly to a decrease in cropland resulting from reforestation (Ellison 1985f; Brauning 1992a). The acreage of land in Québec used as pasture has also been decreasing steadily since the beginning of the 1970s (Falardeau and DesGranges 1991). Erskine et al. (1990) and Falardeau and DesGranges (1991) hold that blackbird control programs targeting wintering populations in the United States have had repercussions on Brown-headed Cowbird populations.

Christian Houle and Line Choinière

Northern Oriole*

Oriole du nord

Icterus galbula

Christiane Girard

Only when the trees begin to leaf out and apple blossoms scent the air does the Northern Oriole make its appearance here. Males two years or older, in brilliant orange and black livery, rival each other in song from the topmost branches of tall shade trees.

They arrive from Mexico and Central America, where they spend the winter. Orioles sometimes winter in the southeastern United States (Root 1988) and occasionally as far north as New York State (Bull 1985), Nova Scotia (Tufts 1961), and New Brunswick (Squires 1976).

Orioles can be attracted to feeders—similar in appearance to hummingbird feeders—containing coloured sugar water or by orange sections nailed to feeders. Bits of string for nest making may be hung from branches or a clothes line.

Habitat and behaviour

The preferred habitat is an open or semi-open area dominated by tall deciduous trees such as elm, maple, willow, or poplar. Orioles also frequent orchards, tall hedgerows, and open forest near water, in city and country alike.

On arriving in early May, males establish territories, from which they sing. Neighbouring territories often adjoin where tall trees and food are plentiful (Pleasants 1979; Sealy 1980b; Beletsky 1982a). The males sing incessantly before the females arrive, but frequency of singing decreases once pair bonds are established and as the nesting season progresses (Beletsky 1982a). The female also gives song and calls similar to those of the male (Beletsky 1982b). Pairs form quickly, often on the morning the female arrives, and the female may be the instigator in pair formation (Butcher 1991). Courtship behaviour is relatively unaggressive and of short duration. Other birds that intrude on the male's territory are pursued, however (Sealy 1980b; Butcher 1991). Intruding males will sometimes try to copulate with the female (Flood 1985; Edinger 1988).

The female builds the nest alone, often outside the territory defended by the male (Edinger 1988; Butcher 1991). A skilful weaver, she uses hundreds or even thousands of plant fibres to construct the pouchlike structure over a period of 4–8 days (Tyler 1958a; Orians 1985a). This seemingly fragile structure, which can vary depending on the nature of the local climate, will withstand high winds and inclement weather. Often it is hung from the tip of a bough where it is concealed by leaves and is out of the reach of predators, such as crows, owls, and squirrels (Orians 1985a). When the inside and outside are completed, the female lays the eggs which she alone incubates. Nests are occasionally parasitized by the Brown-headed Cowbird, but this is fairly rare (Peck and James 1987), as this oriole is a known rejecter species that disposes of parasitic eggs by puncturing them with its sharply pointed bill and levering them out (Rothstein 1977; Rohwer et al. 1989).

Notes on territory

On arriving on the breeding grounds, the male establishes and defends a territory from which it sings (Sealy 1980a; Butcher 1991). The nest, however, is often sited outside the territory (Butcher 1991). Sealy (1980b) reported that the immediate vicinity of the nest (≤ 5 m) is vigorously defended by the male, and occasionally by the female, against intruding conspecifics. Sealy also stated that pair density varies as a function of the food supply. Density increased between 1976 and 1977 from 5.4 to 10.5 breeding pairs/ha when an outbreak of forest tent caterpillars occurred, reverting to 5.3 breeding pairs/ha in 1978 once the outbreak died down.

Pleasants (1979) reported that in some riparian habitat in the southwestern United States, the western subspecies (Icterus galbula bullockii) may nest either in colonies or in solitary pairs, depending on the abundance of food and availability of nesting habitat.

* The Northern Oriole (*Icterus galbula*) has recently been split into three species (AOU 1995). Two occur in Québec: the Baltimore Oriole (*Icterus galbula*, Oriole de Baltimore) is a migrant breeder, and Bullock's Oriole (*Icterus bullockii*, Oriole de Bullock) is a visitant.

Northern Orioles particularly favour tall trees beside rivers or in open woodland near water. Saint-François Bay, Lake Saint-Pierre. *October 1993, Pierre Normand.*

Northern Oriole

Breeding distribution
1984–1989

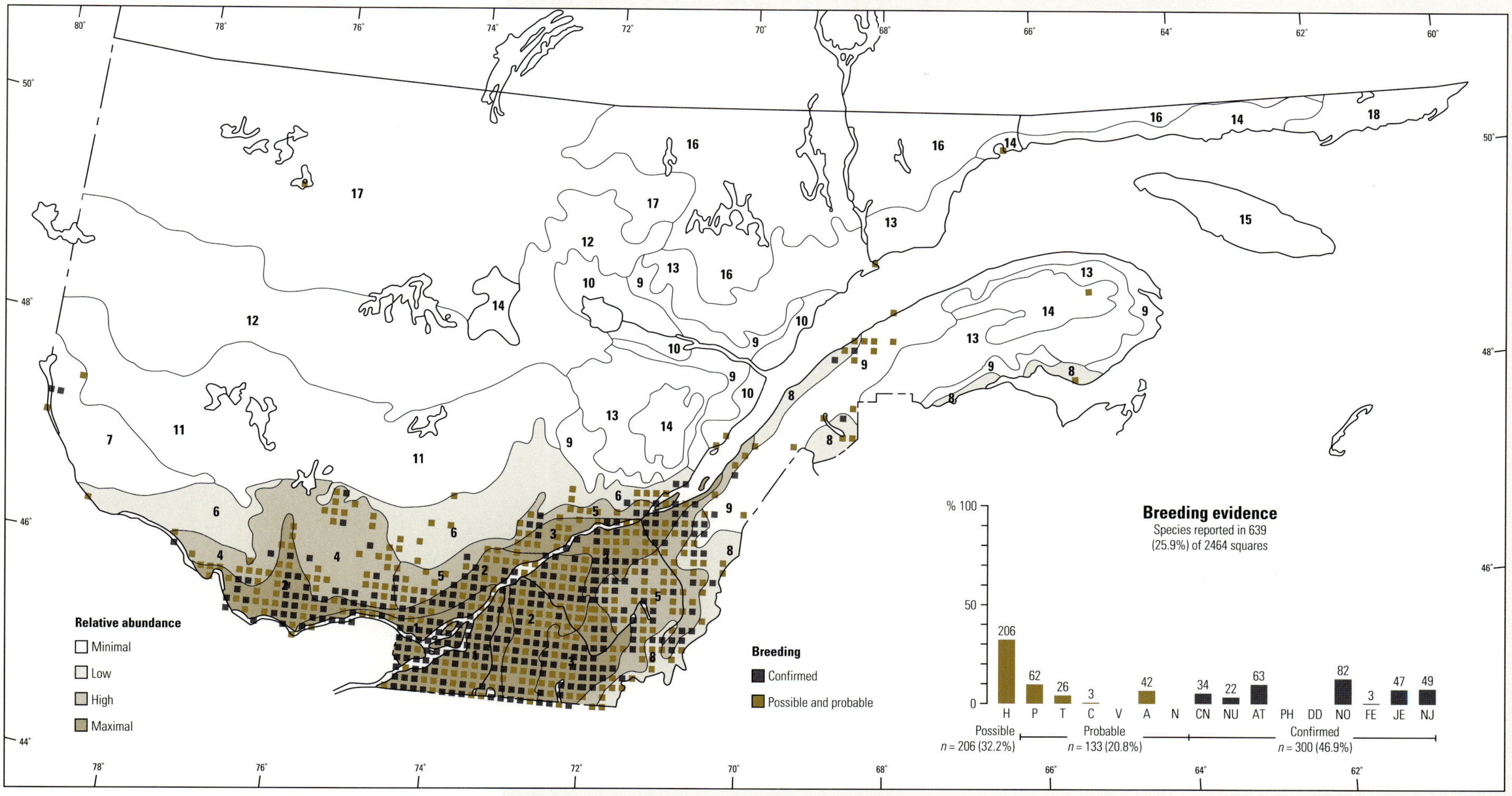

World breeding range

Species profile

Clutch size	usually 4–5 eggs (2–6)[1,2,4]
Laying interval	?
Start of incubation	?
Incubation period	12–15 days[3]
Maturity at hatching	altricial
Care of young	by female and male[5]
Nestling period	11–14 days[11]
Age at first flight	12–14 days[4]
Dependence of young	probably until 10 days after leaving nest[1]
Broods per year	1[4]
Breeding age	1 year or more[2]
Mating system	monogamy, sometimes polygyny[5,8]
Duration of pair bond	1 season[1,7]
Total length	18.0–22.8 cm[3]
Wing span	28.6–31.1 cm[9]
Weight	M 34.3 g (22.3–41.5) / F 33.2 g (28.1–41.3)[10]
Longevity record	11 years 7 months[6]

References:
1. Sealy 1979
2. Labedz 1984
3. Godfrey 1986
4. Ehrlich et al. 1988
5. Flood 1985
6. Klimkiewicz and Futcher 1987
7. Butcher 1991
8. Edinger 1988
9. Terres 1980
10. Dunning 1984
11. DeGraaf and Rudis 1987

Chronology $n = 197$

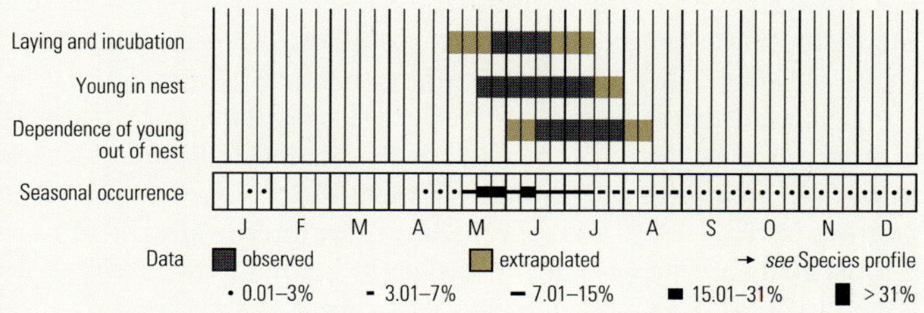

Frequency of occurrence by ecological region

Ecological regions** (forest regions)	Adequately covered squares	Squares in which reported	% of squares in which reported	Relative abundance (%)*
1. Sugar maple-hickory and sugar maple-basswood	97	96	99.0 ± 1.4	100.0
2. Sugar maple-basswood	153	147	96.1 ± 1.8	97.1
3. Sugar maple-basswood and sugar maple-yellow birch	101	95	94.1 ± 3.0	95.0
4. Sugar maple-yellow birch-basswood	56	37	66.1 ± 10.9	66.8
5. Sugar maple-yellow birch-beech and sugar maple-yellow birch-basswood	113	85	75.2 ± 4.5	76.0
6. Sugar maple-yellow birch-beech	71	27	38.0 ± 10.3	38.4
7. Sugar maple-yellow birch-balsam fir	19	3	15.8 ± 17.9	16.0
8. Sugar maple-yellow birch and balsam fir-yellow birch	47	22	46.8 ± 12.8	47.3
9. Balsam fir-yellow birch	87	8	9.2 ± 5.8	9.3
10. Balsam fir-white birch or balsam fir-red maple	46	2	4.3 ± 5.7	4.4
11. Yellow birch-balsam fir	52	1	1.9 ± 4.5	1.9
12. Balsam fir-white birch (west)	59	0	0	0
13. Balsam fir-white birch (east)	76	1	1.3 ± 3.0	1.3
14. Balsam fir-black spruce	28	1	3.6 ± 8.3	3.6
15. White spruce-balsam fir and balsam fir-white spruce	22	0	0	0
16. Black spruce-balsam fir-moss	18	0	0	0
17. Black spruce-moss	32	0	0	0
18.			*not analysed*	

* minimal: 0–29.9%; low: 30–59.9%; high: 60–89.9%; maximal: 90–100%

** Deciduous forest: 1–7; mixed forest: 8–11; coniferous forest: 12–17

A Northern Oriole might nest in this tall poplar growing beside the water. Sainte-Foy. *August 1991, Yves Aubry.*

Both parents feed the young caterpillars of all sorts, as well as ants, aphids, Colorado potato beetles, and other insects gleaned mostly from the leaves, bark, and branches of trees (Sealy 1980b). Adults also eat fruit and nectar. When the time comes for them to leave the nest, the young cling to the outside and climb in and out several times before taking flight. According to Sealy (1979), the young are probably dependent at least until they are capable of sustained flight, 10 or so days after leaving the nest. The female, however, leaves her mate and fledglings a few days before the male leaves the young. Females roam widely and begin their moult earlier than males; from this point on, both sexes grow increasingly silent and secretive (Sealy 1979).

The male's loud singing and brilliant plumage, the open habitat it frequents, and the characteristic shape of the nest—highly visible in autumn after the leaves have fallen—all make the oriole an easy bird to detect. The large number of sightings and high rate of breeding confirmation (46.9%) show how easy the species is to locate during the breeding season.

Sealy (1979) reported that Northern Orioles are territorial on the wintering grounds. Schemske (1975) observed an oriole defending a territory containing a bed of nectar-producing flowers.

Distribution

The Northern Oriole breeds from southern Canada to northern Mexico (AOU 1983). In Québec, it is widely distributed in the Central St. Lawrence Lowland as far east as Île d'Orléans, but distribution is patchier in the Laurentians and Appalachians. Forest fragmentation and thinning of riparian woodlands in these regions and in the Ottawa Valley provide favourable nesting habitat.

Distribution is spotty in Témiscamingue and along the St. Lawrence east of Québec City, at Baie-Saint-Paul, Baie-Comeau, Sept-Îles, near Kamouraska, and near Murdochville and Bonaventure in the Gaspé. Local populations around Rimouski and Lake Témiscouata are proof that the species is gradually becoming established in areas from which it once was absent. These areas represent an extension of the range from the Maritimes (Erskine 1992). Occurrence in all of these regions indicates the species' partiality for habitat that has been modified by humans.

History and trends

In the days before European settlement, the Northern Oriole, a canopy-dwelling species, was probably confined to shoreline scrub and forest clearings in the southern part of the province. Openings in the continuous forest and edge habitat created by the gradual spread of urban development have been favourable to the species.

In the past, Mousley (1916a) considered this oriole fairly common at Hatley in the Eastern Townships, as did Lloyd (1923, 1944) in the Ottawa–Hull region, and Wintle (1896) described it as common in the Montréal area. The situation was apparently different in the Québec City region, where Dionne (1906) reported having seen the bird only twice. These items of information would seem to suggest that the Northern Oriole has been extending its range eastward during the 20th century. In the southeast of the province, the species is still uncommon in the Lower St. Lawrence (Gendron and Gauthier 1984) and rare in the Gaspé (Girard 1988). Two nests have been found in the Rouyn-Noranda region, the first at Lake Pelletier during the breeding season and the second at Lake Beauchastel in autumn (J. Lapointe, pers. comm.). There are several summer sightings from the

The female oriole may hang her nest from the tip of a drooping branch of elm or weeping willow, well out of reach of predators. When the leaves disappear in the fall, nests can be spotted in the topmost branches of these tall trees. Cap Tourmente National Wildlife Area, northeast of Québec City. *August 1991, Léo-Guy de Repentigny.*

Northern Orioles are also found in orchards. Saint-Augustin-de-Desmaures, west of Québec City. *Summer 1993, Jean Gauthier.*

An oriole sitting on its nest in a white elm. The nest, tightly woven by the female, is a deep pouch, hung from the tip of a bough. Warwick, south of Victoriaville. *July 1986, Denis Faucher.*

Saguenay–Lac-Saint-Jean region, although breeding has not yet been confirmed there (Breton 1980; Club des ornithologues amateurs du Saguenay–Lac-Saint-Jean 1993a).

Population trends are described differently in the various sources: Robbins et al. (1986) reported an increase in the eastern part of the continent and for the United States and Canada overall; Falardeau and DesGranges (1991) also noted an increase in open habitat in southern Québec. On the other hand, Erskine et al. (1992), Cyr and Larivée (1993a), and Hill and Hagan (1991) all reported a decline. The magnitude of the increases or decreases cited were generally small.

Certain factors, such as abundance of food, may have an effect on local populations. A study by Sealy (1980b) showed that outbreaks of the forest tent caterpillar did not produce any increase in clutch size or productivity, although the number of nesting pairs may double in an infested area. Compared with normal years, a greater number of pairs of yearlings, males and females of similar plumage, was observed. These young pairs begin nesting somewhat later in the season than 2-year-old birds and appear to produce fewer eggs (Sealy 1980b; Labedz 1984). After an infestation, pair density returns to pre-epidemic levels (Sealy 1980b). Adults may benefit from the abundance of prey, especially if it coincides with the nestling phase (Edinger 1988). DeGraaf (1987), however, reported that surveys in oak forests in Pennsylvania during an outbreak of the gypsy moth, a defoliator, showed a decline in the numbers of singing orioles.

Desneige Perreault and Yves Aubry

Pine Grosbeak

Durbec des pins

Durbec des sapins

Pinicola enucleator

Christiane Girard

IN SOUTHERN QUÉBEC, THE BEST CHANCES OF SEEING THE PINE GROSBEAK, a secretive bird of the boreal forest, are in winter, when it is sometimes driven south in search of better feeding grounds. The Pine Grosbeak is gregarious in winter and travels in flocks (Terres 1980; Cramp 1994a). At this time of the year, it frequents conifers and occasionally deciduous trees, particularly fruit trees, where it feeds on the seeds, buds, and fruits (Wintle 1896; Taverner 1922; Terres 1980; Cramp 1994a). The Pine Grosbeak is quite tame and is usually easily approached (Bent 1968).

This large finch is about the same size as an American Robin. The male is easy to recognize with his pinkish-red plumage. Females and immature birds are both greyish, although the female has a dull yellow tinge and the immature birds have a reddish tinge.

During the breeding season, the Pine Grosbeak is one of the most difficult species to find, mainly because of its secretive behaviour and relative scarcity (Erskine 1977). In North America, the species winters in the southern part of its breeding range in two distinct areas: from southern Alaska and Yukon south along the Rocky Mountains to northern New Mexico, and from central Manitoba to the Atlantic provinces, including southern Ontario and southern and eastern Québec (National Geographic Society 1987).

Habitat and behaviour

During the breeding season, the Pine Grosbeak is mainly associated with coniferous woodland and, to a lesser extent, mixed forest (AOU 1983; Godfrey 1986). It is found on the edges of open places such as clearings and small bodies of water (Terres 1980; Godfrey 1986). The species also frequents the margins of hayfields and pastures (Terres 1980).

Pairs may form as early as March (Bent 1968). Courtship displays include the male offering food to the female (Adkisson 1981), which encourages him with soft cries. She builds the loosely woven cup-shaped nest from twigs; it is well hidden on a branch near the trunk or in the fork of a tree (French 1954), generally a spruce or other conifer or a shrub (DeGraaf and Rudis 1987), between 0.6 and 10.0 m up depending on the type of habitat (French 1954).

The clutch of 2–5 eggs is incubated by the female for 13–15 days (Pulliainen 1979). The male feeds her while she is incubating and for a few days after the eggs hatch (Pulliainen and Hakanen 1972). When the female is disturbed on the nest, she rarely flushes until the intruder approaches to within 30 cm or less (Pulliainen 1979), and some birds even let themselves be taken by hand (Pulliainen and Hakanen 1972; Pulliainen 1979).

Both male and female take turns caring for the young, removing fecal sacs from the nest and feeding the young by regurgitation (French 1954; Pulliainen and Hakanen 1972). During the brief brood-rearing period both parents develop a pair of gular pouches in the floor of the mouth that extend for a full centimetre or more along the sides of the tongue and are used to carry food for the young (Ehrlich et al. 1988). The sacs develop at the same time as the gonads, and their development may be controlled by a similar physiological mechanism (French 1954). Examination of these sacs in three specimens collected in Utah found that they all contained seeds of Engelmann spruce; one specimen contained primarily seeds of moss-campion along with some spruce seeds and insect parts, and in another specimen, tender new growth from the tips of spruce branches was also present (French 1954).

The young remain in the nest for roughly 14 days (Pulliainen 1979). French (1954) reported adults feeding newly fledged young in mid-July in Utah. According to DeGraaf and Rudis (1987), the Pine Grosbeak raises only one brood a year.

The species' song is a varied and musical warble, like a softer version of the Purple Finch's song (Bent 1968). Both in summer and winter, the male mimics other species occurring in the same habitat, and its vast repertoire includes imitations of the Hairy Woodpecker, Gray Jay, American Robin, and Greater Yellowlegs (Taylor 1979; Wells 1981).

Notes on territory

Territoriality in the Pine Grosbeak is hard to characterize. According to Erskine (1977), the cardueline finches, including the Pine Grosbeak, are largely non-territorial. French (1954) estimated the species' territory to be about 360 m in diameter, but qualified this by commenting that pairs, particularly the males, did not defend the territory actively and allowed intruders to come within a few metres of the nest.

In the Great Whale River region, the density of breeding pairs was found to be highest in black spruce-lichen woodland (Cons. G. G. – GREBE 1991a). In the Eastmain River area, Tremblay and Belles-Isles (1991) found the highest densities in shoreline scrub (a very unusual finding), jack pine stands, and recent burns.

The Pine Grosbeak nests mainly in coniferous woodland, like that on the summits of the Gaspé Peninsula's rugged interior. Gaspé Conservation Park. *July 1991, Yves Aubry.*

Pine Grosbeak

Breeding distribution
1984–1989

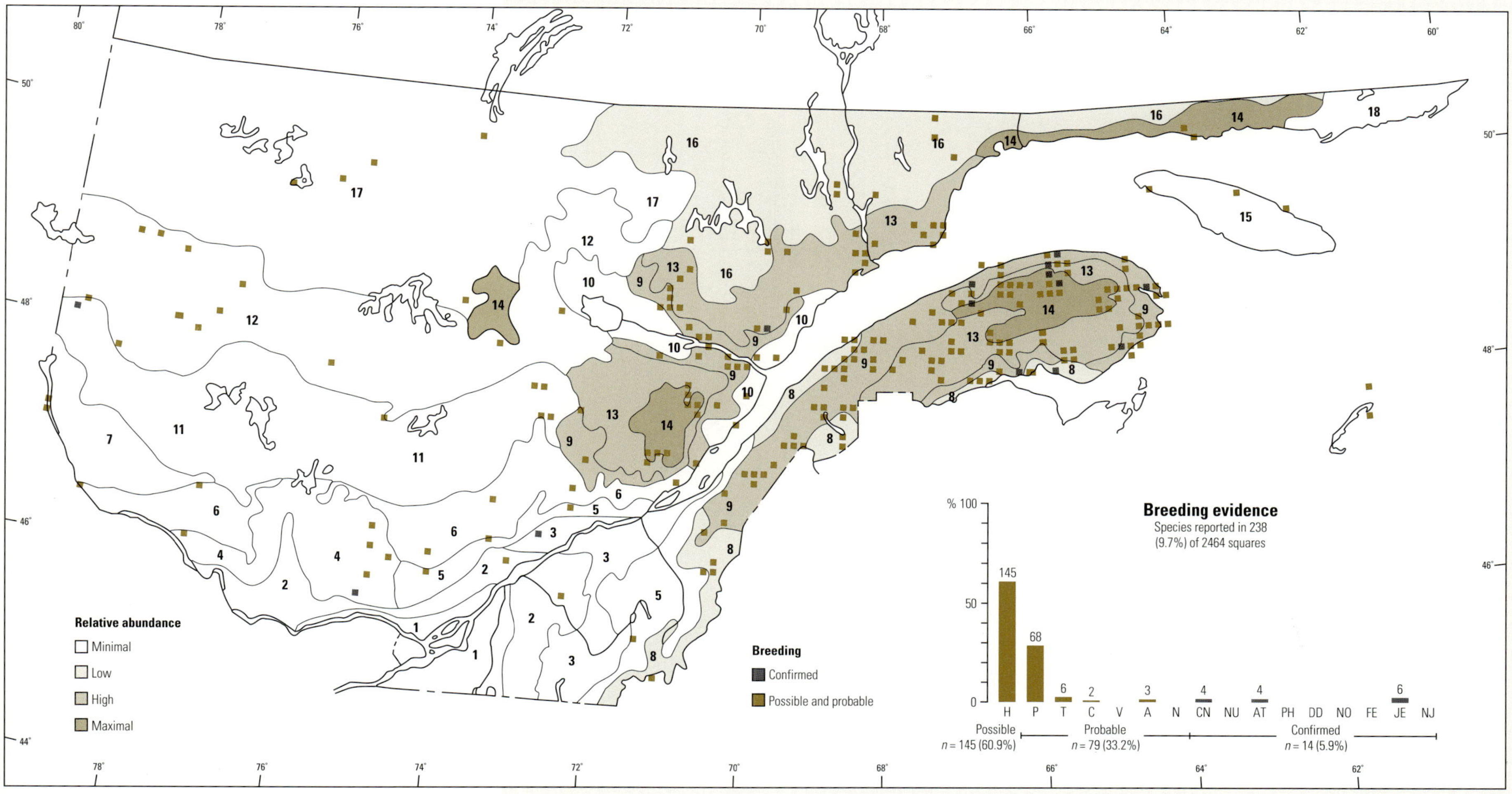

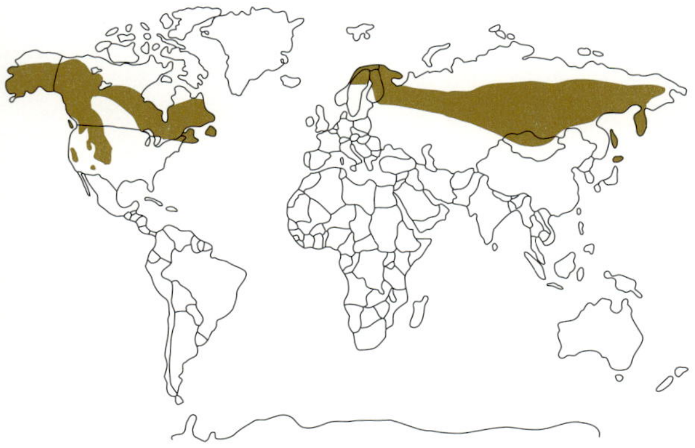

World breeding range

Species profile

Clutch size	usually 4 eggs (2–5)[6,7]
Laying interval	?
Start of incubation	?
Incubation period	13–15 days[5,6,7]
Maturity at hatching	altricial
Care of young	by female and male[5,6]
Nestling period	14–18 days[5,6,7]
Age at first flight	18–23 days[7]
Dependence of young	until ≈ 3 weeks after leaving nest[7]
Broods per year	1[5]
Breeding age	1 year[8]
Mating system	monogamy[5,6]
Duration of pair bond	?
Total length	23.0–24.5 cm[1]
Wing span	30.5–33.0 cm[2]
Weight	56.4 g (52.0–62.0)[3]
Longevity record	9 years 9 months[4]

References:
1. Godfrey 1986
2. Terres 1980
3. Dunning 1984
4. Klimkiewicz and Futcher 1987
5. Ehrlich et al. 1988
6. Pulliainen and Hakanen 1972
7. Adkisson 1977
8. DeGraaf and Rudis 1987

Chronology $n = 4$

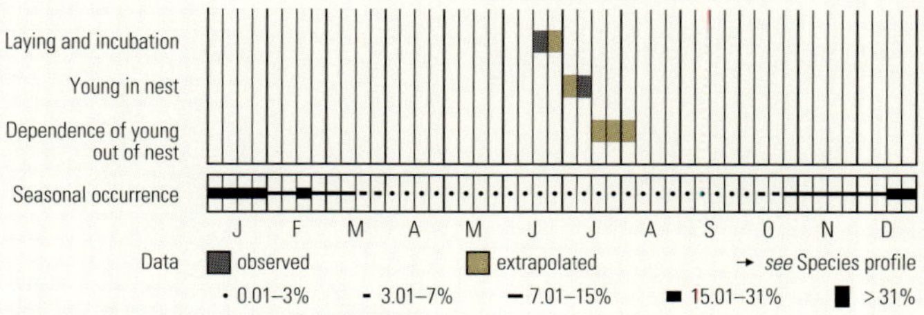

Frequency of occurrence by ecological region

Ecological regions** (forest regions)	Adequately covered squares	Squares in which reported	% of squares in which reported	Relative abundance (%)*
1. Sugar maple-hickory and sugar maple-basswood	97	0	0	0
2. Sugar maple-basswood	153	2	1.3 ± 1.2	1.9
3. Sugar maple-basswood and sugar maple-yellow birch	101	1	1.0 ± 1.6	1.5
4. Sugar maple-yellow birch-basswood	56	5	8.9 ± 6.9	13.2
5. Sugar maple-yellow birch-beech and sugar maple-yellow birch-basswood	113	1	0.9 ± 1.3	1.3
6. Sugar maple-yellow birch-beech	71	5	7.0 ± 5.8	10.4
7. Sugar maple-yellow birch-balsam fir	19	2	10.5 ± 15.5	15.5
8. Sugar maple-yellow birch and balsam fir-yellow birch	47	10	21.3 ± 10.7	31.4
9. Balsam fir-yellow birch	87	53	60.9 ± 9.5	89.8
10. Balsam fir-white birch or balsam fir-red maple	46	4	8.7 ± 7.5	12.8
11. Yellow birch-balsam fir	52	4	7.7 ± 7.9	11.3
12. Balsam fir-white birch (west)	59	5	8.5 ± 7.7	12.5
13. Balsam fir-white birch (east)	76	39	51.3 ± 10.9	75.6
14. Balsam fir-black spruce	28	19	67.9 ± 18.2	100.0
15. White spruce-balsam fir and balsam fir-white spruce	22	2	9.1 ± 13.3	13.4
16. Black spruce-balsam fir-moss	18	6	33.3 ± 24.8	49.1
17. Black spruce-moss	32	3	9.4 ± 11.7	13.8
18.			*not analysed*	

* minimal: 0–29.9%; low: 30–59.9%; high: 60–89.9%; maximal: 90–100%

** Deciduous forest: 1–7; mixed forest: 8–11; coniferous forest: 12–17

The species is found on the margins of open areas in coniferous forests, including clearings and the banks of streams and rivers. The nest is usually built in a spruce tree or shrub. Gaspé Conservation Park. *July 1991, Sylvain St-Onge.*

This bird of the north country breeds in the boreal forest up to the tree line. De Pas River, northeast of Schefferville. *July 1981, Jean-Luc DesGranges.*

In northern Québec, the species is found mainly in wooded valleys, forest islands, and other parts of the forest spared by fire. Caribou River, Hudson Bay region. *August 1979, Pierre Lamothe.*

The Pine Grosbeak was reported in 9.7% (238/2464) of Atlas squares surveyed. Breeding is difficult to confirm because of the species' secretive nature. Confirmations were obtained in only 5.9% (14/238) of the squares in which the species was observed, on the basis of records of adults carrying nesting materials or food, or of downy young incapable of sustained flight. No nests were found.

Distribution

The Pine Grosbeak has a holarctic distribution ranging across North America, Europe, and Russia (Terres 1980). In North America, it breeds along the Rockies from Alaska to New Mexico and from Yukon east to Newfoundland; its breeding range excludes the Prairies and ends at the tree line (National Geographic Society 1987).

The species was atlassed mainly in the northeastern part of southern Québec, particularly where Balsam fir is dominant. Frequency of occurrence was very high in the Balsam fir-black spruce forest region, the forest type found on the mountaintops of the Gaspé Peninsula, the Laurentide Wildlife Reserve, the Lake Marceau region west of Lake Saint-Jean, and on the Middle North Shore. Occurrence was also high in the Balsam fir-yellow birch and Balsam fir-white birch (east) forest regions on the Gaspé Peninsula and in the Lower St. Lawrence, Appalachian, Québec City, and Upper North Shore regions. However, frequency of occurrence was much lower in the western part of southern Québec and on Anticosti Island, and the species was rare or absent in the Central St. Lawrence Lowland, Eastern Townships, and the Chaudière River area. Nevertheless, the Pine Grosbeak was atlassed in a few locations in southern Québec, such as Notre-Dame-du-Bon-Conseil, near Drummondville, and Sainte-Marguerite-de-Lingwick in the Eastern Townships, where it was found in conifer-dominated stands. According to David (1980a), the Pine Grosbeak does not breed in the Central St. Lawrence Lowland or the southern Appalachians.

North of the Atlas area, the Pine Grosbeak breeds to the tree line (Todd 1963). Over the years, the species has been observed sporadically near Gagnon and Fermont, near Lake Mistassini, in the Nottaway, Broadback, and Rupert river watersheds, near the Eastmain River, near Lac à l'Eau Claire and Lake Guillaume Delisle in the Great Whale River region, and near Kuujjuaq (Godfrey 1949; Hildebrand 1950; Bordage and Aubry 1982; Cons. G. G. – GREBE 1991a, 1992o; ÉPOQ).

Pine Grosbeaks do not migrate in the strict sense of the word, although in winter they may irrupt out of the boreal forest to seek better foraging opportunities farther south. They have even been observed on occasion in the southern United States, in Virginia and the Carolinas (AOU 1983). According to Bock and Lepthien (1976b), boreal seed-eating birds irrupt south in winter when the cone crop is poor in the north; such irruptions are unpredictable and vary in intensity. This no doubt applies to the Pine Grosbeak, which feeds mainly on conifer seeds (Erskine 1977).

History and trends

The Pine Grosbeak has been reported in eastern Québec since the turn of the century, when large numbers of birds apparently summered there, particularly in the Magdalen Islands, nesting in stands of white spruce (Macoun and Macoun 1915; Fradette 1992). The species was also fairly common on Anticosti Island in summer (Dionne 1920), as it still was in the 1960s (Ouellet 1969). At the turn of the century, the Pine Grosbeak was a common winter visitant in the Montréal area (Wintle 1896), and in the 1960s, was still considered to be a relatively common, though irregular, winter visitor in the Montréal and Montérégie regions (Ouellet 1974a).

In general, the Pine Grosbeak is considered uncommon, even on its breeding grounds (French 1954; Todd 1963; Erskine 1977). According to Erskine (1978b), the species' poor showing in Breeding Bird Surveys (BBS) from 1966 to 1975 cannot be taken as an indication of overall population trends in Canada. An analysis of ÉPOQ summer records from 1970 to 1989, however, shows a decline in Québec (Cyr and Larivée 1993b). The species' secretive behaviour during the breeding season (Helleiner 1987b), its scarcity (Erskine 1978b), and the remoteness of its preferred habitat (northern and mountainous regions) make it difficult to census adequately in breeding bird surveys.

Louis Breton

Purple Finch

Roselin pourpré

Carpodacus purpureus

Christiane Girard

MANY A WALK IN THE WOODS IS ENLIVENED BY THE MELODIOUS SONG OF the Purple Finch. Delivered from a treetop or in flight, the lively sustained warbling is an aid to detecting the songster himself. Care should be taken, though, not to confuse it with the song of the Warbling Vireo, Pine Grosbeak, or particularly the House Finch. The male's dull rose-red plumage on the sides, head, neck, and back distinguishes it from the male House Finch, whose plumage is brighter red except on the back, which is brownish-grey, though often washed with red. The Purple Finch is a common migrant breeder in southern Québec. In winter, it is uncommon or scarce, depending on the area, and may even be absent in some years (David 1980a; David and Gosselin 1984).

The Purple Finch generally winters from Manitoba east to Nova Scotia and south to Texas and Florida. On the Pacific coast, the wintering grounds extend from British Columbia to Baja California (National Geographic Society 1987).

The Purple Finch nests mainly in coniferous and mixed forest, but also in deciduous woodland provided it can find sufficiently mature conifers. It prefers relatively open stands and edges of clearings, bodies of water, or logging roads to deep, dense forests. Sainte-Anne River, Gaspé Conservation Park. *July 1991, Yves Aubry.*

Habitat and behaviour

During the breeding season, the Purple Finch is found in various types of habitat containing conifers, from fairly open coniferous and mixed forest to even deciduous woodland containing at least a few conifers. It is also found in forest edges, woodlands bordering waterways and roads, marshes, city parks, the ornamental evergreens of urban and suburban residential areas, farm woodlots, and conifer plantations (Terres 1980; Godfrey 1986; DeGraaf and Rudis 1987; Peck and James 1987). The Purple Finch appears to prefer stands of conifers with a diameter (dbh) of at least 23 cm (DeGraaf and Chadwick 1987).

In courtship display, the male hops along the ground with breast puffed out and wings drooping. It beats its wings faster and faster, cocking its tail like a wren, and chippering all the while. Then, with wings blurred like a hummingbird's, it rises straight into the air, about 30 cm off the ground, and drops down next to the female. Sometimes the courtship display takes the form of a song duel between two males which, in the female's presence, sing energetically with crown feathers raised, neck out-stretched, and wings and tail spread (Bent 1968).

The nest is rather difficult to find. It is usually placed at a height of 1.5–18 m, almost always in a conifer, in the densest part of the tree (Godfrey 1986; DeGraaf and Rudis 1987; Peck and James 1987). Peck and James (1987) found nests in both the uppermost and lower parts of trees, although most were in the upper third. Both parents build the nest and feed the young, but only the female incubates, though the male feeds her while she is incubating (Bent 1968). Some Purple Finches produce 2 broods per season (Yunick 1983a; Bull 1985; DeGraaf and Rudis 1987); this may occur in Ontario (Peck and James 1987) and thus perhaps also in Québec. The noisy young are fed by the parents for a fair length of time after leaving the nest (Bent 1968).

The Purple Finch is occasionally parasitized by the Brown-headed Cowbird. Of 16 nests examined in southern Québec, only one (6.2%) was parasitized (Terrill 1961), and of 88 nests examined in Ontario, 35 (39.8%) were found to have been parasitized (Peck and James 1987).

The Purple Finch is mainly granivorous, as suggested by its conical bill. It consumes the seeds of grasses, weeds, and trees such as balsam fir, spruce, maple, birch, and ash. The winter and spring diet also includes small quantities of apple, poplar, maple, and birch buds. In late spring, it catches insects such as beetles and caterpillars, and in summer it eats large amounts of fruit, including strawberries and raspberries, both cultivated and wild, and the fruit of dogwood, elder, and viburnum (Bent 1968; DeGraaf and Witman 1979; Ehrlich et al. 1988). It is easily attracted to bird feeders, where it prefers sunflower seeds, millet, thistle, and canary-grass seed (David and Duquette 1982).

In some winters when food is scarce, irruptions occur (Bent 1968; Bock and Lepthien 1976b; Terres 1980; Yunick 1983b), at which times large numbers may appear at bird feeders (Dunn 1992).

The Purple Finch was reported in 60.1% (1481/2464) of Atlas squares surveyed, and breeding was confirmed in 14.7% (217/1481) of these, mostly on the basis of records of newly fledged young, easily detected by their noisy behaviour.

Distribution

In Canada, the Purple Finch breeds in Yukon, the Mackenzie District of the Northwest Territories, and all 10 provinces (Godfrey 1986). In the United States, it nests in the northeastern states, the Great Lakes region, and the Pacific coast states (National Geographic Society 1987). The species was atlassed as a breeder throughout southern Québec. Frequency of occurrence is high everywhere, although somewhat lower in the Montréal and Montérégie regions, where conifers are less abundant. The low frequency of occurrence on Anticosti Island is, however, difficult to explain.

History and trends

At the turn of the century, the Purple Finch was common in summer in the Montréal and Québec City regions (Wintle 1896; Dionne 1906). In recent years, BBS data from Québec show an increase in population between 1966 and 1989, though ÉPOQ summer data do not show any significant trend for the same period (1970–89); if 1980–89 BBS data are analysed separately, they reveal a decline for Québec.

Wootton (1987) reported that the breeding population of Purple Finches had decreased in the northeastern United States between 1966 and 1979. He also established a correlation between this decline and the increase in density of House Finch populations during the same period. Apparently, these two species compete in summer, as their diets are similar and both nest in coniferous trees. This competition may one day oust the Purple Finch from urban and suburban areas and restrict it to forest habitat. Shedd (1990) noted conflict between the two species in winter at feeding stations, where both are frequent visitors, with the House Finch usually dominant. He concluded that expansion in House Finch numbers could cause a decline in Purple Finch populations, especially during severe winters, although it was not yet demonstrated that the introduction of the House Finch in the eastern United States may be causing a decline in Purple Finch populations.

Bent (1968) reported an apparent decline in the Purple Finch population as a result of the introduction of the House Sparrow. Wootton (1987) tried to determine whether there was any correlation in the occurrence of these two species in areas where the House Finch was present and where it was absent, but found none in the latter case.

GILLES PAUZÉ AND SIMON BARRETTE

Notes on territory

In Québec, McNeil (1964 in Cyr 1974) recorded territories of 0.5–1.4 ha, averaging 0.9 ha. An average of 3.2 ha was obtained by Cyr (1974) for populations nesting in hemlock stands at Mirabel, and an average of 2.1 ha was recorded by Sabo (1980) in subalpine habitat in New Hampshire.

In Québec, densities of 4.9 pairs/10 ha were reported in a balsam fir-black spruce stand at Sainte-Anne-des-Monts (Erskine 1980a), 2.5 pairs/10 ha in mature sugar maple-yellow birch stands at Woburn and north of Beauport (Erskine 1984), and an average density of 1.4 pairs/10 ha in hemlock stands at Mirabel (Cyr 1974). In La Mauricie National Park, DesGranges (1980b) recorded densities of 2.7 pairs/10 ha in a balsam fir stand, 2.2 pairs/10 ha in a white spruce stand, 1.1 pairs/10 ha in a mixed stand and also in a stand of red spruce, and 0.5 pair/10 ha in a white birch stand.

Purple Finch

Breeding distribution
1984–1989

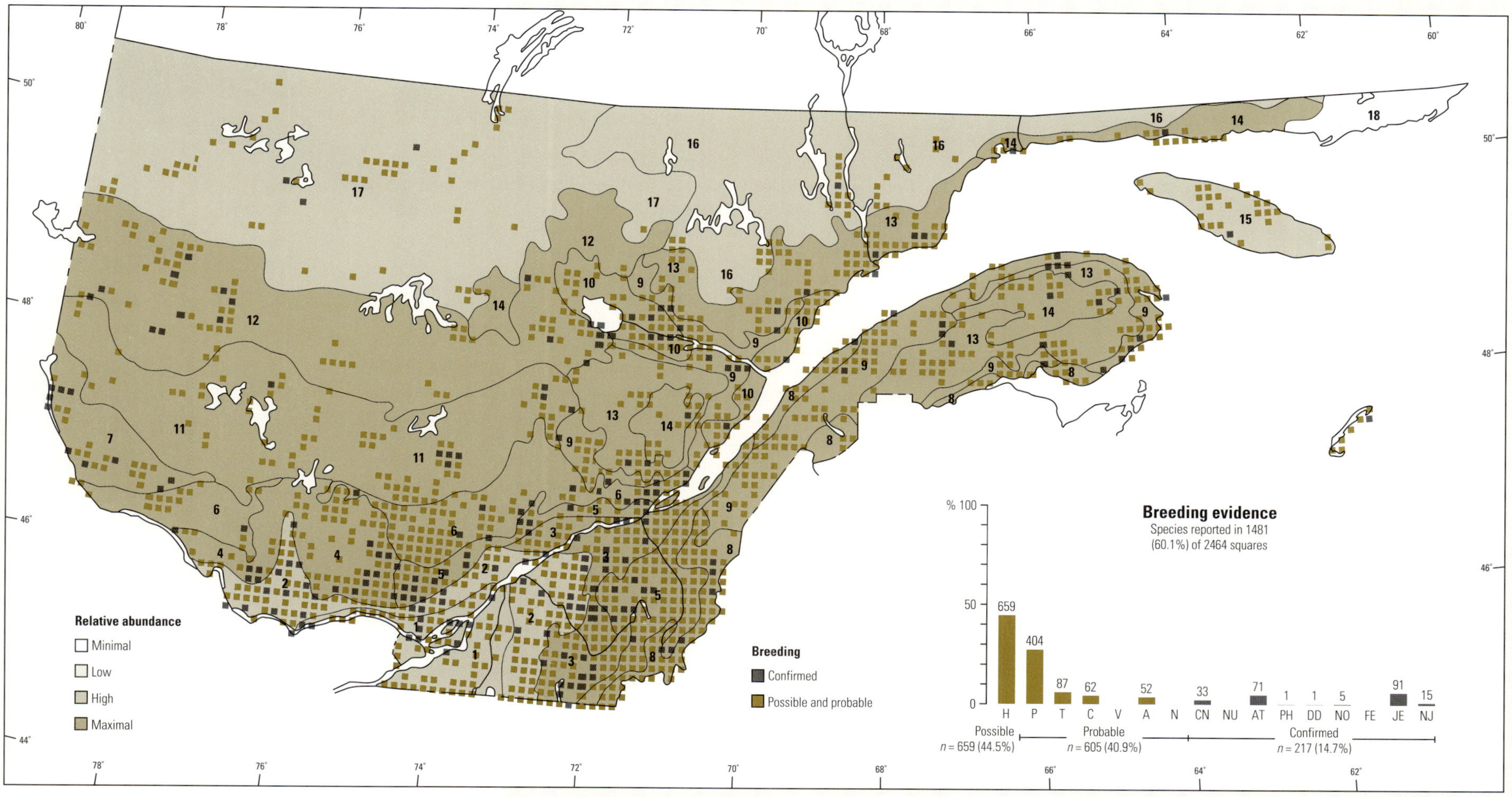

Species profile

Clutch size	usually 4–5 eggs (3–6)[1,2,4,9,10]
Laying interval	?
Start of incubation	?
Incubation period	≈ 13 days[1,2,3,4,9,10]
Maturity at hatching	altricial
Care of young	by female and male[1,2,9]
Nestling period	≈ 14 days[1,2,4,9]
Age at first flight	≈ 14 days or more[8]
Dependence of young	until after first flight[9]
Broods per year	usually 1 (1–2)[1,2,5,10,11]
Breeding age	1 year or more[5]
Mating system	probably monogamy[2]
Duration of pair bond	?
Total length	14–16 cm[3]
Wing span	23.5–26.7 cm[4]
Weight	24.9 g (18.1–35.3)[7]
Longevity record	12 years 8 months[6]

References:
1. Harrison 1978
2. Ehrlich et al. 1988
3. Godfrey 1986
4. Terres 1980
5. Yunick 1983a
6. Kennard 1975
7. Dunning 1984
8. according to nestling period
9. Bent 1968
10. DeGraaf and Rudis 1987
11. Bull 1985

Chronology $n = 44$

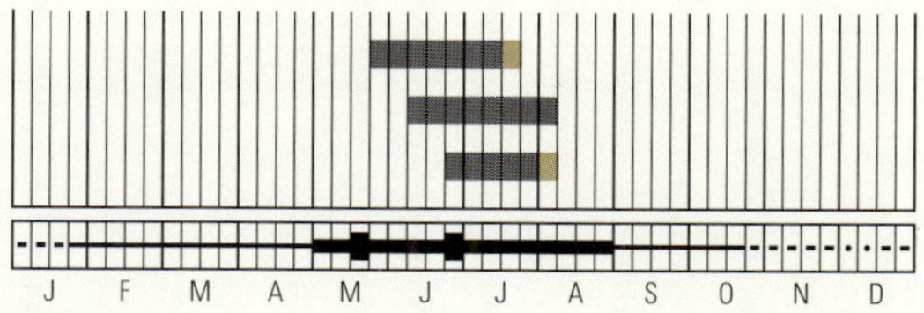

World breeding range

Frequency of occurrence by ecological region

Ecological regions** (forest regions)	Adequately covered squares	Squares in which reported	% of squares in which reported	Relative abundance (%)*
1. Sugar maple-hickory and sugar maple-basswood	97	63	64.9 ± 4.8	64.9
2. Sugar maple-basswood	153	133	86.9 ± 2.9	86.9
3. Sugar maple-basswood and sugar maple-yellow birch	101	95	94.1 ± 3.0	94.1
4. Sugar maple-yellow birch-basswood	56	54	96.4 ± 4.8	96.4
5. Sugar maple-yellow birch-beech and sugar maple-yellow birch-basswood	113	111	98.2 ± 1.7	98.2
6. Sugar maple-yellow birch-beech	71	69	97.2 ± 4.0	97.2
7. Sugar maple-yellow birch-balsam fir	19	18	94.7 ± 12.0	94.7
8. Sugar maple-yellow birch and balsam fir-yellow birch	47	47	100.0 ± 1.1	100.0
9. Balsam fir-yellow birch	87	86	98.9 ± 2.5	98.9
10. Balsam fir-white birch or balsam fir-red maple	46	45	97.8 ± 4.4	97.8
11. Yellow birch-balsam fir	52	49	94.2 ± 7.0	94.2
12. Balsam fir-white birch (west)	59	55	93.2 ± 7.0	93.2
13. Balsam fir-white birch (east)	76	76	100.0 ± 0.7	100.0
14. Balsam fir-black spruce	28	27	96.4 ± 8.3	96.4
15. White spruce-balsam fir and balsam fir-white spruce	22	16	72.7 ± 19.4	72.7
16. Black spruce-balsam fir-moss	18	16	88.9 ± 17.4	88.9
17. Black spruce-moss	32	28	87.5 ± 13.0	87.5
18.			*not analysed*	

* minimal: 0–29.9%; low: 30–59.9%; high: 60–89.9%; maximal: 90–100%

** Deciduous forest: 1–7; mixed forest: 8–11; coniferous forest: 12–17

House Finch

Roselin familier

Carpodacus mexicanus

THE HOUSE FINCH IS NATIVE TO THE WESTERN AND SOUTHERN UNITED States and was introduced accidentally into eastern North America. In the early 1940s, several birds were released on Long Island in New York State by a pet dealer who had been holding them for sale illegally (Elliott and Arbib 1953; Hansrote and Hansrote 1992). Over the next 20 years, small colonies became established, and flocks of House Finches began to be reported in winter, mainly around New York City, on Long Island, and on the Connecticut coast.

This new population increased rapidly after the mid-1960s, at which time a virtual population explosion occurred. Between 1966 and 1979, numbers increased at the rate of 24% a year (Wootton 1986). The House Finch also began expanding its range, not only toward the south and west, but also northward, notably along the New England coast (Nichols 1985a). By 1982, the eastern population had spread to Nova Scotia and Ontario in the north, Tennessee and Georgia in the south, and Michigan, Illinois, and Iowa in the west (Nichols 1985a). Today the House Finch is a common breeder throughout much of the eastern United States and southeastern Canada (Hill 1993).

In Québec, the species was rare before the 1980s, but is now relatively abundant in many locations in the southernmost part of the province.

Unlike native populations in the west, House Finches in the east have undergone major morphological changes, including a significant reduction in size of the wings, tail, and feet (Aldrich and Weske 1978). Moreover, it appears that the harsher climatic conditions in the east have made the species partially migratory, even though the original population is sedentary (Belthoff and Gauthreaux 1991). For example, part of the population in the northeastern United States and the Great Lakes region was found to be moving south in the winter (Hill 1993). In Québec, birds have been banded every summer since the late 1980s in Châteauguay, and results show that some birds remain in the area throughout the winter while others move south. Banded birds were sighted or captured around Coaticook and in the Kinderhook region in New York State and Millerton area in Pennsylvania (R. McNeil, U. of Montréal, unpubl. data).

Habitat and behaviour

Certain aspects of the species' breeding biology, notably its fecundity, have contributed to its success in the northeast. A very early nester—a nest containing 5 eggs was found in early April near Montréal (Yank and Aubry 1985b)—the House Finch can easily produce 2 broods in one season, and up to 4 have been reported in the Montréal region (Yank and Aubry 1984). As each clutch generally contains 4 or 5 eggs, a triple-brooded pair could well produce 8–15 young a year. At this rate, the population could increase by 20% annually and double in a little over 3 years (Kozlovic 1987; Meade 1988b; Paquin 1989).

In eastern North America, the House Finch is normally seen in urban habitat, near human dwellings, and is less common in rural areas. It frequents mostly residential areas where open lawn is interspersed with ornamental trees and shrubs. It prefers nesting sites well sheltered from the wind, such as evergreens and vines growing on buildings, and avoids deciduous trees (Nichols 1985a; Graham 1988a). Nests have been reported built in a variety of sites, including rose trellises, hanging flower baskets, shutters, and overhangs on buildings, an indication of its eclecticism. It occasionally uses bird houses or the abandoned nests of such species as the Cliff Swallow, American Robin, and orioles (Bergtold 1913; Godfrey 1986; Ehrlich et al. 1988; Hill, J. R. III 1988; Paquin 1989).

The House Finch is highly tolerant of conspecifics during the breeding season and may nest in isolated pairs or in loose colonies (Létourneau 1983; Bull 1985; Hill 1993). The nest is built by the female of such flexible materials as grass, twigs, rootlets, plant down, and even string. It is usually found 1.5–6 m up (Nichols 1985a; DeGraaf and Rudis 1987). Typical clutch size is 4 or 5 eggs, laid at the rate of one a day (Bergtold 1913). The nest is often reused for a second clutch (Nichols 1985a).

Christiane Girard

Notes on territory

Both male and female House Finches sing during the breeding season, although females sing less energetically and less often (Bergtold 1913). The male House Finch is a tireless singer and can usually be observed perched conspicuously on a tree branch, making it easy to spot (Kozlovic 1987).

The House Finch nests in loose colonies, but may also nest singly (Hill 1993).

The House Finch nests near urban dwellings in areas with tall shade trees. View of the area around Saint-Lambert (background), south of Montréal, where a pair was found in the spring of 1978, the third sight record for the species in Québec. *September 1978, Robert Parent, Ministère du Loisir, de la Chasse et de la Pêche.*

House Finch

Breeding distribution
1984–1989

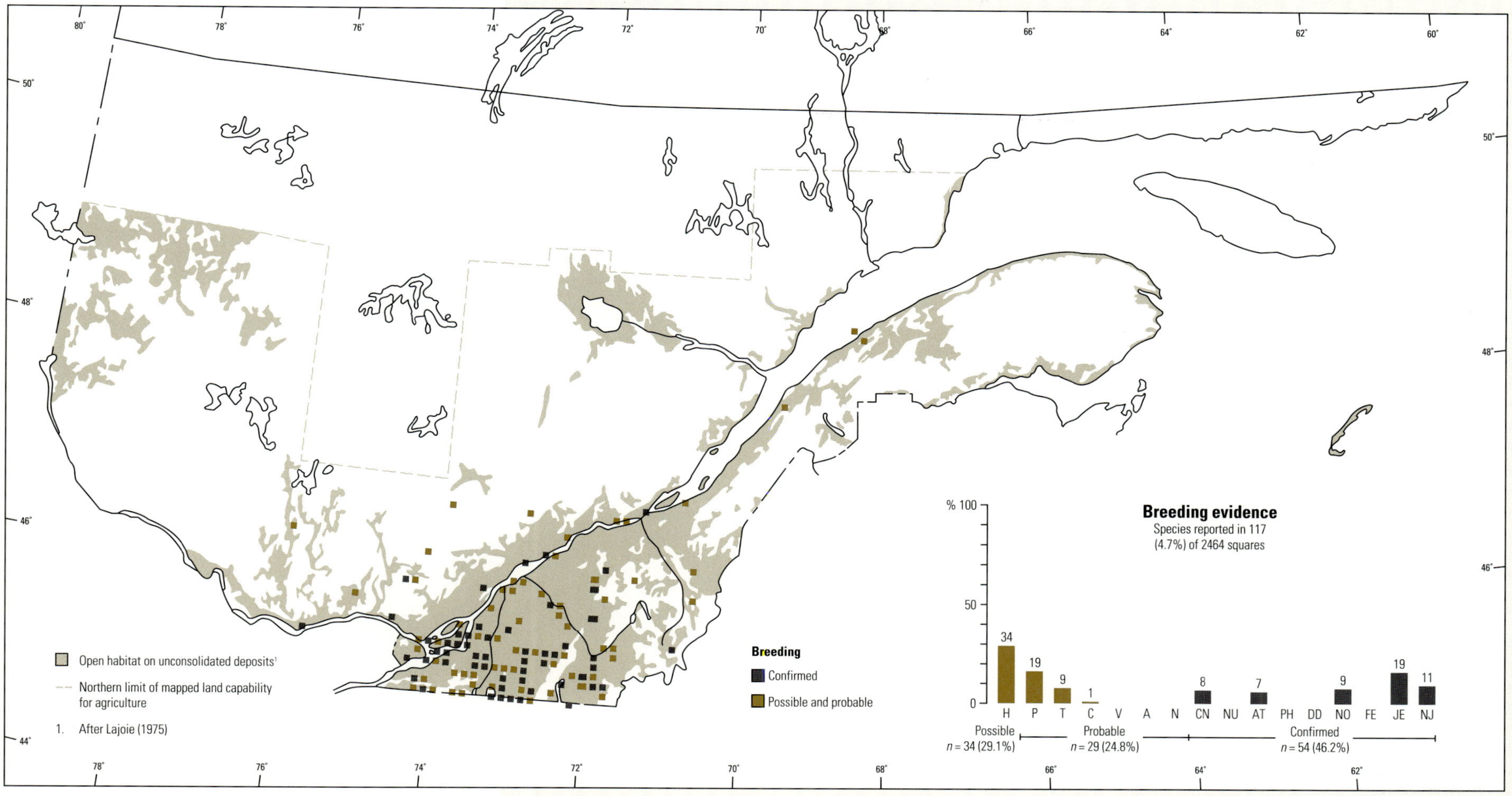

Species profile

Clutch size	usually 4–5 eggs (2–7)[1,4,6,7,11]
Laying interval	1 egg per day[7,12]
Start of incubation	with last egg, sometimes before[14]
Incubation period	12–17 days[1,2,7,9,11,14]
Maturity at hatching	altricial
Care of young	by female and male[7,9,14]
Nestling period	11–19 days[7,9,11]
Age at first flight	11–19 days or more[13]
Dependence of young	?
Broods per year	usually 2 (2–4)[3,4,7,10,11,14,16]
Breeding age	1 year[11,14,15]
Mating system	monogamy, sometimes polygyny[9,14]
Duration of pair bond	usually more than 1 season[5,14]
Total length	12.5–14.0 cm[1]
Wing span	?
Weight	21.4 g (19.0–25.5)[12]
Longevity record	11 years 7 months[8]

References:
1. Godfrey 1986
2. Nichols 1985a
3. Kozlovic 1987
4. Meade 1988b
5. Hooge 1990
6. Wootton 1986
7. Bergtold 1913
8. Klimkiewicz and Futcher 1989
9. Ehrlich et al. 1988
10. Harrison 1975
11. DeGraaf and Rudis 1987
12. Dunning 1984
13. according to nestling period
14. Hill 1993
15. Gill and Lanyon 1965
16. Yank and Aubry 1984

World breeding range

Chronology $n = 22$

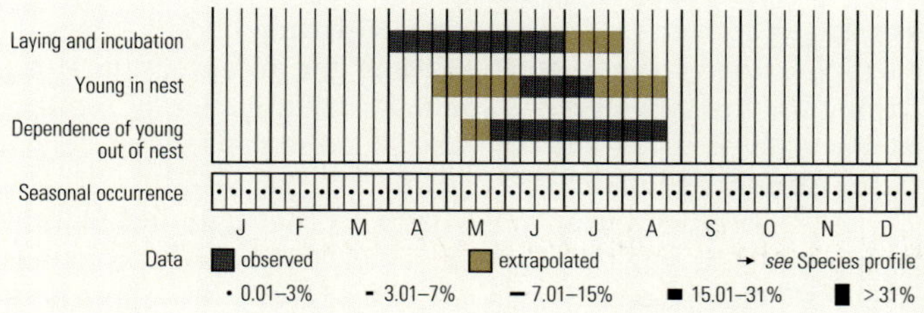

The species inhabits residential areas with lawns and tall trees, where it shows a preference for evergreens. Saint-Lambert, near Montréal. *June 1991, Gaétan Duquette.*

Incubation, by the female, lasts 12–17 days (DeGraaf and Rudis 1987; Hill 1993). Before the onset of incubation, the male remains constantly at his mate's side, energetically driving off any males that attempt to approach her. When courting, the male feeds the female. During the first few days, he simulates regurgitation of food, but later actually feeds her and continues to do so regularly throughout the incubation period (Hill 1993).

The nestlings remain in the nest for 11–19 days and are fed by both parents (Bergtold 1913; Ehrlich et al. 1988). The species is granivorous in all seasons, even though it does feed on buds and fruit at certain times of the year. The young are fed a diet composed almost exclusively of seeds. They may also be fed insects, but these comprise only a negligible portion of the diet (Bergtold 1913; Ehrlich et al. 1988; Hill 1993). Throughout its range the House Finch is a rare cowbird host (Hill 1993), though in Ontario 42.2% of 64 nests examined were found to have been parasitized. However, cowbirds nestlings invariably did not survive, probably because the seed diet fed to them was unsuitable (Peck and James 1987).

During the breeding season, the male sings continually from a conspicuous perch and is easily located, either alone or with the female.

The species was reported in 4.7% (117/2464) of Atlas squares surveyed. In 11 of the 54 squares in which breeding was confirmed, evidence was based on discovery of nests containing eggs or young. In 19 other squares, breeding was confirmed through sightings of newly fledged young, several of which were found near birdfeeding stations, where they were being fed by adults.

Birdfeeders have contributed to the spread within Québec of this almost exclusively granivorous species. Montréal. *July 1991, Pierre Bannon.*

A House Finch nest in an evergreen—a shallow cup built mainly of grasses and twigs. Saint-Basile-le-Grand, east of Montréal. *June 1989, François Morneau.*

The House Finch is partial to the shelter of tall spruce trees, in which it can build its nest. Saint-Lambert, near Montréal. *June 1991, Gaétan Duquette.*

Distribution

The breeding range of the House Finch extends from British Columbia down to Mexico in the west, across most of the United States, and from Ontario to Nova Scotia in the east (Hill 1993).

The House Finch is well established in southernmost Québec, breeding south of the St. Lawrence River, and locally a little way north of the Ottawa and St. Lawrence rivers to the Québec City area.

At present, the House Finch has a rather extensive range in Québec, although it was first observed in the province only in the fall of 1976, at a feeding station in Philipsburg (David and Gosselin 1977a), and the first nest was found as recently as 1983 in Pointe-Claire (Gosselin 1983). The present extent of the range shows how quickly the species has established itself in Québec, just as it has throughout the northeastern part of the continent (Mundinger and Hope 1982; Nichols 1985a; Robbins et al. 1986; Dunn 1987c; Kozlovic 1987; Meade 1988b).

History and trends

Between 1976 and 1983, the House Finch was mainly observed in the southwesternmost part of Québec. The dramatic population increase in the province began in 1983, when the first nests were found in Pointe-Claire, Sherbrooke, and Philipsburg (David 1983; Gosselin 1983). Between 1984 and 1986, the range expanded eastward, along both shores of the St. Lawrence River, and the species is now found in places such as Hull and Joliette and even Québec City and Rivière-du-Loup (Yank and Aubry 1985b, 1986a; Yank et al. 1987a). Since 1987, the House Finch has continued its northward and eastward expansion, and birds have been sighted in the Lower St. Lawrence, Gaspé and Saguenay–Lac-Saint-Jean regions, on the North Shore to Sept-Îles, and on the Magdalen Islands (Gosselin et al. 1988b, 1989; Yank et al. 1988, 1990a, 1991b; Aubry et al. 1991). The species was found breeding for the first time at Mont-Laurier in 1991 (Yank et al. 1991b).

Feeding stations provide an easy source of food for House Finches in winter and have most certainly contributed to the establishment of the species in Québec, as they have elsewhere in northeastern North America (Nichols 1985a; Paquin 1989). Other factors have aided its rapid expansion. Since it nests near human habitation, it occupies an ecological niche that is relatively unexploited by indigenous species, which reduces competition (Paquin 1989). In fact, its main potential competitor is the House Sparrow. In the northeastern United States, while the House Finch population increased, House Sparrow numbers declined; these phenomena appear to be at least partially linked (Kricher 1983). It has not yet been shown, however, that the introduction of the House Finch in the east might cause a decline in populations of the Purple Finch (Shedd 1990).

Jean Paquin

Red Crossbill

Bec-croisé rouge

Bec-croisé des sapins

Loxia curvirostra

Christiane Girard

AMONG QUÉBEC'S BREEDING BIRDS, THE RED CROSSBILL IS NOTORIOUS FOR its irregular distribution and movements. These birds travel in small nomadic flocks that may appear suddenly in an area, breed, and then disappear as abruptly as they came. Breeding occurs at any time of the year, provided there are abundant food supplies. Red Crossbills are occasionally seen feeding on the cones of ornamental conifers such as mugho and Scots pine, and may even make an appearance at bird feeders. However, the species' distribution in northeastern North America appears to be mainly related to the presence of abundant cone crops of white spruce, Eastern white pine, and Eastern hemlock (Benkman 1985), trees which are all native to Québec.

The different forms of the species are distinguished chiefly by their size and by the size and shape of their bills and, to a lesser extent, by the colour of their plumage (Phillips 1981; Dickerman 1987). Two subspecies are found in Québec: *Loxia curvirostra pusilla* and *Loxia curvirostra minor*, which some authors have renamed respectively *Loxia curvirostra percna* and *Loxia curvirostra neogaea* (Phillips 1981; Dickerman 1987; Benkman 1993). *Minor* is paler and smaller than *pusilla* (Godfrey 1986). However, their status as subspecies has been called into question for several reasons, among them the apparent absence of crosses between these two forms where they are sympatric; this suggests that they may be separate species (Groth 1988, 1993).

The Red Crossbill may breed at any time of the year. Outside the breeding season, the birds wander erratically mainly within their breeding range. In some winters, however, they may irrupt into the midwestern, eastern, and southern United States, as far south as the northern portions of the Gulf states (AOU 1983).

Habitat and behaviour

The Red Crossbill has a widespread distribution in the Northern Hemisphere. There are a number of different subspecies with a variety of bill shapes. These variations allow each population to feed on the dominant species of conifer in its range (Dickerman 1987). In Europe, the irruptions noted during some years are thought to result from population increases, although they could also be linked to the size of the cone crop. When the crossbill population is high, the birds no doubt irrupt as soon as seed becomes scarce, while a good cone crop may induce them to stay put (Newton 1970).

Crossbills are highly specialized foragers. The tip of the lower mandible is curved to the left or right for extracting seeds from conifer cones, their main source of food. Benkman (1985, 1987, 1992) provides a detailed account of how this works. First, the crossbill pushes the tip of its bill between two scales

This crossbill breeds mainly in the boreal coniferous forest, but may also breed in Eastern white pine plantations farther south in the deciduous forest. Its stout, powerful bill allows it to open the pine cones, which are larger than other cones, when they are ripening. Franklin, south of Montréal. *Summer of 1984, Léo-Guy de Repentigny.*

Notes on territory

Just before the breeding season, Red Crossbill flocks break up into solitary pairs or small groups of 2 or 3 pairs (Lawrence 1949). Subsequently, males establish small breeding territories, usually limited to the nest tree (Lawrence 1949; DeGraaf et al. 1980; Benkman 1985). In Ontario, nests were observed only 23 m apart (Peck and James 1987).

The Red Crossbill competes with the White-winged Crossbill and the redpolls, particularly when cones are ripening. Yet, the White-winged Crossbill and the larger subspecies of the Red Crossbill sometimes nest only 60 m apart, even feeding their young within a few metres of each other (Benkman 1985).

In Manitoba, densities of 2.5 pairs/10 ha were observed in a mixed spruce-aspen stand with planted jack pine in openings (Erskine 1976a).

Red Crossbill

Breeding distribution

1984–1989

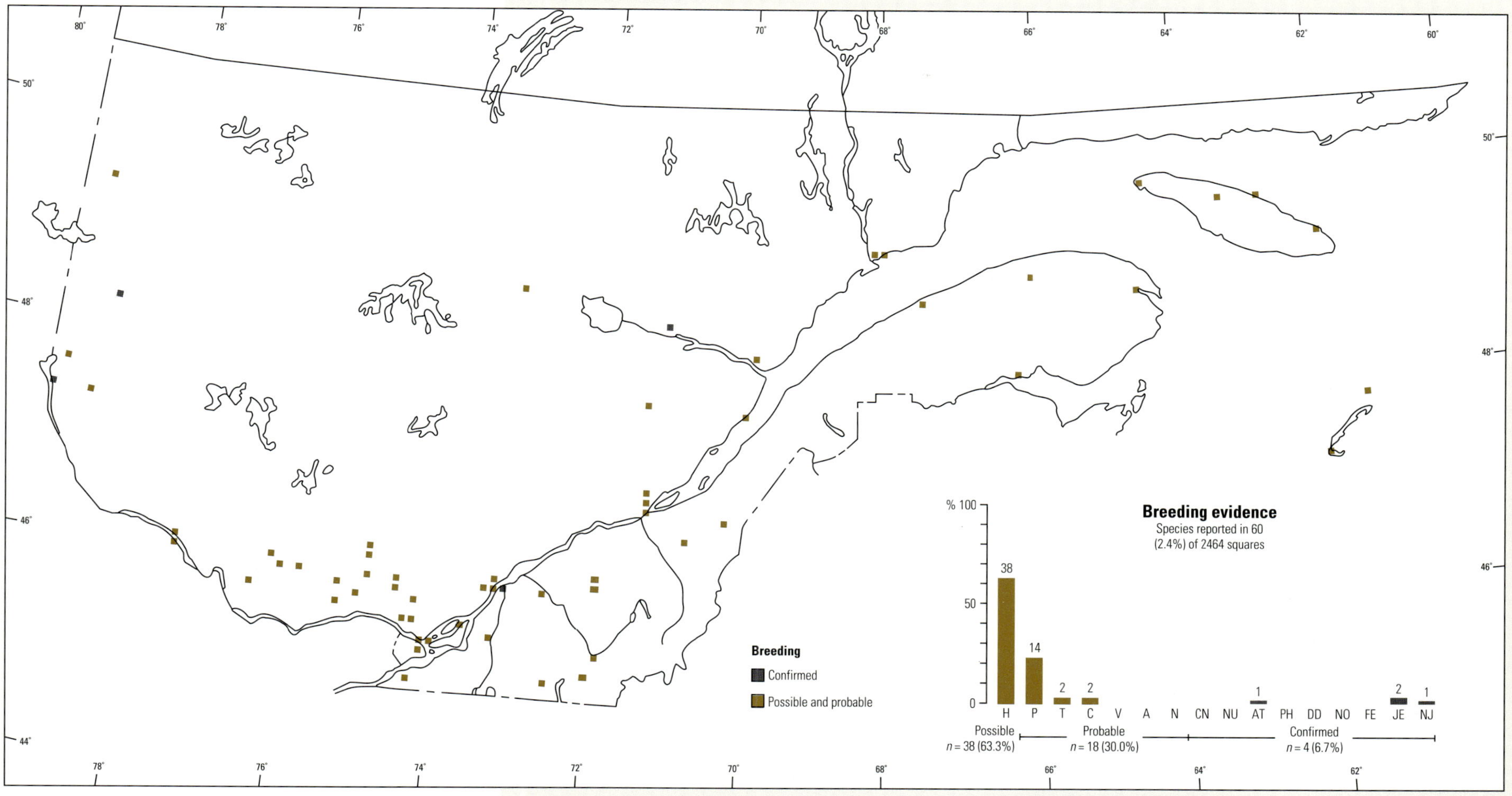

Species profile

Clutch size	usually 3–4 eggs (2–5)[2,3,4,6]
Laying interval	?
Start of incubation	sometimes with first egg[11]
Incubation period	12–18 days[1,3,4,5,6,14]
Maturity at hatching	altricial
Care of young	by female and male[3]
Nestling period	15–24 days[15]
Age at first flight	15–24 days or more[12]
Dependence of young	until at least 33 days after leaving nest[13]
Broods per year	1, probably more[10,11]
Breeding age	1 year or less[10,11]
Mating system	monogamy[6]
Duration of pair bond	?
Total length	14.0–16.5 cm[4]
Wing span	25.4–27.3 cm[7]
Weight	36.5 g (29.2–44.9)[8]
Longevity record	4 years 2 months[9]

Chronology *n* = 1

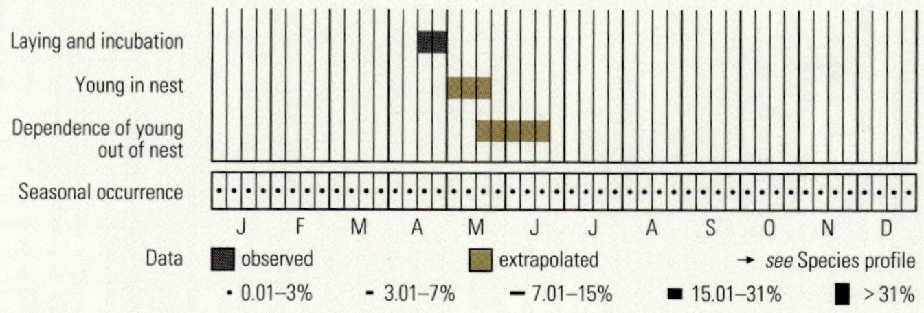

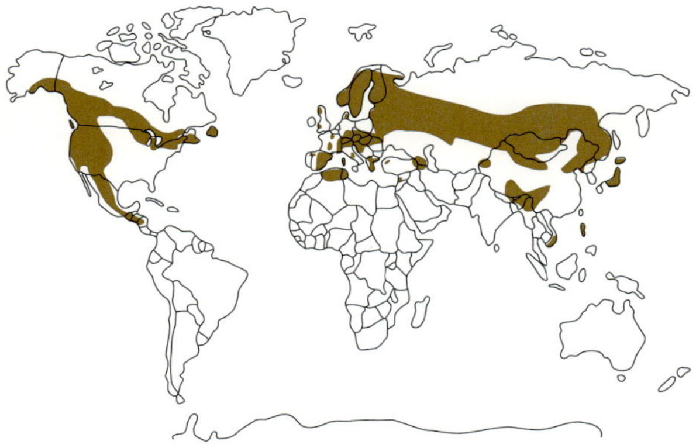

World breeding range

References:

1. Peck and James 1987
2. Bull 1985
3. Bent 1968
4. Godfrey 1986
5. Harrison 1975
6. Ehrlich et al. 1988
7. Terres 1980
8. Dunning 1984
9. Klimkiewicz and Futcher 1989
10. DeGraaf and Rudis 1987
11. Benkman 1985
12. according to nestling period
13. Newton 1973
14. Lawrence 1949
15. Skutch 1976

and, with a lateral movement of the lower mandible, spreads the scales apart to expose the seed. It then scoops the seed out with its tongue and removes the seed coat with its bill. When feeding, crossbills usually orient themselves so that their lower mandible is directed toward the distal side of the gap and they push their upper mandible against the basal scale to widen the gap (Benkman 1985, 1987, 1992). They may also use their bills in other ways, particularly when the cone is already open (Cramp 1994a).

Crossbills are very agile, even acrobatic in their movements when foraging, hanging from the cones, sometimes upside down, and using their feet and bill to manoeuvre among the branches parrot-fashion. Their large feet are superbly adapted to this type of foraging (Newton 1973; Cramp 1994a).

The Red Crossbill usually frequents coniferous forests (Godfrey 1986), but may also be found in conifer plantations in deciduous forest regions farther south (Peck and James 1987). Its diet consists almost exclusively of conifer seeds, particularly Eastern white pine (Benkman 1985), but it also feeds occasionally on berries, the seeds of deciduous trees, buds, and insects, with insects sometimes making up a significant part of its diet in summer (Lawrence 1949; Bent 1968; Benkman 1985; Godfrey 1986). Having a wider and more powerful bill than the White-winged Crossbill, the Red Crossbill is more adept at extracting seeds from pine cones, which have larger, thicker scales than, for example, spruce cones, and it is more efficient at husking pine and hemlock seeds than the smaller seeds of spruce (Benkman 1985, 1987). Red Crossbills generally travel in search of prime food sites in small flocks numbering 8 individuals on average (Benkman 1988a).

The Red Crossbill breeds at all times of the year, whenever and wherever it finds an abundant source of ripening cones (McCabe and McCabe 1933; Bent 1968; Benkman 1985; Bull 1985; Lumsden and Smith 1987). In eastern North America, according to Benkman (1985), breeding has been observed mainly between January and May, when Eastern white pine and Eastern hemlock cones are available, but also at the end of summer, when white spruce, red spruce, and Eastern hemlock cones are plentiful. Breeding may also occur in June and July, when the species feeds on jack pine cones (Benkman 1985).

Lawrence (1949) was one of the first to describe the Red Crossbill's courtship behaviour. The male approaches the female while she is nibbling cone seeds at the top of a tree. He rises above the female on vibrating wings and circles her, uttering loud whistled notes that run into an "enraptured melody." The performance usually ends with pursuits. Courtship feeding is also observed in the species (Lawrence 1949; COQ 1988).

The female usually builds the nest 1.5–24 m up in a conifer (Harrison 1975; DeGraaf et al. 1980; Peck and James 1987), on a branch well out from the trunk. The nest, which is often difficult to find, is made of conifer twigs, bark strips, moss, lichen, and grasses and is lined with hair, feathers, lichen, moss, and sometimes pine needles (Bent 1968; Harrison 1975; Godfrey 1986; Peck and James 1987).

Typical clutch size is 3 or 4 (less frequently 2 or 5) eggs, incubated solely by the female (Bent 1968; Harrison 1975; Godfrey 1986) for 12–18 days (Lawrence 1949; Ehrlich et al. 1988).

During courtship feeding, the female crouches and flutters her wings to beg food from the male. She and the male also call back and forth. The male may give her 3–30 regurgitated boli of food on each visit to the nest (Lawrence 1949). The newly hatched young are covered with down, which allows them to withstand temperatures below 0°C for short periods (Leahy 1982).

This nomadic species may nest at varying times of the year, depending on the region and availability and abundance of ripe cones. In jack pine stands, nesting usually occurs in June and July. James Bay region. *August 1982, Vincent Létourneau.*

The Red Crossbill is a real acrobat when foraging. To remove the seeds from cones, it must place its bill at the optimum angle to the cone, which sometimes requires it to hang upside down. Drummondville. *June 1990, Jean Paquin.*

During the first few days after hatching, the female broods the young while the male continues to bring her food, which she regurgitates into the mouths of her offspring (Cramp 1994a). After about 5 days, she stops brooding during the day and joins the male in bringing food to the nest; at this time, the birds become very quiet, making the nest more difficult to find (Lawrence 1949). When it is very cold, the young may become torpid and must be brooded to restore their metabolism before they can be fed (Ternovski 1954 *in* Cramp 1994a).

The mandibles of the young are not yet crossed when they leave the nest. During the next few weeks, the horny surface layer of the bill curves progressively until the bill resembles the adult's (Newton 1973; Leahy 1982; Ehrlich et al. 1988). Once the nesting period is over, family groups gather and invade new territories in search of food (Lawrence 1949; Bent 1968). The newly fledged young may remain in these family groups and be fed by their parents for at least 33 days (Bailey et al. 1953 *in* Newton 1973). Red Crossbills may reach sexual maturity when they are less than a year old and still have their immature plumage, a great advantage in a nomadic species like the Red Crossbill which may encounter food sources suitable to sustain breeding several times a year (Benkman 1985).

Distribution

The Red Crossbill has a circumboreal distribution (Leahy 1982), nesting primarily in the northern coniferous forest of Eurasia and North America (Godfrey 1986). In Canada, it is found in the southern boreal forest and in temperate regions from Newfoundland to southern Yukon, including all of British Columbia. In the United States, it occurs in southern Alaska as well as in the northeastern and Great Lakes states, and the Western Cordilleras. It also extends south to the Central American highlands, from northern Mexico to Nicaragua (AOU 1983).

Since the Red Crossbill prefers conifers with large cones, it is generally found farther south than the White-winged Crossbill, particularly in coniferous stands in the mixed and deciduous forest zones. The Red Crossbill was found in only 2.4% (60/2464) of Atlas squares surveyed. Most records were from the southernmost part of Québec (Ottawa Valley, Laurentians, Montérégie, Eastern Townships, and the Bois-Francs region), but the species was also observed in more northerly regions such as Abitibi–Témiscamingue, Saguenay–Lac-Saint-Jean, the Upper North Shore, the Québec City area, and the Gaspé Peninsula. Breeding was confirmed in only 6.7% (4/60) of the squares in which the species was reported, the 4 records involving an adult carrying food near Sorel, pairs with recently fledged young in the Témiscamingue and Saguenay regions, and a nest in the Rouyn-Noranda area.

The Red Crossbill was also seen in conifer stands on the Magdalen Islands, where it is considered an irregular nomadic visitant and probable breeder (Fradette 1992), as well as in Anticosti Island's coniferous forests, where it was reported for the first time in 1963 (Ouellet 1969).

Before the Atlas project, only one nest had ever been found in Québec, in 1977, on a golf course in Saint-Bruno-de-Montarville with a small belt of red pines edging a maple stand. The female was observed between 10 and 13 April building her nest in a pine, on a branch 1.2 m up and incubating her clutch of 4 eggs from 19 to 28 April. On 3 May, however, the nest was found to be empty and abandoned.

The Red Crossbill is usually seen only on rare occasions and in small numbers in Québec. It is classified as a rare permanent resident in the southern part of the province (David 1980a). ÉPOQ data for the period 1960–89 show fewer than 20 records a year on average; usually only 1–3 individuals were observed, rarely more than 20. In addition, during this 30-year period, two main irruptions occurred in Québec, from June to December 1969 and from March to May 1988 (40% of Atlas records were obtained in 1988). During these irruptions, frequency of occurrence increased dramatically, between five and sevenfold, although the size of the flocks did not rise substantially. Irruptions in Ontario and New York have also been recorded in 1947–48, 1961, 1963–64, 1970, and 1984–85, when breeding was confirmed or breeding attempts were observed (Lawrence 1949; Cringan et al. 1971; Raynor 1983; Dickerman 1987).

History and trends

At the turn of the century, Wintle (1896) described the Red Crossbill as a common "transient visitant" in the Montréal area, while Caulfield (1890 *in* Ouellet 1974a) stated that it was an irregular visitant although large flocks were recorded in some winters. Subsequently, populations declined sharply in Québec, as they did elsewhere in northeastern North America (Dickerman 1987). The decline may have been caused by the intensive logging of large stands of Eastern white pines and Eastern hemlocks during the 19th and early 20th centuries. Eastern white pine was one of the dominant species in the forests of southern Québec in the early 1800s, as was Eastern hemlock in the mid-1800s (Bouchard et al. 1989).

Since then, populations appear to have stabilized, and in North America as a whole, numbers may even have increased slightly over the past few decades (Robbins et al. 1986).

Vincent Létourneau

White-winged Crossbill

Bec-croisé à ailes blanches

Bec-croisé bifascié
Loxia leucoptera

Christiane Girard

THE WHITE-WINGED CROSSBILL IS A DENIZEN OF THE CONIFEROUS FOREST. IN Québec, it breeds as far north as the Subarctic and is also found sporadically farther south in regions of mixed and deciduous forest, generally in winter. Like the Red Crossbill, it is quite tame, allowing observers to get close enough to see its remarkable bill. When foraging, it uses its bill and feet to hoist itself among the branches like a little parrot. The crossed mandibles are perfectly adapted for extracting seeds from conifer cones, particularly spruce and tamarack (Benkman 1985).

Of the two species of crossbills found in Québec, the White-winged is the most common. It is usually seen only occasionally, in small flocks; however, during irruption years, the everpresent flocks may comprise hundreds of birds. The White-winged Crossbill feeds mainly on conifer seeds (Newton 1972), but sometimes supplements its diet with birch and alder seeds, fruit, and insects (Taber 1968; DeGraaf et al. 1980). It is also strongly attracted to salt spread on roads in winter and, because of its tameness, is sometimes killed by passing cars (Meade 1942; Ehrlich et al. 1988).

Habitat and behaviour

The White-winged Crossbill is a boreal species that occasionally nests in mixed woodland and islands of conifers in deciduous forests farther south. Among the four species of crossbills found worldwide, the White-winged Crossbill has the slenderest bill, which explains its liking for conifers with small cones. Therefore, it is found most often in stands dominated by spruce and tamarack, particularly when the ripening cones begin to open (Benkman 1985, 1987, 1992). The White-winged Crossbill is less fond of balsam fir, with its sticky cones and resinous seed coats (Benkman 1992). However, farther south, it feeds readily on the cones of Eastern hemlock and Eastern white-cedar (Benkman 1985, 1987; Smith and Lumsden 1987).

Breeding takes place when and where seed production is adequate for the high energy demands of producing eggs and feeding young (Newton 1972; Benkman 1985, 1990). Since seeds may be abundant at different times of the year, breeding times also vary. In fact, breeding may occur in any month of the year (Newton 1972; Godfrey 1986).

Benkman (1985) has found that breeding usually coincides with three main periods in which conifer seeds are most available: from July to October for white spruce and tamarack; from January to April for white spruce; and from March to May for black spruce. In red spruce areas, the cones open at the same time as those of white spruce. A crop of balsam fir cones may also occasionally sustain a breeding population, as was seen in the Jacques Cartier Conservation Park in the late summer of 1988 (pers. obs.).

Apparently, the pair bond is maintained year-round, enabling a pair to rapidly commence nesting when suitable conditions arise (Benkman 1985). The male's song consists of a succession of loud trills on different pitches, sometimes interspersed with warbling or harsh, telegraph-like notes.

The nest is usually built on a horizontal branch of a spruce, although sometimes in a pine or other conifer, 1–21 m up (Harrison 1975; DeGraaf et al. 1980). It is made of twigs, lichens, and bark strips, and lined with grasses, hair, and feathers (Taber 1968; Harrison 1975; Godfrey 1986). Only the female incubates, but she is fed frequently on the nest by the male. He usually announces his arrival with loud whistles, alighting on a nearby tree and calling loudly for a moment before going directly to the nest. This behaviour is a good indication that a nest is nearby (Taber 1968; Harrison 1975). The average clutch consists of 2–4 eggs, rarely 5 (Benkman 1992).

The down-covered young (Bond 1938) are fed by both parents on regurgitated milky seed pulp (Taber 1968; Ehrlich et al. 1988). Feeding continues after the young leave the nest (Benkman 1992). The young can fly fairly well about a week after leaving the nest, when they follow their parents, or sometimes the male only, in their wanderings. Their bills do not begin to cross until they are around 30 days old (Newton 1973; James et al. 1987), or the second week after they have left

Notes on territory

Territorial defence is limited mainly to the immediate vicinity of the nest. The White-winged Crossbill does not defend a feeding territory, since it forages in flocks even during the breeding period (Benkman 1985, 1988b). In fact, given the abundance of cones in suitable nesting areas, it would be difficult for individuals to keep other birds out of their territories and, even if it were possible, there would be few benefits. Feeding in flocks, on the other hand, does have advantages, since the birds can spend more time foraging and less time watching out for predators (Benkman 1985).

This crossbill mainly frequents the boreal forest, where it may breed at any time of the year. Montmorency forest, Laurentide Wildlife Reserve, north of Québec City. *May 1991, Réginald Ouellet.*

White-winged Crossbill

Breeding distribution

1984–1989

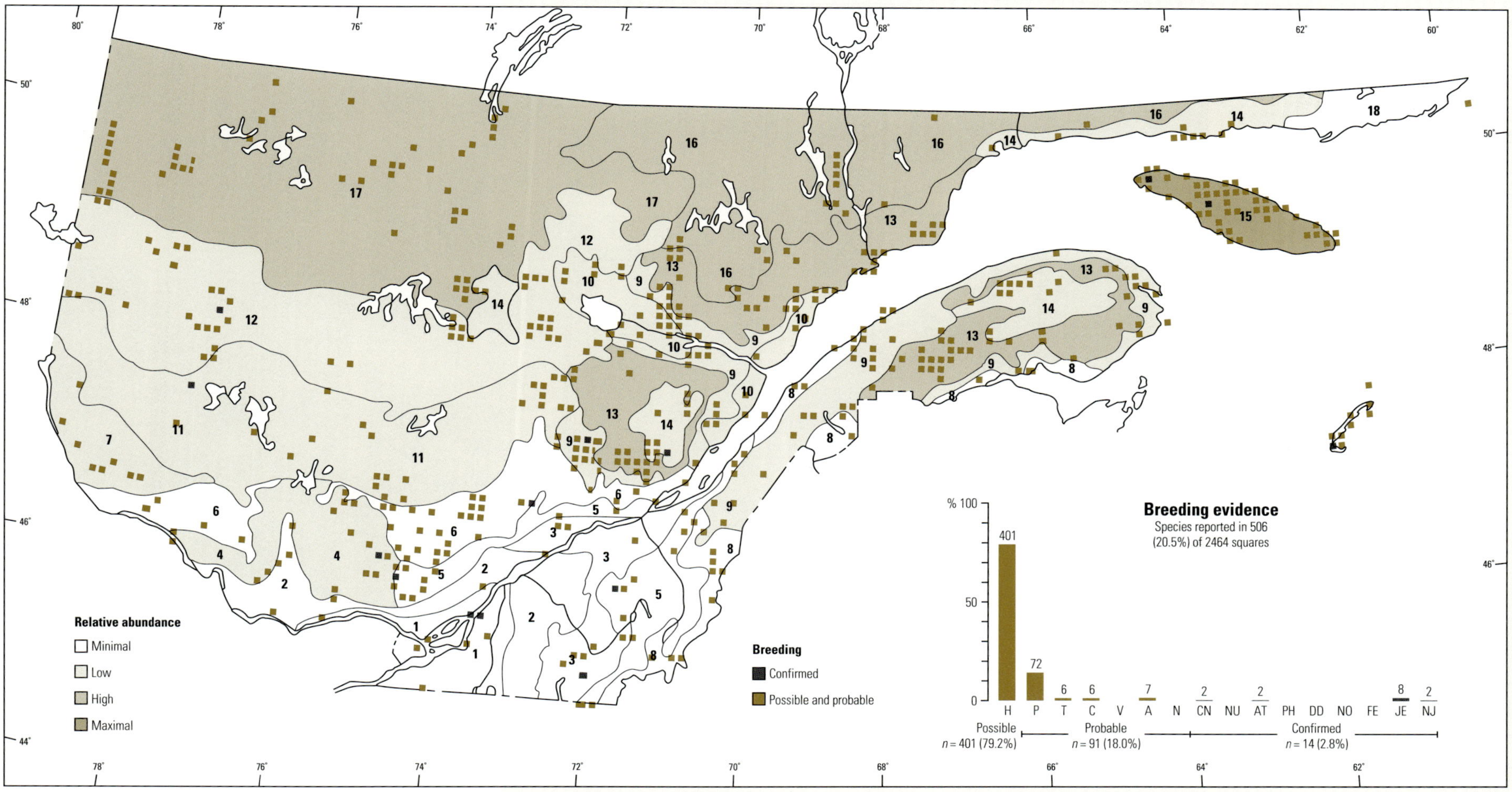

Species profile

Clutch size	2–4 eggs, rarely 5[1,3,4,5]
Laying interval	probably 1 egg per day[5]
Start of incubation	sometimes with first egg[5]
Incubation period	probably 12–14 days[5]
Maturity at hatching	altricial
Care of young	by female and male[5]
Nestling period	?
Age at first flight	≈ 1 week after leaving nest[5]
Dependence of young	until several days after leaving nest[5]
Broods per year	probably 2–4[5]
Breeding age	probably less than 5 months[5]
Mating system	monogamy, probably polyandry[5]
Duration of pair bond	?
Total length	15–17 cm[2]
Wing span	?
Weight	M 27.9 g (?) / F 25.3 g (?)[6]
Longevity record	4 years[5]

References:
1. Taber 1968
2. Godfrey 1986
3. Ehrlich et al. 1988
4. DeGraaf and Rudis 1987
5. Benkman 1992
6. Dunning 1984

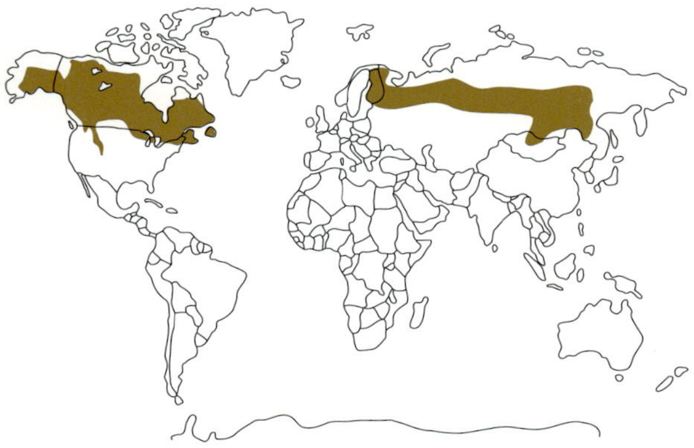
World breeding range

Chronology $n = 9$

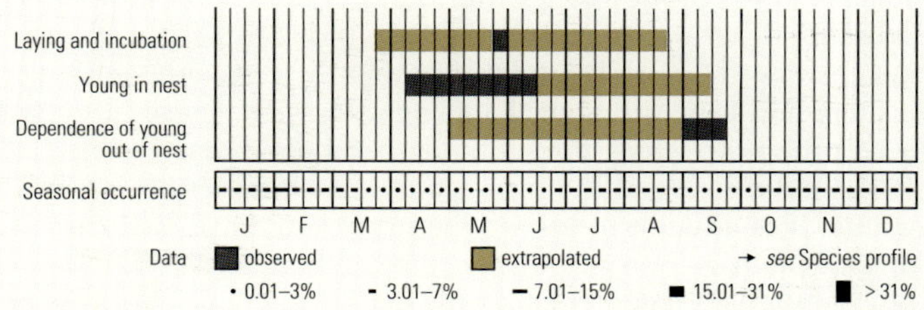

Frequency of occurrence by ecological region

Ecological regions** (forest regions)	Adequately covered squares	Squares in which reported	% of squares in which reported	Relative abundance (%)*
1. Sugar maple-hickory and sugar maple-basswood	97	8	8.2 ± 3.0	8.2
2. Sugar maple-basswood	153	7	4.6 ± 1.9	4.6
3. Sugar maple-basswood and sugar maple-yellow birch	101	11	10.9 ± 3.9	10.9
4. Sugar maple-yellow birch-basswood	56	17	30.4 ± 10.6	30.4
5. Sugar maple-yellow birch-beech and sugar maple-yellow birch-basswood	113	17	15.0 ± 3.8	15.0
6. Sugar maple-yellow birch-beech	71	19	26.8 ± 9.5	26.8
7. Sugar maple-yellow birch-balsam fir	19	7	36.8 ± 22.8	36.8
8. Sugar maple-yellow birch and balsam fir-yellow birch	47	12	25.5 ± 11.3	25.5
9. Balsam fir-yellow birch	87	39	44.8 ± 9.6	44.8
10. Balsam fir-white birch or balsam fir-red maple	46	21	45.7 ± 12.5	45.7
11. Yellow birch-balsam fir	52	17	32.7 ± 13.2	32.7
12. Balsam fir-white birch (west)	59	33	55.9 ± 13.0	55.9
13. Balsam fir-white birch (east)	76	47	61.8 ± 10.7	61.8
14. Balsam fir-black spruce	28	13	46.4 ± 19.3	46.4
15. White spruce-balsam fir and balsam fir-white spruce	22	22	100.0 ± 2.3	100.0
16. Black spruce-balsam fir-moss	18	11	61.1 ± 25.5	61.1
17. Black spruce-moss	32	28	87.5 ± 13.0	87.5
18.			*not analysed*	

* minimal: 0–29.9%; low: 30–59.9%; high: 60–89.9%; maximal: 90–100%

** Deciduous forest: 1–7; mixed forest: 8–11; coniferous forest: 12–17

The White-winged Crossbill has the slenderest bill of all the crossbills and feeds mainly on conifers with small cones. It was atlassed in tamarack peatlands and white spruce stands on Anticosti Island. *July 1987, Jean Beaubien and Guy Simard, Canadian Forest Service.*

the nest (Benkman 1992). The crossing of the bill involves the gradual lengthening and curving of the horny surface layer (Newton 1973). As their bills cross, the young begin to manipulate the cones, awkwardly at first, and then extract the seeds (Benkman 1992).

The White-winged Crossbill was reported in 20.5% (506/2464) of Atlas squares surveyed, 10 times more often than the Red Crossbill. Breeding was confirmed in only 2.8% (14/506) of the squares in which the species was found. A number of factors make breeding behaviour extremely difficult to discern, especially as the White-winged Crossbill sings throughout the year, although singing is more prevalent during the breeding season. The female also rarely leaves the nest, which is well hidden and often inaccessible (Taber 1968), and because of the precocious wandering tendencies of the young, it is difficult to know whether young birds with streaked juvenal plumage were born where they are seen (Todd 1963). In years of exceptionally good cone crops, the female may start a second brood, letting the male finish raising the first brood (Benkman 1988a, 1989). It is not known if, in those instances when a female lays a second clutch, she mates with another male or the same one (Benkman 1992).

In both North American crossbill species, the orientation of the lower mandible varies from individual to individual and can be deflected to either the left or right. Although all White-winged Crossbills use the same technique to extract seeds from cones, those with the lower mandible deflected to the left attack the cone in the opposite direction from those with the mandible deflected to the right. To remove a seed, the bird inserts the tip of its bill between two scales (on closed cones, crossbills first bite between the overlapping scales to form a gap). The bird then places its lower mandible against the distal scale and its upper mandible against the basal scale (the scale closer to the base of the cone) and spreads the two scales apart with a lateral movement of its lower mandible. Finally, it scoops the seed out with its tongue and removes the seed coat with its bill. It can also use its upper mandible, which is particularly long, as a hook to dislodge the seed (Benkman 1985, 1987, 1992). Crossbills usually forage on cones while they are still attached to branches, though they sometimes wrench cones off with their bills and carry them to a horizontal branch, where they clamp the cone between their foot and the perch. The foot used to hold the cone is usually the one opposite the side which the lower mandible crosses (Knox 1983 *in* Benkman 1992; James et al. 1987).

Distribution

The White-winged Crossbill is a circumboreal species that nests in the coniferous forests of Eurasia from Scandinavia to Siberia and northern North America. It is found in Alaska and throughout Canada, except on the Prairies and in the Arctic and High Subarctic. It also occurs in some of the northern U.S. states, and may range farther south during irruption years. Surprisingly, the mountains of Hispaniola in the Caribbean also support an isolated resident population (AOU 1983; Godfrey 1986).

The White-winged Crossbill was atlassed virtually throughout southern Québec. However, the species may breed at any time of the year, while atlassing was carried out mainly during June and July.

Frequency of occurrence was highest on Anticosti Island, where the forests are dominated by black and white spruce, balsam fir, and tamarack; breeding was also confirmed on the island. Frequency of occurrence was fairly high also in all regions with extensive mature conifer stands, particularly the area north of the Gouin and Pipmuacan reservoirs, the Laurentide Wildlife Reserve, and the interior of the Gaspé Peninsula. This corresponds roughly to the area demarcated by the mean July isotherm of 15°C. The species is observed much less regularly in the deciduous and mixed forests in the southern and central parts of southern Québec, including the Lac-Saint-Jean region.

Breeding was confirmed on the Magdalen Islands, despite a low frequency of occurrence. The species is considered a common nomadic visitant to the islands during most of the year (Fradette 1992). Breeding confirmation was also obtained in the Eastern Townships and the Lanaudière and Chaudière–Appalaches regions. In the Montréal area, most of the records were obtained after the major irruption of 1984–85. The two breeding records northeast of Montréal appear to be the first ever obtained in the Central St. Lawrence Lowland, where the species occurs irregularly.

Breeding in the White-winged Crossbill coincides with the ripening of the cones of various types of conifers; therefore, White-winged Crossbills may nest at different times of the year depending on the type of stand frequented. In tamarack stands, nesting usually occurs between July and October. Saint-Charles, east of Lévis. *July 1991, Guy Lemieux.*

In southern Québec, this nomadic species frequents plantations of white spruce when there is a good cone crop. Saint-Joachim-de-Courval, north of Drummondville. *June 1991, Daniel Jauvin.*

History and trends

In years of poor cone crops in the boreal forest, crossbills migrate well south of their usual range (Bock and Lepthien 1976b; Forster 1990; Janssen 1990; Plunkett 1990).

The last major irruption was during the winter of 1984–85, when the cone crops of several conifer species, including white spruce, were exceptionally good in the portions of southern Québec south of the boreal forest (Grenier and Masse 1990). Presumably the irruption was caused by poor crops in the boreal forest. In fact, in the early autumn of 1984, Benkman (1985) observed an early and massive drop in numbers in the Laurentide Wildlife Reserve, where neither the tamarack nor the spruce crop was sufficient to allow the birds to overwinter.

At the turn of the century, this species was sometimes seen in considerable numbers in the Montréal area, where it was considered a common "transient visitant" (Caulfield 1890a; Wintle 1896). More contemporary authors (Ouellet 1974a; David 1980a) have suggested that the species is less common today in this region than it was in the past. This may be due to the razing of extensive stands of mature conifers in the southern boreal forest for agriculture and logging. Even though BBS data show a downward trend for 1965–79, the evidence is insufficient to prove a significant modification in the population (Robbins et al. 1986). In addition, BBS data for Québec between 1966 and 1989 do not show any significant change in numbers.

Over the last few decades, land that used to be farmed or where the forest cover was deciduous has been planted with conifers. As these stands mature, they will provide new nest sites and sources of food for crossbills in regions not traditionally frequented by the birds. However, forestry practices that promote the rapid growth and particularly the early felling of trees, before they are old enough to produce cones, may be detrimental to the species.

VINCENT LÉTOURNEAU

Common Redpoll

Sizerin flammé

Carduelis flammea

In Québec, as elsewhere, the Common Redpoll is a casual breeder south of the 50th parallel, though winter irruptions of this circumpolar species (Clement 1968a; Godfrey 1986; Knox 1988) occur irregularly, some years on a massive scale, throughout southern Québec (Clement 1968a; Godfrey 1986; Larson and Bock 1986). Most birders and ornithologists are thus more familiar with the winter habits of the Common Redpoll than with its summer behaviour.

The sometimes massive winter irruptions of redpolls and several other northern cardueline finches, such as the Pine Siskin, Purple Finch, Evening Grosbeak, and the crossbills, appear to be closely linked to seed crop failure in coniferous and some deciduous trees, such as birch, in northern forests (Newton 1985b).

In winter, as in summer, the species is typified by the restlessness of foraging flocks. The birds never stay still for more than a few seconds, flitting constantly from one plant, shrub or tree to another, so that the flock appears to be in constant motion (Clement 1968a).

Habitat and behaviour

During the breeding season in the Arctic, the Common Redpoll frequents ravines, rocky slopes with some shrubby ground cover, as well as drier, well-drained barrens (Godfrey 1986; Watson 1986). In subarctic regions, it is found in patches of stunted spruce or scrub birch, in willow and alder thickets, on the taiga, and in tamarack stands (Godfrey 1986; Nyström and Nyström 1987; pers. obs.).

Nesting typically extends from the end of May, when the gonads are fully expanded (Watson 1986), until the end of July. It is likely that some females produce 2 broods per year, especially in the southern parts of the breeding range. Scandinavian observers have even noted that some birds nest early in the southern part of the range and then migrate north to renest (Peiponen 1957). In Nain, Labrador, an early nest with 4 eggs was found on 28 April (Clement 1968a).

Little work has been done on pair formation. Dilger (1960), however, in studies of captive birds, noted the existence of a well-defined, linear social hierarchy for each sex. Outside the breeding season, males dominate females. With the approach of the breeding season, females become increasingly aggressive until dominance is reversed. Each female then singles out a male, which she aggressively defends. Once a pair is formed, the male sings to keep away other males until the flock breaks up. Monogamy is assumed to be the general rule, but an instance of sequential polyandry has been reported, with one female producing two successive broods from different males (Seutin et al. 1991).

The secretive behaviour of the adults during the nesting period can make nests difficult to find (Clement 1968a). However, before incubation starts, a careful observer can easily detect the early signs of breeding. The most obvious sign is a female, heedless of the observer, carrying nesting material to a clump of conifers or willows. Although they build the nest alone, females are constantly followed by their mates. The nest is made of grasses and twigs and lined with cottony plant down, feathers (mainly ptarmigan and grouse), and sometimes hair or even fur, including that of lemmings (Clement 1968a; Godfrey 1986). In the High Arctic, the nest may be located on the ground in scrub, a sedge tussock, or a rock crevice. Farther south, it is usually built 1–2 m up in a shrub or stunted tree that sometimes is still surrounded by snow at the base (Clement 1968a; Godfrey 1986; pers. obs.). Early nests are typically built in conifers, but later ones are usually found in deciduous trees.

Christiane Girard

Notes on territory

Cardueline finches nest as solitary pairs or in small groups within which each pair defends only a small territory around the nest. Foraging occurs outside this territory (Newton 1985a). In the Common Redpoll, territorial behaviour in the male is not very pronounced, as indicated by the irregular spacing of nests, which are sometimes very close together (Clement 1968a). Pairs are formed before the territory is established (Newton 1985a), indicating that the females do not select a mate on the basis of the quality of the territory.

In the Ungava Peninsula, as here in the New Québec Crater region, Common Redpolls may nest on the ground in scrub, a sedge tussock, or a rock crevice. *September 1977, Pierre Lamothe.*

Common Redpoll

Breeding distribution

1984–1989

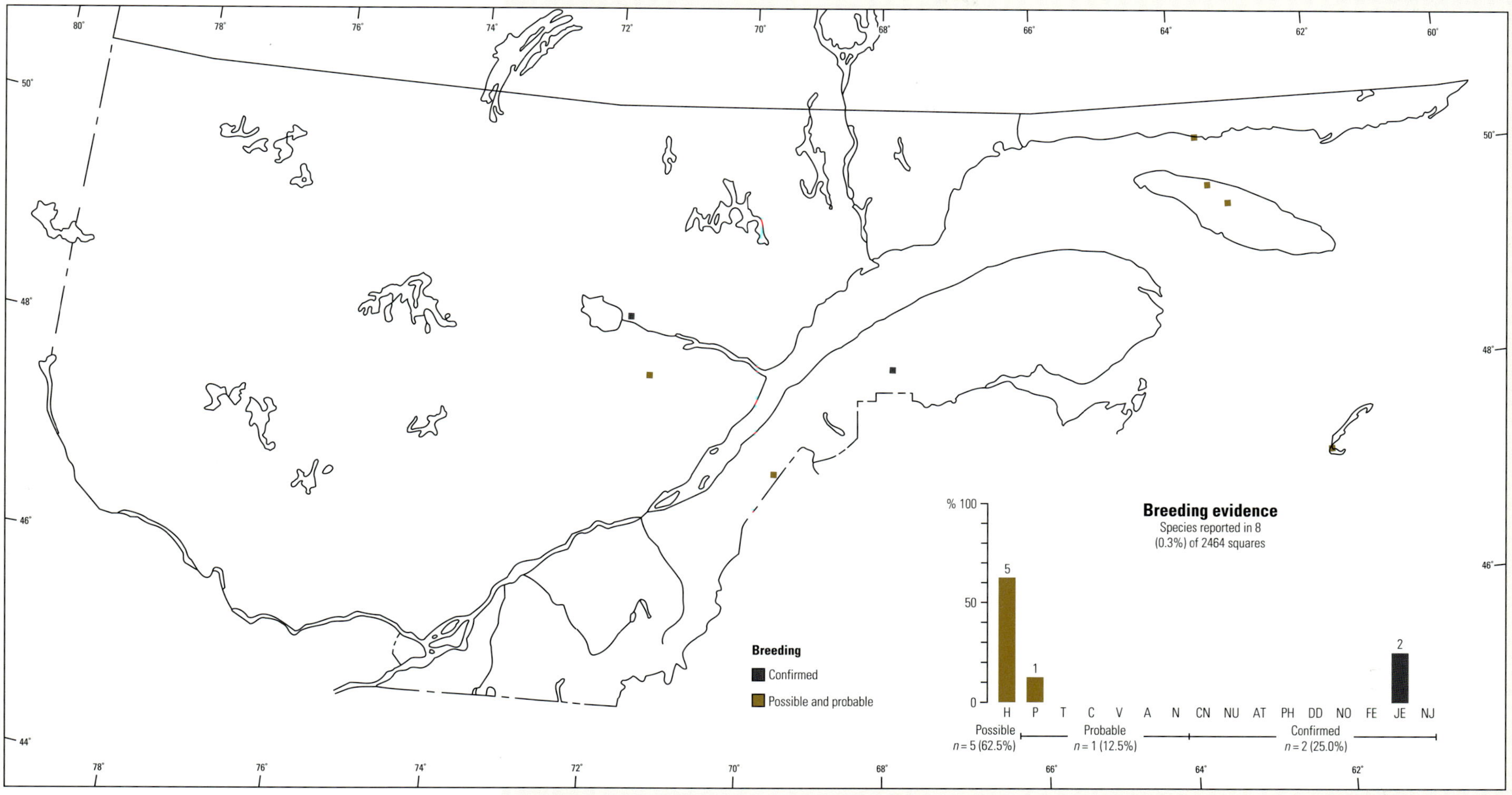

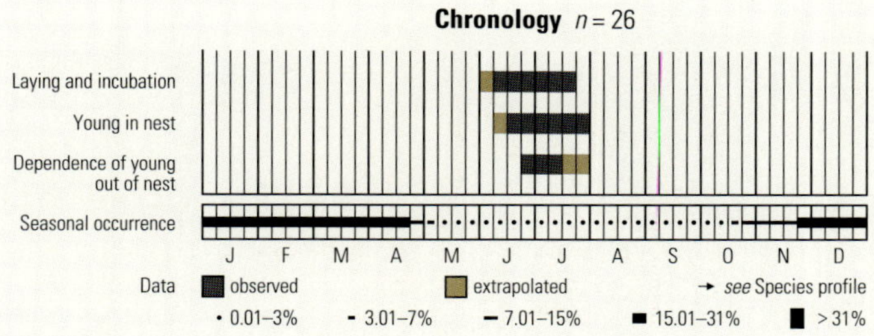

Species profile

Clutch size	usually 3–5 eggs (3–7)[1,2,6]
Laying interval	1 egg per day[6]
Start of incubation	usually with penultimate egg[6]
Incubation period	10–11 days[1,2,6]
Maturity at hatching	altricial
Care of young	by female and male[2,6]
Nestling period	10–12 days[2,6]
Age at first flight	10–12 days[6]
Dependence of young	probably less than 2 weeks after leaving nest[6]
Broods per year	often 2, 1 in the Arctic[6]
Breeding age	probably 1 year[6]
Mating system	probably monogamy[5], *see* text
Duration of pair bond	?
Total length	11–15 cm[1]
Wing span	21.0–22.2 cm[7]
Weight	13.0 g (10.1–15.0)[4]
Longevity record	5 years 9 months[3]

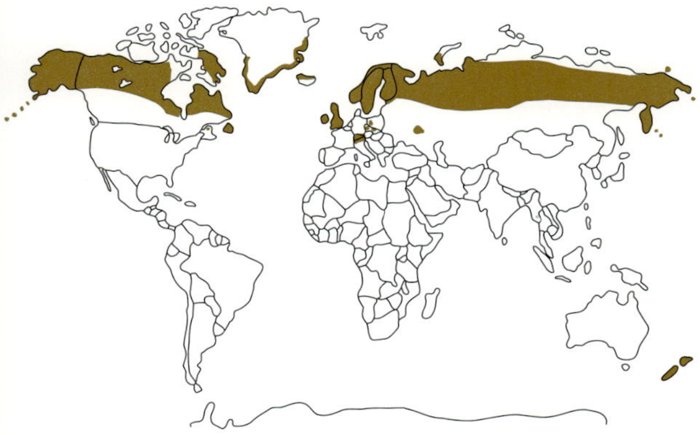

World breeding range

References:
1. Godfrey 1986
2. Clement 1968a
3. Klimkiewicz and Futcher 1987
4. Dunning 1984
5. Ehrlich et al. 1988
6. G. Seutin, unpubl. data
7. Terres 1980

Clutch size is usually 3–5 eggs, which are blue with faint reddish-brown speckles and blotches. Incubation lasts 10 or 11 days, starting when the penultimate egg has been laid. The males feed the sitting females regularly, but also spend a good deal of time singing from an exposed perch and pursuing one another. The reason for the latter behaviour is unclear, as the species is not territorial, but when it occurs regularly and in the same location, it is indicative of nesting.

Both parents feed the nestlings. During this period, which lasts 10–12 days, the adults are generally silent and difficult to observe, but the begging calls of the young can often be heard from a distance of a few metres whenever an adult arrives at the nest to feed them. When females initiate a second clutch, the males seem to take over the entire responsability of feeding the fledglings (Seutin et al. 1991).

In winter, in northern areas, Common Redpolls feed mainly on the seeds of birch, willow, and alder; farther south, the seeds of forbs and grasses are added to the diet (Clement 1968a; Godfrey 1986). During the nesting period, insects may comprise a sizable proportion of the diet of both adults and nestlings (Clement 1968a).

Distribution

The Common Redpoll is one of the most abundant birds of the Québec tundra and taiga. Breeding has been confirmed up to the northern tip of the Ungava Peninsula (Todd 1963; Cons. G. G. – GREBE 1991a), but the southern limit of the breeding range has not been as clearly defined. Todd (1963) reported nesting sites south of the southern limit described by Godfrey (1986) and there are summer records from around the Missisicabi, Nottaway, and Broadback rivers near James Bay (Cons. G. G. – GREBE 1992o). The southern limit of the breeding range probably lies just south of 51° N in the west and 50°30' N on the Lower North Shore.

Breeding is sporadic farther south and the species was atlassed in only 8 squares. In 1989, breeding was confirmed at Alma and at Lake Mistigougèche, near the Rimouski Wildlife Reserve. Possible and probable breeding were reported from various locations: the Laurentide Wildlife Reserve in 1989; Lac de l'Est, southwest of Kamouraska, where a pair was sighted in a black spruce stand in 1984; around Havre-Saint-Pierre in 1986; and at Havre-Aubert on the Magdalen Islands in 1989. Between 3 and 15 July 1989, young and flocks of adults and first-year birds were sighted on Anticosti Island, suggesting that the species may have bred there that year.

The Common Redpoll was probably such a scarce local breeder in southern Québec in the past that it was overlooked.

On the northeast coast of James Bay, Common Redpolls nest in shrubbery both on the islands and along the shoreline. Bay of Many Islands.
July 1990, Yves Aubry.

Although it breeds very far north, the Common Redpoll also finds the tundra habitat it prefers farther south on mountaintops in the Gaspé Peninsula, as here on Mount Albert. Gaspé Conservation Park.
October 1988, Alain Desrosiers, Ministère du Loisir, de la Chasse et de la Pêche.

The Common Redpoll nests sporadically on the Magdalen Islands in the clumps of stunted trees that are typical of part of the islands' forest cover. Pointe de l'Est National Wildlife Area. *June 1991, Desneige Perreault.*

Perched on the rim of its nest, this Common Redpoll is bringing food to its newly hatched chicks. In the Arctic, nests may be placed in a shrub, such as this scrub birch, a tree that grows in sheltered areas. Ferguson Lake, Northwest Territories. *July 1990, Yves Aubry.*

History and trends

In southern Québec, the Common Redpoll is generally considered a winter visitor, although it is thought to breed irregularly on the Magdalen Islands, where adults and first-year birds have been reported on several occasions in June and July (Fradette 1992). Breeding was confirmed on La Grosse Île, where a nest containing four young 4 or 5 days old was discovered on 29 June 1905 (Callender 1906). On the Magdalen Islands, breeding evidence has been obtained at least once every 10 years over the last 40 years (Fradette 1992).

Presently available data is insufficient to assess the dynamics of the breeding population in northern Québec and on the North Shore. It is clear, however, that numbers in a given area fluctuate significantly from year to year, as is the case elsewhere in North America as well as in Europe and Asia (see Knox 1988).

Gilles Seutin

Pine Siskin

Chardonneret des pins

Tarin des pins
Carduelis pinus

Christiane Girard

THE PINE SISKIN IS A GREGARIOUS BIRD THAT WOULD EASILY GO UNNOTICED were it not for its sporadic visits to feeding stations. Even its plumage, heavily streaked with dark brown with no apparent distinctive traits in either sex, makes it rather unobstrusive, although the yellow at the base of the tail and on the wings does make it easier to identify.

This fast-flying bird travels in flocks of varying size all year round. In winter, it is often found in the company of redpolls and other finches, including the American Goldfinch. In Canada, it winters all along the West Coast, as well as irregularly in the southern parts of its breeding range. It is found throughout the United States in winter, except southern Florida. It also winters in northern Mexico (AOU 1983; Godfrey 1986).

Habitat and behaviour

Distinctive traits of the Pine Siskin are its almost constant nomadic tendencies and its erratic comings and goings. In summer, it may be common in a given locality one year and totally absent the next (Todd 1963; Bock and Lepthien 1976b; R. Rivard, unpubl. data). According to Yunick (1981), this unpredictable species is rather opportunistic; it has no real attachment to a given region during the breeding season, but instead prospects for suitable conditions. Populations on the wintering grounds may also fluctuate wildly in size from year to year. According to Bock and Lepthien (1976b), fall or early winter southward irruptions seem to be related to reduced availability of the seeds of various plants and trees in the northern part of its range. The fact that some tree species do not produce seeds every year makes the movements of the Pine Siskin unpredictable.

The Pine Siskin is found in coniferous and mixed woods, in rural areas as well as on the outskirts of urban centres with abundant sources of food and suitable nest sites. It can sometimes find suitable conditions in ornamental conifers in towns and cities (Erskine 1977; Godfrey 1986; McNair 1988).

It also takes advantage of winter bird feeders in residential areas (Palmer 1968; McNair 1988), but again, it does not frequent the same places every year. Nonetheless, in any given winter it may become sedentary, returning regularly to the same feeding stations, as demonstrated by Aubry and Laporte (1990): between January and May 1989, 23.3% (926/3966) of birds banded in the Charlesbourg area were recaptured the same season, and 77.2% (715/926) of them were taken at the station where they had been banded.

The diet of the Pine Siskin varies somewhat with the season and region. It eats mainly seeds of conifers such as spruce, pine, cedar, tamarack, and hemlock, but also those of deciduous trees such as birch, alder, maple, and elm, as well as the seeds of a number of herbaceous plants. It also sometimes eats insects (Roland et al. 1986), buds, and berries (Barrow 1912 in Higgins 1991; Forbush and May 1939; Palmer 1968; Terres 1980; Roland et al. 1986). Flocks of siskins have been observed pecking at the bare ground or mortar in the stonework of the wall of a house, and this behaviour has been attributed to a need for minerals (Flaxman 1983; Bennetts and Hutto 1985). The Pine Siskin is one of the few songbirds that has been observed during winter at a bird feeder at night (Prescott 1985).

As the breeding season approaches, flocks of Pine Siskins making their way to suitable areas gradually break up into smaller groups. The males then begin to engage in courtship feeding (Palmer 1968). Once pairs have formed within the flock, the female selects a nest site, to which the male will then follow her. The pair establishes a small territory around the nest, but remains gregarious outside the territory (Newton 1972).

Notes on territory

Two or three days before laying commences, both mates stake out and defend a territory 0.9–1.8 m in diameter around the nest (Weaver and West 1943). Once laying has begun, the male leaves the area for short periods to find food for his mate. His absences grow longer as incubation progresses. The female then becomes very aggressive, chasing off any conspecifics or other songbirds that attempt to approach the nest (Weaver and West 1943).

The density of Pine Siskins in an Ontario spruce-fir forest was estimated at 4.7 pairs/10 ha (Erskine 1971a).

The Pine Siskin is a resident of extensive coniferous forests like this fir stand, where it can find an abundant supply of the seeds it eats. Laurentide Wildlife Reserve, north of Québec City. *July 1991, Guy Lemieux.*

Pine Siskin

Breeding distribution

1984–1989

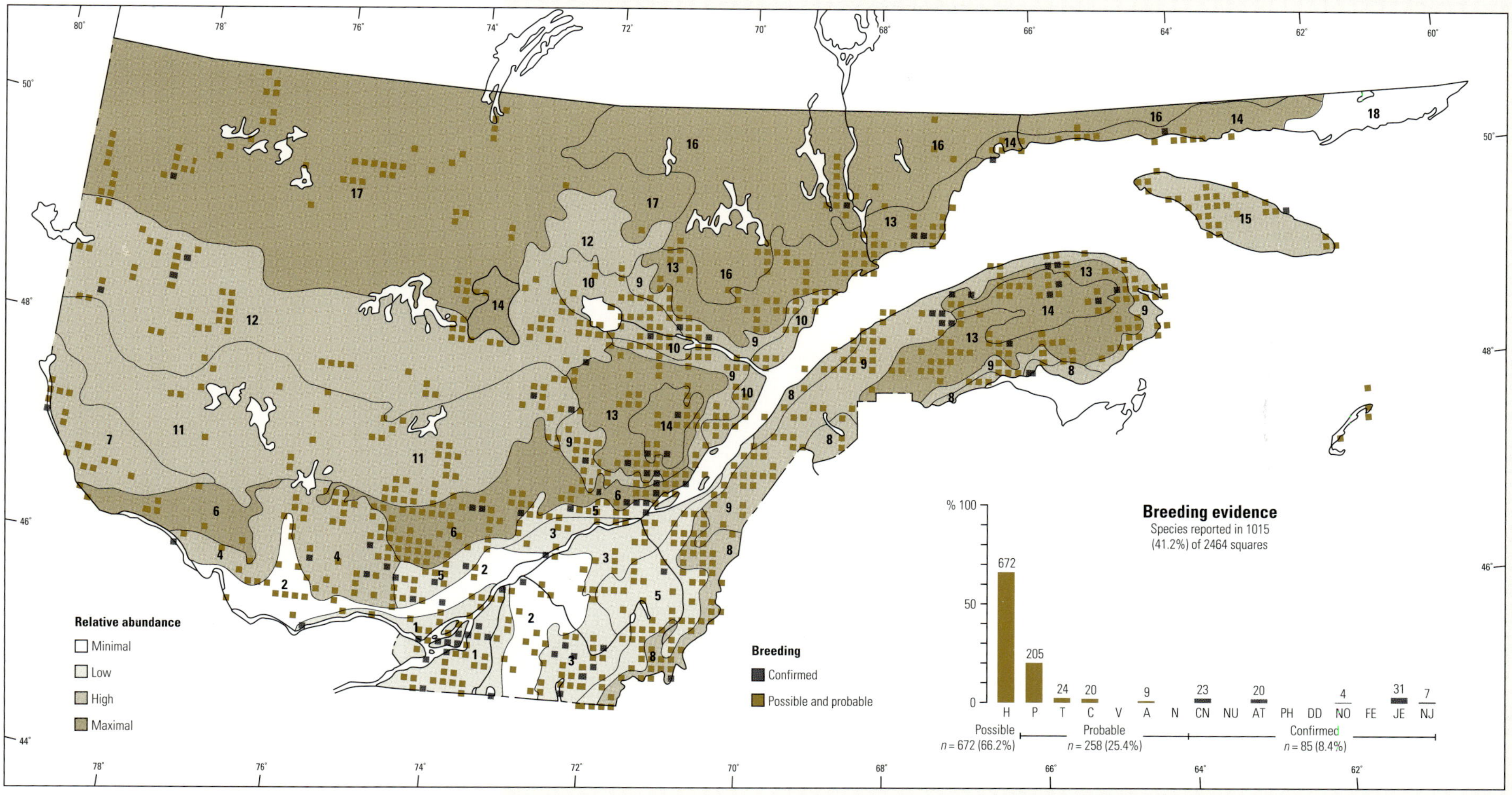

World breeding range

Species profile

Clutch size	usually 3–4 eggs (2–6)[1,2,3,4]
Laying interval	1 egg per day[1]
Start of incubation	with first egg[1,2]
Incubation period	13–14 days[1,2,4]
Maturity at hatching	altricial
Care of young	by female and male[1,2]
Nestling period	14–15 days[1,2,4]
Age at first flight	14–15 days or more[8]
Dependence of young	probably until a few weeks after leaving nest[2]
Broods per year	usually 1 (1–2)[1,4]
Breeding age	1 year[1]
Mating system	monogamy[1]
Duration of pair bond	?
Total length	11–13 cm[7]
Wing span	?
Weight	14.6 g (10.8–20.1)[5]
Longevity record	9 years 5 months[6]

References:
1. Palmer 1968
2. Weaver and West 1943
3. Perry 1965
4. Terres 1980
5. Dunning 1984
6. Klimkiewicz and Futcher 1987
7. Godfrey 1986
8. according to nestling period

Chronology $n = 153$

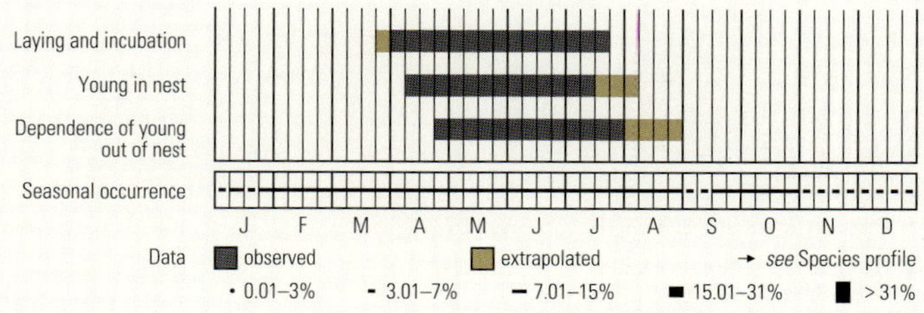

Frequency of occurrence by ecological region

Ecological regions** (forest regions)	Adequately covered squares	Squares in which reported	% of squares in which reported	Relative abundance (%)*
1. Sugar maple-hickory and sugar maple-basswood	97	42	43.3 ± 4.9	44.7
2. Sugar maple-basswood	153	36	23.5 ± 3.5	24.3
3. Sugar maple-basswood and sugar maple-yellow birch	101	45	44.6 ± 5.9	46.0
4. Sugar maple-yellow birch-basswood	56	34	60.7 ± 11.2	62.7
5. Sugar maple-yellow birch-beech and sugar maple-yellow birch-basswood	113	64	56.6 ± 5.1	58.5
6. Sugar maple-yellow birch-beech	71	63	88.7 ± 7.0	91.6
7. Sugar maple-yellow birch-balsam fir	19	13	68.4 ± 22.1	70.6
8. Sugar maple-yellow birch and balsam fir-yellow birch	47	33	70.2 ± 11.8	72.5
9. Balsam fir-yellow birch	87	74	85.1 ± 7.1	87.8
10. Balsam fir-white birch or balsam fir-red maple	46	34	73.9 ± 11.1	76.3
11. Yellow birch-balsam fir	52	35	67.3 ± 13.2	69.5
12. Balsam fir-white birch (west)	59	50	84.7 ± 9.7	87.5
13. Balsam fir-white birch (east)	76	72	94.7 ± 5.3	97.8
14. Balsam fir-black spruce	28	27	96.4 ± 8.3	99.5
15. White spruce-balsam fir and balsam fir-white spruce	22	16	72.7 ± 19.4	75.1
16. Black spruce-balsam fir-moss	18	17	94.4 ± 13.5	97.5
17. Black spruce-moss	32	31	96.9 ± 7.6	100.0
18.			*not analysed*	

* minimal: 0–29.9%; low: 30–59.9%; high: 60–89.9%; maximal: 90–100%

** Deciduous forest: 1–7; mixed forest: 8–11; coniferous forest: 12–17

The variable rate of seed production in various tree species from year to year forces the Pine Siskin to be nomadic. It may even leave the boreal forest and settle in mixed woods in which birch and other deciduous trees will meet its dietary needs. Laurentide Wildlife Reserve, north of Québec City. *June 1991, Réginald Ouellet.*

Pairs of Pine Siskins most often nest close together, forming loose colonies; on occasion some pairs may nest singly, however (Palmer 1968; Newton 1972; Ehrlich et al. 1988). The female alone builds the nest, which is customarily concealed in a mature conifer, well out from the trunk at the tip of a branch (Elderkin 1989). It can be found anywhere from 1 to 15 m up (DeGraaf et al. 1980), but is usually located at mid-height (Harrison 1978; Terres 1980; Peck and James 1987; Ehrlich et al. 1988). Construction takes approximately 4 days (Weaver and West 1943). The flat structure, which is usually not attached to the branch, is made of twigs, rootlets, and grasses and is lined with hair and feathers. The female has been known to dismantle an old American Goldfinch nest and use the materials for her own nest (Palmer 1968). Though nests are usually found in conifers (Peck and James 1987), they are sometimes built in a deciduous tree (Palmer 1968).

Incubation, which begins with the laying of the first egg, is by the female only and lasts about 2 weeks. During this time, the male feeds his mate by regurgitation, enabling her to remain on the nest for long periods. For the first few days after hatching, the female feeds the young frequently, every 10–15 minutes, with food brought by the male; the feedings are gradually spaced further apart toward the end of the nestling period. Once the young are 10 days old, they are fed directly by both the male and the female, and she begins to forage for herself (Palmer 1968).

The young leave the nest at about 15 days of age, but appear to remain nearby for another few weeks (Weaver and West 1943). When the breeding season is over, young and adults gather in flocks of 50–200 or more to search the surrounding area for food (Palmer 1968). Where Pine Siskins nest in solitary pairs, as in urban areas, families of 4–6 can be seen at feeding stations for part of the summer.

The Pine Siskin was reported in 41.2% (1015/2464) of Atlas squares surveyed. This high observation rate is probably partially due to the spectacular invasions that occurred in the province in the winters of 1984–85 and 1986–87 (Yank et al. 1987b; R. Rivard, unpubl. data) and that persisted into the following breeding season. Breeding

This finch is not averse to seeds of herbaceous plants growing in open areas bordering mixed or coniferous woodland. Forillon National Park, Gaspé Peninsula. *July 1991, Yves Aubry.*

The species typically nests in evergreen conifers, but may occasionally use a tamarack. Sainte-Foy. *September 1991, Yves Aubry.*

was confirmed in 8.4% of squares (85/1015) in which the species was reported. It is known that siskins frequenting a particular feeding station all winter will often breed nearby the following spring (Aubry and Laporte 1990). This may explain the high number of reports of confirmed breeding in the inhabited regions of southern Québec; on the other hand, the low percentage of confirmations obtained for the province as a whole can probably be attributed to the early breeding and secretive behaviour of the species on the nest. It is worth noting that most confirmations were based on records of young outside the nest, easily recognized by their darker plumage and downy heads and necks.

Distribution

The Pine Siskin breeds across central and southern Canada from Newfoundland to southern Alaska, including the Great Lakes region; it also breeds in the western United States, in the Mexican highlands, and as far south as Guatemala (AOU 1983). The breeding range of the species is hard to define, since its presence in a given region is irregular from one year to the next and seems to be mainly determined by the abundance of food. In Québec, the Pine Siskin breeds irregularly throughout the southern part of the province from James Bay and southern Labrador, particularly in the mixed and coniferous forest regions south of the 53rd parallel (David 1980a). Frequency of occurrence is higher in the forest regions dominated by black spruce or balsam fir, though it is still quite high in the mixed forest zone in the forest regions dominated by balsam fir or yellow birch. In the deciduous forest zone, the sometimes high frequency of occurrence in some forest regions is associated with islands of conifers scattered throughout agricultural, wooded, and urban areas.

The Pine Siskin nests on Anticosti Island, but breeding on the Magdalen Islands has not yet been confirmed. The sighting of pairs between May and August, as well as various historical accounts, suggests, however, that a small population breeds on the islands, at least irregularly (Fradette 1992).

The most northerly report of this species comes from near Great Whale River, where Savile (1950) told of observing a flock of 18 birds in August. Todd (1963), however, casts doubt on this observation and feels that they were Common Redpolls. Other, more recent reports mention sightings of Pine Siskins at Blanc-Sablon and Brador on the Lower North Shore, at Fermont, and around the La Grande Rivière (Dryade 1992; ÉPOQ).

The Pine Siskin winters throughout most of the southern part of its breeding range except Anticosti Island (David 1980a). It is rare in winter on the Magdalen Islands, where it has been observed in small numbers only during the winters of 1983–84 and 1990–91 (Fradette 1992).

Confirmations of breeding south of the 47th parallel, in particular the Central St. Lawrence Lowland upriver from Québec City and in the Eastern Townships, are probably partially the result of the great abundance of the species during its southern incursions in the springs of 1985 and 1987 (Yank et al. 1987b). It nested in the Greater Montréal area in 1984, 1985, and 1987. Adults and nestlings were recorded at Roxboro, Cartierville, and Saint-Lambert in 1984 (Yank and Aubry 1984). At the end of March the following year, a pair building a nest near Huntingdon, another pair carrying twigs in Hampstead, and adults feeding each other at a feeder in Boucherville were sighted (Bannon 1985a). In addition, in 1987, recently fledged young were reported at 3 different locations on the Island of Montréal (Yank et al. 1987b).

History and trends

Based on his personal observations, Todd (1963) claimed that the Pine Siskin originated from the western part of the continent and expanded its range eastward.

Not much is known about the Québec population; as it is nomadic and not always observed in the same areas from year to year, it is hard to determine whether its numbers have changed over the years. According to Ouellet (1974a), this siskin probably nested in the Montréal area in the last century and was probably more numerous there in winter then. Elsewhere in Québec, it was abundant on the Magdalen Islands (Fradette 1992) and considered a common summer resident in the eastern part of the province, on Anticosti Island (Schmitt 1904 *in* Ouellet 1969) and at Godbout on the North Shore (Dionne 1906). A few individuals, but sometimes also large flocks, used to be seen in summer on the Gaspé Peninsula (Demille 1926; Ball 1938).

Today, the Pine Siskin is considered a fairly common permanent resident in the Gaspé (Girard 1988). Overall, it is quite likely that the status of the Pine Siskin in Québec as a whole is similar today to what it was, except perhaps in the Central St. Lawrence Lowland, where its decline may be due to loss of local conifer stands (Ouellet 1974a) and peatlands (A. Bouchard, pers. comm.), where it probably used to dwell.

Reynald Rivard and Manon Bombardier

American Goldfinch

Chardonneret jaune

Carduelis tristis

Christiane Girard

The American Goldfinch provides birders with a spectacle of sounds and colours, with its undulating, rhythmic flight and lively, high-pitched *perchicoree* call.

The American Goldfinch moults twice a year and is the only cardueline finch to undergo a prenuptial moult, during which all its body feathers are replaced, except for the flight feathers (Middleton 1977b, 1978). The female's spring and fall plumages are identical, a drab olive-green. In fall, the male resembles the female, but as the time of courtship approaches, he dons a bright yellow livery with contrasting black forehead patch and black and white wings and tail. At this time, the male also enlarges his repertoire of songs and engages in some fascinating courtship behaviour (Drum 1939).

The American Goldfinch is migratory in the northern part of its breeding range but is generally a permanent resident in the southern part. Males seem to winter slightly north of the females (Prescott and Middleton 1990). In winter, goldfinches are sometimes found in the southern and central parts of southern Québec (David 1980a). Middleton (1977a) believes that the increase in the overwintering population in southern Ontario since 1915 is due to the proliferation of feeders, which ensure an abundant food supply.

However, most of the population winters in northern Mexico and the United States, except for the Great Lakes and northern Plains regions (National Geographic Society 1987).

Habitat and behaviour

The American Goldfinch breeds in a variety of semi-open habitat, such as roadsides, woodland edges, riparian areas, open forest, parks, second growth, orchards, gardens, cultivated or abandoned fields, and marshes (Nickell 1951; Godfrey 1986; Peck and James 1987). It is usually not found in mature forest. In Ontario, Peck and James (1987) reported that goldfinch nests were found most often in hawthorn bushes, while in Wisconsin, Worley (1987) found nests more often in dogwoods than in any other kind of shrub. In fact, the species occurs almost anywhere it can procure thistle seed, its favourite food, as well as the seeds of dandelions, ragweed, mullein, goat's-beard, sunflowers, and alders (Stokes and Stokes 1986; Hayakawa 1987; Ehrlich et al. 1988).

The species frequents mainly open areas with abundant food and preferably a few trees or shrubs for perching and nesting. It also adapts very well to humans.

This small bird's behaviour changes with the seasons. In winter, the males and females form flocks and are rather quiet; their repertoire of songs and flight patterns is quite limited. In spring, the males don their bright nuptial plumage, sing sweet, warbling songs from prominent perches, engage in spirited chases with other males, and so on; to escape, females adopt an elusive zigzagging flight. Altercations between males may result in dramatic confrontations in which both males, facing each other, flutter together high into the air. In summer, at the beginning of the nesting season, the male performs two types of courtship displays, each accompanied by calls. Autumn is ushered in by the characteristic *chipee, chipee, chipee* call of the young, which is heard until they become independent of their parents (Stokes 1979).

Unlike most other songbirds, American Goldfinches begin their courtship activities up to 2 months before nesting (Stokes 1950). Interestingly, dominant females appear to select subdominant males for mates, and vice versa (Watt 1991).

The female alone builds the nest, usually in the fork of a shrub or tree, though occasionally in non-woody plants 0.3–10 m up (Nickell 1951; Ehrlich et al. 1988). The cup-shaped nest is woven very tightly of plant fibres, grasses, and rootlets. The rim is reinforced with bark strips held in place with sticky materials such as spider silk and caterpillar tents. The interior, with a diameter of roughly 6.5 cm (4.5–8.0 cm), is lined with thistle, milkweed, or cattail down (Harrison 1975; Hayakawa 1987; Peck and James 1987). The female is silent when transporting nesting materials (Stokes 1979).

A week before nest building starts, the male establishes a territory, which encompasses a 30-m area around the nest, and defends it until the nest has been completed (Stokes 1950). The species is not strongly territorial, and once incubation has begun, the area defended begins to shrink until, by the time the eggs hatch, it consists of only the immediate area around the nest (Stokes 1979). Although other

The American Goldfinch nests on the edges of forests and abandoned fields. Saint-Lambert, near Montréal. *September 1930, Henry Mousley, Laval University Collection.*

Notes on territory

This species is gregarious outside the breeding season, and pairs are formed while the birds are still in flocks. Although birds separate from the flock around early May as soon as pairing occurs, the pairs do not establish a territory until a few days before nest building begins (Drum 1939; Stokes 1950).

The male establishes a small territory about 30 m in diameter (or 660 m² on average) in an appropriate site (Stokes 1950). During the breeding season, the male displays both his typical undulating flight and an exaggerated version used in courtship. In the latter, termed "deep-loop-flight," he climbs upward, makes a steep dive with wings held in to gain momentum, and then spreads his wings to coast upward again (Stokes 1979; Hayakawa 1987).

When incubation starts, territorial boundaries begin to dissolve until just the immediate area around the nest is defended, mainly by the female (Middleton 1979, 1987). Like most cardueline finches, the American Goldfinch may nest in small, loose colonies, within which each pair defends a small territory (Middleton 1979). However, this type of behaviour may reflect the uneven distribution of resources and adequate nesting sites rather than an innate tendency to nest colonially (Miller 1978).

American Goldfinch

Breeding distribution
1984–1989

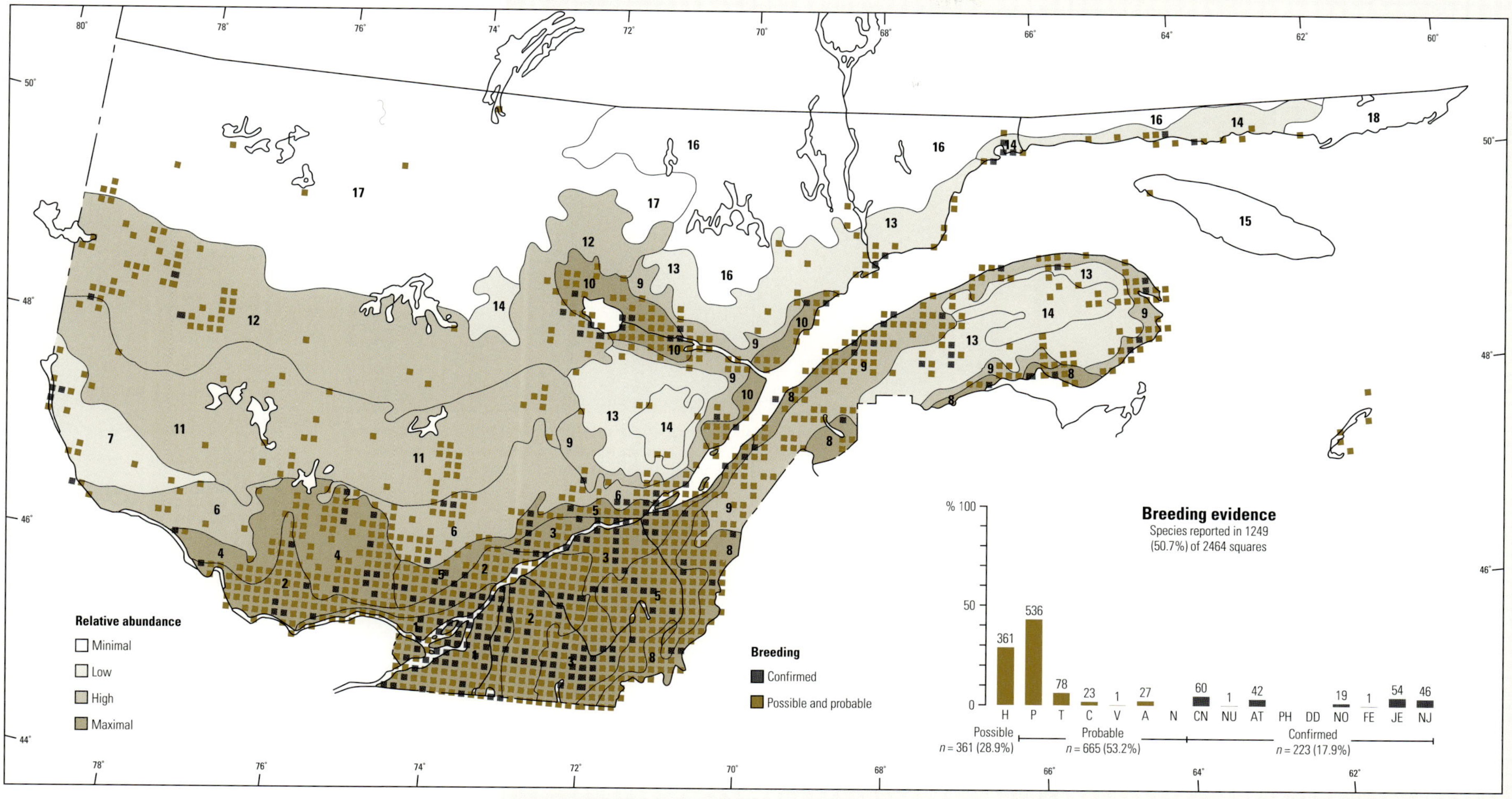

Species profile

Clutch size	usually 4–6 eggs (2–7)[1,2,6,7]
Laying interval	usually 1 egg per day[5,12]
Start of incubation	with second or third egg[13], before last egg[12,14]
Incubation period	10–14 days[7,8,9,10]
Maturity at hatching	altricial
Care of young	by female and male[1,5,7]
Nestling period	11–17 days[1,6,7,14]
Age at first flight	≈ 11–17 days[14]
Dependence of young	until ≈ 3 weeks after leaving nest[1,14]
Broods per year	usually 1 (1–2)[2,6,7,10]
Breeding age	1 year[9]
Mating system	monogamy, sometimes polyandry[4,7]
Duration of pair bond	1 season[2]
Total length	11.4–14.0 cm[8]
Wing span	22.2–22.9 cm[9]
Weight	M 13.2 g (8.6–20.7) / F 12.6 g (10.0–17.1)[11]
Longevity record	9 years 3 months[3]

References:
1. Hayakawa 1987
2. Middleton 1979
3. Klimkiewicz and Futcher 1987
4. Middleton 1988a
5. Stokes and Stokes 1986
6. Stokes 1950
7. Ehrlich et al. 1988
8. Godfrey 1986
9. Terres 1980
10. DeGraaf and Rudis 1987
11. Dunning 1984
12. Peck and James 1987
13. Stokes 1979
14. Middleton 1993

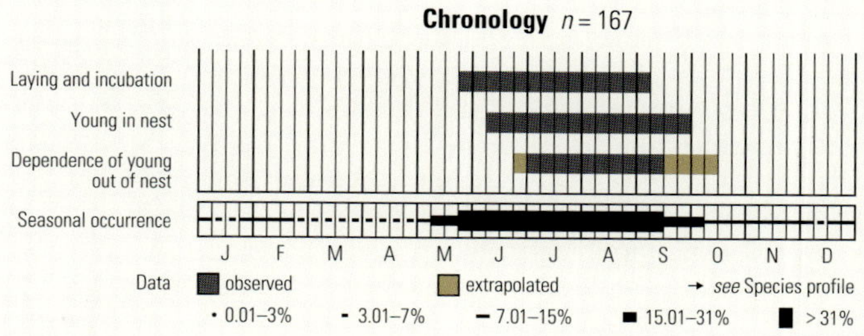

Frequency of occurrence by ecological region

Ecological regions** (forest regions)	Adequately covered squares	Squares in which reported	% of squares in which reported	Relative abundance (%)*
1. Sugar maple-hickory and sugar maple-basswood	97	97	100.0 ± 0.5	100.0
2. Sugar maple-basswood	153	153	100.0 ± 0.3	100.0
3. Sugar maple-basswood and sugar maple-yellow birch	101	101	100.0 ± 0.5	100.0
4. Sugar maple-yellow birch-basswood	56	54	96.4 ± 4.8	96.4
5. Sugar maple-yellow birch-beech and sugar maple-yellow birch-basswood	113	113	100.0 ± 0.4	100.0
6. Sugar maple-yellow birch-beech	71	52	73.2 ± 9.5	73.2
7. Sugar maple-yellow birch-balsam fir	19	9	47.4 ± 23.5	47.4
8. Sugar maple-yellow birch and balsam fir-yellow birch	47	45	95.7 ± 5.8	95.7
9. Balsam fir-yellow birch	87	78	89.7 ± 6.1	89.7
10. Balsam fir-white birch or balsam fir-red maple	46	42	91.3 ± 7.5	91.3
11. Yellow birch-balsam fir	52	32	61.5 ± 13.6	61.5
12. Balsam fir-white birch (west)	59	42	71.2 ± 11.9	71.2
13. Balsam fir-white birch (east)	76	43	56.6 ± 10.9	56.6
14. Balsam fir-black spruce	28	15	53.6 ± 19.3	53.6
15. White spruce-balsam fir and balsam fir-white spruce	22	1	4.5 ± 10.3	4.5
16. Black spruce-balsam fir-moss	18	1	5.6 ± 13.5	5.6
17. Black spruce-moss	32	6	18.8 ± 15.1	18.8
18.			*not analysed*	

* minimal: 0–29.9%; low: 30–59.9%; high: 60–89.9%; maximal: 90–100%

** Deciduous forest: 1–7; mixed forest: 8–11; coniferous forest: 12–17

World breeding range

This species mainly frequents open areas with an abundant supply of food and a few trees and shrubs for perching and nesting. Since it nests late, it can take advantage of the abundant supplies of plant down then available to line its nest. Mount Saint-Hilaire. *August 1991, Yves Aubry.*

goldfinches flying over or foraging inside the territory are tolerated, a male taking up a prominent position is immediately driven off, and female intruders are also expelled, usually by the female (Stokes 1950).

Once the nest has been completed, the pair leaves the area for a few days, sometimes for up to 2 weeks, as if they had abandoned it (Stokes 1979; Stokes and Stokes 1986). When they return, the female begins egg laying. A good way to locate the nest is to listen for the *bearbee, bearbee, bee, bearbee* or *sweet, sweet, sweet* calls given by the male and female when the nest is threatened (Stokes 1979).

Clutch size can range from 2 to 7 eggs, but the average is 4–6 (Stokes 1950; Middleton 1979; Ehrlich et al. 1988). The smooth, unmarked, ovate eggs are a very pale bluish or greenish-blue hue (Harrison 1975; Harrison 1978). The female is very attentive and incubates almost without a break for 10–14 days (Godfrey 1986; Ehrlich et al. 1988). During this time she is fed by the male. At periodic intervals ranging from 45 minutes to over 2 hours, he flies over the nest, in either the flat-flight or deep-loop-flight display (Stokes 1979); if the female is hungry she calls *teeteetee*, and this induces the male to land close to the nest and regurgitate partially digested seeds (Stokes 1979; Stokes and Stokes 1986; Middleton 1987).

Both parents feed the young by regurgitation, mainly seeds (Terres 1980; Stokes and Stokes 1986; Middleton 1987), although sometimes insects are given as well (Miller 1978; Ehrlich et al. 1988). Nesting is late, timed to coincide with the abundant food available at the end of summer, when supplies of plant seeds are greater than in May or June. Because the goldfinch lays its eggs at the end of the Brown-headed Cowbird's breeding season, nests are rarely parasitized (Terres 1980); the cowbird's and goldfinch's most active laying periods are about a month apart (see Peck and James 1987).

The young leave the nest 11–17 days after hatching (Stokes 1950; Hayakawa 1987; Ehrlich et al. 1988; Middleton 1993) and follow their parents (Stokes 1979). The female sometimes lays a second clutch, building the nest while the male is taking care of the first brood (Middleton 1979; Hayakawa 1987), sometimes taking them to bird feeders.

American Goldfinch nest solidly attached to a shrub. Before starting to lay, the female leaves the area for a few days. Clutches may consist of up to 7 eggs. Saint-Lambert, near Montréal. *September 1933, Henry Mousley, Laval University Collection.*

Notes on territory (continued)

Given its small size, the territory does not usually provide all the water, food, and nesting material required (Stokes 1950), and the birds use common foraging areas often within 90 m of the nest site (Drum 1939; Nickell 1951).

Erskine (1971a, 1976a, 1980a, 1984) reported different densities of territorial males in southern Ontario, depending on habitat: 2.3–6.5 males/10 ha in pine forest or plantations; 3.6 and 6.2 males/10 ha in Eastern white-cedar stands located near openings; 0.5-5.5 males/10 ha in pastureland, brushy areas, and partially reforested fields; 1.6 and 3.5 males/10 ha in deciduous forest; 3.3 males/10 ha in an urban park; and 3.2 males/km in hedgerows alongside old fields.

Nesting mostly in late summer, goldfinches take advantage of abundant supplies of plant seeds. After the young leave the nest, the lively and brightly coloured family groups can be seen foraging in herbaceous vegetation. Saint-Augustin-de-Desmaures, west of Québec City. *Summer 1992, Jean Gauthier.*

Young goldfinch perched just above the nest, ready to make his first flight. The late date, 11 September, suggests that he is from a second brood. Saint-Lambert, near Montréal. *September 1900, Henry Mousley, Laval University Collection.*

Distribution

The American Goldfinch is found only in North America, nesting from southern Canada to the south-central United States (Godfrey 1986; National Geographic Society 1987).

In Québec, the American Goldfinch is found primarily in urban and agricultural areas, although it also frequents regions of extensively fragmented forest. Therefore, the highest frequencies of occurrence were obtained in the Central St. Lawrence Lowland, Ottawa Valley, and the Saguenay–Lac-Saint-Jean region, as well as in the lower Laurentians, the Chaudière–Appalaches region, and the Eastern Townships.

The species is slightly less common in densely wooded regions, where it is generally found only in clearings, burns, and cutovers and along logging roads, transmission line rights-of-way, and shores of lakes and rivers.

The American Goldfinch was also atlassed on Anticosti Island and the Magdalen Islands, where it had been recorded previously (Ouellet 1969; Bergeron 1977; Fradette 1992).

North of the Atlas area, there are records as far north as the Pontax River and Lake Mesgouez, east of James Bay (Cons. G. G. – GREBE 1992o), and up to Lake Plétipi, west of the Manicouagan Reservoir (ÉPOQ), suggesting that the northern limit of its breeding range is around the 52nd parallel.

History and trends

The American Goldfinch is probably more common today than it was before European settlement, when it was probably largely limited to marshes, lakeshores, riverbanks, burns, and other natural openings (Nickell 1951).

BBS data for 1966–83 indicate a decline in breeding populations in the Maritimes, Ontario, Québec, and the central Prairies; this could be partly due to the loss of potential breeding habitat, changing agricultural practices, and increased urbanization (Hayakawa 1987). In Québec, breeding season data for 1980–89 also suggest a decline (BBS).

Maryse Lepage and Benoît Limoges

Evening Grosbeak

Gros-bec errant

Coccothraustes vespertinus

Christiane Girard

THE EVENING GROSBEAK, ORIGINALLY NATIVE TO WESTERN NORTH America, was sighted in southern Québec for the first time at the turn of the century. As the years passed, the species was seen with increasing frequency, and today the Evening Grosbeak can be observed both during the breeding season in its habitat and in winter at feeders. The Evening Grosbeak is noisy and quarrelsome in winter, when it travels in flocks, but becomes less gregarious and more secretive during the breeding season, when it is found in pairs (Scott and Bekoff 1991). The species is now considered a resident breeder in southern Québec (David 1980a).

During the winter, the Evening Grosbeak either remains in its breeding range or irrupts to the south, sometimes as far as the Gulf states and Mexico (AOU 1983; Aubry and Laporte 1990).

Habitat and behaviour

During the breeding season, the Evening Grosbeak frequents mainly second-growth and mature coniferous woodland, particularly stands of balsam fir and spruce, as well as mixed forest (Erskine 1977; Godfrey 1986; Peck and James 1987); it is also seen occasionally in deciduous woodland, parks, and orchards (Peck 1987c). Winter flocks frequent a wide range of forest and other types of wooded habitat, and the species is often seen in inhabited areas during this season.

The Evening Grosbeak's feeding habits are quite variable, depending on the season. Before and during the breeding season, insects—including beetles, cankerworms (Geometridae), and spruce budworms—make up a significant part of its diet (Terres 1980). The Evening Grosbeak, along with the Tennessee Warbler, is one of the best biological indicators of the presence of the spruce budworm, and infested stands have a high density of nesting grosbeaks (Blais and Parks 1964; Erskine 1977). Despite its fondness for insects at certain times of the year, the Evening Grosbeak is primarily a seed eater that feeds on the buds and seeds of many types of herbaceous plants, shrubs, and trees, particularly maples and conifers. With its powerful bill, it cracks open the pits of fruits like choke cherries to get to the seed, but rejects the fleshy part of the fruit; when grosbeaks are eating berries, you can sometimes hear the cracking sound as they live up to their nickname of "berry-breaker" (Speirs 1968; Giroux and Laporte 1988). In winter, they are particularly fond of the seeds of Manitoba maple and also gorge on sunflower seeds, as feeder owners can attest (Speirs 1968; Aubry and Laporte 1990). At this time of year, flocks can often be seen along roadsides eating road salt, and in summer, grosbeaks may even be observed feeding on the calcium chloride spread on dirt roads to keep dust down (Speirs 1968; Giroux and Laporte 1988).

Evening Grosbeaks arrive on their nesting grounds already paired (Scott and Bekoff 1991). The breeding season begins in mid-May in the south, but may not start until the end of June in the northern part of the species' range (Harrison 1978). Shortly after the pairs arrive, nest building begins. The female does most of the work, although the male may sometimes bring nest materials. There is no evidence that pairs attempt to establish a territory (Scott and Bekoff 1991).

During courtship, pairs are very secretive, and even if a bird is spotted, it is practically impossible to find the nest by following it (Scott and Bekoff 1991). The most frequent courtship behaviour includes wing fluttering, calling from perches, wing drooping, and courtship feeding by the male. Sometimes courtship dancing can be observed just prior to copulation (Scott and Bekoff 1991). In this display, the male faces the female, and with crest raised, wings drooped, and bill and tail pointed upward, he slowly pivots back and forth (Downs 1958 *in* Speirs 1968).

Notes on territory

Evening Grosbeaks arrive on their breeding grounds already paired and do not appear to establish territories. While the female incubates the eggs, her mate generally perches nearby, ready to defend the nest. During this period, he forages around the nest site, though occasionally venturing up to 1 km from the nest (Scott and Bekoff 1991).

The Evening Grosbeak appears to select its habitat mainly on the basis of availability of food rather than its structure or local competition (Bekoff et al. 1987).

The Evening Grosbeak is a denizen of mixed and coniferous forest. Primarily a seed eater, it feeds on the seeds and buds of herbaceous plants, shrubs, and trees, particularly maples and conifers. Tantaré Ecological Reserve, near Québec City. *Summer 1990, Ministère de l'Environnement du Québec.*

Evening Grosbeak

Breeding distribution 1984–1989

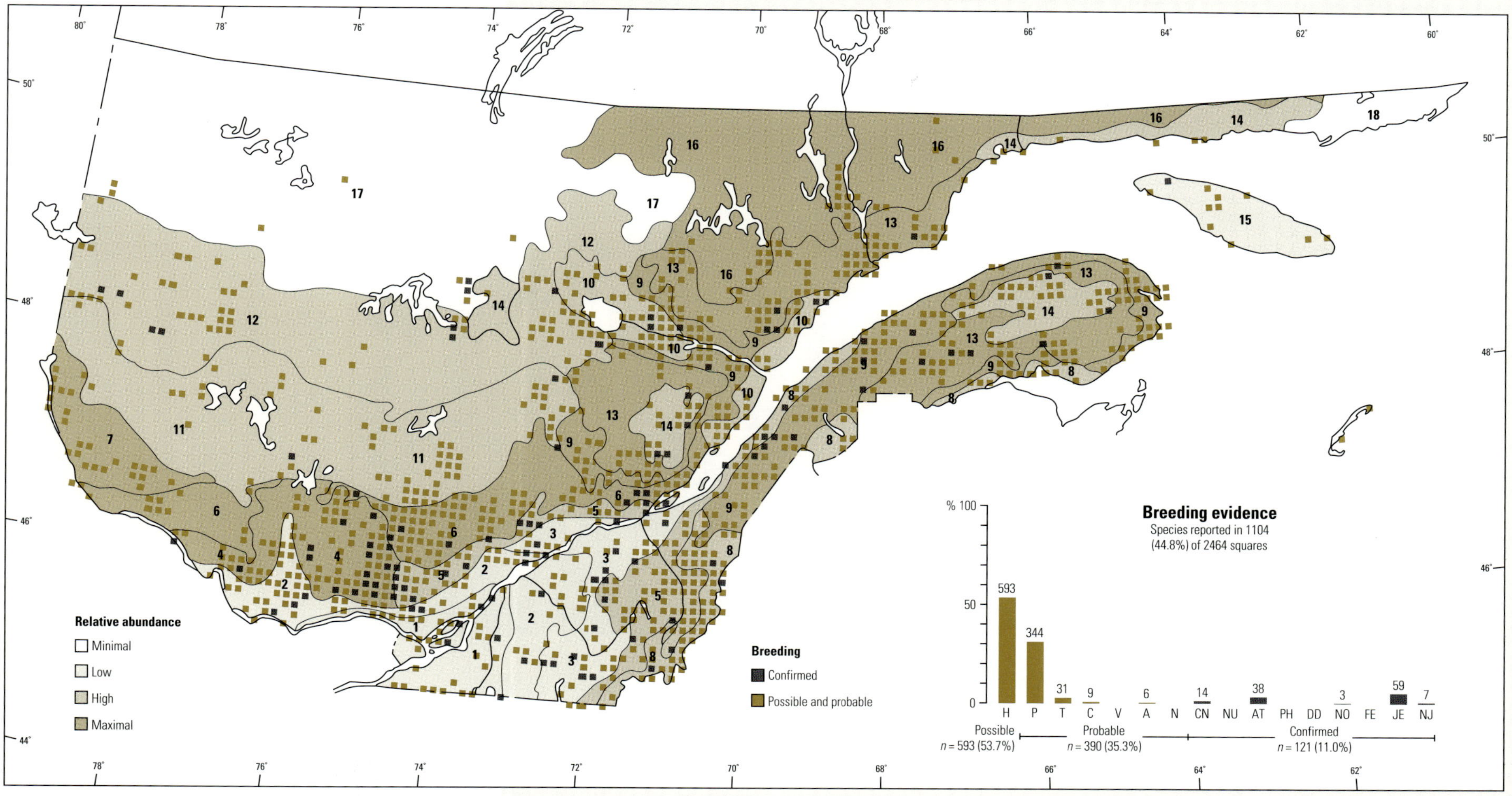

Species profile

Clutch size	usually 3–4 eggs (2–5)[1]
Laying interval	1 egg per day[1]
Start of incubation	?
Incubation period	usually 11–14 days[1,2]
Maturity at hatching	altricial
Care of young	by female and male[1,2]
Nestling period	usually 13–14 days[1,2]
Age at first flight	usually 13–14 days or more[7]
Dependence of young	until autumn[6]
Broods per year	usually 1 (1–2)[2]
Breeding age	?
Mating system	monogamy, sometimes polygyny[2]
Duration of pair bond	?
Total length	17.5–21.5 cm[3]
Wing span	33.0–34.9 cm[4]
Weight	M 60.1 g (38.7–86.1) / F 58.7 g (43.2–73.5)[5]
Longevity record	15 years 3 months[8]

References:
1. Speirs 1968
2. Scott and Bekoff 1991
3. Godfrey 1986
4. Terres 1980
5. Dunning 1984
6. Bekoff and Scott 1989
7. according to nestling period
8. Klimkiewicz and Futcher 1987

World breeding range

Chronology $n = 32$

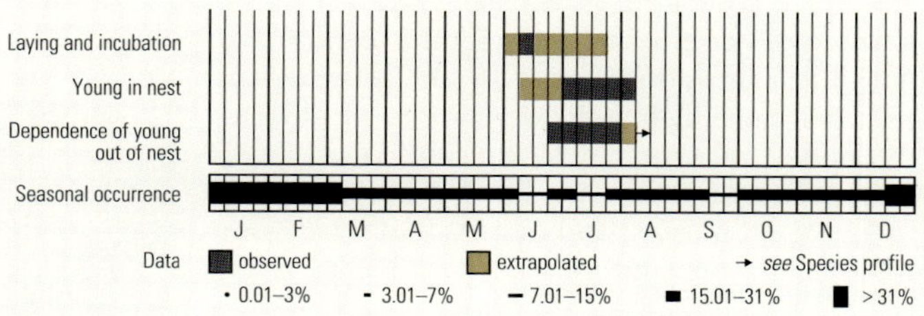

Frequency of occurrence by ecological region

Ecological regions** (forest regions)	Adequately covered squares	Squares in which reported	% of squares in which reported	Relative abundance (%)*
1. Sugar maple-hickory and sugar maple-basswood	97	33	34.0 ± 4.7	34.0
2. Sugar maple-basswood	153	70	45.8 ± 4.1	45.8
3. Sugar maple-basswood and sugar maple-yellow birch	101	45	44.6 ± 5.9	44.6
4. Sugar maple-yellow birch-basswood	56	51	91.1 ± 6.9	91.1
5. Sugar maple-yellow birch-beech and sugar maple-yellow birch-basswood	113	82	72.6 ± 4.6	72.6
6. Sugar maple-yellow birch-beech	71	67	94.4 ± 5.3	94.4
7. Sugar maple-yellow birch-balsam fir	19	18	94.7 ± 12.0	94.7
8. Sugar maple-yellow birch and balsam fir-yellow birch	47	41	87.2 ± 8.9	87.2
9. Balsam fir-yellow birch	87	84	96.6 ± 3.9	96.6
10. Balsam fir-white birch or balsam fir-red maple	46	40	87.0 ± 8.8	87.0
11. Yellow birch-balsam fir	52	44	84.6 ± 10.4	84.6
12. Balsam fir-white birch (west)	59	44	74.6 ± 11.5	74.6
13. Balsam fir-white birch (east)	76	76	100.0 ± 0.7	100.0
14. Balsam fir-black spruce	28	24	85.7 ± 14.1	85.7
15. White spruce-balsam fir and balsam fir-white spruce	22	7	31.8 ± 20.2	31.8
16. Black spruce-balsam fir-moss	18	17	94.4 ± 13.5	94.4
17. Black spruce-moss	32	8	25.0 ± 16.6	25.0
18.			*not analysed*	

* minimal: 0–29.9%; low: 30–59.9%; high: 60–89.9%; maximal: 90–100%

** Deciduous forest: 1–7; mixed forest: 8–11; coniferous forest: 12–17

The species seeks coniferous forests with an abundant supply of insects, particularly spruce budworm infested balsam fir stands, like this one on the Gaspé Peninsula. The infested area appears grey in the centre of the photograph. *August 1992, Lina Breton, Ministère des Forêts.*

Grosbeaks prefer to nest in open forest, selecting a shrub or tree, usually a conifer, that provides adequate concealment and stability for the nest (Bekoff et al. 1989). The open habitat allows pairs to detect approaching predators, while the nest site, being very difficult of access, makes it harder for predators to locate the nest and provides protection against inclement weather; these two factors combined were found to improve breeding success (Bekoff et al. 1987, 1989). The nest is usually located high in the tree (6–30 m up), close to the trunk, and generally facing south; the flimsy, loosely constructed cup is formed of interwoven twigs, buds, mosses, and lichens and lined with rootlets, grasses, and pine needles (Harrison 1975; Peck and James 1987; Ehrlich et al. 1988).

Typical clutch size is 3 or 4 eggs, which are incubated by the female alone for 11–14 days (Speirs 1968). While the female is on the nest, the male feeds her by regurgitation and remains nearby, probably to defend the nest (Scott and Bekoff 1991). After the eggs hatch, both parents take care of the young, feeding them masticated insects and other prey and keeping the nest clean by removing the fecal sacs. The young remain in the nest for around 13 or 14 days (Speirs 1968) and, after fledging, stay with their parents in the area around the nest for another 2–5 days (Scott and Bekoff 1991).

Subsequently, the parents and young depart on their annual wanderings. The young continue to be fed by their parents until August or even September (Bekoff and Scott 1989). In winter, major population movements occur, and thus the same grosbeaks rarely return to the same bird feeder. The Evening Grosbeak is a very lively, quarrelsome, nervous bird, particularly at feeders, where it tends to flee at the slightest movement or hint of danger (Aubry and Laporte 1990).

The Evening Grosbeak prefers to build its nest at the top of a conifer, where it has an unobstructed view of its surroundings. In summer, these birds are sometimes seen eating gravel at the side of the road, which helps them digest seeds and fruit pits in their gizzards. Matapédia River, Gaspé Peninsula. *August 1991, Yves Aubry.*

Grosbeak eating unripe fruit. Its stout, powerful bill is used to crack open the pits of the fruits of many trees and shrubs year-round. Philipsburg, south of Saint-Jean-sur-Richelieu. *May 1994, André Bilodeau.*

The species was reported in 44.8% (1104/2464) of Atlas squares surveyed, but was confirmed as breeding in only 11.0% (121/1104) of these. The low rate of breeding confirmations was due in part to the difficulty of finding nests, which are usually placed high in trees, mainly conifers, where they are well hidden in the dense foliage; nests were found only in 0.9% (10/1104) of the squares in which the species was reported. Pairs' secretive behaviour during the breeding season probably also played a role in limiting breeding confirmations in some squares.

Distribution

The Evening Grosbeak's breeding range lies mainly in central and southern Canada, from British Columbia to the Atlantic provinces, although it also extends south to the northern Great Lakes states and parts of New England. On the West Coast, the species nests mainly in mountainous regions, including high-elevation areas in Mexico (AOU 1983; Montevecchi and Tuck 1987; Erskine 1992).

The Evening Grosbeak was atlassed as a breeder throughout southern Québec except the Magdalen Islands, though it was reported there. The first sight record from the Magdalen Islands dates only from 1966 (Fradette 1992), and the sole breeding record, from August 1972; the dearth of breeding records is probably due to the species' irregular presence in very small numbers during the breeding season (Fradette 1992).

In the southern part of the province, the Evening Grosbeak appears to be associated mainly with the mixed and coniferous forest zones. The highest frequencies of occurrence for the Atlas correspond with the principal areas of spruce budworm infestation during the Atlas period 1984–89 (Bordeleau 1988; Bordeleau and Boulet 1989, 1990). The species' association with forests infested with the spruce budworm has also been reported by Erskine (1977). During the breeding season, the Evening Grosbeak is less common in regions where its preferred habitat is limited, such as Montérégie, the Eastern Townships, and the Central St. Lawrence Lowland between the Ottawa Valley and Québec City.

The relatively low frequency of occurrence in the Black spruce-moss forest region, located in the northwest part of the Atlas area, can be explained in part by the grosbeak's association with spruce budworm infestations, which are less likely to affect black spruce (Morris 1963).

The species' distribution extends slightly north of the Atlas area, with records from Eastmain and Gagnon (ÉPOQ).

History and trends

Originally from the coniferous forests of northwestern North America, the Evening Grosbeak was first observed in the Montréal and Québec City areas during the 1889–90 winter invasion (Caulfield 1890b; Butler 1892; Dionne 1906). Until then, it had been unknown to eastern naturalists (Aubry and Laporte 1990), and in 1906, Dionne wrote that the first Black-crowned Grosbeak (*sic*) that, to his knowledge, had appeared in the Québec City region, had been shot on 11 March 1890 at La Jeune Lorette (Loretteville). Wintle (1896) reported it at Montréal in late January of the same year. Subsequently, the species was reported fairly regularly in small numbers, at least until 1916, when Terrill (1916) mentioned a few records in the Montréal region.

According to Taverner (1921) and Todd (1963), the species' eastward expansion may have been prompted by the planting of Manitoba maples in parks, cities, towns, and villages and along the roadsides of the eastern provinces. Blais and Parks (1964) observed that, between the 1940s and 1960s, grosbeak populations reached exceptional numbers in areas infested by the spruce budworm. They reported seeing large numbers of these birds between 1960 and 1962 in the Patapedia River watershed, southeast of Rimouski, which was severely affected by the budworm at the time. This predator-prey relationship was also reported by Erskine (1977). Ouellet (1974a), too, believes that recurrent spruce budworm epidemics may have contributed to the species' eastward expansion.

The Evening Grosbeak's summer range appears to have expanded eastward in gradual stages, resulting in part from the species' winter movements (Todd 1963). Breeding was confirmed in Québec for the first time on 3 May 1940 at L'Assomption, where a male was seen transporting nesting materials (straw and horsehair) in his bill. Another male carrying rootlets was seen two years later on 13 June 1942 in the same place (Cayouette 1946; Gosselin 1989a). The first breeding record in the Montréal area was obtained during the Atlas survey, at Saint-Sulpice, in 1984 (Yank and Aubry 1984).

Population trends are difficult to assess in the species, since it is observed irregularly throughout its breeding range. BBS data for 1968–79 do not show any significant trends in the North American population as a whole (Robbins et al. 1986). The Québec population appears to have remained fairly stable during the same period, according to BBS (1966–89) and ÉPOQ (summer, 1970–89) data. However, BBS data for 1980–89 show a strong downward trend. Nevertheless, the species' range appears to be still expanding in eastern North America (National Geographic Society 1987).

JACQUELINE VINCENT

House Sparrow

Moineau domestique

Passer domesticus

Christiane Girard

THE FEISTY HOUSE SPARROW, THE VERITABLE URCHIN OF CITY, SUBURB, AND country, is related to the weavers of Africa, Eurasia, and Oceania that are sold in pet shops. The large Québec population is resident throughout the year. The House Sparrow, a bird with a long history, was introduced into North America in the mid-19th century.

Few species have been as widely and abundantly documented as the House Sparrow. Very early on, ornithologists were intrigued by the bird's ability to learn and adapt (Porter 1904, 1906, 1910), which enabled it to be successfully introduced into many countries. In New Zealand in 1990, it was found that House Sparrows were entering a bus station by flying in front of the sensor that operated the automatic doors, an example of how wily the bird can be in gaining access to a source of food (Breitwisch and Breitwisch 1991). The House Sparrow has long aroused passions among defenders and detractors alike (see Anonymous 1901, 1920; Ingram 1992; Dion 1993). Morency (1992:90) describes how at the turn of the century two members of the clergy took sides in the controversy, one for, the other against.

Habitat and behaviour

In Europe, the House Sparrow gradually moved from the agrarian ecosystem to urban habitat (Summers-Smith 1963). In the 19th and 20th centuries, it spread throughout Northern Europe and Asia as well as the countries where it had been introduced, by following expanding road and rail networks (Summers-Smith 1988), and in North America birds were transplanted to other places once they were well established in the localities where they had been introduced (Long 1981).

House Sparrows are generally social, foraging and roosting, taking water and dust baths communally (Summers-Smith 1963; Lowther and Cink 1992), though not always; for example, Cyr (1980) made the unusual observation of a lone female House Sparrow taking a dust bath in an old robin's nest.

Nest sites are usually located near a food source in an open area such as a park, garden, or green space, or near a dwelling or building. House Sparrows feed on the ground on seeds, insects, and waste grain around farms, and on undigested grain gleaned from the droppings of cattle and horses. Generally, the nest is placed on or very near an inhabited building, whether in city or country (Summers-Smith 1963), the nooks and crannies in buildings being favoured sites. When these are lacking, the sparrow nests in a vine, a tall conifer or, more rarely, in a deciduous tree. In urban habitat, it occupies a wide variety of nest sites, proving itself to be even more eclectic than the European Starling (Savard and Falls 1981). In rural habitat, it nests mainly under the eaves of buildings and in nooks in barns. Nests were once found in holes in bales of hay in a barn loft (Werler and Franks 1975).

Nests in trees are located rather high up (3–9 m), where the foliage is densest (Savard and Falls 1981). Several pairs may occupy the same tree, in which case they tolerate each other, but drive away any House Sparrows not nesting there (McGillivray 1980a). The nest is a globular structure with a side entrance, built of dried grasses and diverse materials such as feathers, string, and paper (Lowther and Cink 1992) found nearby or occasionally filched from other nests (McGillivray 1980a). McGillivray (1980b) reported finding a Colorado spruce near Calgary that contained 14 nests, 10 of which formed a sphere 1 m in radius, within which individual nests were joined into a single communal structure. When the House Sparrow uses a nest box, it either fills it with a variety of materials or it makes a simple cup of dried grasses (Lowther and Cink 1992).

The number of broods per season varies with latitude (Murphy 1978), from as many as 4 in Panama to as few as one in Churchill, northern Manitoba (Kendeigh 1976), though broods tend to be larger in the north than in the south (Murphy 1978). When several broods are raised in one season, the second brood often has better fledging success (McGillivray 1983). The time interval between broods is greater when trees rather than nest boxes are used. McGillivray (1981, 1983) considers nest boxes to be superior sites, being warmer and generally more conducive to nesting.

Monogamy is the norm, although instances of polygyny have been recorded (Lowther and Cink 1992; Veiga 1992). A polygynous male feeds its young as if it were monogamous and tending only one brood. Females assisted by males have higher breeding success than do unassisted females, which have a tendancy to kill the young of assisted females (Veiga 1990b). Females of a polygynous male produce fewer broods in a season than do those mated with a monogamous male (Veiga 1990a).

Recent studies have shown that some or all of the nestlings in a brood may be genetically different from the birds that tend the nest (Wetton et al. 1987; Wetton and Parkin 1991). In one colony in England, Wetton and Parkin (1991) found that 13.6% of the young in 183 broods were not related to the attendant male. Furthermore, some females lay in the nests of other House Sparrows, thereby increasing host clutch size (Kendra et al. 1988).

The female incubates alone, leaving the eggs only for brief periods to feed. The first two or three eggs hatch at about the same time and the others on the next day, or two days after at most (Weaver 1943). In most cases, both parents feed the young (Kendeigh 1976). At first, the nestlings are fed by regurgitation, and the older nestlings may receive more food than the younger ones (Weaver 1942; Lowther and Cink 1992). The adults continue to feed the young for up to 2 weeks after they have left the nest (Weaver 1942). Insects make up 50–70% of the diet of the young (Lowther and Cink 1992), sourced possibly from a local infestation such as Dutch elm disease (Wilmot 1917), or from spider's webs (Rossetti 1983). The adults' diet is mainly vegetable matter (Bent 1958).

Notes on territory

The House Sparrow defends a small territory around its nest (Will 1973). Males and females defend the territory only against birds of their own sex. The pair may frequent the breeding site throughout the year, but generally only defend it actively from the onset of nesting, though defence of nest boxes used as winter roosts has been reported (Lowther and Cink 1992).

After the breeding season, the birds gather in flocks to spend the night in trees and shrubs or in the vines growing on the walls of buildings in the city or country. They congregate at an initial site one or two hours before sundown, then move to the main roost just before sunset. Most of the sparrows that flock in this manner are young birds that have not adopted a permanent breeding site, adult birds tending to remain near the nest site throughout the year (Lowther and Cink 1992).

The House Sparrow has learned to adapt to urban habitat, where it finds food year-round in streets, parks, and at bird feeders. Montréal. *March 1991, Yves Aubry.*

House Sparrow

Breeding distribution

1984–1989

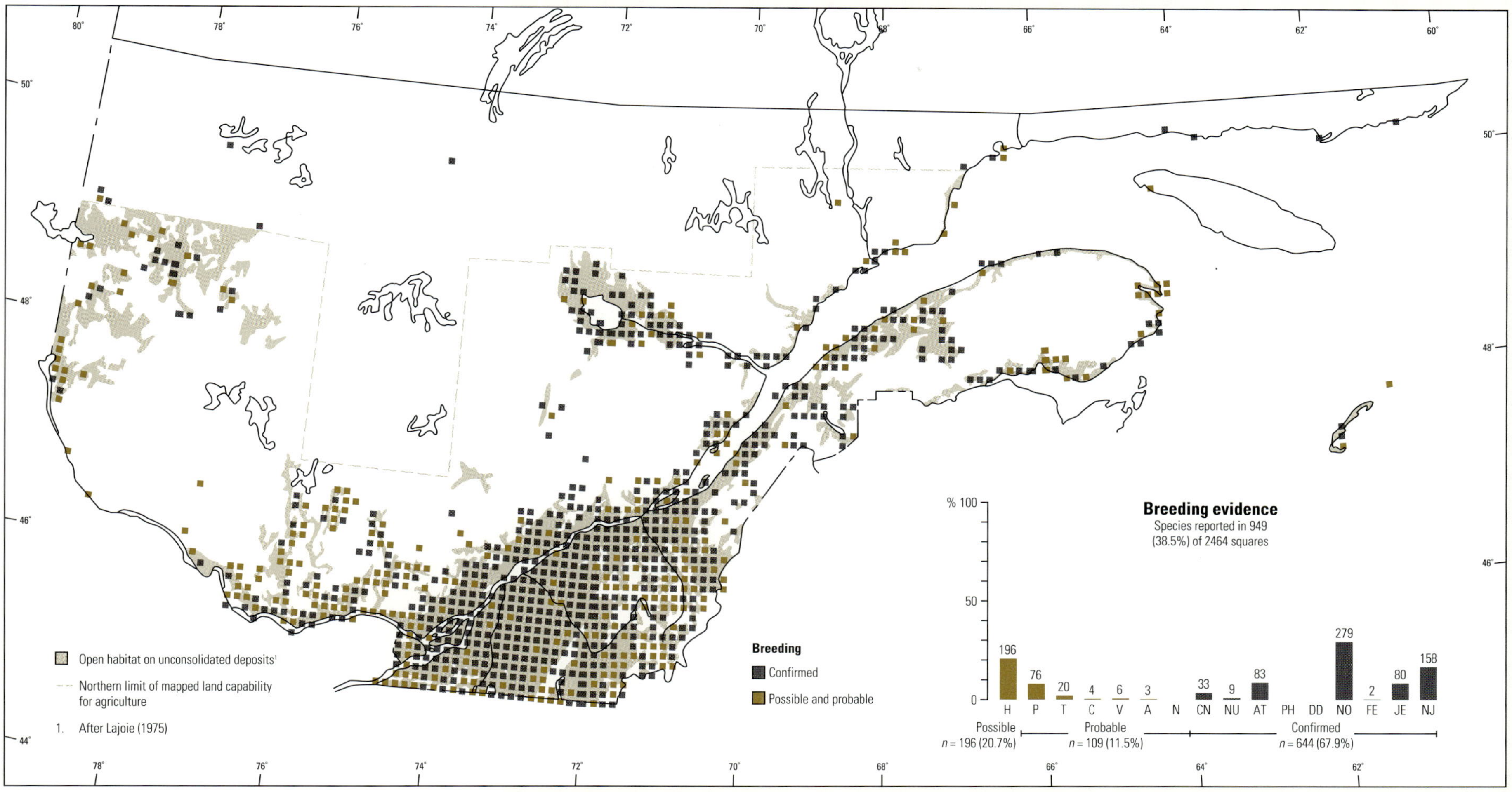

Species profile

Clutch size	usually 4–6 eggs (1–8)[1]
Laying interval	1 egg per day[1]
Start of incubation	usually with penultimate egg[1]
Incubation period	usually 11 days (10–14)[1]
Maturity at hatching	altricial
Care of young	by female and male[1]
Nestling period	usually 14 days (10–14)[1]
Age at first flight	≈ 14 days[1]
Dependence of young	until ≈ 7–15 days after leaving nest[1,9]
Broods per year	usually 2–3 (1–4)[5,6,7]
Breeding age	1 year[5]
Mating system	monogamy, sometimes polygyny[1,8]
Duration of pair bond	more than 1 season[1]
Total length	14.7–17.0 cm[4]
Wing span	24.1–25.4 cm[3]
Weight	M 28.6 g (?) / F 28.4 g (?)[1]
Longevity record	13 years 4 months[2]

References:
1. Lowther and Cink 1992
2. Klimkiewicz and Futcher 1987
3. Terres 1980
4. Godfrey 1986
5. DeGraaf and Rudis 1987
6. Murphy 1978
7. Kendeigh 1976
8. Veiga 1992
9. Weaver 1942

Chronology n = 116

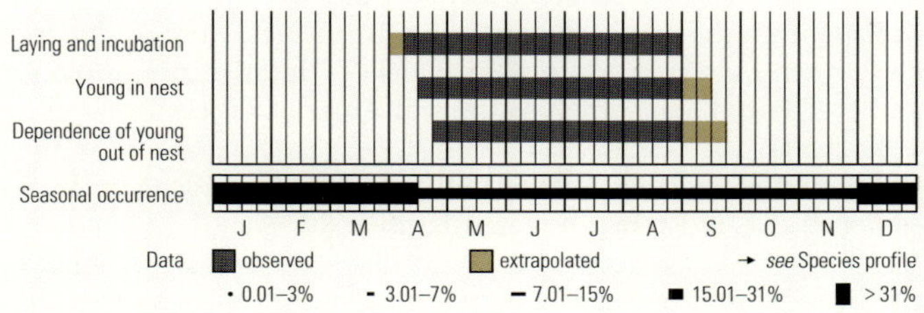

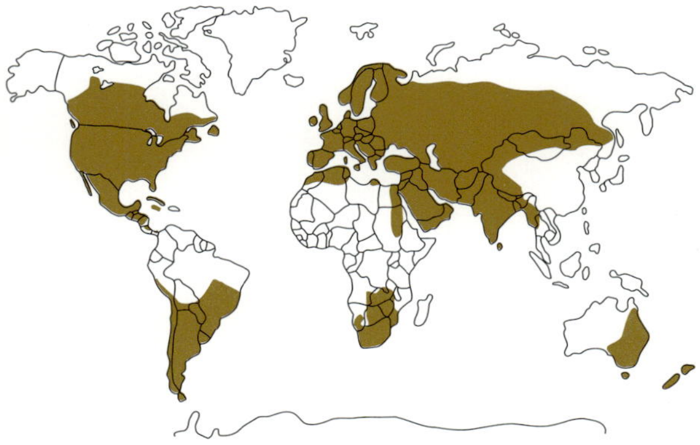

World breeding range

Sometimes birds other than the parents, either young of the year or other adults, may assist in feeding a brood. This behaviour may be fairly common locally, as indicated by a study by Sappington (1975) in Mississippi. The only reported effect of this was earlier fledging of the young. Shortly after leaving the nest, the young flock together and disperse, sometimes over considerable distances, especially in late summer and early autumn (Summers-Smith 1963; Will 1969, 1973; Lowther and Cink 1992). Young females tend to range farther afield than young males (Fleisher et al. 1984). In late summer, flocks of House Sparrows in rural areas are attracted to cultivated fields, which are a major food source. There they congregate along the edges, near protective cover (Summers-Smith 1963).

In agricultural habitat, House Sparrows feed on waste grain in the fields or around silos and other farm buildings. It also nests in the nooks and crannies of such buildings. Beloeil, east of Montréal. *April 1991, Yves Aubry.*

In North America, House Sparrow size varies with latitude. Birds in the desert regions of the southwestern United States are smaller than those in more northerly locations, such as Montréal and Edmonton in Canada (Johnston and Selander 1964, 1971; Lowther and Cink 1992). Moreover, birds that must endure severe winter cold need to accumulate more fat and water to survive the long, cold winter nights; this extra weight may have a bearing on their survival rate (Blem 1975).

Distribution

In North America, the House Sparrow breeds throughout Canada from the southern Northwest Territories, through the western provinces, Ontario, southern Québec, the Maritime provinces, and Newfoundland, down to Central America (Montevecchi and Tuck 1987; Erskine 1992; Lowther and Cink 1992).

House Sparrow distribution is highly dependent on human settlement, as illustrated by the Atlas map.

The St. Lawrence and Ottawa valleys, Eastern Townships, and the Beauce, predominantly agricultural and urban areas, all have abundant populations of House Sparrows. Occurrence is local in Abitibi–Témiscamingue, east of Baie-Comeau, and on Anticosti Island and the Magdalen Islands, where inhabited areas are more dispersed and farms are smaller. The House Sparrow was even sighted at Gagnon, from 9 to 11 July 1976 (ÉPOQ) and at Blanc-Sablon on 25 July 1981 (David and Gosselin 1982a).

History and trends

The House Sparrow was first imported from England in 1850 when an unsuccessful attempt was made to introduce the species into Brooklyn, New York. A second attempt, in 1852, proved to be more successful, and between 1852 and 1870, large numbers of birds were brought from England and Germany to be released. Birds from the initial release points were subsequently transplanted to other cities in the United States and Canada. High prices set by dealers brought about massive importation of birds for release, and this is how the species came to colonize the cities and countryside of North America (Barrows 1889; Long 1981).

Barrows (1889) credits these introductions to N. Pike, Director of the Brooklyn Institute, among others. Chapman (1925), however, maintains that E. Scheifflin, who introduced the European Starling, was responsible. It is highly probable that both men were involved in some of the many attempts to introduce the species.

Some supporters of the House Sparrow extolled it as the foe of insects injurious to ornamental trees (Barrows 1889; Robbins 1973). Others aspired to introduce all of the species mentioned in the works of Shakespeare. Some immigrants wanted a bird that would remind them of their homeland. Only a few naturalists and immigrants voiced any reservations in the face of this mass enthusiasm (Robbins 1973).

The species was afforded extensive protection. Food was provided, nest boxes installed, and predators such as cats, falcons, hawks, and shrikes were controlled; most importantly, legislation was passed to protect it. By 1870, however, public opinion began to change, and legal protection ceased. House Sparrows were accused of attacking grain and fruit crops, defacing the houses and buildings where they roosted and nested, and competing with indigenous

House Sparrows occasionally nest in deciduous trees. This globular nest has a side entrance. Sainte-Foy. *February 1994, Yves Aubry.*

View of the Montcalm Market, Québec City around 1900. At the turn of the century horses contributed to the spread of House Sparrows, which fed on grain spilled on the ground or found in horse droppings. After automobiles began replacing horses, House Sparrow numbers began steadily declining. St. Jean Street, Québec City. *Québec City Archives, Iconographic Collection, Valentine and Sons Pub., Negative No. 019295.*

species. Furthermore, it was discovered that the birds were insectivorous only when the young were in the nest. But by then the House Sparrow was well on its way to colonizing the whole continent, spreading across North America much faster than the European Starling (Wing 1943). In 40 years, it had already occupied all suitable regions (Lowther and Cink 1992).

In the face of such rapid proliferation, drastic measures were taken to reduce House Sparrow numbers in certain parts of the world where it had been introduced. In Australia, a price was set on its head and eggs (Long 1981), and in Montréal and elsewhere in Canada, advertisements were placed in newspapers soliciting live sparrows for target practice (Anonymous 1889).

The species was introduced into Canada by Colonel William Rhodes (*c.* 1821–92). After several vain attempts (Rhodes 1877; Wintle 1896), he succeeded in Québec City, in the Jardin des Gouverneurs, near the Château Frontenac, on 10 June 1868 (Anonymous 1868; Rhodes 1868), his purpose being to add spice to the lives of the citizens and garrison, especially during the long winters. Rhodes had also contributed to the releases of 1854 in the United States by bringing the birds across the Atlantic (Rhodes 1877; Barrows 1889). By the spring of 1872, the House Sparrows released by Colonel Rhodes had reportedly become acclimatized to Québec City and had been sighted at Cap-Rouge the previous summer (Anonymous 1872). According to Wintle, the species appeared in Montréal around 1870 (Barrows 1889:231) and by 1896 was abundant (Wintle 1896).

House Sparrow populations peaked around 1915 and then began to decline (Bergtold 1921; Bent 1958). Abundance remained high, nonetheless. The decline is thought to have been prompted mainly by the displacement of the horse by the automobile in cities, beginning around 1905 (Bergtold 1921). Until then, millions of horses had plied city and country, and the House Sparrow gleaned spilled and undigested grain from their abundant droppings. Other causes were the introduction of mechanized farming, higher sanitation standards on dairy farms and in cities, and less ornate architectural fashions, providing fewer nest sites. Severe winter weather was another constraint on their spread in the northern part of the range.

Although it declined in North America, the House Sparrow continued to spread throughout most of Central and South America (Smith 1980; Dott 1986; Thurber 1986; Summers-Smith 1988), down to Puerto Williams, in Chile, the southernmost town in the world (Wallace 1991).

Overall, the eastern North American population declined significantly between 1966 and 1979 (Robbins et al. 1986), a trend echoed also in Québec in Breeding Bird Survey data for 1966–89 and 1980–89 (BBS) and in ÉPOQ data for 1969–88 (Larivée 1991).

With increased urbanization and the proliferation of bird feeders, House Sparrow populations might have been expected to stabilize or increase. Such has not been the case, however. The House Sparrow now must share its habitat in the cities, towns, and villages of southern Québec with the House Finch, which, since its introduction in Long Island, New York in 1940, has continued to colonize new areas (Bosakowski 1986). The ecological niches occupied by both species have several points in common. Both species frequent urban and suburban habitat; both are granivorous throughout most of their life cycle; both visit bird feeding stations; and both nest in tall, ornamental trees and on or around buildings. Some authors have posited a link between the decrease in House Sparrow abundance and the increase in House Finch populations (Kricher 1983; Wise and Walls 1988), although Bennett (1990) discounts the effect of interspecific competition at the local and regional levels.

Yves Aubry

LA PRESSE

CINQUIÈME ANNÉE—No. 106 MONTRÉAL, MERCREDI 20 FÉVRIER 1889 UN CENT

Mort aux moineaux

Les municipalités voisines, à la requête des jardiniers et des cultivateurs, ont autorisé dernièrement le massacre des moineaux, et la tête de ces derniers est mise à prix, pourvu que l'oiseau soit en vie. On donne dix centins pour chaque douzaine de moineaux livrés en vie dans une cage ou une boîte grillée, au No 147, rue des Carrières, chez M. le conseiller Joseph Paquette, qui les recevra tous les samedis, de 2 p.m. à 6 p.m. Les cages seront échangées ou rendues. On achètera aussi les moineaux aux mêmes conditions, au No 1184, rue Saint Laurent, le samedi, aux mêmes heures.

Les oiseaux captifs sont ensuite revendus 20 centins la douzaine aux membres des différents clubs de tir, qui en font une consommation extraordinaire. Ils tirent les moineaux au vol après les avoir lancés d'une trappe à 26 verges de distance. Au dernier concours, les bons tireurs tuaient 26 moineaux sur 30. Disons qu'il est beaucoup plus difficile de tuer un moineau qu'un pigeon au vol. La semaine dernière, plus de 600 moineaux ont été tués de la sorte.

Introduced for aesthetic and sentimental reasons, House Sparrows soon became so abundant that they were hunted and killed, as illustrated by this article which appeared in *La Presse* on 20 February 1889. *Reprinted with the kind permission of La Presse Ltée.*

Clipping from *La Presse*
Newspaper article – 20 February 1889

Death to English [House] Sparrows

Neighbouring municipalities, at the request of gardeners and farmers, have recently authorized the extermination of English Sparrows, and a bounty is offered on live specimens. Ten cents will be paid for every dozen sparrows delivered live in a cage or screened box to Councillor Joseph Paquette, 147 Rue des Carrières, on Saturdays between 2 p.m. and 6 p.m. Cages will be exchanged or returned. Sparrows will also be purchased under the same terms at 1184 St. Lawrence Street, on Saturdays at the same times.

The captured birds are resold for 20 cents a dozen to various shooting clubs, which have a tremendous demand for them. Sharpshooters target birds in flight at a distance of 26 yards as they are released through a trap door. In the last competition, crack shots downed 26 birds out of 30. Sparrows are much harder to hit in flight than pigeons. Last week, over 600 English Sparrows were killed in this way.

OTHER SPECIES OF RECORD IN QUÉBEC

Extinct species,
northern breeders,
and other species…

Pacific Loon

Huart du Pacifique

Plongeon du Pacifique
Gavia pacifica

Christiane Girard

VERY FEW PEOPLE IN THE PROVINCE EVER HEAR THE PACIFIC LOON'S bloodcurdling cry, because it breeds high in northern Québec and is rarely seen in the south. It is probably much more common in Québec than the five known breeding records suggest. Since 1985, the Pacific Loon has been considered distinct from the Arctic Loon, which breeds in Eurasia (AOU 1985), and for which much more information is available (Lehtonen 1970; Nilsson 1977; Sjölander 1978).

Habitat and behaviour

Pacific Loons breed on ponds and lakes on the arctic tundra and taiga. Nests are built at the water's edge on islands, vegetation platforms, or on the mainland. The young are first led from the nest when they are 1 or 2 days old and are brooded occasionally at various locations along the shore until they are 12 days old. Young chicks are fed small items, probably insects, whereas older ones are fed fish and insects until they fledge at 50–55 days of age. The young abandon the ponds when they are 57–64 days old (Bergman and Derksen 1977; Petersen 1989).

Pacific Loons are often conspicuous and can be seen from far away in open tundra habitat. Nests can be found by combing the shoreline on foot or by canoe. Distraction displays are further evidence of breeding, as are family groups, which stay together until autumn (Peck 1987b). An adult flying with a fish in its bill, presumably to feed a chick on the nesting pond, is also a good indication of breeding.

Distribution

Pacific Loons breed from northeastern Siberia and Alaska east across northern Canada to Baffin Island, northern Ontario, and some islands in Hudson Bay and James Bay (Godfrey 1986). In Québec, Manning (1949) collected 2 adults and 1 downy chick from a lake 40 km north of Inukjuak and 16 km inland from Hudson Bay, on 21–22 July 1944. Two breeding records come from the west coast of Ungava Bay: on 29 July 1966, a pair with a downy chick was seen 1 km south of De Villiers Cove on a small shallow lake with rocky islets (R. Ouellet, pers. comm.), while on 14 July 1980, a nest with 2 eggs was found on an islet in a lake south of Hall Bay (G. Chapdelaine, pers. comm.). In 1992, the known breeding range was extended south to James Bay, where a pair with 2 flightless young were seen on a lake 2 km east of Attikuan Point; a pair was also seen 10 km east of Point Louis XIV the same year (Yank and Aubry 1992a). The last known breeding record is from the Akulivik airport, where a nest containing one egg was discovered on the edge of a nearby pond on 15 July 1993 (ÉPOQ). Since 1980, there have also been sightings during the nesting season from Kangirsuk, Ikattok Bay, and Inukjuak (ÉPOQ). Gaston et al. (1985) found the Pacific Loon to be an uncommon breeder on West Digges Island and Nuvuk Island near Cape Wolstenholme, and noted that it probably breeds on the Québec mainland. All this evidence, along with the fact that the Pacific Loon nests on a number of other islands off Québec in the Northwest Territories (Todd 1963; Godfrey 1986), suggests that it may breed more or less continuously along Québec's northern coast from Ungava Bay westward around Cape Wolstenholme at least as far south as Attikuan Point. How far inland it breeds in Québec is unknown, but in Ontario and Manitoba it nests at least 100 km inland from Hudson Bay (Jehl and Smith 1970; Peck 1987b).

History and trends

Very little is known of the Pacific Loon's distribution and status in Québec, and more data are needed regarding the coastal limits of its breeding range and how far inland it breeds.

Thousands of loons—Pacific, Red-throated, and Common combined—are taken every year by Natives on Québec's west coast (James Bay and Northern Quebec Native Harvesting Research Committee 1982).

ROBERT ALVO

Notes on territory

Pacific Loons have been found nesting on ponds and lakes ranging in size from 0.3 to 41 ha (Bergman and Derksen 1977; Petersen, M. R. 1979; Campbell et al. 1985), that are often used in successive years, possibly by the same pair (Jehl and Smith 1970; Bergman and Derksen 1977). Adults often feed elsewhere (Jehl and Smith 1970), but may fly back to the pond regularly to check their eggs from the air (McNicholl 1975).

Adults usually feed chicks with food captured in the breeding pond, but they may also ferry fish from the sea to the breeding pond (Bent 1919; Taverner and Sutton 1934; Bergman and Derksen 1977). Families often move by land to nearby ponds (Grinell and Palmer 1940; Bergman and Derksen 1977).

Nest density in Alaska was 0.8 nest/km² (Bergman and Derksen 1977), while pair density in northern Ontario was only 0.02–0.1 pair/km² (Peck 1987b).

The Pacific Loon breeds mainly in Quebec's Far North and prefers open sites such as a rocky island in a pond or lake. South of Hall Bay, coast of Ungava Bay. *July 1981, Gilles Chapdelaine.*

The Pacific Loon's nest, often at the water's edge, is a depression in the ground. South of Hall Bay, coast of Ungava Bay. *July 1981, Gilles Chapdelaine.*

World breeding range

Species profile	
Clutch size	usually 2 eggs (1–2)[2,3,4]
Laying interval	1 egg per 2–3 days[5]
Start of incubation	with first egg[1,5]
Incubation period	23–25 days[2,5]
Maturity at hatching	precocial
Care of young	by female and male[1,2,3]
Nestling period	less than 48 hours[1]
Age at first flight	50–55 days[1]
Dependence of young	until ≈ 57–64 days old[1]
Broods per year	1[2]
Breeding age	probably 2 years[9]
Mating system	monogamy[2]
Duration of pair bond	more than 1 season[5]
Total length	58–74 cm[6]
Wing span	109.2–127.0 cm[7]
Weight	1659 g (990–2450)[8]
Longevity record	?

References:
1. Petersen 1989
2. Ehrlich et al. 1988
3. Petersen, M. R. 1979
4. Peck and James 1983
5. Jehl and Smith 1970
6. Godfrey 1986
7. Terres 1980
8. Dunning 1984
9. Palmer 1962

Chronology $n = 5$

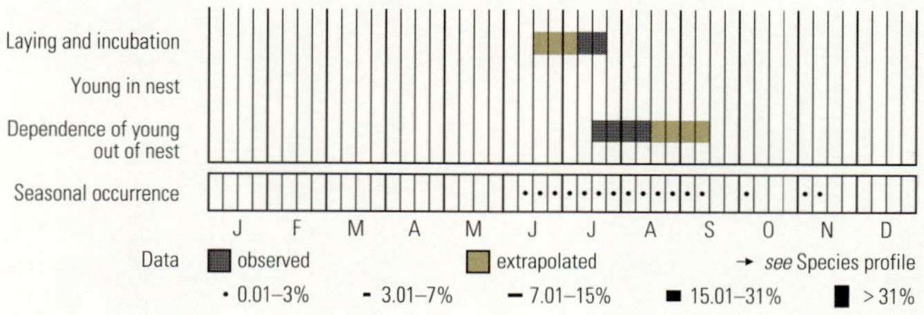

Tundra Swan

Cygne siffleur
Cygnus columbianus

Christiane Girard

THE TUNDRA SWAN IS ONE OF THE FEW SPECIES THAT HAS BEEN MAKING A comeback after probably being eliminated as a breeder from Québec (Lumsden 1984). Recovery started when the Migratory Birds Convention Act was passed in 1917 (S.C. 1917, c.18), and was aided by the bird's habit of nesting early, when the snow is deep and too soft for travelling, as well as by its wary nature. It breeds along the Hudson Bay coast and is transient in southern Québec, seen only during migration.

Habitat and behaviour

Tundra Swans usually breed north of the tree line in the tundra. Nesting starts very early in the season, and the swans select snow-free areas that are slightly elevated, often commanding a sweeping view. The same site is frequently used in subsequent years. Nests are usually situated on the main shore of a lake or pond, or on small islands or points in lakes, but also in marshes, tidal meadows, or tundra not always close to water. The cob (male) stands guard close by while the pen (female) incubates, but the cob may also incubate. Soon after the young hatch, the parents go through a flightless period (Palmer 1976; Cramp and Simmons 1977). The downy young start diving at 10 days; older flightless young and flightless parents dive well when pursued. Cygnets are at first helped by the parents to find food (Palmer 1976; Cramp and Simmons 1977).

Tundra Swans are easy to see from the air from a distance of up to 1.1 km, and their voice carries up to 6.5 km (Lumsden 1987h). They feed largely on the leaves, stems, and tubers of aquatic and marsh plants, but also on aquatic invertebrates. On the breeding grounds the families feed along the margins of lakes and occasionally on upland vegetation (Cramp and Simmons 1977).

Distribution

Tundra Swans breed in the arctic regions of Eurasia and North America. The Québec population is at the eastern edge of the North American breeding range. Heyland et al. (1970) found them breeding along the east coast of Hudson Bay from Cape Wolstenholme to Inukjuak. Twenty individuals were tallied between Povungnituk and Ivujivik in a survey along the coast between 28 June and 2 July 1978 (Chapdelaine and Tremblay 1979). In June 1954, Cooch (1954) saw two Tundra Swans north of Inukjuak between the Nauberakvik River and Cape Gertrude. Recently, several individuals were reported around Rivière au Phoque, Roggan River, and Kakassituq Point (Benoit et al. 1993b), as were two adults with five young near Kakassituq Point (Yank et al. 1990b). This suggests that the known

Notes on territory

Tundra Swans are gregarious except during the breeding season, when they are territorial. All breeding activities except pair formation occur on the territory (Cramp and Simmons 1977). The male and female chase conspecifics by aerial flight, and the pair have a triumph display with wings spread and shaking (Palmer 1976). Broods stay within about 100–400 m of the nest for considerable periods of time (Bellrose 1976), and are led to larger waters as the parents begin to moult (Ehrlich et al. 1988), probably to avoid early freeze-up (Palmer 1976).

In the Yukon Delta, pairs disperse soon after arrival, while nonbreeders remain in flocks along coastal estuaries, and only 40–60% of the swans are paired and hold territories (Bellrose 1976). Some 2-year-old birds hold territories, but do not nest (Palmer 1976).

Breeding density in the former USSR averaged 0.05 pair/km² (Cramp and Simmons 1977), while in Alaska and northwestern Canada there were 0.22–3.70 potential

The breeding range of the Tundra Swan in Québec is confined to the Hudson Bay coast and northern part of James Bay. Here, a pair escorts its brood of 5 cygnets on a lake near the coast of James Bay. Bay of Many Islands.
August 1990, Austin Reed.

breeders/km² (Wilk 1988). In winter flocks, the whole family defends the area around it, acting as a unit in threatening other birds and joining in the subsequent triumph ceremony (Cramp and Simmons 1977).

breeding range now extends north of James Bay. Farther south, in the La Grande Rivière area, one individual was sighted on the Opinaca River in June 1981 and a second in May 1984 (Bordage and Aubry 1982; Bordage 1985b). Breeding was not confirmed in this area, however. On 24 June 1993, the species was reported near Kuujjuaq (Y. Gauthier, pers. comm.).

Tundra Swans are rare in southern Québec (David 1980a). The ÉPOQ database contains only 63 reports from 1969 to 1990, mostly between March and June. The only record for the Magdalen Islands involves an adult with two juveniles in migration killed in October 1977 (Fradette 1992).

Chesapeake Bay is the most important North American wintering ground, with, in some years, nearly half of the North American population congregating there (Stewart and Manning 1958). The 222 sites in the United States with winter observations are along the east coast from South Carolina to New Jersey, around Great Salt Lake in Utah, and on the California coast (Root 1988).

History and trends

Archaeological records suggest that Tundra Swans may have been more common in eastern Canada in the 16th century and earlier, breeding around Hudson Bay on both the Ontario and Québec sides, possibly at Ungava Bay, and perhaps locally in Labrador and northern Newfoundland (Lumsden 1984). However, most of these populations were probably destroyed during the 19th century, victims of the large-scale trade in swan skins by the Hudson's Bay Company (Bent 1925).

Their recent appearance as a breeding bird in Manitoba, Ontario, and Québec indicates a return to a former breeding range after an absence of over 150 years (Lumsden 1984). Winter surveys have shown an increase in the North American population, from 78 000 in the 1950s to 133 000 during 1970–74, with the increase apparently concentrated in Canadian breeding areas. An estimated 5000 adults occur in eastern arctic Canada (Bellrose 1976). A subsistence harvest takes place in Canada, but its size is not known (USFWS and CWS 1986).

Robert Alvo

World breeding range

Species profile

Clutch size	usually 4 eggs (3–5)[1]
Laying interval	1 egg per 2 days[8]
Start of incubation	with last egg[8], 1–2 days after first egg is laid[10]
Incubation period	usually 30–32 days (29–40)[1,2,4,5,7,8,9]
Maturity at hatching	precocial
Care of young	by female and male[2,8]
Nestling period	less than 24 hours[4]
Age at first flight	60–75 days[1,2,3,4,9]
Dependence of young	until ≈ 1 year old[8]
Broods per year	1[2,5,8]
Breeding age	2–4 years[3,9]
Mating system	monogamy[2,8]
Duration of pair bond	probably for life[2,3,8]
Total length	M 116–145 cm / F 115–147 cm[1,6]
Wing span	182.8–213.4 cm[4]
Weight	M 8590 g (7400–9580) / F 7290 g (6480–8800)[6]
Longevity record	21 years[11]

References:
1. Bellrose 1976
2. Ehrlich et al. 1988
3. Palmer 1976
4. Terres 1980
5. Harrison 1978
6. Miller et al. 1988
7. Godfrey 1986
8. Cramp and Simmons 1977
9. del Hoyo et al. 1992
10. Hawkins 1986
11. Limpert and Earnst 1994

Chronology $n = 16$

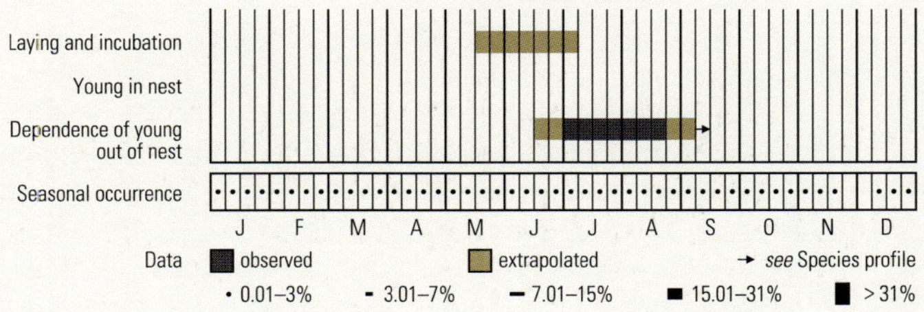

Mute Swan

Cygne tuberculé

Cygnus olor

The Mute Swan, a native of Eurasia, was introduced into the United States in the last century to ornament lakes and ponds. Its orange bill with a black knob sets it apart from other North American swans. This graceful bird often holds its neck in an S-curve and arches its wings during displays.

In North America, Mute Swans are mainly kept in captivity or semi-captivity on private property, in city parks or zoos. Nonetheless, as a result of introductions and a number of escapes, they are now well established in the wild. The Mute Swan is considered accidental in Québec (Robert 1989b).

Habitat and behaviour

The Mute Swan likes the calm, shallow water of marshes, ponds, lakes, and rivers, as well as coastal bays, where the emergent vegetation is dense and abundant (DeGraaf and Rudis 1987; Ehrlich et al. 1988).

Pairs form throughout the fall and winter. Age at first breeding is usually between 3 and 4 years of age (Perrins and Reynolds 1967), although the pens (females), more precocious than the cobs (males), may mate as early as their second year. In any case, a pair will stay together for at least a year before breeding (Wilmore 1974; Johnsgard 1975b; Ehrlich et al. 1988) and generally bonds for life (Bellrose 1976; Palmer 1976).

The courtship display of Mute Swans is a graceful spectacle. Before copulating, they exhibit a series of behaviours ranging from ruffling the neck feathers to mutual preening to rising in the water breast to breast. The entire ritual may resemble a dance.

From the onset of breeding, which may be as early as March, these swans defend a large territory, which often includes an expanse of water big enough to provide food and a runway. Breeding pairs are more aggressive than immatures or non-breeders and may thus acquire the best nest sites (Wilmore 1974).

Most of the time, pairs nest solitarily (Terres 1980). Nest building is a joint effort, with the pen doing most of the work (Johnsgard 1975b; Palmer 1976). The nest is generally a platform on the ground near water, on an island or on emergent vegetation in shallow water. Made of cattails, reeds, and roots of aquatic plants and lined with finer vegetation, feathers, and down (Reese 1975; Palmer 1976; Ehrlich et al. 1988), it can measure approximately 1 or even 2 m in diameter and 0.6 to 0.8 m in height (Johnsgard 1975b; Harrison 1978). A pair reuses the same nest for several years (Reese 1975; Palmer 1976; Cramp and Simmons 1977). In places where the water does not freeze over, Mute Swans winter near their nest site (Perrins and Reynolds 1967).

Typical clutch size is 4–8 eggs, laid at approximate 48-h intervals (Perrins and Reynolds 1967; Wilmore 1974; Reese 1975). Incubation, generally by the pen, begins when the last egg has been laid or a little before (Johnsgard 1975b; Harrison 1978). The cygnets are covered with pale grey down and lack the knob on the bill; they remain in the nest less than 48 hours. The cob takes care of the first hatchlings and leads them to water while the pen finishes incubating (Palmer 1976).

Mortality among cygnets is greatest in the second week (Reynolds 1965). It is estimated that a pair of swans successfully raises only 2 or 3 young a year, since the others fall victim to predators, such as snapping turtles, pike, foxes, and crows. Human harassment and egging, along with bad weather and nest flooding, also limit breeding success (Perrins and Reynolds 1967; Wilmore 1974; Reese 1975).

Their diet consists mainly of submerged and emergent aquatic vegetation (Johnsgard 1975b). Although the young can feed themselves, the parents may pull up underwater plants for them (Perrins and Reynolds 1967; Cramp and Simmons 1977; Lumsden 1987c). Cygnets also eat insects and small invertebrates (Bellrose 1976). The Mute Swan usually feeds in deeper water than do most dabbling ducks, and some dabblers, like the American Wigeon, eat the uprooted vegetation left floating after the swans have moved on (Terres 1980).

The family does not break up until autumn or winter, and the young may remain with their parents until the next breeding season (Reese 1975; Cramp and Simmons 1977). Although they do not migrate far in spring or fall, Mute Swans leave waterways that freeze over in winter, heading for expanses of fresh or salt water that remain open. They form flocks including both breeding and non-breeding birds (Perrins and Reynolds 1967; Reese 1975; Terres 1980). Non-breeders may also flock in large numbers all year long (Cramp and Simmons 1977).

Distribution

The only record of breeding in Québec concerns a pair that was released in 1978 at Lake Édouard, in Rouyn (COQ 1978). The semi-captive birds—a gift of the City of Ottawa, where they had been released and caught again in the summer of 1977—had built a nest and were incubating 5 eggs in the month of June. In Ottawa, pairs of Mute Swans are released on the Rideau River each spring. They breed during the summer and are recaptured with their offspring in fall.

Between 1965 and the spring of 1993, at least 10 sightings of Mute Swans were reported in Québec. In most cases, the birds were of unknown origin, but had probably escaped from farms or were part of introduced populations now well established in some eastern American states and Ontario. In spring, Mute Swans may also occasionally be seen among flights of Canada Geese staging through the St. Lawrence.

To date, winter sightings of this species have all been around Montréal, probably because more water remains open here at this time of year. Spring sightings have extended farther east, up to the Québec City region (David and Gosselin 1979b; David 1980a; PQSPB 1983; Aubry et al. 1987, 1988; Gosselin et al. 1990a; Yank et al. 1991b).

The Mute Swan is now established on the eastern seaboard of the United States, from New Hampshire and Massachusetts down to Virginia, as well as in Michigan (Long 1981; DeGraaf and Rudis 1987). Canadian distribution is more irregular (Lumsden 1987c); the species is now a permanent resident locally in southern British Columbia and southern Saskatchewan (Regina) (Godfrey 1986), and in Ontario, where it nests chiefly in the marshes of Lake Ontario, between Bowmanville and Hamilton (Lumsden 1987c).

Christiane Girard

Notes on territory

Throughout entire breeding season, from March to October, the Mute Swan aggressively defends a huge territory (Reese 1975; DeGraaf and Rudis 1987) of extremely variable size. On Rhode Island, Willey found 12 territories varying in area between 0.2 and 4.7 ha (Palmer 1976). In Michigan, Wood and Geltson (1972) calculated that territories varied in size between 1.6 and 4.2 ha. Perrins and Reynolds (1967) noted that territories near Oxford, England, were generally quite far apart, with average inter-nest distance between 2.4 and 3.2 km.

History and trends

A domestic bird for centuries in Europe, raised for its down, meat, and beauty (Wilmore 1974), the Mute Swan was introduced into North America in the late 19th century (Long 1981).

In the United States, individuals of this species were released in New York State just before the turn of the century (Bump 1941). Early in this century, other birds were released in some northeastern states (Long 1981). The first Mute Swans in Canada were introduced near Victoria, British Columbia, around 1889 (Warren 1970).

In some places, such as Long Island, New York, the populations are of considerable size, as a result of large releases (Sibley 1988e), and populations have increased rapidly in the northeastern United States in recent decades. In New York, the population has risen from 700 to 2000 individuals since the 1970s. In New Jersey, there are 1100 birds and in Connecticut, over 1800 (The Gazette 1993).

Due to our harsh winters, the species is not very abundant in Canada. Since they are fairly sedentary, migrating very little or not at all, Mute Swans are confined to habitats where there is open water all year long (Peck and James 1983). The Ontario population is about 120 feral birds (Lumsden 1987c), while there are only a few hundred in British Columbia (Campbell et al. 1990a). In Québec, the lack of ice-free areas and the often severe winters are limiting factors. Generally speaking, it is difficult to characterize and quantify the wild populations of Mute Swans in North America, since they include captive and semi-captive birds.

The large increase in the size of North American populations is of concern to biologists, as this highly territorial swan occupies large areas and may drive out other species of waterfowl (Reese 1975). During the breeding season, it is also extremely aggressive toward people who approach.

Christine Langevin, Josée Tardif, and Derin Henderson

World breeding range

Species profile

Clutch size	usually 4–8 eggs[4,5,6,7,8]
Laying interval	1 egg per 2 days[5,6,7]
Start of incubation	with last egg, sometimes before[5,10,11]
Incubation period	usually 35 days (34–41)[5,6,7]
Maturity at hatching	precocial
Care of young	by female and male[3,5,6,7]
Nestling period	less than 48 hours[3,5,7]
Age at first flight	100–155 days[3,7,8]
Dependence of young	until ≈ 1 year old[6,7,8]
Broods per year	1[3,5,6]
Breeding age	M 3–4 years / F 2–3 years[4,5]
Mating system	monogamy[5,6]
Duration of pair bond	for life[6]
Total length	130–160 cm[9]
Wing span	213.4–243.8 cm[3]
Weight	M 11 800 g (9200–14 300) / F 9670 g (7600–10 600)[2]
Longevity record	26 years 9 months[1]

References:
1. Klimkiewicz and Futcher 1989
2. Dunning 1984
3. Terres 1980
4. Perrins and Reynolds 1967
5. Cramp and Simmons 1977
6. Palmer 1976
7. Wilmore 1974
8. Reese 1975
9. Long 1981
10. Johnsgard 1975b
11. Harrison 1978

Chronology $n = 1$

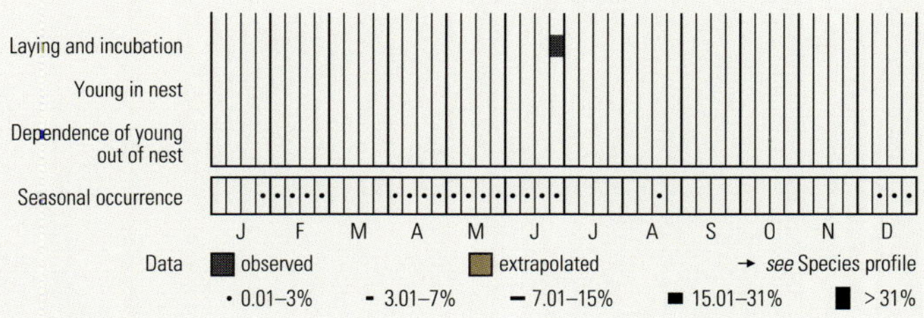

Snow Goose

Oie des neiges

Chen caerulescens

Pierre Dupuis

In the St. Lawrence estuary in late March, while winter still holds a firm grip and the intertidal mudflats are almost completely ice-covered, the first of these hardy migrants make their spring appearance. Increasingly through early April, large clamorous skeins of Greater Snow Geese arrive from their wintering grounds on the east coast of the United States as almost the entire world population of this subspecies assembles in the St. Lawrence. They remain until late May, when they continue their lengthy migration to the breeding grounds in the eastern Canadian Arctic, returning again in the fall for an extended stopoff on the flight back to the wintering grounds. The spectacle of these immense flocks of geese with their strikingly white plumage and vibrant and vociferous activity, set against the magnificent scenery of the St. Lawrence Valley as a backdrop, is clearly one of the most impressive wildlife displays in the world.

By the end of May, Greater Snows are back on their Arctic nesting grounds and seeking nest sites in meadows, on gently rolling terrain, or on the exposed slopes of ravines. Bylot Island, Northwest Territories. *July 1990, Austin Reed.*

Each year these geese cover more than 8000 km on their migrations, and the St. Lawrence estuary constitutes the most important staging area for them along that lengthy route. Because of its geography, the St. Lawrence estuary is attractive to the geese; it is situated directly on their path almost 1000 km north of the starting point on spring migration, and, because of the influence of the maritime climate which extends inland up the valley, the estuarine wetlands thaw while the mountainous terrain to the north and to the south is still in the grip of winter. The estuary is biologically attractive too because of the ecological richness of its intertidal marshes, which provide an abundant supply of nutritious plant foods. In recent years, the geese have also discovered an additional food supply in the form of spilt cereal grains and herbaceous forage on adjacent farmland. These migratory stopoffs in the St. Lawrence are of utmost importance to the geese in allowing them to build up the energy reserves they need to continue their migration. Also, in spring, the reserves they accumulate during staging help sustain them, and contribute to the formation of their eggs during the critical first weeks on their Arctic breeding grounds when new vegetative growth is minimal.

Greater Snows, unlike Lesser Snows, however, feed intensively on arrival on the breeding grounds after their long journey (Gauthier and Tardif 1991). At that time they sustain themselves on smartweed bulbs and oxytropis roots because the spring growth of green foliage has not yet begun (Gauthier 1993b).

As the only North American goose subspecies that normally breeds entirely north of the Arctic Circle, the Greater Snow Goose is clearly adapted to breeding in the High Arctic. The discovery, during the summers of 1990–93, of several nests on islands in the St. Lawrence estuary, 2500 km south of the southern boundary of the known breeding range, was therefore greeted with considerable surprise. Only time will tell whether these recent records represent the start of a new breeding subpopulation.

Habitat and behaviour

Anyone who has watched the huge dense flocks at Cap Tourmente or Montmagny in October needs no reminding that Greater Snow Geese, like other geese, are gregarious. But within these large milling masses the individual geese are acting neither individually, randomly, nor chaotically. The basis of their gregariousness is a social structure centred on the family unit—two parents and their young of the year. Geese are monogamous, mate for life (they usually mate and breed for the first time at 3 years old), and remain with their young throughout almost a full year until they begin nesting to produce the following year's brood (Owen 1980). For the keen observer, it is especially fascinating to watch the behaviour of individual families of Greater Snow Geese as both parents, identifiable by their white plumage, and goslings, with grey plumage, squabble and

The nest, usually built on slightly elevated ground, is a scrape in the ground, lined with grasses and roots and down from the female's breast. Bylot Island. *July 1990, Austin Reed.*

Notes on territory

Snow Geese establish exclusive territories for nesting. The territories are defended, most vigorously by the male, against intruding pairs and most predators. Most territorial conflicts are resolved by posturing and threat, but fighting may occur when nests are close together. Lesser Snow Geese characteristically nest in high densities, which leads to frequent territorial squabbling and occasionally to rape of nesting females by neighbouring ganders. Such manifestations of social disruption are less frequent among Greater Snow Geese, which nest in lower densities. The territory always includes the nest site, but may or may not include feeding areas. For most Greater Snow Geese breeding on Bylot Island, Northwest Territories, the territory is usually larger than 0.25 ha and includes the nest site and the feeding area for the adults during nesting. Exclusive territories are not used during brood rearing; individual broods range over areas exceeding 3 km² which overlap with those of other broods.

Another form of territoriality consists of protecting a small moving area around the brood (both parents participate in its defence) or around the adult female before and during nesting (protection afforded by the

Just a few hours after hatching, the goslings are led by their parents to feed intensively on the tundra. Bylot Island. *July 1990, Austin Reed.*

male). This behaviour affords the young geese or the nesting female increased feeding opportunities by preventing disturbance from other geese (Lemieux 1959a,b; Mineau and Cooke 1979; Gauthier and Tardif 1991; Hughes 1992; Hughes et al. 1994a,b; A. Reed, CWS, unpubl. data).

jockey for position with other families and non-breeding adults in huge feeding flocks (Turcotte and Bédard 1989a,b).

On the St. Lawrence prior to the 1960s, the geese fed almost exclusively in freshwater tidal marshes, dominated by the American bulrush, of the traditional range, mainly exploiting the bulrush rhizomes, and making only occasional use of a few areas of reclaimed unimproved tidal grassland (Reed 1976; Anonymous 1981). Through the 1960s and 1970s, bulrush marshes continued to be the main feeding grounds, but some flocks moved increasingly onto arable farmland, gradually extending their daily foraging flights farther inland. The downstream expansion of their spring staging area brought them into contact with cord-grass salt marshes. In the mid-1970s, geese in this area divided their feeding activities between the cord-grass marshes and adjacent farmland, but a decade later they were concentrating heavily on agricultural land, the marshes being used principally as roost sites (Gauthier et al. 1988); decreased feeding on marshland in this area is explained by the absence of bulrushes and the high cost of extraction and low energy return associated with cord-grass roots (Bédard and Gauthier 1989). The recently occupied Lake Saint-Pierre staging area is characterized by freshwater non-tidal wetlands in which bulrush is not an important component; there the geese feed almost exclusively on agricultural land where, in addition to the typical hay crops found in the traditional range, they have access to large acreages of corn (Bergeron 1992; DeKoster 1992; A. Reed, CWS, unpubl. data). A firm tradition of field feeding has now become established throughout the expanded spring range, even though bulrush rhizomes continue to be a dietary mainstay.

In the fall, presumably in reponse to heavy hunting pressure, the goose flock has not expanded much beyond the bulrush marshes of the traditional range and rarely frequents farmland. An exception is the establishment over the past two decades of a regular pattern of field feeding at a restricted non-hunting area in the Cap Tourmente National Wildlife Area.

Intensive feeding in fall on the bulrush marshes can result in complete denudation. Some observers claim this accelerates the erosion of sediments (Dionne 1985). Studies showed that heavy grazing in a marsh near Montmagny lowered the production of plant biomass (Giroux and Bédard 1987), but an equilibrium was achieved and no long-term decline in plant abundance was found (A. Reed, CWS, and J.-F. Giroux, UQAM, unpubl. data). Similarly, at Cap Tourmente between 1971 and 1984, no long-term decline in the abundance of bulrush was evident (Reed 1989).

World breeding range

Species profile

Clutch size	usually 3–5 eggs (2–9)[1,2,5]
Laying interval	1 egg per 1.5 days[2]
Start of incubation	with last egg[2]
Incubation period	24 days[1,2]
Maturity at hatching	precocial
Care of young	by female and male[1,2]
Nestling period	24–48 hours[1,2]
Age at first flight	42 days[1,2]
Dependence of young	until ≈ 1 year old[2]
Broods per year	1[1,2]
Breeding age	3 years[1,2]
Mating system	monogamy[1,2]
Duration of pair bond	for life[1,2]
Total length	M 71.1–83.8 cm / F 71.1–81.3 cm[5]
Wing span	134.6–152.4 cm[6]
Weight	M 3118 g (1900–4050) / F 2828 g (1840–3850)[3]
Longevity record	26 years 7 months[4]

References:
1. Lemieux 1959b
2. G. Gauthier, Laval U., and A. Reed, CWS, unpubl. data
3. Reed and Plante (in prep.)
4. Clapp et al. 1982
5. Bellrose 1980
6. Terres 1980

Chronology *n* = 16

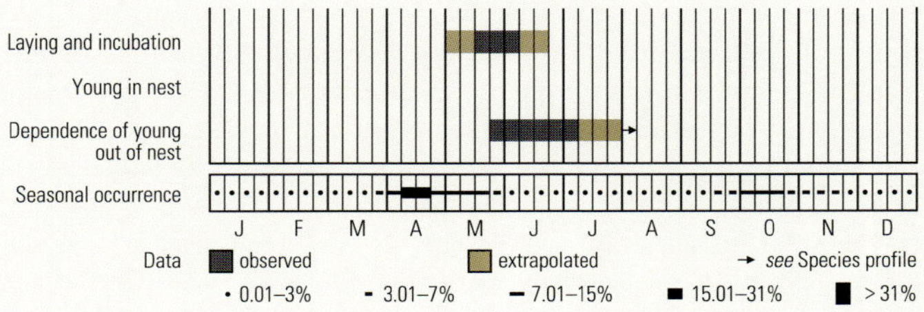

An incubating female near a research camp on Bylot Island. Northwest Territories. *July 1990, Austin Reed.*

Greater Snow Geese arrive on their High Arctic breeding grounds in late May or early June and begin egg laying 2–3 weeks later (Lemieux 1959b; Gauthier and Tardif 1991). Most nesting occurs in loose colonies of a few pairs to several hundred pairs, usually with nests 20–200 m apart. The colonies may be situated on rolling terrain, on the exposed slopes of ravines, or, more usually, in low-lying wet meadows characterized by polygon formations and an abundance of grasses, sedges, and cotton-grass. Nests are usually on slightly elevated ground often partially protected by low mounds; the low tundra vegetation offers little protection against weather and predators. Clutch size is typically 3–5 eggs, which are incubated for about 24 days by the female while the male keeps guard nearby. Peak hatch is usually around 7–12 July (Lemieux 1959b; Gauthier 1993b; G. Gauthier, Laval U., and A. Reed, CWS, unpubl. data).

The parents lead the goslings from the nest 1–2 days after hatching. The family roams over adjacent tundra, feeding regularly and intensively on protein-rich graminoid vegetation (Manseau and Gauthier 1993) through the 24-h daylight period. The goslings grow so rapidly that most are almost adult size and capable of flight at the age of 6 weeks (Lemieux 1959a; Giroux et al. 1984, 1986; Manseau 1991; Hughes 1992; Manseau and Gauthier 1993; Hughes et al. 1994a,b; G. Gauthier, Laval U., and A. Reed, CWS, unpubl. data).

Nesting by Greater Snow Geese in the St. Lawrence estuary was first confirmed on 16 May 1990 when 5 nests were found, and a sixth suspected on Battures aux Loups Marins, a small island in the estuary. None of the breeding adults observed was crippled. Clutch sizes were 3, 3, 5, 6, and 7 (G. Chapdelaine *in* Aubry et al. 1990b). On 21 and 23 May 1991, partial searches of the same island (G. Chapdelaine and J. Rodrigue, CWS, and J. Bédard, Laval U., unpubl. data) revealed at least 4 nests; by the next visit on 6 June, one of these nests had hatched, and weights and measurements of eggs in another unhatched nest indicated an expected hatch date (based on egg density) of 19 June (A. Reed, CWS, unpubl. data). On the next visit, 2 July, a group of approximately 8 adults and 10 goslings was observed (J. Rosa, CWS, unpubl. data).

On 25 May 1992, 3 nests were found in a partial search of the island (J. Bédard, Laval U., unpubl. data). On 27 May 1993, 8 nests were found on Battures aux Loups Marins (J. Bédard, pers. comm.), 3 containing 4 eggs and 3 containing 6 eggs. Another nest was empty and had apparently been destroyed by a predator. The last nest contained one intact egg, but an adult accompanied by a gosling was observed nearby. In one of the nests containing 6 eggs, 2 of the eggs were in the process of hatching. On 16 July of the same year on the same island, 13 flightless goslings were seen (P. Laporte, CWS, pers. comm.). Elsewhere in the St. Lawrence estuary, nesting was suspected on Île Brûlée (Kamouraska) and Île aux Lièvres in 1991, the latter case being confirmed by the subsequent observation of a family group (Gosselin et al. 1991b). In 1968, a pair nested unsuccessfully in Connecticut (Zeranski and Baptist 1990), but none has been reported since.

According to the hatching dates recorded in 1991 and 1993, the geese that nested in the St. Lawrence did so a full month earlier than their Arctic-nesting counterparts and had in fact begun nesting before the main flocks had left the St. Lawrence for the north. In the St. Lawrence, the occurrence of several nests in four consecutive years is clearly more than an accidental or isolated event; might it indicate the early stages of the establishment of a new breeding subpopulation? In a population of Barnacle Geese, which winters in The Netherlands and breeds near the Barents Sea in the European Arctic, a few individuals began nesting on a spring staging area in Gotland, Sweden, during the early 1970s; that small nucleus of pioneering breeders had increased to 450 pairs by 1985 (Larsson et al. 1988).

Distribution

Snow Geese occur from northwest Greenland across North America to Alaska and into easternmost Siberia (Wrangel Island). Two subspecies are recognized: the Lesser Snow Goose (*Chen caerulescens caerulescens*) and the Greater Snow Goose (*Chen caerulescens atlantica*) (Palmer 1976). The Greater Snow Goose is on average about 15% larger and has a somewhat heavier bill than the Lesser. Both subspecies occur in Québec.

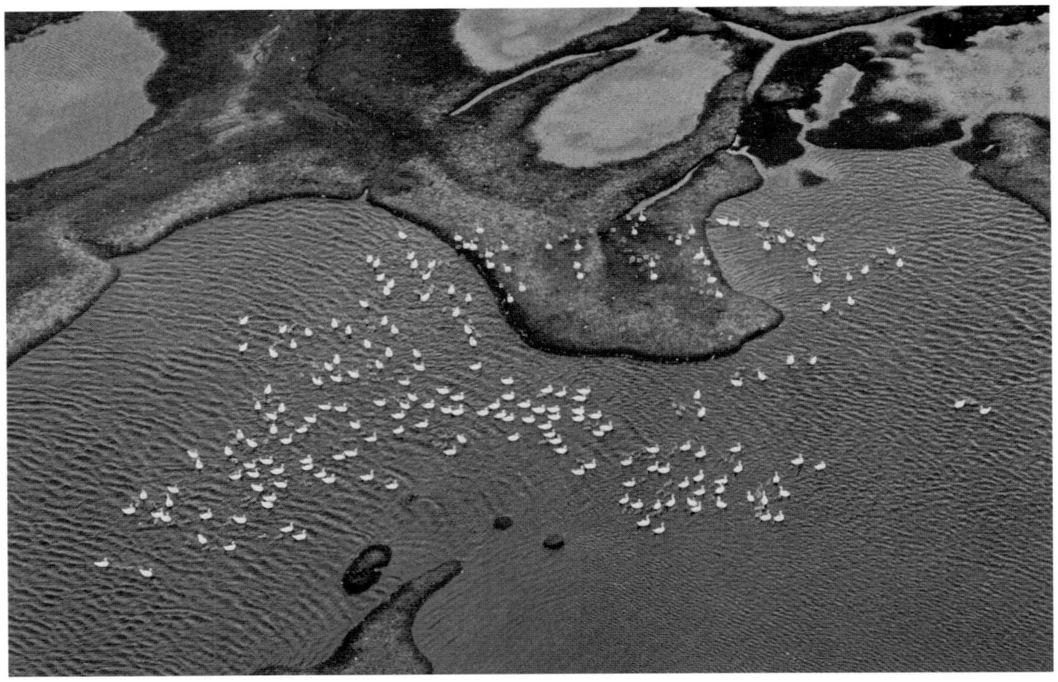

Families start gathering in July, when the young are still flightless. Bylot Island. *July 1982, Austin Reed.*

The Greater subspecies nests principally on Bylot Island and northern Baffin Island (Lemieux 1959a,b; Reed and Chagnon 1987; Reed et al. 1992a), but the range extends northward throughout the Canadian Arctic Archipelago and to northwest Greenland, and southward to central Baffin Island and the northern half of Foxe Basin (Palmer 1976; Bellrose 1980; Reed et al. 1980; Anonymous 1981). On migration they funnel through an important staging area in the St. Lawrence estuary and winter along the Atlantic coast from New Jersey to South Carolina (Bellrose 1980; Anonymous 1981).

Lesser Snow Geese breed mainly from southwest Baffin Island and the west and south coasts of Hudson Bay, across the northern mainland coast of the Northwest Territories, and the southern portions of Victoria and Banks islands to Alaska and Wrangel Island (Bellrose 1980; Boyd et al. 1982; Kerbes 1975, 1983; Reed et al. 1980, 1987a). They follow various migration pathways southward to main wintering areas in the southern mid-continent states near the Gulf of Mexico, in California and Mexico, and in Washington State and British Columbia (Palmer 1976; Bellrose 1980).

Most, if not all, Greater Snow Geese migrate through the St. Lawrence Valley, mainly between Lake Saint-Pierre and L'Isle-Verte. They also stop off, probably briefly in most years, in the vast tundra area of the Ungava Peninsula (A. Reed, J. D. Heyland, P. Dupuis, J.-F. Giroux, D. LeHenaff, unpubl. data). Lesser Snow Geese from eastern Arctic colonies also use the same tundra area of Ungava on migration, concentrated however near the Hudson Bay coast (A. Reed, CWS, unpubl. data); farther south their migration path follows the Hudson and James bay coasts of Québec, with an important staging area in Rupert Bay (Curtis and Allen 1976; Palmer 1976; Bellrose 1980; Cons. G. G. – GREBE 1992q). A few Lesser Snow Geese mingle with Greaters on the

Since 1990, there have been several records of confirmed breeding of Greater Snow Geese on Battures aux Loups Marins in the St. Lawrence estuary. *May 1991, Pierre Brousseau.*

The vegetation on Battures aux Loups Marins is more abundant than in the Arctic, affording the nest better protection against bad weather and predators. *May 1991, Pierre Brousseau.*

St. Lawrence staging area, but they undoubtedly represent less than 5% of the staging flock (Anonymous 1981; A. Reed, CWS, unpubl. data). A few pairs of Lesser Snow Geese were recorded nesting near the Hudson Bay coast of the Ungava Peninsula (near Povungnituk) in 1966–68 (J. D. Heyland *in* Palmer 1976), but during an aerial survey on 30 June–1 July 1978, fewer than 100 Snow Geese were seen and no evidence of nesting was obtained (Chapdelaine and Tremblay 1979). A pair of adult Snow Geese (subspecies not indicated) accompanied by 3 flightless young were observed near Déception Bay during the summer of 1983 (Gariépy 1983).

The existence of two colour morphs in Lesser Snow Geese has been well documented (Cooch 1963; Cooke and Cooch 1968), a typical white-plumaged phase, and a blue phase with head and part of neck white, most of body dark greyish-blue. Flocks of Lesser Snows from colonies in the eastern Canadian Arctic (those that migrate through James Bay) typically have a high proportion of blue-phase individuals—as high as 64% on southwest Baffin Island (Reed et al. 1987a) and 70–75% at Cape Henrietta Maria on the northwest corner of James Bay, Ontario (Lumsden 1987g)—but the proportion grades to almost nil in flocks from the western Arctic and Wrangel Island (Palmer 1976). Until recently it was thought that there was no blue phase of the Greater Snow Goose, but in 1973 J. D. Heyland (*in* Palmer 1976) examined two blue-phase geese of typical Greater Snow Goose size; since then a substantial number of apparent blue-phase Greaters have been measured in the St. Lawrence staging area and on Bylot Island (J. D. Heyland and A. Reed, CWS, unpubl. data). Although the existence of blue-phase Greater Snow Geese now appears to be established, their proportion in the flock is low: less than 1% are blue phase among breeding and moulting flocks on Bylot Island (Reed et al. 1992a) and in staging flocks along the St. Lawrence (A. Reed, CWS, unpubl. data), and some of these may be Lessers.

Starting in the 1960s, expansion of staging and wintering ranges of the Greater Snow Geese became evident. The extensions to the St. Lawrence staging area are well documented. Up until the mid-1950s, spring and fall staging occurred in the same small sector extending along approximately 80 km of the St. Lawrence estuary (Dionne 1883; Lewis 1921b; White and Lewis 1937; Lemieux 1959b). That area, which includes Cap Tourmente and Montmagny, continues to harbour the bulk of the population during fall migration, but increasing numbers are appearing to the southwest in Lake Saint-Pierre and northern Lake Champlain; use of areas to the east of the traditional range is still irregular and only small numbers are involved.

Important changes have, however, occurred in spring. By 1965, the spring range had extended 40 km farther downstream; by 1975, additional sites another 45 km downstream and as far as 100 km upstream were being used (Anonymous 1981). The spring range now extends along more than 400 km of the St. Lawrence (mainly between Lake Saint-Pierre and Rimouski), and small pioneering groups are appearing along two main tributaries, the Saguenay and Richelieu rivers. Of special interest is the emergence over the last decade of an important staging area at Lake Saint-Pierre (Anonymous 1981; DeKoster 1992; Reed 1992); large numbers of geese use the area in April before joining the main flock in the Cap Tourmente/Montmagny/Kamouraska area for the final three to four weeks of the staging period.

History and trends

Since the turn of the century, the Greater Snow Goose population has soared from about 3000 to more than 400 000 (Lemieux 1959a; Reed 1990, 1992). The most recent spring surveys (1992–93) by the Canadian Wildlife Service in the St. Lawrence indicate a population of about 420 000 (A. Reed, CWS, unpubl. data); in the fall of 1993, following an exceptionally good breeding year, the population was estimated by modelling (see Gauvin and Reed 1987) to exceed half a million geese.

The increase in population is best explained by the combined effects of several factors. Legislation outlawing spring hunting of migratory birds in both Canada and the United States undoubtedly had a strong effect in the early decades of this century. Fall hunting of Greater Snow Geese was also abolished in the United States from 1931 to 1975. Refuges and sanctuaries were established in both countries (Anonymous 1981). More important in the population increase of recent decades is the consistently good breeding success the geese have achieved and an apparently increased survival rate. Prior to 1965, weather conditions in the High Arctic prevented successful nesting in about one in every three years; an apparent improvement in the climate has allowed the geese to nest successfully in all but two of the last 19 years (1975–93), during which they have maintained an average of 27% young in the fall flights (Reed 1992; A. Reed, CWS, unpubl. data). It is speculated that their recent adoption of agricultural crops to supplement the natural food supply in winter has reduced natural mortality of young birds (Reed 1990). Although the hunting take has increased, the percentage of the population shot by hunters has decreased greatly in recent years (Reed 1990). Despite its abundance, the population remains vulnerable to the threat of oil pollution and disease because it assembles in such massive concentrations on staging and wintering locations.

AUSTIN REED

The family bond is evident here as one of the parent birds stands guard while the rest of the family feeds in a tidal marsh. With their strong neck and many lamellae on the mandibles, tongue, and palate, the geese can uproot nutritious bulrush rhizomes from depths of up to 20 cm. Cap Tourmente National Wildlife Area, northeast of Québec City. *October 1985, Austin Reed.*

Eurasian Wigeon

Canard siffleur d'Europe

Canard siffleur

Anas penelope

The Eurasian Wigeon is an Old World species, and is in many ways the Eurasian counterpart of the American Wigeon. It is now sighted regularly in North America, mainly along the Atlantic and Pacific coasts. In Québec, the Eurasian Wigeon is a casual visitor (Edgell 1984; Robert 1989b).

Habitat and behaviour

In Europe, the Eurasian Wigeon breeds in boreal and subarctic regions, around shallow lakes with abundant floating and submerged vegetation. For nesting, it appears to prefer partially wooded shorelines (Hildén 1964) rather than those choked with emergent vegetation and also shuns fast-flowing streams and rivers (Cramp and Simmons 1977).

Pairs are formed in the autumn and courtship behaviour in the species is similar to that in the American Wigeon (Johnsgard 1965). The nest, built by the female not far from the water, is a shallow scrape in the ground lined with grasses, leaves, and, after the usual clutch of 8–9 eggs has been laid, the female's down. The hen chooses a site that is well concealed by the surrounding vegetation, whether in open habitat or in deciduous or coniferous forest. The female alone incubates the eggs and rears the young, which are independent at 40–45 days of age, when they are fully feathered (Cramp and Simmons 1977).

The females of both wigeon species are very similar, which makes it extremely difficult to confirm breeding in the species. When sighted alone, female Eurasians are often mistaken for female Americans, and even when they are accompanied by the easily identified Eurasian male, it is not always possible to determine whether the two birds are a true breeding pair.

Distribution

In Québec, as in most other provinces, the Eurasian Wigeon is seen mainly in migration, especially in spring. The 11 Atlas records collected from the Montréal area, Lake Saint-Pierre, Saint-Fulgence, Rimouski, the Mingan area, and the Magdalen Islands did not provide evidence of breeding. The other Québec records have come mainly from the North Shore, Lac-Saint-Jean, the Lower St. Lawrence, the Central St. Lawrence Lowland, and the Ottawa Valley (David 1980a).

Although there still is no evidence of the species breeding in North America, it is highly probable that the Eurasian Wigeon does nest here, since its Old World boreal and subarctic breeding habitats closely resemble corresponding habitats in North America. Moreover, the species hybridizes with the American Wigeon (Aubry 1981; Merrifield 1993). Birds banded in Iceland have been recovered in the Atlantic provinces and along the eastern seaboard of the United States, which suggests that some of the Iceland population winters in eastern North America (Palmer 1976; Bellrose 1980). Winter records have also been obtained in British Columbia, especially in the Vancouver area (Campbell et al. 1990a).

History and trends

Most information on population trends in the species has been obtained from studies in Europe, where either declines or increases have been reported, depending on the region (Cramp and Simmons 1977).

The species was no doubt rarer in North America during the 19th century than it is now (Palmer 1976). The first Québec records are from the North Shore and date from the beginning of the century (Todd 1963). The Eurasian Wigeon has been reported with increasing frequency over the last few decades, and between 1980 and 1992 was seen a dozen or so times a year. In recent years, it has even been sighted in the summer (R. Yank, pers. comm.). This wigeon shows up in very small numbers every year among duck wings mailed in to the Canadian Wildlife Service for the Species Composition Survey (SCS) of waterfowl hunters (P. Dupuis, CWS, unpubl. data).

Simon Barrette and Michel Robert

Jean-Pierre Vallée

Notes on territory

The Eurasian Wigeon is not a colony nester, although nests may sometimes be as close as 5 m. At a site in Scotland containing 40 or so pairs, the distance between nests was 8–34 m (Cramp and Simmons 1977).

In Québec, the Eurasian Wigeon is a casual visitor that is being reported with increasing frequency. It can be seen during migration at Baie-du-Febvre, notably in the settling pond. South shore of Lake Saint-Pierre. *June 1991, Daniel Jauvin.*

World breeding range

Species profile

Clutch size	usually 8–9 eggs (6–12)[1]
Laying interval	1 egg per day[1]
Start of incubation	with last egg[1]
Incubation period	22–25 days[1,2]
Maturity at hatching	precocial
Care of young	by female[1]
Nestling period	probably less than 48 hours
Age at first flight	40–45 days[1,2]
Dependence of young	until first flight or a few days before[1]
Broods per year	1[1]
Breeding age	1–2 years[1]
Mating system	monogamy[1]
Duration of pair bond	probably 1 season[1]
Total length	43–51 cm[3]
Wing span	75–86 cm[1]
Weight	M 726 g (544–953) / F 635 g (499–816)[4]
Longevity record	8 years 7 months[5]

References:
1. Cramp and Simmons 1977
2. Palmer 1976
3. Godfrey 1986
4. Bellrose 1980
5. Clapp et al. 1982

Chronology $n = 0$

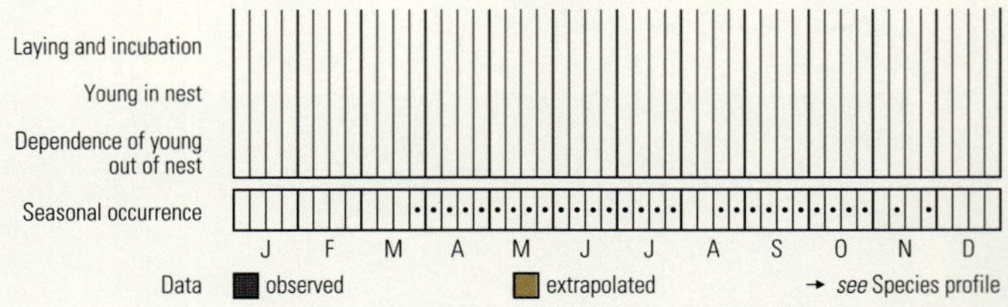

Canvasback

Morillon à dos blanc

Fuligule à dos blanc
Aythya valisineria

The Canvasback is a diving duck exclusive to North America that breeds chiefly in the West. Its scientific name *valisineria* stems from its fondness for the aquatic plant *Vallisneria americana*, or wild-celery, whose rhizome is especially relished by the species (Terres 1980). The Canvasback closely resembles the Redhead, as both have rust-red heads and black breasts. The Canvasback, however, can be distinguished by its sloping forehead, all-black (rather than tricoloured) bill, and lighter coloured back and sides (Alison 1980; Godfrey 1986; National Geographic Society 1987).

Canvasbacks winter mainly along the Atlantic and Pacific coasts of the United States and to a lesser extent inland, from the northern and central states to the Gulf states and Mexico (Alison 1980; Bellrose 1980; Terres 1980; Godfrey 1986). In Canada, the species winters in southwestern British Columbia and southern Ontario in varying numbers (Godfrey 1986).

The Canvasback is a rare migrant in southern Québec, although it is abundant some autumns (Lehoux et al. 1985). It is casual in winter (David 1980a).

Habitat and behaviour

The Canvasback nests in open habitat, such as freshwater marshes or the edges of ponds, lakes, rivers, or bays fringed with lush emergent vegetation (Macoun and Macoun 1915; Palmer 1976; Bellrose 1980; Godfrey 1986; del Hoyo et al. 1992). The nest is generally built at water level, in emergent vegetation a few metres from the shore or near an island, or occasionally on dry ground near a small body of water. It is a bulky, basket-shaped structure made of reeds and sedges and lined with finer materials and down (Macoun and Macoun 1915; Alison 1980; Surprenant 1993).

Typical clutch size is 7–10 eggs. Sometimes the nest is parasitized by conspecifics, Redheads, or Ruddy Ducks, in which case the host female lays fewer eggs (Palmer 1976; Bellrose 1980; Terres 1980; Godfrey 1986; Bluhm 1992; del Hoyo et al. 1992)—or she may even desert the nest altogether if she is overly disturbed (Bluhm 1992). The female incubates the eggs for 23–29 days. After tending her brood for a few weeks, she abandons the unfledged young, to flock with other adult birds for the moult (Palmer 1976). The male leaves the nest site soon after the onset of laying to join other moulting males rafting on large bodies of water (Palmer 1976; Alison 1980).

Canvasbacks may breed as yearlings, although it seems females younger than 4 or 5 have less success at rearing young. Populations with a high proportion of young females are therefore not very productive (Alison 1980).

Canvasbacks forage mainly by diving but may also feed on the surface, like dabbling ducks. Their diet consists primarily of the seeds, roots, tubers, leaves, and stems of aquatic plants (Palmer 1976; Alison 1980), complemented by aquatic invertebrates (insects and molluscs) and small fish (Palmer 1976; Alison 1980; del Hoyo et al. 1992). The ducklings' diet is composed mainly of vegetable matter, especially floating plants (Palmer 1976; Alison 1980).

Distribution

In the United States, the Canvasback breeds in Alaska and some of the northwestern states; in Canada, it ranges from British Columbia to Manitoba and the southernmost part of Ontario (Alison 1980; Peterson 1984; Godfrey 1986; National Geographic Society 1987; Sandilands 1987d; del Hoyo et al. 1992; Surprenant 1993).

In Québec, Canvasbacks are mainly observed in late autumn, when they are most abundant along the St. Lawrence River, especially on the riverine lakes in the Montréal area and on Lake Saint-Pierre. They are also seen in spring, but in fewer numbers (Ouellet 1974a; Lehoux et al. 1985). The species is seldom observed at other times of the year (David 1980a).

There are two pre-Atlas breeding records in Québec, the first from Lake Chicobi, Abitibi, in August 1973 (Trépanier 1975) and the second from Lac du Milieu, in Chibougamau Park (now Ashuapmushuan Wildlife Reserve), in late summer in 1980 (COQ 1980b). Both were sightings of females with broods. These reports are highly unusual for Québec, and since there is little substantiating evidence such as photographs, breeding has not been officially confirmed in the province.

History and trends

For a long time, the Canvasback was threatened by overhunting, both for sport and market, as the species is highly esteemed as a table bird (Ouellet 1974a; Alison 1980). At the end of the 19th century, Wintle (1896) reported that, of all ducks sold in Montréal markets, Canvasbacks brought the highest prices; birds imported from the West fetched anywhere from $2.50 to $7.50 a brace. Numbers have declined severely over this century because of overhunting, periodic droughts on the breeding grounds, and degradation of summer and winter habitat (Alison 1980; Environment Canada 1990; del Hoyo et al. 1992). The species was on the Blue List between 1975 and 1981, and was designated a species of Special Concern in 1982 and 1986 (Tate and Tate 1982; Tate 1986). As a result, stringent protective measures have been implemented to ensure its continued survival (Ouellet 1974a; Environment Canada 1990; Surprenant 1993).

The Canvasback is one of the least abundant North American game birds (Semenchuk 1992). Over the last few years, Canvasbacks have accounted for less than 0.6% of the total duck harvest in Canada (Lévesque et al. 1993). The harvest of Canvasbacks in Québec between 1973 and 1992 was regular but negligible (Appendix II).

Josée Tardif and Nicolas Gagnon

Carole Bérubé

Notes on territory

A pair's home range includes areas for loafing, nesting, and feeding. The home range of one pair may overlap those of several other pairs, whether Canvasbacks or other species of ducks, without any serious territorial behaviour occurring. Territorial behaviour in the male varies; in some circumstances, a male may defend a territory of about 4.5 m around the hen but in other circumstances, he will tolerate the presence of other males nearby.

A male's home range may exceed 525 ha (Palmer 1976; Bellrose 1980).

Species profile

Clutch size	usually 7–10 eggs (7–12)(1,3,4,6,10)
Laying interval	1 egg per day(6)
Start of incubation	≈ 1–3 days before last egg is laid(3,6)
Incubation period	usually 24 days (23–29)(1,2,3,6,10)
Maturity at hatching	precocial
Care of young	by female(1,3,6,11)
Nestling period	less than 24 hours(5)
Age at first flight	63–80 days(2,4,11)
Dependence of young	from a few days to a few weeks before first flight(6)
Broods per year	1(9)
Breeding age	1 year(3,4,6)
Mating system	monogamy(4,6,11)
Duration of pair bond	1 season(6,10)
Total length	48–66 cm(1)
Wing span	71–91 cm(2)
Weight	M 1248 g (?) / F 1190 g (?)(8)
Longevity record	16 years 11 months(7)

References:
1. Godfrey 1986
2. Terres 1980
3. Bellrose 1980
4. del Hoyo et al. 1992
5. Surprenant 1993
6. Palmer 1976
7. Clapp et al. 1982
8. Dunning 1984
9. Harrison 1978
10. Ehrlich et al. 1988
11. Alison 1980

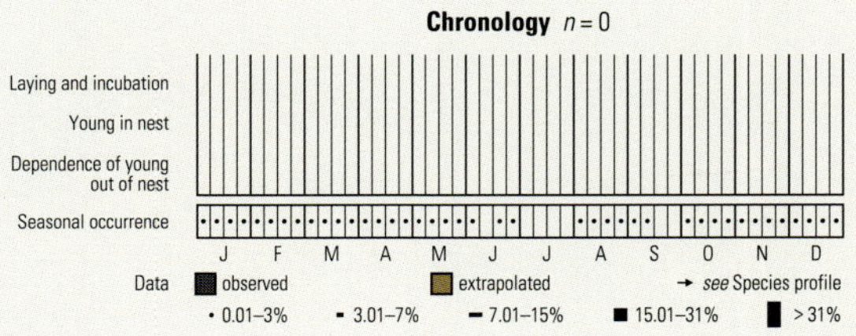

World breeding range

King Eider

Eider à tête grise
Somateria spectabilis

Pierre Dupuis

The King Eider, a bird of the Far North, nests on the arctic and subarctic tundra. It is the most abundant of the four species of eiders in the world, but because of the remoteness of its breeding grounds, which few people ever visit, birders generally tend to believe that it is scarcer than it actually is. The handful of confirmed breeding records in Québec all come from north of the 58th parallel, and sightings of single birds and flocks most often occur along the coasts of Ungava and Hudson bays.

This majestic bird can be seen in southern Québec during spring and fall migration and in winter, when it flocks with the Common Eider in the Gulf of St. Lawrence. The male's orange frontal shield, outlined in black, and bluish-grey head and black lower back readily distinguish it from the Common Eider. Female King and Common eiders are very similar, and the two species may in fact hybridize (Bardarson 1986; R. S. Palmer, pers. comm.).

According to ancient belief, the King Eider was an old male Common Eider, grown grey with age, who reigned over younger eiders by virtue of his venerability (Palmer 1976).

Habitat and behaviour

In Québec, the King Eider probably breeds on wet tundra dotted with freshwater ponds, generally within 10 km of the coast. According to Cramp and Simmons (1977), sightings farther than 50 km inland are exceptional.

Age at first breeding is usually 3 years. Immature birds often linger on the wintering grounds, and most do not migrate as far as the breeding areas (Salomonsen 1967), but go directly to the moulting areas (Salomonsen 1968), which explains why immature birds have been sighted on the North Shore as late as 15 June (Beetz 1916 in Palmer 1976).

Pairs are normally formed prior to spring migration. Shortly after arriving in the north, they gradually move inland from marine areas to the tundra lakes and ponds where they mate. During the fairly elaborate courtship displays, only the drake vocalizes, uttering a low, dovelike cooing.

The female, often accompanied by the male, selects the nest site (Lamothe 1973; Johnson and Herter 1989; R. C. Cotter, pers. comm.). The male aggressively repels any other males that attempt to court his mate. Although King Eiders are gregarious for most of the year, they generally nest in widely dispersed pairs rather than in colonies. The nest is usually located on dry ground, generally in habitats with less than 30% plant cover, mainly saxifrage (Lamothe 1973), and a hummock or even a bare rocky slope may be used (Johnsgard 1975b). Males leave their mates once incubation has begun—after the laying of the last egg (Lamothe 1973)—and join flocks of immature birds at moulting sites, one area being off the west coast of Greenland (Salomonsen 1967). During the 22- to 24-day incubation period, females rarely leave the nest and consequently eat very little. With their habit of sitting tight and their cryptic plumage they can easily be overlooked, even in areas where there is no plant cover. The young leave the nest about 16 hours after hatching, remaining for a dozen days or so on inland ponds and lakes before making for the sea. Although two broods will often share a tundra pond, large crèches with several females and many young only occur in marine habitats (Lamothe 1973).

Notes on territory

Male King Eiders are very aggressive toward other eiders that venture too close to their mate or the nest. Such aggression occurs infrequently, however, as the species nests mostly in solitary pairs. Since the King Eider has been little studied in Québec, there are no breeding density estimates for the province. Densities of 5–10 nests/km² were reported on Southampton Island in the Northwest Territories (Abraham and Finney 1986).

King Eiders often nest on the edges of tundra ponds. This photo was taken in early June, just before the birds disperse to breed in isolated pairs. Bathurst Island, Northwest Territories. *June 1971, Pierre Lamothe.*

Female King Eider incubating eggs on the barren tundra of Bathurst Island. Her protective coloration enables her to blend in with her surroundings. Northwest Territories. *July 1969, Pierre Lamothe.*

Distribution

The few breeding records in Québec are from the coast of the Ungava Peninsula. In Ungava Bay, at Rivière au Chien Rouge, 10 juveniles and 4 females were sighted in July 1966 (R. Ouellet, pers. comm.). More recently, on 4 July 1980, a female incubating 6 eggs was seen on a small island connected to the mainland at low tide, at the mouth of the Rivière aux Feuilles (COQ 1980a). Two other records come from Kogaluc River and around Cap du Prince-de-Galles (Todd 1963).

Rafts of nearly 900 birds have been reported in early June near the Gyrfalcon Islands along Ungava Bay (Lamothe 1978) and Chapdelaine and Tremblay (1979) saw several birds in late June between Bélanger Island in Hudson Bay and De Rozière Bay in Ungava Bay. In July 1985, a bird was seen in Ivujivik (ÉPOQ). Hildebrand (1950) reported that summer sightings of the species at the mouth of the Koksoak River were not unusual.

Between 1963 and 1976, several drakes were seen every year in May and June among rafts of Common Eiders in the St. Lawrence estuary between Kamouraska and Rimouski. Although only males were reported, it is hard to believe that females were totally absent (A. Reed, pers. comm.). In May 1973, a male King Eider was seen with a female eider (*Somateria* sp.) on Bicquette Island in the estuary (J. Gauthier, pers. comm.). There are 4 summer Atlas records—estuary, Grandes-Bergeronnes, Rimouski, and the south shore of Anticosti Island—but no evidence of breeding was obtained.

Species profile

Clutch size	usually 5 eggs (2–6)[1]
Laying interval	1 egg per day[4]
Start of incubation	with last egg[4]
Incubation period	22–24 days[2,4]
Maturity at hatching	precocial
Care of young	by female[2,3,4]
Nestling period	≈ 16 hours[8]
Age at first flight	30–50 days[2]
Dependence of young	until sometime before first flight[7]
Broods per year	1[2,3,4]
Breeding age	2 or possibly 3 years[1,2,4]
Mating system	monogamy[2,4]
Duration of pair bond	1 season[1,3,4]
Total length	M 55.1–63.5 cm / F 47.0–57.2 cm[1]
Wing span	86–102 cm[3,5]
Weight	M 1655 g (1367–1934) / F 1569 g (1213–1923)[4]
Longevity record	18 years 11 months[6]

References:
1. Bellrose 1980
2. Ehrlich et al. 1988
3. Harrison 1978
4. Cramp and Simmons 1977
5. Terres 1980
6. Clapp et al. 1982
7. Palmer 1976
8. Lamothe 1973

Chronology *n* = 4

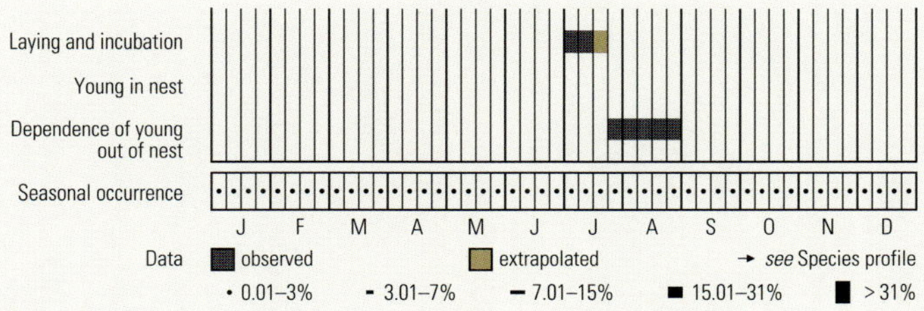

World breeding range

The chicks hatch after an incubation period of 22–24 days. A few hours after hatching, the hen leads the young to a brood-rearing pond. Bathurst Island. Northwest Territories. *July 1971, Pierre Lamothe.*

Nests are usually lined with a thick layer of down from the female's breast. This is the down used in real eiderdown comforters. Bathurst Island, Northwest Territories. *June 1971, Pierre Lamothe.*

The winter population in the gulf has been estimated at 2500 birds, or about 1.6% of the total eider population there. King Eiders winter in small flocks with Common Eiders, mainly around the Mingan Archipelago and Anticosti Island (Bourget et al. 1986). In winter, a flock of about 75 King Eiders—males, females, and immatures—was also seen in Manitounuk Sound off the coast of Hudson Bay (Bider and Lamothe 1982). Populations breeding in eastern North America generally winter in the North Atlantic from Labrador and Greenland to New England (AOU 1983).

History and trends

The King Eider's breeding grounds in Québec are very remote, thinly populated, and seldom visited. Consequently, confirmed sightings from this area during the breeding season are few and far between and there are insufficient data to describe the species breeding history in Québec.

Since this eider is an arctic species, it is hunted mainly by the Inuit, although Canadian Wildlife Service questionnaires indicate that eiders are occasionally shot in fall by sport hunters in Québec (see Appendix II) and elsewhere in Canada. This harvest is small, considering that the North American population has been estimated at over 1 million birds (Palmer 1976). Oil spills are still the main threat to the species in areas where it rafts in large numbers (several thousand birds), such as off the coasts of Greenland and Alaska.

PIERRE LAMOTHE AND LINE CHOINIÈRE

Labrador Duck

Canard du Labrador

Eider du Labrador

Camptorhynchus labradorius

Michel Gosselin

The Labrador Duck became extinct during the second half of the 19th century, before anything was recorded of its natural history. Its nearest living relatives were believed to be the eiders (Zusi and Bentz 1978), but a recent study finds them to be most closely related to the scoters (Livezey 1995).

Distribution

The exact breeding range of the Labrador Duck is not known. It probably bred in Labrador and on the Québec North Shore, formerly known as the Canadian Labrador. This is uncertain, however, as the exact origin of the few specimens collected during the breeding season is not known (Todd 1963). The only Québec specimen relates to an adult male shot in the spring of 1862, in the La Prairie Basin, south of Montréal (Wintle 1896). This specimen is now in the American Museum of Natural History (No. 734023).

Some early authors reported that the Labrador Duck bred on the north shore of the Gulf of St. Lawrence. The American naturalist J. J. Audubon, for example, who passed through Brador on the Lower North Shore in 1833, wrote in his journal:

> The Pied Duck breeds here at the top of the low bushes, but the season is so far advanced we have not found its nest (*in* Audubon and Coues 1898).

"Pied Duck" was how the Labrador Duck was known at that time. Not having seen the duck himself, Audubon's assertion must have been based on local hearsay and has been challenged by some authors such as Todd (1963), who considered it unlikely that the bird, an "eider," would have nested up in bushes.

The Labrador Duck became extinct during the last century. This stuffed specimen, which constitutes Québec's only record, was shot at La Prairie, near Montréal, in the spring of 1862. Originally part of the Rothschild collection, it was acquired by the American Museum of Natural History in 1932.
Richard Sloss, American Museum of Natural History, New York, No. 734023.

W. R. King, in *The Sportsman and Naturalist in Canada*, published in 1866, claimed:

> The pied duck or Labrador duck is common in the Gulf of St. Lawrence, and breeds on its northern shore, a short distance inland (*in* Todd 1963).

Here again, it is not known whether the author's claim was based on actual observation or mere speculation. Thus, it is not known whether the Labrador Duck did in fact breed in Québec.

In winter and during migration, the bird ranged along the Atlantic coast from Nova Scotia and New Brunswick to Chesapeake Bay in the United States. It was apparently fairly regular on the south coast of Long Island, New York and in New Jersey (Greenway 1967; Bull 1974). It frequented bays, sandy shores, and estuaries, and fed on small molluscs (Bent 1925; Godfrey 1986).

History and trends

The last specimen of the Labrador Duck was taken in the fall of 1875 on the coast of Long Island. Another bird, thought to have been a Labrador Duck, was shot in Elmira, New York, in December 1878 (Bull 1974). The last specimen in Canada was shot in April 1871 on Grand Manan Island, New Brunswick (Greenway 1967).

The Labrador Duck apparently was never abundant; only small flocks visited the Atlantic coast in autumn and winter. Birds were sometimes killed by market hunters in New England, but like most sea ducks, Labrador Ducks were not very good eating and the slaughtered birds often did not sell. The species gradually disappeared from its wintering grounds and the markets between 1850 and 1870 (Greenway 1967). The breeding grounds of the Labrador Duck may have been very localized, leaving it vulnerable to extermination by eggers and feather hunters, who were very active on the North Shore and the Labrador coast during the last century. It has also been suggested that the bird's main prey may have grown scarce (Bent 1925; Greenway 1967; Godfrey 1986; Goudie *in* Kirk 1994a). In any case, the reasons for the species' extinction remain unknown.

Michel Robert and Michel Gosselin

Species profile

Clutch size	?
Laying interval	?
Start of incubation	?
Incubation period	?
Maturity at hatching	?
Care of young	?
Nestling period	?
Age at first flight	?
Dependence of young	?
Broods per year	?
Breeding age	?
Mating system	?
Duration of pair bond	?
Total length	probably up to 57 cm, male larger than female[1]
Wing span	probably ≈ 80 cm[2]
Weight	?
Longevity record	?

References:
1. Palmer 1976
2. M. Gosselin, CMN, unpubl. data

Oldsquaw

Canard kakawi

Harelde kakawi

Clangula hyemalis

THROUGHOUT ITS ANNUAL CYCLE, THE OLDSQUAW CAN BE FOUND IN FRESH or salt water habitat. In southern Québec, it is seen mainly during spring and fall migration. It breeds in northern Québec and, in winter, is the province's most abundant duck after the Common Eider (Lehoux et al. 1985). During most of the year, the male's elongated tail (the two central feathers are about 25 cm long), is diagnostic. In flight, the Oldsquaw can be confused only with the Northern Pintail, but pintails are bigger and have longer necks. Oldsquaws in their first fall may be difficult to distinguish from young Harlequin Ducks, however (Palmer 1976). The Oldsquaw is a loquacious bird and, on the breeding grounds, the male's loud distinctive calls can be heard far and wide.

Habitat and behaviour

Pair formation occurs on the wintering grounds and is a gradual process. At first, the drakes gather in all-male flocks and only display to each other. They then display to female groups, but the hens respond aggressively to these advances until the beginning of February. Later, the displays become less frequent but almost always end with the formation of pair bonds (Alison 1975). Courtship displays are very elaborate and are performed mainly on the water, accompanied by spectacular aerial acrobatics. Ritualized behaviours maintain the pair bond (Palmer 1976). On arrival on the breeding grounds, the male becomes even more garrulous. The unmistakable call carries a great distance over the tundra and, once heard, it cannot be forgotten.

The female chooses the nest site, while the male remains on a nearby body of water, usually a tundra pond. Territoriality is limited to the defence of this pond. The female does not necessarily nest within the territory defended by the male, however (Alison 1972).

The nest is built on dry ground, in an existing depression, generally no more than a few hundred metres from water (Bellrose 1976). The hen selects a spot where her plumage will blend in with the surroundings. On islands, whether in salt or fresh water, nests are often found grouped together (Bellrose 1976).

At the onset of laying, the hen covers the eggs with vegetation from around the nest, such as moss or sedges; then she gradually adds down from her breast. Even in the absence of plant cover, the nest is difficult to find because of the hen's cryptic plumage, posture, and tendency to sit tight in the presence of intruders.

The brood is led to water within 24 hours of hatching (Alison 1972). If there is salt water nearby, the brood does not linger long on fresh water, whereas inland—on the Ungava Peninsula, for example—the young are reared entirely on large stretches of fresh water. Age at first flight is 35 days (Alison 1972).

The males abandon the females during incubation and go to the coasts of Hudson Bay or Ungava Bay to moult, forming rafts of up to 200 birds, but rarely more.

Chick mortality on the breeding grounds is considerable (Alison 1972), but is counterbalanced by high annual adult survival rates around 72% (Palmer 1976), which help maintain populations at healthy levels.

Distribution

The Oldsquaw breeds along the coasts of northern Québec. It is abundant in Ungava Bay but less common along Hudson Bay (Todd 1963). Breeding has been confirmed on James Bay, on the islands between Point Louis XIV and Attikuan Point (Benoit et al. 1993b), and as far south as the Twin Islands (Manning 1981). Intensive aerial surveys of the Ungava Peninsula showed that almost 50% of the ducks nesting on the tundra to be Oldsquaws (Eklund 1956).

Breeding was not confirmed in the Atlas area, despite the potentially suitable tundra habitat along the Lower North Shore. Some authors indeed suggest that the species may once have nested along the Gulf of St. Lawrence. Comeau (1945) reported that the Oldsquaw was tolerably common and supposed to breed along the North Shore at the turn of the century, while Todd (1963) classed it as a transient and winter visitant along the North Shore, though he

Pierre Dupuis

Notes on territory

The male defends the territory, which is roughly 0.5 ha on average and is used for courtship and feeding, but not necessarily for nesting.

The Oldsquaw is not aggressive toward either immature birds or other species of ducks. When another pair of Oldsquaws trespasses on an occupied area, the resident male will first chase away the female, so that the male accompanying her will be more easily persuaded to leave (Alison 1975). *The female never engages in any kind of territorial behaviour: for example, should her mate be captured, she will continue to lay or incubate, even though another pair will often stake out a territory nearby and try to drive her away. Should the female be captured, the male will stop defending the territory after a certain period* (Alison 1975).

In Québec, the Oldsquaw breeds on both salt water and on inland tundra lakes and ponds, often on islands. Ungava Bay region. *September 1977, Pierre Lamothe.*

The female chooses the nest site, which is not necessarily in the territory defended by the male. The nest may be on a tundra pond like this one. Bathurst Island, Northwest Territories. *June 1969, Pierre Lamothe.*

World breeding range

Species profile

Clutch size	usually 7 eggs (2–11)[7]
Laying interval	1 egg per day[5,6]
Start of incubation	with last egg[6,7]
Incubation period	usually 26 days (24–29)[1,2,6,7]
Maturity at hatching	precocial
Care of young	by female[1,2,3,4,6,7]
Nestling period	less than 24 hours[7]
Age at first flight	35–40 days[4,6]
Dependence of young	until first flight or shortly afterward[6]
Broods per year	1[6,7]
Breeding age	2 years[4,5,6,7]
Mating system	monogamy[4,6]
Duration of pair bond	sometimes more than 1 season[5,6,7]
Total length	M 48.3–57.4 cm / F 37.6–43.7 cm[7]
Wing span	66.0–78.7 cm[2]
Weight	M 913.9 g (?) / F 797.2 g (?)[9]
Longevity record	15 years 4 months[8]

References:
1. Godfrey 1986
2. Terres 1980
3. Harrison 1978
4. Ehrlich et al. 1988
5. Johnsgard 1975b
6. Cramp and Simmons 1977
7. Bellrose 1980
8. Clapp et al. 1982
9. Palmer 1976

Chronology $n = 9$

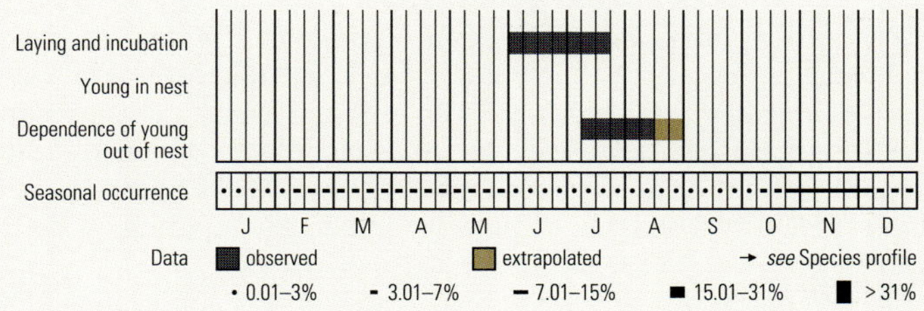

did cite a breeding record from Kegaska River near Natashquan, at the northern limit of the Atlas area. Verrill (1862) reported that the Oldsquaw was very common as a breeder on Anticosti Island, without giving any other details. He also reported aggregations of males during the summer of 1861. One Atlas record was obtained on the south shore of Anticosti Island, confirming that the species is present there occasionally in summer, as previously reported by Verrill (1862), Brewster (1884), and Lewis (1924).

Outside the breeding season, the Oldsquaw is especially abundant in the estuary and the gulf. In spring it is found mainly from the upper estuary to the Middle North Shore. Flocks of over 1000 birds have been sighted along the Gaspé coast and the Upper North Shore (Lehoux et al. 1985).

A few Oldsquaws are seen in spring in southwestern Québec on the riverine lakes in the St. Lawrence. These birds are probably part of the population that winters on the Great Lakes and migrates to the coasts of James and Hudson bays in late spring. Some individuals may also cross northern Québec to reach their breeding grounds on the Ungava Peninsula and the Arctic Archipelago, as was probably the case for the hundreds of Oldsquaws seen resting and feeding on insect larvae at the outlet of Lake Bienville during the third week of May in 1977 and 1978 (Bider and Lamothe 1982). The birds only stay in these ice-free areas for a few days, however.

In the fall, beginning in mid-October, part of the population stages along the freshwater course and estuary of the St. Lawrence. Small flocks are regularly sighted between Lake Saint-François and Québec City, while several thousand birds can often be seen in the lower estuary at this time (Lehoux et al. 1985). Oldsquaws winter along the Middle and Lower North Shore, where they are particularly common in December, January, and February (Comeau 1945). Many birds spend the whole winter in the estuary, at the mouth of the Saguenay River (Savard 1990), where nearly 27 500 Oldsquaws were reported on 28 December 1987 (LeBaron 1987). On 1 January 1992, over 60 000 birds were sighted off Cap-d'Espoir, in the gulf (Yank and Aubry 1992b). In winter, the ducks tend to remain fairly far offshore, in large rafts that are sexually segregated to some extent (Palmer 1976).

History and trends

The Oldsquaw has a circumpolar distribution, breeding in the Low and High Arctic. The world population is estimated at about 15 million (Gooders and Boyer 1986), while the North American early summer population is believed to be roughly 3–4 million, with about 1.5 million in the eastern Canadian Arctic (Bellrose 1976). Roughly 65 000 pairs breed in northern Québec (CWS and MLCP 1986).

The Oldsquaw breeds in remote areas far from major industrial centres, and the Québec population is relatively unaffected by human activity. This fishy-tasting duck is also not a favourite of hunters and only accounts for around 1% of the annual harvest of migratory game birds in Québec (Appendix II). Although mortality from entanglement in commercial gill nets was significant in the Great Lakes in the early 1950s (Bellrose 1976), it has declined considerably in recent years. The greatest threat to the species now is rather a massive oil spill, which would endanger wintering birds rafting in inshore areas.

Pierre Lamothe

The nest is often built on the open tundra, which makes the incubating hen very vulnerable but, because of her plumage, she is able to blend in with her surroundings. Bathurst Island, Northwest Territories. *July 1972, Pierre Lamothe.*

Black Scoter

Macreuse à bec jaune

Macreuse noire
Melanitta nigra

Pierre Dupuis

THE BLACK SCOTER IS UNDOUBTEDLY THE MOST VOCAL OF THE THREE species of scoters, and probably of any of the ducks found in Québec. The male's prolonged plaintive whistle, given during courtship on the wintering grounds and in migration, is unmistakable. In many respects, the Black Scoter is one of North America's least-known ducks, and until recently, even the location of its breeding grounds in the eastern part of the continent was a mystery. The extensive field research of recent years in the watersheds east of James Bay and southeast of Hudson Bay has documented the species more fully and confirmed it as a common breeder in northern Québec.

On the taiga, the Black Scoter nests on the margins of ponds, small lakes, and less frequently good-sized lakes. Three adults on a lake in the Lake Bienville region. North-central Québec. *June 1976, Jean-Pierre L. Savard.*

Habitat and behaviour

Staging along the St. Lawrence between mid-April and mid-May on their way north (Lehoux et al. 1985), Black Scoters begin arriving in mid-May in the heart of their eastern breeding grounds, in the watersheds of the Great Whale River, Little Whale River, and Lake Guillaume Delisle; the main wave arrives from late May to early June (Cons. G. G. – GREBE 1992u). Although flocks can be seen on rivers or along the coast of James Bay and Hudson Bay at this time (Bider and Lamothe 1982; Cons. G. G. – GREBE 1991b, 1992p), the main contingent apparently travels to their inland breeding areas, stopping on lakes and ponds along the way (Cons. G. G. – GREBE 1992u, 1993f). As a result, large numbers of birds seem to converge directly on the nesting grounds, while the rest remain on ice-free water on rivers, large lakes, or the coast, waiting for the ice to break up on their breeding ponds and lakes (Savard 1977; Cons. G. G. – GREBE 1992u).

In Iceland, the nest is usually built in dense shrubby cover, especially willows or birch, 1–30 m from water (Bengtson 1970). In the watersheds southeast of Hudson Bay, the onset of incubation is generally between the first and third weeks of June (Cons. G. G. – GREBE 1990a; Savard and Lamothe 1991); two or three days later, the drakes leave the hens to incubate the eggs and rear the young alone, and head for the coast to moult (Bengtson 1970; Bider and Lamothe 1982; Savard and Lamothe 1991), congregating in rafts mainly off the coast of James and Hudson bays (Ross 1983; Benoit et al. 1991, 1992). Although some early clutches may hatch at the beginning of July, hatching peaks between the second and last weeks (Cons. G. G. – GREBE 1990a; Savard and Lamothe 1991).

In the Great Whale, Little Whale, and Nastapoka river watersheds, broods are mainly found on ponds or lakes smaller than 30 ha, although they may be seen on lakes as large as 100 ha or so (Cons. G. G. – GREBE 1990a, 1993e). They frequent bodies of water between 1 and 5 m deep, usually with a rocky bottom, and with clear, mineral-poor water that ranges from alkaline to slightly acidic (Savard 1977; Bider and Lamothe 1982; Cons. G. G. – GREBE 1990b, 1993e; Savard and Lamothe 1991). These ponds often contain no fish (Savard 1988c) and little or no aquatic vegetation (Cons. G. G. – GREBE 1990b, 1993e), but are richer in benthic fauna than unutilized lakes (Bider and Lamothe 1982; Cons. G. G. – GREBE 1990b). In Scotland, in late summer, broods frequent clear, sandy-bottomed lakes with a neutral or alkaline pH (Fox et al. 1989).

As in the White-winged Scoter (Brown and Brown 1981; Kehoe 1989), brood amalgamation may occur (Cramp and Simmons 1977). However, the causes and consequences of this behaviour are still unknown.

Moulting Black Scoters are highly gregarious, rafting in thousands along the coast (Bellrose 1976; Ross 1982, 1983; Benoit et al. 1991). On the shores of James and Hudson bays, the scoters feed and rest on open water between coastal islands or farther offshore, unlike other species of ducks that frequent the coastal marshes or littoral zones (Benoit et al. 1991, 1992; Cons. G. G. – GREBE 1991b, 1992p, 1993g). The birds can be found along the Hudson Bay coast until at least the end of October (Cons. G. G. – GREBE 1991b, 1993g).

Around Lake Amichinatwayach, at the headwaters of the Little Whale River, hens and broods feed largely on insects and molluscs, as well as on plant matter (Cons. G. G. – GREBE 1993e). Moulting flocks

Abundance of Black Scoters in southern Québec. Aerial surveys conducted in May 1990, 1991, and 1992 *(D. Bordage, CWS, unpubl. data).*

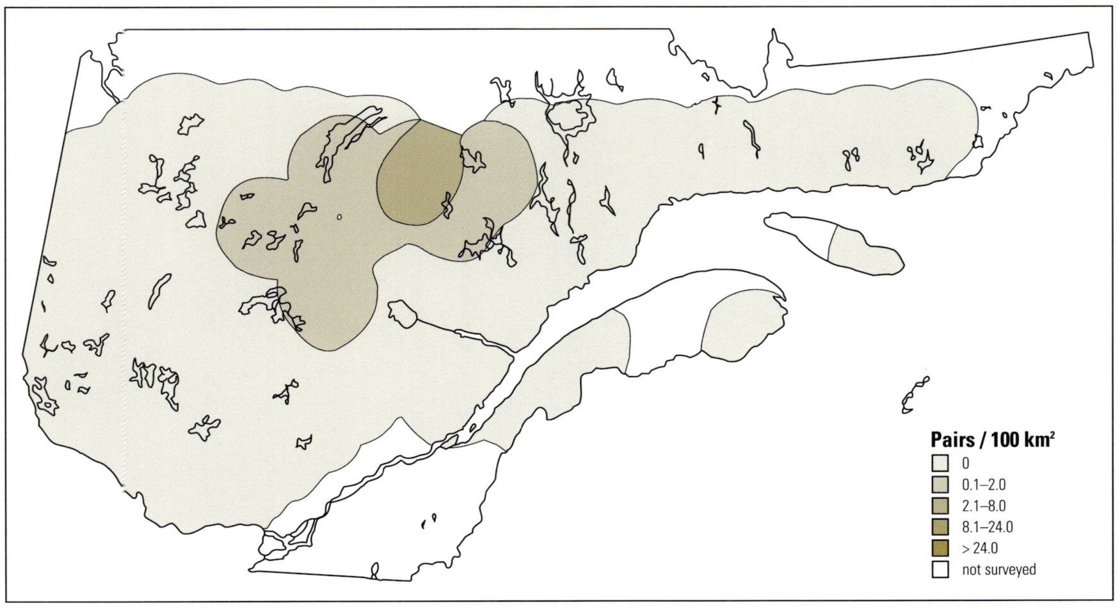

Notes on territory

There is virtually no information on territoriality in the species or on the size of its home range. In Iceland, males have been observed defending an area around their mates in early spring (Cramp and Simmons 1977).

During the summer of 1989, the highest breeding densities in northwestern Québec —16.5 females/100 km² —were found near Lake Amichinatwayach, at the headwaters of the Little Whale River (Cons. G. G. – GREBE 1990a). In southern Québec, the highest densities of presumed breeding pairs, 4–8 pairs/100 km², were observed between Lakes Manouane and Mistassini (D. Bordage and M. Grenier, CWS, unpubl. data).

on the east coast of James and Hudson bays consume mainly molluscs, particularly blue mussels, but also aquatic insects, small crustaceans, and sometimes plants (Cons. G. G. – GREBE 1990d, 1991b; Benoit et al. 1993b). On the wintering grounds on the Atlantic and Pacific coasts, mussels, snails, crustaceans, and occasionally barnacles make up its diet (Stott and Olson 1973; Vermeer and Levings 1977; Vermeer and Bourne 1984; Goudie and Ankney 1986).

Distribution

There are two subspecies of Black Scoter: *Melanitta nigra nigra* breeds in Iceland and northern Europe and Asia, while *Melanitta nigra americana* breeds in eastern Siberia and North America (Palmer 1976; Madge and Burn 1988). The American race has two disjunct populations: one in Alaska (Palmer 1976), and the other in Québec and Labrador (Erskine 1987; Savard and Lamothe 1991; Cons. G. G. – GREBE 1992t), with some of this population occasionally breeding in Newfoundland (Godfrey 1986; Goudie 1987). The winter ranges are also distinct, with western birds wintering in the Aleutian Islands and along the Pacific coast, and eastern birds wintering on the Atlantic coast from Newfoundland down to Georgia, and in small numbers on the Great Lakes (Bellrose 1976; Goudie and Ankney 1986).

The Black Scoter's migration routes are still poorly known. In Québec, tens of thousands of birds can be seen on the St. Lawrence in spring (Savard et al. 1995). Given the location of its breeding grounds, this is the best time to observe the species in southern Québec; the vast majority of birds stage along the Upper North Shore, the estuary, and Chaleur Bay. In fall, migrants are more widely scattered, staging in these same areas, as well as on Lac des Deux Montagnes and Lake Saint-Pierre (Lehoux et al. 1985).

The limits of the Black Scoter's breeding range in Québec have still not been determined, since several regions in northern Québec are unexplored or have not yet been surveyed in spring. Broods have been reported inland from Lake Minto in the north to La Grande-3 Reservoir in the south, to the Caniapiscau River in the east, and to the Hudson Bay coastal plain in the west (but not in the coastal area itself) (Bider and Lamothe 1982; Savard and Lamothe 1991; Cons. G. G. – GREBE 1992t; D. Bordage, CWS, unpubl. data); in terms of both abundance and density, this region is probably the heart of the species' breeding range in Québec (Savard and Lamothe 1991; Cons. G. G. – GREBE 1992t). There is also a single record from Baie aux Feuilles in southwestern Ungava Bay (Godfrey 1986).

In the summer of 1978, the lake in the background was home to a brood of Black Scoters and five broods of Surf Scoters. Near Lake Bienville, north-central Québec. *July 1978, Jean-Pierre L. Savard.*

The Black Scoter's breeding range in the province is probably more extensive than that just described. Isolated pairs have been seen during the breeding season in suitable habitat to the east and south (Cons. G. G. – GREBE 1992t); densities were generally very low, however. Pairs observed to the east (DesGranges 1989; Cons. G. G. – GREBE 1992t; Dryade 1992, 1993; J.-L. DesGranges, CWS, unpubl. data) and in most of Labrador numbered barely 2100 (Gillespie and Wetmore 1974; Erskine 1987; Goudie and Whitman 1987). Pairs have also been reported to the south and southeast, to the Gouin, Pipmuacan, and Manicouagan reservoirs (Bordage 1987, 1988a,b; Tardif and Morneau 1991; Cons. G. G. – GREBE 1992a; D. Bordage, CWS, unpubl. data).

Unlike the other two scoters, no Black Scoter nest or brood has been observed in the Hudson or James Bay coastal plains, nor on any of the coastal islands, despite the many studies conducted there during the brood-rearing period (Bordage 1985b; Cons. G. G. – GREBE 1990a,d, 1992a,p; Benoit et al. 1991, 1992, 1993b, 1994; Savard and Lamothe 1991). One exception was a brood sighted at Long Island, off Point Louis XIV, in 1991 (Cons. G. G. – GREBE 1993g).

The Great Whale River watershed, with its wealth of suitable habitat, is the heart of the species' breeding range in Québec. Hudson Bay region. *July 1976, Pierre Lamothe.*

Rafts of Black Scoters on Chaleur Bay during migration. Carleton, Gaspé Peninsula. *April 1992, Réginald Ouellet.*

The importance of the moulting areas along the James and Hudson bay coasts is well established (Chapdelaine and Tremblay 1979; Ross 1982, 1983; Savard and Lamothe 1991), particularly the mouth of the Piagochioui River, Point Louis XIV (Ross 1983), Bay of Many Islands, Dead Duck Bay, Attikuan Point (Benoit et al. 1991, 1992), and Long Island (Cons. G. G. – GREBE 1993g).

History and trends

Until recently, knowledge of this species in North America was limited to some facets of courtship behaviour and wintering ecology; distribution, abundance, and density were described in only very general terms (Johnsgard 1975b; Bellrose 1976; Palmer 1976; Savard and Lamothe 1991). Ornithologists have known since the beginning of the century, however, that rafts of scoters moult along the coasts of James and Hudson bays (Todd 1963). On the basis of the greater number of birds observed on the Ontario side, Bellrose (1976) concluded that the species probably bred in northern Ontario. Ross (1983) counted close to 20 000 birds along the east coast of James Bay in 1974 and presumed that the species nested at various locations inland in Ontario and Québec.

The Black Scoter has still not been confirmed as a breeder in Ontario, but scattered pairs have been seen in suitable habitat in the Hudson Bay Lowland, where the species probably breeds in very small numbers (K. Ross, pers. comm.). On the Québec side, breeding was confirmed in 1953 near Baie aux Feuilles, on southwestern Ungava Bay (Godfrey 1986), and in 1976 in the watersheds of the Great Whale and Little Whale rivers (Savard 1977). All the field research done in this region has confirmed its great importance for the species (Savard and Lamothe 1991; Cons. G. G. – GREBE 1992t); brood density may be as high as 11 broods/100 km^2 locally, mainly in the area around Lake Louet, east of Lake Bienville (Savard and Lamothe 1991).

It is difficult to assess the size of the breeding population in Québec. Ross (1983) observed close to 90 000 moulting males in 1977 in northwestern James Bay and, from this and sightings elsewhere, extrapolated a fall flight from James and Hudson bays of 200 000–320 000 birds. Moreover, there is not enough data to ascertain trends in the province's population. The Black Scoter is not considered a good table bird and therefore is little hunted. Most hunting along the St. Lawrence is done west of Québec City (Lehoux et al. 1985), and the harvest, though quite variable, has declined significantly in the past 20 years (Appendix II).

ALAIN MORRIER

World breeding range

Species profile

Clutch size	usually 7–9 eggs (5–11)[7,9]
Laying interval	1 egg per 1–2 days[2]
Start of incubation	with last egg[8,9]
Incubation period	27–33 days[7,8]
Maturity at hatching	precocial
Care of young	by female[1]
Nestling period	probably less than 48 hours
Age at first flight	45–50 days[1,2,3]
Dependence of young	until first flight or before[2,9]
Broods per year	1[1,2,8,9]
Breeding age	2–3 years[2,3]
Mating system	monogamy[1,2,9]
Duration of pair bond	1 season[2], ?[9]
Total length	43–54 cm[5]
Wing span	76.2–88.9 cm[4]
Weight	M 1134 g (998–1270) / F 998 g (862–1089)[6], M 1117.0 g (?) / F 987.4 g (?)[9]
Longevity record	8 years[9]

References:
1. Ehrlich et al. 1988
2. Cramp and Simmons 1977
3. del Hoyo et al. 1992
4. Terres 1980
5. Godfrey 1986
6. Bellrose 1976
7. Palmer 1976
8. Harrison 1978
9. Bordage and Savard 1995

Chronology *n* = 71

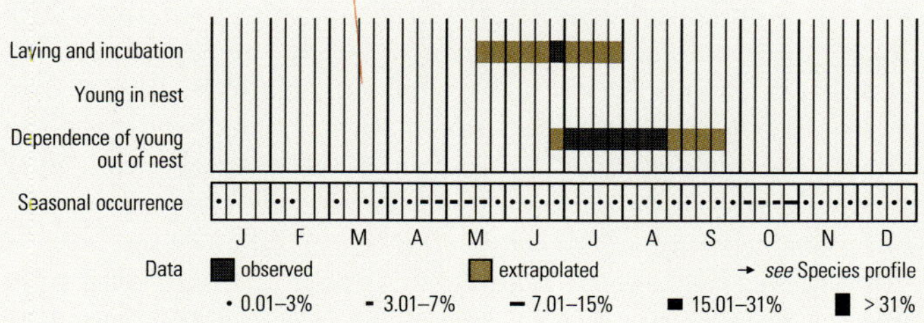

White-winged Scoter

Macreuse à ailes blanches

Macreuse brune

Melanitta fusca

Pierre Dupuis

THE WHITE-WINGED SCOTER IS THE LARGEST OF THE THREE SCOTERS AND also the one whose distribution and breeding biology is best known. Godfrey (1986), in the absence of satisfactory breeding evidence, still classified it as a transient in Québec. The dozens of broods reported since 1989 in northwestern Québec, mainly from the James Bay coastal plain, and survey results (Cons. G. G. – GREBE 1990a, 1992t, 1993e; Benoit et al. 1991, 1992, 1993b, 1994) prove that, on the contrary, the species is well established here.

There are three subspecies of the circumpolar White-winged Scoter: *Melanitta fusca fusca*, which breeds in Europe and western Asia; *Melanitta fusca stejnegeri*, which breeds in eastern Asia; and *Melanitta fusca deglandi*, which breeds in North America (Cramp and Simmons 1977).

Habitat and behaviour

In the James Bay region of Québec, breeding pairs and broods frequent freshwater lakes 20–100 ha in area and less than 4 m deep; these lakes have fairly clear, alkaline water, floating or submerged aquatic vegetation, and scrub cover around the edges (Benoit et al. 1993b). Similar brood-rearing habitat is used by White-wings in Saskatchewan (Brown and Brown 1981; Brown and Fredrickson 1986).

The staple summer diet of adults and young at Redberry Lake, Saskatchewan is the scud (Brown 1981; Brown and Fredrickson 1986), an amphipod which also occurs in lakes on the James Bay coastal plain, where broods of White-wings have been reported (Benoit et al. 1994). The birds feed at depths of 1–3 m, with adults tending to forage deeper than ducklings (Brown 1981; Brown and Fredrickson 1986). Average diving time during a feeding session is about 28 seconds; a duck fleeing danger may stay under as long as 78 seconds (Grosz 1966).

The chief component of the White-winged Scoter's winter diet on the New Hampshire coast is bivalves (Stott and Olson 1973), while in British Columbia, their main winter foods are bivalves, gastropods, and, to a lesser degree, barnacles and crustaceans (Vermeer and Levings 1977; Sanger and Jones 1984; Vermeer and Bourne 1984). Grosz (1966) reported that on the California coast, this scoter also feeds on plants in addition to the molluscs and crustaceans that make up the bulk of its diet. From fall until spring, it forages in water with a gravelly or sandy bottom where molluscs and crustaceans are abundant.

Pairs usually form on the wintering grounds. In California, groups of 6–8 and sometimes up to 20 males may court the same female. The male approaches the female, bobbing his head up and down and stretching his neck towards her; the female responds in kind. Displaying males are often attacked by other males, however. Most females have already found mates by the end of December, and once pairs have formed, they keep away from the courting groups (Grosz 1966).

At the start of the breeding season, the male does not stake out a fixed territory, but rather defends an area around the female, following her constantly (Brown and Fredrickson 1987a). Protected from the harassment of other males, the hen is able to feed for long periods and build up the energy reserves needed for laying and incubation, when she may lose up to 23% of her body weight (Brown 1981; Brown and Fredrickson 1987b).

The nest, a shallow depression lined with twigs and leaves, is usually concealed in dense shrubs, grasses, or forbs (Bellrose 1976; Brown and Brown 1981; Brown and Fredrickson 1989). It is generally located near water, but nests up to 800 m away have been found (Palmer 1976). In one Saskatchewan study, the mean distance between nests and the nearest shoreline was 96 m (Brown and Brown 1981). On occasion, intra- and interspecific parasitism may occur (Brown and Brown 1981). When laying is complete, the hen adds some of her thick, dark down to the nest to keep the eggs warm when she leaves to feed.

The drake abandons the hen at the onset of incubation (Brown and Fredrickson 1989) and heads to a large lake or to the nearest coast, to moult in the company of other adult males and sometimes immature birds (Palmer 1976). The native peoples used to hunt White-winged Scoters when they congregated in their large moulting flocks. The moulting birds were easy picking and natives used to surround the frightened flightless birds in their canoes, forcing them to dive until they drowned and floated belly-up on the surface (Todd 1963).

After hatching, the ducklings may remain up to a full day in the nest (Brown and Brown 1981) before the hen leads them to water. During the first 24 hours, the ducklings are vulnerable to intense predation from gulls (Brown and Brown 1981). When several broods share the same body of water, crèches may form, although the mechanisms behind this have not yet been documented (Kehoe 1989). Many young are fully independent at 2 weeks, before they have even fledged: the crèches are then led by the older ducklings (Brown and Brown 1981; Brown and Fredrickson 1989). After taking leave of the ducklings, the females begin their fall migration, with the young following suit shortly afterward.

Distribution

Until recently, it was thought that the breeding range of the subspecies *Melanitta fusca deglandi* extended from Alaska south to North Dakota, and only as far east as northwestern Ontario (Godfrey 1986; Ross 1987). Although White-wings had been observed in summer along the coast of northern Québec and Labrador, from James Bay through Ungava Bay and Hudson Strait to the Gulf of St. Lawrence (Manning 1949; Todd 1963; Curtis and Allen 1976; Chapdelaine and Tremblay 1979), breeding had not been confirmed. Audubon's breeding record from Brador, at the eastern tip of the Lower North Shore, was generally discounted as dubious (Todd 1963) and only Dionne (1906) extended the species' breeding range as far east as Labrador.

Breeding in Québec was finally confirmed in 1976 with the sighting of a brood near the Saindon Lakes, in the Little Whale River

Notes on territory

According to Brown and Fredrickson (1987a), the drake does not defend a territory per se, but rather an area around the hen so that she can spend all her time building up energy reserves for laying and incubation. Hens show extremely strong nest site tenacity from year to year (Brown and Brown 1981). There is practically no information available on the size of the home range, but Brown and Fredrickson (1987a) noted brood movements of over 1 km.

The White-winged Scoter has been confirmed as a breeder on the Québec side of James Bay since the early 1990s, and many nest in the area around the Roggan River. East of Kakassituq Point.
June 1995, Austin Reed.

The Lake Bienville area offers suitable breeding habitat for the species; pairs nest in shallow parts of the lake. North-central Québec. *June 1989, Jean-Luc DesGranges.*

area (Savard 1977). Recently, 5 broods were observed in the headwaters of the Little Whale River, where the species seems to be well established (Cons. G. G. – GREBE 1990a, 1993e). A brood was also sighted near the Caniapiscau River (56°32' N, 69°25' W; Cons. G. G. – GREBE 1992t), and a nest was discovered on an island in Manitounuk Sound, south of the mouth of the Little Whale River (Cons. G. G. – GREBE 1990d).

Another very important breeding area has recently been discovered along a 20–40 km wide strip of the James Bay coastal plain, from the Rivière au Castor to Point Louis XIV: between 1990 and 1993, over 60 broods were spotted there (Benoit et al. 1991, 1992, 1993b, 1994). The Point Louis XIV and Roggan River areas are especially productive, yielding the highest breeding pair densities —80 pairs/100 km² and 65 pairs/100 km², respectively—in Québec to date (Benoit et al. 1993b).

A part of the White-winged Scoter population stages in the estuary and the Gulf of St. Lawrence in spring and fall (Lehoux et al. 1985); at these times, large numbers can be seen in southern Québec all along the St. Lawrence (David 1980a), especially in the Lower St. Lawrence region between Sainte-Luce-sur-Mer and Matane (J. Gauthier, pers. comm.). In North America, White-wings winter along both the Atlantic and Pacific coasts: in the east, from Newfoundland to North Carolina and sometimes as far south as the Gulf of Mexico, and in the west, from Alaska to California (Bellrose 1976; Houston and Brown 1983).

History and trends

On the Atlantic seaboard, the species was hunted extensively, particularly in the United States, before the Migratory Birds Convention was signed in 1916, and in fact, population declines had already been reported by then (Palmer 1976). In Québec, Ouellet (1974a), based on Hall (1862) and Wintle (1896), who both classed the White-winged Scoter as a common "transient visitant," ascribed the decline this century to water pollution and habitat degradation; in the eastern part of the continent, oil pollution on marine waters may have surpassed hunting as the major cause of mortality in recent years (Palmer 1976). The population decline in the heart of the species' breeding range in northern Alberta may be attributed to the opening of the hunting season two or three weeks before the young are able to fly (Ehrlich et al. 1988). Throughout Québec and the rest of Canada, harvests of the various species of scoters are small. The number of White-winged Scoters taken by sport hunters has ranged between 2000 and 14 000 since 1973, which is barely 1% of the annual harvest of ducks in the province (Appendix II).

It is premature to identify population trends for this species in Québec.

Benoît Limoges and Alain Morrier

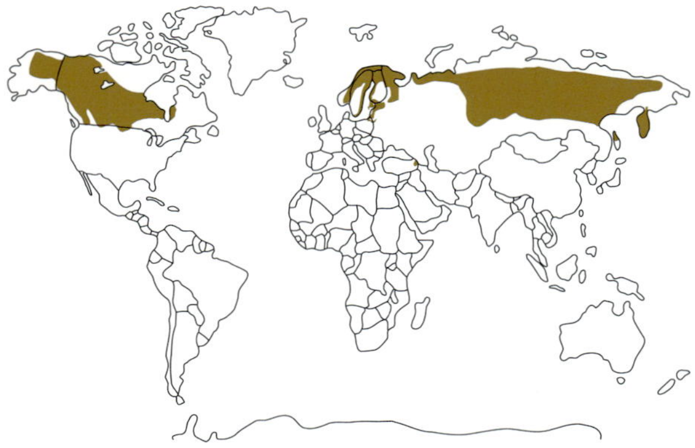

World breeding range

Species profile

Clutch size	usually 9 eggs (5–12)[3]
Laying interval	1 egg per 1–2 days[1]
Start of incubation	?
Incubation period	25–30 days[2,3]
Maturity at hatching	precocial
Care of young	by female[3]
Nestling period	12–24 hours[2]
Age at first flight	63–77 days[3,5]
Dependence of young	usually until 2 weeks old[2,9]
Broods per year	1[3]
Breeding age	2 years[3,4,5]
Mating system	monogamy[3]
Duration of pair bond	1 season[4]
Total length	48–58 cm[7]
Wing span	85.1–105.4 cm[8]
Weight	M 1588 g (1361–1769) / F 1179 g (952–1406)[1]
Longevity record	18 years[6]

References:
1. Bellrose 1976
2. Brown and Brown 1981
3. Ehrlich et al. 1988
4. Johnsgard 1975b
5. Palmer 1976
6. Kehoe et al. 1989
7. Godfrey 1986
8. Terres 1980
9. Brown and Fredrickson 1989

Chronology *n* = 13

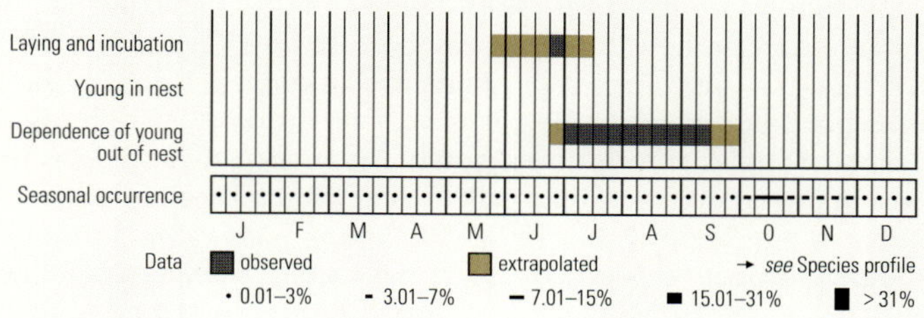

Rough-legged Hawk

Buse pattue

Buteo lagopus

THE ROUGH-LEGGED HAWK IS THE MOST COMMON HAWK OF THE NORTH American Arctic (Zarn 1975). It owes its name to its fully feathered legs, or tarsi, which presumably keeps them warm in cold arctic conditions. This hawk is large, but relatively light, so it can easily hover in search of prey. Individual coloration varies widely, from light brown with only scattered darker markings on a whitish ventral surface and wide whitish edges on the dorsal feathers to a nearly uniform dark brown plumage (Cade 1955b). A northern nester, this buteo is a rare winter visitor in southern Québec, but a common transient during spring and fall migration (David 1980a).

Habitat and behaviour

The Rough-legged Hawk breeds in the Arctic and Subarctic, mainly on the open tundra, and builds its nest on cliff ledges, boulders, escarpments, shelves on steep riverbanks, and occasionally, in trees or on flat ground (Godfrey 1986). The nest is composed of sticks, twigs, weeds, and stalks, and lined with grasses, sedges, feathers, or fur. Accumulated droppings and food remains around the nest turn to soil, providing nutrients for orange-coloured lichens and for grasses, which are nest markers easily spotted from a distance (Zarn 1975; Palmer 1988). The Rough-leg's strong site tenacity means that nests may be repaired and reused over many years. One breeding pair may have several alternative sites within a territory, especially since nests may be taken over by the more aggressive Peregrine Falcon, which competes for the same kind of nest site (White and Cade 1971).

Rough-legs form monogamous pair bonds, and some pairs may stay together for several seasons, even migrating together (Townsend 1937; Brown and Amadon 1968). Courtship displays may consist of high circling above the nesting territory by a single bird or a pair, mutual calling, and sky dancing, in which a soaring male partially closes his wings, dives abruptly downward, climbs back up, stalls, and then stoops back down again (Johnsgard 1990).

Incubation is primarily by the female, while the male hunts or guards the territory, but sometimes the male may incubate briefly. Breeding densities and clutch size fluctuate locally with the available food supply. Up to 7 eggs are laid when food is abundant, but the female deserts her clutch if prey populations crash (Zarn 1975). Outside the breeding season, humans can often get quite close to this bird, but when nesting it is very territorial. On sighting an intruder, one or both parents may fly at the unwanted visitor, screaming an almost catlike call, thus revealing their presence and the nest's location (Palmer 1988). The young hatch asynchronously after an incubation period of 28–31 days. Both parents hunt for and guard the young, which begin to tear up their own food at about 15–20 days (Cramp and Simmons 1980). The males fledge at about 34–36 days, while their larger female siblings may remain in the nest 40–45 days. The young remain dependent on their parents for food for another 20–35 days (Palmer 1988).

Rough-legged Hawks hunt by quartering over wet meadows, peatlands, marshlands, and other open areas, and often hover in one place while scanning the ground. They prey mainly on lemmings and voles, which often make up 80–90% of their diet, and more rarely on larger mammals such as arctic ground squirrels, young arctic hares, and weasels (Springer 1975; Palmer 1988). Having relatively small, weak feet, they are opportunistic, sometimes taking frogs from marshy areas on their wintering grounds, small birds on the ground and fish from the water's surface. When prey is scarce, they have been known to usurp the prey of Red-tailed Hawks (Hogan 1983) or feed upon carrion (Smith 1975).

Distribution

The Rough-legged Hawk is a panboreal species, breeding mainly between latitudes 61° and 76° N in the tundra and taiga regions of Europe, Asia, and North America (Brown and Amadon 1968). For hunting and nesting, it prefers open terrain with low-growing vegetation for good long-distance visibility (Cramp and Simmons 1980). In Canada, its northern limits extend far into the High Arctic, where it nests on such islands as Prince Patrick, Melville, Bathurst, and Bylot; this range coincides with the distribution of native rodents (Murie 1959; Palmer 1988). In good prey years, its southern limits may extend to the fringe of the boreal forest, where it may nest in trees. In Québec, this species breeds throughout virtually the entire Ungava Bay region and south to about 52° N, and on the outer Lower North Shore of the Gulf of St. Lawrence from Wolf Bay eastward (Godfrey 1986).

The Rough-legged Hawk is very common along the Hudson Bay coast. In 1990, 52 nests were found there from the Great Whale River to north of Lake Guillaume Delisle (Cons. G. G. – GREBE 1990c). In 1993, breeding in the James Bay region was confirmed by discovery of a nest on the wall of the La Grande-2 spillway (Aubry and Yank 1993c).

During the Atlas survey period, on 2 August 1989, a cliff nest was sighted on Île du Petit Mécatina near Harrington Harbour, Duplessis County, on the Middle North Shore (QNRCP). The nest

In the Subarctic, this hawk sometimes nests on cliff faces in the taiga. Schefferville. *August 1987, Christian Marcotte.*

Carole Bérubé

Notes on territory

During the breeding season, this species can be very territorial. In a study in northern Sweden, human intruders approaching within 600 m of a Rough-legged Hawk nest were met and then circled first by the male, then at 400 m by the female. If the person persisted and came within 100–200 m, the male dived at and then chased the intruder until he retreated to at least 500–600 m away from the nest (Anderson and Wiklund 1985). *Rough-legged Hawks often nested within 50 m of Peregrines and Gyrfalcons, and within 20 m of ravens, in years of high local prey abundance in the central Canadian Arctic* (Poole and Bromley 1988b). *Aggressive interactions were frequently observed among close conspecific neighbours, but little contact was made.*

Although this is a common breeding species in many areas, breeding densities fluctuate locally depending on food supplies (Brown and Amadon 1968; White and Cade 1971). *Along the Colville River in Alaska, distances between pairs ranged from 0.4 km to 22 km, averaging 3.8 km* (White and Cade 1971). *A nesting raptor survey conducted on 16 rivers in southwestern Alaska found pairs nesting 1.0–12.9 km from other Rough-legged Hawks* (Mindell and Dotson 1982). *In Québec, along the*

The tundra is the main habitat of the Rough-legged Hawk. Ivujivik, Ungava Peninsula. *August 1982, Gaëtan Guertin.*

contained 2 young about to fledge, so breeding in the area was confirmed, although the site was just beyond the Atlas boundary.

In the summer of 1993, Rough-legged Hawks nested on Île Nue de Mingan, where 2 dark-phase birds were spotted on a bare cliff on 18 June (J. Ibarzabal, pers. comm.).

Sightings of Rough-legged Hawks far south of their breeding range, such as those reported at Saint-Bruno and Métabetchouan, Lac-Saint-Jean; Val-Paradis, Abitibi; Bic and Cacouna along the Lower St. Lawrence; and Tracy (R. Yank, pers. comm.) well into July, may possibly have been sexually immature yearlings or perhaps adults foregoing breeding due to low prey availability or poor body condition.

Although usually solitary, these birds often migrate in large flocks, profiting from favourable winds and temperatures. On their wintering grounds, they will roost communally with other raptors at night near abundant food supplies on prairies, open fields, marshes, and peatlands.

World breeding range

Species profile

Clutch size	usually 3–5 eggs (2–7)[2,3,4,5]
Laying interval	1 egg per 2 days[3,6]
Start of incubation	with first egg[4,6]
Incubation period	28–31 days[2,4,6]
Maturity at hatching	semi-altricial
Care of young	by female and male[1,3,4,6]
Nestling period	usually 40 days (31–45)[1]
Age at first flight	M 34–36 days / F 40–45 days[1,5]
Dependence of young	until 20–35 days after first flight[1,3,6]
Broods per year	1[1,4,6]
Breeding age	probably 1 year (1–2)[1,6]
Mating system	monogamy[1,3,6]
Duration of pair bond	probably sometimes more than 1 season[1,4,5]
Total length	M 50–56 cm / F 54–59 cm[2]
Wing span	121.9–142.2 cm[7]
Weight	M 828 g (779–900) / F 1130 g (1055–1227)[1]
Longevity record	18 years 1 month[1]

References:
1. Palmer 1988
2. Godfrey 1986
3. Johnsgard 1990
4. Brown and Amadon 1968
5. Townsend 1937
6. Cramp and Simmons 1980
7. Terres 1980

Chronology *n* = 79

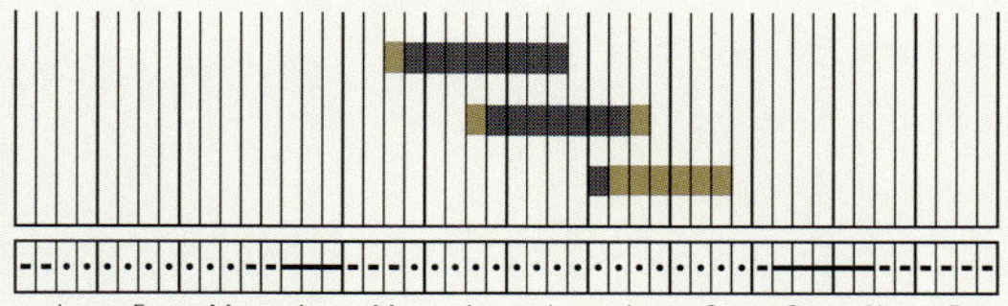

In 1993, this opportunistic pair of Rough-legged Hawks built its nest on the side of the La Grande-2 spillway. James Bay region. *August 1993, Robert Alvo.*

Notes on territory (continued)

Hudson Bay coast, distances between active nests ranged from 250 m to 11.8 km in 1990, averaging 5.0 km (Brodeur et al. 1994; F. Morneau, pers. comm.). In the Northwest Territories, nesting densities averaged 1 pair/13 km² to 1 pair/26 km² (Sealy 1966). According to Brown and Amadon (1968), an average density may be 1 pair/7.8 km² to 1 pair/10.4 km², with a maximum density of about 1 pair/4 km², although in some areas it is only 1 pair/78 km².

History and trends

In the early 1900s, Rough-legged Hawks were seen frequently in the province during the fall migration (Dionne 1906) and were reported breeding on "nearly all the high cliffs all the way across Ungava Bay District from James Bay to Ungava Bay" (Macoun 1900). It is still considered to be a fairly common breeding bird on the arctic tundra. Population numbers and productivity in the breeding range are influenced most strongly by local fluctuations in prey availability (Palmer 1988).

As they eat mostly non-migratory mammals, Rough-legs have been little affected by pesticide contamination. Thanks to the remoteness of their nest sites, their populations have remained relatively unaffected by human activities in the north (Ouellet 1974a; Johnsgard 1990). This situation is rapidly changing, however, and the greatest threat to Rough-legged Hawks is now the development of mineral and energy reserves in the Subarctic and Arctic.

DERIN HENDERSON AND DAVID M. BIRD

Wide shot of a ledge with a nest containing a Rough-leg chick. La Grande-2 spillway, James Bay region. *August 1993, Robert Alvo.*

Gyrfalcon

Faucon gerfaut

Falco rusticolus

Christiane Girard

THE GYRFALCON IS THE LARGEST AND MOST NORTHERLY OF ALL FALCONS. Individuals vary from virtually pure white to almost black morphs, with many variations in between. The morphs interbreed; a nest in Ungava Bay contained 3 young, 1 white, 1 grey, and 1 dark brown (D. M. Bird, McGill U., unpubl. data). Occasionally, a bird has mixed characteristics; for example, a dark bird may have areas of whitish feathers (Palmer 1988). Females are larger and heavier than males by as much as 40% (Brown and Amadon 1968). Gyrfalcons have a holarctic distribution and are permanent residents in northern Québec, although every year some wanderers visit southern Québec in winter (Brown and Amadon 1968; Platt 1976a).

Adult and young on an old raven or Rough-legged Hawk nest in a fissure sheltered by an overhang. North of the Nastapoka River, Hudson Bay region. *June 1992, Serge Brodeur.*

Habitat and behaviour

In the breeding season, the Gyrfalcon inhabits open country in the Arctic and Subarctic with cliffs, rock outcrops or mountains, both inland and on rocky seacoasts and islands. It has also been found in forested or partially wooded areas in the northern boreal forest (Brown and Amadon 1968; Godfrey 1986). Nests, or eyries, are typically a scrape on a cliff ledge 1–2 m wide and deep, most often under a sheltering overhang and usually snow-free year-round. They use the same territory each year, but within this territory will commonly use alternative nest sites (Poole and Bromley 1988a). They occasionally appropriate old stick nests of ravens, Rough-legged Hawks, and Golden Eagles on cliffs, but rarely those in trees (Palmer 1988). Gyrfalcons are difficult to study because they can be extremely shy in some areas, but certain habits during the breeding season are telltale signs of their presence. The unmated male makes himself conspicuous to females by perching at more open and higher sites than usual and attempts to attract a mate by displaying at the nest ledge. During courtship display flights, the male flies in a repeated figure eight, with the central crossing in front of the eyrie, giving an almost constant wailing call. Once occupied, nest sites are marked by droppings along the side of the cliff, or the mewing call of the begging nestlings may reveal its location (Palmer 1988).

Age at first breeding has been reported as 2 years by some researchers (Cramp and Simmons 1980), but Platt (1977) maintained that they are not fully sexually mature and able to breed successfully until the age of 4 years. They appear to be monogamous and generally form a lifelong attachment to a particular nesting territory. They begin nesting in April or early May throughout their range (Brown and Amadon 1968; Tommeraas 1989), sometimes competing with ravens for cliff sites (Poole and Bromley 1988b). Gyrfalcons can share cliffs with Peregrine Falcons simultaneously and successfully (White and Cade 1971). Despite the northerly latitudes of their breeding range, renesting has been documented in the Northwest Territories and Yukon (Platt 1976a; Poole 1988). Throughout the incubation period, the male is the sole hunter, but will relieve the sitting female at the nest for 2–3 hours after bringing in food. Not until the young are 18–25 days old does the female participate in hunting forays (Poole and Boag 1988). Both sexes will defend the nest against predators, such as foxes and ravens. Male nestlings fledge at 45–47 days, 2–3 days earlier than the larger females. The young falcons may remain within several hundred metres of the nest for the first 7–10 days after fledging, relying on the parents for food, but are independent after 3–4 weeks (Poole and Bromley 1988a).

Notes on territory

Gyrfalcons have a large home range, but only the area around the eyrie is defended, usually by both sexes. Gyrfalcons have killed Peregrine Falcons, Common Ravens, and Rough-legged Hawks that came too close to the nest (Cade 1960; White and Cade 1971). The area defended is variable, perhaps in relation to the food supply (Newton 1979). Observations by Platt (1977) in Yukon suggest

Gyrfalcon perched near its nest on a rocky cliff ledge. George River, Ungava Bay region. *June 1977, Pierre Lamothe.*

Using her powerful talons and hooked beak, this female Gyrfalcon is tearing her prey into pieces to feed to her 4 young. The eldest of these hatchlings is 16 days old. Fosheim Peninsula, near Eureka, Ellesmere Island, Northwest Territories. *July 1973, David M. Bird.*

Unlike Peregrines, Gyrfalcons generally capture prey by surprise—flying fast, hugging the contours of the terrain, and using small rocks and trees as temporary perches. They will soar high in search of prey, however (White and Weeden 1966). These opportunistic birds prey on a variety of animals, mostly birds and especially ptarmigan. In some areas, ptarmigan account for 90% of their diet by weight, and in years of scarcity, many Gyrfalcons do not breed (Gudmundsson 1972; Barichello and Mossop 1983). Other prey species include waterfowl, shorebirds, seabirds, arctic ground squirrels, arctic hares, lemmings, and voles (Cade 1960; Muir and Bird 1984).

Distribution

Gyrfalcons inhabit the arctic and subarctic regions of the Northern Hemisphere (Cade 1982). In Québec, they have been found breeding at suitable sites throughout the Ungava Bay region (Godfrey 1986). Along Hudson Bay, a nest was found as far south as Point Louis XIV (J.-L. DesGranges, pers. comm.). Inland, the breeding range extends south to the Nastapoka River, where 2 nests were found in 1992 (Cons. G. G. – GREBE 1992v; GREBE 1992), and probably reaches the Little Whale River, where a pair was seen 30 km from the sea in August 1990 (Cons. G. G. – GREBE 1990c).

Pairs are generally well spaced, as they need large hunting ranges, and do not nest in relatively flat areas nor on wet tundra lacking elevated nesting sites (Johnsgard 1990). Most sites, but not all, are located beside water, such as rivers, lakes, and seacoasts.

David (1980a) listed the Gyrfalcon as a casual winter visitor to southern Québec, but it is now regularly sighted in small numbers, particularly along the St. Lawrence. Spring sightings are rarer; some of the latest spring records during migration include Château-Richer (18 May in 1962 and in 1972), Portneuf (5 May 1983), Cacouna (7 May 1983), and Saint-Gabriel-de-Brandon (11 May 1987) (ÉPOQ).

Three weeks later, only 3 of the brood are left in the nest. Because of its smaller size, the youngest sibling often fails in the competition for food. Fosheim Peninsula, near Eureka, Ellesmere Island, Northwest Territories. *July 1973, David M. Bird.*

History and trends

Historically, the Gyrfalcon has been considered rare over almost all of its range (Freemark 1978), except very locally in areas of abundant prey (Dionne 1906), where nests may contain 4 or 5 young. Gyrfalcon populations in northern Québec have remained rather stable at least in the last 25 years (J. Weaver and D. M. Bird, McGill U., unpubl. data), with minor annual fluctuations due to changes in the prey base. The Gyrfalcon is among the most coveted species for falconry throughout the world. As it is listed in Appendix I of the Convention on International Trade in Endangered Species (CITES), poaching of the birds and their eggs is punishable by heavy fines (Mossop et al. 1985). At present, the greatest threat to their populations lies in increased industrial exploration and development on their breeding grounds, and increasing tourism along the northern rivers. Pesticides have had little impact on their populations (Cade et al. 1971; Bromley 1987). Low-level military flights east of Ungava Bay and in Labrador could seriously disturb Gyrfalcons during the breeding season (Lepage *in* Bromley 1987).

Derin Henderson and David M. Bird

Notes on territory (continued)

that pairs defend an oval area around the nest measuring a little under 1352 m on its long axis. Poole and Bromley (1988b), on the other hand, reported that ravens and gulls were not attacked until they were within 100 m of the nest. In their study, the minimum distance between a Gyrfalcon's nest and that of another bird was 50 m for the Rough-legged Hawk, 65 m for the Common Raven, 700 m for the Peregrine Falcon, 1000 m for other Gyrfalcons, and 1800 m for Golden Eagles. Agressive encounters were seen regularly between Gyrfalcons and the other species, but those with Golden Eagles were the most vigorous, and eagles were often chased up to 2 km from the nest sites. This is probably due to the limited variety of available prey for these two species early in the nesting season, when the resident population of prey animals is low and migratory species have not yet arrived. As a result, Gyrfalcons and Golden Eagles have a forced dietary overlap during nest initiation, and competition is intense (Poole and Bromley 1988b).

Gyrfalcon densities are extremely low everywhere, even near seacoasts, where the highest breeding densities are usually found (Brown and Amadon 1968). A relatively high nesting density of 0.8 pair/100 km^2 was estimated during a study at Kilgavik, Northwest Territories, with an annual mean inter-nest distance of 9.5–11.4 km (Poole and Bromley 1988a). In the Northwest Territories and Yukon, and along the Colville River in

Alaska, other reported breeding densities have ranged from a minimum of 0.07 up to 0.6 pair/100 km² (Cade 1960; Platt 1977; Bromley 1983; Bromley and McLean 1986). Iceland has the highest recorded breeding density, with up to 1.6 pairs/100 km² (Wayre and Jolly 1958).

This falcon is a bird of the vast open tundra of northern Québec. Here, a watchful adult perches on a rock. George River, Ungava Bay region. *June 1977, Pierre Lamothe.*

World breeding range

Species profile

Clutch size	usually 3–4 eggs (2–7)[1,2,3,4,5,8]
Laying interval	1 egg per 2–3 days[4,5]
Start of incubation	with penultimate egg[9]
Incubation period	usually 34–36 days[4,5,7]
Maturity at hatching	semi-altricial
Care of young	by female and male[1,3,4,5,6,7], *see* text
Nestling period	M 45–47 days / F 47–50 days[6]
Age at first flight	M 45–47 days / F 47–50 days[6]
Dependence of young	until at least 3 or 4 weeks after first flight[1,2,3,4,5,6,7]
Broods per year	usually 1[4,5]
Breeding age	2–4 years[3,4,5,9]
Mating system	monogamy[3,4,5,7]
Duration of pair bond	for life[1,4,5]
Total length	M 53–57 cm / F 56–63 cm[2]
Wing span	M 101–124 cm / F 120–135 cm[5]
Weight	M 1115–1313 g (805–1325) / F 1735–1760 g (1396–2100)[8]
Longevity record	12 years[8]

References:
1. Brown and Amadon 1968
2. Godfrey 1986
3. Johnsgard 1990
4. Cramp and Simmons 1980
5. Palmer 1988
6. Poole and Bromley 1988a
7. Ehrlich et al. 1988
8. Clum and Cade 1994
9. Platt 1977

Chronology *n* = 3

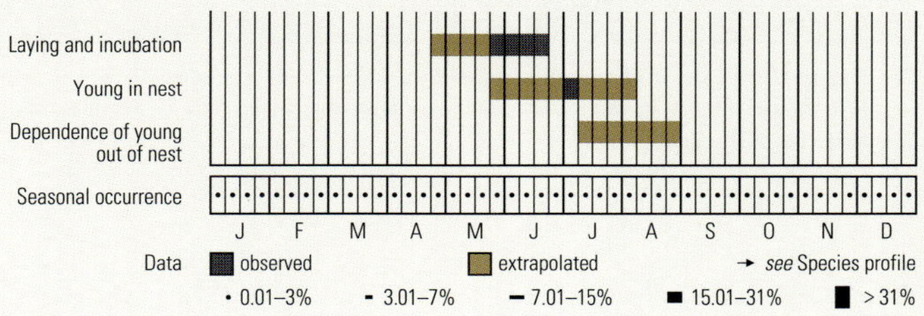

Chukar

Perdrix choukar
Alectoris chukar

Christiane Girard

THIS CHICKEN-SIZED PARTRIDGE IS NATIVE TO THE ROCKY, MOUNTAINOUS regions of central and southern Eurasia. It was successfully introduced into several countries, including the United States, Mexico, and Canada, chiefly with a view to hunting. Attempts to introduce it into Québec have failed, although commercial farming of Chukars for their meat is very successful, and they are often raised along with Barbary Partridge and Red-legged Partridge. Partridge farms are extremely large in area, and birds occasionally escape. Chukars sighted in Québec come from these farms (David 1980a).

The sexes have identical plumage. The upperparts are greyish brown with a pinkish wash. A black band runs across the forehead and through the eye, down the sides of the neck and across the upper breast. The throat and abdomen are buff and the flanks are barred with black and chestnut. The colour of the bill, feet and eyelids varies from red to pink (Godfrey 1986).

The Red-legged Partridge is the species that most resembles the Chukar, but it has a white eyebrow and its black collar is not as well defined as the Chukar's. The Barbary Partridge has no black stripe through the eye, neck and breast, and has a grey face (Cramp and Simmons 1980).

The Chukars in southern Québec have escaped from captivity and live in open areas. In British Columbia, they nest in the shelter of rocks or shrubs, preferring uneven terrain. Kamloops, British Columbia. *June 1971, Pierre Lamothe.*

Habitat and behaviour

This sedentary species lives in rugged mountainous areas where only a few shrubs grow, a habitat in which they can exercise their main defence against predators, running away (Johnsgard 1973). When the peaks are covered with snow, Chukars go down into the valleys, where it is milder. In the northern parts of the Chukar's range, depth of snow cover may have a significant impact on the survival of the species. According to a study by Christensen, as reported by Johnsgard (1973), birds can stand temperatures as cold as –30°C, but a few centimetres of snow on the ground for several weeks running may result in high mortality.

Roosting in winter, Chukars sometimes nestle together in a cave on a rocky face of a cliff or in the shelter of a shrub, forming a circle with their tails inward and their heads outward (Terres 1980).

Chukars generally forage on the ground from midmorning until early afternoon, mostly for graminaceous seeds, but also clover and dandelion leaves, and a few insects (Cramp and Simmons 1980). On hot days, they may feed early in the morning and late in the afternoon, spending the rest of the day in the shade.

In spring, coveys of 5–150 Chukar disperse as birds pair up. Those over a year old are the first to mate, followed by first year birds in late spring (Mackie and Buechner 1963). The cock's often-aggressive courtship display includes standing in an erect, stiff posture, and then turning sideways to expose his barred flanks (Johnsgard 1973).

The hen makes her nest in the shelter of a rock or shrub, a simple scrape lined with dry grass and feathers. She covers the first eggs with straw and leaves. Sometimes three weeks go by between the end of laying and the beginning of incubation. Hatching is synchronous and virtually simultaneous. In captivity, the hen may lay again while the cock incubates the first clutch (Cramp and Simmons 1980).

The hatchlings soon leave the nest, following the hen; the male may or may not help with rearing the young (Mackie and Buechner 1963; Johnsgard 1973; Cramp and Simmons 1980). Cramp and Simmons (1980) reported that if there are two broods, each parent accompanies one of them. The covey cackles constantly, except when a predator is sighted. The young can forage independently from their earliest days, but the hen regularly broods them for another few weeks. They remain together until the following spring (Cramp and Simmons 1980).

Notes on territory

According to Mackie and Buechner (1963), the Chukar cock does not really defend a territory, but fiercely protects the hen from the onset of nesting until the start of incubation.

In Utah, these investigators reported up to 20 breeding pairs/6 ha or 1 pair/3000 m^2.

Distribution

The Chukar was introduced into North America for the first time in 1893 (Long 1981). Since then, attempts at introducing it have been made in 42 American states, and the species is now established in 8 western states. Data from Nevada show that 968 000 birds were taken by hunters there between 1946 and 1967.

In Canada, the Chukar was introduced into British Columbia some time between 1950 and 1955. It was then introduced into Alberta, Saskatchewan, and Ontario, but unsuccessfully (Godfrey 1986).

In Québec, unsuccessful attempts at introducing it were made in Lennoxville (David 1980a) and Rigaud (PQSPB 1982). On the other hand, single birds and sometimes coveys are occasionally seen around Montréal (PQSPB 1980, 1982, 1983, 1984) and Québec City (F. Shaffer, pers. comm.) and are either escapes or releases (David 1980a).

After 19 birds were released in 1980 at Rigaud, at least one brood was seen the following summer at the same site. In the spring of 1982, 2 birds were still there (David and Gosselin 1983). This sighting is the only evidence that the species breeds or has bred in Québec.

History and trends

The Chukar is mainly associated with dry western regions. It cannot thrive in eastern North America because essential features of its habitat—low precipitation, no cover in winter—are missing. Escaped birds do not appear to be able to survive very long in Québec.

BENOÎT LIMOGES AND MARIE-RENÉE DOYON

Chronology *n* = 0

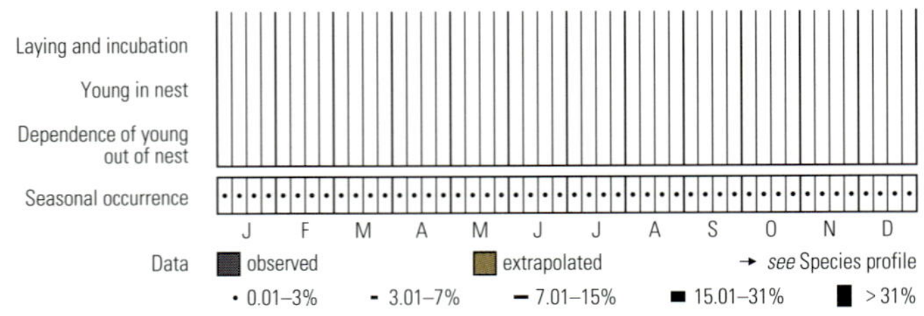

World breeding range

Species profile

Clutch size	usually 9–16 eggs (7–20)[4]
Laying interval	1 egg per 1–2 days[1,3]
Start of incubation	with last egg[1]
Incubation period	usually 22–24 days[1]
Maturity at hatching	precocial
Care of young	usually by female[6], by female and male[1]
Nestling period	probably less than 24 hours
Age at first flight	7–10 days[1]
Dependence of young	until ≈ 1 year old[1]
Broods per year	1, 2 possible[1]
Breeding age	1 year[1]
Mating system	monogamy, polygyny possible[6]
Duration of pair bond	usually until start of incubation[6]
Total length	33.0–39.5 cm[2]
Wing span	47–52 cm[1]
Weight	M 619 g (?) / F 537 g (?)[5]
Longevity record	?

References:
1. Cramp and Simmons 1980
2. Godfrey 1986
3. Johnsgard 1973
4. Ehrlich et al. 1988
5. Dunning 1984
6. Mackie and Buechner 1963

Common Quail

Caille des blés

Coturnix coturnix

Christiane Girard

IN THE 19TH AND 20TH CENTURIES, SEVERAL ATTEMPTS WERE MADE AT introducing the Common Quail, a Eurasian species, into North America. Although breeding has been reported in Québec and in the United States, the species has never become established on the continent. Today the Common Quail is raised by many Québec poultry farmers; quail and quail eggs are sold in restaurants and food markets. Common Quail have speckled, mottled brown plumage and are smaller than Northern Bobwhites (Terres 1980), from which they also differ in that they are strong fliers able to fly long distances (Wallace 1881).

Habitat and behaviour

The Common Quail is primarily a migratory species that originally inhabited the grassy, shrubby steppes. Today it is found on several continents, where it occupies a variety of mostly open habitats, including woodlands, grasslands, crop fields—cornfields in particular—hilly pastures, and seashores with grassy areas (Cramp and Simmons 1980; Long 1981; Robbins 1981; Grzimek 1984a).

Males generally arrive on the breeding grounds before females. Pairs form on the breeding territories or sometimes during spring migration. The nature of the pair bond varies—males may be monogamous, bigamous, successively polygamous, or even sexually promiscuous—and appears to be a function of the ratio of males to females in a given area. In the breeding season, males defend their territories against other males, which may give rise to violent combat (Cramp and Simmons 1980; Grzimek 1984a).

The nest is a slight depression in the ground lined with fine grasses. Clutch size typically ranges from 6–14 eggs, which are incubated for 16–20 days. The female alone incubates the eggs and rears the young (Orcutt and Orcutt 1976; Cramp and Simmons 1980; Long 1981; Robbins 1981; Grzimek 1984a). The young leave the nest a few hours after hatching and are able to forage for themselves, although they may be fed by the female (Orcutt and Orcutt 1976; Cramp and Simmons 1980). They begin making short flights at 11–14 days of age, but are incapable of sustained flight until 19 days old (Cramp and Simmons 1980; Grzimek 1984a). They are sexually mature by the age of 1 year (Grzimek 1984a), and sometimes even by their first autumn if hatched in early summer (Cramp and Simmons 1980).

The Common Quail is omnivorous and feeds mainly on seeds, grain, shoots, and invertebrates found on the ground, especially insects (Cramp and Simmons 1980; Long 1981).

Distribution and history

In the wild, the Common Quail is found mainly in Europe, Asia, and Africa, but attempts have been made to introduce the species elsewhere. These introductions were successful in the Hawaiian Islands and on Réunion, but failed in other parts of the world, notably in North America, Australia, New Zealand, and Tahiti (Long 1981).

With the purpose of establishing a quail population for sport hunting in North America (Wallace 1881), numerous releases were made in 20 continental states from 1870 to about 1925 and from 1955 to 1960. Birds were also released in Canada in the 1880s, in New Brunswick and Québec (Long 1981). Quail were released in Québec in 1880, 1881, and 1882 (Wallace 1881; LeMoine 1882). In 1881, several nests were found in Lachine, where quail had been introduced that spring (Wallace 1881). Despite a handful of breeding records, the attempts to introduce the species into the province ended in failure. The quail migrated south in the fall and did not return (Gosselin 1977b). In short, all attempts at introducing the species into North America have failed (Long 1981).

NICOLAS GAGNON AND JOSÉE TARDIF

Notes on territory

The male fiercely defends his territory against other males (Cramp and Simmons 1980; Grzimek 1984a). In Europe, minimal defended territory size ranges from 0.7 to 1.5 ha. In Great Britain, a mean density of 0.3 male/10 ha has been observed in good breeding years (Cramp and Simmons 1980).

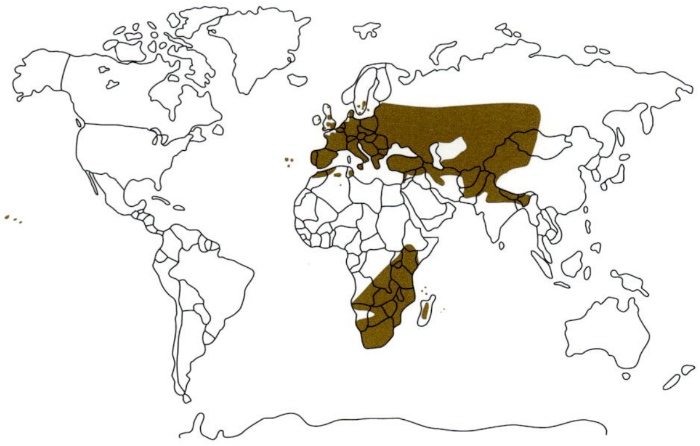

World breeding range

Species profile

Clutch size	usually 6–14 eggs (3–18)[1,2,3,4,5]
Laying interval	1 egg per day[1,5]
Start of incubation	with last egg[1,5]
Incubation period	16–20 days[1,3,4,5]
Maturity at hatching	precocial
Care of young	by female[1,4,5]
Nestling period	less than 24 hours[1,5]
Age at first flight	11–14 days[1,4]
Dependence of young	until 49–69 days old[1,4]
Broods per year	1, sometimes 2[1]
Breeding age	1 year[1,4]
Mating system	monogamy, polygyny, promiscuity[1]
Duration of pair bond	?
Total length	12.3–20.0 cm[2]
Wing span	?
Weight	64.8–125.5 g[2]
Longevity record	?

References:
1. Cramp and Simmons 1980
2. Long 1981
3. Robbins 1981
4. Grzimek 1984a
5. Orcutt and Orcutt 1976

Rock Ptarmigan

Lagopède des rochers

Lagopède alpin

Lagopus mutus

Christiane Girard

DENIZEN OF THE TUNDRA FOUND SOUTH OF THE TREE LINE ONLY IN winter, the Rock Ptarmigan is one of the most seldom seen grouse in Québec. It is, however, quite tame and easily observed in its breeding range on the rocky tundra, especially from mid-May to late June.

The Rock Ptarmigan is one of three ptarmigan species found worldwide, with two—the Rock and the Willow Ptarmigan—breeding in northern Québec. All three species are distinguishable from other grouse by their fully feathered toes, white winter plumage and brown summer plumage, and a monogamous mating system (Bergerud and Mossop 1984; Bernard-Laurent 1987).

Habitat and behaviour

Both sexes are easy to spot on their breeding territories during courtship and egg laying, especially the male with its spectacular flight-song displays (MacDonald 1970). The male Rock Ptarmigan retains his white plumage until the end of incubation, and so is easily distinguished from its close relative, the Willow Ptarmigan. The males in this species acquires a rusty red plumage on the head and neck at the onset of the breeding season (Godfrey 1986). Males and females moult from their white plumage into their brown plumage during incubation—females at the onset of incubation and males near the end. After hatching, Rock Ptarmigan can be very difficult to spot until adults and juveniles begin flocking in early autumn (MacDonald 1970).

Typical breeding terrain consists of moderately sloping ground in tundra or alpine regions, often with scattered rock outcrops. Males use rocks, knolls, or other high points for territorial displays. Nesting habitat is usually dry and rocky; nests are typically located in well-concealed depressions in clumps of grass or on mossy hummocks on open tundra, or in grassy patches on rock outcrops. Rock Ptarmigan are mostly monogamous, and begin establishing territories at the onset of snowmelt (Johnsgard 1983a; pers. obs.).

In northern Canada, clutch size of first nests range from 7 to 11 eggs, most commonly 8 or 9 (MacDonald 1970; Brodsky 1988; Cotter 1991). A hen may lay a second, though smaller, clutch if the first is lost early (Cotter 1991). The female alone incubates, generally for 21 days. The pair bond breaks down in the last week of incubation, after which the male remains on his territory, but provides no parental care of the young. As in many other grouse species, Rock Ptarmigan chicks can fly short distances by 10–12 days after hatching, and are strong fliers by 3 weeks of age (Johnsgard 1983a).

Rock Ptarmigan have a varied diet, although they rely heavily on the buds and leaves of willow and scrub birch throughout the year. In summer, the leaves and flowers of herbaceous plants, such as mountain-avens, horsetail, saxifrage, and vaccinium are an important part of the diet (Weeden 1969; Gardarsson and Moss 1970; Moss and Watson 1984; Thomas 1984).

Distribution

The Rock Ptarmigan is a circumpolar species, breeding primarily on the tundra, but also into the taiga where suitable habitat occurs (Godfrey 1986). It breeds throughout northern Québec south to the tree line and into the taiga around Schefferville (Godfrey 1986). At one time the species probably bred on Anticosti Island (Ouellet 1969), which is at the same latitude and not too far from Newfoundland, where there is a breeding population (Godfrey 1986). Comeau (1923a) claimed to have observed some young Rock Ptarmigan in July and August on the higher ranges east and west of the Washikuti River on the North Shore. Although this sighting was questioned by Todd (1963), he mentioned corroborating evidence provided by later records from elevated points in the region of Knob Lake. The confirmed southern limits descends to at least the 55th parallel in Québec (Todd 1963).

While some populations overwinter on the James Bay coast, within their breeding range, others migrate south into the taiga and boreal forest (Godfrey 1986). Rock Ptarmigan often overwinter in central Québec, specifically in the barren uplands 240–320 km inland from James Bay, as well as inland along the Lower North Shore (Godfrey 1986).

This grouse of the tundra likes to breed on bare rocky or pebbly terrain, preferably high up, where its nest will be well camouflaged. Akpatok Island, Ungava Bay, Northwest Territories. *August 1987, Lyne St-Jacques.*

This female Rock Ptarmigan in her cryptic brown plumage blends right in with a landscape of patchy snow. In females, the moult from winter white to summer brown occurs near the onset of incubation. Bathurst Island, Northwest Territories. *June 1969, Pierre Lamothe.*

Notes on territory

Soon after their arrival on the breeding grounds, the previously gregarious males begin dispersing and become increasingly aggressive toward other males. Their aggressiveness is marked by more frequent vocalizations from the ground, short runs and head bowing, and exposure of their combs. They compete as well for favourable feeding, roosting, and sheltering areas (MacDonald 1970). Territorial behaviour varies among females. Some may threaten and even attack intruding females (MacDonald 1970), while others show little or no aggression (Olpinski 1986). Females gradually associate themselves with a specific male and his territory, building their nest within his territorial boundaries (Johnsgard 1983a).

A study on the arctic tundra near Déception Bay, in northern Québec, found that the size of the male's territory depended on his age and number of mates. The largest territories were defended by adult males or those paired polygynously, while yearlings, monogamous males, or unmated but territorial males held significantly smaller territories (Olpinski 1986). In this 2-year study, the mean overall territory size for male Rock Ptarmigan was 5.00 ha in 1982 and 7.24 ha in 1983, with breeding densities of 8.5 pairs/km² (0.85 pair/10 ha) and 9.7 pairs/km² (0.97 pair/10 ha), respectively

(Olpinski 1986). MacDonald (1970) estimated that males on Bathurst Island, Northwest Territories, defended areas of about 2.6 km^2. Population densities from various studies in the Northwest Territories were summarized by Weeden (1963); they ranged from 3 adults/km^2 (0.3 adult/10 ha) on Baffin Island (Soper 1940) to 4000 adults/32 500 km^2 (0.01 adult/10 ha) on Prince of Wales Island (Manning and MacPherson 1961).

The farther north the male Rock Ptarmigan goes, the longer it keeps its winter plumage in spring. As soon as the snow begins to melt, the male starts to stake out a territory. Bathurst Island, Northwest Territories. *June 1969, Pierre Lamothe.*

History and trends

In many areas, breeding populations undergo regular cyclical fluctuations in density, peaking approximately every 10 years and fluctuating three- to fourfold (Weeden and Theberge 1972; Cotter 1991). In the Arctic, ptarmigan are killed for food in all seasons, with little impact on the population as a whole (Weeden 1963). At present, most sport hunting is of migrant ptarmigan that move southward late in the fall (Weeden 1963). In Québec, these cyclical invasions involve mixed flocks of both Rock and Willow ptarmigan, but the initial wave is made up primarily of Willow Ptarmigan. The Willow Ptarmigan's preference for boreal forest habitat in winter brings it farther south and into closer contact with human settlement, and as a result it is killed in far greater numbers than is the Rock Ptarmigan.

However, Rock Ptarmigan may face increasing hunting pressure as arctic and subarctic regions are exploited for mining, military operations, hydro-electric projects, and other types of commercial development. In particular, the new roads required for northern development will open up formerly inaccessible areas of the ptarmigan's range. Rock Ptarmigan could be susceptible to overhunting, because of their preference for open habitat, their tendency to form large flocks, and their unwariness, especially during the courtship period (Weeden 1963).

Richard C. Cotter

World breeding range

Species profile

Clutch size	usually 6–9 eggs (3–13)[2,4,6,7]
Laying interval	usually 1 egg per 1–2 days[4,8]
Start of incubation	with last or penultimate egg[4,5,8]
Incubation period	usually 21 days (20–26)[3,4,5,6,7,8]
Maturity at hatching	precocial
Care of young	mainly by female[4,5,8]
Nestling period	less than 24 hours[5]
Age at first flight	usually 10–12 days (7–15)[2,3,4,7,8]
Dependence of young	until 10–12 weeks old[7,8]
Broods per year	1[2,3,8]
Breeding age	1 year[1,2,8]
Mating system	monogamy, sometimes polygyny[1,2,5,8,9]
Duration of pair bond	less than 1 season[1,7,8]
Total length	32.5–39.5 cm[6]
Wing span	54–60 cm[8]
Weight	M 498.9–600.2 g (441–645) / F 474.9–550.3 g (454–660)[9]
Longevity record	9 years[10]

References:
1. Olpinski 1986
2. Johnsgard 1983a
3. Weeden 1963
4. Harrison 1978
5. MacDonald 1970
6. Godfrey 1986
7. Ehrlich et al. 1988
8. Cramp and Simmons 1980
9. Holder and Montgomerie 1993
10. extrapolated from Holder and Montgomerie 1993

Chronology $n = 63$

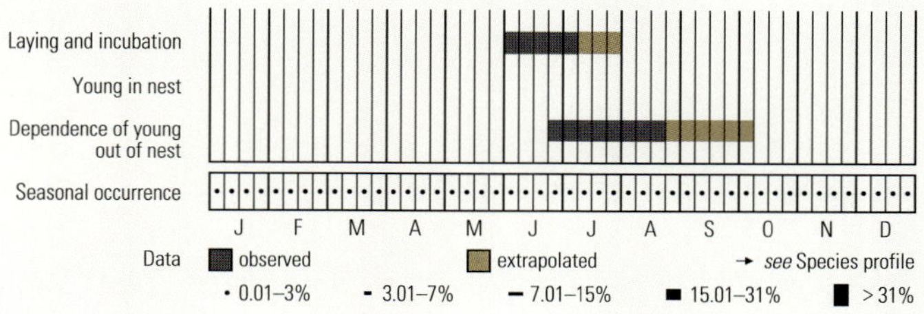

Northern Bobwhite

Colin de Virginie

Colinus virginianus

THE NORTHERN BOBWHITE IS SMALLER THAN ITS RELATIVES, THE partridges, grouse, and pheasants. This quail inhabits the central and southern United States, and those observed on rare occasions in Québec have usually escaped from captivity or been introduced.

The species was named for its characteristic whistling call, *bobwhite*. The male and female have similar plumage, except for the throat, chin, and stripe over the eye, which are white in the male and buffy in the female.

Habitat and behaviour

The Northern Bobwhite is found in wooded areas with low shrubs, grainfields, and abandoned farmland overgrown with weeds. Unlike the Gray Partridge, coveys of which are often observed in open areas, the Northern Bobwhite prefers brushy fields and the edges of woodlots. It usually tries to remain hidden and runs quickly for cover from bush to bush in more open areas (Rosene 1969).

In spring, at the start of the breeding season, the males make their distinctive whistling calls from the ground, a fence post, or a tree to proclaim their territories and challenge other males. The pair bond is formed once competing males have been driven away. The simple nest, which is built by both parents, is concealed in a tuft of high grass and contains 12–18 eggs. The male and female incubate the eggs for a period of just over three weeks. The chicks leave the nest soon after hatching and can fly in short spurts when they are only a week old. In the southern part of its range, the Northern Bobwhite often produces a second and sometimes even a third clutch when earlier clutches are destroyed by predation. Eggs can be found in nests as late as September in places like South Carolina and Georgia (Bent 1932).

The Northern Bobwhite has an extremely varied diet, ranging from buds, seeds, and herbaceous plants to fruit and insects. It is welcome in farming areas because it both eats large quantities of insect pests and is prized as a game bird. Since it is plentiful in many accessible areas, the Northern Bobwhite is the main game bird in many parts of the United States (Bent 1932; Dziepak 1991).

Distribution

The Northern Bobwhite inhabits parts of Central America, the eastern and central United States as far west as New Mexico, and the southern tip of Ontario (Johnsgard 1973; Long 1981; Page and Austen 1993).

In Québec, the species has only been spotted on a few occasions over the last 50 years (David 1980a). A small release on Mount Royal in 1943 was unsuccessful, and the birds were never seen again (Ouellet 1974a). The species has been sighted only 9 times in the last dozen years, in eastern Québec (Lake Beauport, 1980; Saint-Ferréol-les-Neiges, 1982) and in the Montérégie region (La Prairie, 1980; Montréal, 1983; Dorval, 1984; Nuns' Island, 1984; Brossard, 1985; Repentigny, 1986; and Saint-Hyacinthe, 1991) (ÉPOQ). Since these were most likely escaped or introduced birds, other observations have no doubt remained unreported. To the 4 ÉPOQ records in the Atlas period 1984–89 can be added a single Atlas record, involving a bird seen during the summer of 1986 in Havelock, near the American border.

History and trends

The Northern Bobwhite spread rapidly through southern Ontario after European settlement, when the land was cleared. By 1840, the species had reached southern Georgian Bay, the northern limit of its range in the province (Gosselin 1977b; Godfrey 1986; Lumsden 1987d). Since then, the population has declined considerably and is limited to farmland that is not being intensively cultivated.

The Northern Bobwhite does not appear to be able to gain a foothold in southern Québec or other parts of southern Canada. Intensive agriculture and the long months of heavy snow cover probably limit its distribution (Lumsden 1987d).

ALAIN LANOUE AND MARIE-RENÉE DOYON

Ghislain Caron

Notes on territory

After the coveys disperse in spring, males defend territories of approximately 0.4 ha (DeGraaf and Rudis 1987), where they make their distinctive whistling calls. In Maryland, Stewart and Robbins (1958) observed 0.4 pair/10 ha on cultivated farmland, compared with 1.25 pairs/10 ha in fields and along the edges of woodlots.

World breeding range

Species profile

Clutch size	usually 14–16 eggs (12–18)[1,2,3]
Laying interval	1 egg per 1–2 days[4,10]
Start of incubation	with last egg[11]
Incubation period	23–24 days[1,3]
Maturity at hatching	precocial
Care of young	by female and male[3,4]
Nestling period	less than 6 hours[3]
Age at first flight	≈ 2 weeks[4]
Dependence of young	until following spring[4]
Broods per year	1–2[1,9]
Breeding age	1 year[1,9]
Mating system	monogamy[5,6]
Duration of pair bond	sometimes more than 1 season[6]
Total length	24–27 cm[7]
Wing span	35.6–40.6 cm[3]
Weight	? g (148.6–256.0)[2]
Longevity record	6 years 5 months[8]

References:
1. Rosene 1969
2. Long 1981
3. Terres 1980
4. Harrison 1978
5. Ehrlich et al. 1988
6. Bent 1932
7. Godfrey 1986
8. Clapp et al. 1982
9. DeGraaf and Rudis 1987
10. Johnsgard 1973
11. Stokes and Stokes 1990

Chronology *n* = 0

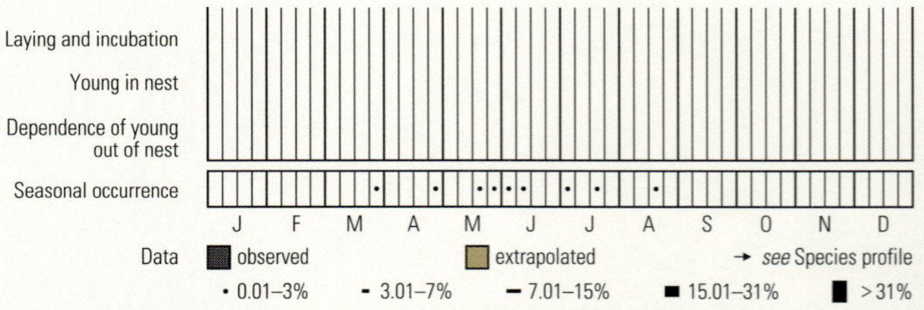

The Northern Bobwhite, a species rarely seen in Québec, prefers brushy habitat to open ground. This male, which escaped from captivity or was released, was spotted along the edges of fields and woods in the Cap Tourmente National Wildlife Area during the winter of 1994–95. Northeast of Québec City. *December 1994, Stéphane Turgeon.*

Sandhill Crane

Grue du Canada

Grus canadensis

Ghislain Caron

THIS TALL, STATELY BIRD, WITH A WINGSPREAD OF UP TO 2.10 m, IS ONE of the largest birds breeding in Québec. The Sandhill Crane breeds north of the Abitibi and in the Eastmain Lowland, but is rarely encountered farther south. Breeding was not confirmed for the Atlas despite a pre-Atlas breeding record from July 1981 of a family group at Hall Cove, north of Waskaganish, on Rupert Bay.

Sandhill Crane downy running through the tall grasses of Boatswain Bay, northeast of Rupert Bay. This photo is the first proof that the Sandhill Crane breeds in Québec. *August 1981, Yves Aubry.*

With its impressive size and long legs and neck, the Sandhill Crane resembles the Great Blue Heron, which is often wrongly called a crane, but the Sandhill Crane is heavier. The adult plumage is a uniform grey, often tinged with rust, while the young birds are brown. The crane also has dull red skin on its crown and lores, but no breeding plumes (Godfrey 1986).

Cranes fly with their necks outstretched, whereas herons in flight hold their necks tucked in an S-shape. Their wingbeat is slow, with a characteristic rapid upstroke. They are able to fly at high altitudes and soar in wide circles within thermal updrafts (Salt and Wilk 1966). Their loud *garooo-a-a-a*, a shrill, rolling call that carries over long distances, is often heard when the birds are in flight.

Habitat and behaviour

The Sandhill Crane is an omnivorous and opportunistic feeder. In winter, it eats primarily waste grain and roots in cultivated fields. In summer, young grass shoots, berries, insects, small fish, frogs, larvae, and spiders are also part of its usual diet (Walkinshaw 1949a).

The species is monogamous, and pairs remain together for several years, probably until one of the mates dies (Tacha 1988). When they arrive on the breeding grounds, pairs may be accompanied by the young of the previous year, which are driven away when nesting begins (Terres 1980; Tacha 1988). In the Hudson Bay Lowland of Ontario, egg laying probably starts at the beginning of May (Lumsden 1987e), but little is known about the breeding cycle of the species in Québec. By the end of April, cranes are already back around Cabbage Willows Bay on Rupert Bay, and adults with young have been seen there as early as 19 June (Cons. G. G. – GREBE 1992j).

Sandhill Cranes nest on vast open plains with shallow bodies of water, generally quite remote from human activity (Stephen 1990). In Manitoba, they are typically found in peatlands (Melvin et al. 1990), similar to their habitat in Ontario, where they live in peaty wetlands abounding in sphagnum, cattails, sedges, leatherleaf, and sweet gale (Riley 1982; Lumsden 1987e). In Québec, they mainly inhabit peatlands, ponds, streams (Cons. G. G. – GREBE 1992j), and coastal marshes (Gosselin and David 1981; Benoit et al. 1991; Cons. G. G. – GREBE 1992j). Flocks of non-breeding cranes appear to prefer open habitat, such as coastal meadows (Riley 1982) like those at Cabbage Willows Bay (Cons. G. G. – GREBE 1992j).

Cranes are known for their courtship dances. During these displays, they may leap as high as 2 m into the air, with their wings half spread. Most of them bounce like rubber balls, calling loudly, while others bow (Terres 1980; Tacha 1988).

Notes on territory

In winter, Sandhill Cranes live in groups (Tacha 1988), but in the breeding season, pairs defend individual territories, to which they return year after year (Stephen 1990). Breeding pair densities of 1.3–77.8/100 km² have been recorded (Reed 1988; Melvin et al. 1990). Territory size appears to vary considerably from one region to another. There are no data for Québec, where territories appear to be occupied from the beginning of May until the end of August (Cons. G. G. – GREBE 1992j).

At the end of April, the cranes start to return to the vast coastal meadows of Cabbage Willows Bay, on northwestern Rupert Bay. *August 1991, Michel Robert.*

Nests discovered in the Hudson Bay Lowland of Ontario were rudimentary and made of a few sticks and stems. Nests in cattail marshes are much bulkier (Lumsden 1971, 1987e). These are placed within 2 m of the water's edge and are surrounded by vegetation high enough to conceal them. Nests built of aquatic plants or sticks and measuring up to 1.5 m in diameter have been found in other regions (Harrison 1978; Stephen 1990). Sandhill Cranes typically lay 2-egg clutches throughout their breeding range (Salt and Wilk 1966; Miller 1973; Melvin et al. 1990; Dwyer and Tanner 1992).

The pair takes turns incubating, alternating for periods ranging from 30 minutes to nearly 11 hours (Skutch 1976; Reed 1988). Incubation lasts 27–32 days (Lumsden 1971). Both adults also rear the brood (Ehrlich et al. 1988). The downy young are able to follow their parents soon after hatching and can fly by 65–90 days of age (Terres 1980; Ehrlich et al. 1988). Shortly after hatching, the young run nimbly about, and the adults sometimes appear to leave them alone temporarily (Y. Aubry, pers. comm.).

Sandhill Cranes are extremely secretive near the nest, although a call at dawn or dusk will sometimes betray their presence. They are responsive to playback recordings when on their breeding territories. If disturbed, a crane will leave its nest only as a last resort and may show agitated behaviour when flushed (Lumsden 1987e).

Beginning in August, the cranes gather before migrating to their winter quarters (Stephen 1990).

Distribution

There are six subspecies of Sandhill Cranes distributed across Canada, the United States (including Alaska), Cuba, and Siberia (Terres 1980; Tacha et al. 1985). Three of them breed in Canada, including the subspecies *Grus canadensis rowani*, which is found in northern Ontario and Québec (Aldrich 1978; Godfrey 1986).

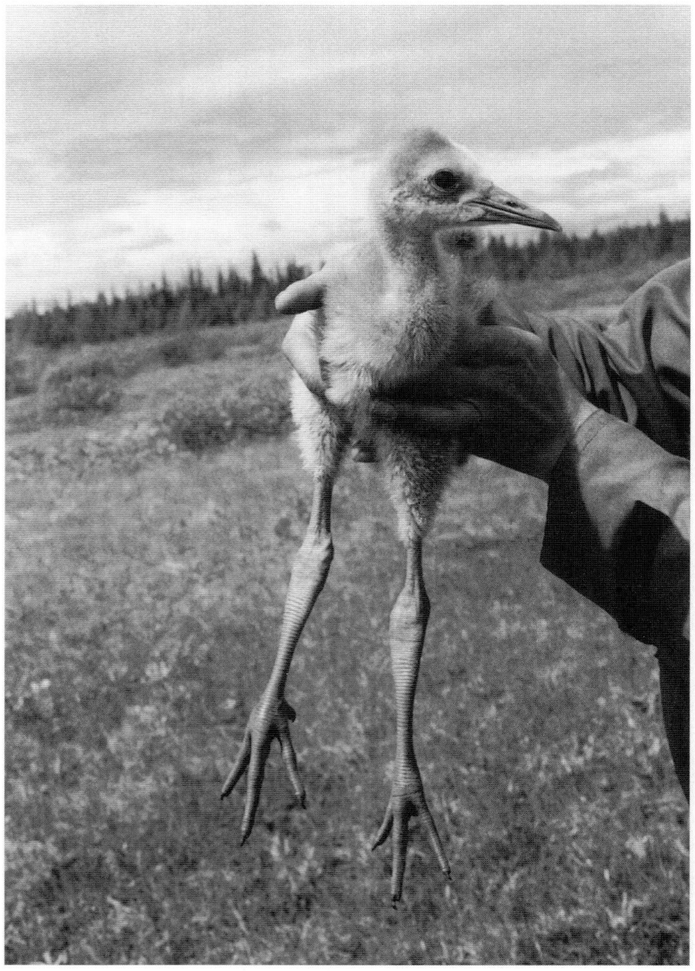

Young crane caught at the Bay of Many Islands. Notice the bird's highly developed legs. Running is key to the young's survival. James Bay. *July 1990, Yves Aubry.*

The Sandhill Crane breeds in the Bay of Many Islands area on northern James Bay. *July 1990, Yves Aubry.*

The species was reported in only one Atlas square, south of Matagami. Nevertheless, it has been known since 1972 to occur around Rupert Bay (Ouellet and Bourget 1975). Sightings over about 10 years indicate that numbers are increasing there, with a single brood recorded in 1981 (Gosselin and David 1981) and 5 others in 1991 (Cons. G. G. – GREBE 1992j). Flocks of up to 30 cranes are frequently seen at Cabbage Willows Bay, on northwestern Rupert Bay (Cons. G. G. – GREBE 1992j). Sandhill Cranes have also been sighted in various places along the James Bay coast between Rupert Bay and the Bay of Many Islands (Ouellet and Bourget 1975; Bordage and Aubry 1982; Benoit et al. 1991; Cons. G. G. – GREBE 1992p); breeding has been confirmed at the Bay of Many Islands, 35 km north of La Grande Rivière (Benoit et al. 1991).

Three broods have been found inland: the first near Lake Grasset, the second near the Missisicabi River, south of Rupert Bay, and the third near Rivière à la Truite, 30 km east of Rupert Bay. Adults have been observed from Lebel-sur-Quévillon north to the Opinaca River and east to Lake Nemiscau (Bordage 1984, 1985b; Cons. G. G. – GREBE 1992j). In July 1993, a bird was heard in a peatland at L'Ascension in the Saguenay–Lac-Saint-Jean region (Club des ornithologues amateurs du Saguenay–Lac-Saint-Jean 1993b). Already considered a

In the Bay of Many Islands, the crane occupies this vast coastal salt marsh edged with shrubbery. James Bay. *July 1990, Yves Aubry.*

regular migrant in the Abitibi region (Aubry et al. 1992), the Sandhill Crane now seems to be a well-established breeder north of this region and in the western Eastmain Lowland (Cons. G. G. – GREBE 1992j), where there are many peatlands and ponds, its preferred breeding habitat. Recently, on 9 July 1994, an agitated adult was seen in a peatland near Launay, in Abitibi; its behaviour suggested that there were young nearby (L. Rochefort, Laval U., pers. comm.).

This member of the family Gruidae has for some years been a regular migrant on the ponds and peatlands in the Abitibi. Saint-Dominique-du-Rosaire. *August 1991, André-Guy Bernier.*

History and trends

According to Godfrey (1986), the Sandhill Crane was extirpated from almost all its former breeding range in the inhabited areas of Canada. Despite this, the crane populations in British Columbia, the Prairies, and northern Ontario seem to have been increasing since the 1940s.

The Sandhill Crane was apparently very common in northern Ontario in the 17th century, but was probably extirpated in the 19th century (Lumsden 1971, 1987e). It was rarely observed in aerial surveys of aquatic birds on the west side of James Bay and Hudson Bay between 1950 and 1970 (Lumsden 1987e), but a sudden increase was subsequently noted. Today it is found throughout most of Ontario north of the 51st parallel (Peck and James 1983; Lumsden 1987e).

There were only a few records from Québec prior to 1980 (David 1980a). The first sight record from the Québec side of James Bay dates from 1972 (Ouellet and Bourget 1975), but the first breeding record in the province was not obtained until 1981, at Hall Cove on Rupert Bay (Y. Aubry, pers. comm.). Recent observations in the Eastmain Lowland indicate that the Sandhill Crane's Québec range is expanding (Cons. G. G. – GREBE 1992j).

Vincent Létourneau and Alain Morrier

World breeding range

Species profile

Clutch size	usually 2 eggs (1–3)[2,3,5,6,12]
Laying interval	usually 1 egg per 2 days[12]
Start of incubation	with first egg[12]
Incubation period	usually 30 days (27–32)[2,3,12,13]
Maturity at hatching	precocial
Care of young	by female and male[2,3,12]
Nestling period	≈ 24 hours [8,12]
Age at first flight	65–90 days[3,12]
Dependence of young	until 9–10 months old[12]
Broods per year	1[2,12]
Breeding age	2–7 years[12]
Mating system	monogamy[1,2,12]
Duration of pair bond	probably for life[1,12]
Total length	86–122 cm[10]
Wing span	183–213 cm[3]
Weight	5089 g (4426–5788)[4]
Longevity record	18 years[7]

References:
1. Tacha 1988
2. Ehrlich et al. 1988
3. Terres 1980
4. Dunning 1984
5. Miller 1973
6. Melvin et al. 1990
7. Clapp et al. 1982
8. Reed 1988
9. Stephen 1990
10. Godfrey 1986
11. Harrison 1978
12. Tacha et al. 1992
13. Lumsden 1971

Chronology *n* = 13

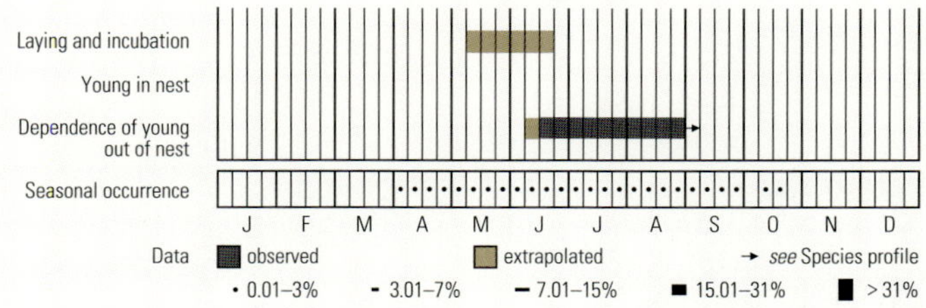

Lesser Yellowlegs

Petit Chevalier

Tringa flavipes

Clodin Roy

The Lesser Yellowlegs is one of the most common medium-sized shorebirds in North America. In Canada, it breeds from Yukon to Lake Bienville in north-central Québec (Johnsgard 1981; Godfrey 1986; Gosselin et al. 1991b). It is a rather scarce and unfamiliar bird in the province, although it can be common in suitable breeding habitat. Like its close relative, the Greater Yellowlegs, it is very noisy at the approach of intruder (Bent 1927).

In southern Québec, the Lesser Yellowlegs is an uncommon transient in spring, but a common transient in fall (Fortin et al. 1978; David 1980a), when the extensive St. Lawrence River system, especially the gulf, becomes an important staging area (Maisonneuve et al. 1990). It winters along the coast of the southern United States, through Mexico to Chile and Argentina (Godfrey 1986), as well as inland to Paraguay (Hayes and Fox 1991).

Habitat and behaviour

Lesser Yellowlegs nest in the subarctic muskeg and boreal forest, preferring woodland clearings and burnt-over areas of forest with nearby ponds, lakes, and wetlands (Johnsgard 1981). The nest is a depression in the ground lined with leaves and grass on a slope or ridge, usually among charred tree trunks or similar burnt debris, fallen logs, or at the base of a small tree, stump, or shrub (Bent 1927; Johnsgard 1981). Lesser Yellowlegs are thought to be monogamous (Johnsgard 1981; Ehrlich et al. 1988). Typical clutch size is 4 eggs, and both parents incubate, probably for 22–23 days (Johnsgard 1981; Godfrey 1986). The young are precocial, with the whole brood leaving the nest for the water soon after hatching (Johnsgard 1981).

Lesser Yellowlegs actively search for food while wading in shallow water, pecking their prey from the water's surface or mud (Michaud and Ferron 1986). On their wintering grounds, they also forage at night, possibly to take advantage of the increased availability of prey, which they find by employing a tactile side-sweeping technique (Robert and McNeil 1989). They feed predominantly on terrestrial and aquatic insects and their larvae, crustaceans, and worms, as well as a few small fish and gastropods (Bent 1927; Robert and McNeil 1989).

Distribution

Lesser Yellowlegs breed in the boreal forest of North America, ranging from Alaska and British Columbia in the west to the east coast of James Bay in the east. Both Todd (1963) and Godfrey (1986) support the observations of Manning and MacPherson (1952) that the species nests along the east coast of James Bay. In 1950, Manning and MacPherson (1952) observed 3 pairs in a small marsh near Moar Bay on their visits between 27 June and 10 July. Although they found no nests, they deduced from the birds' behaviour that Lesser Yellowlegs were nesting in the vicinity. Furthermore, on a canoe trip on the Roggan River (north of Fort George) between 12 and 14 July of that same year, they observed 3 Lesser Yellowlegs, one of which was a juvenile female that they collected.

A report of adults and young in a palsa bog near Kuujjuarapik in July 1982 (Rochefort 1983) is probably the first breeding record of Lesser Yellowlegs in Québec. The only confirmed nest of this species in the province was found on 13 June 1991, near Lake Bienville (Gosselin et al. 1991b; Cons. G. G. – GREBE 1993c), quite far inland at the head of the Great Whale River, which empties into Hudson Bay. A single Lesser Yellowlegs was observed at Lake Bienville the previous summer, and others were observed that same summer on the Great Whale River, on the Mollet Lakes, and at the La Grande-1 Dam (Yank et al. 1990b).

No evidence of breeding or possible breeding of Lesser Yellowlegs was found in the Atlas area. Birds were seen throughout the summer along the St. Lawrence, but these were identified as transients.

Notes on territory

Lesser Yellowlegs are territorial in the breeding season, and soon after their arrival on the breeding grounds, they disperse over the available habitat (Johnsgard 1981). Information regarding the size of the area defended around the nest is lacking, but Randall (1961) maintained that 2 nests are rarely closer than 800 m apart, except in areas where nesting sites are limited, such as on the outskirts of Edmonton, where 3 or 4 pairs may nest in an area of 40 ha or less. Myers and Myers (1979) reported that on their wintering grounds, many birds defend feeding territories of 0.1–0.5 ha, depending on the size of the wetland, and that on the edges of marshes, defended areas may be up to 100 m long.

Lesser Yellowlegs like large ponds in extensive peatlands. Only a few records exist of this species breeding in Québec. Northwest of Lake Bienville, north-central Québec. *June 1991, Serge Brodeur.*

This Lesser Yellowlegs nest in the grass beside a rock near Lake Bienville is the only one found to date in the province. North-central Québec. *June 1991, Renée Lemieux.*

The Lesser Yellowlegs is particularly abundant along the shores of the St. Lawrence River and in the Magdalen Islands during fall migration (Mousseau et al. 1976; Brousseau 1981; Maisonneuve 1982; Bourget 1990; Maisonneuve et al. 1990). Adults migrate mainly from the end of July into August whereas juveniles move south from the end of August to mid-September (Mousseau et al. 1976).

History and trends

The available data suggest that the current and possibly historical breeding range of the Lesser Yellowlegs may extend well into the boreal forest of central Québec.

Current population trends are difficult to evaluate because of the lack of information on the historical and current size of breeding populations, but by the early 20th century, they had undoubtedly already been decimated by hunters (Bent 1927). Fortunately for the species, hunting was banned in 1927 under the Migratory Birds Convention Act (Migratory Birds Regulations 1927).

RICHARD C. COTTER AND DERIN HENDERSON

World breeding range

Species profile

Clutch size	usually 4 eggs (3–5)[1,3,4]
Laying interval	?
Start of incubation	with last egg[1]
Incubation period	probably 22–23 days[1,4]
Maturity at hatching	precocial
Care of young	by female and male[1,3,4]
Nestling period	less than 24 hours[1,3,4]
Age at first flight	probably ≈ 23–25 days[1]
Dependence of young	?
Broods per year	probably 1[2]
Breeding age	?
Mating system	probably monogamy[1,2]
Duration of pair bond	?
Total length	23–28 cm[5]
Wing span	49.5–54.6 cm[6]
Weight	81 g (69–94)[7]
Longevity record	5 years 11 months[8]

References:
1. Johnsgard 1981
2. Ehrlich et al. 1988
3. Bent 1927
4. Randall 1961
5. Godfrey 1986
6. Terres 1980
7. Dunning 1984
8. Kennard 1975

Chronology *n* = 2

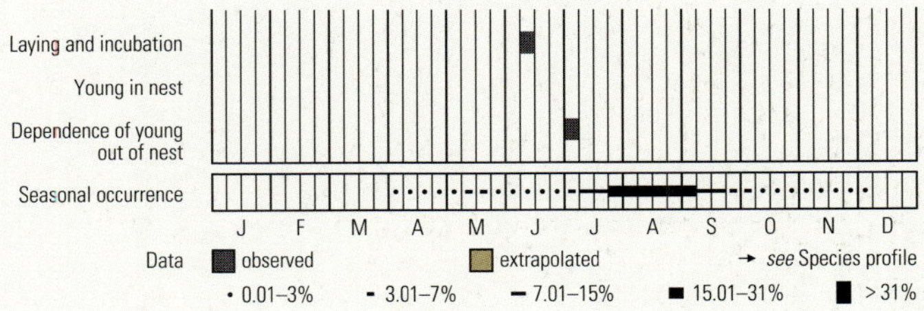

Willet

Chevalier semipalmé

Catoptrophorus semipalmatus

Clodin Roy

The Willet is a large shorebird named after its call, a musical *pill-will-willet*. The black-and-white wing pattern visible in flight and its distinctive call make the Willet easy to identify (Hayman et al. 1986).

In Québec, the Willet is classed as a very rare visitor (David 1980a). Its breeding status is unclear. The only probable breeding record is an Atlas report of a pair exhibiting typical courtship behaviour from the Magdalen Islands.

Habitat and behaviour

There are two subspecies of Willets. The eastern race, which may breed in the Magdalen Islands, nests in salt marshes and beaches along the Atlantic coast (Johnsgard 1981; Godfrey 1986). Salt marshes dominated by salt meadow cord-grass, particularly those that are burned or cropped periodically—by wild geese, for example—are preferred for nesting (Johnsgard 1981). Atypical of this species is a small colony found recently in Corea, Maine, nesting in a 30–50 ha sphagnum bog characterized by hummocks covered in lichen and sedges (Wells and Vickery 1990).

The Willet is a monogamous, semicolonial shorebird (Johnsgard 1981; Ehrlich et al. 1988). The nest is a simple depression in the ground lined with grass, and may be well concealed by vegetation, or completely exposed on open sand (Johnsgard 1981; Ehrlich et al. 1988). Typical clutch size is 4 eggs; the female incubates during the day and the male at night (Ehrlich et al. 1988). The precocial chicks hatch after an incubation period of 21–29 days, and a few hours later, the brood is led by both parents to the rearing site—a large shallow pond in a marshy area, for example (Howe 1982). Initially, both parents brood the young and defend them from predators such as raccoons and foxes, but the female abandons the brood 2–4 weeks after hatching (Hansen 1979) and the male abandons them about 2 weeks later (Howe 1982; Ehrlich et al. 1988).

Willets stalk fiddler crabs, mole crabs, shore crabs, and small fish by probing and pecking at the water's surface. They also forage by "mowing," that is, walking through the mud or shallow water, rapidly opening and closing the bill while moving the head up and down, preying on small molluscs, fish fry, marine worms, and aquatic insects (Johnsgard 1981).

Distribution

The eastern race, *Catoptrophorus semipalmatus semipalmatus*, nests in Canada along coastal areas of Prince Edward Island, Newfoundland, and Nova Scotia, including Cape Breton Island. It is considered a probable breeder in the Magdalen Islands in Québec (Fradette 1992). This race also nests along the Atlantic coast of the United States from Maine to Florida, along the coast of Texas, and in the West Indies (Terres 1980; Godfrey 1986). The western subspecies, *Catoptrophorus semipalmatus inornatus*, nests on the Canadian Prairies and in adjacent states (Godfrey 1986).

Willets were observed in only one Atlas square, at Baie du Havre aux Basques, Magdalen Islands. On 17 June 1989, a pair was seen performing courtship displays, and one bird was calling, apparently to its mate. Also, in the last few years, Willets have been regularly observed in summer at Grande-Entrée (Yank et al. 1987b), as well as near Rimouski (Yank et al. 1985) and at La Pocatière (Yank et al. 1990b) in the estuary. It is difficult to assess whether these observations indicate a possible range extension, or whether the birds seen were non-breeders.

Notes on territory

This aggressive territorial species defends multipurpose territories throughout the breeding season. Beginning in the pre-nesting period, the male defends the area around the female from other males (Tomkins 1965). In coastal populations, shore territories used for feeding and rearing the young are defended vigorously by both members of the pair, but more so by the male (Tomkins 1965; Howe 1982). The much smaller nesting territories, whether adjacent to or separate from the shore territories, are defended by both adults during the pre-laying, laying, and incubation periods (Tomkins 1965; Hansen 1979; Howe 1982).

Inter-nest distance varies. In South Carolina and Georgia, most nests were at least 60 m apart (Tomkins 1965). In Nova Scotia, nests were 75–265 m apart, with a density of 7.3 nests/10 ha (Hansen 1979). On Wallops Island, Virginia, the mean distance between nests was 21.1 m for marsh sites (density 24.8 nests/ha) and 45.1 m on natural sand dunes (density 4.8 nests/ha) (Howe 1982).

The Willet inhabits the salt marshes and beaches of the Atlantic coast. Courtship displays have been observed in the marshes of the Magdalen Islands, where the species is a probable breeder. Grosse-Île. *June 1991, Desneige Perreault.*

In Nova Scotia, the mean area defended was largest while young were on the marsh (1.22 ha for females and 1.31 ha for males) and smallest during incubation (0.19 ha for females and 0.58 ha for males) (Hansen 1979). On the Great Plains in North Dakota, Willet densities are much lower than on the Atlantic coast, and adults defend large feeding and nesting territories averaging 44.3 ha (Ryan and Renken 1987).

Pair of Willets feeding in a lagoon at Havre-aux-Maisons on 5 June 1989. On 17 June, a pair, possibly the same one, was seen engaged in courtship activities in Baie du Havre aux Basques, about 10 km away. Magdalen Islands.
June 1989, Normand David.

History and trends

Populations of both subspecies currently appear quite healthy (Johnsgard 1981). In fact, the eastern population seems to be expanding its breeding range northward after being severely depleted in the late 19th and early 20th centuries by market hunters and egg collectors (Terres 1980; Ehrlich et al. 1988). During that period the Willet almost disappeared along the Atlantic coast north of Virginia, but with protection and unmolested nesting, these populations have rebounded (Terres 1980).

Richard C. Cotter and Derin Henderson

World breeding range

Species profile

Clutch size	4 eggs, rarely 5[2,4,5,6]
Laying interval	1 egg per 1–4 days[3,5,6]
Start of incubation	with last egg[2,6]
Incubation period	21–29 days[2,3,5,7]
Maturity at hatching	precocial
Care of young	by female and male[2,4]
Nestling period	a few hours[2,5,6]
Age at first flight	28–29 days[2,6]
Dependence of young	until 4–6 weeks old[2,6,7]
Broods per year	1[2,5,6]
Breeding age	2 years[3,4]
Mating system	monogamy[2,6,7]
Duration of pair bond	usually for life[2]
Total length	35.5–43.0 cm[1]
Wing span	61.0–78.7 cm[9]
Weight	215 g (?)[8]
Longevity record	?

References:
1. Godfrey 1986
2. Howe 1982
3. Johnsgard 1981
4. Palmer 1967
5. Tomkins 1965
6. Hansen 1979
7. Ehrlich et al. 1988
8. Dunning 1984
9. Terres 1980

Chronology $n = 0$

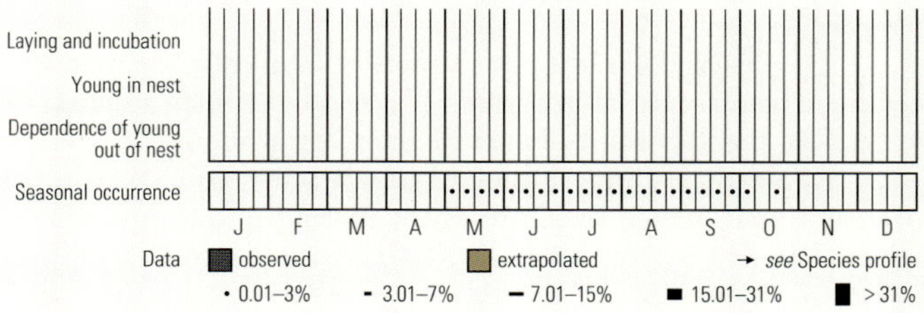

Marbled Godwit

Barge marbrée

Limosa fedoa

Christiane Girard

UNTIL LATELY, THIS LARGE PRAIRIE SHOREBIRD WAS CONSIDERED A CASUAL summer visitor in Québec (David 1980a). Recent sightings indicate, however, that it probably nests around Rupert Bay.

The Marbled Godwit is identifiable by its size and its long, straight or slightly upturned bill. It is a long-legged bird with rich buffy brown plumage. Its flight is direct and swift; it alights in an elegant manner, momentarily holding its wings upright over its back as its feet touch the ground, revealing its cinnamon-coloured wing linings.

The broad plains of Cabbage Willows Bay are probably the only breeding grounds for this member of the family Scolopacidae in Québec. Records of various types of breeding behaviour in this area suggest that the species nests here. Northwestern Rupert Bay. *August 1991, Michel Robert.*

Habitat and behaviour

With its bill, which may be as long as 14 cm, it probes the shallow waters of pools and shorelines for insects, molluscs, and other aquatic invertebrates. It is also found on low-lying prairies, where it feeds on grasshoppers and many other kinds of insects (Bent 1927).

The Marbled Godwit has a variety of loud calls, the most characteristic being *kerreck, kreck, kreck, kerreck*, and *eradica radica, radica, radica*, heard continually on the breeding grounds (Bent 1927).

In North Dakota, godwits arrive from migration in flocks (Johnsgard 1981), which break up as the males begin to fly over their selected territories on nearby fields. They fly up to an altitude of 20–90 m, then slowly circle downward, vocalizing all the while. Then they return to the flock and pursue the females, on the ground and in the air, sometimes even pecking them, urging them to select a breeding territory. During pursuit flights, the birds fly almost side by side, with shallow wingbeats and tails fanned.

On the Prairies, the Marbled Godwit's breeding habitat is characterized by sloughs and a variety of ponds and lakes. It avoids tall, dense grasslands, being found most often in grass 15–60 cm high (Ryan et al. 1984; Morrison 1987). It prefers a habitat peculiar to the lower courses of some streams—broad, alluvial, permanently damp plains that are flat, low, and regularly flooded during periods of high water; they are covered with short, thick grass only a few centimetres high (Bent 1927). Several pairs may nest in the same area (Harrison 1978).

Clutch size is 4 eggs, laid in the open on a dry hummock, sometimes far from water (Ryan et al. 1984; Colwell and Oring 1990). The nest is a shallow depression lined with a few strands of dead vegetation. In Saskatchewan, nests are located in tall, dense vegetation (taller than surrounding vegetation) dominated by upland grasses (Colwell and Oring 1990). Egg laying there occurs from mid-May to the end of June, but mostly between 30 May and 9 June (Bent 1927). The dates of the few records from the west side of James Bay, Ontario, are similar (Morrison et al. 1976; Peck and James 1983). The Ontario data and sightings at Rupert Bay, Québec, in 1991 suggest that egg laying could occur in early to mid-June around Rupert Bay (Cons. G. G. – GREBE 1992g).

The chicks, on hatching, already have the characteristic silhouette of the species. They are covered with dull down that enables them to blend in with their surroundings. Both parents tend the young (Ehrlich et al. 1988). They soon acquire their juvenal plumage, similar to the winter plumage of the adult, and are able to fly by the end of July (Bent 1927).

Marbled Godwits gather in small flocks at the end of the breeding season before the start of migration (Wishart and Sealy 1980). On the west coast of James Bay, the adults begin to migrate south by the end of July, the young in mid-August (Morrison et al. 1976).

Distribution

There are three disjunct populations of Marbled Godwits in North America (Gibson and Kessel 1989). The largest is found in the southern part of the Canadian Prairies and in Montana, North and South Dakota, and Minnesota. The other two are small, isolated populations, one in southern Alaska and the other on the west coast of James Bay, where the Marbled Godwit has been confirmed breeding at North Point and on southern Akimiski Island, and may possibly be breeding along the southern coast of Hudson Bay (Morrison 1987).

Recent sightings indicate that the western James Bay population may extend into Québec and that Marbled Godwits may possibly be breeding at various inland sites north of the Missisicabi River and along western Rupert Bay; the species is considered a probable breeder at Cabbage Willows Bay (Cons. G. G. – GREBE 1992g). A bird was sighted on 26 June 1990 a few kilometres east of the Ontario border, just north of the Missisicabi River. Godwits have also been heard in 3 places, 15 km inland, west of the mouth of the Nottaway River.

The most interesting records, however, come from Cabbage Willows Bay, on northwestern Rupert Bay (Cons. G. G. – GREBE 1992g). Single birds, pairs, and flocks were sighted in the area between May and mid-August in 1990 and 1991. One bird was observed on 9 June 1990 harassing a Golden Eagle: this was possibly territorial behaviour. Several isolated pairs were also seen. As many as 80 Marbled Godwits were observed in mid-June 1991. These godwits displayed noisy, active behaviour similar to that of the flocks of breeding birds arriving from migration in North Dakota (Johnsgard 1981). Two birds were also seen copulating (P. Bannon, D. Daigneault, V. Létourneau, pers. comm.). On 21 July 1990, two observers were dive-bombed for nearly an hour when they flushed a very noisy, aggressive pair (A. Morrier and M. Robert, pers. obs.). It is thought that these birds were defending a brood.

Notes on territory

In the heart of its range, the Prairie provinces and the northern midwestern states, the Marbled Godwit often breeds in loose colonies. Territorial borders are ill defined, and nests are sometimes quite far apart (Johnsgard 1981; Hayman et al. 1986). In North Dakota, Ryan et al. (1984) recorded densities of 2–7 breeding pairs/10 km². Also in North Dakota, nests have been found as close as 60 m, and as many as 4 nests within a 0.15 km² (15 ha) area (Johnsgard 1981). Parents and young leave the nest together soon after hatching, but the distance they then travel is not known.

In 1990 and 1991, several Marbled Godwits were observed in the breeding season on the alluvial plain of Cabbage Willows Bay. Northwestern Rupert Bay. *June 1991, Vincent Létourneau.*

The Marbled Godwit uses its long bill to probe the shallow waters of wet fields for insects and aquatic invertebrates. Near Beaverhill Lake, Alberta. *June 1970, Pierre Lamothe.*

Cabbage Willows Bay is an alluvial plain a few kilometres long and several hundred metres wide at the mouth of the Novide River. This habitat seems to be ideally suited to the breeding requirements of the Marbled Godwit.

History and trends

The Marbled Godwit was first described and named in 1758, based on a specimen taken from the Hudson Bay region (Hall 1960). At the beginning of the 20th century, Bent (1927) considered that its numbers were declining sharply.

The Marbled Godwit, being a large and fairly tame bird, was easy game, especially because of its curious habit of circling and recircling over other godwits that had been shot. Overhunting was probably the reason for the loss of a large population that migrated along the Atlantic coast, from New England southward (Bent 1927).

The main cause of its decline, however, was and will always be habitat loss due to the encroachment of agriculture on the western plains (Bent 1927; Ryan et al. 1984).

The first sight records of Marbled Godwits at James Bay date back to the middle of the 18th century (Manning 1952). Although the species had been regularly sighted at several locations since then (Manning 1952; Todd 1963), breeding was not confirmed in the region until 1975 (Morrison et al. 1976). Further breeding evidence was later obtained, but always confined to Ontario (Peck and James 1983; Morrison 1987). This population probably numbers at least a few hundred birds. Hagar, for example, observed a flock of 300–400 godwits on the Ontario side of James Bay in 1966 (Morrison et al. 1976).

Present data do not enable us to confirm whether this population extends into Québec or even whether the bird ever nested here. The few summer visits by ornithologists to Cabbage Willows Bay have not produced conclusive evidence (Bourget 1973b; Ouellet and Bourget 1975).

Rupert Bay appears to be a major migratory stopover in Québec, and, in all likelihood, is a breeding site of the Marbled Godwit. Any activity that might affect this vast ecosystem could have repercussions on the ecology of the species.

Vincent Létourneau and Alain Morrier

World breeding range

Species profile

Clutch size	usually 4 eggs (3–5)[1,2,7]
Laying interval	probably 1 egg per 1–3 days[2]
Start of incubation	probably before last egg[2]
Incubation period	probably 21–23 days[1,2]
Maturity at hatching	precocial
Care of young	by female and male[1,2]
Nestling period	probably less than 24 hours
Age at first flight	probably 21–22 days[1,2]
Dependence of young	?
Broods per year	?
Breeding age	?
Mating system	probably monogamy[1]
Duration of pair bond	?
Total length	42.5–50.5 cm[3]
Wing span	81.3 cm[5]
Weight	371 g (277–420)[6]
Longevity record	8 years 7 months[4]

References:
1. Ehrlich et al. 1988
2. Johnsgard 1981
3. Godfrey 1986
4. Clapp et al. 1982
5. Terres 1980
6. Dunning 1984
7. Bent 1927

Chronology $n = 0$

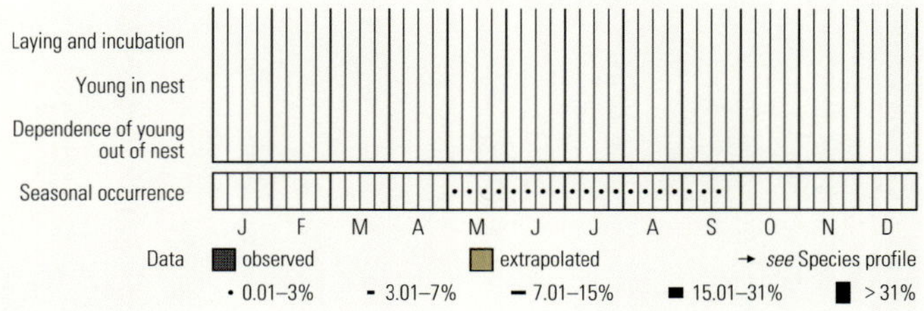

Semipalmated Sandpiper

Bécasseau semipalmé

Calidris pusilla

Christiane Girard

THE SEMIPALMATED SANDPIPER IS THE MOST COMMON SMALL SHOREBIRD breeding in eastern North America. It nests on the arctic tundra in northern Québec, and is a common fall transient in southern Québec (Godfrey 1986). This sparrow-sized sandpiper is one of the smallest shorebirds and is part of a group of other small *Calidris* species commonly referred to as peeps. It is the only peep nesting in Québec that has partially webbed toes (Godfrey 1986).

Habitat and behaviour

These small shorebirds favour coastal and low inland tundra for nesting (Johnsgard 1981; Godfrey 1986). They particularly like the well-vegetated shorelines of rivers, ponds, and lakes, as well as sand dunes with scattered small pools (Johnsgard 1981; Godfrey 1986). Although they are most abundant on moist or wet sedge-grass meadows near marshy tundra ponds, they sometimes breed on drier and more exposed peaty areas (Johnsgard 1981).

Semipalmated Sandpipers are monogamous. Until the start of incubation, males perform elaborate courtship displays, flying 5–10 m above the tundra and then hovering in midair, and all the while giving out a complex song (Johnsgard 1981; Miller 1983b). The nest is a depression in a grassy hummock or knoll, usually close to water on wet tundra or in sandy areas (Johnsgard 1981; Godfrey 1986). Typical clutch size is 4 eggs, with both sexes sharing incubation duties (Ashkenazie and Safriel 1979). The precocial chicks hatch after 20 days, and within a few hours are foraging for themselves on midge larvae and the adults of other flies (Holmes and Pitelka 1968). The female usually departs within 11 days of hatching, leaving brood rearing to the male. The male then deserts the brood shortly before or after fledging, which is at approximately 16–19 days of age (Gratto-Trevor 1991, 1992).

Adults feed by pecking and probing in the soft mud along shorelines or lake and marsh edges, searching continuously for benthic invertebrates, such as small arthropods, molluscs, and annelids, as well as some terrestrial invertebrates, such as insects and spiders (Gratto-Trevor 1992). During fall migration near L'Isle-Verte, on the south shore of the St. Lawrence estuary, Semipalmated Sandpipers feed largely on clam worms, a small polychaete found in abundance there (Michaud and Ferron 1986).

Distribution

The Semipalmated Sandpiper breeds only in North America, principally in the tundra regions of northern Alaska, the central Canadian Arctic, southern Baffin Island, and along coastal areas of Hudson Bay, northern Québec, and Labrador (Godfrey 1986). A nest found on an island near the coast of the Bay of Many Islands provided evidence of breeding along James Bay. South of 51°30' N, there is a single breeding record of this species from the Magdalen Islands: on 27 June 1907, a well-known naturalist, Mr. Philipp, observed a pair of adults with a nest containing 4 eggs (Fradette 1992).

No evidence of breeding or even of possible breeding was found for the Atlas in southern Québec. Birds were seen, but because of the time of year (late summer), they were considered to be in migration. They are common fall and spring transients throughout southern Québec, particularly along the estuary and Gulf of St. Lawrence (Cramp and Simmons 1983; Bourget 1990). Semipalmated Sandpipers usually migrate in enormous flocks, often thousands-strong. Particularly large numbers stage in coastal areas from New Brunswick to Maryland. Several hundred thousand regularly stop over at Mary's Point, on the upper Bay of Fundy, New Brunswick (Morrison 1984), where they feed on beach fleas at low tide, both day and night (Hicklin and Smith 1984; Manseau and Ferron 1991). Though not so spectacular, flocks of more than 10 000 can be seen regularly in the Magdalen Islands (Mousseau et al. 1976) and even near Québec City at Montmagny Bay (Bourget 1990). The St. Lawrence River system,

The Semipalmated Sandpiper nests and feeds in grassy and pebbly areas along the shores of James Bay. Bay of Many Islands. *June 1990, Yves Aubry.*

Notes on territory

In the breeding season, this territorial species defends a nesting territory, primarily against conspecifics (Gratto-Trevor 1992), and forages in undefended areas up to 2–3 km away (Ashkenazie and Safriel 1979).

The males arrive on the breeding grounds a few days before the females, most often returning to the same area as the previous year, and begin almost immediately to defend their territories through flight displays and chases. While intraspecific aggression is more common, they may also show aggression toward morphologically similar species like the Least Sandpiper (Gratto-Trevor 1992).

Females usually pair with their previous mates if they return to the same territory (Gratto et al. 1985), and both sexes defend the territory during incubation. Although there are no published data on home-range size, the mean size of nesting territories at La Pérouse Bay, near Churchill, Manitoba, was 1.0 ha (Gratto et al. 1985).

Nesting appears to be patchy, even in suitable habitat (Norton 1973). Breeding densities range from 1.1 pairs/10 ha in Barrow, Alaska, to 10 pairs/10 ha at La Pérouse Bay, Manitoba (Gratto and Cooke 1987).

This sandpiper nests in short grass. This nest was photographed on an island near the shore in the Bay of Many Islands. James Bay. *June 1990, Yves Aubry.*

especially between Québec City and Matane, is a major staging area for fall-migrating shorebirds, of which the most numerous is the Semipalmated Sandpiper (Maisonneuve et al. 1990).

The Semipalmated Sandpiper winters on both the Atlantic and Pacific coasts of South America, with the largest concentrations occurring in Surinam and French Guiana (Morrison 1984).

History and trends

It is difficult to estimate the size of the breeding population of the Semipalmated Sandpiper. It is, however, considered the most abundant small shorebird along the Atlantic coast during fall migration (Harrington and Morrison 1979). This was not always the case, for between 1890 and 1916, large numbers of Semipalmated Sandpipers, along with other shorebirds, were shot by market hunters. Populations recovered after hunting was banned in both the United States and Canada (Townsend 1927b) under the Migratory Birds Convention of 1916.

Richard C. Cotter

World breeding range

Species profile

Clutch size	4 eggs, rarely 3[2,3,4]
Laying interval	usually 1 egg per day[2,4]
Start of incubation	a few hours before last egg is laid, sporadic before[4]
Incubation period	usually 20 days (18–22)[2,3,4]
Maturity at hatching	precocial
Care of young	mainly by male[2,4]
Nestling period	a few hours[4]
Age at first flight	usually 16–19 days[4,7]
Dependence of young	usually until 14 days old (9–20)[4,7]
Broods per year	1[4]
Breeding age	M 1–4 years / F usually 2 years (1–3)[4]
Mating system	monogamy[2,4]
Duration of pair bond	sometimes more than 1 season[4]
Total length	14–17 cm[3]
Wing span	27.9–33.0 cm[5]
Weight	28.1 g (20.0–41.0)[6]
Longevity record	12 years 2 months[1]

References:
1. Klimkiewicz and Futcher 1989
2. Johnsgard 1981
3. Godfrey 1986
4. Gratto-Trevor 1992
5. Terres 1980
6. Dunning 1984
7. Gratto-Trevor 1991

Chronology *n* = 11

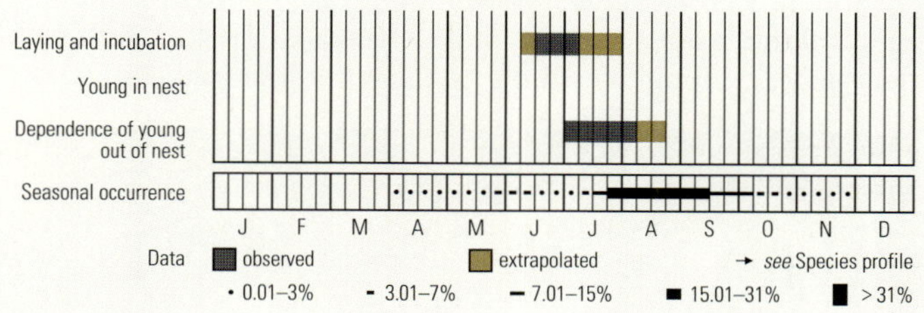

Red-necked Phalarope

Phalarope hyperboréen

Phalarope à bec étroit
Phalaropus lobatus

PHALAROPES ARE SMALL SANDPIPERS, DIFFERENT FROM OTHER SHOREBIRDS by their lobed toes and distinctive swimming habit; they often feed while spinning on the water, picking up food items as they go (Godfrey 1986). Like a typical phalarope, this species is one of the few birds to exhibit reversed sex roles and sexual dimorphism, meaning the larger and more brightly coloured females compete for males, and the males provide all or most parental care (Reynolds 1987).

The Red-necked Phalarope, like the Red Phalarope, which also nests in northern Québec, overwinters on the ocean, and like other truly pelagic birds, has salt glands (Terres 1980). These paired, enlarged nasal glands are located at the base of the bill and regulate the salt content of the blood. After a drink of sea water, these glands remove the excess salt from the water and excrete the concentrated salt solution (Faaborg 1988).

Habitat and behaviour

The Red-necked Phalarope is a tundra-nesting species. It nests in low-lying wet areas, including marshes with small ponds and islands in river deltas (Hussell 1987b). The nest is a shallow cup lined with grass and leaves, built solely by the male (Terres 1980; Cramp and Simmons 1983). Nests are usually placed close to water in tussocks or sunk in moss, and often sheltered by low vegetation (Hildén and Vuolanto 1972; Johnsgard 1981; Cramp and Simmons 1983; Reynolds 1987). The species may also nest on small coastal islands with little vegetation, where it feeds in pools of water, among the rocks in small sheltered bays or off shore (Y. Aubry, pers. comm.).

Red-necked Phalaropes are non-territorial, and are either monogamous or, less commonly, serially polyandrous. Typically a single clutch of 4 eggs is laid, but the female may renest if the first clutch is lost or if a second male is available (Hildén and Vuolanto 1972; Reynolds 1987). The incubation period averages 18–19 days, and only the male incubates.

Before the eggs hatch, the adults are easy to observe as they feed on small ponds, and the male greets intruders with a mild but agitated *prek* call (Cramp and Simmons 1983). The precocial young can swim immediately after hatching (Bent 1927). Their diet consists of aquatic insects, including larvae, as well as crustaceans, molluscs, zooplankton, a few small fish, and some seeds (Wetmore 1925), which they gather by spinning rapidly on the water (Johnsgard 1981).

Distribution

The Red-necked Phalarope has a circumpolar distribution, nesting throughout the Low Arctic and Subarctic (Cramp and Simmons 1983). It breeds across northern Québec, including the Lake Bienville area (Cons. G. G. – GREBE 1993c), islands in Ungava Bay, Lake Bérard, the Schefferville area, and on the east coast of Hudson and James bays and their offshore islands south to Rupert Bay (Todd 1963; Godfrey 1986). In 1990, nests and birds exhibiting breeding behaviour were found at Bay of Many Islands (Benoit et al. 1991). This species winters mainly at sea off the west coast of South America, in the eastern Pacific, and off the west coast of Africa (Johnsgard 1981).

In spring and autumn, these phalaropes migrate principally along either the Pacific or Atlantic coast. Huge aggregations of 2–3 million birds have been observed in late August along the coast of Maine up to lower Passamaquoddy Bay of New Brunswick and Maine (Finch et al. 1978). The Red-necked Phalarope is a rare transient in the interior of southern Québec (Godfrey 1986). It is listed as a rare transient in the Montréal area (Bannon 1991a) and throughout the St. Lawrence Valley and Plain, where it is mainly seen during fall migration (Fortin et al. 1978; David 1980b).

No evidence was found for the Atlas that the Red-necked Phalarope breeds in southern Québec. Birds were sighted, but were identified as fall transients.

History and trends

Little is known of the historical population trends or breeding distribution of the Red-necked Phalarope due to the remoteness of both its breeding and wintering areas. Because this species has such a vast breeding range that extends well beyond the range of most human activity, and because its wintering areas are primarily oceanic, it probably has a very large total population (Johnsgard 1981).

RICHARD C. COTTER

Clodin Roy

Notes on territory

Adults of both sexes have a strong tendency to return to the same breeding site year after year, although the return rate to the natal nesting area is higher among males than females (Schamel and Tracy 1991). Neither sex defends a territory during the breeding season, and foraging areas may overlap with those of other birds (Reynolds 1987).

Although pairs may nest solitarily, far from local feeding pools, some form loose colonies (Cramp and Simmons 1983), with inter-nest distances as little as 2 m (Hildén and Vuolanto 1972). A study at Cape Espenberg, Alaska, estimated nesting densities at 8.8–24 nests/10 ha (Schamel and Tracy 1991). The highest nesting density reported was on Victoria Island, Northwest Territories, with an estimated 66–88 nests/10 ha (Johnsgard 1981).

The nest is a depression lined with dry grass. The male incubates the 4 eggs alone. Bay of Many Islands, James Bay. *June 1990, Yves Aubry.*

The Red-necked Phalarope breeds across the tundra of northern Québec. Here a male has built its nest in a clump of grass among the boulders on an island in Bay of Many Islands.
June 1990, Yves Aubry.

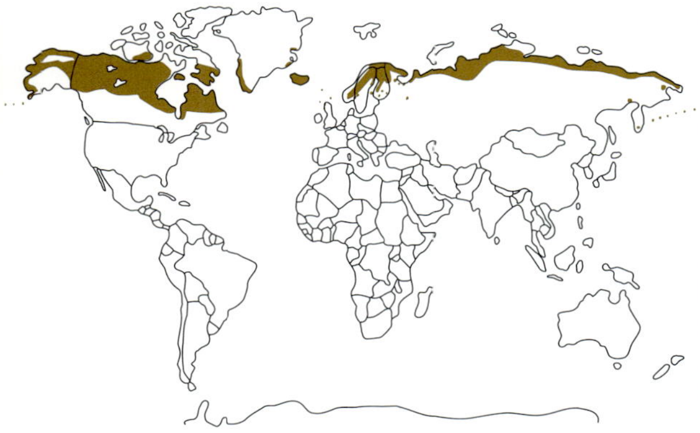

World breeding range

Species profile

Clutch size	4 eggs, sometimes 3[3,4,5]
Laying interval	1 egg per 1–2 days[1,3]
Start of incubation	with third or fourth egg, sporadic before[1,3,5]
Incubation period	usually 18–19 days (17–21)[1,3,5]
Maturity at hatching	precocial
Care of young	by male[1,5,10]
Nestling period	less than 24 hours[1,3]
Age at first flight	≈ 20 days[1,3,5]
Dependence of young	until 14–17 days old[1,5]
Broods per year	1–2[1,3,5]
Breeding age	1 year[1,5]
Mating system	monogamy, often polyandry[1,5,9,10]
Duration of pair bond	8–17 days[1,5]
Total length	16.5–20.0 cm[2]
Wing span	34.3–36.8 cm[7]
Weight	M 32 g (29–35) / F ? g (29–33)[8]
Longevity record	5 years[6]

References:
1. Cramp and Simmons 1983
2. Godfrey 1986
3. Johnsgard 1981
4. Bent 1927
5. Reynolds 1987
6. Clapp et al. 1982
7. Terres 1980
8. Dunning 1984
9. Schamel and Tracy 1991
10. Ehrlich et al. 1988

Chronology $n = 6$

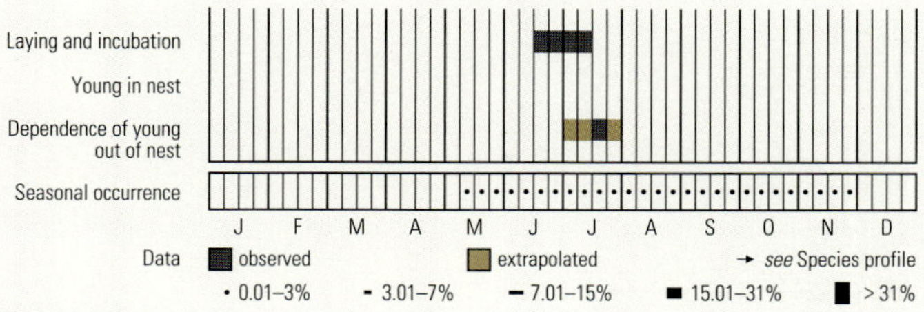

Red Phalarope

Phalarope roux

Phalarope à bec large
Phalaropus fulicaria

Christiane Girard

THE RED PHALAROPE BREEDS IN THE ARCTIC REGIONS OF EURASIA AND North America; it is a more pelagic species than the Wilson's and Red-necked phalaropes, nesting near the coast and spending the rest of the year at sea. It also migrates at sea, where it sometimes forms flocks of several thousand. Most of the Nearctic population overwinters off the west coast of Africa (Cramp and Simmons 1983).

In Québec, the Red Phalarope appears to breed only in the most northwesterly part of the province. The rare sight records from southern Québec are mainly from along the St. Lawrence in autumn (ÉPOQ). An unusual historical record is of several thousand migrating Red Phalaropes off the Magdalen Islands in the spring of 1890 (Fradette 1992).

Habitat and behaviour

Red Phalaropes are semicolonial, several pairs breeding together in suitable habitat without defending individual territories (Cramp and Simmons 1983). They nest on the ground on wet, grassy tundra, often near small freshwater pools. It forages in these pools or at sea, where it find insects, molluscs, crustaceans, or small fish that it takes on or near the surface of the water. One foraging technique consists of spinning like a top on the water's surface, usually clockwise at about 40 revolutions per minute, and capturing organisms that are stirred up by the whirlpool it creates (Cramp and Simmons 1983).

As in the other two phalarope species, the usual roles of the male and female birds are reversed. The female Red Phalarope is more brightly coloured and more aggressive than the male, which she courts, and the male alone incubates the eggs and tends the young. The female usually has only one mate, but is occasionally polyandrous. Typical clutch size is 4 eggs, which are usually laid in June or July, and are incubated for 18–20 days. The young phalaropes are precocial and leave the nest less than 24 hours after hatching. Accompanied by the adult male, they sometimes roam hundreds of metres, moving from one pool to another, where they forage for themselves (Bent 1927; Cramp and Simmons 1983; Godfrey 1986).

Distribution and history

In Québec, the Red Phalarope's breeding range is apparently confined to the coastal regions of the northwestern part of the province, from Inukjuak northward (Todd 1963; Godfrey 1986). This is only speculative, of course, as no nests have ever been found in the province. An accurate description of the Québec breeding range of this phalarope will require much more field work.

Notes on territory

The Red Phalarope is gregarious most of the year. During the breeding season it is semicolonial and it does not defend a breeding territory, as nests are usually fairly distant from each other. A small pond, for example, may be used for foraging, resting, and courtship displays by a number of pairs of phalaropes without any of them exhibiting territorial behaviour.

Concentrations of 7–10 pairs/ha have been reported in areas where suitable nesting habitat is limited (Parmelee et al. 1967). Pairs are more widely dispersed where larger expanses of good breeding habitat are found.

Female Red Phalarope foraging on a tundra pond. Bathurst Island, Northwest Territories. *June 1971, Pierre Lamothe.*

The Red Phalarope nests on wet grassy tundra near small freshwater ponds. This male phalarope is incubating on a nest that blends into its surroundings. Bathurst Island, Northwest Territories. *July 1971, Pierre Lamothe.*

For now, the strongest breeding evidence was obtained at Cusson Point on the east coast of Hudson Bay, where on 29 July 1930, Todd (1963) captured 3 male Red Phalaropes, 2 of which had brood patches. The birds were very agitated, calling and running toward him, which suggests they were nesting nearby. More recently, on 20 June 1986 and 21 and 24 June 1987, the species was observed at Inukjuak, also on the Hudson Bay coast (ÉPOQ). A female was captured at Kuujjuaq, south of Ungava Bay, on 18 June 1884 (Gabrielson and Wright 1951), and 4 birds were sighted at Kangiqsujuaq on Hudson Strait on 9 June 1963 (Brochu 1963). Interesting as these records may be, they do not constitute evidence that the species breeds on the coast of northern Québec, since any of the birds seen could have been migrating (Gaston et al. 1985) or nesting on islands off Québec belonging to the Northwest Territories. Only one Red Phalarope nest has ever been found in these regions to date, on West Digges Island, Northwest Territories, just a few kilometres off Cape Wolstenholme, on 23 July 1886 (Todd 1963).

Michel Robert

World breeding range

Species profile

Clutch size	4 eggs, sometimes 3[1]
Laying interval	1 egg per day[1]
Start of incubation	with last egg[1]
Incubation period	18–20 days[1]
Maturity at hatching	precocial
Care of young	by male[1]
Nestling period	less than 24 hours[1]
Age at first flight	16–18 days[1]
Dependence of young	until first flight[1]
Broods per year	usually 1 (1–2)[1]
Breeding age	1 year[1]
Mating system	monogamy, sometimes polyandry[1,5]
Duration of pair bond	usually until start of incubation[1]
Total length	19.5–23.0 cm[2]
Wing span	35.6–40.6 cm[3]
Weight	M 50.2 g (?) / F 61.1 g (?)[4]
Longevity record	?

References:
1. Cramp and Simmons 1983
2. Godfrey 1986
3. Terres 1980
4. Dunning 1984
5. Schamel and Tracy 1991

Chronology $n = 0$

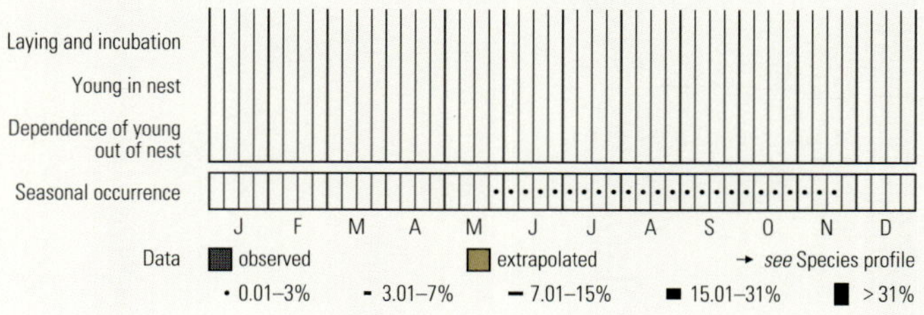

Pomarine Jaeger

Labbe pomarin

Stercorarius pomarinus

Christiane Girard

LARGER AND HEAVIER THAN THE PARASITIC AND LONG-TAILED JAEGERS, the Pomarine Jaeger is identified by its elongated central tail streamers with their twisted ends. Like other jaegers, it pirates food by aggressively harrying other larids, even the large gulls, in flight (Bent 1921). In Québec, it breeds only in the northwesternmost part of the province, and is a very rare, though regularly reported, transient in southern Québec (David 1980a).

Habitat and behaviour

The Pomarine Jaeger nests on tundra interspersed with lakes, ponds, and sluggish streams, usually near the coast (Godfrey 1986).

The two eggs are laid in a slight depression in the moss, sometimes lined with grass. Both parents incubate the eggs and feed the young (Cramp and Simmons 1983). The semiprecocial chicks leave the nest at 2–4 days (Ehrlich et al. 1988), but remain nearby until they fledge, at about 5 weeks. The young are tended by the adults for up to 2 weeks after fledging (Cramp and Simmons 1983).

Pomarine Jaegers defend their nests and young against predators, including humans, through aerial displays, including dive-attacks, and alarm calls. This species is considered less aggressive than the other two jaegers toward human intruders (Cramp and Simmons 1983).

Pomarine Jaegers are normally silent, but during the breeding season they produce a variety of piercing whistles and high-pitched calls (Cramp and Simmons 1983). At this time, they prey mostly on lemmings (Maher 1970a); for the rest of the year they are pelagic, feeding mainly on fish (Cramp and Simmons 1983). They are sometimes seen following fishing vessels, foraging on offal and garbage thrown overboard (Bent 1921; Terres 1980).

Distribution

The Pomarine Jaeger breeds in the Arctic in North America and the former USSR. It winters at sea off the coasts of the United States to the West Indies and Peru and off the coasts of Africa and the east coast of Australia (AOU 1983).

In Québec, the Pomarine Jaeger breeds along the northern part of the Hudson Bay coast. In 1930, Todd (1963) found two nests, one containing 2 eggs near Inukjuak on 21 June, and a second one, also containing 2 eggs, one of which was on the point of hatching, at Povungnituk on 17 July. A few days later, as Todd travelled northward, he sighted other Pomarine Jaegers at the mouth of the Povungnituk River, Cusson Point, and Mosquito Bay. During a later trip, in 1945, he again sighted Pomarines at Povungnituk and at the first rapid on the Kogaluc River. He commented that he had hoped to find the species more common in this area, but the pairs were isolated and scattered. The Pomarine Jaeger may also nest on the west coast of Ungava Bay. On 16 July 1966, at Lake Voltz, a pair was observed displaying agitated behaviour, and another pair, behaving similarly, was sighted the next day on the Rivière au Chien Rouge (R. Ouellet, pers. comm.).

In southern Québec, the Pomarine Jaeger is mainly seen in migration, in the estuary and the Gulf of St. Lawrence, and is rare in the interior. Non-breeding birds may wander south of the breeding grounds in summer (Godfrey 1986); this explains the handful of summer sight records for the gulf and estuary.

History and trends

Because the Pomarine Jaeger frequents remote and isolated areas throughout its life cycle, there are insufficient data to assess trends in distribution or abundance, or to identify potential threats to the species. We do know that in Alaska, breeding pair density is strongly affected by the cyclical fluctuations in lemming populations, which are its preferred summer prey (Maher 1970a; Custer and Pitelka 1987).

SIMON BARRETTE

Notes on territory

The Pomarine Jaeger is more territorial than its two congeners and will drive away both conspecifics and other jaegers from its territory (Parmelee et al. 1967; Cramp and Simmons 1983). Breeding displays, copulation, nesting, and foraging all occur within the territory, which may range in size from 0.06 to 1.21 km², depending on habitat and prey availability (Cramp and Simmons 1983). When the young reach the age of 2 or 3 weeks, the territory is expanded to increase the foraging area. Pair density in the Pomarine Jaeger is more closely linked to cyclical fluctuations in rodent numbers than it is for the other jaeger species (Cramp and Simmons 1983). In Alaska, pair density can reach 7.9/km², depending on the abundance of brown lemmings (Custer and Pitelka 1987).

In Québec, the Pomarine Jaeger nests on the tundra in the coastal region of northern Hudson Bay. Arctic breeding grounds include Bathurst Island. Northwest Territories. *July 1971, Pierre Lamothe.*

The great variety of waterbirds found in the river deltas and along the west coast of Ungava Bay attracts the Pomarine Jaeger, an opportunistic predator. Aupaluk region. *July 1978, Pierre Lamothe.*

World breeding range

Species profile

Clutch size	usually 2 eggs (1–3)[1,5]
Laying interval	1 egg per 2 days[1]
Start of incubation	with first egg[1]
Incubation period	24–28 days[1,2,4]
Maturity at hatching	semiprecocial
Care of young	by female and male[1,4,5]
Nestling period	2–4 days[5]
Age at first flight	31–37 days[1,5]
Dependence of young	for up to 14 days after first flight[1,5]
Broods per year	1[1,4,5]
Breeding age	?
Mating system	monogamy[1]
Duration of pair bond	?
Total length	51.0–58.5 cm[2]
Wing span	121.9 cm[3]
Weight	M 648 g (542–797) / F 740 g (576–917)[6]
Longevity record	?

References:
1. Cramp and Simmons 1983
2. Godfrey 1986
3. Terres 1980
4. Harrison 1978
5. Ehrlich et al. 1988
6. Dunning 1984

Chronology *n* = 2

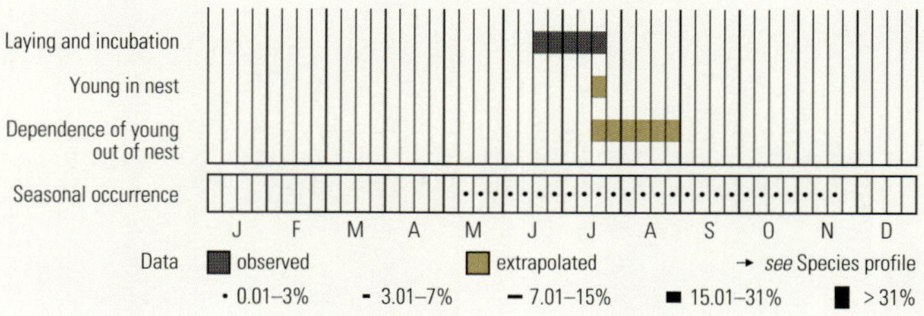

Parasitic Jaeger

Labbe parasite

Stercorarius parasiticus

Christiane Girard

THE PARASITIC JAEGER'S NAME IS WELL DESERVED, SINCE IT OFTEN HARASSES terns and gulls in order to steal their prey. Its agile falconlike flight makes it easy to distinguish from other larids (Townsend 1921d). In Québec, it breeds solely on the northern tundra, usually not far inland from the coast (Todd 1963; Godfrey 1986), and is a regular transient in southern Québec (David 1980a).

Habitat and behaviour

The Parasitic Jaeger breeds on the rocky or mossy coastal tundra of the Arctic and Subarctic (Godfrey 1986). Expanses of fresh water seem to be more important to this species than to Pomarine and Long-tailed jaegers (Maher 1974). Pairs are usually solitary (Parmelee et al. 1967; Maher 1974), although they sometimes nest in colonies (Godfrey 1986).

The eggs, usually 2, are laid in a depression in the ground and incubated by both parents. The young leave the nest a few days after hatching and hide separately in nearby vegetation; they are not seen together again until they are 3 weeks old (Cramp and Simmons 1983). Chicks over 10 days old are often left unattended while the parents go off to hunt together (Cramp and Simmons 1983). The young fledge at around 25–33 days (Harrison 1978; Cramp and Simmons 1983). They are fed by both adults until 7–10 days after fledging and remain with them another 1–4 weeks at least (Cramp and Simmons 1983).

In Sweden, Elmberg (1991) observed three birds on the same territory during two breeding seasons. When humans or predators approached too close, all three birds actively defended the nest with calls, aerial attacks, and distraction displays to draw the intruders away (Cramp and Simmons 1983; Elmberg 1991). It was not possible, however, to determine whether all three birds were feeding the young, or if there was some sort of kinship between the third bird and the parents of the young.

Although the Parasitic Jaeger is known for food piracy (Birt and Cairns 1987), it does forage for itself, taking lemmings, mice, fish, invertebrates, berries, passerines, shorebirds, eggs, and chicks (Martin and Barry 1978; Cramp and Simmons 1983). According to Pruett-Jones (1980), Parasitic Jaegers practise a sort of co-operative hunting never seen in Pomarine and Long-tailed jaegers. This typically involves a pair of birds, with the first flying a few metres off the ground and the second following behind about 10 m up.

Distribution

The Parasitic Jaeger breeds in the Arctic and Subarctic in North America and Eurasia. It spends the winter at sea, from southern California to southern Chile and eastern Australia and New Zealand in the Pacific, and offshore from Maine and the British Isles south to Argentina and west Africa in the Atlantic (AOU 1983). It also winters in the Mediterranean, the Persian Gulf, the Arabian Sea, and the Indian Ocean.

In Québec, the Parasitic Jaeger breeds only in the Far North. A nest containing 2 eggs was discovered on 8 July 1945 at the mouth of the Kogaluc River. Another nest, also containing 2 eggs, was found just across the border in Labrador at Cape Chidley in June 1902 (Todd 1963). Todd (1963) reported several observations of Parasitic Jaegers along the Hudson Bay coast, between Cape Jones and Povungnituk. According to Todd, the species probably nests all along the coast from Lake Guillaume Delisle northward. He also cited an observation at Fort George in 1882. There have been sightings in Ungava Bay, as well, where 4 birds were seen at the mouth of the False River, on 22 August 1948 (Gabrielson and Wright 1951). A pair exhibiting agitated behaviour was also observed near Rivière au Chien Rouge on 17 July 1966 (R. Ouellet, pers. comm.). A report of another 2 agitated birds carrying food in their bills and diving at the observers on 18 and 22 July 1966 at Payne Lake (R. Ouellet, pers. comm.) indicates that the species may nest fairly far inland, as is the

Notes on territory

Parasitic Jaegers nest in isolated pairs or in small, loose aggregations. Solitary pairs use their territories for copulation, nesting, and feeding, chasing off all other jaegers (Cramp and Simmons 1983). In Alaska, the average territory size of a pair nesting solitarily was 1.3 km^2 (Maher 1974). When the species nests in loose groups, the nests may be as little as 30 m apart, and the territory is used solely for copulation and nesting. Under such conditions, pair density may reach 100 pairs/km^2, or even, in one documented case near a colony of Arctic Terns, 211 pairs/km^2 (Cramp and Simmons 1983).

On the coastal tundra of Hudson Bay, this Parasitic Jaeger has made its nest on a hillock in the midst of a colony of Lesser Snow Geese. La Pérouse Bay, Manitoba. *June 1980, Yves Aubry.*

case in the Northwest Territories. Recently, during aerial surveys conducted on the Ungava Peninsula, isolated observations of Parasitic Jaegers were made near the Chien Rouge and Kogaluc rivers on 17 and 21 June 1993, and around Rivière à la Baleine, east of Kuujjuaq, on 15 June 1993 (B. Harvey, USFWS and D. Bordage, CWS, pers. comm.).

In the southern part of the province, the Parasitic Jaeger is the jaeger most often seen in migration. It is sighted mainly in the spring, in May and June, and in the fall, from August to November, in the estuary and the Gulf of St. Lawrence. It is unusual inland (ÉPOQ). Non-breeders may wander south of the breeding grounds in summer (Godfrey 1986), which may explain the few July sightings in the gulf.

History and trends

Population trends for this species, as with the other jaegers, are not well known, since it spends almost the entire year in remote regions rarely visited by people. We can only surmise that, like many other species breeding essentially in northern regions, it is spared most of the negative effects of human presence (Robert 1989b).

Simon Barrette

The rudimentary nest of the Parasitic Jaeger is a simple depression, in which it lays its eggs. La Pérouse Bay, Manitoba. *June 1980, Yves Aubry.*

World breeding range

Species profile

Clutch size	usually 2 eggs (1–2)[2,3,4,5,6]
Laying interval	1 egg per 1–2 days[1]
Start of incubation	with first egg[1,2]
Incubation period	23–28 days[1,5]
Maturity at hatching	semiprecocial
Care of young	by female and male[1,2,3]
Nestling period	a few days[2,6]
Age at first flight	25–33 days[1,2]
Dependence of young	until at least 2–5 weeks after first flight[1,2]
Broods per year	1[1,2,3]
Breeding age	3–5 years[1,4]
Mating system	monogamy[1,3]
Duration of pair bond	for life[1]
Total length	46.0–53.5 cm[5]
Wing span	110–125 cm[1]
Weight	M 421 g (301–540) / F 508 g (346–644)[7]
Longevity record	30 years[4]

References:
1. Cramp and Simmons 1983
2. Harrison 1978
3. Ehrlich et al. 1988
4. Terres 1980
5. Godfrey 1986
6. Martin and Barry 1978
7. Dunning 1984

Chronology $n = 1$

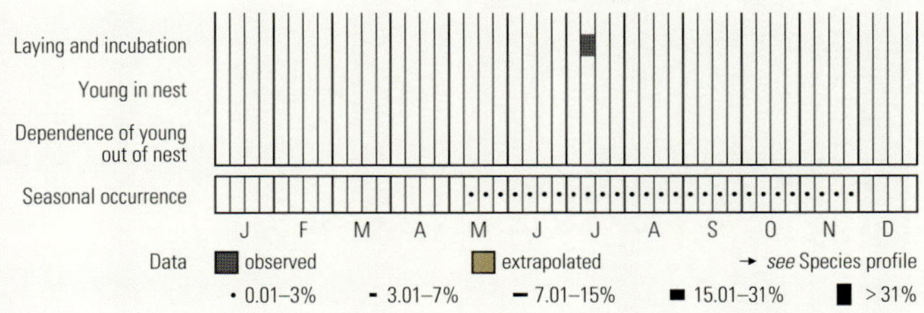

Long-tailed Jaeger

Labbe à longue queue

Stercorarius longicaudus

Christiane Girard

THE LONG-TAILED JAEGER, THE SMALLEST OF OUR JAEGERS, CAN BE identified by its very long, slender, tail streamers, slim body, and graceful flight. In Québec, it breeds only on the Ungava Peninsula. For southern Québec it is classified as a very rare, though regularly reported, transient (David 1980a).

Habitat and behaviour

The Long-tailed Jaeger nests on the tundra farther inland and in drier areas with fewer large expanses of fresh water than the Parasitic Jaeger (Parmelee et al. 1967; Maher 1974). The nest is a slight depression in moss or peat, which may or may not be lined with available vegetation (Godfrey 1986). Hatching is asynchronous, with anywhere from 1 to 3 days between the hatching of the first chick and the last. The roles of the male and female birds are more clearly differentiated than in the other jaegers of the genus *Stercorarius*. Nest building is probably carried out for the most part by the female, which also incubates and feeds the young, while the male hunts and defends the territory (Cramp and Simmons 1983). The male brings back food to share with the female, which in turn feeds the young by regurgitation. The young usually leave the nest about 2 days after hatching, remaining nearby until they fledge. Throughout the nesting period, the parents are territorial, attacking and attempting to drive intruders away from the nest and young.

During the breeding season, Long-tailed Jaegers prey mostly on lemmings, voles, and other small mammals. When these are scarce, they eat a larger proportion of birds, eggs, invertebrates, berries, carrion, and refuse (Maher 1970b; Andersson 1976; Cramp and Simmons 1983). They hover, kestrellike, to detect their prey, and being agile fliers take more flying insects than Parasitic and Pomarine jaegers (Maher 1974). Outside the breeding season, they eat more fish, carrion, and refuse (Drury 1960; Cramp and Simmons 1983). The Long-tailed Jaeger is less piratical than the Pomarine and Parasitic jaegers (Cramp and Simmons 1983).

Although apparently silent when at sea, Long-tailed Jaegers utter a whole range of calls and whistles during the breeding season. At sea, they tend to fly at a higher altitude than the other jaegers and are less likely to follow ships (Cramp and Simmons 1983).

Distribution

The Long-tailed Jaeger breeds in the Arctic in North America and Eurasia. It migrates chiefly at sea and is rarely observed in inland areas (Cramp and Simmons 1983; Godfrey 1986). It winters in the Southern Hemisphere off the coasts of South America, the west coast of Africa, and in the central Pacific (Cramp and Simmons 1983).

Todd (1963) considered the Long-tailed the most common of the three jaegers in northern Québec. In 1930, he found nests at Inukjuak on 23 June and 4 July, and at Povungnituk on 14 July. He also found nests at the mouth of the Kogaluc River on 16 and 18 June and 5 July in 1945. The species was also sighted south of the Nastapoka River and at Cusson Point, on Hudson Bay. The sighting of a pair on the Broadback River on 14 June 1914 is virtually the only known record south of Hudson Bay during the breeding season (Todd 1963). The species has also been seen inland on the Ungava Peninsula. Eklund (1956), for example, sighted it twice at Payne Lake and once at Lake Maryland in July 1949. Todd (1963) believed that it nests throughout the Ungava Peninsula, north of the tree line. More recently, R. Ouellet (pers. comm.) observed a pair acting aggressively while carrying food in their bills at Baie aux Feuilles on 30 July 1966.

Notes on territory

It is primarily the male that defends the territory from intrusion by conspecifics and other jaegers. The male Long-tailed Jaeger constantly defends a central core 400–600 m in diameter. It does not regularly patrol the rest of the territory, but will drive off any jaegers encountered when foraging. Territory size is usually 0.68–1.38 km^2 (Cramp and Simmons 1983). In Sweden, breeding pair density varied between 0.02 and 0.63 pair/km^2, depending upon the abundance of small mammals (Andersson 1976). In the North American Arctic, pair densities range from 0.07 to 0.94 pair/km^2 (Maher 1970b, 1974; Erskine 1984).

The smallest of our three jaegers nests in dry areas on the tundra, including substrates such as this stony plateau on Bathurst Island. Eggs are laid in a shallow depression in the ground. Northwest Territories. *July 1971, Pierre Lamothe.*

In southern Québec Long-tailed Jaegers are observed mostly in migration in May, June, August, and September in the estuary and the Gulf of St. Lawrence. There are some records of inland sightings, from the southern Appalachians, the Laurentians, and the St. Lawrence Plain (Ouellet 1974a; David 1980a). This jaeger is found farther out to sea in the Atlantic (Lee 1989) and is rarer in the Gulf of St. Lawrence (David 1980a; Tufts 1986) than the other two jaegers.

History and trends

As with the Parasitic and Pomarine jaegers, there is a lack of data with which to assess trends in abundance and distribution of the Long-tailed Jaeger, which frequents very remote areas. In Eurasia, the abundance of breeding Long-tails appears to vary with fluctuations in food availability (Cramp and Simmons 1983).

Simon Barrette

The Long-tailed Jaeger breeds in the Arctic in North America and Eurasia; in Québec, the Ungava Peninsula appears to be its main breeding area. Shown here is a nesting bird on Bathurst Island. Northwest Territories. *July 1971, Pierre Lamothe.*

World breeding range

Species profile

Clutch size	usually 2 eggs (1–3)[2,4,5]
Laying interval	1 egg per 1–3 days[5]
Start of incubation	with first egg[2,5]
Incubation period	23–25 days[1,2,3,4,5]
Maturity at hatching	semiprecocial
Care of young	by female and male[1,2,5], *see* text
Nestling period	1–2 days, sometimes more[2,5]
Age at first flight	21–27 days[1,2,4,5]
Dependence of young	until 10–21 days after first flight[1,2,5]
Broods per year	1[1,5]
Breeding age	?
Mating system	monogamy[1,5]
Duration of pair bond	more than 1 season[1,5]
Total length	50.7–58.5 cm[3]
Wing span	105–117 cm[5]
Weight	M 280 g (236–343) / F 313 g (258–358)[6]
Longevity record	8 years 1 month[7]

References:
1. Ehrlich et al. 1988
2. Harrison 1978
3. Godfrey 1986
4. Terres 1980
5. Cramp and Simmons 1983
6. Dunning 1984
7. Clapp et al. 1982

Chronology *n* = 7

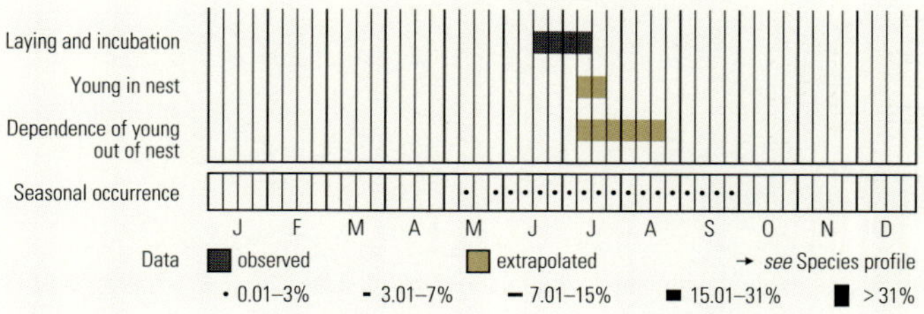

Iceland Gull

Goéland arctique

Larus glaucoides

Only about 160 pairs of Iceland Gulls breed in Québec. The two known colonies are located on the province's northwest tip, on cliffs overlooking Digges Sound near several large Thick-billed Murre colonies (Gaston et al. 1985). This migrant breeder is seen regularly in winter along the St. Lawrence.

Habitat and behaviour

Iceland Gulls breed in either single- or mixed-species colonies on steep arctic cliffs, often well inland (Ehrlich et al. 1988; Gaston and Elliot 1990). Nests are usually built on a ledge 100–200 m up facing a sound or fjord (Cramp and Simmons 1983) and are made of moss, grass, and seaweed (Godfrey 1986). The typical clutch of 2–3 eggs is often laid while the waters below are still covered with ice (Townsend 1921c). The semiprecocial chicks are fed by regurgitation. On Digges Sound, adults prey on Thick-billed Murre eggs, but apparently to a lesser extent than Glaucous Gulls, with which the Iceland Gulls at the two Québec colonies nest in association. Iceland Gulls also feed on small organisms taken close to the surface of the water (Gaston et al. 1985).

Distribution

Iceland Gulls breed on Greenland and in the Canadian Arctic (Godfrey 1966; Gaston and Decker 1985; Gaston and Elliot 1990). The two Canadian subspecies are known as Kumlien's Gull (*Larus glaucoides kumlieni*) and Thayer's Gull (*Larus glaucoides thayeri*). On Digges Sound, Iceland Gulls breed in seven colonies, two of which are on the mainland and therefore in Québec (Gaston et al. 1985). The gulls' presence is probably related to the two large Thick-billed Murre colonies nearby, which contain roughly 300 000 pairs, and to the general high productivity of the waters. Atlas records on the St. Lawrence River probably involved non-breeders.

Iceland Gulls winter on the Atlantic coast (Godfrey 1986) and along the St. Lawrence, particularly the gulf (David 1980a). They can be seen in harbours, at dumps, or loafing on the ice. Iceland Gulls look like miniature Glaucous Gulls, but have a proportionately smaller head, neck, and bill and noticeably faster wingbeats in calm air (Thomas and Andresen 1986). Kumlien's Gulls are sometimes difficult to distinguish from the Thayer's Gull, which breeds farther west and north (Godfrey 1966) and is observed much less frequently in southern Québec (Steeves et al. 1989). A number of field marks must be carefully checked to differentiate the two subspecies (Gosselin and David 1975; Kaufman 1990a).

History and trends

Of the estimated 1000–3000 pairs breeding in colonies in the Hudson Strait area (Gaston 1990), only about 160 nest on Québec soil (Gaston et al. 1985). Iceland Gulls apparently did not breed in great numbers on Digges Sound in the 19th century (Todd 1963), and there is no evidence that any change has occurred in the size of the population since then.

There is still much taxonomic confusion regarding Kumlien's and Thayer's gulls and other closely related gulls (Rand 1942; Snell 1991). Apparently Kumlien's and Thayer's gulls interbreed on southeastern Southampton Island, about 200 km west of the Québec colonies (Gaston and Decker 1985), and on Baffin Island (AOU 1983). Godfrey (1986) and Sibley and Monroe (1990) consider the Kumlien's and Thayer's gulls as subspecies, whereas the AOU treats them as separate species since 1973 (AOU 1973, 1983).

Robert Alvo

Clodin Roy

Notes on territory

Iceland Gulls often travel long distances from the colony to forage. The Digges Sound birds fly several kilometres to the Thick-billed Murre colonies, where they prey on eggs and chicks. They also pick food from the water and scavenge on beaches (Cramp and Simmons 1983; Gaston et al. 1985).

On Digges Sound from 1980 to 1983, about 350 pairs nested in seven colonies (including 2 in Québec) of 10–126 pairs each. While two colonies were about 55 km apart, five were within 11 km of each other. Some pairs also nested solitarily (Gaston et al. 1985).

The cliffs along Digges Sound, at the extreme northwestern tip of the Ungava Peninsula, mark the southern limit of the Iceland Gull's breeding range in Canada. View of Cape Wolstenholme from Digges Islands. *July 1981. Gilles Chapdelaine.*

South of Cape Wolstenholme, the Iceland Gull nests on the ledges of steep coastal and island cliffs, sometimes close to Thick-billed Murre colonies, which provide the gulls with an abundant food source. Northern tip of the Ungava Peninsula. *August 1982, Gaétan Guertin.*

World breeding range

Species profile

Clutch size	usually 2–3 eggs[1,2,3,4]
Laying interval	probably 1 egg per 2 days[4]
Start of incubation	?
Incubation period	24–26 days[4]
Maturity at hatching	semiprecocial
Care of young	probably by female and male[1]
Nestling period	?
Age at first flight	?
Dependence of young	?
Broods per year	1[1,2]
Breeding age	probably ≈ 4 years[1]
Mating system	monogamy[1]
Duration of pair bond	?
Total length	57.0–63.5 cm[6]
Wing span	127–150 cm[3,5]
Weight	557 g (?)[7]
Longevity record	?

References:
1. Ehrlich et al. 1988
2. Harrison 1978
3. Terres 1980
4. Gaston et al. 1985
5. Cramp and Simmons 1983
6. Godfrey 1986
7. Dunning 1984

Chronology*

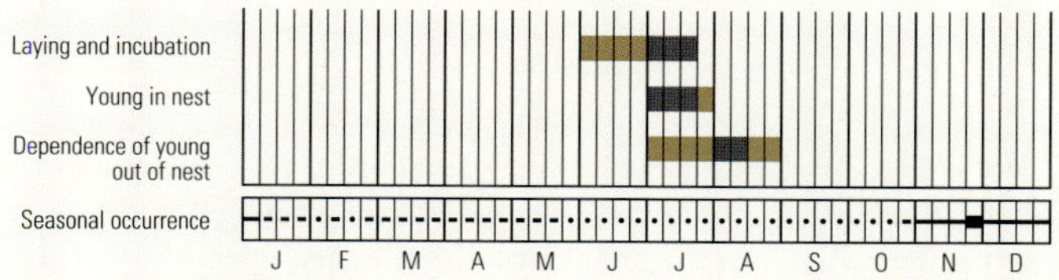

Data ■ observed ■ extrapolated → *see* Species profile

• 0.01–3% ‐ 3.01–7% – 7.01–15% ▬ 15.01–31% █ > 31%

* Data from the two Québec colonies are insufficient to establish a precise breeding chronology. This table is therefore based on the work of Gaston et al. (1985), who monitored 182 breeding pairs in a number of colonies on the Digges Islands and Staffe Islet, Northwest Territories, both within 6 km of the Québec coast.

Glaucous Gull

Goéland bourgmestre

Larus hyperboreus

"THE NAME BURGOMASTER IS A FITTING NAME FOR THIS CHIEF MAGISTRATE of the feathered tribes of the Arctic seas, where it reigns supreme over all the lesser water fowl." This is how Bent (1921) described the holarctic Glaucous Gull, which breeds in large numbers along the coasts of northern Québec (Todd 1963). It is easily observed in winter along the St. Lawrence River east of Tadoussac, but can also be seen farther upstream (David 1980a; Steeves et al. 1989).

Pair of Glaucous Gulls on the nest. Bathurst Island, Northwest Territories. *June 1971, Pierre Lamothe.*

Habitat and behaviour

Glaucous Gulls usually nest in colonies on high rocky cliffs, often near larid or murre colonies, where they prey on the eggs and young (Bent 1921). Some pairs breed solitarily on islets in inland lakes, rivers, river mouths, or bays, on tundra slopes, or on grassy clifftop slopes. Nests are built on rock outcrops, pinnacles, cliff edges, or sloping or flat ground.

Glaucous Gulls eat invertebrates and fish taken from the intertidal zone, the sea surface, and from rivers and lakes. They also prey on adult birds (Stempniewicz 1983a) and their eggs and chicks (Gaston and Nettleship 1981), insects, berries, carrion washed up on shore, seal placentas and feces, and items pirated from other birds (Stempniewicz 1983b; Cramp and Simmons 1983).

A study conducted on Digges Sound, southwest of Cape Wolstenholme, showed that Glaucous Gulls often nest in mixed colonies with Iceland Gulls on cliff ledges, but also nest solitarily on small offshore islands (Gaston et al. 1985). An estimated 170 pairs nested in the area, about 70 of these within 1 km of a Thick-billed Murre colony. Peak hatching dates in 1981 and 1982 were 22–26 June. Survival of chicks from hatching to fledging was 76–81% in 1981 and 68% in 1982. Gulls breeding near the murres apparently obtained most of their food by scavenging around the colonies or by preying on murre eggs and chicks, with some pairs and individuals specializing in taking murre eggs. Conversely, birds nesting about 6 km from the murres seemed to find their food elsewhere, perhaps at sea. The rate at which broods were fed was relatively constant over the season, averaging between 6 and 10 meals daily, but the amount of food regurgitated at each feeding may have increased as the chicks got older. Young chicks stimulate regurgitation by pecking at the red spot on the adult's bill (Cramp and Simmons 1983). Most food deliveries, where divisible, were shared by the whole brood. Breeding individuals generally roost in the colony at night (Gaston et al. 1985).

Clodin Roy

Distribution

Glaucous Gulls breed along many of the arctic coasts and islands of both the Eastern and Western hemispheres. In Québec, this gull's breeding range extends from roughly the mouth of the Nastapoka River to Cape Wolstenholme (Gaston et al. 1985), and around Ungava Bay to the northern tip of Labrador; in Labrador, the species breeds along the coast down to Hamilton Inlet (Todd 1963; Godfrey 1986). Bent's (1921) record on the east side of James Bay may have been based on incorrectly identified Herring Gull eggs (Manning and MacPherson 1952; Todd 1963).

Glaucous Gulls tend to breed along the coastlines, with a few individuals selecting sites a short distance inland. Stretches of coast with cliffs have much higher concentrations. For example, very few birds nest between Cape Wolstenholme and Inukjuak, while they are abundant between Inukjuak and Lake Guillaume Delisle, the southernmost breeding area along the west coast of Québec. On the Labrador coast, the centre of abundance is between Hamilton Inlet and Nachvak (Todd 1963).

In Québec, Glaucous Gulls winter within their breeding range and farther south. Birders have reported them frequently along the St. Lawrence River, particularly at Bergeronnes, Pointe-au-Père, Rimouski, Baie-Comeau, and Les Escoumins, and to a lesser extent at points along the gulf, the Saguenay River, the Ottawa River, and in the Lac-Saint-Jean region (ÉPOQ). In winter, they are generally seen in coastal ice-free bays and harbours, but some are seen inland at garbage dumps throughout the province. Wintering birds are generally silent.

Notes on territory

Glaucous Gull colonies generally consist of up to 30 pairs, with inter-nest distances ranging from 5 m to 165 m. Colonies often have clubs nearby, where males defend territories 2–5 m in diameter for courtship and copulation.

In Alaska, the mean distance between solitary pairs was 322 m on the coast and 1150 m inland (Cramp and Simmons 1983). Nest density was 5.9 nests/km² along the coast (along a 5-km section of coastline) and 0.15 nest/km² inland. Solitary pairs defend a small territory used for copulation, nesting,

The Glaucous Gull is common along the coasts of northern Québec, where it nests on clifftops and ledges. Cape Wolstenholme, northern tip of the Ungava Peninsula. *September 1992, Tony Gaston.*

and the early stages of brood rearing. Two to four days after the chicks hatch, the family shifts to a rearing territory 1–10 m from the nearest body of water larger than 0.5 ha, though distances up to 200 m are not unknown. This territory is defended against conspecific adults and older chicks. The young remain close to the rearing territory for up to 2 weeks after they fledge and are fed and cared for by the parents, but may make short flights, accompanied or not by their parents, up to 3 km away. The family later moves to the sea or to a river, where the parents continue to feed the young, which achieve independence gradually (Cramp and Simmons 1983).

Staffe Islet, the islet to the left, is in Digges Sound, off the northwest tip of the Ungava Peninsula. It is home to some 10 pairs of Glaucous Gulls, which nest in a colony in association with a much larger number of Iceland Gulls. Northwest Territories. *August 1982, Gaétan Guertin.*

History and trends

People living along Hudson Bay take gull eggs, and this may have an impact on local Glaucous Gull populations. However, in Ontario there are no data to suggest a significant change in distribution during the past 100 years (Weseloh 1987b), and the same is probably true for Québec.

ROBERT ALVO

World breeding range

Species profile

Clutch size	usually 3 eggs (2–4)[1,3,6,7]
Laying interval	1 egg per 2 days[3]
Start of incubation	with first egg[3]
Incubation period	27–30 days[1,3,4,6]
Maturity at hatching	semiprecocial
Care of young	by female and male[3,4,5]
Nestling period	a few days[1,3]
Age at first flight	probably 45–50 days[3]
Dependence of young	until first flight, but may be as much as 2 or more weeks longer[3]
Broods per year	1[1,3]
Breeding age	?
Mating system	monogamy[3]
Duration of pair bond	1 season[3]
Total length	66–76 cm[6]
Wing span	152.4–167.6 cm[2]
Weight	M 1576 g (1280–1820) / F 1249 g (1070–1430)[8]
Longevity record	22 years 1 month[9]

References:
1. Bent 1921
2. Terres 1980
3. Cramp and Simmons 1983
4. Harrison 1978
5. Ehrlich et al. 1988
6. Godfrey 1986
7. Todd 1963
8. Dunning 1984
9. Clapp et al. 1982

Chronology *n* = 11

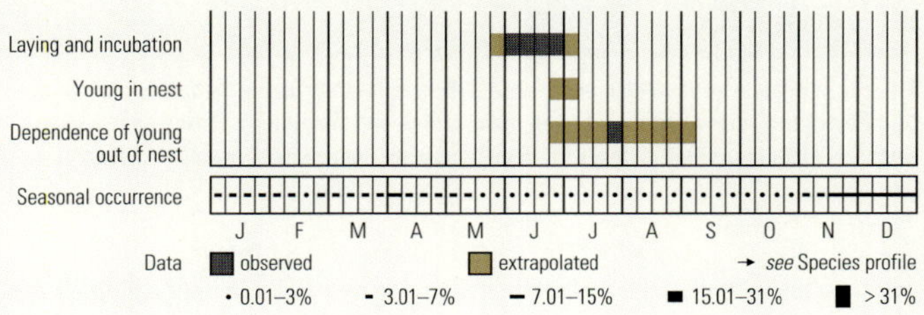

Great Auk

Grand Pingouin

Pinguinus impennis

Michel Gosselin

THE GREAT AUK, WHICH BECAME EXTINCT IN THE MIDDLE OF THE 19th century, was the largest alcid of the modern era. It measured nearly 70 cm long, twice the length of a murre, and weighed about 5 kg (Bédard 1969a). Unlike the other alcids, but in common with the penguins (Spheniscidae) of the Southern Hemisphere, the Great Auk was flightless. Like penguins, the Great Auk was an expert swimmer and used its wings, much reduced in size, for underwater "flight" (Bengtson 1984).

Habitat and behaviour

The gregarious Great Auk was a seabird of the cold-temperate parts of the North Atlantic. Unlike most other seabirds, however, it could not nest on steep cliffs but was constrained by its flightlessness to breed on sloping islands onto which it could drag itself out of the water; its nest sites also had to have a sufficient inshore food supply and be free of land predators. A single pear-shaped egg was laid, which averaged 12 cm in length and weighed about 300 g. Incubation lasted about 40 days (Kirk 1994c). The young apparently took to the water at 9 or 10 days. Age at first breeding is thought to have been 4–7 years (Bengtson 1984; Harris and Birkhead 1985a).

A victim of ruthless exploitation, the Great Auk became extinct during the last century. It lived on Bird Rocks and other nearby rocks and reefs, which due to erosion over the centuries are less extensive today than when Jacques Cartier visited in the 16th century. Magdalen Islands. *July 1983, Yves Aubry.*

Distribution and history

Little is known about the historical breeding range of the Great Auk. In Europe, it bred on Westmann, Geirfuglasker, and Eldey islands off the coast of Iceland, and on the Orkney Islands off the coast of Scotland, and perhaps also on the Faeroes and Shetland (Bengtson 1984; Nettleship and Evans 1985). Funk Island, off the east coast of Newfoundland, is the only authenticated breeding site in North America (Bengston 1984), although it may well have nested on other islands in the Western Atlantic, between Labrador and Maine (Nettleship and Evans 1985). Despite Greenway's (1967) assertions, there is every likelihood that it bred on Bird Rocks in the Magdalen Islands (Hardy 1888). It was here that Jacques Cartier captured several birds which he called *Apponatz*. We know from his description of birds encountered during his visit to Funk Island in 1534 that the *Apponatz* was indeed the Great Auk:

> In the air and round about are an hundred times as many more as on the island itself. Some of these birds are as large as geese, being black and white with a beak like a crow's[1]. They are always in the water, not being able to fly in the air, inasmuch as they have only small wings about the size of half one's hand, with which however they move as quickly along the water as the other birds fly through the air. And these birds are so fat that it is marvellous. We call them apponatz[2]; and our two long-boats were laden with them as with stones, in less than half an hour. Of these, each of our ships salted four or five casks, not counting those we were able to eat fresh (Jacques Cartier, translated by Biggar 1924).

A few weeks later, at Bird Rocks, Cartier wrote:

> We landed on the lower part of the smaller island and killed more than a thousand murres[3] and great auks[2], of which we took away as many as we wished in our long-boats. One might have loaded in an hour thirty such long-boats (*ibid.*).

Cartier visited Bird Rocks toward the end of June in 1534; hence it seems reasonable to assume that the birds were nesting there at that time.

In 1623, the missionary Gabriel Sagard, en route to New France, stopped at the same place and noted:

> When there is wind the birds rise easily from the ground, but otherwise certain species of them are hardly able to get on the wing, and can be easily knocked over with sticks, as the sailors from another ship had done, who, before we got there, had loaded their shallop with them and filled several barrels with the eggs found on the nests... These birds for the most part live only on fish, and though there are different species, some big, others small, they do not usually form several flocks... Their nests are so arranged in the island according to their species that there is no confusion, in fact excellent order. The large birds are placed near one another, and those less big or of other species along with those that match them, and all in such numbers that hardly could you ever persuade a person to believe it who had not seen them. I ate one of them, which the sailors call *Guillaume*, and the natives *Apponath*, with white and black plumage, as big as a hen with a short tail and small wings; it was not inferior to any game we have. There is another species smaller than the rest, called *Godets* (Sagard 1939).

These appear to be the only records we have of the occurrence of the Great Auk on Bird Rocks and in Québec (see also Fradette 1992).

Man drove the Great Auk to extinction. When the Europeans rediscovered North America in the 15th century, the colony on Funk Island was the largest. It was estimated at 100 000 pairs, while the other colonies certainly were smaller (Nettleship and Evans 1985). The species was so common on the Grand Banks of Newfoundland that ships' pilots used the birds as a marker for the banks. The Newfoundland colony, like all the other known colonies, was subjected to relentless exploitation from the time it was discovered at the beginning of the 16th century until its total extermination at the end of the 18th century. Being flightless, the Great Auk was more vulnerable than other species. The many vessels that came to fish on the Grand Banks—380 in the year 1578 alone (Nettleship and Evans 1985)—often stopped off at the Great Auk colonies to take on a supply of meat and eggs. Between 1770 and 1780, the colony began to be harvested for its feathers, which were used to make mattresses and pillows (Nettleship and Evans 1985). Birds were thus slaughtered every year until, by 1800, the colony had disappeared.

In the 19th century, the last surviving pairs nested on a few skerries southwest of Iceland; they were much coveted by collectors. The very last pair was taken at Eldey Rock in June 1844 (Townsend 1919; Nettleship and Evans 1985).

MICHEL ROBERT AND MICHEL GOSSELIN

1 The original term, *corbin*, is more properly translated as raven (Livory 1985:253).
2 The original term, *Apponatz*, refers to the Great Auk.
3 The original term, *Godez*, could refer to murres, razorbills, or puffins (Livory 1985:253).

Species profile

Clutch size	1 egg[2,3,5,6]
Laying interval	–
Start of incubation	?
Incubation period	probably ≈ 43 days (30–65)[2,3,6]
Maturity at hatching	probably semiprecocial[2]
Care of young	probably by female and male[4]
Nestling period	no nest, took to the sea at 9–10 days[6]
Age at first flight	flightless, took to the sea at 9–10 days[6]
Dependence of young	?
Broods per year	1[2,5]
Breeding age	probably 4–7 years[6]
Mating system	probably monogamy[4]
Duration of pair bond	?
Total length	≈ 76 cm[3]
Wing span	probably ≈ 70 cm[7]
Weight	probably ≈ 5000 g[1]
Longevity record	probably 20–25 years[6]

References:
1. Bédard 1969a
2. Harris and Birkhead 1985a
3. Greenway 1967
4. Ehrlich et al. 1988
5. Townsend 1919
6. Bengtson 1984
7. M. Gosselin, CMN, unpubl. data

Ringed Turtle-Dove

Tourterelle rieuse

Streptopelia risoria

THE RINGED TURTLE-DOVE CAN OFTEN BE SEEN IN PET SHOPS. ROUGHLY the same size as the Mourning Dove, the Ringed Turtle-Dove is a paler, cream to buff colour, has a rounded rather than pointed tail, and a black half-collar on the back of its neck. Even free-ranging, the Ringed Turtle-Dove is usually very tame and can be easily approached.

Linnaeus described the Ringed Turtle-Dove and gave it the scientific name *Columba risoria* (now *Streptopelia risoria*) in the 10th edition of his *Systema naturae*, published in 1758. His account of the species was based on domestic specimens that he assumed had come from India (AOU 1983; Smith 1987). Recently, it has been shown, however, that the Ringed Turtle-Dove is in fact a pale domesticated form of the African Collared-Dove (*Streptopelia roseogrisea*), which was discovered much later in Africa in the wild. Some authors have also confused the Ringed Turtle-Dove with the closely related Eurasian Collared-Dove (*Streptopelia decaocto*) (Cramp 1985; Smith 1987).

Habitat and behaviour

Little is known about the species' reproductive behaviour in the wild and most of the data collected has been based on birds in captivity. The Ringed Turtle-Dove builds its nest, a flimsy platform made of twigs, about 4–5 m up in a shrub or tree (Cramp 1985). The 2 white eggs are incubated by both parents for 2 weeks (Miller and Miller 1958; Goodwin 1983). For their first 12 days of life, the chicks are fed pigeon's milk, a whitish liquid high in fat and protein that is produced in the parents' crops. Later, regurgitated vegetable matter is gradually introduced into the chicks' diet. The young leave the nest 16–19 days after hatching (ten Cate and Hilbers 1991).

Distribution and history

Throughout the 20th century, small local populations of escaped Ringed Turtle-Doves have become established on and off in various parts of the United States, mainly Florida and California (Smith 1987), but it is difficult to determine whether these are truly wild populations because many owners keep their birds at semi-liberty, while providing them with food and shelter. Birds observed in Québec are no doubt part of such semidomesticated flocks, or accidental or intentional releases. There have been sight records of these doves almost every year since 1981, mainly from urban areas. The only breeding records are from Hull (1981, 1982, and 1988), and involve birds kept at semi-liberty by a breeder; these birds apparently nested in an abandoned American Robin's nest (Chabot 1986, *contra* Yank and Aubry 1985c; Gosselin et al. 1988b). A pair observed mating in Boucherville in the summer of 1983 (David 1983), and a failed attempt at nesting in Dorval in 1984 (Yank and Aubry 1984) are the only other breeding records for the province. Winter sightings of birds kept at semi-liberty in Hull (Yank and Aubry 1985c; Chabot 1986) and a bird seen near a feeder in Bromptonville in 1989 (Gosselin et al. 1989) show that the species can sometimes survive the winter in the wild.

So far, the Ringed Turtle-Dove is the only member of the genus *Streptopelia* that has been observed in Québec, both in its light (domestic) and dark (wild) forms.

MICHEL GOSSELIN AND MICHEL ROBERT

Jean-Marc Giroux

Notes on territory

The biology of the Ringed Turtle-Dove in the wild is poorly understood and most information on the species is derived from birds in captivity.

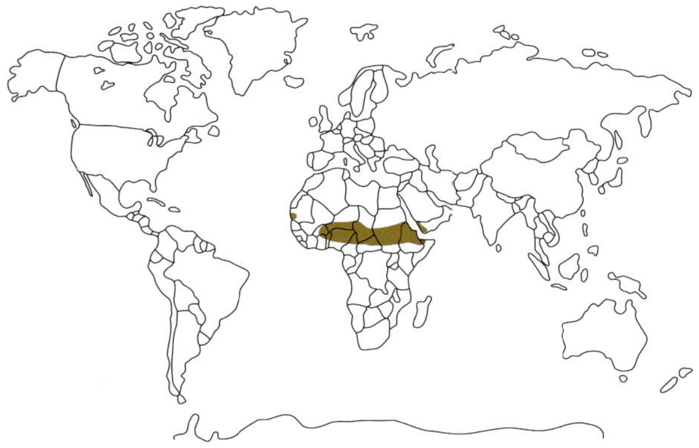

World breeding range

Species profile	
Clutch size	2 eggs (1–2)[1,4]
Laying interval	1 egg per 1–2 days[3,4]
Start of incubation	?
Incubation period	≈ 14 days[3,4]
Maturity at hatching	altricial
Care of young	by female and male[2,4]
Nestling period	16–19 days[2]
Age at first flight	16–19 days[2,4]
Dependence of young	probably 4–10 months[4]
Broods per year	several in captivity[2]
Breeding age	probably 4–10 months[4]
Mating system	?
Duration of pair bond	?
Total length	29–30 cm[1]
Wing span	45–50 cm[1]
Weight	M 157.3 g (135.0–172.0) / F 148.0 g (?)[1]
Longevity record	?

References:
1. Cramp 1985
2. ten Cate and Hilbers 1991
3. Goodwin 1983
4. Miller and Miller 1958

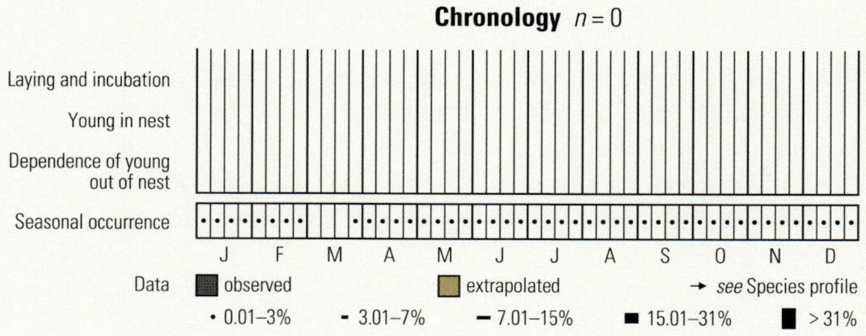

Passenger Pigeon

Tourte

Tourte voyageuse
Ectopistes migratorius

Michel Gosselin

TWO OR THREE HUNDRED YEARS AGO, THE NOW EXTINCT PASSENGER Pigeon was one of the most abundant birds in the world, with a total population of three to five billion birds. Immense flocks roamed the forests in eastern North America in search of food, in numbers so huge, sometimes into the millions, that they shut out the sunlight for hours (Schorger 1955; Bucher 1992).

Habitat and behaviour

The Passenger Pigeon, though closely related to the Mourning Dove, was larger and a more powerful flier, in keeping with its endless wanderings. Unlike the Mourning Dove, however, the Passenger Pigeon was highly sexually dimorphic, the plumage of the male and female differing quite significantly.

The Passenger Pigeon's breeding habitat was the vast beech and oak forests that formerly covered southeastern Canada and the northeastern United States, where it found the food it needed to sustain its immense colonies (Schorger 1955; Kirk 1994b). It nested early in spring to take advantage of the acorns and beechnuts left on the ground from the previous autumn before they germinated. The nesting cycle was brief, with only 30 days between building of the nest, a rudimentary affair, and chick abandonment (Schorger 1955).

Most of the population nested in a few central sites. Some forests in the eastern and central United States, in Pennsylvania and Michigan for example, supported huge colonies. The largest colony observed was in Wisconsin; it occupied an area of almost 2200 km^2 and housed 135 million birds (Blockstein and Tordoff 1985). Numerous other small colonies were also found. The species relied on its colonial and migratory behaviour to ensure nesting success; the locations of the colonies changed from year to year, and local predators, small in number compared to the pigeons, had little effect on the birds. Passenger Pigeons laid only a single egg each year and abandoned the squabs when they were 2 weeks old, before they had even made their first flight. This behaviour, which was useful when the birds nested in huge colonies, was disastrous for pairs nesting solitarily (Schorger 1955; Blockstein and Tordoff 1985).

In colonies in the northern United States, the breeding season was probably over by the end of May (Schorger 1955). After nesting, adult and juvenile birds travelled in separate flocks until autumn. Indeed, most of the Québec records involve these post-breeding movements, when the birds would travel north to feast on the abundant and varied supply of seeds and berries (Schorger 1955). Several smaller colonies, established late in the breeding season, could also be found in Canada.

In Québec, the Passenger Pigeon was observed mainly in summer; the earliest arrival date recorded was 30 April 1861, at Sillery (Le Moine 1861), and the latest fall record was 9 December 1888, at Lachine (Wintle 1896). Boucher (1664), who lived in the Trois-Rivières area, reported that the species could be observed between May and September, which seems fairly consistent with what is known about the species' annual occurrence in the province.

Distribution and history

Although the Passenger Pigeon was reported widely in southern Québec, there are very few breeding records and a relatively small number of individuals appear to have nested in the province. On this subject, Schorger (1955) cites Pehr Kalm, a Swedish naturalist who visited Québec in 1749 and provided the best description of the species' breeding behaviour in Québec. In July 1749, Kalm wrote from Saint-Jean-sur-Richelieu:

> Wild pigeons. The French call this bird *Tourte*, claiming that in summer an infinite number of them can be found in the large tracts of forest and virgin lands here, where they build their nests in the trees; the flocks often darken the sky; when it begins to get cold, they go south and are not seen again until the following spring...(Kalm 1759 *in* Rousseau et al. 1977 [Translation]).

Elsewhere, Kalm gives a detailed account of this pigeon in North America:

> The birds spend the entire summer in Canada, and particularly do they nest in the vast wild forests and wastes which abound there, where no men are to be found and where seldom any human being ventures. When in summer a person travels through these forests he might easily become terrified by the enormous number of these birds, which in some places almost entirely cover the branches of the trees and, when taking wing, obscure the sky. These Pigeons have, however, their distinct boundaries, outside of which they do not often venture; as for example, somewhat south of Bay St. Paul, which is 20 French miles north of Quebec, not very many of them nest in the woods; and the cause of this is said to be that the oak and beech tree, which supply them with their principal food, are here arrested in their growth, and grow no further north.
>
> In forests where there are human settlements or where the country is inhabited, only a few are to be seen; and as the land is being gradually cultivated by man, the Pigeons move further away into the wilderness. It is maintained that the cause of this is, partly, that their nests and young are disturbed by boys, partly their own sense of a lack of safety, and finally that during a great part of the year their food is shared by the swine.
>
> They build their nests in high trees, pine trees as well [as] deciduous ones; often as many as 40 or 50 nests are to be found in the same tree.
>
> Some maintain that they raise two broods of young every summer.
>
> In places where they nest in abundance the ground is often covered with their droppings to a thickness of one or two feet.
>
> While these birds are hatching their young, or while the latter are not yet able to fly, the savages or Indians in North America are in the habit of never shooting or killing them, nor of allowing others to do so, pretending that it would be a great pity on their young, which would in that case have to starve to death. Some of the Frenchmen told me that they had set out with the intention of shooting some of them at that season of the year, but that the savages had at first with kindness endeavoured to dissuade them from such a purpose, and later added threats to their entreaties when the latter were of no avail.
>
> In Canada it is almost everywhere the custom for young farmhands and boys to investigate where the Pigeons have their nests, and as soon as the young are able to fly they are taken from the nest and brought to the farm, where they are afterward kept in suitable quarters and industriously fed, whereupon they are killed and eaten. To make doubly sure that they do not escape, one of their wings is generally cut short so that even in case they do get out, they cannot fly away. Such nestlings have a good appetite, thrive comfortably, become quite tame and within a short time, if well taken care of, accumulate so much fat that they afford a most palatable dish (Kalm 1759, translated by Cronberger *in* Kalm 1911).

According to Kalm (*in* Rousseau et al. 1977), some probably fairly small colonies could still be found in the 18th century in the St. Lawrence Valley, at least in the Saint-Jean-sur-Richelieu and Baie-Saint-Paul areas. These colonies were most likely satellites of ones farther south.

Although Jacques Cartier himself spoke of the *turtre* (Laporte 1974a), most written reports of the Passenger Pigeon in Québec date from the second half of the 19th century, when the huge colonies in the northern United States were being increasingly decimated. Breeding records of the time show that its breeding range in Québec extended north to Belleterre in the Témiscamingue region to Lake Abitibi (Mitchell 1935), and to Fort George on the coast of James Bay (Oological Collection, Canadian Museum of Nature). The species had been driven to the northern limit of its range in what were probably its final attempts to breed.

The decline of the Passenger Pigeon was caused by market hunting, which started before 1800 but did not gather momentum until around 1840, when the development of the railroad allowed the professional pigeoners to access the colonies and the telegraph kept them informed of the birds' whereabouts (Blockstein and Tordoff 1985). Hundreds of thousands of young and adult birds were taken and sold at market every year. This appalling trade has been described in detail by many authors (Townsend 1932; Schorger 1955; McKinley 1960; Greenway 1967). According to Blockstein and Tordoff (1985), the taking of chicks and the disturbance of the colonies in general prevented the adults from breeding; the species probably declined very rapidly between 1871 and 1880 before finally disappearing altogether. However, Lambert (1814 *in* Kirk 1994b) had already mentioned a decline in population by the beginning of the 19th century. Other factors, particularly logging, may have also contributed to the Passenger Pigeon's decline (Schorger 1955; Blockstein and Tordoff 1985; Bucher 1992).

Comeau (1911) is the last author to report the species' presence in Québec: a male was observed in mid-May 1911 by the keeper of the Pointe des Monts lighthouse on the North Shore. Records from 1913 and 1925 from near Oka (Dionne 1914) and L'Ange-Gardien (Yourans 1930) are suspect and most probably involved Mourning Doves.

The last Passenger Pigeon died in the Cincinnati Zoo on 1 September 1914.

Michel Gosselin and Michel Robert

Species profile

Clutch size	1 egg[1]
Laying interval	–
Start of incubation	after laying of egg[1]
Incubation period	13 days[1]
Maturity at hatching	altricial
Care of young	by female and male[1]
Nestling period	≈ 14–16 days[1]
Age at first flight	≈ 17–18 days[1]
Dependence of young	until 2 weeks old[1]
Broods per year	1, rarely 2[1,2]
Breeding age	?
Mating system	monogamy[2]
Duration of pair bond	?
Total length	M 41.2 cm (36.1–45.9) / F 35.2 cm (29.0–42.8)[1]
Wing span	58.4–62.5 cm[1]
Weight	? g (255.6–340.8)[1]
Longevity record	at least 17 years[1]

References:
1. Schorger 1955
2. Ehrlich et al. 1988

Chronology $n = 0$

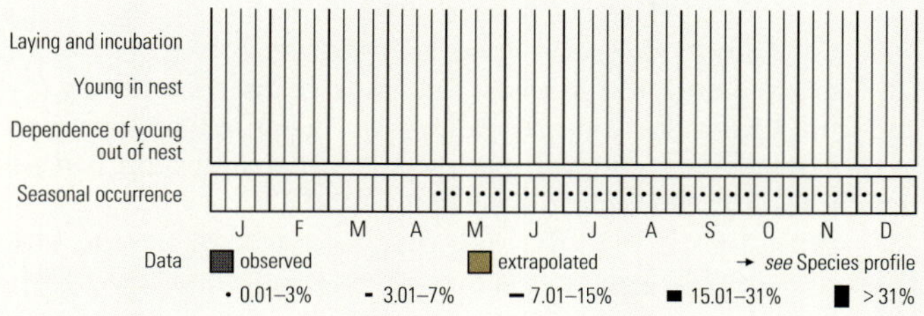

Monk Parakeet

Conure veuve

Myiopsitta monachus

Christiane Girard

IF, ONE DAY, YOU SEE A SMALL GREEN PARAKEET WITH A BUFF BAND ACROSS its belly at your feeder, don't worry, you are not hallucinating. The Monk Parakeet, a South American species, is not a migrant, but rather a cage bird that escaped or was let go. Its name comes from the grey plumage on its head and breast, which gives it a monkish air. Gentle and intelligent, the Monk Parakeet is a valued pet, but some bird owners, initially captivated by its striking coloration and ability to imitate the human voice, grow weary of its squawking and decide to set it free.

Monk Parakeets with their communal nest in an urban area. This nest, in a pine tree, is made of twigs and branches. Any birds of this species in Québec have either escaped or been released from captivity. North Carolina, United States. *January 1993, Denis Faucher.*

Habitat and behaviour

In Québec, the species forages in gardens, orchards, and farmland, where it finds thistle seeds, berries, large insects, ripening corn, and fruit. It also likes to feast on sunflower seeds in bird feeders and, in the United States, is more often seen in urban and suburban areas than in rural areas (Neidermyer and Hickey 1977).

Although it may fly long distances to forage, the Monk Parakeet is highly sedentary. Once it has established a territory, it starts to make a nest, regardless of the time of year. The spherical or domed nest is built in a treetop, on a telephone pole, or on some other artificial support (Bull 1974). The large, bulky structure of twigs, often thorny ones, may be up to 1 m in diameter and contain several chambers, each with its own entrance tunnel below. Although nests with a single chamber are not unusual, enormous communal nests with as many as 20 chambers have been found (Forshaw 1978). Nests occupied by up to 12 pairs have been reported in South America (Terres 1980).

Nest are used for brood rearing as well as for roosting in winter. The same ones are used year after year and are continually reinforced with twigs. In South America, pairs may produce two broods a year (Forshaw 1978).

Distribution and history

The Monk Parakeet is native to South America, where it is found from Bolivia to Argentina. In the early 1960s, it was introduced into the United States as a cage bird and, between 1968 and 1972, more than 64 000 of these parakeets were brought into the country (Banks 1976). Many were set free or escaped from pet shops, zoos or private collections. For example, in New York City, a large number of birds apparently gained their freedom when the crate they were in was accidentally dropped at the John F. Kennedy Airport.

The first evidence of breeding in North America was found in 1969 in Miami (Owre 1973). The species spread rapidly through the northeastern United States and the population peaked in 1972. The next year, the U.S. Fish and Wildlife Service and various state governments launched an active control program against it to ensure that it did not transmit psittacosis, damage crops, or compete with native species (Neidermyer and Hickey 1977). In Argentina, the Monk Parakeet is said to destroy 2–15% of corn crops every year, and sometimes up to 45% locally (Long 1981). It is also blamed for crop damage in Great Britain (Bump 1971). By the time the control program was initiated in the United States, the species appeared to have become acclimatized to winter conditions in the northeastern part of the country, which are similar to those in its native South America. In fact, the population was growing and many juvenile birds had already been observed. By 1975, however, it had been virtually wiped out in the United States.

In Québec, a Monk Parakeet was first sighted in the wild in May 1980 in Montréal (Gosselin and David 1980b). In 1984, a pet shop closed and 6 birds were released in Dollard-des-Ormeaux (Yank and Aubry 1984). Over the next few years, they were spotted in several places in and around Montréal and apparently attempted to breed, but without success. In May 1987, parakeets nested successfully in Laval, producing 4 young (Yank et al. 1987b); this is the first record of breeding in Canada. One parakeet was seen at a feeder in Aylmer on 10 November 1990 (Gosselin et al. 1991a), and another was seen in Tadoussac on 19 June 1991 (Gosselin et al. 1991b).

BENOÎT LIMOGES

Notes on territory

The Monk Parakeet does not defend a territory. In its native South America, nests can contain up to 20 chambers, each pair having its own chamber, although pairs often nest singly (Forshaw 1978).

It has been observed successfully defending a nest site against a House Sparrow (Freeland 1973).

Chronology *n* = 0

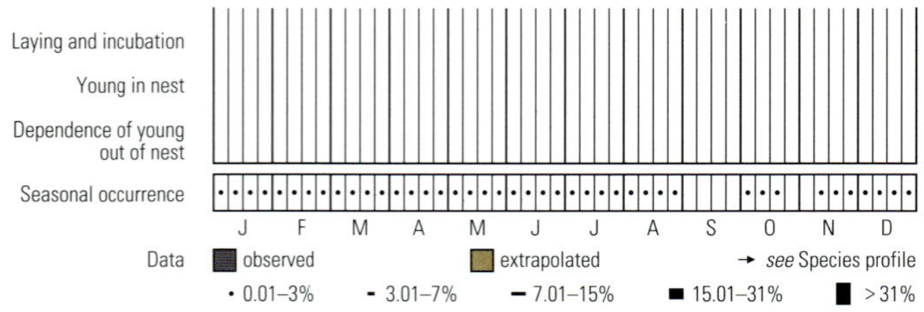

World breeding range

Species profile	
Clutch size	4–8 eggs[1,4]
Laying interval	?
Start of incubation	with first egg[2]
Incubation period	24–27 days[4,5]
Maturity at hatching	altricial
Care of young	?
Nestling period	6–7 weeks[1,2,4,5]
Age at first flight	6–7 weeks or more[6]
Dependence of young	?
Broods per year	1–2[1]
Breeding age	?
Mating system	?
Duration of pair bond	?
Total length	29 cm[1,3]
Wing span	?
Weight	140 g (?)[3]
Longevity record	?

References:
1. Forshaw 1978
2. Neidermyer and Hickey 1977
3. Terres 1980
4. Urban 1990
5. Vriends 1984
6. according to nestling period

Snowy Owl

Harfang des neiges

Nyctea scandiaca

AS THE PROVINCIAL BIRD, THE SNOWY OWL SYMBOLIZES QUÉBEC'S WHITE winters and northern character. The species was chosen from among 17 candidates and gained official status in December 1987 (Aubry et al. 1980; Simard 1988).

This large arctic owl has no prominent ear tufts and is easily identified. Adult males are almost pure white, usually with some scattered dark brown spots or barring, while the larger and heavier female has more of the same markings on the body and wings. Immatures are even darker and more heavily barred than adult females. Both sexes become lighter with age. These birds are well adapted to their northerly range; the dense underlayer of down overlaid with thick lightweight feathering to the base of the talons insulates them even in temperatures of –50°C.

Snowy Owls breed on the arctic tundra, including northern Québec, but may be seen in the south during their periodic winter invasions and are regular winter visitors to the southern part of the province.

Habitat and behaviour

The Snowy Owl breeds on the tundra in the High and Low Arctic, usually among hummocks or rocks (Godfrey 1986). Outside the nesting season, the birds are shy and quiet, but during breeding they are more vocal, especially near the nest (Johnsgard 1988). During courtship, unpaired males make long, slow flights in search of females, periodically uttering a loud hooting sound, which is answered by receptive females. The most spectacular aspect of courtship is an aerial display performed by the male, in which he glides with the wings in a high V position alternating with rapid wingbeats in a silent, mothlike flight. The courting male, often with a lemming clamped in his beak, then lands near the female, and keeping his wings partly spread, with the wrists held high, walks, bows, and turns in front of her. The owl may repeat this courtship display as he flies from perch to perch (Taylor 1973).

The nest site is chosen as a vantage point from which to survey the surrounding terrain, usually on top of large boulders or hummocks, but sometimes on a gravel bank or in a coastal marshy area. The nest is a shallow depression, usually unlined but occasionally lined with feathers, moss, or grass and constructed solely by the female (Parmelee 1992). Both parents will actively defend the nest, sometimes simulating a broken wing to lure a predator away (Murie 1929). If the predator persists, it risks an attack to its head in a flurry of talons and beak (Karalus and Eckert 1974). Incubation and then feeding of the young is performed entirely by the female while the male hunts and delivers food to her at the nest. Clutch size varies greatly, depending on local or temporal fluctuations in the food supply, with as many as 14 eggs being laid in years when lemmings, their staple food, are abundant (Robinson and Becker 1986). In years when local lemming populations crash, owls will not nest or will move elsewhere (Parker 1974). After hatching, the young are brooded for 14–26 days by the female (Parmelee 1992), after which they disperse to find safety among the nearby plants and rocks. The scattered young are fed by the male, which continues to supply the female as she incubates eggs that have not yet hatched and broods the younger chicks. At about 50 days the young can fly, and when 60 days old they hunt by themselves (Johnsgard 1988; Voous 1989). The family tends to remain loosely together until the end of the summer, when the young begin to disperse (Karalus and Eckert 1974).

Most Snowy Owls are monogamous, but cases of polygamy involving one male and two females have been reported (Watson 1957). While pair bonds appear to last for only one season, exceptions have been recorded: some pairs will remate for many successive breeding seasons (Robinson and Becker 1986). In the wild, they probably do not breed until they are at least 2 years old, but captive birds have bred at 1 year of age (Johnsgard 1988).

On the breeding grounds, these owls prey almost exclusively on lemmings, but they will also hunt mice, rats, hares, ground squirrels, and birds such as ducks, young geese, grouse, crows, ravens, and ptarmigan. They will even fish for their meal. Their hunting methods are varied: watching for prey from a perch, flying low and silently over the ground and then diving upon the prey, or walking and hopping over the snow and then listening for prey immediately below (Karalus and Eckert 1974; Chamberlin 1980). They have also been known to kill and eat mammals such as foxes caught in leg-hold

Ghislain Caron

Notes on territory

Breeding territories are defended by the males with vocalizations and threat postures (Parmelee 1992).

Breeding densities vary considerably, depending on the abundance of prey. The smallest area occupied by a breeding pair reported for North America is just under 1 km² (Pitelka et al. 1955).

The Snowy Owl breeds on the arctic tundra. It perches on whatever low vantage point it can find to watch for any movement by rodents or birds. Bathurst Island, Northwest Territories. *June 1969, Pierre Lamothe.*

The nest is no more than a slight depression in a low mound. Clutch size depends on the abundance of prey, but heavy snowfall does not prevent the species from nesting. Bathurst Island, Northwest Territories. *July 1969, Pierre Lamothe.*

traps. They owe their remarkable hearing to their facial discs, which act as parabolic reflectors, collecting and focussing sound waves, allowing them to zero in on the exact location of their prey (Payne 1962). Unlike many other owls, Snowy Owls are highly diurnal, hunting by daylight, which is almost continuous within the Arctic Circle throughout the summer breeding season.

Distribution

Snowy Owls have a circumpolar distribution, and in Québec, they breed above the tree line around Povungnituk, the Kogaluc River, Kuujjuaq, and Ungava Bay (Godfrey 1986). The Snowy Owl is not considered a migratory species and will winter on its arctic breeding territory as long as there is sufficient food. However, when there is a decline in the local availability of lemmings, large numbers do make irregular incursions into southern Canada and the northern United States (Kerlinger and Lein 1988b). These movements vary considerably in their intensity, their location, and the total amount of territory affected (Johnsgard 1988). Areas in which rodents are both numerous and easily caught, such as stubblefields and hayfields, appear to be favoured winter habitats (Lein and Webber 1979). While these irruptions generally take place every 3–5 years, some have been as much as eight years apart (Karalus and Eckert 1974) while others have occured in two successive winters (Godfrey 1986). According to Gross (1947), these invasions have been recorded since 1833–34, and the winter of 1945–46 saw one of the largest southward movements noted since the first records. These owls remain territorial even in winter, and both adults and immatures may defend territories for several months (Evans 1980; Boxall and Lein 1982).

Recent studies have shown that there are major age- and sex-related differences in winter movements and distribution among Snowy Owls in North America (Kerlinger and Lein 1986). Immature males were found to winter the farthest south, adult females the farthest north, and the immature females and adult males in between. Adults appear to winter south to the northern Great Plains regularly, while in Québec and points east, and on the west coast, most winter migrants represent the periodic and irregular incursions of immatures (Kerlinger et al. 1985a; Kerlinger and Lein 1988b). In the St. Lawrence Valley, population fluctuations peaked every three or four years between 1960 and 1981, but these fluctuations cannot be considered regular cycles (David and Larivée 1991). Although some Snowy Owls may be nomadic in winter, many become winter residents on the Prairies of Canada and the north-central United States (Keith 1960, 1964; Boxall and Lein 1982), and some return to the same place each winter (Bent 1938; Oeming 1958, 1964; M. R. Lein, U. of Calgary, unpubl. data).

Summer sightings have occurred in southern Québec, although they are rare and are usually of non-breeding birds that have remained in the south when prey such as rodents are plentiful. Individual owls have been observed on several occasions in the Magdalen Islands in summer (Fradette 1992). A possible breeding pair was atlassed on Brion Island on 21 June 1987, where two adults were seen with another of indeterminate age but which was heavily spotted.

Notes on territory (continued)

Several studies on Snowy Owls have been done in the Northwest Territories. On Baffin Island, breeding territories ranged from 1.2–3.9 km^2 (mean 2.6 km^2) in years of abundant prey to 8–10 km^2 in years of scarcity (Watson 1957). On Southampton Island, territories were as large as 22 km^2, and mean distance between nests was 4.5 km (Sutton 1932). On the Dundas Peninsula on eastern Melville Island, the density was 1 owl/69 km^2 (Miller and Russell 1973).

In winter, territories were as small as 0.8 km^2 in northwestern Michigan (Chamberlin 1980), and on Wolf Island, Ontario, the highest density found was 2.3 owls/100 ha (Weir 1973). Winter ranges are smaller than summer territories because each supports only one bird instead of a growing family (Chamberlin 1980).

Areas close to open water are good nesting habitat for the Snowy Owl, enabling it to vary its diet with the many species of waterbirds and small rodents to be found there. Rankin Inlet, Northwest Territories. *June 1980, Yves Aubry.*

Like the young of the Short-eared Owl, young Snowy Owls disperse into the area around the nest at the age of 2–4 weeks. Bathurst Island, Northwest Territories. *August 1969, Pierre Lamothe.*

History and trends

Since most of the Snowy Owl's breeding range is in remote areas well away from human activity, it is believed that its general status has changed little in North America. The species is neither evenly distributed nor permanently abundant anywhere, however, because its population densities are so interlinked with the periodic fluctuations of its food supply (Bent 1938).

On the breeding grounds, Snowy Owls have been trapped and eaten by humans for thousands of years (Parmelee 1992). Food shortages are the greatest threat to Snowy Owl populations, particularly immatures. On their southern wintering grounds, humans are the chief enemy (Duhamel and Gaudreau 1983). Although owls are protected by provincial regulations in Québec, as in the rest of Canada, trophy hunters continue to shoot many winter visitors in the southern part of the province. Approximately a third of the Snowy Owls brought in to the clinic of the Union québécoise de réhabilitation des oiseaux de proies at the University of Montréal Veterinary College in Saint-Hyacinthe in the winters of 1991–92 and 1992–93 had been wounded by gunshot (G. Fitzgerald, pers. comm.). On the Canadian Prairies, traumatic injuries—including collisions with automobiles and utility lines, as well as gunshot wounds—appear to be the major cause of mortality (Kerlinger and Lein 1988a). Because these owls are attracted to wide, relatively undisturbed fields, they have recently become a problem at airports in Canada, particularly in Toronto and Vancouver, where they have collided with airplanes (Voous 1989).

Derin Henderson

World breeding range

Species profile

Clutch size	usually 5–9 eggs (3–14)[1,2,4,5,6,8]
Laying interval	1 egg per 2 days[1,2,3]
Start of incubation	with first egg[1,2,5]
Incubation period	30–37 days[1,2,4,5,7]
Maturity at hatching	semi-altricial
Care of young	by female and male[2,3], *see* text
Nestling period	14–26 days[2]
Age at first flight	≈ 50 days (43–63)[1,2,3,7]
Dependence of young	until 5–7 weeks after leaving nest[2]
Broods per year	1[2,3,5]
Breeding age	probably more than 2 years[1,2]
Mating system	monogamy, rarely polygamy[2]
Duration of pair bond	usually 1 season[2,3,8]
Total length	M 53.1–70.7 cm / F 63.0–76.7 cm[5]
Wing span	M 157.7 cm (131.7–165.6) / F 164.4 cm (156.7–181.6)[5]
Weight	M 1612.9 g (1448.0–1839.9) / F 1706.7 g (1593.0–2002.6)[5]
Longevity record	17 years[5]

References:
1. Mebs 1989
2. Parmelee 1992
3. Voous 1989
4. Taylor 1973
5. Karalus and Eckert 1974
6. Godfrey 1986
7. Harrison 1978
8. Robinson and Becker 1986

Chronology *n* = 2

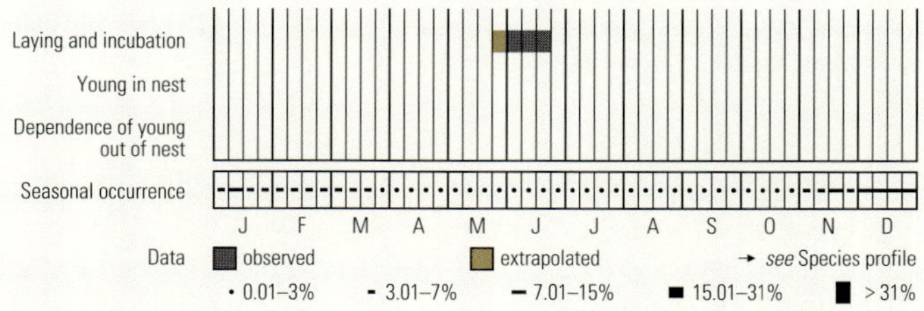

Red-bellied Woodpecker

Pic à ventre roux

Melanerpes carolinus

Christiane Girard

ANY BIRDER WANTING TO FIND A RED-BELLIED WOODPECKER SHOULD FIRST become familiar with the breeding calls of amphibians, as one of its calls is suggestive of the voice of the tetraploid gray treefrog (Godfrey 1986). Another of its calls is similar to that of the Red-headed Woodpecker. The Red-bellied is rarely seen in Québec and there is no evidence of its ever having bred here.

Habitat and behaviour

In New York, the Red-bellied Woodpecker breeds in flooded woodlands and in deciduous forest near clearings, pastures, and roads (Meade 1988d). The few nests that have been found in Ontario were in deciduous forests and wooded residential areas (Peck and James 1983).

The breeding cycle begins at the end of January, when at dawn on cold mornings the male begins uttering his characteristic *churr* near his winter roosting cavity. Later in the season, he will start drumming, beating at a rhythmic rate of three taps a second. A female may then join him and the pair drums together in synchrony (Kilham 1958a).

A Red-bellied Woodpecker spent most of the summer of 1989 in an oak forest at the foot of Mount Saint-Hilaire. *August 1991, Yves Aubry.*

The pair excavates the nest cavity in a dead tree, somewhere between 2 and 15 m above the ground (Bent 1939). It is not unusual for the more aggressive European Starling to take over the nest hole on completion. The pair defends the territory successfully, however, against intruding conspecifics (Kilham 1961). Both parents incubate the eggs for 12–14 days. They also take turns brooding the young throughout most of their first week. Both adults feed the young, which may follow them for up to 2½ months after they leave the nest (Kilham 1983).

The species feeds on wood-boring insects and their larvae as well as acorns, beechnuts, pine seeds, juniper berries, and the fruits of other shrubs. It is sometimes seen foraging for ants, beetles, crickets, and grasshoppers on the ground (Terres 1980).

Although the Red-bellied Woodpecker is extremely vocal at the start of the breeding period, it grows silent and secretive from May onward when the eggs are being incubated and the young raised. This means that the species was less likely to be found by atlassers, who were usually just beginning their fieldwork at that time of the year.

Distribution

The Red-bellied Woodpecker is resident in the eastern United States, and a partial migrant in the northern part of its range. Most Québec records are winter ones, however (David 1980a). The only Atlas record is from Mount Saint-Hilaire in 1989, where one bird was seen in an oak forest throughout most of the summer. There has since been a post-Atlas sight record of a pair on 7 May 1992 at Mount Saint-Bruno (Aubry and Yank 1992a; D. Daigneault, pers. comm.). The Red-bellied Woodpecker is probably at the northern limit of its range in New York State and in the southernmost part of Ontario, where it mainly occupies the Carolinian Forest zone (Terres 1980; Woodliffe 1987c), a habitat not found in Québec.

History and trends

Since the first record in 1944 (David 1980a), 30 sightings have been reported in Québec, 20 of them in the 1980s, so the range of the Red-bellied Woodpecker appears to be expanding northward. Although Ontario populations have fluctuated in the past, an increase in numbers has also been observed there over the last two decades (Woodliffe 1987c). The same thing has been happening in New York since the 1950s. This is due to both the increased popularity of bird feeders, which supplement traditional sources of food, and the effects of Dutch elm disease, which first occurred then. The dead trees attract insects and are thus both food sources and nest sites for the woodpecker (Bull 1985).

SIMON BARRETTE

Notes on territory

The male and female defend the territory against intruding conspecifics of either sex (Kilham 1961) by means of calls, drumming, displays, and even physical combat (Kilham 1983). The territory, which is about 2 ha (Graber et al. 1977), is used for mating and nesting, but not always for feeding (Stickel 1965).

Adults maintain individual territories of about 1.7 ha over the winter. The male may remain within the nesting territory throughout the year (Kilham 1961, 1983).

In New York State, territory densities were 1.2/10 ha (Maxwell 1993), 2.0/10 ha (Künstler 1993), and 3.5/10 ha (Jaslowitz 1993) at 3 mature deciduous woodland sites, and 2.8/10 ha in the Theodore Roosevelt Memorial Bird Sanctuary (Richard 1993).

Species profile

Clutch size	usually 4–5 eggs (3–8)[1,2]
Laying interval	1 egg per day[7]
Start of incubation	with third egg[7]
Incubation period	12–14 days[1,2]
Maturity at hatching	altricial
Care of young	by female and male[2]
Nestling period	24–27 days[2]
Age at first flight	24–27 days or more[10]
Dependence of young	until 42–75 days after leaving nest[1,3]
Broods per year	1–2[1,3]
Breeding age	1 year[8]
Mating system	monogamy[2]
Duration of pair bond	1 season[3,6]
Total length	22.2–25.4 cm[4]
Wing span	38.1–45.7 cm[1]
Weight	M 67.2 g (?) / F 56.2 g (?)[9]
Longevity record	20 years 8 months[5]

References:

1. Terres 1980
2. Ehrlich et al. 1988
3. Kilham 1983
4. Godfrey 1986
5. Kennard 1975
6. Kilham 1958a
7. Stickel 1965
8. DeGraaf and Rudis 1987
9. Dunning 1984
10. according to nestling period

Chronology *n* = 0

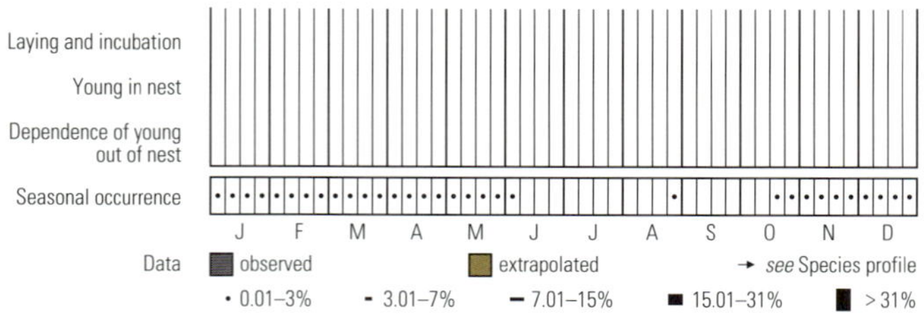

World breeding range

Acadian Flycatcher

Moucherolle vert

Empidonax virescens

Christiane Girard

LIKE MOST OF THE *EMPIDONAX* FLYCATCHERS, THE ACADIAN FLYCATCHER is identified by its song, an explosive *pit-see* or *wee-see*, with the first note accented and a rising inflection. It is accidental in Québec (David 1980a).

Habitat and behaviour

The Acadian Flycatcher is elusive and difficult to detect because of its habit of frequenting the middle level of mature deciduous woodlands (Godfrey 1986). In Ontario, it favours maple and beech forests, and is sometimes found on the edges of wooded ravines (Peck and James 1987; Woodliffe 1987a). In New York State, it also frequents groves of eastern hemlock, and is reportedly associated with rocky ravines and swift-moving streams (Eaton 1988a).

Pairs often nest in the same site year after year (Walkinshaw 1966a). The nest, located 2–9 m up, is a rather untidy structure suspended from a fork at the end of a horizontal limb (Peck and James 1987). Built by the female, it is held together with twigs, plant debris, and spider webs, and can be identified by the nesting materials dangling a few centimetres below the nest (Christy 1942; Newman 1958).

Both the male and female feed the young, which leave the nest when they are 13 days old and remain dependent on their parents for food for an additional 12 days or so (Harrison 1978).

The Acadian Flycatcher hawks for insects from a perch on a dead or leafless branch well beneath the canopy. Darting after its prey, it seizes it adeptly with its bill and returns to its branch. Its varied diet includes wasps, bees, flies, ants, caterpillars, moths, and sometimes crane flies and dragonflies. Prey that is too large to be swallowed whole is brought back to the perch, where it is held in the feet and decapitated (Terres 1980).

Distribution

The Acadian Flycatcher winters from northern South America to Central America and breeds from southernmost Ontario south to Texas and Florida (Godfrey 1986), arriving on its breeding grounds in mid-May (Christy 1942). It is an accidental visitor to Québec and has been recorded only twice in the province. One bird was seen at Natashquan on 2 June 1928 (Todd 1963). A male was atlassed singing in a willow swamp at Lake Sergent, Portneuf County from 22 July to 4 August 1984.

History and trends

In Ontario, habitat loss to agriculture has probably caused a decline in the breeding population and perhaps even reduced the species' breeding range in the province, where it is now listed as rare (Woodliffe 1987a). However, in New York State, populations appear to be increasing in response to forest regeneration on abandoned land logged at the turn of the century (Eaton 1988a).

SIMON BARRETTE

Notes on territory

The male defends a territory slightly larger than 1 ha against conspecifics (Christy 1942; Newman 1958).

In a Michigan study, territories ranged between 0.7 ha and 1.7 ha, with 1.2 ha the average size. The territory is used for nesting (Walkinshaw 1966a) and probably also for foraging.

In Pennsylvania, in a wooded suburb with deciduous trees, 3 territories were found within a 10-ha area (Robertson and Carey 1993), while in Connecticut, in an oak and hemlock forest, density was 0.9 territory/10 ha (1.0/10.5 ha) (Braunfield 1993).

World breeding range

Species profile

Clutch size	usually 3 eggs (2–4)[1,2,3,4]
Laying interval	1 egg per day[1,4]
Start of incubation	?
Incubation period	13–15 days[1,2,3,4]
Maturity at hatching	altricial
Care of young	by female and male[1,3,4]
Nestling period	13–15 days[1,3,4]
Age at first flight	13–15 days[1,2]
Dependence of young	until ≈ 10–12 days after first flight[1,4]
Broods per year	1[5]
Breeding age	1 year[5]
Mating system	monogamy[3]
Duration of pair bond	?
Total length	14–16 cm[7]
Wing span	21–24 cm[9]
Weight	12.9 g (9.9–16.1)[8]
Longevity record	10 years 11 months[6]

References:
1. Walkinshaw 1966a
2. Terres 1980
3. Ehrlich et al. 1988
4. Harrison 1978
5. DeGraaf and Rudis 1987
6. Clapp et al. 1983
7. Godfrey 1986
8. Dunning 1984
9. Cramp 1988

Tufted Titmouse

Mésange bicolore

Parus bicolor

Christiane Girard

THE TUFTED TITMOUSE—ACTIVE, VIGOROUS, CURIOUS, AND FEARLESS—IS very much like the more familiar Black-capped Chickadee. In the eastern United States, this little grey bird, easily identified by its tufted crest, dark forehead, and rusty flanks, is often seen in gardens, where it is extremely popular, as it is quite tame and will eat right out of your hand. It is a fairly recent winter visitor in southern Québec.

Habitat and behaviour

The Tufted Titmouse usually nests in mature hardwood forest, or sometimes in orchards, parks, or ornamental trees in residential areas. It may also make use of nest boxes, but it does seem to prefer natural tree cavities (0.9–27.5 m up), such as knotholes or old woodpecker holes. It does not usually do any excavating itself and reuses the same cavity year after year if it is available (Bent 1946).

The female may make her nest any time from the end of April onward (Bent 1946). It takes her 6–10 days to gather the various materials needed (Brackbill 1970), which may often include cast snakeskins. She also uses hair, sometimes taken from dead animals, but sometimes boldly plucked from live animals, even a human being seated quietly near the nest (Bent 1946).

Typical clutch size is 5 or 6 eggs, which the female incubates for 12–14 days (Ehrlich et al. 1988; Grubb and Pravosudov 1994). The altricial young hatch naked and blind. Their eyes open at about 4 days, and at 10 days they are fully feathered and already look like the adults (Bent 1946). They leave the nest 15–18 days after hatching, but may remain with their parents until the following spring. Pairs generally seem to bond for life (Laskey 1957; Brackbill 1970).

The Tufted Titmouse is mainly insectivorous, eating primarily caterpillars and wasps, although acorns, beechnuts, and smaller seeds are also part of its diet. Beechnuts, which may account for 23% of its diet, are its chief source of vegetable food between August and February (Bent 1946).

The song of the Tufted Titmouse is heard especially in March and April (Currie 1987). It consists of two notes repeated from 3 to 11 times: *peto, peto, peto . . .* It is a clear, strong whistle, and its timbre is reminiscent of the whistled *tew* or *fee-bee* of the Black-capped Chickadee, but on a lower note. From late spring to early fall, the Tufted Titmouse tends to disappear and becomes unusually silent (Bent 1946).

Distribution

This titmouse is a permanent resident throughout almost the entire eastern United States, even in the most northerly reaches of its breeding range (Bent 1946; Bull 1974). Since the winter of 1978–79, small numbers have been sighted in southern Québec almost every winter except those of 1983–84 to 1985–86, especially between the Richelieu River and Lake Memphrémagog (Aubry et al. 1991; R. Yank, pers. comm.). These titmice probably come from Vermont. In the winter of 1990–91, 50 or so Tufted Titmice were reported (often near lakes or rivers) in various municipalities in the southernmost part of the province, some of them close to the Vermont border (Aubry et al. 1991).

Notes on territory

Territory size is 0.8–1.2 ha, and the territory is defended, chiefly by the male, from late winter until the end of the breeding season. Conspecifics are warned of the presence of a territorial male mainly by his song. The territory is used for mating, nesting, and feeding (Stokes and Stokes 1983). According to Offutt (1965), the territory begins 5 m from the ground and goes up to the treetops.

The family group occupies a winter territory of 6–8 ha (Stokes and Stokes 1983). A density of 3.2 breeding pairs/10 ha was found in Maryland forests (Stewart and Robbins 1958). In two mature hardwood stands in urban areas of New York State, 4.5 and 5.0 territories/10 ha were counted (Jaslowitz 1993; Künstler 1993).

The Tufted Titmouse has been seen at feeders almost every winter since 1978 in southern Québec. Lennoxville, near Sherbrooke. *February 1982, Normand David.*

There are also three summer records: one bird was seen on 4 June 1988 at Saint-Bruno-de-Montarville, another on 12 and 18 June 1988 at Victoriaville (about 130 km from Vermont), and the third on 27 May 1989 at Philipsburg.

Banding statistics for the species in the United States have shown that fewer than 1% of birds venture more than 30 km from their place of capture (Elder 1985). As a result, considering that several sightings have been made more than 30 km from the American border and that the species is fairly hard to spot during the summer, the Tufted Titmouse may already be breeding in Québec, despite the low number of summer records and the lack of breeding evidence. It is doubtless only a question of time until breeding is confirmed.[1]

History and trends

The Tufted Titmouse has definitely been making incursions northward in the last few decades. It spread throughout the state of New York in the mid-1950s (Meade 1988e), southern Ontario in the early 1960s (Currie 1987), then Vermont in the late 1970s (Nichols 1985c).

This expansion is probably not linked with the growing prevalence of winter bird feeders, since providing extra food at that time of year does not seem to have any effect on the size of Tufted Titmouse flocks (Samson and Lewis 1979). Various authors cite instead climatic factors or the regrowth of mature forests on abandoned farmland (Eaton 1959; Boyd 1962).

In Vermont, the Tufted Titmouse seems to prefer to nest in valleys near rivers (Nichols 1985c). The first breeding record in Québec could therefore come from one of those lovely villages in the southern Eastern Townships or Montérégie, in a park or garden bordering deciduous woodlands dominated by hardwoods such as the American beech.

VINCENT LÉTOURNEAU

1 On 7 July 1995, an expected event occurred with the sighting of a pair of Tufted Titmice feeding three recently fledged young at Saint-Armand-Ouest. This represents the first confirmed breeding record for the province (Aubry and Bannon 1995).

World breeding range

Species profile

Clutch size	usually 5–6 eggs (3–9)[1,2,4,8]
Laying interval	1 egg per day[3]
Start of incubation	?
Incubation period	usually 13 days (12–14)[2,3,4,9]
Maturity at hatching	altricial
Care of young	by female and male[2,3]
Nestling period	15–18 days[2,3]
Age at first flight	15–18 days[2,3]
Dependence of young	until at least 46–47 days old[3]
Broods per year	1[1,2,3]
Breeding age	1 year[1]
Mating system	monogamy[1]
Duration of pair bond	for life[2,3,4]
Total length	15.3–16.5 cm[5]
Wing span	23.5–27.3 cm[4]
Weight	21.6 g (17.5–26.1)[7]
Longevity record	13 years 3 months[6]

References:

1. DeGraaf and Rudis 1987
2. Laskey 1957
3. Brackbill 1970
4. Terres 1980
5. Godfrey 1986
6. Clapp et al. 1983
7. Dunning 1984
8. Grubb and Pravosudov 1994
9. Bent 1946

Chronology $n = 1$

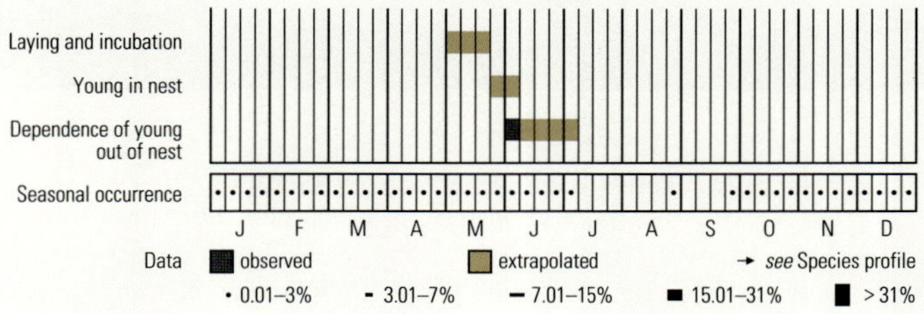

Northern Wheatear

Traquet motteux

Oenanthe oenanthe

Christiane Girard

THE NORTHERN WHEATEAR IS PRIMARILY AN OLD WORLD SPECIES. The birds that breed in the northeastern Canadian Arctic are of the subspecies *Oenanthe oenanthe leucorhoa*, which is also found in Greenland and Iceland. These wheatears migrate an incredible distance between their northern breeding grounds and their wintering grounds in western Africa, south of the Sahara, making an annual round trip of about 16 000 km! They migrate across Western Europe and the Atlantic Ocean to Greenland, a distance of up to 2400 km (Cramp 1988; Sandberg et al. 1991). Other than accidentals, the birds that breed in northern Québec do not pass through the southern part of the province in migration, and the species is rarely sighted there.

The Northern Wheatear nests on the tundra and coastal areas of the Québec Arctic. Salluit, northern Ungava Peninsula. *June 1993, Yves Gauthier.*

Habitat and behaviour

In North America, the Northern Wheatear, a bird of open terrain, nests on the arctic tundra (Godfrey 1986). Like the other members of its family, it maintains a well-defined breeding territory, which it establishes wherever it can find suitable rocks or cavities for nesting and wide open areas for hunting. In Greenland, for example, it frequents stony moraines, valleys, dry river beds, and heathland with scattered boulders or large rocks (Cramp 1988). The Northern Wheatear is a ground feeder, foraging in the short turf for insects, which it finds mostly by running along rocks or sitting on a rock or small bush and keeping a sharp eye out. The female builds the nest under rocks, or in a burrow, crevice, or other cavity. The eggs, usually numbering 5–7, are incubated mostly by the female, while both parents share the tasks of caring for the young (Moreno 1984; Cramp 1988). The birds that breed in the Canadian Arctic are probably single-brooded as is generally the case in Iceland and Scandinavia (Bent 1949; Cramp 1988).

Distribution

Our knowledge of the breeding distribution of the Northern Wheatear in northern Québec is limited. Along the shores of Hudson Bay, on 1 August 1981, a pair with young was seen near Ivujivik, about 30 km southwest of Erik Cove (Gaston et al. 1985). Farther south along the Hudson Bay shore, breeding was confirmed near Akulivik and Inukjuak. At Akulivik, an adult wheatear accompanied by young was seen on 13 and 14 July 1993, and at Inukjuak wheatears were spotted on 9 and 20 June 1986 and on 3 to 8 June 1988. At Inukjuak 2 adults accompanied by one of that year's young were seen on 9 July 1993 (ÉPOQ; Y. Gauthier, pers. comm.). The species was also observed near Lake Guillaume Delisle on 27 June 1981 (ÉPOQ). East of Cape Wolstenholme, Erik Cove is a known breeding site (Godfrey 1986). A young bird in postjuvenal moult was captured there on 1 September 1926 among a pile of rocks, and a family group was seen there on 14 August 1938 (Todd 1963). The most recent breeding record comes from Aupaluk, on Ungava Bay, where a chick still with a few tufts of down was seen in the company of an adult on 12 July 1994, as well as 2 adults seen on 11 and 13 July 1994 (Y. Gauthier, pers. comm.). Further records from the coastal region of northern Québec include one made by Hantzsch (1929), who reported on a clutch that had

Notes on territory

The Northern Wheatear almost always defends a territory, whether on its breeding grounds, on the sites where it stages for a few days during migration, or on its wintering grounds. Wheatears are often highly aggressive to same-sex conspecifics and have a whole range of behaviour to express their displeasure to intruders (Cramp 1988). The size of the territory defended by the pair over the summer is quite variable and depends, among other things, on the number of wheatears present in the area. In Greenland, territories were found to range from 12 to 16 ha. In Europe, however, as well as on the wintering grounds, territories are usually much smaller (Cramp 1988).

The species nests in rocky areas covered with short turf around Ivujivik. Northwest tip of the Ungava Peninsula. *August 1982, Gaétan Guertin.*

The species has long been reported as a breeder along the coast of Erik Cove. The surrounding valleys and vast expanses of tundra in the region are perfect for it. Northwest tip of the Ungava Peninsula. *August 1982, Gaétan Guertin.*

been found near Killiniq (N.W.T.), near the province's northeastern tip. He also observed a family group in northeastern Ungava Bay. A bird was spotted at Lake Pio about 50 km south of Aupaluk, in mid-July 1982, and 2 others were seen at Tasiujaq, about 100 km south of Aupaluk, on 30 May 1984 (David and Gosselin 1983; ÉPOQ). On 8 June 1993, 2 wheatears engaged in what seemed to be a territorial struggle were observed at Salluit (ÉPOQ). In the light of this information, the species probably also nests elsewhere along the coasts of northern Québec, from Hudson Bay to Labrador.

History and trends

Some authors have reported that the Northern Wheatear used to breed on the North Shore (Merriam 1885b; Comeau 1890, 1923b; Todd 1963). In the last century, Comeau collected several specimens in the area around Godbout; although these reports date mainly from migration periods, some details are indicative of possible breeding there. For example, on 9 June 1885, Comeau killed a pair, the female of which had fairly well developed eggs in the oviduct (Merriam 1885b). He also observed 2 birds on 2 July 1891 and 7 or 8 on 7 August of the same year. Also at Godbout, he later made other interesting sightings, notably of 7 birds on 6 August 1898 and of juveniles in August 1922 (Comeau 1923b). Lewis (*in* Todd 1963) observed an individual at Sept-Îles on the North Shore during the breeding season, on 11 June 1949. It is therefore not impossible that the species breeds or has occasionally bred on the North Shore, since it nests on the Labrador coast, at least as far down as Gready, just a few hundred kilometres to the north of Blanc-Sablon. The Northern Wheatear has also been observed at Menihek Lakes, south of Schefferville, in early June (Smith 1969).

Michel Robert

World breeding range

Species profile

Clutch size	usually 5–7 eggs (2–9)[1,2,3,4,5]
Laying interval	1 egg per day[2]
Start of incubation	with last or penultimate egg[2]
Incubation period	usually 13–14 days (10–16)[1,2,4,5]
Maturity at hatching	altricial
Care of young	mainly by female[1,2]
Nestling period	usually 10 days[2]
Age at first flight	usually 15 days (10–21)[2]
Dependence of young	until 28–32 days old[2]
Broods per year	usually 1 (1–2)[2]
Breeding age	1 year[2]
Mating system	monogamy, sometimes polygyny[2]
Duration of pair bond	1 season[2]
Total length	14.0–15.9 cm[5]
Wing span	26–32 cm[2]
Weight	24.6 g (19.9–35.5)[6]
Longevity record	7 years[4]

References:
1. Harrison 1978
2. Cramp 1988
3. Bent 1949
4. Terres 1980
5. Godfrey 1986
6. Dunning 1984

Chronology $n = 5$

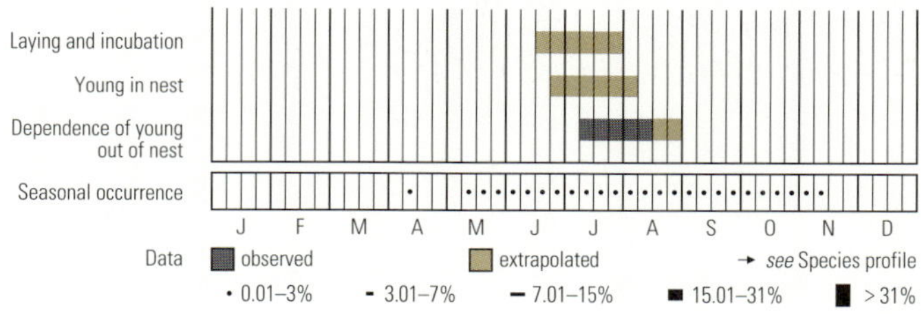

Gray-cheeked Thrush

Grive à joues grises

Catharus minimus

Christiane Girard

The Gray-cheeked Thrush breeds farther north than any of the other thrushes that occur in Québec; it is also the only North American thrush whose breeding range extends beyond the North American continent. It ranges as far as northeastern Siberia (Stepanyan 1990), although most of the population is found in North America, in the coniferous forests of northern Canada and Alaska. The species is not easy to observe, whether on the breeding grounds or in migration. In spring, it is the last thrush to return from the wintering grounds, with most birds arriving only after mid-May.

The Gray-cheeked Thrush and Bicknell's Thrush used to be considered conspecific, and both were known as the Gray-cheeked Thrush (AOU 1957), but recent research has shown that they are two distinct species (Ouellet 1993). Bicknell's Thrush is found in the southern part of Québec, whereas the Gray-cheeked Thrush occurs farther north (Ouellet 1993).

The Gray-cheeked Thrush is difficult to distinguish from Bicknell's Thrush in the field, and sight records south of its breeding range could well be of either species. Moreover, inexperienced observers may confuse the Gray-cheeked Thrush with Swainson's Thrush.

Habitat and behaviour

The Gray-cheeked Thrush is the typical thrush of the boreal forest and the taiga (Erskine 1977; unpubl. data). During the breeding season, it occurs in coniferous woods and in denser stands near the tree limit (Godfrey 1986; unpubl. data), and prefers a fairly thick, mossy ground cover and rather moist subsoil. The dominant tree species in these stands are black spruce, white spruce, and, to a lesser extent, tamarack (unpubl. data). In the boreal forest and the taiga, it is also found along rivers and streams, lakes, and ponds in clumps of alders or willows (Bent 1949; Todd 1963; unpubl. data). Harper (1958) reported it in areas of regeneration after a forest fire, in muskeg, and in spruce and fir stands. In migration, it may be sighted in any treed or shrubby area, even in city parks and suburban gardens (unpubl. data).

The species is highly secretive throughout the breeding period, from the middle or end of May to mid-August. The male sings mainly at dawn or dusk, but sometimes early in the breeding season it can be heard singing during the day perched near the top of a conifer. It is rare to find high densities of nesting pairs, as territories appear to be fairly spread out and distant from one another (unpubl. data). The loud, musical song differs from that of Bicknell's Thrush in that it drops toward the end. The Gray-cheeked Thrush generally forages on the ground and feeds mainly on insects (Bent 1949). It also eats fruit, especially during fall migration. In the coniferous forest, the nest is usually built in a tree less than 2 m up (unpubl. data), though sometimes higher (Bent 1949). As with other thrushes, the female builds the nest and incubates, although the male helps in rearing the young (Bent 1949; Godfrey 1986; unpubl. data).

This migrant species is found in southern Québec in spring, from mid-May to early June. In the fall, it migrates to its wintering grounds in Central and South America, beginning in mid-August in the northernmost part of the range and continuing until the end of September and possibly later. Like the other thrushes, Gray-cheeks migrate mainly during the night and are seldom seen in large numbers during migration. Most birds nesting in Québec appear to migrate through the eastern part of the continent, between the Mississippi Valley and the Atlantic coast (Wallace 1939; Bent 1949; AOU 1983).

Notes on territory

High breeding densities of Gray-cheeked Thrushes have not been reported in Québec, and information on territory size is still fragmentary and incomplete.

Counts made in 1975 in a mature fir stand near Candlestick Lake in Newfoundland yielded 5.1 singing birds/10 ha (Erskine 1980a). In 1966, in open black spruce forest near Rae, not far from Great Slave Lake in the Northwest Territories, 2.0 singing males/10 ha were reported (Erskine 1971a). In 1974, at Porter Lake, in the same region, 1.5 singing males/10 ha were recorded in an upland lichen woodland (Erskine 1976a).

On the taiga, the Gray-cheeked Thrush favours damp, mossy coniferous habitat. Lake Vincelotte, James Bay region. *July 1983, Henri Ouellet.*

Of all the North American thrushes, the Gray-cheeked has the most northerly breeding range. It nests primarily on the taiga, the transition area between boreal forest and tundra. A nest was found in this muskeg, north of Schefferville. *July 1958, Henri Ouellet.*

Distribution

The breeding range of the Gray-cheeked Thrush extends across a broad band of the boreal forest up to the tree line from Newfoundland to Alaska and eastern Siberia (AOU 1983; Godfrey 1986; Stepanyan 1990). In Québec, it occupies a vast area between the Lower North Shore of the Gulf of St. Lawrence and the tree line, sometimes beyond where ecological conditions are suitable, as in the forested enclaves of sheltered valleys (Harper 1958; Todd 1963; unpubl. data).

The Gray-cheeked Thrush is distinguished from Bicknell's Thrush, which has a more southerly range, by its morphological characteristics and its song. The Newfoundland race (*Catharus minimus minimus*), the nominate form, is found on the Lower North Shore of the Gulf of St. Lawrence, in the Strait of Belle Isle (Blanc-Sablon area), and somewhat farther west. Farther north, in the Goose Bay area of Labrador, it gives way to the subspecies *Catharus minimus aliciae*, which is found throughout the remaining portion of the boreal forest in Québec (AOU 1957; Todd 1963; Godfrey 1986; unpubl. data). In the western part of Québec, southernmost nesting sites are found at the southern limit of Hudson Bay (Cons. G. G. – GREBE 1991a) and in the northern part of the James Bay region (unpubl. data).

History and trends

Information is too scant to compare present and past abundance and assess changes in the breeding population. The species does not appear to be much affected by human activity on the breeding grounds. Data collected on the wintering grounds, however, indicate that Gray-cheeked Thrush populations may have suffered a decline, the extent of which is difficult to determine, again because of the paucity of data (Faaborg and Arendt 1985; Terborgh 1989; Askins 1992; unpubl. data).

Henri Ouellet

World breeding range

Species profile

Clutch size	usually 4 eggs (3–5)[2]
Laying interval	probably 1 egg per day[3]
Start of incubation	probably with third egg[3]
Incubation period	13–14 days[4]
Maturity at hatching	altricial
Care of young	by female and male[3]
Nestling period	11–13 days[2]
Age at first flight	11–13 days[2]
Dependence of young	until a few days after first flight[5]
Broods per year	1[2]
Breeding age	1 year[5]
Mating system	monogamy[2]
Duration of pair bond	?
Total length	M 18.9 cm (18.2–19.7) / F 18.1 cm (17.2–18.8)[6]
Wing span	26.7–34.3 cm[1]
Weight	M 34.2 g (28.2–36.8) / F 33.0 g (27.5–41.8)[6]
Longevity record	?

References:
1. Cramp 1988
2. Bent 1949
3. Wallace 1939
4. Godfrey 1986
5. H. Ouellet, CMN, unpubl. data
6. Ouellet 1993

Chronology *n* = 5

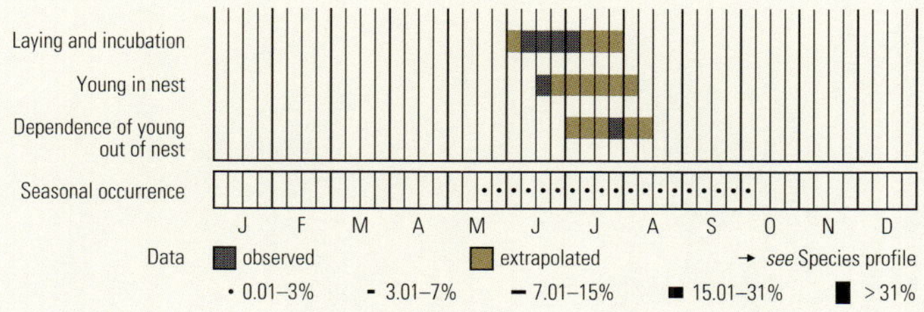

Bohemian Waxwing

Jaseur boréal

Bombycilla garrulus

Christiane Girard

In Québec, the Bohemian Waxwing is mainly an irregular winter visitor that can be observed here from late October to late April (David 1980a). It has never been confirmed as a breeder, but summer sightings of individual birds and small flocks in various northern boreal areas suggest that one day breeding will be documented in Québec.

The bird is very similar to the Cedar Waxwing, but is larger and has greyer plumage. In addition, it has white wing patches and rusty undertail coverts. It is sometimes located by its sibilant, buzzy call, lower in pitch than the Cedar Waxwing's, with notes somewhat more separated.

Habitat and behaviour

Throughout most of the year, the Bohemian Waxwing feeds on small fruit and berries, such as raspberries, choke cherries, haws, and rose hips. It seems to prefer mainly the berries of the common juniper and those of the American and the European mountain-ash (Bent 1950), the latter a naturalized ornamental tree found throughout southern Québec (Marie-Victorin 1964). In spring and early summer, the Bohemian Waxwing expertly hawks insects, which it takes in midair from an exposed perch (Terres 1980; Cramp 1988).

Fruit is digested rapidly (16–40 minutes) and extremely efficiently (Dorst 1974). Since it contains a great deal of water, fruit must be consumed in large quantities, 2 or 3 times the weight of the bird per day, to meet its energy requirements (Cramp 1988). A 57-g waxwing has been estimated to eat about 170 g of fruit a day (Gibb and Gibb 1951).

The male performs his courtship display near the female and may even touch her. The tail is lowered to a vertical position, the feathers on the belly and rump are fluffed out, the wings are drooped and the crest raised, making the bird appear larger. This behaviour is accompanied by a slight rotation of the head (Cramp 1988). The female responds by fluffing out her feathers as well. The male solemnly takes an object in its bill and proffers it to the female, which accepts it and then gives it back. This exchange may be repeated several times. The ritual may also occur in winter, as has been observed in mid-March in Manitoba (Cramp 1988). Such behaviour observed at this time of year should not, of course, be considered evidence of breeding.

The onset of nesting occurs between late May and the end of June (Terres 1980). In Scandinavia, it begins most often in mid-June (Cramp 1988). The Bohemian Waxwing nests in pine and spruce forest sometimes with an admixture of birch. The nest is often located at the forest edge, near a lake or stream, or in a peatland or burn (Harrison 1978; Godfrey 1986; Cramp 1988). It is built by both the male and female (Bent 1950; Cramp 1988), and is made of twigs, grass, mosses, and lichens (Harrison 1978; Godfrey 1986). It is generally located in an isolated conifer on a horizontal branch, well out from the trunk, about 1.2–15.2 m up (Terres 1980). The 4–7 eggs are incubated solely by the female for 13–15 days (Godfrey 1986; Cramp 1988). The young remain in the nest for 13–17 days and are fed a diet of insects and fruit by both parents (Harrison 1978; Cramp 1988). After the young fledge, the family group may remain together for a long period, until the end of January in North America (Cramp 1988).

As the species nests locally and in small numbers within its breeding range, it is easily overlooked (Cadman 1987a).

Notes on territory

Little is known about territoriality in the Bohemian Waxwing. It is a nomadic species that shows no inclination to nest in the same area from year to year (Dorst 1974). It is often considered a colonial species, in the sense that two or more pairs may nest within a few metres of each other (Bent 1950; Cramp 1988). The birds rarely show any signs of aggression, toward either other waxwings or different species (Cramp 1988).

The Bohemian Waxwing is mainly a fruit-eating species of the northern coniferous forest, often found near lakes and streams. Breeding has not yet been confirmed in Québec. Southwest of Schefferville. *August 1987, Christian Marcotte.*

Distribution

The Bohemian Waxwing is circumpolar in its distribution. Two subspecies occur in Eurasia, *Bombycilla garrulus garrulus*, which nests in northern Eurasia, and *Bombycilla garrulus centralasiae*, which is found in central and eastern Siberia (Bent 1950; Cramp 1988). The birds that breed in North America belong to a distinct subspecies, *Bombycilla garrulus pallidiceps*, documented to range across Alaska, the western Northwest Territories, British Columbia, and the northern parts of the Prairie provinces and Ontario (Godfrey 1986). Summer observations had long led ornithologists to suspect that the species bred in northern Ontario, but breeding was only confirmed there in 1984 (Peck and James 1987), on the Hudson Bay Lowland, and Cadman (1987a) believes that it may breed farther east, perhaps as far as James Bay.

There was only one pre-Atlas summer record in Québec, from Lac à la Loutre, in the southern Ottawa Valley on 6 August 1978 (David 1980a). Between 1984 and 1989, the Bohemian Waxwing was atlassed as a possible or probable breeder in 3 squares in Abitibi and near Matagami. A flock and a single bird in flight were also observed on the North Shore in early July 1986. A single bird was sighted on 29 June 1987 near Weedon Centre in the Eastern Townships.

Summer occurrence of the species in the boreal regions of Québec was again confirmed in 1991 during intensive surveys conducted in the James Bay region: several flocks were sighted in June, in an area between Rupert Bay and a point 100 km east of Nemiscau, and between the Broadback and Pontax rivers (Cons. G. G. – GREBE 1992o). In late July of the same year, 2 birds were sighted about 40 km east of the mouth of the Broadback River, and 2 others were sighted midway between the Rupert and Pontax rivers, 90 km west of Nemiscau (M. Robert, D. Daigneault, and A. Morrier, pers. obs.). These birds were in flight, and their status is uncertain. Also in 1991, sightings in the Great Whale and Little Whale river watersheds established the species as a probable breeder there (Cons. G. G. – GREBE 1991a). The species was also reported in the Schefferville region in August 1987 (C. Marcotte, pers. comm.).

History and trends

In North America, as in Europe, Bohemian Waxwings appear in sometimes massive invasions, as well as on their normal migratory movements. Dorst (1974) has associated increased abundance and expansion of the breeding range with some of these invasions. In North America, major invasions were reported during the winters of 1958–59, 1961–62, and 1968–69 (James 1962; Bagg 1969; Terres 1980).

Given the number and extent of summer sightings, it is probable that the Bohemian Waxwing breeds in certain parts of Québec. Better field coverage and more intensive monitoring of the species in northern regions will be necessary, however, to determine its status and evaluate the stability of summer populations.

VINCENT LÉTOURNEAU

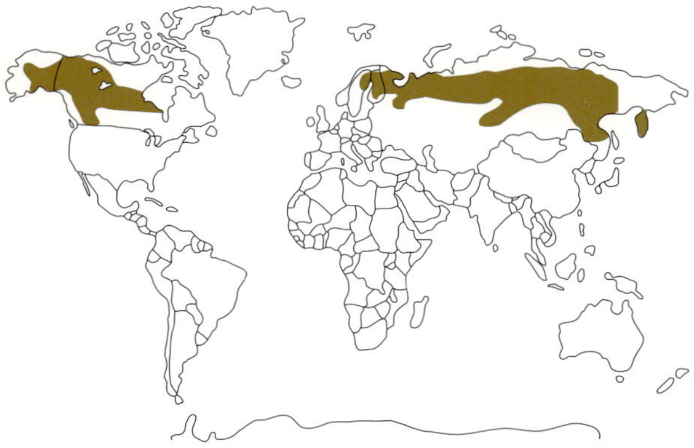

World breeding range

Species profile

Clutch size	usually 5–6 eggs (4–7)[5,6]
Laying interval	?
Start of incubation	?
Incubation period	usually 14 days (13–15)[1,3,4,5,6]
Maturity at hatching	altricial
Care of young	by female and male[5,6]
Nestling period	usually 14–15 days (13–17)[1,5,6]
Age at first flight	usually 14–15 days (13–17) or more[8]
Dependence of young	possibly until January[9]
Broods per year	probably 1[1]
Breeding age	?
Mating system	monogamy[1]
Duration of pair bond	?
Total length	19–22 cm[3]
Wing span	33.0–35.6 cm[4]
Weight	56.4 g (46.5–69.0)[7]
Longevity record	5 years 10 months[2]

References:
1. Ehrlich et al. 1988
2. Klimkiewicz et al. 1983
3. Godfrey 1986
4. Terres 1980
5. Harrison 1978
6. Cramp 1988
7. Dunning 1984
8. according to nestling period
9. Cameron 1908

Chronology $n = 0$

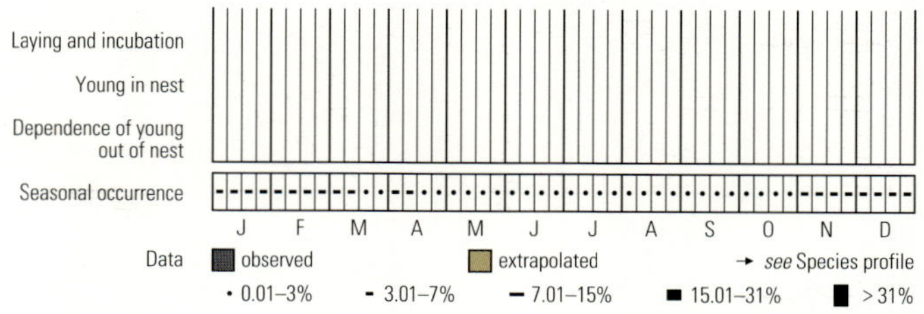

Northern Shrike

Pie-grièche grise

Lanius excubitor

Christiane Girard

Few Québec birdwatchers have been lucky enough to see the Northern Shrike in its northern breeding habitat, but most have seen it in the southern part of the province in winter, where, like the Snowy Owl, it is numerous in some years and scarce in others (Davis 1974). During these winter incursions, it is often observed perched at the top of a tall tree or pole or sitting on a telephone wire as it scans for prey. It also occurs around bird feeders, where it terrorizes smaller birds, much to the consternation of feeder watchers, though it does not take too great a toll and makes an interesting winter sighting.

Habitat and behaviour

Found in Eurasia, North Africa, and North America, the Northern Shrike is a rather ubiquitous species, occurring in a variety of settings, from the palm trees and deserts of North Africa to subarctic taiga (Parrot 1980). In North America, it breeds only in the north, within the broad belt of coniferous forest stretching from Alaska to Labrador. It is not, however, associated with unbroken forest, preferring instead open coniferous woodland and the outer edges of denser forest. It can be found, for example, near peatlands, burns, lakes, and rivers. It also frequents areas of low, shrubby vegetation near the tree line (Cade 1967; Godfrey 1986).

The rather bulky nest is built in a tree or shrub, generally at a height of 1–10 m. Some nests found in Québec and Labrador were built in spruce trees, 4–6 m up (Bent 1950; Todd 1963); another was located in an alder thicket about 1 m from the ground (H. Ouellet, pers. comm.). Both parents build the nest and feed the nestlings, but only the female incubates the eggs (Grönlund et al. 1970; Ehrlich et al. 1988). The Northern Shrike feeds mainly on small rodents, birds, and insects; juveniles eat mostly invertebrates during the first weeks after fledging. After the prey—small mammals and birds, most often small passerines—is killed, the shrike impales it or wedges it into the fork of a tree before devouring it (Bent 1950; Cade 1967). Prey is not always consumed immediately; some is cached in "larders" to be eaten later by the adults or young (Cade 1967; Grönlund et al. 1970).

Distribution

In Québec, according to Godfrey (1986), the Northern Shrike breeds from the James Bay and Hudson Bay coastal plains to Labrador between Schefferville and the southern part of Ungava Bay. Godfrey also reported breeding in the areas around the Rivière aux Feuilles, Koroc River, Schefferville, Great Whale River, and locally south to Cabbage Willows Bay. Breeding birds have been observed at the mouth of the Great Whale River, at Kuujjuaq, at Lac de la Hutte Sauvage (a widening of the George River), and near Cabbage Willows Bay, in Rupert Bay (Todd 1963; H. Ouellet, pers. comm.). More recently, adults with young were sighted on the De Pas River and on the George River, near the mouth of the Falcoz River and Kangiqsualujjuaq Bay (D'Amours 1989). The species probably breeds in the Lake Chavigny area, about 200 km east of Inukjuak (Rochefort 1983). Individuals were also sighted on the North Shore, at Harrington Harbour, in June 1976 (ÉPOQ).

The Northern Shrike was first confirmed as breeding in Ontario in 1981, in the Hudson Bay Lowland. A female with enlarged ova was collected at Moosonee in May 1972 (Cadman 1987b), suggesting that the species probably also breeds on the Ontario side of James Bay, just as it does on the Québec side.

Little is known about the Québec distribution of the Northern Shrike. Very little work on passerines has been done in central and northern Québec, and breeding evidence for this species is scant. The northern and southern limits of its breeding range will no doubt be established in the future.

Notes on territory

In Alaska, pairs are so widely dispersed that territories are not well defined. The birds do not always react aggressively when a conspecific approaches the nest. In his Alaska study, Cade (1967) estimated the foraging area of the male to be about 125 ha, but noted that the birds are rarely found more than 800 m from the nest.

The Northern Shrike lives in open coniferous habitat with sparse, shrubby vegetation, as shown here along the McPhadyen River, near Schefferville. Labrador, Newfoundland. *August 1985, Christian Marcotte.*

A small rodent impaled by a Northern Shrike. Rivière-Ouelle, southwest of Rivière-du-Loup.
February 1990, Denis Faucher.

History and trends

Data collected on the Northern Shrike in winter are insufficient to indicate any definite trends either in Québec or North America in general, as the species is irregular in its movements. This is also true of the fragmentary breeding range data.

Breeding density in northern Ontario is generally low (Cadman 1987b), whereas in Churchill, Manitoba and in Alaska, abundance varies from year to year (Jehl and Smith 1970; Cade and Swem 1993).

Michel Robert

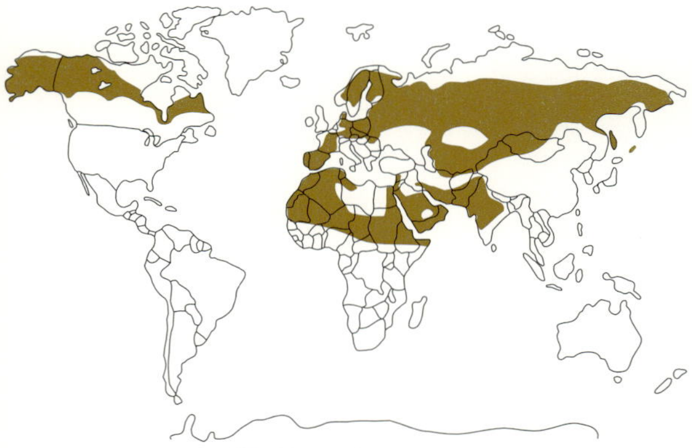

World breeding range

Species profile

Clutch size	usually 5–8 eggs (4–9)[1,7,12]
Laying interval	?
Start of incubation	before last egg[6]
Incubation period	usually 16 days (14–18)[1,6]
Maturity at hatching	altricial
Care of young	by female and male[7,8]
Nestling period	17–20 days[1,6,12]
Age at first flight	17–20 days or more[10]
Dependence of young	until ≈ 10–14 days after first flight[1,6,8]
Broods per year	1[8]
Breeding age	2 years[1]
Mating system	monogamy, sometimes polygyny[3,11]
Duration of pair bond	1 season[2]
Total length	22.5–27.5 cm[4]
Wing span	34.3–41.9 cm[5]
Weight	65.6 g (61.8–69.0)[9]
Longevity record	12 years[5]

References:
1. Degen et al. 1992
2. Yosef 1992
3. Yosef and Pinshow 1988
4. Godfrey 1986
5. Terres 1980
6. Cade 1967
7. Ehrlich et al. 1988
8. Harrison 1978
9. Dunning 1984
10. according to nestling period
11. Cramp and Perrins 1993
12. Cade and Swem 1993

Chronology *n* = 6

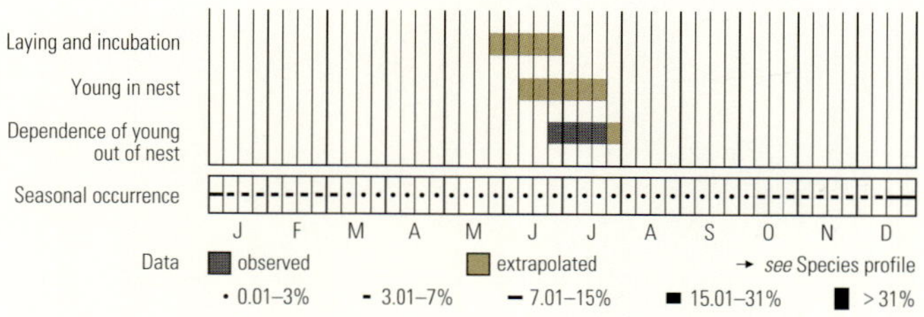

Blue-winged Warbler

Paruline à ailes bleues

Vermivora pinus

Christiane Girard

THIS DELIGHTFUL BUT RARE WARBLER, MUCH SOUGHT AFTER BY QUÉBEC birdwatchers, is currently listed as a casual visitor in the province (Robert 1989b). However, its status may well change in the next few years, if breeding evidence is reported.

The main hindrance to observing the species in Québec is, of course, its rarity. Once found, it can be fairly easily identified by its striking plumage: the narrow, black stripe through the eye, the greyish-blue wings and tail contrasting with the bright yellow face and underparts, and white wing bars. Its distinctive, two-syllable *bee buzz* song is easy to remember and recognize but is rather faint and can only be heard at fairly close range. The wheezy notes are similar to those of the Golden-winged Warbler and the hybrids of the two species (Gill and Murray 1972b; DeBenedictis 1984), Brewster's and Lawrence's warblers. For this reason, singing males should always be identified visually. The similarity of the Blue-winged and Golden-winged songs was noted during the first 2 years that the species was observed at Fitch Bay in the Eastern Townships (pers. obs.).

Habitat and behaviour

The Blue-winged Warbler frequents young deciduous woodlands, alder thickets around swamps and streams, old fields, and forest edges and clearings (Ewert 1981; Frech and Confer 1987). It is less of a habitat specialist than the Golden-winged Warbler and will use a wider variety of sites (Confer and Knapp 1981). The overlapping of the species' breeding ranges and their mutual tolerance favours hybridization.

Courtship behaviour occurs over a brief period of 1 or 2 days (Ficken and Ficken 1968d), whereas similar behaviour can be observed in other warblers over several days (Morse 1989a). Males begin establishing territorial limits on arriving at the nesting sites (Murray and Gill 1976). Perched at the top of a small tree, they sing until the females arrive, 1 or 2 days later. Pair formation occurs at this point, and the females begin nest building shortly after (Ficken and Ficken 1968d).

The nest is usually on or near the ground, on a mat of dead leaves, among ferns, or in tussocks. It is built of strips of bark, leaves, and grasses and lined with fine grasses and bits of bark (Murray and Gill 1976; Terres 1980; Godfrey 1986; DeGraaf and Rudis 1987). The female alone incubates the eggs for a period of 10–12 days. The young are fed by both the male and female (Godfrey 1986; DeGraaf and Rudis 1987; Ehrlich et al. 1988).

The birds leave for their wintering grounds, located from the south of Mexico to Panama, after the breeding season (Godfrey 1986).

Distribution

The Blue-winged Warbler has not yet been confirmed as breeding in Québec. The few Atlas reports were from the Ottawa Valley and the Eastern Townships in the southernmost part of Québec. It was sighted 3 years in a row on the same territory in Fitch Bay, in the Eastern Townships, from 1986 to 1988, but only one bird was seen each year. The species is considered a possible breeder because of territorial behaviour observed for over 7 consecutive days. During the same period, the Blue-winged Warbler was seen elsewhere in the region, on the west shore of Lake Memphrémagog. In 1984, the species was observed in Cantley, in the Ottawa Valley.

Notes on territory

Upon arrival at the breeding site, the male begins singing to establish his territory. All or almost all of his subsequent activity occurs within the boundaries of this territory (Murray and Gill 1976). The nest, however, is often built at the periphery. In Michigan, Murray and Gill (1976) estimated that Blue-winged Warblers defend an average area of 1.8 ha. They noted, however, that territory size may vary according to habitat. In New York State, Ficken and Ficken (1968c) estimated territory size to be 0.4–0.8 ha/pair.

In Ontario, 4.5 territories/10 ha were found in a habitat of fields and shrubs and 3.4 territories/10 ha in habitat consisting of shrubs and a forest (Clark 1993a,b).

One of the habitat types favoured by the Blue-winged Warbler, a casual visitor in the southernmost part of Québec, is forest regrowth in cut overs. Gowan Road, Herdman, southwest of Montréal. *June 1991, Pierre Bannon.*

History and trends

In the 19th century, the North American breeding range of the Blue-winged Warbler was restricted mainly to west of the Appalachians and south of the Great Lakes. Isolated populations were also found along the Atlantic seaboard. For about 150 years, the species has been expanding its range northeast-ward (Gill 1980). The first breeding evidence for the southernmost part of Ontario was obtained in 1956, and since then breeding has been increasingly reported in that region. The Atlas of the breeding birds of Ontario confirmed breeding of the species in 28 squares in the south of the province (Mills 1987a). The first breeding evidence in Vermont was obtained in 1976, which was the year atlassing began in the state. During the next 4 years, 3 other breeding confirmations were obtained (Clark 1985b). These examples show how rapidly the range is expanding.

The first sighting in Québec was at Nantes, northwest of Lake Mégantic, in 1963 (ÉPOQ). Since then about 15 sightings have been reported; one of these was a recent sighting of at least 2 birds—possibly 3, including a pair—between 17 and 23 May 1992 in the Huntingdon region (ÉPOQ; David 1980a; Aubry and Yank 1992a). Thus, there is reason to believe that the species is continuing its northeastward expansion, and breeding evidence for Québec may soon be forthcoming.

At one time, the breeding ranges of the Blue-winged and Golden-winged warblers were almost totally exclusive of each other. Today, however, the Blue-winged is increasingly invading the range of the Golden-winged, which has even become outnumbered in many regions in New England. There are many possible explanations for this expansion, one of them being changes in land use, such as abandonment of agricultural land. Early successional vegetation on abandoned farmland first favours the Golden-winged Warbler, which frequents areas of relatively early deciduous regrowth. As succession progresses, the habitat becomes more suitable for the Blue-winged Warbler, which prefers older growth and uses a wider variety of habitat (Confer and Knapp 1981). Hybridization is favoured by the fact that the species are closely related and mutually tolerant when inhabiting the same area. They also have similar courtship behaviour and feeding habits (Ficken and Ficken 1968b,c). With each generation, hybridization brings a small reduction in Golden-winged numbers and may eventually contribute to its decline, if populations of the Blue-winged, a more competitive species, should increase more rapidly (Gill 1980). All of these factors are contributing to the expansion of the breeding range of the Blue-winged Warbler.

François Shaffer

World breeding range

Species profile

Clutch size	usually 5–6 eggs (4–7)[1,2,4,5,7]
Laying interval	1 egg per day[6,7]
Start of incubation	with last egg[6,7]
Incubation period	10–12 days[1,2,3,4,7]
Maturity at hatching	altricial
Care of young	by female and male[5,7]
Nestling period	8–10 days[1,6,7]
Age at first flight	8–10 days or more[10]
Dependence of young	?
Broods per year	1[1]
Breeding age	1 year[1]
Mating system	monogamy[5]
Duration of pair bond	?
Total length	11.4–12.7 cm[2]
Wing span	17.2–19.1 cm[4]
Weight	8.4 g (7.2–11.0)[9]
Longevity record	7 years 11 months[8]

References:
1. DeGraaf and Rudis 1987
2. Godfrey 1986
3. Griscom and Sprunt 1979
4. Terres 1980
5. Ehrlich et al. 1988
6. Harrison 1978
7. Bent 1953
8. Klimkiewicz et al. 1983
9. Dunning 1984
10. according to nestling period

Chronology $n = 0$

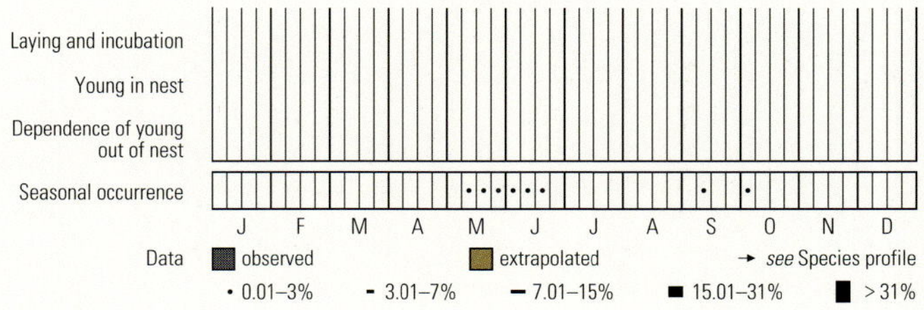

Prairie Warbler

Paruline des prés

Dendroica discolor

Christiane Girard

MORE IS KNOWN ABOUT THE ECOLOGY AND BEHAVIOUR OF THE PRAIRIE Warbler than of any other passerine in North America. The wealth of information contained in Nolan's (1978) monograph highlights the paucity of information available on most of our avifauna.

The Prairie Warbler is a species of eastern North America that winters chiefly in the Caribbean and the Bahamas (Lambert and Smith 1985). As yet, it is a casual visitor in southern Québec (Robert 1989b). Many birders would love to be the first to find it breeding in the province.

Habitat and behaviour

Wilson found this species in an area in Kentucky locally called "Prairie," which created the misnomer in English (Ehrlich 1988) that was probably later adopted in French. In fact, Prairie Warblers breed in dry scrubby areas, such as brushy second growths, fallow land, abandoned pastures, forest margins and, in sand dune country, areas with pine and ground juniper (Nolan 1978; Lambert and Smith 1985; Godfrey 1986). The most important habitat in Ontario, the only Canadian province where breeding has been confirmed, is rocky pine-oak-juniper scrub (Lambert 1987).

Nests are built by the female in low woody vegetation, usually less than 3 m high (Nolan 1978). In Ontario, most nests are built in junipers, but many other trees and shrubs are used (Lambert and Smith 1985; Peck and James 1987). Only the female incubates. She also broods and shades the nestlings, but both adults feed them invertebrates, mostly caterpillars. Families separate into two units after fledging, with each parent taking part of the brood. However, if the female renests, the male takes over care of all dependent young (Nolan 1978).

The two sexes forage at different levels within the vegetation, doubtless as a result of the constraints imposed by their respective activities. The male sings high up in the bushes to defend his territory and feeds mainly at a height of over 3 m, whereas the female, busy with nest building and incubation, forages lower down, mostly below 1.5 m (Morimoto and Wasserman 1991b).

Prairie Warblers are best located by the buzzy ascending song, which is sung throughout the nesting season.

Distribution

In the United States, the Prairie Warbler breeds from Michigan south to Texas and east to the Atlantic seaboard states (National Geographic Society 1987). In Canada, the species has been reported in Ontario, Québec, New Brunswick, Nova Scotia, and Newfoundland, but breeds only in Ontario (Lambert and Smith 1985). It was first observed in Québec from 30 May to 1 June 1966 in Montréal (Carleton 1966). Most subsequent records have come from the Island of Montréal; others are from Aylmer (Ottawa Valley); Sainte-Foy, Cap Tourmente (Québec City area); La Pocatière, Rimouski (Lower St. Lawrence); Métabetchouan (Lac-Saint-Jean); Baie-Comeau (North Shore); and Barachois (Gaspé) (Aubry et al. 1990b, 1992; Aubry and Yank 1993b, 1994; ÉPOQ; R. Yank, pers. comm.). The only Atlas record of a Prairie Warbler was of a singing male on 13 June 1988 at Saint-Édouard-de-Lotbinière. Rumours of possible breeding in Gatineau Park have so far remained unsubstantiated (Lambert and Smith 1985).

History and trends

The Prairie Warbler was likely rare and local in North America before European settlement. While its range probably has not changed much since that time, human activity tending to create suitable habitat (e.g. farming, logging) has encouraged it to invade many previously forested areas within its range (Nolan 1978). More recently, BBS data show a continent-wide decline in numbers from 1968 to 1979 (Robbins et al. 1986). Despite this, in the past 20 years its range has expanded in Ohio, Maine, and New York (Lambert and Smith 1985), but not in Ontario (Lambert 1987); Clark (1985c) does not mention any northward expansion for Vermont.

Though the Prairie Warbler does not yet nest in Québec, it may soon do so, as it is already breeding in Ontario and the U.S. states bordering Québec.

ROBERT ALVO

Notes on territory

In Indiana, breeding territories were 0.4–3.5 ha (mean 1.6 ha), but experienced males probably have larger ones than yearlings. Territory size also likely decreases as population density increases. Breeding densities in 32 habitats in 14 U.S. states were 0.5–21.3 pairs/10 ha, the highest values being in dense brush with openings in the canopy (Nolan 1978).

World breeding range

Species profile

Clutch size	usually 4 eggs (3–5)[1,7]
Laying interval	1 egg per day[1,7]
Start of incubation	with last egg[1]
Incubation period	usually 12 days (10–15)[1]
Maturity at hatching	altricial
Care of young	by female and male[1]
Nestling period	usually 9–10 days (8–11)[1]
Age at first flight	usually 9–10 days (8–11)[1]
Dependence of young	usually until 41 days old (34–55)[1,7]
Broods per year	1–2[1]
Breeding age	1 year[2,7]
Mating system	monogamy, sometimes polygyny[1,7]
Duration of pair bond	sometimes more than 1 season[1]
Total length	10.9–13.2 cm[6]
Wing span	15.9–18.4 cm[5]
Weight	M 8.0 g (6.1–10.1) / F 7.3 g (5.7–10.8)[3]
Longevity record	10 years 3 months[4]

References:
1. Nolan 1978
2. DeGraaf and Rudis 1987
3. Dunning 1984
4. Kennard 1975
5. Terres 1980
6. Godfrey 1986
7. Lambert and Smith 1985

Chronology *n* = 0

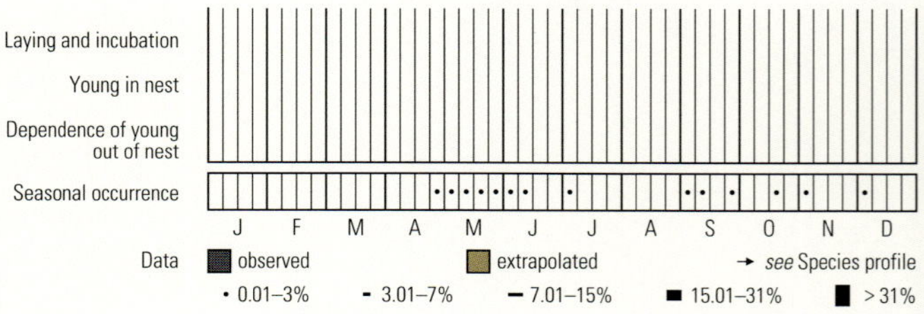

Prothonotary Warbler

Paruline orangée

Protonotaria citrea

Christiane Girard

THIS IS THE ONLY WARBLER BREEDING IN CANADA THAT NESTS ALMOST exclusively in tree cavities. With fewer than 80 pairs, largely restricted to the Carolinian deciduous swamps of southern Ontario (McCracken 1987b), it was listed as vulnerable in 1984 (COSEWIC 1993). Only a few records exist for Québec, where it does not breed (Robert 1989b).

Habitat and behaviour

The Prothonotary Warbler has highly specific habitat requirements, breeding in flooded or swampy deciduous woodlands; in Ontario, the species is particularly associated with sandspits along the north shore of Lake Erie (McCracken 1987b). According to a Michigan study, nests are built over standing water, over or along the edge of running water, or over dry land 0.6–138 m from water at a readily flooded spot. Shade for the nest is important to prevent overheating (Walkinshaw 1953).

Canopy cover near nest sites at Long Point, Ontario averaged 34% (McCracken 1981), and silver maple was the dominant living tree species near nests, though most nests were in rotting dead trees or stubs. Nest trees were small. Cavities were about 10 cm long, and entrance diameters were roughly 6 cm. Of the 20 nests observed, nine cavities had been excavated by Downy Woodpeckers or Black-capped Chickadees, but others had formed naturally. Nests had an average elevation of 1.7 m (0.7–3.2 m) and were constructed of moss outside, lined with grasses and fine strips of shredded bark; moss is a critical nesting material (McCracken 1981). Nest boxes are often used when available (Petit et al. 1987; Blem and Blem 1991).

Incubation is by the female. Both parents feed the young. Southern birds typically raise two broods, while those in the northern states and Ontario typically raise only one (Walkinshaw 1941; McCracken 1981); nevertheless, one pair raised two broods at Point Pelee, Ontario (Goodwin 1978; McCracken 1981). Major causes of egg or nestling loss are predation by raccoons, mice (notably the white-footed mouse), and snakes; cowbird parasitism; nest competition with House Wrens, which pierce and remove eggs; and flooding (Walkinshaw 1941, 1991b; Blem and Blem 1991; Petit 1991a).

Although its wet habitat is often difficult to access, the male's rather Ovenbird-like song (Walkinshaw 1991b) facilitates detection. Nests are easy to find because the male often feeds the female on the nest and builds one or more dummy nests that betray the territory (McCracken 1987b). It should be remembered, though, that on the fringe of the species' range, unmated males also defend territories and build nests (Leberman 1992g).

Distribution

The Prothonotary Warbler breeds in the eastern United States from Texas to New York State, and in southern Ontario. It is absent from Vermont, New Hampshire, and Maine (McCracken 1981; AOU 1983; Laughlin and Kibbe 1985; Adamus 1987). The closest breeding area to Québec is central New York State (Eaton 1988l).

Only six records exist for Québec, but none involves birds in their breeding habitat during the breeding season. The first individual was seen near Saint-Charles-de-Mandeville on 17 and 18 June 1977 (COQ 1977). Single birds were seen in mid-May on Mount Royal (David and Gosselin 1981a), at Sherbrooke (David and Gosselin 1982b), and at Sainte-Pétronille (Gosselin 1983). A bird was carried to La Grosse Île, in the Magdalen Islands, on 24 August 1989 during a storm (Yank et al. 1990a). A male was atlassed visiting a nest box in a rural area close to human habitation near Drummondville on 21 and 26 May 1988.

History and trends

The species was considered an accidental visitant in New York State at the beginning of the century (Eaton 1988l). In Canada, the Prothonotary Warbler was first seen in the spring of 1888 at Hamilton, Ontario, but breeding was confirmed only in 1929 at Rondeau, Ontario (McCracken 1981; McCracken et al. 1981), though it is possible that populations breeding in inaccessible locations had not been discovered (McCracken 1987b). In the early 1930s, some 100 pairs were breeding at Rondeau, but the removal of dead timber in or about 1935 led to a steep decline in that population.

BBS data show a strong increase in the southeastern United States from 1968 to 1979, with the highest density in the Mississippi flood plain, while Louisiana recorded the greatest number of individuals (Robbins et al. 1986). The species has been increasing in Pennsylvania (Leberman 1992g), but has been recommended for Special Concern status in Michigan (Walkinshaw 1991b). In Canada, Prothonotary Warbler numbers, though small, appear to be stable. The two largest populations are at Rondeau and Long Point on the north shore of Lake Erie (McCracken 1981).

If the species expands its breeding range northeastward, we may expect to see this warbler breeding in the swampy woodlands of southern Québec.

ROBERT ALVO

Notes on territory

Males arrive on the breeding grounds before the females and immediately establish territories, select nest sites, and even build nests (Bent 1953). They are strongly territorial, attempting to drive off conspecifics and other cavity nesters, such as the House Wren and Black-capped Chickadee (Walkinshaw 1941). They chase all birds that come within 1 m or so of the nest (Walkinshaw 1953). Territory size was 0.1–1.2 ha in Tennessee (Petit 1989) and 0.8–2.6 ha (mean 1.5 ha) in southern Michigan (Walkinshaw 1953).

Chronology *n* = 0

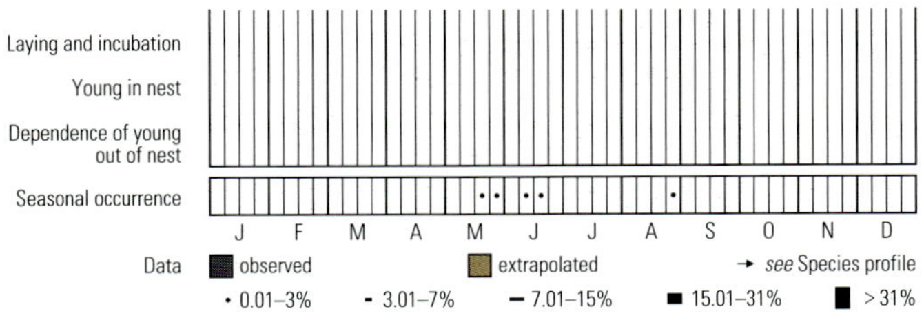

Species profile

Clutch size	usually 4–6 eggs (3–8)[1,2,3,6,7]
Laying interval	1 egg per day[6]
Start of incubation	before last egg[6,8]
Incubation period	12–14 days[1,2,3,4,5]
Maturity at hatching	altricial
Care of young	by female and male[1,3]
Nestling period	10–11 days[1,4,7,9]
Age at first flight	10–11 days[8]
Dependence of young	?
Broods per year	1, sometimes 2[1,3,4,8]
Breeding age	1 year[4]
Mating system	monogamy, sometimes polygyny[3,12]
Duration of pair bond	sometimes more than 1 season[8]
Total length	13.4–14.7 cm[5]
Wing span	?
Weight	M 15.0 g (13.6–15.8) / F 17.4 g (13.6–20.0)[10]
Longevity record	5 years 11 months[11]

References:
1. Harrison 1978
2. Terres 1980
3. Ehrlich et al. 1988
4. DeGraaf and Rudis 1987
5. Godfrey 1986
6. Bent 1953
7. Griscom and Sprunt 1979
8. Petit 1989
9. Walkinshaw 1953
10. Dunning 1984
11. Kennard 1975
12. Petit 1991b

World breeding range

Louisiana Waterthrush

Paruline hochequeue

Seiurus motacilla

Christiane Girard

THE LOUISIANA WATERTHRUSH IS EXCEEDINGLY RARE IN QUÉBEC. THE first records for the province are as recent as the mid-1970s, mostly from the Ottawa Valley region. Although no nests have yet been found, there is a great deal of evidence to suggest that the species breeds in the province.

During the breeding season, the Louisiana Waterthrush is found in wooded ravines with streams, or in woodland swamps (Godfrey 1986). Like the Northern Waterthrush, which it resembles a great deal, it is often seen walking on the ground tilting its tail (Bent 1953). The Louisiana Waterthrush can be distinguished from the Northern Waterthrush by its white, rather than yellowish, underparts and eyebrow stripe, and by its unspotted throat (Godfrey 1986).

In addition, the Louisiana Waterthrush's song is quite different from that of the Northern Waterthrush. It usually consists of two parts: three clear slurred whistles, followed by a characteristic jumble of twittering notes dropping in pitch (Peterson 1980). The song often mingles with the roar of the stream, making it difficult to hear.

The Louisiana Waterthrush, which is very rare in Québec, builds its nest near streams in extensive deciduous forests. It was atlassed as a probable breeder in the hilly regions of Gatineau Park.
October 1991, Daniel St-Hilaire.

Habitat and behaviour

The Louisiana Waterthrush occupies a very specialized ecological niche: nest sites must have water and at least 75% canopy cover (McCracken 1990). The species favours large, predominantly deciduous forest tracts, usually over 100 ha, situated on ravine slopes, but is also occasionally found in swamps with mature trees (Eagles 1987c; McCracken 1990), habitat more typical of the Northern Waterthrush. Because of its specific habitat requirements, the Louisiana Waterthrush has a scattered distribution (McCracken 1990).

Both partners participate in nest construction. The nest is built near streams or rivers, particularly rushing streams with cold, clear water (Eagles 1987c; McCracken 1990). According to studies from New York State, pairs occupy territories approximately 400 m long along the water (Eaton 1958).

The well-hidden nest is generally located between 0.5 and 4 m above the waterline in an old stump, the upturned roots of a tree, or a cavity in the stream bank (Eaton 1958; McCracken 1990). In Ontario, nests were found to be made of leaves and mosses, and lined with fine grasses and animal hair (Peck and James 1987).

Clutch size is 4–6 eggs, incubated by the female alone (Bent 1953). The young leave the nest when they are 10 days old, but are still fed by their parents for another month, during which time they remain near the stream where they were born (Eaton 1958).

The Louisiana Waterthrush is also a specialist in terms of its diet, which consists mainly of aquatic and terrestrial insects and spiders. However, it occasionally feeds on molluscs, fish, crustaceans, and amphibians (Bent 1953; Eaton 1958; Craig 1984), which is very unusual among North American passerines (McCracken 1990). The Louisiana Waterthrush spends most of its time on or near the ground, usually walking along the margins of streams or pools, where it can often be seen foraging in the shallow water (Bent 1953).

The Louisiana Waterthrush is often parasitized by the Brown-headed Cowbird, and will usually desert its nest when 3 or more cowbird eggs are laid (Bent 1953). According to McCracken (1990), parasitism may have a significant effect on the species' reproductive success.

Distribution

Most pre-Atlas records of the Louisiana Waterthrush in Québec come from the hilly areas of Gatineau Park: a specimen was collected on 24 May 1974 at Cascades (Godfrey 1986), individuals were sighted on 20–25 May and 4–9 July 1974 near Meech Creek (Club des ornithologues de l'Outaouais 1982), and another was sighted on 25 May 1975 at Aylmer (Godfrey 1986). In addition, in 1982, the species was present at Kingsmere from 30 May to 23 June, with as many as 3 birds being sighted at any one time, which suggests that the species may indeed breed in Gatineau Park (Gosselin and David 1982; Yank and Aubry 1984; Club des ornithologues de l'Outaouais 1985). The Atlas records come from Kingsmere and Mount Hereford in the Eastern

Notes on territory

When males arrive on their breeding grounds, they proclaim their newly established territories with their wild, rambling song. The females arrive up to 9 days later (Eaton 1958).

Territories tend to be long and narrow and are located on the banks of streams or rivers. In New York State, territories were approximately 400 m long (Eaton 1958). *In Connecticut, they averaged 0.67 ha* (Craig 1984). *The nest is usually located in the core of the territory* (Eaton 1958).

Townships. At Kingsmere, an individual was seen on 14 July 1984, in the same place as in 1982. The bird at Mount Hereford was observed on 22 June 1988 in an alder thicket on the edge of a lake, habitat which is somewhat atypical for the species. More recently, the species was sighted on 10 May 1992 near the Champlain lookout in Gatineau Park (Yank and Aubry 1992a), on 12 and 18 May 1993 at Saint-Ignace-de-Loyola, west of Lake Saint-Pierre (R. Ouellet, pers. comm.), between 23 June and 1 July 1993 at Lac la Blanche in the Ottawa Valley (Yank and Aubry 1993a), from 7 to 15 May 1994 at Philipsburg, on 12 May 1994 in downtown Montréal, and on 14 May 1994 in the Mount Royal Cemetery (Aubry and Bannon 1994a).

The hills of Gatineau Park have many small streams with waterfalls in its deciduous forests, the species' preferred breeding habitat. In fact, these sites are common on the entire plateau overlooking the Ottawa Valley. These areas are often difficult to get to, and little intensive field work has been done to confirm nesting. Mature forests in the Appalachian Mountains in the Eastern Townships also contain suitable nesting habitat for the species.

The Louisiana Waterthrush breeds exclusively in North America. In Canada, its known breeding range is limited to southern Ontario, where the breeding population probably numbers somewhere from 150 to a little over 300 pairs (McCracken 1990). Although the population is considered stable (Eagles 1987c), the Louisiana Waterthrush has been designated as vulnerable in Canada because of its scarcity and limited distribution (McCracken 1990).

The Louisiana Waterthrush is considered a common breeding bird in the United States, however, where it is abundant and widespread (Eagles 1987c). South of the border, it breeds as far north as Minnesota and New England and as far south as Texas and Florida (AOU 1983). The Louisiana Waterthrush winters mainly in southern Florida, the West Indies, Mexico, Central America, and northern South America (McCracken 1990).

Recent data suggest that the species is expanding northward and eastward from its traditional breeding range (Kibbe 1985g; Eagles 1987c; Eaton 1988j). However, this may reflect the increased number of observers in the field rather than an actual range expansion (Eagles 1987c). Nevertheless, the sightings in Gatineau Park are 100 km north of the species' known breeding range in Ontario, which has traditionally been considered to extend only as far north as Kingston (Eagles 1987c). Similarly, the sighting in the Eastern Townships is less than 100 km north of the northernmost record in Vermont (Kibbe 1985g).

History and trends

Ornithologists at the beginning of the 19th century confused the Louisiana and Northern waterthrushes, believing they were one species (Bent 1953). Both Wilson and Nuttall described a single species possessing the characteristics of both the Louisiana Waterthrush and the Northern Waterthrush, but as early as 1841, Audubon had recognized the existence of two separate species (Bent 1953). The Louisiana Waterthrush was considered rare in Canada at the turn of the century and thought to be limited to southwestern Ontario (Taverner 1919). The first records in Québec date from the mid-1970s.

Because it is a bird with very specific ecological requirements, the Louisiana Waterthrush is sensitive to activities that affect its habitat and food sources, such as clearcutting or other logging practices that raze the forest cover or create openings in the canopy (McCracken 1990). Logging may also affect the aquatic life and alter the physical characteristics of the streams and rivers flowing through cutover areas as well as habitats farther downstream. Fortunately, the Louisiana Waterthrush frequents sites that are often inaccessible, which protects the species to a certain extent (McCracken 1990).

Daniel St-Hilaire and Diane Dauphin

World breeding range

Species profile

Clutch size	usually 5 eggs (4–6)[1,2,6,7]
Laying interval	1 egg per day[9]
Start of incubation	before last egg[1,2]
Incubation period	12–14 days[1,2,6,7]
Maturity at hatching	altricial
Care of young	by female and male[1,2,7]
Nestling period	usually 10 days (9–11)[1,2,9]
Age at first flight	16 days[2]
Dependence of young	until 30 days after leaving nest[1]
Broods per year	1[6]
Breeding age	1 year[8]
Mating system	monogamy[9]
Duration of pair bond	probably 1 season[9]
Total length	15–16 cm[5]
Wing span	?
Weight	M 19.8 g (17.4–22.7) / F 20.8 g (17.7–26.0)[4]
Longevity record	7 years 11 months[3]

References:
1. Eaton 1958
2. Harrison 1978
3. Klimkiewicz et al. 1983
4. Dunning 1984
5. Godfrey 1986
6. Harrison 1984
7. Ehrlich et al. 1988
8. McCracken 1990
9. Robinson 1995

Chronology $n = 0$

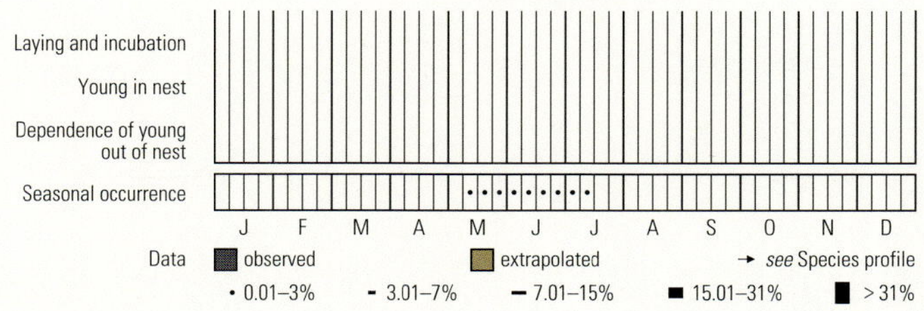

American Tree Sparrow

Bruant hudsonien

Spizella arborea

The musical song of the American Tree Sparrow is heard more often in northern Québec than in the inhabited areas of the southern part of the province. This sparrow breeds in subarctic regions as far as the tree line and is only seen in the south during migration and in winter (David 1980a). As winter approaches, scattered groups are often observed rustling about in the half-bare shrubbery, uttering their tinkling call notes (Godfrey 1986).

Tree Sparrows winter throughout southern Canada, except Newfoundland, and in most of the northern and central states of the United States (AOU 1983; Godfrey 1986).

Habitat and behaviour

The American Tree Sparrow spends the summer in forest and shrub tundra and open lichen woodland, as well as above timberline in the western mountains (Godfrey 1986). It owes its Latin specific name *arborea*, which means "tree," to a vague resemblance to the Eurasian Tree Sparrow, *Passer montanus* (Naugler 1993), but Baumgartner (1960) does not consider it particularly appropriate. In fact, the species prefers shrubby habitats of willow, scrub birch, alder, and sometimes stunted conifers along streams and in peatlands (Godfrey 1986).

In spring, migrating flocks break up just before arriving on the nesting grounds, where pairs are formed. The male's courtship behaviour is not very elaborate; it consists for the most part simply of singing, although he is sometimes seen next to the female, puffing out his feathers and spreading his wings. Territory size decreases as the breeding season progresses. Territories have an inner core where breeding activities occur, and an outer zone that is used less (Weeden 1965; Naugler 1993). In the morning, territorial behaviour is concentrated in the inner core, with the male proclaiming dominance by singing from the highest branches. Later in the day, there is more activity in the outer area (Weeden 1965; Naugler 1993). The female has a smaller activity space, located wholly within the male's territory (Weeden 1965). The female may on occasion participate in defending the territory (Baumgartner 1968). Tree Sparrows are not particularly aggressive toward other species that invade their territories but will drive away intruders that come too close to the nest (Baumgartner 1968).

The female builds the nest, usually on the ground but sometimes in a stunted shrub or tree, rarely higher than 1.5 m. The outer cup consists of blades of grass, stems, rootlets, mosses, lichens, and strips of bark, and the inside is lined with feathers (usually ptarmigan or duck), lemming fur, and sometimes moss. Only one clutch of 3–5 eggs is laid per season, which is incubated by the female (Weeden 1965; Baumgartner 1968).

The young are fed insects by both parents. After they leave the nest at the age of 8–10 days, the male plays a greater role in feeding. The young can barely fly on leaving the nest, learning to do so only after a few days (Baumgartner 1968). The American Tree Sparrow feeds almost exclusively on plant seeds in autumn, winter, and spring. During the breeding period the diet diversifies to include insects, as well as flowers and fruit of such plants as crowberry and cranberry (West 1973). Occasionally it catches insects on the wing.

In September, after the breeding season, the birds congregate in flocks to migrate south, where they also occupy open, shrubby habitat, such as vacant lots, roadsides, gardens, and marshes. These migrating flocks do not appear to have any social structure, and individuals do not belong to any flock in particular (West and Peyton 1972) as they sometimes do in winter (Baumgartner 1968). Females migrate farther south, while males remain in more northerly areas (Baumgartner 1968), including southern Québec. Throughout the winter, they scratch the ground or dig through the snow for dry grass, looking for seeds. Baumgartner (1968) reported finding 982 seeds in the stomach of one Tree Sparrow. This species is also a familiar visitor at bird feeders.

Christiane Girard

Notes on territory

The territory of the American Tree Sparrow is not well defined but is rather an area of activity of varying dimensions. The boundaries are imprecise and vary according to daily territorial activity and the stage of the breeding season (Weeden 1965). Confrontations appear to be over females rather than over territories (Weeden 1965).

A study in Alaska showed territory size to range between 0.6 and 3.8 ha and density to be 5.2–8.6 pairs/10 ha (Weeden 1965). Territory size tends to decrease as the season progresses (Naugler 1993). In Canada, studies in the Northwest Territories showed densities to range from 1.2 pairs/10 ha in riverside habitat with herbaceous growth and low willows to 5.6 pairs/10 ha in upland alder thickets (Erskine 1976a).

The American Tree Sparrow nests in shrubby stands of willow and alder in tundra barrens such as those found along the James Bay coast. Bay of Many Islands, James Bay. *August 1990, Yves Aubry.*

Distribution

In summer, the American Tree Sparrow breeds in Alaska, Yukon, northwestern British Columbia, and in subarctic regions from the Northwest Territories to Newfoundland. In Québec, it is found throughout the province north of about the 53rd parallel (Godfrey 1986). Todd (1963) even reported a specimen taken at Korak Bay near Akulivik in northern Québec, well beyond the tree limit. This is the northernmost record for the species in Québec. A pair whose agitated behaviour denoted the probable presence of a nest or young, was reported on 13, 14, and 21 July 1987 near Lake Preissac, in the Abitibi region. They were seen in a flooded stand of alders, bordered by an abandoned meadow and a young deciduous forest (R. Roy, pers. comm.). This instance is likely accidental.

Although Tree Sparrows are common in migration throughout southern Québec, it is only in southernmost Québec, the Central St. Lawrence Lowland, and the southern Appalachians that they are seen with any frequency in winter. They occur less often in more northerly areas (David 1980a; Godfrey 1986). The northern limit of the species' winter range corresponds roughly to the –23°C isotherm for January, and includes southern Canada. The southern limit extends to North Carolina, northern Texas, and Nevada (Root 1988).

History and trends

The status of the American Tree Sparrow appears to have changed little since the end of the last century. The species was common during spring and fall migrations in the south of the province, and it was reported in winter in the Québec City region (Wintle 1896; Dionne 1906; Macoun and Macoun 1909).

There is little information on the population dynamics of the American Tree Sparrow. It does not appear to have been much affected by European settlement. Its breeding grounds are not particularly subject to human disturbance, and the shrubby, open habitat of its wintering grounds has no doubt increased as a result of the deforestation that accompanied agricultural and urban development. Moreover, the Tree Sparrow has probably also benefited from the proliferation of winter feeding stations throughout North America.

Benoît Limoges and Simon Barrette

World breeding range

Species profile

Clutch size	usually 3–5 eggs (3–7)[1,2,4]
Laying interval	1 egg per day[1,8]
Start of incubation	with last egg[1,8]
Incubation period	10–13 days[1,2,3,4,8]
Maturity at hatching	altricial
Care of young	by female and male[1,2,8]
Nestling period	8–10 days[1,4,8]
Age at first flight	14–16 days[1,8]
Dependence of young	until 2–3 weeks after leaving nest[8]
Broods per year	1[5,8]
Breeding age	?
Mating system	monogamy[2,4,8]
Duration of pair bond	?
Total length	14.5–16.4 cm[3]
Wing span	21.6–24.8 cm[4]
Weight	20.1 g (?)[7]
Longevity record	10 years 9 months[6]

References:
1. Harrison 1978
2. Ehrlich et al. 1988
3. Godfrey 1986
4. Terres 1980
5. Baumgartner 1968
6. Klimkiewicz and Futcher 1987
7. Dunning 1984
8. Naugler 1993

Chronology *n* = 33

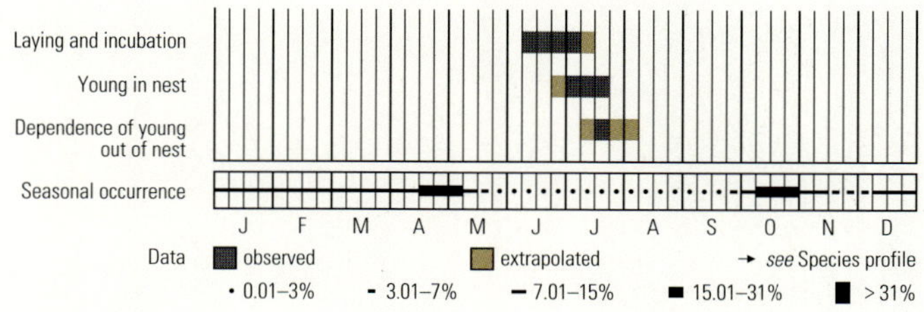

Lapland Longspur

Bruant lapon

Calcarius lapponicus

Christiane Girard

In summer the Lapland Longspur, a characteristic bird of the tundra, is seen only in the northern part of the province. In migration, some birds may be observed in mixed-species flocks, usually with Snow Buntings and Horned Larks, from which the longspurs can be told by their smaller size and darker plumage (Godfrey 1986; National Geographic Society 1987). In winter, the Lapland Longspur is found in the Atlantic provinces and the extreme south of the other Canadian provinces, notably in southwestern Québec, though only in small numbers (Godfrey 1986). It also winters throughout most of the United States.

Habitat and behaviour

In spring, the males are the first to arrive on the breeding grounds and are joined by the females a few days later (Hussell and Holroyd 1974; Fox et al. 1987). The birds remain in flocks until the snow melts (Mikheev 1939 *in* Cramp 1994b; Drury 1961); they then begin their courtship displays, which consist of aerial pursuits and displays on the ground by both sexes (Drury 1961).

The Lapland Longspur nests on the tundra, preferably in wet areas with tussocks of grasses or sedges and possibly clumps of shrubs (Williamson 1968; Godfrey 1986). As Erskine (1971a) indicated, the Lapland Longspur and the Snow Bunting are the most common and widespread land birds of the Canadian tundra.

Occurring as it does in high latitudes, the Lapland Longspur has a short breeding season, and nesting is usually timed for the young to leave the nest when adult flies, their main source of food, are most abundant (Cramp 1994b). Breeding is thus synchronous among birds in the same area (Williamson 1968; Fox et al. 1987). Seastedt and MacLean (1979), for example, in a study in Alaska, noted that all males that succeeded in attracting mates had established their territories at about the same time, within a period of 8 days. Custer and Pitelka (1977) noted that in 75% of the nests studied in Alaska over a 7-year period, the females laid their first egg within 7 days of each other.

The nest is built in a slight depression in the ground, on a hummock, or in a grass tussock, and is often sheltered by a shrub (Harrison 1978; Godfrey 1986). Made of mosses and grasses, it is completed in 3 or 4 days (Ehrlich et al. 1988). Typical clutch size is 4–6 eggs. Incubation lasts for 10–14 days; within any given population, hatching is fairly synchronous, occurring over a period of 1–4 days (Drury 1961; Williamson 1968; Hussell 1972). Insects, brought by both the male and the female, comprise 90% of the nestlings' diet (Hussell 1972). After 8–10 days, the young leave the nest not yet able to fly (Harrison 1978; McLaughlin and Montgomerie 1989). Within a day, the adults divide the brood into two units of 2 or 3 young each. These groups remain stable until the young are independent, at about the age of 23 days (McLaughlin and Montgomerie 1985, 1989). Each parent raises its group separately.

Distribution

The Lapland Longspur is a circumpolar species that breeds solely on the tundra, ranging across northern Eurasia, Greenland, Alaska, and northern Canada, specifically northern Yukon, the Northwest Territories, and the northern parts of Manitoba, Ontario, Québec, and Labrador (Godfrey 1986). In Québec, it nests on the Ungava Peninsula, and along the Ungava Bay and Hudson Bay coasts south to southern James Bay (Godfrey 1986; National Geographic Society 1987; Cons. G. G. – GREBE 1992o). Except on the coast, the southern limit of the breeding range is still unknown.

Notes on territory

The territory is generally used for mating, nesting, and some foraging (Cramp 1994b). In Alaska, Seastedt and MacLean (1979) recorded territories of between 1.1–2.8 ha (mean 1.8 ha), while Drury (1961) reported between 0.1 and 0.2 ha on Bylot Island. The female selects the nest site and thereby directly influences the size of the territory; thus, if she selects a site outside the area defended by the male, he must expand the territory to include the nest. The greater the abundance of prey, the smaller the territory (Seastedt and MacLean 1979).

Seastedt and MacLean (1979) reported densities of 3.3 pairs/10 ha in Alaska, while Lyon and Montgomerie (1987 in Cramp 1994b) found up to 2.5 pairs/10 ha in the Northwest Territories.

A small tundra passerine, the Lapland Longspur nests in open habitat and uses rocks or shrubs as singing posts or for surveying the territory. Lake Brisson, near the George River, Ungava Bay region.
July 1981, Benoît Houde.

An incubating female. The nest, surrounded by grasses and lichens, is placed in a depression in the shelter of a small shrub. Rankin Inlet, Northwest Territories. *July 1980, Yves Aubry.*

History and trends

The abundance and distribution of the Lapland Longspur do not appear to have changed appreciably since the beginning of the century, when breeding was generally widespread throughout Greenland and northern Canada. It was abundant in Labrador and northern Hudson Bay, and was known to nest at James Bay (Macoun and Macoun 1915). As David (1980a) states, the Lapland Longspur is a casual winter visitor in southern Québec, where in most places it is seen only in migration, and then rarely.

Alain Lanoue and Marie-Renée Doyon

World breeding range

Species profile

Clutch size	usually 4–6 eggs (2–8)[1,2,4,5,6,8,11,14]
Laying interval	1 egg per day[1,15]
Start of incubation	?
Incubation period	10–14 days[1,2,8,14]
Maturity at hatching	altricial
Care of young	by female and male[8,9]
Nestling period	8–10 days[1,10]
Age at first flight	11–15 days[1,3,9]
Dependence of young	until ≈ 23 days old[10]
Broods per year	1, rarely 2[2,3,4,11]
Breeding age	1 year[12]
Mating system	monogamy, sometimes polygyny[2,3,4,13]
Duration of pair bond	?
Total length	15.2–17.6 cm[5]
Wing span	26.7–29.8 cm[6]
Weight	27.3 g (23.5–32.5)[7]
Longevity record	6 years[4]

References:
1. Harrison 1978
2. Ehrlich et al. 1988
3. Montgomerie 1988
4. Custer and Pitelka 1977
5. Godfrey 1986
6. Terres 1980
7. Dunning 1984
8. Drury 1961
9. McLaughlin and Montgomerie 1985
10. McLaughlin and Montgomerie 1989
11. Fox et al. 1987
12. Gierow and Gierow 1991 *in* Cramp 1994b
13. Alekseeva et al. 1993
14. Peck and James 1987
15. Cramp 1994b

Chronology *n* = 8

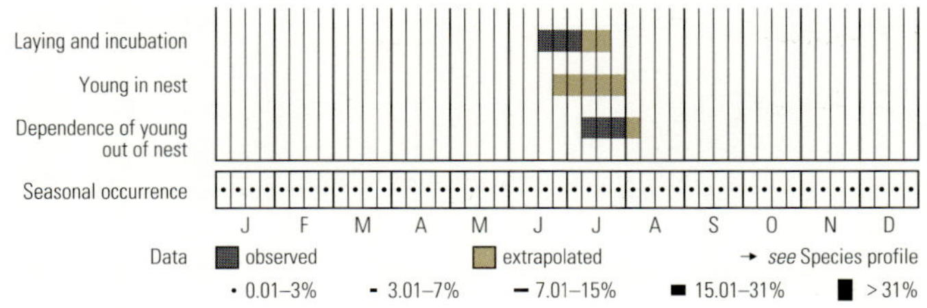

Young in a nest, a cuplike structure made of grasses and mosses. Rankin Inlet, Northwest Territories. *July 1980, Yves Aubry.*

Snow Bunting

Bruant des neiges

Plectrophenax nivalis

Christiane Girard

The aptly named Snow Bunting, a tiny white irregular winter visitor from the far north, arrives in southern Québec with the first October snow squalls, enlivening the winter with its distinctive plumage and musical twittering, given in flight.

Snow Buntings frequent mainly abandoned fields, tidal flats along the St. Lawrence, treeless banks and shores of lakes and rivers, and roads and railway cuts. They feed on seeds blown over the snow by the wind. They can also be seen in flocks at feeders in open areas. Sometimes they will elect to stay for a week or two in suitable habitat, where stalks of goldenrod and aster protrude through the snow. At such times, they can be observed feeding regularly, at the same places and times, retiring half an hour before sunset to their nocturnal roosts (pers. obs.). The birds are highly gregarious in winter and fly about in flocks of a hundred or so during the day. At night, however, they divide into smaller groups of 4–20 birds in roosts located in an open field or in the shelter of a snow drift. Snow Buntings sleep on the ground and are light sleepers (Parmelee 1968; pers. obs.). Contrary to what Salomonsen (1951) wrote, the birds sleep alone, even in the coldest conditions (–27°C in winds of 45 km/h: Vincent and Bédard 1976). They are hardy birds that do not huddle together to keep warm (Parmelee 1968; Vincent and Bédard 1976), but maintain a distance from each other of 2.5 m on average, and rarely less than 30 cm (Vincent and Bédard 1976).

Beginning in mid-March, the males congregate in flocks and begin eating more so as to better contend with their imminent migration and especially the difficult foraging conditions they will face upon arriving on their breeding grounds. This means accumulating enormous reserves of fat, and some birds will even double their body weight (Vincent and Bédard 1976; Banks et al. 1989). Huge premigratory flocks of several thousand birds are often seen in the Lower St. Lawrence region. In April, some males already have their bold black-and-white breeding plumage, and by listening carefully a birdwatcher might even hear their crystalline song, which is given mostly in flight. One or two nights before their departure for the north, they will often perch in trees; several hundred birds were once seen in late April in Saint-André, Kamouraska County, perched on the branches of a small stand of birches, a host of white balls against an immaculate background (pers. obs.). Snow Buntings migrate by both day and night and are sometimes joined by small flocks of Lapland Longspurs (Parmelee 1968) and occasionally Horned Larks (Terres 1980; pers. obs.). In North America, Snow Buntings winter within a broad belt extending from southern Alaska to Newfoundland and south to the central United States (AOU 1983).

A bird of the tundra, the Snow Bunting also inhabits outposts and villages, concealing its nest under houses and piles of debris of all kinds. Kangiqsualujjuaq, east coast of Ungava Bay. *June 1977, Pierre Lamothe.*

During the breeding season, the Snow Bunting seeks out rocky terrain with cavities and crannies in which it can build its nest. Bathurst Island, Northwest Territories. *July 1971, Pierre Lamothe.*

Notes on territory

When they arrive on the breeding grounds, the males begin establishing and defending their territories (Tinbergen 1939; Parmelee 1968) by singing from rock perches and driving away conspecifics (Drury 1961). The territory is always confined to land, even though in April the snow makes land indistinguishable from streams, lakes, or the sea (Tinbergen 1961). Parmelee (1968) reported that the territory is usually 300–400 m in diameter. In Scotland, Milsom and Watson (1984) estimated it to be 3–40 ha in area. In Greenland, Meltofte (1983) found Snow Bunting density to be over 0.5 pair/10 ha in suitable habitat. However, in barren

A male Snow Bunting feeding its young on the nest. The nest is generally in a sheltered spot on the ground, such as this one that was found under a panel of wood. Rankin Inlet, Northwest Territories. *July 1980, Yves Aubry.*

habitat without rocky outcroppings, or in areas that are late in thawing, densities are very low or even nil. Erskine (1971a, 1972a, 1984) reported densities ranging between 0.1 and 1.0 territorial male/10 ha in various tundra habitats of the continental Northwest Territories and the Arctic Archipelago north of the 62nd parallel.

Habitat and behaviour

Some males arrive on the breeding grounds as early as late March (Tinbergen 1939; Parmelee 1968; Meltofte 1983), but most arrive around mid-April. For the Inuit, the return of the *Amauligaks*, arriving in waves "with every warm, wet storm from the east" (Tinbergen 1961), signifies the end of winter (Parmelee 1968). On arrival on their breeding grounds, the males begin defining and defending their territories (Tinbergen 1939; Parmelee 1968), singing from rock perches. Snow Buntings breed farther north than any other land bird (Terres 1980) and are the first passerines to return in spring (Wynne-Edwards 1952), occupying rocky terrain (Godfrey 1986) on hillsides or mountain slopes. Females and yearlings arrive 3–4 weeks after the males, the longest time lag recorded for any passerine species (Parmelee 1968). On arriving, newcomers are attacked by territorial males. The females, however, adopt behaviour that enables them to be recognized. This provokes a marked change in the behaviour of the males, which at this point begin their courtship displays. During courtship, the male soars to a height of about 10 m then glides back toward the female, all the while singing its crystalline song. Sometimes it turns its back to the female with wings and tail outspread, then runs rapidly off, stops, and repeats the performance, flashing the bold black and white of its breeding plumage. Courtship may last for several weeks, and the female will copulate only when she is ready to begin building the nest (Tinbergen 1939). Before mating, the female investigates cavities and crevices around the territory, looking for suitable nest sites. She then begins carrying moss in her bill and allows the male to approach. The male copulates with her no more than 2–5 times a day (Tinbergen 1939). Nest construction begins shortly afterward and is performed almost solely by the female, with the male sporadically bringing nesting material to her. Preferred nest sites are rocky outcroppings or rock piles located near damp areas of vegetation that thaw (Meltofte 1983) or are snow-free early on. Nests are almost always placed in a cavity or crack, often 30 cm deep. On Baffin Island, nests have been found up to an elevation of 2000 m (Parmelee 1968). Snow Buntings may nest in northern villages, using cracks in the foundations of houses or any other structure where the nest can be hidden (Drury 1961; Tinbergen 1961; Parmelee 1968; Hussell and Holroyd 1974; Y. Aubry, pers. comm.).

The nest is a bulky, loosely built structure made of grasses, mosses, and lichens (Hussell 1972). The female lines it with dry grasses, rootlets, sometimes downy willow seeds, and invariably with feathers or fur. Occasionally, she will reuse a nest from the previous year, adding a new lining (Parmelee 1968). Sutton and Parmelee found a nest on Baffin Island that had been started on 15 June and was ready to be lined on 19 June. By 20 June, it was well furnished with hair, and by 22 June feathers had been added and the first egg laid (Parmelee 1968). The female generally incubates the 4–7 eggs for about 12 days, during which time she is fed regularly by the male (Lyon and Montgomerie 1985, 1987).

World breeding range

Species profile

Clutch size	usually 4–7 eggs (3–9)[1,2,4,6,11]
Laying interval	1 egg per day[3]
Start of incubation	with last egg[3]
Incubation period	usually 12–13 days (10–15)[2]
Maturity at hatching	altricial
Care of young	by female and male[7]
Nestling period	usually 10–12 days (8–17)[4,5,10,11]
Age at first flight	13–14 days[2,5]
Dependence of young	until ≈ 12 days after leaving nest[2]
Broods per year	usually 1 (1–2)[3,4]
Breeding age	usually 2 years (1–2)[5]
Mating system	usually monogamy[3]
Duration of pair bond	sometimes more than 1 season[10]
Total length	15.0–18.5 cm[1]
Wing span	30.5–33.0 cm[4]
Weight	42.2 g (34.0–56.0)[8]
Longevity record	8 years 9 months[9]

References:

1. Godfrey 1986
2. Parmelee 1968
3. Tinbergen 1939
4. Terres 1980
5. Meltofte 1983
6. Hussell and Holroyd 1974
7. Ehrlich et al. 1988
8. Dunning 1984
9. Klimkiewicz and Futcher 1987
10. Nethersole-Thompson 1966
11. Drury 1961

Chronology *n* = 7

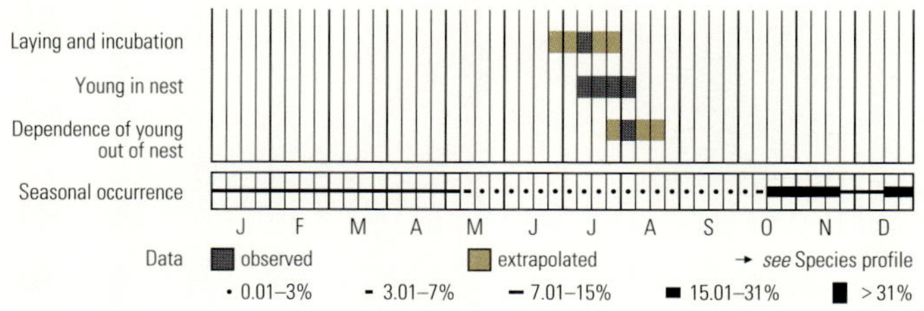

Even though there may be annual or geographic variations, Snow Bunting populations breed highly synchronously. Most eggs of a given population hatch within a one-week period (Meltofte 1977 *in* Meltofte 1983).

Within individual broods, hatching is partially asynchronous, which, according to Hussell (1985), is advantageous to the species. The young remain in the nest for 10–12 days. The male assists the female in feeding the nestlings a diet composed entirely of insects (Lepidoptera and Diptera) and spiders (Nethersole-Thompson 1966). Some young birds leave the nest before being able to fly. The parents continue feeding them for another 10 days or so, even though they may disperse onto adjacent territories, where they are tolerated by other adults (Tinbergen 1939; Nethersole-Thompson 1966). Young Snow Buntings grow at the same rate as Lapland Longspurs, although the latter leave the nest much sooner. Their growth pattern is similar to that of some of the less northerly emberizine species such as the White-crowned, Chipping, and Vesper sparrows (Maher 1964). In August, some of the young travel to the coast, where they feed in large flocks until the fall migration. On Ellesmere Island, most birds head south around mid-September, but some birds linger in arctic areas until December, and some even overwinter (Parmelee 1968).

Distribution

The Snow Bunting is a circumpolar species that nests in the northernmost areas of Québec, around Ungava Bay and on the Ungava Peninsula (Godfrey 1986). There are also a few summer records for the Kuujjuarapik region (ÉPOQ), though nesting has not yet been confirmed there. The breeding range elsewhere in Canada covers the northern and northeastern parts of the continental Northwest Territories, all of the Arctic Archipelago, the Belcher and Nastapoka islands along the east coast of Hudson Bay, and the northern part of the Labrador coast. It nests locally in Yukon and extreme northwestern British Columbia (Campbell and Van Der Raay 1985; Godfrey 1986). The Snow Bunting is also found in Alaska and Greenland (Godfrey 1986).

A male Snow Bunting in breeding plumage, standing guard in alarm posture at the entrance to its nest cavity. Bathurst Island, Northwest Territories. *July 1971, Pierre Lamothe.*

This male, still in its winter plumage, is feasting on seeds and insects uncovered by the melting snow. Québec City. *March 1987, Jean Giroux.*

History and trends

Meltofte (1983) views the Snow Bunting as probably the most common bird in the High Arctic and the only passerine that breeds there abundantly. There have been no specific studies, however, on its abundance or population trends. Because the Snow Bunting is highly nomadic and gregarious on its wintering grounds, Christmas Bird Count data cannot be used to detect any continent-wide population trends (Root 1988). Several factors, however, are known to affect this species' abundance. Snow Buntings were once hunted and trapped in late winter by the *habitants* of the St. Lawrence Valley, for the pot or for sale in local markets. They were caught with slip-noose traps on Île d'Orléans and the south shore of the estuary and in live-decoy door traps in the Charlevoix region and on the North Shore. Although hunting of the species has been prohibited in Québec since 1963, trapping was still being carried out here and there in the 1980s (Hamann 1985; pers. obs.). A study on Devon Island in the Northwest Territories by Hussell and Holroyd (1974) showed brood predation rates to be over 50%, the predators for the most part being arctic fox and, to a lesser extent, ermine.

In northern Scotland, which is at the southern limit of the Snow Bunting's range, Nethersole-Thompson (*in* Sharrock 1976) noted an increase in the breeding population since the mid-1960s. Some birds may have been prompted to nest farther south because of unusually hard winters during those years in Iceland and Scandinavia (Sharrock 1976).

Jacqueline Vincent

Yellow-headed Blackbird

Carouge à tête jaune

Xanthocephalus xanthocephalus

THIS LARGE MARSH-NESTING BLACKBIRD IS COMMON IN WESTERN NORTH America. Some of its nesting colonies are huge, and the marshes may throb with the male's distinctive song, described as a "cacophonous scraping or strangling noise" by Kaufman (1983). Adult males are easily identified by their bright yellow head, neck, and breast in stark contrast to the black body, wings, and tail. They have a white patch on the primary coverts of their wings, which is highly visible in flight. Females are smaller and dusky brown overall except for their yellow face and upper breast. Breeding east of the Canadian Prairies is sporadic, and Godfrey (1986) classifies this species as casual in Québec.

Habitat and behaviour

This gregarious bird travels in large flocks and may mix with grackles, Red-winged Blackbirds, or cowbirds outside the breeding season. It nests in small to large colonies in deep marshes or sloughs, or among the cattails, bulrushes, and reeds of the marshy areas of lakes and rivers (Bent 1958; Peck and James 1987). The male arrives on the breeding grounds first in the spring and defends his selected territory from rival males. The Yellow-headed Blackbird may share some marshes with the Common Grackle (Minock 1980) and the more widely distributed Red-winged Blackbird, but the larger Yellow-head occupies the deeper areas and will usually evict any already established Red-wings from these parts of the marsh (Orians 1985a). The male is polygynous and will actively pursue females throughout the marsh, displaying to them by spreading his tail and half opening his wings to show the white patches (Bent 1958). Harems average 3 or 4 females per male, but 10 or more females on one male's territory is not uncommon (Orians and Wittenberger 1991).

The nest is built by the female over water that is at least 0.6 m deep, which protects the nest and young from swimming predators. It is a deep, basketlike structure of dead grasses, reeds, and cattails woven around emergent herbaceous plants or, occasionally, wound around shrubs and small trees growing in the water (Orians 1985a). Incubation is performed entirely by the female, but the male usually perches near the nest, warning the female of possible dangers and aggressively defending the nest against intruding hawks, crows, and even humans when the young are present. Males may also feed the nestlings of the first nest to hatch on his territory, unless his harem is large, in which case his courting duties come first (Orians 1985a). The young are fed by regurgitation during the first two days, mainly on aquatic insects. The adults will also eat grain, seeds, and other vegetable matter. Nestlings leave the nest at 9–12 days of age but they remain hidden among the dense vegetation until they can fly. At about 3 weeks of age, they may make short flights and soon after are pursuing their parents, begging for food (Fautin 1941). The Yellow-headed Blackbird is rarely parasitized by the Brown-headed Cowbird, probably due to a combination of factors. Nest initiation is earlier than in the cowbird, the incubation period is shorter than in many other cowbird hosts, and it has a high rate of nest abandonment at the end of the season, compromising the young cowbird's survival; it usually has a high nesting density and the colony is defended communally, and the lack of woody vegetation where cowbirds can perch adjacent to their nest sites makes blackbird nests less accessible to searching cowbirds (Ortega and Cruz 1991).

Christiane Girard

Notes on territory

In southwestern Manitoba, male territory size ranges from 0.02 ha to 0.2 ha, with up to 7 females on one territory (Lightbody and Weatherhead 1987b). Males that were more aggressive and were therefore able to secure large territories had larger harems (Lightbody and Weatherhead 1987a).

World breeding range

Species profile

Clutch size	usually 4 eggs (3–5)[1,3,4,5]
Laying interval	1 egg per day[1]
Start of incubation	with second egg or afterward[1]
Incubation period	10–13 days[1,2,3,4,5]
Maturity at hatching	altricial
Care of young	mainly by female[1,2,4,9]
Nestling period	9–12 days[1,2,4,5]
Age at first flight	≈ 21 days[1,2,4]
Dependence of young	until at least 28–35 days old[2]
Broods per year	usually 1 (1–2)[6]
Breeding age	M 2 years / F 1 year[1]
Mating system	polygyny[1,5,9,10,11]
Duration of pair bond	1 season[1]
Total length	21.5–28.0 cm[3]
Wing span	?
Weight	M 79.7 g (72.5–85.5) / F 49.3 g (42.4–56.0)[8]
Longevity record	10 years 11 months[7]

References:

1. Bent 1958
2. Fautin 1941
3. Godfrey 1986
4. Terres 1980
5. Ehrlich et al. 1988
6. Orians and Wittenberger 1991
7. Klimkiewicz and Futcher 1987
8. Dunning 1984
9. Searcy and Yasukawa 1989
10. Lightbody and Weatherhead 1987a
11. Lightbody and Weatherhead 1987b

Chronology $n = 0$

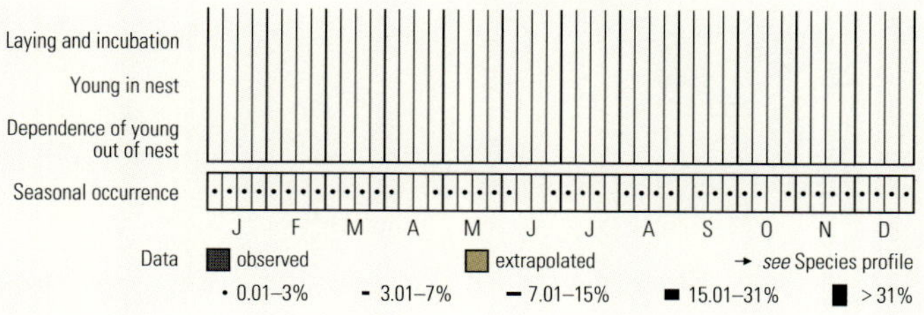

Distribution

The Yellow-headed Blackbird is confined to the North American continent, breeding from central British Columbia east throughout the Canadian Prairies to Ontario, and south to southern California, Missouri, and northwestern Ohio. It winters from the southern United States to south-central Mexico (Godfrey 1986).

The easternmost confirmed breeding record of this species in Canada is from the southern part of northwestern Ontario (Flood 1987c). Robert (1989b) lists the Yellow-headed Blackbird as a casual visitor in Québec. There have been single summer sightings in June and July: at Montréal in 1894, and between 1984 and 1994 at Granby, Bishopton, Saint-Augustin-de-Desmaures, Alma, Radisson, Pointe-Lebel, Harrington Harbour, Rivière-Ouelle, Cacouna, and Grande-Cascapédia and on the Magdalen Islands at La Vernière and L'Étang-du-Nord (N. David and R. Yank, pers. comm.). A pair was recorded at Radisson, near La Grande-2 for several weeks in the spring of 1980 (Côté 1980).

The sightings of single individuals, such as the lone bird atlassed on the Madgalen Islands, could in some cases be construed as possible breeding. Although Yellow-headed Blackbirds usually nest colonially, occasionally solitary pairs have been recorded breeding in the Great Lakes region (Peterjohn and Rice 1991). Breeding pairs have yet to be documented for the province of Québec.

History and trends

Historically, the Yellow-headed Blackbird has been considered a rare or accidental visitor in Québec (Dionne 1906; Taverner 1919). The winter range of this species has greatly expanded in recent decades due to modern harvesting practices, which leave a substantial quantity of seeds in the fields, either from the crop plant or from weeds. These fields along with cattle feedlots, where there is plenty of grain, are ideal foraging places and have greatly increased the birds' overwinter survival (Orians 1985a). This is reflected in the BBS data, which shows a significant continental increase in their population, particularly in the central region of the continent (Robbins et al. 1986).

Since 1980 in Québec, this species is being observed with increasing regularity in winter and during fall migration, often in mixed flocks and occasionally at feeders. Three were sighted among a flock of about 2000 Brown-headed Cowbirds near Melocheville during the winter of 1980 (Gosselin and David 1980a). A single male at Jonquière was the first local occurrence in the Saguenay–Lac-Saint-Jean region in the spring of 1981 (David and Gosselin 1981a). The most popular location appears to be Saint-Étienne-de-Beauharnois, where winter sightings were made from 1986 to 1991 (Yank and Aubry 1986b; Gosselin et al. 1987, 1990b; Yank et al. 1989b; Aubry et al. 1991).

Derin Henderson

Notes on territory (continued)

A study in eastern Washington State found that marshes with the highest densities of females had the highest rate of emergence of dragonflies and damselflies (Odonata), the staple diet of nestlings (Orians and Wittenberger 1991). The females are not territorial, and although most foraging is done outside the territory on which they nest, usually along undefended shorelines and in nearby fields, they will also forage regularly on their own and neighbouring territories.

Orchard Oriole

Oriole des vergers

Icterus spurius

Christiane Girard

THE ORCHARD ORIOLE IS A CASUAL VISITOR IN QUÉBEC. SINCE THE EARLY 1960s, only 25 sightings have been recorded in the province (R. Yank, pers. comm.). Although the species may very well nest in Québec, there is no evidence to this effect. Sightings are usually of single birds and usually do not occur more than once in the same area.

Besides being very rare in Québec, the Orchard Oriole usually remains for only a brief period on its breeding grounds, preparing as early as July to head back to the wintering grounds, which lie between southeastern Mexico and northern Colombia (AOU 1983). The brevity of the Orchard Oriole's sojourn reduces chances of observing it. Smaller than the Northern Oriole, from which it is distinguished by its less conspicuous dark chestnut plumage, the male is recognized mainly by its song. Nearly two thirds of Québec sightings were made during the last weeks of May.

Habitat and behaviour

Throughout its range, the Orchard Oriole, as its name indicates, is found in orchards or other places where fruit trees are growing. A species typical of open habitat, it is likely to be found near farms, in open forest, fields containing a few trees, and sometimes suburban parks.

The Québec sightings were probably of wandering yearlings that had little chance of finding a mate. Young males apparently establish their territories north of the normal breeding range (Sealy 1980a; Enstrom 1992a,b). The olive-green and yellowish plumage of young but sexually mature males differs markedly from that of older males and may easily be confused with that of the female. The only distinguishing field marks are the black on the throat, which extends lower on the breast to form a bib, and an occasional few chestnut feathers (Orians 1985a). It was first thought that this difference in colouring worked to the advantage of the young males, as their resemblance to the females would protect them from aggression by territory holders and enable them to compete more easily for high quality food supplies (Flood 1984). A study by Enstrom (1992a), however, does not support this hypothesis, finding rather that adult males are more aggressive toward young males than toward females and other adult males and suggesting that young males will try to establish themselves in new areas to avoid intense competition.

Generally, the male arrives on the territory a few days before the female, which he attracts and courts through his singing (Sealy 1980a). The nest, an open pouch about as deep as it is wide, is built by the female, which skilfully weaves plant fibres to a tree fork 2–20 m above the ground (Peck and James 1987). After fledging, the young remain near the nest, concentrating their activities within a 20–30 m radius for about a week. Unlike females and yearlings, breeding males appear to be fairly faithful to the nest site (Sealy 1980a). Peck and James (1987) reported that in Ontario a pair of Orchard Orioles nested in the same tree for three consecutive years.

Distribution

The Orchard Oriole breeds from southern Saskatchewan, Manitoba, and Ontario to northern Mexico and from the central United States to the Atlantic coast (National Geographic Society 1987).

The centre of abundance is in the southern states bordering the Mississippi Valley (Bent 1958; Robbins et al. 1986). Orchard Orioles may nest in the extreme south of Manitoba (Sealy 1980a), Saskatchewan (Shadick 1980), and Ontario (Flood 1987a), and in the U.S. states bordering Québec breeding is rare or uncommon (Ellison 1985r; Connor 1988c).

In Québec, the Orchard Oriole has been recorded on the North Shore and on Bonaventure Island, but is most often sighted in the Central St. Lawrence Lowland, especially in the Greater Montréal area (David 1980a). During the Atlas survey, it was sighted in 10 different places: Hudson, Sainte-Marthe, Nuns' Island, Westmount, Saint-Esprit, the Taureau Reservoir, Saint-Benoît-du-Lac, Thetford Mines, in the Batiscan-Neilson ZEC, and at Moisie River.

History and trends

The first sight records in Québec date from 1919 and 1920. No other sightings appear to have been recorded until the 1960s (R. Yank, pers. comm.). Since 1984, the Orchard Oriole has been reported more consistently in the south of the province. It is difficult to determine, however, whether occurrence is more frequent or whether more birds are being sighted because there are more birdwatchers.

Robbins et al. (1986) reported a declining trend in the south-central United States. Although the decline has not become generalized throughout the range, the situation was deemed precarious enough to warrant inclusion of the species on the Blue List as of 1982 (Tate 1986). In the northern part of its range, populations are tending to increase (Robbins et al. 1986), which may explain the recent range expansion into Manitoba (Sealy 1980a) and possibly Ontario (Flood 1987a).

VINCENT LÉTOURNEAU

Notes on territory

According to Enstrom (1992a), the male defends an area about 15 m in diameter around the nest. The female defends this area as well, but less vigorously. When threatened, the bird will sing to advertise its presence or launch a direct, repeated attack on the intruder, especially if it is a young male (Enstrom 1992a).

On agricultural land and in residential suburban areas in Maryland, Stewart and Robbins (1958) recorded average densities of 4.5 territorial males/10 ha.

Chronology *n* = 0

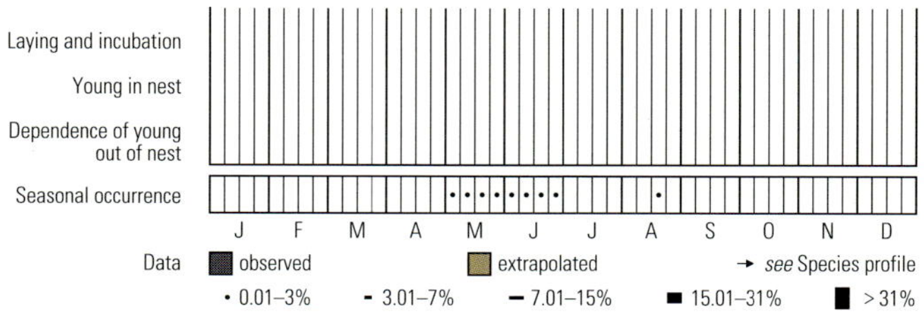

Species profile

Clutch size	usually 4–5 eggs (3–7)[1,2,3]
Laying interval	?
Start of incubation	?
Incubation period	usually 14 days (12–15)[3]
Maturity at hatching	altricial
Care of young	by female and male[4,5]
Nestling period	11–14 days[3,4]
Age at first flight	11–14 days or more[9]
Dependence of young	until at least fall migration[10]
Broods per year	1[5]
Breeding age	1 year[5]
Mating system	monogamy[5]
Duration of pair bond	1 season[5]
Total length	15.0–18.5 cm[2]
Wing span	23.5–26.0 cm[6]
Weight	19.6 g (16.0–25.1)[8]
Longevity record	9 years 7 months[7]

References:
1. Bull 1985
2. Godfrey 1986
3. Dennis 1948
4. Harrison 1978
5. Sealy 1980a
6. Terres 1980
7. Klimkiewicz and Futcher 1987
8. Dunning 1984
9. according to nestling period
10. Sealy 1979

World breeding range

Hoary Redpoll

Sizerin blanchâtre

Carduelis hornemanni

Christiane Girard

The Hoary Redpoll breeds in the Arctic and Subarctic regions of North America and Eurasia, including northernmost Québec. Some winters, a few individuals can be seen among the flocks of Common Redpolls that wander all over the province. More rarely, small flocks consisting solely of Hoary Redpolls can be seen.

The Hoary Redpoll is very similar to the Common Redpoll, which is much more common in Québec at all times of year. In winter, a Hoary Redpoll in a flock of Common Redpolls will often attract attention because of its paler body and hoary, or frosted, appearance. The coloration of both species varies markedly from one individual to another. Depending on its age and sex, a dark Hoary Redpoll may easily be confused with a pale Common Redpoll (Troy 1985; Knox 1988; David 1991b; Seutin et al. 1992).

In North America, different subspecies of Hoary Redpoll breed in the High and Low Arctic (Knox 1988): *Carduelis hornemanni exilipes*, the only subspecies that breeds in Québec, is smaller and has a shorter bill than *Carduelis hornemanni hornemanni*, which is found at higher latitudes.

Habitat and behaviour

In the Arctic, the Hoary Redpoll usually seeks out ravines and rocky slopes (Godfrey 1986), with a microclimate suitable for nesting; further south where trees and shrubs are available, it is found in the same habitats as the Common Redpoll, although it typically prefers more open areas. Like the Common Redpoll (Peiponen 1957), the Hoary Redpoll may nest twice in a summer, at least in the southern part of its breeding range. This is made possible, even during the short northern summer, by the fact that nest building takes as little as 4 days, incubation a minimum of 10 days, and the chicks sometimes leave the nest after only 9 days.

Nest building begins in late May or early June. Early nests are usually set in conifers, while in July they are more often built a little above the ground in small deciduous trees or shrubs, such as willows, alders, or birches, or sometimes even on the ground, sheltered by a rock. The nest, 6 cm deep with an outside diameter of about 10 cm, is lined with ptarmigan feathers, plant down, and sometimes caribou or lemming fur, and usually contains 4 or 5 pale blue or greenish eggs, spotted with brown (Macoun and Macoun 1909).

Hoary Redpolls do not defend a territory during the breeding season, and nests are often clustered in the same thicket. Only the female incubates, while the male feeds her frequently on the nest, continuing to do so for a few days after hatching.

According to Baldwin (1968), the female broods the young alone, but both parents feed them until they are independent. In July, most chicks have usually left the nest and are feeding alongside the adults. The diet of the Hoary Redpoll is composed of the seeds and buds of birches and willows, the seeds of herbaceous plants such as cotton-grass, and some insects.

As in the Common Redpoll, individual Hoary Redpolls have little fidelity to their wintering sites (Troy 1983), and thus the species is rarely observed at the same location several winters in a row.

Distribution

The Hoary Redpoll is a holarctic species. In North America, it breeds in the Arctic and Subarctic from Alaska to Greenland (Godfrey 1986). In Québec, breeding has been confirmed only at Kuujjuaq (Hildebrand 1950; Gabrielson and Wright 1951; Todd 1963), although the species has been observed in May, June, and July in the Inukjuak and Tasiujaq regions (ÉPOQ).

David (1980a) listed the Hoary Redpoll as a casual winter visitor throughout southern Québec, except on Anticosti Island and the Magdalen Islands, where it had never been reported. According to Fradette (1992), however, there have been some sporadic winter sightings in the Magdalen Islands.

The Hoary Redpoll winters down to southern Canada (Root 1988) and the northern U.S. states in small numbers (AOU 1983). Southern Québec is thus at the southern limit of its winter range, while the northern part of the province is the southern limit of its breeding range.

History and trends

The Hoary Redpoll had already been reported as a rare visitor several times in winter south of the Québec border by the turn of the century (Forbush 1929). Around the same time, Macoun and Macoun (1909) stated that the species bred abundantly near Fort Chimo (Kuujjuaq), and it was reported there as early as 1882 (Todd 1963). Ouellet (1974a) also cited several records in the Montréal region in the 1950s, suggesting that the Hoary Redpoll was probably more common in winter than reported sightings would indicate, given that it is easily overlooked in flocks of Common Redpolls.

Notes on territory

Hoary Redpolls breed in solitary pairs or in small groups. The species is not territorial. Even when nests are close together, males exhibit little aggressive behaviour (Baldwin 1968), either to each other or toward the Common Redpoll.

This redpoll breeds in northernmost Québec, near Ungava Bay. Compared to the Common Redpoll, it prefers more open areas. Rivière aux Feuilles. *September 1977, Pierre Lamothe.*

The Hoary Redpoll may raise two broods on the shrub tundra; early in the season, it usually nests in a conifer, while later, in July, it will often nest in a small deciduous tree, in a shrub, or even on the ground. Lake Bérard, near Baie aux Feuilles, Ungava Bay region. *June 1977, Pierre Lamothe.*

The irregular occurrence of the species in the southern part of the province, where it is seen only in winter, does not provide enough information to determine population trends. The growing number of records in recent years certainly has more to do with the increased experience of the expanding ranks of birders and the popularity of bird feeders than any real growth in populations.

No significant change in the habitat of the Hoary Redpoll is foreseen, so its numbers should remain fairly stable in the near future.

Alain Lanoue
and Gilles Seutin

World breeding range

Species profile

Clutch size	usually 4–5 eggs (3–6)[1,2,3,5,9]
Laying interval	1 egg per day[9]
Start of incubation	with second or third egg[9]
Incubation period	10–15 days[1,2,4,5,9]
Maturity at hatching	altricial
Care of young	by female and male[1,5,9]
Nestling period	9–14 days[1,2,4,5]
Age at first flight	10–14 days[9]
Dependence of young	probably less than 2 weeks after leaving nest[9]
Broods per year	1, 2 possible[3,10], *see* text
Breeding age	probably 1 year[9]
Mating system	probably monogamy[4]
Duration of pair bond	?
Total length	11–15 cm[6]
Wing span	?
Weight	12.7 g (10.4–16.1)[7]
Longevity record	4 years 7 months[8]

References:
1. Baldwin 1968
2. Terres 1980
3. Forbush 1929
4. Ehrlich et al. 1988
5. Harrison 1978
6. Godfrey 1986
7. Dunning 1984
8. Klimkiewicz and Futcher 1989
9. G. Seutin, unpubl. data
10. Peiponen 1957

Chronology *n* = 6

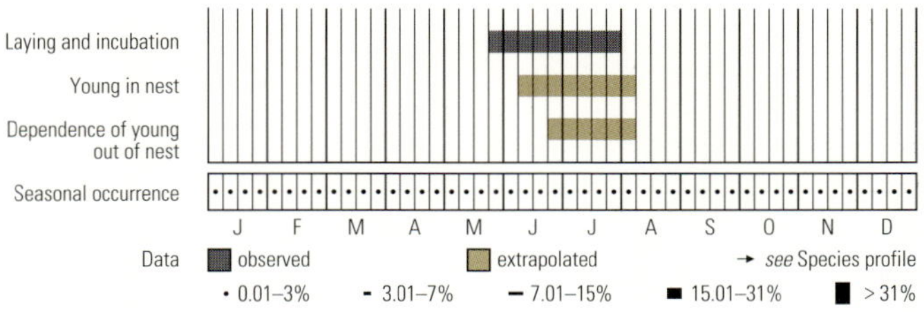

AVIAN DIVERSITY IN QUÉBEC

BIODIVERSITY

Biodiversity can be generally defined as the variety of life in a given area. It encompasses genetic composition and species richness (number of species) as well as the distribution and abundance of communities and ecosystems, and the processes whereby living organisms interact with one another and their environment (Salwasser 1990; Mosquin and Whiting 1992; Trauger and Hall 1992; World Conservation Monitoring Centre 1992). Biodiversity is thus much more than a simple inventory of species in a given area: it is a concept that embraces the range of variation in life on the Earth. Administratively, conservation of biodiversity can be defined as management of the interaction of human beings with the variety of life forms and ecosystems so they yield the greatest benefit to present generations yet maintain their potential to meet the needs and aspirations of future generations. In other words, the aim is to reverse the process of biological impoverishment at all levels of biological organization. This is the main thrust of the current movement to conserve our resources.

The concept of biodiversity covers a number of temporal and spatial scales, and this is a source of confusion (West 1993). One can speak of the biodiversity of a habitat, a complex of habitats, a region, a province, a country, a continent, or even the entire Earth. Questions about measuring, monitoring, value, and causes of biodiversity have different answers depending on the spatial or temporal scale (day, season, year, or decade) considered. Bird diversity in a given habitat, for example, is often proportional to the structural diversity of the habitat, whereas bird diversity in a given region depends on habitat diversity within the region. Relations also differ depending on the spatial scale on which they are considered. The Least Flycatcher and the American Redstart are a case in point. On the scale of the habitat, these two birds do not share the same forest segment, since the flycatcher excludes the redstart from its territory. On the scale of the region, however, the two species are associated, since they prefer the same type of habitat (Sherry and Holmes 1988; Wiens 1989). The perception of a phenomenon depends on the scale on which it is observed: the levels of biological organization (gene, population, species, and community) correspond to different spatial scales. In addition, the spatial needs of an organism depend on its size: a mouse or a warbler requires only a few square metres, but a wolf or an eagle needs several square kilometres. It is often difficult to decide on an appropriate scale when studying any given phenomenon. This is why a multispatial approach is essential when discussing biodiversity (Noss and Harris 1986). Almost all facets of biodiversity are affected by factors of scale—the scale on which the phenomenon takes place and the one on which it is observed. The impact of a housing development on bird diversity may be significant on a local scale (loss of one or more habitats), for example, yet negligible on the regional scale.

Characterizing biodiversity is no simple matter, especially considering the multitude of methods, levels of biological organization, and spatial scales for classifying life (Frontier and Pichot-Viale 1993). Noss (1990) recognizes three types of diversity (composition diversity, structural diversity, and functional diversity) that apply to all levels of organization, from gene to ecosystem. Though the emphasis is often placed on composition diversity (species richness, for example), structural and functional diversity are nonetheless important for maintaining composition diversity and essential for preserving ecosystem integrity. Diversity includes, besides species diversity, genetic (subspecies and races, for example), taxonomic (evolutionary distance or kinship between organisms), class (size, age, and sex), and habitat diversity (microhabitat, habitat, and ecosystem). All of these facets contribute in their own ways to a region's biodiversity.

Why conserve biodiversity?

There are intrinsic, legal, and utilitarian reasons for conserving biodiversity (Norton 1986, 1987; World Conservation Monitoring Centre 1992; World Resources Institute 1992). The intrinsic reasons are based on philosophical, moral, or religious grounds—the individual's or society's decision to attribute moral value to the existence of other forms of life (Callicott 1986). The legal reasons are laws adopted by society to enforce respect for certain components of nature—laws such as the Québec Act respecting threatened or vulnerable species. Utilitarian reasons fall into three main categories:

1. Economics. Human beings derive economic benefits from the diversity of natural environments (logging, agriculture, hunting, fishing, and so forth).
2. Public health. Human beings derive direct benefits from numerous "naturally" produced substances. Most of our medicines come from plants and animals, and many of them cannot be produced synthetically.
3. Future generations. Since it is impossible to predict scientific knowledge in the future, and since only a very small portion of living organisms have been studied for their direct usefulness to human beings, prudence demands that biodiversity be maintained for future generations. Many species once considered worthless are now important components of our economy. Only a few years ago, trembling aspen and Western yew were considered valueless. Today, the former supplies Alberta's forest industry and the latter provides a substance (taxol) that fights cancer. The functioning of ecosystems that support life on the Earth is complex and poorly understood. We must therefore conserve as many of their components as possible.

How can biodiversity be measured?

The variety of ways of measuring biodiversity is proportional to the complexity of the concept. Because of the spatial and temporal scales involved as well as the multitude of levels of biological organization, this question has many answers, and they vary depending on the spatial scale considered (local, regional, national, and so forth), the level of organization targeted, and the reasons for measuring biodiversity.

There are many indexes for measuring diversity (Whittaker 1977; Magurran 1988). These indexes take two factors into account: richness and relative abundance (of species, habitats, and so forth). The indexes fall into two categories: those that measure diversity within a spatial unit (a habitat, for example) and those that measure the degree of difference between two or more spatial units (two habitats, for example). The indexes also differ according to the spatial scale to which they apply: *alpha* diversity indexes measure diversity within a given habitat—bird diversity in a maple stand, a marsh, or a field; *gamma* diversity indexes measure diversity in an area that comprises a complex of habitats—a watershed or a county, for example; and *epsilon* diversity indexes measure the biodiversity of many regions or of a biogeographic region such as the boreal forest. *Beta* diversity indexes, on the other hand, measure differences in diversity between two or more habitats (Magurran 1988). The *beta* diversity index for two habitats is the inverse of their similarity index; the more the bird community in two habitats differs, the higher the *beta* index. One other index that is sometimes used is the *delta* diversity index, which measures the difference between two or more habitat mosaics.

To conserve biodiversity, efforts must focus on maximizing diversity between habitats (*beta* diversity) rather than diversity within a habitat (*alpha* diversity). The importance of spatial scale, then, is clear. Habitat A with three unique species contributes much more to regional diversity than habitat B with 10 common species found in many other habitats (C, D, and E). Transforming habitat A to accommodate 10 common species could result in loss of the three unique ones. The diversity of habitat A would, of course, triple, but on the regional level there would be a decline of three species. In other words, a hierarchical approach is important for sound management of biodiversity. Such an approach implies that higher levels of organization encompass and restrict behaviour at lower levels. Intervention on a local scale (that is, maximizing *alpha* diversity) must accordingly be weighed in terms of the consequences on a larger scale. The concept of hierarchy, which cannot be dissociated from biodiversity, demands that biodiversity be monitored on a number of spatial and temporal scales. The impact of environmental stress on biodiversity is expressed in different ways depending on level of biological organization. For example, global warming will undoubtedly provoke changes in the distribution of certain species, modify the dynamics (growth, mortality, and reproduction rates) of many populations, cause physiological changes in many individuals, and promote selection of genes best adapted to the new conditions. A comprehensive approach to biodiversity must take variability at all levels into account—from gene to ecosystem.

The value of a species

One of the problems with diversity indexes is the assumption that all species are equivalent—an assumption that is not generally well founded, especially when the goal is conservation of biodiversity. Species vary in distribution, abundance, and ecology as well as in the role they play within the community and the ecosystem. Certain species sometimes play a crucial role in the functioning of a particular community or ecosystem (Bond 1993; Daily et al. 1993; Johnsson 1993). The disappearance of such species can trigger a whole chain of events that may affect many other species. A forest without woodpeckers, for example, is also a forest without nest holes for secondary cavity nesters that cannot make their own holes, and such species accordingly decline and sometimes even disappear.

Rarity

The concept of rarity is more complex than it seems. There are several types of rarity (Rabinowitz et al. 1986). An understanding of the mechanisms and circumstances responsible for the rarity of an organism is important if judgments are to be made about the value of a rare species.

Most wildlife communities are composed of a few dominant species and several less abundant species. The rarer species are often representative of a population adapted to conditions that prevailed in the past. Small populations of species associated with young forests, for example, can inhabit clearings in mature forest. In the event of a fire, an epidemic, or clearcutting, these species become dominant in the new community that takes over the regenerating habitat. The species associated with mature forests, on the other hand, are relegated to the forest fragments that remain, and they become rare. As the forest matures, however, these species gradually come to dominate the local bird community. Rarity of a species does not necessarily mean the species is vital for maintaining an ecosystem (Noss 1991). In fact, the most abundant species in an ecosystem are often those that contribute most to the proper functioning of the ecosystem. There are also species of variable abundance that play a central role in the functioning of an ecosystem (keystone species). A decline in the population of a keystone species often has a much greater impact on the functioning of an ecosystem than the disappearance of a rare species. It is thus important not to focus solely on pinpointing and protecting rare or threatened species.

Identifying keystone species in a community or an ecosystem is no easy task (Walker 1992; Bond 1993). Often, the role of a species differs from one ecosystem to the next, from one year to the next, or even from one season to the next. The identification of keystone species thus requires close observation over many years. Caution is accordingly required, especially since we know little about the ecological rules governing most species.

A comprehensive approach

Our knowledge of biodiversity is very limited; we know little about insects and even less about smaller organisms (World Conservation Monitoring Centre 1992). One approach recommended for biodiversity conservation given this lack of knowledge is what is known as the coarse- and fine-filter approach. The idea is to first protect entire ecosystems because of their differences, on the assumption that they contain most species of a region. Likewise, by protecting the habitat of species at the top of the food chain and those that need large home ranges, the survival of most species with smaller home ranges that use the same habitat will also be protected (Noss 1991). Some species nonetheless escape this coarse filter, and their conservation requires a more specific approach—what is known as the fine-filter approach. This approach is often reserved for high-profile species (game and threatened species).

Gap analysis is one technique used to plan the protection of an ecosystem. A variety of habitat and animal distribution maps are superposed to determine what is or is not protected by the reserve system. This method can be used to identify key unprotected zones (Scott et al. 1993).

Sustainable development

Though designation of protected areas is a major component of biodiversity conservation strategy, the creation of protected areas alone cannot guarantee conservation of biodiversity. To ensure maintenance of biodiversity, we will have to focus on the "semi-natural matrix," that is, on areas designated for human use and activity. It is here that the concept of biodiversity intersects with that of sustainable development. Sustainable development can be defined as positive socioeconomic change that does not weaken the ecological and social systems on which communities and society are based (Rees 1989). The message conveyed by the concept of sustainable development is that a long-term perspective is required, that natural resources must be used in a way that does not jeopardize the possibility of maintaining or improving their use in the future. This principle implies as well that exploitation of any one resource must not compromise use of another resource. In short, human activity must not jeopardize global biodiversity.

In areas that constitute the semi-natural matrix, conservation of local biodiversity is a tremendous challenge, but it is crucial if we are to progress without compromising the use of natural resources by future generations. One essential and urgent activity is accurate measurement of the effects of human activity on fauna and flora as well as on ecosystems in general. This knowledge will lay the essential groundwork for design and formulation of development strategies that can ensure the conservation of biodiversity. The conservation of biodiversity in Québec demands a commitment to sustainable development in all fields of endeavour, public as well as private.

Threatened species

Efforts must also be devoted to species at risk. This is a "reactionary" (action should have been taken long before the species became at risk), yet necessary measure. It is clear that with the growing number of threatened species, it is becoming more and more difficult to work to save them all. This means more energy must be dedicated to prevention to reduce the number of species that become threatened.

Maintaining the biodiversity of ecosystems and species is crucial, but the importance of genetic diversity must not be forgotten either. One need only think of domesticated animal and plant species to understand the benefits of genetic diversity. Among other things, genetic diversity increases a species' ability to adapt to environmental change as well as its resistance to disease. The phenomenal productivity of most of our grain and vegetable species derives from crossing of diverse subspecies.

The loss of a subspecies, a race, or any other genetic entity composing a species is a first step toward disappearance of that species, besides being a potential loss for human use.

Habitat fragmentation

One factor that contributes to loss of biodiversity is the fragmentation of natural habitats (Wilcox and Murphy 1985; Temple and Wilcox 1986; Harris and Silva-Lopez 1992; Morrison et al. 1992). This is a global phenomenon that turns up on all spatial scales and affects all organisms (Lord and Norton 1990). Habitat fragmentation increases what is known as the edge effect (Leopold 1933; Yahner 1988; Hunter 1990); that is, sometimes transition zones between two habitats are particularly attractive to many animal species. In fact, certain transition zones support a greater diversity than either of the two adjacent habitats because they incorporate elements of both.

Some species, on the other hand, shun edges, preferring forest interiors (Ovenbird, Wood Thrush, and Northern Goshawk). Likewise, some species nest in grassland and avoid forest edges (Savannah Sparrow, Le Conte's Sparrow, and Horned Lark). Such species are often affected by excessive fragmentation of their habitat.

The impact of habitat fragmentation on birds has been particularly well researched (Wilcove and Robinson 1990; Faaborg et al. 1993; Andrén 1994; Villard et al. 1995). Small fragments are not generally suitable for species with large home ranges or those that do not tolerate edges (Brown and Dinsmore 1986; Freemark and Collins 1992; Warner 1992). In addition, it is harder for the males of a number of migratory species to attract females in small habitat fragments that support only one or two territories (Gibbs and Faaborg 1990; Villard et al. 1992, 1993). Also, breeding success sometimes declines in small fragments because of high parasitism and predation rates (Wilcove 1985; Yahner and Scott 1988; Robinson et al. 1993.).

The problem of fragmentation affects all habitats, forests as well as other types of habitat (Herkert 1991; Askins 1993). The problem is the same for species that live in mature forest, natural grassland, marshland, or anywhere else: the smaller the fragment, the greater the influence of adjacent habitats, and the more the fragment comes to differ from the original habitat. When the fragments of a habitat become isolated from one another, the wildlife populations also become fragmented and isolated. The separation of a population into subpopulations is one of the steps that can lead to extinction of the population. When many subpopulations continue to interact through dispersal and exchange of individuals, we speak of a metapopulation (Fritz 1979; Temple and Cary 1988; Freemark 1989; Kalkhoven 1993). In some metapopulations, subpopulations survive only because of immigration of individuals from other subpopulations (Gilpin and Hanski 1991). Besides maintaining the subpopulations, these exchanges of individuals also ensure genetic integrity by preventing the generally harmful effects of inbreeding. In other words, it is important to maintain genetic variability, that is, an appropriate combination of genes corresponding to an adaptation to a given habitat. The introduction of another individual is that much more beneficial if the individual adds to the genetic diversity (in the sense of increasing the genetic variability) of the population.

For many types of organisms (plants, small mammals, reptiles, and amphibians), though less frequently for birds, the presence of corridors between fragments facilitates dispersal of individuals and recolonization of the fragments. With weak fliers such as the Ruffed Grouse, for example, forest corridors can facilitate dispersal of young, especially in agricultural mosaics, and maintain genetic diversity.

Population viability

Another concept often associated with biodiversity is minimum viable population, that is, a critical "floor" below which the population cannot survive for long (Gilpin and Soulé 1986; Soulé 1987). Population preservation is an important aspect of biodiversity conservation. In certain tropical countries, species loss is a major concern, but in temperate countries such as Canada, gradual extinction of local populations of certain species is the most telling sign of the deterioration of overall biodiversity (Ehrlich and Daily 1993). Population extinction affects biodiversity on the local scale first, but it also undermines its conservation on a regional and even national scale.

Population survival depends on a variety of factors. There is therefore no set size for population viability. Minimum population size varies greatly from one species to the next and from one area to the next (Simberloff 1988). This is because there are two components to any judgment about population viability: a scientific component and an intuitive component. Science makes it possible to understand genetic and demographic aspects of the "natural" survival of a population in a given area. However, the length of time a population is to survive and the level of probability that it will indeed survive during this period depend on value judgments made by society (Shaffer 1981, 1987). For example, depending on development options selected, a population could have a 95% chance of surviving for the next 20 years or an 80% chance of surviving for the next 100 years. Each of these possibilities is associated with a different population size. The concept of a minimum viable population should be considered a limit to stay as far away from as possible rather than a goal.

Conclusion

In Québec, as elsewhere, bird diversity can be measured in different ways, at different levels of biological organization (gene, population, and species) and on a variety of spatial scales. Two factors contribute to the vulnerability of a species: its abundance and its distribution. As a general rule, the more abundant a species is and the more extensive its range, the less vulnerable it is. Habitat specialization is also an important factor; species that are habitat specialists are entirely dependent on what happens to their habitat.

Clearly, efforts to conserve biodiversity should focus increasingly on ecosystems (Franklin 1993). Development of a protected area system is only one aspect of a conservation strategy. Appropriate handling of the semi-natural matrix is equally, if not more, important. In fact, the conservation capacity of protected areas depends largely on the use that is made of adjacent spaces.

The concept of biodiversity conservation is complex. Maurer (1993) lists five major principles of habitat management for biodiversity: first, species diversity per se is not necessarily the best criterion upon which to build management objectives (Samson and Knopf 1982); second, the spatial structure of habitats in a landscape is important (Martin 1992); third, the geographic context of a management situation must influence the management approach (Samson 1992); fourth, management objectives must take account of natural levels of environmental variability (Landres 1992); and fifth, wildlife management (conservation and preservation) must become a global discipline.

In the urban ecosystem, developments on the scale of the regional county municipality, the city, the town, the residential neighbourhood, or the individual residential property all in their own way affect biodiversity. Every citizen can contribute to conserving biodiversity on each of these scales—by fixing up his or her backyard or by becoming involved in developing a network of urban parks in his or her neighbourhood or town. The concepts of hierarchy, scale, and level of biological organization are integral parts of the concept of biodiversity. The conservation of biodiversity must be viewed holistically, encompassing the diverse spatial scales and levels of biological organization.

Jean-Pierre L. Savard

SPECIES DIVERSITY IN ECOLOGICAL REGIONS

One of the most useful features of an atlas of this kind is that the collected data can be used to establish biogeographical relationships, that is, relationships between bird distribution and such geographical features as vegetation, climate, and topography. Although individual species accounts discuss distribution in varying degrees of detail, they say little about interspecific relationships. In fact, it would be hard to obtain an overall picture of bird distribution throughout southern Québec without examining every map. This section presents a comprehensive approach for identifying those regions in southern Québec with the greatest species richness for birds, the species composition of the various ecological regions, species with similar distributions, and the most common and unusual species in the major regions of southern Québec.

Comparisons among ecological regions

The dataset in the frequency of occurrence file, presented in Appendix I, was subjected to a principal component analysis with ecological regions treated as variables and bird species as observations. The purpose of this analysis was to obtain a broad graphical representation of inter-regional variations in birdlife, as well as to make species comparisons with regard to their relative abundance in all ecological regions. Figure 1 compares ecological regions with respect to species reported within them. Two component axes (explaining 87% of the variability in the data) were retained. In principal component analysis, the direction of variables (in this case, ecological regions) with respect to the origin are interpretable; in other words, the direction of the arrows is important information, although their length is not. Regions separated by a small angle are highly correlated; those intersecting at right angles are uncorrelated (r = 0); and those pointing in opposite directions are negatively correlated. Similarly, observations (bird species) are interpreted according to their position in space; in general, species plotted close to one another have similar distributions and frequencies of occurrence.

Thibault (1985) ranked his forest regions (in Atlas terminology, ecological regions) in decreasing order by degree-days, ranging from pure deciduous growth to pure coniferous growth. Adjoining regions have very similar climates (degree-days, aridity index, and so forth) and vegetation. Our analysis reproduces Thibault's classification almost exactly, in that each region is separated from its "neighbour" by a fairly sharp angle (Figure 1). Avian communities in regions of similar vegetation and climate also tend to be highly similar in species composition. Interestingly, no 2 regions point in opposite directions: the regions at either end of the gradient for vegetation (Sugar maple-hickory and sugar maple-basswood, and Black spruce-moss) are separated by an angle of no more than about 90°. A zero correlation between two regions indicates that the most abundant species in one region tend not to be more abundant in the other, and vice versa. The matrix of correlation coefficients between ecological regions (Table 1) reaffirms the results of the principal component analysis.

Arrows are numbered to correspond with ecological regions, as follows:

Deciduous forest zone

1. Sugar maple-hickory and sugar maple-basswood
2. Sugar maple-basswood
3. Sugar maple-basswood and sugar maple-yellow birch
4. Sugar maple-yellow birch-basswood
5. Sugar maple-yellow birch-beech and sugar maple-yellow birch-basswood
6. Sugar maple-yellow birch-beech
7. Sugar maple-yellow birch-balsam fir

Mixed forest zone

8. Sugar maple-yellow birch and balsam fir-yellow birch
9. Balsam fir-yellow birch
10. Balsam fir-white birch or balsam fir-red maple
11. Yellow birch-balsam fir

Coniferous forest zone

12. Balsam fir-white birch (west)
13. Balsam fir-white birch (east)
14. Balsam fir-black spruce
15. White spruce-balsam fir and balsam fir-white spruce
16. Black spruce-balsam fir-moss
17. Black spruce-moss

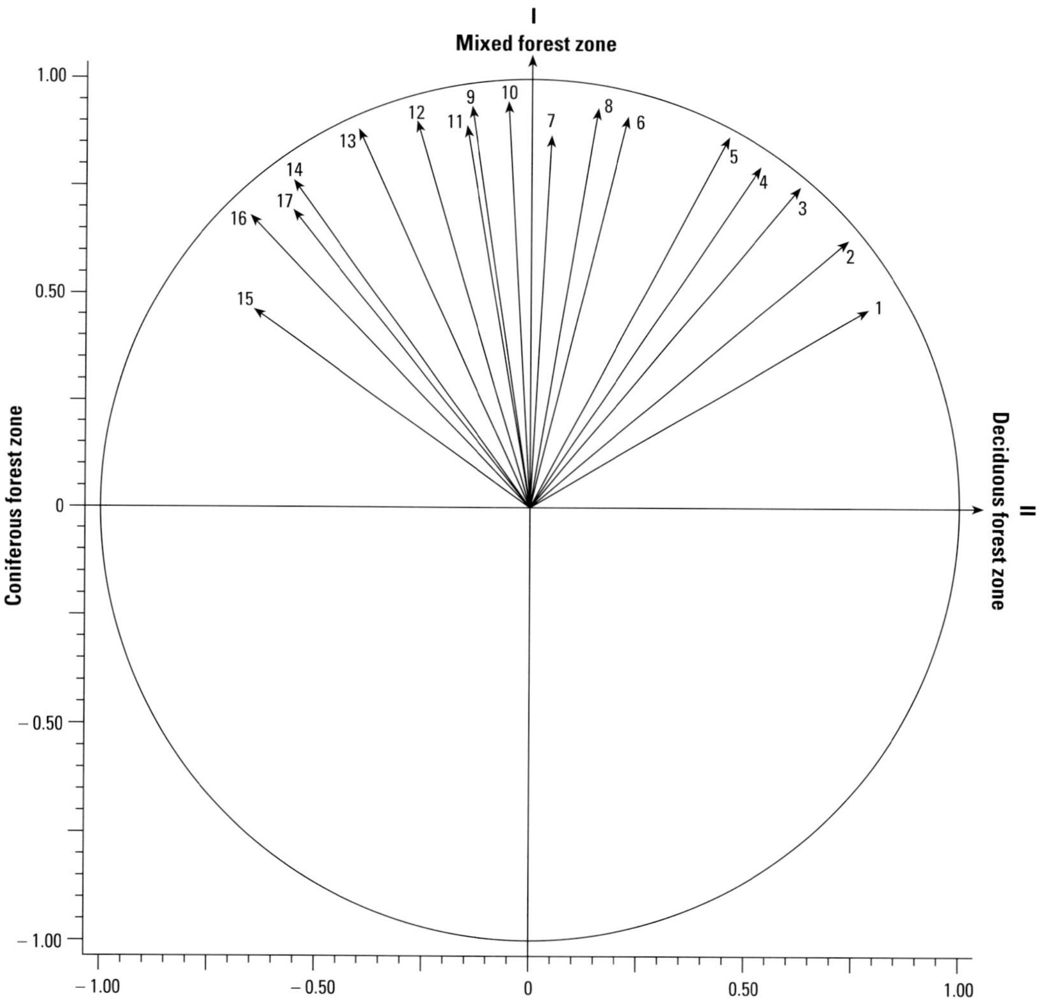

Figure 1
Plot of ecological regions with respect to the two principal component axes

The arrows illustrate the relationship of bird communities among the ecological regions. Note that ecological regions characterized by deciduous growth clustered to the right (with the exception of Region 7; Sugar maple-yellow birch-balsam fir) (see text), regions characterized by coniferous growth clustered to the left, and regions characterized by mixed forests plotted out in between the two.

Comparisons among bird species

Even more interesting is the two-dimensional plot of bird species (Figure 2). The first component axis represents abundance and ubiquity, on an increasing scale toward the top of the graph; in other words, species further up on the graph are more common and more widely distributed over more regions than those further down. The second axis represents a deciduous-coniferous gradient, from right to left. A loop can be drawn around the periphery of the plot in Figure 2, starting at the bottom centre with the points representing the Rufous-sided Towhee and Northern Cardinal. These fairly rare, local species are found almost exclusively in the Central St. Lawrence Lowland. Proceeding counter-clockwise, species such as the House Wren and Indigo Bunting, more common and not as local, are encountered; these species, like the preceding ones, are more abundant in the Central St. Lawrence Lowland while not as scarce elsewhere in the deciduous forest zone. Several edge species that primarily inhabit the Central St. Lawrence Lowland (such as the four just mentioned) are found in this part of the graph. The multiplicity of open habitats and forest fragmentation in the Central Lowland are ideal for these species, which are also often at the northern limit of their North American range in southern Québec.

Continuing around the loop in Figure 2, we encounter species such as the Wood Thrush and Great Crested Flycatcher. These species are common throughout the deciduous forest zone, not only in the Central St. Lawrence Lowland. The next species encountered (such as the Veery and Blue Jay) are more generalist, common to the deciduous and mixed forest zones alike. Next come species (such as the Red-eyed Vireo) that are almost as common in the coniferous forest zone as in the deciduous and mixed forest zones, followed by the most ubiquitous and evenly distributed species in southern Québec according to our data (American Robin, White-throated Sparrow, Northern Flicker).

Then (still moving counter-clockwise) comes the Yellow-rumped Warbler, common everywhere in southern Québec except the Montréal area. The next species (Swainson's Thrush, Dark-eyed Junco, Ruby-crowned Kinglet) are less and less common in the Central St. Lawrence Lowland. Note that these species are plotted directly opposite those found solely in the Central Lowland. Next come species (such as the Tennessee Warbler, Lincoln's Sparrow, and Bay-breasted Warbler) that are even less common in the deciduous forest zone, followed by species (such as the Gray Jay and Rusty Blackbird) more typical of the boreal forest. The next species (Blackpoll Warbler, Fox Sparrow, and Pine Grosbeak) are even more confined to the boreal forest; interestingly, these species are much more common in the eastern part of southern Québec than in the western part. The loop terminates with three species (Greater Yellowlegs, Red-breasted Merganser, and Canada Goose) that are confined to just a few regions of the boreal forest; in fact, all three species are most common on Anticosti Island.

It is also interesting to observe how species within a group are distributed. For example, the Wood Thrush, Veery, Hermit Thrush, and Swainson's Thrush fall along the deciduous-coniferous gradient in that order (Figure 2). Among woodpeckers, the Pileated Woodpecker is the most closely associated with the deciduous forest zone; the Yellow-bellied Sapsucker, Downy Woodpecker, and Hairy Woodpecker cluster together in the group of birds common to both the deciduous and mixed forest zones. The Northern Flicker is one of the most widespread and common species throughout southern Québec. The preference of the Black-backed Woodpecker and Three-toed Woodpecker for the boreal forest is clear; the Black-backed Woodpecker, however, is less confined than the Three-toed Woodpecker to the boreal forest.

Table 1

Spearman's correlation coefficients between ecological regions

Ecological region	1	2	3	4	5	6	7	8	9	10	11	12	13	14	15	16	17
1	1.00	0.91	0.84	0.67	0.71	0.44	0.24	0.47	0.34	0.40	0.19	0.32	0.14	0.07	–0.13	–0.04	0.03
2		1.00	0.96	0.89	0.89	0.72	0.46	0.71	0.55	0.58	0.42	0.50	0.30	0.14	–0.06	0.04	0.20
3			1.00	0.90	0.95	0.79	0.57	0.81	0.65	0.67	0.53	0.58	0.41	0.23	–0.02	0.10	0.27
4				1.00	0.92	0.89	0.70	0.83	0.72	0.70	0.66	0.62	0.47	0.23	0.01	0.08	0.27
5					1.00	0.90	0.67	0.91	0.79	0.79	0.68	0.67	0.57	0.36	0.12	0.19	0.34
6						1.00	0.85	0.90	0.87	0.84	0.87	0.79	0.73	0.49	0.26	0.35	0.47
7							1.00	0.74	0.73	0.70	0.86	0.68	0.64	0.41	0.33	0.28	0.43
8								1.00	0.90	0.88	0.79	0.77	0.75	0.54	0.29	0.35	0.46
9									1.00	0.91	0.83	0.84	0.89	0.74	0.41	0.51	0.50
10										1.00	0.79	0.87	0.83	0.64	0.28	0.48	0.55
11											1.00	0.85	0.82	0.62	0.44	0.49	0.64
12												1.00	0.86	0.74	0.44	0.63	0.79
13													1.00	0.90	0.59	0.79	0.73
14														1.00	0.56	0.83	0.72
15															1.00	0.55	0.53
16																1.00	0.79
17																	1.00

Table 2 lists the 30 commonest species in the Central St. Lawrence Lowland and tables 3, 4, and 5 show the 30 commonest species in the three forest zones—deciduous, mixed, and coniferous. Note that five species (American Robin, White-throated Sparrow, Northern Flicker, Tree Swallow, and Common Yellowthroat), all of which fall in the upper top half (ubiquitous species) of the plot produced by the principal component analysis, are among the 12 most common species in the three forest zones (see tables 3, 4, and 5). In addition, species plotted to the right in Figure 2 are more common in the deciduous forest zone, while those plotted to the left are more common in the coniferous forest zone. Interestingly, no bird species are unique to the mixed forest zone; those common in the mixed forest zone are also common in the deciduous or coniferous forest zones, or in both at once, perhaps because of the wealth of both deciduous and coniferous stands in the mixed forest zone.

Species found in the Central St. Lawrence Lowland reflect the habitat diversity found there. Many of the most common species (such as the American Goldfinch, Eastern Kingbird, Barn Swallow, Killdeer, and Bobolink; see Table 2) are associated with open habitats of all types, accurately reflecting the predominance of such habitat in the Central St. Lawrence Lowland. Even more interesting is the complex of species that breed almost exclusively in the Central Lowland, as described below in the section on species richness.

Table 2

The 30 commonest species in the Central St. Lawrence Lowland

Species	Number of squares in which reported	% of squares in which reported
Cedar Waxwing	351	100 %
Song Sparrow	351	100 %
Red-winged Blackbird	351	100 %
American Goldfinch	351	100 %
Northern Flicker	350	99.7 %
Eastern Kingbird	350	99.7 %
Tree Swallow	350	99.7 %
Barn Swallow	350	99.7 %
American Robin	350	99.7 %
American Crow	349	99.4 %
Black-capped Chickadee	349	99.4 %
Veery	349	99.4 %
Common Yellowthroat	349	99.4 %
Chipping Sparrow	348	99.1 %
White-throated Sparrow	348	99.1 %
Blue Jay	347	98.9 %
European Starling	347	98.9 %
Rose-breasted Grosbeak	347	98.9 %
Red-eyed Vireo	346	98.6 %
Common Grackle	346	98.6 %
Killdeer	345	98.3 %
Bobolink	341	97.2 %
Brown-headed Cowbird	341	97.2 %
Eastern Wood-Pewee	340	96.9 %
Gray Catbird	340	96.9 %
Least Flycatcher	338	96.3 %
Yellow Warbler	338	96.3 %
Northern Oriole	338	96.3 %
Ovenbird	337	96.0 %
Savannah Sparrow	336	95.7 %

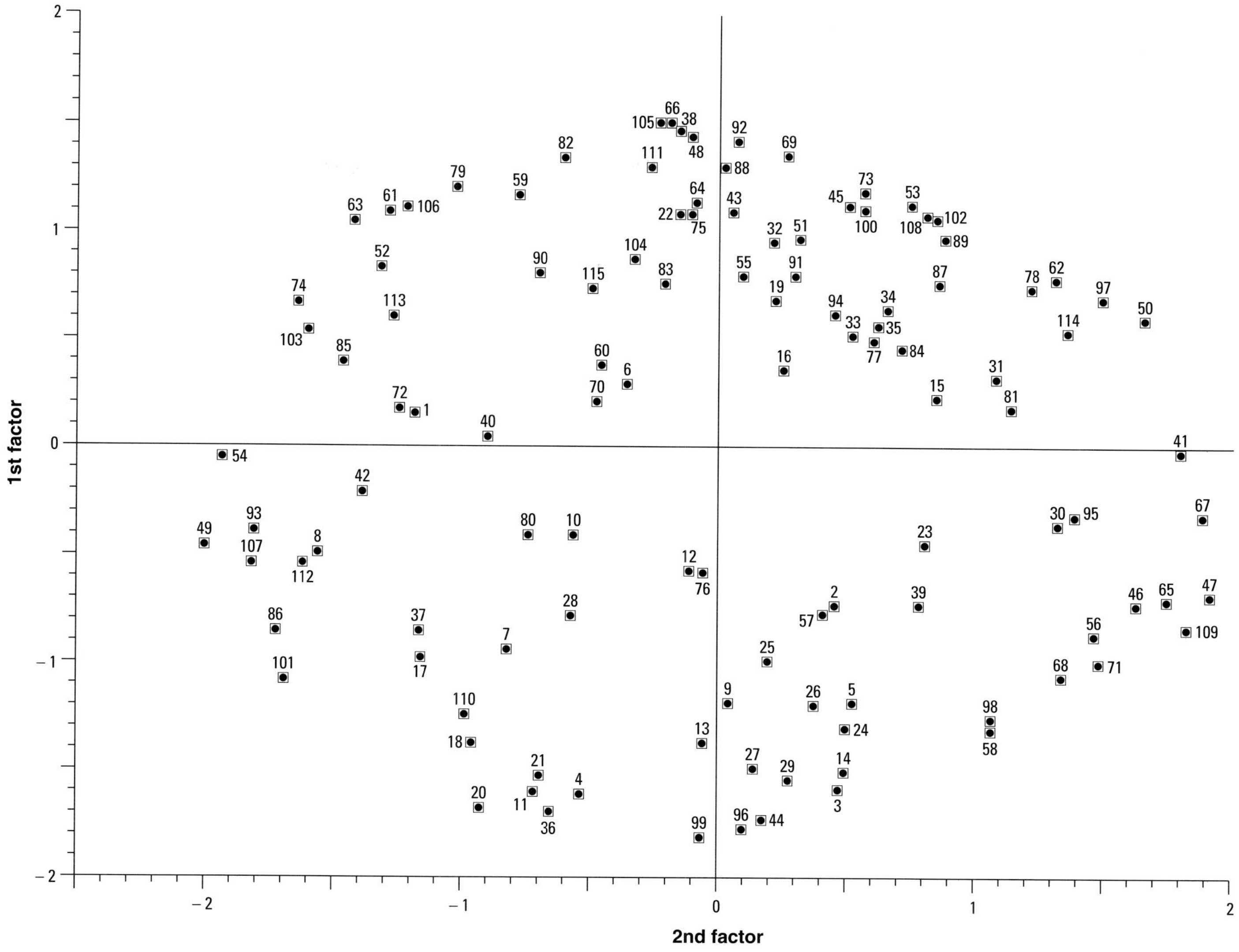

Figure 2
Plot of bird species with respect to the two principal component axes
Principal component analysis of the frequency of species occurrence, by ecological region.
Restricted to the 115 species included in the analysis.

Species code

1 Common Loon
2 American Bittern
3 Green Heron
4 Canada Goose
5 Wood Duck
6 American Black Duck
7 Ring-necked Duck
8 Common Goldeneye
9 Hooded Merganser
10 Common Merganser
11 Red-breasted Merganser
12 Sharp-shinned Hawk
13 Northern Goshawk
14 Red-shouldered Hawk
15 Broad-winged Hawk
16 American Kestrel
17 Merlin
18 Spruce Grouse
19 Ruffed Grouse
20 Greater Yellowlegs
21 Solitary Sandpiper
22 Spotted Sandpiper
23 American Woodcock
24 Black-billed Cuckoo
25 Great Horned Owl
26 Barred Owl
27 Northern Saw-whet Owl
28 Common Nighthawk
29 Whip-poor-will
30 Chimney Swift
31 Ruby-throated Hummingbird
32 Belted Kingfisher
33 Yellow-bellied Sapsucker
34 Downy Woodpecker
35 Hairy Woodpecker
36 Three-toed Woodpecker
37 Black-backed Woodpecker
38 Northern Flicker
39 Pileated Woodpecker
40 Olive-sided Flycatcher
41 Eastern Wood-Pewee
42 Yellow-bellied Flycatcher
43 Alder Flycatcher
44 Willow Flycatcher
45 Least Flycatcher
46 Eastern Phoebe
47 Great Crested Flycatcher
48 Tree Swallow
49 Gray Jay
50 Blue Jay
51 American Crow
52 Common Raven
53 Black-capped Chickadee
54 Boreal Chickadee
55 Red-breasted Nuthatch
56 White-breasted Nuthatch
57 Brown Creeper
58 House Wren
59 Winter Wren
60 Golden-crowned Kinglet
61 Ruby-crowned Kinglet
62 Veery
63 Swainson's Thrush
64 Hermit Thrush
65 Wood Thrush
66 American Robin
67 Gray Catbird
68 Brown Thrasher
69 Cedar Waxwing
70 Solitary Vireo
71 Warbling Vireo
72 Philadelphia Vireo
73 Red-eyed Vireo
74 Tennessee Warbler
75 Nashville Warbler
76 Northern Parula
77 Yellow Warbler
78 Chestnut-sided Warbler
79 Magnolia Warbler
80 Cape May Warbler
81 Black-throated Blue Warbler
82 Yellow-rumped Warbler
83 Black-throated Green Warbler
84 Blackburnian Warbler
85 Bay-breasted Warbler
86 Blackpoll Warbler
87 Black-and-white Warbler
88 American Redstart
89 Ovenbird
90 Northern Waterthrush
91 Mourning Warbler
92 Common Yellowthroat
93 Wilson's Warbler
94 Canada Warbler
95 Scarlet Tanager
96 Northern Cardinal
97 Rose-breasted Grosbeak
98 Indigo Bunting
99 Rufous-sided Towhee
100 Chipping Sparrow
101 Fox Sparrow
102 Song Sparrow
103 Lincoln's Sparrow
104 Swamp Sparrow
105 White-throated Sparrow
106 Dark-eyed Junco
107 Rusty Blackbird
108 Common Grackle
109 Northern Oriole
110 Pine Grosbeak
111 Purple Finch
112 White-winged Crossbill
113 Pine Siskin
114 American Goldfinch
115 Evening Grosbeak

The list of the commonest species in the deciduous forest zone (Table 3) is almost the same as for the Central St. Lawrence Lowland (Table 2), not surprising given that more than half (351 out of 610) of the squares in the deciduous forest zone are located in the Central Lowland, and because forest stands remain fairly similar. The major difference can be accounted for by open habitat, which is more extensive in the Central Lowland. Consequently, forest species tend to rank higher (be more common) than on the Central Lowland list; on the other hand, open-country species tend to rank somewhat lower.

The list of commonest species in the mixed forest zone (Table 4) is somewhat different from the list for the deciduous forest zone, with the appearance of species (such as the Magnolia Warbler, Swainson's Thrush, Yellow-rumped Warbler, Dark-eyed Junco, Ruby-crowned Kinglet, Winter Wren, Hermit Thrush, and Nashville Warbler) that are mostly associated with the coniferous forest. This trend is even more marked on the list for the coniferous forest zone itself, with the appearance of a few newcomers (Tennessee Warbler, Lincoln's Sparrow, Pine Siskin, Common Raven, Northern Waterthrush, Swamp Sparrow, Bay-breasted Warbler, Philadelphia Vireo, and Boreal Chickadee) among the 30 commonest species (Table 5).

Table 3
The 30 commonest species in the deciduous forest zone

Species	Number of squares in which reported	% of squares in which reported
American Robin	608	99.7 %
Song Sparrow	608	99.7 %
Northern Flicker	607	99.5 %
Cedar Waxwing	606	99.3 %
Common Yellowthroat	606	99.3 %
Tree Swallow	605	99.2 %
Blue Jay	605	99.2 %
Black-capped Chickadee	605	99.2 %
White-throated Sparrow	605	99.2 %
Common Grackle	603	98.9 %
Veery	602	98.7 %
Red-winged Blackbird	602	98.7 %
Red-eyed Vireo	601	98.5 %
Rose-breasted Grosbeak	601	98.5 %
Barn Swallow	599	98.2 %
Eastern Kingbird	597	97.9 %
Chipping Sparrow	596	97.7 %
Ovenbird	589	96.6 %
Least Flycatcher	582	95.4 %
American Crow	581	95.2 %
American Redstart	580	95.1 %
American Goldfinch	579	94.9 %
Chestnut-sided Warbler	568	93.1 %
Killdeer	561	92.0 %
European Starling	557	91.3 %
Eastern Wood-Pewee	555	91.0 %
Black-and-white Warbler	554	90.8 %
Brown-headed Cowbird	551	90.3 %
Alder Flycatcher	549	90.0 %
Purple Finch	543	89.0 %

Species richness

The total number of species reported by ecological region, counting all species and all squares visited, is presented in Table 6, and is graphically illustrated in Figure 3. A general decrease in diversity can be observed from south to north. The Central St. Lawrence Lowland is the most diverse region for birds in southern Québec. As noted above, this region offers a wide variety of habitats, among them forest, open and fragmented areas, woodland edges and hedgerows, farmland, suburban areas, marshes, and riparian and other wetlands.

Table 4
The 30 commonest species in the mixed forest zone

Species	Number of squares in which reported	% of squares in which reported
American Robin	232	100 %
White-throated Sparrow	232	100 %
Northern Flicker	230	99.1 %
Magnolia Warbler	230	99.1 %
Common Yellowthroat	230	99.1 %
Tree Swallow	229	98.7 %
Swainson's Thrush	229	98.7 %
Yellow-rumped Warbler	229	98.7 %
Black-capped Chickadee	228	98.3 %
Cedar Waxwing	228	98.3 %
Dark-eyed Junco	227	97.8 %
Common Grackle	227	97.8 %
Purple Finch	227	97.8 %
Least Flycatcher	226	97.4 %
American Redstart	226	97.4 %
Ovenbird	226	97.4 %
Ruby-crowned Kinglet	225	97.0 %
Song Sparrow	225	97.0 %
Red-winged Blackbird	225	97.0 %
Red-eyed Vireo	223	96.1 %
Chipping Sparrow	222	95.7 %
Rose-breasted Grosbeak	219	94.4 %
Barn Swallow	217	93.5 %
Winter Wren	216	93.1 %
Hermit Thrush	213	91.8 %
Alder Flycatcher	212	91.4 %
Nashville Warbler	212	91.4 %
Chestnut-sided Warbler	212	91.4 %
Spotted Sandpiper	211	90.9 %
Veery	211	90.9 %

Table 5
The 30 commonest species in the coniferous forest zone

Species	Number of squares in which reported	% of squares in which reported
White-throated Sparrow	235	100 %
American Robin	233	99.1 %
Swainson's Thrush	232	98.7 %
Dark-eyed Junco	230	97.9 %
Ruby-crowned Kinglet	229	97.4 %
Northern Flicker	227	96.6 %
Tree Swallow	226	96.2 %
Magnolia Warbler	226	96.2 %
Winter Wren	225	95.7 %
Yellow-rumped Warbler	225	95.7 %
Purple Finch	218	92.8 %
Common Yellowthroat	217	92.3 %
Tennessee Warbler	215	91.5 %
Lincoln's Sparrow	213	90.6 %
Pine Siskin	213	90.6 %
Common Raven	209	88.9 %
Cedar Waxwing	209	88.9 %
American Redstart	208	88.5 %
Spotted Sandpiper	199	84.7 %
Alder Flycatcher	199	84.7 %
Hermit Thrush	198	84.3 %
Northern Waterthrush	198	84.3 %
Nashville Warbler	197	83.8 %
Swamp Sparrow	191	81.3 %
Bay-breasted Warbler	189	80.4 %
Red-eyed Vireo	187	79.6 %
Philadelphia Vireo	181	77.0 %
Boreal Chickadee	179	76.2 %
Chipping Sparrow	179	76.2 %
American Crow	177	75.3 %

The species richness of the Central Lowland is explained by the fact that it represents the northern limit of the range in eastern North America of many species, which in Québec breed there exclusively. These species can be classified into three major groups: those that breed in the forests of southern Québec (Yellow-throated Vireo, Eastern Screech-Owl, Wild Turkey, Blue-gray Gnatcatcher, and Cerulean Warbler); those that breed in man-made and fragmented environments such as fields, scrub, woodland edges, hedgerows, thickets, and suburbs (including the House Finch, Field Sparrow, Gray Partridge, Northern Cardinal, Rufous-sided Towhee, Red-headed Woodpecker, Loggerhead Shrike,

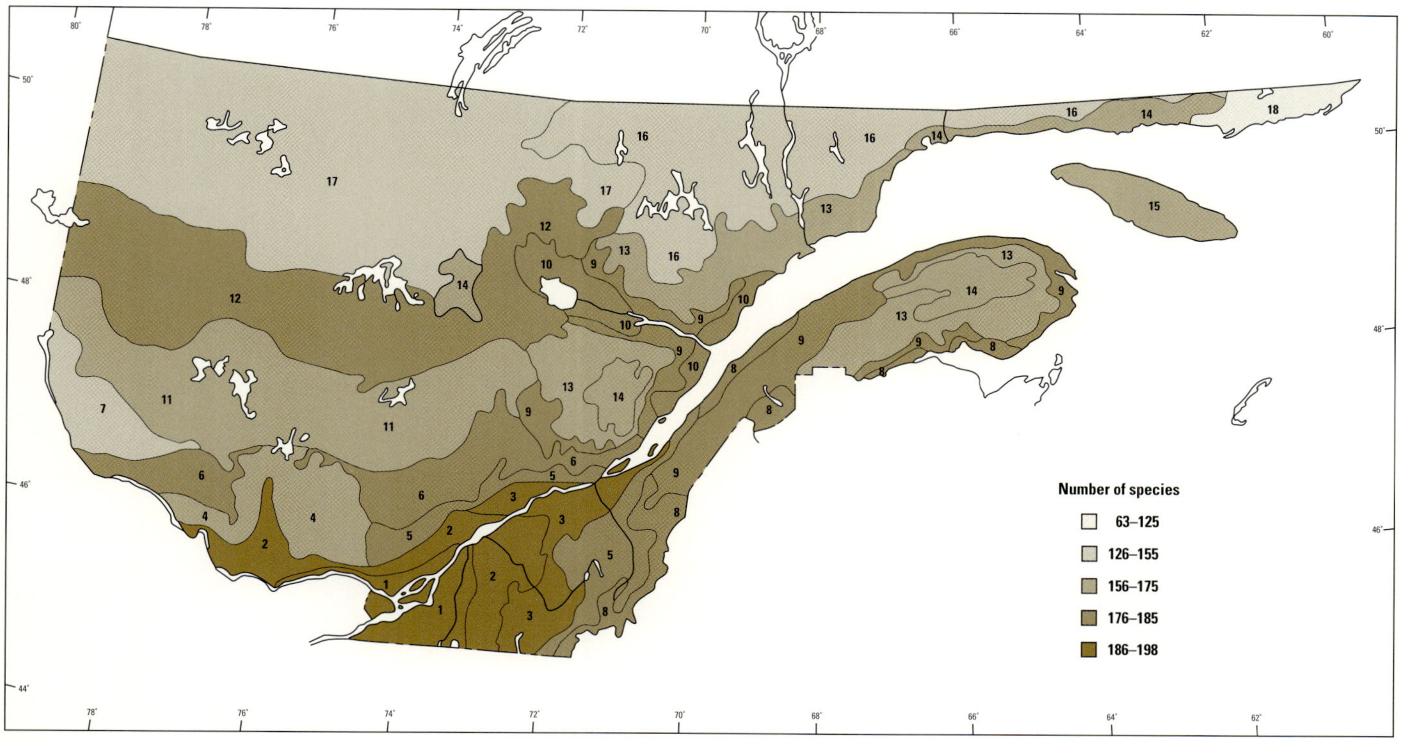

Figure 3
Number of species reported in the ecological regions

Golden-winged Warbler, Clay-colored Sparrow, Grasshopper Sparrow, Yellow-billed Cuckoo, and Carolina Wren); and those, equally impressive in number, that breed exclusively (or almost exclusively) in marshes and other wetlands in the Central Lowland (Willow Flycatcher, Marsh Wren, Common Moorhen, Least Bittern, Sedge Wren, Redhead, Ruddy Duck, Little Gull, and Great Egret). Region 1 (Sugar maple-hickory and sugar maple-basswood) is the least rich ecological region in the Central Lowland, as a number of species highly typical of the coniferous forest (Merlin, Spruce Grouse, Three-toed and Black-backed woodpeckers, Gray Jay) were not reported during the breeding season in view of the rarity of this habitat here; these species were, however, reported (albeit very rarely) elsewhere in the Central Lowland.

Table 6
Number of species reported in the ecological regions

Ecological region (forest region)	Number of species	Squares visited	Field hours
Deciduous forest zone			
1. Sugar maple-hickory and sugar maple-basswood	186	120	7598
2. Sugar maple-basswood	198	193	6742
3. Sugar maple-basswood and sugar maple-yellow birch	191	144	7235
4. Sugar maple-yellow birch-basswood	171	124	3916
5. Sugar maple-yellow birch-beech and sugar maple-yellow birch-basswood	180	149	7587
6. Sugar maple-yellow birch-beech	180	164	5087
7. Sugar maple-yellow birch-balsam fir	145	55	714
Mixed forest zone			
8. Sugar maple-yellow birch and balsam fir-yellow birch	177	97	2811
9. Balsam fir-yellow birch	181	216	5846
10. Balsam fir-white birch or balsam fir-red maple	183	96	2671
11. Yellow birch-balsam fir	157	152	2223
Coniferous forest zone			
12. Balsam fir-white birch (west)	176	228	2928
13. Balsam fir-white birch (east)	169	216	3613
14. Balsam fir-black spruce	164	93	2288
15. White spruce-balsam fir and balsam fir-white spruce	173	78	1438
16. Black spruce-balsam fir-moss	126	68	953
17. Black spruce-moss	150	256	2156
18. Tundra and forests	63	15	368

Species richness is not quite as great elsewhere in the deciduous and mixed forest zones. Region 7 (Sugar maple-yellow birch-balsam fir) is the least rich ecological region in the deciduous forest zone (Figure 3). A small region that received little coverage, it is characterized by the fact that many species associated with the coniferous forest (such as the Merlin, Spruce Grouse, Solitary Sandpiper, Black-backed Woodpecker, Yellow-bellied Flycatcher, Gray Jay) are more common there than in the other ecological regions of the deciduous and mixed forest zones. Furthermore, the principal component analysis classified this region within the mixed forest zone instead of in the deciduous forest zone as classified by Thibault (1985) (see Figure 1). Note, however, that the region contains very little marshland and open country, as reflected in the smaller number of species that breed in such habitats.

After the ecological regions that cover the Central St. Lawrence Lowland and the Ottawa River valley (regions 1, 2, and 3), the two regions demonstrating the greatest species diversity are Region 10 (Balsam fir-white birch or balsam fir-red maple) and Region 9 (Balsam fir-yellow birch). Region 9 has the advantage that it is divided into three ecologically rather distinct parts, including both forest bird habitats in the Saguenay and Appalachian regions and waterbird and seabird habitats along the coast of the Gaspé Peninsula. Region 10 supports a diverse birdlife thanks to the variety of habitats around Lake Saint-Jean as well as the contribution it receives from the birds that nest along the shore of the St. Lawrence in Charlevoix and on part of the Upper North Shore.

There is even less avian diversity in the coniferous forest zone. Region 12 (Balsam fir-white birch [west]) is the most diverse for birds of the ecological regions in the boreal forest, because it includes the open and aquatic habitats in the Abitibi region that support many species (such as the Red-necked Grebe, Sharp-tailed Grouse, and Le Conte's Sparrow) with very local distributions. In this analysis, Region 15 (White spruce-balsam fir and balsam fir-white spruce), since it is comprised of Anticosti Island and the Magdalen Islands—which have fairly distinct bird communities—, hosts a wide variety of species; in fact, Anticosti Island alone supports 139 species, the Magdalen Islands, 158. A few of the species that breed in the Magdalen Islands (Horned Grebe, Piping Plover, Roseate Tern, and Thick-billed Murre) are not known to breed regularly elsewhere in Québec. Other species that breed in the Magdalen Islands (Semipalmated Plover, Least Sandpiper, and Common Black-headed Gull) also nest (in southern Québec) only on Anticosti Island and in a few places on the North Shore.

Apart from Region 7 (Sugar maple-yellow birch-balsam fir), the regions with the lowest species diversity—Regions 16 (Black spruce-balsam fir-moss), 17 (Black spruce-moss), and 18 (Tundra and forests)—tend to be those farther north. Although Region 18 accounts for the smallest number of reported species in southern Québec, it was also the least well covered. The landscape of this North Shore region is characterized by tundra in exposed areas in high elevations and along the shore, and by wooded areas in valleys or other low-lying areas; therefore, it can only support a limited number of species. It is, however, not without interest because it provides habitat for large colonies of seabirds as well as tundra species such as the Willow Ptarmigan and the American Pipit.

Gilles Falardeau

CLIFF-NESTING BIRDS

Of southern Québec's breeding birds, 36 species nest on cliffs and escarpments. Whether of rock or unconsolidated deposits, such habitat is often inaccessible to terrestrial predators. Birds using rocky cliffs build their nests in holes or crevices, or on ledges. Some species construct a nest of twigs or branches, or of mud applied directly to the cliff face, while others lay their eggs on the bare rock. Some ledge-nesting alcids such as the Common Murre and Thick-billed Murre lay pyriform eggs, which are less likely to roll off. In unconsolidated material, birds may dig a cavity or use another species' cavity. Some species also nest in microhabitats associated with cliffs, such as talus and the veneer of loose material that accumulates at the base or top of the cliff or on cliff ledges.

Of the 36 cliff-nesting species, 30 use rocky cliffs. The Great Cormorant, Iceland Gull, Black-legged Kittiwake, and the alcids nest only on sea cliffs; other species like the Great Black-backed Gull, Glaucous Gull, and Herring Gull nest in both coastal and inland environments, although the only gull found breeding on cliffs in freshwater environments is the Herring Gull. The Razorbill and Black Guillemot often select crevices or talus at the cliff base. The Double-crested Cormorant also nests on cliffs with wide enough ledges. The Atlantic Puffin may sometimes nest in cavities in cliffs, as on Bonaventure Island, although it usually digs a burrow in the ground on an island, like Leach's Storm-Petrel.

Cliffs along the coast provide nesting habitat for many species of seabirds, while those in remote inland areas are used by species like the Golden Eagle, Peregrine Falcon, and Common Raven. Moisie River, northeast of Sept-Îles. *July 1994, François Morneau.*

Stratified cliffs provide excellent nesting habitat for many seabird species. Forillon National Park, Gaspé Peninsula. *July 1992, Yves Aubry.*

Many diurnal birds of prey build their nests on rocky cliffs inland or overlooking salt or fresh water, where they can survey their hunting territory. The Peregrine Falcon, Gyrfalcon, Golden Eagle, and Rough-legged Hawk prefer such habitat, while the Bald Eagle, Merlin, and Red-tailed Hawk nest less often on cliffs, generally preferring trees. The Common Raven is likely to be found sharing a cliff with these raptors.

Some species such as the Bank Swallow and Belted Kingfisher nest in burrows in steep banks of unconsolidated material. This sand pit houses a colony of Bank Swallows. Martin-Valin ZEC, north of the Saguenay River. *August 1991, Jean Gauthier.*

The Turkey Vulture usually builds its nest in a crevice or between boulders in a talus slope where trees are growing, while the Osprey sometimes nests on the top of a large boulder near a cliff when there are no suitable trees nearby.

Of the nocturnal birds of prey, the Great Horned Owl occasionally nests on cliffs, as does the Barn Owl, which may use a crevice or even a mine shaft.

Six other species that usually nest on man-made structures also occasionally use rocky cliffs, including the Rock Dove, Chimney Swift, Eastern Phoebe, Cliff Swallow, Barn Swallow, and European Starling.

In northern areas, such as the Ungava Peninsula, two passerines nest in crevices in cliffs and talus slopes: the Snow Bunting, which prefers this habitat, and the Northern Wheatear, which generally nests in burrows in the loose organic deposits at the base of cliffs or, on flatter terrain, in holes under rocks.

Three species nest in escarpments of unconsolidated deposits, in steep banks of earth, sand, gravel, or clay: the Belted Kingfisher and Bank Swallow, which are seldom seen nesting elsewhere, and the Northern Rough-winged Swallow, which uses artificial cavities in addition to natural ones, including digging its own burrows in banks, usually along a river or stream.

Louis Messely

PEATLAND BIRDS

In Québec, more than a hundred breeding bird species inhabit or frequent peatlands (Erskine 1977; Rochefort and Quinty 1994; Stockwell 1994). Species richness and abundance may vary within this habitat type, however, the main factor appearing to be the diversity of vegetation cover. For instance, peatlands with a mixture of wooded, shrubby, and open areas support the greatest number of species and individuals (Rochefort and Quinty 1994; Stockwell 1994). A number of studies have also concluded that species diversity increases with the size of the peatland (Hakala 1971; Boström and Nilsson 1983; Cons. G. G. – GREBE 1993c).

In Sweden, the density and diversity of passerine species in peatlands declines from south to north, while the converse has been observed in other groups of birds such as anatids, larids, and sandpipers (Boström and Nilsson 1983). The same general pattern is found in southern Québec, with the number of passerine species in bogs and fens decreasing from the St. Lawrence Plain to the Saguenay–Lac-Saint-Jean region (Rochefort and Quinty 1994). Yet the taiga peatlands, especially the muskeg of the Lake Plateau physiographic region, supports more species and individuals than the muskeg of the Eastmain Lowland (Cons. G. G. – GREBE 1991a, 1992o; Tremblay and Belles-Isles 1991). Among Eastmain Lowland peatlands, the vast tamarack peatlands around Rupert Bay support the greatest species diversity, in part because these tamarack peatlands and most of the peatlands in the Lake Plateau region are fens, and thus more productive than the bogs of the Eastmain Lowland (Grondin and Ouzilleau 1980).

Three groups of birds—passerines, waterfowl, and shorebirds—dominate peatland bird communities in terms of abundance and species richness. The Savannah Sparrow is probably the most widespread passerine in Québec peatlands (Cons. G. G. – GREBE 1991a, 1992o; Tremblay and Belles-Isles 1991; Rochefort and Quinty 1994). In the Atlas area, the most common passerines in peatlands are the Palm Warbler, Common Yellowthroat, Lincoln's Sparrow, and White-throated Sparrow (Erskine 1977; Cons. G. G. – GREBE 1992o; Rochefort and Quinty 1994; Stockwell 1994).

Waterfowl can be found in the lakes, ponds, and pools that are scattered throughout peatlands. In the Atlas area, the Ring-necked Duck and American Black Duck are common peatland inhabitants. The American Black Duck is particularly abundant in peatlands without pools near Rivière-du-Loup (Reed 1970; Erskine 1977; Cons. G. G. – GREBE 1990a, 1992a). However, it generally uses peatlands only for nesting, preferring to raise its brood in adjacent habitats such as rivers and coastal marshes. North of the Atlas area, the Green-winged Teal and Canada Goose are common in muskeg, while the Northern Pintail and the Black Scoter may be locally abundant (Cons. G. G. – GREBE 1990a, 1992a). The Canada Goose often builds its nest on an island in a peatland pond (Tardif and Reed 1991).

Among the shorebirds, the most widespread species in both bogs and fens is undoubtedly the Common Snipe (Erskine 1977; Tremblay and Belles-Isle 1991; Cons. G. G. – GREBE 1992g, 1993c; Stockwell 1994). In northern Québec, muskeg provides the primary breeding habitat for shorebirds. On the Lake Plateau, a single fen may support as many as eight shorebird species (Cons. G. G. – GREBE 1993c). The utilization of peatlands by a species depends on a number of factors, including the size of the habitat, water, and nutrient characteristics (minerotrophic or ombrotrophic) and the number, shape, and location of the water bodies present (Cons. G. G. – GREBE 1993c). For example, the Least Sandpiper and Short-billed Dowitcher appear to exclusively inhabit string fens (minerotrophic peatlands dotted with pools), with the latter species being restricted to large fens of this type.

Although more than 10 species of raptors hunt in peatlands, the Northern Harrier and Short-eared Owl are the only ones that breed there (Cons. G. G. – GREBE 1992m; Stockwell 1994). In the Atlas area, the Northern Harrier is one of the commonest birds of prey in this type of habitat.

The vast peatlands of Anticosti Island are breeding habitat for various species with a northern affinity such as the Greater Yellowlegs and Canada Goose. *June 1978, Gilles Chapdelaine.*

Many species of birds including Bonaparte's Gull, Arctic Tern, and Sharp-tailed Grouse, limit their use of peatlands, mainly to breeding. Bonaparte's Gull and the Arctic Tern nest in peatlands but forage for some of their food in adjacent lakes (Cons. G. G. – GREBE 1990a, 1992h, 1992y). Similarly, the Red-throated Loon nests on a hummock in the middle of a peatland pond but fishes on an adjoining lake or out at sea (Douglas and Reimchen 1988a; Eriksson et al. 1990), as on the Lower North Shore.

Most species that use peatlands also frequent other habitats. For example, the Yellow-rumped Warbler, a common peatland species, is also found in most types of mixed and coniferous forest (Erskine 1977). The Great Gray Owl hunts in peatlands with trees for perches, but can also be found in shoreline scrub (Bull and Dun can 1993).

Several species—generally shorebirds such as the Short-billed Dowitcher, Least Sandpiper, Greater Yellowlegs, and Lesser Yellowlegs—are exclusive to peatlands, spending the breeding season in the peatlands of northern Québec (Cons. G. G. – GREBE 1993c).

François Morneau

WATERFOWL AND SHOREBIRDS OF THE ST. LAWRENCE

Habitat for waterbirds

Québec has over 8000 km of coastline, including approximately 4000 km on the St. Lawrence and just over 4000 km along James Bay, Hudson Bay, Hudson Strait, and Ungava Bay in the north. A total of 90 000 ha of wetlands (including wet meadows and upper freshwater marshes with shrubs and trees) are found along these coasts, nearly 80 000 ha on the St. Lawrence alone. Approximately 40% of these wetlands consist of beds of submergent vegetation, which are found almost exclusively upstream of Trois-Rivières. Wetlands become scarcer as the river flows into the estuary and the Gulf of St. Lawrence, decreasing from 63 000 ha in the fluvial section to 12 000 ha in the estuary and to around 5000 ha in the gulf, where the rocky shores largely exclude this type of habitat (Lands Directorate 1986; Gratton and Dubreuil 1990).

Unfortunately, there is little information on the importance of these wetlands for breeding waterfowl and shorebirds. Most of the studies done in the St. Lawrence during the last 20 years have focussed mainly on staging areas for migrating birds. This discussion attempts to fill in some of these gaps, although it also covers staging and wintering regions. Because of the lack of data in some areas, we were obliged to make extrapolations, which will have to be validated at some point in the future.

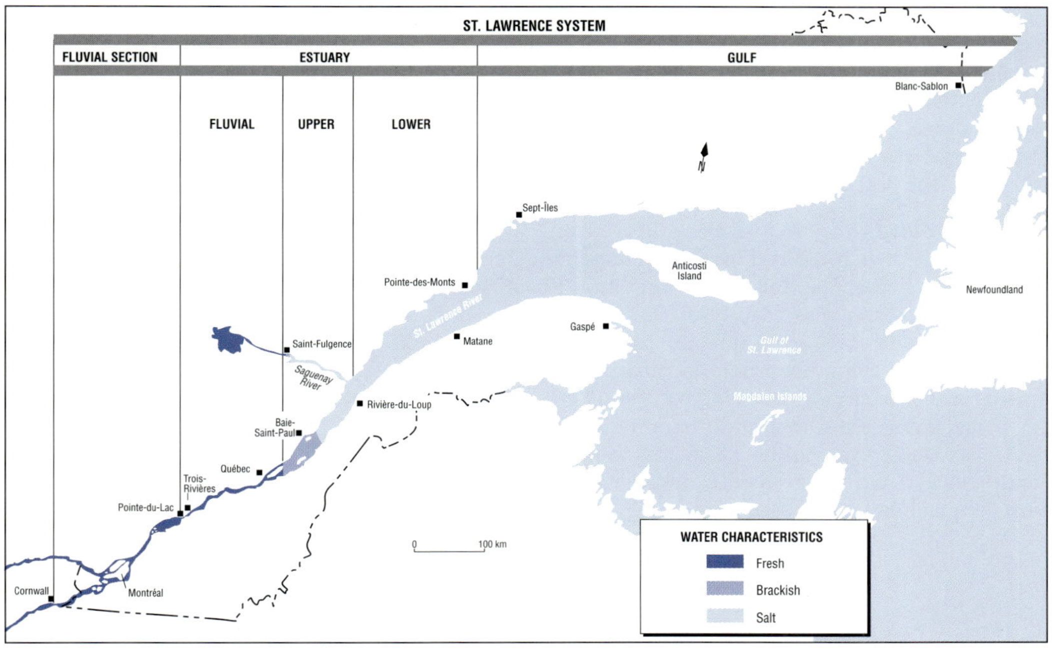

Hydrographic Divisions of the St. Lawrence

Source: Bouchard and Millet 1993, St. Lawrence Centre, Environment Canada

The importance of the St. Lawrence for staging and wintering waterfowl and shorebirds

Waterfowl

The St. Lawrence River system is one of the major migration routes for waterfowl in eastern North America. Every year, an estimated one million waterfowl stage along this corridor, including roughly 500 000 geese, 200 000 sea ducks, 150 000 diving ducks, and 125 000 dabbling ducks (Lehoux et al. 1985). Numbers often vary significantly from season to season and from year to year, however. These staging areas are of crucial importance to waterfowl, particularly during the critical few weeks in spring before the breeding season begins in earnest. At this time of year, the birds must have access to an abundant supply of energy-rich food before they travel to their breeding grounds farther north or on the river itself.

Waterfowl in the fluvial section

More staging waterfowl use the estuary (≈700 000) than the river proper (≈250 000), which, however, is favoured by pochards, goldeneyes, and Canada Geese. In fall, pochards and goldeneyes (≈125 000) stage mainly in areas with extensive beds of submergent vegetation—water-milfoil, pondweed, and tape-grass—typical of the riverine lakes around Montréal and in Lake Saint-Pierre (Dimension Environnement 1982; Dolan and Bourgeois 1985; Lehoux et al. 1985; Bourgeois et al. 1994). At this time of year, these ducks feed mainly on the seeds of submergents, as well as on the associated arthropods and molluscs (Anderson 1959; Rogers and Korschgen 1966; Cronan and halla 1968; Bartonek and Hickey 1969; Dirschl 1969; Nilsson 1970; Thompson 1973). Pochards are rarely found in water over 6 m deep. Diving ducks are gregarious and usually gather in large rafts during the day to feed; they may also feed at night if disturbed during the day by recreational activities such as pleasure boating, fishing, or hunting (Paris 1985).

In spring, Lake Saint-Pierre is also a favourite haunt of migrating Canada Geese, which are attracted by the region's vast floodplains and grainfields (Lehoux et al. 1983; Dolan and Bourgeois 1985; Lehoux et al. 1985; Bourgeois et al. 1994). The goose population, which was estimated at 58 000 individuals in 1988 (J.-C. Bourgeois, MEF, pers. comm.), usually ranges between 50 000 and 70 000 birds (Benoit et al. 1987; Bourgeois 1994). During the day, the geese forage in the fields near the river (Reed et al. 1977; Bourgeois et al. 1983), feeding on corn kernels and the seeds of smartweed and horsetails and grazing on young, tender plants such as timothy. At night, the geese gather primarily in shallow flooded areas (≤45 cm deep), which provide plentiful food and protection from terrestrial predators. At night, they spend on average 40% of the time feeding (Dolan et al. 1983).

Dabbling ducks are also abundant in the Lake Saint-Pierre region, particularly in spring (Benoit et al. 1987; Bourgeois et al. 1994), when they are much more numerous than in the riverine lakes of the Montréal region (Léveillé et al., in prep.). Like Canada Geese, ducks gather on the floodplains in flocks of up to 17 000 individuals. Lake Saint-Pierre is the second largest staging area for dabbling ducks in the St. Lawrence system, supporting close to 30% of the spring population (Benoit et al. 1987). The ducks feed mainly on the seeds of wild plants such as smartweed and barnyard-grass (CWS, unpubl. data), generally in the shallower parts of the floodplain, where the water is less than 30–40 cm deep (Bourgeois et al. 1983). Weed seeds are particularly abundant here at this time of year (5 kg/m^3) (Savignac et al. 1982).

Snow Geese, which were uncommon in the fluvial section during the 1970s (Lehoux et al. 1985), are now a major part of the region's avifauna. Numbers have increased enormously since the early 1980s. In the late 1970s, only around 300 geese were found around Lake Saint-Pierre in spring, but by 1984, this had risen to 35 000 and by 1987 to 60 000. Between 1990 and 1993, the number of geese using the Baie-du-Febvre floodplain nearly tripled from 100 000 to 250 000 (Benoit et al. 1987; Cazelais 1992; DeKoster 1992; Bourgeois et al. 1994).

In winter, stretches of open water can be found in both the river itself and the estuary, with the former attracting 10 000–12 000 ducks (Reed and Bourget 1977; Lehoux et al. 1985). This winter population consists mainly of diving ducks such as Common Mergansers and Common Goldeneyes, and a few hundred American Black Ducks, found mostly near Pointe au Foin and Cornwall on Lake Saint-François and Dixie Island and Pointe du Domaine on Lake Saint-Louis.

Waterfowl in the estuary

The estuary undoubtedly attracts the largest number of ducks and geese in the St. Lawrence, with an estimated 700 000 individuals staging in this area. These flocks consist mainly of Snow Geese and various species of dabbling ducks. Although Snow Geese are feeding more and more in farmers' fields in spring (Bédard et al. 1986a; Reed 1992), they still rely largely on bulrush marshes, particularly in autumn, when the rhizomes of this plant provide most of their food (Reed 1977a, 1984, 1989; Gauthier et al. 1988). The main staging areas in the estuary for geese and dabbling ducks have changed little over the years and are found mainly downstream of Québec City, in the North Channel of Île d'Orléans, at Cap Tourmente, Bellechasse Cove, the Montmagny Islands, the Montmagny flats, and, in the case of dabbling ducks, the Alouettes flats (Lehoux et al. 1985; Banville and St-Onge 1990a,b).

The estuary also attracts some 60 000 staging scoters, eiders, Oldsquaws, and mergansers, particularly in autumn. These species frequent mainly maritime habitats, although scoters are found regularly in Lake Saint-Pierre in the fall (Dolan 1983). The scoters tend to be limited to the areas between Tadoussac and Pointe-des-Monts on the north side of the estuary and between Rivière-du-Loup and Matane on the south side, mainly in areas with sandy and pebbly bottoms. Their favourite prey are molluscs, which they dive for at fairly shallow depths, generally under 20 m (Madsen 1954; Grosz and Yocom 1972; Sanger and Jones 1984; Vermeer and Bourne 1984). In contrast, Common Eiders are associated chiefly with muddy and rocky substrates. In spring, their favourite foods are herring roe and nereids, while at other times of the year, they dive for periwinkles, sea urchins, blue mussels, and gammarids (Cantin et al. 1974; Guillemette and Himmelman 1988; Guillemette et al. 1992). Oldsquaw are found in areas with sandy or muddy bottoms. They feed generally at depths between 5 and 10 m, although they have been found at depths of over 30 m (Madsen 1954). This highly opportunistic species has a diverse diet, composed of up to 61 different prey items (Madsen 1954; Nilsson 1970; Vermeer and Levings 1977; Johnson 1984; Sanger and Jones 1984). In spring, fairly large numbers of Red-breasted Mergansers are also seen regularly off Rivière-du-Loup (J. Bédard, Laval U., pers. comm.).

In winter, significant numbers of ducks (≈40 000) remain in the estuary, mainly Oldsquaws, goldeneyes, and American Black Ducks (Reed and Bourget 1977; Lehoux et al. 1985; Savard 1990). They gather mainly off the north shore, where the tides and currents keep the water open all winter, particularly between the Saguenay and Betsiamites rivers.

Waterfowl in the gulf

The Gulf of St. Lawrence, with its rocky and sandy shores, is inhospitable to marshland and thus provides relatively little habitat for waterfowl such as geese and dabbling ducks that use marshes during migration. The 150 000 ducks found here in spring are mainly sea ducks like the Common Eider, Oldsquaw, and scoters, as well as mergansers (Lehoux et al. 1985). Habitat use by Common Eiders and Oldsquaws is roughly the same as in the estuary, with the greatest concentrations occurring between Pointe-des-Monts and Sept-Îles and around the Mingan Islands. There is not much information on the use of Anticosti Island during migration.

The Red-breasted Merganser and the Common Merganser also occur regularly in the gulf during migration, though in smaller numbers than the sea ducks. Mergansers are more numerous here than anywhere else in the St. Lawrence at this time, although they are found in much smaller numbers (≈5000 in fall) than the Common Eider. Since mergansers are primarily fish eaters (Munro and Clemens 1939; Madsen 1957; Latta and Sharkey 1966; Cronan and Halla 1968; Timken and Anderson 1969), they tend to be found at river mouths, for example, in Gaspé Bay, where they feast on prey less than 10 cm long, including eel, stickleback, cod, flounder, carp, and even salmon. They are particularly fond of sheltered bays with sandy bottoms, and dive for food in relatively shallow water, usually less than 5 m.

In winter, close to 150 000 ducks frequent the gulf (Reed and Bourget 1977; Lehoux et al. 1985; Bourget et al. 1986; Bordage et al., in prep.). The Common Eider, with a population around 100 000, is unquestionably the dominant species, followed by the Oldsquaw (≈40 000). Indeed, roughly 30% of all the eiders overwintering on the east coast of North America concentrate in the gulf, mostly off the Mingan Islands, Gaspé Bay, and the western tip of Anticosti Island.

Shorebirds

No less than 160 000 shorebirds representing 22 species stop along the St. Lawrence during fall migration (Brousseau 1981; Maisonneuve 1982; Maisonneuve et al. 1990). The greatest numbers, over 100 000 individuals, are found in the estuary. Semipalmated Sandpipers are the most common, making up close to 70% of records in the estuary.

The gulf is the second most important shorebird staging area (40 000 individuals). Among the 17 species found, the Semipalmated Sandpiper and the Short-billed Dowitcher are the most common, with 36% and 24% of records respectively.

Table 1
Estimates of migrating and wintering waterfowl populations in the St. Lawrence system

	FLUVIAL			ESTUARY			GULF			TOTAL		
	Spring	Fall	Winter	Spring	Fall	Winter	Spring	Fall	Winter	Spring	Fall	Winter
Snow Goose	100 000	1 000		300 000	500 000					400 000	>500 000	
Brant				15 000	300		3 000			18 000	<500	
Canada Goose	60 000	1 000		35 000	4 000		3 000	2 000		98 000	7 000	
Total geese	**160 000**	**2 000**		**350 000**	**504 300**		**6 000**	**2 000**		**516 000**	**508 300**	
Total dabbling ducks	**25 000**	**25 000**	**2 000**	**25 000**	**100 000**	**4 000**	**1 000**	**4 000**		**51 000**	**129 000**	**6 000**
Pochards	60 000	100 000		3 000	5 000					63 000	105 000	
Goldeneyes	4 000	25 000	3 000	3 000	10 000	12 000	3 000	2 000	1 500	10 000	37 000	16 500
Mergansers	1 000	500	7 000	2 000	500	500	1 000	5 000	500	4 000	6 000	8 000
Total diving ducks	**65 000**	**125 500**	**10 000**	**8 000**	**15 500**	**15 500**	**4 000**	**7 000**	**2 000**	**77 000**	**148 000**	**24 500**
Eiders				10 000	10 000		100 000	30 000	100 000	110 000	40 000	100 000
Oldsquaw				500	7 000	25 000	2 000	3 000	40 000	2 500	10 000	65 000
Scoters		2 000		25 000	40 000		50 000	3 000		75 000	45 000	
Total sea ducks		**2 000**		**35 500**	**57 000**	**25 000**	**152 000**	**36 000**	**140 000**	**187 500**	**95 000**	**165 000**
TOTAL	**250 000**	**154 500**	**12 000**	**418 500**	**676 800**	**41 500**	**163 000**	**49 000**	**142 000**	**831 500**	**880 300**	**195 500**

See text for documents consulted.

Greater Yellowlegs foraging in the lagoons and ponds of the Magdalen Islands. These prime feeding sites attract not only local breeders but also birds in migration. Baie du Have aux Basques. *August 1983, Yves Aubry.*

Upstream of Lake Saint-Pierre, however, fewer shorebirds (≈4000) are found, because of the lack of tides, which would leave the shores regularly exposed, and the absence of extensive sedimentation, which also creates mudflats where shorebirds can feed. In the fluvial section, shorebirds are found mainly in small areas of sedimentation. The Semipalmated Sandpiper is the dominant species.

Île Rouge and the Portneuf sandbank in the estuary and Baie du Havre aux Basques in the Magdalen Islands are among the most important staging areas in eastern North America for the Black-bellied Plover, Short-billed Dowitcher, and White-rumped Sandpiper, respectively.

The importance of the St. Lawrence for breeding waterfowl

Since no systematic surveys have been carried out on waterfowl in the St. Lawrence system during the breeding season, there are little reliable data on the number of nests and broods produced every year in the system as a whole. Individual studies done here and there in the region suggest that probably 5000–10 000 nests or broods of dabbling ducks and a few dozen nests or broods of diving ducks, excluding the Common Eider, are produced annually (Reed 1975b; Pilon et al. 1981; Dimension Environnement 1982; Bélanger 1989; Bourgeois et al. 1994).

Although these numbers are only estimates, we believe that the figure of 5000–10 000 broods is fairly accurate. Contrary to widespread belief, most of the broods produced in Québec do not come from the St. Lawrence, but from its various tributaries and the boreal forest.

Excluding the Common Eider, the five main species breeding in the St. Lawrence are, in descending order, the Mallard (20%), American Black Duck (19%), Northern Pintail (17%), Gadwall (16%), and American Wigeon (10%). These species account for 80% of the duck population present during the breeding season.

One point to be noted is that the distribution and abundance of waterfowl appear to be closely linked to the distribution and abundance of wetlands. Therefore, the greatest number of broods are produced in the stretches of the river farthest upstream. The following section provides detailed data by region.

Waterfowl in the fluvial section

Close to 75% of the wetlands in the St. Lawrence system are located in the fluvial section. This section is dominated by floodplain forests and wet meadows and contains no fewer than 10 separate groups of islands comprising some 400 islands in all. These two factors explain in large part why over 75% of the duck broods produced in the system come from this part of the river. In addition, species richness is quite high in the fluvial section due to the variety of wetlands it offers, and almost all of the species of waterfowl known to nest along the St. Lawrence system can be found here, except for species like the Common Eider, Red-breasted Merganser, and Snow Goose.

Lake Saint-François

Few exhaustive studies have been carried out on Lake Saint-François despite its great potential for supporting breeding waterfowl. Dimension Environnement (1982) estimated that roughly 1300 nests can be found every year in the various habitats around the lake, in line with Alliston's (1979a) estimate of 1000 broods.

Brood surveys by the Canadian Wildlife Service (unpubl. data) between 1977 and 1992 in the Lac Saint-François National Wildlife Area show that the four main species associated with this lake are, in descending order: Mallard (25%), Redhead (22%), American Wigeon (19%), and Gadwall (18%). The presence of large numbers of Redheads is typical of the type of habitat found around the lake, and in the 1970s, Alliston (1979a) found 172 nests of this species. The numerous marshes and islands, mainly in the western part, make Lake Saint-François particularly attractive to waterfowl. Potential nesting habitat includes 367 ha of wet meadows and 279 ha of floodplain forests (Dimension Environnement 1982).

Table 2

Estimates of waterfowl abundance and density (except for Common Eider) during the breeding season in different sections of the St. Lawrence

	Lake Saint-François	**Lake Saint-Louis**	**Montréal –Sorel**	**Pierre**[1]	**Lake Saint- estuary**	**Fluvial estuary**	**Upper estuary**	**Gulf Lower Islands)**	**Gulf (Magdalen Islands)**	**(Mingan TOTAL**
Mallard	25 %	67 %[2]	13 %	24 %	13 %	9%	<1 %	<1 %		20 %
American Black Duck	3 %		2 %	9 %	54 %	15%	98 %	18 %	50 %	19 %
Northern Pintail	7 %	11 %	23 %	39 %	22 %	21 %	1 %	31 %	3 %	17 %
Gadwall	18 %	2 %	28 %	12 %	1 %	13 %			3 %	16 %
American Wigeon	19 %	9 %	10 %	5 %	1 %	1 %		<1 %		10 %
Redhead	22 %									6 %
Blue-winged Teal	2 %	2 %	8 %	5 %	6 %	6 %		3 %		4 %
Green-winged Teal		2 %	<5 %		<1 %	25 %		16 %	23 %	3 %
Northern Shoveler	1 %		5 %	2 %	1 %	7 %		<1 %		2 %
Wood Duck	1 %	5 %	<5 %	2 %	<1 %	<1 %				2 %
Other[3]	2 %	2 %	<1 %	<1 %		<3 %		29 %	16 %	1 %
TOTAL PAIRS, NESTS, OR BROODS	**1300**	**350**	**1500**	**350**	**400**	**200**	**500**	**?**	**?**	**4600**

1 Excluding Berthier–Sorel Islands.

2 Including a certain percentage of American Black Ducks.

3 Including one or more of the following species: Red-breasted Merganser, Common Goldeneye, Ring-necked Duck, Greater Scaup, Lesser Scaup, Canada Goose.

The Lac Saint-François National Wildlife Area is a vast mosaic of islands, channels, and pools, including one of the largest freshwater marshes in southern Québec. The region attracts a number of aquatic species. **Southwest of Montréal.** *June 1974, Léo-Guy de Repentigny.*

Broods are raised on and around the islands and in beds of submergent and emergent vegetation in freshwater marshes (Thompson 1974; Gervais et al. 1984). The islands and islets are used most intensively for brood rearing by early-nesting species—Mallard, Northern Pintail, and American Black Duck—whose broods hatch when the emergent plants are not yet developed. As the season progresses, however, broods tend to forage more in areas with emergent plants. Other species of dabbling ducks, mainly Gadwall and American Wigeon, make greater use of emergents for brood rearing, but avoid overly dense areas such as dense stands of cattails, bur-reed, and horsetails, particularly later in the season. However, Gadwall and American Wigeon broods frequent beds of river bulrush throughout the brood-rearing period. The presence of openings in the vegetation (≈50%) also attracts broods (Thompson 1974).

Lake Saint-Louis

Lake Saint-Louis has much less potential for breeding waterfowl than Lake Saint-François because of its heavily industrialized and urbanized shoreline and fewer islands and wetlands. However, according to Dimension Environnement (1982), around 333 nests are found annually in the area, 180 of which hatch successfully. The nests are located in the 640 ha of floodplain forests and 354 ha of wet meadows around the lake.

Species composition varies somewhat with habitat. Floodplain forests are home to mainly (over 80%) American Black Ducks/Mallards, Wood Ducks, Northern Pintails, and occasionally American Wigeons, Green-winged Teals, Blue-winged Teals, and Hooded Mergansers (Laperle 1974; Dimension Environnement 1982). Although the Dimension Environnement study did not differentiate between Mallards and American Black Ducks, Mallards were apparently assumed to clearly predominate. If this is true, the Mallard has taken less than 10 years to replace the American Black Duck in this part of the St. Lawrence, since Laperle (1974) reported a ratio of 1.47 Black Ducks to 1 Mallard in the early 1970s in his study on the Îles de la Paix from 1968 to 1972.

Densities of approximately 0.33 nest/ha can be found in flooded riparian forests, both on islands and around the lake. The highest densities occur in forests with dead trees (0.45 nest/ha) and the lowest in healthy forests (0.06 nest/ha). This can be explained by the lack, in younger forests, of cavities, crotches, and multiple trunks, which are all abundant in degraded stands. Although most species of ducks prefer to nest on the ground, in the forest, during years of severe flooding, they nest mainly (up to 90%) in trees (Laperle 1974). In these years, species which do not nest in crotches or cavities like teals, pintails, and wigeons are less abundant or even absent altogether.

Around Lake Saint-Louis, nest density in wet meadows (reed canary-grass dominant) is usually nearly twice as high as in forests (0.53 nest/ha compared with 0.33 nest/ha) (Dimension Environnement 1982). However, around the Lachine Rapids, Dimension Environnement (1985a) found much higher densities in floodplain forests (2.0 nests/ha compared with 1.4 nests/ha). The species mix in these wet meadows is different from that in floodplain forests, with the Gadwall replacing the Wood Duck and Hooded Merganser. However, the American Black Duck and Mallard still dominate, with close to 45% of the population.

The areas with the highest potential productivity include the Îles de la Paix, Saint-Jean Creek, the islands and islets in the Lachine Rapids and the St. Lawrence Seaway, and Dixie and Bushy islands.

Various dabbling ducks breed in island swamps and then move their broods into areas with beds of emergent vegetation. Other species, like the Tree Swallow, take advantage of the cavities in the dead trees to nest. Such dead trees are a common feature of these swamps. **Îles de la Paix National Wildlife Area, near Montréal.** *June 1987, Léo-Guy de Repentigny.*

Montréal–Sorel section

The most outstanding feature of the fluvial section between Montréal and Sorel is the presence of six groups of islands containing 234 islands in all. The islands range from very small (0.1 ha) to very large (over 1800 ha). The vegetation is also very diverse, consisting of trees and shrubs (40%) and wet meadows (38% with tall grasses [>50 cm] and 14% with short grasses [<50 cm], used mainly as pastureland and farmland) (Bélanger 1989). Around these islands are over 3000 ha of emergent and submergent aquatic vegetation, which provide ideal brood-rearing habitat (Dryade 1980).

Along with Lake Saint-François, this part of the river is the most productive in the entire St. Lawrence system. Every year, close to 1500 nests are found (Bélanger 1989). Ten species nest on the islands (Cantin et al. 1976; Pilon et al. 1980; Bélanger 1989; Giroux et al. 1992; Bélanger and Lehoux 1995), including Gadwall and Northern Pintail, which alone make up close to 50% of the breeding duck population. The presence of such a large Gadwall population in this area is particularly noteworthy, since the species was considered rare in Québec as recently as the 1960s. Indeed, the first breeding record in Québec was obtained on Lake Saint-François in 1968 (Reed 1969), during a general survey of breeding waterfowl populations in the St. Lawrence Valley. This suggests that the species colonized the area southwest of Montréal between 1962 and 1967. The almost complete absence of the American Black Duck in this area (only around 2% of the population) should also be noted.

Nearly 75% of nests can be found in the tall grass on the islands, compared with only 11% in the short grass and 14% in forested areas. Most nests (50%) are built in reed canary-grass communities. On average, each island contains 8.9 nests and the mean potential density is 0.7 nest/ha. The actual density varies depending on the size of the island, with the higher densities occurring on islands less than 0.5 ha (Bélanger et al. 1989; Bélanger and Lehoux 1995).

Several factors may influence nesting density on islands in this stretch of the river, including human disturbance, agriculture, and predation. Human disturbance is a constant problem on many islands and close to 30% have at least one house or summer cottage (Bélanger 1989); human presence is particularly great on the Berthier–Sorel Islands, where 75% of islands are inhabited. In addition, 43% of the total area of these islands is used for agriculture, 5186 ha for crops and 1685 ha for grazing (Bélanger 1991). Breeding density in these agricultural habitats is on average two to three times lower than in natural habitats, due mainly to the ensuing changes to the plant cover, including a reduction in its height and abundance, rather than the direct destruction of nests from machinery or trampling by cattle. Predation is also a significant limiting factor. A study by Giroux et al. (1992) shows that predation by mammals results in the loss of almost 50% of the nests on the Contrecoeur Islands.

Particularly rich breeding habitat for waterfowl can be found on the Berthier–Sorel Islands, where slightly over 750 nests have been counted in a year, or close to 50% of the total production in the sector. Densities as high as 4 broods/10 ha have been recorded in the marshes of some of the islands (Bourgeois et al. 1994).

Lake Saint-Pierre

The main feature distinguishing Lake Saint-Pierre from the rest of the St. Lawrence system (we do not consider the Berthier–Sorel Islands to be part of the lake) is its vast floodplains. In spring, 7000 ha of land, including farmland, riparian forests, and scrubland, are flooded. The 4000 ha of flooded farmland provide ideal staging habitat for waterfowl because of the rich food resources present. These areas allow the birds to rest, rebuild the fat reserves they require for migration, and obtain protein for egg development. The area also contains close to 10 000 ha of marshes (Benoit et al. 1987). Although marshes are found all along the shoreline of this huge body of water, the most extensive ones occur on the south shore between Nicolet and Baie de Lavallière (Dryade 1980).

In 1978, the Québec Department of Recreation, Fish and Game (MLCP) estimated breeding density in mainland meadows around the lake to be 0.25 nest/ha. Dimension Environnement (1982) found a total of 344 nests in these meadows. At least 10 species of waterfowl breed around Lake Saint-Pierre. A brood survey by the MLCP in the summer of 1984 revealed that the two main species are the Northern Pintail and Mallard, with 39% and 24% of broods respectively (Bourgeois et al. 1994). The Gadwall, which is the commonest species in the area immediately upstream of the lake, is third in importance on the lake itself, undoubtedly because of the absence of nesting islands.

On the Berthier–Sorel Islands, freshwater marshes and fields lie side by side. Waterfowl that nest in the fields, shrubby areas or woods move into the marshes to raise their broods. In the centre of this island is a marsh where Black Terns, Marsh Wrens, Hooded Mergansers, Wood Ducks, and other species nest. Île du Mitan, near Berthierville. *June 1988, M. Letendre, Ministère du Loisir, de la Chasse et de la Pêche.*

Natural areas (excluding sites managed for duck production) with the highest brood densities were Saint-François Bay (2 broods/10 ha) and the north side of the lake between Maskinongé Bay and Pointe-du-Lac (2 broods/10 ha). The least productive sector is on the south side between the Nicolet and Saint-François rivers, where the density was only 1 brood/10 ha (Bourgeois et al. 1994).

Aerial view of the southern part of Saint-François Bay, where the many small ponds are used by several species of waterfowl. Saint-François Bay, Lake Saint-Pierre. *September 1993, Luc Bélanger.*

Waterfowl in the estuary

The fluvial estuary forms a narrow corridor over 200 km in length. The tidal influence is evident here, with the shoreline gently sloping in the west and more precipitous in the east, as the estuary flows toward the gulf. The distribution of marshes, which cover over 5000 ha, reflects this relief, and these wetlands tend to be concentrated on the southwest side of the estuary between Port-Saint-François and Saint-Pierre-les-Becquets, and on the north shore of Île d'Orléans (Dryade 1980). The dominant plant species in these wetlands is American bulrush. Submergent aquatic vegetation is less prevalent than in the fluvial section, although there are extensive tidal flats (approximately 2000 ha) opposite Bécancour.

The only data available on breeding waterfowl in the fluvial estuary are from studies by Banville and St-Onge (1988) and the Canadian Wildlife Service (D. Lehoux, unpubl. data) on the north shore of Île d'Orléans, and by Lehoux and Bourget (1981) in the area between Trois-Rivières and Québec City. According to these studies, the bulrush marshes in the region can support densities of 0.87 brood/km on average, which suggests a total production of 400–500 broods. Of the nine species that breed in this part of the estuary, American Black Duck (54%) and Northern Pintail (22%) are the most common. The Gadwall represents no more than 1% of the total duck population and is limited, like the Blue-winged Teal and Wood Duck, to the upstream portion of the fluvial estuary. This may very well be due again to the lack of small islands suitable for breeding, since greater numbers of Gadwall are found in areas farther downstream, where there are more islands. Mallards are fairly abundant, representing 13% of the broods observed.

The two areas used most extensively by waterfowl during the breeding season are the section between Nicolet and the Rivière aux Orignaux, and the north shore of Île d'Orléans.

The 80-km-long upper estuary is shaped like a huge funnel. Here, the freshwater flow mingles with salt water from the gulf. This stretch of the St. Lawrence also has the greatest tidal variations and level of sedimentation in the entire system, with close to 2.7 million tonnes (dry weight) of material deposited annually (Sérodes 1980). The upper estuary also has some 4000 ha of marshes, including 3000 ha of the lushest bulrush marshes in Québec (Dryade 1980).

There is very little information on breeding waterfowl in the upper estuary. Surveys by Ducks Unlimited (unpubl. data) in the Montmagny Islands give a general idea of species composition; among the five species of ducks found, the most common are the American Black Duck (49%) and the Northern Pintail (19%). The Mallard is also fairly abundant (15%).

Unfortunately, there are no data on the total number of broods produced annually in the upper estuary. However, extrapolating the densities obtained in fairly similar habitat upstream (Lehoux and Bourget 1981; Banville and St-Onge 1988; 0.85 brood/ha) to the bulrush marshes of the upper estuary, which extend over approximately 200 km of shoreline, yields an estimate of slightly under 200 broods per year. Although the most productive sites have not yet been identified, the Montmagny Islands and Cap Tourmente area are thought to be important since they account for 40% of all the bulrush marshes in the estuary, and are being managed for waterfowl production by Ducks Unlimited.

Downstream from the upper estuary is the lower estuary, which is approximately 500 km long. It is characterized by cord-grass marshes, instead of bulrush marshes, downstream of La Pocatière. Cord-grass marshes cover 4500 ha in the lower estuary, with 70% of this habitat found on the south shore.

During the breeding season, the lower estuary is a favourite haunt of the American Black Duck. Banding data from the Fondation Les oiseleurs between 1987 and 1990 (unpubl. data) show the species accounting for over 98% of the population in this region. A few Northern Pintails and Mallards are also found.

The American Black Duck breeds on islands, in the upper level of tidal salt marshes, and on the nearby mainland (Reed and Moisan 1971; Reed 1975a,b). According to Bélanger et al. (1994), peatlands have also become one of the species' primary nesting habitats in the estuary, particularly those on L'Isle-Verte. Nest density in peatlands (22.5 nests/100 ha) is significantly higher than in the upper level of tidal salt marshes (2.1 nests/100 ha) and farmland (0.4 nest/100 ha). These high densities can probably be explained by the dense shrub cover, which hinders terrestrial predators and provides effective camouflage against avian predators. Bélanger et al. (1994) report 47% breeding success in this habitat.

Brood rearing occurs mainly in cord-grass marshes (Reed and Moisan 1971; L. Bélanger, CWS, unpubl. data), particularly the middle marsh dominated by salt-meadow cord-grass, which provides ducklings with good escape cover and plentiful food in the form of seeds and invertebrates (Lehoux 1972). Broods forage in small, shallow pools in the marsh.

Annual productivity in the estuary's cord-grass marshes has yet to be determined. Banding data collected by the Fondation Les oiseleurs (unpubl. data) in various marshes on the south shore of the lower estuary show a density of 2.1 broods/km of marsh. Therefore, by extrapolation, the number of broods produced annually in the lower estuary can be estimated at around 250, a fairly small number given the size of this area. Even using the highest densities recorded in the sector does not give a figure much over 500 broods.

Herons and ducks can frequently be seen on this man-made impoundment in the Cap Tourmente National Wildlife Area. Northeast of Québec City. *August 1987, Léo-Guy de Repentigny.*

The most productive site in the lower estuary is probably the Baie de L'Isle-Verte National Wildlife Area (4.7 broods/km), although the density is almost half that reported by Reed and Moisan (1971) 15 years earlier. This decrease is probably due to several factors: because of a major change in the plant cover, some islands like the Île aux Pommes that were previously used extensively by the American Black Duck only have a few nests; furthermore, some peatlands have grown up over the years, becoming more wooded and hence less attractive to this species. Lastly, new agricultural practices have led to the extensive and ongoing draining of some parts of marshes, particularly the salt-meadow cord-grass portion. Draining transforms the vegetation so that it provides less effective escape cover. In addition, in the hottest part of the summer, the tidal pools where ducklings forage dry up (up to 100 tidal pools are affected for each drainage canal created) (Bachand 1992).

The cord-grass marshes in the Baie de L'Isle-Verte National Wildlife Area are among the most extensive salt marshes in the St. Lawrence system, attracting ducks, herons, and shorebirds. Northeast of Rivière-du-Loup. *August 1978, Léo-Guy de Repentigny.*

Waterfowl in the gulf

The Gulf of St. Lawrence is characterized by sandy deposits, extensive rocky shores, and innumerable islands, islets, rocks, and reefs. Most of these islands, except for Bonaventure and Anticosti, are sparsely vegetated. Because of the nature of the coastline, wetlands are not only small, but few and far between. They are generally limited to river mouths and small sheltered embayments; marshes can often be found in the region's barachois, which are lagoons protected by a sand spit (Dryade 1980; Gratton and Dubreuil 1990).

As in the upper estuary, data on breeding waterfowl in the gulf are very scarce. The small number of existing studies are limited mainly to a description of the status of each species and potential nesting sites (Bernard 1984; Fradette 1992), or a quantitative analysis of the avifauna at very specific sites (Chapdelaine and Blais 1979; Parks Canada, unpubl. data). Therefore, we were not able to assess annual productivity in the gulf overall, but were limited to describing the probable species composition in two regions, the Magdalen Islands and the Mingan Islands, based on breeding pair surveys.

A study by Chapdelaine and Blais (1979) at Pointe de l'Est on the Magdalen Islands, in the summer of 1977, found 10 species of ducks; 90% of pairs recorded were Northern Pintail, Red-breasted Mergansers, American Black Duck, and Green-winged Teal. An aerial survey of breeding pairs by Parks Canada (unpubl. data) in the spring of 1990 in the Mingan Islands found the dominant species (73%) to be the American Black Duck and Green-winged Teal, as on the Magdalen Islands. Here, too, Common Eiders were excluded. Substantial numbers of Canada Geese (5%), Common Goldeneyes, and Gadwalls were also found. The Northern Pintail appears to be less abundant in this part of the gulf than on the Magdalen Islands.

Denis Lehoux, Luc Bélanger, Léo-Guy de Repentigny, and Jean-Claude Bourgeois

SEABIRDS OF THE ST. LAWRENCE

Seabird surveys provide a way of determining the health of seabird populations in marine ecosystems.

Data obtained from seabird surveys in Québec are stored in a computer data bank (Banque informatisée des oiseaux marins du Québec, or BIOMQ) managed by the Canadian Wildlife Service, Québec Region. The bank contains, not only data from modern quantitative surveys, but also information on distribution and abundance provided by early naturalists. Information on survey methodology is also included, which allows ornithologists to assess the value of surveys and determine whether they can be compared with previous data. Survey data supplied by other government and nongovernmental agencies involved in wildlife and habitat conservation are also included. The Société Duvetnor, for example, not only harvests eider down from the Common Eider population in the estuary, but conducts research, organizes nature interpretation activities, and co-operates closely with the CWS in providing up-to-date information for the BIOMQ on various seabird species breeding in the estuary. Parks Canada and Québec Ministry of Environment and Wildlife (MEF) also contribute data to the BIOMQ.

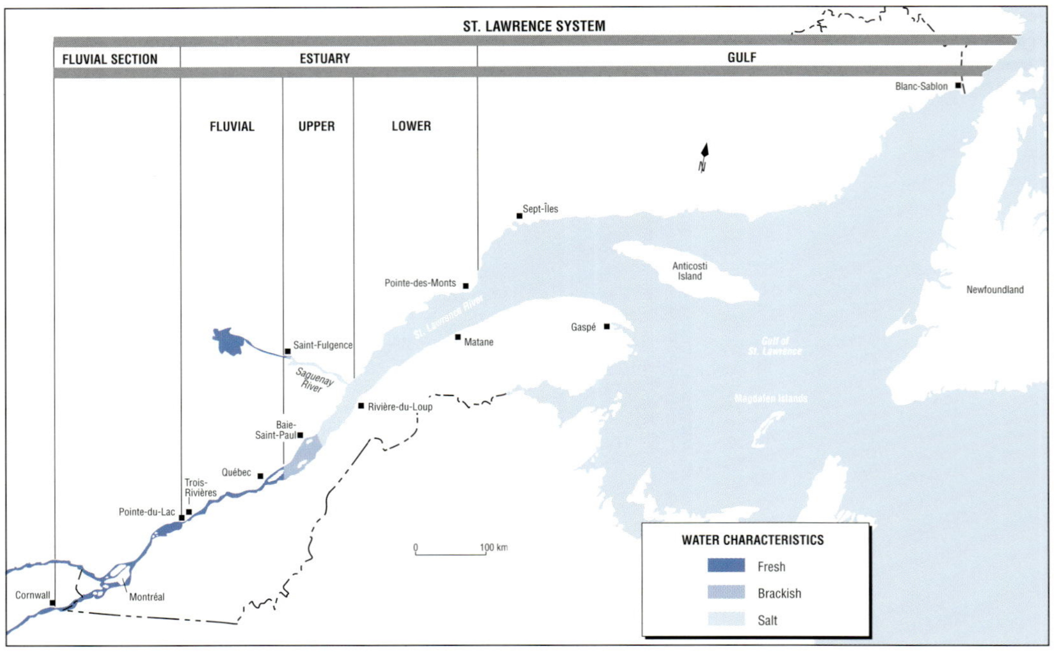

Hydrographic Divisions of the St. Lawrence

Source: Bouchard and Millet 1993, St. Lawrence Centre, Environment Canada

Corossol Island, near Sept-Îles on the North Shore, houses several seabird colonies. On the cliff face in the foreground, Black-legged Kittiwakes can be seen on their nests. Herring Gulls and a few Great Black-backed Gulls occupy the rocky terraces nearest the water and the flat top of the island. Razorbills can be seen on the right, and a few Common Murres (barely visible) occupy the rocky point on the far left. *May 1985, Gilles Chapdelaine.*

Note: In 1996, this island was home to the first confirmed breeding attempt of the Northern Fulmar in Québec (G. Chapdelaine, P. Brousseau, J.-F. Rail, and J. Thibault, pers. obs.). Observed by several biologists, a female was seen incubating a single egg, the typical clutch size for this species (del Hoyo et al. 1992). This breeding attempt was not unexpected: in 1993, a pair of fulmars was seen on this island, probably prospecting for a future nest site (Chapdelaine 1995), which is characteristic of the species (Palmer 1962). Further breeding records are expected as a second prospecting pair was observed in 1996.

This photograph, taken in 1976, of the Double-crested Cormorant colony on Corossol Island (visible in the trees in the foreground) shows the impact of the birds' droppings on the vegetation. The trees die, creating openings that attract mainly the larger gulls. With the tree cover gone, the soil erodes, uncovering ledges and crevices in the rock, which provide nest sites for Black-legged Kittiwakes and Razorbills. By 1994, the healthy forest had almost disappeared, leaving shrubs, raspberry bushes, and tall ferns (2 m), where a small number of Common Eiders nested. *May 1976, Gilles Chapdelaine.*

Québec is surrounded by several bodies of water: in the south, the St. Lawrence River and estuary and the Gulf of St. Lawrence, and in the north, James Bay, Hudson Bay, Hudson Strait, and Ungava Bay. Between 1988 and 1991, seabird surveys were carried out in five sections of the estuary and gulf: (1) the seabird sanctuaries on the North Shore, (2) the lower and upper estuary, (3) the Gaspé Peninsula, (4) the Magdalen Islands (not completed), and (5) the upper St. Lawrence between Cornwall and Québec City. The results of some of these surveys have already been published, including the censuses of the North Shore seabird sanctuaries (Chapdelaine and Brousseau 1991) and the survey of seabirds on the Gaspé Peninsula (Chapdelaine and Brousseau 1992b). This text and accompanying tables summarize the findings of these surveys, and also discuss the population trends observed in the estuary, on the Magdalen Islands, and the upper St. Lawrence (Montréal to Québec City), although in the last three cases, the results have yet to be analysed in detail.

Readers should also note that the results presented here only apply in the areas surveyed rather than to the total population in the region. For example, on the North Shore, there are many more tern colonies outside the sanctuaries than inside, and the trends discussed here must not be construed as representing population trends outside the sanctuaries. Conversely, over 95% of the North Shore alcid population is found inside the sanctuaries, and we believe the trends discussed here are an accurate reflection of the region's alcid population as a whole. On the Gaspé Peninsula and the upper St. Lawrence between Montréal and Québec City, comprehensive surveys were undertaken and, therefore, the results represent the overall populations. The surveys on the Magdalen Islands and upper and lower estuary did not provide enough information to determine total seabird populations in these regions. However, we believe that the data for the estuary are fairly accurate given the large number of sites ($n = 60$) that were surveyed (Table 4).

Migratory bird sanctuaries on the North Shore of the Gulf of St. Lawrence

Table 1 summarizes trends in the populations of seabirds breeding in the nine sanctuaries along the North Shore of the Gulf of St. Lawrence. For more details about survey methods and species distribution in the sanctuaries, see Brousseau and Chapdelaine (1990) and Chapdelaine and Brousseau (1991). The seabird colonies outside the sanctuaries were not taken into account in this analysis.

On Île aux Perroquets in the Mingan Islands, Black-legged Kittiwakes nest on the cliff ledges, Razorbills and Black Guillemots occupy crevices in the cliff face and talus slopes at the bottom, and Atlantic Puffins dig their burrows in the ground on the grassy flat top. [The island is named for the puffin, which somewhat resembles a parrot (*perroquet* in French).] In the 19th century, before the lighthouse was built, Northern Gannets and Common Murres also nested here. Middle North Shore. *August 1983, Gilles Chapdelaine.*

Table 1
Trends in seabird populations (individuals) in the North Shore sanctuaries, Gulf of St. Lawrence, Québec, 1977–93

Species	1977	1982	1988	1993	Trend[1]
Red-throated Loon	72	68	76	66	=
Leach's Storm-Petrel	?	234	1 744	1 840	+
Great Cormorant	214	134	86	78	–
Double-crested Cormorant	472	1 353	4 558	3 472	+
Common Eider	2 965	2 410	8 536	14 548	+
Great Black-backed Gull	1 413	1 722	2 201	2 284	+
Herring Gull	11 358	18 843	22 409	4 879	–
Ring-billed Gull	1 716	945	288	104	–
Black-legged Kittiwake	3 256	7 506	8 536	6 294	+
Caspian Tern	3	7	15	0	–
Common Tern/Arctic Tern	1 470	1 935	1 982	545	–
Common Murre	10 165	14 615	26 049	30 829	+
Razorbill	3 597	3 572	7 036	8 389	+
Black Guillemot	514	484	521	411	–
Atlantic Puffin	15 223	30 466	35 142	46 684	+

1 Population: + increasing; = stable; – decreasing

After Chapdelaine and Brousseau (1991) and Chapdelaine (1995).

A total of 16 species were surveyed. Between 1977 and 1993, the populations of 7 species declined—the Great Cormorant, Herring Gull, Ring-billed Gull, Caspian Tern, Common Tern, Arctic Tern, and Black Guillemot—although in some species, including the Herring Gull and terns, these declines are fairly recent. Conversely, significant and indeed rather spectacular increases were observed during this period in populations of Leach's Storm-Petrel, Double-crested Cormorant, Common Eider, Common Murre, and Atlantic Puffin. The Common Murre population is the largest it has been since the network of migratory bird sanctuaries was established on the North Shore in 1925. The Razorbill population also rose significantly, particularly in the Sainte-Marie Islands sanctuary.

Double-crested Cormorant colonies on Île de l'Est in the Sainte-Marie Islands. *August 1992, Gilles Chapdelaine.*

View of the Audubon Islets and the mainland of the Lower North Shore in the background. These islands were named after naturalist and artist John James Audubon, who visited the area in 1833. Black Guillemots and Herring Gulls nest here in small numbers. Lower North Shore. *June 1983, Kathleen Blanchard.*

Typical club of alcids, composed of Common Murres, Razorbills, and Atlantic Puffins. Sainte-Marie Islands, Lower North Shore. *July 1991, Claude Nadeau.*

Small numbers of cormorants or Black-legged Kittiwakes nests here and there along the cliffs on the eastern and southern coast of the Gaspé Peninsula. A few Black Guillemots are always present, however, taking advantage of the crevices and boulder piles at the bottom. Cap d'Espoir, south of Percé. *July 1985, Gilles Chapdelaine.*

Gaspé Peninsula

Table 2 summarizes and compares the results of surveys on the Gaspé coast in 1979 and 1989. More detailed information about the survey methods can be found in Chapdelaine and Brousseau (1992b). As on the North Shore, the results are presented by species.

Table 2
Trends in seabird populations (pairs) on the Gaspé Peninsula, 1979–89

Species	1979	1989	Trend[1]
Leach's Storm-Petrel	10	21	+
Northern Gannet[2]	18 245	24 125	+
Common Eider	47	256	+
Great Cormorant	16	35	+
Double-crested Cormorant	2 443	4 666	+
Ring-billed Gull	0	21	+
Herring Gull	9 796	7 718	–
Great Black-backed Gull	229	1 762	+
Black-legged Kittiwake	21 827	43 077	+
Common Tern	459	2 407	+
Common Murre	11 866	27 857	+
Razorbill	312[3]	587[3]	+
Black Guillemot	2 008	3 078	+
Atlantic Puffin	5	*pr*	=

1 Population: + increasing; = stable; – decreasing
2 1994 survey: 32 048 pairs (BIOMQ)
3 Individuals

pr Species present but not censused

After Chapdelaine and Brousseau (1992b).

Part of the Northern Gannet colony on Bonaventure Island. On the top, you can see the close, evenly spaced arrangement of the birds on their nests. Non-breeding individuals, on the other hand, like those on the left, are irregularly spaced and often occupy areas on the periphery of the colony. Leach's Storm-Petrels also nest in burrows in the island's meadows and coniferous forests, only returning to their burrows at night. The presence of the burrows is revealed by a strong musky odour. The red fox is the petrel's main predator on Bonaventure Island. The wings on the ground give a good idea of how extensive predation is. *July 1994, Pierre Brousseau.*

Renowned for their diverse avifauna (11 colony-nesting species) and large colonies of many species, Bonaventure Island (in the background) and Percé Rock (rocky island facing the village of Percé) are among the world's natural wonders. *February 1975, Réal Bisson.*

In all, 14 species of seabirds breed along the Gaspé coast, with the total seabird population in 1989 consisting of roughly 230 000 adult individuals. The main species are the Black-legged Kittiwake (37.4%), Common Murre (24.2%), and Northern Gannet (20.9%). Between 1979 and 1989, the populations of almost all the species increased, except the Herring Gull. The greatest increases were in the Great Black-backed Gull and Common Tern populations, although the latter has recently experienced a significant decline in numbers. Furthermore, the Northern Gannet population has increased steadily since 1979, and if the current annual growth rate of 3.1% continues (Chapdelaine and Brousseau 1992b), the Bonaventure Island population will soon exceed 30 000.

This photograph of Île aux Loups Marins in the Magdalen Islands is dramatic testimony to the impact that cormorant and heron guano can have on the landscape. Herring Gulls nest on the ground in the clearings. *June 1987, Jean-Pierre le Bel.*

On Falaise aux Goélands on Anticosti Island, seabirds only nest in the sites that cannot be reached by foxes, so there are no colonies at the top. In the centre is a colony of Northern Gannets; Black-legged Kittiwakes have built their nests on narrow ledges here and there on the cliff face. *July 1994, Pierre Brousseau.*

Magdalen Islands

A total of 16 species of seabirds nest on the cliffs, sand dunes, and beaches of the Magdalen Islands and nearby islands. Table 3 summarizes data on the seabird populations on the main islands, as well as on Bird Rocks and Brion Island. Because of logistical problems and the difficulties in censusing some species, particularly the alcids, population trends could only be analysed in 9 species (Table 3).

On Bird Rocks (including North Bird Rock), the Northern Gannet population is increasing by 3% a year, a rate similar to that on Bonaventure Island. Great and Double-crested cormorant populations are also on the rise. Among larids, significant increases have occurred in the Herring Gull, Great Black-backed Gull, and Black-legged Kittiwake populations. Given the difficulties of distinguishing between Common and Arctic terns, which often nest side by side in the same colonies, they were combined. Therefore, it is impossible to determine to what amount each species contributed to the increase observed on the islands.

Table 3
Trends in seabird populations (*p* = pairs, *i* = individuals) on the Magdalen Islands, including Bird Rocks and Brion Island, 1976–90

Species	1976–77	1990	Trend[1]
Leach's Storm-Petrel	*pr*	*pr*	?
Northern Gannet[2]	4 453 *p*	7 640 *p*	+
Great Cormorant	1 032 *i*	2 106 *i*	+
Double-crested Cormorant	728 *p*	1 601 *p*	+
Common Black-headed Gull	0	5 *p*	+
Herring Gull	2 313 *i*	3 070 *i*	+
Great Black-backed Gull	1 586 *i*	2 388 *i*	+
Black-legged Kittiwake	5 590 *p*	7 433 *p*	+
Roseate Tern	3 *i*	5 *i*	=
Common Tern/Arctic Tern	634 *p*	2 872 *p*	+
Common Murre	500–1 000 *i*[3]	?	?
Thick-billed Murre	500–1 000 *i*[3]	?	?
Razorbill	662 *p*	>445 *i*[4]	?
Black Guillemot	267 *i*	665 *i*[4]	?
Atlantic Puffin	113 *i*	500 *i*[4]	?

1 Population: + increasing; = stable; – decreasing
2 1994 survey: 9 868 pairs (BIOMQ)
3 According to Brown et al. (1975)
4 Incomplete data

pr Species present but not censused

Source: BIOMQ

Great Cormorant colony on Shag Island. Shag is one of the colloquial names for these birds. Magdalen Islands. *June 1988, Jean-Pierre le Bel.*

Many Northern Gannets still nest on Great Bird Rock, although the population is greatly diminished from what it was in 1830, when it was estimated to consist of over 100 000 pairs. When the lighthouse was built in 1869, the birds were driven off the top of the island. After the lighthouse was automated (the last lighthouse keeper left in 1987), the gannets returned to the top and the colony has gradually increased since then. Great Bird also houses the only colony of Thick-billed Murres in the gulf. Other alcids such as the Atlantic Puffin and Common Murre also nest here. Magdalen Islands. *July 1994, Pierre Brousseau.*

Northern Gannets and murres return year after year to nest on storm-tossed North Bird Rock, which is located near Great Bird Rock and actually consists of two islands. The sea is gradually wearing away the islands and a third island present during Cartier's time has completely vanished, reducing the number of nest sites. Magdalen Islands. *July 1994, Pierre Brousseau.*

Upper and lower estuary

A total of 9 species of seabirds (Table 4) breed on the islands and cliffs of the estuary.

The Common Eider colonies in the estuary are particularly noteworthy, not only because of their size (there were an estimated 10 000 pairs on Bicquette Island in 1990), but because of the quantity of down produced. Although populations have declined in places such as Île Blanche, they have increased on other islands, such as Îlet aux Alouettes, Battures aux Loups Marins, and islands near Ragueneau (Île de la Mine, Les Trois Soeurs); with the exception of Battures aux Loups Marins, these sites are along the north shore of the St. Lawrence estuary, where breeding populations were considered marginal 20 years ago.

Table 4

Trends in seabird populations (p = pairs, i = individuals) in the St. Lawrence estuary, between 1971–72 and 1990

Species	1971–72[1] (36 sites)	1990[2] (36 sites)	1990[3] (60 sites)	Trend[4]
Double-crested Cormorant	2 695 *p*	6 123 *p*	10 366 *p*[5]	+
Common Eider	19 150 *p*	22 665 *p*	25 407 *p*	+
Ring-billed Gull	900 *p*	6 578 *p*	7 592 *p*	+
Herring Gull	12 255 *p*	8 617 *p*	10 274 *p*	–
Great Black-backed Gull	1 293 *p*	2 636 *p*	2 907 *p*	+
Black-legged Kittiwake	0	530 *p*	710 *p*	+
Common Tern	0	0	19 *p*	=
Razorbill	1 090 *i*	1 930 *i*	1 934 *i*	+
Black Guillemot	2 600 *i*	588 *i*	645 *i*	–

1 Based on results from Reed (1975a)
2 The same sites visited by Reed (1975a)
3 The column shows estimates for almost all the islands in the estuary surveyed in 1990. Some islands between Baie-Comeau and Port-Cartier still have not been surveyed; historically, this section of the estuary was of little importance but today may house significant populations of seabirds, which are expanding out of and spilling over from their traditional areas.
4 Population: + increasing; = stable; – decreasing
5 Information not available on some colonies. Therefore, this is not the total population in the estuary.

Source: BIOMQ

The species that has experienced the greatest increase during the past 20 years, however, is the Double-crested Cormorant, although numbers should stabilize in the future because of a control program recently implemented by the provincial government. Ring-billed Gull, Great Black-backed Gull, and Razorbill populations are also on the rise. Also noteworthy is the fact that the Black-legged Kittiwake has recently managed to establish itself as a regular breeder in this area. Herring Gull and Black Guillemot populations, on the other hand, have decreased.

The Brandypot Islands in the estuary are home to no less than 9 colonial species. Each species occupies its own niche on the islands: Razorbills in crevices in the cliffs, Black-legged Kittiwakes on the narrow cliff ledges, Black Guillemots in the talus at the cliff base, Herring and Great Black-backed gulls on flat, bare terraces on the edge of the islands, Common Eiders in the forest undergrowth, Double-crested Cormorants in the treetops and on exposed rocks, Great Blue Herons in the treetops, and Black-crowned Night-Herons lower down in the trees. *August 1985, Bernard Filion, Ducks Unlimited Canada.*

The exposed rocks on Île aux Pommes, an island in the estuary off L'Isle-Verte, attract Double-crested Cormorants, while the grassy meadows on the island provide cover for nesting Common Eiders and American Black Ducks. *June 1971, Austin Reed.*

Le Petit Pèlerin, an island situated in the estuary, is home to several seabird colonies. In the upper left-hand corner, Double-crested Cormorants can be seen nesting on the trees and rocks. Razorbills, whose white breasts can be seen, are occupying a ledge on the left. In the centre, Black-legged Kittiwakes cling to the narrow ledges where they have built their nests. Herring Gulls and Great Black-backed Gulls, which are difficult to distinguish from the kittiwakes in the photo, occupy the widest ledges, while Black Guillemots nest in crevices in the rock. *May 1988, Gilles Chapdelaine.*

Cormorant colony on Le Pilier de Bois. Three other colony-nesting birds also breed on this islet: Herring Gulls, Great Black-backed Gulls, and Razorbills. The colony of 140 Razorbills is the most westerly one in the St. Lawrence (1995 survey, CWS, unpubl. data). *September 1991, Jean-Luc DesGranges.*

The largest Ring-billed Gull colony in the Québec City region is at the mouth of the Saint-Charles River, in a dump on the grounds of a paper mill. The species has prospered here, because of the peacefulness of the site and the abundant food in the Saint-Charles and St. Lawrence rivers and nearby urban areas. A recent study by the Canadian Wildlife Service on the diet of not yet fledged Ring-billed Gull chicks reveals that they are fed mainly insects, earthworms, and domestic waste. *May 1994. Pierre Brousseau.*

Table 5
Trends in seabird populations (pairs) in the Cornwall–Québec City section of the St. Lawrence, 1979–91

Species	1979	1991	Trend[1]
Double-crested Cormorant	0	4	+
Ring-billed Gull	36 417	115 756	+
Herring Gull	2	52	+
Great Black-backed Gull	2	8	+
Common Tern	41	192	+

1 Population: + increasing; = stable; – decreasing

Source: BIOMQ

Upper St. Lawrence River (Cornwall–Québec City)

This part of the river is dominated by the Ring-billed Gull (Table 5). Ring-billed populations increased spectacularly here between 1979 and 1991, as they did elsewhere in Canada (Ryder 1993). Another significant event is the arrival in the region of the Double-crested Cormorant, which frequented the Lake Saint-Pierre area between 1952 and 1965. The Common Tern population in the Lake Saint-Pierre region also increased and in 1991 the colony was home to 150 pairs.

Île de la Couvée is a dredge spoil island created during the construction of the St. Lawrence Seaway, which opened in 1958. Ring-billed Gulls colonized the island in the early 1970s. The island, which is just off Saint-Lambert and opposite Montréal, can be seen from the Champlain Bridge (shown in the background). In 1994 the colony housed close to 30 000 nesting pairs. *August 1993, Pierre Brousseau.*

Deslauriers Island, located off Varennes near the northeastern tip of the Island of Montréal on the edge of the St. Lawrence Seaway, has the largest Ring-billed Gull colony in the province. *May 1991, Pierre Brousseau.*

Discussion

In general, seabird populations along the St. Lawrence are increasing, with a few exceptions such as Herring Gulls in the estuary and the Gaspé, and Black Guillemots in the estuary. It is difficult to determine why some species are increasing while others are decreasing since all species cannot be studied everywhere and at the same time.

However, a study on Northern Gannets on Bonaventure Island (Chapdelaine et al. 1987) has provided a credible explanation for population fluctuations during the last 20 years. Between 1969 and 1976, the gannet population dropped from 20511 to 16400 pairs and, then, between 1976 and 1989, began to rebound, reaching 24125 pairs in 1989. Parallel studies on productivity and the presence of contaminants revealed that, between 1969 and 1976, hatching success was only 36–40%, coinciding with high concentrations of DDT and its derivatives in eggs at this time. The poor hatching rates had an effect on net productivity, which must be at least 67% for a Northern Gannet population to maintain itself; from 1966 to 1974 productivity was only 29.6–44.7%. Beginning in 1976, researchers observed a strong improvement in net productivity (68.6%) and a marked drop in the concentrations of DDT and its derivatives in gannet eggs. In 1979, the population began to increase and has continued to do so since.

The overall increases in seabird populations in the North Shore sanctuaries are probably due in part to the strengthened protection program by the Canadian Wildlife Service and Parks Canada on the Mingan Islands and to efforts by the Quebec–Labrador Foundation to educate people in the North Shore communities. However, the main reason is no doubt the abundance of small prey fish such as sand lance and capelin, which in turn probably resulted from a decrease in the populations of large predatory species such as the Atlantic cod (Québec Bureau of Statistics 1977–87). However, protection and education efforts no doubt played an important role in reducing poaching, which was quite prevalent in North Shore communities in the past, thus allowing the seabirds to take advantage of the abundant food (Chapdelaine 1987). Ongoing studies on the diet of the Razorbill show that food is abundant and that net productivity reached record levels between 1988 and 1991.

An abundant food resource may also explain overall increases in seabird populations in the Gaspé, Magdalen Islands, and estuary, but no specific studies have been carried out to confirm this hypothesis.

The most important factors affecting fluctuations in seabird populations are probably the quality and abundance of food, which influence breeding success, the number of young that reach adulthood, and annual survival rates among adults. When these parameters are known, population increases and decreases can be explained. However, few long-term studies have been carried out in Canada to provide a clear understanding of what is happening in specific seabird populations and we know very little about the mechanisms that control the fish populations on which these seabirds depend.

Gilles Chapdelaine and Pierre Brousseau

A Great Cormorant colony has been established on this small island in the Magdalen Islands. The cormorants share the island with Herring Gulls. Île aux Goélands. *June 1988, Jean-Pierre le Bel.*

In 1914, Percy A. Taverner photographed this group of 22 Atlantic Puffins on the cliffs of Bonaventure Island. Today, only a few individuals remain. *P. A. Taverner. Photo No. 32295, by permission of the Canadian Museum of Nature, Ottawa.*

Le Pilier de Pierre is a small rocky islet where, in 1995, 6799 pairs of Ring-billed Gulls nested along with Herring Gulls and a few Great Black-backed Gulls (CWS, unpubl. data). St. Lawrence estuary. *June 1991, Gilles Chapdelaine.*

BREEDING BIRDS OF AQUATIC AND RIPARIAN HABITATS OF THE ST. LAWRENCE SYSTEM

The objective of this study is not so much to explain the distribution of individual riverine species (species associated with aquatic and riparian habitats throughout the St. Lawrence system), but rather to determine the geographical patterns associated with the avifauna on a broader scale. The patterns revealed will, of course, depend on the criteria used to select species and squares for analysis. Since this approach is exploratory, the results obtained may differ somewhat from those presented in other parts of the Atlas. Furthermore, the terms used here, which are defined in the text, may have a more limited meaning than elsewhere in the Atlas.

Since the main purpose of a zoological atlas is to map the geographic distribution of the wildlife in the particular atlas area, the data that underpin it are generally analysed from the standpoint of individual species. This, however, is somewhat at cross-purposes with nature itself, which is composed of assemblages of species interconnected by invisible links in the food web. The animal and plant populations in a given area constantly interact, forming complex groups or communities that evolve in response to common factors. The closer communities are geographically, the more they resemble one another. This intimate relation between living organisms and geography is the very basis of biogeography.

One of the potential applications of the Québec Atlas data is the characterization of the main avian assemblages in Québec in terms of their distribution and similarities, thus directly contributing to the knowledge of Québec's biogeography. This is the purpose of this study, which focuses on the subset of Atlas data gathered along the St. Lawrence River.

Dataset (species)

Although 233 species were confirmed as breeders in squares along the St. Lawrence River, we have limited our analysis to strictly riverine species, defined as those that commonly breed in the riparian and aquatic habitats associated with the St. Lawrence or in the freshwater habitats in squares bordering it (river mouths, lakes, ponds, and peatland/pond complexes). Consequently, species that are summer visitors in the St. Lawrence system (shearwaters and scoters) and species usually associated with terrestrial habitats such as forests and cropland (including a number of diurnal and nocturnal birds of prey) are excluded, unless they are clearly attracted by nesting sites and food sources that are enhanced because of the river system's proximity. For example, the Rock Dove and House Sparrow nest in high numbers near the river, taking advantage of the abundance of food in ports in which grain is transshipped. A species must be a probable or confirmed breeder in a square to be included in the analysis.

Spatial units

For the purposes of analysis, we divided the St. Lawrence into 9 regions (Figure 1). In most cases, these regions correspond to the ornithological zones of the St. Lawrence defined by Ghanimé et al. (1990). Some adjoining zones were combined, however, to obtain a large enough sample size for statistical purposes. We also subdivided some of the zones into separate regions (Anticosti Island and the Magdalen Islands). The 9 regions were subsequently grouped into 4 main sections based on their physical, chemical, and physiographic characteristics: the freshwater course, estuary, southern gulf, and northern gulf.

Frequency of occurrence of riverine species used to select study squares for analysis (% of study squares), dividing the river system into upstream (freshwater and brackish water) and downstream (saltwater) portions

	Occurrence upstream (%)	Occurrence downstream (%)
Red-winged Blackbird	100.0	–
Killdeer	96.3	–
Common Yellowthroat	95.1	–
Common Grackle	95.1	–
European Starling	93.8	–
American Black Duck	85.2	–
American Crow	100.0	95.0
Tree Swallow	97.5	89.2
Savannah Sparrow	92.6	87.5
Spotted Sandpiper	92.6	85.0
Common Tern	–	57.5
Black Guillemot	–	62.5
Common Eider	–	64.2
Double-crested Cormorant	–	68.3
Herring Gull	–	95.0
Great Black-backed Gull	–	95.0

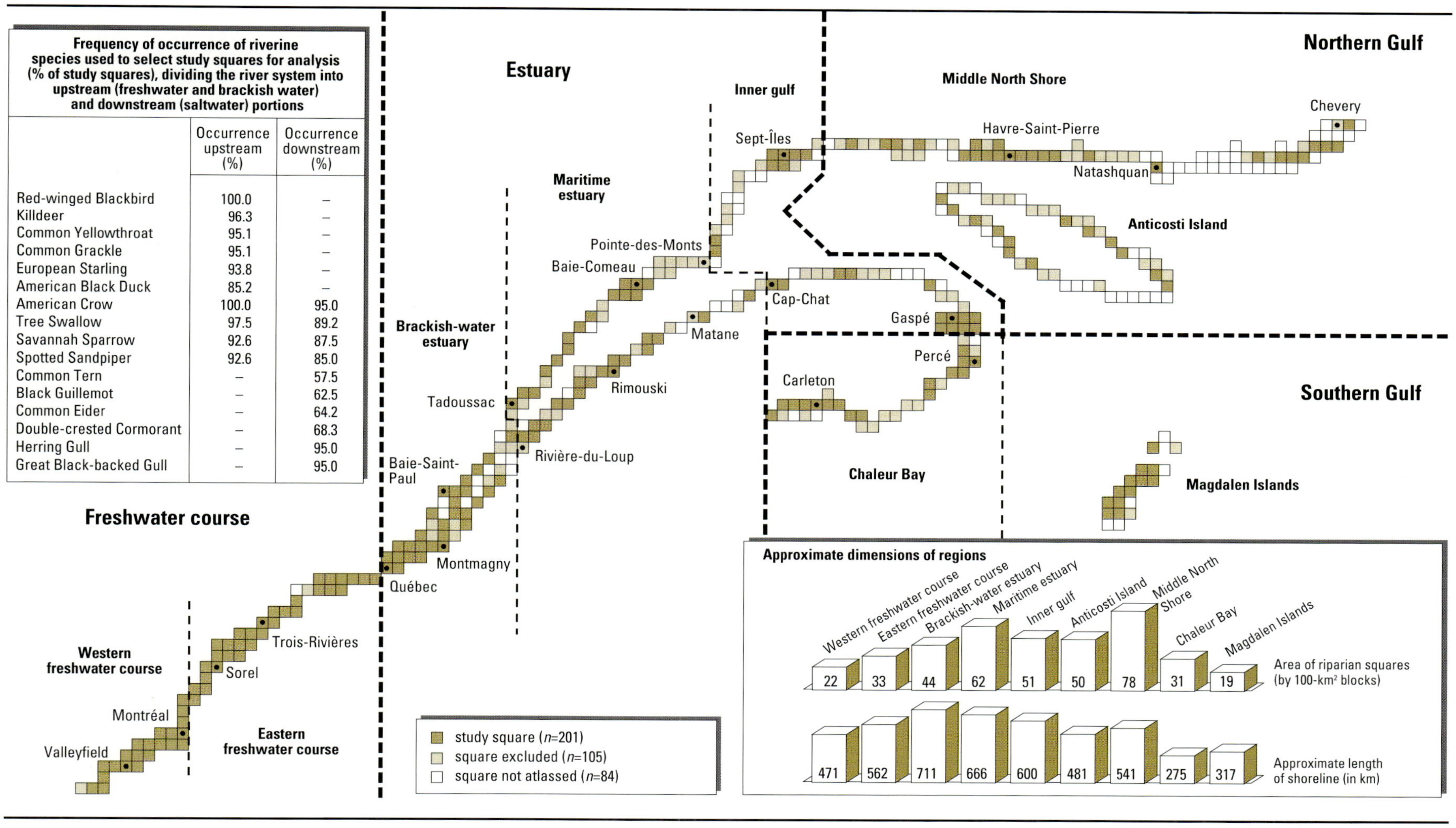

Figure 1
Geographical subdivisions of the St. Lawrence system (4 sections and 9 regions) showing the squares surveyed for the Atlas

Study squares

The Québec portion of the shoreline of the St. Lawrence River can be divided into 390 squares of 100 km², 306 of which were surveyed for the Atlas (Figure 1). However, 105 of these squares received insufficient coverage or have very little wetland habitat and had to be excluded from the statistical calculations of such things as average number of species per square and occurrence to avoid skewing the data. This leaves 201 representative squares, statistically speaking. Such a large number of squares had to be excluded since the number of species observed is a function of atlassing effort. Squares retained for analysis are those where at least 6 of the 10 commonest species in the section were present (shown in Figure 1). Statistics were compiled separately for the upstream portion (encompassing the freshwater course and the brackish-water estuary) and the downstream portion (from the maritime estuary to the gulf). Of the 201 final study squares, 45% supported the 10 species selected. Of the remaining squares, 21% had 9 species, 16% had 8 species, 12% had 7 species, and 5% had 6 species.

Occurrence and rarity

Frequency of occurrence was measured by calculating the percentage of study squares that supported the various riverine species. Frequency of occurrence must be distinguished from abundance, or the number of individuals censused of each species, which was not available. A species was considered common in one of the 9 regions, 4 sections, or the entire river system if it was present in 50% or more of the study squares, while rare species were defined as those found in 10% or less of squares. A regular species was defined as one with a frequency of occurrence between 10 and 50%. It should be noted that although frequency of occurrence and abundance, which are two different ways of measuring rarity, are correlated to some extent, they do not necessarily lead to the same conclusions (Schoener 1987): a species may be very abundant in a particular square (locally), but have a low frequency of occurrence regionally, as is the case with colonial species such as the Northern Gannet, which, while it breeds in large colonies, is concentrated in only 3 colonies in the St. Lawrence. However, the species frequents a number of sites in the gulf and estuary to feed.

Main sections of the St. Lawrence

Although all 4 sections of the river support rare species which are exclusive to that section, they are also home to a number of common or regular species that are typical of the section.

Freshwater course

Unlike the other sections, the freshwater course has a very uniform mix of species in all the squares. Although 4 rare species breed exclusively here—the Great Egret, Lesser Scaup, Little Gull, and Red-headed Woodpecker—the freshwater course is distinctive by its community of common and regular species (present in over 10% of study squares) that frequent freshwater habitats. Many are birds of freshwater marshes, including dabbling ducks, such as Wood Ducks, Northern Shovelers, and Blue-winged Teal, and small waders, such as Green Herons, Least Bitterns, Common Moorhens, and Virginia Rails. Willow Flycatchers and Warbling Vireos frequent seasonally flooded treed and shrub swamps, and Purple Martins, Northern Rough-winged Swallows, Sedge Wrens, and Marsh Wrens also show a marked preference for the freshwater course.

Estuary

Unlike the freshwater course, the estuary supports very few species that are exclusive to it. Only the Snow Goose (which has become established only very recently on the tidal flats of Battures aux Loups Marins in the Montmagny Islands) and Le Conte's Sparrow (a probable breeder) breed exclusively in this section of the river. Common estuarine species include the Black-crowned Night-Heron, which occurs more frequently here than anywhere else in the St. Lawrence system. The estuary is best characterized, however, by its unique combination of species typical of either freshwater (Gadwall, American Woodcock, and Eastern Phoebe) or saltwater habitats (Common Eider, Red-breasted Merganser, and Black Guillemot). Consequently, the estuary forms an ecotone, or transitional zone, between the freshwater course upstream and the maritime sections downstream, which is also the reason such a large proportion of rare species are found here.

Gulf of St. Lawrence

Species that are rare or absent from the freshwater course and estuary—the Northern Gannet, Great Cormorant, Common Murre, and Razorbill—give the gulf its distinctive character. Although very large colonies of some species are found, generally the number of nesting sites is quite limited.

Lake Saint-Pierre, with its cluster of islands broken by a complex network of channels and extensive floodplains, contains vast areas of freshwater habitat that are among the richest in Québec in terms of the riverine species they attract. In the photo, the spring-flooded wet (sedge) meadows and flat farmland stretch out as far as the eye can see, interspersed with treed swamps, shrub swamps, vast marshes and immense beds of floating and submergent vegetation. This mosaic of wetlands is a crucial staging area for waterfowl, marsh birds, and shorebirds and also contains the northernmost breeding populations of such southern species as the Redhead, Common Moorhen, Black Tern, and Marsh Wren. *October 1993, Pierre Normand.*

Southern gulf

Although the southern gulf, particularly the Magdalen Islands, has a very diverse birdlife, very few species are truly typical of the area. Particularly noteworthy species include the Piping Plover and 3 rare species in Québec that breed only on the Magdalen Islands —the Horned Grebe, Willet (probable breeder), and Roseate Tern.

Northern gulf

This region has few common species. Aside from the typical seabirds that breed throughout the gulf, the most characteristic species of this section are the Bald Eagle, Semipalmated Plover, and Atlantic Puffin. Although they are not exclusive to the northern gulf, they are commoner here than in the other sections. In addition, the Red-throated Loon, Willow Ptarmigan, and American Pipit breed exclusively in this section of the river, usually near freshwater ponds on sparsely vegetated islands.

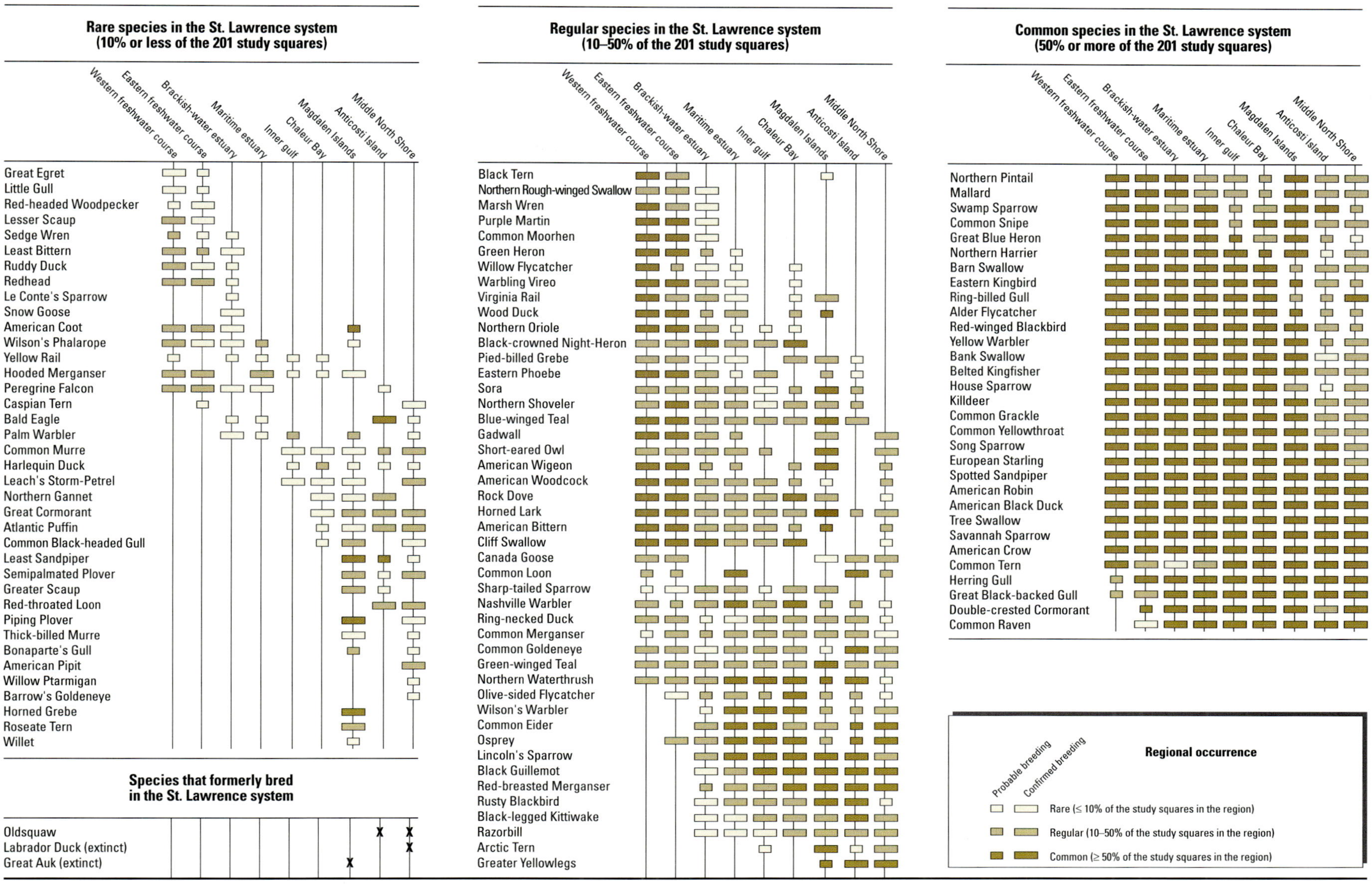

Figure 2
Frequency of occurrence of species in the 9 regions of the St. Lawrence system

Species richness

Species richness in the river system overall

Among the 233 species of birds that occur along the river, 107 can be classified as riverine breeders (Table 1 and Figure 2). To these must be added 8 other riverine species that apparently defend territories along the St. Lawrence but have not been confirmed as breeders: the Harlequin Duck, Barrow's Goldeneye, Willow Ptarmigan, Yellow Rail, Willet, Bonaparte's Gull, Sedge Wren, and Le Conte's Sparrow. In addition, 3 other species, 2 of which are now extinct, used to be part of the river's birdlife. The Oldsquaw, which is abundant in salt water outside the breeding season used to breed on the Lower North Shore (Todd 1963) and Anticosti Island (Verrill 1862), while the Great Auk, which once colonized Bird Rocks in the Magdalen Islands (Laporte 1974a), and the Labrador Duck, which bred all along the Lower North Shore according to Todd (1963), are now extinct.

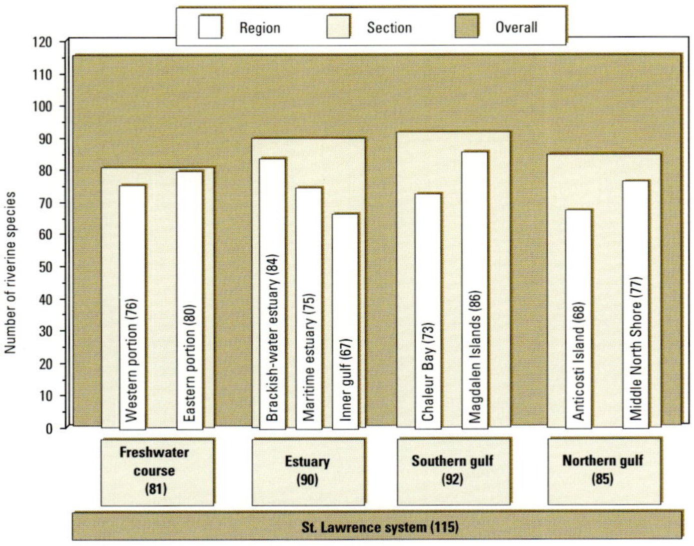

Figure 3
Number of riverine species breeding in the different geographical subdivisions of the St. Lawrence system, according to 3 levels of analysis

Of the 115 riverine species, only 18 are present year-round in significant numbers (Table 1). The 115 species can be divided into 30 families (15 of which are represented by a single species) and 7 subfamilies. The family Anatidae (ducks and geese) is the most diverse, with 23 species (Table 1).

Species richness by section and region

The number of riverine species found varies according to section and region (Figure 3). The freshwater course has 81 species, the estuary, 90, and the gulf, 100 (92 in the southern part and 85 in the northern part). At the regional level, the Magdalen Islands and brackish-water estuary have the highest species diversity, with 86 and 84 species respectively, while the inner gulf and Anticosti Island have the lowest, with 67 and 84 species respectively. No significant relation was found between the total area of the squares in a region and the number of species present. Therefore, other factors must come into play, most likely a significant change in the avian assemblages found. In other words, it is not so much the number of species that varies, but rather the composition of the avian community itself, since only 52 of the 115 riverine species breed throughout the entire river system. Therefore, many species are associated with only some sections or regions of the river and a few species with only one section or region, thus giving each region and section its special character. As species give way to different ones along the river, the cumulative number of species continuously increases (Figure 4). In fact, as one proceeds along the river in either direction, species richness increases by one species for every 120 km of shoreline on average (or 60 km in terms of actual distance), although the rate of increase is not constant. Whether heading upstream (western freshwater course) or downstream (Middle North Shore), cumulative species richness increases rapidly at first, then flattens out in the middle of the system, then increases again at the end of the gradient (Figure 4).

Table 1

Ecological characteristics, geographic affinity, and sensitivity of riverine birds in the St. Lawrence system

Species preceded by an asterisk (*) have resident populations remaining year-round on the St. Lawrence; those preceded by the symbol ¥ are colonial breeders. Species in which breeding is probable but not confirmed are shown in italics. The primary characteristic used in the statistical analyses is shown in darker shading with a dot in the middle, while secondary characteristics are in lighter shading.

Family / Subfamily / Species[1]	Habitat[2]: Lentic						Habitat: Lotic	Habitat: Marine			Diet[3]			Size[4]			Breeding strategy[5]		Geographical affinity[6]					Sensitivity[7]		
	Wet meadows	Peatlands	Freshwater and brackish-water marshes	Shrub swamp	Treed swamp	Lakes	St. Lawrence and tributaries	Salt marshes	Beaches	Coastal habitat	Herbivore/granivore	Aquatic carnivore	Terrestrial carnivore	Small	Medium	Large	r-strategist	K-strategist	North	East	Cosmopolitan	West	South	Not sensitive	Not very sensitive	Sensitive
Gaviidae (2)																										
Red-throated Loon										•		•				•		•	•						•	
Common Loon						•						•				•		•			•				•	
Podicipedidae (2)																										
Pied-billed Grebe			•									•			•		•						•	•		
Horned Grebe						•						•			•		•					•				•
Hydrobatidae (1)																										
(¥) Leach's Storm-Petrel										•		•		•				•		•						•
Sulidae (1)																										
(¥) Northern Gannet										•		•				•		•		•						•
Phalacrocoracidae (2)																										
(*¥) Great Cormorant										•		•				•		•		•					•	
(¥) Double-crested Cormorant										•		•				•		•			•			•		
Ardeidae (6)																										
American Bittern			•									•			•		•				•			•		
Least Bittern			•									•		•			•						•		•	
(¥) Great Blue Heron			•									•				•		•					•	•		
(¥) Great Egret			•									•			•			•					•			•
Green Heron					•							•			•		•						•		•	
(¥) Black-crowned Night-Heron								•				•			•			•					•		•	
Anatidae (23)																										
Anserinae (2)																										
(¥) Snow Goose			•								•					•		•	•							•
Canada Goose			•								•					•		•	•						•	
Anatinae (21)																										
Wood Duck					•						•				•		•						•		•	
Green-winged Teal			•								•				•		•				•			•		
(*) American Black Duck			•									•				•	•		•					•		
(*) Mallard			•								•					•	•					•		•		
Northern Pintail			•								•					•	•				•			•		
Blue-winged Teal			•								•				•		•						•	•		
Northern Shoveler			•									•			•		•					•		•		
Gadwall			•								•				•		•					•		•		
American Wigeon						•					•				•		•					•		•		
Redhead			•								•				•		•					•				•
Ring-necked Duck						•					•				•		•					•		•		
(¥) Greater Scaup						•						•			•		•		•							•
Lesser Scaup			•									•			•		•					•				•
(*¥) Common Eider										•		•				•	•		•						•	
Harlequin Duck							•					•			•			•	•							•
(*) Common Goldeneye						•						•			•		•		•					•		
(*) *Barrow's Goldeneye*						•						•			•		•		•							•
Hooded Merganser						•						•			•		•						•		•	
(*) Common Merganser							•					•				•	•				•			•		
(*) Red-breasted Merganser						•						•			•		•		•						•	
Ruddy Duck			•								•				•		•					•				•
Accipitridae (3)																										
Pandioninae (1)																										
Osprey										•		•				•		•			•			•		
Accipitrinae (2)																										
(*) Bald Eagle									•			•				•		•			•				•	
Northern Harrier			•										•		•		•					•		•		
Falconidae (1)																										
Peregrine Falcon									•				•		•			•					•		•	
Phasianidae (1)																										
Tetraoninae (1)																										
(*) *Willow Ptarmigan*				•							•				•		•		•							•
Rallidae (5)																										
Yellow Rail	•											•		•			•					•				•
Virginia Rail			•									•		•			•					•			•	
Sora			•									•		•			•					•		•		
Common Moorhen			•									•			•		•						•		•	
American Coot			•									•			•		•					•			•	
Charadriidae (3)																										
Semipalmated Plover									•			•		•			•		•						•	
Piping Plover									•			•		•			•						•		•	
Killdeer									•				•		•		•					•		•		
Scolopacidae (7)																										
Scolopacinae (6)																										
Greater Yellowlegs		•										•			•			•	•						•	
Willet								•				•			•			•					•			•
Spotted Sandpiper									•			•		•			•				•			•		
Least Sandpiper		•										•		•			•		•						•	
Common Snipe			•									•			•		•				•			•		
American Woodcock				•									•		•		•						•	•		
Phalaropodinae (1)																										
Wilson's Phalarope			•									•		•			•					•			•	
Laridae (12)																										
Larinae (7)																										
Little Gull							•					•			•			•		•						•
(¥) Common Black-headed Gull										•		•			•			•		•						•
(¥) *Bonaparte's Gull*								•				•			•			•				•				•
(¥) Ring-billed Gull							•						•		•			•				•		•		
(*¥) Herring Gull										•		•				•		•			•			•		
(*¥) Great Black-backed Gull										•		•				•		•		•				•		
(¥) Black-legged Kittiwake										•		•			•			•		•					•	
Sterninae (5)																										
(¥) Caspian Tern										•		•			•			•					•			•
(¥) Roseate Tern										•		•			•			•					•			•
(¥) Common Tern										•		•			•			•			•			•		
(¥) Arctic Tern										•		•			•			•	•						•	
(¥) Black Tern			•									•		•				•				•			•	
Alcidae (5)																										
(¥) Common Murre										•		•			•			•		•					•	
(¥) Thick-billed Murre										•		•			•			•		•						•
(¥) Razorbill										•		•			•			•		•					•	
(*¥) Black Guillemot										•		•		•				•		•					•	
(¥) Atlantic Puffin										•		•			•			•		•					•	
Columbidae (1)																										
(*¥) Rock Dove							•				•				•		•				•			•		
Strigidae (1)																										
Short-eared Owl								•					•		•		•				•			•		
Alcedinidae (1)																										
Belted Kingfisher							•					•			•		•				•			•		
Picidae (1)																										
Red-headed Woodpecker					•								•	•			•						•			•
Tyrannidae (5)																										
Fluvicolinae (4)																										
Olive-sided Flycatcher		•											•	•			•				•			•		
Alder Flycatcher				•								•		•			•				•			•		
Willow Flycatcher					•								•	•			•						•		•	
Eastern Phoebe					•								•	•			•						•	•		
Tyranninae (1)																										
Eastern Kingbird				•									•	•			•						•	•		
Alaudidae (1)																										
(*) Horned Lark									•				•	•			•				•			•		
Hirundinidae (6)																										
(¥) Purple Martin							•					•		•			•						•		•	
Tree Swallow							•					•		•			•				•			•		
(¥) N. Rough-winged Swallow							•					•		•			•						•		•	
(¥) Bank Swallow							•					•		•			•				•			•		
(¥) Cliff Swallow							•					•		•			•				•			•		
Barn Swallow							•					•		•			•				•			•		
Corvidae (2)																										
(*) American Crow									•				•			•		•			•			•		
(*) Common Raven									•				•			•		•			•			•		
Troglodytidae (2)																										
Sedge Wren	•												•	•			•						•			•
Marsh Wren			•										•	•			•						•		•	
Muscicapidae (1)																										
American Robin					•								•	•			•				•			•		
Motacillidae (1)																										
American Pipit	•												•	•			•		•							•
Sturnidae (1)																										
(*) European Starling					•								•	•			•				•			•		
Vireonidae (1)																										
Warbling Vireo					•								•	•			•						•		•	
Emberizidae (16)																										
Parulinae (6)																										
Nashville Warbler		•											•	•			•					•		•		
Yellow Warbler				•									•	•			•				•			•		
Palm Warbler		•											•	•			•				•					•
Northern Waterthrush				•								•		•			•				•			•		
Common Yellowthroat				•									•	•			•				•			•		
Wilson's Warbler		•											•	•			•		•						•	
Emberizinae (6)																										
Savannah Sparrow	•												•	•			•				•			•		
Le Conte's Sparrow	•												•	•			•					•				•
Sharp-tailed Sparrow								•					•	•			•			•					•	
Song Sparrow				•									•	•			•				•			•		
Lincoln's Sparrow		•											•	•			•				•				•	
Swamp Sparrow			•										•	•			•				•			•		
Icterinae (4)																										
Red-winged Blackbird			•										•	•			•					•		•		
Rusty Blackbird		•										•		•			•		•						•	
(¥) Common Grackle					•							•		•			•					•		•		
Northern Oriole					•								•	•			•						•	•		
Passeridae (1)																										
(*) House Sparrow							•				•			•			•				•			•		

1 This list consists only of riverine species, or those directly associated with the aquatic and riparian ecosystems of the St. Lawrence.

2 Habitats frequented most during foraging, taken mainly from Godfrey (1967) and DeGraaf and Rudis (1987) and by consulting the experts at the Canadian Wildlife Service.

3 Taken mainly from Martin et al. (1951) and a review of the most recent literature on the species.

4 Small: ≤100 g; Medium: 101–1000 g; Large: >1000 g.

5 Most of the information on breeding biology used to determine breeding strategies was taken from the Atlas species profiles. When information was not available, we used data on a related species.

6 Distribution maps in the *Field Guide to the Birds of North America* (National Geographic Society 1987) were used to determine the species' distribution centre in North America for the St. Lawrence population.

7 Not sensitive: species that breeds in 33 squares or more; present in all sections;

Not very sensitive: species that breeds in 11–100 squares; absent from at least one section;

Sensitive: species that breeds in 10 squares or fewer; absent from at least one section.

The Pointe-au-Père salt marsh, located 5 km east of Rimouski, comprises approximately 20 ha of cord-grass marsh dotted with tidal pools. This site, which became a National Wildlife Area in 1986, is a staging ground for thousands of shorebirds and, on the regional level, a major foraging and brood-raising site for a large number of aquatic species such as the Great Blue Heron and Common Eider. It is an important site in the maritime estuary for nesting riverine birds because of its great biological productivity, diverse saltwater habitats, and adjoining agricultural, forest, and urban habitats. *June 1990, Léo-Guy de Repentigny.*

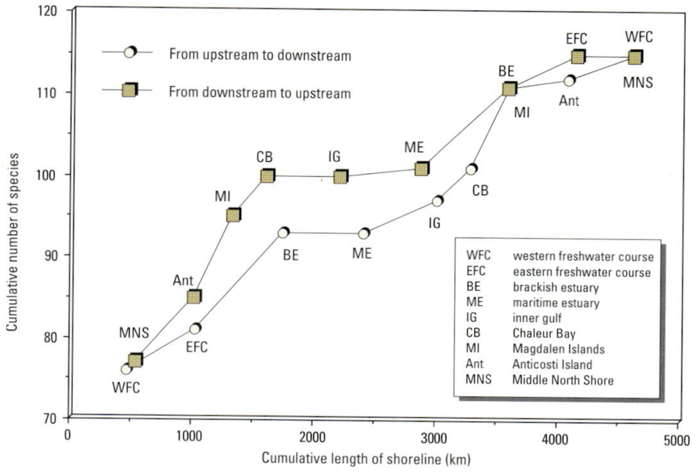

Figure 4
Cumulative number of breeding bird species in the St. Lawrence system

Species richness at the local level

The number of breeding bird species present in a square is indicative of the diversity of the local avifauna. Squares in the freshwater course have an average of 45 species, while squares of the same size in the northern part of the gulf have only 26 (Figure 5). On average, squares in the freshwater course and the southern gulf contain 56% and 44% of the species present in their regions respectively, while squares in the estuary and northern gulf have 41% and 30% respectively.

In general, squares in the northern gulf have fewer breeding birds than squares in other parts of the system (Figure 5), despite overall levels of species richness that are similar to or even greater than in other sections (Figure 3). This suggests that habitats in the northern gulf, and therefore the avian assemblages associated with them, are not very diverse locally (at the square level), but vary considerably from one square to another, giving this section a heterogeneous appearance. Conversely, the freshwater course is more uniform in terms of the avian assemblages found locally, reflecting its continuous mix of habitats throughout the section. Indeed, almost all the breeding birds in this section can be found in some squares on the shores of Lake Saint-Pierre.

Ecological structure of local avian communities

Habitat

The distribution of riverine species varies, depending on whether they frequent freshwater ecosystems—which may be either lentic (the calm waters of lakes, freshwater marshes, peatlands, wet meadows, and treed and shrub swamps) or lotic (the faster moving waters of streams and rivers, including the St. Lawrence itself)—or maritime ecosystems (salt marshes, beaches, and coastal habitats) (Table 1). Since the prevalent habitats change as the river flows down from the freshwater course into the estuary and gulf, a corresponding change occurs in the local avifauna (Environment Canada 1996; Lehoux et al., this publication: 1207–1213). Therefore, the proportion of species in each square that usually frequent habitats associated with fresh water decreases progressively from the upstream sections to the downstream sections of the river (Figure 6), from an average of 86% in the freshwater course to 54% in the northern gulf. Along this gradient, the main change that occurs is the replacement of species that frequent lentic habitats by species that frequent maritime habitats, which, in turn, reflects the river's progression from fresh to salt water.

Diet

Food resources along the river vary considerably with type of habitat. In the freshwater course, the 3 large riverine lakes—Lake Saint-François, Lake Saint-Louis, and Lake Saint-Pierre—which represent the prime lentic habitat here, have vast beds of aquatic vegetation and abundant plant cover on their shores. These highly productive habitats support a profusion of plant and animal life. In the northern gulf, where the rugged shoreline precludes salt marsh development, biological productivity is concentrated in the marine environment. Consequently, as the river flows downstream, birds using the resources found in marine habitats (aquatic carnivores) gradually replace the herbivores and terrestrial carnivores that are prevalent farther upstream (Table 1, Figure 6). The average percentage of aquatic carnivores increases from 38% in the freshwater course to 65% in the northern gulf.

Surrounded on the west by a row of dunes and on the east by a tombolo with an outlet to the sea, Baie du Havre aux Basques in the Magdalen Islands is the best site for aquatic birds in the Gulf of St. Lawrence. In the middle of the lagoon is a large sandy shoal that is uncovered at low tide, helping to isolate the salt meadows and intertidal marshes with pools in the northern part of the lagoon from the pond of brackish water in the south. Beyond the bay, silting and better drainage have allowed crowberry heaths to become established. The value of the habitat is due not only to the rare species that frequent it, such as the Greater Scaup, Piping Plover, Common Black-headed Gull, and Roseate Tern, but more importantly to its significance as a staging area for thousands of shorebirds, which stop here to feed during post-breeding migration. *Summer 1990, Pierre Fradette.*

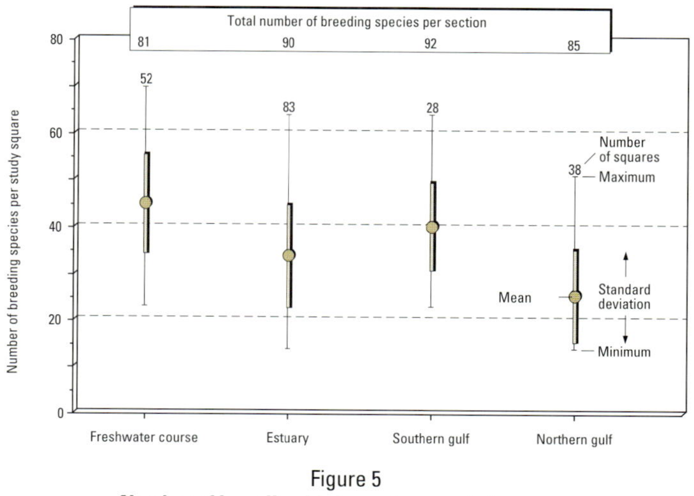

Figure 5
Number of breeding bird species per study square (local species richness) in the 4 sections of the St. Lawrence system

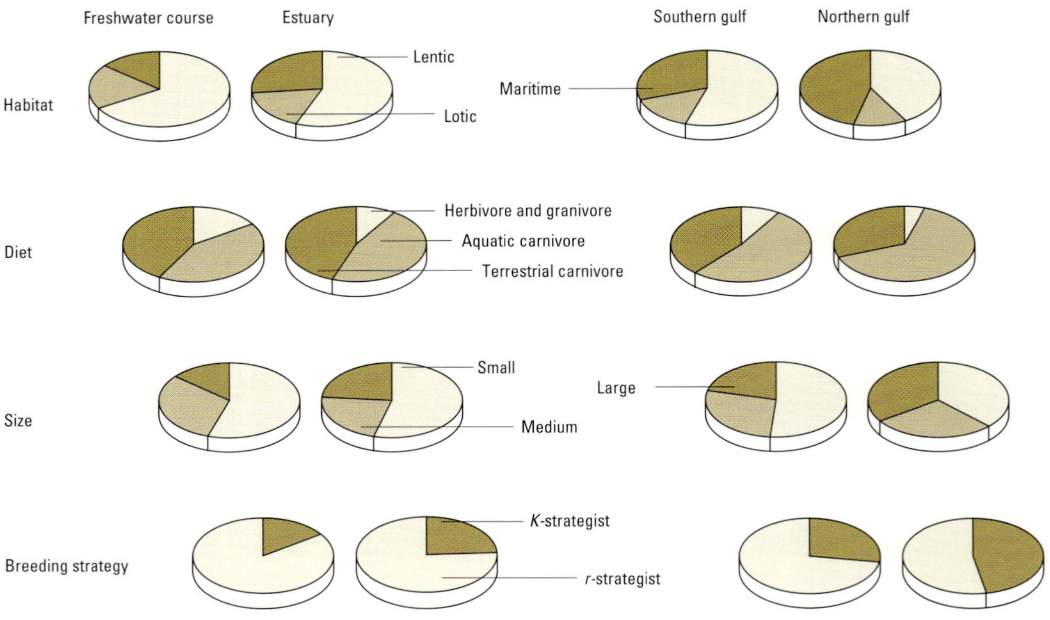

Figure 6
Characteristics of the local avifauna in the 4 sections of the St. Lawrence (calculations based on study squares)

Size of birds

A bird's size is usually related to the size of the organisms on which it feeds. Since the species that breed in the freshwater course feed mainly on plant seeds or aquatic or emergent insects, large birds make up a small proportion (14%) of the region's avifauna (Table 1, Figure 6). In marine habitats, large fish are abundant and larger birds that are able to catch and swallow them make up most of the avifauna. Therefore, there is a greater proportion of large birds in the marine environment, particularly the northern gulf (34%).

Breeding strategies

Animals adopt breeding strategies that give them an advantage in the environment in which they live. Species using the *r*-strategy have high reproductive rates (early sexual maturity, multiple broods, and large clutches), which is generally associated with a short lifespan. These species are usually able to colonize new areas fairly easily, but because individuals are short-lived, populations may fluctuate considerably from year to year. Conversely, species using the *K*-strategy take several years to reach sexual maturity, produce a small number of eggs every year, but spend more time and energy raising their young. Therefore, the progeny of *K*-strategist species are able to more easily face the pressures of the environment, giving them greater longevity and population stability (Smith 1990). Because of the complexity of the selective pressures that occur throughout the yearly life cycle, some species use a mixture of the two strategies. However, these species can generally be associated with one of the two strategies on the basis of the number and size of clutches, age at sexual maturity, and lifespan (Table 1).

The proportion of *r*- and *K*-strategist species varies from the downstream to the upstream portions (Figure 6). While *r*-strategist species dominate the local avifauna in the freshwater course, making up 85% of the species on average in each study square, this strategy is used by fewer and fewer species as the river flows downstream, and *r*-strategist species make up only 53% of the species in the northern gulf. Indeed, the avifauna in the northern gulf consists mainly of large seabirds that take several years to reach sexual maturity and usually produce few offspring, notwithstanding their longer lifespan.

Geographical affinities of regional avifauna

On the basis of the North American distribution (National Geographic Society 1987) of the 115 riverine species, we have classified each species according to its geographical affinity (Table 1). Southern species are defined as those having their distribution centre (during breeding season) in the southern part of the continent, or south of the coniferous forest. Northern species have their distribution centre mainly north of the mixed forest. Similarly, eastern and western species are those having most of their breeding range either in the central part of the continent or along the Atlantic seaboard respectively. Species that could not be assigned to a specific category were classified as cosmopolitan. Cosmopolitan species are found throughout the river system, although they do not necessarily breed in all regions.

Cosmopolitan species dominate the avifauna along the entire length of the system (Figure 7). Indeed, cosmopolitan species comprise roughly 30% of the river's avifauna regardless of section, which reflects, in part, our definition of this element. Species with more specific distribution centres—the southern (23% of species), western (20%), northern (16%), and eastern (11%, including 2 Eurasian species, the Little Gull and the Common Black-headed Gull) elements—exhibit a clinal pattern of variation along the St. Lawrence. The northern and eastern elements gradually replace the southern and western elements as one moves farther downstream (Figure 7), reflecting the southwest to northeast direction of the river itself. Furthermore, the southern and western elements have managed to expand farther east and north (toward the gulf) than the northern and eastern elements have been able to expand south and west (toward the freshwater section). This suggests that, at the beginning of the Holocene (roughly 18000 years ago), many of the species that were confined to the southern and western part of the continent began to colonize areas left uncovered by the retreating glaciers and receding oceans (Pielou 1991).

Furthermore, only 3 riverine species are not native to North America: the Rock Dove, European Starling, and House Sparrow. All three are Old World species deliberately introduced into the continent by citizens nostalgic for the birds of the old country. In addition, 5 North American species have been re-introduced in the northeast (Gosselin 1977b). Young Peregrine Falcons were released in the St. Lawrence Valley between 1976 and 1992 to help the species get re-established, while 4 game species—the Canada Goose, Mallard, Redhead, and Gadwall—have been released near the river.

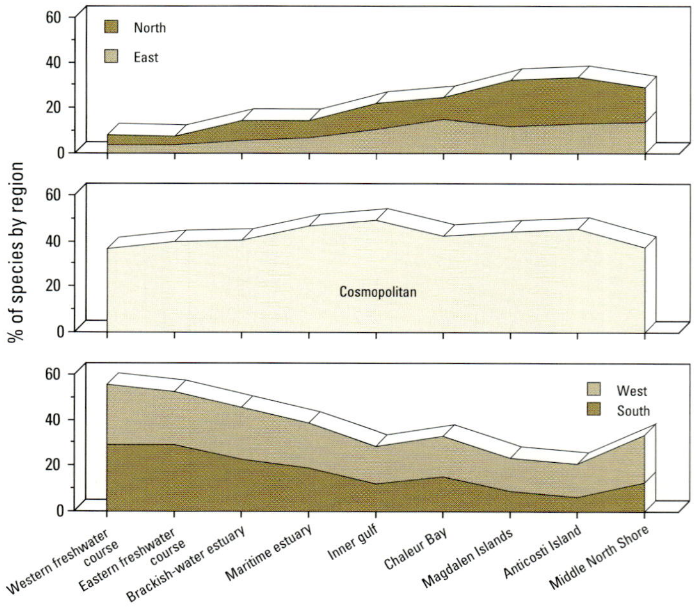

Figure 7
Geographical affinity of the avifauna of the 9 regions of the St. Lawrence (calculations based on all squares)

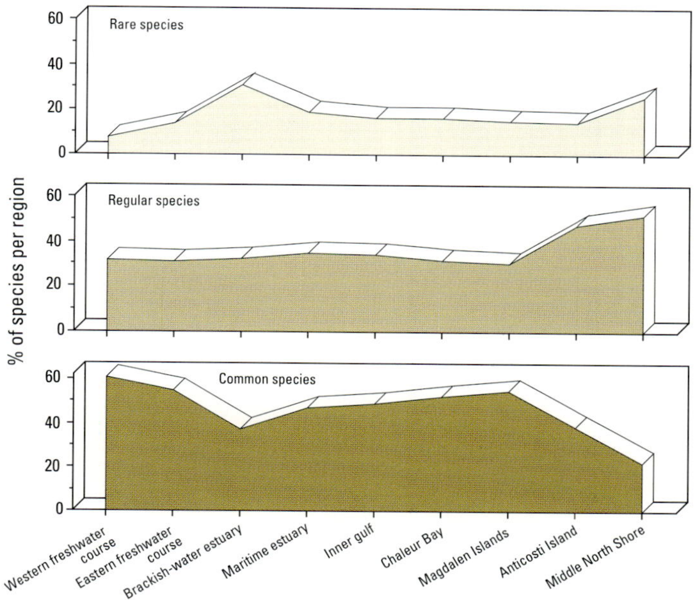

Figure 8
Relative importance of rare, regular, and common species in the 9 regions of the St. Lawrence

Frequency of occurrence (1984–89)

Most common species

Of the 115 riverine species, 31 are common along the river, meaning that they breed in over half the study squares (Figure 2). Only 5 of the species that are common in all 9 regions breed in 80% or more of the squares: the American Crow (97% of squares), American Robin (93%), Tree Swallow (93%), Savannah Sparrow (90%), and Spotted Sandpiper (88%). Six species, though common in the river system as a whole, have significant gaps in their distribution: the Double-crested Cormorant and the Common Raven are common in most regions but did not breed in the western freshwater course between 1984 and 1989; the Common Tern was rare in the brackish-water estuary; the Northern Harrier and Bank Swallow were rare on Anticosti Island; and the Great Blue Heron was rare on the Middle North Shore (Figure 2).

Table 2
Frequency of occurrence, sensitivity, and status in Québec of 38 rare species that breed along the St. Lawrence
Status as an endangered, threatened or vulnerable species in Canada and the United States is also shown for species that, according to the sources consulted, are considered to be at risk.

	Occurrence in the St. Lawrence system			Sensitivity			Status in Québec[1]				Status in Canada[2] ● and North America[3] ■			
Species	Freshwater course	Estuary	Gulf	Sensitive	Not very sensitive	Common	Threatened	Vulnerable	Endangered	Other	Not classified[4]	Threatened[4]	Vulnerable[4]	Endangered[4]
Red-throated Loon			Regular		•	•								
Horned Grebe			Rare	•				•						
Leach's Storm-Petrel		Rare	Rare	•						a				
Northern Gannet			Rare	•						a				
Great Cormorant			Regular		•					a				
Least Bittern	Regular	Rare			•			•					•	
Great Egret	Rare			•						c				
Snow Goose		Rare		•						c				
Redhead	Regular	Rare		•						c				
Greater Scaup			Rare	•						b				
Lesser Scaup	Rare			•						b				
Harlequin Duck		Rare	Rare	•						b				•
Barrow's Goldeneye			Rare	•						f				
Hooded Merganser	Regular	Rare	Rare		•	•								
Ruddy Duck	Rare	Rare		•						c				
Bald Eagle		Rare	Regular		•		•				•			■
Peregrine Falcon	Regular	Rare	Rare		•			•						•■
Willow Ptarmigan			Rare	•		•								
Yellow Rail	Rare	Rare	Rare	•				•						
American Coot	Regular	Rare	Rare		•	•					•			
Semipalmated Plover			Regular		•	•								
Piping Plover			Regular		•				•			■		•
Willet				•						e				
Least Sandpiper			Regular		•	•								
Wilson's Phalarope	Regular				•					c				
Little Gull	Rare			•						d				
Common Black-headed Gull			Rare	•						c				
Bonaparte's Gull			Rare	•						b				
Caspian Tern	Rare		Rare	•					•				•	
Roseate Tern			Rare	•			•					•		■
Common Murre		Rare	Regular		•					a				
Thick-billed Murre			Rare	•						b				
Atlantic Puffin			Regular		•					a				
Red-headed Woodpecker	Rare			•				•						
Sedge Wren	Regular	Rare		•				•			•			
American Pipit			Rare	•		•								
Palm Warbler		Rare	Rare	•		•								
Le Conte's Sparrow		Rare		•						a				

1 Robert (1989b)
For the "Other" category:
a - Breeder confined to southern Québec
b - Breeder confined to central and northern Québec
c - Breeder that has recently appeared and/or is expanding
d - Casual or accidental breeder
e - Casual visitor
f - Transient (breeding unconfirmed)

2 Committee on the Status of Endangered Wildlife in Canada (1993)
3 United States Fish and Wildlife Service (1993)
4 Not classified: species studied but of unknown status because of a lack of information
Threatened: species that is likely to become endangered if factors affecting its vulnerability are not reversed
Vulnerable: species that is particularly at risk because of its limited breeding range or because of low or declining numbers
Endangered: species in danger of immediate extinction or extirpation through all or part of its breeding range

Rare species

Rare species at the level of the entire system

One third (38/115) of the riverine species were atlassed in 10% or less of the study squares and were therefore considered rare (Figure 2). Four of these species breed exclusively in the freshwater course (Great Egret, Lesser Scaup, Little Gull, and Red-headed Woodpecker), while two others (Snow Goose and Le Conte's Sparrow, a probable breeder) are limited to the estuary (Table 2). Almost half of the rare species (17/38) were found only in the gulf, which demonstrates its importance in the biodiversity of the river system as a whole. The remaining 15 species that were rare in the river system overall but were not found exclusively in a particular section were all rare (14/15) or absent (Caspian Tern) in the estuary.

Rare species at the regional level

Birds are not distributed uniformly throughout the river system, and their frequency of occurrence varies depending on the level of analysis (region, section, or entire system). For example, the Horned Grebe is common in the Magdalen Islands but is not found elsewhere in the system, making it rare overall. Frequency of occurrence also depends on scale, on the size of the geographic area under consideration; the larger the area, the greater the number of rare species, as the species that are exclusive to the various geographic divisions add up.

Consequently, rare species make up a smaller proportion of the avifauna when the scale of analysis is reduced. Only 11 of the 38 species considered rare in the river system overall are rare in every region in which they occur (Figure 2). The western and eastern parts of the freshwater course contain only 8% and 14% of rare species respectively, while other regions have 20% on average (Figure 8). The greatest proportion of rare species (31%) is found in the brackish-water estuary. This is probably due to the fact that many species typical of the adjacent freshwater and maritime regions are able to breed in the estuary but occur there with lesser frequency than in their usual breeding areas. Furthermore, 4 species common in one region and one species common in 2 regions are rare overall: the Bald Eagle and the Least Sandpiper on Anticosti Island, and the Horned Grebe, American Coot, Piping Plover, and Least Sandpiper on the Magdalen Islands. Although these species have established populations in one region (or 2 regions in the case of the Least Sandpiper), they do not seem to be able to colonize other parts of the system and may in fact be remnant populations which formerly occupied a much larger range.

Sensitive species, exclusive to one section

Horned Grebe 46, 47, 49, 50, 51, 52
Great Egret 1, 13, 16
Snow Goose 23
Lesser Scaup 2, 5, 7, 13, 15
Barrow's Goldeneye 61, 62, 64
Willow Ptarmigan 76
Willet 44
Little Gull 6, 8, 9, 16
Roseate Tern 45, 48
Red-headed Woodpecker 2, 3, 11, 13
American Pipit 74, 77, 78, 79
Le Conte's Sparrow 19, 20, 26

Sensitive species, not exclusive to one section

Leach's Storm-Petrel 30, 36, 52, 74, 78, 79
Northern Gannet 36, 53, 59, 60
Redhead 1, 2, 7, 8, 9, 10, 13, 14, 16, 20
Greater Scaup 41, 42, 43, 44, 51, 58
Harlequin Duck 34, 36, 37, 38, 39, 52, 56, 63, 68, 70
Ruddy Duck 4, 7, 8, 16, 24
Yellow Rail 1, 20, 21, 22, 25, 28, 31, 32, 35
Common Black-headed Gull 39, 41, 43, 44, 45, 66
Bonaparte's Gull 44, 47, 76
Caspian Tern 13, 73
Thick-billed Murre 53, 78
Sedge Wren 1, 2, 3, 4, 12, 13, 16, 20
Palm Warbler 18, 27, 29, 31, 33, 34, 44, 52, 65, 72

Species that are not sensitive and not exclusive to one section, but are present in only one square in a section

Pied-billed Grebe 55
Least Bittern 20
Peregrine Falcon 54
Common Moorhen 20
American Coot 20
Piping Plover 80
Wilson's Phalarope 44
Black Tern 44
Willow Flycatcher 36
Eastern Phoebe 54
Northern Rough-winged Swallow 17
Cliff Swallow 71
Warbling Vireo 40

Species that are not sensitive but are exclusive to one section

Red-throated Loon 57, 59, 60, 66, 67, 69, 70, 73, 74, 75, 76

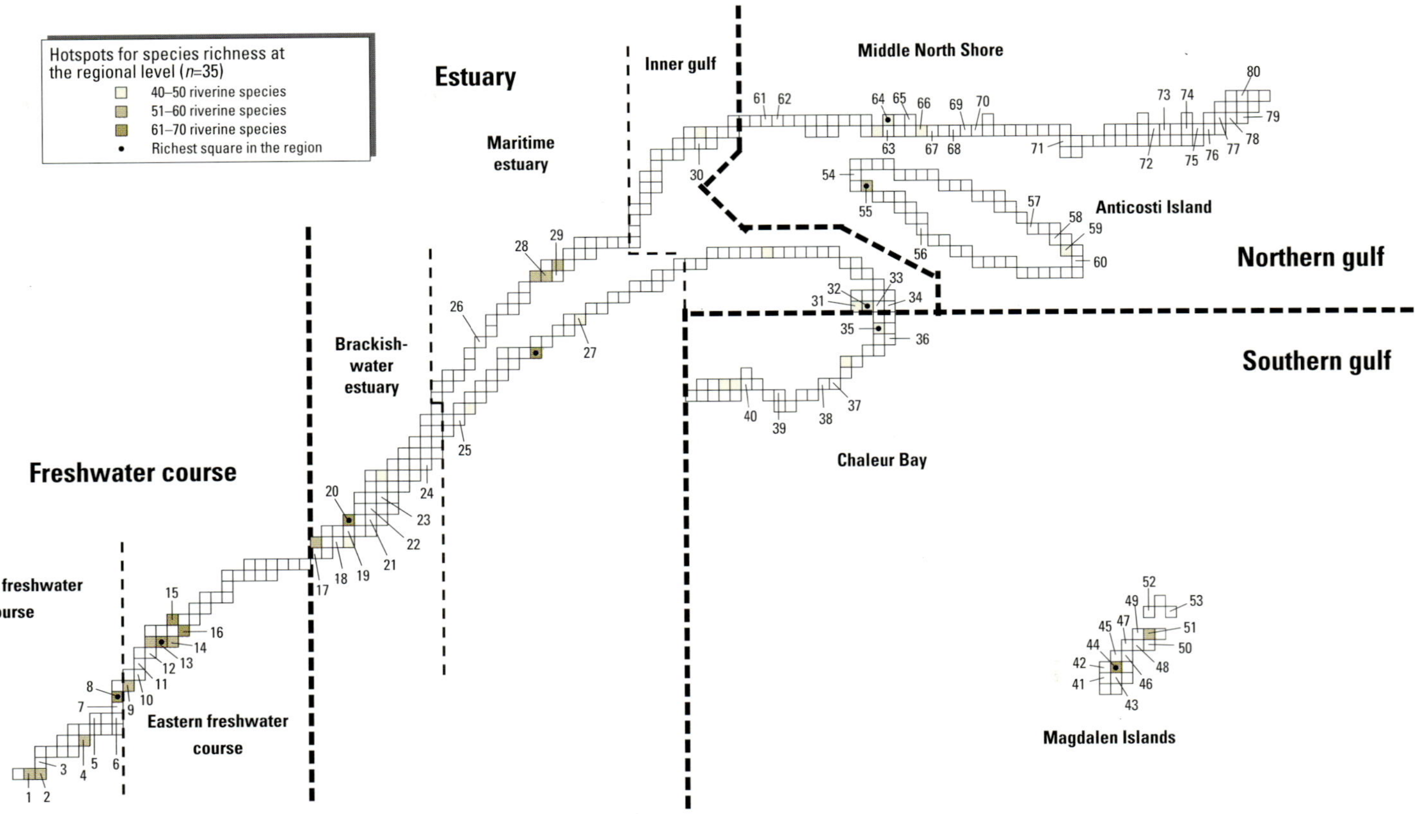

Figure 9
Ornithologically significant sites for breeding birds in the St. Lawrence system

Sensitive species

Each species was also assigned a sensitivity rating. Sensitivity is a notion akin to vulnerability that takes account of both frequency of occurrence and geographic distribution (Table 1). By definition, the 52 species that breed in at least 33 of the 201 study squares and that are present in all 4 sections of the river are not sensitive. They all have a widespread distribution in the river system and in all likelihood pose no conservation problem at the present time. Conversely, the 25 species that breed in 10 squares or fewer and have a much smaller range have the highest sensitivity rating. Although some of these species, such as colonial seabirds, are very abundant locally, the limited availability of suitable nest sites means that there is much greater risk of their extirpation from the St. Lawrence. The distribution of these 25 sensitive species is shown in Figure 9. Robert (1989b) classified 10 of these species as endangered (Piping Plover, Caspian Tern), threatened (Bald Eagle and Roseate Tern), or vulnerable (Horned Grebe, Least Bittern, Sedge Wren, Peregrine Falcon, Yellow Rail, and Red-headed Woodpecker) (Table 2). Six of these species—the Least Bittern, Bald Eagle, Peregrine Falcon, Piping Plover, Caspian Tern, and Roseate Tern—are considered to be at risk either in Canada (COSEWIC 1993) or in North America (USFWS 1993) or at both levels simultaneously (Table 2). As with frequency of occurrence, the sensitivity level of a species depends on the scale of the analysis.

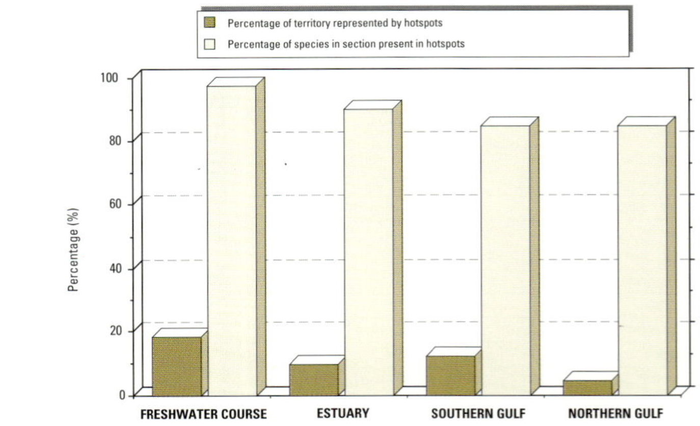

	FRESHWATER COURSE	ESTUARY	SOUTHERN GULF	NORTHERN GULF
Number of hotspots	10	14	6	5
Number of species in section that breed in hotspots	79/81	81/90	78/92	72/85

Figure 10
Relative importance of squares with high species diversity (regional hotspots) and percentage of avifauna found there, in the 4 sections of the St. Lawrence

Ornithologically significant sites and species conservation

Figure 9 shows the location of sites that have an unusually high number of species. These hotspots are defined as squares with at least the average number of breeding species found in the region, plus the standard deviation. The high species diversity in these squares makes them ideal birdwatching spots in the region and excellent sites for bird sanctuaries or conservation areas. However, the farther downstream the section in the river system, the smaller the area covered by hotspots (18% in the freshwater course compared with 5% in the northern gulf). The proportion of the avifauna found in hotspots also decreases progressively as one moves downstream (98% in the freshwater course compared with 85% in the northern gulf: Figure 10). Furthermore, the distribution of rare species, particularly the most sensitive ones, does not always correspond with the distribution of hotspots, especially in the gulf (Figure 9). Consequently, efforts to maintain species diversity in the St. Lawrence system must also take account of squares that support a more limited avifauna. Like hotspots, these squares, which make up 73% (58/79) of the sites of ornithological interest shown in Figure 9, are important for maintaining bird diversity overall since they may support species that are not found in the richest squares. Therefore, in establishing bird sanctuaries to protect the avifauna of the St. Lawrence, a balance must be struck between hotspots and other squares (Tardif and DesGranges, in prep.).

Jean-Luc DesGranges and Bernard Tardif

APPENDICES

Appendix I

Frequency of occurrence by ecological region

In each column and for each species, the first line gives the percentage of squares adequately covered (selected for analysis) in which the species was reported; the line below gives the 95% confidence interval. Only those species reported in squares with adequate coverage are shown here, so that some species, such as the Brant, may not appear, even though they were observed during atlassing.

	Region 1	Region 2	Region 3	Region 4	Region 5	Region 6	Region 7	Region 8	Region 9	Region 10	Region 11	Region 12	Region 13	Region 14	Region 15	Region 16	Region 17
Red-throated Loon* (R,A)													2.6 0.0-6.6	7.1 0.0-18.0	4.5 0.0-14.8		
Common Loon	7.2 4.4-10.0	25.5 21.9-29.1	17.8 13.2-22.4	71.4 61.0-81.8	34.5 29.6-39.4	71.8 62.2-81.5	94.7 82.8-100.0	48.9 36.1-61.8	66.7 57.5-75.8	41.3 29.0-53.6	98.1 93.5-100.0	78.0 67.0-89.0	68.4 58.2-78.6	57.1 38.0-76.3	100.0 97.7-100.0	55.6 29.6-81.5	56.2 37.5-75.0
Pied-billed Grebe (A)	25.8 21.4-30.2	18.3 15.0-21.6	9.9 6.2-13.6	10.7 3.3-18.1	11.5 8.1-14.9	15.5 7.6-23.4		19.1 8.8-29.5	18.4 10.8-26.0	10.9 2.7-19.1	5.8 0.0-12.8	13.6 4.3-22.8	3.9 0.0-8.6		4.5 0.0-14.8		
Red-necked Grebe (R,A)												3.4 0.0-8.7					6.2 0.0-16.2
Northern Gannet (R,A,C)															4.5 0.0-14.8		
Great Cormorant (R,A,C)									1.1 0.0-3.7						13.6 0.0-29.1		
Double-crested Cormorant (A,C)	8.2 5.3-11.2	5.2 3.2-7.2	12.9 8.8-17.0	3.6 0.0-8.4	5.3 2.8-7.9	1.4 0.0-4.5	5.3 0.0-17.2	8.5 0.9-16.1	24.1 15.8-32.5	30.4 18.8-42.0	1.9 0.0-6.5	6.8 0.0-13.8	13.2 5.5-20.8	17.9 2.6-33.1	18.2 1.1-35.3		
American Bittern	53.6 48.6-58.6	44.4 40.4-48.5	36.6 30.9-42.3	53.6 42.2-65.0	38.1 33.1-43.1	46.5 35.9-57.1	21.1 1.4-40.7	44.7 31.9-57.5	42.5 33.0-52.1	39.1 26.9-51.4	25.0 12.8-37.2	47.5 34.4-60.5	22.4 13.1-31.6	17.9 2.6-33.1		11.1 0.0-28.6	12.5 0.0-25.5
Least Bittern (R,A)	21.6 17.5-25.8	7.8 5.5-10.2	2.0 0.0-4.0		0.9 0.0-2.2	1.4 0.0-4.5		2.1 0.0-6.6									
Great Blue Heron (C)	60.8 56.0-65.7	60.1 56.1-64.2	55.4 49.6-61.3	85.7 77.4-94.0	50.4 45.3-55.6	64.8 54.6-75.0	100.0 97.4-100.0	53.2 40.4-66.0	47.1 37.5-56.8	37.0 24.9-49.1	73.1 60.6-85.6	66.1 53.7-78.5	31.6 21.4-41.8	42.9 23.7-62.0	18.2 1.1-35.3	22.2 0.0-44.4	62.5 44.2-80.8
Great Egret (R,A,C)	1.0 0.0-2.4	1.3 0.1-2.5															
Green Heron	58.8 53.9-63.7	34.6 30.7-38.6	30.7 25.2-36.2	10.7 3.3-18.1	13.3 9.6-16.9	9.9 3.2-16.5		10.6 2.3-19.0	2.3 0.0-5.6			1.7 0.0-5.7					
Black-crowned Night-Heron (A,C)	20.6 16.5-24.7	5.2 3.2-7.2	5.9 2.9-9.0		3.5 1.4-5.7	7.0 1.3-12.8		12.8 3.8-21.7	32.2 23.1-41.3	39.1 26.9-51.4		10.2 1.9-18.4	9.2 2.6-15.8	3.6 0.0-11.9			
Canada Goose	14.4 10.8-18.1	7.8 5.5-10.2	3.0 0.6-5.3	10.7 3.3-18.1	5.3 2.8-7.9	7.0 1.3-12.8			3.4 0.0-7.3	4.3 0.0-10.1	3.8 0.0-9.8	8.5 0.8-16.1	2.6 0.0-6.6	14.3 0.2-28.4	54.5 33.1-75.9	5.6 0.0-19.0	18.8 3.7-33.8
Wood Duck	45.4 40.4-50.3	36.6 32.6-40.6	32.7 27.1-38.2	44.6 33.3-56.0	25.7 21.1-30.2	35.2 25.0-45.4	31.6 9.5-53.6	14.9 5.4-24.3	10.3 4.2-16.5	8.7 1.2-16.2	21.2 9.5-32.8	22.0 11.0-33.0	7.9 1.7-14.1				6.2 0.0-16.2
Green-winged Teal (A)	22.7 18.4-26.9	11.1 8.4-13.8	14.9 10.5-19.2		8.8 5.7-12.0	11.3 4.3-18.3	10.5 0.0-26.0	8.5 0.9-16.1	24.1 15.8-32.5	15.2 5.9-24.5	7.7 0.0-15.6	28.8 16.9-40.8	14.5 6.6-22.4	25.0 8.0-42.0	22.7 4.4-41.1	11.1 0.0-28.6	15.6 1.5-29.8
American Black Duck	60.8 56.0-65.7	51.6 47.5-55.7	63.4 57.7-69.1	50.0 38.6-61.4	58.4 53.3-63.5	69.0 59.1-78.9	89.5 74.0-100.0	68.1 56.0-80.1	60.9 51.5-70.4	69.6 58.0-81.2	75.0 62.8-87.2	62.7 50.0-75.4	57.9 47.1-68.7	46.4 27.1-65.7	77.3 58.9-95.6	61.1 35.6-86.6	65.6 47.6-83.6
Mallard (A)	85.6 81.9-89.2	54.9 50.8-59.0	67.3 61.8-72.9	26.8 16.6-37.0	40.7 35.7-45.8	33.8 23.7-43.9	42.1 18.8-65.4	25.5 14.2-36.9	21.8 13.7-29.9	32.6 20.8-44.4	34.6 21.3-48.0	52.5 39.5-65.6	11.8 4.5-19.1	14.3 0.2-28.4	18.2 1.1-35.3	16.7 0.0-36.8	21.9 6.0-37.8
Northern Pintail (A)	41.2 36.3-46.1	16.3 13.2-19.5	9.9 6.2-13.6	3.6 0.0-8.4	9.7 6.5-13.0	5.6 0.3-10.9		6.4 0.0-13.2	11.5 5.1-17.9	21.7 11.2-32.2		8.5 0.8-16.1	7.9 1.7-14.1	10.7 0.0-23.4	22.7 4.4-41.1	5.6 0.0-19.0	
Blue-winged Teal (A)	44.3 39.4-49.3	28.8 25.0-32.5	17.8 13.2-22.4	8.9 2.0-15.8	9.7 6.5-13.0	12.7 5.4-20.0	15.8 0.0-33.7	12.8 3.8-21.7	24.1 15.8-32.5	21.7 11.2-32.2	1.9 0.0-6.5	27.1 15.4-38.9	9.2 2.6-15.8	7.1 0.0-18.0	22.7 4.4-41.1		9.4 0.0-21.0
Northern Shoveler (A)	21.6 17.5-25.8	10.5 7.8-13.1	2.0 0.0-4.0	1.8 0.0-5.5	2.7 0.7-4.6	4.2 0.0-8.9	5.3 0.0-17.2	2.1 0.0-6.6	3.4 0.0-7.3	10.9 2.7-19.1	1.9 0.0-6.5	10.2 1.9-18.4	1.3 0.0-4.3	3.6 0.0-11.9	9.1 0.0-22.4		
Gadwall (A)	28.9 24.3-33.4	9.8 7.2-12.4	4.0 1.4-6.6		2.7 0.7-4.6	2.8 0.0-6.8		4.3 0.0-10.1		4.3 0.0-10.1		3.4 0.0-8.7	3.9 0.0-8.6				
Eurasian Wigeon (R,A)	4.1 1.8-6.4	1.3 0.1-2.5						2.1 0.0-6.6		2.2 0.0-6.6							
American Wigeon (A)	35.1 30.3-39.8	11.8 9.0-14.5	5.0 2.1-7.8		3.5 1.4-5.7	2.8 0.0-6.8	5.3 0.0-17.2	2.1 0.0-6.6	3.4 0.0-7.3	6.5 0.0-13.2	5.8 0.0-12.8	20.3 9.6-31.0	3.9 0.0-8.6	3.6 0.0-11.9		5.6 0.0-19.0	
Redhead (R,A)	7.2 4.4-10.0	2.0 0.6-3.3				1.4 0.0-4.5											
Ring-necked Duck	8.2 5.3-11.2	7.2 4.9-9.5	9.9 6.2-13.6	10.7 3.3-18.1	11.5 8.1-14.9	36.6 26.3-46.9	31.6 9.5-53.6	34.0 21.8-46.3	40.2 30.7-49.7	8.7 1.2-16.2	48.1 34.1-62.1	44.1 31.1-57.1	38.2 27.5-48.8	35.7 17.1-54.3	27.3 7.9-46.7	22.2 0.0-44.4	37.5 19.2-55.8
Greater Scaup (R,A)												1.7 0.0-5.7			4.5 0.0-14.8		3.1 0.0-10.7
Lesser Scaup (R,A)	5.2 2.7-7.6	1.3 0.1-2.5								4.3 0.0-10.1	1.9 0.0-6.5	5.1 0.0-11.3					
Common Eider (A)								6.4 0.0-13.2	9.2 3.4-15.0	23.9 13.1-34.7			3.9 0.0-8.6	17.9 2.6-33.1	27.3 7.9-46.7		
Harlequin Duck (R,A)									1.1 0.0-3.7					7.1 0.0-18.0			
Surf Scoter (R,A)													1.3 0.0-4.3			5.6 0.0-19.0	3.1 0.0-10.7
Common Goldeneye	8.2 5.3-11.2	6.5 4.3-8.7	5.0 2.1-7.8	14.3 6.0-22.6	5.3 2.8-7.9	42.3 31.7-52.8	63.2 40.4-86.0	23.4 12.4-34.4	43.7 34.1-53.3	37.0 24.9-49.1	84.6 74.2-95.0	45.8 32.7-58.8	59.2 48.4-70.0	46.4 27.1-65.7	63.6 42.9-84.4	83.3 63.2-100.0	62.5 44.2-80.8
Barrow's Goldeneye (R,A)														3.6 0.0-11.9			
Bufflehead (R,A)							5.3 0.0-17.2				1.9 0.0-6.5						6.2 0.0-16.2
Hooded Merganser	12.4 8.9-15.8	19.0 15.7-22.3	10.9 7.0-14.7	39.3 28.1-50.5	11.5 8.1-14.9	45.1 34.5-55.7	68.4 46.4-90.5	12.8 3.8-21.7	20.7 12.7-28.6	4.3 0.0-10.1	40.4 26.6-54.1	11.9 3.1-20.6	10.5 3.6-17.5	3.6 0.0-11.9	4.5 0.0-14.8		6.2 0.0-16.2
Common Merganser	6.2 3.5-8.8	24.2 20.6-27.8	19.8 15.0-24.6	67.9 57.1-78.6	31.9 27.0-36.7	71.8 62.2-81.5	73.7 52.6-94.7	19.1 8.8-29.5	41.4 31.8-50.9	26.1 15.0-37.2	76.9 65.0-88.9	52.5 39.5-65.6	36.8 26.3-47.4	50.0 30.7-69.3	45.5 24.1-66.9	33.3 8.6-58.1	31.3 13.6-48.9
Red-breasted Merganser		2.6 1.1-4.2	1.0 0.0-2.6	7.1 0.8-13.5	3.5 1.4-5.7	5.6 0.3-10.9		4.3 0.0-10.1	17.2 9.8-24.7	15.2 5.9-24.5	1.9 0.0-6.5	1.7 0.0-5.7	14.5 6.6-22.4	14.3 0.2-28.4	68.2 48.0-88.4	5.6 0.0-19.0	3.1 0.0-10.7
Ruddy Duck (R,A)	4.1 1.8-6.4	0.7 0.0-1.6	1.0 0.0-2.6							2.2 0.0-6.6		1.7 0.0-5.7					
Turkey Vulture (L)	26.8 22.3-31.3	30.1 26.3-33.9	24.8 19.6-29.9	46.4 35.0-57.8	8.8 5.7-12.0	5.6 0.3-10.9											
Osprey (Ca)	5.2 2.7-7.6	20.3 16.9-23.6	22.8 17.7-27.8	33.9 23.1-44.8	26.5 22.0-31.1	45.1 34.5-55.7	42.1 18.8-65.4	38.3 25.8-50.8	43.7 34.1-53.3	41.3 29.0-53.6	57.7 43.9-71.5	39.0 26.2-51.8	51.3 40.4-62.3	50.0 30.7-69.3	36.4 15.6-57.1	72.2 48.5-95.9	34.4 16.4-52.4
Bald Eagle (A,L)			1.0 0.0-2.6	5.4 0.0-11.0		7.0 1.3-12.8		2.1 0.0-6.6		4.3 0.0-10.1	5.8 0.0-12.8	3.4 0.0-8.7	2.6 0.0-6.6		59.1 37.9-80.3		3.1 0.0-10.7
Northern Harrier (O)	92.8 90.0-95.6	74.5 70.9-78.1	86.1 81.9-90.4	62.5 51.4-73.6	76.1 71.7-80.5	38.0 27.7-48.4	10.5 0.0-26.0	74.5 63.1-85.8	47.1 37.5-56.8	65.2 53.3-77.2	23.1 11.1-35.0	55.9 42.9-68.9	19.7 10.9-28.6	32.1 14.0-50.3	4.5 0.0-14.8	5.6 0.0-19.0	12.5 0.0-25.5
Sharp-shinned Hawk	24.7 20.4-29.1	32.7 28.8-36.6	43.6 37.7-49.4	48.2 36.8-59.6	34.5 29.6-39.4	47.9 37.3-58.5	26.3 5.3-47.4	46.8 34.0-59.6	47.1 37.5-56.8	34.8 22.8-46.7	55.8 41.9-69.7	37.3 24.6-50.0	40.8 30.0-51.6	42.9 23.7-62.0	13.6 0.0-29.1	16.7 0.0-36.8	31.3 13.6-48.9
Cooper's Hawk (R)	12.4 8.9-15.8	4.6 2.7-6.5	5.0 2.1-7.8	12.5 4.6-20.4	3.5 1.4-5.7			6.4 0.0-13.2									
Northern Goshawk	14.4 10.8-18.1	15.0 12.0-18.1	20.8 15.9-25.7	17.9 8.9-26.8	12.4 8.9-15.9	18.3 9.9-26.7	31.6 9.5-53.6	27.7 16.1-39.3	16.1 8.8-23.4	10.9 2.7-19.1	11.5 2.2-20.8	8.5 0.8-16.1	13.2 5.5-20.8	17.9 2.6-33.1	4.5 0.0-14.8		3.1 0.0-10.7
Red-shouldered Hawk	45.4 40.4-50.3	30.1 26.3-33.9	33.7 28.1-39.3	25.0 15.0-35.0	20.4 16.1-24.6	28.2 18.5-37.8		6.4 0.0-13.2		6.5 0.0-13.2	3.8 0.0-9.8		1.3 0.0-4.3				
Broad-winged Hawk	51.5 46.6-56.5	65.4 61.4-69.3	79.2 74.3-84.1	89.3 81.9-96.7	82.3 78.3-86.3	80.3 71.7-88.9	78.9 59.3-98.6	87.2 78.3-96.2	67.8 58.7-76.9	76.1 65.3-86.9	69.2 56.2-82.2	45.8 32.7-58.8	44.7 33.8-55.6	35.7 17.1-54.3	4.5 0.0-14.8	16.7 0.0-36.8	21.9 6.0-37.8
Red-tailed Hawk (Ca)	63.9 59.1-68.7	43.8 39.7-47.9	37.6 31.9-43.3	42.9 31.5-54.2	35.4 30.5-40.3	56.3 45.8-66.9	52.6 29.1-76.1	48.9 36.1-61.8	27.6 18.9-36.3	10.9 2.7-19.1	46.2 32.2-60.1	23.7 12.5-35.0	42.1 31.3-52.9	67.9 49.7-86.0		100.0 97.2-100.0	59.4 40.8-78.0
Golden Eagle (R)									2.3 0.0-5.6	2.2 0.0-6.6			1.3 0.0-4.3	7.1 0.0-18.0			
American Kestrel	91.8 88.8-94.7	90.2 87.6-92.8	94.1 91.0-97.1	67.9 57.1-78.6	79.6 75.4-83.9	45.1 34.5-55.7	15.8 0.0-33.7	85.1 75.7-94.6	73.6 65.0-82.2	73.9 62.8-85.0	30.8 17.8-43.8	71.2 59.2-83.1	61.8 51.2-72.5	60.7 41.8-79.6	50.0 28.5-71.5	66.7 41.9-91.4	81.2 66.2-96.3
Merlin		0.7 0.0-1.6	5.9 2.9-9.0	12.5 4.6-20.4	8.0 5.0-10.9	18.3 9.9-26.7	31.6 9.5-53.6	25.5 14.2-36.9	37.9 28.5-47.3	45.7 33.2-58.1	25.0 12.8-37.2	42.4 29.4-55.3	36.8 26.3-47.4	35.7 17.1-54.3	77.3 58.9-95.6	44.4 18.5-70.4	9.4 0.0-21.0
Peregrine Falcon (R)	6.2 3.5-8.8	1.3 0.1-2.5	5.0 2.1-7.8		0.9 0.0-2.2	2.8 0.0-6.8		4.3 0.0-10.1	1.1 0.0-3.7	2.2 0.0-6.6			1.3 0.0-4.3		9.1 0.0-22.4		
Gray Partridge (O)	38.1 33.3-43.0	17.6 14.4-20.9	8.9 5.3-12.5		3.5 1.4-5.7												
Ringed-necked Pheasant (R,O)	8.2 5.3-11.2	0.7 0.0-1.6	2.0 0.0-4.0		0.9 0.0-2.2	1.4 0.0-4.5											
Spruce Grouse			2.0 0.0-4.0	1.8 0.0-5.5	1.8 0.1-3.4	4.2 0.0-8.9	21.1 1.4-40.7	2.1 0.0-6.6	11.5 5.1-17.9	6.5 0.0-13.2	32.7 19.5-45.9	32.2 19.9-44.5	14.5 6.6-22.4	42.9 23.7-62.0	13.6 0.0-29.1	27.8 4.1-51.5	40.6 22.0-59.2
Ruffed Grouse	74.2 69.8-78.6	78.4 75.0-81.9	86.1 81.9-90.4	85.7 77.4-94.0	88.5 85.1-91.9	90.1 83.5-96.8	68.4 46.4-90.5	74.5 63.1-85.8	80.5 72.7-88.2	67.4 55.6-79.2	78.8 67.2-90.5	81.4 71.0-91.7	73.7 64.0-83.4	75.0 58.0-92.0	63.6 42.9-84.4	38.9 13.4-64.4	56.2 37.5-75.0
Sharp-tailed Grouse (R,O)												5.1 0.0-11.3					3.1 0.0-10.7
Wild Turkey (R)	8.2 5.3-11.2		5.9 2.9-9.0														
Yellow Rail (R,A)	1.0 0.0-2.4					1.4 0.0-4.5			2.3 0.0-5.6	4.3 0.0-10.1		3.4 0.0-8.7	2.6 0.0-6.6				
Virginia Rail (A)	36.1 31.3-40.9	17.0 13.8-20.2	15.8 11.4-20.3	12.5 4.6-20.4	9.7 6.5-13.0	4.2 0.0-8.9		6.4 0.0-13.2	4.6 0.2-9.0	10.9 2.7-19.1	3.8 0.0-9.8	10.2 1.9-18.4	2.6 0.0-6.6				
Sora (A)	21.6 17.5-25.8	10.5 7.8-13.1	4.0 1.4-6.6	7.1 0.8-13.5	8.8 5.7-12.0	5.6 0.3-10.9		4.3 0.0-10.1	13.8 6.9-20.6	21.7 11.2-32.2	1.9 0.0-6.5	20.3 9.6-31.0	7.9 1.7-14.1	3.6 0.0-11.9	18.2 1.1-35.3		3.1 0.0-10.7
Common Moorhen (A)	38.1 33.3-43.0	11.1 8.4-13.8	2.0 0.0-4.0		0.9 0.0-2.2	1.4 0.0-4.5			1.1 0.0-3.7	2.2 0.0-6.6							
American Coot (R,A)	11.3 8.0-14.7	4.6 2.7-6.5	3.0 0.6-5.3			1.4 0.0-4.5				2.2 0.0-6.6	1.9 0.0-6.5	5.1 0.0-11.3					
Sandhill Crane (R,A)												1.7 0.0-5.7					
Semipalmated Plover (R)															4.5 0.0-14.8		
Killdeer (O)	100.0 99.5-100.0	96.1 94.3-97.9	100.0 99.5-100.0	89.3 81.9-96.7	98.2 96.6-99.9	66.2 56.1-76.3	42.1 18.8-65.4	93.6 86.8-100.0	80.5 72.7-88.2	93.5 86.8-100.0	23.1 11.1-35.0	66.1 53.7-78.5	34.2 23.8-44.6	35.7 17.1-54.3	9.1 0.0-22.4	11.1 0.0-28.6	43.7 25.0-62.5
Greater Yellowlegs			1.0 0.0-2.6								1.9 0.0-6.5	10.2 1.9-18.4		7.1 0.0-18.0	95.5 85.2-100.0		34.4 16.4-52.4
Solitary Sandpiper		0.7 0.0-1.6				5.6 0.3-10.9	36.8 14.0-59.6	6.4 0.0-13.2	5.7 0.9-10.6	4.3 0.0-10.1	17.3 6.5-28.1	16.9 6.9-27.0	7.9 1.7-14.1	3.6 0.0-11.9	4.5 0.0-14.8	22.2 0.0-44.4	50.0 31.1-68.9
Spotted Sandpiper	94.8 92.4-97.3	79.1 75.7-82.5	91.1 87.5-94.7	75.0 65.0-85.0	90.3 87.0-93.5	84.5 76.6-92.4	89.5 74.0-100.0	95.7 89.9-100.0	88.5 82.1-94.9	91.3 83.8-98.8	90.4 81.7-99.0	86.4 77.2-95.7	86.8 79.2-94.5	89.3 76.6-100.0	59.1 37.9-80.3	94.4 81.0-100.0	84.4 70.2-98.5
Upland Sandpiper (O)	67.0 62.3-71.7	50.3 46.2-54.4	55.4 49.6-61.3	3.6 0.0-8.4	29.2 24.5-33.9	4.2 0.0-8.9		14.9 5.4-24.3	1.1 0.0-3.7	6.5 0.0-13.2		1.7 0.0-5.7					
Least Sandpiper (R)															27.3 7.9-46.7		
Common Snipe (A,O)	77.3 73.1-81.6	85.0 81.9-88.0	95.0 92.2-97.9	55.4 44.0-66.7	80.5 76.4-84.7	47.9 37.3-58.5	47.4 23.9-70.9	89.4 81.0-97.7	66.7 57.5-75.8	60.9 48.6-73.1	48.1 34.1-62.1	64.4 51.8-77.0	40.8 30.0-51.6	39.3 20.4-58.2	59.1 37.9-80.3	38.9 13.4-64.4	40.6 22.0-59.2
American Woodcock	67.0 62.3-71.7	74.5 70.9-78.1	64.4 58.7-70.0	58.9 47.7-70.2	54.0 48.9-59.1	49.3 38.7-59.9	26.3 5.3-47.4	42.6 29.8-55.3	42.5 33.0-52.1	56.5 44.1-68.9	36.5 23.0-50.1	44.1 31.1-57.1	35.5 25.0-46.0	25.0 8.0-42.0	4.5 0.0-14.8	11.1 0.0-28.6	9.4 0.0-21.0
Wilson's Phalarope (R,A)	6.2 3.5-8.8	1.3 0.1-2.5			0.9 0.0-2.2	1.4 0.0-4.5		2.1 0.0-6.6		2.2 0.0-6.6			1.3 0.0-4.3				
Little Gull (R,A)	3.1 1.0-5.2	0.7 0.0-1.6															
Common Black-headed Gull (R,A)														3.6 0.0-11.9			
Bonaparte's Gull (R)								2.1 0.0-6.6	1.1 0.0-3.7	8.7 1.2-16.2		3.4 0.0-8.7					6.2 0.0-16.2
Ring-billed Gull (A,C)	29.9 25.3-34.5	39.2 35.2-43.2	34.7 29.0-40.3	28.6 18.2-39.0	24.8 20.3-29.3	18.3 9.9-26.7	26.3 5.3-47.4	10.6 2.3-19.0	20.7 12.7-28.6	30.4 18.8-42.0	26.9 14.4-39.4	13.6 4.3-22.8	13.2 5.5-20.8	14.3 0.2-28.4	31.8 11.6-52.0		3.1 0.0-10.7

	Region 1	Region 2	Region 3	Region 4	Region 5	Region 6	Region 7	Region 8	Region 9	Region 10	Region 11	Region 12	Region 13	Region 14	Region 15	Region 16	Region 17
Herring Gull **(A,C)**	11.3 8.0-14.7	21.6 18.1-25.0	21.8 16.8-26.7	33.9 23.1-44.8	12.4 8.9-15.9	36.6 26.3-46.9	73.7 52.6-94.7	17.0 7.1-26.9	40.2 30.7-49.7	41.3 29.0-53.6	61.5 47.9-75.2	59.3 46.4-72.2	38.2 27.5-48.8	32.1 14.0-50.3	54.5 33.1-75.9	83.3 63.2-100.0	46.9 28.0-65.7
Great Black-backed Gull **(A,C)**	8.2 5.3-11.2	5.2 3.2-7.2	6.9 3.7-10.2		0.9 0.0-2.2	4.2 0.0-8.9		10.6 2.3-19.0	24.1 15.8-32.5	23.9 13.1-34.7			10.5 3.6-17.5	14.3 0.2-28.4	72.7 53.3-92.1		
Black-legged Kittiwake **(A,C)**								2.1 0.0-6.6	5.7 0.9-10.6				2.6 0.0-6.6	10.7 0.0-23.4	36.4 15.6-57.1		
Caspian Tern **(R,A,C)**	1.0 0.0-2.4	0.7 0.0-1.6															
Common Tern **(A,C)**	28.9 24.3-33.4	4.6 2.7-6.5	4.0 1.4-6.6	3.6 0.0-8.4	0.9 0.0-2.2	4.2 0.0-8.9	15.8 0.0-33.7	2.1 0.0-6.6	11.5 5.1-17.9	17.4 7.7-27.1	13.5 3.6-23.3	28.8 16.9-40.8	5.3 0.0-10.5	10.7 0.0-23.4	68.2 48.0-88.4		18.8 3.7-33.8
Arctic Tern **(R,A,C)**									1.1 0.0-3.7					10.7 0.0-23.4	4.5 0.0-14.8		
White-winged Tern **(R,A,C)**										2.2 0.0-6.6							
Black Tern **(A,C)**	30.9 26.3-35.6	8.5 6.1-10.9	3.0 0.6-5.3		1.8 0.1-3.4			2.1 0.0-6.6	1.1 0.0-3.7	2.2 0.0-6.6	3.8 0.0-9.8	8.5 0.8-16.1					
Common Murre **(R,A,C)**									1.1 0.0-3.7					3.6 0.0-11.9	9.1 0.0-22.4		
Razorbill **(R,A,C)**									2.3 0.0-5.6					3.6 0.0-11.9	4.5 0.0-14.8		
Black Guillemot **(A,C)**								4.3 0.0-10.1	12.6 6.0-19.3	8.7 1.2-16.2			2.6 0.0-6.6	10.7 0.0-23.4	31.8 11.6-52.0		
Atlantic Puffin **(R,A,C)**														3.6 0.0-11.9	4.5 0.0-14.8		
Rock Dove **(O)**	95.9 93.6-98.2	86.9 84.0-89.8	93.1 89.8-96.3	53.6 42.2-65.0	71.7 67.0-76.4	29.6 19.8-39.3	26.3 5.3-47.4	44.7 31.9-57.5	25.3 16.8-33.8	43.5 31.1-55.9		20.3 9.6-31.0	14.5 6.6-22.4	3.6 0.0-11.9			3.1 0.0-10.7
Mourning Dove **(O)**	100.0 99.5-100.0	90.2 87.6-92.8	97.0 94.7-99.4	57.1 45.8-68.5	77.0 72.6-81.4	38.0 27.7-48.4	15.8 0.0-33.7	66.0 53.7-78.2	31.0 22.0-40.0	56.5 44.1-68.9	5.8 0.0-12.8	23.7 12.5-35.0	17.1 8.7-25.5	14.3 0.2-28.4	18.2 1.1-35.3		
Black-billed Cuckoo	53.6 48.6-58.6	41.2 37.1-45.2	33.7 28.1-39.3	21.4 11.9-31.0	21.2 17.0-25.5	11.3 4.3-18.3	63.2 40.4-86.0	10.6 2.3-19.0	2.3 0.0-5.6	6.5 0.0-13.2	9.6 1.0-18.3	16.9 6.9-27.0	5.3 0.0-10.5		4.5 0.0-14.8		
Yellow-billed Cuckoo **(R)**	4.1 1.8-6.4	3.9 2.1-5.7	1.0 0.0-2.6	1.8 0.0-5.5													
Barn Owl **(R)**			1.0 0.0-2.6														
Eastern Screech-Owl **(R)**	25.8 21.4-30.2	4.6 2.7-6.5	2.0 0.0-4.0		0.9 0.0-2.2												
Great Horned Owl	54.6 49.7-59.6	40.5 36.5-44.6	46.5 40.7-52.4	28.6 18.2-39.0	32.7 27.9-37.6	22.5 13.5-31.5	15.8 0.0-33.7	25.5 14.2-36.9	17.2 9.8-24.7	23.9 13.1-34.7	34.6 21.3-48.0	25.4 13.9-36.9	17.1 8.7-25.5	10.7 0.0-23.4	40.9 19.7-62.1	22.2 0.0-44.4	9.4 0.0-21.0
Northern Hawk Owl **(R)**									1.1 0.0-3.7	2.2 0.0-6.6			2.6 0.0-6.6	3.6 0.0-11.9			3.1 0.0-10.7
Barred Owl	25.8 21.4-30.2	28.8 25.0-32.5	35.6 30.0-41.3	37.5 26.4-48.6	31.9 27.0-36.7	33.8 23.7-43.9	36.8 14.0-59.6	17.0 7.1-26.9	21.8 13.7-29.9	13.0 4.3-21.8	23.1 11.1-35.0	6.8 0.0-13.8	9.2 2.6-15.8	3.6 0.0-11.9		5.6 0.0-19.0	
Great Gray Owl **(R)**												1.7 0.0-5.7					3.1 0.0-10.7
Long-eared Owl **(R)**	5.2 2.7-7.6	5.9 3.8-8.0	3.0 0.6-5.3	3.6 0.0-8.4	2.7 0.7-4.6	4.2 0.0-8.9		2.1 0.0-6.6	5.7 0.9-10.6			5.1 0.0-11.3	2.6 0.0-6.6				
Short-eared Owl **(O,L)**	15.5 11.7-19.2	4.6 2.7-6.5	5.0 2.1-7.8	1.8 0.0-5.5	2.7 0.7-4.6	4.2 0.0-8.9		4.3 0.0-10.1	3.4 0.0-7.3	23.9 13.1-34.7		6.8 0.0-13.8	5.3 0.0-10.5	14.3 0.2-28.4			3.1 0.0-10.7
Boreal Owl **(R)**		0.7 0.0-1.6							2.3 0.0-5.6				6.6 0.8-12.3	10.7 0.0-23.4			
Northern Saw-whet Owl	13.4 9.9-17.0	17.6 14.4-20.9	29.7 24.3-35.1	21.4 11.9-31.0	18.6 14.5-22.7	16.9 8.8-25.1		10.6 2.3-19.0	13.8 6.9-20.6	17.4 7.7-27.1	23.1 11.1-35.0	3.4 0.0-8.7	1.3 0.0-4.3				
Common Nighthawk	28.9 24.3-33.4	22.2 18.7-25.7	20.8 15.9-25.7	26.8 16.6-37.0	16.8 12.9-20.8	26.8 17.3-36.3	26.3 5.3-47.4	23.4 12.4-34.4	24.1 15.8-32.5	47.8 35.3-60.3	42.3 28.5-56.1	61.0 48.2-73.8	46.1 35.1-57.0	25.0 8.0-42.0	4.5 0.0-14.8	55.6 29.6-81.5	37.5 19.2-55.8
Whip-poor-will	19.6 15.5-23.6	29.4 25.6-33.2	21.8 16.8-26.7	35.7 24.7-46.7	10.6 7.3-14.0	22.5 13.5-31.5	10.5 0.0-26.0			4.3 0.0-10.1	11.5 2.2-20.8	3.4 0.0-8.7					
Chimney Swift	78.4 74.2-82.5	58.2 54.1-62.2	63.4 57.7-69.1	76.8 67.0-86.6	75.2 70.7-79.7	90.1 83.5-96.8	68.4 46.4-90.5	68.1 56.0-80.1	47.1 37.5-56.8	26.1 15.0-37.2	44.2 30.3-58.1	1.7 0.0-5.7	26.3 16.6-36.0	10.7 0.0-23.4			
Ruby-throated Hummingbird	78.4 74.2-82.5	79.1 75.7-82.5	87.1 83.0-91.2	83.9 75.3-92.6	86.7 83.1-90.4	80.3 71.7-88.9	94.7 82.8-100.0	91.5 83.9-99.1	80.5 72.7-88.2	56.5 44.1-68.9	63.5 49.9-77.0	37.3 24.6-50.0	53.9 43.0-64.9	60.7 41.8-79.6	13.6 0.0-29.1	5.6 0.0-19.0	3.1 0.0-10.7
Belted Kingfisher	87.6 84.2-91.1	88.9 86.2-91.6	93.1 89.8-96.3	98.2 94.5-100.0	78.8 74.5-83.0	94.4 89.1-99.7	84.2 66.3-100.0	80.9 70.5-91.2	93.1 87.9-98.3	82.6 72.9-92.3	86.5 76.7-96.4	83.1 73.0-93.1	75.0 65.4-84.6	78.6 62.4-94.8	50.0 28.5-71.5	83.3 63.2-100.0	62.5 44.2-80.8
Red-headed Woodpecker **(R)**	10.3 7.1-13.5	6.5 4.3-8.7	3.0 0.6-5.3		0.9 0.0-2.2												
Red-bellied Woodpecker **(R)**	1.0 0.0-2.4																
Yellow-bellied Sapsucker	52.6 47.6-57.5	80.4 77.1-83.7	90.1 86.4-93.8	92.9 86.5-99.2	92.0 89.1-95.0	90.1 83.5-96.8	94.7 82.8-100.0	85.1 75.7-94.6	80.5 72.7-88.2	43.5 31.1-55.9	90.4 81.7-99.0	49.2 36.1-62.2	59.2 48.4-70.0	21.4 5.2-37.6	77.3 58.9-95.6	44.4 18.5-70.4	37.5 19.2-55.8
Downy Woodpecker	92.8 90.0-95.6	81.7 78.4-85.0	89.1 85.3-93.0	89.3 81.9-96.7	88.5 85.1-91.9	80.3 71.7-88.9	94.7 82.8-100.0	97.9 93.4-100.0	82.8 75.3-90.2	67.4 55.6-79.2	82.7 71.9-93.5	57.6 44.7-70.6	71.1 61.1-81.0	28.6 10.9-46.2	86.4 70.9-100.0	38.9 13.4-64.4	25.0 8.4-41.6
Hairy Woodpecker	81.4 77.5-85.4	82.4 79.1-85.6	85.1 80.8-89.5	85.7 77.4-94.0	93.8 91.1-96.5	76.1 66.9-85.2	100.0 97.4-100.0	85.1 75.7-94.6	78.2 70.1-86.3	65.2 53.3-77.2	82.7 71.9-93.5	52.5 39.5-65.6	65.8 55.4-76.2	42.9 23.7-62.0	72.7 53.3-92.1	22.2 0.0-44.4	37.5 19.2-55.8
Three-toed Woodpecker		0.7 0.0-1.6	1.0 0.0-2.6			5.6 0.3-10.9		6.4 0.0-13.2	2.3 0.0-5.6			1.7 0.0-5.7	9.2 2.6-15.8	21.4 5.2-37.6	13.6 0.0-29.1	22.2 0.0-44.4	21.9 6.0-37.8
Black-backed Woodpecker		0.7 0.0-1.6	5.9 2.9-9.0	7.1 0.8-13.5	19.5 15.3-23.6	32.4 22.4-42.4	42.1 18.8-65.4	36.2 23.8-48.6	46.0 36.3-55.6	23.9 13.1-34.7	42.3 28.5-56.1	15.3 5.6-24.9	40.8 30.0-51.6	64.3 45.7-82.9	36.4 15.6-57.1	44.4 18.5-70.4	46.9 28.0-65.7
Northern Flicker	100.0 99.5-100.0	100.0 99.7-100.0	99.0 97.4-100.0	96.4 91.6-100.0	100.0 99.6-100.0	100.0 99.3-100.0	100.0 97.4-100.0	100.0 98.9-100.0	97.7 94.4-100.0	100.0 98.9-100.0	100.0 99.0-100.0	98.3 94.3-100.0	94.7 89.5-100.0	92.9 82.0-100.0	95.5 85.2-100.0	100.0 97.2-100.0	100.0 98.4-100.0
Pileated Woodpecker	48.5 43.5-53.4	47.7 43.6-51.8	46.5 40.7-52.4	67.9 57.1-78.6	46.9 41.8-52.0	60.6 50.1-71.0	57.9 34.6-81.2	23.4 12.4-34.4	24.1 15.8-32.5	10.9 2.7-19.1	67.3 54.1-80.5	30.5 18.4-42.6	9.2 2.6-15.8	3.6 0.0-11.9		5.6 0.0-19.0	
Olive-sided Flycatcher	5.2 2.7-7.6	19.6 16.3-22.9	35.6 30.0-41.3	58.9 47.7-70.2	49.6 44.4-54.7	81.7 73.3-90.1	73.7 52.6-94.7	57.4 44.7-70.2	74.7 66.2-83.2	41.3 29.0-53.6	76.9 65.0-88.9	32.2 19.9-44.5	71.1 61.1-81.0	82.1 66.9-97.4	45.5 24.1-66.9	66.7 41.9-91.4	50.0 31.1-68.9
Eastern Wood-Pewee	100.0 99.5-100.0	96.1 94.3-97.9	95.0 92.2-97.9	87.5 79.6-95.4	92.0 89.1-95.0	73.2 63.7-82.7	52.6 29.1-76.1	93.6 86.8-100.0	70.1 61.2-79.0	52.2 39.7-64.7	34.6 21.3-48.0	11.9 3.1-20.6	34.2 23.8-44.6	14.3 0.2-28.4			
Yellow-bellied Flycatcher	4.1 1.8-6.4	4.6 2.7-6.5	22.8 17.7-27.8	19.6 10.4-28.9	33.6 28.8-38.5	40.8 30.4-51.3	73.7 52.6-94.7	76.6 65.6-87.6	73.6 65.0-82.2	47.8 35.3-60.3	50.0 36.0-64.0	37.3 24.6-50.0	60.5 49.8-71.2	60.7 41.8-79.6	63.6 42.9-84.4	77.8 55.6-100.0	71.9 54.7-89.0
Alder Flycatcher	88.7 85.3-92.0	90.2 87.6-92.8	95.0 92.2-97.9	82.1 73.2-91.1	97.3 95.4-99.3	81.7 73.3-90.1	78.9 59.3-98.6	95.7 89.9-100.0	92.0 86.4-97.5	95.7 89.9-100.0	82.7 71.9-93.5	89.8 81.6-98.1	89.5 82.5-96.4	89.3 76.6-100.0	22.7 4.4-41.1	94.4 81.0-100.0	96.9 89.3-100.0
Willow Flycatcher	52.6 47.6-57.5	10.5 7.8-13.1	14.9 10.5-19.2	5.4 0.0-11.0	5.3 2.8-7.9	5.6 0.3-10.9		4.3 0.0-10.1	1.1 0.0-3.7								
Least Flycatcher	95.9 93.6-98.2	97.4 95.8-98.9	95.0 92.2-97.9	92.9 86.5-99.2	96.5 94.3-98.6	90.1 83.5-96.8	100.0 97.4-100.0	95.7 89.9-100.0	98.9 96.3-100.0	100.0 98.9-100.0	94.2 87.2-100.0	84.7 75.1-94.4	90.8 84.2-97.4	75.0 58.0-92.0	9.1 0.0-22.4	88.9 71.4-100.0	53.1 34.3-72.0
Eastern Phoebe	74.2 69.8-78.6	90.8 88.3-93.4	87.1 83.0-91.2	94.6 89.0-100.0	81.4 77.3-85.5	40.8 30.4-51.3	10.5 0.0-26.0	34.0 21.8-46.3	10.3 4.2-16.5	6.5 0.0-13.2	25.0 12.8-37.2	11.9 3.1-20.6	7.9 1.7-14.1	10.7 0.0-23.4	4.5 0.0-14.8		
Great Crested Flycatcher	96.9 94.8-99.0	96.1 94.3-97.9	91.1 87.5-94.7	85.7 77.4-94.0	72.6 67.9-77.2	57.7 47.2-68.3	52.6 29.1-76.1	40.4 27.8-53.0	9.2 3.4-15.0	8.7 1.2-16.2	5.8 0.0-12.8		2.6 0.0-6.6				
Eastern Kingbird **(O)**	100.0 99.5-100.0	99.3 98.4-100.0	100.0 99.5-100.0	98.2 94.5-100.0	100.0 99.6-100.0	87.3 80.0-94.6	89.5 74.0-100.0	89.4 81.0-97.7	86.2 79.4-93.1	80.4 70.3-90.6	69.2 56.2-82.2	62.7 50.0-75.4	50.0 39.1-60.9	25.0 8.0-42.0	18.2 1.1-35.3	27.8 4.1-51.5	28.1 11.0-45.3
Horned Lark **(O)**	79.4 75.3-83.5	60.1 56.1-64.2	60.4 54.6-66.2	8.9 2.0-15.8	50.4 45.3-55.6	19.7 11.1-28.3		27.7 16.1-39.3	13.8 6.9-20.6	8.7 1.2-16.2	3.8 0.0-9.8	5.1 0.0-11.3	9.2 2.6-15.8	7.1 0.0-18.0	9.1 0.0-22.4		
Purple Martin **(O)**	70.1 65.5-74.7	32.0 28.2-35.9	21.8 16.8-26.7	8.9 2.0-15.8	7.1 4.2-9.9	5.6 0.3-10.9			1.1 0.0-3.7								
Tree Swallow	100.0 99.5-100.0	99.3 98.4-100.0	100.0 99.5-100.0	98.2 94.5-100.0	100.0 99.6-100.0	97.2 93.2-100.0	94.7 82.8-100.0	100.0 98.9-100.0	98.9 96.3-100.0	100.0 98.9-100.0	96.2 90.2-100.0	96.6 91.3-100.0	98.7 95.7-100.0	100.0 98.2-100.0	86.4 70.9-100.0	100.0 97.2-100.0	90.6 79.0-100.0
Northern Rough-winged Swallow **(A,O)**	50.5 45.5-55.5	45.1 41.0-49.2	26.7 21.5-32.0	25.0 15.0-35.0	14.2 10.4-17.9	2.8 0.0-6.8		6.4 0.0-13.2									
Bank Swallow **(O,A,C)**	76.3 72.0-80.6	87.6 84.8-90.4	86.1 81.9-90.4	75.0 65.0-85.0	70.8 66.1-75.5	62.0 51.6-72.3	31.6 9.5-53.6	70.2 58.4-82.0	82.8 75.3-90.2	89.1 80.9-97.3	21.2 9.5-32.8	50.8 37.8-63.9	63.2 52.6-73.7	60.7 41.8-79.6	4.5 0.0-14.8	61.1 35.6-86.6	53.1 34.3-72.0
Cliff Swallow **(O,A,C)**	75.3 70.9-79.6	72.5 68.8-76.3	91.1 87.5-94.7	66.1 55.2-76.9	88.5 85.1-91.9	49.3 38.7-59.9	36.8 14.0-59.6	72.3 60.7-83.9	47.1 37.5-56.8	30.4 18.8-42.0	13.5 3.6-23.3	37.3 24.6-50.0	13.2 5.5-20.8	7.1 0.0-18.0			15.6 1.5-29.8
Barn Swallow **(O,A)**	100.0 99.5-100.0	99.3 98.4-100.0	100.0 99.5-100.0	96.4 91.6-100.0	99.1 97.8-100.0	95.8 91.1-100.0	78.9 59.3-98.6	97.9 93.4-100.0	98.9 96.3-100.0	97.8 93.4-100.0	76.9 65.0-88.9	84.7 75.1-94.4	88.2 80.9-95.5	60.7 41.8-79.6	13.6 0.0-29.1	83.3 63.2-100.0	46.9 28.0-65.7
Gray Jay		0.7 0.0-1.6	4.0 1.4-6.6	10.7 3.3-18.1	8.0 5.0-10.9	29.6 19.8-39.3	78.9 59.3-98.6	25.5 14.2-36.9	34.5 25.3-43.7	8.7 1.2-16.2	75.0 62.8-87.2	64.4 51.8-77.0	52.6 41.7-63.6	82.1 66.9-97.4	100.0 97.7-100.0	72.2 48.5-95.9	84.4 70.2-98.5
Blue Jay	97.9 96.2-99.7	98.7 97.5-99.9	100.0 99.5-100.0	100.0 99.1-100.0	99.1 97.8-100.0	100.0 99.3-100.0	100.0 97.4-100.0	95.7 89.9-100.0	98.9 96.3-100.0	67.4 55.6-79.2	92.3 84.4-100.0	47.5 34.4-60.5	52.6 41.7-63.6	17.9 2.6-33.1	9.1 0.0-22.4		3.1 0.0-10.7
American Crow	100.0 99.5-100.0	98.7 97.5-99.9	100.0 99.5-100.0	91.1 84.2-98.0	99.1 97.8-100.0	85.9 78.3-93.5	42.1 18.8-65.4	97.9 93.4-100.0	97.7 94.4-100.0	100.0 98.9-100.0	59.6 45.9-73.4	79.7 69.0-90.4	75.0 65.4-84.6	89.3 76.6-100.0	63.6 42.9-84.4	88.9 71.4-100.0	56.2 37.5-75.0
Common Raven	11.3 8.0-14.7	33.3 29.4-37.2	49.5 43.6-55.4	85.7 77.4-94.0	56.6 51.5-61.7	87.3 80.0-94.6	94.7 82.8-100.0	89.4 81.0-97.7	95.4 91.0-99.8	82.6 72.9-92.3	88.5 79.2-97.8	93.2 86.2-100.0	78.9 69.9-88.0	92.9 82.0-100.0	90.9 77.6-100.0	94.4 81.0-100.0	96.9 89.3-100.0
Black-capped Chickadee	100.0 99.5-100.0	98.7 97.5-99.9	100.0 99.5-100.0	100.0 99.1-100.0	98.2 96.6-99.9	98.6 95.5-100.0	100.0 97.4-100.0	97.9 93.4-100.0	98.9 96.3-100.0	95.7 89.9-100.0	100.0 99.0-100.0	93.2 86.2-100.0	92.1 85.9-98.3	67.9 49.7-86.0	40.9 19.7-62.1	27.8 4.1-51.5	50.0 31.1-68.9
Boreal Chickadee		1.3 0.1-2.5	28.7 23.3-34.1	8.9 2.0-15.8	35.4 30.5-40.3	28.2 18.5-37.8	42.1 18.8-65.4	80.9 70.5-91.2	73.6 65.0-82.2	47.8 35.3-60.3	82.7 71.9-93.5	55.9 42.9-68.9	80.3 71.4-89.1	92.9 82.0-100.0	100.0 97.7-100.0	88.9 71.4-100.0	65.6 47.6-83.6
Tufted Titmouse **(R)**	1.0 0.0-2.4	0.7 0.0-1.6															
Red-breasted Nuthatch	47.4 42.5-52.4	76.5 72.9-80.0	88.1 84.1-92.1	100.0 99.1-100.0	85.8 82.1-89.6	93.0 87.2-98.7	100.0 97.4-100.0	91.5 83.9-99.1	83.9 76.6-91.2	69.6 58.0-81.2	92.3 84.4-100.0	62.7 50.0-75.4	65.8 55.4-76.2	57.1 38.0-76.3	81.8 64.7-98.9	55.6 29.6-81.5	62.5 44.2-80.8
White-breasted Nuthatch	86.6 83.0-90.1	75.8 72.2-79.4	66.3 60.7-71.9	73.2 63.0-83.4	60.2 55.1-65.2	59.2 48.7-69.6	36.8 14.0-59.6	21.3 10.6-32.0	4.6 0.2-9.0		19.2 8.0-30.5	5.1 0.0-11.3	7.9 1.7-14.1	3.6 0.0-11.9			
Brown Creeper	56.7 51.8-61.6	42.5 38.4-46.6	52.5 46.6-58.4	46.4 35.0-57.8	54.9 49.8-60.0	36.6 26.3-46.9	57.9 34.6-81.2	31.9 19.9-44.0	14.9 7.9-22.0	15.2 5.9-24.5	44.2 30.3-58.1	15.3 5.6-24.9	9.2 2.6-15.8	21.4 5.2-37.6	40.9 19.7-62.1	5.6 0.0-19.0	31.3 13.6-48.9
Carolina Wren **(R)**	4.1 1.8-6.4																
House Wren	84.5 80.8-88.3	71.2 67.5-75.0	57.4 51.6-63.3	32.1 21.4-42.9	29.2 24.5-33.9	15.5 7.6-23.4	5.3 0.0-17.2	4.3 0.0-10.1	2.3 0.0-5.6	8.7 1.2-16.2	3.8 0.0-9.8	5.1 0.0-11.3					
Winter Wren	49.5 44.5-54.5	60.1 56.1-64.2	84.2 79.7-88.6	85.7 77.4-94.0	89.4 86.0-92.7	94.4 89.1-99.7	94.7 82.8-100.0	97.9 93.4-100.0	95.4 91.0-99.8	76.1 65.3-86.9	100.0 99.0-100.0	93.2 86.2-100.0	97.4 93.4-100.0	89.3 76.6-100.0	95.5 85.2-100.0	100.0 97.2-100.0	100.0 98.4-100.0
Sedge Wren **(R,A)**	10.3 7.1-13.5	3.9 2.1-5.7	2.0 0.0-4.0	3.6 0.0-8.4		2.8 0.0-6.8											
Marsh Wren **(A)**	32.0 27.3-36.6	13.7 10.8-16.7	8.9 5.3-12.5		4.4 2.1-6.8	1.4 0.0-4.5				2.2 0.0-6.6							
Golden-crowned Kinglet	21.6 17.5-25.8	44.4 40.4-48.5	69.3 63.8-74.8	80.4 71.1-89.6	73.5 68.9-78.0	70.4 60.7-80.2	94.7 82.8-100.0	80.9 70.5-91.2	50.6 40.9-60.2	58.7 46.4-71.0	80.8 69.5-92.0	67.8 55.5-80.1	44.7 33.8-55.6	46.4 27.1-65.7	86.4 70.9-100.0	55.6 29.6-81.5	87.5 74.5-100.0
Ruby-crowned Kinglet	14.4 10.8-18.1	40.5 36.5-44.6	63.4 57.7-69.1	78.6 69.0-88.1	86.7 83.1-90.4	90.1 83.5-96.8	100.0 97.4-100.0	93.6 86.8-100.0	100.0 99.4-100.0	93.5 86.8-100.0	98.1 93.5-100.0	94.9 88.7-100.0	97.4 93.4-100.0	100.0 98.2-100.0	95.5 85.2-100.0	100.0 97.2-100.0	100.0 98.4-100.0
Blue-gray Gnatcatcher **(R)**	7.2 4.4-10.0	2.0 0.6-3.3			0.9 0.0-2.2	1.4 0.0-4.5											
Eastern Bluebird **(O)**	42.3 37.4-47.2	53.6 49.5-57.7	51.5 45.6-57.4	73.2 63.0-83.4	33.6 28.8-38.5	32.4 22.4-42.4	5.3 0.0-17.2	23.4 12.4-34.4	9.2 3.4-15.0	47.8 35.3-60.3	1.9 0.0-6.5	11.9 3.1-20.6	6.6 0.8-12.3	3.6 0.0-11.9			9.4 0.0-21.0
Veery	97.9 96.2-99.7	100.0 99.7-100.0	100.0 99.5-100.0	100.0 99.1-100.0	98.2 96.6-99.9	94.4 89.1-99.7	100.0 97.4-100.0	100.0 98.9-100.0	89.7 83.5-95.8	97.8 93.4-100.0	78.8 67.2-90.5	61.0 48.2-73.8	59.2 48.4-70.0	60.7 41.8-79.6	4.5 0.0-14.8	27.8 4.1-51.5	9.4 0.0-21.0
Bicknell's Thrush **(L)**					0.9 0.0-2.2	1.4 0.0-4.5		2.1 0.0-6.6	10.3 4.2-16.5		1.9 0.0-6.5		23.7 14.3-33.1	28.6 10.9-46.2	4.5 0.0-14.8	5.6 0.0-19.0	
Swainson's Thrush	15.5 11.7-19.2	30.1 26.3-33.9	57.4 51.6-63.3	67.9 57.1-78.6	84.1 80.2-87.9	90.1 83.5-96.8	89.5 74.0-100.0	100.0 98.9-100.0	97.7 94.4-100.0	100.0 98.9-100.0	98.1 93.5-100.0	100.0 99.2-100.0	98.7 95.7-100.0	100.0 98.2-100.0	100.0 97.7-100.0	100.0 97.2-100.0	93.7 83.8-100.0
Hermit Thrush	52.6 47.6-57.5	85.0 81.9-88.0	92.1 88.7-95.5	92.9 86.5-99.2	95.6 93.2-97.9	95.8 91.1-100.0	94.7 82.8-100.0	100.0 98.9-100.0	83.9 76.6-91.2	93.5 86.8-100.0	96.2 90.2-100.0	94.9 88.7-100.0	86.8 79.2-94.5	82.1 66.9-97.4	18.2 1.1-35.3	94.4 81.0-100.0	100.0 98.4-100.0
Wood Thrush	85.6 81.9-89.2	86.3 83.3-89.2	80.2 75.4-85.0	82.1 73.2-91.1	77.0 72.6-81.4	53.5 42.9-64.1	15.8 0.0-33.7	76.6 65.6-87.6	18.4 10.8-26.0	10.9 2.7-19.1	1.9 0.0-6.5	3.4 0.0-8.7	5.3 0.0-10.5				
American Robin	100.0 99.5-100.0	99.3 98.4-100.0	100.0 99.5-100.0	98.2 94.5-100.0	100.0 99.6-100.0	100.0 99.3-100.0	100.0 97.4-100.0	100.0 98.9-100.0	100.0 99.4-100.0	100.0 98.9-100.0	100.0 99.0-100.0	100.0 99.2-100.0	98.7 95.7-100.0	100.0 98.2-100.0	100.0 97.7-100.0	100.0 97.2-100.0	96.9 89.3-100.0
Gray Catbird	95.9 93.6-98.2	95.4 93.5-97.3	100.0 99.5-100.0	83.9 75.3-92.6	91.2 88.0-94.3	64.8 54.6-75.0	26.3 5.3-47.4	89.4 81.0-97.7	49.4 39.8-59.1	32.6 20.8-44.4	13.5 3.6-23.3	10.2 1.9-18.4	13.2 5.5-20.8	3.6 0.0-11.9	9.1 0.0-22.4		
Northern Mockingbird **(Ca)**	34.0 29.3-38.8	19.0 15.7-22.3	30.7 25.2-36.2	7.1 0.8-13.5	16.8 12.9-20.8	8.5 2.2-14.7		4.3 0.0-10.1	3.4 0.0-7.3	26.1 15.0-37.2		6.8 0.0-13.8	6.6 0.8-12.3	7.1 0.0-18.0	4.5 0.0-14.8		
Brown Thrasher	68.0 63.4-72.7	78.4 75.0-81.9	70.3 64.9-75.7	62.5 51.4-73.6	60.2 55.1-65.2	25.4 16.0-34.7	5.3 0.0-17.2	36.2 23.8-48.6	3.4 0.0-7.3	2.2 0.0-6.6	5.8 0.0-12.8	8.5 0.8-16.1	2.6 0.0-6.6				
American Pipit **(R)**														7.1 0.0-18.0			
Bohemian Waxwing **(R)**												3.4 0.0-8.7					

	Region 1	Region 2	Region 3	Region 4	Region 5	Region 6	Region 7	Region 8	Region 9	Region 10	Region 11	Region 12	Region 13	Region 14	Region 15	Region 16	Region 17
Cedar Waxwing	100.0 99.5-100.0	100.0 99.7-100.0	100.0 99.5-100.0	98.2 94.5-100.0	97.3 95.4-99.3	100.0 99.3-100.0	100.0 97.4-100.0	100.0 98.9-100.0	97.7 94.4-100.0	97.8 93.4-100.0	98.1 93.5-100.0	96.6 91.3-100.0	100.0 99.3-100.0	85.7 71.6-99.8	18.2 1.1-35.3	94.4 81.0-100.0	96.9 89.3-100.0
Loggerhead Shrike (R)	10.3 7.1-13.5	4.6 2.7-6.5	5.0 2.1-7.8	1.8 0.0-5.5	1.8 0.1-3.4	1.4 0.0-4.5				2.2 0.0-6.6							
European Starling (O)	100.0 99.5-100.0	97.4 95.8-98.9	100.0 99.5-100.0	87.5 79.6-95.4	97.3 95.4-99.3	66.2 56.1-76.3	21.1 1.4-40.7	93.6 86.8-100.0	92.0 86.4-97.5	93.5 86.8-100.0	26.9 14.4-39.4	64.4 51.8-77.0	38.2 27.5-48.8	46.4 27.1-65.7	36.4 15.6-57.1	16.7 0.0-36.8	21.9 6.0-37.8
Solitary Vireo	9.3 6.2-12.4	25.5 21.9-29.1	61.4 55.6-67.1	62.5 51.4-73.6	77.0 72.6-81.4	73.2 63.7-82.7	78.9 59.3-98.6	85.1 75.7-94.6	70.1 61.2-79.0	43.5 31.1-55.9	80.8 69.5-92.0	44.1 31.1-57.1	63.2 52.6-73.7	71.4 53.8-89.1	27.3 7.9-46.7	72.2 48.5-95.9	46.9 28.0-65.7
Yellow-throated Vireo (L)	15.5 11.7-19.2	14.4 11.4-17.4	14.9 10.5-19.2	3.6 0.0-8.4	0.9 0.0-2.2			2.1 0.0-6.6									
Warbling Vireo	92.8 90.0-95.6	88.2 85.5-91.0	73.3 68.0-78.5	58.9 47.7-70.2	46.9 41.8-52.0	23.9 14.8-33.1	21.1 1.4-40.7	36.2 23.8-48.6	12.6 6.0-19.3		13.5 3.6-23.3		3.9 0.0-8.6				
Philadelphia Vireo	18.6 14.6-22.5	28.1 24.4-31.8	26.7 21.5-32.0	33.9 23.1-44.8	30.1 25.3-34.8	50.7 40.1-61.3	57.9 34.6-81.2	61.7 49.2-74.2	80.5 72.7-88.2	91.3 83.8-98.8	75.0 62.8-87.2	81.4 71.0-91.7	90.8 84.2-97.4	75.0 58.0-92.0		94.4 81.0-100.0	81.2 66.2-96.3
Red-eyed Vireo	97.9 96.2-99.7	98.7 97.5-99.9	99.0 97.4-100.0	98.2 94.5-100.0	98.2 96.6-99.9	100.0 99.3-100.0	94.7 82.8-100.0	100.0 98.9-100.0	93.1 87.9-98.3	97.8 93.4-100.0	96.2 90.2-100.0	96.6 91.3-100.0	93.4 87.7-99.2	64.3 45.7-82.9	22.7 4.4-41.1	61.1 35.6-86.6	78.1 62.2-94.0
Blue-winged Warbler (R)		0.7 0.0-1.6	2.0 0.0-4.0														
Brewster's Warbler (R)	1.0 0.0-2.4	2.0 0.6-3.3	1.0 0.0-2.6														
Golden-winged Warbler (R)	5.2 2.7-7.6	7.2 4.9-9.5	5.0 2.1-7.8	10.7 3.3-18.1		2.8 0.0-6.8											
Tennessee Warbler	14.4 10.8-18.1	28.1 24.4-31.8	30.7 25.2-36.2	55.4 44.0-66.7	54.9 49.8-60.0	77.5 68.5-86.5	89.5 74.0-100.0	72.3 60.7-83.9	92.0 86.4-97.5	95.7 89.9-100.0	82.7 71.9-93.5	78.0 67.0-89.0	97.4 93.4-100.0	96.4 88.1-100.0	95.5 85.2-100.0	100.0 97.2-100.0	90.6 79.0-100.0
Orange-crowned Warbler (R)										2.2 0.0-6.6		1.7 0.0-5.7	1.3 0.0-4.3	3.6 0.0-11.9	4.5 0.0-14.8	5.6 0.0-19.0	6.2 0.0-16.2
Nashville Warbler	51.5 46.6-56.5	83.7 80.5-86.8	85.1 80.8-89.5	92.9 86.5-99.2	92.9 90.1-95.8	95.8 91.1-100.0	89.5 74.0-100.0	95.7 89.9-100.0	87.4 80.7-94.0	95.7 89.9-100.0	90.4 81.7-99.0	91.5 83.9-99.2	89.5 82.5-96.4	78.6 62.4-94.8	31.8 11.6-52.0	83.3 63.2-100.0	96.9 89.3-100.0
Northern Parula	1.0 0.0-2.4	6.5 4.3-8.7	27.7 22.4-33.1	33.9 23.1-44.8	53.1 48.0-58.2	80.3 71.7-88.9	42.1 18.8-65.4	70.2 58.4-82.0	75.9 67.5-84.2	30.4 18.8-42.0	67.3 54.1-80.5	13.6 4.3-22.8	40.8 30.0-51.6	25.0 8.0-42.0		5.6 0.0-19.0	3.1 0.0-10.7
Yellow Warbler	97.9 96.2-99.7	95.4 93.5-97.3	96.0 93.4-98.6	87.5 79.6-95.4	91.2 88.0-94.3	57.7 47.2-68.3	47.4 23.9-70.9	70.2 58.4-82.0	58.6 49.1-68.2	80.4 70.3-90.6	46.2 32.2-60.1	84.7 75.1-94.4	51.3 40.4-62.3	53.6 34.3-72.9	18.2 1.1-35.3	83.3 63.2-100.0	71.9 54.7-89.0
Chestnut-sided Warbler	83.5 79.7-87.3	94.1 92.0-96.2	95.0 92.2-97.9	96.4 91.6-100.0	94.7 92.1-97.2	94.4 89.1-99.7	100.0 97.4-100.0	100.0 98.9-100.0	83.9 76.6-91.2	89.1 80.9-97.3	98.1 93.5-100.0	88.1 79.4-96.9	52.6 41.7-63.6	21.4 5.2-37.6		22.2 0.0-44.4	31.3 13.6-48.9
Magnolia Warbler	20.6 16.5-24.7	60.1 56.1-64.2	66.3 60.7-71.9	91.1 84.2-98.0	93.8 91.1-96.5	97.2 93.2-100.0	100.0 97.4-100.0	100.0 98.9-100.0	100.0 99.4-100.0	95.7 89.9-100.0	100.0 99.0-100.0	88.1 79.4-96.9	100.0 99.3-100.0	96.4 88.1-100.0	95.5 85.2-100.0	100.0 97.2-100.0	100.0 98.4-100.0
Cape May Warbler	1.0 0.0-2.4	11.1 8.4-13.8	25.7 20.5-31.0	41.1 29.8-52.3	42.5 37.4-47.6	62.0 51.6-72.3	5.3 0.0-17.2	53.2 40.4-66.0	60.9 51.5-70.4	63.0 50.9-75.1	69.2 56.2-82.2	35.6 23.0-48.2	61.8 51.2-72.5	78.6 62.4-94.8	18.2 1.1-35.3	50.0 23.9-76.1	6.2 0.0-16.2
Black-throated Blue Warbler	41.2 36.3-46.1	67.3 63.4-71.2	86.1 81.9-90.4	94.6 89.0-100.0	92.0 89.1-95.0	95.8 91.1-100.0	89.5 74.0-100.0	91.5 83.9-99.1	60.9 51.5-70.4	56.5 44.1-68.9	71.2 58.4-83.9	42.4 29.4-55.3	36.8 26.3-47.4	10.7 0.0-23.4	4.5 0.0-14.8	22.2 0.0-44.4	
Yellow-rumped Warbler	47.4 42.5-52.4	85.0 81.9-88.0	87.1 83.0-91.2	100.0 99.1-100.0	97.3 95.4-99.3	98.6 95.5-100.0	100.0 97.4-100.0	100.0 98.9-100.0	98.9 96.3-100.0	95.7 89.9-100.0	100.0 99.0-100.0	91.5 83.9-99.2	94.7 89.5-100.0	96.4 88.1-100.0	100.0 97.7-100.0	100.0 97.2-100.0	100.0 98.4-100.0
Black-throated Green Warbler	37.1 32.3-41.9	70.6 66.8-74.4	88.1 84.1-92.1	82.1 73.2-91.1	92.0 89.1-95.0	78.9 70.1-87.7	84.2 66.3-100.0	100.0 98.9-100.0	89.7 83.5-95.8	71.7 60.4-83.1	76.9 65.0-88.9	62.7 50.0-75.4	85.5 77.6-93.4	78.6 62.4-94.8	100.0 97.7-100.0	44.4 18.5-70.4	53.1 34.3-72.0
Blackburnian Warbler	34.0 29.3-38.8	66.7 62.8-70.6	75.2 70.1-80.4	91.1 84.2-98.0	89.4 86.0-92.7	93.0 87.2-98.7	94.7 82.8-100.0	97.9 93.4-100.0	89.7 83.5-95.8	78.3 67.8-88.8	86.5 76.7-96.4	52.5 39.5-65.6	61.8 51.2-72.5	35.7 17.1-54.3	4.5 0.0-14.8	22.2 0.0-44.4	12.5 0.0-25.5
Pine Warbler (L)	10.3 7.1-13.5	27.5 23.7-31.2	8.9 5.3-12.5	19.6 10.4-28.9	2.7 0.7-4.6	9.9 3.2-16.5	26.3 5.3-47.4										
Palm Warbler (L)		2.6 1.1-4.2	7.9 4.5-11.3	3.6 0.0-8.4		2.8 0.0-6.8		6.4 0.0-13.2	9.2 3.4-15.0	8.7 1.2-16.2		6.8 0.0-13.8	3.9 0.0-8.6				21.9 6.0-37.8
Bay-breasted Warbler	9.3 6.2-12.4	14.4 11.4-17.4	30.7 25.2-36.2	30.4 19.8-40.9	59.3 54.2-64.3	71.8 62.2-81.5	73.7 52.6-94.7	80.9 70.5-91.2	83.9 76.6-91.2	84.8 75.5-94.1	76.9 65.0-88.9	72.9 61.1-84.6	90.8 84.2-97.4	85.7 71.6-99.8	50.0 28.5-71.5	100.0 97.2-100.0	75.0 58.4-91.6
Blackpoll Warbler		0.7 0.0-1.6	4.0 1.4-6.6	3.6 0.0-8.4	8.0 5.0-10.9	14.1 6.5-21.7		42.6 29.8-55.3	63.2 53.9-72.6	23.9 13.1-34.7	11.5 2.2-20.8	8.5 0.8-16.1	82.9 74.5-91.3	89.3 76.6-100.0	77.3 58.9-95.6	100.0 97.2-100.0	3.1 0.0-10.7
Cerulean Warbler (R)	4.1 1.8-6.4	0.7 0.0-1.6		1.8 0.0-5.5													
Black-and-white Warbler	75.3 70.9-79.6	89.5 86.9-92.2	93.1 89.8-96.3	100.0 99.1-100.0	95.6 93.2-97.9	94.4 89.1-99.7	100.0 97.4-100.0	97.9 93.4-100.0	77.0 68.8-85.2	82.6 72.9-92.3	96.2 90.2-100.0	74.6 63.1-86.1	57.9 47.1-68.7	28.6 10.9-46.2	54.5 33.1-75.9	22.2 0.0-44.4	40.6 22.0-59.2
American Redstart	88.7 85.3-92.0	93.5 91.3-95.7	97.0 94.7-99.4	96.4 91.6-100.0	99.1 97.8-100.0	95.8 91.1-100.0	100.0 97.4-100.0	97.9 93.4-100.0	97.7 94.4-100.0	97.8 93.4-100.0	96.2 90.2-100.0	93.2 86.2-100.0	97.4 93.4-100.0	89.3 76.6-100.0	81.8 64.7-98.9	88.9 71.4-100.0	62.5 44.2-80.8
Ovenbird	92.8 90.0-95.6	96.1 94.3-97.9	99.0 97.4-100.0	92.9 86.5-99.2	97.3 95.4-99.3	100.0 99.3-100.0	100.0 97.4-100.0	100.0 98.9-100.0	95.4 91.0-99.8	100.0 98.9-100.0	96.2 90.2-100.0	83.1 73.0-93.1	71.1 61.1-81.0	53.6 34.3-72.9	45.5 24.1-66.9	11.1 0.0-28.6	46.9 28.0-65.7
Northern Waterthrush	39.2 34.3-44.0	58.8 54.8-62.9	58.4 52.6-64.2	80.4 71.1-89.6	79.6 75.4-83.9	83.1 74.9-91.2	89.5 74.0-100.0	83.0 73.1-92.9	93.1 87.9-98.3	93.5 86.8-100.0	61.5 47.9-75.2	84.7 75.1-94.4	92.1 85.9-98.3	82.1 66.9-97.4	72.7 53.3-92.1	94.4 81.0-100.0	68.7 51.1-86.4
Louisiana Waterthrush (R)		0.7 0.0-1.6															
Connecticut Warbler (R)							5.3 0.0-17.2			6.5 0.0-13.2	1.9 0.0-6.5	6.8 0.0-13.8					15.6 1.5-29.8
Mourning Warbler	69.1 64.4-73.7	83.0 79.8-86.2	80.2 75.4-85.0	82.1 73.2-91.1	89.4 86.0-92.7	88.7 81.7-95.7	94.7 82.8-100.0	89.4 81.0-97.7	79.3 71.4-87.3	97.8 93.4-100.0	76.9 65.0-88.9	83.1 73.0-93.1	68.4 58.2-78.6	39.3 20.4-58.2	9.1 0.0-22.4	88.9 71.4-100.0	78.1 62.2-94.0
Common Yellowthroat	100.0 99.5-100.0	98.7 97.5-99.9	100.0 99.5-100.0	96.4 91.6-100.0	100.0 99.6-100.0	100.0 99.3-100.0	100.0 97.4-100.0	100.0 98.9-100.0	97.7 94.4-100.0	100.0 98.9-100.0	100.0 99.0-100.0	96.6 91.3-100.0	97.4 93.4-100.0	96.4 88.1-100.0	45.5 24.1-66.9	100.0 97.2-100.0	96.9 89.3-100.0
Wilson's Warbler		2.6 1.1-4.2	5.9 2.9-9.0	10.7 3.3-18.1	13.3 9.6-16.9	25.4 16.0-34.7	21.1 1.4-40.7	53.2 40.4-66.0	67.8 58.7-76.9	58.7 46.4-71.0	30.8 17.8-43.8	69.5 57.4-81.6	80.3 71.4-89.1	92.9 82.0-100.0	18.2 1.1-35.3	88.9 71.4-100.0	84.4 70.2-98.5
Canada Warbler	29.9 25.3-34.5	64.7 60.8-68.7	83.2 78.6-87.7	91.1 84.2-98.0	86.7 83.1-90.4	93.0 87.2-98.7	100.0 97.4-100.0	100.0 98.9-100.0	94.3 89.4-99.1	97.8 93.4-100.0	69.2 56.2-82.2	57.6 44.7-70.6	76.3 66.9-85.7	46.4 27.1-65.7		50.0 23.9-76.1	28.1 11.0-45.3
Scarlet Tanager	63.9 59.1-68.7	81.0 77.7-84.3	66.3 60.7-71.9	89.3 81.9-96.7	62.8 57.9-67.8	74.6 65.3-84.0	94.7 82.8-100.0	72.3 60.7-83.9	26.4 17.8-35.0	32.6 20.8-44.4	53.8 39.9-67.8	18.6 8.3-29.0	3.9 0.0-8.6	7.1 0.0-18.0	9.1 0.0-22.4		
Northern Cardinal	43.3 38.4-48.2	13.7 10.8-16.7	8.9 5.3-12.5	5.4 0.0-11.0	4.4 2.1-6.8	1.4 0.0-4.5											
Rose-breasted Grosbeak	97.9 96.2-99.7	100.0 99.7-100.0	98.0 96.0-100.0	94.6 89.0-100.0	100.0 99.6-100.0	97.2 93.2-100.0	100.0 97.4-100.0	97.9 93.4-100.0	94.3 89.4-99.1	89.1 80.9-97.3	96.2 90.2-100.0	37.3 24.6-50.0	68.4 58.2-78.6	42.9 23.7-62.0		11.1 0.0-28.6	
Indigo Bunting	73.2 68.7-77.7	74.5 70.9-78.1	35.6 30.0-41.3	69.6 59.1-80.2	23.9 19.5-28.3	22.5 13.5-31.5	15.8 0.0-33.7	6.4 0.0-13.2		2.2 0.0-6.6	1.9 0.0-6.5	1.7 0.0-5.7	1.3 0.0-4.3				3.1 0.0-10.7
Rufous-sided Towhee	19.6 15.5-23.6	16.3 13.2-19.5	4.0 1.4-6.6	1.8 0.0-5.5	1.8 0.1-3.4			4.3 0.0-10.1									
American Tree Sparrow (R)												1.7 0.0-5.7					
Chipping Sparrow	96.9 94.8-99.0	100.0 99.7-100.0	100.0 99.5-100.0	98.2 94.5-100.0	100.0 99.6-100.0	90.1 83.5-96.8	84.2 66.3-100.0	100.0 98.9-100.0	100.0 99.4-100.0	93.5 86.8-100.0	86.5 76.7-96.4	84.7 75.1-94.4	85.5 77.6-93.4	89.3 76.6-100.0	36.4 15.6-57.1	33.3 8.6-58.1	78.1 62.2-94.0
Clay-colored Sparrow (R,O)	4.1 1.8-6.4	4.6 2.7-6.5	7.9 4.5-11.3	3.6 0.0-8.4						6.5 0.0-13.2							
Field Sparrow (O)	27.8 23.3-32.3	31.4 27.5-35.2	18.8 14.1-23.5	19.6 10.4-28.9	6.2 3.5-8.9	5.6 0.3-10.9			1.1 0.0-3.7								
Vesper Sparrow (O)	53.6 48.6-58.6	58.8 54.8-62.9	35.6 30.0-41.3	53.6 42.2-65.0	18.6 14.5-22.7	23.9 14.8-33.1	5.3 0.0-17.2	23.4 12.4-34.4	25.3 16.8-33.8	43.5 31.1-55.9		16.9 6.9-27.0	17.1 8.7-25.5	3.6 0.0-11.9			
Savannah Sparrow (O)	97.9 96.2-99.7	93.5 91.3-95.7	97.0 94.7-99.4	82.1 73.2-91.1	89.4 86.0-92.7	47.9 37.3-58.5	26.3 5.3-47.4	93.6 86.8-100.0	82.8 75.3-90.2	95.7 89.9-100.0	23.1 11.1-35.0	55.9 42.9-68.9	44.7 33.8-55.6	32.1 14.0-50.3	50.0 28.5-71.5	22.2 0.0-44.4	43.7 25.0-62.5
Grasshopper Sparrow (R,O)	7.2 4.4-10.0	5.2 3.2-7.2	1.0 0.0-2.6	3.6 0.0-8.4													
Henslow's Sparrow (R,O)	1.0 0.0-2.4	0.7 0.0-1.6															
Le Conte's Sparrow (R,O)			1.0 0.0-2.6							6.5 0.0-13.2		10.2 1.9-18.4					9.4 0.0-21.0
Sharp-tailed Sparrow (R,A)	2.1 0.3-3.8	0.7 0.0-1.6			0.9 0.0-2.2	1.4 0.0-4.5		6.4 0.0-13.2	5.7 0.9-10.6	4.3 0.0-10.1			3.9 0.0-8.6				
Fox Sparrow				1.8 0.0-5.5		2.8 0.0-6.8		27.7 16.1-39.3	52.9 43.2-62.5	2.2 0.0-6.6	13.5 3.6-23.3	10.2 1.9-18.4	68.4 58.2-78.6	78.6 62.4-94.8	100.0 97.7-100.0	61.1 35.6-86.6	15.6 1.5-29.8
Song Sparrow	100.0 99.5-100.0	100.0 99.7-100.0	100.0 99.5-100.0	100.0 99.1-100.0	100.0 99.6-100.0	97.2 93.2-100.0	100.0 97.4-100.0	100.0 98.9-100.0	96.6 92.7-100.0	95.7 89.9-100.0	96.2 90.2-100.0	86.4 77.2-95.7	69.7 59.6-79.8	71.4 53.8-89.1	40.9 19.7-62.1	27.8 4.1-51.5	50.0 31.1-68.9
Lincoln's Sparrow	3.1 1.0-5.2	17.6 14.4-20.9	67.3 61.8-72.9	30.4 19.8-40.9	71.7 67.0-76.4	46.5 35.9-57.1	47.4 23.9-70.9	97.9 93.4-100.0	81.6 74.0-89.2	95.7 89.9-100.0	76.9 65.0-88.9	84.7 75.1-94.4	93.4 87.7-99.2	78.6 62.4-94.8	90.9 77.6-100.0	100.0 97.2-100.0	100.0 98.4-100.0
Swamp Sparrow	84.5 80.8-88.3	75.2 71.6-78.8	67.3 61.8-72.9	85.7 77.4-94.0	69.9 65.2-74.7	83.1 74.9-91.2	89.5 74.0-100.0	87.2 78.3-96.2	69.0 60.0-78.0	76.1 65.3-86.9	96.2 90.2-100.0	84.7 75.1-94.4	77.6 68.4-86.9	71.4 53.8-89.1	68.2 48.0-88.4	100.0 97.2-100.0	90.6 79.0-100.0
White-throated Sparrow	97.9 96.2-99.7	99.3 98.4-100.0	100.0 99.5-100.0	98.2 94.5-100.0	99.1 97.8-100.0	100.0 99.3-100.0	100.0 97.4-100.0	100.0 98.9-100.0	100.0 99.4-100.0	100.0 98.9-100.0	100.0 99.0-100.0	100.0 99.2-100.0	100.0 99.3-100.0	100.0 98.2-100.0	100.0 97.7-100.0	100.0 97.2-100.0	100.0 98.4-100.0
White-crowned Sparrow (R)												1.7 0.0-5.7		10.7 0.0-23.4			
Dark-eyed Junco	25.8 21.4-30.2	37.9 33.9-41.9	64.4 58.7-70.0	82.1 73.2-91.1	86.7 83.1-90.4	94.4 89.1-99.7	89.5 74.0-100.0	95.7 89.9-100.0	98.9 96.3-100.0	97.8 93.4-100.0	98.1 93.5-100.0	94.9 88.7-100.0	98.7 95.7-100.0	96.4 88.1-100.0	100.0 97.7-100.0	100.0 97.2-100.0	100.0 98.4-100.0
Bobolink (O)	99.0 97.6-100.0	95.4 93.5-97.3	98.0 96.0-100.0	73.2 63.0-83.4	86.7 83.1-90.4	40.8 30.4-51.3	21.1 1.4-40.7	87.2 78.3-96.2	72.4 63.7-81.1	67.4 55.6-79.2	7.7 0.0-15.6	45.8 32.7-58.8	19.7 10.9-28.6	3.6 0.0-11.9	4.5 0.0-14.8	5.6 0.0-19.0	3.1 0.0-10.7
Red-winged Blackbird	100.0 99.5-100.0	100.0 99.7-100.0	100.0 99.5-100.0	96.4 91.6-100.0	100.0 99.6-100.0	97.2 93.2-100.0	78.9 59.3-98.6	100.0 98.9-100.0	100.0 99.4-100.0	95.7 89.9-100.0	90.4 81.7-99.0	72.9 61.1-84.6	84.2 76.0-92.4	71.4 53.8-89.1	27.3 7.9-46.7	33.3 8.6-58.1	28.1 11.0-45.3
Eastern Meadowlark (O)	95.9 93.6-98.2	89.5 86.9-92.2	86.1 81.9-90.4	58.9 47.7-70.2	49.6 44.4-54.7	19.7 11.1-28.3		23.4 12.4-34.4	1.1 0.0-3.7	8.7 1.2-16.2	3.8 0.0-9.8	6.8 0.0-13.8					
Western Meadowlark (R,O)		0.7 0.0-1.6						2.1 0.0-6.6	1.1 0.0-3.7								
Rusty Blackbird		2.6 1.1-4.2	9.9 6.2-13.6	10.7 3.3-18.1	9.7 6.5-13.0	38.0 27.7-48.4	31.6 9.5-53.6	10.6 2.3-19.0	55.2 45.5-64.8	19.6 9.4-29.7	71.2 58.4-83.9	54.2 41.2-67.3	67.1 56.8-77.4	71.4 53.8-89.1	63.6 42.9-84.4	88.9 71.4-100.0	68.7 51.1-86.4
Common Grackle	100.0 99.5-100.0	97.4 95.8-98.9	99.0 97.4-100.0	98.2 94.5-100.0	100.0 99.6-100.0	98.6 95.5-100.0	100.0 97.4-100.0	100.0 98.9-100.0	98.9 96.3-100.0	100.0 98.9-100.0	92.3 84.4-100.0	76.3 65.0-87.5	82.9 74.5-91.3	82.1 66.9-97.4	18.2 1.1-35.3	44.4 18.5-70.4	37.5 19.2-55.8
Brown-headed Cowbird	99.0 97.6-100.0	95.4 93.5-97.3	98.0 96.0-100.0	92.9 86.5-99.2	96.5 94.3-98.6	60.6 50.1-71.0	31.6 9.5-53.6	89.4 81.0-97.7	81.6 74.0-89.2	82.6 72.9-92.3	7.7 0.0-15.6	40.7 27.8-53.6	38.2 27.5-48.8	28.6 10.9-46.2	4.5 0.0-14.8		3.1 0.0-10.7
Northern Oriole	99.0 97.6-100.0	96.1 94.3-97.9	94.1 91.0-97.1	66.1 55.2-76.9	75.2 70.7-79.7	38.0 27.7-48.4	15.8 0.0-33.7	46.8 34.0-59.6	9.2 3.4-15.0	4.3 0.0-10.1	1.9 0.0-6.5		1.3 0.0-4.3	3.6 0.0-11.9			
Pine Grosbeak		1.3 0.1-2.5	1.0 0.0-2.6	8.9 2.0-15.8	0.9 0.0-2.2	7.0 1.3-12.8	10.5 0.0-26.0	21.3 10.6-32.0	60.9 51.5-70.4	8.7 1.2-16.2	7.7 0.0-15.6	8.5 0.8-16.1	51.3 40.4-62.3	67.9 49.7-86.0	9.1 0.0-22.4	33.3 8.6-58.1	9.4 0.0-21.0
Purple Finch	64.9 60.2-69.7	86.9 84.0-89.8	94.1 91.0-97.1	96.4 91.6-100.0	98.2 96.6-99.9	97.2 93.2-100.0	94.7 82.8-100.0	100.0 98.9-100.0	98.9 96.3-100.0	97.8 93.4-100.0	94.2 87.2-100.0	93.2 86.2-100.0	100.0 99.3-100.0	96.4 88.1-100.0	72.7 53.3-92.1	88.9 71.4-100.0	87.5 74.5-100.0
House Finch (O)	42.3 37.4-47.2	20.3 16.9-23.6	22.8 17.7-27.8	1.8 0.0-5.5	8.0 5.0-10.9	2.8 0.0-6.8		2.1 0.0-6.6									
Red Crossbill (Ca)	7.2 4.4-10.0	5.9 3.8-8.0	4.0 1.4-6.6	17.9 8.9-26.8	2.7 0.7-4.6	2.8 0.0-6.8			4.6 0.2-9.0	4.3 0.0-10.1	1.9 0.0-6.5	1.7 0.0-5.7	2.6 0.0-6.6		13.6 0.0-29.1		
White-winged Crossbill	8.2 5.3-11.2	4.6 2.7-6.5	10.9 7.0-14.7	30.4 19.8-40.9	15.0 11.2-18.8	26.8 17.3-36.3	36.8 14.0-59.6	25.5 14.2-36.9	44.8 35.2-54.5	45.7 33.2-58.1	32.7 19.5-45.9	55.9 42.9-68.9	61.8 51.2-72.5	46.4 27.1-65.7	100.0 97.7-100.0	61.1 35.6-86.6	87.5 74.5-100.0
Common Redpoll (R)									1.1 0.0-3.7					3.6 0.0-11.9	9.1 0.0-22.4		
Pine Siskin	43.3 38.4-48.2	23.5 20.0-27.1	44.6 38.7-50.4	60.7 49.5-71.9	56.6 51.5-61.7	88.7 81.7-95.7	68.4 46.4-90.5	70.2 58.4-82.0	85.1 78.0-92.1	73.9 62.8-85.0	67.3 54.1-80.5	84.7 75.1-94.4	94.7 89.5-100.0	96.4 88.1-100.0	72.7 53.3-92.1	94.4 81.0-100.0	96.9 89.3-100.0
American Goldfinch	100.0 99.5-100.0	100.0 99.7-100.0	100.0 99.5-100.0	96.4 91.6-100.0	100.0 99.6-100.0	73.2 63.7-82.7	47.4 23.9-70.9	95.7 89.9-100.0	89.7 83.5-95.8	91.3 83.8-98.8	61.5 47.9-75.2	71.2 59.2-83.1	56.6 45.7-67.4	53.6 34.3-72.9	4.5 0.0-14.8	5.6 0.0-19.0	18.8 3.7-33.8
Evening Grosbeak	34.0 29.3-38.8	45.8 41.6-49.9	44.6 38.7-50.4	91.1 84.2-98.0	72.6 67.9-77.2	94.4 89.1-99.7	94.7 82.8-100.0	87.2 78.3-96.2	96.6 92.7-100.0	87.0 78.2-95.7	84.6 74.2-95.0	74.6 63.1-86.1	100.0 99.3-100.0	85.7 71.6-99.8	31.8 11.6-52.0	94.4 81.0-100.0	25.0 8.4-41.6
House Sparrow (O)	100.0 99.5-100.0	88.9 86.2-91.6	96.0 93.4-98.6	62.5 51.4-73.6	86.7 83.1-90.4	38.0 27.7-48.4	10.5 0.0-26.0	78.7 68.0-89.4	74.7 66.2-83.2	84.8 75.5-94.1	7.7 0.0-15.6	45.8 32.7-58.8	23.7 14.3-33.1	14.3 0.2-28.4	4.5 0.0-14.8	5.6 0.0-19.0	9.4 0.0-21.0

* The relative abundance by ecological region of species shown with any of the following letter codes in parentheses were not mapped, for reasons that are self-explanatory:

R : rare species reported in a small number of squares, i.e., no more than 50
O : species living in open country
A : species living solely on or near water
L : localized species with highly specific habitat requirements
C : colonial species
Ca : species whose distribution cannot be interpreted from analysis results

Appendix II
Québec migratory game bird harvest estimates, 1973–92

The Canadian Wildlife Service co-ordinates the annual National Harvest Survey (NHS) to estimate the number of migratory game birds harvested in Canada by sport hunters holding a Migratory Game Bird Hunting Permit (MGBHP). NHS teams in each province administer two types of stratified sample surveys: the Harvest Questionnaire Survey and the Species Composition Survey, which together make up the National Harvest Survey.

The Canadian Wildlife Service establishes each year a new list of hunters to be contacted for each survey based on a random selection of names from the data file of MGBHP.

The Harvest Questionnaire Survey is sent to 3000 Québec hunters, as well as hunters living in the rest of Canada and the United States. The questionnaire is designed to assess sport hunting activities and total kills. Hunters are asked to provide information on their hunting activities and kills for nine species groups. The sample is constructed to include hunters in all regions with varying levels of experience.

The Species Composition Survey is administered separately to a second random sample of 3100 permit holders in Québec. Hunters are asked to return in a special envelope a part of each bird they kill during the hunting season, a wing for a duck or a woodcock, and the tail feathers for a goose. This allows estimated harvests to be partitioned by species for each hunting period in each regional subdivision surveyed.

The data from respondents in both surveys allow harvesting levels to be assessed by providing indices on the number of birds harvested by species. These figures are fairly accurate for species harvested in greatest numbers. However, in species that are harvested much less often (generally representing less than 2% of the total harvest, marked with an asterisk in the table), these indices merely provide a measure of relative abundance either because of the small numbers of birds involved or because the species frequents mainly remote regions. Finally, estimates for species harvested only on rare occasions (generally representing less than 1% of the total harvest) are only indicative of the species' presence in the harvest and accordingly are shown as *p* in the table. These estimates were, however, used to calculate the totals in the table.

The table shows estimates of annual harvests by sport hunters in Québec from 1973 to 1992. As the table indicates, harvests for several species dropped significantly during this period. However, these data cannot be used in isolation as a universal indicator of the status of waterfowl populations or of trends in these populations.

Estimates are in fact closely tied to the number of hunters with permits and the activities they report. The largest harvests were recorded at the end of the 1970s, a period when the number of waterfowl hunters in the province also peaked. Since then, the number of permit holders has decreased by about a third. At the same time, the average total harvest reported by successful hunters has increased in recent years, indicating, perhaps, that less skilful and less committed hunters have dropped out of the sport.

In general, indices derived from these surveys cannot be used on their own to explain annual variations in migratory bird populations or to draw conclusions on the abundance of game species. For example, although Common Eider breeding populations are thriving in Québec, harvests of this sea duck have dropped significantly in the province. The case of the Greater Snow Goose also illustrates this paradox in a somewhat different way: the spectacular increase in the population, which has been well documented in photographic surveys of the spring flight, has not necessarily translated into an increase in sport hunting harvests. In the case of other species, like the Blue-winged Teal, harvesting levels seem to be more in line with other indices, such as a drop in numbers in spring.

Species harvested but not included in the table

Some species taken during the hunting season in Québec do not appear in the table because of their irregular occurrence and small showing in the Species Composition Survey.

Hybrids and subspecies fall into this category. Hybrids observed include crosses between the American Black Duck and Mallard, as well as domesticated forms of the Mallard. Hybrid individuals are observed every year, but have only been quantified systematically since 1982. In the calculation of total harvests, hybrids are grouped with one of the parental types, depending on the dominant plumage.

A small number of duck wings and goose tail feathers are also received—albeit sporadically and irregularly from year to year—for the Eurasian Wigeon, Greater White-fronted Goose, and Ross' Goose. Harvest estimates for these accidental species are not broken out separately in the Québec figures but combined with the corresponding species (e.g., the Eurasian Wigeon is lumped in with the American Wigeon).

The Harlequin Duck is now on the list of threatened species and hunting of this duck has been banned since 1990. It has only shown up six times in the Species Composition Survey during the period.

Barrow's Goldeneye, the Ruddy Duck, and the King Eider also appear regularly in the National Harvest Survey. Because of their small showing in most years, they are represented by *p* in the table. The use of new identification criteria developed in the 1980s (CWS, Québec Region, unpubl. data) for Barrow's Goldeneye and the King Eider has allowed researchers to monitor both harvesting levels and indices of relative abundance for these species more closely since then.

Figures in table

Percentage (%) figures for ducks and geese were calculated as follows:

[estimated annual species harvest/estimated total annual harvest for all ducks and geese × 100].

The National Harvest Survey data are derived from surveys of hunters designed to provide a statistically reliable picture of harvests of the main species of migratory game birds in each province. The figures shown here are estimates based on data provided by sport hunters. Generally, the reliability of these estimates depends on an adequate response rate in both surveys.

In the case of species that are rarely harvested, estimates have been replaced by *p* in the table but were taken into account in calculating the totals (harvests and %).

The table is based on estimates published in the Canadian Wildlife Service's *Progress Notes* (Cooch et al. 1975; Wendt et al. 1978; Wendt and Hyslop 1981; Métras 1985; Dickson 1989; Lévesque et al. 1993; Lévesque and Collins, CWS, unpubl. data).

In conclusion, interpreting information from harvest data is a complex task. Such data directly reflect hunters' activities, but only indirectly reflect population trends among migratory game birds. However, the results of these surveys, when combined with the data from research projects and various other surveys carried out during the breeding season (breeding pairs, breeding success, and habitats), can be used to monitor the status of waterfowl populations and to ensure conservation by providing information for regulatory decision making in such areas as the migratory bird hunt.

Every autumn, thousands of hunters contribute to the effort to manage and conserve migratory game bird populations by participating in these surveys.

Pierre Dupuis, André Bourget, and Hélène Lévesque

Species	1973 Harvest (%)	1974 Harvest (%)	1975 Harvest (%)	1976 Harvest (%)	1977 Harvest (%)	1978 Harvest (%)	1979 Harvest (%)	1980 Harvest (%)	1981 Harvest (%)	1982 Harvest (%)	1983 Harvest (%)	1984 Harvest (%)	1985 Harvest (%)	1986 Harvest (%)	1987 Harvest (%)	1988 Harvest (%)	1989 Harvest (%)	1990 Harvest (%)	1991 Harvest (%)	1992 Harvest (%)	Average 1988–92 Harvest	(%)
Wood Duck	6 487 (1.3)	22 879 (4.0)	16 291 (3.5)	17 322 (2.9)	22 476 (3.6)	18 019 (3.4)	18 513 (4.0)	17 610 (3.5)	25 214 (5.0)	20 997 (4.1)	24 964 (5.5)	22 497 (5.1)	23 598 (4.8)	15 194 (3.5)	22 820 (4.9)	25 746 (5.8)	24 988 (5.8)	23 402 (5.1)	25 801 (7.2)	23 690 (6.9)	24 725	(6.2)
Green-winged Teal	38 243 (7.8)	56 335 (9.9)	40 323 (8.6)	68 781 (11.4)	83 396 (13.2)	60 573 (11.4)	35 616 (7.7)	45 733 (9.1)	47 914 (9.4)	66 131 (12.9)	43 584 (9.6)	40 724 (9.3)	60 543 (12.2)	55 129 (12.7)	40 219 (8.6)	52 088 (11.7)	68 800 (15.9)	89 355 (19.6)	36 895 (10.3)	30 500 (8.8)	55 528	(13.3)
American Black Duck	120 668 (24.5)	110 139 (19.3)	91 394 (19.5)	118 438 (19.6)	131 457 (20.8)	130 843 (24.7)	113 058 (24.6)	120 399 (24.1)	105 548 (20.8)	117 847 (23.0)	101 908 (22.4)	107 624 (24.5)	110 541 (22.3)	114 703 (26.4)	129 920 (27.7)	127 134 (28.5)	99 674 (23.0)	105 277 (23.1)	91 820 (25.7)	84 170 (24.4)	101 615	(24.9)
Mallard	72 282 (14.7)	74 116 (13.0)	58 237 (12.4)	73 667 (12.2)	83 600 (13.2)	62 449 (11.8)	68 250 (14.8)	81 304 (16.3)	92 834 (18.3)	94 808 (18.5)	87 107 (19.2)	66 849 (15.2)	97 157 (19.6)	84 445 (19.5)	116 793 (24.9)	83 748 (18.8)	79 419 (18.3)	86 525 (19.0)	77 668 (21.7)	90 102 (26.1)	83 492	(20.8)
Northern Pintail	19 355 (3.9)	21 043 (3.7)	21 999 (4.7)	29 010 (4.8)	40 558 (6.4)	20 550 (3.9)	15 313 (3.3)	17 099 (3.4)	17 462 (3.4)	20 294 (4.0)	16 283 (3.6)	9 469 (2.2)	16 519 (3.3)	13 012 (3.0)	11 921 (2.5)	12 159 (2.7)	15 461 (3.6)	19 568 (4.3)	7 593 (2.1)	6 367 (1.8)	12 230	(2.9)
Blue-winged Teal	44 315 (9.0)	44 737 (7.9)	33 766 (7.2)	55 053 (9.1)	44 381 (7.0)	36 008 (6.8)	13 984 (3.0)	10 324 (2.1)	8 748 (1.7)	14 768 (2.9)	16 489 (3.6)	10 390 (2.4)	9 495 (1.9)	9 448 (2.2)	7 997 (1.7)	5 941 (1.3)	6 029 (1.4)	4 160 (0.9)	3 228 (0.9)	3 517 (1.0)	4 575	(1.1)
Northern Shoveler*	7 204 (1.5)	5 451 (1.0)	3 578 (0.8)	2 638 (0.4)	6 669 (1.1)	2 754 (0.5)	2 530 (0.5)	5 135 (1.0)	5 567 (1.1)	3 133 (0.6)	3 368 (0.7)	4 339 (1.0)	3 457 (0.7)	2 361 (0.5)	2 000 (0.4)	2 712 (0.6)	2 677 (0.6)	1 874 (0.4)	2 417 (0.7)	3 206 (0.9)	2 577	(0.6)
Gadwall*	1 894 (0.4)	3 681 (0.6)	3 907 (0.8)	6 392 (1.1)	6 710 (1.1)	7 484 (1.4)	9 759 (2.1)	10 703 (2.1)	6 118 (1.2)	7 406 (1.4)	7 844 (1.7)	5 232 (1.2)	9 133 (1.8)	5 250 (1.2)	4 765 (1.0)	4 806 (1.1)	3 633 (0.8)	3 579 (0.8)	5 780 (1.6)	6 499 (1.9)	4 859	(1.2)
American Wigeon*	6 256 (1.3)	9 631 (1.7)	5 970 (1.3)	11 633 (1.9)	9 143 (1.4)	6 870 (1.3)	9 124 (2.0)	7 213 (1.4)	6 152 (1.2)	5 688 (1.1)	8 611 (1.9)	4 331 (1.0)	5 753 (1.2)	6 929 (1.6)	5 222 (1.1)	2 942 (0.7)	5 520 (1.3)	5 422 (1.2)	2 072 (0.6)	6 653 (1.9)	4 522	(1.1)
Canvasback	*p*	*p*	*p*	*p*	*p*	*p*	*p*	*p*	*p*	*p*	*p*	*p*	*p*	*p*	*p*	*p*	*p*	*p*	*p*	*p*	*p*	
Redhead	*p*	*p*	*p*	*p*	*p*	*p*	*p*	*p*	*p*	*p*	*p*	*p*	*p*	*p*	*p*	*p*	*p*	*p*	*p*	*p*	*p*	
Ring-necked Duck	17 477 (3.5)	19 415 (3.4)	16 143 (3.5)	16 477 (2.7)	19 739 (3.1)	24 558 (4.6)	17 928 (3.9)	21 197 (4.2)	20 674 (4.1)	21 352 (4.2)	18 687 (4.1)	26 255 (6.0)	23 096 (4.7)	28 513 (6.6)	23 652 (5.0)	29 549 (6.6)	24 866 (5.7)	19 521 (4.3)	24 112 (6.7)	15 874 (4.6)	22 784	(5.6)
Greater Scaup	48 067 (9.8)	29 829 (5.2)	25 071 (5.4)	28 000 (4.6)	20 266 (3.2)	18 577 (3.5)	20 414 (4.4)	19 284 (3.9)	22 603 (4.5)	15 516 (3.0)	13 912 (3.1)	20 128 (4.6)	17 656 (3.6)	11 727 (2.7)	7 183 (1.5)	13 622 (3.1)	9 380 (2.2)	9 284 (2.0)	6 089 (1.7)	4 705 (1.4)	8 616	(2.1)
Lesser Scaup	24 331 (4.9)	32 166 (5.6)	29 538 (6.3)	33 782 (5.6)	32 944 (5.2)	22 021 (4.2)	26 566 (5.8)	28 746 (5.7)	33 368 (6.6)	19 690 (3.8)	20 243 (4.5)	17 260 (3.9)	24 798 (5.0)	23 467 (5.4)	12 191 (2.6)	22 430 (5.0)	26 711 (6.2)	24 047 (5.3)	17 545 (4.9)	15 026 (4.3)	21 152	(5.1)
Common Eider*	4 824 (1.0)	4 792 (0.8)	7 901 (1.7)	9 322 (1.5)	6 154 (1.0)	7 802 (1.5)	5 268 (1.1)	6 842 (1.4)	9 430 (1.9)	3 485 (0.7)	9 026 (2.0)	7 498 (1.7)	4 727 (1.0)	3 034 (0.7)	5 351 (1.1)	1 777 (0.4)	2 062 (0.5)	3 458 (0.8)	3 503 (1.0)	3 442 (0.9)	2 700	(0.7)
King Eider	*p*	*p*	*p*	*p*	*p*	*p*	*p*	*p*	*p*	*p*	*p*	*p*	*p*	*p*	*p*	*p*	*p*	*p*	*p*	*p*	*p*	
Harlequin Duck	*p*	*p*	*p*	*p*	*p*	*p*	*p*	*p*	*p*	*p*	*p*	*p*	*p*	*p*	*p*	*p*	*p*	*p*	*p*	*p*	*p*	
Oldsquaw	9 781 (2.0)	9 263 (1.6)	6 103 (1.3)	11 649 (1.9)	4 575 (0.7)	7 394 (1.4)	11 471 (2.5)	7 937 (1.6)	7 769 (1.5)	10 290 (2.0)	8 882 (2.0)	12 690 (2.9)	6 405 (1.3)	5 522 (1.3)	5 647 (1.2)	2 382 (0.5)	6 174 (1.4)	1 761 (0.4)	3 074 (0.9)	2 171 (0.6)	3 112	(0.8)
Black Scoter	6 178 (1.3)	8 254 (1.4)	11 516 (2.5)	18 000 (3.0)	26 470 (4.2)	6 512 (1.2)	10 819 (2.4)	12 271 (2.5)	11 380 (2.2)	5 895 (1.2)	4 684 (1.0)	7 666 (1.7)	6 400 (1.3)	2 060 (0.5)	7 435 (1.6)	3 457 (0.8)	5 000 (1.2)	3 896 (0.9)	3 294 (0.9)	1 474 (0.4)	3 424	(0.8)
Surf Scoter	6 925 (1.4)	14 348 (2.5)	15 945 (3.4)	20 278 (3.4)	18 219 (2.9)	8 373 (1.6)	4 155 (0.9)	10 253 (2.0)	12 403 (2.4)	15 451 (3.0)	4 184 (0.9)	7 965 (1.8)	6 407 (1.3)	1 982 (0.5)	6 736 (1.4)	7 370 (1.7)	5 086 (1.2)	5 195 (1.1)	1 671 (0.5)	3 469 (1.0)	4 558	(1.1)
White-winged Scoter	5 306 (1.1)	14 122 (2.5)	5 067 (1.1)	8 704 (1.4)	10 352 (1.6)	4 802 (0.9)	8 432 (1.8)	10 811 (2.2)	8 023 (1.6)	7 627 (1.5)	6 528 (1.4)	11 185 (2.5)	7 650 (1.5)	2 416 (0.6)	7 165 (1.5)	7 082 (1.6)	8 079 (1.9)	5 319 (1.2)	2 418 (0.7)	5 045 (1.5)	5 589	(1.4)
Common Goldeneye	31 783 (6.5)	50 647 (8.9)	29 700 (6.3)	24 652 (4.1)	26 375 (4.2)	39 910 (7.5)	26 892 (5.8)	28 775 (5.8)	28 396 (5.6)	20 996 (4.1)	21 850 (4.8)	19 428 (4.4)	23 329 (4.7)	16 058 (3.7)	20 611 (4.4)	17 065 (3.8)	9 508 (2.2)	17 666 (3.9)	10 155 (2.8)	11 938 (3.5)	13 266	(3.2)
Barrow's Goldeneye	*p*	*p*	*p*	*p*	*p*	*p*	*p*	*p*	*p*	*p*	*p*	*p*	*p*	*p*	*p*	*p*	*p*	*p*	*p*	*p*	*p*	
Bufflehead*	3 835 (0.8)	9 733 (1.7)	10 290 (2.2)	8 466 (1.4)	4 742 (0.8)	9 082 (1.7)	7 735 (1.7)	5 085 (1.0)	8 982 (1.8)	4 155 (0.8)	5 683 (1.3)	5 275 (1.2)	3 733 (0.8)	2 183 (0.5)	4 936 (1.1)	5 364 (1.2)	4 726 (1.1)	3 811 (0.8)	2 310 (0.6)	8 157 (2.4)	4 874	(1.2)
Hooded Merganser	6 370 (1.3)	8 706 (1.5)	9 155 (2.0)	13 669 (2.3)	11 457 (1.8)	11 235 (2.1)	10 347 (2.2)	8 658 (1.7)	10 079 (2.0)	9 804 (1.9)	8 260 (1.8)	10 961 (2.5)	11 256 (2.3)	7 494 (1.7)	8 979 (1.9)	5 619 (1.3)	11 894 (2.7)	7 672 (1.7)	13 622 (3.8)	10 244 (3.0)	9 810	(2.5)
Common Merganser	6 035 (1.2)	8 998 (1.6)	13 063 (2.8)	12 710 (2.1)	10 545 (1.7)	14 389 (2.7)	11 887 (2.6)	15 811 (3.2)	13 499 (2.7)	16 175 (3.2)	14 364 (3.2)	14 757 (3.4)	12 226 (2.5)	11 431 (2.6)	10 979 (2.3)	7 576 (1.7)	7 277 (1.7)	11 290 (2.5)	10 968 (3.1)	7 539 (2.2)	8 930	(2.2)
Red-breasted Merganser*	2 986 (0.6)	2 869 (0.5)	5 647 (1.2)	4 575 (0.8)	6 031 (1.0)	1 533 (0.3)	3 335 (0.7)	3 438 (0.7)	3 717 (0.7)	7 426 (1.4)	4 835 (1.1)	4 119 (0.9)	6 407 (1.3)	6 920 (1.6)	3 078 (0.7)	3 216 (0.7)	3 786 (0.9)	3 235 (0.7)	4 144 (1.2)	2 277 (0.7)	3 332	(0.8)
Ruddy Duck	*p*	*p*	*p*	*p*	*p*	*p*	*p*	*p*	*p*	*p*	*p*	*p*	*p*	*p*	*p*	*p*	*p*	*p*	*p*	*p*	*p*	
Total – Ducks:	**493 081 (100.0)**	**569 457 (100.0)**	**467 894 (100.0)**	**602 901 (100.0)**	**631 240 (100.0)**	**529 082 (100.0)**	**460 354 (100.0)**	**500 245 (100.0)**	**507 924 (100.0)**	**512 452 (100.0)**	**454 006 (100.0)**	**439 566 (100.0)**	**494 931 (100.0)**	**434 145 (100.0)**	**468 881 (100.0)**	**445 554 (100.0)**	**433 786 (100.0)**	**456 329 (100.0)**	**357 425 (100.0)**	**348 913 (100.0)**	**407 713**	**(100.0)**
Snow Goose	25 863 (47.9)	9 826 (18.3)	35 711 (68.4)	27 061 (50.8)	20 606 (28.6)	43 297 (39.9)	26 557 (33.3)	68 549 (55.7)	31 051 (55.6)	43 865 (59.6)	46 316 (55.3)	49 426 (66.6)	25 431 (46.3)	9 932 (19.8)	39 058 (38.8)	45 259 (68.8)	44 698 (44.5)	62 940 (53.4)	49 869 (49.4)	27 601 (48.3)	46 073	(52.3)
Brant	*p*	*p*	*p*	*p*	*p*	*p*	*p*	*p*	*p*	*p*	*p*	*p*	*p*	*p*	*p*	*p*	*p*	*p*	*p*	*p*	*p*	
Canada Goose	28 101 (52.1)	43 470 (80.8)	16 494 (31.6)	26 186 (49.2)	51 372 (71.4)	65 266 (60.1)	52 694 (66.1)	51 386 (41.8)	24 801 (44.4)	29 364 (39.9)	37 465 (44.7)	24 220 (32.6)	29 202 (53.1)	39 577 (78.9)	60 407 (60.1)	20 424 (31.1)	55 852 (55.5)	54 740 (46.5)	50 716 (50.2)	28 012 (49.0)	41 949	(46.5)
Total – Geese:	**53 964 (100.0)**	**53 796 (100.0)**	**52 205 (100.0)**	**53 247 (100.0)**	**71 978 (100.0)**	**108 563 (100.0)**	**79 667 (100.0)**	**123 028 (100.0)**	**55 852 (100.0)**	**73 608 (100.0)**	**83 781 (100.0)**	**74 210 (100.0)**	**54 945 (100.0)**	**50 171 (100.0)**	**100 543 (100.0)**	**65 684 (100.0)**	**100 550 (100.0)**	**117 679 (100.0)**	**100 940 (100.0)**	**57 150 (100.0)**	**88 401**	**(100.0)**
Common Snipe	38 035	46 624	38 843	47 298	26 968	41 027	30 117	26 973	20 376	30 901	25 362	28 916	21 791	23 567	25 544	15 747	20 373	16 059	15 171	15 119	16 493	
American Woodcock	33 751	20 094	39 921	56 112	30 610	47 011	32 112	47 175	27 891	36 560	36 855	37 279	30 094	33 995	45 866	43 967	33 501	37 432	29 167	29 250	34 663	
No. of permit sales in Québec	57 247	58 345	63 768	66 453	72 828	74 745	73 209	76 133	75 178	72 850	67 700	65 308	60 823	59 685	55 124	57 206	54 605	54 700	53 739	49 262	53 904	
No. of permit sales in Canada	452 841	434 162	471 320	484 771	520 530	524 946	504 459	498 916	465 352	465 820	443 332	414 958	375 743	379 821	341 263	323 324	315 300	311 714	295 777	277 006	304 624	

Appendix III
Species status in Québec

The following is a list of the 429 species of birds in the wild that have been observed to date in Québec and have been documented by a specimen or photograph, or at the very least, a sight record with an acceptable published description. The status of each species in Québec is indicated by a code, with some species having a second code qualifying the first, applicable to only a small proportion of the individuals that frequent Québec, as follows:

SB **(sedentary breeder)**: A year-round resident species that breeds in Québec and usually does not undertake an annual migration. The adults' post-breeding movements, if they occur at all, are strictly local (Spruce Grouse).

RB **(resident breeder)**: A year-round resident species that breeds in Québec and all or most of the population winters within the Québec portion of its breeding range. Pre- and post-breeding movements usually involve only short distances and may be annual (Blue Jay, Black-capped Chickadee), cyclical (Northern Goshawk), irregular or unpredictable (Black-backed Woodpecker).

MB **(migrant breeder)**: A summer resident species that breeds in Québec and all or most of the population winters outside the Québec portion of its breeding range. Pre- and post-breeding movements are migrations in the true sense (Bobolink, Song Sparrow).

T **(transient)**: Species seen regularly in migration between its breeding and wintering grounds, both of which are located outside Québec (Black-bellied Plover).

V **(visitant)**: Species observed irregularly, being found outside its usual range because of various dispersal-related phenomena: vagrants, movements outside usual migration routes, pre- and post-breeding dispersal, etc. (Snowy Egret).

IN **(introduced)**: According to current knowledge, species that has been released deliberately or that has escaped accidentally from captivity, that is not acclimated but which may have bred on one or more occasions in Québec (Monk Parakeet).

PC **(probably from captivity)**: Species that probably originated from captivity but may have been born in the wild (Northern Bobwhite).

E **(extinct)**: Species that no longer exists in Québec or anywhere else in the world.

?: status has not been confirmed but is likely.

ENGLISH NAME	FRENCH NAME	LATIN NAME	STATUS
Red-throated Loon	Huart à gorge rousse	*Gavia stellata*	MB
Pacific Loon	Huart du Pacifique	*Gavia pacifica*	MB
Common Loon	Huart à collier	*Gavia immer*	MB
Yellow-billed Loon	Huart à bec blanc	*Gavia adamsii*	V
Pied-billed Grebe	Grèbe à bec bigarré	*Podilymbus podiceps*	MB
Horned Grebe	Grèbe cornu	*Podiceps auritus*	MB
Red-necked Grebe	Grèbe jougris	*Podiceps grisegena*	MB
Eared Grebe	Grèbe à cou noir	*Podiceps nigricollis*	V
Western Grebe	Grèbe élégant	*Aechmophorus occidentalis*	V
Yellow-nosed Albatross	Albatros à nez jaune	*Diomedea chlororhynchos*	V
Northern Fulmar	Fulmar boréal	*Fulmarus glacialis*	T
Cory's Shearwater	Puffin cendré	*Calonectris diomedea*	V
Greater Shearwater	Puffin majeur	*Puffinus gravis*	T
Sooty Shearwater	Puffin fuligineux	*Puffinus griseus*	T
Manx Shearwater	Puffin des Anglais	*Puffinus puffinus*	T
Wilson's Storm-Petrel	Pétrel océanite	*Oceanites oceanicus*	T
Leach's Storm-Petrel	Pétrel cul-blanc	*Oceanodroma leucorhoa*	MB
Northern Gannet	Fou de Bassan	*Morus bassanus*	MB
American White Pelican	Pélican blanc d'Amérique	*Pelecanus erythrorhynchos*	V
Brown Pelican	Pélican brun	*Pelecanus occidentalis*	V
Great Cormorant	Grand Cormoran	*Phalacrocorax carbo*	MB
Double-crested Cormorant	Cormoran à aigrettes	*Phalacrocorax auritus*	MB
American Bittern	Butor d'Amérique	*Botaurus lentiginosus*	MB
Least Bittern	Petit Butor	*Ixobrychus exilis*	MB
Great Blue Heron	Grand Héron	*Ardea herodias*	MB
Great Egret	Grande Aigrette	*Casmerodius albus*	MB
Little Egret	Aigrette garzette	*Egretta garzetta*	V
Snowy Egret	Aigrette neigeuse	*Egretta thula*	V
Little Blue Heron	Aigrette bleue	*Egretta caerulea*	V
Tricolored Heron	Aigrette tricolore	*Egretta tricolor*	V
Cattle Egret	Héron garde-boeufs	*Bubulcus ibis*	V
Green Heron	Héron vert	*Butorides virescens*	MB
Black-crowned Night-Heron	Bihoreau à couronne noire	*Nycticorax nycticorax*	MB
Yellow-crowned Night-Heron	Bihoreau violacé	*Nyctanassa violacea*	V
White Ibis	Ibis blanc	*Eudocimus albus*	V
Glossy Ibis	Ibis falcinelle	*Plegadis falcinellus*	V
Greater Flamingo	Flamant rose	*Phoenicopterus ruber*	PC
Fulvous Whistling-Duck	Dendrocygne fauve	*Dendrocygna bicolor*	V
Black-bellied Whistling-Duck	Dendrocygne à ventre noir	*Dendrocygna autumnalis*	PC
Tundra Swan	Cygne siffleur	*Cygnus columbianus*	MB
Whooper Swan	Cygne chanteur	*Cygnus cygnus*	V
Trumpeter Swan	Cygne trompette	*Cygnus buccinator*	V
Mute Swan	Cygne tuberculé	*Cygnus olor*	V (MB)
Bean Goose	Oie des moissons	*Anser fabalis*	V
Pink-footed Goose	Oie à bec court	*Anser brachyrhynchus*	V
Greater White-fronted Goose	Oie rieuse	*Anser albifrons*	T
Snow Goose	Oie des neiges	*Chen caerulescens*	T (MB)
Ross' Goose	Oie de Ross	*Chen rossii*	T
Brant	Bernache cravant	*Branta bernicla*	T (MB)
Barnacle Goose	Bernache nonnette	*Branta leucopsis*	PC
Canada Goose	Bernache du Canada	*Branta canadensis*	MB
Ruddy Shelduck	Tadorne casarca	*Tadorna ferruginea*	PC
Wood Duck	Canard branchu	*Aix sponsa*	MB
Green-winged Teal	Sarcelle à ailes vertes	*Anas crecca*	MB
American Black Duck	Canard noir	*Anas rubripes*	MB
Mallard	Canard colvert	*Anas platyrhynchos*	MB
Northern Pintail	Canard pilet	*Anas acuta*	MB
Garganey	Sarcelle d'été	*Anas querquedula*	PC
Blue-winged Teal	Sarcelle à ailes bleues	*Anas discors*	MB
Cinnamon Teal	Sarcelle cannelle	*Anas cyanoptera*	PC
Northern Shoveler	Canard souchet	*Anas clypeata*	MB
Gadwall	Canard chipeau	*Anas strepera*	MB
Eurasian Wigeon	Canard siffleur d'Europe	*Anas penelope*	T
American Wigeon	Canard siffleur d'Amérique	*Anas americana*	MB
Canvasback	Morillon à dos blanc	*Aythya valisineria*	T
Redhead	Morillon à tête rouge	*Aythya americana*	MB
Ring-necked Duck	Morillon à collier	*Aythya collaris*	MB
Tufted Duck	Fuligule Morillon	*Aythya fuligula*	PC
Greater Scaup	Grand Morillon	*Aythya marila*	MB
Lesser Scaup	Petit Morillon	*Aythya affinis*	MB
Common Eider	Eider à duvet	*Somateria mollissima*	MB
King Eider	Eider à tête grise	*Somateria spectabilis*	MB
Steller's Eider	Eider de Steller	*Polysticta stelleri*	V
Labrador Duck	Canard du Labrador	*Camptorhynchus labradorius*	E
Harlequin Duck	Canard arlequin	*Histrionicus histrionicus*	MB
Oldsquaw	Canard kakawi	*Clangula hyemalis*	MB
Black Scoter	Macreuse à bec jaune	*Melanitta nigra*	MB
Surf Scoter	Macreuse à front blanc	*Melanitta perspicillata*	MB
White-winged Scoter	Macreuse à ailes blanches	*Melanitta fusca*	MB
Common Goldeneye	Garrot à oeil d'or	*Bucephala clangula*	MB
Barrow's Goldeneye	Garrot de Barrow	*Bucephala islandica*	T (MB?)
Bufflehead	Petit Garrot	*Bucephala albeola*	MB
Smew	Harle piette	*Mergellus albellus*	PC
Hooded Merganser	Bec-scie couronné	*Lophodytes cucullatus*	MB
Common Merganser	Grand Bec-scie	*Mergus merganser*	MB
Red-breasted Merganser	Bec-scie à poitrine rousse	*Mergus serrator*	MB
Ruddy Duck	Canard roux	*Oxyura jamaicensis*	MB
Black Vulture	Urubu noir	*Coragyps atratus*	V
Turkey Vulture	Urubu à tête rouge	*Cathartes aura*	MB
Osprey	Balbuzard	*Pandion haliaetus*	MB
American Swallow-tailed Kite	Milan à queue fourchue	*Elanoides forficatus*	V
Bald Eagle	Pygargue à tête blanche	*Haliaeetus leucocephalus*	MB

ENGLISH NAME	FRENCH NAME	LATIN NAME	STATUS
Northern Harrier	Busard Saint-Martin	*Circus cyaneus*	MB
Sharp-shinned Hawk	Épervier brun	*Accipiter striatus*	MB
Cooper's Hawk	Épervier de Cooper	*Accipiter cooperii*	MB
Northern Goshawk	Autour des palombes	*Accipiter gentilis*	RB
Red-shouldered Hawk	Buse à épaulettes	*Buteo lineatus*	MB
Broad-winged Hawk	Petite Buse	*Buteo platypterus*	MB
Swainson's Hawk	Buse de Swainson	*Buteo swainsoni*	V
Red-tailed Hawk	Buse à queue rousse	*Buteo jamaicensis*	MB
Rough-legged Hawk	Buse pattue	*Buteo lagopus*	MB
Golden Eagle	Aigle royal	*Aquila chrysaetos*	MB
American Kestrel	Crécerelle d'Amérique	*Falco sparverius*	MB
Merlin	Faucon émerillon	*Falco columbarius*	MB
Peregrine Falcon	Faucon pèlerin	*Falco peregrinus*	MB
Gyrfalcon	Faucon gerfaut	*Falco rusticolus*	MB
Gray Partridge	Perdrix grise	*Perdix perdix*	SB
Chukar	Perdrix choukar	*Alectoris chukar*	IN
Common Quail	Caille des blés	*Coturnix coturnix*	IN
Ring-necked Pheasant	Faisan de chasse	*Phasianus colchicus*	IN
Spruce Grouse	Tétras du Canada	*Dendragapus canadensis*	SB
Willow Ptarmigan	Lagopède des saules	*Lagopus lagopus*	RB
Rock Ptarmigan	Lagopède des rochers	*Lagopus mutus*	RB
Ruffed Grouse	Gélinotte huppée	*Bonasa umbellus*	SB
Sharp-tailed Grouse	Gélinotte à queue fine	*Tympanuchus phasianellus*	SB
Wild Turkey	Dindon sauvage	*Meleagris gallopavo*	SB
Northern Bobwhite	Colin de Virginie	*Colinus virginianus*	IN
Yellow Rail	Râle jaune	*Coturnicops noveboracensis*	MB
King Rail	Râle élégant	*Rallus elegans*	V
Virginia Rail	Râle de Virginie	*Rallus limicola*	MB
Sora	Râle de Caroline	*Porzana carolina*	MB
Purple Gallinule	Gallinule violacée	*Porphyrula martinica*	V
Common Moorhen	Poule-d'eau	*Gallinula chloropus*	MB
American Coot	Foulque d'Amérique	*Fulica americana*	MB
Sandhill Crane	Grue du Canada	*Grus canadensis*	MB
Northern Lapwing	Vanneau huppé	*Vanellus vanellus*	V
Black-bellied Plover	Pluvier argenté	*Pluvialis squatarola*	T
Greater Golden-Plover	Pluvier doré d'Eurasie	*Pluvialis apricaria*	V
American Golden-Plover	Pluvier doré d'Amérique	*Pluvialis dominica*	T
Common Ringed Plover	Grand Gravelot	*Charadrius hiaticula*	V
Semipalmated Plover	Pluvier semipalmé	*Charadrius semipalmatus*	MB
Piping Plover	Pluvier siffleur	*Charadrius melodus*	MB
Killdeer	Pluvier kildir	*Charadrius vociferus*	MB
American Oystercatcher	Huîtrier d'Amérique	*Haematopus palliatus*	V
Black-necked Stilt	Échasse d'Amérique	*Himantopus mexicanus*	V
American Avocet	Avocette d'Amérique	*Recurvirostra americana*	V
Common Greenshank	Chevalier aboyeur	*Tringa nebularia*	V
Greater Yellowlegs	Grand Chevalier	*Tringa melanoleuca*	MB
Lesser Yellowlegs	Petit Chevalier	*Tringa flavipes*	MB
Solitary Sandpiper	Chevalier solitaire	*Tringa solitaria*	MB
Willet	Chevalier semipalmé	*Catoptrophorus semipalmatus*	V (MB?)
Spotted Sandpiper	Chevalier branlequeue	*Actitis macularia*	MB
Upland Sandpiper	Maubèche des champs	*Bartramia longicauda*	MB
Eskimo Curlew	Courlis esquimau	*Numenius borealis*	E ?
Whimbrel	Courlis corlieu	*Numenius phaeopus*	T
Long-billed Curlew	Courlis à long bec	*Numenius americanus*	V
Black-tailed Godwit	Barge à queue noire	*Limosa limosa*	V
Hudsonian Godwit	Barge hudsonienne	*Limosa haemastica*	T
Bar-tailed Godwit	Barge rousse	*Limosa lapponica*	V
Marbled Godwit	Barge marbrée	*Limosa fedoa*	V (MB?)
Ruddy Turnstone	Tournepierre à collier	*Arenaria interpres*	T
Red Knot	Bécasseau maubèche	*Calidris canutus*	T
Sanderling	Bécasseau sanderling	*Calidris alba*	T
Semipalmated Sandpiper	Bécasseau semipalmé	*Calidris pusilla*	MB
Western Sandpiper	Bécasseau d'Alaska	*Calidris mauri*	T
Temminck's Stint	Bécasseau de Temminck	*Calidris temminckii*	V

ENGLISH NAME	FRENCH NAME	LATIN NAME	STATUS
Least Sandpiper	Bécasseau minuscule	*Calidris minutilla*	MB
White-rumped Sandpiper	Bécasseau à croupion blanc	*Calidris fuscicollis*	T
Baird's Sandpiper	Bécasseau de Baird	*Calidris bairdii*	T
Pectoral Sandpiper	Bécasseau à poitrine cendrée	*Calidris melanotos*	T
Sharp-tailed Sandpiper	Bécasseau à queue fine	*Calidris acuminata*	V
Purple Sandpiper	Bécasseau violet	*Calidris maritima*	T
Dunlin	Bécasseau variable	*Calidris alpina*	T
Curlew Sandpiper	Bécasseau cocorli	*Calidris ferruginea*	V
Stilt Sandpiper	Bécasseau à échasses	*Calidris himantopus*	T
Buff-breasted Sandpiper	Bécasseau roussâtre	*Tryngites subruficollis*	T
Ruff	Bécasseau combattant	*Philomachus pugnax*	V
Short-billed Dowitcher	Bécasseau roux	*Limnodromus griseus*	MB
Long-billed Dowitcher	Bécasseau à long bec	*Limnodromus scolopaceus*	V
Common Snipe	Bécassine des marais	*Gallinago gallinago*	MB
Eurasian Woodcock	Bécasse des bois	*Scolopax rusticola*	V
American Woodcock	Bécasse d'Amérique	*Scolopax minor*	MB
Wilson's Phalarope	Phalarope de Wilson	*Phalaropus tricolor*	MB
Red-necked Phalarope	Phalarope hyperboréen	*Phalaropus lobatus*	MB
Red Phalarope	Phalarope roux	*Phalaropus fulicaria*	MB
Pomarine Jaeger	Labbe pomarin	*Stercorarius pomarinus*	MB
Parasitic Jaeger	Labbe parasite	*Stercorarius parasiticus*	MB
Long-tailed Jaeger	Labbe à longue queue	*Stercorarius longicaudus*	MB
Great Skua	Grand Labbe	*Catharacta skua*	V
South Polar Skua	Labbe antarctique	*Catharacta maccormicki*	V
Laughing Gull	Mouette à tête noire	*Larus atricilla*	V
Franklin's Gull	Mouette de Franklin	*Larus pipixcan*	V
Little Gull	Mouette pygmée	*Larus minutus*	V (MB)
Common Black-headed Gull	Mouette rieuse	*Larus ridibundus*	MB
Bonaparte's Gull	Mouette de Bonaparte	*Larus philadelphia*	MB
Mew Gull	Goéland cendré	*Larus canus*	V
Ring-billed Gull	Goéland à bec cerclé	*Larus delawarensis*	MB
California Gull	Goéland de Californie	*Larus californicus*	V
Herring Gull	Goéland argenté	*Larus argentatus*	MB
Yellow-legged Gull	Goéland leucophée	*Larus cachinnans*	V
Thayer's Gull (see Iceland Gull)			
Iceland Gull	Goéland arctique	*Larus glaucoides*	MB
Lesser Black-backed Gull	Goéland brun	*Larus fuscus*	V
Slaty-backed Gull	Goéland à manteau ardoisé	*Larus schistisagus*	V
Glaucous Gull	Goéland bourgmestre	*Larus hyperboreus*	MB
Great Black-backed Gull	Goéland à manteau noir	*Larus marinus*	MB
Black-legged Kittiwake	Mouette tridactyle	*Rissa tridactyla*	MB
Ross' Gull	Mouette rosée	*Rhodostethia rosea*	V
Sabine's Gull	Mouette de Sabine	*Xema sabini*	T
Ivory Gull	Mouette blanche	*Pagophila eburnea*	V
Caspian Tern	Sterne caspienne	*Sterna caspia*	MB
Sandwich Tern	Sterne caugek	*Sterna sandvicensis*	V
Roseate Tern	Sterne de Dougall	*Sterna dougallii*	MB
Common Tern	Sterne pierregarin	*Sterna hirundo*	MB
Arctic Tern	Sterne arctique	*Sterna paradisaea*	MB
Forster's Tern	Sterne de Forster	*Sterna forsteri*	V
Sooty Tern	Sterne fuligineuse	*Sterna fuscata*	V
White-winged Tern	Guifette leucoptère	*Chlidonias leucopterus*	V (MB)
Black Tern	Guifette noire	*Chlidonias niger*	MB
Black Skimmer	Bec-en-ciseaux noir	*Rynchops niger*	V
Dovekie	Mergule nain	*Alle alle*	T
Common Murre	Marmette de Troïl	*Uria aalge*	MB
Thick-billed Murre	Marmette de Brünnich	*Uria lomvia*	MB
Razorbill	Petit Pingouin	*Alca torda*	MB
Great Auk	Grand Pingouin	*Pinguinus impennis*	E
Black Guillemot	Guillemot à miroir	*Cepphus grylle*	MB
Marbled Murrelet	Alque marbrée	*Brachyramphus marmoratus*	V
Atlantic Puffin	Macareux moine	*Fratercula arctica*	MB
Rock Dove	Pigeon biset	*Columba livia*	SB

ENGLISH NAME	FRENCH NAME	LATIN NAME	STATUS
Laughing Dove	Tourterelle maillée	*Streptopelia senegalensis*	PC
Ringed Turtle-Dove	Tourterelle rieuse	*Streptopelia risoria*	IN
White-winged Dove	Tourterelle à ailes blanches	*Zenaida asiatica*	V
Mourning Dove	Tourterelle triste	*Zenaida macroura*	MB
Passenger Pigeon	Tourte	*Ectopistes migratorius*	E
Monk Parakeet	Conure veuve	*Myiopsitta monachus*	IN
Black-billed Cuckoo	Coulicou à bec noir	*Coccyzus erythropthalmus*	MB
Yellow-billed Cuckoo	Coulicou à bec jaune	*Coccyzus americanus*	V (MB)
Mangrove Cuckoo	Coulicou manioc	*Coccyzus minor*	V
Barn Owl	Effraie des clochers	*Tyto alba*	V (MB)
Eastern Screech-Owl	Petit-duc maculé	*Otus asio*	SB
Great Horned Owl	Grand-duc d'Amérique	*Bubo virginianus*	SB
Snowy Owl	Harfang des neiges	*Nyctea scandiaca*	RB
Northern Hawk Owl	Chouette épervière	*Surnia ulula*	RB
Burrowing Owl	Chouette des terriers	*Athene cunicularia*	V
Barred Owl	Chouette rayée	*Strix varia*	SB
Great Gray Owl	Chouette lapone	*Strix nebulosa*	RB
Long-eared Owl	Hibou moyen-duc	*Asio otus*	MB
Short-eared Owl	Hibou des marais	*Asio flammeus*	MB
Boreal Owl	Nyctale boréale	*Aegolius funereus*	RB
Northern Saw-whet Owl	Petite Nyctale	*Aegolius acadicus*	MB
Common Nighthawk	Engoulevent d'Amérique	*Chordeiles minor*	MB
Chuck-will's-widow	Engoulevent de Caroline	*Caprimulgus carolinensis*	V
Whip-poor-will	Engoulevent bois-pourri	*Caprimulgus vociferus*	MB
Chimney Swift	Martinet ramoneur	*Chaetura pelagica*	MB
Ruby-throated Hummingbird	Colibri à gorge rubis	*Archilochus colubris*	MB
Belted Kingfisher	Martin-pêcheur d'Amérique	*Ceryle alcyon*	MB
Red-headed Woodpecker	Pic à tête rouge	*Melanerpes erythrocephalus*	MB
Red-bellied Woodpecker	Pic à ventre roux	*Melanerpes carolinus*	V
Yellow-bellied Sapsucker	Pic maculé	*Sphyrapicus varius*	MB
Downy Woodpecker	Pic mineur	*Picoides pubescens*	RB
Hairy Woodpecker	Pic chevelu	*Picoides villosus*	RB
Three-toed Woodpecker	Pic tridactyle	*Picoides tridactylus*	RB
Black-backed Woodpecker	Pic à dos noir	*Picoides arcticus*	RB
Northern Flicker	Pic flamboyant	*Colaptes auratus*	MB
Pileated Woodpecker	Grand Pic	*Dryocopus pileatus*	SB
Olive-sided Flycatcher	Moucherolle à côtés olive	*Contopus borealis*	MB
Eastern Wood-Pewee	Pioui de l'Est	*Contopus virens*	MB
Yellow-bellied Flycatcher	Moucherolle à ventre jaune	*Empidonax flaviventris*	MB
Acadian Flycatcher	Moucherolle vert	*Empidonax virescens*	V
Alder Flycatcher	Moucherolle des aulnes	*Empidonax alnorum*	MB
Willow Flycatcher	Moucherolle des saules	*Empidonax traillii*	MB
Least Flycatcher	Moucherolle tchébec	*Empidonax minimus*	MB
Eastern Phoebe	Moucherolle phébi	*Sayornis phoebe*	MB
Say's Phoebe	Moucherolle à ventre roux	*Sayornis saya*	V
Great Crested Flycatcher	Tyran huppé	*Myiarchus crinitus*	MB
Tropical Kingbird	Tyran mélancolique	*Tyrannus melancholicus*	V
Western Kingbird	Tyran de l'Ouest	*Tyrannus verticalis*	V
Eastern Kingbird	Tyran tritri	*Tyrannus tyrannus*	MB
Scissor-tailed Flycatcher	Tyran à longue queue	*Tyrannus forficatus*	V
Fork-tailed Flycatcher	Tyran à queue fourchue	*Tyrannus savana*	V
Horned Lark	Alouette cornue	*Eremophila alpestris*	MB
Purple Martin	Hirondelle noire	*Progne subis*	MB
Tree Swallow	Hirondelle bicolore	*Tachycineta bicolor*	MB
Northern Rough-winged Swallow	Hirondelle à ailes hérissées	*Stelgidopteryx serripennis*	MB
Bank Swallow	Hirondelle de rivage	*Riparia riparia*	MB
Cliff Swallow	Hirondelle à front blanc	*Hirundo pyrrhonota*	MB
Barn Swallow	Hirondelle des granges	*Hirundo rustica*	MB
Gray Jay	Geai du Canada	*Perisoreus canadensis*	RB
Steller's Jay	Geai de Steller	*Cyanocitta stelleri*	V
Blue Jay	Geai bleu	*Cyanocitta cristata*	RB
Black-billed Magpie	Pie bavarde	*Pica pica*	V
Eurasian Jackdaw	Choucas des tours	*Corvus monedula*	V

ENGLISH NAME	FRENCH NAME	LATIN NAME	STATUS
American Crow	Corneille d'Amérique	*Corvus brachyrhynchos*	MB
Pied Crow	Corbeau pie	*Corvus albus*	PC
Common Raven	Grand Corbeau	*Corvus corax*	SB
Black-capped Chickadee	Mésange à tête noire	*Parus atricapillus*	RB
Boreal Chickadee	Mésange à tête brune	*Parus hudsonicus*	RB
Tufted Titmouse	Mésange bicolore	*Parus bicolor*	V (MB ?)
Red-breasted Nuthatch	Sittelle à poitrine rousse	*Sitta canadensis*	RB
White-breasted Nuthatch	Sittelle à poitrine blanche	*Sitta carolinensis*	SB
Brown Creeper	Grimpereau brun	*Certhia americana*	MB
Carolina Wren	Troglodyte de Caroline	*Thryothorus ludovicianus*	V (SB)
Bewick's Wren	Troglodyte de Bewick	*Thryomanes bewickii*	V
House Wren	Troglodyte familier	*Troglodytes aedon*	MB
Winter Wren	Troglodyte des forêts	*Troglodytes troglodytes*	MB
Sedge Wren	Troglodyte à bec court	*Cistothorus platensis*	MB
Marsh Wren	Troglodyte des marais	*Cistothorus palustris*	MB
Golden-crowned Kinglet	Roitelet à couronne dorée	*Regulus satrapa*	MB
Ruby-crowned Kinglet	Roitelet à couronne rubis	*Regulus calendula*	MB
Blue-gray Gnatcatcher	Gobemoucheron gris-bleu	*Polioptila caerulea*	MB
Northern Wheatear	Traquet motteux	*Oenanthe oenanthe*	MB
Eastern Bluebird	Merle-bleu de l'Est	*Sialia sialis*	MB
Mountain Bluebird	Merle-bleu azuré	*Sialia currucoides*	V
Townsend's Solitaire	Solitaire de Townsend	*Myadestes townsendi*	V
Veery	Grive fauve	*Catharus fuscescens*	MB
Gray-cheeked Thrush	Grive à joues grises	*Catharus minimus*	MB
Bicknell's Thrush	Grive de Bicknell	*Catharus bicknelli*	MB
Swainson's Thrush	Grive à dos olive	*Catharus ustulatus*	MB
Hermit Thrush	Grive solitaire	*Catharus guttatus*	MB
Wood Thrush	Grive des bois	*Hylocichla mustelina*	MB
Eurasian Blackbird	Merle noir	*Turdus merula*	PC
Fieldfare	Grive litorne	*Turdus pilaris*	V
American Robin	Merle d'Amérique	*Turdus migratorius*	MB
Varied Thrush	Grive à collier	*Ixoreus naevius*	V
Gray Catbird	Moqueur chat	*Dumetella carolinensis*	MB
Northern Mockingbird	Moqueur polyglotte	*Mimus polyglottos*	MB
Brown Thrasher	Moqueur roux	*Toxostoma rufum*	MB
American Pipit	Pipit d'Amérique	*Anthus rubescens*	MB
Bohemian Waxwing	Jaseur boréal	*Bombycilla garrulus*	V (MB ?)
Cedar Waxwing	Jaseur des cèdres	*Bombycilla cedrorum*	MB
Northern Shrike	Pie-grièche grise	*Lanius excubitor*	RB
Loggerhead Shrike	Pie-grièche migratrice	*Lanius ludovicianus*	MB
European Starling	Étourneau sansonnet	*Sturnus vulgaris*	MB
White-eyed Vireo	Viréo aux yeux blancs	*Vireo griseus*	V
Solitary Vireo	Viréo à tête bleue	*Vireo solitarius*	MB
Yellow-throated Vireo	Viréo à gorge jaune	*Vireo flavifrons*	MB
Warbling Vireo	Viréo mélodieux	*Vireo gilvus*	MB
Philadelphia Vireo	Viréo de Philadelphie	*Vireo philadelphicus*	MB
Red-eyed Vireo	Viréo aux yeux rouges	*Vireo olivaceus*	MB
Blue-winged Warbler	Paruline à ailes bleues	*Vermivora pinus*	V
Golden-winged Warbler	Paruline à ailes dorées	*Vermivora chrysoptera*	MB
Tennessee Warbler	Paruline obscure	*Vermivora peregrina*	MB
Orange-crowned Warbler	Paruline verdâtre	*Vermivora celata*	MB
Nashville Warbler	Paruline à joues grises	*Vermivora ruficapilla*	MB
Northern Parula	Paruline à collier	*Parula americana*	MB
Yellow Warbler	Paruline jaune	*Dendroica petechia*	MB
Chestnut-sided Warbler	Paruline à flancs marron	*Dendroica pensylvanica*	MB
Magnolia Warbler	Paruline à tête cendrée	*Dendroica magnolia*	MB
Cape May Warbler	Paruline tigrée	*Dendroica tigrina*	MB
Black-throated Blue Warbler	Paruline bleue à gorge noire	*Dendroica caerulescens*	MB
Yellow-rumped Warbler	Paruline à croupion jaune	*Dendroica coronata*	MB
Black-throated Gray Warbler	Paruline grise à gorge noire	*Dendroica nigrescens*	V
Hermit Warbler	Paruline à tête jaune	*Dendroica occidentalis*	V
Black-throated Green Warbler	Paruline verte à gorge noire	*Dendroica virens*	MB
Blackburnian Warbler	Paruline à gorge orangée	*Dendroica fusca*	MB
Yellow-throated Warbler	Paruline à gorge jaune	*Dendroica dominica*	V

ENGLISH NAME	FRENCH NAME	LATIN NAME	STATUS
Pine Warbler	Paruline des pins	*Dendroica pinus*	MB
Kirtland's Warbler	Paruline de Kirtland	*Dendroica kirtlandii*	V
Prairie Warbler	Paruline des prés	*Dendroica discolor*	V
Palm Warbler	Paruline à couronne rousse	*Dendroica palmarum*	MB
Bay-breasted Warbler	Paruline à poitrine baie	*Dendroica castanea*	MB
Blackpoll Warbler	Paruline rayée	*Dendroica striata*	MB
Cerulean Warbler	Paruline azurée	*Dendroica cerulea*	MB
Black-and-white Warbler	Paruline noir et blanc	*Mniotilta varia*	MB
American Redstart	Paruline flamboyante	*Setophaga ruticilla*	MB
Prothonotary Warbler	Paruline orangée	*Protonotaria citrea*	V
Worm-eating Warbler	Paruline vermivore	*Helmitheros vermivorus*	V
Ovenbird	Paruline couronnée	*Seiurus aurocapillus*	MB
Northern Waterthrush	Paruline des ruisseaux	*Seiurus noveboracensis*	MB
Louisiana Waterthrush	Paruline hochequeue	*Seiurus motacilla*	V
Kentucky Warbler	Paruline du Kentucky	*Oporornis formosus*	V
Connecticut Warbler	Paruline à gorge grise	*Oporornis agilis*	MB
Mourning Warbler	Paruline triste	*Oporornis philadelphia*	MB
Common Yellowthroat	Paruline masquée	*Geothlypis trichas*	MB
Hooded Warbler	Paruline à capuchon	*Wilsonia citrina*	V
Wilson's Warbler	Paruline à calotte noire	*Wilsonia pusilla*	MB
Canada Warbler	Paruline du Canada	*Wilsonia canadensis*	MB
Yellow-breasted Chat	Paruline polyglotte	*Icteria virens*	V
Hepatic Tanager	Tangara orangé	*Piranga flava*	V
Summer Tanager	Tangara vermillon	*Piranga rubra*	V
Scarlet Tanager	Tangara écarlate	*Piranga olivacea*	MB
Western Tanager	Tangara à tête rouge	*Piranga ludoviciana*	V
Northern Cardinal	Cardinal rouge	*Cardinalis cardinalis*	SB
Rose-breasted Grosbeak	Cardinal à poitrine rose	*Pheucticus ludovicianus*	MB
Black-headed Grosbeak	Cardinal à tête noire	*Pheucticus melanocephalus*	V
Blue Grosbeak	Passerin bleu	*Guiraca caerulea*	V
Indigo Bunting	Passerin indigo	*Passerina cyanea*	MB
Painted Bunting	Passerin nonpareil	*Passerina ciris*	V
Dickcissel	Dickcissel	*Spiza americana*	V
Green-tailed Towhee	Tohi à queue verte	*Pipilo chlorurus*	V
Rufous-sided Towhee	Tohi à flancs roux	*Pipilo erythrophthalmus*	MB
American Tree Sparrow	Bruant hudsonien	*Spizella arborea*	MB
Chipping Sparrow	Bruant familier	*Spizella passerina*	MB
Clay-colored Sparrow	Bruant des plaines	*Spizella pallida*	V (MB)
Field Sparrow	Bruant des champs	*Spizella pusilla*	MB
Vesper Sparrow	Bruant vespéral	*Pooecetes gramineus*	MB
Lark Sparrow	Bruant à joues marron	*Chondestes grammacus*	V
Black-throated Sparrow	Bruant à gorge noire	*Amphispiza bilineata*	V

ENGLISH NAME	FRENCH NAME	LATIN NAME	STATUS
Lark Bunting	Bruant noir et blanc	*Calamospiza melanocorys*	V
Savannah Sparrow	Bruant des prés	*Passerculus sandwichensis*	MB
Grasshopper Sparrow	Bruant sauterelle	*Ammodramus savannarum*	MB
Henslow's Sparrow	Bruant de Henslow	*Ammodramus henslowii*	V (MB ?)
Le Conte's Sparrow	Bruant de Le Conte	*Ammodramus leconteii*	MB
Sharp-tailed Sparrow	Bruant à queue aiguë	*Ammodramus caudacutus*	MB
Fox Sparrow	Bruant fauve	*Passerella iliaca*	MB
Song Sparrow	Bruant chanteur	*Melospiza melodia*	MB
Lincoln's Sparrow	Bruant de Lincoln	*Melospiza lincolnii*	MB
Swamp Sparrow	Bruant des marais	*Melospiza georgiana*	MB
White-throated Sparrow	Bruant à gorge blanche	*Zonotrichia albicollis*	MB
White-crowned Sparrow	Bruant à couronne blanche	*Zonotrichia leucophrys*	MB
Harris' Sparrow	Bruant à face noire	*Zonotrichia querula*	V
Dark-eyed Junco	Junco ardoisé	*Junco hyemalis*	MB
Lapland Longspur	Bruant lapon	*Calcarius lapponicus*	MB
Smith's Longspur	Bruant de Smith	*Calcarius pictus*	V
Snow Bunting	Bruant des neiges	*Plectrophenax nivalis*	MB
Bobolink	Goglu	*Dolichonyx oryzivorus*	MB
Red-winged Blackbird	Carouge à épaulettes	*Agelaius phoeniceus*	MB
Eastern Meadowlark	Sturnelle des prés	*Sturnella magna*	MB
Western Meadowlark	Sturnelle de l'Ouest	*Sturnella neglecta*	V (MB)
Yellow-headed Blackbird	Carouge à tête jaune	*Xanthocephalus xanthocephalus*	V
Rusty Blackbird	Quiscale rouilleux	*Euphagus carolinus*	MB
Brewer's Blackbird	Quiscale de Brewer	*Euphagus cyanocephalus*	V
Common Grackle	Quiscale bronzé	*Quiscalus quiscula*	MB
Brown-headed Cowbird	Vacher à tête brune	*Molothrus ater*	MB
Orchard Oriole	Oriole des vergers	*Icterus spurius*	V
Northern Oriole	Oriole du Nord	*Icterus galbula*	MB
Brambling	Pinson du Nord	*Fringilla montifringilla*	V
Gray-crowned Rosy-Finch	Roselin à tête grise	*Leucosticte tephrocotis*	V
Pine Grosbeak	Durbec des pins	*Pinicola enucleator*	RB
Purple Finch	Roselin pourpré	*Carpodacus purpureus*	MB
House Finch	Roselin familier	*Carpodacus mexicanus*	RB
Red Crossbill	Bec-croisé rouge	*Loxia curvirostra*	RB
White-winged Crossbill	Bec-croisé à ailes blanches	*Loxia leucoptera*	RB
Common Redpoll	Sizerin flammé	*Carduelis flammea*	RB
Hoary Redpoll	Sizerin blanchâtre	*Carduelis hornemanni*	RB
Pine Siskin	Chardonneret des pins	*Carduelis pinus*	RB
American Goldfinch	Chardonneret jaune	*Carduelis tristis*	MB
European Goldfinch	Chardonneret élégant	*Carduelis carduelis*	PC
Evening Grosbeak	Gros-bec errant	*Coccothraustes vespertinus*	RB
House Sparrow	Moineau domestique	*Passer domesticus*	SB

Appendix IV
Computerizing the Atlas distribution maps

Inevitably in this day and age, the Atlas maps were produced using computer graphics, thanks to the collaboration of the federal Department of the Environment, State of the Environment Reporting Branch, Environmental Information Systems Division. The following information on how these maps were created may be of technical interest.

In the early 1960s, the federal government wanted to make a nationwide land inventory to meet land management needs. Thus was born the Canada Geographic Information System (CGIS). The designers of this system were pioneers in the creation of many concepts, technical terms, and algorithms now used in other systems. CGIS's strength lies in the way it creates large thematic databases. It is made up of several subsystems which analyse and model data in various forms for specific applications, as for example, the plotting in raster format of images from cartographic databases. This entails a complex process of selecting and converting data to create a file whose format is compatible with the plotter.

This is precisely the kind of technology that the Canadian Wildlife Service was looking for to make maps showing the breeding distribution of each species in Québec. Existing expertise and technology were harnessed to produce maps on plate-ready film.

The specifications had to allow for the overriding constraints of document size and good image definition. Tests were run to set the parameters that would give the desired quality. As a result, the perimeter of the Atlas squares had to be reduced to leave a visible boundary between them. The graphics output was generated by an OPTRONICS X4040 plotter hooked up to a minicomputer. This high-precision drum plotter achieves resolution of anything from 25 to 200 micrometres. The resulting image consists of square pixels projected onto photographic film by laser. The maps were plotted to a resolution of 50 micrometres to give the squares a linear aspect.

Using the OPTRONICS plotter imposed its own set of requirements. First, the software available on the market was analysed to establish the appropriate methodology. A processing system similar to the CGIS system had to be used to speed up implementation of the data conversion process. Then, other software programs were developed to rework the data and produce maps in accordance with the plotting specifications.

From Atlas survey data to printed map, the whole process involves three computing environments (see flowchart).

First, the raw data file is processed on a microcomputer. Each record in this file includes a UTM co-ordinate, a breeding code, and the species' taxonomic number. The computer sorts the entries by species number and orders the data. The file is then validated so that only the strongest entry, i.e., the one with the highest breeding code, is kept from among identical combinations of species number and UTM co-ordinates. Then, the records are rearranged by species (for a total of 242 species files) and copied onto magnetic tape, for transfer to the next environment.

The process of data conversion then begins to produce the plot files. This consists of three main batch processing stages run on each of the 242 species files in the Atlas database. First, the contours of the squares are vectorized using the UTM co-ordinate as the point of origin to create polygons with a perimeter of a few millimetres. A special process reduces those polygons straddling UTM zone boundaries. Breeding codes and species numbers are also represented in the form of vectors and shown on the maps to provide a guide for further processing. Next, the vector file is converted to raster format, consistent with the desired scale and pixel resolution of the output. Each pixel has an attribute code to identify the various elements in the map. In the final stage, the data are sorted by attribute code to yield two plot files: one for those squares with confirmed breeding, and another for those with possible or probable breeding. Each map requires two plot files in the form of encoded rasters which can be "read" by the plotter. To optimize the plotting process and keep tape requirements to a minimum, the files are copied onto magnetic tape at the rate of six maps per tape.

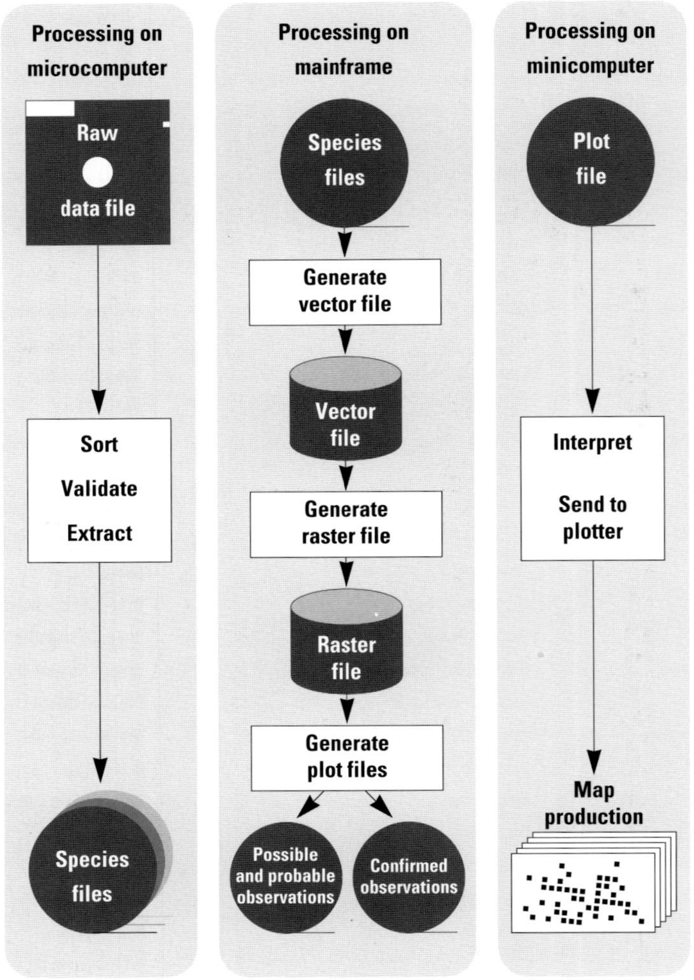

Flowchart

The final process involves reproducing the maps on photographic film. There is a procedure for automatically positioning the six maps on each tape onto the same film. Plotting takes some 20 minutes. The negative is then developed using a large format camera specially designed for reversal development. In all, nearly 500 maps were made: 242 maps of possible and probable breeding, 242 maps of confirmed breeding, 5 maps showing the number of field hours spent in each square surveyed, and 5 maps showing the number of species recorded in each of the squares surveyed.

This project took about eight months, the first five of which were devoted to developing the software and procedures, and benchmarking. It yielded an interesting product with the development of a system designed for general use applicable to raster file plotting. It also fostered research resulting in refinement of existing tools. Production of these maps was made possible by the collaboration of a team of analysts, programmers, and support staff who pooled their expertise.

Johanne Baril

Appendix V
Names of living things mentioned in the Atlas

Plants (Abbreviation)

Nonvascular (algae, fungi, lichens, mosses, and liverworts) (NP)
Vascular (VP)

Publications consulted:
Benoit 1993; Bois 1986; Borror and White 1991; Borror et al. 1989; Chagnon and Robert 1962; Desrosiers et al. 1995; Devillers et al. 1993; Farrar 1995; Fleurbec 1978, 1981, 1983, 1985, 1987, 1993; Lamoureux et al. 1996; Macdonald 1984; Marie-Victorin 1995; McAllister 1990; Mulligan 1992; Ouellet et al. 1990; Porsild and Cody 1980; Reynolds 1976, 1977b; Scoggan 1978; Scott and Scott 1988.

Animals

Vertebrates (Abbreviation)
- Fish (F)
- Amphibians (A)
- Reptiles (R)
- Birds (B)
- Mammals (M)

Invertebrates (Abbreviation)
- Arthropods
  - Insects (I)
  - Crustaceans (C)
  - Other arthropods (OA)
- Annelids (An)
- Molluscs (Mo)
- Echinoderms (E)

English	Latin	French	Type of organism
Alder	*Alnus* spp.	Aulne	(VP)
Alder, Green	*Alnus viridis crispa*	Aulne crispé	(VP)
Alder, Speckled	*Alnus incana rugosa*	Aulne rugueux	(VP)
Alewife	*Alosa pseudoharengus*	Gaspareau	(F)
Alfalfa	*Medicago sativa*	Luzerne (or Luzerne cultivée)	(VP)
Alligator, American	*Alligator mississipiensis*	Alligator	(R)
Amphipods	Amphipoda (order)	Amphipodes	(C)
Ant, Boreal carpenter	*Camponotus herculeanus*	Fourmi bicolore (or Fourmi charpentière)	(I)
Ants	Formicidae (family)	Fourmis	(I)
Aphids	Aphidoidea (superfamily)	Pucerons	(I)
Apple	*Malus* spp.	Pommier	(VP)
Arachnids	Arachnida (class)	Arachnides	(OA)
Arrowhead, Broad-leaved	*Sagittaria latifolia*	Sagittaire à larges feuilles	(VP)
Arrowhead, Erect	*Sagittaria rigida*	Sagittaire dressée	(VP)
Ash, Black	*Fraxinus nigra*	Frêne noir	(VP)
Ash, Blue	*Fraxinus quadrangulata*	Frêne anguleux (or Frêne bleu)	(VP)
Ash, Red	*Fraxinus pennsylvanica*	Frêne rouge (or Frêne de Pennsylvanie)	(VP)
Ash, White	*Fraxinus americana*	Frêne blanc (or Frêne d'Amérique)	(VP)
Aspen, Largetooth	*Populus grandidentata*	Peuplier à grandes dents	(VP)
Aspen, Trembling	*Populus tremuloides*	Peuplier faux-tremble (or Tremble)	(VP)
Aster	*Aster* spp.	Aster	(VP)
Aster, Simple	*Aster simplex*	Aster simple	(VP)
Atriplex, Hastate (see Spearscale)			
Barley	*Hordeum vulgare*	Orge (or Orge vulgaire)	(VP)
Barley, Foxtail	*Hordeum jubatum*	Orge agréable (or Orge queue-d'écureuil)	(VP)
Barnacle	*Balanus* spp.	Balane	(C)
Barnyard-grass	*Echinocloa* spp.	Echinocloé	(VP)
Bass	*Micropterus* spp.	Achigan	(F)
Bass, Rock	*Ambloplites rupestris*	Crapet de roche	(F)
Basswood	*Tilia americana*	Tilleul d'Amérique	(VP)
Bat, Big brown	*Eptesicus fuscus*	Grande Chauve-souris brune	(M)
Bayberry (see Wax myrtle)			
Beach flea (Amphipod)	*Corophium volutator*		(C)
Beachgrass	*Ammophila breviligulata*	Ammophile à ligule courte	(VP)
Bear, Black	*Ursus americanus*	Ours noir	(M)
Bear, Polar	*Ursus maritimus*	Ours blanc	(M)
Beaver	*Castor canadensis*	Castor	(M)
Beech, American	*Fagus grandifolia*	Hêtre à grandes feuilles	(VP)
Beech family	Fagaceae (family)	Hêtres	(VP)
Bees	Apoidea (superfamily)	Abeilles	(I)
Bees, Bumble	Apinae (subfamily)	Bourdons	(I)
Beetle, Colorado potato	*Leptinotarsa decemlineata*	Doryphore de la pomme de terre	(I)
Beetle, Fourspotted sap	*Glischrochilus quadrisignatus*	Nitidule à quatre points	(I)
Beetle, Native elm bark	*Hylurgopinus rufipes*	Scolyte de l'orme	(I)
Beetle, Spotted cucumber	*Diabrotica undecimpunctata howardi*	Chrysomèle maculée du concombre	(I)
Beetles	Coleoptera (order)	Coléoptères	(I)
Beetles, Bark and wood-boring	Scolytidae (family)	Scolytes	(I)
Beetles, Click	Elateridae (family)	Taupins	(I)
Beetles, Leaf	Chrysomelidae (family)	Chrysomèles (or Vers rongeurs des plantes)	(I)
Beetles, Metallic Wood-boring	Buprestidae (family)	Buprestes	(I)
Beetles, Rove	Staphylinidae (family)	Staphylins	(I)
Beetles, Scarab	Scarabaeidae (family)	Scarabées	(I)
Beetles, Snout	Curculionidae (family)	Charançons	(I)
Birch, Dwarf (see Birch, Swamp)			
Birch, Gray	*Betula populifolia*	Bouleau gris (or Bouleau à feuilles de Peuplier)	(VP)
Birch, Scrub	*Betula glandulosa*	Bouleau glanduleux	(VP)
Birch, Swamp	*Betula pumila*	Bouleau nain	(VP)
Birch, White	*Betula papyrifera*	Bouleau à papier (or Bouleau blanc)	(VP)
Birch, Yellow	*Betula alleghaniensis*	Bouleau jaune	(VP)
Bittern, Eurasian	*Botaurus stellaris*	Butor étoilé	(B)
Bivalves	Pelecypoda (class)	Bivalves	(Mo)
Black flies	Simuliidae (family)	Mouches noires	(I)
Black-grass	*Juncus gerardii*	Jonc de Gérard	(VP)
Bleeding-heart	*Dicentra* spp.	Coeurs-saignants	(VP)
Blennies, Combtooth	Blenniidae (family)	Blennies	(F)
Blue-joint	*Calamagrostis canadensis*	Calamagrostide du Canada (or Foin bleu)	(VP)
Blueberry	*Vaccinium* spp.	Bleuet	(VP)
Bluegrass	*Poa* spp.	Pâturin	(VP)
Bobcat	*Lynx rufus*	Lynx roux	(M)
Bramble	*Rubus* spp.	Ronce	(VP)
Brome-grass	*Bromus* spp.	Brome	(VP)
Buckbean	*Menyanthes trifoliata*	Trèfle d'eau commun	(VP)
Buckwheat	*Fagopyrum esculentum*	Sarrasin	(VP)
Budworm, Jack pine	*Choristoneura pinus*	Tordeuse du Pin gris	(I)
Budworm, Spruce	*Choristoneura fumiferana*	Tordeuse des bourgeons de l'épinette	(I)
Bugloss, Viper's	*Echium vulgare*	Vipérine (or Vipérine vulgaire)	(VP)
Bugs	Hemiptera (order)	Punaises	(I)
Bugs, Stink	Pentatomidae (family)	Punaises à bouclier (or Punaises puantes)	(I)
Bullhead catfishes	Ictaluridae (family)	Barbottes	(F)
Bulrush, American	*Scirpus americanus*	Scirpe d'Amérique	(VP)
Bulrush, Great	*Scirpus lacustris*	Scirpe des étangs	(VP)
Bulrush, River	*Scirpus fluviatilis*	Scirpe des rivières	(VP)
Bulrush, Salt-marsh	*Scirpus maritimus*	Scirpe maritime	(VP)

English	Latin	French	Type of organism
Bulrush, Sea (see Bulrush, Salt-marsh)			
Bur-reed, Branching	*Sparganium androcladum*	Rubanier rameux	(VP)
Bur-reed, Giant	*Sparganium eurycarpum*	Rubanier à gros fruits	(VP)
Burnet, Canada	*Sanguisorba canadensis*	Sanguisorbe du Canada	(VP)
Butterflies	Lepidoptera (order)	Papillons	(I)
Butternut	*Juglans cinerea*	Noyer cendré	(VP)
Buttonbush	*Cephalanthus occidentalis*	Céphalanthe occidental	(VP)
Caddisflies	Trichoptera (order)	Phryganes	(I)
Canary-grass	*Phalaris* spp.	Alpiste	(VP)
Canary-grass, Reed	*Phalaris arundinacea*	Alpiste roseau (or Phalaris roseau)	(VP)
Cankerworm, Fall	*Alsophila pometaria*	Arpenteuse d'automne	(I)
Cankerworm, Spring	*Paleacrita vernata*	Arpenteuse du printemps	(I)
Capelin	*Mallotus villosus*	Capelan	(F)
Caribou	*Rangifer tarandus*	Caribou	(M)
Carp, Common	*Cyprinus carpio*	Carpe	(F)
Cat, Domestic	*Felis catus*	Chat domestique	(M)
Cattail, Common	*Typha latifolia*	Typha à feuilles larges	(VP)
Cattail, Narrow-leaved	*Typha angustifolia*	Typha à feuilles étroites	(VP)
Cedar (see White-cedar, Eastern)			
Centipedes	Chilopoda (class)	Centipèdes	(OA)
Cherry, Black	*Prunus serotina*	Cerisier tardif	(VP)
Cherry, Choke	*Prunus virginiana*	Cerisier de Virginie (or Cerisier à grappes)	(VP)
Cherry, Pin	*Prunus pensylvanica*	Cerisier de Pennsylvanie	(VP)
Chipmunk, Eastern	*Tamias striatus*	Tamia rayé (or Suisse)	(M)
Chironomids (see Midges)			
Chrysomelids (see Beetles, Leaf)			
Cicada, Periodical	*Magicicada septendecim*	Cigale dix-sept ans	(I)
Cicadas	Cicadidae (family)	Cigales	(I)
Cloudberry	*Rubus chamaemorus*	Ronce plaquebière (or Ronce petit-mûrier)	(VP)
Clover	*Trifolium* spp.	Trèfle	(VP)
Club-moss	*Lycopodium* spp.	Lycopode	(VP)
Cod, Atlantic	*Gadus morhua*	Morue franche	(F)
Coleopterans (see Beetles)			
Collared-Dove, African	*Streptopelia roseogrisea*	Tourterelle rose-et-gris	(B)
Collared-Dove, Eurasian	*Streptopelia decaocto*	Tourterelle turque	(B)
Columbine	*Aquilegia* spp.	Ancolie	(VP)
Coontail, Common	*Ceratophyllum demersum*	Cornifle nageante	(VP)
Copepods	Copepoda (subclass)	Copépodes	(C)
Cord-grass, Freshwater	*Spartina pectinata*	Spartine pectinée	(VP)
Cord-grass, Salt-meadow	*Spartina patens*	Spartine étalée	(VP)
Cord-grass, Saltwater	*Spartina alterniflora*	Spartine à fleurs alternes (or Spartine alterniflore)	(VP)
Corn	*Zea mays*	Maïs	(VP)
Corn borer, European	*Ostrinia nubilalis*	Pyrale du maïs	(I)
Cotton-grass	*Eriophorum* spp.	Linaigrette	(VP)
Cottontail, Eastern	*Sylvilagus floridanus*	Lapin à queue blanche	(M)
Cottonwood, Black	*Populus trichocarpa*	Peuplier baumier de l'Ouest	(VP)
Cottonwood, Eastern	*Populus deltoides*	Peuplier deltoïde	(VP)
Couch-grass (see Grass, Quack)			
Cow-parsnip	*Heracleum maximum*	Berce laineuse (or Berce très grande)	(VP)
Coyote	*Canis latrans*	Coyote	(M)
Crab, Fiddler	*Uca* spp.		(C)
Crab, Mole	*Emerita talpoida*		(C)
Crab, Shore	*Sesarma* spp.		(C)

English	Latin	French	Type of organism
Crabs	Decapoda (order)	Crabes	(C)
Cranberry	*Vaccinium* spp.	Airelle	(VP)
Crayfishes	Astacidae (family)	Écrevisses	(C)
Creeper, Virginia	*Parthenocissus quinquefolia*	Vigne vierge (or Parthénocisse à cinq folioles)	(VP)
Crickets	Gryllidae (family)	Grillons	(I)
Crickets, Ground	Nemobiinae (subfamily)	Némobies (Grillons)	(I)
Crickets, Tree	Oecanthinae (subfamily)	Grillons des arbres	(I)
Crow, Carrion	*Corvus corone*	Corneille noire	(B)
Crow, Northwestern	*Corvus caurinus*	Corneille d'Alaska	(B)
Crowberry, Black	*Empetrum nigrum*	Camarine noire	(VP)
Cuckoo, Common	*Cuculus canorus*	Coucou gris	(B)
Cunner	*Tautogolabrus adspersus*	Tanche-tautogue	(F)
Cypresses	Cupressaceae (family)	Cyprès	(VP)
Damselflies	Zygoptera (suborder)	Demoiselles	(I)
Dandelion	*Taraxacum* spp.	Pissenlit	(VP)
Deer, White-tailed	*Odocoileus virginianus*	Cerf de Virginie	(M)
Dipterans (see Flies)			
Ditch-grass	*Ruppia maritima*	Ruppie maritime	(VP)
Dock	*Rumex* spp.	Rumex	(VP)
Dog	*Canis familiaris*	Chien	(M)
Dogwood	*Cornus* spp.	Cornouiller	(VP)
Dogwood, Red-osier	*Cornus stolonifera*	Cornouiller stolonifère	(VP)
Dragonflies and damselflies	Odonata (order)	Odonates	(I)
Duck, Muscovy	*Cairina moschata*	Canard musqué	(B)
Duckweed, Lesser	*Lemna minor*	Lenticule mineure (or Lentille d'eau)	(VP)
Dune grass, American	*Elymus arenarius*	Élyme des sables (or Élyme de mer)	(VP)
Earthworms	Lumbricidae (family)	Lombrics (or Vers de terre)	(An)
Eel, American	*Anguilla rostrata*	Anguille d'Amérique	(F)
Eelgrass	*Zostera marina*	Zostère marine	(VP)
Elder	*Sambucus* spp.	Sureau	(VP)
Elm, Slippery	*Ulmus rubra*	Orme rouge	(VP)
Elm, White	*Ulmus americana*	Orme d'Amérique	(VP)
Ephemeropterans (see Mayflies)			
Ermine	*Mustela erminea*	Hermine	(M)
Euphausiids (see Krill shrimps)			
False Solomon's-seal, Starry	*Smilacina stellata*	Smilacine étoilée	(VP)
Fern, Sensitive	*Onoclea sensibilis*	Onoclée sensible	(VP)
Fescue, Red	*Festuca rubra*	Fétuque rouge	(VP)
Fir, Balsam	*Abies balsamea*	Sapin baumier	(VP)
Fir, Grand	*Abies grandis*	Sapin grandissime (or Sapin de Vancouver)	(VP)
Fisher	*Martes pennanti*	Pékan	(M)
Fleas	Siphonaptera (order)	Puces	(I)
Flies	Diptera (order)	Diptères	(I)
Flies, Crane	Tipulidae (family)	Tipules	(I)
Flies, Deer	Tabanidae (family)	Taons	(I)
Flies, Horse	Tabanidae (family)	Taons	(I)
Flies, Muscid	Muscidae (family)	Muscides	(I)
Flounder, Winter	*Pleuronectes americanus*	Plie rouge	(F)
Flounders, Righteye	Pleuronectidae (family)	Plies	(F)
Fly, House	*Musca domestica*	Mouche domestique	(I)
Fox, Arctic	*Alopex lagopus*	Renard arctique	(M)
Fox, Red	*Vulpes vulpes*	Renard roux	(M)
Foxglove, Purple	*Digitalis purpurea*	Digitale pourpre	(VP)

English	Latin	French	Type of organism
Foxtail	*Setaria* spp.	Sétaire	(VP)
Fresh-water mussels	Pelecypoda (class)	Moules d'eau douce	(Mo)
Gale, Sweet	*Myrica gale*	Myrique baumier	(VP)
Gall fly, Goldenrod	*Eurosta* spp.	Tephrite (or Galle de la Verge d'or or Mouche des fruits)	(I)
Gammarids	Gammaridae (family)	Gammares	(C)
Gaspereau (see Alewife)			
Geometers (see Measuringworms)			
Glasswort	*Salicornia europaea*	Salicorne d'Europe	(VP)
Goat's-beard	*Tragopogon* spp.	Salsifis	(VP)
Goldenrod	*Solidago* spp.	Verge d'or	(VP)
Goose, Magpie	*Anseranas semipalmata*	Canaroie semipalmé	(B)
Goose-grass, Poor	*Puccinellia paupercula*	Puccinellie maigre	(VP)
Grass, Quack	*Agropyron repens*	Chiendent	(VP)
Grass, Squirrel-tail (see Barley, Foxtail)			
Grass, Sweet	*Hierochloe odorata*	Hiérochloé odorante	(VP)
Grasses	Gramineae (family)	Graminées	(VP)
Grasshoppers, Long-horned	Tettigoniidae (family)	Sauterelles	(I)
Grasshoppers, Short-horned	Acrididae (family)	Criquets	(I)
Ground squirrel	*Spermophilus* spp.	Spermophile	(M)
Ground squirrel, Arctic	*Spermophilus parryii*	Spermophile arctique	(M)
Groundsel, Sea beach	*Senecio pseudo-arnica*	Séneçon faux-arnica	(VP)
Grouse, Sage	*Centrocercus urophasianus*	Gélinotte des armoises (or Tétras des armoises)	(B)
Gull, Glaucous-winged	*Larus glaucescens*	Goéland à ailes grises	(B)
Gunnel	*Pholis* spp.	Sigouine	(F)
Hackberry	*Celtis occidentalis*	Micocoulier occidental	(VP)
Hare, Arctic	*Lepus arcticus*	Lièvre arctique	(M)
Hare, Snowshoe	*Lepus americanus*	Lièvre d'Amérique	(M)
Hawthorn	*Crataegus* spp.	Aubépine	(VP)
Hazel	*Corylus* spp.	Noisetier	(VP)
Heath family	Ericaceae (family)	Éricacées	(VP)
Hemipterans (see Bugs)			
Hemlock, Eastern	*Tsuga canadensis*	Pruche du Canada	(VP)
Herring, Atlantic	*Clupea harengus*	Hareng atlantique	(F)
Hickory, Bitternut	*Carya cordiformis*	Caryer cordiforme	(VP)
Hollow green weed	*Enteromorpha* spp.	Entéromorphe	(NP)
Holly	*Ilex* spp.	Houx	(VP)
Homopterans (cicadas, aphids, scale insects, etc.)	Homoptera (order)	Homoptères	(I)
Honeysuckle	*Lonicera* spp.	Chèvrefeuille	(VP)
Hornets	Vespinae (subfamily)	Guêpes	(I)
Hornwort (see Coontail, Common)			
Horse-mint (see Monarda)			
Horsetail	*Equisetum* spp.	Prêle	(VP)
Hymenopterans (ants, bees, wasps, etc.)	Hymenoptera (order)	Hyménoptères	(I)
Ironwood	*Ostrya virginiana*	Ostryer de Virginie	(VP)
Juniper, Common	*Juniperus communis*	Genévrier commun	(VP)
Katydids	Tettigoniidae (family)	Katydides	(I)
Krill shrimps	Euphausiacea (order)	Euphausides	(C)
Labrador-tea, Common	*Ledum groenlandicum*	Lédon du Groenland (or Thé du Labrador)	(VP)
Lagomorphs (hares, rabbits, and pikas)	Lagomorpha (order)	Lagomorphes	(M)
Laurel	*Kalmia* spp.	Kalmia	(VP)
Leafhoppers	Cicadellidae (family)	Cicadelles	(I)
Leather-leaf	*Cassandra calyculata*	Cassandre caliculé	(VP)
Leeches	Hirudinea (class)	Sangsues	(An)
Lemming, Brown	*Lemmus sibiricus*	Lemming brun	(M)
Lemming, Northern bog	*Synaptomys borealis*	Campagnol lemming boréal	(M)
Lemming, Southern bog	*Synaptomys cooperi*	Campagnol lemming de Cooper	(M)
Lemming, Ungava collared	*Dicrostonyx hudsonius*	Lemming d'Ungava	(M)
Lepidopterans (see Butterflies and moths)			
Lichen, Alpine reindeer	*Cladina stellaris*	Cladine étoilée	(NP)
Lichen, Old man's beard	*Usnea* spp.	Usnée	(NP)
Lichen, Reindeer	*Cladina rangiferina*	Cladine rangifère	(NP)
Lichen, Yellow reindeer	*Cladina mitis*	Cladine douce	(NP)
Lobelia	*Lobelia* spp.	Lobélie	(VP)
Locusts (see Grasshoppers, Short-horned)			
Loon, Arctic	*Gavia arctica*	Huart arctique	(B)
Looper, Hemlock	*Lambdina fiscellaria fiscellaria*	Arpenteuse de la pruche	(I)
Loosestrife, Purple	*Lythrum salicaria*	Salicaire	(VP)
Lynx	*Lynx canadensis*	Lynx du Canada (or Loup-cervier)	(M)
Mackerel, Atlantic	*Scomber scombrus*	Maquereau bleu	(F)
Maple, Manitoba	*Acer negundo*	Érable à feuilles composées (or Érable à Giguère)	(VP)
Maple, Red	*Acer rubrum*	Érable rouge	(VP)
Maple, Silver	*Acer saccharinum*	Érable argenté	(VP)
Maple, Striped	*Acer pensylvanicum*	Érable de Pennsylvanie (or Bois barré)	(VP)
Maple, Sugar	*Acer saccharum*	Érable à sucre	(VP)
Marmot, Hoary	*Marmota caligata*	Marmotte des rocheuses	(M)
Marten, American	*Martes americana*	Martre d'Amérique	(M)
Mayflies	Ephemeroptera (order)	Éphéméroptères	(I)
Meadowsweet	*Spiraea* spp.	Spirée	(VP)
Measuringworms	Geometridae (family)	Arpenteuses (or Géomètres)	(I)
Midges	Chironomidae (family)	Chironomes (or Moucherons)	(I)
Milkweed, Common	*Asclepias syriaca*	Asclépiade commune	(VP)
Millet (see Foxtail)			
Millipedes	Diplopoda (class)	Millipèdes	(OA)
Mink	*Mustela vison*	Vison d'Amérique	(M)
Minnows	Cyprinidae (family)	Ménés	(F)
Mites	Acarina (order)	Mites	(OA)
Moles	Talpidae (family)	Taupes	(M)
Monarda	*Monarda* spp.	Monarde	(VP)
Moose	*Alces alces*	Orignal	(M)
Mosquitoes	Culicidae (family)	Moustiques (or Maringouins)	(I)
Moss, Hairy-cap	*Polytrichum juniperinum*		(NP)
Moss, Spanish	*Tillandsia usneoides*	Mousse espagnole	(NP)
Moss-campion	*Silene acaulis*	Silène acaule	(VP)
Mosses, Feather	Hypnaceae (family)	Mousses hypnacées	(NP)
Moth, Gypsy	*Lymantria dispar*	Spongieuse	(I)
Moths	Lepidoptera (order)	Papillons	(I)
Moths, Geometer (see Measuringworms)			
Moths, Sphinx	Sphingidae (family)	Sphinx	(I)
Mountain-ash, American	*Sorbus americana*	Sorbier d'Amérique	(VP)
Mountain-ash, European	*Sorbus aucuparia*	Sorbier des oiseleurs (or Sorbier des oiseaux)	(VP)
Mountain-avens	*Dryas* spp.	Dryas	(VP)
Mouse	*Peromyscus* spp.	Souris	(M)
Mouse, Deer	*Peromyscus maniculatus*	Souris sylvestre	(M)
Mouse, White-footed	*Peromyscus leucopus*	Souris à pattes blanches	(M)
Mullein, Common	*Verbascum thapsus*	Molène vulgaire	(VP)
Muskrat	*Ondatra zibethicus*	Rat musqué	(M)

English	Latin	French	Type of organism
Mussel, Blue	*Mytilus edulis*	Moule bleue	(Mo)
Mussel, Zebra	*Dreissena polymorpha*	Moule zébrée	(Mo)
Mustelids (weasels and allies)	Mustelidae (family)	Mustélidés	(M)
Naiads	Naiadaceae (family)	Naïadacées	(VP)
Nereids (clam worms)	Nereidae (family)	Néréis	(An)
Nettle, Stingless	*Pilea* spp.	Piléa	(VP)
Nettles	Urticaceae (family)	Orties	(VP)
Nightingale	*Luscinia megarhynchos*	Rossignol philomèle	(B)
Nightshade	*Solanum* spp.	Morelle	(VP)
Nightshade family	Solanaceae (family)	Solanacées	(VP)
Nuthatch, Brown-headed	*Sitta pusilla*	Sittelle à tête brune	(B)
Oak, Bur	*Quercus macrocarpa*	Chêne à gros fruits	(VP)
Oak, Chestnut	*Quercus montana*	Chêne Châtaignier	(VP)
Oak, Red	*Quercus rubra*	Chêne rouge	(VP)
Oak, Swamp white	*Quercus bicolor*	Chêne bicolore (or Chêne bleu)	(VP)
Oak, White	*Quercus alba*	Chêne blanc	(VP)
Oat	*Avena* spp.	Avoine	(VP)
Orach (see Spearscale)			
Orthopterans (grasshoppers, crickets, walking sticks, etc.)	Orthoptera (order)	Orthoptères	(I)
Ostracods	Ostracoda (subclass)	Ostracodes	(C)
Ostrich-fern	*Matteuccia struthiopteris*	Matteuccie fougère-à-l'autruche	(VP)
Otter, River	*Lutra canadensis*	Loutre de rivière	(M)
Otter, Sea	*Enhydra lutris*	Loutre de mer	(M)
Owl, Eagle	*Bubo bubo*	Grand-duc d'Europe	(B)
Owl, Spotted	*Strix occidentalis*	Chouette tachetée	(B)
Owl, Tawny	*Strix aluco*	Chouette hulotte	(B)
Oxytropis	*Oxytropis* spp.	Oxytropis	(VP)
Oysters	Pteroconchida (order)	Huîtres	(Mo)
Partridge, Barbary	*Alectoris barbara*	Perdrix gambra	(B)
Partridge, Red-legged	*Alectoris rufa*	Perdrix rouge	(B)
Peat-moss (see Sphagnum)			
Penguins	Spheniscidae (family)	Manchots	(B)
Perch, Yellow	*Perca flavescens*	Perchaude	(F)
Periwinkle	*Littorina* spp.	Littorine	(Mo)
Pickerel	*Esox* spp.	Brochet	(F)
Pike, Northern	*Esox lucius*	Brochet	(F)
Pine, Eastern white	*Pinus strobus*	Pin blanc	(VP)
Pine, Jack	*Pinus banksiana*	Pin gris	(VP)
Pine, Loblolly	*Pinus taeda*	Pin taeda	(VP)
Pine, Mugho	*Pinus mugo*	Pin mugo	(VP)
Pine, Pitch	*Pinus rigida*	Pin rigide	(VP)
Pine, Red	*Pinus resinosa*	Pin rouge	(VP)
Pine, Scots	*Pinus sylvestris*	Pin sylvestre (or Pin d'Écosse)	(VP)
Pine, Scrub	*Pinus virginiana*	Pin de Virginie	(VP)
Pine, Shortleaf	*Pinus echinata*	Pin épineux	(VP)
Pipit, Water	*Anthus spinoletta*	Pipit spioncelle	(B)
Plantain, Seaside	*Plantago maritima*	Plantain maritime	(VP)
Pocket gopher, Northern	*Thomomys talpoides*	Gaufre gris	(M)
Poison-ivy	*Rhus radicans* (or *Toxicodendron rydbergii*)	Herbe-à-puce (or Herbe à la puce)	(VP)
Polychaetes	Polychaeta (class)	Polychètes	(An)
Pond-lily	*Nuphar* spp.	Nénuphar	(VP)
Pondweed	*Potamogeton* spp.	Potamot	(VP)
Pondweed, Sago	*Potamogeton pectinatus*	Potamot pectiné	(VP)

English	Latin	French	Type of organism
Poplar	*Populus* spp.	Peuplier	(VP)
Poplar, Balsam	*Populus balsamifera*	Peuplier baumier	(VP)
Porcupines	Erethizontidae (family)	Porcs-épics	(M)
Prominent, Saddle	*Heterocampa guttivitta*	Hétérocampe de l'érable	(I)
Puccinellia, Poor (see Goose-grass, Poor)			
Pumpkinseed	*Lepomis gibbosus*	Crapet-soleil	(F)
Raccoon	*Procyon lotor*	Raton laveur	(M)
Ragweed	*Ambrosia* spp.	Herbe-à-poux	(VP)
Raspberry, Flowering	*Rubus odoratus*	Ronce odorante	(VP)
Raspberry, Red	*Rubus idaeus*	Framboisier rouge	(VP)
Rat, Norway	*Rattus norvegicus*	Rat surmulot	(M)
Redcedar, Eastern	*Juniperus virginiana*	Genévrier de Virginie (or Cèdre rouge)	(VP)
Redtop	*Agrostis alba*	Agrostide blanche	(VP)
Reed, Common	*Phragmites communis*	Phragmite commun (or Roseau commun)	(VP)
Reedgrass, Slimstem	*Calamagrostide neglecta*	Calamagrostide négligé	(VP)
Rice, Wild	*Zizania aquatica*	Zizanie aquatique (or Folle avoine or Riz sauvage)	(VP)
Rice cut grass	*Leersia oryzoides*	Léersie faux-riz	(VP)
Robin, European	*Erithacus rubecula*	Rougegorge familier	(B)
Rockweeds	Fucaceae (family)	Fucus	(NP)
Rodents	Rodentia (order)	Rongeurs	(M)
Rose, Multiflora	*Rosa multiflora*	Rosier multiflore	(VP)
Rose, Wild	*Rosa blanda*	Églantier	(VP)
Rush	*Juncus* spp.	Jonc	(VP)
Rush, Baltic	*Juncus balticus*	Jonc de la Baltique	(VP)
Rye	*Secale cereale*	Seigle	(VP)
Salmon, Atlantic	*Salmo salar*	Saumon atlantique	(F)
Sand lance, American	*Ammodytes americanus*	Lançon d'Amérique	(F)
Sand-spurrey, Canada	*Spergularia canadensis*	Spergulaire du Canada	(VP)
Sandwort, Seabeach	*Arenaria peploides*	Sabline faux-péplus	(VP)
Sauger	*Stizostedion canadense*	Doré noir	(F)
Sawflies, Common	Tenthredinidae (family)	Tenthrèdes	(I)
Sawflies, Conifer	Diprionidae (family)	Diprions	(I)
Sawfly, Swaine jack pine	*Neodiprion swainei*	Diprion de Swaine	(I)
Sawyer, Whitespotted	*Monochamus scutellatus*	Longicorne noir	(I)
Saxifrage	*Saxifraga* spp.	Saxifrage	(VP)
Scallops	Pectinidae (family)	Pétoncles	(Mo)
Scud (Amphipod)	*Hyalella azteca*		(C)
Sculpins	Cottidae (family)	Chabots	(F)
Sea-grass (see Ditch-grass)			
Sea-lettuce	*Ulva* spp.	Laitue de mer	(NP)
Sea-lungwort	*Mertensia maritima*	Mertensie maritime	(VP)
Sea-milkwort	*Glaux maritima*	Glauce maritime	(VP)
Sea-rocket	*Cakile edentula*	Caquillier édentulé	(VP)
Sea urchins	Echinoidea (class)	Oursins	(E)
Seals	Phocidae (family)	Phoques	(M)
Sedge, Aquatic	*Carex aquatilis*	Carex aquatique	(VP)
Sedge, Chaffy (see Sedge, Scaly)			
Sedge, Lake	*Carex lacustris*	Carex lacustre	(VP)
Sedge, Mackenzie's	*Carex mackenziei*	Carex de Mackenzie	(VP)
Sedge, Moniliform	*Carex hormathodes*	Carex moniliforme	(VP)
Sedge, Scaly	*Carex paleacea*	Carex écailleux (or Carex paléacé)	(VP)
Sedge, Woolly	*Carex lanuginosa*	Carex laineux	(VP)
Sedges	Cyperaceae (family)	Cypéracées	(VP)
Serviceberry	*Amelanchier* spp.	Amélanchier	(VP)

English	Latin	French	Type of organism
Shanny, Arctic	*Stichaeus punctatus*	Stichée arctique	(F)
Shanny, Daubed	*Lumpenus maculatus*	Lompénie tachetée	(F)
Shrews	Soricidae (family)	Musaraignes	(M)
Shrimps	Decapoda (order)	Crevettes	(C)
Silverweed	*Potentilla anserina*	Potentille ansérine	(VP)
Skeletonizer, Birch	*Bucculatrix canadensisella*	Squeletteuse du bouleau	(I)
Skunk, Striped	*Mephitis mephitis*	Moufette rayée	(M)
Slugs	Limacidae (family)	Limaces	(Mo)
Smartweed	*Polygonum* spp.	Renouée	(VP)
Smelt, Rainbow	*Osmerus mordax*	Éperlan arc-en-ciel	(F)
Snails	Gastropoda (class)	Gastéropodes	(Mo)
Snakes, Colubrid	Colubridae (family)	Couleuvres	(R)
Sorghum	*Sorghum vulgare*	Sorgho	(VP)
Sorrel (see Dock)			
Sow-thistle	*Sonchus* spp.	Laiteron	(VP)
Spanworm, Bruce	*Operophtera bruceata*	Arpenteuse de Bruce	(I)
Sparrow, Eurasian Tree	*Passer montanus*	Moineau friquet	(B)
Sparrow, Seaside	*Ammodramus maritimus*	Bruant maritime	(B)
Spearscale	*Atriplex hastata*	Arroche hastée	(VP)
Sphagnum	*Sphagnum* spp.	Sphaigne	(NP)
Spiders	Araneida (order)	Araignées (Arachnides)	(OA)
Spike-rush, Single-scaled	*Eleocharis uniglumis*	Éléocharide uniglume (or Éléocharide halophile)	(VP)
Spruce, Black	*Picea mariana*	Épinette noire	(VP)
Spruce, Colorado	*Picea pungens*	Épinette du Colorado (or Épinette bleue)	(VP)
Spruce, Engelmann	*Picea engelmannii*	Épinette d'Engelmann	(VP)
Spruce, Red	*Picea rubens*	Épinette rouge	(VP)
Spruce, White	*Picea glauca*	Épinette blanche	(VP)
Squirrel, Flying	*Glaucomys* spp.	Polatouche (or Écureuil volant)	(M)
Squirrel, Grey	*Sciurus carolinensis*	Écureuil gris	(M)
Squirrel, Northern flying	*Glaucomys sabrinus*	Grand polatouche	(M)
Squirrel, Red	*Tamiasciurus hudsonicus*	Écureuil roux	(M)
Stickleback, Threespine	*Gasterosteus aculeatus*	Épinoche à trois épines	(F)
Strawberry, Wild	*Fragaria virginiana*	Fraisier des champs	(VP)
Sucker	*Catostomus* spp.	Meunier	(F)
Sumac	*Rhus* spp.	Sumac	(VP)
Sumac, Staghorn	*Rhus typhina*	Sumac vinaigrier (or Vinaigrier)	(VP)
Sunfishes	Centrarchidae (family)	Crapets	(F)
Sunflower	*Helianthus* spp.	Tournesol	(VP)
Sweetflag	*Acorus calamus*	Acorus roseau	(VP)
Tamarack	*Larix laricina*	Mélèze laricin	(VP)
Tape-grass	*Vallisneria americana*	Vallisnérie d'Amérique	(VP)
Tent caterpillar, Eastern	*Malacosoma americanum*	Livrée d'Amérique	(I)
Tent caterpillar, Forest	*Malacosoma disstria*	Livrée des forêts	(I)
Tent caterpillars and lappet moths	Lasiocampidae (family)	Livrées	(I)
Thistle	*Cirsium* spp.	Chardon	(VP)
Threadfoot	*Podostemon ceratophyllum*	Podostémon cératophylle	(VP)
Ticks	Acarina (order)	Tiques	(I)
Timothy	*Phleum pratense*	Mil (or Phléole des prés)	(VP)
Tomcod, Atlantic	*Microgadus tomcod*	Poulamon atlantique	(F)
Tortrix, Large aspen	*Choristoneura conflictana*	Tordeuse du tremble	(I)
Touch-me-not	*Impatiens* spp.	Impatiente	(VP)
Touch-me-not, Spotted	*Impatiens capensis*	Impatiente du Cap	(VP)
Treecreeper, Eurasian	*Certhia familiaris*	Grimpereau des bois	(B)
Treefrog, Tetraploid gray	*Hyla versicolor*	Rainette versicolore	(A)
Trout	Salmoninae (subfamily)	Truites	(F)
Turtle, Snapping	*Chelydra serpentina*	Tortue serpentine	(R)
Umbrella-sedge	*Cyperus* spp.	Souchet	(VP)
Vetch, Tufted	*Vicia cracca*	Vesce jargeau	(VP)
Viburnum	*Viburnum* spp.	Viorne	(VP)
Vines	Vitaceae (family)	Vignes	(VP)
Violet	*Viola* spp.	Violette	(VP)
Vole	*Clethrionomys* spp./*Microtus* spp.	Campagnol	(M)
Vole, Gapper's red-backed	*Clethrionomys gapperi*	Campagnol à dos roux de Gapper	(M)
Vole, Meadow	*Microtus pennsylvanicus*	Campagnol des champs	(M)
Walleye	*Stizostedion vitreum*	Doré	(F)
Walnut	*Juglans* spp.	Noyer	(VP)
Wapiti	*Cervus elaphus*	Wapiti	(M)
Wasps, Wood	Xiphydriidae (family)	Faux-sirex	(I)
Wasps, Yellow-jacket	Vespidae (family)	Guêpes	(I)
Water flea	*Daphnia pulex*	Daphnie	(C)
Water fleas	Cladocera (order)	Cladocères	(C)
Water-lily, Magnolia	*Nymphaea tuberosa*	Nymphée tubéreuse	(VP)
Water-milfoil, Eurasian	*Myriophyllum spicatum*	Myriophylle à épi	(VP)
Water-plantain, American	*Alisma plantago-aquatica*	Alisme plantain-d'eau	(VP)
Water-stargrass	*Heteranthera dubia*	Hétéranthère litigieuse	(VP)
Waterweed, Canada	*Elodea canadensis*	Élodée du Canada	(VP)
Wax myrtle	*Myrica* spp.	Myrique	(VP)
Weasel, Least	*Mustela nivalis*	Belette pygmée	(M)
Weasel, Long-tailed	*Mustela frenata*	Belette à longue queue	(M)
Weavers	*Ploceus* spp.	Tisserins	(B)
Webworm, Fall	*Hyphantria cunea*	Chenille à tente estivale	(I)
Weevils (see Beetles, Snout)			
Whales	Cetacea (order)	Baleines	(M)
Wheat	*Triticum aestivum*	Blé	(VP)
White-cedar, Eastern	*Thuja occidentalis*	Thuya occidental (or Cèdre)	(VP)
Wild-celery (see Tape-grass)			
Willow, Black	*Salix nigra*	Saule noir	(VP)
Willow, Crack	*Salix fragilis*	Saule fragile	(VP)
Willow, Golden weeping	*Salix alba*	Saule pleureur doré	(VP)
Willow, Heartleaf	*Salix eriocephala*	Saule à tête laineuse (or Saule rigide)	(VP)
Willow, Meadow	*Salix petiolaris*	Saule à long pétiole (or Saule pétiolé)	(VP)
Willow, Pussy	*Salix discolor*	Saule discolore	(VP)
Willow, Sandbar	*Salix exigua*	Saule de l'Intérieur	(VP)
Willow, Shining	*Salix lucida lucida*	Saule brillant	(VP)
Willow, Weeping (see Willow, Golden weeping)			
Wolf	*Canis lupus*	Loup	(M)
Wood-nettle	*Laportea canadensis*	Ortie du Canada	(VP)
Wood rat, Bushy-tailed	*Neotoma cinerea*	Rat à queue touffue	(M)
Woodchuck	*Marmota monax*	Marmotte commune	(M)
Worm, Clam	*Nereis virens*		(An)
Worm, Octagonal-tail	*Dendrobaena octaedra*	Ver à queue octogonale	(An)
Worm, Pasture	*Aporrectodea turgida*	Ver de pâturage	(An)
Worm, Québec	*Lumbricus festivus*	Ver québécois	(An)
Worm, Red marsh	*Lumbricus rubellus*	Ver rouge du marécage	(An)
Wracks (see Rockweeds)			
Yellow cress, Amphibious	*Rorippa amphibia*	Rorippa amphibie	(VP)
Yew, Western	*Taxus brevifolia*	If de l'Ouest	(VP)

Appendix VI
Gazetteer

LOCALITY	Lat. (N)	Long. (W)
Aiguebelle Conservation Park	48°30′	78°45′
Akulivik	60°48′	78°12′
Akwesasne	45°00′	74°39′
Alma	48°33′	71°39′
Alouettes flats	48°05′	69°42′
Amos	48°34′	78°07′
Anderson Island	56°19′	76°42′
Anticosti Island	49°30′	63°00′
Antoine Creek	48°36′	79°29′
Aqueduct Canal	45°24′	73°37′
Archipel de L'Isle-aux-Grues	47°04′	70°34′
Arnaud River	59°59′	69°46′
Arntfield	48°12′	79°15′
Ash Island	45°04′	73°20′
Ashuapmushuan River	48°37′	72°20′
Ashuapmushuan Wildlife Reserve	49°05′	73°32′
Asselin Creek (Abitibi)	48°22′	78°47′
Asselin Creek (Témiscouata)	47°48′	68°24′
Attikuan Point	54°16′	79°27′
Audubon Islets	50°12′	60°21′
Aupaluk	59°21′	69°41′
Aylmer	45°24′	75°50′
Baie aux Feuilles (formerly Leaf Bay)	58°55′	69°10′
Baie aux Outardes	49°02′	68°30′
Baie-Comeau	49°13′	68°09′
Baie de Brador Migratory Bird Sanctuary	51°23′	57°11′
Baie de L'Isle-Verte National Wildlife Area	48°01′	69°21′
Baie de Lavallière	46°04′	72°58′
Baie de Mille-Vaches	48°34′	69°13′
Baie-des-Bacon	48°31′	69°16′
Baie des Ha! Ha!	48°21′	70°49′
Baie des Loups (formerly Wolf Bay)	50°14′	60°13′
Baie-des-Loups (see Wolf Bay)		
Baie des Loups Migratory Bird Sanctuary	50°12′	60°15′
Baie des Rochers	47°58′	69°49′
Baie-des-Rochers	47°58′	69°51′
Baie-du-Febvre	46°08′	72°43′
Baie du Havre aux Basques	47°17′	61°57′
Baie du Renard	49°17′	61°50′
Baie-du-Renard	49°17′	61°50′
Baie-Johan-Beetz	50°17′	62°48′
Baie-Saint-Laurent	48°30′	67°29′
Baie-Saint-Paul	47°27′	70°30′
Baie-Sainte-Catherine	48°06′	69°44′
Baie-Sainte-Claire	49°54′	64°30′
Baie-Trinité	49°25′	67°18′
Baldwin Mills	45°02′	71°55′
Barachois	48°37′	64°17′
Barachois-de-Malbaie (see Barachois)		
Barboteau Rock	51°41′	79°00′
Baskatong Reservoir	46°48′	75°50′
Batiscan	46°30′	72°15′
Batiscan-Neilson ZEC	47°11′	71°52′
Battures aux Loups Marins	47°15′	70°25′
Bay of Many Islands	54°08′	79°10′
Bazin River	47°29′	75°22′
Beaconsfield	45°26′	73°52′
Beauharnois	45°19′	73°52′
Beauharnois Dam	45°19′	73°55′
Beauport	46°52′	71°11′
Beauport flats	46°51′	71°11′
Beaupré	47°03′	70°54′
Beaurepaire	45°25′	73°53′
Bec-scie River	49°43′	64°03′
Bécancour	46°20′	72°26′
Bélanger Island	51°01′	73°24′
Bell River	49°49′	77°39′
Bellechasse Cove	46°55′	70°46′
Belleterre	47°23′	78°42′
Beloeil	45°34′	73°12′
Bergeronnes	48°20′	69°35′
Bernierville	46°06′	71°34′
Berry	48°49′	78°16′
Berthier	46°15′	73°25′
Berthier Islands	46°06′	73°06′
Berthierville	46°05′	73°11′
Betchouane	50°14′	63°11′
Betsiamites Point	48°56′	68°39′
Betsiamites River	48°56′	68°38′
Bic (see Le Bic)		

LOCALITY	Lat. (N)	Long. (W)
Bic Conservation Park	48°21′	68°47′
Bicquette Island	48°25′	68°53′
Bird Rocks (see Rochers aux Oiseaux)		
Bishopton	45°35′	71°35′
Blacklands Island	50°13′	60°13′
Blanc Reservoir	47°45′	73°15′
Blanc-Sablon	51°25′	57°08′
Boatswain Bay	51°50′	78°50′
Bogan Creek	48°43′	66°17′
Bois-de-Coulonge Park	46°47′	71°14′
Boischatel (see Saint-Jean-de-Boischatel)		
Boisvert Point	48°34′	69°09′
Bonaventure	48°03′	65°29′
Bonaventure Island	48°30′	64°10′
Bonaventure River	48°02′	65°29′
Boniface River	57°38′	76°50′
Bonne-Espérance	51°30′	57°35′
Boucherville	45°36′	73°27′
Boucherville Islands	45°37′	73°28′
Bourjoli Point	59°24′	77°55′
Boyer River	46°53′	70°52′
Brador (formerly Bradore)	51°28′	57°15′
Brador Bay (formerly Bradore Bay)	51°28′	57°17′
Brandypot Islands (see Îles du Pot à l'Eau-de-Vie)		
Breckenridge	45°29′	75°57′
Brion Island	47°48′	61°28′
Bristol Ridge	45°37′	76°25′
Broadback River	51°21′	78°52′
Brome Lake	45°15′	72°30′
Brome-Missisquoi	45°08′	72°48′
Bromptonville	45°28′	71°57′
Brossard	45°27′	73°28′
Bury	45°28′	71°30′
Bushy Island	45°26′	73°44′
Butte du Vent	47°23′	61°55′
Cabbage Willows Bay	51°34′	79°14′
Cabonga Reservoir	47°20′	76°35′
Cacouna (see Saint-Georges-de-Cacouna)		
Cacouna marsh	47°56′	69°30′
Campbell Bay	45°34′	75°17′
Campbell's Bay	45°44′	76°36′
Caniapiscau	54°52′	69°50′
Caniapiscau Reservoir	54°10′	69°50′
Caniapiscau River	57°40′	69°29′
Cantley	45°34′	75°47′
Cap à l'Aigle (formerly Cape Eagle)	49°47′	64°20′
Cap à l'Orignal	48°22′	68°50′
Cap-aux-Meules	47°23′	61°52′
Cap-Chat	49°06′	66°41′
Cap d'Espoir	48°25′	64°19′
Cap-d'Espoir	48°26′	64°20′
Cap de l'Ours	49°33′	62°27′
Cap du Prince-de-Galles (formerly Cape Prince of Wales)	61°36′	71°30′
Cap Noir	47°14′	62°01′
Cap-Rouge	46°45′	71°22′
Cap Tourmente	47°05′	70°45′
Cap-Tourmente	47°04′	70°48′
Cap Tourmente National Wildlife Area	47°00′	70°54′
Cape Bon Ami	48°48′	64°12′
Cape Gaspé	48°45′	64°10′
Cape Gertrude (see Bourjoli Point)		
Cape Jones (see Point Louis XIV)		
Cape Observation	49°40′	62°42′
Cape Sandtop	49°12′	61°44′
Cape Tunnel	49°23′	62°02′
Cape Vauréal	49°37′	62°37′
Cape Whittle	50°11′	60°08′
Cape Wolstenholme	62°35′	77°30′
Caplan	48°06′	65°41′
Carheil River	52°17′	66°59′
Caribou Lake	47°40′	78°29′
Caribou River	56°17′	76°11′
Carillon	45°34′	74°22′
Carleton	48°06′	66°08′
Carleton Centre	48°06′	66°07′
Cartierville	45°32′	73°43′
Cascades	45°35′	75°52′
Cascapédia River	48°11′	65°54′
Castagnier	48°44′	77°50′
Causapscal	48°22′	67°14′

LOCALITY	Lat. (N)	Long. (W)
Causapscal River	48°21′	67°13′
Cazaville	45°05′	74°22′
Chaleur Bay	48°00′	65°45′
Chalifour River	50°22′	73°44′
Chambly	45°27′	73°17′
Chambly Basin	45°27′	73°17′
Chambord	48°26′	72°04′
Champlain	45°29′	73°29′
Champlain Bridge	45°28′	73°32′
Champlain lookout	45°30′	75°55′
Chandler	48°21′	64°41′
Chapais	49°47′	74°51′
Chapeau	45°55′	77°04′
Charlesbourg	46°51′	71°16′
Charlevoix	47°39′	70°09′
Charron Island	45°35′	73°29′
Château-Richer	46°58′	71°01′
Châteauguay	45°23′	73°45′
Châteauguay River	45°23′	73°45′
Châteauguay-Station	45°22′	73°45′
Chaudière Falls	45°25′	75°43′
Chaudière River	46°45′	71°17′
Chevery	50°28′	59°36′
Chibougamau	49°55′	74°22′
Chicoutimi	48°26′	71°04′
Chisasibi	53°47′	78°53′
Choinière Reservoir	45°25′	72°37′
Chute-des-Passes Dam	49°54′	71°15′
Clarenceville	45°04′	73°15′
Clarke Bay	47°37′	61°27′
Cliff Island	50°19′	59°42′
Cliff Islands	50°18′	59°41′
Clova	48°07′	75°22′
Coacoachou Bay	50°14′	60°17′
Coaticook	45°08′	71°48′
Coats River	55°19′	76°55′
Coin-du-Banc	48°34′	64°18′
Commune marsh	45°23′	73°46′
Como	45°27′	74°07′
Contrecoeur	45°51′	73°14′
Contrecoeur Islands	45°52′	73°15′
Corossol Island	50°05′	66°23′
Côte-de-Beaupré	47°02′	70°56′
Côte-du-Sud	47°22′	70°02′
Côte-Saint-Luc	45°28′	73°40′
Côte-Sainte-Catherine Camp-ground	45°25′	73°35′
Côte-Sainte-Catherine Park (see Côte-Sainte-Catherine Camp-ground)		
Coteau-du-Lac	45°18′	74°11′
Covey Hill	45°01′	73°48′
Cowansville	45°12′	72°45′
Croche River	47°30′	72°46′
Cusson Point (formerly Magnet Point)	60°24′	77°46′
Daniel Johnson Dam	50°39′	68°44′
Daragon Bay	45°35′	75°16′
Dartmouth River	48°53′	64°33′
Dauphiné Point	49°07′	62°30′
De Pas River	55°53′	64°45′
De Rozière Bay	60°32′	69°45′
De Villiers Cove	59°19′	69°19′
Dead Duck Bay	53°35′	79°00′
Decelles Reservoir	47°42′	78°08′
Déception Bay	57°20′	74°33′
Dégelis	47°33′	68°39′
Delisle	48°38′	71°42′
Deslauriers Island	45°43′	73°26′
Desmaraisville	49°30′	76°11′
Destor	48°27′	78°57′
Deuxième Îlet	47°30′	61°44′
Dewittville	45°07′	74°05′
Dickerson Island	45°02′	74°35′
Dieppe Hills	49°31′	79°22′
Digges Sound	62°28′	77°44′
Ditchfield	45°33′	70°50′
Dixie Island	45°26′	73°43′
Dolbeau	48°53′	72°14′
Dollard-des-Ormeaux	45°29′	73°49′
Donnacona	46°40′	71°45′
Dorval	45°27′	73°45′
Dozois Reservoir	47°30′	77°05′
Drummondville	45°53′	72°29′
Drummondville-Sud (see Drummondville)		

LOCALITY	Lat. (N)	Long. (W)
Dubé Bay	45°35′	75°08′
Dubuisson	48°07′	77°57′
Duchesnay Forestry Station	46°54′	71°40′
Dundee	45°00′	74°30′
Dune du Nord	47°31′	61°44′
Dune du Sud	47°30′	61°42′
Dune du Sud beach	47°27′	61°45′
Dunham	45°08′	72°48′
Duparquet	48°30′	79°14′
Duval Island	45°54′	73°13′
Eardley	45°33′	76°06′
East Angus	45°29′	71°40′
Eastmain	52°15′	78°30′
Eastmain River	52°15′	78°35′
Eccles Hill	45°01′	72°54′
Electoral District of Ahuntsic	45°33′	73°40′
Electoral District of Hochelaga	45°32′	73°33′
Electoral District of Snowdon	45°29′	73°37′
Erik Cove	62°35′	77°24′
Étang de l'Est (formerly East Pond)	47°37′	61°26′
Étang-des-Caps	47°14′	62°00′
Fabre	47°12′	79°22′
Falaise aux Goélands (formerly Gullcliff Bay)	49°09′	61°42′
Falardeau (see Saint-David-de-Falardeau)		
Falcoz River	56°56′	65°00′
False River	58°28′	67°50′
Farnham	45°17′	72°59′
Fatima	47°24′	61°53′
Fermont	52°47′	67°05′
Figuery (see Saint-Marc-de-Figuery)		
Fitch Bay	45°07′	72°11′
Forestville	48°44′	69°05′
Forillon	48°50′	64°20′
Forillon National Park	48°54′	64°21′
Forillon Peninsula	48°46′	64°12′
Forks	57°40′	69°30′
Fort-Coulonge	45°51′	76°44′
Fort George	53°50′	79°00′
Fort Mackenzie	56°50′	68°57′
Fort Rupert (see Waskaganish)		
Franklin	45°02′	73°55′
Franklin Centre	45°02′	73°55′
Fraser Point	45°03′	74°28′
Frelighsburg	45°03′	72°50′
Gagnon	51°53′	68°10′
Galibois Island	51°12′	58°30′
Gallichan	48°36′	79°17′
Gaspé	48°50′	64°29′
Gaspé Bay	48°46′	64°17′
Gaspé Conservation Park	48°52′	65°58′
Gaspé Harbour	48°51′	64°27′
Gaspé Peninsula	48°40′	66°00′
Gaspé Provincial Park (see Gaspé Conservation Park)		
Gatineau	45°30′	75°39′
Gatineau Park	45°34′	75°57′
Gatineau River	45°27′	75°42′
Gentilly (see Bécancour)		
George River	58°49′	66°10′
Georgeville	45°08′	72°15′
Gethsémani	50°13′	60°40′
Giffard	46°51′	71°12′
Godbout	49°19′	67°36′
Godbout River	49°19′	67°36′
Gouin Reservoir	48°38′	74°54′
Gould	45°35′	71°22′
Gracefield	46°06′	76°03′
Granby	45°24′	72°44′
Grand Lac à l'Épaule (see Lac à l'Épaule)		
Grand Lac Delay	56°11′	70°52′
Grand Lac Germain	51°12′	66°40′
Grand Lac Jacques-Cartier (see Lake Jacques Cartier)		
Grand Lac Macpès	48°18′	68°29′
Grand Lac Salé	49°20′	63°19′
Grand Lac Victoria	47°31′	77°30′
Grand-Métis	48°38′	68°08′
Grand Pabos Bay	48°20′	64°42′
Grande-Cascapédia	48°15′	65°54′
Grande-Entrée	47°33′	61°33′
Grande Entrée Island (formerly Coffin Island)	47°34′	61°31′

LOCALITY	Lat. (N)	Long. (W)
Grande Entrée Lagoon	47°33′	61°40′
Grande Plée Bleue	46°47′	71°03′
Grande Rivière de la Baleine	55°16′	77°47′
Grande-Vallée	49°13′	65°08′
Grandes-Bergeronnes	48°15′	69°32′
Grands-Jardins Conservation Park	47°41′	70°51′
Great Bird Rock (see Rocher aux Oiseaux)		
Great Whale River (see Grande Rivière de la Baleine)		
Greenfield Park	45°29′	73°29′
Gregory Lake (see Lake Bérard)		
Grondines	46°36′	72°03′
Gros-Morne	49°15′	65°33′
Grosse-Île	47°37′	61°31′
Grosse Île Bay	47°37′	61°32′
Groulx Mountains	51°37′	67°37′
Gulf of St. Lawrence	48°15′	62°00′
Guy Point	49°35′	62°31′
Gyrfalcon Islands	59°03′	69°02′
Hall Bay	60°53′	69°43′
Hall Cove	51°41′	78°57′
Hall River	48°05′	65°26′
Hampstead	45°29′	73°38′
Harricanaw River	50°56′	79°32′
Harrington	45°50′	74°40′
Harrington Harbour	50°30′	59°29′
Harvey Point	49°27′	62°15′
Hatley	45°11′	71°56′
Hauterive (see Baie-Comeau)		
Havelock	45°00′	73°45′
Havre-Aubert	47°14′	61°57′
Havre-aux-Maisons	47°24′	61°49′
Havre-Saint-Pierre	50°14′	63°36′
Heath Point	49°05′	61°42′
Hemmingford	45°03′	73°35′
Henryville	45°08′	73°11′
Herdman	45°02′	74°06′
High Rock Island (see Ispisakas Island)		
Hopkins Point	45°02′	74°32′
Hudson	45°27′	74°09′
Hudson Bay	58°25′	78°10′
Hudson Strait	62°10′	72°55′
Hull	45°26′	75°44′
Huntingdon	45°05′	74°10′
Ikattok Bay	59°12′	69°20′
Île à Calculot (formerly Gull Island)	50°12′	63°37′
Île à Calculot des Betchouanes (formerly Gun Island)	50°12′	63°13′
Île à Durand	47°05′	70°36′
Île à la Brume (formerly Fog Island)	50°10′	60°30′
Île à la Brume Migratory Bird Sanctuary	50°10′	60°30′
Île aux Allumettes	45°51′	77°04′
Île-aux-Allumettes	45°51′	77°04′
Île aux Basques	48°09′	69°15′
Île aux Basques Migratory Bird Sanctuary	48°11′	69°09′
Île aux Cochons	47°24′	61°50′
Île aux Coudres	47°24′	70°23′
Île aux Fermiers	45°40′	73°27′
Île aux Fraises	47°46′	69°48′
Île aux Goélands (formerly Gull Island; Lower North Shore)	50°12′	60°40′
Île aux Goélands (formerly Gull Island; Magdalen Islands)	47°22′	61°58′
Île aux Grues	47°04′	70°33′
Île aux Hérons	45°25′	73°35′
Île aux Lièvres	47°51′	69°44′
Île aux Loups	47°32′	61°43′
Île aux Loups Marins (formerly Seal Island)	47°36′	61°29′
Île aux Noix	45°07′	73°16′
Île aux Oies	47°07′	70°29′
Île aux Perroquets (Lower North Shore)	51°26′	57°14′
Île aux Perroquets (Mingan Islands)	50°13′	64°13′
Île aux Pommes	48°06′	69°19′
Île Blanche	47°56′	69°40′
Île Bonaventure and Rocher Percé Conservation Park	48°30′	64°10′
Île Bonaventure and rocher Percé Migratory Bird Sanctuary	48°30′	64°10′
Île Brion Ecological Reserve	47°48′	61°28′
Île Brûlée	47°36′	69°53′
Île d'Entrée (formerly Entry Island)	47°17′	61°42′
Île d'Orléans	46°55′	70°58′
Île de Bellechasse	46°56′	70°46′
Île de l'Est (formerly East Island; Magdalen Islands)	47°37′	61°26′
Île de l'Est (formerly Eastern Island; Lower North Shore)	50°19′	59°39′
Île de la Couvée	45°29′	73°30′

LOCALITY	Lat. (N)	Long. (W)
Île de la Maison	50°13′	64°12′
Île de la Mine	49°03′	68°33′
Île de la Petite Colonie	45°53′	73°14′
Île de Ouapitagone (formerly Wapitagun Island)	50°13′	60°02′
Île des Barques	46°05′	73°00′
Île des Loups (formerly Wolf Island)	50°10′	60°17′
Île des Soeurs	45°28′	73°33′
Île des Soeurs Grises	45°23′	73°45′
Île du Bic	48°24′	68°52′
Île du Cap aux Meules (formerly Grindstone Island)	47°23′	61°54′
Île du Corossol Migratory Bird Sanctuary	50°05′	66°25′
Île du Fantôme	50°14′	63°41′
Île du Havre Aubert (formerly Amherst Island)	47°14′	61°57′
Île du Havre aux Maisons (formerly Alright Island)	47°25′	61°47′
Île du Lac	50°11′	60°05′
Île du Mitan	46°04′	73°10′
Île du Moine	46°04′	73°00′
Île du Petit Mécatina (formerly Little Mecatina Island)	50°33′	59°20′
Île Jésus	45°35′	73°45′
Île Notre-Dame (formerly Moffat Island)	45°31′	73°32′
Île Nue de Mingan	50°13′	64°07′
Île Perrot	45°22′	73°57′
Île Plate	48°38′	64°09′
Île Razade Nord-Est	48°12′	69°08′
Île Razade Sud-Ouest	48°11′	69°09′
Île Ronde	46°05′	73°05′
Île Rouge	48°04′	69°33′
Île Sainte-Hélène (formerly St. Helen's Island)	45°31′	73°32′
Île Verte	48°02′	69°26′
Îles aux Loups Marins (formerly Seal Islands)	47°14′	70°26′
Îles de Contrecoeur National Wildlife Area	45°52′	73°15′
Îles de l'estuaire National Wildlife Area	47°35′	69°53′
Îles-de-la-Madeleine	47°24′	61°48′
Îles de la Paix	45°20′	73°51′
Îles de la Paix National Wildlife Area	45°20′	73°54′
Îles de Ouapitagone du Large (formerly Outer Wapitagun Islands)	50°12′	60°02′
Îles du Pot à l'Eau-de-Vie	47°52′	69°41′
Îles Sainte-Marie Migratory Bird Sanctuary	50°18′	59°40′
Îlet à Lefebvre	45°53′	73°14′
Îlet aux Alouettes	48°06′	69°41′
Innommée Bay	49°15′	61°48′
Inukjuak	58°27′	78°06′
Island of Montréal	45°30′	73°40′
Ispisakas Island (formerly High Rock Island)	52°26′	78°34′
Ivujivik	62°25′	77°55′
Jack River (see Jolicoeur River)		
Jacques Cartier Bay	51°16′	58°16′
Jacques Cartier Conservation Park	47°20′	71°21′
Jacques Cartier Passage	50°00′	63°30′
Jacques Cartier River	46°40′	71°45′
James Bay	53°40′	79°05′
Jaro ZEC	45°53′	70°22′
Jérémie Islets	48°53′	68°47′
Johan Beetz Bay	50°17′	62°48′
Jolicoeur River (formerly Jack River)	52°01′	78°42′
Joliette	46°01′	73°27′
Joly	46°29′	71°40′
Jones Island	45°31′	74°18′
Jonquière	48°25′	71°13′
Joseph Point	49°24′	62°08′
Joutel	49°27′	78°17′
Kahnawake	45°25′	73°41′
Kakassituq Point	54°11′	79°23′
Kamouraska	47°34′	69°52′
Kamouraska Bay	47°32′	69°54′
Kamouraska Islands	47°36′	69°53′
Kangiqsualujjuaq	58°41′	65°57′
Kangiqsujuaq	61°36′	71°58′
Kangirsuk	60°01′	70°02′
Katevale (see Sainte-Catherine-de-Hatley)		
Katevale marsh	45°16′	72°04′
Kazabazua	45°57′	76°01′
Kécarpoui River	51°04′	58°50′
Kegaska	50°11′	61°16′
Kegaska River	50°11′	61°21′
Kénogami	48°25′	71°15′

LOCALITY	Lat. (N)	Long. (W)
Killiniq (N.W.T.)	60°25′	64°50′
Kingsmere	45°29′	75°50′
Km 105 Camp (Rupert River)	50°22′	76°58′
Knob Lake	54°48′	66°48′
Knowlton	45°13′	72°31′
Kogaluc River	59°40′	77°35′
Koksoak River	58°32′	68°10′
Korak Bay	60°43′	77°45′
Koroc River	58°51′	65°47′
Kuujjuaq (formerly Fort Chimo)	58°06′	68°24′
Kuujjuarapik (formerly Great Whale)	55°17′	77°45′
Kuuraapik River	58°30′	68°04′
L'Ancienne-Lorette	46°48′	71°21′
L'Ange-Gardien	46°55′	71°06′
L'Anse-Saint-Jean	48°14′	70°12′
L'Ascension	48°42′	71°40′
L'Assomption	45°50′	73°25′
L'Étang-du-Nord	47°22′	61°57′
L'Isle-Verte	48°01′	69°20′
L'Islet	47°06′	70°21′
La Baie	48°20′	70°52′
La Barrière	46°25′	73°43′
La Doré	48°43′	72°39′
La Ferme	48°36′	78°13′
La Grande Commune	46°07′	72°54′
La Grande Décharge	48°32′	71°37′
La Grande Île	46°07′	73°01′
La Grande Reservoirs	53°45′	76°00′
La Grande Rivière (formerly Fort George River; Northern Québec)	53°50′	79°00′
La Grande Rivière (Gaspé)	48°24′	64°30′
La Grande-1 Dam	53°44′	78°34′
La Grande-2 Dam	53°47′	77°27′
La Grande-3	53°44′	75°59′
La Grande-3 Dam	53°45′	76°00′
La Grande-4 Generating Station	53°54′	73°23′
La Grosse Île	47°37′	61°31′
La Longue Pointe	52°45′	78°52′
La Malbaie (Bay)	48°35′	64°14′
La Malbaie (Town)	47°39′	70°09′
La Martinique beach	47°19′	61°56′
La Mauricie National Park	46°48′	72°58′
La Morandière	48°37′	77°38′
La Pérade	46°35′	72°12′
La Pocatière	47°22′	70°02′
La Prairie	45°25′	73°30′
La Prairie Basin	45°26′	73°32′
La Romaine	50°13′	60°40′
La Sarre	48°48′	79°12′
La Tuque	47°26′	72°47′
La Vérendrye Park (see La Vérendrye Wildlife Reserve)		
La Vérendrye Wildlife Reserve	47°20′	77°00′
La Vernière	47°22′	61°54′
Lac à l'Eau Claire (formerly Clearwater Lake)	50°10′	75°10′
Lac à l'Empêche	47°39′	70°31′
Lac à l'Épaule	47°15′	71°14′
Lac à la Loutre	46°08′	76°51′
Lac-à-la-Tortue	46°37′	72°38′
Lac à la Truite	46°05′	71°30′
Lac au Chien	46°57′	71°43′
Lac au Goéland (formerly Gull Lake)	49°47′	76°48′
Lac aux Cèdres	52°02′	67°07′
Lac d'Ours (formerly Lake Leslie)	45°51′	76°32′
Lac de l'Est	47°12′	69°34′
Lac de la Hutte Sauvage (formerly Indian House Lake)	56°10′	64°45′
Lac des Cygnes	47°39′	70°35′
Lac des Deux Montagnes (formerly Lake of Two Mountains)	45°27′	74°00′
Lac des Neiges	47°28′	71°02′
Lac du Bec Croche	47°14′	71°12′
Lac du Gros Ours	48°12′	74°21′
Lac du Milieu	48°49′	72°48′
Lac la Blanche	45°44′	75°19′
Lac-Mégantic	45°35′	70°53′
Lac Nouvel	47°26′	71°16′
Lac-Saint-Charles	46°54′	71°23′
Lac Saint-François National Wildlife Area	45°02′	74°29′
Lac-Supérieur	46°12′	74°28′
Lachine	45°26′	73°41′
Lachine Rapids	45°25′	73°36′
Lachute	45°39′	74°20′
Lacolle	45°05′	73°22′
Lafontaine Island	45°44′	76°45′
Laforge Reservoir	54°18′	72°15′
Laforge-1 Hydro-electric Station	54°02′	73°22′
Lake Abitibi	48°40′	79°31′
Lake Albanel	50°54′	73°18′
Lake Amichinatwayach	56°00′	72°40′

LOCALITY	Lat. (N)	Long. (W)
Lake Andou	47°23′	76°45′
Lake Ashuapmushuan	49°10′	73°45′
Lake Aylmer	45°49′	71°21′
Lake Beauchamp	48°35′	78°13′
Lake Beauchastel	48°09′	79°07′
Lake Beauport	46°57′	71°17′
Lake Belle Fontaine	47°24′	71°08′
Lake Bérard (formerly Gregory Lake)	58°28′	70°07′
Lake Bernard	45°45′	75°59′
Lake Bienville	55°05′	72°40′
Lake Bleury	47°19′	71°43′
Lake Boivin	45°24′	72°41′
Lake Boulrisse	46°57′	76°15′
Lake Brisson	56°18′	64°20′
Lake Caopatina	49°27′	74°46′
Lake Carroll	47°33′	71°03′
Lake Cascapédia	48°55′	66°20′
Lake Cawachagamite	51°42′	74°08′
Lake Chabouillé	50°53′	77°50′
Lake Champlain	45°03′	73°09′
Lake Chavigny	58°12′	75°08′
Lake Chevrillon	50°00′	74°29′
Lake Chibougamau	49°50′	74°15′
Lake Chicobi	48°53′	78°30′
Lake Choret	54°43′	72°36′
Lake Colomb	51°02′	77°40′
Lake Corbeau	49°18′	73°08′
Lake Denys	54°55′	76°47′
Lake Desaulniers	53°34′	77°34′
Lake Deyglun	46°39′	78°36′
Lake Dieppe	49°30′	79°20′
Lake Drolet	45°44′	70°53′
Lake Édouard (Québec) (formerly Lake Edward)	47°35′	72°21′
Lake Édouard (Témiscamingue)	48°14′	79°01′
Lake Élizabeth	55°36′	75°28′
Lake Evans	50°55′	77°00′
Lake Fagnant	55°05′	76°49′
Lake Fleuricourt	50°18′	73°05′
Lake Fortier	46°57′	71°21′
Lake Fressel	55°27′	75°12′
Lake Fronsac	48°48′	64°43′
Lake Gaudreault	50°52′	62°15′
Lake Georges	48°36′	78°13′
Lake Giles	46°07′	75°46′
Lake Grasset	49°57′	78°10′
Lake Guilbeault	47°04′	72°22′
Lake Guillaume Delisle (formerly Richmond Gulf)	56°15′	76°17′
Lake Hélène	53°26′	77°31′
Lake Hertel	45°33′	73°09′
Lake Jacques Cartier	47°35′	71°13′
Lake Kakupis	55°43′	75°38′
Lake Kapwakami	53°03′	78°23′
Lake Kegaska	50°20′	61°25′
Lake Kénogami	48°20′	71°23′
Lake La Bardelière	51°24′	74°50′
Lake La Mothe	48°46′	71°09′
Lake La Motte	48°24′	78°03′
Lake Larose	50°35′	74°34′
Lake Leamy	45°27′	75°43′
Lake Leslie (see Lac d'Ours)		
Lake Leverrier	46°47′	69°59′
Lake Louet	55°18′	71°45′
Lake Lucie	50°23′	78°27′
Lake Lusignan	46°41′	74°09′
Lake Lyster	45°02′	71°54′
Lake Macpès (see Grand Lac Macpès)		
Lake Magog	45°18′	72°02′
Lake Malbaie	47°34′	71°00′
Lake Malin	47°21′	71°03′
Lake Manouane	50°41′	70°45′
Lake Marceau	48°44′	73°51′
Lake Maricourt	56°34′	70°50′
Lake Maryland	60°27′	73°20′
Lake Matagami	49°53′	77°30′
Lake Matane	48°42′	66°59′
Lake Matemace	54°54′	66°49′
Lake Mégantic	45°32′	70°53′
Lake Memphrémagog	45°08′	72°16′
Lake Mesgouez	51°24′	75°05′
Lake Minishtuk	49°17′	72°59′
Lake Minto	57°13′	75°00′
Lake Mistassini	51°00′	73°37′
Lake Mistigougèche	48°08′	67°58′
Lake Montauban	46°52′	72°10′
Lake Moquachéa	50°22′	75°57′
Lake Nemiscau	51°26′	76°43′
Lake Nixon	48°07′	71°16′
Lake Nominingue	46°26′	74°59′
Lake Nudant	54°10′	76°34′
Lake Olga	49°47′	77°15′
Lake Opocopa	52°38′	66°34′

LOCALITY	Lat. (N)	Long. (W)
Lake Ouescapis	50°14′	77°00′
Lake Paimpont	54°56′	73°20′
Lake Pau	54°52′	69°54′
Lake Pelletier	48°13′	79°03′
Lake Perceval	46°55′	71°33′
Lake Péribonka	50°07′	71°15′
Lake Philippe	45°36′	76°00′
Lake Pio	58°55′	69°35′
Lake Plétipi	51°41′	70°06′
Lake Poncheville	50°10′	76°55′
Lake Preissac	48°20′	78°20′
Lake Pythonga	46°23′	76°26′
Lake Rimouski	48°19′	68°28′
Lake Rouvray	49°18′	70°49′
Lake Rouyn	48°14′	78°58′
Lake Saint-Augustin	46°45′	71°23′
Lake Saint-Charles	46°56′	71°23′
Lake Saint-François (formerly Lake St. Francis, Frontenac)	45°55′	71°10′
Lake Saint-François (Huntingdon)	45°10′	74°22′
Lake Saint-Jean (formerly Lake St. John)	48°35′	72°05′
Lake Saint-Louis (formerly Lake St. Louis)	45°24′	73°48′
Lake Saint-Paul	46°18′	72°29′
Lake Saint-Pierre (formerly Lake St. Peter)	46°12′	72°52′
Lake Saint-Pierre islets	46°12′	72°50′
Lake Sainte-Anne	50°05′	67°50′
Lake Sakami	53°15′	76°45′
Lake Sergent	46°52′	71°43′
Lake Simard	47°37′	78°41′
Lake Soscumica	50°15′	77°27′
Lake Tantaré	47°04′	71°33′
Lake Témiscouata	47°40′	68°50′
Lake Timiskaming	47°10′	79°25′
Lake Trois Saumons	47°08′	70°11′
Lake Villebon	47°58′	77°19′
Lake Vincelotte	54°18′	72°15′
Lake Voltz	59°11′	69°43′
Lake Waswanipi	49°34′	76°29′
Lake Wayagamac	47°22′	72°39′
Landrienne	48°33′	77°57′
Languedoc	48°48′	78°42′
Lanoraie	45°58′	73°13′
LaSalle	45°26′	73°38′
Latulipe	47°26′	79°02′
Launay	48°39′	78°32′
Laurentide Provincial Park (see Laurentide Wildlife Reserve)		
Laurentide Wildlife Reserve	47°45′	71°15′
Laval	45°35′	73°45′
Laval Island	48°45′	69°02′
Le Bic	48°22′	68°42′
Le Cormoran (formerly Seal House Cove)	50°16′	65°53′
Le Corps-Mort (formerly Deadman's Island)	47°16′	62°12′
Le Gardeur	45°45′	73°28′
Le Gros Cap	47°13′	62°00′
Le Gros Pèlerin (formerly Great Pilgrim Island)	47°45′	69°42′
Le Petit Pèlerin (formerly Long Pilgrim Island)	47°42′	69°45′
Le Pilier de Bois (formerly Wood Pillar Islet)	47°12′	70°24′
Le Pilier de Pierre (formerly Stone Pillar Islet)	47°12′	70°22′
Le Pinacle (formerly Pinnacle Mountain)	45°03′	72°44′
Lebel-sur-Quévillon	49°03′	76°59′
Leclercville	46°34′	72°00′
Lennoxville	45°22′	71°51′
Léry	45°21′	73°48′
Les Éboulements	47°29′	70°19′
Les Escoumins	48°21′	69°24′
Les Grands Jardins	47°36′	71°00′
Les Îlots	48°15′	64°45′
Les Pèlerins (formerly Pilgrim Islands)	47°44′	69°44′
Les Trois Soeurs	49°03′	68°32′
Lévis	46°48′	71°11′
Little Mecatina River	50°40′	59°25′
Little Whale River (see Petite Rivière de la Baleine)		
Long Portage	51°55′	67°03′
Longue-Pointe	50°16′	64°09′
Longue-Pointe-de-Mingan (see Longue-Pointe)		
Longueuil	45°32′	73°30′
Loretteville	46°51′	71°21′
Lotbinière	46°32′	71°38′
Louis-Hippolyte LaFontaine Bridge-Tunnel	45°35′	73°30′
Louise-Gosford ZEC	45°31′	70°51′

LOCALITY	Lat. (N)	Long. (W)
Louiseville	46°15′	72°57′
Lourdes-de-Blanc-Sablon	51°25′	57°12′
Louvicourt	48°05′	77°28′
Luskville	45°32′	76°01′
Madeleine River	49°15′	65°19′
Magdalen Islands (see Îles-de-la-Madeleine)		
Magog	45°16′	72°08′
Magpie River	50°19′	64°27′
Maizerets	46°50′	71°13′
Malartic	48°08′	78°08′
Malbaie	48°38′	64°23′
Malbaie River	47°39′	70°09′
Manic-1 Dam	49°12′	68°20′
Manic-2 Dam	49°19′	68°21′
Manic-5	50°38′	68°44′
Manic-5 Dam (see Daniel Johnson Dam)		
Manicouagan	49°13′	68°09′
Manicouagan Peninsula	49°08′	68°20′
Manicouagan Point	49°06′	68°12′
Manicouagan Reservoir	51°24′	68°44′
Manicouagan River	49°11′	68°13′
Manitounuk Sound	55°30′	77°25′
Maniwaki	46°23′	75°58′
Manowin Island	50°06′	66°24′
Marcel Raymond Ecological Reserve	45°07′	73°15′
Maria	48°10′	65°59′
Maricourt Camp	56°30′	71°10′
Marsoui	49°13′	66°04′
Martin-Valin ZEC	48°40′	70°37′
Mascouche	45°45′	73°36′
Maskinongé	46°14′	73°01′
Maskinongé Bay	46°12′	73°00′
Maskinongé River	45°49′	74°40′
Masson	45°32′	75°25′
Masta Island	45°39′	73°28′
Mastigouche Wildlife Reserve	46°40′	73°30′
Matagami	49°45′	77°38′
Matamec	50°17′	65°58′
Matamec River	50°17′	65°58′
Matane	48°51′	67°32′
Matane River	48°51′	67°32′
Matane Wildlife Reserve	48°44′	66°55′
Matapédia River	47°58′	66°56′
Matapédia Valley	48°10′	67°10′
Matchiatic Island	50°16′	59°55′
McNab Rocks	51°45′	79°06′
Meech Creek	45°36′	75°53′
Melocheville	45°19′	73°56′
Métabetchouan	48°26′	71°52′
Métis-sur-Mer	48°40′	68°00′
Milton	45°30′	72°45′
Mingan	50°18′	64°02′
Mingan Archipelago	50°13′	63°35′
Mingan Archipelago National Park Reserve	50°13′	63°35′
Mingan Islands	50°13′	63°50′
Mingan River	50°18′	63°59′
Minganie	50°14′	63°36′
Miquelon	49°24′	76°27′
Mirabel	45°39′	74°05′
Missisicabi River	51°15′	79°41′
Missisquoi	45°10′	73°00′
Missisquoi Bay	45°03′	73°08′
Mistassini (see Mistissini)		
Mistassini River	48°42′	72°20′
Mistissini	50°25′	73°53′
Moar Bay	52°48′	78°47′
Moisie	50°11′	66°05′
Moisie Bay	50°16′	65°56′
Moisie River	50°12′	66°04′
Mollet Lakes	55°37′	74°25′
Mont-Brun	48°22′	78°43′
Mont-Carmel	47°26′	69°52′
Mont de la Table	48°59′	66°00′
Mont du Lac à l'Empêche	47°42′	70°35′
Mont du Lac des Cygnes	47°40′	70°36′
Mont-Joli	48°35′	68°11′
Mont-Joli Airport	48°36′	68°12′
Mont-Laurier	46°33′	75°30′
Mont-Orford Provincial Park (see Mont-Orford Recreation Park)		
Mont-Orford Recreation Park	45°20′	72°13′
Mont-Saint-Bruno Conservation Park	45°33′	73°19′
Mont-Saint-Hilaire	45°34′	73°12′
Mont-Saint-Hilaire Biosphere Reserve	45°33′	73°10′
Mont-Saint-Pierre	49°13′	65°49′
Mont-Tremblant Provincial Park (see Mont-Tremblant Recreation Park)		
Mont-Tremblant Recreation Park	46°26′	74°21′

LOCALITY	Lat. (N)	Long. (W)
Montagne Blanche	51°35′	67°11′
Montagne Noire	46°14′	74°18′
Monteregian Hills	45°28′	73°03′
Montmagny	46°59′	70°33′
Montmagny flats	47°00′	70°35′
Montmagny Islands (see Archipel de L'Isle-aux-Grues)		
Montmorency Demonstration and Research Forest	47°17′	71°08′
Montmorency forest (see Montmorency Demonstration and Research Forest)		
Montmorency River	46°53′	71°09′
Montréal	45°30′	73°36′
Montréal Botanical Gardens	45°33′	73°33′
Montréal-Est	45°38′	73°31′
Montréal-Nord	45°36′	73°38′
Montréal-Ouest	45°27′	73°39′
Montréal Urban Community	45°30′	73°41′
Monts Chic-Chocs	48°55′	66°00′
Mosquito Bay	60°46′	78°00′
Mount Albert	48°55′	66°11′
Mount Bleu	47°18′	70°48′
Mount D'Iberville	58°53′	63°43′
Mount Gosford	45°18′	70°52′
Mount Hereford	45°05′	71°36′
Mount Jacques Cartier	48°59′	65°57′
Mount Jauffret	51°37′	68°07′
Mount L.-A. Richard	47°48′	71°15′
Mount Mégantic	45°28′	71°09′
Mount Orford	45°19′	72°15′
Mount Owl's Head	45°04′	72°18′
Mount Raoul Blanchard	47°19′	70°50′
Mount Rougemont	45°28′	73°03′
Mount Royal	45°30′	73°36′
Mount Royal Cemetery	45°30′	73°36′
Mount Royal Park	45°30′	73°36′
Mount Saint-Alban	48°49′	64°14′
Mount Saint-Bruno	45°33′	73°19′
Mount Saint-Grégoire	45°21′	73°09′
Mount Saint-Hilaire	45°33′	73°10′
Mount Sir Wilfrid	46°41′	75°36′
Mount Veyrier	51°32′	68°05′
Mount Yamaska	45°27′	72°52′
Mount Yapeitso	52°19′	70°27′
Murdochville	48°58′	65°30′
Nantes	45°38′	71°02′
Napierville	45°11′	73°24′
Nastapoka Islands (N.W.T.)	56°40′	76°40′
Nastapoka River	56°55′	76°33′
Natashquan	50°12′	61°49′
Natashquan River	50°07′	61°48′
Nauberakvik River	58°56′	78°20′
Nemiscau	51°42′	76°15′
Neuville	46°42′	71°35′
New Québec Crater	61°17′	73°40′
New Richmond	48°10′	65°52′
Newport	48°16′	64°45′
Newport Point	48°17′	64°43′
Niapiskau Island	50°13′	63°45′
Nicolet	46°13′	72°37′
Nicolet Migratory Bird Sanctuary	46°08′	72°44′
Nicolet River	46°15′	72°39′
Noranda (see Rouyn-Noranda)		
North Bird Rock (see Rocher aux Margaux)		
North Hatley	45°17′	71°58′
Notre-Dame Mountains	48°10′	68°00′
Notre-Dame-de-Lourdes	46°19′	71°49′
Notre-Dame-de-Montauban	46°52′	72°20′
Notre-Dame-de-Pierreville	46°06′	72°53′
Notre-Dame-de-Portneuf	46°44′	71°55′
Notre-Dame-de-Stanbridge	45°10′	73°02′
Notre-Dame-des-Sept-Douleurs	48°00′	69°27′
Notre-Dame-du-Bon-Conseil	46°00′	72°21′
Nottaway River	51°22′	78°55′
Novide River	51°32′	79°15′
Nuns' Island (see Île des Soeurs)		
Nuvuk Island (N.W.T.)	62°23′	77°55′
Oka	45°28′	74°05′
Old Chelsea	45°30′	75°49′
Old Harry Point	47°34′	61°28′
Old Port of Montréal	45°30′	73°33′
Olomane River	50°14′	60°38′
Opinaca Reservoir	52°39′	76°20′
Opinaca River	52°14′	78°02′
Oster Island	47°24′	79°32′
Otish Mountains	52°19′	70°35′
Ottawa River	45°20′	73°55′
Otter Lake	45°51′	76°26′
Ouapitagone Archipelago (formerly Wapitagun Islands)	50°12′	60°17′
Outaouais Regional Community	45°35′	75°49′
Outardes-3 Dam	49°34′	68°48′
Pabos	48°22′	64°37′

LOCALITY	Lat. (N)	Long. (W)
Paspébiac	48°02′	65°15′
Patapedia River	47°50′	67°22′
Paul Bay	54°00′	79°00′
Payne Lake	59°25′	74°00′
Percé	48°32′	64°13′
Percé Rock	48°31′	64°12′
Péribonka	48°46′	72°03′
Péribonka River	48°45′	72°05′
Perroquets Islands	50°17′	59°44′
Petit Lac Saint-Germains	48°26′	70°39′
Petit Lac Wayagamac	47°23′	72°31′
Petit-Mécatina	51°30′	60°00′
Petit rocher aux Oiseaux (see Rocher aux Margaux)		
Petite Rivière Cascapédia	48°09′	65°51′
Petite Rivière de la Baleine	56°00′	76°45′
Petite-Rivière-Saint-François	47°18′	70°34′
Philipsburg	45°02′	73°05′
Philipsburg Migratory Bird Sanctuary	45°01′	73°04′
Piagochioui River	54°02′	79°02′
Pic de l'Aurore	48°32′	64°14′
Piedmont	45°54′	74°08′
Pierrefonds	45°29′	73°52′
Pierreville	46°04′	72°49′
Pikauba River	48°20′	71°27′
Pin-Rigide Ecological Reserve	45°06′	73°52′
Pintendre	46°45′	71°08′
Pipmuacan Reservoir	49°40′	70°20′
Plains of Abraham	46°48′	71°13′
Plaisance	45°37′	75°07′
Plaisance Wildlife Reserve	45°36′	75°08′
Plée de Saint-Charles	46°46′	71°00′
Plessisville	46°13′	71°47′
Point Louis XIV	54°38′	79°45′
Pointe-à-Boisvert	48°34′	69°11′
Pointe à la Batterie (formerly Battery Point)	49°25′	62°13′
Pointe-à-la-Croix	48°01′	66°41′
Pointe au Foin	45°13′	74°17′
Pointe-au-Père (see Sainte-Anne-de-la-Pointe-au-Père)		
Pointe-au-Père National Wildlife Area	48°31′	68°28′
Pointe-aux-Loups	47°32′	61°43′
Pointe-aux-Outardes	49°03′	68°26′
Pointe-aux-Trembles	45°39′	73°30′
Pointe-Claire	45°26′	73°50′
Pointe de l'Est (formerly East Point)	47°37′	61°23′
Pointe de l'Est National Wildlife Area	47°38′	61°27′
Pointe de la Tour	49°30′	62°25′
Pointe des Monts	49°19′	67°23′
Pointe-des-Monts	49°19′	67°23′
Pointe du Domaine	45°23′	73°54′
Pointe du Gouvernement	45°08′	73°15′
Pointe-du-Lac	46°17′	72°42′
Pointe du Renard	49°19′	61°50′
Pointe-Lebel	49°10′	68°12′
Pontax River	51°37′	78°50′
Port-aux-Quilles	47°55′	69°52′
Port-Cartier	50°01′	66°52′
Port-Daniel	48°11′	64°58′
Port-Daniel River	48°11′	64°58′
Port-Menier	49°49′	64°21′
Port of Gros-Cacouna	47°56′	69°31′
Port of Montréal	45°30′	73°33′
Port of Québec	46°49′	71°13′
Port-Saint-François	46°16′	72°37′
Portage Lake	45°56′	70°16′
Portage-du-Fort	45°36′	76°40′
Portneuf	46°42′	71°53′
Portneuf sandbank	48°37′	69°05′
Portneuf Wildlife Reserve	47°09′	72°17′
Poste-de-la-Baleine (see Kuujjuarapik)		
Povungnituk	60°02′	77°17′
Povungnituk River	60°03′	77°13′
Princeville	46°10′	71°53′
Quaqtaq	61°02′	69°37′
Québec City	46°49′	71°14′
Racine Point	48°44′	72°19′
Radisson	53°47′	77°37′
Ragueneau	49°04′	68°32′
Rapides-des-Joachims	46°12′	77°41′
Repentigny	45°44′	73°28′
Restigouche	48°01′	66°42′
Restigouche River	48°04′	66°20′
Richard Island	47°26′	76°42′
Richelieu	46°00′	73°00′
Richelieu River	46°03′	73°07′
Richelieu Valley	45°27′	73°16′
Richmond	45°40′	72°09′
Rigaud	45°29′	74°18′

LOCALITY	Lat. (N)	Long. (W)
Rigaud Mountain	45°27′	74°18′
Rimouski	48°26′	68°33′
Rimouski Wildlife Reserve	48°03′	68°15′
Rivière à la Baleine (formerly Whale River)	58°15′	67°35′
Rivière à la Truite (formerly Trout River)	51°38′	78°52′
Rivière-à-Pierre	46°59′	72°11′
Rivière au Castor (formerly Beaver River)	53°24′	78°58′
Rivière au Chien Rouge	59°19′	69°45′
Rivière au Phoque (formerly Seal River)	54°31′	79°30′
Rivière aux Feuilles (formerly Leaf River)	58°47′	70°04′
Rivière aux Orignaux	46°26′	72°14′
Rivière aux Outardes	49°04′	68°28′
Rivière-aux-Outardes	50°15′	69°00′
Rivière aux Saumons	45°02′	74°31′
Rivière-aux-Saumons	49°25′	62°14′
Rivière-Blanche ZEC	47°18′	72°05′
Rivière de la Montagne Blanche	51°24′	67°01′
Rivière des Caps	47°43′	69°40′
Rivière des Mille Îles	45°42′	73°32′
Rivière des Prairies	45°42′	73°29′
Rivière du Lièvre	45°31′	75°26′
Rivière-du-Loup	47°50′	69°32′
Rivière du Nord (formerly North River)	45°31′	74°20′
Rivière du Sud (formerly South River)	45°08′	73°15′
Rivière du Sud-Ouest	48°21′	68°46′
Rivière Noire	46°12′	73°34′
Rivière Nouvelle	48°07′	66°17′
Rivière-Nouvelle	48°18′	66°29′
Rivière-Ouelle	47°26′	70°01′
Rivière Rouge	45°39′	74°41′
Roberval	48°31′	72°13′
Rocher aux Margaux	47°50′	61°09′
Rocher aux Oiseaux	47°50′	61°09′
Rochers aux Oiseaux	47°51′	61°09′
Rochers aux Oiseaux Migratory Bird Sanctuary	47°51′	61°00′
Rock Forest	45°21′	71°59′
Rockburn	45°01′	74°01′
Roggan River	54°25′	79°28′
Romaine River	50°18′	63°48′
Routhierville	48°11′	67°09′
Rouyn (see Rouyn-Noranda)		
Rouyn-Noranda	48°14′	79°01′
Roxboro	45°31′	73°48′
Rupert Bay	51°35′	79°00′
Rupert House (see Waskaganish)		
Rupert River	51°30′	78°45′
Sacré-Coeur	48°14′	69°48′
Saguenay Fjord	48°08′	69°44′
Saguenay River	48°08′	69°44′
Saguenay–St. Lawrence Marine Park	48°07′	69°40′
Saindon Lakes	55°36′	73°30′
Saint-Adrien-d'Irlande	46°07′	71°27′
Saint-Ambroise	48°33′	71°20′
Saint-André	47°41′	69°44′
Saint-Anicet	45°07′	74°21′
Saint-Antoine-Abbé	45°03′	73°53′
Saint-Armand-Ouest	45°02′	73°03′
Saint-Armand-Station	45°02′	73°03′
Saint-Athanase	47°26′	69°25′
Saint-Augustin	51°13′	58°39′
Saint-Augustin Migratory Bird Sanctuary	51°08′	58°28′
Saint-Augustin-de-Desmaures	46°45′	71°28′
Saint-Basile-le-Grand	45°32′	73°17′
Saint-Benoît-du-Lac	45°10′	72°16′
Saint-Bernard Island (see Île des Soeurs Grises)		
Saint-Bernard-de-Lacolle	45°05′	73°25′
Saint-Bruno (Chicoutimi)	48°17′	70°49′
Saint-Bruno (Lac-Saint-Jean)	48°28′	71°39′
Saint-Bruno-de-Kamouraska	47°27′	69°45′
Saint-Bruno-de-Montarville	45°32′	73°21′
Saint-Charles	46°46′	70°57′
Saint-Charles River	46°49′	71°13′
Saint-Charles-de-Mandeville	46°22′	73°21′
Saint-Chrysostome	45°06′	73°46′
Saint-Cléophas	46°14′	73°25′
Saint-Coeur-de-Marie (see Delisle)		
Saint-Colomban	45°44′	74°08′
Saint-Constant	45°22′	73°34′
Saint-David-de-Falardeau	48°37′	71°07′
Saint-Denis	47°30′	69°56′
Saint-Denis Point	47°30′	69°59′
Saint-Dominique-du-Rosaire	48°46′	78°07′
Saint-Donat	46°19′	74°13′
Saint-Edmond-de-Grantham	45°53′	72°40′

LOCALITY	Lat. (N)	Long. (W)
Saint-Édouard-de-Lotbinière	46°34′	71°50′
Saint-Elphège	46°03′	72°42′
Saint-Elzéar	48°09′	65°24′
Saint-Esprit	45°54′	73°40′
Saint-Étienne-de-Beauharnois	45°15′	73°55′
Saint-Eusèbe	47°33′	68°55′
Saint-Fabien	48°18′	68°52′
Saint-Fabien-de-Panet	46°39′	70°09′
Saint-Fabien-sur-Mer	48°19′	68°52′
Saint-Félicien	48°39′	72°27′
Saint-Félix-de-Kingsey	45°48′	72°12′
Saint-Ferréol-les-Neiges	47°07′	70°51′
Saint-Fidèle	47°44′	69°59′
Saint-François	47°00′	70°49′
Saint-François Bay	46°05′	72°55′
Saint-François River (formerly St. Francis River)	46°07′	72°55′
Saint-François-du-Lac	46°04′	72°50′
Saint-Fulgence	48°27′	70°54′
Saint-Fulgence marsh	48°27′	70°55′
Saint-Gabriel-de-Brandon	46°16′	73°23′
Saint-Gabriel-de-Gaspé	48°31′	64°32′
Saint-Gédéon	48°30′	71°46′
Saint-Gédéon marsh	48°30′	71°46′
Saint-Georges	46°07′	70°40′
Saint-Georges-de-Cacouna	47°55′	69°30′
Saint-Gérard-des-Laurentides	46°36′	72°49′
Saint-Germain	47°35′	69°48′
Saint-Gilles	46°31′	71°22′
Saint-Godefroi	48°05′	65°07′
Saint-Hilaire-sur-Richelieu (see Mont-Saint-Hilaire)		
Saint-Hippolyte	45°31′	73°41′
Saint-Honoré	48°32′	71°05′
Saint-Hubert	45°30′	73°25′
Saint-Hubert Airport	45°31′	73°25′
Saint-Hyacinthe	45°37′	72°57′
Saint-Ignace-de-Loyola	46°04′	73°08′
Saint-Jacques River	45°26′	73°29′
Saint-Jean Cove	48°15′	70°12′
Saint-Jean Creek	45°22′	73°47′
Saint-Jean River	48°46′	64°26′
Saint-Jean-Chrysostome	46°43′	71°12′
Saint-Jean-de-Boischatel	46°54′	71°09′
Saint-Jean-Port-Joli	47°13′	70°16′
Saint-Jean-sur-Richelieu	45°19′	73°16′
Saint-Jérôme	45°47′	74°00′
Saint-Joachim	47°03′	70°51′
Saint-Joachim-de-Courval	45°58′	72°33′
Saint-Jovite	46°07′	74°36′
Saint-Lambert	45°30′	73°30′
Saint-Lambert-de-Lévis	46°35′	71°13′
Saint-Laurent	45°30′	73°40′
Saint-Lazare (Bellechasse)	46°39′	70°48′
Saint-Lazare (Vaudreuil)	45°24′	74°08′
Saint-Léon-de-Standon	46°29′	70°37′
Saint-Léonard-de-Portneuf	46°53′	71°55′
Saint-Louis	47°34′	69°52′
Saint-Majoric (see Saint-Majorique)		
Saint-Majorique	45°56′	72°35′
Saint-Marc-de-Figuery	48°28′	78°03′
Saint-Mathieu	48°11′	68°59′
Saint-Maurice	47°00′	73°30′
Saint-Maurice River	46°21′	72°31′
Saint-Maurice Wildlife Reserve	47°05′	73°15′
Saint-Maurice-de-Dalquier	48°39′	77°59′
Saint-Méthode	48°44′	72°25′
Saint-Michel-des-Saints	46°41′	73°55′
Saint-Nicolas	46°42′	71°24′
Saint-Omer	48°07′	66°13′
Saint-Ours Island	45°55′	73°13′
Saint-Pascal	47°32′	69°49′
Saint-Paul-du-Nord	48°34′	69°14′
Saint-Pierre	45°27′	73°39′
Saint-Pierre Point	48°38′	64°10′
Saint-Pierre-de-Montmagny (see Saint-Pierre-Montmagny)		
Saint-Pierre-les-Becquets	46°30′	72°12′
Saint-Pierre-Montmagny	46°55′	70°38′
Saint-Placide	45°32′	74°12′
Saint-Prime	48°35′	72°20′
Saint-Prosper	46°37′	72°17′
Saint-Quentin Island	46°21′	72°32′
Saint-Raphaël	46°48′	70°45′
Saint-Raymond	46°54′	71°50′
Saint-Roch-de-Mékinac	46°49′	72°46′
Saint-Roch-des-Aulnaies	47°19′	70°11′
Saint-Romuald	46°45′	71°14′
Saint-Sauveur-des-Monts	45°54′	74°10′
Saint-Siméon	47°50′	69°53′
Saint-Simon	48°12′	69°03′
Saint-Sulpice	45°50′	73°21′
Saint-Tharcisius	48°32′	67°20′
Saint-Tite-des-Caps	47°08′	70°46′

LOCALITY	Lat. (N)	Long. (W)
Saint-Ubalde	46°45′	72°16′
Saint-Valérien-de-Milton	45°34′	72°43′
Saint-Vincent-de-Paul	45°37′	73°39′
Saint-Vital-de-Clermont	48°55′	79°15′
Saint-Zéphirin	46°04′	72°38′
Sainte-Anne Cove	47°20′	70°05′
Sainte-Anne Creek	49°05′	72°21′
Sainte-Anne River	49°08′	66°30′
Sainte-Anne-de-Bellevue	45°25′	73°56′
Sainte-Anne-de-la-Pocatière	47°21′	70°00′
Sainte-Anne-de-la-Pointe-au-Père	48°31′	68°28′
Sainte-Anne-des-Monts	49°08′	66°30′
Sainte-Béatrix	46°12′	73°37′
Sainte-Blandine	48°22′	68°28′
Sainte-Catherine Bay	48°07′	69°43′
Sainte-Catherine-de-Hatley	45°15′	72°03′
Sainte-Dorothée	45°32′	73°49′
Sainte-Émélie-de-l'Énergie	46°19′	73°39′
Sainte-Euphémie	46°46′	70°26′
Sainte-Famille-d'Aumond	46°28′	75°53′
Sainte-Félicité	46°57′	69°56′
Sainte-Foy	46°47′	71°17′
Sainte-Gertrude-Manneville	48°32′	78°22′
Sainte-Hedwidge	48°29′	72°21′
Sainte-Luce	48°33′	68°23′
Sainte-Luce-sur-Mer (see Sainte-Luce)		
Sainte-Marguerite River (Sacré-Coeur)	48°16′	69°57′
Sainte-Marguerite River (Sept-Îles)	50°09′	66°36′
Sainte-Marguerite-de-Lingwick	45°36′	71°21′
Sainte-Marguerite-Marie	48°49′	72°13′
Sainte-Marie Island (formerly St. Mary Island)	50°18′	59°39′
Sainte-Marie Islands (formerly St. Mary Islands)	50°19′	59°39′
Sainte-Marie River	49°40′	63°55′
Sainte-Marthe	45°24′	74°18′
Sainte-Martine	45°15′	73°48′
Sainte-Pétronille	46°51′	71°08′
Sainte-Rita	47°57′	68°55′
Sainte-Rose	45°37′	73°47′
Sainte-Rose-de-Watford	46°19′	70°25′
Sainte-Thérèse Island	45°41′	73°28′
Sainte-Thérèse-de-Gaspé	48°25′	64°25′
Sainte-Thérèse-de-Lisieux	46°54′	71°12′
Salaberry-de-Valleyfield	45°15′	74°08′
Salluit	62°13′	75°39′
Sandy Beach	48°49′	64°26′
Sandy Beach Point	48°50′	64°25′
Sandy Hook Dune	47°15′	61°48′
Schefferville	54°48′	66°50′
Schwartz	45°48′	76°22′
Seal House Cove (see Le Cormoran)		
Senneterre	48°23′	77°14′
Senneville	45°25′	73°57′
Sept-Îles (formerly Seven Islands)	50°12′	66°23′
Sept Îles Bay	50°12′	66°28′
Serpent River (formerly Snake River)	48°12′	79°22′
Shag Island	47°29′	61°42′
Shannon	46°53′	71°31′
Shawinigan	46°33′	72°45′
Shawinigan Falls	46°32′	72°46′
Shefford	45°27′	72°32′
Sheldrake	50°16′	64°54′
Sherbrooke	45°25′	71°54′
Shickshock Mountains (see Monts Chic-Chocs)		
Sillery	46°46′	71°15′
Snake River (see Serpent River)		
Sorel	46°02′	73°07′
Sorel Islands	46°06′	73°01′
St. Lawrence River	49°40′	64°30′
Stadacona Pond	48°14′	79°01′
Stater Pond	46°04′	71°28′
Strait of Belle Isle	51°24′	57°07′
Streit Pond	45°02′	73°04′
Sud-Ouest Point	49°23′	63°36′
Summit Park	45°29′	76°36′
Sutton Mountains	45°05′	72°30′
Table Head	49°21′	61°54′
Table Mountain (see Mont de la Table)		
Tadoussac	48°09′	69°43′
Taillon Point	48°42′	71°58′
Takpangajok Inlet (Ungava Bay; see also Tarpangajok Bay)	58°57′	—
Tantaré Ecological Reserve	47°03′	71°31′
Tarpangajok Bay (see also Takpangajok Inlet)	60°04′	65°04′
Tartigou River	48°45′	67°48′
Tasiujaq	58°42′	69°56′
Taureau Reservoir	46°46′	73°50′

LOCALITY	Lat. (N)	Long. (W)
Témiscamingue	47°33′	79°14′
Témiscamingue Point	47°17′	79°28′
Terrebonne	45°42′	73°38′
Tête-à-la-Baleine	50°42′	59°19′
Thetford Mines	46°05′	71°18′
Thurso	45°36′	75°15′
Ticouapé River	48°41′	72°21′
Torngat Mountains	59°30′	64°25′
Town of Mount Royal	45°31′	73°39′
Tracy	46°01′	73°09′
Trécesson	48°39′	78°19′
Trembling Mountain Park (formerly Montagne Tremblante Provincial Park; see Mont-Tremblant Recreation Park)		
Trois-Pistoles	48°07′	69°10′
Trois-Rivières (formerly Three Rivers)	46°21′	72°33′
Trois-Rivières-Ouest	46°20′	72°36′
Twin Islands	51°13′	58°27′
Ulverton	45°43′	72°14′
Ungava Bay	58°20′	67°20′
Ungava Peninsula	60°00′	74°00′
Val-Alain	46°25′	71°45′
Val-Bélair	46°52′	71°26′
Val-d'Or	48°06′	77°47′
Val-Jalbert (see Chambord)		
Val-Paradis	49°10′	79°17′
Val-Racine	45°29′	71°04′
Valcartier (see Valcartier-Village)		
Valcartier-Village	46°56′	71°28′
Vale Perkins	45°05′	72°18′
Valleyfield (see Salaberry-de-Valleyfield)		
Varennes	45°41′	73°26′
Varennes Islands	45°40′	73°27′
Vaudreuil	45°24′	74°02′
Vaudreuil Bay	45°24′	74°00′
Vaudreuil-Soulanges	45°21′	74°13′
Vauvert	48°45′	72°07′
Verchères	45°47′	73°21′
Verchères Islands	45°48′	73°21′
Verdun	45°27′	73°34′
Victoriaville	46°03′	71°58′
Ville-Marie	47°20′	79°26′
Villebois	49°06′	79°09′
Villemontel	48°38′	78°22′
Wabano River	48°20′	74°03′
Wabouchagamou River	50°29′	60°25′
Wakeham	48°50′	64°34′
Warwick	45°56′	71°59′
Washicoutai Bay	50°13′	60°52′
Washicoutai River	50°15′	60°50′
Washikuti River (see Washicoutai River)		
Waskaganish (formerly Rupert House)	51°29′	78°45′
Waswanipi	49°41′	75°58′
Watshishou Migratory Bird Sanctuary	50°16′	62°29′
Way Rock	51°51′	79°03′
Weedon	45°42′	71°27′
Weedon Centre	45°42′	71°27′
Westmount	45°29′	73°36′
Woburn	45°23′	70°52′
Wolf Bay (hamlet)	50°16′	60°08′
Wotton (see Wottonville)		
Wottonville	45°44′	71°48′
Wreck Island	50°13	64°11′
Wyman	45°32′	76°18′
Yamachiche	46°16′	72°50′
Yamaska Recreation Park	45°25′	72°36′
Yamaska River	46°06′	72°56′
York River	48°49′	64°34′

Glossary

Principal publications consulted:

Anonymous 1992; Ayotte 1994; Bates and Jackson 1984; Buteau et al. 1994; Campbell and Lack 1985; Canadian Forestry Service 1992; Cauboue et al. 1996; Cayouette 1993; Commission de toponymie du Québec 1994a,b; Conseil international de la langue française 1979; del Hoyo et al. 1992; Ehrlich et al. 1988; Fleurbec 1987, 1993; Godfrey 1986; Hayes 1995; Hosie 1978; Johnsgard 1983b; Jones and Collaborators 1990; Kimmins 1992; Leahy 1982; Librairie Larousse 1982; Marchand 1991; Marie-Victorin 1964; Ménard-Lépine 1980; Métro 1975; Moore 1968; National Wetlands Working Group 1981; Palmer 1976; Parent 1990; Payette 1983; Perrins and Middleton 1985; Pettingill 1970; Pielou 1988; Poole et al. 1992; Ramade 1993; Robert 1989b; Smith 1990; Strahler and Strahler 1987; Terres 1980; Welty 1975; Wittenberger 1981.

Adult plumage
Definitive plumage, which varies according to the seasons and the point in the moult cycle. It may consist either of *basic plumage*, which is also called *eclipse plumage* in some species, and *nuptial*, or *alternate*, *plumage*.

Agonistic behaviour
Aggressive interactions between individuals, characterized by a variety of postures, movements, and vocalizations which serve as a threat, to repel or intimidate the opponent without actual combat. The purpose of such interactions may also be to establish a dominance hierarchy.

Air sacs
Air-filled cavities that make up part of a bird's respiratory system, located under the skin between the internal organs and muscles as well as inside some bones. Air sacs are essential in respiration, are used to control buoyancy in aquatic species, and play a role in thermoregulation.

Allopatry
Situation in which two species or subspecies occupy separate geographic ranges, at least during the breeding season (as opposed to *sympatry*). Adj.: Allopatric.

Alternate plumage
See *Nuptial plumage*.

Altricial
Term used to describe the maturity at hatching of young, as in the Emberizidae, that are born naked with little or no down, with their eyes closed, incapable of locomotion, and which must be fed by their parents. See *Nidicolous*.

Anisodactyl(ous)
Having the *hallux* directed backward and three toes directed forward, as in the passerines.

Antelingual pouch
Extensible pouch in mouth cavity which birds can expand at will by distending the floor and placing their tongues in the back of their mouths. This morphological adaptation, found in the corvids and other groups, is used, among other things, to carry food to the young. See *Gular pouch*.

Arena
A display ground where males of the same species congregate and compete with each other for the sole purpose of attracting and courting females ready to mate. The dominant males usually occupy the centre of the arena, relegating other males to the periphery. Arena mating is a form of *polygyny*. Syn.: Lek.

Asynchronous
In which the life cycle and activities are not simultaneous (e.g., asynchronous hatching). Ant.: *Synchronous*.

Avian
Pertaining to birds (e.g., avian community, avian cholera).

Barachois
Sandbar formed by waves and currents at the entrance to a bay. By extension, also applies to the brackish water lagoon and marsh behind the sandbar. Syn.: Baymouth bar.

Basal metabolism
Rate of energy utilization by an animal at rest and under no stress. Expressed in kilocalories per kilogram of body weight per hour. Syn.: Basal metabolic rate.

Basic plumage
Plumage that is usually duller than the *nuptial*, or *alternate*, *plumage*, that some birds acquire at the end of the breeding season and maintain until spring, when they moult into their nuptial plumage. Also used for birds that have similar plumage throughout the entire year. Syn.: Winter plumage, non-breeding plumage, non-nuptial plumage.

Bigamy
Type of social organization in which the male or female mates with two partners. See *Polygamy*.

Biocenosis
Association of living things, both plants and animals, forming an ecological community within a *biotope*.

Biotope
Geographic area characterized by a set of biophysical conditions (climate, soil, relief, etc.). The biotope, *biocenosis*, and their relations form the ecosystem.

Black Duck Joint Venture (BDJV)
Program launched in 1989 under the *North American Waterfowl Management Plan*; its objective is to promote and co-ordinate data gathering for projects to protect and manage American Black Duck populations.

Sexual promiscuity
Mating of the male or female with several partners without a pair bond being formed; practised by approximately 6% of species.

Blue List
List published since 1971 in the National Audubon Society's journal "American Birds" based on sightings in North America compiled by region. The list features species that are undergoing non-cyclical population declines or changes in their breeding ranges. Its purpose is to give the scientific community and government agencies early warning of species potentially at risk.

Bog
Peatland fed exclusively by precipitation and therefore acid and low in nutrients. The dominant vegetation is sphagnum, lichens, and shrubs (heaths); in treed bogs in Québec, the most common species encountered is black spruce. Syn.: Ombrotrophic peatland, ombrotrophic bog.

Boreal
An area where the dominant vegetation consists of closed-canopy coniferous forest or various proportions of coniferous and hardwood forests, found in cool to cold temperate regions of the Northern Hemisphere.

Borrow pit
Sand or gravel pit, usually small in size and easy to access, from which road construction material is taken.

Braided stream
Stream with multiple threads that subdivide and rejoin repeatedly.

Breeding plumage
See *Nuptial plumage*.

Brooding
The act of sitting over the young to keep them warm and protect them from rain, sun, or predators. *Altricial* young must be brooded by the parents almost constantly until they are able to maintain their own body temperature, while *precocial* young require only intermittent brooding. Sometimes also used to refer to the incubation of eggs. See *Incubation*.

Burn
Portion of forest damaged by a fire of natural or accidental origins.

Canada Wildlife Act
Act adopted in 1973 which applies to animals, plants, and their respective habitats. It defines the federal government's powers in acquiring land for wildlife conservation and research; the powers of wildlife officers in enforcing the Act; and penalties and related procedures in the case of contravention of the law. Since 1994, the Act also provides for the creation of protected marine areas and the implementation of regulations on wildlife areas.

Caruncle
Unfeathered appendage most often situated above the bird's eye, but may also be on the throat or head; usually brightly coloured and most often associated with *sexual dimorphism*. Caruncles undoubtedly play a role in display and their state may be subject to hormonal control.

Chaparral
In North America, type of vegetation characterized by low, dense evergreen scrub adapted to dry conditions. Chaparral is found in the southwestern U.S. and northern Mexico in areas where the climate and environmental conditions support neither a desert nor a forest. Known as mattoral in Europe and South America, mallee in Australia, and fynbos in Africa.

Circumboreal
Pertains to regions located within the *boreal* zone, in Asia, Europe, and North America.

Circumpolar
Pertains to the geographic regions located near the poles, by convention between 75° and 90° latitude. In the Atlas, only applies to the Northern Hemisphere.

Clay Belt
Region located south of James Bay and Hudson Bay, where the soil contains a high proportion of glacial clays. The Clay Belt extends through the Abitibi and Témiscamingue regions in Québec across the Ontario border as far west as Lake Nipigon.

Clearcutting
The felling of all trees in a stand, without taking any measures to protect advance regeneration.

Club
A social gathering place in a seabird colony (Northern Gannets, alcids) used by adults and sexually immature prospectors that are visiting the colony.

Co-operative breeding
Phenomenon of non-breeding individuals participating in some or all of the tasks relating to breeding in their species such as nest construction, *incubation*, and feeding the young. This behaviour is most common in birds associated with terrestrial ecosystems, particularly in tropical and subtropical regions. Non-breeding individuals participating in a co-operative breeding system are not necessarily related to the breeding individuals.

Communal roost
Place where one or more species of birds congregate, often in the hundreds or thousands, to sleep or rest.

Community
See *Biocenosis*.

Congener
Term for species of birds that belong to the same genus.

Conspecific
Term for birds of the same species.

Contour feathers
Feathers giving the bird its distinctive outline; they comprise the *tectrices*, *remiges*, and *rectrices* but not the down, *vibrissae*, filoplumes, semiplumes, and powder feathers.

Convergent evolution
Series of transformations resulting in structural or behaviourial similarities in organisms that are not closely related. This phenomenon occurs when different species occupy similar ecological niches.

Courtship display
Set of ritualized behaviour engaged in by some species that brings males and females together as mates and leads to coition.

Coverts
Small contour feathers that overlie or underlie the bases of the main flight feathers and cover the rest of the wing or tail. Also known as the *tectrices*.

Crèche
An assemblage pooling the still dependent young of different broods, accompanied by one or more adults.

Cryptic
Used to describe the colouring of the plumage or eggs of some species which allows them to blend into their surroundings.

DBH
Diameter at breast height; a standardized tree measurement taken at 1.3 m above ground level.

Density
Number of individuals, pairs, or territories per unit of area. The following correspondence can be observed: 1 pair = 1 territory = 1 territorial male = 2 individuals (occupying a territory).

Dump nest
Nest containing one or more eggs laid by several different females, usually of the same species.

Dutch elm disease
Disease caused by a microscopic fungus (*Ceratocystis ulmi*) and propagated by the native elm bark beetle (Scolytidae). In Québec, the disease appeared in 1944, attacking mainly the white elm. It hinders the functioning of specialized tissues for transporting water and causes the wilting, yellowing, and browning of the foliage, then defoliation and ultimately the death of the tree. Also known as elm blight.

Eclipse plumage
A post-nuptial plumage occurring in some species (notably in male ducks) at the end of the breeding season usually when a simultaneous moult of the *remiges* renders them temporarily flightless.

Ecological niche
The status or place of a species within an ecosystem. The notion of niche takes in elements such as habitat, diet, and intra- and interspecific relations. A species with a fairly broad niche is called a generalist, while a species with a narrow niche is called a specialist.

Ecological region
In the Atlas, term adapted from Thibault's (1985) forest region (domaine forestier) and used for statistical analysis in the mapping of relative abundance of many species. For further details, see the sections "Vegetation formations" and "Frequency of occurrence by ecological region and mapping of relative abundance". Syn.: Forest region.

Ecotone
Transition zone between two habitats, where species characteristic of each of the habitats can be found, as well as some plants and animals associated exclusively with the ecotone.

Ectoparasite
Parasite inhabiting exterior of host's body (e.g., ticks and fleas).

Endangered species
A species facing imminent extirpation or extinction.

Endoparasite
Parasite inhabiting interior of host's body (e.g., intestinal worms).

Epiphyte
A plant that grows on the surface of another plant, deriving support but not nutrients from it.

Erratic block
Boulder carried by glacial ice far from its place of origin from a bedrock outcrop and left stranded when the ice melted; it is thus often of a different type from neighboring rocks.

Erythrism
Abnormal colouring of plumage, observed when the normal melanins are replaced by a reddish melanin.

Esker
A long, narrow, steep and sinuous ridge of fluvioglacial origin. Eskers are formed by the accumulation of sediments deposited by a meltwater stream within or under the ice sheet of the inlandsis (continental glacier), and indicate the direction of the drainage at the moment of melting.

Extinct species
A species that no longer exists.

Extirpated species
A species no longer existing in the wild in Canada, but occurring elsewhere.

Extrapair copulation
Occasional copulation between an already paired individual (*monogamy*, *polygyny*, or *polyandry*) and one or more individuals of the opposite sex without formation of a pair bond. This type of behaviour has been reported in most families of birds.

Fecal sac
Whitish, gelatinous sac that contains the *feces* of *altricial* nestlings. The parents may either eat the sacs or transport them from the nest, which contributes to nest sanitation, helps maintain the insulating qualities of the nest, and makes the nest more difficult to spot for predators.

Feces
The excrement formed from digestive remains. Syn.: Droppings, whitewash.

Fen
A peatland fed not only by rainwater but by nutrient-rich and minerotrophic waters from nearby mineral soils. The dominant vegetation is graminoids and shrubs (alder, willow); in treed fens in Québec, the most common species encountered is Eastern white-cedar (southern latitudes) or tamarack (northern latitudes). Syn.: Minerotrophic peatland.

Fledge
Vb. Of a chick, to take its first flight. Coincides with the end of the nestling period in *altricial* species.

Fledgling
Term generally used to describe an altricial chick from the time it leaves the nest until it becomes independent.

Forb
Broadleaved herb, as distinguished from the grasses.

Forest region
See *Ecological region*.

Fovea
Depression in the central area of the retina, containing from 200 000 to one million visual cells, that magnifies images. Most birds have at least one fovea per eye, while others have two: one, called the search fovea, to locate objects and the second, called the pursuit fovea, to follow movement.

Genotype
The set of genetic characteristics possessed and transmitted by each individual, whether morphological, physiological, or behavioural.

Gleaner
Bird that feeds mainly on insects at its feet by picking its prey off leaves, branches, and tree trunks as it moves through the vegetation. Vb.: to glean.

Greater Montréal
In the Atlas, that area of the Central St. Lawrence Lowland in Québec bordered by the Ottawa Valley to the west, the Mauricie–Bois-Francs region to the northeast, and the Eastern Townships to the east.

Gregarious
Term for animal species that habitually gathers in flocks in its habitat. Ant.: *Solitary*.

Gular pouch
Pouch in the upper throat in some species of birds, including cormorants, pigeons and pheasants. Syn.: Throat pouch. See *Antelingual pouch*.

Hallux
The first, or hind, toe in birds (equivalent to humans' big toe), often directed backward. In some species, it is much reduced or vestigial, or even absent altogether.

Hawking
Method of foraging in which the bird sallies forth from a perch to seek moving prey in the air. Vb.: to hawk.

Heterogeneous habitat
Ecological framework consisting of diverse elements such as vegetation of dissimilar ages and structures, various types of soil, etc. Syn.: Mixed habitat, habitat mosaic.

Holarctic
The temperate and cold regions in the Northern Hemisphere, including much of North America, Asia, and Europe.

Home range
The area used by an individual, pair, or group of birds for their activities. A bird's home range is usually larger than its territory, but may be the same as the territory when the home range is defended (in this case it is called the territory). See *Territory*.

Horizon, (soil)
A layer of soil which lies more or less parallel to the surface and has fairly distinctive soil properties, usually designated by the letters A, B, and C extending downward from the surface to the parent material.

Hovering
1.Method of flight, as in hummingbirds and some falcons, such as the American Kestrel, whereby the bird flaps its wings rapidly so as to remain stationary in the air.
2.Method of foraging in some passerines in which the bird, while remaining nearly stationary in the air, takes a prey from a substrate such as the top of a branch, or food such as nectar, insects, or berries.

Hydric
In the Atlas, used to describe environments and habitats with a high degree of moisture in the soil and atmosphere.

Hypothermia
Physiological condition in which the body core temperature falls below that considered normal for a homeothermic animal. When food resources are scarce, some species of birds are able to maintain a type of controlled hypothermia, or torpor, in which their metabolic expenditures and thus amount of energy required to survive are reduced.

Immature
A term for a bird from the time it acquires juvenile plumage (first plumage of true contour feathers) until it acquires full adult plumage. A bird is considered immature until the age of adulthood (i.e., until it is sexually mature, is of adult size, has adult plumage, etc.). It does not connote sexual immaturity unless so specified because, in some species, immatures may be capable of breeding. See *Juvenile*.

Imprinting
A form of learning observed mainly in *nidifugous* birds such as the anatids and rails, as well as some *nidicolous* species, and resulting in the establishment of family bonds. Normally established with parents during a brief period after hatching, but can also be with siblings, humans, and even objects, colours, and sounds.

Incubation
The act by which parent birds sit on the eggs, using their body heat to warm them so that they will develop and hatch. Incubation lasts from laying to hatching. See *Brooding*.

Interspecific
Pertaining to birds of two or more different species and their relations.

Intolerant
Refers to a plant species that cannot grow in the shade. Syn.: Shade intolerant. See *Tolerant*.

Intraspecific
Pertaining to birds of the same species and their relations.

Introduced, (species)
See Appendix III for definition.

Invasion
Movements of a bird species, often in large numbers, into new territory, thus expanding its range. Sometimes used as a synonym for *irruption*.

Irruption
A form of migration that occurs irregularly in response to decreased availability of food in the species' usual range. The proportion of the population that migrates and the distance travelled varies greatly from year to year.

Juvenile
A term for any young bird (in juvenile plumage) that has not attained sexual maturity. Sometimes synonymous with *immature*.

Juvenile plumage
In the general sense, all plumage that is not adult plumage. In the more specific sense, the first plumage in which *contour feathers* (*tectrices*, *remiges*, and *rectrices*) are present. Sometimes used as a synonym for juvenal plumage.

Kame
A mound or hill resulting from the accumulation of various sediments (gravel, loess, etc.) near the front of a glacier, which is visible after the glacier has melted.

Kleptoparasitism
Form of parasitism associated with feeding, whereby a species steals food from another species. Syn.: Food piracy.

Krummholz
Bent, stunted, ground-hugging trees, shaped by the wind and other climatic factors and found near the tree line (alpine or arctic) or along the seashore.

Lek
See *Arena*.

Lentic
Used to describe freshwater ecosystems in which the water circulates and is renewed slowly (e.g., lakes, marshes, *peatlands*), as well as the organisms that live in these ecosystems. Ant.: *Lotic*.

Lobate
Character of the feet, the toes of which are separately fringed by small lobes (round,

fleshy flaps) as in the grebes, an adaptation for swimming. In some species of diving ducks, the feet are *palmate*, but the *hallux* may be lobed.

Lore
Naked area or area covered with feathers or *vibrissae* between the eye and the bill.

Lotic
Used to describe freshwater ecosystems in which the water circulates and is renewed fairly quickly (e.g., rivers, streams, brooks), as well as the organisms that live in these ecosystems. Ant.: *Lentic*.

Mandible
Lower part of the bill. In the plural, it designates both the lower and upper parts of the bill.

Marsh
A mineral wetland or a peatland that is periodically or permanently inundated by standing or slowly moving water and is covered with clumps of low vegetation such as emergent grasses, sedges, rushes, and reeds that are characteristic of muddy substrata.

Mating system
See *Bigamy, Extrapair copulation, Monogamy, Polyandry, Polygamy, Polygyny, and Sexual promiscuity*.

Mature forest
A mature forest that is sufficiently developed to be harvestable.

Mature stand
Stand that is fully developed, particularly in terms of its size and reproductive capacity.

Maxilla
Upper part of the bill.

Mesic
Pertaining to environments and habitats with a moderate degree of moisture in the soil, i.e., neither *hydric* nor *xeric*.

Migrant breeder
See Appendix III for definition.

Migratory Birds Convention
Agreement signed in 1916 between Canada and the United States, which became law in Canada in 1917 with the adoption of the Migratory Birds Convention Act and in the United States in 1918 with the adoption of the Migratory Bird Treaty Act. The Canadian Act regulated the hunting and use of migratory birds and prohibited trafficking and commercialization of them. It also provided for the establishment of migratory bird sanctuaries. The Act was repealed and replaced by the 1994 Migratory Birds Convention Act, which provides enhanced regulatory authority to implement the Convention and updates and strengthens enforcement provisions, among other things.

Mimetic
Pertaining to mimicry.

Mimicry
Aspect of colour pattern, shape, or even behaviour that allows the organism to resemble an element in its environment or another animal species less likely to be attacked by predators. Mimicry is rare in birds.

Mixed forest
Forest in which deciduous and coniferous trees intermingle. In Québec, the mixed forest zone lies generally between the St. Lawrence Lowlands and the coniferous forest zone.

Monogamy
An exclusive relationship between the male and female during at least part of the breeding season; approximately 90% of birds are monogamous. There are two types:
Seasonal: Exclusive pair bond maintained for a year, less than a year, one breeding season, or part of a breeding season. Syn.: Serial monogamy.
Permanent: Exclusive pair bond maintained for the partners' lifetimes. However, mates may change partners if the first breeding attempts fail or if one of the members of the pair dies or disappears.

Moult
The periodic loss and replacement of all or part of the feathers, simultaneous or not. Generally, birds moult once or twice a year, and sometimes three or very rarely four times. Moulting has a number of purposes: camouflage, *sexual dimorphism* for courtship purposes, and insulation from the cold. The duration of the moult ranges from 2 to 4 weeks in the Anseriformes, from 5 to 12 weeks in most passerines, and over two years in some raptors.

Mull
A homogeneous forest humus where organic litter does not accumulate on the surface. Mull is formed by the intense biological activity of the microfauna (earthworms in particular).

Muskeg
Term of Algonquin Indian origin frequently employed for peatland. It refers to a habitat characteristic of boreal regions in North America colonized by acid-loving plants (sphagnum moss, heaths) and, more sporadically, by trees such as tamaracks. Similar to a treed bog.

Nearctic
The cold, temperate regions of North America north of the tropics.

Neotropical migrant
Species of bird that breeds in North America and spends the non-breeding season mainly in the *neotropics*. See Hayes (1995) for a discussion and a different definition of this term.

Neotropics
Name given to the area extending from the West Indies and northern Mexico to Cape Horn in South America. Also called the Neotropical Region.

Nestling
Term generally used to describe an altricial chick while it is still in the nest.

New Québec (Nouveau-Québec)
The former district of Ungava in the Northwest Territories (commonly known as the Ungava Peninsula) lying north of the general line of the Eastmain River corresponding to the 53rd parallel annexed to the province of Québec in 1912 and named New Québec under an Act of the Legislative Assembly of Québec.

Nidicolous
Young that remain in the nest for some time after hatching. See *Altricial*.

Nidifugous
Young that leave the nest immediately or shortly after hatching. See *Precocial*.

Nomadic
Used to describe a species or population whose movements occur in response to certain climatic conditions or the availability of food.

Nord-du-Québec
Since 1987, administrative region encompassing the area north of the 49th parallel (see map, p. 1300). This administrative region is sometimes cited as Northern Québec, not to be confused with the geographical region, northern Québec. See *New Québec*.

North American Waterfowl Management Plan (NAWMP)
Plan signed in 1986, which establishes a framework for managing and conserving waterfowl populations by setting objectives for populations and habitats. The plan also contains general recommendations on developing measures to protect North American waterfowl.

Northern Québec
See *Nord-du-Québec*.

Nuptial plumage
Adult plumage during the breeding season. Brighter and often more colourful than the basic plumage. In some species, including the Bobolink, House Sparrow, and European Starling, the nuptial plumage is acquired by the wearing down of the tips of the feathers rather than by a moult as such. Syn.: Breeding plumage, alternate plumage, summer plumage.

Old-growth forest
A descriptive term attributed to natural forests that have reached an advanced stage in succession relatively uninfluenced by human activity or by frequent natural disturbances. This type of forest has multiple canopy layers created by the crowns of trees of various ages and species. The dominant stratum consists of huge, long-lived trees, and also contains a significant number of standing dead trees and logs lying about the forest floor.

Overmature forest
Stand that has reached maturity and has begun to lose its vigour and its commercial value.

Palearctic
The cold, temperate regions of Europe and northern Asia.

Palmate
Character of the feet, the toes of which are joined by a web. In ducks, the three forward toes are webbed and the *hallux* is free.

Parapatric
Designates species or subspecies that occupy different but contiguous geographical areas, at least during the breeding season.

Parasite
A plant or animal that lives at the expense of another species—called the host—often at the latter's disadvantage but usually not at the expense of its life. Birds are potential hosts to a number of parasites, but rarely constitute parasites themselves in the strict sense of the word. However, in the more general sense, some birds are parasitic in terms of their food-gathering habits (*kleptoparasitism*) or breeding habits (interspecific and intraspecific brood parasitism). In the latter case, a distinction can be made between facultative parasites, that can also build their own nest and raise their own young (e.g., some anatids) and obligate parasites, whose eggs are incubated and young are raised by host species (e.g., Brown-headed Cowbird).

Parasitism
Exploitation of one organism (host) by another organism (parasite). See *Parasite*.

Parkland
Transition zone between the prairies and forest, fairly extensive in area, characterized by tall herbaceous vegetation, scattered large trees, and scrub. It is found in western North America.

Peatland
An ecosystem in which the rate of production of organic matter by living organisms exceeds the rate at which these compounds are degraded. As a result, part of the production accumulates as organic deposits or peat. These habitats are very wet and are often dominated by sphagnum moss and heaths.

Pellet
A small compact mass composed of the undigested or partially digested parts of a bird's food (hair, bones) retained in the stomach or esophagus for a certain time before being regurgitated through the mouth. These pellets, which are most commonly associated with birds of prey, facilitate the study of a number of parameters relating to diet. Syn.: Casting.

Philopatry
The tendency of an individual or pair to return to the same nest site year after year. This behaviour appears to maximize reproductive success. Syn.: Site tenacity.

Piedmont
Land located at the foot of mountains.

Plant down
Plant material such as hairs, the aigret of some fruits, or any other silky part of the plant used in nest construction.

Podzol
Soil of the order Podzolic. For further details, see the "Soil" section in "The Biogeography of Québec".

Polyandry
Mating system in which a female mates with two or more males during the same breeding cycle. In birds, fewer than 1% of species engage in polyandry; these species are often characterized by reverse *sexual dimorphism*, with the female generally brighter plumaged and larger than the male (see species accounts for Spotted Sandpiper, Wilson's Phalarope, and Red Phalarope).

Polygamy
Mating system in which one male or one female mates with more than one partner. In contrast with *sexual promiscuity*, polygamy implies a form of social organization between

the sexes and in the choice of mate. It includes *polyandry* and *polygyny*, but not *extrapair copulation. Bigamy* is also a form of polygamy.

Polygyny
Mating system in which one male mates with two or more females during the same breeding season. There are several different forms: the male may have a harem, maintaining bonds with several females simultaneously, or may mate successively with several females.

Polynya
An expanse of open water surrounded by ice.

Precocial
Term used to describe the maturity at hatching of young born with their eyes open, covered with down, and capable of locomotion. They leave the nest shortly after hatching and follow their parents, which may either feed them (e.g., Rallidae) or lead them to food (e.g., Anatidae). See *Nidifugous.*

Prescribed burn
Portion of a forest where controlled burning has been deliberately carried out for forest management purposes.

Primaries
See *Remiges.*

Pygostyle
Term for the vertebrae at the end of the vertebral column that form the terminal bone at the base of the tail, which is the base of attachment for the *rectrices.*

Race
Population of an animal species that has distinctive morphological characteristics from other members of the same species. In everyday language, race is used for human populations and domesticated animals, but is sometimes used as a synonym of *subspecies.*

Raised beach
An ancient beach occurring above the present shoreline, originally formed by a postglacial sea, then elevated by isostatic lifting.

Rectrices
The tail feathers of a bird, which are attached to the pygostyle or rump and which help birds keep their balance or steer while in flight.

Regeneration
Renewal of a forest stand by natural or artificial means.

Remiges
Large flight feathers present on the wings. In general, there are three kinds: the primaries, which are attached to the hand bones, the secondaries, which are attached to the forearm, and the tertiaries, which are attached to the humerus.

Resident breeder
See Appendix III for definition.

Resource partitioning
Division of the resources of a habitat whereby several species co-exist without competing against each other. It often takes the form of feeding in different parts of the same habitat (as do some closely related warblers) or else taking food of different sizes.

Riparian
Pertaining to the bank of a watercourse (e.g., river or stream).

Sapling
A young, vigorously growing tree 3–5 m high and with a *dbh* under 10 cm.

Second-growth forest
Stand that has developed naturally after a major disturbance (e.g., clearcutting, fire, insect infestation).

Secondaries
See *Remiges.*

Sedentary breeder
See Appendix III for definition.

Sedge meadow
Association of herbaceous plants dominated by one or more species of sedges.

Selective cutting
The felling of trees selected according to their size, species, proportions, etc.

Semi-altricial
Term used to describe the maturity at hatching of young born covered with down, incapable of locomotion, and dependent on their parents for food. Their eyes may be open, as in the herons and diurnal raptors or closed, as in the nocturnal raptors.

Semiprecocial
Term used to decribe the maturity at hatching of young born with their eyes open and covered with down that remain in or near the nest even if they are capable of locomotion and are dependent on their parents for food, as in the larids, alcids, and caprimulgids.

Sexual dimorphism
Term for the difference in structure, size, and appearance between males and females of a species or subspecies at an equal age and during the same season.

Slip-noose trap
Snare made of horsehair, installed in a lattice of string. Used to capture Snow Buntings and banned in Québec by Order in Council in 1963.

Sole
Part of the foot under the tarsus, that supports the weight of the bird when standing or perched.

Solitary
Pertaining to birds that live as isolated individuals or in single pairs.

Speculum
A patch on the outer secondaries, its colour often contrasting with the rest of the wing. Found mainly in ducks.

Strip cutting
The harvesting of a stand by removing the crop in narrow strips. The width of the strips to be removed is adjusted to promote optimal regrowth according to the dominant species, the age classes present, and the rate of growth.

Subspecies
A population, or a group of individuals, found within the species' breeding range that can be morphologically or genetically distinguished from the members of other populations of the same species. Term sanctioned by the International Code of Zoological Nomenclature. Subspecies are designated by a trinomial Latin system.

Summer visitor
In the Atlas, used to refer to individuals of some species that spend the summer in the Atlas area, but do not breed (e.g., Common Loon, scoters, and sandpipers frequenting the St. Lawrence estuary).

Swamp
A mineral wetland or peatland with standing or gently flowing water. The water table is usually at or near the surface. The vegetation is characterized by a dense cover of deciduous or coniferous trees, or shrubs.

Sympatry
Situation in which species or subspecies occupy ranges that overlap partially or entirely, at least during the breeding season (in contrast with *allopatry*). Adj.: Sympatric.

Synchronous
In which the life cycle and activities are simultaneous. Ant.: *Asynchronous.*

Syndactyl(ous)
Having two toes (the third and the fourth) fused for part of their length, as in the Belted Kingfisher.

Taiga
Subarctic forest formation composed mainly of conifers, fairly open or dissected by numerous wetlands and forming a transition zone between the commercial coniferous forest and the tundra. Syn.: Open boreal woodland, subarctic forest, Hudsonian zone.

Talus
Accumulation of loose rock fragments (slide rock) derived by rockfall from a cliff. Syn.: Scree, rock pile.

Tarsometatarsus
The lower part of the leg just above the foot, formed by a fusion of the tarsus and metatarsus. Commonly called tarsus.

Ternery
A tern colony.

Territory
Space that an individual, pair, or a small group of birds of the same species appropriates. The defended space may be used to nest, to ensure exclusive use of a food supply, to protect the space around one or more mates or the young, or to defend a display area or singing or lookout perch.

Thinning
Felling in an immature stand to accelerate the growth of the remaining trees while maintaining the integrity of the canopy (e.g., commercial thinning, precommercial thinning).

Threatened species
A species likely to become endangered if limiting factors are not reversed.

Tidal marsh
Marsh that is affected by tides.

Tolerant
Refers to a plant species that can grow in the shade. See *Intolerant.* Also used to describe a plant species that can survive in an environment with a deficiency (water, nutrients) or excess (salts or organic acids) of an essential element.

Tomium
The edge of the upper or lower mandible, which can be cutting or lamellate (sieve-billed).

Totipalmate
The character of feet in which all four toes are joined by a web, as in the Pelecaniformes (cormorants, gannets, pelicans, etc.).

Transient
See Appendix III for definition.

Tundra
Plant formation typical of very cold environments, characterized by low vegetation (average height 25 cm) and a lack of trees. In Canada, the tundra is found in the Far North (arctic tundra), high mountains (alpine tundra), and sometimes on very cold coastlines (along Hudson Bay and the Lower North Shore).

Uropygial gland
Skin gland located on the rump, at the base of the tail feathers, which secretes an oily substance that birds spread on their feathers to clean and lubricate them, increase their water repellency and flexibility and protect them against bacteria and fungi. Syn.: Oil gland, preen gland.

Vibrissae
Specialized feathers typically located near the bill. Found in certain birds that feed on flying insects. Syn.: Rictal bristles.

Visitant
See Appendix III for definition.

Vulnerable species
A species of special concern because of characteristics that make it particularly sensitive to human activities or natural events.

Wetland
Land that is saturated with water or is periodically or permanently inundated; characterized by hydrophytic vegetation.

Windthrow
Forest or group of trees uprooted by the wind. Syn.: Windfall, blowdown.

Winter plumage
See *Basic plumage.*

Winter visitor
Species occupying a different geographic area from its breeding range outside the breeding season, usually in winter.

Wintering range
Geographic area occupied by the species between two breeding seasons, i.e., where it overwinters. It may be located in regions with a more favourable climate or, in some cases, correspond roughly to the species' breeding range.

Woody
Used to describe a plant in which the root, stem, and branches contain a compact, fibrous substance called lignin.

Xeric
Used to describe environments and habitats with very little water or excessive drainage.

Zygodactyl(ous)
Having the second and third toes directed forward and the first toe (*hallux*) and the fourth toe directed backward, as in the woodpeckers.

Bibliography

Multiple entries by the same author are arranged by single author works, followed by multiauthor entries for which that author is the principal author. Single-author entries are listed in chronological order of date of publication (e.g., Chapdelaine, G. 1977; Chapdelaine, G. 1979a; Chapdelaine, G. 1979b). Joint works are listed next in alphabetical order of the second author and, as applicable, in order of date of publication (e.g., Chapdelaine, G. and P. Blais 1979; Chapdelaine, G. and P. Brousseau 1984; Chapdelaine, G. and P. Brousseau 1989; Chapdelaine, G. and P. Laporte 1982). Multiauthor entries with two or more other authors (cited Chapdelaine et al. in the text) are listed in order of date of publication (e.g., Chapdelaine, G., P. Brousseau, R. Anderson, and R. Marsan 1985; Chapdelaine, G., A.J. Gaston, and P. Brousseau 1986a; Chapdelaine, G., P. Dupuis, and A. Reed 1986b).

Abbott, P.C., J.D. Delafield, and G.H. Montgomery. 1957. Annual report for year ending December 31st, 1957. Province of Quebec Society for the Protection of Birds, Montréal, 40 pp.

Abbott, P.C., J.D. Delafield, and G.H. Montgomery. 1960. Annual report for year ending December 31st, 1960. Province of Quebec Society for the Protection of Birds, Montréal, 39 pp.

Able, K.P. and B.R. Noon. 1976. Avian community structure along elevational gradients in the northeastern United States. Oecologia 26: 275-294.

Abraham, K.F. and C.D. Ankney. 1986. Summer birds of East Bay, Southampton Island, Northwest Territories. Canadian Field-Naturalist 100(2): 180-185.

Abraham, K.F. and G.H. Finney. 1986. Eiders of the eastern Canadian Arctic, pp. 55-73 *in* A. Reed, ed. Eider ducks in Canada/Les eiders au Canada. Canadian Wildlife Service, Ottawa. Report Series no. 47, 177 pp.

Adamcik, R.S. and L.B. Keith. 1978. Regional movements and mortality of Great Horned Owls in relation to Snowshoe Hare fluctuations. Canadian Field-Naturalist 92(3): 228-234.

Adamcik, R.S., A.W. Todd, and L.B. Keith. 1979. Demographic and dietary responses of Red-tailed Hawks during a Snowshoe Hare fluctuation. Canadian Field-Naturalist 93(1): 16-27.

Adams, A.M., J. Bunn, B.S. Davis, A.L. Jones, and L.W. Odne. 1973. A study of the Pig's Eye Lake heron colonies. Loon 45: 32-45.

Adams, R.J. Jr. 1982. The changing season. American Birds 36(5): 820-826.

Adamus, P.R. 1987. The Atlas of breeding birds of Maine, 1978-1983. Maine Department of Inland Fisheries and Wildlife, Augusta, 366 pp.

Addison, E.M., R.D. Strickland, and D.J.H. Fraser. 1989. Gray Jays, *Perisoreus canadensis*, and Common Ravens, *Corvus corax*, as predators of winter ticks, *Dermacentor albipictus*. Canadian Field-Naturalist 103(3): 406-408.

Addy, C.E. and J.D. Heyland. 1968. Canada Goose management in eastern Canada and the Atlantic Flyway, pp. 10-23 *in* R.L. Hine and C. Schoenfeld, eds. Canada Goose Management. Dembar Educational Research Service, Madison, Wisconsin.

Adkisson, C.S. 1977. The breeding and maintenance of Pine Grosbeaks in captivity. Avicultural Magazine 83(4): 195-198.

Adkisson, C.S. 1981. Geographic variation in vocalization and the evolution of North American Pine Grosbeaks. Condor 83(4): 277-288.

Afton, A.D. 1979. Time budget of breeding Northern Shovelers. Wilson Bulletin 91(1): 42-49.

Afton, A.D. 1983. Male and female strategies for reproduction in Lesser Scaup. Ph.D. dissertation, University of North Dakota, Grand Forks, 151 pp.

Afton, A.D. 1984. Influence of age and time on reproductive performance of female Lesser Scaup. Auk 101: 255-265.

Ahlund, M. and F. Gotmark. 1989. Gull predation on eider ducklings: effects of human disturbance. Biological Conservation 48(2): 115-128.

Ailes, I.W. 1976. Ecology of the Upland Sandpiper in central Wisconsin. M.Sc. thesis, University of Wisconsin, Oshkosh, 63 pp.

Ainley, M.[G.] 1978. Prey piracy by Hawk Owl on Common Crow. Annual report, Tchebec, pp. 93-96.

Ainley, M.G. 1981. Henry Mousley and the ornithology of Hatley and Montreal, 1910-1946. Tchebec 11: 113-133.

Ainley, M.G. 1982. Lewis McIver Terrill – Promoter of bird study and conservation in Quebec. Tchebec 12: 72-85.

Alain, G. 1988. Plan tactique de la Gélinotte huppée. Ministère du Loisir, de la Chasse et de la Pêche, Québec, Direction de la faune terrestre, 50 pp.

Albright, D.M. and D.W. Doidge. 1985. Sighting of nesting Peregrine Falcons (*Falco peregrinus*) in eastern Hudson Bay. Unpublished report, MacDonald College of McGill University, 4 pp.

Aldrich, J.W. 1963. Geographic orientation of American Tetraonidae. Journal of Wildlife Management 27(4): 529-545.

Aldrich, J.W. 1978. Status of the Canadian Sandhill Crane, pp. 139-148 *in* J.C. Lewis, ed. Proceedings. 1978 Crane Workshop. Colorado State University, Fort Collins.

Aldrich, J.W. and J.S. Weske. 1978. Origin and evolution of the eastern House Finch population. Auk 95(3): 528-536.

Aldridge, H.D. and R.M. Brigham. 1991. Factors influencing foraging time in two aerial insectivores: the bird *Chordeiles minor* and the bat *Eptesicus fuscus*. Canadian Journal of Zoology 69(1): 62-69.

Alekseeva, N.S., E.A. Polents, and V.K. Ryabitsev. 1993. Population ecology of Lapland Bunting in central Yamal. 1. Nesting density, fertility, reproductive success, polygyny. Soviet Journal of Ecology 23(3): 175-181.

Alexandersson, H. 1979. En hybridisering mellan Vitvingad Tarna och Svarttarna pa Oland 1978. Calidris 8: 151-155.

Alison, R.M. 1972. The breeding biology of Oldsquaw at Churchill, Manitoba. Ph.D. dissertation, University of Toronto, Ontario, 129 pp.

Alison, R.M. 1975. Breeding biology and behavior of the Oldsquaw (*Clangula hyemalis* L.). Ornithological Monographs no. 18, 52 pp.

Alison, R.M. 1976. The history of the Wild Turkey in Ontario. Canadian Field-Naturalist 90: 481-485.

Alison, R.M. 1980. Le Morillon à dos blanc. Hinterland Who's Who. Environment Canada, Canadian Wildlife Service, 4 pp. [Also available in English as: Canvasback.]

Allaby, M. 1977. A dictionary of the environment. The Camelot Press, Southampton, Great Britain, 532 pp.

Allainé, D. 1991. The effects of colony size and breeding stage on colony defense pattern in the Black-headed Gull. Acta Oecologia 12(3): 385-396.

Allaire, P.N. 1972. Field Sparrow uses abandoned nest for August brood. Auk 89(4): 886.

Allaire, P.N. and C.D. Fisher. 1975. Feeding ecology of three resident sympatric sparrows in eastern Texas. Auk 92(2): 260-269.

Allard, H.A. 1939. Vocal mimicry of the Starling and the Mockingbird. Science 90(2338): 370-371.

Allen, A.A. 1915. The behavior of the Least Bittern. Bird-Lore 17(6): 425-430.

Allen, A.A. 1930. Ruby-throat. Bird-Lore 32(3): 223-231.

Allen, A.A. 1948. Sea-bird cities off Audubon's Labrador. National Geographic Magazine 93: 755-774.

Allen, E.G. 1939. Nicolas Denys, a forgotten observer of birds. Auk 56(3): 283-290.

Allen, E.G. 1951. The history of American ornithology before Audubon. Transactions. American Philosophical Society 41: 386-591.

Allen, F.H. 1940. Nicolas Denys' birds. Auk 57(1): 75-82.

Allen, J.A. 1908. Pennant's Indian zoology. Bulletin of the American Museum of Natural History 24: 111-116.

Allen, R.B., P.O. Corr, and J.A. Dorso. 1990. Nesting success and efficiency of waterfowl using nest boxes in central Maine: a management perspective, pp. 291-295 *in* L.H. Fredrickson, G.V. Burger, S.P. Havera, D.A. Graber, R.E. Kirby, and T.S. Taylor, eds. Proceedings. 1988 North American Wood Duck symposium, St. Louis, Missouri, 390 pp.

Allen, R.W. and M.M. Nice. 1952. A study of the breeding biology of the Purple Martin (*Progne subis*). American Midland Naturalist 47(3): 606-665.

Alliston, W.G. 1979a. The population ecology of an isolated nesting population of Redheads (*Aythya americana*). Ph.D. dissertation, Cornell University, Ithaca, New York, 211 pp.

Alliston, W.G. 1979b. Renesting by Redhead Duck. Wildfowl 30: 40-44.

Alvo, R. 1981. Marsh nesting of Common Loons. Canadian Field-Naturalist 95: 357.

Alvo, R. 1985. Determining whether Common Loon eggs have hatched. Wilson Bulletin 97: 242-243.

Alvo, R. and M. Berrill. 1992. Adult Common Loon feeding behaviour is related to food fed to chicks. Wilson Bulletin 104: 184-185.

Alvo, R. and K. Prior. 1986. Using eggshells to determine the year of a Common Loon, *Gavia immer*, nesting attempt. Canadian Field-Naturalist 100(1): 114-115.

Alvo, R., D.J.T. Hussell, and M. Berrill. 1988. The breeding success of Common Loons (*Gavia immer*) in relation to alkalinity and other lake characteristics in Ontario. Canadian Journal of Zoology 66: 746-752.

Amadon, D. 1975. Why are female birds of prey larger than males? Raptor Research 9(1-2): 1-11.

Ambuel, B. and S.A. Temple. 1982. Songbird populations in southern Wisconsin forests: 1954 and 1979. Journal of Field Ornithology 53(2): 149-158.

American Ornithologists' Union (AOU). 1957. Check-list of North American birds, 5th editon. American Ornithologists' Union, Port City Press, Inc., Baltimore, Maryland, 691 pp.

American Ornithologists' Union (AOU). 1973. Thirty-second supplement to the American Ornithologists' Union. Check-list of North American birds. Auk 90(2): 411-419.

American Ornithologists' Union (AOU). 1983. Check-list of North American birds, 6th edition. American Ornithologists' Union, Allen Press, Inc., Lawrence, Kansas, 877 pp.

American Ornithologists' Union (AOU). 1985. Thirty-fifth supplement to the American Ornithologists' Union. Check-list of North American birds. Auk 102(3): 680-686.

American Ornithologists' Union (AOU). 1987. Thirty-sixth supplement to the American Ornithologists' Union. Check-list of North American birds. Auk 104(3): 591-596.

American Ornithologists' Union (AOU). 1989. Thirty-seventh supplement to the American Ornithologists' Union. Check-list of North American birds. Auk 106(3): 532-538.

American Ornithologists' Union (AOU). 1991. Thirty-eighth supplement to the American Ornithologists' Union. Check-list of North American birds. Auk 108(3): 750-754.

American Ornithologists' Union (AOU). 1993. Thirty-ninth supplement to the American Ornithologists' Union. Check-list of North American birds. Auk 110(3): 675-682.

American Ornithologists' Union (AOU). 1995. Fortieth supplement to the American Ornithologists' Union. Check-list of North American birds. Auk 112(3): 819-830.

Ammann, A. 1981. A guide to capturing and banding American Woodcock using pointing dogs. Ruffed Grouse Society, 30 pp.

Ammann, G.A. 1963. Status and management of Sharp-tailed Grouse in Michigan. Journal of Wildlife Management 27(4): 802-809.

Andersen, D.E. 1990. Nest defense behavior of Red-tailed Hawks. Condor 92(4): 991-997.

Anderson, B.W. and S.A. Laymon. 1989. Creating habitat for the Yellow-billed Cuckoo (*Coccyzus americana*), pp. 468-472 *in* D.L. Abell, ed. Proceedings. California Riparian Systems Conference: Protection, Management, and Restoration for the 1990's. United States Department of Agriculture, Berkeley, California. General Technical Report PSW-110, 544 pp.

Anderson, D.W. and J.J. Hickey. 1972. Eggshell changes in certain North American birds, pp. 514-540 *in* K.H. Voous, ed. Proceedings. XVth International Ornithological Congress, The Hague, 30 August-5 September 1970. E.J. Brill, Leiden, 745 pp.

Anderson, H.G. 1959. Food habits of migratory ducks in Illinois. Illinois Natural History Survey Bulletin 27(4): 289-344.

Anderson, K.S. 1980. Least Sandpiper (*Calidris minutilla*) breeding in Massachusetts. American Birds 34(6): 867.

Anderson, S.H. and H.H. Shugart, Jr. 1974. Habitat selection of breeding birds in an east Tennessee deciduous forest. Ecology 55(4): 828-837.

Anderson, W.L. 1964. Survival and reproduction of pheasants released in southern Illinois. Journal of Wildlife Management 28(2): 254-264.

Anderson, W.L. 1975. Lead poisoning in waterfowl at Rice Lake, Illinois. Journal of Wildlife Management 39(2): 264-270.

Andersson, M. 1976. Population ecology of the Long-tailed Skua (*Stercorarius longicaudus* Vieill.). Journal of Animal Ecology 45(2): 537-559.

Andersson, S. and C.G. Wiklund. 1985. Sex role partitioning during offspring protection in the Rough-legged Buzzard *Buteo lagopus*. Ibis 129: 103-107.

Andrén, H. 1992. Corvid density and nest predation in relation to forest fragmentation: a landscape perspective. Ecology 73(3): 794-804.

Andrén, H. 1994. Effects of habitat fragmentation on birds and mammals in landscapes with different proportions of suitable habitat: a review. Oikos 71(3): 355-366.

Andrews, R., G. Atwell, B.G. Blodget, I.C.T. Nisbet, and M. Scheibel. 1989. Roseate Tern Recovery plan – Northeast population. United States Fish and Wildlife Service, Newton Corner, Massachusetts, 78 pp.

Andrle, R.F. 1969. Red-tailed Hawks nesting on cliffs in Ontario. Canadian Field-Naturalist 83(2): 165.

Andrle, R.F. 1971. Range extension of the Golden-crowned Kinglet in New York. Wilson Bulletin 83(3): 313-316.

Aniskowicz, B.T. 1981. Behavior of a male Least Bittern incubating after loss of mate. Wilson Bulletin 93(3): 397-400.

Ankney, C.D. and A.D. Afton. 1988. Bioenergetics of breeding Northern Shovelers: diet, nutrient reserves, clutch size, and incubation. Condor 90(2): 459-472.

Ankney, C.D. and D.M. Scott. 1982. On the mating system of Brown-headed Cowbirds. Wilson Bulletin 94(3): 260-268.

Ankney, C.D., D.G. Dennis, L.N. Wishard, and J.E. Seeb. 1986. Low genic variation between Black Ducks and Mallards. Auk 103(4): 701-709.

Anonymous. 1868. Faits divers – L'acclimatation des oiseaux. Le Courrier du Canada no. 54, Friday edition, 12 June 1868.

Anonymous. 1872. Nos moineaux. Naturaliste canadien 5: 51.

Anonymous. 1889. Mort aux moineaux. La Presse 5(106): 1. Edition 20 February 1889.

Anonymous. 1901. Guêpes et moineaux. Naturaliste canadien 26: 132-134.

Anonymous. 1915. Le faisan dans la province de Québec. Naturaliste canadien 41: 180.

Anonymous. 1920. Encore le moineau – Utile ou nuisible? L'éternelle question. Naturaliste canadien 6(11): 241-247.

Anonymous. 1923. Les faisans dans la province de Québec. Naturaliste canadien 49: 242-243.

Anonymous. 1957. Observations ornithologiques, printemps 1957. Bulletin ornithologique 2(3): 2-18.

Anonymous. 1968. Canadianismes Mieux dire. Bulletin de l'Office de la langue française 6(8): 1. Ministère des Affaires culturelles, Québec.

Anonymous. 1979. Observations ornithologiques. Juin-juillet 1979. Bulletin ornithologique 24(3): 58-69.

Anonymous. 1980a. Observations ornithologiques. Mars, avril et mai 1980. Bulletin ornithologique 25(2): 53-62.

Anonymous. 1980b. Observations ornithologiques sur le territoire de la Baie James (1978-1979). Société de développement de la Baie James, Aménagement régional et Protection de la forêt. Feuillet d'ornithologie, juin 1980, no. 1 [9 pp.].

Anonymous. 1981. Un plan d'aménagement de la Grande Oie blanche. Canadian Wildlife Service, United States Fish and Wildlife Service, and Atlantic Flyway Council. Sainte-Foy, Québec, 76 pp. [Also available in English as: A Greater Snow Goose Management Plan.]

Anonymous. 1985. Guide d'observation des oiseaux de l'Outaouais. Club des ornithologues de l'Outaouais, Hull, 223 pp.

Anonymous. 1990. Breeding biology of Goshawks in lowland Britain. British Birds 83: 527-540.

Anonymous. 1992. Just what is a neotropical migrant? Partners in flight 2(1): 29-30.

Apfelbaum, S. and A. Haney. 1977. Nesting and foraging activity of the Brown Creeper in northeastern Minnesota. Loon 49: 78-80.

Apfelbaum, S. and A. Haney. 1981. Bird populations before and after wildfire in a Great Lakes pine forest. Condor 83(4): 347-354.

Applegate, R.D. 1977. Possible ecological role of food caches of Loggerhead Shrike. Auk 94(2): 391-392.

Aquila, C.D. 1993. 97. Intertidal salt marsh, pp. 93-94 *in* J.D. Lowe, ed. Resident bird counts 1992. Journal of Field Ornithology 64(1)(supplement): 1-128.

Arbib, R. 1977. The Blue List for 1978. American Birds 31(6): 1087-1096.

Arbib, R. 1988a. Common Yellowthroat, pp. 416-417 *in* R.F. Andrle and J.R. Carroll, eds. The Atlas of breeding birds in New York State. Cornell University Press, Ithaca, 551 pp.

Arbib, R. 1988b. Double-crested Cormorant, pp. 30-31 *in* R.F. Andrle and J.R. Carroll, eds. The Atlas of breeding birds in New York State. Cornell University Press, Ithaca, 551 pp.

Arbib, R. 1988c. Eastern Meadowlark, pp. 470-471 *in* R.F. Andrle and J.R. Carroll, eds. The Atlas of breeding birds in New York State. Cornell University Press, Ithaca, 551 pp.

Arbib, R. 1988d. Grasshopper Sparrow, pp. 448-449 *in* R.F. Andrle and J.R. Carroll, eds. The Atlas of breeding birds in New York State. Cornell University Press, Ithaca, 551 pp.

Arbib, R. 1988e. Swamp Sparrow, pp. 460-461 *in* R.F. Andrle and J.R. Carroll, eds. The Atlas of breeding birds in New York State. Cornell University Press, Ithaca, 551 pp.

Arcese, P. 1987. Age, intrusion pressure and defence against floaters by territorial male Song Sparrows. Animal Behaviour 35: 773-784.

Arcese, P. 1989a. Intrasexual competition and the mating system in primarily monogamous birds: the case of the Song Sparrow. Animal Behaviour 38(1): 96-111.

Arcese, P. 1989b. Territory acquisition and loss in male Song Sparrows. Animal Behaviour 37(1): 45-55.

Arcese, P. and J.N.M. Smith. 1988. Effects of population density and supplemental food on reproduction in Song Sparrows. Journal of Animal Ecology 57: 119-136.

Arcese, P., J.N.M. Smith, W.M. Hochachka, C.M. Rogers, and D. Ludwig. 1992. Stability, regulation, and the determination of abundance in an insular Song Sparrow population. Ecology 73(3): 805-822.

Archibald, G.W. 1985. Cranes, pp. 142-145 *in* C.M. Perrins and A.L.A. Middleton, eds. The encyclopedia of birds. Facts on File, New York, 447 pp.

Archibald, H.L. 1975. Temporal patterns of spring space use by Ruffed Grouse. Journal of Wildlife Management 39(3): 472-481.

Argus (Les consultants en environnement). 1989. Création des mares dans les marais à spartines du Saint-Laurent: identification des sites potentiels. Report prepared for the Canadian Wildlife Service, Québec Region, Conservation and Protection, Environment Canada, 47 pp.

Armistead, H.T. 1983. Purple Martin, p. 298 *in* J. Farrand, ed. The Audubon Society master guide to birding, vol. II. Alfred A. Knopf, New York, 398 pp.

Armstrong, E. and D. Euler. 1983. Habitat usage of two woodland *Buteo* species in central Ontario. Canadian Field-Naturalist 97(2): 200-207.

Armstrong, E.A. 1955. The Wren. Collins, London, 312 pp.

Armstrong, E.A. 1956. Territory in the Wren *Troglodytes troglodytes*. Ibis 98: 430-437.

Armstrong, E.A. and H.L.K. Whitehouse. 1977. Behavioural adaptations of the Wren (*Troglodytes troglodytes*). Biological Reviews 52(2): 235-294.

Armstrong, E.R. 1987. Red-necked Grebe, pp. 42-43 *in* M.D. Cadman, P.F.J. Eagles, and F.M. Helleiner, eds. Atlas of the breeding birds of Ontario. University of Waterloo Press, Waterloo, 617 pp.

Armstrong, E.R. and D.L.G. Noakes. 1983. Wintering biology of Mourning Doves, *Zenaida macroura*, in Ontario. Canadian Field-Naturalist 77(4): 434-438.

Armstrong, J.T. 1965. Breeding home range in the nighthawk and other birds; its evolutionary and ecological significance. Ecology 46(5): 619-629.

Artmann, J.W. and E.M. Martin. 1975. Incidence of ingested lead shot in Sora rails. Journal of Wildlife Management 39(3): 514-519.

Ashkenazie, S. and U.N. Safriel. 1979. Breeding cycle and behavior of the Semipalmated Sandpiper at Barrow, Alaska. Auk 96(1): 56-67.

Askins, R.A. 1992. Forest fragmentation and the decline of migratory song birds. Bird Observer 20: 13-21.

Askins, R.A. 1993. Population trends in grassland, shrubland, and forest birds in eastern North America, pp. 1-34 *in* D.M. Power, ed. Current Ornithology, vol. 11. Plenum Press, New York.

Askins, R.A., M.J. Philbrick, and D.S. Sugeno. 1987. Relationship between the regional abundance of forest and the composition of forest bird communities. Biological Conservation 39: 129-152.

Askins, R.A., J.F. Lynch, and R. Greenberg. 1990. Population declines in migratory birds in eastern North America, pp. 1-57 *in* D.M. Power, ed. Current Ornithology, vol. 7. Plenum Press, New York, 402 pp.

Asselin, D. 1991. Septième compte rendu des activités de nidification du Merle-bleu à poitrine rouge au Québec, 1990. Société linnéenne du Québec, Sainte-Foy, 8 pp.

Asselin, D. 1992. Huitième compte rendu des activités de nidification du Merle-bleu de l'Est au Québec, 1991. Société linnéenne du Québec, Sainte-Foy, 8 pp.

Aubry, Y. 1981. Observation d'un hybride probable de *Anas americana x Anas penelope* à Lauzon, Québec; printemps 1981. Bulletin ornithologique 26(2): 41-42.

Aubry, Y. 1984. First nests of the Common Black-headed Gull in North America. American Birds 38(3): 366-367.

Aubry, Y. and P. Bannon. 1994a. The spring season: Quebec region. National Audubon Society Field Notes 48(3): 272-274.

Aubry, Y. and P. Bannon. 1994b. The nesting season: Quebec region. National Audubon Society Field Notes 48(5): 921-923.

Aubry, Y. and P. Bannon. 1994c. The winter season: Quebec region. National Audubon Society Field Notes 48(2): 179-181.

Aubry, Y. and P. Bannon. 1995. The summer season: Quebec region. National Audubon Society Field Notes 49(5): 903-906.

Aubry, Y. and D. Bordage. 1982. Un nid d'Aigle-pêcheur (*Pandion haliaetus*) construit au sommet d'un poteau, Caniapiscau, Nouveau-Québec. Carnets de zoologie 4(4): 58-59.

Aubry, Y. and F. Lambert. 1984. Nidification des oiseaux du Québec: chronologie/Quebec breeding birds: nesting and rearing chronology. Environment Canada, Canadian Wildlife Service, 27 pp.

Aubry, Y. and P. Laporte. 1990. Les gros-becs aux mangeoires: "jamais" les mêmes! QuébecOiseaux 1(3): 8-11.

Aubry, Y. and R. Yank. 1992a. The spring season: Quebec region. American Birds 46(3): 391-393.

Aubry, Y. and R. Yank. 1992b. The winter season: Quebec region. American Birds 46(2): 233-235.

Aubry, Y. and R. Yank. 1993a. Les observations saisonnières: l'été 1992: juin et juillet. QuébecOiseaux 4(3): 27-30.

Aubry, Y. and R. Yank. 1993b. The autumn migration: Quebec region. American Birds 47(1): 60-62.

Aubry, Y. and R. Yank. 1993c. The summer season: Quebec region. American Birds 47(5): 1086-1088.

Aubry, Y. and R. Yank. 1994. The autumn migration: Quebec region. American Birds 48(1): 84-86.

Aubry, Y., N. David, M. Gosselin, and J. Hardy. 1980. Comment choisir l'oiseau emblème du Québec. Bulletin ornithologique 25(1): 34.

Aubry, Y., M. Gosselin, and R. Yank. 1987. The spring migration: Quebec region. American Birds 41(3): 406.

Aubry, Y., M. Gosselin, and R. Yank. 1988. The winter season: Quebec region. American Birds 42(2): 238-240.

Aubry, Y., M. Gosselin, and R. Yank. 1989. The nesting season: Quebec region. American Birds 43(5): 1291-1293.

Aubry, Y., M. Gosselin, and R. Yank. 1990a. Les observations saisonnières; l'été 1989. QuébecOiseaux 1(3): 22-25.

Aubry, Y., M. Gosselin, and R. Yank. 1990b. The spring season: Quebec region. American Birds 44(3): 392-394.

Aubry, Y., M. Gosselin, and R. Yank. 1991. The winter season: Quebec region. American Birds 45(2): 247-249.

Aubry, Y., M. Gosselin, and R. Yank. 1992. The autumn migration: Quebec region. American Birds 46(1): 57-60.

Audet, L.-P. 1949. L'École de la Route. Carnets de zoologie 9(4): 185-187.

Audubon, J.J. 1827-1838. The birds of America (Double elephant folio). London: published at cost to the author.

Audubon, J.J. 1840. The birds of America. [1967, Dover Publications, reprinting of volume published between 1840-44, 7 vol.].

Audubon, J.J. 1990. Birds of America, vol. 2 (Multimedia Birds of America). CMC Research, Portland [1 CD-Rom disk].

Audubon, M.R. and E. Coues. 1898. Audubon and his journals, vol. 1 and 2. J.C. Nimmo, London, 532 pp.

Austin, D.D. and M.L. Perry. 1979. Birds in six communities within a lodgepole pine forest. Journal of Forestry 77(9): 584-586.

Austin, J. 1990. Comparison of activities within families and pairs of wintering Canada Geese. Wilson Bulletin 102(3): 536-542.

Austin, M. and M. Cadman. 1992. An update on the status of the Henslow's Sparrow (*Ammodramus henslowii*) in Canada. Committee on the Status of Endangered Wildlife in Canada (COSEWIC), Ottawa, 22 pp.

Austin, O.L. Jr. 1928. Migration-Routes of the Arctic Tern (*Sterna paradisaea* Brünnich). Northeastern Bird-Banding Association, Bulletin 4: 144-148.

Austin, O.L. Jr. and S.H. Low. 1932. Notes on the breeding of the Tree Swallow. Bird-Banding 3(2): 39-44.

Austin, O.L. Jr. and A. Singer. 1990. Familles d'oiseaux: un guide sur la classification des oiseaux. Éditions Broquet, Laprairie, 211 pp.

Avery, M.L., K.J. Goocher, and M.A. Cone. 1993. Handling efficiency and berry size preferences of Cedar Waxwings. Wilson Bulletin 105(4): 604-611.

Axtel, H.H. 1947. Distribution of the vertebrate animals of Chemung and Schuyler counties, New York. Ph.D. dissertation, Cornell University, Ithaca, 350 pp.

Ayotte, G. 1994. Glossaire de botanique, autoformation. Éditions MultiMondes, Sainte-Foy, 616 pp.

Bachand, J. 1992. Inventaire des canaux de drainage dans les marais de la baie de L'Isle-Verte. Société de conservation de la baie de L'Isle-Verte, 29 pp.

Baekken, B., J.O. Nybo, and G.A. Sonerud. 1987. Home-range size of Hawk Owls: dependence on calculation method, number of tracking days, and number of plotted perchings, pp. 145-148 *in* R.W. Nero, R.J. Clark, R.J. Knapton, and R.H. Hamre, eds. Proceedings. Biology and conservation of northern forest owls. Symposium, 3-7 February, Winnipeg. United States Department of Agriculture, Forest Service, Rocky Mountain Forest and Range Experiment Station, Fort Collins, Colorado. General Technical Report RM-142, 309 pp.

Bagg, A.M. 1969. The changing seasons. Audubon Field Notes 23(1): 4-12.

Bagg, A.M. and H.M. Parker. 1951. The Turkey Vulture in New England and eastern Canada up to 1950. Auk 68: 315-333.

Bagg, A.M. and H.M. Parker. 1953. Further information on the Turkey Vulture (*Cathartes aura*) in New England and eastern Canada. Auk 70: 490-491.

Baggerman, B., G.P. Baerends, H.S. Heikens, and J.H. Mook. 1956. Observations on the behaviour of the Black Tern (*Chlidonias n. niger* L.), in the breeding area. Ardea 44: 1-71.

Bailey, R.O. and R.D. Titman. 1984. Habitat use and feeding ecology of postbreeding Redheads. Journal of Wildlife Management 48(4): 1144-1155.

Baillie, J.L. 1955. The Golden Eagle nesting in the Gaspé Peninsula, Québec. Canadian Field-Naturalist 69: 13-14.

Baillie, J.L. and P. Harrington. 1936. Distribution of breeding birds in Ontario, part 1. Transactions. Royal Canadian Institute 21: 1-50.

Baird, J. 1968. Eastern Savannah Sparrow, pp. 678-696 *in* A.C. Bent. Life histories of North American cardinal, grosbeaks, buntings, towhees, finches, sparrows, and allies. (O.L. Austin, Jr., ed.). United States National Museum, Washington. [1968, Dover Publications, New York]. Bulletin no. 237, vol. 1, pp. 1-602; vol. 2, pp. 603-1248; vol. 3, pp. 1249-1889.

Baird, J.C. 1987. La Bécasse d'Amérique. Hinterland Who's Who. Environment Canada, Canadian Wildlife Service, 4 pp. [Also available in English as: American Woodcock.]

Baird, P.H. 1994. Black-legged Kittiwake (*Rissa tridactyla*), *in* A. Poole and F. Gill, eds. The birds of North America, no. 92. The Academy of Natural Sciences, Philadelphia. The American Ornithologists' Union, Washington, D.C., 28 pp.

Baird, S.F., T.M. Brewer, and R. Ridgway. 1905. A history of North American birds, vol. 1. Little, Brown and Co., Boston, 596 pp.

Baker, J.A. and R.J. Brooks. 1981. Raptor and vole populations at an airport. Journal of Wildlife Management 45(2): 390-396.

Baker, M.C. and A.E.M. Baker. 1973. Niche relationships among six species of shorebirds on their wintering and breeding ranges. Ecological Monographs 43: 193-212.

Baker, M.C., G.L. Sherman, T.C. Theimer, and D.C. Bradley. 1982. Population biology of White-crowned Sparrows: residence time and local movements of juveniles. Behavioral Ecology and Sociobiology 11: 133-137.

Baldwin, P.H. 1968. Hoary Redpoll, pp. 400-407 *in* A.C. Bent. Life histories of North American cardinals, grosbeaks, buntings, towhees, finches, sparrows, and allies. (O.L. Austin, Jr., ed.). United States National Museum, Washington. [1968, Dover Publications, New York]. Bulletin no. 237, vol. 1, pp. 1-602; vol. 2, pp. 603-1248; vol. 3, pp. 1249-1889.

Balgooyen, T.G. 1976. The behaviour and ecology of the American Kestrel (*Falco sparverius* L.) in the Sierra Nevada of California. University of California. Publications in Zoology 103: 1-83.

Ball, G.F. 1983. Functional incubation in male Barn Swallows. Auk 100(4): 997-998.

Ball, I.J., D.S. Gilmer, L.M. Cowardin, and J.H. Riechmann. 1975. Survival of Wood Duck and Mallard broods in north-central Minnesota. Journal of Wildlife Management 39(4): 776-780.

Ball, S.C. 1938. Summer birds of the Forillon, Gaspé County, Québec. Canadian Field-Naturalist 52(7): 95-103, 52(8): 120-122.

Ball, S.C. 1943. Further bird notes from Gaspé, Québec. Canadian Field-Naturalist 57(1): 1-4.

Ball, S.C. 1945. The European Starling in Gaspé. Auk 62: 79-97.

Ball, S.C. 1952. Fall migration on the Gaspé Peninsula. Peabody Museum of Natural History, Yale University. Bulletin no. 7, 211 pp.

Ball, S.C. 1954. Additional birds from eastern Gaspé. Canadian Field-Naturalist 68: 103-108.

Baltz, D.M. and G.V. Morejohn. 1977. Food habits and niche overlap of seabirds wintering on Monterey Bay, California. Auk 94(3): 526-534.

Bancroft, J. 1984. Further observations of Great Crested Flycatchers. Blue Jay 42(2): 114-116.

Bancroft, J. 1986. A further report on Great Crested Flycatchers. Blue Jay 44(2): 114-116.

Banks, K.W., H. Clark, I.R.K. Mackay, S.G. Mackay, and R.M. Sellers. 1989. Biometrics and pre-migratory fattening in the Snow Bunting *Plectrophenax nivalis*. Ringing and Migration 10: 141-157.

Banks, R.C. 1976. Wildlife importation into the United States, 1900-1972. Special Scientific Report. Wildlife. United States Fish and Wildlife Service 200: 1-18.

Bannerman, D.A. 1956. The birds of the British Isles. Oliver and Boyd, Edinburgh, Scotland, vol. 5, 350 pp.

Bannerman, D.A. 1961. The birds of the British Isles. Oliver and Boyd, Edinburgh, Scotland, 398 pp.

Bannon, P. 1983. First nesting of the Little Gull (*Larus minutus*) in Quebec. American Birds 37(5): 838-839.

Bannon, P. 1984. Summary of sightings in the Montreal area. Tchebec 14: 6-27.

Bannon, P. 1985a. Bird views, March 17 through April 24. Province of Quebec Society for the Protection of Birds. Newsletter 27(8): 4-6.

Bannon, P. 1985b. Bird Views. Province of Quebec Society for the Protection of Birds. Newsletter 28(3): 4-5.

Bannon, P. 1986. Brewster's Warbler, *Vermivora chrysoptera x pinus* backcross, breeding in Huntingdon County, Quebec. Canadian Field-Naturalist 100(1): 118-119.

Bannon, P. 1991a. Birdfinding in the Montreal area. Province of Quebec Society for the Protection of Birds and Centre de conservation de la faune ailée de Montréal, Montréal, 362 pp.

Bannon, P. 1991b. Où et quand observer les oiseaux dans la région de Montréal. Province of Quebec Society for the Protection of Birds and Centre de conservation de la faune ailée de Montréal, Montréal, 361 pp.

Bannon, P. 1991c. Parlons d'oiseaux. Province of Quebec Society for the Protection of Birds. Newsletter 34(1): 10-11.

Bannon, P. 1993. Birds views. Province of Quebec Society for the Protection of Birds. Newsletter 36(1): 10.

Bannon, P. and P. Laporte. 1985. Réapparition du Bruant de Henslow au Québec. Bulletin ornithologique 30(4): 158-163.

Banville, D. and S. St-Onge. 1988. Inventaire des couvées de canards dans le bras nord de l'île d'Orléans en 1984 et 1985. Ministère du Loisir, de la Chasse et de la Pêche, 14 pp.

Banville, D. and S. St-Onge. 1990a. Inventaire aérien de la sauvagine sur le fleuve Saint-Laurent entre Grondines/Leclercville et Saint-Roch-des-Aulnaies/Baie-Sainte-Catherine au printemps 1989. Service de l'aménagement et de l'exploitation de la faune, région de Québec, 79 pp.

Banville, D. and S. St-Onge. 1990b. Inventaire aérien de la sauvagine sur le fleuve Saint-Laurent entre Grondines/Leclercville et Saint-Roch-des-Aulnaies/Baie-Sainte-Catherine à l'automne 1988. Service de l'aménagement et de l'exploitation de la faune, région de Québec, 73 pp.

Banville, D., S. St-Onge, and N. Desrosiers. 1995. Inventaire des couvées de canards dans le bras nord de l'île d'Orléans en 1991. Ministère du Loisir, de la Chasse et de la Pêche, région de Québec. Preliminary report.

Barclay, R.M.R., M.L. Leonard, and G. Friesen. 1985. Nocturnal singing by Marsh Wrens. Condor 87(3): 418-422.

Bardarson, H.R. 1986. Birds of Iceland. H.R. Bardarson, ed. Reykjavik, 336 pp.

Barichello, N. and D. Mossop. 1983. Productivity of Gyrfalcons (*Falco rusticolus*) as a response to changes in the ptarmigan (*Lagopus* spp.) abundance over a five year period. Paper presented at the 34th Alaska Science Conference, Whitehorse.

Barklow, W.E. 1979. Graded frequency variations of the tremolo call of the Common Loon (*Gavia immer*). Condor 81: 53-64.

Barlow, J.C. and W.B. McGillivray. 1983. Foraging and habitat relationships of the sibling species Willow Flycatcher (*Empidonax traillii*) and Alder Flycatcher (*E. alnorum*) in southern Ontario. Canadian Journal of Zoology 61(7): 1510-1516.

Barlow, J.C. and D.M. Power. 1970. An analysis of character variation in Red-eyed and Philadelphia vireos (Aves: Vireonidae) in Canada. Canadian Journal of Zoology 48: 673-694.

Barlow, J.C. and J.C. Rice. 1977. Aspects of the comparative behaviour of Red-eyed and Philadelphia vireos. Canadian Journal of Zoology 55: 528-542.

Barnhurst, B. 1979. Saw-whet Owl banding at île Perrot. Tchebec, pp. 94-97.

Barnhurst, B. and M.[W.] McIntosh. 1976. Hawk-watching in Montreal. Tchebec, pp. 67-75.

Barnhurst, R.J. and M.W. McIntosh. 1993. Hawk migration studies, spring and fall 1980-1993. Hawk Migration Association of North America. Biannual Newsletters.

Barr, J.F. 1973. Feeding biology of the Common Loon (*Gavia immer*) in oligotrophic lakes of the Canadian Shield. Ph.D. dissertation, University of Guelph, Ontario.

Barr, J.F. 1986. Population dynamics of the Common Loon (*Gavia immer*) associated with mercury-contaminated waters in northwestern Ontario. Canadian Widlife Service. Occasional Paper no. 56, 28 pp.

Barrett, R.T. 1985. Further changes in the breeding distribution and numbers of cliff-breeding seabirds in Sør-Varanger, North Norway, 1975-1983. Fauna Norvegica Series C, Cinclus 8: 35-39.

Barrows, W.B. 1889. The English Sparrow (*Passer domesticus*) in North America, especially in its relation to agriculture. United States Department of Agriculture, Division of Economic Ornithology and Mammalogy. Bulletin no. 1, 405 pp.

Barry, T.W. 1956. Observations of a nesting colony of American Brant. Auk 73(2): 193-202.

Barry, T.W. 1962. Effects of late seasons on Atlantic Brant reproduction. Journal of Wildlife Management 26(1): 19-26.

Barske, P. 1968. The Black Duck. Evaluation, management and research: a symposium. Atlantic Waterfowl Council and Wildlife Management Institute, Washington, 193 pp.

Bart, J., R.A. Stehn, J.A. Herrick, N.A. Heaslip, T.A. Bookhout, and J.R. Stenzel. 1984. Survey methods for breeding Yellow Rails. Journal of Wildlife Management 48(4): 1382-1386.

Bartholomew, G.A. 1986. The role of natural history in contemporary biology. BioScience 36(5): 324-329.

Bartonek, J.C. and J.J Hickey. 1969. Food habits of Canvasbacks, Redheads and Lesser Scaups in Manitoba. Condor 71(3): 280-290.

Bartonek, J.C., R.J. Blohm, R.K. Brace, F.D. Caswell, K.E. Gamble, H.W. Miller, R.S. Pospahala, and M.M. Smith. 1984. Status and needs of the Mallard. Transactions. North American Wildlife and Natural Resources Conference 49: 501-517.

Bateman, B. 1953. Observations on the natural history of the Leaf Bay-Fort Chimo region, Ungava, Quebec. Intermediate Naturalist 8: 1-7.

Bates, R.L. and J.A. Jackson. 1984. Dictionary of geological terms, 3rd edition. Prepared under the direction of the American Geological Institute. Anchor Press/Doubleday, Garden City, New York, 571 pp.

Batt, B.D.J., A.D. Afton, M.G. Anderson, C.D. Ankney, D.H. Johnson, J.A. Kadlec, and G.L. Krapu. 1992. Ecology and management of breeding waterfowl. University of Minnesota Press, Minneapolis, 635 pp.

Baudvin, H. 1983. Le régime alimentaire de la Chouette effraie (*Tyto alba*), *in* Le Jean Le Blanc, Bulletin du Centre d'études ornithologiques de Bourgogne, vol. 22, 108 pp.

Baumgartner, A.M. 1968. Tree Sparrow, pp. 1137-1165 *in* A.C. Bent. Life histories of North American cardinals, grosbeaks, buntings, towhees, finches, sparrows, and allies. (O.L. Austin, Jr., ed.). United States National Museum, Washington. [1968, Dover Publications, New York]. Bulletin no. 237, vol. 1, pp. 1-602; vol. 2, pp. 603-1248; vol. 3, pp. 1249-1889.

Baumgartner, F.M. 1939. Territory and population in the Great Horned Owl. Auk 56(3): 274-282.

Bautz, L. and P. Nye. 1987. Spring movement of an adult Bald Eagle from southeastern New York to central Ontario. Eyas 10(1): 32-38.

Baydack, R.K. 1988. Characteristics of Sharp-tailed Grouse, *Tympanuchus phasianellus*, leks in the parklands of Manitoba. Canadian Field-Naturalist 102(2): 39-44.

Baydack, R.K. and D.A. Hein. 1987. Tolerance of Sharp-tailed Grouse to lek disturbance. Wildlife Society Bulletin 15: 535-539.

Beal, F.E.L. 1897. The Blue Jay and its food. United States Department of Agriculture, Yearbook for 1896.

Beal, F.E.L. 1898. The food of cuckoos, pp. 7-14 *in* C.H. Merriam, ed. Cuckoos and Shrikes in their relation to agriculture. United States Department of Agriculture, Division of biological survey, Washington, Bulletin no. 9, 26 pp.

Beal, F.E.L. 1911. Food of the woodpeckers of the United States. United States Department of Agriculture. Biological Survey Bulletin no. 37, Washington, 64 pp.

Beard, E.B. 1964. Duck brood behaviour at the Seney National Wildlife Refuge. Journal of Wildlife Management 28(3): 492-521.

Beason, R.C. 1984. Altruism in the Horned Lark. Journal of Field Ornithology 55(4): 489-490.

Beason, R.C. and L.L. Trout. 1984. Cooperative breeding in the Bobolink. Wilson Bulletin 96(4): 709-710.

Beauchamp, G. 1992. Diving behavior in Surf Scoters and Barrow's Goldeneyes. Auk 109(4): 819-827.

Beauchesne, P. 1991. Effets de la largeur de la bande riveraine de protection sur l'abondance et la diversité de la faune dans la sapinière boréale. M.Sc. thesis, Université Laval, Sainte-Foy.

Beaudoin, P. 1978. Des Dindons sauvages élevés en captivité survivraient en forêt. Le Peuple-Courrier de la Côte-du-Sud. Wednesday edition, 12 April 1978, p. 27.

Beaudoin, P. 1984. Réintroduction du Dindon sauvage à Saint-Fabien: un ornithologue autodidacte gagne son pari haut la main. Le Peuple-Courrier de la Côte-du-Sud. Wednesday edition, 9 May 1984, p. A-3.

Beaulieu, H. 1992. Liste des espèces de la faune vertébrée susceptibles d'être désignées menacées ou vulnérables. Ministère du Loisir, de la Chasse et de la Pêche, Gouvernement du Québec, 107 pp.

Becker, D.M. and C.H. Sieg. 1987. Home range and habitat utilization of breeding male Merlins, *Falco columbarius*, in southeastern Montana. Canadian Field-Naturalist 101(3): 398-403.

Beckwith, R.C. and E.L. Bull. 1985. Scat analysis of the arthropod component of Pileated Woodpecker diet. Murrelet 66(3): 90-92.

Bédard, J. 1969a. Adaptative radiation in Alcidae. Ibis 111: 190-198.

Bédard, J. 1969b. Histoire naturelle du Gode, *Alca torda* L., dans le golfe Saint-Laurent, province de Québec, Canada. Canadian Wildlife Service, Ottawa. Report Series no. 7, 79 pp. [An abstract is available in English.]

Bédard, J. 1985. Evolution and characteristics of the Atlantic Alcidae, pp. 1-51 *in* D.N. Nettleship and T.R. Birkhead, eds. The Atlantic Alcidae: the evolution, distribution and biology of the auks inhabiting the Atlantic Ocean and adjacent water areas. Academic Press, London, 574 pp.

Bédard, J. 1988. Gestion des populations de Cormoran à aigrettes dans l'estuaire du Saint-Laurent: résultats de l'étude pilote et recommandations d'intervention. Final report submitted to Ducks Unlimited (Canada).

Bédard, J. and G. Gauthier. 1989. Comparative energy budgets of Greater Snow Geese *Chen caerulescens atlantica* staging in two habitats in spring. Ardea 77: 3-20.

Bédard, J. and G. LaPointe. 1984a. Banding returns, arrival times, and site fidelity in the Savannah Sparrow. Wilson Bulletin 96(2): 196-205.

Bédard, J. and G. LaPointe. 1984b. The Savannah Sparrow territorial system: can habitat features be related to breeding success? Canadian Journal of Zoology 62: 1819-1828.

Bédard, J. and G. LaPointe. 1985. Influence of parental age and season on Savannah Sparrow reproductive success. Condor 87: 106-110.

Bédard, J. and M. Meunier. 1983. Parental care in the Savannah Sparrow. Canadian Journal of Zoology 61(12): 2836-2843.

Bédard, J. and J. Munro. 1972. Fourth Annual Quebec Waterfowl Seminar.

Bédard, J. and J. Munro. 1977. Brood and crèche stability in the Common Eider of the St. Lawrence estuary. Behaviour 60: 221-236.

Bédard, J. and A. Nadeau. 1994. L'Eider à duvet dans l'estuaire du Saint-Laurent: un plan de gestion (révisé: 1994). Société Duvetnor Ltée, report submitted to Ducks Unlimited (Canada), 64 pp.

Bédard, J., A. Nadeau, and G. Gauthier. 1986a. Effects of grazing by Greater Snow Geese on hay production. Journal of Applied Ecology 23(1): 65-75.

Bédard, J., J. Gauthier, and J. Munro. 1986b. La distribution de l'Eider à duvet durant l'élevage des canetons dans l'estuaire du Saint-Laurent, pp. 12-19 *in* A. Reed, ed. Eider ducks in Canada/Les eiders au Canada. Canadian Wildlife Service, Ottawa. Report Series no. 47, 177 pp.

Bédard, J., J. Huot, and J.-F. Giroux. 1987. Île Blanche. Programme de suivi environnemental. Report submitted to Ducks Unlimited (Canada), 25 pp.

Beddall, B.G. 1963. Range expansion of the Cardinal and other birds in the northeastern states. Wilson Bulletin 75(2): 140-158.

Bednarz, J.C. and J.J. Dinsmore. 1981. Status, habitat use, and management of Red-shouldered Hawks in Iowa. Journal of Wildlife Management 45(1): 236-241.

Bednarz, J.C. and J.J. Dinsmore. 1982. Nest-sites and habitat of Red-shouldered and Red-tailed hawks in Iowa. Wilson Bulletin 94(1): 31-45.

Bednarz, J.C., D. Kleim, Jr., L.J. Goodrich, and S.E. Senner. 1990. Migration counts of raptors at Hawk Mountain, Pennsylvania, as indicators of population trends, 1934-1986. Auk 107(1): 96-109.

Beecher, M.D. and I.M. Beecher. 1979. Sociobiology of Bank Swallows: reproductive strategy of the male. Science 205(21): 1282-1285.

Beecher, M.D., M.B. Medvin, P.K. Stoddard, and P. Loesche. 1986. Acoustic adaptations for parent-offspring recognition in swallows. Experimental Biology 45(3): 179-183.

Beecher, W.J. 1978. Feeding adaptations and evolution in the starlings. Bulletin of the Chicago Academy of Science 11: 269-298.

Beedy, E.C. 1981. Bird communities and forest structure in the Sierra Nevada of California. Condor 83(2): 97-105.

Bégin, P., F. Brousseau, F. Demers, B. Michaud, S. Plourde, and M. Thériault. 1988. Étude de la population de Gélinottes huppées (*Bonasa umbellus*) sur la réserve faunique de Rimouski. Automne 1988. Université du Québec à Rimouski for Ministère du Loisir, de la Chasse et de la Pêche, 26 pp.

Béique, R. 1980. Notions générales d'entomologie forestière. Service d'entomologie et de pathologie, Ministère de l'Énergie et des Ressources, Laboratoire d'entomologie forestière, Sainte-Foy, 20 pp.

Bekoff, M. and A.C. Scott. 1989. Aggression, dominance, and social organization in Evening Grosbeaks. Ethology 83: 177-194.

Bekoff, M., A.C. Scott, and D.A. Conner. 1987. Nonrandom nest-site selection in Evening Grosbeaks. Condor 89(4): 819-829.

Bekoff, M., A.C. Scott, and D.A. Conner. 1989. Ecological analyses of nesting success in Evening Grosbeaks. Oecologia 81(1): 67-74.

Bélanger, F., S. Berberi, J.R. Breton, D. Carrier, and R. Lessard. 1990. La Beauce et les Beaucerons: portrait d'une région 1737-1987. Société du patrimoine des Beaucerons et Corporation du 250e anniversaire de la Beauce, Saint-Joseph-de-Beauce, 380 pp.

Bélanger, L. 1989. Potentiel des îles du Saint-Laurent dulcicole pour la sauvagine et plan de protection. Environment Canada, Canadian Wildlife Service, Québec Region, 71 pp.

Bélanger, L. 1991. Programme de valorisation des îles du Saint-Laurent pour la nidification de la sauvagine et des autres espèces aviennes (Montréal à Trois-Rivières). I. Gestion intégrée des activités agricoles. Environment Canada, Canadian Wildlife Service, Québec Region, 61 pp.

Bélanger, L. and R. Couture. 1988. Use of man-made ponds by dabbling duck broods. Journal of Wildlife Management 52(4): 718-723.

Bélanger, L. and R. Couture. 1989. Pintail, *Anas acuta*, use of man-made ponds in Quebec. Canadian Field-Naturalist 103(4): 547-550.

Bélanger, L. and D. Lehoux. 1995. Use of various habitat types by nesting ducks on islands in the St. Lawrence River between Montréal and Trois-Rivières. Canadian Wildlife Service. Occasional Paper no. 87, 26 pp.

Bélanger, L. and S. Tremblay. 1989. Distribution et caractéristiques forestières des héronnières du Québec. Technical report, Ministère du Loisir, de la Chasse et de la Pêche, 52 pp.

Bélanger, L., D. Lehoux, and C. Grenier. 1989. Propositions d'aménagement des îles du fleuve Saint-Laurent pour la sauvagine à partir de matériaux de dragage. Environment Canada, Canadian Wildlife Service, Québec Region for the St. Lawrence Centre, 125 pp.

Bélanger, L., L. Gratton, and D. Lehoux. 1994. Étude de la nidification du Canard noir dans divers habitats côtiers de l'estuaire du Saint-Laurent et perspectives de conservation. Environment Canada, Canadian Wildlife Service, Québec Region. Technical Report Series no. 210, 37 pp. [An abstract is available in English.]

Belant, J.L. 1991. Multiple lake defense by Common Loons. Passenger Pigeon 53(2): 187-188.

Belant, J.L. and R.K. Anderson. 1991a. Winter recoveries and territorial affinity of Common Loons banded in Wisconsin. Wilson Bulletin 103(1): 141-142.

Belant, J.L. and R.K. Anderson. 1991b. Common Loon, *Gavia immer*, brood habitat use in northern Wisconsin. Canadian Field-Naturalist 105(3): 372-375.

Beletsky, L.D. 1982a. Vocal behavior of the Northern Oriole. Wilson Bulletin 94(3): 372-381.

Beletsky, L.D. 1982b. Vocalizations of female Northern Orioles. Condor 84(4): 445-447.

Beletsky, L.D. and G.H. Orians. 1989. Territoriality among male Red-winged Blackbirds. III: Testing hypotheses of territorial dominance. Behavioral Ecology and Sociobiology 24(5): 333-339.

Beletsky, L.D. and G.H. Orians. 1990. Male parental care in a population of Red-winged Blackbird, 1983-1988. Canadian Journal of Zoology 68(3): 606-609.

Beletsky, L.D. and G.H. Orians. 1991. Effects of breeding experience and familiarity on site fidelity in female Red-winged Blackbirds. Ecology 72(3): 787-796.

Belles-Isles, J.-C. and J. Picman. 1986a. House Wren nest-destroying behaviour. Condor 88(2): 190-193.

Belles-Isles, J.-C. and J. Picman. 1986b. Nesting losses and nest site preferences in House Wrens. Condor 88(4): 483-486.

Bellrose, F.C. 1976. Ducks, geese and swans of North America, 2nd edition. Stackpole Books, Harrisburg, Pennsylvania, 543 pp.

Bellrose, F.C. 1980. Ducks, geese and swans of North America, 3rd edition. Stackpole Books, Harrisburg, Pennsylvania, 540 pp.

Bellrose, F.C. 1990. The history of Wood Duck management, pp. 13-20 *in* L.H. Fredrickson, G.V. Burger, S.P. Havera, D.A. Graber, R.E. Kirby, and T.S. Taylor, eds. Proceedings. 1988 North American Wood Duck symposium, St. Louis, Missouri, 390 pp.

Belthoff, J.R. and S.A. Gauthreaux, Jr. 1991. Migration and differential winter distribution of House Finches in the eastern United States. Condor 93(2): 374-382.

Belthoff, J.R. and G. Ritchison. 1989. Natal dispersal of Eastern Screech-Owls. Condor 91(2): 254-265.

Belthoff, J.R. and G. Ritchison. 1990. Nest-site selection by Eastern Screech-Owls in central Kentucky. Condor 92(4): 982-990.

Bendell, B.E. 1980. The role of the Red-winged Blackbird, *Agelaius phoeniceus*, as predator of insects. M.Sc. thesis, McGill University, Montréal, 90 pp.

Bendell, B.E., P.J. Weatherhead, and R.K. Stewart. 1981. The impact of predation by Red-winged Blackbirds on European Corn Borer populations. Canadian Journal of Zoology 59(8): 1535-1538.

Bengtson, S.-A. 1966. Field studies on the Harlequin Duck in Iceland. Wildfowl Trust Annual Report 17: 79-94.

Bengtson, S.-A. 1970. Location of nest-sites of ducks in Lake Myvatn area, north-east Iceland. Oikos 21(2): 218-229.

Bengtson, S.-A. 1971a. Breeding success of the Arctic Tern *Sterna paradisea* (Pontoppidan) in the Kongsfjord area, Spitsbergen in 1976. Norwegian Journal of Zoology 19: 77-82.

Bengtson, S.-A. 1971b. Habitat selection of duck broods in Lake Myvatn area, north-east Iceland. Ornis Scandinavica 2: 17-26.

Bengtson, S.-A. 1972. Breeding ecology of the Harlequin Duck *Histrionicus histrionicus* (L.) in Iceland. Ornis Scandinavica 3: 1-19.

Bengtson, S.-A. 1984. Breeding ecology and extinction of the Great Auk (*Pinguinus impennis*): anecdotal evidence and conjectures. Auk 101(1): 1-12.

Bengtson, S.-A. and S. Ulfstrand. 1971. Food resources and breeding frequency of the Harlequin Duck *Histrionicus histrionicus* in Iceland. Oikos 22: 235-239.

Benkman, C.W. 1985. The foraging ecology of crossbills in eastern North America. Ph.D. dissertation, State University of New York, Albany, 182 pp.

Benkman, C.W. 1987. Crossbill foraging behavior, bill structure, and patterns of food profitability. Wilson Bulletin 99(3): 351-368.

Benkman, C.W. 1988a. Flock size, food dispersion, and the feeding behavior of crossbills. Behavioral Ecology and Sociobiology 23: 167-175.

Benkman, C.W. 1988b. Why White-winged Crossbills do not defend feeding territories. Auk 105(3): 578-579.

Benkman, C.W. 1989. Breeding opportunities, foraging rates, and parental care in White-winged Crossbills. Auk 106(3): 483-485.

Benkman, C.W. 1990. Intake rates and timing of crossbill reproduction. Auk 107(2): 376-386.

Benkman, C.W. 1992. White-winged Crossbill (*Loxia leucoptera*), *in* A. Poole, P. Stettenheim, and F. Gill, eds. The birds of North America, no. 27. The Academy of Natural Sciences, Philadelphia. The American Ornithologists' Union, Washington, 20 pp.

Benkman, C.W. 1993. The evolution, ecology, and decline of the Red Crossbill of Newfoundland. American Birds 47(2): 225-229.

Bennett, G. 1987. Great Crested Flycatcher, pp. 266-267 *in* M.D. Cadman, P.F.J. Eagles, and F.M. Helleiner, eds. Atlas of the breeding birds of Ontario. University of Waterloo Press, Waterloo, 617 pp.

Bennett, W.A. 1990. Scale of investigation and the detection of competition: an example from the House Sparrow and House Finch introductions in North America. American Naturalist 135(6): 725-747.

Bennetts, R.E. and R.L. Hutto. 1985. Attraction of social Fringillids to mineral salts: an experimental study. Journal of Field Ornithology 56(2): 187-189.

Benoit, J., R. Bergeron, J.-C. Bourgeois, S. Desjardins, and J. Picard. 1987. Les habitats et la faune de la région du lac Saint-Pierre. Synthèse des connaissances. Québec, Ministère du Loisir, de la Chasse et de la Pêche, 123 pp.

Benoit, P. 1993. *Common names of insects in Canada / Noms communs des insectes du Canada.* E.M. Belton and D.C. Eidt, eds. Based on *Nomenclatura Insectorum Canadensis*. The Entomological Society of Canada [1 disk].

Benoit, R. 1991. Axes de vol et activité diurne et nocturne du Grand Héron (*Ardea herodias*) au lac Saint-Louis, Québec. M.Sc. thesis, Université de Montréal, Québec, 63 pp.

Benoit, R., A. Reed, R. Lalumière, and G. Morissette. 1991. Utilisation par la sauvagine des habitats côtiers de la baie de Many Islands, baie James. Technical report for Direction ingénierie et environnement, Société d'Énergie de la Baie James, by Groupe Environnement Shooner, Québec, 62 pp.

Benoit, R., A. Reed, and R. Lalumière. 1992. Utilisation par la sauvagine des habitats de la côte nord-est de la baie James, été 1991. Report for Service écologie, Direction ingénierie et environnement, Société d'Énergie de la Baie James, by Groupe Environnement Shooner, Québec, 62 pp.

Benoit, R., J.-L. DesGranges, and R. McNeil. 1993a. Flight path of Great Blue Herons at a colony near Montreal. Canadian Journal of Zoology 71(11): 2250-2257.

Benoit, R., R. Lalumière, and A. Reed. 1993b. Étude de la sauvagine sur la côte nord-est de la baie James – 1992. Report for Service écologie, Société d'Energie de la Baie James, by Groupe Environnement Shooner, Québec, 91 pp.

Benoit, R., A. Reed, and R. Lalumière. 1994. Étude de la sauvagine sur la côte nord-est de la baie James – 1993. Report for Service écologie, Direction ingénierie et environnement, Société d'Énergie de la Baie James, by Groupe Environnement Shooner, Québec, 113 pp.

Bent, A.C. 1919. Life histories of North American diving birds. United States National Museum, Washington. [1963, Dover Publications, New York]. Bulletin no. 107, 239 pp.

Bent, A.C. 1921. Life histories of North American gulls and terns. United States National Museum, Washington. [1963, Dover Publications, New York]. Bulletin no. 113, 337 pp.

Bent, A.C. 1922. Life histories of North American petrels and pelicans and their allies. United States National Museum, Washington. [1964, Dover Publications, New York]. Bulletin no. 121, 327 pp.

Bent, A.C. 1923. Life histories of North American wild fowl. United States National Museum, Washington. [1962, Dover Publications, New York]. Bulletin no. 126, vol. 1, 244 pp.

Bent, A.C. 1925. Life histories of North American wild fowl. United States National Museum, Washington. [1962, Dover Publications, New York]. Bulletin no. 130, vol. 2, 316 pp.

Bent, A.C. 1926. Life histories of North American marsh birds. United States National Museum, Washington. [1963, Dover Publications, New York]. Bulletin no. 135, 392 pp.

Bent, A.C. 1927. Life histories of North American shore birds. United States National Museum, Washington. [1962, Dover Publications, New York]. Bulletin no. 142, vol. 1, 420 pp.

Bent, A.C. 1929. Life histories of North American shore birds. United States National Museum, Washington. [1962, Dover Publications, New York]. Bulletin no. 146, vol. 2, 412 pp.

Bent, A.C. 1932. Life histories of North American gallinaceous birds. United States National Museum, Washington. [1963, Dover Publications, New York]. Bulletin no. 162, 490 pp.

Bent, A.C. 1937. Life histories of North American birds of prey. United States National Museum, Washington. [1961, Dover Publications, New York]. Bulletin no. 167, vol. 1, 409 pp.

Bent, A.C. 1938. Life histories of North American birds of prey. United States National Museum, Washington. [1961, Dover Publications, New York]. Bulletin no. 170, vol. 2, 482 pp.

Bent, A.C. 1939. Life histories of North American woodpeckers. United States National Museum, Washington. [1964, Dover Publications, New York]. Bulletin no. 174, 334 pp.

Bent, A.C. 1940. Life histories of North American cuckoos, goatsuckers, hummingbirds and their allies. United States National Museum, Washington. [1964, Dover Publications, New York]. Bulletin no. 176, vol. 1, pp. 1-244; vol. 2, pp. 244-506.

Bent, A.C. 1942. Life histories of North American flycatchers, larks, swallows, and their allies. United States National Museum, Washington. [1963, Dover Publications, New York]. Bulletin no. 179, 555 pp.

Bent, A.C. 1946. Life histories of North American jays, crows and titmice. United States National Museum, Washington. [1964, Dover Publications, New York]. Bulletin no. 191, vol. 1, pp. 1-214; vol. 2, pp. 215-495.

Bent, A.C. 1948. Life histories of North American nuthatches, wrens, thrashers, and their allies. United States National Museum, Washington [1964, Dover Publications, New York]. Bulletin no. 195, 475 pp.

Bent, A.C. 1949. Life histories of North American thrushes, kinglets, and their allies. United States National Museum, Washington. [1964, Dover Publications, New York]. Bulletin no. 196, 452 pp.

Bent, A.C. 1950. Life histories of North American wagtails, shrikes, vireos, and their allies. United States National Museum, Washington. [1965, Dover Publications, New York]. Bulletin no. 197, 411 pp.

Bent, A.C. 1953. Life histories of North American wood warblers. United States National Museum, Washington. [1963, Dover Publications, New York]. Bulletin no. 203, vol. 1, pp. 1-367; vol. 2, pp. 367-734.

Bent, A.C. 1958. Life histories of North American blackbirds, orioles, tanagers, and allies. United States National Museum, Washington. [1965, Dover Publications, New York]. Bulletin no. 211, 549 pp.

Bent, A.C. 1968. Life histories of North American cardinals, grosbeaks, buntings, towhees, finches, sparrows, and allies. (O.L. Austin, Jr., ed.). United States National Museum, Washington. [1968, Dover Publications, New York]. Bulletin no. 237, vol. 1, pp. 1-602; vol. 2, pp. 603-1248; vol. 3, pp. 1249-1889.

Benton, A.H. 1968. An unusual nesting situation of the Tree Swallow *Iridoprocne bicolor*. Wilson Bulletin 80(2): 233.

Benton, A.H. and H. Tucker. 1968. Weather and Purple Martin mortality in western New York. Kingbird 18(1): 71-75.

Benyus, J.M. 1989. Northwoods wildlife: a watcher's guide to habitats. United States Department of Agriculture, Forest Service, North Central Forest Experiment Station. Northwood Press, Inc., St. Paul, Minnesota, 453 pp.

Berger, A.J. 1951. Nesting density of Virginia and Sora rails in Michigan. Condor 53(4): 202.

Berger, A.J. 1968. Eastern Vesper Sparrow, pp. 868-882 *in* A.C. Bent. Life histories of North American cardinals, grosbeaks, buntings, towhees, finches, sparrows, and allies. (O.L. Austin, Jr., ed.). United States National Museum, Washington. [1968, Dover Publications, New York]. Bulletin no. 237, vol. 1, pp. 1-602; vol. 2, pp. 603-1248; vol. 3, pp. 1249-1889.

Berger, A.J. and D.F. Parmelee. 1952. The Alder Flycatcher in Washtenaw County, Michigan: breeding distribution and cowbird parasitism. Wilson Bulletin 64(1): 33-38.

Berger, D.D., D.W. Anderson, J.D. Wheaver, and R.W. Risebrough. 1970. Shell thinning in eggs of Ungava Peregrines. Canadian Field-Naturalist 84(3): 265-267.

Berger, M. and K. Johansen. 1989. The respiratory pattern and expiratory gas concentrations in tropic hummingbirds *Colibri coruscans*, pp. 179-186 *in* C. Bech and R.E. Reinertsen, eds. Proceedings. NATO Advanced Research Workshop on Physiology of Cold Adaptation in Birds, Norway. Life Sciences, vol. 173. Plenum Press, New York.

Bergeron, M. 1977. Mes observations ornithologiques sur l'île Anticosti du 30 juin au 9 juillet 1976. Club des ornithologues du Québec, Feuille de contact, janvier 1977, no. 1, pp. 6-7.

Bergeron, R. 1992. Régime alimentaire de la Grande Oie blanche et de la Bernache du Canada dans la région du Lac Saint-Pierre au printemps. M.Sc. thesis, Université du Québec à Montréal, Montréal, 81 pp.

Bergeron, R., R.J. Hughes, and A. Reed. 1993. Projet de Laforge 1: étude de la sauvagine et caractérisation de ses habitats, été 1992. Submitted to Société d'Énergie de la Baie James by Le Groupe Dryade, 87 pp.

Bergerud, A.T. 1970. Population dynamics of Willow Ptarmigan *Lagopus lagopus alleni* L. in Newfoundland 1955 to 1965. Oikos 21: 299-325.

Bergerud, A.T. and M.W. Gratson. 1988. Adaptive strategies and population ecology of northern grouse. University of Minnesota Press, Minneapolis, vol. 1: Population studies, pp. 3-419; vol. 2: Theory and synthesis, pp. 423-809.

Bergerud, A.T. and D.S. Huxter. 1969. Breeding season, habitat utilisation and movement of Newfoundland Willow Ptarmigan. Journal of Wildlife Management 33(4): 967-974.

Bergerud, A.T. and D.H. Mossop. 1984. The pair bond in ptarmigan. Canadian Journal of Zoology 62: 2129-2141.

Bergman, R.D. 1973. Use of southern boreal lakes by postbreeding Canvasbacks and Redheads. Journal of Wildlife Management 37(2): 160-170.

Bergman, R.D. and D.V. Derksen. 1977. Observations on Arctic and Red-throated loons at Storkersen Point, Alaska. Arctic 30(1): 41-51.

Bergman, R.D., P. Swain, and M.W. Weller. 1970. A comparative study of nesting Forster's and Black terns. Wilson Bulletin 82(4): 435-444.

Bergtold, W.H. 1913. A study of the House Finch. Auk 30(1): 40-73.

Bergtold, W.H. 1921. The English Sparrow (*Passer domesticus*) and the motor vehicle. Auk 38(2): 244-250.

Bernard, M.A. 1984. Aménagement en milieu naturel. Rivière Petite-Cascapédia. Environment Canada, Canadian Wildlife Service, Québec Region, 184 pp.

Bernard-Laurent, A. 1987. Le régime alimentaire du Lagopède alpin (*Lagopus mutus*; Montin 1776): synthèse bibliographique. Gibier Faune Sauvage 4: 321-347.

Bertin, R.I. 1977. Breeding habitats of the Wood Thrush and Veery. Condor 79(3): 303-311.

Bertin, R.I. 1982. The Ruby-throated Hummingbird and its major food plants: ranges, flowering phenology, and migration. Canadian Journal of Zoology 60(2): 210-219.

Berube, P. and R. Couture. 1986. Étude de la composition structurale des habitats de nidification et d'élevage de la Bécasse d'Amérique (*Scolopax minor*), dans le centre sud du Québec. Report submitted to Ministère du Loisir, de la Chasse et de la Pêche, Direction générale de la faune, 158 pp.

Best, L.B. 1974. An unusual case of nesting persistence in a female Field Sparrow. Condor 76(3): 349.

Best, L.B. 1977a. Patterns of feeding Field Sparrow young. Wilson Bulletin 89(4): 625-627.

Best, L.B. 1977b. Nesting biology of the Field Sparrow. Auk 94(2): 308-319.

Best, L.B. 1977c. Territory quality and mating success in the Field Sparrow (*Spizella pusilla*). Condor 79: 192-204.

Best, L.B. 1978. Field Sparrow reproductive success and nesting ecology. Auk 95(1): 9-22.

Best, L.B. 1979. Effects of fire on a Field Sparrow population. American Midland Naturalist 101(2): 434-442.

Best, L.B. and N.L. Rodenhouse. 1984. Territory preference of Vesper Sparrows in cropland. Wilson Bulletin 96(1): 72-82.

Bider, J.R. and P. Lamothe. 1982. Complexe Grande-Baleine. Études d'environnement. Les Canards: synthèse des études sur les populations de canards menées entre les années 1975 et 1980 dans les bassins de la Grande rivière de la Baleine et de la Petite rivière de la Baleine. Hydro-Québec, Document de synthèse no. 7, 30 pp.

Bielefeldt, J., R.N. Rosenfield, and J.M. Papp. 1992. Unfounded assumptions about diet of the Cooper's Hawk. Condor 94(2): 427-436.

Biggar, H.P. 1924. The voyages of Jacques Cartier. (Published from the originals with translations, notes and appendices). Publications of the Public Archives of Canada, no. 11, Ottawa, 330 pp.

Binford, L.C. 1991. Black-throated Blue Warbler, pp. 402-403 *in* R. Brewer, G.A. McPeek, and R.J. Adams, Jr., eds. The Atlas of breeding birds of Michigan. Michigan State University Press, East Lansing, 594 pp.

Bird, C.C. and E.G. Bird. 1935. The birds of Jan Mayen Island. Ibis 77: 837-855.

Bird, D.M. and Y. Aubry. 1982. Reproductive and hunting behaviour in Peregrine Falcons, *Falco peregrinus*, in southern Quebec. Canadian Field-Naturalist 96(2): 167-171.

Bird, D.M. and D.T. Brown. 1991. Étude de faisabilité pour l'implantation de la fauconnerie et de l'autourserie au Québec. Ministère du Loisir, de la Chasse et de la Pêche, Québec, 138 pp.

Bird, D.M. and J.D. Weaver. 1988. Peregrine Falcon populations in Ungava Bay, Quebec, 1980-1985, pp. 45-49 *in* T.J. Cade, J.H. Enderson, C.G. Thelander, and C.M. White, eds. Peregrine Falcon populations: their management and recovery. The Peregrine Fund, Inc., Boise, Idaho.

Bird, D.M., I. Ritchie, J.D. Weaver, and R. Bowman. 1990. Impact of forced renesting in Ungava Bay Peregrine Falcons. Canadian Field-Naturalist 104(2): 219-221.

Birkhead, T.R. 1985. Coloniality and social behaviour in the Atlantic Alcidae, pp. 355-382 *in* D.N. Nettleship and T.R. Birkhead, eds. The Atlantic Alcidae: the evolution, distribution and biology of the auks inhabiting the Atlantic Ocean and adjacent water areas. Academic Press, London, 574 pp.

Birkhead, T.R. and M.P. Harris. 1985. Ecological adaptations for breeding in the Atlantic Alcidae, pp. 205-231 *in* D.N. Nettleship and T.R. Birkhead, eds. The Atlantic Alcidae: the evolution, distribution and biology of the auks inhabiting the Atlantic Ocean and adjacent water areas. Academic Press, London, 574 pp.

Birkhead, T.R. and P.J. Hudson. 1977. Population parameters for the Common Guillemot (*Uria aalge*). Ornis Scandinavica 8: 145-154.

Birkhead, T.R. and D.N. Nettleship. 1980. Census methods for murres, *Uria* species: a unified approach. Canadian Wildlife Service. Occasional Paper no 43, 23 pp.

Birkhead, T.R. and D.N. Nettleship. 1987. Ecological relationships between Common Murres, *Uria aalge*, and Thick-billed Murres, *Uria lomvia*, at Gannet Islands, Labrador. III. Feeding ecology of the young. Canadian Journal of Zoology 65: 1638-1649.

Biro, M. 1987. Wood Duck, pp. 66-67 *in* M.D. Cadman, P.F.J. Eagles, and F.M. Helleiner, eds. Atlas of the breeding birds of Ontario. University of Waterloo Press, Waterloo, 617 pp.

Birt, V.L. and D.K. Cairns. 1987. Kleptoparasitic interactions of Arctic Skuas *Stercorarius parasiticus* and Black Guillemots *Cepphus grylle* in northeastern Hudson Bay, Canada. Ibis 129(2): 190-196.

Bishop, L.B. 1889. Notes on the birds of the Magdalen Islands. Auk 6(2): 144-150.

Bishop, P.G. Jr. 1980. The history and recent breeding of the Common Raven in New York State. M.Sc. thesis, State University of New York, Albany.

Blackburn, Y. 1981a. Résultats des observations du printemps 1980 au Saguenay. Le Harfang 4(2): 14-31.

Blackburn, Y. 1981b. Résumé des observations de l'été 1980 au Saguenay. Le Harfang 4(6): 201-207.

Black, J.M. and M. Owen. 1989. Agonistic behaviour in Barnacle Goose flocks: assessment, investment and reproductive success. Animal Behaviour 37: 199-209.

Blair, H.M.S. 1962. Studies of less familiar birds. 119. Great Gray Owl. British Birds 55: 414-18.

Blais, B. 1992. Suivi des nids de Faucons pèlerins dans le sud du Québec, été 1992. Report submitted to Ministère du Loisir, de la Chasse et de la Pêche and Canadian Wildlife Service, 68 pp.

Blais, B. 1993. Suivi des nids de Faucons pèlerins dans le sud du Québec, été 1993. Report submitted to Ministère du Loisir, de la Chasse et de la Pêche and Canadian Wildlife Service, 50 pp.

Blais, J.R. 1983. Trends in frequency, extent and severity of Spruce Budworm outbreaks in eastern Canada. Canadian Journal of Forest Research 13(4): 539-547.

Blais, J.R. and G.H. Parks. 1964. Interaction of Evening Grosbeak (*Hesperiphona vespertina*) and Spruce Budworm (*Choristoneura fumiferana*) [Clem.] in a localized bud outbreak treated with DDT in Quebec. Canadian Journal of Zoology 42(5): 1017-1024.

Blake, C.H. 1953. Notes on the Rough-winged Swallow. Bird-Banding 24(3): 107-108.

Blake, E.R. 1977. Manual of neotropical birds, vol. 1: Spheniscidae (penguins) to Laridae (gulls and allies). University of Chicago Press, Chicago, 674 pp.

Blake, E.R. and G.H. Orians. 1985. Orioles, pp. 412-413 *in* B. Campbell and E. Lack, eds. A dictionary of birds. British Ornithologists' Union, Buteo Books, Vermillion, South Dakota, 670 pp.

Blakers, M., S.J.J.F. Davies, and P.N. Reilley. 1984. The Atlas of Australian birds. Royal Australasian Ornithologists Union. Melbourne University Press, Carleton, 738 pp.

Blanchard, K.A. 1984. Seabird harvest and the importance of education in seabird management on the North Shore of the Gulf of St. Lawrence. Ph.D. dissertation, Cornell University, Ithaca, New York, 242 pp.

Blanchard, K.A. 1994. Culture and seabird conservation: the North Shore of the Gulf of St. Lawrence, Canada, pp. 294-310 *in* D.N. Nettleship, J. Burger, and M. Gochfeld, eds. Seabirds on islands: threats, case studies and action plans. Birdlife Conservation Series no. 1. Birdlife International, Cambridge, 318 pp.

Blancher, P.J. and D.K. McNicol. 1991. Tree Swallow diet in relation to wetland acidity. Canadian Journal of Zoology 69(10): 2629-2637.

Blancher, P.J. and R.J. Robertson. 1982. A double-brooded Eastern Kingbird. Wilson Bulletin 94(2): 212-213.

Blancher, P.J., C.L. Furlonger, and D.K. McNicol. 1987. Diet of nestling Tree Swallows (*Tachycineta bicolor*) near Sudbury, Ontario, summer 1986. Canadian Wildlife Service, Ottawa. Technical Report Series no. 31.

Blanchet, B. 1963. Les sciences naturelles au Camp-École Trois-Saumons. Carnets de zoologie 23(1): 4-8.

Blanchette, P. and P. Larue. 1993. Développement d'un indice de qualité de l'habitat pour la Paruline couronnée (*Seiurus aurocapillus* L.) au Québec. Gouvernement du Québec, Ministère du Loisir, de la Chasse et de la Pêche, Direction générale de la ressource faunique, Gestion intégrée des ressources, document technique 93/2, 20 pp.

Bleakney, S. 1953. Starlings in the Ungava district, Quebec province. Canadian Field-Naturalist 67: 44-45.

Bledsoe, A.H. 1988. A hybrid *Oporornis philadelphia x Geothlypis trichas* with comments on the taxonomic interpretation and evolutionary significance of intergeneric hybridization. Wilson Bulletin 100(1): 1-8.

Blem, C.R. 1972. Stomach capacity in the Common Nighthawk. Wilson Bulletin 84(4): 492-493.

Blem, C.R. 1975. Geographic variation in wing-loading of the House Sparrow. Wilson Bulletin 87(4): 543-549.

Blem, C.R. and L.B. Blem. 1991. Nest-box selection by Prothonotary Warblers. Journal of Field Ornithology 62(3): 299-307.

Blockstein, D.E. 1989. Crop milk and clutch size in Mourning Doves. Wilson Bulletin 101(1): 11-25.

Blockstein, D.E. and H.B. Tordoff. 1985. Gone forever: a contemporary look at the extinction of the Passenger Pigeon. American Birds 39(5): 845-851.

Blokpoel, H. 1976. Bird hazards to aircraft. Clarke Irwin, Toronto, 235 pp.

Blokpoel, H. 1977. Gulls and terns nesting in northern Lake Ontario and the upper St. Lawrence River. Canadian Wildlife Service. Progress Notes no. 75, 12 pp.

Blokpoel, H. 1987. Bonaparte's Gull, p. 538 *in* M.D. Cadman, P.F.J. Eagles, and F.M. Helleiner, eds. Atlas of the breeding birds of Ontario. University of Waterloo Press, Waterloo, 617 pp.

Blokpoel, H. and W.C. Scharf. 1991. Status and conservation of seabirds nesting in the Great Lakes of North America, pp. 17-41 *in* J.P. Croxall, ed. Seabird status and conservation: a supplement. International Council for Bird Preservation, Technical Publication no. 11, Cambridge, Great Britain.

Blokpoel, H. and B. Smith. 1988. First record of roof-nesting by Ring-billed Gulls and Herring Gulls in Ontario. Ontario Birds 6: 15-18.

Blokpoel, H. and G.D. Tessier. 1986. Le Goéland à bec cerclé en Ontario: une nouvelle espèce problème. Canadian Wildlife Service, Québec Region. Occasional Paper no. 57, 38 pp. [Also available in English as: The Ring-billed Gull in Ontario: a review of a new problem species.]

Blokpoel, H. and G.D. Tessier. 1991. Distribution and abundance of colonial water birds nesting in the Canadian portions of the lower Great Lakes system in 1990. Canadian Wildlife Service, Ontario Region. Technical Report Series no. 117, 16 pp.

Blomme, C. 1987a. Brown Creeper, pp. 302-303 *in* M.D. Cadman, P.F.J. Eagles, and F.M. Helleiner, eds. Atlas of the breeding birds of Ontario. University of Waterloo Press, Waterloo, 617 pp.

Blomme, C. 1987b. Common Raven, pp. 290-291 *in* M.D. Cadman, P.F.J. Eagles, and F.M. Helleiner, eds. Atlas of the breeding birds of Ontario. University of Waterloo Press, Waterloo, 617 pp.

Bloom, P.H., M.D. McCrary, and M.J. Gibson. 1993. Red-shouldered Hawk home-range and habitat use in southern California. Journal of Wildlife Management 57(2): 258-265.

Bluhm, C.K. 1992. Chapter 10. Environmental and endocrine control of waterfowl reproduction, pp. 323-360 *in* B.D.J. Batt, A.D. Afton, M.G. Anderson, C.D. Ankney, D.H. Johnson, J.A. Kadlec, and G.L. Krapu, eds. Ecology and management of breeding waterfowl. University of Minnesota Press, Minneapolis, 635 pp.

Blume, D. 1985. Woodpeckers, pp. 297-301 *in* C.M. Perrins and A.L.A. Middleton, eds. The encyclopedia of birds. Facts on File, New York, 447 pp.

Boag, D.A. and M.A. Schroeder. 1992. Spruce Grouse (*Dendragapus canadensis*), *in* A. Poole, P. Stettenheim, and F. Gill, eds. The birds of North America, no. 5. The Academy of Natural Sciences, Philadelphia. The American Ornithologists' Union, Washington, 28 pp.

Bock, C.E. and J.H. Bock. 1974. On the geographical ecology and evolution of the Three-toed Woodpeckers, *Picoides tridactylus* and *P. arcticus*. American Midland Naturalist 92(2): 397-405.

Bock, C.E. and L.W. Lepthien. 1972. Winter eruptions of Red-breasted Nuthatches in North America 1950-1970. American Birds 26(3): 558-561.

Bock, C.E. and L.W. Lepthien. 1975. A Christmas count analysis of woodpecker abundance in the United States. Wilson Bulletin 87(3): 355-366.

Bock, C.E. and L.W. Lepthien. 1976a. Changing winter distribution and abundance of the Blue Jay, 1962-1971. American Midland Naturalist 96(1): 232-236.

Bock, C.E. and L.W. Lepthien. 1976b. Synchronous eruptions of boreal seed-eating birds. American Naturalist 110(974): 559-571.

Bock, C.E. and R.B. Smith. 1971. An analysis of Colorado Christmas counts. American Birds 25(6): 945-947.

Boersma, D.C. 1990. Prise d'oiseaux migrateurs au Canada, pendant la saison de chasse de 1988. Environment Canada, Canadian Wildlife Service. Progress Notes no. 190, 42 pp. [Also available in English as: Migratory game birds harvested in Canada during the 1988 hunting season.]

Bohm, R.T. 1985. Use of artificial nests by Great Gray Owls, Great Horned Owls, and Red-tailed Hawks in northeastern Minnesota. Loon 57(4): 150-152.

Bohm, R.T. 1988. Use of nestboxes by Great Gray Owls in northern Minnesota. Loon 60(3): 121.

Boileau, G. and M. Dumont. 1979. L'Abitibi-Témiscamingue. Collection Études et Dossiers, Série Études régionales, Ministère des Communications. Éditeur officiel du Québec, Québec, 232 pp.

Boily, P. 1978. Liste annotée d'espèces d'oiseaux observées dans la région méridionale des Appalaches du Québec (Cantons de l'Est et Beauce). Preliminary report on the regional avifauna. Le Musée du Séminaire de Sherbrooke, 34 pp.

Boily, P. 1983. Distribution et abondance relative des anatidés et autres oiseaux aquatiques dans l'Estrie. Cahiers d'ornithologie Victor-Gaboriault no. 5. Canadian Wildlife Service, Québec, 91 pp.

Boily, P., P. Fragnier, V. Létourneau, F. Shaffer, and J. Shaffer. 1988. Les oiseaux en Estrie: feuillet d'abondance saisonnière. Société de loisir ornithologique de l'Estrie, Sherbrooke, 2 pp.

Bois, D. 1986. Espèces marines de l'archipel des Sept-Îles. Comité touristique, Corporation de promotion industrielle commerciale et touristique de Sept-Îles, Sept-Îles, 70 pp.

Boisclair, J. 1980. Les oiseaux familiers du Québec. Stanké, Montréal, 183 pp.

Bollinger, E.K. and T.A. Gavin. 1989. The effects of site quality on breeding-site fidelity in Bobolinks. Auk 106(4): 584-594.

Bollinger, E.K. and T.A. Gavin. 1991. Patterns of extra-pair fertilizations in Bobolinks. Behavioral Ecology and Sociobiology 29: 1-7.

Bollinger, E.K. and T.A. Gavin. 1992. Eastern Bobolink populations: ecology and conservation in an agricultural landscape, pp. 497-506 *in* J.M. Hagan and D.W. Johnston, III, eds. Ecology and conservation of neotropical migrant landbirds. Smithsonian Institution Press, Washington, 609 pp.

Bollinger, E.K., T.A. Gavin, C.J. Hibbard, and J.T. Wootton. 1986. Two male Bobolinks feed young at the same nest. Wilson Bulletin 98(1): 154-156.

Bollinger, E.K., P.B. Bollinger, and T.A. Gavin. 1990. Effects of hay-cropping on eastern populations of the Bobolink. Wildlife Society Bulletin 18(2): 142-150.

Bond, J. 1938. Nesting of the White-winged Crossbill. Canadian Field-Naturalist 52(1): 3-5.

Bond, W.J. 1993. Keystone species, pp. 237-253 *in* E.D. Schulze and H.A. Mooney, eds. Biodiversity and ecosystem function. Springer-Verlag, New York.

Bondrup-Nielsen, S. 1976. First Boreal Owl nest for Ontario, with notes on development of the young. Canadian Field-Naturalist 90(4): 477-479.

Bondrup-Nielsen, S. 1978. Vocalizations, nesting and habitat preferences of the Boreal Owl (*Aegolius funereus*) in North America. M.Sc. thesis, University of Toronto, Ontario, 158 pp.

Bondrup-Nielsen, S. 1984. Vocalizations of the Boreal Owl, *Aegolius funereus richardsoni*, in North America. Canadian Field-Naturalist 98(2): 191-197.

Bonn, F. and G. Rochon. 1992. Précis de télédétection, vol. 1, Principes et méthodes. Presses de l'Université du Québec, Québec, 485 pp.

Bonney, R.E. Jr. 1988a. Warbling Vireo, pp. 348-349 *in* R.F. Andrle and J.R. Carroll, eds. The Atlas of breeding birds in New York State. Cornell University Press, Ithaca, 551 pp.

Bonney, R.E. Jr. 1988b. Black-throated Blue Warbler, pp. 376-377 *in* R.F. Andrle and J.R. Carroll, eds. The Atlas of breeding birds in New York State. Cornell University Press, Ithaca, 551 pp.

Bonney, R.E. Jr. 1988c. European Starling, pp. 340-341 *in* R.F. Andrle and J.R. Carroll, eds. The Atlas of breeding birds in New York State. Cornell University Press, Ithaca, 551 pp.

Bonney, R.E. Jr. 1988d. House Wren, pp. 302-303 *in* R.F. Andrle and J.R. Carroll, eds. The Atlas of breeding birds in New York State. Cornell University Press, Ithaca, 551 pp.

Bonney, R.E. Jr. 1988e. Prairie Warbler, pp. 388-389 *in* R.F. Andrle and J.R. Carroll. The Atlas of breeding birds in New York State. Cornell University Press, Ithaca, 551 pp.

Bonney, R.E. Jr. 1988f. Rose-breasted Grosbeak, pp. 430-431 *in* R.F. Andrle and J.R. Carroll, eds. The Atlas of breeding birds in New York State. Cornell University Press, Ithaca, 551 pp.

Bonney, R.E. Jr. 1988g. White-breasted Nuthatch, pp. 296-297 *in* R.F. Andrle and J.R. Carroll, eds. The Atlas of breeding birds in New York State. Cornell University Press, Ithaca, 551 pp.

Bonney, R.E. Jr. 1988h. Yellow Warbler, pp. 368-369 *in* R.F. Andrle and J.R. Carroll, eds. The Atlas of breeding birds in New York State. Cornell University Press, Ithaca, 551 pp.

Bonney, R.E. Jr. 1988i. American Crow, pp. 282-283 *in* R.F. Andrle and J.R. Carroll, eds. The Atlas of breeding birds in New York State. Cornell University Press, Ithaca, 551 pp.

Bonney, R.E. Jr. and J.L. Burrill. 1988. Wood Thrush, pp. 326-327 *in* R.F. Andrle and J.R. Carroll, eds. The Atlas of breeding birds in New York State. Cornell University Press, Ithaca, 551 pp.

Bookhout, T.A. 1995. Yellow Rail (*Coturnicops noveboracensis*), *in* A. Poole and F. Gill, eds. The birds of North America, no. 139. The Academy of Natural Sciences, Philadelphia. The American Ornithologists' Union, Washington, 16 pp.

Bookhout, T.A. and J.R. Stenzel. 1987. Habitat and movements of breeding Yellow Rails. Wilson Bulletin 99(3): 441-447.

Booth, C.J. 1982. Fledging success of some Red-throated Divers in Orkney. Scottish Birds 12(2): 33-38.

Bordage, D. 1984. Distribution et abondance des Anatidés le long des rivières Eastmain, Opinaca et Petite rivière Opinaca en aval des ouvrages de dérivation. Report for Société d'Énergie de la Baie James, Canadian Wildlife Service, Québec Region, Progress Report no. 3, 1983, 46 pp.

Bordage, D. 1985a. Bilan de surveillance des populations de sauvagine des régions sud-ouest et est du Nouveau-Québec (1981-1984). Report for Société d'Énergie de la Baie James, Canadian Wildlife Service, Québec Region, 92 pp.

Bordage, D. 1985b. Distribution et abondance des Anatidés dans la région nord-ouest du Nouveau-Québec. Report for Société d'Énergie de la Baie James, Canadian Wildlife Service, Québec Region, Progress Report no. 4, 1984, 43 pp.

Bordage, D. 1987. Suivi des couples nicheurs de Canard noir en forêt boréale – 1985. Canadian Wildlife Service, Québec Region. Technical Report Series no. 18, 29 pp. [An abstract is available in English.]

Bordage, D. 1988a. Suivi des couples nicheurs de Canard noir en forêt boréale – 1986. Canadian Wildlife Service, Québec Region. Technical Report Series no. 36, 47 pp. [An abstract is available in English.]

Bordage, D. 1988b. Suivi des couples nicheurs de Canard noir en forêt boréale – 1987. Canadian Wildlife Service, Québec Region. Technical Report Series no. 35, 42 pp. [An abstract is available in English.]

Bordage, D. 1991. Utilisation par les oiseaux migrateurs de la région du complexe Grande-Baleine. Programme d'étude conjoint, Canadian Wildlife Service, Québec Region and Hydro-Québec, 1990-1992, unpublished data.

Bordage, D. and Y. Aubry. 1982. Inventaire de la distribution et de l'abondance des oiseaux aquatiques dans les régions sud-ouest et est du Nouveau-Québec. Environment Canada, Canadian Wildlife Service, Progress Report no. 1, 1981, 93 pp.

Bordage, D. and Y. Aubry. 1983. Distribution et abondance des Anatidés dans les régions sud-ouest et est du Nouveau-Québec, région du Québec. Report for Société d'Énergie de la Baie James, Environment Canada, Canadian Wildlife Service, Progress Report no. 2, 1982, 64 pp.

Bordage, D. and J.-P.L. Savard. 1995. Black Scoter (*Melanitta nigra*), *in* A. Poole and F. Gill, eds. The birds of North America, no. 177. The Academy of Natural Sciences, Philadelphia. The American Ornithologists' Union, Washington, 20 pp.

Bordage, D., N. Plante, S. Paradis, and A. Bourget. (In preparation). Estimating the size of Common Eider populations in winter using ratio estimates. Environment Canada, Canadian Wildlife Service, Québec Region and Parks Canada, 24 pp.

Bordeleau, C. 1988. Programme de surveillance en forêt naturelle: Tordeuse des bourgeons de l'épinette, pp. 3-7 *in* Insectes et maladies des arbres, Québec 1987. Supplément de Forêt Conservation 54(10): 1-32.

Bordeleau, C. and M. Auger. 1991. Programme de surveillance en forêt naturelle: Tordeuse des bourgeons de l'épinette, pp. 3-5 *in* Insectes et maladies des arbres, Québec 1990. Ministère des Forêts du Québec, 34 pp.

Bordeleau, C. and M. Auger. 1992. Programme de surveillance en forêt naturelle: Tordeuse des bourgeons de l'épinette, pp. 3-5 *in* Insectes et maladies des arbres, Québec 1991. Ministère des Forêts, 38 pp.

Bordeleau, C. and B. Boulet. 1989. Programme de surveillance en forêt naturelle: Tordeuse des bourgeons de l'épinette, pp. 3-11 *in* Insectes et maladies des arbres, Québec 1988. Supplément de Forêt Conservation 56(1): 1-32.

Bordeleau, C. and B. Boulet. 1990. Programme de surveillance en forêt naturelle: Tordeuse des bourgeons de l'épinette. pp. 3-8 *in* Insectes et maladies des arbres, Québec 1989. Supplément de Forêt Conservation 57(1): 1-40.

Borror, D.J. 1965. Song variation in Maine Song Sparrows. Wilson Bulletin 77(1): 5-37.

Borror, D.J. 1977. Rufous-sided Towhees mimicking Carolina Wren and Field Sparrow. Wilson Bulletin 89(3): 477-480.

Borror, D.J. and R.E. White. 1991. Les insectes de l'Amérique du Nord (au nord du Mexique). Les Guides Peterson, Éditions Broquet, Laprairie, 408 pp.

Borror, D.J., C.A. Triplehorn, and N.F. Johnson. 1989. An introduction to the study of insects, 6th edition. Harcourt Brace College Publishers.

Bortolotti, G.R. 1987. Bald Eagle, pp. 110-111 *in* M.D. Cadman, P.F.J. Eagles, and F.M. Helleiner, eds. Atlas of the breeding birds of Ontario. University of Waterloo Press, Waterloo, 617 pp.

Bosakowski, T. 1986. Winter population trends of the House Finch and ecologically similar species in northeastern New Jersey. American Birds 40(4): 1105-1110.

Bosakowski, T., R. Speiser, and J. Benzinger. 1987. Distribution, density, and habitat relationships of the Barred Owl in northern New Jersey, pp. 135-143 *in* R.W. Nero, R.J. Clark, R.J. Knapton, and R.H. Hamre, eds. Biology and conservation of northern forest owls: symposium proceedings. 3-7 February 1987, Winnipeg, Manitoba. United States Department of Agriculture, Forest Service, Rocky Mountain Forest and Range Experiment Station, Fort Collins, Colorado. General Technical Report RM-142, 309 pp.

Bosakowski, T., R. Kane, and D.G. Smith. 1989a. Decline of the Long-eared Owl in New Jersey. Wilson Bulletin 101(3): 481-485.

Bosakowski, T., R. Speiser, and D.G. Smith. 1989b. Nesting ecology of forest dwelling Great Horned Owls, *Bubo virginianus*, in the eastern deciduous forest biome. Canadian Field-Naturalist 103(1): 65-69.

Boström, U. and S. Nilsson. 1983. Latitudinal gradients and local variations in species richness and structure of bird communities on raised peat-bogs in Sweden. Ornis Scandinavica 14: 213-226.

Bouchard, A., S. Dyrda, Y. Bergeron, and A. Meilleur. 1989. The use of notary deeds to estimate the changes in the composition of 19th century forests, in Haut-Saint-Laurent, Québec. Canadian Journal of Forest Research 19(9): 1146-1150.

Bouchard, H. and P. Millet. 1993. Le Saint-Laurent: milieux de vie diversifiés. Environment Canada, Conservation and Protection, Québec Region, St. Lawrence Centre. St. Lawrence UPDATE collection, thematic report on the state of the St. Lawrence River. Montréal, 97 p. [Also available in English as: The St. Lawrence River: Diversified Environments.]

Boucher, P. 1664. Histoire véritable et naturelle des moeurs et productions du pays de la Nouvelle France, vulgairement dite le Canada. Paris: chez Florentin Lambert, rue Saint-Jacques, vis-à-vis Saint-Yves, à l'image Saint-Paul [1964, Garneau, Laval], 168 pp.

Boucher, P.R. 1992. Les milieux naturels protégés au Québec. Transcript of the conference held on the occasion of the 21st Congress of the Fédération canadienne de la nature in Québec on 14 August 1992, 46 pp.

Boucher, R. and R. Lalonde. 1980. Inventaires des faunes terrestre et avienne dans la vallée Desaulniers en 1976, 1977 et 1978. Société d'Énergie de la Baie James, direction Environnement, Service de l'écologie.

Bouffard, S.H. 1983. Redhead egg parasitism of Canvasback nests. Journal of Wildlife Management 47(1): 214-216.

Boulva, J. and P. Browne. 1964. Le Pinson de Le Conte, *Passerherbulus caudacutus* (Latham), au Saguenay. Naturaliste canadien 91(5): 159-164.

Bourdages, J.-L., C. Gauvin, M. Robert, G. Domon, A. Bouchard, and P. Drapeau. 1988. Étude des ressources et des potentiels du Parc régional du Cap-Saint-Jacques. Report for Communauté Urbaine de Montréal by the Centre de recherches écologiques de Montréal, 227 pp.

Bourdages, J.-L., G. Domon, L. Cloutier, P. Drapeau, and A. Bouchard. 1990. Caractéristiques écologiques et potentiels du Parc régional de l'Anse-à-l'Orme. Report for Communauté Urbaine de Montréal by the Centre de recherches écologiques de Montréal, 235 pp.

Bourgeois, A. 1977. Quantitative analysis of American Woodcock nest and brood habitat. Proceedings. Woodcock Symposium 6: 109-118.

Bourgeois, J.-C. 1994. La halte migratoire du lac Saint-Pierre: un habitat d'importance internationale pour la sauvagine. QuébecOiseaux 5(3): 18-22.

Bourgeois, J.-C. and R. Couture. 1979. Contribution à l'étude des problèmes relatifs à l'interprétation des recensements de populations de Bécasses d'Amérique (*Philohela minor*). Ministère du Loisir, de la Chasse et de la Pêche, Direction générale de la faune, Québec, 65 pp.

Bourgeois, J.-C., J. Bourgeois, D. Lehoux, and M. Darveau. 1983. Bilan d'activité diurne de la sauvagine et sélection des types de culture pour son alimentation lors de la halte migratoire printanière dans le secteur Nicolet–Longue Pointe, lac Saint-Pierre. Ministère du Loisir, de la Chasse et de la Pêche, Direction générale de Trois-Rivières and Environment Canada, Canadian Wildlife Service, Québec Region, 99 pp.

Bourgeois, J.-C., D. Dolan, and L. Houde. 1992. Rapport d'inventaire de couvées de canards barboteurs au lac Saint-Pierre en 1984 et 1985. Ministère du Loisir, de la Chasse et de la Pêche, Direction régionale Mauricie–Bois-Francs, Service de l'aménagement et de l'exploitation de la faune, Trois-Rivières, Programme informatique [Micro-diskette, 90 mm].

Bourgeois, J.-C., D. Dolan, and L. Houde. 1994. Dénombrement de la sauvagine sur le fleuve Saint-Laurent entre le lac Saint-Pierre et Grondines (1982-1983). Ministère du Loisir, de la Chasse et de la Pêche, Direction régionale Mauricie–Bois-Francs, Service de l'aménagement et de l'exploitation de la faune, Trois-Rivières, Programme informatique [Micro-diskette, 90 mm].

Bourget, A. 1973a. Relation of eiders and gulls nesting in mixed colonies in Penobscot Bay, Maine. Auk 90(4): 809-820.

Bourget, A. 1973b. Études à la baie James, Québec, 1972. Unpublished report, Canadian Wildlife Service, Québec Region, 92 pp.

Bourget, A. 1990. Programme d'inventaire de limicoles, rapport sommaire 1989. Unpublished report, Canadian Wildlife Service, Québec Region, 17 pp.

Bourget, A. 1991. Programme d'inventaire de limicoles: rapport sommaire 1990. Unpublished report, Canadian Wildlife Service, Québec Region, 18 pp.

Bourget, A., P. Dupuis, and W.R. Whitman. 1986. Les eiders hivernant dans le golfe du Saint-Laurent: effectifs et distribution, pp. 94-99 *in* A. Reed, ed. Eider ducks in Canada/Les eiders au Canada. Canadian Wildlife Service, Ottawa. Report Series no. 47, 177 pp.

Bourget, E. 1976. Inventaire des ressources, littoral de la rive sud de l'estuaire maritime du Saint-Laurent. Institut national de recherche scientifique – Océanologie, Laboratoire océanographique de Rimouski. Mimeographic report, 42 pp.

Bourne, W.R.P. 1985. Petrel, pp. 451-456 *in* B. Campbell and E. Lack, eds. A dictionary of birds. British Ornithologists' Union, Buteo Books, Vermillion, South Dakota, 670 pp.

Bousfield, M.A., I.R. Kirkham, and R.D. McRae. 1986. Breeding of Wilson's Phalarope, *Phalaropus tricolor*, at Churchill, Manitoba. Canadian Field-Naturalist 100(3): 392-393.

Bouvier, J.M. 1974a. Breeding biology of the Hooded Merganser in southwestern Quebec, including interactions with Common Goldeneyes and Wood Ducks. Canadian Field-Naturalist 88(3): 323-330.

Bouvier, J.M. 1974b. Avifauna of the CIP Nature Centre, Harrington, Quebec. Unpublished report, 63 pp.

Bowdish, B.S. and P.B. Philipp. 1916. The Tennessee Warbler in New Brunswick. Auk 33(1): 1-8.

Bowman, R. and D.M. Bird. 1985. Reproductive performance of American Kestrels laying replacement clutches. Canadian Journal of Zoology 63: 2590-2593.

Bowman, R. and D.M. Bird. 1986. Ecological correlates of mate replacement in American Kestrels. Condor 88(4): 440-445.

Boxall, P.C. 1981. Ruby-crowned Kinglets (*Regulus calendula*) feeding a Brown-headed Cowbird (*Molothrus ater*). Canadian Field-Naturalist 95(1): 99-100.

Boxall, P.C. 1983. Observations suggesting parental division of labor by American Redstarts. Wilson Bulletin 95(4): 673-674.

Boxall, P.C. and M.R. Lein. 1982. Territoriality and habitat selection of female Snowy Owls (*Nyctea scandiaca*) in winter. Canadian Journal of Zoology 60(10): 2344-2350.

Boyajian, N.R. 1970. The Nesting season – Hudson–St. Lawrence region. Audubon Field Notes 24(5): 665-668.

Boyce, M.S. 1980. Possible competitive displacement between Common Nighthawks and bats in southwestern Oregon. Murrelet 61(1): 35.

Boyd, E.M. 1962. A half-century's changes in the birdlife around Springfield, Massachusetts. Bird-Banding 33(3): 137-148.

Boyd, H. 1979. Management of *Branta bernicla* in Canada, pp. 181-199 *in* Proceedings. First technical meeting on western Palearctic migration. Bird Management, International Waterfowl Research Bureau.

Boyd, H. and L.S. Maltby. 1979. The Brant of the Western Queen Elizabeth Islands, Northwest Territories, pp. 5-21 *in* R.L. Jarvis and J.C. Bartonek, eds. Management and biology of Pacific Flyway geese, a symposium. Oregon State University Press, Book Stores, Corvallis, Oregon.

Boyd, H., G.E.J. Smith, and F.G. Cooch. 1982. The Lesser Snow Geese of the eastern Canadian Arctic: their status during 1974-79 and their management from 1981 to 1990. Canadian Wildlife Service. Occasional Paper no. 46, 25 pp.

Boyd, H., L.S. Maltby, and A. Reed. 1988. Variations du plumage des Bernaches cravants nichant dans l'Arctique canadien. Canadian Wildlife Service. Progress Notes no. 174, 9 pp. [Also available in English as: Differences in the plumage patterns of Brant breeding in High Arctic Canada.]

Braaten, D.J. 1975. Observations at three Brown Creeper nests in Itasca State Park in 1973. Loon 47: 110-113.

Brackbill, H. 1947. Period of dependency in the American Robin. Wilson Bulletin 59(2): 114-115.

Brackbill, H. 1958. Nesting behavior of the Wood Thrush. Wilson Bulletin 70(1): 70-89.

Brackbill, H. 1970. Tufted Titmouse breeding behaviour. Auk 87(3): 522-536.

Brackney, A.W. and T.A. Bookhout. 1982. Population ecology of Common Gallinules in southwestern Lake Erie marshes. Ohio Journal of Science 82(5): 229-237.

Bradstreet, M.S.W. 1980. Thick-billed Murres and Black Guillemots in the Barrow Strait area, Northwest Territories, during spring: diets and food availability along ice edges. Canadian Journal of Zoology 58(11): 2120-2140.

Bradstreet, M.S.W. and R.G.B. Brown. 1985. Feeding ecology of the Atlantic Alcidae, pp. 263-313 *in* D.N. Nettleship and T.R. Birkhead, eds. The Atlantic Alcidae: the evolution, distribution and biology of the auks inhabiting the Atlantic Ocean and adjacent water areas. Academic Press, London, 574 pp.

Brand, A.R. 1938. Vibration frequencies of passerine bird song. Auk 55: 263-268

Brassard, L. 1959. Pour vos rapports quotidiens. Le Jeune Naturaliste 9(6): 143.

Brassard, L. 1961. Grottes d'intérêt historique sur le fjord Saguenay. Saguenayensia Mai-juin-juillet-août: 51-75.

Brassard, L. 1962. Le Camp des jeunes Explorateurs. Carnets de zoologie 22(3): 36-39.

Braun, B.M., P.A. Heinz, and G.H. Heinz. 1980. Herring Gull predation on Red-breasted Merganser ducklings. Wilson Bulletin 92(3): 403.

Braun, C.E., J.H. Enderson, C.J. Henny, H. Meng, and A.G. Nye, Jr. 1977. Falconry: effects on raptor populations and management in North America. Wilson Bulletin 89(2): 360-369.

Braund, F.W. and E.P. McCullagh. 1940. The birds of Anticosti Island, Quebec. Wilson Bulletin 52(2): 96-123.

Braune, B.M. 1987a. Seasonal aspects of the diet of Bonaparte's Gull (*Larus philadelphia*) in the Quoddy region, New Brunswick, Canada. Auk 104(2): 167-172.

Braune, B.M. 1987b. Body morphometrics and molt of Bonaparte's Gulls in the Quoddy region, New Brunswick, Canada. Condor 89(1): 150-157.

Braune, B.M. 1989. Autumn migration and comments on the breeding range of Bonaparte's Gull, *Larus philadelphia*, in eastern North America. Canadian Field-Naturalist 103(4): 524-530.

Braune, B.M. and D.E. Gaskin. 1982a. Feeding methods and diving rates of migrating larids off Deer Island, New Brunswick. Canadian Journal of Zoology 60(9): 2190-2197.

Braune, B.M. and D.E. Gaskin. 1982b. Feeding ecology of nonbreeding populations of larids off Deer Island, New Brunswick. Auk 99(1): 67-76.

Braunfield, R.A. 1993. 66. Upland oak and hemlock forest, p. 74 *in* J.D. Lowe, ed. Resident bird counts 1992. Journal of Field Ornithology 64(1)(supplement): 1-128.

Brauning, D.W. 1992a. Brown-headed Cowbird, pp. 406-407 *in* D.W. Brauning, ed. Atlas of breeding birds in Pennsylvania. University of Pittsburgh Press, Pittsburgh, 484 pp.

Brauning, D.W. 1992b. Common Nighthawk, pp. 168-169 *in* D.W. Brauning, ed. Atlas of breeding birds in Pennsylvania. University of Pittsburgh Press, Pittsburgh, 484 pp.

Brauning, D.W. 1992c. Solitary Vireo, pp. 290-291 *in* D.W. Brauning, ed. Atlas of breeding birds in Pennsylvania. University of Pittsburgh Press, Pittsburgh, 484 pp.

Brauning, D.W. 1992d. Spotted Sandpiper, pp. 136-137 *in* D.W. Brauning, ed. Atlas of breeding birds in Pennsylvania. University of Pittsburgh Press, Pittsburgh, 484 pp.

Breitwisch, R. and M. Breitwisch. 1991. House Sparrows open an automatic door. Wilson Bulletin 103(4): 725-726.

Breitwisch, R., P.G. Merritt, and G.H. Whitesides. 1986. Parental investment by the Northern Mockingbird: male and female roles in feeding nestlings. Auk 103(1): 152-159.

Breton, N. 1980. Calendrier des migrations des oiseaux du Saguenay–Lac-Saint-Jean, 2nd edition. Club des ornithologues amateurs du Saguenay–Lac-Saint-Jean, Arvida, 54 pp.

Brewer, R. 1991. Black Tern, pp. 226-227 *in* R. Brewer, G.A. McPeek, and R.J. Adams, Jr., eds. The Atlas of breeding birds of Michigan. Michigan State University Press, East Lansing, 594 pp.

Brewer, R., G.A. McPeek, and R.J. Adams, Jr. 1991. The Atlas of breeding birds of Michigan. Michigan State University Press, East Lansing, 594 pp.

Brewster, W. 1884. Notes on the birds observed during a summer cruise in the Gulf of St. Lawrence. Proceedings. Boston Society of Natural History 22: 364-412.

Briggs, S.A. 1993. 1. Mixed upland broadleaf forest, p. 7 *in* J.D. Lowe, ed. Resident bird counts 1992. *Journal of Field Ornithology* 64(1)(supplement): 1-128.

Brigham, R.M. 1989. Roost and nest sites of Common Nighthawks: are gravel roofs important? Condor 91(4): 722-724.

Brigham, R.M. 1990. Prey selection by Big Brown Bats (*Eptesicus fuscus*) and Common Nighthawks (*Chordeiles minor*). American Midland Naturalist 124: 73-80.

Brigham, R.M. 1991. Foraging and the use of torpor by free-ranging goatsuckers. No. 230. Abstracts of scientific papers presented at the 109th stated meeting of the AOU, held at McGill University, Montréal. Institute of Wetland and Waterfowl Research, Ducks Unlimited.

Brigham, R.M. and M.B. Fenton. 1991. Convergence in foraging strategies by two morphologically and philogenetically distinct nocturnal aerial insectivores. Journal of Zoology 223(3): 475-489.

Briskie, J.V. 1994. Least Flycatcher (*Empidonax minimus*), *in* A. Poole and F. Gill, eds. The birds of North America, no. 99. The Academy of Natural Sciences, Philadelphia. The American Ornithologists' Union, Washington, 24 pp.

Briskie, J.V. and S.G. Sealy. 1987. Polygyny and double-brooding in the Least Flycatcher. Wilson Bulletin 99(3): 492-494.

Briskie, J.V. and S.G. Sealy. 1989a. Changes in nest defense against a brood parasite over the breeding cycle. Ethology 82: 61-67.

Briskie, J.V. and S.G. Sealy. 1989b. Nest-failure and the evolution of hatching asynchrony in the Least Flycatcher. Journal of Animal Ecology 58(2): 653-665.

Brisson, J., Y. Bergeron, and A. Bouchard. 1988. Les successions secondaires sur sites mésiques dans le Haut-Saint-Laurent. Canadian Journal of Botany 66: 1192-1203.

Brisson, J., Y. Bergeron, and A. Bouchard. 1992. The history and tree stratum of an old-growth forest of Haut-Saint-Laurent region, Quebec. Natural Areas Journal 12(1): 3-9.

Brisson, J., Y. Bergeron, A. Bouchard, and A. Leduc. 1994. Beech-maple dynamics in an old-growth forest in southern Québec, Canada. Écoscience 1(1): 40-46.

Brisson, M.J. 1760. Ornithologie ou Méthode contenant la division des oiseaux en ordres, sections, genres, espèces et leurs variétés, à laquelle on a joint une description exacte de chaque espèce, avec les citations des auteurs qui en ont traité, les noms qu'ils leur ont donnés, ceux que leur ont donné les différentes nations et les noms vulgaires, 6 vol. Paris, quai des Augustins chez Cl. Jean-Baptiste Bauche, libraire, à l'Image Sainte Geneviève et S. Jean dans le désert.

Brittingham, M.C. and S.A. Temple. 1983. Have cowbirds caused forest songbirds to decline? BioScience 33(1): 31-35.

Brittingham, M.C. and S.A. Temple. 1988. Impacts of supplemental feeding on survival rates of Black-capped Chickadees. Ecology 69: 581-589.

Brochu, M. 1963. Observations ornithologiques à Maricourt, Nouveau-Québec, du 7 mai au 2 août 1963. Bulletin ornithologique 8(4): 7-8.

Brodeur, S. and F. Morneau. 1991. Statut de l'Aigle royal (*Aquila chrysaetos canadensis*) au Québec. Preliminary report. Direction générale de la ressource faunique, Ministère du Loisir, de la Chasse et de la Pêche, Québec, 36 pp.

Brodeur, S., F. Morneau, R. Décarie, J.J. Negro, and D.M. Bird. 1994. Breeding density and brood size of Rough-legged Hawks in northwestern Québec. Journal of Raptor Research 28(4): 259-262.

Brodeur, S., F. Morneau, R. Décarie, J.-L. DesGranges, and J.[J.] Negro. 1995. Southern extension to the breeding range of the Gyrfalcon, *Falco rusticolus*, in eastern North America. Arctic 48(1): 94-95.

Brodsky, L.M. 1988. Mating tactics of male Rock Ptarmigan (*Lagopus mutus*): a conditional mating strategy. Animal Behaviour 36: 335-342.

Broley, C.L. 1947. Migration and nesting of Florida Bald Eagles. Wilson Bulletin 59: 3-20.

Bromley, M. 1991. Update report on the status of Tundra Peregrine Falcons (*Falco peregrinus tundrius*) in Canada. COSEWIC, D.M. Albright and D.W. Doidge, 1985. Sighting of nesting Peregrine Falcons (*Falco peregrinus*) in eastern Hudson Bay. Unpublished report, MacDonald College of McGill University, Sainte-Anne-de-Bellevue, 4 pp.

Bromley, R.G. 1983. 1982 Raptor survey. Government of Northwest Territories, Department of Renewable Resources, File Report no. 35, 47 pp.

Bromley, R.G. 1987. Update report on the status of the Gyrfalcon (*Falco rusticolus*) in Canada. Committee on the Status of Endangered Wildlife in Canada (COSEWIC), Ottawa, 21 pp.

Bromley, R.G. and B. McLean. 1986. Raptor surveys in the Kitikmeot and Baffin regions, Northwest Territories, 1983 and 1984. Government of Northwest Territories, Department of Renewable Resources, File Report no. 65, 63 pp.

Brooks, A. 1941. The loon as a duck killer. Condor 43: 197.

Brooks, E.W. 1980. Interspecific nesting of Clay-colored and Field sparrows. Wilson Bulletin 92(2): 264-265.

Broun, M. 1979. Golden-winged Warbler, pp. 42-43 *in* L. Griscom and A. Sprunt, Jr., eds. The warblers of America. Doubleday and Company, Garden City, New York, 302 pp.

Brousseau, P. 1981. Distribution et abondance des oiseaux de rivage le long du Saint-Laurent, section Cornwall–La Pocatière. Environment Canada, Canadian Wildlife Service, Québec Region, 132 pp.

Brousseau, P. and G. Chapdelaine. 1983. Techniques d'inventaires détaillées utilisées dans le cadre du douzième inventaire des oiseaux marins des refuges de la Côte-Nord du golfe du Saint-Laurent. Canadian Wildlife Service, Québec Region. Unpublished report, 152 pp.

Brousseau, P. and G. Chapdelaine. 1987. Les oiseaux marins de l'archipel des Sept-Îles (golfe du Saint-Laurent, Québec). Naturaliste canadien 114: 177-186.

Brousseau, P. and G. Chapdelaine. 1989. Inventaire des colonies d'oiseaux marins de la région de la baie des Chaleurs et de la Gaspésie. Canadian Wildlife Service, Québec Region. Unpublished report, 47 pp.

Brousseau, P. and G. Chapdelaine. 1990. Treizième inventaire des oiseaux marins dans les refuges de la Côte-Nord: techniques et résultats détaillés. Canadian Wildlife Service, Québec Region. Technical Report Series no. 96, 159 pp. [An abstract is available in English.]

Brousseau, P., J. Lefebvre, and J.-F. Giroux. 1996. Diet of Ring-billed Gull chicks in urban and non-urban colonies in Quebec. Colonial Waterbirds 19(1): 22-30.

Brower, A.E. 1969. Birds observed in the Schefferville area of Labrador, Québec: June 25 to August 20, 1968. Field research in Labrador-Ungava. McGill Sub-arctic Research Paper 24: 183-184.

Brown, B.A., J.O. Whitaker, T.W. French, and C. Maser. 1986. Note on food habits of the Screech Owl and the Burrowing Owl of southeastern Oregon. Great Basin Naturalist 46(3): 421-426.

Brown, C.R. 1975. Polygamy in the Purple Martin. Auk 92(3): 602-604.

Brown, C.R. 1978a. On early spring arrival of Purple Martin. Bird-Banding 49(2): 130-133.

Brown, C.R. 1978b. Post-fledging behavior of Purple Martin. Wilson Bulletin 90(3): 376-385.

Brown, C.R. 1979. Territoriality in the Purple Martin. Wilson Bulletin 91(4): 583-591.

Brown, C.R. 1985. The costs and benefits of coloniality in the Cliff Swallow. Ph.D. dissertation, Princeton University, Princeton, New Jersey.

Brown, C.R. 1988. Social foraging in Cliff Swallows: local enhancement, risk sensitivity, competition and the avoidance of predators. Animal behaviour 36(3): 780-792.

Brown, C.R. 1991. Food-sharing signals among socially foraging Cliff Swallows. Animal Behaviour 42(4): 551-564.

Brown, C.R. and E.J. Bitterbaum. 1980. Implications of juvenile harassment in Purple Martins. Wilson Bulletin 92(4): 452-457.

Brown, C.R. and M.B. Brown. 1988. The costs and benefits of egg destruction by conspecifics in colonial Cliff Swallows. Auk 105(4): 737-748.

Brown, C.R. and M.B. Brown. 1989. Behavioural dynamics of intraspecific brood parasitism in colonial Cliff Swallows. Animal Behaviour 37(5): 777-796.

Brown, C.R. and M.B. Brown. 1995. Cliff Swallow (*Hirundo pyrrhonota*), *in* A. Poole and F. Gill, eds. The birds of North America, no. 149. The Academy of Natural Sciences, Philadelphia. The American Ornithologists' Union, Washington, 32 pp.

Brown, J.L. 1969. Territorial behavior and population regulation in birds. Wilson Bulletin 81(3): 293-329.

Brown, L.H. 1976. Birds of prey: their biology and ecology. Hamlyn, 256 pp.

Brown, L.H. and D. Amadon. 1968. Eagles, hawks and falcons of the world, 2 vol. McGraw-Hill Book Company, New York, 945 pp.

Brown, M. and J.J. Dinsmore. 1986. Implications of marsh size and isolation for marsh bird management. Journal of Wildlife Management 50(3): 392-397.

Brown, P.W. 1981. Reproductive ecology and productivity of White-winged Scoters. Ph.D. dissertation, University of Missouri-Columbia, Missouri, 175 pp.

Brown, P.W. and M.A. Brown. 1981. Nesting biology of the White-winged Scoter. Journal of Wildlife Management 45(1): 38-45.

Brown, P.W. and L.H. Fredrickson. 1986. Food habits of breeding White-winged Scoters. Canadian Journal of Zoology 64(8): 1652-1654.

Brown, P.W. and L.H. Fredrickson. 1987a. Time budget and incubation behavior of breeding White-winged Scoters. Wilson Bulletin 99(1): 50-55.

Brown, P.W. and L.H. Fredrickson. 1987b. Body and organ weights and carcass composition of breeding White-winged Scoters. Wildfowl 38: 103-107.

Brown, P.W. and L.H. Fredrickson. 1989. White-winged Scoter, *Melanitta fusca*, populations and nesting on Redberry Lake, Saskatchewan. Canadian Field-Naturalist 103(2): 240-247.

Brown, R.G.B. 1986. Atlas révisé des oiseaux de mer de l'Est du Canada. 1. Relevés effectués à bord de bateaux. Canadian Wildlife Service, 115 pp.

Brown, R.G.B. and D.N. Nettleship. 1984a. Capelin and seabirds in the north west Atlantic, pp. 184-194 *in* D.N. Nettleship, G.A. Sanger, and P.F. Springer, eds. Marine birds: their feeding ecology and commercial fisheries relationships. Canadian Wildlife Service, Special Publication, 220 pp.

Brown, R.G.B. and D.N. Nettleship. 1984b. The seabirds of northeastern North America: their present status and conservation requirements, pp. 85-100 *in* J.P. Croxall, P.G. Evans, and R.W. Schreiber, eds. Status and conservation of the world's seabirds. International Council for Bird Preservation, Technical Publication no. 2, Cambridge, 778 pp.

Brown, R.G.B., D.N. Nettleship, P. Germain, C.E. Tull, and T. Davis. 1975. The Atlas of eastern Canadian seabirds. Canadian Wildlife Service, 220 pp.

Brown, W. 1971. Winter population trends in the Red-shouldered Hawks. American Birds 25: 813-817.

Browne, P. 1967. Status of birds, Lake St. John region, Quebec. Canadian Field-Naturalist 81: 50-62.

Brownell, V.R. and M.J. Oldham. 1983. Status report on the Bald Eagle (*Haliaetus leucocephalus*) in Canada. Ontario Ministry of Natural Resources, Toronto, 80 pp.

Browning, M.R. 1989. The nomenclatural status and type locality of the Yellow Warbler subspecies *Dendroica petechia aestiva* (Gmelin) (Aves: Parulinae). Canadian Field-Naturalist 103(4): 597-599.

Brunton, D.F. and W.J. Crins. 1975. Status and habitat preference of the Yellow-bellied Flycatcher in Algonquin Park, Ontario. Ontario Field Biologist 29(1): 25-28.

Brunton, D.F. and R. Pittaway, Jr. 1971. Observations of the Great Gray Owl on winter range. Canadian Field-Naturalist 85: 315-322.

Brunton, D.H. 1986. Fatal antipredator behavior of a Killdeer. Wilson Bulletin 98(4): 605-607.

Brunton, D.H. 1988. Sequential polyandry by a female Killdeer. Wilson Bulletin 100(4): 670-672.

Brush, T. and E.W. Stiles. 1990. Habitat use by breeding birds in the New Jersey pine barrens. Bulletin of the New Jersey Academy of Science 35(2): 13-16.

Bryant, A. 1987. Spotted Sandpiper, pp. 168-169 *in* M.D. Cadman, P.F.J. Eagles, and F.M. Helleiner, eds. Atlas of the breeding birds of Ontario. University of Waterloo Press, Waterloo, 617 pp.

Bryant, A.A. 1986. Influence of selective logging on Red-shouldered Hawks, *Buteo lineatus*, in Waterloo region, Ontario, 1953-1978. Canadian Field-Naturalist 100(4): 520-525.

Bryant, D.M. 1985. Swallow, pp. 571-572 *in* B. Campbell and E. Lack, eds. A dictionary of birds. British Ornithologists' Union, Buteo Books, Vermillion, South Dakota, 670 pp.

Bryant, H. 1861. Remarks on some of the birds that breed in the Gulf of St. Lawrence. Proceedings. Boston Society of Natural History 8: 65-75.

Bryson, R. 1966. Air masses, streamlines and the boreal forest. Geographical Bulletin 8: 228-269.

Bub, H. and H. Oelke. 1989. The history of bird marking till the inception of scientific bird ringing. Ring 12: 141-163.

Buchanan, J.B. 1988a. The effect of kleptoparasitic pressure on hunting behavior and performance of host Merlins. Journal of Raptor Research 22(2): 63-64.

Buchanan, J.B. 1988b. North American Merlin populations: an analysis using Christmas Bird Count data. American Birds 42: 1178-1180.

Buchanan, J.B. 1988c. Migration and winter populations of Greater Yellowlegs, *Tringa melanoleuca*, in western Washington. Canadian Field-Naturalist 102(4): 611-616.

Bucher, E.H. 1992. The causes of extinction of the Passenger Pigeon. Current Ornithology 9: 1-36.

Buckley, P.A. and F.G. Buckley. 1981. The endangered status of North American Roseate Terns. Colonial Waterbirds 4: 166-173.

Buech, R.R. 1982. Nesting ecology and cowbird parasitism of Clay-colored, Chipping, and Field sparrows in a Christmas tree plantation. Journal of Field Ornithology 53(4): 363-369.

Buehler, D.A., T.J. Mersmann, J.D. Fraser, and J.K.D. Seegar. 1991. Effects of human activity on Bald Eagle distribution on the northern Chesapeake Bay. Journal of Wildlife Management 55(2): 282-290.

Buhnerkempe, J.E. and R.L. Westemeier. 1984. Nest-sites of Turkey Vultures in southeastern Illinois. Wilson Bulletin 98(3): 495-496.

Buhnerkempe, J.E. and R.L. Westemeier. 1988. Breeding biology and habitat of Upland Sandpiper on prairie-chicken sanctuaries in Illinois. Transactions. Illinois State Academy of Science 81(1-2): 153-162.

Bull, E.L. 1975. Habitat utilization of the Pileated Woodpecker, Blue Mountains, Oregon. M.Sc. thesis, Oregon State University, Corvallis, 58 pp.

Bull, E.L. and J.R. Duncan. 1993. Great Gray Owl (*Strix nebulosa*), *in* A. Poole and F. Gill, eds. The birds of North America, no. 41. The Academy of Natural Sciences, Philadelphia. The American Ornithologists' Union, Washington, 16 pp.

Bull, E.L. and J.E. Jackson. 1995. Pileated Woodpecker (*Dryocopus pileatus*), *in* A. Poole and F. Gill, eds. The birds of North America, no. 148. The Academy of Natural Sciences, Philadelphia. The American Ornithologists' Union, Washington, 24 pp.

Bull, E.L. and E.C. Meslow. 1977. Habitat requirements of the Pileated Woodpecker in northeastern Oregon. Journal of Forestry 75(6): 335-337.

Bull, E.L. and E.C. Meslow. 1988. Breeding biology of the Pileated Woodpecker – management implications. United States, Department of Agriculture, Forest Service, Pacific Northwest Research Station. Research note, PNW-RN-474, 8 pp.

Bull, E.L., S.R. Peterson, and J.W. Thomas. 1986. Resource partitioning among woodpeckers in northeastern Oregon. United States Department of Agriculture, Forest Service, Pacific Northwest Research Station. Research note, PNW-444, 19 pp.

Bull, E.L., M.G. Henjum, and R.S. Rohweder. 1988a. Nesting and foraging habitat of Great Gray Owls. Journal of Raptor Research 22: 107-115.

Bull, E.L., M.G. Henjum, and R.S. Rohweder. 1988b. Home range and dispersal of Great Gray Owls in northeastern Oregon. Journal of Raptor Research 22: 101-106.

Bull, E.L., M.G. Henjum, and R.S. Rohweder. 1989a. Reproduction and mortality of Great Gray Owls in Oregon. Northwest Science 63: 38-43.

Bull, E.L., M.G. Henjum, and R.S. Rohweder. 1989b. Diet and optimal foraging of Great Gray Owls. Journal of Wildlife Management 53: 47-50.

Bull, E.L., A.L. Wright, and M.G. Henjum. 1989c. Nesting and diet of Long-eared Owls in conifer forests, Oregon. Condor 91(4): 908-912.

Bull, J. 1974. Birds of New York State. Doubleday, Garden City, New York, 655 pp.

Bull, J. 1985. Birds of New York State, including the 1976 supplement. Comstock Publishing Associates, Cornell University Press, Ithaca, 703 pp.

Bull, J. and J. Farrand. 1977. The Audubon Society field guide to North American birds, eastern region. Alfred A. Knopf, New York, 775 pp.

Bump, G. 1941. The introduction and transplantation of game birds and mammals into the state of New York. Transactions. North American Wildlife and Natural Resources Conference 5: 409-420.

Bump, G. 1971. The South American Monk, Quaker, or Gray-headed Parakeet. United States Department of Interior, Washington. Wildlife Leaflet no. 496.

Bump, G., R.W. Darrow, F.C. Edminster, and W.F. Crissey. 1947. The Ruffed Grouse: its life history and management. New York State Conservation Department, 915 pp.

Bunch, K.G. and D. Tomback. 1986. Bolus recovery by Gray Jays: an experimental analysis. Animal Behaviour 34: 754-762.

Bundy, G. 1976. Breeding biology of the Red-throated Diver. Bird Study 23: 249-256.

Bunn, D.S., A.B. Warburton, and R.D.S. Wilson. 1982. The Barn Owl. Buteo Books, Vermillion, South Dakota, 164 pp.

Burger, J. 1978. The pattern and mechanism of nesting in mixed-species heronries, pp. 45-48 *in* A. Sprunt IV, J.C. Ogden, and S. Wincklerer, eds. Wading birds. Research Report. National Audubon Society, no. 7, New York, 381 pp.

Burger, J. 1979. Resource partitioning: nest site selection in mixed species colonies of herons, egrets and ibises. American Midland Naturalist 101(4): 191-209.

Burger, J. 1984. Abiotic factors affecting migrant shorebirds, pp. 17-73 *in* J. Burger and B.L. Olla, eds. Behavior of marine animals, vol. 5. Shorebirds: breeding behavior and populations. Plenum Press, New York, 437 pp.

Burger, J. 1987. Physical and social determinants of nest-site selection in Piping Plover in New Jersey. Condor 89(4): 811-818.

Burger, J. 1988. Foraging behavior in gulls: differences in method, prey, and habitat. Colonial Waterbirds 11(1): 9-23.

Burger, J. and M. Gochfeld. 1988. Nest-site selection and temporal patterns in habitat use of Roseate and Common terns. Auk 105: 433-438.

Burger, J. and M. Howe. 1975. Notes on winter feeding behavior and molt in Wilson's Phalaropes. Auk 92(3): 442-451.

Burger, J., M.A. Howe, D.C. Hahn, and J. Chase. 1977. Effects of tide cycle on habitat selection and habitat partitioning by migrating shorebirds. Auk 94(4): 743-759.

Burgham, M.C.J. and J. Picman. 1989. Effect of Brown-headed Cowbirds on the evolution of Yellow Warbler anti-parasite strategies. Animal Behaviour 38(2): 298-308.

Burns, F.L. 1911. A monograph of the Broad-winged Hawk (*Buteo platypterus*). Wilson Bulletin 23(3-4): 139-320.

Burns, J.T. 1982. Nests, territories and reproduction of Sedge Wrens (*Cistothorus platensis*). Wilson Bulletin 94(3): 338-349.

Burton, J. 1978. L'alimentation estivale du Fou de Bassan, du Grand Cormoran et du Cormoran à aigrettes aux Îles-de-la-Madeleine en relation avec les pêches commerciales. Report for Department of Fisheries and Oceans, Environment Canada, Centre de recherches écologiques de Montréal, 35 pp.

Burton, J.A. 1973. Owls of the world: their evolution, structure, and ecology. E.P. Dutton and Co., Inc., New York, 216 pp.

Burton, J.A. 1984. Owls of the world: their evolution, structure, and ecology, 2nd edition. Tanager Books, Dover, New Hampshire, 208 pp.

Busby, D.G. and S.G. Sealy. 1979. Feeding ecology of a population of nesting Yellow Warblers. Canadian Journal of Zoology 57(8): 1670-1681.

Busby, D.G., S.B. Holmes, P.A. Pearce, and R.A. Fleming. 1987. The effect of aerial application of Zectran® on brain cholinesterase activity in forest songbirds. Archives of Environmental Contamination and Toxicology 16: 623-629.

Busby, D.G., L.M. White, and P.A. Pearce. 1990. Effects of aerial spraying of fenitrothion on breeding White-throated Sparrows. Journal of Applied Ecology 27(2): 743-755.

Bush, E.F. 1986. A history of hydro-electric development in Canada. Bureau des Monuments et Sites historiques.

Buskirk, J.V. 1984. Vocal mimicry of Nashville Warblers by Yellow-rumped Warblers. Wilson Bulletin 96(3): 477-482.

Buss, I.O. 1942. A managed Cliff Swallow colony in southern Wisconsin. Wilson Bulletin 54(3): 153-161.

Buss, I.O., R.K. Meyer, and C. Kabat. 1951. Wisconsin pheasant reproduction studies based on ovulated follicle technique. Journal of Wildlife Management 15(1): 32-46.

Bussières, B. 1992. La déforestation subalpine au mont du lac à l'Empêche, Charlevoix, Québec. Ph.D. dissertation, Faculté des Sciences de l'Agriculture et de l'Alimentation, Université Laval, Sainte-Foy.

Butcher, G.S. 1991. Mate choice in female Northern Orioles with a consideration of the role of the black male coloration in female choice. Condor 93(1): 82-88.

Buteau, P., N. Dignard, and P. Grondin. 1994. Système de classification des milieux humides du Québec. Department of Energy, Mines and Resources (Canada), and Ministère des Ressources naturelles du Québec, 25 pp.

Butler, A.W. 1892. Some notes concerning the Evening Grosbeak. Auk 9: 238-247.

Butler, R.G. and W. Trivelpiece. 1981. Nest spacing, reproductive success and behaviour of the Great Black-backed Gull. Auk 98(2): 99-107.

Butler, R.W. 1988. Population dynamics and migration routes of Tree Swallows, *Tachycinela bicolor*, in North America. Journal of Field Ornithology 59(4): 395-402.

Butler, R.W. 1992. Great Blue Heron (*Ardea herodias*), *in* A. Poole, P. Stettenheim, and F. Gill, eds. The birds of North America, no. 25. The Academy of Natural Sciences, Philadelphia. The American Ornithologists' Union, Washington, 20 pp.

Butler, R.W. 1995. The patient predator: foraging and population ecology of the Great Blue Heron *Ardea herodias* in British Columbia. Canadian Wildlife Service. Occasional Paper no. 86, 44 pp.

Butts, W.K. 1931. A study of the Chickadee and White-breasted Nuthatch by means of marked individuals. Bird-Banding 2: 59-76.

Cabe, P.R. 1993. European Starling (*Sturnus vulgaris*), *in* A. Poole and F. Gill, eds. The birds of North America, no. 48. The Academy of Natural Sciences, Philadelphia. The American Ornithologists' Union, Washington, 24 pp.

Caccamise, D.F. 1974. Competitive relationship of the Common and Lesser nighthawks. Condor 76: 1-20.

Caccamise, D.F. and J. Fischl. 1985. Patterns of association of secondary species in roosts of European Starlings and Common Grackles. Wilson Bulletin 97(2): 173-182.

Cade, T.J. 1955a. Experiments on the winter territoriality of the American Kestrel, *Falco sparverius*. Wilson Bulletin 67: 5-17.

Cade, T.J. 1955b. Variation of the common Rough-legged Hawks in North America. Condor 57: 313-346.

Cade, T.J. 1960. Ecology of the Peregrine and Gyrfalcon populations in Alaska. University of California. Publications in Zoology 63(3): 151-290.

Cade, T.J. 1967. Ecological and behavioral aspects of predation by the Northern Shrike. Living Bird 6: 43-86.

Cade, T.J. 1982. The falcons of the world. Comstock and Cornell University Press, Ithaca, New York, 188 pp.

Cade, T.J. and D.M. Bird. 1990. Peregrine Falcons, *Falco peregrinus*, nesting in an urban environment: a review. Canadian Field-Naturalist 104(2): 209-218.

Cade, T.J. and R.W. Fyfe. 1970. The North American Peregrine survey, 1970. Canadian Field-Naturalist 84(3): 231-245.

Cade, T.J. and T. Swem. 1993. Ecology of Northern Shrikes nesting in Arctic Alaska. International Shrike symposium abstracts, 11-15 January, Archbold Biological Station, Lake Placid, Florida, 21 pp.

Cade, T.J., J.L. Lincer, C.M. White, D.G. Rosenesu, and L.G. Swartz. 1971. DDE residues and eggshell changes in Alaskan falcons and hawks. Science 172: 955-957.

Cadman, M.D. 1986. Status report of the Loggerhead Shrike (*Lanius ludovicianus*) in Canada. Committee on the Status of Endangered Wildlife in Canada (COSEWIC), Ottawa, 95 pp.

Cadman, M.D. 1987a. Bohemian Waxwing, p. 548 *in* M.D. Cadman, P.F.J. Eagles, and F.M. Helleiner, eds. Atlas of the breeding birds of Ontario. University of Waterloo Press, Waterloo, 617 pp.

Cadman, M.D. 1987b. Northern Shrike, p. 549 *in* M.D. Cadman, P.F.J. Eagles, and F.M. Helleiner, eds. Atlas of the breeding birds of Ontario. University of Waterloo Press, Waterloo, 617 pp.

Cadman, M.D. 1987c. Turkey Vulture, pp. 106-107 *in* M.D. Cadman, P.F.J. Eagles, and F.M. Helleiner, eds. Atlas of the breeding birds of Ontario. University of Waterloo Press, Waterloo, 617 pp.

Cadman, M.D. 1990. Update status report on the Loggerhead Shrike (*Lanius ludovicianus*) in eastern Canada. Committee on the Status of Endangered Wildlife in Canada (COSEWIC), Ottawa, 33 pp.

Cadman, M.D. 1991. Status of the Northern Harrier in Canada. Committee on the Status of Endangered Wildlife in Canada (COSEWIC), unpublished report.

Cadman, M.D. and A.M. Page. 1993. Status of the Short-eared Owl (*Asio flammeus*) in Canada. Report prepared for the Ontario Ministry of Natural Resources and the Committee on the Status of Endangered Wildlife in Canada (COSEWIC), 55 pp.

Cadman, M.D., P.F.J. Eagles, and F.M. Helleiner. 1987. Atlas of the breeding birds of Ontario. University of Waterloo Press, Waterloo, 617 pp.

Caine, L.A. and W.R. Marion. 1991. Artificial addition of snags and nest boxes to slash pine plantations. Journal of Field Ornithology 62(1): 97-106.

Cairns, D.K. 1978a. Seabird sanctuaries ignored. Nature Canada 7(2): 31-32.

Cairns, D.K. 1978b. Some aspects of the biology of the Black Guillemot *Cepphus grylle* in the estuary and the Gulf of St. Lawrence. M.Sc. thesis, Université Laval, Sainte-Foy, 85 pp.

Cairns, D.K. 1979. Censusing hole-nesting auks by visual counts. Bird-Banding 50: 358-364.

Cairns, D.K. 1980. Nesting density, habitat structure and human disturbance as factors in Black Guillemot reproduction. Wilson Bulletin 92(3): 352-361.

Cairns, D.K. 1981. Breeding, feeding, and chick growth of the Black Guillemot (*Cepphus grylle*) in southern Quebec. Canadian Field-Naturalist 95(3): 312-318.

Cairns, D.K. 1987. Diet and foraging ecology of Black Guillemots in northeastern Hudson Bay. Canadian Journal of Zoology 65: 1257-1263.

Cairns, D.K., G. Chapdelaine, and W.A. Montevecchi. 1991. Prey exploitation by seabirds in the Gulf of St. Lawrence, pp. 277-291 *in* J.-C. Therriault, ed. The Gulf of St. Lawrence: small ocean or big estuary. Department of Fisheries and Oceans, Canadian Special Publication of Fisheries and Aquatic Sciences 113, Ottawa, 359 pp.

Cairns, W.E. 1977. Breeding biology and behaviour of the Piping Plover (*Charadrius melodus*) in southern Nova Scotia. M.Sc. thesis, Dalhousie University, Halifax.

Cairns, W.E. 1982. Breeding biology and behaviour of Piping Plover. Wilson Bulletin 94(4): 531-545.

Cairns, W.E. and I.A. McLaren. 1980. Status of the Piping Plover on the east coast of North America. American Birds 34(2): 206-208.

Caldwell, P.D. and J.S. Lindzey. 1974. The behavior of adult female woodcock in central Pennsylvania, pp. 1-9 *in* J.W. Artmann and S.R. Pursglove, eds. Proceedings. Fifth American Woodcock Workshop, Athens.

Calef, R.T. 1953. Avian populations of the funk forest natural area in McLean Co. Illinois. Transactions. Illinois State Academy of Science 46: 240-257.

Callender, J.P. 1906. The Redpoll breeding on the Magdalen Islands. Auk 23: 102-103.

Callicott, J.B. 1986. On the intrinsic value of nonhuman species, pp. 138-172 *in* B.G. Norton, ed. The preservation of species: the value of biological diversity. Princeton University Press, Princeton, New Jersey.

Cameron, E.S. 1908. The birds of Custer and Dawson counties, Montana. Auk 25(1): 39-56.

Campbell, B. and E. Lack. 1985. A dictionary of birds. The British Ornithologists' Union, Buteo Books, Vermillion, South Dakota, 670 pp.

Campbell, E.C. and R.W. Campbell. 1983. Status report on the Common Barn-Owl (*Tyto alba*) in Canada, 1982. Committee on the Status of Endangered Wildlife in Canada (COSEWIC), Ottawa, 71 pp.

Campbell, J.M. 1965. An impressive Gray Jay migration. Canadian Field-Naturalist 79(2): 157-158.

Campbell, R.W. and B.M. Van Der Raay. 1985. First breeding record of the Snow Bunting for British Columbia. Wilson Bulletin 97(1): 128-129.

Campbell, R.W., B.M. Van Der Raay, I. Robertson, and B.J. Petrar. 1985. Spring and summer distribution, status, and nesting ecology of the Arctic Loon, *Gavia arctica*, in interior British Columbia. Canadian Field-Naturalist 99(3): 337-442.

Campbell, R.W., N.K. Dawe, I. McTaggart-Cowan, J.M. Cooper, G.W. Kaiser, and M.C.E. McNall. 1990a. The birds of British Columbia, vol. 1, Nonpasserines. Introduction, loons through waterfowl. Royal British Columbia Museum and Environment Canada, Canadian Wildlife Service, Victoria, 514 pp.

Campbell, R.W., N.K. Dawe, I. McTaggart-Cowan, J.M. Cooper, G.W. Kaiser, and M.C.E. McNall. 1990b. The birds of British Columbia, vol. 2, Nonpasserines. Diurnal birds of prey through woodpeckers. Royal British Columbia Museum and Environment Canada, Canadian Wildlife Service, Victoria, 636 pp.

Campbell, T.R. 1970. Loon concentration on Mille Lacs Lake. Loon 42: 36-37.

Canada Committee on Ecological Land Classification. 1989. Régions écoclimatiques du Canada. Environment Canada, Lands Directorate. Ecological Land Classification Series, no. 23, 118 pp. [Also available in English as: Ecoclimatic regions of Canada.]

Canadian Forestry Service. 1992. Silvicultural terms in Canada/Terminologie de la sylviculture au Canada. Policy and Economics Directorate, 63 pp.

Canadian Wildlife Service (CWS). 1972. Canadian bird names – French English and scientific/Noms des oiseaux du Canada: noms français, anglais et scientifiques, 3rd edition. Canadian Wildlife Service. Occasional Paper no. 2, 32 pp.

Canadian Wildlife Service (CWS). 1973. Faisan de chasse. Hinterland Who's Who. Environment Canada, Canadian Wildlife Service, 4 pp. [Also available in English as: Ring-necked Pheasant.]

Canadian Wildlife Service (CWS). 1988. *Anatum* Peregrine Falcon recovery plan. Prepared by Western Raptor Technical Committee. Environment Canada, 52 pp.

Canadian Wildlife Service (CWS). 1992. Breeding waterfowl survey in eastern Canada and the state of Maine: 1992 Progress Report. A component of the Black Duck Joint Venture; surveys conducted by the Canadian Wildlife Service and the Maine Department of Inland Fisheries and Wildlife, 10 pp.

Canadian Wildlife Service (CWS) and Ministère du Loisir, de la Chasse et de la Pêche (MLCP). 1986. Plan de gestion de la sauvagine au Québec. Environment Canada, Canadian Wildlife Service, and Ministère du Loisir, de la Chasse et de la Pêche, Québec City, 108 pp.

Canadian Wildlife Service (CWS) and Statistics Canada. 1993. L'importance de la faune pour les Canadiens: données statistiques pour le Québec en 1991. Environment Canada, Canadian Wildlife Service and Statistics Canada, Ottawa.

Cannings, R.A., R.J. Cannings, and S.G. Cannings. 1987. Birds of the Okanagan valley, British Columbia. Royal British Columbia Museum, Victoria, 420 pp.

Cannings, R.J. 1987. The breeding biology of Northern Saw-whet Owls in southern British Columbia, pp. 193-198 *in* R.W. Nero, R.J. Clark, R.J. Knapton, and R.H. Hamre, eds. Proceedings. Biology and conservation of northern forest owls. Symposium, 3-7 February, Winnipeg. United States Department of Agriculture, Forest Service, Rocky Mountain Forest and Range Experiment Station, Fort Collins, Colorado. General Technical Report RM-142, 309 pp.

Cannings, R.J. 1993. Northern Saw-whet Owl (*Aegolius acadicus*), *in* A. Poole and F. Gill, eds. The birds of North America, no. 42. The Academy of Natural Sciences, Philadelphia. The American Ornithologists' Union, Washington, 20 pp.

Cantin, M. 1974. Kamouraska, réserve nationale de faune, projet d'implantation. Environment Canada, Canadian Wildlife Service, 54 pp. [Also available in English as: Proposal for a national wildlife area at Kamouraska.]

Cantin, M., J. Bédard, and H. Milne. 1974. The food and feeding of Common Eiders in the St. Lawrence estuary in summer. Canadian Journal of Zoology 52(3): 319-334.

Cantin, M., A. Bourget, G. Chapdelaine, and W.G. Alliston. 1976. Distribution et écologie de la reproduction du Canard chipeau (*Anas strepera*) au Québec. Naturaliste canadien 103(5): 469-481.

Capen, D.E. 1979. Management of northeastern pine forests for nongame birds, pp. 90-109 *in* R.M. DeGraaf and K.E. Evans, eds. Proceedings. Management of northcentral and northeastern forests for nongame birds. Workshop, 23-25 January, Minneapolis, Minnesota. United States Department of Agriculture, Forest Service, North Central Forest Experiment Station. General Technical Report NC-51, 268 pp.

Capp, M.S. 1992. Tests of the function of the song repertoire in Bobolinks. Condor 94(2): 468-479.

Carey, M. 1990. Effects of brood size and nestling age on parental care by male Field Sparrows (*Spizella pusilla*). Auk 107(3): 580-586.

Carey, M. and V. Nolan, Jr. 1979. Population dynamics of Indigo Buntings and the evolution of avian polygyny. Evolution 33(4): 1180-1192.

Carey, M., D.E. Burhans, and D.A. Nelson. 1994. Field Sparrow (*Spizella pusilla*), *in* A. Poole and F. Gill, eds. The birds of North America, no. 103. The Academy of Natural Sciences, Philadelphia. The American Ornithologists' Union, Washington, 20 pp.

Carleton, G. 1966. Hudson – St.Lawrence. Audubon Field Notes 20: 488-490.

Carlson, S. 1991. Common Grackle, pp. 506-507 *in* R. Brewer, G.A. McPeek, and R.J. Adams, Jr., eds. The Atlas of breeding birds of Michigan. Michigan State University Press, East Lansing, 594 pp.

Carlsson, B.-G. and B. Hörnfeldt. 1989. Trigyny in Tengmalm's Owl *Aegolius funereus* induced by supplementary feeding? Ornis Scandinavica 20(2): 155-156.

Caron, M., R. Ouellet, and M. Lepage. 1986. La situation de la Tourterelle triste au Québec. Ministère du Loisir, de la Chasse et de la Pêche, Direction de la faune terrestre, Service de la petite faune, 37 pp.

Carpenter, B. 1973. Deciduous-coniferous second-growth north woods. American Birds 27(6): 975.

Carpentier, A.G. 1986. Little Gull nesting on the James Bay lowlands, Ontario. Ontario Birds 4(3): 112-113.

Carroll, J.P. 1993. Gray Partridge (*Perdix perdix*), *in* A. Poole and F. Gill, eds. The birds of North America, no. 58. The Academy of Natural Sciences, Philadelphia. The American Ornithologists' Union, Washington, 20 pp.

Carroll, J.R. 1988a. Osprey, pp. 98-99 *in* R.F. Andrle and J.R. Carroll, eds. The Atlas of breeding birds in New York State. Cornell University Press, Ithaca, 551 pp.

Carroll, J.R. 1988b. Status and breeding ecology of Black Tern (*Chlidonias niger*) in New York. Kingbird 38(3): 159-172.

Carroll, J.R. 1988c. Palm Warbler, pp. 390-391 *in* R.F. Andrle and J.R. Carroll, eds. The Atlas of breeding birds in New York State. Cornell University Press, Ithaca, 551 pp.

Carter, B.C. 1958. The American Goldeneye in central New Brunswick. Canadian Wildlife Service, Wildlife Management Bulletin 2(9): 47.

Carter, J.W. 1992. Upland Sandpiper, *Bartramia longicauda*, pp. 235-252 *in* K.J. Schneider and D.M. Pence, eds. Migratory nongame birds of management concern in the northeast. United States Department of the Interior, Fish and Wildlife Service, Newton Corner, Massachusetts, 400 pp.

Cartier, J. 1981. Voyages au Canada. François Maspero, ed., Paris, 272 pp.

Cartwright, J. 1987. Black-and-White Warbler, pp. 398-399 *in* M.D. Cadman, P.F.J. Eagles, and F.M. Helleiner, eds. Atlas of the breeding birds of Ontario. University of Waterloo Press, Waterloo, 617 pp.

Casebere, L.A. 1979. Nesting record for the Canada Warbler in Lagrange County. Indiana Audubon Quarterly 57(2): 113-117.

Cassirer, R.E.F. and C.R. Groves. 1992. Ecology of Harlequin Duck in northern Idaho. Progress Report 1991. Heritage Section, Nongame and Endangered Wildlife Program, Bureau of Wildlife, Idaho Department of Fish and Game, 74 pp.

Cassirer, [R.]E.F., G. Shirato, F. Sharpe, C.R. Groves, and R.N. Anderson. 1993. Cavity nesting by Harlequin Ducks in the Pacific northwest. Wilson Bulletin 105(4): 691-694.

Catesby, M. 1771. The Natural History of Carolina, Florida, and the Bahama Islands. London, B. White.

Catling, P.M. 1972. A study of the Boreal Owl in southern Ontario with particular reference to the irruption of 1968-1969. Canadian Field-Naturalist 86(3): 223-232.

Cauboue, M., W.L. Strong, L. Archambault, and R.A. Sims. 1996. Terminology of ecological land classification in Canada. Information report LAU-X-114E. Canadian Forest Service – Québec, Natural Resources Canada, 62 pp.

Caulfield, F.B. 1890a. Our winter birds. Canadian Record of Science 4(3): 143-151.

Caulfield, F.B. 1890b. Notes on a bird new to the province of Quebec. Canadian Record of Science 4(2): 109-111.

Causey, M.K. and J.G. Graves. 1969. Insecticide residues in Least Bittern eggs. Wilson Bulletin 81(3): 340-341.

Cayouette, R. 1946. Summer records of the Eastern Evening Grosbeak (*Hesperiphona v. vespertina*) from Québec. Canadian Field-Naturalist 60: 134.

Cayouette, R. 1957. Oiseaux d'été 1957: observations ornithologiques. Club des ornithologues de Québec (COQ). Bulletin ornithologique 2(4): 1-19.

Cayouette, R. 1966. Le peintre animalier Audubon. Son oeuvre. Ses séjours au Québec. Carnets de zoologie 26(1): 10-13.

Cayouette, R. 1977. Oiseaux sauvages reçus au Jardin zoologique de Québec, à Charlesbourg, de 1937 à 1969, dont l'origine est connue. Unpublished report.

Cayouette, R. 1978a. La Gélinotte à queue fine au Québec en 1932. Bulletin ornithologique 23(1): 17.

Cayouette, R. 1978b. Nichoirs d'oiseaux. Illustrations of Jean-Luc Grondin. Société zoologique de Québec, Charlesbourg, 36 pp.

Cayouette, R. 1993. Huart ou Huard. Bulletin ornithologique 37(4): 144.

Cayouette, R. and J.-L. Grondin. 1972. Les oiseaux du Québec. Société zoologique de Québec, Orsainville, 117 pp.

Cayouette, R. and J.-L. Grondin. 1977. Les oiseaux du Québec, 2e édition revue et corrigée. Société zoologique de Québec, Orsainville, 117 pp.

Cazelais, S. 1992. Écologie de la Bernache du Canada lors de sa halte migratoire printanière dans la région du lac Saint-Pierre, Québec. M.Sc. thesis, Université du Quebec à Montréal, Montréal, 86 pp.

Cazelais, S. 1994. Historique de la réglementation sur les oiseaux de proie au Québec. Bulletin de l'Union québécoise de réhabilitation des oiseaux de proie, L'Envol 5(1): 8-9.

Centre de télédétection du Québec and Ministère de l'Énergie et des Ressources du Québec. 1992. Image du Québec par satellite.

Chabot, J. 1981. Recensement des colonies nicheuses de Laridés sur la rivière des Outaouais, entre Grenville et Rapide-des-Joachims en 1981. Ministère du Loisir, de la Chasse et de la Pêche, Service de l'aménagement et de l'exploitation de la faune, Région de l'Outaouais, 41 pp.

Chabot, J. 1986. La Tourterelle rieuse (*Streptopelia risoria*) au Québec. Ornitaouais 8(3): 107-108.

Chabot, J. and S. Barrette. 1989. Nidification de l'Urubu à tête rouge (*Cathartes aura*) et son statut au Québec. Naturaliste canadien 116(3): 205-209.

Chabot, J., R. McNeil, and J. Burton. 1979. Une étude de la Perdrix grise (*Perdix perdix*) au Québec. Final report for Ministère du Tourisme, de la Chasse et de la Pêche by Centre de recherches écologiques de Montréal, 186 pp.

Chabot, J., R. McNeil, and J. Burton. 1980. Histoire et dispersion de la Perdrix grise (*Perdix perdix*) au Québec. Naturaliste canadien 107: 243-257.

Chabot, J., J. Lapointe, and E. Langevin. 1982. Nidification du Grèbe jougris au Québec. Naturaliste canadien 109: 135-137.

Chabreck, R.H. 1963. Breeding habits of the Pied-billed Grebe in an impounded coastal marsh in Louisiana. Auk 80(4): 447-452.

Chabrzyk, G. and J.C. Coulson. 1976. Survival and recruitment in the Herring Gull *Larus argentatus*. Journal of Animal Ecology 45: 187-203.

Chagnon, G. and A. Robert. 1962. Principaux coléoptères de la province de Québec. Les Presses de l'Université de Montréal, Montréal, 440 pp.

Chalifour, J. 1982. Alimentation, croissance et reproduction des Sternes commune (*Sterna hirundo* L.) et arctique (*S. paradisaea* Pontopp.) aux Îles-de-la-Madeleine, Québec. M.Sc. thesis, University of Montréal, Montréal, 135 pp.

Chamberlain, M. 1887. A catalogue of Canadian birds, with notes on the distribution of the species. J. and A. McMillan, eds. Saint John, New Brunswick, 143 pp.

Chamberlain, M. 1888. A systematic table of Canadian birds. Saint John, New Brunswick.

Chamberlain, M. 1891. A popular handbook of the ornithology of the United States and Canada, based on Nuttall's manual. Vol. 1: The land birds; vol. 2: Game and water birds. Boston: Little, Brown and Co., vol. 1, 473 pp.; vol. 2, 431 pp.

Chamberlain, M. 1894. A popular handbook of the ornithology of eastern North America by Thomas Nuttall, revised and annotated by M. Chamberlain. Vol. 1: The land birds; vol. 2: Game and water birds. London: Truslove and Hauson. Vol. 1, 473 pp.; vol. 2, 431 pp.

Chamberlain, M. 1895. Some Canadian birds; a brief account of some of the common birds of eastern Canada. First series – birds of field and grove. Toronto: The Coop, Clark Co., 96 pp.

Chamberlain, M. 1903. A popular handbook of the birds of Canada and the United States by Thomas Nuttall. New revised and annotated edition by Montague Chamberlain. Toronto, The Musson Book Co., vol. 1, 473 pp.; vol. 2, 431 pp.

Chamberlain-Auger, J.A., P.J. Auger, and E.G. Strauss. 1990. Breeding biology of American Crows. Wilson Bulletin 102(4): 615-622.

Chamberlin, M.L. 1977. Observations on the Red-necked Grebe nesting in Michigan. Wilson Bulletin 89(1): 33-46.

Chamberlin, M.L. 1980. Winter hunting behavior of a Snowy Owl in Michigan. Wilson Bulletin 92(1): 116-120.

Champagne, J. and M. Melançon. 1985. Les milieux humides de la région de Montréal, 1966-1981. Environment Canada, Lands Directorate. Working Paper no. 39, 21 pp. [Also available in English as: Wetlands in the Montreal Region 1966-1981.]

Champagne, P. 1982. Morphologie littorale de la baie de Rupert. Naturaliste canadien 109(3): 375-384.

Champagne, Y. and M. Surprenant. 1981. Une mention de nidification du Canard roux (*Oxyura jamaicensis*) au Québec. Bulletin ornithologique 26(4): 131-132.

Chancellor, J. 1978. Audubon. London, Weidenfeld and Nicolson, 224 pp.

Chapdelaine, G. 1977. Le Grand Cormoran en hiver, le long des côtes de la péninsule de Gaspésie, Québec. Canadian Field-Naturalist 91: 184.

Chapdelaine, G. 1979a. Situation du Pétrel cul-blanc (*Oceanodroma leucorhoa*) à l'île Bonaventure. Environment Canada, Canadian Wildlife Service, Québec Region. Unpublished report, 6 pp.

Chapdelaine, G. 1979b. Situation du Canard huppé (*Aix sponsa*) au Québec. Environment Canada, Canadian Wildlife Service, Québec Region. Unpublished report, 12 pp.

Chapdelaine, G. 1987. Seabirds nesting in the estuary and gulf. Nexus 9(2): 9-13.

Chapdelaine, G.. 1995. Fourteenth census of seabird populations in the sanctuaries of the North Shore of the Gulf of St. Lawrence, 1993. Canadian Field-Naturalist 109(2): 220-226.

Chapdelaine, G. and P. Blais. 1979. Inventaire de l'avifaune de la réserve de l'Est en 1977. Environment Canada, Canadian Wildlife Service, Québec Region, 35 pp.

Chapdelaine, G. and A. Bourget. 1981. Distribution, abondance et fluctuations des populations d'oiseaux marins de l'archipel de Mingan (golfe du Saint-Laurent, Québec). Naturaliste canadien 108: 219-227.

Chapdelaine, G. and P. Brousseau. 1984. Douzième inventaire des populations d'oiseaux marins dans les refuges de la côte nord du golfe Saint-Laurent. Canadian Field-Naturalist 98: 178-183.

Chapdelaine, G. and P. Brousseau. 1989. Size and trends of Black-legged Kittiwake (*Rissa tridactyla*) populations in the Gulf of St. Lawrence (Quebec) 1974-1985. American Birds 43(1): 21-24.

Chapdelaine, G. and P. Brousseau. 1991. Thirteenth census of seabird populations in the sanctuaries of the North Shore of the Gulf of St. Lawrence, 1982-1988. Canadian Field-Naturalist 105(1): 60-66.

Chapdelaine, G. and P. Brousseau. 1992a. Tendances observées chez les populations d'oiseaux marins du fleuve Saint-Laurent, de l'estuaire et du golfe, pp. 7-9 *in* C. Hyslop and J. Kennedy, eds. Canadian Wildlife Service. Bird Trends no. 2, 20 pp.

Chapdelaine, G. and P. Brousseau. 1992b. Distribution, abundance and changes of seabird population of the Gaspé Peninsula, Québec, 1979 to 1989. Canadian Field-Naturalist 106(4): 427-434.

Chapdelaine, G. and P. Brousseau. (In press). Diet of Razorbill (*Alca torda*) chicks and breeding success in the St. Mary's Islands, Gulf of St. Lawrence, Quebec, Canada, 1990-1992, pp. 27-36 *in* W.A. Montevecchi, ed. Studies of high-latitude seabirds. 4. Trophic relationships and energetics of endotherms in cold ocean systems. Canadian Wildlife Service, Occasional Paper Series no.91, 73 pp. [1996.]

Chapdelaine, G. and P. Laporte. 1982. Population, reproduction et analyse des contaminants chez le Gode (*Alca torda*) dans l'estuaire et le golfe du Saint-Laurent (Québec). Canadian Wildlife Service, Québec Region. Progress Notes no. 129, 10 pp. [Also available in English as: Population, reproductive success, and analysis of contaminants in Razorbills (*Alca torda*) in the estuary and Gulf of St. Lawrence, Quebec.]

Chapdelaine, G. and G. Tremblay. 1979. Indices de distribution et de l'abondance de l'Eider à duvet (*Somateria mollissima sedentaria* et *S. m. borealis*) le long de la côte est de la baie d'Hudson, du détroit d'Hudson et de la baie d'Ungava. Environment Canada, Canadian Wildlife Service, Québec Region. Unpublished report, 21 pp.

Chapdelaine, G., P. Brousseau, R. Anderson, and R. Marsan. 1985. Breeding ecology of Common and Arctic terns in the Mingan Archipelago, Québec. Colonial Waterbirds 8(2): 166-177.

Chapdelaine, G., A.J. Gaston, and P. Brousseau. 1986a. Recensement des colonies de Marmettes de Brünnich de l'île Akpatok, Territoires du Nord-Ouest. Canadian Wildlife Service, Québec Region. Progress Notes no. 163, 9 pp. [Also available in English as: Censusing the Thick-billed Murre colonies of Akpatok Island, NWT.]

Chapdelaine, G., P. Dupuis, and A. Reed. 1986b. Distribution, abondance et fluctuation des populations d'Eider à duvet dans l'estuaire et le golfe du Saint-Laurent, pp. 6-11 *in* A. Reed, ed. Eider ducks in Canada/Les eiders au Canada. Canadian Wildlife Service, Ottawa. Report Series no. 47, 177 pp.

Chapdelaine, G., A. Bourget, W.B. Kemp, D.J. Nakashima, and D.J. Murray. 1986c. Population d'Eider à duvet près des côtes du Québec septentrional, pp. 39-50 *in* A. Reed, ed. Eider ducks in Canada/Les eiders au Canada. Canadian Wildlife Service, Ottawa. Report Series no. 47, 177 pp.

Chapdelaine, G., P. Laporte, and D.N. Nettleship. 1987. Population, productivity, and DDT contamination trends of Northern Gannets (*Sula bassanus*) at Bonaventure Island, Quebec, 1967-1984. Canadian Journal of Zoology 65: 2922-2926.

Chapin, E.A. 1925. Food habits of the vireos, a family of insectivorous birds. United States Department of Agriculture, Washington. Bulletin no. 1335, 43 pp.

Chapman, B.-A. and L.S. Forbes. 1984. Observations on detrimental effects of Great Blue Herons on breeding Black Terns. Journal of Field Ornithology 55(2): 251-252.

Chapman, F.M. 1925. The European Starling as an American citizen. Natural History 25: 480-485.

Chapman, F.M. 1932. Handbook of birds of eastern North America. D. Appleton and Company, New York, 421 pp.

Chapman, F.M. 1966. Handbook of birds of eastern North America: with introductory chapters on the study of birds in nature, 2nd edition. Dover Publications, New York, 581 pp.

Chapman, F.M. 1968. The warblers of North America. Dover Publications, New York, 295 pp.

Chapman, L.B. 1955. Studies of a Tree Swallow colony. Bird-Banding 26: 45-70.

Chapman Mosher, B. 1986. Factors influencing reproductive success and nesting strategies in Black Terns. Ph.D. dissertation, Simon Fraser University, Burnaby, British Columbia.

Charland, J.-P. 1990. Les pâtes et papiers au Québec 1880-1980: technologies, travail et travailleurs. Institut québécois de recherche sur la culture, Québec, 447 pp.

Charlevoix, F.X. 1744. Histoire et description générale de la Nouvelle France avec le journal historique d'un voyage fait par ordre du Roi dans l'Amérique septentrionale. (6 volumes). Paris, Didot.

Charlevoix, F.X. 1761. Journal of a voyage to North America. 2 vol. R. and J. Dodsley, London.

Chelazzi, G. and F. Pineschi. 1974. Recoveries after displacement in pigeons belonging to an urban population. Monitore Zoologico Italiano 8(3): 152-157.

Cheskey, E.D. 1990. Red-breasted Nuthatch nesting in residential Waterloo. Ontario birds 8(2): 71-75.

Cheskey, T. 1987. Olive-sided Flycatcher, pp. 250-251 *in* M.D. Cadman, P.F.J. Eagles, and F.M. Helleiner, eds. Atlas of the breeding birds of Ontario. University of Waterloo Press, Waterloo, 617 pp.

Chew, L. 1981. Geographic and individual variation in the morphology and sequential organization of the song of the Savannah Sparrow (*Passerculus sandwichensis*). Canadian Journal of Zoology 59(4): 702-713.

Choate, E.A. 1973. The dictionary of American bird names. Gambit, Boston, 261 pp.

Christiansen, D.A. Jr. and S.E. Reinert. 1990. Habitat use of the Northern Harrier in a coastal Massachusetts shrubland with notes on population trends in southeastern New England. United States of America. Journal of Raptor Research 24(4): 84-90.

Christy, B.H. 1939. Northern Pileated Woodpecker, pp. 171-189 *in* A.C. Bent. Life histories of North American woodpeckers. United States National Museum, Washington. [1964, Dover Publications, New York]. Bulletin no. 174, 334 pp.

Christy, B.H. 1942. Acadian Flycatcher, pp. 183-197 *in* A.C. Bent. Life histories of North American flycatchers, larks, swallows, and their allies. United States National Museum, Washington. [1963, Dover Publications, New York]. Bulletin no. 179, 555 pp.

Church, K.E. and W.F. Porter. 1990. Winter and spring habitat use by Gray Partridge in New York. Journal of Wildlife Management 54(4): 653-657.

Clapp, R.B. 1976. Review of longevity records by R. Rydzewski. Bird-Banding 47(3): 279-281.

Clapp, R.B., M.K. Klimkiewicz, and J.H. Kennard. 1982. Longevity records of North American birds: Gaviidae through Alcidae. Journal of Field Ornithology 53(2): 81-124.

Clapp, R.B., M.K. Klimkiewicz, and A.G. Futcher. 1983. Longevity records of North American birds: Columbidae through Paridae. Journal of Field Ornithology 54(2): 123-137.

Clark, D.B. 1985a. Blue-gray Gnatcatcher, pp. 236-237 *in* S.B. Laughlin and D.P. Kibbe, eds. The Atlas of breeding birds of Vermont. Vermont Institute of Natural Science, University Press of New England, Hanover, 456 pp.

Clark, D.B. 1985b. Blue-winged Warbler, pp. 274-275 *in* S.B. Laughlin and D.P. Kibbe, eds. The Atlas of breeding birds of Vermont. Vermont Institute of Natural Science, University Press of New England, Hanover, 456 pp.

Clark, D.B. 1985c. Prairie Warbler, pp. 302-303 *in* S.B. Laughlin and D.P. Kibbe, eds. The Atlas of breeding birds of Vermont. Vermont Institute of Natural Science, University Press of New England, Hanover, 456 pp.

Clark, G.C. and R.D. Ohmart. 1985. Spread-winged posture of Turkey Vultures: single or multiple function? Condor 87(3): 350-355.

Clark, K., D. Euler, and E. Armstrong. 1983. Habitat associations of breeding birds in cottage and natural areas of central Ontario. Wilson Bulletin 95(1): 77-96.

Clark, K., D. Euler, and E. Armstrong. 1984. Predicting avian community response to lakeshore cottage development. Journal of Wildlife Management 48(4): 1239-1247.

Clark, K.L. and R.J. Robertson. 1981. Cowbird parasitism and evolution of anti-parasite strategies in the Yellow Warbler. Wilson Bulletin 93(2): 249-258.

Clark, L. and J.R. Mason. 1985. Use of nest material as insecticidal and anti-pathogenic agents by the European Starling. Oecologia 67: 169-176.

Clark, M.F.G. 1991. 67. Field, ridge, and shrubby trees, p. 72 *in* R.M. Marshall, ed. Resident bird counts 1990. Journal of Field Ornithology 62(1)(supplement): 1-99.

Clark, M.F.G. 1992. 94. Field, ridge, and shrubby trees, pp. 92-93 *in* R.M. Marshall, ed. Resident bird counts 1991. Journal of Field Ornithology 63(1)(supplement): 1-123.

Clark, M.F.G. 1993a. 89. Field, ridge, and shrubby trees, pp. 88-89 *in* J.D. Lowe, ed. Resident bird counts 1992. Journal of Field Ornithology 64(1)(supplement): 1-128.

Clark, M.F.G. 1993b. 90. Shrubby trees, ridge, and woods, p. 89 *in* J.D. Lowe, ed. Resident bird counts 1992. Journal of Field Ornithology 64(1)(supplement): 1-128.

Clark, R.G. 1985d. Effects of agricultural land use of the biology of the Red-winged Blackbird (*Agelaius phoeniceus*). Ph.D. dissertation, McGill University, Montréal, 165 pp.

Clark, R.G. and P.J. Weatherhead. 1986. The effect of fine scale variations in agricultural land use on the abundance of Red-winged Blackbirds. Canadian Journal of Zoology 64: 1951-1955.

Clark, R.G. and P.J. Weatherhead. 1987. Influence of population size on habitat use by territorial male Red-winged Blackbirds in agricultural landscapes. Auk 104: 311-315.

Clark, R.G., R.D. Titman, J.R. Bider, and M. Prévost. 1982. Aspects of the socio-economic impact of blackbirds and other vertebrates on farm production in Québec. Phytoprotection 63: 56-67.

Clark, R.G., R.D. Titman, J.R. Bider, and H. Greenwood. 1988. Influence of agricultural land-use practices on bird damage and control, pp. 464-473 *in* H. Ouellet, ed. Acta XIX Congressus Internationalis Ornithologici, Ottawa, 22-29 June 1986. Ottawa University Press, vol. I, pp. 1-1404; vol. II, pp. 1405-2815.

Clark, R.J. 1972. Observations of nesting Marsh Hawks in Manitoba. Blue Jay 30: 43-48.

Clark, R.J. 1975. A field study of the Short-eared Owl, *Asio flammeus* (Pontoppidan), in North America. Wildlife Monographs no. 47, 67 pp.

Clark, R.J. and J.G. Ward. 1974. Interspecific competition in two species of open country raptors *Circus cyaneus* and *Asio flammeus*. Proceedings. Pennsylvania Academy of Science 48: 79-87.

Clark, R.J., D.G. Smith, and L. Kelso. 1987. Distributional status and literature of northern forest owls, pp. 47-55 *in* R.W. Nero, R.J. Clark, R.J. Knapton, and R.H. Hamre, eds. Proceedings. Biology and conservation of northern forest owls. Symposium, 3-7 February, Winnipeg. United States Department of Agriculture, Forest Service, Rocky Mountain Forest and Range Experiment Station, Fort Collins, Colorado. General Technical Report RM-142, 309 pp.

Clark, W.S. and B.K. Wheeler. 1987. A field guide to hawks of North America. Houghton Mifflin Company, Boston, 198 pp.

Clarke, R.G. 1982. Nest site selection by Sharp-shinned Hawks in interior Alaska, pp. 155-162 *in* W.N. Ladd and P.F. Schempf, eds. Raptor management and biology in Alaska and western Canada. United States Fish and Wildlife Service, Anchorage. Research Report no. FWS/AK/PROC-82, 335 pp.

Clawson, R.L., G.W. Hartman, and L.H. Fredrickson. 1979. Dump nesting in a Missouri Wood Duck population. Journal of Wildlife Management 43: 347-355

Cleary, L. 1977. Succès de reproduction du Cormoran à aigrettes, *Phalacrocorax auritus a.* sur trois îles du Saint-Laurent, en 1975 et 1976. M.Sc. thesis, Université Laval, Sainte-Foy.

Cleave, A. 1989. Hummingbirds. Hamlyn, London, 128 pp.

Clement, R.C. 1968a. Common Redpoll, pp. 407-421 *in* A.C. Bent. Life histories of North American cardinals, grosbeaks, buntings, towhees, finches, sparrows, and allies. (O.L. Austin, Jr., ed.). United States National Museum, Washington. [1968, Dover Publications, New York]. Bulletin no 237, vol. 1, pp. 1-602; vol. 2, pp. 603-1248; vol. 3, pp. 1249-1889.

Clement, R.C. 1968b. Eastern White-crowned Sparrow, pp. 1273-1291 *in* A.C. Bent. Life histories of North American cardinals, grosbeaks, buntings, towhees, finches, sparrows, and allies. (O.L. Austin, Jr., ed.). United States National Museum, Washington. [1968, Dover Publications, New York]. Bulletin no. 237, vol. 1, pp. 1-602; vol. 2, pp. 603-1248; vol. 3, pp. 1249-1889.

Cline, K.W. 1986. 1986 Chesapeake Bay eagle banding project report. National Wildlife Federation, Washington.

Cloutier, D. and C. Houle. 1987. Les pics ont-ils des maux de tête? Jaseur 7(1): 4-5.

Club des ornithologues amateurs du Saguenay–Lac-Saint-Jean. 1993a. Les observations ornithologiques au Saguenay–Lac-Saint-Jean: printemps 1992. Le Harfang 16(1): 19-35.

Club des ornithologues amateurs du Saguenay–Lac-Saint-Jean. 1993b. Les observations ornithologiques au Saguenay–Lac-Saint-Jean: l'été 1993. Le Harfang 16(4): 13-27.

Club des ornithologues de l'Outaouais. 1982. Liste annotée des oiseaux de l'Outaouais (Québec). Hull, 115 pp.

Club des ornithologues de l'Outaouais. 1985. Guide d'observation des oiseaux de l'Outaouais. Hull, 223 pp.

Club des ornithologues de Québec (COQ). 1971. Les observations ornithologiques de l'été 1971. Bulletin ornithologique 16(3): 49-70.

Club des ornithologues de Québec (COQ). 1972. Observations ornithologiques. Bulletin ornithologique 17(3): 30-47.

Club des ornithologues de Québec (COQ). 1974. Observations ornithologiques. Bulletin ornithologique 19(3): 38-67.

Club des ornithologues de Québec (COQ). 1975. La saison de nidification 1975: observations ornithologiques. Bulletin ornithologique 20(3): 55-77.

Club des ornithologues du Québec (COQ). 1976. Observations ornithologiques, été 1976 (juin, juillet et août). Bulletin ornithologique 21(3): 65-96.

Club des ornithologues du Québec (COQ). 1977. Observations ornithologiques, été 1977, juin, juillet. Bulletin ornithologique 22(3): 92-106.

Club des ornithologues du Québec (COQ). 1978. Observations ornithologiques: juin et juillet 1978. Bulletin ornithologique 23(3): 82-91.

Club des ornithologues du Québec (COQ). 1980a. Observations ornithologiques, juin et juillet 1980. Bulletin ornithologique 25(3): 85-94.

Club des ornithologues du Québec (COQ). 1980b. Observations ornithologiques, août, septembre et octobre 1980. Bulletin ornithologique 25(4): 122-130.

Club des ornithologues du Québec (COQ). 1985. Les observations extrarégionales. Bulletin ornithologique 30(3): 100-101.

Club des ornithologues du Québec (COQ). 1986. Observations ornithologiques. Bulletin ornithologique 31(3): 75-89.

Club des ornithologues du Québec (COQ). 1987. Observations ornithologiques. Bulletin ornithologique 32(3): 101-112.

Club des ornithologues du Québec (COQ). 1988. Observations ornithologiques. Bulletin ornithologique 33(2): 56-69.

Club des ornithologues du Québec (COQ). 1989. Observations ornithologiques. Bulletin ornithologique 34(3): 92-105.

Club des ornithologues du Québec (COQ). 1992. Observations ornithologiques. Bulletin ornithologique 37(3): 70-80.

Club d'ornithologie de la Manicouagan. 1992. Guide des sites de la Côte-Nord. Club d'ornithologie de la Manicouagan, Baie-Comeau, Québec, 48 pp.

Clum, N.J. and T.J. Cade. 1994. Gyrfalcon (*Falco rusticolus*), *in* A. Poole and F. Gill, eds. The birds of North America, no. 114. The Academy of Natural Sciences, Philadelphia. The American Ornithologists' Union, Washington, 28 pp.

Cobb, S. 1959. On the angle of the cerebral axis in the American Woodcock. Auk 76(1): 55-59.

Cochet, G. 1980. Nidification de la Mouette rieuse, *Larus ridibundus*, en Haute Loire à 900 m d'altitude. Nos Oiseaux 35: 288.

Cochran, W.G. 1977. Sampling techniques, 3rd edition. John Wiley and Sons, New York, 428 pp.

Cogbill, C.V. 1985. Dynamics of the boreal forests of the Laurentian Highlands, Canada. Canadian Journal of Forest Research 15: 252-261.

Cohen, R.R. 1980. Color versus age in female Tree Swallows. Journal of Colorado-Wyoming Academy of Science 12: 44-45.

Cohen, R.R. 1985. First-come, first-served mate and nest-site selection by returning successful breeders of a nest-box Tree Swallow population. Abstract, Cooper Ornithological Society, Wilson Ornithological Society Meeting, Boulder, Colorado.

Cohen, R.R. 1987. Spring arrival and nest-site acquisition patterns in Tree Swallows (*Tachycineta bicolor*). Abstract, American Ornithology Union Management, San Francisco.

Cole Harris, R. 1987. Atlas historique du Canada, vol. I. Des origines à 1800. Presses de l'Université de Montréal, Montréal, 198 pp.

Coleman, J.S. and J.D. Fraser. 1989. Habitat use and home ranges of Black and Turkey vultures. Journal of Wildlife Management 53(3): 782-792.

Coles, V. 1938. The life history of the Turkey Vulture (*Cathartes aura septentrionalis* Wied). Ph.D. dissertation, Cornell University, Ithaca, New York, 617 pp.

Collar, N.J. 1985. Crane, pp. 117-118 *in* B. Campbell and E. Lack, eds. A dictionary of birds. British Ornithologists' Union, Buteo Books, Vermillion, South Dakota, 670 pp.

Collias, N.E. and E.C. Collias. 1984. Nest building and bird behavior. Princeton University Press, Princeton, New Jersey, 336 pp.

Collias, N.E. and L.R. Jahn. 1959. Social behavior and breeding success in Canada Geese (*Branta canadensis*) confined under semi-natural conditions. Auk 76(4): 478-509.

Collins, C.T. 1970. The Black-crowned Night-Heron as a predator of tern chicks. Auk 87(3): 584-586.

Collins, C.T. 1991. Diurnal body temperature cycle in the Northern Hawk-Owl (*Surnia ulula*). Journal of Raptor Research 25(3): 92-93.

Collins, H.H. Jr. and N.R. Boyajian. 1965. Familiar garden birds of America. Harper and Row, New York, 309 pp.

Collins, S.L. 1981. A comparison of nest-site and perch-site vegetation structure for seven species of warblers. Wilson Bulletin 93(4): 542-547.

Collins, S.L. 1983a. Geographic variation in habitat structure for the wood warblers in Maine and Minnesota. Oecologia 59: 246-252

Collins, S.L. 1983b. Geographic variation in habitat structure of the Black-throated Green Warbler (*Dendroica virens*). Auk 100(2): 382-389.

Collins, S.L., F.C. James, and P.G. Risser. 1982. Habitat relationships of wood warblers (Parulidae) in northern central Minnesota. Oikos 39(1): 50-58.

Collopy, M.W. 1984. Parental care and feeding ecology of Golden Eagle nestlings. Auk 101(4): 753-760.

Colston, P. and P. Burton. 1988. A field guide to the waders of Britain and Europe, with North Africa and the Middle East. Hodder and Stoughton, London, 224 pp.

Colston, P.R. 1990. Les Becs-de-cire et les Tisserins, pp. 134-139 *in* C.M. Perrins and A.L.A. Middleton, eds. Les oiseaux des plaines et des forêts. Solar, Paris, 143 pp.

Colwell, M.A. 1986. The first documented case of polyandry for Wilson's Phalarope (*Phalaropus tricolor*). Auk 103(3): 611-612.

Colwell, M.A. and J.R. Jehl, Jr. 1994. Wilson's Phalarope (*Phalaropus tricolor*), *in* A. Poole and F. Gill, eds. The birds of North America, no. 83. The Academy of Natural Sciences, Philadelphia. The American Ornithologists' Union, Washington, 20 pp.

Colwell, M.A. and L.W. Oring. 1988a. Wing fluttering display by incubating male Wilson's Phalaropes. Canadian Journal of Zoology 66: 2315-2317.

Colwell, M.A. and L.W. Oring. 1988b. Breeding biology of Wilson's Phalarope in southcentral Saskatchewan. Wilson Bulletin 100(4): 567-582.

Colwell, M.A. and L.W. Oring. 1990. Nest-site characteristics of prairie shorebirds. Canadian Journal of Zoology 68(2): 297-302.

Comeau, N.-A. 1890. Additional notes on the probable breeding of *Saxicola oenanthe* near Godbout, province of Quebec, Canada. Auk 7: 294.

Comeau, N.-A. 1909. Life and sport on the North Shore of the Lower St. Lawrence and Gulf. Daily Telegraph Printing House, Québec, 440 pp.

Comeau, N.-A. 1911. Passenger Pigeon Notes. Forest and Stream 76(23): 899.

Comeau, N.-A. 1923a. Life and sport on the North Shore of the Lower St. Lawrence and Gulf, 2nd edition. Telegraph Printing Company, Québec, 440 pp.

Comeau, N.-A. 1923b. Wheatear at Godbout, Quebec. Auk 40: 544-545.

Comeau, N.-A. 1945. La vie et le sport sur la Côte-Nord du Bas-Saint-Laurent et du golfe. Garneau, Québec, 372 pp.

Comeau, N.-A. 1954. Life and sport on the North Shore of the Lower St. Lawrence and Gulf, 3rd edition. Telegraph Printing Company, Québec, 440 pp.

Comité de recherche sur l'exploitation par les autochtones de la Baie-James et du Nord québécois. 1982. Terre d'abondance. Étude sur l'exploitation de la faune par les Cris de la Baie James, de 1972 à 1979. Québec, 860 pp.

Comité des parcs en forêt. 1974. Parcs, territoires et zones analogues: plein air et conservation au Québec. Corporation des ingénieurs forestiers de la province de Québec, Québec, 274 pp.

Commission de toponymie du Québec. 1987. Répertoire toponymique du Québec 1987. Publications du Québec, Québec, 1900 pp.

Commission de toponymie du Québec. 1994a. Noms et lieux du Québec: dictionnaire illustré. Publications du Québec, Sainte-Foy, 925 pp.

Commission de toponymie du Québec. 1994b. Répertoire toponymique du Québec 1987: supplément cumulatif 1993. Publications du Québec, Sainte-Foy, 324 pp.

Commission de toponymie du Québec. 1995. Fichier informatique TOPOS – Population du Québec en 1991.

Commission internationale des noms français des oiseaux. 1993. Noms français des oiseaux du monde. Éditions MultiMondes, Sainte-Foy, 452 pp.

Commission sur la protection des forêts. 1991. Des forêts en santé. Rapport d'enquête et d'audience publique sur la stratégie de protection des forêts. Bureau d'audiences publiques sur l'environnement, Les Publications du Québec, Sainte-Foy, 277 pp.

Committee on the Status of Endangered Wildlife in Canada (COSEWIC). 1993. Espèces canadiennes en péril [Canadian species in peril]. Ottawa, 11 pp.

Committee on the Status of Endangered Wildlife in Canada (COSEWIC). 1996. Espèces canadiennes en péril [Canadian species at risk]. Ottawa, 18 pp.

Confer, J.[L.]. 1988. Golden-winged Warbler, pp. 356-357 *in* R.F. Andrle and J.R. Carroll, eds. The Atlas of breeding birds in New York State. Cornell University Press, Ithaca, 551 pp.

Confer, J.L. 1992a. Golden-winged Warbler (*Vermivora chrysoptera*), *in* A. Poole, P. Stettenheim, and F. Gill, eds. The birds of North America, no. 20. The Academy of Natural Sciences, Philadelphia. The American Ornithologists' Union, Washington, 16 pp.

Confer, J.L. 1992b. Golden-winged Warbler (*Vermivora chrysoptera*), pp. 369-383 *in* K.J. Schneider and D.M. Pence, eds. Migratory nongame birds of management concern in the northeast. United States Department of the Interior, Fish and Wildlife Service, Newton Corner, Massachusetts, 400 pp.

Confer, J.L. and K. Knapp. 1981. Golden-winged Warblers and Blue-winged Warblers: the relative success of a habitat specialist and a generalist. Auk 98(1): 108-114.

Conner, R.N. 1974a. A note on aerial courtship of Red-tailed Hawks. Bird-Banding 45(3): 269.

Conner, R.N. 1974b. Eastern Bluebirds nesting in clearcuts. Journal of Wildlife Management 38(4): 934-935.

Conner, R.N. 1980. Foraging habitats of woodpeckers in southwestern Virginia. Journal of Field Ornithology 51: 119-127.

Conner, R.N. and C.S. Adkisson. 1976. Discriminant function analysis: a possible aid in determining the impact of forest management on woodpecker nesting habitat. Forest Science 22: 122-127.

Conner, R.N., R.G. Hooper, H.S. Crawford, and H.S. Mosby. 1975. Woodpecker nesting habitat in cut and uncut woodlands in Virginia. Journal of Wildlife Management 39(1): 144-150.

Conner, R.N., O.K. Miller, Jr., and C.S. Adkisson. 1976. Woodpecker dependence on trees infected by fungal heart rots. Wilson Bulletin 88(4): 575-581.

Conner, R.N., J.G. Dickson, B.A. Locke, and C.A. Segelquist. 1983. Vegetation characteristics important to common songbirds in east Texas. Wilson Bulletin 95(3): 349-361.

Conner, R.N., M.E. Anderson, and J.G. Dickson. 1986. Relationships among territory size, habitat, song and nesting success of Northern Cardinals. Auk 103(1): 23-31.

Connor, P.F. 1988a. Cerulean Warbler, pp. 396-397 *in* R.F. Andrle and J.R. Carroll, eds. The Atlas of breeding birds in New York State. Cornell University Press, Ithaca, 551 pp.

Connor, P.F. 1988b. Marsh Wren, pp. 308-309 *in* R.F. Andrle and J.R. Carroll, eds. The Atlas of breeding birds in New York State. Cornell University Press, Ithaca, 551 pp.

Connor, P.F. 1988c. Orchard Oriole, pp. 482-483 *in* R.F. Andrle and J.R. Carroll, eds. The Atlas of breeding birds in New York State. Cornell University Press, Ithaca, 551 pp.

Connor, P.F. 1988d. Pied-billed Grebe, pp. 28-29 *in* R.F. Andrle and J.R. Carroll, eds. The Atlas of breeding birds in New York State. Cornell University Press, Ithaca, 551 pp.

Connor, P.F. 1988e. Sedge Wren, pp. 306-307 *in* R.F. Andrle and J.R. Carroll, eds. The Atlas of breeding birds in New York State. Cornell University Press, Ithaca, 551 pp.

Connor, P.F. 1988f. White-throated Sparrow, pp. 462-463 *in* R.F. Andrle and J.R. Carroll, eds. The Atlas of breeding birds in New York State. Cornell University Press, Ithaca, 551 pp.

Connors, P.G., J.P. Myers, and F.A. Pitelka. 1979. Seasonal habitat use by arctic Alaskan shorebirds. Studies in Avian Biology 2: 101-111.

Conover, M.R. and D.E. Miller. 1980. Rictal bristle function in Willow Flycatcher. Condor 82: 469-471.

Conrad, K.F. and R.J. Robertson. 1992. Intraseasonal effects of clutch manipulation on parental provisioning and residual reproductive value of Eastern Phoebes (*Sayornis phoebe*). Oecologia 89: 356-364.

Conseil international de la langue française. 1979. Vocabulaire de la géomorphologie. Hachette, Paris, 218 pp.

Consortium Gauthier et Guillemette – GREBE. 1989. Complexe Grande-Baleine, Avant-projet, phase II. Étude de l'avifaune et du Castor: Répartition et abondance du Balbuzard (*Pandion haliaetus*) sur le territoire du complexe Grande-Baleine en 1989. Report for Hydro-Québec, vice-présidence Environnement. Preliminary version, Montréal, 32 pp.

Consortium Gauthier et Guillemette – GREBE. 1990a. Complexe Grande-Baleine, Avant-projet, phase II. Étude de l'avifaune et du Castor: Écologie de la sauvagine (été 1989). Report for Hydro-Québec, vice-présidence Environnement. Preliminary version, Montréal, 214 pp.

Consortium Gauthier et Guillemette – GREBE. 1990b. Complexe Grande-Baleine, Avant-projet, phase II. Étude de l'avifaune et du Castor: Écologie des Macreuses à bec jaune (*Melanitta nigra*) et à front blanc (*M. perspicillata*) en période de reproduction sur le territoire du complexe Grande-Baleine. Report for Hydro-Québec, vice-présidence Environnement. Preliminary version, Montréal, 70 pp.

Consortium Gauthier et Guillemette – GREBE. 1990c. Complexe Grande-Baleine, Avant-projet, phase II. Étude de l'avifaune et du Castor: Répartition et abondance des oiseaux de proies. Report for Hydro-Québec, vice-présidence Environnement. Preliminary version, Montréal, 55 pp.

Consortium Gauthier et Guillemette – GREBE. 1990d. Complexe Grande-Baleine, Avant-projet, phase II. Étude de l'avifaune et du Castor: Utilisation estivale et automnale du détroit de Manitounuk par la sauvagine (1989). Report for Hydro-Québec, vice-présidence Environnement. Preliminary version, Montréal, 42 pp.

Consortium Gauthier et Guillemette – GREBE. 1991a. Complexe Grande-Baleine, Avant-projet, phase II. Habitat et abondance de l'avifaune terrestre. Report for Hydro-Québec, vice-présidence Environnement, Montréal, 94 pp.

Consortium Gauthier et Guillemette – GREBE. 1991b. Complexe Grande-Baleine, Avant-projet, phase II. Étude de l'avifaune et du Castor: Habitats et utilisation du passage de Manitounuk par la sauvagine, 1990. Report for Hydro-Québec, vice-présidence Environnement. Preliminary version, Saint-Romuald, Québec, 117 pp.

Consortium Gauthier et Guillemette – GREBE. 1991c. Complexe Grande-Baleine, Avant-projet, phase II. Utilisation des arbustaies sur le complexe La Grande et régime alimentaire des lagopèdes. Report for Hydro-Québec, vice-présidence Environnement. Preliminary version, Saint-Romuald, Québec, 84 pp.

Consortium Gauthier et Guillemette – GREBE. 1992a. Complexe Nottaway-Broadback-Rupert. La sauvagine, vol. 1: Densité, abondance et habitats de la sauvagine. Report for Hydro-Québec, vice-présidence Environnement, Montréal, 238 pp.

Consortium Gauthier et Guillemette – GREBE. 1992b. Complexe Nottaway-Broadback-Rupert. La sauvagine, vol. 2: Comparaisons des densités de la sauvagine entre 1990 et 1991. Report for Hydro-Québec, vice-présidence Environnement, Montréal, 37 pp.

Consortium Gauthier et Guillemette – GREBE. 1992c. Complexe Nottaway-Broadback-Rupert. La sauvagine, vol. 3: Fréquentation des rivières par la sauvagine au printemps et en été. Report for Hydro-Québec, vice-présidence Environnement, Montréal, 72 pp.

Consortium Gauthier et Guillemette – GREBE. 1992d. Complexe Nottaway-Broadback-Rupert. La sauvagine, vol. 4: Utilisation spatio-temporelle des milieux humides lors de la migration automnale 1990. Report for Hydro-Québec, vice-présidence Environnement, Montréal, 57 pp.

Consortium Gauthier et Guillemette – GREBE. 1992e. Complexe Nottaway-Broadback-Rupert. La sauvagine, vol. 5: Revue des techniques d'aménagement des habitats de la sauvagine. Report for Hydro-Québec, vice-présidence Environnement, Montréal, 81 pp.

Consortium Gauthier et Guillemette – GREBE. 1992f. Complexe Nottaway-Broadback-Rupert. Oiseaux aquatiques, vol. 1: Habitats et répartition du Grand Héron (*Ardea herodias*). Report for Hydro-Québec, vice-présidence Environnement, Montréal, 69 pp.

Consortium Gauthier et Guillemette – GREBE. 1992g. Complexe Nottaway-Broadback-Rupert. Oiseaux aquatiques, vol. 2: Habitats et répartition des limicoles en période de reproduction. Report for Hydro-Québec, vice-présidence Environnement, Montréal, 55 pp.

Consortium Gauthier et Guillemette – GREBE. 1992h. Complexe Nottaway-Broadback-Rupert. Oiseaux aquatiques, vol. 3: Habitats, abondance et répartition de la Mouette de Bonaparte (*Larus philadelphia*). Report for Hydro-Québec, vice-présidence Environnement, Montréal, 33 pp.

Consortium Gauthier et Guillemette – GREBE. 1992i. Complexe Nottaway-Broadback-Rupert. Oiseaux aquatiques, vol. 4: Habitats, abondance et répartition de la Sterne pierregarin (*Sterna hirundo*). Report for Hydro-Québec, vice-présidence Environnement, Montréal, 29 pp.

Consortium Gauthier et Guillemette – GREBE. 1992j. Complexe Nottaway-Broadback-Rupert. Oiseaux aquatiques, vol. 5: Habitats, abondance et répartition des huarts, des râles, de la Grue du Canada et des autres oiseaux aquatiques. Report for Hydro-Québec, vice-présidence Environnement, Montréal, 51 pp.

Consortium Gauthier et Guillemette – GREBE. 1992k. Complexe Nottaway-Broadback-Rupert. Les oiseaux terrestres, vol. 1: Habitats, abondance et répartition du Balbuzard (*Pandion haliaetus*). Report for Hydro-Québec, vice-présidence Environnement, Montréal, 33 pp.

Consortium Gauthier et Guillemette – GREBE. 1992l. Complexe Nottaway-Broadback-Rupert. Les oiseaux terrestres, vol. 2: Habitats, abondance et répartition du Pygargue à tête blanche (*Haliaeetus leucocephalus*). Report for Hydro-Québec, vice-présidence Environnement, Montréal, 39 pp.

Consortium Gauthier et Guillemette – GREBE. 1992m. Complexe Nottaway-Broadback-Rupert. Les oiseaux terrestres, vol. 3: Habitats et répartition des autres espèces d'oiseaux de proie. Report for Hydro-Québec, vice-présidence Environnement, Montréal, 54 pp.

Consortium Gauthier et Guillemette – GREBE. 1992n. Complexe Nottaway-Broadback-Rupert. Les oiseaux terrestres, vol. 4: Habitats, abondance et répartition de la Chouette lapone (*Strix nebulosa*). Report for Hydro-Québec, vice-présidence Environnement, Montréal, 34 pp.

Consortium Gauthier et Guillemette – GREBE. 1992o. Complexe Nottaway-Broadback-Rupert. Les oiseaux terrestres, vol. 6: Habitats, abondance et répartition des passereaux et des pics. Report for Hydro-Québec, vice-présidence Environnement, Montréal, 2 vol., 142 pp.

Consortium Gauthier et Guillemette – GREBE. 1992p. Complexe Nottaway-Broadback-Rupert. Avifaune de la baie de Rupert, vol. 1: Utilisation estivale par la sauvagine des habitats côtiers de l'est de la baie James, 1991. Report for Hydro-Québec, vice-présidence Environnement, Saint-Romuald, 55 pp.

Consortium Gauthier et Guillemette – GREBE. 1992q. Complexe Nottaway-Broadback-Rupert. Avifaune de la baie de Rupert, vol. 2: Utilisation des baies de Rupert et Boatswain par la sauvagine, printemps et automne 1991. Report for Hydro-Québec, vice-présidence Environnement, Montréal, 108 pp.

Consortium Gauthier et Guillemette – GREBE. 1992r. Complexe Nottaway-Broadback-Rupert. Avifaune de la baie de Rupert, vol. 3: Régime alimentaire de la sauvagine dans la région de la baie de Rupert au printemps et à l'automne 1991. Report for Hydro-Québec, vice-présidence Environnement, Saint-Romuald, 39 pp.

Consortium Gauthier et Guillemette – GREBE. 1992s. Complexe Nottaway-Broadback-Rupert. Avifaune de la baie de Rupert, vol. 4: Abondance et composition des limicoles durant la migration automnale dans les baies de Rupert et Boatswain, 1991. Report for Hydro-Québec, vice-présidence Environnement, Montréal, 49 pp.

Consortium Gauthier et Guillemette – GREBE. 1992t. Complexe Grande-Baleine, Avant-projet, phase II. Étude de l'avifaune et du Castor: Aire de reproduction des macreuses dans la péninsule Québec-Labrador. Report for Hydro-Québec, vice-présidence Environnement, Montréal, 35 pp.

Consortium Gauthier et Guillemette – GREBE. 1992u. Complexe Grande-Baleine, Avant-projet, phase II. Étude de l'avifaune et du Castor: Stratégies d'accès des macreuses aux lieux de nidification. Report for Hydro-Québec, vice-présidence Environnement, Montréal, 19 pp.

Consortium Gauthier et Guillemette – GREBE. 1992v. Complexe Grande-Baleine, Avant-projet, phase II. Étude de l'avifaune et du Castor: Compte rendu de l'inventaire du Canard arlequin (*Histrionicus histrionicus*) réalisé dans les régions des complexes La Grande et Grande-Baleine en juin 1992 et de la visite de nids d'Aigle royal (*Aquila chrysaetos*) au complexe Grande-Baleine; mission G 5A. Report for Hydro-Québec, vice-présidence Environnement. Preliminary version, Montréal, 9 pp.

Consortium Gauthier et Guillemette – GREBE. 1992w. Complexe Grande-Baleine, Avant-projet, phase II. Étude de l'avifaune et du Castor: Habitat de nidification de la Macreuse à front blanc et de la Macreuse à bec jaune. Report for Hydro-Québec, vice-présidence Environnement, Montréal, 32 pp.

Consortium Gauthier et Guillemette – GREBE. 1992x. Complexe Grande-Baleine, Avant-projet, phase II. Validation des modèles de prédiction du potentiel pour la sauvagine et le Balbuzard dans la zone d'étude du complexe Grande-Baleine. Report for Hydro-Québec, vice-présidence Environnement, Montréal, 9 pp.

Consortium Gauthier et Guillemette – GREBE. 1992y. Complexe Nottaway-Broadback-Rupert. Les oiseaux terrestres, vol. 5: Habitats hivernaux, abondance et répartition des Tétraoninés. Report for Hydro-Québec, vice-présidence Environnement, Montréal, 78 pp.

Consortium Gauthier et Guillemette – GREBE. 1993a. Complexe Grande-Baleine, Avant-projet, phase II. Étude de l'avifaune et du Castor: Acquisition de connaissances sur l'écologie du Canard arlequin. Report for Hydro-Québec, vice-présidence Environnement, Montréal, 38 pp.

Consortium Gauthier et Guillemette – GREBE. 1993b. Complexe Grande-Baleine, Avant-projet, phase II. Étude de l'avifaune et du Castor: dénombrement de l'effectif reproducteur du Canard arlequin, 1992. Report for Hydro-Québec, vice-présidence Environnement, Montréal, 22 pp.

Consortium Gauthier et Guillemette – GREBE. 1993c. Complexe Grande-Baleine, Avant-projet, phase II. Étude de l'avifaune et du Castor: Habitat et répartition des limicoles. Report for Hydro-Québec, vice-présidence Environnement, Montréal, 87 pp.

Consortium Gauthier et Guillemette – GREBE. 1993d. Complexe Grande-Baleine, Avant-projet, phase II. Étude télémétrique de l'Aigle royal. Report for Hydro-Québec, vice-présidence Environnement, Montréal, 56 pp.

Consortium Gauthier et Guillemette – GREBE. 1993e. Complexe Grande-Baleine, Avant-projet, phase II. Étude de l'avifaune et du Castor: Description et utilisation de l'habitat d'élevage des macreuses à l'été 1990. Final report for Hydro-Québec, vice-présidence Environnement, Saint-Romuald, 54 pp.

Consortium Gauthier et Guillemette – GREBE. 1993f. Complexe Grande-Baleine, Avant-projet, phase II. Étude de l'avifaune et du Castor: Fréquentation et description des habitats de prénidification des macreuses, 1990. Final report for Hydro-Québec, vice-présidence Environnement, Saint-Romuald, 49 pp.

Consortium Gauthier et Guillemette – GREBE. 1993g. Complexe Grande-Baleine, Avant-projet, phase II. Étude de l'avifaune et du Castor: Fréquentation par la sauvagine des habitats côtiers situés à l'est de la baie d'Hudson, 1991. Final report for Hydro-Québec, vice-présidence Environnement, Saint-Romuald, 77 pp.

Cooch, F.G. 1954. Eider duck survey, eastern Arctic 1954. Environment Canada, Canadian Wildlife Service, 20 pp.

Cooch, F.G. 1963. Recent changes in the distribution of colour phases of *Chen caerulescens caerulescens*. Congressus internationalis ornithologici XII: 1182-1194.

Cooch, F.G. 1965. The breeding biology and management of the Northern Eider (*Somateria mollissima borealis*) in the Cape Dorset area, Northwest Territories. Canadian Wildlife Service. Wildlife Management Bulletin 2(10): 68.

Cooch, F.G. and K. Newell. 1977. Species of waterfowl and age and sex ratios of ducks and geese harvested in Canada during the 1975 season. Canadian Wildlife Service. Progress Notes no. 71, 43 pp.

Cooch, F.G., K. Newell, and H. Raible. 1975. Species of waterfowl and age and sex ratios of ducks harvested in Canada during the 1974 season. Canadian Wildlife Service. Progress Notes no. 57, 53 pp.

Cooke, F. 1981. 200 years of ornithology in Canada. Canadian Field-Naturalist 95(1): 1-5.

Cooke, F. and F.G. Cooch. 1968. The genetics of polymorphism in the Snow Goose, *Anser caerulescens*. Evolution 22: 289-300.

Coon, R.A., B.K. Williams, J.S. Lindzey, and J.L. George. 1982. Examination of woodcock nest sites in central Pennsylvania, pp. 55-62 *in* T.J. Dwyer and G.L. Storm, eds. Woodcock Ecology and Management. Papers from the seventh woodcock symposium held at the Pennsylvania State University, University Park, Pennsylvania, 28-30 October 1980. United States Fish and Wildlife Service. Research Report no. 14, Washington, 191 pp.

Cooper, J.A. 1978. The history and breeding biology of the Canada Geese of Marshy Point, Manitoba. Wildlife Monographs no. 61, 87pp.

Cooper, J.M. 1994. Least Sandpiper (*Calidris minutilla*), *in* A. Poole and F. Gill, eds. The birds of North America, no. 115. The Academy of Natural Sciences, Philadelphia. The American Ornithologists' Union, Washington, 28 pp.

Cooper, R.J., K.M. Dodge, P.J. Martinat, S.B. Donahoe, and R.C. Whitmore. 1990. Effect of diflubenzuron application on eastern deciduous forest birds. Journal of Wildlife Management 54(3): 486-493.

Cooper, S. 1984. Habitat and size of the Le Conte's Sparrow's territory. Loon 56(3): 162-165.

Coote, F. 1916. Faune ornithologique des îles Pèlerin. Naturaliste canadien 43: 2-8.

Cormier, C. and N. Breton. 1984. Liste annotée des oiseaux du Saguenay–Lac-Saint-Jean. Le Harfang 7(4): 589.

Cornwell, G.W. 1963. Observations on the breeding biology and behavior of a nesting population of Belted Kingfisher. Condor 65: 426-431.

Côté, R. 1980. Réseau de surveillance écologique du complexe La Grande: faune avienne. Compte rendu des observations de 1980. Société d'Énergie de la Baie James, 5 pp.

Côté, R. 1981. Réseau de surveillance écologique du complexe La Grande: faune avienne. Compte rendu des observations de 1981. Société d'Énergie de la Baie James, 8 pp.

Côté, R. 1982. Réseau de surveillance écologique du complexe La Grande: faune avienne. Compte rendu des observations de 1982. Société d'Énergie de la Baie James.

Cottam, C. 1939. Food habits of North American diving ducks. United States Department of Agriculture. Technical Bulletin no. 643, 139 pp.

Cottam, C., J.J. Lynch, and A.L. Nelson. 1944. Food habits and management of American Sea Brant. Journal of Wildlife Management 8: 36-56.

Cotter, R.C. 1991. Population attributes and reproductive biology of Rock Ptarmigan (*Lagopus mutus*) in the central Canadian Arctic. M.Sc. thesis, University of Alberta, Edmonton, 103 pp.

Cotter, R.C. 1992. A literature review of the food habits of Rock Ptarmigan (*Lagopus mutus*) and Willow Ptarmigan (*L. lagopus*), pp. 32-103 *in* A. Gunn, ed. Experimental commercial harvest of ptarmigan, Kitikmeot Region, 1989. Government of Nortwest Territories, Department of Renewable Resources. Manuscript Report no. 47, 103 pp.

Cottrell, S.D., H.H. Prince, and P.I. Padding. 1990. Nest success, duckling survival, and brood habitat selection of Wood Ducks in a Tennessee riverine system, pp. 191-197 *in* L.H. Fredrickson, G.V. Burger, S.P. Havera, D.A. Graber, R.E. Kirby, and T.S. Taylor, eds. Proceedings. 1988 North American Wood Duck symposium, St. Louis, Missouri, 390 pp.

Coues, E. 1903. Key to North American birds, 5th edition. Dana Estes and Company, Boston.

Couillard, L. and P. Grondin. 1986. La végétation des milieux humides du Québec. Gouvernement du Québec, Ministère de l'Environnement, Direction du patrimoine écologique, Les Publications du Québec, 400 pp.

Coulson, J.C. 1972. The significance of the pair-bond in the Kittiwake, pp. 424-433 *in* K.H. Voous, ed. Proceedings. XVth International Ornithological Congress, The Hague, 30 August-5 September 1970. E.J. Brill, Leiden, 745 pp.

Coulson, J.C. 1983. The changing status of the Kittiwake *Rissa tridactyla* in the British Isles, 1969-1979. Bird Study 30: 9-11.

Coulson, J.C. 1985. Changes in the biology of the Kittiwake *Rissa tridactyla*: a 31-year study of a breeding colony. Journal of Animal Ecology 54: 9-26.

Coulson, J.C. and J. Horobin. 1976. The influence of age on the breeding biology and survival of the Arctic Tern (*Sterna paradisaea*). Journal of Zoology (London) 178: 247-260.

Coulter, M.W. and J.C. Baird. 1982. Changing forest land uses and opportunities for woodcock management in New England and the Maritime provinces, pp. 75-85 *in* T.J. Dwyer and G.L. Storm, eds. Woodcock Ecology and Management. Papers from the seventh woodcock symposium held at the Pennsylvania State University, University Park, Pennsylvania, 28-30 October 1980. United States Fish and Wildlife Service. Research Report no. 14, Washington, 191 pp.

Coulter, M.W. and W.R. Miller. 1968. Nesting biology of Black Ducks and Mallards in northern New England. Vermont Fish and Game Department. Bulletin no. 68-2, 74 pp.

Couper, W. 1881. The American Goshawk. Canadian Sportsman and Naturalist 1(3): 18.

Court, G.S., D.M. Bradley, C.C. Gates, and D.A. Boag. 1988. The population biology of Peregrine Falcons in the Keewatin district of the Northwest Territories, Canada, pp. 729-739 *in* T.J. Cade, J.H. Enderson, C.G. Thelander, and C.M. White, eds. Peregrine Falcon populations: their management and recovery. The Peregrine Fund, Boise, Idaho.

Courville, S. 1980. La crise agricole du Bas-Canada: Éléments d'une réflexion géographique (parties 1 et 2). Cahiers de géographie du Québec 24(62): 193-223 and 24(63): 385-428.

Couture, R. 1990. Coup d'oeil sur les habitats de la Bécasse d'Amérique. Proceedings. Prepared for Canadian Wildlife Service by Fondation Les oiseleurs du Québec. 1er colloque provincial sur la Bécasse d'Amérique [1st provincial symposium on the American Woodcock], Rivière-du-Loup.

Couture, R. and J.-C. Bourgeois. 1974. Woodcock studies at Trois-Rivières, Québec, pp. 1-8 *in* J.W. Artmann and S.R. Pursglove, eds. Proceedings. Fifth American Woodcock Workshop, Athens.

Couture, R. and J.-C. Bourgeois. 1977. American Woodcock density variations during the breeding season in central Quebec. Proceedings. Woodcock Symposium 6: 93-99.

Cowardin, L.M., G.E. Cummings, and P.E. Reed, Jr. 1967. Stump and tree nesting by Mallards and Black Ducks. Journal of Wildlife Management 31(2): 229-235

Cox, G.W. 1960. A life history of the Mourning Warbler. Wilson Bulletin 72(1): 5-28.

Coy, P.L., W. Prem, and L. Baertlein. 1993. 70. Young aspen forest with scattered mature pines, pp. 76-77 *in* J.D. Lowe, ed. Resident bird counts 1992. Journal of Field Ornithology 64(1)(supplement): 1-128.

Crabtree, R.L., L.S. Broome, and M.L. Wolfe. 1989. Effects of habitat characteristics on Gadwall nest predation and nest-site selection. Journal of Wildlife Management 53(1): 129-137.

Craig, E.H., T.H. Craig, and L.R. Powers. 1988. Activity patterns and home-range use of nesting Long-eared Owls. Wilson Bulletin 100(2): 204-213.

Craig, R.B. 1978. An analysis of the predatory behavior of the Loggerhead Shrike. Auk 95(2): 221-234.

Craig, R.J. 1984. Comparative foraging ecology of Louisiana and Northern waterthrushes. Wilson Bulletin 96(2): 173-183.

Craig, R.J. 1985. Comparative habitat use by Louisiana and Northern waterthrushes. Wilson Bulletin 93(3): 347-355.

Craig, R.J. 1987. Divergent prey selection in two species of waterthrushes (*Seiurus*). Auk 104: 180-187.

Craig, T.H., J.W. Connelly, E.H. Craig, and T.L. Parker. 1990. Lead concentrations in Golden and Bald eagles. Wilson Bulletin 102(1): 130-133.

Craighead, J.[J.] and F.[C.] Craighead. 1940. Nesting Pigeon Hawks. Wilson Bulletin 52: 241-248.

Craighead, J.J. and F.C. Craighead, Jr. 1956. Hawks, owls and wildlife. Stackpole Co. and Wildlife Management Institute, Harrisburg, Pennsylvania [1969, Dover Publications, New York], 443 pp.

Craighead, J.J. and D.S. Stockstad. 1964. Breeding age of Canada Geese. Journal of Wildlife Management 28(1): 57-64.

Cramp, S. 1985. Handbook of the birds of Europe, the Middle East and North Africa: the birds of the Western Palearctic. Vol. 4: terns to woodpeckers. Oxford University Press, Oxford, 960 pp.

Cramp, S. 1988. Handbook of the birds of Europe, the Middle East and North Africa: the birds of the Western Palearctic. Vol. 5: tyrant flycatchers to thrushes. Oxford University Press, Oxford, 1063 pp.

Cramp, S. 1992. Handbook of the birds of Europe, the Middle East and North Africa: the birds of the Western Palearctic. Vol. 6: warblers. Oxford University Press, Oxford, 728 pp.

Cramp, S. 1994a. Handbook of the birds of Europe, the Middle East and North Africa: the birds of the Western Palearctic. Vol. 8: crows to finches. Oxford University Press, Oxford, 899 pp.

Cramp, S. 1994b. Handbook of the birds of Europe, the Middle East and North Africa: the birds of the Western Palearctic. Vol. 9: buntings and new world warblers. Oxford University Press, Oxford, 488 pp.

Cramp, S. and C. M. Perrins. 1993. Handbook of the birds of Europe, the Middle East and North Africa: the birds of the Western Palearctic. Vol. 7: flycatchers to shrikes. Oxford University Press, Oxford, 577 pp.

Cramp, S. and K.E.L. Simmons. 1977. Handbook of the birds of Europe, the Middle East and North Africa: the birds of the Western Palearctic. Vol. 1: ostrich to ducks. Oxford University Press, Oxford, 722 pp.

Cramp, S. and K.E.L. Simmons. 1980. Handbook of the birds of Europe, the Middle East and North Africa: the birds of the Western Palearctic. Vol. 2: hawks to bustards. Oxford University Press, Oxford, 695 pp.

Cramp, S. and K.E.L. Simmons. 1983. Handbook of the birds of Europe, the Middle East and North Africa: the birds of the Western Palearctic. Vol. 3: waders to gulls. Oxford University Press, Oxford, 913 pp.

Craven, S.R. 1984. Food habits of Canada Geese on the coast of Hudson Bay. Journal of Wildlife Management 48(2): 567-569.

Crawford, H.S. and D.T. Jennings. 1989. Predation by birds on Spruce Budworm *Choristoneura fumiferana*: functional, numerical, and total responses. Ecology 70(1): 152-163.

Crawford, H.S., R.G. Hooper, and R.W. Titterington. 1981. Songbird population response to silvicultural practices in central appalachian hardwoods. Journal of Wildlife Management 45(3): 680-692.

Crawford, H.S., R.W. Titterington, and D.T. Jennings. 1983. Bird predation and Spruce Budworm populations. Journal of Forestry 81(7): 433-435, 478.

Crawford, R.D. 1977. Polygynous breeding of Short-billed Marsh Wren. Auk 94(2): 359-362.

Cringan, A.T., A. Salvadori, and R.H. Manske. 1971. Red Crossbill breeding in Wellington County, Ontario. Canadian Field-Naturalist 85(3): 258.

Crins, W.J. 1987a. Black-throated Green Warbler, pp. 382-383 *in* M.D. Cadman, P.F.J. Eagles, and F.M. Helleiner, eds. Atlas of the breeding birds of Ontario. University of Waterloo Press, Waterloo, 617 pp.

Crins, W.J. 1987b. Yellow-rumped Warbler, pp. 380-381 *in* M.D. Cadman, P.F.J. Eagles, and F.M. Helleiner, eds. Atlas of the breeding birds of Ontario, University of Waterloo Press, Waterloo, 617 pp.

Crocker-Bedford, D.C. and B. Chaney. 1988. Characteristics of Goshawk nesting stands, pp. 210-217 *in* R.L. Glinski, B.G. Pendleton, M.B. Moss, M.N. Lefranc, Jr., B.A. Millsap, and S.W. Hoffman, eds. Proceedings. Southwest raptor management symposium and workshop. Institute for Wildlife Research, National Wildlife Federation, Washington. National Wildlife Federation Scientific and Technical Series no. 11, 395 pp.

Crockett, A.B. and P.L. Hansley. 1978. Apparent response of *Picoides* woodpeckers to outbreaks of pine bark beetle. Western Birds 9: 67-70.

Crocoll, S.T. 1994. Red-shouldered Hawk (*Buteo lineatus*), *in* A. Poole and F. Gill, eds. The birds of North America, no. 107. The Academy of Natural Sciences, Philadelphia. The American Ornithologists' Union, Washington, 20 pp.

Crocoll, S.T. and J.W. Parker. 1989. The breeding biology of Broad-winged and Red-shouldered hawks in western New York. Journal of Raptor Research 23(4): 125-139.

Cronan, J.M. and B.F. Halla. 1968. Fall and winter foods of Rhode Island waterfowl. Department of National Wildlife, pamphlet no. 7, p. 40.

Crooks, M.P. and G.O. Hendrickson. 1953. Field Sparrow life history in central Iowa. Bird Life 23: 10-13.

Croskery, P.R. 1988. Reoccupation of Common Loon, *Gavia immer*, territories following removal of the resident pair. Canadian Field-Naturalist 102(2): 264-265.

Croskery, P.R. 1991. Common Loon, *Gavia immer*, nesting success and young survival in northwestern Ontario. Canadian Field-Naturalist 105(1): 45-48.

Crouch, J.E. 1936. Nesting habits of the Cedar Waxwing (*Bombycilla cedrorum*). Auk 53(1): 1-8.

Croxall, J.P. and A.J. Gaston. 1988. Patterns of reproduction in high-latitude Northern- and Southern-Hemisphere seabirds, pp. 1176-1194 *in* H. Ouellet, ed. Acta XIX Congressus Internationalis Ornithologici, Ottawa, 22-29 June 1986. Ottawa University Press, vol. I, pp. 1-1404; vol. II, pp. 1405-2815.

Croxall, J.P., P.G.H. Evans, and R.W. Schreiber. 1984. Status and conservation of the world's seabirds. International Council for Bird Preservation. Technical Publication no. 2, Cambridge, 778 pp.

Cruikshank, A.D. 1979. American Redstart (*Setophaga ruticilla*), pp. 139-141 *in* L. Griscom and A. Sprunt, Jr., eds. The warblers of America, 2nd edition. Doubleday and Company, Garden City, New York, 302 pp.

Cummings, J.L., J.L. Guarino, and C.E. Knittle. 1989. Chronology of blackbird damage to sunflowers. Wildlife Society Bulletin 17(1): 50-52.

Curley, E.M., R. Bowman, and D.M. Bird. 1987. Nest site characteristics of boxes occupied by starlings and kestrels. Raptor Research 6: 160-164.

Currie, H.G. 1987. Tufted Titmouse, pp. 296-297 *in* M.D. Cadman, P.F.J. Eagles, and F.M. Helleiner, eds. Atlas of the breeding birds of Ontario. University of Waterloo Press, Waterloo, 617 pp.

Curry, R. 1987. Gray Catbird, pp. 332-333 *in* M.D. Cadman, P.F.J. Eagles, and F.M. Helleiner, eds. Atlas of the breeding birds of Ontario. University of Waterloo Press, Waterloo, 617 pp.

Curtis, S. and L. Allen. 1976. The waterfowl ecology of the Quebec coast of James Bay. Canadian Wildlife Service, unpublished report, Ottawa, 72 pp.

Custer, T.W. and W.E. Davis, Jr. 1982. Nesting by one-year-old Black-crowned Night-Herons on Hope Island, Rhode Island. Auk 99: 784-786.

Custer, T.W. and D.W. Peterson, Jr. 1991. Growth rates of Great Egret, Snowy Egret and Black-crowned Night-Heron chicks. Colonial Waterbirds 14(1): 46-50.

Custer, T.W. and F.A. Pitelka. 1977. Demographic features of a Lapland Longspur population near Barrow, Alaska. Auk 94(3): 505-525.

Custer, T.W. and F.A. Pitelka. 1987. Nesting by Pomarine Jaegers near Barrow, Alaska, 1971. Journal of Field Ornithology 58(2): 225-230.

Cuthbert, N.L. 1954. A nesting study of the Black Tern in Michigan. Auk 71: 36-63.

Cuvier, G. 1826-1828. Oeuvres complètes de Buffon. (Vol. 20-28). Delangle Frères, ed., Paris.

Cyr, A. 1974. L'avifaune nicheuse de deux types d'associations forestières dans la zone du nouvel aéroport international de Montréal à Mirabel, Québec. M.Sc. thesis, Université de Montréal, Montréal, 237 pp.

Cyr, A. 1975. Méthode de description cartographique de l'habitat d'oiseaux forestiers. Alauda 43(4): 417-426.

Cyr, A. 1980. Un bain de poussière inusité. Bulletin ornithologique 25(3): 83.

Cyr, A. 1988. Planter des fruits, jardiner des oiseaux. Franc-Nord, hors série 2: 28-33.

Cyr, A. and D. Lacombe. 1988. La stérilisation est-elle une avenue fertile pour contrôler les populations d'oiseaux nuisibles, pp. 484-492 *in* H. Ouellet, ed. Acta XIX Congressus Internationalis Ornithologici, Ottawa, 22-29 June 1986. Ottawa University Press, vol. I, pp. 1-1404; vol. II, pp. 1405-2815.

Cyr, A. and J. Larivée. 1993a. A check-list approach for monitoring neotropical migrant birds: twenty-year trends in birds of Québec using ÉPOQ, pp. 229-236 *in* D.M. Finch and P.W. Stangel, eds. Status and management of neotropical migratory birds. United States Department of Agriculture, Forest Service, Rocky Mountain Forest and Range Experiment Station, Fort Collins, Colorado. General Technical Report RM-229, 422 pp.

Cyr, A. and J. Larivée. 1993b. Monitoring bird population changes with ÉPOQ (Étude des populations d'oiseaux du Québec/Studies of bird populations in Québec): twenty-year trends. Presented at the Bird Monitoring Symposium: the Canadian experience. Society of Canadian Ornithologists, Guelph, Ontario, April 1993, 18 pp.

Cyr, A. and J. Larivée. 1995. Atlas saisonnier des oiseaux du Québec. Les Presses de l'Université de Sherbrooke and Société de loisir ornithologique de l'Estrie, Sherbrooke, 711 pp.

Daily, G.C., P.R. Ehrlich, and N.M. Haddad. 1993. Double keystone bird in a keystone species complex. Proceedings. National Academy of Science 90: 592-594.

Dale, F.H. 1956. Pheasants and pheasant populations, *in* D.L. Allen, ed. Pheasants in North America. The Stackpole Company and the Wildlife Management Institute, Harrisburg, Pennsylvania, 490 pp.

D'Amours, P. 1989. Distribution géographique de 50 espèces d'oiseaux dans le bassin hydrographique de la rivière George (Nouveau-Québec). Unpublished report, 52 pp.

Dance, K. 1987. Pileated Woodpecker, pp. 248-249 *in* M.D. Cadman, P.F.J. Eagles, and F.M. Helleiner, eds. Atlas of the breeding birds of Ontario. University of Waterloo Press, Waterloo, 617 pp.

Danchin, E. 1987a. Contexte social et comportements reproducteurs dans les colonies de Mouette tridactyle (*Rissa tridactyla*). Alauda 55(2): 93-111.

Danchin, E. 1987b. Les comportements liés à l'occupation du site de reproduction chez la Mouette tridactyle (*Rissa tridactyla*). Les comportements de pré-atterrissage. Behaviour 100: 226-246.

Dane, C. 1966. Some aspects of breeding biology of the Blue-winged Teal. Auk 83(3): 389-402.

Darley, J.A. 1982. Territoriality and mating behavior of the male Brown-headed Cowbird. Condor 84(1): 15-21.

Darley, J.A. 1983. Territorial behavior of the female Brown-headed Cowbird (*Molothrus ater*). Canadian Journal of Zoology 61: 65-69.

Darley, J.A., D.M. Scott, and N.K. Taylor. 1971. Territorial fidelity of catbirds. Canadian Journal of Zoology 49(11): 1465-1478.

Darley, J.A., D.M. Scott, and N.K. Taylor. 1977. Effects of age, sex and breeding success on site fidelity of Gray Catbirds. Bird-Banding 48(2): 145-151.

Darveau, M. 1988. Le Vacher à tête brune ou l'art de ne pas mettre tous ses oeufs dans le même panier. Carnets de zoologie 48: 67-71.

Darveau, M. 1993. Notes sur les oiseaux de l'île aux Basques. Bulletin ornithologique 37(4): 128-130.

Darveau, M., P. Bellefleur, and B. Houde. 1982. Modèle de simulation de la dynamique des communautés aviennes en fonction de la succession de l'érablière à Bouleau jaune. Naturaliste canadien 109: 51-62.

Darveau, M., J.-L. DesGranges, Y. Mauffette, and G. Gauthier. 1990. Impacts du dépérissement des érablières sur trois oiseaux insectivores. Canadian Wildlife Service, Québec Region. Technical Report Series no. 106, 65 pp. [An abstract is available in English.]

Darveau, M., J.-L. DesGranges, and G. Gauthier. 1992. Habitat use by three breeding insectivorous birds in declining maple forests. Condor 94(1): 72-82.

Darveau, M., G. Gauthier, J.-L. DesGranges, and Y. Mauffette. 1993. Nesting success, nest sites, and parental care of the Least Flycatcher in declining maple forests. Canadian Journal of Zoology 71(8): 1592-1601.

Darveau, M., P. Beauchesne, L. Bélanger, J. Huot, and P. Larue. 1995. Riparian forest strips as habitat for breeding birds in boreal forest. Journal of Wildlife Management 59(1): 67-78.

D'Astous, N. 1994. Fréquentation de la Basse-Côte-Nord par le Canard arlequin selon les témoignages des résidents innus et non-autochtones. Technical report for Environment Canada, Canadian Wildlife Service, Québec Region, by GREBE, Montréal, 11 pp.

Dau, C.P. 1992. The fall migration of Pacific Flyway Brant in relation to climatic conditions. Wildfowl 43: 80-95.

Dauphin, D. 1985. Évaluation des propriétés de la méthode de dénombrement ponctuel d'oiseaux chanteurs: D.R.L. – I.P.A. M.Sc. thesis, Université du Québec à Montréal, Montréal, 84 pp.

David, F. and J.M. Parmelee. 1979. Nest switchings among American Robins and Eastern Phoebes at Itasca State Park. Loon 51(4): 169-170.

David, N. 1976a. Historique de l'établissement du cardinal au Québec. Bulletin ornithologique 21(3): 90-92.

David, N. 1976b. Précisions sur la fréquence de l'hivernage des oiseaux au Québec. Bulletin ornithologique 21(4): 124-133.

David, N. 1978a. Bref aperçu de la distribution de l'Épervier de Cooper au Québec. Bulletin ornithologique 23(2): 64-70.

David, N. 1978b. Victor Gaboriault, ornithologue. Cahiers d'ornithologie Victor-Gaboriault no. 1. Club des ornithologues du Québec, Charlesbourg, 46 pp.

David, N. 1979. Barn Owls in Quebec. Canadian Field-Naturalist 93: 323-324.

David, N. 1980a. État et distribution des oiseaux du Québec méridional. Cahiers d'ornithologie Victor-Gaboriault, no. 3. Club des ornithologues du Québec, Charlesbourg, 213 pp.

David, N. 1980b. Status and distribution of birds in southern Quebec. Cahiers d'ornithologie Victor-Gaboriault no. 4. Club des ornithologues du Québec, Charlesbourg, 213 pp.

David, N. 1983. The nesting season: Quebec region. American Birds 37(6): 970-971.

David, N. 1990. Les meilleurs sites d'observation des oiseaux au Québec. Québec Science Éditeur, Sillery, 311 pp.

David, N. 1991a. Du tétras au Goglu. QuébecOiseaux 3(1): 18-22.

David, N. 1991b. Reconnaître le Sizerin blanchâtre. QuébecOiseaux 3(2): 16-17.

David, N. and G. Duquette. 1982. Comment nourrir les oiseaux autour de chez soi. Éditions Québec Science, Presses de l'Université du Québec, Sillery, 70 pp.

David, N. and G. Duquette. 1983. Comment nourrir les oiseaux autour de chez soi, 2nd edition. Éditions Québec Science, Presses de l'Université du Québec, Sillery, 70 pp.

David, N. and M. Gosselin. 1975. The nesting season: Quebec region. American Birds 29(5): 945-947.

David, N. and M. Gosselin. 1976a. The spring migration: Quebec region. American Birds 30(4): 815-817.

David, N. and M. Gosselin. 1976b. The nesting season: Quebec region. American Birds 30(5): 931-932.

David, N. and M. Gosselin. 1977a. The autumn migration: Quebec region. American Birds 31(2): 153-155.

David, N. and M. Gosselin. 1977b. The nesting season: Quebec region. American Birds 31(6): 1114-1116.

David, N. and M. Gosselin. 1978. The spring migration: Quebec region. American Birds 32(5): 981-983.

David, N. and M. Gosselin. 1979a. The autumn migration: Quebec region. American Birds 33(2): 158-159.

David, N. and M. Gosselin. 1979b. The spring migration, Quebec region. American Birds 33(5): 753-754.

David, N. and M. Gosselin. 1981a. The spring migration: Quebec region. American Birds 35(5): 803-804.

David, N. and M. Gosselin. 1981b. The autumn migration: Quebec region. American Birds 35(2): 160-162.

David, N. and M. Gosselin. 1982a. The fall season: Quebec region. American Birds 36(2): 155-157.

David, N. and M. Gosselin. 1982b. The spring migration: Quebec region. American Birds 36(5): 831-832.

David, N. and M. Gosselin. 1983. The autumn migration: Quebec region. American Birds 37(2): 158-160.

David, N. and M. Gosselin. 1984. Observer les oiseaux au Québec, 2nd edition. Collection Faire, Éditions Québec Science, Presses de l'Université du Québec, Sillery, 261 pp.

David, N. and J. Larivée. 1991. Le Harfang des neiges en hiver: invasions cycliques ou non? QuébecOiseaux 3(2): 12-15.

David, N., M. Gosselin, and G. Seutin. 1990. Pattern of colonization by the Northern Mockingbird in Quebec. Journal of Field Ornithology 61(1): 1-8.

Davidson, F.S. 1944. Nesting of the Least Bittern. Flicker 16: 19-21.

Davies, C. 1990. Migration, pp. 590-600 *in* D. McFarland, ed. Dictionnaire du comportement animal. Éditions Robert Lafont, Paris, 1013 pp.

Davis, A.H. 1982. Terns showing mixed characters of Black and White-winged Black terns. British Birds 75(12): 579-580.

Davis, C.M. 1978. A nesting study of the Brown Creeper. Living Bird 17: 237-263.

Davis, D.E. 1961. Principles for population control by gametocides. Transactions. North American Wildlife and Natural Resources Conference 26: 160-167.

Davis, D.E. 1974. Emigrations of Northern Shrikes 1959-1970. Auk 91(4): 821-825.

Davis, E. 1943. A study of wild and hand reared Killdeers. Wilson Bulletin 55: 223-233.

Davis, J.W.F. 1975. Specialization in feeding location by Herring Gulls. Journal of Animal Ecology 44: 795-804.

Davis, M.B. 1981. Quaternary history and the stability of forest communities, pp. 132-153 *in* D.C. West, H.H. Shugart, and D.B. Botkin, eds. Forest succession, concept and application. Springer-Verlag, New York, 517 pp.

Davis, M.B. 1986. Climatic instability, time lags and community disequilibrium, pp. 269-285 *in* J.M. Diamond and T.J. Case, eds. Community Ecology. Harper and Row, New York.

Davis, R.A. 1972. A comparative study of the use of habitat by Arctic Loons and Red-throated Loons. Ph.D. dissertation, University of Western Ontario, London, 290 pp.

Davis, T.A. and R.A. Ackerman. 1985. Adaptations of Black Tern (*Chlidonias niger*) eggs for water loss in a moist nest. Auk 102(3): 640-643.

Davis, W.E. Jr. 1993. Black-crowned Night-Heron (*Nycticorax nycticorax*), *in* A. Poole and F. Gill, eds. The birds of North America, no. 74. The Academy of Natural Sciences, Philadelphia. The American Ornithologists' Union, Washington, 20 pp.

Davis, W.E. Jr. and J.A. Kushlan. 1994. Green Heron (*Butorides virescens*), *in* A. Poole and F. Gill, eds. The birds of North America, no. 129. The Academy of Natural Sciences, Philadelphia. The American Ornithologists' Union, Washington, 24 pp.

Davis, W.H., W.C. McComb, and P.N. Allaire. 1986. Nest box use by starlings: does it inhibit bluebird production? Transactions. Kentucky Academy of Science 47(3-4): 133-136.

Davis, W.J. 1980. The Belted Kingfisher (*Megaceryle alcyon*): its ecology and territoriality. M.Sc. thesis, University of Cincinnati, Ohio.

Davis, W.J. 1982. Territory size in *Megaceryle alcyon* along a stream habitat. Auk 99(2): 353-362.

Dawson, W.R., R.L. Marsh, W.A. Buttemer, and C. Carrey. 1983. Seasonal and geographic variation of cold resistance in House Finches *Carpodacus mexicanus*. Physiological Zoology 56(30): 353-369.

Dean, W.R.J. 1985. Starlings, orioles and drongos, pp. 428-430 *in* C.M. Perrins and A.L.A. Middleton, eds. The encyclopedia of birds. Facts on File, New York, 447 pp.

Deane, R. 1910. Audubon's Labrador trip of 1833. Auk 27(1): 42-52.

DeBenedictis, P.A. 1984. Gleanings from the technical literature: Blue-winged/Golden-winged Warbler complex. Birding 16(5): 205-207.

DeBenedictis, P.A. 1990. ABA Checklist: birds of the continental United States and Canada. American Birding Association, 91 pp.

Décarie, R., J.-L. DesGranges, C. Lépine, and F. Morneau. 1993. Impact of insecticides on American Robin (*Turdus migratorius*) in a suburban environment. Environmental Pollution 80: 231-238.

Degen, A.A., B. Pinshow, R. Yosef, M. Kam, and K.A. Nagy. 1992. Energetics and growth rate of Northern Shrike (*Lanius excubitor*) nestlings. Ecology 73(6): 2273-2283.

DeGraaf, R.M. 1987. Breeding birds and gypsy moth defoliation: short-term responses of species and guilds. Wildlife Society Bulletin 15(2): 217-221.

DeGraaf, R.M. 1989. Territory sizes of Song Sparrows, *Melospiza melodia*, in rural and suburban habitats. Canadian Field-Naturalist 103(1): 43-47.

DeGraaf, R.M. and N.L. Chadwick. 1987. Forest type, timber size class, and New England breeding birds. Journal of Wildlife Management 51(1): 212-217.

DeGraaf, R.M. and D.D. Rudis. 1987. New England wildlife: habitat, natural history, and distribution. United States Department of Agriculture, Forest Service, Northeastern Forest Experiment Station, Broomall, Pennsylvania. General Technical Report NE-108, 491 pp.

DeGraaf, R.M. and G.M. Witman. 1979. Trees, shrubs, and vines for attracting birds. A manual for the northeast. University of Massachusetts Press, Amherst, 194 pp.

DeGraaf, R.M., G.M. Witman, J.W Lanier, B.J. Hill, and J.M. Keniston. 1980. Forest habitat for birds of the northeast. United States Department of Agriculture, Forest Service, Eastern Region, Washington, 589 pp.

DeGraaf, R.M., V.E. Scott, R.H. Hamre, L. Ernst, and S.H. Anderson. 1991. Forest and rangeland birds of the United States; natural history and habitat use. United States Department of Agriculture, Forest Service, Agriculture Handbook 688, 625 pp.

De Grazio, J.W., J.F. Besser, T.J. DeCino, J.L. Guarino, and R.I. Starr. 1971. Use of 4-aminopyridine to protect ripening corn from blackbirds. Journal of Wildlife Management 35(3): 565-569.

De Grazio, J.W., J.F. Besser, T.J. DeCino, J.L. Guarino, and E.W. Schafer, Jr. 1972. Protecting ripening corn from blackbirds by broadcasting 4-aminopyridine baits. Journal of Wildlife Management 36(4): 1316-1320.

De Kay, J.E. 1844. Zoology of New York or the New York fauna, vol. 2: Birds. Carroll and Cook, Albany, New York, 141 pp.

DeKoster, R. 1992. Sélection de l'habitat et bilan d'activité de la Grande Oie des neiges au printemps dans la région du lac Saint-Pierre, Québec. M.Sc. thesis, Université du Québec à Montréal, Montréal, 77 pp.

Delacour, J. 1975. The waterfowl of the world. Arco Publishing, New York, vol. 1, 284 pp.; vol. 2, 231 pp.

Delacour, J.T. and M.W. Ridley. 1985. Pheasant, pp. 457-460 *in* B. Campbell and E. Lack, eds. A dictionary of birds. British Ornithologists' Union, Buteo Books, Vermillion, South Dakota, 670 pp.

Delannoy, C.A. and A. Cruz. 1988. Breeding biology of Puerto Rican Sharp-shinned Hawk (*Accipiter striatus venator*). Auk 105(4): 649-662.

Delany, M.F., H.M. Stevenson, and R. McCracken. 1985. Distribution, abundance, and habitat of the Florida Grasshopper Sparrow. Journal of Wildlife Management 49(3): 626-631.

Delaunois, A. 1990. Les oiseaux de chez nous. Collection Nos richesses. Héritage, Saint-Lambert, Québec, 223 pp.

del Hoyo, J., A. Elliot, and J. Sargatal. 1992. Handbook of the birds of the world. Vol. 1: ostrich to ducks. Lynx Edicions, Barcelona, 696 pp.

DellaSala, D.A. 1986. Polygyny in the Yellow Warbler. Wilson Bulletin 98(1): 152-154.

DellaSala, D.A. and D.L. Rabe. 1987. Response of Least Flycatcher *Empidonax minimus* to forest disturbances. Biological Conservation 41: 291-299.

Dement'ev, G.P. and N.A. Gladkov. 1951a. Birds of the Soviet Union, vol. I. English translation from Russian, Israel Program for Scientific Translations, Jerusalem 1966, 704 pp.

Dement'ev, G.P. and N.A. Gladkov. 1951b. Birds of the Soviet Union, vol. II. English translation from Russian, Israel Program for Scientific Translations, Jerusalem 1968, 553 pp.

Dement'ev, G.P. and N.A. Gladkov. 1951c. Birds of the Soviet Union, vol. III. English translation from Russian, Israel Program for Scientific Translations, Jerusalem 1969, 756 pp.

Dement'ev, G.P. and N.A. Gladkov. 1952. Birds of the Soviet Union, vol. IV. English translation from Russian, Israel Program for Scientific Translations, Jerusalem 1967, 683 pp.

Dement'ev, G.P. and N.A. Gladkov. 1954a. Birds of the Soviet Union, vol. V. English translation from Russian, Israel Program for Scientific Translations, Jerusalem 1970, 957 pp.

Dement'ev, G.P. and N.A. Gladkov. 1954b. Birds of the Soviet Union, vol. VI. English translation from Russian, Israel Program for Scientific Translations, Jerusalem 1968, 879 pp.

Demers, A. and P. Laporte. 1988. Inventaire des habitats du Pluvier siffleur sur la Côte-Nord et en Gaspésie. Environment Canada, Canadian Wildlife Service, Québec Region. Unpublished report, 45 pp.

Demille, J.B. 1926. Birds of Gaspé County, Quebec. Auk 43: 508-527.

Dennis, C.J. 1969. The Yellow-shafted Flicker (*Colaptes auratus*) on Nantucket Island, Massachusetts. Bird-Banding 40(4): 290-308.

Dennis, D.G. 1990. Status of Wood Ducks in Canada, pp. 135-137 *in* L.H. Fredrickson, G.V. Burger, S.P. Havera, D.A. Graber, R.E. Kirby, and T.S. Taylor, eds. Proceedings. 1988 North American Wood Duck symposium, St. Louis, Missouri, 390 pp.

Dennis, J.V. 1948. Observations on the Orchard Oriole in the lower Mississippi delta. Bird-Banding 19: 12-20.

Dennis, R.H. and H. Dow. 1984. The establishment of a population of Goldeneyes *Bucephala clangula* breeding in Scotland. Bird Study 31: 217-222.

Denys, N. 1908. The description and natural history of the coast of North America (Acadia). [Translated from the original of 1672] Champlain Society, Toronto, 625 pp.

Derleth, E.L. and G.F. Sepik. 1990. Summer-fall survival of American Woodcock in Maine. Journal of Wildlife Management 54(1): 97-106.

Derleth, E.L., D.G. McAuley, and T.J. Dwyer. 1989. Avian community response to small-scale habitat disturbance in Maine. Canadian Journal of Zoology 67(2): 385-390.

Derrickson, K.C. and R. Breitwisch. 1992. Northern Mockingbird (*Mimus polyglottos*), *in* A. Poole, P. Stettenheim, and F. Gill, eds. The birds of North America, no. 7. The Academy of Natural Sciences, Philadelphia. The American Ornithologists' Union, Washington, 26 pp.

Déry, D.A. 1925. In memoriam: Charles Eusèbe Dionne. Canadian Field-Naturalist 39(3): 61-63.

Déry, D.A. 1933. Premières mentions et descriptions originales de quelques oiseaux de la province de Québec. Naturaliste canadien 60(1): 16-34.

DeSante, D. and P. Pyle. 1986. Distributional checklist of North American birds, vol. 1. United States and Canada, Artemisia Press, Lee Vining, California, 442 pp.

DesGranges, J.-L. 1978a. Adaptive value of social behaviour in the Great Blue Heron (*Ardea herodias*). Proceedings. 1978 meeting of the Colonial Waterbird Group. Colonial Waterbirds 2: 192-201.

DesGranges, J.-L. 1978b. Un couple d'Aigles pêcheurs (*Pandion haliaetus*) niche au sommet d'un tepee indien. Bulletin ornithologique 23: 18-19.

DesGranges, J.-L. 1979. Le Grand Héron au Québec. Carnets de zoologie 39: 20-23.

DesGranges, J.-L. 1980a. Compétition entre le Cormoran à aigrettes et le Grand Héron au moment de la nidification. Naturaliste canadien 107: 199-200.

DesGranges, J.-L. 1980b. Des communautés aviennes du Parc national de la Mauricie, Québec. Canadian Wildlife Service. Occasional Paper no. 41, 34 pp. [Also available in English as: Avian community structure of six forest stands in La Mauricie National Park, Quebec.]

DesGranges, J.-L. 1981a. Comportement reproducteur du Cormoran à aigrettes et du Grand Héron dans l'estuaire du Saint-Laurent. Carnets de zoologie 41: 5-7.

DesGranges, J.-L. 1981b. Observations sur l'alimentation du Grand Héron *Ardea herodias* au Québec, Canada. Alauda 49: 25-34.

DesGranges, J.-L. 1981c. Existe-t-il des nids de débarras chez les goélands? Naturaliste canadien 108: 191-193.

DesGranges, J.-L. 1982. Weight growth of young Double-crested Cormorants in the St. Lawrence estuary, Québec. Colonial Waterbirds 5: 79-86.

DesGranges, J.-L. 1989. Étude des effets de l'acidification sur la faune aquatique au Canada: les oiseaux lacustres et leurs habitats au Québec. Environment Canada, Canadian Wildlife Service. Occasional Paper no. 67, 73 pp. [Also available in English as: Studies of the effects of acidification on aquatic wildlife in Canada: Lacustrine birds and their habitats in Quebec.]

DesGranges, J.-L. and R. Angers. 1981. Considérations écologiques sur les goélands de l'archipel de Montmagny en 1979. Canadian Wildlife Service, Québec Region. Unpublished report, 22 pp.

DesGranges, J.-L. and M. Darveau. 1985. Effect of lake acidity and morphometry on the distribution of aquatic birds in southern Quebec. Holarctic Ecology 8(13): 181-190.

DesGranges, J.-L. and M. Darveau. 1988. Fréquentation des lacs du Québec méridional par les oiseaux aquatiques à la période de reproduction. Naturaliste canadien 115(1): 1-7.

DesGranges, J.-L. and P. Laporte. 1979. Aperçu préliminaire de la situation des huarts (Gaviidae) au Québec. Unpublished report for Le groupe de travail sur la sauvagine du Comité conjoint de l'entente de la baie James et du Nord québécois, by Canadian Wildlife Service, Québec Region, 35 pp. [Also available in English as: Preliminary considerations on the status of loons (Gaviidae) in Quebec.]

DesGranges, J.-L. and P. Laporte. 1983. Fourth and fifth tours of inspection of Quebec heronries 1980-1981. Canadian Wildlife Service, Quebec Region. Progress Notes no. 139, 11 pp.

DesGranges, J.-L. and A. Reed. 1981. Disturbance and control of selected colonies of Double-crested Cormorants in Quebec. Colonial Waterbirds 4: 12-19.

DesGranges, J.-L. and G. Rondeau. 1993. Forest bird response to natural perturbations and silvicultural practices: does logging mimic nature? pp. 80-91 *in* D.H. Kuhnke, ed. Proceedings. Birds in the boreal forest. Workshop, 10-12 March 1992. Prince Albert, Saskatchewan. Forestry Canada, Northwest Region, Northern Forestry Centre, Edmonton, Alberta.

DesGranges, J.-L. and G. Rondeau. 1995. Modifications des communautés d'oiseaux de la sapinière suite au ravage d'une épidémie d'insectes. Forestry Chronicle 71(2): 201-210.

DesGranges, J.-L., G. Chapdelaine, and P. Dupuis. 1984. Sites de nidification et dynamique des populations du Cormoran à aigrettes au Québec. Canadian Journal of Zoology 62: 1260-1267.

DesGranges, J.-L., Y. Mauffette, and G. Gagnon. 1987. Sugar maple forest decline and implications for forest insects and birds. Transactions. North American Wildlife and Natural Resources Conference 52: 677-689.

DesGranges, J.-L., J. Rodrigue, B. Tardif, and M. Laperle. 1994. Exposition au mercure de Balbuzards nichant sur les territoires de la baie James et de la baie d'Hudson. Canadian Wildlife Service, Québec Region. Technical Report Series no. 220, 152 pp. [An abstract is available in English.]

Desloges, C. and E. Émond. 1974. Les grandes unités naturelles de l'île d'Anticosti. Ministère des Terres et Forêts, Service de l'aménagement des terres, 95 pp.

De Smet, K.[D.] 1984. Status report on the Merlin (*Falco columbarius*) in North America. Committee on the Status of Endangered Wildlife in Canada (COSEWIC), Ottawa, 32 pp.

De Smet, K.D. 1987a. Organochlorines, predators, and reproductive success of the Red-necked Grebe in southern Manitoba. Condor 89(3): 460-467.

De Smet, K.D. 1987b. Status report on the Golden Eagle (*Aquila chrysaetos*) in Canada. Committee on the Status of Endangered Wildlife in Canada (COSEWIC), Ottawa, 37 pp.

De Smet, K.[D.] and R.D. James. 1987. Golden Eagle, p. 519 *in* M.D. Cadman, P.F.J. Eagles, and F.M. Helleiner, eds. Atlas of the breeding birds of Ontario. University of Waterloo Press, Waterloo, 617 pp.

Desrochers, A. and P. Fragnier. 1983. Étude de l'avifaune nicheuse et de la végétation de la Réserve nationale de faune du lac Saint-François, été 1983. Canadian Wildlife Service, Québec Region. Unpublished report, 160 pp.

Desrochers, A., S.J. Hannon, and K.E. Nordin. 1988. Winter survival and territory acquisition in a northern population of Black-capped Chickadees. Auk 105(4): 727-736.

Desrochers, B.A. and C.D. Ankney. 1986. Effect of brood size and age on the feeding behavior of adult and juvenile American Coots (*Fulica americana*). Canadian Journal of Zoology 64(7): 1400-1406.

Desrosiers, A. 1993. Inventaire des héronnières du Québec – Été 1992. Technical report, Ministère du Loisir, de la Chasse et de la Pêche, 31 pp.

Desrosiers, A., F. Caron, and R. Ouellet. 1995. Liste de la faune vertébrée du Québec, new edition. Ministère de l'Environnement et de la Faune, Les publications du Québec, Sainte-Foy, 122 pp.

Dessecker, D.R. and R.H. Yahner. 1987. Breeding bird communities associated with Pennsylvania northern hardwood clearcut stands. Proceedings. Pennsylvania Academy of Science 61(2): 170-173.

DeStefano, S. and D.H. Rusch. 1986. Harvest rates of Ruffed Grouse in northeastern Wisconsin. Journal of Wildlife Management 50(3): 361-367.

De Steven, D. 1978. The influence of age on the breeding biology of the Tree Swallow, *Iridoprocne bicolor*. Ibis 120(4): 516-523.

Devereux, J.G. and J.A. Mosher. 1984. Breeding ecology of Barred Owls in the central Appalachians. Raptor Research 18(2): 49-52.

Devillers, P., H. Ouellet, E. Benito-Espinal, R. Beudels, R. Cruon, N. David, C. Érard, M. Gosselin, and G. Seutin. 1993. Noms français des oiseaux du monde. Commission internationale des noms français des oiseaux. Éditions MultiMondes, Sainte-Foy, 452 pp.

Devitt, O.E. 1939. The Yellow Rail in Ontario. Auk 56: 238-243.

De Vos, G.J. 1979. Adaptedness of arena behaviour in Black Grouse (*Tetraotetrix*) and other grouse species (Tetraoninae). Behaviour 68: 277-314.

Dexter, R.W. 1957. Additional recoveries of banded crows, hawks, and owls. Bird-Banding 28: 99.

Dexter, R.W. 1974. Unusually large numbers of Chimney Swifts at a nest. Bird-Banding 45(4): 365.

Dexter, R.W. 1981. Nesting success of Chimney Swifts related to age and the number of adults at the nest, and the subsequent fate of the visitors. Journal of Field Ornithology 52(3): 228-232.

Dexter, R.W. 1992. Sociality of Chimney Swifts (*Chaetura pelagica*) nesting in a colony. North American Bird Bander 17(2): 61-64.

Diamond, A.W. 1986. An evaluation of the vulnerability of Canadian migratory birds to changes in neotropical forest habitats. Unpublished report to Latin American Program. Environment Canada, Canadian Wildlife Service, Ottawa, 93 pp.

Diamond, A.W. 1991. Assessment of the risks from tropical deforestation to Canadian songbirds. Transactions. North American Wildlife and Natural Resources Conference 56: 177-194.

Dickerman, R.W. 1987. The "Old Northeastern" subspecies of Red Crossbill. American Birds 41(2): 189-194.

Dickinson, J.C. Jr. 1968. Rufous-sided Towhee: eastern United States subspecies, pp. 562-580 *in* A.C. Bent. Life histories of North American cardinals, grosbeaks, buntings, towhees, finches, sparrows, and allies. United States National Museum, Washington. [1968, Dover Publications, New York]. Bulletin no. 237, vol. 1, pp. 1-602; vol. 2, pp. 603-1248; vol. 3, pp. 1249-1889.

Dickson, D.L. 1992. The Red-throated Loon as an indicator of environmental quality. Canadian Wildlife Service, Ottawa. Occasional Paper no. 73, 16 pp.

Dickson, D.L. 1993. Breeding biology of Red-throated Loon in the Canadian Beaufort Sea region. Arctic 46: 1-7.

Dickson, K. and L. Métras. 1987. Les oiseaux migrateurs abattus au Canada pendant la saison de chasse de 1986. Environment Canada, Canadian Wildlife Service, Québec Region. Progress Notes no. 171, 42 pp. [Also available in English as: Migratory birds killed in Canada during the 1986 season.]

Dickson, K.M. 1989. Les oiseaux migrateurs abattus au Canada pendant la saison de chasse de 1987. Environment Canada, Canadian Wildlife Service. Progress Notes no. 179, 41 pp. [Also available in English as: Migratory game birds killed in Canada during the 1987 hunting season.]

Dickson, K.M. and A.M. Scheuhammer. 1993. Concentrations of lead in wing bones of three species of ducks in Canada, pp. 6-28 *in* J.A. Kennedy and S. Nadeau, eds. Lead shot contamination of waterfowl and their habits in Canada. Canadian Wildlife Service, Ottawa. Technical Report Series no. 164, 109 pp.

Didiuk, A.B. 1980. Movements and distribution of Canada Geese near Cape Churchill, Manitoba. M.Sc. thesis, University of Wisconsin, Madison, 45 pp.

Dignard, N. 1990. Distribution du Scirpe américain au Québec et localisation des principaux marais, pp. 9-15 *in* L. Bélanger. Compte rendu du minicolloque sur les marais à scirpe du Québec, Sainte-Foy, mars 1990. Environment Canada, Canadian Wildlife Service, Québec Region.

Di Labio, B.M. and P.R. Martin. 1989. Common Raven nesting in eastern Ontario. Ontario Birds 7(2): 62-64.

Dilger, W.C. 1956a. Adaptative modifications and ecological isolating mechanisms in the thrush genera *Catharus* and *Hylocichla*. Wilson Bulletin 68(3): 170-199.

Dilger, W.C. 1956b. Hostile behavior and reproductive isolating mechanisms in the avian genera *Catharus* and *Hylocichla*. Auk 73(3): 313-353.

Dilger, W.C. 1960. Agonistic and social behavior of captive redpolls. Wilson Bulletin 72: 114-132.

Dimension Environnement Ltée. 1982. Étude de la sauvagine du lac Saint-Louis et du bassin de Laprairie, Québec. Project proposal for Direction Environnement d'Hydro-Québec, 182 pp.

Dimension Environnement Ltée. 1985a. Étude de la nidification de la sauvagine et des espèces nichant en colonie dans les îles des rapides de Lachine et le secteur de la vieille centrale à ville LaSalle. Technical report for Direction Environnement d'Hydro-Québec, 36 pp.

Dimension Environnement Ltée. 1985b. Étude des aires d'alimentation du Bihoreau à couronne noire dans la zone sud-est de l'archipel de Montréal et de la sensibilité de l'espèce aux dérangements humains. Technical Report for Direction Environnement d'Hydro-Québec, 91 pp.

Dingle, E. 1942. Rough-winged Swallow, *Stelgidopteryx rufficolis serripennis*, pp. 424-433 *in* A.C. Bent. Life histories of North American flycatchers, larks, swallows, and their allies. United States National Museum, Washington. [1963, Dover Publications, New York]. Bulletin no. 179, 555 pp.

DION, A.[P.]. 1988. Les jardins d'oiseaux: Quoi planter pour avoir des oiseaux en toute saison. Recherche photographique André Cyr. Québec Agenda and Brimar, Beauceville, 191 pp.

DION, A.P. 1993. Nos oiseaux en péril: pouvons-nous encore les sauver? Éditions de l'Homme, Montréal, 175 pp.

DIONNE, A. and M. ST-GEORGES. 1993. Évaluation et suppression des inconvénients causés par les oiseaux au Casino de Montréal. Report for Provancher, Roy et associés, Architectes, by GREBE, 57 pp.

DIONNE, C.-E. 1883. Les oiseaux du Canada. Imprimerie de P.G. Delisle, Québec, 284 pp.

DIONNE, C.-E. 1889. Catalogue des oiseaux de la province de Québec, avec des notes sur leur distribution géographique. J. Dussault, Québec, 119 pp.

DIONNE, C.-E. 1906. Les oiseaux de la province de Québec. Dussault et Proulx, Québec, 414 pp.

DIONNE, C.-E. 1914. Le Pigeon voyageur existe-t-il encore? Naturaliste canadien 41: 1-2.

DIONNE, C.-E. 1918. Le Hibou maculé. Naturaliste canadien 44: 181.

DIONNE, C.-E. 1920. Liste des oiseaux de l'île d'Anticosti. Naturaliste canadien 47(2): 25-29.

DIONNE, J.C. 1972. La dénomination des mers du post-glaciaire au Québec. Cahiers de géographie de Québec 16(39): 483-487.

DIONNE, J.C. 1985. Tidal marsh erosion by geese, St. Lawrence estuary, Québec. Géographie physique et quaternaire 39(1): 99-105.

DIRSCHL, H.J. 1969. Foods of Lesser Scaup and Blue-winged Teal in the Saskatchewan River delta. Journal of Wildlife Management 33(1): 77-87.

DIXON, C.L. 1978. Breeding biology of the Savannah Sparrow on Kent Island. Auk 95(2): 235-246.

DOEPKER, R.[V.] and J.J. OZOGA. 1991a. Magnolia Warbler, pp. 398-399 *in* R. Brewer, G.A. McPeek, and R.J. Adams, Jr., eds. The Atlas of breeding birds of Michigan. Michigan State University Press, East Lansing, 594 pp.

DOEPKER, R.[V.] and J.J. OZOGA. 1991b. Black-throated Green Warbler, pp. 406-407 *in* R. Brewer, G.A. McPeek, and R.J. Adams, Jr., eds. The Atlas of breeding birds of Michigan. Michigan State University Press, East Lansing, 594 pp.

DOEPKER, R.V., R.D. EARLE, and J.J. OZOGA. 1992. Characteristics of Blackburnian Warbler, *Dendroica fusca*, breeding habitat in Upper Michigan. Canadian Field-Naturalist 106: 366-371.

DOERR, P.D., L.B. KEITH, D.H. RUSCH, and C.A. FISCHER. 1974. Characteristics of winter feeding aggregations of Ruffed Grouse in Alberta. Journal of Wildlife Management 38(4): 601-615.

DOLAN, D. 1983. Dénombrement aérien de la sauvagine en migration d'automne sur le fleuve Saint-Laurent entre Berthier et Grondines, 1982. Ministère du Loisir, de la Chasse et de la Pêche, Service de l'aménagement et de l'exploitation de la faune, Direction régionale de Trois-Rivières, 35 pp.

DOLAN, D. 1985. Calcul des superficies de marais et densité des couvées de canards inventoriées au lac Saint-Pierre à l'été 1984. Ministère du Loisir, de la Chasse et de la Pêche, Service de l'aménagement et de l'exploitation de la faune, Direction générale de Trois-Rivières, 28 pp.

DOLAN, D. and J.-C. BOURGEOIS. 1985. Abondance et distribution de la sauvagine dans la région du lac Saint-Pierre aux printemps de 1984, 1985. Ministère du Loisir, de la Chasse et de la Pêche, Service de l'aménagement et de l'exploitation de la faune, Direction générale de Trois-Rivières, 24 pp.

DOLAN, D., J.-C. BOURGEOIS, J. BOURGEOIS, and D. LEHOUX. 1983. Bilan d'activité nocturne de la Bernache du Canada lors de la halte migratoire printanière dans le secteur Nicolet–Longue Pointe, lac Saint-Pierre. Ministère du Loisir, de la Chasse et de la Pêche, Direction générale de Trois-Rivières and Environment Canada, Canadian Wildlife Service, 53 pp.

DOLBEER, R.A. 1990. Ornithology and integrated pest management: Red-winged Blackbirds *Agelaius phoeniceus* and corn. Ibis 132: 309-322.

DOLBEER, R.A. and R.A. STEHN. 1979. Population trends of blackbirds and starlings in North America, 1966-1976. United States Fish and Wildlife Service. Research Report – Wildlife, 214 pp.

DOLBEER, R.A. and R.A. STEHN. 1983. Population status of blackbirds and starlings in North America, 1966-1981. United States Fish and Wildlife Service. Research Report – Bird Damage, 294 pp.

DOLBEER, R.A., P.P. WORONECKI, and R.L. BRUGGERS. 1986. Reflecting tapes repel blackbirds from millet, sunflower, and sweet corn. Wildlife Society Bulletin 14(4): 418-425.

DONAGHEY, R.H. 1975. Spacing behaviour of breeding Bufflehead (*Bucephala albeola*) on ponds in the southern boreal forest. M.Sc. thesis, University of Alberta, Edmonton, 216 pp.

DONOVAN, L.G. and H. OUELLET. 1993. Dictionnaire étymologique des noms d'oiseaux du Canada. Guérin, Montréal, 349 pp.

DORST, J. 1974. The life of birds, vol. 1. Weidenfeld and Nicolson, London, 349 pp.

DOTT, H.E.M. 1986. The spread of the House Sparrow *Passer domesticus* in Bolivia. Ibis 128: 132-137.

DOUGLAS, S.D. and T.E. REIMCHEN. 1988a. Habitat characteristics and population estimate of breeding Red-throated Loons, *Gavia stellata*, on the Queen Charlotte Islands, British Columbia. Canadian Field-Naturalist 102(4): 679-684.

DOUGLAS, S.D. and T.E. REIMCHEN. 1988b. Reproductive phenology and early survivorship in Red-throated Loons, *Gavia stellata*. Canadian Field-Naturalist 102(4): 701-704.

DOUVILLE, R. and J.D. CASANOVA. 1982. La vie quotidienne des Indiens du Canada à l'époque de la colonisation française. Hachette, Paris, 306 pp.

DOW, D.D. 1965. The role of saliva in food storage by the Gray Jay. Auk 82(2): 139-154.

DOW, D.D. 1969. Home range and habitat of the Cardinal in peripheral and central populations. Canadian Journal of Zoology 47: 103-114.

DOW, H. and S. FREDGA. 1983. Breeding and natal dispersal of the Goldeneye, *Bucephala clangula*. Journal of Animal Ecology 52: 681-695.

DOW, H. and S. FREDGA. 1984. Factors affecting reproductive output of the Goldeneye Duck *Bucephala clangula*. Journal of Animal Ecology 53: 679-692.

DOWNS, E.H. 1958. Evening Grosbeaks of South Londonderry, Vermont: 1956. Bird-Banding 29(1): 27-31.

DOYON, M.-R. 1992. Évaluation de l'introduction de la Gélinotte huppée (*Bonasa umbellus*) sur l'île aux Lièvres dans l'estuaire du Saint-Laurent. M.Sc. thesis, Université Laval, Sainte-Foy, 50 pp.

DRAPEAU, P. 1981. État et distribution des populations de Grand Héron et d'Aigle pêcheur au parc de la Gatineau. Commission de la capitale nationale. Technical report, 90 pp.

DRAPEAU, P. 1990. L'avifaune, pp. 116-147 *in* G. Domon, G. Vincent, and A. Bouchard, eds. Le bois de Saraguay, histoire et caractéristiques. Jardin botanique, Ville de Montréal.

DRAPEAU, P. 1993. Structure des communautés d'oiseaux et analyse distributionnelle des espèces le long d'un gradient d'habitats de la forêt du sud du Québec: déterminismes écologiques et variations stochastiques. Ph.D. dissertation, Université de Montréal, Montréal, 279 pp.

DRAPEAU, P. and A. LEDUC. 1990. Analyses géostatistiques des données de l'Atlas des oiseaux nicheurs du Québec. Association québécoise des groupes d'ornithologues, 30 pp.

DRENNAN, M.P., D.C. FOLGER, and C. TREYBALL. 1987. Common Black-headed Gulls on Petit Manan Island, Maine. American Birds 41(2): 195-196.

DREWIEN, R.C. and P.F. SPRINGER. 1969. Ecological relationships of breeding Blue-winged Teal to prairies potholes, pp. 102-105 *in* Canadian Wildlife Service, Saskatoon Wetlands Seminar. Department of Indian Affairs and Northern Development, Ottawa. Report Series no. 6, 262 pp.

DRILLING, N.E. and C.F. THOMPSON. 1988. Natal and breeding dispersal in House Wrens. Auk 105(3): 480-491.

DRILLING, N.E. and C.F. THOMPSON. 1991. Mate switching in multibrooded House Wrens. Auk 108(1): 60-70.

DROZDEK, K. 1991. Some notes to the breeding biology of *Larus ridibundus*. Acta Scientiarum Naturalium scientiarum Bohemoslovacae – Brno 25(1): 1-37.

DRUM, M. 1939. Territorial studies on the Eastern Goldfinch. Wilson Bulletin 51(2): 69-77.

DRURY, W.H. 1973. Population changes in New England seabirds. Bird-Banding 44(4): 267-313.

DRURY, W.H. 1974. Population changes in New England seabirds. Bird-Banding 45(1): 1-15.

DRURY, W.H. JR. 1960. Breeding activities of Long-tailed Jaeger, Herring Gull, and Arctic Tern on Bylot Island, Northwest Territories, Canada. Bird-Banding 31(2): 63-78.

DRURY, W.H. JR. 1961. Studies of the breeding biology of Horned Lark, Water Pipit, Lapland Longspur, and Snow Bunting on Bylot Island, Northwest Territories, Canada. Bird-Banding 32(1): 1-46.

DRYADE. 1980. Habitats propices aux oiseaux migrateurs le long des rives de la rivière Richelieu, de la rivière Outaouais, du fleuve Saint-Laurent, de l'estuaire du Saint-Laurent, de la côte nord du golfe du Saint-Laurent, de la péninsule gaspésienne, des Îles-de-la-Madeleine. Report for Canadian Wildlife Service, Québec Region, Environment Canada, 66 pp.

DRYADE. 1981a. Analyse des pertes de végétation riveraine le long du Saint-Laurent de Cornwall à Matane. Report for Canadian Wildlife Service, Québec Region, Environment Canada, 25 pp.

DRYADE. 1981b. Habitats propices aux oiseaux migrateurs. Analyse des pertes de végétation riveraine le long du Saint-Laurent entre Cornwall et Matane (1945-1960, 1960-1976). Québec, 25 pp.

DRYADE. 1992. Projet de Laforge 1. Caractérisation de la sauvagine et de son utilisation des milieux humides. Final report for Direction Ingénierie et Environnement, Société d'Énergie de la Baie James, 49 pp.

DRYADE. 1993. Projet de Laforge 1. Étude de la sauvagine et caractérisation de ses habitats, été 1992. Final report for Direction Ingénierie et Environnement, Société d'Énergie de la Baie James, 87 pp.

DUBOWY, P.J. 1985. Feeding ecology and behavior of postbreeding male Blue-winged Teals and Northern Shovelers. Canadian Journal of Zoology 63(6): 1291-1297.

DUCAS, P. 1971. L'Hirondelle pourprée. Carnets de zoologie 31(1): 4-8.

DUCKS UNLIMITED CANADA. 1991. Les hauts et les bas du Canard huppé. Conservationniste 5(1): 10-16.

DUDONES, T. 1983. An American Bittern hunting dragonflies. Kingbird 33(2): 108.

DUEBBERT, H.F. 1966. Island nesting of the Gadwall in North Dakota. Wilson Bulletin 78(1): 12-25.

DUEBBERT, H.F. 1969. High nest density and hatching success of ducks on South Dakota CAP land. Transactions. North American Wildlife and Natural Resources Conference 34: 218-228.

DUFFIN, K. 1973. Barn Swallows use freshwater and marine algae in nest construction. Wilson Bulletin 85(2): 237-238.

DUFTY, A.M. JR. 1982. Movements and activities of radio-tracked Brown-headed Cowbirds. Auk 99(2): 316-327.

DUGAS, R., J. DÉZIEL, L. TARDIF, and R. MOISAN. 1989. Traits géographiques, pp. 233-260 *in* Le Québec statistique. Bureau de la statistique du Québec, Québec, 1028 pp.

DUGGER, B.D., K.M. DUGGER, and L.H. FREDRICKSON. 1994. Hooded Merganser (*Lophodytes cucullatus*), *in* A. Poole and F. Gill, eds. The birds of North America, no. 98. The Academy of Natural Sciences, Philadelphia. The American Ornithologists' Union, Washington, 24 pp.

DUHAMEL, A. and L. GAUDREAU. 1983. Le Harfang des neiges, l'oiseau du Québec? Forêt Conservation 49(9): 24-28.

DUKE, G.E. 1966. Reliability of censuses of singing male woodcocks. Journal of Wildlife Management 30(4): 697-707.

DUMAIS, R. 1962. Le Camp Marie-Victorin. Carnets de zoologie 22(2): 20-23.

DUNCAN, J.R. 1987. Movement strategies, mortality, and behavior of radio-marked Great Gray Owls in southeastern Manitoba and northern Minnesota, pp. 101-107 *in* R.W. Nero, R.J. Clark, R.J. Knapton, and R.H. Hamre, eds. Proceedings. Biology and conservation of northern forest owls. Symposium, 3-7 February, Winnipeg. United States Department of Agriculture, Forest Service, Rocky Mountain Forest and Range Experiment Station, Fort Collins, Colorado. General Technical Report RM-142, 309 pp.

DUNCAN, S. 1980. An analysis of the stomach contents of some Sharp-shinned Hawks (*Accipiter striatus*). Journal of Field Ornithology 51(2): 178.

DUNFORD, R.D. and R.B. OWEN, JR. 1973. Summer behavior of immature radio-equipped woodcock in central Maine. Journal of Wildlife Management 37(4): 462-469.

DUNLOP, C.L., H. BLOCKPOEL, and S. JARVIE. 1991. Nesting rafts as a management tool for a declining Common Tern (*Sterna hirundo*) colony. Colonial Waterbirds 14(2): 116-120.

DUNN, E.H. 1979. Nesting biology and development of young in Ontario Black Terns. Canadian Field-Naturalist 93(3): 276-281.

DUNN, E.H. 1987a. Common Loon, pp. 36-37 *in* M.D. Cadman, P.F.J. Eagles, and F.M. Helleiner, eds. Atlas of the breeding birds of Ontario. University of Waterloo Press, Waterloo, 617 pp.

DUNN, E.H. 1987b. Great Blue Heron, pp. 50-51 *in* M.D. Cadman, P.F.J. Eagles, and F.M. Helleiner, eds. Atlas of the breeding birds of Ontario. University of Waterloo Press, Waterloo, 617 pp.

DUNN, E.H. 1987c. Using atlas data to monitor changes in House Finch distribution, pp. 574-575 *in* M.D. Cadman, P.F.J. Eagles, and F.M. Helleiner, eds. Atlas of the breeding birds of Ontario. University of Waterloo Press, Waterloo, 617 pp.

DUNN, E.H. 1991. Population trends in Canadian songbirds. Canadian Wildlife Service. Bird Trends 1: 2-11.

DUNN, E.H. 1992. Annual report: Project FeederWatch 1991-92. FeederWatch News 5(2): 1-4.

DUNN, E.H. and D.J. AGRO. 1995. Black Tern (*Chlidonias niger*), *in* A. Poole and F. Gill, eds. The birds of North America, no. 147. The Academy of Natural Sciences, Philadelphia. The American Ornithologists' Union, Washington, 24 pp.

DUNN, E.H. and C.D. MACINNES. 1987. Geographic variation in clutch size and body size of Canada Geese. Journal of Field Ornithology 58(3): 355-371.

DUNN, E.H. and D.L. TESSAGLIA. 1991. Predation at bird feeders in winter. Paper presented at the 109th meeting, American Ornithologists' Union, Montréal, 14-17 August.

DUNN, G. 1975. Les forts de l'Outaouais. Éditions du Jour, Montréal, 162 pp.

DUNN, P.O. 1989. The maintenance of monogamy in Black-billed Magpies and Tree Swallows. Ph.D. dissertation, University of Alberta, Edmonton.

DUNN, P.O. and S.J. HANNON. 1991. Intraspecific competition and the maintenance of monogamy in Tree Swallows. Behavioral Ecology 2(3): 258-266.

DUNNING, J.B. JR. 1984. Body weights of 686 species of North American birds. Western Bird Banding Association. Monograph no. 1, 34 pp.

DUNNING, J.B. JR. 1992. Significant encounters. North American Bird-Bander 17: 29.

DUNNING, J.S. 1982. South American land birds: a photographic aid to identification. Harrowood Books, Newtown Square, Pennsylvania, 364 pp.

DUNSTAN, T.C. and S.D. SAMPLE. 1972. Biology of Barred Owls in Minnesota. Loon 44(4): 111-115.

DUPUIS, P. 1990. La Bécasse d'Amérique au Québec: bilan et perspective, un commentaire. Proceedings. Prepared for Canadian Wildlife Service by Fondation Les oiseleurs du Québec. 1er colloque provincial sur la Bécasse d'Amérique [1st provincial symposium on the American Woodcock], Rivière-du-Loup.

DUQUETTE, G. and N. LÉPINE. 1981. Prédation d'une Corneille d'Amérique sur un Pigeon biset. Bulletin ornithologique 26(4): 133.

D'URBAN, W.S.M. 1857. Notes on the land birds observed round Montreal during the winter of 1856-57. Canadian Naturalist and Geologist 2(2): 138-145.

DUSI, J.L. 1977. Impact of Cattle Egrets on an upland colony area. Proceedings. Colonial Waterbird Group [1]: 128-130.

DUSI, J.L. 1979. Stabilities of heron colonies in swamp and upland sites. Proceedings. Colonial Waterbird Group (1978) 2: 38-40.

DWYER, N.C. and G.W. TANNER. 1992. Nesting success in Florida Sandhill Cranes. Wilson Bulletin 104(1): 22-31.

DWYER, T.J. 1974. Social behavior of breeding Gadwall in North Dakota. Auk 91(2): 375-386.

DWYER, T.J., E.L. DERLETH, and D.G. MCAULEY. 1982. Woodcock brood ecology in Maine, pp. 63-70 *in* T.J. Dwyer and G.L. Storm, eds. Woodcock ecology and management. Papers from the Seventh Woodcock Symposium held at the Pennsylvania State University, University Park, Pennsylvania, 28-30 October 1980. United States Fish and Wildlife Service. Research Report no. 14, Washington, 191 pp.

DWYER, T.J., D.G. MCAULEY, and E.L. DERLETH. 1983. Woodcock singing-ground counts and habitat changes in the northeastern United States. Journal of Wildlife Management 47(3): 772-779.

DYBBRO, T. 1976. De danske ynglefugles udbredelse. Resultaterne of Atlas-projektet, kortloegningen af Danmarks ynglefugle 1971-74. Dansk Ornithologisk Forening, København, 293 pp.

DYER, M.I., D.B. SINIFF, S.G. CURTIS, and J.S. WEBB. 1972. Distribution of Red-winged Blackbird (*Agelaius phoeniceus* L.) breeding populations in the Lake Erie region of the United States and Canada, pp. 213-234 *in* S.C. Kendeigh and J. Pinowski, eds. Proceedings. Productivity, population dynamics and systematics of granivorous birds, 6-8 September 1970. General Meeting for the Working Group on Granivorous Birds, IBP, PT Section, The Hague, Holland.

DYKE, A.S. and V.K. PREST. 1987. Lake Wisconsinan and Holocene retreat of the Laurentide Ice Sheet. Geological Survey of Canada, Map 1702A.

DZIEPAK, P. 1991. Northern Bobwhite, pp. 190-191 *in* R. Brewer, G.A. McPeek, and R.J. Adams, Jr., eds. The Atlas of breeding birds of Michigan. Michigan State University Press, East Lansing, 594 pp.

DZUBIN, A. 1955. Some evidence of home range in waterfowl. Transactions. North American Wildlife and Natural Resources Conference 20: 278-298.

DZUBIN, A. 1957. Pairing display and spring and summer flights of the Mallard. Blue Jay 15(1): 10-13

DZUBIN, A. 1969. Assessing breeding populations of ducks by ground counts, pp. 178-230 *in* Canadian Wildlife Service, Saskatoon Wetlands Seminar. Department of Indian Affairs and Northern Development, Ottawa. Report Series no. 6, 262 pp.

EADIE, J. MCA. 1989. Alternative reproductive tactics in a precocial bird: the ecology and evolution of brood parasitism in goldeneye. Ph.D. dissertation, Department of Zoology, University of British Columbia, Vancouver, 202 pp.

EADIE, J. MCA. and G. GAUTHIER. 1985. Prospecting for nest sites by cavity-nesting ducks of the genus *Bucephala*. Condor 87(4): 528-534.

EAGLES, P.F.J. 1987a. Cerulean Warbler, pp. 396-397 *in* M.D. Cadman, P.F.J. Eagles, and F.M. Helleiner, eds. Atlas of the breeding birds of Ontario. University of Waterloo Press, Waterloo, 617 pp.

EAGLES, P.F.J. 1987b. Horned Grebe, pp. 40-41 *in* M.D. Cadman, P.F.J. Eagles, and F.M. Helleiner, eds. Atlas of the breeding birds of Ontario. University of Waterloo Press, Waterloo, 617 pp.

EAGLES, P.F.J. 1987c. Louisiana Waterthrush, pp. 408-409 *in* M.D. Cadman, P.F.J. Eagles, and F.M. Helleiner, eds. Atlas of the breeding birds of Ontario. University of Waterloo Press, Waterloo, 617 pp.

EAGLES, P.F.J. 1987d. Pine Warbler, pp. 386-387 *in* M.D. Cadman, P.F.J. Eagles, and F.M. Helleiner, eds. Atlas of the breeding birds of Ontario. University of Waterloo Press, Waterloo, 617 pp.

EAGLES, P.F.J. 1987e. Western Meadowlark, pp. 472-473 *in* M.D. Cadman, P.F.J. Eagles, and F.M. Helleiner, eds. Atlas of the breeding birds of Ontario. University of Waterloo Press, Waterloo, 617 pp.

EAGLES, P.F.J. and M.D. CADMAN. 1983. Ontario breeding bird atlas: guide for participants, revised edition. Ontario Breeding Bird Atlas Management Committee, Toronto, 28 pp.

EARHART, C.M. and N.K. JOHNSON. 1970. Size dimorphism and food habits of North American owls. Condor 72(3): 251-264.

EARLEY, C.G. 1991. Brown-headed Cowbird, *Molothrus ater*, seen removing a Chipping Sparrow, *Spizella passerina*, egg. Canadian Field-Naturalist 105(2): 281-282.

EASTMAN, J. 1991. Upland Sandpiper, pp. 210-211 *in* R. Brewer, G.A. McPeek, and R.J. Adams, Jr., eds. The Atlas of breeding birds of Michigan. Michigan State University Press, East Lansing, 594 pp.

EATON, E.H. 1914. Birds of New York, vol. 2: General chapters, land birds. New York State Museum. Memoir no. 12, State of New York University, Albany, 719 pp.

EATON, S.W. 1957. A life history study of *Seiurus noveboracensis*. St-Bonaventure University Science Studies 19: 7-36.

EATON, S.W. 1958. A life history study of the Louisiana Waterthrush. Wilson Bulletin 70: 210-236.

EATON, S.W. 1959. The Tufted Titmouse invades New York. Kingbird 9: 59-62.

EATON, S.W. 1968. Northern Slate-colored Junco, pp. 1029-1043 *in* A.C. Bent. Life histories of North American cardinals, grosbeaks, buntings, towhees, finches, sparrows, and their allies. (O.L. Austin, Jr., ed.). United States National Museum, Washington. [1968, Dover Publications, New York]. Bulletin no. 237, vol. 1, pp. 1-602; vol. 2, pp. 603-1248; vol. 3, pp. 1249-1889.

EATON, S.W. 1988a. Acadian Flycatcher, pp. 250-251 *in* R.F. Andrle and J.R. Carroll, eds. The Atlas of breeding birds in New York State. Cornell University Press, Ithaca, 551 pp.

EATON, S.W. 1988b. American Coot, pp. 146-147 *in* R.F. Andrle and J.R. Carroll, eds. The Atlas of breeding birds in New York State. Cornell University Press, Ithaca, 551 pp.

EATON, S.W. 1988c. American Redstart, pp. 400-401 *in* R.F. Andrle and J.R. Carroll, eds. The Atlas of breeding birds in New York State. Cornell University Press, Ithaca, 551 pp.

EATON, S.W. 1988d. Barn Swallow, pp. 276-277 *in* R.F. Andrle and J.R. Carroll, eds. The Atlas of breeding birds in New York State. Cornell University Press, Ithaca, 551 pp.

EATON, S.W. 1988e. Black-throated Green Warbler, pp. 380-381 *in* R.F. Andrle and J.R. Carroll, eds. The Atlas of breeding birds in New York State. Cornell University Press, Ithaca, 551 pp.

EATON, S.W. 1988f. Black-and-white Warbler, pp. 398-399 *in* R.F. Andrle and J.R. Carroll, eds. The Atlas of breeding birds in New York State. Cornell University Press, Ithaca, 551 pp.

EATON, S.W. 1988g. Eastern Phoebe, pp. 258-259 *in* R.F. Andrle and J.R. Carroll, eds. The Atlas of breeding birds in New York State. Cornell University Press, Ithaca, 551 pp.

EATON, S.W. 1988h. Henslow's Sparrow, pp. 450-451 *in* R.F. Andrle and J.R. Carroll, eds. The Atlas of breeding birds in New York State. Cornell University Press, Ithaca, 551 pp.

EATON, S.W. 1988i. Killdeer, pp. 150-151 *in* R.F. Andrle and J.R. Carroll, eds. The Atlas of breeding birds in New York State. Cornell University Press, Ithaca, 551 pp.

EATON, S.W. 1988j. Louisiana Waterthrush, pp. 410-411 *in* R.F. Andrle and J.R. Carroll, eds. The Atlas of breeding birds in New York State. Cornell University Press, Ithaca, 551 pp.

EATON, S.W. 1988k. Northern Goshawk, pp. 108-109 *in* R.F. Andrle and J.R. Carroll, eds. The Atlas of breeding birds in New York State. Cornell University Press, Ithaca, 551 pp.

EATON, S.W. 1988l. Prothonotary Warbler, pp. 402-403 *in* R.F. Andrle and J.R. Carroll, eds. The Atlas of breeding birds in New York State. Cornell University Press, Ithaca, 551 pp.

EATON, S.W. 1988m. Wild Turkey, pp. 130-131 *in* R.F. Andrle and J.R. Carroll, eds. The Atlas of breeding birds in New York State. Cornell University Press, Ithaca, 551 pp.

EATON, S.W. 1988n. Yellow-billed Cuckoo, pp. 198-199 *in* R.F. Andrle and J.R. Carroll, eds. The Atlas of breeding birds in New York State. Cornell University Press, Ithaca, 551 pp.

Eaton, S.W. 1988o. Magnolia Warbler, pp. 372-373 *in* R.F. Andrle and J.R. Carroll, eds. The Atlas of breeding birds in New York State. Cornell University Press, Ithaca, 551 pp.

Eaton, S.W. 1988p. Vesper Sparrow, pp. 444-445 *in* R.F. Andrle and J.R. Caroll, eds. The Atlas of breeding birds in New York State. Cornell University Press, Ithaca, 551 pp.

Eaton, S.W. 1992. Wild Turkey (*Meleagris gallopavo*), *in* A. Poole, P. Stettenheim, and F. Gill, eds. The birds of North America, no. 22. The Academy of Natural Sciences, Philadelphia. The American Ornithologists' Union, Washington, 28 pp.

Eckert, C.G. and P.J. Weatherhead. 1987. Owners, floaters and competitive asymmetries among territorial Red-winged Blackbirds. Animal Behaviour 35: 1317-1323.

Eckert, K.[R.] 1978. Invasion of Great Gray and Boreal owls 1977-78. Loon 50:63-68.

Eckert, K.R. and T.L. Savaloja. 1979. First documented nesting of the Boreal Owl south of Canada. American Birds 33(2): 135-137.

Eckhardt, R.C. 1976. Polygyny in the Western Wood Pewee. Condor 78: 561-562.

Eddleman, W.R. 1993. 55. Mature white pine forest, p. 67 *in* J.D. Lowe, ed. Resident bird counts 1992. Journal of Field Ornithology 64(1)(supplement): 1-128.

Edgell, M.C.R. 1984. Trans-hemispheric movements of Holarctic Anatidae: the Eurasian Wigeon (*Anas penelope* L.) in North America. Journal of Biogeography 11: 27-39.

Edinger, B.B. 1988. Extra-pair courtship copulation attempts in Northern Orioles. Condor 90(3): 546-554.

Edminster, F.C. 1954. American game birds of field and forest. Charles Scribner's Sons, New York, 490 pp.

Edwards, G. 1776. A natural history of uncommon birds and of some other rare and undescribed animals. College of Physicians, London, 4 vol.

Edwards, R.Y. 1953. Barrow's Goldeneye using crow nests in British Columbia. Wilson Bulletin 65: 197-198.

Ehrlich, P.R. and G.C. Daily. 1988. Red-naped Sapsuckers feeding at willows: possible keystone herbivores. American Birds 42(3): 357-365.

Ehrlich, P.R. and G.C. Daily. 1993. Population extinction and saving biodiversity. Ambio 22: 64-68.

Ehrlich, P.R., D.S. Dobkin, and D. Wheye. 1988. The birder's handbook: a field guide to the natural history of North American birds. Fireside Book, Simon and Schuster, New York, 785 pp.

Eifrig, C.W.G. 1905. Ornithological results of the Canadian Neptune's expedition to Hudson Bay and northward, 1903-1904. Auk 22(3): 232-241.

Eifrig, C.W.G. 1910-1911. The birds of Ottawa. Ottawa Naturalist 24(9): 152-163; (10): 176-187; (11): 198-206; (12): 222-228.

Eigelis, Y.K. and B.V. Nekrasov. 1967. Morphological peculiarities of the buccal cavity of Corvidae related to food transportation. Zoologicheskii zhurnal 34: 258-263.

Einarsson, A. 1985. Use of space in relation to food in Icelandic Barrow's Goldeneye (*Bucephala islandica*). Ph.D. dissertation, University of Aberdeen, Scotland.

Einarsson, A. 1988. Distribution and movements of Barrow's Goldeneye (*Bucephala islandica*) young in relation to food. Ibis 130: 153-163.

Einarsson, A. 1990. Settlement into breeding habitats by Barrow's Goldeneye (*Bucephala islandica*): evidence for temporary oversaturation of preferred habitat. Ornis Scandinavica 21: 7-16.

Eiserer, L.[A.] 1976. The American Robin. Nelson-Hall, Chicago, 175 pp.

Eiserer, L.A. 1980. Effects of grass length and mowing in foraging behaviour of American Robin (*Turdus migratorius*). Auk 97(3): 576-580.

Eklund, C.R. 1944. Nesting notes on the Arctic Terns. Auk 61: 648.

Eklund, C.R. 1956. Bird and mammal notes from the interior Ungava Peninsula. Canadian Field-Naturalist 70: 69-76.

Elder, W.H. 1985. Survivorship in the Tufted Titmouse. Wilson Bulletin 97(4): 517-524.

Elderkin, M.F. 1987. The breeding and feeding ecology of a Barred Owl, *Strix varia* Barton, population in King's County, Nova Scotia. M.Sc. thesis, Acadia University, Wolfville, Nova Scotia, 203 pp.

Elderkin, M.F. 1989. Habitat supply analysis: forest birds in New Brunswick. New Brunswick Department of Natural Resources and Energy, Fish and Wildlife Branch, Forest Lands Habitats Management Program, 148 pp.

Eldridge, J.L. and G.L. Krapu. 1988. The influence of diet quality on clutch size and laying pattern in Mallards. Auk 105(1): 102-110.

Eliason, B.C. 1986a. Female site fidelity and polygyny in the Blackpoll Warbler (*Dendroica striata*). Auk 103(4): 782-790.

Eliason, B.C. 1986b. Mating system, parental care, and reproductive success in the Blackpoll Warbler (*Dendroica striata*). Ph.D. dissertation, University of Minnesota, Minneapolis, 132 pp.

Elliot, P.F. 1978. Cowbird parasitism in the Kansas tallgrass prairie. Auk 95(1): 161-167.

Elliot, P.F. 1980. Evolution of promiscuity in the Brown-headed Cowbird. Condor 82(2): 138-141.

Elliot, R.D. 1991. The management of the Newfoundland murre hunt, pp. 29-35 *in* A.J. Gaston and R.D. Elliot, eds. Studies of high-latitude seabirds. 2. Conservation biology of Thick-billed Murres in the northwest Atlantic. Environment Canada, Canadian Wildlife Service, Ottawa. Occasional Paper no. 69, 63 pp.

Elliot, R.D. and R.I.G. Morrison. 1979. The incubation period of the Yellow Rail. Auk 96: 422-423.

Elliot, R.D., B.T. Collins, E.C. Hayakawa, and L. Métras. 1991. The harvest of murres in Newfoundland from 1977-78 to 1987-88, pp. 36-44 *in* A.J. Gaston and R.D. Elliot, eds. Studies of high-latitude seabirds. 2. Conservation biology of Thick-billed Murres in the northwest Atlantic. Environment Canada, Canadian Wildlife Service, Ottawa. Occasional Paper no. 69, 63 pp.

Elliott, J.E., K.M. Langelier, A.M. Scheuhammer, P.H. Sinclair, and P.E. Whitehead. 1992. Incidence de l'intoxication au plomb chez le Pygargue à tête blanche et de la présence de grenaille de plomb dans le gésier des oiseaux aquatiques de la Colombie-Britannique, 1988-1991. Canadian Wildlife Service. Progress Notes no. 200, 7 pp. [Also available in English as: Incidence of lead poisoning in Bald Eagles and lead shot in waterfowl gizzards from British Columbia, 1988-91.]

Elliott, J.J. 1958. Regional report (region 10–Marine). Kingbird 7: 104.

Elliott, J.J. and R.S. Arbib, Jr. 1953. Origin and status of the House Finch in the eastern United States. Auk 70: 31-37.

Ellis, P.M. and J.D. Okill. 1990. Breeding ecology of the Merlin *Falco columbarius* in Shetland. Bird Study 37: 101-110.

Ellison, G.F. 1985. European Starling, pp. 262-263 *in* S.B. Laughlin and D.P. Kibbe, eds. The Atlas of breeding birds of Vermont. Vermont Institute of Natural Science, University Press of New England, Hanover, 456 pp.

Ellison, L.N. 1971. Territoriality in Alaska Spruce Grouse. Auk 88: 652-664.

Ellison, L.N. 1973. Seasonal social organization and movements of Spruce Grouse. Condor 75: 375-385.

Ellison, L.N. and L. Cleary. 1978. Effects of human disturbance on breeding of Double-crested Cormorants. Auk 95: 510-517.

Ellison, W.G. 1985a. Rose-breasted Grosbeak, pp. 332-333 *in* S.B. Laughlin and D.P. Kibbe, eds. The Atlas of breeding birds of Vermont. Vermont Institute of Natural Science, University Press of New England, Hanover, 456 pp.

Ellison, W.G. 1985b. American Crow, pp. 206-207 *in* S.B. Laughlin and D.P. Kibbe, eds. The Atlas of breeding birds of Vermont. Vermont Institute of Natural Science, University Press of New England, Hanover, 456 pp.

Ellison, W.G. 1985c. Bobolink, pp. 358-359 *in* S.B. Laughlin and D.P. Kibbe, eds. The Atlas of breeding birds of Vermont. Vermont Institute of Natural Science, University Press of New England, Hanover, 456 pp.

Ellison, W.G. 1985d. Black-and-white Warbler, pp. 310-311 *in* S.B. Laughlin and D.P. Kibbe, eds. The Atlas of breeding birds of Vermont. Vermont Institute of Natural Science, University Press of New England, Hanover, 456 pp.

Ellison, W.G. 1985e. Brown Creeper, pp. 220-221 *in* S.B. Laughlin and D.P. Kibbe, eds. The Atlas of breeding birds of Vermont. Vermont Institute of Natural Science, University Press of New England, Hanover, 456 pp.

Ellison, W.G. 1985f. Brown-headed Cowbird, pp. 368-369 *in* S.B. Laughlin and D.P. Kibbe, eds. The Atlas of breeding birds of Vermont. Vermont Institute of Natural Science, University Press of New England, Hanover, 456 pp.

Ellison, W.G. 1985g. Carolina Wren, pp. 222-223 *in* S.B. Laughlin and D.P. Kibbe, eds. The Atlas of breeding birds of Vermont. Vermont Institute of Natural Science, University Press of New England, Hanover, 456 pp.

Ellison, W.G. 1985h. Cedar Waxwing, pp. 258-259 *in* S.B. Laughlin and D.P. Kibbe, eds. The Atlas of breeding birds of Vermont. Vermont Institute of Natural Science, University Press of New England, Hanover, 456 pp.

Ellison, W.G. 1985i. Cerulean Warbler, pp. 308-309 *in* S.B. Laughlin and D.P. Kibbe, eds. The Atlas of breeding birds of Vermont. Vermont Institute of Natural Science, University Press of New England, Hanover, 456 pp.

Ellison, W.G. 1985j. Common Grackle, pp. 366-367 *in* S.B. Laughlin and D.P. Kibbe, eds. The Atlas of breeding birds of Vermont. Vermont Institute of Natural Science, University Press of New England, Hanover, 456 pp.

Ellison, W.G. 1985k. Common Yellowthroat, pp. 322-323 *in* S.B. Laughlin and D.P. Kibbe, eds. The Atlas of breeding birds of Vermont. Vermont Institute of Natural Science, University Press of New England, Hanover, 456 pp.

Ellison, W.G. 1985l. Dark-eyed Junco, pp. 356-357 *in* S.B. Laughlin and D.P. Kibbe, eds. The Atlas of breeding birds of Vermont. Vermont Institute of Natural Science, University Press of New England, Hanover, 455 pp.

Ellison, W.G. 1985m. Grasshopper Sparrow, pp. 346-347 *in* S.B. Laughlin and D.P. Kibbe, eds. The Atlas of breeding birds of Vermont. Vermont Institute of Natural Science, University Press of New England, Hanover, 456 pp.

Ellison, W.G. 1985n. Horned Lark, pp. 188-189 *in* S.B. Laughlin and D.P. Kibbe, eds. The Atlas of breeding birds of Vermont. Vermont Institute of Natural Science, University Press of New England, Hanover, 456 pp.

Ellison, W.G. 1985o. House Wren, pp. 224-225 *in* S.B. Laughlin and D.P. Kibbe, eds. The Atlas of breeding birds of Vermont. Vermont Institute of Natural Science, University Press of New England, Hanover, 456 pp.

Ellison, W.G. 1985p. Killdeer, pp. 104-105 *in* S.B. Laughlin and D.P. Kibbe, eds. The Atlas of breeding birds of Vermont. Vermont Institute of Natural Science, University Press of New England, Hanover, 456 pp.

Ellison, W.G. 1985q. Northern Goshawk, pp. 78-79 *in* S.B. Laughlin and D.P. Kibbe, eds. The Atlas of breeding birds of Vermont. Vermont Institute of Natural Science, University Press of New England, Hanover, 456 pp.

Ellison, W.G. 1985r. Orchard Oriole, pp. 370-371 *in* S.B. Laughlin and D.P. Kibbe, eds. The Atlas of breeding birds of Vermont. Vermont Institute of Natural Science, University Press of New England, Hanover, 456 pp.

Ellison, W.G. 1985s. Pine Warbler, pp. 300-301 *in* S.B. Laughlin and D.P. Kibbe, eds. The Atlas of breeding birds of Vermont. Vermont Institute of Natural Science, University Press of New England, Hanover, 456 pp.

Ellison, W.G. 1985t. Savannah Sparrow, pp. 344-345 *in* S.B. Laughlin and D.P. Kibbe, eds. The Atlas of breeding birds of Vermont. Vermont Institute of Natural Science, University Press of New England, Hanover, 456 pp.

Ellison, W.G. 1985u. Scarlet Tanager, pp. 328-329 *in* S.B. Laughlin and D.P. Kibbe, eds. The Atlas of breeding birds of Vermont. Vermont Institute of Natural Science, University Press of New England, Hanover, 456 pp.

Ellison, W.G. 1985v. Spotted Sandpiper, pp. 106-107 *in* S.B. Laughlin and D.P. Kibbe, eds. The Atlas of breeding birds of Vermont. Vermont Institute of Natural Science, University Press of New England, Hanover, 456 pp.

Ellison, W.G. 1985w. Vesper Sparrow, pp. 342-343 *in* S.B. Laughlin and D.P. Kibbe, eds. The Atlas of breeding birds of Vermont. Vermont Institute of Natural Science, University Press of New England, Hanover, 456 pp.

Ellison, W.G. 1985x. White-breasted Nuthatch, pp. 218-219 *in* S.B. Laughlin and D.P. Kibbe, eds. The Atlas of breeding birds of Vermont. Vermont Institute of Natural Science, University Press of New England, Hanover, 456 pp.

Ellison, W.G. 1985y. Yellow-rumped Warbler, pp. 294-295 *in* S.B. Laughlin and D.P. Kibbe, eds. The Atlas of breeding birds of Vermont. Vermont Institute of Natural Science, University Press of New England, Hanover, 456 pp.

Ellison, W.G. 1985z. Belted Kingfisher, pp. 152-153 *in* S.B. Laughlin and D.P. Kibbe, eds. The Atlas of breeding birds of Vermont. Vermont Institute of Natural Science, University Press of New England, Hanover, 456 pp.

Ellison, W.G. 1985aa. Black-throated Green Warbler, pp. 296-297 *in* S.B. Laughlin and D.P. Kibbe, eds. The Atlas of breeding birds of Vermont. Vermont Institute of Natural Science, University Press of New England, Hanover, 456 pp.

Ellison, W.G. 1992. Blue-gray Gnatcatcher (*Polioptila caerulea*), *in* A. Poole, P. Stettenheim, and F. Gill, eds. The birds of North America, no. 23. The Academy of Natural Sciences, Philadelphia. The American Ornithologists' Union, Washington, 20 pp.

Elman, R. 1977. The Atlantic Flyway. Winchester Press, New York, 203 pp.

Elmberg, J. 1991. Cooperative nest defence by trios of Arctic Skuas *Stercorarius parasiticus*. Ibis 134(3): 298.

Elson, P.F. 1962. Predator-prey relationships between fish-eating birds and Atlantic Salmon. Fisheries Research Board of Canada. Bulletin no. 133.

Elton, C. 1965. Voles, mice and lemmings. J. Cramer und Wheinheim, Germany, 496 pp.

Emlen, J.M. 1966. The role of time and energy in food preference. American Naturalist 100: 611-617.

Emlen, J.T. 1973. Territorial aggression in wintering warblers at Bahamas Agave blossoms. Wilson Bulletin 85(1): 71-74.

Emlen, J.T. Jr. 1952. Social behavior in nesting Cliff Swallows. Condor 54(4): 177-199.

Emlen, J.T. Jr. 1954. Territory, nest building, and pair formation in the Cliff Swallow. Auk 71(1): 16-35.

Enderson, J.H. 1960. American Kestrel population study. Wilson Bulletin 72: 222-231.

Energy, Mines and Resources Canada. 1973. National Atlas of Canada, 4th edition. Surveys and Mapping Branch, Department of Energy, Mines and Resources, Ottawa.

Enstrom, D.A. 1992a. Breeding season communication hypotheses for delayed plumage maturation in passerines: tests in the Orchard Oriole, *Icterus spurius*. Animal Behaviour 43(3): 463-472.

Enstrom, D.A. 1992b. Delayed plumage maturation in the Orchard Oriole (*Icterus spurius*): test of winter adaptation hypothesis. Behavioral Ecology and Sociobiology 30: 35-42.

Environment Canada. 1984. Climatic Atlas – Canada: A series of maps portraying Canada's climate. Map series 1 – Temperature and degree days/Atlas climatique – Canada : Série de cartes illustrant le climat du Canada. Série de cartes no. 1 – Température et degrés-jours. Atmospheric Environment Service, Ottawa.

Environment Canada. 1986. Climatic Atlas – Canada: A series of maps portraying Canada's climate. Map series 2 – Precipitation/Atlas climatique – Canada : Série de cartes illustrant le climat du Canada. Série de cartes no. 2 – Précipitations. Environment Canada, Atmospheric Environment Service, Ottawa.

Environment Canada. 1990. Compte rendu de la situation des oiseaux migrateurs considérés comme gibier au Canada et propositions relatives au règlement de chasse de 1991. Canadian Wildlife Service, Conservation and Protection, 21 pp.

Environment Canada. 1991. The state of Canada's environment. Government of Canada, Ottawa.

Environment Canada. 1993. Compte rendu de la situation des oiseaux migrateurs considérés comme gibier au Canada et propositions relatives au règlement de chasse de 1994. Canadian Wildlife Service, Conservation and Protection, 36 pp.

Environment Canada. 1994. Rapport synthèse sur l'état du Saint-Laurent, Tome II. Les caractéristiques biologiques. Centre Saint-Laurent, Montréal. [Also available in English.]

Eriksson, M.O.G. 1976. Food and feeding habits of downy Goldeneye *Bucephala clangula* (L.) ducklings. Ornis Scandinavica 7: 159-169.

Eriksson, M.O.G. 1979. Aspects of the breeding biology of the Goldeneye *Bucephala clangula*. Holarctic Ecology 2: 186-194.

Eriksson, M.O.G., D. Blomqvist, M. Hake, and O.C. Johansson. 1990. Parental feeding in the Red-throated Diver *Gavia stellata*. Ibis 132: 1-13.

Erpelding, R., R.O. Kimmel, and D.J. Lockman. 1987. Foods and feeding behavior of young Gray Partridge in Minnesota, pp. 17-30 *in* R.O. Kimmel, J.W. Schulz, and G.J. Mitchell, eds. Perdix IV: Gray Partridge workshop. Farmland Wildlife Populations and Research Group, Minnesota Department of Natural Resources, Madelia, Minnesota, 155 pp.

Errington, P.L. 1930. Territory disputes of three pairs of nesting Marsh Hawks. Wilson Bulletin 42: 237-239.

Erskine, A.J. 1963. The Black-headed Gull (*Larus ridibundus*) in eastern North America. Audubon Field Notes 17(3): 334-338.

Erskine, A.J. 1971a. A preliminary catalogue of bird census studies in Canada. Canadian Wildlife Service. Progress Notes no. 20, 78 pp.

Erskine, A.J. 1971b. Fox Sparrow summering in Abitibi County, Québec. Tchebec 1(4): 97-98.

Erskine, A.J. 1971c. Some new perspectives on the breeding ecology of Common Grackles. Wilson Bulletin 83(4): 352-370.

Erskine, A.J. 1972a. A preliminary catalogue of bird census plot studies in Canada, vol. 2. Canadian Wildlife Service. Progress Notes no. 30, 42 pp.

Erskine, A.J. 1972b. Buffleheads. Canadian Wildlife Service. Monograph Series no. 4, 240 pp.

Erskine, A.J. 1972c. The Great Cormorants of eastern Canada. Canadian Wildlife Service. Occasional Paper no. 14, 23 pp.

Erskine, A.J. 1972d. Populations, movements and seasonal distribution of mergansers in northern Cape Breton Island. Canadian Wildlife Service. Report Series no. 17, 36 pp.

Erskine, A.J. 1976a. A preliminary catalogue of bird census plot studies in Canada, vol. 3. Canadian Wildlife Service. Progress Notes no. 59, 24 pp.

Erskine, A.J. 1976b. Chronology of nesting in urban birds as a guide of timing of census. American Birds 30(3): 667-672.

Erskine, A.J. 1977. Birds in boreal Canada: communities, densities and adaptations. Canadian Wildlife Service. Report Series no. 41, 73 pp.

Erskine, A.J. 1978a. Durability of the tree holes used by Buffleheads. Canadian Field-Naturalist 92(2): 94-95.

Erskine, A.J. 1978b. The first ten years of the co-operative Breeding Bird Survey in Canada. Canadian Wildlife Service. Report Series no. 42, 61 pp.

Erskine, A.J. 1979. Man's influence on potential nesting sites and populations of swallows in Canada. Canadian Field-Naturalist 93(4): 371-377.

Erskine, A.J. 1980a. A preliminary catalogue of bird census plot studies in Canada, part 4. Canadian Wildlife Service. Progress Notes no. 112, 26 pp.

Erskine, A.J. 1980b. Urban birds in the context of Canadian climate and settlement, pp. 1321-1326 *in* Herausgegeben von Rolf Nohring. Actis XVII Congressus Internationalis Ornithologici. Deutschen Ornithologen-Gesellschaft, Berlin, 5-11 June 1978, vol. I, pp. 1-747; vol. II, pp. 756-1463.

Erskine, A.J. 1984. A preliminary catalogue of bird census plot studies in Canada, part 5. Canadian Wildlife Service. Progress Notes no. 144, 34 pp.

Erskine, A.J. 1987. Bilan préliminaire des populations d'oiseaux aquatiques dans les provinces de l'Atlantique, 1978-1985, pp. 72-81 *in* A.J. Erskine, ed. Les dénombrements de populations reproductrices d'oiseaux aquatiques dans les provinces de l'Atlantique. Environment Canada, Canadian Wildlife Service. Occasional Paper no. 60, 89 pp. [Also available in English as: Waterfowl breeding population surveys, Atlantic Provinces.]

Erskine, A.J. 1988. The changing patterns of Brant migration in eastern North America. Journal of Field Ornithology 59(2): 110-119.

Erskine, A.J. 1990. Joint laying in *Bucephala* ducks – "parasitism" or nest-site competition? Ornis Scandinavica 21: 52-56.

Erskine, A.J. 1992. Atlas of breeding birds of the Maritime Provinces. Nimbus and Nova Scotia Museum, 270 pp.

Erskine, A.J. and W.D. McLaren. 1972. Sapsucker nest holes and their use by other species. Canadian Field-Naturalist 86(4): 357-361.

Erskine, A.J. and A.D. Smith. 1986. Status and movements of Common Eiders in the Maritime provinces, pp. 20-29 *in* A. Reed, ed. Eider ducks in Canada/Les eiders au Canada. Canadian Wildlife Service. Report Series no. 47, 177 pp.

Erskine, A.J., B.T. Collins, and J.W. Chardine. 1990. Étude collective des populations d'oiseaux au Canada, 1988. Canadian Wildlife Service, Ottawa. Progress Notes no. 188, 16 pp. [Also available in English as: The cooperative Breeding Bird Survey in Canada, 1988.]

Erskine, A.J., B.T. Collins, E. Hayakawa, and C. Downes. 1992. Étude collective des populations d'oiseaux nicheurs au Canada, 1989-1991. Canadian Wildlife Service. Progress Notes no. 199, 15 pp. [Also available in English as: The cooperative Breeding Bird Survey in Canada, 1989-91.]

Erwin, R.M. 1978. Coloniality in terns: the role of social feeding. Condor 80(2): 211-215.

Erwin, W.G. 1935. Some nesting habits of the Brown Thrasher. Journal of the Tennessee Academy of Science 10: 179-204.

Ettinger, A.O. and J.R. King. 1980. Time and energy budgets of the Willow Flycatcher (*Empidonax traillii*) during the breeding season. Auk 97(3): 533-546.

Evans, D.L. 1980. Vocalizations and territorial behavior of wintering Snowy Owls. American Birds 34(5): 748-749.

Evans, D.L. 1983. Northern Harrier, pp. 224-226 *in* J. Farrand, Jr., ed. The Audubon Society master guide to birding, vol. 1. Alfred A. Knopf, New York, 447 pp.

Evans, E.W. 1978. Nesting responses of Field Sparrows (*Spizella pusilla*) to plant succession on a Michigan old field. Condor 80(1): 34-40.

Evans, K.E. and R.N. Conner. 1979. Snag management, pp. 215-225 *in* R.M. DeGraaf and K.E. Evans, eds. Management of northcentral and northeastern forest for nongame birds. United States Department of Agriculture, Forest Service, St. Paul, Minnesota. General Technical Report NC-51, 268 pp.

Evans, P.G.H. 1984. Status and conservation of seabirds of northwest Europe (excluding Norway and the U.S.S.R.), pp. 293-322 *in* J.P. Croxall, P.G.H. Evans, and R.W. Schreiber, eds. Status and conservation of the world's seabirds. Cambridge, Great Britain. ICBP Technical Publication no. 2, 778 pp.

Evans, P.G.H. and D.N. Nettleship. 1985. Conservation of the Atlantic Alcidae, pp. 427-486 *in* D.N. Nettleship and T.R. Birkhead, eds. The Atlantic Alcidae: the evolution, distribution and biology of the auks inhabiting the Atlantic Ocean and adjacent water areas. Academic Press, London, 574 pp.

Evans, P.R. 1985. Plover, pp. 467-469 *in* B. Campbell and E. Lack, eds. A dictionary of birds. British Ornithologists' Union, Buteo Books, Vermillion, South Dakota, 670 pp.

Evans, R.M. 1969. Territorial stability in Sharp-tailed Grouse. Wilson Bulletin 81(1): 75-78.

Evans, R.M. and M.K. McNicholl. 1972. Variations in the reproductive activities of Arctic Terns at Churchill, Manitoba. Arctic 25: 131-171.

Evans, R.M., D.B. Krindle, and M.E. Mattson. 1970. Caspian Terns nesting near Spruce Island, Lake Winnipegosis, Manitoba. Blue Jay 28: 68-71.

Evarts, S. and C.J. Williams. 1987. Multiple paternity in a wild population of Mallards. Auk 104(4): 597-602.

Evers, D.C. 1991a. Olive-sided Flycatcher, pp. 276-277 *in* R. Brewer, G.A. McPeek, and R.J. Adams, Jr., eds. The Atlas of breeding birds of Michigan. Michigan State University Press, East Lansing, 594 pp.

Evers, D.C. 1991b. Pine Warbler, pp. 412-413 *in* R. Brewer, G.A. McPeek, and R.J. Adams, Jr., eds. The Atlas of breeding birds of Michigan. Michigan State University Press, East Lansing, 594 pp.

Ewald, P.W. and S. Rohwer. 1982. Effects of supplemental feeding on timing of breeding, clutch-size and polygyny in Red-winged Blackbird *Agelaius phoeniceus*. Journal of Animal Ecology 51: 429-450.

Ewaschuk, E. and D.A. Boag. 1972. Factors affecting hatching success of densely nesting Canada Geese. Journal of Wildlife Management 36(4): 1097-1106.

Ewert, D.N. 1981. The occurence of closely related species-pairs in central Michigan: Willow and Alder flycatchers and Golden-winged and Blue-winged warblers. Jack-Pine Warbler 59(3): 95-98.

Faaborg, J. 1976. Habitat selection and territorial behavior of the small grebes of North Dakota. Wilson Bulletin 88(3): 390-399.

Faaborg, J. 1988. Ornithology, an ecological approach. Prentice Hall, Englewood Cliffs, New Jersey, 470 pp.

Faaborg, J. and W.J. Arendt. 1985. Long-term declines in winter resident warblers in a Puerto Rican dry forest. American Birds 43 (4): 1226-1230.

Faaborg, J., M. Brittingham, T. Donovan, and J. Blake. 1993. Neotropical migrant responses to habitat fragmentation in the temperate zone, *in* T. Martin and D. Finch, eds. Ecology and management of neotropical migratory birds: a synthesis and review of the critical issues. Oxford University Press.

Faanes, C.A. 1980. Breeding biology of Eastern Phoebes in northern Wisconsin. Wilson Bulletin 92(1): 107-110.

Falardeau, G. and J.-L. DesGranges. 1991. Sélection de l'habitat et fluctuations récentes des populations d'oiseaux des milieux agricoles du Québec. Canadian Field-Naturalist 105(4): 469-482.

Falls, J.B. and J.G. Kopachena. 1994. White-throated Sparrow (*Zonotrichia albicollis*), *in* A. Poole and F. Gill, eds. The birds of North America, no. 128. The Academy of Natural Sciences, Philadelphia. The American Ornithologists' Union, Washington, 32 pp.

Fannucchi, W.A., G.T. Fannucchi, and L.E. Nauman. 1986. Effects of harvesting Wild Rice, *Zizania aquatica*, on Soras, *Porzana carolina*. Canadian Field-Naturalist 100(4): 533-536.

Farber, P.L. 1977. The development of taxidermy and the history of ornithology. Ibis 68: 550-566.

Farber, P.L. 1982. The emergence of ornithology as a scientific discipline: 1760-1850. D. Reidel Publications, Boston, 191 pp.

Farrand, J. Jr. 1983. The Audubon Society master guide to birding. Alfred A. Knopf, New York, vol. 1, 447 pp.; vol. 2, 398 pp.; vol. 3, 399 pp.

Farrand, J. Jr. 1993. Moments in history: the Connecticut Warbler. American Birds 47(2): 308-310.

Farrar, J.L. 1995. Trees in Canada. Fitzhenry and Whiteside, Canadian Forest Service, Natural Resources Canada and Canada Communication Group Publishing, Supply and Services Canada, Markham, Ontario, 502 pp.

Farrar, R.B. 1973. Rural stream border. American Birds 27(6): 982-983.

Fast, A.H. and L.H. Barnes. 1950. American Kestrel behaviour. Wilson Bulletin 62: 38.

Faucher, D. 1994. Le jeu chez la Corneille d'Amérique. QuébecOiseaux 5(4): 20-21.

Fautin, R.W. 1941. Development of nestling Yellow-headed Blackbirds. Auk 58: 215-232.

Fawks, E. 1937. Bird-Lore's first breeding bird census, second-growth hardwood. Bird-Lore 39(5): 380.

Fawver, B.J. 1947. Bird population of an Illinois flood plain forest. Transactions. Illinois State Academy of Science 40: 178-189.

Faxon, W. 1917. Brewster's Warbler (*Vermivora leucobronchialis*) in Lexington, Massachusetts. Auk 34: 481-482.

Feare, C. 1984. The Starling. Oxford University Press, Oxford, 315 pp.

Feduccia, A. 1980. The age of birds. Harvard University Press, Cambridge, Massachusetts, 196 pp.

Feierabend, J.S. 1984. The Black Duck: an international resource on trial in the United States. Wildlife Society Bulletin 12(2): 128-134.

Feldsine, J.W. and L.W. Oliphant. 1985. Courtship behaviour of Merlins. Raptor Research 19: 60-67.

Ferguson, R.S. 1981. Territorial attachment and mate fidelity by Horned Grebes. Wilson Bulletin 93(4): 560-561.

Ferguson R.S. and S.G. Sealy. 1983. Breeding ecology of the Horned Grebe, *Podiceps auritus*, in southwestern Manitoba. Canadian Field-Naturalist 97(4): 401-408.

Ferrara, J. 1991. Le merle cocu. Science et vie 881: 60-61.

Ferron, J. and Y. Lemay. 1987. Prévisions démographiques pour la population de Tétras des savanes (*Dendragapus canadensis*) introduite à l'île d'Anticosti en 1985 et 1986. Report for Direction générale de la faune, Ministère du Loisir, de la Chasse et de la Pêche, 33 pp.

Fetterolf, P.M. and H. Blokpoel. 1983. Reproductive performance of Caspian Terns at a new colony on Lake Ontario, 1979-1981. Journal of Field Ornithology 54(2): 170-186.

Fichtel, C. 1985a. Common Loon, pp. 30-31 *in* S.B. Laughlin and D.P Kibbe, eds. The Atlas of breeding birds of Vermont. Vermont Institute of Natural Science, University Press of New England, Hanover, 456 pp.

Fichtel, C. 1985b. Red-breasted Merganser, pp. 68-69 *in* S.B. Laughlin and D.P. Kibbe, eds. The Atlas of breeding birds of Vermont. Vermont Institute of Natural Science, University Press of New England, Hanover, 456 pp.

Fichtel, C. 1985c. Olive-sided Flycatcher, pp. 170-171 in S.B. Laughlin and D.P. Kibbe, eds. The Atlas of breeding birds of Vermont. Vermont Institute of Natural Science, University Press of New England, Hanover, 456 pp.

Fichtel, C. 1985d. Osprey, pp. 398-399 *in* S.B. Laughlin and D.P. Kibbe, eds. The Atlas of breeding birds of Vermont. Vermont Institute of Natural Science, University Press of New England, Hanover, 456 pp.

Ficken, M.S. and R.W. Ficken. 1962. The comparative ethology of the wood warblers: a review. Living Bird 1: 103-122.

Ficken, M.S. and R.W. Ficken. 1965a. Comparative ethology of the Chestnut-sided Warbler, Yellow Warbler, and American Redstart. Wilson Bulletin 77(4): 363-375.

Ficken, M.S. and R.W. Ficken. 1965b. Territorial display as a population-regulating mechanism in the Yellow Warbler. Auk 82(2): 274-275.

Ficken, M.S. and R.W. Ficken. 1967. Age-specific differences in the breeding behaviour and ecology of the American Redstart. Wilson Bulletin 79: 188-199.

Ficken, M.S. and R.W. Ficken. 1968a. Ecology of Blue-winged Warblers, Golden-winged Warblers and some other *Vermivora*. American Midland Naturalist 79(2): 311-319.

Ficken, M.S. and R.W. Ficken. 1968b. Reproductive isolating mechanisms in the Blue-winged Warbler – Golden-winged Warbler complex. Evolution 22: 166-179.

Ficken, M.S. and R.W. Ficken. 1968c. Territorial relationships of Blue-winged Warblers, Golden-winged Warblers, and their hybrids. Wilson Bulletin 80(4): 442-451.

Ficken, M.S. and R.W. Ficken. 1968d. Courtship of Blue-winged Warblers, Golden-winged Warblers, and their hybrids. Wilson Bulletin 80(2): 161-172.

Filion, G. 1985. Relevés fauniques dans l'archipel des Pèlerins, été 1985. Ducks Unlimited Canada, 30 pp.

Fillmore, E.R. and R.D. Titman. 1977. Chipping Sparrow hanged. Canadian Field-Naturalist 91(1): 69.

Finch, D.M. 1983. Brood parasitism of the Abert's Towhee: timing, frequency, and effects. Condor 85(3): 355-359.

Finch, D.M. 1989. Relationships of surrounding riparian habitat to nest-box use and reproductive outcome in House Wrens. Condor 91(4): 848-859.

Finch, D.W. 1978. Black-headed Gull (*Larus ridibundus*) in Newfoundland. American Birds 32(3): 312.

Finch, D.W., W.C. Russell, and E.V. Thompson. 1978. Pelagic birds in the Gulf of Maine. American Birds 32(3): 281-284.

Finlay, J.C. 1971. Breeding biology of Purple Martins at the northern limit of their range. Wilson Bulletin 83(3): 255-269.

Finlay, J.C. 1976. Some effects of weather on Purple Martin activity. Auk 93(2): 231-244.

Fischl, J. and D.F. Caccamise. 1985. Influence of habitat and season on foraging flock composition in the European Starling (*Sturnus vulgaris*). Oecologia 67: 532-539.

Fisher, A.K. 1893. Hawks and owls of the United States in their relation to agriculture. United States Bureau of Biological Survey, Department of Agriculture, Washington. Bulletin no. 3, 210 pp.

Fisher, B.M. 1975. Possible intra-specific killing by a Great Gray Owl. Canadian Field-Naturalist 89: 71-72.

Fisher, J. and R.M. Lockley. 1954. Sea-birds. Collins Ltd, London, 320 pp.

Fisher, J. and H.G. Vevers. 1943. The breeding distribution, history and population of the North Atlantic Gannet (*Sula bassana*). Journal of Animal Ecology 12: 173-213.

Fisher, J. and H.G. Vevers. 1944. The breeding distribution, history and population of the North Atlantic Gannet (*Sula bassana*). Journal of Animal Ecology 13: 49-62.

Fisher, R.B. 1958. The breeding biology of the Chimney Swift *Chaetura pelagica* (Linnaeus). New York State Museum and Science Service, Bulletin no. 368, 141 pp.

Fisk, E.J. 1971. Increase of fall Traill's Flycatchers in southern Florida. Bird-Banding 42(2): 121.

Fitch, H.S. 1958. Home ranges, territories and seasonal movements of vertebrates of the Natural History Reservation. University of Kansas Publications, Museum of Natural History 11(3): 63-326.

Fitch, H.S. 1974. Observations on the food and nesting of the Broad-winged Hawk (*Buteo platypterus*) in northeastern Kansas. Condor 76(2): 331-333.

Fitch, H.S., F. Swenson, and D.F. Tillotson. 1946. Behavior and food habits of the Red-tailed Hawk. Condor 48(5): 205-237.

Fitzpatrick, I.W. 1980. Foraging behaviour of neotropical tyrant flycatchers. Condor 82(1): 43-57.

Fjeldså, J. 1973. Feeding and habitat selection of the Horned Grebe, *Podiceps auritus* (Aves), in the breeding season. VidensKabelige Meddelelser fra dansk naturhistorisk 136: 57-95.

Flaxman, E.W. 1983. Communal mineral-eating by siskins. British Birds 76(8): 352.

Fleischer, R.C., P.E. Lowther, and R.F. Johnston. 1984. Natal dispersal in House Sparrows: possible causes and consequences. Journal of Field Ornithology 55(4): 444-456.

Fleischer, R.C., M.T. Murphy, and L.E. Hunt. 1985. Clutch size increase and intraspecific brood parasitism in the Yellow-billed Cuckoo. Wilson Bulletin 97(1): 125-127.

Fleming, T.L. and S.M. Speich. 1988. Ravens in southwestern Washington coniferous forests. Murrelet 69(3): 69-72.

Flemming, S.P., R.D. Chiasson, P.C. Smith, P.J. Austin-Smith, and R.P. Bancroft. 1988. Piping Plover status in Nova Scotia related to its reproductive and behavioral responses to human disturbance. Journal of Field Ornithology 59(4): 321-330.

Flemming, S.P., R.P. Bancroft, and N.R. Seymour. 1991. Enhanced fledging success by colonially nesting Ospreys in Nova Scotia coastal habitat. Wilson Bulletin 103: 668-675.

Fleurbec 1978. Plantes sauvages des villes et des champs, Guide d'identification Fleurbec. Éditions Fleurbec. Saint-Augustin-de-Desmaures, Québec, vol. 1, 288 pp.

Fleurbec 1981. Plantes sauvages comestibles, Guide d'identification Fleurbec. Éditions Fleurbec. Saint-Cuthbert, Québec, 168 pp.

Fleurbec 1983. Plantes sauvages des villes et des champs, Guide d'identification Fleurbec. Éditions Fleurbec. Saint-Augustin-de-Desmaures, Québec, vol. 2, 208 pp.

Fleurbec 1985. Plantes sauvages du bord de la mer, Guide d'identification Fleurbec. Éditions Fleurbec. Saint-Augustin-de-Desmaures, Québec, 288 pp.

Fleurbec 1987. Plantes sauvages des lacs, rivières et tourbières, Guide d'identification Fleurbec. Éditions Fleurbec. Saint-Augustin (Portneuf), 399 pp.

Fleurbec 1993. Fougères, prêles et lycopodes, Guide d'identification Fleurbec. Éditions Fleurbec. Saint-Henri-de-Lévis, 511 pp.

Flint, V.E., R.L. Boehme, Y.V. Kostin, and A.A. Kuznetsov. 1984. A field guide to birds of USSR: including Eastern Europe and central Asia. Princeton University Press, Princeton, New Jersey, 353 pp.

Flood, N.J. 1984. Adaptive significance of delayed plumage maturation in male Northern Orioles. Evolution 38(2): 267-279.

Flood, N.J. 1985. Incidences of polygyny and extra-pair copulation in the Northern Oriole. Auk 102(2): 410-413.

Flood, N.[J.] 1987a. Orchard Oriole, pp. 484-485 *in* M.D. Cadman, P.F.J. Eagles, and F.M. Helleiner, eds. Atlas of the breeding birds of Ontario. University of Waterloo Press, Waterloo, 617 pp.

Flood, N.[J.] 1987b. Rusty Blackbird, pp. 476-477 *in* M.D. Cadman, P.F.J. Eagles, and F.M. Helleiner, eds. Atlas of the breeding birds of Ontario. University of Waterloo Press, Waterloo, 617 pp.

Flood, N.[J.] 1987c. Yellow-headed Blackbird, pp. 474-475 *in* M.D. Cadman, P.F.J. Eagles, and F.M. Helleiner, eds. Atlas of the breeding birds of Ontario. University of Waterloo Press, Waterloo, 617 pp.

Flood, N.J. and G.R. Bortolotti. 1984. Status of the Sharp-shinned Hawk (*Accipiter striatus*) in Canada. Committee on the Status of Endangered Wildlife in Canada (COSEWIC), Ottawa, 86 pp.

Fogarty, M.J., K.A. Arnold, L. McKibben, L.B. Pospichal, and R.J. Tully. 1980. Common Snipe, pp. 189-209 *in* G.C. Sanderson, ed. Management of migratory shore and upland game birds in North America. International Association of Fish and Wildlife Agencies, Washington, 358 pp.

Follen, D.G. Sr. and J.C. Haug. 1981. Saw-whet Owl nest in Wood Duck box. Passenger Pigeon 43(1): 47-48.

Forbes, L.S. 1991. Hunger and food allocation among nestlings of facultatively siblicidal Ospreys. Behavioral Ecology and Sociobiolgy 29: 189-195.

Forbes, L.S., K. Simpson, J.P. Kelsall, and D.R. Flook. 1985. Great Blue Heron colonies in British Columbia. Canadian Wildlife Service – Pacific and Yukon Region, Technical Report, 78 pp.

Forbes, M.R.L. 1983. The nesting ecology and breeding behaviour of the Pied-billed Grebe, *Podilymbus podiceps* (L.), at a national wildlife refuge in Nova Scotia. M.Sc. thesis, Acadia University, Wolfville, 62 pp.

Forbes, M.R.L. 1987. Extrapair feeding in Pied-billed Grebes. Wilson Bulletin 99(1): 109-111.

Forbes, M.R.L. and C.D. Ankney. 1987. Hatching asynchrony and food allocation within broods of Pied-billed Grebes, *Podilymbus podiceps*. Canadian Journal of Zoology 65: 2872-2877.

Forbes, M.R.L. and C.D. Ankney. 1988. Nest attendance by adult Pied-billed Grebes, *Podilymbus podiceps* (L.). Canadian Journal of Zoology 66: 2019-2023.

Forbes, M.R.L., H.P. Barkhouse, and P.C. Smith. 1989. Nest-site selection by Pied-billed Grebes *Podilymbus podiceps*. Ornis Scandinavica 20: 211-218.

Forbush, E.H. 1912. A history of the game birds, wild-fowl, and shorebirds of Massachusetts and adjacent states. Massachusetts State Board of Agriculture, Wright and Potter Printing Company, Boston, 622 pp.

Forbush, E.H. 1927. Birds of Massachusetts and other New England states, vol. 2: Land birds from bob-whites to grackles. Massachusetts Department of Agriculture, Boston, [1978, Arno Press, New York], 461 pp.

Forbush, E.H. 1929. Birds of Massachusetts and other New England states, vol. 3: Land birds from sparrows to thrushes. Massachusetts Department of Agriculture, Boston, [1978, Arno Press, New York], 466 pp.

Forbush, E.H. and J.B. May. 1939. Natural history of the birds of eastern and central North America. Houghton Mifflin Company, Boston, 554 pp.

Forbush, E.H. and J.B. May. 1955. A natural history of American birds of eastern and central North America. Bramhall House, New York, 552 pp.

Foreman, L.D. 1976. Observations of Common Merganser broods in northwestern California. California Fish and Game 62(3): 207-212.

Forestry Canada 1992. Terminologie de la sylviculture au Canada. Direction générale des politiques et de l'économie, Ottawa, 74 pp. [Also available in English as: Silvicultural terms in Canada.]

Forman, R.T.T., A.E. Galli, and C.F. Leck. 1976. Forest size and avian diversity in New Jersey woodlots with some land use implications. Oecologia 26: 1-28.

Forshaw, J.M. 1977. Parrots of the world. T.F.H. Publications, Neptune, New Jersey, 584 pp.

Forshaw, J.M. 1978. Parrots of the world, 2nd edition. Lansdowne Editions, Melbourne, 616 pp.

Forsman, E. and E.L. Bull. 1987. Great Horned, Great Gray, Spotted and Barred owls, pp. 118-123 *in* B.G. Pendleton, ed. Proceedings. Western raptor management symposium and workshop. National Wildlife Federation, Washington. National Wildlife Federation Scientific and Technical Series no. 12, 317 pp.

Forster, J.R. 1771. A catalogue of the animals of North America, or, Faunula americana. P.L. Sclater, ed. [1882, Willughby Society for the Reprinting of scarce Ornithological works, London], 53 pp.

Forster, J.R. 1772. Part XXIX. An account of the birds sent from Hudson's Bay; with observations relative to their natural history and Latin descriptions of some of the most uncommon. Read June 18-25, 1772. Willughby Society, London, 382-433 pp.

Forster, R.A. 1990. An attempt to unravel the mystery of winter finch occurrence. Bird Observer 18(1): 25-31.

Fortin, C. 1985. La chasse contingentée de l'Orignal et de la chasse du petit gibier dans la réserve faunique des Laurentides. Bilan 1985 et recommandations pour 1986-88. Ministère du Loisir, de la Chasse et de la Pêche, 8 pp.

Fortin, L., J.-P.[L.] Savard, and C. Aubert. 1978. Étude des populations d'oiseaux de la région de Québec. Club des ornithologues du Québec, Charlesbourg. Cahiers d'ornithologie Victor-Gaboriault no. 2, 101 pp.

Foster, W.L. and J. Tate, Jr. 1966. The activities and coactions of animals at sapsucker trees. Living Bird 5: 87-113.

Fournier, P. 1932. ... Les voyages naturalistes du clergé français avant la révolution.... Paris, Le Chevalier, 254 pp. (Encyclopédie biologique, vol. X.) Voyages et découvertes scientifiques des missionnaires naturalistes français à travers le monde (XVe au XXe siècles).

Fowle, C.D. 1972. Effects of phosphamidon on forest birds in New Brunswick. Canadian Wildlife Service. Report Series no. 16, 25 pp.

Fox, A.D., I.S. Francis, J. Madsen, and J.M. Stroud. 1987. The breeding biology of the Lapland Bunting *Calcarius lapponicus* in west Greenland during two contrasting years. Ibis 129(4): 541-552.

Fox, A.D., N. Jarrett, H. Gitay, and D. Paynter. 1989. Late summer habitat selection by breeding waterfowl in northern Scotland. Wildfowl 40: 106-114.

Fox, G.A. 1961. A contribution to the life history of the Clay-colored Sparrow. Auk 78(2): 220-224.

Fox, G.A. 1964. Notes on the western race of the Pigeon Hawk. Blue Jay 22(4): 140-147.

Fox, G.A. 1971. Reproductive success of Merlins. Journal of Wildlife Management 35(1): 122-128.

Fradette, P. 1992. Les oiseaux des Îles-de-la-Madeleine: populations et sites d'observation. Attention frag'Îles, Mouvement pour la valorisation du patrimoine naturel des Îles, L'Étang-du-Nord, 292 pp.

Fragnier, P. 1992. Cartes illustrant l'intensité de l'exploitation forestière sur le territoire public québécois entre 1974 et 1993. Environment Canada, Canadian Wildlife Service, Québec Region, 14 pp.

Frakes, R.A. and R.E. Johnson. 1982. Niche convergence in *Empidonax* flycatchers. Condor 84(3): 286-291.

Francis, C.M. and F. Cooke. 1986. Differential timing of spring migration in wood warblers (Parulinae). Auk 103(3): 548-556.

Franklin, A.B. 1988. Breeding biology of the Great Gray Owl in southeastern Idaho and northeastern Wyoming. Condor 90: 689-696.

Franklin, J.F. 1993. Preserving biodiversity: species, ecosystems or landscapes. Ecological Applications 3: 202-205.

Franzblau, M.A. and J.P. Collins. 1980. Test of a hypothesis of territory regulation in an insectivorous bird by experimentally increasing prey abundance. Oecologia 46: 164-170.

Franzreb, K.E. 1984. Foraging habits of Ruby-crowned and Golden-crowned kinglets in an Arizona montane forest. Condor 86(2): 139-145.

Franzreb, K.E. 1985. Foraging ecology of Brown Creepers in a mixed-coniferous forest. Journal of Field Ornithology 56(1): 9-16.

Fraser, D. and N. St-Louis. 1979. Étude de la faune avienne dans les bassins de la Grande rivière de la Baleine et de la Petite rivière de la Baleine (été 1978). Technical report for Direction Environnement of Hydro-Québec by Éco-recherches, 128 pp.

Fraser, D.M. 1987a. Black-throated Blue Warbler, pp. 378-379 *in* M.D. Cadman, P.F.J. Eagles, and F.M. Helleiner, eds. Atlas of the breeding birds of Ontario. University of Waterloo Press, Waterloo, 617 pp.

Fraser, D.M. 1987b. Mourning Warbler, pp. 414-415 *in* M.D. Cadman, P.F.J. Eagles, and F.M. Helleiner, eds. Atlas of the breeding birds of Ontario. University of Waterloo Press, Waterloo, 617 pp.

Fraser, D.M. 1987c. Red-breasted Merganser, pp. 102-103 *in* M.D. Cadman, P.F.J. Eagles, and F.M. Helleiner, eds. Atlas of the breeding birds of Ontario. University of Waterloo Press, Waterloo, 617 pp.

Frawley, B.J. and L.B. Best. 1991. Effects of mowing on breeding bird abundance and species composition in alfalfa fields. Wildlife Society Bulletin 19(2): 134-142.

Frech, M.H. and J.L. Confer. 1987. The Golden-winged Warbler: competition with the Blue-winged Warbler and habitat selection in portions of southern, central and northern New York. Kingbird 37(2): 65-71.

Fredrickson, L.H. 1970. Breeding biology of American Coots in Iowa. Wilson Bulletin 82(4): 445-457.

Fredrickson, L.H. 1971. Common Gallinule breeding biology and development. Auk 88(4): 914-919.

Fredrickson, L.H. 1977. American Coot (*Fulica americana*), pp. 123-147 *in* G.C. Sanderson, ed. Management of migratory shore and upland game birds in North America. International Association of Fish and Wildlife Agencies, University of Nebraska Press, Lincoln, 358 pp.

Fredrickson, L.H. 1990. Wood Duck behavior: fall courtship to egg laying, pp. 35-43 *in* L.H. Fredrickson, G.V. Burger, S.P. Havera, D.A. Graber, R.E. Kirby, and T.S. Taylor, eds. Proceedings. 1988 North American Wood Duck symposium, St. Louis, Missouri, 390 pp.

Fredrickson, L.H. and J.L. Hansen. 1983. Second broods in Wood Ducks. Journal of Wildlife Management 47(2): 320-326.

Fredrickson, L.H. and M.E. Heitmeyer. 1991. Life history strategies and habitat needs of the Northern Pintail. United States Department of the Interior,

Fish and Wildlife Service. Leaflet no. 13.1.3, Washington, 8 pp.

Freedman, B., C. Beauchamp, I.A. McLaren, and S.I. Tingley. 1981. Forestry management practices and populations of breeding birds in a hardwood forest in Nova Scotia. Canadian Field-Naturalist 95(3): 307-311.

Freeland, D.B. 1973. Some food preferences and aggressive behavior by Monk Parakeets. Wilson Bulletin 85(3): 332-335.

Freemark, K.E. 1978. The status of the Gyrfalcon (*Falco rusticolus*) in Canada. Environment Canada, Canadian Wildlife Service, Ottawa, 11 pp.

Freemark, K.E. 1989. Landscape ecology of forest birds in the northeast, pp. 7-12 *in* R.M. DeGraaf and W.M. Healy, eds. Is forest fragmentation a management issue in the northeast? United States Department of Agriculture, Forest Service, Northeastern Forest Experiment Station. General Technical Report NE-140.

Freemark, K.E. and B. Collins. 1992. Landscape ecology of birds breeding in temperate forest fragments, pp. 443-454 *in* J.M. Hagan III and D.W. Johnston, eds. Ecology and conservation of neotropical migrant landbirds. Smithsonian Institution Press, Washington, 609 pp.

Freemark, K.E. and H.G. Merriam. 1986. Importance of area and habitat heterogeneity to bird assemblages in temperate forest fragments. Biological Conservation 36: 115-141.

Frelich, L.E. and C.G. Lorimer. 1991. Natural disturbance regimes in hemlock-hardwood forests of the Upper Great Lakes region. Ecological Monographs 61: 145-164.

French, N.R. 1954. Notes on breeding activities and on gular sacs in the Pine Grosbeak. Condor 56: 83-85.

Frenette, M. 1990. The St. Lawrence, pp. 295-304 *in* M.G. Wolman and H.C. Riggs, eds. Surface water hydrology. The Geological Society of America, Boulder, Colorado, 374 pp.

Friedmann, H. 1929. The cowbirds, a study in the biology of social parasitism. Charles C. Thomas Publisher, Springfield, Illinois, 395 pp.

Friedmann, H. 1963. Host relations of the parasitic cowbirds. United States National Museum, Washington. Bulletin no. 233, 276 pp.

Friedmann, H. 1966. Additional data on the host relations of the parasitic cowbirds. Smithsonian Miscellaneous Collection 149(11): 1-12.

Friedmann, H. 1971. Further information on the host relations of the parasitic cowbirds. Auk 88(2): 239-255.

Friedmann, H., L.F. Kiff, and S.I. Rothstein. 1977. A further contribution to knowledge of the host relations of the parasitic cowbirds. Smithsonian Contributions to Zoology no. 235, 75 pp.

Fritz, R.S. 1979. Consequences of insular population structure: distribution and extinction of Spruce Grouse populations. Oecologia 42: 57-65.

Frontier, S. and D. Pichot-Viale. 1993. Écosystèmes: structure, fonctionnement, évolution. Masson, ed. Paris, 428 pp.

Fry, C.H. 1985. Kingfishers, pp. 266-269 *in* C.M. Perrins and A.L.A. Middleton, eds. The encyclopedia of birds. Facts on File, New York, 447 pp.

Fryer, R. 1937. The Yellow Rail in southern Manitoba. Canadian Field-Naturalist 51: 41-42.

Fuller, A.B. 1938. Yellow Rail at Churchill, Manitoba. Auk 55: 670-671.

Fuller, M.R. 1979. Spatiotemporal ecology of four sympatric raptor species. Ph.D. dissertation, University of Minnesota, Minneapolis, 220 pp.

Fuller, M.R. and J.A. Mosher. 1981. Methods of detecting and counting raptors: a review. Studies in Avian Biology 6: 235-246.

Furness, R.W. and R.T. Barrett. 1985. The food requirements and ecological relationship of a seabird community in North Norway. Ornis Scandinavica 16: 305-313.

Furness, R.W. and R.T. Barrett. 1991. Seabirds and fish declines. National Geographic Research and Exploration 7: 82-95.

Futuyma, D.J. and F. Gould. 1979. Associations of plants and insects in a deciduous forest. Ecological Monographs 49: 33-50.

Fyfe, R.W. 1976. Status of Canadian raptor populations. Canadian Field-Naturalist 90(3): 370-375.

Fyfe, R.W. 1977. Status of Canadian raptor populations, pp. 34-39 *in* R.D. Chancellor, ed. World Conference on Birds of Prey, Vienna, 1975. International Council for Bird Preservation.

Gaboriault, V. 1974. Charles-Eusèbe Dionne, naturaliste. La Société historique de la Côte-du-Sud, La Pocatière. Cahier d'histoire no. 9, 144 pp.

Gaboriault, W. 1961. Les oiseaux aux Îles-de-la-Madeleine. Naturaliste canadien 88: 166-224.

Gabrielson, I.N. 1952. Notes on the birds of the North Shore of the Gulf of St. Lawrence. Canadian Field-Naturalist 66: 44-59.

Gabrielson, I.N. and F.C. Lincoln. 1959. The birds of Alaska. Stackpole Company and Wildlife Management Institution, Harrisburg, 922 pp.

Gabrielson, I.N. and B.S. Wright. 1951. Notes on the birds of Fort Chimo, Ungava district. Canadian Field-Naturalist 65(4): 127-140.

Gaddis, P.R. 1983. Composition and behavior of mixed-species flocks of forest birds in north-central Florida. Florida Field Naturalist 11: 25-34.

Gagné, S. 1993. Création des parcs, plus vite et plus loin. Franc-Vert 10(3): 32-33.

Gagnon, J., G. Savard, and C. Cormier. 1986. Bruant sauterelle. Le Harfang 9(5): 202.

Galati, B. and C.B. Galati. 1985. Breeding of the Golden-crowned Kinglet in northern Minnesota. Journal of Field Ornithology 56(1): 28-40.

Galli, A.E., C.F. Leck, and R.T.T. Forman. 1976. Avian distribution patterns in forest islands of different sizes in central New Jersey. Auk 93(2): 356-364.

Galusha, J.G. Jr. and C.J. Amlaner, Jr. 1978. The effects of diurnal and tidal periodicities in the numbers and activities of Herring Gulls *Larus argentatus* in a colony. Ibis 120(3): 322-328.

Gardarsson, A. 1978. Distribution and numbers of the Barrow's Goldeneye (*Bucephala islandica*) in Iceland. Natturu Fraedin gurinn 48: 162-191.

Gardarsson, A. and R. Moss. 1970. Selection of food by Icelandic Ptarmigan in relation to its availability and nutritive value, pp. 47-71 *in* A. Watson, ed. Animal populations in relation to their food resources. Blackwell Scientific Publications, Oxford, 477 pp.

Gariepy, F. 1980. Les oiseaux de la rivière aux Feuilles. Bulletin ornithologique 25(2): 41-49.

Gariépy, L. 1983. Les oiseaux de la baie Déception. Bulletin ornithologique 28(3): 111-113.

Garland, L.E. 1977. The effects of historical land use practices upon the abundance and distribution of deer in Vermont. Vermont Fish and Game Department, Montpelier.

Garson, P.J. 1985. Dippers and wrens, pp. 60-63 *in* C.M. Perrins and A.L.A. Middleton, eds. Songbirds. Series: All the world's animals. Torstar Books, New York, 160 pp.

Gascon, C. and D. Planas. 1986. Spring pond water chemistry and the reproduction of the wood frog, *Rana sylvatica*. Canadian Journal of Zoology 64: 543-550.

Gaston, A.J. 1985. Development of the young in the Atlantic Alcidae, pp. 319-354 *in* D.N. Nettleship and T.R. Birkhead, eds. The Atlantic Alcidae: the evolution, distribution and biology of the auks inhabiting the Atlantic Ocean and adjacent water areas. Academic Press, London, 574 pp.

Gaston, A.J. 1990. Seabirds: populations, ecology and energy consumption, pp. 127-133 *in* J.A. Percy, ed. Proceedings. Marine ecosystem studies in Hudson Strait. Workshop, Canadian Technical Report of Fisheries and Aquatic Sciences, no. 1770.

Gaston, A.J. and R. Decker. 1985. Interbreeding of Thayer's Gull, *Larus thayeri*, and Kumlien's Gull, *Larus glaucoides kumlieni*, on Southampton Island, Northwest Territories. Canadian Field-Naturalist 99(2): 257-259.

Gaston, A.J. and R.D. Elliot. 1990. Kumlien's Gull (*Larus glaucoides kumlieni*) on Coats Island, Northwest Territories. Canadian Field-Naturalist 104(3): 477-479.

Gaston, A.J. and R.D. Elliot. 1991. Studies of high-latitudes seabirds. 2. Conservation biology of Thick-billed Murres in the northwest Atlantic. Environment Canada, Canadian Wildlife Service. Occasional Paper no. 69, 63 pp.

Gaston, A.J. and M. Malone. 1980. Range extension of Atlantic Puffin and Razorbill in Hudson Strait. Canadian Field-Naturalist 94(3): 328-329.

Gaston, A.J. and D.N. Nettleship. 1981. The Thick-billed Murres of Prince Leopold Island: a study of the breeding ecology of a colonial high arctic seabird. Canadian Wildlife Service. Monograph Series no. 6, 350 pp.

Gaston, A.J., G. Chapdelaine, and D.G. Noble. 1983. The growth of Thick-billed Murre chicks at colonies in Hudson Strait: inter and intra-colony variation. Canadian Journal of Zoology 61(11): 2465-2473.

Gaston, A.J., G. Chapdelaine, and D.G. Noble. 1984. Phenotypic variation among Thick-billed Murres from colonies in Hudson Strait. Arctic 37(3): 284-287.

Gaston, A.J., D.K. Cairns, R.D. Elliot, and D.G. Noble. 1985. A natural history of Digges Sound. Canadian Wildlife Service. Report Series no. 46, 65 pp.

Gaston, A.J., L.N. de Forest, G. Gilchrist, and D.N. Nettleship. 1993. Monitoring Thick-billed Murre populations at colonies in northern Hudson Bay, 1972-92. Canadian Wildlife Service. Occasional Paper no. 80, 16 pp.

Gates, J.M. 1962. Breeding biology of the Gadwall in northern Utah. Wilson Bulletin 74(1): 43-67.

Gates, J.M. 1972. Red-tailed Hawk populations and ecology in east-central Wisconsin. Wilson Bulletin 84(4): 421-433.

Gaudreau, G. 1986. L'exploitation des forêts publiques au Québec 1842-1905. Institut québécois de recherche sur la culture, Québec, 126 pp.

Gauthier, B. 1977. Recherches sur les limites biologiques du Saint-Laurent. Ph.D. dissertation, Université Laval, Sainte-Foy, 233 pp.

Gauthier, B. and M. Goudreau. 1983. Mares glacielles et non glacielles dans les marais de L'Isle-Verte, estuaire du Saint-Laurent, Québec. Géographie Physique et Quaternaire 37(1): 49-56.

Gauthier, G. 1987a. Further evidence of long-term pair bonds in ducks of the genus *Bucephala*. Auk 104: 521-522.

Gauthier, G. 1987b. The adaptive significance of territorial behaviour in breeding Buffleheads: a test of three hypothesis. Animal Behaviour 34: 348-360.

Gauthier, G. 1987c. Brood territories in Buffleheads: determinants and correlates of territory size. Canadian Journal of Zoology 65: 1402-1410.

Gauthier, G. 1988. Factors affecting nest-box use by Buffleheads and other cavity-nesting birds. Wildlife Society Bulletin 16(2): 132-141.

Gauthier, G. 1989. The effect of experience and timing on reproductive performance in Buffleheads. Auk 106: 568-576.

Gauthier, G. 1990. Philopatry, nest-site fidelity, and reproductive performance in Buffleheads. Auk 107: 126-132.

Gauthier, G. 1993a. Bufflehead (*Bucephala albeola*), *in* A. Poole and F. Gill, eds. The birds of North America, no. 67. The Academy of Natural Sciences, Philadelphia. The American Ornithologists' Union, Washington, 24 pp.

Gauthier, G. 1993b. Feeding ecology of nesting Greater Snow Geese. Journal of Wildlife Management 57(2): 216-223.

Gauthier, G. and J.N.M. Smith. 1987. Territorial behaviour, nest-site availability, and breeding density in Buffleheads. Journal of Animal Ecology 56: 171-184.

Gauthier, G. and J. Tardif. 1991. Female feeding and male vigilance during nesting in Greater Snow Geese. Condor 93: 701-711.

Gauthier, G., Y. Bédard, and J. Bédard. 1988. Habitat use and activity budgets of Greater Snow Geese in spring. Journal of Wildlife Management 52(2): 191-201.

Gauthier, J. and Y. Aubry. 1985. Atlas des oiseaux nicheurs du Québec: guide du participant. Environment Canada, Canadian Wildlife Service, 34 pp. [Also available in English as: Guide for participants.]

Gauthier, J. and J. Bédard. 1976. Les déplacements de l'Eider commun (*Somateria mollissima*) dans l'estuaire du Saint-Laurent. Naturaliste canadien 103(4): 261-283.

Gauthier, J., J. Bédard, and A. Reed. 1976. Overland migration by Common Eiders of the St. Lawrence estuary. Wilson Bulletin 88(2): 333-344.

Gauthier, J., D. Lehoux, and J. Rosa. 1980. Les marécages intertidaux dans l'estuaire du Saint-Laurent. Environment Canada, Canadian Wildlife Service, 91 pp.

Gauthier, M. and D.W. Thomas. 1993. Nest site selection and cost of nest building by Cliff Swallow (*Hirundo pyrrhonota*). Canadian Journal of Zoology 71: 1120-1123.

Gauthier, M., D.W. Thomas, J.R. Speakman, and H. Lapierre. 1994. Bénéfices associés à la réutilisation des nids chez l'Hirondelle à front blanc (*Hirundo pyrrhonota*). Ecoscience 1(2): 119-126.

Gauthier, Y. and J. Larivée. 1989. L'Hirondelle des granges (*Hirundo rustica*) dans le Bas-Saint-Laurent. Kakawi 11(2): 2-4.

Gauthreaux, S.A. Jr. 1971. The changing seasons. American Birds 25(5): 821-825.

Gauvin, J. and A. Reed. 1987. A simulation model for the Greater Snow Goose population. Environment Canada, Canadian Wildlife Service. Occasional Paper no. 64, 28 pp.

Gavin, T.A. 1984. Broodedness in Bobolinks. Auk 101(1): 179-181.

Gavin, T.A. and E.K. Bollinger. 1985. Multiple paternity in a territorial passerine: the Bobolink. Auk 102(3): 550-555.

Gavin, T.A. and E.K. Bollinger. 1988. Reproductive correlates of breeding-site fidelity in Bobolinks. Ecology 69(1): 96-103.

Gawlik, D.E. and K.L. Bildstein. 1990. Reproductive success and nesting habitat of Loggerhead Shrike in north-central South Carolina. Wilson Bulletin 102(1): 37-48.

Gehrman, K.H. 1951. An ecological study of the Lesser Scaup duck (*Aythya affinis* Eyton) at West Medical Lake, Spokane County, Washington. M.Sc. thesis, Washington State College, Pullman, 94 pp.

Gélinas, C. 1984. L'exploitation et la conservation forestière au parc national de la Mauricie, 1830-1940. Dossier documentaire (internal report), Parks Canada, Québec, 460 pp.

Gélinas, M. 1990. Colloque provincial le 8 décembre 1990. Proceedings. Prepared for Canadian Wildlife Service by Fondation Les oiseleurs du Québec. 1er colloque provincial sur la Bécasse d'Amérique [1st provincial symposium on the American Woodcock], Rivière-du-Loup.

Gendron, G. and Y. Gauthier. 1984. Les oiseaux du Bas-Saint-Laurent: liste annotée. Club des ornithologues du Bas-Saint-Laurent, Rimouski, 61 pp.

George, P. 1970. Dictionnaire de la géographie. Presses universitaires de France, Paris, 510 pp.

Gérardin, V. 1980. Les régions écologiques et la végétation des sols minéraux. L'inventaire du capital nature du territoire de la Baie-James. Environment Canada, Ottawa, 398 pp.

Gerell, G. 1985. Habitat selection and nest predation in a Common Eider population in southern Sweden. Ornis Scandinavica 16: 129-139.

Géroudet, P. 1984. Les rapaces diurnes et nocturnes d'Europe, 6th edition Delachaux et Niestlé, ed. Neuchâtel, 426 pp.

Gerrard, J.M. 1983. A review of the current status of Bald Eagles in North America, pp. 5-22 *in* D.M. Bird, ed. Biology and management of Bald Eagles and Ospreys. Harpell Press, Sainte-Anne-de-Bellevue, Québec, 325 pp.

Gerrard, J.M. and G.R. Bortolotti. 1988. The Bald Eagle: haunts and habits of a wilderness monarch. Western Producer Prairie Books, Saskatoon, 177 pp.

Gerrard, J.M., D.W.A. Whitfield, P. Gerrard, P.N. Gerrard, and W.J. Maher. 1978. Migratory movements and plumage of subadult Saskatchewan Bald Eagles. Canadian Field-Naturalist 92(4): 375-382.

Gerrard, J.M., P.N. Gerrard, and D.W.A. Whitfield. 1980. Behavior in a non-breeding Bald Eagle. Canadian Field-Naturalist 94(3): 391-397.

Gerrard, P.N., J.M. Gerrard, D.W.A. Whitfield, and W.J. Maher. 1974. Post-fledging movements of juvenile Bald Eagles. Blue Jay 32: 218-226.

Gerrard, P.N., S.N. Wiemeyer, and J.M. Gerrard. 1979. Some observations of the behaviour of captive Bald Eagles before and during incubation. Raptor Research 13: 57-64.

Gerson, H. 1988. Status report on the Black Tern, *Chlidonias niger*, in Canada. Committee on the Status of Endangered Wildlife in Canada (COSEWIC) and Ontario Ministry of Natural Resources, Toronto, Ontario, 51 pp.

Gervais, D., C. Lachance, and C. Robidoux. 1983. Étude préliminaire des marais du sud-ouest du lac Saint-François, Huntingdon, Québec. Projet Relais and Ducks Unlimited Canada, 104 pp.

Gervais, D., M. Loiselle, C. Robidoux, and F. Shaffer. 1984. Étude de la sauvagine et du Merle-bleu de l'Est sur la Réserve nationale de faune du lac Saint-François, 1984. Projet Environnement 2000, 44 pp.

Ghanimé, L., J.-L. DesGranges, S. Loranger, and Collaborators. 1990. Les régions biogéographiques du Saint-Laurent. Technical report prepared for Environment Canada and Fisheries and Oceans Canada, Québec Region, by Lavalin Environnement.

Gibb, J. and C. Gibb. 1951. Waxwings in the winter of 1949-1950. British Birds 44: 158-163.

Gibbs, H.L., P.J. Weatherhead, P.T. Boag, B.N. White, L.M. Tabak, and D.J. Hoysak. 1990. Realized reproductive success of polygynous Red-winged Blackbirds revealed by DNA markers. Science 250(4986): 1394-1397.

Gibbs, J.P. 1991. Spatial relationships between nesting colonies and foraging areas of Great Blue Herons. Auk 108: 764-770.

Gibbs, J.P. and J. Faaborg. 1990. Estimating the viability of Ovenbird and Kentucky Warbler populations in forest fragments. Conservation Biology 4: 193-196.

Gibbs, J.P. and S.M. Melvin. 1992a. Least Bittern (*Ixobrychus exilis*), pp. 71-88 *in* K.J. Schneider and D.M. Pence, eds. Migratory nongame birds of management concern in the northeast. United States Department of the Interior, Fish and Wildlife Service, Newton Corner, Massachusetts, 400 pp.

Gibbs, J.P. and S.M. Melvin. 1992b. Pied-billed Grebe (*Podilymbus podiceps*), pp. 31-49 *in* K.J. Schneider and D.M. Pence, eds. Migratory nongame birds of management concern in the northeast. United States Department of the Interior, Fish and Wildlife Service, Newton Corner, Massachusetts, 400 pp.

Gibbs, J.P. and S.M. Melvin. 1992c. Sedge Wren (*Cistothorus platensis*), pp. 191-209 *in* K.J. Schneider and D.M. Pence, eds. Migratory nongame birds of management concern in the northeast. United States Department of the Interior, Fish and Wildlife Service, Newton Corner, Massachusetts, 400 pp.

Gibbs, J.P., J.R. Longcore, D.G. McAuley, and J.K. Ringelman. 1991a. Use of wetland habitats by selected nongame waterbirds in Maine. United States Fish and Wildlife Service, Fish and Wildlife Research 9.

Gibbs, J.P., W.G. Shriver, and S.M. Melvin. 1991b. Spring and summer records of the Yellow Rail in Maine. Journal of Field Ornithology 62(4): 509-516.

Gibbs, J.P., F.A. Reid, and S.M. Melvin. 1992a. Least Bittern (*Ixobrychus exilis*), *in* A. Poole, P. Stettenheim, and F. Gill, eds. The birds of North America, no. 17. The Academy of Natural Sciences, Philadelphia. The American Ornithologists' Union, Washington, 12 pp.

Gibbs, J.P., S.M. Melvin, and F.A. Reid. 1992b. American Bittern (*Botaurus lentiginosus*), *in* A. Poole, P. Stettenheim, and F. Gill, eds. The birds of North America, no. 18. The Academy of Natural Sciences, Philadelphia. The American Ornithologists' Union, Washington, 12 pp.

Gibbs, R.M. 1962. Juvenile mortality in the Common Goldeneye, *Bucephala clangula americana*, in Maine. Maine Field Naturalist 18(5): 67-68.

Gibson, D.D. and B. Kessel. 1989. Geographic variation in the Marbled Godwit and description of an Alaska subspecies. Condor 91(2): 436-443.

Giguère, G.E. 1973. Oeuvres de Champlain. Éditions du Jour, Montréal, 479 pp.

Gilbert, G. 1981. Le comportement social du Pinson à queue aiguë. M.Sc. thesis, Université Laval, Sainte-Foy, 80 pp.

Gilbert, G., R.G. Hélie, and J.M. Mondoux. 1985. Écorégions et écodistricts du Québec, pp. 1-55 *in* Sensibilité de l'écosystème aux précipitations acides au Québec. Environment Canada, Ottawa. Ecological Land Classification Series no. 20, 96 pp. [Also available in English as: Ecosystem sensitivity to acid precipitation for Quebec.]

Gill, D. 1974. The Gray Jay as a predator of small mammals. Canadian Field-Naturalist 88(3): 370-371.

Gill, D.E. and W.E. Lanyon. 1965. Establishment, growth, and behavior of an extralimital population of House Finches at Huntington, New York. Bird-Banding 36(1): 1-14.

Gill, F.B. 1980. Historical aspects of hybridization between Blue-winged and Golden-winged warblers. Auk 97(1): 1-18.

Gill, F.B. 1985. Birds, pp. 259-351 *in* H.H. Genoways. and F.J. Brenner, eds. Species of special concern in Pennsylvania. Carnegie Museum of Natural History, Pittspurgh Special Publication no. 11.

Gill, F.B. 1990. Ornithology. W.H. Freeman and Company, New York, 660 pp.

Gill, F.B. and B.G. Murray, Jr. 1972a. Discrimination behaviour and hybridization of the Blue-winged and Golden-winged warblers. Evolution 26: 282-293.

Gill, F.B. and B.G. Murray, Jr. 1972b. Song variation in sympatric Blue-winged and Golden-winged warblers. Auk 89(3): 625-643.

Gillespie, D.I. and S.P. Wetmore. 1974. Waterfowl survey in Labrador-Ungava, 1970, 1971, 1972, pp. 8-18 *in* H. Boyd, ed. Canadian Wildlife Service waterfowl studies in eastern Canada, 1969-73/ Études du SCF sur les oiseaux aquatiques dans l'est du Canada, 1969-73. Canadian Wildlife Service. Report Series no. 29, 106 pp.

Gilmer, D.S., P.M. Konrad, and R.E. Stewart. 1983. Nesting ecology of Red-tailed Hawks and Great Horned Owls in central North Dakota and their interactions with other large raptors. Prairie Naturalist 15(3): 133-143.

Gilpin, M. and I. Hanski. 1991. Metapopulation dynamics: empirical and theoretical investigations. Academic Press, New York.

Gilpin, M.E. and M.E. Soulé. 1986. Minimum viable populations: processes of species extinction, pp. 19-34 *in* M.E. Soulé, ed. Conservation biology: the science of scarcity and diversity. Sinauer Associates, Sunderland, Massachusetts.

Gingras, P. 1993. Saint-Cléophas, terre d'urubus et de merles bleus. La Presse, 7 November 1993, p. C-12.

Ginn, H.B. and D.S. Melville. 1983. Moult in birds. The British Trust for Ornithology, England, Guide 19, 112 pp.

Girard, S. 1988. Itinéraire ornithologique de la Gaspésie. Club des ornithologues de la Gaspésie, Percé, 166 pp.

Giroux, J. and P. Laporte. 1988. Le Gros-bec errant: hôte coloré de nos mangeoires. Bulletin ornithologique 33(3): 129.

Giroux, J.-F. 1981. Interspecific nest parasitism by Redheads on islands in southeastern Alberta. Canadian Journal of Zoology 59(11): 2053-2057.

Giroux, J.-F. and J. Bédard. 1987. The effects of grazing by Greater Snow Geese on the vegetation of tidal marshes in the St. Lawrence estuary. Journal of Applied Ecology 24: 773-788.

Giroux, J.-F., J. Bédard, and Y. Bédard. 1984. Habitat use by Greater Snow Geese during the brood-rearing period. Arctic 37: 155-160.

Giroux, J.-F., J. Bédard, and Y. Bédard. 1986. Time budget of Greater Snow Geese during the brood-rearing period. Wildfowl 37: 46-50.

Giroux, J.-F., M. Caron, M. Picard, and S. Carrière. 1992. Effet du contrôle des prédateurs sur la nidification des canards à la Réserve nationale de faune des îles de Contrecoeur. Département des sciences biologiques, Université du Québec à Montréal, Montréal, 36 pp.

Glahn, J.F. 1974. Study of breeding rails with recorded calls in north-central Colorado. Wilson Bulletin 86(3): 206-214.

Glase, J.C. 1973. Ecology of social organization in the Black-capped Chickadee. Living Bird 12: 235-267.

Glenny, F.H. 1939. An anomalous artery in the Kingfisher (*Ceryle alcyon*). Ohio Journal of Science 39(2): 94-96.

Glover, F.A. 1953. Nesting ecology of the Pied-billed Grebe in north-eastern Iowa. Wilson Bulletin 65: 32-39.

Glutz von Blotzheim, U.N. and K.M. Bauer. 1982. Handbuch der Vögel Mitteleuropas [Manuel of birds of central Europe], vol. 8, part 2. Akademische Verlagsgesellschaft, Wiesbaden, Federal Republic of Germany, 1270 pp.

Gobeil, R.E. 1970. Vocalizations of the Savannah Sparrow. Bird-Banding 41: 18-21.

Gochfeld, M. and J. Burger. 1987. Nest-site selection: comparison of Roseate and Common terns (*Sterna dougallii* and *S. hirundo*) in a Long Island, New York, colony. Bird Behaviour 7: 58-66.

Godfrey, W.E. 1949. Birds of Lake Mistassini and Lake Albanel, Quebec. National Museum of Canada, Ottawa. Bulletin no. 114, Biological Series no. 38, 43 pp.

Godfrey, W.E. 1956. Some distributional notes on Canadian birds. Canadian Field-Naturalist 70(3): 136-138.

Godfrey, W.E. 1966. The birds of Canada. National Museum of Canada, Ottawa. Bulletin no. 203, Biological Series no. 73, 428 pp.

Godfrey, W.E. 1967. Les oiseaux du Canada. National Museum of Canada, Ottawa, [1979, Bryant Press, Toronto]. Bulletin no. 203, Biological Series no. 73, 506 pp.

Godfrey, W.E. 1968. Notes on birds of the Amos region, Quebec. National Museum of Canada, Natural History Papers 44: 1-5.

Godfrey, W.E. 1986. Les oiseaux du Canada, éd. révisée. Musée national des sciences naturelles, Musées nationaux du Canada, Ottawa, 650 pp.

Godfrey, W.E. 1986. The birds of Canada, revised edition. National Museum of Natural Sciences, Ottawa, 595 pp.

Godfrey, W.E. and A.L. Wilk. 1948. Birds of the Lake St. John region, Quebec. National Museum of Canada, Ottawa. Bulletin no. 110, Biological Series no. 36, 32 pp.

Goldwasser, S., D. Gaines, and S.R. Wilbur. 1980. The Least Bell's Vireo in California: a de facto endangered race. American Birds 34(5): 742-745.

Gollop, J.B. and W.H. Marshall. 1954. A guide for aging duck broods in the field. Mississippi Flyway Council, 14 pp.

Good, E.E. 1952. The life history of the American Crow, *Corvus brachyrhynchos* Brehm. Ph.D. dissertation, Ohio State University, Columbus, 190 pp.

Gooders, J. 1975. The great book of birds. The Hamlyn Publishing Group Ltd, London, 352 pp.

Gooders, J. and T. Boyer. 1986. Ducks of North America and the Northern Hemisphere. Facts on File, New York, 176 pp.

Goodwin, C.E. 1978. The nesting season: Ontario region. American Birds 32(6): 1153-1156.

Goodwin, C.E. 1987a. Black-crowned Night-Heron, pp. 58-59 *in* M.D. Cadman, P.F.J. Eagles, and F.M. Helleiner, eds. Atlas of the breeding birds of Ontario. University of Waterloo Press, Waterloo, 617 pp.

Goodwin, C.E. 1987b. Green-backed Heron, pp. 56-57 *in* M.D. Cadman, P.F.J. Eagles, and F.M. Helleiner, eds. Atlas of the breeding birds of Ontario. University of Waterloo Press, Waterloo, 617 pp.

Goodwin, C.E. and R.C. Rosche. 1970. The nesting season: Ontario-western New York region. Audubon Field Notes 24(5): 677-680.

Goodwin, C.E. and R.C. Rosche. 1974. The nesting season: Ontario-western New York region. American Birds 28(5): 896-903.

Goodwin, D. 1976. Crows of the world. Comstock Publishing Associates, Cornell University Press, Ithaca, New York, 354 pp.

Goodwin, D. 1977. Pigeons and doves of the world. Cornell University Press, Ithaca, New York, 446 pp.

Goodwin, D. 1983. Pigeons and doves of the world, 3rd edition. Cornell University Press, Ithaca, New York, 363 pp.

Goodyear, D.L. 1993. 54. Tamarack slough, pp. 66-67 *in* J.D. Lower, ed. Resident bird counts 1992. Journal of Field Ornithology 64(1)(supplement): 1-128.

Goossen, J.P. 1989. Le Pluvier siffleur. Hinterland Who's Who. Environment Canada, Canadian Wildlife Service, 4 pp. [Also available in English as: Piping Plover.]

Goossen, J.P. and S.G. Sealy. 1982. Production of young in a dense nesting population of Yellow Warblers, *Dendroica petechia*, in Manitoba. Canadian Field-Naturalist 96(2): 189-199.

Göransson, G., T. von Schantz, I. Fröberg, A. Helgée, and H. Wittzell. 1990. Male characteristics, viability and harem size in the Pheasant *Phasianus colchicus*. Animal Behaviour 40: 89-104.

Gorchov, D.L. 1988. Does asynchronous fruit ripening avoid satiation of seed dispersers? A field test. Ecology 69(5): 1545-1551.

Gorman, M.L. 1970. Behavioural and physiological adaptations for breeding in the Eider duck, *Somateria m. mollissima* (L.). Ph.D. dissertation, University of Aberdeen, Aberdeen, United Kingdom.

Gorman, M.L. and H. Milne. 1972. Crèche behaviour in the Common Eider *Somateria m. mollissima* L. Ornis Scandinavica 3: 21-25.

Gosnell, A.L. 1985. Red-breasted Nuthatch, pp. 216-217 *in* S.B. Laughlin and D.P. Kibbe, eds. The Atlas of breeding birds of Vermont. Vermont Institute of Natural Science, University Press of New England, Hanover, 456 pp.

Gosselin, M. 1975a. Average first spring arrival of birds in the Montréal area. Annual report of The Province of Quebec Society for the Protection of Birds 1974 and 1975, Montréal. Tchebec, pp. 88-93.

Gosselin, M. 1975b. Le Faisan à collier, domestique ou sauvage? Bulletin ornithologique 20(2): 51-52.

Gosselin, M. 1977a. Historique de l'établissement du Phalarope de Wilson (*Steganopus tricolor*) au Québec. Bulletin ornithologique 22(3): 115-119.

Gosselin, M. 1977b. Oiseaux sauvages d'origine captive. Carnets de zoologie 37(3): 36-42.

Gosselin, M. 1983. The spring migration, Quebec region. American Birds 37(5): 849-850.

Gosselin, M. 1989a. Une de perdue, dix de retrouvées. QuébecOiseaux 1(1): 20-23.

Gosselin, M. 1989b. Une famille d'oiseaux méconnus: les Sturnidés. QuébecOiseaux 1(2): 18-21.

Gosselin, M. 1990a. Des oiseaux méconnus: les Bombycillidés. QuébecOiseaux 2(1): 19-21.

Gosselin, M. 1990b. Des oiseaux méconnus: les Mimidés. QuébecOiseaux 2(3): 22-25.

Gosselin, M. 1991. Une famille d'oiseaux méconnus: les Muscicapidés. QuébecOiseaux 3(1): 12-15.

Gosselin, M. 1992a. Une famille d'oiseaux méconnus: les Paridés. QuébecOiseaux 3(3): 20-22.

Gosselin, M. 1992b. Une famille d'oiseaux méconnus: les Podicipédidés. QuébecOiseaux 4(1): 19-21.

Gosselin, M. 1993. Une famille d'oiseaux méconnus: les Sittidés. QuébecOiseaux 4(3): 15-17.

Gosselin, M. and N. David. 1975. Field identification of Thayer's Gull (*Larus thayeri*) in eastern North America. American Birds 29(6): 1059-1066.

Gosselin, M. and N. David. 1980a. The winter season: Quebec region. American Birds 34(3): 256-257.

Gosselin, M. and N. David. 1980b. The nesting season: Quebec region. American Birds 34(6): 877-878.

Gosselin, M. and N. David. 1981. The nesting season: Quebec region. American Birds 35(6): 921-922.

Gosselin, M. and N. David. 1982. The nesting season: Quebec region. American Birds 36(6): 956-958.

Gosselin, M., R. Yank, and Y. Aubry. 1987. The winter season: Quebec region. American Birds 41(2): 258-260.

Gosselin, M., R. Yank, and Y. Aubry. 1988a. The autumn migration: Quebec region. American Birds 42(1): 45-47.

Gosselin, M., R. Yank, and Y. Aubry. 1988b. The nesting season: Quebec region. American Birds 42(5): 1270-1272.

Gosselin, M., R. Yank, and Y. Aubry. 1989. The spring season: Quebec region. American Birds 43(3): 452-454.

Gosselin, M., R. Yank, and Y. Aubry. 1990a. Les observations saisonnières: l'hiver 1989-1990. QuébecOiseaux 2(1): 22-25.

Gosselin, M., R. Yank, and Y. Aubry. 1990b. The winter season: Quebec region. American Birds 44(2): 233-235.

Gosselin, M., R. Yank, and Y. Aubry. 1991a. The spring season: Quebec region. American Birds 45(1): 71-73.

Gosselin, M., R. Yank, and Y. Aubry. 1991b. The summer season: Quebec region. American Birds 45(5): 1090-1092.

Gottfried, B.M. and E.C. Franks. 1975. Habitat use and flock activity of Dark-eyed Juncos in winter. Wilson Bulletin 87(3): 374-383.

Gottlieb, G. 1963. A naturalistic study of imprinting in Wood Ducklings (*Aix sponsa*). Journal of Comparative and Physiological Psychology 56: 86-91.

Gottlieb, G. 1968. Species recognition in ground nesting and hole nesting ducklings. Ecology 49: 87-95.

Goudie, R.I. 1985. Range extension of the American Wigeon, *Anas americana*, to the island of Newfoundland. Canadian Field-Naturalist 99(4): 533.

Goudie, R.I. 1987. Estimation préliminaire des populations d'oiseaux aquatiques reproducteurs de Terre-Neuve, 1978-1979, pp. 38-48 *in* A.J. Erskine, ed. Les dénombrements de populations reproductrices d'oiseaux aquatiques dans les provinces de l'Atlantique. Canadian Wildlife Service. Occasional Paper no. 60, 88 pp. [Also available in English as: Waterfowl breeding population surveys, Atlantic Provinces.]

Goudie, R.I. 1989a. Historical status of Harlequin Duck wintering in eastern North America – a reappraisal. Wilson Bulletin 101(1): 112-114.

Goudie, R.I. 1989b. The status of the Harlequin Duck (*Histrionicus histrionicus*) in eastern North America. Committee on the Status of Endangered Wildlife in Canada (COSEWIC), Ottawa, 53 pp.

Goudie, R.I. and C.D. Ankney. 1986. Body size, activity budgets, and diets of sea ducks wintering in Newfoundland. Ecology 67(6): 1475-1482.

Goudie, R.I. and W.R. Whitman. 1987. Populations d'oiseaux aquatiques au Labrador, 1980-1982, pp. 49-71 *in* A.J. Erskine, ed. Les dénombrements de populations reproductrices d'oiseaux aquatiques dans les provinces de l'Atlantique. Canadian Wildlife Service. Occasional Paper no. 60, 88 pp. [Also available in English as: Waterfowl breeding population surveys, Atlantic Provinces.]

Gould, J. 1987. Sedge Wren, pp. 310-311 *in* M.D. Cadman, P.F.J. Eagles, and F.M. Helleiner, eds. Atlas of the breeding birds of Ontario. University of Waterloo Press, Waterloo, 617 pp.

Government of Canada. 1991. The state of Canada's environment. Minister of Supply and Services Canada, Ottawa.

Gowaty, P.A. 1983. Male parental care and apparent monogamy among Eastern Bluebird (*Sialia sialis*). American Naturalist 121(2): 149-157.

Gowaty, P.A. 1984. House Sparrows kill Eastern Bluebirds. Journal of Field Ornithology 55(3): 378-380.

Gowaty, P.A. and A.A. Karlin. 1984. Multiple maternity and paternity in single broods of apparently monogamous Eastern Bluebirds (*Sialia sialis*). Behavioral Ecology and Sociobiology 15: 91-95.

Gowaty, P.A. and J.H. Plissner. 1987. Association of male and female American Robins (*Turdus migratorius*) during the breeding season: paternity assurance by sexual access or mate-guarding. Wilson Bulletin 99(1): 56-62.

Graber, J.W. 1968. Western Henslow's Sparrow, pp. 779-788 *in* A.C. Bent. Life histories of North American cardinals, grosbeaks, buntings, towhees, finches, sparrows, and their allies. (O.L. Austin, Jr., ed.). United States National Museum, Washington. [1968, Dover Publications, New York]. Bulletin no. 237, vol. 1, pp. 1-602; vol. 2, pp. 603-1248; vol. 3, pp. 1249-1889.

Graber, J.W., R.R. Graber, and E.L. Kirk. 1977. Illinois birds: Picidae. Illinois Natural History Survey, Biological Notes no. 102, 73 pp.

Graber, J.W., R.R. Graber, and E.L. Kirk. 1983. Illinois birds: wood warblers. Illinois Natural History Survey, Biological Notes no. 118.

Graber, R.R. and J.[W.] Graber. 1951. Nesting of the Parula Warbler in Michigan. Wilson Bulletin 63(2): 75-83.

Graber, R.R. and J.W. Graber. 1963. A comparative study of bird populations in Illinois, 1906-1909 and 1956-1958. Illinois Natural History Survey Bulletin 28(3): 383-528.

Graber, R.R., J.W. Graber, and E.L. Kirk. 1970. Illinois birds: Mimidae. Illinois Natural History Survey, Biological Notes no. 68, 38 pp.

Graber, R.R., J.W. Graber, and E.L. Kirk. 1971. Illinois birds: Turdidae. Illinois Natural History Survey, Biological Notes no. 75, 44 pp.

Graber, R.R., J.W. Graber, and E.L. Kirk. 1974. Illinois birds: Tyrannidae. Illinois Natural History Survey, Biological Notes no. 86, 57 pp.

Graham, D.J. and J.-L. DesGranges. 1990. Effets de l'azinphos-méthyl (Guthion) sur les oiseaux fréquentant les champs de pommes de terre et l'influence des facteurs comportementaux. SAGE Ltée for Environment Canada, Canadian Wildlife Service, Québec Region. Technical Report Series no. 103, 46 pp. [An abstract is available in English.]

Graham, D.J. and J.-L. DesGranges. 1993. Effects of organophosphate azinphos-methyl on birds of potato fields and apple orchards in Quebec, Canada. Agriculture, Ecosystems and Environment 43: 183-199.

Graham, D.S. 1988a. House Finch nest-site selection at Guelph, Ontario. Condor 90(1): 58-60

Graham, D.S. 1988b. Responses of five host species to cowbird parasitism. Condor 90(3): 588-591.

Gramza, A.F. 1967. Responses of brooding nighthawks to a disturbance stimulus. Auk 84: 72-86.

Grandtner, M.M. 1966. La végétation forestière du Québec méridional. Les Presses de l'Université Laval, Sainte-Foy, 216 pp.

Grandtner, M.M. 1967. Les ressources végétales des Îles-de-la-Madeleine. Université Laval, Sainte-Foy, 42 pp.

Grandy, J.W. 1983. The North American Black Duck (*Anas rubripes*): A case study of 28 years of failure in American wildlife management. International Journal for the Study of Animal Problems 4(4)(supplement): 1-35.

Grange, W.W. 1948. Wisconsin grouse problems. Wisconsin Conservation Department, Publication no. 338, 318 pp.

Grant, J. 1966. The Barred Owl in British Columbia. Murrelet 47(2): 39-45.

Grant, P.J. 1986. Gulls, a guide to identification, 2nd edition. Buteo Books, Vermillion, South Dakota, 352 pp.

Gratto-Trevor, C.L. 1991. Parental care in Semipalmated Sandpipers *Calidris pusilla*: brood desertion by females. Ibis 133: 394-399.

Gratto-Trevor, C.L. 1992. Semipalmated Sandpiper (*Calidris pusilla*), *in* A. Poole, P. Stettenheim, and F. Gill, eds. The birds of North America, no. 6. The Academy of Natural Sciences, Philadelphia. The American Ornithologists' Union, Washington, 20 pp.

Gratto[-Trevor], C.L. and F. Cooke. 1987. Geographic variation in the breeding biology of the Semipalmated Sandpiper. Ornis Scandinavica 18(3): 233-235.

Gratto[-Trevor], C.L., R.I.G. Morrison, and F. Cooke. 1985. Philopatry, site tenacity, and mate fidelity in the Semipalmated Sandpiper. Auk 102(1): 16-24.

Gratton, L. and C. Dubreuil. 1990. Portrait de la végétation et de la flore du Saint-Laurent. Ministère de l'Environnement, Direction de la conservation et du patrimoine écologique and Environment Canada, Canadian Wildlife Service, 55 pp.

GREBE. 1992. Complexe Grande-Baleine. Avant-projet, phase II. Compte rendu de l'étude télémétrique de l'Aigle royal réalisée dans le complexe Grande-Baleine en juin 1992: mission G4. Report for Hydro-Québec, vice-présidence Environnement. Montréal, 8 pp.

GREBE. 1994. Aménagement hydroélectrique Sainte-Marguerite-3. Dénombrement de l'avifaune aquatique et terrestre dans le bassin de la rivière Sainte-Marguerite. Report for Hydro-Québec, vice-présidence Ingénierie. Montréal, 63 pp.

Green, R. 1976. Breeding behaviour of Ospreys *Pandion haliaetus* in Scotland. Ibis 118: 475-490.

Green, R.E. 1985. Lark, pp. 319-320 *in* B. Campbell and E. Lack, eds. A dictionary of birds. British Ornithologists' Union, Buteo Books, Vermillion, South Dakota, 670 pp.

Greenberg, R. 1979. Body size, breeding habitat, and winter exploitation systems in *Dendroica*. Auk 96(4): 756-766.

Greenberg, R. 1988. Water as a habitat cue for breeding Swamp and Song sparrows. Condor 90(2): 420-427.

Greenberg, R. 1989. Neophobia, aversion to open space, and ecological plasticity in Song and Swamp sparrows. Canadian Journal of Zoology 67(5): 1194-1199.

Greenberg, R. 1990. Feeding neophobia and ecological plasticity: a test of the hypothesis with captive sparrows. Animal Behaviour 39(2): 375-379.

Greenberg, R. and S. Droege. 1990. Adaptations to tidal marshes in breeding populations of the Swamp Sparrow. Condor 92(2): 393-404.

Greenberg, R.S. and J.A. Gradwohl. 1980. Observations of paired Canada Warblers *Wilsonia canadensis* during migration in Panama. Ibis 122(4): 509-512.

Greenlaw, J.S. 1978. The relation of breeding schedule and clutch size to food supply in the Rufous-sided Towhee. Condor 80(1): 24-33.

Greenlaw, J.S. and J.D. Rising. 1994. Sharp-tailed Sparrow (*Ammodramus caudacutus*), *in* A. Poole and F. Gill, eds. The birds of North America, no. 112. The Academy of Natural Sciences, Philadelphia. The American Ornithologists' Union, Washington, 28 pp.

Greenway, J.C. Jr. 1967. Extinct and vanishing birds of the world. Dover Publications, New York, 520 pp.

Greenwood, H. 1985. Sexual selection and delayed plumage maturation in the sub-adult male cohort of the Red-winged Blackbird (*Agelaius phoeniceus*). Ph.D. dissertation, McGill University, Montréal, 133 pp.

Greenwood, P.J. and P.H. Harvey. 1982. The natal breeding dispersal of birds. Annual Review of Ecology and Systematics 13: 1-21.

Gregg, L.E. 1974. Woodcock nesting habitat in northern Wisconsin, *in* J.W. Artmann and S.R. Pursglove, eds. Proceedings. Fifth American Woodcock workshop, Athens, 5 pp.

Gregg, L.[E.] 1984. Population ecology of woodcock in Wisconsin. Department of Natural Resources, Madison. Technical Bulletin no. 144, 51 pp.

Grenier, A. 1989. État des populations de Mouettes tridactyles (*Rissa tridactyla*) de la réserve de parc national de l'Archipel-de-Mingan. Environment Canada, Parks Service, Service de la conservation des ressources naturelles, district de Mingan, 29 pp.

Grenier, A. and C. Kavanagh. 1993. État des populations de goélands (*Larus argentatus*, *Larus delawarensis* et *Larus marinus*) de la réserve de parc national de l'Archipel-de-Mingan, 1991. Environment Canada, Parks Service, Service de la conservation des ressources naturelles, district de Mingan, 35 pp.

Grenier, J. and L. Masse. 1990. La production de semences forestières en 1988-1989, portrait statistique. Gouvernement du Québec, Ministère des Forêts.

Grier, J.W. 1969. Bald Eagle behavior and productivity responses to climbing to nests. Journal of Wildlife Management 33(4): 961-966.

Grier, J.W. 1972. Current status of the Bald and Golden eagles in North America: a summary of information and opinions from field researchers, pp. 76-78. Canadian Wildlife Service, Environment Canada, Ottawa. Transactions. Federal-Provincial Wildlife Conference 36: 108 pp.

Grier, J.W. 1982. Ban of DDT and subsequent recovery of reproduction in Bald Eagles. Science 218: 1232-1235.

Grinell, L.I. and R.S. Palmer. 1940. Notes of bird-life of Churchill, Manitoba. Canadian Field-Naturalist 55(4): 47-55.

Griscom, L. and A. Sprunt, Jr. 1957. The warblers of America. The Devin-Adair Company, New York, 356 pp.

Griscom, L. and A. Sprunt, Jr. 1979. The warblers of America, 2nd edition. Doubleday and Company, Garden City, New York, 302 pp.

Grondin, P. and J. Ouzilleau. 1980. Les tourbières du sud de la Jamésie. Géographie Physique et Quaternaire 34(3): 267-299.

Grönlund, S., J. Itämies, and H. Mikkola. 1970. On the food and feeding habits of the Great Grey Shrike *Lanius excubitor* in Finland. Ornis Fennica 47(4): 167-171.

Gross, A.O. 1935. The life history cycle of Leach's Petrel (*Oceanodroma leucorhoa*) on the outer islands of the Bay of Fundy. Auk 52: 382-399.

Gross, A.O. 1937. Birds of the Bowdoin-MacMillan Arctic Expedition, 1934. Auk 54: 12-42.

Gross, A.O. 1940. Eastern Nighthawk, pp. 206-234 *in* A.C. Bent. Life histories of North American cuckoos, goatsuckers, hummingbirds, and their allies. United States National Museum, Washington. [1964, Dover Publications, New York]. Bulletin no. 176, 2 vol., 506 pp.

Gross, A.O. 1942a. Bank swallow, pp. 400-424 *in* A.C. Bent. Life histories of North American flycatchers, larks, swallows, and their allies. United States National Museum, Washington. [1963, Dover Publications, New York]. Bulletin no. 179, 555 pp.

Gross, A.O. 1942b. Northern Cliff Swallow, pp. 463-484 *in* A.C. Bent. Life histories of North American flycatchers, larks, swallows, and their allies. United States National Museum, Washington. [1963, Dover Publications, New York]. Bulletin no. 179, 555 pp.

Gross, A.O. 1946. Eastern Crow, pp. 226-259 *in* A.C. Bent. Life histories of North American jays, crows and titmice. United States National Museum, Washington. [1964, Dover Publications, New York]. Bulletin no. 191, 2 vol., 495 pp.

Gross, A.O. 1947. Cyclic invasions of the Snowy Owl and the migration of 1945-1946. Auk 64: 584-601.

Gross, A.O. 1948a. Catbird, pp. 320-351 *in* A.C. Bent. Life histories of North American nuthatches, wrens, thrashers, and their allies. United States National Museum, Washington. [1964, Dover Publications, New York]. Bulletin no. 195, 475 pp.

Gross, A.O. 1948b. Eastern House Wren, pp. 113-141 *in* A.C. Bent. Life histories of North American nuthatches, wrens, thrashers, and their allies. United States National Museum, Washington. [1964, Dover Publications, New York]. Bulletin no. 195, 475 pp.

Gross, A.O. 1949. Eastern Hermit Thrush, pp. 143-162 *in* A.C. Bent. Life histories of North American thrushes, kinglets, and their allies. United States National Museum, Washington. [1964, Dover Pulbications, New York]. Bulletin no. 196, 452 pp.

Gross, A.O. 1953. Black-polled Warbler, pp. 389-408 *in* A.C. Bent. Life histories of North American wood warblers. United States National Museum, Washington. [1963, Dover Publications, New York]. Bulletin no. 203, vol. 1, pp. 1-367; vol. 2, pp. 367-734.

Gross, A.O. 1958a. Bronzed Grackle, pp. 395-421 *in* A.C. Bent. Life histories of North American blackbirds, orioles, tanagers, and their allies. United States National Museum, Washington. [1965, Dover Publications, New York]. Bulletin no. 211, 549 pp.

Gross, A.O. 1958b. Eastern Meadowlark, pp. 53-80 *in* A.C. Bent. Life histories of North American blackbirds, orioles, tanagers, and their allies. United States National Museum, Washington. [1965, Dover Publications, New York]. Bulletin no. 211, 549 pp.

Gross, D.A. 1992a. Yellow-rumped Warbler, pp. 316-317 *in* D.W. Brauning, ed. Atlas of breeding birds in Pennsylvania. University of Pittsburgh Press, Pittsburgh, 484 pp.

Gross, D.A. 1992b. American Crow, pp. 230-231 *in* D.W. Brauning, ed. Atlas of breeding birds in Pennsylvania. University of Pittsburgh Press, Pittsburgh, 484 pp.

Gross, D.A. 1992c. Northern Saw-whet Owl, pp. 166-167 *in* D.W. Brauning, ed. Atlas of breeding birds in Pennsylvania. University of Pittsburgh Press, Pittsburgh, 484 pp.

Gross, D.A. 1993. 74. Mixed oak-pine forest II, pp. 79-80 *in* J.D. Lowe, ed. Resident bird counts 1992. Journal of Field Ornithology 64(1)(supplement): 1-128.

Grossman, M.L. and J. Hamlet. 1964. Birds of prey of the world. Clarkson N. Potter, New York, 496 pp.

Grossman, M.L. and J. Hamlet. 1988. Birds of prey of the world, 2nd edition. Bonanza Books, New York, 496 pp.

Grosz, T. 1966. Food habits and parasites of the Pacific White-winged Scoter (*Melanitta fusca dixoni*) in the Humboldt Bay area. M.Sc. thesis, Humboldt State College, Arcata, California, 44 pp.

Grosz, T. and C.F. Yocom. 1972. Food habits of the White-winged Scoter in northwestern California. Journal of Wildlife Management 36(4): 1279-1282.

Groth, J.G. 1988. Resolution of cryptic species in appalachian Red Crossbills. Condor 90(4): 745-760.

Groth, J.G. 1993. Evolutionary differentiation in morphology, vocalizations, and allozymes among nomadic sibling species in the North American Red Crossbill (*Loxia curvirostra*) complex. University of California. Publications in Zoology no. 127, 143 pp.

Groulx, L. 1970. Roland-Michel Barrin de La Galissonière, 1693-1756. Les Presses de l'Université Laval, Sainte-Foy, 102 pp.

Groupe Boréal. 1992. Complexe Nottaway-Broadback-Rupert. Les oiseaux terrestres, vol. 5: Habitats hivernaux, abondance et répartition des tétraoninés. Report for Hydro-Québec, vice-présidence Environnement, Montréal, Québec; Le Groupe Boréal, 78 pp.

Grove, G.W. 1993. 72. Hardwood forest with scattered pine, p. 78 *in* J.D. Lowe, ed. Resident bird counts 1992. Journal of Field Ornithology 64(1)(supplement): 1-128.

Grubb, T.C. Jr. 1973. Absence of "individual distance" in the Tree Swallow during adverse weather. Auk 90(2): 432-433.

Grubb, T.C. Jr. 1982. On sex-specific foraging behavior in the White-breasted Nuthatch. Journal of Field Ornithology 53(4): 305-314.

Grubb, T.C. Jr. and V.V. Pravosudov. 1994. Tufted Titmouse (*Parus bicolor*), *in* A. Poole and F. Gill, eds. The birds of North America, no. 86. The Academy of Natural Sciences, Philadelphia. The American Ornithologists' Union, Washington, 16 pp.

Grubb, T.C. Jr. and T.A. Waite. 1987. Caching by Red-breasted Nuthatches. Wilson Bulletin 99(4): 696-699.

Grzimek, B. 1984a. Grzimek's animal life encyclopedia. Vol. 7, birds I. Van Nostrand Reinhold Company, New York, 579 pp.

Grzimek, B. 1984b. Grzimek's animal life encyclopedia. Vol. 8, birds II. Van Nostrand Reinhold Company, New York, 620 pp.

Grzimek, B. 1984c. Grzimek's animal life encyclopedia. Vol. 9, birds III. Van Nostrand Reinhold Company, New York, 648 pp.

Guay, D. 1983. Histoires vraies de la chasse au Québec. VLB éditeur, Montréal, 258 pp.

Gudlin, M.J. and R.W. Dimmick. 1984. Habitat utilization by Ruffed Grouse transplanted from Wisconsin to Tennessee, pp. 75-88 *in* W.L. Robinson, ed. Ruffed Grouse management: state of the art in the early 1980's. North central section, The Wildlife Society, Maryland, 181 pp.

Gudmundsson, F. 1971. The Harlequin Duck (*Histrionicus histrionicus*) in Iceland. Nátturufraedingurinn 41: 84-98. [Summary in English.]

Gudmundsson, F. 1972. The predator-prey relationship of the Gyrfalcon (*Falco rusticolus*) and the Rock Ptarmigan (*Lagopus mutus*) in Iceland, p. 649 *in* K.H. Voous, ed. Proceedings. XVth International Ornithological Congress, The Hague, 30 August-5 September 1970. E.J. Brill, Leiden, 745 pp.

Guérin, D. 1987. Livrée des forêts, *Malacosoma disstria*, pp. 10-11 *in* C. Bordeleau, D. Guérin, L. Innes, D. Lachance, and R. Picher, eds. Insectes et maladies des arbres. Ministère de l'Énergie et des Ressources, Direction de la conservation, Service de la protection contre les insectes et les maladies, Québec, 33 pp.

Guignion, D.L. 1967. A nesting study of the Common Eider (*Somateria mollissima dresseri*) in the St. Lawrence estuary. M.Sc. thesis, Université Laval, Sainte-Foy, 131 pp.

Guillemette, M. 1991. Sélection de l'habitat et stratégies alimentaires chez l'Eider à duvet (*Somateria mollissima*) en hiver dans l'archipel de Mingan, Golfe du Saint-Laurent. M.Sc. thesis, Université Laval, Sainte-Foy, 116 pp.

Guillemette, M. 1993. État des populations de Sternes pierregarins (*Sterna hirundo*) de la Gaspésie et causes de l'insuccès de la reproduction en 1993. Unpublished report. Prepared for Environment Canada, Canadian Wildlife Service, Québec Region, 81 pp.

Guillemette, M. 1994. Élimination des goélands prédateurs et impact de cette intervention sur la productivité de la Sterne Pierregarin (*Sterna hirundo*) nichant sur le Banc de Carleton. Unpublished report. Prepared for Environment Canada, Canadian Wildlife Service, Québec Region, 41 pp.

Guillemette, M. and V. Guillemette. 1983. Observations d'Aigles à tête blanche et de Gerfauts hivernant à la pointe ouest de l'île d'Anticosti. Bulletin ornithologique 28(2): 77-82.

Guillemette, M. and J.H. Himmelman. 1988. Écologie hivernale de l'Eider à duvet dans l'archipel de Mingan, golfe du Saint-Laurent. Final report submitted to Department of Supply and Services Canada, Canadian Wildlife Service, and Canadian Parks Service, 42 pp.

Guillemette, M., R.C. Ydenberg, and J.H. Himmelman. 1992. The role of energy intake rate in prey and habitat selection of Common Eiders *Somateria mollissima*: a risk-sensitive interpretation. Journal of Animal Ecology 61: 599-610.

Guillemette, M., J.H. Himmelman, C. Barrette, and A. Reed. 1993. Habitat selection by Common Eiders in winter and its interaction with flock size. Canadian Journal of Zoology 71(6): 1259-1266.

Guillemette, M., F. Bolduc, and J.-L. DesGranges. 1994. Stomach contents of diving and dabbling ducks during fall migration in the St. Lawrence River, Québec, Canada. Wildfowl 45: 167-175.

Gullion, G.W. 1953. Territorial behavior of the American Coot. Condor 55(4): 169-186.

Gullion, G.W. 1954. The reproductive cycle of American Coots in California. Auk 71(4): 366-412.

Gullion, G.W. 1967. Selection and use of drumming sites by male Ruffed Grouse. Auk 84(1): 87-112.

Gullion, G.W. 1984. Managing northern forests for wildlife. The Ruffed Grouse Society, Coraopolis, Pennsylvania, 72 pp.

Gullion, G.W. 1990. Ruffed Grouse use of conifer plantations. Wildlife Society Bulletin 18: 183-187.

Gunn, W.W.H. 1979. Tennessee Warbler, pp. 50-52 *in* L. Griscom and A. Sprunt, Jr., eds. The warblers of America, 2nd edition. Doubleday and Company, Garden City, New York, 302 pp.

Gutzwiller, K.J., C.H. Strauss, K.R. Kinsley, J.S. Wakeley, and G.L. Storm. 1982. Relationships between land use and woodcock habitat in Pennsylvania, with guidelines for rangewide research, pp. 86-96 *in* T.J. Dwyer and G.L. Storm. Woodcock ecology and management. Papers from the Seventh Woodcock Symposium held at the Pennsylvania State University, University Park, Pennsylvania, 28-30 October 1980. United States. Fish and Wildlife Service. Research Report no. 14, Washington, 191 pp.

Ha, J.C. and P.N. Lehner. 1990. Notes on Gray Jay demographics in Colorado. Wilson Bulletin 102(4): 698-702.

Haas, C.A. 1987. Eastern subspecies of the Loggerhead Shrike: the need for measurements of live birds. North American Bird Bander 12: 99-102.

Hadley, N. 1969. Breeding biology of the Gray-headed Junco, *Junco caniceps* (Woodhouse) in the Colorado front range. Colorado Field Ornithologist 5: 15-21.

Hagan, J.M. III. 1993. Decline of the Rufous-sided Towhee in the eastern United States. Auk 110(4): 863-874.

Hagan, J.M. III and J.R. Walters. 1990. Foraging behavior, reproductive success, and colonial nesting in Ospreys. Auk 107: 506-521.

Hagar, J.A. 1956. Birds seen at Magdalens in 1956. Manuscript, 15 pp.

Hagen, Y. 1969. Norwegian studies on the reproduction of birds of prey and owls in relation to microrodent population fluctuations. Fauna 22(2): 73-126.

Haig, S.M. 1985. The status of the Piping Plover in Canada. National Museum of Canada, Ottawa.

Haig, S.M. 1992. Piping Plover (*Charadrius melodus*), *in* A. Poole, P. Stettenheim, and F. Gill, eds. The birds of North America, no. 2. The Academy of Natural Sciences, Philadelphia. The American Ornithologists' Union, Washington, 18 pp.

Haig, S.M. and L.W. Oring. 1988a. Distribution and dispersal in the Piping Plover. Auk 105(4): 630-638.

Haig, S.M. and L.W. Oring. 1988b. Mate, site, and territory fidelity in Piping Plovers. Auk 105(2): 268-277.

Haig, S.M. and J.H. Plissner. 1993. Distribution and abundance of Piping Plover: results and implications of the 1991 International Census. Condor 95: 145-156.

Hailman, J.P. 1976. Leaf-scratching in White-crowned Sparrows and Fox Sparrows: test of a model. Wilson Bulletin 88(2): 354-356.

Hainault, P. and D. St-Hilaire. 1989. Inventaire du Bruant sauterelle (*Ammodramus savannarum*) dans le sud-ouest du Québec, en 1989. Club des ornithologues de l'Outaouais, Hull, 47 pp.

Hakala, A. 1971. A quantitative study of the bird fauna of some open peatlands in Finland. Ornis Fennica 41: 1-11.

Hall, A. 1862. On the mammals and birds of the district of Montreal. Canadian Naturalist and Geologist 7(1): 44-78; 7(3): 171-193; 7(4): 289-316; 7(5): 3[illegible]4-376.

Hall, E.R. and K.R. Kelson. 1959. The mammals of North America, vol. 1. Ronald Press, New York, 546 pp.

Hall, G.A. 1969. Breeding range expansion of the Brown Creeper in the middle Atlantic States. Redstart 36: 98-103.

Hall, G.A. 1984. A long-term bird population study in an Appalachian spruce forest. Wilson Bulletin 96(2): 228-240.

Hall, G.A. 1994. Magnolia Warbler (*Dendroica magnolia*), *in* A. Poole and F. Gill, eds. The birds of North America, no. 136. The Academy of Natural Sciences, Philadelphia. The American Ornithologists' Union, Washington, 16 pp.

Hall, G.H. 1970. Great moments in action: the story of the Sun Life falcons. Canadian Field-Naturalist 84(3): 209-230.

Hall, H.M. 1960. A gathering of shore birds. The Devin-Adair Company, New York, 242 pp.

Hamann, J. 1985. Une tradition qui ne veut pas mourir. Franc-Nord 2(1): 6-9.

Hamas, M.J. 1994. Belted Kingfisher (*Ceryle alcyon*), *in* A. Poole and F. Gill, eds. The birds of North America, no. 84. The Academy of Natural Sciences, Philadelphia. The American Ornithologists' Union, Washington, 16 pp.

Hamel, D. 1975. Le nourrissage symbolique (courtship feeding) chez un couple de Martin-pêcheur (*Megaceryle alcyon*). Bulletin ornithologique 20(3): 80-81.

Hamel, F. 1961. Le Club des ornithologues de Québec. Carnets de zoologie 21(3): 36-38.

Hamel, P.B. 1992. Cerulean Warbler (*Dendroica cerulea*), pp. 385-400 *in* K.J. Schneider and D.M. Pence, eds. Migratory nongame birds of management concern in the northeast. United States Department of the Interior, Fish and Wildlife Service, Newton Corner, Massachusetts, 400 pp.

Hamer, T.E., S.G. Seim, and K.R. Dixon. 1989. Northern Spotted Owl and Northern Barred Owl habitat use and home range size in Washington. Preliminary report. Washington Department of Wildlife, Wildlife Management – Nongame, Washington, [illegible]5 pp.

Hamerstrom, F. 1969. A harrier population study, pp. 367-383 *in* J.J. Hickey, ed. Peregrine Falcon populations: their biology and decline. University of Wisconsin Press, Madison, 596 pp.

Hamerstrom, F. 1972. Birds of prey of Wisconsin. Department of Natural Resources, Madison, Wisconsin, 64 pp.

Hamerstrom, F. 1979. Effect of prey on predator: voles and harriers. Auk 96(2): 375-390.

Hamerstrom, F. and M. Kopeny. 1981. Harrier nest-site vegetation. Raptor Research 15: 86-88.

Hamerstrom, F., F.N. Hamerstrom, and J. Hart. 1973. Nest boxes: an effective management tool for kestrels. Journal of Wildlife Management 37: 400-403.

Hamerstrom, F., T. Ray, C.M. White, and C.E. Braun. 1975. Conservation committee report on status of eagles. Conservation Report 87(1): 140-143.

Hamerstrom, F.N. Jr. and F. Hamerstrom. 1951. Mobility of the Sharp-tailed Grouse in relation to its ecology and distribution. American Midland Naturalist 46([illegible]): 174-226.

Hamerstrom, F.N., F. Hamerstrom, and P.A. Johnsgard. 1985. Grouse, pp. 257-259 *in* B. Campbell and E. Lack, eds. A dictionary of birds. British Ornithologists' Union, Buteo Books, Vermillion, South Dakota, 670 pp.

Hamilton, W.I. III and M.E. Hamilton. 1965. Breeding characteristics of Yellow-billed Cuckoos in Arizona. Proceedings. California Academy of Science (4th Series) 32: 405-432.

Hancock, D. 1970. Adventures with eagles. The Wildlife Conservation Centre, Saanichton, British Columbia, 40 pp.

Hancock, J. and J. Kushlan. 1984. The herons handbook. Croom Helm, London, 288 pp.

Hancock, J. and J. Kushlan. 1989. Guide des hérons du monde: aigrettes, bihoreaux, butors, hérons, onorés. Delachaux et Niestlé, Lausanne, 288 pp.

Hands, H.M., R.D. Drobney, and M.R. Ryan. 1989a. Status of the Least Bittern in the northcentral United States. United States Fish and Wildlife Service, Twin Cities, Minnesota, 13 pp.

Hands, H.M., R.D. Drobney, and M.R. Ryan. 1989b. Status of the Common Barn-Owl in the north-central United States. United States Fish and Wildlife Service, Twin Cities, Minnesota, 19 pp.

Haney, J.C. 1990. Winter habitat of Common Loons on the continental shelf of the southeastern United States. Wilson Bulletin 102(2): 253-263.

Hann, H.W. 1937. Life history of the Ovenbird in southern Michigan. Wilson Bulletin 49(3): 145-237.

Hann, H.W. 1940. Polyandry in the Ovenbird in southern Michigan. Wilson Bulletin 52: 69-72.

Hannon, S.J. 1984. Factors limiting polygyny in the Willow Ptarmigan. Animal Behaviour 32: 153-161.

Hannon, S.J. and T.W. Barry. 1986. Demography, breeding biology and predation of Willow Ptarmigan at Anderson River delta, Northwest Territories. Arctic 39: 300-303.

Hannon, S.J. and J. Roland. 1984. Morphology and territory acquisition in Willow Ptarmigan. Canadian Journal of Zoology 62: 1502-1506.

Hansen, G.L. 1979. Territorial and foraging behaviour of the Eastern Willet *Catoptrophorus semipalmatus semipalmatus* (Gmelin). M.Sc. thesis, Acadia University, Wolfville, Nova Scotia, 174 pp.

Hansen, J.L., L. Fredrickson, and J.P. Rogers. 1973. Nesting of the Hooded Merganser in southeastern Missouri. Unpublished manuscript, 29 pp.

Hanson, H.C. 1965. The Giant Canada Goose. Southern Illinois University Press, Carbondale, 226 pp.

Hansrote, C. and M. Hansrote. 1992. Movements of House Finches (*Carpodacus mexicanus*) and Pine Siskins (*Carduelis pinus*) through the Virginia Piedmont. North American Bird Bander 17(3): 102-107.

Hanssen, O.J. 1982. Recent changes in the breeding population and nesting habits of Black-headed Gull *Larus ridibundus* L. in the Ostfold archipelago, SE Norway. Cinclus 5: 10-15.

Hantzsch, B. 1929. Contribution to the knowledge of the avifauna of north-eastern Labrador. Canadian Field-Naturalist 43: 52-59.

Haramis, G.M. 1990. Breeding ecology of the Wood Duck: a review, pp. 45-60 *in* L.H. Fredrickson, G.V. Burger, S.P. Havera, D.A. Graber, R.E. Kirby, and T.S. Taylor, eds. Proceedings. 1988 North American Wood Duck symposium, St. Louis, Missouri, 390 pp.

Haramis, G.M. and D.Q. Thompson. 1985. Density-production characteristics of box-nesting Wood Ducks in a northern greentree impoundment. Journal of Wildlife Management 49: 429-436.

Hardy, B. and J.-L. DesGranges. 1989. Évaluation des effets à moyen terme sur les communautés aviennes de l'entretien des plantations d'Épinettes noires (*Picea mariana*) aux phénoxys (ESTAPROP) et au glyphosate (ROUNDUP). Technical report prepared by Environnement Illimité for Environment Canada, Canadian Wildlife Service, Québec Region, 42 pp.

Hardy, F.P. 1888. Testimony of some early voyagers on the Great Auk. Auk 5: 380-384.

Hardy, J. 1977. Étude de la distribution de la Chouette cendrée au Québec. Bulletin ornithologique 22(1): 14-20.

Hardy, J. 1979a. La Chouette cendrée au Québec durant l'hiver 78-79. Bulletin ornithologique 24(2): 8-10.

Hardy, J. 1979b. Statut de la Sterne rosée au Québec. Bulletin ornithologique 24(3): 70-74.

Hardy, J.D. and T.C. Tacha. 1989. Age-related recruitment of Canada Geese from the Mississippi valley population. Journal of Wildlife Management 53(1): 97-98.

Hardy, J.W. 1961. Studies in behavior and phylogeny of certain New World jays (Garrulinae). University of Kansas Science, Bulletin 42(2): 13-149.

Hardy, R. and N. Séguin. 1983. Forêt et société rurale en Mauricie. Réseau, octobre 1983, 5 pp.

Hardy, R., N. Séguin, A. Gamelin, A. Miville, and G. Trépanier. 1980. L'exploitation forestière en Mauricie. Dossier statistique: 1850-1930. Publication du groupe de recherche sur la Mauricie, Université du Québec à Trois-Rivières, Trois-Rivières. Cahier no. 4, 199 pp.

Harestad, A.S. and D.G. Keisker. 1989. Nest tree use by primary cavity-nesting birds in south central British Columbia. Canadian Journal of Zoology 67(4): 1067-1073.

Harlow, R.C. 1922. The breeding habits of the Northern Raven in Pennsylvania. Auk 39(3): 399-410.

Harlow, R.C., R.G. Hooper, R.D. Chamberlain, and H.S. Crawford. 1975. Some winter and nesting season food of the Common Raven in Virginia. Auk 92: 298-306.

Harper, F. 1958. Birds of the Ungava Peninsula. Museum of Natural History, University of Kansas, Lawrence. Miscellaneous Publications no. 17, 171 pp.

Harrington, B.A. and R.I.G. Morrison. 1979. Semipalmated Sandpiper migration in North America, pp. 83-100 *in* F.A. Pitelka, ed. Studies in avian biology, no. 2. Cooper Ornithological Society, Los Angeles, 261 pp.

Harrington-Tweit, B., P.W. Mattocks, Jr., and E.S. Hunn. 1979. The nesting season: northern Pacific coast region. American Birds 33(6): 892.

Harris, L.D. and G. Silva-Lopez. 1992. Forest fragmentation and the conservation of biological diversity, pp. 197-237 *in* P.L. Fielder and S.K. Jain, eds. Conservation Biology: the theory and practice of nature conservation, preservation, and management. Chapman and Hall, New York, 507 pp.

Harris, M.A. and R.E. Lemon. 1972. Songs of Song Sparrows (*Melospiza melodia*): individual variation and dialects. Canadian Journal of Zoology 50(3): 301-309.

Harris, M.A. and R.E. Lemon. 1974. Songs of Song Sparrows: reactions of males to songs of different localities. Condor 76(1): 33-44.

Harris, M.P. 1984. The Puffin. Poyser, Calton, United Kingdom, 224 pp.

Harris, M.P. and T.R. Birkhead. 1985a. Breeding ecology of the Atlantic Alcidae, pp. 155-204 *in* D.N. Nettleship and T.R. Birkhead, eds. The Atlantic Alcidae: the evolution, distribution and biology of the auks inhabiting the Atlantic Ocean and adjacent water areas. Academic Press, London, 574 pp.

Harris, M.P. and T.R. Birkhead. 1985b. Ecological adaptations for breeding in the Atlantic Alcidae, pp. 205-231 *in* D.N. Nettleship and T.R. Birkhead, eds. The Atlantic Alcidae: the evolution, distribution and biology of the auks inhabiting the Atlantic Ocean and adjacent water areas. Academic Press, London, 574 pp.

Harris, M.P., D.J. Halley, and R.L. Swann. 1994. Age of first breeding in Common Murres. Auk 111(1): 207-209.

Harris, R.E. 1989. Breeding biology of the Short-billed Dowitcher in the Schefferville area, Quebec–Labrador. M.Sc. thesis, University of Ottawa, Ottawa, 72 pp.

Harris, R.N. 1979. Aggression, superterritories and reproductive success in Tree Swallows. Canadian Journal of Zoology 57(10): 2072-2078.

Harris, S.W., C.L. Buechele, and C.F. Yocom. 1954. The status of Barrow's Goldeneye in eastern Washington. Murrelet 35: 33-38.

Harrison, C. 1978. A field guide to the nests, eggs and nestlings of North American birds. Collins, Toronto, 416 pp.

Harrison, C. 1982. An atlas of the birds of the Western Palearctic. Princeton University Press, Princeton, New Jersey, 322 pp.

Harrison, C.J.O. 1978. Bird families of the world. Harryn Abrams, New York, 264 pp.

Harrison, H.H. 1975. A field guide to birds' nests of 285 species found breeding in the United States east of the Mississippi River. The Peterson Field Guide Series no. 21, Houghton Mifflin Company, Boston, 257 pp.

Harrison, H.H. 1984. Wood warblers' world. Simon and Schuster, New York, 335 pp.

Harrison, P. 1983. Seabirds: an identification guide. Houghton Mifflin Company, Boston, 448 pp.

Harrison, P. 1985. Seabirds: an identification guide, revised edition. Houghton Mifflin Company, Boston, 448 pp.

Hartman, F.E. 1992. Wood Duck, pp. 68-69 *in* D.W. Brauning, ed. Atlas of breeding birds in Pennsylvania. University of Pittsburgh Press, Pittsburgh, 484 pp.

Hartshorne, J.M. 1962. Behavior of the Eastern Bluebird at the nest. Living Bird 1: 131-149.

Harvey, W.F. and A. Bourget. 1995. A breeding pair survey of Canada Geese in northern Quebec – 1995. Maryland Department of Natural Resources, Wildlife Division, and Canadian Wildlife Service, Québec Region, 10 pp.

Hasbrouck, M.E. 1944. The status of Barrow's Goldeneye in the eastern United States. Auk 61: 544-554.

Hatch, D.R.M. 1971. Brown-headed Cowbird parasitism on Spotted Sandpiper and Wilson's Phalarope. Blue Jay: 17-18.

Hatch, J.J. 1970. Predation and piracy by gulls at a ternery in Maine. Auk 87(2): 244-254.

Hausman, I.A. 1927. On the winter food of the Tree Swallow (*Iridoprocne bicolor*) and the Myrtle Warbler (*Dendroica coronata*). American Naturalist 61: 379-382.

Havera, S.P. and R.E. Kirby. 1990. Biology of the species, pp. 377-379 *in* L.H. Fredrickson, G.V. Burger, S.P. Havera, D.A. Graber, R.E. Kirby, and T.S. Taylor, eds. Proceedings. 1988 North American Wood Duck symposium, St. Louis, Missouri, 390 pp.

Havera, S.P., R.M. Whitton, and R.T. Sealy. 1992. Blood lead and ingested and embedded shot in diving ducks during spring. Journal of Wildlife Management 56(3): 539-545.

Hawkins, L.L. 1986. Nesting behaviour of male and female Whistling Swans and implications of male incubation. Wildfowl 37: 5-27.

Hawksley, O. 1957. Ecology of a breeding population of Arctic Terns. Bird-Banding 28: 57-92.

Hay, O.P. 1887. The Red-headed Woodpecker: a hoarder. Auk 4: 193-196.

Hayakawa, E. 1987. Le Chardonneret jaune. Hinterland Who's Who. Environment Canada, Canadian Wildlife Service, 4 pp. [Also available in English as: American Goldfinch.]

Hayes, F.E. 1995. Definitions for migrant birds: what is a neotropical migrant? Auk 112(2): 521-523.

Hayes, F.E. and J.A. Fox. 1991. Seasonality, habitat use, and flock sizes of shorebirds at the Bahia de Asuncion, Paraguay. Wilson Bulletin 103(4): 637-649.

Hayes, P.A. and R.J. Robertson. 1989. The impact of male parental care on female Eastern Kingbird reproductive success. Wilson Bulletin 101(3): 462-467.

Hayman, P., J. Marchant, and T. Prater. 1986. Shorebirds: an identification guide to the waders of the world. Houghton Mifflin Company, Boston, 412 pp.

Haymes, G.T. and H. Blokpoel. 1978. Food of Ring-billed Gull chicks at the Eastern Headland of the Toronto Outer Harbour in 1977. Canadian Field-Naturalist 92(4): 392-395.

Hays, H. 1972. Polyandry in the Spotted Sandpiper. Living Bird 11: 43-57.

Hayward, C.L. 1935. The breeding status and migration of the Caspian Tern in Utah. Condor 37: 140-144.

Hayward, G.D. 1983. Resource partitioning among six forest owls in the River of No Return Wilderness, Idaho. M.Sc. thesis, University of Idaho, Moscow, 132 pp.

Hayward, G.D. and E.O. Garton. 1983. First nesting record for Boreal Owl in Idaho. Condor 85(4): 501.

Hayward, G.D. and E.O. Garton. 1984. Roost habitat selection by three small forest owls. Wilson Bulletin 96(4): 690-692.

Hayward, G.D. and P.H. Hayward. 1993. Boreal Owl (*Aegolius funereus*), *in* A. Poole and F. Gill, eds. The birds of North America, no. 63. The Academy of Natural Sciences, Philadelphia. The American Ornithologists' Union, Washington, 20 pp.

Hayward, G.D., P.H. Hayward, and E.O. Garton. 1987. Movements and home range use by Boreal Owls in central Idaho, pp. 175-184 *in* R.W. Nero, R.J. Clark, R.J. Knapton, and R.H. Hamre, eds. Proceedings. Biology and conservation of northern forest owls. Symposium, 3-7 February, Winnipeg. United States Department of Agriculture, Forest Service, Rocky Mountain Forest and Range Experiment Station, Fort Collins, Colorado. General Technical Report RM-142, 309 pp.

Haywood, D.R. and R.D. Ohmart. 1983. Preliminary report on habitat utilization by two pairs of breeding Bald Eagles in Arizona, pp. 87-94 *in* D.M. Bird, ed. Biology and management of Bald Eagles and Ospreys. Harpell Press, Sainte-Anne-de-Bellevue, 325 pp.

Healy, W.M. 1985. Turkey poult feeding activity, invertebrate abundance, and vegetation structure. Journal of Wildlife Management 49(2): 466-472.

Hebrard, J.J. 1978. Habitat selection in two species of *Spizella*: a concurrent laboratory and field study. Auk 95(2): 404-410.

Hecht, W.R. 1951. Nesting of the Marsh Hawk at Delta, Manitoba. Wilson Bulletin 63: 167-176.

Heinrich, B. 1988. Winter foraging at carcasses by three sympatric corvids, with emphasis on recruitment by the raven, *Corvus corax*. Behavioral Ecology and Sociobiology 23: 141-156.

Heinselman, M.L. 1973. Fire in the virgin forests of the Boundary Waters Canoe Area, Minnesota. Quaternary Research 3: 329-382.

Heintzelman, D.S. 1979. Hawks and owls of North America. Universe Books, New York, 197 pp.

Heintzelman, D.S. 1982. Variations in numbers of Sharp-shinned Hawks migrating past Bake Oven Knob, Pennsylvania, in autumn (1961 to 1981). American Hawkwatcher 1: 1-4.

Helleiner, F.M. 1987a. Connecticut Warbler, pp. 412-413 *in* M.D. Cadman, P.F.J. Eagles, and F.M. Helleiner, eds. Atlas of the breeding birds of Ontario. University of Waterloo Press, Waterloo, 617 pp.

Helleiner, F.M. 1987b. Pine Grosbeak, pp. 488-489 *in* M.D. Cadman, P.F.J. Eagles, and F.M. Helleiner, eds. Atlas of the breeding birds of Ontario. University of Waterloo Press, Waterloo, 617 pp.

Helleiner, F.M. 1987c. Wilson's Warbler, pp. 420-421, *in* M.D. Cadman, P.F.J. Eagles, and F.M. Helleiner, eds. Atlas of the breeding birds of Ontario. University of Waterloo Press, Waterloo, 617 pp.

Helleiner, F.M. 1987d. Yellow-billed Cuckoo, pp. 200-201 *in* M.D. Cadman, P.F.J. Eagles, and F.M. Helleiner, eds. Atlas of the breeding birds of Ontario. University of Waterloo Press, Waterloo, 617 pp.

Henderson, A.D. 1915. Nesting of the Great Gray Owl in central Alberta. Oölogist 32(1): 2-6.

Henderson, A.D. 1919. Nesting of the American Hawk Owl. Oölogist 36(4): 59-63.

Henderson, A.D. 1923. Nesting of the Solitary Sandpiper. Oölogist 40(4): 53-56.

Henderson, A.D. 1924. The Common Loon in Alberta. Condor 26: 143-145.

Henderson, A.D. 1925. With the early breeders. Oölogist 42(4): 50-57.

Hendricks, P. 1987a. Foraging patterns of Water Pipits (*Anthus spinoletta*) with nestlings. Canadian Journal of Zoology 65: 1522-1529.

Hendricks, P. 1987b. Habitat use by nesting Water Pipits (*Anthus spinoletta*): a test of the snowfield hypothesis. Arctic and Alpine Research 19(3): 313-320.

Hennemann, W.W. 1984. Spread-winged behaviour of Double-crested and Flightless cormorants (*Phalacrocorax auritus* and *P. harrisi*): wing drying or thermoregulation? Ibis 126: 230-239.

Hennepin, L. 1903. A new discovery of a vast country in America by Father Louis Hennepin; reprinted from the second London issue of 1698. A.C. McClurg and Company, 2 vol., 711 pp.

Henny, C.J. 1972. An analysis of the population dynamics of selected avian species, with special reference to changes during the modern pesticides era. United States Fish and Wildlife Service, Washington, Research Report no. 1, 98 pp.

Henny, C.J. 1988. Osprey, pp. 73-101 *in* R.S. Palmer, ed. Handbook of North American birds. Vol. 4: Diurnal Raptors (part 1). Yale University Press, New Haven, Connecticut, 433 pp.

Henny, C.J. and N.E. Holgersen. 1974. Range expansion and population increase of the Gadwall in eastern North America. Wildfowl 25: 95-101.

Henny, C.J. and W.T. Van Velzen. 1972. Migration patterns and wintering localities of American Ospreys. Journal of Wildlife Management 36: 1133-1141.

Henny, C.J. and H.M. Wight. 1972. Population ecology and environmental pollution: Red-tailed and Cooper's hawks, pp. 229-250 *in* Population ecology of migratory birds: a symposium. United States Fish and Wildlife Service. Research Report no. 2, Washington.

Henny, C.J., F.C. Schmid, E.M. Martin, and L.L. Hood. 1973. Territorial behavior, pesticides, and the population ecology of Red-shouldered Hawks in central Maryland, 1943-1971. Ecology 54(3): 545-554.

Henny, C.J., L.J. Blus, and T.E. Kaiser. 1984. Heptachlor seed treatment contaminates hawks, owls and eagles of Columbia Basin, Oregon. Raptor Research 18: 41-48.

Hensel, R.J. and W.A. Troyer. 1964. Nesting studies of the Bald Eagle in Alaska. Condor 66(4): 282-286.

Hepp, G.R., R.A. Kennamer, and W.F. Harvey, IV. 1989. Recruitment and natal philopatry of Wood Ducks. Ecology 70(4): 897-903.

Heppner, F. 1965. Sensory mechanisms and environmental clues used by the American Robin in locating earthworms. Condor 67: 247-256.

Herkert, J.R. 1991. Prairie birds of Illinois: population response to two centuries of habitat change. Illinois Natural History Survey 34: 393-399.

Herrick, F.H. 1910. Life and behavior of the cuckoo. Journal of Experimental Zoology 9(1): 169-233.

Herrick, F.H. 1932. Daily life of the American Eagle: early phase. Auk 49: 307-323.

Herzog, P.W. and D.M. Keppie. 1980. Migration in a local population of Spruce Grouse. Condor 82(4): 366-372.

Hespenheide, H.A. 1971. Flycatcher habitat selection in the eastern deciduous forest. Auk 88(1): 61-74.

Hess, E.H. 1973. Imprinting, early experience and the developmental psychobiology of attachment. Van Nostrand Reinhold Company, New York. Behavioural Science Series, 472 pp.

Heusmann, H.W. 1975. Several aspects of the nesting biology of yearling Wood Ducks. Journal of Wildlife Management 39(3): 503-507.

Heusmann, H.W. 1991. The history and status of the Mallard in the Atlantic Flyway. Wildlife Society Bulletin 19: 14-22.

Heusmann, H.W., R. Bellville, and R.G. Burrell. 1980. Further observations on dump nesting by Wood Ducks. Journal of Wildlife Management 44(4): 908-915.

Hewitt, O.H. 1967. The Wild Turkey and its management. The Wildlife Society, Washington, 589 pp.

Heyland, J.D. 1970. The Common Starling in Arctic Quebec. Canadian Field-Naturalist 84(4): 397-398.

Heyland, J.D., E.B. Chamberlain, C.F. Kimball, and D.H. Baldwin. 1970. Whistling Swans breeding on the northwest coast of New Quebec. Canadian Field-Naturalist 84(4): 398-399.

Hiam, A. and M. Sutherland. 1979. Ground-nesting and related behaviour of Nighthawks (*Chordeiles minor*) in Massachusetts. Bird Observer 7(3): 98-105.

Hickey, E. and T. Bider. 1981. Étude écologique de la Bernache canadienne (*Branta canadensis interior*) à la Grande rivière de la Baleine. Direction de l'environnement d'Hydro-Québec, Projet d'aménagement du complexe hydro-électrique Grande rivière de la Baleine, Éco-recherches, Pointe-Claire, 112 pp. [Study summer 1978, submitted 1979, published 1981].

Hickey, J.J. 1969. Peregrine Falcon populations: their biology and decline. University of Wisconsin Press, Madison, 596 pp.

Hickey, J.J. and D.W. Anderson. 1969. The Peregrine Falcon: life history and population literature, pp. 3-42 *in* J.J. Hickey, ed. Peregrine Falcon populations: their biology and decline. University of Wisconsin Press, Madison, 596 pp.

Hicklin, P.W. and P.C. Smith. 1984. Selection of foraging sites and invertebrate prey by migrant Semipalmated Sandpipers, *Calidris pusilla* (Pallas), in Minas Basin, Bay of Fundy. Canadian Journal of Zoology 62: 2201-2210.

Hickman, S. 1981. Evidence for aerodynamic advantages of tail keeling in the Common Grackle. Wilson Bulletin 93(4): 500-505.

Hiebert, S.M., P.K. Stoddard, and P. Arcese. 1989. Repertoire size, territory acquisition and reproductive success in the Song Sparrow. Animal Behaviour 37(2): 266-273.

Hiemenz, N. 1981. Northern Green Heron colony. Loon 53(4): 235-236.

Higgins, K.F. and L.M. Kirsch. 1975. Some aspects of the breeding biology of the Upland Sandpiper in North Dakota. Wilson Bulletin 87(1): 96-102.

Higgins, M.J. 1991. Pine Siskin, pp. 522-523 *in* R. Brewer, G.A. McPeek, and R.J. Adams, Jr. eds. The Atlas of breeding birds of Michigan. Michigan State University Press, East Lansing, 594 pp.

Higuchi, H. 1988. Individual differences in bait-fishing by the Green-backed Heron *Ardeola striata* associated with territory quality. Ibis 130(1): 39-44.

Hildebrand, H. 1950. Notes on the birds of the Ungava Bay district. Canadian Field-Naturalist 64(2): 55-67.

Hildén, O. 1964. Ecology of duck populations in the island group of Valassaaret, Gulf of Bothnia. Annales Zoologici Fennici 1: 1-279.

Hildén, O. and T. Solonen. 1987. Status of the Great Gray Owl in Finland, pp. 115-120 *in* R.W. Nero, R.J. Clark, R.J. Knapton, and R.H. Hamre, eds. Proceedings. Biology and conservation of northern forest owls. Symposium, 3-7 February, Winnipeg. United States Department of Agriculture, Forest Service, Rocky Mountain Forest and Range Experiment Station, Fort Collins, Colorado. General Technical Report RM-142, 309 pp.

Hildén, O. and S. Vuolanto. 1972. Breeding biology of the Red-necked Phalarope, *Phalaropus lobatus*, in Finland. Ornis Fennica 49: 57-85.

Hill, D. and P.R. Robertson. 1988a. Breeding success of wild and hand-reared Ring-necked Pheasants. Journal of Wildlife Management 52(3): 446-450.

Hill, D. and P.R. Robertson. 1988b. The pheasant: ecology, management and conservation. BSP Professional Books, Oxford, 281 pp.

Hill, G.E. 1993. House Finch (*Carpodacus mexicanus*), *in* A. Poole and F. Gill, eds. The birds of North America, no. 46. The Academy of Natural Sciences, Philadelphia. The American Ornithologists' Union, Washington, 24 pp.

Hill, J.R. III. 1988. The Eastern House Finch nesting in Purple Martin houses and gourds. American Birds 42(1): 36-38.

Hill, N.P. 1968. Eastern Sharp-tailed Sparrow, pp. 795-819 *in* A.C. Bent. Life histories of North American cardinals, grosbeaks, buntings, towhees, finches, sparrows, and allies. (O.L. Austin, Jr., ed.). United States National Museum, Washington. [1968, Dover Publications, New York]. Bulletin no. 237, vol. 1, pp. 1-602; vol. 2, pp. 603-1248; vol. 3, pp. 1249-1889.

Hill, N.P. and J.M. Hagan, III. 1991. Population trends of some northeastern North American landbirds: a half-century of data. Wilson Bulletin 103(2): 165-182.

Hill, W.L. 1988. The effect of food abundance on the reproduction patterns of Coots. Condor 90(2): 324-331.

Hindman, L.J. and F. Ferrigno. 1990. Atlantic Flyway goose population: status and management. Transactions. North American Wildlife and Natural Resources Conference 55: 293-311.

Hines, J.E. 1977. Nesting and brood ecology of Lesser Scaup at Waterhen marsh, Saskatchewan. Canadian Field-Naturalist 91(3): 248-255.

Hirons, G.J.M. 1985. Owls, pp. 235-245 *in* C.M. Perrins and A.L.A. Middleton, eds. The encyclopedia of birds. Facts on File, New York, 447 pp.

Hitchcock, R.R. and R.E. Mirarchi. 1984. Duration of dependence of wild fledgling Mourning Doves upon parental care. Journal of Wildlife Management 48(1): 99-108.

Hjertaas, D.G., P. Hjertaas, and W.J. Maher. 1988. Colony size and reproductive biology of the Bank Swallow, *Riparia riparia*, in Saskatchewan. Canadian Field-Naturalist 102(3): 465-470.

Hobson, K.A. and S.G. Sealy. 1989a. Mate guarding in the Yellow Warbler *Dendroica petechia*. Ornis Scandinavica 20(4): 241-249.

Hobson, K.A. and S.G. Sealy. 1989b. Female-female aggression in polygynously nesting Yellow Warblers. Wilson Bulletin 101(1): 84-86.

Hobson, K.A. and S.G. Sealy. 1989c. Response of Yellow Warblers to the threat of cowbird parasitism. Animal Behaviour 38(3): 510-519.

Hobson, K.A. and S.G. Sealy. 1989d. Breeding-season aggression of female Yellow Warblers to models of male and female conspecific intruders. Animal Behaviour 39(4): 809-811.

Hobson, K.A. and S.G. Sealy. 1990. Female song in the Yellow Warbler. Condor 92(1): 259-261.

Hochachka, W.M., J.N.M. Smith, and P. Arcese. 1989. Song Sparrow, pp. 135-152 *in* I. Newton, ed. Lifetime reproduction in birds. Academic Press, New York, 479 pp.

Hochbaum, G.S. and E.F. Bossenmaier. 1972. Response of Pintails to improved breeding habitat in southern Manitoba. Canadian Field-Naturalist 86(1): 79-81.

Hochstetler, K.J., E.L. Garza, W. Lohmann, and T.A. Sordahl. 1991. Distribution of Eastern and Western meadowlarks in Winneshiek County, Iowa. Bird Life 61(1): 8-13.

Hocutt, G.E. and R.W. Dimmick. 1971. Summer food habits of juvenile Wood Duck in east Tennessee. Journal of Wildlife Management 35(2): 286-292.

Hogan, G.G. 1983. Interspecific food piracy of Rough-legged Hawk, *Buteo lagopus*, on Red-tailed Hawk, *B. jamaicensis*, and scavenging by Rough-legged Hawk. Canadian Field-Naturalist 97(3): 330-331.

Hogstad, O. 1978. Sexual dimorphism in relation to winter foraging and territorial behaviour of the Three-toed Woodpecker *Picoides tridactylus* and three *Dendrocopos* species. Ibis 120(2): 198-203.

Hohman, W.L. 1985. Feeding ecology of Ring-necked Ducks in northwestern Minnesota. Journal of Wildlife Management 49(3): 546-557.

Höhn, E.O. 1967. Observations on the breeding biology of Wilson's Phalarope (*Steganopus tricolor*) in central Alberta. Auk 84(2): 220-244.

Holcomb, L.C. 1972. Traill's Flycatcher breeding biology. Nebraska Bird Review 40(3): 50-68.

Holcomb, L.C. and M. Jaeger. 1978. Growth and calculation of age in Mourning Dove nestlings. Journal of Wildlife Management 42(4): 843-852.

Holder, K. and R. Montgomerie. 1993. Rock Ptarmigan (*Lagopus mutus*), *in* A. Poole and F. Gill, eds. The birds of North America, no. 51. The Academy of Natural Sciences, Philadelphia. The American Ornithologists' Union, Washington, 24 pp.

Holmes, R.T. 1986. Foraging patterns of forest birds: male-female differences. Wilson Bulletin 98(2): 196-213.

Holmes, R.T. 1990. The structure of temperate deciduous forest bird community: variability in time and space, pp. 121-129 *in* A. Keast, ed. Biogeography and ecology of forest bird communities. SPB Academic Publisher, The Hague, 410 pp.

Holmes, R.T. 1994. Black-throated Blue Warbler (*Dendroica caerulescens*), *in* A. Poole and F. Gill, eds. The birds of North America, no. 87. The Academy of Natural Sciences, Philadelphia. The American Ornithologists' Union, Washington, 24 pp.

Holmes, R.T. and F.A. Pitelka. 1968. Food overlap among coexisting sandpipers on northern Alaskan tundra. Systematic Zoology 17: 305-318.

Holmes, R.T. and S.K. Robinson. 1981. Tree species preferences of foraging insectivorous birds in a northern hardwood forest. Oecologia 48: 31-35.

Holmes, R.T. and S.K. Robinson. 1988. Spatial patterns, foraging tactics, and diets of ground-foraging birds in a northern hardwoods forest. Wilson Bulletin 100(3): 377-394.

Holmes, R.T. and T.W. Sherry. 1988. Assessing population trends of New Hampshire forest birds: local vs regional trends. Auk 105(4): 756-768.

Holmes, R.T. and F.W. Sturges. 1975. Bird community dynamics and energetics in a northern hardwoods ecosystem. Journal of Animal Ecology 44: 175-200.

Holmes, R.T., T.W. Sherry, and S.E. Bennett. 1978. Diurnal and individual variability in the foraging behaviour of the American Redstart (*Setophaga ruticilla*). Oecologia 36: 141-149.

Holmes, R.T., C.P. Black, and T.W. Sherry. 1979. Comparative population bio-energetics of three insectivorous passerines in a northern deciduous forest. Condor 81(1): 9-20.

Holmes, R.T., T.W. Sherry, and F.W. Sturges. 1986. Bird community dynamics in a temperate deciduous forest: long-term trends at Hubbard Brook. Ecological Monographs 56(3): 201-220.

Holmes, R.T., T.W. Sherry, P.P. Marra, and K.E. Petit. 1992. Multiple brooding and productivity of a neotropical migrant, the Black-throated Blue Warbler (*Dendroica caerulescens*), in an unfragmented temperate forest. Auk 109(2): 321-333.

Holt, D.W. 1986. Status Report: the Short-eared Owl in the northeast. Eyas 9(2): 3-5.

Holt, D.W. 1992a. Nestling growth rates of Short-eared Owls. Wilson Bulletin 104(2): 326-333.

Holt, D.W. 1992b. Notes on Short-eared Owl, *Asio flammeus*, nest sites, reproduction, and territory sizes in coastal Massachusetts. Canadian Field-Naturalist 106(3): 352-356.

Holt, D.W. and N.N. Childs. 1991. Non-breeding season diet of Long-eared Owls in Massachusetts. Journal of Raptor Research 25(1): 23-24.

Holt, D.W. and S.M. Leasure. 1993. Short-eared Owl (*Asio flammeus*), *in* A. Poole and F. Gill, eds. The birds of North America, no. 62. The Academy of Natural Sciences, Philadelphia. The American Ornithologists' Union, Washington, 24 pp.

Holt, D.W., J.P. Lortie, B.J. Nikula, and R.C. Humphrey. 1986. First record of Common Black-headed Gulls breeding in the United States. American Birds 40(2): 204-206.

Holt, D.W., S.M. Melvin, and B. Steele. 1992. Nestling growth rates of Short-eared Owls. Wilson Bulletin 104(2): 326-333.

Holthuijzen, A.M.A. and T.L. Sharik. 1985. The avian seed dispersal system of Eastern red cedar (*Juniperus virginiana*). Canadian Journal of Botany 63: 1508-1515.

Holway, D.A. 1991. Nest-site selection and the importance of nest concealment in the Black-throated Blue Warbler. Condor 93(3): 575-581.

Holyoak, D.T. 1990. Les Corvidés, pp. 140-143 *in* C.M. Perrins and A.L.A. Middleton, eds. Les oiseaux des plaines et des forêts. Solar, Paris, 143 pp.

Hooge, P.N. 1990. Maintenance of pair-bonds in the House Finch. Condor 92(4): 1066-1067.

Hoogland, J.L. and P.W. Sherman. 1976. Advantages and disadvantages of Bank Swallow (*Riparia riparia*) coloniality. Ecological Monographs 46(1): 33-58.

Hooper, R.G. 1977. Nesting habitat of Common Ravens in Virginia. Wilson Bulletin 89(2): 233-242.

Hooper, R.G., H.S. Crawford, D.R. Chamberlain, and R.F. Harlow. 1975. Nesting density of Common Ravens in the Ridge-Valley region of Virginia. American Birds 29(5): 931-935.

Hooper, T.D. and M. Nyhof. 1986. Food habits of the Long-eared Owl in south central British Columbia. Murrelet 67(1): 28-30.

Horak, G.J. 1970. A comparative study of the foods of the Sora and Virginia Rail. Wilson Bulletin 82(2): 206-213.

Horn, A.G. and J.B. Falls. 1991. Song switching in mate attraction and territory defense by Western Meadowlarks (*Sturnella neglecta*). Ethology 87: 262-268.

Horton, G.I. and M.K. Causey. 1982. Association among Woodcock brood members after brood breakup, pp. 71-74 *in* T.J. Dwyer and G.L. Storm, eds. Woodcock ecology and management. Papers from the Seventh Woodcock Symposium held at the Pennsylvania State University, University Park, Pennsylvania, 28-30 October 1980. United States Fish and Wildlife Service. Research Report no. 14, Washington, 191 pp.

Hosie, R.C. 1978. Arbres indigènes du Canada. Éditions Fides, Hull, 389 pp.

Houde, A. 1978. Atlas climatologique du Québec. Service météorologique, Ministère des Richesses Naturelles du Québec. Publication no. M-36, 45 pp.

Hounsell, R.G. 1991. Status of the woodcock in Canada 1990. Canadian Wildlife Service, Sackville, 5 pp.

Houston, C.S. 1975. Close proximity of Red-tailed Hawk and Great Horned Owl nests. Auk 92(3): 612-614.

Houston, C.S. 1978. Recoveries of Saskatchewan-banded Great Horned Owls. Canadian Field-Naturalist 92(1): 61-66.

Houston, C.S. 1983. Birds first described from Hudson Bay. Canadian Field-Naturalist 97: 95-98.

Houston, C.S. 1987. Nearly synchronous cycles of the Great Horned Owl and Snowshoe Hare in Saskatchewan, pp. 56-58 *in* R.W. Nero, R.J. Clark, R.J. Knapton, and R.H. Hamre, eds. Proceedings. Biology and conservation of northern forest owls. Symposium, 3-7 February, Winnipeg. United States Department of Agriculture, Forest Service, Rocky Mountain Forest and Range Experiment Station, Fort Collins, Colorado. General Technical Report RM-142, 309 pp.

Houston, C.S. and P.W. Brown. 1983. Recoveries of Saskatchewan-banded White-winged Scoters, *Melanitta fusca*. Canadian Field-Naturalist 97: 454-455.

Houston, D.C. 1986. Scavenging efficiency of Turkey Vultures in tropical forest. Condor 88(3): 318-323.

Howard, D.V. 1967. Variation in the breeding season and clutch-size of the Robin in the northeastern United States and the Maritime provinces of Canada. Wilson Bulletin 79(4): 432-440.

Howe, H.F. 1974. Age-specific differences in habitat selection by the American Redstart. Auk 91(1): 161-162.

Howe, H.F. 1976. Egg size, hatching asynchrony, sex, and brood reduction in the Common Grackle. Ecology 57: 1195-1207.

Howe, H.F. 1978. Initial investment, clutch size, and brood reduction in the Common Grackle (*Quiscalus quiscula* L.). Ecology 59(6): 1109-1122.

Howe, M.A. 1975. Behavioral aspects of the pair bond in Wilson's Phalarope. Wilson Bulletin 87(2): 248-270.

Howe, M.A. 1982. Social organization in a nesting population of Eastern Willets (*Catoptrophorus semipalmatus*). Auk 99(1): 88-102.

Howe, R.H. Jr. 1898. Breeding habits of the American Robin (*Merula migratoria*) in eastern Massachusetts. Auk 15: 162-167.

Howell, J.C. 1942. Notes on the nesting habits of the American Robin (*Turdus migratorius* L.). American Midland Naturalist 28(3): 529-603.

Howes-Jones, D. 1985. Nesting habits and activity patterns of Warbling Vireos, *Vireo gilvus*, in southern Ontario. Canadian Field-Naturalist 99(4): 484-489.

Hoyt, S.F. 1957. The ecology of the Pileated Woodpecker. Ecology 38(2): 246-256.

Hubel, F.C. 1907. Preliminary list of the summer birds of the Cobalt Mining region, Nipissing district, Ontario. Auk 24(1): 48-52.

Hudon, J.M. 1978. Inventaire des oiseaux marins: parc national Forillon, 1977. Parks Canada, Service de la conservation des ressources naturelles, 64 pp.

Hudson, A.V. 1982. Great Black-backed Gulls on Great Saltee Island. Irish Birds 2: 167-185.

Hudson, P.J. 1985. Population parameters for the Atlantic Alcidae, pp. 233-254 *in* D.N. Nettleship and T.R. Birkhead, eds. The Atlantic Alcidae: the evolution, distribution and biology of the auks inhabiting the Atlantic Ocean and adjacent water areas. Academic Press, London, 574 pp.

Hughes, J. 1986. Distribution, abundance, and productivity of Ospreys in interior Alaska. Alaska Department of Fish and Game, Juneau.

Hughes, R.J. 1992. Utilisation de l'habitat par la Grande Oie blanche pendant la période d'élevage des couvées à l'île Bylot, Territoires du Nord-Ouest. M.Sc. thesis, Université Laval, Sainte-Foy, 87 pp.

Hughes, R.J., A. Reed, and G. Gauthier. 1994a. Space and habitat use by Greater Snow Goose broods on Bylot Island, Northwest Territories. Journal of Wildlife Management 58(3): 536-545.

Hughes, R.J., G. Gauthier, and A. Reed. 1994b. Summer habitat use and behaviour of Greater Snow Geese. Wildfowl 45: 49-64.

Hughes, R.J., A. Reed, L. Rancourt, and R. Bergeron. 1994c. Nesting and brood-rearing ecology of Canada Geese in the vicinity of a future hydroelectric reservoir in northern Quebec, 1992-93. Paper presented at the 50th Northeast Fish and Wildlife Conference, Burlington, Vermont, May 1994.

Huhtala, K., E. Korpimäki, and E. Pulliainen. 1987. Foraging activity and growth of nestlings in the Hawk Owl: adaptative strategies under northern conditions, pp. 152-156 *in* R.W. Nero, R.J. Clark, R.J. Knapton, and R.H. Hamre, eds. Proceedings. Biology and conservation of northern forest owls. Symposium, 3-7 February, Winnipeg. United States Department of Agriculture, Forest Service, Rocky Mountain Forest and Range Experiment Station, Fort Collins, Colorado. General Technical Report RM-142, 309 pp.

Hunt, L.B. 1969. Physiological susceptibility of Robins to D.D.T. poisoning. Wilson Bulletin 81(4): 407-418.

Hunter, M.L. Jr. 1990. Wildlife forests and forestry: principles of managing forests for biological diversity. Prentice Hall, Englewood Cliffs, New Jersey.

Hunter, M.L. Jr. 1991. Coping with ignorance: the coarse-filter strategy for maintaining biodiversity, pp. 266-281 *in* K.A. Kohn, ed. Balancing on the brink of extinction. Island Press, Washington.

Hunter, M.L. Jr., G.L. Jacobson, Jr., and W. Thompson, III. 1988. Paleoecology and the coarse-filter approach to maintaining biological diversity. Conservation Biology 2: 375-385.

Huot, G. 1980. L'observation des oiseaux. Éditions Marcel Broquet, LaPrairie, 183 pp.

Hurd, P.L., P.J. Weatherhead, and S.B. McRae. 1991. Parental consumption of nestling feces: good food or sound economics? Behavioral Ecology 2(1): 69-76.

Hurtley, R.J. and E.C. Franks. 1976. Changes in the breeding ranges of two grassland birds. Auk 93(1): 108-115.

Hussell, D.J.T. 1972. Factors affecting clutch size in arctic passerines. Ecological Monographs 42(3): 317-364.

Hussell, D.J.T. 1981. The use of migration counts for monitoring bird population levels. Studies in Avian Biology 6: 92-102.

Hussell, D.J.T. 1982. Migrations of the Yellow-bellied Flycatcher in southern Ontario. Journal of Field Ornithology 53(3): 223-234.

Hussell, D.J.T. 1983. Age and plumage color in female Tree Swallows. Journal of Field Ornithology 54(3): 312-318.

Hussell, D.J.T. 1985. On the adaptive basis for hatching synchrony: brood reduction, nest failure and asynchronous hatching in Snow Buntings. Ornis Scandinavica 16: 205-212.

Hussell, D.J.T. 1987a. Horned Lark, pp. 270-271 *in* M.D. Cadman, P.F.J. Eagles, and F.M. Helleiner, eds. Atlas of the breeding birds of Ontario. University of Waterloo Press, Waterloo, 617 pp.

Hussell, D.J.T. 1987b. Red-necked Phalarope, p. 536 *in* M.D. Cadman, P.F.J. Eagles, and F.M. Helleiner, eds. Atlas of the breeding birds of Ontario. University of Waterloo Press, Waterloo, 617 pp.

Hussell, D.J.T. 1987c. Semipalmated Plover, p. 522 *in* M.D. Cadman, P.F.J. Eagles, and F.M. Helleiner, eds. Atlas of the breeding birds of Ontario. University of Waterloo Press, Waterloo, 617 pp.

Hussell, D.J.T. and G.L. Holroyd. 1974. Birds of the Truelove Lowland and adjacent areas of northeastern Devon Island, Northwest Territories. Canadian Field-Naturalist 88(2): 197-212.

Hutchinson, B.C. 1985. Status report on Beaver (*Castor canadensis*) in Canada. COSEWIC report for Canadian Wildlife Service, 17 pp.

Hyde, A.S. 1939. The life history of the Henslow's Sparrow, *Passerherbulus henslowii*. Museum of Zoology, University of Michigan. Miscellaneous Publications no. 41.

Hydro-Québec. 1993. Complexe Grande-Baleine. Rapport d'avant-projet. Partie 2: Complexe hydroélectrique. Tome 2: Description du milieu. Vol. 2: Milieu naturel marin, 233 pp.

Ibarzabal, J. and L. Imbeau. 1992. Les observations ornithologiques au Saguenay/Lac Saint-Jean – Été 1991. Le Harfang 15(1): 36-47.

Ickes, R. 1992. Great Crested Flycatcher, pp. 210-211 *in* D.W. Brauning, ed. Atlas of breeding birds in Pennsylvania. University of Pittsburgh Press, Pittsburgh, 484 pp.

Ignatiuk, J.B. and R.G. Clark. 1991. Breeding biology of American Crows in Saskatchewan parkland habitat. Canadian Journal of Zoology 69: 168-175.

Imhof, T.A. 1993. 79. Suburban cemetery, p. 82 *in* J.D. Lowe, ed. Resident bird counts 1992. Journal of Field Ornithology 64(1)(supplement): 1-128.

Ingold, J.L. and G.E. Wallace. 1994. Ruby-crowned Kinglet (*Regulus calendula*), *in* A. Poole and F. Gill, eds. The birds of North America, no. 119. The Academy of Natural Sciences, Philadelphia. The American Ornithologists' Union, Washington, 24 pp.

Ingolfsson, A. and B.T. Estrella. 1978. The development of the shell-cracking behavior in Herring Gulls. Auk 95(3): 577-579.

Ingram, C. 1959. The importance of juvenile cannibalism in the breeding biology of certain birds of prey. Auk 76(2): 218-266.

Ingram, J. 1992. La science du quotidien. Éditions MultiMondes, 240 pp.

Inskipp, T.P. and J.T.R. Sharrock. 1992. English names of West Palearctic birds. British Birds 85(6): 263-290.

Isenmann, P. 1977. Données sur la biologie de reproduction de la Mouette rieuse en Camargue. Nos Oiseaux 34(4): 143-154.

Isler, M.L. and P.R. Isler. 1987. The tanagers: natural history, distribution and identification. R.W. Spiegel, ed. Smithsonian Institution Press, Washington, 404 pp.

Ivey, G.[L.] and C. Baars. 1990. A second Semipalmated Plover nest in Oregon. Oregon birds 16(3): 207-208.

Ivey, G.L., K.A. Fothergill, and K.L. Yates-Mills. 1988. A Semipalmated Plover nest in Oregon. Western Birds 19: 35-36.

Jackson, H.D. 1985. Commentary and observations on the alleged transportation of eggs and young by caprimulgids. Wilson Bulletin 97(3): 381-385.

Jackson, J.A. 1970. Quantitative study of the foraging ecology of Downy Woodpeckers. Ecology 51: 318-323.

Jackson, J.A. 1983. Nesting phenology, nest site selection and reproductive success of Black and Turkey vultures, pp. 245-270 *in* S.R. Wilbur and J.A. Jackson, eds. Vulture biology and management. University of California Press, Los Angeles, 550 pp.

Jackson, J.A. and J. Tate, Jr. 1974. An analysis of nest box use by Purple Martins, House Sparrows, and Starlings in eastern North America. Wilson Bulletin 86(4): 435-449.

Jackson, N.H. and D.D. Roby. 1992. Fecundity and egg-laying patterns of captive yearling Brown-headed Cowbirds. Condor 94(3): 585-589.

Jalava, J. 1993. Status report on the Sedge Wren (*Cistothorus platensis stellaris*) in Canada. Committee on the Status of Endangered Wildlife in Canada (COSEWIC), Ottawa, 50 pp.

James Bay and Northern Quebec Native Harvesting Research Committee. 1982. The wealth of the land: wildlife harvests by the James Bay Cree, 1972-73 to 1978-79. Québec, 811 pp.

James, C.P., T.W. Barry, A.R. Smith, and S.J. Barry. 1987. Bill crossover ratios in Canadian crossbills *Loxia* spp. Ornis Scandinavica 18(4): 310-312.

James, D. 1962. The changing seasons. Audubon Field Notes 16(3): 306-307.

James, F.C., F.R. Johnston, N.O. Warner, G.J. Niemi, and W.J. Boecklen. 1984. The Grinnellian niche of the Wood Thrush. American Naturalist 124(1): 17-47.

James, R.D. 1976. Foraging behavior and habitat selection of three species of vireos in southern Ontario. Wilson Bulletin 88(1): 62-75.

James, R.D. 1978. Pairing and nest site selection in Solitary and Yellow-throated vireos with a description of a ritualized nest building display. Canadian Journal of Zoology 56(5): 1163-1169.

James, R.D. 1984. Structure, frequency of usage, and apparent learning in the primary song of the Yellow-throated Vireo, with comparative notes on Solitary Vireos (Aves: Vireonidae). Canadian Journal of Zoology 62: 468-472.

James, R.D. 1987a. Greater Scaup, p. 512 *in* M.D. Cadman, P.F.J. Eagles, and F.M. Helleiner, eds. Atlas of the breeding birds of Ontario. University of Waterloo Press, Waterloo, 617 pp.

James, R.D. 1987b. Greater Yellowlegs, pp. 524-525 *in* M.D. Cadman, P.F.J. Eagles, and F.M. Helleiner, eds. Atlas of the breeding birds of Ontario. University of Waterloo Press, Waterloo, 617 pp.

James, R.D. 1987c. Solitary Vireo, pp. 346-347 *in* M.D. Cadman, P.F.J. Eagles, and F.M. Helleiner, eds. Atlas of the breeding birds of Ontario. University of Waterloo Press, Waterloo, 617 pp.

James, R.D. 1987d. Warbling Vireo, pp. 350-351 *in* M.D. Cadman, P.F.J. Eagles, and F.M. Helleiner, eds. Atlas of the breeding birds of Ontario. University of Waterloo Press, Waterloo, 617 pp.

James, R.D. 1987e. Yellow-throated Vireo, pp. 348-349 *in* M.D. Cadman, P.F.J. Eagles, and F.M. Helleiner, eds. Atlas of the breeding birds of Ontario. University of Waterloo Press, Waterloo, 617 pp.

James, R.D. 1987f. Philadelphia Vireo, pp. 352-353 *in* M.D. Cadman, P.F.J. Eagles, and F.M. Helleiner, eds. Atlas of the breeding birds of Ontario. University of Waterloo Press, Waterloo, 617 pp.

James, R.D. 1988. Blue Jays mimic the calls of the Red-shouldered and Broad-winged hawks. Ontario Birds 6(1): 11-14.

James, R.D. and M.K. Peck. 1986. Are Red-eyed and Philadelphia vireos always territorial? Ontario Birds 4: 101-103.

Jamieson, I., N. Seymour, and R.P. Bancroft. 1982. Time and activity budgets of Ospreys nesting in northeastern Nova Scotia. Condor 84: 439-441.

Janssen, R.B. 1990. A major invasion of White-winged Crossbills, 1989-1990. Loon 62(2): 69-71.

Jansson, C., J. Ekman, and A. von Bromssen. 1981. Winter mortality and food supply in tits *Parus* spp. Oikos 37: 313-322.

Jarvis, R.L. and J.H. Noyes. 1986. Foods of Canvasbacks and Redheads in Nevada: paired males and ducklings. Journal of Wildlife Management 50(2): 199-203.

Jaslowitz, C.A. 1993. 24. Mature urban deciduous forest II, p. 48 *in* J.D. Lowe, ed. Resident bird counts 1992. Journal of Field Ornithology 64(1)(supplement): 1-128.

Jauvin, D. 1981. Phalarope de Wilson. Bulletin ornithologique 26(2): 1981.

Jean, B. 1985. Agriculture et développement dans l'est du Québec. Presses de l'Université du Québec, Québec, 431 pp.

Jean, M. and A. Bouchard. 1991. Temporal changes in wetland landscapes of a section of the St. Lawrence River, Canada. Environmental Management 15: 241-250.

Jefferson, B. 1989. Observations of Common Raven in metropolitan Toronto. Ontario Birds 7(1): 15-20.

Jefferson, B. 1991. Evidence of pair bonding between Common Raven (*Corvus corax*) and American Crow (*Corvus brachyrhynchos*). Ontario Birds 9(2): 45-48.

Jehl, J.R. Jr. 1963. An investigation of fall-migrating dowitchers in New Jersey. Wilson Bulletin 75: 250-261.

Jehl, J.R. Jr. and D.T. Hussell. 1966. Incubation periods of some subarctic birds. Canadian Field-Naturalist 80(3): 179-180.

Jehl, J.R. Jr. and B.A. Smith. 1970. Birds of the Churchill region, Manitoba. Manitoba Museum of Man and Nature, Winnipeg. Special Publication no. 1, 87 pp.

Jenness, D. 1958. The Indians of Canada, 4th edition. Anthropological Bulletin: National Museum of Canada, Series no. 15, 452 pp.

Job, H.K. 1902. Among the waterfowl. The Musson Book Co. Limited, Toronto, 224 pp.

Job, H.K. 1906. Some bird-notes from the Magdalens. Bird-Lore 2: 43-48.

Jobin, B., J.-L. DesGranges, N. Plante, and C. Boutin. 1994. Relations entre la modification du paysage rural, les changements de pratique agricole et les fluctuations des populations d'oiseaux champêtres du sud du Québec (plaine du Saint-Laurent). Gauthier et Guillemette Consultants for Environment Canada, Canadian Wildlife Service, Québec Region. Technical Report Series no. 191, 95 pp. [An abstract is available in English.]

Joe, S. and L. Duchesne. 1989. Populations et ménages, pp. 294-323 *in* Le Québec statistique. Bureau de la statistique du Québec, Québec, 1028 pp.

John, R. 1987. American Crow, pp. 288-289 *in* M.D. Cadman, P.F.J. Eagles, and F.M. Helleiner, eds. Atlas of the breeding birds of Ontario. University of Waterloo Press, Waterloo, 617 pp.

John, R.D. 1991. Observations on soil requirements for nesting Bank Swallows, *Riparia riparia*. Canadian Field-Naturalist 105(2): 251-254.

Johns, J.E. 1969. Field studies of Wilson's Phalarope. Auk 86(4): 660-670.

Johnsgard, P.A. 1961. The sexual behavior and systematic position of the Hooded Merganser. Wilson Bulletin 73(3): 226-236.

Johnsgard, P.A. 1965. Handbook of waterfowl behavior. Cornell University Press, Ithaca, New York, 378 pp.

Johnsgard, P.A. 1973. Grouse and quails of North America. University of Nebraska Press, Lincoln, 553 pp.

Johnsgard, P.A. 1975a. North American game birds of upland and shoreline. University of Nebraska Press, Lincoln, 183 pp.

Johnsgard, P.A. 1975b. Waterfowl of North America. Indiana University Press, Bloomington, 575 pp.

Johnsgard, P.A. 1978. Ducks, geese, and swans of the world. University of Nebraska Press, Lincoln, 404 pp.

Johnsgard, P.A. 1980. Copulatory behavior of the American Bittern. Auk 97(4): 868-869.

Johnsgard, P.A. 1981. The plovers, sandpipers, and snipes of the world. University of Nebraska Press, Lincoln, 493 pp.

Johnsgard, P.A. 1983a. Grouse of the World. University of Nebraska Press, Lincoln, 413 pp.

Johnsgard, P.A. 1983b. The hummingbirds of North America. Smithsonian Institution Press, Washington, 303 pp.

Johnsgard, P.A. 1986. The pheasants of the world. Oxford University Press, Oxford, 300 pp.

Johnsgard, P.A. 1988. North American owls: biology and natural history. Smithsonian Institution Press, Washington, 295 pp.

Johnsgard, P.A. 1989. Courtship and mating, pp. 112-117 *in* S. Atwater and J. Schnell, eds. Ruffed Grouse. Stackpole Books, Harrisburg, 370 pp.

Johnsgard, P.A. 1990. Hawks, eagles, and falcons of North America: biology and natural history. Smithsonian Institution Press, Washington, 403 pp.

Johnsgard, P.A. and S.J. Maxson. 1989. Nesting, pp. 130-137 *in* S. Atwater and J. Schnell, eds. Ruffed grouse. Stackpole Books, Harrisburg, 370 pp.

Johnsgard, P.A. and R.W. Wood. 1968. Distributional changes and interaction between Prairie chickens and Sharp-tailed Grouse in the Midwest. Wilson Bulletin 80: 173-188.

Johnsgard, P.A., R.D. Kimmel, S.J. Maxson, J.G. Scott, R. Small, and G.L. Storm. 1989. The young grouse, pp. 140-159 *in* S. Atwater and J. Schnell, eds. Ruffed Grouse. Stackpole Books, Harrisburg, 370 pp.

Johnson, B.R. and W.J. Adams. 1977. The Little Gull (*Larus minutus*) in Arctic North America. Canadian Field-Naturalist 91(3): 294-296.

Johnson C.M. and G.A. Baldassarre. 1988. Aspects of the wintering ecology of Piping Plovers in coastal Alabama. Wilson Bulletin 100(2): 214-223.

Johnson, D.H. 1980. Barred Owls use nest box. Loon 52(4): 193-194.

Johnson, D.H. and D.G. Fullen, Sr. 1984. Barred Owls and nest boxes. Raptor Research 18(1): 34-35.

Johnson, E.J. and L.B. Best. 1982. Factors affecting feeding and brooding of nestlings. Auk 99(1): 148-156.

Johnson, E.V., L.M. Guilford, and D.Q. Thompson. 1976. The effects of orchard pesticides applications on breeding robins. Wilson Bulletin 88(1): 16-35.

Johnson, H.S. 1933. Notes on the family life of a pair of American Pipits. Wilson Bulletin 45: 114-117.

Johnson, L.L. 1967. The Common Goldeneye duck and the role of nesting boxes in its management in north-central Minnesota. Journal of the Minnesota Academy of Science 34: 129-132.

Johnson, L.S. and L.H. Kermott. 1989. Territorial intrusions in the House Wren *Troglodytes aedon*: evidence for the sperm competition hypothesis. Ornis Scandinavica 20: 89-92.

Johnson, L.S. and L.H. Kermott. 1990. Possible causes of territory takeovers in a north-temperate population of House Wrens. Auk 107(4): 781-784.

Johnson, L.S. and L.H. Kermott. 1991. Effect of nest-site supplementation on polygynous behavior in the House Wren (*Troglodytes aedon*). Condor 93(3): 784-787.

Johnson, P.H. and J.W. Grier. 1988. Determinants of breeding distributions of ducks. Wildlife Monographs no. 100, 37 pp.

Johnson, R.A. 1940. Present range, migration and abundance of the Atlantic Murre in North America. Bird-Banding 11: 1-17.

Johnson, R.A. and H.S. Johnson. 1935. A study of the nesting and family life of the Red-throated Loon. Wilson Bulletin 47: 98-103.

Johnson, R.K., R.R. Roth, S.E. Kleiner, and C.R. Bartlett. 1991. A case of polygyny in the Wood Thrush. Wilson Bulletin 103(3): 509-510.

Johnson, R.R. and J.J. Dinsmore. 1985. Brood-rearing and postbreeding habitat use by Virginia Rails and Soras. Wilson Bulletin 97(4): 551-554.

Johnson, R.R. and J.J. Dinsmore. 1986a. The use of tape-recorded calls to count Virginia Rails and Soras. Wilson Bulletin 98(2): 303-306.

Johnson, R.R. and J.J. Dinsmore. 1986b. Habitat use by breeding Virginia Rails and Soras. Journal of Wildlife Management 50(3): 387-392.

Johnson, R.R., L.T. Haight, and J.M. Simpson. 1979. Owl populations and species status in the southwestern United States, pp. 40-49 *in* P.P. Schaeffer and S.M. Ehlers, eds. Proceedings. National Audubon's Society symposium on owls of the west: their ecology and conservation. National Audubon Society Western Education Center, Tiburon, California, 97 pp.

Johnson, S.J. 1973. Post-fledging activity of the Red-tailed Hawk. Raptor Research 7(2): 43-48.

Johnson, S.R. 1984. Prey selection by Oldsquaws (*Clangula hyemalis* L.) in a Beaufort Sea lagoon, pp. 12-19 *in* Marine birds: their feeding ecology and commercial fisheries relationship. Canadian Wildlife Service Special Publications, Ottawa.

Johnson, S.R. and D.R. Herter. 1989. The birds of the Beaufort Sea. BP Exploration (Alaska) Inc., Anchorage, Alaska, 372 pp.

Johnson, W.N. and P.W. Brown. 1990. Avian use of a lakeshore buffer strip and an undisturbed lakeshore in Maine. Northern Journal of Applied Forestry 7: 114-117.

Johnsson, K. 1993. The Black Woodpecker (*Dryocopus martius*) as a keystone species in forest. Sveriges Lantbruksuniversitat, Institutionen for Viltekologi Rapport 24. Swedish University of Agricultural Sciences, Department of Wildlife Ecology, Uppsala.

Johnston, D.W. 1971. Niche relationships among some deciduous forest flycatchers. Auk 88(4): 796-804.

Johnston, D.W. and E.P. Odum. 1956. Breeding bird populations in relation to plant succession on the Piedmont of Georgia. Ecology 37: 50-62.

Johnston, R.F. 1958. Breeding of the Brown Thrasher in Kansas. Kansas Ornithological Society Bulletin 9(3): 17-19.

Johnston, R.F. 1967. Seasonal variation in the food of the Purple Martin *Progne subis* in Kansas. Ibis 109(1): 8-13.

Johnston, R.F. 1992. Rock Dove (*Columba livia*), *in* A. Poole, P. Stettenheim, and F. Gill, eds. The birds of North America, no. 13. The Academy of Natural Sciences, Philadelphia. The American Ornithologists' Union, Washington, 16 pp.

Johnston, R.F. and R.K. Selander. 1964. House Sparrows: rapid evolution of races in North America. Science 144: 548-550.

Johnston, R.F. and R.K. Selander. 1971. Evolution in the House Sparrow. II. Adaptive differentiation in North American populations. Evolution 25(1): 1-28.

Johnstone, K. 1991. Timber and trauma: 75 years with the federal forestry service, 1899-1974. Forestry Canada, 194 pp.

Jones, E.T. 1987. Observations of the Northern Hawk Owls in Alberta, pp. 149-151 *in* R.W. Nero, R.J. Clark, R.J. Knapton, and R.H. Hamre, eds. Proceedings. Biology and conservation of northern forest owls. Symposium, 3-7 February, Winnipeg. United States Department of Agriculture, Forest Service, Rocky Mountain Forest and Range Experiment Station, Fort Collins, Colorado. General Technical Report RM-142, 309 pp.

Jones, G. and Collaborators. 1990. Collins reference dictionnary: environmental science. Collins, London, 473 pp.

Jones, J.C. 1940. Food habits of the American Coot with notes on distribution. United States Department of the Interior. Wildlife Research Bulletin no. 2, 52 pp.

Jones, R.E. and A.S. Leopold. 1967. Nesting interference in a dense population of Wood Ducks. Journal of Wildlife Management 31: 221-228.

Jones, S. 1979. Habitat management series for unique or endangered species. Report no. 17: the Accipiters – Goshawk, Cooper's Hawk, Sharp-shinned Hawk. United States Department of the Interior, Bureau of Land Management Technical Note no. 335, 25 pp.

Joyner, D.E. 1969. A survey of the ecology and behavior of the Ruddy Duck (*Oxyura jamaicensis*) in northern Utah. M.Sc. thesis, University of Utah, Salt Lake City, 82 pp.

Joyner, D.E. 1976. Effects of interspecific nest parasitism by Redheads and Ruddy Ducks. Journal of Wildlife Management 40(1): 33-38.

Joyner, D.E. 1983. Parasitic egg laying in Redheads and Ruddy Ducks in Utah: incidence and success. Auk 100(3): 717-725.

Judd, S.D. 1898. The food of shrikes, pp. 15-26 *in* C.H. Merriam, ed. Cuckoos and Shrikes in their relation to agriculture. United States Department of Agriculture, Division of Biological Survey, Washington. Bulletin no. 9, 26 pp.

Judd, S.D. 1901. The relation of sparrows to agriculture. United States Department of Agriculture, Division of Biological Survey. Bulletin no. 15, 98 pp.

Juillard, M. and J. Beuret. 1983. L'aménagement de sites de nidification et son influence sur une population de Chouettes effraies, *Tyto alba*, dans le nord-ouest de la Suisse. Nos Oiseaux 37(1): 1-20.

Julin, K.R. 1986. Decline of second-growth Douglas-fir in relation to Great Blue Heron nesting. Northwest Science 60: 201-205.

Jurdant, M., J.-L. Belair, V. Gerardin, and J.-P. Ducruc. 1977. L'inventaire du Capital-Nature: méthode de classification et de cartographie écologique du territoire (3e approximation) [Inventory of Capital-Nature: a method of ecological land classification and mapping]. Lands Directorate, Environmental Management Service, Fisheries and Environment Canada. Ecological Land Classification Series no. 2, 202 pp.

Kaczynski, C.F. and E.B. Chamberlain. 1968. Aerial surveys of Canada Geese and Black Ducks in eastern Canada. United States Fish and Wildlife Service, Special Scientific Report – Wildlife no. 118, 29 pp.

Kadlec, J.A. and W.H. Drury. 1968. Structure of the New England Herring Gull population. Ecology 49: 644-676.

Kaiser, S.M. and E.K. Fritzell. 1984. Effects of river recreationists on Green-backed Heron behaviour. Journal of Wildlife Management 48(2): 561-567.

Kaiser, S.M. and F.A. Reid. 1987. A comparison of Green-backed Heron nesting in two freshwater ecosystems. Colonial Waterbirds 10(1): 78-83.

Kale, H.W. II. 1967. Agressive behavior by a migrating Cape May Warbler. Auk 84(1): 120-121.

Kale, H.W. II. 1968. The relationship of Purple Martins to mosquito control. Auk 85(4): 654-661.

Kalkhoven, J.T.R. 1993. Survival of populations and the scale of the fragmented agricultural landscape, pp. 83-90 *in* R.G.M. Bunes, L. Ryzlowski, and M.G. Paoletti, eds. Landscape Ecology and Agroecosystems. Lewis Publishers, Ann Arbor, Michigan.

Kalm, P. 1770-1771. Travels into North America. William Eyres, Warrington, England, 3 vol.

Kalm, P. 1911. A description of the Wild Pigeons which visit the southern English colonies in North America, during certain years, in incredible multitudes. Translated by S.M. Gronberger. Auk 28(1): 53-66.

Kane, R. 1982. Natural Bluebird nest sites in New Jersey. Audubon 8(3): 38-39.

Karalus, K.E. and A.W. Eckert. 1974. The owls of North America, north of Mexico. Doubleday, Garden City, New York, [1987, Weathervane Books, New York], 278 pp.

Karalus, K.E. and A.W. Eckert. 1987. (See Karalus, K.E. and A.W. Eckert 1974.)

Karlin, A.A., G.S. Kimberly, C.S. Micheline, and R.A. Barnhill. 1990. Additional evidence of multiple parentage in Eastern Bluebirds. Condor 92: 520-521.

Kaufman, K. 1983. Yellow-headed Blackbird, p. 296 *in* J. Farrand, ed. The Audubon Society master guide to birding, vol. 3. Alfred A. Knopf, New York, 399 pp.

Kaufman, K. 1990a. A field guide to advanced birding: birding challenges and how to approach them. Houghton Mifflin Company, Boston, 299 pp.

Kaufman, K. 1990b. The practiced eye: Common Merganser and Red-breasted Merganser. American Birds 44(5): 1203-1205.

Kaufmann, G.W. 1983. Displays and vocalizations of the Sora and the Virginia Rail. Wilson Bulletin 95(1): 42-59.

Kaufmann, G.W. 1987. Growth and development of Sora and Virginia Rail chicks. Wilson Bulletin 99(3): 432-440.

Kaufmann, G.W. 1989. Breeding ecology of the Sora, *Porzana carolina*, and the Virginia Rail, *Rallus limicola*. Canadian Field-Naturalist 103(2): 270-282.

Keast, A. 1973. The utilization of the Churchill Falls reservoir area by wild geese and ducks and a breeding density survey, July 15-21, 1973. Manuscript Report, Department of Biology, Queen's University, Kingston, Ontario, 53 pp.

Keast, A. and E.S. Morton. 1980. Migrant birds in the neotropics: ecology, behavior, distribution and conservation. Symposium, October 1977, Conservation and Research Center, National Zoological Park, Smithsonian Institution. Smithsonian Institution Press, Washington, 576 pp.

Keast, A. and S. Saunders. 1991. Ecomorphology of the North American Ruby-crowned (*Regulus calendula*) and Golden-crowned (*R. satrapa*) kinglets. Auk 108(4): 880-888.

Keeton, W.T. 1969. Orientation by pigeons: is the sun necessary? Science 165: 922-928.

Keeton, W.T. 1977. Pigeon navigation, pp. 1-20 *in* A.M. Granda and J.H. Maxwell, eds. Neural mechanisms of behaviour in the pigeon. Plenum Press, New York, 436 pp.

Kehoe, F.P. 1986. The adaptive significance of crèching behavior in the White-winged Scoter (*Melanitta fusca deglandi*). M.Sc. thesis, University of Guelph, Guelph, Ontario, 33 pp.

Kehoe, F.P. 1989. The adaptive significance of crèching behaviour in the White-winged Scoter (*Melanitta fusca deglandi*). Canadian Journal of Zoology 67(2): 406-411.

Kehoe, [F.]P. 1993. Status of sea ducks in the Atlantic Flyway with strategies towards improved management. Ad Hoc Sea Ducks Committee Atlantic Flyway Technical Section, 15 pp.

Kehoe, F.P., P.W. Brown, and C.S. Houston. 1989. Survival and longevity of White-winged Scoters nesting in central Saskatchewan. Journal of Field Ornithology 60(2): 133-136.

Keith, L.B. 1960. Observations on Snowy Owls at Delta, Manitoba. Canadian Field-Naturalist 74: 106-112.

Keith, L.B. 1961. A study of waterfowl ecology on small impoundments in southeastern Alberta. Wildlife Monographs no. 6, 88 pp.

Keith, L.B. 1964. Territoriality among Snowy Owls. Canadian Field-Naturalist 78: 17-24.

Keith, L.B. 1987. Le Grand Duc. Hinterland Who's Who. Environment Canada, Canadian Wildlife Service, 4 pp. [Also available in English as: Great Horned Owl.]

Keith, L.B. and D.H. Rusch. 1988. Predation's role in the cyclic fluctuations of Ruffed Grouse, pp. 699-732 *in* H. Ouellet, ed. Acta XIX Congressus Internationalis Ornithologici, Ottawa, 22-29 June 1986. Ottawa University Press, vol. I; pp. 1-1404; vol. II, pp. 1405-2815.

Keller, R.C. 1993. 132. Conventional dairy farm, p. 114 *in* J.D. Lowe, ed. Resident bird counts 1992. Journal of Field Ornithology 64(1)(supplement): 1-128.

Kellner, C. 1990. Post-fledging behavior of American Kestrels in central Kentucky. Journal of Raptor Research 24(3): 56-58.

Kelso, J.R.M., C.K. Minns, J.E. Gray, and M.L. Jones. 1986. Acidification of surface waters in eastern Canada and its relationship to aquatic biota. Canadian Special Publication of Fisheries and Aquatic Sciences, 87 pp.

Kendeigh, S.C. 1941. Territorial and mating behaviour of the House Wren. Illinois Biological Monographs 18(3): 1-120.

Kendeigh, S.C. 1945. Nesting behavior of wood warblers. Wilson Bulletin 57(3): 145-164.

Kendeigh, S.C. 1947. Bird population studies in the coniferous forest biome during a spruce budworm outbreak. Ontario Department of Lands and Forests. Biological Bulletin 1: 1-100.

Kendeigh, S.C. 1976. Latitudinal trends in the metabolic adjustments of the House Sparrow. Ecology 57: 509-519.

Kendra, P.E., R.R. Roth, and D.W. Tallamy. 1988. Conspecific broods parasitism in the House Sparrow. Wilson Bulletin 100(1): 80-90.

Kennamer, J.E. and M.C. Kennamer. 1990. Current status and distribution of the Wild Turkey 1989, pp. 1-12 *in* W.M. Healy and G.B. Healy, eds. Proceedings. Sixth National Wild Turkey symposium. National Wild Turkey Federation, Edgefield, South Carolina, 228 pp.

Kennamer, R.A. and G.R. Hepp. 1987. Frequency and timing of second broods in Wood Ducks. Wilson Bulletin 99(4): 655-662.

Kennard, F.H. 1920. Notes on the breeding habits of the Rusty Blackbird in northern New England. Auk 32: 412-422.

Kennard, J.H. 1975. Longevity records of North American birds. Bird-Banding 46(1): 55-73.

Kennedy, E.D. and D.W. White. 1992. Nest building in House Wrens. Journal of Field Ornithology 63(1): 35-42.

Kenward, R.E. and P. Widén. 1989. Do Goshawks *Accipiter gentilis* need forests? Some conservation lessons from radio tracking, pp. 561-567 *in* B.-U. Meyburd and R.D. Chancellor, eds. Raptors in the modern world. Proceedings. Third world conference on birds of prey and owls, 22-27 March 1987, Eilat, Israël. World Working Group on Birds of Prey and Owls, Berlin, 611 pp.

Kenyon, K.W. 1942. Hunting strategy of Pigeon Hawks. Auk 59(3): 443-444.

Keppie, D.M. 1987. Impact of demographic parameters upon a population of Spruce Grouse in New Brunswick. Journal of Wildlife Management 51(4): 771-777.

Keppie, D.M. and P.W. Herzog. 1978. Nest site characteristics and nest success of Spruce Grouse. Journal of Wildlife Management 42(3): 628-632.

Keppie, D.M. and R.M. Whiting, Jr. 1994. American Woodcock (*Scolopax minor*), *in* A. Poole and F. Gill, eds. The birds of North America, no. 100. The Academy of Natural Sciences, Philadelphia. The American Ornithologists' Union, Washington, 28 pp.

Keran, D. 1978. Nest site selection by the Broad-winged Hawk in north central Minnesota and Wisconsin. Raptor Research 12: 15-20.

Keran, D. 1981. The incidence of man-caused and natural mortalities to raptors. Raptor Research 15(4): 108-112.

Kerbes, R.H. 1975. Lesser Snow Geese in the eastern Canadian Arctic: a photographic inventory of June 1973. Canadian Wildlife Service, Ottawa. Technical Report Series no 35, 47 pp.

Kerbes, R.H. 1983. Lesser Snow Goose colonies in the western Canadian Arctic. Journal of Wildlife Management 47(2): 523-526.

Kerlinger, P. and M.R. Lein. 1986. Differences in winter range among age-sex classes of Snowy Owls (*Nyctea scandiaca*) in North America. Ornis Scandinavica 17(1): 1-7.

Kerlinger, P. and M.R. Lein. 1988a. Causes of mortality, fat condition, and weights of wintering Snowy Owls. Journal of Field Ornithology 59(1): 7-12.

Kerlinger, P. and M.R. Lein. 1988b. Population ecology of Snowy Owls during winter on the Great Plains of North America. Condor 90(4): 866-874.

Kerlinger, P., M.R. Lein, and B.J. Sevick. 1985a. Distribution and population fluctuations of wintering Snowy Owls (*Nyctea scandiaca*) in North America. Canadian Journal of Zoology 63: 1829-1834.

Kerlinger, P., V.P. Bingman, and K.P. Able. 1985b. Comparative flight behaviour of migrating hawks studied with tracking radar during autumn in central New York. Canadian Journal of Zoology 63: 755-761.

Kerns, J.M. and J.F. Howe. 1967. Factors determining Great Blue Heron rookery movement. Journal of Minnesota Academy of Science 34: 80-83.

Kertell, K. 1986. Reproductive biology of Northern Hawk Owls in Denali National Park, Alaska. Raptor Research 20(3-4): 91-101.

Kessel, B. 1953. Distribution and migration of the European Starling in North America. Condor 55(2): 49-67.

Kessel, B. 1957. A study of the breeding biology of the European Starling (*Sturnus vulgaris* L.) in North America. American Midland Naturalist 58: 257-331.

Ketterson, E.D. and V. Nolan, Jr. 1976. Geographic variation and its climatic correlates in the sex-ratio of eastern-wintering Dark-eyed Juncos (*Junco hyemalis hyemalis*). Ecology 57(4): 679-693.

Kibbe, D.P. 1985a. Black-crowned Night-Heron, pp. 44-45 *in* S.B. Laughlin and D.P. Kibbe, eds. The Atlas of breeding birds of Vermont. Vermont Institute of Natural Science, University Press of New England, Hanover, 456 pp.

Kibbe, D.P. 1985b. Black-throated Blue Warbler, pp. 292-293 *in* S.B. Laughlin and D.P. Kibbe, eds. The Atlas of breeding birds of Vermont. Vermont Institute of Natural Science, University Press of New England, Hanover, 456 pp.

Kibbe, D.P. 1985c. Blackpoll Warbler, pp. 306-307 *in* S.B. Laughlin and D.P. Kibbe, eds. The Atlas of breeding birds of Vermont. Vermont Institute of Natural Science, University Press of New England, Hanover, 456 pp.

Kibbe, D.P. 1985d. Common Nighthawk, pp. 144-145 *in* S.B. Laughlin and D.P. Kibbe, eds. The Atlas of breeding birds of Vermont. Vermont Institute of Natural Science, University Press of New England, Hanover, 456 pp.

Kibbe, D.P. 1985e. Gray Partridge, pp. 88-89 *in* S.B. Laughlin and D.P. Kibbe, eds. The Atlas of breeding birds of Vermont. Vermont Institute of Natural Science, University Press of New England, Hanover, 456 pp.

Kibbe, D.P. 1985f. Least Bittern, pp. 36-37 *in* S.B. Laughlin and D.P. Kibbe, eds. The Atlas of breeding birds of Vermont. Vermont Institute of Natural Science, University Press of New England, Hanover, 456 pp.

Kibbe, D.P. 1985g. Louisiana Waterthrush, pp. 318-319 *in* S.B. Laughlin and D.P. Kibbe, eds. The Atlas of breeding birds of Vermont. Vermont Institute of Natural Science, University Press of New England, Hanover, 456 pp.

Kibbe, D.P. 1985h. Magnolia Warbler, pp. 288-289 *in* S.B. Laughlin and D.P. Kibbe, eds. The Atlas of breeding birds of Vermont. Vermont Institute of Natural Science, University Press of New England, Hanover, 456 pp.

Kibbe, D.P. 1985i. Philadelphia Vireo, pp. 270-271 *in* S.B. Laughlin and D.P. Kibbe, eds. The Atlas of breeding birds of Vermont. Vermont Institute of Natural Science, University Press of New England, Hanover, 456 pp.

Kibbe, D.P. 1985j. Pied-billed Grebe, pp. 32-33 *in* S.B. Laughlin and D.P. Kibbe, eds. The Atlas of breeding birds of Vermont. Vermont Institute of Natural Science, University Press of New England, Hanover, 456 pp.

Kibbe, D.P. 1985k. Solitary Vireo, pp. 264-265 *in* S.B. Laughlin and D.P. Kibbe, eds. The Atlas of breeding birds of Vermont. Vermont Institute of Natural Science, University Press of New England, Hanover, 456 pp.

Kibbe, D.P. and S.B. Laughlin. 1985. Double-crested Cormorant, pp. 391-392 *in* S.B. Laughlin and D.P. Kibbe, eds. The Atlas of breeding birds of Vermont. Vermont Institute of Natural Science, University Press of New England, Hanover, 456 pp.

Kibler, L.F. 1969. The establishment and maintenance of a Bluebird nest-box project. Bird-Banding 40(2): 114-129.

Kiff, L.F. 1988. Commentary – Changes in status of the Peregrine in North America: an overview, pp. 123-129 *in* T.J. Cade, J.H. Enderson, C.G. Thelander, and C.M. White, eds. Peregrine Falcon populations: their management and recovery. The Peregrine Fund, Boise, Idaho, 949 pp.

Kilgore, D.L. Jr. and K.L. Knudsen. 1977. Analysis of materials in Cliff and Barn Swallow nests: relationship between mud selection and nest architecture. Wilson Bulletin 89(4): 562-571.

Kilham, L. 1958a. Pair formation, mutual tapping, nest hole selection of Red-bellied Woodpeckers. Auk 75(3): 318-329.

Kilham, L. 1958b. Territorial behavior of wintering Red-headed Woodpeckers. Wilson Bulletin 70(4): 347-358.

Kilham, L. 1959. Behavior and methods of communication of Pileated Woodpeckers. Condor 61(6): 377-387.

Kilham, L. 1961. Reproductive behavior of Red-bellied Woodpeckers. Wilson Bulletin 73(3): 237-254.

Kilham, L. 1962a. Breeding behavior of Yellow-bellied Sapsuckers. Auk 79(1): 31-43.

Kilham, L. 1962b. Reproductive behavior of Downy Woodpeckers. Condor 64(2): 126-133.

Kilham, L. 1968. Reproductive behavior of White-breasted Nuthatches. I. Distraction display, bill-sweeping and nest hole defense. Auk 85(3): 477-492.

Kilham, L. 1971a. Use of blister beetle in bill-sweeping by White-breasted Nuthatch. Auk 88(1): 175-176.

Kilham, L. 1971b. Roosting habits of White-breasted Nuthatches. Condor 73(1): 113-114.

Kilham, L. 1972. Reproductive behavior of White-breasted Nuthatches. II. Courtship. Auk 89(1): 115-129.

Kilham, L. 1973a. Colonial-type nesting in Yellow-shafted Flickers as related to staggering of nesting times. Bird-Banding 44(4): 317-318.

Kilham, L. 1973b. Reproductive behavior of the Red-breasted Nuthatch. I. Courtship. Auk 90(3): 597-609.

Kilham, L. 1979a. Courtship and the pair-bond of Pileated Woodpeckers. Auk 96(3): 587-594.

Kilham, L. 1979b. Three-week vs 4-week nestling periods in *Picoides* and other woodpeckers. Wilson Bulletin 91(2): 335-338.

Kilham, L. 1981. Agonistic behavior of the White-breasted Nuthatch. Wilson Bulletin 93(2): 271-274.

Kilham, L. 1983. Life history studies of woodpeckers of eastern North America. Nuttall Ornithological Club, no. 20, Cambridge, Massachusetts, 231 pp.

Kilham, L. 1984a. Cooperative breeding of American Crows. Journal of Field Ornithology 55(3): 349-356.

Kilham, L. 1984b. Intra- and extrapair copulatory behavior of American Crows. Wilson Bulletin 96(4): 716-717.

Kilham, L. 1985. Territorial behavior of American Crows. Wilson Bulletin 97(3): 389-390.

Kilham, L. 1989. The American Crow and the Common Raven. The W.L. Moody, Jr., Natural History Series. Texas A and M University Press, College Station, 255 pp.

Kilpatrick, H.J., T.P. Husband, and C.A. Pringle. 1988. Winter roost site characteristics of eastern Wild Turkeys. Journal of Wildlife Management 52(3): 461-463.

Kimmins, J.P. 1992. Balancing act: environmental issues in forestry. University of British Columbia Press, Vancouver, 244 pp.

King, J. 1955. Notes on the life history of Traill's Flycatcher (*Empidonax traillii*) in southeastern Washington. Auk 72: 148-173.

King, J.R. and J.D. Hubbard. 1981. Comparative patterns of nestling growth in White-crowned Sparrows. Condor 83(4): 362-369.

Kinsley, K.R., S.A. Liscinsky, and G.L. Storm. 1982. Changes in habitat structure on woodcock singing grounds in central Pennsylvania, pp. 40-50 *in* T.J. Dwyer and G.L. Storm, eds. Woodcock ecology and management. Papers from the Seventh woodcock Symposium held at the Pennsylvania State University, University Park, Pennsylvania, 28-30 October 1980. United States Fish and Wildlife Service. Research Report no. 14, Washington, 191 pp.

Kirby, R.E. 1976. Breeding chronology and interspecific relations of Pied-billed Grebes in northern Minnesota. Wilson Bulletin 88(3): 493-495.

Kirby, R.E. 1990. Wood Duck nonbreeding ecology: fledging to spring migration, pp. 61-76 *in* L.H. Fredrickson, G.V. Burger, S.P. Havera, D.A. Graber, R.E. Kirby, and T.S. Taylor, eds. Proceedings. 1988 North American Wood Duck symposium, St. Louis, Missouri, 390 pp.

Kirby, R.E., P. Dupuis, and G.L. Hensler. 1987. Extent and consequences of recent changes in numbers of the American Black Duck and in hybridization between Black Ducks and Mallards. Review Draft. United States Department of the Interior, United States Fish and Wildlife Service, Washington, 194 pp.

Kirk, D.A. 1993. Status report of the Passenger Pigeon *Ectopistes migratorius* in Canada. Second draft, December 1993. Environment Canada, Canadian Wildlife Service, Ontario Region, Nepean, 18 pp.

Kirk, D.A. 1994a. Status report on the Labrador Duck *Camptorynchus labradorius* in Canada. Committee on the Status of Endangered Wildlife in Canada (COSEWIC), Ottawa, 24 pp.

Kirk, D.A. 1994b. Status report on the Passenger Pigeon *Ectopistes migratorius* in Canada. Committee on the Status of Endangered Wildlife in Canada (COSEWIC), Ottawa, 18 pp.

Kirk, D.A. 1994c. Status report on the Great Auk *Pinguinus impennis* in Canada. Committee on the Status of Endangered Wildlife in Canada (COSEWIC), Ottawa, 27 pp.

Kirkham, I.R. and D.N. Nettleship. 1986. Status report on the Roseate Tern in Canada (*Sterna dougallii*). Committee on the Status of Endangered Wildlife in Canada (COSEWIC), Ottawa, 44 pp.

Kirkham, I.R. and D.N. Nettleship. 1987. Status of Roseate Tern in Canada. Journal of Field Ornithology 58(4): 505-515.

Kitchen, D.W. and G.S. Hunt. 1969. Brood habitat of the Hooded Merganser. Journal of Wildlife Management 33(3): 605-609.

Klett, A.T., H.F. Duebbert, C.A. Faanes, and K.F. Higgins. 1986. Techniques for studying nest success of ducks in upland habitats in the prairie pothole region. United States Fish and Wildlife Service. Resource Publication no. 158, Washington, 24 pp.

Klimkiewicz, M.K. and A.G. Futcher. 1987. Longevity records of North American birds: Coerebinae through Estrildidae. Journal of Field Ornithology 58(3): 318-333.

Klimkiewicz, M.K. and A.G. Futcher. 1989. Longevity records of North American birds, supplement 1. Journal of Field Ornithology 60(4): 469-494.

Klimkiewicz, M.K., R.B. Clapp, and A.G. Futcher. 1983. Longevity records of North American birds: Remizidae through Parulinae. Journal of Field Ornithology 54(3): 287-294.

Klimstra, W.D. and W.O. Stieglitz. 1957. Notes on reproductive activities of Robins in Iowa and Illinois. Wilson Bulletin 69(4): 333-337.

Klingensmith, C.W. 1970. Breeding bird census no. 21: maturing upland red pine – white pine plantation. Audubon Field Notes 24: 754-755.

Knapton, R.W. 1979a. Breeding ecology of the Clay-colored Sparrow. Living Bird 17: 137-158.

Knapton, R.W. 1979b. Optimal size of territory in the Clay-colored Sparrow, *Spizella pallida*. Canadian Journal of Zoology 57(7): 1358-1370.

Knapton, R.W. 1980. Nestling foods and foraging patterns in the Clay-colored Sparrow. Wilson Bulletin 92(4): 458-465.

Knapton, R.W. 1982. The Henslow's Sparrow (*Ammodramus henslowii*) in Canada: a status report. Committee on the Status of Endangered Wildlife in Canada (COSEWIC), Ottawa, 77 pp.

Knapton, R.W. 1984a. Parental feeding of nestling Nashville Warblers: the effects of food type, brood-size, nestling age, and time of day. Wilson Bulletin 96(4): 594-602.

Knapton, R.W. 1984b. The Henslow's Sparrow in Ontario: a historical perspective. Ontario Birds 2: 70-74.

Knapton, R.W. 1987a. Clay-colored Sparrow, pp. 442-443 *in* M.D. Cadman, P.F.J. Eagles, and F.M. Helleiner, eds. Atlas of the breeding birds of Ontario. University of Waterloo Press, Waterloo, 617 pp.

Knapton, R.W. 1987b. Henslow's Sparrow, pp. 452-453 *in* M.D. Cadman, P.F.J. Eagles, and F.M. Helleiner, eds. Atlas of the breeding birds of Ontario. University of Waterloo Press, Waterloo, 617 pp.

Knapton, R.W. 1988a. Nesting success is higher for polygynously mated females than for monogamously mated females in the Eastern Meadowlark. Auk 105(2): 325-329.

Knapton, R.W. 1988b. Unusual Bank and Barn swallow nesting on Lake Winnipegosis, Manitoba. Blue Jay 46(1): 37.

Knapton, R.W. 1994. Clay-colored Sparrow (*Spizella pallida*), *in* A. Poole and F. Gill, eds. The birds of North America, no. 120. The Academy of Natural Sciences, Philadelphia. The American Ornithologists' Union, Washington, 16 pp.

Knapton, R.W. and J.B. Falls. 1982. Polymorphism in the White-throated Sparrow: habitat occupancy and nest-site selection. Canadian Journal of Zoology 60(3): 452-459.

Knapton, R.W. and J.B. Falls. 1983. Differences in parental contribution among pair types in the polymorphic White-throated Sparrow. Canadian Journal of Zoology 61: 288-292.

Knapton, R.W., R.V. Cartar, and J.B. Falls. 1984. A comparison of breeding ecology and reproductive success between morphs of the White-throated Sparrow. Wilson Bulletin 96(1): 60-71.

Knight, R.L. 1986. Responses of nesting ravens to people in areas of different human densities. Condor 86(3): 345-346.

Knight, R.L., D.J. Grout, and S.A. Temple. 1987. Nest-defense behavior of the American Crow in urban and rural areas. Condor 89: 175-177.

Knopf, F.L. and B.A. Knopf. 1983. Flocking pattern of foraging American Crows in Oklahoma. Wilson Bulletin 95(1): 153-155.

Knox, A.C. 1988. The taxonomy of redpolls. Ardea 76: 1-26.

Koenig, W.D. 1986. Geographical ecology of clutch size variation in North American woodpeckers. Condor 88(4): 499-504.

Kohler, L.S. 1915. Home life of the Scarlet Tanager. Oriole 3: 4-8.

Kopachena, J.G. and J.B. Falls. 1993. Postfledging parental care in the White-throated Sparrow (*Zonotrichia albicollis*). Canadian Journal of Zoology 71(2): 227-232.

Korpimäki, E. 1981. On the ecology and biology of Tengmalm's Owls (*Aegolius funereus*) in southern Ostrobothnia and Suomenselkä, western Finland. Acta Univertatis Ouluensis. Series A, no. 118, Biol. 13: 1-84.

Korpimäki, E. 1986. Niche relationships and life history tactics of three sympatric *Strix* owl species in Finland. Ornis Scandinavica 17: 126-132.

Korpimäki, E. 1987. Timing of breeding of Tengmalm's Owl *Aegolius funereus* in relation to vole dynamics in western Finland. Ibis 129: 58-68.

Korpimäki, E. 1989. Mating system and mate choice of Tengmalm's Owls *Aegolius funereus*. Ibis 131: 41-50.

Korpimäki, E. 1992. Diet composition, prey choice and breeding success of Long-eared Owls: effects of multiannual fluctuations in food abundance. Canadian Journal of Zoology 70(12): 2373-2381.

Korpimäki, E. and K. Norrdahl. 1991. Numerical and functional responses of kestrels, Short-eared Owls, and Long-eared Owls to vole densities. Ecology 72(3): 814-826.

Kortegaard, L. 1968. Studies on the breeding biology of the Red-breasted Merganser (*Mergus serrator*) in Vejlerne, North Jutland. Dansk Ornithologisk Forenings Tiddsskrift 62: 37-67.

Kowalski, M.P. 1983. Factors affecting the performance of flight songs and perch songs in the Common Yellowthroat. Wilson Bulletin 95(1): 140-142.

Kozlovic, D.R. 1987. House Finch, pp. 492-493 *in* M.D. Cadman, P.F.J. Eagles, and F.M. Helleiner, eds. Atlas of the breeding birds of Ontario. University of Waterloo Press, Waterloo, 617 pp.

Kramer, H.G., R.E. Lemon, and M.J. Morris. 1985. Song switching and agonistic stimulation in the Song Sparrow (*Melospiza melodia*): five tests. Animal Behaviour 33(1): 135-149.

Krapu, G.L. 1974. Feeding ecology of pintail hens during reproduction. Auk 91(2): 278-290.

Krapu, G.L. and G.A. Swanson. 1975. Some nutritional aspects of reproduction in prairie nesting pintails. Journal of Wildlife Management 39(1): 156-162.

Krauth, S. 1972. The breeding biology of the Common Gallinule. M.Sc. thesis, University of Wisconsin, Oshkosh, 74 pp.

Krebs, J.R. 1978. Colonial nesting in birds, with special reference to the Ciconiiformes, pp. 299-314 *in* A. Sprunt IV, J.C. Ogden, and S. Winckler, eds. Wading birds. Research Report. National Audubon Society, no. 7, New York, 381 pp.

Kricher, J.C. 1983. Correlation between House Finch increase and House Sparrow decline. American Birds 37(4): 358-360.

Kricher, J.C. 1995. Black-and-White Warbler (*Mniotilta varia*), *in* A. Poole and F. Gill, eds. The birds of North America, no. 158. The Academy of Natural Sciences, Philadelphia. The American Ornithologists' Union, Washington, 20 pp.

Kridelbaugh, A. 1983. Nesting ecology of the Loggerhead Shrike in central Missouri. Wilson Bulletin 95(2): 303-308.

Krohn, W.B. 1970. Woodcock feeding habits as related to summer field usage in central Maine. Journal of Wildlife Management 34(4): 769-775.

Krohn, W.B. 1971. Some patterns of woodcock activities on Maine summer fields. Wilson Bulletin 83(4): 396-407.

Krohn, W.B., P.O. Corr, and A.E. Hutchinson. 1992. Status of the American Eider with special reference to northern New England. United States Department of the Interior, Fish and Wildlife Service. Research Report no. 12, 12 pp.

Kroodsma, D.E. 1984. Songs of the Alder Flycatcher (*Empidonax alnorum*) and Willow Flycatcher (*Empidonax traillii*) are innate. Auk 101(1): 13-24.

Kroodsma, D.E. and L.D. Parker. 1977. Vocal virtuosity in the Brown Thrasher. Auk 94(4): 783-785.

Kroodsma, D.E., R.C. Bereson, B.E. Byers, and E. Minear. 1989. Use of song types by the Chesnut-sided Warbler: evidence for both intra- and inter-sexual functions. Canadian Journal of Zoology 67(2): 447-456.

Kroodsma, R.L. 1984. Effect of edge on breeding forest bird species. Wilson Bulletin 96(3): 426-436.

Krueger, H.O. 1981. Breeding adaptations of the Vesper Sparrow (*Pooecetes gramineus*) in a fire-altered ecosystem. M.Sc. thesis, Central Michigan University, Mount Pleasant, 71 pp.

Kruuk, H. 1964. Predators and anti-predator behaviour of the Black-headed Gull (*Larus ridibundus* L.). E.J. Brill, Leiden, 129 pp.

Kuerzi, R.G. 1941. Life history studies of the Tree Swallow. Proceedings. Linnean Society of New York 52-53: 1-52.

Kuitunen, M. and P. Helle. 1988. Relationship of the Common Treecreeper (*Certhia familiaris*) to edge effect and forest fragmentation. Ornis Fennica 65: 150-155.

Kuitunen, M. and T. Törmälä. 1983. The food of Treecreeper *Certhia f. familiaris* nestlings in southern Finland. Ornis Fennica 60: 42-44.

Künstler, D.S. 1993. 23. Mature urban deciduous forest I, pp. 47-48 *in* J.D. Lowe, ed. Resident bird counts 1992. Journal of Field Ornithology 64(1)(supplement): 1-128.

Kurzejeski, E.W. and J.B. Lewis. 1990. Home ranges, movements, and habitat use of Wild Turkey hens in northern Missouri, pp. 67-71 *in* W.M. Healy and G.B. Healy. Proceedings. Sixth national Wild Turkey symposium. National Wild Turkey Federation, Edgefield, 228 pp.

Kushlan, J.A. 1973. Least Bittern nesting colonially. Auk 90(3): 685-686.

Kushlan, J.A. 1976. Feeding behavior of North American herons. Auk 93(1): 86-94.

Labedz, T.E. 1984. Age and reproductive success in Northern Orioles. Wilson Bulletin 96(2): 303-305.

Lack, D. 1944. Early references to territory in bird-life. Condor 46(1): 108-111.

Lack, D. 1945. Possible life-pairing of nuthatch. British Birds 38(15): 297.

Lack, D. 1968. Ecological adaptations for breeding in birds. Chapman and Hall, London, 409 pp.

Lacombe, D., A. Cyr, and J.-M. Bergeron. 1986. Effects of the chemosterilant Ornitrol on the nesting success of Red-winged Blackbirds. Journal of Applied Ecology 23: 773-779.

Lacombe, D., A. Cyr, and P. Matton. 1990. Spermatogenesis in photostimulated Red-winged Blackbirds (*Agelaius phoeniceus*) chronically treated with a potent gonadotropin releasing hormone analogue. Canadian Journal of Zoology 68(12): 2518-2523.

Lafontaine, P. 1988. Évaluation des propriétés de la méthode parcellaire (6,25 ha). M.Sc. thesis, Université du Québec à Montréal, 29 pp.

Lagarec, D. 1974. Le relief du Québec, pp. 7-14 *in* Annuaire du Québec. Ministère de l'Industrie et du Commerce and Bureau de la Statistique du Québec, 1224 pp.

Lagrenade, M.-C. and P. Mousseau. 1981. Alimentation des poussins de Goélands à bec cerclé de l'île de la Couvée, Québec. Naturaliste canadien 108(2): 131-138.

Lahontan, L.A. 1703. Nouveaux voyages de Mr le baron de Lahontan, dans l'Amérique Septentrionale. Chez les Frères l'Honoré, La Haye, vol. 1, 279 pp.; vol. 2, 220 pp.

Laine, H. 1981. Male participation in incubation and brooding in the Blue Jay. Auk 98(3): 622-623.

Laine, H. 1983. Parental chores distribution in the monogamous Blue Jay. American Zoologist 23(4): 930.

Lajoie, P.G. 1975. Les terres agricoles du Québec méridional: distribution, étendue et qualité. Centre de recherche sur les terres, Direction générale de la recherche, Agriculture Canada, Ottawa. Publication no. 1556, 62 pp. + 1 map, scale 1:2 500 000.

Lalumière, R. and C. Lemieux. 1993. Guide de transplantation de la Zostère marine. Report for Canadian Wildlife Service, Environment Canada, 16 pp.

Lalumière, R., L. Belzile, and C. Lemieux. 1992. Étude de la Zostère marine le long de la côte nord-est de la baie James. Report for Société d'Énergie de la Baie James, by Groupe Environnement Shooner, Montréal, 31 pp.

Lambert, A.B. 1987. Prairie Warbler, pp. 390-391 *in* M.D. Cadman, P.F.J. Eagles, and F.M. Helleiner, eds. Atlas of the breeding birds of Ontario. University of Waterloo Press, Waterloo, 617 pp.

Lambert, A.B. and R.B.H. Smith. 1985. Status report of the Prairie Warbler (*Dendroica discolor*). Committee on the Status of Endangered Wildlife in Canada (COSEWIC), Ottawa, 318 pp.

Lamontagne, R. 1960a. Les échanges scientifiques entre Roland-Michel Barrin de La Galissonière et les chercheurs contemporains. Revue d'Histoire de l'Amérique Française 13(1): 25-33.

Lamontagne, R. 1960b. La contribution scientifique de La Galissonière au Canada. Revue d'Histoire de l'Amérique Française 13(4): 509-524.

Lamothe, P. 1973. Biology of King Eider (*Somateria spectabilis*) in a fresh water breeding area on Bathurst Island, Northwest Territories. M.Sc. thesis, University of Alberta, Edmonton, 125 pp.

Lamothe, P. 1978. Étude préliminaire des rivières de l'Ungava. Description des populations de sauvagine. Hydro-Québec, Direction de l'environnement, 24 pp.

Lamothe, P. 1982. Synthèse des études sur la Bernache du Canada des bassins de la Grande rivière de la Baleine et de la Petite rivière de la Baleine (1975-1980) et notes sur les autres espèces d'Ansérinés. Hydro-Québec Direction Environnement, technical report, 55 pp.

Lamoureux, G. and Collaborators. 1996. Plantes sauvages printanières, Guide d'identification Fleurbec, 2e éd. Éditions Fleurbec. Saint-Henri-de-Lévis, Québec.

Land Resource Research Centre. 1990. Pédopaysages du Canada. Québec-Sud-Ouest. Contribution 89-02, Research Branch, Agriculture Canada, Ottawa, 37 pp.

Landres, P.B. 1992. Temporal scale perspectives in managing biological diversity. Transactions. North American Wildlife and Natural Resources Conference 57: 292-307.

Landry, B. and M. Mercier. 1983. Notions de géologie. Éditions Modulo, Outremont, 437 pp.

Lands Directorate. 1986. Milieux humides le long du fleuve Saint Laurent, 1950-1978/Wetlands of the St. Lawrence River region, 1950-1978. Environment Canada, Environmental Conservation Service, Québec Region. Working Paper no. 45, 29 pp.

Lane, J. 1962. Nesting of the Yellow Rail in southwestern Manitoba. Canadian Field-Naturalist 76: 189-191.

Lane, P. 1980. Les oiseaux d'hiver au Québec. Héritage, Montréal, 110 pp.

Lane, P. 1987. L'alimentation des oiseaux. Éditions Broquet, Ottawa, 181 pp.

Lane, P.A. and J.R. Duncan. 1987. Observations of Northern Hawk Owls nesting in Roseau County. Loon 59(4): 165-174.

Lang, A. 1990. Status report on the American Coot (*Fulica americana*). Committee on the Status of Endangered Wildlife in Canada (COSEWIC), Ottawa, 23 pp.

Lanyon, W.E. 1956. Ecological aspects of the sympatric distribution of meadowlarks in north-central states. Ecology 37: 98-108.

Lanyon, W.E. 1957. The comparative biology of the meadowlarks (*Sturnella*) in Wisconsin. Nuttall Ornithological Club, no. 1, Cambridge, Massachusetts, 67 pp.

Lanyon, W.E. 1966. Hybridization in meadowlarks. American Museum of Natural History. Bulletin 134: 1-25.

Lanyon, W.E. 1979. Hybrid sterility in meadowlarks. Nature 279: 557-558.

Lanyon, W.E. 1994. Western Meadowlark (*Sturnella neglecta*), *in* A. Poole and F. Gill, eds. The birds of North America, no. 104. The Academy of Natural Sciences, Philadelphia. The American Ornithologists' Union, Washington, 20 pp.

Lanyon, W.E. 1995. Eastern Meadowlark (*Sturnella magna*), *in* A. Poole and F. Gill, eds. The birds of North America, no. 160. The Academy of Natural Sciences, Philadelphia. The American Ornithologists' Union, Washington, 24 pp.

Laperle, M. 1969. A waterfowl ecological nesting study at Îles-de-la-Paix, Lake St. Louis, Québec. Canadian Wildlife Service, Annual Progress Report, Project no. 360, 65 pp.

Laperle, M. 1974. Effects of water level fluctuations on duck breeding success, pp. 18-30 *in* H. Boyd, ed. Canadian Wildlife Service waterfowl studies in eastern Canada, 1969-73/Études du SCF sur les oiseaux aquatiques dans l'est du Canada, 1969-73. Canadian Wildlife Service, Ottawa. Report Series no. 29, 106 pp.

Lapierre, B. 1991. Granby "dortoir" de 20 000 corneilles, du jamais vu au Québec. La Voix de l'Est, samedi 2 mars 1991, p. 4.

Laplante, G. 1979. Inventaire de la nidification du Grand Héron (*Ardea herodias*), du Bihoreau à couronne noire (*Nycticorax nycticorax*) et du Goéland argenté (*Larus argentatus*) sur l'île Steamboat, comté de Laviolette. Ministère du Loisir, de la Chasse et de la Pêche du Québec, 29 pp.

Laplante, J.-P. 1985. Papillons et chenilles du Québec et de l'est du Canada. Éditions France-Amérique, Montréal, 280 pp.

LaPointe, G. and J. Bédard. 1986. Savannah Sparrow, *Passerculus sandwichensis*, reproductive success. Canadian Field-Naturalist 100(2): 264-267.

Lapointe, J. 1986. Présence de projectiles dans des gésiers de différentes espèces de canards en migration dans la région de Rouyn-Noranda, Québec. Ministère du Loisir, de la Chasse et de la Pêche, Service de l'Aménagement et de l'Exploitation de la Faune, Rouyn-Noranda, 9 pp.

Laporte, P. 1974a. L'ornithologie au Canada à l'époque de Jacques Cartier 1500-1600. Carnets de zoologie 34(4): 52-58.

Laporte, P. 1974b. Common Grackle kills a Barn Swallow. Wilson Bulletin 86(4): 477-478.

Laporte, P. 1982. Organochlorine residues and eggshell measurements of Great Blue Heron eggs from Québec. Colonial Waterbirds 5: 95-103.

Laporte, P. and F. Shaffer. 1994. Recensement du Pluvier siffleur au Québec, en 1991, pp. 24-29 *in* S.P. Flemming, ed. The 1991 International Piping Plover Census in Canada. Canadian Wildlife Service. Occasional Paper no. 82, 59 pp.

Larivée, J. 1989. Variation des observations d'oiseaux du Québec méridional de 1969 à 1988. Carnets de zoologie 49(4): 83-90.

Larivée, J. 1991. Le Moineau domestique est de moins en moins observé au Québec. QuébecOiseaux 2(4): 24-25.

Larivée, J. 1993. Chronobiologie des oiseaux du Bas-Saint-Laurent: migration et reproduction. Club des ornithologues du Bas-Saint-Laurent, 160 pp.

L'Arrivée, L.P. and H. Blokpoel. 1990. Seasonal distribution and site tenacity of Black-crowned Night-Herons, *Nycticorax nycticorax*, banded in Canada. Canadian Field-Naturalist 104(4): 534-539.

Larson, D.L. and C.E. Bock. 1986. Eruptions of some North American boreal seed-eating birds, 1901-1980. Ibis 128: 137-140.

Larson, J.S. 1991. North America, pp. 57-84 *in* M. Finlays and M. Moser, eds. Wetlands. Toucan Books, London, 224 pp.

Larsson, K., P. Forslund, L. Gustafsson, and B.S. Ebbinge. 1988. From the high Arctic to the Baltic: the successful establishment of a Barnacle Goose *Branta leucopsis* population on Gotland, Sweden. Ornis Scandinavica 19: 182-189.

Lasiewski, R.C. 1962. The energetics of migrating hummingbirds. Condor 64: 324.

Laskey, A.R. 1939. A study of nesting Bluebirds. Bird-Banding 10: 23-32.

Laskey, A.R. 1944. A study of the Cardinal in Tennessee. Wilson Bulletin 56(1): 27-44.

Laskey, A.R. 1948. Some nesting data on the Carolina Wren at Nashville, Tennessee. Bird-Banding 19(3): 101-121.

Laskey, A.R. 1950. Cowbird behavior. Wilson Bulletin 62: 157-174.

Laskey, A.R. 1957. Some Tufted Titmouse life history. Bird-Banding 28: 135-145.

Laskey, A.R. 1958. Blue Jays at Nashville, Tennessee: movements, nesting, age. Bird-Banding 29(4): 211-218.

Laskey, A.R. 1962. Breeding biology of mockingbirds. Auk 79(4): 596-606.

Lasley, G.W. and C. Sexton. 1993. The autumn migration: Texas region. American Birds 47(1): 115-121.

Latham, J. 1781-1785. A general synopsis of birds. Leigh and Sotheby, London. Vol. I, parts 1, 2; vol. II, parts 1, 2; vol. III, parts 1, 2.

Latham, J. 1809. *J. Lathami* Systema Ornithologiae, sive index ornithiligicus (editio nova). Paris, 444 pp.

Latta, C.W. and R.F. Sharkey. 1966. Feeding behavior of the American Merganser in captivity. Journal of Wildlife Management 30(1): 17-23.

Laughlin, S.B. 1985a. Eastern Screech-Owl, pp. 132-133 *in* S.B. Laughlin and D.P. Kibbe, eds. The Atlas of breeding birds of Vermont. Vermont Institute of Natural Science, University Press of New England, Hanover, 456 pp.

Laughlin, S.B. 1985b. Great Horned Owl, pp. 134-135 *in* S.B. Laughlin and D.P. Kibbe, eds. The Atlas of breeding birds of Vermont. Vermont Institute of Natural Science, University Press of New England, Hanover, 456 pp.

Laughlin, S.B. 1985c. Wild Turkey, pp. 94-95 *in* S.B. Laughlin and D.P. Kibbe, eds. The Atlas of breeding birds of Vermont. Vermont Institute of Natural Science, University Press of New England, Hanover, 456 pp.

Laughlin, S.B. and D.P. Kibbe. 1985. The Atlas of breeding birds of Vermont. Vermont Institute of Natural Science, University Press of New England, Hanover, 456 pp.

Laurenzi, A.W., B.W. Anderson, and R.D. Ohmart. 1982. Wintering biology of Ruby-crowned Kinglets in the lower Colorado River valley. Condor 84(4): 385-398.

Lauriault, J. 1987. Guide d'identification des arbres du Canada. National Museums of Canada. Éditions Marcel Broquet, La Prairie, 551 pp.

Laurin, S. 1989. Histoire des Laurentides. Institut québécois de recherche sur la culture, Québec, 892 pp.

Lawrence, L. de K. 1948. Comparative study of the nesting behavior of Chestnut-sided and Nashville warblers. Auk 65: 204-219.

Lawrence, L. de K. 1949. The Red Crossbill at Pimisi Bay, Ontario. Canadian Field-Naturalist 63: 147-160.

Lawrence, L. de K. 1952. New sight records of three species at Pimisi Bay, Ontario. Canadian Field-Naturalist 66(2): 67.

Lawrence, L. de K. 1953a. Nesting life and behavior of the Red-eyed Vireo. Canadian Field-Naturalist 67: 46-77.

Lawrence, L. de K. 1953b. Notes on the nesting behavior of the Blackburnian Warbler. Wilson Bulletin 65(3): 135-144.

Lawrence, L. de K. 1955. Is the sapsucker destructive? Nature Magazine 48(9): 487-488, 500.

Lawrence, L. de K. 1967. A comparative life-history study of four species of woodpeckers. The American Ornithologists' Union, Washington. Ornithological Monographs no. 5, 156 pp.

Lawrence, L. de K. 1974. Whip-poor-will. Hinterland Who's Who. Environment Canada, Canadian Wildlife Service, 4 pp.

Laymon, S.A. and M.D. Halterman. 1989. A proposed habitat management plan for Yellow-billed Cuckoos in California, pp. 272-277 *in* D.L. Abell, ed. Proceedings. California riparian systems conference: protection, management, and restoration for the 1990's. United States Department of Agriculture, Berkeley, California. General Technical Report PSW-110, 544 pp.

Lea, R.B. 1942. A study of the nesting habits of the Cedar Waxwing. Wilson Bulletin 54(4): 225-227.

Leahy, C. 1982. The birdwatcher's companion. An Encyclopedic Handbook of North American Birdlife. Hill and Wang, New York, 917 pp.

Leak, W.B. 1987. Fifty years of compositional change in deciduous and coniferous forest types in New Hampshire. Canadian Journal of Forest Research 17: 388-393.

LeBaron, G.S. 1987. The 88th Christmas Bird Count. American Birds 42(4): 503-1198.

le Bel, J.-P. 1990. Quelques observations du Canard arlequin (*Histrionicus histrionicus*) sur les rivières gaspésiennes, été 1990. Unpublished report, Ministère du Loisir, de la Chasse et de la Pêche, direction régionale du Bas-Saint-Laurent-Gaspésie–Îles-de-la-Madeleine, Rimouski, 6 pp.

Leberman, R.C. 1992a. American Bittern, pp. 46-47 *in* D.W. Brauning, ed. Atlas of breeding birds in Pennsylvania. University of Pittsburgh Press, Pittsburgh, 484 pp.

Leberman, R.C. 1992b. Black Tern, pp. 144-145 *in* D.W. Brauning, ed. Atlas of breeding birds in Pennsylvania. University of Pittsburgh Press, Pittsburgh, 484 pp.

Leberman, R.C. 1992c. Mourning Warbler, pp. 348-349 *in* D.W. Brauning, ed. Atlas of breeding birds in Pennsylvania. University of Pittsburgh Press, Pittsburgh, 484 pp.

Leberman, R.C. 1992d. Ovenbird, pp. 340-341 *in* D.W. Brauning, ed. Atlas of breeding birds in Pennsylvania. University of Pittsburgh Press, Pittsburgh, 484 pp.

Leberman, R.C. 1992e. Pileated Woodpecker, pp. 192-193 *in* D.W. Brauning, ed. Atlas of breeding birds in Pennsylvania. University of Pittsburgh Press, Pittsburgh, 484 pp.

Leberman, R.C. 1992f. Swamp Sparrow, pp. 390-391 *in* D.W. Brauning, ed. Atlas of breeding birds in Pennsylvania. University of Pittsburgh Press, Pittsburgh, 484 pp.

Leberman, R.C. 1992g. Prothonotary Warbler, pp. 334-335 *in* D.W. Brauning, ed. Atlas of breeding birds in Pennsylvania. University of Pittsburgh Press, Pittsburgh, 484 pp.

Leck, C. 1989. Sunbathing in the Brown Creeper. Western Birds 20: 91-92.

LeCroy, M. and C.T. Collins. 1972. Growth and survival of Roseate and Common tern chicks. Auk 89: 595-611.

Lee, D. and J.M. Speirs. 1977. Breeding bird censuses at Big Trout Lake, 1975. Ontario Field Biologist 31(2): 48-54.

Lee, D.S. 1989. Jaegers and skuas in the western North Atlantic: some historical misconceptions. American Birds 43(1): 18-20.

Lee, D.S. 1990. Nesting of Golden Eagles in the central and southern Appalachians. Wilson Bulletin 102(4): 693-698.

Lefebvre, J. and J.-F. Giroux. 1993. Étude du régime alimentaire du Goéland à bec cerclé à la Réserve nationale de faune des îles de Contrecoeur. Université du Québec à Montréal, Montréal, 17 pp.

Lefebvre, L. 1985. Stability of flock composition in urban pigeons. Auk 102(4): 886-888.

Lefebvre, P.W. and J.L. Seubert. 1970. Surfactants as blackbird stressing agents. Proceedings. Vertebrate Pest Conference 4: 156-161.

Leffelaar, D. and R.J. Robertson. 1984. Do male Tree Swallows guard their mates? Behavioral Ecology and Sociobiology 16: 73-79.

Leffelaar, D. and R.J. Robertson. 1986. Equality of feeding roles and the maintenance of monogamy in Tree Swallows. Behavioral Ecology and Sociobiology 18: 199-206.

Léger, C. and R. McNeil. 1987. Brood size and chick position as factors influencing feeding frequency, growth, and survival of nestling Double-crested Cormorants, *Phalacrocorax auritus*. Canadian Field-Naturalist 101(3): 351-361.

Legris, A.M. and H. Lévesque. 1991a. Prises d'oiseaux migrateurs, au Canada, pendant la saison de chasse de 1990. Canadian Wildlife Service. Progress Notes no. 197, 41 pp. [Also available in English as: Migratory game birds harvested in Canada during the 1990 hunting season.]

Legris, A.M. and H. Lévesque. 1991b. Prises d'oiseaux migrateurs, au Canada, pendant la saison de chasse de 1989. Canadian Wildlife Service. Progress Notes no. 194, 43 pp. [Also available in English as: Migratory game birds harvested in Canada during the 1989 hunting season.]

Lehoux, D. 1972. La nourriture disponible et la préférence des jeunes Canards noirs. Travail de fin d'études, Université Laval, Sainte-Foy, 38 pp.

Lehoux, D. 1978. La situation de la Sarcelle à ailes vertes (*Anas crecca carolinensis*) au Québec. Environment Canada, Canadian Wildlife Service, Québec Region, 14 pp.

Lehoux, D. and A. Bourget. 1981. La sauvagine et le développement industriel dans la région Gentilly–Bécancour. Environment Canada, Canadian Wildlife Service, Québec Region, 19 pp.

Lehoux, D. and A. Bourget. 1987. Le Saint-Laurent: un lieu de rencontre pour la sauvagine. Environment Canada, Canadian Wildlife Service, Québec Region, 5 pp.

Lehoux, D., R. Ouellet, and P. Laporte. 1981. Étude comparative des différentes méthodes de surveillance des oiseaux forestiers suite à des épandages expérimentaux d'insecticides [A comparative study of various methods for monitoring forest bird populations following experimental insecticide sprays]. Environment Canada, Canadian Wildlife Service, and Ministère du Loisir, de la Chasse et de la Pêche, 108 pp.

Lehoux, D., A. Bourget, M. Darveau, J. Bourgeois, and J.-C. Bourgeois. 1983. Abondance, distribution et chronologie de migration des oiseaux aquatiques au lac Saint-Pierre. Environment Canada, Canadian Wildlife Service, and Ministère du Loisir, de la Chasse et de la Pêche, 77 pp.

Lehoux, D., A. Bourget, P. Dupuis, and J. Rosa. 1985. La sauvagine dans le système du Saint-Laurent (fleuve, estuaire, golfe). Environment Canada, Canadian Wildlife Service, Québec Region, vol. 1, 76 pp.; vol. 2 (appendix), 72 pp. [An abstract is available in English.]

Lehtonen, L. 1970. Biology of the Black-throated Diver, *Gavia a. arctica* (L.). Annales Zoologici Fennici 7: 25-60.

Leighton, F.A., J.M. Gerrard, P. Gerrard, D.W.A. Whitfield, and W.J. Maher. 1979. An aerial survey of Bald Eagles in Saskatchewan. Journal of Wildlife Management 43(1): 61-69.

Lein, M.R. 1978. Song variation in a population of Chestnut-sided Warblers (*Dendroica pensylvanica*): its nature and suggested significance. Canadian Journal of Zoology 56(6): 1266-1283.

Lein, M.R. 1981. Display behavior of Ovenbirds (*Seiurus aurocapillus*). II. Song variation and singing behavior. Wilson Bulletin 93(1): 21-41.

Lein, M.R. and G.A. Webber. 1979. Habitat selection by wintering Snowy Owls (*Nyctea scandiaca*). Canadian Field-Naturalist 93(2): 176-178.

Lemay, A.B., R. McNicoll, and R. Ouellet. 1989. Incidence de la grenaille de plomb dans les gésiers de canards, d'oies et de bernaches récoltés au Québec. Ministère du Loisir, de la Chasse et de la Pêche, Direction de la gestion des espèces et des habitats, 87 pp.

Lemay, Y. 1989. Caractérisation de l'habitat de reproduction du Tétras du Canada (*Dendragapus canadensis*) sur l'île d'Anticosti. M.Sc. thesis, Université du Québec à Trois-Rivières, Trois-Rivières, 133 pp.

Lemieux, L. 1959a. Histoire naturelle et aménagement de la Grande Oie blanche, *Chen hyperborea atlantica*. Naturaliste canadien 86(8-9): 133-192.

Lemieux, L. 1959b. The breeding biology of the Greater Snow Goose on Bylot Island, Northwest Territories. Canadian Field-Naturalist 73: 117-128.

Lemmetyinen, R. 1972. Growth and mortality in the chicks of Arctic Terns in the Kangsford area, Spitsbergen in 1970. Ornis Fennica 49: 45-53.

Lemmetyinen, R. 1973. Feeding ecology of *Sterna paradisaea* Pontop. and *S. hirundo* L. in the archipelago of southwestern in Finland. Annales Zoologici Fennici 10: 507-525.

Le Moine, J.M. 1860. Ornithologie du Canada; quelques groupes d'après la nomenclature de Baird. E.R. Fréchette, Québec, vol. 1, 100 pp.; vol. 2, 398 pp.

Le Moine, J.M. 1861. Ornithologie du Canada d'après la nomenclature du Smithsonian Institution de Washington, 2nd edition, 2nd part. Atelier typographique J.T. Brousseau, Québec.

Le Moine, J.M. 1864. Tableau synoptique de l'ornithologie du Canada: classification et nomenclature du Smithsonian Institution de Washington. Des Presses L. Brousseau, Québec, 24 pp.

Le Moine, J.M. 1866. On the birds of Canada. Transactions. Quebec Literary and Historical Society, pp. 1-27, pp. 121-145.

Le Moine, J.M. 1874. Ornithologie du Canada. Conférence populaire lue devant l'Institut canadien de Québec, le 20 novembre 1874, 27 pp.

LeMoine, J.M. 1882. Birds notes. Canadian Sportsman and Naturalist 1: 143-144.

Le Moine, J.M. 1889. Lettres familières sur les oiseaux du Canada. R. Renault, Lowell, Massachusetts.

Le Moine, J.M. 1892. The birds of Quebec. Canadian Record of Science, pp. 157-164.

Le Moine, J.M. 1901. The avi-fauna of the province of Quebec. Transactions. Royal Society of Canada, 2nd Series, 1901-1902, 7: 73-76.

Lemon, R.E. and C. Chatfield. 1973. Organization of song of Rose-breasted Grosbeaks. Animal Behaviour 21(1): 28-44.

Lemon, R.E., S. Monette, and D. Roff. 1987. Song repertoires of American warblers (Parulinae): honest advertising or assessment? Ethology 74: 265-284.

Lenington, S. 1980. Bi-parental care in Killdeer: an adaptive hypothesis. Wilson Bulletin 92(1): 8-20.

Lenington, S. and T. Mace. 1975. Mate fidelity and nesting site tenacity in the Killdeer. Auk 92(1): 149-151.

Lent, R.A. 1992. Variation in Gray Catbird nest morphology. Journal of Field Ornithology 63(4): 411-419.

Leonard, M.L. and J. Picman. 1987a. Nesting mortality and habitat selection by Marsh Wrens. Auk 104(3): 491-495.

Leonard, M.L. and J. Picman. 1987b. The adaptive significance of multiple nest building by male Marsh Wrens. Animal Behaviour 35: 271-277.

Leonard, M.L., A.G. Horn, and S.F. Eden. 1988. Parent-offspring aggression in Moorhens. Behavioral Ecology and Sociobiology 23(4): 265-270.

Leopold, A. 1933. Game management. Charles Scribner's Sons, New York, 481 pp.

Leopold, F. 1951. A study of nesting Wood Ducks in Iowa. Condor 53(5): 209-220.

Lepage, D. 1993. L'observation des oiseaux en Estrie: les meilleurs sites, les périodes favorables. Société de loisir ornithologique, Sherbrooke, 290 pp.

Lepage, M. 1973. Survie et croissance du Canard noir (*Anas rubripes*) durant la période d'élevage. M.Sc. thesis, Université Laval, Sainte-Foy, 128 pp.

Lepage, M. 1975. Les oiseaux migrateurs dans le couloir fluvial, entre Montréal et le lac Saint-Pierre. Technical report. Service de l'Aménagement de la Faune, district de Montréal, Ministère du Tourisme, de la Chasse et de la Pêche, Québec, 22 pp.

Lepage, M. 1976. Utilisation des îles de Boucherville par les oiseaux migrateurs. Technical Report. Service de l'Aménagement de la Faune, district de Montréal, Ministère du Tourisme, de la Chasse et de la Pêche, Québec, 46 pp.

Lepthien, L.W. and C.E. Bock. 1976. Winter abundance patterns of North American kinglets. Wilson Bulletin 88(3): 483-485.

Lescarbot, M. 1928. Nova Francia, a description of Acadia, 1606. Translated by P. Erondelle, 1609. George Routledge and Sons, London, 346 pp.

Lester, A.N. 1980. Numerical response of woodpeckers and their effect on mortality of mountain pine beetles in lodgepole pine in northwestern Montana. M.Sc. thesis, University of Montana, Missoula, 103 pp.

Létourneau, V. 1983. Tilu-tilu...tio-tio-tio-tio-tio...dribriure. Jaseur 3(3): 149-152.

Lett, D.W. and D.M. Bird. 1987. Postfledging behavior of American Kestrels in southwestern Quebec. Wilson Bulletin 99(1): 77-82.

Levasseur, J.-M. and M. Letendre. 1979. Observations ornithologiques dans le territoire de la Baie James (1974-1977) et le long du Caniapiscau-Koksoak (1977). Société de Développement de la Baie James, Environnement et Aménagement du territoire, 52 pp.

Léveillé, M., L.M. Soyez, J. Brisebois, J. Leclerc, R. Ouellet, and R. McNicoll. (In preparation). Inventaire aérien de la sauvagine au printemps 1990 sur le fleuve Saint-Laurent, la rivière des Outaouais, le lac des Deux-Montagnes, la rivière Richelieu et la Baie Missisquoi. Ministère du Loisir, de la Chasse et de la Pêche, Service de l'aménagement et de l'exploitation de la faune, Direction régionale de Montréal.

Lévesque, H. and R. McNeil. 1985. Abondance et activités du Pigeon biset, *Columba livia*, dans le port de Montréal, Québec. Canadian Field-Naturalist 99(3): 343-355.

Lévesque, H. and R. McNeil. 1986. Déplacements du Pigeon biset (*Columba livia*) dans le Vieux-Port de Montréal. Naturaliste canadien 113: 47-54.

Lévesque, H., B. Collins, and A.M. Legris. 1993. Prises d'oiseaux migrateurs au Canada, pendant la saison de chasse de 1991. Canadian Wildlife Service. Progress Notes no. 204, 42 pp. Also available in English as: Migratory game birds harvested in Canada during the 1991 hunting season.

Levine, E. 1988a. Brown Creeper, pp. 298-299 *in* R.F. Andrle and J.R. Carroll, eds. The Atlas of breeding birds in New York State. Cornell University Press, Ithaca, 551 pp.

Levine, E. 1988b. Great Crested Flycatcher, pp. 260-261 *in* R.F. Andrle and J.R. Carroll, eds. The Atlas of breeding birds in New York State. Cornell University Press, Ithaca, 551 pp.

Levine, E. 1988c. Solitary Vireo, pp. 344-345 *in* R.F. Andrle and J.R. Carroll, eds. The Atlas of breeding birds in New York State. Cornell University Press, Ithaca, 551 pp.

Lewis, H.F. 1920. Changes in the status of certain birds in the vicinity of Quebec, P.Q. Canadian Field-Naturalist 34: 132-136.

Lewis, H.F. 1921a. Nesting of the Philadelphia Vireo. Auk 38: 26-44.

Lewis, H.F. 1921b. The Greater Snow Goose. Canadian Field-Naturalist 39: 35.

Lewis, H.F. 1922. Notes on some Labrador birds. Auk 39: 507-516.

Lewis, H.F. 1924. List of birds recorded from the island of Anticosti, Québec. Canadian Field-Naturalist 38(3): 43-46; (4): 72-75; (5): 88-90; (7): 125-127; (8): 146-147.

Lewis, H.F. 1925a. Notes on birds of the Labrador Peninsula in 1923. Auk 42: 74-86.

Lewis, H.F. 1925b. The new bird sanctuaries in the Gulf of St. Lawrence. Canadian Field-Naturalist 39: 177-179.

Lewis, H.F. 1927. Notes. Canadian Field-Naturalist 41: 189-190.

Lewis, H.F. 1929. The natural history of the Double-crested Cormorant. Ph.D. dissertation, Cornell University, Ithaca, 94 pp.

Lewis, H.F. 1931. The relation of Canada Geese and Brant to commercial gathering of eel-grass in the St. Lawrence estuary. Canadian Field-Naturalist 45(2): 57-62.

Lewis, H.F. 1934a. Notes on some details of the explorations by Jacques Cartier in the Gulf of St. Lawrence. Transactions. Royal Society of Canada, Section 2: 117-148.

Lewis, H.F. 1934b. Some observations indicating the northeastward extension of the range of the starling. Auk 51: 88-89.

Lewis, H.F. 1937. Migrations of the American Brant (*Branta bernicla hrota*). Auk 54: 73-95.

Lewis, H.F. 1941a. Breeding European cormorants of North America. Auk 58(3): 360-363.

Lewis, H.F. 1941b. Ring-billed Gulls of the Atlantic coast. Wilson Bulletin 53(1): 22-30.

Lewis, H.F. 1942. Fourth census of non-passerine birds in the bird sanctuaries of the North Shore of the Gulf of St. Lawrence. Canadian Field-Naturalist 56: 5-8.

Lewis, H.F. and C. Cottam. 1954. Eelgrass status and environmental relations. Journal of Wildlife Management 18(4): 449-460.

Lewis, R.A. 1975a. Reproductive biology of the White-crowned Sparrow (*Zonotrichia leucophrys pugetensis* Grinnell). I. Temporal organization of reproductive and associated cycles. Condor 77: 46-59.

Lewis, R.A. 1975b. Reproductive biology of the White-crowned Sparrow. II. Environmental control of reproductive and associated cycles. Condor 77: 111-124.

Librairie Larousse. 1982. Grand dictionnaire encyclopédique Larousse. Librairie Larousse, ed., Paris, 10 vol., 11038 pp.

Lien, J. 1975. Aggression between Great Black-backed Gulls and Bald Eagles. Auk 92(3): 584-585.

Lightbody, J.P. and P.J. Weatherhead. 1987a. Polygyny in the Yellow-headed Blackbird: female choice versus male competition. Animal Behaviour 35(6): 1670-1680.

Lightbody, J.P. and P.J. Weatherhead. 1987b. Interactions among females in polygynous Yellow-headed Blackbirds. Behavioral Ecology and Sociobiology 21(1): 23-30.

Limpert, R.J. and S.L. Earnst. 1994. Tundra Swan (*Cygnus columbianus*), *in* A. Poole and F. Gill, eds. The birds of North America, no. 89. The Academy of Natural Sciences, Philadelphia. The American Ornithologists' Union, Washington, 20 pp.

Lincer, J.L. 1975. DDE – Induced eggshell-thinning in the American Kestrel: a comparison of the field situation and laboratory results. Journal of Applied Ecology 12: 781-793.

Lincoln, F.C. 1924. A "Territory" note on the Belted Kingfisher. Wilson Bulletin 36: 113-115.

Link, M. 1980. A northern Green Heron colony. Loon 52(3): 99-101.

Livezey, B.C. 1981a. Locations and success of duck nests evaluated through discriminant analysis. Wildfowl 32: 23-37.

Livezey, B.C. 1981b. Duck nesting in retired croplands at Horicon National Wildlife Refuge, Wisconsin. Journal of Wildlife Management 45(1): 27-37.

Livezey, B.C. 1995. Phylogeny and evolutionary ecology of modern seaducks (Anatidae: Mergini). Condor 97: 233-255.

Livory, A. 1985. Essai sur les noms français des oiseaux d'Europe et sur leur étymologie. Groupe ornithologique Normand, Université de Caen, France, 330 pp.

Lloyd, C., M.L. Tasker, and K. Partridge. 1991. The Status of seabirds in Britain and Ireland. T. and A.D. Poyser, eds., London, 355 pp.

Lloyd, C.S. 1977. The ability of the Razorbill *Alca torda* to raise an additional chick to fledging. Ornis Scandinavica 8(2): 155-159.

Lloyd, C.S. 1979. Factors affecting breeding of Razorbills *Alca torda* on Skokholm. Ibis 121(2): 165-176.

Lloyd, H. 1923. The birds of Ottawa, 1923: revised to March 20, 1923. Canadian Field-Naturalist 37(6): 101-105; (7): 125-127; (8): 151-156.

Lloyd, H. 1924. The birds of Ottawa, 1923: revised to March 20, 1923. Canadian Field-Naturalist 38(1): 10-16.

Lloyd, H. 1944. The birds of Ottawa, 1944. Canadian Field-Naturalist 58(5): 143-175.

Lock, A.R. 1986. A census of Common Eiders breeding in Labrador and the Maritime provinces, pp. 30-38 *in* A. Reed, ed. Eider ducks in Canada/Les eiders au Canada. Canadian Wildlife Service. Report Series no. 47, 177 pp.

Lock, A.R. and R.K. Ross. 1973. The nesting of the Great Cormorant (*Phalacrocorax carbo*) and the Double-crested Cormorant (*Phalacrocorax auritus*) in Nova Scotia in 1971. Canadian Field-Naturalist 87: 43-49.

Lock, T. 1991. A new breeding gull for Nova Scotia. Nova Scotia Birds 33(1): 48.

Lockie, J.D. 1955. The breeding habits and food of Short-eared Owls after a vole plague. Bird Study (2): 53-69.

Löfgren, L. 1984. Oceans birds: their breeding, biology and behaviour. Croom Helm, Sydney, 240 pp.

Loft, B., R.K. Murton, and N.J. Westwood. 1966. Gonadal cycles and the evolution of breeding seasons in British Columbidae. Journal of Zoology 150: 249-272.

Loftin, H. and E.L. Tyson. 1965. Stylized behavior in the Turkey Vulture's courtship dance. Auk 77(2): 193.

Logan, C.A. 1991. Mate switching and mate choice in female Northern Mockingbirds: facultative monogamy. Wilson Bulletin 103(2): 277-281.

Logan, C.A. and L.E. Hyatt. 1991. Mate attraction by autumnal song in the Northern Mockingbird (*Mimus polyglottos*). Auk 108(2): 429-432.

Löhrl, H. 1985. Nuthatches, pp. 94-95 *in* C.M. Perrins and A.L.A. Middleton, eds. Songbirds, Series: all the World's animals. Torstar Books, New York, 160 pp.

Lokemoen, J.T. 1966. Breeding ecology of the Redhead duck in western Montana. Journal of Wildlife Management 30(4): 668-681.

Lokemoen, J.T., H.F. Duebbert, and D.E. Sharp. 1990. Homing and reproductive habits of Mallards, Gadwalls, and Blue-winged Teal. Wildlife Monographs no. 106, 28 pp.

Lombardo, M.P. 1983. Female Tree Swallow lays three clutches during one breeding season. Wilson Bulletin 95(1): 146.

Lombardo, M.P. 1985. Mutual restraint in the Tree Swallows: a test of the TIT FOR TAT model of reciprocity. Science 227(4692): 1363-1365.

Lombardo, M.P. 1987. Attendants at Tree Swallow nests. II. The exploratory-dispersal hypothesis. Condor 89(1): 138-149.

Lombardo, M.P., H.W. Power, P.C. Stouffer, L.C. Romagnano, and A.S. Hoffenbert. 1989. Egg removal and intraspecific brood parasitism in the European Starling (*Sturnus vulgaris*). Behavioral Ecology and Sociobiology 24: 217-223.

Loncke, D.J. and J.B. Falls. 1973. An attempted third brood in the White-throated Sparrow. Auk 90:904.

Long, J.L. 1981. Introduced birds of the world: the worldwide history, distribution and influence of birds introduced to new environments. Universe Books, New York, 528 pp.

Lord, J.M. and D.A. Norton. 1990. Scale and the spatial concept of fragmentation. Conservation Biology 4:197-202.

Lortie, M. 1977. Arbres, forêts et perturbations naturelles au Québec. Les Presses de l'Université Laval, Sainte-Foy, 170 pp.

Low, J.B. 1940. Production of the Redhead (*Nyroca americana*) in Iowa. Wilson Bulletin 52(3): 153-164.

Lowe, C. 1978. Certain life history aspects of the Red-tailed Hawk in central Oklahoma and interior Alaska. M.Sc. thesis, University of Alaska, Fairbanks.

Lowther, J.K. 1961. Polymorphism in the White-throated Sparrow, *Zonotrichia albicollis* (Gmelin). Canadian Journal of Zoology 38: 281-292.

Lowther, J.K. 1977. Nesting biology of the Sora at Vermillion, Alberta. Canadian Field-Naturalist 91(1): 63-67.

Lowther, J.K. and J.B. Falls. 1968. White-throated Sparrow, pp. 1364-1392 *in* A.C. Bent. Life histories of North American cardinals, grosbeaks, buntings, towhees, finches, sparrows, and allies. (O.L. Austin, Jr., ed.). United States National Museum, Washington. [1968, Dover Publications, New York]. Bulletin no. 237, vol. 1, pp. 1-602; vol. 2, pp. 603-1248; vol. 3, pp. 1249-1889.

Lowther, P.E. 1993. Brown-headed Cowbird (*Molothrus ater*), *in* A. Poole and F. Gill, eds. The birds of North America, no. 47. The Academy of Natural Sciences, Philadelphia. The American Ornithologists' Union, Washington, 24 pp.

Lowther, P.E. and C.L. Cink. 1992. House Sparrow (*Passer domesticus*), *in* A. Poole, P. Stettenheim, and F. Gills, eds. The birds of North America, no. 12. The Academy of Natural Sciences, Philadelphia. The American Ornithologists' Union, Washington, 20 pp.

Lucas, F.A. 1891. Explorations in Newfoundland and Labrador in 1887, made in connection with the cruise of the United States Fish Commission Schooner Grampus. Report of the United States National Museum for 1888-89, pp. 709-728.

Ludwig, J.P. 1965. Biology and structure of the Caspian Tern (*Hydroprogne caspia*) population of the Great Lakes from 1896-1964. Bird-Banding 26(4): 217-233.

Lumsden, H.G. 1971. The status of the Sandhill Crane in Northern Ontario. Canadian Field-Naturalist 85(4): 285-293.

Lumsden, H.G. 1980. Starling nest sites and cleared land. Journal of Field Ornithology 51(2): 178-179.

Lumsden, H.G. 1984. The pre-settlement breeding distribution of Trumpeter and Tundra swans in eastern Canada. Canadian Field-Naturalist 98(4): 415-424.

Lumsden, H.G. 1986. Choice of nest boxes by Tree Swallows, *Tachycineta bicolor*, House Wrens, *Troglodytes aedon*, Eastern Bluebirds, *Sialia sialis*, and European Starlings, *Sturnus vulgaris*. Canadian Field-Naturalist 100(3): 343-349.

Lumsden, H.G. 1987a. Brant, p. 563 *in* M.D. Cadman, P.F.J. Eagles, and F.M. Helleiner, eds. Atlas of the breeding birds of Ontario. University of Waterloo Press, Waterloo, 617 pp.

Lumsden, H.G. 1987b. Gray Partridge, pp. 132-133 *in* M.D. Cadman, P.F.J. Eagles, and F.M. Helleiner, eds. Atlas of the breeding birds of Ontario. University of Waterloo Press, Waterloo, 617 pp.

Lumsden, H.G. 1987c. Mute Swan, pp. 62-63 *in* M.D. Cadman, P.F.J. Eagles, and F.M. Helleiner, eds. Atlas of the breeding birds of Ontario. University of Waterloo Press, Waterloo, 617 pp.

Lumsden, H.G. 1987d. Northern Bobwhite, pp. 144-145 *in* M.D. Cadman, P.F.J. Eagles, and F.M. Helleiner, eds. Atlas of the breeding birds of Ontario. University of Waterloo Press, Waterloo, 617 pp.

Lumsden, H.G. 1987e. Sandhill Crane, pp. 160-161 *in* M.D. Cadman, P.F.J. Eagles, and F.M. Helleiner, eds. Atlas of the breeding birds of Ontario. University of Waterloo Press, Waterloo, 617 pp.

Lumsden, H.G. 1987f. Sharp-tailed Grouse, p. 140 *in* M.D. Cadman, P.F.J. Eagles, and F.M. Helleiner, eds. Atlas of the breeding birds of Ontario. University of Waterloo Press, Waterloo, 617 pp.

Lumsden, H.G. 1987g. Snow Goose, p. 510 *in* M.D. Cadman, P.F.J. Eagles, and F.M. Helleiner, eds. Atlas of the breeding birds of Ontario. University of Waterloo Press, Waterloo, 617 pp.

Lumsden, H.G. 1987h. Tundra Swan, p. 509 *in* M.D. Cadman, P.F.J. Eagles, and F.M. Helleiner, eds. Atlas of the breeding birds of Ontario. University of Waterloo Press, Waterloo, 617 pp.

Lumsden, H.G. and R.B.H. Smith. 1987. Red Crossbill, pp. 494-495 *in* M.D. Cadman, P.F.J. Eagles, and F.M. Helleiner, eds. Atlas of the breeding birds of Ontario. University of Waterloo Press, Waterloo, 617 pp.

Lunk, W.A. 1962. The Rough-winged Swallow, *Stelgidopteryx ruficollis* (Vieillot), a study based on its breeding biology in Michigan. Nuttall Ornithological Club, no. 4, Cambridge, 155 pp.

Luther, D.H. 1974. Observations at a Carolina Wren nest from which Brown-headed Cowbirds fledged. Wilson Bulletin 86(1): 51-57.

Lutz, R.S. and J.A. Crawford. 1987. Reproductive success and nesting habitat of Merriam's Wild Turkeys in Oregon. Journal of Wildlife Management 51(4): 783-787.

Luukkonen, D.R. 1987. Status and breeding ecology of the Loggerhead Shrike in Virginia. M.Sc. thesis, Virginia Polytechnic Institute and State University, Blacksburg, 78 pp.

Lynch, J.M. 1981. Status of the Cerulean Warbler in the Roanoke River Basin of North Carolina. Chat 45(2): 29-35.

Lynch, P.G. and D.G. Smith. 1984. Census of Eastern Screech-Owls (*Otus asio*) in urban open-space areas using tape-recorded song. American Birds 38(4): 388-391.

Lyon, B. 1978. Eye to eye with the Red-shouldered Hawk. Nature Canada 7(2): 5-9.

Lyon, B.E. and R.D. Montgomerie. 1985. Incubation feeding in Snow Buntings: female manipulation or indirect male parental care? Behavioral Ecology and Sociobiology 17: 279-284.

Lyon, B.E. and R.D. Montgomerie. 1987. Ecological correlates of incubation feeding: a comparative study of high arctic finches. Ecology 68(3): 713-722.

Lyons, D.M., K. Titus, and J.A. Mosher. 1986. Sprig delivery by Broad-winged Hawks. Wilson Bulletin 98(3): 469-471.

Mabbott, D.C. 1920. Food habits of seven species of American shoal-water ducks. United States Department of Agriculture, Bulletin 862: 17-22.

MacAloney, R.K. 1973. Breeding ecology of the Common Eider (*Somateria mollissima dresseri*) in the Liscombe area of Nova Scotia. M.Sc. thesis, Acadia University, Wolfville, Nova Scotia.

MacArthur, R.H. 1958. Population ecology of some warblers of northeasthern coniferous forest. Ecology 39(4): 599-619.

Macdonald, D. 1984. The encyclopedia of mammals. Facts on File Publications, New York, 895 pp.

MacDonald, S.D. 1970. The breeding behaviour of the Rock Ptarmigan. Living Bird 9: 195-238.

MacInnes, C.D. 1962. Nesting of small Canada Geese near Eskimo Point, Northwest Territories. Journal of Wildlife Management 26(3): 247-256.

MacInnes, C.D. and E.H. Dunn. 1988a. Estimating proportion of an age class nesting in Canada Geese. Journal of Wildlife Management 52(3): 421-423.

MacInnes, C.D. and E.H. Dunn. 1988b. Components of clutch size variation in Arctic-nesting Canada Geese. Condor 90(1): 83-89.

Macior, L.W. 1959. Predation by Gray Jay on the young of the Hoary Bat. Flicker 31(3): 100.

Mackey, W.J. Jr. 1945. American bird decoys. Bonanza Books, New York, 256 pp.

Mackie, R.J. and H.K. Buechner. 1963. Reproductive cycle of the Chukar. Journal of Wildlife Management 27(2): 246-260.

MacKinnon, D. 1989. Effects of sylvicultural herbicide use on the breeding birds of regenerating forest clearcuts in Nova Scotia. M.Sc. thesis, Dalhousie University, Halifax, Nova Scotia, 156 pp.

MacNamara, C. 1926. Champlain as a naturalist. Canadian Field-Naturalist 40: 125-133.

Macoun, J. 1900. Catalogue of Canadian birds. Vol. I. Geological Survey of Canada. Department of Mines, Ottawa. Bulletin no. 692, pp.1-218.

Macoun, J. 1900-1904. Catalogue of Canadian birds. Geological Survey of Canada, 3 vol.

Macoun, J. 1903. Catalogue of Canadian birds. Vol. II. Geological Survey of Canada. Department of Mines, Ottawa. Bulletin no. 822, pp. 219-413.

Macoun, J. 1979. Autobiography of John Macoun, Canadian explorer and naturalist, 1831-1920, 2nd edition. Ottawa, Ottawa Field-Naturalist Club, 361 pp.

Macoun, J. and J.M. Macoun. 1909. Catalogue of Canadian birds, 2nd edition. Geological Survey of Canada. Department of Mines, Ottawa. Bulletin no. 973, 761 pp.

Macoun, J. and J.M. Macoun. 1915. Catalogue des oiseaux du Canada. Commission géologique, Canada, no. 1529, Ministère des Mines, Ottawa, 909 pp.

Macoun, W.T. 1920. Obituary: John Macoun, 1831-1920. Canadian Field-Naturalist 34: 110-114.

MacQueen, P.M. 1950. Territory and song in the Least Flycatcher. Wilson Bulletin 62(4): 194-205.

Madge, S. and H. Burn. 1988. Waterfowl: an identification guide to the ducks, geese and swans of the world. Houghton Mifflin Co., Boston, 298 pp.

Madsen, F.J. 1954. On the food habits of the diving ducks in Denmark. Danish Review of Game Biology 2: 155-266.

Madsen, F.J. 1957. On the food habits of some fish-eating birds in Denmark. Danish Review of Game Biology 3: 19-83.

Magee, A. 1993a. 7. Second-growth hardwood forest, pp. 37-38 *in* J.D. Lowe, ed. Resident bird counts 1992. Journal of Field Ornithology 64(1)(supplement): 1-128.

Magee, A. 1993b. 96. Shrubby swamp and sedge hummocks, p. 93 *in* J.D. Lowe, ed. Resident bird counts 1992. Journal of Field Ornithology 64(1)(supplement): 1-128.

Magoun, A.J. 1976. Summer scavenging in north-eastern Alaska. M.Sc. thesis, University of Alaska, Fairbanks.

Magurran, A.E. 1988. Ecological diversity and its measurement. Princeton University Press, Princeton, New Jersey.

Maher, W.J. 1964. Growth rate and development of endothermy in the Snow Bunting (*Plectrophenax nivalis*) and Lapland Longspur (*Calcarius lapponicus*) at Barrow, Alaska. Ecology 45(3): 520-528.

Maher, W.J. 1970a. The Pomarine Jaeger as a Brown Lemming predator in northern Alaska. Wilson Bulletin 82(2): 130-151.

Maher, W.J. 1970b. Ecology of the Long-tailed Jaeger at Lake Hazen, Ellesmere Island. Arctic 23(2): 112-129.

Maher, W.J. 1974. Ecology of Pomarine, Parasitic, and Long-tailed jaegers in northern Alaska. Pacific Coast Avifauna no. 37, Los Angeles, California, 148 pp.

Mahoney, S.A. and J.R. Jehl. 1985. Adaptations of migratory shorebirds to highly saline and alkaline lakes: Wilson's Phalarope and American Avocet. Condor 87(4): 520-527.

Maisonneuve, C. 1982. Distribution et abondance des oiseaux de rivage le long du Saint-Laurent: estuaire moyen, estuaire maritime, Gaspésie, baie des Chaleurs, Côte-Nord et Îles-de-la-Madeleine. Environment Canada, Canadian Wildlife Service, Québec Region, 81 pp.

Maisonneuve, C., P. Brousseau, and D. Lehoux. 1990. Critical fall staging sites for shorebirds migrating through the St. Lawrence system, Québec. Canadian Field-Naturalist 104(3): 372-378.

Maisonneuve, C., R. McNicoll, and R. Ouellet. 1993. Utilisation du bassin versant de la rivière Boyer par la sauvagine en 1992: rapport d'étape. Service de la faune terrestre, Direction de la faune et des habitats, Ministère du Loisir, de la Chasse et de la Pêche, Québec, 27 pp.

Majka, C.G., B.L. Roscoe, and M.V. MacKinnon. 1976. The first nest record of the Greater Yellowlegs (*Tringa melanoleuca*) in Nova Scotia. Canadian Field-Naturalist 90(2): 200-201.

Malecki, R.A. and R.E. Trost. 1990. A breeding ground survey of Atlantic Flyway Canada Geese, *Branta canadensis*, in northern Quebec. Canadian Field-Naturalist 104(4): 575-578.

Mallory, E.P. and D.C. Schneider. 1979. Agonistic behavior in Short-billed Dowitchers feeding on a patchy resource. Wilson Bulletin 91(2): 271-278.

Malmborg, P.K. and M.F. Willson. 1988. Foraging ecology of avian frugivores and some consequences for seed dispersal in an Illinois woodlot. Condor 90(1): 173-186.

Maltby, F. 1915. Nesting of the Yellow Rail in North Dakota. Oölogist 32: 122-124.

Manci, K.M. and D.H. Rusch. 1988. Indices to distribution and abundance of some inconspicuous waterbirds at Horicon marsh. Journal of Field Ornithology 59: 67-75.

Mannan, R.W. 1984. Summer area requirements of Pileated Woodpeckers in western Oregon. Wildlife Society Bulletin 12: 265-268.

Mannan, R.W. and E.C. Meslow. 1984. Bird populations and vegetation characteristics in managed and old-growth forests, northeastern Oregon. Journal of Wildlife Management 48(4): 1219-1238.

Mannan, R.W., E.C. Meslow, and H.M. Wight. 1980. Use of snags by birds in Douglas-fir forests, western Oregon. Journal of Wildlife Management 44(4): 787-797.

Manning, T.H. 1946. Bird and mammal notes from the east side of Hudson Bay. Canadian Field-Naturalist 60(4): 71-85.

Manning, T.H. 1949. The birds of north-western Ungava, pp. 153-224 *in* T.[E.W.] Manning, ed. A summer on Hudson Bay. Hodder and Stoughton, London, 224 pp.

Manning, T.H. 1952. Birds of the west James Bay and southern Hudson Bay coasts. National Museum of Canada. Bulletin 125: 1-114.

Manning, T.H. 1981. Birds of the Twin Islands, James Bay, Northwest Territories. National Museum of Canada. Syllogeus no. 30, 50 pp.

Manning, T.H. and D.F. Coates. 1952. Notes on the birds of some James Bay Islands. National Museum of Canada. Bulletin 126: 195-207.

Manning, T.H. and A.H. MacPherson. 1952. Birds of the east James Bay coast between Long Point and Cape Jones. Canadian Field-Naturalist 66(1): 1-35.

Manning, T.H. and A.H. MacPherson. 1961. A biological investigation of Prince of Wales Island, Northwest Territories. Transactions. Royal Canadian Institute 33(2): 116-239.

Manseau, M. 1991. Habitat d'élevage des oisons de la Grande Oie des neiges, *Chen caerulescens atlantica*: une approche comparative incluant la perception de l'animal. M.Sc. thesis, Université Laval, Sainte-Foy, 67 pp.

Manseau, M. and J. Ferron. 1991. Activité alimentaire nocturne des Bécasseaux semipalmés (*Calidris pusilla*) lors d'une halte migratoire dans la baie de Fundy. Canadian Journal of Zoology 69(2): 380-384.

Manseau, M. and G. Gauthier. 1993. Interactions between Greater Snow Geese and their rearing habitat. Ecology 74(7): 2045-2055.

Marchand, P.J. 1991. Life in the cold: an introduction to winter ecology, 2nd edition. University Press of New England, Hanover, New Hampshire, 239 pp.

Marie-Victorin, Fr. 1964. Flore laurentienne, 2e édition. Les Presses de l'Université de Montréal, Montréal, 925 pp.

Marie-Victorin, Fr. 1995. Flore laurentienne, 3e édition. Les Presses de l'Université de Montréal, Montréal, 1083 pp.

Marks, J.S. 1985. Yearling male Long-eared Owls breed near natal nest. Journal of Field Ornithology 56(2): 181-182.

Marks, J.S. 1986. Nest-site characteristics and reproductive success of Long-eared Owls in southwestern Idaho. Wilson Bulletin 98(4): 547-560.

Marks, J.S., D.L. Evans, and D.W. Holt. 1994. Long-eared Owl (*Asio otus*), *in* A. Poole and F. Gill, eds. The birds of North America, no. 133. The Academy of Natural Sciences, Philadelphia. The American Ornithologists' Union, Washington, 24 pp.

Marler, P. and S. Peters. 1988. Sensitive periods for song acquisition from tape recording and live tutors in the Swamp Sparrow, *Melospiza georgiana*. Ethology 77: 76-84.

Marshall, J.T. 1948. Ecologic rates of Song Sparrows in the San Francisco Bay region, part 1. Habitat and abundance. Condor 50(5): 193-215.

Marshall, R.M. [ed.] 1992. Resident bird counts 1991. Journal of Field Ornithology 63(1)(supplement): 1-123.

Marsi, H.T. 1980. Are Golden-winged Warblers selective in choosing a nesting habitat? North American Bird Bander 5(4): 144-145.

Martell, M., P. Redig, J. Nibe, G. Buhl, and D. Frenzel. 1991. Survival and movements of released rehabilitated Bald Eagles. Journal of Raptor Research 25(3): 72-76.

Marti, C.D. 1988. A long-term study of food-niche dynamics in the Common Barn Owl: comparisons within and between populations. Canadian Journal of Zoology 66: 1803-1812.

Marti, C.D. 1989. The secretive Common Barn-Owl difficult to study and protect. Eyas 12(1): 7-9.

Marti, C.D. 1992. Barn Owl (*Tyto alba*), *in* A. Poole, P. Stettenheim, F. Gill, eds. Birds of North America, no. 1. The Academy of Natural Sciences, Philadelphia. The American Ornithologists' Union, Washington, 16 pp.

Marti, C.D. and J.S. Marks. 1989. Medium-sized owls, pp. 124-133 *in* B.G. Pendleton, ed. Proceedings. Western Raptor Management Symposium and Workshop. National Wildlife Federation Scientific and Technical Series no. 12, Washington, D.C., 320 pp.

Marti, C.D., P.M. Wagner, and K.W. Benne. 1979. Nests boxes for the management of Barn Owls. Wildlife Society Bulletin 7: 145-148.

Martin, A.C., H.S. Zim, and A.L. Nelson. 1951. American wildlife and plants – A guide to wildlife food habits. McGraw-Hill Book Co. [1961, Dover Publications, New York], 500 pp.

Martin, D.J. 1980. Response by male Fox Sparrows to broadcast of particular conspecific songs. Wilson Bulletin 92(1): 21-32.

Martin, G.R. 1985a. Eye, pp. 311-373 *in* A.S. King and J. McLelland, ed. Form and function in birds, vol. 3. Academic Press, London, 522 pp.

Martin, K. and S.J. Hannon. 1988. Early pair and extra-pair copulations in Willow Ptarmigan. Condor 90(1): 245-246.

Martin, M.L. 1977. Flocking and roosting activities of Red-winged Blackbird in southern Quebec. M.Sc. thesis, McGill University, Montréal, 102 pp.

Martin, M.[L.] and T.W. Barry. 1978. Nesting behavior and food habits of Parasitic Jaegers at Anderson River delta, Northwest Territories. Canadian Field-Naturalist 92(1): 45-50.

Martin, M.[L.], R.[D.] Titman, and [J.]R. Bider. 1976. Protection du maïs-grain contre les oiseaux noirs. Ministère de l'Agriculture, Québec.

Martin, N.D. 1960. An analysis of bird populations in relation to forest succession in Algonquin Provincial Park, Ontario. Ecology 41(1): 126-140.

Martin, N.L. 1985b. Turkey Vulture, pp. 70-71 *in* S.B. Laughlin and D.P. Kibbe, eds. The Atlas of breeding birds of Vermont. Vermont Institute of Natural Science, University Press of New England, Hanover, 456 pp.

Martin, P.-L. 1990. La chasse au Québec. Éditions du Boréal, La Prairie, 409 pp.

Martin, S.G. 1973. Longevity surprise: the Bobolink. Bird-Banding 44(1): 157-158.

Martin, S.G. 1974. Adaptation for polygynous breeding in the Bobolink, *Dolichonyx oryzivorus*. American Zoologist 14: 109-119.

Martin, T.E. 1992. Landscape considerations for viable populations and biological diversity. Transactions. North American Wildlife and Natural Resources Conference 57: 283-291.

Martini, I.P., R.I.G. Morrison, W.A. Glooschenko, and R. Protz. 1980. Coastal studies in James Bay, Ontario. Geoscience Canada 7: 11-21.

Martz, G.F. 1967. Effects of nesting cover removal on breeding puddle ducks. Journal of Wildlife Management 31(2): 236-247.

Marvil, R.E. and A. Cruz. 1989. Impact of Brown-Headed Cowbird parasitism on the reproductive success of the Solitary Vireo. Auk 106(3): 476-480.

Mason, C.F. and S.M. Macdonald. 1976. Aspects of the breeding ecology of the Snipe. Bird Study 23(1): 33-38.

Masse, D. and M. Raymond. 1988. La nidification de la sauvagine dans le marécage de la Rivière-du-Sud et la zone agricole environnante. Canadian Journal of Zoology 66(5): 1160-1167.

Mathieu, J. and S. Courville. 1987. Peuplement colonisateur aux XVIIe et XVIIIe siècles. Université Laval, Sainte-Foy, November 1987. Cahiers du CELAT, no. 8, 292 pp.

Matray, P.F. 1974. Broad-winged Hawk nesting and ecology. Auk 91(2): 307-324.

Matthews, G.V.T. 1968. Bird navigation. University Press, Cambridge.

Matthysen, E., D. Cimprich, and T.C. Grubb, Jr. 1992. Is social organization in winter determined by short – or long-term benefits? A case on migrant Red-breasted Nuthatches *Sitta canadensis*. Ornis Scandinavica 23(1) 43-48.

Maunder, J.E. and W. Threlfall. 1972. The breeding biology of the Black-legged Kittiwake in Newfoundland. Auk 89: 789-816.

Maurer, B.A. 1993. Biological diversity, ecological integrity, and neotropical migrants: new perspectives for wildlife management, pp. 24-31 *in* D.M. Finch and P.W. Stangel, eds. Status and management of neotropical migratory birds. United States Department of Agriculture, Forest Service. General Technical Report RM-229.

Maurer, B.A. and R.C. Whitmore. 1981. Foraging of five bird species in two forests with different vegetation structure. Wilson Bulletin 93: 478-490.

Maxson, S.J. 1978. Spring home range and habitat use by female Ruffed Grouse. Journal of Wildlife Management 42(1): 61-71.

Maxson, S.J. and L.W. Oring. 1980. Breeding season time and energy budgets of the polyandrous Spotted Sandpiper. Behaviour 74: 200-263.

Maxwell, G.R. 1993. 22. Mature beech-maple forest on Lake Ontario shore, p. 47 *in* J.D. Lowe, ed. Resident bird counts 1992. Journal of Field Ornithology 64(1)(supplement): 1-128.

Maxwell, G.R. II. 1970. Pair formation, nest building, and egg laying of the Common Grackle in northern Ohio. Ohio Journal of Science 70(5): 284-291.

Maxwell, G.R. [II] and L.S. Putnam. 1972. Incubation, care of young, and nest success of the Common Grackle (*Quiscalus quiscula*) in northern Ohio. Auk 89(2): 349-359.

Mayfield, H. 1965. The Brown-headed Cowbird, with old and new hosts. Living Bird 4: 13-28.

Mayfield, H. 1977. Brown-headed Cowbird: agent of extermination? American Birds 31(2): 107-113.

Mayr, E. 1975. Materials for a history of American ornithology, pp. 365-396 *in* E. Stresemann, ed. Ornithology from Aristotle to the present. Harvard University Press, Cambridge, Massachusetts, 432 pp.

Mayr, E. and P.D. Ashlock. 1991. Principles of systematic zoology. McGraw-Hill, New York, 475 pp.

McAllister, D.E. 1990. A list of the fishes of Canada / Liste des poissons du Canada. National Museum of Natural Sciences, Ottawa. Syllogeus no. 64, 310 pp.

McAloney, R.K. 1973. Brood ecology of the Common Eider (*Somateria mollissima dresseri*) in the Liscombe area of Nova Scotia. M.Sc. thesis, Acadia University, Wolfville, Nova Scotia.

McAlpine, D.F., J. Finne, M. Phinney, S. Gilliland, and S. Makepeace. 1988a. Breeding records for the Gadwall (*Anas strepera*) in New Brunswick. Naturaliste canadien 115: 95-96.

McAlpine, D.F., M. Phinney, and S. Makepeace. 1988b. New Brunswick breeding of Wilson's Phalarope, *Phalaropus tricolor*, confirmed. Canadian Field-Naturalist 102(1): 77-78.

McAtee, W.L. 1908. Food habits of the grosbeaks. United States Department of Agriculture, Government Printing Office, Washington. Biological Survey Bulletin no. 32, 92 pp.

McAtee, W.L. 1911. Woodpeckers in relation to trees and wood products. United States Department of Agriculture, Government Printing Office, Washington. Biological Survey Bulletin no. 39, 99 pp.

McAtee, W.L. 1918. Food habits of the Mallard ducks of the United States. United States Department of Agriculture, Government Printing Office, Washington. Bulletin no. 720, 36 pp.

McAtee, W.L. 1926. The relation of birds to woodlots in New York State. Roosevelt Wild Life Bulletin 4: 7-152.

McAtee, W.L. 1949. Ornithological publications of Percy Algernon Taverner, 1895-1945. Annual reports 1939-1947. National Museum of Canada Bulletin 112: 102-113.

McAuley, D.G. 1986. Ring-necked Duck productivity in relation to wetland acidity: nest success, duckling foods and survival. M.Sc. thesis, University of Maine, Orono, 71 pp.

McAuley, D.G. and J.R. Longcore. 1988. Foods of juvenile Ring-necked Ducks: relationships to wetland pH. Journal of Wildlife Management 52(2): 177-185.

McAuley, D.G., J.R. Longcore, and G.F. Sepik. 1990. Renesting by American Woodcocks (*Scolopax minor*) in Maine. Auk 107: 407-410.

McCabe, R.A. 1965. Nest construction by House Wrens. Condor 67: 229-234.

McCabe, R.A. and A.S. Hawkins. 1946. The Hungarian Partridge in Wisconsin. American Midland Naturalist 36: 1-75.

McCabe, T.T. and E.B. McCabe. 1933. Notes on the anatomy and breeding habits of crossbills. Condor 35: 136-137.

McClelland, B.R. 1979. The Pileated Woodpecker in forests of the northern Rocky Mountains, pp. 283-299 *in* J.G. Dickson, R.N. Conner, R.R. Fleet, J.C. Kroll, and J.A. Jackson, eds. The role of insectivorous birds in forest ecosystems. Academic Press, New York, 381 pp.

McClure, E. 1984. Bird Banding. The Boxwood Press, Pacific Grove, California, 341 pp.

McCourt, K.H., D.A. Boag, and D.M. Keppie. 1973. Female Spruce Grouse activities during laying and incubation. Auk 90: 619-623.

McCracken, J.D. 1981. Status report on the Prothonotary Warbler (*Protonotaria citrea*) in Canada. Committee on the Status of Endangered Wildlife in Canada (COSEWIC), Ottawa, 27 pp.

McCracken, J.D. 1987a. Marsh Wren, pp. 312-313 *in* M.D. Cadman, P.F.J. Eagles, and F.M. Helleiner, eds. Atlas of the breeding birds of Ontario. University of Waterloo Press, Waterloo, 617 pp.

McCracken, J.D. 1987b. Prothonotary Warbler, pp. 402-403 *in* M.D. Cadman, P.F.J. Eagles, and F.M. Helleiner, eds. Atlas of the breeding birds of Ontario. University of Waterloo Press, Waterloo, 617 pp.

McCracken, J.D. 1990. Status report on the Louisiana Waterthrush (*Seiurus motacilla*) in Canada. Committee on the Status of Endangered Wildlife in Canada (COSEWIC), Ottawa, 30 pp.

McCracken, J.D. 1992. Status report on the Cerulean Warbler (*Dendroica cerulea*) in Canada. Committee on the Status of Endangered Wildlife in Canada (COSEWIC), Ottawa, 31 pp.

McCracken, J.D., M.S.W. Bradstreet, and G.L. Holroyd. 1981. Breeding birds of Long Point, Lake Erie: a study in community succession. Canadian Wildlife Service, Ottawa. Technical Report Series no. 44, 74 pp.

McGahan, J. 1966. Ecology of the Golden Eagle. M.Sc. thesis, University of Montana, Missoula, 78 pp.

McGahan, J. 1968. Ecology of the Golden Eagle. Auk 85(1): 1-12.

McGarigal, K. and J.D. Fraser. 1984. The effect of forest stand age on owl distribution in southwestern Virginia. Journal of Wildlife Management 48(4): 1393-1398.

McGhee, R. 1989. Le Canada au temps des envahisseurs. Libre Expression, Musée Canadien des Civilisations, 175 pp.

McGillivray, W.B. 1980a. Communal nesting in the House Sparrow. Journal of Field Ornithology 51(4): 371-372.

McGillivray, W.B. 1980b. Nest grouping and productivity in the House Sparrow. Auk 97: 396-399.

McGillivray, W.B. 1981. The breeding ecology of House Sparrows. Ph.D. dissertation, University of Kansas, Lawrence, 83 pp.

McGillivray, W.B. 1983. Intraseasonal reproductive costs for the House Sparrow (*Passer domesticus*). Auk 100(1): 25-32.

McGillivray, W.B. 1988. Breeding of the Rock Dove, *Columba livia*, in January at Edmonton, Alberta. Canadian Field-Naturalist 102(1): 76-77.

McGilvrey, F.B. 1966. Nesting of Hooded Mergansers on the Patuxent Wildlife Research Center, Laurel, Maryland. Auk 83(3): 477-479.

McGilvrey, F.B. 1969. Survival in Wood Duck broods. Journal of Wildlife Management 33(1): 73-76.

McGowan, K.J. and G.E. Woolfenden. 1986. Aerial rain bathing by Common Nighthawks. Wilson Bulletin 98(4): 612-613.

McHenry, M.G. 1971. Breeding and post-breeding movements of Blue-winged Teal (*Anas discors*) in southwestern Manitoba. Ph.D. dissertation, University of Oklahoma, Norman, 67 pp.

McIlroy, M. 1961. Possible hybridization between a Clay-colored Sparrow and a Chipping Sparrow in Ithaca. Kingbird 11: 7-10.

McIlveen, W.D. 1987. Three-toed Woodpecker, p. 242 *in* M.D. Cadman, P.F.J. Eagles, and F.M. Helleiner, eds. Atlas of the breeding birds of Ontario. University of Waterloo Press, Waterloo, 617 pp.

McIntyre, J.W. 1974. Territorial affinity of a Common Loon. Bird-Banding 45: 178.

McIntyre, J.W. 1975. Biology and behavior of the Common Loon (*Gavia immer*) with reference to its adaptability in a man-altered environment. Ph.D. dissertation, University of Minnesota, Minneapolis.

McIntyre, J.W. 1978. Wintering behavior of Common Loons. Auk 95: 396-403.

McIntyre, J.W. 1988. The Common Loon: spirit of northern lakes. Fitzhenry and Whiteside, Markham, Ontario, 229 pp.

McInvaille, W.B. and L.B. Keith. 1974. Predator-prey relations and breeding biology of the Great Horned Owl and Red-tailed Hawk in central Alberta. Canadian Field-Naturalist 88(1): 1-20.

McKenzie, P.M. 1991. A Ruby-throated Hummingbird (*Archilochus colubris*) trapped in a spider-web. Journal of Louisiana Ornithology 1(2): 54-58.

McKinley, D. 1960. A history of the Passenger Pigeon in Missouri. Auk 77: 399-420.

McKinney, F. 1965. The displays of the American Green-winged Teal. Wilson Bulletin 77(2): 112-121.

McKinney, F. 1970. Displays of four species of blue-winged ducks. Living Bird 9: 29-64.

McKinney, F. 1973. Ecoethological aspects of reproduction, pp. 6-21 *in* D.S. Farmer, ed. Breeding biology of birds. National Academy of Science, Washington, 515 pp.

McKinney, F. and P. Stolen. 1982. Extra-pair-bond courtship and forced copulation among captive Green-winged Teal (*Anas crecca carolinensis*). Animal Behaviour 30: 461-474.

McKinney, F., L.G. Sorenson, and M. Hart. 1990. Multiple functions of courtship displays in dabbling ducks (Anatini). Auk 107(1): 188-191.

McLaren, M.A. 1975. Breeding biology of the Boreal Chickadee. Wilson Bulletin 87(3): 344-354.

McLaren, P.L. 1987. Common Yellowthroat, pp. 416-417 *in* M.D. Cadman, P.F.J. Eagles, and F.M. Helleiner, eds. Atlas of the breeding birds of Ontario. University of Waterloo Press, Waterloo, 617 pp.

McLaughlin, R.L. and R.D. Montgomerie. 1985. Brood division by Lapland Longspurs. Auk 102(4): 687-695.

McLaughlin, R.L. and R.D. Montgomerie. 1989. Brood dispersal and multiple central place foraging by Lapland Longspur parents. Behavioral Ecology and Sociobiology 25(3): 207-216.

McLean, I.G., J.N.M. Smith, and K.G. Stewart. 1986. Mobbing behavior, nest exposure, and breeding success in the American Robin. Behaviour 96(1-2): 171-186.

McLeod, J.M. 1974. Bird populations studies in the Swaine Jack Pine Sawfly life system. Rapport d'inf. LAU-X-10, Environment Canada, Centre de recherches forestières des Laurentides, Sainte-Foy, 86 pp.

McNair, D.B. 1988. Review of breeding records of Red Crossbill and Pine Siskin in the southern Appalachian Mountains and adjacent regions. Migrant 59(4): 105-113.

McNeil, R. and J. Burton. 1973. Dispersal of some southbound migrating North American shorebirds away from the Magdalen Islands, Gulf of St. Lawrence, and Sable Island, Nova Scotia. Caribbean Journal of Science 13(3-4): 257-278.

McNeil, R. and J. Burton. 1977. Southbound migration of shorebirds from the Gulf of St. Lawrence. Wilson Bulletin 89(1): 167-171.

McNeil, R. and F. Cadieux. 1972. Fat content and flight-range capabilities of some adult spring and fall migrant North American shorebirds in relation to migration routes on the Atlantic coast. Naturaliste canadien 99(6): 589-606.

McNeil, R. and C. Léger. 1987. Nest-site quality and reproductive success of early- and late-nesting Double-crested Cormorants. Wilson Bulletin 99: 262-267.

McNeil, R., J. Boulva, W. Gaboriault, and J.G. Strauch, Jr. 1973. Observations récentes sur les oiseaux aux Îles-de-la-Madeleine, Québec. Revue de Géographie de Montréal 27(2): 157-171.

McNeil, R., R. Benoit, and J.-L. DesGranges. 1993. Day-time and night-time foraging of Great Blue Herons in a non-tidal environment. Canadian Journal of Zoology 71: 1075-1078.

McNicholl, M.K. 1975. Arctic Loon "checking" nest. Auk 92(2): 393-394.

McNicholl, M.K. 1987. Northern Cardinal, pp. 430-431 *in* M.D. Cadman, P.F.J. Eagles, and F.M. Helleiner, eds. Atlas of the breeding birds of Ontario. University of Waterloo Press, Waterloo, 617 pp.

McNicol, D.K. 1979. Seasonal, habitat and sex-specific patterns of food utilization by Red-winged Blackbirds (*Agelaius phoeniceus*) in eastern Ontario and their economic importance. M.Sc. thesis, Queen's University, Kingston, Ontario, 162 pp.

McNicol, D.K., R.J. Robertson, and P.J. Weatherhead. 1982. Seasonal, habitat, and sex-specific food habits of Red-winged Blackbirds: implication for agriculture. Canadian Journal of Zoology 60(12): 3282-3289.

McNicol, D.K., B.E. Bendell, and R.K. Ross. 1987. Studies of the effects of acidification on aquatic wildlife in Canada: waterfowl and trophic relationships in small lakes in northern Ontario. Canadian Wildlife Service. Occasional Paper no. 62, 80 pp.

McNicoll, R., R. Ouellet, and M. Lepage. 1991. Cinquième inventaire quinquennal du Faucon pèlerin au Québec. Ministère du Loisir, de la Chasse et de la Pêche, Direction de la gestion des espèces et des habitats, Service de la faune terrestre, 30 pp.

McPherson, J.M. 1988. Preferences of Cedar Waxwings in the laboratory for fruit species, color and size: a comparison with field observations. Animal Behaviour 36(4): 961-969.

Meade, G.M. 1942. Calcium chloride – a death lure for crossbills. Auk 59: 439-440.

Meade, G.M. 1988a. Blue-gray Gnatcatcher, p. 314 *in* R.F. Andrle and J.R. Carroll, eds. The Atlas of breeding birds in New York State. Cornell University Press, Ithaca, 551 pp.

Meade, G.M. 1988b. House Finch, pp. 488-489 *in* R.F. Andrle and J.R. Carroll, eds. The Atlas of breeding birds in New York State. Cornell University Press, Ithaca, 551 pp.

Meade, G.M. 1988c. Northern Cardinal, pp. 428-429 *in* R.F. Andrle and J.R. Carroll, eds. The Atlas of breeding birds in New York State. Cornell University Press, Ithaca, 551 pp.

Meade, G.M. 1988d. Red-bellied Woodpecker, pp. 228-229 *in* R.F. Andrle and J.R. Carroll, eds. The Atlas of breeding birds in New York State. Cornell University Press, Ithaca, 551 pp.

Meade, G.M. 1988e. Tufted Titmouse, pp. 292-293 *in* R.F. Andrle and J.R. Carroll, eds. The Atlas of breeding birds in New York State. Cornell University Press, Ithaca, 551 pp.

Meade, G.M. 1988f. Turkey Vulture, pp. 96-97 *in* R.F. Andrle and J.R. Carroll, eds. The Atlas of breeding birds in New York State. Cornell University Press, Ithaca, 551 pp.

Mebs, T. 1989. Les rapaces nocturnes: chouettes et hiboux. Delachaux et Niestlé, Neuchâtel, France, 123 pp.

Mech, L.D. 1970. The wolf: the ecology and behavior of an endangered species. Natural History Press, Garden City, New York, 384 pp.

Meehan, R.H. 1980. Behavioral significance of Boreal Owl vocalizations during the breeding season. M.Sc. thesis, University of Alaska, Fairbanks, 58 pp.

Meehan, R.H. and R.J. Ritchie. 1982. Habitat requirements of Boreal and Hawk owls in Interior Alaska, pp. 188-199 *in* W.N. Ladd and P.F. Schempf, eds. Proceedings. Symposium workshop on raptor management and biology in Alaska and western Canada, Anchorage, Alaska, 335 pp.

Mélançon, C. 1940. Charmants voisins. Librairie Granger frères, Montréal, 281 pp.

Mélançon, C. 1964. Charmants voisins. Éditions du Jour, Montréal, 255 pp.

Mellen, T.K., E.C. Meslow, and R.W. Mannan. 1992. Summertime home range and habitat use of Pileated Woodpeckers in western Oregon. Journal of Wildlife Management 56(1): 96-103.

Meltofte, H. 1983. Arrival and pre-nesting period of the Snow Bunting *Plectrophenax nivalis* in east Greenland. Polar Research 1: 185-198.

Melvin, S.M., W.J.D. Stephen, and S.A. Temple. 1990. Population estimates, nesting biology, and habitat preferences of Interlake, Manitoba, Sandhill Cranes, *Grus canadensis*. Canadian Field-Naturalist 104(3): 354-361.

Ménard-Lépine, L. 1980. Vocabulaire de l'environnement. Publication de la Société de l'Énergie de la Baie James, by Services linguistiques, relations publiques and Direction de l'environnement, Montréal, 48 pp.

Mendall, H.L. 1936. The home-life and economic status of the Double-crested Cormorant. University of Maine Studies, Second Series, no. 38. Maine Bulletin 39: 1-159.

Mendall, H.L. 1937. Nesting of the Bay-breasted Warbler. Auk 54(4): 429-439.

Mendall, H.L. 1944. Food of hawks and owls in Maine. Journal of Wildlife Management 8(3): 204.

Mendall, H.L. 1958. The Ring-necked Duck in the northeast. University of Maine, Bulletin, vol. 60, no. 16, 317 pp.

Mendall, H.L. 1968. An inventory of Maine's breeding eider ducks. Transactions. Northeast Section Wildlife Society 25: 95-104.

Mendelssohn, H. and U. Paz. 1977. Mass mortality of birds of prey caused by azodrin, an organophosphorus insecticide. Biological Conservation 11: 163-170.

Mendenhall, V.M. and H. Milne. 1985. Factors affecting duckling survival of Eiders *Somateria mollissima* in northeast Scotland. Ibis 127: 148-158.

Meng, H. 1959. Food habits of nesting Cooper's Hawks and Goshawks in New York and Pennsylvania. Wilson Bulletin 71(2): 169-174.

Mengel, R.M. 1964. The probable history of species formation in some northern wood warblers (Parulidae). Living Bird 3: 9-43.

Menkens, G.E. Jr. and R.A. Malecki. 1991. Winter sightings of Canada Geese, *Branta canadensis*, banded in northern Quebec and James Bay. Canadian Field-Naturalist 105(3): 350-353.

Mercier, Y. 1992. L'intendance des terres privées: un nouveau concept. Habitats 3(1): 1-2.

Meredith, R. 1935. The Yellow Rail in the province of Quebec. Canadian Field-Naturalist 49: 58-59.

Merilees, W.J. 1967. Spotted Sandpiper chicks returning to the nest. Canadian Field-Naturalist 81(3): 223.

Merriam, C.H. 1882. List of birds ascertained to occur within ten miles from Point de Monts, province of Quebec, Canada; based chiefly upon the notes of Napoleon A. Comeau. Nuttall Ornithological Club, Bulletin 7(4): 233-242.

Merriam, C.H. 1883a. Addendum to the list of birds ascertained to occur within ten miles from Point de Monts, province of Quebec, Canada; based chiefly upon the notes of Napoleon A. Comeau. Nuttall Ornithological Club, Bulletin 8(4): 244.

Merriam, C.H. 1883b. On a bird new to northern North America (*Vireo flavoviridis*). Nuttall Ornithological Club, Bulletin 8(4): 213.

Merriam, C.H. 1884. Second addendum to the list of birds ascertained to occur within ten miles from Point de Monts, province of Quebec, Canada; based chiefly upon the notes of Napoleon A. Comeau. Auk 1(3): 295.

Merriam, C.H. 1885a. Third addendum to the list of birds ascertained to occur within ten miles from Point de Monts, province of Quebec, Canada; based chiefly upon the notes of Napoleon A. Comeau. Auk 2(1): 113.

Merriam, C.H. 1885b. Probable breeding of the Wheatear (*Saxicola oenanthe*) on the North Shore of the Gulf of St. Lawrence. Auk 2(3): 305.

Merriam, C.H. 1885c. Fourth addendum to the list of birds ascertained to occur within ten miles from Point de Monts, province of Quebec, Canada; based chiefly upon the notes of Napoleon A. Comeau. Auk 2(3): 315.

Merrie, T.D.H. 1977. Relationship between spatial distribution of breeding divers and the availability of fishing waters. Bird Study 25: 119-122.

Merrie, T.D.H. 1979. Success of artificial island nest-sites divers. British Birds 72(1): 32-33.

Merrifield, K. 1993. Eurasian x American Wigeons in western Oregon. Western Birds 24: 105-107.

Metcalf, L.N. 1977. Coniferous forest, pp. 53-54 *in* W.T. Van Velzen, ed. Fourtieth breeding bird census, no. 66. American Birds 31(1): 24-93.

Métras, L. 1983. Les oiseaux migrateurs abattus au Canada pendant la saison de chasse de 1981. Canadian Wildlife Service. Progress Notes no. 137, 39 pp. [Also available in English as: Migratory birds killed in Canada during the 1981 season.]

Métras, L. 1985. Les oiseaux migrateurs abattus au Canada pendant la saison de chasse de 1983. Canadian Wildlife Service. Progress Notes no. 154, 39 pp. [Also available in English as: Migratory birds killed in Canada during the 1983 season.]

Métras, L. 1986. Les oiseaux migrateurs abattus au Canada pendant la saison de chasse de 1985. Canadian Wildlife Service. Progress Notes no. 166, 42 pp. [Also available in English as: Migratory birds killed in Canada during the 1985 season.]

Métras, L. and S. Wendt. 1986. Les oiseaux migrateurs abattus au Canada pendant la saison de chasse de 1984. Canadian Wildlife Service. Progress Notes no. 161, 40 pp. [Also available in English as: Migratory birds killed in Canada during the 1984 season.]

Métro, A. 1975. Dictionnaire forestier multilingue, terminologie forestière: sciences forestières, technologie, pratiques et produits forestiers. Collection de terminologie forestière multilingue no. 2, Association française des eaux et forêts, Conseil international de la langue française, 432 pp.

Metz, K.J. 1991. The enigma of multiple nest building by male Marsh Wrens. Auk 108(1): 170-173.

Meuller, H. and S. Hall. 1991. Floodplain forest. Journal of Field Ornithology 62(1): 42.

Meyer, R.L. and T.G. Balgooyen. 1987. A study and implications of habitat separation by sex of wintering American Kestrels (*Falco sparverius* L.), pp. 107-123 *in* R. Bowman and D.M. Bird, eds. The ancestral Kestrel. Raptor Research Foundation and Macdonald Raptor Research Centre of McGill University, Sainte-Anne-de-Bellevue. Raptor Research Report no. 6, 178 pp.

Meyerriecks, A.J. 1960. Comparative breeding behavior of four species of North American herons. Publication of the Nuttall Ornithological Club, no. 2. Cambridge, 158 pp.

Michaud, G. and J. Ferron. 1986. Étude comparative des techniques de quête alimentaire de quatre espèces d'oiseaux limicoles. Naturaliste canadien 113: 281-292.

Middleton, A.L.A. 1977a. Increase in overwintering by the American Goldfinch, *Carduelis tristis*, in Ontario. Canadian Field-Naturalist 91(2): 165-172.

Middleton, A.L.A. 1977b. The molt of the American Goldfinch. Condor 79(4): 440-444.

Middleton, A.L.A. 1978. The annual cycle of the American Goldfinch. Condor 80(4): 401-406.

Middleton, A.L.A. 1979. Influence of age and habitat on reproduction by the American Goldfinch. Ecology 60(2): 418-432.

Middleton, A.L.A. 1984. Longevity of the American Goldfinch. Journal of Field Ornithology 55(3): 383-386.

Middleton, A.L.A. 1987. American Goldfinch, pp. 500-501 *in* M.D. Cadman, P.F.J. Eagles, and F.M. Helleiner, eds. Atlas of the breeding birds of Ontario. University of Waterloo Press, Waterloo, 617 pp.

Middleton, A.L.A. 1988a. Polyandry in the mating system of the American Goldfinch, *Carduelis tristis*. Canadian Journal of Zoology 66: 296-299.

Middleton, A.L.A. 1988b. The urban environment and its potential as a refuge from brood parasitism by the Brown-headed Cowbird, pp. 1802-1808 *in* H. Ouellet, ed. Acta XIX Congressus Internationalis Ornithologici, Ottawa, 22-29 June 1986. Ottawa University Press, vol. I, pp. 1-1404; vol. II, pp. 1405-2815.

Middleton, A.L.A. 1991. Failure of Brown-headed Cowbird parasitism in nests of the American Goldfinch. Journal of Field Ornithology 62(2): 200-203.

Middleton, A.L.A. 1993. American Goldfinch (*Carduelis tristis*), *in* A. Poole and F. Gill, eds. The birds of North America, no. 80. The Academy of Natural Sciences, Philadelphia. The American Ornithologists' Union, Washington, 24 pp.

Middleton, A.L.A. and D.R.C. Prescott. 1989. Polygyny, extra-pair copulations, and nest helpers in the Chipping Sparrow, *Spizella passerina*. Canadian Field-Naturalist 103(1): 61-64.

Middleton, D.S. 1949. Close proximity of two nests of American Bitterns. Wilson Bulletin 61: 113

Migratory Birds Regulations. 1927. The Migratory Birds Convention Act, R.S.C. 1927, c. 130. Department of Resources and Development, Canadian Wildlife Service, Ottawa, 33 pp.

Migratory Birds Regulations. 1950. The Migratory Birds Convention Act, R.S.C. 1927, c. 130. Department of Resources and Development, Canadian Wildlife Service, Ottawa, 33 pp.

Mikkola, H. 1983. Owls of Europe. T. Toyser and A.D. Toyser, Calton [Buteo Books, Vermillion, South Dakota], 397 pp.

Mikkola, H. and S. Sulkava. 1970. Food of Great Gray Owls in Fenno-Scandia. British Birds 63: 23-27.

Milko, R. 1986. Potential ecological effects of the proposed Grand Canal diversion project on Hudson and James bays. Arctic 39(4): 316-326.

Miller, A.H. 1931. Systematic revision and natural history of the American shrikes (*Lanius* spp.). University of California. Publications in Zoology 38: 11-242.

Miller, A.H. 1956. Ecologic factors that accelerate formation of races and species of terrestrial vertebrates. Evolution 10(3): 262-277.

Miller, A.H. and J.L. Gulledge. 1985. Mockingthrush, pp. 357-358 *in* B. Campbell and E. Lack, eds. A dictionary of birds. British Ornithologists' Union, Buteo Books, Vermillion, South Dakota, 670 pp.

Miller, E.H. 1979. Functions of display flights by males of the Least Sandpiper, *Calidris minutilla* (Vieill.), on Sable Island, Nova Scotia. Canadian Journal of Zoology 57: 876-893.

Miller, E.H. 1983a. Habitat and breeding cycle of the Least Sandpiper (*Calidris minutilla*) on Sable Island, Nova Scotia. Canadian Journal of Zoology 61(12): 2880-2898.

Miller, E.H. 1983b. The structure of aerial displays in three species of Calidridinae (Scolopacidae). Auk 100: 440-451.

Miller, E.H. 1985. Parental behaviour in the Least Sandpiper (*Calidris minutilla*). Canadian Journal of Zoology 63: 1593-1601.

Miller, E.H. 1986. Components of variation in nuptial calls of the Least Sandpiper (*Calidris minutilla*, Aves, Scolopacidae). Systematic Zoology 35(3): 400-413.

Miller, E.H., W.W.H. Gunn, and R.E. Harris. 1983. Geographic variation in the aerial song of the Short-billed Dowitcher (Aves, Scolopacidae). Canadian Journal of Zoology 61(10): 2191-2198.

Miller, F.L. and R.H. Russell. 1973. Unequal distribution of Snowy Owls on eastern Melville Island, Northwest Territories. Canadian Field-Naturalist 87(2): 180-181.

Miller, J.R. and J.T. Miller. 1948. Nesting of the Spotted Sandpiper at Detroit, Michigan. Auk 65: 558-567.

Miller, K.J. 1976. Activity patterns, vocalizations, and site selection in nesting Blue-winged Teal. Wildfowl 27: 33-43.

Miller, L.J. 1978. The spatial and temporal dispersion of nests as an adaptation to food exploitation and nest predation in the American Goldfinch. Ph.D. dissertation, Graduate College of the University of Illinois, Urbana-Champaign, 98 pp.

Miller, R.S. 1973. The brood size of cranes. Wilson Bulletin 85(4): 436-441.

Miller, R.S. and R.W. Nero. 1983. Hummingbird-sapsucker associations in northern climates. Canadian Journal of Zoology 61: 1540-1546.

Miller, R.S., S. Tamm, G.D. Sutherland, and C.L. Gass. 1985. Cues for orientation in hummingbird foraging: color and position. Canadian Journal of Zoology 63: 18-21.

Miller, S.L., M.A. Gregg, A.R. Kuritsubo, S.M. Combs, M.K. Murdock, J.A. Nilsson, B.R. Noon, and R.G. Botzler. 1988. Morphometric variation in Tundra Swans: relationships among sex and age classes. Condor 90(4): 802-815.

Miller, W.J. and L.S. Miller. 1958. Synopsis of behaviour traits of the Ring Neck Dove. Animal Behaviour 6(1-2): 3-8.

Mills, A. 1987a. Blue-winged Warbler, pp. 356-357 *in* M.D. Cadman, P.F.J. Eagles, and F.M. Helleiner, eds. Atlas of the breeding birds of Ontario. University of Waterloo Press, Waterloo, 617 pp.

Mills, A. 1987b. Common Nighthawk, pp. 220-221 *in* M.D. Cadman, P.F.J. Eagles, and F.M. Helleiner, eds. Atlas of the breeding birds of Ontario. University of Waterloo Press, Waterloo, 617 pp.

Mills, A. 1987c. Golden-winged Warbler, pp. 358-359 *in* M.D. Cadman, P.F.J. Eagles, and F.M. Helleiner, eds. Atlas of the breeding birds of Ontario. University of Waterloo Press, Waterloo, 617 pp.

Mills, A. 1987d. Northern Saw-whet Owl, pp. 218-219 *in* M.D. Cadman, P.F.J. Eagles, and F.M. Helleiner, eds. Atlas of the breeding birds of Ontario. University of Waterloo Press, Waterloo, 617 pp.

Mills, A. 1987e. Red-breasted Nuthatch, pp. 298-299 *in* M.D. Cadman, P.F.J. Eagles, and F.M. Helleiner, eds. Atlas of the breeding birds of Ontario. University of Waterloo Press, Waterloo, 617 pp.

Mills, A. 1987f. Whip-poor-will, pp. 224-225 *in* M.D. Cadman, P.F.J. Eagles, and F.M. Helleiner, eds. Atlas of the breeding birds of Ontario. University of Waterloo Press, Waterloo, 617 pp.

Mills, A. 1987g. White-breasted Nuthatch, pp. 300-301 *in* M.D. Cadman, P.F.J. Eagles, and F.M. Helleiner, eds. Atlas of the breeding birds of Ontario. University of Waterloo Press, Waterloo, 617 pp.

Mills, A.M. 1986. The influence of moonlight on the behaviour of goatsuckers (Caprimulgidae). Auk 103(2): 370-378.

Millsap, B.A. 1986. Status of wintering Bald Eagles in the conterminous 48 states. Wildlife Society Bulletin 14: 433-440.

Millsap, B.A. 1987. Introduction to Federal laws and raptor management, chap. 3, pp. 23-33 *in* B.A.G. Pendleton, B.A. Millsap, K.W. Cline, and D.M. Bird, eds. Raptor management techniques manual. National Wildlife Federation, Washington. Scientific and Technical Series no. 10, 420 pp.

Milne, H. 1968. The eider-mussel food link in the community of the Ythan estuary. Report Challenger Society 3: 31.

Milne, H. and A. Reed. 1974. Annual production of fledged young from the eider colonies of the St. Lawrence estuary. Canadian Field-Naturalist 88(2): 163-169.

Milonoff, M. 1991. Renesting ability and clutch size in precocial birds. Oikos 62: 189-194.

Milosevich, J. and J. Olson. 1981. Nesting Canada Warblers in Joliet, Illinois. Audubon Bulletin 197: 23-26.

Milsom, T.P. and A. Watson. 1984. Numbers and spacing of summering Snow Buntings and snow cover in the Cairngorms. Scottish Birds 13(1): 19-23.

Mindell, D.P. and R.A. Dotson. 1982. Distribution and abundance of nesting raptors in southwestern Alaska, pp. 112-137 *in* W.N. Ladd and P.F. Schempf, eds. Proceedings. Symposium and workshop on raptor management and biology in Alaska and western Canada. United States Fish and Wildlife Service, Alaska Regional Office, Anchorage, 335 pp.

Mineau, P. and F. Cooke. 1979. Territoriality in Snow Geese or the protection of parenthood. Wildfowl 30: 16-90.

Ministère de l'Environnement du Québec (MENVIQ). 1983. Les réserves écologiques au Québec (ou la protection intégrale et permanente d'échantillons du patrimoine naturel du Québec), 37 pp.

Ministère de l'Environnement du Québec (MENVIQ). 1987. Climatologie du Québec méridional. Prepared by H. Proulx, G. Jacques, A.-M. Lamothe, and J. Litynski. Ministère de l'Environnement du Québec, Direction de la météorologie, 198 pp.

Ministère de l'Environnement du Québec (MENVIQ). 1988. L'environnement au Québec: Un premier bilan: Document technique. Gouvernement du Québec, Québec, 427 pp.

Ministère de l'Environnement du Québec (MENVIQ). 1989. Bassins versants et régions hydrographiques, p. 254 (map) *in* Le Québec statistique. Bureau de la Statistique du Québec, 1028 pp.

Ministère de l'Environnement du Québec (MENVIQ). 1991. Plan de conservation de la Réserve écologique de l'île Brion. Direction de la conservation et du patrimoine écologique and Direction régionale du Bas-Saint-Laurent, Gaspésie, Îles-de-la-Madeleine, 66 pp.

Ministère de l'Environnement du Québec (MENVIQ). 1993. État de l'environnement au Québec, 1992. Guérin, Montréal, 560 pp.

Ministère des Terres et Forêts du Québec (MTFQ). 1977. Carte synthèse des forêts du Québec. Prepared by Darveau, Grenier, Lussier et associés.

Ministère du Loisir, de la Chasse et de la Pêche (MLCP). 1992. Liste des espèces de la faune vertébrée susceptibles d'être désignées menacées ou vulnérables. Gouvernement du Québec, Ministère du Loisir, de la chasse et de la Pêche, publication SP-2124-11-92, 107 pp.

Minock, M.E. 1980. Some relationships between Common Grackles and Yellow-headed Blackbirds nesting in the same marsh. Passenger Pigeon 42(4): 125-127.

Minot, E.O. 1980. Tidal, diurnal and habitat influences on Common Eider rearing activities. Ornis Scandinavica 11: 165-172.

Mirarchi, R.E. and T.S. Baskett. 1994. Mourning Dove (*Zenaida macroura*), *in* A. Poole and F. Gill, eds. The birds of North America, no. 117. The Academy of Natural Sciences, Philadelphia. The American Ornithologists' Union, Washington, 32 pp.

Mitchell, M.H. 1935. The Passenger Pigeon in Ontario. Royal Ontario Museum. Zoological Contribution no. 7, 181 pp.

Mock, D.W. 1978. Pair-formation displays of the Great Egret. Condor 80(2): 159-172.

Mock, D.W. 1985. Siblicidal brood reduction: the prey-size hypothesis. American Naturalist 125: 327-343.

Mock, D.W., T.C. Lamey, C.F. Williams, and A. Pelletier. 1987a. Flexibility in the development of heron sibling aggression: an intraspecific nest of the prey-size hypothesis. Animal behaviour 35: 1386-1393.

Mock, D.W., T.C. Lamey, and B.J. Ploger. 1987b. Proximate and ultimate roles of food amount in regulating egret sibling aggression. Ecology 68(6): 1760-1772.

Moilanen, P. and M. Kuitunen. 1977. Nesting boxes for tree creepers. Suomen Luonto 36(1): 16-17.

Moisan, G. 1966. The Green-winged Teal in the Atlantic Flyway. Naturaliste canadien 93(2): 69-88.

Moisan, G. and R.W. Fyfe. 1967. Ninth census of non-passerine birds in the sanctuaries of the North Shore of the Gulf of St. Lawrence. Canadian Field-Naturalist 81(1): 67-70.

Moisan G. and J.-M. Poulin. 1967. Déplacements saisonniers et taux de mortalité du Goéland argenté *Larus argentatus* des îles Razades, Québec. Naturaliste canadien 94: 421-435.

Moisan, G., R.I. Smith, and R.K. Martinson. 1967. The Green-winged Teal: its distribution, migration and population dynamics. United States Fish and Wildlife Service. Special Scientific Report, Washington, no. 100, 248 pp.

Møller, A.P. 1990a. Male tail length and female mate choice in the monogamous swallow, *Hirundo rustica*. Animal Behaviour 39(3): 458-465.

Møller, A.P. 1990b. Changes in the size of avian breeding territories in relation to the nesting cycle. Animal Behaviour 40(6): 1070-1079.

Møller, A.P. 1991a. The effect of feather nest lining on reproduction in the swallow *Hirundo rustica*. Ornis Scandinavica 22(4): 396-400.

Møller, A.P. 1991b. Defence of offspring by male swallows, *Hirundo rustica*, in relation to participation in extra-pair copulations by their mates. Animal Behaviour 42(2): 261-267.

Monaghan, P., J.D. Uttley, M.D. Burns, C. Thaine, and J. Blackwood. 1989. The relationship between food supply, reproductive effort and breeding success in Arctic Terns *Sterna paradisaea*. Journal of Animal Ecology 58: 261-274.

Mondloch, C.J. and W. Timberlake. 1991. The effect of parental food supply on parental feeding and squab growth in Pigeons, *Columba livia*. Ethology 88: 236-248.

Montagna, W. 1943. Weights and plumages of the Horned Lark of central New York. Auk 60: 210-215.

Montevecchi, W.A. 1996. Studies of high-latitude seabirds. 4. Trophic relationships and energetics of endotherms in cold ocean systems. Canadian Wildlife Service. Occasional Paper no. 91, 73 pp.

Montevecchi, W.A. and L.M. Tuck. 1987. Newfoundland birds: exploitation, study, conservation. Nuttall Ornithological Club, Cambridge, Massachusetts, 273 pp.

Montevecchi, W.A., D.K. Cairns, A.E. Burger, R.D. Elliot, and J. Wells. 1987. The status of the Common Black-headed Gull in Newfoundland and Labrador. American Birds 41(2): 197-203.

Montevecchi, W.A., R. Myers, J. Neuman, and J.W. Chardine. 1992. Seabird production and capelin harvest in the northeast Atlantic in 1991: ice, anomally, ignorance, p. 14 *in* Studies on high-latitude marine homeotherms: energetics, trophic interactions, and multi-species bioindication. Conference abstracts. Workshop 3-7 April 1992, Memorial University, St. John's, Newfoundland. Canadian Wildlife Service and Memorial University, 17 pp.

Montgomerie, R.D. 1988. Seasonal patterns of mate guarding in Lapland Longspurs, pp. 442-453 *in* H. Ouellet, ed. Acta XIX Congressus Internationalis Ornithologici, Ottawa, 22-29 June 1986. Ottawa University Press, vol. I, pp. 1-1404; vol. II, pp. 1405-2815.

Montgomery, G.H. Jr. and W.H. Rawlings. 1947. Report on birds, pp. 6-22 *in* L.M. Terrill, ed. Thirtieth annual report. Province of Quebec Society for the Protection of Birds, Montréal, 27 pp.

Moore, F.R. and P.A. Simm. 1985. Migratory disposition and choice of diet by the Yellow-rumped Warbler (*Dendroica coronata*). Auk 102(4): 820-826.

Moore, H. 1992. 4. Upland second-growth oak-hickory forest, p. 9 *in* R.M. Marshall, ed. Resident bird counts 1991. Journal of Field Ornithology 63(1)(supplement): 1-123.

Moore, P.D. and D.J. Bellamy. 1974. Peatlands. Springer-Verlag, New York, 221 pp.

Moore, R.T. 1912. The Least Sandpiper during the nesting season in the Magdalen Islands. Auk 24: 210-223.

Moore, R.T. 1913. The Fox Sparrow as a songster. Auk 30(2): 177-187.

Moore, W.G. 1968. A Dictionary of geography: definitions and explanations of terms used in physical geography, 4th edition. Penguin, Harmondsworth, 234 pp.

Morehouse, E.L. and R. Brewer. 1968. Feeding of nestling and fledging Eastern Kingbirds. Auk 85(1): 44-54.

Morency, P. 1992. Lumière des oiseaux, histoires naturelles du Nouveau Monde. Boréal, Seuil, 331 pp.

Moreno, J. 1984. Parental care of fledged young, division of labor, and the development of foraging techniques in the Northern Wheatear (*Oenanthe oenanthe* L.). Auk 101(4): 741-752.

Morgan, F.D. 1976. Nesting studies of the Indigo Bunting (*Passerina cyanea*) at Thornhill, Indiana. Indiana Academy of Science 86: 461-465.

Morgan, J.H. and P.A. Pearce. 1969. An adult White-winged Black Tern in New Brunswick. Canadian Field-Naturalist 83(4): 394-395.

Morgan, K. and B. Freedman. 1986. Breeding bird communities in a hardwood forest succession in Nova Scotia. Canadian Field-Naturalist 100(4): 506-519.

Morimoto, D.C. and F.E. Wasserman. 1991a. Dispersion patterns and habitat associations of Rufous-sided Towhees, Common Yellowthroats, and Prairie Warblers in the southeastern Massachusetts pine barrens. Auk 108(2): 264-276.

Morimoto, D.C. and F.E. Wasserman. 1991b. Intersexual and interspecific differences in the foraging behavior of Rufous-sided Towhees, Common Yellowthroats and Prairie Warblers in the pine barrens of southeastern Massachusetts. Journal of Field Ornithology 62(4): 436-449.

Morneau, F., R. Décarie, and J.-L. DesGranges. 1990. Effet du traitement des arbres ornementaux au diazinon (Basudin) et à l'acéphate (Orthene) sur le Merle d'Amérique (*Turdus migratorius*) en milieu urbain. Report for Environment Canada by GREBE. Canadian Wildlife Service, Québec Region. Technical Report Series no. 104, 42 pp. [An abstract is available in English.]

Morneau, F., S. Brodeur, R. Décarie, S. Carrière, and D.M. Bird. 1994. Abundance and distribution of nesting Golden Eagles in Hudson Bay, Québec. Journal of Raptor Research 28(4): 220-225.

Morris, M.M.J. and R.E. Lemon. 1983. Characteristics of vegetation and topography near Red-shouldered Hawk nests in southwestern Québec. Journal of Wildlife Management 47(1): 138-145.

Morris, M.M.J. and R.E. Lemon. 1988. Mate choice in American Redstarts: by territory quality? Canadian Journal of Zoology 66: 2255-2261.

Morris, M.M.J., B.L. Penak, R.E. Lemon, and D.M. Bird. 1982. Characteristics of Red-shouldered Hawk nests in southwestern Québec. Canadian Field-Naturalist 96(2): 139-146.

Morris, R.D. and J.S. Quinn. 1979. Reproductive and chick feeding ecology of Caspian Terns (*Sterna caspia*) on South Limestone Island, Georgian Bay. Unpublished report.

Morris, R.F. 1963. The dynamics of epidemic Spruce Budworm populations. Memoirs of the Entomological Society of Canada, no. 31, 270 pp.

Morris, R.F., W.F. Cheshire, C.A. Miller, and D.G. Mott. 1958. The numerical response of avian and mammalian predators during a gradation of the Spruce Budworm. Ecology 39: 487-494.

Morris, R.O. 1905. The Gadwall and Yellow Rail near Springfield, Massachusetts. Auk 22: 207-208.

Morrison, M.L. 1980. Seasonal aspects of the predatory behavior of Loggerhead Shrikes. Condor 82(3): 296-300.

Morrison, M.L. 1981. The structure of western warbler assemblages: analysis of foraging behavior and habitat selection in Oregon. Auk 98(3): 578-588.

Morrison, M.L. 1986. Bird populations as indicators of environmental change, pp. 429-451 *in* R.F. Johnston, ed. Current Ornithology, vol. 3. Plenum Press, New York.

Morrison, M.L. and E.C. Meslow. 1983. Bird community structure on early-growth clearcuts in western Oregon. American Midland Naturalist 110: 129-137.

Morrison, M.L., B.G. Marcot, and R.W. Mamman. 1992. Wildlife-habitat relationships: concepts and applications. University of Wisconsin Press, Madison, 343 pp.

Morrison, R.I.G. 1983. A hemispheric perspective on the distribution and migration of some shorebirds in North and South America, pp. 84-94 *in* H. Boyd, ed. Proceedings. First western hemisphere waterfowl and waterbird symposium, 25-28 May 1982, Edmonton. International Waterfowl Research Bureau. Canadian Wildlife Service, 147 pp.

Morrison, R.I.G. 1984. Migration systems of some New World shorebirds, pp. 134-167 *in* J. Burger and B.L. Olla, eds. Shorebirds: migration and foraging behavior. Plenum Press, New York, 437 pp.

Morrison, R.I.G. 1987. Marbled Godwit, p. 528 *in* M.D. Cadman, P.F.J. Eagles, and F.M. Helleiner, eds. Atlas of the breeding birds of Ontario. University of Waterloo Press, Waterloo, 617 pp.

Morrison, R.I.G. and R.K. Ross. 1989. Atlas of Nearctic shorebirds on the coast of South America. Canadian Wildlife Service, Ottawa, vol. 1, pp. 1-128; vol. 2, pp. 131-325.

Morrison, R.I.G., T.H. Manning, and J.A. Hagar. 1976. Breeding of the Marbled Godwit, *Limosa fedoa*, in James Bay. Canadian Field-Naturalist 90(4): 487-490.

Morrison, R.I.G., R.W. Butler, H.L. Dickson, A. Bourget, P.W. Hicklin, and J.P. Goossen. 1991. Potential Western Hemisphere shorebird reserve network. Sites for migrant shorebirds in Canada. Canadian Wildlife Service, Ottawa. Technical Report Series no. 144, 98 pp.

Morse, D.H. 1966. The context of songs in the Yellow Warbler. Wilson Bulletin 78(4): 444-445.

Morse, D.H. 1967a. Competitive relationships between Parula Warblers and other species during the breeding season. Auk 84(4): 490-502.

Morse, D.H. 1967b. Foraging relationships of Brown-headed Nuthatches and Pine Warblers. Ecology 48: 94-193.

Morse, D.H. 1967c. The contexts of songs in the Black-throated Green and Blackburnian warblers. Wilson Bulletin 79: 62-72.

Morse, D.H. 1968. A quantitative study of foraging of male and female spruce-wood warblers. Ecology 49(4): 779-784

Morse, D.H. 1970. Ecological aspects of some mixed-species foraging flocks of birds. Ecological Monographs 40: 119-168.

Morse, D.H. 1971a. Effects of the arrival of a new species upon habitat utilization by two forest thrushes in Maine. Wilson Bulletin 83(1): 57-65.

Morse, D.H. 1971b. The foraging of warblers isolated on small islands. Ecology 52(2): 216-228.

Morse, D.H. 1973. The foraging of small populations of Yellow Warblers and American Redstarts. Ecology 54(2): 346-355.

Morse, D.H. 1974. Foraging of Pine Warblers allopatric and sympatric to Yellow-throated Warbler. Wilson Bulletin 86: 474-477.

Morse, D.H. 1976a. Hostile encounters among spruce-wood warblers (*Dendroica*: Parulidae). Animal Behaviour 24: 764-771.

Morse, D.H. 1976b. Variables affecting the density and territory size of breeding spruce-woods warblers. Ecology 57: 290-301.

Morse, D.H. 1977. The occupation of small islands by passerine birds. Condor 79(4): 399-412

Morse, D.H. 1978. Populations of Bay-breasted and Cape May warblers during an outbreak of the Spruce Budworm. Wilson Bulletin 90(3): 404-413.

Morse, D.H. 1979. Habitat use by the Blackpoll Warbler. Wilson Bulletin 91(2): 234-243.

Morse, D.H. 1980. Foraging and coexistence of spruce-woods warblers. Living Bird 18: 7-25.

Morse, D.H. 1985. Habitat selection in North American parulid warblers, pp. 131-157 *in* M.L. Cody, ed. Habitat selection in birds. Academic Press, New York, 558 pp.

Morse, D.H. 1989a. American warblers, an ecological and behavioral perspective. Harvard University Press, Cambridge, Massachusetts, 406 pp.

Morse, D.H. 1989b. Song patterns of warblers at dawn and dusk. Wilson Bulletin 101(1): 26-35.

Morse, D.H. 1993. Black-throated Green Warbler (*Dendroica virens*), *in* A. Poole and F. Gill, eds. The birds of North America, no. 55. The Academy of Natural Sciences, Philadelphia. The American Ornithologists' Union, Washington, 20 pp.

Morse, D.H. 1994. Blackburnian Warbler (*Dendroica fusca*), *in* A. Poole and F. Gill, eds. The birds of North America, no. 102. The Academy of Natural Sciences, Philadelphia. The American Ornithologists' Union, Washington, 20 pp.

Morse, T.E. and H.M. Wight. 1969. Dump nesting and its effect on production in Wood Ducks. Journal of Wildlife Management 33(2): 284-293.

Morse, T.E., J.L. Jakabosky, and V.P. McCrow. 1969. Some aspects of the breeding biology of the Hooded Merganser. Journal of Wildlife Management 33(3): 596-604.

Morton, E.S. 1980. Adaptations to seasonal changes by migrant land birds in the Panama Canal zone, pp. 347-453 *in* A. Keast and E.S. Morton. Migrant birds in the neotropics: ecology, behavior, distribution, and conservation. Smithsonian Institution Press, Washington, 576 pp.

Morton, E.S. 1982. Grading, discreteness, redundancy, and movitation-structural rules, pp. 183-212 *in* D. Kroodsma and E.H. Miller, eds. Acoustic communication in birds, vol. 1. Academic Press, New York.

Morton, E.S. 1987. Variation in mate guarding intensity by male Purple Martins. Behaviour 101: 211-224.

Morton, E.S. and K.C. Derrickson. 1990. The biological significance of age-specific return schedules in breeding Purple Martins. Condor 92(4): 1040-1050.

Morton, E.S. and M.D. Shalter. 1977. Vocal response to predators in pair-bonded Carolina Wrens. Condor 79(2): 222-227.

Morton, E.S., L. Forman, and M. Braun. 1990. Extrapair fertilizations and the evolution of colonial breeding in Purple Martins. Auk 107(2): 275-283.

Mosby, H.S. 1973. The changed status of the Wild Turkey over the past three decades, pp. 71-76 *in* G.C. Sanderson and H.C. Schultz, eds. Wild Turkey management: current problems and programs. The Missouri Chapter of the Wildlife Society and University of Missouri Press, Columbia.

Mosher, J.A. and P.F. Matray. 1974. Size dimorphism: a factor in energy savings for Broad-winged Hawks. Auk 91(2): 325-341.

Moskoff, W. 1995a. Solitary Sandpiper (*Tringa solitaria*), *in* A. Poole and F. Gill, eds. The birds of North America, no. 156. The Academy of Natural Sciences, Philadelphia. The American Ornithologists' Union, Washington, 16 pp.

Moskoff, W. 1995b. Veery (*Catharus fuscescens*), *in* A. Poole and F. Gill, eds. The birds of North America, no. 142. The Academy of Natural Sciences, Philadelphia. The American Ornithologists' Union, Washington, 16 pp.

Moskovits, D. 1978. Winter territorial and foraging behavior of Red-headed Woodpeckers in Florida. Wilson Bulletin 90(4): 521-535.

Mosquin, T. and P.G. Whiting. 1992. Canada country study of biodiversity: taxonomic and ecological census, economic benefits, conservation costs and unmet needs. Canadian Centre for Biodiversity, Ottawa, 282 pp.

Moss, R. and A. Watson. 1984. Maternal nutrition, egg quality and breeding success of Scottish Ptarmigan (*Lagopus mutus*). Ibis 126(2): 212-220.

Mossop, D.H. 1988. Winter survival and spring breeding strategies of Willow Ptarmigan, pp. 330-378 *in* A.T. Bergerud and M.W. Gratson, eds. Adaptive strategies and population ecology of northern Grouse, vol. 1. Population studies. University of Minnesota Press, Minneapolis, 419 pp.

Mossop, D.H., R.G. Bromley, and M. Shoesmith. 1985. The status and management of the Gyrfalcon (*Falco rusticolus*) in Canada. A Canadian position paper for the 1985 meeting of the parties in the Convention of the International Trade in Endangered Species of Wild Fauna and Flora (CITES), Buenos Aires, Argentina.

Mountjoy, D.J. and R.J. Robertson. 1988a. Why are waxwings waxy? Delayed plumage maturation in the Cedar Waxwing. Auk 105(1): 61-69.

Mountjoy, D.J. and R.J. Robertson. 1988b. Nest-construction tactics in the Cedar Waxwing. Wilson Bulletin 100(1): 128-130.

Mousley, H. 1916a. Five years personal notes and observations on the birds of Hatley, Stanstead County, Quebec 1911-1915. Auk 33(1): 57-73; (2): 168-186.

Mousley, H. 1916b. The breeding of the prairie Horned Lark at Hatley, Stanstead County, Quebec. Auk 33(3): 281-286.

Mousley, H. 1917a. The breeding of the Black-throated Blue Warbler at Hatley, Stanstead County, Quebec 1916. Auk 34(2): 187-191.

Mousley, H. 1917b. The Cape May and other rare warblers at Hatley, Stanstead County, Quebec. Auk 34(2): 214.

Mousley, H. 1917c. Hudsonian Chickadees at Hatley, Stanstead County, Quebec. Auk 34(2): 215.

Mousley, H. 1917d. Winter birds at Hatley, Stanstead County, Quebec. Auk 34(2): 217.

Mousley, H. 1917e. A study of subsequent nestings after the loss of the first. Auk 34(4): 381-393.

Mousley, H. 1917f. Evening Grosbeaks at Hatley, Stanstead County, Quebec. Auk 34(4): 479.

Mousley, H. 1917g. Rare warblers at Hatley, Stanstead County, Quebec. Auk 34(4): 484.

Mousley, H. 1917h. Labrador Chickadees at Hatley, Stanstead County, Quebec. Auk 34(4): 486-487.

Mousley, H. 1917i. Sap drinking habits of warblers. Auk 34(4): 484-485.

Mousley, H. 1918a. Labrador and Acadian chickadees at Hatley, Stanstead County, Quebec. Auk 35(1): 83.

Mousley, H. 1918b. Uncommon birds at Hatley, Stanstead County, Quebec. Auk 35(1): 84-85.

Mousley, H. 1918c. Subsequent nestings. Auk 35(2): 237-238.

Mousley, H. 1918d. Further notes and observations on the birds of Hatley, Stanstead County, Quebec, 1916-1917. Auk 35(3): 289-310.

Mousley, H. 1919a. The singing tree, or how near to the nest do the male birds sing? Auk 36(3): 339-348.

Mousley, H. 1919b. Further notes and observations on the birds of Hatley, Stanstead County, Quebec, 1918. Auk 36(4): 472-487.

Mousley, H. 1921a. The Horned Grebe (*Colymbus auritus*) at Hatley, Stanstead County, Quebec. Auk 38(1): 108.

Mousley, H. 1921b. Three important records from Hatley, Stanstead County, Quebec. Auk 38(1): 126-127.

Mousley, H. 1921c. Further notes and observations on the birds of Hatley, Stanstead County, Quebec, 1919. Auk 38(1): 51-59.

Mousley, H. 1921d. Which sex selects the nesting locality? Auk 38(3): 321-328.

Mousley, H. 1921e. Goldfinches and Purple Finches wintering at Hatley, Stanstead County, Quebec. Auk 38(4): 606.

Mousley, H. 1921f. The Philadelphia Vireo (*Vireosylva philadelphica*) in the province of Quebec. Auk 38(4): 607.

Mousley, H. 1921g. Early spring records at Hatley, Stanstead County, Quebec. Auk 38(4): 609.

Mousley, H. 1922. Palm Warbler (*Dendroica palmarum palmarum*) at Hatley, Quebec. Canadian Field-Naturalist 36(7): 137-138.

Mousley, H. 1923a. The Starling (*Sturnus vulgaris*) breeding at Hatley, Quebec. Auk 40(3): 539.

Mousley, H. 1923b. Further notes on the breeding of the Starling (*Sturnus vulgaris*) at Hatley, Quebec. Auk 40(4): 694-695.

Mousley, H. 1923c. The Orange-crowned Warbler (*Vermivora c. celata*) in the province of Quebec. Canadian Field-Naturalist 37(9): 168-169.

Mousley, H. 1924a. Further notes on the Starling (*Sturnus vulgaris*) in Canada. Auk 41(1): 158-159.

Mousley, H. 1924b. A study of the home life of the Northern Parula and other warblers at Hatley, Stanstead County, Quebec, 1921-1922. Auk 41(2): 263-288.

Mousley, H. 1924c. Further notes and observations on the birds of Hatley, Stanstead County, Quebec, 1921-1923. Auk 41(4): 572-589.

Mousley, H. 1925a. Further notes on the breeding of the Starling (*Sturnus vulgaris*) at Hatley, Quebec, 1924. Auk 42(2): 273-275.

Mousley, H. 1925b. Further notes on the birds of Hatley, Quebec, 1924. Auk 42(2): 281-282.

Mousley, H. 1926. A further study of the home life of the Northern Parula and of the Yellow Warbler and Ovenbird. Auk 43(2): 184-197.

Mousley, H. 1927. Further notes on the birds of Hatley, Quebec, 1925-1926. Auk 44(4): 520-523.

Mousley, H. 1928. A further study of the home life of the Northern Parula Warbler (*Compsothlypis americana usneae*), 1925. Auk 45(4): 475-479.

Mousley, H. 1930a. The home life of the American Goldfinch. Canadian Field-Naturalist 44(8): 177-179.

Mousley, H. 1930b. A further study of the home life of the American Goldfinch. Canadian Field-Naturalist 44(9): 204-207.

Mousley, H. 1931a. A study of the home life of the Alder Flycatcher (*Empidonax traillii traillii*). Auk 48(4): 547-552.

Mousley, H. 1931b. Notes on the home life of the Virginia Rail. Canadian Field-Naturalist 45(3): 65-66.

Mousley, H. 1931c. Reminiscences of the home life of the Black-billed Cuckoo. Canadian Field-Naturalist 45(4): 78-79.

Mousley, H. 1932a. Further notes on the birds, orchids, ferns and butterflies of the province of Quebec, 1931. Canadian Field-Naturalist 46(8): 171-173.

Mousley, H. 1932b. A third study of the home life of the Eastern Goldfinch (*Spinus tristis tristis*). Canadian Field-Naturalist 46(9): 200-203.

Mousley, H. 1934a. A study of the home life of the Short-billed Marsh Wren (*Cistothorus stellaris*). Auk 51(2): 439-445.

Mousley, H. 1934b. A study of the home life of the Northern Crested Flycatcher (*Myiarchus crinitus boreus*). Auk 51(1): 207-216.

Mousley, H. 1935. The birth of a Snipe family (*Capella delicata*). Auk 52(4): 408-411.

Mousley, H. 1936. Birth of a Black Duck family (*Anas rubripes tristis*). Auk 53(4): 377-380.

Mousley, H. 1937a. Nesting habits of the Spotted Sandpiper. Auk 54(4): 445-451.

Mousley, H. 1937b. A study of a Virginia Rail and Sora Rail at their nests. Wilson Bulletin 49(2): 80-84.

Mousley, H. 1938. A study of the home life of the Eastern Belted Kingfisher. Wilson Bulletin 50(1): 1-12.

Mousley, H. 1939a. Nesting behavior of Wilson's Snipe and Spotted Sandpiper. Auk 56(2): 129-133.

Mousley, H. 1939b. Home life of the American Bittern. Wilson Bulletin 51(2): 83-85.

Mousley, H. 1940. Further notes on the nesting habits of the Virginia Rail. Wilson Bulletin 52(2): 87-90.

Mousley, H. 1945a. The nesting of the Eastern Green Heron *Butorides virescens virescens* (Linn.) in the province of Quebec, May 1942. Canadian Field-Naturalist 59(2): 51-52.

Mousley, H. 1945b. The Pied-billed Grebe, Florida Gallinule, and Eastern Mourning Dove in the Montreal district. Annual report, Provancher Society of Natural History of Canada, pp. 48-58.

Mousley, H. 1946. The Eastern Kingbird (*Tyrannus tyrannus* L.). Annual report, Provancher Society of Natural History of Canada, pp. 34-40.

Mousseau, P. 1983. Écologie de la colonie du Bihoreau à couronne noire (*Nycticorax nycticorax*) de l'île aux Hérons, rapides de Lachine, Québec. Report for Direction des réserves écologiques et des sites naturels, Ministère de l'Environnement du Québec, by Centre de recherches écologiques de Montréal, 68 pp.

Mousseau, P. 1984a. Établissement du Goéland à bec cerclé, *Larus delawarensis*, au Québec. Canadian Field-Naturalist 98(1): 29-37.

Mousseau, P. 1984b. L'avifaune nicheuse de la Réserve écologique du Pin-rigide, Saint-Antoine-Abbé, Québec. Centre de recherches écologiques de Montréal, 227 pp.

Mousseau, P. 1984c. Synthèse des données écologiques de l'île Brion, Îles-de-la-Madeleine, Québec. Ministère de l'Environnement du Québec, Direction des réserves écologiques et des sites naturels, 173 pp.

Mousseau, P. 1985. État de la colonie du Bihoreau à couronne noire (*Nycticorax nycticorax*) de Newport-Point après la mise en place d'un centre de service pour bateaux. Report for Ministère de l'Agriculture, des Pêcheries et de l'Alimentation du Québec, by Centre de recherches écologiques de Montréal, 47 pp.

Mousseau, P. and J.-P. Beaumont. 1982. Distribution, caractéristiques biophysiques et de l'utilisation, et facteurs limitant des aires d'alimentation du Grand Héron (*Ardea herodias*) et du Bihoreau à couronne noire (*Nycticorax nyticorax*) au lac Saint-Louis et au bassin de La Prairie, Québec. Report for Direction Environnement, Hydro-Québec, by Centre de recherches écologiques de Montréal, 191 pp.

Mousseau, P. and M.C. Lagrenade. 1979. Étude du succès de reproduction de la colonie de Goéland à bec cerclé du pont Champlain, Montréal. Report for Canadian Wildlife Service, by Centre de recherches écologiques de Montréal, 72 pp.

Mousseau, P. and M.C. Lagrenade. 1980. Succès de reproduction et contaminants présents chez le Goéland à bec cerclé du sud-ouest du Québec. Report for Canadian Wildlife Service, by Centre de recherches écologiques de Montréal, 128 pp.

Mousseau, P., N. David, R. McNeil, and J. Burton. 1976. Les Îles-de-la-Madeleine: sites de nidification et d'alimentation des oiseaux aquatiques. Report for Office de planification et de développement du Québec. Centre de recherches écologiques de Montréal (CREM) and Centre de Recherche en Sciences de l'Environnement (CERSE), Montréal, 204 pp.

Moyles, D.L.J. and D.A. Boag. 1981. Where, when, and how male Sharp-tailed Grouse establish territories on arenas. Canadian Journal of Zoology 59(8): 1576-1581.

Moynihan, M. 1955. Some aspects of reproductive behavior in the Black-headed Gull (*Larus ridibundus ridibundus* L.) and related species. E.J. Brill, Leiden, 201 pp.

Mueller, H.C. 1971. Oddity and specific searching image more important than conspicuousness in prey selection. Nature 233: 345-346.

Mueller, H.C. and D.D. Berger. 1970. Prey preferences in the Sharp-shinned Hawk: the roles of sex, experience and motivation. Auk 87(3): 452-457.

Mueller, H.C., D.D. Berger, and G. Allez. 1979. Age and sex differences in size of Sharp-shinned Hawks. Bird-Banding 50(1): 33-44.

Mueller, H.C., N.S. Mueller, and P.G. Parker. 1981. Observation of brood of Sharp-shinned Hawk in Ontario, with comments on the functions of sexual dimorphism. Wilson Bulletin 93(1): 85-92.

Muir, D. and D.M. Bird. 1984. Food of Gyrfalcons at a nest on Ellesmere Island. Wilson Bulletin 96(3): 464-467.

Muldal, A., H.L. Gibbs, and R.J. Robertson. 1985. Preferred nest spacing of an obligate cavity-nesting bird, the Tree Swallow. Condor 87(3): 356-363.

Mulligan, G.A. 1992. Common and botanical names of weeds in Canada/Noms populaires et scientifiques des plantes nuisibles du Canada. Centre de recherches sur les terres et les ressources biologiques, Direction de la recherche, Agriculture Canada, Ottawa. Publication 1397/B, 131 pp.

Mulligan, J.A. 1966. Singing behavior and its development in the Song Sparrow *Melospiza melodia*. University of California. Publications in Zoology 81: 1-76.

Mulvihill, R.S. 1992a. American Redstart, pp. 332-333 *in* D.W. Brauning, ed. Atlas of breeding birds in Pennsylvania. University of Pittsburgh Press, Pittsburgh, 484 pp.

Mulvihill, R.S. 1992b. Chimney Swift, pp. 174-175 *in* D.W. Brauning, ed. Atlas of breeding birds in Pennsylvania. University of Pittsburgh Press, Pittsburgh, 484 pp.

Mulvihill, R.S. 1992c. Dark-eyed Junco, pp. 394-395 *in* D.W. Brauning, ed. Atlas of breeding birds in Pennsylvania. University of Pittsburgh Press, Pittsburgh, 484 pp.

Mulvihill, R.S. 1992d. Magnolia Warbler, pp. 312-313 *in* D.W. Brauning, ed. Atlas of breeding birds in Pennsylvania. University of Pittsburgh Press, Pittsburgh, 484 pp.

Mundinger, P.C. and S. Hope. 1982. Expansion of the winter range of the House Finch: 1947-79. American Birds 36(4): 347-353.

Munoff, J.A. 1963. Food habits, growth and mortality in nestling Marsh Hawks. Kingbird 13: 67-74.

Munro, D.A. 1979. Tribute to Hoyes Lloyd, 1888-1978. Canadian Field-Naturalist 93: 331-336.

Munro, H.L. and R.C. Rounds. 1985. Selection of artificial nest sites by five sympatric passerines. Journal of Wildlife Management 49(1): 264-276.

Munro, J. and J. Bédard. 1977a. Crèche formation in the Common Eider. Auk 94: 759-771.

Munro, J. and J. Bédard. 1977b. Gull predation and crèching behaviour in the Common Eider. Journal of Animal Ecology 46: 799-810.

Munro, J.A. 1935. Barrow's Golden-eye nesting in marmot's burrow. Condor 37: 82-83.

Munro, J.A. 1939. Studies of waterfowl in British Columbia: Barrow's Golden-eye, American Golden-eye. Transactions. Royal Canadian Institute 48: 259-318.

Munro, J.A. 1941a. Studies of waterfowl in British Columbia: Greater Scaup duck, Lesser Scaup duck. Canadian Journal of Research 19: 113-138.

Munro, J.A. 1941b. Studies of waterfowl in British Columbia: the grebes. British Columbia Provincial Museum, Victoria. Occasional Paper no. 3, 71 pp.

Munro, J.A. 1942. Studies of waterfowl in British Columbia: Buffle-head. Canadian Journal of Research 20(6): 133-160.

Munro, J.A. 1945. Observation of the loon in the Cariboo parklands, British Columbia. Auk 62: 38-49.

Munro, J.A. 1949. Studies of waterfowl in British Columbia: Green-winged Teal. Canadian Journal of Research 27(3): 149-178.

Munro, J.A. 1956. Two records of Barred Owls in southern British Columbia. Murrelet 37(3): 37.

Munro, J.A. and W.A. Clemens. 1939. Waterfowl in relation to the spawning of herring in British Columbia. Biological Board of Canada, 46 pp.

Murie, O.J. 1929. Nesting of the Snowy Owl. Condor 31(1): 3-12.

Murie, O.J. 1959. Fauna of the Aleutian Islands and Alaska Peninsula. United States Fish and Wildlife Service. North American Fauna, no. 61.

Murphy, E.C. 1978. Breeding ecology of House Sparrows: spatial variation. Condor 80: 180-193.

Murphy, M.T. 1983a. Nest success and nesting habits of Eastern Kingbirds and other flycatchers. Condor 85(2): 208-219.

Murphy, M.T. 1983b. Ecological aspects of the reproductive biology of Eastern Kingbirds: geographic comparisons. Ecology 64(4): 914-928.

Murphy, M.T. 1986a. Brood parasitism of Eastern Kingbirds by Brown-headed Cowbirds. Auk 103(3): 626-628.

Murphy, M.T. 1986b. Temporal components of reproductive variability in Eastern Kingbirds (*Tyrannus tyrannus*). Ecology 67(6): 1483-1492.

Murphy, M.T. 1987. The impact of weather on kingbird foraging behavior. Condor 89(4): 721-730.

Murphy, M.T. and R.C. Fleischer. 1986. Body size, nest predation, and reproductive patterns in Brown Thrashers and other mimids. Condor 88(4): 446-455.

Murphy, R.K. 1992. Long-eared Owl ingests nestlings' feces. Wilson Bulletin 104(1): 192-193.

Murphy, R.K., M.W. Gratson, and R.N. Rosenfield. 1988. Activity and habitat use by a breeding male Cooper's Hawk in a suburban area. Journal of Raptor Research 22(4): 97-100.

Murray, B.G. Jr. 1969. A comparative study of Le Conte's and Sharp-tailed sparrows. Auk 86: 199-231.

Murray, B.G. Jr. and F.B. Gill. 1976. Behavioral interactions of Blue-winged and Golden-winged warblers. Wilson Bulletin 88(2): 231-254.

Murray, B.G. Jr. and J.R. Jehl, Jr. 1964. Weights of autumn migrants from coastal New Jersey. Bird-Banding 35(4): 253-263.

Musselman, T.E. 1941. Bluebird mortality in 1940. Auk 58(3): 409-410.

Myers, J.P. and L.P. Myers. 1979. Shorebirds of coastal Buenos Aires province, Argentina. Ibis 121: 186-200.

Myres, M.T. 1959a. The behaviour of the sea ducks and its value in the systematics of the tribes Mergini and Somateriini of the family Anatidae. Ph.D. dissertation, University of British Columbia, Vancouver.

Myres, M.T. 1959b. Display behaviour of Bufflehead, scoters and goldeneyes at copulation. Wilson Bulletin 71: 159-168.

Nagy, A.C. 1977. Population trend indices based on 40 years of autumn counts at Hawk Mountain Sanctuary in north-eastern Pennsylvania, pp. 243-253 *in* R.D. Chancellor, ed. Proceedings. World conference on birds of prey, 1-3 October, Vienna. International Council for Bird Preservation, Hampshire, Great Britain, 442 pp.

Naiman, R.J., J.M. Melillo, and J.E. Hobbie. 1986. Ecosystem alteration of boreal forest streams by beaver (*Castor canadensis*). Ecology 67: 1254-1269.

Nakashima, D.J. and D.J. Murray. 1988. The Common Eider (*Somateria mollissima sedentaria*) of eastern Hudson Bay: a survey of nest colonies and Inuit ecological knowledge. Environmental Studies Revolving Funds Report, no. 102, Ottawa, 174 pp.

Nash, S.V. and J.A. Dick. 1981. First documentation of Greater Yellowlegs breeding in Ontario. Ontario Field Biologist 35: 48.

National Geographic Society. 1987. A field guide to the birds of North America, 2nd edition. National Geographic Society, Washington, 464 pp.

National Museum of Man. 1983. Le bois et la main-d'oeuvre au XIXe siècle. National Museum of Canada, no. 45, 7 pp.

National Wetlands Working Group. 1981. Wetlands of Canada (maps)/Terres humides du Canada (cartes). Ecological Land Classification Series no. 14. Minister of Supply and Services, Hull.

National Wetlands Working Group. 1988. Wetlands of Canada. Ecological Land Classification Series, no. 24. Sustainable Development Branch, Environment Canada, Ottawa, Ontario, and Polyscience Publications Inc., Montréal, Québec, 452 pp.

Natural Resources Canada. 1994. Données altimétriques et bathymétriques intégrées. Image tirée de la base de données GÉOSCOPE [Données d'orign: Integrated elevation and bathymetry. Digital data NOAA/NGDC/WD-A, Boulder, Colorado], adapted by Service des technologies à référence spatiale du MRN.

Naugler, C.T. 1993. American Tree Sparrow (*Spizella arborea*), *in* A. Poole, P. Stettenheim, and F. Gill, eds. The birds of North America, no. 37. The Academy of Natural Sciences, Philadelphia. The American Ornithologists' Union, Washington, 12 pp.

Naylor, B.J. and J.F. Bendell. 1989. Clutch size and egg size of Spruce Grouse in relation to spring diet, food supply, and endogenous reserves. Canadian Journal of Zoology 67(4): 969-980.

Nazar, W., M. Nazar, and I.F. Keymer. 1988. Barn Owls nesting in wells in Portugal. British Birds 81(5): 237-238.

Neidermyer, W.J. and J.J. Hickey. 1977. The Monk Parakeet in the United States, 1970-75. American Birds 31(3): 273-278.

Nelson, A.L. 1934. Some early summer food preferences of the American Raven in southeastern Oregon. Condor 36: 10-15.

Nelson, D.A. and L.J. Croner. 1991. Song categories and their functions in the Field Sparrow (*Spizella pusilla*). Auk 108(1): 42-52.

Nelson, J.B. 1978a. The Gannet. Buteo Books, Vermillion, South Dakota, 336 pp.

Nelson, J.B. 1978b. The Sulidae: gannets and boobies. Aberdeen University Studies Series no. 154. Oxford University Press, Oxford, 1012 pp.

Nelson, R.W. 1968. Nest-robbing by Cooper's Hawks. Auk 85(4): 696-697.

Nero, R.W. 1956a. A behavior study of the Red-winged Blackbird. I. Mating and nesting activities. Wilson Bulletin 68(1): 5-37.

Nero, R.W. 1956b. A behavior study of the Red-winged Blackbird. II. Territoriality. Wilson Bulletin 68(2): 129-150.

Nero, R.W. 1963. Birds of the Lake Athabasca region, Saskatchewan. Saskatchewan Natural History Society, Regina. Special Publication no. 5, 143 pp.

Nero, R.W. 1972. Further records of summer flocking of Common Loon. Blue Jay 30: 85-86.

Nero, R.W. 1980. The Great Gray Owl – Phantom of the northern forest. Smithsonian Nature Book, 167 pp.

Nero, R.W. 1988. Le Geai bleu. Hinterland Who's Who. Environment Canada, Canadian Wildlife Service, 4 pp. [Also available in English as: Blue Jay.]

Nero, R.W. and H.W. Copland. 1981. High mortality of Great Gray Owls in Manitoba–winter 1980-81. Blue Jay 39:158-165.

Nethersole-Thompson, D. 1966. The Snow Bunting. Oliver and Boyd, London, 316 pp.

Nethersole-Thompson, D. 1976. Recent distribution, ecology and breeding of Snow Buntings in Scotland. Scottish Birds 9: 147-162.

Nettleship, D.N. 1972. Breeding success of the Common Puffin (*Fratercula arctica*) on different habitats at Great Island, Newfoundland. Ecological Monographs 42: 239-268.

Nettleship, D.N. and G. Chapdelaine. 1988. Population size and status of the Northern Gannet *Sula bassanus* in North America, 1984. Journal of Field Ornithology 59(2): 120-127.

Nettleship, D.N. and P.G.H. Evans. 1985. Distribution and status of the Atlantic Alcidae, pp. 53-154 *in* D.N. Nettleship and T.R. Birkhead, eds. The Atlantic Alcidae: the evolution, distribution and biology of the auks inhabiting the Atlantic Ocean and adjacent water areas. Academic Press, London, 574 pp.

Nettleship, D.N. and A.R. Lock. 1973. Tenth census of seabirds in the sanctuaries of the North Shore of the Gulf of St. Lawrence. Canadian Field-Naturalist 87(4): 395-402.

Neuman, J., J.M. Carter, and J.W. Chardine. 1992. Breeding failure in Newfoundland colony of Black-legged Kittiwakes, pp. 14-15 *in* Studies on high-latitude marine homeotherms energetics, trophic interactions and multispecies bio-indication. Conferences abstracts. Workshop 3-7 April 1992, Memorial University, St. John's, Canadian Wildlife Service and Memorial University, 17 pp.

Newman, D.L. 1958. A nesting of the Acadian Flycatcher. Wilson Bulletin 70(2): 130-144.

Newton, I. 1970. Irruption of crossbills in Europe, pp. 337-357 *in* A. Watson, ed. Animal populations in relation to their food resources. Symposium of the British Ecological Society, Blackwell Scientific Publications, Oxford, 477 pp.

Newton, I. 1972. Finches. New Naturalist, A survey of British natural history, no. 55, Collins, London, 288 pp.

Newton, I. 1973. Finches. Taplinger Publishing Company, New York, 288 pp.

Newton, I. 1979. Population ecology of raptors. Buteo Books, Vermillion, South Dakota, 399 pp.

Newton, I. 1985a. Finch, pp. 214-215 *in* B. Campbell and E. Lack, eds. A dictionary of birds. British Ornithologists' Union, Buteo Books, Vermillion, South Dakota, 670 pp.

Newton, I. 1985b. Irruption, pp. 307-309 *in* B. Campbell and E. Lack, eds. A dictionary of birds. British Ornithologists' Union, Buteo Books, Vermillion, South Dakota, 670 pp.

Newton, I., E. Meek, and B. Little. 1978. Breeding ecology of the Merlin in Northumberland. British Birds 71: 376-398.

Nice, M.M. 1937. Studies in the life history of the Song Sparrow, part I. Transactions. Linnaean Society of New York, no. 4.

Nice, M.M. 1943. Studies in the life history of the Song Sparrow, part II. Transactions. Linnaean Society of New York, no. 6.

Nice, M.M. 1961. The belligerency of a pair of Wood Pewees. Audubon Bulletin 118: 1-7.

Nice, M.M. and R.H. Thomas. 1948. A nesting of the Carolina Wren. Wilson Bulletin 60: 139-158.

Nicholls, T.H. and D.W. Warner. 1972. Barred Owl habitat use as determined by radiotelemetry. Journal of Wildlife Management 36(2): 213-224.

Nichols, W. 1985a. House Finch, pp. 376-377 *in* S.B. Laughlin and D.P. Kibbe, eds. The Atlas of breeding birds of Vermont. Vermont Institute of Natural Science, University Press of New England, Hanover, 456 pp.

Nichols, W. 1985b. Mourning Warbler, pp. 320-321 *in* S.B. Laughlin and D.P. Kibbe, eds. The Atlas of breeding birds of Vermont. Vermont Institute of Natural Science, University Press of New England, Hanover, 456 pp.

Nichols, W. 1985c. Tufted Titmouse, pp. 214-215 *in* S.B. Laughlin and D.P. Kibbe, eds. The Atlas of breeding birds of Vermont. Vermont Institute of Natural Science, University Press of New England, Hanover, 456 pp.

Nichols, W.D. 1985d. Lincoln's Sparrow, pp. 350-351 *in* S.B. Laughlin and D.P. Kibbe, eds. The Atlas of breeding birds of Vermont. Vermont Institute of Natural Science, University Press of New England, Hanover, 456 pp.

Nichols, W.D. 1985e. Rusty Blackbird, pp. 364-365 *in* S.B. Laughlin and D.P. Kibbe, eds. The Atlas of breeding birds of Vermont. Vermont Institute of Natural Science, University Press of New England, Hanover, 456 pp.

Nicholson, S.A., J.T. Scott, and A.R. Breisch. 1979. Structure and succession in the tree stratum at Lake George, New York. Ecology 60: 1240-1254.

Nickell, W.P. 1943. Observations on the nesting of the Killdeer. Wilson Bulletin 55: 23-28.

Nickell, W.P. 1951. Studies of habitats, territory, and nests of the Eastern Goldfinch. Auk 68: 447-470.

Nickell, W.P. 1957. Robins use same nest for three sets of eggs in one season. Auk 74(1): 95.

Nickell, W.P. 1965. Habitats, territory and nesting of the catbird. American Midland Naturalist 73(2): 433-478.

Niemi, G.J. and J.M. Hanowski. 1984. Effect of a transmission line on bird populations in the Red Lake peatland, northern Minnesota. Auk 101(3): 487-498.

Nikula, B. 1991. The winter season: New England region. American Birds 45(2): 249-253.

Nilsson, L. 1965. Observations on the spring behaviour of the Red-breasted Merganser. Vår Fågelvärld 24(3): 255-256.

Nilsson, L. 1970. Food-seeking activity of south Swedish diving ducks in the non-breeding season. Oikos 21: 145-154.

Nilsson, S.G. 1977. Adult survival rate of the Black-throated Diver *Gavia arctica*. Ornis Scandinavica 8: 193-195.

Nisbet, I.C.T. 1973. Courtship-feeding, egg-size and breeding success in Common Terns. Nature 241: 141-142.

Nisbet, I.C.T. 1981. Biological characteristics of the Roseate Tern *Sterna dougallii*. United States Fish and Wildlife Service, Newton Corner, Massachusetts, and Massachusetts Audubon Society, Lincoln, 136 pp.

Nisbet, I.C.T. 1984. Migration and winter quarters of North American Roseate Terns as shown by banding recoveries. Journal of Field Ornithology 55(1): 1-17.

Nisbet, I.C.T. 1989. Status and biology of the northeastern population of the Roseate Tern *Sterna dougallii* – a literature survey and update: 1981-89. United States Fish and Wildlife Service, Newton Corner, Massachusetts, 74 pp.

Nisbet, I.C.T. 1992. A closer look: Roseate Tern. Birding 24(5): 304-308.

Nisbet, I.C.T. and W.M. Drury. 1972. Measuring breeding success in Common and Roseate terns. Bird-Banding 43:97-106.

Nisbet, I.C.T. and M.J. Welton. 1984. Seasonal variation in breeding success of Common Terns: consequences of predation. Condor 86(1): 53-60.

Noble, D.G. and J.E. Elliott. 1990. Levels of contaminants in Canadian raptors, 1966 to 1988: effects and temporal trends. Canadian Field-Naturalist 104(2): 222-243.

Noble, G.K. 1936. Courtship and sexual selection of the Flicker (*Colaptes auratus luteus*). Auk 53: 269-282.

Nol, E. 1987. Killdeer, pp. 164-165 *in* M.D. Cadman, P.F.J., Eagles, and F.M. Helleiner, eds. Atlas of the breeding birds of Ontario. University of Waterloo Press, Waterloo, 617 pp.

Nolan, V. Jr. 1968. Eastern Song Sparrow, pp. 1492-1512 *in* A.C. Bent. Life histories of North American cardinals, grosbeaks, buntings, towhees, finches, sparrows, and allies. (O.L. Austin, Jr., ed.). United States National Museum, Washington. [1968, Dover Publications, New York]. Bulletin no. 237, vol. 1, pp. 1-602; vol. 2, pp. 603-1248; vol. 3, pp. 1249-1889.

Nolan, V. Jr. 1975. External differences between newly hatched cuckoos (*Coccyzus americanus* and *C. erythropthalmus*). Condor 77(3): 341.

Nolan, V. [Jr.]. 1978. The ecology and behavior of the Prairie Warbler (*Dendroica discolor*). American Ornithologists' Union. Allen Press, Lawrence. Monograph no. 26, 595 pp.

Nolan, V. Jr. and E.D. Ketterson. 1990. Timing of autumn migration and its relation to winter distribution in Dark-eyed Juncos. Ecology 71(4): 1267-1278.

Nolan, V. Jr. and C.F. Thompson. 1975. The occurrence and significance of anomalous reproductive activities in two North American non-parasitic cuckoos *Coccyzus* spp. Ibis 117(4): 496-503.

Noon, B.R. 1981. The distribution of an avian guild along a temperate elevational gradient: the importance and expression of competition. Ecological Monographs 51(1): 105-124.

Noon, B.R., D.K. Dawson, D.B. Inkley, C.S. Robbins, and S.H. Anderson. 1980. Consistency in habitat preference of forest bird species. Transactions. North American Wildlife and Natural Resources Conference 45: 226-244.

Norberg, R.A. 1987. Evolution, structure and ecology of northern forest owls, pp. 9-43 *in* R.W. Nero, R.J. Clark, R.J. Knapton, and R.H. Hamre, eds. Proceedings. Biology and conservation of northern forest owls. Symposium, 3-7 February, Winnipeg. United States Department of Agriculture, Forest Service, Rocky Mountain Forest and Range Experiment Station, Fort Collins, Colorado. General Technical Report RM-142, 309 pp.

Norberg, R.A. and U.M. Norberg. 1976. Size of fish carried by flying Red-throated Divers *Gavia stellata* (Pont.) to nearby fledged young in nesting tarn. Ornis Fennica 53: 92-95.

Norse, W.J. 1985. Pileated Woodpecker, pp. 168-169 *in* S.B. Laughlin and D.P. Kibbe, eds. The Atlas of breeding birds of Vermont. Vermont Institute of Natural Science, University Press of New England, Hanover, 456 pp.

Norse, W.J. and C. Fichtel. 1985a. Common Merganser, pp. 66-67 *in* S.B. Laughlin and D.P. Kibbe, eds. The Atlas of breeding birds of Vermont. Vermont Institute of Natural Science, University Press of New England, Hanover, 456 pp.

Norse, W.J. and C. Fichtel. 1985b. Great Blue Heron, pp. 38-39 *in* S.B. Laughlin and D.P. Kibbe, eds. The Atlas of breeding birds of Vermont. Vermont Institute of Natural Science, University Press of New England, Hanover, 456 pp.

Norse, W.J. and C. Fichtel. 1985c. Eastern Phoebe, pp. 182-183 *in* S.B. Laughlin and D.P. Kibbe, eds. The Atlas of breeding birds of Vermont. Vermont Institute of Natural Science, University Press of New England, Hanover, 456 pp.

Norton, B.G. 1986. The preservation of species, the value of biological diversity. Princeton University Press, Princeton, New Jersey.

Norton, B.G. 1987. Why preserve natural variety? Princeton University Press, Princeton, New Jersey.

Norton, D.W. 1973. Ecological energetics of Calidrine sandpipers breeding in northern Alaska. Ph.D. dissertation, University of Alaska, Fairbanks.

Norvell, J.R. and P.D. Creighton. 1990. Foraging of Horned Larks and Water Pipits in alpine communities. Journal of Field Ornithology 61(4): 434-440.

Noss, R.F. 1987. From plant communities to landscapes in conservation inventories: a look at the nature conservancy (U.S.A.). Biological Conservation 41: 11-37.

Noss, R.F. 1990. Indicators for monitoring biodiversity: a hierarchical approach. Conservation Biology 4: 355-364.

Noss, R.F. 1991. From endangered species to biodiversity, pp. 227-246 *in* K.A. Kohan, ed. Balancing on the brink of extinction. Island Press, Washington.

Noss, R.F. 1992. Issues of scale in conservation biology, pp. 240-250 *in* P.L. Fidler and S.K. Jain, eds. Conservation Biology. Chapman and Hall, New York.

Noss, R.F. and L.D. Harris. 1986. Nodes, networks and MUMs: preserving diversity at all scales. Environmental Management 10: 299-309.

Novak, P.G. 1988. Clay-colored Sparrow, pp. 440-441 *in* R.F. Andrle and J.R. Carroll, eds. The Atlas of breeding birds in New York State. Cornell University Press, Ithaca, New York, 551 pp.

Novak, P.G. 1989. Breeding ecology and status of the Loggerhead Shrike (*Lanius ludovicianus*) in New York State. M.Sc. thesis, Cornell University, Ithaca, New York 156 pp.

Novak, P.G. 1992. Black Tern, pp. 149-169 *in* K.J. Schneider and D.M. Pence, eds. Migratory nongame birds of management concern in the northeast. United States Department of the Interior, Fish and Wildlife Service, Newton Corner, Massachusetts, 400 pp.

NOVE Environment. 1990. Identification des peuplements forestiers d'intérêt phyto-sociologique. Report for Service de Recherches en environnement et en santé publique, vice-présidence Environnement, Hydro-Québec, 133 pp.

Noyes, J.H. and R.L. Jarvis. 1985. Diet and nutrition of breeding female Redhead and Canvasback ducks in Nevada. Journal of Wildlife Management 49(1): 203-211.

Nudds, T.D. and R.W. Cole. 1991. Changes in populations and breeding success of boreal forest ducks. Journal of Wildlife Management 55(4): 569-573.

Nybo J.O. and G.A. Sonerud. 1990. Seasonal changes on diet of Hawk Owls *Surnia ulula*: importance of snow cover. Ornis Fennica 67: 45-51.

Nyström, B. and H. Nyström. 1987. Biotopval och häckning hos gråsiskor *Carduelis flammea* och snösiskor *C. hornemanni* i Ammarnäsområdet, södra Lappland. [Habitat preferences and breeding biology of Common Redpoll (*Carduelis flammea*) and Arctic Redpoll (*C. hornemanni*) in the area of Ammarnas, southern Lapland; unedited translation, Secretary of State, Canada.]. Vår Fågelvärld 46: 119-128.

Oatman, G.F. 1985a. Common Raven, pp. 208-209 *in* S.B. Laughlin and D.P. Kibbe, eds. The Atlas of breeding birds of Vermont. Vermont Institute of Natural Science, University Press of New England, Hanover, 456 pp.

Oatman, G.F. 1985b. Tennessee Warbler, pp. 278-279 *in* S.B. Laughlin and D.P. Kibbe, eds. The Atlas of breeding birds of Vermont. Vermont Institute of Natural Science, University Press of New England, Hanover, 456 pp.

Oatman, G.F. 1985c. Three-toed Woodpecker, pp. 162-163 *in* S.B. Laughlin and D.P. Kibbe, eds. The Atlas of breeding birds of Vermont. Vermont Institute of Natural Science, University Press of New England, Hanover, 456 pp.

Oatman, G.F. 1985d. Wilson's Warbler, pp. 324-325 *in* S.B. Laughlin and D.P. Kibbe, eds. The Atlas of breeding birds of Vermont. Vermont Institute of Natural Science, University Press of New England, Hanover, 456 pp.

O'Connor, R.J. 1978. Growth strategies of nestling passerines. Living Bird 16: 209-238.

O'Donoghue, C.H. and J.N. Gowanlack. 1919. Notes on the Caspian Tern and the Parasitic Jaeger in Manitoba. Canadian Field-Naturalist 33: 1-6.

Odum, E.P. and E.J. Kuenzler. 1955. Measurement of territory and home range size in birds. Auk 72: 128-137.

Oeming, A.F. 1958. Recoveries in Alberta of banded Snowy Owls. Canadian Field-Naturalist 72: 171-172.

Oeming, A.F. 1964. Banding recovery of Great Gray Owl. Blue Jay 22: 10.

Offutt, G.C. 1965. Behavior of the Tufted Titmouse before and during the nesting season. Wilson Bulletin 77(3): 382-387.

Ogden, J.C. 1978. Recent population trends of colonial wading birds on the Atlantic and Gulf coastal plains, pp. 137-159 *in* A. Sprunt IV, J.C. Ogden, and S. Winckler, eds. Wading birds. Research Report. National Audubon Society no 7. New York, 381 pp.

Ogilvie, M.A. 1976. The winter birds. Michael Joseph Ltd, London, 224 pp.

Ohanjanian, I.A. 1989. Food flights of Red-necked Grebes during the breeding season. Journal of Field Ornithology 60(2): 143-153.

Ohlendorf, H.M., E.E. Klass, and T.E. Kaiser. 1978. Environmental pollutants and eggshell thinning in the Black-crowned Night-Heron, pp. 63-72 *in* A. Sprunt IV, J.C. Ogden, and S. Winckler, eds. Wading birds. Research Report. National Audubon Society no. 7, New York, 381 pp.

Ohlendorf, H.M., D.M. Swineford, and L.N. Locke. 1981. Organochlorine residues and mortality of herons. Pesticides Monitoring Journal 14: 125-135.

Olendorff, R.R. 1976. The food habits of North American Golden Eagles. American Midland Naturalist 95: 231-236.

Oliphant, L.W. 1974. Merlins in Saskatoon. Blue Jay 33: 140-147.

Oliphant, L.W. 1985. North American Merlin breeding survey. Raptor Research 19: 37-41.

Oliphant, L.W. and E. Haug. 1985. Population changes of Saskatoon Merlins. Raptor Research 19: 56-59.

Oliphant, L.W. and W.J.P. Thompson. 1976. Food-caching in Merlins. Canadian Field-Naturalist 90(3): 364-365.

Olpinski, S.C. 1986. Breeding ecology, habitat and morphometrics of Rock Ptarmigan (*Lagopus mutus*) in Nouveau-Québec. M.Sc. thesis, McGill University, Montréal, 40 pp.

Olson, S.T. and W.M. Marshall. 1952. The Common Loon in Minnesota. Minnesota Museum of Natural History. Occasional Paper 5: 1-77.

Orcutt, F.S. Jr. and A.B. Orcutt. 1976. Nesting and parental behavior in domestic Common Quail. Auk 93(1): 135-141.

Orde, C.J. and B.E. Harrell. 1977. Hunting techniques and predators efficiency of nesting Red-tailed Hawks. Raptor Research 11(4): 82-85.

Ordre des ingénieurs forestiers du Québec and Union des municipalités régionales de comté et des municipalités locales du Québec. 1994. La forêt: un enjeu pour le Québec des régions. Supplément de la revue L'Aubelle, no. 104, octobre 1994.

Orians, G.H. 1961. The ecology of blackbird (*Agelaius*) social systems. Ecological Monographs 31(3): 285-312.

Orians, G.H. 1969. On the evolution of mating systems in birds and mammals. American Naturalist 103: 589-603.

Orians, G.H. 1980. Some adaptations of marsh-nesting blackbirds. Princeton University Press, Princeton, New Jersey, 295 pp.

Orians, G.H. 1985a. Blackbirds of the Americas. University of Washington Press, Seattle, 163 pp.

Orians, G.H. 1985b. American blackbirds, pp. 118-119 *in* C.M. Perrins and A.L.A. Middleton, eds. Songbirds, Series: all the world's animals. Torstar Books, New York, 160 pp.

Orians, G.H. and G.M. Christman. 1968. A comparative study of the behavior of Red-winged, Tricolored, and Yellow-headed blackbirds. University of California. Publications in Zoology no. 84, 81 pp.

Orians, G.[H.] and F. Kuhlman. 1956. The Red-tailed Hawk and Great Horned Owl population in Wisconsin. Condor 58(5): 371-385.

Orians, G.H. and J.F. Wittenberger. 1991. Spatial and temporal scales in habitat selection. American Naturalist 137: 529-549.

Oring, L.W. 1964. Behavior and ecology of certain ducks during the postbreeding period. Journal of Wildlife Management 28(2): 223-233.

Oring, L.W. 1968. Vocalizations of the Green and Solitary sandpipers. Wilson Bulletin 80(4): 395-420.

Oring, L.W. 1969. Summer biology of the Gadwall at Delta, Manitoba. Wilson Bulletin 81(1): 44-54.

Oring, L.W. 1973. Solitary Sandpiper early reproductive behavior. Auk 90(3): 652-663.

Oring, L.W. and M.L. Knudson. 1972. Monogamy and polyandry in the Spotted Sandpiper. Living Bird 11: 59-73.

Oring, L.W. and D.B. Lank. 1982. Sexual selection, arrival times, philopatry and site fidelity in the polyandrous Spotted Sandpiper. Behavioral Ecology and Sociobiology 10: 185-191.

Oring, L.W. and D.B. Lank. 1984. Breeding area fidelity, natal philopatry, and the social systems of sandpipers, pp. 125-147 *in* J. Burger and B.L. Olla, eds. Behavior of marine animals, vol. 5. Shorebirds: breeding behavior and populations. Plenum Press, New York, 437 pp.

Oring, L.W. and S.J. Maxson. 1978. Instances of simultaneous polyandry by a Spotted Sandpiper *Actitis macularia*. Ibis 120: 349-353.

Oring, L.W. and J.M. Reed. 1991. Spotted Sandpiper opportunism and reconnaissance for breeding opportunities. Abstract of paper presented to the AOU's 109th stated meeting, Montréal, August 14-17, 1991.

Oring, L.W., D.B. Lank, and S.J. Maxson. 1983. Population studies of the polyandrous Spotted Sandpiper. Auk 100(2): 272-285.

Ortega, C.P. and A. Cruz. 1991. A comparative study of cowbird parasitism in Yellow-headed Blackbirds and Red-winged Blackbirds. Auk 108(1): 16-24.

Osborne, D.R. and A.T. Peterson. 1984. Decline of the Upland Sandpiper (*Bartramia Longicauda*) in Ohio: an endangered species. Ohio Journal of Science 84(1)8-10.

Osborne, P.E. and B.J. Tigar. 1992. Interpreting bird atlas using logistic models: an example from Lesotho, southern Africa. Journal of Applied Ecology 29: 55-62.

Osborne, T.O. 1987. Biology of the Great Gray Owl in interior Alaska, pp. 91-195 *in* R.W. Nero, R.J. Clark, R.J. Knapton, and R.H. Hamre, eds. Proceedings. Biology and conservation of northern forest owls. Symposium, 3-7 February, Winnipeg. United States Department of Agriculture, Forest Service, Rocky Mountain Forest and Range Experiment Station, Fort Collins, Colorado. General Technical Report RM-142, 309 pp.

Otis, P., L. Messely, and D. Talbot. 1993. Guide des sites ornithologiques de la grande région de Québec. Club des ornithologues de Québec, 300 pp.

Otto, H.E. 1988. Antagonistic behaviour of Horned Grebes. Blue Jay 46(2): 92.

Ouellet, F. 1966. Histoire économique et sociale du Québec 1760-1850. Structures et conjoncture. Fides, Montréal, 640 pp.

Ouellet, F. 1971. Histoire économique et sociale du Québec 1760-1850. Structures et conjoncture. Fides, Montréal, 2 vol., 639 pp.

Ouellet, H. 1966. Histoire et dispersion de la Fauvette azurée *Dendroica cerulea* (Wilson) dans la province de Québec, Canada. Naturaliste canadien 93: 335-337.

Ouellet, H. 1967. The distribution of the Cerulean Warbler in the province of Quebec, Canada. Auk 84(2): 272-274.

Ouellet, H. 1969. Les oiseaux de l'île Anticosti, province de Québec, Canada. National Museums of Canada, National Museum of Natural Sciences, Ottawa. Publications in Zoology no. 1, 79 pp.

Ouellet, H. 1970a. Changes in the bird fauna of the Montreal region, Canada. Canadian Field-Naturalist 84(1): 27-34.

Ouellet, H. 1970b. Further observations on the food and predatory habits of the Gray Jay. Canadian Journal of Zoology 48(2): 327-330.

Ouellet, H. 1974a. Les oiseaux des collines montérégiennes et de la région de Montréal, Québec, Canada. National Museums of Canada, National Museum of Natural Sciences, Ottawa. Publications in Zoology no. 5, 167 pp.

Ouellet, H. 1974b. An intergeneric grouse hybrid (*Bonasa x Canachites*). Canadian Field-Naturalist 88(2): 183-186.

Ouellet, H. 1987. Profile of a pioneer: P.A. Taverner. American Birds 41: 20-26.

Ouellet, H. 1988. Ornithology in Canada from the beginning to 1950/L'ornithologie au Canada depuis les origines jusqu'à 1950, pp. 109-123 *in* H. Ouellet, ed. Acta Congressus Internationalis Ornithologici, Ottawa, 22-29 June 1986. Ottawa University Press, vol. 1, pp. 1-1404; vol. II, pp. 1405-2815.

Ouellet, H. 1991. Is Bicknell's Thrush a Gray-cheeked Thrush? Paper presented at the 109th stated meeting of the American Ornithologists' Union, Montréal, August 14-17 1991.

Ouellet, H. 1993. Bicknell's Thrush: taxonomic position and distribution. Wilson Bulletin 105(4): 545-572.

Ouellet, H. 1995. Ornithology at Canada's National Museum. Chapter 12. Pp. 303-322 *in* W.E. Davis, Jr. and J.A. Jackson, eds. Contributions to the history of North American ornithology. Memoirs of the Nuttall Ornithological Club, no. 12, Cambridge, Massachusetts.

Ouellet, H. and J.R. Bider. 1973. White-crowned Sparrow breeding at Deception Bay, Nouveau-Québec. Canadian Field-Naturalist 87(3): 321.

Ouellet, H. and A. Bourget. 1975. The Sandhill Crane in Quebec. Canadian Field-Naturalist 89(1): 73-74.

Ouellet, H. and W.E. Godfrey. 1980. Ornithology in Canada in the 20th century: a capsule overview. British Ornithologists' Club. Bulletin 100: 331-336.

Ouellet, H. and M. Gosselin. 1983. Les noms français des oiseaux d'Amérique du Nord. National Museum of Natural Sciences. Syllogeus no. 43, 36 pp.

Ouellet, H. and R. Ouellet. 1963. Bird notes from lac Ste. Anne, Saguenay County, Quebec. Canadian Field-Naturalist 77: 146-153.

Ouellet, H., M. Gosselin, and J.-P. Artigau. 1990. Nomenclature française des oiseaux d'Amérique du Nord. Canadian Museum of Nature and Secretary of State of Canada. Terminology Bulletin no. 202, 127 pp.

Owen, M. 1980. Wild geese of the world. Batsford Ltd, London, 236 pp.

Owen, R.B. Jr. and W.J. Galbraith. 1989. Earthworm biomass in relation to forest types, soil, and land use: implications for woodcock management. Wildlife Society Bulletin 17(2): 130-136.

Owen, R.B. Jr. and J.W. Morgan. 1975. Summer behavior of adult radio-equipped woodcock in central Maine. Journal of Wildlife Management 39(1): 179-182.

Owen, R.B. Jr., J.M. Anderson, J.W. Artmann, E.R. Clark, T.G. Dilworth, L.E. Gregg, F.W. Martin, J.D. Newsom, and S.R. Pursglove. 1977. American Woodcock, pp. 149-186 *in* G.C. Sanderson, ed. Management of migratory shore and upland game birds in North America. International Association of Fish and Wildlife Agencies, Washington, 358 pp.

Owre, O.T. 1973. A consideration of the exotic avifauna of southeastern Florida. Wilson Bulletin 85(4): 491-500.

Page, A.M. and M.J. Austen. 1993. Status report on the Northern Bobwhite (*Colinus virginianus*) in Canada. Committee on the Status of Endangered Wildlife in Canada (COSEWIC), Ottawa, 36 pp.

Palmer, D.A. 1986. Habitat selection, movements and activity of Boreal and Saw-whet owls. M.Sc. thesis, Colorado State University, Fort Collins, 101 pp.

Palmer, D.A. 1987. Annual, seasonal and nightly variation in calling activity of Boreal and Northern Saw-whet owls, pp. 162-168 *in* R.W. Nero, R.J. Clark, R.J. Knapton, and R.H. Hamre, eds. Proceedings. Biology and conservation of northern forest owls. Symposium, 3-7 February, Winnipeg. United States Department of Agriculture, Forest Service, Rocky Mountain Forest and Range Experiment Station, Fort Collins, Colorado. General Technical Report RM-142, 309 pp.

Palmer, D.A. and R.A. Ryder. 1984. The first documented breeding of the Boreal Owl in Colorado. Condor 86(2): 215-217.

Palmer, R.S. 1962. Handbook of North American birds, vol. 1: loons through flamingos. Yale University Press, New Haven, Connecticut, 567 pp.

Palmer, R.S. 1967. The shorebirds of North America. Viking Press, New York, 270 pp.

Palmer, R.S. 1968. Pine Siskin, pp. 424-447 *in* A.C. Bent. Life histories of North American cardinals, grosbeaks, buntings, towhees, finches, sparrows, and allies. (O.L. Austin, Jr., ed.). United States National Museum, Washington. [1968, Dover Publications, New York]. Bulletin no. 237, vol. 1, pp. 1-602; vol. 2, pp. 603-1248; vol. 3, pp. 1249-1889.

Palmer, R.S. 1976. Handbook of North American birds: waterfowl (parts 1 and 2). Yale University Press, New Haven, Connecticut, vol. 2, 521 pp.; vol. 3, 560 pp.

Palmer, R.S. 1985. Cardinal, p. 79 *in* B. Campbell and E. Lack, eds. A dictionary of birds. British Ornithologists' Union, Buteo Books, Vermillion, South Dakota, 670 pp.

Palmer, R.S. 1988. Handbook of North American birds: diurnal raptors (parts 1 and 2). Yale University Press, New Haven, Connecticut, vol. 4, 433 pp.; vol. 5, 465 pp.

Papi, F. 1989. Pigeons use olfactory cues to navigate. Ethology, Ecology and Evolution 1(3): 219-231.

Paquin, J. 1989. Un roselin de plus en plus... familier. Franc-Nord 6(2): 20-23.

Paquin, J. 1990a. Affluence au nichoir de l'Hirondelle bicolore. QuébecOiseaux 1(4): 8-11.

Paquin, J. 1990b. Des tourterelles de plus en plus nombreuses durant l'hiver. QuébecOiseaux 1(3): 12-15.

Paquin, J. 1993a. Labbes et bécasseaux: poursuites sur les rives du Saint-Laurent. QuébecOiseaux 4(4): 19-21.

Paquin, J. 1993b. Une première nidification du Goéland à bec cerclé sur le toit d'un édifice au Québec. QuébecOiseaux 5(2): 11.

Parent, B. 1989. Ressource et industrie forestières: portrait statistique édition 1988. Ministère de l'Énergie et des Ressources, Sainte-Foy, 55 pp.

Parent, G. and J. Lebrun. 1988. Un sucre au goût acide. Québec Science, avril 1988, pp. 18-23.

Parent, S. 1990. Dictionnaire des sciences de l'environnement. Éditions Broquet, Laprairie, 748 pp.

Paris, B. 1985. Utilisation du lac Saint-Louis et de la partie est du lac des Deux-Montagnes par le Grand Morillon (*Aythya marila*) et le Petit Morillon (*Aythya affinis*) lors de la migration d'automne 1984. Research report for Université du Québec à Montréal, Montréal, 38 pp.

Parker, G.R. 1974. A population peak and crash of lemmings and Snowy Owls on Southampton Island, Northwest Territories. Canadian Field-Naturalist 88(2): 151-156.

Parker, G.R. and W.R. Barrow. 1987. The status of the Mallard (*Anas platyrhynchos*) in Atlantic Canada. Canadian Wildlife Service, Sackville, New Brunswick. Unpublished Report.

Parker, K.E. 1985. Observations of a flying Common Loon carrying a fish. Journal of Field Ornithology 56(4): 412-413.

Parker, K.E. 1988. Common Loon reproduction and chick feeding on acidified lakes in the Adirondack Park, New York. Canadian Journal of Zoology 66: 804-810.

Parkes, K.C. 1951. The genetics of the Golden-winged–Blue-winged warbler complex. Wilson Bulletin 63(1): 5-15.

Parkes, K.C. 1983. "Brewster's" Warbler and "Lawrence's" Warbler, p. 112 *in* J. Farrand, Jr., ed. The Audubon Society Master Guide to birding. Alfred A. Knopf, New York, vol. 3, 399 pp.

Parkin, D.T. 1988. Genetic variation in the House Sparrow (*Passer domesticus*), pp. 1652-1657 *in* H. Ouellet, ed. Acta XIX Congressus Internationalis Ornithologici, Ottawa, 22-29 June 1986. Ottawa University Press, vol. 1, pp. 1-1404; vol. II, pp. 1405-2815.

Parmelee, D.F. 1968. Snow Bunting, pp. 1652-1675 *in* A.C. Bent. Life histories of North American cardinals, grosbeaks, buntings, towhees, finches, sparrows, and allies. (O.L. Austin, Jr., ed.) United States National Museum, Washington. [1968, Dover Publications, New York]. Bulletin no. 237, vol. 1, pp. 1-602; vol. 2, pp. 603-1248; vol. 3, pp. 1249-1889.

Parmelee, D.F. 1992. Snowy Owl (*Nyctea scandiaca*), *in* A. Poole, P. Stettenheim and F. Gill, eds. The birds of North America, no. 10. The Academy of Natural Sciences, Philadelphia. The American Ornithologists' Union, Washington, 20 pp.

Parmelee, D.F., H.A. Stephens, and R.H. Schmidt. 1967. The birds of southeastern Victoria Island and adjacent small islands. National Museum of Canada. Bulletin no. 222.

Parren, S.G. 1991. Evaluation of nest-box sites selected by Eastern Bluebirds, Tree Swallows, and House Wrens. Wildlife Society Bulletin 19: 270-277.

Parrish, J., R. Benjamin, and R. Smith. 1981. Near-ultraviolet light reception in the Mallard. Auk 98(3): 627-628.

Parrot, J. 1980. Frugivory by Great Grey Shrikes *Lanius excubitor*. Ibis 122: 532-533.

Pasquier, R.F. 1977. Watching birds: an introduction to ornithology. Houghton Mifflin Company, Boston, 301 pp.

Paszkowski, C.A. 1982. Vegetation, ground, and frugivorous foraging of the American Robin. Auk 99(4): 701-709.

Patterson, C.B. 1991. Relative parental investment in the Red-winged Blackbird. Journal of Field Ornithology 62(1): 1-18.

Patterson, I.J. 1965. Timing and spacing of broods in the Black-headed Gull (*Larus ridibundus*). Ibis 107(4): 433-459.

Paulus, S.L. 1982. Feeding ecology of Gadwalls in Louisiana in winter. Journal of Wildlife Management 46(1): 71-79.

Payette, S. 1975. La limite septentrionale des forêts sur la côte orientale de la Baie d'Hudson, Nouveau-Québec. Naturaliste canadien 102: 317-329.

Payette, S. 1983. The forest tundra and the present tree-line of the northern Québec–Labrador Peninsula, pp. 3-24 *in* P. Morisset and S. Payette, eds. Tree-line ecology. Proceedings. Northern Québec tree-line conference. Nordicana no. 47.

Payne, R. 1962. How the Barn Owl locates its prey by hearing. Living Bird 1: 151-159.

Payne, R.B. 1965. Clutch size and numbers of eggs laid by Brown-headed Cowbirds. Condor 67: 44-60.

Payne, R.B. 1969. Breeding seasons and reproductive physiology of Tricolored Blackbirds and Red-winged Blackbirds. University of California. Publications in Zoology 90: 1-137.

Payne, R.B. 1982. Ecological consequences of song matching: breeding success and intraspecific song mimicry in Indigo Buntings. Ecology 63(2): 401-411.

Payne, R.B. 1985. Cuckoo, pp. 123-126 *in* B. Campbell and E. Lack, eds. A dictionary of birds. British Ornithologists' Union, Buteo Books, Vermillion, South Dakota, 670 pp.

Payne, R.B. 1992. Indigo Bunting (*Passerina cyanea*), *in* A. Poole, P. Stettenheim, and F. Gill, eds. The birds of North America, no. 4. The Academy of Natural Sciences, Philadelphia. The American Ornithologists' Union, Washington, 24 pp.

Payne, R.B. and L.L. Payne. 1989. Heritability estimates and behaviour observations: extra-pair matings in Indigo Buntings. Animal Behaviour 38: 457-467.

Peabody, P.B. 1922. Haunts and breeding habits of the Yellow Rail. Journal of the Museum of Comparative Oology 2: 33-44.

Peakall, D.B. and A.C. Kemp. 1980. Organochlorine levels in owls in Canada and South Africa. Ostrich 51(3): 186-187.

Pearce, P.A. 1973. A White-winged Black Tern specimen from Canada. Auk 90(2): 420.

Pearce, P.A., D.B. Peakall, and A.J. Erskine. 1979. Impact on forest birds of the 1976 Spruce Budworm spray operation in New Brunswick. Canadian Wildlife Service. Progress Notes no. 97, 15 pp.

Peck, G.K. 1987a. Great Egret, pp. 52-53 *in* M.D. Cadman, P.F.J. Eagles, and F.M. Helleiner, eds. Atlas of the breeding birds of Ontario. University of Waterloo Press, Waterloo, 617 pp.

Peck, G.K. 1987b. Pacific Loon, p. 507 *in* M.D. Cadman, P.F.J. Eagles, and F.M. Helleiner, eds. Atlas of the breeding birds of Ontario. University of Waterloo Press, Waterloo, 617 pp.

Peck, G.K. 1987c. Evening Grosbeak, pp. 502-503 *in* M.D. Cadman, P.F.J. Eagles, and F.M. Helleiner, eds. Atlas of the breeding birds of Ontario. University of Waterloo Press, Waterloo, 617 pp.

Peck, G.K. and R.D. James. 1983. Breeding birds of Ontario: nidiology and distribution, vol. 1: Nonpasserines. Life Sciences Miscellaneous Publications, Royal Ontario Museum, Toronto, 321 pp.

Peck, G.K. and R.D. James. 1987. Breeding birds of Ontario: nidiology and distribution, vol. 2: Passerines. Life Sciences Miscellaneous Publications, Royal Ontario Museum, Toronto, 387 pp.

Pehrsson, O. 1991. Egg and clutch size in the Mallard as related to food quality. Canadian Journal of Zoology 69(1): 156-162.

Peiponen, V. 1957. Wechsell der birkenzeisig, *Carduelis flammea* (L.), sein brutgebiet während des sommers? [Does the redpoll change its breeding area during the summer?]. Ornis Fennica 34: 41-64.

Pelletier, M., R. Vaillancourt, S. Hébert, R. Greendale, and Y. Vigneault. 1990. Habitats côtiers perturbés dans le réseau Saint-Laurent en aval de l'île d'Orléans. Canadian Technical Report of Fisheries and Aquatic Sciences no. 1754, 61 pp.

Pelletier, R. and B. Scherrer. 1983. Étude de l'avifaune du parc de la Gatineau: caractéristiques des communautés d'oiseaux aquatiques et relations avec leur milieu. Final report for National Capital Commission. Laboratoire d'ornithologie et de biostatistique. Université du Québec à Montréal, Montréal, 80 pp.

Pelzl, H.W. 1971. Nest parasitism by Red-breasted Mergansers in Wisconsin. Auk 88(1): 184-185.

Penak, B.L. 1979. The Red-shouldered Hawks of Vaudreuil County. Tchebec [9]: 89-90.

Penak, B.L. 1981. The status of the Cooper's Hawk (*Accipiter cooperii*) in Ontario, with an overview of the status in Canada. Committee on the Status of Endangered Wildlife in Canada (COSEWIC), Ottawa, 70 pp.

Penak, B.L. 1985. The status of the Eastern Screech-Owl (*Otus asio*) in Canada with an overview of the status in North America. World Wildlife Fund (WWF) and Committee on the Status of Endangered Wildlife in Canada (COSEWIC). Unpublished report.

Pennant, T. 1785. Arctic zoology. Vol II. Class II birds. Henry Hughs, London, pp. 188-586.

Peppard, W.P. 1963. Timberdoodlings. Maine Fish and Game, vol. 5 [3 pp.].

Pepper, G.W. 1972. The ecology of Sharp-tailed Grouse during spring and summer in the aspen parklands of Saskatchewan. Saskatchewan Department of Natural Resources, Wildlife Report no. 1, 56 pp.

Perez, D.E. 1982. Parental foraging in Chipping Sparrows (*Spizella passerina*). M.Sc. thesis, Western Michigan University, Kalamazoo, 32 pp.

Perreault, P. 1983. Colonie de Laridés à Limoilou. Bulletin ornithologique 28(3): 127-132.

Perrins, C.M. 1985. Tits, pp. 382-387 *in* C.M. Perrins and A.L.A. Middleton, eds. The encyclopedia of birds. Facts on File, New York, 447 pp.

Perrins, C.M. and A.L.A. Middleton. 1985. The encyclopedia of birds. Facts of File, New York, 447 pp.

Perrins, C.M. and C.M. Reynolds. 1967. A preliminary study of the Mute Swan (*Cygnus olor*). Wildfowl Trust Annual Report 18: 74-84.

Perritt, J.E. and L.B. Best. 1989. Effects of weather on the breeding ecology of Vesper Sparrows in Iowa crop fields. American Midland Naturalist 121(2): 355-360.

Perry, A.E. 1965. The nesting of the Pine Siskin in Nebraska. Wilson Bulletin 77(3): 243-250.

Perry, J.H. 1986. L'industrie forestière: la plus grande ressource du Canada. Deuxième partie. Le Banquier, avril 1986: 6-16.

Peterjohn, B.G. and D.L. Rice. 1991. The Ohio breeding bird Atlas. Ohio Department of Natural Resources, Columbus, 416 pp.

Petersen, A.J. 1955. The breeding cycle in the Bank Swallow. Wilson Bulletin 67(4): 235-286.

Petersen, J.D. 1993. 1. Northern oak woodland, p. 33 *in* J.D. Lowe, ed. Resident bird counts 1992. Journal of Field Ornithology 64(1)(supplement): 1-128.

Petersen, L.R. 1979. Ecology of Great Horned Owls and Red-tailed Hawks in southeastern Wisconsin. Wisconsin Department of Natural Resources, Madison. Technical Bulletin no. 111, 63 pp.

Petersen, M.R. 1973. Breeding biology of Arctic and Red-throated loons. M.Sc. thesis, University of California, Davis, 55 pp.

Petersen, M.R. 1979. Nesting ecology of Arctic Loons. Wilson Bulletin 91(4): 608-617.

Petersen, M.R. 1989. Nesting biology of Pacific Loons, *Gavia pacifica*, on the Yukon-Kuskokwim Delta, Alaska. Canadian Field-Naturalist 103(2): 265-269.

Peterson, A. 1983. Observations on habitat selection by Henslow's Sparrow in Broome County, New York. Kingbird 33: 155-164.

Peterson, B. and G. Gauthier. 1985. Nest site use by cavity-nesting birds of the Cariboo parkland, British Columbia. Wilson Bulletin 97(3): 319-331.

Peterson, D.M. 1988. Great Egret, pp. 38-39 *in* R.F. Andrle and J.R. Carroll, eds. The Atlas of breeding birds in New York State. Cornell University Press, Ithaca, 551 pp.

Peterson, J.M. and S.T. Crocoll. 1992. Red-shouldered Hawk, pp. 333-351 *in* K.J. Schneider and D.M. Pence, eds. Migratory nongame birds of management concern in the northeast. United States Department of the Interior, Fish and Wildlife Service, Newton Corner, Massachusetts, 400 pp.

Peterson, J.M.C. 1988a. Great Blue Heron, pp. 36-37 *in* R.F. Andrle and J.R. Carroll, eds. The Atlas of breeding birds in New York State. Cornell University Press, Ithaca, 551 pp.

Peterson, J.M.C. 1988b. Cape May Warbler, pp. 374-375 *in* R.F. Andrle and J.R. Carroll, eds. The Atlas of breeding birds in New York State. Cornell University Press, Ithaca, 551 pp.

Peterson, J.M.C. 1988c. Blackpoll Warbler, pp. 394-395 *in* R.F. Andrle and J.R. Carroll, eds. The Atlas of breeding birds in New York State. Cornell University Press, Ithaca, 551 pp.

Peterson, J.M.C. 1988d. Common Grackle, pp. 478-479 *in* R.F. Andrle and J.R. Carroll, eds. The Atlas of breeding birds in New York State. Cornell University Press, Ithaca, 551 pp.

Peterson, J.M.C. 1988e. Common Loon, pp. 26-27 *in* R.F. Andrle and J.R. Carroll, eds. The Atlas of breeding birds in New York State. Cornell University Press, Ithaca, 551 pp.

Peterson, J.M.C. 1988f. Common Raven, pp. 286-287 *in* R.F. Andrle and J.R. Carroll, eds. The Atlas of breeding birds in New York State. Cornell University Press, Ithaca, 551 pp.

Peterson, J.M.C. 1988g. Gray Partridge, pp. 122-123 *in* R.F. Andrle and J.R. Carroll, eds. The Atlas of breeding birds in New York State. Cornell University Press, Ithaca, 551 pp.

Peterson, J.M.C. 1988h. Olive-sided Flycatcher, pp. 244-245 *in* R.F. Andrle and J.R. Carroll, eds. The Atlas of breeding birds in New York State. Cornell University Press, Ithaca, 551 pp.

Peterson, J.M.C. 1988i. Philadelphia Vireo, pp. 350-351 *in* R.F. Andrle and J.R. Carroll, eds. The Atlas of breeding birds in New York State. Cornell University Press, Ithaca, 551 pp.

Peterson, J.M.C. 1988j. Pine Warbler, pp. 386-387 *in* R.F. Andrle and J.R. Carroll, eds. The Atlas of breeding birds in New York State. Cornell University Press, Ithaca, 551 pp.

Peterson, J.M.C. 1988k. Red-breasted Nuthatch, pp. 294-295 *in* R.F. Andrle and J.R. Carroll, eds. The Atlas of breeding birds in New York State. Cornell University Press, Ithaca, 551 pp.

Peterson, J.M.C. 1988l. Rusty Blackbird, pp. 474-475 *in* R.F. Andrle and J.R. Carroll, eds. The Atlas of breeding birds in New York State. Cornell University Press, Ithaca, 551 pp.

Peterson, J.M.C. 1988m. Savannah Sparrow, pp. 446-447 *in* R.F. Andrle and J.R. Carroll, eds. The Atlas of breeding birds in New York State. Cornell University Press, Ithaca, 551 pp.

Peterson, J.M.C. 1988n. Tennessee Warbler, pp. 362-363 *in* R.F. Andrle and J.R. Carroll, eds. The Atlas of breeding birds in New York State. Cornell University Press, Ithaca, 551 pp.

Peterson, J.M.C. 1988o. Wilson's Warbler, p. 421 *in* R.F. Andrle and J.R. Carroll, eds. The Atlas of breeding birds in New York State. Cornell University Press, Ithaca, 551 pp.

Peterson, J.M.C. 1988p. Yellow-bellied Flycatcher, pp. 248-249 *in* R.F. Andrle and J.R. Carroll, eds. The Atlas of breeding birds in New York State. Cornell University Press, Ithaca, 551 pp.

Peterson, J.M.C. 1988q. Yellow-rumped Warbler, pp. 378-379 *in* R.F. Andrle and J.R. Carroll, eds. The Atlas of breeding birds in New York State. Cornell University Press, Ithaca, 551 pp.

Peterson, J.M.C. and C. Fichtel. 1992. Olive-sided Flycatcher, pp. 353-367 *in* K.J. Schneider and D.M. Pence, eds. Migratory nongame birds of management concern in the northeast. United States Department of the Interior, Fish and Wildlife Service, Newton Corner, Massachusetts, 400 pp.

Peterson, R.T. 1969. Les oiseaux, 2e éd. Life, Le monde vivant, Collections Time-Life, 192 pp.

Peterson, R.T. 1980. Field guide to the birds east of the Rockies. Houghton Mifflin Co., Boston, 384 pp.

Peterson, R.T. 1984. Guide des oiseaux de l'Amérique du Nord à l'est des Rocheuses. Éditions France-Amérique, Montréal, 384 pp.

Pethon, P. 1967. Food and feeding habits of the Common Eider (*Somateria mollissima*). Nytt Magasin foer Zoology 15: 97-111.

Petit, D.R., L.J. Petit, and K.E. Petit. 1989. Winter caching ecology of deciduous woodland birds and adaptations for protection of stored food. Condor 91(4): 766-776.

Petit, D.R., K.E. Petit, and L.J. Petit. 1990. Geographic variation in foraging ecology of North American insectivorous birds. Studies in Avian Biology 13: 254-263.

Petit, K.E., D.R. Petit, and L.J. Petit. 1988a. On measuring vegetation characteristics in bird territories: nest sites *vs* perch sites and the effect of plot size. American Midland Naturalist 119(1): 209-215.

Petit, K.E., M.D. Dixon, and R.T. Holmes. 1988b. A case of polygyny in the Black-throated Blue Warbler. Wilson Bulletin 100(1): 132-143.

Petit, L.J. 1989. Breeding biology of Prothonotary Warblers in riverine habitat in Tennessee. Wilson Bulletin 101(1): 51-61.

Petit, L.J. 1991a. Adaptive tolerance of cowbird parasitism by Prothonotary Warblers: a consequence of nest-site limitation. Animal Behaviour 41: 425-432.

Petit, L.J. 1991b. Experimentally induced polygyny in a monogamous bird species: Prothonotary Warblers and the polygyny threshold. Behavioral Ecology and Sociobiology 29: 177-187.

Petit, L.J., W.J. Fleming, K.E. Petit, and D.R. Petit. 1987. Nest-box use by Prothonotary Warblers (*Protonotaria citrea*) in riverine habitat. Wilson Bulletin 99(3): 485-488.

Petrie, M. 1984. Territory size in the Moorhen (*Gallinula chloropus*), an outcome of RHP asymmetry between neighbours. Animal Behaviour 32(3): 861-870.

Petrie, M. 1986. Mating decisions by female Common Moorhens (*Gallinula chloropus*), pp. 947-955 *in* H. Ouellet, ed. Acta XIX Congressus Internationalis Ornithologici, Ottawa, 22-29 June 1986. Ottawa University Press, vol. I, pp. 1-1404; vol. II, pp. 1405-2815.

Petrinovich, L. and L.F. Baptista. 1987. Song development in the White-crowned Sparrow: modification of learned song. Animal Behaviour 35(4): 961-974.

Petter, S.C., D.B. Miles, and M.M. White. 1990. Genetic evidence of mixed reproductive strategy in a monogamous bird. Condor 92(3): 702-708.

Pettingill, O.S. Jr. 1936. The American Woodcock. Memoirs of the Boston Society of Natural History 9(2): 169-391

Pettingill, O.S. Jr. 1970. Ornithology in laboratory and field, 4th edition. Burgess Publishing Co., Minneapolis, Minnesota, 524 pp.

Pettingill, O.S. Jr. 1985. Ornithology in laboratory and field, 5th edition. Academic Press, Orlando, Florida, 403 pp.

Philipp, P.B. 1925. Notes on some summer birds of the Magdalen Islands. Canadian Field-Naturalist 39: 75-78.

Philippe, L. 1945. L'alimentation hivernale de la Gélinotte à fraise, *Bonasa umbellus Togota* L., dans la région de Montréal. Naturaliste canadien 72(9): 235-240.

Phillips, A.R. 1981. The races of the Red Crossbill, *Loxia curvirostra*, in Arizona, pp. 223-230 *in* G. Monson and A.R. Phillips. Annotated Checklist of the Birds of Arizona, 2nd edition. University of Arizona Press, Tucson, 240 pp.

Phillips, A.R., M.A. Howe, and W.E. Lanyon. 1966. Identification of the flycatchers of eastern North America, with special emphasis on the genus *Empidonax*. Bird-Banding 37: 153-171.

Phillips, J.C. 1986. A natural history of the ducks, vol. III. Dover Publications, New York, 383 pp.

Piatt, J.F. and D.N. Nettleship. 1985. Diving depths of four Alcids. Auk 102(2): 293-297.

Piatt, J.F., D.N. Nettleship, and W. Threlfall. 1984. Net-mortality of Common Murres and Atlantic Puffins in Newfoundland, 1951-1981, pp. 196-206 *in* D.N. Nettleship, G.A. Sanger, and P.F. Springer, eds. Marine birds: their feeding ecology and commercial fisheries relationships. Proceedings. Pacific Seabird Group Symposium, Canadian Wildlife Service, Ottawa, 220 pp.

Pickwell, G. 1925. The nesting of the Killdeer. Auk 42: 485-496.

Picman, J. 1977a. Intraspecific nest destruction in the Long-billed Marsh Wren (*Telmatodytes palustris palustris*). Canadian Journal of Zoology 55: 1997-2003.

Picman, J. 1977b. Destruction of eggs by the Long-billed Marsh Wren (*Telmatodytes palustris palustris*). Canadian Journal of Zoology 55: 1914-1920.

Picman, J. 1980. Impact of Marsh Wrens on reproductive strategy of Red-winged Blackbirds. Canadian Journal of Zoology 58: 337-350.

Picman, J. and J.-C. Belles-Isles. 1988. Evidence for intraspecific brood parasitism in the House Wren. Condor 90(2): 513-514.

Picman, J. and A.K. Picman. 1980. Destruction of nests by the Short-billed Marsh Wren. Condor 82(2): 176-179.

Pielou, E.C. 1988. The world of northern evergreens. Comstock Publishers Associates, Ithaca, New York, 174 pp.

Pielou, E.C. 1991. After the Ice Age. The return of life to glaciated North America. University of Chicago Press, Chicago, Illinois, 366 pp.

Pierce, B.A. 1985. Acid tolerance in amphibians. BioScience 35: 239-243.

Pierotti, R.J. and T.P. Good. 1994. Herring Gull (*Larus argentatus*), *in* A. Poole and F. Gill, eds. The birds of North America, no. 124. The Academy of Natural Sciences, Philadelphia. The American Ornithologists' Union, Washington, 28 pp.

Pierson, T.A. 1978. Distraction display by male Canada Warbler. Raven 49(2): 62-63.

Pietz, P.J. and J.R. Tester. 1982. Habitat selection by sympatric Spruce and Ruffed grouse in north central Minnesota. Journal of Wildlife Management 46(2): 391-403.

Pilon, C., J.M. Boisvert, D. Carrière, J. Champagne, P. Chevalier, D. Le Quéré, V. Sicard, and G. Sylvain. 1980. Les îles du Saint-Laurent de Boucherville à Contrecoeur: environnement biophysique. Centre de recherches écologiques de Montréal, Université de Montréal, Montréal, 292 pp.

Pilon, C., J. Champagne, and P. Chevalier. 1981. Environnement biophysique des îles de Berthier-Sorel. Centre de recherches écologiques de Montréal, 203 pp.

Pilon, C., J. Burton, and R. McNeil. 1983a. Reproduction du Grand Cormoran (*Phalacrocorax carbo*) et du Cormoran à aigrettes (*P. auritus*) aux Îles-de-la-Madeleine, Québec. Canadian Journal of Zoology 61: 524-530.

Pilon, C., J. Burton, and R. McNeil. 1983b. Summer food of the Great and Double-crested cormorants on the Magdalen Islands, Quebec. Canadian Journal of Zoology 61(12): 2733-2739.

Pinkowski, B.C. 1975. Yearling male Eastern Bluebird assists parents in feeding young. Auk 92(4): 801-802.

Pinkowski, B.C. 1976. Use of tree cavities by nesting Eastern Bluebirds. Journal of Wildlife Management 40(3): 556-563.

Pinkowski, B.C. 1977. Foraging behavior of the Eastern Bluebird. Wilson Bulletin 89(3): 404-414.

Pinkowski, B.C. 1978. Two successive male Eastern Bluebirds tending the same nest. Auk 95: 606-608.

Pitelka, F.[A.]. 1940. Breeding behaviour of the Black-throated Green Warbler. Wilson Bulletin 53: 3-18.

Pitelka, F.A. 1950. Geographic variation and the species problem in the shorebird genus *Limnodromus*. University of California. Publication in Zoology 50: 1-108.

Pitelka, F.A., P.Q. Tomich, and G.W. Treichel. 1955. Ecological relations of jaegers and owls as lemming predators near Barrow, Alaska. Ecological Monographs 25: 85-117.

Pitocchelli, J. 1993. Mourning Warbler (*Oporornis philadelphia*), *in* A. Poole and F. Gill, eds. The birds of North America, no. 72. The Academy of Natural Sciences, Philadelphia. The American Ornithologists' Union, Washington, 16 pp.

Pittaway, R. 1988. Wing-tail flicking as a means of distinguishing crows from ravens. Ontario Birds 6(2): 74-75.

Pitzrick, J.M. 1993. 125. Abandonned upland pasture, *in* J.D. Lowe, ed. Resident bird counts 1992. Journal of Field Ornithology 64(1)(supplement): 1-128.

Place, A.R. and E.W. Stiles. 1992. Living off the wax of the land: bayberries and Yellow-rumped Warblers. Auk 109(2): 334-345.

Platt, J.B. 1973. Habitat and time utilization of a pair of nesting Sharp-shinned Hawks (*Accipiter striatus velox*): a telemetry study. Ph.D. dissertation, Brigham Young University, Provo, Utah.

Platt, J.B. 1976a. Gyrfalcon nest sites selection and winter activity in the western Canadian Arctic. Canadian Field-Naturalist 90(3): 338-345.

Platt, J.B. 1976b. Sharp-shinned Hawk nesting and nest-site selection in Utah. Condor 78: 102-103.

Platt, J.B. 1977. The breeding behavior of wild and captive Gyrfalcons in relation to their environment and human disturbance. Ph.D. dissertation, Cornell University, Ithaca, New York, 173 pp.

Pleasants, B.Y. 1979. Adaptive significance of the variable dispersion pattern of breeding Northern Orioles. Condor 81(1): 28-34.

Plissner, J.H. and P.A. Gowaty. 1988. Evidence of reproductive error in adoption of nestling Eastern Bluebirds (*Sialia sialis*). Auk 105(3): 575-578.

Plunkett, A.M. 1990. Pine Grosbeaks invade – winter 1989-1990. Loon 62(3): 123-125.

Podos, J., S. Peters, T. Rudnicky, P. Marler, and S. Nowicky. 1992. The organization of song repertories in Song Sparrows: themes and variations. Ethology 90(2): 89-106.

Polunin, N. and C.R. Eklund. 1953. Notes on food habits of waterfowl in the interior of Ungava Peninsula. Canadian Field-Naturalist 67: 134-137.

Pomerleau, R. 1991. Expérience sur les mécanismes du dépérissement des forêts de feuillus au Québec. Forest Canada. Information report LAU-X-96F, 48 pp.

Pool, W. 1961. Feeding habits of the Harlequin Duck. Wildfowl Trust Annual Report 13: 126-129.

Poole, A., P. Stettenheim, and F. Gill. 1992. Editorial Guidelines and Glossary, pp. 6-15 *in* The birds of North America. The Academy of Natural Sciences, Philadelphia. The American Ornithologists' Union, Washington, 16 pp.

Poole, A.F. 1989. Ospreys: a natural and unnatural history. Cambridge University Press, Cambridge, Massachusetts, 246 pp.

Poole, K.G. 1988. A replacement clutch in wild Gyrfalcons, *Falco rusticolus*, in the Northwest Territories. Canadian Field-Naturalist 102(1): 62-64.

Poole, K.G. and D.A. Boag. 1988. Ecology of Gyrfalcons (*Falco rusticolus*) in the central Canadian Arctic: diet and feeding behaviour. Canadian Journal of Zoology 66: 334-344.

Poole, K.G. and R.G. Bromley. 1988a. Natural history of the Gyrfalcon in the central Canadian Arctic. Arctic 41: 31-38.

Poole, K.G. and R.G. Bromley. 1988b. Interrelationships within a raptor guild in the central Canadian Arctic. Canadian Journal of Zoology 66(10): 2275-2282.

Porsild, A.E. and W.J. Cody 1980. Vascular plants of continental Northwest Territories, Canada. National Museum of Natural Sciences, National Museums of Canada, 667 pp.

Porter, D.K., M.A. Strong, J.B. Giezentanner, and R.A. Ryder. 1975. Nest ecology, productivity, and growth of the Loggerhead Shrike on the shortgrass prairie. Southwestern Naturalist 19: 429-436.

Porter, J.P. 1904. A preliminary study of the psychology of the English Sparrow. American Journal of Psychology 15(3): 313-346.

Porter, J.P. 1906. Further study of the English Sparrow and other birds. American Journal of Psychology 17: 248-271.

Porter, J.P. 1910. Intelligence and imitation in birds; a criterious of imitation. American Journal of Psychology 21(1): 1-71.

Porter, R.D. and S.N. Wiemeyer. 1969. Dieldrin and DDT: effects on Sparrow Hawk eggshells and reproduction. Science 165: 199-200.

Porter, W.G. 1983. Effects of winter conditions on reproduction in a northern Wild Turkey population. Journal of Wildlife Management 47(2): 281-290.

PORTNOY, J.W. and W.E. DODGE. 1979. Red-shouldered Hawk nesting ecology and behavior. Wilson Bulletin 91(1): 104-117.

POST, W. and J.S. GREENLAW. 1982. Comparative costs of promiscuity and monogamy: a test of reproductive effort theory. Behavioral Ecology and Sociobiology 10: 101-107.

POSTON, H.J. 1969. Relationship between the Shoveler and its breeding habitat at Strathmore, Alberta, pp. 132-137 *in* Canadian Wildlife Service. Saskatoon Wetlands Seminar. Canadian Wildlife Service Report Series, no. 6, 262 pp.

POTTER, E.F. 1980. Notes on nesting Yellow-billed Cuckoos. Journal of Field Ornithology 51(1): 17-28.

POTTER, E.F. 1985. Antings Antics. Living Bird Quarterly 4(3): 12-15.

POTTER, E.F. and D.C. HAUSER. 1974. Relationship of anting and sunbathing to molting in wild birds. Auk 91(3): 537-563.

POTTER, J.G. 1965. Snow cover. Climatological studies, no. 3, Metropolitan Board, Department of Transport, Toronto, 69 pp.

POTTS, G.R. 1972. The cause of the decline of the Partridge (*Perdix perdix*), p. 679 *in* K.H. Voous, ed. Proceedings. XVth International Ornithological Congress, The Hague, 30 August-5 September 1970. E.J. Brill, Leiden, 745 pp.

POTVIN, D. 1952. Naturalistes et sciences naturelles. Carnets de zoologie 12(4): 141-153.

POTVIN, N. 1977. Régime alimentaire d'oiseaux noirs s'attaquant au maïs fourrager et comparaison de méthodes de répression. M.Sc. thesis, Université de Sherbrooke, Sherbrooke, 75 pp.

POTVIN, N. and J.-M. BERGERON. 1981. Different modifications in the use of the acetylene cannon as a deterrent against blackbird damage to forage corn. Phytoprotection 62(1): 44-52.

POTVIN, N., J.-M. BERGERON, and C. FERNET. 1976. Régime alimentaire d'oiseaux fréquentant un agrosystème. Canadian Journal of Zoology 54(11): 1992-2000.

POTVIN, N., J.-M. BERGERON, and J. GENEST. 1978. Comparaison des méthodes de répression d'oiseaux s'attaquant au maïs fourrager. Canadian Journal of Zoology 56(1): 40-47.

POTVIN, N., J.-M. BERGERON, M. NORMAN, and A. CYR. 1982. Evaluating the sterile male method on Red-winged Blackbirds: clinical evaluation of thiotepa as a sterilant. Canadian Journal of Zoology 60(3): 460-465.

POUGH, R.H. 1949. Audubon land bird guide: birds of eastern and central North America from southern Texas to central Greenland. Doubleday and Co., Garden City, New York, 312 pp.

POUGH, R.H. 1951. Audubon water bird guide: water, game, and large land birds of eastern and central North America from southern Texas to central Greenland, most of which range to the Pacific coast. Doubleday and Co., Garden City, New York, 352 pp.

POULIN, J.M. 1968. Reproduction du Fou de Bassan (*Sula bassana*), île Bonaventure (Québec). M.Sc. thesis, Université Laval, Sainte-Foy, 110 pp.

POULIOT, J.C. 1934. La grande aventure de Jacques Cartier: relations de 1534 et 1535-1536. Québec, 328 pp.

POWE, N.N. 1972. The climate of the Saguenay–Lake St. John region, 10 pp.

POWELL, V.N. and H.L. JONES. 1978. An observation of polygyny in the Common Yellowthroat. Wilson Bulletin 90(4): 656-657.

POWER, H.W., E. LITOVICH, and M.P. LOMBARDO. 1981. Male starlings delay incubation to avoid being cuckolded. Auk 98(2): 386-389.

POWER, H.W., E.D. KENNEDY, L.C. ROMAGNANO, M.P. LOMBARDO, A.S. HOFFENBERG, P.C. STOUFFER, and T.R. MCGUIRE. 1989. The parasitism insurance hypothesis: why starlings leave space for parasitic eggs. Condor 91(4): 753-765.

POWERS, K.D. and J. CHERRY. 1983. Loon migrations off the coast of the northeastern United States. Wilson Bulletin 95(1): 125-132.

POWERS, K.D. and W.T. RUMAGE. 1978. Effect of the Argo Merchant oil spill on the bird populations of the New England coast, pp. 142-148 *in* Proceedings. In the wake of the Argo Merchant. Symposium, 11-13 January 1978. Center for Ocean Management Studies, University of Rhode Island, Kingston, 181 pp.

PRATT, H.M. 1972. Nesting success of Common Egrets and Great Blue Herons in the San Francisco Bay region. Condor 74: 447-453.

PRATT, H.M. and D.W. WINKLER. 1985. Clutch size, timing of laying, and reproductive success in a colony of Great Blue Herons and Great Egrets. Auk 102(1): 49-63.

PRAVOSUDOV, V.V. and T.C. GRUBB, JR. 1993. White-breasted Nuthatch (*Sitta carolinensis*), *in* A. Poole and F. Gill, eds. The birds of North America, no. 54. The Academy of Natural Sciences, Philadelphia. The American Ornithologists' Union, Washington, 16 pp.

PREBLE, N.A. 1957. Nesting habits of the Yellow-billed Cuckoo. American Midland Naturalist 57(2): 474-482.

PRENDERGAST, J.R., R.M. QUINN, J.H. LAWTON, B.C. EVERSHAM, and D.W. GIBBONS. 1993. Rare species, the coincidence of diversity hotspots and conservation strategies. Nature 365: 335-337.

PRESCOTT, D.R.C. 1985. Feeding at night by wintering Pine Siskins. Journal of Field Ornithology 56(4): 419.

PRESCOTT, D.R.C. 1986. Polygyny in the Willow Flycatcher. Condor 88(3): 385-386.

PRESCOTT, D.R.C. 1987a. Territorial responses to song playback in allopatric and sympatric populations of Alder (*Empidonax alnorum*) and Willow (*E. traillii*) flycatchers. Wilson Bulletin 99(4): 611-619.

PRESCOTT, D.R.C. 1987b. Yellow Rail, pp. 146-147 *in* M.D. Cadman, P.F.J. Eagles, and F.M. Helleiner, eds. Atlas of the breeding birds of Ontario. University of Waterloo Press, Waterloo, 617 pp.

PRESCOTT, D.R.C. 1987c. Yellow-bellied Flycatcher, pp. 254-255 *in* M.D. Cadman, P.F.J. Eagles, and F.M. Helleiner, eds. Atlas of the breeding birds of Ontario. University of Waterloo Press, Waterloo, 617 pp.

PRESCOTT, D.R.C. and A.L.A. MIDDLETON. 1988. Feeding-time minimization and the territorial behavior of the Willow Flycatcher (*Empidonax traillii*). Auk 105(1): 17-28.

PRESCOTT, D.R.C. and A.L.A. MIDDLETON. 1990. Age and sex differences in winter distribution of American Goldfinches in eastern North America. Ornis Scandinavica 21: 99-104.

PRESCOTT, K.W. 1965. The Scarlet Tanager. New Jersey State Museum Investigations, vol. 2. Department of Education of New Jersey, Trenton, 159 pp.

PREST, V.K., D.R. GRANT, and V.N. RAMPTON. 1970. Carte glaciaire du Canada. Geological Survey of Canada, Department of Energy, Mines and Resources. Map, scale 1:500000.

PRESTON, C.R. 1990. Distribution of raptor foraging in relation to prey biomass and habitat structure. Condor 92(1): 107-112.

PRESTON, C.R. and R.D. BEANE. 1993. Red-tailed Hawk (*Buteo jamaicensis*), *in* A. Poole and F. Gill, eds. The birds of North America, no. 52. The Academy of Natural Sciences, Philadelphia. The American Ornithologists' Union, Washington, 24 pp.

PREVETT, J.P., H.G. LUMSDEN, and F.C. JOHNSON. 1983. Waterfowl kill by Cree hunters of the Hudson Bay Lowland, Ontario. Arctic 36: 185-192.

PREVETT, J.P., I.F. MARSHALL, and V.G. THOMAS. 1985. Spring foods of Snow and Canada geese at James Bay. Journal of Wildlife Management 49(3): 558-563.

PRÉVOST, Y.A. 1977. Feeding ecology of Ospreys in Antigonish County, Nova Scotia. M.Sc. thesis, McGill University, Montréal, 111 pp.

PRIBIL, S. and J. PICMAN. 1991. Why House Wrens destroy clutches of other birds: a support for the nest site competition hypothesis. Condor 93(1): 184-185.

PRICE, I.M. 1976. Black-crowned Night-Heron reproduction on Pigeon Island, Lake Ontario, pp. 248-274 *in* Proceedings. Fish-eating birds of the Great Lakes and environmental contaminants. Symposium, 2-3 December.

PRINCE, H.H. 1968. Nest sites used by Wood Ducks and Common Goldeneyes in New Brunswick. Journal of Wildlife Management 32(3): 489-500.

PRINCE, H.H. 1991. Wood Duck, pp. 122-123 *in* R. Brewer, G.A. McPeek, and R.J. Adams, Jr., eds. The Atlas of breeding birds of Michigan. Michigan State University Press, East Lansing, 594 pp.

PRINDIVILLE GAINES, E. and M.R. RYAN. 1988. Piping Plover habitat use and reproductive success in North Dakota. Journal of Wildlife Management 52(2): 266-273.

PRINSENBERG, S.J. 1992. L'effet des aménagements hydroélectriques sur le milieu marin et glaciel de la baie d'Hudson. Services d'information Vent du Nord, Montréal, 11 pp.

PRIOR, K.A. 1990. Turkey Vulture food habits in southern Ontario. Wilson Bulletin 102(4): 706-710.

PRIOR, K.A. and P.J. WEATHERHEAD. 1991a. Turkey Vultures foraging at experimental food patches: a test of information transfer at communal roosts. Behavioral Ecology and Sociobiology 28: 385-390.

PRIOR, K.A. and P.J. WEATHERHEAD. 1991b. Competition at the carcass: opportunities for social foraging by Turkey Vultures in southern Ontario. Canadian Journal of Zoology 69(6): 1550-1556.

PROCTER-GRAY, E. 1991. Female-like plumage of subadult male American Redstarts does not reduce aggression from other males. Auk 108(4): 872-879.

PROULX, H., G. JACQUES, A.-M. LAMOTHE, and J.K. LITYNSKI. 1987. Climatologie du Québec méridional. Ministère de l'Environnement du Québec, Québec. Publication no. M.P. 45, 198 pp.

PROVANCHER, L. 1870. Faune canadienne. Naturaliste canadien 2: 193-196.

PROVINCE OF QUEBEC SOCIETY FOR THE PROTECTION OF BIRDS (PQSPB). 1950. Annual report, Montréal, 59 pp.

PROVINCE OF QUEBEC SOCIETY FOR THE PROTECTION OF BIRDS (PQSPB). 1951. Report on birds. Annual report, Montréal, pp. 9-43.

PROVINCE OF QUEBEC SOCIETY FOR THE PROTECTION OF BIRDS (PQSPB). 1952. Annual report, Montréal, 35 pp.

PROVINCE OF QUEBEC SOCIETY FOR THE PROTECTION OF BIRDS (PQSPB). 1958. Annual report, Montréal, 46 pp.

PROVINCE OF QUEBEC SOCIETY FOR THE PROTECTION OF BIRDS (PQSPB). 1959. Annual report, Montréal, 36 pp.

PROVINCE OF QUEBEC SOCIETY FOR THE PROTECTION OF BIRDS (PQSPB). 1971. PQSPB News. Newsletter 14(2): 2.

PROVINCE OF QUEBEC SOCIETY FOR THE PROTECTION OF BIRDS (PQSPB). 1975. Summary of sightings in the Montreal area 1974-1975. Annual report. Tchebec 5: 15-76.

PROVINCE OF QUEBEC SOCIETY FOR THE PROTECTION OF BIRDS (PQSPB). 1980. Summary of sightings in the Montreal area 1980. Annual report. Tchebec 10: 4-82.

PROVINCE OF QUEBEC SOCIETY FOR THE PROTECTION OF BIRDS (PQSPB). 1982. Summary of sightings in the Montreal area. Annual report. Tchebec 12: 6-66.

PROVINCE OF QUEBEC SOCIETY FOR THE PROTECTION OF BIRDS (PQSPB). 1983. Summary of sightings in the Montreal area. Annual report. Tchebec 13: 5-68.

PROVINCE OF QUEBEC SOCIETY FOR THE PROTECTION OF BIRDS (PQSPB). 1984. Summary of sightings in the Montreal area. Annual report. Tchebec 14: 6-27.

PROVOST, M.W. 1947. Nesting of birds in the marshes of northwest Iowa. American Midland Naturalist 38: 485-503.

PRUD'HOMME-CYR, J. 1975. Expérimentation de l'α-chloralose en vue du contrôle de quelques oiseaux dans la zone du nouvel aéroport international de Montréal à Mirabel, Québec. M.Sc. thesis, Université de Montréal, Montréal, 107 pp.

PRUETT-JONES, S.G. 1980. Team-hunting and food sharing in Parasitic Jaegers. Wilson Bulletin 92(4): 524-526.

PRY-JONES, R.P. 1985. Bunting, pp. 72-74 *in* B. Campbell and E. Lack, eds. A dictionary of birds. British Ornithologists' Union, Buteo Books, Vermillion, South Dakota, 670 pp.

PULLIAINEN, E. 1979. On the breeding of the Pine Grosbeak *Pinicola enucleator* in northeast Finland. Ornis Fennica 56(4):156-162.

PULLIAINEN, E. and R. HAKANEN. 1972. Parental care of a pair of Pine Grosbeaks *Pinicola enucleator* during the nestling period. Ornis Fennica 49(3-4): 86-90.

PULLIAM, H.R. and J.B. DUNNING. 1987. The influence of food supply on local density and diversity of sparrows. Ecology 68(4): 1009-1014.

PULLIAM, H.R. and F. ENDERS. 1971. The feeding ecology of five sympatric finch species. Ecology 52(4): 557-566.

PUTNAM, L.S. 1949. The life history of the Cedar Waxwing. Wilson Bulletin 61(3): 141-182.

QUAY, W.B. 1985. Sperm release in migrating wood-warblers (Parulinae) nesting at higher latitude. Wilson Bulletin 97(3): 283-295.

QUAY, W.B. 1989. Insemination of Tennessee Warblers during spring migration. Condor 91(3): 660-670.

QUENNEVILLE, R. 1990. Monitoring des oiseaux marins du parc national Forillon 1989. Parks Canada. Service de la conservation des ressources naturelles, parc national Forillon, 28 pp.

QUINLAN, R.W., W.A. HUNT, K. WILSON, and J. KERR. 1990. Habitat requirements of selected wildlife species in the Weldwood Forest Management Agreement Area. Final report submitted to the Weldwood Forest Management Agreement Area, Integrated Resource Management Steering Committee, 189 pp.

QUINN, M.S. and G.L. HOLROYD. 1989. Nestling and egg destruction by House Wrens. Condor 91(1): 206-207.

QUINNEY, T.E. 1983. Tree Swallows cross a polygyny threshold. Auk 100(3): 750-754.

QUINNEY, T.E. 1986. Male and female parental care in Tree Swallows. Wilson Bulletin 98(1): 147-150.

QUINTON, M.S. 1994. Alone with the Northern Goshawk. National Geographic 186(1): 117-128.

RABE, D. 1977. Structural analysis of woodcock diurnal habitat in northern Michigan. Proceedings. Woodcock Symposium 6: 125-134.

RABENOLD, P.P. and M.D. DECKER. 1989. Black and Turkey vultures expand their ranges northward. Eyas 12(2): 11-15.

RABINOWITZ, D., S. CAIRNS, and T. DILLON. 1986. Seven forms of rarity and their frequency in the flora of the British Isles, pp. 182-204 *in* M.E. Soulé, ed. Conservation biology: the science of scarcity and diversity. Sinauer Associates, Sunderland, Massachusetts.

RACINE, R.N. and N.S. THOMPSON. 1983. Social organization of wintering Blue Jays. Behaviour 87(3-4): 237-255.

RÅD, O. 1980. Breeding distribution and habitat selection of Red-breasted Mergansers in freshwater in western Norway. Wildfowl 31: 53-56.

RAE, J. 1888. Notes on some of the birds and mammals of the Hudson's Bay Co's Territories and the Arctic coast. Canadian Record of Science 3: 125-136.

RAIL, J.-F., G. CHAPDELAINE, P. BROUSSEAU, and J.-P.L. SAVARD. 1996. Utilisation des oiseaux marins comme bioindicateurs de l'écosystème marin de l'estuaire et du golfe Saint-Laurent. Canadian Wildlife Service, Québec Region, Environment Conservation Direction. Technical Report Series no. 254, 113 pp.

RAITASUO, D. 1964. Social behaviour of the Mallard in the course of the annual cycle. Papers on Game Research (Riistatieteelisiä Julkaisuja) 24: 1-62.

RAMADE, F. 1993. Dictionnaire encyclopédique de l'écologie et des sciences de l'environnement. Édiscience international, Paris, 822 pp.

RAMSAY, A.D. 1987. Thermal adaptions of the Carolina Wren. North American Bird Bander 12(2): 64.

RAND, A.L. 1942. *Larus kumlieni* and its allies. Canadian Field-Naturalist 56: 123-126.

RAND, A.L. 1945. Hungarian Partridge in the Ottawa–Montreal area. Canadian Field-Naturalist 59: 26-27.

RANDALL, T.E. 1961. Lesser Yellowlegs: life history in Canada, pp. 88-92 *in* D.A. Bannerman, ed. The birds of the British Isles, vol. 10. Oliver and Boyd, Edinburgh, 320 pp.

RANDS, M.R.W. 1986. The survival of gamebird (Galliformes) chicks in relation to pesticide use on cereals. Ibis 128: 57-64.

RAPHAEL, M.G. and M. WHITE. 1984. Use of snags by cavity-nesting birds in the Sierra Nevada. Wildlife Monographs no. 86, 66 pp.

RAPPOLE, J.H. and D.W. WARNER. 1976. Relationships between behavior, physiology and weather in avian transients at migration stopover sites. Oecologia 26: 193-212.

RATCLIFFE, D.A. 1962. Breeding density of the Peregrine Falcon *Falco peregrinus* and Raven *Corvus corax*. Ibis 104(1): 13-39.

RATCLIFFE, D.A. 1980. The Peregrine Falcon. Buteo Books, Vermillion, South Dakota, 416 pp.

RAVELING, D.G. 1969. Social classes of Canada Geese in winter. Journal of Wildlife Management 33(2): 304-318.

RAVELING, D.G. 1970. Dominance relationships and agonistic behavior of Canada Geese in winter. Behavior 37: 291-319.

RAVELING, D.G. and H.G. LUMSDEN. 1977. Nesting ecology of Canada Geese in the Hudson Bay Lowlands of Ontario: evolution and population regulation. Fish and Wildlife Resource, Ontario Ministry of Natural Resources, Report no. 98, 77 pp.

RAYNOR, G. 1983. Nesting of the Red Crossbill on Long Island. Kingbird 33(1): 15-18.

RAZUREL, É. 1994. Niveau de contamination de la Sterne pierregarin (*Sterna hirundo*) dans le système hydrographique du Saint-Laurent et son utilisation possible comme sonde bioanalytique. M.Sc. thesis, Université du Québec à Rimouski, Rimouski, 96 pp.

REA, A.M. 1983. Cathartid affinities: a brief overview, pp. 26-54 *in* S.R. Wilbur and J.A. Jackson, eds. Vulture biology and management. University of California Press, Berkeley, 550 pp.

REAM, C.H. 1976. Loon productivity, human disturbance, and pesticide residues in northern Minnesota. Wilson Bulletin 88: 427-432.

REAUME, T. 1987. Nest building by American Crows. Ontario Birds 5(3): 102-105.

RECHER, H.F. and J.A. RECHER. 1969. Some aspects of the ecology of migrant shorebirds. II. Aggression. Wilson Bulletin 81(2): 140-154.

REDMOND, G.W. and D.M. KEPPIE. 1988. Variation in occupancy of coniferous territories by male American Woodcock. Canadian Journal of Zoology 66: 2029-2035.

REED, A. 1964. Observations on the birds of Laurentides Park, Quebec. Naturaliste canadien 40(4): 97-116.

REED, A. 1969. Waterfowl breeding reconnaissance 1968. Quebec Wildlife Service Program Report (unpublished report), 9 pp.

REED, A. 1970. The breeding ecology of the Black Duck in the St. Lawrence estuary. Ph.D. dissertation, Université Laval, Sainte-Foy, 175 pp.

REED, A. 1971. Pre-dusk rafting flights of wintering Goldeneyes and other ducks in the province of Quebec. Wildfowl 22: 61-62.

REED, A. 1973. Aquatic bird colonies in the St. Lawrence estuary. Ministère du Tourisme, de la Chasse et de la Pêche, Gouvernement du Québec. Bulletin no. 18, 54 pp.

REED, A. 1975a. Les colonies d'oiseaux aquatiques dans l'estuaire du Saint-Laurent. Ministère du Tourisme, de la Chasse et de la Pêche, Québec, Faune du Québec. Bulletin no. 19, 76 pp.

REED, A. 1975b. Reproductive output of Black Ducks in the St. Lawrence estuary. Journal of Wildlife Management 39(2): 243-255.

REED, A. 1975c. Migration, homing, and mortality of breeding female Eiders *Somateria mollissima dresseri* of the St. Lawrence estuary, Quebec. Ornis Scandinavica 6: 41-47.

REED, A. 1976. Geese, nutrition and farmland. Wildfowl 27: 153-156.

REED, A. 1977a. L'écologie de l'alimentation de la Grande Oie blanche dans l'estuaire du Saint-Laurent: un rapport d'étape. Symposium sur l'écologie de l'alimentation organisé par le Bureau international de recherche sur la sauvagine suisse, 5 pp.

REED, A. 1977b. Spring surveys of Atlantic Brant in eastern North America, 1977. Canadian Wildlife Service, Quebec Region, unpublished report, 10 pp.

REED, A. 1978. Canadian Waterfowl Management Plan – Quebec. Canadian Wildlife Service, Quebec Region, unpublished report, 18 pp.

REED, A. 1984. Production of *Scirpus americanus* and its use by Greater Snow Geese at the Cap Tourmente National Wildlife Area. Draft of paper presented at the 5th North American Snow Goose Conference, Quebec, 16 pp.

REED, A. 1989. Use of a freshwater tidal marsh in the St. Lawrence estuary by Greater Snow Geese, pp. 605-616 *in* R.R. Sharitz and J.W. Gibbons, eds. Freshwater wetlands and wildlife, CONF-8603101. Department of Energy Symposium no. 61. United States Department of Energy, Office of Scientific and Technical Information, Oak Ridge, Tennessee.

REED, A. 1990. Population dynamics in a successful species: challenges in managing the increasing population of Greater Snow Geese. Transactions. International Union of Game Biologist Congress, Trondheim 19: 136-142.

REED, A. 1991. Subsistence harvesting of waterfowl in northern Quebec: goose hunting and the James Bay Cree. Transactions. North American Wildlife and Natural Resources Conference 56: 344-349.

REED, A. 1992. Management of Greater Snow Geese in relation to crop damage: the need for a diversified and integrated approach, pp. 93-100 *in* M.V. Roomen and J. Madsen, eds. Waterfowl and agriculture: review and future perspective of the crop damage conflict in Europe. International Waterfowl Research Bureau. Special Publication no. 21, 184 pp.

REED, A. 1993. Duration of family bonds of Brant geese *Branta bernicla* on the Pacific coast of North America. Wildfowl 44: 33-38.

REED, A. and A. BOURGET. 1977. Distribution and abundance of waterfowl wintering in southern Quebec. Canadian Field-Naturalist 91(1): 1-7.

REED, A. and H. BOYD. 1974. The impact of opening weekend hunting on local Black Ducks breeding in the St. Lawrence estuary and on transient ducks, pp. 84-91 *in* H. Boyd, ed. Canadian Wildlife Service waterfowl studies in eastern Canada, 1969-73/Études du SCF sur les oiseaux aquatiques dans l'est du Canada, 1969-73. Canadian Wildlife Service. Report Series no. 29, 106 pp.

REED, A. and P. CHAGNON. 1987. Greater Snow Geese on Bylot Island, Northwest Territories, 1983. Journal of Wildlife Management 51(1): 128-131.

REED, A. and A.J. ERSKINE. 1986. La taille et l'état des populations d'Eider à duvet dans l'est de l'Amérique du Nord, pp. 163-170 *in* A. Reed, ed. Eider ducks in Canada/Les eiders au Canada. Canadian Wildlife Service. Report Series no. 47, 177 pp.

REED, A. and R. JOLY. 1977. Situation de la Bernache du Canada (*Branta canadensis*) au Québec. Canadian Wildlife Service, Québec Region, unpublished report, 25 pp.

REED, A. and G. MOISAN. 1971. The Spartina tidal marshes of the St. Lawrence estuary and their importance to aquatic birds. Naturaliste canadien 98: 905-922.

REED, A. and N. PLANTE. (In preparation). Longterm decline in body mass, size, and condition of Greater Snow Geese at a fall staging area.

REED, A., G. CHAPDELAINE, and P. DUPUIS. 1977. Use of farmland in spring by migrating Canada Geese in the St. Lawrence valley, Québec. Journal of Applied Ecology 14: 667-680.

REED, A., P. DUPUIS, K. FISCHER, and J. MOSER. 1980. An aerial survey of breeding geese and other wildlife in Foxe Basin and northern Baffin Island, Northwest Territories, July 1979. Canadian Wildlife Service. Progress Notes no. 114, 21 pp.

REED, A., P. DUPUIS, A. BOURGET, and H.L. MENDALL. 1986. Sous-espèces d'Eider à duvet hivernant dans le golfe du Saint-Laurent, pp. 89-91 *in* A. Reed, ed. Eider ducks in Canada/Les eiders au Canada. Canadian Wildlife Service. Report Series no. 47, 177 pp.

Reed, A., P. Dupuis, and G.E.J. Smith. 1987a. A survey of Lesser Snow Geese on Southampton and Baffin islands, Northwest Territories, 1979. Canadian Wildlife Service. Occasional Paper no. 61, 24 pp.

Reed, A., T.W. Barry, L.S. Maltby, H. Boyd, and M. O'Briain. 1987b. Migration of Brant from the Canadian Arctic, Abstract, pp. 83-84 *in* C.R. Harington, ed. The Canadian Arctic Islands: Canada's missing dimension. Ottawa, 112 pp.

Reed, A., D. Goyette, and G. Lameboy. 1990. Observations préliminaires sur le régime alimentaire de la Bernache du Canada sur la côte est de la baie James, au printemps. Canadian Wildlife Service. Progress Notes no. 191, 5 pp. [Also available in English as: Preliminary observations on the spring diet of Canada Geese on the east coast of James Bay.]

Reed, A., H. Boyd, P. Chagnon, and J. Hawkings. 1992a. The numbers and distribution of Greater Snow Geese on Bylot Island and near Jungersen Bay, Baffin Island, in 1988 and 1983. Arctic 45(2): 115-119.

Reed, A., R. Lalumière, and R. Benoit. 1992b. Use of eelgrass meadows by Brant on the northeast coast of James Bay. Abstract of poster presented at 7th North American Arctic Goose Conference, January 1992, Vallejo, California, 123 pp.

Reed, A., Y. Aubry, and E. Reed. 1994. Surf Scoter, *Melanitta perspicillata*, nesting in southern Québec. Canadian Field-Naturalist 108(3): 364-365.

Reed, J.M. 1985. A comparison of the "flush" and spot-map methods for estimating the size of Vesper Sparrow territories. Journal of Field Ornithology 56(2): 131-137.

Reed, J.M. 1986. Vegetation structure and Vesper Sparrow territory location. Wilson Bulletin 98(1): 144-147.

Reed, J.R. 1988. Arctic adaptations in the breeding biology of Sandhill Cranes, *Grus canadensis*, on Banks Island, Northwest Territories. Canadian Field-Naturalist 102(4): 643-648.

Rees, W.E. 1989. Planning for sustainable developments: a resource book. University of British Columbia Centre for Human Settlements, Vancouver.

Reese, J.G. 1975. Productivity and management of feral Mute Swans in Chesapeake Bay. Journal of Wildlife Management 39(2): 280-286.

Reese, K.P., J.A. Kadlec, and L.M. Smith. 1987. Characteristics of islands selected by nesting Canada Geese *Branta canadensis*. Canadian Field-Naturalist 101(4): 539-542.

Régnière, J. and D. Welsh. 1988. Bird predation in a declining population of the Spruce Budworm (Lepidoptera: Tortricidae). Presented at the National Meeting of the Entomological Society of America, Louisville, Kentucky, December 1988 and at the Easter Spruce Budworm Research Work Conference, Fredericton, New Brunswick, January 1989, 19 pp.

Reid, W. 1992a. Black-throated Blue Warbler, pp. 314-315 *in* D.W. Brauning, ed. Atlas of breeding birds in Pennsylvania. University of Pittsburgh Press, Pittsburgh, 484 pp.

Reid, W. 1992b. Horned Lark, pp. 214-215 *in* D.W. Brauning, ed. Atlas of breeding birds in Pennsylvania. University of Pittsburgh Press, Pittsburgh, 484 pp.

Reilly, E.M. Jr. 1968. The Audubon illustrated handbook of American birds. McGraw-Hill Book Company, New York.

Reimann, E.J. 1942. Some studies on the Eastern Crow. Proceedings. Delaware Valley Ornithological Club of Philadelphia, 1938-1941. Cassinia 31: 19-24.

Reimchen, T.E. and S. Douglas. 1984. Feeding schedule and daily food consumption in Red-throated Loons (*Gavia stellata*) over the prefledging period. Auk 101(3): 593-599.

Reimchen, T.E. and S. Douglas. 1985. Differential contribution of the sexes to prefledged young in Red-throated Loons. Auk 102(1): 198-201.

Remsen, J.V. Jr., M. Ellerman, and J. Cole. 1989. Dead-leaf-searching by the Orange-crowned Warbler in Louisiana in winter. Wilson Bulletin 101(4): 645-648.

Renaud, W.E. and M.S.W. Bradstreet. 1980. Late winter distribution of Black Guillemots in northern Baffin Bay and the Canadian High Arctic. Canadian Field-Naturalist 94(4): 421-425.

Rendell, W.B. and R.J. Robertson. 1990. Influence of forest edge on nest-site selection by Tree Swallows. Wilson Bulletin 102(4): 634-644.

Rendell, W.B. and R.J. Robertson. 1991. Competition for cavities among Great Crested Flycatchers, *Myiarchus crinitus*, Northern Flickers, *Colaptes auratus*, and Tree Swallows, *Tachycineta bicolor*. Canadian Field-Naturalist 105(1): 113-114.

Renken, R.B. and E.P. Wiggers. 1989. Forest characteristics related to Pileated Woodpecker territory size in Missouri. Condor 91(3): 642-652.

Rever, M. and R.S. Miller. 1973. Common Goldeneyes and the Emma Lake nest boxes. Blue Jay 31(1): 27-30.

Reynolds, C.M. 1965. The survival of Mute Swan cygnets. Bird Study 12: 128-129.

Reynolds, J.D. 1987. Mating system and nesting biology of the Red-necked Phalarope *Phalaropus lobatus*: what constrains polyandry? Ibis 129: 225-242.

Reynolds, J.D. and R.W. Knapton. 1984. Nest-site selection and breeding biology of the Chipping Sparrow. Wilson Bulletin 96(3): 488-493.

Reynolds, J.D., M.A. Colwell, and F. Cooke. 1986. Sexual selection and spring arrival times of Red-necked and Wilson's phalaropes. Behavioral Ecology and Sociobiology 18: 303-310.

Reynolds, J.W. 1976. Catalogue et clé d'identification des Lombricidés du Québec. Naturaliste canadien 103: 21-27.

Reynolds, J.W. 1977a. Earthworms utilized by the American Woodcock. Proceedings. Woodcock Symposium 6: 161-169.

Reynolds, J.W. 1977b. The earthworms (Lumbricidae and Sparganophilidae) of Ontario. Life Sciences Miscellaneous Publications, Royal Ontario Museum.

Reynolds, J.W., W.B. Krohn, and G.A. Jordan. 1977. Earthworm populations as related to woodcock habitat usage in central Maine. Proceedings. Woodcock Symposium 6: 135-146.

Reynolds, R.E. and J.R. Sauer. 1991. Changes in Mallard breeding population in relation to production and harvest rates. Journal of Wildlife Management 55(3): 483-487.

Reynolds, R.S. 1985. Atlantic waterfowl council, technical section minutes, pp. 52-61.

Reynolds, R.T. and E.C. Meslow. 1984. Partitioning of food and niche characteristics of coexisting *Accipiter* during breeding. Auk 101(4): 761-779.

Reynolds, R.T. and H.M. Wight. 1978. Distribution, density and productivity of *Accipiter* hawks breeding in Oregon. Wilson Bulletin 90(2): 182-196.

Reynolds, R.T., E.C. Meslow, and H.M. Wight. 1982. Nesting habitat of coexisting *Accipiter* in Oregon. Journal of Wildlife Management 46(1): 124-138.

Rheinwald, G. 1977. Atlas der brutverbreitung westdeutscher vogelarten. Dachverband Deutscher Avifaunisten, Verlag, 72 pp.

Rhodes, W. 1868. House Sparrows. Morning Chronicle and Commercial, vol. 22, 12 June edition.

Rhodes, W. 1877. Imported birds for our woods and parks. Forest and Stream 8: 165.

Rice, J.C. 1978a. Behavioural interactions of two interspecifically territorial vireos. I. Song discrimination and natural interactions. Animal Behaviour 26: 527-538.

Rice, J.C. 1978b. Ecological relationships of two interspecifically territorial vireos. Ecology 59: 526-538.

Rice, O.O. 1969. Record of female cardinals sharing nest. Wilson Bulletin 81(2): 216.

Rice, W.R. 1982. Acoustical location of the prey by the Marsh Hawk: adaptation to concealed prey. Auk 99(3): 403-413.

Richard, A. 1991. Grèbe cornu – 1991. Unpublished report. Association québécoise des groupes d'ornithologues, 13 pp.

Richard, M.A. 1987. Carolina Wren, pp. 304-305 *in* M.D. Cadman, P.F.J. Eagles and F.M. Helleiner, eds. Atlas of the breeding birds of Ontario. University of Waterloo Press, Waterloo, 617 pp.

Richard, M.E. 1993. 87. Suburban bird sanctuary, p. 87 *in* J.D. Lowe, ed. Resident bird counts 1992. Journal of Field Ornithology 64(1)(supplement): 1-128.

Richard, P.J.H. 1987. Le couvert végétal du Québec-Labrador et son histoire postglaciaire. Université de Montréal, Montréal. Notes et documents du Département de géographie, no. 87-1, 74 pp.

Ricklefs, R.E. 1972. Latitudinal variation in breeding productivity of the Rough-winged Swallow. Auk 89(4): 826-836.

Ridgway, R. 1882. Descriptions of two new thrushes from the United States. Proceedings. United States National Museum, 1881, 4: 374-379.

Riley, J.L. 1982. Habitats of Sandhill Cranes in the southern Hudson Bay Lowland, Ontario. Canadian Field-Naturalist 96(1): 51-55.

Ringuet, I. and L.-G. de Repentigny. 1982. Un profil des réserves nationales de faune au Québec. Canadian Wildlife Service, Québec Region, unpublished report, 79 pp.

Ripley, S.D. 1976. Rails of the world: a monograph of the family Rallidae. M.F. Feheley Publishers Limited, Toronto, 406 pp.

Ripley, S.D. 1977. Rails of the world: a monograph of the family Rallidae. David R. Godine, Boston, 406 pp.

Rippin, A.B. and D.A. Boag. 1974. Spatial organization among male Sharp-tailed Grouse on arenas. Canadian Journal of Zoology 52: 591-597.

Rising, J.D. 1987a. Dark-eyed Junco, pp. 464-465 *in* M.D. Cadman, P.F.J. Eagles, and F.M. Helleiner, eds. Atlas of the breeding birds of Ontario. University of Waterloo Press, Waterloo, 617 pp.

Rising, J.D. 1987b. Savannah Sparrow, pp. 448-449 *in* M.D. Cadman, P.F.J. Eagles, and F.M. Helleiner, eds. Atlas of the breeding birds of Ontario. University of Waterloo Press, Waterloo, 617 pp.

Rising, J.D. 1987c. Sharp-tailed Sparrow, p. 554 *in* M.D. Cadman, P.F.J. Eagles, and F.M. Helleiner, eds. Atlas of the breeding birds of Ontario. University of Waterloo Press, Waterloo, 617 pp.

Rising, J.D. 1987d. Song Sparrow, pp. 458-459 *in* M.D. Cadman, P.F.J. Eagles, and F.M. Helleiner, eds. Atlas of the breeding birds of Ontario. University of Waterloo Press, Waterloo, 617 pp.

Rising, J.D. 1987e. Swamp Sparrow, pp. 460-461 *in* M.D. Cadman, P.F.J. Eagles, and F.M. Helleiner, eds. Atlas of the breeding birds of Ontario. University of Waterloo Press, Waterloo, 617 pp.

Rising, J.D. 1987f. Vesper Sparrow, pp. 446-447 *in* M.D. Cadman, P.F.J. Eagles, and F.M. Helleiner, eds. Atlas of the breeding birds of Ontario. University of Waterloo Press, Waterloo, 617 pp.

Rising, J.D. 1987g. White-crowned Sparrow, p. 556 *in* M.D. Cadman, P.F.J. Eagles, and F.M. Helleiner, eds. Atlas of the breeding birds of Ontario. University of Waterloo Press, Waterloo, 617 pp.

Rising, J.D. and J.C. Avise. 1993. Application of genealogical-concordance principles to the taxonomy and evolutionary history of the Sharp-tailed Sparrow (*Ammodramus caudacutus*). Auk 110(4): 844-856.

Rising, T. and J. Rising. 1982. Canadian songbirds and their ways. Tundra Books, Montréal, 176 pp.

Risley, C.J. 1984. Report on the status of the Eastern Bluebird (*Sialia sialis*). Committee on the Status of Endangered Wildlife in Canada (COSEWIC), Ottawa, 46 pp.

Ritchison, G. 1986. The singing behavior of female Northern Cardinal. Condor 88(2): 156-159.

Ritchison, G. 1991. The flight songs of Common Yellowthroats: description and causation. Condor 93(1): 12-18.

Ritchison, G. and M.K. Omer. 1990. Winter behavior of Northern Cardinals (*Cardinalis cardinalis*). Transactions. Kentucky Academy of Science 51(3-4): 145-153.

Ritchison, G., J.R. Belthoff, and E.J. Sparks. 1992. Dispersal restlessness: evidence for innate dispersal by juvenile Eastern Screech-Owl? Animal Behaviour 43(1): 57-65.

Robards, F.C. and J.I. Hodges. 1977. Observations from 2760 Bald Eagle nests in southeast Alaska. United States Fish and Wildlife Service, Juneau, Alaska.

Robards, F.C. and J.G. King. 1966. Nesting and productivity of Bald Eagles, southeast Alaska. United States Fish and Wildlife Service, Juneau, Alaska.

Robbins, C.S. 1973. Introduction, spread, and present abundance of the House Sparrow in North America, pp. 3-9 *in* S.C. Kendeigh, ed. A Symposium of the House Sparrow (*Passer domesticus*) and European Tree Sparrow (*P. montanus*) in North America. Ornithological Monographs no. 14, 121 pp.

Robbins, C.S. 1974. A history of North American hawk-watching, pp. 29-40 *in* M. Harwood, ed. Proceedings. North American hawk migration conference. Hawk Migration Association of North America. Shiver Mountain Press, Washington, 165 pp.

Robbins, C.S., B. Bruun, and H.S. Zim. 1966. A guide to field identification: birds of North America. Golden Press, New York, 340 pp.

Robbins, C.S., B. Bruun, and H.S. Zim. 1980. Guide des oiseaux d'Amérique du Nord: Guide d'identification sur le terrain. Éditions Marcel Broquet, LaPrairie, 351 pp.

Robbins, C.S., D. Bystrak, and P.H. Geissler. 1986. The breeding bird survey: its first fifteen years, 1965-1979. United States Department of the Interior, Fish and Wildlife Service, Washington. Resource Publication no. 157, 196 pp.

Robbins, C.S., D.K. Dawson, and B.A. Dowell. 1989a. Habitat area requirements of breeding forest birds of the middle Atlantic states. Wildlife Monographs no. 103, 34 pp.

Robbins, C.S., J.R. Sauer, R.S. Greenberg, and S. Droege. 1989b. Population declines in North American birds that migrate to the neotropics. Proceedings. National Academy of Science 86: 7658-7662.

Robbins, C.S., J.W. Fitzpatrick, and P.B. Harnel. 1989c. A warbler in trouble *Dendroica cerulea*, pp. 549-562 *in* J.M. Hagan III and D.W. Johnston, eds. Ecology and conservation of neotropical migrant landbirds. Smithsonian Institution Press, Washington, 609 pp.

Robbins, G.E.S. 1981. Quail, their breeding and management. Worlds Pheasant Association, Payn Essex Printers Ltd, Sudbury, Suffolk, Great Britain, 108 pp.

Robbins, S.D. Jr. 1991. Wisconsin bird life: population and distribution, past and present. University of Wisconsin Press, Madison, 720 pp.

Robert, M. 1988. Abondance et répartition du Dindon sauvage dans le sud-ouest du Québec. Unpublished report. Association québécoise des groupes d'ornithologues, 40 pp.

Robert, M. 1989a. L'odyssée du Dindon sauvage. Franc-Nord 6(4): 20-23.

Robert, M. 1989b. Les oiseaux menacés du Québec. Association québécoise des groupes d'ornithologues and Canadian Wildlife Service, Québec Region, Environment Canada, 109 pp. [Also available in English as: The threatened birds of Quebec].

Robert, M. 1990. Statut du Canard arlequin (*Histrionicus histrionicus*) sur le territoire québécois. Unpublished report. Canadian Wildlife Service, Québec Region, 10 pp.

Robert, M. 1991. L'identification des pies-grièches: oubliez les critères magiques. QuébecOiseaux 2(3): 16-17.

Robert, M. 1993. Le Râle jaune (*Coturnicops noveboracensis*) à l'île aux Grues et à Cacouna. Unpublished research report (summer period) for Canadian Wildlife Service, Québec Region, 37 pp.

Robert, M. and P. Laporte. 1991a. Plan d'action pour le rétablissement de la Pie-grièche migratrice au Québec. Unpublished report. Canadian Wildlife Service, Québec Region, 46 pp.

Robert, M. and P. Laporte. 1991b. Situations historique et actuelle de la Pie-grièche migratrice au Québec. Canadian Wildlife Service, Québec Region. Progress Notes no. 196, 7 pp. [Also available in English as: History and current status of the Loggerhead Shrike in Quebec.]

Robert, M. and R. McNeil. 1989. Comparative day and night feeding strategies of shorebird species in a tropical environnement. Ibis 131: 69-79.

Robert, M., P. Laporte, and A. Demers. 1991. La Pie-grièche migratrice au Québec: un siècle pour s'établir... et disparaître. QuébecOiseaux 2(4): 21-23.

Robertson, D. and R. Carey. 1993. 34. Suburban deciduous woodlot, p. 54 *in* J.D. Lowe, ed. Resident bird counts 1992. Journal of Field Ornithology 64(1)(supplement): 1-128.

Robertson, R.J. and H.L. Gibbs. 1982. Superterritoriality in Tree Swallows: a reexamination. Condor 84(3): 313-316.

Robertson, R.J., H.L. Gibbs, and B.J. Stutchbury. 1986. Spitefulness, altruism and the cost of aggression: evidence against superterritoriality in Tree Swallows. Condor 88(1): 123-124.

Robertson, R.J., B.J. Stutchbury, and R.R. Cohen. 1992. Tree Swallow (*Tachycineta bicolor*), *in* A. Poole, P. Stettenheim, and F. Gill, eds. The birds of North America, no. 11. The Academy of Natural Sciences, Philadelphia. The American Ornithologists' Union, Washington, 28 pp.

Robidoux, Y.P. and A. Cyr. 1989. Sélection granulométrique pour la construction du nid chez l'Hirondelle à front blanc, *Hirundo pyrrhonota*. Canadian Field-Naturalist 3(4): 577-583.

Robins, J.D. 1971. A study of Henslow's Sparrow in Michigan. Wilson Bulletin 83(1): 39-48.

Robinson, M. and C.D. Becker. 1986. Snowy Owls on Fetlar. British Birds 79(5): 228-242.

Robinson, S.K. 1981. Ecological relations and social interactions of Philadelphia and Red-eyed vireos. Condor 83(1): 16-26.

Robinson, S.K. and R.T. Holmes. 1982. Foraging behavior of forest birds: the relationships among search tactics, diet, and habitat structure. Ecology 63(6): 1918-1931.

Robinson, S.K. and R.T. Holmes. 1984. Effects of plant species and foliage structure on the foraging behaviour of forest birds. Auk 101(4): 672-684.

Robinson, S.K., J.A. Grzybowski, S.I. Rothstein, M.C. Brittingham, L.J. Petit, and F.R. Thompson. 1993. Management implications of cowbird parasitism on neotropical migrant songbirds, pp. 93-102 *in* D.M. Finch and P.W. Stangel, eds. Status and management of neotropical migratory birds. United States Department of Agriculture, Forest Service. General Technical Report RM-229.

Robinson, W.D. 1995. Louisiana Waterthrush (*Seiurus motacilla*), *in* A. Poole and F. Gill, eds. The birds of North America, no. 151. The Academy of Natural Sciences, Philadelphia. The American Ornithologists' Union, Washington, 20 pp.

Robinson, W.L. 1980. Fool hen: the Spruce Grouse on the Yellow Dog Plains. University of Wisconsin Press, Madison, 224 pp.

Robitaille, L. 1985. Le dépérissement des érablières au Québec: problématique et état des recherches. Ministère des Mines et des Ressources, Service de la recherche forestière, 3 pp.

Rochefort, L. 1983. Observations ornithologiques au Nouveau-Québec. Bulletin ornithologique 28(2): 83-84.

Rochefort, L. and L. Quinty. 1994. La restauration des tourbières exploitées: le développement d'une stratégie intégrée au Québec. Report for Direction de la recherche et des techniques environnementales, service des milieux agricoles, urbains et naturels, Ministère de l'Environnement et de la Faune.

Rodenhouse, N.L. 1992. Potential effects of climatic change on a neotropical migrant landbird. Conservation Biology 6: 263-272.

Rodenhouse, N.[L.] and L.[B.] Best. 1981. Nesting ecology of Vesper Sparrows breeding on Iowa cropland. Proceedings. Iowa Academy of Science 88(1): 20.

Rodenhouse, N.L. and L.B. Best. 1983. Breeding ecology of Vesper Sparrows in corn and soybean fields. American Midland Naturalist 110: 265-275.

Rodenhouse, N.L. and R.T. Holmes. 1992. Results of experimental and natural food reductions for breeding Black-throated Blue Warblers. Ecology 73(1): 357-372.

Rogers, J.P. 1979. *Branta bernicla hrota* in the United States of America – A management review, pp. 198-205 *in* M. Smart, ed. Proceedings. First technical meeting on western Palearctic migratory bird management *Branta bernicla bernicla*. International Waterfowl Research Bureau, Gloucester, England, 228 pp.

Rogers, J.P. and L.J. Korschgen. 1966. Foods of Lesser Scaups on breeding, migration and wintering areas. Journal of Wildlife Management 30(2): 258-264.

Rogers, J.P. and J.H. Patterson. 1984. The Black Duck population and its management. Transactions. North American Wildlife and Natural Resources Conference 49: 527-534.

Rohner, C. and F.I. Doyle. 1992. Methods of locating Great Horned Owl nests in the boreal forest. Journal of Raptor Research 26(1): 33-35.

Rohwer, F.C. and M.G. Anderson. 1988. Female based philopatry, monogamy and the timing of pair formation in waterfowl. Current Ornithology 5: 187-221.

Rohwer, F.C. and H.W. Heusmann. 1991. Effects of brood size and age on survival of female Wood Ducks. Condor 93(4): 819-824.

Rohwer, S., S.D. Fretwell, and D.M. Niles. 1980. Delayed maturation in passerine plumages and the deceptive acquisition of resources. American Naturalist 115: 400-437.

Rohwer, S., C.D. Spaw, and E. Roskaft. 1989. Costs to Northern Orioles of puncture-ejecting parasitic cowbird eggs from their nests. Auk 106(4): 734-738.

Rohwer, S.A. 1972. Multivariate assessment of interbreeding between the meadowlarks, *Sturnella*. Systematic Zoology 21: 313-338.

Roland, J., S.J. Hannon, and M.A. Smith. 1986. Foraging pattern of Pine Siskins and its influence on winter moth survival in an apple orchard. Ecologia 69(1): 47-52.

Rolando, A. and M. Zunino. 1992. Observations of tool use in corvids. Ornis Scandinavica 23(2): 201-202.

Rolfinke, B.F., R.H. Yahner, and J.S. Wakeley. 1990. Effects of forest irrigation on long-term trends in breeding-bird communities. Wilson Bulletin 102(2): 264-278.

Romagnano, L., A.S. Hoffenberg, and H.W. Power. 1990. Intraspecific brood parasitism in the European Starling. Wilson Bulletin 102(2): 279-291.

Romanow, J. 1993. 126. Abandoned pasture scrubland, pp. 109-110 *in* J.D. Lowe, ed. Resident bird counts 1992. Journal of Field Ornithology 64(1)(supplement): 1-128.

Root, O.M. 1968. Clay-colored Sparrow, pp. 1186-1208 *in* A.C. Bent. Life histories of North American cardinals, grosbeaks, buntings, towhees, finches, sparrows, and allies. (O.L. Austin, Jr., ed.). United States National Museum, Washington. [1968, Dover Publications, New York]. Bulletin no. 287, vol. 1, pp. 1-602; vol. 2, pp. 603-1248; vol. 3, pp. 1249-1889.

Root, R.B. 1967. The niche exploitation pattern of the Blue-gray Gnatcatcher. Ecological Monographs 37: 317-350.

Root, T. 1988. Atlas of wintering North American birds. An analysis of Christmas Bird Count data. University of Chicago Press, Chicago, 312 pp.

Roseberry, J.L. and W.D. Klimstra. 1970. The nesting ecology and reproductive performance of the Eastern Meadowlark. Wilson Bulletin 82(3): 243-267.

Rosene, W. 1969. The Bobwhite Quail: its life and management. Rutgers University Press, New Brunswick, New Jersey, 418 pp.

Rosenfield, R.N. 1984. Nesting biology of Broad-winged Hawks in Wisconsin. Raptor Research 18(1): 6-9.

Rosenfield, R.N. 1990. Preincubation behavior and paternity assurance in the Cooper's Hawk (*Accipiter cooperii* [Bonaparte]). Ph.D. dissertation, North Dakota State University, Fargo, 90 pp.

Rosenfield, R.N. and J. Bielefeldt. 1993. Cooper's Hawk (*Accipiter cooperii*), *in* A. Poole and F. Gill, eds. The birds of North America, no. 75. The Academy of Natural Sciences, Philadelphia. The American Ornithologists' Union, Washington, 24 pp.

Rosenfield, R.N., J. Bielefeldt, and J. Cary. 1991. Copulatory and other pre-incubation behaviors of Cooper's Hawks. Wilson Bulletin 103(4): 656-660.

Rosenfield, R.N., J. Bielefeldt, and J. Cary. 1992. Natal dispersal and inbreeding in the Cooper's Hawk. Wilson Bulletin 104(1): 182-184.

Rosenfield, R.N., M.W. Gratson, and L.B. Carson. 1984. Food brought by Broad-winged Hawks to a Wisconsin nest. Journal of Field Ornithology 55(2): 246-247.

Roskaft, E. and S. Rohwer. 1987. An experimental study of the function of the red epaulettes and the black body colour of male Red-winged Blackbirds. Animal Behaviour 35: 1070-1077.

Roskaft, E., G.H. Orians, and L.D. Beletsky. 1990. Why do Red-winged Blackbirds accept eggs of Brown-headed Cowbird? Evolutionary Ecology 4(1): 35-42.

Ross, R.K. 1982. Duck distribution along the James and Hudson Bay coasts of Ontario. Naturaliste canadien 109: 927-932.

Ross, R.K. 1983. An estimate of the Black Scoter, *Melanitta nigra*, population moulting in James and Hudson bays. Canadian Field-Naturalist 97(2): 147-150.

Ross, R.K. 1987. White-winged Scoter, p. 517 *in* M.D. Cadman, P.F.J. Eagles, and F.M. Helleiner, eds. Atlas of the breeding birds of Ontario. University of Waterloo Press, Waterloo, 617 pp.

Rossetti, K. 1983. House Sparrow taking insects from spiders' webs. British Birds 76(9): 412.

Roth, R.R. 1987. Assessment of habitat quality for Wood Thrush in a residential area, pp. 139-149 *in* L.W. Adams and D.L. Leedy, eds. Proceedings. National symposium on urban wildlife. National Institute for Urban Wildlife, Columbia, Maryland.

Roth, R.R. and R.I. Johnson. 1989. Abundance and return rates of Wood Thrush in a forest fragment, 1974-1939, *in* Symposium Abstracts. Ecology and conservation of neotropical migrant landbirds. Manomet Bird Observatory, December 1989.

Rothstein, S.I. 1971. High nest density and non-random nest placement in the Cedar Waxwing. Condor 73(4): 483-485.

Rothstein, S.I. 1973. Extreme overlap between first and second nesting in the Rose-breasted Grosbeak. Wilson Bulletin 85(2): 242-243.

Rothstein, S.I. 1974. Mechanism of avian egg recognition: possible learned and innate factors. Auk 91(1): 796-807.

Rothstein, S.I. 1975. An experimental and teleonomic investigation of avian brood parasitism. Condor 77: 250-271.

Rothstein, S.I. 1976a. Cowbird parasitism of the Cedar Waxwing and its evolutionary implications. Auk 93(3): 498-509.

Rothstein, S.I. 1976b. Experiments on defenses Cedar Waxwings use against cowbird parasitism. Auk 93(4): 675-691.

Rothstein, S.I. 1977. Cowbird parasitism and egg recognition of the Northern Oriole. Wilson Bulletin 89(1): 21-32.

Rousseau, C. 1974. Géographie floristique du Québec/Labrador. Distribution des principales espèces vasculaires. Travaux et documents du Centre d'études nordiques. Les Presses de l'Université Laval, Sainte-Foy, 978 pp.

Rousseau, J. and V. Legendre. 1966. Les animaux du Canada en 1664. La faune indigène. Canadian Wildlife Service, Ministère du Tourisme, de la Chasse et de la Pêche du Québec. Bulletin 8: 303-335.

Rousseau, J., G. Béthune, and P. Morisset. 1977. Voyage de Pehr Kalm au Canada en 1749. Traduction annotée du journal de route. Pierre Tisseyre, Montréal, 674 pp.

Rowdon, L.F. 1969. Winter birds on the Magdalen Islands, additions to the list. Mimeographic document.

Rowe, J.S. 1972. Forest regions of Canada. Canadian Forestry Service, Environment Canada, Ottawa. Publication no. 1300, 172 pp.

Roy, C. 1994. Inventaire des sites de reproduction du Pluvier siffleur aux Îles-de-la-Madeleine. Report presented to Environment Canada, Conservation and Protection, by Attention FragÎles, Cap-aux-Meules, Îles-de-la-Madeleine.

Royall, W.C. Jr. 1966. Breeding of the Starling in central Arizona. Condor 68: 196-205.

Rumsey, R.L. 1970. Woodpecker nest failures in creosoted utility poles. Auk 87(2): 367-369.

Rusch, D.H. and L.B. Keith. 1971. Seasonal and annual trends in numbers of Alberta Ruffed Grouse. Journal of Wildlife Management 35(4): 803-821.

Rusch, D.H., E.C. Meslow, P.D. Doerr, and L.B. Keith 1972. Response of Great Horned Owl populations to changing prey densities. Journal of Wildlife Management 36(2): 282-296.

Rusch, D.H., C.D. Ankney, H. Boyd, J.R. Longcore, F. Montalbano, III, J.K. Ringelman, and V.D. Stotts. 1989. Population ecology and harvest of the American Black Duck: a review. Wildlife Society Bulletin 17: 379-406.

Rushmore, F.M. 1969. Sapsucker damage varies with tree species and seasons. United States Forest Research Paper NE-136: 1-19.

Rust, H.J. 1947. Migration and nesting of nighthawks in northern Idaho. Condor 49(5): 177-188.

Rutter, R.J. 1969. A contribution to the biology of the Gray Jay (*Perisoreus canadensis*). Canadian Field-Naturalist 83(4): 300-316.

Ryan, M.R. and J.J. Dinsmore. 1979. A quantitative study of the behavior of breeding American Coots. Auk 96(4): 704-713.

Ryan, M.R. and R.B. Renken. 1987. Habitat use by breeding Willets in the Northern Great Plains. Wilson Bulletin 99(2): 175-189.

Ryan, M.R., R.B. Renken, and J.J. Dinsmore. 1984. Marbled Godwit habitat selection in the northern prairie region. Journal of Wildlife Management 48(4): 1206-1218.

Ryder, J.A., D.A. Palmer, and J.J. Rawinski. 1987. Distribution and status of the Boreal Owl in Colorado, pp. 169-174 *in* R.W. Nero, R.J. Clark, R.J. Knapton, and R.H. Hamre, eds. Proceedings. Biology and conservation of northern forest owls. Symposium, 3-7 February, Winnipeg. United States Department of Agriculture, Forest Service, Rocky Mountain Forest and Range Experiment Station, Fort Collins, Colorado. General Technical Report RM-142, 309 pp.

Ryder, J.P. 1978. Possible origins and adaptive value of female-female pairing in gulls, pp. 138-145 *in* W.E. Southern, ed. Proceedings. 1978 Conference of the Colonial Waterbird Group, Northern Illinois University, DeKalb, Illinois, 213 pp.

Ryder, J.P. 1993. Ring-billed Gull (*Larus delawarensis*), *in* A. Poole, P. Stettenheim, and F. Gill, eds. The birds of North America, no 33. The Academy of Natural Sciences, Philadelphia. The American Ornithologists' Union, Washington, 28 pp.

Ryder, J.P. and P.L. Somppi. 1979. Female-female pairing in Ring-billed Gulls. Auk 96: 1-5.

Sabo, S.R. 1980. Niche and habitat relations in subalpine bird communities of the White Mountains of New Hampshire. Ecological Monographs 50(2): 241-259.

Sabo, S.R. and R.T. Holmes. 1983. Foraging niches and the structure of forest bird communities in contrasting montane habitats. Condor 85: 121-138.

Sadler, D.C. 1987. Wood Thrush, pp. 328-329 *in* M.D. Cadman, P.F.J. Eagles, and F.M. Helleiner, eds. Atlas of the breeding birds of Ontario. University of Waterloo Press, Waterloo, 617 pp.

Sagard, G. 1865. Le Grand voyage au pays des Hurons, situé en l'Amérique vers la Mer douce, ès derniers confins de la Nouvelle-France, dite Canada. Première et seconde parties. Nouvelle édition. Librairie Tross, Paris, 268 pp. [originally published in 1632].

Sagard, G. 1939. The long journey to the country of the Hurons (Translation of Le Grand voyage au pays des Hurons, situé en l'Amérique vers la Mer douce, ès derniers confins de la Nouvelle-France, dite Canada. By Father Gabriel Sagard, first published in 1632, new edition 1865. Ed. G.M. Wrong, Translated by H. Langton). The Champlain Society, Toronto.

Salomonsen, F. 1951. The birds of Greenland. E. Munksgaard, Copenhagen, 3 parts, 607 pp.

Salomonsen, F. 1967. The water birds of Greenland. (English translation, 1982). Studies of northern seabirds of the Canadian Wildlife Service. Environment Canada. Report no. 124, 85 pp.

Salomonsen, F. 1968. The moult migration. Wildfowl 19: 5-24.

Salt, W.R. 1973. Alberta vireos and wood warblers – Families Vireonidae and Parulidae – Distribution and breeding. Provincial Museum and Archives of Alberta. Publication 3: 7-9.

Salt, W.R. and A.L. Wilk. 1966. The birds of Alberta, 2nd edition. Department of Industry and Development, Queen's Printer, Edmonton, 511 pp.

Salter, R. 1991. Aggregations of Barn Owls in abandoned desert mines. American Birds 45(1): 56-57.

Salwasser, H. 1990. Conserving biological diversity: a perspective on scope and approaches. Forest Ecology and Management 35(1): 75-90.

Salzman, E. 1993. Sibley's classification of birds. Birding 25(6): 446-455.

Samson, F.B. 1992. Conserving biological diversity in sustainable ecological systems. Transactions. North American Wildlife and Natural and Resources Conference 57: 308-320.

Samson, F.B. and F.L. Knopf. 1982. In search of a diversity ethic for wildlife management. Transactions. North American Wildlife and Natural Resources Conference 47: 421-431.

Samson, F.B. and S.J. Lewis. 1979. Experiments on population regulation in two North American parids. Wilson Bulletin 91(2): 222-233.

Samson, P. 1995. Écoguide d'Anticosti. La Chambre de Commerce de l'île d'Anticosti and la Pourvoirie du Lac Geneviève, 21 pp.

Samuel, D.E. 1971. The breeding biology of Barn and Cliff swallows in West Virginia. Wilson Bulletin 83(3): 284-301.

Sandberg, R., U. Ottosson, and J. Pettersson. 1991. Magnetic orientation of migratory Wheatears (*Oenanthe oenanthe*) in Sweden and Greenland. Journal of Experimental Biology 155: 51-64.

Sanders, C.J. 1970. Populations of breeding birds in the spruce-fir forests of northwestern Ontario. Canadian Field-Naturalist 84: 131-135.

Sanders, S.D. and M.A. Flett. 1989. Montane riparian habitat and Willow Flycatchers: threats to a sensitive environment and species, pp. 262-266 *in* D.L. Abell, ed. Proceedings. California riparian systems conference: protection, management, and restoration for the 1990's. United States Department of Agriculture, Pacific Southwest Forest and Range Experiment Station, Forest Service, Davis, California. General Technical Report PSW-110, 544 pp.

Sanderson, G.C. and F.C. Bellrose. 1986. Lead poisoning in waterfowl. Illinois Natural History Survey, Special Publications no. 4, 34 pp.

Sandilands, A.[D.] 1987a. American Wigeon, pp. 84-85 *in* M.D. Cadman, P.F.J. Eagles, and F.M. Helleiner, eds. Atlas of the breeding birds of Ontario. University of Waterloo Press, Waterloo, 617 pp.

Sandilands, A.[D.] 1987b. Green-winged Teal, pp. 68-69 *in* M.D. Cadman, P.F.J. Eagles, and F.M. Helleiner, eds. Atlas of the breeding birds of Ontario. University of Waterloo Press, Waterloo, 617 pp.

Sandilands, A.[D.] 1987c. Northern Shoveler, pp. 80-81 *in* M.D. Cadman, P.F.J. Eagles, and F.M. Helleiner, eds. Atlas of the breeding birds of Ontario. University of Waterloo Press, Waterloo, 617 pp.

Sandilands, A.[D.] 1987d. Canvasback, pp. 86-87 *in* M.D. Cadman, P.F.J. Eagles, and F.M. Helleiner, eds. Atlas of the breeding birds of Ontario. University of Waterloo Press, Waterloo, 617 pp.

Sandilands, A.[D.] and C.[A.] Campbell. 1987a. Le Conte's Sparrow, pp. 454-455 *in* M.D. Cadman, P.F.J. Eagles, and F.M. Helleiner, eds. Atlas of the breeding birds of Ontario. University of Waterloo Press, Waterloo, 617 pp.

Sandilands, A.D. and C.A. Campbell. 1987b. Status report on the Least Bittern (*Ixobrychus exilis*) in Canada. Committee on the Status of Endangered Wildlife in Canada (COSEWIC), Ottawa, 34 pp.

Sanger, G.A. and R.J. Jones, Jr. 1984. Winter feeding ecology and trophic relationships of Oldsquaws and White-winged Scoters on Kachemak Bay, Alaska, pp. 20-28 *in* D.N. Nettleship, G.A. Sanger, and P.F. Springer, eds. Proceedings. Pacific seabird group. Symposium, 6-8 January 1982, Seattle, Washington. Canadian Wildlife Service, Ottawa, 220 pp.

Santana, C.E., R.L. Knigh, and S.A. Temple. 1986. Parental care at a Red-tailed Hawk nest tended by three adults. Condor 88(1): 109-110.

Santner, S. 1992a. Black-throated Green Warbler, pp. 318-319 *in* D.W. Brauning, ed. Atlas of breeding birds in Pennsylvania. University of Pittsburgh Press, Pittsburgh, 484 pp.

Santner, S. 1992b. Killdeer, pp. 132-133 *in* D.W. Brauning, ed. Atlas of breeding birds in Pennsylvania. University of Pittsburgh Press, Pittsburgh, 484 pp.

Santner, S. 1992c. Pine Warbler, pp. 324-325 *in* D.W. Brauning, ed. Atlas of breeding birds in Pennsylvania. University of Pittsburgh Press, Pittsburgh, 484 pp.

Santner, S. 1992d. Vesper Sparrow, pp. 380-381 *in* D.W. Brauning, ed. Atlas of breeding birds in Pennsylvania. University of Pittsburgh Press, Pittsburgh, 484 pp.

Sappington, J.N. 1975. Cooperative breeding in the House Sparrow (*Passer domesticus*). Ph.D. dissertation, Mississippi State University, Mississippi State, 102 pp.

Sarrazin, R., M. Cantin, A. Gagnon, C. Gauthier, and G. Lefebvre. 1983. La protection des habitats fauniques au Québec, Annexe B. Les activités d'altérations des milieux et leurs répercussions sur la faune. Ministère du Loisir, de la Chasse et de la Pêche, Direction générale de la faune, Québec, 175 pp.

Saunders, A.A. 1929. Bird Song. State Museum Handbook no. 7, 220 pp.

Saunders, R.M. 1935. The first introduction of European plants and animals into Canada. Canadian Historical Review 16: 388-406.

Savage, C. 1985. Ces merveilleux oiseaux du Canada. Éditions La Presse, Ottawa, 209 pp.

Savaloja, T. 1981. Yellow Rail. Birding 13(3): 80-85.

Savard, G. and C. Cormier. 1995. Liste annotée des oiseaux du Saguenay–Lac-Saint-Jean. Club des ornithologues amateurs du Saguenay–Lac-Saint-Jean, Jonquière, 175 pp.

Savard, J.-P.L. 1977. Étude de la faune avienne dans les bassins de la Grande rivière de la Baleine et de la Petite rivière de la Baleine (été 1976). Technical report for Hydro-Québec, Direction Environnement. Éco-recherches, 132 pp.

Savard, J.-P.L. 1978. Birds in metropolitan Toronto: distribution relationships with habitat features and nesting sites. M.Sc. thesis, University of Toronto, Toronto, 138 pp.

Savard, J.-P.L. 1982a. Intra and inter-specific competition between Barrow's Goldeneye (*Bucephala islandica*) and Bufflehead (*Bucephala albeola*). Canadian Journal of Zoology 60(12): 3439-3446.

Savard, J.-P.L. 1982b. Barrow's Goldeneye nest box utilization in the Cariboo parkland, British Columbia: year 1. Canadian Wildlife Service. Progress Notes no. 131.

Savard, J.-P.L. 1984. Territorial behaviour of Common Goldeneye, Barrow's Goldeneye and Bufflehead in areas of sympatry. Ornis Scandinavica 15: 211-216.

Savard, J.-P.L. 1985. Evidence of long-term pair bonds in Barrow's Goldeneye. Auk 102(2): 389-391.

Savard, J.-P.L. 1986a. Territorial behaviour, nesting success and brood survival in Barrow's Goldeneye and its congeners. Ph.D. dissertation, University of British Columbia, Vancouver, 219 pp.

Savard, J.-P.L. 1986b. Polygyny in Barrow's Goldeneye. Condor 88(2): 250-252.

Savard, J.-P.L. 1987a. Status report on Barrow's Goldeneye. Canadian Wildlife Service, Pacific and Yukon Region, British Columbia. Technical Report Series no. 23, 57 pp.

Savard, J.-P.L. 1987b. Causes and functions of brood amalgamation in Barrow's Goldeneye and Bufflehead. Canadian Journal of Zoology 65: 1548-1553.

Savard, J.-P.L. 1988a. Winter, spring and summer territoriality in Barrow's Goldeneye: characteristics and benefits. Ornis Scandinavica 19: 119-128.

Savard, J.-P.L. 1988b. Use of nest boxes by Barrow's Goldeneyes: nesting success and effect on the breeding population. Wildlife Society Bulletin 16(2): 125-132.

Savard, J.-P.L. 1988c. Memorandum sur les développements hydro-électriques à la baie James, et leurs impacts possibles sur l'avifaune: identification du problème et des lacunes dans nos connaissances. Environment Canada, Canadian Wildlife Service, Québec Region, 55 pp.

Savard, J.-P.L. 1990. Population de sauvagine hivernant dans l'estuaire du Saint-Laurent: écologie, distribution et abondance. Canadian Wildlife Service, Québec Region. Technical Report Series no. 89, 91 pp. [An abstract is available in English.]

Savard, J.-P.L. and J. McA. Eadie. 1989. Survival and breeding philopatry in Barrow's and Common goldeneyes. Condor 91(1): 198-203.

Savard, J.-P.L. and J.B. Falls. 1981. Influence of habitat structure on the nesting height of birds in urban areas. Canadian Journal of Zoology 59(6): 924-932.

Savard, J.-P.L. and P. Lamothe. 1991. Distribution, abundance, and aspects of breeding ecology of Black Scoter, *Melanitta nigra*, and Surf Scoter, *M. perspicillata*, in northern Quebec. Canadian Field-Naturalist 105: 488-496.

Savard, J.-P.L. and J.N.M. Smith. 1987. Interspecific aggression by Barrow's Goldeneye: a descriptive and functional analysis. Behaviour 102: 168-184.

Savard, J.-P.L., G.E.J. Smith, and J.N.M. Smith. 1991. Duckling mortality in Barrow's Goldeneye and Bufflehead broods. Auk 108: 568-577.

Savard, J.-P.L., J. Bédard, and A. Nadeau. 1995. Spring and early summer distribution of seaducks (scoters and eiders) in the St. Lawrence River estuary. Canadian Wildlife Service, Québec Region, unpublished report.

Savard, M. 1984a. Présence d'une Mouette rieuse (*Larus ridibundus*) sur la rivière Mistassini au Lac-Saint-Jean. Le Harfang 7(5): 111-112.

Savard, M. 1984b. Le Chevalier solitaire (*Tringa solitaria*), important prédateur de larves d'odonates en Sagamie. Fabreries 11(1): 1-6.

Savignac, R., L. Baribeau, and J. Bourgeois. 1982. Étude préliminaire du zooplancton et des graines en suspension dans la plaine de débordement du lac Saint-Pierre. Report for Environment Canada, Canadian Wildlife Service and Ministère du Loisir, de la Chasse et de la Pêche, Direction générale de Trois-Rivières, by G.D.G. Environnement Mauricie, 46 pp.

Savile, D.B.O. 1950. Bird notes from Great Whale River, Québec. Canadian Field-Naturalist 64(3): 95-99.

Schaadt, C.P. 1989. Intraspecific comparisons of sexual and geographic variation in the growth of migratory and sedentary Ospreys. Ph.D. dissertation, McGill University, Montréal, 122 pp.

Schaefer, J.M. 1978. Observations at Turkey Vulture nest in southern Iowa. Bird Life 48(3): 87-93.

Schaefer, J.M. and J.J. Dinsmore. 1992. Movement of a nestling between American Crow nests. Wilson Bulletin 104(1): 185-187

Schaeff, C. and J. Picman. 1988. Destruction of eggs by Western Meadowlarks. Condor 90(4): 935-937.

Schaeffer, F.S. 1973. Tactile bristles of Saw-whet Owls are sensitive to touch. Bird-Banding 44(2): 125.

Schafer, E.W. Jr. and R.B. Brunton. 1971. Chemicals as bird repellents: two promising agents. Journal of Wildlife Management 35(3): 569-572.

Schamel, D. and D.M. Tracy. 1985. Replacement clutches in the Red-throated Loon. Journal of Field Ornithology 56(3): 282-283.

Schamel, D. and D.M. Tracy. 1991. Breeding site fidelity and natal philopatry in the sex role-reversed Red and Red-necked phalaropes. Journal of Field Ornithology 62(3): 390-398.

Schardien, B.J. and J.A. Jackson. 1982. Killdeers feeding on frogs. Wilson Bulletin 94(1): 85-87.

Scharf, W.C. 1989. Coastal Great Blue Heron and Great Egret colonies of the Michigan Great Lakes. Jack-Pine Warbler 67: 52-65.

Scharf, W.C. 1991. Great Egret, pp. 110-111 *in* R. Brewer, G.A. McPeek, and R.J. Adams, Jr., eds. The Atlas of breeding birds of Michigan. Michigan State University Press, East Lansing, 594 pp.

Schemske, D.W. 1975. Territoriality in a nectar feeding Northern Oriole in Costa Rica. Auk 92(3): 594-595.

Schenkeveld, L.E. and R.C. Ydenberg. 1985. Synchronous diving by Surf Scoters flocks. Canadian Journal of Zoology 63: 2516-2519.

Scherrer, B. 1982. Étude de l'avifaune du parc de la Gatineau: méthodologie générale. Université du Québec à Montréal, Laboratoire d'ornithologie et de biostatistique, Montréal, 242 pp.

Scherrer, B. and F. Morneau. 1984. Étude de l'avifaune du parc de la Gatineau: synthèse. Laboratoire d'ornithologie et de biostatistique, Département des Sciences biologiques et CERSE, Université du Québec à Montréal, Montréal, 105 pp.

Schieck, J.O. and S.J. Hannon. 1989. Breeding site fidelity in Willow Ptarmigan: the influence of previous reproductive success and familiarity with partner and territory. Oecologia 81(4): 465-472.

Schifferli, A., E. Géroudet, and R. Winkler. 1980. Atlas des oiseaux nicheurs de Suisse. Station ornithologique suisse de Sempach, 462 pp.

Schlichter, L. 1978. Winter predation by Black-capped Chickadees and Downy Woodpeckers on inhabitants of the Goldenrod Ball Gall. Canadian Field-Naturalist 92(1): 71-74.

Schmitt, J. 1904. Monographie de l'île d'Anticosti (golfe du Saint-Laurent). Hermann, Paris, 370 pp.

Schnell, J.H. 1958. Nesting behavior and food habits of Goshawks in the Sierra Nevada of California. Condor 60(6): 377-403.

Schoener, T.W. 1987. The geographical distribution of rarity. Oecologia 74: 161-173.

Schorger, A.W. 1947. The deep diving of the loon and Oldsquaw and its mechanism. Wilson Bulletin 59: 151-159.

Schorger, A.W. 1952. Introduction of the Domestic Pigeon. Auk 69(4): 462-463.

Schorger, A.W. 1955. The Passenger Pigeon, its natural history and extinction. University of Wisconsin Press, Madison, 424 pp.

Schorger, A.W. 1966. The Wild Turkey: its history and domestication. University of Oklahoma Press, Norman, 625 pp.

Schrantz, F.G. 1943. Nest life of the Eastern Yellow Warbler. Auk 60(3): 367-387.

Schroeder, M.A. 1986. The fall phase of dispersal in juvenile Spruce Grouse. Canadian Journal of Zoology 64(1): 16-20.

Schroeder, M.A. and D.A. Boag. 1985. Behavior of Spruce Grouse broods in the field. Canadian Journal of Zoology 63: 2494-2500.

Schroeder, M.A. and D.A. Boag. 1988. Dispersal in Spruce Grouse: is inheritance involved? Animal Behaviour 36(1): 305-307.

Schroeder, R.L. 1982. Habitat suitability index models: Downy Woodpecker. United States Department of Interior, Fish and Wildlife Service, 10 pp.

Schwalbe, P.W. 1992. Black-and-white Warbler, pp. 330-331 *in* D.W. Brauning, ed. Atlas of breeding birds in Pennsylvania, University of Pittsburgh Press, Pittsburgh, 484 pp.

Scoggan, H.J. 1978. The flora of Canada. National Museum of Natural Sciences, National Museums of Canada, Ottawa, 4 vol., 1711 pp.

Scott, A.C. and M. Bekoff. 1991. Breeding behavior of Evening Grosbeaks. Condor 93(1): 71-81.

Scott, D.M. and C.D. Ankney. 1980. Fecundity of the Brown-headed Cowbird in southern Ontario. Auk 97(4): 677-683.

Scott, D.M. and C.D. Ankney. 1983. The laying cycle of Brown-headed Cowbirds: passerine chickens? Auk 100: 583-592.

Scott, D.M., J.A. Darley, and A.V. Newsome. 1988. Length of the laying season and clutch size of Gray Catbirds at London, Ontario, Canada. Journal of Field Ornithology 59(4): 355-360.

Scott, D.M., P.J. Weatherhead, and C.D. Ankney. 1992. Egg-eating by female Brown-headed Cowbirds. Condor 94(3): 579-584.

Scott, G.A. 1963. First nesting of the Little Gull (*Larus minutus*) in Ontario and the New World. Auk 80(4): 548-549.

Scott, J.M., F. Davis, B. Csuti, R. Noss, B. Butterfield, C. Groves, H. Anderson, S. Caicco, F. D'Erchia, T.C. Edwards, Jr., J. Ulliman, and G. Wright. 1993. Gap analysis: a geographic approach to protection of biological diversity. Wildlife Monographs no. 123, 41 pp.

Scott, L. 1970. Great Horned Owls occupy artificial nesting site. Blue Jay 28(3): 123.

Scott, M.P. 1985. Duck, pp. 157-161 *in* B. Campbell and E. Lack, eds. A dictionary of birds. British Ornithologists' Union, Buteo Books, Vermillion, South Dakota, 670 pp.

Scott, T.A. and M.L. Morrison. 1990. Natural history and management of the San Clemente Loggerhead Shrike. Proceedings. Western Foundation of Vertebrate Zoology 4(2): 23-57.

Scott, V.E. and G.L. Crouch. 1987. Response of breeding birds to commercial clearcutting of aspen in southwestern Colorado. United States Forest Service Resource, Note RM 475, 5 pp.

Scott, W.B. and E.J. Crossman. 1973. Freshwater fishes of Canada. Environment Canada, Fisheries Research Board of Canada, Ottawa, 1026 pp.

Scott, W.B. and E.J. Crossman. 1978. Poissons d'eau douce du Canada. Environment Canada, Fisheries and Marine Service, Ottawa. Bulletin no. 184, 1026 pp.

Scott, W.B. and M.G. Scott. 1988. Atlantic fishes of Canada. Canadian Bulletin of Fisheries and Aquatic Sciences no. 219, 731 pp.

Sealy, S.G. 1966. Notes on the Rough-legged Hawk in the Perry River region, Northwest Territories. Blue Jay 24(3): 127-128.

Sealy, S.G. 1979. Prebasic molt of the Northern Oriole. Canadian Journal of Zoology 57: 1473-1478.

Sealy, S.G. 1980a. Breeding biology of Orchard Orioles in a new population in Manitoba. Canadian Field-Naturalist 94(2): 154-158.

Sealy, S.G. 1980b. Reproductive responses of Northern Orioles to a changing food supply. Canadian Journal of Zoology 58(2): 221-227.

Sealy, S.G. 1984. Extra-pair copulation in the Yellow Warbler (*Dendroica petechia*). Animal Behaviour 32(1): 295-296.

Sealy, S.G. 1985. Erect posture of the young Black-billed Cuckoo: an adaptation for early mobility in a nomadic species. Auk 102(4): 889-892.

Sealy, S.G. 1988. Aggressiveness in migrating Cape May Warblers: defense of an aquatic food source. Condor 90(1): 271-274.

Sealy, S.G. and G.C. Biermann. 1983. Timing of breeding and migrations in a population of Least Flycatchers in Manitoba. Journal of Field Ornithology 54(2): 113-122.

Sealy, S.G., K.A. Hobson, and J.V. Briskie. 1989. Responses of Yellow Warblers to experimental intraspecific brood parasitism. Journal of Field Ornithology 60(2): 224-229.

Searcy, W.A. 1988. Do female Red-winged Blackbirds limit their own breeding densities? Ecology 69(1): 85-95.

Searcy, W.A. and P. Marler. 1981. A test for responsiveness to song structure and programming in female sparrows. Science 213: 926-928.

Searcy, W.A. and K. Yasukawa. 1989. Alternative models of territorial polygyny in birds. American Naturalist 134(3): 323-343.

Seastedt, T.R. and S.F. MacLean. 1979. Territory size and composition in relation to resource abundance in Lapland Longspurs breeding in arctic Alaska. Auk 96(1): 131-142.

Secunda, R.C. and T.W. Sherry. 1991. Polyterritorial polygyny in the American Redstart. Wilson Bulletin 103(2): 190-203.

Sedgwick, J.A. 1987. Avian habitat relationships in pinyon-juniper woodland. Wilson Bulletin 99(3): 413-431.

Sedgwick, J.A. and F.L. Knopf. 1989. Regionwide polygyny in Willow Flycatchers. Condor 91(2): 473-475.

Sedgwick, J.A. and F.L. Knopf. 1992. Describing Willow Flycatcher habitats: scale perspectives and gender differences. Condor 94: 720-733.

Sedinger, J.S. and D.G. Raveling. 1990. Parental behavior of Cackling Canada Geese during brood rearing: division of labor within pairs. Condor 92(1): 174-181.

Séguin, N. 1977. La conquête du sol au 19e siècle. Éditions du Boréal Express, Québec.

Séguin, N. 1980. Agriculture et colonisation au Québec. Aspects historiques. Éditions du Boréal Express, Montréal, 220 pp.

Semenchuk, G.P. 1992. The Atlas of breeding birds of Alberta. Federation of Alberta Naturalists, Edmonton, 391 pp.

Senner, S.E. and M.A. Howe. 1984. Conservation of Nearctic shorebirds, pp. 379-421 *in* J. Burger and B.L. Olla, eds. Behavior of marine animals, vol. 5. Shorebirds: breeding behavior and populations. Plenum Press, New York, 437 pp.

Sepik, G., R.B. Owen, Jr., and M.W. Coulter. 1981. A landowner's guide to woodcock management in the northeast. Moosehorn National Wildlife Refuge, United States Fish and Wildlife Service, Life Sciences and Agriculture Experiment Station, University of Maine, Orono, 23 pp.

Serie, J.R. 1987. Results of the 1987 midwinter waterfowl survey in the Atlantic Flyway states. United States Fish and Wildlife Service, Laurel, Maryland, 8 pp.

Serie, J.R. 1989. Results of the 1989 midwinter waterfowl survey in the Atlantic Flyway states. United States Fish and Wildlife Service, Laurel, Maryland, 7 pp.

Serie, J.R. 1991. Results of the 1991 midwinter waterfowl survey in the Atlantic Flyway states. United States Fish and Wildlife Service, Laurel, Maryland, 5 pp.

Serie, J.R. 1992. Results of the 1992 midwinter waterfowl survey in the Atlantic Flyway states. Preliminary report, United States Fish and Wildlife Service, Laurel, Maryland, 5 pp.

Serie, J.R. 1993. Results of the 1993 midwinter waterfowl survey in the Atlantic Flyway states. Draft report, United States Fish and Wildlife Service, Office of Migratory Bird Management, Laurel, Maryland, 3 pp.

Serie, J.R. and G.A. Swanson. 1976. Feeding ecology of breeding Gadwalls on saline wetlands. Journal of Wildlife Management 40(1): 69-81.

Sérodes, J.B. 1980. Étude de la sédimentation intertidale de l'estuaire moyen du Saint-Laurent. Report for Environment Canada by Université Laval, Sainte-Foy, 28 pp.

Serpell, J.A. 1985. Parrots, lories and cockatoos, pp. 220-227 *in* C.M. Perrins and A.L.A. Middleton, eds. The encyclopedia of birds. Facts on File, New York, 447 pp.

Serrentino, P. 1992. Northern Harrier, *Circus cyaneus*, pp. 89-117 *in* K.J. Schneider and D.M. Pence, eds. Migratory nongame birds of management concern in the northeast. United States Department of the Interior, Fish and Wildlife Service, Newton Corner, Massachusetts, 400 pp.

Serventy, D.L. 1971. Biology of desert birds, pp. 287-339 *in* D.S. Farner and J.R. King, eds. Avian Biology, vol. 1. Academic Press, New York, 586 pp.

Servos, M.C. 1987. Summer habitat use by Great Gray Owls in southeastern Manitoba, pp. 108-114 *in* R.W. Nero, R.J. Clark, R.J. Knapton, and R.H. Hamre, eds. Proceedings. Biology and conservation of northern forest owls. Symposium, 3-7 February, Winnipeg. United States Department of Agriculture, Forest Service, Rocky Mountain Forest and Range Experiment Station, Fort Collins, Colorado. General Technical Report RM-142, 309 pp.

Seutin, G. 1987. Female song in Willow Flycatchers (*Empidonax traillii*). Auk 104(2): 329-330.

Seutin, G. 1991a. The Gray-cheeked Thrush: one species or two? An analysis of mtDNA. Paper presented at the 109th stated meeting of the AOU, Montréal, August 14-17 1991.

Seutin, G. 1991b. Morphometric identification of Traill's Flycatchers: an assessment of Stein's formula. Journal of Field Ornithology 62(3): 308-313.

Seutin, G. and J.P. Simon. 1988a. Genetic variation in sympatric Willow Flycatchers (*Empidonax traillii*) and Alder Flycatchers (*E. alnorum*). Auk 105(2): 235-243.

Seutin, G. and J.P. Simon. 1988b. Protein and enzyme uniformity in a new, isolated population of the Sharp-tailed Sparrow. Biochemical Systematics and Ecology 16(2): 233-236.

Seutin, G., P.T. Boag, B.N. White, and L.M. Ratcliffe. 1991. Sequential polyandry in the Common Redpoll (*Carduelis flammea*). Auk 108(1): 166-170.

Seutin, G., P.T. Boag, and L.M. Ratcliffe. 1992. Plumage variability in redpolls from Churchill, Manitoba. Auk 109(4): 771-785.

Sexton, D.A. 1979. Off-lek copulation in Sharp-tailed Grouse. Wilson Bulletin 91(1): 150-151.

Seymour, N.R. 1974a. Territorial behaviour of wild Shovelers at Delta, Manitoba. Wildfowl 25: 49-55.

Seymour, N.R. 1974b. Site attachment in the Northern Shoveler. Auk 91(2): 423-427.

Seymour, N.R. and R.D. Titman. 1978. Changes in activity patterns, agonistic behaviour and territoriality of Black Ducks (*Anas rubripes*) in a Nova Scotia tidal marsh. Canadian Journal of Zoology 56(8): 1773-1785.

Shadick, S.J. 1980. Saskatchewan breeding records for Red Crossbill, Orchard Oriole and Red-headed Woodpecker. Blue Jay 38(4): 247-249.

Shaffer, F. 1990. État de la population du Bruant (Pinson) à queue aiguë (*Ammodramus caudacutus subvirgatus*) au Québec et au Canada. Report for Ministère du Loisir, de la Chasse et de la Pêche. Association québécoise des groupes d'ornithologues, 45 pp.

Shaffer, F. 1992. Rapport d'inventaire du Grèbe cornu (*Podiceps auritus*) aux Îles-de-la-Madeleine – 1992. Unpublished report. Canadian Wildlife Service, Québec Region, 13 pp.

Shaffer, F. 1995. Suivi des populations de sternes et de goélands aux Îles-de-la-Madeleine en 1995. Unpublished report. Canadian Wildlife Service, Québec Region, 13 pp.

Shaffer, F. and P. Laporte. 1989. Rapport de recherches sur le Pluvier siffleur (*Charadrius melodus*) aux Îles-de-la-Madeleine, été 1989. Association québécoise des groupes d'ornithologues and Canadian Wildlife Service, Québec Region, 42 pp.

Shaffer, F. and P. Laporte. 1990. Rapport de recherches sur le Pluvier siffleur (*Charadrius melodus*) aux Îles-de-la Madeleine, été 1990. Association québécoise des groupes d'ornithologues and Canadian Wildlife Service, Québec Region, 70 pp.

Shaffer, F. and P. Laporte. 1992a. Rapport synthèse des recherches relatives au Pluvier siffleur (*Charadrius melodus*) effectuées aux Îles-de-la-Madeleine de 1987 à 1991. Association québécoise des groupes d'ornithologues and Canadian Wildlife Service, Québec Region, 78 pp.

Shaffer, F. and P. Laporte. 1992b. Rapport de recherches relatives au Pluvier siffleur (*Charadrius melodus*) effectuées aux Îles-de-la-Madeleine, été 1992. Association québécoise des groupes d'ornithologues and Canadian Wildlife Service (Internal report), 36 pp.

Shaffer, F. and P. Laporte. 1993. Inventaires de la Sterne de Dougall (*Sterna dougallii*), des colonies de sternes et de goélands aux Îles-de-la-Madeleine en 1993. Unpublished report. Canadian Wildlife Service, Québec Region, 51 pp.

Shaffer, F. and P. Laporte. 1994. Diet of Piping Plovers on the Magdalen Islands, Quebec. Wilson Bulletin 106(3): 531-536.

Shaffer, F. and C. Pineau. 1987. État du Pluvier siffleur (*Charadrius melodus*) aux Îles-de-la-Madeleine. Association québécoise des groupes d'ornithologues, 15 pp.

Shaffer, M.L. 1981. Minimum population sizes for species conservation. BioScience 31: 131-134.

Shaffer, M.L. 1987. Minimum viable populations: coping with uncertainty, pp. 69-86 *in* M.E. Soulé, ed. Viable populations for conservation. Cambridge University Press, Cambridge, 189 pp.

Sharp, M.J. and C.A. Campbell. 1982. Breeding ecology and status of Red-shouldered Hawks (*Buteo l. lineatus*) in Waterloo region. Ontario Field Biologist 36: 1-10.

Sharrock, J.T.R. 1976. The Atlas of breeding birds in Britain and Ireland. Poyser, Berkhamsted, 479 pp.

Shealer, D.A. and S.W. Kress. 1991. Nocturnal abandonment response to Black-crowned Night-Heron disturbance in a Common Tern colony. Colonial Waterbirds 14(1): 51-56.

Shedd, D.H. 1990. Aggressive interactions in wintering House Finches and Purple Finches. Wilson Bulletin 102(1): 174-178.

Sheldon, W.G. 1961. Summer crepuscular flights of American Woodcocks in central Massachusetts. Wilson Bulletin 73(2): 126-139.

Sheldon, W.G. 1971. The book of the American Woodcock, 2nd edition. University of Massachusetts Press, Amherst, 227 pp.

Shepherd, P., T. Crockett, T.L. de Santo, and K.L. Bildstein. 1991. The impact of Hurricane Hugo on the breeding ecology of wading birds at Pumpkinseed Island, Hobcan Barony, South Carolina. Colonial Waterbirds 14(2): 150-157.

Sherman, A.R. 1910. At the sign of the Northern Flicker. Wilson Bulletin 22: 135-171.

Sherman, A.R. 1911. Nest life of the Screech-Owl. Auk 28: 155-168.

Sherman, A.R. 1913. The nest life of the Sparrow Hawk. Auk 30: 406-418.

Sherrod, S.K., C.M. White, and F.S.L. Williamson. 1976. Biology of the Bald Eagle on Amchitka Island, Alaska. Living Bird 15: 143-182.

Sherry, T.S. 1979. Competitive interactions and adaptive strategies of American Redstarts and Least Flycatchers in a northern hardwood forest. Auk 96(2): 265-283.

Sherry, T.W. and R.T. Holmes. 1985. Dispersion patterns and habitat responses of birds in northern hardwood forests, pp. 283-309 *in* M.L. Cody, ed. Habitat selection in birds. Academic Press, New York, 558 pp.

Sherry, T.W. and R.T. Holmes. 1988. Habitat selection by breeding American Redstarts in response to a dominant competitor, the Least Flycatcher. Auk 105(2): 350-364.

Sherry, T.W. and R.T. Holmes. 1989. Age-specific social dominance affects habitat use by breeding American Redstarts (*Setophaga ruticilla*): a removal experiment. Behavioral Ecology and Sociobiology 25: 327-333.

Shields, W.M. 1984. Factors affecting nest and site fidelity in Adirondack Barn Swallows (*Hirundo rustica*). Auk 101(4): 780-789.

Shields, W.M. and K.L. Bildstein. 1979. Birds versus bats: behavioral interactions at a localized food source. Ecology 60: 468-474.

Shields, W.M. and J.R. Crook. 1987. Barn Swallow coloniality: a net cost for group breeding in the Adirondacks? Ecology 68(5): 1373-1386.

Short, L.L. 1979. Burdens of the picid hole-excavating habit. Wilson Bulletin 91(1): 16-28.

Short, L.L. 1982. Woodpeckers of the world. Delaware Museum of Natural History, Greenville. Monograph Series no. 4, 676 pp.

Shugart, G.W., W.C. Scharf, and F.J. Cuthbert. 1978. Status and reproductive success of the Caspian Tern (*Sterna caspia*) in the United States Great Lakes. Proceedings. Colonial Waterbird Group [1978]: 146-156.

Shy, E. 1984. Habitat shift and geographical variation in North American Tanagers (Thraupinae: *Piranga*). Oecologia 63: 281-285.

Sibley, C.G. and J.E. Ahlquist. 1984. The relationships of the starlings (Sturnidae: sturnini) and the mockingbirds (Sturnidae: mimini). Auk 101(2): 230-243.

Sibley, C.G. and J.E. Ahlquist. 1990. Phylogeny and classification of birds: a study in molecular evolution. Yale University Press, New Haven, Connecticut, 976 pp.

Sibley, C.G. and B.L. Monroe. 1990. Distribution and taxonomy of birds of the world. Yale University Press, New Haven, Connecticut, 1111 pp.

Sibley, S.C. 1988a. Common Barn-Owl, pp. 200-201 *in* R.F. Andrle and J.R. Carroll, eds. The Atlas of breeding birds in New York State. Cornell University Press, Ithaca, 551 pp.

Sibley, S.C. 1988b. Common Nighthawk, pp. 214-215 *in* R.F. Andrle and J.R. Carroll, eds. The Atlas of breeding birds in New York State. Cornell University Press, Ithaca, 551 pp.

Sibley, S.C. 1988c. Horned Lark, pp. 264-265 *in* R.F. Andrle and J.R. Carroll, eds. The Atlas of breeding birds in New York State. Cornell University Press, Ithaca, 551 pp.

Sibley, S.C. 1988d. Spotted Sandpiper, pp. 156-157 *in* R.F. Andrle and J.R. Carroll, eds. The Atlas of breeding birds in New York State. Cornell University Press, Ithaca, 551 pp.

Sibley, S.C. 1988e. Mute Swan, pp. 56-57 *in* R.F. Andrle and J.R. Carroll, eds. The Atlas of breeding birds in New York State. Cornell University Press, Ithaca, 551 pp.

Sibley, S.C. 1988f. Purple Martin, pp. 266-267 *in* R.F. Andrle and J.R. Carroll, eds. The Atlas of breeding birds in New York State. Cornell University Press, Ithaca, 551 pp.

Sibley, S.C. 1988g. Red-breasted Merganser, pp. 92-93 *in* R.F. Andrle and J.R. Carroll, eds. The Atlas of breeding birds in New York State. Cornell University Press, Ithaca, 551 pp.

Sidle, J. 1985. Endangered and threatened wildlife and plants; determination of endangered and threatened status for the Piping Plover. United States Federal Register 50: 50726-50734.

Sieg, C.H. and D.M. Becker. 1990. Nest-site habitat selected by Merlins in southeastern Montana. Condor 92(3): 688-694.

Siegfried, W.R. 1973. Summer food and feeding of the Ruddy Duck in Manitoba. Canadian Journal of Zoology 51(12): 1293-1297.

Siegfried, W.R. 1976a. Breeding biology and parasitism in the Ruddy Duck. Wilson Bulletin 88(4): 566-574.

Siegfried, W.R. 1976b. Segregation in feeding behavior of four diving ducks in southern Manitoba. Canadian Journal of Zoology 54(5): 730-736.

Siegfried, W.R. 1977. Notes on behaviour of Ruddy Ducks during the brood period. Wildfowl 28: 126-128.

Simard, C. 1988. Historique de la démarche pour l'emblème aviaire du Québec. Bulletin ornithologique 32(4): 135-136.

Simard, G. 1986. Inventaire des populations de Sternes pierregarin et arctique, Réserve de Parc national de l'Archipel-de-Mingan. Parks Canada, Québec Region.

Simberloff, D. 1988. The contribution of population and community biology to conservation science. Annual Review of Ecological Systematics 19: 473-511.

Simmons, G.A. and N.F. Sloan. 1974. Consumption of Jack-pine budworm, *Choristoneura pinus* (Freeman), by the Eastern Chipping Sparrow, *Spizella passerina* (Bechstein). Canadian Journal of Zoology 52(7): 817-821.

Simmons, K.E.L. 1956. Feather-eating and pellet-formation in the Great Crested Grebe. British Birds 49: 432-435.

Simmons, R.E. 1989. How polygynous female Northern Harriers *Circus cyaneus* choose their mates and why they are deceived, pp. 243-246 *in* B.U. Meyburg and R.D. Chancellor, eds. Raptors in the modern world. World Working Group on Birds of Prey and Owls, Berlin, Germany, 611 pp.

Simmons, R.E. and P.C. Smith. 1985. Northern Harrier nest-site selection. Canadian Journal of Zoology 63: 494-498.

Simmons, R.E., P.C. Smith, and R.B. MacWhirter. 1986a. Hierarchies among Northern Harrier (*Circus cyaneus*) harems and the costs of polygyny. Journal of Animal Ecology 55: 755-772.

Simmons, R.E., R.B. MacWhirter, P.E. Barnard, and G.L. Hanse. 1986b. The influence of microtines on polygyny, productivity, age and provisioning of Northern Harriers: a 5-year study. Canadian Journal of Zoology 64: 2447-2456.

Simon, S. and D. Simon. 1980. Observations sur la reproduction de la Chouette épervière *Surnia ulula*. Alauda 48(2-3): 152-153.

Simons, T., S.K. Herrod, M.W. Collopy, and M.A. Jenkins. 1988. Restoring the Bald Eagle. American Scientist 76: 253-260.

Simpson, B.S. 1985. Effects of location in territory and distance from neighbors on the use of song repertoires by Carolina Wrens. Animal Behaviour 33(3): 793-804.

Simpson, K.J., J.N.M. Smith, and J.P. Kelsall. 1987. Correlates and consequences of coloniality in Great Blue Herons. Canadian Journal of Zoology 65: 572-577.

Simpson, M.B. Jr. 1978. Breeding season distribution and ecology of the Vesper Sparrow in the Southern Blue Ridge Mountain province. Chat 42(1): 1-2.

Sirén, M. 1952. Studies on the breeding biology of the Goldeneye, *Bucephala clangula*. Riistatieteellisia julkaisuja 8: 101-111.

Sjöberg, K. 1988. Food selection, food-seeking patterns and hunting success of captive Goosanders *Mergus merganser* and Red breasted Mergansers, *M. serrator* in relation to the behaviour of their prey. Ibis 130(1): 79-93.

Sjöberg, K. and K. Danell. 1982. Feeding activity of ducks in relation to diel emergence of chironomids. Canadian Journal of Zoology 60: 1383-1387.

Sjölander, S. 1978. Reproductive behaviour of the Black-throated Diver *Gavia arctica*. Ornis Scandinavica 9: 51-65.

Sjölander, S. and G. Ågren. 1972. Reproductive behavior of the Common Loon. Wilson Bulletin 84(3): 296-308.

Sklepkovych, B.O. and W.A. Montevecchi. 1989. The world's largest known nesting colony of Leach's Storm-Petrel on Baccalieu Island, Newfoundland. American Birds 43: 38-42.

Skutch, A.F. 1976. Parent birds and their young. University of Texas Press, Austin, 503 pp.

Skutch, A.F. 1985a. Buntings and tanagers, p. 409 *in* C.M. Perrins and A.L.A. Middleton, eds. The encyclopedia of birds. Facts on File, New York, 447 pp.

Skutch, A.F. 1985b. Life of the woodpecker. Ibis Publishing, Santa Monica, California, 136 pp.

Skutch, A.F. 1985c. Tanager, pp. 582-584 *in* B. Campbell and E. Lack, eds. A dictionary of birds. British Ornithologists' Union, Buteo Books, Vermillion, South Dakota, 670 pp.

Slack, R.D. 1976. Nest guarding behavior by male Gray Catbirds. Auk 93(2): 292-300.

Sloan, N.F. and G.A. Simmons. 1973. Foraging behavior of the Chipping Sparrow in response to high populations of Jack pine budworm. American Midland Naturalist 90(1): 210-215.

Smeenk, C. 1972. Ökologische vergleiche zwischen waldkauz *Strix aluco* und waldohreule *Asio otus*. Ardea 60: 1-71.

Smith, A.G. 1964. The 1964 waterfowl survey on the Lousana study area, Alberta, Canada. Unpublished Progress Report. United States Fish and Wildlife Service, Wildlife Resource Laboratory, Denver, Colorado, 25 pp.

Smith, C.R. 1988a. Mourning Warbler, pp. 414-416 *in* R.F. Andrle and J.R. Carroll, eds. The Atlas of breeding birds in New York State. Cornell University Press, Ithaca, 551 pp.

Smith, C.R. 1988b. Scarlet Tanager, pp. 426-427 *in* R.F. Andrle and J.R. Carroll, eds. The Atlas of breeding birds in New York State. Cornell University Press, Ithaca, 551 pp.

Smith, C.R. 1992. Henslow's Sparrow, *Ammodramus henslowii*, pp. 315-330 *in* K.J. Schneider and D.M. Pence, eds. Migratory nongame birds of management concern in the northeast. United States Department of the Interior, Fish and Wildlife Service, Newton Corner, Massachusetts, 400 pp.

Smith, D.A. 1970. Observations on nesting Hawk Owls at the Mer Bleue, near Ottawa, Canada. Canadian Field-Naturalist 84(4): 377-383.

Smith, D.G. 1972. The role of the epaulets in the Red-winged Blackbird (*Agelaius phoeniceus*) social system. Behaviour 41: 251-268.

Smith, D.G. and R. Gilbert. 1984. Eastern Screech-Owl home range and use of suburban habitats in southern Connecticut. Journal of Field Ornithology 55(3): 322-329.

Smith, D.G. and E. Hiestand. 1990. Alloparenting at an Eastern Screech-Owl nest. Condor 92(1): 246-247.

Smith, D.G. and J.R. Murphy. 1982. Nest site selection in raptor communities of the Eastern Great Basin Desert. Great Basin Naturalist 42(4): 395-404.

Smith, D.G., C.R. Wilson, and H.H. Frost. 1972. The biology of the American Kestrel in central Utah. Southwest Naturalist 17(1): 73-83.

Smith, D.J. and C.R. Smith. 1992. Henslow's Sparrow and Grasshopper Sparrow: a comparison of habitat use in Finger Lakes National Forest, New York. Bird Observer 20(4): 187-194.

Smith, G.A. 1988a. Least Bittern, pp. 34-35 *in* R.F. Andrle and J.R. Carroll, eds. The Atlas of breeding birds in New York State. Cornell University Press, Ithaca, 551 pp.

Smith, G.A. 1988b. Red-tailed Hawk, pp. 114-115 *in* R.F. Andrle and J.R. Carroll, eds. The Atlas of breeding birds in New York State. Cornell University Press, Ithaca, 551 pp.

Smith, H.G. and R. Montgomerie. 1991. Nestling American Robins compete with siblings by begging. Behavioral Ecology and Sociobiology 29(4): 307-312.

Smith, H.G. and R. Montgomerie. 1992. Male incubation in Barn Swallows: the influence of nest temperature and sexual selection. Condor 94: 750-759.

Smith, J.N.[M.]. 1978. Division of labor by Song Sparrows feeding fledged young. Canadian Journal of Zoology 56(2): 187-191.

Smith, J.N.M., Y. Yom-Tov, and R. Moses. 1982. Polygyny, male parental care, and sex ratio in Song Sparrows: an experimental study. Auk 99(3): 555-564.

Smith, K.G. 1978. Range extension of the Blue Jay into western North America. Bird-Banding 49(3): 208-214.

Smith, K.G. and D.C. Andersen. 1982. Food, predation, and reproductive ecology of the Dark-eyed Junco in northern Utah. Auk 99(4): 650-661.

Smith, L.M., L.D. Vangilder, and R.A. Kennamer. 1985. Foods of wintering Brant in eastern North America. Journal of Field Ornithology 56(3): 286-289.

Smith, N.J.H. 1980. Further advance of House Sparrows into the Brazilian amazon. Condor 82: 109-111.

Smith, N.N. 1969. Greenland Wheatears observed at Esker, Labrador, May 20 - June 2, 1967. Canadian Field-Naturalist 83(1): 56.

Smith, P.W. 1987. The Eurasian Collared-Dove arrives in the Americas. American Birds 41(5): 1371-1379.

Smith, R.B.H. and H.G. Lumsden. 1987. White-winged Crossbill, pp. 496-497 *in* M.D. Cadman, P.F.J. Eagles and F.M. Helleiner, eds. Atlas of the breeding birds of Ontario. University of Waterloo Press, Waterloo, 617 pp.

Smith, R.I. 1968. The social aspects of reproductive behavior in the Pintail. Auk 85(3): 381-396.

Smith, R.I. 1970. Response of Pintail breeding populations to drought. Journal of Wildlife Management 34(4): 943-946.

Smith, R.L. 1968. Grasshopper Sparrow, pp. 725-745 *in* A.C. Bent. Life histories of North American cardinals, grosbeaks, buntings, towhees, finches, sparrows, and allies. (O.L. Austin, Jr., ed.). United States National Museum, Washington. [1968, Dover Publications, New York]. Bulletin no. 237, vol. 1, pp. 1-602; vol. 2, pp. 603-1248; vol. 3, pp. 1249-1889.

Smith, R.L. 1990. Ecology and field biology, 4th edition. Harper and Row Publishers, New York, 922 pp.

Smith, S.M. 1972. The ontogeny of impaling behaviour in the Loggerhead Shrike, *Lanius ludovicianus*. Behaviour 42: 232-247.

Smith, S.M. 1991. The Black-capped Chickadee. Cornell University Press, Ithaca, New York, 362 pp.

Smith, S.M. 1992. Pairbond persistence and "divorce" in Black-capped Chickadees. Wilson Bulletin 104(2): 338-342.

Smith, S.M. 1993. Black-capped Chickadee (*Parus atricapillus*), *in* A. Poole, P. Stettenheim, and F. Gill, eds. The birds of North America, no. 39. The Academy of Natural Sciences, Philadelphia. The American Ornithologists' Union, Washington, 20 pp.

Smith, T.G. 1975. Rough-legged Hawks, *Buteo lagopus* (Pontoppidan), as carrion feeders in the Arctic. Canadian Field-Naturalist 89: 190.

Smith, T.M. and H.H. Shugart. 1987. Territory size variation in the Ovenbird: the role of habitat structure. Ecology 68: 695-704.

Smith, W.J. 1957. Birds of the Clay Belt of Northern Ontario and Quebec. Canadian Field-Naturalist 71(4): 163-181.

Smith, W.J. 1988. Patterned daytime singing of the Eastern Wood-Pewee, *Contopus virens*. Animal Behaviour 36(4): 1111-1123.

Smith, W.P. 1942. Nesting habits of the Eastern Phoebe. Auk 59(3): 410-417.

Snell, R.R. 1991. Interspecific allozyme differentiation among North Atlantic white-headed larid gulls. Auk 108(2): 319-328.

Snow, B.K. 1985. Hummingbirds, pp. 258-262 *in* C.M. Perrins and A.L.A. Middleton, eds. The encyclopedia of birds. Facts on File, New York, 447 pp.

Snow, C. 1973. Golden Eagle. Habitat management series for unique or endangered species. United States Department of the Interior, Bureau of Land Management, Denver, Colorado. Report no. 7, 52 pp.

Snyder, L.L. 1951. Ontario Birds. Clarke, Irwin and Co., Toronto, 248 pp.

Snyder, L.L. 1957. Changes in the avifauna of Ontario, pp. 26-42 *in* F.A. Urquhart, ed. Changes in the fauna of Ontario. University of Toronto Press, Toronto, 75 pp.

Snyder, N.F.R. and H.A. Snyder. 1973. Experimental study of feeding rates of nesting Cooper's Hawks. Condor 75(4): 461-463.

Snyder, N.F.R. and J.W. Wiley. 1976. Sexual size dimorphism in hawks and owls of North America. American Ornithologists' Union, Battle Creek. Ornithological Monographs no. 20, 96 pp.

Snyder, N.F.R., H.A. Snyder, J.L. Lincer, and R.T. Reynolds. 1973. Organochlorines, heavy metals and the biology of North American accipiters. BioScience 23(5): 300-305.

Société Duvetnor. 1990. Inventaire d'oiseaux aquatiques, Bas-Saint-Laurent. Mai-juin 1990. Preliminary report for Canadian Wildlife Service, Québec Region, 10 pp.

Société ornithologique du centre du Québec (SOCQ). 1988. L'observation des oiseaux au lac Saint-Pierre (guide des sites). Société ornithologique du centre du Québec, Drummondville, 242 pp.

Société zoologique de Québec. 1983. Noms français des oiseaux du Canada et des États-Unis continentaux. Société zoologique de Québec, 80 pp.

Sodhi, N.S. 1989. Attempted polygyny by a Merlin. Wilson Bulletin 101(3): 506-507.

Sodhi, N.S. 1991. House Sparrow, *Passer domesticus*, flock size in relation to the proximity of Merlin, *Falco columbarius*, nests. Canadian Field-Naturalist 105(2): 278-279.

Sodhi, N.S. and L.W. Oliphant. 1992. Hunting ranges and habitat use and selection of urban-breeding Merlins. Condor 94: 743-749.

Sodhi, N.S., P.C. James, I.G. Warkentin, and L.W. Oliphant. 1992. Breeding ecology of urban Merlins (*Falco columbarius*). Canadian Journal of Zoology 70(8): 1477-1483.

Sodhi, N.S., L.W. Oliphant, P.C. James, and I.G. Warkentin. 1993. Merlin (*Falco columbarius*), *in* A. Poole and F. Gill, eds. The birds of North America, no. 44. The Academy of Natural Sciences, Philadelphia. The American Ornithologists' Union, Washington, 20 pp.

Sogge, M.K., W.M. Gilbert, and C. van Riper III. 1994. Orange-crowned Warbler (*Vermivora celata*), *in* A. Poole and F. Gill, eds. The birds of North America, no. 101. The Academy of Natural Sciences, Philadelphia. The American Ornithologists' Union, Washington, 20 pp.

Soil Survey of Canada. 1987. Le système canadien de classification des sols, 2nd edition. Agriculture Canada, Ottawa. Publication no. 1646F. [Also available in English as: The Canadian System of Soil Classification.]

Solheim, R. 1983. Bigyny and biandry in the Tengmalm's Owl *Aegolius funereus*. Ornis Scandinavica 14(1): 51-57.

Solman, V.E.F. 1945. The ecological relations of pike (*Esox lucius*) and waterfowl. Ecology 26: 157-170.

Solman, V.E.F. 1974. Harrison Flint Lewis, 1893-1974. Canadian Field-Naturalist 88: 507-516.

Sonerud, G.A. 1985. Risk of nest predation in three species of hole nesting owls: influence on choice of nesting habitat and incubation behaviour. Ornis Scandinavica 16: 261-269.

Sonerud, G.A., J.O. Nybo, P.E. Fjeld, and C. Knoff. 1987. A case of bigyny in the Hawk Owl *Surnia ulula*: spacing of nests and allocation of male feeding effort. Ornis Fennica 64: 144-148.

Sootes, R.F. Jr. 1975. Ecological succession of breeding birds in relation to plant succession on dredge islands in North Carolina estuaries. Ph.D. dissertation, North Carolina State University, Raleigh, 91 pp.

Soper, J.D. 1940. Local distribution of eastern Canadian Arctic birds. Auk 57(1): 13-21.

Soucy, L.J. Jr. 1976. Barred Owl nest. North American Bird Bander 1(2): 68-69.

Soucy, L.J. Jr. 1980. Three long distance recoveries of banded New Jersey Barn Owls. North American Bird Bander 5: 97.

Soulé, M.E. 1987. Viable populations for conservation. Cambridge University Press, Cambridge, 189 pp.

Soulliere, G.J. 1990. Regional and site-specific trends in Wood Duck use of nest boxes, pp. 235-244 *in* L.H. Fredrickson, G.V. Burger, S.P. Havera, D.A. Graber, R.E. Kirby, and T.S. Taylor, eds. Proceedings. 1988 North American Wood Duck symposium, St. Louis, Missouri, 390 pp.

Southern, W.E. 1958. Nesting of the Red-eyed Vireo in the Douglas Lake region, Michigan. Jack-Pine Warbler 36: 105-130.

Sowls, L.K. 1955. Prairie ducks: a study of their behavior, ecology and management. A Wildlife Management Institute Publication, University of Nebraska Press, Lincoln, 193 pp.

Spahn, R.G. 1988. Red-headed Woodpecker, pp. 226-227 *in* R.F. Andrle and J.R. Carroll, eds. The Atlas of breeding birds in New York State. Cornell University Press, Ithaca, 551 pp.

Speich, S.M., H.L. Jones, and E.M. Benedict. 1986. Review of natural nesting of the Barn Swallow in North America. American Midland Naturalist 115(2): 248-254.

Speirs, D.H. 1968. Eastern Evening Grosbeak, pp. 206-237 *in* A.C. Bent. Life histories of North American cardinals, grosbeaks, buntings, towhees, finches, sparrows, and allies. (O.L. Austin, Jr., ed.). United States National Museum, Washington. [1968, Dover Publications, New York]. Bulletin no. 237, vol. 1, pp. 1-602; vol. 2, pp. 603-1248; vol. 3, pp. 1249-1889.

Speirs, J.M. 1985. Birds of Ontario. National Heritage/Natural History, Toronto, 2 vol., 1524 pp.

Speirs, J.M. and D.H. Speirs. 1968. Lincoln's Sparrow, pp. 1434-1467 *in* A.C. Bent. Life histories of North American cardinals, grosbeaks, buntings, towhees, finches, sparrows, and allies. (O.L. Austin, Jr., ed.). United States National Museum, Washington. [1968, Dover Publications, New York]. Bulletin no. 237, vol. 1, pp. 1-602; vol. 2, pp. 603-1248; vol. 3, pp. 1249-1889.

Speiser, R. and T. Bosakowski. 1988. Nest-site preferences of Red-tailed Hawks in the highlands of southeastern New York and northern New Jersey. Journal of Field Ornithology 59(4): 361-368.

Speiser, R. and T. Bosakowski. 1989. Nest trees selected by Northern Goshawks along the New York–New Jersey border. Kingbird 39(3): 132-141.

Speiser, R. and T. Bosakowski. 1991. Nesting phenology, site fidelity, and defense behavior of Northern Goshawks in New York and New Jersey. Journal of Raptor Research 25(4): 132-135.

Spencer, O.R. 1943. Nesting habits of the Black-billed Cuckoo. Wilson Bulletin 55(1): 11-22.

Spendelow, J.A. and S.R. Patton. 1988. National atlas of coastal waterbird colonies in the contiguous United States: 1976-82. United States Fish and Wildlife Service, Biological Report 88(5): 326 pp.

Spendelow, J.A., R.M. Erwin, and B.K. Williams. 1989. Patterns of species co-occurrence of nesting colonial Ciconiiformes in Atlantic coast estuarine areas. Colonial Waterbirds 12(1): 51-59.

Sperry, C.C. 1940. Food habits of a group of shorebirds: woodcock, snipe, knot, and dowitcher. United States Department of the Interior, Bureau of Biological Survey, Washington. Wildlife Research Bulletin no. 1, 37 pp.

Spofford, W.R. 1959. Nesting of the Golden Eagle at Sugluk Inlet, Ungava, Quebec. Canadian Field-Naturalist 73: 177.

Spofford, W.R. 1971. The breeding status of the Golden Eagle in the Appalachians. American Birds 25(1): 3-7.

Sprague, R.T. and R.D. Weir. 1984. The birds of Prince Edward County, 2nd edition. Kingston Field Naturalists, Kingston, 190 pp.

Spreyer, M.F. 1987. A floristic analysis of Great Gray Owl habitat in Aitkin County, Minnesota, pp. 96-100 *in* R.W. Nero, R.J. Clark, R.J. Knapton, and R.H. Hamre, eds. Proceedings. Biology and conservation of northern forest owls. Symposium, 3-7 February, Winnipeg. United States Department of Agriculture, Forest Service, Rocky Mountain Forest and Range Experiment Station, Fort Collins, Colorado. General Technical Report RM-142, 309 pp.

Springer, A.M. 1975. Observations on the summer diet of Rough-legged Hawks from Alaska. Condor 77(3): 338.

Sprunt, A. Jr. 1942. Purple Martin, pp. 489-509 *in* A.C. Bent. Life histories of North American flycatchers, larks, swallows and their allies. (O.L. Austin, Jr., ed.). United States National Museum, Washington. [1963, Dover Publications, New York]. Bulletin no. 179, 555 pp.

Sprunt, A., W.B. Robertson, Jr., S. Postupalsky, R.J. Hensel, C.E. Knoder, and F.J. Ligas. 1973. Comparative productivity of six Bald Eagle populations. Transactions. North American Wildlife and Natural Resources Conference 38: 96-106.

Squires, W.A. 1976. The birds of New Brunswick, 2nd edition. New Brunswick Museum, St. John. Monograph Series no. 7, 221 pp.

St-Hilaire, D. 1990. Deuxième inventaire du Bruant sauterelle (*Ammodramus savannarum*) dans le sud-ouest du Québec, saison 1990. Club des ornithologues de l'Outaouais, Hull, 15 pp.

St-Hilaire, D. 1991. Liste annotée des oiseaux du Témiscamingue. Société d'ornithologie du Témiscamingue in collaboration with Association québécoise des groupes d'ornithologues, 12 pp.

St. Lawrence Centre and Laval University. 1991. Atlas environnemental du Saint-Laurent – Les milieux humides: des habitats au contact de la terre et de l'eau. St. Lawrence UPDATE series. St. Lawrence Centre, Conservation and Protection, Québec Region, Environment Canada, Montréal. [Also available in English as: Wetlands: habitats on the edge of land and water, environmental atlas of the St. Lawrence.]

St-Louis, N. 1978. Étude de la faune avienne dans les bassins de la Grande rivière de la Baleine et de la Petite rivière de la Baleine (été 1977). Report for Hydro-Québec, Direction de l'Environnement. Éco-recherches, Pointe-Claire, 159 pp. [Published in 1982.]

St-Louis, N. 1979. Étude de la faune avienne dans les bassins de la Grande rivière de la Baleine et de la Petite rivière de la Baleine (été 1978). Report for Hydro-Québec, Direction de l'Environnement. Éco-recherches, Pointe-Claire, 128 pp. [Published in 1982.]

Stalheim, P.S. 1974. Behavior and ecology of the Yellow Rail (*Coturnicops noveboracensis*). M.Sc. thesis, University of Minnesota, Minneapolis, 83 pp.

Stalheim, P.S. 1975. Breeding behaviour of captive Yellow Rails. Avicultural Magazine 81(3): 133-141.

Stalmaster, M.V. 1987. The Bald Eagle. Universe Books, New York, 227 pp.

Stamp, D. 1966. A glossary of normal geographical terms. British Association for the Advancement of Science, Longman Group Ltd, London, 539 pp.

Stanwood, C.J. 1913. The Olive-backed Thrush (*Hylocichla ustulata swainsoni*) at his summer home. Wilson Bulletin 25: 118-137.

Stauffer, D.F. and L.B. Best. 1980. Habitat selection by birds of riparian communities: evaluating effects of habitat alterations. Journal of Wildlife Management 44(1): 1-15.

Stauffer, D.F. and L.B. Best. 1986. Nest-site characteristics of open-nesting birds in riparian habitats in Iowa. Wilson Bulletin 98(2): 231-242.

Steeves, J.B. and S. Holohan. 1975. Wilson's Phalarope nesting range extended to the St. Lawrence valley in Quebec. Canadian Field-Naturalist 89(2): 185.

Steeves, J.[B.], S. Holohan, and R. Barnhurst. 1989. Migration of larids at Beauharnois, Quebec, 1967-1980. Canadian Field-Naturalist 103(1): 23-28.

Stefanski, R.A. 1967. Utilization of the breeding territory in the Black-capped Chickadee. Condor 69(3): 259-267.

Stefanski, R.A. and J.B. Falls. 1972a. A study of distress calls of Song, Swamp, and White-throated sparrows (Aves: Fringillidae). I. Intraspecific responses and functions. Canadian Journal of Zoology 50(12): 1501-1512.

Stefanski, R.A. and J.B. Falls. 1972b. A study of distress calls of Song, Swamp, and White-throated sparrows (Aves: Fringillidae). II. Interspecific responses and properties used in recognition. Canadian Journal of Zoology 50(12): 1513-1525.

Stein, R.C. 1958. The behavioral, ecological and morphological characteristics of two populations of the Alder Flycatcher (*Empidonax traillii*). New York State Museum and Science Service. Bulletin no. 371, 63 pp.

Stein, R.C. 1963. Isolating mechanisms between populations of Traill's Flycatchers. Proceedings. American Philosophical Society 107(1): 21-50.

Steiner, A.J. 1984. Mid-winter waterfowl inventory: Atlantic Flyway, 1954-1984 trend analysis. United States Fish and Wildlife Service, Region 5, Massachusetts, 284 pp.

Stempniewicz, L. 1983a. Hunting methods of the Glaucous Gull and escape manoeuvers of its prey, the Dovekie. Journal of Field Ornithology 54(3): 329-331.

Stempniewicz, L. 1983b. Glaucous Gulls stealing spoil from Parasitic Jaegers. Journal of Field Ornithology 54(3): 331.

Stendell, R.C., J.W. Artmann, and E. Martin. 1980. Lead residues in Sora rails from Maryland. Journal of Wildlife Management 44(2): 525-527.

Stenzel, J.R. 1982. Ecology of breeding Yellow Rails at Seney National Wildlife Refuge. M.Sc. thesis, Ohio State University, Columbus, 96 pp.

Stepanyan, L.S. 1990. Conspectus of the ornithological fauna of the USSR. Nauka, Moscow.

Stephen, W.J.D. 1990. Status report on the Greater Sandhill Crane. Committee on the Status of Endangered Wildlife in Canada (COSEWIC), Ottawa, 16 pp.

Stevens, O.A. 1936. The first descriptions of North American birds. Wilson Bulletin 47: 203-215.

Stewart, G.R. and R.D. Titman. 1980. Territorial behaviour by prairie pothole Blue-winged Teal. Canadian Journal of Zoology 58: 639-649.

Stewart, P.A. 1952. Dispersal, breeding behaviour and longevity of banded Barn Owls in North America. Auk 99: 227-245.

Stewart, P.A. 1972. Mortality of Purple Martins from adverse weather. Condor 74(4): 480.

Stewart, P.A. 1982. Migration of Blue Jays in eastern North America. North American Bird Bander 7(3): 107-112.

Stewart, R.E. 1953. A life history study of the Yellow-throat. Wilson Bulletin 65(2): 99-115.

Stewart, R.E. and H.A. Kantrud. 1972. Population estimates of breeding birds in North Dakota. Auk 89(4): 766-788.

Stewart, R.E. and J.H. Manning. 1958. Distribution and ecology of Whistling Swans in the Chesapeake Bay region. Auk 75(2): 203-212.

Stewart, R.E. and C.S. Robbins. 1958. Birds of Maryland and the District of Columbia. United States Department of the Interior, Fish and Wildlife Service, Washington. North American Fauna, no. 62, 401 pp.

Stewart, R.M. 1973. Breeding behaviour and life history of the Wilson's Warbler. Wilson Bulletin 85(1): 21-29.

Stewart, R.M., R.P. Henderson, and K. Darling. 1977. Breeding ecology of the Wilson's Warbler in the High Sierra Nevada, California. Living Bird 16: 83-102

Stickel, D.W. 1965. Territorial and breeding habits of Red-bellied Woodpeckers. American Midland Naturalist 74(1): 110-118.

Stickney, G.P. 1923. Correspondence with Prof. F.E.L. Beal. Wilson Bulletin 25(2): 125-126.

Stiehl, R.B. 1978. Aspects of the ecology of the Common Raven in Harney Basin, Oregon. Ph.D. dissertation, Portland State University, Portland, Oregon, 189 pp.

Stiehl, R.B. 1985. Brood chronology of the Common Raven. Wilson Bulletin 97(1): 78-87.

Stiles, E.W. 1980. Patterns of fruit presentation and seed dispersal in bird-disseminated woody plants in the eastern deciduous forest. American Naturalist 116(5): 570-688.

Stiles, E.W. 1982. Expansions of Mockingbird and multiflora rose in the northeastern United States and Canada. American Birds 36(4): 358-364.

Stiles, F.G. and A.F. Skutch. 1989. A guide to the birds of Costa Rica. Comstock Publishing Associates, Ithaca, 511 pp.

Still, H.R. Jr. and D.P. Baumann, Jr. 1990. Wild Turkey nesting ecology on the Francis Marion National Forest, pp. 13-17 *in* W.M. Healy and G.B. Healy, eds. Proceedings. Sixth national Wild Turkey symposium. National Wild Turkey Federation, Edgefield, South Carolina, 228 pp.

Stocek, A.W. 1950. Breeding behavior of the Goldfinch. Wilson Bulletin 62(3): 107-127.

Stocek, R.F. 1970. Observations on the breeding ecology of the Tree Swallow. Cassimia 52: 3-20.

Stocek, R.F. 1979. Decline of summering Bald Eagles in central New Brunswick. Canadian Field-Naturalist 93(4): 443-445.

Stocek, R.F. 1986. Spring weather and local movements of Tree Swallows, *Tachycineta bicolor*. Canadian Field-Naturalist 100(1): 134-136.

Stocek, R.F. 1992. Le Pygargue à tête blanche. Hinterland Who's Who. Environment Canada, Canadian Wildlife Service, 4 pp. [Also available in English as: Bald Eagle.]

Stockwell, S.S. 1994. Habitat selection and community organization of birds in eight peatlands of Maine. Ph.D. dissertation, University of Maine, Orono, 115 pp.

Stokes, A.W. 1950. Breeding behaviour of the Goldfinch. Wilson Bulletin 62(3): 107-127.

Stokes, D.W. 1979. A guide to bird behavior, vol. 1. Stokes Nature Guides, Little, Brown and Company, Toronto, 336 pp.

Stokes, D.W. 1989. Nos oiseaux, tous les secrets de leur comportement, vol. 1. Les guides Stokes de la nature, Éditions de l'Homme, Montréal, 360 pp.

Stokes, D.W. and L.Q. Stokes. 1983. A guide to bird behavior, vol. 2. Stokes Nature Guides, Little, Brown and Company, Boston, 334 pp.

Stokes, D.[W.] and L.[Q.] Stokes. 1986. American Goldfinches. Living Bird Quarterly 5(3): 22.

Stokes, D.W. and L.Q. Stokes. 1989a. A guide to bird behavior, vol. 3. Stokes Nature Guides, Little, Brown and Company, Boston, 397 pp.

Stokes, D.W. and L.Q. Stokes. 1989b. Nos oiseaux, tous les secrets de leur comportement, vol. 2. Les guides Stokes de la nature, Éditions de l'Homme, Montréal, 366 pp.

Stokes, D.W. and L.Q. Stokes. 1990. Nos oiseaux, tous les secrets de leur comportement, vol. 3. Les guides Stokes de la nature, Éditions de l'Homme, Montréal, 415 pp.

Stone, W. 1965. Bird studies at Old Cape May: an ornithology of coastal New Jersey (2 vol.), 2nd edition. Dover Publications, New York, 941 pp.

Stoner, D. 1936. Studies on the Bank Swallow *Riparia riparia riparia* (Linnaeus) in the Oneida Lake region. Roosevelt Wild Life Annals 4(2): 126-233.

Storer, R.W. 1954. A hybrid between the Chipping and Clay-colored sparrows. Wilson Bulletin 66: 143-144.

Storer, R.W. 1960. The classification of birds, pp. 57-93 *in* A.J. Marshall, ed. Biology and comparative physiology of birds, vol. 1. Academic Press, New York, 518 pp.

Storer, R.W. 1961. Observations of pellet-casting by Horned and Pied-billed grebes. Auk 78(1): 90-92.

Storer, R.W. 1966. Sexual dimorphism and food habits in three North American accipiters. Auk 83(3): 423-436.

Storer, R.W. 1969. The behavior of the Horned Grebe in spring. Condor 71: 180-205.

Storer, R.W. 1971. Adaptation radiation of birds, pp. 149-188 *in* D.S. Farner and J.R. King, eds. Avian Biology, vol. 1. Academic Press, New York, 586 pp.

Storer, R.W. and T.R. Birkhead. 1985. Auk, pp. 27-28 in B. Campbell and E. Lack, eds. A dictionary of birds. British Ornithologists' Union, Buteo Books, Vermillion, South Dakota, 670 pp.

Stott, R.S. and D.P. Olson. 1973. Food-habitat relationship of sea ducks on the New Hampshire coastline. Ecology 54(5): 996-1007.

Stotts, V.D. and D.E. Davis. 1960. The Black Duck in the Chesapeake Bay of Maryland: breeding behaviour and biology. Chesapeake Science 1: 127-154.

Stoudt, J.H. 1969. Relationships between waterfowl and water areas on the Redvers waterfowl study area, pp. 123-131 *in* Saskatoon Wetlands Seminar. Canadian Wildlife Service. Report Series no. 6, 262 pp.

Stouffer, P.C. and D.F. Caccamise. 1991. Roosting and diurnal movements of radio-tagged American Crows. Wilson Bulletin 103(3): 387-400.

Stouffer, P.C., E.D. Kennedy, and H.W. Power. 1987. Recognition and removal of intraspecific parasite eggs by starlings. Animal Behaviour 35(5): 1583-1584.

Stout, J. and G.W. Cornwell. 1976. Nonhunting mortality of fledged North American waterfowl. Journal of Wildlife Management 40(4): 681-693.

Strahler, A.N. and A.H. Strahler. 1987. Modern physical geography, 3rd edition. John Wiley and Sons, Toronto, 544 pp.

Strain, J.G. and R.L. Mumme. 1988. Effects of food supplementation, song playback, and temperature on vocal territorial behavior of Carolina Wrens. Auk 105(1): 11-16.

Strauch, J.G. 1971. Killdeer breeding range extension. Auk 88(1): 171.

Straw, J.A. Jr. 1992. American Woodcock harvest and breeding population status, 1992. United States Department of the Interior, Fish and Wildlife Service, 15 pp.

Street, J.F. 1923. On the nesting grounds of the Solitary Sandpiper and the Lesser Yellow-legs. Auk 40(4): 577-583.

Stresemann, E. 1975. Ornithology from Aristotle to the present. Harvard University Press, Cambridge, Massachusetts, 432 pp.

Strickland, D. 1969. Écologie, comportement social et nidification du Geai Gris (*Perisoreus canadensis*). M.Sc. thesis, Université de Montréal, Montréal.

Strickland, D. 1991. Juvenile dispersal in Gray Jays: dominant brood member expels siblings from natal territory. Canadian Journal of Zoology 69: 2935-2945.

Strickland, D. and H. Ouellet. 1993. Gray Jay (*Perisoreus canadensis*), *in* A. Poole, P. Stettenheim, and F. Gill, eds. The birds of North America, no. 40. The Academy of Natural Sciences, Philadelphia. The American Ornithologists' Union, Washington, 24 pp.

Strong, P.I.V. and J.A. Bissonette. 1987. Effects of nest-site loss on Common Loons, *Gavia immer*. Canadian Field-Naturalist 101(4): 581-583.

Strong, P.I.V. and J.A. Bissonette. 1989. Common Loons, *Gavia immer*, nesting in deadwater streams. Canadian Field-Naturalist 103(1): 94-96.

Studd, M.V. and R.J. Robertson. 1985. Evidence for reliable badges of status in territorial Yellow Warblers (*Dendroica petechia*). Animal Behaviour 33(4): 1102-1113.

Stull, W.D. 1968. Eastern and Canadian Chipping Sparrows, pp. 1166-1184 *in* A.C. Bent. Life histories of North American cardinals, towhees, finches, sparrows, and allies. (O.L. Austin, Jr., ed.). United States National Museum, Washington. [1968, Dover Publications, New York]. Bulletin no. 237, vol. 1, pp. 1-602; vol. 2, pp. 603-1248; vol. 3, pp. 1249-1889.

Sturm, L. 1945. A study of the nesting activities of the American Redstart. Auk 62: 189-206.

Stutchbury, B.J. 1988. Evidence that Bank Swallow colonies do not function as information centers. Condor 90(4): 953-955.

Stutchbury, B.J. 1991. Floater behaviour and territory acquisition in male Purple Martins. Animal Behaviour 42(3): 435-443.

Stutchbury, B.J. and R.J. Robertson. 1985. Floating populations of female Tree Swallows. Auk 102(3): 651-654.

Stutchbury, B.J. and R.J. Robertson. 1987. Signaling subordinate and female status: two hypotheses for the adaptive significance of subadult plumage in female Tree Swallows. Auk 104(4): 717-723.

Sugden, L.G. 1963. Barrow's Goldeneye using crow nests. Condor 65: 330.

Sugden, L.G. 1977. Horned Grebe breeding habitat in Saskatchewan parklands. Canadian Field-Naturalist 91(4): 372-376.

Sugden, L.G. 1979. Habitat use by nesting American Coots in Saskatchewan parklands. Wilson Bulletin 91(4): 599-607.

Sugden, L.G. and G.W. Beyersbergen. 1986. Effect of density and concealment on American Crow predation of simulated duck nests. Journal of Wildlife Management 50(1): 9-14.

Sullivan, B.D. and J.J. Dinsmore. 1990. Factors affecting egg predation by American Crows. Journal of Wildlife Management 54(3): 433-437.

Sullivan, B.D. and J.J. Dinsmore. 1992. Home range and foraging habitat of American Crow, *Corvus brachyrhynchos*, in a waterfowl breeding area in Manitoba. Canadian Field-Naturalist 106(2): 181-184.

Summers-Smith, [J.]D. 1963. The House Sparrow. Collins, London, 269 pp.

Summers-Smith, J.D. 1988. The sparrows – A study of the genus *Passer*. T. and A.D. Poyser, Calton, England, 342 pp.

Surprenant, M. 1993. Les oiseaux aquatiques du Québec, de l'Ontario et des Maritimes. Éditions Michel Quintin, Waterloo, Québec, 285 pp.

Sutherland, C.A. 1963. Notes on the behaviour of Common Nighthawks in Florida. Living Bird 2: 31-39.

Sutherland, D.A. and M.E. Gartshore. 1987. Blue-gray Gnatcatcher, pp. 318-319 *in* M.D. Cadman, P.F.J. Eagles, and F.M. Helleiner, eds. Atlas of the breeding birds of Ontario. University of Waterloo Press, Waterloo, 617 pp.

Sutherland, J.M. and W.J. Maher. 1987. Nest-site selection of the American Coot in the aspen parklands of Saskatchewan. Condor 89(4): 804-810.

Sutton, C.C. and P.T. Sutton. 1985. The status and distribution of Barred Owl and Red-shouldered Hawk in southern New Jersey. Cassinia 61: 20-29.

Sutton, G.M. 1928. The birds of Pymatuning Swamp and Conneaut Lake, Crawford County, Pennsylvania. Annals of the Carnegie Museum 18: 19-239.

Sutton, G.M. 1930. Notes on the Northern Pileated Woodpecker in Pennsylvania. The Cardinal 2(8): 207-217.

Sutton, G.M. 1932. The birds of Southampton Island. Memoirs of the Carnegie Museum no. 12, 275 pp.

Sutton, G.M. 1960. Nesting fringillids of the George Reserve, southeastern Michigan. Jack-Pine Warbler 38: 46-65, 125-139.

Swales, B.H. 1912. Yellow Rail (*Coturnicops noveboracensis*). Auk 29: 100-101.

Swallow, S.K., R.A. Howard, Jr., and R.J. Gutiérrez. 1988. Snag preferences of woodpeckers foraging in a northeastern hardwood forest. Wilson Bulletin 100(2): 236-246.

Swanson, G.A. and M.I. Meyer. 1977. Impact of fluctuating water levels on feeding ecology of breeding Blue-winged Teal. Journal of Wildlife Management 41(3): 426-432.

Swanson, G.A., M.I. Meyer, and J.R. Serie. 1974. Feeding ecology of breeding Blue-winged Teals. Journal of Wildlife Management 38(3): 396-407.

Swengel, S.R. and A.B. Swengel. 1987. Study of a Northern Saw-whet Owl population in Sauk County, Wisconsin, pp. 199-208 *in* R.W. Nero, R.J. Clark, R.J. Knapton, and R.H. Hamre, eds. Proceedings. Biology and conservation of northern forest owls. Symposium, 3-7 February, Winnipeg. United States Department of Agriculture, Forest Service, Rocky Mountain Forest and Range Experiment Station, Fort Collins, Colorado. General Technical Report RM-142, 309 pp.

Swengel, S.R. and A.B. Swengel. 1992. Diet of Northern Saw-whet Owls in southern Wisconsin. Condor 94(4): 707-711.

Swenson, J.E. 1983. Is the northern interior Bald Eagle population in North America increasing?, pp. 23-34 *in* D.M. Bird, ed. Biology and management of Bald Eagles and Ospreys. Harpell Press, Sainte-Anne-de-Bellevue, 325 pp.

Swenson, J.E. 1985. Seasonal habitat use by Sharp-tailed Grouse, *Tympanuchus phasianellus*, on mixed-grass prairie in Montana. Canadian Field-Naturalist 99(1): 40-46.

Taber, R.D. 1949. Observations on the breeding behavior of the Ring-necked Pheasant. Condor 51(4): 153-175.

Taber, W. 1968. White-winged Crossbill, pp. 527-544 *in* A.C. Bent. Life histories of North American cardinals, grosbeaks, buntings, towhees, finches, sparrows, and allies. (O.L. Austin, Jr., ed.). United States National Museum, Washington. [1968, Dover Publications, New York]. Bulletin no. 237, vol. 1, pp. 1-602; vol. 2, pp. 603-1248; vol. 3, pp. 1249-1889.

Taber, W. and D.W. Johnston. 1968. Indigo Bunting, pp. 80-111 *in* A.C. Bent. Life histories of North American cardinals, grosbeaks, buntings, towhees, finches, sparrows and allies. (O.L. Austin, Jr., ed.) United States National Museum, Washington. [1968, Dover Publications, New York]. Bulletin no. 237, vol. 1, pp. 1-602; vol. 2, pp. 603-1248; vol. 3, pp. 1249-1889.

Tacha, T.C. 1988. Social organization of Sandhill Cranes from mid-continental North America. Wildlife Monographs no. 99, 37 pp.

Tacha, T.C., P.A. Vohs, and W.D. Warde. 1985. Morphometric variation of Sandhill Cranes from mid-continental North America. Journal of Wildlife Management 49(1): 246-250.

Tacha, T.C., S.A. Nesbitt, and P.A. Vohs. 1992. Sandhill Crane (*Grus canadensis*), *in* A. Poole, P. Stettenheim, and F. Gill, eds. The birds of North America, no. 31. The Academy of Natural Sciences, Philadelphia. The American Ornithologists' Union, Washington, 24 pp.

Tailleux, I. and S. Marchand. 1987. Influence des perturbations forestières sur l'abondance des couples nicheurs de Canard noir (*Anas rubripes*) et de Garrot commun (*Bucephala clangula*) en milieu boréal. Report for Canadian Wildlife Service by Université Laval, Sainte-Foy, 22 pp.

Talent, L.B., B.L. Krapu, and R.L. Jarvis. 1981. Effects of Redhead nest parasitism on Mallards. Wilson Bulletin 93(4): 562-563.

Tamisier, A. 1972. Etho-écologie des Sarcelles d'hiver (*Anas crecca* L.) pendant leur hivernage en Camargue. Ph.D. dissertation, Université des Sciences et Techniques du Languedoc, Académie de Montpellier, France, 157 pp.

Tamisier, A. 1976. Diurnal activities of Green-winged Teal and Pintail wintering in Louisiana. Wildfowl 27: 19-32.

Tanguay, R. 1964. Les oiseaux des comtés de Kamouraska, L'Islet et Montmagny, P.Q. Naturaliste canadien 91: 309-331.

Tanguay, R. 1965. Les oiseaux des comtés de Kamouraska, L'Islet et Montmagny (province de Québec), partie 2. Naturaliste canadien 92(1): 8-48; 92(2): 49-58.

Tanguay, S. 1988. Guide des sites naturels du Québec. Éditions Michel Quintin, Waterloo, Québec, 251 pp.

Tanner, W.D. Jr. and G.O. Hendrickson. 1956. Ecology of the Sora in Clay County, Iowa. Iowa Bird Life 26(4): 78-81.

Tardif, B. and J.-L. DesGranges. (In preparation). Characterization of breeding birds and rare vascular plants associated with the St. Lawrence River, to optimize the conservation of biodiversity. Biological Conservation.

Tardif, J. and F. Morneau. 1991. Avifaune aquatique. Centrale Eastmain 1. Avant-projet, phase 2. Rapport sectoriel no. 16. Report for vice-présidence Environnement, Hydro-Québec by Le Groupe Roche-Boréal, 173 pp.

Tardif, J. and A. Reed. 1991. Caractérisation préliminaire des habitats de nidification de la Bernache du Canada (*Branta canadensis*) dans la région du Lac Bienville. Report for Hydro-Québec, direction de l'Environnement by Environment Canada, 33 pp.

Tate, G.R. 1992. Short-eared Owl, pp. 171-189 *in* K.J. Schneider and D.M. Pence, eds. Migratory nongame birds of management concern in the northeast. United States Department of the Interior, Fish and Wildlife Service, Newton Corner, Massachusetts, 400 pp.

Tate, J. Jr. 1981. The Blue List for 1981 – the first decade. American Birds 35(1): 3-10.

Tate, J. Jr. 1986. The Blue List for 1986. American Birds 40(2): 227-236.

Tate, J. Jr. and D.J. Tate. 1982. The Blue List for 1982. American Birds 36(2): 126-135.

Taverner, P.A. 1919. Birds of Eastern Canada. Memoir no. 104, Geological Survey, Canadian Department of Mines. Biological Series no. 1563 [1974, Coles, Toronto, 290 pp.], 297 pp.

Taverner, P.A. 1921. The Evening Grosbeak in Canada. Canadian Field-Naturalist 35(3): 41-45.

Taverner, P.A. 1922. Les oiseaux de l'Est du Canada, 2e édition. Memoir no. 104, Geological Survey, Canadian Department of Mines. Biological Series no. 3, 308 pp.

Taverner, P.A. 1925. Montague Chamberlain: in memoriam. Canadian Field-Naturalist 39: 20.

Taverner, P.A. 1929. Bird notes from the Canadian Labrador, 1928. Canadian Field-Naturalist 43: 74-79.

Taverner, P.A. 1934. Birds of Canada. National Museums of Canada, Department of Mines, Ottawa. [1938, 1945, 1953, Musson Book Company, Toronto]. Bulletin no. 72, Biological Series no. 19, 445 pp.

Taverner, P.A. and G.M. Sutton. 1934. The birds of Churchill, Manitoba. Annals of the Carnegie Museum 23: 1-83.

Taverner, P.A. and B.H. Swales. 1911. Notes on the migration of the Saw-whet Owl. Auk 28(3): 329-334.

Taylor, A.L. Jr. and E.D. Forsman. 1976. Recent range extensions of the Barred Owl in western North America, including the first records for Oregon. Condor 78(4): 560-561.

Taylor, D.M. and C.D. Littlefield. 1986. Willow Flycatcher and Yellow Warbler response to cattle grazing. American Birds 40(5): 1169-1173.

Taylor, P. 1979. Interspecific vocal mimicry by Pine Grosbeaks. Canadian Field-Naturalist 93(4): 436-437.

Taylor, P.D. 1987. Purple Martin, pp. 272-273 *in* M.D. Cadman, P.F.J. Eagles, and F.M. Helleiner, eds. Atlas of the breeding birds of Ontario. University of Waterloo Press, Waterloo, 617 pp.

Taylor, P.S. 1973. Breeding behaviour of the Snowy Owl. Living Bird 12: 137-154.

Taylor, P.S. and D.N. Nettleship. 1974. The seabirds of Bonaventure Island, Percé Rock and area, Québec, in summer 1974. Canadian Wildlife Service – Studies on northern Seabirds. Manuscript Report no. 29, 154 pp.

Taylor, W.K. 1972. Analysis of Ovenbirds killed in central Florida. Bird-Banding 43: 15-19.

Taylor, W.K. and M.A. Kershner. 1991. Breeding biology of the Great Crested Flycatcher in central Florida. Journal of Field Ornithology 62(1): 28-39.

Taylor, W.K., R.L. Crawford, M.[A.] Kershner, and S. Gravel. 1983. House Wren migration compared with other wrens: an emphasis on Florida. Journal of Field Ornithology 54(1): 17-28.

Teather, K.L. and R.J. Robertson. 1986. Pair bonds and factors influencing the diversity of mating systems in Brown-headed Cowbirds. Condor 88(1): 63-69.

Teixeira, R.M. 1979. Atlas van de Nederlandse broedvogels. Vereniging tot Behoud van Natuurmonumenten in Nederland, 431 pp.

Telfer, E.S. 1992. Habitat change as a factor in the decline of the western Canadian Loggerhead Shrike, *Lanius ludovicianus*, population. Canadian Field-Naturalist 106(3): 321-326.

Telfer, E.S. 1993. La Pie-grièche migratrice. Hinterland Who's Who. Environment Canada, Canadian Wildlife Service, 3 pp. [Also available in English as: Loggerhead Shrike.]

Temeles, E.J. 1990. Interspecific territoriality of Northern Harriers: the role of kleptoparasitism. Animal Behaviour 40(2): 361-366.

Temminck, C.J. 1807. Catalogue systématique du cabinet d'ornithologie et de la collection de quadrumanes. C. Sepp Janez, Amsterdam, 270 pp.

Temminck, C.J. 1813-1815. Histoire naturelle générale des pigeons et des gallinacées (3 tomes). H.O. Brouwer, Amsterdam, vol. 1, 499 pp.; vol. 2, 477 pp.; vol. 3, 457 pp.

Temminck, C.J. 1817. Observations sur la classification méthodique des oiseaux et remarques sur l'analyse d'une nouvelle ornithologie élémentaire par L.P. Vieillot. G. Dufour, Amsterdam, 60 pp.

Temple, S.A. 1972. Sex and age characteristics of North American Merlins. Bird-Banding 43: 191-196.

Temple, S.A. and J.R. Cary. 1988. Modeling dynamics of habitat-interior bird populations in fragmented landscapes. Conservation Biology 2(4): 340-347.

Temple, S.A. and B.L. Temple. 1976. Avian population trends in central New York State, 1935-1972. Bird-Banding 47: 238-257.

Temple, S.A. and B.A. Wilcox. 1986. Introduction: predicting effects of habitat patchiness and fragmentation, pp. 261-262 *in* J. Verner, M.L. Morrison, and C.J. Ralph, eds. Wildlife 2000: modeling habitat relationships of terrestrial vertebrates. University of Wisconsin Press, Madison.

ten Cate, C. and J. Hilbers. 1991. Effects of brood size on inter-clutch intervals, offspring development and male-female interactions in the Ring Dove *Streptopelia risoria*. Animal Behaviour 41(1): 27-36.

Terborgh, J.W. 1980. The conservation status of neotropical migrants: present and future, pp. 21-30 *in* A. Keast and E.S. Morton, eds. Migrant birds in the neotropics: ecology, behavior, distribution and conservation. Smithsonian Institution Press, Washington, 576 pp.

Terborgh, J.[W.]. 1989. Where have all the birds gone? Princeton University Press, Princeton, New Jersey, 207 pp.

Terborgh, J.[W.]. 1992. Why American songbirds are vanishing. Scientific American 266(5): 98-104.

Terrasse, J.F. and M. Terrasse. 1977. Le Balbuzard pêcheur *Pandion haliaetus* (L.) en Méditerrannée occidentale, essai de recensement, reproducteur, avenir. Nos Oiseaux 34: 111-127.

Terres, J.K. 1980. The Audubon Society encyclopedia of North American birds. [1991, Wings Books, New York.] Alfred A. Knopf, New York, 1109 pp.

Terrill, L.M. 1910a. Notes on winter birds in Hochelaga and Compton counties. Ottawa Naturalist 24(2): 38-40.

Terrill, L.M. 1910b. Notes on the White-throated Sparrow. Ottawa Naturalist 24(8): 135-138.

Terrill, L.M. 1911a. Blue Jay imitating Red-shouldered Hawk: absence of cones on evergreen trees. Ottawa Naturalist 25(1): 20.

Terrill, L.M. 1911b. Changes in the status of certain birds in the vicinity of Montreal, P.Q. Ottawa Naturalist 25(4): 57-63.

Terrill, L.M. 1913. Early winter bird notes: 1912-13. Ottawa Naturalist 27(3-4): 43-46.

Terrill, L.M. 1915. Notes from the Laurentian Hills. Wilson Bulletin 37(2): 302-309.

Terrill, L.M. 1916. Unusual bird records at Montreal during the fall and winter of 1915-1916. Canadian Field-Naturalist 30(1): 15-16.

Terrill, L.M. 1917. Occurrence of the Ring-necked Pheasant in the vicinity of Montreal. Ottawa Naturalist 30(10): 132.

Terrill, L.M. 1921a. The Grasshopper Sparrow in the Montreal district. Auk 38(1): 115-116.

Terrill, L.M. 1921b. Unusual migration records in the vicinity of Montreal - March 1921. Canadian Field-Naturalist 35(7): 140-141.

Terrill, L.M. 1922. The Short-billed Marsh Wren in the Montreal district. Auk 39: 112-115.

Terrill, L.M. 1923. Song Sparrow wintering in the Montreal district. Canadian Field-Naturalist 37(2): 30-31.

Terrill, L.M. 1924. Occurrence of the Starling in the Montreal district. Canadian Field-Naturalist 38(3): 58.

Terrill, L.M. 1929. The Brown Creeper and other birds wintering on St. Helen's Island (Montreal). Canadian Field-Naturalist 43(3): 50-51.

Terrill, L.M. 1931a. Nesting of the Saw-whet Owl (*Cryptoglaux acadica acadica*) in the Montreal district. Auk 48(2): 169-174.

Terrill, L.M. 1931b. Occurrence of the Pomarine Jaeger in the Montreal district. Canadian Field-Naturalist 45(6): 142.

Terrill, L.M. 1938. Whip-poor-will and Nighthawk in the province of Quebec. Provancher Society of Natural History of Canada, Annual report, pp. 150-157.

Terrill, L.M. 1942. Report on birds. Annual report. Province of Quebec Society for the Protection of Birds 25: 9-25.

Terrill, L.M. 1943a. Report on birds. Annual report. Province of Quebec Society for the Protection of Birds 26: 8-31.

Terrill, L.M. 1943b. Nesting habits of the Yellow Rail in Gaspé County, Quebec. Auk 60(2): 171-180.

Terrill, L.M. 1944. Report on birds. Annual report. Province of Quebec Society for the Protection of Birds 27: 5-26.

Terrill, L.M. 1945. Report on birds. Annual report. Province of Quebec Society for the Protection of Birds 28: 7-31.

Terrill, L.M. 1946. Report on birds. Annual report. Province of Quebec Society for the Protection of Birds 29: 6-38.

Terrill, L.M. 1947. Report on birds. Annual report. Province of Quebec Society for the Protection of Birds 30: 6-22.

Terrill, L.M. 1951. Shore bird migration at Montreal. Canadian Field-Naturalist 65(3): 87-98.

Terrill, L.M. 1961. Cowbird hosts in southern Quebec. Canadian Field-Naturalist 75(1): 2-11.

Terrill, L.M. 1968. Eastern Fox Sparrow, pp. 1395-1415 *in* A.C. Bent. Life histories of North American cardinals, grosbeaks, buntings, towhees, finches, sparrows, and allies. (O.L. Austin, Jr., ed.). United States National Museum, Washington. [1968, Dover Publications, New York]. Bulletin no. 237, vol. 1, pp. 1-602; vol. 2, pp. 603-1248; vol. 3, pp. 1249-1889.

Terrill, L.M. and N. Smith. 1930. Birds of the Razades and Basque islands. The Provancher Society of Natural History of Canada, Annual report, pp. 27-35.

Terrill, S.B. and R.L. Crawford. 1988. Additional evidence of nocturnal migration by Yellow-rumped Warblers in winter. Condor 90(1): 261-263.

Terrill, S.B. and R.D. Ohmart. 1984. Facultative extension of fall migration by Yellow-rumped Warblers (*Dendroica coronata*). Auk 101(3): 427-438.

Test, H. 1969. Relation of wing and tail color of the woodpeckers *Colaptes auratus* and *C. cafer* to their food. Condor 71: 206-211.

The Gazette. 1993. New York State struggles with swan nuisance. February 13, 1993, Montréal, p. K-8.

Thibault, J.C. and F. Bouvet. 1983. Les caractéristiques du nid du Balbuzard pêcheur, *Pandion haliaetus*, en Corse. Nos Oiseaux 37: 65-73.

Thibault, J.C. and O. Patrimonio. 1991. Some aspects of breeding success of the Osprey *Pandion haliaetus* in Corsica, West Mediterranean. Bird Study 38: 98-102.

Thibault, M. 1985. Les régions écologiques du Québec méridional, deuxième approximation. Service de la cartographie, Ministère de l'Énergie et des Ressources du Québec, Québec. Map, scale 1:1 250 000.

THIBAULT, M. 1988. Description du milieu forestier du Québec méridional par domaine et sous-domaine climatique. Service de la recherche appliquée, Direction de la recherche et du développement, Ministère de l'Énergie et des Ressources du Québec, Québec, rapport interne no. 309, 31 pp.

THIBAULT, M.T.H. 1989. Environnement, pp. 261-291 *in* Le Québec statistique. Bureau de la statistique du Québec, Québec, 1028 pp.

THIBODEAU, J.C., M. GAUDREAU, and J. BERGERON. 1986. Le zonage agricole, un bilan positif; les effets de la Loi 90 dans la région sud de Montréal. Institut national de la recherche scientifique [INRS] – Urbanisation. Rapport de recherche no. 9.

THOMAS, G. 1985a. Plovers and sandpipers, pp. 162-166 *in* C.M. Perrins and A.L.A. Middleton, eds. The encyclopedia of birds. Facts on File, New York, 447 pp.

THOMAS, G. 1985b. Plovers and sandpipers, pp. 166-170 *in* C.M. Perrins and A.L.A. Middleton, eds. The encyclopedia of birds. Facts on File, New York, 447 pp.

THOMAS, G. 1990. Les pluviers, les courlis et les avocettes, pp. 98-111 *in* C.M. Perrins and A.L.A. Middleton, eds. Les oiseaux des mers et des rivières: les autruches et les kiwis. Solar, Paris, 146 pp.

THOMAS, J.W., R. ANDERSON, C. MASER, and E. BULL. 1979. Snags, pp. 60-67 *in* J.W. Thomas, ed. Wildlife habitats in managed forests – the Blue Mountains of Oregon and Washington. United States Department of Agriculture, Forest Service, Washington. Agriculture Handbook no. 533, 512 pp.

THOMAS, L. and K. ANDRESEN. 1986. Wingbeat rates of Glaucous and Iceland gulls. British Birds 79(1): 42.

THOMAS, V.G. 1984. Winter diet and intestinal proportions of Rock and Willow ptarmigan and Sharp-tailed Grouse in Ontario. Canadian Journal of Zoology 62(11): 2258-2263.

THOMPSON, D. 1973. Feeding ecology of diving ducks on Keokuk Pool, Mississippi River. Journal of Wildlife Management 37(3): 367-381.

THOMPSON, D.H. 1978. Declines in populations of Great Blue Herons and Great Egrets in five mid-western States. Proceedings. Colonial Waterbird Group 2: 114-123.

THOMPSON, F.R. III and D.E. CAPEN. 1988. Avian assemblages in seral stages of a Vermont forest. Journal of Wildlife Management 52(4): 771-777.

THOMPSON, F.R. III and E.K. FRITZELL. 1988. Ruffed Grouse winter roost site preference and influence on energy demands. Journal of Wildlife Management 52(3): 454-460.

THOMPSON, F.R. III and E.K. FRITZELL. 1989. Habitat use, home range, and survival of territorial male Ruffed Grouse. Journal of Wildlife Management 53(1): 15-21.

THOMPSON, I.D. 1974. Movement and use of habitat by brood-rearing ducks at Lake St. Francis, Québec. M.Sc thesis, York University, Toronto, 123 pp.

THRELFALL, W. and J.R. BLACQUIERE. 1982. Breeding biology of the Fox Sparrow in Newfoundland. Journal of Field Ornithology 53(3): 235-239.

THURBER, W.A. 1986. Range expansion of the House Sparrow through Guatemala and El Salvador. American Birds 40(2): 341-350.

THWAITES, R.G. 1895-1901. The Jesuit relations and allied documents: travels and explorations of the Jesuit Missionaries in New France, 1610-1791. Cleveland, Burrows Brothers, 73 vol.

TILGHMAN, N.G. 1987. Characteristics of urban woodlands affecting breeding bird diversity and abundance. Landscape and Urban Planning 14: 481-495.

TIMKEN, R.L. and B.W. ANDERSON. 1969. Food habits of Common Mergansers in the northcentral United States. Journal of Wildlife Management 33(1): 87-91.

TINBERGEN, N. 1939. The behavior of the Snow Bunting in spring. Transactions. Linnean Society of New York 5: 1-94.

TINBERGEN, N. 1956. On the functions of territory in gulls. Ibis 98: 401-411.

TINBERGEN, N. 1958. Curious naturalists. Country Life, London, 280 pp.

TINBERGEN, N. 1961. Carnets d'un naturaliste. Hachette, Paris, 296 pp.

TITMAN, R.D. and N.R. SEYMOUR. 1981. A comparison of pursuit flights by six North American ducks of the genus *Anas*. Wildfowl 32: 11-18.

TITTERINGTON, R.W., H.S. CRAWFORD, and B.N. BURGASON. 1979. Songbird responses to commercial clear-cutting in Maine spruce-fir forests. Journal of Wildlife Management 43(3): 602-609.

TITUS, J.R. and L.W. VANDRUFF. 1981. Response of the Common Loon to recreational pressure in the Boundary Waters Canoe Area, northeastern Minnesota. Wildlife Monographs no. 79, 60 pp.

TITUS, K. and M.R. FULLER. 1990. Recent trends in counts of migrant hawks from northeastern North America. Journal of Wildlife Management 54(3): 463-470.

TITUS, K. and J.A. MOSHER. 1981. Nest-site habitat selected by woodland hawks in the central Appalachians. Auk 98(2): 270-281.

TITUS, K. and J.A. MOSHER. 1987. Selection of nest tree species by Red-Shouldered and Broad-winged hawks in two temperate forest regions. Journal of Field Ornithology 58(3): 274-283.

TODD, W.E.C. 1940. Birds of western Pennsylvania. University of Pittsburgh Press, Pittsburgh, 710 pp.

TODD, W.E.C. 1963. Birds of the Labrador Peninsula and Adjacent Areas. A distributional list. University of Toronto Press, Toronto, 819 pp.

TOLAND, B.R. 1990. Nesting ecology of Red-tailed Hawks in central Missouri. Transactions. Missouri Academy of Science 24: 1-16.

TOLFREY, F. 1845. The sportsman in Canada. T.C. Newby, London.

TOMKINS, I.R. 1965. The Willets of Georgia and South Carolina. Wilson Bulletin 77(2): 151-167.

TOMMERAAS, P.J. 1989. A time-lapse nest study of a pair of Gyrfalcons (*Falco rusticolus*) from their arrival at the nesting ledge to the completion of egg-laying. Fauna Norvegica Series C 12(2): 52-64.

TOMPA, F.S. 1964. Factors determining the numbers of Song Sparrows, *Melospiza melodia* (Wilson), on Mandarte Island, B.C., Canada. Acta Zoologica Fennica 109(1): 3-73.

TOWNSEND, C.W. 1913a. Some more Labrador notes. Auk 30(1): 1-10.

TOWNSEND, C.W. 1913b. A short trip into the Labrador Peninsula by way of the Natashquan River. Geographical Society, Philadelphia. Bulletin 11(3): 170-182.

TOWNSEND, C.W. 1914. A plea for the conservation of the Eider. Auk 31(1): 14-21.

TOWNSEND, C.W. 1916. A new subspecies of Hudsonian Chickadee from the Labrador Peninsula. Auk 33(1): 74.

TOWNSEND, C.W. 1917a. In Audubon's Labrador. Auk 34(2): 133-146.

TOWNSEND, C.W. 1917b. The Labrador Chickadee (*Penthestes hudsonicus nigricans*) in a southward migration. Auk 34(2): 160-163.

TOWNSEND, C.W. 1918. In Audubon's Labrador. Houghton Mifflin, Boston and New York, 354 pp.

TOWNSEND, C.W. 1919. Great Auk, pp. 208-215 *in* A.C. Bent. Life histories of North American diving birds. United States National Museum, Washington. [1963, Dover Publications, New York]. Bulletin no. 107, 239 pp.

TOWNSEND, C.W. 1920. Notes on the summer birds of the Gaspé Peninsula, province of Quebec. Canadian Field-Naturalist 34(4): 78-80; 34(5): 86-95.

TOWNSEND, C.W. 1921a. Arctic Tern, pp. 249-255 *in* A.C. Bent. Life histories of North American gulls and terns. United States National Museum, Washington. [1963, Dover Publications, New York]. Bulletin no. 113, 337 pp.

TOWNSEND, C.W. 1921b. Herring Gull, pp. 102-120 *in* A.C. Bent. Life histories of North American gulls and terns. United States National Museum, Washington. [1963, Dover Publications, New York]. Bulletin no. 113, 337 pp.

TOWNSEND, C.W. 1921c. Iceland Gull, pp. 62-65 *in* A.C. Bent. Life histories of North American gulls and terns. United States National Museum, Washington. [1963, Dover Publications, New York]. Bulletin no. 113, 337 pp.

TOWNSEND, C.W. 1921d. Parasitic Jaeger, pp. 14-20 *in* A.C. Bent. Life histories of North American gulls and terns. United States National Museum, Washington. [1963, Dover Publications, New York]. Bulletin no. 113, 337 pp.

TOWNSEND, C.W. 1923a. A breeding station of the Horned Lark and Pipit on the Gaspé Peninsula. Auk 40(1): 85-87.

TOWNSEND, C.W. 1923b. Additional notes on the birds of the Gaspé Peninsula. Canadian Field-Naturalist 37(1): 6-11.

TOWNSEND, C.W. 1923c. Red-breasted Merganser, pp. 13-22 *in* A.C. Bent. Life histories of North American wild fowl. United States National Museum, Washington. [1962, Dover Publications, New York]. Bulletin no. 126, vol. 1, 244 pp.

TOWNSEND, C.W. 1924. Diving of grebes and loons. Auk 41(1): 29-41.

TOWNSEND, C.W. 1926. Virginia Rail, pp. 292-301 *in* A.C. Bent. Life histories of North American marsh birds. United States National Museum, Washington. [1963, Dover Publications, New York]. Bulletin no. 135, 392 pp.

TOWNSEND, C.W. 1927a. Least Sandpiper, pp. 202-213 *in* A.C. Bent. Life histories of North American shore birds. United States National Museum, Washington. [1962, Dover Publications, New York]. Bulletin no. 142, vol. 1, 420 pp.

TOWNSEND, C.W. 1927b. Semipalmated Sandpiper, pp. 244-255 *in* A.C. Bent. Life histories of North American shore birds. United States National Museum, Washington. [1962, Dover Publications, New York]. Bulletin no. 142, vol. 1, 420 pp.

TOWNSEND, C.W. 1929a. Killdeer, pp. 202-217 *in* A.C. Bent. Life histories of North American shore birds. United States National Museum, Washington. [1962, Dover Publications, New York]. Bulletin no. 146, vol. 2, 412 pp.

TOWNSEND, C.W. 1929b. Semipalmated Plover, pp. 217-229 *in* A.C. Bent. Life histories of North American shore birds. United States National Museum, Washington. [1962, Dover Publications, New York]. Bulletin no. 146, vol. 2, 412 pp.

TOWNSEND, C.W. 1932. Passenger Pigeon, pp. 379-402 *in* A.C. Bent. Life histories of North American gallinaceous birds. United States National Museum, Washington. [1963, Dover Publications, New York]. Bulletin no. 162, 490 pp.

TOWNSEND, C.W. 1937. American Rough-legged Hawk, pp. 269-284 *in* A.C. Bent. Life histories of North American birds of prey. United States National Museum, Washington. [1961, Dover Publications, New York]. Bulletin no. 167, vol. 1, 409 pp.

TOWNSEND, C.W. 1938a. Richardson's Owl, pp. 220-227 *in* A.C. Bent. Life histories of North American birds of prey. United States National Museum, Washington. [1961, Dover Publications, New York]. Bulletin no. 170, vol. 2, 482 pp.

TOWNSEND, C.W. 1938b. Short-eared Owl, pp. 169-182 *in* A.C. Bent. Life histories of North American birds of prey. United States National Museum, Washington. [1961, Dover Publications, New York]. Bulletin no. 170, vol. 2, 482 pp.

TOWNSEND, C.W. and G.M. ALLEN. 1907. Birds of Labrador. Proceedings. Boston Society of Natural History 23: 277-428.

TOWNSEND, C.W. and A.C. BENT. 1910. Additional notes on the birds of Labrador. Auk 27(1): 1-18.

TOZER, R.G. and J.M. RICHARDS. 1974. Birds of the Oshawa-Lake Scugog region, Ontario. Alger Press, 384 pp.

TRAMER, E.J. and F.E. TRAMER. 1977. Feeding response of fall migrants to prolonged inclement weather. Wilson Bulletin 89(1): 166.

TRAUGER, D.L. 1971. Population ecology of Lesser Scaup (*Aythya affinis*) in subarctic taiga. Ph.D. dissertation, Iowa State University, Ames, 118 pp.

TRAUGER, D.L. and R.J. HALL. 1992. The challenge of biological diversity: professional responsibilities, capabilities and realities. Transactions. North American Wildlife and Natural Resources Conference 57: 20-36.

TREMBLAY, C. and J.-C. BELLES-ISLES. 1991. Faune avienne terrestre. Aménagement hydroélectrique d'Eastmain. 1. Étude d'impact sur l'environnement. Avant-projet. Rapport sectoriel no. 15. Report for vice-présidence Environnement, Hydro-Québec by Le Groupe Roche-Boréal, 43 pp.

TREMBLAY, G. 1974. Géologie du Quaternaire: région de Rouyn-Noranda et d'Abitibi, comtés d'Abitibi-est et d'Abitibi-ouest. Ministère des richesses naturelles du Québec, DP-236, 100 pp.

TREMBLAY, J. and L.N. ELLISON. 1979. Effects of human disturbance on breeding of Black-crowned Night-Herons. Auk 96(2): 364-369.

TREMBLAY, J. and L.N. ELLISON. 1980a. Breeding success of the Black-crowned Night-Heron in the St. Lawrence estuary. Canadian Journal of Zoology 58: 1259-1263.

TREMBLAY, J. and L.N. ELLISON. 1980b. Le régime alimentaire des jeunes Bihoreaux à couronne noire dans l'estuaire du Saint-Laurent. Naturaliste canadien 107: 15-20.

TREMBLAY, J.-L. and R. GAUDRY. 1936. Décimation des zostères (herbe à bernaches), dans la région de l'île Verte. Naturaliste canadien 63: 257-262.

TRÉPANIER, J.P. 1975. Synthèse ornithologique 1966-75. Camp-école Chicobi, 26 pp.

TRIMBLE, S.A. 1975. Habitat management series for unique or endangered species: Merlin. United States Department of the Interior, Bureau of Land Management. Technical Notes no. 271, 41 pp.

TROY, D.M. 1983. Recaptures of redpolls: movement of an irruptive species. Journal of Field Ornithology 54(2): 146-151.

TROY, D.M. 1985. A phenetic analysis of the redpolls *Carduelis flammea flammea* and *C. hornemanni exilipes*. Auk 102(1): 82-96.

TRUSLOW, F.K. 1966. Ground-nesting Great Horned Owl: a photographic study. Living Bird 5: 177-186.

TRUSLOW, F.K. 1967. Egg-carrying by the Pileated Woodpecker. Living Bird 6: 227-236.

TSUJI, L.J.S., D.R. KOZLOVIC, and M.B. SOKDOWSKI. 1992. Territorial position in Sharp-tailed Grouse leks: the probability of fertilization. Condor 94: 1030-1031.

TUCK, G.S. 1978. A field guide to the seabirds of Britain and the world. W.M. Collins' Sons and Co., Ltd, London, 292 pp.

TUCK, L.M. 1965. Les marmettes: leur répartition, leurs populations et leurs particularités biologiques – Étude du genre *Uria*. Canadian Wildlife Service. Monograph Series no. 1, 284 pp. [Also available in English as: The murres: their distribution, populations and biology – a study of the genus *Uria*.]

TUCK, L.M. 1972. The snipes: A study of the genus *Capella*. Canadian Wildlife Service. Monograph Series no. 5, 429 pp.

TUCKER, J.W. JR. 1990. Male Eastern Bluebird rears four broods during one nesting season. Wilson Bulletin 102(4): 726-728.

TUFTS, R.W. 1961. The birds of Nova Scotia. Nova Scotia Museum, Halifax, 481 pp.

TUFTS, R.W. 1986. Birds of Nova Scotia, 3rd edition. Nimbus Publishing Limited and Nova Scotia Museum, Halifax, 478 pp.

TURCEK, F.J. and L. KELSO. 1968. Ecological aspects of good transportation and storage in the Corvidae. Communication in Behavioral Biology, part A 1: 277-297.

TURCOTTE, F., R. COUTURE, J. FERRON, and R. COURTOIS. 1993. Caractérisation des habitats essentiels du Tétras du Canada (*Dendragapus canadensis*) dans la région de l'Abitibi-Témiscamingue. Ministère de l'Environnement et de la Faune, Direction de la faune et des habitats, Service de la faune terrestre, Québec, 77 pp.

TURCOTTE, F., R. COUTURE, R. COURTOIS, and J. FERRON. 1994. Réactions du Tétras du Canada (*Dendragapus canadensis*) face à l'exploitation forestière en forêt boréale. Ministère de l'Environnement et de la Faune, Direction de la faune et des habitats, Service de la faune terrestre, Québec, 77 pp.

TURCOTTE, Y. and J. BÉDARD. 1989a. Prolonged parental care and foraging of Greater Snow Goose juveniles. Wilson Bulletin 101(3): 500-503.

TURCOTTE, Y. and J. BÉDARD. 1989b. Shared parental investment, parent-offspring conflict and brood size in Greater Snow Geese. Animal Behaviour 38: 703-706.

TURGEON, P. 1992. La Radissonie, le pays de la baie James. Éditions Libre Expression [Montréal], 191 pp.

TURNER, A. 1989. A handbook to the swallows and martins of the world. Christopher Helm, London, 258 pp.

TURNER, L.M. 1885. List of the birds of Labrador, including Ungava, East Main, Moose, and Gulf districts of the Hudson Bay Company, together with the island of Anticosti. Proceedings. United States National Museum 8: 233-254.

TWEDT, D.J. 1984. Pellet casting by Common Grackles. Wilson Bulletin 96(3): 470-471.

TYLER, W.M. 1929a. Piping Plover, pp. 236-246 *in* A.C. Bent. Life histories of North American shore birds. United States National Museum, Washington. [1962, Dover Publications, New York]. Bulletin no. 146, vol. 2, 412 pp.

TYLER, W.M. 1929b. Spotted Sandpiper, pp. 78-97 *in* A.C. Bent. Life histories of North American shore birds. United States National Museum, Washington. [1962, Dover Publications, New York]. Bulletin no. 146, vol. 2, 412 pp.

TYLER, W.M. 1939. Yellow-bellied Sapsucker, pp. 126-141 *in* A.C. Bent. Life histories of North American woodpeckers. United States National Museum, Washington. [1964, Dover Publications, New York]. Bulletin no. 174, 334 pp.

TYLER, W.M. 1940a. Chimney Swift, pp. 271-293 *in* A.C. Bent. Life histories of North American cuckoos, goatsuckers, hummingbirds, and their allies. United States National Museum, Washington. [1964, Dover Publications, New York]. Bulletin no. 176, vol. 1, pp. 1-244; vol. 2, pp. 245-506.

TYLER, W.M. 1940b. Eastern Whip-poor-will, pp. 163-183 *in* A.C. Bent. Life histories of North American cuckoos, goatsuckers, hummingbirds, and their allies. United States National Museum, Washington. [1964, Dover Publications, New York]. Bulletin no. 176, vol. 1, pp. 1-244; vol. 2, pp. 245-506.

TYLER, W.M. 1940c. Ruby-throated Hummingbird, pp. 332-352 *in* A.C. Bent. Life histories of North American cuckoos, goatsuckers, hummingbirds, and their allies. United States National Museum, Washington. [1964, Dover Publications, New York]. Bulletin no. 176, vol. 1, pp. 1-244; vol. 2, pp. 245-506.

TYLER, W.M. 1942a. Eastern Kingbird, pp. 11-29 *in* A.C. Bent. Life histories of North American flycatchers, larks, swallows, and their allies. United States National Museum, Washington. [1963, Dover Publications, New York]. Bulletin no. 179, 555 pp.

TYLER, W.M. 1942b. Eastern Phoebe, pp. 140-154 *in* A.C. Bent. Life histories of North American flycatchers, larks, swallows, and their allies. United States National Museum, Washington. [1963, Dover Publications, New York]. Bulletin no. 179, 555 pp.

TYLER, W.M. 1942c. Tree Swallow, pp. 384-400 *in* A.C. Bent. Life histories of North American flycatchers, larks, swallows, and their allies. United States National Museum, Washington. [1963, Dover Publications, New York]. Bulletin no. 179, 555 pp.

TYLER, W.M. 1946. Northern Blue Jay, pp. 32-52 *in* A.C. Bent. Life histories of North American Jays, Crows, and Titmice. United States National Museum, Washington. [1964, Dover Publications, New York]. Bulletin no. 195, vol. 1, pp. 1-214; vol. 2, pp. 215-495.

TYLER, W.M. 1948a. Brown Creeper, pp. 56-71 *in* A.C. Bent. Life histories of North American nuthatches, wrens, thrashers, and their allies. United States National Museum, Washington. [1964, Dover Publications, New York]. Bulletin no. 195, 495 pp.

TYLER, W.M. 1948b. Red-breasted Nuthatch, pp. 22-35 *in* A.C. Bent. Life histories of North American nuthatches, wrens, thrashers, and their allies. United States National Museum, Washington. [1964, Dover Publications, New York]. Bulletin no. 195, 475 pp.

TYLER, W.M. 1948c. White-breasted Nuthatch, pp. 1-12 *in* A.C. Bent. Life histories of North American nuthatches, wrens, thrashers, and their allies. United States National Museum, Washington. [1964, Dover Publications, New York]. Bulletin no. 195, 475 pp.

TYLER, W.M. 1949. Veery, pp. 217-231 *in* A.C. Bent. Life histories of North American thrushes, kinglets, and their allies. United States National Museum, Washington. [1964, Dover Publications, New York]. Bulletin no. 196, 452 pp.

TYLER, W.M. 1950. Eastern Warbling Vireo, pp. 362-373 *in* A.C. Bent. Life histories of North American wagtails, shrikes, vireos, and their allies. United States National Museum, Washington. [1964, Dover Publications, New York]. Bulletin no. 197, 411 pp.

TYLER, W.M. 1953a. Golden-winged Warbler, pp. 47-57 *in* A.C. Bent. Life histories of North American wood warblers. United States National Museum, Washington. [1963, Dover Publications, New York]. Bulletin no. 203, vol. 1, pp. 1-367; vol. 2, pp. 367-734.

TYLER, W.M. 1953b. Wilson's Pileolated Warbler, pp. 626-639 *in* A.C. Bent. Life histories of North American wood warblers. United States National Museum, Washington. [1963, Dover Publications, New York]. Bulletin no. 203, vol. 1, pp. 1-367; vol. 2, pp. 367-734.

TYLER, W.M. 1953c. Yellow Palm Warbler, pp. 450-456 *in* A.C. Bent. Life histories of North American wood warblers. United States National Museum, Washington. [1963, Dover Publications, New York]. Bulletin no. 203, vol. 1, pp. 1-367; vol. 2, pp. 367-734.

TYLER, W.M. 1958a. Baltimore Oriole, pp. 247-270 *in* A.C. Bent. Life histories of North American blackbirds, orioles, tanagers, and allies. United States National Museum, Washington. [1965, Dover Publications, New York]. Bulletin no. 211, 549 pp.

TYLER, W.M. 1958b. Scarlet Tanager, pp. 479-491 *in* A.C. Bent. Life histories of North American blackbirds, orioles, tanagers and allies. United States National Museum, Washington. [1965, Dover Publications, New York]. Bulletin no. 211, 549 pp.

UNITED STATES FISH AND WILDLIFE SERVICE (USFWS). 1990. American Woodcock Management Plan. United States Department of the Interior, Fish and Wildlife Service, Washington, D.C., 11 pp.

UNITED STATES FISH AND WILDLIFE SERVICE (USFWS). 1993. Endangered and threatened wildlife and plants. Title 50 CFR 17.11 and 17.12, 40 pp.

UNITED STATES FISH AND WILDLIFE SERVICE (USFWS) and CANADIAN WILDLIFE SERVICE (CWS). 1986. Plan nord-américain de gestion de la sauvagine: stratégie de collaboration. U.S. Department of the Interior, Fish and Wildlife Service, Washington, D.C. and Environment Canada, Canadian Wildlife Service, Ottawa, 21 pp. [Also available in English as: North American Waterfowl Management Plan: a strategy for cooperation.]

UNITED STATES FISH AND WILDLIFE SERVICE (USFWS) and CANADIAN WILDLIFE SERVICE (CWS) [1991 ?]. Le Canard noir: une intervention bilatérale est urgente. Projet conjoint sur le Canard noir, Plan nord-américain de gestion de la sauvagine, Prospectus pour les commanditaires du projet conjoint sur le Canard noir. United States Department of the Interior, Fish and Wildlife Service, Washington, D.C. and Environment Canada, Canadian Wildlife Service, Ottawa, 8 pp.

URBAN, J. 1990. Élevage des oiseaux d'ornement – Étude systématique. Éditions J. et D. Pau, 447 pp.

UTTLEY, J., P. MONAGHAN, and J. BLACKWOOD. 1990. Hedgehog *Erinaceous europaeus* predation on Arctic Tern *Sterna paradisaea* eggs: the impact on breeding success. Seabird 13: 2-6.

VAN BUSKIRK, J.V. 1984. Vocal mimicry of Nashville Warblers by Yellow-rumped Warblers. Wilson Bulletin 96(3): 477-482.

VANDER HAEGEN, W.M., W.E. DODGE, and M.W. SAYRE. 1988. Factors affecting productivity in a northern Wild Turkey population. Journal of Wildlife Management 52(1): 127-133.

VANDER HAEGEN, W.M., M.W. SAYRE, and W.E. DODGE. 1989. Winter use of agricultural habitats by Wild Turkeys in Massachusetts. Journal of Wildlife Management 53(1): 30-33.

VAN DIJK, B. 1986. The breeding biology of eiders at île aux Pommes, Québec, pp. 119-126 *in* A. Reed, ed. Eider ducks in Canada/Les eiders au Canada. Canadian Wildlife Service. Report Series no. 47, 177 pp.

VANGILDER, L.D., L.M. SMITH, and R.K. LAWRENCE. 1986. Nutrient reserves of premigratory Brant during spring. Auk 103(1): 237-241.

Van Horn, M.A. and T. Donovan. 1994. Ovenbird (*Seiurus aurocapillus*), *in* A. Poole and F. Gill, eds. The birds of North America, no. 88. The Academy of Natural Sciences, Philadelphia. The American Ornithologists' Union, Washington, 24 pp.

Van Horne, B. and A. Bader. 1990. Diet of nestling Winter Wrens in relationship to food availability. Condor 92(2): 413-420.

van IJzendoorn, E.J. 1980. Breeding record of Black Tern, *Chlidonias niger* x White-winged Black Tern, *C. leucopterus*. Dutch Birding 2: 17-18.

Van Rhijn, J.G. and T. Groothuis. 1985a. Biparental care and the basis of alternative associations among gulls, with special reference to Black-headed Gulls. Ardea 73(2): 159-174.

Van Rhijn, J.G. and T. Groothuis. 1985b. On the mechanism of mate selection in Black-headed Gulls. Behaviour 100: 134-169.

Van Scheik, W.J. and V. Lewin. 1986. Double-crested Cormorant, *Phalacrocorax auritus*, egg ejection. Canadian Field-Naturalist 100(4): 561-562.

Van Tyne, J. and A.J. Berger. 1966. Fundamentals of ornithology. Science Publishing, 624 pp.

Van Velzen, W.T. and C.S. Robbins. 1969. The Breeding Bird Survey, 1969. United States Fish and Wildlife Service, Migratory Bird Populations Station Administrative Report, Laurel, Maryland.

Varland, D.E. and E.E. Klaas. 1991. Development of foraging behavior in the American Kestrel. Journal of Raptor Research 25(1): 9-17.

Vaudry, R. 1994. Suivi des populations des Sternes pierregarins, des Sternes arctiques, des Mouettes tridactyles et des Macareux moine dans la réserve de parc national de l'Archipel-de-Migan, réalisé en 1992-1993. Parks Canada, Service de la conservation des ressources naturelles, district de Mingan, 27 pp.

Veen, J. 1980. Breeding behaviour and breeding success of a colony of Little Gulls *Larus minutus* in the Netherlands. Limosa 53: 73-83.

Vehrencamp, S.L. 1977. Relative fecundity and parental effort in communally nesting Anis, *Crotophaga sulcirostris*. Science 197: 403-405.

Veiga, J.P. 1990a. A comparative study of reproductive adaptations in House and Tree sparrows. Auk 107: 45-59.

Veiga, J.P. 1990b. Infanticide by male and female House Sparrows. Animal Behaviour 39: 496-502.

Veiga, J.P. 1990c. Sexual conflict in the House Sparrow: interference between polygynously mated females versus asymmetric male investment. Behavioral Ecology and Sociobiology 27: 345-350.

Veiga, J.P. 1992. Why are House Sparrows predominantly monogamous? A test of hypotheses. Animal Behaviour 43: 361-370.

Veillette, J.J. 1988. Déglaciation et évolution des lacs proglaciaires post-Algonquin et Barlow au Témiscamingue, Québec et Ontario. Géographie physique et quaternaire 42: 7-31.

Vennor, H.G. Jr. 1860. Notes on birds wintering in and around Montreal. From observations taken during the winters of 1856-57-58-59-60. Canadian Naturalist and Geologist 5(6): 425-430.

Verbeek, N.A.M. 1967. Breeding biology and ecology of the Horned Lark in alpine tundra. Wilson Bulletin 79(2): 208-218.

Verbeek, N.A.M. 1970. Breeding ecology of the Water Pipit. Auk 87(3): 425-451.

Verbeek, N.A.M. 1981. Nesting success and orientation of Water Pipits *Anthus spinoletta* nests. Ornis Scandinavica 12(1): 37-39.

Verbeek, N.A.M. and R.W. Butler. 1981. Cooperative breeding of the Northwestern Crow, *Corvus caurinus*, in British Columbia. Ibis 123: 183-189.

Verbeek, N.A.M. and P. Hendricks. 1994. American Pipit (*Anthus rubescens*), *in* A. Poole and F. Gill, eds. The birds of North America, no. 95. The Academy of Natural Sciences, Philadelphia. The American Ornithologists' Union, Washington, 24 pp.

Vermeer, K. 1970. Some aspects of the nesting of ducks on islands in Lake Newell, Alberta. Journal of Wildlife Management 34(1): 126-129.

Vermeer, K. 1973. Some aspects of the breeding and mortality of Common Loons in east-central Alberta. Canadian Field-Naturalist 87: 403-408.

Vermeer, K. 1981. Food and population of Surf Scoters in British Columbia. Wildfowl 32: 107-116.

Vermeer, K. 1982. Food and distribution of the three *Bucephala* species in British Columbia waters. Wildfowl 33: 22-30.

Vermeer, K. and N. Bourne. 1984. The White-winged Scoter diet in British Columbia waters: resource partitioning with other scoters, pp. 30-38 *in* D.N. Nettleship, G.A. Sanger, and P.F. Springer, eds. Marine birds: their feeding ecology and commercial fisheries relationships. Canadian Wildlife Service, Ottawa, 200 pp.

Vermeer, K. and C.D. Levings. 1977. Populations, biomass and food habits of ducks on the Fraser delta intertidal area, British Columbia. Wildfowl 28: 49-60.

Verner, J. 1965. Breeding biology of the Long-billed Marsh Wren. Condor 67: 6-30.

Verner, J. and G.H. Engelsen. 1970. Territories, multiple nest building and polygyny in the Long-billed Marsh Wren. Auk 87(3): 557-567.

Verner, J. and M.F. Willson. 1966. The influence of habitats on mating systems of North American passerine birds. Ecology 47(1): 143-144.

Verner, J. and M.F. Willson. 1969. Mating system, sexual dimorphism and the role of male North American passerine birds in the nesting cycle. The American Ornithologists' Union, Washington. Ornithological Monographs no. 9, 76 pp.

Verrill, A.E. 1862. Notes on the natural history of Anticosti [and] Catalogue of the birds observed at Anticosti and vicinity. Proceedings. Boston Society of Natural History 9: 132-143.

Vesall, D.B. 1940. Notes on nesting habits of the American Bittern. Wilson Bulletin 52(3): 207-208.

Vézina, B. 1975. Écologie de la Gélinotte huppée (*Bonasa umbellus*) au Québec. M.Sc. thesis, Université Laval, Sainte-Foy, 126 pp.

Via, J.W. 1979. Foraging tactics of flycatchers in southwestern Virginia, pp. 191-202 *in* J.G. Dickson, R.N. Conner, R.R. Fleet, J.A. Jackson, and J.C. Kroll, eds. The role of insectivorous birds in forest ecosystems. Academic Press, London, 381 pp.

Vickery, P.D. 1988. Distribution and population status of Harlequin Duck (*Histrionicus histrionicus*) wintering in eastern North America. Wilson Bulletin 100(1): 119-126.

Vieillot, L.J.P. 1807. Histoire naturelle des oiseaux de l'Amérique septentrionale, contenant un grand nombre d'espèces décrites ou figurées pour la première fois, tomes 1 et 2. Desray Libraire, Paris.

Vieillot, L.J.P. 1816. Analyse d'une nouvelle ornithologie élémentaire. Deterville, Paris.

Vieillot, L.J.P. 1817. Observations sur la classification méthodique des oiseaux et remarques sur l'analyse d'une nouvelle ornithologie élémentaire. G. Dufour, Amsterdam.

Village, A. 1987. Numbers, territory-size and turnover of Short-eared Owls *Asio flammeus* in relation to vole abundance. Ornis Scandinavica 18(3): 198-204.

Village, A. 1990. The Kestrel. Poyser, London, 352 pp.

Villard, M.-A. 1990. Les oiseaux forestiers du parc de la Gatineau: une perspective régionale. QuébecOiseaux 2(4): 14-18.

Villard, M.-A. 1991. Spatio-temporal dynamics of forest bird patch populations in agricultural landscapes. Ph.D. dissertation, Carleton University, Ottawa, 170 pp.

Villard, M.-A. and R.A. Bracken. 1989. First confirmed breeding record of Le Conte's Sparrow, *Ammodramus leconteii*, in eastern Ontario. Canadian Field-Naturalist 103(1): 89-90.

Villard, M.-A., K.E. Freemark, and G. Merriam. 1992. Metapopulation theory and neotropical migrant birds in temperate forests: an empirical investigation, pp. 474-482 *in* J.M. Hagan III and D.W. Johnston, eds. Ecology and conservation of neotropical migrant landbirds. Smithsonian Institution Press, Washington.

Villard, M.-A., P.R. Martin, and C.G. Drummond. 1993. Habitat fragmentation and pairing success in the Ovenbird (*Seiurus aurocapillus*). Auk 110(4): 759-768.

Villard, M.-A., G. Merriam, and B.A. Maurer. 1995. Dynamics in subdivided populations of neotropical migratory birds in a fragmented temperate forest. Ecology 76(1): 27-40.

Villard, P. and C.W. Beninger. 1993. Foraging behavior of male Black-backed and Hairy woodpeckers in a forest burn. Journal of Field Ornithology 64(1): 71-76.

Vincent, J. and J. Bédard. 1976. Fat reserves in Snow Buntings. Canadian Journal of Zoology 54: 1051-1063.

Vincent, J.-S. 1989. Le Quaternaire du sud-est du Bouclier canadien, pp. 266-295 *in* R.J. Fulton, ed. Le Quaternaire du Canada et du Groenland, vol. 1. Geological Survey of Canada, Geology of Canada, 907 pp.

Vincent, J.-S. and L. Hardy. 1979. The evolution of glacial Lakes Barlow and Ojibway, Quebec and Ontario. Geological Survey of Canada. Bulletin no. 316, 18 pp.

Vinicombe, K. 1980. Tern showing mixed characters of Black Tern and White-winged Black Tern. British Birds 73(5): 223-225.

Viverette, C.B. 1993. 33. Oak-maple slope forest, p. 54 *in* J.D. Lowe, ed. Resident bird counts 1992. Journal of Field Ornithology 64(1)(supplement): 1-128.

Voous, K.H. 1989. Owls of the Northern Hemisphere. MIT Press, Cambridge, Massachusetts, 320 pp.

Voronetsky, V.I. 1987. Some features of Long-eared Owl ecology and behavior: mechanisms maintaining territoriality, pp. 229-230 *in* R.W. Nero, R.J. Clark, R.J. Knapton, and R.H. Hamre, eds. Proceedings. Biology and conservation of northern forest owls. Symposium, 3-7 February, Winnipeg. United States Department of Agriculture, Forest Service, Rocky Mountain Forest and Range Experiment Station, Fort Collins, Colorado. General Technical Report RM-142, 309 pp.

Vriends, M.M. 1984. Guide to pet birds. Simon and Schuster, New York, 319 pp.

Waite, T.A. 1987. Vigilance in the White-breasted Nuthatch: effects of dominance and sociality. Auk 104(3): 429-434.

Waite, T.A. 1988. A field test of density-dependent survival of simulated Gray Jay caches. Condor 90(1): 247-249.

Waite, T.A. 1991. Economics and consequences of scatterhoarding in Gray Jays (*Perisoreus canadensis*). Ph.D. dissertation, Ohio State University, Columbus.

Waite, T.A. 1992. Winter fattening in Gray Jays: seasonal, diurnal and climatic correlates. Ornis Scandinavica 23(4): 499-503.

Waite, T.A. and T.C. Grubb, Jr. 1988. Diurnal caching rhythm in captive White-breasted Nuthatches *Sitta carolinensis*. Ornis Scandinavica 19(1): 68-70.

Walker, B.H. 1992. Biodiversity and ecological redundancy. Conservation Biology 6: 18-23.

Walkinshaw, L.H. 1935. Studies of the Short-billed Marsh Wren (*Cistothorus stellaris*) in Michigan. Auk 52: 362-369.

Walkinshaw, L.H. 1939. The Yellow Rail in Michigan. Auk 56: 227-237.

Walkinshaw, L.H. 1940. Summer life of the Sora rail. Auk 57(2): 153-168.

Walkinshaw, L.H. 1941. The Prothonotary Warbler, a comparison of nesting conditions in Tennessee and Michigan. Wilson Bulletin 53(1): 3-21.

Walkinshaw, L.H. 1949a. The Sandhill Cranes. Cranbrook Institute of Science, Bloomfield Hills, Michigan, 202 pp.

Walkinshaw, L.H. 1949b. Twenty-five eggs apparently laid by a cowbird. Wilson Bulletin 61(2): 82-85.

Walkinshaw, L.H. 1953. Life-history of the Prothonotary Warbler. Wilson Bulletin 65(3): 152-168.

Walkinshaw, L.H. 1957a. Incubation period of the Sora rail. Auk 74(4): 496.

Walkinshaw, L.H. 1957b. Louisiana Waterthrush, pp. 148-149 *in* L. Griscom and A. Sprunt, Jr., eds. The warblers of America. Doubleday and Company, New York, 302 pp.

Walkinshaw, L.H. 1966a. Studies of the Acadian Flycatcher in Michigan. Bird-Banding 37(4): 227-257.

Walkinshaw, L.H. 1966b. Summer biology of Traill's Flycatcher. Wilson Bulletin 78(1): 31-46.

Walkinshaw, L.H. 1968a. Eastern Field Sparrow, pp. 1217-1235 *in* A.C. Bent. Life histories of North American cardinals, grosbeaks, buntings, towhees, finches, sparrows, and allies. (O.L. Austin Jr., ed.). United States National Museum, Washington. [1968, Dover Publications, New York]. Bulletin no. 237, vol. 1, pp. 1-602; vol. 2, pp. 603-1248; vol. 3, pp. 1249-1889.

Walkinshaw, L.H. 1968b. Le Conte's Sparrow, pp. 765-776 *in* A.C. Bent. Life histories of North American cardinals, grosbeaks, buntings, towhees, finches, sparrows, and allies. (O.L. Austin Jr., ed.). United States National Museum, Washington. [1968, Dover Publications, New York]. Bulletin no. 237, vol. 1, pp. 1-602; vol. 2, pp. 603-1248; vol. 3, pp. 1249-1889.

Walkinshaw, L.H. 1991a. Yellow Rail, pp. 536-537 *in* R. Brewer, G.A. McPeek, and R.J. Adams, Jr., eds. The Atlas of breeding birds of Michigan. Michigan State University Press, East Lansing, 594 pp.

Walkinshaw, L.H. 1991b. Prothonotary Warbler, pp. 430-431 *in* R. Brewer, G.A. McPeek, and R.J. Adams, Jr., eds. The Atlas of breeding birds of Michigan. Michigan State University Press, East Lansing, 594 pp.

Walkinshaw, L.H. and C.J. Henry. 1957. Yellow-bellied Flycatcher nesting in Michigan. Auk 74(3): 293-304.

Wallace, [?]. 1881. The migratory quail. Canadian Sportsman and Naturalist 1: 75.

Wallace, G.E. 1991. Noteworthy bird records from southernmost Chile. Condor 93: 175-176.

Wallace, G.J. 1939. Bicknell's Thrush: its taxonomy, distribution and life history. Proceedings. Boston Society of Natural History 41(6): 211-402.

Wallace, G.J. 1949. Bicknell's Thrush, pp. 199-217 *in* A.C. Bent. Life histories of North American thrushes, kinglets, and their allies. United States National Museum, Washington. [1964, Dover Publications, New York]. Bulletin no. 196, 452 pp.

Wallace, G.J. 1969. Endangered and declining species of Michigan birds. Jack-Pine Warbler 47: 70-75.

Wallace, G.J. and H.D. Mahan. 1975. An introduction to ornithology, 3rd edition. MacMillan Publishing, New York, 546 pp.

Wallace, G.J., W.P. Nickell, and R.F. Bernard. 1961. Bird mortality in the Dutch Elm disease program in Michigan. Cranbrook Institute of Science, Bloomfield Hills, Michigan. Bulletin 41, 44 pp.

Walley, W.J. 1989. Breeding of the Scarlet Tanager, *Piranga olivacea*, in western Manitoba. Canadian Field-Naturalist 103(4): 572-576.

Wander, S.A. 1985. Observations at a Northern Waterthrush nest. Journal of Field Ornithology 56(1): 69.

Wardle, J.S. 1974. Alpine timberlines, pp. 371-402 *in* J.D. Ives and R. Barry, eds. Arctic and alpine environments. Methuen, London.

Warkentin, I.G., P.C. James, and L.W. Oliphant. 1991. Influence of site fidelity on mate switching in urban-breeding Merlins (*Falco columbarius*). Auk 108(2): 294-302.

Warne, F.L. 1926. Crows is Crows. Bird-Lore 28: 110-116.

Warner, R.E. 1992. Nest ecology of grassland passerines on road rights-of-way in central Illinois. Biological Conservation 59: 1-7.

Warren, B.H. 1888. Report on the birds of Pennsylvania. Edwin K. Meyers, State Printer, Harrisburg, 260 pp.

Warren, W.H. 1970. Mute Swans in Victoria. Victoria Naturalist 27: 11.

Washington, D. 1986. Treecreeper attaching faecal sacs to pole. British Birds 79(1): 43.

Wasserman, F.E. 1982. The spatial distribution of singing and foraging in the territories of White-throated Sparrows. Canadian Journal of Zoology 60(11): 2757-2760.

Wasserman, F.E. 1983. Territories of Rufous-sided Towhees contain more than minimal food resources. Wilson Bulletin 95(4): 664-667.

Watson, A. 1957. The behaviour, breeding and food ecology of the Snowy Owl, *Nyctea scandiaca*. Ibis 99: 419-462.

Watson, A. 1986. Notes on the sexual cycle of some Baffin Island birds. Canadian Field-Naturalist 100(3): 396-397.

Watson, M.D., G.J. Robertson, and F. Cooke. 1993. Egg-laying time and laying interval in the Common Eider. Condor 95: 869-878.

Watt, D.J. 1983. Plumage coloration and dominance behavior in three species of sparrows of the genus *Zonotrichia*. Ph.D. dissertation, University of Oklahoma, Norman, 131 pp.

Watt, D.J. 1991. Correlates to mate pairing and breeding success in the American Goldfinch (*Carduelis tristis*). Abstracts of scientific papers presented at the 109th stated meeting of the American Ornithologists' Union, Montréal, no. 38.

Wayre, P. and G.F. Jolly. 1958. Notes on the breeding of the Iceland Gyrfalcon. British Birds 51(8): 285-290.

Weatherhead, P.J. 1980a. Study of the ecology and behavioural patterns of the Red-winged Blackbird populations in the Beauharnois roost. Report presented to Agriculture Canada, 254 pp.

Weatherhead, P.J. 1980b. An investigation to assess alternatives for reducing damage by Red-winged Blackbirds to agricultural crops. Final Report, Department of Renewable Resources, MacDonald Campus of McGill University, Sainte-Anne-de-Bellevue, 134 pp.

Weatherhead, P.J. 1987a. Common Grackle, pp. 480-481 *in* M.D. Cadman, P.F.J. Eagles, and F.M. Helleiner, eds. Atlas of the breeding birds of Ontario. University of Waterloo Press, Waterloo, 617 pp.

Weatherhead, P.J. 1987b. Red-winged Blackbird, pp. 468-469 *in* M.D. Cadman, P.F.J. Eagles, and F.M. Helleiner, eds. Atlas of the breeding birds of Ontario. University of Waterloo Press, Waterloo, 617 pp.

Weatherhead, P.J. 1987c. Bobolink, pp. 466-467 *in* M.D. Cadman, P.F.J. Eagles, and F.M. Helleiner, eds. Atlas of the breeding birds of Ontario. University of Waterloo Press, Waterloo, 617 pp.

Weatherhead, P.J. and J.R. Bider. 1979. Management options for blackbird problems in agriculture. Phytoprotection 60(3): 145-155.

Weatherhead, P.J. and R.J. Robertson. 1980. Sexual recognition and anticuckoldry behavior in Savannah Sparrows. Canadian Journal of Zoology 58(6): 991-996.

Weatherhead, P.J., R.G. Clark, J.R. Bider, and R.D. Titman. 1980. Movements of blackbirds and starlings in southwestern Quebec and eastern Ontario in relation to crop damage and control. Canadian Field-Naturalist 94(1): 75-79.

Weatherhead, P.J., H. Greenwood, and R.D. Titman. 1985. Numerical responses of Red-winged Blackbird populations to changes in regional land-use patterns. Canadian Journal of Zoology 64: 1944-1950.

Weaver, F.G. 1949. Wood Thrush, pp. 101-123 *in* A.C. Bent. Life histories of North American thrushes, kinglets, and their allies. United States National Museum, Washington. [1964, Dover Publications, New York]. Bulletin no. 196, 452 pp.

Weaver, R.L. 1942. Growth and development of English Sparrows. Wilson Bulletin 54(3): 183-191.

Weaver, R.L. 1943. Reproduction in English Sparrows. Auk 60(1): 62-74.

Weaver, R.L. and F.H. West. 1943. Notes on the breeding of the Pine Siskin. Auk 60(4): 492-504.

Webb, R. and J.R. King. 1984. Nests and microclimates of the White-crowned Sparrow: a thermal analysis on geographic gradient. American Zoologist 24(3): 142a.

Webb, W.L., D.F. Behrend, and B. Saisorn. 1977. Effect of logging on songbird populations in a northern hardwood forest. Wildlife Monographs no. 55, 35 pp.

Webster, J.M. 1964. Fall foods of Soras from two habitats in Connecticut. Journal of Wildlife Management 28(1): 163-165.

Weeden, J.S. 1965. Territorial behavior of the Tree Sparrow. Condor 67(3): 193-209.

Weeden, R.B. 1963. Management of ptarmigan in North America. Journal of Wildlife Management 27(4): 673-683.

Weeden, R.B. 1969. Foods of Rock and Willow ptarmigan in central Alaska with comments on interspecific competition. Auk 86(2): 271-281.

Weeden, R.B. and J.B. Theberge. 1972. The dynamics of a fluctuating population of Rock Ptarmigan in Alaska, pp. 90-106 *in* K.W. Voous, ed. Proceedings. XVth International Ornithological Congress, The Hague, 30 August-5 September 1970. E.J. Brill, Leiden, 745 pp.

Weeks, H.P. Jr. 1977. Nest reciprocity in Eastern Phoebes and Barn Swallows. Wilson Bulletin 89(4): 632-635.

Weeks, H.P. Jr. 1978. Clutch size variation in the Eastern Phoebe in southern Indiana. Auk 95(4): 656-666.

Weeks, H.P. Jr. 1979. Nesting ecology of the Eastern Phoebe in southern Indiana. Wilson Bulletin 91(3): 441-454.

Weeks, H.P. Jr. 1994. Eastern Phoebe (*Sayornis phoebe*), *in* A. Poole and F. Gill, eds. The birds of North America, no. 94. The Academy of Natural Sciences, Philadelphia. The American Ornithologists' Union, Washington, 20 pp.

Wein, R.W. and D.A. MacLean. 1983. An overview of fire in northern ecosystems, pp. 1-18 *in* R.W. Wein and D.A. MacLean, eds. The role of fire in northern circumpolar ecosystems. Scope, John Wiley and Sons.

Weir, D. and A. Hanson. 1989. Food habits of Great Horned Owls, *Bubo virginianus*, in the northern taiga of the Yukon Territory and Alaska. Canadian Field-Naturalist 103(1): 12-17.

Weir, R.D. 1973. Snowy Owl invasion of Wolf Island, winter 72. Ontario Field Biologists 27: 3-17.

Weir, R.D. 1987a. Eastern Screech-Owl, pp. 204-205 *in* M.D. Cadman, P.F.J. Eagles, and F.M. Helleiner, eds. Atlas of the breeding birds of Ontario. University of Waterloo Press, Waterloo, 617 pp.

Weir, R.D. 1987b. Northern Goshawk, pp. 118-119 *in* M.D. Cadman, P.F.J. Eagles, and F.M. Helleiner, eds. Atlas of the breeding birds of Ontario. University of Waterloo Press, Waterloo, 617 pp.

Weir, R.D. 1987c. Osprey, pp. 108-109 *in* M.D. Cadman, P.F.J. Eagles, and F.M. Helleiner, eds. Atlas of the breeding birds of Ontario. University of Waterloo Press, Waterloo, 617 pp.

Weir, R.D. 1987d. Red-tailed Hawk, pp. 122-123 *in* M.D. Cadman, P.F.J. Eagles, and F.M. Helleiner, eds. Atlas of the breeding birds of Ontario. University of Waterloo Press, Waterloo, 617 pp.

Weir, R.D. 1987e. Short-eared Owl, pp. 214-215 *in* M.D. Cadman, P.F.J. Eagles, and F.M. Helleiner, eds. Atlas of the breeding birds of Ontario. University of Waterloo Press, Waterloo, 617 pp.

Weir, R.D. 1987f. Northern Hawk-Owl, p. 542 *in* M.D. Cadman, P.F.J. Eagles, and F.M. Helleiner, eds. Atlas of the breeding birds of Ontario. University of Waterloo Press, Waterloo, 617 pp.

Weir, R.D., F. Cooke, M.H. Edwards, and R.B. Stewart. 1980. Fall migration of Saw-whet Owls at Prince Edward Point, Ontario. Wilson Bulletin 92(4): 475-488.

Weller, M.W. 1958. Observations on the incubation behaviour of a Common Nighthawk. Auk 75: 48-59.

Weller, M.W. 1959. Parasitic egg laying in the Redhead (*Aythya americana*) and other North American Anatidae. Ecological Monographs 29: 333-365.

Weller, M.W. 1961. Breeding biology of the Least Bittern. Wilson Bulletin 73(1): 11-35.

Weller, M.W. 1964. Distribution and migration of the Redhead. Journal of Wildlife Management 28(1): 64-103.

Weller, M.W. 1965a. Bursa regression, gonad cycle, and molt of the Great Horned Owl. Bird-Banding 38(2): 102-112.

Weller, M.W. 1965b. Chronology of pair formation in some nearctic *Aythya* (Anatidae). Auk 82(2): 227-235.

Weller, M.W. 1967. Courtship of the Redhead (*Aythya americana*). Auk 84(4): 544-559.

Weller, M.W. 1971. Robin mortality in relation to Dutch Elm disease control programs on the Iowa State University campus. Iowa State Journal of Science 45(3): 471-475.

WELLER, M.W. 1979. Density and habitat relationships of Blue-winged Teal nesting in northwestern Iowa. Journal of Wildlife Management 43(2): 367-374.

WELLER, M.W. and L.H. FREDRICKSON. 1973. Avian ecology of a managed glacial marsh. Living Bird 12: 269-291.

WELLER, M.W., D.L. TRAUGER, and G.L. KRAPU. 1969. Breeding birds of the West Mirage Islands, Great Slave Lake, N.W.T. Canadian Field-Naturalist 83(4): 344-360.

WELLS, J. 1981. An observation of vocal mimicry in Pine Grosbeak. Osprey 12(3): 82-84.

WELLS, J.V. and P.D. VICKERY. 1990. Willet nesting in sphagnum bog in eastern Maine. Journal of Field Ornithology 61(1): 73-75.

WELSH, D.A. 1971. Breeding and territoriality of the Palm Warbler in a Nova Scotia bog. Canadian Field-Naturalist 85(1): 31-37.

WELSH, D.A. 1975. Savannah Sparrows breeding and territoriality on a Nova Scotia dune beach. Auk 92(1): 235-251.

WELSH, D.A. 1987a. Blackpoll Warbler, p. 551 *in* M.D. Cadman, P.F.J. Eagles, and F.M. Helleiner, eds. Atlas of the breeding birds of Ontario. University of Waterloo Press, Waterloo, 617 pp.

WELSH, D.A. 1987b. Cape May Warbler, pp. 376-377 *in* M.D. Cadman, P.F.J. Eagles, and F.M. Helleiner, eds. Atlas of the breeding birds of Ontario. University of Waterloo Press, Waterloo, 617 pp.

WELSH, D.A. 1987c. Magnolia Warbler, pp. 374-375 *in* M.D. Cadman, P.F.J. Eagles, and F.M. Helleiner, eds. Atlas of the breeding birds of Ontario. University of Waterloo Press, Waterloo, 617 pp.

WELSH, D.A. 1987d. Palm Warbler, pp. 392-393 *in* M.D. Cadman, P.F.J. Eagles, and F.M. Helleiner, eds. Atlas of the breeding birds of Ontario. University of Waterloo Press, Waterloo, 617 pp.

WELSH, D.A. 1987e. Tennessee Warbler, pp. 364-365 *in* M.D. Cadman, P.F.J. Eagles, and F.M. Helleiner, eds. Atlas of the breeding birds of Ontario. University of Waterloo Press, Waterloo, 617 pp.

WELSH, D.A. and F.M. HELLEINER. 1987. Orange-crowned Warbler, p. 550 *in* M.D. Cadman, P.F.J. Eagles, and F.M. Helleiner, eds. Atlas of the breeding birds of Ontario. University of Waterloo Press, Waterloo, 617 pp.

WELTER, W.A. 1935. The natural history of the Long-billed Marsh Wren. Wilson Bulletin 47: 3-34.

WELTY, J.C. 1975. The life of birds, 2nd edition. W.B. Saunders Company, Philadelphia, 623 pp.

WENDT, J.S. and C. HYSLOP. 1981. Migratory birds killed in Canada during the 1980 season. Canadian Wildlife Service. Progress Notes no. 126, 42 pp.

WENDT, J.S. and E. SILIEFF. 1986. The kill of eiders and other sea ducks by hunters in eastern Canada, pp. 147-154 *in* A. Reed, ed. Eider ducks in Canada/Les eiders au Canada. Canadian Wildlife Service. Report Series no. 47, 177 pp.

WENDT, J.S., F.G. COOCH, and K.L. NEWELL. 1978. Species of waterfowl killed in Canada during the 1976 and 1977 seasons. Canadian Wildlife Service. Progress Notes no. 92, 27 pp.

WENGER, J.D. 1991. American Crow, p. 316 *in* R. Brewer, G.A. McPeek, and R.J. Adams, Jr., eds. The Atlas of breeding birds of Michigan. Michigan State University Press, East Lansing, 594 pp.

WENSTROM, W.P. 1974. Habitat selection by brood rearing American Woodcock, pp. 1-19 *in* J.W. Artmann and S.R. Pursglove, eds. Proceedings. Fifth American Woodcock workshop, Athens.

WERLER, E. and E.C. FRANKS. 1975. Some unusual nest sites of the House Sparrow. Wilson Bulletin 87(1): 113.

WERSCHKUL, D.F., E. MCMAHON, and M. LEIDSCHUH. 1976. Some effects of human activities on the Great Blue Heron in Oregon. Wilson Bulletin 88: 660-662.

WESELOH, D.V. 1987a. Double-crested Cormorant, pp. 44-45 *in* M.D. Cadman, P.F.J. Eagles, and F.M. Helleiner, eds. Atlas of the breeding birds of Ontario. University of Waterloo Press, Waterloo, 617 pp.

WESELOH, D.V. 1987b. Glaucous Gull, p. 539 *in* M.D. Cadman, P.F.J. Eagles, and F.M. Helleiner, eds. Atlas of the breeding birds of Ontario. University of Waterloo Press, Waterloo, 617 pp.

WESELOH D.V. 1987c. Little Gull, pp. 178-179 *in* M.D. Cadman, P.F.J. Eagles, and F.M. Helleiner, eds. Atlas of the breeding birds of Ontario. University of Waterloo Press, Waterloo, 617 pp.

WESELOH, D.V. and R.T. BROWN. 1971. Plant distribution within a heron rookery. American Midland Naturalist 86: 57-64.

WESELOH, D.V. and P. MINEAU. 1986. Apparent hybrid Common Black-headed Gull nesting in Lake Ontario. American Birds 40(1): 18-20.

WEST, G.C. 1973. Foods eaten by Tree Sparrows in relation to availability during summer in northern Manitoba. Arctic 26(1): 7-22.

WEST, G.C. and L.J. PEYTON. 1972. The spring migration of the Tree Sparrow through southern Yukon Territory. Bird-Banding 43(4): 241-246.

WEST, J.D. and J.M. SPEIRS. 1959. The 1956-1957 invasion of Three-toed Woodpeckers. Wilson Bulletin 71(4): 348-352.

WEST, N.E. 1993. Biodiversity of rangelands. Journal of Range Management 46: 2-13.

WESTNEAT, D.F. 1987a. Extra-pair copulations in a predominantly monogamous bird: observations of behaviour. Animal Behaviour 35: 865-876.

WESTNEAT, D.F. 1987b. Extra-pair fertilizations in a predominantly monogamous bird: genetic evidence. Animal Behaviour 35: 877-886.

WESTNEAT, D.F. 1988a. Male parental care and extra-pair copulations in the Indigo Bunting. Auk 105(1): 149-160.

WESTNEAT, D.F. 1988b. The relationships among polygyny, male parental care and female breeding success in the Indigo Bunting. Auk 105(2): 372-374.

WETHERBEE, D.K. 1968. Southern Swamp Sparrow, pp. 1475-1490 *in* A.C. Bent. Life histories of North American cardinals, grosbeaks, buntings, towhees, finches, sparrows, and allies. (O.L. Austin, Jr., ed.). United States National Museum, Washington. [1968, Dover Publications, New York]. Bulletin no. 237, vol. 1, pp. 1-602; vol. 2, pp. 603-1248; vol. 3, pp. 1249-1889.

WETMORE, A. 1916. Birds of Puerto Rico. United States Department of Agriculture, Washington. Technical Bulletin no. 326, 140 pp.

WETMORE, A. 1925. Food of American phalaropes, avocets, and stilts. United States Department of Agriculture, Washington. Department Bulletin no. 1359, 20 pp.

WETMORE, A. 1935. Shadowy birds of the night. National Geographic Magazine 67: 217-240.

WETMORE, S.P. and D.I. GILLESPIE. 1976. Osprey and Bald Eagle populations in Labrador and northeastern Quebec, 1969-1973. Canadian Field-Naturalist 90(3): 330-337.

WETTON, J.H. and D.T. PARKIN. 1991. An association between fertility and cuckoldry in the House Sparrow, *Passer domesticus*. Proceedings. Royal Society of London. Series B. Biological Sciences, pp. 227-233.

WETTON, J.H., R.E. CARTER, D.T. PARKIN, and D. WALTERS. 1987. Demographic study of a wild House Sparrow population by DNA fingerprinting. Nature 327: 147-149.

WHEELWRIGHT, N.T. 1986. The diet of American Robins: an analysis of United States Biological Survey records. Auk 103(4): 710-725.

WHEELWRIGHT, N.T. 1988. Seasonal changes in food preferences of American Robins in captivity. Auk 105(2): 374-378.

WHEELWRIGHT, N.T. and J.D. RISING. 1993. Savannah Sparrow (*Passerculus sandwichensis*), *in* A. Poole and F. Gill, eds. The birds of North America, no. 45. The Academy of Natural Sciences, Philadelphia. The American Ornithologists' Union, Washington, 28 pp.

WHELAN, C.J. 1989. An experimental test of prey distribution learning in two paruline warblers. Condor 91(1): 113-199.

WHITE, C.M. 1968. Diagnosis and relationships of the North American tundra-inhabiting Peregrine Falcons. Auk 85(2): 179-191.

WHITE, C.M. and D.A. BOYCE. 1988. An overview of Peregrine Falcon subspecies, pp. 789-810 *in* T.J. Cade, J.H. Enderson, C.G. Thelander, and C.M. White, eds. Peregrine Falcon populations: their management and recovery. The Peregrine Fund, Boise, Idaho, 949 pp.

WHITE, C.M. and T.J. CADE. 1971. Cliff-nesting raptors and Ravens along the Colville River in Arctic Alaska. Living Bird 10: 107-150.

WHITE, C.M. and R.B. WEEDEN. 1966. Hunting methods of Gyrfalcons and behaviour of their prey (ptarmigan). Condor 68: 517-519.

WHITE, D.W. and E.W. STILES. 1991. Fruit harvesting by American Robins: influence of fruit size. Wilson Bulletin 103(4): 690-692.

WHITE, E.F.G. and H.F. LEWIS. 1937. The Greater Snow Goose in Canada. Auk 54: 440-444.

WHITE, H.C. 1939. Change in gastric digestion of kingfishers with development. American Naturalist 73: 188-190.

WHITE, H.C. 1953. The Eastern Belted Kingfisher in the Maritime provinces. Fisheries Research Board of Canada, Ottawa. Bulletin no. 97.

WHITE, R.P. 1988. Wintering grounds and migration patterns of the Upland Sandpiper. American Birds 42(5): 1247-1253.

WHITEAVES, J.F. 1870. Notes on Canadian birds. Canadian Naturalist 5: 230-231.

WHITFIELD, D.W.A., J.M. GERRARD, W.J. MAHER, and D.W. DAVIS. 1974. Bald Eagle nesting habitat, density, and reproduction in central Saskatchewan and Manitoba. Canadian Field-Naturalist 88(4): 399-407.

WHITMORE, R.C. 1979. Short-term change in vegetation structure and its effect on Grasshopper Sparrows in West Virginia. Auk 96: 621-625.

WHITMORE, R.C. 1981. Structural characteristics of Grasshopper Sparrow habitat. Journal of Wildlife Management 45(3): 811-814.

WHITNEY, C.L. 1991. Avoidance of song matching in the Wood Thrush: a field experiment. Wilson Bulletin 103(1): 96-101.

WHITNEY, C.L. 1992. Temporal stability of song in a local population of Wood Thrushes. Wilson Bulletin 104(3): 516-520.

WHITTAKER, R.H. 1977. Evolution of species diversity in land communities, pp. 1-67 *in* M.K. Hecht, W.C. Steere, and B. Wallace, eds. Evolutionary biology, vol. 10. Plenum, New York.

WHITTAM, R.F. 1971. Intraspecific relationships of *Melospiza georgiana* and *M. melodia*. M.Sc. thesis, Bishops University, Lennoxville, 37 pp.

WHITTINGHAM, L.A. and R.J. ROBERTSON. 1993. Nestling hunger and parental care in Red-winged Blackbirds. Auk 110(2): 240-246.

WHITTLE, C.L. 1922. Additional data regarding the famous Arnold Arboretum Mockingbird. Auk 39: 496-506.

WIEMEYER, S.N. and D.W. SPARLING. 1991. Acute toxicity of four anticholinesterase insecticides to American Kestrels, Eastern Screech-Owls and Northern Bobwhites. Environmental Toxicology and Chemistry 10(9): 1139-1148.

WIENS, J.A. 1969. An approach to the study of ecological relationships among grassland birds. Ornithological Monographs no. 8, 93 pp.

WIENS, J.A. 1973. Interterritorial habitat variation in Grasshopper and Savannah sparrows. Ecology 54(4): 877-884.

WIENS, J.A. 1989. Spatial scaling in ecology. Functional Ecology 3:385-397.

WIENS, J.A., J.T. ROTENBERRY, and B.V. HORNE. 1986. A lesson in the limitations of field experiments: shrubsteppe birds and habitat alterations. Ecology 67(2): 365-376.

WIENS, T.P. and F.J. CUTHBERT. 1988. Nest-site tenacity and mate retention of Piping Plover. Wilson Bulletin 100(4): 545-553.

WIESE, J.H. 1978. Heron nest-site selection and its ecological effects, pp. 27-34 *in* A. Sprunt IV, J.C. Ogden, and S. Winckler, eds. Wading birds. Naturalist Audubon Society, New York. Research report no. 7.

WIGGINS, D.A. 1987. L'Hirondelle pourprée. Hinterland Who's Who. Environment Canada, Canadian Wildlife Service, 4 pp. [Also available in English as: Purple Martin.]

WIGGINS, D.A. 1991. Foraging success and aggression in solitary and group-feeding Great Egrets (*Casmerodius albus*). Colonial Waterbirds 14(2): 176-179.

WILBUR, S.R. 1983. The status of vultures in the Western Hemisphere, pp. 113-123 *in* S.R. Wilbur and J.A. Jackson, eds. Vulture biology and management. University of California Press, Los Angeles, 550 pp.

WILCOVE, D.S. 1985. Nest predation in forest tracts and the decline of migratory songbirds. Ecology 66(4): 1211-1214.

WILCOVE, D.S. and S.K. ROBINSON. 1990. The impact of forest fragmentation on bird communities in eastern North America, pp. 319-331 *in* A. Keast, ed. Biogeography and ecology of forest bird communities. SPB Academic Publishers, The Hague, Netherlands.

WILCOX, B.A. and D.D. MURPHY. 1985. Conservation strategy: the effects of fragmentation on extinction. American Naturalist 125: 879-887.

WILCOX, L. 1959. A twenty year banding study of the Piping Plover. Auk 76: 129-152.

WILDLIFE HABITAT CANADA. 1986. Le statut des habitats fauniques au Canada: problèmes-questions et possibilités. Wildlife Habitat Canada, Ottawa, 72 pp. [Also available in English as: The status of wildlife habitat in Canada: problems, issues and opportunities.]

WILEY, R.H. 1976. Communication and spatial relationships in a colony of Common Grackles. Animal Behaviour 24(3): 570-584.

WILK, R.J. 1988. Distribution, abundance, population structure and productivity of Tundra Swans in Bristol Bay, Alaska. Arctic 41(4): 288-292.

WILL, R.L. 1969. Fecundity, density, and movements of a House Sparrow population in southern Illinois. Ph.D. dissertation, University of Illinois, Urbana.

WILL, R.L. 1973. Breeding success, numbers, and movements of House Sparrows at McLeansboro, Illinois, pp. 60-78 *in* S.C. Kendeigh, ed. A symposium of the House Sparrow (*Passer domesticus*) and European Tree Sparrow (*P. montanus*) in North America. Ornithological Monographs no. 14, 121 pp.

WILLARD, D.E. 1977. The feeding ecology and behavior of five species of heron in southeastern New Jersey. Condor 79: 462-470.

WILLIAMS, G. 1978. Andrew Graham and Thomas Hutchins – collaboration and plagiarism in 18th century natural history. Beaver (spring 1978): 6-15.

WILLIAMS, G.P. 1954. Population fluctuations in some Northern Hemisphere game birds (Tetraonidae). Journal of Animal Ecology 23(1): 1-34.

WILLIAMS, J.B. 1975. Habitat utilization by four species of woodpeckers in a central Illinois woodland. American Midland Naturalist 93: 354-367.

WILLIAMS, J.B. and G.O. BATZLI. 1979. Interference competition and niche shifts in the bark-foraging guild in central Illinois. Wilson Bulletin 91(3): 400-411.

WILLIAMSON, F.S.L. 1968. Common Lapland Longspur, pp. 1597-1598 *in* A.C. Bent. Life histories of North American cardinals, grosbeaks, buntings, towhees, finches, sparrows, and allies. (O.L. Austin, Jr., ed.). United States National Museum, Washington. [1968, Dover Publications, New York]. Bulletin no. 237, vol. 1, pp. 1-602; vol. 2, pp. 603-1248; vol. 3, pp. 1249-1889.

WILLIAMSON, P. 1971. Feeding ecology of the Red-eyed Vireo (*Vireo olivaceus*) and associated foliage-gleaning birds. Ecological Monographs 41(2): 129-152.

WILLOUGHBY, E.J. and T.J. CADE. 1964. Breeding behaviour of the American Kestrel (Sparrow Hawk). Living Bird 3: 75-96.

WILLOUGHBY SOCIETY 1882. Forster's animals of Hudson's Bay. Philip Lutley Sclater, London.

WILMERS, T.J. 1987. Competition between starlings and kestrels for nest boxes: a review. Raptor Research 6: 156-159.

WILMORE, S.B. 1974. Swans of the world. David and Charles, London, 229 pp.

WILMOT, N.E. 1917. English Sparrow (*Passer domesticus*) feeding on the larva of the elm tree beetle. Auk 34: 479-480.

WILSON, A. 1808-1814. American ornithology or the natural history of birds of the United States. Bradford and Inskeep, Philadelphia, 9 vol.

WILSON, A. and C.L. BONAPARTE. 1831. American Ornithology, vol. 2. Edinburgh, 334 pp.

WILSON, C.V. 1971. Le climat du Québec, première partie – Atlas climatique [The climate of Québec, Part 1 (Climatic Atlas)]. Atmospheric Environment Service, Environment Canada. Climatological Studies no. 11, 8 pp.

WILTSCHKO, R. and W. WILTSCHKO. 1990. Zur Entwicklung der Sonnenkompaßorientierung bei jungen [The development of sun compass orientation in young homing pigeons]. Journal Für Ornithologie 131(1): 1-19.

WILTZ, K.J. and V. GIAMPA. 1978. Habitat use by Yellow-rumped Warblers at the northern extremities of their winter range. Wilson Bulletin 90(4): 566-574.

WING, L. 1943. Spread of the Starling and English Sparrow. Auk 60: 74-87.

WINKER, K., J.H. RAPPOLE, and M.A. RAMOS. 1990a. Population dynamics of the Wood Thrush in southern Veracruz, Mexico. Condor 92(2): 444-460.

WINKER, K., J.H. RAPPOLE, and M.A. RAMOS. 1990b. Within-forest preferences of Wood Thrushes wintering in the rainforest of southern Veracruz. Wilson Bulletin 102(4): 715-720.

WINKLER, D.W. 1992. Causes and consequences of variation in parental defense behavior by Tree Swallows. Condor 94(4): 502-520.

WINKLER, D.W. 1993. Use and importance of feathers as nest lining in Tree Swallows (*Tachycineta bicolor*). Auk 110(1): 29-36.

WINTER, T.C. and M.-K. WOO. 1990. Hydrology of lakes and wetlands, pp. 159-187 *in* M.G. Wolman and H.C. Riggs, eds. Surface water hydrology. Geological Society of America, Boulder, Colorado.

WINTLE, E.D. 1882. Ornithology of the island of Montreal. Canadian Sportsman and Naturalist 2(2): 108-110.

WINTLE, E.D. 1893. The birds of Montreal. Desbarats and Co., Montréal.

WINTLE, E.D. 1896. The birds of Montreal. W. Drysdale and Co., Montréal, 227 pp.

WISE, C.D. and A. WALLS. 1988. A preliminary study of the impact of the House Finch on populations of the House Sparrow in Indiana. Indiana Audubon Quarterly 66(1): 26-29.

WISHART, R.A. and J.R. BIDER. 1976. Habitat preferences of woodcock in southwestern Quebec. Journal of Wildlife Management 40(3): 523-531.

WISHART, R.A. and J.R. BIDER. 1977. Aspects of woodcock nocturnal activity in southwestern Quebec. Canadian Field-Naturalist 91(2): 141-147.

WISHART, R.A. and S.G. SEALY. 1980. Late summer time budget and feeding behaviour of Marbled Godwit (*Limosa fedoa*) in southern Manitoba. Canadian Journal of Zoology 58(7): 1277-1282.

WITHERBY, H., F.C.R. JOURDAIN, N.F. TICEHURST, and B.W. TUCKER. 1940. The handbook of British birds. H.F. and G. Witherby, London, 5 vol.

WITTENBERGER, J.F. 1978. The breeding biology of an isolated Bobolink population in Oregon. Condor 80(4): 355-371.

WITTENBERGER, J.F. 1980a. Feeding of secondary nestling by polygynous male Bobolinks in Oregon. Wilson Bulletin 92(3): 330-340.

WITTENBERGER, J.F. 1980b. Vegetation structure, food supply, and polygyny in Bobolinks (*Dolichonyx oryzivorus*). Ecology 61(1): 140-150.

WITTENBERGER, J.F. 1981. Animal social behavior. Duxbury Press, Boston, 722 pp.

WITTENBERGER, J.F. 1982. Factors affecting how male and female Bobolinks apportion parental investments. Condor 84(1): 22-39.

WOJNOWSKI, J.K. 1993. 31. Red Oak–Ironwood savannah, pp. 52-53 *in* J.D. Lowe, ed. Resident bird counts 1992. Journal of Field Ornithology 64(1)(supplement): 1-128.

WOJNOWSKI, J.K. and D.L. GOODYEAR. 1993. 30. Red Ash–Red Oak savannah, p. 52 *in* J.D. Lowe, ed. Resident bird counts 1992. Journal of Field Ornithology 64(1)(supplement): 1-128.

WOLDOW, N. 1972. Pigeon and man: a spotty old friendship. Natural History 81: 26-36.

WOLF, E.D. 1970. The impact of seasonnal flowering on the biology of some tropical hummingbirds. Condor 72: 1-14.

WOLF, L., E.D. KETTERSON, and V. NOLAN, JR. 1988. Paternal influence on growth and survival of Dark-eyed Junco young: do parental males benefit? Animal Behaviour 36(6): 1601-1618.

WOLFORD, J.W. and D.A. BOAG. 1971a. Distribution and biology of Black-crowned Night-Herons in Alberta. Canadian Field-Naturalist 85(1): 13-19.

WOLFORD, J.W. and D.A. BOAG. 1971b. Food habits of Black-crowned Night-Herons in southern Alberta. Auk 88(2): 435-437.

WOOD, B. 1985. Wagtails and pipits, pp. 338-340 *in* C.M. Perrins and A.L.A. Middleton, eds. The encyclopedia of birds. Facts on File, New York, 447 pp.

WOOD, N.A. 1974. The breeding behaviour and biology of the Moorhen. British Birds 67(3-4): 104-157.

WOOD, R. and W.L. GELTSON. 1972. Preliminary report: the Mute Swans of Michigan's Grand Traverse Bay region. Michigan Department of Natural Resources, Wildlife Division. Report no. 2683, 6 pp.

WOODIN, M.C. and G.A. SWANSON. 1989. Foods and dietary strategies of prairie-nesting Ruddy Ducks and Redheads. Condor 91(2): 280-287.

WOODLIFFE, P.A. 1987a. Acadian Flycatcher, pp. 256-257 *in* M.D. Cadman, P.F.J. Eagles, and F.M. Helleiner, eds. Atlas of the breeding birds of Ontario. University of Waterloo Press, Waterloo, 617 pp.

WOODLIFFE, P.A. 1987b. Least Bittern, pp. 48-49 *in* M.D. Cadman, P.F.J. Eagles, and F.M. Helleiner, eds. Atlas of the breeding birds of Ontario. University of Waterloo Press, Waterloo, 617 pp.

WOODLIFFE, P.A. 1987c. Red-bellied Woodpecker, pp. 234-235 *in* M.D. Cadman, P.F.J. Eagles, and F.M. Helleiner, eds. Atlas of the breeding birds of Ontario. University of Waterloo Press, Waterloo, 617 pp.

WOODLIFFE, P.A. 1987d. Red-headed Woodpecker, pp. 232-233 *in* M.D. Cadman, P.F.J. Eagles, and F.M. Helleiner, eds. Atlas of the breeding birds of Ontario. University of Waterloo Press, Waterloo, 617 pp.

WOODREY, M.S. 1991. Caching behavior in free-ranging White-breasted Nuthatches: the effects of social dominance. Ornis Scandinavica 22: 160-166.

WOODS, R.S. 1968. House Finch, pp. 290-314 *in* A.C. Bent. Life histories of North American cardinals, grosbeaks, buntings, towhees, finches, sparrows, and allies. (O.L. Austin, Jr., ed.). United States National Museum, Washington. [1968, Dover Publications, New York]. Bulletin no. 237, vol. 1, pp. 1-602; vol. 2, pp. 603-1248; vol. 3, pp. 1249-1889.

WOODWARD, J.C. 1976. Successful parasitism of the Gray Catbird by the Brown-headed Cowbird. Wilson Bulletin 88(3): 504-505.

WOODWARD, P.W. 1983. Behavioral ecology of fledging Brown-headed Cowbirds and their hosts. Condor 85(2): 151-163.

WOOLF, A., R. NORRIS, and J. KUBE. 1984. Evaluation of Ruffed Grouse reintroduction in southern Illinois, pp. 59-74 *in* W.L. Robinson, ed. Ruffed Grouse management: state of the art in the early 1980's. North Central Section of the Wildlife Society, Maryland, 181 pp.

WOOLFENDEN, G.E. 1956. Comparative breeding behaviour of *Ammospiza caudacutus* and *A. maritima*. University of Kansas Publications. Museum of Natural History 10(2): 45-75.

WOOTTON, J.T. 1986. Clutch-size in western and introduced eastern populations of House Finches: patterns and hypothesis. Wilson Bulletin 98(3): 459-462.

WOOTTON, J.T. 1987. Interspecific competition between introduced House Finch populations and two associated passerine species. Oecologia 71: 325-331.

Wootton, J.T., E.K. Bollinger, and C.J. Hibbard. 1986. Mating systems in homogeneous habitats: the effects of female uncertainty, knowledge costs, and random settlement. American Naturalist 128(4): 499-512.

Work, T.H. and A.J. Wool. 1942. The nest life of the Turkey Vulture. Condor 44: 145-149.

World Conservation Monitoring Centre. 1992. Global biodiversity: status of the earth's living resources. Chapman and Hall, London, 585 pp.

World Resources Institute. 1992. Global biodiversity strategy: guidelines for action to save, study and use earth's biotic wealth sustainably and equitably. World Resources Institute, World Conservation Union and United Nations Environment Programme.

Worley, C. 1987. Nest site selectivity by the goldfinch in old field habitat. Passenger Pigeon 49(1): 19-21.

Wray, T. and R.C. Whitmore. 1979. Effects of vegetation on nesting success of Vesper Sparrows. Auk 96(4): 802-805.

Wray, T., K.A. Strait, and R.C. Whitmore. 1982. Reproductive success of grassland sparrows on a reclaimed surface mine in West Virginia. Auk 99(1): 157-164.

Wright, B.S. 1954. High tide and east wind. Stackpole, Harrisburg, 162 pp.

Wright, H.F. 1921. The Mockingbird in the Boston region and in New England and Canada. Auk 38: 382-432.

Wunderle, J.M. Jr. 1978. Territorial defense of a nectar source by a Palm Warbler. Wilson Bulletin 90(2): 297-299.

Wyllie, I. 1981. The Cuckoo. Universe Books, New York, 176 pp.

Wynne-Edwards, V.C. 1932. Notes on some rare birds in Montreal district. Canadian Field-Naturalist 46(8): 181-183.

Wynne-Edwards, V.C. 1952. Zoology of the Baird expedition (1950). I. The birds observed in central and south-east Baffin Island. Auk 69(4): 353-391.

Yahner, R.H. 1983. Site-related nesting success of Mourning Doves and American Robins in shelterbelts. Wilson Bulletin 95(4): 573-580.

Yahner, R.H. 1988. Changes in wildlife communities near edges. Conservation Biology 2: 333-339.

Yahner, R.H. and C.A. DeLong. 1992. Avian predation and parasitism on artificial nests and eggs in two fragmented landscapes. Wilson Bulletin 104(1): 162-168.

Yahner, R.H. and D.P. Scott. 1988. Effects of forest fragmentation on depredation of artificial nests. Journal of Wildlife Management 52(1): 158-161.

Yahner, R.H. and A.L. Wright. 1985. Depredation on artificial ground nests: effects of edge and plotage. Journal of Wildlife Management 49(2): 508-513.

Yank, R. 1982. Summary of sightings in the Montreal area. Tchebec 12: 6-63.

Yank, R. and Y. Aubry. 1984. The nesting season: Quebec region. American Birds 38(6): 1000-1001.

Yank, R. and Y. Aubry. 1985a. The nesting season: Quebec region. American Birds 39(5): 889-890.

Yank, R. and Y. Aubry. 1985b. The spring migration: Quebec region. American Birds 39(3): 274-276.

Yank, R. and Y. Aubry. 1985c. The winter season: Quebec region. American Birds 39(2): 148-150.

Yank, R. and Y. Aubry. 1986a. The nesting season: Quebec region. American Birds 40(5): 1180-1182.

Yank, R. and Y. Aubry. 1986b. The winter season: Quebec region. American Birds 40(2): 260-261.

Yank, R. and Y. Aubry. 1992a. The summer season: Quebec region. American Birds 46(5): 1116-1118.

Yank, R. and Y. Aubry. 1992b. The winter season: Quebec region. American Birds 46(2): 233-235.

Yank, R. and Y. Aubry. 1993a. The summer season: Quebec region. American Birds 47(5): 1086-1088.

Yank, R. and Y. Aubry. 1993b. The winter season: Quebec region. American Birds 47(2): 237-239.

Yank, R., Y. Aubry, and M. Gosselin. 1985. The nesting season: Quebec region. American Birds 39(5): 889-890.

Yank, R., Y. Aubry, and M. Gosselin. 1987a. The autumn migration: Quebec region. American Birds 41(1): 61-63.

Yank, R., Y. Aubry, and M. Gosselin. 1987b. The nesting season: Quebec region. American Birds 41(5): 1411-1413.

Yank, R., Y. Aubry, and M. Gosselin. 1988. The spring migration: Quebec region. American Birds 42(3): 410-412.

Yank, R., Y. Aubry, and M. Gosselin. 1989a. Les observations saisonnières: l'hiver 1988-1989. QuébecOiseaux 1(1): 24-26.

Yank, R., Y. Aubry, and M. Gosselin. 1989b. The winter season: Quebec region. American Birds 43(2): 287-289.

Yank, R., Y. Aubry, and M. Gosselin. 1990a. The autumn migration: Quebec region. American Birds 44(1): 51-53.

Yank, R., Y. Aubry, and M. Gosselin. 1990b. The nesting season: Quebec region. American Birds 44(5): 1110-1112.

Yank, R., Y. Aubry, and M. Gosselin. 1991a. L'été 1990: juin et juillet. QuébecOiseaux 2(3): 27-30.

Yank, R., Y. Aubry, and M. Gosselin. 1991b. The spring season: Quebec region. American Birds 45(3): 412-414.

Yasukawa, K., J.L. McClure, R.A. Boley, and J. Zanocco. 1990. Provisioning of nestlings by male and female Red-winged Blackbirds, *Agelaius phoeniceus*. Animal Behaviour 40(1): 153-166.

Yeatman, L. 1976. Atlas des oiseaux nicheurs de France de 1970 à 1975. Ministère de la qualité de la vie – Environnement et Société ornithologique de France, Paris, 281 pp.

Yeaton, R.J. and M.L. Cody. 1974. Competitive release in island Song Sparrow populations. Theoretical Population Biology 5: 42-48.

Yokel, D.A. 1986. Monogamy and brood parasitism: an unlikely pair. Animal Behaviour 34: 1348-1358.

Yokel, D.A. 1989. Intrasexual aggression and the mating behavior of Brown-headed Cowbirds: their relation to population densities and sex ratios. Condor 91: 43-51.

Yosef, R. 1992. Male-biased breeding site fidelity in a population of Northern Shrikes. Condor 94(4): 1025-1027.

Yosef, R. and B. Pinshow. 1988. Polygyny in the Northern Shrike (*Lanius excubitor*) in Israel. Auk 105(3): 581-582.

Young, A.D. 1985. Costs and benefits to Red-breasted Mergansers nesting in Tern and Gull colonies. M.Sc. thesis, McGill University, Montréal, 36 pp.

Young, A.D. and R.D. Titman. 1988. Intraspecific nest parasitism in Red-breasted Mergansers. Canadian Journal of Zoology 66: 2454-2458.

Young, A.M. 1982. Population biology of tropical insects. Plenum Press, New York, 511 pp.

Young, C.J. 1897. Notes on the birds of the Magdalen Islands. Ottawa Naturalist 11(8): 145-158.

Young, C.M. 1967. Overland migration of duck broods in a drought-free area. Canadian Journal of Zoology 45: 249-251.

Young, H. 1951. Territorial behavior of the Eastern Robin. Proceedings. Linnean Society of New York 58-62: 1-37.

Yourans, E. 1930. Chez les oiseaux. Naturaliste canadien 57: 115-116.

Yunick, R.P. 1981. Some observations on the breeding status of the Pine Siskin. Kingbird 31(4): 219-225.

Yunick, R.P. 1983a. Age and sex determination of Purple Finches during the breeding season. North American Bird Bander 8(2): 48-51.

Yunick, R.P. 1983b. Winter site fidelity of some northern finches (Fringillidae). Journal of Field Ornithology 54(3): 254-258.

Yunick, R.P. 1985. A review of recent irruptions of the Black-backed Woodpecker and Three-toed Woodpecker in eastern North America. Journal of Field Ornithology 56(2): 138-152.

Zach, R. 1982. Hatching asynchrony, egg size, growth, and fledging in Tree Swallows. Auk 99(4): 695-700.

Zach, R. and J.B. Falls. 1975. Response of the Ovenbird (Aves: Parulidae) to an outbreak of the Spruce Budworm. Canadian Journal of Zoology 53: 1669-1672.

Zach, R. and J.B. Falls. 1979. Foraging and territoriality of male Ovenbirds (Aves: Parulidae) in a heterogenous habitat. Journal of Animal Ecology 48: 33-52.

Zammuto, R.M. and E.C. Franks. 1978. Forty adult Chimney Swifts at an active nest. Bird-Banding 49(3): 278-279.

Zammuto, R.M. and E.C. Franks. 1981. Environmental effects on roosting behavior of Chimney Swifts. Wilson Bulletin 93(1): 77-84.

Zarn, M. 1975. Rough-legged Hawk (*Buteo lagopus sanctijohannis*). Habitat management series for unique or endangered species. United States Department of the Interior, Bureau of Land Management, Denver. Report no. 14, 23 pp.

Zarnowitz, J.E. and D.A. Manuwal. 1985. The effects of forest management on cavity-nesting birds in northwestern Washington. Journal of Wildlife Management 49(1): 255-263.

Zeleny, L. 1976. The Bluebird: how you can help its fight for survival. Indiana University Press, Bloomington, 170 pp.

Zeranski, J.D. and T.R. Baptist. 1990. Connecticut birds. University Press of New England, Hanover, 328 pp.

Zicus, M.C. 1990. Nesting biology of Hooded Mergansers using nest boxes. Journal of Wildlife Management 54(4): 637-643.

Zicus, M.C. and S.K. Hennes. 1990. Renesting by a Common Goldeneye. Journal of Field Ornithology 6(2): 245-248.

Zicus, M.C., R.H. Hier, and S.J. Maxson. 1983. A Common Loon nest from Minnesota containing four eggs. Wilson Bulletin 95(4): 671-672.

Zimmerman, J.L. 1980. Virginia Rail (*Rallus limicola*), pp. 46-56 *in* G.C. Sanderson, ed. Management of migratory shore and upland game birds in North America. University of Nebraska Press, Lincoln, 358 pp.

Zimmerman, J.L. 1983. Cowbird parasitism of Dickcissels in different habitats and at different nest densities. Wilson Bulletin 95(1): 7-22.

Zimmerman, J.L. 1988. Breeding season habitat selection by the Henslow's Sparrow (*Ammodramus henslowii*) in Kansas. Wilson Bulletin 100(1): 17-24.

Zink, R.M. 1991. The geography of mtDNA variation in the Fox Sparrow. Paper presented at the 109th stated meeting of the AOU, Montréal, August 14-17 1991.

Zink, R.M. and B.A. Fall. 1981. Breeding distribution, song and habitat of the Alder Flycatcher and Willow Flycatcher in Minnesota. Loon 53: 208-214.

Ziswiler, V. and D.S. Farner. 1972. Digestion and the digestive system, pp. 343-430 *in* D.S. Farner and J.R. King, eds. Avian Biology, vol. 2. Academic Press, New York, 612 pp.

Zoltai, S.C. 1983. An overview of the wetlands of Canada, pp. 45-49 *in* H. Boyd, ed. First western hemisphere waterfowl and waterbird symposium. Canadian Wildlife Service, Ottawa, 147 pp.

Zusi, R.L. and G.D. Bentz. 1978. The appendicular myology of the Labrador Duck (*Camptorhynchus labradorius*). Condor 80(4): 407-418.

Index

Where a species' international French name as recommended by the Commission internationale des noms français (1993) is different from the French common name given by Ouellet et al. (1990), it is marked with an asterisk (*).

See page 1238 for an explanation of the species status letter codes after species name.

Errata

P. 93
Taxonomic List of Birds Described in Atlas
Table: column 5 (Family), Tytonidae
Amend to read: (17, 1, 1, 1, 1)
Table: column 9 (Common Name), Barn Owl
Delete '+' following 'Barn Owl'

P. 95
Taxonomic List of Birds Described in Atlas
Table: column 8 (Scientific Name), line 9
Amend to read: *Dendroica pensylvanica*

P. 111
Anatidae
Text: column 2, line 8
Amend to read: extent constituting the taxonomic link...

P. 145
Tytonidae
Text: column 2, last 2 lines
Amend to read: It has often been reported in Québec, but has been positively confirmed as breeding only once in the province.

P. 146
Strigidae
Text: column 1, paragraph 3, line 11
Amend to read: 270°

P. 891
Black-throated Green Warbler
Text: column 2, *History and trends*, paragraph 1, last sentence
Amend to read: (Elliott 1958)

Administrative Regions of Québec

Source: Ministère de l'Énergie et des Ressources du Québec (1990)

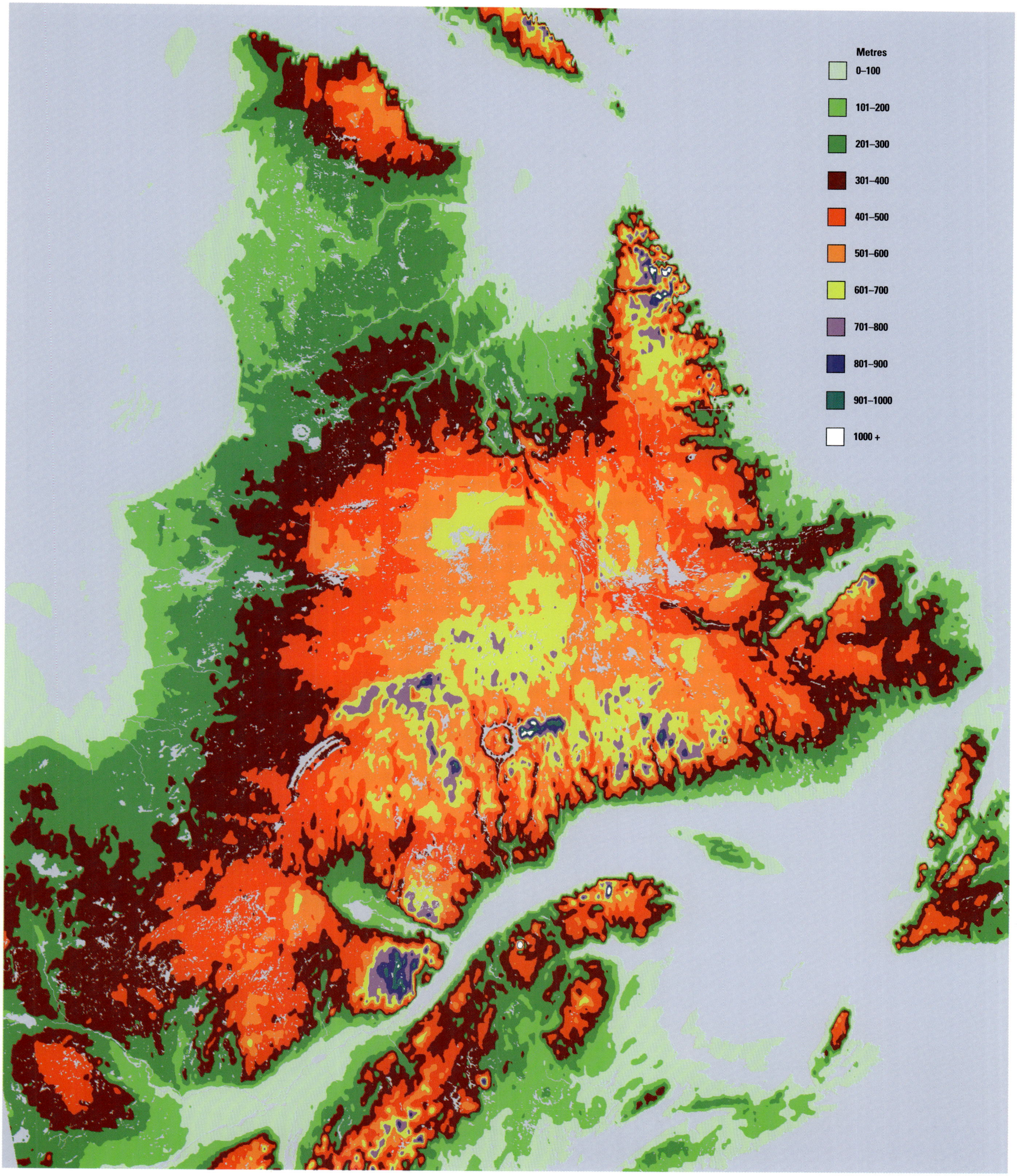

Québec in relief

Source: Image taken from the digital elevation model, GEOSCOPE database (original data: Integrated Global Elevation and Bathymetry. Digital Data NOAA/NGDC/WD-A, Boulder, Colorado). Québec portion adapted by the GIS Technologies Section, Ministère des Ressources naturelles du Québec.

Adapted from Hydro-Québec and Energy, Mines and Resources Canada, 1994.

Typeset in ITC Berkeley Oldstyle
and Univers by Compélec Inc.
and printed in October 1996
by Métropole Litho Inc.

Printed in Canada

Piping Plover
Photograph: Daniel Jauvin